Disha's

Errorless

183 New Syllabus Chapter-wise + Topic-wise

24 Years JEE Main PHYSICS

(2025 - 2002) Previous Year Solved Papers

DISHA™
Publication Inc

DISHA Publication Inc.
A-23 FIEE Complex, Okhla Phase II
New Delhi-110020
Tel: 49842349/ 49842350

Edited By
Sanjeev Kumar Jha
Goving Thakur

Typeset By
DISHA DTP Team

Write To Us At
feedback_disha@aiets.co.in

www.dishapublication.com

DISHA™
Publication Inc

CONTENTS

CLASS XI

A1 - A186

1. **Units and Measurements** A1 – A14
   - **Topic 1 :** Unit of Physical Quantities
   - **Topic 2 :** Dimensions of Physical Quantities
   - **Topic 3 :** Errors in Measurements
2. **Motion in a Straight Line** A15 – A24
   - **Topic 1 :** Distance, Displacement & Uniform Motion
   - **Topic 2 :** Non-uniform Motion **Topic 3 :** Relative Velocity
   - **Topic 4 :** Motion Under Gravity
3. **Motion in a Plane** A25 – A36
   - **Topic 1 :** Vectors
   - **Topic 2 :** Motion in a Plane with Constant Acceleration
   - **Topic 3 :** Projectile Motion
   - **Topic 4 :** Relative Velocity in Two Dimensions & Uniform Circular Motion
4. **Laws of Motion** A37 – A54
   - **Topic 1 :** Ist, IInd & IIIrd Laws of Motion
   - **Topic 2 :** Motion of Connected Bodies, Pulley & Equilibrium of Forces
   - **Topic 3 :** Friction **Topic 4 :** Circular Motion, Banking of Road
5. **Work, Energy and Power** A55 – A68
   - **Topic 1 :** Work **Topic 2 :** Energy
   - **Topic 3 :** Power **Topic 4 :** Collisions
6. **System of Particles and Rotational Motion** A69 – A90
   - **Topic 1 :** Centre of Mass, Centre of Gravity & Principle of Moments
   - **Topic 2 :** Angular Displacement, Velocity and Acceleration
   - **Topic 3 :** Torque, Couple and Angular Momentum
   - **Topic 4 :** Moment of Inertia and Rotational K.E.
   - **Topic 5 :** Rolling Motion
7. **Gravitation** A91 – A104
   - **Topic 1 :** Kepler's Laws of Planetary Motion
   - **Topic 2 :** Newton's Universal Law of Gravitation
   - **Topic 3 :** Acceleration due to Gravity
   - **Topic 4 :** Gravitational Field and Potential Energy
   - **Topic 5 :** Motion of Satellites, Escape Speed and Orbital Velocity
8. **Mechanical Properties of Solids** A105 – A112
   - **Topic 1 :** Hooke's Law & Young's Modulus
   - **Topic 2 :** Bulk and Rigidity Modulus and Work Done in Stretching a Wire
9. **Mechanical Properties of Fluids** A113 – A126
   - **Topic 1 :** Pressure, Density, Pascal's Law and Archimedes' Principle
   - **Topic 2 :** Fluid Flow and Bernoulli's Principle
   - **Topic 3 :** Viscosity and Terminal Velocity
   - **Topic 4 :** Surface Tension, Surface Energy and Capillarity
10. **Thermal Properties of Matter** A127 – A134
    - **Topic 1 :** Thermometer & Thermal Expansion
    - **Topic 2 :** Calorimetry and Heat Transfer
11. **Thermodynamics** A135 – A146
    - **Topic 1 :** First Law of Thermodynamics
    - **Topic 2 :** Specific Heat Capacity and Thermodynamical Processes
    - **Topic 3 :** Second Law of Thermodynamics
12. **Kinetic Theory** A147 – A158
    - **Topic 1 :** Kinetic Theory of an Ideal Gas and Gas Laws
    - **Topic 2 :** Speed of Gas, Pressure and Kinetic Energy
    - **Topic 3 :** Degree of Freedom, Specific Heat Capacity, and Mean Free Path
13. **Oscillations** A159 – A172
    - **Topic 1 :** Displacement, Phase, Velocity and Acceleration in S.H.M.
    - **Topic 2 :** Energy in Simple Harmonic Motion
    - **Topic 3 :** Time Period, Frequency, Simple Pendulum and Spring Pendulum
14. **Waves** A173 – A180
    - **Topic 1 :** Basic of Mechanical Waves, Progressive and Stationary Waves
    - **Topic 2 :** Vibration of String and Organ Pipe
    - **Topic 3 :** Beats, Interference and Superposition of Waves
15. **Experimental Skills–XI** A181 – A186
    - **Topic 1 :** Vernier calipers - its use to measure the internal & external diameter and depth of a vessel; To determine thickness/ diameter of thin sheet/wire with the help of screw gauge.
    - **Topic 2 :** Measure the mass of a given object by the principle of moments using metre scale, Determination of Young's modulus of elasticity of the material of a metallic wire. Calculation of surface tension of water by capillary rise and effect of detergents. Calculation of co-efficient of viscosity of a given viscous liquid, Measurement of specific heat using method of mixtures, Calculation of dissipation of energy by plotting a graph between the square of amplitude and time, Determination of speed of sound using resonance tube,

Hints & Solutions (Class XI)

A187 - A394

1. Units and Measurements A187 – A199
2. Motion in a Straight Line A200 – A211
3. Motion in a Plane A212 – A225
4. Laws of Motion A226 – A243
5. Work, Energy and Power A244 – A263
6. System of Particles and Rotational Motion A264 – A291
7. Gravitation A292 – A308
8. Mechanical Properties of Solids A309 – A316
9. Mechanical Properties of Fluids A317 – A331
10. Thermal Properties of Matter A332 – A339
11. Thermodynamics A340 – A350
12. Kinetic Theory A351 – A361
13. Oscillations A362 – A377
14. Waves A378 – A388
15. Experimental Skills-XI A389 – A394

CLASS XII B1 - B214

1. **Electric Charges and Fields** B1 – B18
   - **Topic 1 :** Electric Charges and Coulomb's Law
   - **Topic 2 :** Electric Field and Electric Field Lines
   - **Topic 3 :** Electric Dipole, Electric Flux and Gauss's Law
2. **Electrostatic Potential and Capacitance** B19 – B38
   - **Topic 1 :** Electrostatic Potential and Equipotential Surfaces
   - **Topic 2 :** Electric Potential Energy and Work Done in Carrying a Charge
   - **Topic 3 :** Capacitors, Grouping of Capacitors and Energy Stored in a Capacitor
3. **Current Electricity** B39 – B62
   - **Topic 1 :** Electric Current, Drift of Electrons, Ohm's Law, Resistance and Resistivity
   - **Topic 2 :** Combination of Resistances
   - **Topic 3 :** Kirchhoff's Laws and Cells
   - **Topic 4 :** Heating Effect of Current
   - **Topic 5 :** Wheatstone Bridge and Different Measuring Instruments
4. **Moving Charges and Magnetism** B63 – B86
   - **Topic 1 :** Motion of Charged Particle in Magnetic Field
   - **Topic 2 :** Magnetic Field Lines, Biot-Savart's Law and Ampere's Circuital Law
   - **Topic 3 :** Force and Torque on Current Carrying Conductor
   - **Topic 4 :** Galvanometer and its Conversion into Ammeter and Voltmeter
5. **Magnetism and Matter** B87 – B92
   - **Topic 1 :** Magnetism, Magnetic Moment, Properties of Magnet
   - **Topic 2 :** Magnetic Materials and their Properties
   - **Topic 3 :** Magnetic Equipment
6. **Electromagnetic Induction** B93 – B104
   - **Topic 1 :** Magnetic Flux, Faraday's and Lenz's Law
   - **Topic 2 :** Motional and Static EMI and Application of EMI
7. **Alternating Current** B105 – B118
   - **Topic 1 :** Alternating Current, Voltage and Power
   - **Topic 2 :** AC Circuit, LCR Circuit and Power Factor
   - **Topic 3 :** Transformers and LC Oscillations
8. **Electromagnetic Waves** B119 – B130
   - **Topic 1 :** Electromagnetic Waves, Conduction and Displacement Current
   - **Topic 2 :** Electromagnetic Spectrum
9. **Ray Optics and Optical Instruments** B131 – B148
   - **Topic 1 :** Plane Mirror, Spherical Mirror and Reflection of Light
   - **Topic 2 :** Refraction of Light at Plane Surface and Total Internal Reflection
   - **Topic 3 :** Refraction at Curved Surface, Lenses and Power of Lens
   - **Topic 4 :** Refraction of Light through a Prism
   - **Topic 5 :** Optical Instruments
10. **Wave Optics** B149 – B160
    - **Topic 1 :** Wavefront, Interference of Light, Coherent and Incoherent Sources
    - **Topic 2 :** Young's Double Slit Experiment
    - **Topic 3 :** Diffraction, Polarisation of Light
11. **Dual Nature of Radiation and Matter** B161 – B174
    - **Topic 1 :** Matter Waves, Cathode and Positive Rays
    - **Topic 2 :** Photon, Photoelectric Effect
12. **Atoms** B175 – B184
    - **Topic 1 :** Atomic Structure and Rutherford's Nuclear Model
    - **Topic 2 :** Bohr's Model and the Spectra of the Hydrogen Atom
13. **Nuclei** B185 – B190
    - **Topic 1 :** Composition and Size of the Nuclei
    - **Topic 2 :** Mass-Energy Equivalence and Nuclear Reactions
14. **Semiconductor Electronics : Materials, Devices and Simple Circuits** B191 – B210
    - **Topic 1 :** Semiconductors and P-N Junction Diode
    - **Topic 2 :** Digital Electronics and Logic Gates
15. **Experimental Skills-XII** B211 – B214
    - **Topic 1 :** Determination of resistivity of the material of a given wire using a Meter bridge
      To determine the resistance of a given wire using Ohm's law
      To determine the resistance and figure of merit of a galvanometer by half deflection method.
    - **Topic 2 :** Determination of focal length of a (i) convex mirror (ii) concave mirror and (iii) convex lens using the parallax method, To plot angle of deviation versus angle of incidence for a triangular prism using parallax method
      To determine the refractive index of a glass slab using a travelling microscope.
    - **Topic 3 :** To draw the characteristic curves of a p-n junction diode in forward and reverse bias
      To draw the characteristic curves of a Zener diode and finding reverse break down voltage
      Identification of diode, LED, resistor, capacitor from mixed collection of such items.

Hints & Solutions (Class XII) B215 - B438

1. Electric Charges and Fields B215 – B235
2. Electrostatic Potential and Capacitance B236 – B256
3. Current Electricity B257 – B286
4. Moving Charges and Magnetism B287 – B308
5. Magnetism and Matter B309 – B312
6. Electromagnetic Induction B313 – B323
7. Alternating Current B324 – B338
8. Electromagnetic Waves B339 – B348
9. Ray Optics and Optical Instruments B349 – B373
10. Wave Optics B374 – B387
11. Dual Nature of Radiation and Matter B388 – B402
12. Atoms B403 – B415
13. Nuclei B416 – B421
14. Semiconductor Electronics : Materials, Devices and Simple Circuits B422 – B435
15. Experimental Skills-XII B436 – B438

Disha's SMART PREP EdGe

CRACK JEE
Deep JEE Analysis
ANALYZING THE EXAM

Jan Attempt:
Total Registered- 13,11,544
Total Appeared- 12,58,136

April Attempt:
Total Registered- 10,61,840
Total Appeared- 9,92,350

Total Registered (Unique): 15.39 lakhs
Total Appeared (Unique): 14.75 lakhs

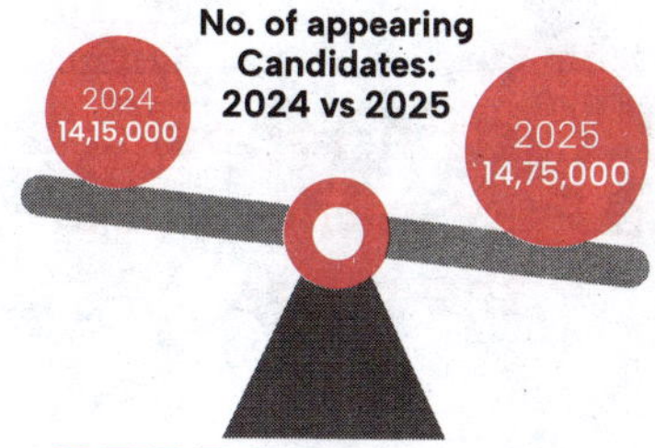

Slight increase from 2024

No. of students who scored in 100 Percentile

2022	2023	2024	2025
24	42	56	24

State-Wise Distribution of Toppers (2025)

State	Session 1	Session 2	Total
Rajasthan	5	2	7
Uttar Pradesh	2	1	3
Delhi (NCT)	2	0	2
Maharashtra	1	2	3
Gujarat	1	1	2
Telangana	1	2	3
Karnataka	1	2	3
Andhra Pradesh	1	0	1
Total	14	10	24

Total qualifying candidates for JEE (Advanced) 2,50,000

Category-wise no. of Qualified Candidates for JEE Adv.

Category	Number of "Top" candidates	
OPEN	96187	101250
OPEN-PwD	5063	
GEN-EWS	23750	25000
GEN-EWS-PwD	1250	
OBC-NCL	64125	67500
OBC-NCL-PwD	3375	
SC	35625	37500
SC-PwD	1875	
ST	17812	18750
ST-PwD	938	

Cut off for writing JEE (Adv) (in percentile)

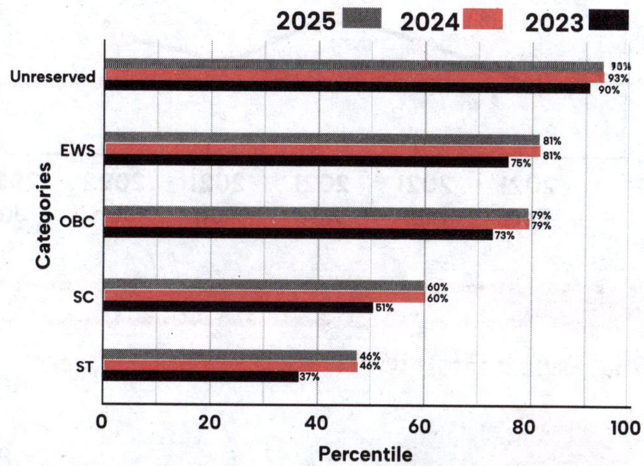

Percentile Vs Marks JEE Main 2025

Marks Range	Percentile Range
290 – 300	99.9989 – 100
280 – 289	99.9961 – 99.9989
270 – 279	99.9903 – 99.9961
260 – 269	99.9772 – 99.9903
250 – 259	99.9601 – 99.9772
240 – 249	99.9350 – 99.9601
230 – 239	99.9011 – 99.9350
220 – 229	99.8516 – 99.9011
210 – 219	99.7950 – 99.8516
200 – 209	99.7108 – 99.7950

Marks Range	Percentile Range
190 – 199	99.5973 – 99.7108
180 – 189	99.4569 – 99.5973
170 – 179	99.2720 – 99.4569
160 – 169	99.0286 – 99.2720
150 – 159	98.7323 – 99.0286
140 – 149	98.3174 – 98.7323
130 – 139	97.8112 – 98.3174
120 – 129	97.1429 – 97.8112
110 – 119	96.2045 – 97.1429
100 – 109	94.9985 – 96.2045

JEE Main cut-off from the past 5 Yrs

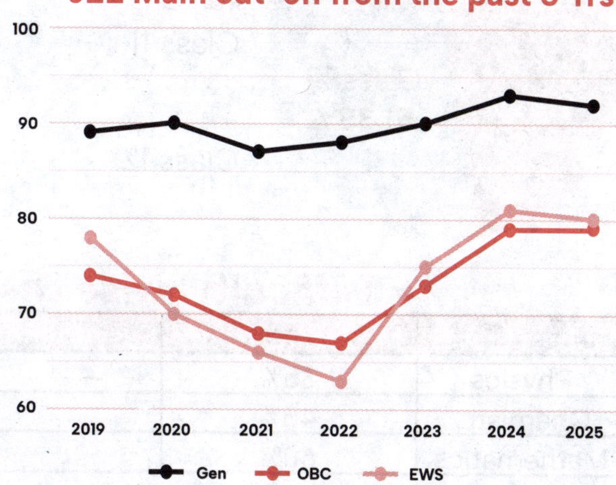

Percentile–Marks–Rank JEE 2025

Percentile Range	Expected Rank Range
100	1 – 24
99.999 – 99.997	25 – 50
99.997 – 99.969	51 – 400
99.969 – 99.946	401 – 700
99.946 – 99.915	701 – 1,100
99.915 – 99.866	1,101 – 1,800
99.866 – 99.807	1,801 – 2,600
99.807 – 99.731	2,601 – 3,600
99.731 – 99.5	3,601 – 6,000
99.5 – 99.4	6,001 – 7,500
99.4 – 99.3	7,501 – 9,000
99.3 – 99.2	9,001 – 10,500
99.2 – 99.1	10,501 – 12,000
99.1 – 99.0	12,001 – 13,500
99.0 – 98.9	13,501 – 15,000
98.9 – 98.8	15,001 – 16,500
98.8 – 98.7	16,501 – 18,000
98.7 – 98.6	18,001 – 19,500
98.6 – 98.5	19,501 – 21,000
98.5 – 98.4	21,001 – 22,500
98.4 – 98.3	22,501 – 24,000
98.3 – 98.2	24,001 – 25,500
98.2 – 98.1	25,501 – 27,000
98.1 – 98.0	27,001 – 28,500
98.0 – 97.9	28,501 – 30,000
97.9 – 97.8	30,001 – 31,500
97.8 – 97.7	31,501 – 33,000
97.7 – 97.6	33,001 – 34,500
97.6 – 97.5	34,501 – 36,000
97.5 – 97.4	36,001 – 37,500
97.4 – 97.3	37,501 – 39,000
97.3 – 97.2	39,001 – 40,500
97.2 – 97.1	40,501 – 42,000
97.1 – 97.0	42,001 – 43,500
97.0 – 96.9	43,501 – 45,000
96.9 – 96.8	45,001 – 46,500
96.8 – 96.7	46,501 – 48,000

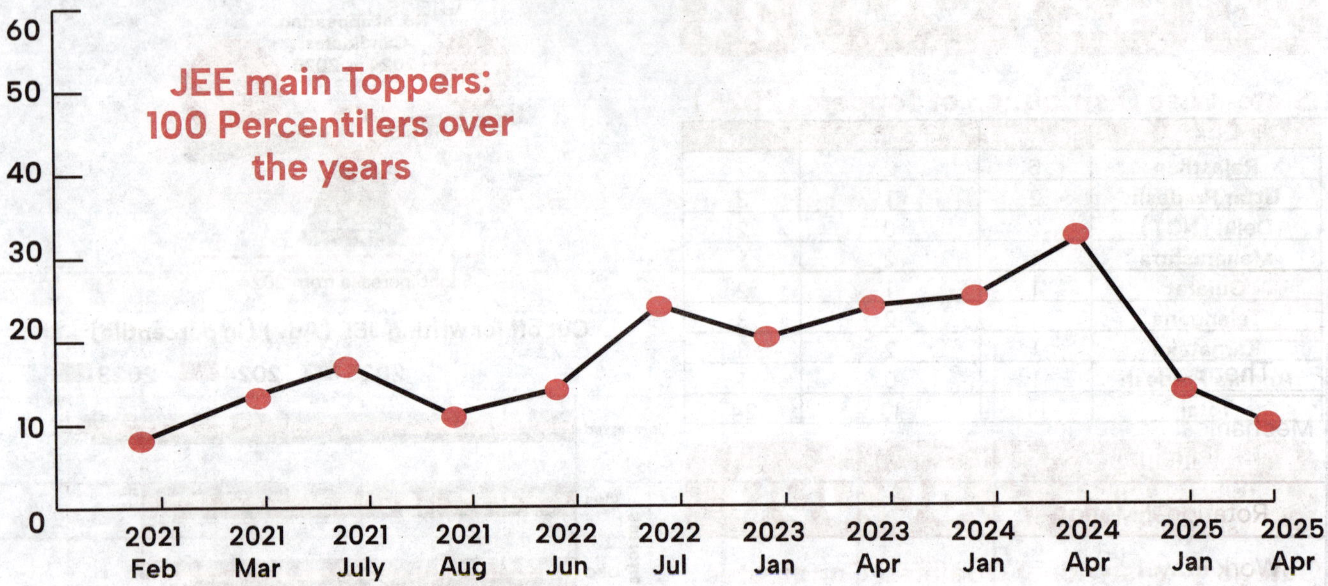

Overall Analysis of JEE Main 2025 Shift Papers

Class-wise Distribution of Overall Paper

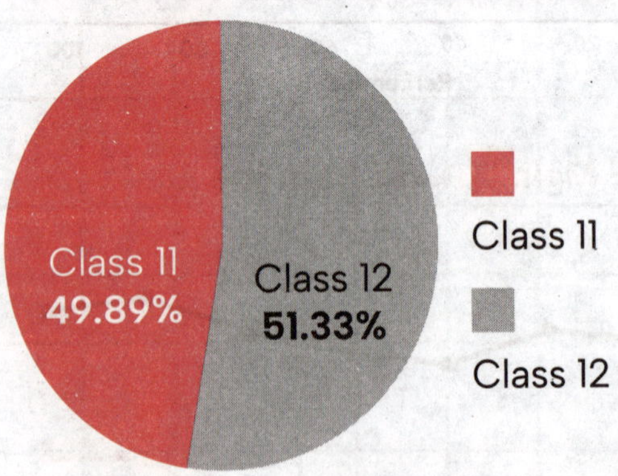

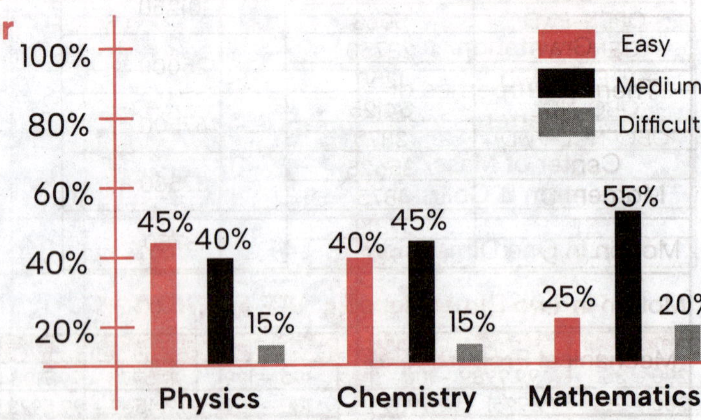

Difficulty-wise Distribution of Overall Paper

Subject level	Moderate, tricky	Easy to moderate, NCERT based Qns	Moderate to difficult, lengthy
Physics	35%	50%	15%
Chemistry	45%	50%	5%
Mathematics	60%	25%	15%

PHYSICS

Class-wise Analysis of JEE Main 2025 Shift Papers

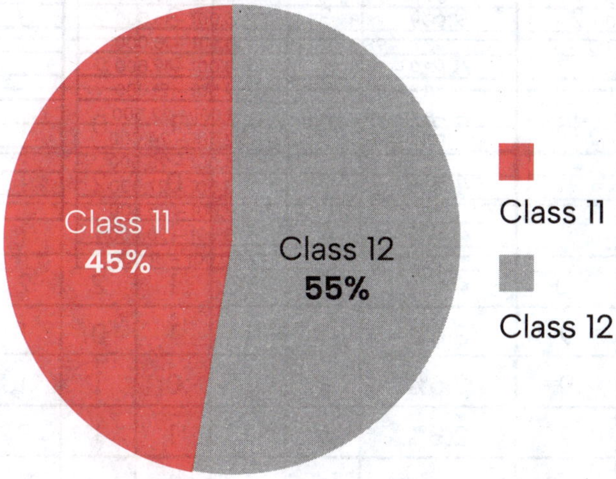

Chapter-wise Analysis of JEE Main 2025 Shift Papers

Class XI

Chapter Name	Total no. of Qns	Chapter Weightage	22S1	22S2	23S1	23S2	24S1	24S2	28S1	28S2	29S1	29S2
Thermodynamics	15	6.0%	0	2	2	2	2	2	1	0	2	2
Mechanical Properties of Fluids	14	5.6%	1	3	1	2	2	0	2	2	1	0
Rotational Motion	13	5.2%	2	1	1	1	2	2	2	0	1	1
Work Power Energy	11	4.4%	1	2	1	1	1	0	2	1	1	1
Units and Dimensions	11	4.4%	1	1	2	1	0	0	1	1	2	2
Oscillations	7	2.8%	0	1	1	0	1	1	0	1	1	1
Gravitation	7	2.8%	1	0	0	2	1	1	0	1	0	1
Thermal Properties of Matter	7	2.4%	3	0	1	0	0	2	0	0	0	1
Center of Mass, Momentum & Collision	6	1.6%	0	0	1	0	0	0	1	2	1	1
Motion In One Dimension	4	1.6%	0	0	1	0	0	0	0	1	1	1
Motion In Two Dimensions	4	1.6%	1	1	0	0	0	1	0	0	1	0
Mechanical Properties of Solids	3	1.2%	0	0	0	0	0	1	0	1	1	0
Laws of Motion	3	1.2%	0	0	0	1	1	1	0	0	0	0
Waves and Sound	3	1.2%	1	0	0	1	0	0	1	0	0	0
Experimental Physics	3	1.2%	1	0	0	0	1	0	1	0	0	0
Kinetic Theory of Gases	3	1.2%	0	0	0	0	0	0	1	2	0	0

Class XII

Chapter Name	Total no. of Qns	Chapter Weightage	22S1	22S2	23S1	23S2	24S1	24S2	28S1	28S2	29S1	29S2
Ray Optics	24	9.6%	3	2	3	2	3	1	2	2	3	3
Electrostatics	17	6.8%	2	1	2	2	1	2	2	1	2	2
Magnetic Effects of Current	13	5.2%	0	3	1	0	1	4	1	1	1	1
Current Electricity	11	4.4%	3	1	0	2	1	0	2	1	0	1
Dual Nature of Matter	11	4.4%	2	1	1	1	1	2	1	0	1	1
Capacitance	10	4.0%	1	0	1	2	2	0	1	1	0	2
Semiconductors	9	3.6%	1	1	1	0	1	1	1	1	1	1
Wave Optics	9	3.6%	1	1	0	1	1	3	1	1	0	0
Electromagnetic Waves	7	2.8%	0	1	1	1	0	0	1	1	1	1
Electromagnetic Induction	7	2.8%	0	1	2	0	0	0	0	2	2	0
Atomic Physics	5	2.0%	0	0	0	0	1	1	0	1	1	1
Mathematics in Physics	5	2.0%	0	1	1	1	1	0	0	0	0	1
Alternating Current	4	1.6%	0	1	0	1	1	0	0	0	1	0
Nuclear Physics	3	1.2%	0	0	1	1	0	0	1	0	0	0
Magnetic Properties of Matter	1	0.4%	0	0	0	0	0	0	0	1	0	0
Total	250	100.0%	25	25	25	25	25	25	25	25	25	25

Number of the Special Type of Questions per Shift

Type of Question	22S1	22S2	23S1	23S2	24S1	24S2	28S1	28S2	29S1	29S2
Statement Based	2	1	1	0	0	0	0	0	0	0
Multi-Statement	1	0	4	0	3	2	2	0	0	0
Assertion-Reason	1	2	0	1	0	2	2	1	4	2
Graphical	1	1	1	0	0	4	2	2	2	1
Incorrect Type	0	0	0	0	0	0	0	1	0	0
Matrix Match	0	0	1	1	0	0	0	1	1	2
Numerical (How many correct/ incorrect)	0	0	0	0	0	0	0	0	0	0
Total Special Questions	**5**	**4**	**7**	**2**	**3**	**8**	**6**	**5**	**7**	**5**

Difficulty-wise Distribution of JEE Main 2025 Shift Papers

Chapter Name	Easy	Medium	Difficult	Total
Electrostatics	11	8	0	19
Electricity & Magnetism	0	63	7	70
Semiconductors	0	9	0	9
Fluid Mechanics	0	17	0	17
Mechanics	0	35	13	48

Chapter Name	Easy	Medium	Difficult	Total
Dual Nature, Atoms & Nuclei	0	19	0	19
Optics	24	9	0	33
Thermal Physics	7	18	0	25
Waves & Oscillations	0	10	0	10
Total	42	188	20	250

CLASS XI

1 Units and Measurements

TREND ANALYSIS
JEE MAIN (2025-2021)

	2025				2024				2023				2022				2021			
	Session II		Session I		Session II		Session I		Session II		Session I		Session II		Session I		Session II		Session I	
	MCQs	NVQs	MCQs	NVQs	MCQs	NVQs	MCQs	NVQs	MCQs	NVQs	MCQs	NVQs	MCQs	NVQs	MCQs	NVQs	MCQs	NVQs	MCQs	NVQs
Topic 1	2	–	1	–	–	–	–	–	–	–	2	–	3	–	–	–	1	–	–	–
Topic 2	10	–	9	1	9	–	7	–	6	–	9	–	6	–	8	–	9	–	2	–
Topic 3	2	1	3	1	2	–	6	–	3	–	–	–	3	2	2	–	3	1	1	2
Topic 4																				
Total	14	1	13	2	11	–	13	–	9	–	11	–	12	2	10	–	13	1	3	2

How to Code Qns for Better & Faster Revision →

Type 1: Able to solve (No code Req)
Type 2: Solved with struggle or by seeing solution
(Use code NR means Needs Revision or orange stickers)
Type 3: Tricky/ Could not solve on my own. Need help from Teacher/ Friend.
(Use code TQ means Tricky Question or red stickers)

You can either hand write the code in boxes or else paste coloured stickers given at end of book.

Topic 1 – Unit of Physical Quantities

1. The unit of $\sqrt{\dfrac{2I}{\epsilon_0 c}}$ is :
(I = intensity of an electromagnetic wave, c : speed of light) **[NCERT : P L-2, 3 | April 7, 2025 (II)]**
(a) Vm (b) NC (c) Nm (d) NC^{-1}

2. Match **List-I** with **List-II**. **[NCERT : P L-2, 3 | April 2, 2025 (II)]**

List - I	List - II
(A) Heat capacity of body	(I) $J\,kg^{-1}$
(B) Specific heat capacity of body	(II) JK^{-1}
(C) Latent heat	(III) $J\,kg^{-1}\,K^{-1}$
(D) Thermal conductivity	(IV) $Jm^{-1}\,K^{-1}\,s^{-1}$

Choose the **correct** answer from the options given below:
(a) (A)–(III), (B)–(I), (C)–(II), (D)–(IV)
(b) (A)–(IV), (B)–(III), (C)–(II), (D)–(I)
(c) (A)–(III), (B)–(IV), (C)–(I), (D)–(II)
(d) (A)–(II), (B)–(III), (C)–(I), (D)–(IV)

3. Match List-I with List-II. **[NCERT : P L-2, 3 | Jan. 29, 2025 (II)]**

List-I	List-II
(A) Magnetic induction	(I) Ampere-metre^2
(B) Magnetic intensity	(II) Weber
(C) Magnetic flux	(III) Gauss
(D) Magnetic moment	(IV) Ampere/metre

Choose the correct answer from the options given below :
(a) (A)-(III), (B)-(IV), (C)-(I), (D)-(II)
(b) (A)-(III), (B)-(IV), (C)-(II), (D)-(I)
(c) (A)-(I), (B)-(II), (C)-(III), (D)-(IV)
(d) (A)-(III), (B)-(II), (C)-(I), (D)-(IV)

4. Electric field in a certain region is given by $\vec{E} = \left(\dfrac{A}{x^2}\hat{i} + \dfrac{B}{y^3}\hat{j}\right)$. The SI unit of A and B are: **[NCERT : P L-2, 3 | Jan. 30, 2023 (I)]**
(a) $Nm^3C^{-1}; Nm^2C^{-1}$ (b) $Nm^2C^{-1}; Nm^3C^{-1}$
(c) $Nm^3C; Nm^2C$ (d) $Nm^2C; Nm^3C$

5. Match **List-I** with **List-II**

[NCERT : P L-2, 3 | Jan. 25, 2023 (I)]

	List-I		List-II
A	Surface tension	I.	$Kg\,m^{-1}\,s^{-1}$
B	Pressure	II.	$Kg\,ms^{-1}$
C	Viscosity	III.	$Kg\,m^{-1}\,s^{-2}$
D	Impulse	IV.	$Kg\,s^{-2}$

Choose the correct answer from the options given below:

(a) A-IV, B-III, C- II, D-I (b) A-IV, B-III, C-I, D-II
(c) A-III, B-IV, C-I, D-II (d) A-II, B-I, C-III, D-IV

6. Match **List-I** with **List-II**.

[NCERT : P L-2, 3 | July 29, 2022 (II)]

List-I	List-II
A. Torque	I. Nms^{-1}
B. Stress	II. $J\,kg^{-1}$
C. Latent Heat	III. Nm
D. Power	IV. Nm^{-2}

Choose the correct answer from the options given below:

(a) A-III, B-II, C-I, D-IV
(b) A-III, B-IV, C-II, D-I
(c) A-IV, B-I, C-III, D-II
(d) A-II, B-III, C-I, D-IV

7. Velocity (v) and acceleration (a) in two systems of units 1 and 2 are related as $v_2 = \frac{n}{m^2}v_1$ and $a_2 = \frac{a_1}{mn}$ respectively. Here m and n are constants. The relations for distance and time in two systems respectively are:

[NCERT : P L-2, 3 | June 28, 2022 (II)]

(a) $\frac{n^3}{m^3}L_1 = L_2$ and $\frac{n^2}{m}T_1 = T_2$

(b) $L_1 = \frac{n^4}{m^2}L_2$ and $T_1 = \frac{n^2}{m}T_2$

(c) $L_1 = \frac{n^2}{m}L_2$ and $T_1 = \frac{n^4}{m^2} = T_2$

(d) $\frac{n^2}{m}L_1 = L_2$ and $\frac{n^4}{m^2}T_1 = T_2$

8. The unit of the coefficient of viscosity in S.I. system is

[NCERT : P L-2, 3 | June 27, 2022 (II)]

(a) m/kg-s (b) m-s/kg^2
(c) kg/m-s^2 (d) kg/m-s

9. If E and H represents the intensity of electric field and magnetising field respectively, then the unit of E/H will be:

[NCERT : P L-2, 3 | Aug. 27, 2021 (I)]

(a) ohm (b) mho
(c) joule (d) newton

10. The density of a material in SI unit is 128 kg m^{-3}. In certain units in which the unit of length is 25 cm and the unit of mass is 50 g, the numerical value of density of the material is: **[NCERT : P L-2, 3 | Jan. 10, 2019 (I)]**

(a) 40 (b) 16 (c) 640 (d) 410

11. A metal sample carrying a current along X-axis with density J_x is subjected to a magnetic field B_z (along z-axis). The electric field E_y developed along Y-axis is directly proportional to J_x as well as B_z. The constant of proportionality has SI unit

[NCERT : P L-2, 3 | Online April 25, 2013]

(a) $\frac{m^2}{A}$ (b) $\frac{m^3}{As}$

(c) $\frac{m^2}{As}$ (d) $\frac{As}{m^3}$

Topic 2 - Dimensions of Physical Quantities

12. The dimension of $\sqrt{\frac{\mu_0}{\epsilon_0}}$ is equal to that of :

(μ_0 = Vacuum permeability and ϵ_0 = Vacuum permittivity)

[NCERT : P L-8 | April 7, 2025 (II)]

(a) Voltage (b) Capacitance
(c) Inductance (d) Resistance

13. Match **List-I** with **List-II**.

List-I	List-II
(A) Mass density	(I) $[ML^2T^{-3}]$
(B) Impulse	(II) $[MLT^{-1}]$
(C) Power	(III) $[ML^2T^0]$
(D) Moment of inertia	(IV) $[ML^{-3}T^0]$

Choose the **correct** answer from the options given below:

[NCERT : P L-7 | April 7, 2025 (II)]

(a) (A)-(IV), (B)-(II), (C)-(III), (D)-(I)
(b) (A)-(I), (B)-(III), (C)-(IV), (D)-(II)
(c) (A)-(IV), (B)-(II), (C)-(I), (D)-(III)
(d) (A)-(II), (B)-(III), (C)-(IV), (D)-(I)

14. If ϵ_0 denotes the permittivity of free space and Φ_E is the flux of the electric field through the area bounded by the closed surface, then dimension of $\left(\epsilon_0 \frac{d\Phi_E}{dt}\right)$ are that of :

[NCERT : P L-8 | April 7, 2025 (I)]

(a) Electric field (b) Electric potential
(c) Electric charge (d) Electric current

15. In an electromagnetic system, a quantity defined as the ratio of electric dipole moment and magnetic dipole moment has dimension of $[M^PL^QT^RA^S]$. The value of P and Q are :

[NCERT : P L-7, 8 | April 4, 2025 (II)]

(a) –1, 0 (b) –1, 1 (c) 1, –1 (d) 0, –1

16. Match the **LIST-I** with **LIST-II**

[NCERT : P L-7 | April 3, 2025 (II)]

LIST-I	**LIST-II**
A. Boltzmann constant	I. ML^2T^{-1}
B. Coefficient of viscosity	II. $MLT^{-3}K^{-1}$
C. Planck's constant	III. $ML^2T^{-2}K^{-1}$
D. Thermal conductivity	IV. $ML^{-1}T^{-1}$

Choose the **correct** answer from the options given below:

(a) A-III, B-IV, C-I, D-II (b) A-II, B-III, C-IV, D-I
(c) A-III, B-II, C-I, D-IV (d) A-III, B-IV, C-II, D-I

17. Match the **LIST-I** with **LIST-II**

[NCERT : P L-7 | April 3, 2025 (I)]

LIST-I	**LIST-II**
A. Gravitational constant	I. $[LT^{-2}]$
B. Gravitational potential energy	II. $[L^2T^{-2}]$
C. Gravitational potential	III. $[ML^2T^{-2}]$
D. Acceleration due to gravity	IV. $[M^{-1}L^3T^{-2}]$

Choose the **correct** answer from the options given below:

(a) A-IV, B-III, C-II, D-I (b) A-III, B-II, C-I, D-IV
(c) A-II, B-IV, C-III, D-I (d) A-I, B-III, C-IV, D-II

18. Given a charge q, current I and permeability of vacuum μ_0. Which of the following quantity has the dimension of momentum? **[NCERT : P L-7, 8 | April 2, 2025 (II)]**

(a) qI/μ_0 (b) $q\mu_0 I$ (c) $q^2\mu_0 I$ (d) $q\mu_0/I$

19. If μ_0 and ε_0 are the permeability and permittivity of free space, respectively, then the dimension of $\left(\frac{1}{\mu_0\varepsilon_0}\right)$ is :

[NCERT : P L-7 | April 2, 2025 (II)]

(a) L/T^2 (b) L^2/T^2 (c) T^2/L (d) T^2/L^2

20. The equation for real gas is given by $\left(P+\frac{a}{V^2}\right)(V-b)=RT$, where P,V,T and R are the pressure, volume, temperature and gas constant, respectively. The dimension of ab^{-2} is equivalent to that of :

[NCERT : P L-8 | April 2, 2025 (I)]

(a) Plankc's constant
(b) Compressibility
(c) Strain
(d) Energy density

21. Match **List-I** with **List-II**.

[NCERT : P L-7 | April 2, 2025 (I)]

List - I	**List - II**
(A) Coefficient of viscosity	(I) $[ML^0T^{-3}]$
(B) Intensity of wave	(II) $[ML^{-2}T^{-2}]$
(C) Pressure gradient	(III) $[ML^{-1}LT^2]$
(D) Compressibility	(IV) $[ML^{-1}T^{-1}]$

Choose the **correct** answer from the options given below:

(a) (A)–(I), (B)–(IV), (C)–(III), (D)–(II)
(b) (A)–(IV), (B)–(I), (C)–(II), (D)–(III)
(c) (A)–(IV), (B)–(II), (C)–(I), (D)–(III)
(d) (A)–(II), (B)–(III), (C)–(IV), (D)–(I)

22. Match List-I with List-II.

[NCERT : P L-7 | Jan. 29, 2025 (II)]

List-I	**List-II**
(A) Young's Modulus	(I) $ML^{-1}T^{-1}$
(B) Torque	(II) $ML^{-1}T^{-2}$
(C) Coefficient of viscosity	(III) $M^{-1}L^3T^{-2}$
(D) Gravitational constant	(IV) ML^2T^{-2}

Choose the correct answer from the options given below :

(a) (A)-(I), (B)-(III), (C)-(II), (D)-(IV)
(b) (A)-(II), (B)-(I), (C)-(IV), (D)-(III)
(c) (A)-(IV), (B)-(II), (C)-(III), (D)-(I)
(d) (A)-(II), (B)-(IV), (C)-(I), (D)-(III)

23. The pair of physical quantities not having same dimensions is : **[NCERT : P L-7 | Jan. 29, 2025 (I)]**

(a) Torque and energy
(b) Surface tension and impulse
(c) Angular momentum and Planck's constant
(d) Pressure and Young's modulus

24. The expression given below shows the variation of velocity (v) with time (t), $v=At^2+\frac{Bt}{C+t}$. The dimension of ABC is: **[NCERT : P L-8 | Jan. 29, 2025 (I)]**

(a) $[M^0L^2T^{-3}]$ (b) $[M^0L^1T^{-3}]$
(c) $[M^0L^1T^{-2}]$ (d) $[M^0L^2T^{-2}]$

25. Match List-I with List-II

[NCERT : P L-7 | Jan. 28, 2025 (II)]

List-I	**List-II**
(A) Angular Impulse	(I) $[M^0L^2T^{-2}]$
(B) Latent Heat	(II) $[ML^2T^{-3}A^{-1}]$
(C) Electrical resistivity	(III) $[ML^2T^{-1}]$
(D) Electromotive force	(IV) $[ML^3T^{-3}A^{-2}]$

Choose the correct answer from the options given below :

(a) (A)-(III), (B)-(I), (C)-(IV), (D)-(II)
(b) (A)-(I), (B)-(III), (C)-(IV), (D)-(II)
(c) (A)-(III), (B)-(I), (C)-(II), (D)-(IV)
(d) (A)-(II), (B)-(I), (C)-(IV), (D)-(III)

26. In a measurement, it is asked to find modulus of elasticity per unit torque applied on the system. The measured quantity has dimension of $[M^aL^bT^c]$. If $b=3$, the value of c is ______.

[NCERT : P L-7 | Jan. 28, 2025 (I)]

27. Match the List-I with List-II

[NCERT : P L-7 | Jan. 23, 2025 (II)]

List-I	**List-II**
A. Permeability of free space	(I) $[ML^2T^{-2}]$
B. Magnetic field	(II) $[MT^{-2}A^{-1}]$
C. Magnetic moment	(III) $[MLT^{-2}A^{-2}]$
D. Torsional constant	(IV) $[L^2A]$

Choose the correct answer from the options given below:

(a) (A)-(I), (B)-(IV), (C)-(II), (D)-(III)
(b) (A)-(II), (B)-(I), (C)-(III), (D)-(IV)
(c) (A)-(IV), (B)-(III), (C)-(I), (D)-(II)
(d) (A)-(III), (B)-(II), (C)-(IV), (D)-(I)

28. The position of a particle moving on x-axis is given by $x(t) = A \sin t + B \cos^2 t + Ct^2 + D$, where t is time. The dimension of $\frac{ABC}{D}$ is

[NCERT : P L-8 | Jan. 23, 2025 (I)]

(a) L (b) L^3T^{-2} (c) L^2T^{-2} (d) L^2

29. The electric flux is $\phi = \alpha\sigma + \beta\lambda$ where λ and σ are linear and surface charge density, respectively, $\left(\frac{\alpha}{\beta}\right)$ represents.

[NCERT : P L-7, 8 | Jan. 23, 2025 (I)]

(a) charge (b) electric field
(c) displacement (d) area

30. If B is magnetic field and μ_o is permeability of free space, then the dimensions of (B/μ_o) is

[NCERT : P L-7 | Jan. 22, 2025 (I)]

(a) MT^2A^{-1} (b) $L^{-1}A$
(c) LT^2A^{-1} (d) $ML^2T^{-2}A^{-1}$

31. Which one of the following is the correct dimensional formula for the capacitance in F ? M, L, T and C stand for unit of mass, length, time and charge,

[NCERT : P L-7 | Jan. 22, 2025 (II)]

(a) $[F] = [C^2M^{-2}L^2T^2]$ (b) $[F] = [CM^{-2}L^{-2}T^{-2}]$
(c) $[F] = [CM^{-1}L^{-2}T^2]$ (d) $[F] = [C^2M^{-1}L^{-2}T^2]$

32. The dimensional formula of latent heat is:

[NCERT : P L-7 | April 9, 2024 (I)]

(a) $[M^0LT^{-2}]$ (b) $[MLT^{-2}]$
(c) $[M^0L^2T^{-2}]$ (d) $[ML^2T^{-2}]$

33. The de-Broglie wavelength associated with a particle of mass m and energy E is $\frac{H}{\sqrt{2mE}}$. The dimensional formula for Planck's constant is :

[NCERT : P L-7 | April 9, 2024 (II)]

(a) $[ML^{-1}T^{-2}]$ (b) $[ML^2T^{-1}]$
(c) $[MLT^{-2}]$ (d) $[M^2L^2T^{-2}]$

34. If $\in_0$ is the permittivity of free space and E is the electric field, then $\in_0 E^2$ has the dimensions :

[NCERT : P L-7 | April 8, 2024 (II)]

(a) $[M^0 L^{-2} T A]$ (b) $[M L^{-1} T^{-2}]$
(c) $[M^{-1} L^{-3} T^4 A^2]$ (d) $[M L^2 T^{-2}]$

35. Match **List-I** with **List-II** :

[NCERT : P L-7 | April 6, 2024 (I)]

List-I	List-II
A. Torque	I. $[M^1 L^1 T^{-2} A^{-2}]$
B. Magnetic field	II. $[L^2 A^1]$
C. Magnetic moment	III. $[M^1 T^{-2} A^{-1}]$
D. Permeability of free space	IV. $[M^1 L^2 T^{-2}]$

Choose the **correct** answer from the options given below:

(a) A-I, B-III, C-II, D-IV (b) A-IV, B-III, C-II, D-I
(c) A-III, B-I, C-II, D-IV (d) A-IV, B-II, C-III, D-I

36. Given below are two statements :

Statement (I) : Dimensions of specific heat is $[L^2T^{-2}K^{-1}]$

Statement (II) : Dimensions of gas constant is $[M L^2T^{-1}K^{-1}]$

[NCERT : P L-7 | April 6, 2024 (II)]

(a) Statement (I) is incorrect but statement (II) is correct
(b) Both statement (I) and statement (II) are incorrect
(c) Statement (I) is correct but statement (II) is incorrect
(d) Both statement (I) and statement (II) are correct

37. If G be the gravitational constant and u be the energy density then which of the following quantity have the dimension as that the $\sqrt{uG}$:

[NCERT : P L-8 | April 5, 2024 (I)]

(a) Pressure gradient per unit mass
(b) Force per unit mass
(c) Gravitational potential
(d) Energy per unit mass

38. What is the dimensional formula of ab^{-1} in the equation $\left(P + \frac{a}{V^2}\right)(V - b) = RT$, where letters have their usual meaning.

[NCERT : P L-8 | April 5, 2024 (II)]

(a) $[M^0L^3T^{-2}]$ (b) $[ML^2T^{-2}]$
(c) $[M^{-1}L^5T^3]$ (d) $[M^6L^7T^4]$

39. The equation of stationary wave is :

$$y = 2a \sin\left(\frac{2\pi nt}{\lambda}\right) \cos\left(\frac{2\pi x}{\lambda}\right)$$

Which of the following is NOT correct **[April 4, 2024 (I)]**

(a) The dimensions of nt is [L]
(b) The dimensions of n is $[LT^{-1}]$
(c) The dimensions of n/λ is [T]
(d) The dimensions of x is [L]

40. Applying the principle of homogeneity of dimensions, determine which one is correct.
Where T is time period, G is gravitational constant, M is mass, r is radius of orbit.

[NCERT : P L-8 | April 4, 2024 (II)]

(a) $T^2 = \frac{4\pi^2 r}{GM^2}$ (b) $T^2 = 4\pi^2 r^3$
(c) $T^2 = \frac{4\pi^2 r^3}{GM}$ (d) $T^2 = \frac{4\pi^2 r^2}{GM}$

41. The dimensional formula of angular impulse is;

[NCERT : P L-7 | Feb. 1, 2024 (I)]

(a) $[ML^{-2}T^{-1}]$ (b) $[ML^2T^{-2}]$
(c) $[MLT^{-1}]$ (d) $[ML^2T^{-1}]$

42. A force is represented by $F = ax^2 + bt^{\frac{1}{2}}$ where x = distance and t = time. The dimensions of b^2/a are:

[NCERT : P L-8 | Jan. 31, 2024 (I)]

(a) $[ML^2T^{-3}]$ (b) $[ML^3T^{-3}]$
(c) $[ML^{-1}T^{-1}]$ (d) $[MLT^{-2}]$

43. Consider two physical quantities A and B related to each other as $E = \frac{B - x^2}{At}$ where E, x and t have dimensions of energy, length and time respectively. The dimension of AB is **[NCERT : P L-8 | Jan. 31, 2024 (II)]**

(a) $L^{-2}M^1T^0$ (b) $L^2M^{-1}T^1$
(c) $L^{-2}M^{-1}T^1$ (d) $L^0M^{-1}T^1$

44. Match **List - I** with **List - II** **[NCERT P : L-7 | Jan. 30, 2024 (I)]**

List - I	List - II
(A) Coefficient of viscosity	(I) $[ML^2T^{-2}]$
(B) Surface tension	(II) $[ML^2T^{-1}]$
(C) Angular momentum	(III) $[ML^{-1}T^{-1}]$
(D) Rotational kinetic energy	(IV) $[ML^0T^{-2}]$

Choose the correct answer from the options given below:
(a) (A) - (II), (B) - (I), (C) - (IV), (D) - (III)
(b) (A) - (I), (B) - (II), (C) - (III), (D) - (IV)
(c) (A) - (III), (B) - (IV), (C) - (II), (D) - (I)
(d) (A) - (IV), (B) - (III), (C) - (II), (D) - (I)

45. If mass if written as $m = k c^P G^{-1/2} h^{1/2}$ then the value of P will be : (Constants have their usual meaning with k a dimensionless constant). **[NCERT : P L-9 | Jan. 30, 2024 (II)]**

(a) 1/2 (b) 1/3 (c) 2 (d) –1/3

46. Given below are two statements :
Statement (I) : Planck's constant and angular momentum have same dimensions.
Statement (II) : Linear momentum and moment of force have same dimensions.
In the light of the above statements, choose the correct answer from the options given below : **[NCERT : P L-7 | Jan. 27, 2024 (I)]**

(a) Statement I is true but Statement II is false
(b) Both Statement I and Statement II are false
(c) Both Statement I and Statement II are true
(d) Statement I is false but Statement II is true

47. The equation of state of a real gas is given by $\left(P + \frac{a}{V^2}\right)(V - b) = RT$, where P, V and T are pressure, volume and temperature respectively and R is the universal gas constant. The dimensions of $\frac{a}{b^2}$ is similar to that of : **[NCERT : P L-8 | Jan. 27, 2024 (II)]**

(a) PV (b) P (c) RT (d) R

48. The speed of a wave produced in water is given by $\upsilon = \lambda^a g^b \rho^c$. Where, λ, g and ρ are wavelength of wave, acceleration due to gravity and density of water respectively. The values of a, b and c respectively, are: **[NCERT : P L-9 | April 15, 2023 (I)]**

(a) $\frac{1}{2}, \frac{1}{2}, 0$ (b) 1, 1, 0 (c) 1, –1, 0 (d) $\frac{1}{2}, 0, \frac{1}{2}$

49. In the equation $\left[X + \frac{a}{Y^2}\right][Y - b] = RT$, X is pressure, Y is volume, R is universal gas constant and T is temperature. The physical quantity equivalent to the ratio $\frac{a}{b}$ is : **[NCERT : P L-8 | April 13, 2023 (II)]**

(a) Energy (b) Impulse
(c) Pressure gradient (d) Coefficient of viscosity

50. Match **List-I** with **List-II** **[NCERT : P L-7 | April 12, 2023 (I)]**

List-I	List-II
A. Spring constant	I. (T^{-1})
B. Angular speed	II. (MT^{-2})
C. Angular momentum	III. (ML^2)
D. Moment of Inertia	IV. (ML^2T^{-1})

Choose the correct answer from the options given below:
(a) A-II, B-I, C-IV, D-III
(b) A-IV, B-I, C-III, D-II
(c) A-II, B-III, C-I, D-IV
(d) A-I, B-III, C-II, D-IV

51. If force (F), velocity (V) and time (T) are considered as fundamental physical quantity, then dimensional formula of density will be: **[NCERT : P L-9 | April 11, 2023 (II)]**

(a) $FV^{-2}T^2$ (b) $FV^{-4}T^{-2}$
(c) FV^4T^{-6} (d) $F^2V^{-2}T^6$

52. Match List I with List II **[NCERT : P L-7 | April 8, 2023 (II)]**

List-I	List-II
A. Torque	I. $ML^{-2}T^{-2}$
B. Stress	II. ML^2T^{-2}
C. Pressure gradient	III. $ML^{-1}T^{-1}$
D. Coefficient of viscosity	IV. $ML^{-1}T^{-2}$

Choose the correct answer from the options given below:
(a) A-III, B-IV, C-I, D-II (b) A-IV, B-II, C-III, D-I
(c) A-II, B-IV, C-I, D-III (d) A-II, B-I, C-IV, D-III

53. $\left(P + \frac{a}{V^2}\right)(V - b) = RT$ represents the equation of state of some gases. Where P is the pressure, V is the volume, T is the temperature and a, b, R are the constants. The physical quantity, which has dimensional formula as that of $\frac{b^2}{a}$, will be: **[NCERT : P L-8 | Feb. 1, 2023 (I)]**

(a) Bulk modulus
(b) Modulus of rigidity
(c) Energy density
(d) Compressibility

54. If the velocity of light c, universal gravitational constant G and Planck's constant h are chosen as fundamental quantities. The dimensions of mass in the new system is: **[NCERT : P L-9 | Feb. 1, 2023 (II)]**

(a) $[h^{1/2} c^{1/2} G^{-1/2}]$ (b) $[h^{-1/2} c^{1/2} G^{1/2}]$
(c) $[h^1 c^1 G^{-1}]$ (d) $[h^{1/2} c^{-1/2} G^1]$

55. If R, X_L and X_C represent resistance, inductive reactance and capacitive reactance. Then which of the following is dimensionless: **[NCERT : P L-7 | Jan. 31, 2023 (I)]**

(a) $R X_L X_C$ (b) $\frac{R}{\sqrt{X_L X_C}}$

(c) $\frac{R}{X_L X_C}$ (d) $R\frac{X_L}{X_C}$

56. Match List-I with List-II. **[NCERT : P L-7 | Jan. 31, 2023 (II)]**

List-I		List-II	
A.	Angular momentum	I.	$[ML^2T^{-2}]$
B.	Torque	II.	$[ML^{-2}T^{-2}]$
C.	Stress	III.	$[ML^2T^{-1}]$
D.	Pressure gradient	IV.	$[ML^{-1}T^{-2}]$

Choose the correct answer from the options given below:

(a) A-I, B-IV, C-III, D-II
(b) A-III, B-I, C-IV, D-II
(c) A-II, B-III, C-IV, D-I
(d) A-IV, B-II, C-I, D-III

57. Match **List-I** with **List-II** : **[NCERT : P L-7 | Jan. 29, 2023 (I)]**

List-I (Physical Quantity)		List-II (Dimensional Formula)	
A.	Pressure gradient	I.	$[M^0L^2T^{-2}]$
B.	Energy density	II.	$[M^1L^{-1}T^{-2}]$
C.	Electric Field	III.	$[M^1L^{-2}T^{-2}]$
D.	Latent heat	IV.	$[M^1L^1T^{-3}A^{-1}]$

Choose the correct answer from the options given below:

(a) A-III, B-II, C-I, D-IV (b) A-II, B-III, C-IV, D-I
(c) A-III, B-II, C-IV, D-I (d) A-II, B-III, C-I, D-IV

58. The equation of a circle is given by $x^2 + y^2 = a^2$, where a is the radius. If the equation is modified to change the origin other than (0, 0), then find out the correct dimensions of A and B in a new equation : $(x - At)^2 + \left(y - \frac{t}{B}\right)^2 = a^2$

The dimensions of t is given as $[T^{-1}]$. **[NCERT : P L-8 | Jan. 29, 2023 (II)]**

(a) $A = [L^{-1}T], B = [LT^{-1}]$ (b) $A = [LT], B = [L^{-1}T^{-1}]$
(c) $A = [L^{-1}T^{-1}], B = [LT^{-1}]$ (d) $A = [L^{-1}T^{-1}], B = [LT]$

59. Match **List-I** with **List-II** **[NCERT : P L-7 | Jan. 25, 2023 (II)]**

List-I		List-II	
A.	Young's Modulus (Y)	I.	$[ML^{-1}T^{-1}]$
B.	Co-efficient of Viscosity (η)	II.	$[ML^2T^{-1}]$
C.	Planck's Constant (h)	III.	$[ML^{-1}T^{-2}]$
D.	Work Function (ϕ)	IV.	$[ML^2T^{-2}]$

Choose the correct answer from the options given below:

(a) A-II, B-III, C-IV, D-I
(b) A-III, B-I, C-II, D-IV
(c) A-I, B-III, C-IV, D-II
(d) A-I, B-II, C-III, D-IV

60. Match **List-I** with **List-II** **[NCERT : P L-7 | Jan. 24, 2023 (I)]**

List-I		List-II	
A.	Planck's Constant (h)	I.	$[M^1L^2T^{-2}]$
B.	Stopping Potential (V_s)	II.	$[M^1L^1T^{-1}]$
C.	Work function (ϕ)	III.	$[M^1L^2T^{-1}]$
D.	Momentum (p)	IV.	$[M^1L^2T^{-3}A^{-1}]$

(a) A-III, B-I, C-II, D-IV (b) A-III, B-IV, C-I, D-II
(c) A-II, B-IV, C-III, D-I (d) A-I, B-III, C-IV, D-II

61. The frequency (v) of an oscillating liquid drop may depend upon radius (r) of the drop, density (ρ) of liquid and the surface tension (s) of the liquid as : $v = r^a \rho^b s^c$. The values of a, b and c respectively are **[NCERT : P L-9 | Jan. 24, 2023 (II)]**

(a) $\left(-\frac{3}{2}, -\frac{1}{2}, \frac{1}{2}\right)$ (b) $\left(\frac{3}{2}, -\frac{1}{2}, \frac{1}{2}\right)$

(c) $\left(\frac{3}{2}, \frac{1}{2}, -\frac{1}{2}\right)$ (d) $\left(-\frac{3}{2}, \frac{1}{2}, \frac{1}{2}\right)$

62. Given below are two statements: One is labelled as Assertion (A) and other is labelled as Reason (R).

Assertion (A): Time period of oscillation of a liquid drop depends on surface tension (S), if density of the liquid is ρ and radius of the drop is r, then $T = K\sqrt{\frac{\rho r^3}{S^{3/2}}}$ is dimensionally correct, where K is dimensionless.

Reason (R): Using dimensional analysis we get R.H.S. having different dimension than that of time period.

In the light of above statements, choose the correct answer from the options given below. **[NCERT : P L-9 | July 29, 2022 (I)]**

(a) Both (A) and (R) are true and (R) is the correct explanation of (A)
(b) Both (A) and (R) are true but (R) is not the correct explanation of (A)
(c) (A) is true but (R) is false
(d) (A) is false but (R) is true

63. The dimensions of $\left(\frac{B^2}{\mu_0}\right)$ will be : (if μ_0 : permeability of free space and B : magnetic field) **[NCERT : P L-7 | July 28, 2022 (I)]**

(a) $[ML^2T^{-2}]$ (b) $[MLT^{-2}]$
(c) $[ML^{-1}T^{-2}]$ (d) $[ML^2T^{-2}A^{-1}]$

64. Consider the efficiency of carnot's engine is given by $\eta = \frac{\alpha\beta}{\sin\theta}\log e\frac{\beta x}{kT}$, where α and β are constants. If T is temperature, k is Boltzmann constant, θ is angular displacement and x has the dimensions of length. Then, choose the incorrect option : **[NCERT : P L-8 | July 28, 2022 (II)]**

(a) Dimensions of β is same as that of force.
(b) Dimensions of $\alpha^{-1}x$ is same as that of energy.
(c) Dimensions of $\eta^{-1}\sin\theta$ is same as that of αβ.
(d) Dimensions of α is same as that of β.

65. An expression of energy density is given by $u = \frac{\alpha}{\beta}\sin\left(\frac{\alpha x}{kt}\right)$. where α, β are constants, x is displacement, k is Boltzmann constant and t is the temperature. The dimensions of β will be : **[NCERT : P L-8 | July 27, 2022 (II)]**

(a) $[ML^2T^{-2}\theta^{-1}]$ (b) $[M^0L^2T^{-2}]$
(c) $[M^0L^0T^0]$ (d) $[M^0L^2T^0]$

66. If momentum [P], area [A] and time [T] are taken as fundamental quantities, then the dimensional formula for coefficient of viscosity is: **[NCERT : P L-9 | July 25, 2022 (I)]**

(a) $[PA^{-1}T^0]$ (b) $[PAT^{-1}]$
(c) $[PA^{-1}T]$ (d) $[PA^{-1}T^{-1}]$

67. Which of the following physical quantities have the same dimensions? **[NCERT : P L-7 | July 25, 2022 (I)]**

(a) Electric displacement $(\vec{D})$ and surface charge density
(b) Displacement current and electric field
(c) Current density and surface charge density
(d) Electric potential and energy

68. In Vander Waals equation $\left[P + \frac{a}{V^2}\right][V - b] = RT$; P is pressure, V is volume, R is universal gas constant and T is temperature. The ratio of constants is $\frac{a}{b}$ dimensionally equal to **[NCERT : P L-8 | June 29, 2022 (I)]**

(a) $\frac{P}{V}$ (b) $\frac{V}{P}$
(c) PV (d) PV^3

69. Given below are two statements : One is labelled as Assertion A and the other is labelled as Reason R.
Assertion A : Product of Pressure (P) and time (t) has the same dimension as that of coefficient of viscosity.
Reason R: Coefficient of viscosity $= \frac{\text{Force}}{\text{Velocity gradient}}$.
Choose the correct answer from the options given below : **[NCERT : P L-7 | June 28, 2022 (I)]**

(a) Both A and R true, and R is correct explanation of A.
(b) Both A and R are true but R is NOT the correct explanation of A.
(c) A is true but R is false.
(d) A is false but R is true.

70. If L, C and R are the self inductance, capacitance and resistance respectively, which of the following does not have the dimension of time ? **[NCERT : P L-7 | June 27, 2022 (II)]**

(a) RC (b) $\frac{L}{R}$ (c) $\sqrt{LC}$ (d) $\frac{L}{C}$

71. The SI unit of a physical quantity is pascal-second. The dimensional formula of this quantity will be **[NCERT : P L-7 | June 27, 2022 (II)]**

(a) $[ML^{-1}T^{-1}]$ (b) $[ML^{-1}T^{-2}]$
(c) $[ML^2T^{-1}]$ (d) $[M^{-1}L^3T^0]$

72. An expression for a dimensionless quantity P is given by $P = \frac{\alpha}{\beta}\log_e\left(\frac{kt}{\beta x}\right)$; where α and β are constants, x is distance; k is Boltzmann constant and t is the temperature. Then the dimensions of α will be : **[NCERT : P L-8 | June 26, 2022 (I)]**

(a) $[M^0L^{-1}T^0]$ (b) $[ML^0T^{-2}]$
(c) $[MLT^{-2}]$ (d) $[ML^2T^{-2}]$

73. The dimension of mutual inductance is : **[NCERT : P L-7 | June 26, 2022 (II)]**

(a) $[ML^2T^{-2}A^{-1}]$ (b) $[ML^2T^{-3}A^{-1}]$
(c) $[ML^2T^{-2}A^{-2}]$ (d) $[ML^2T^{-3}A^{-2}]$

74. Identify the pair of physical quantities which have different dimensions : **[NCERT : P L-7 | June 24, 2022 (I)]**

(a) Wave number and Rydberg's constant
(b) Stress and Coefficient of elasticity
(c) Coercivity and Magnetisation
(d) Specific heat capacity and Latent heat

75. Identify the pair of physical quantities that have same dimensions : **[NCERT : P L-7 | June 24, 2022 (II)]**

(a) velocity gradient and decay constant
(b) wien's constant and Stefan constant
(c) angular frequency and angular momentum
(d) wave number and Avogadro number

76. Match **List-I** with **List-II**. **[NCERT : P L-7 | Aug. 31, 2021 (I)]**

List-I	List-II
(A) Torque	(i) MLT^{-1}
(B) Impulse	(ii) MT^{-2}
(C) Tension	(iii) ML^2T^{-2}
(D) Surface Tension	(iv) MLT^{-2}

Choose the **most appropriate** answer from the option given below :

(a) (A)–(iii), (B)–(i), (C)–(iv), (D)–(ii)
(b) (A)–(ii), (B)–(i), (C)–(iv), (D)–(iii)
(c) (A)–(i), (B)–(iii), (C)–(iv), (D)–(ii)
(d) (A)–(iii), (B)–(iv), (C)–(i), (D)–(ii)

77. Which of the following equations is dimensionally incorrect ? Where t = time, h = height, s = surface tension, θ = angle, ρ = density, r = radius, g = acceleration due to gravity, V = volume, P = pressure, W = work done, ∈ = permittivity, E = electric field, J = current density, L = length, τ = torque **[NCERT : P L-8 | Aug. 31, 2021 (I)]**

(a) $v = \frac{\pi p a^4}{8\eta L}$ (b) $h = \frac{2s\cos\theta}{\rho r g}$
(c) $J = \in\frac{\partial E}{\partial t}$ (d) $W = \tau\theta$

78. If velocity [V], time [T] and force [F] are chosen as the base quantities, the dimensions of the mass will be : **[NCERT : P L-9 | Aug. 31, 2021 (II)]**

(a) $[FT^{-1}V^{-1}]$ (b) $[FTV^{-1}]$
(c) $[FT^2V]$ (d) $[FVT^{-1}]$

79. Which of the following is not a dimensionless quantity ? **[NCERT : P L-8 | Aug. 27, 2021 (I)]**
(a) Relative magnetic permeability (μ_r)
(b) Power factor
(c) Permeability of free space (μ_0)
(d) Quality factor

80. If force (F), length (L) and time (T) are taken as the fundamental quantities. Then what will be the dimension of density : **[NCERT : P L-9 | Aug. 27, 2021 (II)]**
(a) $[FL^{-4}T^2]$ (b) $[FL^{-3}T^2]$
(c) $[FL^{-5}T^2]$ (d) $[FL^{-3}T^3]$

81. If E, L, M and G denote the quantities as energy, angular momentum, mass and constant of gravitation respectively, then the dimensions of P in the formula $P = EL^2M^{-5}G^{-2}$ are :- **[NCERT : P L-7 | Aug. 26, 2021 (I)]**
(a) $[M^0L^1T^0]$ (b) $[M^{-1}L^{-1}T^2]$
(c) $[M^1L^1T^{-2}]$ (d) $[M^0L^0T^0]$

82. Match **List–I** with **List–II**.

List-I	List-II
(A) Magnetic Induction	(i) $ML^2T^{-2}A^{-1}$
(B) Magnetic Flux	(ii) $M^0L^{-1}A$
(C) Magnetic Permeability	(iii) $MT^{-2}A^{-1}$
(D) Magnetization	(iv) $MLT^{-2}A^{-2}$

Choose the most appropriate answer from the options given below : **[NCERT : P L-7 | Aug. 26, 2021 (II)]**
(a) (A)-(ii), (B)-(iv), (C)-(i), (D)-(iii)
(b) (A)-(ii), (B)-(i), (C)-(iv), (D)-(iii)
(c) (A)-(iii), (B)-(ii), (C)-(iv), (D)-(i)
(d) (A)-(iii), (B)-(i), (C)-(iv), (D)-(ii)

83. Match **List-I** with **List-II**.

List-I	List-II
(A) Capacitance, C	(i) $M^1L^1T^{-3}A^{-1}$
(B) Permittivity of free space, ε_0	(ii) $M^{-1}L^{-3}T^4A^2$
(C) Permeability of free space, μ_0	(iii) $M^{-1}L^{-2}T^4A^2$
(D) Electric field, E	(iv) $M^1L^1T^{-2}A^{-2}$

Choose the correct answer from the options given below **[NCERT : P L-7 | July 27, 2021 (II)]**
(a) (A) - (iii), (B) - (ii), (C) - (iv), (D) - (i)
(b) (A) - (iii), (B) - (iv), (C) - (ii), (D) - (i)
(c) (A) - (iv), (B) - (ii), (C) - (iii), (D) - (i)
(d) (A) - (iv), (B) - (iii), (C) - (ii), (D) - (i)

84. The force is given in terms of time t and displacement x by the equation
$F = A\cos Bx + C\sin Dt$
The dimensional formula of $\frac{AD}{B}$ is : **[NCERT : P L-8 | July 25, 2021 (II)]**
(a) $[M^0LT^{-1}]$ (b) $[ML^2T^{-3}]$
(c) $[M^1L^1T^{-2}]$ (d) $[M^2L^2T^{-3}]$

85. Match **List-I** with **List-II**:

List-I	List-II
(A) h (Planck's constant)	(I) $[MLT^{-1}]$
(B) E (kinetic energy)	(II) $[ML^2T^{-1}]$
(C) V (electric potential)	(III) $[ML^2T^{-2}]$
(D) P (linear momentum)	(IV) $[ML^2A^{-1}T^{-3}]$

[NCERT : P L-7 | Feb. 25, 2021 (I)]
Choose the correct answer from the options given below:
(a) (A)→(I), (B)→(II), (C)→(IV), (D)→(III)
(b) (A)→(II), (B)→(III), (C)→(IV), (D)→(I)
(c) (A)→(III), (B)→(II), (C)→(IV), (D)→(I)
(d) (A)→(III), (B)→(IV), (C)→(II), (D)→(I)

86. If e is the electronic charge, c is the speed of light in free space and h is Planck's constant, the quantity $\frac{1}{4\pi\varepsilon_0}\frac{|e|^2}{hc}$ has dimensions of: **[NCERT : P L-7 | Feb. 25, 2021 (II)]**
(a) $[MLT^{-1}]$ (b) $[MLT^0]$
(c) $[LC^{-1}]$ (d) $[M^0L^0T^0]$

87. In a typical combustion engine the work done by a gas molecule is given by $W = \alpha^2\beta e^{-\frac{\beta x^2}{kT}}$, where x is the displacement, k is the Boltzmann constant and T is the temperature, If α and β are constants, dimensions of α will be **[NCERT : P L-8 | Feb. 26, 2021 (I)]**
(a) $[MLT^{-1}]$ (b) $[M^0LT^0]$
(c) $[MLT^{-2}]$ (d) $[M^2LT^{-2}]$

88. The quantities $x = \frac{1}{\sqrt{\mu_0\varepsilon_0}}$, $y = \frac{E}{B}$ and $z = \frac{l}{CR}$ are defined where *C*-capacitance, *R*-Resistance, *l*-length, *E*-Electric field, *B*-magnetic field and ε_0, μ_0, - free space permittivity and permeability respectively. Then : **[NCERT : P L-7 | Sep. 05, 2020 (II)]**
(a) *x*, *y* and *z* have the same dimension.
(b) Only *x* and *z* have the same dimension.
(c) Only *x* and *y* have the same dimension.
(d) Only *y* and *z* have the same dimension.

89. Dimensional formula for thermal conductivity is (here *K* denotes the temperature) : **[NCERT : P L-7 | Sep. 04, 2020 (I)]**
(a) $MLT^{-2}K$ (b) $MLT^{-2}K^{-2}$
(c) $MLT^{-3}K$ (d) $MLT^{-3}K^{-1}$

90. A quantity *x* is given by (IFv^2/WL^4) in terms of moment of inertia *I*, force *F*, velocity *v*, work *W* and Length *L*. The dimensional formula for *x* is same as that of: **[NCERT : P L-7 | Sep. 04, 2020 (II)]**
(a) Planck's constant (b) force constant
(c) coefficient of viscosity (d) energy density

91. Amount of solar energy received on the earth's surface per unit area per unit time is defined as solar constant. Dimension of solar constant is : **[NCERT : P L-7 | Sep. 03, 2020 (II)]**

(a) ML^2T^{-2} (b) ML^0T^{-3}
(c) $M^2L^0T^{-1}$ (d) MLT^{-2}

92. If speed V, area A and force F are chosen as fundamental units, then the dimension of Young's modulus will be :
[NCERT : P L-9 | Sep. 02, 2020 (I)]
(a) FA^2V^{-1} (b) FA^2V^{-3}
(c) FA^2V^{-2} (d) $FA^{-1}V^0$

93. If momentum (P), area (A) and time (T) are taken to be the fundamental quantities then the dimensional formula for energy is :
[NCERT : P L-9 | Sep. 02, 2020 (II); Jan. 11, 2019(II)]
(a) $[P^2AT^{-2}]$ (b) $[PA^{-1}T^{-2}]$
(c) $[PA^{1/2}T^{-1}]$ (d) $[P^{1/2}AT^{-1}]$

94. Which of the following combinations has the dimension of electrical resistance (ϵ_0 is the permittivity of vacuum and μ_0 is the permeability of vacuum)?
[NCERT : P L-7 | Online 11 April 2014, 12 April 2019 (I)]
(a) $\sqrt{\frac{\mu_0}{\epsilon_0}}$ (b) $\frac{\mu_0}{\epsilon_0}$ (c) $\sqrt{\frac{\epsilon_0}{\mu_0}}$ (d) $\frac{\epsilon_0}{\mu_0}$

95. In the formula $X = 5YZ^2$, X and Z have dimensions of capacitance and magnetic field, respectively. What are the dimensions of Y in SI units ?
[NCERT : P L-7 | 10 April 2019 (II)]
(a) $[M^{-3} L^{-2} T^8 A^4]$ (b) $[M^{-1} L^{-2} T^4 A^2]$
(c) $[M^{-2} L^0 T^{-4} A^{-2}]$ (d) $[M^{-2} L^{-2} T^6 A^3]$

96. Let l, r, c and v represent inductance, resistance, capacitance and voltage, respectively. The dimension of $\frac{\ell}{rcv}$ in SI units will be : **[NCERT : P L-7 | 12 Jan. 2019 II]**
(a) $[LA^{-2}]$ (b) $[A^{-1}]$ (c) $[LTA]$ (d) $[LT^2]$

97. A quantity f is given by $f = \sqrt{\frac{hc^5}{G}}$ where c is speed of light, G universal gravitational constant and h is the Planck's constant. Dimension of f is same as that of :
[NCERT : P L-7 | 9 Jan. 2019 I]
(a) area (b) energy
(c) momentum (d) volume

98. The dimensions of stopping potential V_0 in photoelectric effect in units of Planck's constant 'h', speed of light 'c' and Gravitational constant 'G' and ampere A is:
[NCERT : P L-9 | 8 Jan. 2019 I]
(a) $h^{1/3}G^{2/3}c^{1/3} A^{-1}$ (b) $h^{2/3}c^{5/3}G^{1/3} A^{-1}$
(c) $h^{-2/3} e^{-1/3} G^{4/3} A^{-1}$ (d) $h^2G^{3/2}C^{1/3} A^{-1}$

99. The dimensions of $\frac{B^2}{2\mu_0}$, where B is magnetic field and μ_0 is the magnetic permeability of vacuum, is:
[NCERT : P L-7 | 8 Jan. 2019 II]
(a) MLT^{-2} (b) ML^2T^{-1}
(c) ML^2T^{-2} (d) $ML^{-1}T^{-2}$

100. The characteristic distance at which quantum gravitational effects are significant, the Planck length, can be determined from a suitable combination of the fundamental physical constants G, h and c. Which of the following correctly gives the Planck length? **[Online April 15, 2018]**
(a) G^2hc (b) $\left(\frac{Gh}{c^3}\right)^{\frac{1}{2}}$ (c) $G^{\frac{1}{2}}h^2c$ (d) Gh^2c^3

101. Time (T), velocity (C) and angular momentum (h) are chosen as fundamental quantities instead of mass, length and time. In terms of these, the dimensions of mass would be : **[NCERT : P L-9 | Online April 8, 2017]**
(a) $[M]=[T^{-1} C^{-2} h]$ (b) $[M]=[T^{-1} C^2 h]$
(c) $[M]=[T^{-1} C^{-2} h^{-1}]$ (d) $[M]=[T C^{-2} h]$

102. A, B, C and D are four different physical quantities having different dimensions. None of them is dimensionless. But we know that the equation AD = C ln (BD) holds true. Then which of the combination is not a meaningful quantity?
[NCERT : P L-8 | Online April 10, 2016]
(a) $\frac{C}{BD} - \frac{AD^2}{C}$ (b) $A^2 - B^2C^2$
(c) $\frac{A}{B} - C$ (d) $\frac{(A - C)}{D}$

103. In the following 'I' refers to current and other symbols have their usual meaning, Choose the option that corresponds to the dimensions of electrical conductivity :
[NCERT : P L-7 | Online April 9, 2016]
(a) $M^{-1}L^{-3}T^3I$ (b) $M^{-1}L^{-3}T^3I^2$
(c) $M^{-1}L^3T^3I$ (d) $ML^{-3}T^{-3}I^2$

104. If electronic charge e, electron mass m, speed of light in vacuum c and Planck's constant h are taken as fundamental quantities, the permeability of vacuum μ_0 can be expressed in units of : **[NCERT : P L-9 | Online April 11, 2015]**
(a) $\left(\frac{h}{me^2}\right)$ (b) $\left(\frac{hc}{me^2}\right)$
(c) $\left(\frac{h}{ce^2}\right)$ (d) $\left(\frac{mc^2}{he^2}\right)$

105. If the capacitance of a nanocapacitor is measured in terms of a unit 'u' made by combining the electric charge 'e', Bohr radius 'a_0', Planck's constant 'h' and speed of light 'c' then: **[Online April 10, 2015]**
(a) $u = \frac{e^2h}{a_0}$ (b) $u = \frac{hc}{e^2a_0}$
(c) $u = \frac{e^2c}{ha_0}$ (d) $u = \frac{e^2a_0}{hc}$

106. From the following combinations of physical constants (expressed through their usual symbols) the only combination, that would have the same value in different systems of units, is:
[NCERT : P L-7 | Online April 12, 2014]

(a) $\frac{ch}{2\pi\varepsilon_0^2}$ (b) $\frac{e^2}{2\pi\varepsilon_0 G m_e^2}$ (m_e = mass of electron)

(c) $\frac{\mu_0\varepsilon_0}{c^2}\frac{G}{he^2}$ (d) $\frac{2\pi\sqrt{\mu_0\varepsilon_0}}{ce^2}\frac{h}{G}$

107. Let $[\epsilon_0]$ denote the dimensional formula of the permittivity of vacuum. If M = mass, L = length, T = time and A = electric current, then: **[NCERT : P L-7 | 2013]**

(a) $\epsilon_0 = [M^{-1} L^{-3} T^2 A]$ (b) $\epsilon_0 = [M^{-1} L^{-3} T^4 A^2]$

(c) $\epsilon_0 = [M^1 L^2 T^1 A^2]$ (d) $\epsilon_0 = [M^1 L^2 T^1 A]$

108. If the time period t of the oscillation of a drop of liquid of density d, radius r, vibrating under surface tension s is given by the formula $t = \sqrt{r^{2b} s^c d^{a/2}}$. It is observed that the time period is directly proportional to $\sqrt{\frac{d}{s}}$. The value of b should therefore be : **[NCERT : P L-9 | Online April 23, 2013]**

(a) $\frac{3}{4}$ (b) $\sqrt{3}$ (c) $\frac{3}{2}$ (d) $\frac{2}{3}$

109. The dimensions of angular momentum, latent heat and capacitance are, respectively. **[NCERT : P L-7 | Online April 22, 2013]**

(a) $ML^2T^1A^2, L^2T^{-2}, M^{-1}L^{-2}T^2$

(b) $ML^2T^{-2}, L^2T^2, M^{-1}L^{-2}T^4A^2$

(c) $ML^2T^{-1}, L^2T^{-2}, ML^2TA^2$

(d) $ML^2T^{-1}, L^2T^{-2}, M^{-1}L^{-2}T^4A^2$

110. Given that K = energy, V = velocity, T = time. If they are chosen as the fundamental units, then what is dimensional formula for surface tension? **[Online May 7, 2012]**

(a) $[KV^{-2}T^{-2}]$ (b) $[K^2V^2T^{-2}]$

(c) $[K^2V^{-2}T^{-2}]$ (d) $[KV^2T^2]$

111. The dimensions of magnetic field in M, L, T and C (coulomb) is given as **[NCERT : P L-7 | 2008]**

(a) $[MLT^{-1} C^{-1}]$ (b) $[MT^2 C^{-2}]$

(c) $[MT^{-1} C^{-1}]$ (d) $[MT^{-2} C^{-1}]$

112. Which of the following units denotes the dimension $\frac{ML^2}{Q^2}$, where Q denotes the electric charge? **[NCERT : P L-7 | 2006]**

(a) Wb/m^2 (b) Henry (H)

(c) H/m^2 (d) Weber (Wb)

113. Out of the following pair, which one does NOT have identical dimensions ? **[NCERT : P L-7 | 2005]**

(a) Impulse and momentum

(b) Angular momentum and planck's constant

(c) Work and torque

(d) Moment of inertia and moment of a force

114. Which one of the following represents the correct dimensions of the coefficient of viscosity? **[NCERT : P L-7 | 2004]**

(a) $[ML^{-1}T^{-1}]$ (b) $[MLT^{-1}]$

(c) $[ML^{-1}T^{-2}]$ (d) $[ML^{-2}T^{-2}]$

115. Dimensions of $\frac{1}{\mu_0\varepsilon_0}$, where symbols have their usual meaning, are **[NCERT : P L-7 | 2003; April 8, 2023 (I)]**

(a) $[L^{-1}T]$ (b) $[L^{-2}T^2]$ (c) $[L^2T^{-2}]$ (d) $[LT^{-1}]$

116. The physical quantities not having same dimensions are **[NCERT : P L-7 | 2003]**

(a) torque and work

(b) momentum and planck's constant

(c) stress and young's modulus

(d) speed and $(\mu_0\varepsilon_0)^{-1/2}$

117. Identify the pair whose dimensions are equal **[NCERT : P L-7 | 2002]**

(a) torque and work (b) stress and energy

(c) force and stress (d) force and work

Topic 3 - Errors in Measurements

118. A quantity Q is formulated as $X^{-2} Y^{+\frac{3}{2}} Z^{-\frac{2}{5}}$, X, Y and Z are independent parameters which have fractional errors of 0.1, 0.2 and 0.5, respectively in measurement. The maximum fractional error of Q is **[April 8, 2025 (II)]**

(a) 0.1 (b) 0.8 (c) 0.7 (d) 0.6

119. A physical quantity C is related to four other quantities p, q, r and s as follows

$$C = \frac{pq^2}{r^3\sqrt{s}}$$

The percentage errors in the measurement of p, q, r and s are 1%, 2% 3% and 2% respectively. The percentage error in the measurement of C will be ________%. **[April 3, 2025 (II)]**

120. A person measures mass of 3 different particles as 435.42 g, 226.3 g and 0.125 g. According to the rules for arithmetic operations with significant figures, the additions of the masses of 3 particles will be. **[NCERT : P L-4, 5 | April 3, 2025 (I)]**

(a) 661.845 g (b) 662 g

(c) 661.8 g (d) 661.84 g

121. A physical quantity Q is related to four observables a, b, c, d as follows : $Q = \frac{ab^4}{cd}$ where, $a = (60 \pm 3)$Pa; $b = (20 \pm 0.1)$m; $c = (40 \pm 0.2)\ Nsm^{-2}$ and $d = (50 \pm 0.1)$m, then the percentage error in Q is $\frac{x}{1000}$, where x = ________. **[Jan. 29, 2025 (II)]**

122. For an experimental expression $y = \frac{32.3 \times 1125}{27.4}$, where all the digits are significant. Then to report the value of y we should write **[NCERT : P L-4, 5 | Jan. 24, 2025 (I)]**

(a) y = 1326.2 (b) y = 1326.19
(c) y = 1326.186 (d) y = 1330

123. The energy of a system is given as $E(t) = \alpha^3 e^{-\beta t}$, where t is the time and $\beta = 0.3\ s^{-1}$. The errors in the measurement of α and t are 1.2% and 1.6%, respectively. At t = 5s, maximum percentage error in the energy is **[Jan. 23, 2025 (II)]**

(a) 4% (b) 11.6%
(c) 6% (d) 8.4%

124. The maximum percentage error in the measurement of density of a wire is
[Given, mass of wire = (0.60 ± 0.003)g
radius of wire = (0.50 ± 0.01)cm
length of wire = (10.00 ± 0.05)cm] **[Jan. 22, 2025 (II)]**

(a) 4% (b) 5%
(c) 8% (d) 7%

125. In an expression $a \times 10^b$: **[April 8, 2024 (I)]**

(a) a is order of magnitude for $b \le 5$
(b) b is order of magnitude for $a \le 5$
(c) b is order of magnitude for $5 < a \le 10$
(d) b is order of magnitude for $a \ge 5$

126. To find the spring constant (k) of a spring experimentally, a student commits 2% positive error in the measurement of time and 1% negative error in measurement of mass. The percentage error in determining value of k is : **[April 6, 2024 (I)]**

(a) 3% (b) 1%
(c) 4% (d) 5%

127. The radius (r), length (*l*) and resistance (R) of a metal wire was measured in the laboratory as
r = (0.35 ± 0.05) cm
R = (100 ± 10) ohm
l = (15 ± 0.2)cm
The percentage error in resistivity of the material of the wire is: **[Feb. 1, 2024 (I)]**

(a) 25.6% (b) 39.93%
(c) 37.3% (d) 35.6%

128. Match List - I with List - II **[NCERT : P L-4 | Feb. 1, 2024 (II)]**

List - I (Number)	List - II (Significant figure)
(A) 1001	(I) 3
(B) 010.1	(II) 4
(C) 100.100	(III) 5
(D) 0.0010010	(IV) 6

Choose the correct answer from the options given below:

(a) (A)- (II), (B) - (I), (C)- (IV), (D)- (III)
(b) (A) - (IV), (B) - (III), (C) - (I), (D) - (II)
(c) (A) - (I), (B) - (II), (C) - (III), (D) - (IV)
(d) (A) - (III), (B) - (IV), (C) - (II), (D) - (I)

129. If the percentage errors in measuring the length and the diameter of a wire are 0.1% each. The percentage error in measuring its resistance will be : **[Jan. 31, 2024 (I)]**

(a) 0.2% (b) 0.144%
(c) 0.3% (d) 0.1%

130. The measured value of the length of a simple pendulum is 20 cm with 2 mm accuracy. The time for 50 oscillations was measured to be 40 seconds with 1 second resolution. From these measurements, the accuracy in the measurement of acceleration due to gravity is N%. The value of N is: **[Jan. 31, 2024 (II)]**

(a) 4 (b) 8 (c) 6 (d) 5

131. The resistance $R = \frac{V}{I}$ where V = (200 ± 5) V and I = (20 ± 0.2) A, the percentage error in the measurement of R is: **[Jan. 29, 2024 (I)]**

(a) 3.5% (b) 5.5%
(c) 7% (d) 3%

132. A physical quantity Q is found to depend on quantities a, b, c by the relation $Q = \frac{a^4 b^3}{c^2}$. The percentage error in a, b and c are 3%, 4% and 5% respectively. Then, the percentage error in Q is : **[Jan. 29, 2024 (II)]**

(a) 66% (b) 14%
(c) 34% (d) 43%

133. A body of mass (5 ± 0.5) kg is moving with a velocity of (20 ± 0.4) m/s. Its kinetic energy will be **[April 13, 2023 (I)]**

(a) (1000 ± 140) J (b) (1000 ± 0.14) J
(c) (500 ± 0.14) J (d) (500 ± 140) J

134. A cylindrical wire of mass (0.4 ± 0.01)g has length (8 ± 0.04) cm and radius (6 ± 0.03) mm. **[April 8, 2023 (I)]**
The maximum error in its density will be

(a) 1% (b) 3.5%
(c) 4% (d) 5%

135. Two resistances are given as $R_1 = (10 \pm 0.5)\Omega$ and $R_2 = (15 \pm 0.5)\Omega$. The percentage error in the measurement of equivalent resistance when they are connected in parallel is **[April 6, 2023 (I)]**

(a) 6.33 (b) 2.33
(c) 4.33 (d) 5.33

136. In an experiment to find acceleration due to gravity (g) using simple pendulum, time period of 0.5 s is measured from time of 100 oscillation with a watch of 1s resolution. If measured value of length is 10 cm known to 1 mm accuracy. The accuracy in the determination of g is found to be x%. The value of x is **[July 28, 2022 (II)]**

137. A torque meter is calibrated to reference standards of mass, length and time each with 5% accuracy. After calibration, the measured torque with this torque meter will have net accuracy of : **[July 27, 2022 (I)]**

(a) 15% (b) 25%
(c) 75% (d) 5%

138. The maximum error in the measurement of resistance, current and time for which current flows in an electrical circuit are 1%, 2% and 3% respectively. The maximum percentage error in the detection of the dissipated heat will be: **[July 25, 2022 (II)]**

(a) 2 (b) 4 (c) 6 (d) 8

139. A silver wire has mass (0.6 ± 0.006) g, radius (0.5 ± 0.005) mm and length (4 ± 0.04) cm. The maximum percentage error in the measurement of its density will be : **[June 27, 2022 (I)]**

(a) 4% (b) 3% (c) 6% (d) 7%

140. For $z = a^2 x^3 y^{1/2}$, where 'a' is a constant. If percentage error in measurement of 'x' and 'y' are 4% and 12%, respectively, then the percentage error for 'z' will be %. **[June 25, 2022 (II)]**

141. A student determined Young's Modulus of elasticity using the formula $Y = \frac{MgL^3}{4bd^3\delta}$. The value of g is taken to be 9.8 m/s^2, without any significant error, his observation are as following.

Physical Quantity	Least count of the Equipment used for measurement	Observed value
Mass (M)	1 g	2 kg
Length of bar (L)	1 mm	1 m
Breadth of bar (b)	0.1 mm	4 cm
Thickness of bar (d)	0.01 mm	0.4 cm
Depression (δ)	0.01 m	5 mm

Then the fractional error in the measurement of Y is : **[Sep. 1, 2021 (II)]**

(a) 0.0083 (b) 0.0155
(c) 0.155 (d) 0.083

142. Two resistors $R_1 = (4 \pm 0.8)\ \Omega$ and $R_2 = (4 \pm 0.4)\ \Omega$ are connected in parallel. The equivalent resistance of their parallel combination will be : **[Sep. 1, 2021 (II)]**

(a) $(4 \pm 0.4)\Omega$ (b) $(2 \pm 0.4)\Omega$
(c) $(2 \pm 0.3)\Omega$ (d) $(4 \pm 0.3)\Omega$

143. If the length of the pendulum in pendulum clock increases by 0.1%, then the error in time per day is: **[Aug. 26, 2021 (II)]**

(a) 86.4 s (b) 4.32 s
(c) 43.2 s (d) 8.64 s

144. Three students S_1, S_2 and S_3 perform an experiment for determining the acceleration due to gravity (g) using a simple pendulum. They use different lengths of pendulum and record time for different number of oscillations. The observations are as shown in the table.

Student No.	Length of pendulum (cm)	No. of oscillations (n)	Total time for n oscillations	Time period (s)
1.	64.0	8	128.0	16.0
2.	64.0	4	64.0	16.0
3.	20.0	4	36.0	9.0

(Least count of length = 0.1 m
least count for time = 0.1 s)

If E_1, E_2 and E_3 are the percentage errors in 'g' for students 1, 2 and 3 respectively, then the minimum percentage error is obtained by student no. ____. **[NA, July 22, 2021 (II)]**

145. The radius of a sphere is measured to be (7.50 ± 0.85) cm. Suppose the percentage error in its volume is x. The value of x, to the nearest x, is ________. **[March 18, 2021 (II)]**

146. The resistance $R = \frac{V}{I}$, where $V = (50 \pm 2)V$ and $I = (20 \pm 0.2)A$. The percentage error in R is 'x'%. The value of 'x' to the nearest integer is ________. **[March 16, 2021 (I)]**

147. The period of oscillation of a simple pendulum is $T = 2\pi\sqrt{\frac{L}{g}}$. Measured value of '$L$' is 1.0 m from meter scale having a minimum division of 1 mm and time of one complete oscillation is 1.95 s measured from stopwatch of 0.01 s resolution. The percentage error in the determination of 'g' will be : **[Feb. 24, 2021 (II)]**

(a) 1.13% (b) 1.03%
(c) 1.33% (d) 1.30%

148. The density of a solid metal sphere is determined by measuring its mass and its diameter. The maximum error in the density of the sphere is $\left(\frac{x}{100}\right)$%. If the relative errors in measuring the mass and the diameter are 6.0% and 1.5% respectively, the value of x is ______. **[NA Sep. 06, 2020 (I)]**

149. A physical quantity z depends on four observables a, b, c and d, as $z = \frac{a^2 b^{\frac{2}{3}}}{\sqrt{c} d^3}$. The percentages of error in the measurement of a, b, c and d are 2%, 1.5%, 4% and 2.5% respectively. The percentage of error in z is : **[Sep. 05, 2020 (I)]**

(a) 12.25% (b) 16.5%
(c) 13.5% (d) 14.5%

150. For the four sets of three measured physical quantities as given below. Which of the following options is correct? **[NCERT : P L-5 | 9 Jan. 2020 (II)]**

(A) $A_1 = 24.36, B_1 = 0.0724, C_1 = 256.2$
(B) $A_2 = 24.44, B_2 = 16.082, C_2 = 240.2$
(C) $A_3 = 25.2, B_3 = 19.2812, C_3 = 236.183$
(D) $A_4 = 25, B_4 = 236.191, C_4 = 19.5$

(a) $A_4+B_4+C_4 < A_1+B_1+C_1 < A_3+B_3+C_3 < A_2+B_2+C_2$
(b) $A_1+B_1+C_1 = A_2+B_2+C_2 = A_3+B_3+C_3 = A_4+B_4+C_4$
(c) $A_4+B_4+C_4 < A_1+B_1+C_1 = A_2+B_2+C_2 = A_3+B_3+C_3$
(d) $A_1+B_1+C_1 < A_3+B_3+C_3 < A_2+B_2+C_2 < A_4+B_4+C_4$

151. A simple pendulum is being used to determine the value of gravitational acceleration g at a certain place. The length of the pendulum is 25.0 cm and a stop watch with 1 s resolution measures the time taken for 40 oscillations to be 50 s. The accuracy in g is: **[8 Jan. 2020 (II)]**

(a) 5.40% (b) 3.40%
(c) 4.40% (d) 2.40%

152. In the density measurement of a cube, the mass and edge length are measured as (10.00 ± 0.10) kg and (0.10 ± 0.01) m, respectively. The error in the measurement of density is: **[9 April 2019 I]**

(a) 0.01 kg/m^3 (b) 0.10 kg/m^3
(c) 0.013 kg/m^3 (d) 0.07 kg/m^3

153. The area of a square is 5.29 cm^2. The area of 7 such squares taking into account the significant figures is: **[NCERT : P L-5 | 9 April 2019 II]**

(a) 37cm^2 (b) 37.030 cm^2
(c) 37.03 cm^2 (d) 37.0 cm^2

154. The density of a material in the shape of a cube is determined by measuring three sides of the cube and its mass. If the relative errors in measuring the mass and length are respectively 1.5% and 1%, the maximum error in determining the density is: **[2018]**

(a) 2.5% (b) 3.5%
(c) 4.5% (d) 6%

155. The relative uncertainty in the period of a satellite orbiting around the earth is 10^{-2}. If the relative uncertainty in the radius of the orbit is negligible, the relative uncertainty in the mass of the earth is **[Online April 16, 2018]**

(a) 3×10^{-2} (b) 10^{-2}
(c) 2×10^{-2} (d) 6×10^{-2}

156. The relative error in the determination of the surface area of a sphere is α. Then the relative error in the determination of its volume is **[Online April 15, 2018]**

(a) $\frac{2}{3}\alpha$ (b) $\frac{4}{3}\alpha$ (c) $\frac{3}{2}\alpha$ (d) α

157. A student measures the time period of 100 oscillations of a simple pendulum four times. The data set is 90 s, 91 s, 95 s, and 92 s. If the minimum division in the measuring clock is 1 s, then the reported mean time should be: **[2016]**

(a) 92 ± 1.8 s (b) 92 ± 3s
(c) 92 ± 2 s (d) 92 ± 5.0 s

158. The period of oscillation of a simple pendulum is $T = 2\pi\sqrt{\frac{L}{g}}$. Measured value of L is 20.0 cm known to 1 mm accuracy and time for 100 oscillations of the pendulum is found to be 90 s using a wrist watch of 1s resolution. The accuracy in the determination of g is : **[2015]**

(a) 1% (b) 5% (c) 2% (d) 3%

159. The current voltage relation of a diode is given by $I = \left(e^{1000V/T} - 1\right)$ mA, where the applied voltage V is in volts and the temperature T is in degree kelvin. If a student makes an error measuring ± 0.01 V while measuring the current of 5 mA at 300 K, what will be the error in the value of current in mA? **[2014]**

(a) 0.2mA (b) 0.02 mA
(c) 0.5 mA (d) 0.05 mA

160. Match List - I (Event) with List-II (Order of the time interval for happening of the event) and select the correct option from the options given below the lists: **[Online April 19, 2014]**

List - I		List - II	
(1)	Rotation period of earth	(i)	10^5 s
(2)	Revolution period of earth	(ii)	10^7 s
(3)	Period of light wave	(iii)	10^{-15} s
(4)	Period of sound wave	(iv)	10^{-3} s

(a) (1)-(i), (2)-(ii), (3)-(iii), (4)-(iv)
(b) (1)-(ii), (2)-(i), (3)-(iv), (4)-(iii)
(c) (1)-(i), (2)-(ii), (3)-(iv), (4)-(iii)
(d) (1)-(ii), (2)-(i), (3)-(iii), (4)-(iv)

161. An experiment is performed to obtain the value of acceleration due to gravity g by using a simple pendulum of length L. In this experiment time for 100 oscillations is measured by using a watch of 1 second least count and the value is 90.0 seconds. The length L is measured by using a meter scale of least count 1 mm and the value is 20.0 cm. The error in the determination of g would be: **[Online April 9, 2014]**

(a) 1.7% (b) 2.7%
(c) 4.4% (d) 2.27%

162. Resistance of a given wire is obtained by measuring the current flowing in it and the voltage difference applied across it. If the percentage errors in the measurement of the current and the voltage difference are 3% each, then error in the value of resistance of the wire is **[2012]**

(a) 6% (b) zero
(c) 1% (d) 3%

163. The respective number of significant figures for the numbers 23.023, 0.0003 and 2.1×10^{-3} are **[NCERT : P L-4 | 2010]**

(a) 5, 1, 2 (b) 5, 1, 5
(c) 5, 5, 2 (d) 4, 4, 2

164. A body of mass m = 3.513 kg is moving along the x-axis with a speed of $5.00\ ms^{-1}$. The magnitude of its momentum is recorded as **[NCERT : P L-5 | 2008]**

(a) $17.6\ kg\ ms^{-1}$ (b) $17.565\ kg\ ms^{-1}$
(c) $17.56\ kg\ ms^{-1}$ (d) $17.57\ kg\ ms^{-1}$

ANSWER KEYS

1	(d)	18	(b)	35	(b)	52	(c)	69	(c)	86	(d)	103	(b)	120	(c)	137	(b)	154	(c)
2	(d)	19	(b)	36	(c)	53	(d)	70	(d)	87	(b)	104	(c)	121	(77)	138	(d)	155	(c)
3	(b)	20	(d)	37	(b)	54	(a)	71	(a)	88	(a)	105	(d)	122	(d)	139	(a)	156	(c)
4	(b)	21	(b)	38	(b)	55	(b)	72	(c)	89	(d)	106	(b)	123	(c)	140	(18)	157	(c)
5	(b)	22	(d)	39	(c)	56	(b)	73	(c)	90	(d)	107	(b)	124	(b)	141	(b)	158	(d)
6	(b)	23	(b)	40	(c)	57	(c)	74	(d)	91	(b)	108	(c)	125	(b)	142	(c)	159	(a)
7	(a)	24	(a)	41	(d)	58	(b)	75	(a)	92	(d)	109	(d)	126	(d)	143	(c)	160	(a)
8	(d)	25	(a)	42	(b)	59	(b)	76	(a)	93	(c)	110	(a)	127	(b)	144	(1)	161	(b)
9	(a)	26	(0)	43	(b)	60	(b)	77	(a)	94	(a)	111	(c)	128	(a)	145	(34)	162	(a)
10	(a)	27	(d)	44	(c)	61	(a)	78	(b)	95	(a)	112	(b)	129	(c)	146	(5)	163	(a)
11	(b)	28	(c)	45	(a)	62	(d)	79	(c)	96	(b)	113	(d)	130	(c)	147	(a)	164	(a)
12	(d)	29	(c)	46	(a)	63	(c)	80	(a)	97	(b)	114	(a)	131	(a)	148	(1050)		
13	(c)	30	(b)	47	(b)	64	(d)	81	(d)	98	(None)	115	(c)	132	(c)	149	(d)		
14	(d)	31	(d)	48	(a)	65	(d)	82	(d)	99	(d)	116	(b)	133	(a)	150	(None)		
15	(d)	32	(c)	49	(a)	66	(a)	83	(a)	100	(b)	117	(a)	134	(c)	151	(c)		
16	(a)	33	(b)	50	(a)	67	(a)	84	(b)	101	(a)	118	(c)	135	(c)	152	(None)		
17	(a)	34	(b)	51	(b)	68	(c)	85	(b)	102	(d)	119	(15)	136	(5)	153	(d)		

2 Motion in a Straight Line

TREND ANALYSIS
JEE MAIN (2025-2021)

	2025				2024				2023				2022				2021			
	Session II		Session I		Session II		Session I		Session II		Session I		Session II		Session I		Session II		Session I	
	MCQs	NVQs	MCQs	NVQs	MCQs	NVQs	MCQs	NVQs	MCQs	NVQs	MCQs	NVQs	MCQs	NVQs	MCQs	NVQs	MCQs	NVQs	MCQs	NVQs
Topic 1	2	1	–	–	3	–	–	–	3	—	4	1	1	–	–	2	—	—	—	—
Topic 2	2	–	2	1	2	3	4	3	5	—	3	1	2	2	2	–	1	2	1	—
Topic 3	–		–	–	1	–	11	–	3	—	–	–	–	–	–	–	1	—	—	—
Topic 4	–	–	–	–	–	–	–	1	1	—	–	1	5	1	2	3	3	—	2	—
Total	4	1	2	1	6	3	5	4	12	—	7	3	8	3	4	5	5	2	3	—

How to Code Qns for Better & Faster Revision →

Type 1: Able to solve (No code Req)
Type 2: Solved with struggle or by seeing solution
(Use code NR means Needs Revision or orange stickers)

Type 3: Tricky/ Could not solve on my own. Need help from Teacher/ Friend.
(Use code TQ means Tricky Question or red stickers)

You can either hand write the code in boxes or else paste coloured stickers given at end of book.

Topic 1 - Distance, Displacement & Uniform Motion

1. Which of the following curves possibly represent one-dimensional motion of a particle ? **[April 3, 2025 (I)]**

(A)
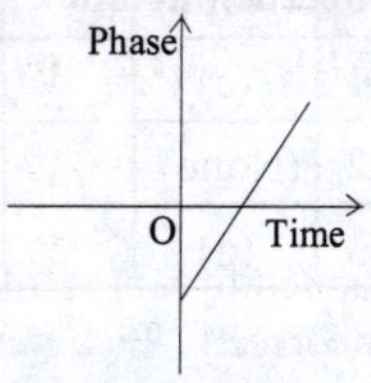

(B)
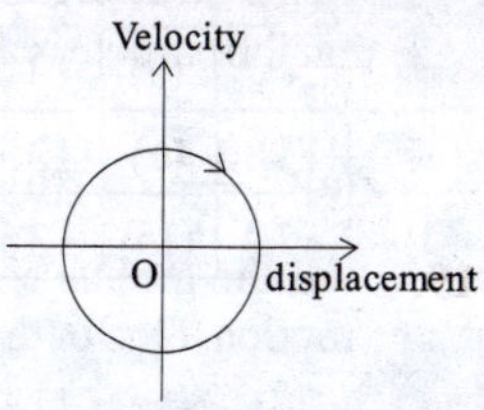

(C)
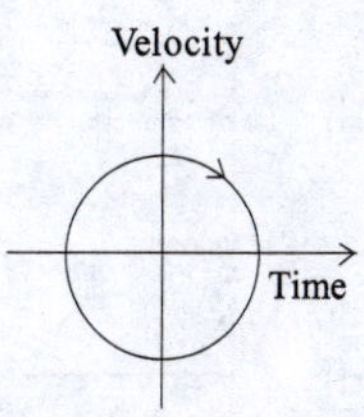

(D)
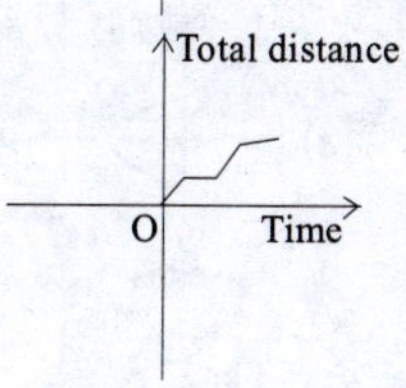

Choose the correct answer from the options given below:

(a) A, B and D only (b) A, B and C only
(c) A and B only (d) A, C and D only

2. A sportsman runs around a circular track of radius r such that he traverses the path ABAB. The distance travelled and displacement, respectively, are **[April 2, 2025 (II)]**

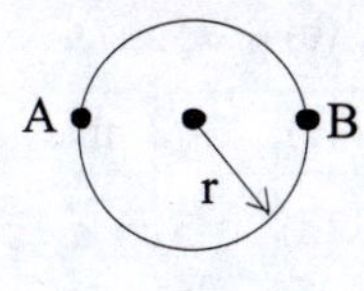

(a) $2r, 3\pi r$
(b) $3\pi r, \pi r$
(c) $\pi r, 3r$
(d) $3\pi r, 2r$

3. A person travelling on a straight line moves with a uniform velocity v_1 for a distance x and with a uniform velocity v_2 for the next $\frac{3}{2}x$ distance. The average velocity in this motion is $\frac{50}{7}$ m/s. If v_1 is 5 m/s then v_2 = ______ m/s. **[April 2, 2025 (I)]**

4. A particle moving in a straight line covers half the distance with speed 6 m/s. The other half is covered in two equal time intervals with speeds 9 m/s and 15 m/s respectively. The average speed of the particle during the motion is : **[April 9, 2024 (I)]**

(a) 8.8 m/s (b) 10 m/s (c) 9.2 m/s (d) 8 m/s

5. A train starting from rest first accelerates uniformly up to a speed of 80 km/h for time t, then it moves with a constant speed for time 3t. The average speed of the train for this duration of journey will be (in km/h) : **[April 6, 2024 (I)]**

(a) 80 (b) 70 (c) 30 (d) 40

6. A cyclist starts from the point P of a circular ground of radius 2 km and travels along its circumference to the point S. The displacement of a cyclist is : **[April 4, 2024 (II)]**

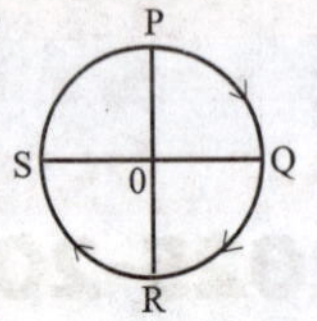

(a) 6km (b) $\sqrt{8}$ km (c) 4km (d) 8km

7. A person travels x distance with velocity v_1 and then x distance with velocity v_2 in the same direction. The average velocity of the person is v, then the relation between v, v_1 and v_2 will be : **[April 10, 2023 (II)]**

(a) $v = v_1 + v_2$ (b) $v = \frac{v_1 + v_2}{2}$

(c) $\frac{2}{v} = \frac{1}{v_1} + \frac{1}{v_2}$ (d) $\frac{1}{v} = \frac{1}{v_1} + \frac{1}{v_2}$

8. A car travels a distance of 'x' with speed V_1 and then same distance 'x' with speed V_2 in the same direction. The average speed of the car is: **[Jan. 25, 2023 (I)]**

(a) $\frac{v_1 v_2}{2(v_1 + v_2)}$ (b) $\frac{v_1 + v_2}{2}$

(c) $\frac{2x}{v_1 + v_2}$ (d) $\frac{2v_1 v_2}{v_1 + v_2}$

9. A particle is moving with constant speed in a circular path. When the particle turns by an angle 90°, the ratio of instantaneous velocity to its average velocity is $\pi : x\sqrt{2}$. The value of x will be **[NCERT : P L-14 | April 6, 2023 (I)]**

(a) 2 (b) 5 (c) 1 (d) 7

10. As shown in the figure, a particle is moving with constant speed π m/s. Considering its motion from A to B, the magnitude of the average velocity is: **[April 6, 2023 (II)]**

(a) π m/s

(b) $\sqrt{3}$ m/s

(c) $2\sqrt{3}$ m/s

(d) $1.5\sqrt{3}$ m/s

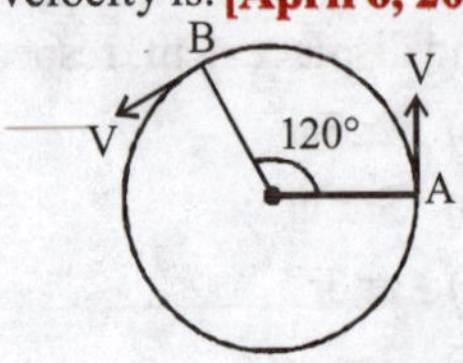

11. An object moves with speed v_1, v_2 and v_3 along a line segment AB, BC and CD respectively as shown in figure. Where AB = BC and AD = 3AB, then average speed of the object will be: **[Feb. 1, 2023 (I)]**

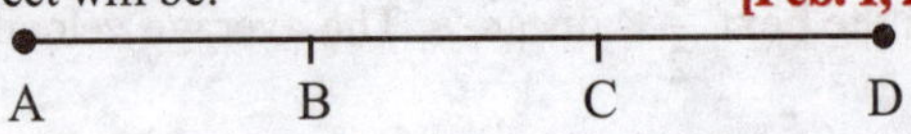

(a) $\frac{v_1 v_2 v_3}{3(v_1 v_2 + v_2 v_3 + v_3 v_1)}$ (b) $\frac{3 v_1 v_2 v_3}{(v_1 v_2 + v_2 v_3 + v_3 v_1)}$

(c) $\frac{(v_1 + v_2 + v_3)}{3}$ (d) $\frac{(v_1 + v_2 + v_3)}{3 v_1 v_2 v_3}$

12. A body is moving with constant speed, in a circle of radius 10 m. The body completes one revolution in 4 s. At the end of 3rd second, the displacement of body (in m) from its starting point is: **[Jan. 31, 2023 (II)]**

(a) 30 (b) 15π (c) 5π (d) $10\sqrt{2}$

13. A horse rider covers half the distance with 5 m/s speed. The remaining part of the distance was travelled with speed 10 m/s for half the time and with speed 15 m/s for other half of the time. The mean speed of the rider averaged over the whole time of motion is x/7 m/s. The value of x is **[Jan. 30, 2023 (I)]**

14. A vehicle travels 4 km with speed of 3 km/h and another 4 km with speed of 5 km/h, then its average speed is : **[Jan. 30, 2023 (II)]**

(a) 4.25 km/h (b) 3.50 km/h

(c) 4.00 km/h (d) 3.75 km/h

15. A person moved from A to B on a circular path as shown in figure. If the distance travelled by him is 60 m, then the magnitude of displacement would be : **[July 25, 2022 (I)]**

(Given cos135° = – 0.7)

(a) 42m

(b) 47m

(c) 19m

(d) 40m

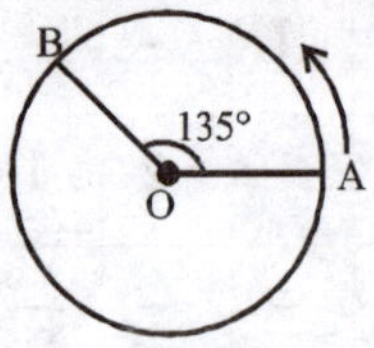

16. A person starts his journey from centre 'O' of the park and comes back to the same position following path OPQO as shown in the figure. The radius of path taken by the person is 200 m and he takes 3 min 58 sec to complete his journey. The average speed of the person is ____ ms^{-1}. (taken π = 3.14) **[June 30, 2022 (I)]**

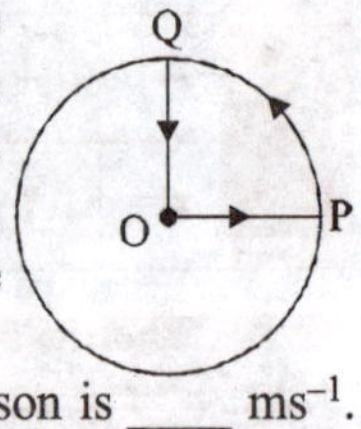

17. A car covers *AB* distance with first one–third at velocity v_1 ms^{-1}, second one–third at v_2 ms^{-1} and last one–third at v_3 ms^{-1}. If $v_3 = 3v_1$, $v_2 = 2v_1$ and $v_1 = 11$ ms^{-1} then the average velocity of the car is ________ ms^{-1}.

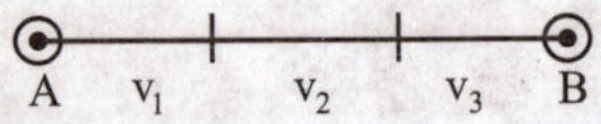

[June 28, 2022 (II)]

18. A particle is moving with speed $v = b\sqrt{x}$ along positive *x*-axis. Calculate the speed of the particle at time $t = \tau$ (assume that the particle is at origin at t = 0). **[12 Apr. 2019 (II)]**

(a) $\frac{b^2\tau}{4}$ (b) $\frac{b^2\tau}{2}$ (c) $b^2\tau$ (d) $\frac{b^2\tau}{\sqrt{2}}$

19. All the graphs below are intended to represent the same motion. One of them does it incorrectly. Pick it up. **[2018]**

(a) 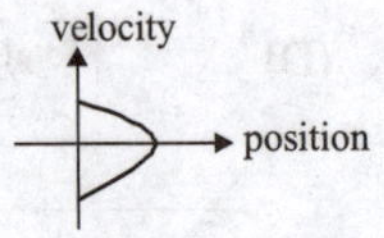

(b) 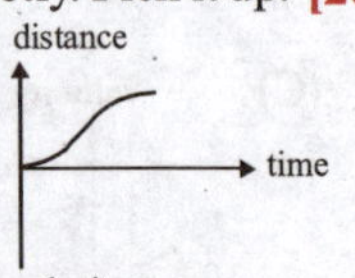

(c) 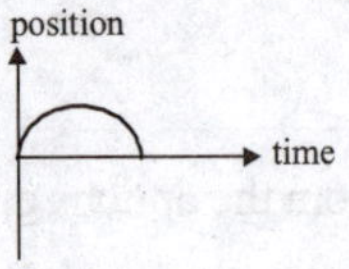

(d) 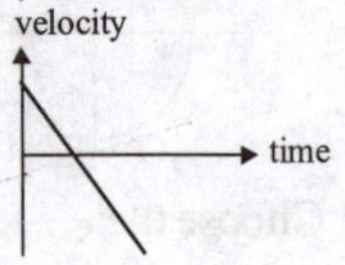

20. A car covers the first half of the distance between two places at 40 km/h and other half at 60 km/h. The average speed of the car is **[Online May 7, 2012]**
(a) 40 km/h (b) 45 km/h (c) 48 km/h (d) 60 km/h

Topic 2 - Non-uniform Motion

21. The displacement x versus time graph is shown below.

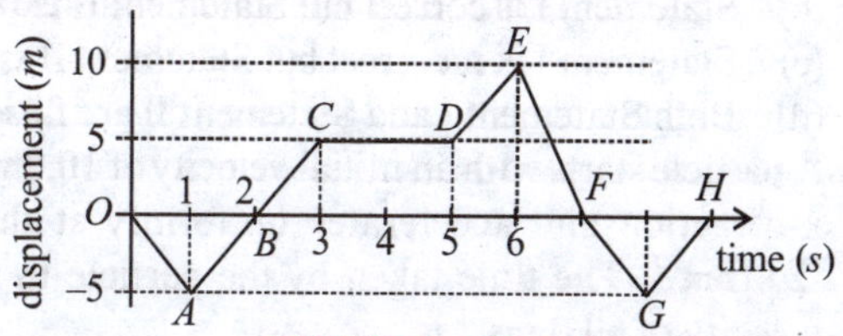

[NCERT : P L-14 | April 4, 2025 (II)]
(A) The average velocity during 0 to 3 s is 10 m/s
(B) The average velocity during 3 to 5 s is 0 m/s
(C) The instantaneous velocity at $t = 2$ s is 5 m/s
(D) The average velocity during 5 to 7 s and instantaneous velocity at $t = 6.5$ s are equal
(E) The average velocity from $t = 0$ to $t = 9$ s is zero
Choose the **correct** answer from the options given below:
(a) (A), (D), (E) only (b) (B), (C), (D) only
(c) (B), (D), (E) only (d) (B), (C), (E) only

22. A particle moves along the x-axis and has its displacement x varying with time t according to the equation
$x = c_0(t^2 - 2) + c(t - 2)^2$
where c_0 and c are constants of appropriate dimensions. Then, which of the following statements is correct? **[NCERT : P L-16 | April 3, 2025 (II)]**
(a) the acceleration of the particle is $2c_0$
(b) the acceleration of the particle is 2c
(c) the initial velocity of the particle is 4c
(d) the acceleration of the particle is $2(c + c_0)$

23. Two cars P and Q are moving on a road in the same direction. Acceleration of car P increases linearly with time whereas car Q moves with a constant acceleration. Both cars cross each other at time $t = 0$, for the first time. The maximum possible number of crossing(s) (including the crossing at $t = 0$) is ________.
[NCERT : P L-16 | Jan. 29, 2025 (II)]

24. The velocity-time graph of an object moving along a straight line is shown in figure. What is the distance covered by the object between t = 0 to t = 4s ?
[NCERT : P L-17 | Jan. 28, 2025 (II)]

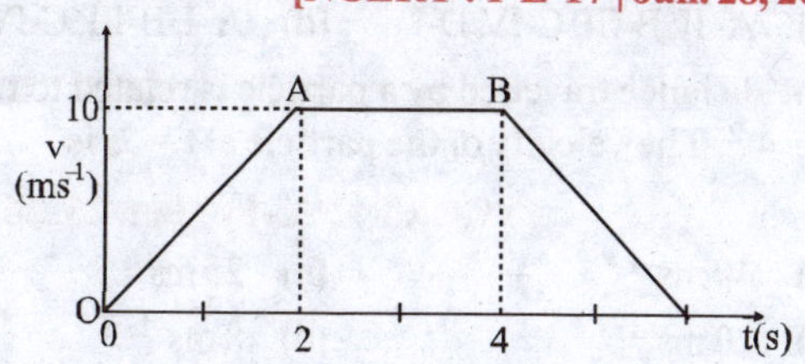

(a) 30 m (b) 10 m (c) 13 m (d) 11 m

25. The motion of an aeroplane is represented by velocity-time graph as shown below. The distance covered by aeroplane in the first 30.5 second is ______ km.
[NCERT : P L-17 | Jan. 23, 2025 (I)]

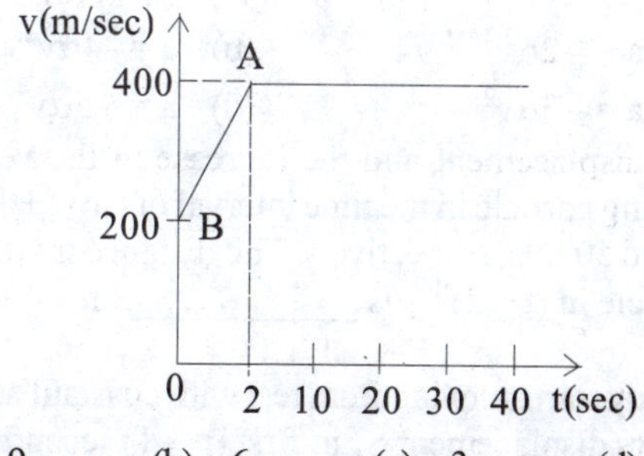

(a) 9 (b) 6 (c) 3 (d) 12

26. A particle is moving in a straight line. The variation of position 'x' as a function of time 't' is given as $x = (t^3 - 6t^2 + 20t + 15)$ m. The velocity of the body when its acceleration becomes zero is : **[NCERT : P L-16 | Jan. 29, 2024 (II)]**
(a) 6 m/s (b) 10 m/s (c) 8 m/s (d) 4 m/s

27. Two cars are travelling towards each other at speed of 20 m s^{-1} each. When the cars are 300 m apart, both the drivers apply brakes and the cars retard at the rate of 2 m s^{-2}. The distance between them when they come to rest is :
[NCERT : P L-18 | April 9, 2024 (II)]
(a) 200 m (b) 50 m (c) 100 m (d) 25 m

28. A particle moves in a straight line so that its displacement x at any time t is given by $x^2 = 1 + t^2$. Its acceleration at any time t is x^{-n} where n = ______.
[NCERT : P L-16 | April 6, 2024 (II)]

29. A body moves on a frictionless plane starting from rest. If S_n is distance moved between t = n – 1 and t = n and S_{n-1} is distance moved between t = n –2 and t = n – 1, then the ratio $\frac{S_{n-1}}{S_n}$ is $\left(1 - \frac{2}{x}\right)$ for n = 10. The value of x is .... **[NCERT : P L-18 | April 5, 2024 (I)]**

30. A body travels 102.5 m in n^{th} second and 115.0 m in $(n + 2)^{th}$ second. The acceleration is :
[NCERT : P L-18 | April 4, 2024 (I)]
(a) 9 m/s^2 (b) 6.25 m/s^2 (c) 12.5 m/s^2 (d) 5 m/s^2

31. A bus moving along a straight highway with speed of 72 km/h is brought to halt within 4s after applying the brakes. The distance travelled by the bus during this time (Assume the retardation is uniform) is ______ m.
[NCERT : P L-18 | April 4, 2024 (II)]

32. A particle is moving in one dimension (along x axis) under the action of a variable force. It's initial position was 16 m right of origin. The variation of its position (x) with time (t) is given as $x = -3t^3 + 18t^2 + 16t$, where x is in m and t is in s. The velocity of the particle when its acceleration becomes zero is ______ m/s.
[NCERT : P L-16 | Feb. 1, 2024 (I)]

33. A particle initially at rest starts moving from reference point x = 0 along x- axis, with velocity v that varies as $v = 4\sqrt{x}$ m/s. The acceleration of the particle is ___ ms^{-2}.
[NCERT : P L-16 | Feb. 1, 2024 (II)]

34. The relation between time 't' and distance 'x' is $t = \alpha x^2 + \beta x$, where α and β are constants. The relation between acceleration (a) and velocity (v) is :
[NCERT : P L-16 | Jan. 31, 2024 (I)]
(a) $a = -2\alpha v^3$ (b) $a = -4\alpha v^4$
(c) $a = -3\alpha v^2$ (d) $a = -5\alpha v^5$

35. The displacement and the increase in the velocity of a moving particle in the time interval of t to (t + 1) s are 125 m and 50 m/s, respectively. The distance travelled by the particle in $(t+2)^{th}$ s is ________ m.
[NCERT : P L-18 | Jan. 30, 2024 (I)]

36. A body starts moving from rest with constant acceleration covers displacement S_1 in first (p – 1) seconds and S_2 in first p seconds. The displacement $S_1 + S_2$ will be made in time: **[NCERT : P L-18 | Jan. 29, 2024 (I)]**
(a) $\sqrt{(2p^2 - 2p + 1)}\,s$ (b) $(2p^2 - 2p + 1)$ s
(c) $(2p + 1)$ s (d) $(2p - 1)$ s

37. A bullet is fired into a fixed target looses one third of its velocity after travelling 4 cm. It penetrates further $D \times 10^{-3}$ m before coming to rest. The value of D is :
[NCERT : P L-18 | Jan. 27, 2024 (II)]
(a) 2 (b) 5 (c) 3 (d) 4

38. The velocity of a particle is $v = v_0 + gt + ft^2$. If its position is $x = 0$ at $t = 0$, then its displacement after unit time ($t = 1$) is **[NCERT : P L-14 | 2007]**
(a) $v_0 + g/2 + f$ (b) $v_0 + 2g + 3f$
(c) $v_0 + g/2 + f/3$ (d) $v_0 + g + f$

39. A particle located at $x = 0$ at time $t = 0$, starts moving along with the positive x-direction with a velocity 'v' that varies as $v = \alpha\sqrt{x}$. The displacement of the particle varies with time as **[NCERT : P L-14 | 2006]**
(a) t^2 (b) t (c) $t^{1/2}$ (d) t^3

40. The position of a particle related to time is given by $x = (5t^2 - 4t + 5)$ m. The magnitude of velocity of the particle at t = 2s will be : **[NCERT : P L-14 | April 15, 2023 (I)]**
(a) $10\ ms^{-1}$ (b) $14\ ms^{-1}$ (c) $16\ ms^{-1}$ (d) $06\ ms^{-1}$

41. The distance travelled by an object in time t is given by $s = (2.5)t^2$. The instantaneous speed of the object at t = 5 s will be : **[NCERT : P L-14 | April 13, 2023 (II)]**
(a) $12.5\ ms^{-1}$ (b) $62.5\ ms^{-1}$
(c) $5\ ms^{-1}$ (d) $25\ ms^{-1}$

42. From the v - t graph shown. the ratio of distance to displacement in 25 s of motion
[NCERT : P L-17 | April 11, 2023 (I)]
(a) $\frac{3}{5}$
(b) $\frac{1}{2}$
(c) $\frac{5}{3}$
(d) 1

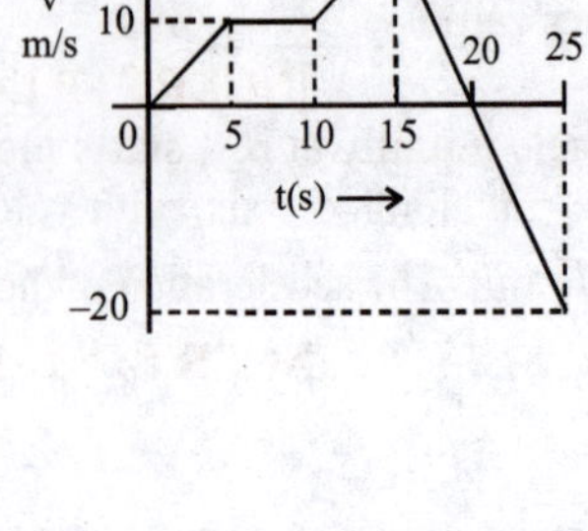

43. Given below are two statements:
Statement I: Area under velocity- time graph gives the distance travelled by the body in a given time.
Statement II: Area under acceleration- time graph is equal to the change in velocity- in the given time.
In the light of given statements, choose the correct answer from the options given below.
[NCERT : P L-17 | April 8, 2023 (II)]
(a) Both Statement I and Statement II are true.
(b) Statement I is correct but Statement II is false.
(c) Statement I is incorrect but Statement II is true.
(d) Both Statement I and Statement II are false.

44. A particle starts with an initial velocity of $10.0\ ms^{-1}$ along x-direction and accelerates uniformly at the rate of $2.0\ ms^{-2}$. The time taken by the particle to reach the velocity of $60.0\ ms^{-1}$ is______.
[NCERT : P L-18 | April 6, 2023 (II)]
(a) 6s (b) 3s (c) 30s (d) 25s

45. For a train engine moving with speed of $20\ ms^{-1}$, the driver must apply brakes at a distance of 500 m before the station for the train to come to rest at the station. If the brakes were applied at half of this distance, the train engine would cross the station with speed $\sqrt{x}\,ms^{-1}$. The value of x is ______. **[NCERT : P L-18 | Feb. 1, 2023 (II)]**
(Assuming same retardation is produced by brakes)

46. Match Column-I with Column-II :

Column-I (x-t graphs)	Column-II (v-t graphs)
A. 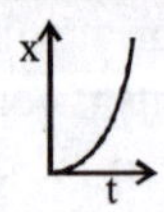	I. 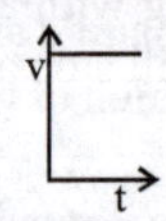
B. 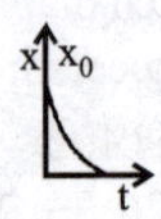	II. 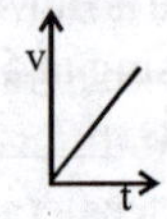
C. 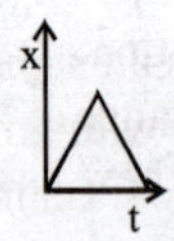	III. 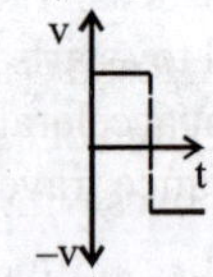
D. 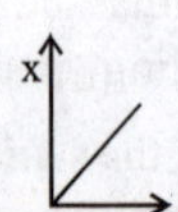	IV. 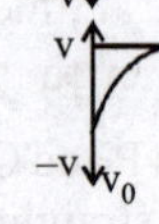

Choose the correct answer from the options given below:
[NCERT : P L-14 | Jan. 30, 2023 (I)]
(a) A-II, B-IV, C-III, D-I (b) A-I, B-II, C-III, D-IV
(c) A-II, B-III, C-IV, D-I (d) A-I, B-III, C-IV, D-II

47. The distance travelled by a particle is related to time t as $x = 4t^2$. The velocity of the particle at t = 5s is .
[NCERT : P L-14 | Jan. 25, 2023 (II)]
(a) $40\ ms^{-1}$ (n) $25\ ms^{-1}$
(c) $20\ ms^{-1}$ (d) $8\ ms^{-1}$

48. The velocity time graph of a body moving in a straight line is shown in figure. [NCERT : P L-17 | Jan. 24, 2023 (II)]

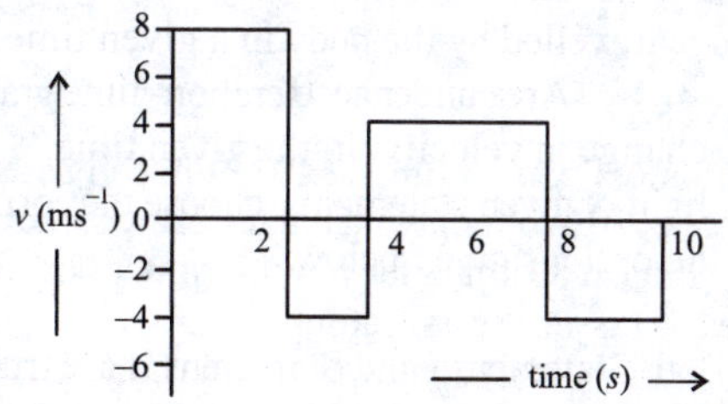

The ratio of displacement to distance travelled by the body in time 0 to 10s is

(a) 1 : 1 (b) 1 : 4 (c) 1 : 2 (d) 1 : 3

49. If $t = \sqrt{x} + 4$, then $\left(\frac{dx}{dt}\right)_{t=4}$ is: [NCERT : P L-14 | July 29, 2022 (I)]

(a) 4 (b) Zero (c) 8 (d) 16

50. The velocity of the bullet becomes one third after it penetrates 4 cm in a wooden block. Assuming that bullet is facing a constant resistance during its motion in the block. The bullet stops completely after travelling at (4 + x) cm inside the block. The value of x is: [NCERT : P L-18 | July 27, 2022 (II)]

(a) 2.0 (b) 1.0 (c) 0.5 (d) 1.5

51. A car is moving with speed of 150 km/h and after applying the break it will move 27 m before it stops. If the same car is moving with a speed of one third the reported speed then it will stop after travelling ________ m distance. [NCERT : P L-18 | July 25, 2022 (I)]

52. A particle is moving in a straight line such that its velocity is increasing at 5 ms^{-1} per meter. The acceleration of the particle is _____ ms^{-2} at a point where its velocity is 20 ms^{-1}. [NCERT : P L-16 | July 25, 2022 (II)]

53. A small toy starts moving from the position of rest under a constant acceleration. If it travels a distance of 10 m in t s, the distance travelled by the toy in the next *t* s will be : [NCERT : P L-18 | June 29, 2022 (II)]

(a) 10m (b) 20m (c) 30m (d) 40m

54. Two buses P and Q start from a point at the same time and move in a straight line and their positions are represented by $X_P(t) = \alpha t + \beta t^2$ and $X_Q(t) = ft - t^2$. At what time, both the buses have same velocity ? [NCERT : P L-14 | June 25, 2022 (II)]

(a) $\frac{\alpha - f}{1+\beta}$ (b) $\frac{\alpha + f}{2(\beta - 1)}$ (c) $\frac{\alpha + f}{2(1+\beta)}$ (d) $\frac{f - \alpha}{2(1+\beta)}$

55. A particle is moving with constant acceleration 'a'. Following graph shows v^2 versus x(displacement) plot. The acceleration of the particle is ___m/s^2. [NCERT : P L-18 | NA, Aug. 31, 2021 (II)]

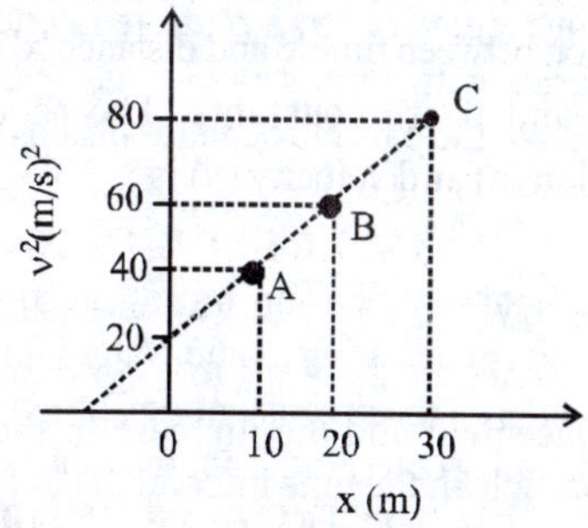

56. If the velocity of a body related to displacement x is given by $\upsilon = \sqrt{5000 + 24x}$ m/s, then the acceleration of the body is ________ m/s^2. [NCERT : P L-16 | NA, Aug. 27, 2021 (I)]

57. The instantaneous velocity of a particle moving in a straight line is given as v = $\alpha t + \beta t^2$, where α and β are constants. The distance travelled by the particle between 1s and 2s is: [NCERT : P L-14 | July 25, 2021 (II)]

(a) $3\alpha + 7\beta$ (b) $\frac{3}{2}\alpha + \frac{7}{3}\beta$

(c) $\frac{\alpha}{2} + \frac{\beta}{3}$ (d) $\frac{3}{2}\alpha + \frac{7}{2}\beta$

58. A scooter accelerates from rest for time t_1 at constant rate u_1 and then retards at constant rate a_2 for time t_2 and comes to rest. The correct value of $\frac{t_1}{t_2}$ will be [NCERT : P L-18 | Feb. 26, 2021 (II)]

(a) $\frac{a_1 + a_2}{a_1}$ (b) $\frac{a_1}{a_2}$ (c) $\frac{a_2}{a_1}$ (d) $\frac{a_1 + a_2}{a_2}$

59. The velocity (*v*) and time (*t*) graph of a body in a straight line motion is shown in the figure. The point *S* is at 4.333 seconds. The total distance covered by the body in 6 s is: [NCERT : P L-17 | 05 Sep. 2020 (II)]

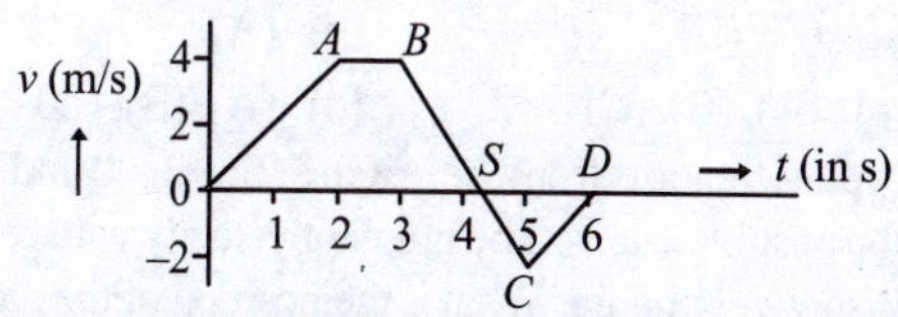

(a) $\frac{37}{3}$ m (b) 12m (c) 11m (d) $\frac{49}{4}$ m

60. The speed verses time graph for a particle is shown in the figure. The distance travelled (in m) by the particle during the time interval *t* = 0 to *t* = 5 s will be ________. [NCERT : P L-17 | NA 4 Sep. 2020 (II)]

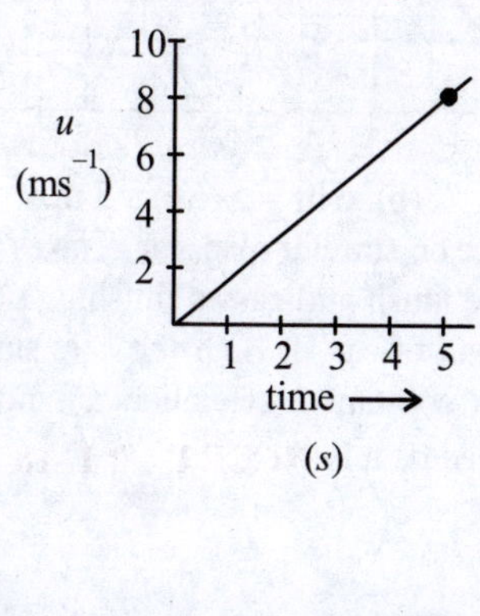

61. The distance x covered by a particle in one dimensional motion varies with time t as $x^2 = at^2 + 2bt + c$. If the acceleration of the particle depends on x as x^{-n}, where n is an integer, the value of n is ______.
[NCERT : P L-16 | NA 9 Jan 2020 I]

62. A bullet of mass 20g has an initial speed of 1 ms^{-1}, just before it starts penetrating a mud wall of thickness 20 cm. If the wall offers a mean resistance of 2.5×10^{-2} N, the speed of the bullet after emerging from the other side of the wall is close to : **[NCERT : P L-18 | 10 Apr. 2019 II]**
(a) 0.1 ms^{-1} (b) 0.7 ms^{-1} (c) 0.3 ms^{-1} (d) 0.4 ms^{-1}

63. The position of a particle as a function of time t, is given by $x(t) = at + bt^2 - ct^3$ where, a, b and c are constants. When the particle attains zero acceleration, then its velocity will be: **[NCERT : P L-16 | 9 Apr. 2019 II]**
(a) $a+\frac{b^2}{4c}$ (b) $a+\frac{b^2}{3c}$ (c) $a+\frac{b^2}{c}$ (d) $a+\frac{b^2}{2c}$

64. A particle starts from origin O from rest and moves with a uniform acceleration along the positive x-axis. Identify all figures that correctly represents the motion qualitatively (a = acceleration, v = velocity, x = displacement, t = time) **[NCERT : P L-16 | 8 Apr. 2019 II]**

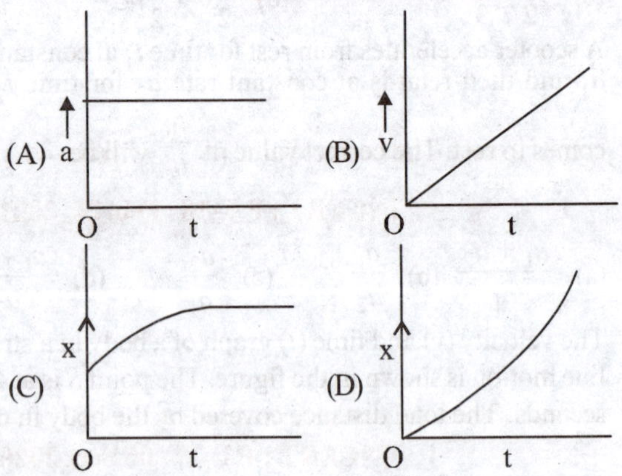

(a) (B), (C) (b) (A)
(c) (A), (B), (C) (d) (A), (B), (D)

65. A particle starts from the origin at time t = 0 and moves along the positive x-axis. The graph of velocity with respect to time is shown in figure. What is the position of the particle at time t = 5s? **[NCERT : P L-17 | 10 Jan. 2019 II]**

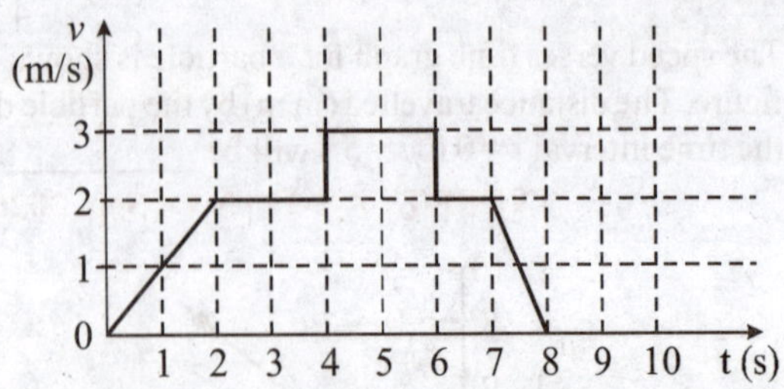

(a) 10 m (b) 6 m (c) 3 m (d) 9 m

66. In a car race on straight road, car A takes a time t less than car B at the finish and passes finishing point with a speed 'v' more than of car B. Both the cars start from rest and travel with constant acceleration a_1 and a_2 respectively. Then 'v' is equal to: **[NCERT : P L-18 | 9 Jan. 2019 II]**
(a) $\frac{2a_1 a_2}{a_1 + a_2} t$ (b) $\sqrt{2a_1 a_2}\, t$
(c) $\sqrt{a_1 a_2}\, t$ (d) $\frac{a_1 + a_2}{2} t$

67. An automobile, travelling at 40 km/h, can be stopped at a distance of 40 m by applying brakes. If the same automobile is travelling at 80 km/h, the minimum stopping distance, in metres, is (assume no skidding)
[NCERT : P L-18 | Online April 15, 2018]
(a) 75m (b) 160m
(c) 100m (d) 150m

68. The velocity-time graphs of a car and a scooter are shown in the figure. (i) the difference between the distance travelled by the car and the scooter in 15 s and (ii) the time at which the car will catch up with the scooter are, respectively **[NCERT : P L-18 | Online April 15, 2018]**

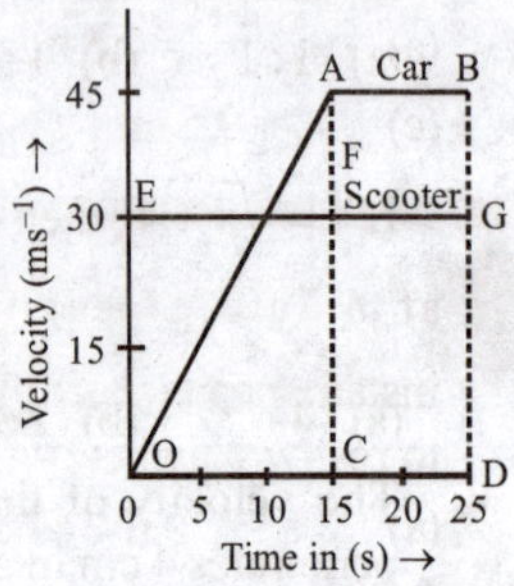

(a) 337.5m and 25s (b) 225.5m and 10s
(c) 112.5m and 22.5s (d) 11.2.5m and 15s

69. A man in a car at location Q on a straight highway is moving with speed v. He decides to reach a point P in a field at a distance d from highway (point M) as shown in the figure. Speed of the car in the field is half to that on the highway. What should be the distance RM, so that the time taken to reach P is minimum?

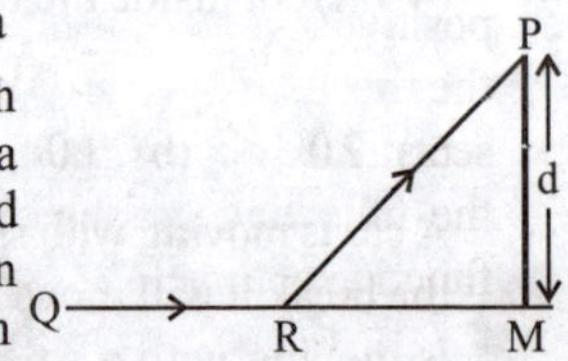

[NCERT : P L-14 | Online April 15, 2018]
(a) $\frac{d}{\sqrt{3}}$ (b) $\frac{d}{2}$ (c) $\frac{d}{\sqrt{2}}$ (d) d

70. The distance travelled by a body moving along a line in time t is proportional to t^3.
The acceleration-time (a, t) graph for the motion of the body will be **[NCERT : P L-16 | Online May 12, 2012]**

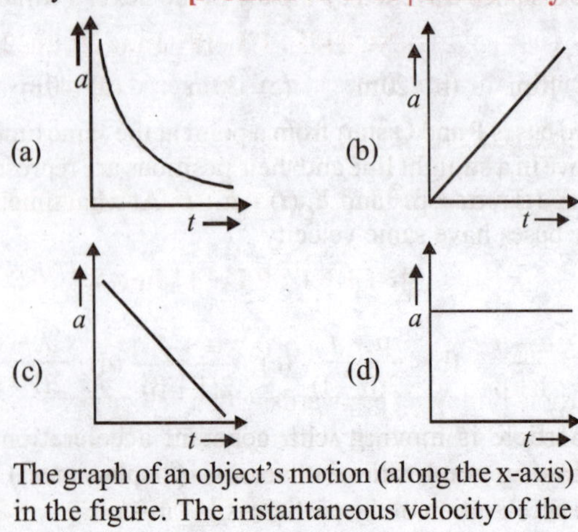

71. The graph of an object's motion (along the x-axis) is shown in the figure. The instantaneous velocity of the object at points A and B are v_A and v_B respectively. Then
[NCERT : P L-14 | Online May 7, 2012]

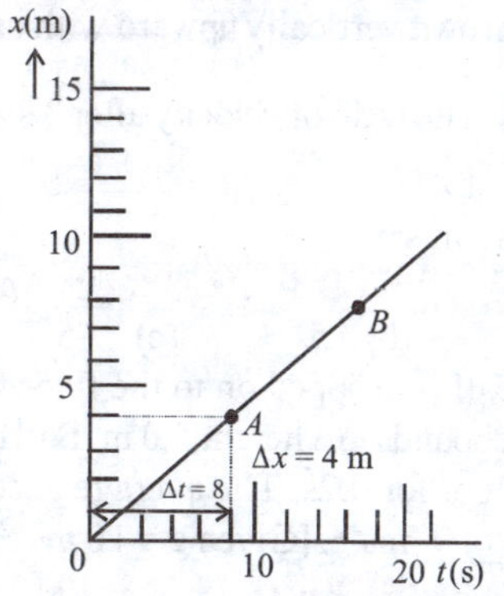

(a) $v_A = v_B = 0.5$ m/s (b) $v_A = 0.5$ m/s $< v_B$

(c) $v_A = 0.5$ m/s $> v_B$ (d) $v_A = v_B = 2$ m/s

72. An object, moving with a speed of 6.25 m/s, is decelerated at a rate given by $\frac{dv}{dt} = -2.5\sqrt{v}$ where v is the instantaneous speed. The time taken by the object, to come to rest, would be: **[NCERT : P L-16 | 2011]**

(a) 2 s (b) 4 s (c) 8 s (d) 1 s

73. A body is at rest at $x = 0$. At $t = 0$, it starts moving in the positive x-direction with a constant acceleration. At the same instant another body passes through $x = 0$ moving in the positive x-direction with a constant speed. The position of the first body is given by $x_1(t)$ after time 't'; and that of the second body by $x_2(t)$ after the same time interval. Which of the following graphs correctly describes $(x_1 - x_2)$ as a function of time 't'? **[NCERT : P L-18 | 2008]**

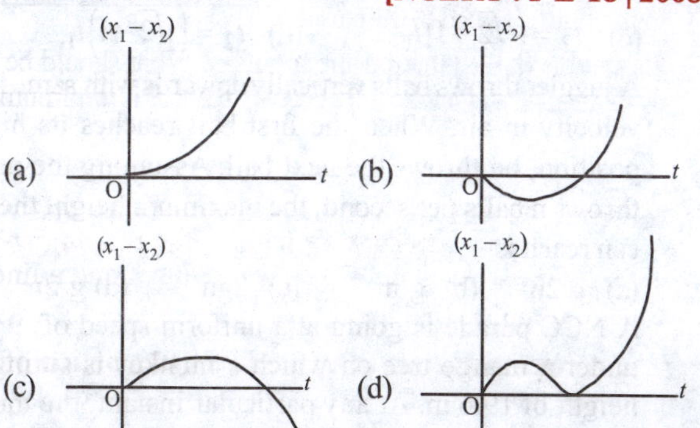

74. A car, starting from rest, accelerates at the rate f through a distance S, then continues at constant speed for time t and then decelerates at the rate $\frac{f}{2}$ to come to rest. If the total distance traversed is 15 S, then **[NCERT : P L-18 | 2005]**

(a) $S = \frac{1}{6}ft^2$ (b) $S = ft$

(c) $S = \frac{1}{4}ft^2$ (d) $S = \frac{1}{72}ft^2$

75. A particle is moving eastwards with a velocity of 5 ms^{-1}. In 10 seconds the velocity changes to 5 ms^{-1} northwards. The average acceleration in this time is **[NCERT : P L-16 | 2005]**

(a) $\frac{1}{2}$ ms^{-2} towards north

(b) $\frac{1}{\sqrt{2}}$ ms^{-2} towards north - east

(c) $\frac{1}{\sqrt{2}}$ ms^{-2} towards north - west

(d) zero

76. The relation between time t and distance x is $t = ax^2 + bx$ where a and b are constants. The acceleration is **[NCERT : P L-16 | 2005]**

(a) $2bv^3$ (b) $-2abv^2$ (c) $2av^2$ (d) $-2av^3$

77. If a body looses half of its velocity on penetrating 3 cm in a wooden block, then how much will it penetrate more before coming to rest? **[NCERT : P L-18 | 2002]**

(a) 1 cm (b) 2 cm (c) 3 cm (d) 4 cm.

Topic 3 - Relative Velocity

78. A body projected vertically upwards with a certain speed from the top of a tower reaches the ground in t_1. If it is projected vertically downwards from the same point with the same speed, it reaches the ground in t_2. Time required to reach the ground, if it is dropped from the top of the tower, is : **[April 6, 2024 (II)]**

(a) $\sqrt{t_1 t_2}$ (b) $\sqrt{t_1 - t_2}$ (c) $\sqrt{\frac{t_1}{t_2}}$ (d) $\sqrt{t_1 + t_2}$

79. Train A is moving along two parallel rail tracks towards north with speed 72 km/h and train B is moving towards south with speed 108 km/h. Velocity of train B with respect to A and velocity of ground with respect to B are (in ms^{-1}) : **[Feb. 1, 2024 (II)]**

(a) – 50 and – 30 (b) – 50 and 30

(c) – 30 and 50 (d) 50 and – 30

80. Two trains 'A' and 'B' of length 'l' and '$4l$' are travelling into a tunnel of length 'L' in parallel tracks from opposite directions with velocities 108 km/h and 72 km/h, respectively. If train 'A' takes 35 s less time than train 'B' to cross the tunnel then, length 'L' of tunnel is: (Given L = 60l) **[April 13, 2023 (I)]**

(a) 1200m (b) 2700m (c) 1800m (d) 900m

81. A passenger sitting in a train A moving at 90 km/h observes another train B moving in the opposite direction for 8 s. If the velocity of the train B is 54 km/h, then length of train B is : **[April 13, 2023 (II)]**

(a) 80m (b) 200m (c) 120m (d) 320m

82. The position-time graphs for two students A and B returning from the school to their homes are shown in figure : **[April 10, 2023 (I)]**

(A) A lives closer to the school

(B) B lives closer to the school

(C) A takes lesser time to reach home

(D) A travels faster than B

(E) B travels faster than A

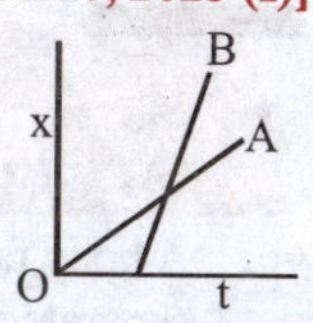

Choose the correct answer from the options given below:

(a) (A) and (E) only (b) (B) and (E) only

(c) (A), (C) and (E) only (d) (A), (C) and (D) only

83. A boy reaches the airport and finds that the escalator is not working. He walks up the stationary escalator in time t_1. If he remains stationary on a moving escalator then the escalator takes him up in time t_2. The time taken by him to walk up on the moving escalator will be : **[July 20, 2021 (II)]**

(a) $\frac{t_1 t_2}{t_2 - t_1}$ (b) $\frac{t_1 + t_2}{2}$ (c) $\frac{t_1 t_2}{t_2 + t_1}$ (d) $t_2 - t_1$

84. Train A and train B are running on parallel tracks in the opposite directions with speeds of 36 km/hour and 72 km/hour, respectively. A person is walking in train A in the direction opposite to its motion with a speed of 1.8 km/hour w.r.t train A. Speed (in ms^{-1}) of this person as observed from train B will be close to : (take the distance between the tracks as negligible) **[Sep. 2, 2020 (I)]**

(a) 29.5 ms^{-1} (b) 28.5 ms^{-1}
(c) 31.5 ms^{-1} (d) 30.5 ms^{-1}

85. A passenger train of length 60 m travels at a speed of 80 km/hr. Another freight train of length 120 m travels at a speed of 30 km/h. The ratio of times taken by the passenger train to completely cross the freight train when: (i) they are moving in same direction, and (ii) in the opposite directions is: **[Jan. 12, 2019 II]**

(a) $\frac{11}{5}$ (b) $\frac{5}{2}$ (c) $\frac{3}{2}$ (d) $\frac{25}{11}$

86. A person standing on an open ground hears the sound of a jet aeroplane, coming from north at an angle 60° with ground level. But he finds the aeroplane right vertically above his position. If v is the speed of sound, speed of the plane is: **[Jan. 12, 2019 II]**

(a) $\frac{\sqrt{3}}{2}v$ (b) $\frac{2v}{\sqrt{3}}$ (c) v (d) $\frac{v}{2}$

87. A car is standing 200 m behind a bus, which is also at rest. The two start moving at the same instant but with different forward accelerations. The bus has acceleration 2 m/s^2 and the car has acceleration 4 m/s^2. The car will catch up with the bus after a time of : **[Online April 9, 2017]**

(a) $\sqrt{110}$ s (b) $\sqrt{120}$ s (c) $10\sqrt{2}$ s (d) 15 s

88. A person climbs up a stalled escalator in 60 s. If standing on the same but escalator running with constant velocity he takes 40 s. How much time is taken by the person to walk up the moving escalator? **[Online April 12, 2014]**

(a) 37 s (b) 27 s (c) 24 s (d) 45 s

89. A goods train accelerating uniformly on a straight railway track, approaches an electric pole standing on the side of track. Its engine passes the pole with velocity u and the guard's room passes with velocity v. The middle wagon of the train passes the pole with a velocity. **[Online May 19, 2012]**

(a) $\frac{u+v}{2}$ (b) $\frac{1}{2}\sqrt{u^2+v^2}$ (c) $\sqrt{uv}$ (d) $\sqrt{\left(\frac{u^2+v^2}{2}\right)}$

Topic 4 - Motion Under Gravity

90. A body falling under gravity covers two points A and B separated by 80 m in 2s. The distance of upper point A from the starting point is ________ m (use g = 10 ms^{-2})

[NCERT : P L-18, 19 | Jan. 27, 2024 (II)]

91. A ball is thrown vertically upward with an initial velocity of 150 m/s. The ratio of velocity after 3 s and 5 s is $\frac{x+1}{x}$. The value of x is ______.
Take (g = 10 m/s^2).

[NCERT : P L-18, 19 | April 12, 2023 (I)]

(a) 6 (b) 5 (c) –5 (d) 10

92. A tennis ball is dropped on to the floor from a height of 9.8 m. It rebounds to a height 5.0 m. Ball comes in contact with the floor for 0.2s. The average acceleration during contact is _____ ms^{-2}. [Given g = 10 ms^{-2}]

[NCERT : P L-18, 19 | Jan. 29, 2023 (I)]

93. A ball is thrown up vertically with a certain velocity so that, it reaches a maximum height h. Find the ratio of the times in which it is at height $\frac{h}{3}$ while going up and coming down respectively. **[NCERT : P L-18, 19 | July 29, 2022 (I)]**

(a) $\frac{\sqrt{2}-1}{\sqrt{2}+1}$ (b) $\frac{\sqrt{3}+\sqrt{2}}{\sqrt{3}-\sqrt{2}}$
(c) $\frac{\sqrt{3}-1}{\sqrt{3}+1}$ (d) $\frac{1}{3}$

94. A ball is released from a height h. If t_1 and t_2 be the time required to complete first half and second half of the distance respectively. Then, choose the correct relation between t_1 and t_2. **[NCERT : P L-18, 19 | July 29, 2022 (II)]**

(a) $t_1 = (\sqrt{2})t_2$ (b) $t_1 = (\sqrt{2}-1)t_2$
(c) $t_2 = (\sqrt{2}+1)t_1$ (d) $t_2 = (\sqrt{2}-1)t_1$

95. A juggler throws balls vertically upwards with same initial velocity in air. When the first ball reaches its highest position, he throws the next ball. Assuming the juggler throws n balls per second, the maximum height the balls can reach is **[NCERT : P L-18, 19 | July 29, 2022 (II)]**

(a) g/2n (b) g/n (c) 2gn (d) $g/2n^2$

96. A NCC parade is going at a uniform speed of 9 km/h under a mango tree on which a monkey is sitting at a height of 19.6 m. At any particular instant, the monkey drops a mango. A cadet will receive the mango whose distance from the tree at time of drop is :
(Given g = 9.8 m/s^2)

[NCERT : P L-18, 19 | July 28, 2022 (I)]

(a) 5m (b) 10m (c) 19.8m (d) 24.5m

97. A ball is thrown vertically upwards with a velocity of 19.6 ms^{-1} from the top of a tower. The ball strikes the ground after 6s. The height from the gound up to which the ball can rise will be $\left(\frac{k}{5}\right)$ m. The value of k is ______
(use g = 9.8 m/s^2) **[NCERT : P L-18, 19 | July 28, 2022 (II)]**

98. A bullet is shot vertically downwards with an initial velocity of 100 m/s from a certain height. Within 10 s, the bullet reaches the ground and instantaneously comes to rest due to the perfectly inelastic collision. The velocity-time curve for total time $t = 20$ s will be : (Take $g = 10$ m/s^2)

[NCERT : P L-18, 19 | July 27, 2022 (I)]

(a)

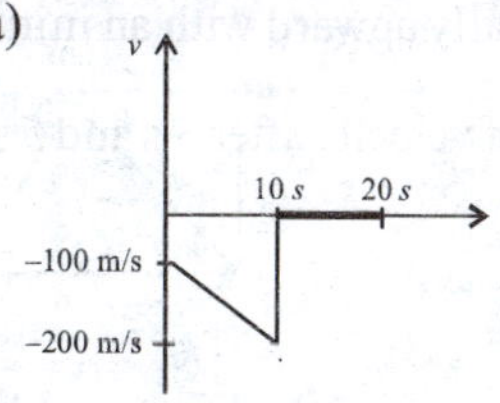

(b)

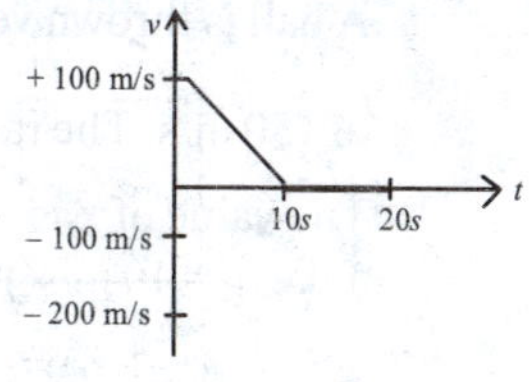

(c)

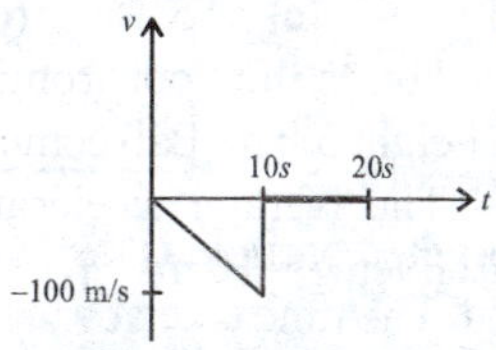

(d)

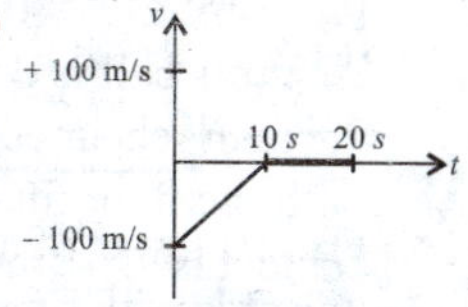

99. Two balls A and B are placed at the top of 180 m tall tower. Ball A is released from the top at $t = 0$ s. Ball B is thrown vertically down with an initial velocity 'u' at $t = 2$ s. After a certain time, both balls meet 100 m above the ground. Find the value of 'u' in ms^{-1}. [use $g = 10\ ms^{-2}$] :

[NCERT : P L-18, 19 | June 29, 2022 (I)]

(a) 10 (b) 15 (c) 20 (d) 30

100. A ball of mass 0.5 kg is dropped from the height of 10m. The height, at which the magnitude of velocity becomes equal to the magnitude of acceleration due to gravity, is .... m. (Use $g = 10\ m/s^2$). **[NCERT : P L-18, 19 | June 26, 2022 (I)]**

101. A ball is projected vertically upward with an initial velocity of $50\ ms^{-1}$ at $t = 0s$. At $t = 2s$. Another ball is projected vertically upward with same velocity. At t = ________s, second ball will meet the first ball ($g = 10\ ms^{-2}$).

[NCERT : P L-18, 19 | June 26, 2022 (II)]

102. Water droplets are coming from an open tap at a particular rate. The spacing between a droplet observed at 4th second after its fall to the next droplet is 34.3 m. At what rate the droplets are coming from the tap ?
(Take $g = 9.8\ m/s^2$) **[NCERT : PL-18, 19 | June 25, 2022 (I)]**

(a) 3 drops / 2 seconds (b) 2 drops / second
(c) 1 drop / second (d) 1 drop / 7 seconds

103. From the top of a tower, a ball is thrown vertically upward which reaches the ground in 6s. A second ball thrown vertically downward from the same position with the same speed reaches the ground in 1.5 s. A third ball released, from the rest from the same location, will reach the ground in ______ s. **[NCERT : P L-18, 19 | June 24, 2022 (I)]**

104. Water drops are falling from a nozzle of a shower onto the floor, from a height of 9.8 m. The drops fall at a regular interval of time. When the first drop strikes the floor, at that instant, the third drop begins to fall. Locate the position of second drop from the floor when the first drop strikes the floor.

[NCERT : P L-18, 19 | Aug. 27, 2021 (II)]

(a) 4.18m (b) 2.94m (c) 2.45m (d) 7.35m

105. Water droplets are coming from an open tap at a particular rate. The spacing between a droplet observed at 4th second after its fall to the next droplet is 34.3 m. At what rate the droplets are coming from the tap ? (Take $g = 9.8\ m/s^2$)

[NCERT : P L-18, 19 | July 25, 2021 (I)]

(a) 3 drops / 2 seconds (b) 2 drops / second
(c) 1 drop / second (d) 1 drop / 7 seconds

106. A balloon was moving upwards with a uniform velocity of 10 m/s. An object of finite mass is dropped from the balloon when it was at a height of 75 m from the ground level. The height of the balloon from the ground when object strikes the ground was around:
(takes the value of g as $10\ m/s^2$)

[NCERT : P L-18, 19 | July 25, 2021 (II)]

(a) 300m (b) 200m (c) 125m (d) 250m

107. The position, velocity and acceleration of a particle moving with a constant acceleration can be represented by:

[NCERT : P L-18, 19 | March 18, 2021 (I)]

(a)

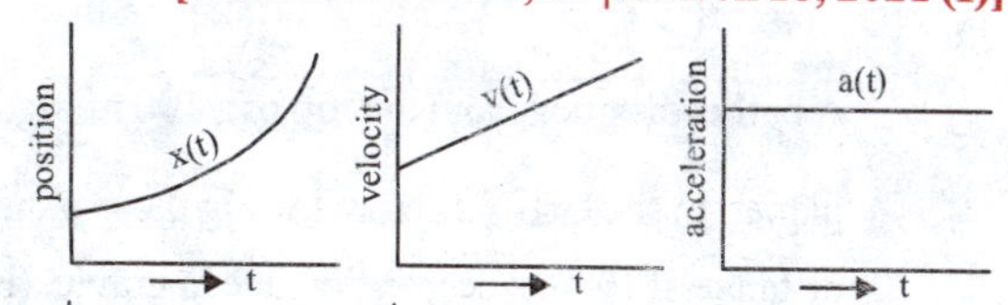

(b)

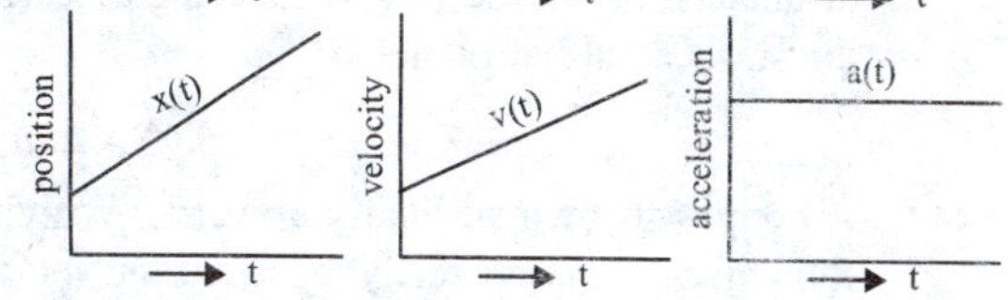

(c)

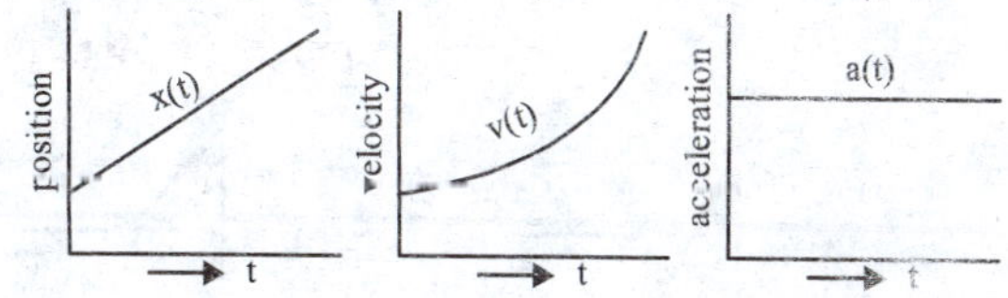

(d)

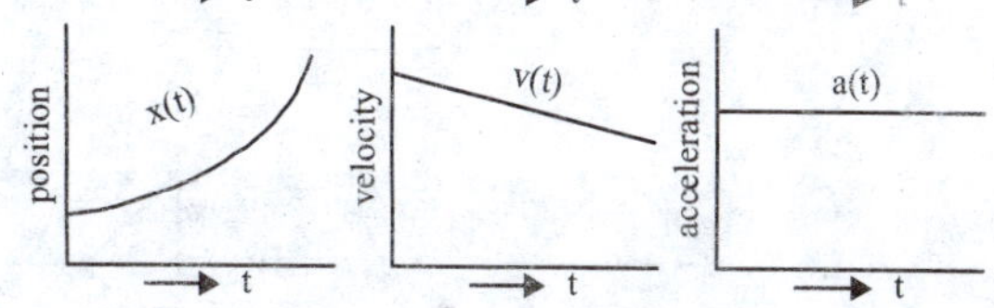

108. A stone is dropped from the top of a building. When it crosses a point 5 m below the top, another stone starts to fall from a point 25 m below the top. Both stones reach the bottom of building simultaneously. The height of the building is **[NCERT : P L-18, 19 | Feb. 25, 2021 (II)]**

(a) 35m (b) 45m (c) 25m (d) 50m

109. A helicopter rises from rest on the ground vertically upwards with a constant acceleration g. A food packet is dropped from the helicopter when it is at a height h. The time taken by the packet to reach the ground is close to [g is the accelertion due to gravity] :

[NCERT : P L-18, 19 | Sep. 5, 2020 (I)]

(a) $t = \frac{2}{3}\sqrt{\left(\frac{h}{g}\right)}$ (b) $t = 1.8\sqrt{\frac{h}{g}}$

(c) $t = \sqrt{\frac{2h}{3g}}$ (d) $t = 3.4\sqrt{\left(\frac{h}{g}\right)}$

110. A Tennis ball is released from a height h and after freely falling on a wooden floor it rebounds and reaches height $\frac{h}{2}$. The velocity versus height of the ball during its motion may be represented graphically by :
(graph are drawn schematically and on not to scale)

[NCERT : P L-18, 19 | Sep. 4, 2020 (I)]

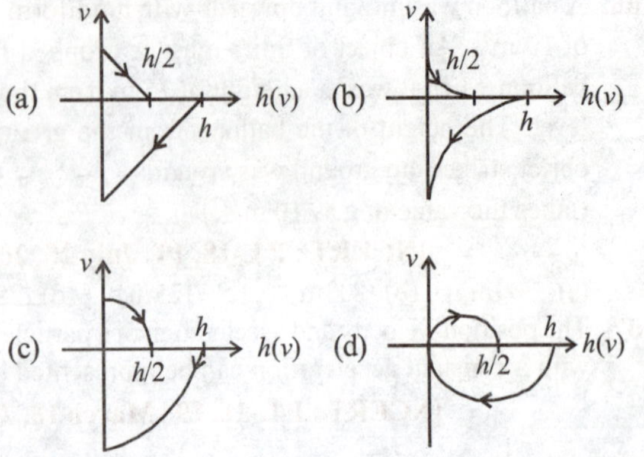

111. A ball is dropped from the top of a 100 m high tower on a planet. In the last $\frac{1}{2}$ s before hitting the ground, it covers a distance of 19 m. Acceleration due to gravity (in ms^{-2}) near the surface on that planet is ______.

[NCERT : P L-18, 19 | NA Jan. 8, 2020 II]

112. A body is thrown vertically upwards. Which one of the following graphs correctly represent the velocity vs time?

[NCERT : P L-18, 19 | 2017]

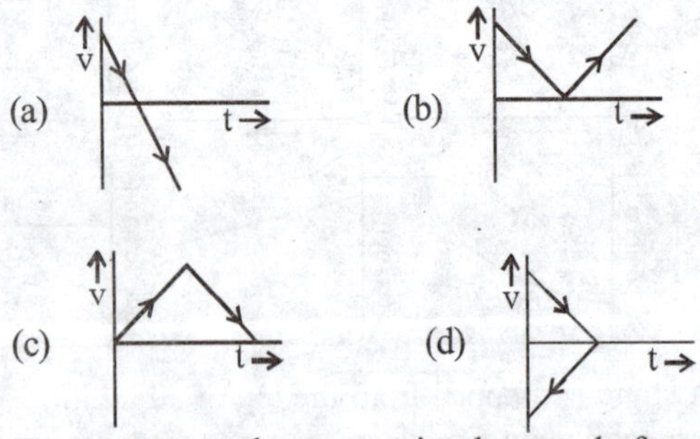

113. Two stones are thrown up simultaneously from the edge of a cliff 240 m high with initial speed of 10 m/s and 40 m/s respectively. Which of the following graph best represents the time variation of relative position of the second stone with respect to the first ?
(Assume stones do not rebound after hitting the ground and neglect air resistance, take $g = 10\ m/s^2$)
(*The figures are schematic and not drawn to scale*)

[NCERT : P L-18, 19 | 2015]

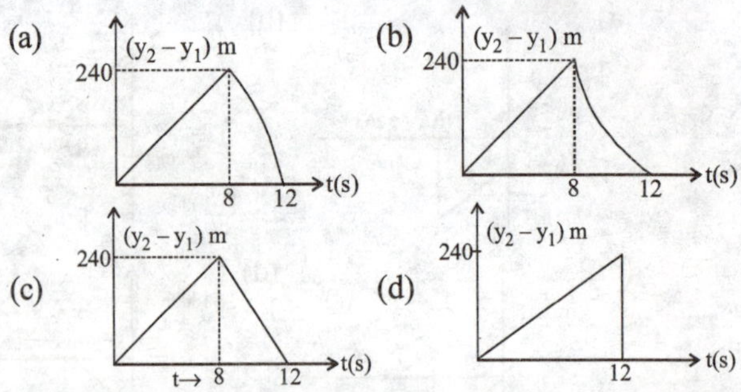

114. From a tower of height H, a particle is thrown vertically upwards with a speed u. The time taken by the particle, to hit the ground, is n times that taken by it to reach the highest point of its path. The relation between H, u and n is: **[NCERT : P L-18, 19 | 2014]**

(a) $2gH = n^2u^2$ (b) $gH = (n-2)^2 u^2d$
(c) $2gH = nu^2 (n-2)$ (d) $gH = (n-2)u^2$

115. A parachutist after bailing out falls 50 m without friction. When parachute opens, it decelerates at $2\ m/s^2$. He reaches the ground with a speed of 3 m/s. At what height, did he bail out ? **[NCERT : P L-18, 19 | 2005]**

(a) 182m (b) 91m (c) 111m (d) 243m

116. A ball is released from the top of a tower of height h meters. It takes *T* seconds to reach the ground. What is the position of the ball at $\frac{T}{3}$ second **[NCERT : P L-18, 19 | 2004]**

(a) $\frac{8h}{9}$ meters from the ground
(b) $\frac{7h}{9}$ meters from the ground
(c) $\frac{h}{9}$ meters from the ground
(d) $\frac{17h}{18}$ meters from the ground

117. From a building two balls *A* and *B* are thrown such that *A* is thrown upwards and *B* downwards (both vertically) with same speed. If v_A and v_B are their respective velocities on reaching the ground, then **[NCERT : P L-18, 19 | 2002]**

(a) $v_B > v_A$
(b) $v_A = v_B$
(c) $v_A > v_B$
(d) their velocities depend on their masses.

ANSWER KEYS

1	(a)	13	(50)	25	(d)	37	(None)	49	(b)	61	(3)	73	(b)	85	(a)	97	(392)	109	(c)
2	(d)	14	(d)	26	(c)	38	(c)	50	(c)	62	(b)	74	(d)	86	(d)	98	(a)	110	(c)
3	(10)	15	(b)	27	(c)	39	(a)	51	(3)	63	(b)	75	(c)	87	(c)	99	(d)	111	(08.00)
4	(d)	16	(3)	28	(3)	40	(c)	52	(100)	64	(d)	76	(d)	88	(c)	100	(5)	112	(a)
5	(b)	17	(18)	29	(19)	41	(d)	53	(c)	65	(d)	77	(a)	89	(d)	101	(6)	113	(a)
6	(b)	18	(b)	30	(b)	42	(c)	54	(d)	66	(c)	78	(a)	90	(45)	102	(c)	114	(c)
7	(c)	19	(b)	31	(40)	43	(a)	55	(1)	67	(b)	79	(b)	91	(b)	103	(3)	115	(d)
8	(d)	20	(c)	32	(52)	44	(d)	56	(12)	68	(c)	80	(c)	92	(120)	104	(d)	116	(a)
9	(a)	21	(d)	33	(8)	45	(200)	57	(b)	69	(a)	81	(d)	93	(b)	105	(c)	117	(b)
10	(d)	22	(d)	34	(a)	46	(a)	58	(c)	70	(b)	82	(a)	94	(d)	106	(c)		
11	(b)	23	(3)	35	(175)	47	(a)	59	(a)	71	(a)	83	(c)	95	(d)	107	(a)		
12	(d)	24	(a)	36	(a)	48	(d)	60	(20)	72	(a)	84	(a)	96	(a)	108	(b)		

Motion in a Plane

3

TREND ANALYSIS
JEE MAIN (2025-2021)

	2025				2024				2023				2022				2021			
	Session II		Session I		Session II		Session I		Session II		Session I		Session II		Session I		Session II		Session I	
	MCQs	NVQs	MCQs	NVQs	MCQs	NVQs	MCQs	NVQs	MCQs	NVQs	MCQs	NVQs	MCQs	NVQs	MCQs	NVQs	MCQs	NVQs	MCQs	NVQs
Topic 1	–	–	–	1	–	4	1	1	2	–	1	2	–	2	3	–	5	1	1	1
Topic 2	1	–	1	–	1	1	1	1	–	–	1	1	1	–	2	1	2	–	1	–
Topic 3	4	–	1	1	1	1	1	1	6	1	4	1	4	3	5	2	2	–	1	1
Topic 4	1	–	–	1	–	–	1	1	–	–	1	2	–	–	1	–	–	1	–	2
Total	6	–	2	3	2	6	4	4	8	1	7	6	5	5	11	3	9	2	3	4

How to Code Qns for Better & Faster Revision

Type 1: Able to solve (No code Req)

Type 2: Solved with struggle or by seeing solution **(Use code NR means Needs Revision or orange stickers)**

Type 3: Tricky/ Could not solve on my own. Need help from Teacher/ Friend. **(Use code TQ means Tricky Question or red stickers)**

You can either hand write the code in boxes or else paste coloured stickers given at end of book.

Topic 1 - Vectors

1. Two particles are located at equal distance from origin. The position vectors of those are represented by $\vec{A} = 2\hat{i} + 3n\hat{j} + 2\hat{k}$ and $\vec{B} = 2\hat{i} - 2\hat{j} + 4p\hat{k}$, respectively. If both the vectors are at right angle to each other, the value of n^{-1} is ______.

[NCERT : P L-72 | Jan. 23, 2025 (I)]

2. If $\vec{a}$ and $\vec{b}$ makes an angle $\cos^{-1}\left(\frac{5}{9}\right)$ with each other, then $|\vec{a} + \vec{b}| = \sqrt{2}|\vec{a} - \vec{b}|$ for $|\vec{a}| = n|\vec{b}|$. The integer value of n is ______.

[NCERT : P L-34 | April 9, 2024 (I)]

3. The resultant of two vectors $\vec{A}$ and $\vec{B}$ is perpendicular to $\vec{A}$ and its magnitude is half that of $\vec{B}$. The angle between vectors $\vec{A}$ and $\vec{B}$ is .... degree.

[NCERT : P L-34 | April 9, 2024 (II)]

4. For three vectors $\vec{A} = \left(-x\hat{i} - 6\hat{j} - 2\hat{k}\right)$, $\vec{B} = \left(-\hat{i} + 4\hat{j} + 3\hat{k}\right)$ and $\vec{C} = \left(-8\hat{i} - \hat{j} + 3\hat{k}\right)$, if $\vec{A}.\left(\vec{B} \times \vec{C}\right) = 0$, them value of x is ______.

[NCERT : P L-102, 103 | April 6, 2024 (I)]

5. Two forces $\vec{F}_1$ and $\vec{F}_2$ are acting on a body. One force has magnitude thrice that of the other force and the resultant of the two forces is equal to the force of larger magnitude.

The angle between $\vec{F}_1$ and $\vec{F}_2$ is $\cos^{-1}\left(\frac{1}{n}\right)$. Th e value of $|n|$ is ______.

[NCERT : P L-34 | April 4, 2024 (I)]

6. If two vectors $\vec{A}$ and $\vec{B}$ having equal magnitude R are inclined at an angle θ, then

[NCERT : P L-34 | Jan. 31, 2024 (II)]

(a) $|\vec{A} - \vec{B}| = \sqrt{2}R\sin\left(\frac{\theta}{2}\right)$
(b) $|\vec{A} + \vec{B}| = 2R\sin\left(\frac{\theta}{2}\right)$
(c) $|\vec{A} + \vec{B}| = 2R\cos\left(\frac{\theta}{2}\right)$
(d) $|\vec{A} - \vec{B}| = 2R\cos\left(\frac{\theta}{2}\right)$

7. A vector has magnitude same as that of $\vec{A} = 3\hat{i} + 4\hat{j}$ and is parallel to $\vec{B} = 4\hat{i} + 3\hat{j}$. The x and y components of this vector in first quadrant are x and 3 respectively where x = ______. [NCERT : P L-32 | Jan. 30, 2024 (II)]

8. A vector in x-y plane makes an angle of 30° with y-axis. The magnitude of y-component of vector is $2\sqrt{3}$. The magnitude of x-component of the vector will be : [NCERT : P L-32 | April 15, 2023 (I)]

(a) $\frac{1}{\sqrt{3}}$ (b) 6 (c) $\sqrt{3}$ (d) 2

9. When vector $\vec{A} = 2\hat{i} + 3\hat{j} + 2\hat{k}$ is subtracted from vector $\vec{B}$, it gives a vector equal to $2\hat{j}$. Then the magnitude of vector $\vec{B}$ will be: [NCERT : P L-33 | April 11, 2023 (II)]

(a) $\sqrt{13}$ (b) 3 (c) $\sqrt{6}$ (d) $\sqrt{5}$

10. If $\vec{P} = 3\hat{i} + \sqrt{3}\hat{j} + 2\hat{k}$ and $\vec{Q} = 4\hat{i} + \sqrt{3}\hat{j} + 2.5\hat{k}$ then, The unit vector in the direction of $\vec{P} \times \vec{Q}$ is $\frac{1}{x}\left(\sqrt{3}\hat{i} + \hat{j} - 2\sqrt{3}\hat{k}\right)$. The value of x is [NCERT : P L-102, 103 | Jan. 25, 2023 (I)]

11. Vectors $a\hat{i} + b\hat{j} + \hat{k}$ and $2\hat{i} - 3\hat{j} + 4\hat{k}$ are perpendicular to each other when $3a + 2b = 7$, the ratio of a to b is $\frac{x}{2}$. The value of x is _____. [NCERT : P L-72 | Jan. 24, 2023 (I)]

12. If two vectors $\vec{P} = \hat{i} + 2m\hat{j} + m\hat{k}$ and $\vec{Q} = 4\hat{i} - 2\hat{j} + m\hat{k}$ are perpendicular to each other. Then, the value of m will be : [NCERT : P L-72 | Jan. 24, 2023 (II)]

(a) 1 (b) −1 (c) −3 (d) 2

13. If the projection of $2\hat{i} + 4\hat{j} - 2\hat{k}$ on $\hat{i} + 2\hat{j} + \alpha\hat{k}$ is zero. Then, the value of α will be [NCERT : P L-72 | July 28, 2022 (I)]

14. If $\vec{A} = \left(2\hat{i} + 3\hat{j} - \hat{k}\right)$ m and $\vec{B} = \left(\hat{i} + 2\hat{j} + 2\hat{k}\right)$ m. The magnitude of component of vector $\vec{A}$ along vector $\vec{B}$ will be______ m. [NCERT : P L-72 | July 26, 2022 (II)]

15. Two vectors $\vec{A}$ and $\vec{B}$ have equal magnitudes. If magnitude of $\vec{A} + \vec{B}$, is equal to two times the magnitude of $\vec{A} - \vec{B}$, then the angle between $\vec{A}$ and $\vec{B}$ will be : [NCERT : P L-34 | June 29, 2022 (I)]

(a) $\sin^{-1}\left(\frac{3}{5}\right)$ (b) $\sin^{-1}\left(\frac{1}{3}\right)$

(c) $\cos^{-1}\left(\frac{3}{5}\right)$ (d) $\cos^{-1}\left(\frac{1}{3}\right)$

16. Which of the following relations is true for two unit vectors $\hat{A}$ and $\hat{B}$ making an angle θ to each other? [NCERT : P L-32 | June 25, 2022 (I)]

(a) $|\hat{A}+\hat{B}| = |\hat{A}-\hat{B}|\tan\frac{\theta}{2}$ (b) $|\hat{A}-\hat{B}| = |\hat{A}+\hat{B}|\tan\frac{\theta}{2}$

(c) $|\hat{A}+\hat{B}| = |\hat{A}-\hat{B}|\cos\frac{\theta}{2}$ (d) $|\hat{A}-\hat{B}| = |\hat{A}+\hat{B}|\cos\frac{\theta}{2}$

17. $\vec{A}$ is a vector quantity such that $|\vec{A}|$ = non-zero constant. Which of the following expressions is true for $\vec{A}$? [NCERT : P L-32 | June 25, 2022 (I)]

(a) $\vec{A}.\vec{A} = 0$ (b) $\vec{A} \times \vec{A} < 0$

(c) $\vec{A} \times \vec{A} = 0$ (d) $\vec{A} \times \vec{A} > 0$

18. **Statement I :** Two forces $\left(\vec{P} + \vec{Q}\right)$ and $\left(\vec{P} - \vec{Q}\right)$ where $\vec{P} \perp \vec{Q}$, when act at an angle θ_1 to each other, the magnitude of their resultant is $\sqrt{3\left(P^2 + Q^2\right)}$, when they act at an angle θ_2, the magnitude of their resultant becomes $\sqrt{2\left(P^2 + Q^2\right)}$. This is possible only when $\theta_1 < \theta_2$.

Statement II : In the situation given above.

$\theta_1 = 60°$ and $\theta_2 = 90°$

In the light of the above statements, choose the most appropriate answer from the options given below :- [NCERT : P L-34 | Aug. 31, 2021 (II)]

(a) Statement-I is false but Statement-II is true
(b) Both Statement-I and Statement-II are true
(c) Statement-I is true but Statement-II is false
(c) Both Statement-I and Statement-II are false.

19. The magnitude of vectors $\overrightarrow{OA}$, $\overrightarrow{OB}$ and $\overrightarrow{OC}$ in the given figure are equal. The direction of $\overrightarrow{OA} + \overrightarrow{OB} - \overrightarrow{OC}$ with x-axis will be :- [Aug. 26, 2021 (I)]

(a) $\tan^{-1}\frac{\left(1-\sqrt{3}-\sqrt{2}\right)}{\left(1+\sqrt{3}+\sqrt{2}\right)}$

(b) $\tan^{-1}\frac{\left(\sqrt{3}-1+\sqrt{2}\right)}{\left(1+\sqrt{3}-\sqrt{2}\right)}$

(c) $\tan^{-1}\frac{\left(\sqrt{3}-1+\sqrt{2}\right)}{\left(1-\sqrt{3}+\sqrt{2}\right)}$

(d) $\tan^{-1}\frac{\left(1+\sqrt{3}-\sqrt{2}\right)}{\left(1-\sqrt{3}-\sqrt{2}\right)}$

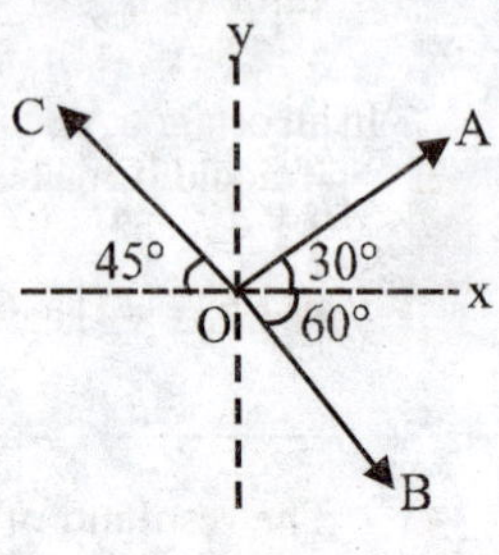

20. The angle between vector $(\vec{A})$ and $(\vec{A}-\vec{B})$ is :

[NCERT : P L-72 | Aug. 26, 2021 (II)]

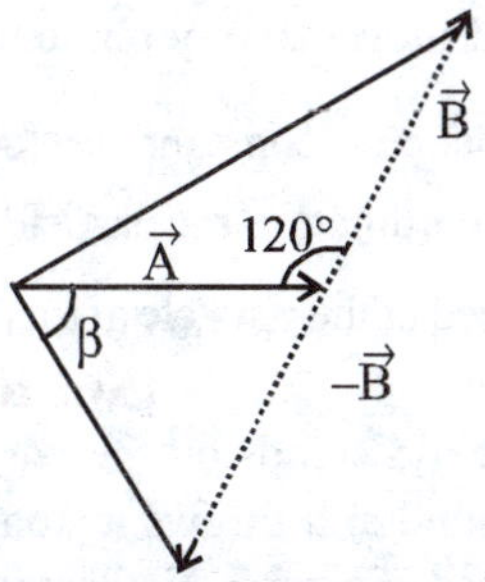

(a) $\tan^{-1}\left(\dfrac{-\frac{B}{2}}{A-B\frac{\sqrt{3}}{2}}\right)$

(b) $\tan^{-1}\left(\dfrac{A}{0.7B}\right)$

(c) $\tan^{-1}\left(\dfrac{\sqrt{3}B}{2A-B}\right)$

(d) $\tan^{-1}\left(\dfrac{B\cos\theta}{A-B\sin\theta}\right)$

21. Two vectors $\vec{X}$ and $\vec{Y}$ have equal magnitude. The magnitude of $\vec{X}-\vec{Y}$ is n times the magnitude of $(\vec{X}+\vec{Y})$. The angle between $\vec{X}$ and $\vec{Y}$ is :

[NCERT : P L-34 | July 25, 20, 2021 (II)]

(a) $\cos^{-1}\left(\dfrac{1-n^2}{1+n^2}\right)$ (b) $\cos^{-1}\left(\dfrac{n^2-1}{-n^2-1}\right)$

(c) $\cos^{-1}\left(\dfrac{n^2+1}{-n^2-1}\right)$ (d) $\cos^{-1}\left(\dfrac{n^2+1}{n^2-1}\right)$

22. What will be the projection of vector $\vec{A}=\hat{i}+\hat{j}+\hat{k}$ on vector $\vec{B}=\hat{i}+\hat{j}$? [NCERT : P L-72 | July 22, 2021 (II)]

(a) $\sqrt{2}(\hat{i}+\hat{j}+\hat{k})$ (b) $2(\hat{i}+\hat{j}+\hat{k})$

(c) $\sqrt{2}(\hat{i}+\hat{j})$ (d) $(\hat{i}+\hat{j})$

23. Three particles P, Q and R are moving along the vectors $\vec{A}=\hat{i}+\hat{j}$, $\vec{B}=\hat{j}+\hat{k}$ and $\vec{C}=-\hat{i}+\hat{j}$ respectively. They strike on a point and start to move in different directions. Now particle P is moving normal to the plane which contains vector $\vec{A}$ and $\vec{B}$. Similarly particle Q is moving normal to the plane which contains vector $\vec{A}$ and $\vec{C}$. The angle between the direction of motion of P and Q is $\cos^{-1}\left(\dfrac{1}{\sqrt{x}}\right)$. Then the value of x is _______.

[NCERT : P L-102, 103 | NA, July 22, 2021 (II)]

24. In an octagon ABCDEFGH of equal side, what is the sum of $\overrightarrow{AB}+\overrightarrow{AC}+\overrightarrow{AD}+\overrightarrow{AE}+\overrightarrow{AF}+\overrightarrow{AG}+\overrightarrow{AH}$,

if $\overrightarrow{AO}=2\hat{i}+3\hat{j}-4\hat{k}$

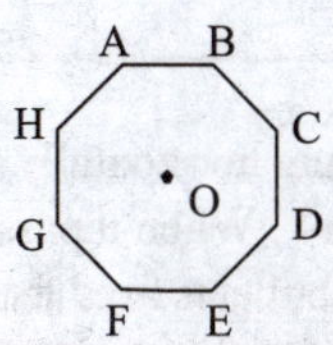

[NCERT : P L-29, 30 | Feb. 25, 2021 (I)]

(a) $-16\hat{i}-24\hat{j}+32\hat{k}$ (b) $16\hat{i}+24\hat{j}+32\hat{k}$

(c) $16\hat{i}+24\hat{j}-32\hat{k}$ (d) $16\hat{i}-24\hat{j}+32\hat{k}$

25. If $\vec{P}\times\vec{Q}=\vec{Q}\times\vec{P}$, the angle between $\vec{P}$ and $\vec{Q}$ is θ $(0° < 30° < 360°)$. The value of θ will be _______°.

[NCERT : P L-102, 103 | Feb. 25, 2021 (II)]

26. A force $\vec{F}=(\hat{i}+2\hat{j}+3\hat{k})$ N acts at a point $(4\hat{i}+3\hat{j}-\hat{k})$ m. Then the magnitude of torque about the point $(\hat{i}+2\hat{j}+\hat{k})$ m will be $\sqrt{x}$ N-m. The value of x is _______.

[NCERT : P L-102, 103 | NA Sep. 05, 2020 (I)]

27. The sum of two forces $\vec{P}$ and $\vec{Q}$ is $\vec{R}$ such that $|\vec{R}|=|\vec{P}|$. The angle θ (in degrees) that the resultant of $2\vec{P}$ and $\vec{Q}$ will make with $\vec{Q}$ is _______.

[NCERT : P L-34 | NA 7 Jan. 2020 II]

28. Let $|\vec{A_1}|=3$, $|\vec{A_2}|=5$ and $|\vec{A_1}+\vec{A_2}|=5$. The value of $(2\vec{A_1}+3\vec{A_2})\cdot(3\vec{A_1}-2\vec{A_2})$ is :

[NCERT : P L-34 | 8 April 2020 II]

(a) – 106.5 (b) –99.5 (c) –112.5 (d) –118.5

29. In the cube of side 'a' shown in the figure, the vector from the central point of the face ABOD to the central point of the face BEFO will be:

[NCERT : P L-33 | 10 Jan. 2019 I]

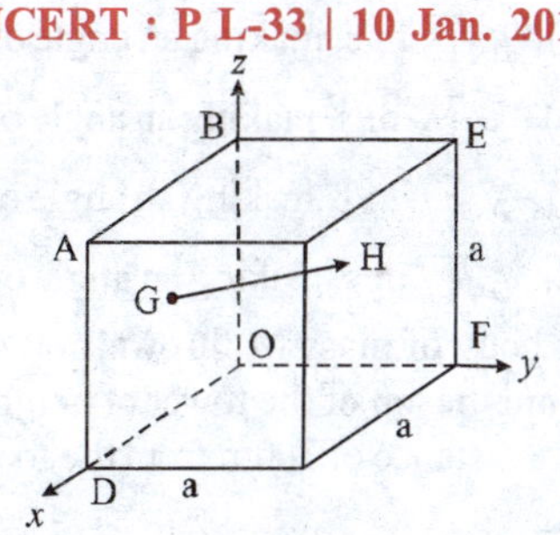

(a) $\dfrac{1}{2}a(\hat{k}-\hat{i})$

(b) $\dfrac{1}{2}a(\hat{i}-\hat{k})$

(c) $\dfrac{1}{2}a(\hat{j}-\hat{i})$

(d) $\dfrac{1}{2}a(\hat{j}-\hat{k})$

30. Two forces P and Q, of magnitude 2F and 3F, respectively, are at an angle θ with each other. If the force Q is doubled, then their resultant also gets doubled. Then, the angle θ is: [NCERT : P L-34 | 10 Jan. 2019 II]

(a) 120° (b) 60° (c) 90° (d) 30°

31. Let $\vec{A}=(\hat{i}+\hat{j})$ and $\vec{B}=(2\hat{i}-\hat{j})$. The magnitude of a coplanar vector $\vec{C}$ such that $\vec{A}.\vec{C}=\vec{B}.\vec{C}=\vec{A}.\vec{B}$ is given by [NCERT : P L-72 | Online April 16, 2018]

(a) $\sqrt{\dfrac{5}{9}}$ (b) $\sqrt{\dfrac{10}{9}}$

(c) $\sqrt{\dfrac{20}{9}}$ (d) $\sqrt{\dfrac{9}{12}}$

32. A vector $\vec{A}$ is rotated by a small angle Δθ radian (Δθ << 1) to get a new vector $\vec{B}$. In that case $|\vec{B}-\vec{A}|$ is :

[Online April 11, 2015]

(a) $|\vec{A}|\Delta\theta$ (b) $|\vec{B}|\Delta\theta - |\vec{A}|$

(c) $|\vec{A}|\left(1-\frac{\Delta\theta^2}{2}\right)$ (d) 0

33. If $\vec{A}\times\vec{B}=\vec{B}\times\vec{A}$, then the angle between A and B is **[NCERT : P L-102, 103 | 2004]**

(a) $\frac{\pi}{2}$ (b) $\frac{\pi}{3}$ (c) π (d) $\frac{\pi}{4}$

Topic 2 - Motion in a Plane with Constant Acceleration

34. A helicopter flying horizontally with a speed of 360 km/h at an altitude of 2 km, drops an object at an instant. The object hits the ground at a point O, 20 s after it is dropped. Displacement of 'O' from the position of helicopter where the object was released is :
(use acceleration due to gravity $g = 10\ m/s^2$ and neglect air resistance) **[NCERT : P L-37 | April 7, 2025 (II)]**

(a) $2\sqrt{5}$ km (b) 4 km (c) 7.2 km (d) $2\sqrt{2}$ km

35. The position vector of a moving body at any instant of time is given as $\vec{r}=\left(5t^2\hat{i}-5t\hat{j}\right)m$. The magnitude and direction of velocity at t = 2s is **[NCERT : P L-36 | Jan. 24, 2025 (II)]**

(a) $5\sqrt{15}$ m/s, making an angle of $\tan^{-1}4$ with –ve Y axis

(b) $5\sqrt{15}$ m/s, making an angle of $\tan^{-1}4$ with +ve X axis

(c) $5\sqrt{17}$ m/s, making an angle of $\tan^{-1}4$ with –ve Y axis

(d) $5\sqrt{17}$ m/s, making an angle of $\tan^{-1}4$ with +ve X axis

36. A body of mass M thrown horizontally with velocity v from the top of the tower of height H touches the ground at a distance of 100m from the foot of the tower. A body of mass 2M thrown at a velocity $\frac{v}{2}$ from the top of the tower of height 4H will touch the ground at a distance of ....m. **[April 8, 2024 (II)]**

37. The co-ordinates of a particle moving in x-y plane are given by :
$x = 2 + 4t,\ y = 3t + 8t^2$.
The motion of the particle is : **[NCERT : P L-36 | April 4, 2024 (I)]**

(a) non-uniformly accelerated.
(b) uniformly accelerated having motion along a straight line.
(c) uniform motion along a straight line.
(d) uniformly accelerated having motion along a parabolic path.

38. Position of an ant (S in metres) moving in Y-Z plane is given by $S = 2t^2\hat{j}+5\hat{k}$ (where t is in second). The magnitude and direction of velocity of the ant at t = 1 s will be : **[NCERT : P L-36 | Jan. 27, 2024 (I)]**

(a) 16 m/s in y-direction (b) 4 m/s in x-direction
(c) 9 m/s in z-direction (d) 4 m/s in y-direction

39. A particle starts from origin at t = 0 with a velocity $5\hat{i}$ m/s and moves in x-y plane under action of a force which produces a constant acceleration of $(3\hat{i}+2\hat{j})m/s^2$. If the x-coordinate of the particle at that instant is 84 m, then the speed of the particle at this time is $\sqrt{\alpha}$ m/s. The value of α is ______. **[NCERT : P L-38 | Jan. 27, 2024 (I)]**

40. A child stands on the edge of the cliff 10 m above the ground and throws a stone horizontally with an initial speed of 5 ms^{-1}. Neglecting the air resistance, the speed with which the stone hits the ground will be ____ ms^{-1} (given, $g = 10\ ms^{-2}$). **[NCERT : P L-36 | Feb. 1, 2023 (I)]**

(a) 15 (b) 20 (c) 30 (d) 25

41. The speed of a swimmer is 4 km h^{-1} in still water. If the swimmer makes his strokes normal to the flow of river of width 1 km, he reaches a point 750 m down the stream on the opposite bank. The speed of the river water is ______ km h^{-1}. **[Jan 31, 2023 (I)]**

42. At time t = 0 a particle starts travelling from a height. $7\hat{z}$ cm in a plane keeping z coordinate constant. At any instant of time it's position along the $\hat{x}$ and $\hat{y}$ directions are defined as 3t and $5t^3$ respectively. At t = 1s acceleration of the particle will be **[NCERT : P L-36 | July 28, 2022 (II)]**

(a) $-30\hat{y}$ (b) $30\hat{y}$
(c) $3\hat{x}+15\hat{y}$ (d) $3\hat{x}+15\hat{y}+7\hat{z}$

43. At t = 0, truck, starting from rest, moves in the positive x-direction at uniform acceleration of 5 ms^{-2}. At t = 20 s, a ball is released from the top of the truck. The ball strikes the ground in 1 s after the release. The velocity of the ball, when it strikes the ground, will be: (Given $g = 10\ ms^{-2}$) **[NCERT : P L-37 | June 30, 2022 (I)]**

(a) $100\hat{i}-10\hat{j}$ (b) $10\hat{i}-100\hat{j}$
(c) $100\hat{i}$ (d) $-10\hat{j}$

44. A girl standing on road holds her umbrella at 45° with the vertical to keep the rain away. If she starts running without umbrella with a speed of $15\sqrt{2}kmh^{-1}$, the rain drops hit her head vertically. The speed of rain drops with respect to the moving girl is: **[June 27, 2022 (I)]**

(a) $30\ kmh^{-1}$ (b) $\frac{25}{\sqrt{2}}kmh^{-1}$
(c) $\frac{30}{\sqrt{2}}kmh^{-1}$ (d) $25\ kmh^{-1}$

45. A fighter jet is flying horizontally at a certain altitude with a speed of 200 ms^{-1}. When it passes directly overhead an anti-aircraft gun, bullet is fired from the gun, at an angle θ with the horizontal, to hit the jet. If the bullet speed is 400 m/s, the value of θ will be .... °. **[June 26, 2022 (I)]**

46. A helicopter is flying horizontally with a speed 'v' at an altitude 'h' has to drop a food packet for a man on the ground. What is the distance of helicopter from the man when the food packet is dropped? **[Aug. 31, 2021 (I)]**

(a) $\sqrt{\frac{2ghv^2+1}{h^2}}$ (b) $\sqrt{2ghv^2+h^2}$

(c) $\sqrt{\frac{2v^2h}{g}+h^2}$ (d) $\sqrt{\frac{2gh}{v^2}+h^2}$

47. A bomb is dropped by fighter plane flying horizontally. To an observer sitting in the plane, the trajectory of the bomb is a : **[NCERT : P L-37 | Aug. 26, 2021 (II)]**

(a) hyperbola

(b) parabola in the direction of motion of plane

(c) straight line vertically down the plane

(d) parabola in a direction opposite to the motion of plane

48. A mosquito is moving with a velocity $\vec{v}=0.5t^2\hat{i}+3t\hat{j}+9\hat{k}$ m/s and accelerating in uniform conditions. What will be the direction of mosquito after 2 s?

[NCERT : P L-36 | March 16, 2021 (II)]

(a) $\tan^{-1}\left(\frac{2}{3}\right)$ from y-axis (b) $\tan^{-1}\left(\frac{5}{2}\right)$ from y-axis

(c) $\tan^{-1}\left(\frac{5}{2}\right)$ from x-axis (d) $\tan^{-1}\left(\frac{2}{3}\right)$ from x-axis

49. A balloon is moving up in air vertically above a point A on the ground. When it is at a height h_1, a girl standing at a distance d (point B) from A (see figure) sees it at an angle 45° with respect to the vertical. When the balloon climbs up a further height h_2, it is seen at an angle 60° with respect to the vertical if the girl moves further by a distance 2.464 d (point C). Then the height h_2 is (given tan 30° = 0.5774): **[Sep. 05, 2020 (I)]**

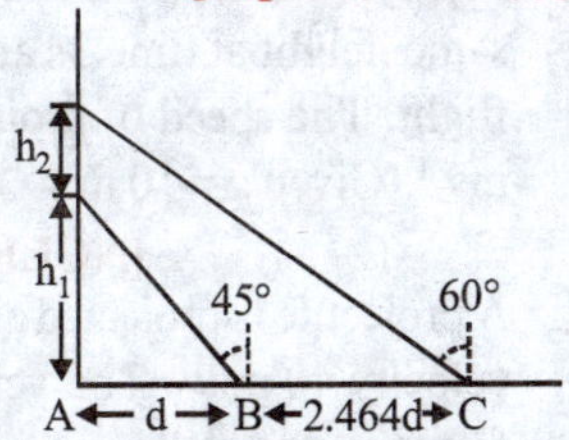

(a) 1.464 d

(b) 0.732 d

(c) 0.464 d

(d) d

50. Starting from the origin at time $t = 0$, with initial velocity $5\hat{j}$ ms^{-1}, a particle moves in the x–y plane with a constant acceleration of $(10\hat{i}+4\hat{j})$ ms^{-2}. At time t, its coordiantes are (20 m, y_0 m). The values of t and y_0 are, respectively: **[NCERT : P L-38 | Sep. 04, 2020 (I)]**

(a) 4 s and 52 m (b) 2 s and 18 m

(c) 2 s and 24 m (d) 5 s and 25 m

51. The position vector of a particle changes with time according to the relation $\vec{r}(t)=15t^2\hat{i}+(4-20t^2)\hat{j}$. What is the magnitude of the acceleration at $t = 1$? **[NCERT : P L-36 | 9 April 2019 II]**

(a) 40 (b) 25 (c) 100 (d) 50

52. A particle moves from the point $(2.0\hat{i}+4.0\hat{j})$m, at t = 0, with an initial velocity $(5.0\hat{i}+4.0\hat{j})$ms^{-1}. It is acted upon by a constant force which produces a constant acceleration $(4.0\hat{i}+4.0\hat{j})$ms^{-2}. What is the distance of the particle from the origin at time 2s? **[NCERT : P L-38 | 11 Jan. 2019 II]**

(a) 15m (b) $20\sqrt{2}$m (c) 5m (d) $10\sqrt{2}$m

53. A particle is moving with a velocity $\vec{v}$ = K $(y\hat{i}+x\hat{j})$, where K is a constant. The general equation for its path is: **[NCERT : P L-36 | 2010, 9 Jan. 2019 I]**

(a) $y = x^2$ + constant (b) $y^2 = x$ + constant

(c) $y^2 = x^2$ + constant (d) xy = constant

54. A particle is moving along the x-axis with its coordinate with time 't' given by $x(t) = 10 + 8t - 3t^2$. Another particle is moving along the y-axis with its coordinate as a function of time given by $y(t) = 5 - 8t^3$. At $t = 1$ s, the speed of the second particle as measured in the frame of the first particle is given as $\sqrt{v}$ m/s. Then v is____

[NCERT : P L-36 | NA 8 Jan. 2020 I]

55. A particle moves such that its position vector $\vec{r}(t)$ = cos ω$t\,\hat{i}$ + sin ω$t\,\hat{j}$ where ω is a constant and t is time. Then which of the following statements is true for the velocity $\vec{v}(t)$ and acceleration $\vec{a}(t)$ of the particle: **[NCERT : P L-36 | 8 Jan. 2020 II]**

(a) $\vec{v}$ is perpendicular to $\vec{r}$ and $\vec{a}$ is directed away from the origin

(b) $\vec{v}$ and $\vec{a}$ both are perpendicular to $\vec{r}$

(c) $\vec{v}$ and $\vec{a}$ both are parallel to $\vec{r}$

(d) $\vec{v}$ is perpendicular to $\vec{r}$ and $\vec{a}$ is directed towards the origin

56. A particle has an initial velocity of $3\hat{i}+4\hat{j}$ and an acceleration of $0.4\hat{i}+0.3\hat{j}$. Its speed after 10 s is : **[NCERT : P L-37 | 2009]**

(a) $7\sqrt{2}$ units (b) 7 units

(c) 8.5 units (d) 10 units

57. The co-ordinates of a moving particle at any time 't' are given by $x=\alpha t^3$ and $y=\beta t^3$. The speed of the particle at time 't' is given by **[NCERT : P L-36 | 2003]**

(a) $3t\sqrt{\alpha^2+\beta^2}$ (b) $3t^2\sqrt{\alpha^2+\beta^2}$

(c) $t^2\sqrt{\alpha^2+\beta^2}$ (d) $\sqrt{\alpha^2+\beta^2}$

Topic 3 - Projectile Motion

58. Two balls with same mass and initial velocity, are projected at different angles in such a way that maximum height reached by first ball is 8 times higher than that of the second ball. T_1 and T_2 are the total flying times of first and second ball, respectively, then the ratio of T_1 and T_2 is : **[NCERT : P L-39 | April 8, 2025 (II)]**

(a) $2\sqrt{2}:1$ (b) $2:1$ (c) $\sqrt{2}:1$ (d) $4:1$

59. Two projectiles are fired from ground with same initial speeds from same point at angles $(45^\circ+\alpha)$ and $(45^\circ+\alpha)$ with horizontal direction. The ratio of their times of flights is **[NCERT : P L-39 | April 7, 2025 (I)]**

(a) 1 (b) $\dfrac{1-\tan\alpha}{1+\tan\alpha}$

(c) $\dfrac{1+\sin 2\alpha}{1-\sin 2\alpha}$ (d) $\dfrac{1+\tan\alpha}{1-\tan\alpha}$

60. A particle is projected with velocity u so that its horizontal range is three times the maximum height attained by it. The horizontal range of the projectile is given as $\dfrac{nu^2}{25g}$, where value of n is : **[NCERT : P L-40 | April 3, 2025 (II)]** (Given 'g' is the acceleration due to gravity).

(a) 6 (b) 18 (c) 12 (d) 24

61. The angle of projection of a particle is measured from the vertical axis as ϕ and the maximum height reached by the particle is h_m. Here h_m as function of ϕ can be presented as **[NCERT : P L-39 | April 3, 2025 (I)]**

(a)

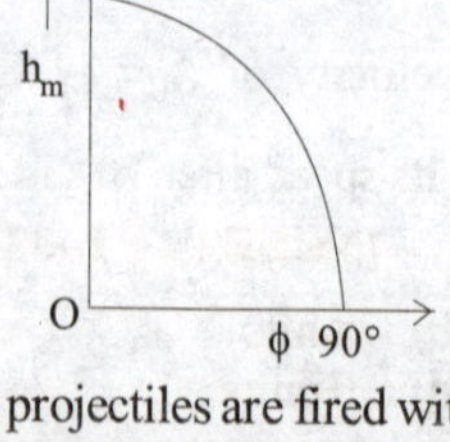

(b)

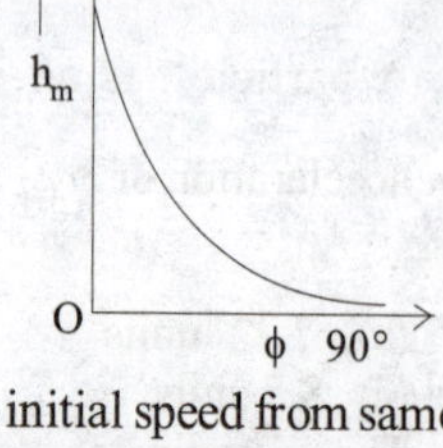

(c)

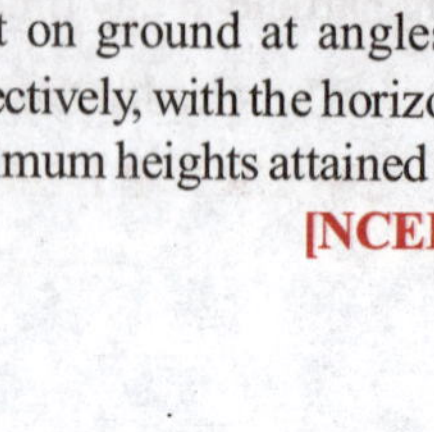

(d)

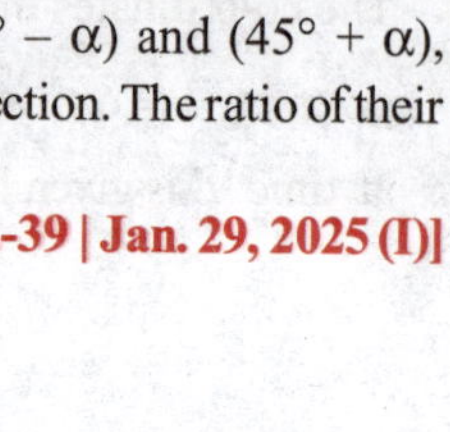

62. Two projectiles are fired with same initial speed from same point on ground at angles of $(45^\circ - \alpha)$ and $(45^\circ + \alpha)$, respectively, with the horizontal direction. The ratio of their maximum heights attained is : **[NCERT : P L-39 | Jan. 29, 2025 (I)]**

(a) $\dfrac{1-\tan\alpha}{1+\tan\alpha}$ (b) $\dfrac{1+\sin\alpha}{1-\sin\alpha}$

(c) $\dfrac{1-\sin 2\alpha}{1+\sin 2\alpha}$ (d) $\dfrac{1+\sin 2\alpha}{1-\sin 2\alpha}$

63. A particle is projected at an angle of 30° from horizontal at a speed of 60 m/s. The height traversed by the particle in the first second is h_0 and height traversed in the last second, before it reaches the maximum heights, is h_1. The ratio $h_0 : h_1$ is _____.
[Take, $g = 10\ m/s^2$] **[NCERT : P L-39 | Jan. 22, 2025 (I)]**

64. The angle of projection for a projectile to have same horizontal range and maximum height is : **[NCERT : P L-39, 40 | April 8, 2024 (II)]**

(a) $\tan^{-1}(2)$ (b) $\tan^{-1}(4)$

(c) $\tan^{-1}\left(\dfrac{1}{4}\right)$ (d) $\tan^{-1}\left(\dfrac{1}{2}\right)$

65. The maximum height reached by a projectile is 64 m. If the initial velocity is halved, the new maximum height of the projectile is ___ m. **[NCERT : P L-39 | April 5, 2024 (II)]**

66. Projectiles A and B are thrown at angles of 45° and 60° with vertical respectively from top of a 400 m high tower. If their ranges and times of flight are same, the ratio of their speeds of projection $v_A : v_B$ is:
[Take $g = 10\ ms^{-2}$]
[NCERT : P L-39, 40 | Jan. 30, 2024 (II)]

(a) $1:\sqrt{3}$ (b) $\sqrt{2}:1$

(c) $1:2$ (d) $1:\sqrt{2}$

67. A ball rolls off the top of a stairway with horizontal velocity u. The steps are 0.1 m high and 0.1 m wide. The minimum velocity u with which that ball just hits the step 5 of the stairway will be $\sqrt{x}\ ms^{-1}$ where x = ______
[use $g = 10\ m/s^2$]. **[NCERT : P L-40 | Jan. 29, 2024 (I)]**

68. A projectile fired at 30° to the ground is observed to be at same height at time 3s and 5s after projection, during its flight. The speed of projection of the projectile is _____ ms^{-1}.(Given $g = 10\ m\ s^{-2}$)
[NCERT : P L-39 | April 11, 2023 (I)]

69. A projectile is projected at 30° from horizontal with initial velocity $40\ ms^{-1}$. The velocity of the projectile at t = 2 s from the start will be: (Given $g = 10\ m/s^2$)
[NCERT : P L-39 | April 11, 2023 (II)]

(a) $20\sqrt{3}\ ms^{-1}$ (b) $40\sqrt{3}\ ms^{-1}$

(c) $20\ ms^{-1}$ (d) Zero

70. The range of the projectile projected at an angle of 15° with horizontal is 50 m. If the projectile is projected with same velocity at an angle of 45° with horizontal, then its range will be : **[NCERT : P L-40 | April 10, 2023 (I)]**

(a) 50m (b) $50\sqrt{2}$ m (c) 100m (d) $100\sqrt{2}$ m

71. Two projectiles are projected at 30° and 60° with the horizontal with the same speed. The ratio of the maximum height attained by the two projectiles respectively is:

[NCERT : P L-39 | April 10, 2023 (II)]

(a) $2:\sqrt{3}$
(b) $\sqrt{3}:1$
(c) 1 : 3
(d) $1:\sqrt{3}$

72. Two projectiles A and B are thrown with initial velocities of 40 m/s and 60 m/s at angles 30° and 60° with the horizontal respectively. The ratio of their ranges respectively is ($g = 10\ m/s^2$)

[NCERT : P L-40 | April 8, 2023 (I)]

(a) $\sqrt{3}:2$
(b) $2:\sqrt{3}$
(c) 1:1
(d) 4:9

73. The trajectory of projectile, projected from the ground is given by $y = x - \frac{x^2}{20}$. Where x and y are measured in meter. The maximum height attained by the projectile will be.

[NCERT : P L-39 | April 8, 2023 (II)]

(a) 5m
(b) $10\sqrt{2}$m
(c) 200m
(d) 10m

74. Given below are two statements : one is labelled as **Assertion A** and the other is labelled as **Reason R.**

Assertion A : When a body is projected at an angle 45°, it's range is maximum.

Reason R : For maximum range, the value of sin2θ should be equal to one.

In the light of the above statements, choose the **correct** answer from the options given below:

[NCERT : P L-40 | April 6, 2023 (I)]

(a) Both **A** and **R** are correct but **R** is **NOT** the correct explanation of **A**
(b) Both **A** and **R** are correct **R** is the correct explanation of **A**
(c) **A** is true but **R** is false
(d) **A** is false but **R** is true

75. The initial speed of a projectile fired from ground is u. At the highest point during its motion, the speed of projectile is $\frac{\sqrt{3}}{2}$u. The time of flight of the projectile is:

[NCERT : P L-39 | Jan. 31, 2023 (I)]

(a) $\frac{u}{2g}$
(b) $\frac{u}{g}$
(c) $\frac{2u}{g}$
(d) $\frac{\sqrt{3}u}{g}$

76. Two bodies are projected from ground with same speeds 40 ms^{-1} at two different angles with respect to horizontal. The bodies were found to have same range. If one of the body was projected at an angle of 60°, with horizontal then sum of the maximum heights, attained by the two projectiles, is _______ m. (Given $g = 10ms^{-2}$)

[NCERT : P L-39, 40 | Jan. 31, 2023 (II)]

77. A stone is projected at angle 30° to the horizontal. The ratio of kinetic energy of the stone at point of projection to its kinetic energy at the highest point of flight will be :

[NCERT : P L-39 | Jan. 29, 2023 (I)]

(a) 1 : 2
(b) 1 : 4
(c) 4 : 1
(d) 4 : 3

78. Two objects are projected with same velocity 'u' however at different angles α and β with the horizontal. If α + β = 90°, the ratio of horizontal range of the first object to the 2nd object will be : [NCERT : P L-40 | Jan. 25, 2023 (II)]

(a) 4 : 1
(b) 2 : 1
(c) 1 : 2
(d) 1 : 1

79. The maximum vertical height to which a man can throw a ball is 136 *m*. The maximum horizontal distance upto which he can throw the same ball is

[NCERT : P L-40 | Jan. 24, 2023 (I)]

(a) 192 *m*
(b) 136 *m*
(c) 272 *m*
(d) 68 *m*

80. An object is projected in the air with initial velocity u at an angle θ. The projectile motion is such that the horizontal range R, is maximum.

Another object is projected in the air with a horizontal range half of the range of first object. The initial velocity remains same in both the case. The value of the angle of projection, at which the second object is projected, will be _____ degree.

[NCERT : P L-40 | July 29, 2022 (I)]

81. A ball of mass m is thrown vertically upward. Another ball of mass 2 m is thrown an angle θ with the vertical. Both the balls stay in air for the same period of time. The ratio of the heights attained by the two balls respectively is $\frac{1}{x}$. The value of x is _____.

[NCERT : P L-39 | July 27, 2022 (I)]

82. A body of mass 10 kg is projected at an angle of 45° with the horizontal. The trajectory of the body is observed to pass through a point (20, 10). If T is the time of flight, then its momentum vector, at time $t = \frac{T}{\sqrt{2}}$, is _____

[Take $g = 10\ m/s^2$] [NCERT : P L-39 | July 27, 2022 (II)]

(a) $100\hat{i} + (100\sqrt{2} - 200)\hat{j}$
(b) $100\sqrt{2}\,\hat{i} + (100 - 200\sqrt{2})\hat{j}$
(c) $100\hat{i} + (100 - 200\sqrt{2})\hat{j}$
(d) $100\sqrt{2}\,\hat{i} + (100\sqrt{2} - 200)\hat{j}$

83. If the initial velocity in horizontal direction of a projectile is unit vector $\hat{i}$ and the equation of trajectory is $y = 5x(1 - x)$. The y component vector of the initial velocity is _________ $\hat{j}$. (Take $g = 10\ m/s^2$)

[NCERT : P L-39 | July 26, 2022 (I)]

84. Two projectiles thrown at 30° and 45° with the horizontal respectively, reach the maximum height in same time. The ratio of their intial velocities is :

[NCERT : P L-39 | July 26, 2022 (I)]

(a) $1:\sqrt{2}$ (b) $2:1$ (c) $\sqrt{2}:1$ (d) $1:2$

85. Two projectiles are thrown with same initial velocity making an angle of 45° and 30° with the horizontal respectively. The ratio of their respective ranges will be

[NCERT : P L-40 | July 26, 2022 (II)]

(a) $1:\sqrt{2}$ (b) $\sqrt{2}:1$

(c) $2:\sqrt{3}$ (d) $\sqrt{3}:2$

86. A ball is projected from the ground with a speed 15 ms^{-1} at an angle θ with horizontal so that its range and maximum height are equal, then tan θ will be equal to

[NCERT : P L-39, 40 | July 25, 2022 (II)]

(a) $\frac{1}{4}$ (b) $\frac{1}{2}$

(c) 2 (d) 4

87. Two projectiles P_1 and P_2 thrown with speed in the ratio $\sqrt{3}:\sqrt{2}$, attain the same height during their motion. If P_2 is thrown at an angle of 60° with the horizontal, the angle of projection of P_1 with horizontal will be:

[NCERT : P L-39 | June 30, 2022 (I)]

(a) 15° (b) 30°

(c) 45° (d) 60°

88. A person can throw a ball upto a maximum range of 100 m. How high above the ground he can throw the same ball?

[NCERT : P L-40 | June 29, 2022 (II)]

(a) 25m (b) 50m

(c) 100m (d) 200m

89. A projectile is launched at an angle 'α' with the horizontal with a velocity 20 ms^{-1}. After 10 s, its inclination with horizontal is 'β'. The value of tan β will be: (g = 10 ms^{-2})

[NCERT : P L-39 | June 27, 2022 (I)]

(a) $\tan\alpha + 5\sec\alpha$ (b) $\tan\alpha - 5\sec\alpha$

(c) $2\tan\alpha - 5\sec\alpha$ (d) $2\tan\alpha + 5\sec\alpha$

90. A fighter jet is flying horizontally at a certain altitude with a speed of 200 ms^{-1}. When it passes directly overhead an anti-aircraft gun, bullet is fired from the gun, at an angle θ with the horizontal, to hit the jet. If the bullet speed is 400 m/s, the value of θ will be .... °.

NCERT : P L-38 | [June 26, 2022 (I)]

91. Given below are two statements. One is labelled as **Assertion A** and the other is labelled as **Reason R.**

Assertion A : Two identical balls A and B thrown with same velocity 'u' at two different angles with horizontal attained the same range R. If A and B reached the maximum height h_1 and h_2 respectively, then $R = 4\sqrt{h_1h_2}$

Reason R: Product of said heights.

$$h_1h_2 = \left(\frac{u^2\sin^2\theta}{2g}\right)\cdot\left(\frac{u^2\cos^2\theta}{2g}\right)$$

Choose the CORRECT answer :

[NCERT : P L-39, 40 | June 25, 2022 (II)]

(a) Both A and R are true and R is the correct explanation of A.

(b) Both A and R are true but R is NOT the correct explanation of A.

(c) A is true but R is false

(d) A is flase but R is true

92. A projectile is projected with velocity of 25 m/s at an angle θ with the horizontal. After *t* seconds its inclination with horizontal becomes zero. If *R* represents horizontal range of the projectile, the value of θ will be: [use g = 10 m/s^2] **[NCERT : P L-40 | June 24, 2022 (I)]**

(a) $\frac{1}{2}\sin^{-1}\left(\frac{5t^2}{4R}\right)$ (b) $\frac{1}{2}\sin^{-1}\left(\frac{4R}{5t^2}\right)$

(c) $\tan^{-1}\left(\frac{4t^2}{5R}\right)$ (d) $\cot^{-1}\left(\frac{R}{20t^2}\right)$

93. A body is projected from the ground at an angle of 45° with the horizontal. Its velocity after 2*s* is 20 ms^{-1}. The maximum height reached by the body during its motion is ________ m. (use g = 10ms^{-2})

[NCERT : P L-39 | June 24, 2022 (II)]

94. The ranges and heights for two projectiles projected with the same initial velocity at angles 42° and 48° with the horizontal are R_1, R_2 and H_1, H_2 respectively. Choose the correct option : **[NCERT : P L-39, 40 | Sep. 1, 2021 (II)]**

(a) $R_1 > R_2$ and $H_1 = H_2$ (b) $R_1 = R_2$ and $H_1 < H_2$

(c) $R_1 < R_2$ and $H_1 < H_2$ (d) $R_1 = R_2$ and $H_1 = H_2$

95. A player kicks a football with an initial speed of 25 ms^{-1} at an angle of 45° from the ground. What are the maximum height and the time taken by the football to reach at the highest point during motion ? (Take g = 10 ms^{-2})

[NCERT : P L-39 | Aug. 27, 2021 (II)]

(a) h_{max} = 10 m T = 2.5 s

(b) h_{max} = 15.625 m T = 3.54 s

(c) h_{max} = 15.625 m T = 1.77 s

(d) h_{max} = 3.54 m T = 0.125 s

96. The projectile motion of a particle of mass 5 g is shown in the figure.

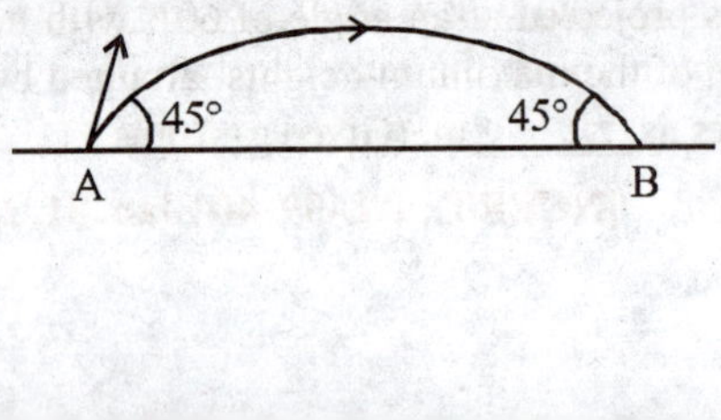

The initial velocity of the particle is $5\sqrt{2}$ ms^{-1} and the air resistance is assumed to be negligible. The magnitude of the change in momentum between the points A and B is $x \times 10^{-2}$ kgms^{-1}.

[NCERT : P L-38, 39 | March 18, 2021 (II)]

The value of x, to the nearest integer, is _______.

97. The trajectory of a projectile in a vertical plane is $y = \alpha x - \beta x^2$, where α and β are constants and x & y are respectively the horizontal and vertical distances of the projectile from the point of projection. The angle of projection θ and the maximum height attained H are respectively given by

[NCERT : P L-39 | Feb. 26, 2021 (II)]

(a) $\tan^{-1}\alpha, \frac{4\alpha^2}{\beta}$ (b) $\tan^{-1}\beta, \frac{\alpha^2}{2\beta}$

(c) $\tan^{-1}\left(\frac{\beta}{\alpha}\right), \frac{\alpha^2}{\beta}$ (d) $\tan^{-1}\alpha, \frac{\alpha^2}{4\beta}$

98. A particle of mass m is projected with a speed u from the ground at an angle $\theta = \frac{\pi}{3}$ w.r.t. horizontal (x-axis). When it has reached its maximum height, it collides completely inelastically with another particle of the same mass and velocity $u\hat{i}$. The horizontal distance covered by the combined mass before reaching the ground is:

[NCERT : P L-40 | 9 Jan. 2020 II]

(a) $\frac{3\sqrt{3}}{8}\frac{u^2}{g}$ (b) $\frac{3\sqrt{2}}{4}\frac{u^2}{g}$

(c) $\frac{5}{8}\frac{u^2}{g}$ (d) $2\sqrt{2}\frac{u^2}{g}$

99. The trajectory of a projectile near the surface of the earth is given as $y = 2x - 9x^2$. If it were launched at an angle θ_0 with speed v_0 then ($g = 10$ ms^{-2}):

[NCERT : P L-39 | 12 April 2019 I]

(a) $\theta_0 = \sin^{-1}\frac{1}{\sqrt{5}}$ and $v_0 = \frac{5}{3}$ ms^{-1}

(b) $\theta_0 = \cos^{-1}\left(\frac{2}{\sqrt{5}}\right)$ and $v_0 = \frac{3}{5}$ ms^{-1}

(c) $\theta_0 = \cos^{-1}\left(\frac{1}{\sqrt{5}}\right)$ and $v_0 = \frac{5}{3}$ ms^{-1}

(d) $\theta_0 = \sin^{-1}\left(\frac{2}{\sqrt{5}}\right)$ and $v_0 = \frac{3}{5}$ ms^{-1}

100. A shell is fired from a fixed artillery gun with an initial speed u such that it hits the target on the ground at a distance R from it. If t_1 and t_2 are the values of the time taken by it to hit the target in two possible ways, the product t_1t_2 is : **[NCERT : P L-39 | 12 April 2019 I]**

(a) R/4g (b) R/g (c) R/2g (d) 2R/g

101. Two particles are projected from the same point with the same speed u such that they have the same range R, but different maximum heights, h_1 and h_2. Which of the following is correct ? **[NCERT : P L-39, 40 | 12 April 2019 II]**

(a) $R^2 = 4h_1h_2$ (b) $R^2 = 16h_1h_2$

(c) $R^2 = 2h_1h_2$ (d) $R^2 = h_1h_2$

102. A plane is inclined at an angle α = 30° with respect to the horizontal. A particle is projected with a speed u = 2 ms^{-1}, from the base of the plane, as shown in figure. The distance from the base, at which the particle hits the plane is close to: (Take g=10 ms^{-2}) **[NCERT : P L-39 | 10 April 2019 II]**

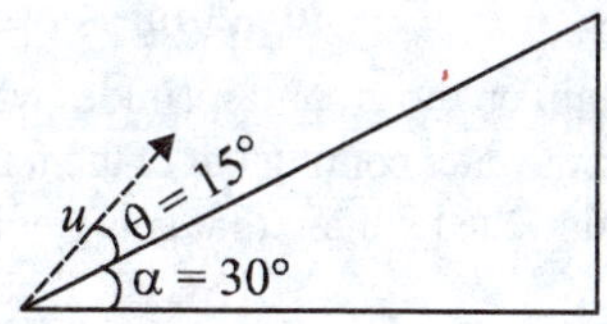

(a) 20 cm (b) 18 cm (c) 26 cm (d) 14 cm

103. A body is projected at t = 0 with a velocity 10 ms^{-1} at an angle of 60° with the horizontal. The radius of curvature of its trajectory at t = 1s is R. Neglecting air resistance and taking acceleration due to gravity g = 10 ms^{-2}, the value of R is: **[NCERT : P L-38, 39 | 11 Jan. 2019 I]**

(a) 10.3 m (b) 2.8 m (c) 2.5 m (d) 5.1 m

104. Two guns A and B can fire bullets at speeds 1 km/s and 2 km/s respectively. From a point on a horizontal ground, they are fired in all possible directions. The ratio of maximum areas covered by the bullets fired by the two guns, on the ground is:

[NCERT : P L-40 | 10 Jan. 2019 I]

(a) 1 : 16 (b) 1 : 2 (c) 1 : 4 (d) 1 : 8

105. The initial speed of a bullet fired from a rifle is 630 m/s. The rifle is fired at the centre of a target 700 m away at the same level as the target. How far above the centre of the target the rifle must be aimed in order to hit the target ?

[NCERT : P L-40 | Online April 11, 2014]

(a) 1.0 m (b) 4.2 m (c) 6.1 m (d) 9.8 m

106. The position of a projectile launched from the origin at $t = 0$ is given by $\vec{r} = (40\hat{i} + 50\hat{j})$ m at $t = 2$s. If the projectile was launched at an angle θ from the horizontal, then θ is (take g = 10 ms^{-2}) **[NCERT : P L-39 | Online April 9, 2014]**

(a) $\tan^{-1}\frac{2}{3}$ (b) $\tan^{-1}\frac{3}{2}$

(c) $\tan^{-1}\frac{7}{4}$ (d) $\tan^{-1}\frac{4}{5}$

107. A projectile is given an initial velocity of $(\hat{i} + 2\hat{j})$ m/s, where $\hat{i}$ is along the ground and $\hat{j}$ is along the vertical. If g = 10 m/s^2, the equation of its trajectory is :

[NCERT : P L-39 | 2013]

(a) $y = x - 5x^2$ (b) $y = 2x - 5x^2$

(c) $4y = 2x - 5x^2$ (d) $4y = 2x - 25x^2$

108. A ball projected from ground at an angle of 45° just clears a wall in front. If point of projection is 4 m from the foot of wall and ball strikes the ground at a distance of 6 m on the other side of the wall, the height of the wall is :
[NCERT : P L-39, 40 | Online April 22, 2013]
(a) 4.4m (b) 2.4m
(c) 3.6m (d) 1.6m

109. A boy can throw a stone up to a maximum height of 10 m. The maximum horizontal distance that the boy can throw the same stone up to will be **[NCERT : P L-39, 40 | 2012]**
(a) $20\sqrt{2}$ m (b) 10m
(c) $10\sqrt{2}$ m (d) 20m

110. A water fountain on the ground sprinkles water all around it. If the speed of water coming out of the fountain is v, the total area around the fountain that gets wet is:
[NCERT : P L-40 | 2011]
(a) $\pi\frac{v^4}{g^2}$ (b) $\frac{\pi}{2}\frac{v^4}{g^2}$ (c) $\pi\frac{v^2}{g^2}$ (d) $\pi\frac{v^2}{g}$

111. A projectile can have the same range 'R' for two angles of projection. If 'T_1' and 'T_2' to be time of flights in the two cases, then the product of the two time of flights is directly proportional to. **[NCERT : P L-39 | 2004]**
(a) R (b) $\frac{1}{R}$ (c) $\frac{1}{R^2}$ (d) R^2

112. A ball is thrown from a point with a speed 'v_0' at an elevation angle of θ. From the same point and at the same instant, a person starts running with a constant speed $\frac{v_0}{2}$ to catch the ball. Will the person be able to catch the ball? If yes, what should be the angle of projection θ?
[NCERT : P L-38 | 2004]
(a) No (b) Yes, 30°
(c) Yes, 60° (d) Yes, 45°

113. A boy playing on the roof of a 10 m high building throws a ball with a speed of 10m/s at an angle of 30° with the horizontal. How far from the throwing point will the ball be at the height of 10 m from the ground ?
[NCERT : P L-40 | 2003]
$[g = 10\text{m/s}^2, \sin 30^o = \frac{1}{2}, \cos 30^o = \frac{\sqrt{3}}{2}]$
(a) 5.20m (b) 4.33m (c) 2.60m (d) 8.66m

Topic 4 - Relative Velocity in Two Dimensions & Uniform Circular Motion

114. A river is flowing from west to east direction with speed of 9 km h^{-1}. If a boat capable of moving at a maximum speed of 27 km h^{-1} in still water, crosses the river in half a minute, while moving with maximum speed at an angle of 150° to direction of river flow, then the width of the river is: **[April 2, 2025 (I)]**
(a) 300m (b) 112.5m
(c) 75m (d) $112.5 \times \sqrt{3}$ m

115. The maximum speed of a boat in still water is 27 km/h. Now this boat is moving downstream in a river flowing at 9 km/h. A man in the boat throws a ball vertically upwards with speed of 10 m/s. Range of the ball as observed by an observer at rest on the river bank, is ______ cm. (Take g = 10 m/s^2) **[Jan. 29, 2025 (I)]**

116. A particle moving in a circle of radius R with uniform speed takes time T to complete one revolution. If this particle is projected with the same speed at an angle θ to the horizontal, the maximum height attained by it is equal to 4R. The angle of projection θ is then given by:
[NCERT : P L-39, 40, 42 | Feb. 1, 2024 (I)]
(a) $\sin^{-1}\left[\frac{2gT^2}{\pi^2 R}\right]^{\frac{1}{2}}$ (b) $\sin^{-1}\left[\frac{\pi^2 R}{2gT^2}\right]^{\frac{1}{2}}$
(c) $\cos^{-1}\left[\frac{2gT^2}{\pi^2 R}\right]^{\frac{1}{2}}$ (d) $\cos^{-1}\left[\frac{\pi R}{2gT^2}\right]^{\frac{1}{2}}$

117. A particle is moving in a circle of radius 50 cm in such a way that at any instant the normal and tangential components of it's acceleration are equal. If its speed at t = 0 is 4m/s, the time taken to complete the first revolution will be $\frac{1}{\alpha}[1 - e^{-2\pi}]$ s, where α = ______.
[NCERT : P L-42 | Jan. 29, 2024 (II)]

118. The speed of a swimmer is 4 km h^{-1} in still water. If the swimmer makes his strokes normal to the flow of river of width 1 km, he reaches a point 750 m down the stream on the opposite bank. The speed of the river water is ______ km h^{-1}. **[Jan. 31, 2023 (II)]**

119. A stone tied to 180 cm long string at its end is making 28 revolutions in horizontal circle in every minute. The magnitude of acceleration of stone is $\frac{1936}{x}$ ms^{-2}. The value of x ______. $\left(\text{Take } \pi = \frac{22}{7}\right)$
[NCERT : P L-42 | Jan. 30, 2023 (II)]

120. An object moves at a constant speed along a circular path in a horizontal plane with centre at the origin. When the object is at x = +2m, its velocity is $-4\hat{j}$ m/s. The object's velocity (v) and acceleration (a) at x = –2 m will be
[NCERT : P L-42 | Jan. 29, 2023 (II)]

(a) $v = 4\hat{i}\ m/s, a = 8\hat{j}\ m/s^2$

(b) $v = 4\hat{j}\ m/s, a = 8\hat{i}\ m/s^2$

(c) $v = -4\hat{j}\ m/s, a = 8\hat{i}\ m/s^2$

(d) $v = -4\hat{i}\ m/s, a = -8\hat{j}\ m/s^2$

121. A girl standing on road holds her umbrella at 45° with the vertical to keep the rain away. If she starts running without umbrella with a speed of $15\sqrt{2}kmh^{-1}$, the rain drops hit her head vertically. The speed of rain drops with respect to the moving girl is: **[June 27, 2022 (I)]**

(a) $30\ kmh^{-1}$ (b) $\frac{25}{\sqrt{2}}kmh^{-1}$

(c) $\frac{30}{\sqrt{2}}kmh^{-1}$ (d) $25\ kmh^{-1}$

122. A swimmer wants to cross a river from point A to point B. Line AB makes an angle of 30° with the flow of river. Magnitude of velocity of the swimmer is same as that of the river. The angle θ with the line AB should be ____°, so that the swimmer reaches point B. **[NA, July 27, 2021 (II)]**

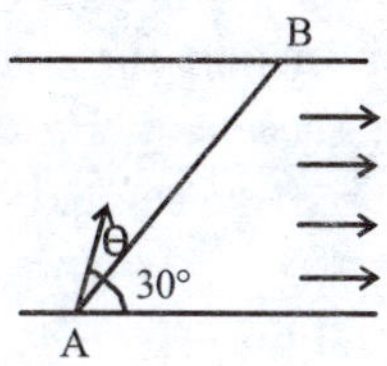

123. A person is swimming with a speed of 10 m/s at an angle of 120° with the flow and reaches to a point directly opposite on the other side of the river. The speed of the flow is 'x' m/s. The value of 'x' to the nearest integer is ______. **[March 18, 2021 (I)]**

124. A swimmer can swim with velocity of 12 km/h in still water. Water flowing in a river has velocity 6 km/h. The direction with respect to the direction of flow of river water he should swim in order to reach the point on the other bank just opposite to his starting point is _____°. (Round off to the Nearest Integer) **[March 16, 2021 (II)]**
(Find the angle in degrees).

125. A clock has a continuously moving second's hand of 0.1 m length. The average acceleration of the tip of the hand (in units of ms^{-2}) is of the order of:

[NCERT : P L-42 | Sep. 06, 2020 (I)]

(a) 10^{-3} (b) 10^{-4} (c) 10^{-2} (d) 10^{-1}

126. When a car is at rest, its driver sees raindrops falling on it vertically. When driving the car with speed v, he sees that raindrops are coming at an angle 60° from the horizontal. On further increasing the speed of the car to $(1 + \beta)v$, this angle changes to 45°. The value of β is close to:

[Sep. 06, 2020 (II)]

(a) 0.73 (b) 0.41 (c) 0.37 (d) 0.50

127. The stream of a river is flowing with a speed of 2 km/h. A swimmer can swim at a speed of 4 km/h. What should be the direction of the swimmer with respect to the flow of the river to cross the river straight? **[9 April 2019 I]**

(a) 90° (b) 150° (c) 120° (d) 60°

128. Ship A is sailing towards north-east with velocity $\vec{v} = 30\hat{i} + 50\hat{j}$ km/hr where $\hat{i}$ points east and $\hat{j}$ north. Ship B is at a distance of 80 km east and 150 km north of Ship A and is sailing towards west at 10 km/hr. A will be at minimum distance from B in: **[8 April 2019 I]**

(a) 4.2 hrs. (b) 2.6 hrs.

(c) 3.2 hrs. (d) 2.2 hrs.

129. Two particles A, B are moving on two concentric circles of radii R_1 and R_2 with equal angular speed ω. At t = 0, their positions and direction of motion are shown in the figure :

[NCERT : P L-42 | 12 Jan. 2019 II]

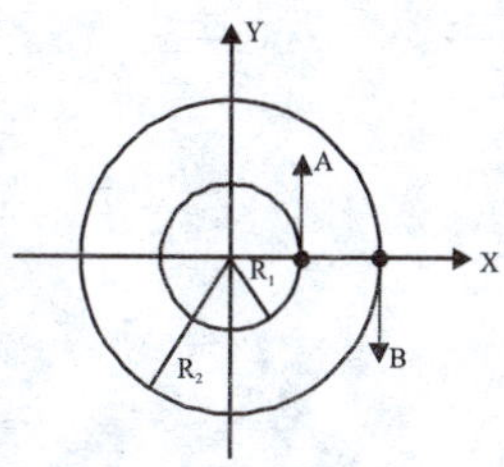

The relative velocity $\vec{v_A} - \vec{v_B}$ at $t = \frac{\pi}{2\omega}$ is given by:

(a) $\omega(R_1 + R_2)\hat{i}$ (b) $-\omega(R_1 + R_2)\hat{i}$

(c) $\omega(R_2 - R_1)\hat{i}$ (d) $\omega(R_1 - R_2)\hat{i}$

130. A particle is moving along a circular path with a constant speed of 10 ms^{-1}. What is the magnitude of the change in velocity of the particle, when it moves through an angle of 60° around the centre of the circle?

[NCERT : P L-42 | Online April 10, 2015]

(a) $10\sqrt{3}$m/s (b) zero

(c) $10\sqrt{2}$m/s (d) 10 m/s

131. If a body moving in circular path maintains constant speed of 10 ms^{-1}, then which of the following correctly describes relation between acceleration and radius?

[NCERT : P L-42 | Online April 10, 2015]

(a) 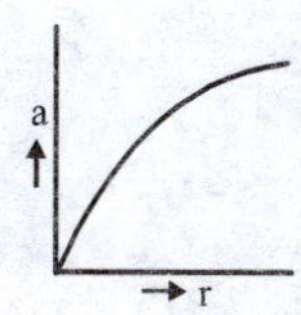

(b) 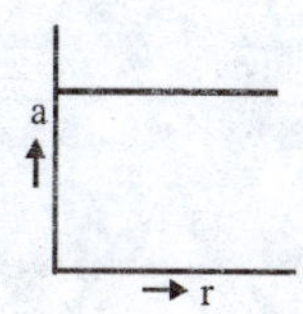

(c) 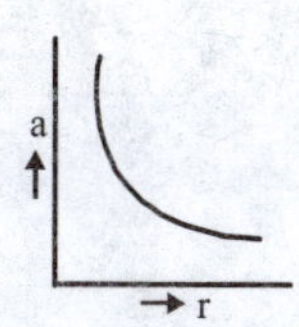

(d) 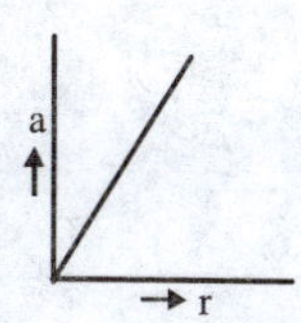

ANSWER KEYS

1	(3)	**15**	(c)	**29**	(c)	**43**	(a)	**57**	(b)	**71**	(c)	**85**	(c)	**99**	(c)	**113**	(d)	**127**	(c)
2	(3)	**16**	(b)	**30**	(a)	**44**	(c)	**58**	(a)	**72**	(d)	**86**	(d)	**100**	(d)	**114**	(b)	**128**	(b)
3	(150)	**17**	(c)	**31**	(a)	**45**	(60)	**59**	(d)	**73**	(a)	**87**	(c)	**101**	(b)	**115**	(2000)	**129**	(c)
4	(4)	**18**	(b)	**32**	(a)	**46**	(c)	**60**	(d)	**74**	(b)	**88**	(a)	**102**	(a)	**116**	(a)	**130**	(d)
5	(6)	**19**	(a)	**33**	(c)	**47**	(c)	**61**	(c)	**75**	(b)	**89**	(b)	**103**	(b)	**117**	(8)	**131**	(c)
6	(c)	**20**	(c)	**34**	(d)	**48**	(None)	**62**	(c)	**76**	(80)	**90**	(60)	**104**	(a)	**118**	(3)		
7	(4)	**21**	(a)	**35**	(c)	**49**	(d)	**63**	(5)	**77**	(d)	**91**	(a)	**105**	(c)	**119**	(125)		
8	(d)	**22**	(d)	**36**	(100)	**50**	(b)	**64**	(b)	**78**	(d)	**92**	(d)	**106**	(c)	**120**	(b)		
9	(None)	**23**	(3)	**37**	(d)	**51**	(d)	**65**	(16)	**79**	(c)	**93**	(20)	**107**	(b)	**121**	(c)		
10	(4)	**24**	(c)	**38**	(d)	**52**	(b)	**66**	(d)	**80**	(15 or 75)	**94**	(b)	**108**	(b)	**122**	(30)		
11	(1)	**25**	(180)	**39**	(673)	**53**	(c)	**67**	(2)	**81**	(1)	**95**	(c)	**109**	(d)	**123**	(5)		
12	(d)	**26**	(195)	**40**	(a)	**54**	(580)	**68**	(80)	**82**	(d)	**96**	(5)	**110**	(a)	**124**	(120)		
13	(5)	**27**	(90)	**41**	(13)	**55**	(d)	**69**	(a)	**83**	(5)	**97**	(d)	**111**	(a)	**125**	(a)		
14	(2)	**28**	(d)	**42**	(b)	**56**	(a)	**70**	(c)	**84**	(c)	**98**	(a)	**112**	(c)	**126**	(a)		

4 Laws of Motion

TREND ANALYSIS
JEE MAIN (2025-2021)

	2025				2024				2023				2022				2021			
	Session II		Session I		Session II		Session I		Session II		Session I		Session II		Session I		Session II		Session I	
	MCQs	NVQs	MCQs	NVQs	MCQs	NVQs	MCQs	NVQs	MCQs	NVQs	MCQs	NVQs	MCQs	NVQs	MCQs	NVQs	MCQs	NVQs	MCQs	NVQs
Topic 1	2	–	1	–	4	–	3	–	5	–	4	–	3	–	3	3	3	–	–	1
Topic 2	2	–	1	–	3	1	2	–	–	–	4	–	8	–	3	3	3	1	–	2
Topic 3	2	–	–	–	3	–	6	–	2	–	6	–	2	–	5	1	–	3	–	5
Topic 4	–	–	1	–	2	1	4	–	2	–	2	1	1	–	4	1	1	–	2	1
Total	6	–	3	–	12	2	15	–	9	–	16	1	14	–	15	8	7	4	2	9

How to Code Qns for Better & Faster Revision

Type 1: Able to solve (No code Req)
Type 2: Solved with struggle or by seeing solution
(Use code NR means Needs Revision or orange stickers)
Type 3: Tricky/ Could not solve on my own. Need help from Teacher/ Friend.
(Use code TQ means Tricky Question or red stickers)

You can either hand write the code in boxes or else paste coloured stickers given at end of book.

Topic 1 - Ist, IInd & IIIrd Laws of Motion

1. An object with mass 500 g moves along x-axis with speed $v = 4\sqrt{x}$ m/s. The force acting on the object is :
[NCERT : P L-54 | April 7, 2025 (II)]
(a) 8 N (b) 5 N (c) 6 N (d) 4 N

2. A body of mass 2 kg moving with velocity of $\vec{v}_{\text{in}} = 3\hat{i} + 4\hat{j}\,ms^{-1}$ enters into a constant force field of 6N directed along positive z-axis. If the body remains in the field for a period of $\frac{5}{3}$ seconds, then velocity of the body when it emerges from force field is
[NCERT : P L-54 | April 8, 2025 (II)]
(a) $4\hat{i} + 3\hat{j} + 5\hat{k}$ (b) $3\hat{i} + 4\hat{j} + 5\hat{k}$
(c) $3\hat{i} + 4\hat{j} - 5\hat{k}$ (d) $3\hat{i} + 4\hat{j} + \sqrt{5}\hat{k}$

3. A balloon and its content having mass M is moving up with an acceleration 'a'. The mass that must be released from the content so that the balloon starts moving up with an acceleration '$3a$' will be : (Take 'g' as acceleration due to gravity)
[NCERT : P L-54 | Jan. 28, 2025 (II)]
(a) $\frac{3Ma}{2a-g}$ (b) $\frac{3Ma}{2a+g}$ (c) $\frac{2Ma}{3a+g}$ (d) $\frac{2Ma}{3a-g}$

4. A player caught a cricket ball of mass 150 g moving at a speed of 20 m/s. If the catching process is completed in 0.1 s, the magnitude of force exerted by the ball on the hand of the player is:
[NCERT : P L-54 | April 8, 2024 (I)]
(a) 150 N (b) 3 N (c) 30 N (d) 300 N

5. A wooden block of mass 5kg rests on soft horizontal floor. When an iron cylinder of mass 25 kg is placed on the top of the block, the floor yields and the block and the cylinder together go down with an acceleration of 0.1 ms^{-2}. The action force of the system on the floor is equal to:
[NCERT : P L-54 | April 5, 2024 (I)]
(a) 297 N (b) 294 N
(c) 291 N (d) 196 N

6. A particle moves in x-y plane under the influence of a force $\vec{F}$ such that its linear momentum is $\vec{P}(t) = \hat{i}\cos(kt) - \hat{j}\sin(kt)$. If k is constant, the angle between $\vec{F}$ and $\vec{P}$ will be:

[NCERT : P L-54 | April 5, 2024 (II)]

(a) $\frac{\pi}{2}$ (b) $\frac{\pi}{6}$ (c) $\frac{\pi}{4}$ (d) $\frac{\pi}{3}$

7. A wooden block, initially at rest on the ground, is pushed by a force which increases linearly with time t. Which of the following curve best describes acceleration of the block with time: **[NCERT : P L-54 | April 4, 2024 (I)]**

(a)
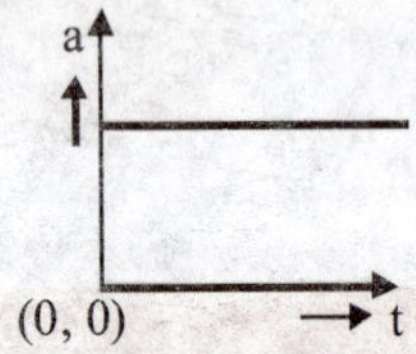

(b)
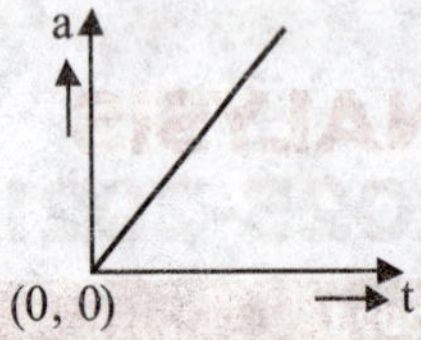

(c)
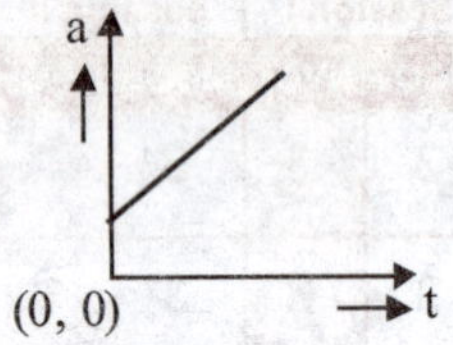

(d)
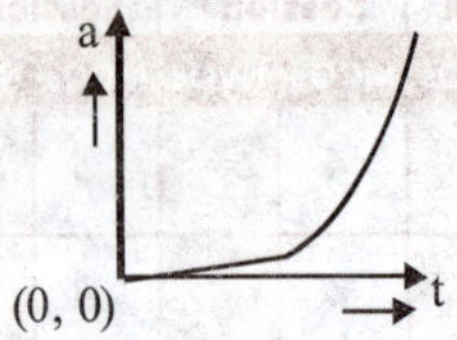

8. A cricket player catches a ball of mass 120g moving with 25m/s speed. If the catching process is completed in 0.1 s then the magnitude of force exerted by the ball on the hand of player will be (in SI unit) :

[NCERT : P L-54 | Feb. 1, 2024 (II)]

(a) 30 (b) 24 (c) 12 (d) 25

9. A spherical body of mass 100 g is dropped from a height of 10m from the ground. After hitting the ground, the body rebounds to a height of 5 m. The impulse of force imparted by the ground to the body is given by :
(given, $g = 9.8\ m/s^2$) **[NCERT : P L-55 | Jan. 30, 2024 (I)]**

(a) $4.32\ kg\ ms^{-1}$ (b) $43.2\ kg\ ms^{-1}$
(c) $23.9\ kg\ ms^{-1}$ (d) $2.39\ kg\ ms^{-1}$

10. A body of mass 1000 kg is moving horizontally with a velocity 6 m/s. If 200 kg extra mass is added, the final velocity (in m/s) is: **[NCERT : P L-57, 58 | Jan. 27, 2024 (I)]**

(a) 6 (b) 2 (c) 3 (d) 5

11. A bullet 10 g leaves the barrel of gun with a velocity of 600 m/s. If the barrel of gun is 50 cm long and mass of gun is 3 kg, then value of impulse supplied to the gun will be:

[NCERT : P L-57, 58 | April 13, 2023 (I)]

(a) 12 Ns (b) 6 Ns (c) 36 Ns (d) 3 Ns

12. Three force $F_1 = 10\ N$, $F_2 = 8\ N$, $F_3 = 6\ N$ are acting on a particle of mass 5 kg. The forces F_2 and F_3 are applied perpendicular so that particle remains at rest. If the force F_1 is removed, then the acceleration of the particle is:

[NCERT : P L-54 | April 12, 2023 (I)]

(a) $2\ ms^{-2}$ (b) $0.5\ ms^{-2}$ (c) $4.8\ ms^{-2}$ (d) $7\ ms^{-2}$

13. An average force of 125 N is applied on a machine gun firing bullets each of mass 10 g at the speed of 250 m/s to keep it in position. The number of bullets fired per second by the machine gun is : **[NCERT : P L-54 | April 11, 2023 (I)]**

(a) 5 (b) 50 (c) 100 (d) 25

14. A body of mass 500 g moves along x-axis such that it's velocity varies with displacement x according to the relation $v = 10\sqrt{x}$ m/s the force acting on the body is:

[NCERT : P L-54 | April 11, 2023 (II)]

(a) 166 N (b) 25 N
(c) 125 N (d) 5 N

15. At any instant the velocity of a particle of mass 500 g is $\left(2t\,\hat{i} + 3t^2\hat{j}\right) ms^{-1}$. If the force acting on the particle at $t = 1 s$ is $\left(\hat{i} + x\hat{j}\right) N$. Then the value of x will be:

[NCERT : P L-54 | April 8, 2023 (I)]

(a) 3 (b) 4 (c) 6 (d) 2

16. Figures (a), (b), (c) and (d) show variation of force with time. **[NCERT : P L-55 | Feb. 1, 2023 (II)]**

(A)
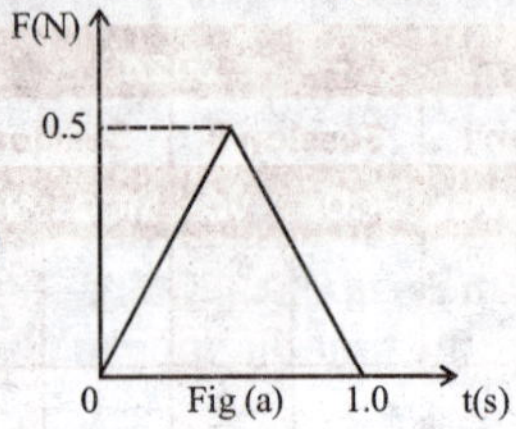

(B)
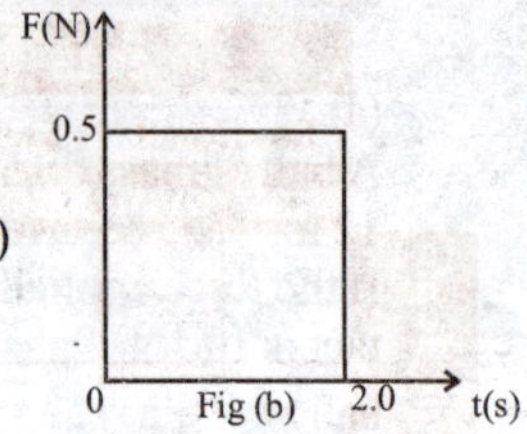

(C)
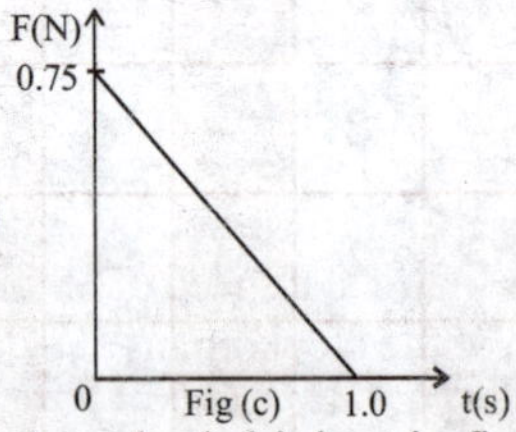

(D)
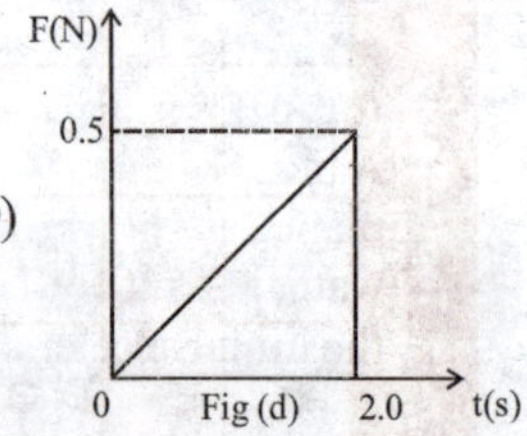

The impulse is highest in figure.

(a) Fig (A) (b) Fig (D) (c) Fig (C) (d) Fig (B)

17. The figure represents the momentum time (p-t) curve for a particle moving along an axis under the influence of the force. Identify the regions on the graph where the magnitude of the force is maximum and minimum respectively ? If $(t_3 - t_2) < t_1$.

[NCERT : P L-54 | Jan. 30, 2023 (I)]

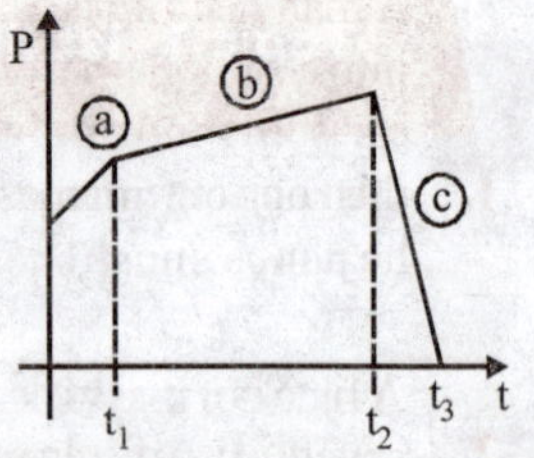

(a) c and a
(b) b and c
(c) c and b
(d) a and b

18. A machine gun of mass 10 kg fires 20 g bullets at the rate of 180 bullets per minute with a speed of $100\ m\ s^{-1}$ each. The recoil velocity of the gun is :

[NCERT : P L-57, 58 | Jan. 30, 2023 (II)]

(a) 0.02 m/s (b) 2.5 m/s
(c) 1.5 m/s (d) 0.6 m/s

19. Force acts for 20 s on a body of mass 20 kg, starting from rest, after which the force ceases and then body describes 50 m in the next 10 s. The value of force will be :

[NCERT : P L-54 | Jan. 29, 2023 (II)]

(a) 40 N (b) 5 N (c) 20 N (d) 10 N

20. In two different experiments, an object of mass 5 kg moving with a speed of 25 ms^{-1} hits two different walls and comes to rest within (i) 3 second, (ii) 5 seconds, respectively. **[NCERT : P L-55 | July 28, 2022 (I)]**
Choose the correct option out of the following :
(a) Impulse and average force acting on the object will be same for both the cases.
(b) Impulse will be same for both the cases but the average force will be different.
(c) Average force will be same for both the cases but the impulse will be different.
(d) Average force and impulse will be different for both the cases.

21. A balloon has mass of 10 g in air. The air escapes from the balloon at a uniform rate with velocity 4.5 cm/s. If the balloon shrinks in 5 s completely. Then, the average force acting on that balloon will be (in dyne). **[NCERT : P L-54 | July 28, 2022 (I)]**
(a) 3 (b) 9 (c) 12 (d) 18

22. A ball of mass 0.15 kg hits the wall with its initial speed of 12 ms^{-1} and bounces back without changing its initial speed. If the force applied by the wall on the ball during the contact is 100 N. Calculate the time duration of the contact of ball with the wall. **[NCERT : P L-55 | July 26, 2022 (II)]**
(a) 0.018 s (b) 0.036 s
(c) 0.009 s (d) 0.072 s

23. A block of metal weighing 2 kg is resting on a frictionless plane (as shown in figure). It is struck by a jet releasing water at a rate of 1 kgs^{-1} and at a speed of 10 ms^{-1}. Then, the initial acceleration of the block, in ms^{-2}, will be : **[NCERT : P L-54 | June 29, 2022 (I)]**

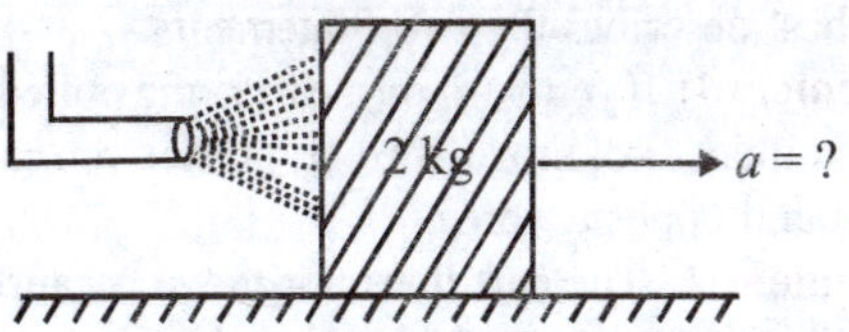

(a) 3 (b) 6 (c) 5 (d) 4

24. A man of 60 kg is running on the road and suddenly jumps into a stationary trolly car of mass 120 kg. Then the trolly car starts moving with velocity 2 ms^{-1}. The velocity of the running man was ________ ms^{-1}. When he jumps into the car. **[NCERT : P L-57, 58 | June 28, 2022 (I)]**

25. A block of mass 2 kg moving on a horizontal surface with speed of 4 ms^{-1} enters a rough surface ranging from $x = 0.5$ m to $x = 1.5$ m. The retarding force in this range of rough surface is related to distance by $F = -kx$ where $k = 12\ Nm^{-1}$. The speed of the block as it just crosses the rough surface will be: **[NCERT : P L-54 | June 28, 2022 (II)]**
(a) Zero (b) 1.5 ms^{-1}
(c) 2.0 ms^{-1} (d) 2.5 ms^{-1}

26. A batsman hits back a ball of mass 0.4 kg straight in the direction of the bowler without changing its initial speed of 15 ms^{-1}. The impulse imparted to the ball is ________ Ns. **[NCERT : P L-55 | June 26, 2022 (II)]**

27. A force on an object of mass 100g is $(10\hat{i}+5\hat{j})$ N. The position of that object at t = 2s is $(a\hat{i}+b\hat{j})$ m after starting from rest. The value of $\frac{a}{b}$ will be ________ **[NCERT : P L-54 | June 26, 2022 (I)]**

28. An object of mass 5 kg is thrown vertically upwards from the ground. The air resistance produces a constant retarding force of 10 *N* throughout the motion. The ratio of time of ascent to the time of descent will be equal to :
[Use $g = 10\ ms^{-2}$] **[NCERT : P L-54 | June 24, 2022 (II)]**
(a) 1 : 1 (b) $\sqrt{2} : \sqrt{3}$
(c) $\sqrt{3} : \sqrt{2}$ (d) 2 : 3

29. The initial mass of a rocket is 1000 kg. Calculate at what rate the fuel should be burnt so that the rocket is given an acceleration of 20 ms^{-2}. The gases come out at a relative speed of 500 ms^{-1} with respect to the rocket :
[Use $g = 10\ m/s^2$] **[NCERT : P L-54 | Aug. 26, 2021 (I)]**
(a) $6.0 \times 10^2\ kg\ s^{-1}$ (b) 500 $kg\ s^{-1}$
(c) 10 $kg\ s^{-1}$ (d) 60 $kg\ s^{-1}$

30. A particle of mass M originally at rest is subjected to a force whose direction is constant but magnitude varies with time according to the relation

$$F = F_0\left[1-\left(\frac{t-T}{T}\right)^2\right]$$

Where F_0 and T are constants. The force acts only for the time interval 2T. The velocity v of the particle after time 2T is: **[NCERT : P L-54 | July 27, 2021 (II)]**
(a) $2F_0T/M$ (b) $F_0T/2M$
(c) $4F_0T/3M$ (d) $F_0T/3M$

31. A force $\vec{F} = (40\hat{i}+10\hat{j})$ N acts on a body of mass 5 kg. If the body starts from rest, its position vector $\vec{r}$ at time t = 10 s, will be: **[NCERT : P L-54 | July 25, 2021 (II)]**
(a) $(100\hat{i}+400\hat{j})$ m (b) $(100\hat{i}+100\hat{j})$ m
(c) $(400\hat{i}+100\hat{j})$ m (d) $(400\hat{i}+400\hat{j})$ m

32. A boy pushes a box of mass 2 kg with a force $\vec{F} = (20\hat{i}+10\hat{j})$N on a frictionless surface. If the box was initially at rest, then ________ m is displacement along the x-axis after 10 s. **[NCERT : P L-54 | Feb. 26, 2021 (I)]**

33. A particle moving in the *xy* plane experiences a velocity dependent force $\vec{F} = k(v_y\hat{i} + v_x\hat{j})$, where v_x and v_y are *x* and *y* components of its velocity $\vec{v}$. if $\vec{a}$ is the acceleration of the particle, then which of the following statements is true for the particle? **[NCERT : P L-54 | Sep. 06, 2020 (II)]**
(a) quantity $\vec{v} \times \vec{a}$ is constant in time
(b) $\vec{F}$ arises due to a magnetic field
(c) kinetic energy of particle is constant in time
(d) quantity $\vec{v} \cdot \vec{a}$ is constant in time

34. A spaceship in space sweeps stationary interplanetary dust. As a result, its mass increases at a rate $\frac{dM(t)}{dt} = bv^2(t)$, where $v(t)$ is its instantaneous velocity. The instantaneous acceleration of the satellite is : **[NCERT : P L-54 | Sep. 05, 2020 (II)]**

(a) $-bv^3(t)$ (b) $-\frac{bv^3}{M(t)}$

(c) $-\frac{2bv^3}{M(t)}$ (d) $-\frac{bv^3}{2M(t)}$

35. A small ball of mass m is thrown upward with velocity u from the ground. The ball experiences a resistive force mkv^2 where v is its speed. The maximum height attained by the ball is : **[NCERT : P L-54 | Sep. 04, 2020 (II)]**

(a) $\frac{1}{2k}\tan^{-1}\frac{ku^2}{g}$ (b) $\frac{1}{k}\ln\left(1+\frac{ku^2}{2g}\right)$

(c) $\frac{1}{k}\tan^{-1}\frac{ku^2}{2g}$ (d) $\frac{1}{2k}\ln\left(1+\frac{ku^2}{g}\right)$

36. A ball is thrown upward with an initial velocity v_0 from the surface of the earth. The motion of the ball is affected by a drag force equal to $m\gamma v^2$ (where m is mass of the ball, v is its instantaneous velocity and γ is a constant). Time taken by the ball to rise to its zenith is : **[NCERT : P L-54 | 10 April 2019 I]**

(a) $\frac{1}{\sqrt{\gamma g}}\tan^{-1}\left(\sqrt{\frac{\gamma}{g}}v_0\right)$ (b) $\frac{1}{\sqrt{\gamma g}}\sin^{-1}\left(\sqrt{\frac{\gamma}{g}}v_0\right)$

(c) $\frac{1}{\sqrt{\gamma g}}ln\left(1+\sqrt{\frac{\gamma}{g}}v_0\right)$ (d) $\frac{1}{\sqrt{2\gamma g}}\tan^{-1}\left(\sqrt{\frac{2\gamma}{g}}v_0\right)$

37. A ball is thrown vertically up (taken as + z-axis) from the ground. The correct momentum-height (p-h) diagram is: **[NCERT : P L-53 | 9 April 2019 I]**

(a) 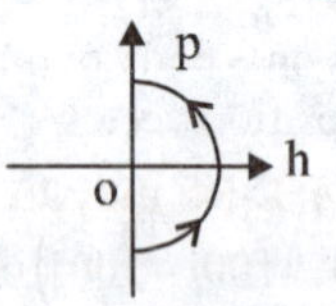

(b) 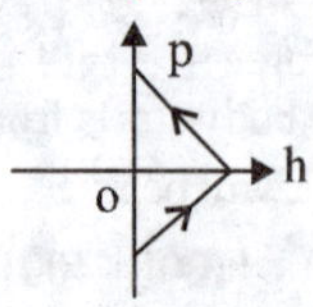

(c) 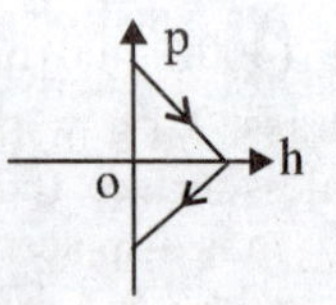

(d) 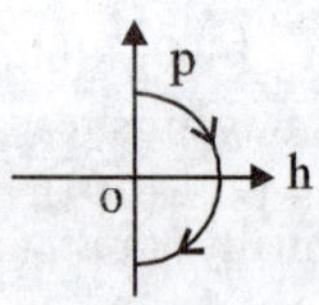

38. A particle of mass m is moving in a straight line with momentum p. Starting at time t = 0, a force F = kt acts in the same direction on the moving particle during time interval T so that its momentum changes from p to 3p. Here k is a constant. The value of T is : **[NCERT : P L-54 | 11 Jan. 2019 II]**

(a) $2\sqrt{\frac{k}{p}}$ (b) $2\sqrt{\frac{p}{k}}$ (c) $\sqrt{\frac{2k}{p}}$ (d) $\sqrt{\frac{2p}{k}}$

39. A particle of mass m is acted upon by a force F given by the empirical law $F=\frac{R}{t^2}v(t)$. If this law is to be tested experimentally by observing the motion starting from rest, the best way is to plot : **[NCERT : P L-54 | Online April 10, 2016]**

(a) log v(t) against $\frac{1}{t}$ (b) v(t) against t^2

(c) log v(t) against $\frac{1}{t^2}$ (d) log v(t) against t

40. A body of mass 5 kg under the action of constant force $\vec{F}=F_x\hat{i}+F_y\hat{j}$ has velocity at t = 0 s as $\vec{v}=(6\hat{i}-2\hat{j})$ m/s and at t = 10s as $\vec{v}=+6\hat{j}$ m/s. The force $\vec{F}$ is: **[NCERT : P L-54 | Online April 11, 2014]**

(a) $(-3\hat{i}+4\hat{j})$N (b) $\left(-\frac{3}{5}\hat{i}+\frac{4}{5}\hat{j}\right)$N

(c) $(3\hat{i}-4\hat{j})$N (d) $\left(\frac{3}{5}\hat{i}-\frac{4}{5}\hat{j}\right)$N

41. A particle of mass m is at rest at the origin at time $t=0$. It is subjected to a force $F(t)=F_0e^{-bt}$ in the x direction. Its speed $v(t)$ is depicted by which of the following curves? **[NCERT : P L-54 | 2012]**

(a) 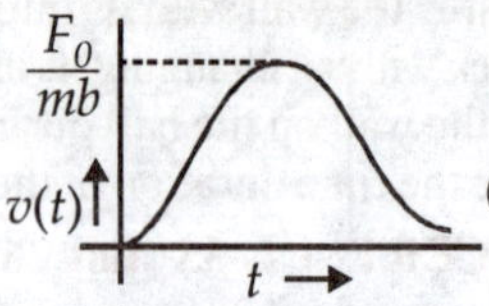

(b) 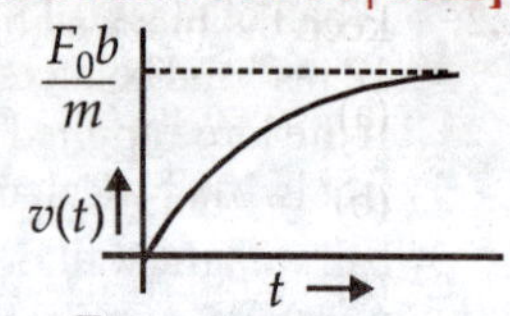

(c) 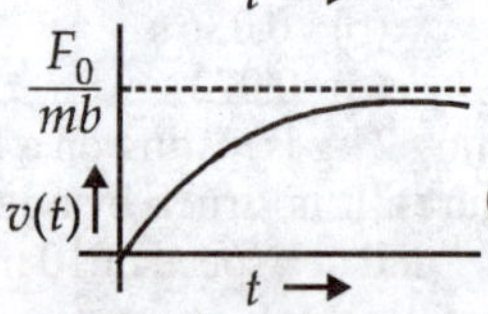

(d) 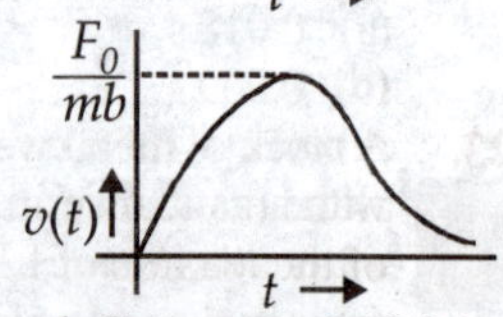

42. This question has Statement 1 and Statement 2. Of the four choices given after the Statements, choose the one that best describes the two Statements.

Statement 1: If you push on a cart being pulled by a horse so that it does not move, the cart pushes you back with an equal and opposite force.

Statement 2: The cart does not move because the force described in statement 1 cancel each other.

[NCERT : P L-56 | Online May 26, 2012]

(a) Statement 1 is true, Statement 2 is true, Statement 2 is the correct explanation of Statement 1.
(b) Statement 1 is false, Statement 2 is true.
(c) Statement 1 is true, Statement 2 is false.
(d) Statement 1 is true, Statement 2 is true, Statement 2 is not the correct explanation of Statement 1.

43. Two fixed frictionless inclined planes making an angle 30° and 60° with the horizontal are shown in the figure. Two blocks A and B are placed on the two planes. What is the relative vertical acceleration of A with respect to B? **[2010]**

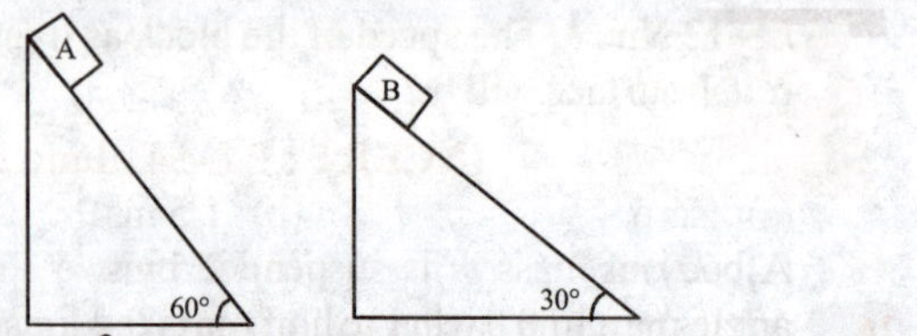

(a) 4.9 ms^{-2} in horizontal direction
(b) 9.8 ms^{-2} in vertical direction
(c) Zero
(d) 4.9 ms^{-2} in vertical direction

44. A ball of mass 0.2 kg is thrown vertically upwards by applying a force by hand. If the hand moves 0.2 m while applying the force and the ball goes upto 2 m height further, find the magnitude of the force. (Consider $g = 10\ m/s^2$). **[NCERT : P L-54 | 2006]**

(a) 4 N (b) 16 N (c) 20 N (d) 22 N

45. A player caught a cricket ball of mass 150 g moving at a rate of 20 m/s. If the catching process is completed in 0.1s, the force of the blow exerted by the ball on the hand of the player is equal to **[NCERT : P L-54 | 2006]**

(a) 150 N (b) 3 N (c) 30 N (d) 300 N

46. A particle of mass 0.3 kg subject to a force $F = -kx$ with $k = 15$ N/m . What will be its initial acceleration if it is released from a point 20 cm away from the origin ? **[NCERT : P L-54 | 2005]**

(a) 15 m/s^2 (b) 3 m/s^2 (c) 10 m/s^2 (d) 5 m/s^2

47. A block is kept on a frictionless inclined surface with angle of inclination 'α'. The incline is given an acceleration 'a' to keep the block stationary. Then a is equal to **[2005]**

(a) g cosec α

(b) g / tan α

(c) g tan α

(d) g

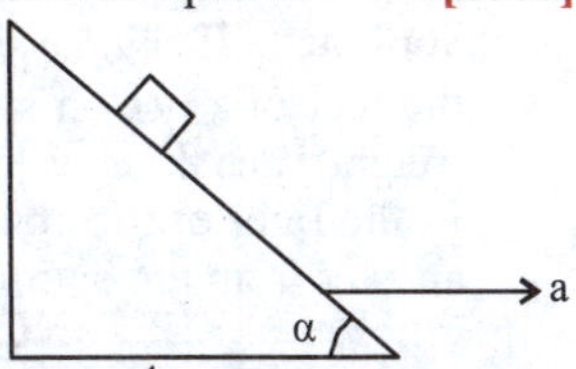

48. A rocket with a lift-off mass 3.5×10^4 kg is blasted upwards with an initial acceleration of $10 m/s^2$. Then the initial thrust of the blast is **[NCERT : P L-54 | 2003]**

(a) 3.5×10^5 N (b) 7.0×10^5 N

(c) 14.0×10^5 N (d) 1.75×10^5 N

49. Three forces start acting simultaneously on a particle moving with velocity, $\vec{v}$. These forces are represented in magnitude and direction by the three sides of a triangle ABC. The particle will now move with velocity **[NCERT : P L-54 | 2003]**

(a) less than $\vec{v}$

(b) greater than $\vec{v}$

(c) $|\vec{v}|$ in the direction of the largest force BC

(d) $\vec{v}$, remaining unchanged

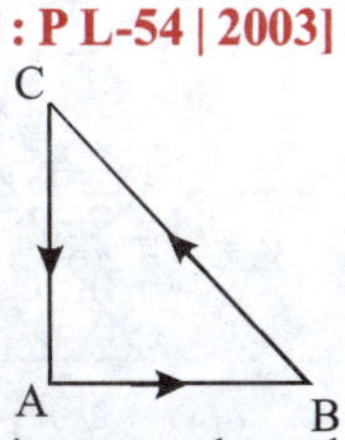

50. A solid sphere, a hollow sphere and a ring are released from top of an inclined plane (frictionless) so that they slide down the plane. Then maximum acceleration down the plane is for (no rolling) **[2002]**

(a) solid sphere (b) hollow sphere

(c) ring (d) all same

Topic 2 - Motion of Connected Bodies, Pulley & Equilibrium of Forces

51. A body of mass m is suspended by two strings making angles θ_1 and θ_2 with the horizontal ceiling with tensions T_1 and T_2 simultaneously. T_1 and T_2 are related by $T_1 = \sqrt{3}T_2$, the angles θ_1 and θ_2 are **[NCERT : P L-65 | April 4, 2025 (I)]**

(a) $\theta_1 = 30° \ \theta_2 = 60°$ with $T_2 = \frac{3mg}{4}$

(b) $\theta_1 = 60° \ \theta_2 = 30°$ with $T_2 = \frac{mg}{2}$

(c) $\theta_1 = 45° \ \theta_2 = 45°$ with $T_2 = \frac{3mg}{4}$

(d) $\theta_1 = 30° \ \theta_2 = 60°$ with $T_2 = \frac{4mg}{5}$

52. A body of mass 1kg is suspended with the help of two strings making angles as shown in figure. Magnitude of tensions T_1 and T_2, respectively, are (in N) : **[NCERT : P L-65 | April 2, 2025 (II)]**

(a) $5, 5\sqrt{3}$

(b) $5\sqrt{3}, 5$

(c) $5\sqrt{3}, 5\sqrt{3}$

(d) 5, 5

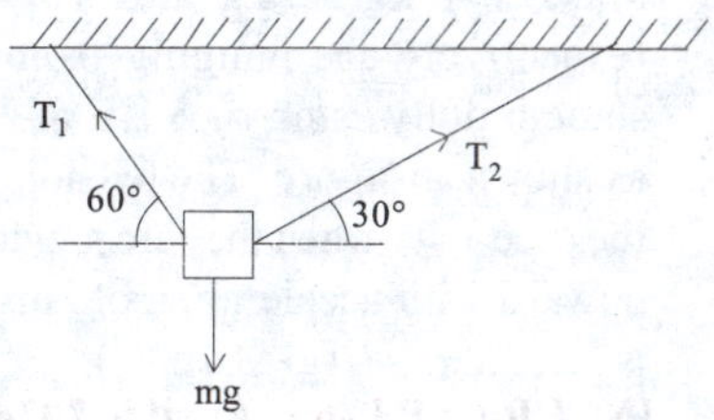

53. A massless spring gets elongated by amount x_1 under a tension of 5N. Its elongation is x_2 under the tension of 7N. For the elongation of $(5x_1 - 2x_2)$, the tension in the spring will be **[NCERT : P L-65 | Jan. 23, 2025 (II)]**

(a) 15 N (b) 20 N

(c) 11 N (d) 39 N

54. A light unstretchable string passing over a smooth light pulley connects two blocks of masses m_1 and m_2. If the acceleration of the system is $\frac{g}{8}$, then the ratio of the masses $\frac{m_2}{m_1}$ is: **[NCERT : P L-65 | April 9, 2024 (I)]**

(a) 9 : 7 (b) 4 : 3

(c) 5 : 3 (d) 8 : 1

55. A 1 kg mass is suspended from the ceiling by a rope of length 4m. A horizontal force 'F' is applied at the mid point of the rope so that the rope makes an angle of 45° with respect to the vertical axis as shown in figure. The magnitude of F is : **[NCERT : P L-58, 59 | April 9, 2024 (II)]**

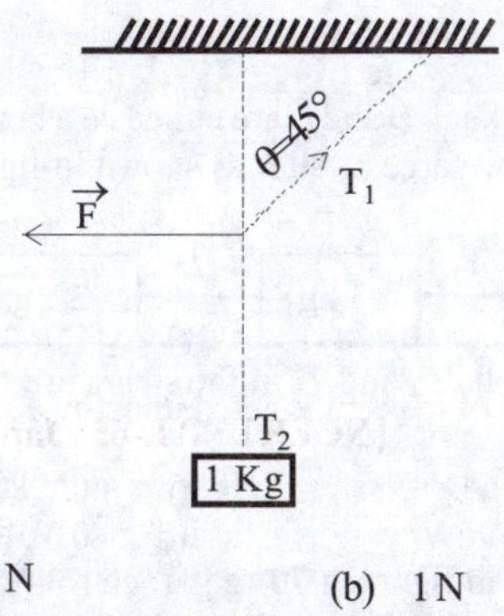

(a) $\frac{10}{\sqrt{2}}$ N (b) 1 N

(c) $\frac{1}{10 \times \sqrt{2}}$ N (d) 10 N

56. A light string passing over a smooth light pulley connects two blocks of masses m_1 and m_2 (where $m_2 > m_1$). If the acceleration of the system is $\frac{g}{\sqrt{2}}$, then the ratio of the masses $\frac{m_1}{m_2}$ is: **[NCERT : P L-65 | April 6, 2024 (I)]**

(a) $\frac{\sqrt{2}-1}{\sqrt{2}+1}$ (b) $\frac{1+\sqrt{5}}{\sqrt{5}-1}$

(c) $\frac{1+\sqrt{5}}{\sqrt{2}-1}$ (d) $\frac{\sqrt{3}+1}{\sqrt{2}-1}$

57. Three blocks M_1, M_2, M_3 having masses 4 kg, 6 kg and 10 kg respectively are hanging from a smooth pully using rope 1, 2 and 3 as shown in figure. The tension in the rope 1, T_1 when they are moving upward with acceleration of $2ms^{-2}$ is .... N (if $g = 10 m/s^2$)

[NCERT : P L-65 | April 5, 2024 (I)]

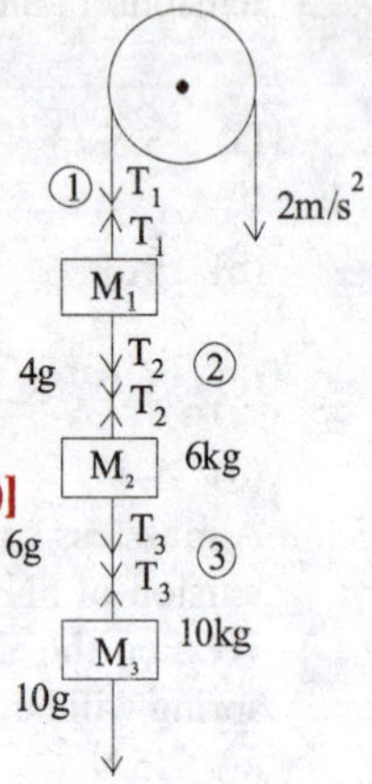

58. All surfaces shown in figure are assumed to be frictionless and the pulleys and the string are light. The acceleration of the block mass 2 kg is:

[NCERT : P L-65 | Jan. 30, 2024 (I)]

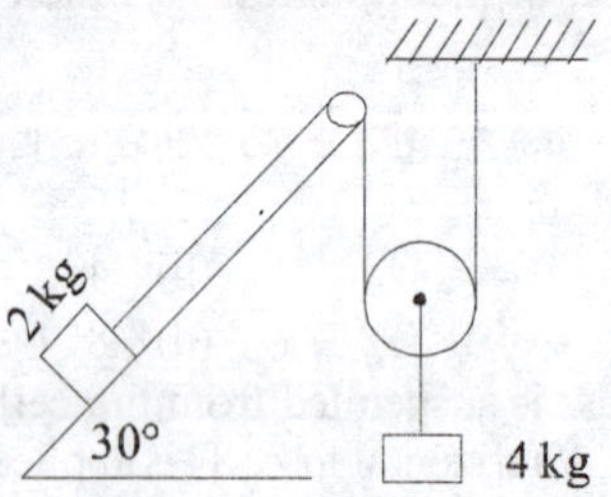

(a) g (b) $\frac{g}{3}$

(c) $\frac{g}{2}$ (d) $\frac{g}{4}$

59. Three blocks *A, B* and *C* are pulled on a horizontal smooth surface by a force of 80 *N* as shown in figure

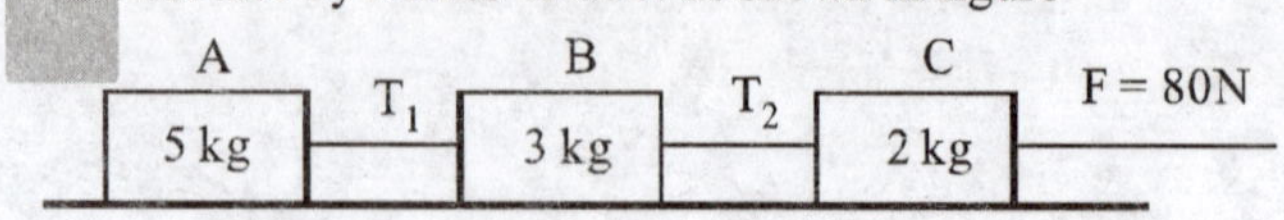

The tensions T_1 and T_2 in the string are respectively:

[NCERT : P L-65 | Jan. 30, 2024 (II)]

(a) 40 *N*, 64*N* (b) 60*N*, 80 *N*
(c) 88 *N*, 96 *N* (d) 80 *N*, 100 *N*

60. As shown in figure, a 70 kg garden roller is pushed with a force of F = 200N at an angle of 30° with horizontal. The normal reaction on the roller is
(Given $g = 10 m s^{-2}$) **[NCERT : P L-65 | Jan. 31, 2023 (I)]**

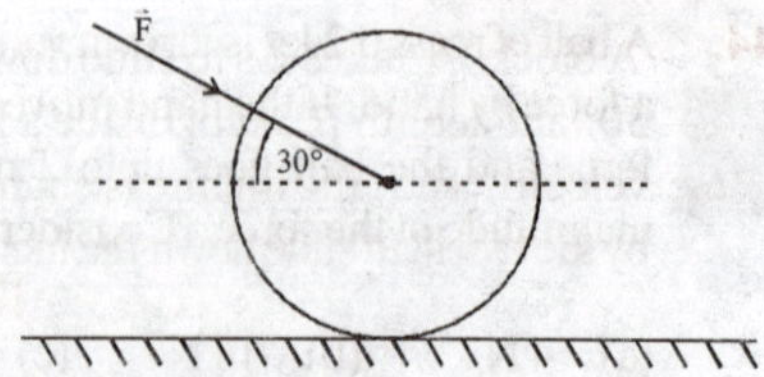

(a) $800\sqrt{2}$ N
(b) 600N
(c) 800N
(d) $200\sqrt{3}$ N

61. A block of $\sqrt{3}$kg is attached to a string whose other end is attached to the wall. An unknown force F is applied so that the string makes an angle of 30° with the wall. The tension T is : (Given $g = 10 ms^{-2}$)

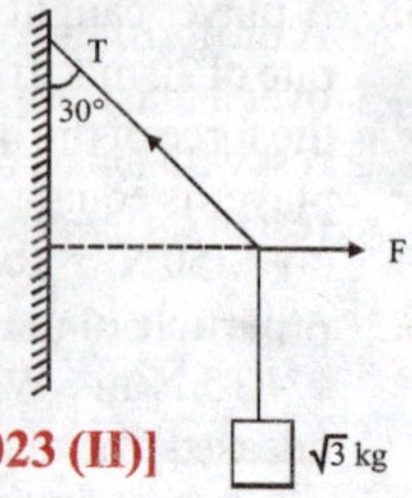

[NCERT : P L-58, 59 | Jan. 30, 2023 (II)]

(a) 20N (b) 25N (c) 10N (d) 15N

62. Given below are two statements :
Statement-I : An elevator can go up or down with uniform speed when its weight is balanced with the tension of its cable.
Statement-II : Force exerted by the floor of an elevator on the foot of a person standing on it is more than his/her weight when the elevator goes down with increasing speed. In the light of the above statements, choose the correct answer from the options given below :

[NCERT : P L-65 | Jan. 24, 2023 (I)]

(a) Both statement I and statement II are false
(b) Statement I is true but Statement II is false
(c) Both Statement I and Statement II are true
(d) Statement I is false but Statement II is true

63. As per given figure, a weightless pulley *P* is attached on a double inclined frictionless surface. The tension in the string (massless) will be (if g = 10 m/s^2)

[NCERT : P L-65 | Jan. 24, 2023 (I)]

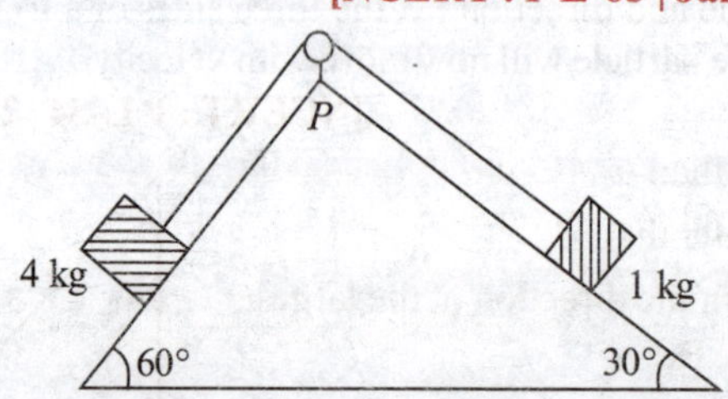

(a) $(4\sqrt{3}+1)$N (b) $4(\sqrt{3}+1)$N

(c) $4(\sqrt{3}-1)$N (d) $(4\sqrt{3}-1)$N

64. Two bodies of masses $m_1 = 5$ kg and $m_2 = 3$ kg are connected by a light string going over a smooth li_ht pulley on a smooth inclined plane as shown in the figure. The system is at rest. The force exerted by the inclined plane on the body of mass m_1 will be : [Take $g = 10 ms^{-2}$]

[NCERT : P L-65 | July 29, 2022 (II)]

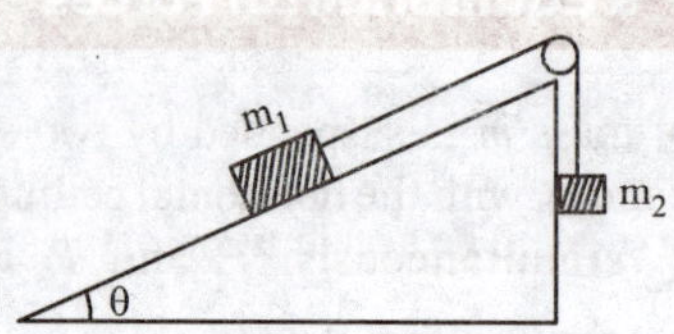

(a) 30N (b) 40N (c) 50N (d) 60N

65. A block 'A' takes 2 s to slide down a frictionless incline of 30° and length 'l', kept inside a lift going up with uniform velocity 'v'. If the incline is changed to 45°, the time taken by the block, to slide down the incline, will be approximately:

[NCERT : P L-65 | July 27, 2022 (II)]

(a) 2.66 s (b) 0.83 s (c) 1.68 s (d) 0.70 s

66. A uniform metal chain of mass m and length 'L' passes over a massless and frictionless pulley. It is released from rest with a part of its length 'ℓ' is hanging on one side and rest of its length 'L – ℓ' is hanging on the other side of the pully. At a certain point of time, when $\ell = \frac{L}{x}$, the acceleration of the chain is $\frac{g}{2}$. The value of x is ____.

[NCERT : P L-65 | July 28, 2022 (II)]

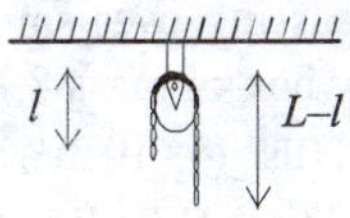

(a) 6
(b) 2
(c) 1.5
(d) 4

67. A monkey of mass 50 kg climbs on a rope which can withstand the tension (T) of 350 N. If monkey initially climbs down with an acceleration of 4 m/s² and then climbs up with an acceleration of 5m/s². Choose the corret option (g = 10 m/s²).

[NCERT : P L-65 | July 26, 2022 (I)]

(a) T = 700 N while climbing upward
(b) T = 350 N while goint downward
(c) Rope will break while climbing upward
(d) Rope will break while going downward

68. Three masses M = 100 kg, m_1 = 10 kg and m_2 = 20 kg are arranged in a system as shown in figure. All the surfaces are frictionless and strings are inextensible and weightless. The pulleys are also weightless and frictionless. A force F is applied on the system so that the mass m_2 moves upward with an acceleration of 2 ms^{-2}. The value of F is : (Take g = 10 ms^{-2})

[NCERT : P L-65 | July 26, 2022 (I)]

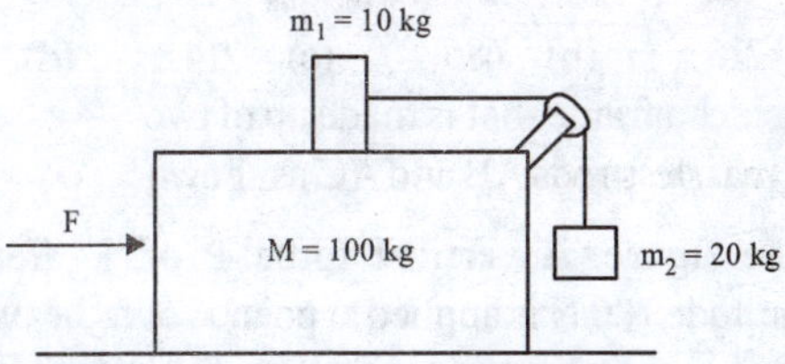

(a) 3360N
(b) 3380N
(c) 3120N
(d) 3240N

69. Two masses M_1 and M_2 are tied together at the two ends of a light inextensible string that passes over a frictionless pulley. When the mass M_2 is twice that of M_1 the acceleration of the system is a_1. When the mass M_2 is thrice that of M_1. The acceleration of the system is a_2. The ratio $\frac{a_1}{a_2}$ will be: **[NCERT : P L-65 | July 26, 2022 (II)]**

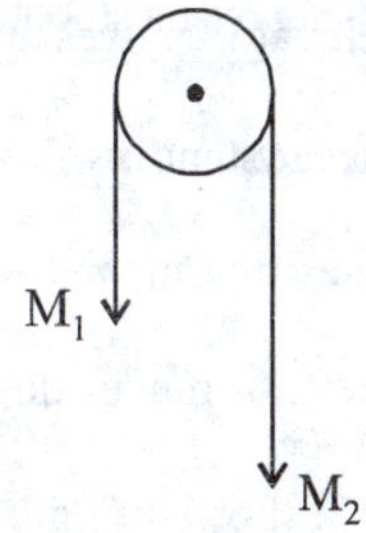

(a) $\frac{1}{3}$

(b) $\frac{2}{3}$

(c) $\frac{3}{2}$

(d) $\frac{1}{2}$

70. Four forces are acting at a point P in equilibrium as shown in figure. The ratio of forcer F_1 to F_2 is 1 : x where x = ______. **[NCERT : P L-58 | July 25, 2022 (I)]**

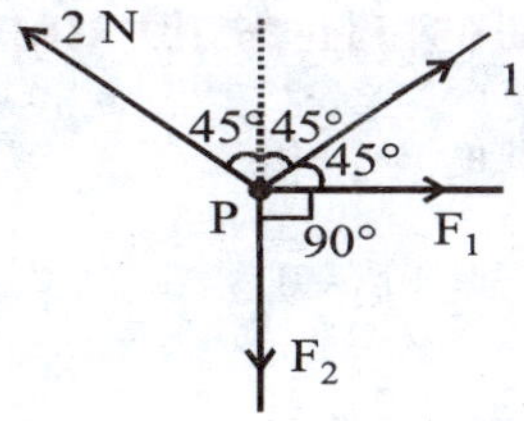

71. For a free body diagram shown in the figure, the four forces are applied in the 'x' and 'y' directions. What additional force must be applied and at what angle with positive x-axis so that the net acceleration of body is zero?

[NCERT : P L-58 | July 25, 2022 (II)]

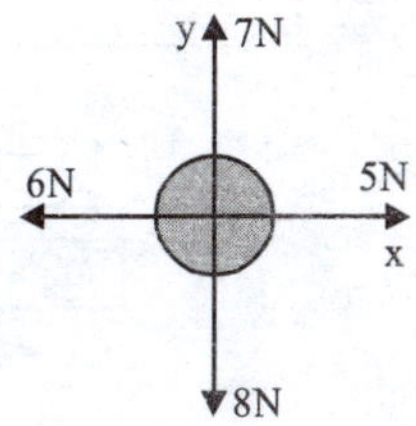

(a) $\sqrt{2}$ N, 45°

(b) $\sqrt{2}$ N, 135°

(c) $\frac{2}{\sqrt{3}}$ N, 30°

(d) 2N, 45°

72. A block of mass M placed inside a box descends vertically with acceleration 'a'. The block exerts a force equal to one-fourth of its weight on the floor of the box. The value of 'a' will be : **[NCERT : P L-65 | June 29, 2022 (II)]**

(a) $\frac{g}{4}$ (b) $\frac{g}{2}$

(c) $\frac{3g}{4}$ (d) g

73. A mass of 10 kg is suspended vertically by a rope of length 5m from the roof. A force of 30 N is applied at the middle point of rope in horizontal direction. The angle made by upper half of the rope with vertical is $\theta = \tan^{-1}(x \times 10^{-1})$. The value of x is ________.
(Given g = 10 m/s²)

[NCERT : P L-58, 59 | June 27, 2022 (II)]

74. A person is standing in an elevator. In which situation, he experiences weight loss ?

[NCERT : P L-65 | June 26, 2022 (I)]

(a) When the elevator moves upward with constant acceleration
(b) When the elevator moves downward with constant acceleration
(c) When the elevator moves upward with uniform velocity
(d) When the elevator moves downward with uniform velocity

75. In the arrangement shown in figure a_1, a_2, a_3 and a_4 are the accelerations of masses m_1, m_2, m_3 and m_4 respectively. Which of the following relation is true for this arrangement?

[NCERT : P L-65 | June 26, 2022 (II)]

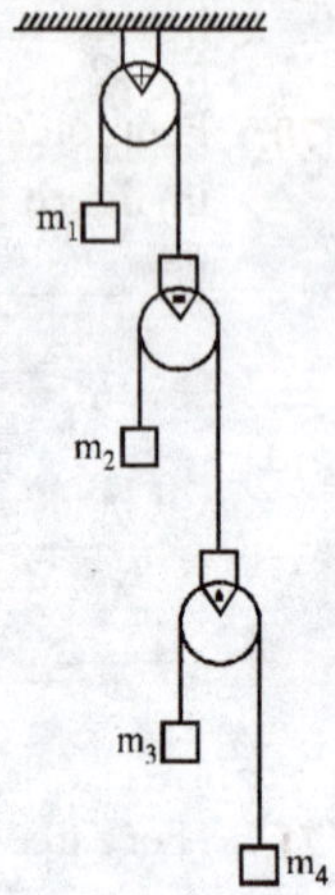

(a) $4a_1 + 2a_2 + a_3 + a_4 = 0$

(b) $a_1 + 4a_2 + 3a_3 + a_4 = 0$

(c) $a_1 + 4a_2 + 3a_3 + 2a_4 = 0$

(d) $2a_1 + 2a_2 + 3a_3 + a_4 = 0$

76. A system of 10 balls each of mass 2 kg are connected via massless and stretchable string. The system is allowed to slip over the edge of a smooth table as shown in figure. Tension on the string between the 7th and 8th ball is____________N when 6th ball just leaves the table.

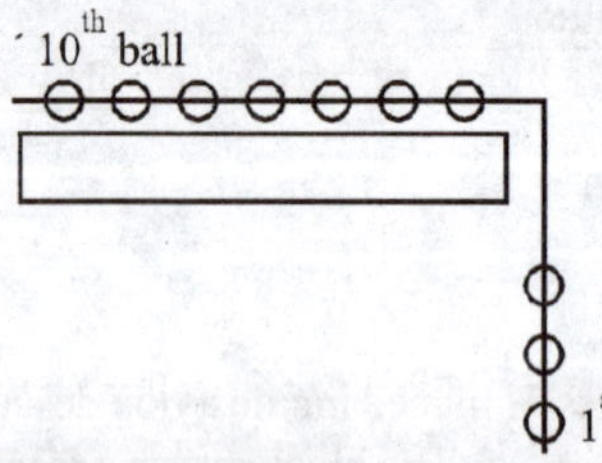

[NCERT : P L-65 | June 26, 2022 (II)]

77. A block of mass 200 g is kept stationary on a smooth inclined plane by applying a minimum horizontal force $F = \sqrt{x}N$ as shown in figure. The value of x = ________.

[NCERT : P L-65 | June 25, 2022 (II)]

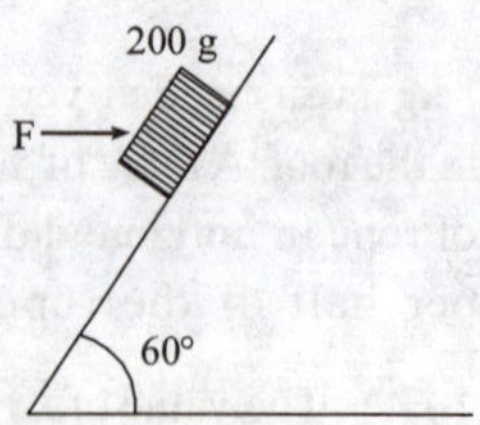

78. A block of mass m slides on the wooden wedge, which in turn slides backward on the horizontal surface. The acceleration of the block with respect to the wedge is: Given m = 8 kg, M = 16 kg.
Assume all the surfaces shown in the figure to be frictionless. **[NCERT : P L-65 | Sep. 1, 2021 (II)]**

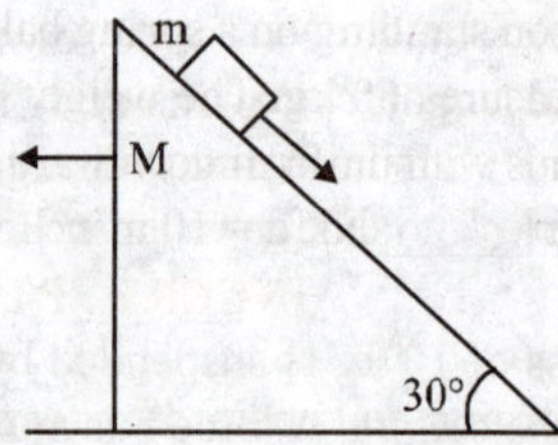

(a) $\frac{4}{3}g$

(b) $\frac{6}{5}g$

(c) $\frac{3}{5}g$

(d) $\frac{2}{3}g$

79. A car is moving on a plane inclined at 30° to the horizontal with an acceleration of 10 ms^{-2} parallel to the plane upward. A bob is suspended by a string from the roof of the car. The angle in degrees which the string makes with the vertical is_______. (Take g = 10 ms^{-2})

[NCERT : P L-58, 59 | NA, Aug. 31, 2021 (I)]

80. The boxes of masses 2 kg and 8 kg are connected by a massless string passing over smooth pulleys. Calculate the time taken by box of mass 8 kg to strike the ground starting from rest. (use g = 10 m/s^2)

[NCERT : P L-65 | Aug. 27, 2021 (II)]

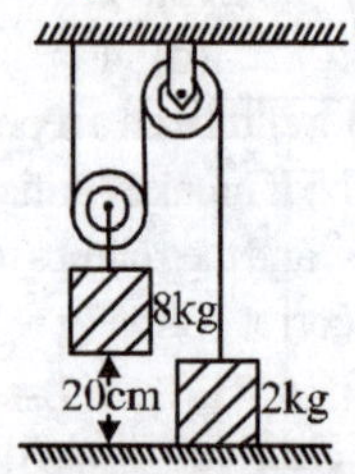

(a) 0.34 s

(b) 0.2 s

(c) 0.25 s

(d) 0.4 s

81. A steel block of 10 kg rests on a horizontal floor as shown. When three iron cylinders are placed on it as shown, the block and cylinders go down with an acceleration 0.2 m/s^2. The normal reaction R_2 by the floor if mass of the iron cylinders are equal and of 20 kg each, is ______N. [Take g = 10 m/s^2] **[NCERT : P L-65 | July 20, 2021 (I)]**

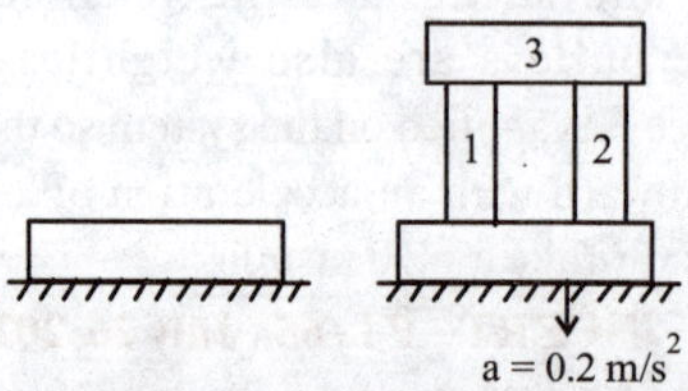

(a) 716 (b) 686 (c) 714 (d) 684

82. Consider a frame that is made up of two thin massless rods AB and AC as shown in the figure. A vertical force $\vec{P}$ of magnitude 100 N is applied at point A of the frame.

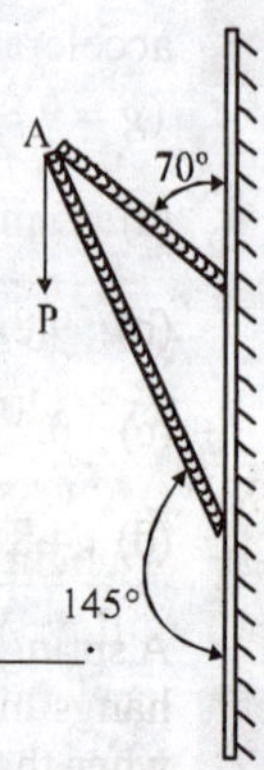

Suppose the force is $\vec{P}$ resolved parallel to the arms AB and AC of the frame. The magnitude of the resolved component along the arm AC is xN.
The value of x, to the nearest integer, is ______.
[Given : sin(35°) = 0.573, cos(35°) = 0.819
sin(110°) = 0.939,
cos(110°) = –0.342]

[NCERT : P L-65 | March 16, 2021 (I)]

83. A person standing on a spring balance inside a stationary lift measures 60 kg. The weight of that person if the lift descends with uniform downward acceleration of 1.8 m/s^2 will be ______ N. [g = 10 m/s^2]

[NCERT : P L-65 | Feb. 26, 2021 (I)]

84. A mass of 10 kg is suspended by a rope of length 4 m, from the ceiling. A force F is applied horizontally at the mid-point of the rope such that the top half of the rope makes an angle of 45° with the vertical. Then F equals: (Take $g = 10$ ms^{-2} and the rope to be massless)

[NCERT : P L-58, 59 | 9 Jan. 2019 II, 7 Jan. 2020 II]

(a) 100 N (b) 90 N
(c) 70 N (d) 75 N

85. Two blocks of mass M_1 = 20 kg and M_2 = 12 kg are connected by a metal rod of mass 8 kg. The system is pulled vertically up by applying a force of 480 N as shown. The tension at the mid-point of the rod is :

[NCERT : P L-65 | Online April 22, 2013]

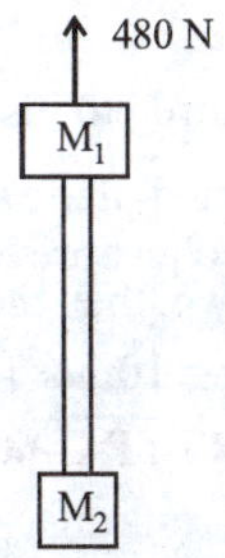

(a) 144 N
(b) 96 N
(c) 240 N
(d) 192 N

86. A block of mass m is connected to another block of mass M by a spring (massless) of spring constant k. The blocks are kept on a smooth horizontal plane. Initially the blocks are at rest and the spring is unstretched. Then a constant force F starts acting on the block of mass M to pull it. Find the force on the block of mass m. **[NCERT : P L-65 | 2007]**

(a) $\dfrac{MF}{(m+M)}$ (b) $\dfrac{mF}{M}$
(c) $\dfrac{(M+m)F}{m}$ (d) $\dfrac{mF}{(m+M)}$

87. Two masses $m_1 = 5$ kg and $m_2 = 4.8$ kg tied to a string are hanging over a light frictionless pulley. What is the acceleration of the masses when left free to move ? $(g = 9.8 \text{m/s}^2)$ **[NCERT : P L-65 | 2004]**

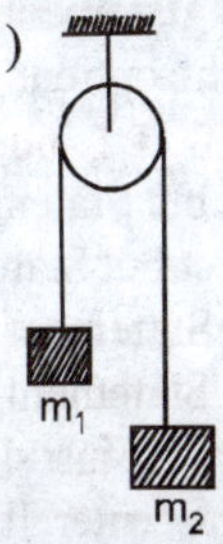

(a) 5 m/s^2
(b) 9.8 m/s^2
(c) 0.2 m/s^2
(d) 4.8 m/s^2

88. A spring balance is attached to the ceiling of a lift. A man hangs his bag on the spring and the spring reads 49 N, when the lift is stationary. If the lift moves downward with an acceleration of 5 m/s^2, the reading of the spring balance will be **[NCERT : P L-65 | 2003]**

(a) 24 N (b) 74 N (c) 15 N (d) 49 N

89. A block of mass M is pulled along a horizontal frictionless surface by a rope of mass m. If a force P is applied at the free end of the rope, the force exerted by the rope on the block is **[NCERT : P L-65 | 2003]**

(a) $\dfrac{Pm}{M+m}$ (b) $\dfrac{Pm}{M-m}$ (c) P (d) $\dfrac{PM}{M+m}$

90. A light spring balance hangs from the hook of the other light spring balance and a block of mass M kg hangs from the former one. Then the true statement about the scale reading is **[NCERT : P L-65 | 2003]**

(a) both the scales read M kg each
(b) the scale of the lower one reads M kg and of the upper one zero
(c) the reading of the two scales can be anything but the sum of the reading will be M kg
(d) both the scales read M/2 kg each

91. A lift is moving down with acceleration a. A man in the lift drops a ball inside the lift. The acceleration of the ball as observed by the man in the lift and a man standing stationary on the ground are respectively

[NCERT : P L-65 | 2002]

(a) g, g (b) $g-a, g-a$
(c) $g-a, g$ (d) a, g

92. When forces F_1, F_2, F_3 are acting on a particle of mass m such that F_2 and F_3 are mutually perpendicular, then the particle remains stationary. If the force F_1 is now removed then the acceleration of the particle is

[NCERT : P L-58 | 2002]

(a) F_1/m (b) F_2F_3/mF_1
(c) $(F_2 - F_3)/m$ (d) F_2/m.

93. Three identical blocks of masses $m = 2$ kg are drawn by a force F = 10. 2 N with an acceleration of 0. 6 ms^{-2} on a frictionless surface, then what is the tension (in N) in the string between the blocks B and C?

[NCERT : P L-65 | 2002]

(a) 9.2 (b) 3.4
(c) 4 (d) 9.8

94. One end of a massless rope, which passes over a massless and frictionless pulley P is tied to a hook C while the other end is free. Maximum tension that the rope can bear is 360 N. With what value of maximum safe acceleration (in ms^{-2}) can a man of 60 kg climb on the rope? **[NCERT : P L-65 | 2002]**

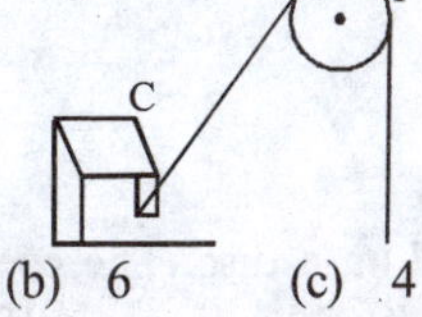

(a) 16 (b) 6 (c) 4 (d) 8

Topic 3 - Friction

95. A cubic block of mass m is sliding down on an inclined plane at 60^0 with an acceleration of $\dfrac{g}{2}$, the value of coefficient of kinetic friction is

[NCERT : P L-60, 61 | April 7, 2025 (I)]

(a) $\sqrt{3}-1$ (b) $\frac{\sqrt{3}}{2}$

(c) $\frac{\sqrt{2}}{3}$ (d) $1-\frac{\sqrt{3}}{2}$

96. A block of mass 25 kg is pulled along a horizontal surface by a force at an angle 45° with the horizontal. The friction coefficient between the block and the surface is 0.25. The displacement of 5 m of the block is:

[NCERT : P L-60, 61 | April 4, 2025 (II)]

(a) 970 J (b) 735 J (c) 245 J (d) 490 J

97. A given object takes n times the time to slide down 45° rough inclined plane as it takes the time to slide down an identical perfectly smooth 45° inclined plane. The coefficient of kinetic friction between the object and the surface of inclined plane is :

[NCERT : P L-60, 61 | April 8, 2024 (II)]

(a) $1-\frac{1}{n^2}$ (b) $1-n^2$ (c) $\sqrt{1-\frac{1}{n^2}}$ (d) $\sqrt{1-n^2}$

98. A heavy box of mass 50 kg is moving on a horizontal surface. If co-efficient of kinetic friction between the box and horizontal surface is 0.3 then force of kinetic friction is :

[NCERT : P L-60, 61 | April 5, 2024 (II)]

(a) 14.7 N (b) 147 N (c) 1.47 N (d) 1470 N

99. A 2 kg brick begins to slide over a surface which is inclined at an angle of 45° with respect to horizontal axis. The co-efficient of static friction between their surfaces is :

[NCERT : P L-60, 61 | April 4, 2024 (II)]

(a) 1 (b) $\frac{1}{\sqrt{3}}$ (c) 0.5 (d) 1.7

100. Consider a block and trolley system as shown in figure. If the coefficient of kinetic friction between the trolley and the surface is 0.04, the acceleration of the system in ms^{-2} is:

[NCERT : P L-60, 61 | Feb. 1, 2024 (I)]

(Consider that the string is massless and unstretchable and the pulley is also massless and frictionless) :

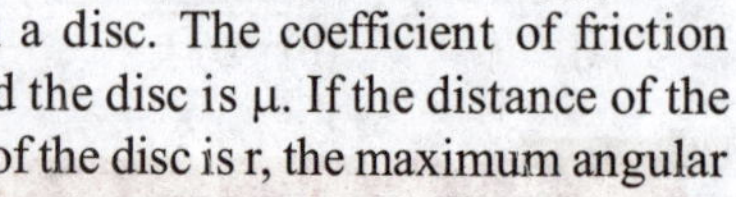

(a) 3

(b) 4

(c) 2

(d) 1.2

101. A coin is placed on a disc. The coefficient of friction between the coin and the disc is μ. If the distance of the coin from the center of the disc is r, the maximum angular velocity which can be given to the disc, so that the coin does not slip away, is :

[NCERT : P L-60, 61 | Jan. 31, 2024 (I)]

(a) $\frac{\mu}{\sqrt{rg}}$ (b) $\sqrt{\frac{\mu g}{r}}$ (c) $\frac{\mu g}{r}$ (d) $\sqrt{\frac{r}{\mu g}}$

102. In the given arrangement of a doubly inclined plane two blocks of masses M and m are placed. The blocks are connected by a light string passing over an ideal pulley as shown. The coefficient of friction between the surface of the plane and the blocks is 0.25. The value of m, for which M = 10 kg will move down with an acceleration of 2 m/s^2, is : (take g = 10 m/s^2 and tan 37° = 3/4)

[NCERT : P L-60, 61 | Jan. 31, 2024 (I)]

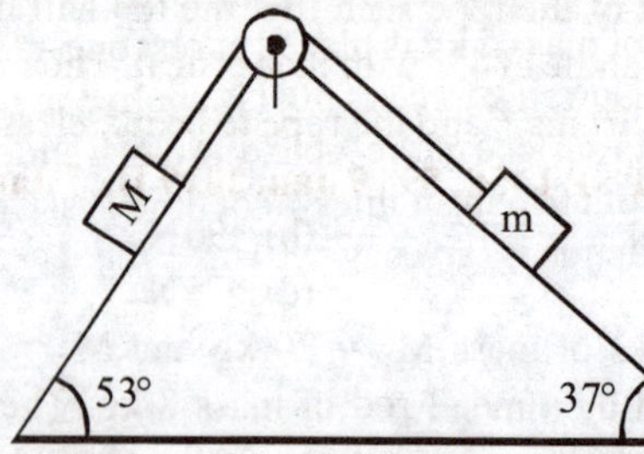

(a) 6.5 kg (b) 4.5 kg

(c) 2.25 kg (d) 9 kg

103. A block of mass 5 kg is placed on a rough inclined surface as shown in the figure.

If $\vec{F_1}$ is the force required to just move the block up the inclined plane and $\vec{F_2}$ is the force required to just prevent the block from sliding down, then the value of $|\vec{F_1}|-|\vec{F_2}|$ is: [Use g = 10m/s^2]

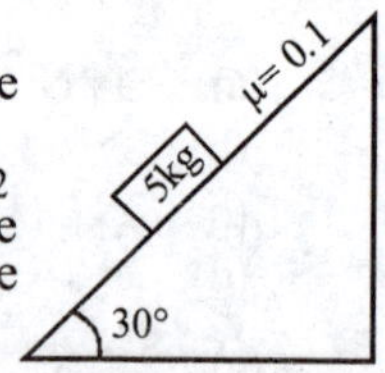

[NCERT : P L-60, 61 | Jan. 31, 2024 (II)]

(a) $25\sqrt{3}$ N (b) $5\sqrt{3}$ N (c) $\frac{50\sqrt{3}}{2}$ N (d) 10 N

104. A block of mass *m* is placed on a surface having vertical crossection given by $y = x^2/4$. If coefficient of friction is 0.5, the maximum height above the ground at which block can be placed without slipping is:

[NCERT : P L-60, 61 | Jan. 30, 2024 (II)]

(a) 1/4 *m* (b) 1/2*m* (c) 1/6 *m* (d) 1/3*m*

105. Given below are two statements :

Statement (I) : The limiting force of static friction depends on the area of contact and independent of materials.

Statement (II) : The limiting force of kinetic friction is independent of the area of contact and depends on materials.

In the light of the above statements, choose the most appropriate answer from the options given below :

[NCERT : P L-60 | Jan. 27, 2024 (II)]

(a) Statement I is correct but Statement II is incorrect

(b) Statement I is incorrect but Statement II is correct

(c) Both Statement I and Statement II are incorrect

(d) Both Statement I and Statement II are correct

106. A coin placed on a rotating table just slips when it is placed at a distance of 1 cm from the center. If the angular velocity of the table in halved, it will just slip when placed at a distance of ______ from the centre:

[NCERT : P L-60, 61 | April 11, 2023 (I)]

(a) 2 cm (b) 1 cm (c) 8 cm (d) 4 cm

107. A bullet of mass 0.1 kg moving horizontally with speed 400 ms^{-1} hits a wooden block of mass 3.9 kg kept on a horizontal rough surface. The bullet gets embedded into the block and moves 20 m before coming to rest. The coefficient of friction between the block and the surface is______. (Given $g = 10\ ms^2$)

[NCERT : P L-60, 61 | April 8, 2023 (II)]

(a) 0.50 (b) 0.90 (c) 0.65 (d) 0.25

108. A block of mass 5 kg is placed at rest on a table of rough surface. Now, if a force of 30N is applied in the direction parallel to surface of the table, the block slides through a distance of 50 m in an interval of time 10s. Coefficient of kinetic friction is (given, $g = 10\ ms^{-2}$):

[NCERT : P L-60 | Feb. 1, 2023 (I)]

(a) 0.50 (b) 0.60 (c) 0.75 (d) 0.25

109. As shown in the figure a block of mass 10 kg lying on a horizontal surface is pulled by a force F acting at an angle 30°, with horizontal. For $\mu_s = 0.25$, the block will just start to move for the value of F : [Given $g = 10\ ms^{-2}$]

[NCERT : P L-60 | Feb. 1, 2023 (II)]

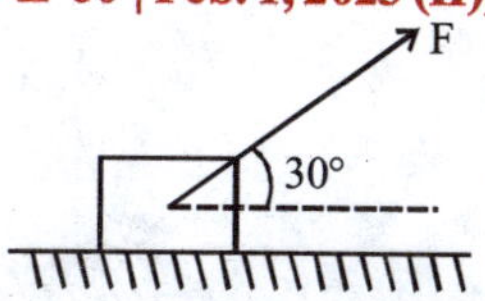

(a) 20N
(b) 33.3N
(c) 25.2N
(d) 35.7N

110. A body of mass 10 kg is moving with an initial speed of 20 m/s. The body stops after 5 s due to friction between body and the floor. The value of the coefficient of friction is: (Take acceleration due to gravity $g = 10\ ms^{-2}$)

[NCERT : P L-60 | Jan. 31, 2023 (II)]

(a) 0.2 (b) 0.3 (c) 0.5 (d) 0.4

111. A block of mass m slides down the plane inclined at angle 30° with an acceleration $\frac{g}{4}$. The value of coefficient of kinetic friction will be :

[NCERT : P L-60, 61 | Jan. 29, 2023 (I)]

(a) $\frac{2\sqrt{3}+1}{2}$ (b) $\frac{1}{2\sqrt{3}}$

(c) $\frac{\sqrt{3}}{2}$ (d) $\frac{2\sqrt{3}-1}{2}$

112. The time taken by an object to slide down 45° rough inclined plane is n times as it takes to slide down a perfectly smooth 45° incline plane. The coefficient of kinetic friction between the object and the incline plane is

[NCERT : P L-60, 61 | Jan. 29, 2023 (II)]

(a) $\sqrt{\frac{1}{1-n^2}}$ (b) $\sqrt{1-\frac{1}{n^2}}$ (c) $1+\frac{1}{n^2}$ (d) $1-\frac{1}{n^2}$

113. Consider a block kept on an inclined plane (inclined at 45°) as shown in the figure. If the force required to just push it up the incline is 2 times the force required to just prevent it from sliding down, the coefficient of friction between the block and inclined plane (μ) is equal to :

[NCERT : P L-60, 61 | Jan. 25, 2023 (II)]

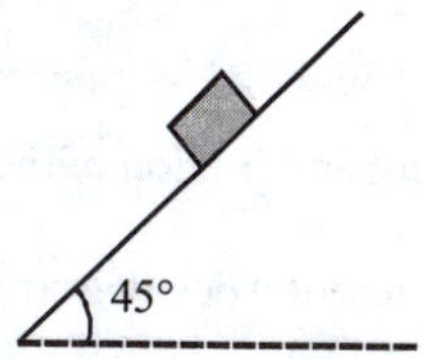

(a) 0.33 (b) 0.60 (c) 0.25 (d) 0.50

114. A bag is gently dropped on a conveyor belt moving at a speed of 2 m/s. The coefficient of friction between the conveyor belt and bag is 0.4. Initially, the bag slips on the belt before it stops due to friction. The distance travelled by the bag on the belt during slipping motion is: [Take $g = 10\ m/s^{-2}$] **[NCERT : P L-60, 61 | July 27, 2022 (I)]**

(a) 2m (b) 0.5m (c) 3.2m (d) 0.8m

115. A block of mass M slides down on a rough inclined plane with constant velocity. The angle made by the incline plane with horizontal is θ. The magnitude of the contact force will be : **[July 27, 2022 (II)]**

(a) Mg
(b) Mg cos θ
(c) $\sqrt{Mg\sin\theta + Mg\cos\theta}$
(d) $Mg\sin\theta\sqrt{1+\mu}$

116. A 2 kg block is pushed against a vertical wall by applying a horizontal force of 50 N. The coefficient of static friction between the block and the wall is 0.5. A force F is also applied on the block vertically upward (as shown in figure). The maximum value of F applied, so that the block does not move upward, will be:

[NCERT : P L-60 | June 30, 2022 (I)]

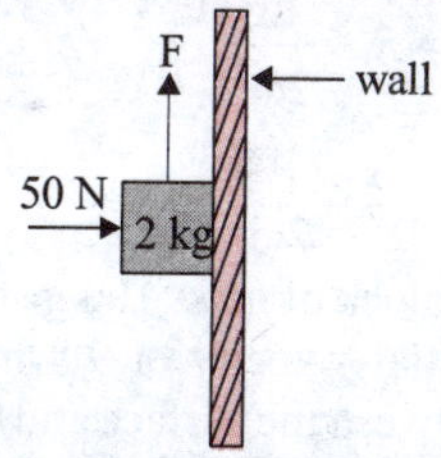

(a) 10N
(b) 20N
(c) 25N
(d) 45N

117. A block of mass 40 kg slides over a surface, when a mass of 4 kg is suspended through an inextensible massless string passing over frictionless pulley as shown below. The coefficient of kinetic friction between the surface and block is 0.02. The acceleration of block is. (Given $g = 10\ ms^{-2}$.)

[NCERT : P L-60 | June 29, 2022 (II)]

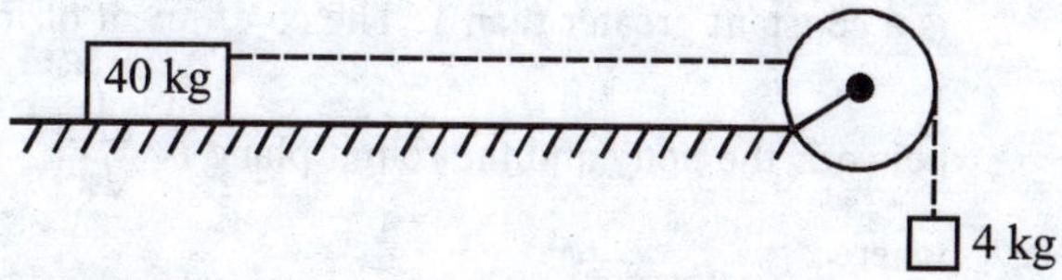

(a) 1 ms^{-2} (b) 1/5 ms^{-2}
(c) 4/5 ms^{-2} (d) 8/11 ms^{-2}

118. A hanging mass M is connected to a four times bigger mass by using a string-pulley arrangement, as shown in the figure. The bigger mass is placed on a horizontal ice-slab and being pulled by 2 Mg force. In this situation.

tension in the string is $\frac{x}{5}$ Mg for $x=$ ________. Neglect mass of the string and friction of the block (bigger mass) with ice slab.
(Given $g=$ acceleration due to gravity)

[June 28, 2022 (I)]

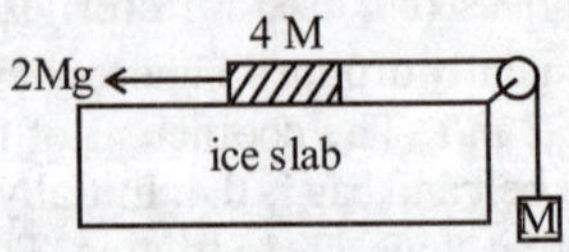

119. A system of two blocks of masses m = 2 kg and M = 8 kg is placed on a smooth table as shown in figure. The coefficient of static friction between two blocks is 0.5. The maximum horizontal force F that can be applied to the block of mass M so that the blocks move together will be:

[NCERT : P L-60 | June 27, 2022 (I)]

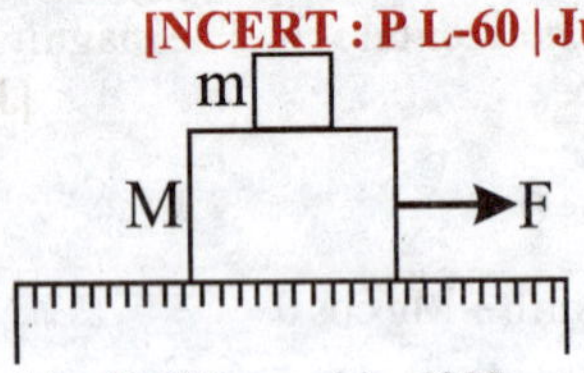

(a) 9.8 N (b) 39.2 N (c) 49 N (d) 78.4 N

120. A disc with a flat small bottom beaker placed on it at a distance R from its center is revolving about an axis passing through the center and perpendicular to its plane with an angular velocity ω. The coefficient of static friction between the bottom of the beaker and the surface of the disc is μ. The breaker will revolve with the disc if:

[NCERT : P L-60 | June 25, 2022 (II)]

(a) $R \le \frac{\mu g}{2\omega^2}$ (b) $R \le \frac{\mu g}{\omega^2}$

(c) $R \ge \frac{\mu g}{2\omega^2}$ (d) $R \ge \frac{\mu g}{\omega^2}$

121. A block of mass 10 kg starts sliding on a surface with an initial velocity of 9.8 ms^{-1}. The coefficient of friction between the surface and bock is 0.5. The distance covered by the block before coming to rest is :
[use $g = 9.8\ ms^{-2}$] **[NCERT : P L-61 | June 24, 2022 (I)]**

(a) 4.9 m (b) 9.8 m (c) 12.5 m (d) 19.6 m

122. When a body slides down from rest along a smooth inclined plane making an angle of 30° with the horizontal, it takes time T. When the same body slides down from the rest along a rough inclined plane making the same angle and through the same distance, it takes time αT, where α is a constant greater than 1. The co-efficient of friction between the body and the rough plane is $\frac{1}{\sqrt{x}}\left(\frac{\alpha^2-1}{\alpha^2}\right)$ where x = ________.

[NCERT : P L-60, 61 | NA, Sep. 1, 2021 (II)]

123. The coefficient of static friction between two blocks is 0.5 and the table is smooth. The maximum horizontal force that can be applied to move the blocks together is ________ N. (take $g = 10\ ms^{-2}$)

[NCERT : P L-61 | NA, Aug. 26, 2021 (II)]

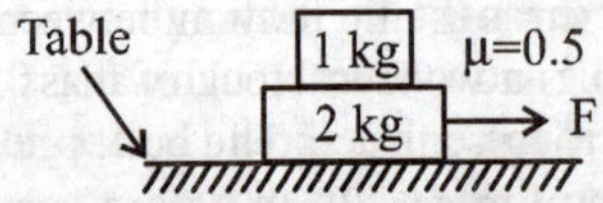

124. A body of mass 'm' is launched up on a rough inclined plane making an angle of 30° with the horizontal. The coefficient of friction between the body and plane is $\frac{\sqrt{x}}{5}$ if the time of ascent is half of the time of descent. The value of x is ________.

[NCERT : P L-60, 61 | NA, July 20, 2021 (II)]

125. A boy of mass 4 kg is standing on a piece of wood having mass 5 kg. If the coefficient of friction between the wood and the floor is 0.5, the maximum force that the boy can exert on the rope so that the piece of wood does not move from its place is ________ N. (Round off to the Nearest integer) [Take $g = 10\ ms^{-2}$]

[NCERT : P L-60, 61 | March 17, 2021 (II)]

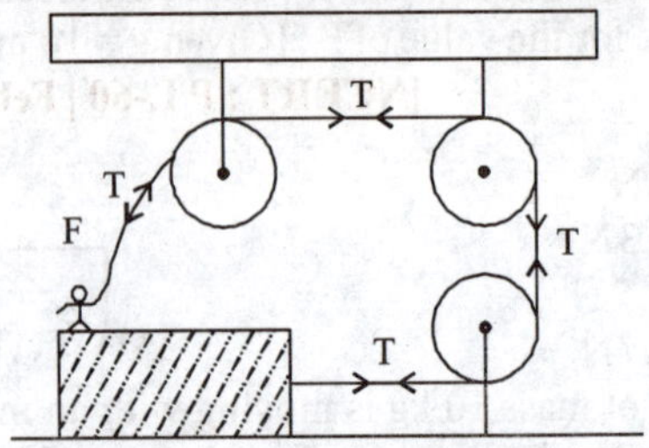

126. A body of mass 1 kg rests on a horizontal floor with which it has a coefficient of static friction $\frac{1}{\sqrt{3}}$. It is desired to make the body move by applying the minimum possible force F N. The value of F will be ________. (Round off to the Nearest integer) [Take $g = 10\ ms^{-2}$]

[NCERT : P L-60, 61 | March 17, 2021 (II)]

127. As shown in the figure, a block of mass $\sqrt{3}$ kg is kept on a horizontal rough surface of coefficient of friction $\frac{1}{3\sqrt{3}}$. The critical force to be applied on the vertical surface as shown at an angle 60° with horizontal such that it does not move, will be 3x. The value of x will be ________.

[$g = 10\ m/s^2$; $\sin 60° = \frac{\sqrt{3}}{2}$; $\cos 60° = \frac{1}{2}$]

[NCERT : P L-60, 61 | Feb. 26, 2021 (I)]

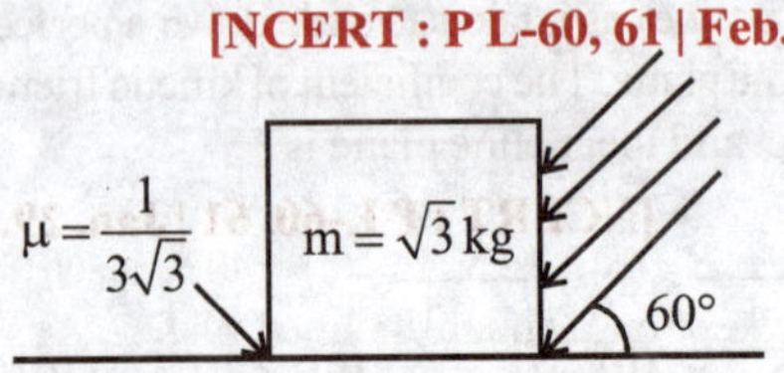

128. An inclined plane is bent in such a way that the vertical cross-section is given by $y = \frac{x^2}{4}$ where y is in vertical and x in horizontal direction. If the upper surface of this curved plane is rough with coefficient of friction μ = 0.5, the maximum height in cm at which a stationary block will not slip downward is ________ cm.

[NCERT : P L-60, 61 | Feb. 24, 2021 (I), 2003 S]

129. The coefficient of static friction between a wooden block of mass 0.5 kg and a vertical rough wall is 0.2. The magnitude of horizontal force that should be applied on the block to keep it adhere to the wall will be ________ N. [$g = 10\, ms^{-2}$]

[NCERT : P L-60 | Feb. 24, 2021 (I)]

130. An insect is at the bottom of a hemispherical ditch of radius 1 m. It crawls up the ditch but starts slipping after it is at height h from the bottom. If the coefficient of friction between the ground and the insect is 0.75, then h is : ($g = 10\, ms^{-2}$)

[NCERT : P L-60, 61 | Sep. 06, 2020 (I)]

(a) 0.20m (b) 0.45m (c) 0.60m (d) 0.80m

131. A block starts moving up an inclined plane of inclination 30° with an initial velocity of v_0. It comes back to its initial position with velocity $\frac{v_0}{2}$. The value of the coefficient of kinetic friction between the block and the inclined plane is close to $\frac{I}{1000}$. The nearest integer to I is ________.

[NCERT : P L-60, 61 | NA Sep. 03, 2020 (II)]

132. A block of mass 5 kg is (i) pushed in case (A) and (ii) pulled in case (B), by a force F=20 N, making an angle of 30° with the horizontal, as shown in the figures. The coefficient of friction between the block and floor is $\mu = 0.2$. The difference between the accelerations of the block, in case (B) and case (A) will be : ($g = 10\, ms^{-2}$)

[NCERT : P L-61 | 12 April 2019 II]

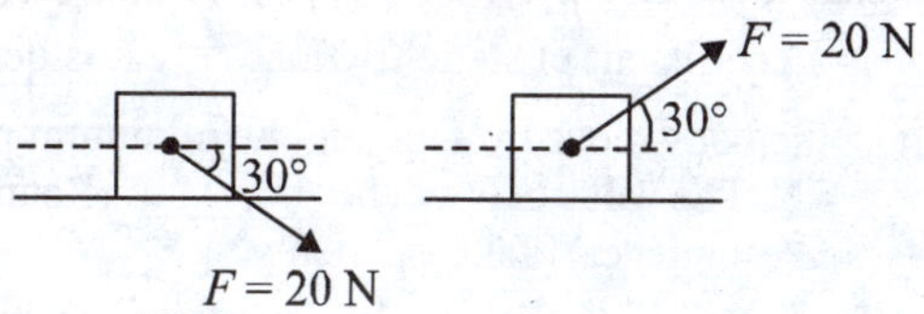

(A) (B)

(a) $0.4\, ms^{-2}$ (b) $3.2\, ms^{-2}$
(c) $0.8\, ms^{-2}$ (d) $0\, ms^{-2}$

133. Two blocks A and B masses m_A=1 kg and m_B = 3 kg are kept on the table as shown in figure. The coefficient of friction between A and B is 0.2 and between B and the surface of the table is also 0.2. The maximum force F that can be applied on B horizontally, so that the block A does not slide over the block B is : [Take $g = 10\, m/s^2$]

[NCERT : P L-61 | 10 April 2019 II]

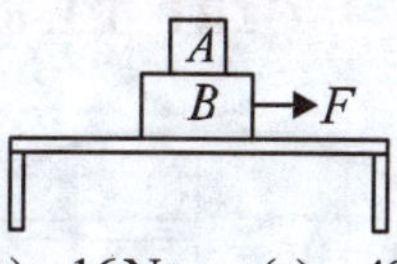

(a) 8 N (b) 16 N (c) 40 N (d) 12 N

134. A block kept on a rough inclined plane, as shown in the figure, remains at rest upto a maximum force 2 N down the inclined plane. The maximum external force up the inclined plane that does not move the block is 10 N. The coefficient of static friction between the block and the plane is : [Take $g = 10\, m/s^2$]

[NCERT : P L-60, 61 | 12 Jan. 2019 II]

(a) $\frac{\sqrt{3}}{2}$ (b) $\frac{\sqrt{3}}{4}$
(c) $\frac{1}{2}$ (d) $\frac{2}{3}$

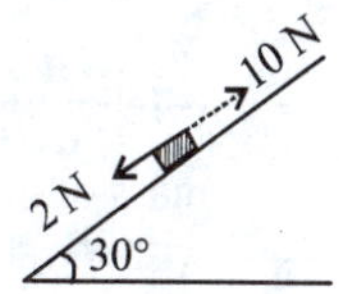

135. Two masses m_1 = 5 kg and m_2 = 10 kg, connected by an inextensible string over a frictionless pulley, are moving as shown in the figure. The coefficient of friction of horizontal surface is 0.15. The minimum weight m that should be put on top of m_2 to stop the motion is:

[NCERT : P L-60, 61 | 2018]

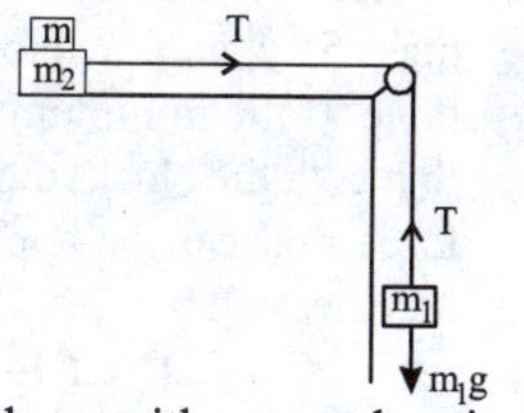

(a) 18.3 kg
(b) 23.3 kg
(c) 43.3 kg
(d) 10.3 kg

136. A body of mass 2kg slides down with an acceleration of $3 m/s^2$ on a rough inclined plane having a slope of 30°. The external force required to take the same body up the plane with the same acceleration will be: ($g = 10 m/s^2$)

[Online April 15, 2018]

(a) 4N (b) 14N
(c) 6N (d) 20N

137. A rocket is fired vertically from the earth with an acceleration of 2g, where g is the gravitational acceleration. On an inclined plane inside the rocket, making an angle θ with the horizontal, a point object of mass m is kept. The minimum coefficient of friction μ_{min} between the mass and the inclined surface such that the mass does not move is :

[NCERT : P L-60 | Online April 9, 2016]

(a) tan2θ (b) tanθ
(c) 3 tanθ (d) 2 tan θ

138. Given in the figure are two blocks A and B of weight 20 N and 100 N, respectively. These are being pressed against a wall by a force F as shown. If the coefficient of friction between the blocks is 0.1 and between block B and the wall is 0.15, the frictional force applied by the wall on block B is:

[NCERT : P L-60 | 2015]

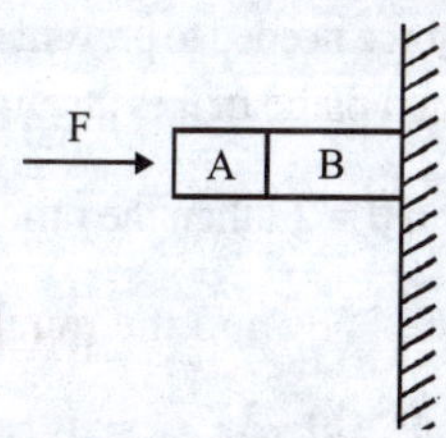

(a) 120 N
(b) 150 N
(c) 100 N
(d) 80 N

139. A small ball of mass m starts at a point A with speed v_0 and moves along a frictionless track AB as shown. The track BC has coefficient of friction μ. The ball comes to stop at C after travelling a distance L which is:

[NCERT : P L-61 | Online April 11, 2014]

(a) $\frac{2h}{\mu}+\frac{v_o^2}{2\mu g}$

(b) $\frac{h}{\mu}+\frac{v_o^2}{2\mu g}$

(c) $\frac{h}{2\mu}+\frac{v_o^2}{\mu g}$

(d) $\frac{h}{2\mu}+\frac{v_o^2}{2\mu g}$

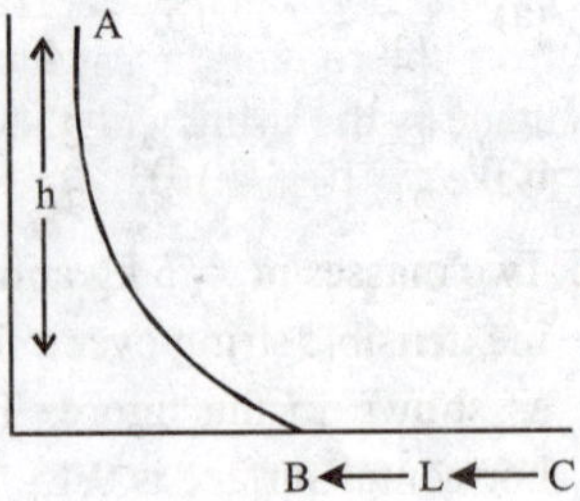

140. A block A of mass 4 kg is placed on another block B of mass 5 kg, and the block B rests on a smooth horizontal table. If the minimum force that can be applied on A so that both the blocks move together is 12 N, the maximum force that can be applied to B for the blocks to move together will be:

[NCERT : P L-61 | Online April 9, 2014]

(a) 30 N (b) 25 N (c) 27 N (d) 48 N

141. A block is placed on a rough horizontal plane. A time dependent horizontal force F = kt acts on the block, where k is a positive constant. The acceleration - time graph of the block is : **[NCERT : P L-61 | Online April 25, 2013]**

(a) 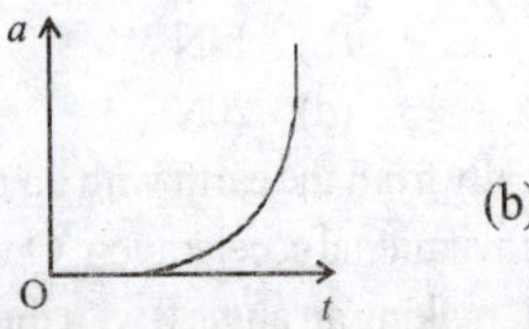

(b) 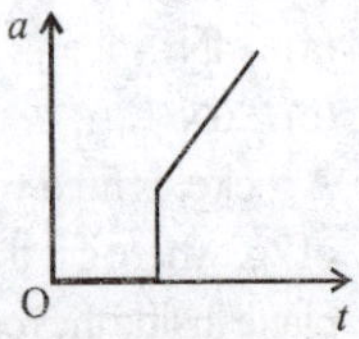

(c) 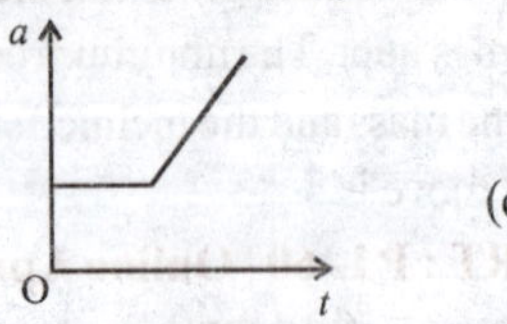

(d) 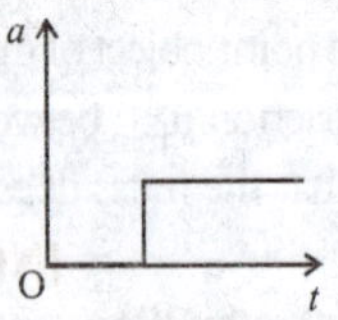

142. A body starts from rest on a long inclined plane of slope 45°. The coefficient of friction between the body and the plane varies as $\mu = 0.3\,x$, where x is distance travelled down the plane. The body will have maximum speed (for $g = 10\ m/s^2$) when $x =$

[NCERT : P L-61 | Online April 22, 2013]

(a) 9.8 m (b) 27 m (c) 12 m (d) 3.33 m

143. The minimum force required to start pushing a body up rough (frictional coefficient μ) inclined plane is F_1 while the minimum force needed to prevent it from sliding down is F_2. If the inclined plane makes an angle θ from the horizontal such that $\tan\theta = 2\mu$ then the ratio of $\frac{F_1}{F_2}$ is

[NCERT : P L-61 | 2011 RS]

(a) 1 (b) 2 (c) 3 (d) 4

144. The upper half of an inclined plane with inclination ϕ is perfectly smooth while the lower half is rough. A body starting from rest at the top will again come to rest at the bottom if the coefficient of friction for the lower half is given by **[NCERT : P L-61 | 2005]**

(a) $2\cos\phi$ (b) $2\sin\phi$

(c) $\tan\phi$ (d) $2\tan\phi$

145. Consider a car moving on a straight road with a speed of 100 m/s. The distance at which car can be stopped is $[\mu_k = 0.5]$ **[NCERT : P L-61 | 2005]**

(a) 1000 m (b) 800 m

(c) 400 m (d) 100 m

146. A block rests on a rough inclined plane making an angle of 30° with the horizontal. The coefficient of static friction between the block and the plane is 0.8. If the frictional force on the block is 10 N, the mass of the block (in kg) is (take $g = 10\ m/s^2$) **[NCERT : P L-61 | 2004]**

(a) 1.6 (b) 4.0

(c) 2.0 (d) 2.5

147. A marble block of mass 2 kg lying on ice when given a velocity of 6 m/s is stopped by friction in 10 s. Then the coefficient of friction is **[NCERT : P L-61 | 2003]**

(a) 0.02 (b) 0.03

(c) 0.04 (d) 0.06

Topic 4 - Circular Motion, Banking of Road

148. A car of mass 'm' moves on a banked road having radius 'r' and banking angle θ. To avoid slipping from banked road, the maximum permissible speed of the car is v_0. The coefficient of friction μ between the wheels of the car and the banked road is **[NCERT : P L-63, 64 | Jan. 24, 2025 (I)]**

(a) $\mu = \frac{v_0^2 + rg\tan\theta}{rg - v_0^2\tan\theta}$ (b) $\mu = \frac{v_0^2 + rg\tan\theta}{rg + v_0^2\tan\theta}$

(c) $\mu = \frac{v_0^2 - rg\tan\theta}{rg + v_0^2\tan\theta}$ (d) $\mu = \frac{v_0^2 - rg\tan\theta}{rg - v_0^2\tan\theta}$

149. A circular table is rotating with an angular velocity of ω rad/s about its axis (see figure). There is a smooth groove along a radial direction on the table. A steel ball is gently placed at a distance of 1m on the groove. All the surface are smooth. If the radius of the table is 3 m, the radial velocity of the ball w.r.t. the table at the time ball leaves the table is $x\sqrt{2}\omega$ m/s, where the value of x is..........

[NCERT : P L-63 | April 8, 2024 (II)]

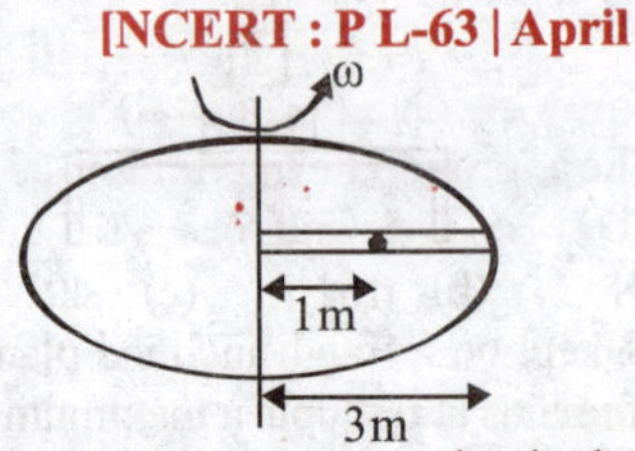

150. A car of 800 kg is taking turn on a banked road of radius 300 m and angle of banking 30°. If coefficient of static friction is 0.2 then the maximum speed with which car can negotiate the turn safely : ($g = 10\ m/s^2$, $\sqrt{3} = 1.73$)

[NCERT : P L-64 | April 6, 2024 (II)]

(a) 70.4 m/s (b) 51.4 m/s

(c) 264 m/s (d) 102.8 m/s

151. A man carrying a monkey on his shoulder does cycling smoothly on a circular track of radius 9m and completes 120 revolutions in 3 minutes. The magnitude of centripetal acceleration of monkey is (in m/s^2) : **[NCERT : P L-63 | April 5, 2024 (II)]**

(a) zero (b) $16\pi^2$ ms^{-2}
(c) $4\pi^2$ ms^{-2} (d) $57600\pi^2$ ms^{-2}

152. A ball of mass 0.5 kg is attached to a string of length 50 cm. The ball is rotated on a horizontal circular path about its vertical axis. The maximum tension that the string can bear is 400 N. The maximum possible value of angularly velocity of the ball is rad/s is, : **[NCERT : P L-63 | Feb. 1, 2024 (I)]**

(a) 1600 (b) 40 (c) 1000 (d) 20

153. If the radius of curvature of the path of two particles of same mass are in the ratio 3 : 4, then in order to have constant centripetal force, their velocities will be in the ratio of : **[NCERT : P L-63 | Jan. 29, 2024 (I)]**

(a) $1:\sqrt{3}$ (b) $\sqrt{3}:1$ (c) $\sqrt{3}:2$ (d) $2:\sqrt{3}$

154. A train is moving with a speed of 12 m/s on rails which are 1.5 m apart. To negotiate a curve radius 400 m, the height by which the outer rail should be raised with respect to the inner rail is **[NCERT : P L-64 | Jan. 27, 2024 (I)]**
(Given, g = 10 m/s^2) :

(a) 6.0 cm (b) 5.4 cm
(c) 4.8 cm (d) 4.2 cm

155. A stone of mass 900g is tied to a string and moved in a vertical circle of radius 1m making 10 rpm. The tension in the string, when the stone is at the lowest point is (if $\pi^2 = 9.8$ and g = 9.8 m/s^2) : **[NCERT : P L-63 | Jan. 29, 2024 (II)]**

(a) 17.8 N (b) 8.82 N (c) 97 N (d) 9.8 N

156. A vehicle of mass 200 kg is moving along a levelled curved road of radius 70 m with angular velocity of 0.2 rad/s. The centripetal force acting on the vehicle is : **[NCERT : P L-63 | April 13, 2023 (II)]**

(a) 560 N (b) 2800 N
(c) 14 N (d) 2240 N

157. A child of mass 5 kg is going round a merry-go-round that makes 1 rotation in 3.14 s. The radius of the merry-go-round is 2 m. The centrifugal force on the child will be **[NCERT : P L-63 | April 6, 2023 (II)]**

(a) 80 N (b) 50 N (c) 100 N (d) 40 N

158. A car is moving on a horizontal curved road with radius 50 m. The approximate maximum speed of car will be, if friction between tyres and road is 0.34. [Take g = 10 ms^{-2}] **[NCERT : P L-63 | Jan. 29, 2023 (I)]**

(a) 3.4ms^{-1} (b) 22.4ms^{-1}
(c) 13ms^{-1} (d) 17ms^{-1}

159. A car is moving on a circular path of radius 600 m such that the magnitudes of the tangential acceleration and centripetal acceleration are equal. The time taken by the car to complete first quarter of revolution, if it is moving with an initial speed of 54 km/hr is $t\left(1-e^{-\pi/2}\right)$s. The value of t is ______. **[NCERT : P L-63 | Jan. 29, 2023 (II)]**

160. A car is moving with a constant speed of 20 m/s in a circular horizontal track of radius 40 m. A bob is suspended from the roof of the car by a massless string. The angle made by the string with the vertical will be :
(Take g = 10 m/s^2) **[NCERT : P L-64 | Jan. 25, 2023 (I)]**

(a) $\frac{\pi}{6}$ (b) $\frac{\pi}{2}$ (c) $\frac{\pi}{4}$ (d) $\frac{\pi}{3}$

161. A smooth circular groove has a smooth vertical wall as shown in figure. A block of mass m moves against the wall with a speed v. Which of the following curve represents the correct relation between the normal reaction on the block by the wall (N) and speed of the block (v)? **[NCERT : P L-63 | July 29, 2022 (I)]**

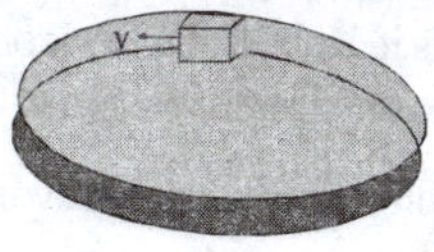

(a)
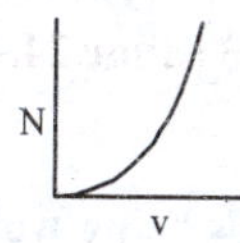

(b)
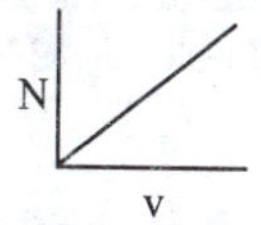

(c)
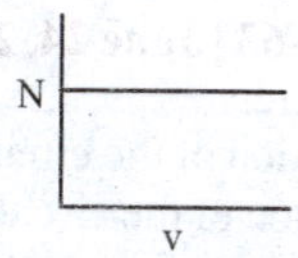

(d)
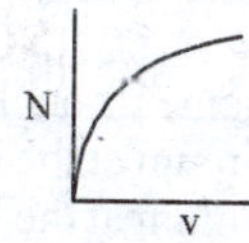

162. A stone tide to a string of length L is whirled in a vertical circle with the other end of the string at the centre. At a certain instant of time, the stone is at its lowest position and has a speed u. The magnitude of change in its velocity, as it reaches a position where the string is horizontal, is $\sqrt{x(u^2-gL)}$. The value of x is **[June 27, 2022 (II)]**

(a) 3 (b) 2 (c) 1 (d) 5

163. A ball is released from rest from point *P* of a smooth semi-spherical vessel as shown in figure. The ratio of the centripetal force and normal reaction on the ball at point *Q* is *A* while angular position of point *Q* is α with respect to point *P*. Which of the following graphs represent the correct relation between *A* and α when ball goes from *Q* to *R*? **[NCERT : P L-63 | June 26, 2022 (I)]**

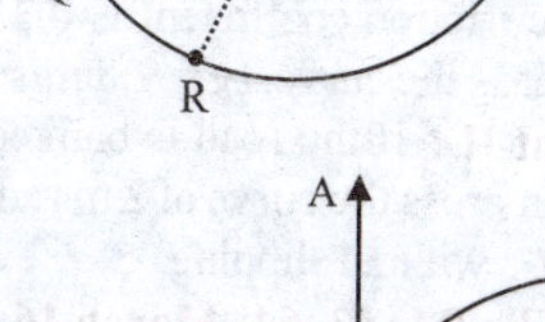

(a)
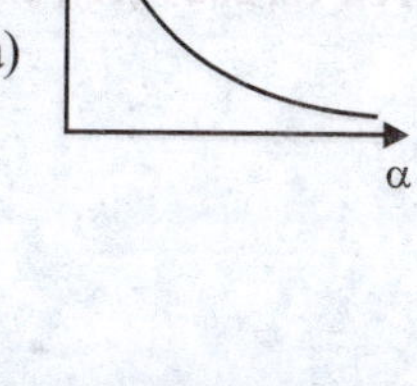

(b)
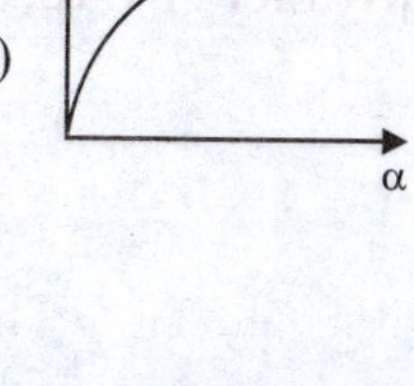

(c) 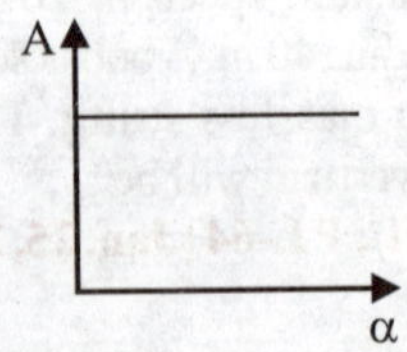

(d) 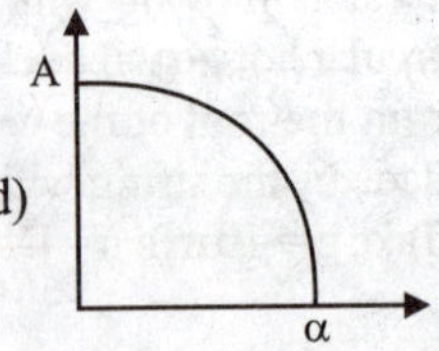

164. A curved in a level road has a radius 75m. The maximum speed of a car turning this curved road can be 30 m/s without skidding. If radius of curved road is changed to 48 m and the coefficient of friction between the tyres and the road remains same, then maximum allowed speed would be _______ m/s.
[NCERT : P L-63 | June 25, 2022 (II)]

165. A boy ties a stone of mass 100 g to the end of a 2 m long string and whirls it around in a horizontal plane. The string can withstand the maximum tension of 80 *N*. If the maximum speed with which the stone can revolve is $\frac{K}{\pi}$ rev. / min. The value of *K* is : (Assume the string is massless and unstretchable) **[NCERT : P L-63 | June 24, 2022 (I)]**

(a) 400 (b) 300
(c) 600 (d) 800

166. A stone of mass m, tied to a string is being whirled in a vertical circle with a uniform speed. The tension in the string is : **[NCERT : P L-63 | June 24, 2022 (II)]**

(a) the same throughout the motion
(b) minimum at the highest position of the circular path
(c) minimum at the lowest position of the circular path
(d) minimum when the rope is in the horizontal position

167. A particle of mass m is suspended from a ceiling through a string of length L. The particle moves in a horizontal circle of radius r such that $r = \frac{L}{\sqrt{2}}$. The speed of particle will be :
[NCERT : P L-64 | Aug. 26, 2021 (II)]

(a) $\sqrt{rg}$ (b) $\sqrt{2rg}$
(c) $2\sqrt{rg}$ (d) $\sqrt{\frac{rg}{2}}$

168. A block of 200 g mass moves with a uniform speed in a horizontal circular groove, with vertical side walls of radius 20 cm. If the block takes 40 s to complete one round, the normal force by the side walls of the groove is :
[NCERT : P L-63 | March 16, 2021 (I)]

(a) 9.859×10^{-2} N (b) 9.859×10^{-4} N
(c) 6.28×10^{-3} N (d) 0.0314 N

169. Statement I : A cyclist is moving on an unbanked road with a speed of 7 kmh^{-1} and takes a sharp circular turn along a path of radius of 2 m without reducing the speed. The static friction coefficient is 0.2. The cyclist will not slip and pass the curve. ($g = 9.8$ m/s^2)
Statement II : If the road is banked at an angle of 45°, cyclist can cross the curve of 2 m radius with the speed of 18.5 kmh^{-1} without slipping.
[NCERT : P L-63, 64 | March 16, 2021 (II); 2002 (S)]

In the light of the above statements, choose the correct answer from the options given below.

(a) Statement I is correct and statement II is incorrect
(b) Statement I is incorrect and statement II is correct
(c) Both statement I and statement II are true
(d) Both statement I and statement II are false

170. A small bob tied at one end of a thin string of length 1 m is describing a vertical circle so that the maximum and minimum tension in the string are in the ratio 5 : 1. The velocity of the bob at the highest position is _______ m/s. (Take $g = 10$ m/s^2) **[Feb. 25, 2021 (I)]**

171. A disc rotates about its axis of symmetry in a hoizontal plane at a steady rate of 3.5 revolutions per second. A coin placed at a distance of 1.25cm from the axis of rotation remains at rest on the disc. The coefficient of friction between the coin and the disc is ($g = 10$m/s^2)
[NCERT : P L-63 | Online April 15, 2018]

(a) 0.5 (b) 0.7 (c) 0.3 (d) 0.6

172. A conical pendulum of length 1 m makes an angle $\theta = 45°$ w.r.t. Z-axis and moves in a circle in the XY plane. The radius of the circle is 0.4 m and its centre is vertically below O. The speed of the pendulum, in its circular path, will be : (Take $g = 10$ ms^{-2})
[NCERT : P L-64 | Online April 9, 2017]

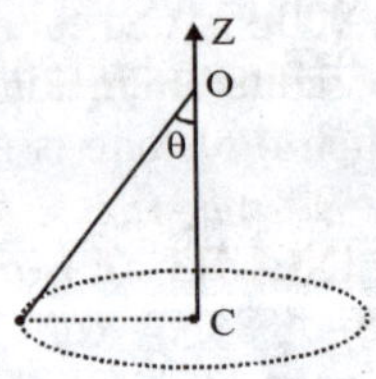

(a) 0.4 m/s (b) 4 m/s
(c) 0.2 m/s (d) 2 m/s

173. A particle is released on a vertical smooth semicircular track from point X so that OX makes angle θ from the vertical (see figure). The normal reaction of the track on the particle vanishes at point Y where OY makes angle ϕ with the horizontal. Then: **[Online April 19, 2014]**

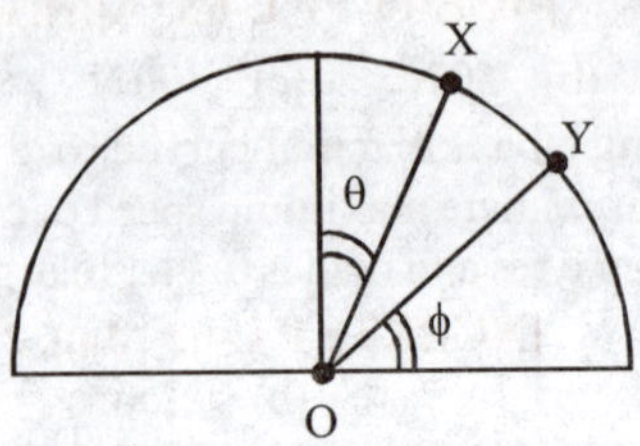

(a) $\sin\phi = \cos\phi$ (b) $\sin\phi = \frac{1}{2}\cos\theta$
(c) $\sin\phi = \frac{2}{3}\cos\theta$ (d) $\sin\phi = \frac{3}{4}\cos\theta$

174. A body of mass '*m*' is tied to one end of a spring and whirled round in a horizontal plane with a constant angular velocity. The elongation in the spring is 1 cm. If the angular velocity is doubled, the elongation in the spring is 5 cm. The original length of the spring is :

[NCERT : P L-63 | Online April 23, 2013]

(a) 15 cm (b) 12 cm

(c) 16 cm (d) 10 cm

175. A point P moves in counter-clockwise direction on a circular path as shown in the figure. The movement of 'P' is such that it sweeps out a length $s = t^3 + 5$, where s is in metres and t is in seconds. The radius of the path is 20 m. The acceleration of 'P' when $t = 2$ s is nearly.

[NCERT : P L-63 | 2010]

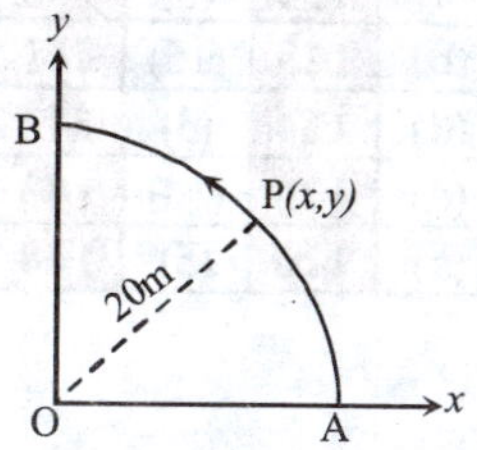

(a) $13 m/s^2$

(b) $12 m/s^2$

(c) $7.2 ms^2$

(d) $14 m/s^2$

176. For a particle in uniform circular motion, the acceleration $\vec{a}$ at a point P(R,θ) on the circle of radius R is (Here θ is measured from the x-axis)

[NCERT : P L-63 | 2010, June 25, 2022 (II)]

(a) $-\frac{v^2}{R}\cos\theta\ \hat{i} + \frac{v^2}{R}\sin\theta\ \hat{j}$

(b) $-\frac{v^2}{R}\sin\theta\ \hat{i} + \frac{v^2}{R}\cos\theta\ \hat{j}$

(c) $-\frac{v^2}{R}\cos\theta\ \hat{i} - \frac{v^2}{R}\sin\theta\ \hat{j}$

(d) $\frac{v^2}{R}\hat{i} + \frac{v^2}{R}\hat{j}$

177. An annular ring with inner and outer radii R_1 and R_2 is rolling without slipping with a uniform angular speed. The ratio of the forces experienced by the two particles situated on the inner and outer parts of the ring, $\frac{F_1}{F_2}$ is

[NCERT : P L-63 | 2005]

(a) $\left(\frac{R_1}{R_2}\right)^2$ (b) $\frac{R_2}{R_1}$

(c) $\frac{R_1}{R_2}$ (d) 1

178. Which of the following statements is **FALSE** for a particle moving in a circle with a constant angular speed?

[NCERT : P L-63 | 2004]

(a) The acceleration vector points to the centre of the circle

(b) The acceleration vector is tangent to the circle

(c) The velocity vector is tangent to the circle

(d) The velocity and acceleration vectors are perpendicular to each other.

ANSWER KEYS

1	(d)	19	(b)	37	(d)	55	(d)	73	(3)	91	(c)	109	(c)	127	(3.33)	145	(a)	163	(c)
2	(b)	20	(b)	38	(b)	56	(a)	74	(b)	92	(a)	110	(d)	128	(25)	146	(c)	164	(24)
3	(c)	21	(b)	39	(a)	57	(240)	75	(a)	93	(b)	111	(b)	129	(25)	147	(d)	165	(c)
4	(c)	22	(b)	40	(a)	58	(d)	76	(36)	94	(c)	112	(d)	130	(a)	148	(c)	166	(b)
5	(c)	23	(c)	41	(c)	59	(a)	77	(12)	95	(a)	113	(a)	131	(346)	149	(2)	167	(a)
6	(a)	24	(6)	42	(c)	60	(c)	78	(d)	96	(c)	114	(b)	132	(c)	150	(b)	168	(b)
7	(b)	25	(c)	43	(d)	61	(a)	79	(30)	97	(a)	115	(a)	133	(b)	151	(b)	169	(c)
8	(a)	26	(12)	44	(d)	62	(b)	80	(d)	98	(b)	116	(d)	134	(a)	152	(b)	170	(5)
9	(d)	27	(2)	45	(c)	63	(b)	81	(b)	99	(a)	117	(d)	135	(b)	153	(c)	171	(d)
10	(d)	28	(b)	46	(c)	64	(b)	82	(82)	100	(c)	118	(6)	136	(d)	154	(b)	172	(d)
11	(b)	29	(d)	47	(c)	65	(c)	83	(492)	101	(b)	119	(c)	137	(b)	155	(d)	173	(c)
12	(a)	30	(c)	48	(b)	66	(d)	84	(a)	102	(b)	120	(b)	138	(a)	156	(a)	174	(a)
13	(b)	31	(c)	49	(d)	67	(c)	85	(d)	103	(b)	121	(b)	139	(b)	157	(d)	175	(d)
14	(b)	32	(500)	50	(d)	68	(a)	86	(d)	104	(a)	122	(3)	140	(c)	158	(c)	176	(c)
15	(a)	33	(a)	51	(b)	69	(b)	87	(c)	105	(b)	123	(15)	141	(b)	159	(40)	177	(c)
16	(b)	34	(b)	52	(b)	70	(3)	88	(a)	106	(d)	124	(3)	142	(d)	160	(c)	178	(b)
17	(c)	35	(d)	53	(c)	71	(a)	89	(d)	107	(d)	125	(30)	143	(c)	161	(a)		
18	(d)	36	(a)	54	(a)	72	(c)	90	(a)	108	(a)	126	(5)	144	(d)	162	(b)		

5

Work, Energy and Power

TREND ANALYSIS
JEE MAIN (2025-2021)

	2025				2024				2023				2022				2021			
	Session II		Session I		Session II		Session I		Session II		Session I		Session II		Session I		Session II		Session I	
	MCQs	NVQs	MCQs	NVQs	MCQs	NVQs	MCQs	NVQs	MCQs	NVQs	MCQs	NVQs	MCQs	NVQs	MCQs	NVQs	MCQs	NVQs	MCQs	NVQs
Topic 1	–	–	2	1	2	1	2	–	–	2	1	2	1	–	1	–	1	2	–	–
Topic 2	3	–	7	1	7	–	7	–	3	4	2	5	5	2	4	–	2	3	–	4
Topic 3	1	–	2	–	1	–	1	–	1	2	2	–	1	–	1	–	–	–	–	–
Topic 4	–	–	2	–	1	–	–	1	1	–	2	2	2	–	2	–	2	–	3	2
Total	4	–	13	2	11	1	10	1	5	8	7	9	9	2	8	–	5	5	3	6

How to Code Qns for Better & Faster Revision

Type 1: Able to solve (No code Req)
Type 2: Solved with struggle or by seeing solution **(Use code NR means Needs Revision or orange stickers)**
Type 3: Tricky/ Could not solve on my own. Need help from Teacher/ Friend. **(Use code TQ means Tricky Question or red stickers)**

You can either hand write the code in boxes or else paste coloured stickers given at end of book.

Topic 1 - Work

1. A force $F = \alpha + \beta x^2$ acts on an object in the x-direction. The work done by the force is 5J when the object is displaced by 1 m. If the constant $\alpha = 1N$ then β will be
[NCERT : P L-75 | Jan. 24, 2025 (I)]
(a) $15\ N/m^2$ (b) $10\ N/m^2$ (c) $12\ N/m^2$ (d) $8\ N/m^2$

2. A force $f = x^2 y\hat{i} + y^2\hat{j}$ acts on a particle in a plane $x + y = 10$. The work done by this force during a displacement from (0, 0) to (4m, 2m) is _______ joule (round off to the nearest integer) **[NCERT : P L-75 | Jan. 23, 2025 (I)]**

3. A force $\vec{F} = 2\hat{i} + b\hat{j} + \hat{k}$ is applied on a particle and it undergoes a displacement $\hat{i} - 2\hat{j} - \hat{k}$. What will be the value of b, if work done on the particle is zero
[NCERT : P L-74 | Jan. 22, 2025 (II)]
(a) 0 (b) $\frac{1}{2}$ (c) $\frac{1}{3}$ (d) 2

4. A particle of mass m moves on a straight line with its velocity increasing with distance according to the equation $v = \alpha\sqrt{x}$, where α is a constant. The total work done by all the forces applied on the particle during its displacement from x = 0 to x = d, will be:
[NCERT : P L-73 | April 9, 2024 (I)]
(a) $\frac{m}{2\alpha^2 d}$ (b) $\frac{md}{2\alpha^2}$ (c) $\frac{m\alpha^2 d}{2}$ (d) $2m\alpha^2 d$

5. A force $(3x^2 + 2x - 5)$ N displaces a body from x = 2 m to x = 4m. Work done by this force is ....J.
[April 9, 2024 (II)]

6. A body of mass 50 kg is lifted to a height of 20 m from the ground in the two different ways as shown in the figures. The ratio of work done against the gravity in both the respective cases, will be: **[NCERT : P L-74 | April 5, 2024 (I)]**

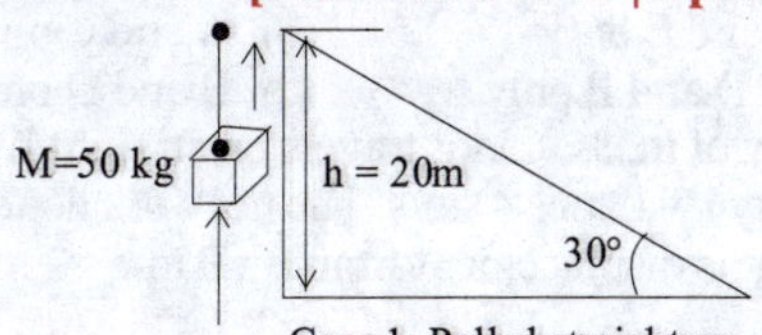

Case-1: Pulled straight up

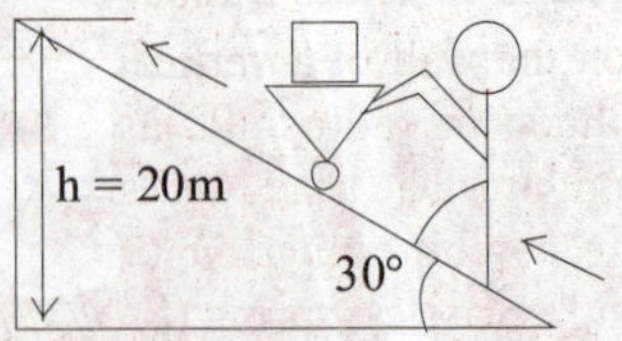

Case-2: Along the ramp

(a) 1 : 1 (b) 2 : 1 (c) $\sqrt{3}:2$ (d) 1 : 2

7. A block of mass 1 kg is pushed up a surface inclined to horizontal at an angle of 60° by a force of 10 N parallel to the inclined surface as shown in figure. When the block is pushed up by 10 m along inclined surface, the work done against frictional force is : $[g = 10\ m/s^2]$ **[NCERT : P L-74 | Jan. 30, 2024 (II)]**

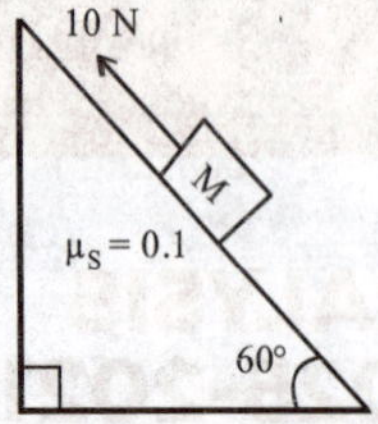

(a) $5\sqrt{3}$J (b) 5 J (c) 5×10^3 J (d) 10 J

8. A block of mass 100 kg slides over a distance of 10 m on a horizontal surface. If the co-efficient of friction between the surfaces is 0.4, then the work done against friction (in J) is: **[NCERT : P L-74 | Jan. 29, 2024 (I)]**

(a) 4000 (b) 3900 (c) 4200 (d) 4500

9. A block of mass 10 kg is moving along x-axis under the action of force F = 5x N. The work done by the force in moving the block from x = 2 m to 4 m will be _______ J. **[NCERT : P L-75 | April 15, 2023 (I)]**

10. To maintain a speed of 80 km/h by a bus of mass 500 kg on a plane rough road for 4 km distance, the work done by the engine of the bus will be _______ KJ. [The coefficient of friction between tyre of bus and road is 0.04]. **[NCERT : P L-74 | April 12, 2023 (I)]**

11. A small particle moves to position $5\hat{i}-2\hat{j}+\hat{k}$ from its initial position $2\hat{i}+3\hat{j}-4\hat{k}$ under the action of force $5\hat{i}+2\hat{j}+7\hat{k}$ N. The value of work done will be _______ J. **[NCERT : P L-74 | Feb. 1, 2023 (I)]**

12. A force $F = (5 + 3y^2)$ acts on a particle in the y-direction, where F is Newton and y is in meter. The work done by the force during a displacement from y = 2m to y = 5m is _______ J. **[NCERT : P L-75 | Feb. 1, 2023 (II)]**

13. Identify the correct statements from the following:

(A) Work done by a man in lifting a bucket out of a well by means of a rope tied to the bucket is negative.

(B) Work done by gravitational force in lifting a bucket out of a well by a rope tied to the bucket is negative.

(C) Work done by friction on a body sliding down an inclined plane is positive.

(D) Work done by an applied force on a body moving on a rough horizontal plane with uniform velocity in zero.

(E) Work done by the air resistance on an oscillating pendulum in negative.

Choose the correct answer from the options given below: **[NCERT : P L-74 | Jan. 29, 2023 (II)]**

(a) B and E only (b) A and C only
(c) B, D and E only (d) B and D only

14. A body of mass 0.5 kg travels on straight line path with velocity $v = (3x^2 + 4)$ m/s. The net work done by the force during its displacement from $x = 0$ to $x = 2$ m is : **[NCERT : P L-75 | July 25, 2022 (I)]**

(a) 64 J (b) 60 J
(c) 120 J (d) 128 J

15. A particle of mass 500 gm is moving in a straight line with velocity $\upsilon = bx^{5/2}$. The work done by the net force during its displacement from $x = 0$ to $x = 4$ m is : (Take $b = 0.25\ m^{-3/2}\ s^{-1}$). **[NCERT : P L-75 | June 29, 2022 (I)]**

(a) 2 J (b) 4 J (c) 8 J (d) 16 J

16. Two persons A and B perform same amount of work in moving a body through a certain distance d with application of forces acting at angle 45° and 60° with the direction of displacement respectively. The ratio of force applied by person A to the force applied by person B is $\frac{1}{\sqrt{x}}$. The value of x is _______ . **[NCERT : P L-74 | NA, Aug. 27, 2021 (I)]**

17. A force of $F = (5y + 20)\hat{j}$ N acts on a particle. The work done by this force when the particle is moved from y = 0 m to y = 10 m is _______ J. **[NCERT : P L-75 | NA, July 25, 2021 (II)]**

18. A porter lifts a heavy suitcase of mass 80 kg and at the destination lowers it down by a distance of 80 cm with a constant velocity. Calculate the work done by the porter in lowering the suitcase. (take $g = 9.8\ ms^{-2}$) **[NCERT : P L-73 | July 22, 2021 (II)]**

(a) –62720.0 J (b) –627.2 J
(c) +627.2 J (d) 784.0 J

19. A person pushes a box on a rough horizontal platform surface. He applies a force of 200 N over a distance of 15 m. Thereafter, he gets progressively tired and his applied force reduces linearly with distance to 100 N. The total distance through which the box has been moved is 30 m. What is the work done by the person during the total movement of the box ? **[NCERT : P L-76 | 4 Sep. 2020 (II)]**

(a) 3280 J (b) 2780 J (c) 5690 J (d) 5250 J

20. A small block starts slipping down from a point B on an inclined plane AB, which is making an angle θ with the horizontal section BC is smooth and the remaining section CA is rough with a coefficient of friction μ. It is found that the block comes to rest as it reaches the bottom (point A) of the inclined plane. If $BC = 2AC$, the coefficient of friction is given by $\mu = k \tan\theta$. The value of k is _______. **[NCERT : P L-74 | NA 2 Sep. 2020 (I)]**

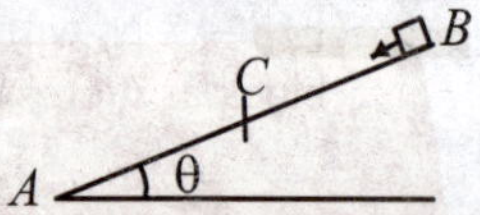

21. Consider a force $\vec{F} = -x\hat{i} + y\hat{j}$. The work done by this force in moving a particle from point A(1, 0) to B(0, 1) along the line segment is: (all quantities are in SI units) **[NCERT : P L-75 | 9 Jan. 2020 I]**

(a) 2J
(b) $\frac{1}{2}$J
(c) 1J
(d) $\frac{3}{2}$J

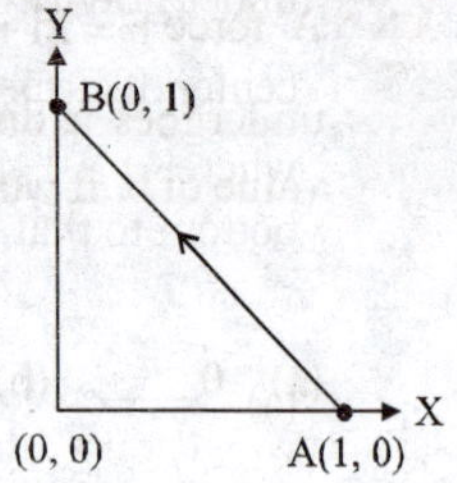

22. A block of mass m is kept on a platform which starts from rest with constant acceleration g/2 upward, as shown in fig. work done by normal reaction on block in time t is: **[NCERT : P L-74 | 10 Jan. 2019 I]**

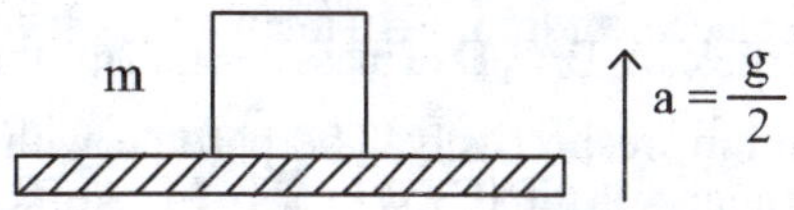

(a) $-\frac{m g^2 t^2}{8}$ (b) $\frac{m g^2 t^2}{8}$

(c) 0 (d) $\frac{3m g^2 t^2}{8}$

23. A uniform chain of length 2 m is kept on a table such that a length of 60 cm hangs freely from the edge of the table. The total mass of the chain is 4 kg. What is the work done in pulling the entire chain on the table ?

[NCERT : P L-74 | 2004]

(a) 12 J (b) 3.6 J (c) 7.2 J (d) 1200 J

24. A spring of spring constant 5×10^3 N/m is stretched initially by 5cm from the unstretched position. Then the work required to stretch it further by another 5 cm is

[NCERT : P L-81 | 2003]

(a) 12.50 N-m (b) 18.75 N-m
(c) 25.00 N-m (d) 6.25 N-m

Topic 2 - Energy

25. Which one of the following forces cannot be expressed in terms of potential energy? **[NCERT : P L-77 | April 7, 2025 (II)]**

(a) Coulomb's force (b) Gravitational force
(c) Frictional force (d) Restoring force

26. A block of mass 1 kg, moving along x with speed $v_i = 10$ m/s enters a rough region ranging from x = 0.1 m to x = 1.9 m. The retarding force acting on the block in this range is $F_r = -kx$ N, with k = 10 N/m. Then the final speed of the block as it crosses rough region is

[NCERT : P L-74 | April 3, 2025 (II)]

(a) 10 m/s (b) 4 m/s (c) 6 m/s (d) 8 m/s

27. A particle is released from height S above the surface of the earth. At certain height its kinetic energy is three times its potential energy. The height from the surface of the earth and the speed of the particle at that instant are respectively. **[NCERT : P L-74, 77 | April 3, 2025 (I)]**

(a) $\frac{S}{2}, \sqrt{\frac{3gS}{2}}$ (b) $\frac{S}{2}, \frac{3gS}{2}$ (c) $\frac{S}{4}, \frac{3gS}{2}$ (d) $\frac{S}{4}, \sqrt{\frac{3gS}{2}}$

28. A body of mass 'm' connected to a massless and unstretchable string goes in verticle circle of radius 'R' under gravity g. The other end of the string is fixed at the center of circle. If velocity at top of circular path is $n\sqrt{gR}$, where, $n \geq 1$, then ratio of kinetic energy of the body at bottom to that at top of the circle is

[NCERT : P L-79 | Jan. 29, 2025 (I)]

(a) $\frac{n}{n+4}$ (b) $\frac{n+4}{n}$ (c) $\frac{n^2}{n^2+4}$ (d) $\frac{n^2+4}{n^2}$

29. A bead of mass 'm' slides without friction on the wall of a vertical circular hoop of radius 'R' as shown in figure. The bead moves under the combined action of gravity and a massless spring (k) attached to the bottom of the hoop. The equilibrium length of the spring is 'R'. If the bead is released from top of the hoop with (negligible) zero initial speed, velocity of bead, when the length of spring becomes 'R' would be (spring constant is 'k', g is acceleration due to gravity)

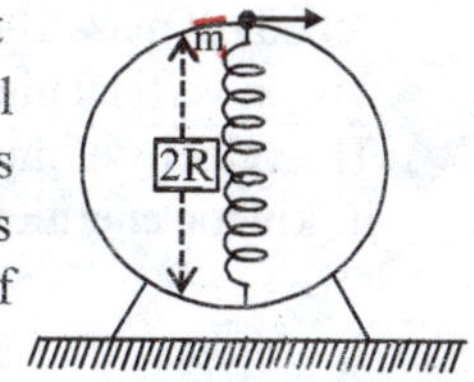

[NCERT : P L-80 | Jan. 28, 2025 (I)]

(a) $2\sqrt{gR + \frac{kR^2}{m}}$ (b) $\sqrt{2Rg + \frac{4kR^2}{m}}$

(c) $\sqrt{2Rg + \frac{kR^2}{m}}$ (d) $\sqrt{3Rg + \frac{kR^2}{m}}$

30. Given below are two statements: one is labelled as **Assertion (A)** and the other is labelled as **Reason (R)**

Assertion (A): In a central force field, the work done is independent of the path chosen.

Reason (R): Every force encountered in mechanics does not have an associated potential energy.

In the light of the above statements, choose the most appropriate answer from the options given below

[NCERT : P L-77 | Jan. 28, 2025 (I)]

(a) A is true but R is false
(b) Both A and R are true but R is NOT the correct explanation of A
(c) Both A and R are true and R is the correct explanation of A
(d) A is false but R is true

31. A string of length L is fixed at one end and carries a mass of M at the other end. The mass makes $\left(\frac{3}{\pi}\right)$ rotations per second about the vertical axis passing through end of the string as shown. The tension in the string is ____ ML.

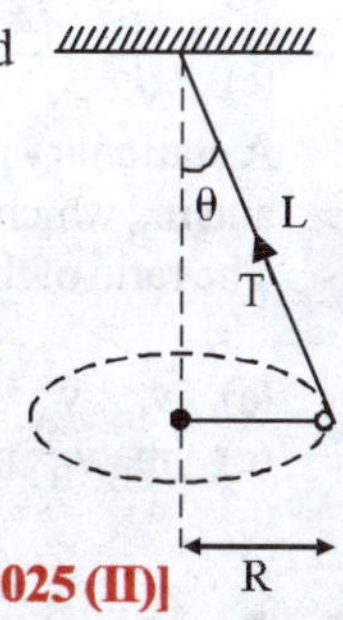

[NCERT : P L-79 | Jan. 24, 2025 (II)]

32. A ball having kinetic energy KE, is projected at an angle of 60° from the horizontal. What will be the kinetic energy of ball at the highest point of its flight?

[NCERT : P L-74 | Jan. 23, 2025 (II)]

(a) $\frac{(KE)}{8}$ (b) $\frac{(KE)}{4}$ (c) $\frac{(KE)}{16}$ (d) $\frac{(KE)}{2}$

33. A ball of mass 100 g is projected with velocity 20 m/s at 60° with horizontal. The decrease in kinetic energy of the ball during the motion from point of projection to highest point is **[NCERT : P L-74 | Jan. 22, 2025 (II)]**

(a) 20 J (b) 15 J (c) zero (d) 5 J

34. A body of mass 100 g is moving in circular path of radius 2 m on vertical plane as shown in figure.
The velocity of the body at point A is 10 m/s. The ratio of its kinetic energies at point B and C is :
[NCERT : P L-79 | Jan. 22, 2025 (II)]

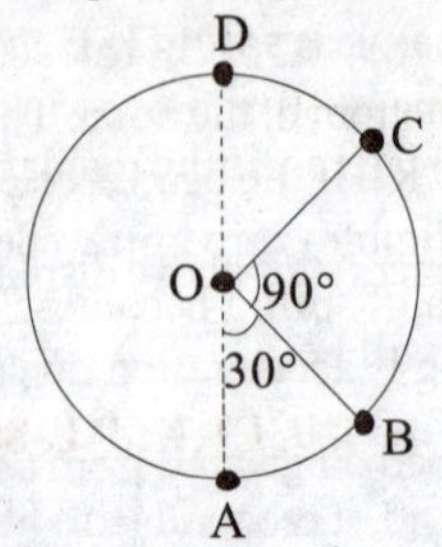

(Take acceleration due to gravity as 10 m/s^2)

(a) $\frac{2+\sqrt{3}}{3}$ (b) $\frac{2+\sqrt{2}}{3}$ (c) $\frac{3+\sqrt{3}}{2}$ (d) $\frac{3-\sqrt{2}}{2}$

35. A bob of mass m is suspended at a point O by a light string of length l and left to perform vertical motion (circular) as shown in figure. Initially, by applying horizontal velocity v_o at the point 'A', the string becomes slack when, the bob reaches at the point 'D'. The ratio of the kinetic energy of the bob at the points B and C is ____.

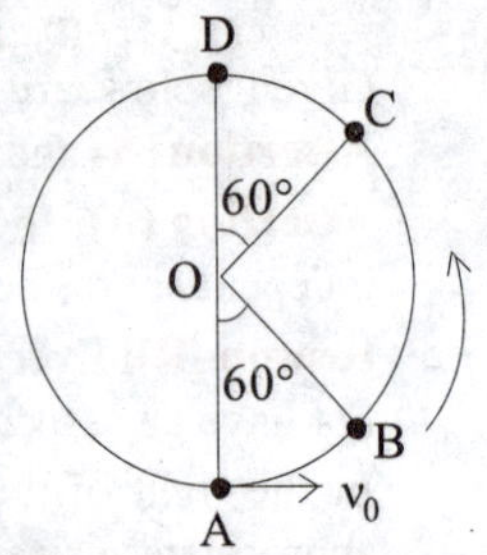

[NCERT : P L-79 | Jan. 22, 2025 (I)]

(a) 2 (b) 1 (c) 4 (d) 3

36. Three bodies A, B and C have equal kinetic energies and their masses are 400 g, 1.2 kg and 1.6 kg respectively. The ratio of their linear momenta is :
[NCERT : P L-74 | April 8, 2024 (I)]

(a) $1:\sqrt{3}:2$ (b) $1:\sqrt{3}:\sqrt{2}$
(c) $\sqrt{2}:\sqrt{3}:1$ (d) $\sqrt{3}:\sqrt{2}:1$

37. A stationary particle breaks into two parts of masses m_A and m_B which move with velocities v_A and v_B respectively. The ratio of their kinetic energies $(K_B : K_A)$ is :
[NCERT : P L-74 | April 8, 2024 (I)]

(a) $v_B : v_A$ (b) $m_B : m_A$
(c) $m_B v_B : m_A v_A$ (d) $1 : 1$

38.

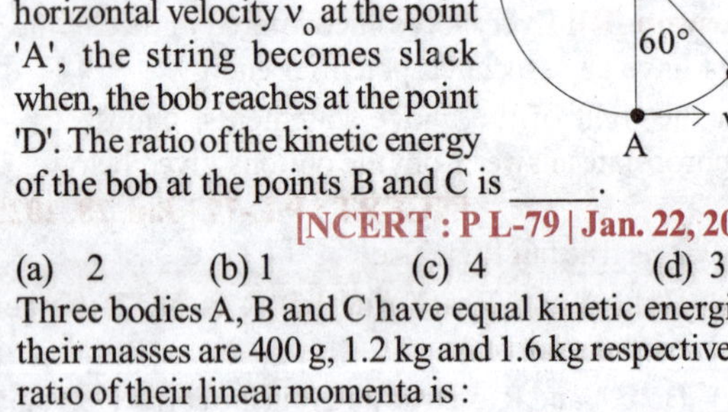

A block is simply released from the top of an inclined plane as shown in the figure above. The maximum compression in the spring when the block hits the spring is : **[NCERT : P L-80 | April 8, 2024 (II)]**

(a) $\sqrt{6}$m (b) 2m (c) 1m (d) $\sqrt{5}$m

39. A bullet of mass 50 g is fired with a speed 100 m/s on a plywood and emerges with 40 m/s. The percentage loss of kinetic energy is : **[NCERT : P L-74 | April 6, 2024 (I)]**

(a) 32% (b) 44% (c) 16% (d) 84%

40. Four particles A, B, C, D of mass $\frac{m}{2}$, m, 2m, 4m, have same momentum, respectively. The particle with maximum kinetic energy is : **[NCERT : P L-74 | April 6, 2024 (I)]**

(a) D (b) C (c) A (d) B

41. When kinetic energy of a body becomes 36 times of its original value, the percentage increase in the momentum of the body will be : **[NCERT : P L-74 | April 6, 2024 (II)]**

(a) 500% (b) 600% (c) 6% (d) 60%

42. A body of m kg slides from rest along the curve of vertical circle from point A to B in friction less path. The velocity of the body at B is :
[NCERT : P L-74, 77 | April 4, 2024 (II)]

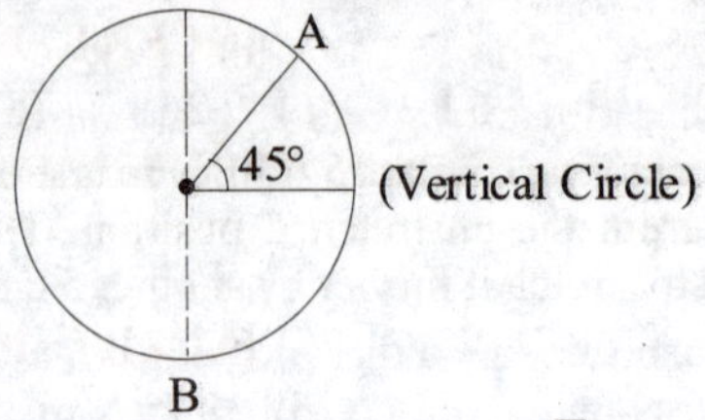

(given, R = 14 m, g = 10 m/s^2 and $\sqrt{2}$ = 1.4)

(a) 19.8 m/s (b) 21.9 m/s (c) 16.7 m/s (d) 10.6m/s

43. A simple pendulum of length 1 m has a wooden bob of mass 1kg. It is struck by a bullet of mass 10^{-2} kg moving with a speed of 2×10^2 ms^{-1}. The bullet gets embedded into the bob. The height to which the bob rises before swinging back is. (use g = 10m/s^2)
[NCERT : P L-78 | Feb. 1, 2024 (I)]

(a) 0.30 m (b) 0.20 m (c) 0.35 m (d) 0.40m

44. An artillery piece of mass M_1 fires a shell of mass M_2 horizontally. Instantaneously after the firing, the ratio of kinetic energy of the artillery and that of the shell is :
[NCERT : P L-74 | Jan. 31, 2024 (I)]

(a) $\frac{M_1}{(M_1+M_2)}$ (b) $\frac{M_2}{(M_1+M_2)}$

(c) $\frac{M_1}{M_2}$ (d) $\frac{M_2}{M_1}$

45. A particle is placed at the point A of a frictionless track ABC as shown in figure. It is gently pushed towards right. The speed of the particle when it reaches the point B is: (Take g = 10 m/s^2). **[NCERT : P L-78 | Jan. 30, 2024 (I)]**

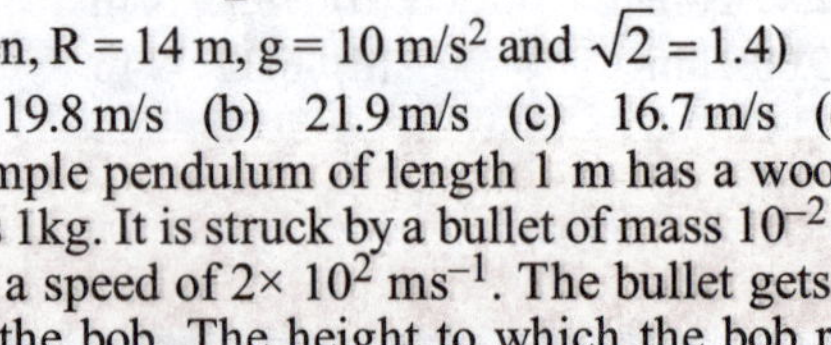

(a) 20 m/s (b) $\sqrt{10}$m/s
(c) $2\sqrt{10}$m/s (d) 10 m/s

46. The potential energy function (in J) of a particle in a region of space is given as $U = (2x^2 + 3y^3 + 2z)$. Here x, y and z are in metre. The magnitude of x - component of force (in N) acting on the particle at point P (1, 2, 3) m is :
[NCERT : P L-78 | Jan. 29, 2024 (I)]

(a) 6 (b) 2 (c) 8 (d) 4

47. A bob of mass 'm' is suspended by a light string of length 'L'. It is imparted a minimum horizontal velocity at the lowest point A such that it just completes half circle reaching the top most position B. The ratio of kinetic energies $\frac{(K.E.)_A}{(K.E.)_B}$ is: **[NCERT : P L-79 | Jan. 29, 2024 (II)]**

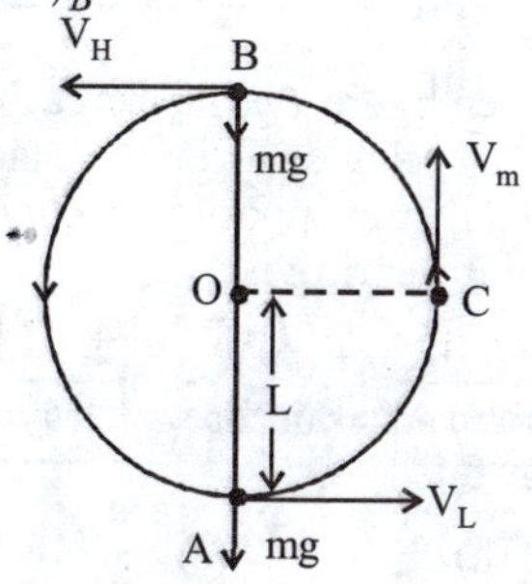

(a) 3 : 2 (b) 5 : 1 (c) 2 : 5 (d) 1 : 5

48. Two bodies of mass 4 g and 25 g are moving with equal kinetic energies. The ratio of magnitude of their linear momentum is : **[NCERT : P L-74 | Jan. 27, 2024 (I)]**

(a) 3 : 5 (b) 5 : 4 (c) 2 : 5 (d) 4 : 5

49. A ball suspended by a thread swings in a vertical plane so that its magnitude of acceleration in the extreme position and lowest position are equal. The angle (θ) of thread deflection in the extreme position will be : **[NCERT : P L-79 | Jan. 27, 2024 (II)]**

(a) $\tan^{-1}(\sqrt{2})$ (b) $2\tan^{-1}\left(\frac{1}{2}\right)$

(c) $\tan^{-1}\left(\frac{1}{2}\right)$ (d) $2\tan^{-1}\left(\frac{1}{\sqrt{5}}\right)$

50. Two bodies are having kinetic energies in the ratio 16 : 9. If they have same linear momentum, the ratio of their masses respectively is: **[NCERT : P L-74 | April 13, 2023 (I)]**

(a) 4 : 3 (b) 3 : 4 (c) 16 : 9 (d) 9 : 16

51. Given below are two statements:

Statement I : A truck and a car moving with same kinetic energy are brought to rest by applying brakes which provide equal retarding forces. Both come to rest in equal distance.

Statement II : A car moving towards east takes a turn and moves towards north, the speed remains unchanged. The acceleration of the car is zero. **[NCERT : P L-74 | April 12, 2023 (I)]**

In the light of given statements, choose the most appropriate answer from the options given below.

(a) Statement I is correct but Statement II is incorrect
(b) Statement I is incorrect but Statement II is correct
(c) Both Statement I is correct but Statement II are incorrect
(d) Both Statement I is correct but Statement II are correct

52. The momentum of a body is increased by 50%. The percentage increase in the kinetic energy of the body is ______%. **[NCERT : P L-74 | April 8, 2023 (I)]**

53. A body of mass 5 kg is moving with a momentum of 10 kg ms^{-1}. Now a force of 2 N acts on the body in the direction of its motion for 5 s. The increase in the kinetic energy of the body is ______ J. **[NCERT : P L-74 | April 8, 2023 (II)]**

54. A small block of mass 100 g is tied to a spring of spring constant 7.5 N/m and length 20 cm. The other end of spring is fixed at a particular point A. If the block moves in a circular path on a smooth horizontal surface with constant angular velocity 5 rad/s about point A, then tension in the spring is **[NCERT : P L-80 | April 6, 2023 (I)]**

(a) 1.5 N (b) 0.75 N (c) 0.25 N (d) 0.50 N

55. A particle of mass 10 g moves in a straight line with retardation 2x, where x is the displacement in SI unts. Its loss of kinetic energy for above displacement is $\left(\frac{10}{x}\right)^{-n}$ J. The value of n will be ______. **[NCERT : P L-76 | April 6, 2023 (I)]**

56. A body is dropped on ground from a height 'h_1' and after hitting the ground, it rebounds to a height 'h_2'. If the ratio of velocities of the body just before and after hitting ground is 4, then percentage loss in kinetic energy of the body is $\frac{x}{4}$. The value of x is ______. **[NCERT : P L-74 | April 6, 2023 (II)]**

57. For a body projected at an angle with the horizontal from the ground, choose the correct statement. **[NCERT : P L-74, 77 | Feb. 1, 2023 (II)]**

(a) The vertical component of momentum is maximum at the highest point.
(b) The Kinetic Energy (K.E.) is zero at the highest point of projectile motion.
(c) Gravitational potential energy is maximum at the highest point.
(d) The horizontal component of velocity is zero at the highest point.

58. A block is fastened to a horizontal spring. The block is pulled to a distance x = 10 cm from its equilibrium position (at x = 0) on a frictionless surface from rest. The energy of the block at x = 5 cm is 0.25 J. The spring constant of the spring is ______ Nm^{-1}. **[NCERT : P L-78 | Feb. 1, 2023 (II)]**

59. A lift of mass M = 500 kg is descending with speed of 2 ms^{-1}. Its supporting cable begins to slip thus allowing it to fall with a constant acceleration of 2 ms^{-2}. The kinetic energy of the lift at the end of fall through to a distance of 6 m will be ____ kJ. **[NCERT : P L-74 | Jan. 31, 2023 (I)]**

60. A 0.4 kg mass takes 8s to reach ground when dropped from a certain height 'P' above surface of earth. The loss of potential energy in the last second of fall is ____ J. [Take g = 10 m/s^2] **[NCERT : P L-77 | Jan. 29, 2023 (I)]**

61. An object of mass 'm' initially at rest on a smooth horizontal plane starts moving under the action of force F = 2N. In the process of its linear motion, the angle θ (as shown in figure) between the direction of force and horizontal varies as θ = kx, where k is a constant and x is the distance covered by the object from its initial position. The expression of kinetic energy of the object will be $E = \frac{n}{k}\sin\theta$. The value of n is ____. **[NCERT : P L-74 | Jan. 25, 2023 (I)]**

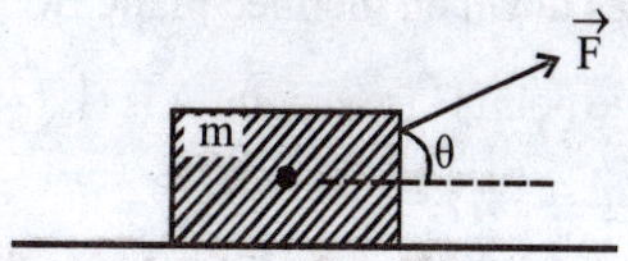

Smooth horizontal surface

62. A spherical body of mass 2 kg starting from rest acquires a kinetic energy of 10000 J at the end of 5th second. The force acted on the body is ______ N.
[NCERT : P L-74 | Jan. 24, 2023 (I)]

63. A body of mass 200 g is tied to a spring of spring constant 12.5 N/m, while the other end of spring is fixed at point O. If the body moves about O in a circular path on a smooth horizontal surface with constant angular speed 5 rad/s, then the ratio of extension in the spring to its natural length will be : **[NCERT : P L-80 | Jan. 24, 2023 (II)]**
(a) 1 : 2 (b) 1 : 1 (c) 2 : 3 (d) 2 : 5

64. A ball is projected with kinetic energy E, at an angle of 60° to the horizontal. The kinetic energy of this ball at the highest point of its flight will become :
[NCERT : P L-74 | July 29, 2022 (I)]
(a) Zero (b) $\frac{E}{2}$ (c) $\frac{E}{4}$ (d) E

65. If momentum of a body is increased by 20%, then its kinetic energy increases by :
[NCERT : P L-74 | July 29, 2022 (II)]
(a) 36% (b) 40% (c) 44% (d) 48%

66. A block of mass 'm' (as shown in figure) moving with kinetic energy E compresses a spring through a distance 25 cm when, its speed is halved. The value of spring constant of used spring will be nE Nm^{-1} for n = ______. **[NCERT : P L-80 | July 28, 2022 (I)]**

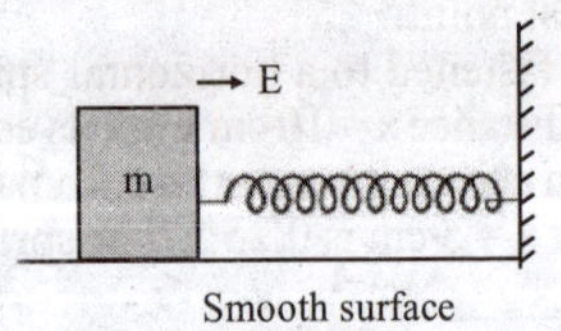

67. A bullect of mass 200 g having initial kinetic energy 90 J is shot inside a long swimming pool as shown in the figure. If it's kinetic energy reduce to 40 J within 1 s, the minimum length of the pool, the bullect has to travel so that it completely comes to rest is
[NCERT : P L-74 | July 28, 2022 (II)]

(a) 45 m (b) 90 m (c) 125 m (d) 25 m

68. Two inclined planes are placed as shown in figure. A block is projected from the Point A of inclined plane AB along its surface with a velocity just sufficient to carry it to the top Point B at a height 10 m. After reaching the Point B the block slides down on inclined plane BC. Time it takes to reach to the point C from point A is $t(\sqrt{2}+1)$ s. The value of t is ______ (use g = 10 m/s²)
[NCERT : P L-77 | July 27, 2022 (II)]

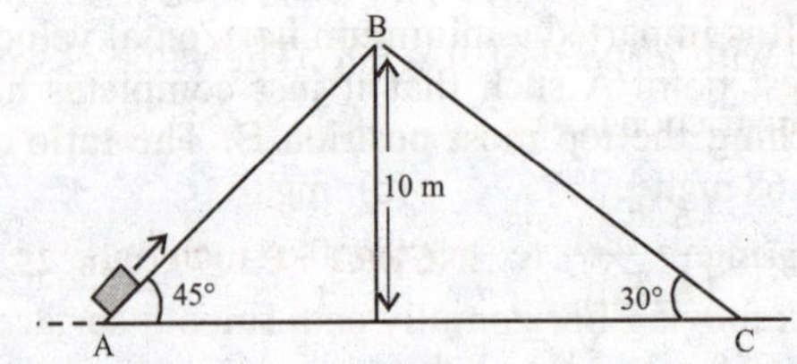

69. As per the given figure, two blocks each of mass 250 g are connected to a spring of spring constant 2 Nm^{-1}. If both are given velocity v in opposite directions, then maximum elongation of the spring is :
[NCERT : P L-74, 81 | July 26, 2022 (I)]

(a) $\frac{v}{2\sqrt{2}}$ (b) $\frac{v}{2}$ (c) $\frac{v}{4}$ (d) $\frac{v}{\sqrt{2}}$

70. A body of mass 8 kg and another of mass 2 kg are moving with equal kinetic energy. The ratio of their respective momenta will be : **[NCERT : P L-74 | July 26, 2022 (II)]**
(a) 1 : 1 (b) 2 : 1 (c) 1 : 4 (d) 4 : 1

71. Two bodies A and B of masses 5 kg and 8 kg are moving such that the momentum of body B is twice that of the body A. The ratio of their kinetic energies will be:
[NCERT : P L-74 | June 30, 2022 (I)]
(a) 4 : 5 (b) 2 : 5 (c) 5 : 4 (d) 5 : 2

72. In the given figure, the block of mass m is dropped from the point 'A'. The expression for kinetic energy of block when it reaches point 'B' is :
[NCERT : P L-78 | June 29, 2022 (II)]

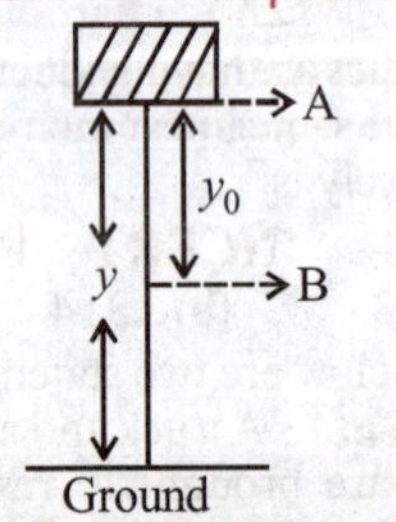

(a) $\frac{1}{2}mgy_0^2$
(b) $\frac{1}{2}mgy^2$
(c) $mg(y-y_0)$
(d) mgy_0

73. An object is thrown vertically upwards. At its maximum height, which of the following quantity becomes zero ?
[NCERT : P L-77 | June 26, 2022 (I)]
(a) Momentum (b) Potential energy
(c) Acceleration (d) Force

74. A particle experiences a variable force $\vec{F} = (4x\hat{i} + 3y^2\hat{j})$ in a horizontal x-y plane. Assume distance in meters and force is newton. If the particle moves from point (1, 2) to point (2, 3) in the x–y plane, the kinetic energy changes by
[NCERT : P L-76 | June 24, 2022 (I)]
(a) 50.0 J (b) 12.5 J (c) 25.0 J (d) 0 J

75. An engine is attached to a wagon through a shock absorber of length 1.5 m. The system with a total mass of 40,000 kg is moving with a speed of 72 kmh^{-1} when the brakes are applied to bring it to rest. In the process of the system being brought to rest, the spring of the shock absorber gets compressed by 1.0 m. If 90% of energy of the wagon is lost due to friction, the spring constant is ______ × 10^5 N/m. **[NCERT : P L-80 | NA, Sep. 1, 2021 (II)]**

76. A body of mass 'm' dropped from a height 'h' reaches the ground with a speed of $0.8\sqrt{gh}$. The value of workdone by the air-friction is : **NCERT : P L-73, 77 | Sep. 1, 2021 (II)]**

(a) –0.68 mgh (b) mgh

(c) 1.64 mgh (d) 0.64 mgh

77. A block moving horizontally on a smooth surface with a speed of 40 ms^{-1} splits into two equal parts. If one of the parts moves at 60 ms^{-1} in the same direction, then the fractional change in the kinetic energy will be x : 4 where x = ________. **[NCERT : P L-74 | NA, Aug. 31, 2021 (I)]**

78. A uniform chain of length 3 meter and mass 3 kg overhangs a smooth table with 2 meter laying on the table. If k is the kinetic energy of the chain in joule as it completely slips off the table, then the value of k is ________.

(Take g = 10 m/s²)

[NCERT : P L-77 | NA, Aug. 26, 2021 (I)]

79. If the Kinetic energy of a moving body becomes four times its initial Kinetic energy, then the percentage change in its momentum will be :

[NCERT : P L-74 | July 20, 2021 (II)]

(a) 100% (b) 200% (c) 300% (d) 400%

80.

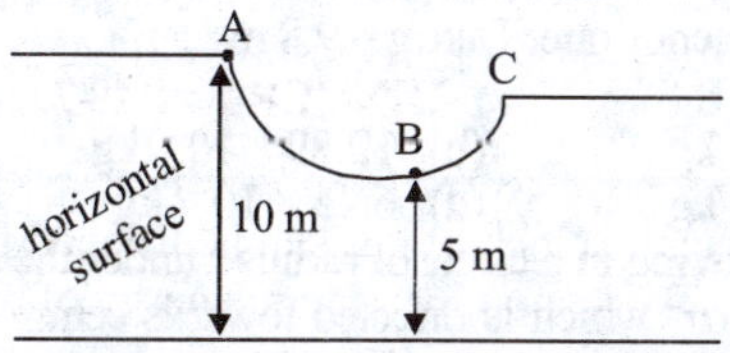

As shown in the figure, a particle of mass 10 kg is placed at a point A. When the particle is slightly displaced to its right, it starts moving and reaches the point B. The speed of the particle at B is x m/s. (Take g = 10 m/s²)

The value of 'x' to the nearest integer is ________.

[NCERT : P L-78 | March 18, 2021 (I)]

81. A ball of mass 4 kg, moving with a velocity of 10ms^{-1}, collides with a spring of length 8 m and force constant 100 Nm^{-1}. The length of the compressed spring is x m. The value of x, to the nearest integer, is ________.

[NCERT : P L-80 | March 18, 2021 (II)]

82. Two particles having masses 4 g and 16 g respectively are moving with equal kinetic energies. The ratio of the magnitudes of their momentum is n : 2. The value of n will

[NCERT : P L-74 | Feb. 25, 2021 (II)]

83. Two solids *A* and *B* of mass 1 kg and 2 kg respectively are moving with equal linear momentum. The ratio of their kinetic energies $(K.E.)_A : (K.E.)_B$ will be $\frac{A}{1}$, so the value of *A* will be ________.

[NCERT : P L-74 | Feb. 24, 2021 (II)]

84. A cricket ball of mass 0.15 kg is thrown vertically up by a bowling machine so that it rises to a maximum height of 20 m after leaving the machine. If the part pushing the ball applies a constant force *F* on the ball and moves horizontally a distance of 0.2 m while launching the ball, the value of *F* (in N) is (g = 10 ms^{-2}) ________. **[NCERT : P L-73 | NA 3 Sep. 2020 (I)]**

85. A particle (m = 1 kg) slides down a frictionless track (AOC) starting from rest at a point *A* (height 2 m). After reaching *C*, the particle continues to move freely in air as a projectile. When it reaching its highest point *P* (height 1 m), the kinetic energy of the particle (in *J*) is: (Figure drawn is schematic and not to scale; take g = 10 ms^{-2})________.

[NCERT : P L-77 | NA 7 Jan. 2020 I]

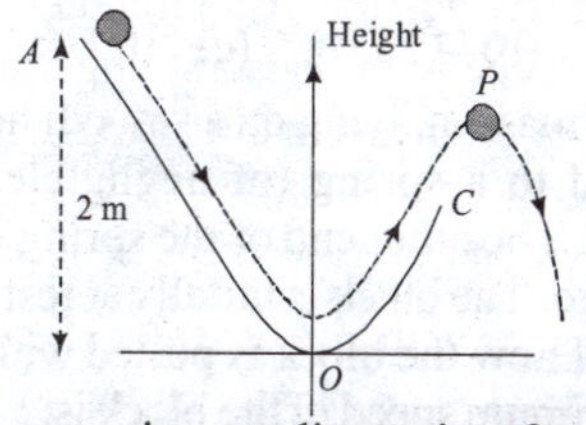

86. A particle moves in one dimension from rest under the influence of a force that varies with the distance travelled by the particle as shown in the figure. The kinetic energy of the particle after it has travelled 3 m is :

[NCERT : P L-76 | 8 April 2019 I, 7 Jan. 2020 II]

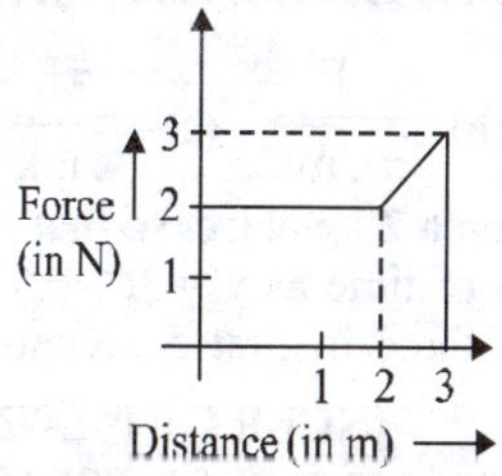

(a) 4 J (b) 2.5 J (c) 6.5 J (d) 5 J

87. A spring whose unstretched length is l has a force constant k. The spring is cut into two pieces of unstretched lengths l_1 and l_2 where, $l_1 = nl_2$ and n is an integer. The ratio k_1/k_2 of the corresponding force constants, k_1 and k_2 will be: **[NCERT : P L-80 | 12 April 2019 II]**

(a) n (b) $\frac{1}{n^2}$ (c) $\frac{1}{n}$ (d) n^2

88. A body of mass 1 kg falls freely from a height of 100m, on a platform of mass 3 kg which is mounted on a spring having spring constant k = 1.25 × 10^6 N/m. The body sticks to the platform and the spring's maximum compression is found to be x. Given that g = 10 ms^{-2}, the value of x will be close to : **[NCERT : P L-80 | 11 April 2019 I]**

(a) 40 cm (b) 4 cm (c) 80 cm (d) 8 cm

89. A uniform cable of mass 'M' and length 'L' is placed on a horizontal surface such that its $\left(\frac{1}{n}\right)^{th}$ part is hanging below the edge of the surface. To lift the hanging part of the cable upto the surface, the work done should be:

[NCERT : P L-77 | 9 April 2019 I]

(a) $\frac{MgL}{2n^2}$ (b) $\frac{MgL}{n^2}$ (c) $\frac{2MgL}{n^2}$ (d) $nMgL$

90. A wedge of mass M = 4m lies on a frictionless plane. A particle of mass m approaches the wedge with speed v. There is no friction between the particle and the plane or between the particle and the wedge. The maximum height climbed by the particle on the wedge is given by:

[NCERT : P L-78 | 9 April 2019 II]

(a) $\frac{v^2}{g}$ (b) $\frac{2v^2}{7g}$ (c) $\frac{2v^2}{5g}$ (d) $\frac{v^2}{2g}$

91. A particle which is experiencing a force, given by $\vec{F} = 3\hat{i} - 12\hat{j}$, undergoes a displacement of $\vec{d} = 4\hat{i}$. If the particle had a kinetic energy of 3 J at the beginning of the displacement, what is its kinetic energy at the end of the displacement? **[NCERT : P L-73 | 10 Jan. 2019 II]**

(a) 9 J (b) 12 J (c) 10 J (d) 15 J

92. A block of mass m, lying on a smooth horizontal surface, is attached to a spring (of negligible mass) of spring constant k. The other end of the spring is fixed, as shown in the figure. The block is initally at rest in its equilibrium position. If now the block is pulled with a constant force F, the maximum speed of the block is: **[NCERT : P L-80 | 9 Jan. 2019 I]**

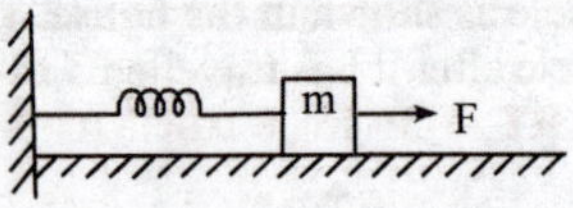

(a) $\frac{2F}{\sqrt{mk}}$ (b) $\frac{F}{\pi\sqrt{mk}}$ (c) $\frac{\pi F}{\sqrt{mk}}$ (d) $\frac{F}{\sqrt{mk}}$

93. A force acts on a 2 kg object so that its position is given as a function of time as $x = 3t^2 + 5$. What is the work done by this force in first 5 seconds? **[NCERT : P L-73 | 9 Jan. 2019 II]**

(a) 850 J (b) 950 J (c) 875 J (d) 900 J

94. A particle is moving in a circular path of radius a under the action of an attractive potential $U = -\frac{k}{2r^2}$. Its total energy is: **[NCERT : P L-78 | 2018]**

(a) $-\frac{k}{4a^2}$ (b) $\frac{k}{2a^2}$ (c) zero (d) $-\frac{3}{2}\frac{k}{a^2}$

95. Two particles of the same mass m are moving in circular orbits because of force, given by $F(r) = \frac{-16}{r} - r^3$.

The first particle is at a distance r = 1, and the second, at r = 4. The best estimate for the ratio of kinetic energies of the first and the second particle is closest to **[NCERT : P L-74 | Online April 16, 2018]**

(a) 10^{-1} (b) 6×10^{-2} (c) 6×10^{2} (d) 3×10^{-3}

96. A body of mass $m = 10^{-2}$ kg is moving in a medium and experiences a frictional force $F = -kv^2$. Its intial speed is $v_0 = 10\ ms^{-1}$. If, after 10 s, its energy is $\frac{1}{8}mv_0^2$, the value of k will be: **[NCERT : P L-74 | 2017]**

(a) 10^{-4} kg m^{-1} (b) 10^{-1} kg m^{-1} s^{-1}
(c) 10^{-3} kg m^{-1} (d) 10^{-3} kg s^{-1}

97. An object is dropped from a height h from the ground. Every time it hits the ground it looses 50% of its kinetic energy. The total distance covered as $t \to \infty$ is **[NCERT : P L-74 | Online April 8, 2017]**

(a) 3h (b) ∞ (c) $\frac{5}{3}$h (d) $\frac{8}{3}$h

98. A time dependent force F = 6t acts on a particle of mass 1 kg. If the particle starts from rest, the work done by the force during the first 1 second will be **[NCERT : P L-73 | 2017]**

(a) 9 J (b) 18 J (c) 4.5 J (d) 22 J

99. A point particle of mass m, moves long the uniformly rough track PQR as shown in the figure. The coefficient of friction, between the particle and the rough track equals μ. The particle is released, from rest from the point P and it comes to rest at a point R. The energies, lost by the ball, over the parts, PQ and QR, of the track, are equal to each other, and no energy is lost when particle changes direction from PQ to QR. The value of the coefficient of friction μ and the distance x (= QR), are, respectively close to : **[2016]**

(a) 0.29 and 3.5 m
(b) 0.29 and 6.5 m
(c) 0.2 and 6.5 m
(d) 0.2 and 3.5 m

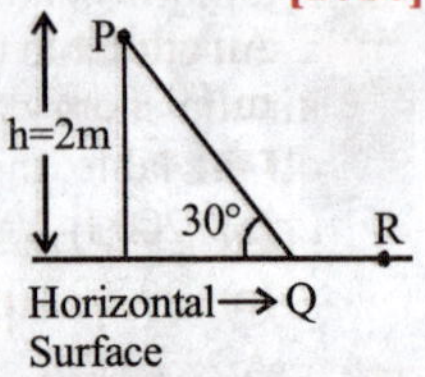

100. A person trying to lose weight by burning fat lifts a mass of 10 kg upto a height of 1 m, 1000 times. Assume that the potential energy lost each time he lowers the mass is dissipated. How much fat will he use up considering the work done only when the weight is lifted up? Fat supplies 3.8×10^7 J of energy per kg which is converted to mechanical energy with a 20% efficiency rate. Take g = 9.8 ms^{-2} : **[NCERT : P L-77 | 2016]**

(a) 9.89×10^{-3} kg (b) 12.89×10^{-3} kg
(c) 2.45×10^{-3} kg (d) 6.45×10^{-3} kg

101. A particle is moving in a circle of radius r under the action of a force $F = \alpha r^2$ which is directed towards centre of the circle. Total mechanical energy (kinetic energy + potential energy) of the particle is (take potential energy = 0 for r = 0) : **[NCERT : P L-78 | Online April 11, 2015]**

(a) $\frac{1}{2}\alpha r^3$ (b) $\frac{5}{6}\alpha r^3$ (c) $\frac{4}{3}\alpha r^3$ (d) αr^3

102. A bullet looses $\left(\frac{1}{n}\right)^{th}$ of its velocity passing through one plank. The number of such planks that are required to stop the bullet can be: **[NCERT : P L-73 | Online April 19, 2014]**

(a) $\frac{n^2}{2n-1}$ (b) $\frac{2n^2}{n-1}$ (c) infinite (d) n

103. Two springs of force constants 300 N/m (Spring A) and 400 N/m (Spring B) are joined together in series. The combination is compressed by 8.75 cm. The ratio of energy stored in A and B is $\frac{E_A}{E_B}$. Then $\frac{E_A}{E_B}$ is equal to : **[NCERT : P L-80 | Online April 9, 2013]**

(a) $\frac{4}{3}$ (b) $\frac{16}{9}$ (c) $\frac{3}{4}$ (d) $\frac{9}{16}$

104. At time $t = 0$ a particle starts moving along the x-axis. If its kinetic energy increases uniformly with time 't', the net force acting on it must be propotional to **[2011 RS]**

(a) constant (b) t
(c) $\frac{1}{\sqrt{t}}$ (d) $\sqrt{t}$

105. The potential energy function for the force between two atoms in a diatomic molecule is approximately given by $U(x)=\frac{a}{x^{12}}-\frac{b}{x^6}$, where a and b are constants and x is the distance between the atoms. If the dissociation energy of the molecule is

$D=\left[U(x=\infty)-U_{\text{at equilibrium}}\right]$, D is **[NCERT : P L-78 | 2010]**

(a) $\frac{b^2}{2a}$ (b) $\frac{b^2}{12a}$ (c) $\frac{b^2}{4a}$ (d) $\frac{b^2}{6a}$

106. An athlete in the olympic games covers a distance of 100 m in 10 s. His kinetic energy can be estimated to be in the range **[NCERT : P L-74 | 2008]**

(a) 200 J - 500 J (b) 2×10^5 J - 3×10^5 J
(c) 20,000 J - 50,000 J (d) 2,000 J - 5,000 J

107. A 2 kg block slides on a horizontal floor with a speed of 4m/s. It strikes a uncompressed spring, and compresses it till the block is motionless. The kinetic friction force is 15N and spring constant is 10,000 N/m. The spring compresses by **[NCERT : P L-80 | 2007]**

(a) 8.5 cm (b) 5.5 cm (c) 2.5 cm (d) 11.0 cm

108. A particle is projected at 60° to the horizontal with a kinetic energy K. The kinetic energy at the highest point is **[NCERT : P L-74 | 2007]**

(a) $K/2$ (b) K (c) Zero (d) $K/4$

109. A particle of mass 100g is thrown vertically upwards with a speed of 5 m/s. The work done by the force of gravity during the time the particle goes up is **[NCERT : P L-73 | 2006]**

(a) –0.5 J (b) –1.25 J (c) 1.25 J (d) 0.5 J

110. The potential energy of a 1 kg particle free to move along the x-axis is given by $V(x)=\left(\frac{x^4}{4}-\frac{x^2}{2}\right)J$.

The total mechanical energy of the particle is 2 J. Then, the maximum speed (in m/s) is **[NCERT : P L-78 | 2006]**

(a) $\frac{3}{\sqrt{2}}$ (b) $\sqrt{2}$ (c) $\frac{1}{\sqrt{2}}$ (d) 2

111. A mass of M kg is suspended by a weightless string. The horizontal force that is required to displace it until the string makes an angle of 45° with the initial vertical direction is **[NCERT : P L-74 | 2006]**

(a) $Mg(\sqrt{2}+1)$ (b) $Mg\sqrt{2}$
(c) $\frac{Mg}{\sqrt{2}}$ (d) $Mg(\sqrt{2}-1)$

112. A spherical ball of mass 20 kg is stationary at the top of a hill of height 100 m. It rolls down a smooth surface to the ground, then climbs up another hill of height 30 m and finally rolls down to a horizontal base at a height of 20 m above the ground. The velocity attained by the ball is **[NCERT : P L-78 | 2005]**

(a) 20 m/s (b) 40 m/s (c) $10\sqrt{30}$ m/s (d) 10 m/s

113. A particle moves in a straight line with retardation proportional to its displacement. Its loss of kinetic energy for any displacement x is proportional to **[NCERT : P L-74 | 2004]**

(a) x (b) e^x (c) x^2 (d) $\log_e x$

114. A particle is acted upon by a force of constant magnitude which is always perpendicular to the velocity of the particle, the motion of the particles takes place in a plane. It follows that **[NCERT : P L-73 | 2004]**

(a) its kinetic energy is constant
(b) its acceleration is constant
(c) its velocity is constant
(d) it moves in a straight line

115. A wire suspended vertically from one of its ends is stretched by attaching a weight of 200N to the lower end. The weight stretches the wire by 1 mm. Then the elastic energy stored in the wire is **[NCERT : P L-80 | 2003]**

(a) 0.2 J (b) 10 J (c) 20 J (d) 0.1 J

116. A ball whose kinetic energy is E, is projected at an angle of 45° to the horizontal. The kinetic energy of the ball at the highest point of its flight will be **[NCERT : P L-74 | 2002]**

(a) E (b) $E/\sqrt{2}$ (c) $E/2$ (d) zero

Topic 3 - Power

117. An object of mass 1000 g experiences a time dependent force $\vec{F}=\left(2t\hat{i}+3t^2\hat{j}\right)N$. The power generated by the force at time t is : **[NCERT : P L-83 | April 7, 2025 (I)]**

(a) $(2t^2+3t^3)W$ (b) $(2t^2+18t^3)W$
(c) $(3t^3+5t^5)W$ (d) $(2t^3+3t^5)W$

118. A sand dropper drops sand of mass $m(t)$ on a conveyer belt at a rate proportional to the square root of speed (v) of the belt, *i.e.* $\frac{dm}{dt}\propto\sqrt{v}$. If P is the power delivered to run the belt at constant speed then which of the following relationship is true ? **[NCERT : P L-83 | Jan. 29, 2025 (II)]**

(a) $P^2\propto v^3$ (b) $P\propto\sqrt{v}$ (c) $P\propto v$ (d) $P^2\propto v^5$

119. A body of mass 4 kg is placed on a plane at a point P having coordinate (3, 4) m. Under the action of force $\vec{F}=(2\hat{i}+3\hat{j})$N, it moves to a new point Q having coordinates (6, 10)m in 4 sec. The average power and instantaneous power at the end of 4 sec are in the ratio of : **[NCERT : P L-83 | Jan. 28, 2025 (II)]**

(a) 13 : 6 (b) 6 : 13
(c) 1 : 2 (d) 4 : 3

120. A body is moving unidirectionally under the influence of a constant power source. Its displacement in time t is proportional to : **[NCERT : P L-83 | April 5, 2024 (II)]**

(a) t^2 (b) $t^{2/3}$ (c) $t^{3/2}$ (d) t

121. A body of mass 2 kg begins to move under the action of a time dependent force given by $\vec{F} = (6t\hat{i} + 6t^2\hat{j})$ N. The power developed by the force at the time t is given by: **[NCERT : P L-83 | Jan. 31, 2024 (II)]**
(a) $(6t^4 + 9t^5)$ W (b) $(3t^3 + 6t^5)$ W
(c) $(9t^5 + 6t^3)$ W (d) $(9t^3 + 6t^5)$ W

122. The ratio of powers of two motors is $\frac{3\sqrt{x}}{\sqrt{x}+1}$, that are capable of raising 300 kg water in 5 minutes and 50 kg water in 2 minutes respectively from a well of 100 m deep. The value of x will be **[NCERT : P L-83 | April 13, 2023 (I)]**
(a) 2 (b) 4 (c) 2.4 (d) 16

123. A block of mass 5 kg starting from rest pulled up on a smooth incline plane making an angle of 30° with horizontal with an affective acceleration of 1 ms^{-2}. The power delivered by the pulling force at t = 10 s from the start is _____ W. [Use g = 10 ms^{-2}] **[NCERT : P L-83 | April 10, 2023 (I)]** (calculate the nearest integer value)

124. If the maximum load carried by an elevator is 1400 kg (600 kg – Passenger + 800 kg – elevator), which is moving up with a uniform speed of 3 ms^{-1} and the frictional force acting on it is 2000 N, then the maximum power used by the motor is______kW (g = 10 m/s^2).
[NCERT : P L-83 | April 10, 2023 (II)]

125. A body of mass 2 kg is initially at rest. It starts moving unidirectionally under the influence of a source of constant power P. Its displacement in 4s is $\frac{1}{3}\alpha^2\sqrt{P}$m. The value of α will be ....
[NCERT : P L-83 | Jan. 30, 2023 (II)]

126. A body of mass 1 kg begins to move under the action of a time dependent force $\vec{F} = (t\hat{i} + 3t^2\hat{j})N$ where $\hat{i}$ and $\hat{j}$ are the unit vectors along x and y axis. The power developed by above force, at the time t = 2s. will be ________ W.
[NCERT : P L-83 | Jan. 24, 2023 (II)]

127. Sand is being dropped from a stationary dropper at a rate of 0.5 kgs^{-1} on a conveyor belt moving with a velocity of 5 ms^{-1}. The power needed to keep belt moving with the same velocity will be : **[NCERT : P L-83 | July 27, 2022 (I)]**
(a) 1.25 W (b) 2.5 W (c) 6.25 W (d) 12.5 W

128. A particle of mass m is moving in a circular path of constant radius r such that its centripetal acceleration (a) is varying with time t as a = k^2rt^2 where k is a constant. The power delivered to the particle by the force acting on it is given as **[NCERT : P L-83 | June 28, 2022 (I)]**
(a) zero (b) $mk^2r^2t^2$ (c) mk^2r^2t (d) mk^2rt

129. A body of mass 2 kg is driven by an engine delivering a constant power of 1 J/s. The body starts from rest and moves in a straight line. After 9 seconds, the body has moved a distance (in m) ________.
[NCERT : P L-83 | 5 Sep. 2020 (II)]

130. A particle is moving unidirectionally on a horizontal plane under the action of a constant power supplying energy source. The displacement (s) - time (t) graph that describes the motion of the particle is (graphs are drawn schematically and are not to scale) :
[NCERT : P L-83 | 3 Sep. 2020 (II)]

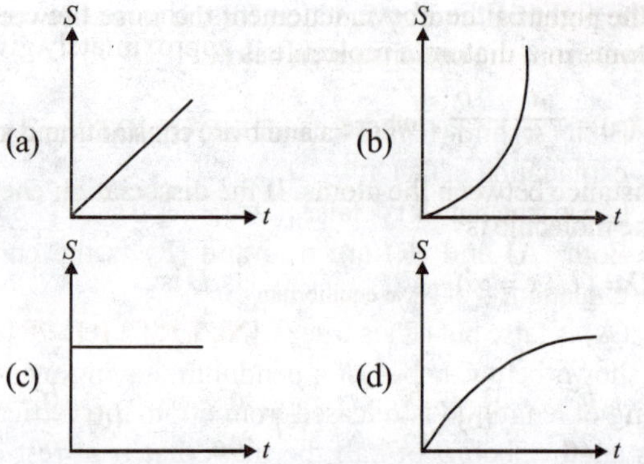

131. A 60 HP electric motor lifts an elevator having a maximum total load capacity of 2000 kg. If the frictional force on the elevator is 4000 N, the speed of the elevator at full load is close to: (1 HP = 746 W, g = 10 ms^{-2})
[NCERT : P L-83 | 7 Jan. 2020 I]
(a) 1.7 ms^{-1} (b) 1.9 ms^{-1} (c) 1.5 ms^{-1} (d) 2.0 ms^{-1}

132. A particle of mass M is moving in a circle of fixed radius R in such a way that its centripetal acceleration at time t is given by n^2Rt^2 where n is a constant. The power delivered to the particle by the force acting on it, is :
[NCERT : P L-83 | Online April 10, 2016]
(a) $\frac{1}{2}Mn^2R^2t^2$ (b) Mn^2R^2t
(c) MnR^2t^2 (d) MnR^2t

133. A wind-powered generator converts wind energy into electrical energy. Assume that the generator converts a fixed fraction of the wind energy intercepted by its blades into electrical energy. For wind speed v, the electrical power output will be most likely proportional to
[NCERT : P L-83 | Online April 25, 2013]
(a) v^4 (b) v^2 (c) v^3 (d) v

134. A body of mass 'm', accelerates uniformly from rest to 'v_1' in time 't_1'. The instantaneous power delivered to the body as a function of time 't' is **[NCERT : P L-83 | 2004]**
(a) $\frac{mv_1t^2}{t_1}$ (b) $\frac{mv_1^2t}{t_1^2}$ (c) $\frac{mv_1t}{t_1}$ (d) $\frac{mv_1^2t}{t_1}$

Topic 4 - Collisions

135. Given below are two statements. One is labelled as **Assertion (A)** and the other is labelled as **Reason (R)**.

Assertion (A) :
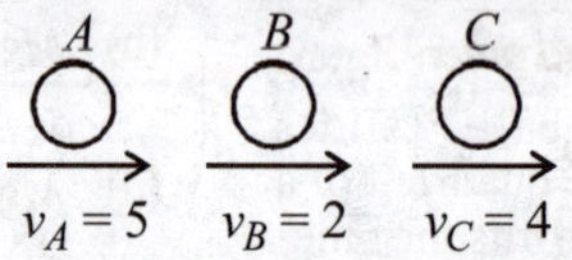

Three identical spheres of same mass undergo one dimensional motion as shown in figure with initial velocities v_A = 5 m/s, v_B = 2 m/s, v_C = 4 m/s. If we wait sufficiently long for elastic collision to happen, then v_A = 4 m/s, v_B = 2 m/s, v_C = 5 m/s will be the final velocities.
Reason (R) : In an elastic collision between identical masses, two objects exchange their velocities.

In the light of the above statements, choose the correct answer from the options given below :

[NCERT : P L-84 | Jan. 29, 2025 (II)]

(a) Both (A) and (R) are true but (R) is NOT the correct explanation of (A)
(b) (A) is true but (R) is false
(c) Both (A) and (R) are true and (R) is the correct explanation of (A).
(d) (A) is false but (R) is true

136. As shown below, bob *A* of a pendulum having massless string of length '*R*' is released from 60° to the vertical. It hits another bob *B* of half the mass that is at rest on a frictionless table in the centre. Assuming elastic collision, the magnitude of the velocity of bob *A* after the collision will be (take *g* as acceleration due to gravity)

[NCERT : P L-84 | Jan. 29, 2025 (I)]

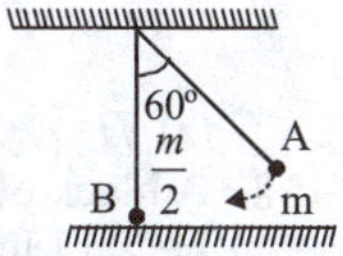

(a) $\frac{1}{3}\sqrt{Rg}$ (b) $\sqrt{Rg}$
(c) $\frac{4}{3}\sqrt{Rg}$ (d) $\frac{2}{3}\sqrt{Rg}$

137. If a rubber ball falls from a height h and rebounds upto the height of h/2. The percentage loss of total energy of the initial system as well as velocity ball before it strikes the ground, respectively, are :

[NCERT : P L-84 | April 4, 2024 (I)]

(a) 50%, $\sqrt{\frac{gh}{2}}$ (b) 50%, $\sqrt{gh}$
(c) 40%, $\sqrt{2gh}$ (d) 50%, $\sqrt{2gh}$

138. A body starts falling freely from height H hits an inclined plane in its path at height h. As a result of this perfectly elastic impact, the direction of the velocity of the body becomes horizontal. The value of $\frac{H}{h}$ for which the body will take the maximum time to reach the ground is ______.

[NCERT : P L-84 | Jan. 31, 2024 (I)]

139. A particle of mass m moving with velocity v collides with a stationary particle of mass 2m. After collision, they stick together and continue to move together with velocity

[NCERT : P L-84 | April 10, 2023 (I)]

(a) v (b) $\frac{v}{2}$ (c) $\frac{v}{3}$ (d) $\frac{v}{4}$

140. A ball is dropped from a height of 20 m. If the coefficient of restitution for the collision between ball and floor is 0.5, after hitting the floor, the ball rebounds to a height of______m. **[Jan. 31, 2023 (II)]**

141. As per the given figure, a small ball P slides down the quadrant of a circle and hits the other ball Q of equal mass which is initially at rest. Neglecting the effect of friction and assume the collision to be elastic, the velocity of ball Q after collision will be : (g = 10 m/s^2) **[NCERT : P L-84 | Jan. 30, 2023 (I)]**

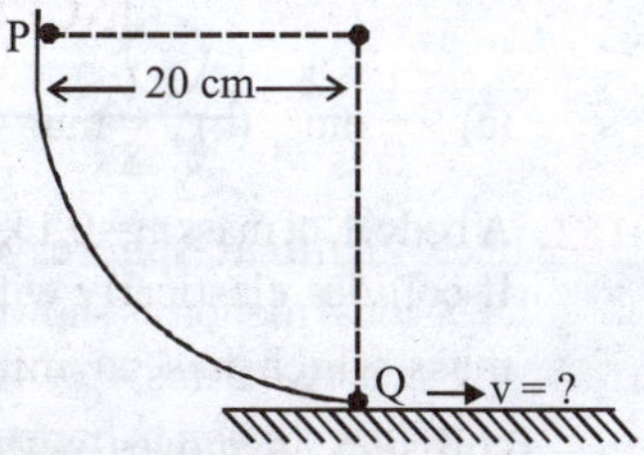

(a) 0 (b) 0.25 m/s (c) 2 m/s (d) 4 m/s

142. A ball of mass 200 g rests on a vertical post of height 20 m. A bullet of mass 10 g, travelling in horizontal direction, hits the centre of the ball. After collision both travels independently. The ball hits the ground at a distance 30 m and the bullet at a distance of 120 m from the foot of the post. The value of initial velocity of the bullet will be (if g = 10 m/s^2) : **[NCERT : P L-83 | Jan. 30, 2023 (I)]**

(a) 120 m/s (b) 60 m/s (c) 400 m/s (d) 360 m/s

143. A body of mass 1 kg collides head on elastically with a stationary body of mass 3 kg. After collision, the smaller body reverses its direction of motion and moves with a speed of 2m/s. The initial speed of the smaller body before collision is _______ ms^{-1}.

[NCERT : P L-84 | Jan. 25, 2023 (II)]

144. A bag of sand of mass 9.8 kg is suspended by a rope. A bullet of 200 g travelling with speed 10 ms^{-1} gets embedded in it, then loss of kinetic energy will be

[NCERT : P L-84 | July 25, 2022 (II)]

(a) 4.9 J (b) 9.8 J (c) 14.7 (d) 19.6 J

145. Two billiard balls of mass 0.05 kg each moving in opposite directions with 10 ms^{-1} collide and rebound with the same speed. If the time duration of contact is t = 0.005 s, then what is the force exerted on the ball due to each other?

[NCERT : P L-83, 84 | July 25, 2022 (II)]

(a) 100 N (b) 200 N (c) 300 N (d) 4000 N

146. A body of mass M at rest explodes into three pieces, in the ratio of masses 1 : 1 : 2. Two smaller pieces fly off perpendicular to each other with velocities of 30 ms^{-1} and 40 ms^{-1} respectively. The velocity of the third piece will be : **[NCERT : P L-84 | June 29, 2022 (I)]**

(a) 15 ms^{-1} (b) 25 ms^{-1} (c) 35 ms^{-1} (d) 50 ms^{-1}

147. What percentage of kinetic energy of a moving particle is transferred to a stationary particle when it strikes the stationary particle of 5 times its mass? (Assume the collision to be head-on elastic collision) **[NCERT : P L-84 | June 27, 2022 (I)]**

(a) 50.0% (b) 66.6% (c) 55.5% (d) 33.3%

148. A body of mass M moving at speed V_0 collides elastically with a mass 'm' at rest. After the collision, the two masses move at angles θ_1 and θ_2 with respect to the initial direction of motion of the body of mass M. The largest possible value of the ratio M/m, for which the angles θ_1 and θ_2 will be equal, is : **[NCERT : P L-85 | Aug. 31, 2021 (I)]**

(a) 4 (b) 1 (c) 3 (d) 2

149. A block moving horizontally on a smooth surface with a speed of 40 m/s splits into two parts with masses in the ratio of 1:2. If the smaller part moves at 60 m/s in the same direction, then the fractional change in kinetic energy is :

[NCERT : P L-84 | Aug. 31, 2021 (II)]

(a) $\frac{1}{3}$ (b) $\frac{2}{3}$ (c) $\frac{1}{8}$ (d) $\frac{1}{4}$

150. A ball of mass 10 kg moving with a velocity $10\sqrt{3}$ m/s along the x-axis, hits another ball of mass 20 kg which is at rest. After the collision, first ball comes to rest while the second ball disintegrates into two equal pieces. One piece starts moving along y-axis with a speed of 10 m/s. The second piece starts moving at an angle of 30° with respect to the x-axis.

The velocity of the ball moving at 30° with x-axis is x m/s. The configuration of pieces after collision is shown in the figure below.

The value of x to the nearest integer is ______.

[NCERT : P L-85 | March 18, 2021 (I)]

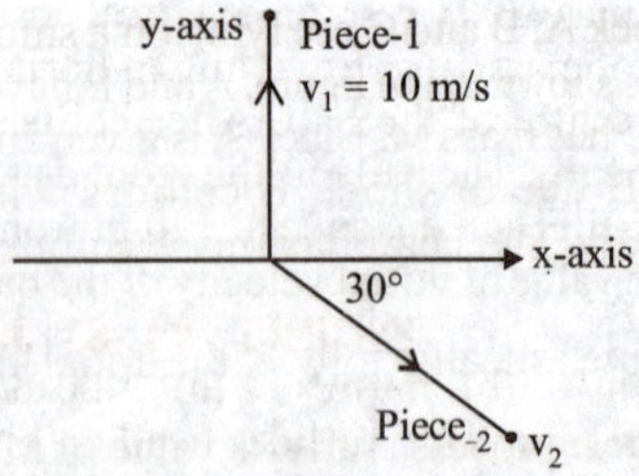

151. A object of mass m_1 collides with another object of mass m_2, which is at rest. After the collision the objects move with equal speeds in opposite direction. The ratio of the masses $m_2 : m_1$ is:
[NCERT : P L-84 | March 18, 2021 (II)]
(a) 2 : 1 (b) 3 : 1 (c) 1 : 2 (d) 1 : 1

152. A large block of wood of mass M = 5.99 kg is hanging from two long massless cords. A bullet of mass m = 10 g is fired into the block and gets embedded in it. The (block + bullet) then swing upwards, their centre of mass rising a vertical distance h = 9.8 cm before the (block + bullet) pendulum comes momentarily to rest at the end of its arc. The speed of the bullet just before collision is :
(take g = 9.8 ms^{-2})
[NCERT : P L-84 | March 16, 2021 (II)]
(a) 821.4 m/s
(b) 831.4 m/s
(c) 841.4 m/s
(d) 811.4 m/s

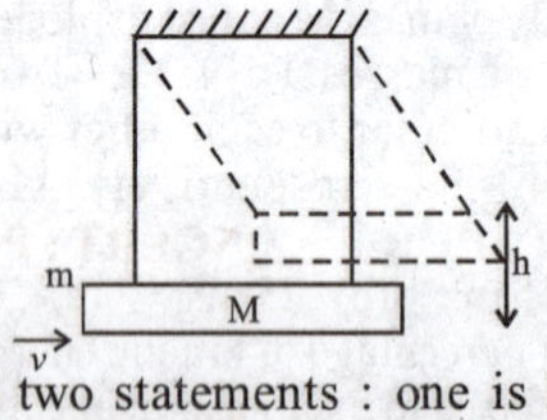

153. Given below are two statements : one is labelled as Assertion A and the other is labelled as Reason R.
Assertion A : Body 'P' having mass M moving with speed 'u' has head-on collision elastically with another body 'Q' having mass 'm' initially at rest. If m << M, body 'Q' will have a maximum speed equal to '2u' after collision.
Reason R : During elastic collision, the momentum and kinetic energy are both conserved.
In the light of the above statements, choose the most appropriate answer from the options given below :
[NCERT : P L-84 | Feb. 24, 2021 (II)]
(a) Both A and R are correct and R is the correct explanation of A
(b) A is not correct but R is correct
(c) A is correct but R is not correct
(d) Both A and R are correct but R is NOT the correct explanation of A

154. A ball with a speed of 9 m/s collides with another identical ball at rest. After the collision, the direction of each ball makes an angle of 30° with the original direction. The ratio of velocities of the balls after collision is $x : y$, where x is ________. **[NCERT : P L-85 | Feb. 24, 2021 (I)]**

155. Two bodies of the same mass are moving with the same speed, but in different directions in a plane. They have a completely inelastic collision and move together thereafter with a final speed which is half of their initial speed. The angle between the initial velocities of the two bodies (in degree) is ______.
[NCERT : P L-85 | NA 6 Sep. 2020 (I)]

156. Particle A of mass m_1 moving with velocity $(\sqrt{3}\hat{i} + \hat{j})$ ms^{-1} collides with another particle B of mass m_2 which is at rest initially. Let $\vec{V}_1$ and $\vec{V}_2$ be the velocities of particles A and B after collision respectively. If $m_1 = 2m_2$ and after collision $\vec{V}_1 = (\hat{i} + \sqrt{3}\hat{j})\, ms^{-1}$, the angle between $\vec{V}_1$ and $\vec{V}_2$ is :
[NCERT : P L-84, 85 | 6 Sep. 2020 (II)]
(a) 15° (b) 60° (c) –45° (d) 105°

157. Blocks of masses m, $2m$, $4m$ and $8m$ are arranged in a line on a frictionless floor. Another block of mass m, moving with speed v along the same line (see figure) collides with mass m in perfectly inelastic manner. All the subsequent collisions are also perfectly inelastic. By the time the last block of mass $8m$ starts moving the total energy loss is p% of the original energy. Value of 'p' is close to :
[NCERT : P L-84 | 4 Sep. 2020 (I)]

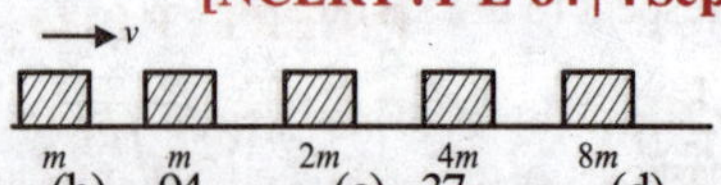

(a) 77 (b) 94 (c) 37 (d) 87

158. A block of mass 1.9 kg is at rest at the edge of a table, of height 1 m. A bullet of mass 0.1 kg collides with the block and sticks to it. If the velocity of the bullet is 20 m/s in the horizontal direction just before the collision then the kinetic energy just before the combined system strikes the floor, is [Take $g = 10$ m/s^2. Assume there is no rotational motion and losss of energy after the collision is negligiable.]
[NCERT : P L-84 | 3 Sep. 2020 (II)]
(a) 20 J (b) 21 J (c) 19 J (d) 23 J

159. A particle of mass m with an initial velocity $u\hat{i}$ collides perfectly elastically with a mass 3 m at rest. It moves with a velocity $v\hat{j}$ after collision, then, v is given by :
[NCERT : P L-84, 85 | 2 Sep. 2020 (I)]
(a) $v = \sqrt{\frac{2}{3}}u$ (b) $v = \frac{u}{\sqrt{3}}$ (c) $v = \frac{u}{\sqrt{2}}$ (d) $v = \frac{1}{\sqrt{6}}u$

160. A particle of mass m is moving along the x-axis with initial velocity $u\hat{i}$. It collides elastically with a particle of mass 10 m at rest and then moves with half its initial kinetic energy (see figure). If $\sin\theta_1 = \sqrt{n}\sin\theta_2$, then value of n is ________.

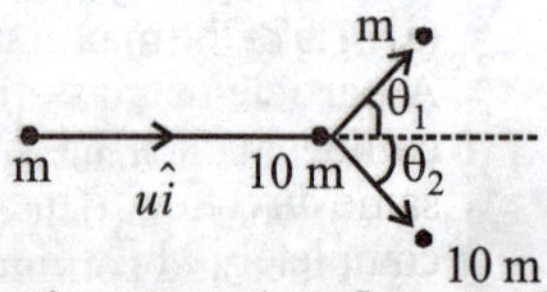

[NCERT : P L-85 | NA 2 Sep. 2020 (II)]

161. Two particles of equal mass m have respective initial velocities $u\hat{i}$ and $u\left(\frac{\hat{i}+\hat{j}}{2}\right)$. They collide completely inelastically. The energy lost in the process is:
[NCERT : P L-84, 85 | 9 Jan. 2020 I]
(a) $\frac{1}{3}mu^2$ (b) $\frac{1}{8}mu^2$ (c) $\frac{3}{4}mu^2$ (d) $\sqrt{\frac{2}{3}}mu^2$

162. A body A, of mass $m = 0.1$ kg has an initial velocity of $3\hat{i}$ ms^{-1}. It collides elastically with another body, B of the same mass which has an initial velocity of $5\hat{j}$ ms^{-1}. After collision, A moves with a velocity $\vec{v} = 4(\hat{i} + \hat{j})$. The energy of B after collision is written as $\frac{x}{10}$ J. The value of x is ________. **[NCERT : P L-84, 85 | NA 8 Jan. 2020 I]**

163. A particle of mass m is dropped from a height h above the ground. At the same time another particle of the same mass is thrown vertically upwards from the ground with a speed of $\sqrt{2gh}$. If they collide head-on completely inelastically, the time taken for the combined mass to reach the ground, in units of $\sqrt{\frac{h}{g}}$ is: **[NCERT : P L-84 | 8 Jan. 2020 (II)]**

(a) $\sqrt{\frac{1}{2}}$ (b) $\sqrt{\frac{3}{4}}$ (c) $\frac{1}{2}$ (d) $\sqrt{\frac{3}{2}}$

164. A man (mass = 50 kg) and his son (mass = 20 kg) are standing on a frictionless surface facing each other. The man pushes his son so that he starts moving at a speed of 0.70 ms^{-1} with respect to the man. The speed of the man with respect to the surface is : **[NCERT : P L-84 | 12 April 2019 I]**

(a) 0.28 ms^{-1} (b) 0.20 ms^{-1} (c) 0.47 ms^{-1} (d) 0.14 ms^{-1}

165. Two particles, of masses M and 2M, moving, as shown, with speeds of 10 m/s and 5 m/s, collide elastically at the origin. After the collision, they move along the indicated directions with speeds v_1 and v_2, respectively. The values of v_1 and v_2 are nearly : **[NCERT : P L-85 | 10 April 2019 I]**

(a) 6.5 m/s and 6.3 m/s

(b) 3.2 m/s and 6.3 m/s

(c) 6.5 m/s and 3.2 m/s

(d) 3.2 m/s and 12.6 m/s

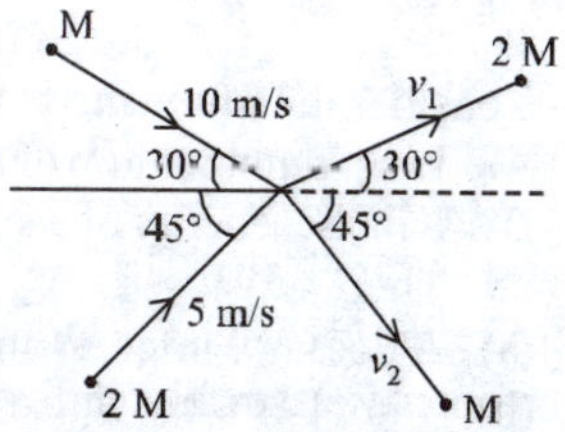

166. A body of mass 2 kg makes an elastic collision with a second body at rest and continues to move in the original direction but with one fourth of its original speed. What is the mass of the second body? **[NCERT : P L-84 | 9 April 2019 I]**

(a) 1.0 kg (b) 1.5 kg (c) 1.8 kg (d) 1.2 kg

167. A particle of mass 'm' is moving with speed '$2v$' and collides with a mass '$2m$' moving with speed 'v' in the same direction. After collision, the first mass is stopped completely while the second one splits into two particles each of mass 'm', which move at angle 45° with respect to the original direction. **[NCERT : P L-85 | 9 April 2019 II]**
The speed of each of the moving particle will be:

(a) $\sqrt{2}\,v$ (b) $2\sqrt{2}\,v$ (c) $v/(2\sqrt{2})$ (d) $v/\sqrt{2}$

168. A body of mass m_1 moving with an unknown velocity of $v_1\hat{i}$, undergoes a collinear collision with a body of mass m_2 moving with a velocity $v_2\hat{i}$. After collision, m_1 and m_2 move with velocities of $v_3\hat{i}$ and $v_4\hat{i}$, respectively.
If $m_2 = 0.5\,m_1$ and $v_3 = 0.5\,v_1$, then v_1 is : **[NCERT : P L-84 | 8 April 2019 II]**

(a) $v_4 - \frac{v_2}{2}$ (b) $v_4 - v_2$ (c) $v_4 - \frac{v_2}{4}$ (d) $v_4 + v_2$

169. An alpha-particle of mass m suffers 1-dimensional elastic collision with a nucleus at rest of unknown mass. It is scattered directly backwards losing, 64% of its initial kinetic energy. The mass of the nucleus is **[NCERT : P L-84 | 12 Jan. 2019 II]**

(a) 2m (b) 3.5m (c) 1.5m (d) 4m

170. There block A, B and C are lying on a smooth horizontal surface, as shown in the figure. A and B have equal masses, m while C has mass M. Block A is given an inital speed v_0 towards B due to which it collides with B perfectly inelastically. The combined mass collides with C, also perfectly inelastically, $\frac{5}{6}$th of the initial kinetic energy is lost in whole process. What is value of M/m? **[NCERT : P L-84 | 9 Jan. 2019 I]**

(a) 5 (b) 2
(c) 4 (d) 3

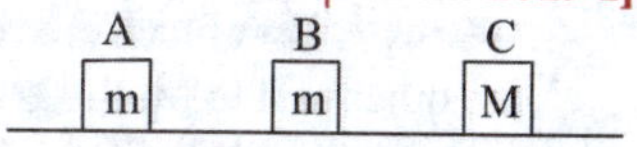

171. In a collinear collision, a particle with an initial speed v_0 strikes a stationary particle of the same mass. If the final total kinetic energy is 50% greater than the original kinetic energy, the magnitude of the relative velocity between the two particles, after collision, is: **[NCERT : P L-84 | 2018]**

(a) $\frac{v_0}{4}$ (b) $\sqrt{2}v_0$ (c) $\frac{v_0}{2}$ (d) $\frac{v_0}{\sqrt{2}}$

172. The mass of a hydrogen molecule is 3.32×10^{-27} kg. If 10^{23} hydrogen molecules strike, per second, a fixed wall of area 2 cm^2 at an angle of 45° to the normal, and rebound elastically with a speed of 10^3 m/s, then the pressure on the wall is nearly: **[NCERT : P L-84, 85 | 2018]**

(a) 2.35×10^3 N/m^2 (b) 4.70×10^3 N/m^2
(c) 2.35×10^2 N/m^2 (d) 4.70×10^2 N/m^2

173. Two particles A and B of equal mass M are moving with the same speed v as shown in the figure. They collide completely inelastically and move as a single particle C. The angle θ that the path of C makes with the X-axis is given by: **[NCERT : P L-85 | Online April 9, 2017]**

(a) $\tan\theta = \frac{\sqrt{3}+\sqrt{2}}{1-\sqrt{2}}$

(b) $\tan\theta = \frac{\sqrt{3}-\sqrt{2}}{1-\sqrt{2}}$

(c) $\tan\theta = \frac{1-\sqrt{2}}{\sqrt{2}(1+\sqrt{3})}$

(d) $\tan\theta = \frac{1-\sqrt{3}}{1+\sqrt{2}}$

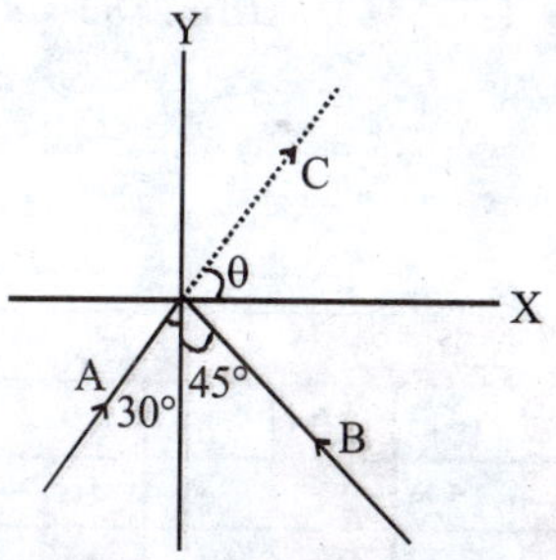

174. A bullet of mass 4 g is fired horizontally with a speed of 300 m/s into 0.8 kg block of wood at rest on a table. If the coefficient of friction between the block and the table is 0.3, how far will the block slide approximately? **[NCERT : P L-84 | Online April 12, 2014]**

(a) 0.19m (b) 0.379m (c) 0.569m (d) 0.758m

175. This question has statement I and statement II. Of the four choices given after the statements, choose the one that best describes the two statements. **[NCERT : P L-84 | 2013]**

Statement - I: A point particle of mass m moving with speed υ collides with stationary point particle of mass M. If the maximum energy loss possible is given as $f\left(\frac{1}{2}mv^2\right)$ then $f = \left(\frac{m}{M+m}\right)$.

Statement - II: Maximum energy loss occurs when the particles get stuck together as a result of the collision.

(a) Statement - I is true, Statment - II is true, Statement - II is the correct explanation of Statement - I.
(b) Statement-I is true, Statment - II is true, Statement - II is **not** the correct explanation of Statement - II.
(c) Statement - I is true, Statment - II is false.
(d) Statement - I is false, Statment - II is true.

176. A projectile moving vertically upwards with a velocity of 200 ms^{-1} breaks into two equal parts at a height of 490 m. One part starts moving vertically upwards with a velocity of 400 ms^{-1}. How much time it will take, after the break up with the other part to hit the ground? **[NCERT : P L-85 | Online May 12, 2012]**

(a) $2\sqrt{10}$ s (b) 5 s (c) 10 s (d) $\sqrt{10}$ s

177. Statement -1 : Two particle moving in the same direction do not lose all their energy in a completely inelastic collision.
Statement -2 : Principle of conservation of momentum holds true for all kinds of collisions. **[NCERT : P L-84 | 2010]**

(a) Statement -1 is true, Statement -2 is true ; Statement - 2 is the correct explanation of Statement -1.
(b) Statement -1 is true, Statement -2 is true; Statement -2 is **not** the correct explanation of Statement -1
(c) Statement -1 is false, Statement -2 is true.
(d) Statement -1 is true, Statement -2 is false.

178. A bomb of mass 16kg at rest explodes into two pieces of masses 4 kg and 12 kg. The velolcity of the 12 kg mass is 4 ms^{-1}. The kinetic energy of the other mass is **[NCERT : P L-84 | 2006]**

(a) 144 J (b) 288 J (c) 192 J (d) 96 J

179. The block of mass M moving on the frictionless horizontal surface collides with the spring of spring constant k and compresses it by length L. The maximum momentum of the block after collision is **[NCERT : P L-84 | 2005]**

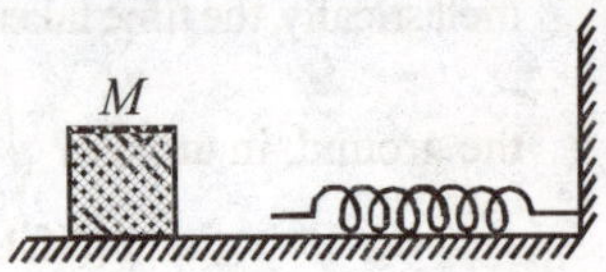

(a) $\frac{kL^2}{2M}$ (b) $\sqrt{Mk}\ L$ (c) $\frac{ML^2}{k}$ (d) zero

180. A mass 'm' moves with a velocity 'v' and collides inelastically with another identical mass. After collision the 1st mass moves with velocity $\frac{v}{\sqrt{3}}$ in a direction perpendicular to the initial direction of motion. Find the speed of the 2nd mass after collision. **[NCERT : P L-85 | 2005]**

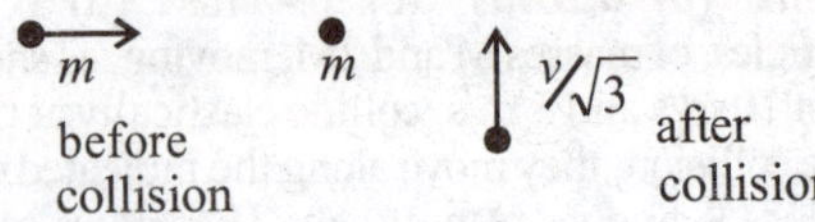

(a) $\sqrt{3}v$ (b) v (c) $\frac{v}{\sqrt{3}}$ (d) $\frac{2}{\sqrt{3}}v$

181. Consider the following two statements : **[2003]**
A. Linear momentum of a system of particles is zero
B. Kinetic energy of a system of particles is zero.
Then
(a) A does not imply B and B does not imply A
(b) A implies B but B does not imply A
(c) A does not imply B but B implies A
(d) A implies B and B implies A

ANSWER KEYS

1	(c)	20	(3)	39	(d)	58	(50)	77	(1)	96	(a)	115	(d)	134	(b)	153	(a)	172	(a)
2	(152)	21	(c)	40	(c)	59	(7)	78	(40)	97	(a)	116	(c)	135	(d)	154	(1)	173	(a)
3	(b)	22	(d)	41	(a)	60	(300)	79	(a)	98	(c)	117	(d)	136	(a)	155	(120)	174	(b)
4	(c)	23	(b)	42	(b)	61	(2)	80	(10)	99	(a)	118	(d)	137	(d)	156	(d)	175	(d)
5	(58)	24	(b)	43	(b)	62	(40)	81	(6)	100	(b)	119	(b)	138	(2)	157	(b)	176	(c)
6	(a)	25	(c)	44	(d)	63	(c)	82	(1)	101	(b)	120	(c)	139	(c)	158	(b)	177	(a)
7	(b)	26	(d)	45	(b)	64	(c)	83	(2)	102	(a)	121	(d)	140	(5)	159	(c)	178	(b)
8	(a)	27	(d)	46	(d)	65	(c)	84	(150)	103	(a)	122	(d)	141	(c)	160	(10.00)	179	(b)
9	(30)	28	(d)	47	(b)	66	(24)	85	(10)	104	(c)	123	(300)	142	(d)	161	(b)	180	(d)
10	(784)	29	(d)	48	(c)	67	(a)	86	(c)	105	(c)	124	(48)	143	(4)	162	(1)	181	(c)
11	(40)	30	(b)	49	(b)	68	(2)	87	(c)	106	(d)	125	(4)	144	(b)	163	(d)		
12	(132)	31	(36)	50	(d)	69	(b)	88	(None)	107	(b)	126	(100)	145	(b)	164	(b)		
13	(a)	32	(b)	51	(a)	70	(b)	89	(a)	108	(d)	127	(d)	146	(b)	165	(a)		
14	(b)	33	(b)	52	(125)	71	(b)	90	(c)	109	(b)	128	(c)	147	(c)	166	(d)		
15	(d)	34	(c)	53	(30)	72	(d)	91	(d)	110	(a)	129	(18)	148	(c)	167	(b)		
16	(2)	35	(a)	54	(b)	73	(a)	92	(d)	111	(d)	130	(b)	149	(c)	168	(b)		
17	(450)	36	(a)	55	(2)	74	(c)	93	(d)	112	(b)	131	(b)	150	(20)	169	(d)		
18	(b)	37	(a)	56	(375)	75	(16)	94	(c)	113	(c)	132	(b)	151	(b)	170	(c)		
19	(d)	38	(b)	57	(c)	76	(a)	95	(b)	114	(a)	133	(c)	152	(b)	171	(b)		

6 System of Particles and Rotational Motion

TREND ANALYSIS
JEE MAIN (2025-2021)

	2025				2024				2023				2022				2021			
	Session II		Session I		Session II		Session I		Session II		Session I		Session II		Session I		Session II		Session I	
	MCQs	NVQs	MCQs	NVQs	MCQs	NVQs	MCQs	NVQs	MCQs	NVQs	MCQs	NVQs	MCQs	NVQs	MCQs	NVQs	MCQs	NVQs	MCQs	NVQs
Topic 1	1	–	2	–	–	2	–	1	–	–	–	–	1	2	1	–	–	1	1	2
Topic 2	–	–	–	–	–	–	–	–	–	–	–	–	–	1	2	–	–	–	–	1
Topic 3	4	3	5	2	2	1	2	2	–	2	1	1	1	–	3	3	1	1	1	2
Topic 4	3	3	1	2	1	2	–	5	–	6	–	4	–	4	2	3	5	3	2	1
Total	8	6	8	4	3	5	2	8	–	8	1	5	2	7	8	6	6	5	4	6

How to Code Qns for Better & Faster Revision

Type 1: Able to solve (No code Req)
Type 2: Solved with struggle or by seeing solution **(Use code NR means Needs Revision or orange stickers)**
Type 3: Tricky/ Could not solve on my own. Need help from Teacher/ Friend. **(Use code TQ means Tricky Question or red stickers)**

You can either hand write the code in boxes or else paste coloured stickers given at end of book.

Topic 1 - Centre of Mass, Centre of Gravity & Principle of Moments

1. A rod of length 5 L is bent right angle keeping one side length as 2 L.

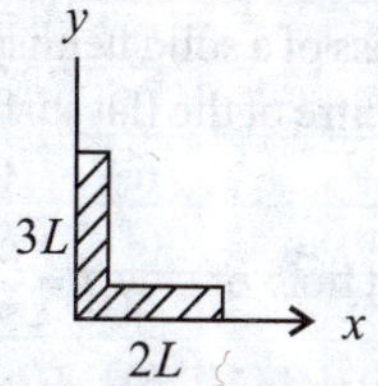

The position of the centre of mass of the system: (Consider $L = 10$ cm) **[NCERT : P L-96 | April 7, 2025 (I)]**

(a) $2\hat{i}+3\hat{j}$ (b) $3\hat{i}+7\hat{j}$ (c) $5\hat{i}+8\hat{j}$ (d) $4\hat{i}+9\hat{j}$

2. The centre of mass of a thin rectangular plate (fig -x) with sides of length a and b, whose mass per unit area (σ) varies as $\sigma = \dfrac{\sigma_0 x}{ab}$ (where σ_0 is a constant), would be______ **[NCERT : P L-96 | Jan. 28, 2025 (I)]**

(a) $\left(\dfrac{2}{3}a, \dfrac{b}{2}\right)$

(b) $\left(\dfrac{2}{3}a, \dfrac{2}{3}b\right)$

(c) $\left(\dfrac{a}{2}, \dfrac{b}{2}\right)$

(d) $\left(\dfrac{1}{3}a, \dfrac{b}{2}\right)$

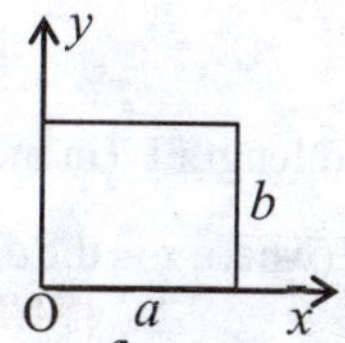

fig-x

3. Consider a circular disc of radius 20 cm with centre located at the origin. A circular hole of a radius 5 cm is cut from this disc in such a way that the edge of the hole touches the edge of the disc. The distance of centre of mass of residual or remaining disc from the origin will be **[NCERT : P L-96 | Jan. 23, 2025 (I)]**

(a) 2.0 cm (b) 0.5 cm (c) 1.5 cm (d) 1.0 cm

4. A uniform thin metal plate of mass 10 kg with dimensions is shown. The ratio of x and y coordinates of center of mass of plate in $\frac{n}{9}$. The value of n is ______. **[NCERT : P L-96 | April 8, 2024 (I)]**

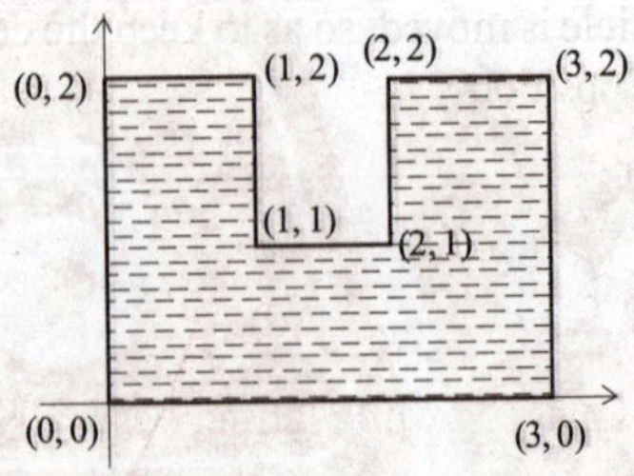

5. In a system two particles of masses $m_1 = 3$kg and $m_2 = 2$kg are placed at certain distance from each other. The particle of mass m_1 is moved towards the center of mass of the system through a distance 2cm. In order to keep the center of mass of the system at the original position, the particle of mass m_2 should move towards the center of mass by the distance ____ cm. **[NCERT : P L-96, 97 | April 4, 2024 (II)]**

6. Two identical spheres each of mass 2M are placed at the corners of a right angled triangle with mutually perpendicular sides equal to 4m each. Taking point of intersection of these two sides as origin, the magnitude of position vector of the centre of mass of the system is $\frac{4\sqrt{2}}{x}$, where the value of x is ____

[NCERT : P L-96 | Feb. 1, 2024 (I)]

7. Two bodies of mass 1 kg and 3 kg have position vectors $\hat{i}+2\hat{j}+\hat{k}$ and $-3\hat{i}-2\hat{j}+\hat{k}$ respectively. The magnitude of position vector of centre of mass of this system will be similar to the magnitude of vector : **[NCERT : P L-96 | July 29, 2022 (I)]**

(a) $\hat{i}-2\hat{j}+\hat{k}$ (b) $-3\hat{i}-2\hat{j}+\hat{k}$

(c) $-2\hat{i}+2\hat{k}$ (d) $-2\hat{i}-\hat{j}+2\hat{k}$

8. The distances of centre of mass from end A of a one dimensional rod (AB) having mass density $\lambda = \lambda_0\left(1-\frac{x^2}{L^2}\right)$ kg/m and length L (in metre) is $\frac{3L}{\alpha}$ m. The value of α is ____. (where x is the distance from end A)

[NCERT : P L-96 | July 28, 2022 (II)]

9. Three identical spheres each of mass M are placed at the corners of a right angled triangle with mutually perpendicular sides equal to 3 m each. Taking point of intersection of mutually perpendicular sides as origin, the magnitude of position vector of centre of mass of the system will be $\sqrt{x}$ m. The value of x is

[NCERT : P L-96 | July 25, 2022 (II)]

10. Two blocks of masses 10 kg and 30 kg are placed on the same straight line with coordinates (0, 0) cm and $(x, 0)$ cm respectively. The block of 10 kg is moved on the same line through a distance of 6 cm towards the other block. The distance through which the block of 30 kg must be moved to keep the position of centre of mass of the system unchanged is: **[NCERT : P L-96 | June 27, 2022 (I)]**

(a) 4 cm towards the 10 kg block
(b) 2 cm away from the 10 kg block
(c) 2 cm towards the 10 kg block
(d) 4 cm away from the 10 kg block

11. The position of the centre of mass of a uniform semi-circular wire of radius 'R' placed in x-y plane with its centre at the origin and the line joining its ends as x-axis is given by $\left(0, \frac{xR}{\pi}\right)$. Then, the value of $|x|$ is __. **[NA, July 22, 2021 (II)]**

12. The disc of mass M with uniform surface mass density σ is shown in the figure. The centre of mass of the quarter disc (the shaded area) is at the position $\frac{x}{3}\frac{a}{\pi}, \frac{x}{3}\frac{a}{\pi}$ where x is. (Round off to the Nearest integer)
[a is an area as shown in the figure]

[NCERT : P L-97 | March 17, 2021 (II)]

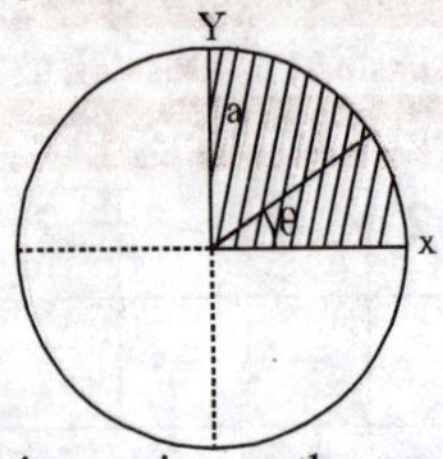

13. A man of 60 kg is running on the road and suddenly jumps into a stationary trolly car of mass 120 kg. Then the trolly car starts moving with velocity 2 ms^{-1}. The velocity of the running man was ________ ms^{-1}. When he jumps into the car.

14. A circular hole of radius $\left(\frac{a}{2}\right)$ is cut out of a circular disc of radius 'a' as shown in figure. The centroid of the remaining circular portion with respect to point 'O' will be: **[NCERT : P L-96 | Feb. 24, 2021 (II)]**

(a) $\frac{1}{6}a$
(b) $\frac{10}{11}a$
(c) $\frac{5}{6}a$
(d) $\frac{2}{3}a$

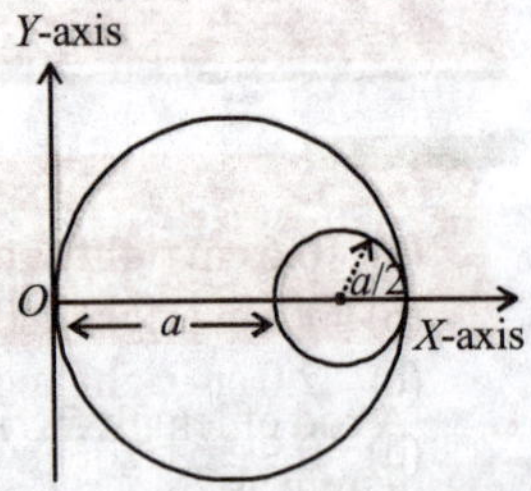

15. The centre of mass of a solid hemisphere of radius 8 cm is x cm from the centre of the flat surface. Then value of x is ____. **[NA Sep. 06, 2020 (II)]**

16. A square shaped hole of side $l = \frac{a}{2}$ is carved out at a distance $d = \frac{a}{2}$ from the centre 'O' of a uniform circular disc of radius a. If the distance of the centre of mass of the remaining portion from O is $-\frac{a}{X}$, value of X (to the nearest integer) is ____.

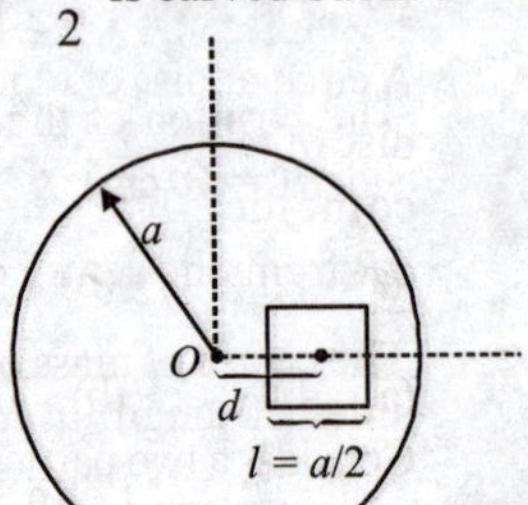

[NCERT : P L-96 | NA Sep. 02, 2020 (II)]

17. The coordinates of centre of mass of a uniform flag shaped lamina (thin flat plale) of mass 4 kg. (The coordinates of the same are shown in figure) are:

[NCERT : P L-96 | 8 Jan. 2020 I]

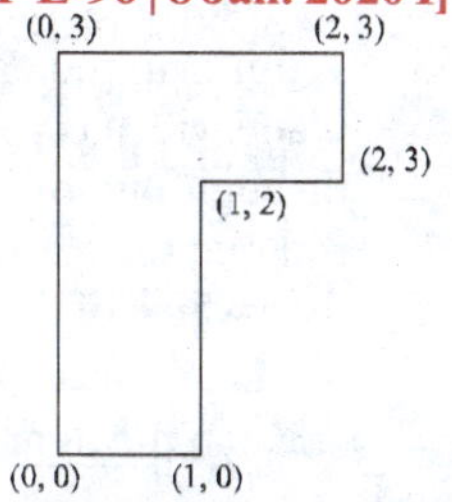

(a) (1.25 m, 1.50 m)

(b) (0.75 m, 1.75 m)

(c) (0.75 m, 0.75 m)

(d) (1 m, 1.75 m)

18. As shown in fig. when a spherical cavity (centred at O) of radius 1 is cut out of a uniform sphere of radius R (centred at C), the centre of mass of remaining (shaded) part of sphere is at G, i.e on the surface of the cavity. R can be determined by the equation:

[NCERT : P L-96 | 8 Jan. 2020 II]

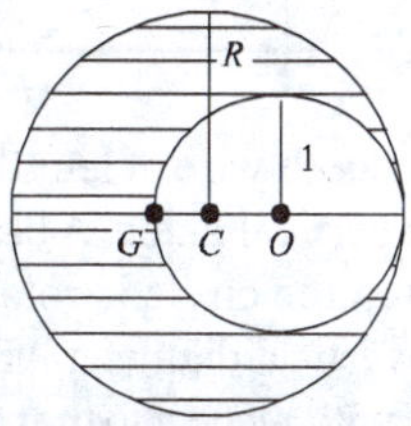

(a) $(R^2+R+1)(2-R)=1$

(b) $(R^2-R-1)(2-R)=1$

(c) $(R^2-R+1)(2-R)=1$

(d) $(R^2+R-1)(2-R)=1$

19. Three point particles of masses 1.0 kg, 1.5 kg and 2.5 kg are placed at three corners of a right angle triangle of sides 4.0 cm, 3.0 cm and 5.0 cm as shown in the figure. The center of mass of the system is at a point:

[NCERT : P L-96 | 7 Jan. 2020 I]

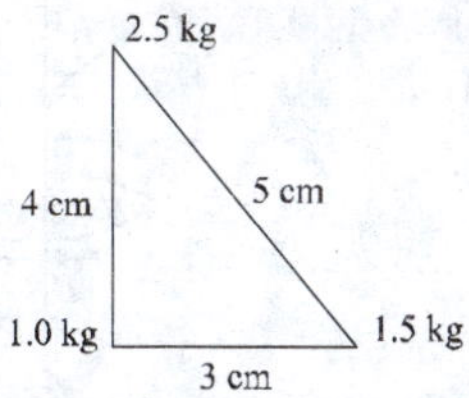

(a) 0.6 cm right and 2.0 cm above 1 kg mass
(b) 1.5 cm right and 1.2 cm above 1 kg mass
(c) 2.0 cm right and 0.9 cm above 1 kg mass
(d) 0.9 cm right and 2.0 cm above 1 kg mass

20. Distance of the centre of mass of a solid uniform cone from its vertex is z_0. If the radius of its base is R and its height is h then z_0 is equal to : **[NCERT : P L-96 | 2015]**

(a) $\frac{5h}{8}$ (b) $\frac{3h^2}{8R}$ (c) $\frac{h^2}{4R}$ (d) $\frac{3h}{4}$

21. A circular disc of radius R is removed from a bigger circular disc of radius $2R$ such that the circumferences of the discs coincide. The centre of mass of the new disc is αR from the centre of the bigger disc. The value of α is

[NCERT : P L-96 | 2007]

(a) 1/4 (b) 1/3 (c) 1/2 (d) 1/6

22. Consider a two particle system with particles having masses m_1 and m_2. If the first particle is pushed towards the centre of mass through a distance d, by what distance should the second particle is moved, so as to keep the centre of mass at the same position? **[NCERT : P L-96 | 2006]**

(a) $\frac{m_2}{m_1}d$ (b) $\frac{m_1}{m_1+m_2}d$

(c) $\frac{m_1}{m_2}d$ (d) d

23. A body A of mass M while falling vertically downwards under gravity breaks into two parts; a body B of mass $\frac{1}{3}M$ and a body C of mass $\frac{2}{3}M$. The centre of mass of bodies B and C taken together shifts compared to that of body A towards **[2005]**

(a) does not shift
(b) depends on height of breaking
(c) body B
(d) body C

24. A 'T' shaped object with dimensions shown in the figure, is lying on a smooth floor. A force '$\vec{F}$' is applied at the point P parallel to AB, such that the object has only the translational motion without rotation. Find the location of P with respect to C. **[NCERT : P L-96 | 2005]**

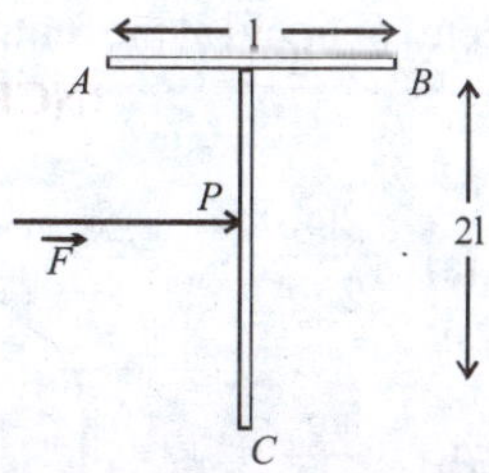

(a) $\frac{3}{2}\ell$

(b) $\frac{2}{3}\ell$

(c) ℓ

(d) $\frac{4}{3}\ell$

Topic 2 - Angular Displacement, Velocity and Acceleration

25. A tube of length 50 cm is filled completely with an incompressible liquid of mass 250 g and closed at both ends. The tube is then rotated in horizontal plane about one of its ends with a uniform angular velocity $x\sqrt{F}$ rad s^{-1}. If F be the force exerted by the liquid at the other end then the value of x will be ______.

[NCERT : P L-104 | July 29, 2022 (II)]

26. A ball is spun with angular acceleration $\alpha = 6t^2 - 2t$ where t is in second and α is in rads^{-2}. At $t = 0$, the ball has angular velocity of 10 rads^{-1} and angular position of 4 rad. The most appropriate expression for the angular position of the ball is:

[NCERT : P L-105 | June 28, 2022 (II)]

(a) $\frac{3}{2}t^4 - t^2 + 10t$ (b) $\frac{t^4}{2} - \frac{t^3}{3} + 10t + 4$

(c) $\frac{2t^4}{3} - \frac{t^3}{6} + 10t + 12$ (d) $2t^4 - \frac{t^3}{2} + 5t + 4$

27. A fly wheel is accelerated uniformly from rest and rotates through 5 rad in the first second. The angle rotated by the flywheel in the next second, will be :
[NCERT : P L-117 | June 24, 2022 (II)]
(a) 7.5 rad (b) 15 rad (c) 20 rad (d) 30 rad

28. The angular speed of truck wheel is increased from 900 rpm to 2460 rpm in 26 seconds. The number of revolutions by the truck engine during this time is ______ : **[NCERT : P L-117 | March 17, 2021 (I)]**

29. A bead of mass m stays at point $P(a, b)$ on a wire bent in the shape of a parabola $y = 4Cx^2$ and rotating with angular speed ω (see figure). The value of ω is (neglect friction) :
[NCERT : P L-104 | Sep. 02, 2020 (I)]

(a) $2\sqrt{2gC}$ (b) $2\sqrt{gC}$

(c) $\sqrt{\frac{2gC}{ab}}$ (d) $\sqrt{\frac{2g}{C}}$

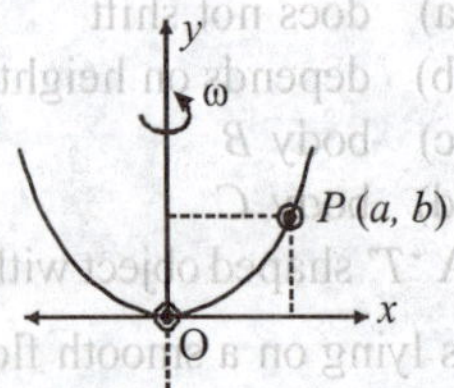

30. A cylindrical vessel containing a liquid is rotated about its axis so that the liquid rises at its sides as shown in the figure. The radius of vessel is 5 cm and the angular speed of rotation is ω rad s^{-1}. The difference in the height, h (in cm) of liquid at the centre of vessel and at the side will be :
[NCERT : P L-104 | Sep. 02, 2020 (I)]

(a) $\frac{2\omega^2}{25g}$

(b) $\frac{5\omega^2}{2g}$

(c) $\frac{25\omega^2}{2g}$

(d) $\frac{2\omega^2}{5g}$

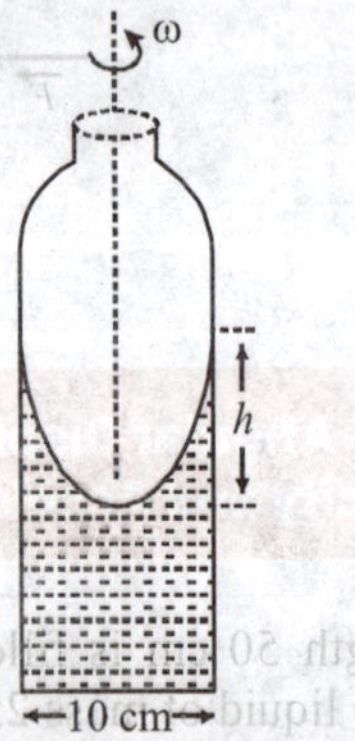

31. A spring mass system (mass m, spring constant k and natural length l) rests in equilibrium on a horizontal disc. The free end of the spring is fixed at the centre of the disc. If the disc together with spring mass system, rotates about it's axis with an angular velocity ω, $(k >> m\omega^2)$ the relative change in the length of the spring is best given by the option: **[NCERT : P L-104 | 9 Jan. 2020 II]**

(a) $\sqrt{\frac{2}{3}}\left(\frac{m\omega^2}{k}\right)$ (b) $\frac{2m\omega^2}{k}$

(c) $\frac{m\omega^2}{k}$ (d) $\frac{m\omega^2}{3k}$

32. A particle of mass m is fixed to one end of a light spring having force constant k and unstretched length l. The other end is fixed. The system is given an angular speed ω about the fixed end of the spring such that it rotates in a circle in gravity free space. Then the stretch in the spring is: **[NCERT : P L-104 | 8 Jan. 2020 I, June 27, 2022 II]**

(a) $\frac{ml\omega^2}{k - \omega m}$ (b) $\frac{ml\omega^2}{k - m\omega^2}$ (c) $\frac{ml\omega^2}{k + m\omega^2}$ (d) $\frac{ml\omega^2}{k + m\omega}$

33. A uniform rod of length l is being rotated in a horizontal plane with a constant angular speed about an axis passing through one of its ends. If the tension generated in the rod due to rotation is $T(x)$ at a distance x from the axis, then which of the following graphs depicts it most closely?
[NCERT : P L-104 | 12 Apr. 2019 II]

(a) 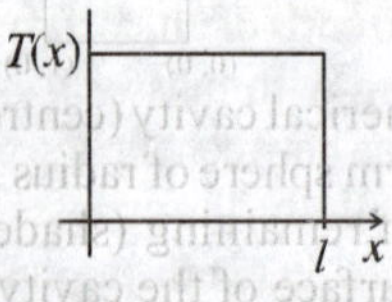

(b) 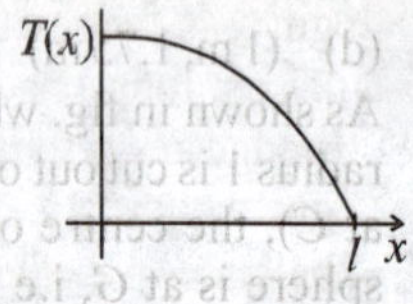

(c) 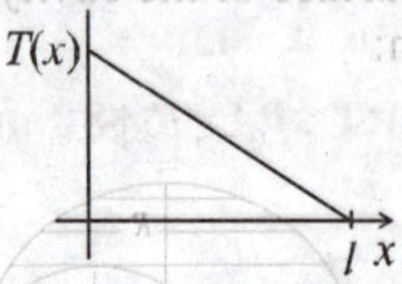

(d) 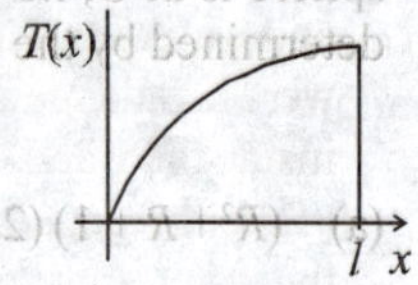

34. A smooth wire of length 2pr is bent into a circle and kept in a vertical plane. A bead can slide smoothly on the wire. When the circle is rotating with angular speed w about the vertical diameter AB, as shown in figure, the bead is at rest with respect to the circular ring at position P as shown. Then the value of w^2 is equal to : **[NCERT : P L-104 | 12 Apr. 2019 II]**

(a) $\frac{\sqrt{3g}}{2r}$

(b) $2g/(r\sqrt{3})$

(c) $(g\sqrt{3})/r$

(d) $2g/r$

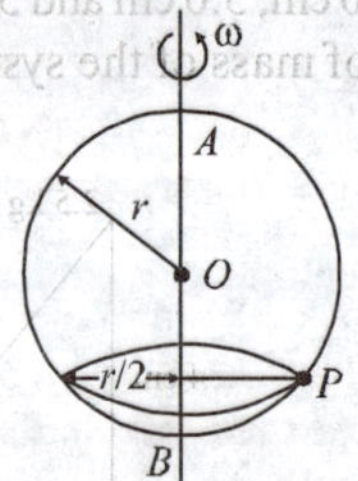

35. A particle is moving with a uniform speed in a circular orbit of radius R in a central force inversely proportional to the nth power of R. If the period of rotation of the particle is T, then: **[NCERT : P L-104 | 2018]**

(a) $T \propto R^{3/2}$ for any n. (b) $T \propto R^{n/2+1}$

(c) $T \propto R^{(n+1)/2}$ (d) $T \propto R^{n/2}$

36. Two point masses of mass $m_1 = fM$ and $m_2 = (1-f)M$ $(f < 1)$ are in outer space (far from gravitational influence of other objects) at a distance R from each other. They move in circular orbits about their centre of mass with angular velocities ω_1 for m_1 and ω_2 for m_2. In that case
[NCERT : P L-104 | Online May 19, 2012]

(a) $(1-f)\,\omega_1 = f\omega$

(b) $\omega_1 = \omega_2$ and independent of f

(c) $f\omega_1 = (1-f)\omega_2$

(d) $\omega_1 = \omega_2$ and depend on f

Topic 3 - Torque, Couple and Angular Momentum

37. A thin solid disk of 1 kg is rotating along its diameter axis at the speed of 1800 rpm. By applying an external torque of 25 π Nm for 40s, the speed increases to 2100 rpm. The diameter of the disk is ____ m.
[NCERT : P L-119 | April 8, 2025 (II)]

38. A cube having a side of 10 cm with unknown mass and 200 gm mass were hung at two ends of an uniform rigid rod of 27 cm long. The rod along with masses was placed on a wedge keeping the distance between wedge point and 200 gm weight as 25 cm. Initially the masses were not at balance. A beaker is placed beneath the unknown mass and water is added slowly to it. At given point the masses were in balance and half volume of the unknown mass was inside the water.
(Take the density of unknown mass is more than that of the water, the mass did not absorb water and water density is 1 gm/cm^3.) The unknown mass is ____ kg.
[NCERT : P L-106, 109 | April 8, 2025 (II)]

39. Which of the following are correct expression for torque acting on a body?

A. $\vec{\tau} = \vec{r} \times \vec{L}$ B. $\vec{\tau} = \frac{d}{dt}(\vec{r} \times \vec{p})$

C. $\vec{\tau} = \vec{r} \times \frac{d\vec{p}}{dt}$ D. $\vec{\tau} = I\vec{\alpha}$

E. $\vec{\tau} = \vec{r} \times \vec{F}$

($\vec{r}$ = position vector; $\vec{p}$ = linear momentum; $\vec{L}$ = angular momentum; $\vec{\alpha}$ = angular acceleration; I = moment of inertia; $\vec{F}$ = force; t = time) **[NCERT : PL-106, 119 | April 4, 2025 (I)]**
Choose the correct answer from the options given below :
(a) B, D and E Only (2) C and D Only
(c) B, C, D and E Only (d) A, B, D and E Only

40. If $\vec{L}$ and $\vec{P}$ represent the angular momentum and linear momentum respectively of a particle of mass 'm' having position vector $\vec{r} = a(\hat{i}\cos\omega t + \hat{j}\sin\omega t)$. The direction of force is **[NCERT : PL-106 | April 4, 2025 (I)]**
(a) Opposite to the direction of $\vec{r}$
(b) Opposite to the direction of $\vec{L}$
(c) Opposite to the direction of $\vec{P}$
(d) Opposite to the direction of $\vec{L} \times \vec{P}$

41. A wheel of radius 0.2 m rotates freely about its center when a string that is wrapped over its rim is pulled by force of 10 N as shown in figure.
The established torque produces an angular acceleration of 2 rad/s^2. Moment of inertia of the wheel is ____ kg m^2.
(Acceleration due to gravity = 10 m/s^2)

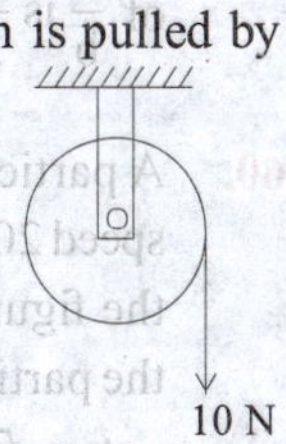

[NCERT : PL-119 | April 2, 2025 (II)]

42. A cord of negligible mass is wound around the rim of a wheel supported by spokes with negligible mass. The mass of wheel is 10 kg and radius is 10 cm and it can freely rotate without any friction. Initially the wheel is at rest. If a steady pull of 20 N is applied on the cord, the angular velocity of the wheel, after the cord is unwound by 1 m, would be : **[NCERT : P L-119 | April 2, 2025 (I)]**

(a) 20 rad/s
(b) 30 rad/s
(c) 10 rad/s
(d) 0 rad/s

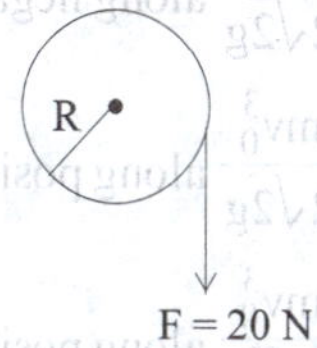

43.

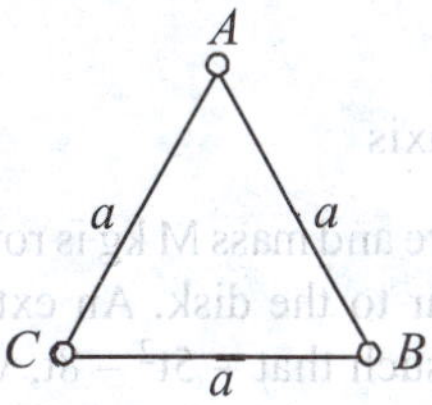

Three equal masses m are kept at vertices (A, B, C) of an equilateral triangle of side a in free space. At $t = 0$, they are given an initial velocity $\vec{V}_A = V_0\overrightarrow{AC}$, $\vec{V}_B = V_0\overrightarrow{BA}$ and $\vec{V}_C - V_0\overrightarrow{CB}$. Here, $\overrightarrow{AC}$, $\overrightarrow{CB}$ and $\overrightarrow{BA}$ are unit vectors along the edges of the triangle. If the three masses interact gravitationally, then the magnitude of the net angular momentum of the system at the point of collision is :
[NCERT : P L-106 | Jan. 29, 2025 (II)]

(a) $\frac{1}{2}amV_0$ (b) $3\ am\ V_0$

(c) $\frac{\sqrt{3}}{2}amV_0$ (d) $\frac{3}{2}amV_0$

44. A square Lamina OABC of length 10 cm is pivoted at 'O'. Forces act at Lamina as shown in figure. If Lamina remains stationary, then the magnitude of F is :
[NCERT : PL-106 | April 2, 2025 (I)]

(a) 20 N
(b) 0 (zero)
(c) 10 N
(d) $10\sqrt{2}$ N

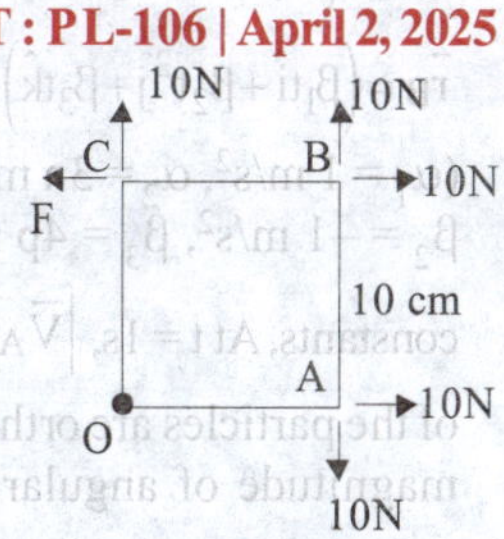

45. The coordinates of a particle with respect to origin in a given reference frame is (1, 1, 1) metres. If a force of $\vec{F} = \hat{i} - \hat{j} + \hat{k}$ acts on the particle, then the magnitude of torque (with respect to origin) in z-direction is ______.
[NCERT : P L-106 | Jan. 29, 2025 (I)]

46. A uniform rod of mass 250 g having length 100 cm is balanced on a sharp edge at 40 cm mark. A mass of 400 g is suspended at 10 cm mark. To maintain the balance of the rod, the mass to be suspended at 90 cm mark, is
[NCERT : P L-106, 109 | Jan. 28, 2025 (II)]
(a) 300 g (b) 190 g (c) 200 g (d) 290 g

47. An object of mass 'm' is projected from origin in a vertical xy plane at an angle 45° with the x-axis with an initial velocity v_0. The magnitude and direction of the angular momentum of the object with respect to origin, when it reaches at the maximum height, will be [g is acceleration due to gravity] **[NCERT : P L-106 | Jan. 24, 2025 (I)]**

(a) $\frac{mv_0^3}{2\sqrt{2}g}$ along negative z-axis

(b) $\frac{mv_0^3}{2\sqrt{2}g}$ along positive z-axis

(c) $\frac{mv_0^3}{4\sqrt{2}g}$ along positive z-axis

(d) $\frac{mv_0^3}{4\sqrt{2}g}$ along negative z-axis

48. A circular disc of radius R metre and mass M kg is rotating around the axis perpendicular to the disk. An external torque is applied to the disc such that = $5t^2 - 8t$, where θ(t) is the angular position of the rotating disc as a function of time t. How much power is delivered by the applied torque, when t = 2s? **[NCERT : P L-119 | Jan. 23, 2025 (II)]**

(a) $60\,MR^2$ (b) $72\,MR^2$
(c) $108\,MR^2$ (d) $8\,MR^2$

49. The torque due to the force $(2\hat{i}+\hat{j}+2\hat{k})$ about the origin, acting on a particle whose position vector is $(\hat{i}+\hat{j}+\hat{k})$, would be **[NCERT : P L-106 | Jan. 22, 2025 (II)]**

(a) $\hat{i}-\hat{j}+\hat{k}$ (b) $\hat{i}+\hat{k}$
(c) $\hat{i}-\hat{k}$ (d) $\hat{j}-\hat{k}$

50. The position vectors of two 1 kg particles, (A) and (B), are given by **[NCERT : P L-106 | Jan. 22, 2025 (I)]**
$\vec{r}_A = \left(\alpha_1 t^2\hat{i}+\alpha_2 t\hat{j}+\alpha_3 t\hat{k}\right)$m and
$\vec{r}_B = \left(\beta_1 t\hat{i}+\beta_2 t^2\hat{j}+\beta_3 t\hat{k}\right)$m, respectively;
($\alpha_1 = 1$ m/s², $\alpha_2 = 3n$ m/s, $\alpha_3 = 2$ m/s, $\beta_1 = 2$ m/s, $\beta_2 = -1$ m/s², $\beta_3 = 4p$ m/s), where t is time, n and p are constants, At t = 1s, $|\vec{V}_A| = |\vec{V}_B|$ and velocities $\vec{V}_A$ and $\vec{V}_B$ of the particles are orthogonal to each other. At t = 1 s, the magnitude of angular momentum of particle (A) with respect to the position of particle (B) is $\sqrt{L}$ kgm²s⁻¹. The value of L is ______.

51. A heavy iron bar, of weight W is having its one end on the ground and the other on the shoulder of a person. The bar makes an angle θ with the horizontal. The weight experienced by the person is: **[NCERT : P L-106, 109 | April 9, 2024 (I)]**

(a) $\frac{W}{2}$ (b) W (c) W cosθ (d) W sinθ

52. A thin circular disc of mass M and radius R is rotating in a horizontal plane about an axis passing through its centre and perpendicular to its plane with angular velocity ω. If another disc of same dimensions but of mass $M/2$ is placed gently on the first disc co-axially, then the new angular velocity of the system is : **[NCERT : P L-122 | April 8, 2024 (II)]**

(a) $\frac{4}{5}\omega$ (b) $\frac{5}{4}\omega$ (c) $\frac{2}{3}\omega$ (d) $\frac{3}{2}\omega$

53. If the radius of earth is reduced to three-fourth of its present value without change in its mass then value of duration of the day of earth will be ______ hours 30 minutes. **[NCERT : P L-122 | April 6, 2024 (I)]**

54. A body of mass 'm' is projected with a speed 'u' making an angle of 45° with the ground. The angular momentum of the body about the point of projection, at the highest point is expressed as $\frac{\sqrt{2}\,mu^3}{Xg}$. The value of 'X' is ____. **[NCERT : P L-106 | Jan. 31, 2024 (II)]**

55. A particle of mass m is projected with a velocity 'u' making an angle of 30° with the horizontal. The magnitude of angular momentum of the projectile about the point of projection when the particle is at its maximum height h is: **[NCERT : P L-106 | Jan. 30, 2024 (I)]**

(a) $\frac{\sqrt{3}}{16}\frac{mu^3}{g}$ (b) $\frac{\sqrt{3}}{2}\frac{mu^2}{g}$ (c) $\frac{mu^3}{\sqrt{2}g}$ (d) zero

56. A body of mass 5 kg moving with a uniform speed $3\sqrt{2}$ ms⁻¹ in X - Y plane along the line y = x + 4. The angular momentum of the particle about the origin will be _____ kg m² s⁻¹. **[NCERT : P L-106 | Jan. 29, 2024 (II)]**

57. A heavy iron bar of weight 12 kg is having its one end on the ground and the other on the shoulder of a man. The rod makes an angle 60° with the horizontal, the weight experienced by the man is : **[NCERT : P L-106, 109 | Jan. 27, 2024 (II)]**

(a) 6 kg (b) 12 kg (c) 3 kg (d) $6\sqrt{3}$kg

58. A light rope is wound around a hollow cylinder of mass 5 kg and radius 70 cm. The rope is pulled with a force of 52.5 N. The angular acceleration of the cylinder will be _____ rad s⁻². **[NCERT : P L-119 | April 13, 2023 (II)]**

59. A force of $-P\hat{k}$ acts on the origin of the coordinate system. The torque about the point (2, –3) is $P(a\hat{i}+b\hat{j})$, the ratio of $\frac{a}{b}$ is $\frac{x}{2}$. The value of x is **[NCERT : P L-106 | April 10, 2023 (II)]**

60. A particle of mass 100 g is projected at time t = 0 with a speed 20 ms⁻¹ at an angle 45° to the horizontal as given in the figure. The magnitude of the angular momentum of the particle about the starting point at time t = 2s is found to be $\sqrt{K}$ kg m²/s. The value of K is ________.

(Take g = 10 ms^{-2}) **[NCERT : P L-106 | Jan. 29, 2023 (II)]**

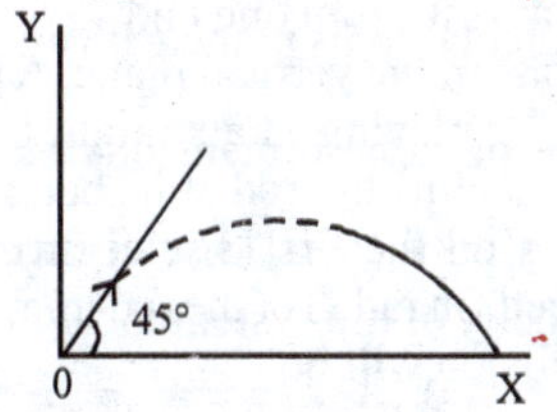

61. An object of mass 8 kg is hanging from one end of a uniform rod CD of mass 2 kg and length 1 m pivoted at its end C on a vertical wall as shown in figure. It is supported by a cable AB such that the system is in equilibrium. The tension in the cable is : (Take g = 10 m/s^2)

[NCERT : P L-106, 109 | Jan. 25, 2023 (I)]

(a) 240 N

(b) 90 N

(c) 300 N

(d) 30 N

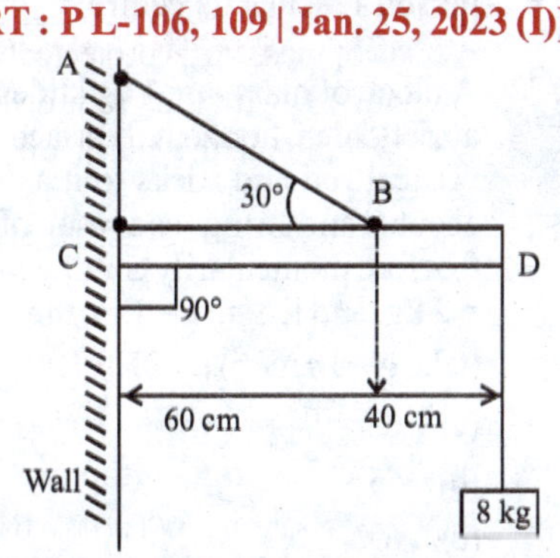

62. The torque of a force $5\hat{i}+3\hat{j}-7\hat{k}$ about the origin is τ. If the force acts on a particle whose position vector is $2\hat{i}+2\hat{j}+\hat{k}$, then the value of τ will be :

[NCERT : P L-106 | July 29, 2022 (II)]

(a) $11\hat{i}+19\hat{j}-4\hat{k}$ (b) $-11\hat{i}+9\hat{j}-16\hat{k}$

(c) $-17\hat{i}+19\hat{j}-4\hat{k}$ (d) $17\hat{i}+9\hat{j}+16\hat{k}$

63. A spherical shell of 1 kg mass and radius R is rolling with angular speed ω on horizontal plane (as shown in figure). The magnitude of angular momentum of the shell about the origin O is $\frac{a}{3}R^2\omega$. The value of a will be :

[NCERT : P L-121 | June 29, 2022 (I)]

(a) 2

(b) 3

(c) 5

(d) 4

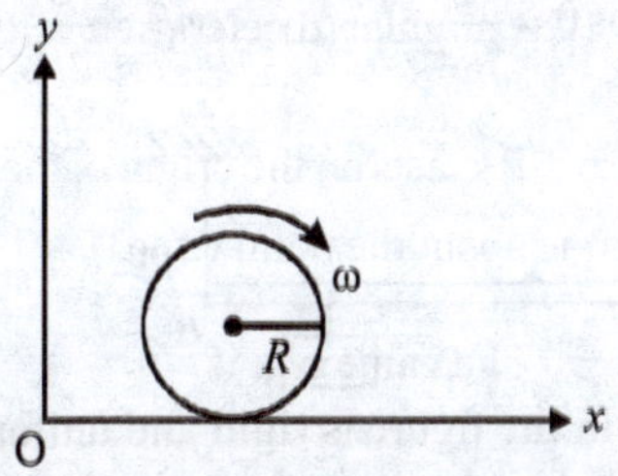

64. The position vector of 1 kg object is $\vec{r}=(3\hat{i}-\hat{j})$ m and its velocity $\vec{v}=(3\hat{j}+\hat{k})\text{ms}^{-1}$. The magnitude of its angular momentum is $\sqrt{x}$ Ns where x is ________.

[NCERT : P L-106 | June 28, 2022 (I)]

65. A $\sqrt{34}$ m long ladder weighing 10 kg leans on a frictionless wall. Its feet rest on the floor 3 m away from the wall as shown in the figure. If F_f and F_w are the reaction forces of the floor and the wall, then ratio of F_w/F_f will be: (Use g = 10 m/s^2)

[NCERT : P L-113, 114 | June 28, 2022 (II)]

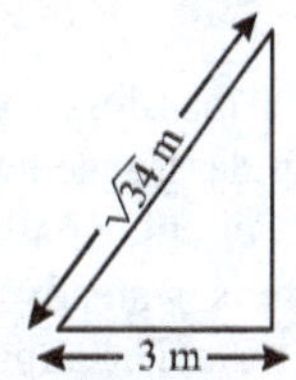

(a) $\frac{6}{\sqrt{110}}$ (b) $\frac{3}{\sqrt{113}}$ (c) $\frac{3}{\sqrt{109}}$ (d) $\frac{2}{\sqrt{109}}$

66. A uniform disc with mass M = 4 kg and radius R = 10 cm is mounted on a fixed horizontal axle as shown in figure. A block with mass m = 2 kg hangs from a massless cord that is wrapped around the rim of the disc. During the fall of the block, the cord does not slip and there is no friction at the axle. The tension in the cord is ________ N.
(Take g = 10 ms^{-2})

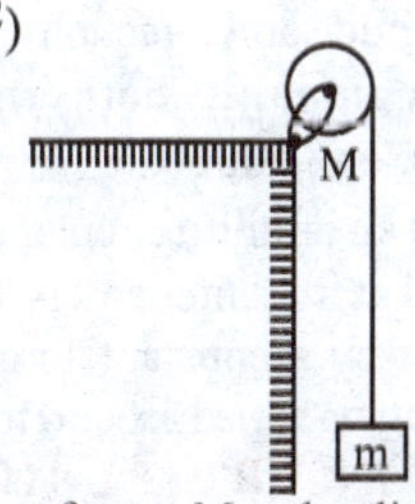

67. A thin circular ring of mass M and radius R is rotating with a constant angular velocity 2 rads^{-1} in a horizontal plane about an axis vertical to its plane and passing through the center of the ring. If two objects each of mass m be attached gently to the opposite ends of a diameter of ring, the ring will then rotate with an angular velocity (in rads^{-1}).

[NCERT : P L-122 | June 26, 2022 (I)]

(a) $\frac{M}{(M+m)}$ (b) $\frac{(M+2m)}{2M}$

(c) $\frac{2M}{(M+2m)}$ (d) $\frac{2(M+2m)}{M}$

68. A particle of mass 'm' is moving in time 't' on a trajectory given by **[NCERT : P L-106 | June 25, 2022 (I)]**

$$\vec{r}=10\,\alpha t^2\hat{i}+5\beta(t-5)\hat{j}$$

Where α and β are dimensional constants.
The angular momentum of the particle becomes the same as it was for $t = 0$ at time t = ________ seconds.

69. A metre scale is balanced on a knife edge at its centre. When two coins, each of mass 10 g are put one on the top of the other at the 10.0 cm mark the scale is found to be balanced at 40.0 cm mark. The mass of the metre scale is found to be $x \times 10^{-2}$ kg. The value of x is

70. Angular momentum of a single particle moving with constant speed along circular path : **[NCERT : P L-106 | Aug. 31, 2021 (I)]**

(a) changes in magnitude but remains same in the direction
(b) remains same in magnitude and direction
(c) remains same in magnitude but changes in the direction
(d) is zero

71. A mass M hangs on a massless rod of length l which rotates at a constant angular frequency. The mass M moves with steady speed in a circular path of constant radius. Assume that the system is in steady circular motion with constant angular velocity ω. The angular momentum of M about point A is L_A which lies in the positive z direction and the angular momentum of M about point B is L_B. The correct statement for this system is : **[NCERT : P L-106 | March 17, 2021 (I)]**

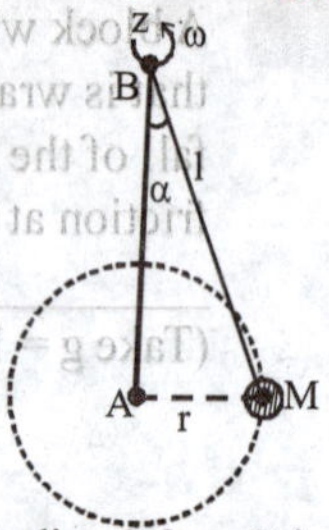

(a) L_A is constant, both in magnitude and direction
(b) L_B is constant in direction with varying magnitude
(c) L_A and L_B are both constant in magnitude and direction
(d) L_B is constant, both in magnitude and direction

72. Consider a 20 kg uniform circular disc of radius 0.2 m. It is pin supported at its center and is at rest initially. The disc is acted upon by a constant force F = 20 N through a massless string wrapped around its periphery as shown in the figure. **[NCERT : P L-119 | March 16, 2021 (I)]**

Suppose the disc makes n number of revolutions to attain an angular speed of 50 rad s^{-1}. The value of n, to the nearest integer, is ______. [Given : In one complete revolution, the disk rotates by 6.28 rad]

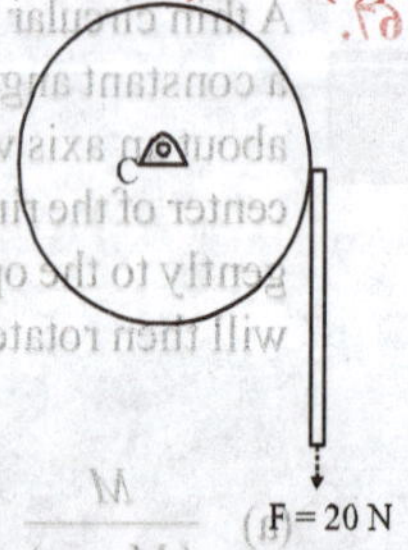

73. A force $\vec{F} = 4\hat{i} + 3\hat{j} + 4\hat{k}$ is applied on an intersection point of $x = 2$ plane and x-axis. The magnitude of torque of this force about a point (2, 3, 4) is ______. (Round off to the Nearest Integer). **[NCERT : P L-106 | March 16, 2021 (II)]**

74. Four point masses, each of mass m, are fixed at the corners of a square of side l. The square is rotating with angular frequency ω, about an axis passing through one of the corners of the square and parallel to its diagonal, as shown in the figure. The angular momentum of the square about this axis is : **[NCERT : P L-119 | Sep. 06, 2020 (I)]**

(a) $ml^2\omega$
(b) $4\,ml^2\omega$
(c) $3\,ml^2\omega$
(d) $2\,ml^2\omega$

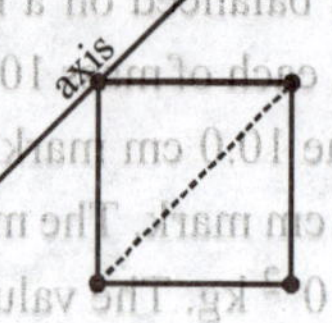

75. A thin rod of mass 0.9 kg and length 1 m is suspended, at rest, from one end so that it can freely oscillate in the vertical plane. A particle of mass 0.1 kg moving in a straight line with velocity 80 m/s hits the rod at its bottom most point and sticks to it (see figure). The angular speed (in rad/s) of the rod immediately after the collision will be ______. **[NCERT : P L-119 | NA Sep. 05, 2020 (II)]**

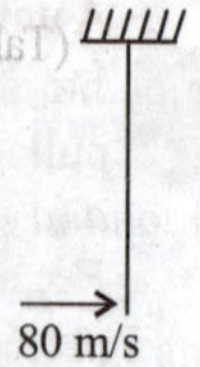

76. A person of 80 kg mass is standing on the rim of a circular platform of mass 200 kg rotating about its axis at 5 revolutions per minute (rpm). The person now starts moving towards the centre of the platform. What will be the rotational speed (in rpm) of the platform when the person reaches its centre ______. **[NCERT : P L-122 | NA Sep. 03, 2020 (I)]**

77. A block of mass $m = 1$ kg slides with velocity $v = 6$ m/s on a frictionless horizontal surface and collides with a uniform vertical rod and sticks to it as shown. The rod is pivoted about O and swings as a result of the collision making angle θ before momentarily coming to rest. If the rod has mass $M = 2$ kg, and length $l = 1$ m, the value of θ is approximately: (take $g = 10\ m/s^2$) **[NCERT : P L-122 | Sep. 03, 2020 (I)]**

(a) 63°
(b) 55°
(c) 69°
(d) 49°

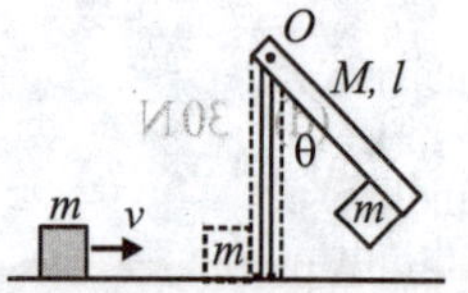

78. A uniform rod of length 'l' is pivoted at one of its ends on a vertical shaft of negligible radius. When the shaft rotates at angular speed ω the rod makes an angle θ with it (see figure). To find θ equate the rate of change of angular momentum (direction going into the paper) $\frac{ml^2}{12}\omega^2 \sin\theta\cos\theta$ about the centre of mass (CM) to the torque provided by the horizontal and vertical forces F_H and F_V about the CM. The value of θ is then such that : **[NCERT : P L-106 | Sep. 03, 2020 (II)]**

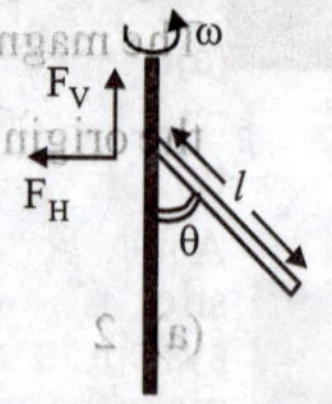

(a) $\cos\theta = \frac{2g}{3l\omega^2}$ (b) $\cos\theta = \frac{g}{2l\omega^2}$

(c) $\cos\theta = \frac{g}{l\omega^2}$ (d) $\cos\theta = \frac{3g}{2l\omega^2}$

79.

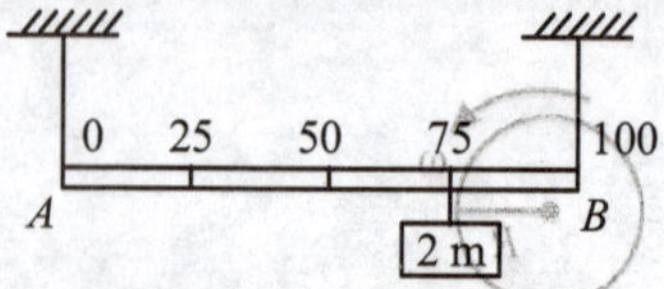

Shown in the figure is rigid and uniform one meter long rod AB held in horizontal position by two strings tied to its ends and attached to the ceiling. The rod is of mass 'm' and has another weight of mass 2 m hung at a distance of 75 cm from A. The tension in the string at A is : **[NCERT : P L-106, 109 | Sep. 02, 2020 (I)]**

(a) 0.5 mg (b) 2 mg (c) 0.75 mg (d) 1 mg

80. A uniform cylinder of mass M and radius R is to be pulled over a step of height a $(a < R)$ by applying a force F at its centre 'O' perpendicular to the plane through the axes of the cylinder on the edge of the step (see figure). The minimum value of F required is :

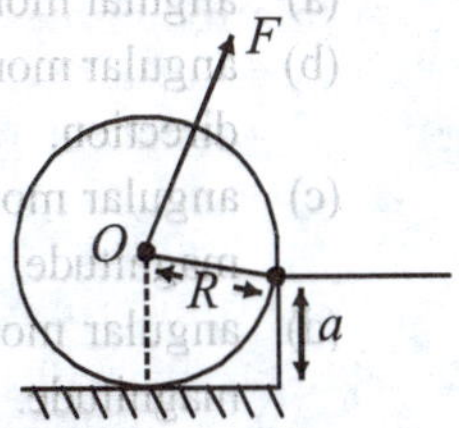

[NCERT : P L-106 | Sep. 02, 2020 (I)]

(a) $Mg\sqrt{1-\left(\frac{R-a}{R}\right)^2}$ (b) $Mg\sqrt{\left(\frac{R}{R-a}\right)^2-1}$

(c) $Mg\frac{a}{R}$ (d) $Mg\sqrt{1-\frac{a^2}{R^2}}$

81. Consider a uniform rod of mass $M = 4m$ and length l pivoted about its centre. A mass m moving with velocity v making angle $\theta = \frac{\pi}{4}$ to the rod's longitudinal axis collides with one end of the rod and sticks to it. The angular speed of the rod-mass system just after the collision is:

[NCERT : P L-122 | 8 Jan. 2020 I]

(a) $\frac{3}{7\sqrt{2}}\frac{v}{l}$ (b) $\frac{3}{7}\frac{v}{l}$ (c) $\frac{3\sqrt{2}}{7}\frac{v}{l}$ (d) $\frac{4}{7}\frac{v}{l}$

82. A particle of mass m is moving along a trajectory given by

$x = x_0 + a\cos\omega_1 t$

$y = y_0 + b\sin\omega_2 t$

The torque, acting on the particle about the origin, at $t = 0$ is: **[NCERT : P L-106 | 10 Apr. 2019 I]**

(a) $m(-x_0 b + y_0 a)\omega_1^2\hat{k}$ (b) $+my_0 a\omega_1^2\hat{k}$

(c) zero (d) $-m(x_0 b\omega_2^2 - y_0 a\omega_1^2)\hat{k}$

83. The time dependence of the position of a particle of mass $m = 2$ kg is given by $\vec{r}(t) = 2t\hat{i} - 3t^2\hat{j}$. Its angular momentum, with respect to the origin, at time t = 2s is :

[NCERT : P L-106 | 10 Apr. 2019 II]

(a) $48(\hat{i}+\hat{j})$ (b) $36\hat{k}$

(c) $-34(\hat{k}-\hat{i})$ (d) $-48\hat{k}$

84. A metal coin of mass 5 g and radius 1 cm is fixed to a thin stick AB of negligible mass as shown in the figure The system is initially at rest. The constant torque, that will make the system rotate about AB at 25 rotations per second in 5s, is close to : **[NCERT : P L-119 | 10 Apr. 2019 II]**

(a) 4.0×10^{-6} Nm

(b) 1.6×10^{-5} Nm

(c) 7.9×10^{-6} Nm

(d) 2.0×10^{-5} Nm

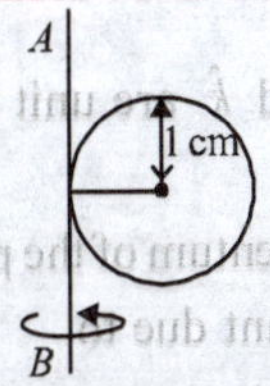

85. A particle of mass 20 g is released with an initial velocity 5 m/s along the curve from the point A, as shown in the figure. The point A is at height h from point B. The particle slides along the frictionless surface. When the particle reaches point B, its angular momentum about O will be : (Take g = 10 m/s²) **[NCERT : P L-106 | 12 Jan. 2019 II]**

(a) 2 kg-m²/s

(b) 8 kg-m²/s

(c) 6 kg-m²/s

(d) 3 kg-m²/s

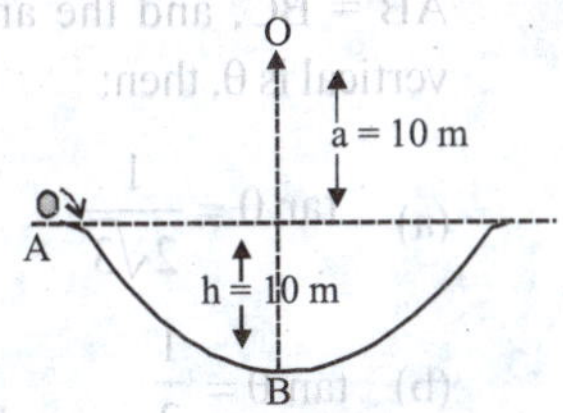

86. A slab is subjected to two' forces $\vec{F_1}$ and $\vec{F_2}$ of same magnitude F as shown in the figure. Force $\vec{F_2}$ is in XY-plane while force F_1 acts along z-axis at the point $(2\vec{i}+3\vec{j})$. The moment of these forces about point O will be : **[NCERT : P L-106 | [11 Jan. 2019 I]**

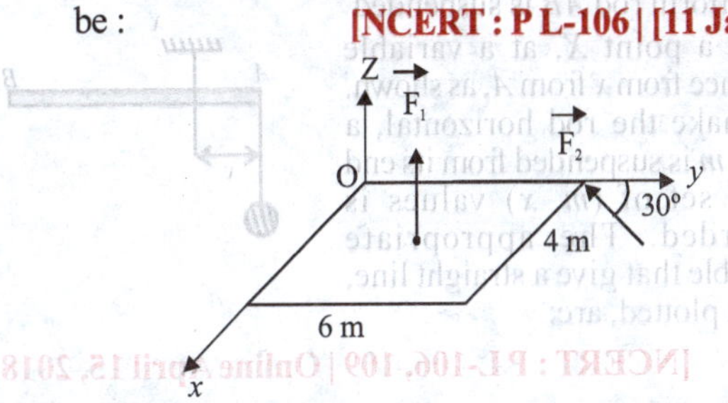

(a) $(3\hat{i}-2\hat{j}+3\hat{k})F$ (b) $(3\hat{i}-2\hat{j}-3\hat{k})F$

(c) $(3\hat{i}+2\hat{j}-3\hat{k})F$ (d) $(3\hat{i}+2\hat{j}+3\hat{k})F$

87. The magnitude of torque on a particle of mass 1 kg is 2.5 Nm about the origin. If the force acting on it is 1 N, and the distance of the particle from the origin is 5m, the angle between the force and the position vector is (in radians):

[NCERT : P L-106 | 11 Jan. 2019 II]

(a) $\frac{\pi}{6}$ (b) $\frac{\pi}{3}$ (c) $\frac{\pi}{8}$ (d) $\frac{\pi}{4}$

88. To mop-clean a floor, a cleaning machine presses a circular mop of radius R vertically down with a total force F and rotates it with a constant angular speed about its axis. If the force F is distributed uniformly over the mop and if coefficient of friction between the mop and the floor is μ, the torque, applied by the machine on the mop is:

[NCERT : P L-106 | 10 Jan. 2019 I]

(a) μ FR/3 (b) μ FR/6 (c) μ FR/2 (d) $\frac{2}{3}$μ FR

89. A rigid massless rod of length $3l$ has two masses attached at each end as shown in the figure. The rod is pivoted at point P on the horizontal axis (see figure). When released from initial horizontal position, its instantaneous angular acceleration will be:

NCERT : P L-119 | 10 Jan. 2019 II]

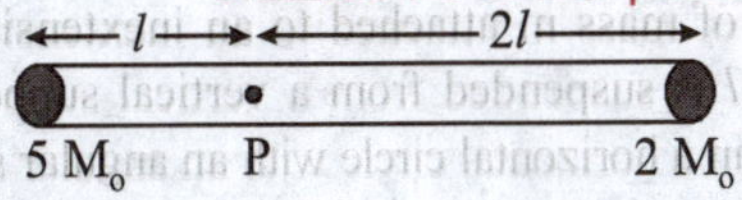

(a) $\frac{g}{13l}$ (b) $\frac{g}{3l}$ (c) $\frac{g}{2l}$ (d) $\frac{7g}{3l}$

90. An L-shaped object, made of thin rods of uniform mass density, is suspended with a string as shown in figure. If AB = BC, and the angle made by AB with downward vertical is θ, then: **[NCERT : P L-106, 109 | 9 Jan. 2019 I]**

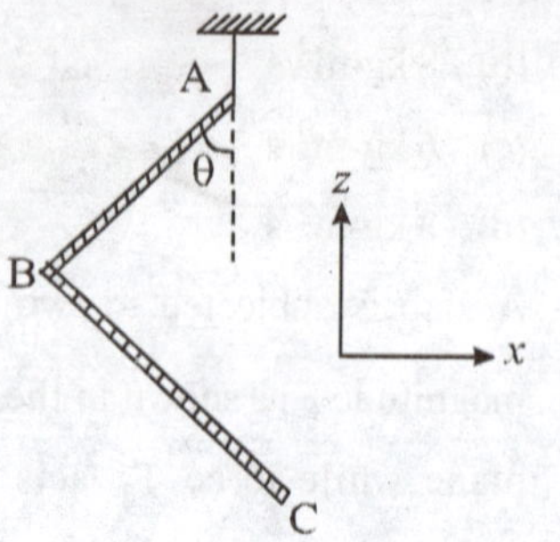

(a) $\tan\theta = \frac{1}{2\sqrt{3}}$

(b) $\tan\theta = \frac{1}{2}$

(c) $\tan\theta = \frac{2}{\sqrt{3}}$

(d) $\tan\theta = \frac{1}{3}$

91. A uniform rod *AB* is suspended from a point *X*, at a variable distance from *x* from *A*, as shown. To make the rod horizontal, a mass *m* is suspended from its end *A*. A set of (*m*, *x*) values is recorded. The appropriate variable that give a straight line, when plotted, are:

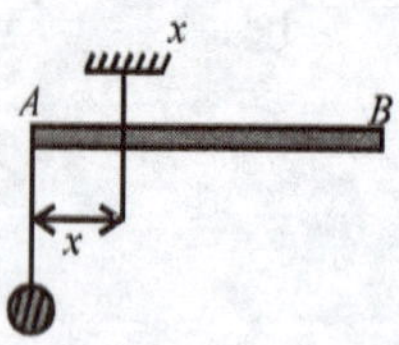

[NCERT : P L-106, 109 | Online April 15, 2018]

(a) $m, \frac{1}{x}$ (b) $m, \frac{1}{x^2}$ (c) m, x (d) m, x^2

92. A particle of mass *m* is moving along the side of a square of side '*a*', with a uniform speed *v* in the *x*-*y* plane as shown in the figure: **[NCERT : P L-106 | 2016]**

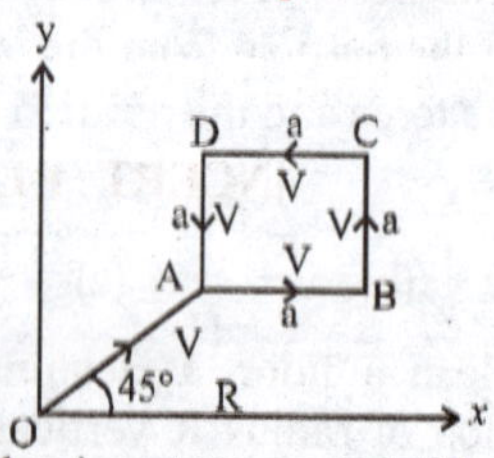

Which of the following statements is false for the angular momentum $\vec{L}$ about the origin?

(a) $\vec{L} = mv\left[\frac{R}{\sqrt{2}} + a\right]\hat{k}$ when the particle is moving from B to C.

(b) $\vec{L} = \frac{mv}{\sqrt{2}}R\hat{k}$ when the particle is moving from D to A.

(c) $\vec{L} = -\frac{mv}{\sqrt{2}}R\hat{k}$ when the particle is moving from A to B.

(d) $\vec{L} = mv\left[\frac{R}{\sqrt{2}} - a\right]\hat{k}$ when the particle is moving from C to D.

93. A bob of mass m attached to an inextensible string of length *l* is suspended from a vertical support. The bob rotates in a horizontal circle with an angular speed ω rad/s about the vertical. About the point of suspension: **[NCERT : P L-106 | 2014]**

(a) angular momentum is conserved.

(b) angular momentum changes in magnitude but not in direction.

(c) angular momentum changes in direction but not in magnitude.

(d) angular momentum changes both in direction and magnitude.

94. A ball of mass 160 g is thrown up at an angle of 60° to the horizontal at a speed of 10 ms^{-1}. The angular momentum of the ball at the highest point of the trajectory with respect to the point from which the ball is thrown is nearly ($g = 10\ ms^{-2}$)

[NCERT : P L-106 | Online April 19, 2014]

(a) 1.73 kg m^2/s (b) 3.0 kg m^2/s

(c) 3.46 kg m^2/s (d) 6.0 kg m^2/s

95. A particle of mass 2 kg is moving such that at time *t*, its position, in meter, is given by $\vec{r}(t) = 5\hat{i} - 2t^2\hat{j}$. The angular momentum of the particle at *t* = 2*s* about the origin in kg m^{-2} s^{-1} is : **[NCERT : P L-106 | Online April 23, 2013]**

(a) $-80\hat{k}$ (b) $(10\hat{i} - 16\hat{j})$

(c) $-40\hat{k}$ (d) $40\hat{k}$

96. A thin horizontal circular disc is rotating about a vertical axis passing through its centre. An insect is at rest at a point near the rim of the disc. The insect now moves along a diameter of the disc to reach its other end. During the journey of the insect, the angular speed of the disc.

[NCERT : P L-122 | 2011]

(a) continuously decreases

(b) continuously increases

(c) first increases and then decreases

(d) remains unchanged

97. A small particle of mass m is projected at an angle θ with the x-axis with an initial velocity v_0 in the x-y plane as shown in the figure. At a time $t < \frac{v_0 \sin\theta}{g}$, the angular momentum of the particle is **[NCERT : P L-106 | 2010]**

(a) $-mg\,v_0 t^2 \cos\theta\,\hat{j}$

(b) $mg\,v_0 t\cos\theta\hat{k}$

(c) $-\frac{1}{2}mg\,v_0 t^2\cos\theta\,\hat{k}$

(d) $\frac{1}{2}mg\,v_0 t^2\cos\theta\hat{i}$

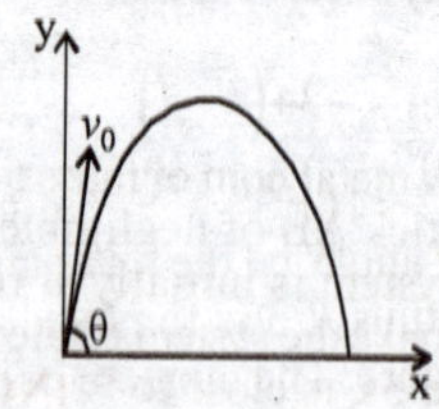

where $\hat{i}, \hat{j}$ and $\hat{k}$ are unit vectors along *x*, *y* and *z*-axis respectively.

98. Angular momentum of the particle rotating with a central force is constant due to **[NCERT : P L-108 | 2007]**

(a) constant torque

(b) constant force

(c) constant linear momentum

(d) zero torque

99. A force of $-F\hat{k}$ acts on O, the origin of the coordinate system. The torque about the point (1, –1) is **[NCERT : P L-106 | 2006]**

(a) $F(\hat{i}-\hat{j})$

(b) $-F(\hat{i}+\hat{j})$

(c) $F(\hat{i}+\hat{j})$

(d) $-F(\hat{i}-\hat{j})$

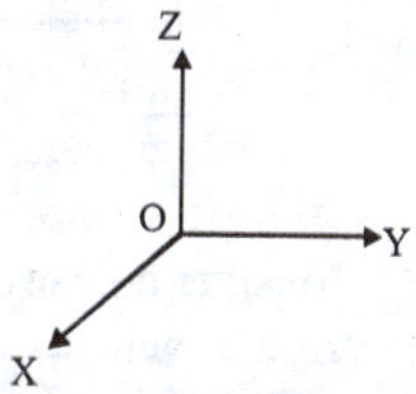

100. A solid sphere is rotating in free space. If the radius of the sphere is increased keeping mass same, which one of the following will not be affected ? **[NCERT : P L-108 | 2004]**

(a) Angular velocity (b) Angular momentum

(c) Moment of inertia (d) Rotational kinetic energy

101. Let $\vec{F}$ be the force acting on a particle having position vector $\vec{r}$, and $\vec{T}$ be the torque of this force about the origin. Then **[NCERT : P L-106 | 2003]**

(a) $\vec{r}.\vec{T}=0$ and $\vec{F}.\vec{T}\neq 0$ (b) $\vec{r}.\vec{T}\neq 0$ and $\vec{F}.\vec{T}=0$

(c) $\vec{r}.\vec{T}\neq 0$ and $\vec{F}.\vec{T}\neq 0$ (d) $\vec{r}.\vec{T}=0$ and $\vec{F}.\vec{T}=0$

102. A particle of mass m moves along line PC with velocity v as shown. What is the angular momentum of the particle about P? **[NCERT : P L-106 | 2002]**

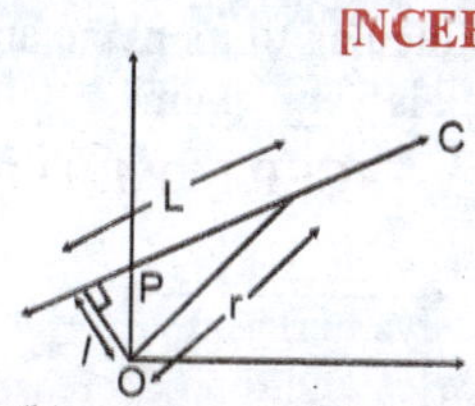

(a) mvL (b) mvl (c) mvr (d) zero.

Topic 4 - Moment of Inertia and Rotational K.E.

103. A rod of linear mass density 'λ' and length 'L' is bent to form a ring of radius 'R'. Moment of inertia of ring about any of its diameter is : **[NCERT : P L-115, 116 | April 8, 2025 (II)]**

(a) $\frac{\lambda L^3}{16\pi^2}$ (b) $\frac{\lambda L^3}{12}$ (c) $\frac{\lambda L^3}{4\pi^2}$ (d) $\frac{\lambda L^3}{8\pi^2}$

104. M and R be the mass and radius of a disc. A small disc of radius $R/3$ is removed from the bigger disc as shown in figure. The moment of inertia of remaining part of bigger disc about an axis AB passing through the centre O and perpendicular to the plane of disc is $\frac{4}{x}$ MR^2. The value of x is_______. **[NCERT : P L-115, 116 | April 7, 2025 (II)]**

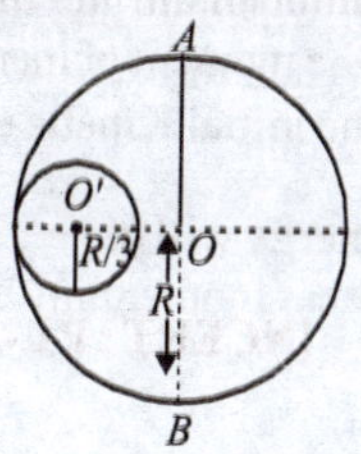

105. A, B and C are disc, solid sphere and spherical shell respectively with same radii and masses. These masses are placed as shown in figure.

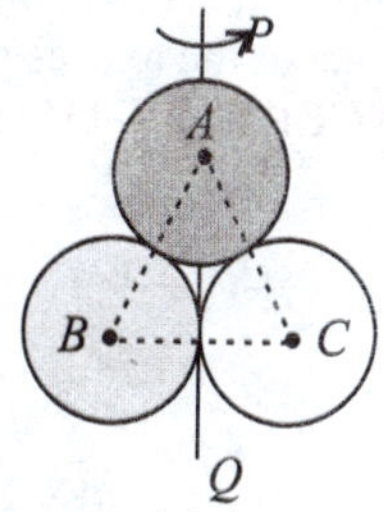

The moment of inertia of the given system about PQ is $\frac{x}{15}$ I, where I is the moment of inertia of the disc about its diameter. The value of x is_____. **[NCERT : P L-115, 116 | April 7, 2025 (I)]**

106. A solid sphere with uniform density and radius R is rotating initially with constant angular velocity (ω_1) about its diameter. After some time during the rotation its starts loosing mass at a uniform rate, with no change in its shape. The angular velocity of the sphere when its radius become $\frac{R}{2}$ is $x\omega_1$. The value of x is _______. **[NCERT : P L-115, 116, 122 | April 4, 2025 (II)]**

107. The moment of inertia of a circular ring of mass M and diameter r about a tangential axis lying in the plane of the ring is : **[NCERT : P L-115, 116 | April 2, 2025 (II)]**

(a) $\frac{1}{2}$Mr^2 (b) $\frac{3}{8}$Mr^2 (c) $\frac{3}{2}$Mr^2 (d) 2 Mr^2

108. Moment of inertia of a rod of mass 'M' and length 'L' about an axis passing through its center and normal to its length is 'α'. Now the rod is cut into two equal parts and these parts are joined symmetrically to form a cross shape. Moment of inertia of cross about an axis passing through its center and normal to plane containing cross is : **[NCERT : P L-115, 116 | April 2, 2025 (I)]**

(a) α (b) α/4 (c) α/8 (d) α/2

109. The moment of inertia of a solid disc rotating along its diameter is 2.5 times higher than the moment of inertia of a ring rotating in similar way. The moment of inertia of a solid sphere which has same radius as the disc and rotating in similar way, is n times higher than the moment of inertia of the given ring. Here, n = _______.
(Consider all the bodies have equal masses.) **[NCERT : P L-115, 116 | Jan. 28, 2025 (I)]**

110. Two iron solid discs of negligible thickness have radii R_1 and R_2 and moment of inertia I_1 and I_2, respectively. For $R_2 = 2R_1$, the ratio of I_1 and I_2 would be $\frac{1}{x}$, where x = _______. **[NCERT : P L-115, 116 | Jan. 28, 2025 (I)]**

111. A uniform circular disc of radius 'R' and mass 'M' is rotating about an axis perpendicular to its plane and passing through its centre. A small circular part of radius R/2 is removed from the original disc as shown in the figure. Find the moment of inertia of the remaining part of the original disc about the axis as given above.
[NCERT : P L-115, 116 | Jan. 22, 2025 (I)]

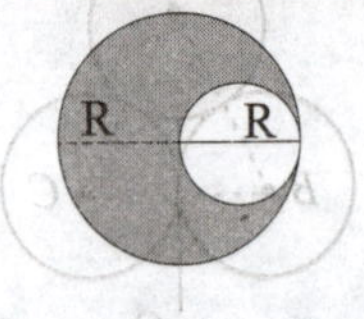

(a) $\frac{7}{32}MR^2$ (b) $\frac{9}{32}MR^2$
(c) $\frac{17}{32}MR^2$ (d) $\frac{13}{32}MR^2$

112. A string is wrapped around the rim of a wheel of moment of inertia 0.40 kgm^2 and radius 10 cm. The wheel is free to rotate about its axis. Initially the wheel is at rest. The string is now pulled by a force of 40 N. The angular velocity of the wheel after 10 s is x rad/s, where x is ______.
[April 9, 2024 (I)]

113. Three balls of masses 2kg, 4kg and 6kg respectively are arranged at centre of the edges of an equilateral triangle of side 2 m. The moment of inertia of the system about an axis through the centroid and perpendicular to the plane of triangle, will be ______ kg m^2.
[NCERT : P L-114 | April 6, 2024 (II)]

114. Ratio of radius of gyration of a hollow sphere to that of a solid cylinder of equal mass, for moment of Inertia about their diameter axis AB as shown in figure is $\sqrt{\frac{8}{x}}$. The value of x is: **[April 5, 2024 (I)]**

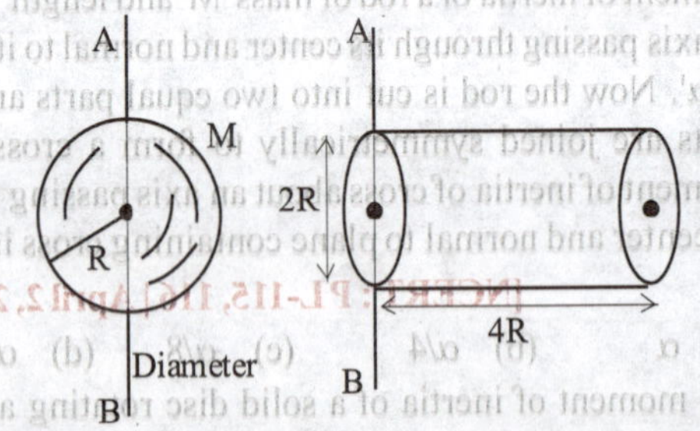

(a) 34 (b) 17 (c) 67 (d) 51

115. A uniform rod AB of mass 2 kg and length 30 cm at rest on a smooth horizontal surface. An impulse of force 0.2 Ns is applied to end B. The time taken by the rod to turn through at right angles will be $\frac{\pi}{x}$ s, where x = ______.
[NCERT : P L-116 | Feb. 1, 2024 (II)]

116. Two identical spheres each of mass 2 kg and radius 50 cm are fixed at the ends of a light rod so that the separation between the centers is 150 cm. Then, moment of inertia of the system about an axis perpendicular to the rod and passing through its middle point is $\frac{x}{20}$ kg m^2, where the value of x is ____. **[NCERT : P L-116 | Jan. 31, 2024 (II)]**

117.

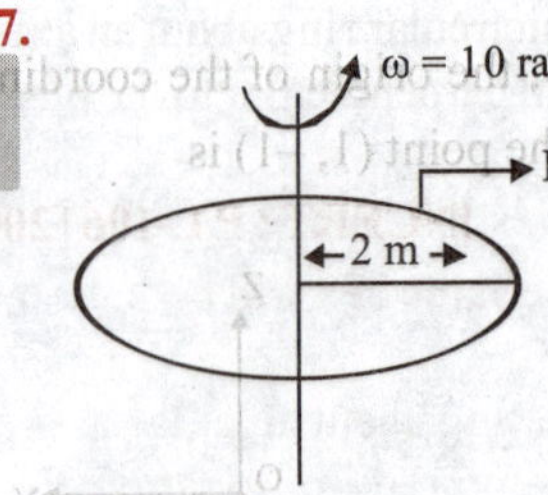

Consider a Disc of mass 5 kg, radius 2 m, rotating with angular velocity of 10 rad/s about an axis perpendicular to the plane of rotation. An identical disc is kept gently over the rotating disc along the same axis. The energy dissipated so that the discs continue to rotate together without slipping is ______ J.
[NCERT : P L-122 | Jan. 30, 2024 (I)]

118. Two discs of moment of inertia $I_1 = 4$ kg m^2 and $I_2 = 2kg\ m^2$, about their central axes & normal to their planes, rotating with angular speeds 10 rad/s & 4 rad/s respectively are brought into contact face to face with their axes of rotation coincident. The loss in kinetic energy of the system in the process is ______ *J*.
[NCERT : P L-122 | Jan. 30, 2024 (II)]

119. Four particles each of mass 1 kg are placed at four corners of a square of side 2 m. Moment of inertia of system about an axis perpendicular to its plane and passing through one of its vertex is _____ kgm^2.
[NCERT : P L-114 | Jan. 27, 2024 (I)]

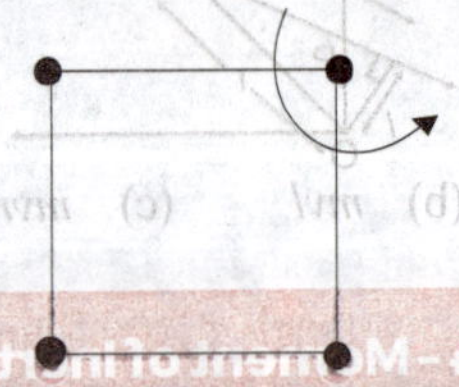

120. A solid sphere and a solid cylinder of same mass and radius are rolling on a horizontal surface without slipping. The ratio of their radius of gyrations respectively (k_{sph} : k_{cyl}) is $2 : \sqrt{x}$, then value of x is ______.
[NCERT : P L-115, 116 | April 15, 2023 (I)]

121. A solid sphere of mass 500 g and radius 5 cm is rotated about one of its diameter with angular speed of 10 rad s^{-1}. If the moment of inertia of the sphere about its tangent is $x \times 10^{-2}$ times its angular momentum about the diameter. Then the value of x will be ______.
[NCERT : P L-116 | April 11, 2023 (I)]

122. A circular plate is rotating in horizontal plane, about an axis passing through its center and perpendicular to the plate, with an angular velocity ω. A person sits at the center having two dumbbells in his hands. When he stretches out his hands, the moment of inertia of the system becomes triple. If E be the initial Kinetic energy of the system, then final Kinetic energy will be $\frac{E}{x}$. The value of x is
[NCERT : P L-119 | April 11, 2023 (II)]

123. The moment of inertia of semicircular ring about an axis, passing through the center and perpendicular to the plane of ring, is $\frac{1}{x}MR^2$, where R is the radius and M is the mass of semicircular ring. The value of x will be

[NCERT : P L-116 | April 8, 2023 (I)]

124. A ring and a solid sphere rotating about an axis passing through their centers have same radii of gyration. The axis of rotation is perpendicular to plane of ring. The ratio of radius of ring to that of sphere is $\sqrt{\frac{2}{x}}$. The value of x is _____. [NCERT : P L-115, 116 | April 6, 2023 (II)]

125. Moment of inertia of a disc of mass M and radius 'R' about any of its diameter is $\frac{MR^2}{4}$. The moment of inertia of this disc about an axis normal to the disc and passing through a point on its edge will be, $\frac{x}{2}MR^2$. The value of x is _____. [NCERT : P L-116 | Feb. 1, 2023 (II)]

126. Two discs of same mass and different radii are made of different materials such that their thicknesses are 1 cm and 0.5 cm respectively. The densities of materials are in the ratio 3:5. The moment of inertia of these discs respectively about their diameters will be in the ratio of $\frac{x}{6}$. The value of x is _____.

[NCERT : P L-114 | Jan. 31, 2023 (II)]

127. I_{CM} is moment of inertia of a circular disc about an axis (CM) passing through its center and perpendicular to the plane of disc. I_{AB} is it's moment of inertia about an axis AB perpendicular to plane and parallel to axis CM at a distance $\frac{2}{3}R$ from center. Where R is the radius of the disc. The ratio of I_{AB} and I_{CM} is x : 9. The value of x is ___.

[NCERT : P L-116 | Jan. 25, 2023 (I)]

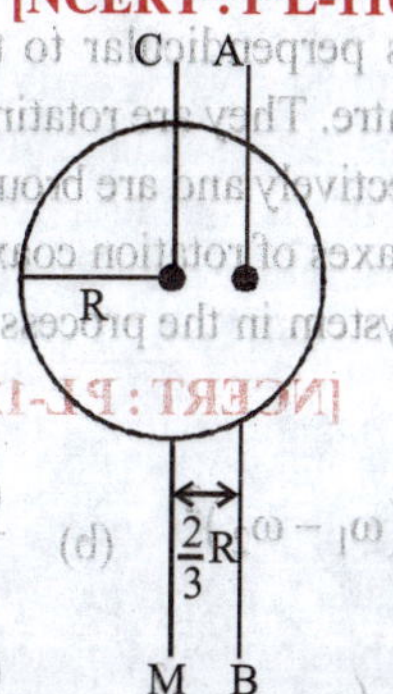

128. If a solid sphere of mass 5 kg and a disc of mass 4 kg have the same radius. Then the ratio of moment of inertia of the disc about a tangent in its plane to the moment of inertia of the sphere about its tangent will be $\frac{x}{7}$. The value of x is _____. [NCERT : P L-116 | Jan. 25, 2023 (II)]

129. Solid sphere A is rotating about an axis PQ. If the radius of the sphere is 5 cm then its radius of gyration about PQ will be $\sqrt{x}$ cm. The value of x is _____ .

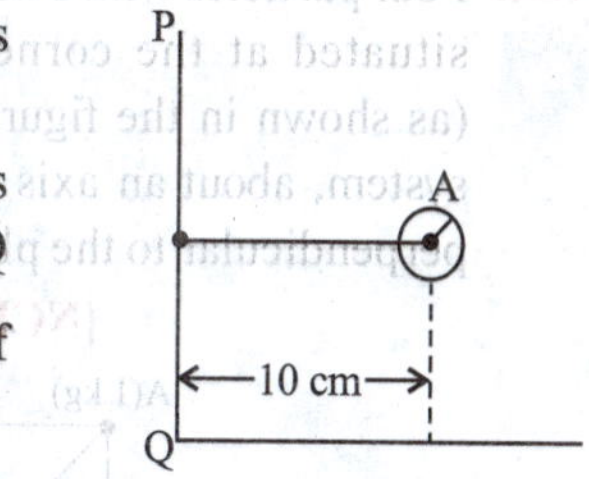

[NCERT : P L-116 | Jan. 24, 2023 (I)]

130. Four identical discs each of mass 'M' and diameter 'a' are arranged in a small plane as shown in figure. If the moment of inertia of the system about OO' is $\frac{x}{4}Ma^2$. Then, the value of x will be _____.

[NCERT : P L-116 | July 28, 2022 (I)]

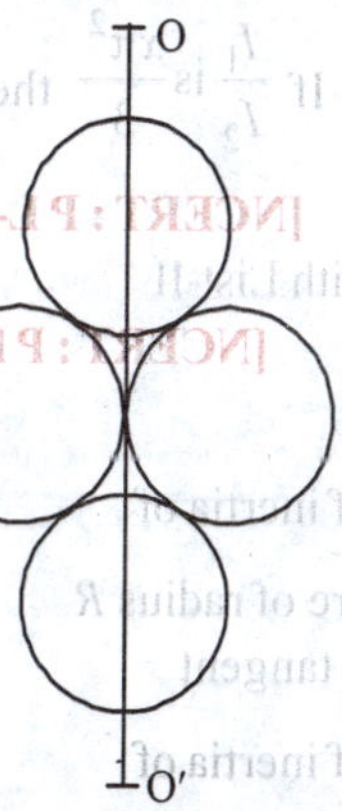

131. A pulley of radius 1.5 m is rotated about its axis by a force $F = (12t - 3t^2)$ N applied tangentially (while t is measured in seconds). If moment of inertia of the pulley about its axis of rotation is 4.5 kg m^2, the number of rotations made by the pulley before its direction of motion is reversed, will be $\frac{K}{\pi}$. The value of K is _____.

[NCERT : P L-119 | July 27, 2022 (I)]

132. A disc of mass 1 kg and radius R is free to rotate about a horizontal axis passing through its centre and perpendicular to the plane of disc. A body of same mass as that of disc is fixed at the highest point of the disc. Now the system is released, when the body comes to the lowest position, its angular speed will be $4\sqrt{\frac{x}{3R}}$ rad s^{-1} where $x =$ _____. $(g = 10 \text{ ms}^{-2})$

[NCERT : P L-116 | July 26, 2022 (I)]

133. The radius of gyration of a cylindrical rod about an axis of rotation perpendicular to its length and passing through the center will be _____ m. Given, the length of the rod is $10\sqrt{3}$ m. [NCERT : P L-116 | July 26, 2022 (II)]

134. Four particles with a mass of 1 kg, 2 kg, 3kg and 4 kg are situated at the corners of a square with side 1 m (as shown in the figure). The moment of inertia of the system, about an axis passing through the point O and perpendicular to the plane of the square, is ____ kgm^2.

[NCERT : P L-114 | June 30, 2022 (I)]

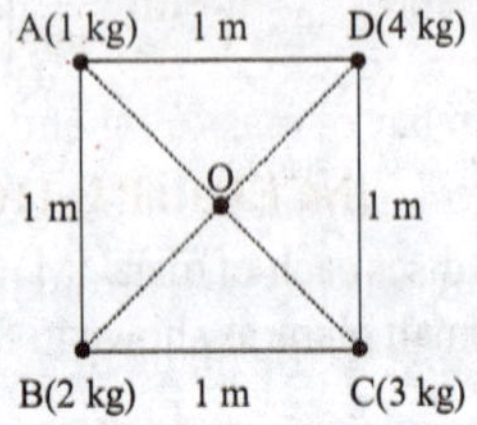

135. The moment of inertia of a uniform thin rod about a perpendicular axis passing through one end is I_1. The same rod is bent into a ring and its moment of inertia about a diameter is I_2. If $\frac{I_1}{I_2}$ is $\frac{x\pi^2}{3}$ then the value of x will be ____.

[NCERT : P L-116 | June 29, 2022 (II)]

136. Match List-I with List-II

[NCERT : P L-116 | June 28, 2022 (I)]

List-I	List-II
A Moment of inertia of solid sphere of radius R about any tangent	I. $\frac{5}{3}MR^2$
B Moment of inertia of hollow sphere of radius (R) about my tangent	II. $\frac{7}{5}MR^2$
C Moment of inertia of circular ring of radius (R) about its diameter.	III. $\frac{1}{4}MR^2$
D Moment of inertia of circular disc of radius (R) about any diameter	IV. $\frac{1}{2}MR^2$

Choose the correct answer from the options given below

(a) A-II, B-I, C-IV, D-III (b) A-I, B-II, C-IV, D-III
(c) A-II, B-I, C-III, D-IV (d) A-I, B-II, C-III, D-IV

137. Given below are two statements : one is labelled as **Assertion A** and the other is labelled as **Reason R**.

Assertion A : Moment of inertia of a circular disc of mass 'M' and radius 'R' about X, Y axes (passing through its plane) and Z-axis which is perpendicular to its plane were found to be I_x, I_y and I_z respectively. The respective radii of gyration about all the three axes will be the same.

Reason R : A rigid body making rotational motion has fixed mass and shape. In the light of the above statements, choose the most appropriate answer from the options given below :

[NCERT : P L-116 | June 25, 2022 (I)]

(a) Both **A** and **R** are correct but **R** is NOT the correct explanation of **A**.
(b) **A** is not correct but **R** is correct.
(c) **A** is correct but **R** is not correct.
(d) Both **A** and **R** are correct and **R** is the correct explanation of **A**.

138. Moment of Inertia (M.I.) of four bodies having same mass 'M' and radius '2R' are as follows:

I_1 = M.I. of solid sphere about its diameter
I_2 = M.I. of solid cylinder about its axis
I_3 = M.I. of solid circular disc about its diameter
I_4 = M.I. of thin circular ring about its diameter

If $2(I_2 + I_3) + I_4 = x.\ I_1$ then the value of x will be ____. **[NCERT : P L-116 | June 25, 2022 (II)]**

139. A 2 kg steel rod of length 0.6 m is clamped on a table vertically at its lower end and is free to rotate in vertical plane. The upper end is pushed so that the rod falls under gravity, Ignoring the friction due to clamping at its lower end, the speed of the free end of rod when it passes through its lowest position is ____ ms^{-1}. (Take $g = 10\ ms^{-2}$)

[NCERT : P L-119 | NA, Sep. 1, 2021 (II)]

140. A system consists of two identical spheres each of mass 1.5 kg and radius 50 cm at the end of light rod. The distance between the centres of the two spheres is 5 m. What will be the moment of inertia of the system about an axis perpendicular to the rod passing through its midpoint?

[NCERT : P L-116 | Aug. 31, 2021 (II)]

(a) $18.75\ kgm^2$ (b) $1.905 \times 10^5\ kgm^2$
(c) $19.05\ kgm^2$ (d) $1.875 \times 10^5\ kgm^2$

141. Moment of inertia of a square plate of side l about the axis passing through one of the corner and perpendicular to the plane of square plate is given by :

[Aug. 27, 2021 (I)]

(a) $\frac{Ml^2}{6}$ (b) Ml^2 (c) $\frac{Ml^2}{12}$ (d) $\frac{2}{3}Ml^2$

142. Two discs have moments of inertia I_1 and I_2 about their respective axes perpendicular to the plane and passing through the centre. They are rotating with angular speeds, ω_1 and ω_2 respectively and are brought into contact face to face with their axes of rotation coaxial. The loss in kinetic energy of the system in the process is given by :

[NCERT : P L-122 | Aug. 27, 2021 (II)]

(a) $\frac{I_1 I_2}{(I_1 + I_2)}(\omega_1 - \omega_2)^2$ (b) $\frac{(I_1 - I_2)^2 \omega_1 \omega_2}{2(I_1 + I_2)}$

(c) $\frac{I_1 I_2}{2(I_1 + I_2)}(\omega_1 - \omega_2)^2$ (d) $\frac{(\omega_1 - \omega_2)^2}{2(I_1 + I_2)}$

143. The solid cylinder of length 80 cm and mass M has a radius of 20 cm. Calculate the density of the material used if the moment of inertia of the cylinder about an axis CD parallel to AB as shown in figure is 2.7 kg m^2.

[NCERT : P L-116 | Aug. 26, 2021 (II)]

(a) 14.9 kg/m^3

(b) $7.5 \times 10^1 \text{ kg/m}^3$

(c) $7.5 \times 10^2 \text{ kg/m}^3$

(d) $1.49 \times 10^2 \text{ kg/m}^3$

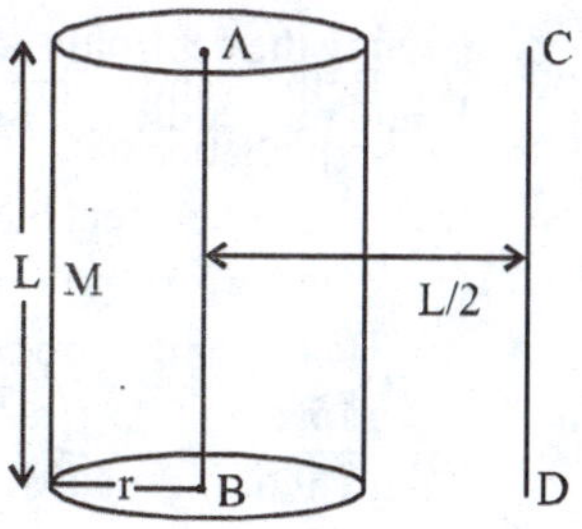

144. The figure shows two solid discs with radius R and r respectively. If mass per unit area is same for both, what is the ratio of MI of bigger disc around axis AB (Which is ⊥ to the plane of the disc and passing through its centre) to MI of smaller disc around one of its diameters lying on its plane? Given 'M' is the mass of the larger disc. (MI stands for moment of inertia) **[NCERT : P L-116 | July 27, 2021 (I)]**

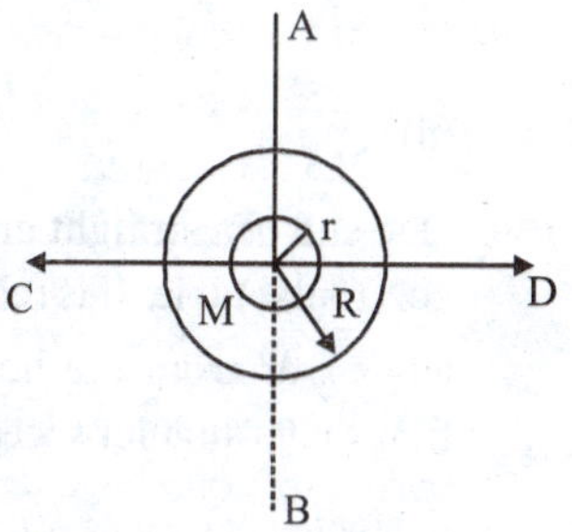

(a) $R^2 : r^2$ (b) $2r^4 : R^4$ (c) $2R^2 : r^2$ (d) $2R^4 : r^4$

145. In the given figure, two wheels P and Q are connected by a belt B. The radius of P is three times as that of Q. In case of same rotational kinetic energy, the ratio of rotational inertias $\left(\frac{I_1}{I_2}\right)$ will be x : 1. The value of x will be _____.

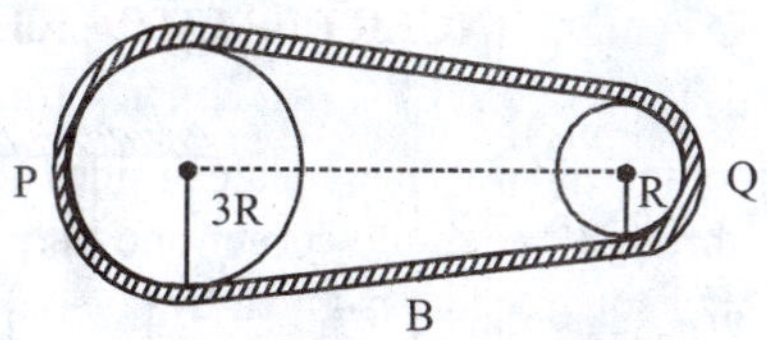

[NCERT : P L-119 | NA, July 27, 2021 (II)]

146. A rod of mass M and length L is lying on a horizontal frictionless surface. A particle of mass 'm' travelling along the surface hits at one end of the rod with a velocity 'u' in a direction perpendicular to the rod. The collision is completely elastic. After collision, particle comes to rest. The ratio of masses $\left(\frac{m}{M}\right)$ is $\frac{1}{x}$. The value of 'x' will be _____. **[NCERT : P L-122 | NA, July, 20, 2021 (I)]**

147. Consider a uniform wire of mass M and length L. It is bent into a semicircle. Its moment of inertia about a line perpendicular to the plane of the wire passing through the centre is : **[NCERT : P L-116 | March 18, 2021 (II), 2005]**

(a) $\frac{2}{3}\frac{ML^2}{\pi^2}$ (b) $\frac{1}{4}\frac{ML^2}{\pi^2}$ (c) $\frac{1}{2}\frac{ML^2}{\pi^2}$ (d) $\frac{ML^2}{\pi^2}$

148. Four identical solid spheres each of mass 'm' and radius 'a' are placed with their centres on the four corners of a square of side 'b'. The moment of inertia of the system about one side of square where the axis of rotation is parallel to the plane of the square is :

[NCERT : P L-116 | Feb. 26, 2021 (I)]

(a) $\frac{4}{5}ma^2$ (b) $\frac{8}{5}ma^2 + mb^2$

(c) $\frac{8}{5}ma^2 + 2mb^2$ (d) $\frac{4}{5}ma^2 + 2mb^2$

149. A uniform thin bar of mass 6 kg and length 2.4 meter is bent to make an equilateral hexagon. The moment of inertia about an axis passing through the centre of mass and perpendicular to the plane of hexagon is _____ $\times 10^{-1}$ kg m^2. **[Feb. 24, 2021 (II)]**

150. Shown in the figure is a hollow icecream cone (it is open at the top). If its mass is *M*, radius of its top, *R* and height, *H*, then its moment of inertia about its axis is :

[NCERT : P L-114 | Sep. 06, 2020 (I)]

(a) $\frac{MR^2}{2}$

(b) $\frac{M(R^2+H^2)}{4}$

(c) $\frac{MH^2}{3}$

(d) $\frac{MR^2}{3}$

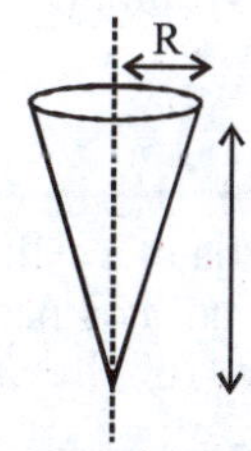

151. A wheel is rotating freely with an angular speed ω on a shaft. The moment of inertia of the wheel is I and the moment of inertia of the shaft is negligible. Another wheel of moment of inertia 3I initially at rest is suddenly coupled to the same shaft. The resultant fractional loss in the kinetic energy of the system is :

[NCERT : P L-122 | Sep. 05, 2020 (I)]

(a) $\frac{5}{6}$ (b) $\frac{1}{4}$ (c) 0 (d) $\frac{3}{4}$

152. *ABC* is a plane lamina of the shape of an equilateral triangle. *D*, *E* are mid points of *AB*, *AC* and *G* is the centroid of the lamina. Moment of inertia of the lamina about an axis passing through *G* and perpendicular to the plane *ABC* is I_0. If part *ADE* is removed, the moment of inertia of the remaining part about the same axis is $\frac{NI_0}{16}$ where *N* is an integer. Value of *N* is _____.

[NCERT : P L-114 | NA Sep. 04, 2020 (I)]

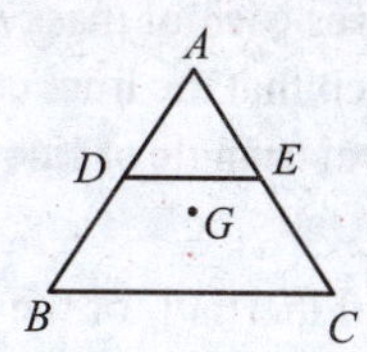

153. A circular disc of mass M and radius R is rotating about its axis with angular speed ω_1. If another stationary disc having radius $\frac{R}{2}$ and same mass M is dropped co-axially on to the rotating disc. Gradually both discs attain constant angular speed ω_2. The energy lost in the process is p% of the initial energy. Value of p is ________.

[NCERT : P L-122 | NA Sep. 04, 2020 (I)]

154. Consider two uniform discs of the same thickness and different radii $R_1 = R$ and $R_2 = \alpha R$ made of the same material. If the ratio of their moments of inertia I_1 and I_2, respectively, about their axes is $I_1 : I_2 = 1 : 16$ then the value of α is :

[NCERT : P L-116 | Sep. 04, 2020 (II)]

(a) $2\sqrt{2}$ (b) $\sqrt{2}$ (c) 2 (d) 4

155. For a uniform rectangular sheet shown in the figure, the ratio of moments of inertia about the axes perpendicular to the sheet and passing through O (the centre of mass) and O' (corner point) is :

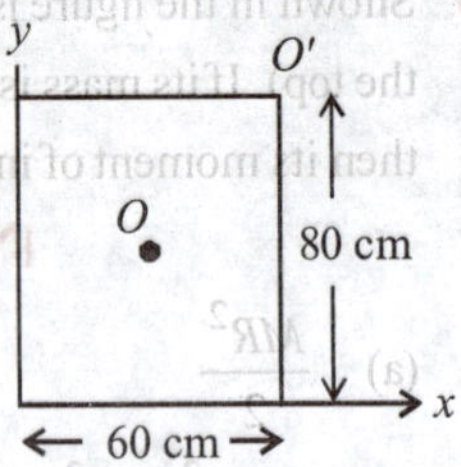

[Sep. 04, 2020 (II)]

(a) 2/3 (b) 1/4 (c) 1/8 (d) 1/2

156. Moment of inertia of a cylinder of mass M, length L and radius R about an axis passing through its centre and perpendicular to the axis of the cylinder is $I = M\left(\frac{R^2}{4} + \frac{L^2}{12}\right)$. If such a cylinder is to be made for a given mass of a material, the ratio L/R for it to have minimum possible I is :

[NCERT : P L-116 | Sep. 03, 2020 (II)]

(a) $\frac{2}{3}$ (b) $\frac{3}{2}$ (c) $\sqrt{\frac{3}{2}}$ (d) $\sqrt{\frac{2}{3}}$

157. A massless equilateral triangle EFG of side 'a' (As shown in figure) has three particles of mass m situated at its vertices. The moment of inertia of the system about the line EX perpendicular to EG in the plane of EFG is $\frac{N}{20}ma^2$ where N is an integer. The value of N is ________.

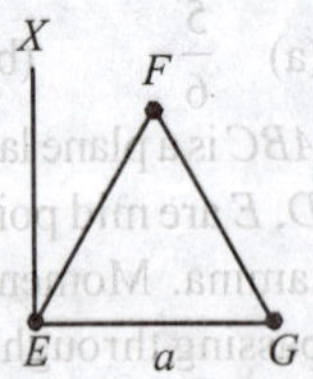

[NCERT : P L-114 | Sep. 03, 2020 (II)]

158. Three solid spheres each of mass m and diameter d are stuck together such that the lines connecting the centres form an equilateral triangle of side of length d. The ratio $\frac{I_0}{I_A}$ of moment of inertia I_0 of the system about an axis passing the centroid and about center of any of the spheres I_A and perpendicular to the plane of the triangle is:

[NCERT : P L-116 | 9 Jan. 2020 I]

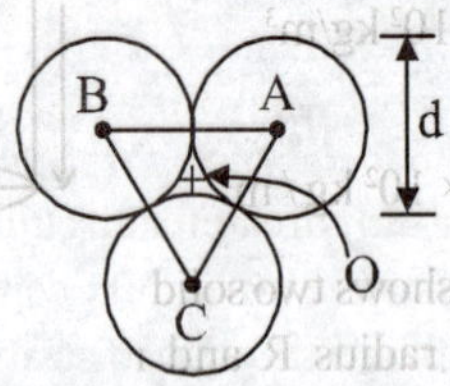

(a) $\frac{13}{23}$ (b) $\frac{15}{13}$ (c) $\frac{23}{13}$ (d) $\frac{13}{15}$

159. One end of a straight uniform 1 m long bar is pivoted on horizontal table. It is released from rest when it makes an angle 30° from the horizontal (see figure). Its angular speed when it hits the table is given as $\sqrt{n}$ rad/s, where n is an integer. The value of n is ________.

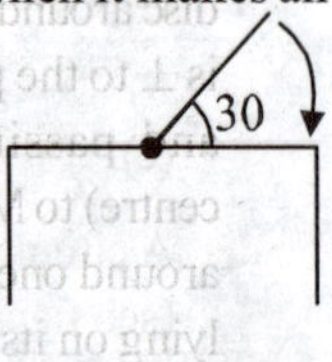

[NCERT : P L-119 | 9 Jan. 2020 I]

160. A uniformly thick wheel with moment of inertia I and radius R is free to rotate about its centre of mass (see fig). A massless string is wrapped over its rim and two blocks of masses m_1 and m_2 ($m_1 > m_2$) are attached to the ends of the string. The system is released from rest. The angular speed of the wheel when m_1 descents by a distance h is:

[NCERT : P L-119 | 9 Jan. 2020 II]

(a) $\left[\frac{2(m_1 - m_2)gh}{(m_1 + m_2)R^2 + I}\right]^{1/2}$

(b) $\left[\frac{2(m_1 + m_2)gh}{(m_1 + m_2)R^2 + I}\right]^{1/2}$

(c) $\left[\frac{(m_1 - m_2)}{(m_1 + m_2)R^2 + I}\right]^{1/2} gh$

(d) $\left[\frac{m_1 + m_2}{(m_1 + m_2)R^2 + I}\right]^{1/2} gh$

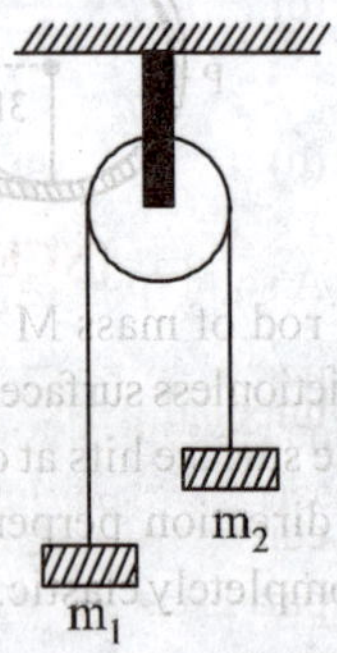

161. As shown in the figure, a bob of mass m is tied by a massless string whose other end portion is wound on a flywheel (disc) of radius r and mass m. When released from rest the bob starts falling vertically. When it has covered a distance of h, the angular speed of the wheel will be:

[NCERT : P L-119 | 7 Jan. 2020 I]

(a) $\frac{1}{r}\sqrt{\frac{4gh}{3}}$ (b) $r\sqrt{\frac{3}{2gh}}$

(c) $\frac{1}{r}\sqrt{\frac{2gh}{3}}$ (d) $r\sqrt{\frac{3}{4gh}}$

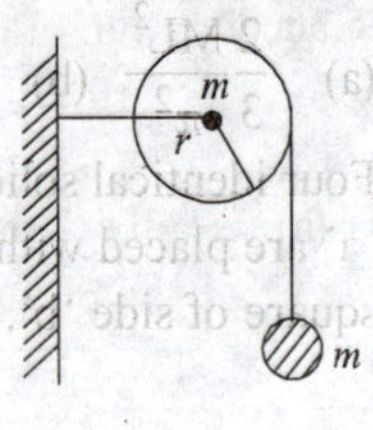

162. The radius of gyration of a uniform rod of length l, about an axis passing through a point $\frac{l}{4}$ away from the centre of the rod, and perpendicular to it, is:

[NCERT : P L-116 | 7 Jan. 2020 I]

(a) $\frac{1}{4}l$ (b) $\frac{1}{8}l$ (c) $\sqrt{\frac{7}{48}}l$ (d) $\sqrt{\frac{3}{8}}l$

163. Mass per unit area of a circular disc of radius a depends on the distance r from its centre as $\sigma(r) = A + Br$. The moment of inertia of the disc about the axis, perpendicular to the plane and passing through its centre is:

[NCERT : P L-114 | 7 Jan. 2020 II]

(a) $2\pi a^4\left(\frac{A}{4}+\frac{aB}{5}\right)$ (b) $2\pi a^4\left(\frac{aA}{4}+\frac{B}{5}\right)$

(c) $\pi a^4\left(\frac{A}{4}+\frac{aB}{5}\right)$ (d) $2\pi a^4\left(\frac{A}{4}+\frac{B}{5}\right)$

164. A circular disc of radius b has a hole of radius a at its centre (see figure). If the mass per unit area of the disc varies as $\left(\frac{\sigma_0}{r}\right)$, then the radius of gyration of the disc about its axis passing through the centre is :

[NCERT : P L-114, 115 | 12 April 2019]

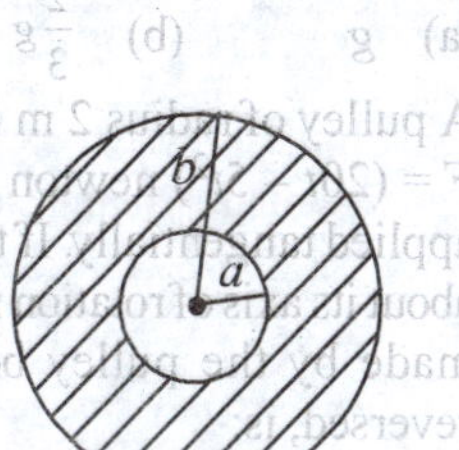

(a) $\sqrt{\frac{a^2+b^2+ab}{2}}$

(b) $\frac{a+b}{2}$

(c) $\sqrt{\frac{a^2+b^2+ab}{3}}$

(d) $\frac{a+b}{3}$

165. A solid sphere of mass M and radius R is divided into two unequal parts. The first part has a mass of $\frac{7M}{8}$ and is converted into a uniform disc of radius 2R. The second part is converted into a uniform solid sphere. Let I_1 be the moment of inertia of the new sphere about its axis. The ratio I_1/I_2 is given by :

[NCERT : P L-116 | 10 Apr. 2019 II]

(a) 185 (b) 140 (c) 285 (d) 65

166. A stationary horizontal disc is free to rotate about its axis. When a torque is applied on it, its kinetic energy as a function of θ, where θ is the angle by which it has rotated, is given as $k\theta^2$. If its moment of inertia is I then the angular acceleration of the disc is:

[NCERT : P L-119 | 9 April 2019 I]

(a) $\frac{k}{4I}\theta$ (b) $\frac{k}{I}\theta$ (c) $\frac{k}{2I}\theta$ (d) $\frac{2k}{I}\theta$

167. Moment of inertia of a body about a given axis is 1.5 kgm². Initially the body is at rest. In order to produce a rotational kinetic energy of 1200 J, the angular acceleration of 20 rad/s² must be applied about the axis for a duration of:

[NCERT : P L-119 | 9 Apr. 2019 II]

(a) 2.5s (b) 2s (c) 5s (d) 3s

168. Let the moment of inertia of a hollow cylinder of length 30 cm (inner radius 10 cm and outer radius 20 cm), about its axis be I. The radius of a thin cylinder of the same mass such that its moment of inertia about its axis is also I, is:

[NCERT : P L-116 | 12 Jan. 2019 I]

(a) 12 cm (b) 16 cm (c) 14 cm (d) 18 cm

169. The moment of inertia of a solid sphere, about an axis parallel to its diameter and at a distance of x from it, is 'I(x)'. Which one of the graphs represents the variation of I(x) with x correctly?

[NCERT : P L-116 | 12 Jan. 2019 II]

(a)
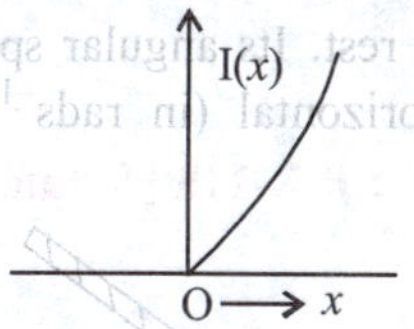

(b)
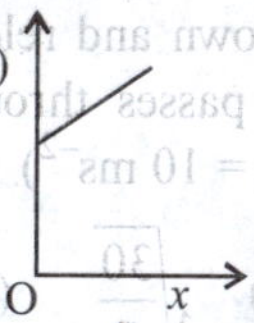

(c)
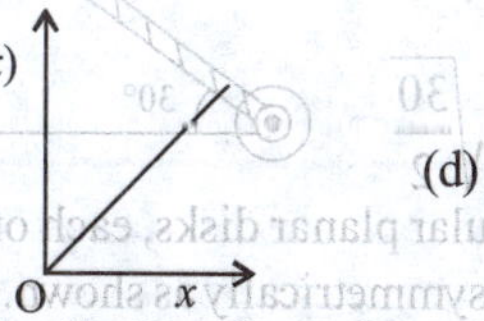

(d)
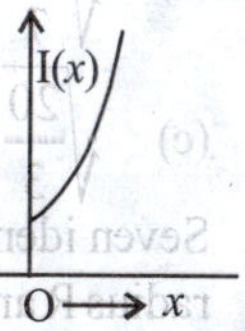

170. A circular disc D_1 of mass M and radius R has two identical discs D_2 and D_3 of the same mass M and radius R attached rigidly at its opposite ends (see figure). The moment of inertia of the system about the axis OO', passing through the centre of D_1, as shown in the figure, will be :

[NCERT : P L-116 | 11 Jan. 2019 II]

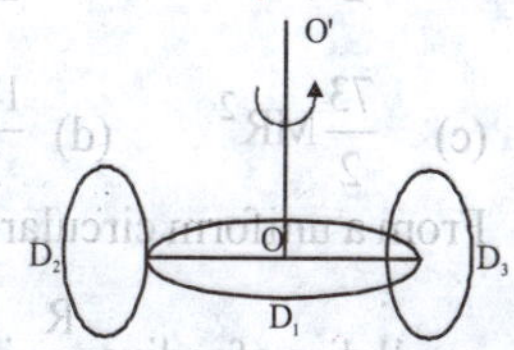

(a) MR^2

(b) $3MR^2$

(c) $\frac{4}{5}MR^2$

(d) $\frac{2}{3}MR^2$

171. Two identical spherical balls of mass M and radius R each are stuck on two ends of a rod of length 2R and mass M (see figure). The moment of inertia of the system about the axis passing perpendicularly through the centre of the rod is:

[NCERT : P L-116 | 10 Jan. 2019 II]

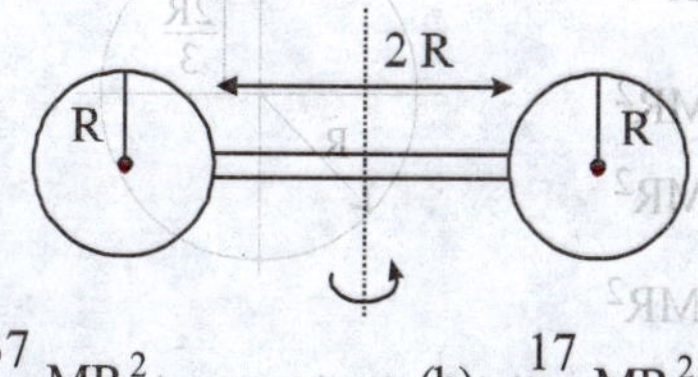

(a) $\frac{137}{15}MR^2$ (b) $\frac{17}{15}MR^2$

(c) $\frac{209}{15}MR^2$ (d) $\frac{152}{15}MR^2$

172. Two masses m and $\frac{m}{2}$ are connected at the two ends of a massless rigid rod of length l. The rod is suspended by a thin wire of torsional constant k at the centre of mass of the rod-mass system (see figure). Because of torsional constant k, the restoring toruque is $\tau = k\theta$ for angular displacement θ. If the rod is rotated by θ_0 and released, the tension in it when it passes through its mean position will be: **[NCERT : P L-119 | 9 Jan. 2019 I]**

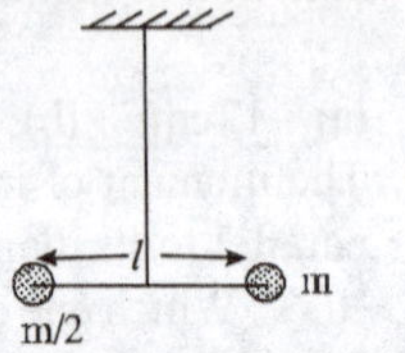

(a) $\frac{3k\theta_0^2}{l}$ (b) $\frac{2k\theta_0^2}{l}$

(c) $\frac{k\theta_0^2}{l}$ (d) $\frac{k\theta_0^2}{2l}$

173. A rod of length 50 cm is pivoted at one end. It is raised such that if makes an angle of 30° from the horizontal as shown and released from rest. Its angular speed when it passes through the horizontal (in $rads^{-1}$) will be $(g = 10\ ms^{-2})$ **[NCERT : P L-119 | 9 Jan. 2019 II]**

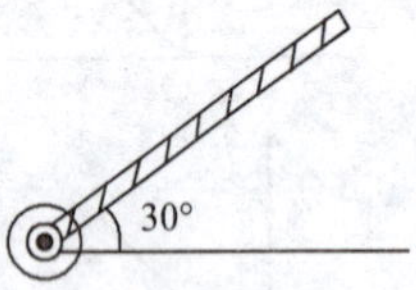

(a) $\sqrt{\frac{30}{7}}$ (b) $\sqrt{30}$

(c) $\sqrt{\frac{20}{3}}$ (d) $\sqrt{\frac{30}{2}}$

174. Seven identical circular planar disks, each of mass M and radius R are welded symmetrically as shown. The moment of inertia of the arrangement about the axis normal to the plane and passing through the point P is: **[NCERT : P L-116 | 2018]**

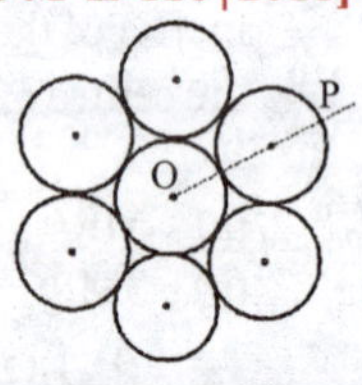

(a) $\frac{19}{2}MR^2$ (b) $\frac{55}{2}MR^2$

(c) $\frac{73}{2}MR^2$ (d) $\frac{181}{2}MR^2$

175. From a uniform circular disc of radius R and mass 9 M, a small disc of radius $\frac{R}{3}$ is removed as shown in the figure. The moment of inertia of the remaining disc about an axis perpendicular to the plane of the disc and passing, through centre of disc is : **[NCERT : P L-116 | Online April 9, 2018]**

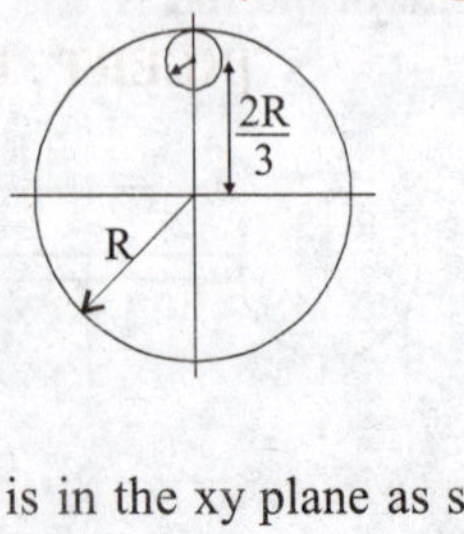

(a) $4\ MR^2$

(b) $\frac{40}{9}\ MR^2$

(c) $10\ MR^2$

(d) $\frac{37}{9}\ MR^2$

176. A thin circular disk is in the xy plane as shown in the figure. The ratio of its moment of inertia about z and z′ axes will be **[NCERT : P L-116 | Online April 16, 2018]**

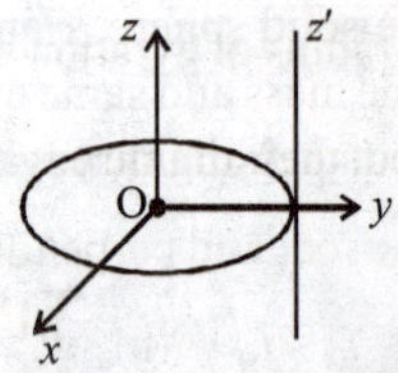

(a) 1 : 2

(b) 1 : 4

(c) 1 : 3

(d) 1 : 5

177. From a solid sphere of mass M and radius R a cube of maximum possible volume is cut. Moment of inertia of cube about an axis passing through its center and perpendicular to one of its faces is : **[2015]**

(a) $\frac{4MR^2}{9\sqrt{3}\pi}$ (b) $\frac{4MR^2}{3\sqrt{3}\pi}$ (c) $\frac{MR^2}{32\sqrt{2}\pi}$ (d) $\frac{MR^2}{16\sqrt{2}\pi}$

178. Consider a thin uniform square sheet made of a rigid material. If its side is 'a' mass m and moment of inertia I about one of its diagonals, then :**[Online April 10, 2015]**

(a) $I > \frac{ma^2}{12}$ (b) $\frac{ma^2}{24} < I < \frac{ma^2}{12}$

(c) $I = \frac{ma^2}{24}$ (d) $I = \frac{ma^2}{12}$

179. A mass m hangs with the help of a string wrapped around a pulley on a frictionless bearing. The pulley has mass m and radius R. Assuming pulley to be a perfect uniform circular disc, the acceleration of the mass m, if the string does not slip on the pulley, is: **[NCERT : P L-119 | 2011]**

(a) g (b) $\frac{2}{3}g$ (c) $\frac{g}{3}$ (d) $\frac{3}{2}g$

180. A pulley of radius 2 m is rotated about its axis by a force $F = (20t - 5t^2)$ newton (where t is measured in seconds) applied tangentially. If the moment of inertia of the pulley about its axis of rotation is 10 kg-m² the number of rotations made by the pulley before its direction of motion is reversed, is: **[NCERT : P L-119 | 2011]**

(a) more than 3 but less than 6

(b) more than 6 but less than 9

(c) more than 9

(d) less than 3

181. A thin uniform rod of length l and mass m is swinging freely about a horizontal axis passing through its end. Its maximum angular speed is ω. Its centre of mass rises to a maximum height of **[NCERT : P L-119 | 2009]**

(a) $\frac{1}{6}\frac{l\omega}{g}$ (b) $\frac{1}{2}\frac{l^2\omega^2}{g}$ (c) $\frac{1}{6}\frac{l^2\omega^2}{g}$ (d) $\frac{1}{3}\frac{l^2\omega^2}{g}$

182. Consider a uniform square plate of side 'a' and mass 'M'. The moment of inertia of this plate about an axis perpendicular to its plane and passing through one of its corners is **[2008]**

(a) $\frac{5}{6}Ma^2$ (b) $\frac{1}{12}Ma^2$ (c) $\frac{7}{12}Ma^2$ (d) $\frac{2}{3}Ma^2$

183. For the given uniform square lamina $ABCD$, whose centre is O, **[2007]**

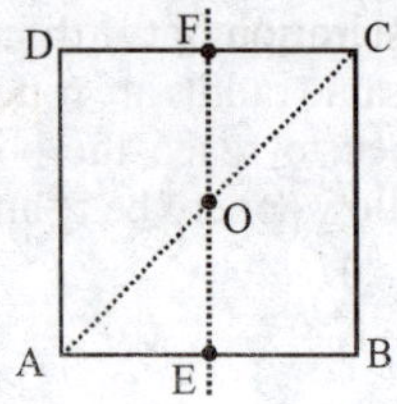

(a) $I_{AC} = \sqrt{2}\ I_{EF}$

(b) $\sqrt{2}I_{AC} = I_{EF}$

(c) $I_{AD} = 3I_{EF}$

(d) $I_{AC} = I_{EF}$

184. One solid sphere A and another hollow sphere B are of same mass and same outer radii. Their moment of inertia about their diameters are respectively I_A and I_B Such that **[NCERT : P L-116 | 2004]**

(a) $I_A < I_B$ (b) $I_A > I_B$ (c) $I_A = I_B$ (d) $\frac{I_A}{I_B} = \frac{d_A}{d_B}$

where d_A and d_B are their densities.

185. A circular disc X of radius R is made from an iron plate of thickness t, and another disc Y of radius $4R$ is made from an iron plate of thickness $\frac{t}{4}$. Then the relation between the moment of inertia I_X and I_Y is **[NCERT : P L-116 | 2003]**

(a) $I_Y = 32 I_X$ (b) $I_Y = 16 I_X$
(c) $I_Y = I_X$ (d) $I_Y = 64 I_X$

186. A particle performing uniform circular motion has angular frequency is doubled & its kinetic energy halved, then the new angular momentum is **[NCERT : P L-119 | 2003]**

(a) $\frac{L}{4}$ (b) $2L$ (c) $4L$ (d) $\frac{L}{2}$

187. Moment of inertia of a circular wire of mass M and radius R about its diameter is **[NCERT : P L-116 | 2002]**

(a) $MR^2/2$ (b) MR^2 (c) $2MR^2$ (d) $MR^2/4$

Topic 5 - Rolling Motion

188. A wheel is rolling on a plane surface. The speed of a particle on the highest point of the rim is 8 m/s. The speed of the particle on the rim of the wheel at the same level as the centre of wheel, will be : **[April 4, 2025 (II)]**

(a) $4\sqrt{2}$ m/s (b) 8 m/s
(c) 4 m/s (d) $8\sqrt{2}$ m/s

189. A force of 49 N acts tangentially at the highest point of a sphere (solid) of mass 20 kg, kept on a rough horizontal plane. If the sphere rolls without slipping, then the acceleration of the center of the sphere is **[April 3, 2025 (I)]**

(a) 3.5 m/s^2
(b) 0.35 m/s^2
(c) 2.5 m/s^2
(d) 0.25 m/s^2

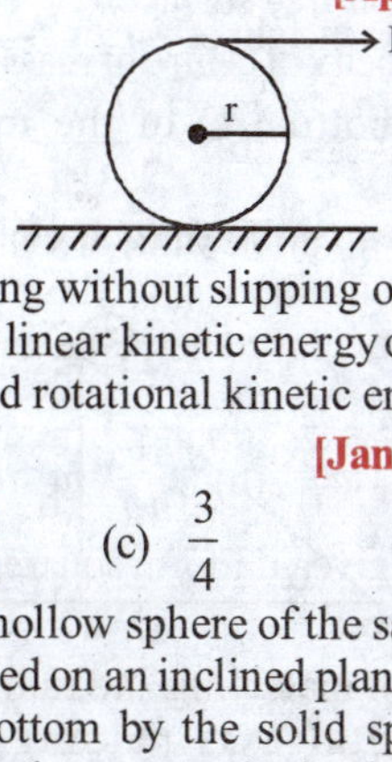

190. A solid sphere is rolling without slipping on a horizontal plane. The ratio of the linear kinetic energy of the centre of mass of the sphere and rotational kinetic energy is **[Jan. 24, 2025 (II)]**

(a) $\frac{2}{5}$ (b) $\frac{5}{2}$ (c) $\frac{3}{4}$ (d) $\frac{4}{3}$

191. A solid sphere and a hollow sphere of the same mass and of same radius are rolled on an inclined plane. Let the time taken to reach the bottom by the solid sphere and the hollow sphere be t_1 and t_2, respectively, then **[Jan. 24, 2025 (II)]**

(a) $t_1 < t_2$ (b) $t_1 = t_2$ (c) $t_1 = 2t_2$ (d) $t_1 > t_2$

192. A uniform solid cylinder of mass 'm' and radius 'r' rolls along an inclined rough plane of inclination 45°. If it starts to roll from rest from the top of the plane then the linear acceleration of the cylinder axis will be **[Jan. 24, 2025 (I)]**

(a) $\frac{1}{\sqrt{2}}g$ (b) $\frac{1}{3\sqrt{2}}g$ (c) $\frac{\sqrt{2}g}{3}$ (d) $\sqrt{2}g$

193. A solid sphere of mass 'm' and radius 'r' is allowed to roll without slipping from the highest point of an inclined plane of length 'L' and makes an angle 30° with the horizontal. The speed of the particle at the bottom of the plane is v_1. If the angle of inclination is increased to 45° while keeping L constant. Then the new speed of the sphere at the bottom of the plane is v_2. The ratio of $v_1^2 : v_2^2$ is **[Jan. 23, 2025 (I)]**

(a) $1:\sqrt{2}$ (b) $1:3$ (c) $1:2$ (d) $1:\sqrt{3}$

194. A circular disc reaches from top to bottom of an inclined plane of length l. When it slips down the plane, if takes t s. When it rolls down the plane then it takes $\left(\frac{\alpha}{2}\right)^{1/2}$ t s, where α is .... **[April 9, 2024 (II)]**

195. A hollow sphere is rolling on a plane surface about its axis of symmetry. The ratio of rotational kinetic energy to its total kinetic energy is $\frac{x}{5}$. The value of x is _____. **[April 5, 2024 (II)]**

196. A solid sphere and a hollow cylinder roll up without slipping on same inclined plane with same initial speed v. The sphere and the cylinder reaches upto maximum heights h_1 and h_2, respectively, above the initial level. The ratio $h_1 : h_2$ is $\frac{n}{10}$. The value of n is _______. **[April 4, 2024 (I)]**

197. A disc of radius R and mass M is rolling horizontally without slipping with speed v. It then moves up an inclined smooth surface as shown in figure. The maximum height that the disc can go up the incline is : **[Feb. 1, 2024 (II)]**

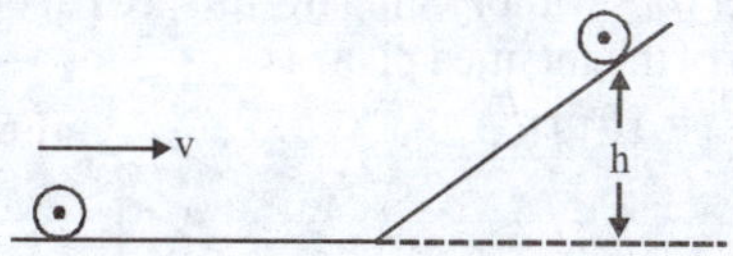

(a) $\frac{3}{4}\frac{v^2}{g}$ (b) $\frac{v^2}{g}$ (c) $\frac{2}{3}\frac{v^2}{g}$ (d) $\frac{1}{2}\frac{v^2}{g}$

198. A solid circular disc of mass 50 kg rolls along a horizontal floor so that its center of mass has a speed of 0.4 m/s. The absolute value of work done on the disc to stop it is _____ J. **[Jan. 31, 2024 (I)]**

199. A cylinder is rolling down on an inclined plane of inclination 60°. It's acceleration during rolling down will be $\frac{x}{\sqrt{3}}$ m/s^2, where x = _____ (use g = 10 m/s^2). **[Jan. 29, 2024 (I)]**

200. A ring and a solid sphere roll down the same inclined plane without slipping. They start from rest. The radii of both bodies are identical and the ratio of their kinetic energies is $\frac{7}{x}$ where x is ________. **[Jan. 27, 2024 (II)]**

201. A disc is rolling without slipping on a surface. The radius of the disc is R. At t = 0, the top most point on the disc is A as shown in figure. When the disc completes half of its rotation, the displacement of point A from its initial position is **[April 13, 2023 (I)]**

(a) $R\sqrt{(\pi^2+4)}$

(b) $R\sqrt{(\pi^2+1)}$

(c) 2R

(d) $2R\sqrt{(1+4\pi^2)}$

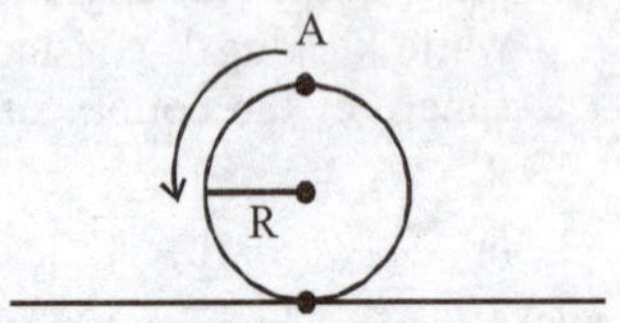

202. A solid sphere is rolling on a horizontal plane without slipping. If the ratio of angular momentum about axis of rotation of the sphere to the total energy of moving sphere is π : 22 then, the value of its angular speed will be ________ rad/s.

203. For a rolling spherical shell, the ratio of rotational kinetic energy and total kinetic energy is $\frac{x}{5}$. The value of x is ________. **[April 12, 2023 (I)]**

204. A hollow spherical ball of uniform density rolls up a curved surface with an initial velocity 3 m/s (as shown in figure). Maximum height with respect to the initial position covered by it will be ______ cm. **[April 8, 2023 (II)]**

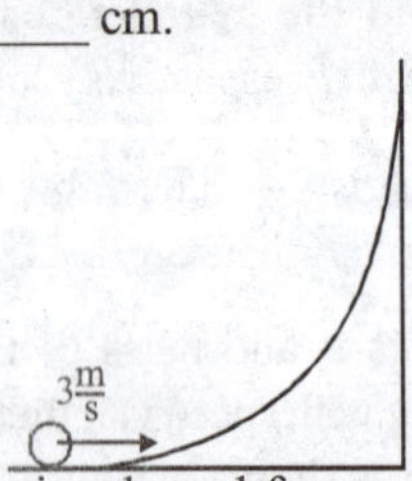

205. A solid cylinder is released from rest from the top of an inclined plane of inclination 30° and length 60 cm. If the cylinder rolls without slipping, its speed upon reaching the bottom of the inclined plane is _____ ms^{-1}.
(Given $g = 10\ ms^{-2}$) **[Feb. 1, 2023 (I)]**

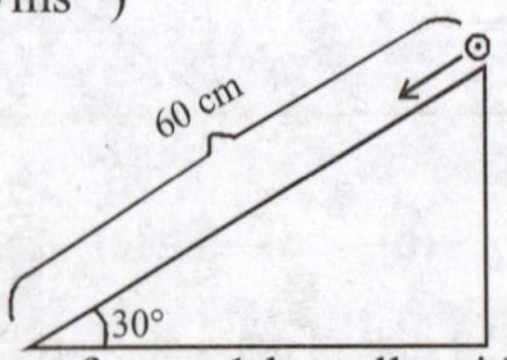

206. A solid sphere of mass 1 kg rolls without slipping on a plane surface. Its kinetic energy is 7×10^{-3} J. The speed of the centre of mass of the sphere is ______ cm s^{-1}. **[Jan. 25, 2023 (I)]**

207. A solid sphere of mass 1 kg rolls without slipping on a plane surface. Its kinetic energy is 7×10^{-3} J. The speed of the centre of mass of the sphere is ______ cm s^{-1}.

208. A thin uniform rod of length 2m. cross sectional area 'A' and density 'd' is rotated about an axis passing through the centre and perpendicular to its length with angular velocity ω. If value of ω in terms of its rotational kinetic energy E is $\sqrt{\frac{\alpha E}{Ad}}$ then the value of α is ________. **[Jan. 30, 2023 (I)]**

209. A uniform disc of mass 0.5 kg and radius r is projected with velocity 18 m/s at t = 0 s on a rough horizontal surface. It starts off with a purely sliding motion at t = 0 s. After 2s it acquires a purely rolling motion (see figure). The total kinetic energy of the disc after 2s will be ________ J (given, coefficient of friction is 0.3 and $g = 10\ m/s^2$). **[Jan. 30, 2023 (II)]**

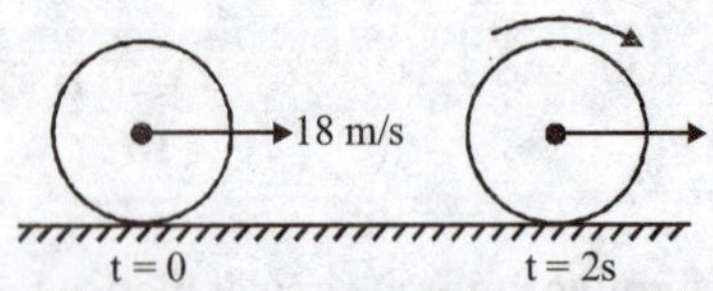

210. A solid sphere of mass 2kg is making pure rolling on a horizontal surface with kinetic energy 2240 J. The velocity of centre of mass of the sphere will be ____ ms^{-1}. **[Jan. 29, 2023 (I)]**

211. A solid cylinder length is suspended symmetrically through two massless strings, as shown in the figure. The distance from the initial rest position, the cylinder should by unbinding the strings to achieve a speed of 4 ms^{-1}, is ______ cm. (take $g = 10\ ms^{-2}$) **[July 27, 2022 (II)]**

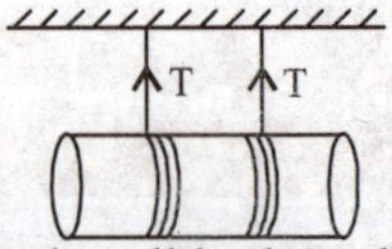

212. A solid cylinder and a solid sphere, having same mass M and radius R, roll down the same inclined plane from top without slipping. They start from rest. The ratio of velocity of the solid cylinder to that of the solid sphere, with which they reach the ground, will be : **[July 25, 2022 (I)]**

(a) $\sqrt{\frac{5}{3}}$ (b) $\sqrt{\frac{4}{5}}$ (c) $\sqrt{\frac{3}{5}}$ (d) $\sqrt{\frac{14}{15}}$

213. A rolling wheel of 12 kg is on an inclined plane at position P and connected to a mass of 3 kg through a string of fixed length and pulley as shown in figure. Consider PR as friction free surface.
The velocity of centre of mass of the wheel when it reaches at the bottom Q of the inclined plane PQ will be $\frac{1}{2}\sqrt{xgh}$ m/s. The value of x is ______. **[June 27, 2022 (II)]**

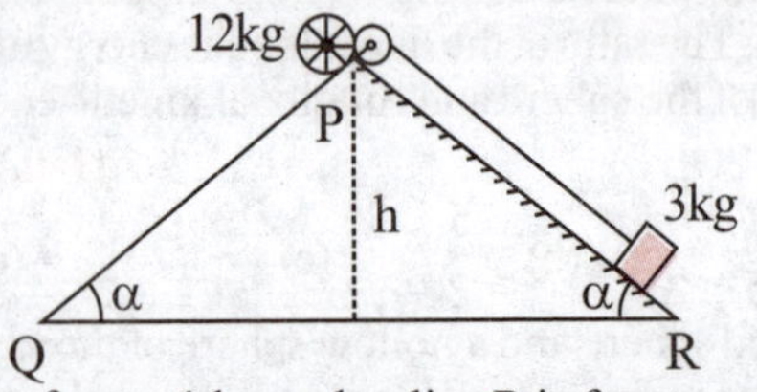

214. A disc of mass 1 kg and radius R is free to rotate about a horizontal axis passing through its centre and perpendicular to the plane of disc. A body of same mass as that of disc is fixed at the highest point of the disc.

Now the system is released, when the body comes to the lowest position, its angular speed will be $4\sqrt{\frac{x}{3R}}$ rad s^{-1} where $x =$ ______. ($g = 10\ ms^{-2}$)

215. Solid spherical ball is rolling on a frictionless horizontal plane surface about is axis of symmetry. The ratio of rotational kinetic energy of the ball to its total kinetic energy is :- **[June 26, 2022 (II)]**

(a) $\frac{2}{5}$ (b) $\frac{2}{7}$ (c) $\frac{1}{5}$ (d) $\frac{7}{10}$

216. The centre of a wheel rolling on a plane surface moves with a speed v_0. A particle on the rim of the wheel at the same level as the centre will be moving at a speed $\sqrt{x}\ v_0$. Then the value of x is ______. **[NA, July 22, 2021 (II)]**

217. A circular disc reaches from top to bottom of an inclined plane of length 'L'. When it slips down the plane, it takes time 't_1'. When it rolls down the plane, it takes time t_2. The value of $\frac{t_2}{t_1}$ is $\sqrt{\frac{3}{x}}$. The value of x will be ______. **[NA, July 20, 2021 (I)]**

218. Two bodies, a ring and a solid cylinder of same material are rolling down without slipping an inclined plane. The radii of the bodies are same. The ratio of velocity of the centre of mass at the bottom of the inclined plane of the ring to that of the cylinder is $\frac{\sqrt{x}}{2}$. Then, the value of x is ______. **[NA, July 20, 2021 (II)]**

219. A body rolls down an inclined plane without slipping. The kinetic energy of rotation is 50% of its translational kinetic energy. The body is : **[July 20, 2021 (II)]**

(a) Solid sphere (b) Solid cylinder
(c) Hollow cylinder (d) Ring

220. The following bodies,
(1) a ring (2) a disc
(3) a solid cylinder (4) a solid sphere,
of same mass 'm' and radius 'R' are allowed to roll down without slipping simultaneously from the top of the inclined plane. The body which will reach first at the bottom of the inclined plane is ______.
[Mark the body as per their respective numbering given in the question] **[March 17, 2021 (I)]**

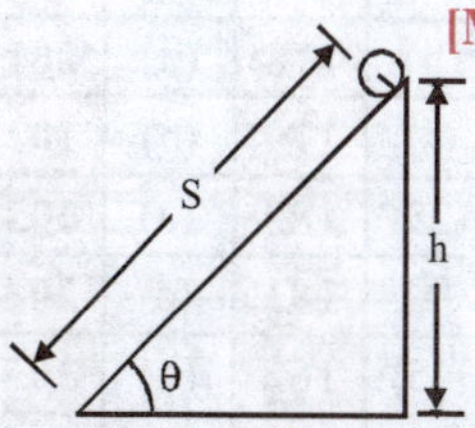

221. A solid disc of radius 'a' and mass 'm' rolls down without slipping on an inclined plane making an angle θ with the horizontal. The acceleration of the disc will be $\frac{2}{b}$ g sin θ where b is ______. (Round off to the Nearest Integer) (g = acceleration due to gravity, θ = angle as shown in figure.) **[March 16, 2021 (II)]**

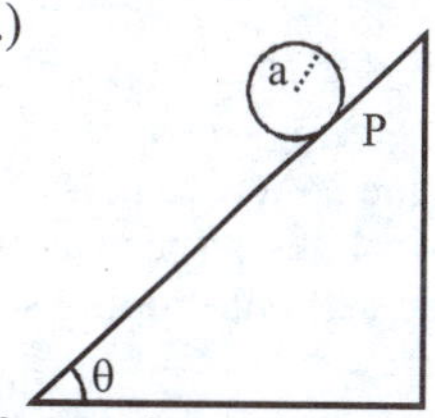

222. A uniform sphere of mass 500 g rolls without slipping on a plane horizontal surface with its centre moving at a speed of 5.00 cm/s. Its kinetic energy is: **[8 Jan. 2020 II]**

(a) 8.75×10^{-4} J (b) 8.75×10^{-3} J
(c) 6.25×10^{-4} J (d) 1.13×10^{-3} J

223.

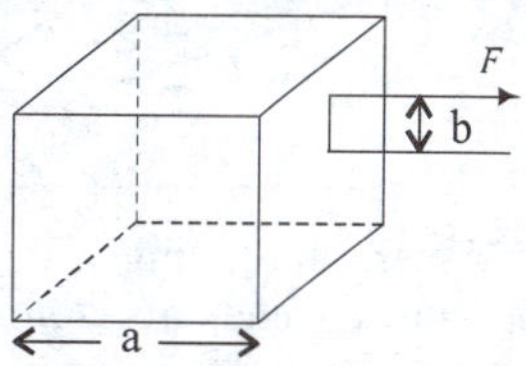

Consider a uniform cubical box of side a on a rough floor that is to be moved by applying minimum possible force F at distance b above its centre of mass (see figure). If the coefficient of friction is $\mu = 0.4$, the maximum possible value of $100 \times \frac{b}{a}$ for box not to topple before moving is ______. **[NA 7 Jan. 2020 II]**

224. A solid sphere and solid cylinder of identical radii approach an incline with the same linear velocity (see figure). Both roll without slipping all throughout. The two climb maximum heights h_{sph} and h_{cyl} on the incline. The ratio $\frac{h_{sph}}{h_{cyl}}$ is given by : **[8 Apr. 2019 II]**

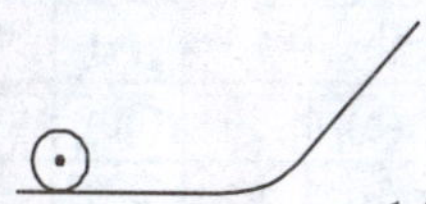

(a) $\frac{2}{\sqrt{5}}$ (b) 1 (c) $\frac{14}{15}$ (d) $\frac{4}{5}$

225. A homogeneous solid cylindrical roller of radius R and mass M is pulled on a cricket pitch by a horizontal force. Assuming rolling without slipping, angular acceleration of the cylinder is: **[10 Jan. 2019 I]**

(a) $\frac{3F}{2mR}$ (b) $\frac{F}{3mR}$ (c) $\frac{F}{2mR}$ (d) $\frac{2F}{3mR}$

226. A roller is made by joining together two cones at their vertices O. It is kept on two rails AB and CD, which are placed asymmetrically (see figure), with its axis perpendicular to CD and its centre O at the centre of line joining AB and CD (see figure). It is given a light push so that it starts rolling with its centre O moving parallel to CD in the direction shown. As it moves, the roller will tend to: **[2016]**

(a) go straight.

(b) turn left and right alternately.

(c) turn left.

(d) turn right.

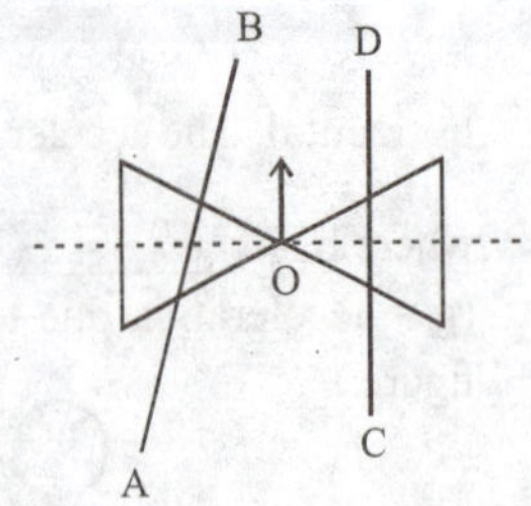

227. A cylinder of mass M_c and sphere of mass M_s are placed at points A and B of two inclines, respectively (See Figure). If they roll on the incline without slipping such that their accelerations are the same, then the ratio $\frac{\sin\theta_c}{\sin\theta_s}$ is: **[Online April 9, 2014]**

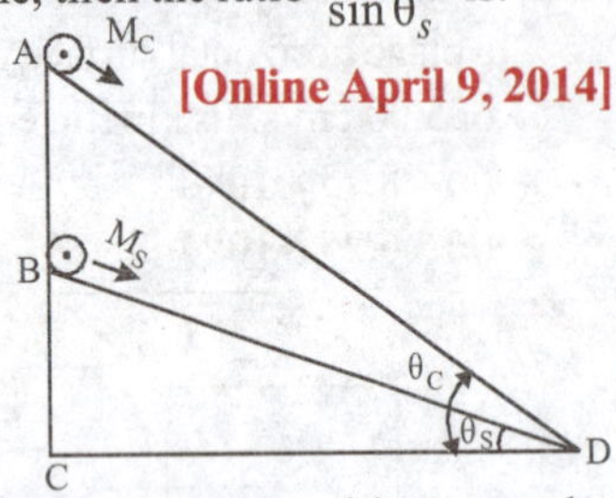

(a) $\sqrt{\frac{8}{7}}$ (b) $\sqrt{\frac{15}{14}}$

(c) $\frac{8}{7}$ (d) $\frac{15}{14}$

228. A loop of radius r and mass m rotating with an angular velocity ω_0 is placed on a rough horizontal surface. The initial velocity of the centre of the hoop is zero. What will be the velocity of the centre of the hoop when it ceases to slip ? **[2013]**

(a) $\frac{r\omega_0}{4}$ (b) $\frac{r\omega_0}{3}$ (c) $\frac{r\omega_0}{2}$ (d) $r\omega_0$

229. A solid sphere is rolling on a surface as shown in figure, with a translational velocity v ms^{-1}. If it is to climb the inclined surface continuing to roll without slipping, then minimum velocity for this to happen is **[Online May 12, 2012]**

(a) $\sqrt{2gh}$

(b) $\sqrt{\frac{7}{5}gh}$

(c) $\sqrt{\frac{7}{2}gh}$

(d) $\sqrt{\frac{10}{7}gh}$

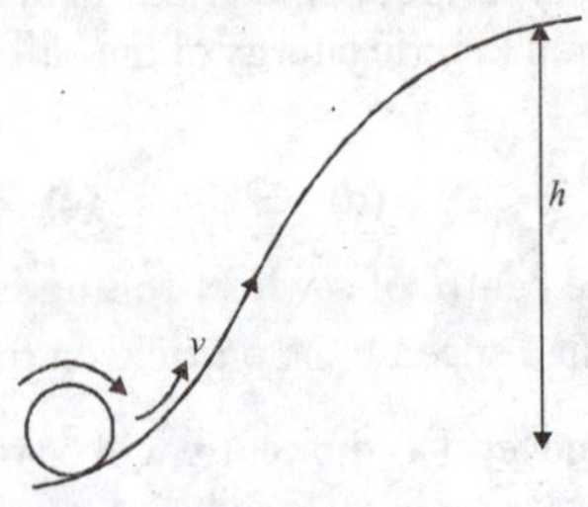

230. A round uniform body of radius R, mass M and moment of inertia I rolls down (without slipping) an inclined plane making an angle θ with the horizontal. Then its acceleration is **[2007]**

(a) $\frac{g\sin\theta}{1-MR^2/I}$ (b) $\frac{g\sin\theta}{1+I/MR^2}$

(c) $\frac{g\sin\theta}{1+MR^2/I}$ (d) $\frac{g\sin\theta}{1-I/MR^2}$

ANSWER KEYS

1	(d)	24	(d)	47	(d)	70	(b)	93	(c)	116	(53)	139	(4.24)	162	(c)	185	(d)	208	(3)
2	(a)	25	(4)	48	(a)	71	(a)	94	(b)	117	(250)	140	(c)	163	(a)	186	(a)	209	(54)
3	(d)	26	(b)	49	(c)	72	(20)	95	(a)	118	(24)	141	(d)	164	(c)	187	(a)	210	(40)
4	(15)	27	(b)	50	(90)	73	(20)	96	(c)	119	(16)	142	(c)	165	(b)	188	(a)	211	(120)
5	(3)	28	(728)	51	(a)	74	(c)	97	(c)	120	(5)	143	(d)	166	(d)	189	(a)	212	(d)
6	(2)	29	(a)	52	(c)	75	(20)	98	(d)	121	(35)	144	(d)	167	(b)	190	(b)	213	(3)
7	(a)	30	(c)	53	(13)	76	(9)	99	(b)	122	(3)	145	(9)	168	(b)	191	(a)	214	(5)
8	(8)	31	(c)	54	(8)	77	(a)	100	(b)	123	(1)	146	(4)	169	(d)	192	(c)	215	(b)
9	(2)	32	(b)	55	(a)	78	(d)	101	(d)	124	(5)	147	(d)	170	(b)	193	(a)	216	(2)
10	(c)	33	(b)	56	(60)	79	(d)	102	(d)	125	(3)	148	(c)	171	(a)	194	(3)	217	(2)
11	(2)	34	(b)	57	(c)	80	(a)	103	(d)	126	(5)	149	(8)	172	(c)	195	(2)	218	(3)
12	(4)	35	(c)	58	(15)	81	(c)	104	(9)	127	(17)	150	(a)	173	(b)	196	(7)	219	(b)
13	(6)	36	(b)	59	(3)	82	(b)	105	(199)	128	(5)	151	(d)	174	(d)	197	(d)	220	(4)
14	(c)	37	(40)	60	(800)	83	(d)	106	(32)	129	(110)	152	(11)	175	(a)	198	(6)	221	(3)
15	(3)	38	(3)	61	(c)	84	(d)	107	(b)	130	(3)	153	(20)	176	(c)	199	(10)	222	(a)
16	(23)	39	(c)	62	(c)	85	(c)	108	(b)	131	(18)	154	(c)	177	(a)	200	(7)	223	(50)
17	(b)	40	(a)	63	(c)	86	(a)	109	(4)	132	(5)	155	(b)	178	(d)	201	(a)	224	(c)
18	(a)	41	(1)	64	(91)	87	(a)	110	(16)	133	(5)	156	(c)	179	(b)	202	(4)	225	(d)
19	(d)	42	(a)	65	(c)	88	(d)	111	(d)	134	(5)	157	(25)	180	(a)	203	(2)	226	(c)
20	(d)	43	(c)	66	(10)	89	(a)	112	(100)	135	(8)	158	(a)	181	(c)	204	(75)	227	(d)
21	(b)	44	(c)	67	(c)	90	(d)	113	(4)	136	(a)	159	(15)	182	(d)	205	(2)	228	(c)
22	(c)	45	(2)	68	(10)	91	(a)	114	(a)	137	(b)	160	(a)	183	(d)	206	(10)	229	(d)
23	(a)	46	(b)	69	(6)	92	(d)	115	(4)	138	(5)	161	(a)	184	(a)	207	(10)	230	(b)

7 Gravitation

TREND ANALYSIS
JEE MAIN (2025-2021)

	2025				2024				2023				2022				2021			
	Session II		Session I		Session II		Session I		Session II		Session I		Session II		Session I		Session II		Session I	
	MCQs	NVQs	MCQs	NVQs	MCQs	NVQs	MCQs	NVQs	MCQs	NVQs	MCQs	NVQs	MCQs	NVQs	MCQs	NVQs	MCQs	NVQs	MCQs	NVQs
Topic 1	1	–	–	–	–	–	2	–	2	1	–	–	4	–	–	–	–	–	2	–
Topic 2	–	1	1	–	1	–	1	–	1	–	–	–	1	–	–	–	1	–	1	–
Topic 3	–	–	–	1	2	–	2	–	7	–	3	–	4	1	5	1	1	–	–	–
Topic 4	1	–	–	–	1	–	1	–	2	–	–	–	1	–	1	–	2	1	1	1
Total	2	1	1	1	4	–	6	–	12	1	3	–	10	1	6	1	4	1	4	1

How to Code Qns for Better & Faster Revision →

Type 1: Able to solve (No code Req)
Type 2: Solved with struggle or by seeing solution **(Use code NR means Needs Revision or orange stickers)**
Type 3: Tricky/ Could not solve on my own. Need help from Teacher/ Friend. **(Use code TQ means Tricky Question or red stickers)**

You can either hand write the code in boxes or else paste coloured stickers given at end of book.

Topic 1 - Kepler's Laws of Planetary Motion

1. Given below are two statements : one is labelled as **Assertion (A)** and the other is labelled as **Reason (R)**.

Assertion (A) : The radius vector from the Sun to a planet sweeps out equal areas in equal intervals of time and thus areal velocity of planet is constant.

Reason (R) : For a central force field the angular momentum is a constant.

In the light of the above statements, choose the **most appropriate answer** from the options given below :

[NCERT : P L-128, 129 | April 7, 2025 (II)]

(a) Both **(A)** and **(R)** are correct and **(R)** is the correct explanation of **(A)**
(b) Both **(A)** and **(R)** are correct but **(R)** is not the correct explanation of **(A)**
(c) **(A)** is correct but **(R)** is not correct
(d) **(A)** is not correct but **(R)** is correct

2. A planet takes 200 days to complete one revolution around the Sun. If the distance of the planet from Sun is reduced to one fourth of the original distance, how many days will it take to complete one revolution :

[NCERT : P L-129 | Jan. 29, 2024 (II)]

(a) 20 (b) 25 (c) 50 (d) 100

3. Given below are two statements : one is labelled as **Assertion (A)** and the other is labelled as **Reason (R)**.

Assertion (A) : The angular speed of the moon in its orbit about the earth is more than the angular speed of the earth in its orbit about the sun.

Reason (R) : The moon takes less time to move around the earth than the time taken by the earth to move around the sun.

In the light of the above statements, choose the most appropriate answer from the options given below:

[NCERT : P L-128-129 | Jan. 27, 2024 (II)]

(a) (A) is correct but (R) is not correct
(b) Both (A) and (R) are correct and (R) is the correct explanation of (A)
(c) Both (A) and (R) are correct but (R) is not the correct explanation of (A)
(d) (A) is not correct but (R) is correct

4. If the earth suddenly shrinks to $\frac{1}{64}$th of its original volume with its mass remaining the same, the period of rotation of earth becomes $\frac{24}{x}$h. The value of x is _____.
[NCERT : P L-129 | April 10, 2023 (I)]

5. The orbital angular momentum of a satellite is L, when it is revolving in a circular orbit at height h from earth surface. If the distance of satellite from the earth centre is increased by eight times to its initial value, then the new angular momentum will be-
[NCERT : P L-129 | April 8, 2023 (II)]
(a) 8L (b) 4L (c) 9L (d) 3L

6. Choose the incorrect statement from the following:
[NCERT : P L-128 | April 6, 2023 (II)]
(a) The speed of satellite in a given circular orbit remains constant.
(b) For a planet revolving around the sun in an elliptical orbit, the total energy of the planet remains constant.
(c) When a body fall towards earth, the displacement of earth towards the body is negligible.
(d) The linear speed of a planet revolving around the sun remains constant.

7. The time period of a satellite revolving around earth in a given orbit is 7 hours. If the radius of orbit is increased to three times its previous value, then approximate new time period of the satellite will be :
[NCERT : P L-129 | June 29, 2022 (II), June 29, 2022 (I)]
(a) 40 hours (b) 36 hours
(c) 30 hours (d) 25 hours

8. Two planets *A* and *B* of equal mass are having their period of revolutions T_A and T_B such that $T_A = 2T_B$. These planets are revolving in the circular orbits of radii r_A and r_B respectively. Which out of the following would be the correct relationship of their orbits?
[NCERT : P L-129 | June 28, 2022 (I)]
(a) $2r_A^2 = r_B^2$ (b) $r_A^3 = 2r_B^3$
(c) $r_A^3 = 4r_B^3$ (d) $T_A^2 - T_B^2 = \frac{\pi^2}{GM}(r_B^3 - 4r_A^3)$

9. The distance between Sun and Earth is *R*. The duration of year if the distance between Sun and Earth becomes 3*R* will be : **[NCERT : P L-129 | June 24, 2022 (II)]**
(a) $\sqrt{3}$ years (b) 3 years
(c) 9 years (d) $3\sqrt{3}$ years

10. The minimum and maximum distances of a planet revolving around the Sun are x_1 and x_2. If the minimum speed of the planet on its trajectory is v_0 then its maximum speed will be:
[NCERT : P L-129 | July 25, 2021 (I), June 25, 2022 (I)]
(a) $\frac{v_0 x_1^2}{x_2^2}$ (b) $\frac{v_0 x_2^2}{x_1^2}$ (c) $\frac{v_0 x_1}{x_2}$ (d) $\frac{v_0 x_2}{x_1}$

11. A planet revolving in elliptical orbit has :
A. a constant velocity of revolution.
B. has the least velocity is when it is nearest to the sun.
C. its areal velocity is directly proportional to its velocity.
D. areal velocity is inversely proportional to its velocity.
E to follow a trajectory such that the areal velocity is constant. **[NCERT : P L-128 | Feb. 26, 2021 (I)]**
Choose the correct answer from the options given below :
(a) A only (b) E only (c) D only (d) C only

12. A particle is moving with uniform speed along the circumference of a circle of radius R under the action of a central fictitious force F which is inversely proportional to R^3. Its time period of revolution will be given by :
[NCERT : P L-129 | Feb. 26, 2021 (I)]
(a) $T \propto R^{\frac{4}{3}}$ (b) $T \propto R^{\frac{3}{2}}$ (c) $T \propto R^2$ (d) $T \propto R^{\frac{5}{2}}$

13. If the angular momentum of a planet of mass m, moving around the Sun in a circular orbit is L, about the center of the Sun, its areal velocity is:
[NCERT : P L-128, 129 | 9 Jan. 2019 I]
(a) $\frac{L}{m}$ (b) $\frac{4L}{m}$ (c) $\frac{L}{2m}$ (d) $\frac{2L}{m}$

14. Figure shows elliptical path abcd of a planet around the sun S such that the area of triangle csa is $\frac{1}{4}$ the area of the ellipse. (See figure) With db as the semimajor axis, and ca as the semiminor axis. If t_1 is the time taken for planet to go over path abc and t_2 for path taken over cda then:
[NCERT : P L-128 | Online April 9, 2016]
(a) $t_1 = 4t_2$
(b) $t_1 = 2t_2$
(c) $t_1 = 3t_2$
(d) $t_1 = t_2$

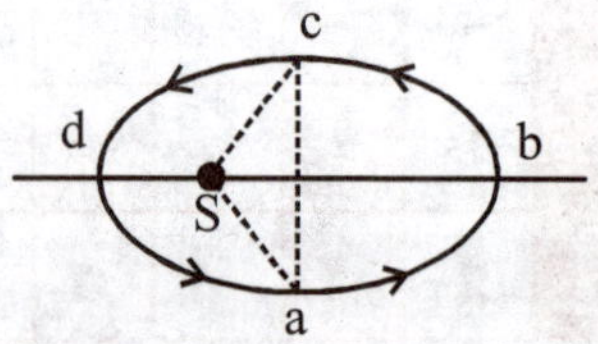

15. The time period of a satellite of earth is 5 hours. If the separation between the earth and the satellite is increased to 4 times the previous value, the new time period will become
[NCERT : P L-129 | 2003]
(a) 10 hours (b) 80 hours (c) 40 hours (d) 20 hours

Topic 2 - Newton's Universal Law of Gravitation

16. Three identical spheres of mass m, are placed at the vertices of an equilateral triangle of length a. When released, they interact only through gravitational force and collide after a time T = 4 seconds. If the sides of the triangle are increased to length 2a and also the masses of the spheres are made 2m, then they will collide after _____ seconds. **[NCERT : P L-131 | April 3, 2025 (I)]**

17. A small point of mass m is placed at a distance 2R from the centre 'O of a big uniform solid sphere of mass M and radius R. The gravitational force on 'm' due to M is F_1. A spherical part of radius R/3 is removed from the big sphere as shown in the figure and the gravitational force on m due to remaining part of M is found to be F_2. The value of ratio $F_1 : F_2$ is **[NCERT : P L-131 | Jan. 22, 2025 (I)]**

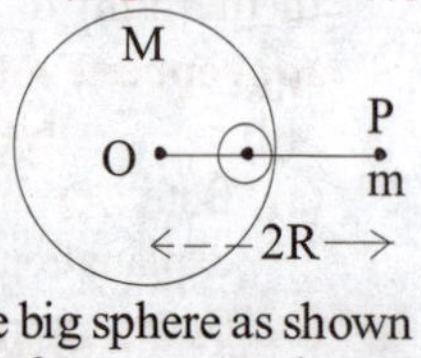

(a) 16:9 (b) 11:10 (c) 12:11 (d) 12:9

18. A metal wire of uniform mass density having length L and mass M is bent to form a semicircular arc and a particle of mass m is placed at the centre of the arc. The gravitational force on the particle by the wire is:

[NCERT : P L-130 | 2003]April 4, 2024 (I)]

(a) $\frac{GMm\pi}{2L^2}$ (b) 0

(c) $\frac{GmM\pi^2}{L^2}$ (d) $\frac{2GmM\pi}{L^2}$

19. Four identical particles of mass m are kept at the four corners of a square. If the gravitational force exerted on one of the masses by the other masses is $\left(\frac{2\sqrt{2}+1}{32}\right)\frac{Gm^2}{L^2}$, the length of the sides of the square is

[NCERT : P L-131 | 2003]Jan. 31, 2024 (I)]

(a) 2L (b) 4L (c) 3L (d) $\frac{L}{2}$

20. Two identical particles each of mass 'm' go round a circle of radius a under the action of their mutual gravitational attraction. The angular speed of each particle will be:

[NCERT : P L-131 | April 15, 2023 (I)]

(a) $\sqrt{\frac{Gm}{2a^3}}$ (b) $\sqrt{\frac{Gm}{8a^3}}$ (c) $\sqrt{\frac{Gm}{4a^3}}$ (d) $\sqrt{\frac{Gm}{a^3}}$

21. Three identical particles A, B and C of mass 100 kg each are placed in a straight line with AB = BC = 13 m. The gravitational force on a fourth particle P of the same mass is F, when placed at a distance 13 m from the particle B on the perpendicular bisector of the line AC. The value of F will be approximately:

[NCERT : P L-131 | July 25, 2022 (I)]

(a) 21 G (b) 100 G (c) 59 G (d) 42 G

22. Four particles each of mass M, move along a circle of radius R under the action of their mutual gravitational attraction as shown in figure. The speed of each particle is :

[NCERT : P L-131 | Sep. 1, 2021 (II), 2014]

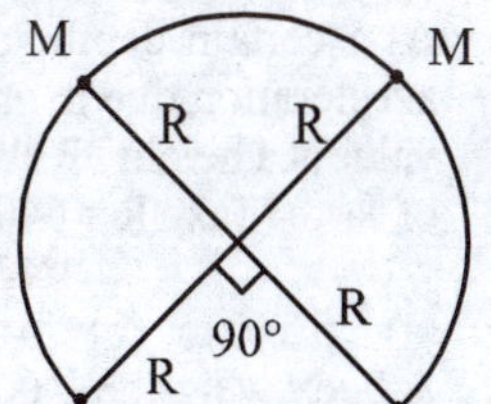

(a) $\frac{1}{2}\sqrt{\frac{GM}{R(2\sqrt{2}+1)}}$

(b) $\frac{1}{2}\sqrt{\frac{GM}{R}(2\sqrt{2}+1)}$

(c) $\frac{1}{2}\sqrt{\frac{GM}{R}(2\sqrt{2}-1)}$

(d) $\sqrt{\frac{GM}{R}}$

23. Find the gravitational force of attraction between the ring and sphere as shown in the diagram, where the plane of the ring is perpendicular to the line joining the centres. If $\sqrt{8}$ R is the distance between the centres of a ring (of mass 'm') and a sphere (mass 'M') where both have equal radius 'R' **[Feb. 26, 2021 (I)]**

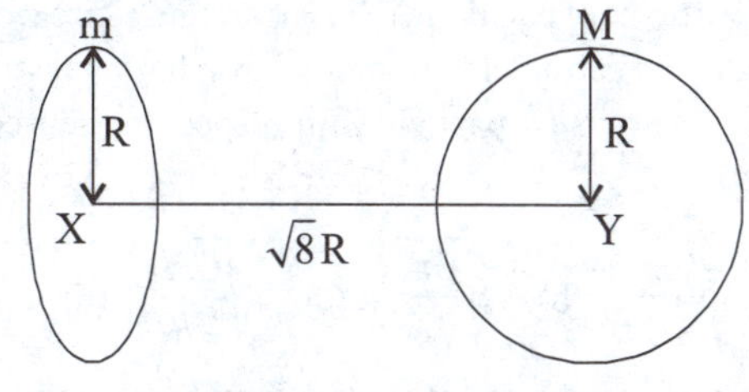

(a) $\frac{1}{3\sqrt{8}}\cdot\frac{GMm}{R^2}$ (b) $\frac{\sqrt{8}}{9}\cdot\frac{GmM}{R}$

(c) $\frac{\sqrt{8}}{27}\cdot\frac{GmM}{R^2}$ (d) $\frac{2\sqrt{2}}{3}\cdot\frac{GMm}{R^2}$

24. A straight rod of length L extends from $x = a$ to $x = L + a$. The gravitational force it exerts on point mass 'm' at $x = 0$, if the mass per unit length of the rod is $A + Bx^2$, is given by: **[12 Jan. 2019 I]**

(a) $Gm\left[A\left(\frac{1}{a+L}-\frac{1}{a}\right)-BL\right]$

(b) $Gm\left[A\left(\frac{1}{a}-\frac{1}{a+L}\right)-BL\right]$

(c) $Gm\left[A\left(\frac{1}{a+L}-\frac{1}{a}\right)+BL\right]$

(d) $Gm\left[A\left(\frac{1}{a}-\frac{1}{a+L}\right)+BL\right]$

25. Take the mean distance of the moon and the sun from the earth to be 0.4×10^6 km and 150×10^6 km respectively. Their masses are 8×10^{22} kg and 2×10^{30} kg respectively. The radius of the earth is 6400 km. Let ΔF_1 be the difference in the forces exerted by the moon at the nearest and farthest points on the earth and ΔF_2 be the difference in the force exerted by the sun at the nearest and farthest points on the earth. Then, the number closest to $\frac{\Delta F_1}{\Delta F_2}$ is:

[Online April 15, 2018]

(a) 2 (b) 6 (c) 10^{-2} (d) 0.6

26. From a sphere of mass M and radius R, a smaller sphere of radius $\frac{R}{2}$ is carved out such that the cavity made in the original sphere is between its centre and the periphery (See figure). For the configuration in the figure where the distance between the centre of the original sphere and the removed sphere is 3R, the gravitational force between the two sphere is:

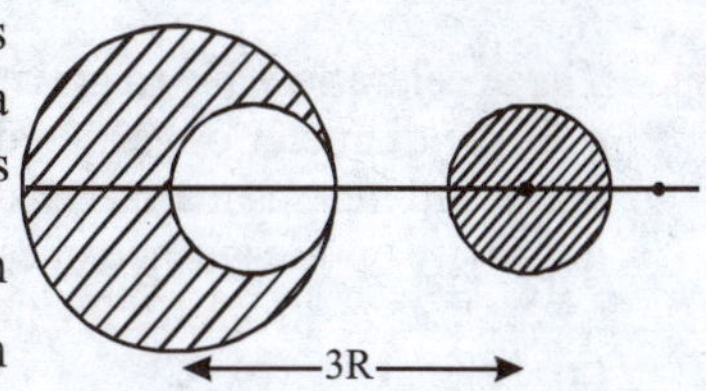

[Online April 11, 2014]

(a) $\frac{41\,GM^2}{3600\,R^2}$ (b) $\frac{41\,GM^2}{450\,R^2}$

(c) $\frac{59\,GM^2}{450\,R^2}$ (d) $\frac{GM^2}{225\,R^2}$

27. Two particles of equal mass 'm' go around a circle of radius R under the action of their mutual gravitational attraction. The speed of each particle with respect to their centre of mass is **[NCERT : P L-131 | 2011 RS]**

(a) $\sqrt{\frac{Gm}{4R}}$ (b) $\sqrt{\frac{Gm}{3R}}$ (c) $\sqrt{\frac{Gm}{2R}}$ (d) $\sqrt{\frac{Gm}{R}}$

28. Two spherical bodies of mass M and $5M$ & radii R & $2R$ respectively are released in free space with initial separation between their centres equal to 12 R. If they attract each other due to gravitational force only, then the distance covered by the smaller body just before collision is **[NCERT : P L-131 | 2003]**

(a) 2.5R (b) 4.5R (c) 7.5R (d) 1.5R

Topic 3 - Acceleration due to Gravity

29. Acceleration due to gravity on the surface of earth is 'g'. If the diameter of earth is reduced to one third of its original value and mass remains unchanged, then the acceleration due to gravity on the surface of the earth is ____g. **[NCERT : P L-133 | Jan. 24, 2025 (II)]**

30. Assuming the earth to be a sphere of uniform mass density, a body weighed 300 N on the surface of earth. How much it would weigh at R/4 depth under surface of earth ? **[NCERT : P L-134 | April 6, 2024 (II)]**

(a) 75 N (b) 375 N (c) 300 N (d) 225 N

31. A 90 kg body placed at 2R distance from surface of earth experiences gravitational pull of :
(R = Radius of earth, g = 10 ms^{-2}) **[NCERT : P L-133 | April 4, 2024 (II)]**

(a) 300 N (b) 225 N (c) 120 N (d) 100 N

32. At what distance above and below the surface of the earth a body will have same weight (take radius of earth as R.) **[NCERT : P L-134 | Jan. 29, 2024 (I)]**

(a) $\frac{R}{2}$ (b) $\sqrt{5}R - R$

(c) $\frac{\sqrt{3}R - R}{2}$ (d) $\frac{\sqrt{5}R - R}{2}$

33. The acceleration due to gravity on the surface of earth is g. If the diameter of earth reduces to half of its original value and mass remains constant, then acceleration due to gravity on the surface of earth would be : **[NCERT : P L-134 | Jan. 27, 2024 (I)]**

(a) g/4 (b) 2g (c) g/2 (d) 4g

34. Two planets A and B of radii R and 1.5 R have densities ρ and ρ/2 respectively. The ratio of acceleration due to gravity at the surface of B to A is : **[NCERT : P L-132, 133 | April 13, 2023 (II)]**

(a) 2 : 3 (b) 2 : 1 (c) 3 : 4 (d) 4 : 3

35. The radii of two planets 'A' and 'B' are 'R' and '4R' and their densities are ρ and ρ/3 respectively. The ratio of acceleration due to gravity at their surfaces ($g_A : g_B$) will be: **[NCERT : P L-132, 133 | April 11, 2023 (I)]**

(a) 1 : 16 (b) 3 : 16 (c) 3 : 4 (d) 4 : 3

36. Assuming the earth to be a sphere of uniform mass density, the weight of a body at a depth $d = \frac{R}{2}$ from the surface of earth, if its weight on the surface of earth is 200 N, will be: **[NCERT : P L-134 | April 10, 2023 (I)]**
(Given R = Radius of earth)

(a) 400 N (b) 500 N (c) 300 N (d) 100 N

37. The weight of a body on the earth is 400 N. Then weight of the body when taken to a depth half of the radius of the earth will be: **[NCERT : P L-133 | April 8, 2023 (I)]**

(a) Zero (b) 300 N (c) 100 N (d) 200 N

38. The acceleration due to gravity at height h above the earth if h <<R (radius of earth) is given by **[NCERT : P L-134 | April 8, 2023 (II)]**

(a) $g' = g\left(1 - \frac{2h}{R}\right)$ (b) $g' = g\left(1 - \frac{2h^2}{R^2}\right)$

(c) $g' = g\left(1 - \frac{h}{2R}\right)$ (d) $g' = g\left(1 - \frac{h^2}{2R^2}\right)$

39. A planet has double the mass of the earth. Its average density is equal to the that of the earth. An object weighing W on earth will weigh on that planet: **[NCERT : P L-132, 133 | April 6, 2023 (I)]**

(a) $2^{2/3}$ W (b) W (c) $2^{1/3}$ W (d) 2 W

40. The weight of a body on the surface of the earth is 100 N. The gravitational force on it when taken at a height, from the surface of earth, equal to one-fourth the radius of the earth is: **[NCERT : P L-133 | April 6, 2023 (II)]**

(a) 100 N (b) 64 N (c) 50 N (d) 25 N

41. Given below are two statements: **[Feb. 1, 2023 (I)]**
Statement I: Acceleration due to gravity is different at different places on the surface of earth.
Statement II: Acceleration due to gravity increases as we go down below the earth's surface.
In the light of the above statements, choose the **correct** answer from the options given below
(a) Both Statement I and Statement II are false
(b) Both Statement I and Statement II are true
(c) Statement I is false but Statement II is true
(d) Statement I is true but Statement II is false

42. At a certain depth "d" below surface of earth. value of acceleration due to gravity becomes four times that of its value at a height 3R above earth surface. Where R is Radius of earth (Take R = 6400 km). The depth d is equal to **[NCERT : P L-134 | Jan. 31, 2023 (I)]**

(a) 5260 km (b) 640 km (c) 2560 km (d) 4800 km

43. A body weight W, is projected vertically upwards from earth's surface to reach a height above the earth which is equal to nine times the radius of earth. The weight of the body at that height will be: **[NCERT : P L-133 | Jan. 31, 2023 (II)]**

(a) $\frac{W}{91}$ (b) $\frac{W}{100}$ (c) $\frac{W}{9}$ (d) $\frac{W}{3}$

44. If the acceleration due to gravity experienced by a point mass at a height h above the surface of earth is same as that of the acceleration due to gravity at a depth αh (h << R_e) from the earth surface. The value of α will be ____.
(use R_e = 6400 km) **[NCERT : P L-134 | July 29, 2022 (I)]**

45. If the radius of earth shrinks by 2% while its mass remains same. The acceleration due to gravity on the earth's surface will approximately :
[NCERT : P L-132 | July 28, 2022 (I)]
(a) decrease by 2% (b) decrease by 4%
(c) increase by 2% (d) increase by 4%

46. The percentage decrease in the weight of a rocket, when taken to a height of 32 km above the surface of earth will, be : (Radius of earth = 6400 km)
[NCERT : P L-133 | July 26, 2022 (I)]
(a) 1% (b) 3% (c) 4% (d) 0.5%

47. An object is taken to a height above the surface of earth at a distance $\frac{5}{4}R$ from the centre of the earth. Where radius of earth, R = 6400 km. The percentage decrease in the weight of the object will be
[NCERT : P L-133 | July 25, 2022 (II)]
(a) 36% (b) 50% (c) 64% (d) 25%

48. The radii of two planets A and B are in the ratio 2 : 3. Their densities are 3ρ and 5ρ respectively. The ratio of their acceleration due to gravity is:
[NCERT : P L-132, 133 | June 30, 2022 (I)]
(a) 9 : 4 (b) 9 : 8 (c) 9 : 10 (d) 2 : 5

49. Given below are two statements :
Statement I : The law of gravitation holds good for any pair of bodies in the universe.
Statement II : The weight of any person becomes zero when the person is at the centre of the earth.
In the light of the above statements, choose the correct answer from the options given below.
[NCERT : P L-132, 134 | June 27, 2022 (I)]
(a) Both Statement I and Statement II are true
(b) Both Statement I and Statement II are false
(c) Statement I is true but Statement II are false
(d) Statement I is false but Statement II is true

50. The variation of acceleration due to gravity (g) with distance (r) from the center of the earth is correctly represented by: (Given R = radius of earth)
[NCERT : P L-133, 134 | June 26, 2022 (I)]

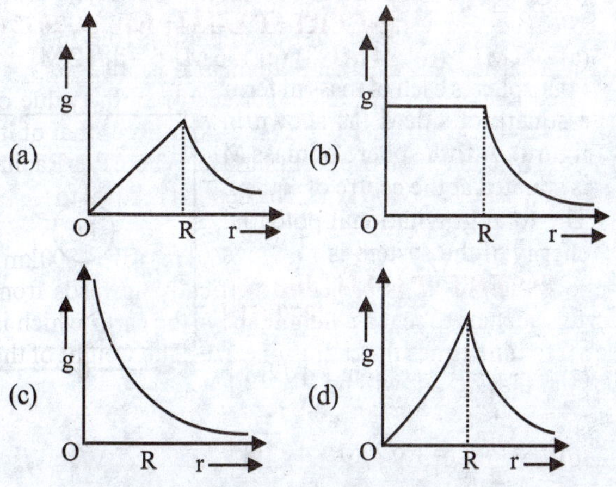

51. The elongation of a wire on the surface of the earth is 10^{-4} m. The same wire of same dimensions is elongated by 6×10^{-5} m on another planet. The acceleration due to gravity on the planet will be .... ms^{-2}. (Take acceleration due to gravity on the surface of earth = 10 m/s^{-2})
[NCERT : P L-132 | June 26, 2022 (I)]

52. Given below are two statements : One is labelled as Assertion A and the other is labelled as Reason R.
Assertion A : If we move from poles to equator, the direction of acceleration due to gravity of earth always points towards the center of earth without any variation in its magnitude.
Reason R : At equator, the direction of acceleration due to the gravity is towards the center of earth.
In the light of above statements, choose the correct answer from the options given below : [June 26, 2022 (II)]
(a) Both A and R are true and R is the correct explanation of A.
(b) Both A and R are true but R is NOT the correct explanation of A.
(c) A is true but R is false
(d) A is false but R is true

53. The approximate height from the surface of earth at which the weight of the body becomes $\frac{1}{3}$ of its weight on the surface of earth is :
[NCERT : P L-133 | June 24, 2022 (I)]
[Radius of earth R = 6400 km and $\sqrt{3}$ = 1.732]
(a) 3840 km (b) 4685 km (c) 2133 km (d) 4267 km

54. Consider a planet in some solar system which has a mass double the mass of earth and density equal to the average density of earth. If the weight of an object on earth is W, the weight of the same object on that planet will be :
[NCERT : P L-132, 133 | July 25, 2021 (II)]
(a) 2W (b) W (c) $2^{\frac{1}{3}}$ W (d) $\sqrt{2}$ W

55. In the reported figure of earth, the value of acceleration due to gravity is same at point A and C but it is smaller than that of its value at point B (surface of the earth). The value of OA : AB will be x : y. The value of x is _______.
[NCERT : P L-133 | Feb. 26, 2021 (II)]

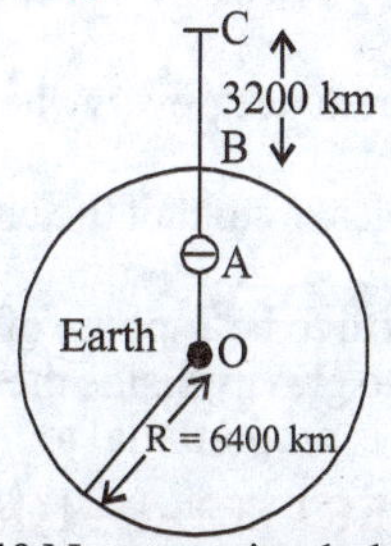

56. A body weighs 49 N on a spring balance at the north pole. What will be its weight recorded on the same weighing machine, if it is shifted to the equator?
(Use and radius of earth, R = 6400 km)
[Feb. 24, 2021 (II)]
(a) 49 N (b) 48.83 N (c) 49.83 N (d) 49.17 N

57. The value of acceleration due to gravity is g_1 at a height $h = \frac{R}{2}$ (R = radius of the earth) from the surface of the earth. It is again equal to g_1 and a depth d below the surface of the earth. The ratio $\left(\frac{d}{R}\right)$ equals :
[NCERT : P L-133 | 5 Sep. 2020 (I)]
(a) $\frac{4}{9}$ (b) $\frac{5}{9}$ (c) $\frac{1}{3}$ (d) $\frac{7}{9}$

58. The acceleration due to gravity on the earth's surface at the poles is g and angular velocity of the earth about the axis passing through the pole is ω. An object is weighed at the equator and at a height h above the poles by using a spring balance. If the weights are found to be same, then h is : ($h<<R$, where R is the radius of the earth) **[5 Sep. 2020 (II)]**

(a) $\frac{R^2\omega^2}{2g}$ (b) $\frac{R^2\omega^2}{g}$ (c) $\frac{R^2\omega^2}{4g}$ (d) $\frac{R^2\omega^2}{8g}$

59. A box weighs 196 N on a spring balance at the north pole. Its weight recorded on the same balance if it is shifted to the equator is close to (Take $g = 10\ ms^{-2}$ at the north pole and the radius of the earth = 6400 km): **[7 Jan. 2020 II]**

(a) 195.66 N (b) 194.32 N (c) 194.66 N (d) 195.32 N

60. The ratio of the weights of a body on the Earth's surface to that on the surface of a planet is 9:4. The mass of the planet is $\frac{1}{9}$th of that of the Earth. If 'R' is the radius of the Earth, what is the radius of the planet ? (Take the planets to have the same mass density).

[NCERT : P L-132, 133 | 2009, 12 April 2019 II]

(a) $\frac{R}{3}$ (b) $\frac{R}{4}$ (c) $\frac{R}{9}$ (d) $\frac{R}{2}$

61. The value of acceleration due to gravity at Earth's surface is 9.8 ms^{-2}. The altitude above its surface at which the acceleration due to gravity decreases to 4.9 ms^{-2}, is close to : (Radius of earth = 6.4×10^6 m)

[NCERT : P L-133 | 10 April 2019 I]

(a) 2.6×10^6 m (b) 6.4×10^6 m
(c) 9.0×10^6 m (d) 1.6×10^6 m

62. Suppose that the angular velocity of rotation of earth is increased. Then, as a consequence.

[Online April 16, 2018]

(a) There will be no change in weight anywhere on the earth
(b) Weight of the object, everywhere on the earth, will decrease
(c) Weight of the object, everywhere on the earth, will increase
(d) Except at poles, weight of the object on the earth will decrease

63. Assuming the earth to be a sphere of uniform density, the acceleration due to gravity inside the earth at a distance of r from the centre is proportional to

[NCERT : P L-134 | Online May 12, 2012]

(a) r (b) r^{-1} (c) r^2 (d) r^{-2}

64. The change in the value of 'g' at a height 'h' above the surface of the earth is the same as at a depth 'd' below the surface of earth. When both 'd' and 'h' are much smaller than the radius of earth, then which one of the following is correct? **[NCERT : P L-133, 134 | 2005]**

(a) $d = \frac{3h}{2}$ (b) $d = \frac{h}{2}$
(c) $d = h$ (d) $d = 2h$

65. Average density of the earth

[NCERT : P L-132, 133 | 2005]

(a) is a complex function of g
(b) does not depend on g
(c) is inversely proportional to g
(d) is directly proportional to g

Topic 4 - Gravitational Field and Potential Energy

66. An astronaut takes a ball of mass m from earth to space. He throws the ball into a circular orbit about earth at an altitude of 318.5 km from earth's surface to the orbit, the change in total mechanical energy of the ball is $x\frac{GM_em}{21R_e}$. The value of x is (take $R_e = 6370$ km):

NCERT : P L-134 | [April 9, 2024 (I)]

(a) 11 (b) 9 (c) 12 (d) 10

67. The gravitational potential at a point above the surface of earth is -5.12×10^7 J/kg and the acceleration due to gravity at that point is 6.4 m/s^2. Assume that the mean radius of earth to be 6400 km. The height of this point above the earth's surface is:

[NCERT : P L-135 | Jan. 30, 2024 (I)]

(a) 1600 km (b) 540 km (c) 1200 km (d) 1000 km

68. A body is released from a height equal to the radius (R) of the earth. The velocity of the body when it strikes the surface of the earth will be :

(Given g = acceleration due to gravity on the earth.)

[NCERT : P L-134 | April 15, 2023 (I)]

(a) $\sqrt{gR}$ (b) $\sqrt{4gR}$ (c) $\sqrt{2gR}$ (d) $\sqrt{\frac{gR}{2}}$

69. If V is the gravitational potential due to sphere of uniform density on it's surface, then it's value at the center of sphere will be: **[NCERT : P L-135 | April 11, 2023 (II)]**

(a) $\frac{3V}{2}$ (b) V (c) $\frac{4}{3}V$ (d) $\frac{V}{2}$

70. An object of mass 1 kg is taken to a height from the surface of earth which is equal to three times the radius of earth. The gain in potential energy of the object will be [If, $g = 10ms^{-2}$ and radius of earth = 6400 km]

[NCERT : P L-134 | July 29, 2022 (II)]

(a) 48 MJ (b) 24 MJ (c) 36 MJ (d) 12 MJ

71. Four spheres each of mass m form a square of side d (as shown in figure). A fifth sphere of mass M is situated at the centre of square. The total gravitational potential energy of the system is :

[NCERT : P L-134 | June 27, 2022 (II)]

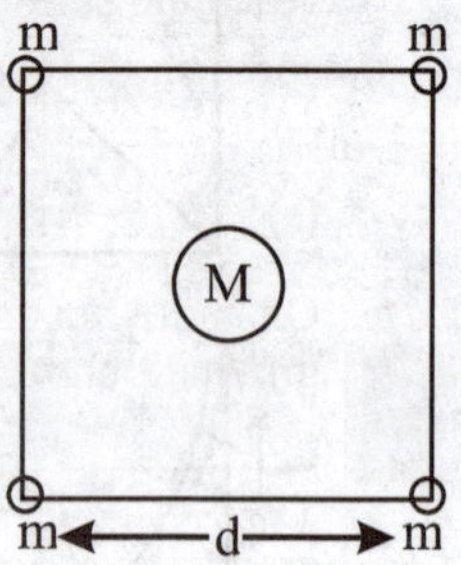

(a) $-\frac{Gm}{d}\left[(4+\sqrt{2})m + 4\sqrt{2}M\right]$

(b) $-\frac{Gm}{d}\left[(4+\sqrt{2})M + 4\sqrt{2}m\right]$

(c) $-\frac{Gm}{d}\left[3m^2 + 4\sqrt{2}M\right]$

(d) $-\frac{Gm}{d}\left[6m^2 + 4\sqrt{2}M\right]$

72. A body of mass (2M) splits into four masses {m, M – m, m, M – m}, which are rearranged to form a square as shown in the figure. The ratio of $\frac{M}{m}$ for which, the gravitational potential energy of the system becomes maximum is x : 1. The value of x is ________.

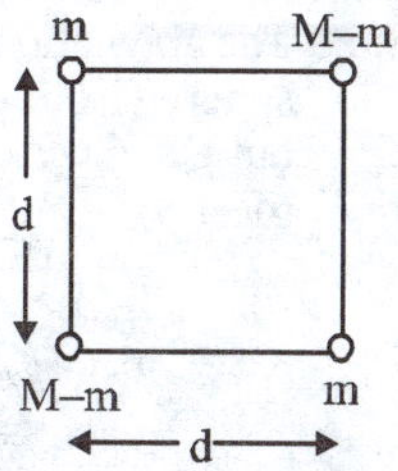

[NCERT : P L-134 | NA, Aug. 27, 2021 (I)]

73. A mass of 50 kg is placed at the centre of a uniform spherical shell of mass 100 kg and radius 50 m. If the gravitational potential at a point, 25 m from the centre is V kg/m. The value of V is : **[NCERT : P L-135 | Aug. 27, 2021 (II)]**

(a) –60 G (b) +2 G (c) –20 G (d) –4 G

74. Inside a uniform spherical shell : **[NCERT : P L-135 | Aug. 26, 2021 (I)]**

(A) the gravitational field is zero
(B) the gravitational potential is zero
(C) the gravitational field is same everywhere
(D) the gravitation potential is same everywhere
(E) all of the above

Choose the most appropriate answer from the options given below :

(a) (A), (C) and (D) only (b) (E) only
(c) (A), (B) and (C) only (d) (B), (C) and (D) only

75. If one wants to remove all the mass of the earth to infinity in order to break it up completely. The amount of energy that needs to be supplied will be $\frac{x}{5}\frac{GM^2}{R}$ where x is ______ (Round off to the Nearest Integer).
(M is the mass of earth, R is the radius of earth, G is the gravitational constant).

[NCERT : P L-134 | March 16, 2021 (II)]

76. Two planets have masses M and 16 M and their radii are a and $2a$, respectively. The separation between the centres of the planets is $10a$. A body of mass m is fired from the surface of the larger planet towards the smaller planet along the line joining their centres. For the body to be able to reach the surface of smaller planet, the minimum firing speed needed is : **[NCERT : P L-134 | 6 Sep. 2020 (II)]**

(a) $2\sqrt{\frac{GM}{a}}$ (b) $4\sqrt{\frac{GM}{a}}$ (c) $\sqrt{\frac{GM^2}{ma}}$ (d) $\frac{3}{2}\sqrt{\frac{5GM}{a}}$

77. On the x-axis and at a distance x from the origin, the gravitational field due to a mass distribution is given by $\frac{Ax}{(x^2+a^2)^{3/2}}$ in the x-direction. The magnitude of gravitational potential on the x-axis at a distance x, taking its value to be zero at infinity, is :

[NCERT : P L-135 | 4 Sep. 2020 (I)]

(a) $\frac{A}{(x^2+a^2)^{1/2}}$ (b) $\frac{A}{(x^2+a^2)^{3/2}}$

(c) $A(x^2+a^2)^{1/2}$ (d) $A(x^2+a^2)^{3/2}$

78. The mass density of a planet of radius R varies with the distance r from its centre as $\rho(r) = \rho_0\left(1-\frac{r^2}{R^2}\right)$. Then the gravitational field is maximum at :

[NCERT : P L-135 | 3 Sep. 2020 (II)]

(a) $r = \sqrt{\frac{3}{4}}R$ (b) $r = R$

(c) $r = \frac{1}{\sqrt{3}}R$ (d) $r = \sqrt{\frac{5}{9}}R$

79. Consider two solid spheres of radii $R_1 = 1m$, $R_2 = 2m$ and masses M_1 and M_2, respectively. The gravitational field due to sphere ① and ② are shown. The value of $\frac{m_1}{m_2}$ is:

[8 Jan. 2020 I]

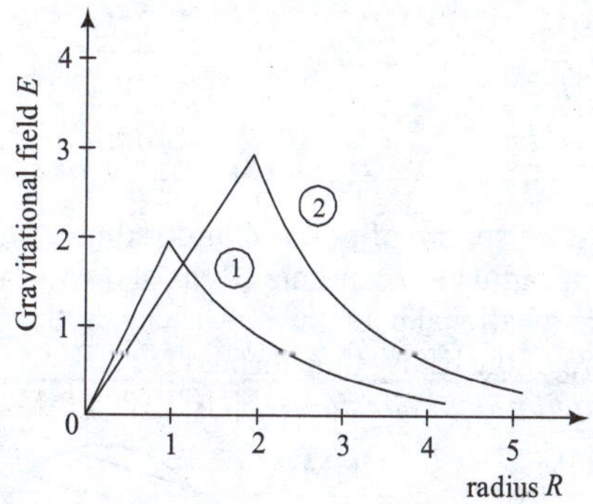

(a) $\frac{2}{3}$ (b) $\frac{1}{6}$ (c) $\frac{1}{2}$ (d) $\frac{1}{3}$

80. An asteroid is moving directly towards the centre of the earth. When at a distance of 10 R (R is the radius of the earth) from the earths centre, it has a speed of 12 km/s. Neglecting the effect of earths atmosphere, what will be the speed of the asteroid when it hits the surface of the earth (escape velocity from the earth is 11.2 km/ s)? Give your answer to the nearest integer in kilometre/s _____.

[NCERT : P L-134 | NA 8 Jan. 2020 II]

81. A solid sphere of mass 'M' and radius 'a' is surrounded by a uniform concentric spherical shell of thickness 2a and mass 2M. The gravitational field at distance '3a' from the centre will be: **[9 April 2019 I]**

(a) $\frac{2GM}{9a^2}$ (b) $\frac{GM}{9a^2}$ (c) $\frac{GM}{3a^2}$ (d) $\frac{2GM}{3a^2}$

82. Four identical particles of mass M are located at the corners of a square of side 'a'. What should be their speed if each of them revolves under the influence of others' gravitational field in a circular orbit circumscribing the square ? **[8 April 2019 I]**

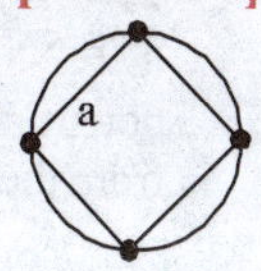

(a) $1.35\sqrt{\frac{GM}{a}}$ (b) $1.16\sqrt{\frac{GM}{a}}$

(c) $1.21\sqrt{\frac{GM}{a}}$ (d) $1.41\sqrt{\frac{GM}{a}}$

83. A test particle is moving in circular orbit in the gravitational field produced by a mass density $\rho(r)=\frac{K}{r^2}$. Identify the correct relation between the radius R of the particle's orbit and its period T: **[8 April 2019 II]**

(a) T/R is a constant (b) T^2/R^3 is a constant
(c) T/R^2 is a constant (d) TR is a constant

84. A body of mass m is moving in a circular orbit of radius R about a planet of mass M. At some instant, it splits into two equal masses. The first mass moves in a circular orbit of radius $\frac{R}{2}$, and the other mass, in a circular orbit of radius $\frac{3R}{2}$. The difference between the final and initial total energies is:
[NCERT : P L-134 | Online April 15, 2018]

(a) $-\frac{GMm}{2R}$ (b) $+\frac{GMm}{6R}$
(c) $-\frac{GMm}{6R}$ (d) $\frac{GMm}{2R}$

85. From a solid sphere of mass M and radius R, a spherical portion of radius R/2 is removed, as shown in the figure. Taking gravitational potential V = 0 at r = ∞, the potential at the centre of the cavity thus formed is :
(*G = gravitational constant*) **[NCERT : P L-135 | 2015]**

(a) $\frac{-2GM}{3R}$ (b) $\frac{-2GM}{R}$
(c) $\frac{-GM}{2R}$ (d) $\frac{-GM}{R}$

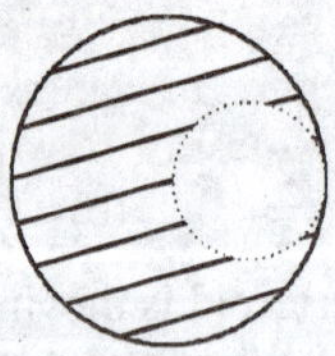

86. Which of the following most closely depicts the correct variation of the gravitational potential V(r) due to a large planet of radius R and uniform mass density ? (figures are not drawn to scale)
[NCERT : P L-135 | Online April 11, 2015]

(a) 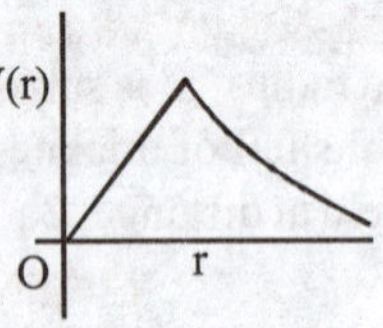

(b) 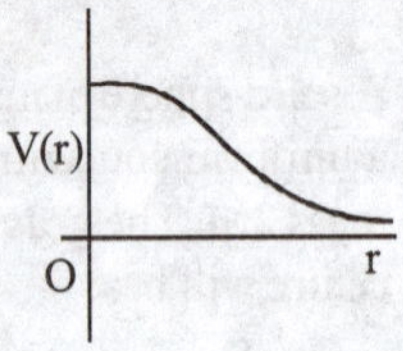

(c) 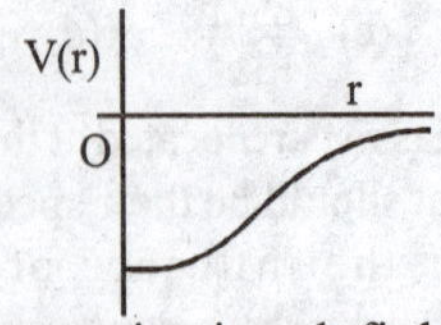

(d) 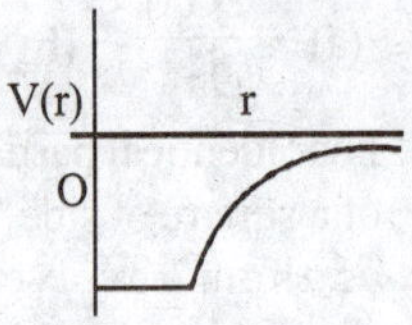

87. The gravitational field in a region is given by $\vec{g}=5\text{N/kg}\,\hat{i}+12\text{N/kg}\,\hat{j}$. The change in the gravitational potential energy of a particle of mass 1 kg when it is taken from the origin to a point (7 m, – 3 m) is:
[Online April 19, 2014]

(a) 71 J (b) $13\sqrt{58}$J (c) – 71 J (d) 1 J

88. The gravitational field, due to the 'left over part' of a uniform sphere (from which a part as shown, has been 'removed out'), at a very far off point, P, located as shown, would be (nearly) : **[Online April 9, 2013]**

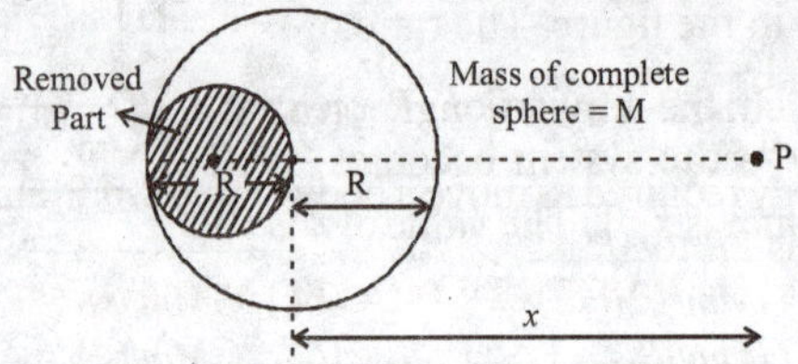

(a) $\frac{5}{6}\frac{GM}{x^2}$ (b) $\frac{8}{9}\frac{GM}{x^2}$ (c) $\frac{7}{8}\frac{GM}{x^2}$ (d) $\frac{6}{7}\frac{GM}{x^2}$

89. The mass of a spaceship is 1000 kg. It is to be launched from the earth's surface out into free space. The value of g and R (radius of earth) are 10 m/s^2 and 6400 km respectively. The required energy for this work will be
[NCERT : P L-134 | 2012]

(a) 6.4×10^{11} Joules (b) 6.4×10^{8} Joules
(c) 6.4×10^{9} Joules (d) 6.4×10^{10} Joules

90. A point particle is held on the axis of a ring of mass m and radius r at a distance r from its centre C. When released, it reaches C under the gravitational attraction of the ring. Its speed at C will be
[NCERT : P L-135 | Online May 26, 2012]

(a) $\sqrt{\frac{2Gm}{r}\left(\sqrt{2}-1\right)}$ (b) $\sqrt{\frac{Gm}{r}}$
(c) $\sqrt{\frac{2Gm}{r}\left(1-\frac{1}{\sqrt{2}}\right)}$ (d) $\sqrt{\frac{2Gm}{r}}$

91. Two bodies of masses m and 4 m are placed at a distance r. The gravitational potential at a point on the line joining them where the gravitational field is zero is: **[2011]**

(a) $-\frac{4Gm}{r}$ (b) $-\frac{6Gm}{r}$ (c) $-\frac{9Gm}{r}$ (d) zero

92. This question contains Statement-1 and Statement-2. Of the four choices given after the statements, choose the one that best describes the two statements. **[2008]**

Statement-1 : For a mass M kept at the centre of a cube of side 'a', the flux of gravitational field passing through its sides 4 π GM. and

Statement-2: If the direction of a field due to a point source is radial and its dependence on the distance 'r' from the source is given as $\frac{1}{r^2}$, its flux through a closed surface depends only on the strength of the source enclosed by the surface and not on the size or shape of the surface.

(a) Statement -1 is false, Statement-2 is true
(b) Statement -1 is true, Statement-2 is true; Statement -2 is a correct explanation for Statement-1
(c) Statement -1 is true, Statement-2 is true; Statement - 2 is not a correct explanation for Statement-1
(d) Statement -1 is true, Statement-2 is false

93. A particle of mass 10 g is kept on the surface of a uniform sphere of mass 100 kg and radius 10 cm. Find the work to be done against the gravitational force between them to take the particle far away from the sphere
(you may take $G=6.67\times10^{-11}\ \text{Nm}^2/\text{kg}^2$) **[2005]**

(a) 3.33×10^{-10} J (b) 13.34×10^{-10} J
(c) 6.67×10^{-10} J (d) 6.67×10^{-9} J

94. If 'g' is the acceleration due to gravity on the earth's surface, the gain in the potential energy of an object of mass 'm' raised from the surface of the earth to a height equal to the radius 'R' of the earth is **[NCERT : P L-134 | 2004]**

(a) $\frac{1}{4}mgR$ (b) $\frac{1}{2}mgR$ (c) $2\,mgR$ (d) mgR

95. Energy required to move a body of mass m from an orbit of radius $2R$ to $3R$ is **[NCERT : P L-134 | 2002]**

(a) $GMm/12R^2$ (b) $GMm/3R^2$
(c) $GMm/8R$ (d) $GMm/6R$.

Topic 5 - Motion of Satellites, Escape Speed and Orbital Velocity

96. An object is kept at rest at a distance of $3R$ above the earth's surface where R is earth's radius. The minimum speed with which it must be projected so that it does not return to earth is :
(Assume M = mass of earth, G = Universal gravitational constant) **[NCERT : P L-136 | April 4, 2025 (II)]**

(a) $\sqrt{\frac{GM}{2R}}$ (b) $\sqrt{\frac{GM}{R}}$ (c) $\sqrt{\frac{3GM}{R}}$ (d) $\sqrt{\frac{2GM}{R}}$

97. A satellite of mass 1000 kg is launched to revolve around the earth in an orbit at a height of 270 km from the earth's surface. Kinetic energy of the satellite in this orbit is ______ $\times 10^{10}$ J. **[NCERT : P L-137 | April 2, 2025 (II)]**
(Mass of earth = 6×10^{24} kg, Radius of earth = 6.4×10^6 m, Gravitational constant = 6.67×10^{-11} Nm^2 kg^{-2})

98. Earth has mass 8 times and radius 2 times that of a planet. If the escape velocity from the earth is 11.2 km/s, the escape velocity in km/s from the planet will be : **[NCERT : P L-136 | Jan. 28, 2025 (II)]**

(a) 11.2 (b) 5.6 (c) 2.8 (d) 8.4

99. A satellite is launched into a circular orbit of radius 'R' around the earth. A second satellite is launched into an orbit of radius 1.03 R. The time period of revolution of the second satellite is larger than the first one approximately by **[NCERT : P L-137 | Jan. 24, 2025 (I)]**

(a) 3% (b) 4.5% (c) 9% (d) 2.5%

100. If a satellite orbiting the Earth is 9 times closer to the Earth than the Moon, what is the time period of rotation of the satellite? (Given rotational time period of Moon = 27 days and gravitational attraction between the satellite and the moon is neglected.) **[NCERT : P L-137 | Jan. 23, 2025 (II)]**

(a) 1 day (b) 81 days (c) 27 days (d) 3 days

101. A satellite of mass $\frac{M}{2}$ is revolving around earth in a circular orbit at a height of $\frac{R}{3}$ from earth surface. The angular momentum of the satellite is $M\sqrt{\frac{GMR}{x}}$. The value of x is ______, where M and R are the mass and radius of earth, respectively.
(G is the gravitational constant)
[NCERT : P L-129 | Jan. 23, 2025 (II)]

102. A satellite of 10^3 kg mass is revolving in circular orbit of radius 2R. If $\frac{10^4 R}{6}$ J energy is supplied to the satellite, it would revolve in a new circular orbit of radius :
(use g = 10m/s^2, R = radius of earth) **[NCERT : P L-138 | April 9, 2024 (II)]**

(a) 2.5 R (b) 3 R (c) 4 R (d) 6 R

103. To project a body of mass m from earth's surface to infinity, the required kinetic energy is (assume, the radius of earth is R_E, g = acceleration due to gravity on the surface of earth) : **[NCERT : P L-135, 136 | April 6, 2024 (I)]**

(a) $2mgR_E$ (b) mgR_E (c) $\frac{1}{2}mgR_E$ (d) $4mgR_E$

104. Match **List - I** with **List-II** :

List - I	List - II
(A) Kinetic energy of planet	(I) $-\frac{GMm}{a}$
(B) Gravitation Potential energy of Sun-planet system.	(II) $\frac{GMm}{2a}$
(C) Total mechanical energy of planet	(III) $\frac{Gm}{r}$
(D) Escape energy at the surface of planet for unit mass object	(IV) $-\frac{GMm}{2a}$

(Where a = radius of planet orbit, r = radius of planet, M = mass of Sun, m = mass of planet)
Choose the correct answer from the options given below: **[NCERT : P L-134, 135, 136 | April 5, 2024 (I)]**

(a) (A) – II, (B) – I, (C) – IV, (D) – III
(b) (A) – III, (B) – IV, (C) – I, (D) – II
(c) (A) – I, (B) – IV, (C) – II, (D) – III
(d) (A) – I, (B) – II, (C) – III, (D) – IV

105. A satellite revolving around a planet in stationary orbit has time period 6 hours. The mass of planet is one-fourth the mass of earth. The radius orbit of planet is :
(Given = Radius of geo-stationary orbit for earth is 4.2×10^4 km) **[NCERT : P L-137 | April 5, 2024 (II)]**

(a) 1.4×10^4 km (b) 8.4×10^4 km
(c) 1.68×10^5 km (d) 1.05×10^4 km

106. Correct formula for height of a satellite from earths surface is : **[NCERT : P L-137 | April 4, 2024 (II)]**

(a) $\left(\frac{T^2R^2g}{4\pi}\right)^{1/2} - R$ (b) $\left(\frac{T^2R^2g}{4\pi^2}\right)^{1/3} - R$

(c) $\left(\frac{T^2R^2}{4\pi^2 g}\right)^{1/3} - R$ (d) $\left(\frac{T^2R^2}{4\pi^2}\right)^{-1/3} + R$

107. A light planet is revolving around a massive star in a circular orbit of radius R with a period of revolution T. If the force of attraction between planet and star is proportional to $R^{-3/2}$ then choose the correct option: **[NCERT : P L-137 | Feb. 1, 2024 (II)]**

(a) $T^2 \propto R^{7/2}$ (b) $T^2 \propto R^3$
(c) $T^2 \propto R^{5/2}$ (d) $T^2 \propto R^{3/2}$

108. The mass of the moon is 1/144 times the mass of a planet and its diameter 1/16 times the diameter of a planet. If the escape velocity on the planet is v, the escape velocity on the moon will be:
[NCERT : P L-136 | Jan. 31, 2024 (II)]

(a) $\frac{v}{3}$ (b) $\frac{v}{4}$ (c) $\frac{v}{12}$ (d) $\frac{v}{6}$

109. Escape velocity of a body from earth is 11.2 *km/s*. If the radius of a planet be one - third the radius of earth and mass be one - sixth that of earth, the escape velocity from the planet is : **[NCERT : P L-136 | Jan. 30, 2024 (II)]**

(a) 11.2 *km/s* (b) 8.4 *km/s*
(c) 4.2 *km/s* (d) 7.9 *km/s*

110. A planet having mass 9 M_e and radius $4R_e$, where M_e and R_e are mass and radius of earth respectively, has escape velocity in km/s given by:
(Given escape velocity on earth $V_e = 11.2 \times 10^3$ m/s)
[NCERT : P L-136 | April 13, 2023 (I)]

(a) 67.2 (b) 16.8 (c) 33.6 (d) 11.2

111. Given below are two statements:
Statement I : For a planet, if the ratio of mass of the planet to its radius increases, the escape velocity from the planet also increases.
Statement II : Escape velocity is independent of the radius of the planet.
In the light of above statements, choose the **most appropriate** answer from the options given below:
[NCERT : P L-136 | April 13, 2023 (II)]

(a) Both Statement I and Statement II are incorrect
(b) Statement I is correct but Statement II is incorrect
(c) Statement I is incorrect but statement II is correct
(d) Both Statement I and Statement II are correct

112. Two satellites A and B move round the earth in the same orbit. The mass of A is twice the mass of B. The quantity which is same for the two satellites will be :
[NCERT : P L-138 | April 12, 2023 (I)]

(a) Potential energy (b) Total energy
(c) Kinetic energy (d) Speed

113. The ratio of escape velocity of a planet to the escape velocity of earth will be:
[NCERT : P L-136 | April 12, 2023 (I)]
Given : Mass of the planet is 16 times mass of earth and radius of the planet is 4 times the radius of earth.

(a) 4 : 1 (b) 2 : 1 (c) $1:\sqrt{2}$ (d) 1 : 4

114. A spaceship of mass 2×10^4 kg is launched into a circular orbit close to the earth surface. The additional velocity to be imparted to the spaceship in the orbit to overcome the gravitational pull will be (if g = 10 m/s^2 and radius of earth = 6400 km) **[NCERT : P L-136 | April 11, 2023 (II)]**

(a) $11.2(\sqrt{2}-1)$ km/s (b) $7.9(\sqrt{2}-1)$ km/s
(c) $8(\sqrt{2}-1)$ km/s (d) $7.4(\sqrt{2}-1)$ km/s

115. Two satellites of masses m and 3m revolve around the earth in circular orbits of radii r & 3r respectively. The ratio of orbital speeds of the satellites respectively is :
[NCERT : P L-137 | April 10, 2023 (I)]

(a) 1 : 1 (b) 3 : 1 (c) $\sqrt{3}:1$ (d) 9 : 1

116. The time period of a satellite, revolving above earth's surface at a height equal to R will be
[NCERT : P L-137 | April 10, 2023 (II)]
(Given $g = \pi^2$ m/s^2, R = radius of earth)

(a) $\sqrt{4R}$ (b) $\sqrt{8R}$ (c) $\sqrt{32R}$ (d) $\sqrt{2R}$

117. Given below are two statements:
Statement I : If E be the total energy of a satellite moving around the earth, then its potential energy will be $\frac{E}{2}$.
Statement II: The kinetic energy of a satellite revolving in an orbit is equal to the half the magnitude of total energy E.
In the light of the above statements, choose the most appropriate answer from the options given below:
[NCERT : P L-138 | April 8, 2023 (I)]

(a) Both Statement I and Statement II are correct
(b) Both Statement I and Statement II are incorrect
(c) Statement I is incorrect but Statement II is correct
(d) Statement I is correct but Statement II is incorrect

118. If earth has a mass nine times and radius twice to that of a planet P. Then $\frac{v_e}{3}\sqrt{x}$ ms^{-1} will be the minimum velocity required by a rocket to pull out of gravitational force of P, where v_e is escape velocity on earth. The value of x is
[NCERT : P L-136 | Feb. 1, 2023 (I)]

(a) 18 (b) 3 (c) 2 (d) 1

119. The escape velocities of two planets A and B are in the ratio 1 : 2. If the ratio of their radii respectively is 1 : 3, then the ratio of acceleration due to gravity of planet A to the acceleration of gravity of planet B will be:
[NCERT : P L-136 | Feb. 1, 2023 (II)]

(a) $\frac{4}{3}$ (b) $\frac{2}{3}$ (c) $\frac{3}{4}$ (d) $\frac{3}{2}$

120. Two satellites A and B having masses in the ratio 4 : 3 are revolving in circular orbits of radii 3*r* and 4*r* respectively around the earth. The ratio of total mechanical energy of A to B is : **[NCERT : P L-138 | July 27, 2022 (I)]**

(a) 9 : 16 (b) 16 : 9 (c) 1 : 1 (d) 4 : 3

121. A body of mass m is projected with velocity λv_e in vertically upward direction from the surface of the earth into space. It is given that v_e is escape velocity and $\lambda < 1$. If air resistance is considered to the negligible, then the maximum height from the centre of earth, to which the body can go, will be (R : radius of earth)
[NCERT : P L-138 | July 27, 2022 (II)]

(a) $\frac{R}{1+\lambda^2}$ (b) $\frac{R}{1-\lambda^2}$ (c) $\frac{R}{1-\lambda}$ (d) $\frac{\lambda^2 R}{1-\lambda^2}$

122. A body is projected vertically upwards from the surface of earth with a velocity equal to one third of escape velocity. The maximum height attained by the body will be: (Take radius of earth = 6400 km and g = 10 ms^{-2})
[NCERT : P L-136 | July 26, 2022 (II)]

(a) 800 km (b) 1600 km (c) 2133 km (d) 4800 km

123. The escape velocity of a body on a planet 'A' is 12 kms^{-1}. The escape velocity of the body on another planet 'B', whose density is four times and radius is half of the planet 'A', is : **[NCERT : P L-136 | June 29, 2022 (I)]**

(a) 12 kms^{-1} (b) 24 kms^{-1}
(c) 36 kms^{-1} (d) 6 kms^{-1}

124. Two satellites S_1 and S_2 are revolving in circular orbits around a planet with radius $R_1 = 3200$ km and $R_2 = 800$ km respectively. The ratio of speed of satellite S_1 to the speed of satellite S_2 in their respective orbits would be $\frac{1}{x}$ where $x =$

[NCERT : P L-137 | June 25, 2022 (II)]

125. Two satellites revolve around a planet in coplanar circular orbits in anticlockwise direction. Their period of revolutions are 1 hour and 8 hours respectively. The radius of the orbit of nearer satellite is 2×10^3 km. The angular speed of the farther satellite as observed from the nearer satellite at the instant when both the satellites are closest is $\frac{\pi}{x}$ rad h^{-1} where x is _______.

[NCERT : P L-137 | NA, Sep. 1, 2021 (II)]

126. The masses and radii of the earth and moon are (M_1, R_1) and (M_2, R_2) respectively. Their centres are at a distance 'r' apart. Find the minimum escape velocity for a particle of mass 'm' to be projected from the middle of these two masses: **[NCERT : P L-138 | Aug. 31, 2021 (I)]**

(a) $V = \frac{1}{2}\sqrt{\frac{4G(M_1 + M_2)}{r}}$ (b) $V = \sqrt{\frac{4G(M_1 + M_2)}{r}}$

(c) $V = \frac{1}{2}\sqrt{\frac{2G(M_1 + M_2)}{r}}$ (d) $V = \frac{\sqrt{2G}(M_1 + M_2)}{r}$

127. The radius in kilometer to which the present radius of earth (R = 6400 km) to be compressed so that the escape velocity is increased 10 times is _______.

[NCERT : P L-136 | March 17, 2021 (I)]

128. The maximum and minimum distances of a comet from the sun are 1.6×10^{12} m and 8.0×10^{10} m respectively. If the speed of the comet at the nearest point is 6×10^4 ms^{-1}, the speed at the farthest point is:

[NCERT : P L-137 | March 16, 2021 (I)]

(a) 6.0×10^3 m/s (b) 3.0×10^3 m/s
(c) 4.5×10^3 m/s (d) 1.5×10^3 m/s

129. Two satellites A and B of masses 200 kg and 400 kg are revolving round the earth at height of 600 km and 1600 km respectively. If T_A and T_B are the time periods of A and B respectively then the value of $T_B - T_A$:

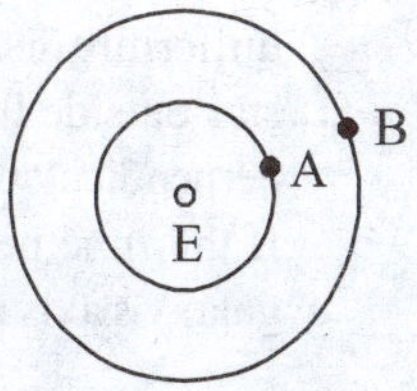

[Given: radius of earth = 6400 km, mass of earth = 6×10^{24} kg]

[NCERT : P L-137 | Feb. 25, 2021 (I)]

(a) 4.24×10^2 s (b) 1.33×10^3 s
(c) 4.24×10^3 s (d) 3.33×10^2 s

130. The initial velocity v_i required to project a body vertically upward from the surface of the earth to reach a height of 10R, where R is the radius of the earth, may be described in terms of escape velocity v_e such that $v_i = \sqrt{\frac{x}{11}} \times v_e$. The value of x will be ______.

[NCERT : P L-136 | Feb. 25, 2021 (II)]

131. Two stars of masses m and $2m$ at a distance d rotate about their common centre of mass in free space. The period of revolution is : **[NCERT : P L-137 | Feb. 24, 2021 (I)]**

(a) $\frac{1}{2\pi}\sqrt{\frac{d^3}{3Gm}}$ (b) $2\pi\sqrt{\frac{d^3}{3Gm}}$

(c) $\frac{1}{2\pi}\sqrt{\frac{3Gm}{d^3}}$ (d) $2\pi\sqrt{\frac{3Gm}{d^3}}$

132. A satellite is in an elliptical orbit around a planet P. It is observed that the velocity of the satellite when it is farthest from the planet is 6 times less than that when it is closest to the planet. The ratio of distances between the satellite and the planet at closest and farthest points is :

[NCERT : P L-137 | NA 6 Sep. 2020 (I)]

(a) 1 : 6 (b) 1 : 3 (c) 1 : 2 (d) 3 : 4

133. A body is moving in a low circular orbit about a planet of mass M and radius R. The radius of the orbit can be taken to be R itself. Then the ratio of the speed of this body in the orbit to the escape velocity from the planet is :

[NCERT : P L-136 | 4 Sep. 2020 (II)]

(a) $\frac{1}{\sqrt{2}}$ (b) 2 (c) 1 (d) $\sqrt{2}$

134. A satellite is moving in a low nearly circular orbit around the earth. Its radius is roughly equal to that of the earth's radius R_e. By firing rockets attached to it, its speed is instantaneously increased in the direction of its motion so that it become $\sqrt{\frac{3}{2}}$ times larger. Due to this the farthest distance from the centre of the earth that the satellite reaches is R. Value of R is: **[NCERT : P L-137 | 3 Sep. 2020 (I)]**

(a) $4R_e$ (b) $2.5R_e$ (c) $3R_e$ (d) $2R_e$

135. The mass density of a spherical galaxy varies as $\frac{K}{r}$ over a large distance 'r' from its centre. In that region, a small star is in a circular orbit of radius R. Then the period of revolution, T depends on R as : **[2 Sep. 2020 (I)]**

(a) $T^2 \propto R$ (b) $T^2 \propto R^3$ (c) $T^2 \propto \frac{1}{R^3}$ (d) $T \propto R$

136. A body A of mass m is moving in a circular orbit of radius R about a planet. Another body B of mass $\frac{m}{2}$ collides with A with a velocity which is half $\left(\frac{\vec{v}}{2}\right)$ the instantaneous velocity $\vec{v}$ of A. The collision is completely inelastic. Then, the combined body:

[NCERT : P L-136, 137 | 9 Jan. 2020 (I)]

(a) continues to move in a circular orbit
(b) Escapes from the Planet's Gravitational field
(c) Falls vertically downwards towards the planet
(d) starts moving in an elliptical orbit around the planet

137. The energy required to take a satellite to a height 'h' above Earth surface (radius of Eareth = 6.4×10^3 km) is E_1 and kinetic energy required for the satellite to be in a circular orbit at this height is E_2. The value of h for which E_1 and E_2 are equal, is:

[NCERT : P L-138 | 9 Jan. 2019 II]

(a) 1.6×10^3 km (b) 3.2×10^3 km
(c) 6.4×10^3 km (d) 28×10^4 km

138. Planet A has mass M and radius R. Planet B has half the mass and half the radius of Planet A. If the escape velocities from the Planets A and B are v_A and v_B, respectively, then $\frac{v_A}{v_B}=\frac{n}{4}$. The value of n is :

[NCERT : P L-136 | 9 Jan. 2020 II]

(a) 4 (b) 1
(c) 2 (d) 3

139. A satellite of mass m is launched vertically upwards with an initial speed u from the surface of the earth. After it reaches height R (R = radius of the earth), it ejects a rocket of mass $\frac{m}{10}$ so that subsequently the satellite moves in a circular orbit. The kinetic energy of the rocket is (G is the gravitational constant; M is the mass of the earth): **[NCERT : P L-136 | 7 Jan. 2020 I]**

(a) $\frac{m}{20}\left(u^2+\frac{113}{200}\frac{GM}{R}\right)$ (b) $5m\left(u^2-\frac{119}{200}\frac{GM}{R}\right)$

(c) $\frac{3m}{8}\left(u+\sqrt{\frac{5GM}{6R}}\right)^2$ (d) $\frac{m}{20}\left(u-\sqrt{\frac{2GM}{3R}}\right)^2$

140. A spaceship orbits around a planet at a height of 20 km from its surface. Assuming that only gravitational field of the planet acts on the spaceship, what will be the number of complete revolutions made by the spaceship in 24 hours around the planet ? [Given : Mass of Planet = 8×10^{22} kg, Radius of planet = 2×10^6 m, Gravitational constant G = 6.67 $\times 10^{-11}$ Nm^2/kg^2] **[NCERT : P L-137 | 10 April 2019 II]**

(a) 9 (b) 17
(c) 13 (d) 11

141. A rocket has to be launched from earth in such a way that it never returns. If E is the minimum energy delivered by the rocket launcher, what should be the minimum energy that the launcher should have if the same rocket is to be launched from the surface of the moon? (Assume that the density of the earth and the moon are equal and that the earth's volume is 64 times the volume of the moon).

[NCERT : P L-136 | 8 April 2019 II]

(a) $\frac{E}{64}$ (b) $\frac{E}{32}$

(c) $\frac{E}{4}$ (d) $\frac{E}{16}$

142. Two satellites, A and B, have masses m and 2m respectively. A is in a circular orbit of radius R, and B is in a circular orbit of radius 2R around the earth. The ratio of their kinetic energies, T_A/T_B, is : **[NCERT : P L-137 | 12 Jan. 2019 II]**

(a) $\frac{1}{2}$ (b) 1

(c) 2 (d) $\sqrt{\frac{1}{2}}$

143. A satellite is revolving in a circular orbit at a height h from the earth surface, such that h << R where R is the radius of the earth. Assuming that the effect of earth's atmosphere can be neglected, the minimum increase in the speed required so that the satellite could escape from the gravitational field of earth is:

[NCERT : P L-136 | 11 Jan. 2019 (I), 2016]

(a) $\sqrt{2gR}$ (b) $\sqrt{gR}$

(c) $\sqrt{\frac{gR}{2}}$ (d) $\sqrt{gR}\left(\sqrt{2}-1\right)$

144. A satellite is moving with a constant speed v in circular orbit around the earth. An object of mass 'm' is ejected from the satellite such that it just escapes from the gravitational pull of the earth. At the time of ejection, the kinetic energy of the object is:

[NCERT : P L-137 | 10 Jan. 2019 I]

(a) $2mv^2$ (b) mv^2

(c) $\frac{1}{2}mv^2$ (d) $\frac{3}{2}mv^2$

145. Two stars of masses 3×10^{31} kg each, and at distance 2×10^{11} m rotate in a plane about their common centre of mass O. A meteorite passes through O moving perpendicular to the star's rotation plane. In order to escape from the gravitational field of this double star, the minimum speed that meteorite should have at O is: (Take Gravitational constant G = 6.6×10^{-11} Nm^2 kg^{-2})

[NCERT : P L-136 | 10 Jan. 2019 II]

(a) 2.4×10^4 m/s (b) 1.4×10^5 m/s
(c) 3.8×10^4 m/s (d) 2.8×10^5 m/s

146. A very long (length L) cylindrical galaxy is made of uniformly distributed mass and has radius R(R << L). A star outside the galaxy is orbiting the galaxy in a plane perpendicular to the galaxy and passing through its centre. If the time period of star is T and its distance from the galaxy's axis is r, then : **[Online April 10, 2015]**

(a) $T \propto r$ (b) $T \propto \sqrt{r}$
(c) $T \propto r^2$ (d) $T^2 \propto r^3$

147. What is the minimum energy required to launch a satellite of mass m from the surface of a planet of mass M and radius R in a circular orbit at an altitude of 2R?

[NCERT : P L-136 | 2013]

(a) $\frac{5GmM}{6R}$ (b) $\frac{2GmM}{3R}$

(c) $\frac{GmM}{2R}$ (d) $\frac{GmM}{2R}$

148. A planet in a distant solar system is 10 times more massive than the earth and its radius is 10 times smaller. Given that the escape velocity from the earth is 11 km s^{-1}, the escape velocity from the surface of the planet would be **[NCERT : P L-136 | 2008]**

(a) 1.1 km s^{-1} (b) 11 km s^{-1}
(c) 110 km s^{-1} (d) 0.11 km s^{-1}

149. Suppose the gravitational force varies inversely as the nth power of distance. Then the time period of a planet in circular orbit of radius 'R' around the sun will be proportional to **[2004]**

(a) R^n (b) $R^{\left(\frac{n-1}{2}\right)}$
(c) $R^{\left(\frac{n+1}{2}\right)}$ (d) $R^{\left(\frac{n-2}{2}\right)}$

150. The time period of an earth satellite in circular orbit is independent of **[NCERT : P L-137 | 2004]**

(a) both the mass and radius of the orbit
(b) radius of its orbit
(c) the mass of the satellite
(d) neither the mass of the satellite nor the radius of its orbit.

151. A satellite of mass m revolves around the earth of radius R at a height x from its surface. If g is the acceleration due to gravity on the surface of the earth, the orbital speed of the satellite is **[NCERT : P L-137 | 2004]**

(a) $\frac{gR^2}{R+x}$ (b) $\frac{gR}{R-x}$
(c) gx (d) $\left(\frac{gR^2}{R+x}\right)^{1/2}$

152. The escape velocity for a body projected vertically upwards from the surface of earth is 11 km/s. If the body is projected at an angle of 45°with the vertical, the escape velocity will be **[NCERT : P L-136 | 2003]**

(a) $11\sqrt{2}$ km/s (b) 22 km/s
(c) 11 km/s (d) $\frac{11}{\sqrt{2}}$ km/s

153. The kinetic energy needed to project a body of mass m from the earth surface (radius R) to infinity is **[NCERT : P L-136 | 2002]**

(a) $mgR/2$ (b) $2mgR$
(c) mgR (d) $mgR/4$.

154. The escape velocity of a body depends upon its mass as **[NCERT : P L-136 | 2002]**

(a) m^0 (b) m^1
(c) m^2 (d) m^3

ANSWER KEYS

1	(a)	17	(c)	33	(d)	49	(a)	65	(d)	81	(c)	97	(3)	113	(b)	129	(b)	145	(d)
2	(b)	18	(d)	34	(c)	50	(a)	66	(a)	82	(b)	98	(b)	114	(c)	130	(10)	146	(a)
3	(b)	19	(b)	35	(c)	51	(6)	67	(a)	83	(a)	99	(b)	115	(c)	131	(b)	147	(a)
4	(16)	20	(c)	36	(d)	52	(d)	68	(a)	84	(c)	100	(a)	116	(c)	132	(a)	148	(c)
5	(d)	21	(b)	37	(d)	53	(b)	69	(a)	85	(d)	101	(3)	117	(b)	133	(a)	149	(c)
6	(d)	22	(b)	38	(a)	54	(c)	70	(a)	86	(c)	102	(d)	118	(c)	134	(c)	150	(c)
7	(b)	23	(c)	39	(c)	55	(4)	71	(a)	87	(d)	103	(b)	119	(c)	135	(a)	151	(d)
8	(c)	24	(d)	40	(b)	56	(b)	72	(2)	88	(c)	104	(a)	120	(b)	136	(d)	152	(c)
9	(d)	25	(a)	41	(d)	57	(b)	73	(d)	89	(d)	105	(d)	121	(b)	137	(b)	153	(c)
10	(d)	26	(a)	42	(d)	58	(a)	74	(a)	90	(c)	106	(b)	122	(a)	138	(a)	154	(a)
11	(b)	27	(a)	43	(b)	59	(d)	75	(3)	91	(c)	107	(c)	123	(a)	139	(b)		
12	(c)	28	(c)	44	(2)	60	(d)	76	(d)	92	(b)	108	(a)	124	(2)	140	(d)		
13	(c)	29	(9)	45	(d)	61	(a)	77	(a)	93	(c)	109	(d)	125	(3)	141	(d)		
14	(c)	30	(d)	46	(a)	62	(d)	78	(d)	94	(b)	110	(b)	126	(b)	142	(b)		
15	(c)	31	(d)	47	(a)	63	(a)	79	(b)	95	(d)	111	(b)	127	(64)	143	(d)		
16	(8)	32	(d)	48	(d)	64	(d)	80	(16)	96	(a)	112	(d)	128	(b)	144	(b)		

Mechanical Properties of Solids

8

TREND ANALYSIS
JEE MAIN (2025-2021)

	2025				2024				2023				2022				2021			
	Session II		Session I		Session II		Session I		Session II		Session I		Session II		Session I		Session II		Session I	
	MCQs	NVQs	MCQs	NVQs	MCQs	NVQs	MCQs	NVQs	MCQs	NVQs	MCQs	NVQs	MCQs	NVQs	MCQs	NVQs	MCQs	NVQs	MCQs	NVQs
Topic 1	2	1	–	–	–	3	4	4	2	4	5	1	4	7	1	–	3	1	1	1
Topic 2	1	3	1	2	1	–	–	2	–	–	–	1	–	1	1	–	–	1	3	–
Topic 3																				
Topic 4																				
Total	3	4	1	2	1	3	4	6	2	4	5	2	4	8	2	-	3	2	4	1

How to Code Qns for Better & Faster Revision →

Type 1: Able to solve (No code Req)
Type 2: Solved with struggle or by seeing solution
(Use code NR means Needs Revision or orange stickers)
Type 3: Tricky/ Could not solve on my own. Need help from Teacher/ Friend.
(Use code TQ means Tricky Question or red stickers)

You can either hand write the code in boxes or else paste coloured stickers given at end of book.

Topic 1 - Hooke's Law & Young's Modulus

1. A 3 m long wire of radius 3 mm shows an extension of 0.1 mm when loaded vertically by a mass of 50 kg in an experiment to determine Young's modulus. The value of Young's modulus of the wire as per this experiment is $P \times 10^{11}\ Nm^{-2}$, where the value of P is : (Take $g = 3\pi\ m/s^2$)
[NCERT : P L-170 | April 8, 2025 (II)]
(a) 5 (b) 10 (c) 25 (d) 2.5

2. Two wires A and B are made of same material having ratio of lengths $\frac{L_A}{L_B} = \frac{1}{3}$ and their diameters ratio $\frac{d_A}{d_B} = 2$. If both the wires are stretched using same force, what would be the ratio of their respective elongations?
[NCERT : P L-170 | April 7, 2025 (I)]
(a) 1 : 6 (b) 1 : 12 (c) 3 : 4 (d) 1 : 3

3. The length of a light string is 1.4 m when the tension on it is 5 N. If the tension increases to 7 N, the length of the string is 1.56 m. The original length of the string is ______ m. **[NCERT : P L-170 | April 2, 2025 (II)]**

4. Two persons pull a wire towards themselves. Each person exerts a force of 200 N on the wire. Young's modulus of the material of wire is $1 \times 10^{11}\ N\ m^{-2}$. Original length of the wire is 2 m and the area of cross section is 2 cm^2. The wire will extend in length by ______ μm.
[NCERT : P L-170 | April 9, 2024 (I)]

5. A wire of cross sectional area A, modulus of elasticity $2 \times 10^{11}\ Nm^{-2}$ and length 2 m is stretched between two vertical rigid supports. When a mass of 2 kg is suspended at the middle it sags lower from its original position making angle $\theta = \frac{1}{100}$ radian on the points of support. The value of A is _____ $\times 10^{-4}\ m^2$ (consider x<<L).
(given : $g = 10\ m/s^2$)
[NCERT : P L-170 | April 6, 2024 (II)]

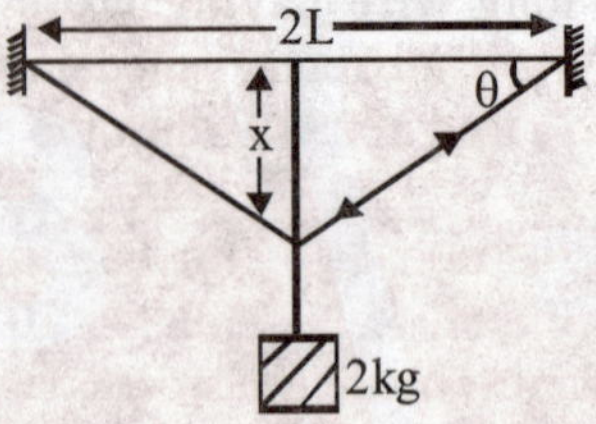

6. The density and breaking stress of a wire are 6×10^4 kg/m^3 and 1.2×10^8 N/m^2 respectively. The wire is suspended from a rigid support on a planet where acceleration due to gravity is $\left(\frac{1}{3}\right)$rd of the value on the surface of earth. The maximum length of the wire with breaking is .... m (take, g = 10 m/s^2) **[NCERT : P L-168 | April 5, 2024 (I)]**

7. With rise in temperature, the Young's modulus of elasticity:
[NCERT : P L-170 | Feb. 1, 2024 (I)]

(a) changes erratically (b) decreases
(c) increases (d) remains unchanged

8. One end of a metal wire is fixed to a ceiling and a load of 2 kg hangs from the other end. A similar wire is attached to the bottom of the load and another load of 1 kg hangs from this lower wire. Then the ratio of longitudinal strain of upper wire to that of the lower wire will be ______.
[Area of cross section of wire = 0.005 cm^2, Y = 2×10^{11} Nm^{-2} and g = 10 ms^{-2}]
[NCERT : P L-170 | Feb. 1, 2024 (II)]

9. Two blocks of mass 2 kg and 4 kg are connected by a metal wire going over a smooth pulley as shown in figure. The radius of wire is 4.0×10^{-5} m and Young's modulus of the metal is 2.0×10^{11} N/m^2.
[NCERT : P L-170 | Jan. 31, 2024 (II)]

The longitudinal strain developed in the wire is $\frac{1}{\alpha\pi}$. The value of α is____.
[Use g = 10 m/s^2)

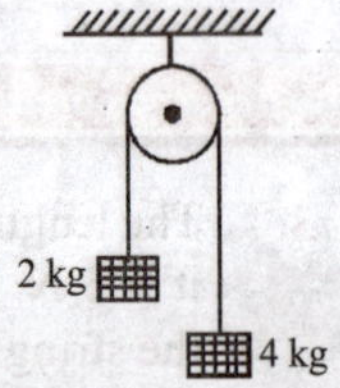

10. Young's modules of materials of a wire of Length 'L' and cross-sectional area A is Y. If the length of the wire is doubled and cross-sectional area is halved then Young's modules will be: **[NCERT : P L-170 | Jan. 30, 2024 (I)]**

(a) $\frac{Y}{4}$ (b) 4Y (c) Y (d) 2Y

11. Each of three blocks P, Q and R shown in figure has a mass of 3kg. Each of the wires A and B has cross - sectional area 0.005 cm^2 and Young's modulus 2×10^{11} Nm^{-2}. Neglecting friction, the longitudinal strain on wire B is ________ $\times 10^{-4}$. (Take g = 10m/s^2)
[NCERT : P L-170 | Jan. 30, 2024 (I)]

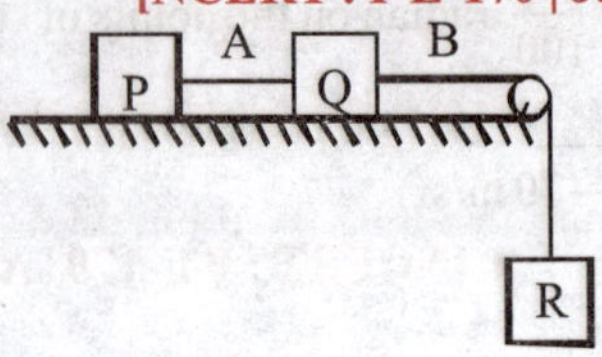

12. A wire of length L and radius r is clamped at one end. If its other end is pulled by a force F, its length increases by l. If the radius of the wire and the applied force both are reduced to half of their original values keeping original length constant, the increase in length will become :
[NCERT : P L-170 | Jan. 29, 2024 (II)]

(a) 3 times (b) 4 times (c) $\frac{3}{2}$ times (d) 2 times

13. Two metallic wires P and Q have same volume and are made up of same material. If their area of cross sections are in the ratio 4 : 1 and force F_1 is applied to P an extension of Δl is produced. The force which is required to produce same extension in Q is F_2.

The value of $\frac{F_1}{F_2}$ is ______.

[NCERT : P L-170 | Jan. 29, 2024 (II)]

14. Given below are two statements : one is labelled as Assertion (A) and the other is labelled as Reason (R)
Assertion (A) : The property of body, by virtue of which it tends to regain its original shape when the external force is removed, is Elasticity.
Reason (R) : The restoring force depends upon the bonded inter atomic and inter molecular force of solid.
In the light of the above statements, choose the correct answer from the options given below :
[NCERT : P L-168, 170 | Jan. 27, 2024 (II)]

(a) (A) is false but (R) is true
(b) (A) is true but (R) is false
(c) Both (A) and (R) are true and (R) is the correct explanation (A)
(d) Both (A) and (R) are true but (R) is not the correct explanation of (A)

15. A wire of length 'L' and radius 'r' is clamped rigidly at one end. When the other end of the wire is pulled by a force f, its length increases by 'ℓ'. Another wire of same material of length '2L' and radius '2r' is pulled by a force '2f'. Then the increase in its length will be :
[NCERT : P L-170 | April 15, 2023 (I)]

(a) 2ℓ (b) ℓ (c) 4ℓ (d) $\ell/2$

16. The length of wire becomes l_1 and l_2 when 100 N and 120 N tensions are applied respectively. If 10 l_2 = 11 l_1, the natural length of wire will be $\frac{1}{x}l_1$. Here the value of x is ____.

[NCERT : P L-170 | April 11, 2023 (I)]

17. Two wires each of radius 0.2 cm and negligible mass, one made of steel and other made of brass are loaded as shown in the figure. The elongation of the steel wire is ____ $\times 10^{-6}$ m. [Young's modulus for steel = 2×10^{11} Nm^{-2} and g = 10ms^{-2}] **[NCERT : P L-170 | April 10, 2023 (I)]**

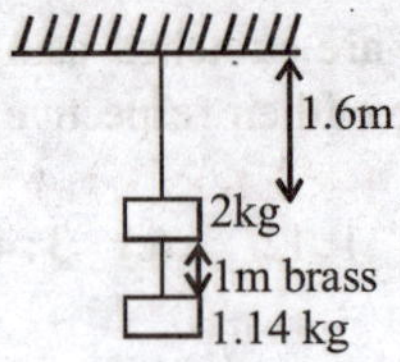

18. Young's moduli of the material of wires A and B are in the ratio of 1 : 4, while its area of cross sections are in the ratio of 1 : 3. If the same amount of load is applied to both the wires, the amount of elongation produced in the wires A and B will be in the ratio of **[NCERT : P L-170 | April 10, 2023 (II)]**

[Assume length of wires A and B are same]

(a) 36 : 1 (b) 12 : 1 (c) 1 : 36 (d) 1 : 12

19. A steel rod has a radius of 20 mm and a length of 2.0 m. A force of 62.8 kN stretches it along its length. Young's modulus of steel is 2.0×10^{11} N/m². The longitudinal strain produced in the wire is ____ $\times 10^{-5}$ **[NCERT : P L-170 | April 6, 2023 (I)]**

20. A metal block of mass m is suspended from a rigid support through a metal wire of diameter 14 mm. The tensile stress developed in the wire under equilibrium state is 7×10^5 Nm^{-2}. The value of mass m is ____ kg.

(Take, $g = 9.8$ ms^{-2} and $\pi = \frac{22}{7}$) **[NCERT : P L-168 | April 6, 2023 (II)]**

21. The Young's modulus of a steel wire of length 6 m and cross-sectional area 3 mm², is 2×10^{11} N/m². The wire is suspended from its support on a given planet. A block of mass 4 kg is attached to the free end of the wire. The acceleration due to gravity on the planet is $\frac{1}{4}$ of its value on the earth. The elongation of wire is (Take g on the earth = 10 m/s²): **[NCERT : P L-170 | Jan. 31, 2023 (I)]**

(a) 1 mm (b) 1 cm (c) 0.1 cm (d) 0.1 mm

22. Under the same load, wire A having length 5.0 m and cross section 2.5×10^{-5} m² stretches uniformly by the same amount as another wire B of length 6.0 m and a cross section of 3.0×10^{-5} m² stretches. The ratio of the Young's modulus of wire A to that of wire B will be: **[NCERT : P L-170 | Jan. 31, 2023 (II)]**

(a) 1 : 4 (b) 1 : 1 (c) 1 : 10 (d) 1 : 2

23. A force is applied to a steel wire 'A', rigidly clamped at one end. As a result elongation in the wire is 0.2 mm. If same force is applied to another steel wire 'B' of double the length and a diameter 2.4 times that of the wire 'A', the elongation in the wire 'B' will be (wires having uniform circular cross sections) **[NCERT : P L-170 | Jan. 30, 2023 (II)]**

(a) 6.06×10^{-2} mm (b) 2.77×10^{-2} mm
(c) 3.0×10^{-2} mm (d) 6.9×10^{-2} mm

24. As shown in the figure, in an experiment to determine Young's modulus of a wire, the extension-load curve is plotted. The curve is a straight line passing through the origin and makes an angle of 45° with the load axis. The length of wire is 62.8 cm and its diameter is 4 mm. The Young's modulus is found to be $x \times 10^4$ Nm^{-2}. The value of x is ____. **[NCERT : P L-170 | Jan. 25, 2023 (I)]**

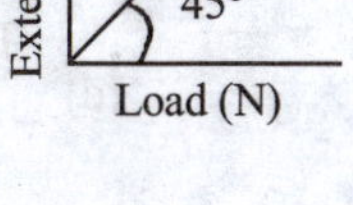

25. A 100 *m* long wire having cross-sectional area 6.25×10^{-4} m^2 and Young's modulus is 10^{10} Nm^{-2} is subjected to a load of 250 *N*, then the elongation in the wire will be : **[NCERT : P L-170 | Jan. 24, 2023 (I)]**

(a) 6.25×10^{-3} m (b) 4×10^{-4} m
(c) 6.25×10^{-6} m (d) 4×10^{-3} m

26. Given below are two statements: one is labelled as Assertion A and the other is labelled as Reason R. **[NCERT : P L-169 | Jan. 24, 2023 (II)]**

Assertion A: Steel is used in the construction of buildings and bridges.

Reason R: Steel is more elastic and its elastic limit is high.

In the light of above statements, choose the most appropriate answer from the options given below:

(a) Both A and R are correct but R is NOT the correct explanation of A.
(b) A is not correct but R is correct.
(c) Both A and R are correct and R is the correct explanation of A.
(d) A is correct but R is not correct.

27. If the length of a wire is made double and radius is halved of its respective values. Then, the Young's modules of the material of the wire will : **[NCERT : P L-170 | July 29, 2022 (I)]**

(a) Remains same
(b) Become 8 times its initial value
(c) Become 1/4th of its initial value
(d) Become 4 times its initial value

28. The speed of a transverse wave passing through a string of length 50 cm and mass 10 g is 60 ms^{-1}. The area of cross-section of the wire is 2.0 mm² and its Young's modulus is 1.2×10^{11} Nm^{-2}. The extension of the wire over its natural length due to its tension will be $x \times 10^{-5}$ m. The value of x is ____. **[NCERT : P L-170 | July 29, 2022 (II)]**

29. A metal wire of length 0.5 m and cross-sectional area 10^{-4} m² has breaking stress 5×10^8 Nm^{-2}. A block of 10 kg is attached at one end of the string and is rotating in a horizontal circle. The maximum linear velocity of block will be ____ ms^{-1}. **[NCERT : P L-168 | July 29, 2022 (II)]**

30. The force required to stretch a wire of cross-section 1 cm² to double its length will be : (Given Young's modulus of the wire = 2×10^{11} N/m²) **[NCERT : P L-170 | July 28, 2022 (I)]**

(a) 1×10^7 N (b) 1.5×10^7 N
(c) 2×10^7 N (d) 2.5×10^7 N

31. A string of area of cross-section 4 mm² and length 0.5 m is connected with a rigid body of mass 2 kg. The body is rotated in a vertical circular path of radius 0.5 m. The body acquires a speed of 5 m/s at the bottom of the circular path. Strain produced in the string when the body is at the bottom of the circle is ____ $\times 10^{-5}$. **[NCERT : P L-170 | July 28, 2022 (II)]**

(use Young's modulus 10^{11} N/m² and $g = 10$ m/s²)

32. In an experiment to determine the Young's modulus, steel wires of five different lengths (1, 2, 3, 4 and 5 m) but of same cross section ($2\ mm^2$) were taken and curves between extension and load were obtained. The slope (extension/load) of the curves were plotted with the wire length and the following graph is obtained. If the Young's modulus of given steel wires is $x \times 10^{11}\ Nm^{-2}$, then the value of x is _______.

[NCERT : P L-170 | July 27, 2022 (II)]

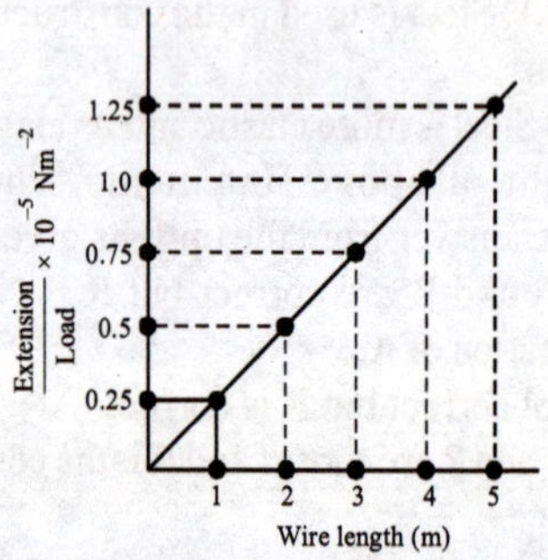

33. A steel wire of length 3.2 m ($Y_S = 2.0 \times 10^{11}\ Nm^{-2}$) and a copper wire of length 4.4 m ($Y_C = 1.1 \times 10^{11}\ Nm^{-2}$), both of radius 1.4 mm are connected end to end. When stretched by a load, the net elongation is found to be 1.4 mm. The load applied, in newton, will be: $\left(\text{Given } \pi = \frac{22}{7}\right)$

[NCERT : P L-170 | July 27, 2022 (II)]

(a) 360 (b) 180 (c) 1080 (d) 154

34. In an experiment to determine the Young's modulus of wire of a length exactly 1 m, the extension in the length of the wire is measured as 0.4 mm with an uncertainty of ± 0.02 mm when a load of 1 kg is applied. The diameter of the wire is measured as 0.4 mm with an uncertainty of ± 0.01 mm. The error in the measurement of Young's modulus (ΔY) is found to be $x \times 10^{10}\ Nm^{-2}$. The value of x is _______. (Take $g = 10\ ms^{-2}$)

[NCERT : P L-170 | July 26, 2022 (I)]

35. The area of cross-section of the rope used to lift a load by a crane is $2.5 \times 10^{-4} m^2$. The maximum lifting capacity of the crane is 10 metric tons. To increase the lifting capacity of the crane to 25 metric tons, the required area of cross-section of the rope should be : (take $g = 10\ ms^{-2}$)

[NCERT : P L-168 | July 26, 2022 (II)]

(a) $6.25 \times 10^{-4} m^2$ (b) $10 \times 10^{-4} m^2$
(c) $1 \times 10^{-4} m^2$ (d) $1.67 \times 10^{-4} m^2$

36. A uniform heavy rod of mass 20 kg, cross-sectional area $0.4\ m^2$ and length 20 m is hanging from a fixed support. Neglecting the lateral contraction, the elongation in the rod due to its own weight is $x \times 10^{-9}$ m. The value of x is____. (Given: Young's modulus $Y = 2 \times 10^{11}\ Nm^{-2}$ and $g = 10\ ms^{-2}$) **[NCERT : P L-170 | July 26, 2022 (II)]**

37. A wire of length L and radius r is clamped rigidly at one end. When the other end of the wire is pulled by a force F, its length increases by 5 cm. Another wire of the same material of length 4L and radius 4r is pulled by a force 4F under same conditions. The increase in length of this wire is _______ cm. **[NCERT : P L-170 | July 25, 2022 (I)]**

38. A wire of length L is hanging from a fixed support. The length changes to L_1 and L_2 when masses 1kg and 2 kg are suspended respectively from its free end. Then the value of L is equal to : **[NCERT : P L-170 | June 29, 2022 (I)]**

(a) $\sqrt{L_1 L_2}$ (b) $\frac{L_1 + L_2}{2}$
(c) $2L_1 - L_2$ (d) $3L_1 - 2L_2$

39. A uniform heavy rod of weight 10 kg, cross-sectional area $100\ cm^2$ and length 20 cm is hanging from a fixed support. Young modulus of the material of the rod is $2 \times 10^{11}\ Nm^{-2}$. Neglecting the lateral contraction, find the elongation of rod due to its own weight.

[NCERT : P L-170 | Aug. 31, 2021 (I)]

(a) 2×10^{-9} m (b) 5×10^{-9} m
(c) 4×10^{-8} m (d) 5×10^{-10} m

40. Four identical hollow cylindrical columns of mild steel support a big structure of mass 50×10^3 kg, the inner and outer radii of each column are 50 cm and 100 cm respectively. Assuming uniform local distribution, calculate the compression strain of each column.
[Use $Y = 2.0 \times 10^{11}$ Pa, $g = 9.8\ m/s^2$]

[NCERT : P L-170 | Aug. 31, 2021 (II)]

(a) 3.60×10^{-8} (b) 2.60×10^{-7}
(c) 1.87×10^{-3} (d) 7.07×10^{-4}

41. Wires W_1 and W_2 are made of same material having the breaking stress of $1.25 \times 10^9\ N/m^2$. W_1 and W_2 have cross-sectional area of $8 \times 10^{-7}\ m^2$ and $4 \times 10^{-7}\ m^2$, respectively. Masses of 20 kg and 10 kg hang from them as shown in the figure. The maximum mass that can be placed in the pan without breaking the wires is ____ kg. (Use $g = 10\ m/s^2$)

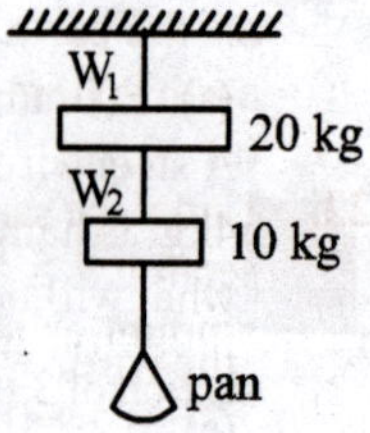

[NCERT : P L-168 | NA, Aug. 27, 2021 (II)]

42. Two blocks of masses 3 kg and 5 kg are connected by a metal wire going over a smooth pulley. The breaking stress of the metal is $\frac{24}{\pi} \times 10^2\ Nm^{-2}$.

What is the minimum radius of the wire?
(Take $g = 10\ ms^{-2}$) **[NCERT : P L-168 | Aug. 26, 2021 (II)]**

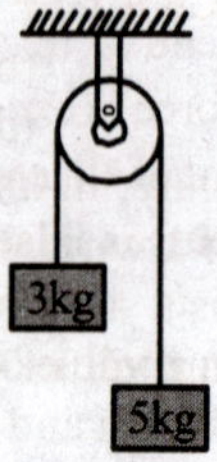

(a) 125 cm
(b) 1250 cm
(c) 12.5 cm
(d) 1.25 cm

43. The length of metallic wire is l_1 when tension in it is T_1. It is l_2 when the tension is T_2. The original length of the wire will be **[NCERT : P L-170 | Feb. 26, 2021 (II)]**

(a) $\frac{T_1 l_1 - T_2 l_2}{T_2 - T_1}$ (b) $\frac{l_1 - l_2}{2}$
(c) $\frac{T_2 l_1 + T_1 l_2}{T_1 + T_2}$ (d) $\frac{T_2 l_1 - T_1 l_2}{T_2 - T_1}$

44. A uniform metallic wire is elongated by 0.04 m when subjected to a linear force *F*. The elongation, if its length and diameter is doubled and subjected to the same force will be ________ cm.
[NCERT : P L-170 | Feb. 24, 2021 (II)]

45. A body of mass m = 10 kg is attached to one end of a wire of length 0.3 m. The maximum angular speed (in rad s^{-1}) with which it can be rotated about its other end in space station is (Breaking stress of wire = 4.8×10^7 Nm^{-2} and area of cross-section of the wire = 10^{-2} cm^2) is ____.
[NCERT : P L-168 | 9 Jan 2020 I]

46. A uniform cylindrical rod of length L and radius *r*, is made from a material whose Young's modulus of Elasticity equals Y. When this rod is heated by temperature T and simultaneously subjected to a net longitudinal compressional force F, its length remains unchanged. The coefficient of volume expansion, of the material of the rod, is (nearly) equal to :
[NCERT : P L-170 | 12 April 2019 (II), 2014]

(a) $9F/(\pi r^2YT)$
(b) $6F/(\pi r^2YT$
(c) $3F/(\pi r^2YT)$
(d) $F/(3\pi r^2YT)$

47. The elastic limit of brass is 379 MPa. What should be the minimum diameter of a brass rod if it is to support a 400 N load without exceeding its elastic limit?
[NCERT : P L-168 | 10 April 2019 II]

(a) 1.00 mm (b) 1.16 mm (c) 0.90 mm (d) 1.36 mm

48. A steel wire having a radius of 2.0 mm, carrying a load of 4kg, is hanging from a ceiling. Given that g = 9.8 m/s^{-2}, what will be the tensile stress that would be developed in the wire? **[NCERT : P L-168 | 8, 9 April 2019 I]**

(a) 6.2×10^6 Nm^{-2}
(b) 5.2×10^6 Nm^{-2}
(c) 3.1×10^6 Nm^{-2}
(d) 4.8×10^6 Nm^{-2}

49. Young's moduli of two wires A and B are in the ratio 7 : 4. Wire A is 2 m long and has radius R. Wire B is 1.5 m long and has radius 2 mm. If the two wires stretch by the same length for a given load, then the value of R is close to :
[NCERT : P L-170 | 8 April 2019 II]

(a) 1.5 mm (b) 1.9 mm (c) 1.7 mm (d) 1.3 mm

50. A thin 1 m long rod has a radius of 5 mm. A force of 50 πkN is applied at one end to determine its Young's modulus. Assume that the force is exactly known. If the least count in the measurement of all lengths is 0.01 mm, which of the following statements is false ?
[NCERT : P L-170 | Online April 10, 2016]

(a) The maximum value of Y that can be determined is 2×10^{14} N/m^2
(b) $\frac{\Delta Y}{Y}$ gets minimum contribution from the uncertainty in the length
(c) $\frac{\Delta Y}{Y}$ gets its maximum contribution from the uncertainty in strain
(d) The figure of merit is the largest for the length of the rod.

51. Two blocks of masses m and M are connected by means of a metal wire of cross-sectional area A passing over a frictionless fixed pulley as shown in the figure. The system is then released. If M = 2 m, then the stress produced in the wire is : **[NCERT : P L-168 | Online April 25, 2013]**

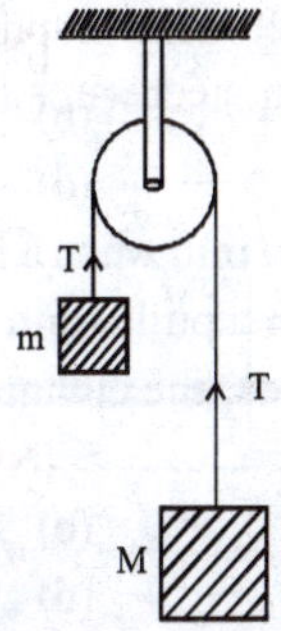

(a) $\frac{2mg}{3A}$
(b) $\frac{4mg}{3A}$
(c) $\frac{mg}{A}$
(d) $\frac{3mg}{4A}$

52. A copper wire of length 1.0 m and a steel wire of length 0.5 m having equal cross-sectional areas are joined end to end. The composite wire is stretched by a certain load which stretches the copper wire by 1 mm. If the Young's modulii of copper and steel are respectively 1.0×10^{11} Nm^{-2} and 2.0×10^{11} Nm^{-2}, the total extension of the composite wire is :
[NCERT : P L-170 | Online April 23, 2013]

(a) 1.75 mm (b) 2.0 mm (c) 1.50 mm (d) 1.25 mm

53. If the ratio of lengths, radii and Young's modull of steel and brass wires in the figure are *a*, *b* and *c* respectively, then the corresponding ratio of increase in their lengths is :
[NCERT : P L-170 | Online April 9, 2013]

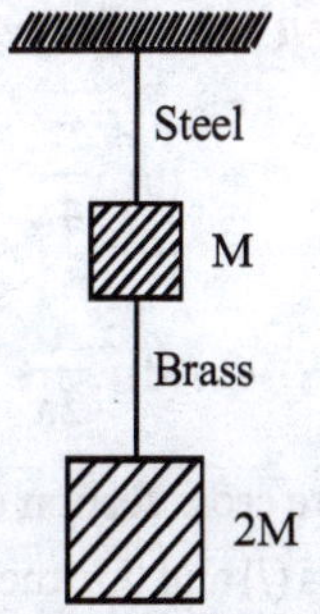

(a) $\frac{3c}{2ab^2}$
(b) $\frac{2a^2c}{b}$
(c) $\frac{3a}{2b^2c}$
(d) $\frac{2ac}{b^2}$

54. A steel wire can sustain 100 kg weight without breaking. If the wire is cut into two equal parts, each part can sustain a weight of **[NCERT : P L-168 | Online May 19, 2012]**

(a) 50 kg
(b) 400 kg
(c) 100 kg
(d) 200 kg

55. The load versus elongation graphs for four wires of same length and made of the same material are shown in the figure. The thinnest wire is represented by the line
[NCERT : P L-170 | Online May 7, 2012]

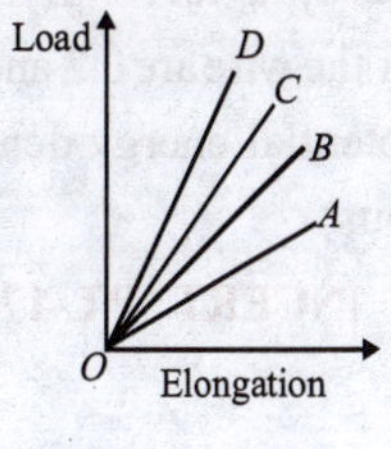

(a) *OA*
(b) *OC*
(c) *OD*
(d) *OB*

56. Two wires are made of the same material and have the same volume. However wire 1 has cross-sectional area A and wire 2 has cross-sectional area $3A$. If the length of wire 1 increases by Δx on applying force F, how much force is needed to stretch wire 2 by the same amount?

[NCERT : P L-170 | 2009]

(a) $4F$ (b) $6F$
(c) $9F$ (d) F

57. A wire elongates by l mm when a load W is hanged from it. If the wire goes over a pulley and two weights W each are hung at the two ends, the elongation of the wire will be (in mm) **[NCERT : P L-170 | 2006]**

(a) l (b) $2l$
(c) zero (d) $l/2$

Topic 2 - Bulk and Rigidity Modulus and Work Done in Stretching a Wire

58. A sample of a liquid is kept at 1 atm. It is compressed to 5 atm which leads to change of volume of 0.8 cm^3. If the bulk modulus of the liquid is 2 GPa, the initial volume of the liquid was ______ litre. (Take 1 atm = 10^5 Pa)

[NCERT : P L-173 | April 8, 2025 (II)]

59. A cylindrical rod of length 1 m and radius 4 cm is mounted vertically. It is subjected to a shear force of 10^5 N at the top. Considering infinitesimally small displacement in the upper edge, the angular displacement θ of the rod axis from its original position would be : (shear moduli, $G = 10^{10}$ N/m^2) **[NCERT : P L-172 | April 4, 2025 (II)]**

(a) $\frac{1}{160\pi}$ (b) $\frac{1}{4\pi}$

(c) $\frac{1}{40\pi}$ (d) $\frac{1}{2\pi}$

60. Two slabs with square cross section of different materials (1, 2) with equal sides (l) and thickness d_1 and d_2 such that $d_2 = 2d_1$ and $l > d_2$. Considering lower edges of these slabs are fixed to the floor, we apply equal shearing force on the narrow faces. The angle of deformation is $\theta_2 = 2\theta_1$. If the shear moduli of material 1 is 4×10^9 N/m^2, then shear moduli of material 2 is $x \times 10^9$ N/m^2, where value of x is ____.

[NCERT : P L-172 | April 4, 2025 (I)]

61. A steel wire of length 2 m and Young's modulus 2.0×10^{11} Nm^{-2} is stretched by a force. If Poisson ratio and transverse strain for the wire are 0.2 and 10^{-3} respectively, then the elastic potential energy density of the wire is ____ $\times 10^5$ (in SI units)

[NCERT : P L-174 | April 2, 2025 (I)]

62. The fractional compression $\left(\frac{\Delta V}{V}\right)$ of water at the depth of 2.5 km below the sea level is ________%. Given, the Bulk modulus of water = 2×10^9 Nm^{-2}, density of water = 10^3 kg m^{-3}, acceleration due to gravity = $g = 10$ ms^{-2}.

[NCERT : P L-173 | Jan. 29, 2025 (I)]

(a) 1.75 (b) 1.0 (c) 1.5 (d) 1.25

63. The volume contraction of a solid copper cube of edge length 10 cm, when subjected to a hydraulic pressure of 7×10^6 Pa, would be _____ mm^3.
(Given bulk modulus of copper = 1.4×10^{11} Nm^{-2})

[NCERT : P L-173 | Jan. 28, 2025 (II)]

64. The increase in pressure required to decrease the volume of a water sample by 0.2% is P $\times 10^5$ Nm^{-2}. Bulk modulus of water is 2.15×10^9 Nm^{-2}. The value of P is ______.

[NCERT : P L-173 | Jan. 24, 2025 (II)]

65. Match **List-I** with **List-II**

List-I	List-II
(A) A force that restores an elastic body of unit area to its original state	(I) Bulk modulus
(B) Two equal and opposite forces parallel to opposite faces	(II) Young's modulus
(C) Forces perpendicular everywhere to the surface per unit area same everywhere	(III) Stress
(D) Two equal and opposite forces perpendicular to opposite faces	(IV) Shear modulus

Choose the correct answer from the options given below:

[NCERT : P L-168, 170, 172, 173 | April 5, 2024 (II)]

(a) (A)-(II), (B)-(IV), (C)-(I), (D)-(III)
(b) (A)-(IV), (B)-(II), (C)-(III), (D)-(I)
(c) (A)-(III), (B)-(IV), (C)-(I), (D)-(II)
(d) (A)-(III), (B)-(I), (C)-(II), (D)-(IV)

66. The depth below the surface of sea to which a rubber ball be taken so as to decrease its volume by 0.02% is ____ m. (Take density of sea water = 10^3 kgm^{-3}, Bulk modulus of rubber = 9×10^8 Nm^{-2}, and g = 10 ms^{-2})

[NCERT : P L-173 | Jan. 31, 2024 (I)]

67. If average depth of an ocean is 4000 m and the bulk modulus of water is 2×10^9 Nm^{-2}, then fractional compression $\frac{\Delta V}{V}$ of water at the bottom of ocean is $\alpha \times 10^{-2}$. The value of α is ______
(Given, g = 10 ms^{-2}, ρ = 1000 kg m^{-3})

[NCERT : P L-173 | Jan. 27, 2024 (I)]

68. A certain pressure 'P' is applied to 1 litre of water and 2 litre of a liquid separately. Water gets compressed to 0.01% whereas the liquid gets compressed to 0.03%. The ratio of Bulk modulus of water to that of the liquid is $\frac{3}{x}$. The value of x is ______. **[NCERT : P L-173 | Feb. 1, 2023 (I)]**

69. A square aluminium (shear modulus is $25 \times 10^9\ Nm^{-2}$) slab of side 60 cm and thickness 15 cm is subjected to a shearing force (on its narrow face) of 18.0×10^4 N. The lower edge is riveted to the floor. The displacement of the upper edg is ____ μm. **[NCERT : P L-172 | July 27, 2022 (I)]**

70. The bulk modulus of a liquid is $3 \times 10^{10}\ Nm^{-2}$. The pressure required to reduce the volume of liquid by 2% is : **[NCERT : P L-173 | June 24, 2022 (I)]**
(a) $3 \times 10^8\ Nm^{-2}$ (b) $9 \times 10^8\ Nm^{-2}$
(c) $6 \times 10^8\ Nm^{-2}$ (d) $12 \times 10^8\ Nm^{-2}$

71. When a rubber ball is taken to a depth of __________ m in deep sea, its volume decreases by 0.5%.
(The bulk modulus of rubber $= 9.8 \times 10^8\ Nm^{-2}$, Density of sea water $= 10^3\ kgm^{-3}$ $g = 9.8\ m/s^2$) **[NCERT : P L-173 | NA, Aug. 31, 2021 (I)]**

72. An object is located at 2 km beneath the surface of the water. If the fractional compression $\frac{\Delta V}{V}$ is 1.36%, the ratio of hydraulic stress to the corresponding hydraulic strain will be ________.
(Given: density of water is $1000\ kgm^{-3}$ and $g = 9.8\ ms^{-2}$) **[NCERT : P L-173 | March 17, 2021 (II)]**
(a) $1.44 \times 10^7\ Nm^{-3}$ (b) $2.26 \times 10^9\ Nm^{-2}$
(c) $1.96 \times 10^7\ Nm^{-2}$ (d) $1.44 \times 10^9\ Nm^{-2}$

73. The normal density of a material is ρ and its bulk modulus of elasticity is K. The magnitude of increase in density of material, when a pressure P is applied uniformly on all sides, will be : **[NCERT : P L-173 | Feb. 26, 2021 (I)]**
(a) $\frac{\rho P}{K}$ (b) $\frac{\rho K}{P}$
(c) $\frac{K}{\rho P}$ (d) $\frac{PK}{\rho}$

74. If Y, K and η are the values of Young's modulus, bulk modulus and modulus of rigidity of any material respectively. Choose the correct relation for these parameters. **[NCERT : P L-170, | Feb. 24, 2021 (I)]**
(a) $Y = \frac{9K\eta}{3K - \eta}\ N/m^2$ (b) $\eta = \frac{3YK}{9K + Y}\ N/m^2$
(c) $Y = \frac{9K\eta}{2\eta + 3K}\ N/m^2$ (d) $K = \frac{Y\eta}{9\eta - 3Y}\ N/m^2$

75. Two steel wires having same length are suspended from a ceiling under the same load. If the ratio of their energy stored per unit volume is 1 : 4, the ratio of their diameters is: **[NCERT : P L-174 | 9 Jan 2020 II]**
(a) $\sqrt{2} : 1$ (b) $1 : 2$ (c) $2 : 1$ (d) $1 : \sqrt{2}$

76. A boy's catapult is made of rubber cord which is 42 cm long, with 6 mm diameter of cross-section and of negligible mass. The boy keeps a stone weighing 0.02 kg on it and stretches the cord by 20 cm by applying a constant force. When released, the stone flies off with a velocity of $20\ ms^{-1}$. Neglect the change in the area of cross-section of the cord while stretched. The Young's modulus of rubber is closest to : **[NCERT : P L-174 | 8 April 2019 I]**
(a) $10^6\ N/m^{-2}$ (b) $10^4\ N/m^{-2}$
(c) $10^8\ N/m^{-2}$ (d) $10^3\ N/m^{-2}$

77. A solid sphere of radius r made of a soft material of bulk modulus K is surrounded by a liquid in a cylindrical container. A massless piston of area a floats on the surface of the liquid, covering entire cross-section of cylindrical container. When a mass m is placed on the surface of the piston to compress the liquid, the fractional decrement in the radius of the sphere $\left(\frac{dr}{r}\right)$, is : **[NCERT : P L-173 | 2018]**
(a) $\frac{Ka}{mg}$ (b) $\frac{Ka}{3mg}$
(c) $\frac{mg}{3Ka}$ (d) $\frac{mg}{Ka}$

78. A bottle has an opening of radius a and length b. A cork of length b and radius $(a + \Delta a)$ where $(\Delta a << a)$ is compressed to fit into the opening completely (see figure). If the bulk modulus of cork is B and frictional coefficient between the bottle and cork is μ then the force needed to push the cork into the bottle is : **[NCERT : P L-173 | Online April 10, 2016]**

(a) $(\pi\mu B b)\, a$
(b) $(2\pi\mu Bb)\, \Delta a$
(c) $(\pi\mu B b)\, \Delta a$
(d) $(4\pi\mu B b)\, \Delta a$

79. In materials like aluminium and copper, the correct order of magnitude of various elastic moduli is: **[NCERT : P L-170, 172, 173 | Online April 9, 2014]**
(a) Young's modulus < Shear modulus < Bulk modulus.
(b) Bulk modulus < Shear modulus < Young's modulus
(c) Shear modulus < Young's modulus < Bulk modulus.
(d) Bulk modulus < Young's modulus < Shear modulus.

ANSWER KEYS

1	(a)	9	(12)	17	(20)	25	(d)	33	(d)	41	(40)	49	(c)	57	(a)	65	(c)	73	(a)
2	(b)	10	(c)	18	(b)	26	(c)	34	(2)	42	(c)	50	(a)	58	(4)	66	(18)	74	(d)
3	(1)	11	(2)	19	(25)	27	(a)	35	(a)	43	(d)	51	(b)	59	(a)	67	(2)	75	(a)
4	(20)	12	(d)	20	(11)	28	(15)	36	(25)	44	(2)	52	(d)	60	(1)	68	(1)	76	(a)
5	(1)	13	(16)	21	(d)	29	(50)	37	(5)	45	(4)	53	(c)	61	(25)	69	(48)	77	(c)
6	(600)	14	(d)	22	(b)	30	(c)	38	(c)	46	(c)	54	(c)	62	(d)	70	(c)	78	(d)
7	(b)	15	(b)	23	(d)	31	(30)	39	(b)	47	(b)	55	(a)	63	(50)	71	(500)	79	(c)
8	(3)	16	(2)	24	(5)	32	(2)	40	(b)	48	(c)	56	(c)	64	(43)	72	(d)		

9 Mechanical Properties of Fluids

TREND ANALYSIS
JEE MAIN (2025-2021)

	2025				2024				2023				2022				2021			
	Session II		Session I		Session II		Session I		Session II		Session I		Session II		Session I		Session II		Session I	
	MCQs	NVQs	MCQs	NVQs	MCQs	NVQs	MCQs	NVQs	MCQs	NVQs	MCQs	NVQs	MCQs	NVQs	MCQs	NVQs	MCQs	NVQs	MCQs	NVQs
Topic 1	1	1	2	1	2	2	–	–	2	–	–	–	2	–	–	–	–	–	–	1
Topic 2	1	–	1	1	2	–	–	3	1	1	1	1	1	–	–	3	1	1	–	1
Topic 3	1	–	1	–	2	1	2	–	3	1	–	1	–	2	3	1	–	–	–	–
Topic 4	2	1	5	2	4	3	4	1	–	2	4	–	2	–	2	–	2	1	2	1
Total	**5**	**2**	**9**	**4**	**10**	**6**	**6**	**4**	**6**	**4**	**5**	**2**	**5**	**2**	**5**	**4**	**3**	**2**	**2**	**3**

How to Code Qns for Better & Faster Revision →

Type 1: Able to solve (No code Req)
Type 2: Solved with struggle or by seeing solution **(Use code NR means Needs Revision or orange stickers)**
Type 3: Tricky/ Could not solve on my own. Need help from Teacher/ Friend. **(Use code TQ means Tricky Question or red stickers)**

You can either hand write the code in boxes or else paste coloured stickers given at end of book.

Topic 1 - Pressure, Density, Pascal's Law and Archimedes' Principle

1. Two cylindrical vessels of equal cross sectional area of $2m^2$ contain water upto height 10m and 6m, respectively. If the vessels are connected at their bottom then the work done by the force of gravity is : (Density of water is 10^3 kg/m^3 and g = 10 m/s^2) **[NCERT : P L-181 | April 3, 2025 (II)]**

(a) 1×10^5 J
(b) 4×10^4 J
(c) 6×10^4 J
(d) 8×10^4 J

2. A vessel with square cross–section and height of 6 m is vertically partitioned. A small window of 100 cm^2 with hinged door is fitted at a depth of 3 m in the partition wall. One part of the vessel is filled completely with water and the other side is filled with the liquid having density 1.5×10^3 kg/m^3. What force one needs to apply on the hinged door so that it does not get opened ?
(Acceleration due to gravity = 10 m/s^2)

[NCERT : P L-183 | April 2, 2025 (I)]

3. In a hydraulic lift, the surface area of the input piston is 6 cm^2 and that of the output piston is 1500 cm^2. If 100 N force is applied to the input piston to raise the output piston by 20 cm, then the work done is ________ kJ.

[NCERT : P L-185 | Jan. 29, 2025 (I)]

4. A 400 g solid cube having an edge of length 10 cm floats in water. How much volume of the cube is outside the water ?
(Given : density of water = 1000 kg m^{-3})

[Jan. 28, 2025 (II)]

(a) 1400 cm^3
(b) 4000 cm^3
(c) 400 cm^3
(d) 600 cm^3

5. Water flows in a horizontal pipe whose one end is closed with a valve. The reading of the pressure gauge attached to the pipe is P_1. The reading of the pressure gauge falls to P_2 when the valve is opened. The speed of water flowing in the pipe is proportional to

[NCERT : P L-183 | Jan. 23, 2025 (II)]

(a) $\sqrt{P_1 - P_2}$
(b) $(P_1 - P_2)^2$
(c) $(P_1 - P_2)^4$
(d) $P_1 - P_2$

6. A sphere of relative density σ and diameter D has concentric cavity of diameter d. The ratio of $\frac{D}{d}$, if it just floats on water surface in a tank is: **[April 9, 2024 (I)]**

(a) $\left(\frac{\sigma}{\sigma-1}\right)^{\frac{1}{3}}$ (b) $\left(\frac{\sigma+1}{\sigma-1}\right)^{\frac{1}{3}}$

(c) $\left(\frac{\sigma-1}{\sigma}\right)^{\frac{1}{3}}$ (d) $\left(\frac{\sigma-2}{\sigma+2}\right)^{\frac{1}{3}}$

7. A cube of ice floats partly in water and partly in kerosene oil. The ratio of volume of ice immersed in water to that in kerosene oil (specific gravity of Kerosene oil = 0.8, specific gravity of ice = 0.9) **[April 8, 2024 (II)]**

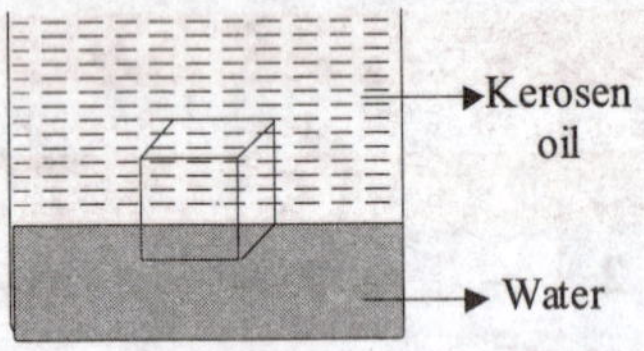

(a) 8 : 9 (b) 5 : 4 (c) 9 : 10 (d) 1 : 1

8. A hydraulic press containing water has two arms with diameters as mentioned in the figure. A force of 10 N is applied on the surface of water in the thinner arm. The force required to be applied on the surface of water in the thicker arm to maintain equilibrium of water is ______N.

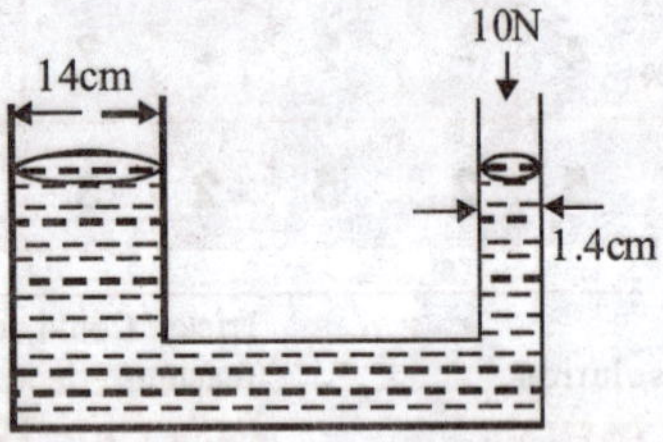

[NCERT : P L-185 | April 5, 2024 (II)]

9. Mercury is filled in a tube of radius 2 cm up to a height of 30 cm. The force exerted by mercury on the bottom of the tube is ______N.

(Given, atmospheric pressure = 10^5 Nm^{-2}, density of mercury = 1.36×10^4 kg m^{-3}, g = 10 ms^{-2}, $\pi = \frac{22}{7}$)

[NCERT : P L-183 | April 4, 2024 (II)]

10. A hydraulic automobile lift is designed to lift vehicles of mass 5000 kg. The area of cross-section of the cylinder carrying the load is 250 cm^2. The maximum pressure the smaller piston would have to bear is [Assume g = 10 m/s^2]: **[NCERT : P L-185 | April 8, 2023 (II)]**

(a) 200×10^6 Pa (b) 20×10^6 Pa
(c) 2×10^6 Pa (d) 2×10^5 Pa

11. Given below are two statements: one is labelled as **Assertion A** and the other is labelled as **Reason R**

Assertion A: When you squeeze one end of a tube to get toothpaste out from the other end, Pascal's principle is observed.

Reason R: A change in the pressure applied to an enclosed incompressible fluid is transmitted undiminished to every portion of the fluid and to the walls of its container.

In the light of the above statements, choose the most appropriate answer from the options given below **[NCERT : P L-182 | April 6, 2023 (II)]**

(a) **A** is not correct but **R** is correct
(b) **A** is correct but **R** is not correct
(c) Both **A** and **R** are correct and **R** is the correct explanation of **A**
(d) Both **A** and **R** are correct but **R** is **NOT** the correct explanation of **A**

12. A pressure-pump has a horizontal tube of cross sectional area 10 cm^2 for the outflow of water at a speed of 20 m/s. The force exerted on the vertical wall just in front of the tube which stops water horizontally flowing out of the tube, is : **[NCERT : P L-181 | July 28, 2022 (II)]**

[given: density of water = 1000 kg/m^3]

(a) 300 N (b) 500 N (c) 250 N (d) 400 N

13. Two cylindrical vessels of equal cross-sectional area 16 cm^2 contain water upto heights 100 cm and 150 cm respectively. The vessels are interconnected so that the water levels in them become equal. The work done by the force of gravity during the process, is [Take density of water = 10^3 kg/m^3 and g = 10 ms^{-2}] **[NCERT : P L-181 | July 27, 2022 (I)]**

(a) 0.25 J (b) 1 J (c) 8 J (d) 12 J

14. A hydraulic press can lift 100 kg when a mass 'm' is placed on the smaller piston. It can lift _____ kg when the diameter of the larger piston is increased by 4 times and that of the smaller piston is decreased by 4 times keeping the same mass 'm' on the smaller piston. **[NCERT : P L-185 | Feb. 24, 2021 (I)]**

15. A hollow spherical shell at outer radius *R* floats just submerged under the water surface. The inner radius of the shell is *r*. If the specific gravity of the shell material is $\frac{27}{8}$ w.r.t water, the value of *r* is : **[NCERT : P L-181 | 5 Sep. 2020 (I)]**

(a) $\frac{8}{9}R$ (b) $\frac{4}{9}R$ (c) $\frac{2}{3}R$ (d) $\frac{1}{3}R$

16. An air bubble of radius 1 cm in water has an upward acceleration 9.8cms^{-2}. The density of water is 1 g/cm^3 and water offers negligible drag force on the bubble. The mass of the bubble is (g = 980 cm/s^2) **[4 Sep. 2020 (I)]**

(a) 4.51 g (b) 3.15 g (c) 4.15 g (d) 1.52 g

17. Two identical cylindrical vessels are kept on the ground and each contain the same liquid of density *d*. The area of the base of both vessels is *S* but the height of liquid in one vessel is x_1 and in the other, x_2. When both cylinders are connected through a pipe of negligible volume very close to the bottom, the liquid flows from one vessel to the other until it comes to equilibrium at a new height. The change in energy of the system in the process is : **[4 Sep. 2020 (II)]**

(a) $gdS(x_2^2 + x_1^2)$ (b) $gdS(x_2 + x_1)^2$

(c) $\frac{3}{4}gdS(x_2 - x_1)^2$ (d) $\frac{1}{4}gdS(x_2 - x_1)^2$

18. A leak proof cylinder of length 1 m, made of a metal which has very low coefficient of expansion is floating vertically in water at 0°C such that its height above the water surface is 20 cm. When the temperature of water is increased to 4°C, the height of the cylinder above the water surface becomes 21 cm. The density of water at $T = 4°C$, relative to the density at $T = 0°C$ is close to: **[8 Jan 2020 (I)]**

(a) 1.26 (b) 1.04
(c) 1.01 (d) 1.03

19. Consider a solid sphere of radius R and mass density $\rho(r) = \rho_0\left(1 - \frac{r^2}{R^2}\right)$, $0 < r \le R$. The minimum density of a liquid in which it will float is: **[8 Jan 2020 (I)]**

(a) $\frac{\rho_0}{3}$ (b) $\frac{\rho_0}{5}$
(c) $\frac{2\rho_0}{5}$ (d) $\frac{2\rho_0}{3}$

20. Two liquids of densities ρ_1 and $\rho_2 (\rho_2 = 2\rho_1)$ are filled up behind a square wall of side 10 m as shown in figure. Each liquid has a height of 5 m. The ratio of the forces due to these liquids exerted on upper part MN to that at the lower part NO is (Assume that the liquids are not mixing):

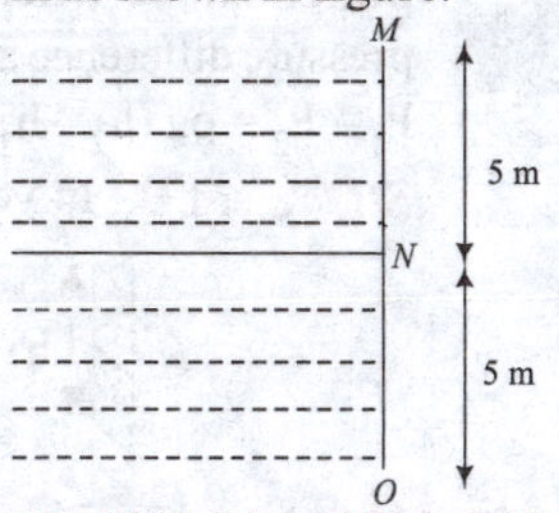

[NCERT : P L-181 | 8 Jan 2020 (II)]

(a) 1/3 (b) 2/3
(c) 1/2 (d) 1/4

21. A cubical block of side 0.5 m floats on water with 30% of its volume under water. What is the maximum weight that can be put on the block without fully submerging it under water? [Take, density of water = 10^3 kg/m³] **[10 April 2019 (II)]**

(a) 46.3 kg (b) 87.5 kg
(c) 65.4 kg (d) 30.1 kg

22. A submarine experiences a pressure of 5.05×10^6 Pa at depth of d_1 in a sea. When it goes further to a depth of d_2, it experiences a pressure of 8.08×10^6 Pa. Then $d_1 - d_2$ is approximately (density of water = 10^3 kg/m³ and acceleration due to gravity = 10 ms⁻²): **[NCERT : P L-183 | 10 April 2019 (II)]**

(a) 300 m (b) 400 m (c) 600 m (d) 500 m

23. A wooden block floating in a bucket of water has $\frac{4}{5}$ of its volume submerged. When certain amount of an oil poured into the bucket, it is found that the block is just under the oil surface with half of its volume under water and half in oil. The density of oil relative to that of water is: **[9 April 2019 (II)]**

(a) 0.5 (b) 0.8
(c) 0.6 (c) 0.7

24. A load of mass M kg is suspended from a steel wire of length 2 m and radius 1.0 mm in Searle's apparatus experiment. The increase in length produced in the wire is 4.0 mm. Now the load is fully immersed in a liquid of relative density 2. The relative density of the material of load is 8.
The new value of increase in length of the steel wire is : **[12 Jan. 2019 (II)]**

(a) 3.0 mm (b) 4.0 mm (c) 5.0 mm (d) Zero

25. A soap bubble, blown by a mechanical pump at the mouth of a tube, increases in volume, with time, at a constant rate. The graph that correctly depicts the time dependence of pressure inside the bubble is given by: **[NCERT : P L-181 | 12 Jan. 2019 (II)]**

(a)
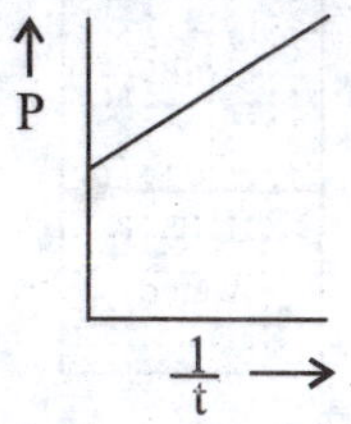

(b)
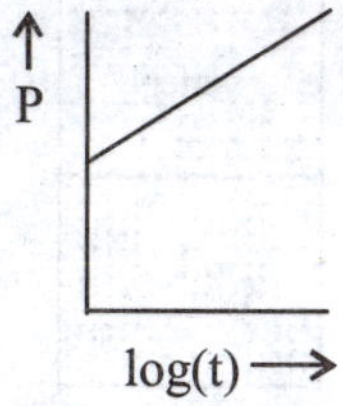

(c)
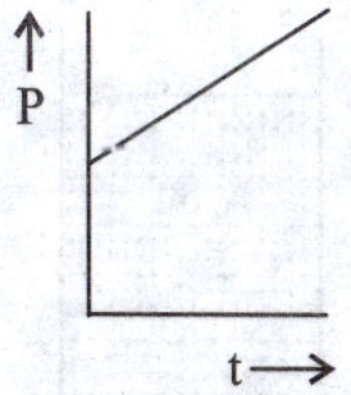

(d)
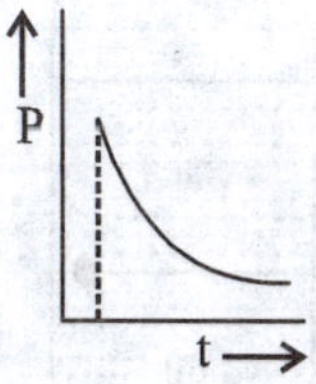

26. A liquid of density ρ is coming out of a hose pipe of radius a with horizontal speed v and hits a mesh. 50% of the liquid passes through the mesh unaffected. 25% looses all of its momentum and 25% comes back with the same speed. The resultant pressure on the mesh will be: **[NCERT : P L-181, 182 | 11 Jan. 2019 (I)]**

(a) $\frac{1}{4}\rho v^2$ (b) $\frac{3}{4}\rho v^2$
(c) $\frac{1}{2}\rho v^2$ (d) ρv^2

27. A thin uniform tube is bent into a circle of radius r in the vertical plane. Equal volumes of two immiscible liquids, whose densities are ρ_1 and ρ_2 $(\rho_1 > \rho_2)$ filled half the circle. The angle θ between the radius vector passing through the common interface and the vertical is **[Online April 15, 2018, 2014 S]**

(a) $\theta = \tan^{-1}\left[\frac{\pi}{2}\left(\frac{\rho_1 - \rho_2}{\rho_1 + \rho_2}\right)\right]$ (b) $\theta = \tan^{-1}\frac{\pi}{2}\left(\frac{\rho_1 - \rho_2}{\rho_1 + \rho_2}\right)$

(c) $\theta = \tan^{-1}\pi\left(\frac{\rho_1}{\rho_2}\right)$ (d) None of above

28. A uniform cylinder of length L and mass M having cross-sectional area A is suspended, with its length vertical, from a fixed point by a massless spring such that it is half submerged in a liquid of density σ at equilibrium position. The extension x_0 of the spring when it is in equilibrium is: **[2013]**

(a) $\frac{Mg}{k}$ (b) $\frac{Mg}{k}\left(1-\frac{LA\sigma}{M}\right)$

(c) $\frac{Mg}{k}\left(1-\frac{LA\sigma}{2M}\right)$ (d) $\frac{Mg}{k}\left(1+\frac{LA\sigma}{M}\right)$

29. A ball is made of a material of density ρ where $\rho_{oil} < \rho < \rho_{water}$ with ρ_{oil} and ρ_{water} representing the densities of oil and water, respectively. The oil and water are immiscible. If the above ball is in equilibrium in a mixture of this oil and water, which of the following pictures represents its equilibrium position ? **[2010, 2008 S]**

(a)
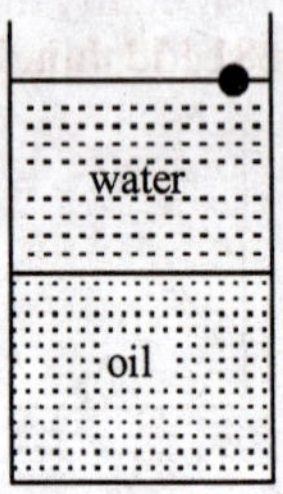

(b)
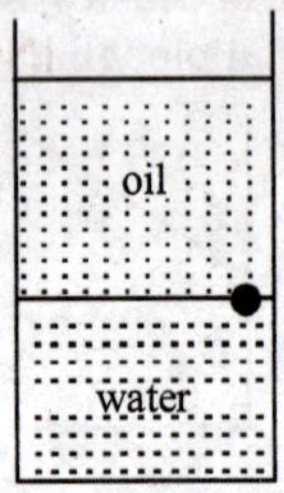

(c)
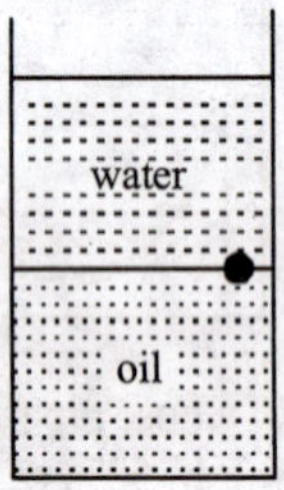

(d)
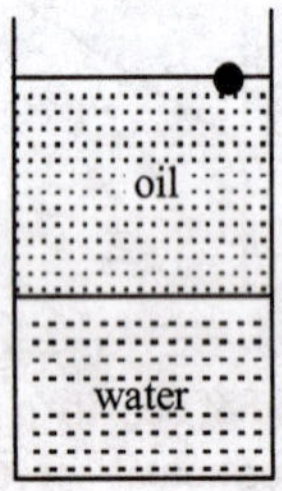

Topic 2 - Fluid Flow and Bernoulli's Principle

30. Consider a completely full cylindrical water tank of height 1.6 m and cross-sectional area 0.5 m^2. It has a small hole in its side at a height 90 cm from the bottom. Assume, the cross-sectional area of the hole to be negligibly small as compared to that of the water tank. If a load 50 kg is applied at the top surface of the water in the tank then the velocity of the water coming out at the instant when the hole is opened is : ($g = 10$ m/s^2) **[NCERT : P L-188 | April 3, 2025 (I)]**

(a) 3 m/s (b) 5 m/s (c) 2 m/s (d) 4 m/s

31.
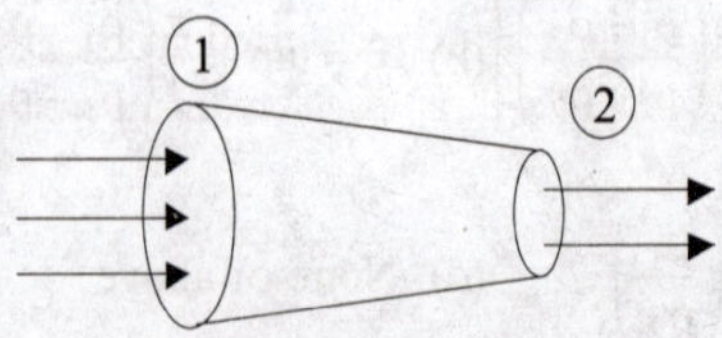

The tube of length L is shown in the figure. The radius of cross section at the point (1) is 2 cm and at the point (2) is 1 cm, respectively. If the velocity of water entering at point (1) is 2 m/s, then velocity of water leaving the point (2) will be **[NCERT : P L-188 | Jan. 22, 2025 (II)]**

(a) 2 m/s (b) 4 m/s
(c) 6 m/s (d) 8 m/s

32. A tube of length 1m is filled completely with an ideal liquid of mass 2M, and closed at both ends.
The tube is rotated uniformly in horizontal plane about one of its ends. If the force exerted by the liquid at the other end is F then angular velocity of the tube is $\sqrt{\frac{F}{\alpha M}}$ in SI unit. The value of α is ______.
[NCERT : P L-? | Jan. 22, 2025 (II)]

33. Correct Bernoulli's equation is (symbols have their usual meaning) : **[NCERT : P L-188 | April 8, 2024 (I)]**

(a) $P + mgh + \frac{1}{2}mv^2 = \text{constant}$

(b) $P + \rho gh + \frac{1}{2}\rho v^2 = \text{constant}$

(c) $P + \rho gh + \rho v^2 = \text{constant}$

(d) $P + \frac{1}{2}\rho gh + \frac{1}{2}\rho v^2 = \text{constant}$

34. Given below are two statements :
Statement I : When speed of liquid is zero everywhere, pressure difference at any two points depends on equation $P_1 - P_2 = \rho g(h_2 - h_1)$
Statement II : In ventury tube shown $2gh = \upsilon_1^2 - \upsilon_2^2$

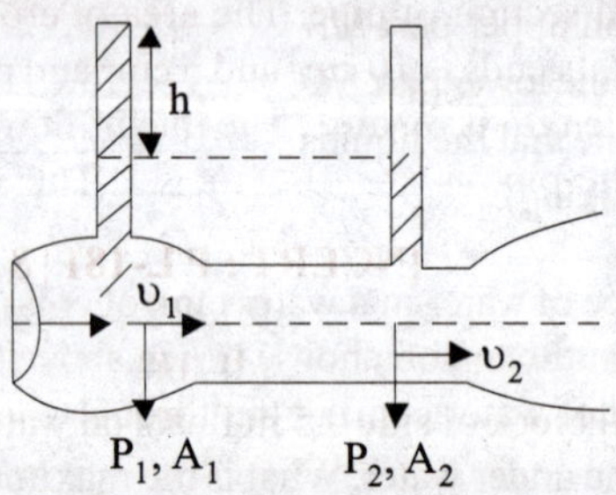

In the light of the above statements, choose the most appropriate answer from the options given below.
[NCERT : P L-189 | April 4, 2024 (I)]

(a) Both Statement I and Statement II are correct.
(b) Statement I is incorrect but Statement II is correct.
(c) Both Statement I and Statement II are incorrect.
(d) Statement I is correct but Statement II is incorrect.

35. A plane is in level flight at constant speed and each of its two wings has an area of 40 m^2. If the speed of the air is 180 km/h over the lower wing surface and 252 km/h over the upper wing surface, the mass of the plane is ______ kg. (Take air density to be 1 kg m^{-3} and g = 10 ms^{-2})
[NCERT : P L-188 | Feb. 1, 2024 (I)]

36. In a test experiment on a model aeroplane in wind tunnel, the flow speeds on the upper and lower surfaces of the wings are 70 ms^{-1} and 65 ms^{-1} respectively. If the wing area is 2m^2, the lift of the wing is ______ N.
(Given density of air = 1.2 kg m^{-3})
[NCERT : P L-188 | Jan. 29, 2024 (I)]

37. The reading of pressure metre attached with a closed pipe is 4.5×10^4 N/m^2. On opening the valve, water starts flowing and the reading of pressure metre falls to 2.0×10^4 N/m^2. The velocity of water is found to be $\sqrt{V}$ m/s. The value of V is ______.
[NCERT : P L-187 | Jan. 27, 2024 (II)]

38.

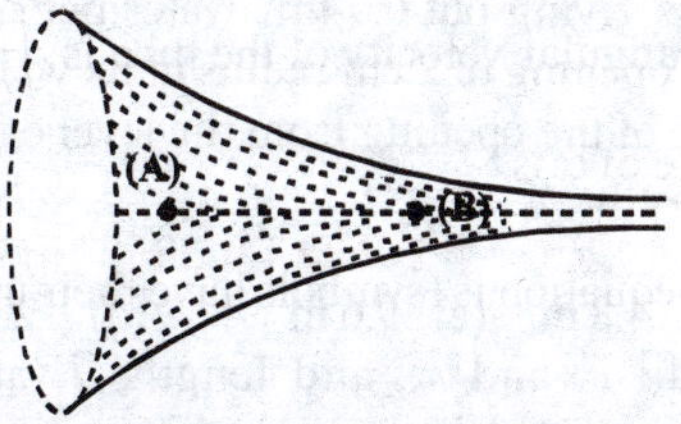

The figure shows a liquid of given density flowing steadily in horizontal tube of varying cross-section. Cross sectional areas at A is 1.5 cm^2, and B is 25 mm^2, if the speed of liquid at B is 60 cm/s then $(P_A - P_B)$ is :
(Given P_A and P_B are liquid pressures at A and B points.)
Density $\rho = 1000$ kg m^{-3}
A and B are on the axis of tube
[NCERT : P L-188 | April 13, 2023 (I)]

(a) 175 Pa (b) 27 Pa
(c) 135 Pa (d) 36 Pa

39. Glycerine of density 1.25×10^3 kg m^{-3} is flowing through the conical section of pipe. The area of cross-section of the pipe at its ends is 10 cm^2 and 5 cm^2 and pressure drop across its length is 3 Nm^{-2}. The rate of flow of glycerine through the pipe is $x \times 10^{-5}$ m^3 s^{-1}. The value of x is ______. **[NCERT : P L-187, 188 | April 12, 2023 (I)]**

40. The surface of water in a water tank of cross section area 750 cm^2 on the top of a house is h m above the tap level. The speed of water coming out through the tap of cross section area 500 mm^2 is 30 cm/s. At that instant, $\frac{dh}{dt}$ is $x \times 10^{-3}$ m/s. The value of x will be ______.
[NCERT : P L-187 | Feb. 1, 2023 (II)]

41. A fully loaded boeing aircraft has a mass of 5.4×10^5 kg. Its total wing area is 500 m^2. It is in level flight with a speed of 1080 km/h. If the density of air ρ is 1.2 kg m^{-3}, the fractional increase in the speed of the air on the upper surface of the wing relative to the lower surface in percentage will be ($g = 10$ m/s^2) **[NCERT : P L-188 | Jan. 29, 2023 (II)]**

(a) 16 (b) 6
(c) 8 (d) 10

42. Consider a cylindrical tank of radius 1m is filled with water. The top surface of water is at 15 m from the bottom of the cylinder. There is a hole on the wall of cylinder at a height of 5 m from the bottom. A force of 5×10^5 N is applied on the top surface of water using a piston. The speed of efflux from the hole will be :
NCERT : P L-188 | July 28, 2022 (II)]

(given atmospheric pressure $P_A = 1.01 \times 10^5$ Pa, density of water $\rho_w = 1000$ kg/m^3 and gravitational acceleration $g = 10$ m/s^2)

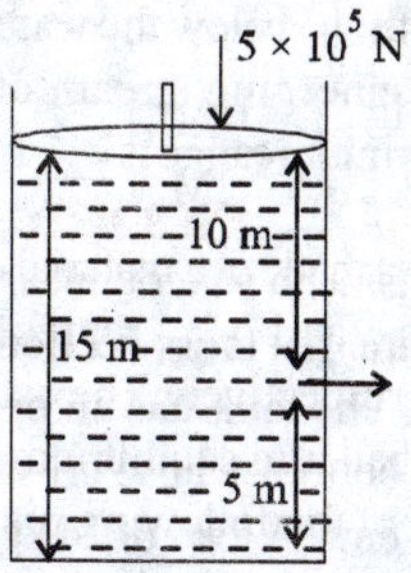

(a) 11.6 m/s
(b) 10.8 m/s
(c) 17.8 m/s
(d) 14.4 m/s

43. A liquid of density 750 kgm^{-3} flows smoothly through a horizontal pipe that tapers in cross–sectional area from $A_1 = 1.2 \times 10^{-2}$ m^2 to $A_2 = \frac{A_1}{2}$. The pressure difference between the wide and narrow sections of the pipe is 4500 Pa. The rate of flow of liquid is ——— $\times 10^{-3}$ m^{3}s^{-1}.
[NCERT : P L-187, 188 | June 28, 2022 (II)]

44. The area of cross-section of a large tank is 0.5 m^2. It has a narrow opening near the bottom having area of cross-section 1 cm^2. A load of 25 kg is applied on the water at the top in the tank. Neglecting the speed of water in the tank, the velocity of the water, coming out of the opening at the time when the height of water level in the tank is 40 cm above the bottom, will be ______ cms^{-1}.
[Take $g = 10$ ms^{-2}] **[NCERT : PL-188 | June 27, 2022 (I)]**

45. An ideal fluid of density 800 kgm^{-3}, flows smoothly through a bent pipe (as shown in figure) that tapers in cross –sectional area from a to $\frac{a}{2}$. The pressure difference between the wide and narrow sections of pipe is 4100 Pa. At wider section, the velocity of fluid is $\frac{\sqrt{x}}{6}$ ms^{-1} for $x =$ .... (Given $g = 10$ m/s^2)

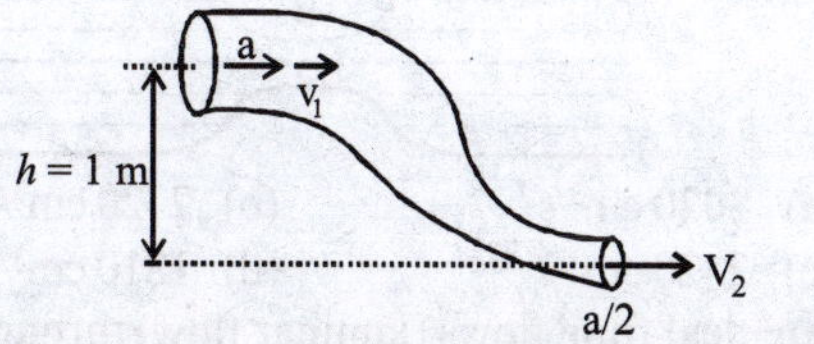

[NCERT : P L-187, 188 | June 26, 2022 (I)]

46. A light cylindrical vessel is kept on a horizontal surface. Area of base is A. A hole of cross-sectional area 'a' is made just at its bottom side. The minimum coefficient of friction necessary to prevent sliding the vessel due to the impact force of the emerging liquid is (a << A): **[NCERT : P L-189 | July 27, 2021 (I)]**

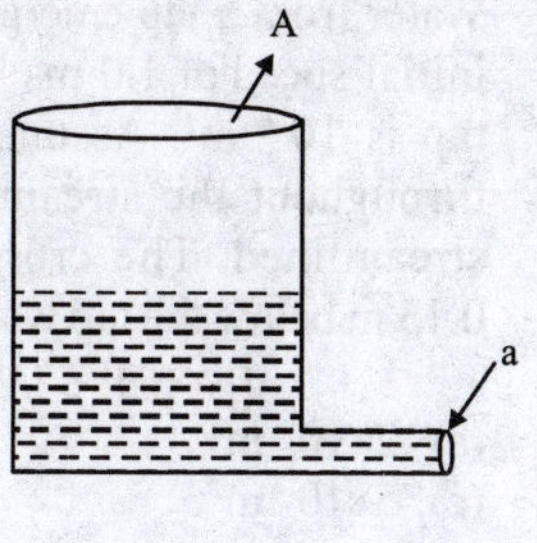

(a) $\frac{A}{2a}$
(b) None of these
(c) $\frac{2a}{A}$
(d) $\frac{a}{A}$

47. The water is filled upto height of 12 m in a tank having vertical sidewalls. A hole is made in one of the walls at a depth 'h' below the water level. The value of 'h' for which the emerging stream of water strikes the ground at the maximum range is ___ m.

[NCERT : P L-189 | NA, July 27, 2021 (II)]

48. Consider a water tank as shown in the figure. It's cross-sectional area is 0.4 m^2. The tank has an opening B near the bottom whose cross-section area is 1 cm^2. A load of 24 kg is applied on the water at the top when the height of the water level is 40 cm above the bottom, the velocity of water coming out the opening B is v ms^{-1}.

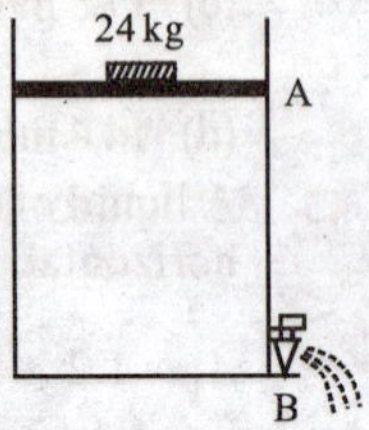

The value of v, to the nearest integer, is ________.
[Take value of g to be 10 ms^{-2}]

[NCERT : P L-188 | March 18, 2021 (II)]

49. A fluid is flowing through a horizontal pipe of varying cross-section, with speed v ms^{-1} at a point where the pressure is P Pascal. At another point where pressure is $\frac{P}{2}$ Pascal its speed is V ms^{-1}. If the density of the fluid is ρ kg m^{-3} and the flow is streamline, then V is equal to :

[NCERT : P L-188 | 6 Sep. 2020 (II)]

(a) $\sqrt{\frac{P}{\rho}+v}$ (b) $\sqrt{\frac{2P}{\rho}+v^2}$ (c) $\sqrt{\frac{P}{2\rho}+v^2}$ (d) $\sqrt{\frac{P}{\rho}+v^2}$

50. Water flows in a horizontal tube (see figure). The pressure of water changes by 700 Nm^{-2} between A and B where the area of cross section are 40 cm^2 and 20 cm^2, respectively. Find the rate of flow of water through the tube.
(density of water = 1000 kgm^{-3})

[NCERT : P L-187, 188 | 9 Jan. 2020 (I)]

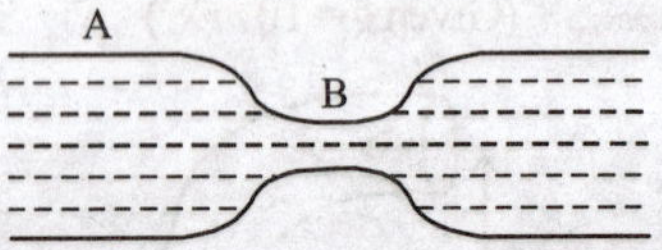

(a) 3020 cm^3/s (b) 2720 cm^3/s
(c) 2420 cm^3/s (d) 1810 cm^3/s

51. An ideal fluid flows (laminar flow) through a pipe of non-uniform diameter. The maximum and minimum diameters of the pipes are 6.4 cm and 4.8 cm, respectively. The ratio of the minimum and the maximum velocities of fluid in this pipe is: [7 Jan. 2020 (II)]

(a) $\frac{9}{16}$ (b) $\frac{\sqrt{3}}{2}$ (c) $\frac{3}{4}$ (d) $\frac{81}{256}$

52. Water from a tap emerges vertically downwards with an initial speed of 1.0 ms^{-1}. The cross-sectional area of the tap is 10^{-4} m^2. Assume that the pressure is constant throughout the stream of water and that the flow is streamlined. The cross-sectional area of the stream, 0.15 m below the tap would be : [Take g = 10 ms^{-2}]

[NCERT : P L-187, 188 | 10 April 2019 (II)]

(a) 2×10^{-5} m^2 (b) 5×10^{-5} m^2
(c) 5×10^{-4} m^2 (d) 1×10^{-5} m^2

53. Water flows into a large tank with flat bottom at the rate of 10^{-4} m^3s^{-1}) Water is also leaking out of a hole of area 1 cm^2 at its bottom. If the height of the water in the tank remains steady, then this height is:

[NCERT : P L-189 | 10 Jan. 2019 I]

(a) 5.1 cm (b) 7 cm (c) 4 cm (d) 9 cm

54. The top of a water tank is open to air and its water level is maintained. It is giving out 0.74m^3 water per minute through a circular opening of 2 cm radius in its wall. The depth of the centre of the opening from the level of water in the tank is close to:

[NCERT : P L-189 | 9 Jan. 2019 (II)]

(a) 6.0 m (b) 4.8 m (c) 9.6 m (d) 2.9 m

55. Two tubes of radii r_1 and r_2, and lengths l_1 and l_2, respectively, are connected in series and a liquid flows through each of them in streamline conditions. P_1 and P_2 are pressure differences across the two tubes. If P_2 is $4P_1$ and l_2 is $\frac{l_1}{4}$, then the radius r_2 will be equal to :

[NCERT : P L-189 | Online April 9, 2017]

(a) r_1 (b) $2r_1$ (c) $4r_1$ (d) $\frac{r_1}{2}$

56. Consider a water jar of radius R that has water filled up to height H and is kept on a stand of height h (see figure). Through a hole of radius r (r << R) at its bottom, the water leaks out and the stream of water coming down towards the ground has a shape like a funnel as shown in the figure. If the radius of the cross–section of water stream when it hits the ground is x. Then :

[NCERT : P L-187, 188 | Online April 9, 2016]

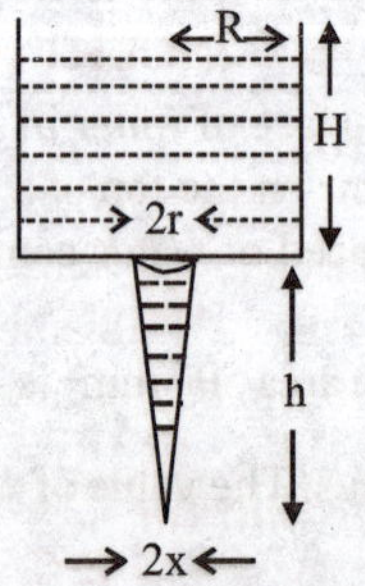

(a) $x = r\left(\frac{H}{H+h}\right)^{\frac{1}{4}}$

(b) $x = r\left(\frac{H}{H+h}\right)$

(c) $x = r\left(\frac{H}{H+h}\right)^2$

(d) $x = r\left(\frac{H}{H+h}\right)^{\frac{1}{2}}$

57. Water is flowing at a speed of 1.5 ms^{-1} through horizontal tube of cross-sectional area 10^{-2} m^2 and you are trying to stop the flow by your palm. Assuming that the water stops immediately after hitting the palm, the minimum force that you must exert should be
(density of water = 10^3 kgm^{-3})

[NCERT : P L-188 | Online April 9, 2014]

(a) 15 N (b) 22.5 N (c) 33.7 N (d) 45 N

58. Air of density 1.2 kg m^{-3} is blowing across the horizontal wings of an aeroplane in such a way that its speeds above and below the wings are 150 ms^{-1} and 100 ms^{-1}, respectively. The pressure difference between the upper and lower sides of the wings, is :

[NCERT : P L-188 | Online April 22, 2013]

(a) 60 Nm^{-2} (b) 180 Nm^{-2}
(c) 7500 Nm^{-2} (d) 12500 Nm^{-2}

59. In a cylindrical water tank, there are two small holes A and B on the wall at a depth of h_1, from the surface of water and at a height of h_2 from the bottom of water tank. Surface of water is at heigh H from the bottom of water tank. Water coming out from both holes strikes the ground at the same point S. Find the ratio of h_1 and h_2

[NCERT : P L-189 | Online May 26, 2012]

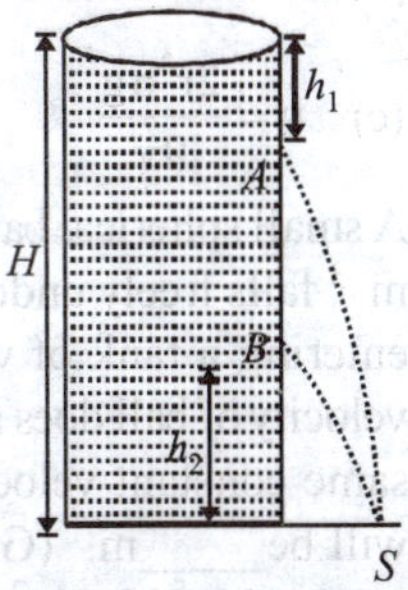

(a) Depends on H

(b) 1 : 1

(c) 2 : 2

(d) 1 : 2

60. A square hole of side length ℓ is made at a depth of h and a circular hole of radius r is made at a depth of 4h from the surface of water in a water tank kept on a horizontal surface. If $\ell << h, r << h$ and the rate of water flow from the holes is the same, then r is equal to

[NCERT : P L-187, 189 | May 7, 2012]

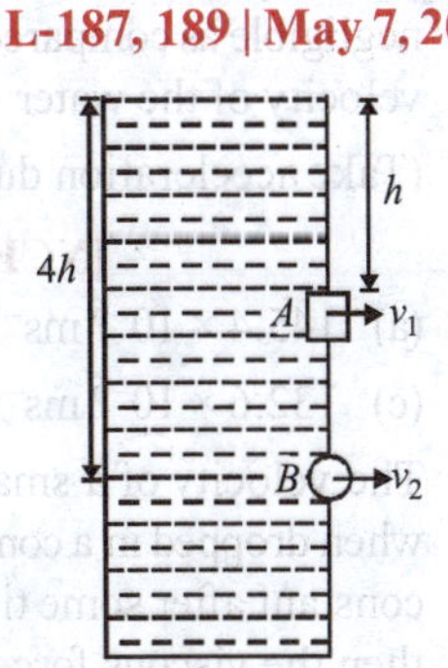

(a) $\frac{\ell}{\sqrt{2\pi}}$

(b) $\frac{\ell}{\sqrt{3\pi}}$

(c) $\frac{\ell}{3\pi}$

(d) $\frac{\ell}{2\pi}$

61. Water is flowing continuously from a tap having an internal diameter 8×10^{-3} m. The water velocity as it leaves the tap is 0.4 ms^{-1}. The diameter of the water stream at a distance 2×10^{-1} m below the tap is close to:

[NCERT : P L-187, 188 | 2011]

(a) 7.5×10^{-3} m (b) 9.6×10^{-3} m

(c) 3.6×10^{-3} m (d) 5.0×10^{-3} m

62. A cylinder of height 20 m is completely filled with water. The velocity of efflux of water (in ms^{-1}) through a small hole on the side wall of the cylinder near its bottom is **[NCERT : P L-189 | 2002]**

(a) 10 (b) 20 (c) 25.5 (d) 5

Topic 3 - Viscosity and Terminal Velocity

63. A solid steel ball of diameter 3.6 mm acquired terminal velocity 2.45×10^{-2} m/s while falling under gravity through an oil of density 925 kg m^{-3}. Take density of steel as 7825 kg m^{-3} and g as 9.8 m/s^2. The viscosity of the oil in SI unit is **[NCERT : P L-192 | April 3, 2025 (II)]**

(a) 2.18 (b) 2.38 (c) 1.68 (d) 1.99

64. A small rigid spherical ball of mass M is dropped in a long vertical tube containing glycerine. The velocity of the ball becomes constant after some time. If the density of glycerine is half of the density of the ball, then the viscous force acting on the ball will be (consider g as acceleration due to gravity)

[NCERT : P L-192 | Jan. 22, 2025 (II)]

(a) $\frac{3}{2}$Mg (b) $\frac{Mg}{2}$ (c) Mg (d) 2 Mg

65. A spherical ball of radius 1×10^{-4} m and density 10^5 kg/m^3 falls freely under gravity through a distance h before entering a tank of water, If after entering in water the velocity of the ball does not change, then the value of h is approximately : (The coefficient of viscosity of water is 9.8×10^{-6} N s/m^2) **[NCERT : P L-192 | April 9, 2024 (II)]**

(a) 2296 m (b) 2249 m (c) 2518 m (d) 2396 m

66. Small water droplets of radius 0.01 mm are formed in the upper atmosphere and falling with a terminal velocity of 10 cm/s. Due to condensation, if 8 such droplets are coalesced and formed a larger drop, the new terminal velocity will be ....cm/s.

[NCERT : P L-192 | April 8, 2024 (II)]

67. A small ball of mass m and density ρ is dropped in a viscous liquid of density ρ_0 . After sometime, the ball falls with constant velocity. The viscous force on the ball is :

[NCERT : P L-192 | April 6, 2024 (I)]

(a) $mg\left(\frac{\rho_0}{\rho}-1\right)$ (b) $mg\left(1+\frac{\rho}{\rho_0}\right)$

(c) $mg(1-\rho\rho_0)$ (d) $mg\left(1-\frac{\rho_0}{\rho}\right)$

68. A small steel ball is dropped into a long cylinder containing glycerine. Which one of the following is the correct representation of the velocity time graph for the transit of the ball ? **[NCERT : P L-192 | Jan. 31, 2024 (I)]**

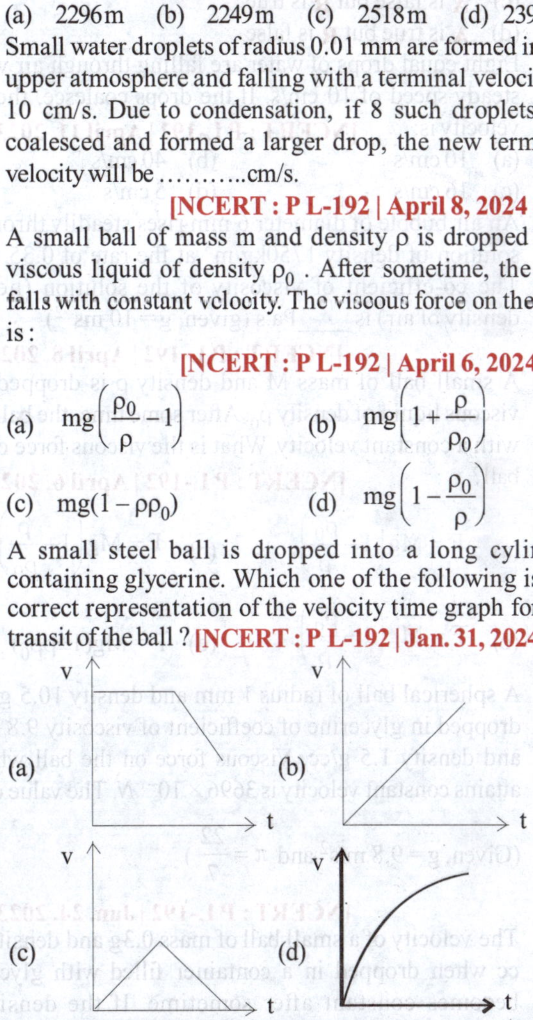

69. A small spherical ball of radius r, falling through a viscous medium of negligible density has terminal velocity 'v'. Another ball of the same mass but of radius 2r, falling through the same viscous medium will have terminal velocity: **[NCERT : P L-192 | Jan. 31, 2024 (II)]**

(a) $\frac{v}{2}$ (b) $\frac{v}{4}$ (c) 4v (d) 2v

70. Given below are two statements: one is labelled as **Assertion A** and the other is labelled as **Reason R.**

Assertion A : A spherical body of radius (5 ± 0.1) mm having a particular density is falling through a liquid of constant density. The percentage error in the calculation of its terminal velocity is 4%.

Reason R : The terminal velocity of the spherical body falling through the liquid is inversely proportional to its radius.

In the light of the above statements, choose the **correct** answer from the options given below

[NCERT : P L-192 | April 13, 2023 (II)]

(a) Both **A** and **R** are true but **R** is **NOT** the correct explanation of **A**
(b) Both **A** and **R** are true but **R** is the correct explanation of **A**
(c) **A** is false but **R** is true
(d) **A** is true but **R** is false

71. Eight equal drops of water are falling through air with a steady speed of 10 cm/s. If the drops coalesce, the new velocity is: **[NCERT : P L-192 | April 11, 2023 (II)]**

(a) 10 cm/s (b) 40 cm/s
(c) 16 cm/s (d) 5 cm/s

72. An air bubble of diameter 6 mm rises steadily through a solution of density 1750kg/m^3 at the rate of 0.35 cm/s. The co-efficient of viscosity of the solution (neglect density of air) is ____Pa.s (given, g = 10 ms^{-2})

[NCERT : P L-192 | April 8, 2023 (I)]

73. A small ball of mass M and density ρ is dropped in a viscous liquid of density ρ_0. After some time, the ball falls with a constant velocity. What is the viscous force on the ball? **[NCERT : P L-192 | April 6, 2023 (I)]**

(a) $F = Mg\left(1-\frac{\rho_0}{\rho}\right)$ (b) $F = Mg\left(1+\frac{\rho}{\rho_0}\right)$

(c) $F = Mg\left(1+\frac{\rho_0}{\rho}\right)$ (d) $F = Mg(1 \pm \rho\rho_0)$

74. A spherical ball of radius 1 mm and density 10.5 g/cc is dropped in glycerine of coefficient of viscosity 9.8 poise and density 1.5 g/cc. Viscous force on the ball when it attains constant velocity is 3696×10^{-x} *N*. The value of *x* is

(Given, g = 9.8 m/s^2 and $\pi = \frac{22}{7}$)

[NCERT : P L-192 | Jan. 24, 2023 (II)]

75. The velocity of a small ball of mass 0.3g and density 8g/cc when dropped in a container filled with glycerine becomes constant after sometime. If the density of glycerine is 1.3 g/cc, then the value of viscous force acting on the ball will be x × 10^{-4} N, the value of x is____. [use g = 10 m/s^2] **[NCERT : P L-192 | July 29, 2022 (II)]**

76. The diameter of an air bubble which was initially 2 mm, rises steadily through a solution of density 1750 kg m^{-3} at the rate of 0.35 cms^{-1}. The coefficient of viscosity of the solution is _______ poise (in nearest integer). (the density of air is negligible). **[NCERT : P L-192 | July 28, 2022 (I)]**

77. An air bubble of negligible weight having radius r rises steadily through a solution of density σ at speed v. The coefficient of viscosity of the solution is given by:

[NCERT : P L-192 | June 30, 2022 (I)]

(a) $\eta = \frac{4r\sigma g}{9v}$ (b) $\eta = \frac{4r^2\sigma g}{9v}$

(c) $\eta = \frac{2r^2\sigma g}{9v}$ (d) $\eta = \frac{2r^2\sigma g}{3\pi v}$

78. A small spherical ball of radius 0.1 mm and density 10^4 kg m^{-3} falls freely under gravity through a distance *h* before entering a tank of water. If after entering the water the velocity of ball does not change and it continue to fall with same constant velocity inside water, then the value of *h* will be_____m. (Given g = 10 ms^{-2}, viscosity of water = 1.0 × 10^{-5} N-sm^{-2}).

[NCERT : P L-192 | June 29, 2022 (II)]

79. A water drop of radius 1 μm falls in a situation where the effect of buoyant force is negligible. Coefficient of viscosity of air is 1.8 × 10^{-5} Nsm^{-2} and its density is negligible as compared to that of water 10^6 gm^{-3}. Terminal velocity of the water drop is:

(Take acceleration due to gravity = 10 ms^{-2})

[NCERT : P L-192 | June 28, 2022 (II)]

(a) 145.4 × 10^{-6} ms^{-1} (b) 118.0 × 10^{-6} ms^{-1}
(c) 132.6 × 10^{-6} ms^{-1} (d) 123.4 × 10^{-6} ms^{-1}

80. The velocity of a small ball of mass 'm' and density d_1, when dropped in a container filled with glycerine, becomes constant after some time. If the density of glycerine is d_2, then the viscous force acting on the ball, will be :

[NCERT : P L-192 | June 27, 2022 (I)]

(a) $mg\left(1-\frac{d_1}{d_2}\right)$ (b) $mg\left(1-\frac{d_2}{d_1}\right)$

(c) $mg\left(\frac{d_1}{d_2}-1\right)$ (d) $mg\left(\frac{d_2}{d_1}-1\right)$

81. A solid sphere, of radius R acquires a terminal velocity v_1 when falling (due to gravity) through a viscous fluid having a coefficient of viscosity **h**. The sphere is broken into 27 identical solid spheres. If each of these spheres acquires a terminal velocity, v_2, when falling through the same fluid, the ratio (v_1/v_2) equals:

[NCERT : P L-192 | 12 April 2019 (II)]

(a) 9 (b) 1/27
(c) 1/9 (d) 27

82. Which of the following option correctly describes the variation of the speed v and acceleration 'a' of a point mass falling vertically in a viscous medium that applies a force F = –kv, where 'k' is a constant, on the body? (Graphs are schematic and not drawn to scale)

[NCERT : P L-192 | Online April 9, 2016]

(a) 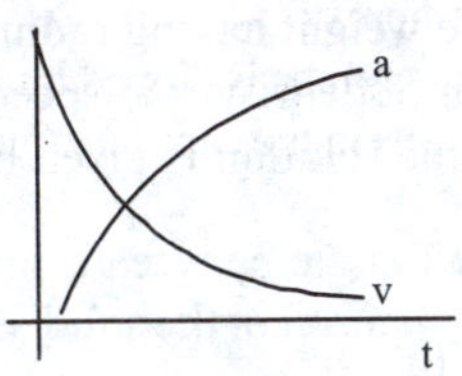

(b) 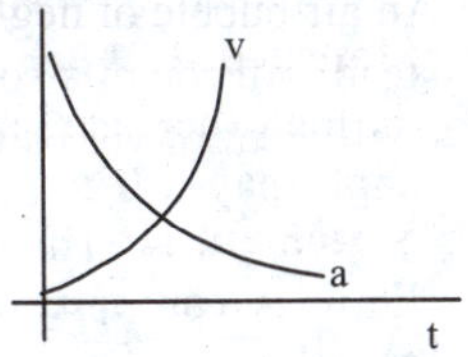

(c) 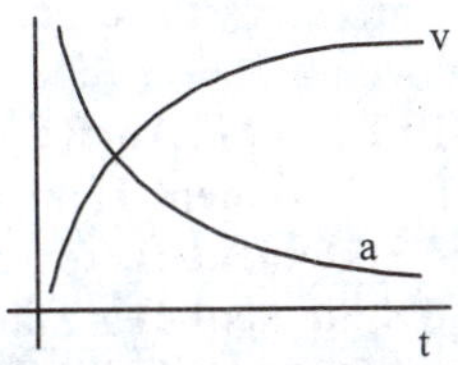

(d) 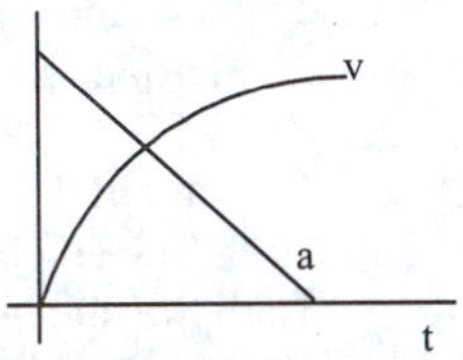

83. The velocity of water in a river is 18 km/hr near the surface. If the river is 5 m deep, find the shearing stress between the horizontal layers of water. The co-efficient of viscosity of water = 10^{-2} poise.

[NCERT : P L-191 | Online April 19, 2014]

(a) 10^{-1} N/m^2 (b) 10^{-2} N/m^2
(c) 10^{-3} N/m^2 (d) 10^{-4} N/m^2

84. In an experiment, a small steel ball falls through a liquid at a constant speed of 10 cm/s. If the steel ball is pulled upward with a force equal to twice its effective weight, how fast will it move upward ?

[NCERT : P L-192 | Online April 25, 2013]

(a) 5 cm/s (b) Zero (c) 10 cm/s (d) 20 cm/s

85. The terminal velocity of a small sphere of radius a in a viscous liquid is proportional to

[NCERT : P L-192 | Online May 26, 2012]

(a) a^2 (b) a^3 (c) a (d) a^{-1}

86. If a ball of steel (density $\rho = 7.8$ g cm^{-3}) attains a terminal velocity of 10 cm s^{-1} when falling in water (Coefficient of viscosity $\eta_{water} = 8.5 \times 10^{-4}$ Pa.s), then, its terminal velocity in glycerine (ρ_ℓ = 1.2 g cm^{-3}, η' = 13.2 Pa.s) would be, nearly

[NCERT : P L-192 | 2011 RS]

(a) 6.25×10^{-4} cm s^{-1} (b) 6.45×10^{-4} cm s^{-1}
(c) 1.5×10^{-5} cm s^{-1} (d) 1.6×10^{-5} cm s^{-1}

87. A spherical solid ball of volume V is made of a material of density ρ_1. It is falling through a liquid of density ρ_2 ($\rho_2 < \rho_1$). Assume that the liquid applies a viscous force on the ball that is proportional to the square of its speed V, i.e., $F_{viscous} = -kv^2$ ($k > 0$). The terminal speed of the ball is

[NCERT : P L-192 | 2008]

(a) $\sqrt{\dfrac{Vg(\rho_1 - \rho_2)}{k}}$ (b) $\dfrac{Vg\rho_1}{k}$

(c) $\sqrt{\dfrac{Vg\rho_1}{k}}$ (d) $\dfrac{Vg(\rho_1 - \rho_2)}{k}$

88. If the terminal speed of a sphere of gold (density = 19.5 kg/m^3) is 0.2 m/s in a viscous liquid (density = 1.5 kg/m^3), find the terminal speed of a sphere of silver (density = 10.5 kg/m^3) of the same size in the same liquid

[NCERT : P L-192 | 2006]

(a) 0.4 m/s (b) 0.133 m/s (c) 0.1 m/s (d) 0.2 m/s

Topic 4 - Surface Tension, Surface Energy and Capillarity

89. A capillary tube of radius 0.1 mm is partly dipped in water (surface tension 70 dyn/cm and glass water contact angle $\simeq 0°$) with 30° inclined with vertical. The length of water risen in the capillary is ______ cm.

[NCERT : P L-197 | April 7, 2025 (II)]

(Take g = 9.8 m/s^2)

(a) $\dfrac{82}{5}$ (b) $\dfrac{57}{2}$ (c) $\dfrac{71}{5}$ (d) $\dfrac{68}{5}$

90. Two liquids A and B have θ_A and θ_B as contact angles in a capillary tube. If $K = \dfrac{\cos\theta_A}{\cos\theta_B}$, then identify the correct statement:

[NCERT : P L-197 | April 4, 2025 (I)]

(a) K is negative, then liquid A and liquid B have convex meniscus.
(b) K is negative, then liquid A and liquid B have concave meniscus.
(c) K is negative, then liquid A has concave meniscus and liquid B has convex meniscus
(d) K is zero, then liquid A has convex meniscus and liquid B has concave meniscus.

91. The excess pressure inside a soap bubble A in air is half the excess pressure inside another soap bubble B in air. If the volume of the bubble A is n times the volume of the bubble B, then, the value of n is ______.

[NCERT : P L-196 | April 3, 2025 (II)]

92. Two water drops each of radius 'r' coalesce to from a bigger drop. If 'T' is the surface tension, the surface energy released in this process is :

[NCERT : P L-196 | April 2, 2025 (II)]

(a) $4\pi r^2 T\left[2 - 2^{\frac{2}{3}}\right]$ (b) $4\pi r^2 T\left[2 - 2^{\frac{1}{3}}\right]$

(c) $4\pi r^2 T\left[1 + \sqrt{2}\right]$ (d) $4\pi r^2 T\left[\sqrt{2} - 1\right]$

93. Consider following statements:

(A) Surface tension arises due to extra energy of the molecules at the interior as compared to the molecules at the surface of a liquid.
(B) As the temperature of liquid rises, the coefficient of viscosity increases.
(C) As the temperature of gas increases, the coefficient of viscosity increases.
(D) The onset of turbulence is determined by Reynold's number.
(E) In a steady flow two streamlines never intersect.

Choose the correct answer from the options given below :

[NCERT : P L-192 | Jan. 28, 2025 (I)]

(a) A, D, E only (b) C, D, E only
(c) B, C, D only (d) A, B, C only

94. An air bubble of radius 0.1 cm lies at a depth of 20 cm below the free surface of a liquid of density 1000 kg/m^3. If the pressure inside the bubble is 2100 N/m^2 greater than the atmospheric pressure, then the surface tension of the liquid in SI unit is (use $g = 10\ m/s^2$)

[NCERT : P L-196 | Jan. 24, 2025 (I)]

(a) 0.02 (b) 0.1 (c) 0.25 (d) 0.05

95. The amount of work done to break a big water drop of radius 'R' into 27 small drops of equal radius is 10 J. The work done required to break the same big drop into 64 small drops of equal radius will be

[NCERT : P L-196 | Jan. 24, 2025 (I)]

(a) 15 J (b) 10 J
(c) 20 J (d) 5 J

96. An air bubble of radius 1.0 mm is observed at a depth of 20 cm below the free surface of a liquid having surface tension 0.095 J/m^2 and density $10^3\ kg/m^3$. The difference between pressure inside the bubble and atmospheric pressure______ N/m^2. (Take $g = 10\ m/s^2$)

[NCERT : P L-196 | Jan. 23, 2025 (II)]

97. Given below are two statements:

Statement-I: The hot water flows faster than cold water.

Statement-II: Soap water has higher surface tension as compared to fresh water.

In the light above statements, choose the correct answer from the options given below.

[NCERT : P L-194 | Jan. 23, 2025 (I)]

(a) Statement-I is false but Statement II is true
(b) Statement-I is true but Statement II is false
(c) Both Statement-I and Statement-II are true
(d) Both Statement-I and Statement-II are false

98. Two soap bubbles of radius 2 cm and 4 cm, respectively, are in contact with each other. The radius of curvature of the common surface, in cm, is______.

[NCERT : P L-196 | Jan. 22, 2025 (I)]

99. The excess pressure inside a soap bubble is thrice the excess pressure inside a second soap bubble. The ratio between the volume of the first and the second bubble is:

[NCERT : P L-196 | April 9, 2024 (II)]

(a) 1 : 9 (b) 1 : 3 (c) 1 : 81 (d) 1 : 27

100. A liquid column of height 0.04 cm balances excess pressure of soap bubble of certain radius. If density of liquid is $8 \times 10^3\ kg\ m^{-3}$ and surface tension of soap solution is $0.28\ Nm^{-1}$, then diameter of the soap bubble is ______cm. (if $g = 10\ ms^{-2}$)

[NCERT : P L-196 | April 8, 2024 (I)]

101. A big drop is formed by coalescing 1000 small droplets of water. The ratio of surface energy of 1000 droplets to that of energy of big drop is $\frac{10}{x}$. The value of x is ______.

[NCERT : P L-196 | April 6, 2024 (I)]

102. Pressure inside a soap bubble is greater than the pressure outside by an amount :
(given : R = Radius of bubble, S = Surface tension of bubble)

[NCERT : P L-196 | April 6, 2024 (II)]

(a) $\frac{4S}{R}$ (b) $\frac{4R}{S}$ (c) $\frac{S}{R}$ (d) $\frac{2S}{R}$

103. Given below are two statements :

Statement-I: When a capillary tube is dipped into a liquid, the liquid neither rises nor falls in the capillary. The contact angle may be 0°.

Statement-II: The contact angle between a solid and a liquid is a property of the material of the solid and liquid as well :

In the light of above statement, choose the **correct** answer from the options given below.

[NCERT : P L-197 | April 5, 2024 (I)]

(a) **Statement-I** is false but **Statement-II** is true.
(b) Both **Statement-I** and **Statement-II** are true.
(c) Both **Statement-I** and **Statement-II** are false.
(d) **Statement-I** is true and **Statement-II** is false.

104. A soap bubble is blown to a diameter of 7 cm. 36960 erg of work is done in blowing it further. If surface tension of soap solution is 40 dyne/cm then the new radius is ____ cm. Take : $\left(\pi = \frac{22}{7}\right)$. **[NCERT : P L-196 | April 4, 2024 (I)]**

105. Given below are two statements :

Statement I : The contact angle between a solid and a liquid is a property of the material of the solid and liquid as well.

Statement II : The rise of a liquid in a capillary tube does not depend on the inner radius of the tube.

In the light of the above statements, choose the correct answer from the options given below :

[NCERT : P L-195, 197 | April 4, 2024 (II)]

(a) Both Statement I and Statement II are false
(b) Statement I is false but Statement II is true.
(c) Statement I is true but Statement II is false.
(d) Both Statement I and Statement II are true.

106. A big drop is formed by coalescing 1000 small droplets of water. The surface energy will become:

[NCERT : P L-196 | Feb. 1, 2024 (II)]

(a) $\frac{1}{100}$th (b) $\frac{1}{10}$th
(c) 100 times (d) 10 times

107. A big drop is formed by coalescing 1000 small identical drops of water. If E_1 be the total surface energy of 1000 small drops of water and E_2 be the surface energy of single big drop of water, then $E_1 : E_2$ is x : 1 where x = ______. **[NCERT : P L-196 | Jan. 30, 2024 (II)]**

108. Given below are two statements:

Statement I : If a capillary tube is immersed first in cold water and then in hot water, the height of capillary rise will be smaller in hot water.

Statement II : If a capillary tube is immersed first in cold water and then in hot water, the height of capillary rise will be smaller in cold water.

[NCERT : P L-197 | Jan. 29, 2024 (I)]

In the light of the above statements, choose the **most appropriate** from the options given below

(a) Statement **I** is false but Statement **II** is true
(b) Both Statement **I** and Statement **II** are false
(c) Statement **I** is true but Statement **II** is false
(d) Both Statement **I** and Statement **II** are true

109. A small liquid drop of radius R is divided into 27 identical liquid drops. If the surface tension is T, then the work done in the process will be : **[NCERT : P L-196 | Jan. 29, 2024 (II)]**

(a) $4\pi R^2 T$ (b) $8\pi R^2 T$ (c) $3\pi R^2 T$ (d) $\frac{1}{8}\pi R^2 T$

110. Given below are two statements :

Statement (I) : Viscosity of gases is greater than that of liquids.

Statement (II) : Surface tension of a liquid decreases due to the presence of insoluble impurities.

In the light of the above statements, choose the most appropriate answer from the options given below : **[NCERT : P L-191, 194 | Jan. 27, 2024 (I)]**

(a) Statement I is correct but statement II is incorrect
(b) Statement I is incorrect but Statement II is correct
(c) Both Statement I and Statement II are incorrect
(d) Both Statement I and Statement II are correct

111. There is an air bubble of radius 1.0 mm in a liquid of surface tension 0.075 Nm^{-1} and density 1000 kg m^{-3} at a depth of 10 cm below the free surface. The amount by which the pressure inside the bubble is greater than the atmospheric pressure is_____ Pa (g = 10 ms^{-2}) **[NCERT : P L-196 | April 15, 2023 (I)]**

112. The surface tension of soap solution is 3.5×10^{-2} Nm^{-1}. The amount of work done required to increase the radius of soap bubble from 10 cm to 20 cm is ____ $\times 10^{-4}$ J. **[NCERT : P L-196 | April 11, 2023 (II)]**

113. A mercury drop of radius 10^{-3} m is broken into 125 equal size droplets. Surface tension of mercury is 0.45 Nm^{-1}. The gain in surface energy is: **[Feb. 1, 2023 (I)]**

(a) 5×10^{-5} J (b) 2.26×10^{-5} J
(c) 17.5×10^{-5} J (d) 28×10^{-5} J

114. If 1000 droplets of water of surface tension 0.07 N/m. having same radius 1 mm each, combine to from a single drop. In the process the released surface energy is- $\left(\text{Take } \pi = \frac{22}{7}\right)$ **[NCERT : P L-196 | Jan. 31, 2023 (I)]**

(a) 7.92×10^{-6} J (b) 7.92×10^{-4} J
(c) 9.68×10^{-4} J (d) 8.8×10^{-5} J

115. The height of liquid column raised in a capillary tube of certain radius when dipped in liquid A vertically is, 5 cm. If the tube is dipped in a similar manner in another liquid B of surface tension and density double the values of liquid A, the height of liquid column raised in liquid B would be_____ m. **[NCERT : P L-197 | Jan. 30, 2023 (I)]**

(a) 0.20 (b) 0.5 (c) 0.05 (d) 0.10

116. Surface tension of a soap bubble is 2.0×10^{-2} Nm^{-1}. Work done to increase the radius of soap bubble from 3.5 cm to 7 cm will be : [Take $\pi = \frac{22}{7}$] **[NCERT : P L-196 | Jan. 29, 2023 (I)]**

(a) 0.72×10^{-4} J (b) 5.76×10^{-4} J
(c) 18.48×10^{-4} J (d) 9.24×10^{-4} J

117. Given below are two statements : One is labelled as Assertion (A) and the other is labelled as Reason (R).

Assertion (A) : Clothes containing oil or grease stains cannot be cleaned by water wash.

Reason (R) : Because the angle of contact between the oil/grease and water is obtuse. In the light of the above statements, choose the correct answer from the option given below. **[NCERT : P L-195 | July 29, 2022 (I)]**

(a) Both (A) and (R) are true and (R) is the correct explanation of (A)
(b) Both (A) and (R) are true but (R) is not the correct explanation of (A)
(c) (A) is true but (R) is false
(d) (A) is true but (R) is true

118. A water drop of radius 1 cm is borken into 729 equal droplets. If surface tension of water is 75 dyne/cm, then the gain in surface energy upto first decimal place will be : (Given $\pi = 3.14$) **[NCERT : P L-196 | July 26, 2022 (I)]**

(a) 8.5×10^{-4} J (b) 8.2×10^{-4} J
(c) 7.5×10^{-4} J (d) 5.3×10^{-4} J

119. The excess pressure inside a liquid drop is 500 Nm^{-2}. If the radius of the drop is 2 mm, the surface tension of liquid is $x \times 10^{-3}$ Nm^{-1}. The value of x is_____. **[NCERT : P L-196 | June 30, 2022 (I)]**

120. A water drop of diameter 2 cm is broken into 64 equal droplets. The surface tension of water is 0.075 N/m. In this process the gain in surface energy will be : **[NCERT : P L-196 | June 28, 2022 (I)]**

(a) 2.8×10^{-4} J (b) 1.5×10^{-3} J
(c) 1.9×10^{-4} J (d) 9.4×10^{-5} J

121. Two narrow bores of diameter 5.0 mm and 8.0 mm are joined together to form a U-shaped tube open at both ends. If this U-tube contains water, what is the difference in the level of two limbs of the tube. [Take surface tension of water T = 7.3×10^{-2} Nm^{-1}, angle of contact = 0°, g = 10 ms^{-2} and density of water = 1.0×10^3 kg m^{-3}] **[NCERT : PL-195 | Aug. 26, 2021 (I)]**

(a) 3.62 mm (b) 2.19 mm (c) 5.34 mm (d) 4.97 mm

122. A soap bubble of radius 3 cm is formed inside the another soap bubble of radius 6 cm. The radius of an equivalent soap bubble which has the same excess pressure as inside the smaller bubble with respect to the atmospheric pressure is _____ cm. **[NCERT : P L-196 | NA, Aug. 26, 2021 (I)]**

123. Two small drops of mercury each of radius R coalesce to form a single large drop. The ratio of total surface energy before and after the change is : **[NCERT : P L-196 | July 20, 2021 (II), 2011 RS]**

(a) $2^{\frac{1}{3}}:1$ (b) $1:2^{\frac{1}{3}}$ (c) 2 : 1 (d) 1 : 2

124. When two soap bubbles of radii a and b (b > a) coalesce, the radius of curvature of common surface is **[NCERT : P L-196 | March 17, 2021 (I)]**

(a) $\frac{ab}{b-a}$ (b) $\frac{ab}{a+b}$ (c) $\frac{b-a}{ab}$ (d) $\frac{a+b}{ab}$

125. Suppose you have taken a dilute solution of oleic acid in such a way that its concentration becomes 0.01 cm^3 of oleic acid per cm^3 of the solution. Then you make a thin film of this solution (monomolecular thickness) of area 4 cm^2 by considering 100 spherical drops of radius $\left(\frac{3}{40\pi}\right)^{\frac{1}{3}} \times 10^{-3}$ cm. Then the thickness of oleic acid layer will be $x \times 10^{-14}$m. Where x is ______.

[NCERT : P L-196 | March 17, 2021 (II)]

126. A large number of water drops, each of radius r, combine to have a drop of radius R. If the surface tension is T and mechanical equivalent of heat is J, the rise in heat energy per unit volume will be

[NCERT : P L-196 | Feb. 26, 2021 (I)]

(a) $\frac{2T}{J}\left(\frac{1}{r}-\frac{1}{R}\right)$ (b) $\frac{3T}{rJ}$

(c) $\frac{3T}{J}\left(\frac{1}{r}-\frac{1}{R}\right)$ (d) $\frac{2T}{rJ}$

127. When a long glass capillary tube of radius 0.015 cm is dipped in a liquid, the liquid rises to a height of 15 cm within it. If the contact angle between the liquid and glass to close to 0°, the surface tension of the liquid, in milliNewton m^{-1}, is $[\rho_{(liquid)} = 900 \text{ kgm}^{-3}, g = 10 \text{ ms}^{-2}]$ (Give answer in closest integer) ______.

[NCERT : P L-197 | NA 3 Sep. 2020 (I)]

128. Pressure inside two soap bubbles are 1.01 and 1.02 atmosphere, respectively. The ratio of their volumes is :

[NCERT : P L-196 | 3 Sep. 2020 (I)]

(a) 4 : 1 (b) 0.8 : 1 (c) 8 : 1 (d) 2 : 1

129. A small spherical droplet of density *d* is floating exactly half immersed in a liquid of density ρ and surface tension T. The radius of the droplet is (take note that the surface tension applies an upward force on the droplet):

[NCERT : P L-196 | 9 Jan. 2020 (II)]

(a) $r=\sqrt{\frac{2T}{3(d+\rho)g}}$ (b) $r=\sqrt{\frac{T}{(d-\rho)g}}$

(c) $r=\sqrt{\frac{T}{(d+\rho)g}}$ (d) $r=\sqrt{\frac{3T}{(2d-\rho)g}}$

130. The ratio of surface tensions of mercury and water is given to be 7.5 while the ratio of their densities is 13.6. Their contact angles, with glass, are close to 135° and 0°, respectively. It is observed that mercury gets depressed by an amount h in a capillary tube of radius r_1, while water rises by the same amount h in a capillary tube of radius r_2. The ratio, (r_1/r_2), is then close to :

[NCERT : P L-197 | 10 April 2019 (I)]

(a) 4/5 (b) 2/5 (c) 3/5 (d) 2/3

131. If 'M' is the mass of water that rises in a capillary tube of radius 'r', then mass of water which will rise in a capillary tube of radius '2r' is : **[NCERT : P L-197 | 9 April 2019 I]**

(a) M (b) $\frac{M}{2}$

(c) 4 M (d) 2 M

132. If two glass plates have water between them and are separated by very small distance (see figure), it is very difficult to pull them apart. It is because the water in between forms cylindrical surface on the side that gives rise to lower pressure in the water in comparison to atmosphere. If the radius of the cylindrical surface is R and surface tension of water is T then the pressure in water between the plates is lower by :

[NCERT : P L-196 | Online April 10, 2015]

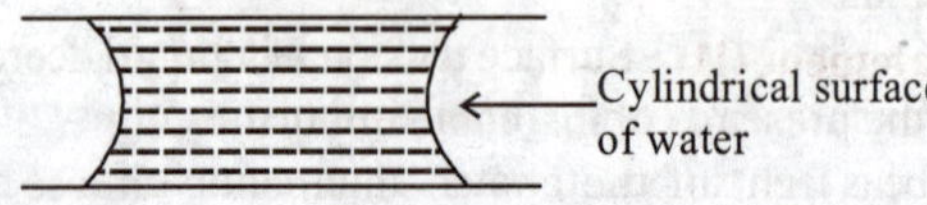

(a) $\frac{2T}{R}$ (b) $\frac{4T}{R}$

(c) $\frac{T}{4R}$ (d) $\frac{T}{R}$

133. On heating water, bubbles being formed at the bottom of the vessel detach and rise. Take the bubbles to be spheres of radius R and making a circular contact of radius r with the bottom of the vessel. If r ≪ R and the surface tension of water is T, value of r just before bubbles detach is: (density of water is ρ_w) **[NCERT : P L-196 | 2014]**

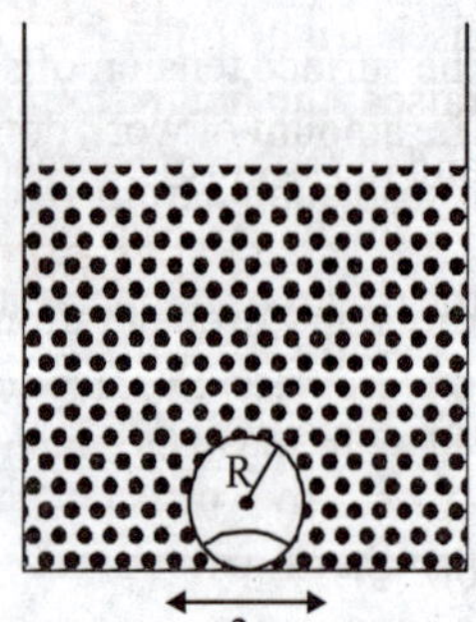

(a) $R^2\sqrt{\frac{2\rho_w g}{3T}}$

(b) $R^2\sqrt{\frac{\rho_w g}{6T}}$

(c) $R^2\sqrt{\frac{\rho_w g}{T}}$

(d) $R^2\sqrt{\frac{3\rho_w g}{T}}$

134. A large number of liquid drops each of radius r coalesce to form a single drop of radius R. The energy released in the process is converted into kinetic energy of the big drop so formed. The speed of the big drop is (given, surface tension of liquid T, density ρ)

[NCERT : P L-196 | Online April 19, 2014, 2012]

(a) $\sqrt{\frac{T}{\rho}\left(\frac{1}{r}-\frac{1}{R}\right)}$ (b) $\sqrt{\frac{2T}{\rho}\left(\frac{1}{r}-\frac{1}{R}\right)}$

(c) $\sqrt{\frac{4T}{\rho}\left(\frac{1}{r}-\frac{1}{R}\right)}$ (d) $\sqrt{\frac{6T}{\rho}\left(\frac{1}{r}-\frac{1}{R}\right)}$

135. An air bubble of radius 0.1 cm is in a liquid having surface tension 0.06 N/m and density 10^3 kg/m^3. The pressure inside the bubble is 1100 Nm^{-2} greater than the atmospheric pressure. At what depth is the bubble below the surface of the liquid? ($g = 9.8 \text{ ms}^{-2}$)

[NCERT : P L-196 | Online April 11, 2014]

(a) 0.1 m (b) 0.15 m (c) 0.20 m (d) 0.25 m

136. A capillary tube is immersed vertically in water and the height of the water column is x. When this arrangement is taken into a mine of depth d, the height of the water column is y. If R is the radius of earth, the ratio $\frac{x}{y}$ is:

[NCERT : P L-197 | Online April 9, 2014]

(a) $\left(1-\frac{d}{R}\right)$ (b) $\left(1-\frac{2d}{R}\right)$

(c) $\left(\frac{R-d}{R+d}\right)$ (d) $\left(\frac{R+d}{R-d}\right)$

137. Wax is coated on the inner wall of a capillary tube and the tube is then dipped in water. Then, compared to the unwaxed capillary, the angle of contact θ and the height *h* upto which water rises change. These changes are :

[NCERT : P L-197 | Online April 23, 2013]

(a) θ increases and *h* also increases
(b) θ decreases and *h* also decreases
(c) θ increases and *h* decreases
(d) θ decreases and *h* increases

138. This question has Statement-1 and Statement-2. Of the four choices given after the Statements, choose the one that best describes the two Statetnents.

Statement-1: A capillary is dipped in a liquid and liquid rises to a height *h* in it. As the temperature of the liquid is raised, the height *h* increases (if the density of the liquid and the angle of contact remain the same).

Statement-2: Surface tension of a liquid decreases with the rise in its temperature.

[NCERT : P L-197 | Online April 9, 2013]

(a) Statement-1 is true, Statement-2 is true; Statement-2 is **not** the correct explanation for Statement-1.
(b) Statement-1 is false, Statement-2 is true.
(c) Statement-1 is true, Statement-2 is false.
(d) Statement-1 is true, Statement-2 is true; Statement-2 is the correct explanation for Statement-1.

139. A thin liquid film formed between a U-shaped wire and a light slider supports a weight of 1.5×10^{-2} N (see figure). The length of the slider is 30 cm and its weight is negligible. The surface tension of the liquid film is

[NCERT : P L-196 | 2012]

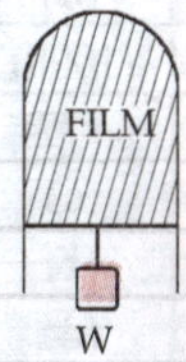

(a) 0.0125 Nm^{-1}
(b) 0.1 Nm^{-1}
(c) 0.05 Nm^{-1}
(d) 0.025 Nm^{-1}

140. Work done in increasing the size of a soap bubble from a radius of 3 cm to 5 cm is nearly (Surface tension of soap solution = 0.03 Nm^{-1}) **[NCERT : P L-196 | 2011]**

(a) 0.2πmJ (b) 2πmJ
(c) 0.4πmJ (d) 4πmJ

141. A 20 cm long capillary tube is dipped in water. The water rises up to 8 cm. If the entire arrangement is put in a freely falling elevator the length of water column in the capillary tube will be **[NCERT : P L-197 | 2005]**

(a) 10 cm (b) 8 cm
(c) 20 cm (d) 4 cm

142. If two soap bubbles of different radii are connected by a tube **[NCERT : P L-196 | 2004]**

(a) air flows from the smaller bubble to the bigger
(b) air flows from bigger bubble to the smaller bubble till the sizes are interchanged
(c) air flows from the bigger bubble to the smaller bubble till the sizes become equal
(d) there is no flow of air.

ANSWER KEYS

1	(d)	**16**	(c)	**31**	(d)	**46**	(c)	**61**	(c)	**76**	(11)	**91**	(8)	**106**	(b)	**121**	(b)	**136**	(a)
2	(150)	**17**	(d)	**32**	(1)	**47**	(6)	**62**	(b)	**77**	(c)	**92**	(a)	**107**	(10)	**122**	(2)	**137**	(c)
3	(5)	**18**	(c)	**33**	(b)	**48**	(3)	**63**	(d)	**78**	(20)	**93**	(b)	**108**	(c)	**123**	(a)	**138**	(b)
4	(d)	**19**	(c)	**34**	(d)	**49**	(d)	**64**	(b)	**79**	(d)	**94**	(d)	**109**	(b)	**124**	(a)	**139**	(d)
5	(a)	**20**	(d)	**35**	(9600)	**50**	(b)	**65**	(c)	**80**	(b)	**95**	(a)	**110**	(b)	**125**	(25)	**140**	(c)
6	(a)	**21**	(b)	**36**	(810)	**51**	(a)	**66**	(40)	**81**	(a)	**96**	(2190)	**111**	(1150)	**126**	(c)	**141**	(c)
7	(d)	**22**	(a)	**37**	(50)	**52**	(b)	**67**	(d)	**82**	(c)	**97**	(b)	**112**	(264)	**127**	(101)	**142**	(a)
8	(1000)	**23**	(c)	**38**	(a)	**53**	(a)	**68**	(d)	**83**	(b)	**98**	(4)	**113**	(b)	**128**	(c)		
9	(177)	**24**	(a)	**39**	(4)	**54**	(b)	**69**	(a)	**84**	(c)	**99**	(d)	**114**	(b)	**129**	(d)		
10	(c)	**25**	(d)	**40**	(2)	**55**	(d)	**70**	(d)	**85**	(a)	**100**	(7)	**115**	(c)	**130**	(b)		
11	(c)	**26**	(b)	**41**	(d)	**56**	(a)	**71**	(b)	**86**	(a)	**101**	(1)	**116**	(c)	**131**	(d)		
12	(d)	**27**	(b)	**42**	(c)	**57**	(b)	**72**	(10)	**87**	(a)	**102**	(a)	**117**	(a)	**132**	(a)		
13	(b)	**28**	(c)	**43**	(24)	**58**	(c)	**73**	(a)	**88**	(c)	**103**	(b)	**118**	(c)	**133**	(a)		
14	(25600)	**29**	(b)	**44**	(300)	**59**	(b)	**74**	(7)	**89**	(a)	**104**	(7)	**119**	(500)	**134**	(d)		
15	(a)	**30**	(d)	**45**	(363)	**60**	(a)	**75**	(25)	**90**	(c)	**105**	(c)	**120**	(a)	**135**	(a)		

Thermal Properties of Matter

10

TREND ANALYSIS
JEE MAIN (2025-2021)

	2025				2024				2023				2022				2021			
	Session II		Session I		Session II		Session I		Session II		Session I		Session II		Session I		Session II		Session I	
	MCQs	NVQs	MCQs	NVQs	MCQs	NVQs	MCQs	NVQs	MCQs	NVQs	MCQs	NVQs	MCQs	NVQs	MCQs	NVQs	MCQs	NVQs	MCQs	NVQs
Topic 1	–	–	1	–	1	–	–	–	1	1	2	3	–	1	2	–	1	2	2	–
Topic 2	2	1	3	1	–	–	1	–	–	–	1	–	2	1	5	2	3	1	2	–
Topic 3																				
Topic 4																				
Total	2	1	4	1	1	–	1	–	1	1	3	3	2	2	7	2	4	3	4	–

How to Code Qns for Better & Faster Revision →

Type 1: Able to solve (No code Req)
Type 2: Solved with struggle or by seeing solution
(Use code NR means Needs Revision or orange stickers)
Type 3: Tricky/ Could not solve on my own. Need help from Teacher/ Friend.
(Use code TQ means Tricky Question or red stickers)

You can either hand write the code in boxes or else paste coloured stickers given at end of book.

Topic 1 – Thermometer & Thermal Expansion

1. Which of the following figure represents the relation between Celsius and Fahrenheit temperatures?
[NCERT : P L-203 | Jan. 24, 2025 (II)]

(a) 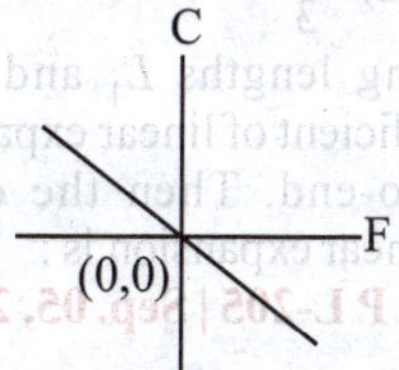

(b) 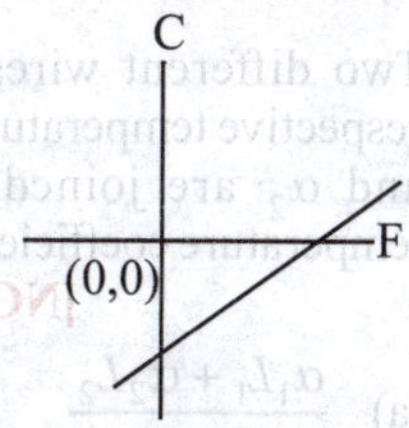

(c) 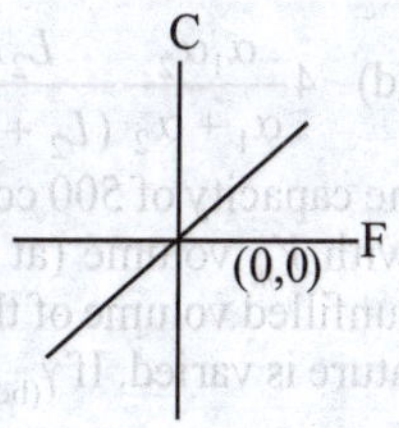

(d) 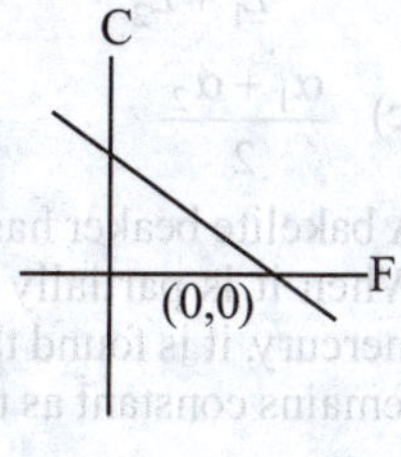

2. On celcius scale the temperature of body increases by 40°C. The increase in temperature on Fahrenheit scale is:
[NCERT : P L-203 | April 4, 2024 (I)]

(a) 70°F (b) 68°F (c) 72°F (d) 75°F

3. On a temperature scale 'X'. The boiling point of water is 65° X and the freezing point is –15° X. Assume that the X scale is linear. The equivalent temperature corresponding to –95° X on the Farenheit scale would be:
[NCERT : P L-203 | April 11, 2023 (I)]

(a) –63° F (b) –112° F (c) –48° F (d) –148° F

4. A steel rod of length 1 m and cross-sectional area 10^{-4} m^2 is heated from 0°C to 200°C without being allowed to extend or bend. The compressive tension produced in the rod is ________ × 10^4 N.
(Given Young's modulus of steel = 2×10^{11} Nm^{-2}, coefficient of linear expansion = 10^{-5}K^{-1}.
[NCERT : P L-205 | April 8, 2023 (II)]

5. A thin rod having a length of 1 m and area of cross-section 3×10^{-6} m^2 is suspended vertically from one end. The rod is cooled from 210°C to 160°C. After cooling, a mass M is attached at the lower end of the rod such that the length of rod again becomes 1 m. Young's modulus and coefficient of linear expansion of the rod are 2×10^{11} Nm^{-2} and

$2 \times 10^{-5}\ K^{-1}$, respectively. The value of M is ______ kg. (Take $g = 10\ m\ s^{-2}$) **[NCERT : P L-205 | Jan. 31, 2023 (I)]**

6. The pressure (P) and temperature (T) relationship of an ideal gas obeys the equation PT^2 = constant. The volume expansion coefficient of the gas will be:
[NCERT : P L-204, 205 | Jan. 30, 2023 (I)]

(a) $3T^2$ (b) $\frac{3}{T^2}$ (c) $\frac{3}{T^3}$ (d) $\frac{3}{T}$

7. A faulty thermometer reads 5°C in melting ice and 95°C in steam. The correct temperature on absolute scale will be......... K when the faulty thermometer reads 41°C.
[NCERT : P L-203 | Jan. 30, 2023 (II)]

8. The graph between two temperature scales P and Q is shown in the figure. Between upper fixed point and lower fixed point there are 150 equal divisions of scale P and 100 divisions on scale Q. The relationship for conversion between the two scales is given by :
[NCERT : P L-203 | Jan. 25, 2023 (II)]

(a) $\frac{t_Q}{150} = \frac{t_P - 180}{100}$

(b) $\frac{t_Q}{100} = \frac{t_P - 30}{150}$

(c) $\frac{t_P}{180} = \frac{t_Q - 40}{100}$

(d) $\frac{t_P}{100} = \frac{t_Q - 180}{150}$

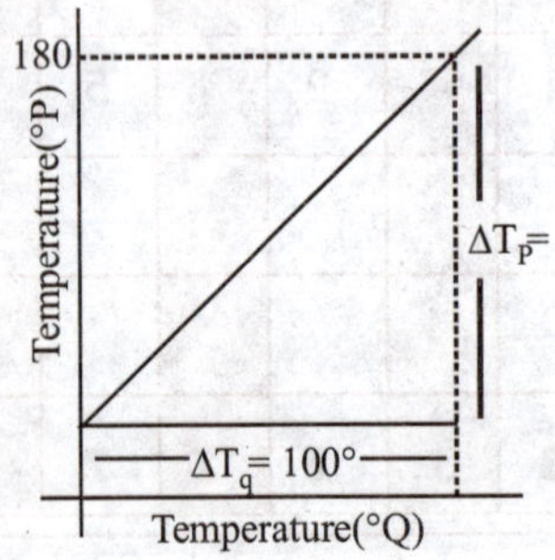

9. A hole is drilled in a metal sheet. At 27°C, the diameter of hole is 5 cm. When the sheet is heated to 177°C, the change in the diameter of hole is $d \times 10^{-3}$ cm. The value of d will be ______ if coefficient of linear expansion of the metal is 1.6×10^{-5}/°C. **[NCERT : P L-205 | Jan. 24, 2023 (I)]**

10. A unit scale is to be prepared whose length does not change with temperature and remains 20 cm, using a bimetallic strip made of brass and iron each of different length. The length of both components would change in such a way that difference between their lengths remains constant. If length of brass is 40 cm and length of iron will be ______ cm. **[NCERT : P L-205 | July 25, 2022 (I)]**
($\alpha_{iron} = 1.2 \times 10^{-5}\ K^{-1}$ and $\alpha_{brass} = 1.8 \times 10^{-5}\ K^{-1}$)

11. At what temperature a gold ring of diameter 6.230 cm be heated so that it can be fitted on a wooden bangle of diameter 6.241 cm? Both the diameters have been measured at room temperature (27 °C).
(Given: coefficient of linear thermal expansion of gold $\alpha_1 = 1.4 \times 10^{-5}\ K^{-1}$)
[NCERT : P L-205 | June 29, 2022 (II)]

(a) 125.7°C (b) 91.7°C
(c) 425.7°C (d) 153.11°C

12. A solid metallic cube having total surface area 24 m^2 is uniformly heated. If its temperature is increased by 10°C, calculate the increase in volume of the cube.
(Given : $\alpha = 5.0 \times 10^{-4}\ °C^{-1}$)
[NCERT : P L-205 | June 25, 2022 (II)]

(a) $2.4 \times 10^6\ cm^3$ (b) $1.2 \times 10^5\ cm^3$
(c) $6.0 \times 10^4\ cm^3$ (d) $4.8 \times 10^5\ cm^3$

13. A steel rod with $Y = 2.0 \times 10^{11}\ Nm^{-2}$ and $\alpha = 10^{-5}\ °C^{-1}$ of length 4 m and area of cross-section 10 cm^2 is heated from 0° C to 400°C without being allowed to extend. The tension produced in the rod is $x \times 10^5$ N where the value of x is ______ **[NCERT : P L-205 | NA, Sep. 1, 2021 (II), Online April 9, 2017(S)]**

14. An ideal gas is expanding such that PT^3 = constant. The coefficient of volume expansion of the gas is :
[NCERT : P L-204, 205 | Aug. 27, 2021 (I)]

(a) $\frac{1}{T}$ (b) $\frac{2}{T}$ (c) $\frac{4}{T}$ (d) $\frac{3}{T}$

15. The area of cross-section of a railway track is 0.01 m^2. The temperature variation is 10°C. Coefficient of linear expansion of material of track is 10^{-5}/°C. The energy stored per meter in the track is ______ J/m.
(Young's modulus of material of track is $10^{11}\ Nm^{-2}$)
[NCERT : P L-205 | NA, July 22, 2021 (II)]

16. Given below are two statements: one is labelled as Assertion A and the other is labelled as Reason R.
Assertion A: When a rod lying freely is heated, no thermal stress is developed in it.
Reason R: On heating, the length of the rod increases.
In the light of the above statements, choose the correct answer from the options given below:
[NCERT : P L-206 | Feb. 25, 2021 (I)]

(a) A is true but R is false
(b) A is false but R is true
(c) Both A and R are true but R is NOT the correct explanation of A
(d) Both A and R are true and R is the correct explanation of A

17. Each side of a box made of metal sheet in cubic shape is 'a' at room temperature 'T', the coefficient of linear expansion of the metal sheet is 'α'. The metal sheet is heated uniformly, by a small temperature ΔT, so that its new temperature is $T + \Delta T$. Calculate the increase in the volume of the metal box.
[NCERT : P L-205 | Feb. 24, 2021 (I)]

(a) $3a^3\alpha\Delta T$ (b) $4a^3\alpha\Delta T$
(c) $3\pi a^3\alpha\Delta T$ (d) $\frac{4}{3}\pi a^3\alpha\Delta T$

18. Two different wires having lengths L_1 and L_2, and respective temperature coefficient of linear expansion α_1 and α_2, are joined end-to-end. Then the effective temperature coefficient of linear expansion is :
[NCERT : P L-205 | Sep. 05, 2020 (II)]

(a) $\frac{\alpha_1 L_1 + \alpha_2 L_2}{L_1 + L_2}$ (b) $2\sqrt{\alpha_1\alpha_2}$

(c) $\frac{\alpha_1 + \alpha_2}{2}$ (d) $4\frac{\alpha_1\alpha_2}{\alpha_1 + \alpha_2}\frac{L_2 L_1}{(L_2 + L_1)^2}$

19. A bakelite beaker has volume capacity of 500 cc at 30°C. When it is partially filled with V_m volume (at 30°C) of mercury, it is found that the unfilled volume of the beaker remains constant as temperature is varied. If $\gamma_{(beaker)} = 6 \times$

$10^{-6}\ °C^{-1}$ and $\gamma_{(mercury)} = 1.5 \times 10^{-4}\ °C^{-1}$, where γ is the coefficient of volume expansion, then V_m (in cc) is close to ______. **[NCERT : P L-205 | NA Sep. 03, 2020 (I)]**

20. When the temperature of a metal wire is increased from 0°C to 10°C, its length increased by 0.02%. The percentage change in its mass density will be closest to : **[NCERT : P L-205 | Sep. 02, 2020 (II)]**

(a) 0.06 (b) 2.3 (c) 0.008 (d) 0.8

21. A non-isotropic solid metal cube has coefficients of linear expansion as: $5 \times 10^{-5}/°C$ along the x-axis and $5 \times 10^{-6}/°C$ along the y and the z-axis. If the coefficient of volume expansion of the solid is $C \times 10^{-6}/°C$ then the value of C is ______. **[NCERT : P L-205 | NA 7 Jan. 2020 I]**

22. At 40°C, a brass wire of 1 mm radius is hung from the ceiling. A small mass, M is hung from the free end of the wire. When the wire is cooled down from 40°C to 20°C it regains its original length of 0.2 m. The value of M is close to: **[NCERT : P L-205 | 12 April 2019 I]**

(Coefficient of linear expansion and Young's modulus of brass are $10^{-5}/°C$ and $10^{11}\ N/m^2$, respectively; $g = 10\ ms^{-2}$)

(a) 9 kg (b) 0.5 kg (c) 1.5 kg (d) 0.9 kg

23. Two rods A and B of identical dimensions are at temperature 30°C. If A is heated upto 180°C and B upto T°C, then the new lengths are the same. If the ratio of the coefficients of linear expansion of A and B is 4 : 3, then the value of T is : **[NCERT : P L-205 | 11 Jan. 2019 II]**

(a) 230°C (b) 270°C (c) 200°C (d) 250°C

24. A thermometer graduated according to a linear scale reads a value x_0 when in contact with boiling water, and $x_0/3$ when in contact with ice. What is the temperature of an object in °C, if this thermometer in the contact with the object reads $x_0/2$? **[NCERT : P L-203 | 11 Jan. 2019 II]**

(a) 25 (b) 60 (c) 40 (d) 35

25. A rod, of length L at room temperature and uniform area of cross section A, is made of a metal having coefficient of linear expansion α/°C. It is observed that an external compressive force F, is applied on each of its ends, prevents any change in the length of the rod, when its temperature rises by ΔT K. Young's modulus, Y, for this metal is: **[NCERT : P L-205 | 8 April 2017 S, 9 Jan. 2019 I]**

(a) $\dfrac{F}{A\alpha\Delta T}$ (b) $\dfrac{F}{A\alpha(\Delta T - 273)}$

(c) $\dfrac{F}{2A\alpha\Delta T}$ (d) $\dfrac{2F}{A\alpha\Delta T}$

26. The ratio of the coefficient of volume expansion of a glass container to that of a viscous liquid kept inside the container is 1 : 4. What fraction of the inner volume of the container should the liquid occupy so that the volume of the remaining vacant space will be same at all temperatures? **[NCERT : P L-205 | Online April 23, 2013]**

(a) 2 : 5 (b) 1 : 4 (c) 1 : 64 (d) 1 : 8

27. On a linear temperature scale Y, water freezes at – 160° Y and boils at – 50° Y. On this Y scale, a temperature of 340 K would be read as : (water freezes at 273 K and boils at 373 K) **[NCERT : P L-203 | Online April 9, 2013]**

(a) –73.7° Y (b) –233.7° Y

(c) –86.3° Y (d) –106.3° Y

28. A wooden wheel of radius R is made of two semicircular part (see figure). The two parts are held together by a ring made of a metal strip of cross sectional area S and length L. L is slightly less than $2\pi R$. To fit the ring on the wheel, it is heated so that its temperature rises by ΔT and it just steps over the wheel. As it cools down to surrounding temperature, it presses the semicircular parts together. If the coefficient of linear expansion of the metal is α, and its Young's modulus is Y, the force that one part of the wheel applies on the other part is : **[NCERT : P L-205 | 2012]**

(a) $2\pi SY\alpha\Delta T$

(b) $SY\alpha\Delta T$

(c) $\pi SY\alpha\Delta T$

(d) $2SY\alpha\Delta T$

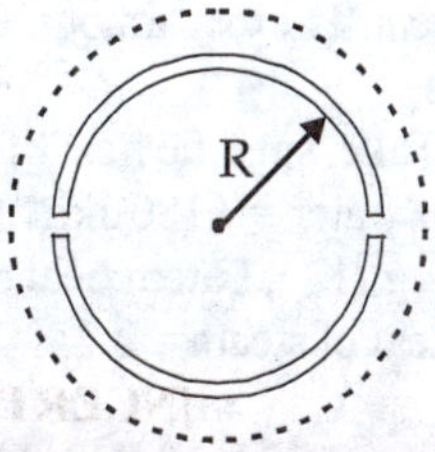

Topic 2 - Calorimetry and Heat Transfer

29. Water falls from a height of 200 m into a pool. Calculate the rise in temperature of the water assuming no heat dissipation from the water in the pool.

(Take $g = 10\ m/s^2$, specific heat of water = 4200 J/(kg K)) **[NCERT : P L-208 | April 8, 2025 (II)]**

(a) 0.23 K (b) 0.36 K (c) 0.14 K (d) 0.48 K

30. Two cylindrical rods A and B made of different materials, are joined in a straight line. The ratio of lengths, radii and thermal conductivities of these rods are :

$\dfrac{L_A}{L_B} = \dfrac{1}{2}, \dfrac{r_A}{r_B} = 2$ and A $\dfrac{K_A}{K_B} = \dfrac{1}{2}$. The free ends of rods A and B are maintained at 400 K, 200 K, respectively. The temperature of rods interface is ______ K, when equilibrium is established. **[NCERT : P L-215, 216 | April 7, 2025 (II)]**

31. Consider a rectangular sheet of solid material of length $l = 9$ cm and width $d = 4$ cm. The coefficient of linear expansion is $\alpha = 3.1 \times 10^{-5}\ K^{-1}$ at room temperature and one atmospheric pressure. The mass of sheet $m = 0.1$ kg and the specific heat capacity $C_v = 900\ J\ kg^{-1}K^{-1}$. If the amount of heat supplied to the material is 8.1×10^2 J then change in area of the rectangular sheet is :- **[NCERT : P L-208, 209 | April 4, 2025 (II)]**

(a) $2.0 \times 10^{-6}\ m^2$ (b) $3.0 \times 10^{-7}\ m^2$

(c) $6.0 \times 10^{-7}\ m^2$ (d) $4.0 \times 10^{-7}\ m^2$

32. A gun fires a lead bullet of temperature 300K into a wooden block. The bullet having melting temperature of 600 K penetrates into the block and melts down. If the total heat required for the process is 625 J, then the mass of the bullet is ___ grams.

(Latent heat of fusion of lead = $2.5 \times 10^4\ JKg^{-1}$ and specific heat capacity of lead = $125\ JKg^{-1}K^{-1}$) **[NCERT : P L-208, 209 | Jan. 23, 2025 (I)]**

(a) 20 (b) 15 (c) 10 (d) 5

33. Two spherical bodies of same materials having radii 0.2 m and 0.8 m are placed in same atmosphere. The temperature of the smaller body is 800 K and temperature of bigger body is 400 K. If the energy radiate from the smaller body is E, the energy radiated from the bigger body is (assume, effect of the surrounding to be negligible)

[NCERT : P L-219 | Jan. 22, 2025 (I)]

(a) 256 E (b) E (c) 64 E (d) 16 E

34. An amount of ice of mass 10^{-3} kg and temperature $-10°C$ is transformed to vapour of temperature 110°C by applying heat. The total amount of work required for this conversion is,

(Take, specific heat of ice = 2100 $Jkg^{-1}K^{-1}$, specific heat of water = 4180 $Jkg^{-1}K^{-1}$, specific heat of steam = 1920 $Jkg^{-1}K^{-1}$, Latent heat of ice = 3.35×10^5 Jkg^{-1} and Latent heat of steam = 2.25×10^6 Jkg^{-1})

[NCERT : P L-208, 209 | Jan. 22, 2025 (I)]

(a) 3022 J (b) 3043 J (c) 3003 J (d) 3024 J

35. Three conductions of same length having thermal conductivity k_1, k_2 and k_3 are connected as shown in figure.

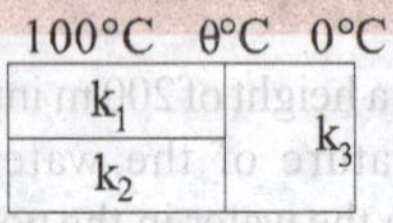

Area of cross sections of 1st and 2nd conductor are same and for 3rd conductor it is double of the 1st conductor. The temperatures are given in the figure. In steady state condition, the value of θ is ______ °C.

(Given : $k_1 = 60$ $Js^{-1}m^{-1}K^{-1}$, $k_2 = 120$ $Js^{-1}m^{-1}K^{-1}$, $k_3 = 135$ $Js^{-1}m^{-1}K^{-1}$) **[NCERT : P L-215, 216 | Jan. 22, 2025 (I)]**

36. A block of ice at $-10°C$ is slowly heated and converted to steam at $100°C$. Which of the following curves represent the phenomenon qualitatively:

[NCERT : P L-210 | Jan. 30, 2024 (II)]

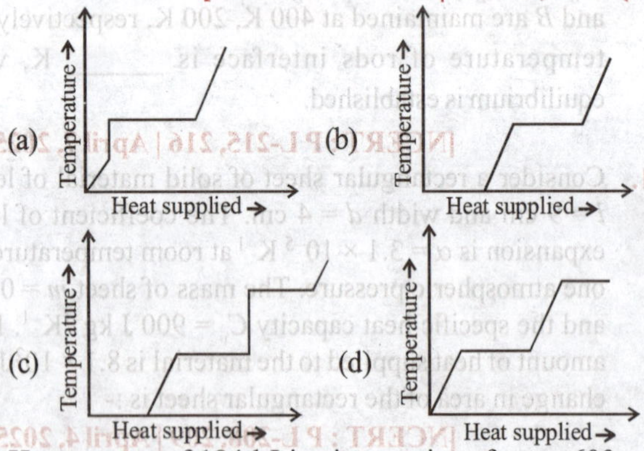

37. Heat energy of 184 kJ is given to ice of mass 600 g at $-12°C$, Specific heat of ice is 2222.3 J kg^{-1}° C^{-1} and latent heat of ice is 336 kJ/kg^{-1}

[NCERT : P L-213 | Jan. 29, 2023 (II)]

(A) Final temperature of system will be 0°C.

(B) Final temperature of the system will be greater than 0°C.

(C) The final system will have a mixture of ice and water in the ratio of 5 : 1.

(D) The final system will have a mixture of ice and water in the ratio of 1 : 5.

(E) The final system will have water only.

Choose the correct answer from the options given below:

(a) A and D only (b) B and D only

(c) A and E only (d) A and C only

38. If K_1 and K_2 are the thermal conductivities, L_1 and L_2 are the lengths, A_1 and A_2 are the cross sectional areas of steel and copper rods respectively such that $\frac{K_2}{K_1} = 9, \frac{A_1}{A_2} = 2, \frac{L_1}{L_2} = 2.$ Then, for the arrangement as shown in the figure, the value of temperature T of the steel – copper junction in the steady state will be :

[NCERT : P L-215 | July 27, 2022 (I)]

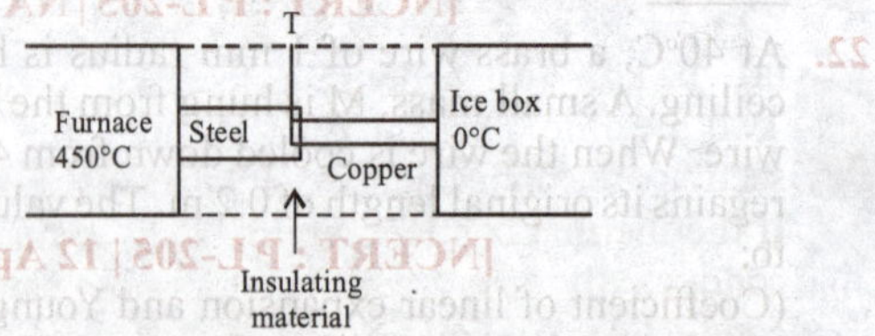

(a) 18 °C (b) 14 °C (c) 45 °C (d) 150 °C

39. An ice cube of dimensions 60 cm × 50 cm × 20 cm is placed in an insulation box of wall thickness 1 cm. The box keeping the ice cube at 0°C of temperature is brought to a room of temperature 40°C. The rate of melting of ice is approximately: (Latent heat of fusion of ice is 3.4×10^5 J kg^{-1} and thermal conducting of insulation wall is 0.05 $Wm^{-1}°C^{-1}$)

[NCERT : P L-213, 215 | July 26, 2022 (II)]

(a) 61×10^{-1} kg s^{-1} (b) 61×10^{-5} kg s^{-1}

(c) 208 kg s^{-1} (d) 30×10^{-5} kg s^{-1}

40. A block of ice of mass 120 g at temperature 0°C is put in 300 gm of water at 25°C. The xg of ice melts as the temperature of the water reaches 0°C. The value of x is

[Use : Specific heat capacity of water = 4200 $Jkg^{-1}K^{-1}$, Latent heat of ice = 3.5×10^5 Jkg^{-1}]

[NCERT : P L-208, 213 | July 25, 2022 (II)]

41. As per the given figure, two plates A and B of thermal conductivity K and $2K$ are joined together to form a compound plate. The thickness of plates are 4.0 cm and 2.5 cm respectively and the area of cross-section is 120 cm^2 for each plate. The equivalent thermal conductivity of the compound plate is $\left(1 + \frac{5}{\alpha}\right)K$, then the value of α will be ______.

[NCERT : P L-216 | June 29, 2022 (I)]

42. A cylinder of fixed capacity of 44.8 litres contains helium gas at standard temperature and pressure. The amount of heat needed to raise the temperature of gas in the cylinder by 20.0°C will be : (Given gas constant R = 8.3 JK^{-1}–mol^{-1})

[NCERT : P L-208 | June 29, 2022 (I)]

(a) 249 J (b) 415 J (c) 498 J (d) 830 J

43. A lead bullet penetrates into a solid object and melts. Assuming that 40% of its kinetic energy is used to heat it, the initial speed of bullet is:

(Given, initial temperature of the bullet = 127°C, Melting point of the bullet = 327°C,

Latent heat of fusion of lead = 2.5×10^4 J Kg^{-1},

Specific heat capacity of lead = 125J/kg K)

[NCERT : P L-208, 213 | June 27, 2022 (II)]

(a) $125\ ms^{-1}$ (b) $500\ ms^{-1}$
(c) $250\ ms^{-1}$ (d) $600\ ms^{-1}$

44. A geyser heats water flowing at a rate of 2.0 kg per minute from 30°C to 70°C. If geyser operates on a gas burner, the rate of combustion of fuel will be ______ $g.\ min^{-1}$
[Heat of combustion = $8 \times 10^3\ Jg^{-1}$
Specific heat of water = $4.2\ Jg^{-1}\ °C^{-1}$] **[NCERT : P L-208 | June 26, 2022 (II)]**

45. A copper block of mass 5.0 kg is heated to a temperature of 500°C and is placed on a large ice block. What is the maximum amount of ice that can melt? [Specific heat of copper : $0.39\ J\ g^{-1}\ °C^{-1}$ and latent heat of fusion of water : $335\ J\ kg^{-1}$]
[NCERT : P L-209, 213 | June 25, 2022 (II)]
(a) 1.5 kg (b) 5.8 kg (c) 2.9 kg (d) 3.8 kg

46. Two metallic blocks M_1 and M_2 of same area of cross-section are connected to each other (as shown in figure). If the thermal conductivity of M_2 is K then the thermal conductivity of M_1 will be : [Assume steady state heat conduction] **[NCERT : P L-216 | June 24, 2022 (I)]**

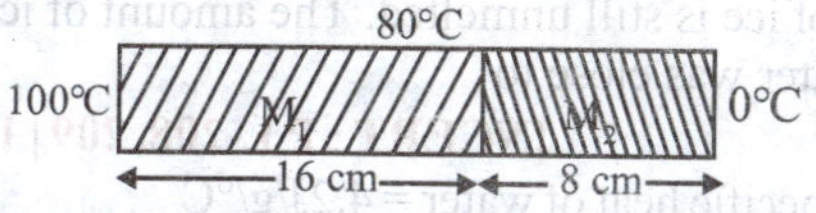

(a) 10K (b) 8K (c) 12.5K (d) 2K

47. A 100 g of iron nail is hit by a 1.5 kg hammer striking at a velocity of $60\ ms^{-1}$. What will be the rise in the temperature of the nail if one fourth of energy of the hammer goes into heating the nail?
[Specific heat capacity of iron = $0.42\ Jg^{-1}\ °C^{-1}$]
[NCERT : P L-208, 209 | June 24, 2022 (II)]
(a) 675°C (b) 1600°C (c) 16.07°C (d) 6.75°C

48. Two thin metallic spherical shells of radii r_1 and r_2 ($r_1 < r_2$) are placed with their centres coinciding. A material of thermal conductivity K is filled in the space between the shells. The inner shell is maintained at temperature θ_1 and the outer shell at temperature θ_2($\theta_1 < \theta_2$). The rate at which heat flows radially through the material is :-
[Aug. 31, 2021 (II), 2005(S)]

(a) $\dfrac{4\pi K r_1 r_2 (\theta_2 - \theta_1)}{r_2 - r_1}$ (b) $\dfrac{\pi r_1 r_2 (\theta_2 - \theta_1)}{r_2 - r_1}$

(c) $\dfrac{K(\theta_2 - \theta_1)}{r_2 - r_1}$ (d) $\dfrac{K(\theta_2 - \theta_1)(r_2 - r_1)}{4\pi r_1 r_2}$

49. A rod CD of thermal resistance $10.0\ KW^{-1}$ is joined at the middle of an identical rod AB as shown in figure, The end A, B and D are maintained at 200°C, 100°C and 125°C respectively. The heat current in CD is P watt. The value of P is ______. **[NA, Aug. 27, 2021 (I)]**

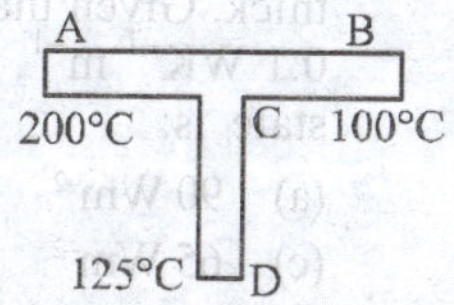

50. The height of victoria falls is 63 m. What is the difference in temperature of water at the top and at the bottom of fall? [Given 1 cal = 4.2 J and specific heat of water = $1\ cal\ g^{-1}\ °C^{-1}$] **[NCERT : P L-208 | Aug. 27, 2021 (II)]**
(a) 0.147°C (b) 14.76°C (c) 1.476°C (d) 0.014°C

51. The temperature of equal masses of three different liquids x,y and z are 10°C, 20°C and 30°C respectively. The temperature of mixture when x is mixed with y is 16°C and that when y is mixed with z is 26°C. The temperature of mixture when x and z are mixed will be :
[NCERT : P L-208, 209 | Aug. 26, 2021 (II)]
(a) 28.32°C (b) 25.62°C (c) 23.84°C (d) 20.28°C

52. Two identical metal wires of thermal conductivities K_1 and K_2 respectively are connected in series. The effective thermal conductivity of the combination is:
[NCERT : P L-216 | March 17, 2021 (I)]

(a) $\dfrac{2K_1K_2}{K_1 + K_2}$ (b) $\dfrac{K_1K_2}{2K_1K_2}$ (c) $\dfrac{K_1 + K_2}{K_1K_2}$ (d) $\dfrac{K_1K_2}{K_1 + K_2}$

53. The temperature θ at the junction of two insulating sheets, having thermal resistances R_1 and R_2 as well as top and bottom temperatures θ_1 and θ_2 (as shown in figure) is given by : **[NCERT : P L-216 | Feb. 26, 2021 (I)]**

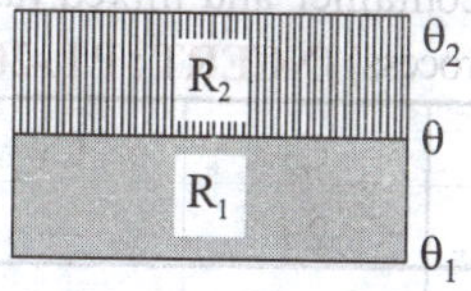

(a) $\dfrac{\theta_1 R_2 - \theta_2 R_1}{R_2 - R_1}$ (b) $\dfrac{\theta_2 R_2 - \theta_1 R_1}{R_2 - R_1}$

(c) $\dfrac{\theta_1 R_1 + \theta_2 R_2}{R_1 + R_2}$ (d) $\dfrac{\theta_1 R_2 + \theta_2 R_1}{R_1 + R_2}$

54. Three rods of identical cross-section and lengths are made of three different materials of thermal conductivity K_1, K_2 and K_3, respectively. They are joined together at their ends to make a long rod (see figure). One end of the long rod is maintained at 100°C and the other at 0°C (see figure). If the joints of the rod are at 70°C and 20°C in steady state and there is no loss of energy from the surface of the rod, the correct relationship between K_1, K_2 and K_3 is:
[NCERT : P L-216 | Sep. 06, 2020 (II)]

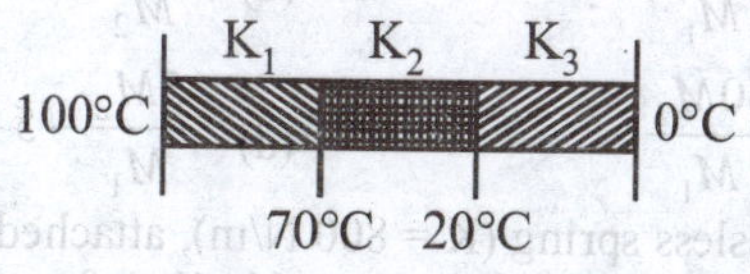

(a) $K_1 : K_3 = 2 : 3, K_2 : K_3 = 2 : 5$
(b) $K_1 < K_2 < K_3$
(c) $K_1 : K_2 = 5 : 2, K_1 : K_3 = 3 : 5$
(d) $K_1 > K_2 > K_3$

55. A bullet of mass 5 g, travelling with a speed of 210 m/s, strikes a fixed wooden target. One half of its kinetics energy is converted into heat in the bullet while the other half is converted into heat in the wood. The rise of temperature of the bullet if the specific heat of its material is 0.030 cal/(g – °C) (1 cal = 4.2×10^7 ergs) close to :
[NCERT : P L-208, 209 | Sep. 05, 2020 (I)]
(a) 87.5°C (b) 83.3°C (c) 119.2°C (d) 38.4°C

56. The specific heat of water = 4200 J kg^{-1} K^{-1} and the latent heat of ice = 3.4×10^5 J kg^{-1}. 100 grams of ice at 0°C is placed in 200 g of water at 25°C. The amount of ice that will melt as the temperature of water reaches 0°C is close to (in grams) : **[NCERT : P L-208 | Sep. 04, 2020 (I)]**

(a) 61.7 (b) 63.8 (c) 69.3 (d) 64.6

57. A calorimter of water equivalent 20 g contains 180 g of water at 25°C. '*m*' grams of steam at 100°C is mixed in it till the temperature of the mixture is 31°C. The value of '*m*' is close to (Latent heat of water = 540 cal g^{-1}, specific heat of water = 1 cal g^{-1} °C^{-1})

[NCERT : P L-208, 213 | Sep. 03, 2020 (II)]

(a) 2 (b) 4 (c) 3.2 (d) 2.6

58. Three containers C_1, C_2 and C_3 have water at different temperatures. The table below shows the final temperature *T* when different amounts of water (given in liters) are taken from each container and mixed (assume no loss of heat during the process) **[NCERT : P L-208, 209 | 8 Jan. 2020 II]**

C_1	C_2	C_3	T
1 *l*	2 *l*	—	60°C
–	1 *l*	2 *l*	30°C
2 *l*	—	1 *l*	60°C
1 *l*	1 *l*	1 *l*	θ

The value of θ (in °C to the nearest integer) is______.

59. M grams of steam at 100°C is mixed with 200 g of ice at its melting point in a thermally insulated container. If it produces liquid water at 40°C [heat of vaporization of water is 540 cal/ g and heat of fusion of ice is 80 cal/g], the value of M is ______.

[NCERT : P L-208, 213 | NA 7 Jan. 2020 II]

60. When M_1 gram of ice at –10 °C (Specific heat = 0.5 cal g^{-1}°C^{-1}) is added to M_2 gram of water at 50°C, finally no ice is left and the water is at 0°C. The value of latent heat of ice, in cal g^{-1} is: **[NCERT : P L-208, 209, 213 | 12 April 2019 I]**

(a) $\frac{50M_2}{M_1} - 5$ (b) $\frac{5M_1}{M_2} - 50$

(c) $\frac{50M_2}{M_1}$ (d) $\frac{5M_2}{M_1} - 5$

61. A massless spring (K = 800 N/m), attached with a mass (500 g) is completely immersed in 1kg of water. The spring is stretched by 2cm and released so that it starts vibrating. What would be the order of magnitude of the change in the temperature of water when the vibrations stop completely? (Assume that the water container and spring receive negligible heat and specific heat of mass = 400 J/kg K, specific heat of water = 4184 J/kg K)

[NCERT : P L-208, 209 | 9 April 2019 II]

(a) 10^{-4} K (b) 10^{-5} K (c) 10^{-1} K (d) 10^{-3} K

62. Two materials having coefficients of thermal conductivity '3*K*' and '*K*' and thickness '*d*' and '3*d*', respectively, are joined to form a slab as shown in the figure. The temperatures of the outer surfaces are 'θ_2' and 'θ_1' respectively, ($\theta_2 > \theta_1$). The temperature at the interface is:

[NCERT : P L-216 | 9 April 2019 II]

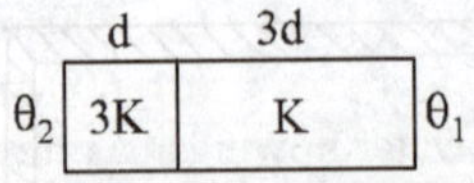

(a) $\frac{\theta_1}{10}+\frac{9\theta_2}{10}$ (b) $\frac{\theta_2+\theta_1}{2}$ (c) $\frac{\theta_1}{6}+\frac{5\theta_2}{6}$ (d) $\frac{\theta_1}{3}+\frac{2\theta_2}{3}$

63. A cylinder of radius R is surrounded by a cylindrical shell of inner radius R and outer radius 2R. The thermal conductivity of the material of the inner cylinder is K_1 and that of the outer cylinder is K_2. Assuming no loss of heat, the effective thermal conductivity of the system for heat flowing along the length of the cylinder is:

[NCERT : P L-216 | 12 Jan. 2019 I]

(a) $\frac{K_1+K_2}{2}$ (b) K_1+K_2

(c) $\frac{2K_1+3K_2}{5}$ (d) $\frac{K_1+3K_2}{4}$

64. Ice at –20°C is added to 50 g of water at 40°C, When the temperature of the mixture reaches 0°C, it is found that 20 g of ice is still unmelted. The amount of ice added to the water was close to

[NCERT : P L-208, 209 | 11 Jan. 2019 I]

(Specific heat of water = 4.2J/g/°C

Specific heat of Ice = 2.1 J/g/°C

Heat of fusion of water at 0°C = 334J/g)

(a) 50g (b) 100 g (c) 60 g (d) 40 g

65. When 100 g of a liquid A at 100°C is added to 50 g of a liquid B at temperature 75°C, the temperature of the mixture becomes 90°C. The temperature of the mixture, if 100 g of liquid A at 100°C is added to 50 g of liquid B at 50°C, will be :

[NCERT : P L-208, 209 | 11 Jan. 2019 II]

(a) 85°C (b) 60°C (c) 80°C (d) 70°C

66. A metal ball of mass 0.1 kg is heated upto 500°C and dropped into a vessel of heat capacity 800 JK^{-1} and containing 0.5 kg water. The initial temperature of water and vessel is 30°C. What is the approximate percentage increment in the temperature of the water? [Specific Heat Capacities of water and metal are, respectively, 4200 $Jkg^{-1}K^{-1}$ and 400 Jkg^{-1} K^{-1}]

[NCERT : P L-208, 209 | 11 Jan. 2019 (II), 10 Jan. 2019 II(S) 2017(S)]

(a) 15% (b) 30% (c) 25% (d) 20%

67. A heat source at T = 10^3 K is connected to another heat reservoir at T = 10^2 K by a copper slab which is 1 m thick. Given that the thermal conductivity of copper is 0.1 WK^{-1} m^{-1}, the energy flux through it in the steady state is: **[NCERT : P L-215 | 10 Jan. 2019 I]**

(a) 90 Wm^{-2} (b) 120 Wm^{-2}

(c) 65 Wm^{-2} (d) 200 Wm^{-2}

68. Temperature difference of 120°C is maintained between two ends of a uniform rod AB of length 2L. Another bent rod PQ, of same cross-section as AB and length $\frac{3L}{2}$, is connected across AB (See figure). In steady state, temperature difference between P and Q will be close to:

[9 Jan. 2019 I]

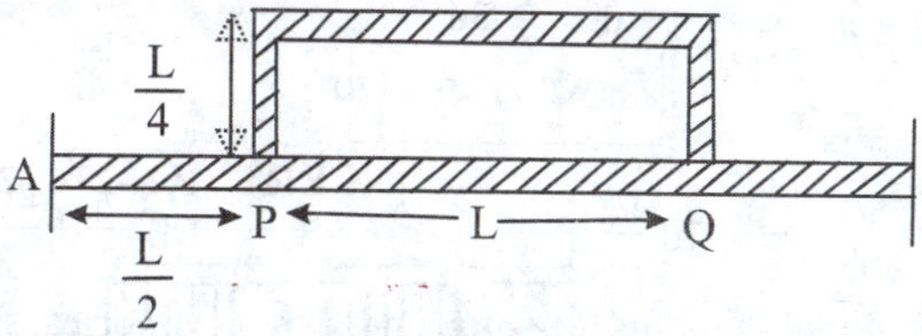

(a) 45°C (b) 75°C (c) 60°C (d) 35°C

69. In an experiment a sphere of aluminium of mass 0.20 kg is heated upto 150°C. Immediately, it is put into water of volume 150 cc at 27°C kept in a calorimeter of water equivalent to 0.025 kg. Final temperature of the system is 40°C. The specific heat of aluminium is :
(take 4.2 Joule=1 calorie)

[NCERT : P L-208, 209 | Online April 8, 2017]

(a) 378 J/kg – °C (b) 315 J/kg – °C
(c) 476 J/kg – °C (d) 434 J/kg – °C

70. An experiment takes 10 minutes to raise the temperature of water in a container from 0°C to 100°C and another 55 minutes to convert it totally into steam by a heater supplying heat at a uniform rate. Neglecting the specific heat of the container and taking specific heat of water to be 1 cal / g °C, the heat of vapourization according to this experiment will come out to be:

[NCERT : P L-209, 213 | Online April 11, 2015]

(a) 560 cal/ g (b) 550 cal/ g
(c) 540 cal/ g (d) 530 cal/ g

71. Three rods of Copper, Brass and Steel are welded together to form a Y shaped structure. Area of cross-section of each rod = 4 cm^2. End of copper rod is maintained at 100°C whereas ends of brass and steel are kept at 0°C. Lengths of the copper, brass and steel rods are 46, 13 and 12 cms respectively. The rods are thermally insulated from surroundings excepts at ends. Thermal conductivities of copper, brass and steel are 0.92, 0.26 and 0.12 CGS units respectively. Rate of heat flow through copper rod is:

[NCERT : P L-215 | 2014]

(a) 1.2 cal/s (b) 2.4 cal/s
(c) 4.8 cal/s (d) 6.0 cal/s

72. Water of volume 2 L in a closed container is heated with a coil of 1 kW. While water is heated, the container loses energy at a rate of 160 J/s. In how much time will the temperature of water rise from 27°C to 77°C? (Specific heat of water is 4.2 kJ/kg and that of the container is negligible).

[NCERT : P L-208 | Online April 9, 2014]

(a) 8 min 20 s (b) 6 min 2 s
(c) 7 min (d) 14 min

73. Assume that a drop of liquid evaporates by decrease in its surface energy, so that its temperature remains unchanged. What should be the minimum radius of the drop for this to be possible? The surface tension is T, density of liquid is ρ and L is its latent heat of vaporization.

[NCERT : P L-213 | 2013]

(a) $\rho L/T$ (b) $\sqrt{T/\rho L}$
(c) $T/\rho L$ (d) $2T/\rho L$

74. A mass of 50g of water in a closed vessel, with surroundings at a constant temperature takes 2 minutes to cool from 30°C to 25°C. A mass of 100g of another liquid in an identical vessel with identical surroundings takes the same time to cool from 30° C to 25° C. The specific heat of the liquid is : (The water equivalent of the vessel is 30g.)

[NCERT : P L-208, 209 | Online April 25, 2013]

(a) 2.0 kcal/kg (b) 7 kcal/kg
(c) 3 kcal/kg (d) 0.5 kcal/kg

75. 500 g of water and 100 g of ice at 0°C are in a calorimeter whose water equivalent is 40 g. 10 g of steam at 100°C is added to it. Then water in the calorimeter is : (Latent heat of ice = 80 cal/g, Latent heat of steam = 540 cal/ g)

[NCERT : P L-208, 209, 213 | Online April 23, 2013]

(a) 580 g (b) 590 g (c) 600 g (d) 610 g

76. The heat radiated per unit area in 1 hour by a furnace whose temperature is 3000 K is ($\sigma = 5.7 \times 10^{-8}\ W m^{-2} K^{-4}$)

[NCERT : P L-219 | Online May 7, 2012]

(a) 1.7×10^{10} J (b) 1.1×10^{12} J
(c) 2.8×10^{8} J (d) 4.6×10^{6} J

77. 100g of water is heated from 30°C to 50°C. Ignoring the slight expansion of the water, the change in its internal energy is (specific heat of water is 4184 J/kg/K):

[NCERT : P L-208 | 2011]

(a) 8.4 kJ (b) 84 kJ (c) 2.1 kJ (d) 4.2 kJ

78. The specific heat capacity of a metal at low temperature (T) is given as

$$C_p (kJK^{-1}kg^{-1}) = 32\left(\frac{T}{400}\right)^3$$

A 100 gram vessel of this metal is to be cooled from 20°K to 4°K by a special refrigerator operating at room temperature (27°C). The amount of work required to cool the vessel is **[NCERT : P L-208 | 2011 RS]**

(a) greater than 0.148 kJ
(b) between 0.148 kJ and 0.028 kJ
(c) less than 0.028 kJ
(d) equal to 0.002 kJ

79. A long metallic bar is carrying heat from one of its ends to the other end under steady–state. The variation of temperature θ along the length x of the bar from its hot end is best described by which of the following figures?

[NCERT : P L-215 | 2009]

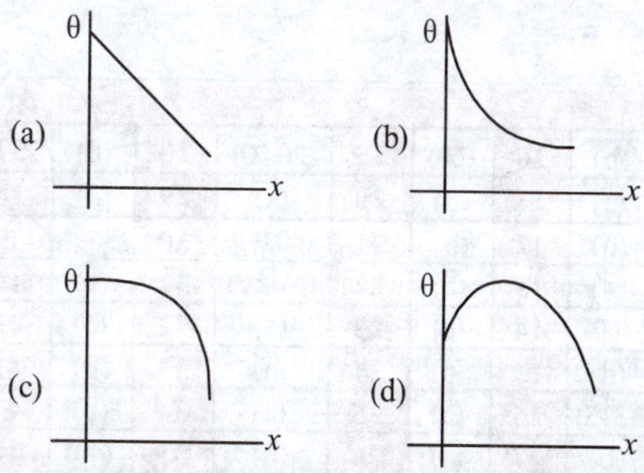

80. One end of a thermally insulated rod is kept at a temperature T_1 and the other at T_2. The rod is composed of two sections of length l_1 and l_2 and thermal conductivities K_1 and K_2 respectively. The temperature at the interface of the two section is

[NCERT : P L-216 | 2007]

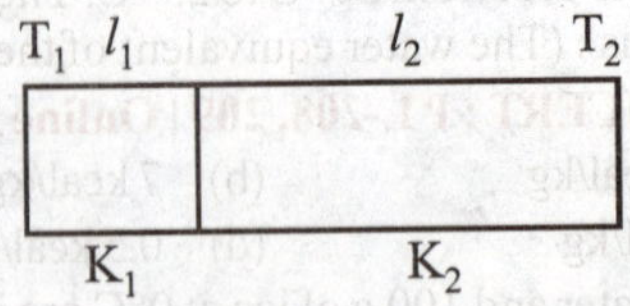

(a) $\frac{(K_1l_1T_1 + K_2l_2T_2)}{(K_1l_1 + K_2l_2)}$ (b) $\frac{(K_2l_2T_1 + K_1l_1T_2)}{(K_1l_1 + K_2l_2)}$

(c) $\frac{(K_2l_1T_1 + K_1l_2T_2)}{(K_2l_1 + K_1l_2)}$ (d) $\frac{(K_1l_2T_1 + K_2l_1T_2)}{(K_1l_2 + K_2l_1)}$

81. Assuming the Sun to be a spherical body of radius R at a temperature of TK, evaluate the total radiant powerd incident of Earth at a distance r from the Sun **[NCERT : P L-219 | 2006]**

(a) $4\pi r_0^2 R^2 \sigma \frac{T^4}{r^2}$ (b) $\pi r_0^2 R^2 \sigma \frac{T^4}{r^2}$

(c) $r_0^2 R^2 \sigma \frac{T^4}{4\pi r^2}$ (d) $R^2 \sigma \frac{T^4}{r^2}$

where r_0 is the radius of the Earth and σ is Stefan's constant.

82. If the temperature of the sun were to increase from T to $2T$ and its radius from R to $2R$, then the ratio of the radiant energy received on earth to what it was previously will be **[NCERT : P L-219 | 2004]**

(a) 32 (b) 16 (c) 4 (d) 64

83. The temperature of the two outer surfaces of a composite slab, consisting of two materials having coefficients of thermal conductivity K and $2K$ and thickness x and $4x$, respectively, are T_2 and $T_1 (T_2 > T_1)$. The rate of heat transfer through the slab, in a steady state is $\left(\frac{A(T_2 - T_1)K}{x}\right) f$, with f equal to **[NCERT : P L-216 | 2004]**

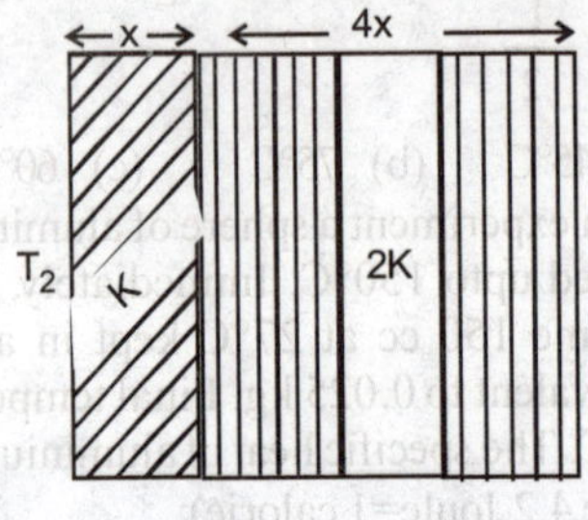

(a) $\frac{2}{3}$

(b) $\frac{1}{2}$

(c) 1

(d) $\frac{1}{3}$

84. Heat given to a body which raises its temperature by 1°C is **[NCERT : P L-208 | 2002]**

(a) water equivalent (b) thermal capacity

(c) specific heat (d) temperature gradient

85. Which of the following is more close to a black body? **[NCERT : P L-219 | 2002]**

(a) black board paint (b) green leaves

(c) black holes (d) red roses

86. If mass-energy equivalence is taken into account, when water is cooled to form ice, the mass of water should **[NCERT : P L-208 | 2002]**

(a) increase

(b) remain unchanged

(c) decrease

(d) first increase then decrease

87. Two spheres of the same material have radii 1 m and 4 m and temperatures 4000 K and 2000 K respectively. The ratio of the energy radiated per second by the first sphere to that by the second is **[NCERT : P L-219 | 2002]**

(a) 1 : 1 (b) 16 : 1 (c) 4 : 1 (d) 1 : 9

ANSWER KEYS

1	(b)	10	(60)	19	(20.00)	28	(d)	37	(a)	46	(b)	55	(a)	64	(d)	73	(d)	82	(d)
2	(c)	11	(d)	20	(a)	29	(d)	38	(c)	47	(c)	56	(a)	65	(c)	74	(d)	83	(d)
3	(d)	12	(b)	21	(60.00)	30	(360)	39	(b)	48	(a)	57	(a)	66	(d)	75	(b)	84	(b)
4	(4)	13	(8)	22	(None)	31	(a)	40	(90)	49	(2)	58	(50.00)	67	(a)	76	(a)	85	(a)
5	(60)	14	(c)	23	(a)	32	(c)	41	(21)	50	(a)	59	(40)	68	(a)	77	(a)	86	(c)
6	(d)	15	(5)	24	(a)	33	(b)	42	(c)	51	(c)	60	(a)	69	(d)	78	(d)	87	(a)
7	(313)	16	(c)	25	(a)	34	(b)	43	(b)	52	(a)	61	(b)	70	(b)	79	(a)		
8	(b)	17	(a)	26	(b)	35	(40)	44	(42)	53	(d)	62	(a)	71	(c)	80	(d)		
9	(12)	18	(a)	27	(c)	36	(d)	45	(c)	54	(a)	63	(d)	72	(a)	81	(b)		

Thermodynamics

11

TREND ANALYSIS
JEE MAIN (2025-2021)

	2025				2024				2023				2022				2021			
	Session II		Session I		Session II		Session I		Session II		Session I		Session II		Session I		Session II		Session I	
	MCQs	NVQs	MCQs	NVQs	MCQs	NVQs	MCQs	NVQs	MCQs	NVQs	MCQs	NVQs	MCQs	NVQs	MCQs	NVQs	MCQs	NVQs	MCQs	NVQs
Topic 1	1	–	1	–	–	–	–	–	3	–	2	–	–	–	1	–	–	–	1	–
Topic 2	8	1	8	2	10	–	6	–	4	–	5	–	4	1	3	2	1	1	6	3
Topic 3	–	–	–	–	–	–	–	–	1	–	–	–	–	–	–	1	1	1	1	1
Topic 4																				
Total	9	1	9	2	10	–	6	–	8	–	7	–	4	1	4	3	2	2	8	4

How to Code Qns for Better & Faster Revision

Type 1: Able to solve (No code Req)
Type 2: Solved with struggle or by seeing solution
(Use code NR means Needs Revision or orange stickers)

Type 3: Tricky/ Could not solve on my own. Need help from Teacher/ Friend.
(Use code TQ means Tricky Question or red stickers)

You can either hand write the code in boxes or else paste coloured stickers given at end of book.

Topic 1 - First Law of Thermodynamics

1. During the melting of a slab of ice at 273 K at atmospheric pressure : **[NCERT : P L-229, 230 | April 3, 2025 (I)]**
   (a) Internal energy of ice-water system remains unchanged.
   (b) Positive work is done by the ice-water system on the atmosphere.
   (c) Internal energy of the ice-water system decreases.
   (d) Positive work is done on the ice-water system by the atmosphere.

2. Given are statements for certain thermodynamic variables,
   (A) Internal energy, volume (V) and mass (M) are extensive variables.
   (B) Pressure (P), temperature (T) and density (ρ) are intensive variables.
   (C) Volume (V), temperature (T) and density (ρ) are intensive variables.
   (D) Mass (M), temperature (T) and internal energy are extensive variables.

   Choose the correct answer from the points given below
   [NCERT : P L-232, 233 | Jan. 22, 2025 (II)]
   (a) (C) and (D) only
   (b) (D) and (A) only
   (c) (A) and (B) only
   (d) (B) and (C) only

3. 1 kg of water at 100°C is converted into steam at 100°C by boiling at atmospheric pressure. The volume of water changes from 1.00×10^{-3} m^3 as a liquid to 1.671 m^3 as steam. Then change in internal energy of the system during the process will be **[NCERT : P L-230 | April 11, 2023 (I)]**
   (Given latent heat of vaporisation = 2257 kJ/kg.
   Atmospheric pressure = 1×10^5 Pa)
   (a) +2090 kJ
   (b) –2090 kJ
   (c) –2426 kJ
   (d) +2476 kJ

4. Given below are two statements:
   Statement I: If heat is added to a system, its temperature must increase.
   Statement II: If positive work is done by a system in a thermodynamic process, its volume must increase.
   In the light of the above statements, choose the correct answer from the options given below:
   [NCERT : P L-230 | April 8, 2023 (I)]

(a) Statement I is true but Statement II is false
(b) Both Statement I and Statement II are true
(c) Both Statement I and Statement II are false
(d) Statement I is false but Statement II is true

5. A source supplies heat to a system at the rate of 1000 W. If the system performs work at a rate of 200 W. The rate at which internal energy of the system increases
[NCERT : P L-230 | April 6, 2023 (I)]

(a) 1200 W (b) 600 W
(c) 500 W (d) 800 W

6. The pressure of a gas changes linearly with volume from A to B as shown in figure. If no heat is supplied to or extracted from the gas then change in the internal energy of the gas will be

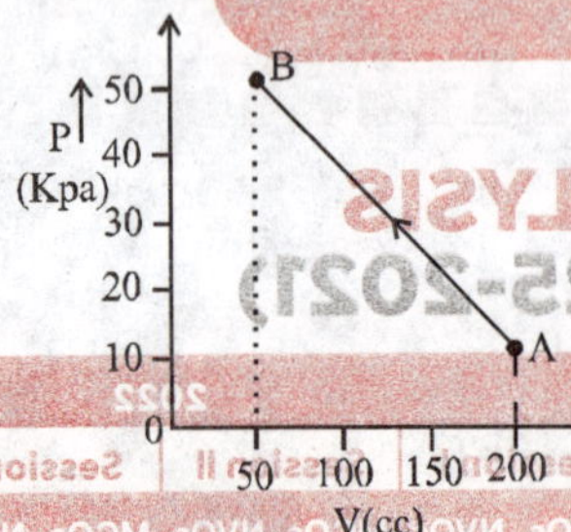

[NCERT : P L-230 | Jan. 31, 2023 (I)]

(a) 6 J (b) Zero
(c) –4.5 J (d) 4.5 J

7. Given below are two statements. One is labelled as Assertion A and the other is labelled as Reason R.
Assertion A : If dQ and dW represent the heat supplied to the system and the work done on the system respectively. Then according to the first law of thermodynamics dQ =dU– dW.
Reason R : First law of thermodynamics is based on law of conservation of energy.
In the light of the above statements, choose the correct answer from the option given below :
[NCERT : P L-230 | Jan. 29, 2023 (I)]

(a) A is correct but R is not correct
(b) A is not correct but R is correct
(c) Both A and R are correct and R is the correct explanation of A
(d) Both A and R are correct but R is not the correct explanation of A

8. A sample of an ideal gas is taken through the cyclic process *ABCA* as shown in figure. It absorbs, 40 J of heat during the part *AB*, no heat during *BC* and rejects 60 J of heat during CA. A work 50 J is done on the gas during the part *BC*. The internal energy of the gas at A is 1560 J. The work done by the gas during the part *CA* is:

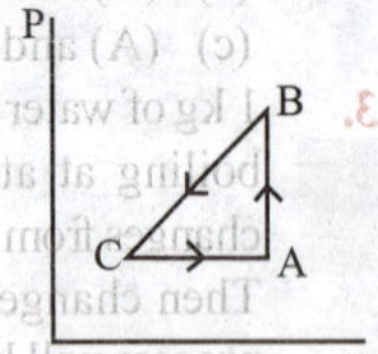

[NCERT : P L-229, 230 | June 28, 2022 (II)]

(a) 20 J (b) 30 J
(c) –30 J (d) –60 J

9. In thermodynamics, heat and work are :
[NCERT : P L-229, 230 | March 16, 2021 (I)]

(a) Point functions
(b) Extensive thermodynamic state variables
(c) Path functions
(d) Intensive thermodynamic state variables

10. A gas can be taken from A to B via two different processes ACB and ADB.
When path ACB is used 60 J of heat flows into the system and 30J of work is done by the system. If path ADB is used work done by the system is 10 J. The heat Flow into the system in path ADB is :

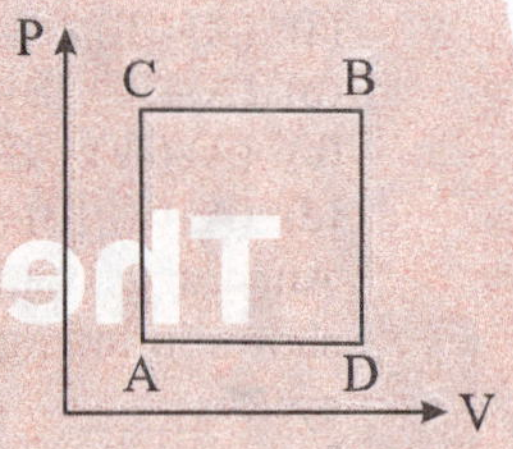

[NCERT : P L-230 | 2007, 9 Jan. 2019 I]

(a) 40 J (b) 80 J (c) 100 J (d) 20 J

11. 200g water is heated from 40°C to 60°C. Ignoring the slight expansion of water, the change in its internal energy is close to (Given specific heat of water = 4184 J/kgK):
[NCERT : P L-230 | Online April 9, 2016]

(a) 167.4 kJ (b) 8.4 kJ (c) 4.2 kJ (d) 16.7 kJ

12. An insulated container of gas has two chambers separated by an insulating partition. One of the chambers has volume V_1 and contains ideal gas at pressure P_1 and temperature T_1. The other chamber has volume V_2 and contains ideal gas at pressure P_2 and temperature T_2. If the partition is removed without doing any work on the gas, the final equilibrium temperature of the gas in the container will be
[NCERT : P L-230 | 2008]

(a) $\dfrac{T_1T_2(P_1V_1+P_2V_2)}{P_1V_1T_2+P_2V_2T_1}$ (b) $\dfrac{P_1V_1T_1+P_2V_2T_2}{P_1V_1+P_2V_2}$

(c) $\dfrac{P_1V_1T_2+P_2V_2T_1}{P_1V_1+P_2V_2}$ (d) $\dfrac{T_1T_2(P_1V_1+P_2V_2)}{P_1V_1T_1+P_2V_2T_2}$

13. Which of the following is incorrect regarding the first law of thermodynamics? **[NCERT : P L-230 | 2005]**

(a) It is a restatement of the principle of conservation of energy
(b) It is not applicable to any cyclic process
(c) It does not introduces the concept of the entropy
(d) It introduces the concept of the internal energy

Topic 2 - Specific Heat Capacity and Thermodynamical Processes

14. A monoatomic gas having $\gamma = \dfrac{5}{3}$ is stored in a thermally insulated container and the gas is suddenly compressed to $\left(\dfrac{1}{8}\right)^{th}$ of its initial volume. The ratio of final pressure and initial pressure is: (γ is the ratio of specific heats of the gas at constant pressure and at constant volume)
[NCERT : P L-235 | April 8, 2025 (II)]

(a) 16 (b) 40 (c) 32 (d) 28

15. Match **List-I** with **List-II**.

List-I	List-II
(A) Isothermal	(I) ΔW (work done) = 0
(B) Adiabatic	(II) ΔQ (supplied heat) = 0
(C) Isobaric	(III) ΔU (change in internal energy) $\neq 0$
(D) Isochoric	(IV) $\Delta U = 0$

Choose the **correct** answer from the options given below:
[NCERT : P L-234, 235 | April 7, 2025 (II)]

(a) (A)-(III), (B)-(II), (C)-(I), (D)-(IV)
(b) (A)-(IV), (B)-(I), (C)-(III), (D)-(II)
(c) (A)-(IV), (B)-(II), (C)-(III), (D)-(I)
(d) (A)-(II), (B)-(IV), (C)-(I), (D)-(III)

16. An ideal gas has undergone through the cyclic process as shown in the figure. Work done by the gas in the entire cycle is_____ $\times 10^{-1}$ J. (Take $\pi = 3.14$)
[NCERT : P L-235 | April 7, 2025 (I)]

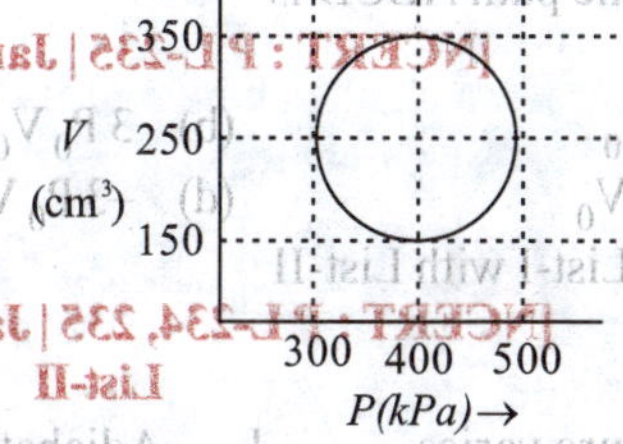

17. Match **List-I** with **List-II**.

List-I	List-II
(A) Isobaric	(I) $\Delta Q = \Delta W$
(B) Isochoric	(II) $\Delta Q = \Delta U$
(C) Adiabatic	(III) ΔQ = zero
(D) Isothermal	(IV) $\Delta Q = \Delta U + P\Delta V$

ΔQ = Heat supplied
ΔW = Work done by the system
ΔU = Change in internal energy
P = Pressure of the system
ΔV = Change in volume of the system

Choose the **correct** answer from the options given below:
[NCERT : P L-234, 235 | April 4, 2025 (II)]

(a) (A)-(IV), (B)-(III), (C)-(II), (D)-(I)
(b) (A)-(IV), (B)-(I), (C)-(III), (D)-(II)
(c) (A)-(IV), (B)-(II), (C)-(III), (D)-(I)
(d) (A)-(II), (B)-(IV), (C)-(III), (D)-(I)

18. An ideal gas exists in a state with pressure P_0, volume V_0. It is isothermally expanded to 4 times of its initial volume (V_0), then isobarically compressed to its original volume. Finally the system is heated isochorically to bring it to its initial state. The amount of heat exchanged in this process is :
[NCERT : P L-235 | April 3, 2025 (II)]

(a) $P_0V_0(2\ln 2 - 0.75)$ (b) $P_0V_0(\ln 2 - 0.75)$
(c) $P_0V_0(\ln 2 - 0.25)$ (d) $P_0V_0(2\ln 2 - 0.25)$

19. A gas is kept in a container having walls which are thermally non-conducting. Initially the gas has a volume of 800 cm^3 and temperature 27°C. The change in temperature when the gas is adiabatically compressed to 200 cm^3 is :
(Take $\gamma = 1.5$, γ is the ratio of specific heats at constant pressure and at constant volume)
[NCERT : P L-235 | April 3, 2025 (I)]

(a) 327 K (b) 600 K (c) 522 K (d) 300 K

20. Identify the characteristics of an adiabatic process in a monoatomic gas. **[NCERT : P L-235 | April 2, 2025 (II)]**
(A) Internal energy is constant.
(B) Work done in the process is equal to the charge in internal energy.
(C) The product of temperature and volume is a constant.
(D) The product of pressure and volume is a constant.
(E) The work done to change the temperature from T_1 to T_2 is proportional to $(T_2 - T_1)$

Choose the **correct** answer from the options given below :
[NCERT : P L-235 | April 2, 2025 (II)]

(a) (A), (C), (D) only (b) (A), (C), (E) only
(c) (B), (E) only (d) (B), (D) only

21. In an adiabatic process, which of the following statements is true ? **[NCERT : P L-235 | April 2, 2025 (I)]**
(a) The molar heat capacity is infinite
(b) Work done by the gas equals the increase in internal energy
(c) The molar heat capacity is zero
(d) The internal energy of the gas decreases as the temperature increases

22. Given below are two statements. One is labelled as **Assertion (A)** and the other is labelled as **Reason (R)**.
Assertion (A) : With the increase in the pressure of an ideal gas, the volume falls off more rapidly in an isothermal process in comparison to the adiabatic process.
Reason (R) : In isothermal process, PV = constant, while in adiabatic process PV^γ = constant. Here γ is the ratio of specific heats, P is the pressure and V is the volume of the ideal gas. **[NCERT : P L-234, 235 | Jan. 29, 2025 (II)]**

In the light of the above statements, choose the correct answer from the options given below :
(a) Both (A) and (R) are true but (R) is NOT the correct explanation of (A)
(b) (A) is true but (R) is false
(c) Both (A) and (R) are true and (R) is the correct explanation of (A).
(d) (A) is false but (R) is true

23.

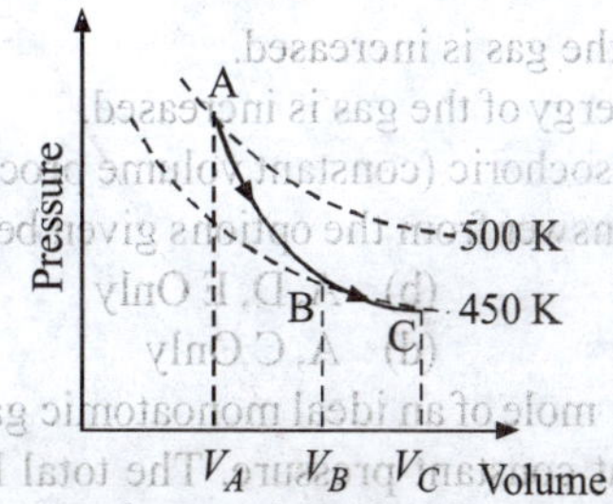

A poly-atomic molecule ($C_V = 3R$, $C_P = 4R$, where R is gas constant) goes from phase space point $A(P_A = 10^5$ Pa, $V_A = 4 \times 10^{-6} m^3)$ to point $B(P_B = 5 \times 10^4$ Pa, $V_B = 6 \times 10^{-6} m^3)$ to point $C(P_C = 10^4$ Pa, $V_C = 8 \times 10^{-6} m^3)$. A to B is an adiabatic path and B to C is an isothermal path.
The net heat absorbed per unit mole by the system is :
[NCERT : P L-234, 235 | Jan. 29, 2025 (II)]

(a) $500R(\ln 3 + \ln 4)$ (b) $450R(\ln 4 - \ln 3)$
(c) $500R \ln 2$ (d) $400R \ln 4$

24. The workdone in an adiabatic change in an ideal gas depends upon only :
[NCERT : P L-235 | Jan. 29, 2025 (I)]

(a) change in its pressure (b) change in its specific heat
(c) change in its volume (d) change in its temperature

25. Given below are two statements. One is labelled as **Assertion (A)** and the other is labelled as **Reason (R)**.

Assertion (A) : In an insulated container, a gas is adiabatically shrunk to half of its initial volume. The temperature of the gas decreases.

Reason (R) : Free expansion of an ideal gas is an irreversible and an adiabatic process.

In the light of the above statement, choose the correct answer from the options given below :

[NCERT : P L-235 | Jan. 24, 2025 (II)]

(a) Both (A) and (R) are true and (R) is the correct explanation of (A)
(b) (A) is true but (R) is false
(c) (A) is false but (R) is true
(d) Both (A) and (R) are true but (R) is NOT the correct explanation of (A)

26. The magnitude of heat exchanged by a system for the given cyclic process ABCA (as shown in figure) is (in SI unit) **[NCERT : P L-235 | Jan. 24, 2025 (II)]**

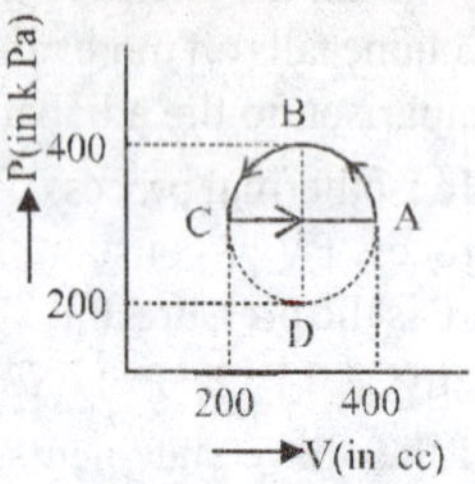

(a) 10π (b) 5π
(c) zero (d) 40π

27. An ideal gas goes from an initial state to final state. During the process, the pressure of gas increases linearly with temperature. **[NCERT : P L-235 | Jan. 24, 2025 (I)]**

A. The work done by gas during the process is zero.
B. The heat added to gas is different from change in its internal energy.
C. The volume of the gas is increased.
D. The internal energy of the gas is increased.
E. The process is isochoric (constant volume process)

Choose the correct answer from the options given below

(a) A, B, C, D Only (b) A, D, E Only
(c) E Only (d) A, C Only

28. The temperature of 1 mole of an ideal monoatomic gas is increased by 50°C at constant pressure. The total heat added and change in internal energy are E_1 and E_2, respectively. If $\frac{E_1}{E_2} = \frac{x}{9}$ then the value of x is ______.

[NCERT : P L-234, 235 | Jan. 24, 2025 (I)]

29. Water of mass m gram is slowly heated to increase the temperature from T_1 to T_2. The change in entropy of the water, given specific heat of water is 1 $Jkg^{-1} K^{-1}$, is

[NCERT : P L-236 | Jan. 23, 2025 (II)]

(a) zero (b) $m(T_2 - T_1)$

(c) $m\ell n\left(\frac{T_1}{T_2}\right)$ (d) $m\ell n\left(\frac{T_2}{T_1}\right)$

30.

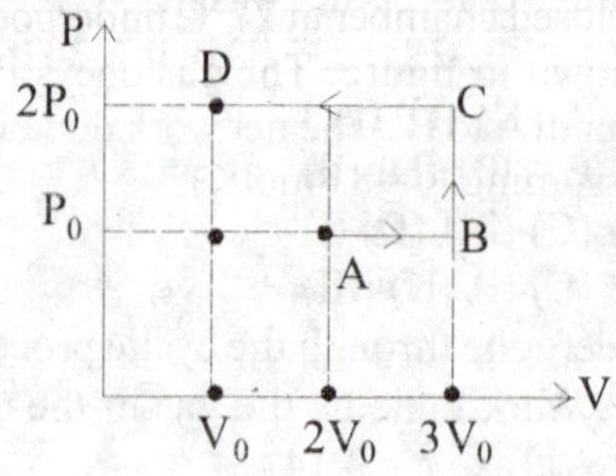

Using the given P-V diagram, the work done by an ideal gas along the path ABCD is

[NCERT : P L-235 | Jan. 23, 2025 (II)]

(a) $4 P_0 V_0$ (b) $3 P_0 V_0$
(c) $-4 P_0 V_0$ (d) $-3 P_0 V_0$

31. Match the List-I with List-II

[NCERT : P L-234, 235 | Jan. 23, 2025 (I)]

	List-I		**List-II**
A.	Pressure varies inversely with volume of an ideal gas.	I.	Adiabatic process
B.	Heat absorbed goes partly to increase internal energy and partly to do work.	II.	Isochoric process
C.	Heat is neither absorbed nor released by a system	III.	Isothermal process
D.	No work is done on or by a gas	IV	Isobaric process

Choose the correct answer from the options given below:

(a) A–I, B–IV, C–II, D–III (b) A–III, B–I, C–IV, D–II
(c) A–I, B–III, C–II, D–IV (d) A–III, B–IV, C–I, D–II

32. An ideal gas initially at 0°C temperature, is compressed suddenly to one fourth of its volume. If the ratio of specific heat at constant pressure to that at constant volume is 3/2, the change in temperature due to the thermodynamics process is ______ K.

[NCERT : P L-235 | Jan. 23, 2025 (I)]

33. The volume of an ideal gas ($\gamma = 1.5$) is changed adiabatically from 5 litres to 4 litres. The ratio of initial pressure to final pressure is:

[NCERT : P L-235 | April 9, 2024 (I)]

(a) $\frac{4}{5}$ (b) $\frac{16}{25}$ (c) $\frac{8}{5\sqrt{5}}$ (d) $\frac{2}{\sqrt{5}}$

34. A sample of 1 mole gas at temperature T is adiabatically expanded to double its volume. If adiabatic constant for the gas is $\gamma = \frac{3}{2}$, then the work done by the gas in the process is: **[NCERT : P L-235 | April 9, 2024 (I)]**

(a) $RT\left[2-\sqrt{2}\right]$ (b) $\frac{R}{T}\left[2-\sqrt{2}\right]$

(c) $RT\left[2+\sqrt{2}\right]$ (d) $\frac{T}{R}\left[2+\sqrt{2}\right]$

35. A real gas within a closed chamber at 27°C undergoes the cyclic process as shown in figure. The gas obeys $PV^3 = RT$ equation for the path A to B. The net work done in the complete cycle is (assuming $R = 8 J/molK$):

[NCERT : P L-235 | April 9, 2024 (II)]

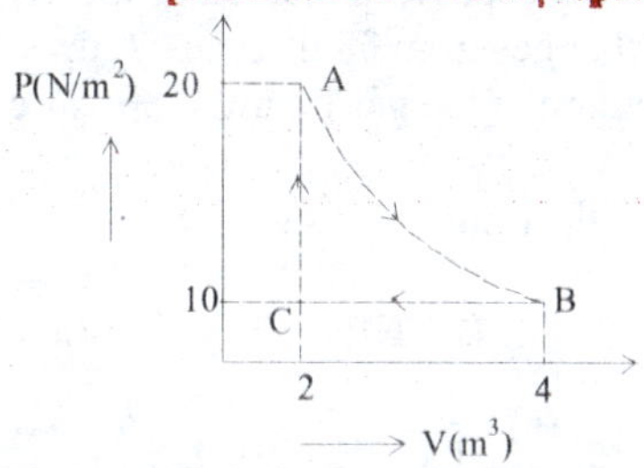

(a) 225 J (b) 205 J (c) 20 J (d) –20 J

36. Two different adiabatic paths for the same gas intersect two isothermal curves as shown in P-V diagram. The relation between the ratio $\frac{V_a}{V_d}$ and the ratio $\frac{V_b}{V_c}$ is:

[NCERT : P L-234, 235 | April 8, 2024 (I)]

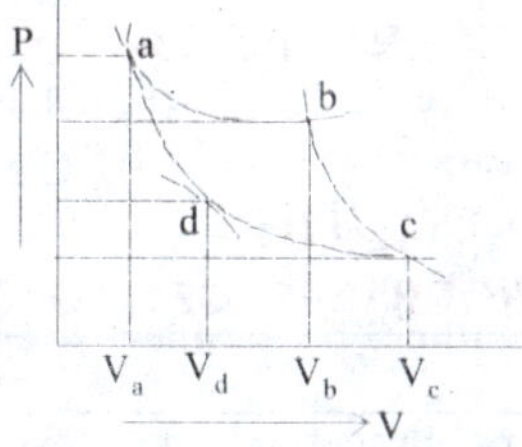

(a) $\frac{V_a}{V_d} = \left(\frac{V_b}{V_c}\right)^{-1}$ (b) $\frac{V_a}{V_d} \neq \frac{V_b}{V_c}$

(c) $\frac{V_a}{V_d} = \frac{V_b}{V_c}$ (d) $\frac{V_a}{V_d} = \left(\frac{V_b}{V_c}\right)^{2}$

37. A diatomic gas ($\gamma = 1.4$) does 100 J of work in an isobaric expansion. The heat given to the gas is :

[NCERT : P L-235 | April 8, 2024 (II)]

(a) 350 J (b) 490 J (c) 150 J (d) 250 J

38. The specific heat at constant pressure of a real gas obeying $PV^2 = RT$ equation is :

[NCERT : P L-231 | April 6, 2024 (I)]

(a) $C_v + R$ (b) $\frac{R}{3} + C_v$ (c) R (d) $C_v + \frac{R}{2V}$

39. A total of 48 J heat is given to one mole of helium kept in a cylinder. The temperature of helium increases by 2°C. The work done by the gas is : (Given, $R = 8.3 J K^{-1} mol^{-1}$.)

[NCERT : P L-234, 235 | April 6, 2024 (II)]

(a) 72.9 J (b) 24.9 J (c) 48 J (d) 23.1 J

40. The heat absorbed by a system in going through the given cyclic process is : **[NCERT : P L-230 | April 5, 2024 (I)]**

(a) 61.6 J

(b) 431.2 J

(c) 616 J

(d) 19.6 J

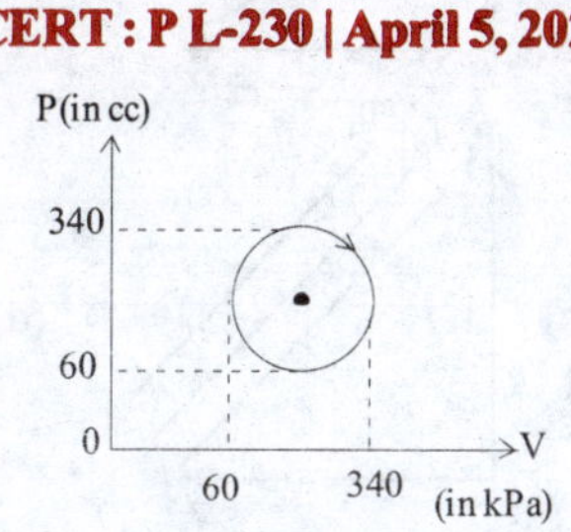

41. During an adiabatic process, if the pressure of a gas is found to be proportional to the cube of its absolute temperature, then the ratio of $\frac{C_P}{C_V}$ for the gas is:

[NCERT : P L-235 | April 5, 2024 (II)]

(a) $\frac{5}{3}$ (b) $\frac{9}{7}$ (c) $\frac{3}{2}$ (d) $\frac{7}{5}$

42. A sample of gas at temperature T is adiabatically expanded to double its volume. Adiabatic constant for the gas is $\gamma = 3/2$. The work done by the gas in the process is : ($\mu = 1$ mole) **[NCERT : P L-235 | April 4, 2024 (II)]**

(a) $RT\left[\sqrt{2} - 2\right]$ (b) $RT\left[1 - 2\sqrt{2}\right]$

(c) $RT\left[2\sqrt{2} - 1\right]$ (d) $RT\left[2 - \sqrt{2}\right]$

43. The pressure and volume of an ideal gas related as $PV^{\frac{3}{2}} = K$ (Constant). The work done when the gas is taken from state A (P_1, V_1, T_1) to state B (P_2, V_2, T_2) is :

[NCERT : P L-235 | Feb. 1, 2024 (I)]

(a) $2(P_1V_1 - P_2V_2)$ (b) $2(P_2V_2 - P_1V_1)$

(c) $2\left(\sqrt{P_1}V_1 - \sqrt{P_2}V_2\right)$ (d) $2\left(P_2\sqrt{V_2} - P_1\sqrt{V_1}\right)$

44. The given figure represents two isobaric processes for the same mass of an ideal gas, then

[NCERT : P L-235 | Jan. 31, 2024 (I)]

(a) $P_1 > P_2$

(b) $P_2 > P_1$

(c) $P_1 = P_2$

(d) $P_2 \geq P_1$

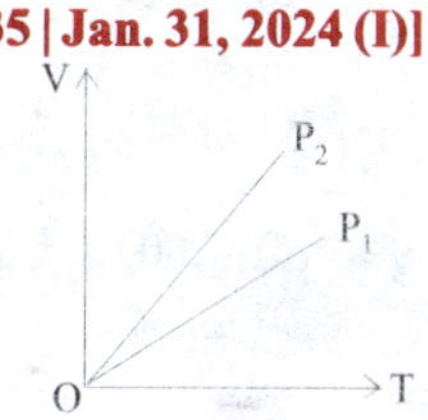

45. Choose the correct statement for processes *A* & *B* shown in figure. **[NCERT : P L-234, 235 | Jan. 30, 2024 (II)]**

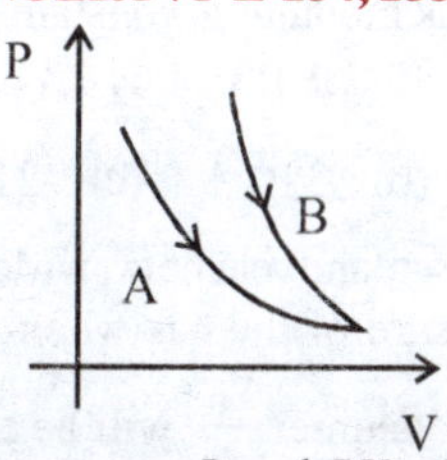

(a) $PV = k$ for process *B* and $PV = k$ for process *A*.

(b) $PV = k$ for process *B* and *A*.

(c) $\frac{P^{\gamma-1}}{T^{\gamma}} = k$ for process *B* and $T = k$ for process *A*.

(d) $\frac{T^{\gamma}}{P^{\gamma-1}} = k$ for process *A* and $PV = k$ for process *B*.

46. A thermodynamic system is taken from an original state A to an intermediate state B by a linear process as shown in the figure. Its volume is then reduced to the original value from B to C by an isobaric process. The total work done by the gas from A to B and B to C would be:

[NCERT : P L-234, 235 | Jan. 29, 2024 (I)]

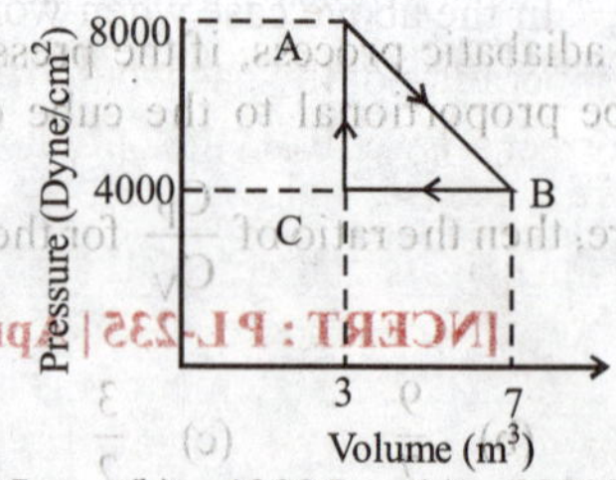

(a) 2200 J (b) 1200 J (c) 33800 J (d) 800 J

47. 0.08 kg air is heated at constant volume through 5°C. The specific heat of air at constant volume is 0.17 kcal/kg°C and J = 4.18 joule/cal. The change in its internal energy is approximately. **[NCERT : P L-231 | Jan. 27, 2024 (I)]**

(a) 318 J (b) 298 J (c) 284 J (d) 142 J

48. During an adiabatic process, the pressure of a gas is found to be proportional to the cube of its absolute temperature. The ratio of $\frac{C_p}{C_v}$ for the gas is: **[NCERT : P L-235 | Jan. 27, 2024 (II)]**

(a) $\frac{5}{3}$ (b) $\frac{3}{2}$ (c) $\frac{7}{5}$ (d) $\frac{9}{7}$

49. A thermodynamic system is taken through cyclic process. The total work done in the process is : **[NCERT : P L-234, 235 | April 15, 2023 (I)]**

(a) 100 J

(b) 300 J

(c) Zero

(d) 200 J

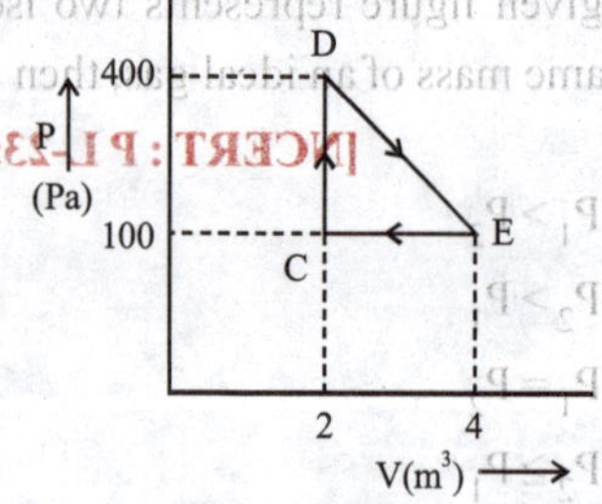

50. Under isothermal condition, the pressure of a gas is given by $P = aV^{-3}$, where a is a constant and V is the volume of the gas. The bulk modulus at constant temperature is equal to **[NCERT : P L-234 | April 13, 2023 (I)]**

(a) $\frac{P}{2}$ (b) 3 P (c) 2 P (d) P

51. The initial pressure and volume of an ideal gas are P_0 and V_0. The final pressure of the gas when the gas is suddenly compressed to volume $\frac{V_0}{4}$ will be : (Given γ = ratio of specific heats at constant pressure and at constant volume) **[NCERT : P L-235 | April 13, 2023 (II)]**

(a) $P_0(4)^{\frac{1}{\gamma}}$ (b) $P_0(4)^{\gamma}$ (c) P_0 (d) $4P_0$

52. The thermodynamic process, in which internal energy of the system remains constant is **[NCERT : P L-234, 235 | April 11, 2023 (II)]**

(a) Isochoric (b) Isothermal
(c) Adiabatic (d) Isobaric

53. A sample of gas at temperature *T* is adiabatically expanded to double its volume. The work done by the gas in the process is $\left(\text{given}, \gamma = \frac{3}{2}\right)$: **[NCERT : P L-235 | Feb. 1, 2023 (I)]**

(a) $W = \frac{T}{R}\left[\sqrt{2} - 2\right]$ (b) $W = TR\left[\sqrt{2} - 2\right]$

(c) $W = RT\left[2 - \sqrt{2}\right]$ (d) $W = \frac{R}{T}\left[2 - \sqrt{2}\right]$

54. A hypothetical gas expands adiabatically such that its volume changes from 08 litres to 27 litres. If the ratio of final pressure of the gas to initial pressure of the gas is $\frac{16}{81}$. Then the ratio of $\frac{C_P}{C_V}$ will be. **[NCERT : P L-235 | Jan. 31, 2023 (II)]**

(a) $\frac{4}{3}$ (b) $\frac{3}{1}$ (c) $\frac{1}{2}$ (d) $\frac{3}{2}$

55. Heat is given to an ideal gas in an isothermal process.

A. Internal energy of the gas will decrease.
B. Internal energy of the gas will increase.
C. Internal energy of the gas will not change.
D. The gas will do positive work.
E. The gas will do negative work.

Choose the **correct** answer from the options given below: **[NCERT : P L-234, 235 | Jan. 30, 2023 (I)]**

(a) A and E only (b) B and D only
(c) C and E only (d) C and D only

56. Match **List I** with **List II** : **[NCERT : P L-234, 235 | Jan. 25, 2023 (II)]**

	List I		List II
A.	Isothermal Process	I.	Work done by the gas decreases internal energy
B.	Adiabatic Process	II.	No change in internal energy
C.	Isochoric Process	III.	The heat absorbed goes partly to increase internal energy and partly to do work
D.	Isobaric Process	IV.	No work is done on or by the gas

Choose the correct answer from the options given below:

(a) A-II, B-I, C-III, D-IV (b) A-II, B-I, C-IV, D-III
(c) A-I, B-II, C-IV, D-III (d) A-I, B-II, C-III, D-IV

57. In an Isothermal change, the change in pressure and volume of a gas can be represented for three different temperature; $T_3 > T_2 > T_1$ as : **[NCERT : P L-235 | Jan. 24, 2023 (II)]**

(a)
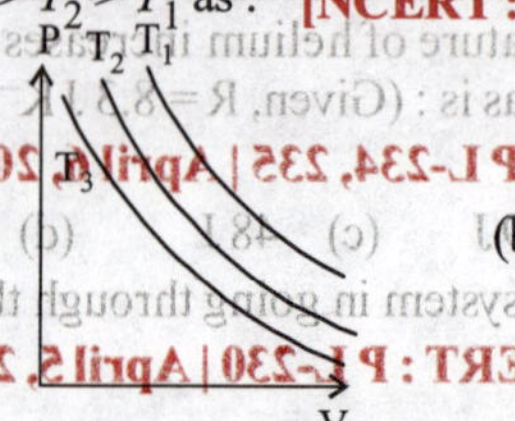

(b)
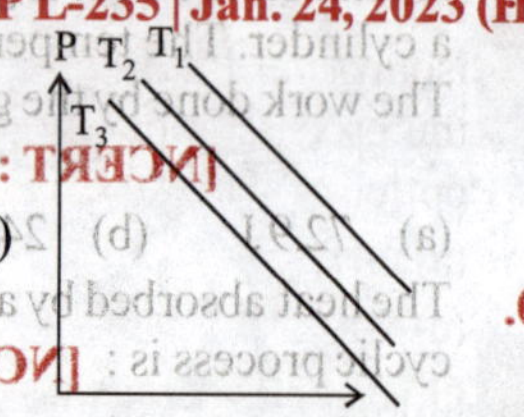

(c)
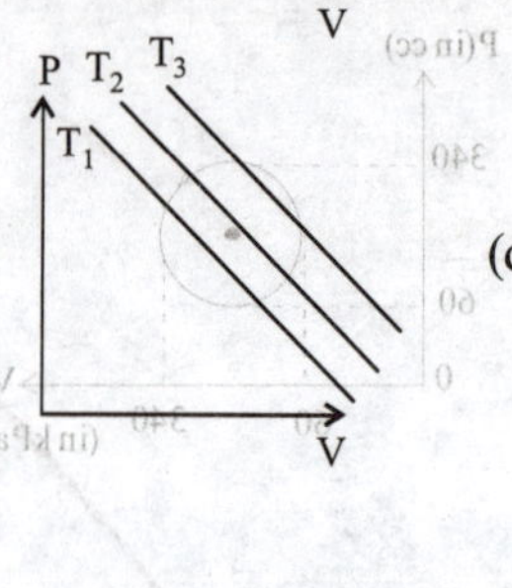

(d)
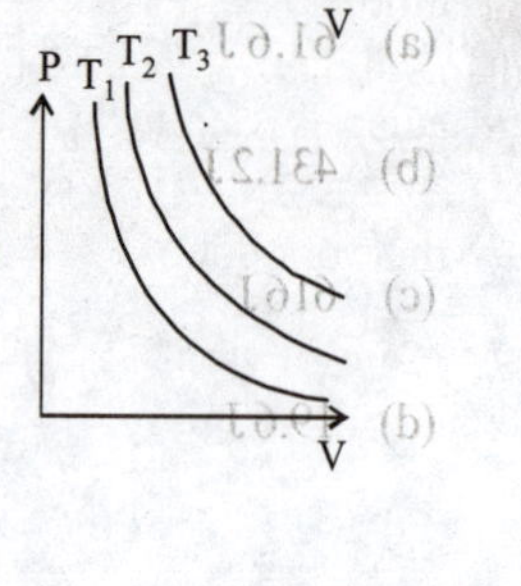

58. The pressure P_1 and density d_1 of diatomic gas $\left(\gamma = \frac{7}{5}\right)$ changes suddenly to $P_2(>P_1)$ and d_2 respectively during an adiabatic process. The temperature of the gas increases and becomes ______ times of its initial temperature.
$\left(\text{given } \frac{d_2}{d_1} = 32\right)$ **[NCERT : P L-235 | July 29, 2022 (I)]**

59. A thermodynamic system is taken from an original state D to an intermediate state E by the linear process shown in the figure. Its volume is then reduced to the original volume from E to F by an isobaric process. The total work done by the gas from D to E to F will be

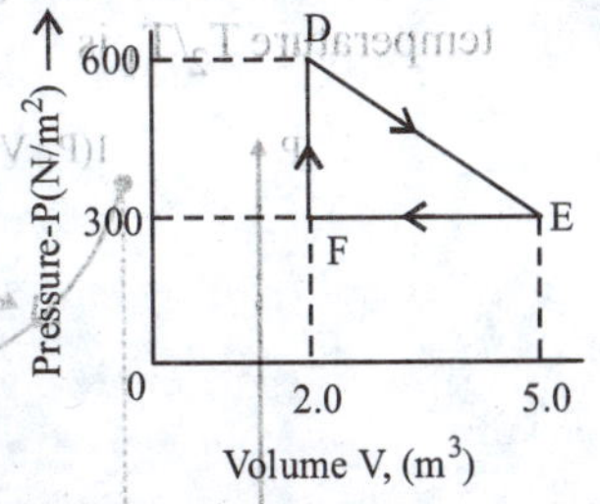

[NCERT : P L-235 | July 29, 2022 (II)]
(a) –450 J (b) 450 J (c) 900 J (d) 1350 J

60. 7 mol of a certain monoatomic ideal gas undegoes a temperature increase of 40 K at constant pressure. The increase in the internal energy of the gas in this process is : (Given R = 8.3 JK^{-1} mol^{-1})
[NCERT : P L-231 | July 26, 2022 (I)]
(a) 5810 J (b) 3486 J (c) 11620 J (d) 6972 J

61. A monoatomic gas at pressure P and volume V is suddenly compressed to one eights of its original volume. The final pressure at constant entropy will be :
[NCERT : P L-235 | July 26, 2022 (I)]
(a) P (b) 8 P (c) 32 P (d) 64 P

62. A certain amount of gas of volume V at 27°C temperature and pressure 2×10^7 Nm^{-2} expands isothermally until its volume gets doubled. Later it expands adiabatically until its volume gets redoubled. The final pressure of the gas will be (Use $\gamma = 1.5$) : **[NCERT : P L-235 | July 25, 2022 (I)]**
(a) 3.536×10^5 Pa (b) 3.536×10^6 Pa
(c) 1.25×10^6 Pa (d) 1.25×10^5 Pa

63. A sample of monoatomic gas is taken at inital pressure of 75 kPa. The volume of the gas is then compressed from 1200 cm^3 to 150 cm^3 adiabatically. In this process, the value of workdone on the gas will be:
[NCERT : P L-235 | June 30, 2022 (I)]
(a) 79 J (b) 405 J (c) 4050 J (d) 9590 J

64. Starting with the same initial conditions, an ideal gas expands from volume V_1 to V_2 in three different ways. The work done by the gas is W_1 if the process is purely isothermal W_2. If the process is purely adiabatic and W_3 if the process is purely isobaric. Then, choose the correct option : **[NCERT : P L-235 | June 29, 2022 (II)]**
(a) $W_1 < W_2 < W_3$ (b) $W_2 < W_3 < W_1$
(c) $W_3 < W_1 < W_2$ (d) $W_2 < W_1 < W_3$

65. Given below are two statement :
Statement-I : What μ amount of an ideal gas undergoes adiabatic change from state (P_1, V_1, T_1) to state (P_2, V_2, T_2) the work done is $W = \frac{\mu R(T_2 - T_1)}{1-\gamma}$, where $\gamma = \frac{C_p}{C_v}$ and R = universal gas constant.
Statement -II: In the above case when work is done on the gas the temperature of the gas would rise.
Choose the correct answer from the options given below:
[NCERT : P L-235 | June 28, 2022 (I)]
(a) Both statement-I and statement-II are true.
(b) Both statement-I and statement-II are false.
(c) Statement-I is true but statement-II is false.
(d) Statement-I is false but statement-II is true.

66. A diatomic gas ($\gamma = 1.4$) does 400 J of work when it is expanded isobarically. The heat given to the gas in the process is ____ J. **[NCERT : P L-235 | June 27, 2022 (II)]**

67. A monoatomic gas performs a work of $\frac{Q}{4}$ where Q is the heat supplied to it. The molar heat capacity of the gas will be ______ R during this transformation. Where R is the gas constant.
[NCERT : P L-231 | June 24, 2022 (II)]

68. A sample of gas with $\gamma = 1.5$ is taken through an adiabatic process in which the volume is compressed from 1200 cm^3 to 300 cm^3. If the initial pressure is 200 kPa. The absolute value of the workdone by the gas in the process = ______J.
[NCERT : P L-235 | NA, Aug. 31, 2021 (II)]

69. In the reported figure, heat energy absorbed by a system in going through a cyclic process is ______ πJ.

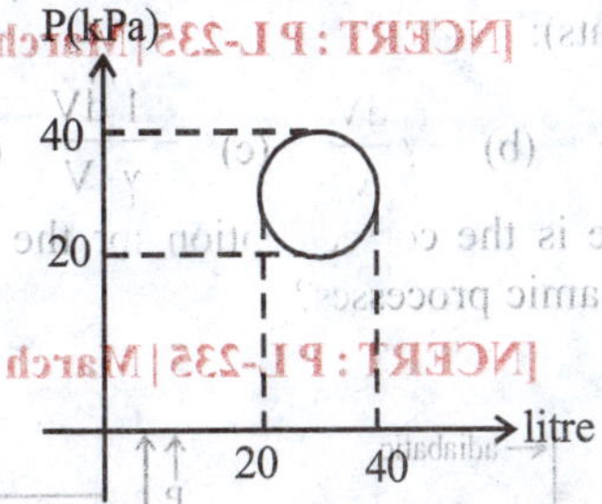

[NCERT : P L-235, 236 | NA, July 20, 2021 (I)]

70. The amount of heat needed to raise the temperature of 4 moles of a rigid diatomic gas from 0°C to 50°C when no work is done is ______. (R is the universal gas constant)
[NCERT : P L-235 | July 20, 2021 (I)]
(a) 250 R (b) 750 R (c) 175 R (d) 500 R

71. One mole of an ideal gas at 27°C is taken from A to B as shown in the given PV indicator diagram. The work done by the system will be ______ $\times 10^{-1}$ J.
[Given : R = 8.3 J / mole K, ln2 = 0.6931]
(Round off to the nearest integer)

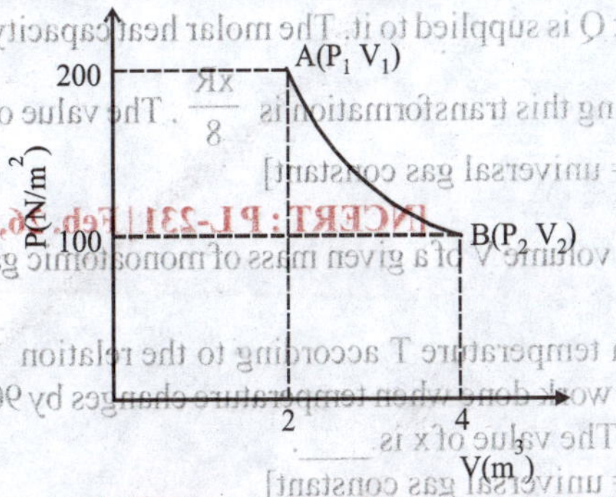

[NCERT : P L-234 | NA, July 20, 2021 (II)]

72. The P-V diagram of a diatomic ideal gas system going under cyclic process as shown in figure. The work done during an adiabatic process CD is (Use $\gamma = 1.4$):

[NCERT : P L-235 | March 18, 2021 (I)]

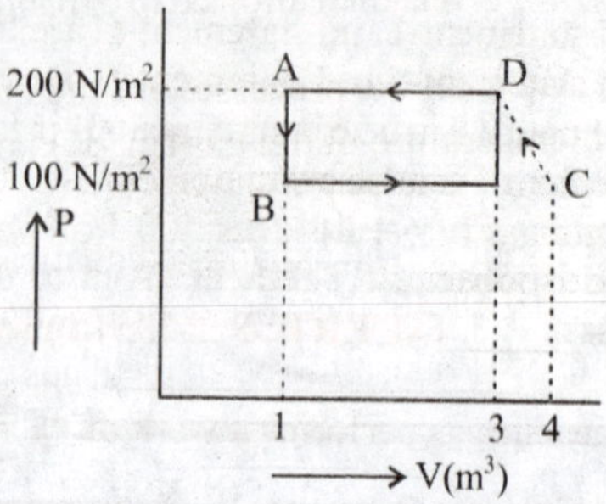

(a) -500 J (b) 200 J (c) -400 J (d) 400 J

73. An ideal gas in a cylinder is separated by a piston in such a way that the entropy of one part is S_1 and that of the other part is S_2. Given that $S_1 > S_2$. If the piston is removed then the total entropy of the system will be:

[March 18, 2021 (II)]

(a) $S_1 - S_2$ (b) $S_1 \times S_2$ (c) $\frac{S_1}{S_2}$ (d) $S_1 + S_2$

74. For an adiabatic expansion of an ideal gas, the fractional change in its pressure is equal to (where γ is the ratio of specific heats): **[NCERT : P L-235 | March 18, 2021 (II)]**

(a) $-\gamma\frac{V}{dV}$ (b) $-\gamma\frac{dV}{V}$ (c) $-\frac{1}{\gamma}\frac{dV}{V}$ (d) $\frac{dV}{V}$

75. Which one is the correct option for the two different thermodynamic processes?

[NCERT : P L-235 | March 17, 2021 (II)]

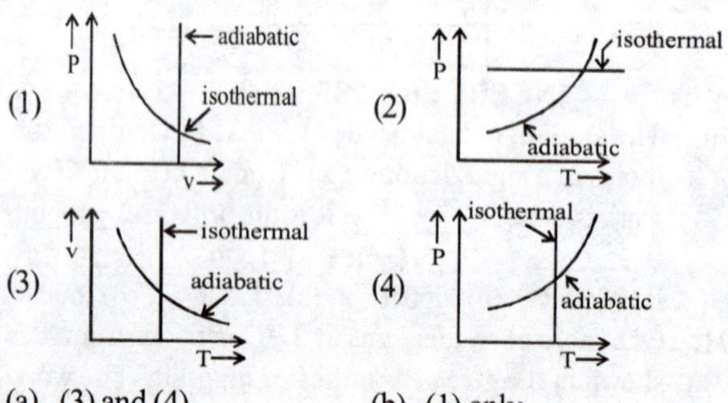

(a) (3) and (4) (b) (1) only

(c) (3) and (1) (d) (2) and (3)

76. 1 mole of rigid diatomic gas performs a work of $\frac{Q}{5}$ when heat Q is supplied to it. The molar heat capacity of the gas during this transformation is $\frac{xR}{8}$. The value of x is____.

[R = universal gas constant]

[NCERT : P L-231 | Feb. 26, 2021 (II)]

77. The volume V of a given mass of monoatomic gas changes with temperature T according to the relation $V = KT^{\frac{2}{3}}$. The work done when temperature changes by 90 K will be xR. The value of x is____.

[R = universal gas constant]

[NCERT : P L-235 | Feb. 26, 2021 (II)]

78. In a certain thermodynamical process, the pressure of a gas depends on its volume as kV^3. The work done when the temperature changes from 100°C to 300°C will be ______ nR, where n denotes number of moles of a gas.

[NCERT : P L-235 | Feb. 25, 2021 (I)]

79. Thermodynamic process is shown below on a P-V diagram for one mole of an ideal gas. If $V_2 = 2V_1$ then the ratio of temperature T_2/T_1 is

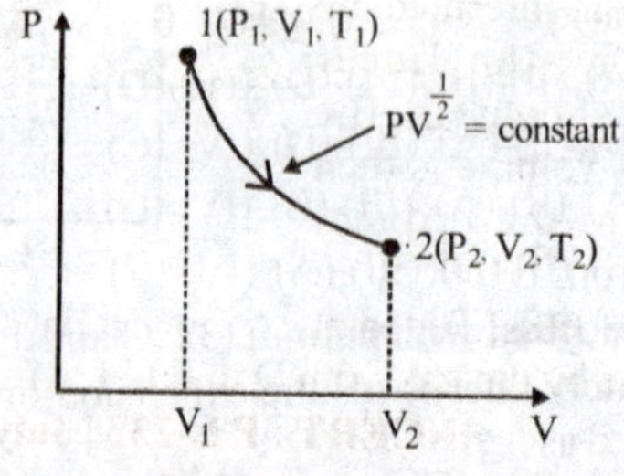

[NCERT : P L-235 | Feb. 25, 2021 (II)]

(a) $\sqrt{2}$ (b) $\frac{1}{2}$ (c) 2 (d) $\frac{1}{\sqrt{2}}$

80. *n* mole a perfect gas undergoes a cyclic process *ABCA* (see (figure) consisting of the following processes.

A → B : Isothermal expansion at temperature *T* so that the volume is doubled from V_1 to $V_2 = 2V_1$ and pressure changes from P_1 to P_2.

B → C : Isobaric compression at pressure P_2 to initial volume V_1.

C → A : Isochoric change leading to change of pressure from P_2 to P_1.

Total work done in the complete cycle *ABCA* is :

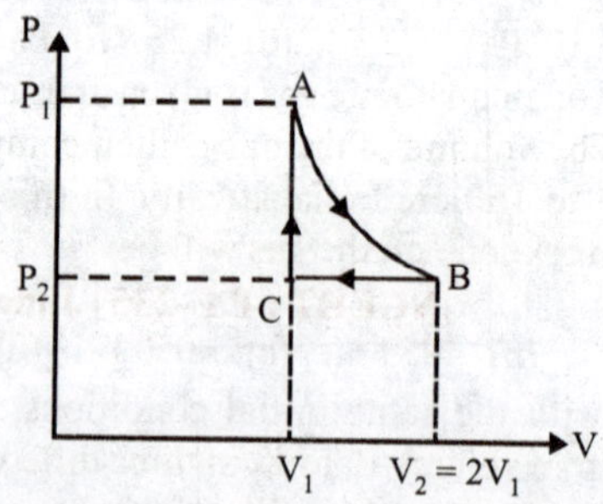

[NCERT : P L-235 | Feb. 24, 2021 (I)]

(a) 0 (b) $nRT\left(\ln 2 + \frac{1}{2}\right)$

(c) $nRT \ln 2$ (d) $nRT\left(\ln 2 - \frac{1}{2}\right)$

81. In an adiabatic process, the density of a diatomic gas becomes 32 times its initial value. The final pressure of the gas is found to be *n* times the initial pressure. The value of *n* is : **[NCERT : P L-235 | 5 Sep. 2020 (II)]**

(a) 32 (b) 326 (c) 128 (d) $\frac{1}{32}$

82. Match the thermodynamic processes taking place in a system with the correct conditions. In the table : ΔQ is the heat supplied, ΔW is the work done and ΔU is change in internal energy of the system.

[NCERT : P L-235 | 4 Sep. 2020 (II)]

Process	Condition
(I) Adiabatic	(A) $\Delta W = 0$
(II) Isothermal	(B) $\Delta Q = 0$
(III) Isochoric	(C) $\Delta U \neq 0, \Delta W \neq 0, \Delta Q \neq 0$
(IV) Isobaric	(D) $\Delta U = 0$

(a) (I)-(A), (II)-(B), (III)-(D), (IV)-(D)
(b) (I)-(B), (II)-(A), (III)-(D), (IV)-(C)
(c) (I)-(A), (II)-(A), (III)-(B), (IV)-(C)
(d) (I)-(B), (II)-(D), (III)-(A), (IV)-(C)

83. A balloon filled with helium (32°C and 1.7 atm.) bursts. Immediately afterwards the expansion of helium can be considered as : **[NCERT : P L-235 | 3 Sep. 2020 (I)]**
(a) irreversible isothermal
(b) irreversible adiabatic
(c) reversible adiabatic
(d) reversible isothermal

84. An engine takes in 5 mole of air at 20°C and 1 atm, and compresses it adiabaticaly to $1/10^{th}$ of the original volume. Assuming air to be a diatomic ideal gas made up of rigid molecules, the change in its internal energy during this process comes out to be X kJ. The value of X to the nearest integer is ______.

[NCERT : P L-235 | NA 2 Sep. 2020 (I)]

85. Which of the following is an equivalent cyclic process corresponding to the thermodynamic cyclic given in the figure? where, $1 \rightarrow 2$ is adiabatic.
(Graphs are schematic and are not to scale)
[NCERT : P L-235 | 9 Jan. 2020 (I), 9 Jan. 2020 (II) (S)]

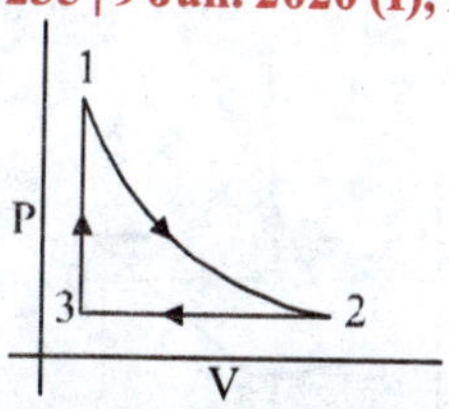

(a)
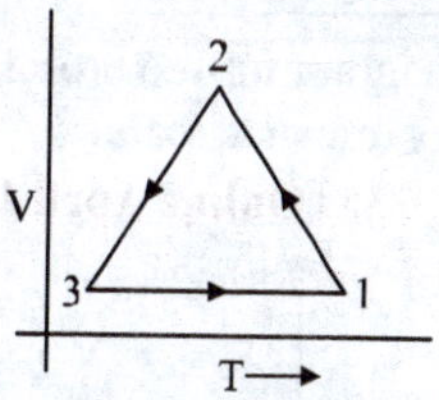

(b)
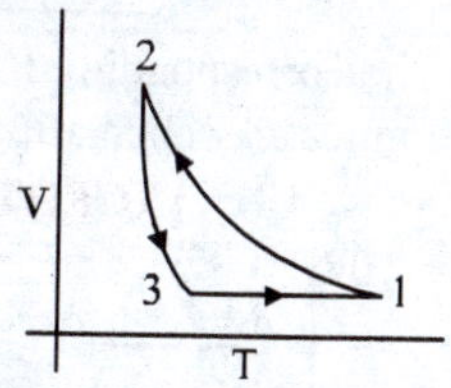

(c)
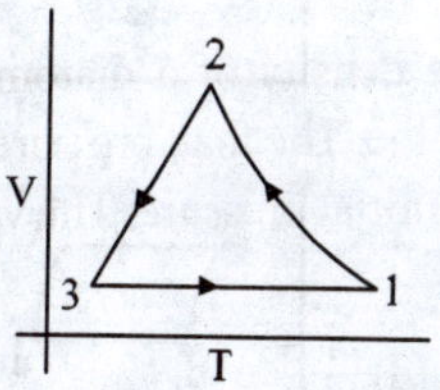

(d)
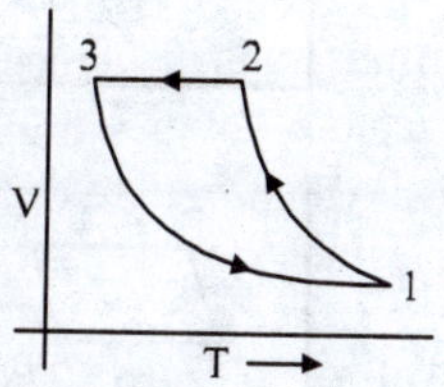

86. Starting at temperature 300 K, one mole of an ideal diatomic gas ($\gamma = 1.4$) is first compressed adiabatically from volume V_1 to $V_2 = \frac{V_1}{16}$. It is then allowed to expand isobarically to volume $2V_2$. If all the processes are the quasi-static then the final temperature of the gas (in °K) is (to the nearest integer) ______. **[NCERT : P L-235 | 9 Jan. 2020 II]**

87. A litre of dry air at STP expands adiabatically to a volume of 3 litres. If $\gamma = 1.40$, the work done by air is: ($3^{1.4} = 4.6555$) [Take air to be an ideal gas]

[NCERT : P L-235 | 7 Jan. 2020 I]

(a) 60.7 J
(b) 90.5 J
(c) 100.8 J
(d) 48 J

88. A sample of an ideal gas is taken through the cyclic process abca as shown in the figure. The change in the internal energy of the gas along the path ca is – 180 J, The gas absorbs 250 J of heat along the path ab and 60 J along the path bc. The work down by the gas along the path abc is: **[NCERT : P L-235 | 12 Apr. 2019 I]**

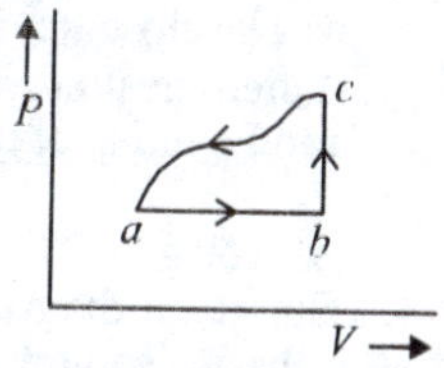

(a) 120 J
(b) 130 J
(c) 100 J
(d) 140 J

89. A cylinder with fixed capacity of 67.2 l contains helium gas at STP. The amount of heat needed to raise the temperature of the gas by 20°C is : [Given that R = 8.31 J $mol^{-1} K^{-1}$] **[NCERT : P L-235 | 10 Apr. 2019 I]**
(a) 350 J
(b) 374 J
(c) 748 J
(d) 700 J

90. n moles of an ideal gas with constant volume heat capacity C_V undergo an isobaric expansion by certain volume. The ratio of the work done in the process, to the heat supplied is: **[NCERT : P L-231 | 10 Apr. 2019 I]**

(a) $\frac{nR}{C_V + nR}$
(b) $\frac{nR}{C_V - nR}$
(c) $\frac{4nR}{C_V - nR}$
(d) $\frac{4nR}{C_V + nR}$

91. One mole of an ideal gas passes through a process where pressure and volume obey the relation $P = P_0\left[1 - \frac{1}{2}\left(\frac{V_0}{V}\right)^2\right]$. Here P_0 and V_0 are constants. Calculate the change in the temperature of the gas if its volume changes from V_0 to $2V_0$.

[NCERT : P L-235 | 10 Apr. 2019 II]

(a) $\frac{1}{2}\frac{P_oV_o}{R}$
(b) $\frac{5}{4}\frac{P_oV_o}{R}$
(c) $\frac{3}{4}\frac{P_oV_o}{R}$
(d) $\frac{1}{4}\frac{P_oV_o}{R}$

92. Following figure shows two processes A and B for a gas. If ΔQ_A and ΔQ_B are the amount of heat absorbed by the system in two cases, and ΔU_A and ΔU_B are changes in internal energies, respectively, then:

[NCERT : P L-230, 235 | 9 April 2019 I]

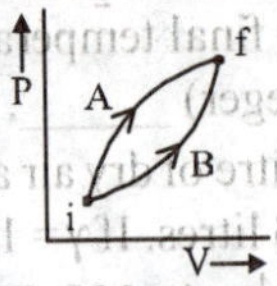

(a) $\Delta Q_A < \Delta Q_B, \Delta U_A < \Delta U_B$

(b) $\Delta Q_A > \Delta Q_B, \Delta U_A > \Delta U_B$

(c) $\Delta Q_A > \Delta Q_B, \Delta U_A = \Delta U_B$

(d) $\Delta Q_A = \Delta Q_B; \Delta U_A = \Delta U_B$

93. A thermally insulated vessel contains 150 g of water at 0°C. Then the air from the vessel is pumped out adiabatically. A fraction of water turns into ice and the rest evaporates at 0°C itself. The mass of evaporated water will be close to:

(Latent heat of vaporization of water = 2.10×10^6 J kg^{-1} and Latent heat of Fusion of water = 3.36×10^5 J kg^{-1})

[NCERT : P L-230 | 8 April 2019 I]

(a) 150 g (b) 20 g (c) 130 g (d) 35 g

94. The given diagram shows four processes *i.e.*, isochoric, isobaric, isothermal and adiabatic. The correct assignment of the processes, in the same order is given by :

[NCERT : P L-235 | 8 Apr. 2019 II]

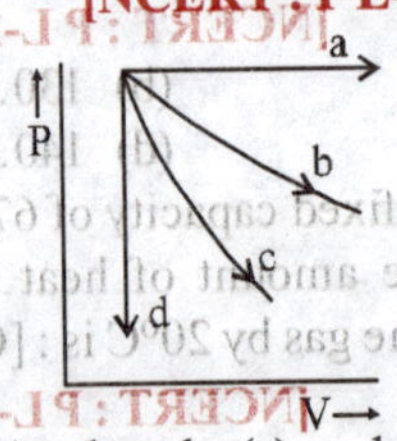

(a) a d b c (b) d a c b (c) a d c b (d) d a b c

95. For the given cyclic process CAB as shown for gas, the work done is: **[NCERT : P L-235, 236 | 12 Jan. 2019 I]**

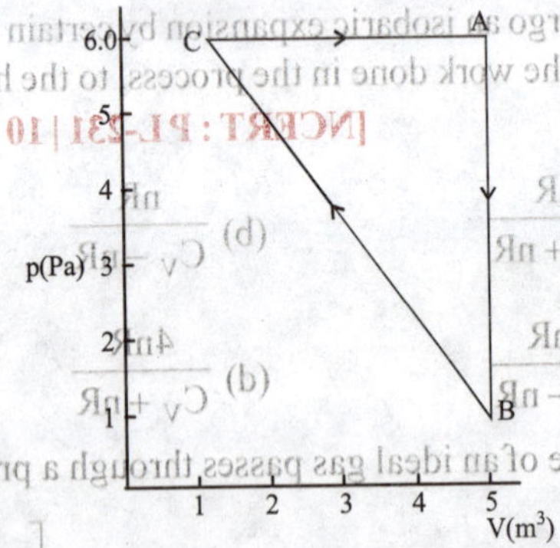

(a) 30 J (b) 10 J (c) 1 J (d) 5 J

96. A rigid diatomic ideal gas undergoes an adiabatic process at room temperature. The relation between temperature and volume for this process is TV^x = constant, then x is:

[NCERT : P L-235 | 11 Jan. 2019 I]

(a) $\frac{3}{5}$ (b) $\frac{2}{5}$ (c) $\frac{2}{3}$ (d) $\frac{5}{3}$

97. Half mole of an ideal monoatomic gas is heated at constant pressure of 1 atm from 20°C to 90°C. Work done by gas is close to: (Gas constant R = 8.31 J/mol-K)

[NCERT : P L-235 | 10 Jan. 2019 II]

(a) 581 J (b) 291 J (c) 146 J (d) 73 J

98. One mole of an ideal monoatomic gas is taken along the path ABCA as shown in the PV diagram. The maximum temperature attained by the gas along the path BC is given by **[NCERT : P L-235 | Online April 16, 2018]**

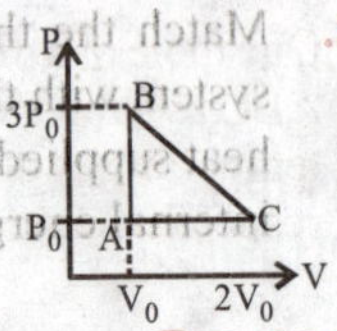

(a) $\frac{25}{8}\frac{P_0V_0}{R}$ (b) $\frac{25}{4}\frac{P_0V_0}{R}$

(c) $\frac{25}{16}\frac{P_0V_0}{R}$ (d) $\frac{5}{8}\frac{P_0V_0}{R}$

99. One mole of an ideal monoatomic gas is compressed isothermally in a rigid vessel to double its pressure at room temperature, 27°C. The work done on the gas will be:

[NCERT : P L-234 | Online April 15, 2018]

(a) 300R ln 6 (b) 300R

(c) 300R ln 7 (d) 300R ln 2

100. The ratio of work done by an ideal monoatomic gas to the heat supplied to it in an isobaric process is :

[NCERT : P L-235 | Online April 9, 2016]

(a) $\frac{2}{5}$ (b) $\frac{3}{2}$ (c) $\frac{3}{5}$ (d) $\frac{2}{3}$

101. Consider a spherical shell of radius R at temperature T. The black body radiation inside it can be considered as an ideal gas of photons with internal energy per unit volume $u = \frac{U}{V} \propto T^4$ and pressure $p = \frac{1}{3}\left(\frac{U}{V}\right)$.

If the shell now undergoes an adiabatic expansion the relation between T and R is : **[NCERT : P L-235 | 2015]**

(a) $T \propto \frac{1}{R}$ (b) $T \propto \frac{1}{R^3}$

(c) $T \propto e^{-R}$ (d) $T \propto e^{-3R}$

102. An ideal gas goes through a reversible cycle a→b→c→d has the V - T diagram shown below. Process d→a and b→c are adiabatic.

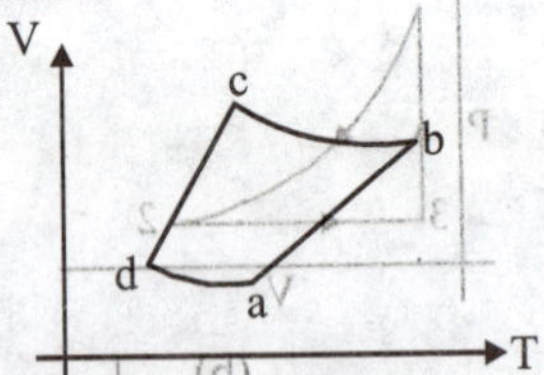

The corresponding P - V diagram for the process is (all figures are schematic and not drawn to scale) :

[NCERT : P L-235 | Online April 10, 2015]

(a)

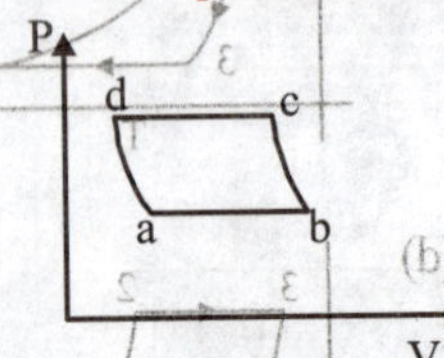

(b)

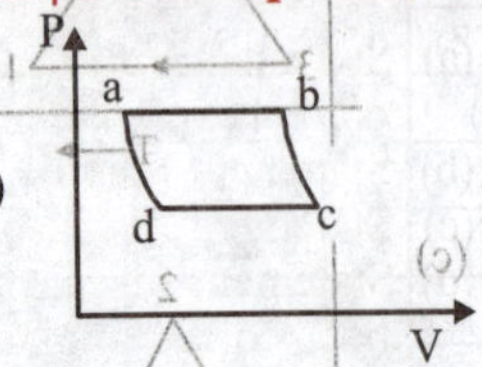

(c)

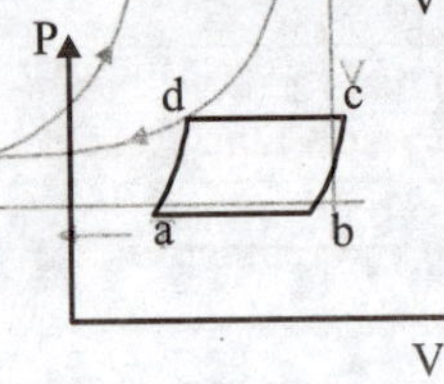

(d)

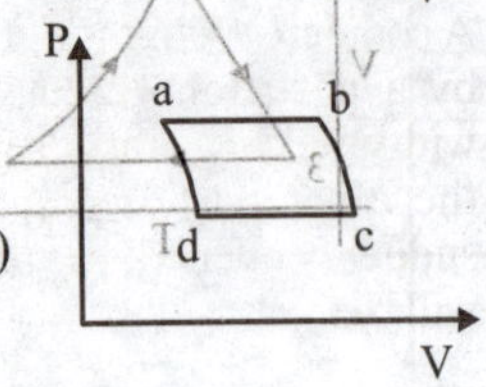

103. One mole of a diatomic ideal gas undergoes a cyclic process ABC as shown in figure. The process BC is adiabatic. The temperatures at A, B and C are 400 K, 800 K and 600 K respectively. Choose the correct statement: **[NCERT : P L-235, 236 | 2014]**

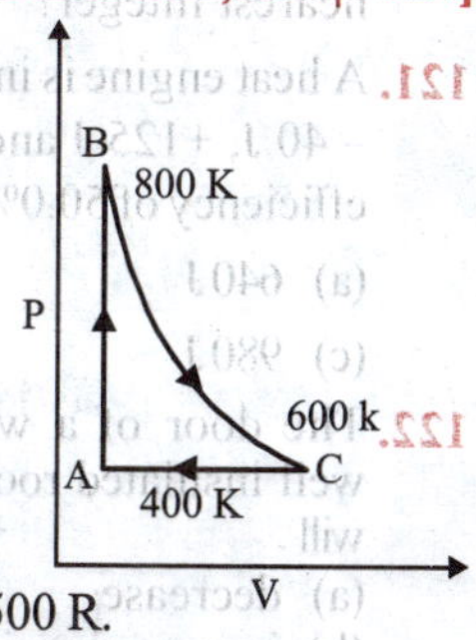

(a) The change in internal energy in whole cyclic process is 250 R.
(b) The change in internal energy in the process CA is 700 R.
(c) The change in internal energy in the process AB is –350 R.
(d) The change in internal energy in the process BC is – 500 R.

104. During an adiabatic compression, 830 J of work is done on 2 moles of a diatomic ideal gas to reduce its volume by 50%. The change in its temperature is nearly:
($R = 8.3\ JK^{-1}\ mol^{-1}$)
[NCERT : P L-235 | Online April 11, 2014]

(a) 40 K (b) 33 K (c) 20 K (d) 14 K

105. An ideal gas at atmospheric pressure is adiabatically compressed so that its density becomes 32 times of its initial value. If the final pressure of gas is 128 atmospheres, the value of 'γ' of the gas is :
[NCERT : P L-235 | Online April 22, 2013]

(a) 1.5 (b) 1.4 (c) 1.3 (d) 1.6

106. The pressure of an ideal gas varies with volume as $P = \alpha V$, where α is a constant. One mole of the gas is allowed to undergo expansion such that its volume becomes '*m*' times its initial volume. The work done by the gas in the process is **[NCERT : P L-235 | Online May 19, 2012]**

(a) $\frac{\alpha V}{2}\left(m^2-1\right)$ (b) $\frac{\alpha^2 V^2}{2}\left(m^2-1\right)$

(c) $\frac{\alpha}{2}\left(m^2-1\right)$ (d) $\frac{\alpha V^2}{2}\left(m^2-1\right)$

107. This question has Statement 1 and Statement 2. Of the four choices given after the Statements, choose the one that best describes the two Statements.

Statement 1: In an adiabatic process, change in internal energy of a gas is equal to work done on/by the gas in the process.

Statement 2: The temperature of a gas remains constant in an adiabatic process.

[NCERT : P L-235 | Online May 7, 2012]

(a) Statement 1 is true, Statement 2 is true, Statement 2 is a correct explanation of Statement 1.
(b) Statement 1 is true, Statement 2 is false.
(c) Statement 1 is false, Statement 2 is true.
(d) Statement 1 is false, Statement 2 is true, Statement 2 is not a correct explanation of Statement 1.

108. A container with insulating walls is divided into equal parts by a partition fitted with a valve. One part is filled with an ideal gas at a pressure *P* and temperature *T*, whereas the other part is completly evacuated. If the valve is suddenly opened, the pressure and temperature of the gas will be : **[NCERT : P L-234 | 2011 RS]**

(a) $\frac{P}{2}, \frac{T}{2}$ (b) P, T (c) $P, \frac{T}{2}$ (d) $\frac{P}{2}, T$

Directions for questions 109 to 113: *Questions are based on the following paragraph.*

Two moles of helium gas are taken over the cycle *ABCDA*, as shown in the *P-T* diagram. **[2009]**

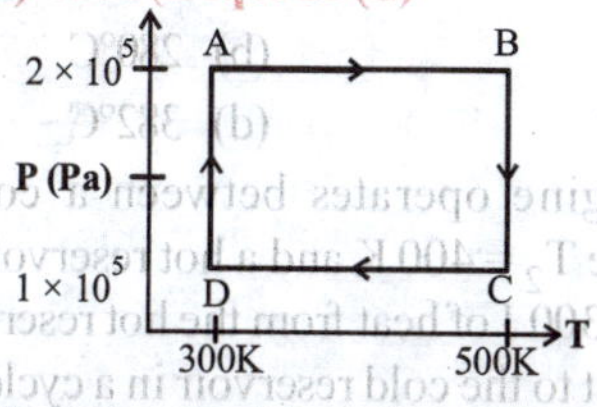

109. Assuming the gas to be ideal the work done on the gas in taking it from *A* to *B* is **[NCERT : P L-235]**

(a) 300 R (b) 400 R
(c) 500 R (d) 200 R

110. The work done on the gas in taking it from *D* to *A* is **[NCERT : P L-235]**

(a) + 414 R (b) – 690 R
(c) + 690 R (d) – 414 R

111. The net work done on the gas in the cycle *ABCDA* is **[NCERT : P L-235]**

(a) 279 R (b) 1076 R
(c) 1904 R (d) zero

112. The work of 146 kJ is performed in order to compress one kilo mole of gas adiabatically and in this process the temperature of the gas increases by 7°C. The gas is
($R = 8.3\ J\ mol^{-1}\ K^{-1}$) **[NCERT : P L-235 | 2006]**

(a) diatomic
(b) triatomic
(c) a mixture of monoatomic and diatomic
(d) monoatomic

113. Which of the following parameters does not characterize the thermodynamic state of matter? **[NCERT : P L-229, 230 | 2003]**

(a) Temperature (b) Pressure
(c) Work (d) Volume

Topic 3 - Second Law of Thermodynamics

114. An engine operating between the boiling and freezing points of water will have **[April 12, 2023 (I)]**

1. efficiency more than 27%
2. efficiency less than the efficiency a Carnot engine operating between the same two temperatures.
3. efficiency equal to 27%
4. efficiency less than 27%

(a) 2, 3 and 4 only
(b) 2 and 3 only
(c) 2 and 4 only
(d) 1 and 2 only

115. 300 cal. of heat is given to a heat engine and it rejects 225 cal. of heat. If source temperature is 227°C, then the temperature of sink will be ___°C. **[June 29, 2022 (I)]**

116. A reversible engine has an efficiency of $\frac{1}{4}$. If the temperature of the sink is reduced by 58°C, its efficiency becomes double. Calculate the temperature of the sink :
[Aug. 31, 2021 (I); July 25, 2021 (II) (S); Feb 25, 2021 (II) (S) 12 April, 2019 (II) (S), 2011 (S)]

(a) 174°C (b) 280°C
(c) 180.4°C (d) 382°C

117. A heat engine operates between a cold reservoir at temperature $T_2 = 400$ K and a hot reservoir at temperature T_1. It takes 300 J of heat from the hot reservoir and delivers 240 J of heat to the cold reservoir in a cycle. The minimum temperature of the hot reservoir has to be ______ K.
[NA, Aug. 27, 2021 (II)]

118. A refrigerator consumes an average 35 W power to operate between temperature –10°C to 25°C. If there is no loss of energy then how much average heat per second does it transfer ? **[Aug. 26, 2021 (II)]**

(a) 263 J/s (b) 298 J/s
(c) 350 J/s (d) 35 J/s

119. An engine operates by taking a monatomic ideal gas through the cycle shown in the figure. The percentage efficiency of the engine is close to ______ **[2012 (S), Online April 8, 2017 (S), NA 6 Sep. 2020 (II), 2013 (S)]**

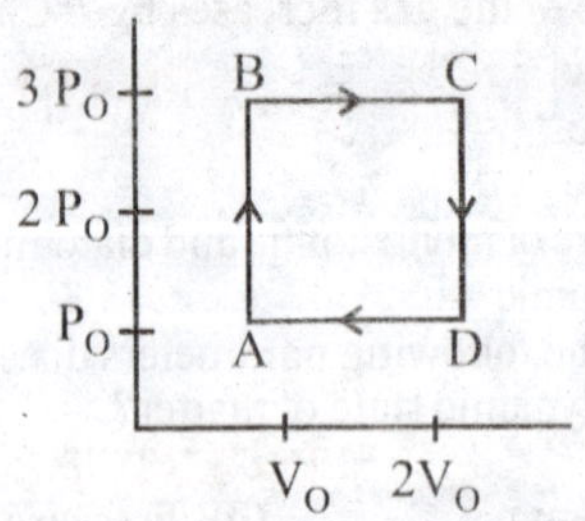

120. If minimum possible work is done by a refrigerator in converting 100 grams of water at 0°C to ice, how much heat (in calories) is released to the surroundings at temperature 27°C (Latent heat of ice = 80 Cal/gram) to the nearest integer? **[NA 3 Sep. 2020 (II)]**

121. A heat engine is involved with exchange of heat of 1915 J, – 40 J, +125 J and – Q J, during one cycle achieving an efficiency of 50.0%. The value of Q is : **[2 Sep. 2020 (II)]**

(a) 640 J (b) 40 J
(c) 980 J (d) 400 J

122. The door of a working refrigerator is left open in a well insulated room. The temperature of air in the room will **[Online May 26, 2012]**

(a) decrease
(b) increase in winters and decrease in summers
(c) remain the same
(d) increase

123. The temperature-entropy diagram of a reversible engine cycle is given in the figure. Its efficiency is **[2005]**

(a) $\frac{1}{4}$ (b) $\frac{1}{2}$
(c) $\frac{2}{3}$ (d) $\frac{1}{3}$

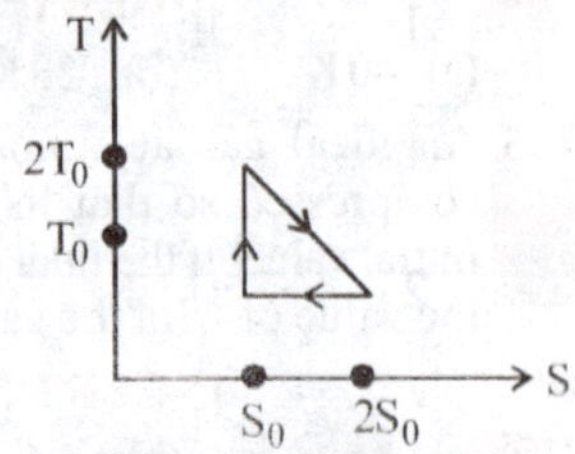

124. Which of the following statements is **correct** for any thermodynamic system ? **[2004]**

(a) The change in entropy can never be zero
(b) Internal energy and entropy are state functions
(c) The internal energy changes in all processes
(d) The work done in an adiabatic process is always zero.

ANSWER KEYS

1	(d)	14	(c)	27	(b)	40	(a)	53	(c)	66	(1400)	79	(a)	92	(c)	105	(b)	118	(a)
2	(c)	15	(c)	28	(15)	41	(c)	54	(a)	67	(2)	80	(d)	93	(b)	106	(d)	119	(19)
3	(a)	16	(314)	29	(d)	42	(d)	55	(d)	68	(480)	81	(c)	94	(d)	107	(b)	120	(8791)
4	(d)	17	(c)	30	(d)	43	(a)	56	(b)	69	(100)	82	(d)	95	(b)	108	(d)	121	(c)
5	(d)	18	(a)	31	(d)	44	(a)	57	(d)	70	(d)	83	(b)	96	(b)	109	(b)	122	(d)
6	(d)	19	(d)	32	(273)	45	(a)	58	(4)	71	(17258)	84	(46)	97	(b)	110	(d)	123	(d)
7	(c)	20	(c)	33	(c)	46	(d)	59	(b)	72	(a)	85	(c)	98	(a)	111	(a)	124	(b)
8	(b)	21	(c)	34	(a)	47	(c)	60	(b)	73	(d)	86	(1818)	99	(d)	112	(a)		
9	(c)	22	(d)	35	(b)	48	(b)	61	(c)	74	(b)	87	(b)	100	(a)	113	(c)		
10	(a)	23	(b)	36	(c)	49	(b)	62	(b)	75	(a)	88	(b)	101	(a)	114	(c)		
11	(d)	24	(d)	37	(a)	50	(b)	63	(b)	76	(25.00)	89	(c)	102	(b)	115	(102)		
12	(a)	25	(c)	38	(d)	51	(b)	64	(d)	77	(60.00)	90	(a)	103	(d)	116	(a)		
13	(b)	26	(b)	39	(d)	52	(b)	65	(a)	78	(50)	91	(b)	104	(c)	117	(500)		

12 Kinetic Theory

TREND ANALYSIS JEE MAIN (2025-2021)

	2025				2024				2023				2022				2021			
	Session II		Session I		Session II		Session I		Session II		Session I		Session II		Session I		Session II		Session I	
	MCQs	NVQs	MCQs	NVQs	MCQs	NVQs	MCQs	NVQs	MCQs	NVQs	MCQs	NVQs	MCQs	NVQs	MCQs	NVQs	MCQs	NVQs	MCQs	NVQs
Topic 1	1	–	1	1	1	–	2	–	–	–	2	–	–	–	4	–	4	–	–	–
Topic 2	1	–	2	–	1	–	4	–	7	–	5	–	6	–	1	–	3	1	1	2
Topic 3	2	2	2	–	7	–	5	–	4	–	4	–	2	2	3	–	2	–	4	1
Topic 4																				
Total	4	2	5	1	9	–	11	–	11	–	11	–	8	2	8	–	9	1	5	3

How to Code Qns for Better & Faster Revision

Type 1: Able to solve (No code Req)

Type 2: Solved with struggle or by seeing solution **(Use code NR means Needs Revision or orange stickers)**

Type 3: Tricky/ Could not solve on my own. Need help from Teacher/ Friend. **(Use code TQ means Tricky Question or red stickers)**

You can either hand write the code in boxes or else paste coloured stickers given at end of book.

Topic 1 - Kinetic Theory of an Ideal Gas and Gas Laws

1. There are two vessels filled with an ideal gas where volume of one is double the volume of other. The large vessel contains the gas at 8 kPa at 1000 K while the smaller vessel contains the gas at 7 kPa at 500 K. If the vessels are connected to each other by a thin tube allowing the gas to flow and the temperature of both vessels is maintained at 600 K, at steady state the pressure in the vessels will be (in kPa). **[NCERT : P L-246 | April 4, 2025 (II)]**
   (a) 4.4 (b) 6
   (c) 24 (d) 18

2. Pressure of an ideal gas, contained in a closed vessel, is increased by 0.4% when heated by 1°C. Its initial temperature must be :
   [NCERT : P L-246 | April 3, 2025 (II)]
   (a) 25° C (b) 2500 K
   (c) 250 K (d) 250° C

3. A container of fixed volume contains a gas at 27ºC. To double the pressure of the gas, the temperature of gas should be raised to ______ ºC.
   [NCERT : P L-246 | Jan. 29, 2025 (I)]

4. P-T diagram of an ideal gas having three different densities ρ_1, ρ_2, ρ_3 (in three different cases) is shown in the figure. Which of the following is correct :
   [NCERT : P L-246 | April 4, 2024 (I)]

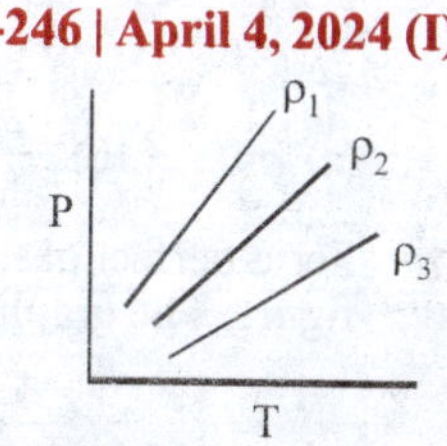

   (a) $\rho_2 < \rho_3$
   (b) $\rho_1 > \rho_2$
   (c) $\rho_1 < \rho_2$
   (d) $\rho_1 = \rho_2 = \rho_3$

5. Two vessels A and B are of the same size and are at same temperature. A contains 1 g of hydrogen and B contains 1 g of oxygen. P_A and P_B are the pressures of the gases in A and B respectively, then $\frac{P_A}{P_B}$ is:
   [NCERT : P L-246 | Jan. 29, 2024 (I)]
   (a) 8 (b) 16 (c) 32 (d) 4

6. The temperature of a gas having 2.0×10^{25} molecules per cubic meter at 1.38 atm. **[NCERT : P L-246 | Jan. 29, 2024 (II)]**
(Given, $k = 1.38 \times 10^{-23}\ JK^{-1}$) as:
(a) 300 K (b) 500 K (c) 100 K (d) 200 K

7. For three low density gases A, B, C pressure versus temperature graphs are plotted while keeping them at constant volume, as shown in the figure. **[NCERT : P L-246 | Feb. 1, 2023 (II)]**

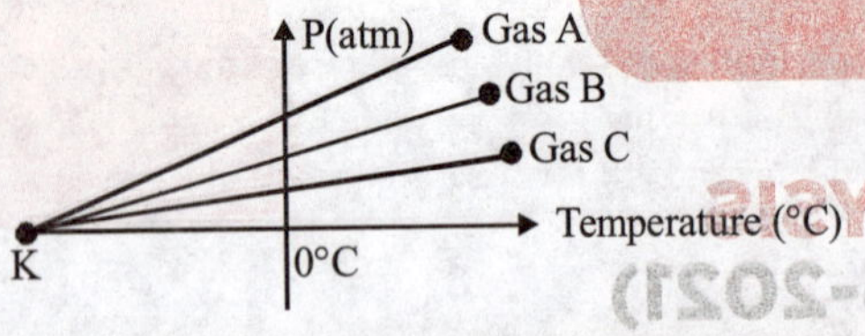

The temperature corresponding to the point 'K' is:
(a) -273° C (b) -100° C
(c) -40° C (d) -373° C

8. A bicycle tyre is filled with air having pressure of 270 kPa at 27°C. The approximate pressure of the air in the tyre when the temperature increases to 36°C is **[NCERT : P L-246 | Jan. 29, 2023 (I)]**
(a) 270 kPa (b) 262 KPa
(c) 278 kPa (d) 360 kPa

9. The pressure of the gas in a constant volume gas thermometer is 100 cm of mercury when placed in melting ice at 1 atm. When the bulb is placed in a liquid, the pressure becomes 180 cm of mercury. Temperature of the liquid is: (Given 0°C = 273 K) **[NCERT : P L-246 | June 30, 2022 (I)]**
(a) 300 K (b) 400 K (c) 600 K (d) 491 K

10. A vessel contains 16 g of hydrogen and 128 g of oxygen at standard temperature and pressure. The volume of the vessel in cm^3 is : **[NCERT : P L-246 | June 29, 2022 (II)]**
(a) 72×10^5 (b) 32×10^5
(c) 27×10^4 (d) 54×10^4

11. A mixture of hydrogen and oxygen has volume 2000 cm^3, temperature 300 K, pressure 100 kPa and mass 0.76 g. The ratio of number of moles of hydrogen to number of moles of oxygen in the mixture will be : **[NCERT : P L-246 | June 27, 2022 (I)]**
(a) $\frac{1}{3}$ (b) $\frac{2}{1}$ (c) $\frac{1}{16}$ (d) $\frac{16}{1}$

12. For a perfect gas, two pressures P_1 and P_2 are shown in figure. The graph shows: **[NCERT : P L-246 | June 27, 2022 (II)]**

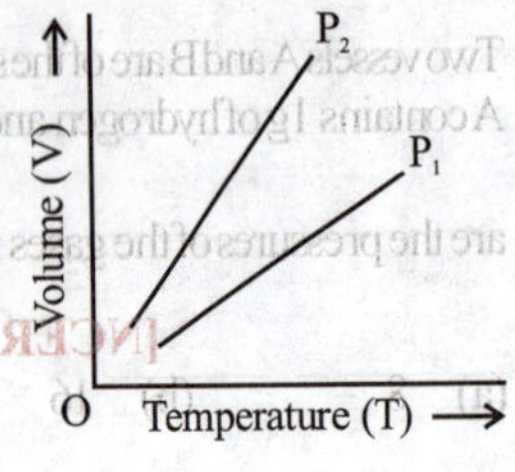

(a) $P_1 > P_2$
(b) $P_1 < P_2$
(c) $P_1 = P_2$
(d) Insufficient data to draw any conclusion

13. According to kinetic theory of gases,
A. The motion of the gas molecules freezes at 0°C
B. The mean free path of gas molecules decreases if the density of molecules is increased.
C. The mean free path of gas molecules increases if temperature is increased keeping pressure constant.
D. Average kinetic energy per molecule per degree of freedom is $\frac{3}{2}k_BT$ (for monoatomic gases)

Choose the most appropriate answer from the options given below: **[NCERT : P L-248 | June 27, 2022 (II)]**
(a) A and C only (b) B and C only
(c) A and B only (d) C and D only

14. For an ideal gas the instantaneous change in pressure 'p' with volume 'v' is given by the equation $\frac{dp}{dv} = -ap$. If $p = p_0$ at $v = 0$ is the given boundary condition, then the maximum temperature one mole of gas can attain is : (Here R is the gas constant) **[NCERT : P L-246 | Aug. 31, 2021 (I)]**
(a) $\frac{p_0}{a\,e\,R}$ (b) $\frac{ap_0}{e\,R}$
(c) infinity (d) 0°C

15. A mixture of hydrogen and oxygen has volume 500 cm^3, temperature 300 K, pressure 400 kPa and mass 0.76 g. The ratio of masses of oxygen to hydrogen will be :- **[NCERT : P L-246 | Aug. 31, 2021 (II)]**
(a) 3 : 8 (b) 3 : 16 (c) 16 : 3 (d) 8 : 3

16. A cylindrical container of volume $4.0 \times 10^{-3}\ m^3$ contains one mole of hydrogen and two moles of carbon dioxide. Assume the temperature of the mixture is 400 K. The pressure of the mixture of gases is : **[NCERT : P L-246 | Aug. 26, 2021 (II)]**
[Take gas constant as 8.3 J mol^{-1} K^{-1}]
(a) 249×10^1 Pa (b) 24.9×10^3 Pa
(c) 24.9×10^5 Pa (d) 24.9 Pa

17. Which of the following graphs represent the behavior of an ideal gas ? Symbols have their usual meaning. **[NCERT : P L-246, 247 | July 20, 2021 (II)]**

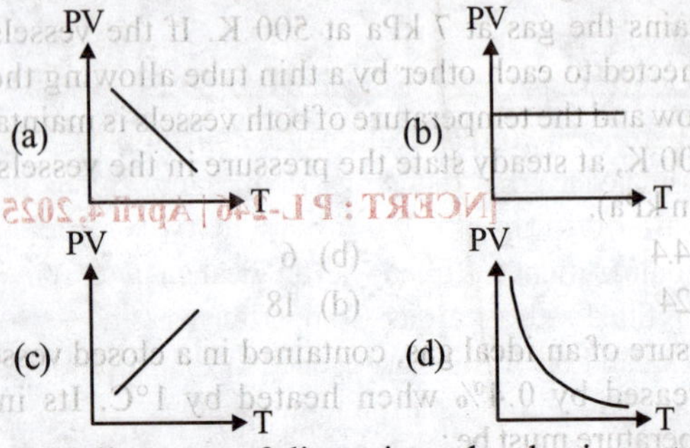

18. Initially a gas of diatomic molecules is contained in a cylinder of volume V_1 at a pressure P_1 and temperature 250 K. Assuming that 25% of the molecules get dissociated causing a change in number of moles. The pressure of the

resulting gas at temperature 2000 K, when contained in a volume $2V_1$ is given by P_2. The ratio P_2/P_1 is ______.
[NCERT : P L-246 | NA Sep. 06, 2020 (I)]

19. The change in the magnitude of the volume of an ideal gas when a small additional pressure ΔP is applied at a constant temperature, is the same as the change when the temperature is reduced by a small quantity ΔT at constant pressure. The initial temperature and pressure of the gas were 300 K and 2 atm. respectively. If $|\Delta T| = C|\Delta P|$, then value of C in (K/atm.) is ______.
[NCERT : P L-246 | NA Sep. 04, 2020 (II)]

20. The number density of molecules of a gas depends on their distance r from the origin as, $n(r) = n_0 e^{-\alpha r^4}$. Then the total number of molecules is proportional to :
[NCERT : P L-246 | 12 April 2019 II]
(a) $n_0\alpha^{-3/4}$ (b) $\sqrt{n_0}\,\alpha^{1/2}$ (c) $n_0\alpha^{1/4}$ (d) $n_0\alpha^{-3}$

21. A vertical closed cylinder is separated into two parts by a frictionless piston of mass m and of negligible thickness. The piston is free to move along the length of the cylinder. The length of the cylinder above the piston is l_1, and that below the piston is l_2, such that $l_1 > l_2$. Each part of the cylinder contains n moles of an ideal gas at equal temperature T. If the piston is stationary, its mass, m, will be given by: (R is universal gas constant and g is the acceleration due to gravity) **[NCERT : P L-246 | 12 Jan. 2019 II]**

(a) $\frac{RT}{ng}\left[\frac{l_1 - 3l_2}{l_1 l_2}\right]$ (b) $\frac{RT}{g}\left[\frac{2l_1 + l_2}{l_1 l_2}\right]$

(c) $\frac{nRT}{g}\left[\frac{1}{l_2} + \frac{1}{l_1}\right]$ (d) $\frac{nRT}{g}\left[\frac{l_1 - l_2}{l_1 l_2}\right]$

22. The temperature of an open room of volume 30 m^3 increases from 17°C to 27°C due to sunshine. The atmospheric pressure in the room remains 1×10^5 Pa. If n_i and n_f are the number of molecules in the room before and after heating, then $n_f - n_i$ will be : **[2017]**
(a) 2.5×10^{25} (b) -2.5×10^{25}
(c) -1.61×10^{23} (d) 1.38×10^{23}

23.

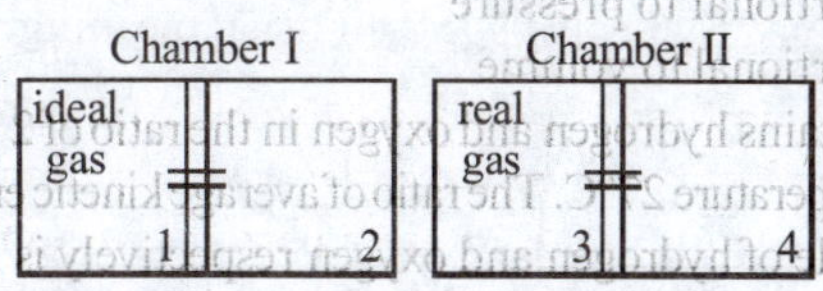

There are two identical chambers, completely thermally insulated from surroundings. Both chambers have a partition wall dividing the chambers in two compartments. Compartment 1 is filled with an ideal gas and Compartment 3 is filled with a real gas. Compartments 2 and 4 are vacuum. A small hole (orifice) is made in the partition walls and the gases are allowed to expand in vacuum.

Statement-1: No change in the temperature of the gas takes place when ideal gas expands in vacuum. However, the temperature of real gas goes down (cooling) when it expands in vacuum.

Statement-2: The internal energy of an ideal gas is only kinetic. The internal energy of a real gas is kinetic as well as potential. **[NCERT : P L-246 | Online April 9, 2013]**
(a) Statement-1 is false and Statement-2 is true.
(b) Statement-1 and Statement-2 both are true. Statement-2 is the correct explanation of Statement-1.
(c) Statement-1 is true and Statement-2 is false.
(d) Statement-1 and Statement-2 both are true. Statement-2 is **not** correct explanation of Statement-1.

24. Cooking gas containers are kept in a lorry moving with uniform speed. The temperature of the gas molecules inside will **[NCERT : P L-246 | 2002]**
(a) increase
(b) decrease
(c) remain same
(d) decrease for some, while increase for others

Topic 2 - Speed of Gas, Pressure and Kinetic Energy

25. The helium and argon are put in the flask at the same room temperature (300 K). The ratio of average kinetic energies (per molecule) of helium and argon is :
(Give : Molar mass of helium = 4 g/mol, Molar mass of argon – 40 g/ mol)
[NCERT : P L-250 | April 7, 2025 (II)]
(a) 1 : 10 (b) 10 : 1
(a) $1 : \sqrt{10}$ (d) 1 : 1

26. The ratio of vapour densities of two gases at the same temperature is $\frac{4}{25}$, then the ratio of r.m.s. velocities will be : **[NCERT : P L-250 | Jan. 28, 2025 (II)]**
(a) $\frac{25}{4}$ (b) $\frac{2}{5}$ (c) $\frac{5}{2}$ (d) $\frac{4}{25}$

27. For a particular ideal gas which of the following graphs represents the variation of mean squared velocity of the gas molecules with temperature ?
[NCERT : P L-250 | Jan. 28, 2025 (I)]

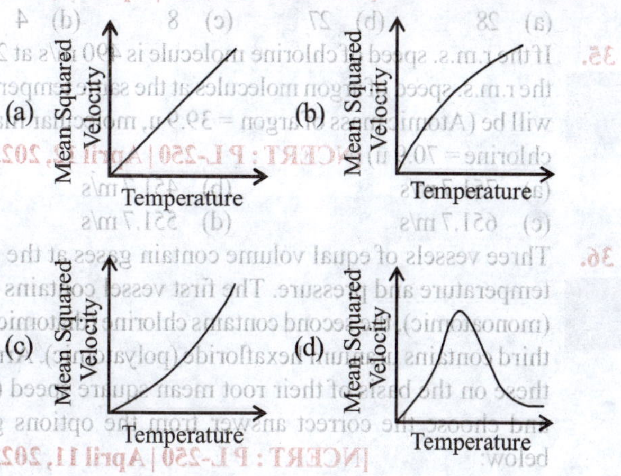

28. A sample contains mixture of helium and oxygen gas. The ratio of root mean square speed of helium and oxygen in the sample, is : [NCERT : P L-250 | April 6, 2024 (I)]

(a) $\frac{1}{32}$ (b) $\frac{2\sqrt{2}}{1}$ (c) $\frac{1}{4}$ (d) $\frac{1}{2\sqrt{2}}$

29. If the root mean square velocity of hydrogen molecule at a given temperature and pressure is 2 km/s, the root mean square velocity of oxygen at the same condition in km/s is: [NCERT : P L-250 | Feb. 1, 2024 (II)]

(a) 1.0 (b) 1.5 (c) 2.0 (d) 0.5

30. At which temperature the r.m.s. velocity of a hydrogen molecule equal to that of an oxygen molecule at 47° C? [NCERT : P L-250 | Jan. 30, 2024 (I)]

(a) 80 K (b) – 73 K (c) 4 K (d) 20 K

31. The average kinetic energy of a monatomic molecule is 0.414 eV at temperature : [NCERT : P L-250 | Jan. 27, 2024 (I)]

(Use $K_B = 1.38 \times 10^{-23}$ J/mol-K)

(a) 3000 K (b) 3200 K (c) 1600 K (d) 1500 K

32. The total kinetic energy of 1 mole of oxygen at 27°C is : [Use universal gas constant (R)= 8.31 J/mole K] [NCERT : P L-253 | Jan. 27, 2024 (II)]

(a) 6845.5 J (b) 5942.0 J
(c) 6232.5 J (d) 5670.5J

33. A flask contains Hydrogen and Argon in the ratio 2 : 1 by mass. The temperature of the mixture is 30°C. The ratio of average kinetic energy per molecule of the two gases (K argon/K hydrogen) is: NCERT : P L-253 | [April 15, 2023 (I)]

(Given: Atomic Weight of Ar = 39.9)

(a) 1 (b) 2 (c) $\frac{39.9}{2}$ (d) 39.9

34. The rms speed of oxygen molecule in a vessel at particular temperature is $\left(1+\frac{5}{x}\right)^{\frac{1}{2}} \upsilon$, where υ is the average speed of the molecule. The value of x will be: (Take $\pi = \frac{22}{7}$) [NCERT : P L-250 | April 13, 2023 (I)]

(a) 28 (b) 27 (c) 8 (d) 4

35. If the r.m.s. speed of chlorine molecule is 490 m/s at 27°C, the r.m.s. speed of argon molecules at the same temperature will be (Atomic mass of argon = 39.9 u, molecular mass of chlorine = 70.9 u) [NCERT : P L-250 | April 12, 2023 (I)]

(a) 751.7 m/s (b) 451.7 m/s
(c) 651.7 m/s (d) 551.7 m/s

36. Three vessels of equal volume contain gases at the same temperature and pressure. The first vessel contains neon (monoatomic), the second contains chlorine (diatomic) and third contains uranium hexafloride (polyatomic). Arrange these on the basis of their root mean square speed (v_{rms}) and choose the correct answer from the options given below: [NCERT : P L-250 | April 11, 2023 (I)]

(a) v_{rms} (mono) = v_{rms} (dia) = v_{rms} (poly)
(b) v_{rms} (mono) > v_{rms} (dia) > v_{rms} (poly)
(c) v_{rms} (dia) < v_{rms} (poly) < v_{rms} (mono)
(d) v_{rms} (mono) < v_{rms} (dia) < v_{rms} (poly)

37. The root mean square speed of molecules of nitrogen gas at 27°C is approximately: [NCERT : P L-250 | April 11, 2023 (II)]

(Given mass of a nitrogen molecule = 4.6×10^{-26} kg and take Boltzmann constant $k_B = 1.4 \times 10^{-23}$ JK^{-1})

(a) 523 m/s (b) 1260 m/s (c) 91 m/s (d) 27.4 m/s

38. The temperature at which the kinetic energy of oxygen molecules becomes double than its value at 27° C is [NCERT : P L-250 | April 8, 2023 (II)]

(a) 1227°C (b) 927°C (c) 327°C (d) 627°C

39. Given below are two statements: one is labelled as **Assertion A** and the other is labelled as **Reason R.**

Assertion A : Earth has atmosphere whereas moon doesn't have any atmosphere.

Reason R : The escape velocity on moon is very small as compared to that on earth.

In the light of the above statement, choose the correct answer from the options given below: [NCERT : P L-250 | April 6, 2023 (I)]

(a) A is true but R is false
(b) A is false but R is true
(c) Both A and R are correct but R is NOT the correct explanation of A
(d) Both A and R are correct and R is correct explanation of A

40. The temperature of an ideal gas is increased from 200 K to 800 K. If r.m.s. speed of gas at 200K is v_0. Then, r.m.s. speed of the gas at 800 K will be: [NCERT : P L-250 | April 6, 2023 (II)]

(a) v_0 (b) $4v_0$ (c) $\frac{v_0}{4}$ (d) $2v_0$

41. The average kinetic energy of a molecule of the gas is [NCERT : P L-250 | Feb. 1, 2023 (I)]

(a) proportional to absolute temperature
(b) dependent on the nature of the gas
(c) proportional to pressure
(d) proportional to volume

42. A flask contains hydrogen and oxygen in the ratio of 2 : 1 by mass at temperature 27°C. The ratio of average kinetic energy per molecule of hydrogen and oxygen respectively is : [NCERT : P L-250 | Jan. 30, 2023 (II)]

(a) 2 : 1 (b) 1 : 1 (c) 1 : 4 (d) 4 : 1

43. At 300 K, the rms speed of oxygen molecules is $\sqrt{\frac{\alpha+5}{\alpha}}$ times to that of its average speed in the gas. Then the value of α will be (used $\pi = \frac{22}{7}$) [NCERT : P L-250 | Jan. 29, 2023 (II)]

(a) 32 (b) 28 (c) 24 (d) 27

44. The root mean square velocity of molecules of gas is [NCERT : P L-250 | Jan. 25, 2023 (I)]
(a) Proportional to square of temperature (T^2).
(b) Inversely proportional to square root of temperature $\sqrt{\frac{1}{T}}$.
(c) Proportional to square root of temperature $\sqrt{T}$.
(d) Proportional to temperature (T).

45. Given below are two statements :
Statements I : The temperature of a gas is – 73°C. When the gas is heated to 527°C, the root mean square speed of the molecules is doubled.
Statement II : The product of pressure and volume of an ideal gas will be equal to translational kinetic energy of the molecules.
In the light of the above statements, choose the correct answer from the options given below :
[NCERT : P L-250, 252 | Jan. 24, 2023 (I)]
(a) Both statement I and Statement II are true
(b) Statement I is true but Statement II is false
(c) Both Statement I and Statement II are false
(d) Statement I is false but Statement II is true

46. The root mean square speed of smoke particles of mass 5×10^{17} kg in their Brownian motion in air at NTP is approximately. [Given $k = 1.38 \times 10^{-23}$ JK^{-1}]
[NCERT : P L-250 | July 29, 2022 (II)]
(a) 60 mm s^{-1} (b) 12 mm s^{-1}
(c) 15 mm s^{-1} (d) 36 mm s^{-1}

47. Given below are two statements :
Statement I : The average momentum of a molecule in a sample of an ideal gas depends on temperature.
Statement II : The rms speed of oxygen molecules in a gas is v. If the temperature is doubled and the oxygen molecules dissociate into oxygen atoms, the rms speed will become 2v.
In the light of the above statements, choose the correct answer from the options given below :
[NCERT : P L-250, 449 | July 28, 2022 (I)]
(a) Both Statement I and Statement II are true
(b) Both Statement I and Statement II are false
(c) Statement I is true but Statement II is false
(d) Statement I is false but Statement II is true

48. A vessel contains 14 g of nitrogen gas at a temperature of 27°C. The amount of heat to be transferred to the gas to double the r.m.s speed of its molecules will be :
Take $R = 8.32$ J mol^{-1} k^{-1}.
[NCERT : P L-250 | July 28, 2022 (II)]
(a) 2229 J (b) 5616 J (c) 9360 J (d) 13.104 J

49. Same gas is filled in two vessels of the same volume at the same temperature. If the ratio of the number of molecules is 1 : 4, then
A. The r.m.s. velocity of gas molecules in two vessels will be the same.
B. The ratio of pressure in these vessels will be 1 : 4
C. The ratio of pressure will be 1 : 1
D. The r.m.s. velocity of gas molecules in two vessels will be in the ratio of 1 : 4
[NCERT : P L-250 | July 27, 2022 (I)]
(a) A and C only (b) B and D only
(c) A and B only (d) C and D only

50. Following statements are given:
[NCERT : P L-250 | July 25, 2022 (I)]
(A) The average kinetic energy of a gas molecule decreases when the temperature is reduced.
(B) The average kinetic energy of a gas molecule increases with increase in pressure at constant temperature.
(C) The average kinetic energy of a gas molecule decreases with increase in volume.
(D) Pressure of a gas increases with increase in temperature at constant pressure.
(E) The volume of gas decreases with increase in temperature.
Choose the correct answer from the options given below:
(a) (A) (b) (A), (B) and (D) only
(c) (B) and (D) only (d) (A), (B) and (E) only

51. Sound travels in a mixture of two moles of helium and n moles of hydrogen. If rms speed of gas molecules in the mixture is $\sqrt{2}$ times the speed of sound, then the value of n will be [NCERT : P L-250 | July 25, 2022 (II)]
(a) 1 (b) 2 (c) 3 (d) 4

52. What will be the effect on the root mean square velocity of oxygen molecules if the temperature is doubled and oxygen molecule dissociates into atomic oxygen?
[NCERT : P L-250 | June 28, 2022 (II)]
(a) The velocity of atomic oxygen remains same
(b) The velocity of atomic oxygen doubles
(c) The velocity of atomic oxygen becomes half
(d) The velocity of atomic oxygen becomes four times

53. The temperature of 3.00 mol of an ideal diatomic gas is increased by 40.0 °C without changing the pressure of the gas. The molecules in the gas rotate but do not oscillate. If the ratio of change in internal energy of the gas to the amount of workdone by the gas is $\frac{x}{10}$. Then the value of x (round off to the nearest integer) is _______.
(Given $R = 8.31$ J mol^{-1} K^{-1})
[NCERT : P L-250 | NA, Sep. 1, 2021 (II)]

54. If the rms speed of oxygen molecules at 0°C is 160 m/s, find the rms speed of hydrogen molecules at 0°C.
[NCERT : P L-250 | Aug. 27, 2021 (II)]
(a) 640 m/s (b) 40 m/s (c) 80 m/s (d) 332 m/s

55. The rms speeds of the molecules of Hydrogen, Oxygen and Carbondioxide at the same temperature are V_H, V_O and V_{CO_2} respectively then :
[NCERT : P L-250 | Aug. 26, 2021 (I)]
(a) $V_H > V_O > V_{CO_2}$ (b) $V_{CO_2} > V_O > V_H$
(c) $V_H = V_O > V_{CO_2}$ (d) $V_H = V_O = V_{CO_2}$

56. The number of molecules in one litre of an ideal gas at 300 K and 2 atmospheric pressure with mean kinetic energy 2×10^{-9} J per molecules is :
[NCERT : P L-250 | July 27, 2021 (I), Sep. 05, 2020 (I)]
(a) 0.75×10^{11} (b) 3×10^{11}
(c) 1.5×10^{11} (d) 6×10^{11}

57. Consider a sample of oxygen behaving like an ideal gas. At 300 K, the ratio of root mean square (rms) velocity to the average velocity of gas molecule would be : (Molecular weight of oxygen is 32 g/mol; $R = 8.3\ JK^{-1}\ mol^{-1}$)
[NCERT : P L-250 | March 18, 2021 (II)]
(a) $\sqrt{\frac{8\pi}{3}}$ (b) $\sqrt{\frac{3}{3}}$ (c) $\sqrt{\frac{8}{3}}$ (d) $\sqrt{\frac{3\pi}{8}}$

58. The root mean square speed of molecules of a given mass of a gas at 27°C and 1 atmosphere pressure is 200 ms^{-1}. The root mean square speed of molecules of the gas at 127°C and 2 atmosphere pressure is $\frac{x}{\sqrt{3}}$ ms^{-1}. The value of x will be ____. **[NCERT : P L-250 | Feb. 24, 2021 (II)]**

59. A monoatomic gas of mass 4.0 u is kept in an insulated container. Container is moving with velocity 30 m/s. If container is suddenly stopped then change in temperature of the gas (R = gas constant) is $\frac{x}{3R}$. Value of x is ____.
[NCERT : P L-250 | Feb. 25, 2021, 2011 (S)]

60. Nitrogen gas is at 300°C temperature. The temperature (in K) at which the rms speed of a H_2 molecule would be equal to the rms speed of a nitrogen molecule, is ____. (Molar mass of N_2 gas 28 g);
[NCERT : P L-250 | NA Sep. 05, 2020 (II)]

61. For a given gas at 1 atm pressure, rms speed of the molecules is 200 m/s at 127°C. At 2 atm pressure and at 227°C, the rms speed of the molecules will be:
[NCERT : P L-250 | 9 April 2019 I]
(a) 100 m/s (b) $80\sqrt{5}$ m/s
(c) $100\sqrt{5}$ m/s (d) 80 m/s

62. If 10^{22} gas molecules each of mass 10^{-26} kg collide with a surface (perpendicular to it) elastically per second over an area 1 m^2 with a speed 10^4 m/s, the pressure exerted by the gas molecules will be of the order of : **[8 April 2019 I]**
(a) $10^4\ N/m^2$ (b) $10^8\ N/m^2$
(c) $10^3\ N/m^2$ (d) $10^{16}\ N/m^2$

63. The temperature, at which the root mean square velocity of hydrogen molecules equals their escape velocity from the earth, is closest to :
[NCERT : P L-250 | 8 April 2019 II]
[Boltzmann Constant $k_B = 1.38 \times 10^{-23}$ J/K
Avogadro Number $N_A = 6.02 \times 10^{26}$ /kg
Radius of Earth : 6.4×10^6 m
Gravitational acceleration on Earth = 10 ms^{-2}]
(a) 800 K (b) 3×10^5 K
(c) 10^4 K (d) 650 K

64. A mixture of 2 moles of helium gas (atomic mass = 4u), and 1 mole of argon gas (atomic mass = 40u) is kept at 300 K in a container. The ratio of their rms speeds $\left[\frac{V_{rms}(\text{helium})}{V_{rms}(\text{argon})}\right]$ is close to :
[NCERT : P L-250 | 9 Jan. 2019 I]
(a) 3.16 (b) 0.32 (c) 0.45 (d) 2.24

65. In an ideal gas at temperature T, the average force that a molecule applies on the walls of a closed container depends on T as T^q. A good estimate for q is:
[Online April 10, 2015]
(a) $\frac{1}{2}$ (b) 2 (c) 1 (d) $\frac{1}{4}$

66. At room temperature a diatomic gas is found to have an r.m.s. speed of 1930 ms^{-1}. The gas is:
[NCERT : P L-250 | Online April 12, 2014]
(a) H_2 (b) Cl_2 (c) O_2 (d) F_2

67. A perfect gas at 27°C is heated at constant pressure so as to double its volume. The final temperature of the gas will be, close to **[NCERT : P L-246 | Online May 7, 2012]**
(a) 327°C (b) 200°C (c) 54°C (d) 300°C

68. Three perfect gases at absolute temperatures T_1, T_2 and T_3 are mixed. The masses of molecules are m_1, m_2 and m_3 and the number of molecules are n_1, n_2 and n_3 respectively. Assuming no loss of energy, the final temperature of the mixture is : **[NCERT : P L-250 | 2011]**
(a) $\frac{n_1T_1 + n_2T_2 + n_3T_3}{n_1 + n_2 + n_3}$ (b) $\frac{n_1T_1^2 + n_2T_2^2 + n_3T_3^2}{n_1T_1 + n_2T_2 + n_3T_3}$
(c) $\frac{n_1^2T_1^2 + n_2^2T_2^2 + n_3^2T_3^2}{n_1T_1 + n_2T_2 + n_3T_3}$ (d) $\frac{(T_1 + T_2 + T_3)}{3}$

69. One kg of a diatomic gas is at a pressure of $8 \times 10^4 N/m^2$. The density of the gas is $4 kg/m^3$. What is the energy of the gas due to its thermal motion?
[NCERT : P L-253 | 2009]
(a) 5×10^4 J (b) 6×10^4 J
(c) 7×10^4 J (d) 3×10^4 J

Topic 3 - Degree of Freedom, Specific Heat Capacity and Mean Free Path

70. Match the List-I with List-II
[NCERT : P L-253 | April 7, 2025 (I)]

	LIST-I		LIST-I
A.	Triatomic rigid gas	I.	$\frac{C_p}{C_v} = \frac{5}{3}$
B.	Diatomic non-rigid gas	II.	$\frac{C_p}{C_v} = \frac{7}{5}$
C.	Monoatomic gas	III.	$\frac{C_p}{C_v} = \frac{4}{3}$
D.	Diatomic rigid gas	IV.	$\frac{C_p}{C_v} = \frac{9}{7}$

Choose the **correct** answer from the options given below :

(a) A-III, B-IV, C-I, D-II (b) A-III, B-II, C-IV, D-I
(c) A-II, B-IV, C-I, D-III (d) A-IV, B-II, C-III, D-I

71. The mean free path and the average speed of oxygen molecules at 300 K and 1 atm are 3×10^{-7} m and 600 m/s, respectively. Find the frequency of its collisions.

[NCERT : P L-255 | April 4, 2025 (I)]

(a) 2×10^{10}/s (b) 9×10^5/s
(c) 2×10^9/s (d) 5×10^8/s

72. The internal energy of air in 4 m × 4 m × 3 m sized room at 1 atmospheric pressure will be ____ $\times 10^6$ J. (Consider air as diatomic molecule)

[NCERT : P L-253 | April 2, 2025 (II)]

73. γ_A is the specific heat ratio of monoatomic gas A having 3 translational degrees of freedom. γ_B is the specific heat ratio of polyatomic gas B having 3 translational, 3 rotational degrees of freedom and 1 vibrational mode. If $\frac{\gamma_A}{\gamma_B} = \left(1 + \frac{1}{n}\right)$, then the value of n is ______.

[NCERT : P L-253 | April 2, ? (I)]

74. The kinetic energy of translation of the molecules in 50g of CO_2 gas at 17°C is :**[NCERT : P L-253 | Jan. 28, 2025 (II)]**

(a) 3986.3 J (b) 4102.8 J
(c) 4205.5 J (d) 3582.7 J

75. For a diatomic gas, if $\gamma_1 = \left(\frac{Cp}{Cv}\right)$ for rigid molecules and $\gamma_2 = \left(\frac{Cp}{Cv}\right)$ for another diatomic molecules, but also having vibrational modes. Then, which one of the following options is correct?

(Cp and Cv are specific heats of the gas at constant pressure and volume)

[NCERT : P L-253 | Jan. 22, 2025 (II)]

(a) $\gamma_2 > \gamma_1$ (b) $\gamma_2 = \gamma_1$
(c) $2\gamma_2 = \gamma_1$ (d) $\gamma_2 < \gamma_1$

76. The temperature of a gas is –78° C and the average translational kinetic energy of its molecules is K. The temperature at which the average translational kinetic energy of the molecules of the same gas becomes 2K is :

[NCERT : P L-250 | April 9, 2024 (II)]

(a) –39°C (b) 117°C (c) 217°C (d) –78°C

77. A mixture of one mole of monoatomic gas and one mole of a diatomic gas (rigid) are kept at room temperature (27°C). The ratio of specific heat of gases at constant volume respectively is: **[NCERT : P L-253 | April 8, 2024 (I)]**

(a) $\frac{7}{5}$ (b) $\frac{3}{2}$ (c) $\frac{3}{5}$ (d) $\frac{5}{3}$

78. Given below are two statements :

Statement (I) : The mean free path of gas molecules is inversely proportional to square of molecular diameter.

Statement (II) : Average kinetic energy of gas molecules is directly proportional to absolute temperature of gas.

In the light of the above statements, choose the correct answer from the option given below:

[NCERT : P L-250 | April 8, 2024 (II)]

(a) **Statement I** is false but **Statement II** is true.
(b) **Statement I** is true but **Statement II** is false.
(c) Both **Statement I** and **Statement II** are false
(d) Both **Statement I** and **Statement II** are true.

79. If the collision frequency of hydrogen molecules in a closed chamber at 27°C is Z, then the collision frequency of the same system at 127° C is :

[NCERT : P L-255 | April 5, 2024 (I)]

(a) $\frac{\sqrt{3}}{2}Z$ (b) $\frac{4}{3}Z$ (c) $\frac{2}{\sqrt{3}}Z$ (d) $\frac{3}{4}Z$

80. If n is the number density and d is the diameter of the molecule, then the average distance covered by a molecule between two successive collisions (i.e. mean free path) is represented by : **[NCERT : P L-255 | April 5, 2024 (II)]**

(a) $\frac{1}{\sqrt{2n\pi d^2}}$ (b) $\sqrt{2n\pi d^2}$
(c) $\frac{1}{\sqrt{2}n\pi d^2}$ (d) $\frac{1}{\sqrt{2}n^2\pi^2d^2}$

81. The translational degrees of freedom (f_t) and rotational degrees of freedom (f_r) of CH_4 molecule are :

[NCERT : P L-253 | April 4, 2024 (II)]

(a) $f_t = 2$ and $f_r = 2$ (b) $f_t = 3$ and $f_r = 3$
(c) $f_t = 3$ and $f_r = 2$ (d) $f_t = 2$ and $f_r = 3$

82. Energy of 10 non rigid diatomic molecules at temperature T is : **[NCERT : P L-253 | April 6, 2024 (II)]**

(a) $\frac{7}{2}RT$ (b) $70\,K_BT$ (c) 35 RT (d) $35\,K_BT$

83. Two moles a monoatomic gas is mixed with six moles of a diatomic gas. The molar specific heat of the mixture at constant volume is: **[NCERT : P L-253 | Feb. 1, 2024 (I)]**

(a) $\frac{9}{4}R$ (b) $\frac{7}{4}R$ (c) $\frac{3}{2}R$ (d) $\frac{5}{2}R$

84. A diatomic gas ($\gamma = 1.4$) does 200 J of work when it is expanded isobarically. The heat given to the gas in the process is: **[NCERT : P L-253 | Feb. 1, 2024 (II)]**

(a) 800 J (b) 600 J (c) 700 J (d) 850 J

85. A gas mixture consists of 8 moles of argon and 6 moles of oxygen at temperature T. Neglecting all vibrational modes, the total internal energy of the system is

[NCERT : P L-253 | Jan. 31, 2024 (II)]

(a) 29 RT (b) 20 RT (c) 27 RT (d) 21 RT

86. If three moles of monoatomic gas $\left(\gamma = \frac{5}{3}\right)$ is mixed with two moles of a diatomic gas $\left(\gamma = \frac{7}{5}\right)$, the value of adiabatic exponent γ for the mixture is

[NCERT : P L-253 | Jan. 30, 2024 (II)]

(a) 1.75 (b) 1.40 (c) 1.52 (d) 1.35

87. N moles of a polyatomic gas (f = 6) must be mixed with two moles of a monoatomic gas so that the mixture behaves as a diatomic gas. The value of N is :

[NCERT : P L-253 | Jan. 29, 2024 (II)]

(a) 4 (b) 3 (c) 6 (d) 2

88. The mean free path of molecules of a certain gas at STP is 1500d, where d is the diameter of the gas molecules. While maintaining the standard pressure, the mean free path of the molecules at 373 K is approximately :

[NCERT : P L-255 | April 13, 2023 (II)]

(a) 1098d (b) 2049d (c) 750d (d) 1500d

89. Match List I with List II :

[NCERT : P L-253 | April 10, 2023 (I)]

List-I	List-II
(A) 3 Translational degrees of freedom	(I) Monoatomic gases
(B) 3 Translational, 2 rotational degrees of freedoms	(II) Polyatomic gases
(C) 3 Translational, 2 rotational and 1 vibrational degrees of freedom	(III) Rigid diatomic gases
(D) 3 Translational, 3 rotational and more than one vibrational degrees of freedom	(IV) Nonrigid diatomic gases

Choose the correct answer from the options given below:

(a) (A) – (IV), (B) – (III), (C) – (II), (D) – (I)
(b) (A) – (IV), (B) – (II), (C) – (I), (D) – (III)
(c) (A) – (I), (B) – (III), (C) – (IV), (D) – (II)
(d) (A) – (I), (B) – (IV), (C) – (III), (D) – (II)

90. A gas mixture consists of 2 moles of oxygen and 4 moles of neon at temperature T. Neglecting all vibrational modes, the total internal energy of the system will be :

[NCERT : P L-253 | April 10, 2023 (II)]

(a) 8 RT (b) 16 RT
(c) 4 RT (d) 11 RT

91. The number of air molecules per cm^3 increased from 3×10^{19} to 12×10^{19}. The ratio of collision frequency of air molecules before and after the increase in number respectively is **[NCERT : P L-253 | April 6, 2023 (I)]**

(a) 1.25 (b) 0.25 (c) 0.75 (d) 0.50

92. The correct relation between $\gamma = \frac{C_p}{C_v}$ and temperature T is : **[NCERT : P L-253 | Jan. 31, 2023 (I)]**

(a) $\gamma \propto \frac{1}{\sqrt{T}}$ (b) $\gamma \propto T^{\circ}$ (c) $\gamma \alpha \frac{1}{T}$ (d) $g \propto T$

93. Heat energy of 735 J is given to a diatomic gas allowing the gas to expand at constant pressure. Each gas molecule rotates around an internal axis but do not oscillate. The increase in the internal energy of the gas will be:

[NCERT : P L-253 | Jan. 31, 2023 (II)]

(a) 525 J (b) 441 J (c) 572 J (d) 735 J

94. According to law of equipartition of energy the molar specific heat of a diatomic gas at constant volume where the molecule has one additional vibrational mode is :-

[NCERT : P L-252 | Jan. 25, 2023 (II)]

(a) $\frac{9}{2}R$ (b) $\frac{5}{2}R$ (c) $\frac{3}{2}R$ (d) $\frac{7}{2}R$

95. γ_1 be the ratio of molar specific heat at constant pressure and molar specific heat at constant volume of a monoatomic gas and γ_2 be the similar ratio of diatomic gas. Considering the diatomic gas molecule as a rigid rotator, the ratio, $\frac{\gamma_1}{\gamma_2}$ is **[NCERT : P L-253 | Jan. 24, 2023 (II)]**

(a) $\frac{27}{35}$ (b) $\frac{35}{27}$ (c) $\frac{25}{21}$ (d) $\frac{21}{25}$

96. One mole of a monoatomic gas is mixed with three moles of a diatomic gas. The molecular specific heat of mixture at constant volume is $\frac{\alpha^2}{4}R$ J/mol K; then the value of α will be ______. (Assume that the given diatomic gas has no vibrational mode). **[NCERT : P L-253 | July 29, 2022 (I)]**

97. At a certain temperature, the degrees of freedom per molecule for gas is 8. The gas performs 150 J of work when it expands under constant pressure. The amount of heat absorbed by the gas will be _____ J.

[NCERT : P L-253 | July 28, 2022 (II)]

98. Which statements are correct about degrees of freedom?

[NCERT : P L-253 | July 27, 2022 (II)]

A. A molecule with n degrees of freedom has n^2 different ways of storing energy.

B. Each degree of freedom is associated with $\frac{1}{2}RT$ average energy per mole.

C. A monoatomic gas molecule has 1 rotational degree of freedom where as diatomic molecule has 2 rotational degrees of freedom

D. CH_4 has a total to 6 degrees of freedom

Choose the correct answer from the option given below:

(a) B and C only (b) B and D only
(c) A and B only (d) C and D only

99. A gas has n degrees of freedom. The ratio of specific heat of gas at constant volume to the specific heat of gas at constant pressure will be :

[NCERT : P L-253 | July 26, 2022 (II)]

(a) $\frac{n}{n+2}$ (b) $\frac{n+2}{n}$ (c) $\frac{n}{2n+2}$ (d) $\frac{n}{n-2}$

100. A thermally insulated vessel contains an ideal gas of molecular mass *M* and ratio of specific heats 1.4. Vessel is moving with speed *v* and is suddenly brought to rest. Assuming no heat is lost to the surrounding and vessel temperature of the gas increases by : (*R* = universal gas constant) **[NCERT : P L-253 | June 26, 2022 (I)]**

(a) $\frac{Mv^2}{7R}$ (b) $\frac{Mv^2}{5R}$ (c) $2\frac{Mv^2}{7R}$ (d) $7\frac{Mv^2}{5R}$

101. A flask contains argon and oxygen in the ratio of 3 : 2 in mass and the mixture is kept at 27°C. The ratio of their average kinetic energy per molecule respectively will be: **[NCERT : P L-250 | June 26, 2022 (II)]**

(a) 3 : 2 (b) 9 : 4 (c) 2 : 3 (d) 1 : 1

102. The ratio of specific heats $\left(\frac{C_P}{C_V}\right)$ in terms of degree of freedom (f) is given by: **[NCERT : P L-253 | June 25, 2022 (II)]**

(a) $\left(1+\frac{f}{3}\right)$ (b) $\left(1+\frac{2}{f}\right)$

(c) $\left(1+\frac{f}{2}\right)$ (d) $\left(1+\frac{1}{f}\right)$

103. One mole of an ideal gas is taken through an adiabatic process where the temperature rises from 27°C to 37°C. If the ideal gas is composed of polyatomic molecule that has 4 vibrational modes, which of the following is true? [$R = 8.314\ J\ mol^{-1}\ k^{-1}$] **[NCERT : P L-253 | July 27, 2021 (II)]**

(a) work done by the gas is close to 332 J
(b) work done on the gas is close to 582 J
(c) work done by the gas is close to 582 J
(d) work done on the gas is close to 332 J

104. What will be the average value of energy for a monoatomic gas in thermal equilibrium at temperature T? **[NCERT : P L-252 | July 22, 2021 (II)]**

(a) $\frac{2}{3}k_BT$ (b) k_BT (c) $\frac{3}{2}k_BT$ (d) $\frac{1}{2}k_BT$

105. Two ideal polyatomic gases at temperatures T_1 and T_2 are mixed so that there is no loss of energy. If F_1 and F_2, m_1 and m_2, n_1 and n_2 be the degrees of freedom, masses, number of molecules of the first and second gas respectively, the temperature of mixture of these two gases is: **[NCERT : P L-253 | March 17, 2021 (I)]**

(a) $\frac{n_1F_1T_1 + n_2F_2T_2}{F_1 + F_2}$ (b) $\frac{n_1T_1 + n_2T_2}{n_1 + n_2}$

(c) $\frac{n_1F_1T_1 + n_2F_2T_2}{n_1 + n_2}$ (d) $\frac{n_1F_1T_1 + n_2F_2T_2}{n_1F_1 + n_2F_2}$

106. Calculate the value of mean free path (λ) for oxygen molecules at temperature 27°C and pressure 1.01×10^5 Pa. Assume the molecular diameter 0.3 nm and the gas is ideal. ($k = 1.38 \times 10^{-23}\ JK^{-1}$) **[NCERT : P L-255 | March 16, 2021 (II)]**

(a) 86 nm (b) 32 nm (c) 58 nm (d) 102 nm

107. A container is divided into two chambers by a partition. The volume of first chamber is 4.5 litre and second chamber is 5.5 litre. The first chamber contain 3.0 moles of gas at pressure 2.0 atm and second chamber contain 4.0 moles of gas at pressure 3.0 atm. After the partition is removed and the mixture attains equilibrium, then, the common equilibrium pressure existing in the mixture is $x \times 10^{-1}$ atm. Value of x is ______. **[NCERT : P L-246 | Feb. 26, 2021 (I)]**

108. A diatomic gas, having $C_p = \frac{7}{2}R$ and $C_v = \frac{5}{2}R$, is heated at constant pressure. The ratio dU : dQ : dW : **[NCERT : P L-253 | Feb. 25, 2021 (I)]**

(a) 5 : 7 : 2 (b) 3 : 5 : 2 (c) 3 : 7 : 2 (d) 5 : 7 : 3

109. The internal energy (U), pressure (P) and volume (V) of an ideal gas are related as U = 3PV + 4. The gas is **[NCERT : P L-253 | Feb. 26, 2021 (II)]**

(a) diatomic only
(b) either monoatomic or diatomic
(c) polyatomic only
(d) monoatomic only

110. In a dilute gas at pressure P and temperature T, the mean time between successive collisions of a molecule varies with T as: **[NCERT : P L-255 | Sep. 06, 2020 (II)]**

(a) T (b) $\frac{1}{\sqrt{T}}$ (c) $\frac{1}{T}$ (d) $\sqrt{T}$

111. Match the C_p/C_v ratio for ideal gases with different type of molecules : **[NCERT : P L-253 | Sep. 04, 2020 (I)]**

Column-I Molecule Type	Column-II C_p/C_v
(A) Monatomic	(I) 7/5
(B) Diatomic rigid molecules	(II) 9/7
(C) Diatomic non-rigid molecules	(III) 4/3
(D) Triatomic rigid molecules	(IV) 5/3

(a) (A)-(IV), (B)-(II), (C)-(I), (D)-(III)
(b) (A)-(III), (B)-(IV), (C)-(II), (D)-(I)
(c) (A)-(IV), (B)-(I), (C)-(II), (D)-(III)
(d) (A)-(II), (B)-(III), (C)-(I), (D)-(IV)

112. A closed vessel contains 0.1 mole of a monatomic ideal gas at 200 K. If 0.05 mole of the same gas at 400 K is added to it, the final equilibrium temperature (in K) of the gas in the vessel will be close to ________. **[NCERT : P L-253 | NA Sep. 04, 2020 (I)]**

113. Consider a gas of triatomic molecules. The molecules are assumed to be triangular and made of massless rigid rods whose vertices are occupied by atoms. The internal energy of a mole of the gas at temperature T is : **[NCERT : P L-253 | Sep. 03, 2020 (I)]**

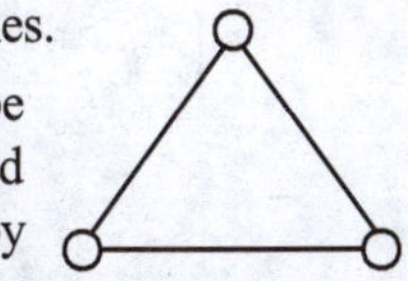

(a) $\frac{5}{2}RT$ (b) $\frac{3}{2}RT$ (c) $\frac{9}{2}RT$ (d) $3RT$

114. To raise the temperature of a certain mass of gas by 50°C at a constant pressure, 160 calories of heat is required. When the same mass of gas is cooled by 100°C at constant volume, 240 calories of heat is released. How many degrees of freedom does each molecule of this gas have (assume gas to be ideal)? **[NCERT : P L-253 | Sep. 03, 2020 (II)]**

(a) 5 (b) 6
(c) 3 (d) 7

115. An ideal gas in a closed container is slowly heated. As its temperature increases, which of the following statements are true? **[NCERT : P L-255 | Sep. 02, 2020 (II)]**

(1) The mean free path of the molecules decreases
(2) The mean collision time between the molecules decreases
(3) The mean free path remains unchanged
(4) The mean collision time remains unchanged

(a) (2) and (3) (b) (1) and (2)
(c) (3) and (4) (d) (1) and (4)

116. Consider two ideal diatomic gases A and B at some temperature T. Molecules of the gas A are rigid, and have a mass m. Molecules of the gas B have an additional vibrational mode, and have a mass $\frac{m}{4}$. The ratio of the specific heats (C_V^A and C_V^B) of gas A and B, respectively is: **[NCERT : P L-253 | 9 Jan 2020 I]**

(a) 7:9 (b) 5:9
(c) 3:5 (d) 5:7

117. Two gases-argon (atomic radius 0.07 nm, atomic weight 40) and xenon (atomic radius 0.1 nm, atomic weight 140) have the same number density and are at the same temperature. The ratio of their respective mean free times is closest to: **[NCERT : P L-255 | 9 Jan 2020 II]**

(a) 3.67 (b) 1.83
(c) 2.3 (d) 4.67

118. The plot that depicts the behavior of the mean free time τ (time between two successive collisions) for the molecules of an ideal gas, as a function of temperature (*T*), qualitatively, is: (Graphs are schematic and not drawn to scale) **[NCERT : P L-255 | 8 Jan. 2020 I]**

(a) 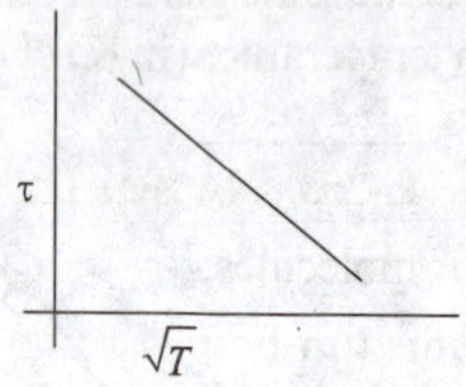

(b) 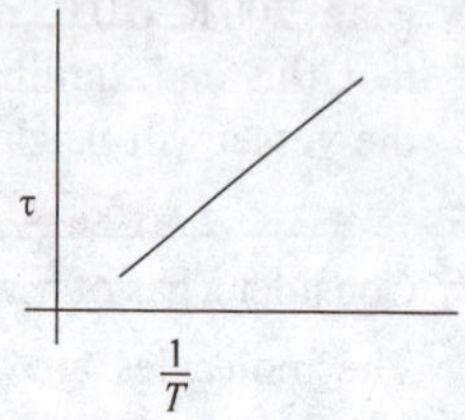

(c) 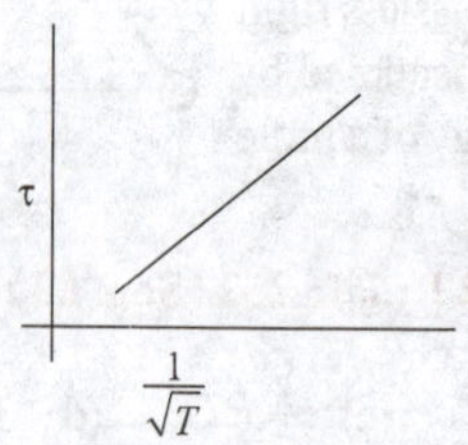

(d) 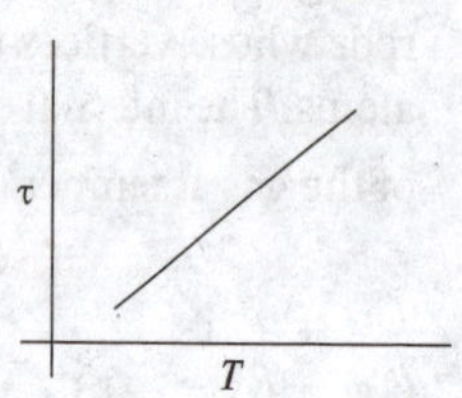

119. Consider a mixture of *n* moles of helium gas and 2*n* moles of oxygen gas (molecules taken to be rigid) as an ideal gas. Its C_P/C_V value will be: **[NCERT : P L-253 | Online April 16, 2018 S, 8 Jan. 2020 II]**

(a) 19/13 (b) 67/45 (c) 40/27 (d) 23/15

120. Two moles of an ideal gas with $\frac{C_P}{C_V}=\frac{5}{3}$ are mixed with 3 moles of another ideal gas with $\frac{C_P}{C_V}=\frac{4}{3}$. The value of $\frac{C_P}{C_V}$ for the mixture is:

[NCERT : P L-253 | 7 Jan. 2020 (I), 12 April 2019 (I)(S), 2004 (S)]

(a) 1.45 (b) 1.50 (c) 1.47 (d) 1.42

121. A diatomic gas with rigid molecules does 10 J of work when expanded at constant pressure. What would be the heat energy absorbed by the gas, in this process ? **[NCERT : P L-253 | 12 April 2019 II]**

(a) 25 J (b) 35 J (c) 30 J (d) 40 J

122. A $25\times10^{-3}\ m^3$ volume cylinder is filled with 1 mol of O_2 gas at room temperature (300 K). The molecular diameter of O_2, and its root mean square speed, are found to be 0.3 nm and 200 m/s, respectively. What is the average collision rate (per second) for an O_2 molecule? **[NCERT : P L-255, 250 | 10 April 2019 I]**

(a) $\sim10^{12}$ (b) $\sim10^{11}$
(c) $\sim10^{10}$ (d) $\sim10^{13}$

123. When heat Q is supplied to a diatomic gas of rigid molecules, at constant volume its temperature increases by ΔT. The heat required to produce the same change in temperature, at a constant pressure is : **[10 April 2019 II]**

(a) $\frac{2}{3}Q$ (b) $\frac{5}{3}Q$ (c) $\frac{7}{5}Q$ (d) $\frac{3}{2}Q$

124. An HCl molecule has rotational, translational and vibrational motions. If the rms velocity of HCl molecules in its gaseous phase is $\bar{v}$, m is its mass and k_B is Boltzmann constant, then its temperature will be: **[NCERT : P L-250 | 9 April 2019 I]**

(a) $\frac{m\bar{v}^2}{6k_B}$ (b) $\frac{m\bar{v}^2}{3k_B}$ (c) $\frac{m\bar{v}^2}{7k_B}$ (d) $\frac{m\bar{v}^2}{5k_B}$

125. The specific heats, C_p and C_v of a gas of diatomic molecules, A, are given (in units of $J\ mol^{-1}\ k^{-1}$) by 29 and 22, respectively. Another gas of diatomic molecules, B, has the corresponding values 30 and 21. If they are treated as ideal gases, then: **[NCERT : P L-253 | 9 April 2019 II]**

(a) A is rigid but B has a vibrational mode.
(b) A has a vibrational mode but B has none.
(c) A has one vibrational mode and B has two.
(d) Both A and B have a vibrational mode each.

126. An ideal gas occupies a volume of $2 \, m^3$ at a pressure of 3×10^6 Pa. The energy of the gas:

[NCERT : P L-253 | 12 Jan. 2019 I]

(a) 9×10^6 J (b) 6×10^4 J
(c) 10^8 J (d) 3×10^2 J

127. An ideal gas is enclosed in a cylinder at pressure of 2 atm and temperature, 300 K. The mean time between two successive collisions is 6×10^{-8} s. If the pressure is doubled and temperature is increased to 500 K, the mean time between two successive collisions will be close to:

[NCERT : P L-255 | 12 Jan. 2019 II]

(a) 2×10^{-7} s (b) 4×10^{-8} s
(c) 0.5×10^{-8} s (d) 3×10^{-6} s

128. A gas mixture consists of 3 moles of oxygen and 5 moles of argon at temperature T. Considering only translational and rotational modes, the total internal energy of the system is : **[NCERT : P L-253 | 11 Jan. 2019 I]**

(a) 15 RT (b) 12 RT
(c) 4 RT (d) 20 RT

129. In a process, temperature and volume of one mole of an ideal monoatomic gas are varied according to the relation VT = K, where K is a constant. In this process the temperature of the gas is increased by ΔT. The amount of heat absorbed by gas is (R is gas constant) :

[NCERT : P L-253 | 11 Jan. 2019 II]

(a) $\frac{1}{2}R\Delta T$ (b) $\frac{1}{2}KR\Delta T$ (c) $\frac{3}{2}R\Delta T$ (d) $\frac{2K}{3}\Delta T$

130. Two kg of a monoatomic gas is at a pressure of 4×10^4 N/m^2. The density of the gas is 8 kg/m^3. What is the order of energy of the gas due to its thermal motion?

[NCERT : P L-250 | 10 Jan 2019 II]

(a) 10^3 J (b) 10^5 J (c) 10^4 J (d) 10^6 J

131. A 15 g mass of nitrogen gas is enclosed in a vessel at a temperature 27°C. Amount of heat transferred to the gas, so that rms velocity of molecules is doubled, is about: [Take R = 8.3 J/K mole]

[NCERT : P L-253 | 9 Jan. 2019 II]

(a) 0.9 kJ (b) 6 kJ (c) 10 kJ (d) 14 kJ

132. Two moles of an ideal monoatomic gas occupies a volume V at 27°C. The gas expands adiabatically to a volume 2 V. Calculate (1) the final temperature of the gas and (2) change in its internal energy. **[NCERT : P L-253 | 2018]**

(a) (1) 189 K (2) 2.7 kJ
(b) (1) 195 K (2) –2.7 kJ
(c) (1) 189 K (2) –2.7 kJ
(d) (1) 195 K (2) 2.7 kJ

133. C_p and C_v are specific heats at constant pressure and constant volume respectively. It is observed that
$C_p - C_v = a$ for hydrogen gas
$C_p - C_v = b$ for nitrogen gas
The correct relation between a and b is :

[NCERT : P L-253 | 2017]

(a) a = 14 b (b) a = 28 b
(c) $a = \frac{1}{14}b$ (d) a = b

134. An ideal gas has molecules with 5 degrees of freedom. The ratio of specific heats at constant pressure (C_p) and at constant volume (C_v) is :

[NCERT : P L-253 | Online April 8, 2017]

(a) 6 (b) $\frac{7}{2}$ (c) $\frac{5}{2}$ (d) $\frac{7}{5}$

135. An ideal gas undergoes a quasi static, reversible process in which its molar heat capacity C remains constant. If during this process the relation of pressure P and volume V is given by PV^n = constant, then n is given by (Here C_P and C_V are molar specific heat at constant pressure and constant volume, respectively) :

[NCERT : P L-253 | 2016]

(a) $n = \frac{C_P - C}{C - C_V}$ (b) $n = \frac{C - C_V}{C - C_P}$

(c) $n = \frac{C_P}{C_V}$ (d) $n = \frac{C - C_P}{C - C_V}$

136. Using equipartition of energy, the specific heat (in $J\,kg^{-1}\,K^{-1}$) of aluminium at room temperature can be estimated to be (atomic weight of aluminium = 27)

[NCERT : P L-254 | Online April 11, 2015]

(a) 410 (b) 25
(c) 1850 (d) 925

137. A given ideal gas with $\gamma = \frac{C_p}{C_v} = 1.5$ at a temperature T. If the gas is compressed adiabatically to one-fourth of its initial volume, the final temperature will be

[NCERT : P L-253 | Online May 12, 2012]

(a) $2\sqrt{2}T$ (b) $4T$
(c) $2T$ (d) $8T$

138. If C_P and C_V denote the specific heats of nitrogen per unit mass at constant pressure and constant volume respectively, then **[NCERT : P L-253 | 2007]**

(a) $C_P - C_V = 28R$
(b) $C_P - C_V = R/28$
(c) $C_P - C_V = R/14$
(d) $C_P - C_V = R$

139. During an adiabatic process, the pressure of a gas is found to be proportional to the cube of its absolute temperature. The ratio C_P/C_V for the gas is

[NCERT : P L-253 | 2003]

(a) $\frac{4}{3}$ (b) 2 (c) $\frac{5}{3}$ (d) $\frac{3}{2}$

ANSWER KEYS																			
1	(b)	15	(c)	29	(d)	43	(b)	57	(d)	71	(c)	85	(c)	99	(a)	113	(d)	127	(b)
2	(c)	16	(c)	30	(d)	44	(c)	58	(400)	72	(12)	86	(c)	100	(b)	114	(b)	128	(a)
3	(327)	17	(c)	31	(b)	45	(b)	59	(3600)	73	(3)	87	(a)	101	(None)	115	(a)	129	(a)
4	(b)	18	(5)	32	(c)	46	(c)	60	(41)	74	(b)	88	(b)	102	(b)	116	(d)	130	(c)
5	(b)	19	(150)	33	(a)	47	(d)	61	(c)	75	(d)	89	(c)	103	(b)	117	(None)	131	(c)
6	(b)	20	(a)	34	(a)	48	(c)	62	(None)	76	(b)	90	(d)	104	(c)	118	(c)	132	(c)
7	(a)	21	(d)	35	(c)	49	(c)	63	(c)	77	(c)	91	(b)	105	(d)	119	(a)	133	(a)
8	(c)	22	(b)	36	(b)	50	(a)	64	(a)	78	(d)	92	(b)	106	(d)	120	(d)	134	(d)
9	(d)	23	(b)	37	(a)	51	(b)	65	(c)	79	(c)	93	(a)	107	(25)	121	(b)	135	(d)
10	(c)	24	(c)	38	(c)	52	(b)	66	(a)	80	(c)	94	(d)	108	(a)	122	(c)	136	(d)
11	(b)	25	(d)	39	(d)	53	(25)	67	(a)	81	(b)	95	(c)	109	(c)	123	(c)	137	(c)
12	(a)	26	(c)	40	(d)	54	(a)	68	(a)	82	(d)	96	(3)	110	(b)	124	(b)	138	(b)
13	(b)	27	(a)	41	(a)	55	(a)	69	(a)	83	(a)	97	(750)	111	(c)	125	(b)	139	(d)
14	(a)	28	(b)	42	(b)	56	(c)	70	(a)	84	(c)	98	(b)	112	(266.67)	126	(a)		

13 Oscillations

TREND ANALYSIS
JEE MAIN (2025-2021)

	2025				2024				2023				2022				2021			
	Session II		Session I		Session II		Session I		Session II		Session I		Session II		Session I		Session II		Session I	
	MCQs	NVQs	MCQs	NVQs	MCQs	NVQs	MCQs	NVQs	MCQs	NVQs	MCQs	NVQs	MCQs	NVQs	MCQs	NVQs	MCQs	NVQs	MCQs	NVQs
Topic 1	1	–	2	–	–	3	–	3	2	–	1	3	1	1	2	2	–	2	1	2
Topic 2	–	–	1	–	1	1	–	1	3	1	1	2	–	–	–	–	2	–	–	1
Topic 3	6	–	3	–	1	1	2	3	3	2	5	3	4	3	4	2	2	–	4	2
Topic 4																				
Total	7	-	6	-	2	5	2	7	8	3	7	8	5	4	6	4	4	2	5	5

How to Code Qns for Better & Faster Revision

Type 1: Able to solve (No code Req)
Type 2: Solved with struggle or by seeing solution **(Use code NR means Needs Revision or orange stickers)**
Type 3: Tricky/ Could not solve on my own. Need help from Teacher/ Friend. **(Use code TQ means Tricky Question or red stickers)**

You can either hand write the code in boxes or else paste coloured stickers given at end of book.

Topic 1 - Displacement, Phase, Velocity and Acceleration in S.H.M.

1. A particle is subjected two simple harmonic motions as : $x_1 = \sqrt{7}\sin 5t$ cm

and $x_2 = 2\sqrt{7}\sin\left(5t + \frac{\pi}{3}\right)$ cm

where x is displacement and t is time in seconds. The maximum acceleration of the particle is $x \times 10^{-2}$ ms^{-2}. The value of x is : **[NCERT : P L-266 | April 2, 2025 (I)]**

(a) 175 (b) $25\sqrt{7}$ (c) $5\sqrt{7}$ (d) 125

2. Given below are two statements. One is labelled as **Assertion (A)** and the other is labelled as **Reason (R)**.

Assertion (A) : Knowing initial position x_0 and initial momentum p_0 is enough to determine the position and momentum at any time t for a simple harmonic motion with a given angular frequency ω.

Reason (R) : The amplitude and phase can be expressed in terms of x_0 an p_0.

In the light of the above statements, choose the correct answer from the options given below :

[NCERT : P L-266 | Jan. 28, 2025 (II)]

(a) Both (A) and (R) are true but (R) is NOT the correct explanation of (A).
(b) (A) is false but (R) is true.
(c) (A) is true but (R) is false.
(d) Both (A) and (R) are true and (R) is the correct explanation of (A).

3. A particle is executing simple harmonic motion with time period 2s and amplitude 1 cm. If D and d are the total distance and displacement covered by the particle in 12.5 s, then $\frac{D}{d}$ is **[NCERT : P L-266 | Jan. 24, 2025 (I)]**

(a) $\frac{15}{4}$ (b) 25

(c) 10 (d) $\frac{16}{5}$

4. The position, velocity and acceleration of a particle executing simple harmonic motion are found to have magnitudes of 4 m, 2 ms^{-1} and 16 ms^{-2} at a certain instant. The amplitude of the motion is $\sqrt{x}$ m where x is ______.
[NCERT : P L-266 | April, 9, 2024 (II)]

5. A particle is doing simple harmonic motion of amplitude 0.06 m and time period 3.14 s. The maximum velocity of the particle is _______ cm/s.
[NCERT : P L-266 | April 6, 2024 (I)]

6. The displacement of a particle executing SHM is given by $x = 10\sin\left(\omega t + \frac{\pi}{3}\right)$ m. The time period of motion is 3.14 s. The velocity of the particle at t = 0 is ______ m/s.
[NCERT : P L-266 | April 4, 2024 (II)]

7. A particle performs simple harmonic motion with amplitude A. Its speed is increased to three times at an instant when its displacement is $\frac{2A}{3}$. The new amplitude of motion is $\frac{nA}{3}$. The value of n is _____.
[NCERT : P L-266 | Jan. 31, 2024 (I)]

8. A simple harmonic oscillator has an amplitude A and time period 6π second. Assuming the oscillation starts from its mean position, the time required by it to travel from $x = A$ to $x = \frac{\sqrt{3}}{2}A$ will be $\frac{\pi}{x}$ s, where x = _____.
[NCERT : P L-266 | Jan. 29, 2024 (II)]

9. A particle executes simple harmonic motion with an amplitude of 4 cm. At the mean position, velocity of the particle is 10 cm/s. The distance of the particle from the mean position when its speed becomes 5 cm/s is $\sqrt{\alpha}$ cm, where α = ______. **[NCERT : P L-266 | Jan. 27, 2024 (I)]**

10. A particle executes S.H.M. of amplitude A along x-axis. At t = 0, the position of the particle is $x = \frac{A}{2}$ and it moves along positive x-axis the displacement of particle in time t is $x = A\sin(\omega t + \delta)$, then the value δ will be :
[NCERT : P L-266 | April 10, 2023 (I)]
(a) $\frac{\pi}{6}$ (b) $\frac{\pi}{3}$ (c) $\frac{\pi}{4}$ (d) $\frac{\pi}{2}$

11. For particle P revolving round the centre O with radius of circular path r and angular velocity ω, as shown in below figure, the projection of OP on the x-axis at time t is
[NCERT : P L-266 | April 8, 2023 (II)]

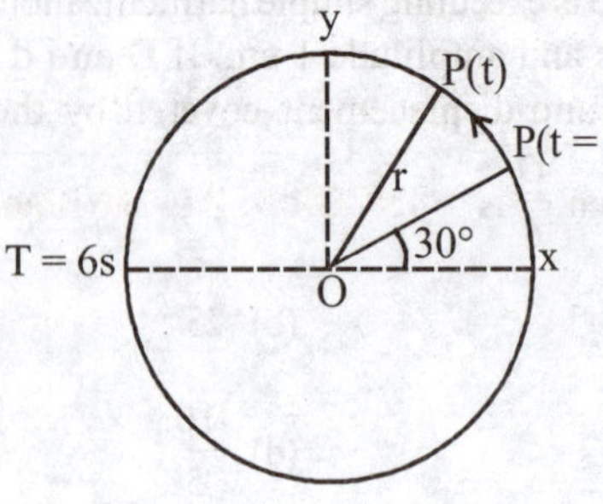

(a) $x(t) = r\cos\left(\omega t + \frac{\pi}{6}\right)$ (b) $x(t) = r\cos(\omega t)$
(c) $x(t) = r\sin\left(\omega t + \frac{\pi}{6}\right)$ (d) $x(t) = r\cos\left(\omega t - \frac{\pi}{6}\omega\right)$

12. The velocity of a particle executing SHM varies with displacement (x) as $4v^2 = 50 - x^2$. The time period of oscillations is $\frac{x}{7}$ s. The value of x is .... $\left(\text{Take } \pi = \frac{22}{7}\right)$
[NCERT : P L-266 | Jan. 30, 2023 (II)]

13. Two simple harmonic waves having equal amplitudes of 8 cm and equal frequency of 10 Hz are moving along the same direction. The resultant amplitude is also 8 cm. The phase difference between the individual waves is _____ degree. **[NCERT : P L-266 | Jan. 29, 2023 (I)]**

14. A particle of mass 250 g executes a simple harmonic motion under a periodic force F = (–25 x) N. The particle attains a maximum speed of 4m/s during its oscillation. The amplitude of the motion is ______cm.
[NCERT : P L-266 | Jan. 29, 2023 (II)]

15. A particle executes simple harmonic motion between x = –A and x = +A. If time taken by particle to go from $x = 0$ to $\frac{A}{2}$ is 2s; then time taken by particle in going from $x = \frac{A}{2}$ to A is: **[NCERT : P L-266 | Jan. 25, 2023 (II)]**
(a) 3 s (b) 2 s (c) 1.5 s (d) 4 s

16. A mass 0.9 kg, attached to a horizontal spring, executes SHM with an amplitude A_1. When this mass passes through its mean position, then a smaller mass of 124 g is placed over it and both masses move together with amplitude A_2. If the ratio $\frac{A_1}{A_2}$ is $\frac{\alpha}{\alpha - 1}$, then the value of α will be _____. **[NCERT : P L-269 | July 27, 2022 (I)]**

17. When a particle executes Simple Harmonic Motion, the nature of graph of velocity as a function of displacement will be : **[NCERT : P L-266, 267 | July 26, 2022 (I)]**
(a) Circular (b) Elliptical
(c) Sinusoidal (d) Straight line

18. A body is performing simple harmonic with an amplitude of 10 cm. The velocity of the body was tripled by air Jet when it is at 5 cm from its mean position. The new amplitude of vibration is $\sqrt{x}$ cm. The value of x is_________.
[NCERT : P L-269 | June 29, 2022 (I)]

19. Motion of a particle in *x-y* plane is described by a set of following equations $x = 4\sin\left(\frac{\pi}{2} - \omega t\right)$ m and $y = 4\sin(\omega t)m$. The path of particle will be –
[NCERT : P L-262 | June 28, 2022 (I)]
(a) circular (b) helical
(c) parabolic (d) elliptical

20. A particle executes simple harmonic motion. Its amplitude is 8 cm and time period is 6s. The time it will take to travel from its position of maximum displacement to the point corresponding to half of its amplitude, is ______ s.
[NCERT : P L-263 | June 27, 2022 (II)]

21. The equation of a particle executing simple harmonic motion is given by $x = \sin \pi \left(t + \frac{1}{3} \right)$ m. At t = 1s, the speed of particle will be (Given : π = 3.14)
[NCERT : P L-262 | June 27, 2022 (II)]
(a) 0 cm s^{-1} (b) 157 cm s^{-1}
(c) 272 cm s^{-1} (d) 314 cm s^{-1}

22. Two simple harmonic motion, are represented by the equations $y_1 = 10 \sin \left(3\pi t + \frac{\pi}{3} \right)$
$y_2 = 5\left(\sin 3\pi t + \sqrt{3} \cos 3\pi t \right)$
Ratio of amplitude of y_1 to y_2 = x : 1. The value of x is ____.
[NCERT : P L-262 | NA, Aug. 27, 2021 (II), 2003 (S); 12 Jan 2019, (II)]

23. A particle executes simple harmonic motion represented by displacement function as
$x(t) = A \sin (\omega t + \phi)$
If the position and velocity of the particle at t = 0 s are 2 cm and 2ω cm s^{-1} respectively, then its amplitude is $x\sqrt{2}$cm where the value of x is ____.
[NCERT : P L-262 | NA, July 27, 2021 (II)]

24. A particle performs simple harmonic motion with a period of 2 second. The time taken by the particle to cover a displacement equal to half of its amplitude from the mean position is $\frac{1}{a}$ s.
The value of 'a' to the nearest integer is ____.
[NCERT : P L-262 | March 18, 2021 (I)]

25. A particle executes S.H.M. with amplitude 'a' and time period 'T'. The displacement of the particle when its speed is half of maximum speed is $\frac{\sqrt{x}\, a}{2}$. The value of x is ____.
[NCERT : P L-266 | Feb. 26, 2021 (II)]

26. The point A moves with a uniform speed along the circumference of a circle of radius 0.36 m and covers 30° in 0.1 s. The perpendicular projection 'P' from 'A' on the diameter MN represents the simple harmonic motion of 'P'. The restoration force per unit mass when P touches M will be **[NCERT : P L-267 | Feb. 25, 2021 (II)]**

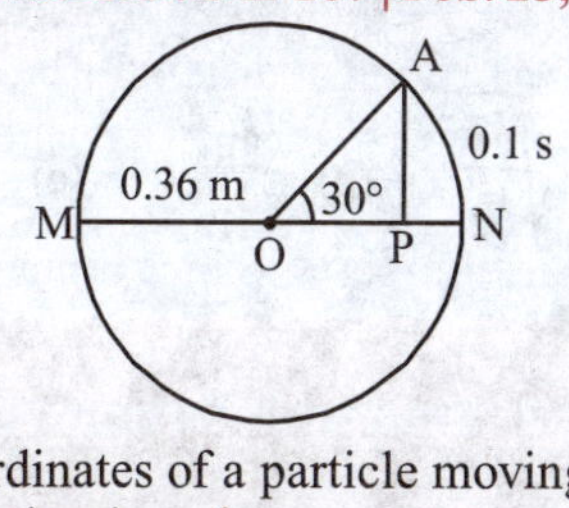

(a) 9.87N
(b) 50N
(c) 100N
(d) 0.49N

27. The position co-ordinates of a particle moving in a 3-D coordinate system is given by
[NCERT : P L-266 | 9 Jan 2019, II]
$x = a \cos\omega t$
$y = a \sin\omega t$
and $z = a\omega t$
The speed of the particle is:
(a) $\sqrt{2}\, a\omega$ (b) $a\omega$ (c) $\sqrt{3}\, a\omega$ (d) $2a\omega$

28. Two simple harmonic motions, as shown, are at right angles. They are combined to form Lissajous figures.
$x(t) = A \sin (at + \delta)$
$y(t) = B \sin (bt)$
Identify the correct match below
[NCERT : P L-265 | Online April 15, 2018]
(a) Parameters: A = B, a = 2b; $\delta = \frac{\pi}{2}$; Curve: Circle
(b) Parameters: A = B, a = b; $\delta = \frac{\pi}{2}$; Curve: Line
(c) Parameters: A $\neq$ B, a = b; $\delta = \frac{\pi}{2}$; Curve: Ellipse
(d) Parameters: A $\neq$ B, a = b; δ = 0; Curve: Parabola

29. The ratio of maximum acceleration to maximum velocity in a simple harmonic motion is 10 s^{-1}. At, t = 0 the displacement is 5 m. What is the maximum acceleration ? The initial phase is $\frac{\pi}{4}$ **[NCERT : P L-266 | Online April 8, 2017]**
(a) 500 m/s^2 (b) 500 $\sqrt{2}$ m/s^2
(c) 750 m/s^2 (d) 750 $\sqrt{2}$ m/s^2

30. A particle performs simple harmonic mition with amplitude A. Its speed is tripled at the instant that it is at a distance $\frac{2A}{3}$ from equilibrium position. The new amplitude of the motion is : **[NCERT : P L-266 | 2016]**
(a) $A\sqrt{3}$ (b) $\frac{7A}{3}$ (c) $\frac{A}{3}\sqrt{41}$ (d) 3A

31. Two particles are performing simple harmonic motion in a straight line about the same equilibrium point. The amplitude and time period for both particles are same and equal to A and T, respectively. At time t = 0 one particle has displacement A while the other one has displacement $\frac{-A}{2}$ and they are moving towards each other. If they cross each other at time t, then t is:
[NCERT : P L-262 | Online April 9, 2016]
(a) $\frac{5T}{6}$ (b) $\frac{T}{3}$ (c) $\frac{T}{4}$ (d) $\frac{T}{6}$

32. x and y displacements of a particle are given as $x(t) = a \sin \omega t$ and $y(t) = a \sin 2\omega t$. Its trajectory will look like :
[NCERT : P L-265 | Online April 10, 2015]

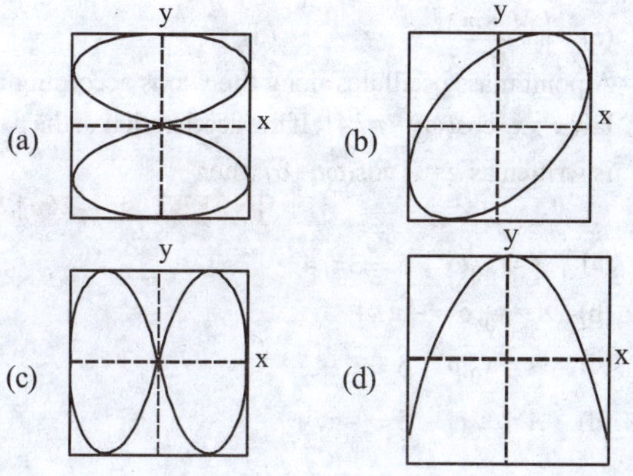

33. Which of the following expressions corresponds to simple harmonic motion along a straight line, where x is the displacement and a, b, c are positive constants? **[NCERT : P L-262 | Online April 12, 2014]**

(a) $a + bx - cx^2$ (b) bx^2
(c) $a - bx + cx^2$ (d) $-bx$

34. The displacement $y(t) = A \sin(\omega t + \phi)$ of a pendulum for $\phi = \frac{2\pi}{3}$ is correctly represented by **[NCERT : P L-267 | Online May 19, 2012]**

(a) 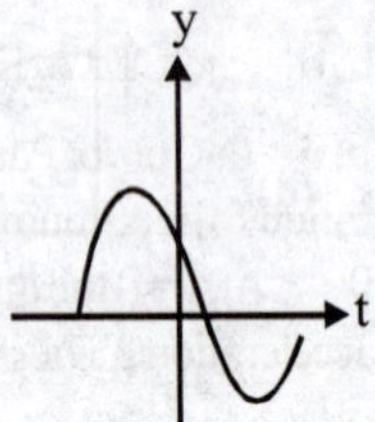

(b) 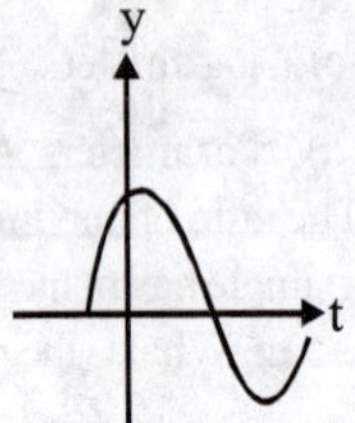

(c) 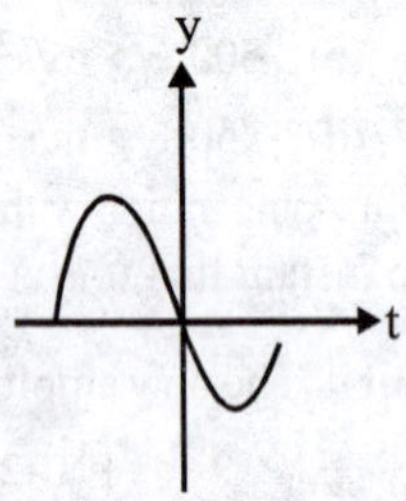

(d) 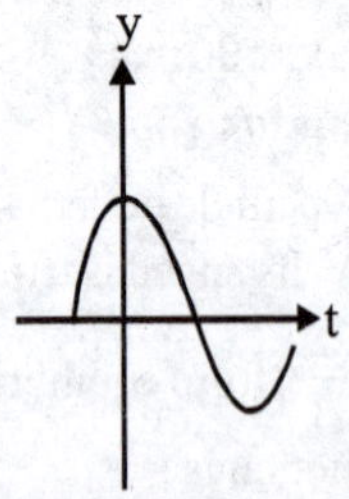

35. Two particles are executing simple harmonic motion of the same amplitude A and frequency ω along the x-axis. Their mean position is separated by distance $X_0 (X_0 > A)$. If the maximum separation between them is $(X_0 + A)$, the phase difference between their motion is: **[NCERT : P L-265 | 2011]**

(a) $\frac{\pi}{3}$ (b) $\frac{\pi}{4}$ (c) $\frac{\pi}{6}$ (d) $\frac{\pi}{2}$

36. A mass M, attached to a horizontal spring, executes S.H.M. with amplitude A_1. When the mass M passes through its mean position then a smaller mass m is placed over it and both of them move together with amplitude A_2. The ratio of $\left(\frac{A_1}{A_2}\right)$ is: **[NCERT : P L-267 | 2011]**

(a) $\frac{M+m}{M}$ (b) $\left(\frac{M}{M+m}\right)^{\frac{1}{2}}$
(c) $\left(\frac{M+m}{M}\right)^{\frac{1}{2}}$ (d) $\frac{M}{M+m}$

37. A point mass oscillates along the x-axis according to the law $x = x_0 \cos(\omega t - \pi/4)$. If the acceleration of the particle is written as $a = A\cos(\omega t + \delta)$, then **[NCERT : P L-266 | 2007]**

(a) $A = x_0\omega^2,\ \delta = 3\pi/4$
(b) $A = x_0,\ \delta = -\pi/4$
(c) $A = x_0\omega^2,\ \delta = \pi/4$
(d) $A = x_0\omega^2,\ \delta = -\pi/4$

38. A coin is placed on a horizontal platform which undergoes vertical simple harmonic motion of angular frequency ω. The amplitude of oscillation is gradually increased. The coin will leave contact with the platform for the first time **[NCERT : P L-267 | 2006]**

(a) at the mean position of the platform
(b) for an amplitude of $\frac{g}{\omega^2}$
(c) for an amplitude of $\frac{g^2}{\omega^2}$
(d) at the highest position of the platform

39. The maximum velocity of a particle, executing simple harmonic motion with an amplitude 7 mm, is 4.4 m/s. The period of oscillation is **[NCERT : P L-267 | 2006]**

(a) 0.01 s (b) 10 s (c) 0.1 s (d) 100 s

40. The function $\sin^2(\omega t)$ represents **[NCERT : P L-262 | 2005]**

(a) a periodic, but not simple harmonic motion with a period $\frac{\pi}{\omega}$
(b) a periodic, but not simple harmonic motion with a period $\frac{2\pi}{\omega}$
(c) a simple harmonic motion with a period $\frac{\pi}{\omega}$
(d) a simple harmonic motion with a period $\frac{2\pi}{\omega}$

41. Two simple harmonic motions are represented by the equations $y_1 = 0.1 \sin\left(100\pi t + \frac{\pi}{3}\right)$ and $y_2 = 0.1\cos \pi t$. The initial phase difference of the velocity of particle 1 with respect to the velocity of particle 2 is **[NCERT : P L-266 | 2005]**

(a) $\frac{\pi}{3}$ (b) $\frac{-\pi}{6}$ (c) $\frac{\pi}{6}$ (d) $\frac{-\pi}{3}$

42. Two particles A and B of equal masses are suspended from two massless springs of spring constants k_1 and k_2, respectively. If the maximum velocities, during oscillation, are equal, the ratio of amplitude of A and B is **[NCERT : P L-267 | 2003]**

(a) $\sqrt{\frac{k_1}{k_2}}$ (b) $\frac{k_2}{k_1}$ (c) $\sqrt{\frac{k_2}{k_1}}$ (d) $\frac{k_1}{k_2}$

Topic 2 - Energy in Simple Harmonic Motion

43. A particle oscillates along the x-axis according to the law, $x(t) = x_0 \sin^2\left(\frac{t}{2}\right)$ where $x_0 = 1$ m. The kinetic energy (K) of the particle as a function of x is correctly represented by the graph. **[NCERT : P L-268 | Jan. 24, 2025 (II)]**

(a) 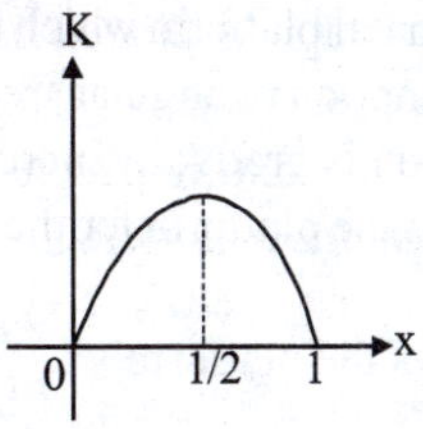
(b) 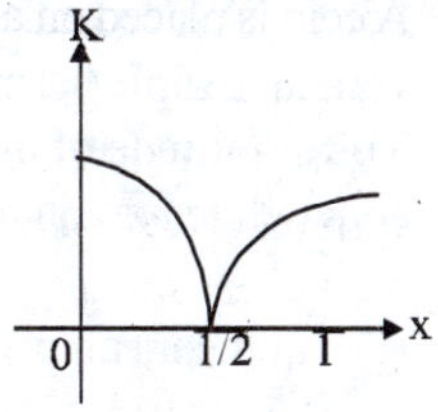
(c) 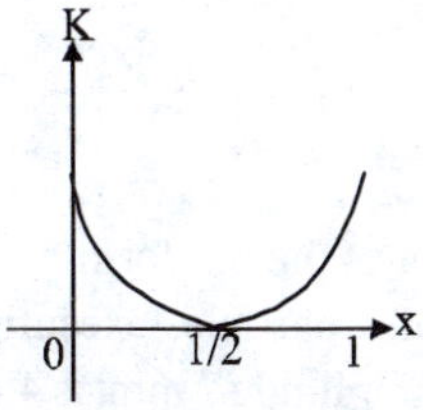
(d) 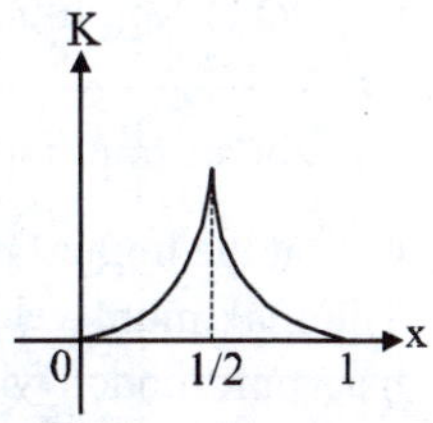

44. An object of mass 0.2 kg executes simple harmonic motion along x axis with frequency of $\left(\frac{25}{\pi}\right)$ Hz . At the position x = 0.04 m the object has kinetic energy 0.5 J and potential energy 0.4 J. The amplitude of oscillation is .... cm. **[NCERT : P L-269 | April 8, 2024 (II)]**

45. In simple harmonic motion, the total mechanical energy of given system is E. If mass of oscillating particle P is doubled then the new energy of the system for same amplitude is : **[NCERT : P L-269 | April 4, 2024 (II)]**

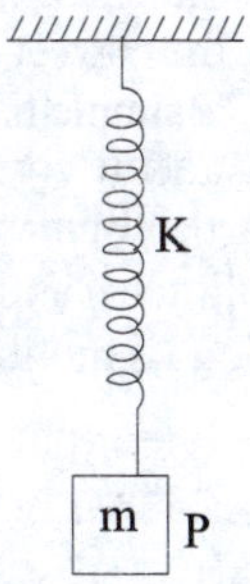

(a) $\frac{E}{\sqrt{2}}$
(b) E
(c) $E\sqrt{2}$
(d) 2E

46. When the displacement of a simple harmonic oscillator is one third of its amplitude, the ratio of total energy to the kinetic energy is $\frac{x}{8}$, where x = ______. **[NCERT : P L-269 | Jan. 29, 2024 (I)]**

47. At a given point of time the value of displacement of a simple harmonic oscillator is given as y = A cos (30°). If amplitude is 40 cm and kinetic energy at that time is 200 J, the value of force constant is $1.0 \times 10^x\ Nm^{-1}$. The value of x is ______. **[NCERT : P L-268 | April 13, 2023 (I)]**

48. A particle executes SHM of amplitude A. The distance from the mean position when it's kinetic energy becomes equal to its potential energy is : **[NCERT : P L-268 | April 13, 2023 (II)]**

(a) $\sqrt{2A}$ (b) 2A (c) $\frac{1}{\sqrt{2}}A$ (d) $\frac{1}{2}A$

49. A particle is executing Simple Harmonic Motion (SHM). The ratio of potential energy and kinetic energy of the particle when its displacement is half of its amplitude will be: **[NCERT : P L-268 | April 12, 2023 (I)]**

(a) 1 : 1 (b) 2 : 1 (c) 1 : 4 (d) 1 : 3

50. The variation of kinetic energy (KE) of a particle executing simple harmonic motion with the displacement (x) starting from mean position to extreme position (A) is given by **[NCERT : P L-268 | April 11, 2023 (I)]**

(a) 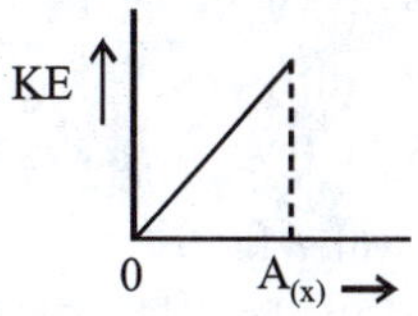
(b) 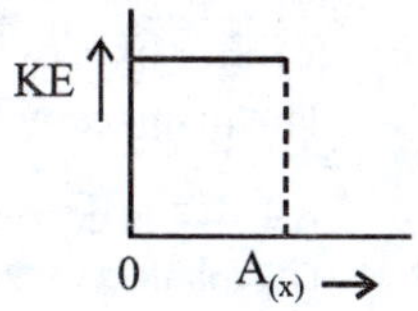
(c) 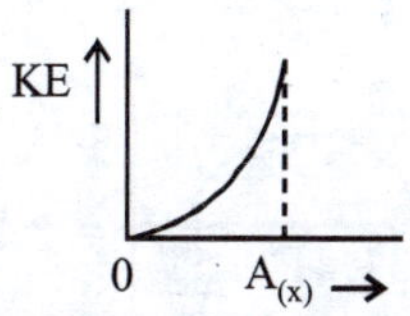
(d) 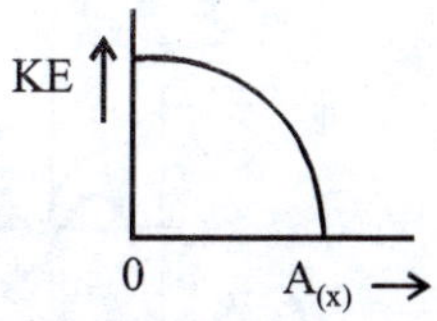

51. The amplitude of a particle executing SHM is 3 cm. The displacement at which its kinetic energy will be 25% more than the potential energy is: ______ cm. **[[NCERT : P L-268 | Feb. 1, 2023 (I)]**

52. The maximum potential energy of a block executing simple harmonic motion is 25 J. A is amplitude of oscillation. At $\frac{A}{2}$, the kinetic energy of the block is **[[NCERT : P L-268 | Jan. 31, 2023 (I)]**

(a) 37.5 J (b) 9.75 J
(c) 18.75 J (d) 12.5 J

53. The general displacement of a simple harmonic oscillator is x = A sin ωt. Let T be its time period. The slope of its potential energy (U) – time (t) curve will be maximum when $t = \frac{T}{\beta}$. The value of β is ______. **[[NCERT : P L-268 | Jan. 30, 2023 (I)]**

54. A particle of mass 1 kg is hanging from a spring of force constant $100\ Nm^{-1}$. The mass is pulled slightly downward and released so that it executes free simple harmonic motion with time period T. The time when the kinetic energy and potential energy of the system will become equal, is $\frac{T}{x}$. The value of x is ______. **[NCERT : P L-268 | March 17, 2021, NA, Aug. 31, 2021 (I), 9 Jan. 2019 (II) (S)]**

55. For a body executing S.H.M. :
(A) Potential energy is always equal to its K.E.
(B) Average potential and kinetic energy over any given time interval are always equal.
(C) Sum of the kinetic and potential energy at any point of time is constant.
(D) Average K.E. in one time period is equal to average potential energy in one time period.

Choose the most appropriate option from the options given below : **[NCERT : P L-268 | Aug. 31, 2021 (II)]**

(a) (C) and (D) (b) only (C)
(c) (B) and (C) (d) only (B)

56. An object of mass 0.5 kg is executing simple harmonic motion. Its amplitude is 5 cm and time period (T) is 0.2 s. What will be the potential energy of the object at an instant $t = \frac{T}{4}$ s starting from mean position. Assume that the initial phase of the oscillation is zero.
[NCERT : P L-268 | July 27, 2021 (II)]
(a) 0.62 J (b) 6.2×10^{-3} J
(c) 1.2×10^{3} J (d) 6.2×10^{3} J

57. The displacement time graph of a particle executing S.H.M. is given in figure : (sketch is schematic and not to scale)

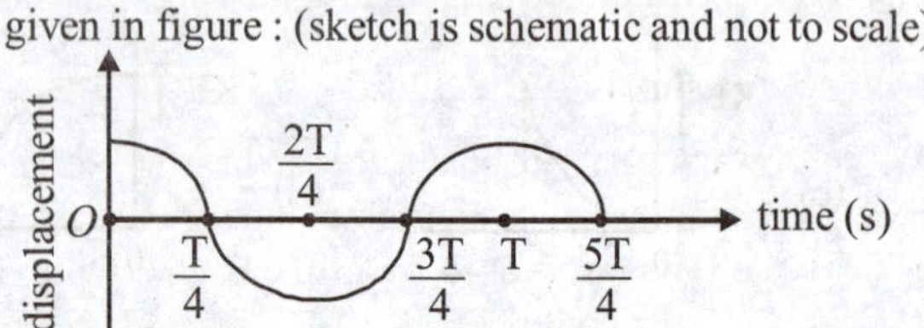

Which of the following statements is/are true for this motion? **[NCERT : P L-266 | Sep. 02, 2020 (II)]**
(1) The force is zero at $t = \frac{3T}{4}$
(2) The acceleration is maximum at $t = T$
(3) The speed is maximum at $t = \frac{T}{4}$
(4) The P.E. is equal to K.E. of the oscillation at $t = \frac{T}{2}$
(a) (1), (2) and (4) (b) (2), (3) and (4)
(c) (1), (2) and (3) (d) (1) and (4)

58. A particle undergoing simple harmonic motion has time dependent displacement given by $x(t) = A\sin\frac{\pi t}{90}$. The ratio of kinetic to potential energy of this particle at t = 210s will be : **[NCERT : P L-268 | 11 Jan 2019, I]**
(a) $\frac{1}{9}$ (b) 1 (c) 2 (d) $\frac{1}{3}$

59. A pendulum is executing simple harmonic motion and its maximum kinetic energy is K_1. If the length of the pendulum is doubled and it performs simple harmonic motion with the same amplitude as in the first case, its maximum kinetic energy is K_2.
[NCERT : P L-268 | 11 Jan 2019, II]
(a) $K_2 = 2K_1$ (b) $K_2 = \frac{K_1}{2}$
(c) $K_2 = \frac{K_1}{4}$ (d) $K_2 = K_1$

60. A particle is executing simple harmonic motion with a time period T. At time t = 0, it is at its position of equilibrium. The kinetic energy-time graph of the particle will look like:
[NCERT : P L-268 | 2017]

(a) 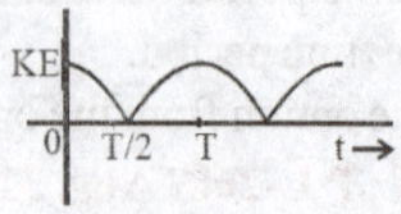

(b) 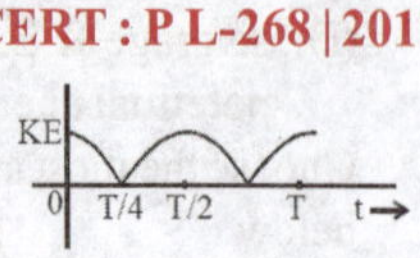

(c) 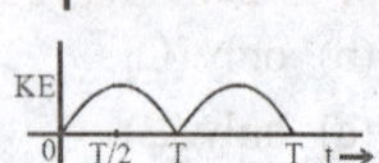

(d) 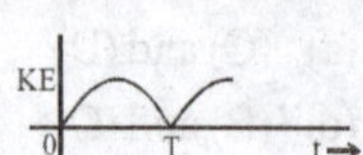

61. For a simple pendulum, a graph is plotted between its kinetic energy (KE) and potential energy (PE) against its displacement d. Which one of the following represents these correctly? (*graphs are schematic and not drawn to scale*)
[NCERT : P L-269 | 2015, April 8, 2023 (I)]

(a) 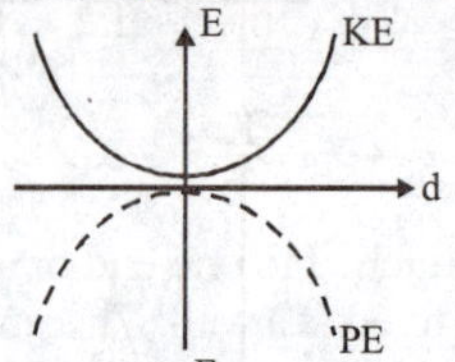

(b) 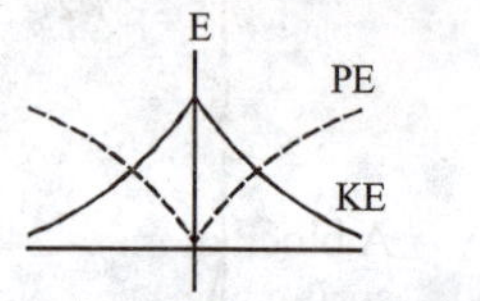

(c) 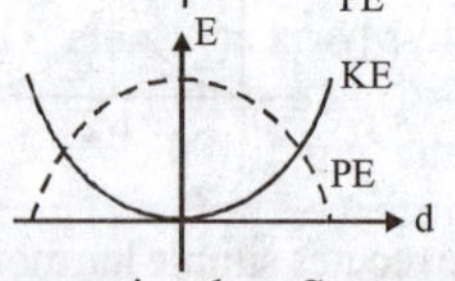

(d) 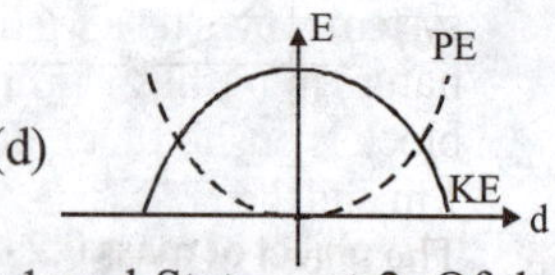

62. This question has Statement 1 and Statement 2. Of the four choices given after the Statements, choose the one that best describes the two Statements.
If two springs S_1 and S_2 of force constants k_1 and k_2 respectively, are stretched by the same force, it is found that more work is done on spring S_1 than on spring S_2.
Statement 1 : If stretched by the same amount work done on S_1, will be more than that on S_2.
Statement 2 : $k_1 < k_2$ **[NCERT : P L-268 | 2012]**
(a) Statement 1 is false, Statement 2 is true.
(b) Statement 1 is true, Statement 2 is false.
(c) Statement 1 is true, Statement 2 is true, Statement 2 is the correct explanation for Statement 1
(d) Statement 1 is true, Statement 2 is true, Statement 2 is not the correct explanation for Statement 1

63. A particle of mass m executes simple harmonic motion with amplitude *a* and frequency ν. The average kinetic energy during its motion from the position of equilibrium to the end is **[NCERT : P L-268 | 2007]**
(a) $2\pi^2 ma^2\nu^2$ (b) $\pi^2 ma^2\nu^2$
(c) $\frac{1}{4}ma^2\nu^2$ (d) $4\pi^2 ma^2\nu^2$

64. Starting from the origin a body oscillates simple harmonically with a period of 2 s. After what time will its kinetic energy be 75% of the total energy?
[NCERT : P L-269 | July 27, 2021 (I), 2006 (S)]
(a) $\frac{1}{6}$s (b) $\frac{1}{4}$s (c) $\frac{1}{3}$s (d) $\frac{1}{12}$s

65. The total energy of a particle, executing simple harmonic motion is **[NCERT : P L-269 | 2004]**
(a) independent of x (b) $\propto x^2$
(c) $\propto x$ (d) $\propto x^{1/2}$
where x is the displacement from the mean position, hence total energy is independent of x.

66. A body executes simple harmonic motion. The potential energy (P.E), the kinetic energy (K.E) and total energy (T.E) are measured as a function of displacement x. Which of the following statements is true ?
[NCERT : P L-268 | 2003]
(a) K.E. is maximum when $x = 0$
(b) T.E is zero when $x = 0$
(c) K.E is maximum when x is maximum
(d) P.E is maximum when $x = 0$

67. In a simple harmonic oscillator, at the mean position **[NCERT : P L-268 | 2002]**

(a) kinetic energy is minimum, potential energy is maximum
(b) both kinetic and potential energies are maximum
(c) kinetic energy is maximum, potential energy is minimum
(d) both kinetic and potential energies are minimum

Topic 3 - Time Period, Frequency, Simple Pendulum and Spring Pendulum

68. A block of mass 2 kg is attached to one end of a massless spring whose other end is fixed at a wall. The spring-mass system moves on a frictionless horizontal table. The spring's natural length is 2 m and spring constant is 200 N/m. The block is pushed such that the length of the spring becomes 1 m and then released. At distance x m (x < 2) from the wall. The speed of the block will be : **[NCERT : P L-267, 268 | April 8, 2025 (II)]**

(a) $10[1-(2-x)]^{\frac{3}{2}}$ m/s (b) $10\left[1-(2-x)^2\right]^{\frac{1}{2}}$ m/s
(c) $10[1-(2-x)^2]$ m/s (d) $10[1-(2-x)^2]^2$ m/s

69. Two simple pendulums having lengths l_1 and l_2 with negligible string mass undergo angular displacements θ_1 and θ_2, from their mean positions, respectively. If the angular accelerations of both pendulums are same, then which expression is correct ? **[NCERT : P L-271 | April 4, 2025 (I)]**

(a) $\theta_1 l_2^2 = \theta_2 l_1^2$ (b) $\theta_1 l_1 = \theta_2 l_2$
(c) $\theta_1 l_2^2 = \theta_2 l_2^2$ (d) $\theta_1 l_2 = \theta_2 l_1$

70.

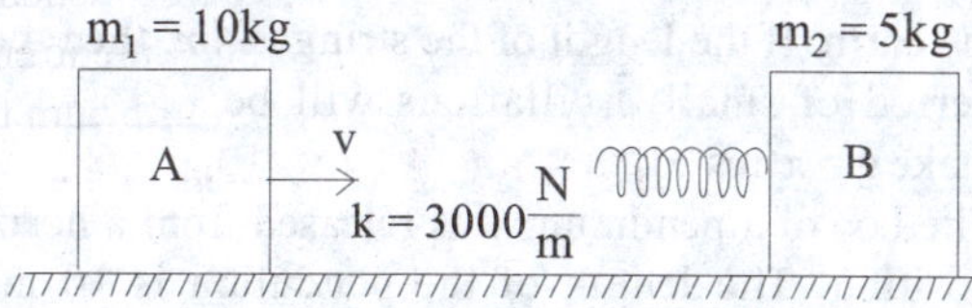

Consider two blocks A and B of masses $m_1 = 10$ kg and $m_2 = 5$ kg that are placed on a frictionless table. The block A moves with a constant speed v = 3 m/s towards the block B kept at rest. A spring with spring constant k = 3000 N/m is attached with the block B as shown in the figure. After the collision, suppose that the blocks A and B, along with the spring in constant compression state, move together, then the compression in the spring is, (Neglect the mass of the spring) **[NCERT : P L-267, 268 | April 3, 2025 (II)]**

(a) 0.2 m (b) 0.4 m (c) 0.1 m (d) 0.3 m

71. Two blocks of masses m and M, (M > m), are placed on a frictionless table as shown in figure. A massless spring with spring constant k is attached with the lower block. If the system is slightly displaced and released then (μ = coefficient of friction between the two blocks) **[NCERT : P L-267, 268 | April 3, 2025 (I)]**

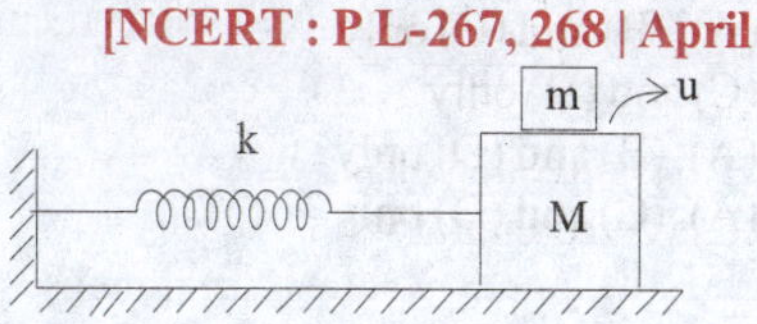

(A) The time period of small oscillation of the two blocks is $T = 2\pi\sqrt{\frac{(m+M)}{k}}$

(B) The acceleration of the blocks is $a = \frac{kx}{M+m}$ (x = displacement of the blocks from the mean position)

(C) The magnitude of the frictional force on the upper block is $\frac{m\mu|x|}{M+m}$

(D) The maximum amplitude of the upper block, if it does not slip, is $\frac{\mu(M+m)g}{k}$

(E) Maximum frictional force can be $\mu(M+m)g$.

Choose the correct answer from the options given below :

(a) A, B, D Only (b) B, C, D Only
(c) C, D, E Only (d) A, B, C Only

72. A piston of mass M is hung from a massless spring whose restoring force law goes as $F = -kx^3$, where k is the spring constant of appropriate dimension. The piston separates the vertical chamber into two parts, where the bottom part is filled with 'n' moles of an ideal gas. An external work is done on the gas isothermally (at a constant temperature T) with the help of a heating filament (with negligible volume) mounted in lower part of the chamber, so that the piston goes up from a height L_0 to L_1, the total energy delivered by the filament is (Assume spring to be in its natural length before heating) **[NCERT : P L-267, 268 | April 3, 2025 (I)]**

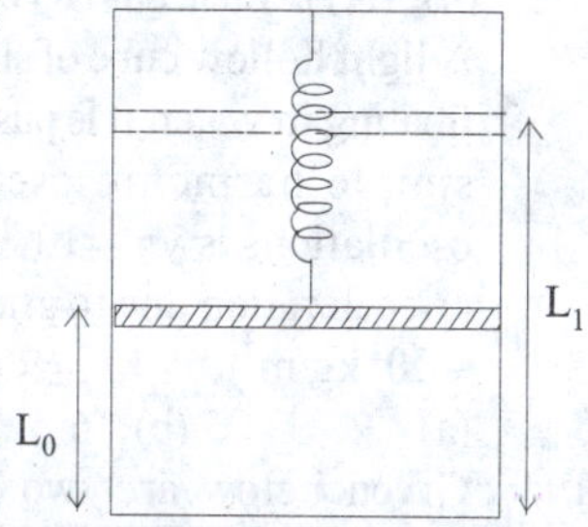

(a) $3nRT\ln\left(\frac{L_1}{L_0}\right) + 2Mg(L_1 - L_0) + \frac{k}{3}(L_1^3 - L_0^3)$

(b) $nRT\ln\left(\frac{L_1^2}{L_0^2}\right) + \frac{Mg}{2}(L_1 - L_0) + \frac{k}{4}(L_1^4 - L_0^4)$

(c) $nRT\ln\left(\frac{L_1}{L_0}\right) + Mg(L_1 - L_0) + \frac{k}{4}(L_1^4 - L_0^4)$

(d) $nRT\ln\left(\frac{L_1}{L_0}\right) + Mg(L_1 - L_0) + \frac{3k}{4}(L_1^4 - L_0^4)$

73. Two bodies *A* and *B* of equal mass are suspended from two massless springs of spring constant k_1 and k_2, respectively. If the bodies oscillate vertically such that their amplitudes are equal, the ratio of the maximum velocity of *A* to the maximum velocity of *B* is **[NCERT : P L-267, 268 | Jan. 29, 2025 (II)]**

(a) $\sqrt{\frac{k_1}{k_2}}$ (b) $\frac{k_1}{k_2}$

(c) $\frac{k_2}{k_1}$ (d) $\sqrt{\frac{k_2}{k_1}}$

74. Given below are two statements : one is labelled as **Assertion (A)** and the other is labelled as **Reason (R)**.
Assertion (A) : Time period of a simple pendulum is longer at the top of a mountain than that at the base of the mountain.
Reason (R) : Time period of a simple pendulum decreases with increasing value of acceleration due to gravity and vice-versa.
In the light of the above statements, choose the most appropriate answer from the options given below:
[NCERT : P L-271 | Jan. 29, 2025 (I)]
(a) Both (A) and (R) are true but (R) is not the correct explanation of (A).
(b) Both (A) and (R) are true and (R) is the correct explanation of (A).
(c) (A) is true but (R) is false.
(d) (A) is false but (R) is true.

75. A light hollow cube of side length 10 cm and mass 10g, is floating in water. It is pushed down and released to execute simple harmonic oscillations. The time period of oscillations is $y\pi \times 10^{-2}$s, where the value of y is
(Acceleration due to gravity, $g = 10$ m/s^2 density of water $= 10^3$ kg/m^3) **[NCERT : P L-267, 268 | Jan. 23, 2025 (I)]**
(a) 2 (b) 6 (c) 4 (d) 1

76. Given below are two statements. One is labelled as **Assertion (A)** and the other is labelled as **Reason (R)**.
Assertion (A): A simple pendulum is taken to a planet of mass and radius, 4 times and 2 times, respectively, than the Earth. The time period of the pendulum remains same on earth and the planet.
Reason (R): The mass of the pendulum remains unchanged at Earth and the other planet. In the light of the above statements, choose the correct answer from the options given below:
[NCERT : P L-271 | Jan. 22, 2025 (II)]
(a) Both (A) and (R) are true but (R) is NOT the correct explanation of (A)
(b) (A) is true but (R) is false
(c) (A) is false but (R) is true
(d) Both (A) and (R) are true and (R) is the correct explanation of (A)

77. A particle of mass 0.50 kg executes simple harmonic motion under force $F = -50(\text{Nm}^{-1})x$. The time period of oscillation is $\frac{x}{35}$ s. The value of x is ....
(Given $\pi = \frac{22}{7}$) **[NCERT : P L-271 | April 9, 2024 (II)]**

78. A simple pendulum doing small oscillations at a place R height above earth surface has time period of $T_1 = 4$ s. T_2 would be it's time period if it is brought to a point which is at a height 2R from earth surface. Choose the correct relation [R = radius of Earth]:
[NCERT : P L-271 | April 5, 2024 (I)]
(a) $T_1 = T_2$ (b) $2T_1 = 3T_2$
(c) $3T_1 = 2T_2$ (d) $2T_1 = T_2$

79. If R is the radius of the earth and the acceleration due to gravity on the surface of earth is $g = \pi^2$m/s^2, then the length of the second's pendulum at a height h = 2R from the surface of earth will be, :
[NCERT : P L-271 | Feb. 1, 2024 (I)]
(a) $\frac{2}{9}$m (b) $\frac{1}{9}$m (c) $\frac{4}{9}$m (d) $\frac{8}{9}$m

80. A mass m is suspended from a spring of negligible mass and the system oscillates with a frequency f_1. The frequency of oscillations if a mass 9 m is suspended from the same spring is f_2. The value of $\frac{f_1}{f_2}$ is ______.
[NCERT : P L-267 | Feb. 1, 2024 (II)]

81. The time period of simple harmonic motion of mass M in the given figure is $\pi\sqrt{\frac{\alpha M}{5K}}$, where the value of α is ______.

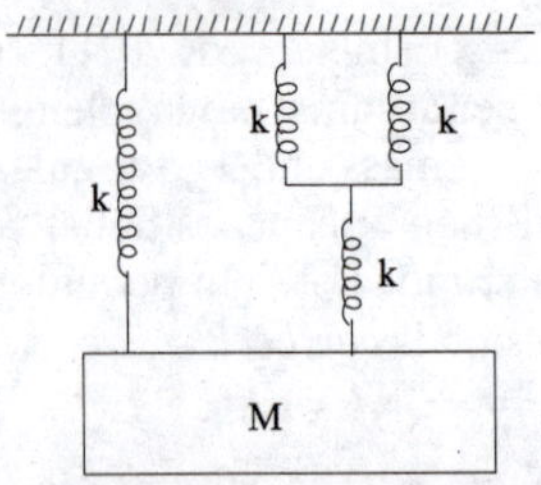

[NCERT : P L-267 | Jan. 31, 2024 (II)]

82. A simple pendulum is placed at a place where its distance from the earth's surface is equal to the radius of the earth. If the lenght of the string is $4m$, then the time period of small oscillations will be ______ s.
[take $g = \pi^2 \text{ms}^{-2}$] **[NCERT : P L-271 | Jan. 30, 2024 (II)]**

83. The bob of a pendulum was released from a horizontal position. The length of the pendulum is 10 m. If it dissipates 10% of its initial energy against air resistance, the speed with which the bob arrives at the lowest point is: **[NCERT : P L-271 | Jan. 29, 2024 (II)]**
[Use, g : 10 ms^{-2}]
(a) $6\sqrt{5}\ ms^{-1}$ (b) $5\sqrt{5}\ ms^{-1}$
(c) $2\sqrt{5}\ ms^{-1}$ (d) $5\sqrt{6}\ ms^{-1}$

84. In a linear simple harmonic motion (SHM)
[NCERT : P L-266 | April 15, 2023 (I)]
(A) Restoring force is directly proportional to the displacement.
(B) The acceleration and displacement are opposite in direction.
(C) The velocity is maximum at mean position.
(D) The acceleration is minimum at extreme points.
Choose the correct answer from the options given below:
(a) (A), (B) and (C) only
(b) (C) and (D) only
(c) (A), (B) and (D) only
(d) (A), (C) and (D) only

85. For a periodic motion represented by the equation $Y = \sin \omega t + \cos \omega t$

[NCERT : P L-262 | April 10, 2023 (II)]

The amplitude of the motion is

(a) 0.5 (b) $\sqrt{2}$ (c) 1 (d) 2

86. A rectangular block of mass 5 kg attached to a horizontal spiral spring executes simple harmonic motion of amplitude 1 m and time period 3.14 s. The maximum force exerted by spring on block is ______ N.

[NCERT : P L-267 | April 10, 2023 (II)]

87. A mass m is attached to two springs as shown in figure. The spring constants of two springs are K_1 and K_2. For the frictionless surface, the time period of oscillation of mass m is **[NCERT : P L-268 | April 6, 2023 (I)]**

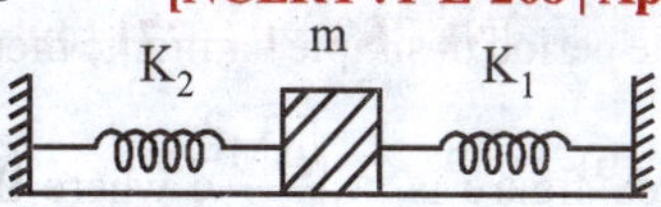

(a) $\frac{1}{2\pi}\sqrt{\frac{K_1+K_2}{m}}$ (b) $\frac{1}{2\pi}\sqrt{\frac{K_1-K_2}{m}}$

(c) $2\pi\sqrt{\frac{m}{K_1+K_2}}$ (d) $2\pi\sqrt{\frac{m}{K_1-K_2}}$

88. A simple pendulum with length 100 cm and bob of mass 250 g is executing S.H.M. of amplitude 10cm. The maximum tension in the string is found to be $\frac{x}{40}$ N. The value of x is______. **[NCERT : P L-271 | April 6, 2023 (II)]**

89. Choose the correct length (L) versus square of time period (T^2) graph for a simple pendulum executing simple harmonic motion. **[NCERT : P L-271 | Feb. 1, 2023 (II); Similar online April 19, 2014]**

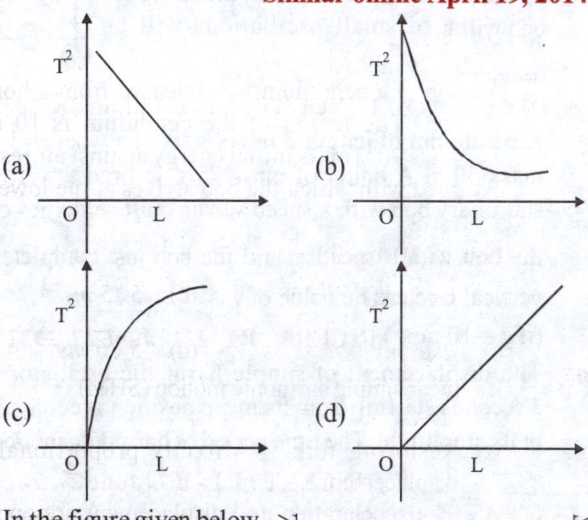

90. In the figure given below. a block of mass M = 490 g placed on a frictionless table is connected with two springs having same spring constant ($K = 2 \text{ N m}^{-1}$). If the block is horizontally displaced through 'X'm then the number of complete oscillations it will make in 14π seconds will be ______.

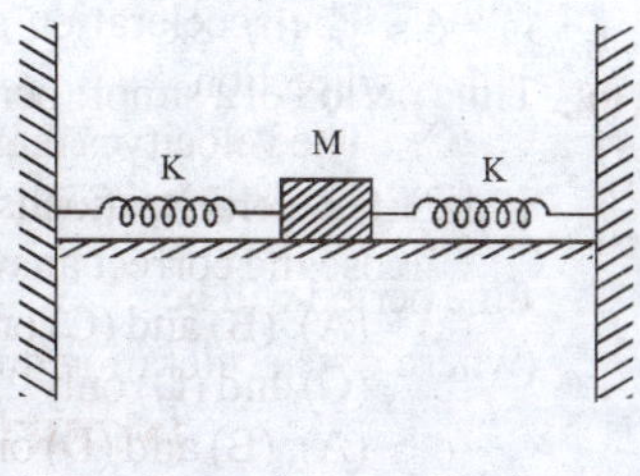

[NCERT : P L-267 | Jan. 31, 2023 (I)]

91. For a simple harmonic motion in a mass spring system shown, the surface is frictionless. When the mass of the block is 1 kg, the angular frequency is ω_1. When the mass block is 2 kg the angular frequency is ω_2. The ratio ω_2/ω_1 is :

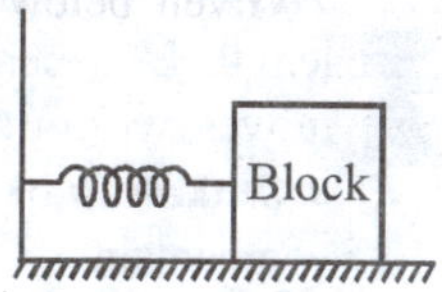

[NCERT : P L-267 | Jan. 30, 2023 (II)]

(a) $\sqrt{2}$ (b) $\frac{1}{\sqrt{2}}$ (c) 2 (d) $\frac{1}{2}$

92. T is the time period of simple pendulum on the earth's surface. Its time period becomes x T when taken to a height R (equal to earth's radius) above the earth's surface. Then, the value of x will be:

[NCERT : P L-271 | Jan. 25, 2023 (I)]

(a) 4 (b) 2 (c) $\frac{1}{2}$ (d) $\frac{1}{4}$

93. Assume that the earth is a solid sphere of uniform density and a tunnel is dug along its diameter throughout the earth. It is found that when a particle is released in this tunnel, it executes a simple harmonic motion. The mass of the particle is 100 g. The time period of the motion of the particle will be (approximately) (take $g = 10 \text{ ms}^{-2}$, radius of earth = 6400 km) **[NCERT : P L-267 | Jan. 25, 2023 (I)]**

(a) 24 hours (b) 1 hour 24 minutes

(c) 1 hour 40 minutes (d) 12 hours

94. A block of mass 2 kg is attached with two identical springs of spring constant 20 N/m each. The block is placed on a frictionless surface and the ends of the springs are attached to rigid supports (see figure). When the mass is displaced from its equilibrium position, it executes a simple harmonic motion. The time period of oscillation is $\frac{\pi}{\sqrt{x}}$ in SI unit. The value of x is ______

[NCERT : P L-267 | Jan. 24, 2023 (I)]

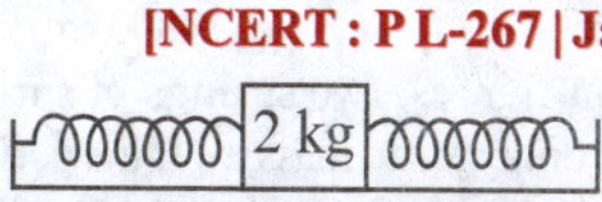

95. Given below are two statements: one is labelled as Assertion A and the other is labelled as Reason R.

[NCERT : P L-271 | Jan. 24, 2023 (II)]

Assertion A: A pendulum clock when taken to Mount Everest becomes fast.

Reason R: The value of g (acceleration due to gravity) is less at Mount Everest than its value on the surface of earth.

In the light of the above statements, choose the most appropriate answer from the options given below:

(a) Both A and R are correct but R is NOT the correct explanation of A.

(b) Both A and R are correct and R is the correct explanation of A.

(c) A is not correct but R is correct.

(d) A is correct but R is not correct.

96. A mass m attached to free end of a spring executes SHM with a period of 1s. If the mass is increased by 3 kg the period of oscillation increases by one second, the value of mass m is ______ kg. **[Jan. 24, 2023 (II)]**

97. The time period of oscillation of a simple pendulum of length L suspended from the roof of a vehicle, which moves without friction down an inclined plane of inclination α, is given by:
[NCERT : P L-271 | July 29, 2022 (I)]

(a) $2\pi\sqrt{\frac{L}{(g\cos\alpha)}}$ (b) $2\pi\sqrt{\frac{L}{(g\sin\alpha)}}$

(c) $2\pi\sqrt{\frac{L}{g}}$ (d) $2\pi\sqrt{\frac{L}{(g\tan\alpha)}}$

98. The metallic bob of simple pendulum has the relative density 5. The time period of this pendulum is 10 s. If the metallic bob is immersed in water, then the new time period becomes $5\sqrt{x}$ s. The value of x will be ________.
[NCERT : P L-271 | July 29, 2022 (II)]

99. The potential energy of a particle of mass 4 kg in motion along the *x*-axis is given by U = 4(1 – cos 4x) J. The time period of the particle for small oscillation $(\sin\theta \simeq \theta)$ is $\left(\frac{\pi}{K}\right)$ *s*. The value of K is _____.
[NCERT : P L-267 | July 28, 2022 (II)]

100. Assume there are two identical simple pendulum clocks. Clock - 1 is placed on the earth and Clock - 2 is placed on a space station located at a height h above the earth surface. Clock – 1 and Clock - 2 operate at time periods 4 s and 6 s respectively. Then the value of h is -
(consider radius of earth R_E = 6400 km and g on earth 10 m/s^2)
[NCERT : P L-271 | July 28, 2022 (II)]

(a) 1200 km (b) 1600 km
(c) 3200 km (d) 4800 km

101. As per given figures, two springs of spring constants K and 2K are connected to mass m. If the period of oscillation in figure (a) is 3s, then the period of oscillation in figure (b) will be $\sqrt{x}$ s. The value of x is________.

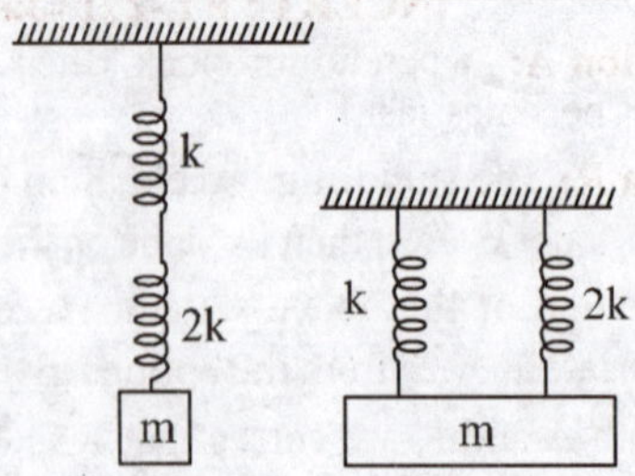

figure (a) figure (b)

[NCERT : P L-267 | July 26, 2022 (II)]

102.

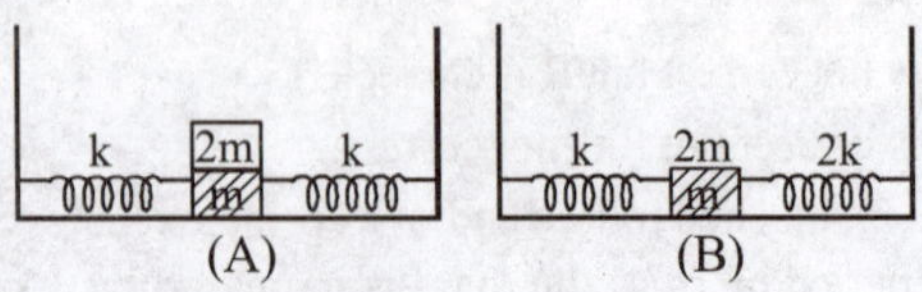

(A) (B)

In figure (A), mass '2 m' is fixed on mass 'm' which is attached to two springs of spring constant k. In figure (B), mass 'm' is attached to two springs of spring constant 'k' and '2k'. If mass 'm' in (A) and in (B) are displaced by distance 'x' horizontally and then released, then time period T_1 and T_2 corresponding to (A) and (B) respectively follow the relation. **[NCERT : P L-267 | July 25, 2022 (I)]**

(a) $\frac{T_1}{T_2} = \frac{3}{\sqrt{2}}$ (b) $\frac{T_1}{T_2} = \sqrt{\frac{3}{2}}$

(c) $\frac{T_1}{T_2} = \sqrt{\frac{2}{3}}$ (d) $\frac{T_1}{T_2} = \frac{\sqrt{2}}{3}$

103. The length of a seconds pendulum at a height h = 2R from earth surface will be:
(Given : R = Radius of earth and acceleration due to gravity at the surface of earth g = π²m/s⁻²)
[NCERT : P L-271 | July 25, 2022 (II)]

(a) $\frac{2}{9}$m (b) $\frac{4}{9}$m (c) $\frac{8}{9}$m (d) $\frac{1}{9}$m

104. The motion of a simple pendulum excuting S.H.M. is represented by following equation.
$Y = A\sin(\pi t + \phi)$ where time is measured in second.
The length of pendulum is : **[June 29, 2022 (II)]**
(a) 97.23 cm (b) 25.3 cm (c) 99.4 cm (d) 406.1 cm

105. A pendulum is suspended by a string of length 250 cm. The mass of the bob of the pendulum is 200 g. The bob is pulled aside until the string is at 60° with vertical as shown in the figure. After releasing the bob the maximum velocity attained by the bob will be ________ ms⁻¹.

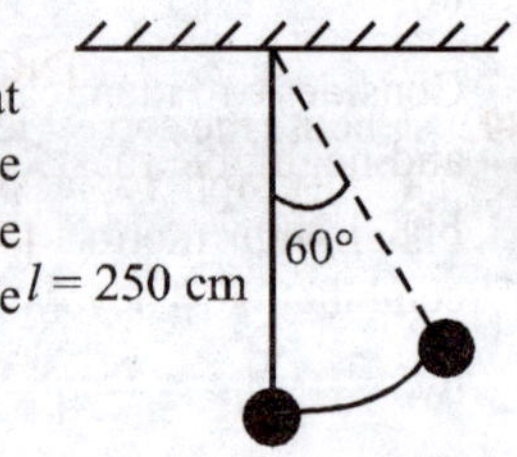

(if g = 10 m/s²) **[NCERT : P L-271 | June 28, 2022 (I)]**

106. A pendulum of length 2 m consists of a wooden bob of mass 50 g. A bullet of mass 75 g is fired towards the stationary bob with a speed v. The bullet emerges out of the bob with a speed $\frac{v}{3}$ and the bob just completes the vertical circle. The value of v is ________ ms⁻¹.
(if g = 10 m/s²) **[NCERT : P L-271 | June 27, 2022 (I)]**

107. The displacement of simple harmonic oscillator after 3 seconds starting from its mean position is equal to half of its amplitude. The time period of harmonic motion is :
[NCERT : P L-267 | June 27, 2022 (I)]
(a) 6 s (b) 8 s (c) 12s (d) 36 s

108. Time period of a simple pendulum in a stationary lift is '*T*'. If the lift accelerates with $\frac{g}{6}$ vertically upwards then the time period will be :
(where *g* = acceleration due to gravity)
[NCERT : P L-271 | June 26, 2022 (I)]

(a) $\sqrt{\frac{6}{5}}T$ (b) $\sqrt{\frac{5}{6}}T$ (c) $\sqrt{\frac{6}{7}}T$ (d) $\sqrt{\frac{7}{6}}T$

109. Two massless springs with spring constants $2k$ and $9k$, carry 50 g and 100 g masses at their free ends. These two masses oscillate vertically such that their maximum velocities are equal. Then, the ratio of their respective amplitudes will be :

[NCERT : P L-267 | June 24, 2022 (II)]

(a) 1 : 2 (b) 3 : 2 (c) 3 : 1 (d) 2 : 3

110. A bob of mass 'm' suspended by a thread of length l undergoes simple harmonic oscillations with time period T. If the bob is immersed in a liquid that has density $\frac{1}{4}$ times that of the bob and the length of the thread is increased by 1/3rd of the original length, then the time period of the simple harmonic oscillations will be :-

[NCERT : P L-271 | Aug. 31, 2021 (II)]

(a) T (b) $\frac{3}{2}$T (c) $\frac{3}{4}$T (d) $\frac{4}{3}$T

111. T_0 is the time period of a simple pendulum at a place. If the length of the pendulum is reduced to $\frac{1}{16}$ times of its initial value, the modified time period is :

[NCERT : P L-271 | July 22, 2021 (II)]

(a) T_0 (b) $8\pi T_0$ (c) $4T_0$ (d) $\frac{1}{4}T_0$

112. Consider two identical springs each of spring constant k and negligible mass compared to the mass M as shown Fig. 1 shows one of them and Fig. 2 shows their series combination. The ratios of time period of oscillation of the two SHM is $T_b / T_a = \sqrt{x}$, where value of x is _____. (Round off to the Nearest integer).

[NCERT : P L-267 | March 17, 2021 (I)]

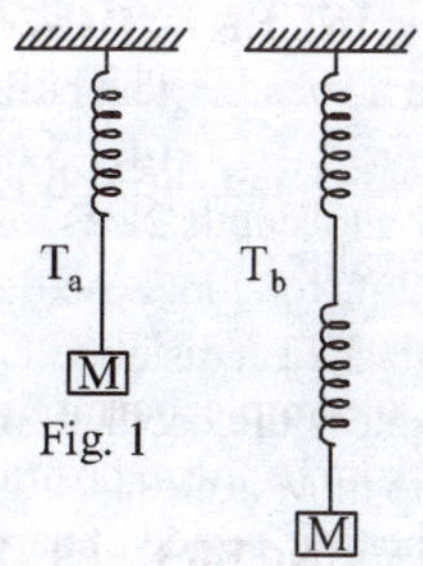

Fig. 1 Fig. 2

113. Time period of a simple pendulum is T inside a lift when the lift is stationary. If the lift moves upwards with an acceleration g/2, the time period of pendulum will be :

[NCERT : P L-271 | March 16, 2021 (I)]

(a) $\sqrt{\frac{2}{3}}T$ (b) $\sqrt{3}T$

(c) $\sqrt{\frac{3}{2}}T$ (d) $\frac{T}{\sqrt{3}}$

114. Assume that a tunnel is dug along a chord of the earth, at a perpendicular distance (R/2) from the earth's centre, where 'R' is the radius of the Earth. The wall of the tunnel is frictionless. If a particle is released in this tunnel, it will execute a simple harmonic motion with a time period :

[NCERT : P L-271 | Feb. 26, 2021 (I)]

(a) $\frac{g}{2\pi R}$ (b) $\frac{2\pi R}{g}$

(c) $2\pi\sqrt{\frac{R}{g}}$ (d) $\frac{1}{2\pi}\sqrt{\frac{g}{R}}$

115. If two similar springs each of spring constant K_1 are joined in series, the new spring constant and time period would be changed by a factor :

[NCERT : P L-267 | Feb. 26, 2021 (I)]

(a) $\frac{1}{4}, 2\sqrt{2}$ (b) $\frac{1}{2}, 2\sqrt{2}$ (c) $\frac{1}{2}, \sqrt{2}$ (d) $\frac{1}{4}, \sqrt{2}$

116. Time period of a simple pendulum is T. The time taken to complete $\frac{5}{8}$ oscillations starting from mean position is $\frac{\alpha}{\beta}$ T. The value of α is_____. **[Feb. 26, 2021 (II)]**

117. In the given figure, a body of mass M is held between two massless springs, on a smooth inclined plane. The free ends of the springs are attached to firm supports. If each spring has spring constant k, the frequency of oscillation of given body is : **[NCERT : P L-267 | Feb. 24, 2021 (II)]**

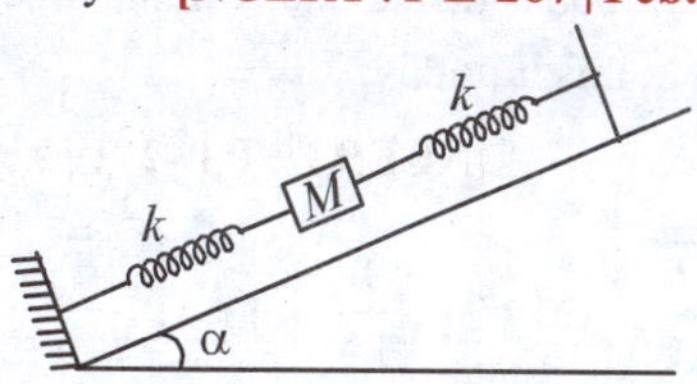

(a) $\frac{1}{2\pi}\sqrt{\frac{k}{2M}}$ (b) $\frac{1}{2\pi}\sqrt{\frac{2k}{Mg\sin\alpha}}$

(c) $\frac{1}{2\pi}\sqrt{\frac{2k}{M}}$ (d) $\frac{1}{2\pi}\sqrt{\frac{k}{Mg\sin\alpha}}$

118. An object of mass m is suspended at the end of a massless wire of length L and area of cross-section, A. Young modulus of the material of the wire is Y. If the mass is pulled down slightly its frequency of oscillation along the vertical direction is:

[NCERT : P L-271 | Sep. 06, 2020 (I)]

(a) $f = \frac{1}{2\pi}\sqrt{\frac{mL}{YA}}$ (b) $f = \frac{1}{2\pi}\sqrt{\frac{YA}{mL}}$

(c) $f = \frac{1}{2\pi}\sqrt{\frac{mA}{YL}}$ (d) $f = \frac{1}{2\pi}\sqrt{\frac{YL}{mA}}$

119. When a particle of mass m is attached to a vertical spring of spring constant k and released, its motion is described by $y(t) = y_0 \sin^2\omega t$, where '$y$' is measured from the lower end of unstretched spring. Then ω is :

[NCERT : P L-267 | Sep. 06, 2020 (II)]

(a) $\frac{1}{2}\sqrt{\frac{g}{y_0}}$ (b) $\sqrt{\frac{g}{y_0}}$

(c) $\sqrt{\frac{g}{2y_0}}$ (d) $\sqrt{\frac{2g}{y_0}}$

120. A block of mass m attached to a massless spring is performing oscillatory motion of amplitude 'A' on a frictionless horizontal plane. If half of the mass of the block breaks off when it is passing through its equilibrium point, the amplitude of oscillation for the remaining system become fA. The value of f is :

[NCERT : P L-268 | Sep. 03, 2020 (II)]

(a) $\frac{1}{\sqrt{2}}$ (b) 1 (c) $\frac{1}{2}$ (d) $\sqrt{2}$

121. A simple pendulum oscillating in air has period T. The bob of the pendulum is completely immersed in a non-viscous liquid. The density of the liquid is $\frac{1}{16}$ th of the material of the bob. If the bob is inside liquid all the time, its period of oscillation in this liquid is :

[NCERT : P L-271 | 9 April 2019 I]

(a) $2T\sqrt{\frac{1}{10}}$ (b) $2T\sqrt{\frac{1}{14}}$ (c) $4T\sqrt{\frac{1}{15}}$ (d) $4T\sqrt{\frac{1}{14}}$

122. Two light identical springs of spring constant k are attached horizontally at the two ends of a uniform horizontal rod AB of length l and mass m. The rod is pivoted at its centre 'O' and can rotate frreely in horizontal plane. The other ends of two springs are fixed to rigid supports as shown in figure. The rod is gently pushed through a small angle and released. The frequency of resulting oscillation is: **[NCERT : P L-267 | 12 Jan 2019, I]**

(a) $\frac{1}{2\pi}\sqrt{\frac{3k}{m}}$

(b) $\frac{1}{2\pi}\sqrt{\frac{2k}{m}}$

(c) $\frac{1}{2\pi}\sqrt{\frac{6k}{m}}$

(d) $\frac{1}{2\pi}\sqrt{\frac{k}{m}}$

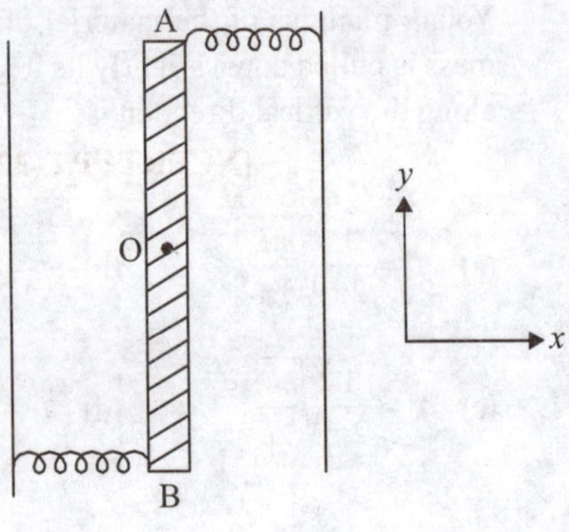

123. A simple pendulum of length 1 m is oscillating with an angular frequency 10 rad/s. The support of the pendulum starts oscillating up and down with a small angular frequency of 1 rad/s and an amplitude of 10^{-2} m. The relative change in the angular frequency of the pendulum is best given by : **[NCERT : P L-271 | 11 Jan 2019, II]**

(a) 10^{-3} rad/s (b) 1 rad/s
(c) 10^{-1} rad/s (d) 10^{-5} rad/s

124. The mass and the diameter of a planet are three times the respective values for the Earth. The period of oscillation of a simple pendulum on the Earth is 2s. The period of oscillation of the same pendulum on the planet would be:

[NCERT : P L-271 | 11 Jan 2019, II]

(a) $\frac{\sqrt{3}}{2}$s (b) $\frac{2}{\sqrt{3}}$s

(c) $\frac{3}{2}$s (d) $2\sqrt{3}$s

125. A particle executes simple harmonic motion with an amplitude of 5 cm. When the particle is at 4 cm from the mean position, the magnitude of its velocity in SI units is equal to that of its acceleration. Then, its periodic time in seconds is: **[NCERT : P L-271 | 10 Jan 2019, II]**

(a) $\frac{4\pi}{3}$ (b) $\frac{3}{8}\pi$

(c) $\frac{8\pi}{3}$ (d) $\frac{7}{3}\pi$

126. A cylindrical plastic bottle of negligible mass is filled with 310 ml of water and left floating in a pond with still water. If pressed downward slightly and released, it starts performing simple harmonic motion at angular frequency ω. If the radius of the bottle is 2.5 cm then ω is close to: (density of water = 10^3 kg/m^3)

[NCERT : P L-267 | 10 Jan 2019, II]

(a) 3.75 rad s^{-1} (b) 1.25 rad s^{-1}
(c) 2.50 rad s^{-1} (d) 5.00 rad s^{-1}

127. A rod of mass 'M' and length '2L' is suspended at its middle by a wire. It exhibits torsional oscillations; If two masses each of 'm' are attached at distance 'L/2' from its centre on both sides, it reduces the oscillation frequency by 20%. The value of ratio m/M is close to :

[NCERT : P L-267 | 9 Jan 2019, II]

(a) 0.77 (b) 0.57
(c) 0.37 (d) 0.17

128. A silver atom in a solid oscillates in simple harmonic motion in some direction with a frequency of 10^{12}/sec. What is the force constant of the bonds connecting one atom with the other? (Mole wt. of silver = 108 and Avagadro number = 6.02×10^{23} gm mole^{-1}) **[NCERT : P L-267 | 2018]**

(a) 6.4 N/m (b) 7.1 N/m
(c) 2.2 N/m (d) 5.5 N/m

129. A particle executes simple harmonic motion and is located at x = a, b and c at times t_0, $2t_0$ and $3t_0$ respectively. The frequency of the oscillation is **[NCERT : P L-267 | Online April 16, 2018]**

(a) $\frac{1}{2\pi t_0}\cos^{-1}\left(\frac{a+b}{2c}\right)$ (b) $\frac{1}{2\pi t_0}\cos^{-1}\left(\frac{a+b}{3c}\right)$

(c) $\frac{1}{2\pi t_0}\cos^{-1}\left(\frac{2a+3c}{b}\right)$ (d) $\frac{1}{2\pi t_0}\cos^{-1}\left(\frac{a+c}{2b}\right)$

130. In an experiment to determine the period of a simple pendulum of length 1 m, it is attached to different spherical bobs of radii r_1 and r_2. The two spherical bobs have uniform mass distribution. If the relative difference in the periods, is found to be 5×10^{-4} s, the difference in radii, $|r_1 - r_2|$ is best given by: **[NCERT : P L-271 | Online April 9, 2017]**

(a) 1 cm (b) 0.1 cm
(c) 0.5 cm (d) 0.01 cm

131. A pendulum clock loses 12 s a day if the temperature is 40°C and gains 4 s a day if the temperature is 20° C. The temperature at which the clock will show correct time, and the co-efficient of linear expansion (α) of the metal of the pendulum shaft are respectively : **[NCERT : P L-271 | 2016]**

(a) 30°C; $\alpha = 1.85 \times 10^{-3}$/°C
(b) 55°C; $\alpha = 1.85 \times 10^{-2}$/°C
(c) 25°C; $\alpha = 1.85 \times 10^{-5}$/°C
(d) 60°C; $\alpha = 1.85 \times 10^{-4}$/°C

132. In an engine the piston undergoes vertical simple harmonic motion with amplitude 7 cm. A washer rests on top of the piston and moves with it. The motor speed is slowly increased. The frequency of the piston at which the washer no longer stays in contact with the piston, is close to : **[NCERT : P L-267 | Online April 10, 2016]**

(a) 0.7 Hz (b) 1.9 Hz
(c) 1.2 Hz (d) 0.1 Hz

133. A pendulum made of a uniform wire of cross sectional area A has time period T. When an additional mass M is added to its bob, the time period changes to T_M. If the Young's modulus of the material of the wire is Y then $\frac{1}{Y}$ is equal to: **[NCERT : P L-271 | 2015]**

(*g = gravitational acceleration*)

(a) $\left[1-\left(\frac{T_M}{T}\right)^2\right]\frac{A}{Mg}$ (b) $\left[1-\left(\frac{T}{T_M}\right)^2\right]\frac{A}{Mg}$

(c) $\left[\left(\frac{T_M}{T}\right)^2-1\right]\frac{A}{Mg}$ (d) $\left[\left(\frac{T_M}{T}\right)^2-1\right]\frac{Mg}{A}$

134. A particle moves with simple harmonic motion in a straight line. In first τ s, after starting from rest it travels a distance a, and in next τ s it travels 2a, in same direction, then: **[NCERT : P L-267 | 2014]**

(a) amplitude of motion is 3a
(b) time period of oscillations is 8τ
(c) amplitude of motion is 4a
(d) time period of oscillations is 6τ

135. An ideal gas enclosed in a vertical cylindrical container supports a freely moving piston of mass M. The piston and the cylinder have equal cross sectional area A. When the piston is in equilibrium, the volume of the gas is V_0 and its pressure is P_0. The piston is slightly displaced from the equilibrium position and released. Assuming that the system is completely isolated from its surrounding, the piston executes a simple harmonic motion with frequency **[NCERT : P L-267 | 2013]**

(a) $\frac{1}{2\pi}\frac{A\gamma P_0}{V_0 M}$ (b) $\frac{1}{2\pi}\frac{V_0 M P_0}{A^2\gamma}$

(c) $\frac{1}{2\pi}\sqrt{\frac{A^2\gamma P_0}{MV_0}}$ (d) $\frac{1}{2\pi}\sqrt{\frac{MV_0}{A\gamma P_0}}$

136. Two simple pendulums of length 1 m and 4 m respectively are both given small displacement in the same direction at the same instant. They will be again in phase after the shorter pendulum has completed number of oscillations equal to : **[NCERT : P L-271 | Online April 9, 2013]**

(a) 2 (b) 7 (c) 5 (d) 3

137. A wooden cube (density of wood 'd') of side 'ℓ' floats in a liquid of density 'ρ' with its upper and lower surfaces horizontal. If the cube is pushed slightly down and released, it performs simple harmonic motion of period 'T' **[NCERT : P L-267 | 2011 RS]**

(a) $2\pi\sqrt{\frac{\ell d}{\rho g}}$ (b) $2\pi\sqrt{\frac{\ell\rho}{dg}}$

(c) $2\pi\sqrt{\frac{\ell d}{(\rho-d)g}}$ (d) $2\pi\sqrt{\frac{\ell\rho}{(\rho-d)g}}$

138. If x, v and a denote the displacement, the velocity and the acceleration of a particle executing simple harmonic motion of time period T, then, which of the following does not change with time? **[NCERT : P L-266 | 2009]**

(a) aT/x (b) $aT+2\pi v$
(c) aT/v (d) $a^2T^2+4\pi^2v^2$

139. Two springs, of force constants k_1 and k_2 are connected to a mass m as shown. The frequency of oscillation of the mass is f. If both k_1 and k_2 are made four times their original values, the frequency of oscillation becomes **[NCERT : P L-267 | 2007]**

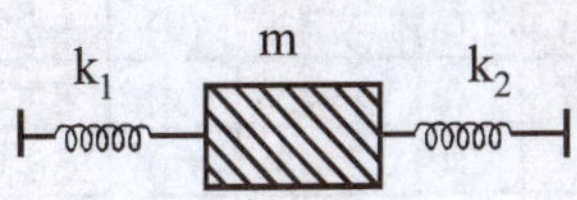

(a) $2f$ (b) $f/2$ (c) $f/4$ (d) $4f$

140. The displacement of an object attached to a spring and executing simple harmonic motion is given by $x = 2 \times 10^{-2} \cos \pi t$ metre. The time at which the maximum speed first occurs is **[NCERT : P L-266 | 2007]**

(a) 0.25 s (b) 0.5 s (c) 0.75 s (d) 0.125 s

141. The bob of a simple pendulum is a spherical hollow ball filled with water. A plugged hole near the bottom of the oscillating bob gets suddenly unplugged. During observation, till water is coming out, the time period of oscillation would **[NCERT : P L-271 | 2005]**

(a) first decrease and then increase to the original value
(b) first increase and then decrease to the original value
(c) increase towards a saturation value
(d) remain unchanged

142. If a simple harmonic motion is represented by $\frac{d^2x}{dt^2} + \alpha x = 0$, its time period is **[2005]**

(a) $\frac{2\pi}{\sqrt{\alpha}}$ (b) $\frac{2\pi}{\alpha}$ (c) $2\pi\sqrt{\alpha}$ (d) $2\pi\alpha$

143. The bob of a simple pendulum executes simple harmonic motion in water with a period t, while the period of oscillation of the bob is t_0 in air. Neglecting frictional force of water and given that the density of the bob is $(4/3) \times 1000 \text{ kg/m}^3$. Which relationship between t and t_0 is true? **[NCERT : P L-271 | 2004]**

(a) $t = 2t_0$ (b) $t = t_0/2$
(c) $t = t_0$ (d) $t = 4t_0$

144. A particle at the end of a spring executes S.H.M with a period t_1. while the corresponding period for another spring is t_2. If the period of oscillation with the two springs in series is T then **[NCERT : P L-267 | 2004]**

(a) $T^{-1} = t_1^{-1} + t_2^{-1}$ (b) $T^2 = t_1^2 + t_2^2$
(c) $T = t_1 + t_2$ (d) $T^{-2} = t_1^{-2} + t_2^{-2}$

145. A mass M is suspended from a spring of negligible mass. The spring is pulled a little and then released so that the mass executes SHM of time period T. If the mass is increased by m, the time period becomes $\frac{5T}{3}$. Then the ratio of $\frac{m}{M}$ is **[NCERT : P L-267 | 2003]**

(a) $\frac{3}{5}$ (b) $\frac{25}{9}$ (c) $\frac{16}{9}$ (d) $\frac{5}{3}$

146. The length of a simple pendulum executing simple harmonic motion is increased by 21%. The percentage increase in the time period of the pendulum of increased length is **[NCERT : P L-271 | 2003]**

(a) 11% (b) 21%
(c) 42% (d) 10%

147. If a spring has time period T, and is cut into n equal parts, then the time period of each part will be **[NCERT : P L-267 | 2002]**

(a) $T\sqrt{n}$ (b) $T/\sqrt{n}$
(c) nT (d) T

148. A child swinging on a swing in sitting position, stands up, then the time period of the swing will **[NCERT : P L-271 | 2002]**

(a) increase
(b) decrease
(c) remains same
(d) increases if the child is long and decreases if the child is short

ANSWER KEYS

1	(a)	16	(16)	31	(d)	46	(9)	61	(d)	76	(a)	91	(b)	106	(10)	121	(c)	136	(a)
2	(d)	17	(b)	32	(c)	47	(4)	62	(b)	77	(22)	92	(b)	107	(d)	122	(c)	137	(a)
3	(b)	18	(700)	33	(d)	48	(c)	63	(b)	78	(c)	93	(b)	108	(c)	123	(a)	138	(a)
4	(17)	19	(a)	34	(a)	49	(d)	64	(a)	79	(b)	94	(5)	109	(b)	124	(d)	139	(a)
5	(12)	20	(1)	35	(a)	50	(d)	65	(a)	80	(3)	95	(c)	110	(d)	125	(c)	140	(b)
6	(10)	21	(b)	36	(c)	51	(2)	66	(a)	81	(12)	96	(1)	111	(d)	126	(None)	141	(b)
7	(7)	22	(1)	37	(a)	52	(c)	67	(c)	82	(8)	97	(a)	112	(2)	127	(c)	142	(a)
8	(2)	23	(2)	38	(b)	53	(8)	68	(b)	83	(a)	98	(5)	113	(a)	128	(b)	143	(a)
9	(12)	24	(6)	39	(a)	54	(8)	69	(d)	84	(a)	99	(2)	114	(c)	129	(d)	144	(b)
10	(a)	25	(03.00)	40	(a)	55	(a)	70	(c)	85	(b)	100	(c)	115	(c)	130	(b)	145	(c)
11	(a)	26	(a)	41	(b)	56	(a)	71	(a)	86	(20)	101	(2)	116	(7)	131	(c)	146	(d)
12	(88)	27	(a)	42	(c)	57	(c)	72	(c)	87	(c)	102	(a)	117	(c)	132	(b)	147	(b)
13	(120)	28	(c)	43	(a)	58	(d)	73	(a)	88	(99)	103	(d)	118	(b)	133	(c)	148	(b)
14	(40)	29	(b)	44	(6)	59	(a)	74	(b)	89	(d)	104	(c)	119	(c)	134	(d)		
15	(d)	30	(b)	45	(b)	60	(b)	75	(a)	90	(20)	105	(5)	120	(d)	135	(c)		

14 Waves

TREND ANALYSIS
JEE MAIN (2025-2021)

	2025				2024				2023				2022				2021			
	Session II		Session I		Session II		Session I		Session II		Session I		Session II		Session I		Session II		Session I	
	MCQs	NVQs	MCQs	NVQs	MCQs	NVQs	MCQs	NVQs	MCQs	NVQs	MCQs	NVQs	MCQs	NVQs	MCQs	NVQs	MCQs	NVQs	MCQs	NVQs
Topic 1	3	–	2	–	1	–	1	–	1	2	3	1	2	–	2	–	–	2	1	–
Topic 2	2	–	1	–	–	3	1	1	–	6	–	–	–	1	–	1	–	2	–	1
Topic 3	3	–	–	–	–	–	–	2	–	–	–	1	–	1	2	2	–	–	1	–
Topic 4																				
Total	8	-	3	-	1	3	2	3	1	8	3	2	2	2	4	3	-	4	2	1

How to Code Qns for Better & Faster Revision

Type 1: Able to solve (No code Req)

Type 2: Solved with struggle or by seeing solution **(Use code NR means Needs Revision or orange stickers)**

Type 3: Tricky/ Could not solve on my own. Need help from Teacher/ Friend. **(Use code TQ means Tricky Question or red stickers)**

You can either hand write the code in boxes or else paste coloured stickers given at end of book.

Topic 1 - Basic of Mechanical Waves, Progressive and Stationary Waves

1. The equation of a wave travelling on a string is $y = \sin[20\pi x + 10\pi t]$, where x and t are distance and time in SI units. The minimum distance between two points having the same oscillating speed is :
[NCERT : P L-283 | April 7, 2025 (II)]
(a) 5.0 cm (b) 20 cm (c) 10 cm (d) 2.5 cm

2. Displacement of a wave is expressed as $x(t) = 5\cos\left(628t + \frac{\pi}{2}\right)$ m. The wavelength of the wave when its velocity is 300 m/s is :
[NCERT : P L-284 | April 4, 2025 (II)]
(a) 5m (b) 3m (c) 0.5m (d) 0.33m

3. A sinusoidal wave of wavelength 7.5 cm travels a distance of 1.2 cm along the x-direction in 0.3 sec. The crest P is at x = 0 at t = 0 sec and maximum displacement of the wave is 2 cm. Which equation correctly represents this wave ?
[NCERT : P L-283| April 2, 2025 (II)]
(a) $y = 2\cos(0.83x - 3.35t)$ cm
(b) $y = 2\sin(0.83x - 3.5t)$ cm
(c) $y = 2\cos(3.35x - 0.83t)$ cm
(d) $y = 2\cos(0.13x - 0.5t)$ cm

4. Given below are two statements: one is labelled as **Assertion (A)** and the other is labelled as **Reason (R)**
Assertion (A): A sound wave has higher speed in solids than gases.
Reason (R): Gases have higher value of Bulk modulus than solids. **[NCERT : P L-286 | Jan. 28, 2025 (I)]**
In the light of the above statements, choose the correct answer from the options given below
(a) Both A and R are true and R is the correct explanation of A
(b) A is false but R is true
(c) Both A and R are true but R is NOT the correct explanation of A
(d) A is true but R is false.

5. The equation of a transverse wave travelling along a string is $y(x, t) = 4.0 \sin [20 \times 10^{-3} x + 600t]$ mm, where x is in the mm and t is in second. The velocity of the wave is **[NCERT : P L-284 | Jan. 23, 2025 (II)]**

(a) +30 m/s (b) – 60 m/s
(c) –30 m/s (d) + 60 m/s

6. A plane progressive wave is given by $y = 2 \cos 2\pi(330t - x)$ m. The frequency of the wave is : **[NCERT : P L-283 | April 8, 2024 (II)]**

(a) 165 Hz (b) 330 Hz (c) 660 Hz (d) 340 Hz

7. The speed of sound in oxygen at S.T.P. will be approximately: **[NCERT : P L-287 | Jan. 31, 2024 (II)]**
(Given, $R = 8.3\ JK^{-1}, \gamma = 1.4$)

(a) 315 m/s (b) 333 m/s (c) 341 m/s (d) 325 m/s

8. The equation of wave is given by
$Y = 10^{-2} \sin 2\pi\left(160t - 0.5x + \frac{\pi}{4}\right)$
Where x and Y are in m and t in s. The speed of the wave is ____ $km\ h^{-1}$. **[NCERT : P L-284 | April 11, 2023 (I)]**

9. A transverse harmonic wave on a string is given by $y(x, t) = 5\sin(6t + 0.003x)$
where x and y are in cm and t in sec. The wave velocity is ______ ms^{-1}. **[NCERT : P L-284 | April 10, 2023 (I)]**

10. The ratio of speed of sound in hydrogen gas to the speed of sound in oxygen gas at the same temperature is: **[NCERT : P L-287 | April 6, 2023 (II)]**

(a) 4 : 1 (b) 1 : 2 (c) 1 : 4 (d) 1 : 1

11. A steel wire with mass per unit length $7.0 \times 10^{-3}\ kg\ m^{-1}$ is under tension of 70N. The speed of transverse waves in the wire will be: **[NCERT : P L-285 | Feb. 1, 2023 (I)]**

(a) 100 m/s (b) 50 m/s (c) 10 m/s (d) 200 π m/s

12. For a solid rod, the Young's modulus of elasticity is $3.2 \times 10^{11}\ Nm^{-2}$ and density is $8 \times 10^3\ kg\ m^{-3}$. The velocity of longitudinal wave in the rod will be. **[NCERT : P L-286 | Jan. 31, 2023 (II)]**

(a) $145.75 \times 10^3\ ms^{-1}$ (b) $3.65 \times 10^3\ ms^{-1}$
(c) $18.96 \times 10^3\ ms^{-1}$ (d) $6.32 \times 10^3\ ms^{-1}$

13. The distance between two consecutive points with phase difference of 60° in a wave of frequency 500 Hz is 6.0 m. The velocity with which wave is traveling, is ______ km/s **[NCERT : P L-284 | Jan. 25, 2023 (I)]**

14. A travelling wave is described by the equation
$y(x, t) = [0.05 \sin (8x - 4t)]\ m$
The velocity of the wave is : [all the quantities are in SI unit] **[NCERT : P L-284 | Jan. 24, 2023 (I)]**

(a) $4\ ms^{-1}$ (b) $2\ ms^{-1}$ (c) $0.5\ ms^{-1}$ (d) $8\ ms^{-1}$

15. In the wave equation $y = 0.5\sin\frac{2\pi}{\lambda}(400t - x)$ m
the velocity of the wave will be : **[NCERT : P L-284 | July 28, 2022 (I)]**

(a) 200 m/s (b) $200\sqrt{2}$ m/s
(c) 400 m/s (d) $400\sqrt{2}$ m/s

16. A transverse wave is represented by $y = 2\sin(\omega t - kx)$ cm. The value of wavelength (in cm) for which the wave velocity becomes equal to the maximum particle velocity, will be ; **[NCERT : P L-284 | July 26, 2022 (II)]**

(a) 4π (b) 2π (c) π (d) 2

17. Which of the following equations correctly represents a travelling wave having wavelength λ = 4.0 cm, frequency ν = 100 Hz and travelling in positive x-axis direction? **[NCERT : P L-283 | June 30, 2022 (I)]**

(a) $y = A \sin[(0.50\pi cm^{-1})x - (100\pi s^{-1})t]$
(b) $y = A \sin 2\pi[(0.25\ cm^{-1})x - (50\ s^{-1})t]$
(c) $y = A\sin\left[\left(\frac{2\pi}{4}cm^{-1}\right)x - \left(\frac{2\pi}{100}s^{-1}\right)t\right]$
(d) $y = A \sin\pi[(0.5cm^{-1})x - (200\ s^{-1})t]$

18. A longitudinal wave is represented by $x = 10 \sin 2\pi\left(nt - \frac{x}{\lambda}\right)$. The maximum particle velocity will be four times the wave velocity if the determined value of wavelength is equal to : **[NCERT : P L-284 | June 29, 2022 (I)]**

(a) 2π (b) 5π (c) π (d) $\frac{\pi}{2}$

19. Two waves are simultaneously passing through a string and their equations are :
$y_1 = A_1 \sin k(x - vt)$, $y_2 = A_2 \sin k(x - vt + x_0)$. Given amplitudes $A_1 = 12$ mm and $A_2 = 5$ mm, $x_0 = 3.5$ cm and wave number k = 6.28 cm^{-1}. The amplitude of resulting wave will be ______ mm. **[NCERT : P L-283 | NA, Aug. 26, 2021 (II)]**

20. The amplitude of wave disturbance propagating in the positive x-direction is given by $y = \frac{1}{(1+x)^2}$ at time t = 0 and $y = \frac{1}{1+(x-2)^2}$ at t = 1s, where x and y are in metres. The shape of wave does not change during the propagation. The velocity of the wave will be ____ m/s. **[NCERT : P L-284 | NA, July 20, 2021 (I)]**

21. A sound wave of frequency 245 Hz travels with the speed of 300 ms^{-1} along the positive x-axis. Each point of the wave moves to and fro through a total distance of 6 cm. What will be the mathematical expression of this travelling wave? **[NCERT : P L-283 | March 17, 2021 (II)]**

(a) $Y(x, t) = 0.06 [\sin 0.8x - (0.5 \times 10^3)t]$
(b) $Y(x, t) = 0.03 [\sin 5.1x - (1.5 \times 10^3)t]$
(c) $Y(x, t) = 0.03 [\sin 5.1x - (0.2 \times 10^3)t]$
(d) $Y(x, t) = 0.06 [\sin 5.1x - (1.5 \times 10^3)t]$

22. Assume that the displacement (*s*) of air is proportional to the pressure difference (Δ*p*) created by a sound wave. Displacement (*s*) further depends on the speed of sound (*v*), density of air (ρ) and the frequency (*f*). If Δ*p* ~ 10Pa, *v* ~ 300 m/s, *ρ* ~ 1 kg/m³ and *f* ~ 1000 Hz, then *s* will be of the order of (take the multiplicative constant to be 1) **[NCERT : P L-287 | Sep. 05, 2020 (I)]**

(a) $\frac{3}{100}$ mm (b) 10mm (c) $\frac{1}{10}$ mm (d) 1mm

23. For a transverse wave travelling along a straight line, the distance between two peaks (crests) is 5 m, while the distance between one crest and one trough is 1.5 m. The possible wavelengths (in m) of the waves are : **[NCERT : P L-283 | Sep. 04, 2020 (I)]**

(a) 1,3,5,..... (b) $\frac{1}{1},\frac{1}{3},\frac{1}{5},.......$

(c) 1,2,3,..... (d) $\frac{1}{2},\frac{1}{4},\frac{1}{6},.......$

24. A progressive wave travelling along the positive x-direction is represented by $y(x,t) = A\sin(kx - \omega t + \phi)$. Its snapshot at $t = 0$ is given in the figure.

[NCERT : P L-283 | 12 April 2019 I]

For this wave, the phase ϕ is :

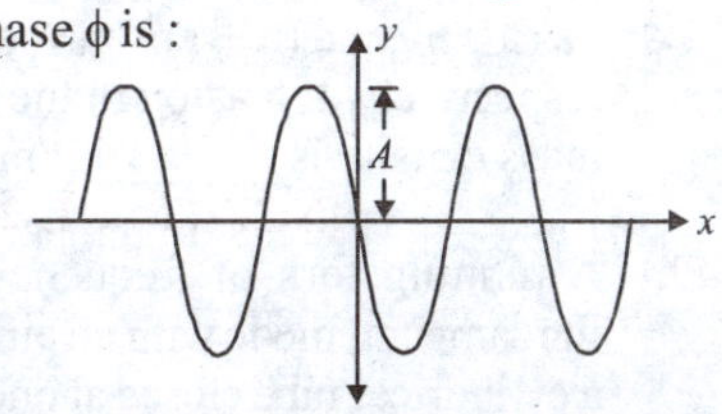

(a) $-\frac{\pi}{2}$ (b) π

(c) 0 (d) $\frac{\pi}{2}$

25. A small speaker delivers 2 W of audio output. At what distance from the speaker will one detect 120 dB intensity sound ? [Given reference intensity of sound as 10^{-12}W/m²]

[12 April 2019 II]

(a) 40 cm (b) 20 cm (c) 10 cm (d) 30 cm

26. The pressure wave, P = 0.01 sin[1000t – 3x] Nm^{-2}, corresponds to the sound produced by a vibrating blade on a day when atmospheric temperature is 0°C. On some other day when temperature is T, the speed of sound produced by the same blade and at the same frequency is found to be 336 ms^{-1}. Approximate value of T is :

[NCERT : P L-287 | 9 April 2019 I]

(a) 4°C (b) 11°C (c) 12°C (d) 15°C

27. A travelling harmonic wave is represented by the equation $y(x, t) = 10^{-3}\sin(50t + 2x)$, where x and y are in meter and t is in seconds. Which of the following is a correct statement about the wave? **[NCERT : P L-284 | 12 Jan. 2019 I]**

(a) The wave is propagating along the negative x-axis with speed 25 ms^{-1}.

(b) The wave is propagating along the positive x-axis with speed 100 ms^{-1}.

(c) The wave is propagating along the positive x-axis with speed 25 ms^{-1}.

(d) The wave is propagating along the negative x-axis with speed 100 ms^{-1}.

28. A transverse wave is represented by

$$y = \frac{10}{\pi}\sin\left(\frac{2\pi}{T}t - \frac{2\pi}{\lambda}x\right)\text{cm}$$

For what value of the wavelength the wave velocity is twice the maximum particle velocity?

[NCERT : P L-284 | Online April 9, 2014]

(a) 40 cm (b) 20 cm (c) 10 cm (d) 60 cm

29. In a transverse wave the distance between a crest and neighbouring trough at the same instant is 4.0 cm and the distance between a crest and trough at the same place is 1.0 cm. The next crest appears at the same place after a time interval of 0.4s. The maximum speed of the vibrating particles in the medium is :

[NCERT : P L-284 | Online April 25, 2013]

(a) $\frac{3\pi}{2}$ cm/s (b) $\frac{5\pi}{2}$ cm/s (c) $\frac{\pi}{2}$ cm/s (d) 2π cm/s

30. The disturbance $y(x, t)$ of a wave propagating in the positive x-direction is given by $y = \frac{1}{1+x^2}$ at time $t = 0$ and by $y = \frac{1}{\left[1+(x-1)^2\right]}$ at $t = 2$ s, where x and y are in meters. The shape of the wave disturbance does not change during the propagation. The velocity of wave in m/s is

[NCERT : P L-282, 283 | Online May 26, 2012]

(a) 2.0 (b) 4.0 (c) 0.5 (d) 1.0

31. The transverse displacement $y(x, t)$ of a wave is given by

$$y(x,t) = e^{-\left(ax^2+bt^2+2\sqrt{ab}\,xt\right)}.$$

This represents a: **[NCERT : P L-283, 284 | 2011]**

(a) wave moving in $-x$ direction with speed $\sqrt{\frac{b}{a}}$

(b) standing wave of frequency $\sqrt{b}$

(c) standing wave of frequency $\frac{1}{\sqrt{b}}$

(d) wave moving in $+x$ direction speed $\sqrt{\frac{a}{b}}$

32. A wave travelling along the x-axis is described by the equation y(x, t) = 0.005 cos (α x – βt). If the wavelength and the time period of the wave are 0.08 m and 2.0s, respectively, then α and β in appropriate units are

[NCERT : P L-283 | 2008]

(a) $\alpha = 25.00\pi, \beta = \pi$ (b) $\alpha = \frac{0.08}{\pi}, \beta = \frac{2.0}{\pi}$

(c) $\alpha = \frac{0.04}{\pi}, \beta = \frac{1.0}{\pi}$ (d) $\alpha = 12.50\pi, \beta = \frac{\pi}{2.0}$

33. The displacement y of a particle in a medium can be expressed as, $y = 10^{-6}\sin\left(100t + 20x + \frac{\pi}{4}\right)m$ where t is in second and x in meter. The speed of the wave is

[NCERT : P L-284 | 2003, 2004]

(a) 20 m/s (b) 5 m/s (c) 2000 m/s (d) 5π m/s

Topic 2 - Vibration of String and Organ Pipe

34. Two strings with circular cross section and made of same material, are stretched to have same amount of tension. A transverse wave is then made to pass through both the strings. The velocity of the wave in the first string having the radius of cross section R is v_1, and that in the other string having radius of cross section $\frac{R}{2}$ is v_2. Then $\frac{v_2}{v_1} =$

[NCERT : P L-291, 292 | April 8, 2025 (II)]

(a) $\sqrt{2}$ (b) 2 (c) 8 (d) 4

35. In an experiment with a closed organ pipe, it is filled with water by $\left(\frac{1}{5}\right)$th of its volume. The frequency of the fundamental note will change by

[NCERT : P L-292, 293 | April 4, 2025 (I)]

(a) 25% (b) 20% (c) –20% (d) –25%

36. A closed organ and an open organ tube filled by two different gases having same bulk modulus but different densities ρ_1 and ρ_2 respectively. The frequency of 9th harmonic of closed tube is identical with 4th harmonic of open tube. If the length of the closed tube is 10 cm and the density ratio of the gases is $\rho_1 : \rho_2 = 1 : 16$, then the length of the open tube is **[NCERT : P L-292, 293 | Jan. 22, 2025 (I)]**

(a) $\frac{20}{7}$ cm (b) $\frac{15}{7}$ cm (c) $\frac{20}{9}$ cm (d) $\frac{15}{9}$ cm

37. A closed and an open organ pipe have same lengths. If the ratio of frequencies of their seventh overtones is $\left(\frac{a-1}{a}\right)$ then the value of a is ______. **[NCERT : P L-292, 293 | April 8, 2024 (I)]**

38. Two open organ pipes of length 60 cm and 90 cm resonate at 6th and 5th harmonics respectively. The difference of frequencies for the given modes is ______ Hz. (Velocity of sound in air = 333 m/s) **[NCERT : P L-292, 293 | April 6, 2024 (II)]**

39. A sonometer wire of resonating length 90 cm has a fundamental frequency of 400 Hz when kept under some tension. The resonating length of the wire with fundamental frequency of 600 Hz under same tension ______ cm. **[NCERT : P L-290, 291 | April 5, 2024 (II)]**

40. The fundamental frequency of a closed organ pipe is equal to the first overtone frequency of an open organ pipe. If length of the open pipe is 60 cm, the length of the closed pipe will be : **[NCERT : P L-292, 293 | Jan. 31, 2024 (I)]**

(a) 30 cm (b) 45 cm (c) 60 cm (d) 15 cm

41. In a closed organ pipe, the frequency of fundamental note is 30 Hz. A certain amount of water is now poured in the organ pipe so that the fundamental frequency is increased to 110 Hz. If the organ pipe has a cross - sectional area of 2 cm^2, the amount of water poured in the organ tube is ________ g. (Take speed of sound in air is 330 m/s) **[NCERT : P L-292, 293 | Jan. 30, 2024 (I)]**

42. The fundamental frequency of vibration of a string stretched between two rigid support is 50 Hz. The mass of the string is 18 g and its linear mass density is 20 g/m. The speed of the transverse waves so produced in the string is ______ ms^{-1}. **[NCERT : P L-290, 291 | April 15, 2023 (I)]**

43. In an experiment with sonometer when a mass of 180 g is attached to the string, it vibrates with fundamental frequency of 30 Hz. When a mass m is attached, the string vibrates with fundamental frequency of 50 Hz. The value of m is ___g. **[NCERT : P L-290, 291 | April 13, 2023 (II)]**

44. For a certain organ pipe, the first three resonance frequencies are in the ratio of 1 : 3 : 5 respectively. If the frequency of fifth harmonic is 405 Hz and the speed of sound in air is 324 ms^{-1} the length of the organ pipe is ________ m. **[NCERT : P L-292, 293 | April 12, 2023 (I)]**

45. A wire of density 8×10^3 kg/m^3 is stretched between two clamps 0.5 m apart. The extension developed in the wire is 3.2×10^{-4} m. If $Y = 8 \times 10^{10}$ N/m^2, the fundamental frequency of vibration in the wire will be ______ Hz. **[NCERT : P L-290, 291 | April 11, 2023 (II)]**

46. An organ pipe 40cm long is open at both ends. The speed of sound in air is 360ms^{-1}. The frequency of the second harmonic is ____Hz. **[NCERT : P L-292, 293 | April 8, 2023 (I)]**

47. A guitar string of length 90 cm vibrates with a fundamental frequency of 120 Hz. The length of the string producing a fundamental frequency of 180 Hz will be ________ cm. **[NCERT : P L-290, 291 | April 8, 2023 (II)]**

48. A wire of length 30 cm, stretched between rigid supports, has it's n^{th} and $(n+1)^{th}$ harmonics at 400 Hz and 450 Hz, respectively. If tension in the string is 2700 N, it's linear mass density is ______ kg/m. **[NCERT : P L-290, 291 | July 27, 2022 (II)]**

49. A tunning fork of frequency 340 Hz resonates in the fundamental mode with an air column of length 125 cm in a cylindrical tube closed at one end. When water is slowly poured in it, the minimum height of water required for observing resonance once again is ______ cm. (Velocity of sound in air is 340 ms^{-1}) **[NCERT : P L-292, 293 | June 28, 2022 (II)]**

50. A wire having a linear mass density 9.0×10^{-4} kg/m is stretched between two rigid supports with a tension of 900 N. The wire resonates at a frequency of 500 Hz. The next higher frequency at which the same wire resonates is 550 Hz. The length of the wire is ______ m. **[NCERT : P L-290, 291 | NA, Aug. 31, 2021 (I)]**

51. A tuning fork is vibrating at 250 Hz. The length of the shortest closed organ pipe that will resonate with the tuning fork will be ______ cm. (Take speed of sound in air as 340 ms^{-1}) **[NCERT : P L-292, 293 | NA, Aug. 27, 2021 (II)]**

52. A closed organ pipe of length L and an open organ pipe contain gases of densities ρ_1 and ρ_2 respectively. The compressibility of gases are equal in both the pipes. Both the pipes are vibrating in their first overtone with same frequency. The length of the open pipe is $\frac{x}{3}L\sqrt{\frac{\rho_1}{\rho_2}}$ where x is ______. (Round off to the Nearest Integer). **[NCERT : P L-292, 293 | March 16, 2021 (II)]**

53. A uniform thin rope of length 12 m and mass 6 kg hangs vertically from a rigid support and a block of mass 2 kg is attached to its free end. A transverse short wave-train of wavelength 6 cm is produced at the lower end of the rope. What is the wavelength of the wavetrain (in cm) when it reaches the top of the rope ? **[NCERT : P L-290, 291 | Sep. 03, 2020 (I)]**

(a) 3 (b) 6 (c) 12 (d) 9

54. Two identical strings X and Z made of same material have tension T_X and T_Z in them. If their fundamental frequencies are 450 Hz and 300 Hz, respectively, then the ratio T_X/T_Z is: **[NCERT : P L-290, 291 | Sep. 02, 2020 (I)]**

(a) 2.25 (b) 0.44 (c) 1.25 (d) 1.5

55. A wire of density 9×10^{-3} kg cm^{-3} is stretched between two clamps 1 m apart. The resulting strain in the wire is 4.9×10^{-4}. The lowest frequency of the transverse vibrations in the wire is (Young's modulus of wire $Y = 9 \times 10^{10}$ Nm^{-2}), (to the nearest integer), ______. **[NCERT : P L-290, 291 | Sep. 02, 2020 (II)]**

56. A one metre long (both ends open) organ pipe is kept in a gas that has double the density of air at STP. Assuming the speed of sound in air at STP is 300 m/s, the frequency difference between the fundamental and second harmonic of this pipe is ______ Hz.
[NCERT : P L-292, 293 | NA 8 Jan. 2020 (I)]

57. A transverse wave travels on a tight steel wire with a velocity of v when tension in it is 2.06×10^4 N. When the tension is changed to T, the velocity changed to $v/2$. The value of T is close to: **[NCERT : P L-290, 291 | 8 Jan. 2020 (II)]**
(a) 2.50×10^4 N (b) 5.15×10^3 N
(c) 30.5×10^4 N (d) 10.2×10^2 N

58. Speed of a transverse wave on a straight wire (mass 6.0 g, length 60 cm and area of cross-section 1.0 mm²) is 90 ms⁻¹. If the Young's modulus of wire is 16×10^{11} Nm⁻² the extension of wire over its natural length is:
[NCERT : P L-290, 291 | 7 Jan. 2020 (I)]
(a) 0.03 mm (b) 0.02 mm (c) 0.04 mm (d) 0.01 mm

59. Equation of travelling wave on a stretched string of linear density 5 g/m is y = 0.03 sin (450 t – 9x) where distance and time are measured in SI units. The tension in the string is:
[NCERT : P L-290, 291 | 11 Jan 2019 (I)]
(a) 10 N (b) 7.5 N (c) 12.5 N (d) 5 N

60. A heavy ball of mass M is suspended from the ceiling of a car by a light string of mass m (m<<M). When the car is at rest, the speeed of transverse waves in the string is 60 ms⁻¹. when the car has acceleration a, the wave-speed increases to 60.5 ms⁻¹. The value of a, in terms of gravitational acceleration g, is closest to:
[NCERT : P L-290, 291 | 9 Jan. 2019 (I)]
(a) $\frac{g}{30}$ (b) $\frac{g}{5}$ (c) $\frac{g}{10}$ (d) $\frac{g}{20}$

61. A wire of length L and mass per unit length 6.0×10^{-3} kgm⁻¹ is put under tension of 540 N. Two consecutive frequencies that it resonates at are: 420 Hz and 490 Hz. Then L in meters is: **[NCERT : P L-290, 291 | 9 Jan. 2020 (II)]**
(a) 2.1 m (b) 1.1 m (c) 8.1 m (d) 5.1 m

62. A string 2.0 m long and fixed at its ends is driven by a 240 Hz vibrator. The string vibrates in its third harmonic mode. The speed of the wave and its fundamental frequency is:
[NCERT : P L-290, 291 | 9 April 2019 (II)]
(a) 180 m/s, 80 Hz (b) 320 m/s, 80 Hz
(c) 320 m/s, 120 Hz (d) 180 m/s, 120 Hz

63. A string is clamped at both the ends and it is vibrating in its 4th harmonic. The equation of the stationary wave is Y = 0.3 sin(0.157x) cos(200πt). The length of the string is: (All quantities are in SI units.)
[NCERT : P L-290, 291 | 9 April 2019 (I)]
(a) 20m (b) 80m (c) 40m (d) 60m

64. A wire of length 2L, is made by joining two wires A and B of same length but different radii r and $2r$ and made of the same material. It is vibrating at a frequency such that the joint of the two wires forms a node. If the number of antinodes in wire A is p and that in B is q then the ratio $p : q$ is: **[NCERT : P L-290, 291 | 8 April 2019 (I)]**
(a) 3 : 5 (b) 4 : 9
(c) 1 : 2 (d) 1 : 4

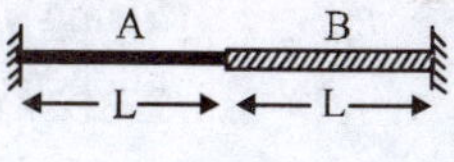

65. A closed organ pipe has a fundamental frequency of 1.5 kHz. The number of overtones that can be distinctly heard by a person with this organ pipe will be: (Assume that the highest frequency a person can hear is 20,000 Hz)
[NCERT : P L-292, 293 | 10 Jan. 2019 (I)]
(a) 6 (b) 4 (c) 7 (d) 5

66. A string of length 1 m and mass 5 g is fixed at both ends. The tension in the string is 8.0 N. The string is set into vibration using an external vibrator of frequency 100 Hz. The separation between successive nodes on the string is close to: **[NCERT : P L-290, 291 | 10 Jan. 2019 (I)]**
(a) 10.0 cm (b) 33.3 cm (c) 16.6 cm (d) 20.0 cm

67. A granite rod of 60 cm length is clamped at its middle point and is set into longitudinal vibrations. The density of granite is 2.7×10^3 kg/m³ and its Young's modulus is 9.27×10^{10} Pa.
What will be the fundamental frequency of the longitudinal vibrations? **[NCERT : P L-290, 291 | 2018]**
(a) 5 kHz (b) 2.5 kHz (c) 10 kHz (d) 7.5 kHz

68. The end correction of a resonance column is 1cm. If the shortest length resonating with the tuning fork is 10cm, the next resonating length should be
[NCERT : P L-290, 291 | Online April 16, 2018]
(a) 32 cm (b) 40 cm (c) 28 cm (d) 36 cm

69. Two wires W_1 and W_2 have the same radius r and respective densities ρ_1 and ρ_2 such that $\rho_2 = 4\rho_1$. They are joined together at the point O, as shown in the figure. The combination is used as a sonometer wire and kept under tension T. The point O is midway between the two bridges. When a stationary waves is set up in the composite wire, the joint is found to be a node. The ratio of the number of antinodes formed in W_1 to W_2 is :
[NCERT : P L-290, 291 | Online April 8, 2017]

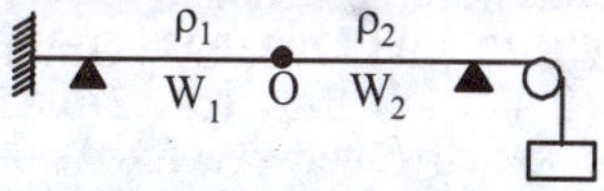

(a) 1 : 1 (b) 1 : 2 (c) 1 : 3 (d) 4 : 1

70. A uniform string of length 20 m is suspended from a rigid support. A short wave pulse is introduced at its lowest end. It starts moving up the string. The time taken to reach the supports is : **[NCERT : P L-290, 291 | 2016]**
(take g = 10 ms⁻²)
(a) $2\sqrt{2}$s (b) $\sqrt{2}$ s (c) $2\pi\sqrt{2}$ s (d) 2 s

71. A pipe open at both ends has a fundamental frequency f in air. The pipe is dipped vertically in water so that half of it is in water. The fundamental frequency of the air column is now : **[NCERT : P L-292, 293 | 2016]**
(a) 2f (b) f (c) $\frac{f}{2}$ (d) $\frac{3f}{4}$

72. A pipe of length 85 cm is closed from one end. Find the number of possible natural oscillations of air column in the pipe whose frequencies lie below 1250 Hz. The velocity of sound in air is 340 m/s. **[NCERT : P L-292, 293 | 2014]**
(a) 12 (b) 8 (c) 6 (d) 4

73. The total length of a sonometer wire between fixed ends is 110 cm. Two bridges are placed to divide the length of wire in ratio 6 : 3 : 2. The tension in the wire is 400 N and the mass per unit length is 0.01 kg/m. What is the minimum common frequency with which three parts can vibrate?

[NCERT : P L-290, 291 | Online April 19, 2014]

(a) 1100 Hz (b) 1000 Hz (c) 166 Hz (d) 100 Hz

74. A sonometer wire of length 1.5 m is made of steel. The tension in it produces an elastic strain of 1%. What is the fundamental frequency of steel if density and elasticity of steel are 7.7×10^3 kg/m^3 and 2.2×10^{11} N/m^2 respectively?

[NCERT : P L-290, 291 | 2013]

(a) 188.5 Hz (b) 178.2 Hz (c) 200.5 Hz (d) 770 Hz

75. A sonometer wire of length 114 cm is fixed at both the ends. Where should the two bridges be placed so as to divide the wire into three segments whose fundamental frequencies are in the ratio 1 : 3 : 4 ?

[NCERT : P L-290, 291 | Online April 23, 2013]

(a) At 36 cm and 84 cm from one end
(b) At 24 cm and 72 cm from one end
(c) At 48 cm and 96 cm from one end
(d) At 72 cm and 96 cm from one end

76. A cylindrical tube, open at both ends, has a fundamental frequency f in air. The tube is dipped vertically in water so that half of it is in water. The fundamental frequency of the air-column is now : **[NCERT : P L-292, 293 | 2012]**

(a) f (b) $f/2$ (c) $3f/4$ (d) $2f$

77. An air column in a pipe, which is closed at one end, will be in resonance wtih a vibrating tuning fork of frequency 264 Hz if the length of the column in cm is (velocity of sound = 330 m/s)

[NCERT : P L-292, 293 | Online May 26, 2012]

(a) 125.00 (b) 93.75 (c) 62.50 (d) 187.50

78. A uniform tube of length 60.5 cm is held vertically with its lower end dipped in water. A sound source of frequency 500 Hz sends sound waves into the tube. When the length of tube above water is 16 cm and again when it is 50 cm, the tube resonates with the source of sound. Two lowest frequencies (in Hz), to which tube will resonate when it is taken out of water, are (approximately).

[NCERT : P L-292, 293 | Online May 19, 2012]

(a) 281, 562 (b) 281, 843 (c) 276, 552 (d) 272, 544

79. A string is stretched between fixed points separated by 75.0 cm. It is observed to have resonant frequencies of 420 Hz and 315 Hz. There are no other resonant frequencies between these two. Then, the lowest resonant frequency for this string is **[NCERT : P L-290, 291 | 2006]**

(a) 105 Hz (b) 1.05 Hz (c) 1050 Hz (d) 10.5 Hz

80. Tube A has both ends open while tube B has one end closed, otherwise they are identical. The ratio of fundamental frequency of tube A and B is

[NCERT : P L-292, 293 | 2002]

(a) 1 : 2 (b) 1 : 4 (c) 2 : 1 (d) 4 : 1

81. A wave $y = a\sin(\omega t - kx)$ on a string meets with another wave producing a node at $x = 0$. Then the equation of the unknown wave is **[NCERT : P L-289 | 2002]**

(a) $y = a\sin(\omega t + kx)$ (b) $y = -a\sin(\omega t + kx)$
(c) $y = a\sin(\omega t - kx)$ (d) $y = -a\sin(\omega t - kx)$

Topic 3 - Beats, Interference and Superposition of Waves

82. The amplitude and phase of a wave that is formed by the superposition of two harmonic travelling waves, $y_1(x, t) = 4\sin(kx - \omega t)$ and $y_2(x, t) = 2\sin(kx - \omega t + \frac{2\pi}{3})$, are : (Take the angular frequency of initial waves same as ω)

[NCERT : P L-288 | April 8, 2025 (II)]

(a) $\left[6, \frac{2\pi}{3}\right]$ (b) $\left[6, \frac{\pi}{3}\right]$ (c) $\left[\sqrt{3}, \frac{\pi}{6}\right]$ (d) $\left[2\sqrt{3}, \frac{\pi}{6}\right]$

83. Two harmonic waves moving in the same direction superimpose to form a wave $x = a\cos(1.5t)\cos(50.5t)$ where t is in seconds. Find the period with which they beat (close to nearest integer)

[NCERT : P L-293, 294 | April 7, 2025 (I)]

(a) 6 s (b) 5 s (c) 1 s (d) 2 s

84. In the resonance experiment, two air columns (closed at one end) of 100 cm and 120 cm long, give 15 beats per second when each one is sounding in the respective fundamental modes. The velocity of sound in the air column is **[NCERT : P L-293, 294 | April 3, 2025 (II)]**

(a) 335 m/s (b) 370 m/s (c) 340 m/s (d) 360 m/s

85. A tuning fork resonated with a sonometer wire of length 1 m stretched with a tension of 6N. When the tension in the wire is changed to 54 N, the same tuning fork produces 12 beats per second with it. The frequency of the tuning fork is ______ Hz. **[NCERT : P L-293 | Feb. 1, 2024 (I)]**

86. Two waves of intensity ratio 1 : 9 cross each other at a point. The resultant intensities at that point, when (a) Waves are incoherent is I_1 (b) Waves are coherent is I_2 and differ in phase by 60°. If $\frac{I_1}{I_2} = \frac{10}{x}$ then x = ______.

87. A closed organ pipe 150 cm long gives 7 beats per second with an open organ pipe of length 350 cm, both vibrating in fundamental mode. The velocity of sound is ______ m/s. **[NCERT : P L-293 | Jan. 27, 2024 (II)]**

88. The displacement equations of two interfering waves are given by

$$y_1 = 10\sin\left(\omega t + \frac{\pi}{3}\right)\text{cm},$$

$$y_2 = 5\left[\sin(\omega t) + \sqrt{3}\cos\omega t\right]\text{cm}$$ respectively. The amplitude of the resultant wave is ______ cm.

[Jan. 31, 2023 (II)]

89. Two waves executing simple harmonic motion travelling in the same direction with same amplitude and frequency are superimposed. The resultant amplitude is equal to the $\sqrt{3}$ times of amplitude of individual motions. The phase difference between the two motions is ___ (degree)

[July 25, 2022 (II)]

90. The velocity of sound in a gas in which two wavelengths 4.08m and 4.16m produce 40 beats in 12s, will be :

[NCERT : P L-293 | June 28, 2022 (I)]

(a) 2.828 ms^{-1} (b) 175.5 ms^{-1}
(c) 353.6 ms^{-1} (d) 707.2 ms^{-1}

91. A set of 20 tuning forks is arranged in a series of increasing frequencies. If each fork gives 4 beats with respect to the preceding fork and the frequency of the last fork is twice the frequency of the first, then the frequency of last fork is______Hz. **[NCERT : P L-293 | June 26, 2022 (II)]**

92. The equations of two waves are given by :

$y_1 = 5\sin 2\pi(x - vt)$ cm
$y_2 = 3\sin 2\pi(x - vt + 1.5)$ cm

These waves are simultaneously passing through a string. The amplitude of the resulting wave is **[June 24, 2022 (I)]**

(a) 2 cm (b) 4 cm (c) 5.8 cm (d) 8 cm

93. Two travelling waves of equal amplitudes and equal frequencies move in opposite directions along a string. They interfere to produce a stationary wave whose equation is given by $y = (10\cos \pi x \sin \frac{2\pi t}{T})$ cm

The amplitude of the particle at $x = \frac{4}{3}$ cm will be _____ cm.

[NCERT : P L-289 | June 24, 2022 (II)]

94. A tuning fork A of unknown frequency produces 5 beats/s with a fork of known frequency 340 Hz. When fork A is filled, the beat frequency decreases to 2 beats/s. What is the frequency of fork A? **[NCERT : P L-293 | Feb. 26, 2021 (II)]**

(a) 335 Hz (b) 338 Hz (c) 345 Hz (d) 342 Hz

95. Three harmonic waves having equal frequency n and same intensity I_0, have phase angles 0, $\frac{\pi}{4}$ and $-\frac{\pi}{4}$ respectively. When they are superimposed the intensity of the resultant wave is close to:

[NCERT : P L-289 | 9 Jan. 2020 I]

(a) $5.8 I_0$ (b) $0.2 I_0$ (c) $3 I_0$ (d) I_0

96. A tuning fork vibrates with frequency 256 Hz and gives one beat per second with the third normal mode of vibration of an open pipe. What is the length of the pipe? (Speed of sound of air is 340 ms^{-1})

[NCERT : P L-293 | Online April 15, 2018]

(a) 190 cm (b) 180 cm (c) 220 cm (d) 200 cm

97. 5 beats/ second are heard when a turning fork is sounded with a sonometer wire under tension, when the length of the sonometer wire is either 0.95m or 1m. The frequency of the fork will be: **[NCERT : P L-293 | Online April 15, 2018]**

(a) 195Hz (b) 251Hz (c) 150Hz (d) 300Hz

98. A standing wave is formed by the superposition of two waves travelling in opposite directions. The transverse displacement is given by

$y(x, t) = 0.5 \sin\left(\frac{5\pi}{4}x\right) \cos(200\pi t)$.

What is the speed of the travelling wave moving in the positive x direction ?

(x and t are in meter and second, respectively.)

[NCERT : P L-289 | Online April 9, 2017]

(a) 160 m/s (b) 90 m/s (c) 180 m/s (d) 120 m/s

99. A wave represented by the equation $y_1 = a\cos(kx - \omega t)$ is superimposed with another wave to form a stationary wave such that the point $x - 0$ is node. The equation for the other wave is **[NCERT : P L-289 | Online May 12, 2012]**

(a) $a\cos(kx - \omega t + \pi)$ (b) $a\cos(kx + \omega t + \pi)$

(c) $a\cos\left(kx + \omega t + \frac{\pi}{2}\right)$ (d) $a\cos\left(kx - \omega t + \frac{\pi}{2}\right)$

100. Following are expressions for four plane simple harmonic waves **[NCERT : P L-289 | Online May 7, 2012]**

(i) $y_1 = A\cos 2\pi\left(n_1 t + \frac{x}{\lambda_1}\right)$

(ii) $y_2 = A\cos 2\pi\left(n_1 t + \frac{x}{\lambda_1} + \pi\right)$

(iii) $y_3 = A\cos 2\pi\left(n_2 t + \frac{x}{\lambda_2}\right)$

(iv) $y_4 = A\cos 2\pi\left(n_2 t - \frac{x}{\lambda_2}\right)$

The pairs of waves which will produce destructive interference and stationary waves respectively in a medium, are

(a) (iii, iv), (i, ii) (b) (i, iii), (ii, iv)

(c) (i, iv), (ii, iii) (d) (i, ii), (iii, iv)

101. A travelling wave represented by $y = A\sin(\omega t - kx)$ is superimposed on another wave represented by $y = A\sin(\omega t + kx)$. The resultant is

[NCERT : P L-289 | 2011 RS]

(a) A wave travelling along + x direction

(b) A wave travelling along – x direction

(c) A standing wave having nodes at $x = \frac{n\lambda}{2}, n = 0, 1, 2....$

(d) A standing wave having nodes at $x = \left(n + \frac{1}{2}\right)\frac{\lambda}{2}$; $n = 0, 1, 2....$

102. Statement - 1 : Two longitudinal waves given by equations : $y_1(x, t) = 2a\sin(\omega t - kx)$ and $y_2(x, t) = a\sin(2\omega t - 2kx)$ will have equal intensity. **[2011 RS]**

Statement - 2 : Intensity of waves of given frequency in same medium is proportional to square of amplitude only.

(a) Statement-1 is true, statement-2 is false.

(b) Statement-1 is true, statement-2 is true, statement-2 is the correct explanation of statement-1

(c) Statement-1 is true, statement-2 is true, statement-2 is not the correct explanation of statement-1

(d) Statement-1 is false, statement-2 is true.

103. Three sound waves of equal amplitudes have frequencies $(\nu - 1)$, ν, $(\nu + 1)$. They superpose to give beats. The number of beats produced per second will be :

[NCERT : P L-289 | 2009]

(a) 3 (b) 2 (c) 1 (d) 4

104. When two tuning forks (fork 1 and fork 2) are sounded simultaneously, 4 beats per second are heard. Now, some tape is attached on the prong of the fork 2. When the tuning forks are sounded again, 6 beats per second are heard. If the frequency of fork 1 is 200 Hz, then what was the original frequency of fork 2? **[NCERT : P L-293 | 2005]**

(a) 202 Hz (b) 200 Hz (c) 204 Hz (d) 196 Hz

105. A tuning fork of known frequency 256 Hz makes 5 beats per second with the vibrating string of a piano. The beat frequency decreases to 2 beats per second when the tension in the piano string is slightly increased. The frequency of the piano string before increasing the tension was **[NCERT : P L-293 | 2003]**

(a) (256 + 2) Hz (b) (256 – 2) Hz

(c) (256 – 5) Hz (d) (256 + 5) Hz

106. A tuning fork arrangement (pair) produces 4 beats/sec with one fork of frequency 288 cps. A little wax is placed on the unknown fork and it then produces 2 beats/sec. The frequency of the unknown fork is **[NCERT : P L-293 | 2002]**

(a) 286 cps (b) 292 cps (c) 294 cps (d) 288 cps

ANSWER KEYS																			
1	(a)	12	(d)	23	(b)	34	(b)	45	(80)	56	(106)	67	(a)	78	(d)	89	(60)	100	(d)
2	(b)	13	(18)	24	(b)	35	(a)	46	(900)	57	(b)	68	(a)	79	(a)	90	(d)	101	(d)
3	(a)	14	(c)	25	(a)	36	(c)	47	(60)	58	(a)	69	(b)	80	(c)	91	(152)	102	(a)
4	(d)	15	(c)	26	(a)	37	(16)	48	(3)	59	(c)	70	(a)	81	(b)	92	(a)	103	(b)
5	(c)	16	(a)	27	(a)	38	(740)	49	(50)	60	(b)	71	(b)	82	(d)	93	(5)	104	(d)
6	(b)	17	(d)	28	(a)	39	(60)	50	(10)	61	(a)	72	(c)	83	(d)	94	(a)	105	(c)
7	(a)	18	(b)	29	(b)	40	(d)	51	(34)	62	(b)	73	(b)	84	(d)	95	(a)	106	(b)
8	(1152)	19	(7)	30	(c)	41	(400)	52	(4)	63	(b)	74	(b)	85	(6)	96	(d)		
9	(20)	20	(2)	31	(a)	42	(90)	53	(c)	64	(c)	75	(d)	86	(13)	97	(a)		
10	(a)	21	(b)	32	(a)	43	(500)	54	(a)	65	(a)	76	(a)	87	(294)	98	(a)		
11	(a)	22	(a)	33	(b)	44	(1)	55	(35.00)	66	(d)	77	(b)	88	(20)	99	(b)		

15 Experimental Skills-XI

TREND ANALYSIS
JEE MAIN (2025-2021)

	2025				2024				2023				2022				2021			
	Session II		Session I		Session II		Session I		Session II		Session I		Session II		Session I		Session II		Session I	
	MCQs	NVQs	MCQs	NVQs	MCQs	NVQs	MCQs	NVQs	MCQs	NVQs	MCQs	NVQs	MCQs	NVQs	MCQs	NVQs	MCQs	NVQs	MCQs	NVQs
Topic 1	–	–	1	2	9	–	4	–	1	–	–	2	4	1	1	5	2	1	2	–
Topic 2	–	–	1	–	–	–	–	–	–	–	–	–	–	2	–	2	–	–	1	–
Topic 3																				
Topic 4																				
Total	–	–	2	2	9	–	4	–	1	–	–	2	4	3	1	7	2	1	3	–

How to Code Qns for Better & Faster Revision

Type 1: Able to solve (No code Req)

Type 2: Solved with struggle or by seeing solution **(Use code NR means Needs Revision or orange stickers)**

Type 3: Tricky/ Could not solve on my own. Need help from Teacher/ Friend. **(Use code TQ means Tricky Question or red stickers)**

You can either hand write the code in boxes or else paste coloured stickers given at end of book.

Topic 1 – Vernier Calipers - its use to measure the internal & external diameter and depth of a vessel; To determine thickness/diameter of thin sheet/wire with the help of screw gauge

1. A tiny metallic rectangular sheet has length and breadth of 5 mm and 2.5mm, respectively. Using a specially designed screw gauge which has pitch of 0.75 mm and 15 divisions in the circular scale, you are asked to find the area of the sheet. In this measurement, the maximum fractional error will be $\frac{x}{100}$ where x is_______. **[Jan. 28, 2025 (I)]**

2. The least count of a screw guage is 0.01 mm. If the pitch is increased by 75% and number of divisions on the circular scale is reduced by 50%, the new least count will be______ $\times 10^{-3}$ mm. **[Jan. 24, 2025 (I)]**

3. Given below are two statements :

Statement I : In a vernier callipers, one vernier scale division is always smaller than one main scale division.

Statement II : The vernier constant is given by one main scale division multiplied by the number of vernier scale division.

In the light of the above statements, choose the correct answer from the options given below. **[Jan. 22, 2025 (I)]**

(a) Both Statement I and Statement II are false.
(b) Statement I is true but Statement II is false.
(c) Both Statement I and Statement II are true.
(d) Statement I is false but Statement II is true.

4. One main scale division of a vernier caliper is equal to m units. If n^{th} division of main scale coincides with $(n+1)^{th}$ division of vernier scale, the least count of the vernier caliper is: **[April 9, 2024 (I)]**

(a) $\frac{n}{(n+1)}$ (b) $\frac{m}{(n+1)}$

(c) $\frac{1}{(n+1)}$ (d) $\frac{m}{n(n+1)}$

5. The diameter of a sphere is measured using a vernier caliper whose 9 divisions of main scale are equal to 10 divisions of vernier scale. The shortest division on the main scale is equal to 1 mm. The main scale reading is 2 cm and second division of vernier scale coincides with a division on main scale. If mass of the sphere is 8.635 g, the density of the sphere is: **[April 8, 2024 (I)]**
(a) $2.5 g/cm^3$ (b) $1.7 g/cm^3$
(c) $2.2 g/cm^3$ (d) $2.0 g/cm^3$

6. There are 100 divisions on the circular scale of a screw gauge of pitch 1 mm. With no measuring quantity in between the jaws, the zero of the circular scale lies 5 divisions below the reference line. The diameter of a wire is then measured using this screw gauge. It is found the 4 linear scale divisions are clearly visible while 60 divisions on circular scale coincide with the reference line. The diameter of the wire is : **[April 8, 2024 (II)]**
(a) 4.65 mm (b) 4.55 mm (c) 4.60 mm (d) 3.35 mm

7. Least count of a vernier caliper is $\frac{1}{20N}$ cm. The value of one division on the main scale is 1 mm. Then the number of divisions of main scale that coincide with N divisions of vernier scale is : **[April 8, 2024 (II)]**
(a) $\left(\frac{2N-1}{20N}\right)$ (b) $\left(\frac{2N-1}{2}\right)$
(c) $(2N-1)$ (d) $\left(\frac{2N-1}{2N}\right)$

8. While measuring diameter of wire using screw gauge the following readings were noted. Main scale reading is 1 mm and circular scale reading is equal to 42 divisions. Pitch of screw gauge is 1 mm and it has 100 divisions on circular scale. The diameter of the wire is $\frac{x}{50}$ mm. The value of x is: **[April 6, 2024 (I)]**
(a) 142 (b) 71 (c) 42 (d) 21

9. In finding out refractive index of glass slab the following observations were made through travelling microscope 50 vernier scale division = 49 MSD; 20 divisions on main scale in each cm
For mark on paper
MSR = 8.45 cm, VC = 26
For mark on paper seen through slab
MSR = 7.12 cm, VC = 41
For powder particle on the top surface of the glass slab
MSR = 4.05 cm, VC = 1
(MSR = Main Scale Reading, VC = Vernier Coincidence)
Refractive index of the glass slab is: **[April 6, 2024 (II)]**
(a) 1.42 (b) 1.52 (c) 1.24 (d) 1.35

10. In a vernier calliper, when both jaws touch each other, zero of the vernier scale shifts towards left and its 4th division coincides exactly with a certain division on main scale. If 50 vernier scale divisions equal to 49 main scale divisions and zero error in the instrument is 0.04 mm then how many main scale divisions are there in 1 cm ? **[April 6, 2024 (II)]**
(a) 40 (b) 5 (c) 20 (d) 10

11. Time periods of oscillation of the same simple pendulum measured using four different measuring clocks were recorded as 4.62 s, 4.632 s, 4.6 s and 4.64 s. The arithmetic mean of these reading in correct significant figure is. **[April 5, 2024 (I)]**
(a) 4.623 s (b) 4.62 s (c) 4.6 s (d) 5 s

12. A vernier callipers has 20 divisions on the vernier scale, which coincides with 19th division on the main scale. The least count of the instrument is 0.1 mm. One main scale division is equal to ____ mm. **[April 5, 2024 (II)]**
(a) 1 (b) 0.5 (c) 2 (d) 5

13. 10 divisions on the main scale of a Vernier calliper coincide with 11 divisions on the Vernier scale. If each division on the main scale is of 5 units, the least count of the instruments is: **[Feb. 1, 2024 (I)]**
(a) $\frac{1}{2}$ (b) $\frac{10}{11}$ (c) $\frac{50}{11}$ (d) $\frac{5}{11}$

14. If 50 Vernier divisions are equal to 49 main scale division of a travelling microscope and one smallest reading of main scale is 0.5 *mm*, the Vernier constant of travelling microscope is **[Jan. 30, 2024 (II)]**
(a) 0.1 *mm* (b) 0.1 *cm* (c) 0.01 *cm* (d) 0.01 *mm*

15. Identify the physical quantity that cannot be measured using spherometer : **[Jan. 27, 2024 (I)]**
(a) Radius of curvature of concave surface
(b) Specific rotation of liquids
(c) Thickness of thin plates
(d) Radius of curvature of convex surface

16. Given below are two statements : one is labelled as Assertion(A) and the other is labelled as Reason (R).
Assertion (A) : In Vernier calliper if positive zero error exists, then while taking measurements, the reading taken will be more than the actual reading.
Reason (R) : The zero error in Vernier Calliper might have happened due to manufacturing defect or due to rough handling.
In the light of the above statements, choose the correct answer from the options given below : **[Jan. 27, 2024 (II)]**
(a) Both (A) and (R) are correct and (R) is the correct explanation of (A)
(b) Both (A) and (R) are correct but (R) is not the correct explanation of (A)
(c) (A) is true but (R) is false
(d) (A) is false but (R) is true

17. In an experiment with Vernier callipers of least count 0.1 mm, when two jaws are joined together the zero of Vernier scale lies right to the zero of the main scale and 6th division of Vernier scale coincides with the main scale division. While measuring the diameter of a spherical bob, the zero of vernier scale lies in between 3.2 cm and 3.3 cm marks, and 4th division of vernier scale coincides with the main scale division. The diameter of bob is measured as : **[April 10, 2023 (II)]**
(a) 3.18 cm (b) 3.25 cm
(c) 3.26 cm (d) 3.22 cm

18. In a screw gauge, there are 100 divisions on the circular scale and the main scale moves by 0.5 mm on a complete rotation of the circular scale. The zero of circular scale lies 6 divisions below the line of graduation when two studs are brought in contact with each other. When a wire is placed between the studs, 4 linear scale divisions are clearly visible while 46th division the circular scale coincide with the reference line. The diameter of the wire is ______ $\times 10^{-2}$ cm. **[Jan. 30, 2023 (I)]**

19. In an experiment of measuring the refractive index of a glass slab using travelling microscope in physics lab, a student measures real thickness of the glass slab as 5.25 mm and apparent thickness of the glass slab at 5.00 mm. Travelling microscope has 20 divisions in one cm on main scale and 50 divisions on Vernier scale is equal to 49 divisions on main scale. The estimated uncertainty in the measurement of refractive index of the slab is $\frac{x}{10} \times 10^{-3}$, where x is _________ **[Jan. 29, 2023 (II)]**

20. A travelling microscope has 20 divisions per cm on the main scale while its Vernier scale has total 50 divisions and 25 Vernier scale divisions are equal to 24 main scale divisions, what is the least count of the travelling microscope? **[July 29, 2022 (I)]**

(a) 0.001 cm (b) 0.002 mm
(c) 0.002 cm (d) 0.005 cm

21. In an experiment to find out the diameter of wire using screw gauge, the following observation were noted :

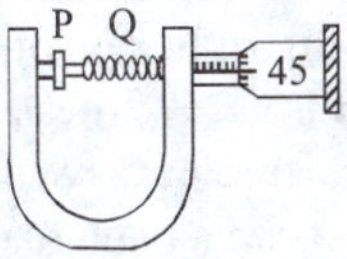

(a) Screw moves 0.5 mm on main scale in one complete rotation.
(b) Total divisions on circular scale = 50
(c) Main scale reading is 2.5 mm
(d) 45th division of circular scale is in the pitch line
(e) Instrument has 0.03 mm negative error

Then the diameter of wire is : **[July 29, 2022 (I)]**

(a) 2.92 mm (b) 2.54 mm (c) 2.98 mm (d) 3.45 mm

22. The one division of main scale of vernier callipers reads 1 mm and 10 divisions of Vernier scale is equal to the 9 divisions on main scale. When the two jaws of the instrument touch each other the zero of the Vernier lies to the right of zero of the main scale and its fourth division coincides with a main scale division. When a spherical bob is tightly placed between the two jaws, the zero of the Vernier scale lies in between 4.1 cm and 4.2 cm and 6th Vernier division coincides with a main scale division. The diameter of the bob will be__ 10^{-2} cm **[July 27, 2022 (I)]**

23. A screw gauge of pitch 0.5 mm is used to measure the diameter of uniform wire of length 6.8 cm, the main scale reading is 1.5 mm and circular scale reading is 7. The calculated curved surface area of wire to appropriate significant figures is : **[July 26, 2022 (I)]**

[Screw gauge has 50 divisions on its circular scale]

(a) 6.8 cm^2 (b) 3.4 cm^2 (c) 3.9 cm^2 (d) 2.4 cm^2

24. In a Vernier Calipers 10 divisions of Vernier scale is equal to the 9 divisions of main scale. When both jaws of Vernier calipers touch each other, the zero of the Vernier scale is shifted to the left of zero of the main scale and 4th Vernier scale division exactly coincides with the main scale reading. One main scale division is equal to 1 mm. While measuring diameter of a spherical body, the body is held between two jaws. It is now observed that zero of the Vernier scale lies between 30 and 31 divisions of main scale reading and 6th Vernier scale division exactly coincides with the main scale reading. The diameter of the spherical body will be : **[July 26, 2022 (II)]**

(a) 3.02 cm (b) 3.06 cm (c) 3.12 cm (d) 3.20 cm

25. If n main scale divisions coincide with (n + 1) vernier scale divisions. The least count of vernier callipers, when each centimetre on the main scale is divided into five equal parts, will be. **[June 30, 2022 (I)]**

(a) $\frac{2n}{n+1}$ mm (b) $\frac{5}{n+1}$ mm
(c) $\frac{1}{2n}$ mm (d) $\frac{1}{5n}$ mm

26. The vernier constant of Vernier callipers is 0.1 mm and it has zero error of (– 0.05) cm. While measuring diameter of a sphere, the main scale reading is 1.7 cm and coinciding vernier division is 5. The corrected diameter will be ____ $\times 10^{-2}$ cm. **[June 29, 2022 (II)]**

27. A student in the laboratory measures thickness of a wire using screw gauge. The readings are 1.22 mm, 1.23 mm, 1.19 mm and 1.20 mm. The percentage error is $\frac{x}{121}$%. The value of x is _________. **[June 28, 2022 (II)]**

28. In a vernier callipers, each cm on the main scale is divided into 20 equal parts. If tenth vernier scale division coincides with nineth main scale division. Then the value of vernier constant will be .... $\times 10^{-2}$ mm. **[June 26, 2022 (I)]**

29. A travelling microscope is used to determine the refractive index of a glass slab. If 40 divisions are there in 1 cm on main scale and 50 Vernier scale divisions are equal to 49 main scale divisions, then least count of the travelling microscope is ______ $\times 10^{-6}$ m. **[June 26, 2022 (II)]**

30. Student A and Student B used two screw gauges of equal pitch and 100 equal circular divisions to measure the radius of a given wire. The actual value of the radius of the wire is 0.322 cm. The absolute value of the difference between the final circular scale readings observed by the students A and B is ________.

[Figure shows position of reference 'O' when jaws of screw gauge are closed]

Given pitch = 0.1 cm.

[NA, July 25, 2021 (I); June 25, 2022 (I)]

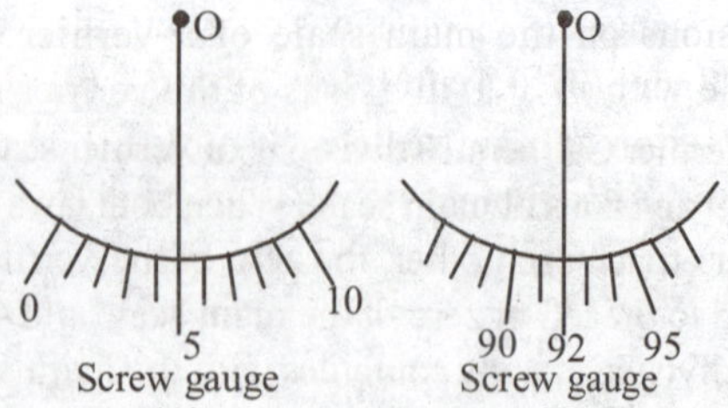

Screw gauge (A) Screw gauge (B)

31. The diameter of a spherical bob is measured using a vernier callipers. 9 divisions of the main scale, in the vernier callipers, are equal to 10 divisions of vernier scale. One main scale division is 1 mm. The main scale reading is 10 mm and 8th division of vernier scale was found to coincide exactly with one of the main scale division. If the given vernier callipers has positive zero error of 0.04 cm, then the radius of the bob is_____ $\times 10^{-2}$ cm. **[Aug. 31, 2021 (II)]**

32. In a Screw Gauge, fifth division of the circular scale coincides with the reference line when the ratchet is closed. There are 50 divisions on the circular scale, and the main scale moves by 0.5 mm on a complete rotation. For a particular observation the reading on the main scale is 5 mm and the 20th division of the circular scale coincides with reference line. Calculate the true reading. **[Aug. 26, 2021 (I)]**

(a) 5.00 mm (b) 5.25 mm (c) 5.15 mm (d) 5.20 mm

33. **Assertion A :** If in five complete rotations of the circular scale, the distance travelled on main scale of the screw gauge is 5 mm and there are 50 total divisions on circular scale, then least count is 0.001 cm.

Reason R : Least count

$$= \frac{\text{Pitch}}{\text{Total divisions on circular scale}}$$

In the light of the above statements, choose the most appropriate answer from the options given below : **[July 27, 2021 (I)]**

(a) A is not correct but R is correct.
(b) Both A and R are correct and R is the correct explanation of A.
(c) A is correct but R is not correct.
(d) Both A and R are correct and R is NOT the correct explanation of A.

34. The vernier scale used for measurement has a positive zero error of 0.2 mm. If while taking a measurement it was noted that '0' on the vernier scale lies between 8.5 cm and 8.6 cm, vernier coincidence is 6, then the correct value of measurement is_____cm. (least count = 0.01 cm) **[March 17, 2021 (I)]**

(a) 8.54 cm (b) 8.36 cm
(c) 8.56 cm (d) 8.58 cm

35. One main scale division of a vernier calipers is 'a' cm and n^{th} division of the vernier scale coincide with $(n-1)^{th}$ division of the main scale. The least count of the calipers in mm is: **[March 16, 2021 (I)]**

(a) $\frac{10a}{(n-1)}$ (b) $\frac{10na}{(n-1)}$ (c) $\left(\frac{n-1}{10n}\right)a$ (d) $\frac{10a}{n}$

36. A screw gauge has 50 divisions on its circular scale. The circular scale is 4 units ahead of the pitch scale marking, prior to use. Upon one complete rotation of the circular scale, a displacement of 0.5 mm is noticed on the pitch scale. The nature of zero error involved, and the least count of the screw gauge, are respectively : **[Sep. 06, 2020 (I)]**

(a) Negative, 2 μm (b) Positive, 10 μm
(c) Positive, 0.1 mm (d) Positive, 0.1 μm

37. A student measuring the diameter of a pencil of circular cross-section with the help of a vernier scale records the following four readings 5.50 mm, 5.55 mm, 5.45 mm, 5.65 mm, The average of these four reading is 5.5375 mm and the standard deviation of the data is 0.07395 mm. The average diameter of the pencil should therefore be recorded as : **[Sep. 06, 2020 (II)]**

(a) (5.5375 ± 0.0739) mm (b) (5.5375 ± 0.0740) mm
(c) (5.538 ± 0.074) mm (d) (5.54 ± 0.07) mm

38. Using screw gauge of pitch 0.1 cm and 50 divisions on its circular scale, the thickness of an object is measured. It should correctly be recorded as : **[Sep. 03, 2020 (I)]**

(a) 2.121 cm (b) 2.124 cm
(c) 2.125 cm (d) 2.123 cm

39. The least count of the main scale of a vernier callipers is 1 mm. Its vernier scale is divided into 10 divisions and coincide with 9 divisions of the main scale. When jaws are touching each other, the 7th division of vernier scale coincides with a division of main scale and the zero of vernier scale is lying right side of the zero of main scale. When this vernier is used to measure length of a cylinder the zero of the vernier scale betwen 3.1 cm and 3.2 cm and 4th VSD coincides with a main scale division. The length of the cylinder is : (VSD is vernier scale division) **[Sep. 02, 2020 (I)]**

(a) 3.2 cm (b) 3.21 cm (c) 3.07 cm (d) 2.99 cm

40. If the screw on a screw-gauge is given six rotations, it moves by 3 mm on the main scale. If there are 50 divisions on the circular scale the least count of the screw gauge is: **[9 Jan. 2020 I]**

(a) 0.001 cm (b) 0.02 mm
(c) 0.01 cm (d) 0.001 mm

41. The least count of the main scale of a screw gauge is 1 mm. The minimum number of divisions on its circular scale required to measure 5 μm diameter of a wire is: **[12 Jan. 2019 I]**

(a) 50 (b) 200 (c) 100 (d) 500

42. The pitch and the number of divisions, on the circular scale for a given screw gauge are 0.5 mm and 100 respectively. When the screw gauge is fully tightened without any object, the zero of its circular scale lies 3 division below the mean line.

The readings of the main scale and the circular scale, for a thin sheet, are 5.5 mm and 48 respectively, the thickness of the sheet is: **[9 Jan. 2019 II]**

(a) 5.755 mm (b) 5.950 mm
(c) 5.725 mm (d) 5.740 mm

43. In a screw gauge, 5 complete rotations of the screw cause it to move a linear distance of 0.25 cm. There are 100 circular scale divisions. The thickness of a wire measured by this screw gauge gives a reading of 4 main scale divisions and 30 circular scale divisions. Assuming negligible zero error, the thickness of the wire is: **[Online April 15, 2018]**
(a) 0.0430 cm (b) 0.3150 cm
(c) 0.4300 cm (d) 0.2150 cm

44. A screw gauge with a pitch of 0.5 mm and a circular scale with 50 divisions is used to measure the thickness of a thin sheet of Aluminium. Before starting the measurement, it is found that when the two jaws of the screw gauge are brought in contact, the 45th division coincides with the main scale line and the zero of the main scale is barely visible. What is the thickness of the sheet if the main scale reading is 0.5 mm and the 25th division coincides with the main scale line? **[2016]**
(a) 0.70 mm (b) 0.50 mm (c) 0.75 mm (d) 0.80 mm

45. Diameter of a steel ball is measured using a Vernier callipers which has divisions of 0.1 cm on its main scale (MS) and 10 divisions of its vernier scale (VS) match 9 divisions on the main scale. Three such measurements for a ball are given as: **[Online April 10, 2015]**

S.No.	MS(cm)	VS divisions
1.	0.5	8
2.	0.5	4
3.	0.5	6

If the zero error is – 0.03 cm, then mean corrected diameter is:
(a) 0.52 cm (b) 0.59 cm (c) 0.56 cm (d) 0.53 cm

46. A student measured the length of a rod and wrote it as 3.50 cm. Which instrument did he use to measure it? **[2014]**
(a) A meter scale.
(b) A vernier calliper where the 10 divisions in vernier scale matches with 9 division in main scale and main scale has 10 divisions in 1 cm.
(c) A screw gauge having 100 divisions in the circular scale and pitch as 1 mm.
(d) A screw gauge having 50 divisions in the circular scale and pitch as 1 mm.

47. N divisions on the main scale of a vernier calliper coincide with $(N+1)$ divisions of the vernier scale. If each division of main scale is 'a' units, then the least count of the instrument is **[Online May 19, 2012]**
(a) a (b) $\frac{a}{N}$ (c) $\frac{N}{N+1}\times a$ (d) $\frac{a}{N+1}$

48. A student measured the diameter of a wire using a screw gauge with the least count 0.001 cm and listed the measurements. The measured value should be recorded as **[Online May 12, 2012]**
(a) 5.3200 cm (b) 5.3 cm
(c) 5.32 cm (d) 5.320 cm

49. A spectrometer gives the following reading when used to measure the angle of a prism.
Main scale reading : 58.5 degree
Vernier scale reading : 09 divisions
Given that 1 division on main scale corresponds to 0.5 degree. Total divisions on the Vernier scale is 30 and match with 29 divisions of the main scale. The angle of the prism from the above data is **[2009, 2012]**
(a) 58.59 degree (b) 58.77 degree
(c) 58.65 degree (d) 59 degree

50. A screw gauge gives the following reading when used to measure the diameter of a wire.
Main scale reading : 0 mm
Circular scale reading : 52 divisions
Given that 1mm on main scale corresponds to 100 divisions of the circular scale. The diameter of wire from the above data is **[2011]**
(a) 0.052 cm (b) 0.026 cm
(c) 0.005 cm (d) 0.52 cm

51. Two full turns of the circular scale of a screw gauge cover a distance of 1mm on its main scale. The total number of divisions on the circular scale is 50. Further, it is found that the screw gauge has a zero error of – 0.03 mm. While measuring the diameter of a thin wire, a student notes the main scale reading of 3 mm and the number of circular scale divisions in line with the main scale as 35. The diameter of the wire is **[2008]**
(a) 3.32 mm (b) 3.73 mm
(c) 3.67 mm (d) 3.38 mm

Topic 2 - Measure the mass of a given object by the principle of moments using metre scale, Determination of Young's modulus of elasticity of the material of a metallic wire. Calculation of surface tension of water by capillary rise and effect of detergents. Calculation of co-efficient of viscosity of a given viscous liquid, Measurement of specific heat using method of mixtures, Calculation of dissipation of energy by plotting a graph between the square of amplitude and time, Determination of speed of sound using resonance tube

52. In the experiment for measurement of viscosity 'η' of given liquid with a ball having radius R, consider following statements. **[Jan. 28, 2025 (I)]**
(A) Graph between terminal velocity V and R will be a parabola
(B) The terminal velocities of different diameter balls are constant for a given liquid.
(C) Measurement of terminal velocity is dependent on the temperature.

(D) This experiment can be utilized to assess the density of a given liquid.

(E) If balls are dropped with some initial speed, the value of η will change.

Choose the correct answer from the options given below:

(a) B, D and E only (b) A, C and D only

(c) C, D and E only (d) A, B and E only

53. In an experiment to determine the Young's modulus, steel wires of five different lengths (1, 2, 3, 4 and 5 m) but of same cross section (2 mm^2) were taken and curves between extension and load were obtained. The slope (extension/load) of the curves were plotted with the wire length and the following graph is obtained. If the Young's modulus of given steel wires is $x \times 10^{11}$ Nm^{-2}, then the value of x is ______. **[July 27, 2022 (II)]**

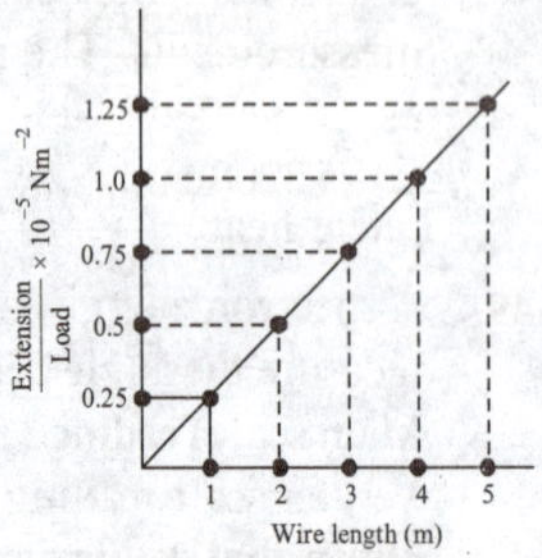

54. In an experiment to determine the Young's modulus of wire of a length exactly 1 m, the extension in the length of the wire is measured as 0.4 mm with an uncertainty of ± 0.02 mm when a load of 1 kg is applied. The diameter of the wire is measured as 0.4 mm with an uncertainty of ± 0.01 mm. The error in the measurement of Young's modulus (ΔY) is found to be $x \times 10^{10}$ Nm^{-2}. The value of x is ______. (Take g = 10 ms^{-2}) **[July 26, 2022 (I)]**

55. In an experiment to determine the velocity of sound in air at room temperature using a resonance is observed when the air column has a length of 20.0 cm for a tuning fork of frequency 400 Hz is used. The velocity of the sound at room temperature is 336 ms^{-1}. The third resonance is observed when the air column has a length of ____cm. **[June 29, 2022 (II)]**

56. A metre scale is balanced on a knife edge at its centre. When two coins, each of mass 10 g are put one on the top of the other at the 10.0 cm mark the scale is found to be balanced at 40.0 cm mark. The mass of the metre scale is found to be $x \times 10^{-2}$ kg. The value of x is **[June 24, 2022 (I)]**

57. A student is performing the experiment of resonance column. The diameter of the column tube is 6 cm. The frequency of the tuning fork is 504 Hz. Speed of the sound at the given temperature is 336 m/s. The zero of the metre scale coincides with the top end of the resonance column tube. The reading of the water level in the column when the first resonance occurs is: **[Feb. 25, 2021 (I)]**

(a) 13 cm (b) 16.6 cm (c) 14.8 cm (d) 18.4 cm

58. In an experiment to verify Stokes law, a small spherical ball of radius r and density ρ falls under gravity through a distance h in air before entering a tank of water. If the terminal velocity of the ball inside water is same as its velocity just before entering the water surface, then the value of h is proportional to : (ignore viscosity of air) **[Sep. 05, 2020 (II)]**

(a) r^4 (b) r (c) r^3 (d) r^2

59. In a resonance tube experiment when the tube is filled with water up to a height of 17.0 cm from bottom, it resonates with a given tuning fork. When the water level is raised the next resonance with the same tuning fork occurs at a height of 24.5 cm. If the velocity of sound in air is 330 m/s, the tuning fork frequency is : **[Sep. 05, 2020 (I)]**

(a) 2200 Hz (b) 550 Hz

(c) 1100 Hz (d) 3300 Hz

60. A capillary tube made of glass of radius 0.15 mm is dipped vertically in a beaker filled with methylene iodide (surface tension = 0.05 Nm^{-1}, density = 667 kg m^{-3}) which rises to height h in the tube. It is observed that the two tangents drawn from liquid-glass interfaces (from opp. sides of the capillary) make an angle of 60° with one another. Then h is close to (g = 10 ms^{-2}). **[2 Sep. 2020 (II)]**

(a) 0.049 m (b) 0.087 m (c) 0.137 m (d) 0.172 m

61. A tuning fork of frequency 480 Hz is used in an experiment for measuring speed of sound (v) in air by resonance tube method. Resonance is observed to occur at two successive lengths of the air column, l_1 = 30 cm and l_2 = 70 cm. Then, v is equal to : **[12 April 2019 (II)]**

(a) 332 ms^{-1} (b) 384 ms^{-1}

(c) 338 ms^{-1} (d) 379 ms^{-1}

62. A resonance tube is old and has jagged end. It is still used in the laboratory to determine velocity of sound in air. A tuning fork of frequency 512 Hz produces first resonance when the tube is filled with water to a mark 11 cm below a reference mark, near the open end of the tube. The experiment is repeated with another fork of frequency 256 Hz which produces first resonance when water reaches a mark 27 cm below the reference mark. The velocity of sound in air, obtained in the experiment, is close to: **[12 Jan. 2019 II]**

(a) 322 ms^{-1} (b) 341 ms^{-1}

(c) 335 ms^{-1} (d) 328 ms^{-1}

63. While measuring the speed of sound by performing a resonance column experiment, a student gets the first resonance condition at a column length of 18 cm during winter. Repeating the same experiment during summer, she measures the column length to be x cm for the second resonance. Then **[2008]**

(a) 18 > x (b) x > 54

(c) 54 > x > 36 (d) 36 > x > 18

ANSWER KEYS

1	(3)	8	(b)	15	(b)	22	(412)	29	(5)	36	(b)	43	(d)	50	(a)	57	(c)
2	(35)	9	(a)	16	(b)	23	(b)	30	(13)	37	(d)	44	(d)	51	(d)	58	(a)
3	(a)	10	(c)	17	(a)	24	(c)	31	(52)	38	(b)	45	(b)	52	(b)	59	(a)
4	(b)	11	(a)	18	(22)	25	(a)	32	(c)	39	(c)	46	(b)	53	(2)	60	(b)
5	(d)	12	(c)	19	(41)	26	(180)	33	(a)	40	(a)	47	(d)	54	(2)	61	(b)
6	(b)	13	(d)	20	(c)	27	(150)	34	(a)	41	(b)	48	(d)	55	(104)	62	(d)
7	(b)	14	(d)	21	(c)	28	(5)	35	(d)	42	(c)	49	(c)	56	(6)	63	(b)

HINTS & SOLUTIONS

Units and Measurements 1

1. **(d)** Intensity of an electromagnetic wave, representing the average power transmitted per unit area

$$I = \frac{1}{2} \epsilon_0 E_0^2 \times C$$

$$E_0 = \sqrt{\frac{2I}{\epsilon_0 C}}$$

Hence, the unit of above expression will be equivalent to electric field.

E_0 : electric field.

2. **(d)** Heat capacity

$$C' = \frac{\Delta Q}{\Delta T} = JK^{-1}$$

Specific heat capacity, $S = \frac{\Delta Q}{m\Delta T} = Jkg^{-1}K^{-1}$

Latent heat $L = \frac{\Delta Q}{m} = Jkg^{-1}$

Thermal conductivity $K = \frac{\Delta Q(L)}{A\Delta T} = Jm^{-1}K^{-1}s^{-1}$

3. **(b)** Magnetic induction → Gauss

Magnetic intensity,

$$H = \frac{I}{\chi} = \frac{M}{\chi V} = \frac{\text{Ampere} \times (\text{metre})^2}{(\text{metre})^3} = \text{Ampere / metre}$$

Magnetic flux → Weber (Wb)

Magnetic moment = iA = Ampere-metre^2

4. **(b)** Electric field in a certain region is given by,

$$\vec{E} = \frac{A}{x^2}\hat{i} + \frac{B}{y^3}\hat{j}$$

$$\left[\frac{A}{x^2}\right] = NC^{-1} \Rightarrow [A] = Nm^2C^{-1}$$

$$\left[\frac{B}{y^3}\right] = NC^{-1} \Rightarrow [B] = Nm^3C^{-1}$$

5. **(b)** (A) Surface Tension $= \frac{F}{\ell} = \frac{MLT^{-2}}{L} = MT^{-2}$
$= Kgs^{-2}$

(B) Pressure $= \frac{F}{A} = \frac{MLT^{-2}}{L^2} = Kgm^{-1}s^{-2}$

(C) Viscosity $= \frac{F}{A\left(\frac{dV}{dz}\right)} = \frac{MLT^{-2}}{L^2\left(\frac{LT^{-1}}{L}\right)}$

$= ML^{-1}T^{-1} = Kg\, m^{-1}\, s^{-1}$

(D) Impulse $= \int Fdt = MLT^{-2} \times T$

$= MLT^{-1} = Kg\, ms^{-1}$

6. **(b)** Torque = F(N) × $r_\perp$(m) → Nm

Stress $= \frac{\text{Force (N)}}{\text{Area}(m^2)}$ → N/m^2

Latent heat $= \frac{\text{Energy (J)}}{\text{Mass (kg)}}$ → J kg^{-1}

Power $= \frac{\text{Work (Nm)}}{\text{Time (s)}}$ → N ms^{-1}

7. **(a)** Given, $v_2 = \frac{n}{m^2} v_1 \Rightarrow \frac{L_2}{T_2} = \frac{n}{m^2} \cdot \frac{L_1}{T_1}$ ...(i)

and, $a_2 = \frac{a_1}{mn} \Rightarrow \frac{L_2}{T_2^2} = \frac{L_1}{T_1^2} \cdot \frac{1}{mn}$ ...(ii)

Dividing (i) by (ii), we get

$$\frac{L_2/T_2}{L_2/T_2^2} = \frac{n/m^2 \cdot L_1/T_1}{L_1/T_1^2 \cdot 1/mm} \Rightarrow T_2 = \frac{n^2}{m} . T_1$$

as, $L_2 = \frac{T_2}{T_1} . L_1 \frac{n}{m^2} = \frac{n^2}{m} L_1 \frac{n}{m^2} = \frac{n^3}{m^3} L_1$

8. **(d)** $F = 6\pi\eta vr$

∴ Unit of coefficient of viscosity in S.I. system kg/m-s.

9. **(a)** $\frac{E}{H} = \frac{\text{Volt /metre}}{\text{Unit of (nI)}} = \frac{\frac{V}{m}}{\frac{1}{m} \cdot A} = \frac{V}{A} = \Omega$ = Resistance

10. **(a)** $n_1 u_1 = n_2 u_2$

$$128 \frac{kg}{m^3} = n_2 \frac{50g}{(25\, cm)^3}$$

$$\Rightarrow 128 \times \frac{1000\, gm}{(100\, cm)^3} = n_2 \frac{50\, gm}{(25\, cm)^3} \Rightarrow n_2 = 40 \text{ units}$$

11. **(b)** $E_y = KJ_x B_Z$ (Given)

∴ Constant of proportionality

$$K = \frac{E_y}{B_Z J_x} = \frac{c}{J_x} = \frac{m^3}{As}$$

[As $\frac{E}{B} = c$ (speed of light) and $J = \frac{I}{\text{Area}}$]

12. **(d)** Vacuum permeability (μ_0) = $[MLT^{-2}A^{-2}]$

Vacuum permittivity (ε_0) = $[M^{-1}L^{-3}T^4A^2]$

Calculating the dimensions of

$$\left(\frac{\mu_0}{\varepsilon_0}\right)^{1/2} = [ML^2T^{-3}A^{-2}]$$

The dimension of (R) = $[ML^2T^{-3}A^{-2}]$.

13. **(c)** Calculating dimensional formula for each quantities

(A) Mass density $= \frac{M}{V} = M^1L^{-3}$ ...(iv)

(B) Impulse $= M \times u = M^1L^1T^{-1}$ ...(ii)

(C) Power $= F \cdot V = M^1L^2T^{-3}$ ...(i)

(D) Moment of inertia $= Mr^2 = M^1L^2$ ...(iii)

14. **(d)** The displacement current is given by

$$I_d = \varepsilon_0 \frac{d\phi_\varepsilon}{dt} \qquad \therefore \left[\varepsilon_0 \frac{d\phi_\varepsilon}{dt}\right] = [\text{Current}].$$

15. **(d)** Electric dipole moment $(\vec{P}) = q \times 2l$

Magnetic dipole moment $(\vec{M}) = IA$

$$\left[\frac{P}{M}\right] = \left[\frac{LTA}{L^2A}\right] = L^{-1}T = M^0L^{-1}T^1A^0$$

Comparing values of P and Q are 0, –1

Correct answer : Option (d).

16. **(a)**

(A) [Boltzmann constant] =

$$[k] = \frac{[PV]}{[NT]} = \frac{ML^2T^{-2}}{K} = ML^2T^{-2}K^{-1}$$

(B) [Coefficient of viscosity] =

$$[\eta] = \frac{[F]}{[6\pi rv]} = \frac{MLT^{-2}}{L^2T^{-1}} = ML^{-1}T^{-1}$$

(C) [Plank's constant] =

$$[h] = \frac{[E]}{[f]} = \frac{ML^2T^{-2}}{T^{-1}} = ML^2T^{-1}$$

(D) $[\text{Thermal conductivity}] = [K] = \left[\frac{Q.x}{A.\theta.t}\right]$

$$= \frac{(ML^2T^{-3})}{L^2.K} L = MLT^{-3}K^{-1}$$

17. **(a)** (A) Gravitational constant, $G = \frac{Fr^2}{m^2}$

$$\therefore [G] = \frac{[MLT^{-2}][L^2]}{[M^2]} = [M^{-1}L^3T^{-2}]$$

(B) $[P.E.] = [mgh] = [MLT^{-2}L] = [ML^2T^{-2}]$

(C) $[\text{Gravitational Potential}] = \left[\frac{GM}{r}\right]$

$$= \frac{[M^{-1}L^3T^{-2}][M]}{[L]} = [M^0L^2T^{-2}] = [L^2T^{-2}]$$

(D) [Acceleration due to gravity] $= [g] = [LT^{-2}]$

18. **(b)** Let momentum P depends as

$P = Q^x \mu_0^y I^z = [AT]^x [MLT^{-2}A^{-2}]^y [A]^z$

$MLT^{-1} = M^yL^yT^{x-2y}A^{-2y+z+x}$

Equating powers y = 1

$x - 2y = -1$

$-2y + z = 0 \qquad \therefore x = y = z = 1$

Therefore dimensions of P = dimensions of $q\mu_0 I$

19. **(b)** Speed of light

$$C = \frac{1}{\sqrt{\mu_0\varepsilon_0}} \Rightarrow \frac{1}{\mu_0\varepsilon_0} = C^2 = L^2T^{-2}$$

20. **(d)** From given real gas equation,

$$\left[P + \frac{a}{V^2}\right](V - b) = RT$$

$\therefore \quad [a] = [P][V^2] = ML^{-1}T^{-2}L^6 = ML^5T^{-2}$

$[b] = [V] = L^3$

$[ab^{-2}] = ML^5T^{-2}T^{-6} = ML^{-1}T^{-2}$ = Dimension of energy density.

21. **(b)** (A) Coefficient of viscosity $\eta = F\frac{dx}{Adv}$

$[\eta] = [ML^{-1}T^{-1}]$

(B) Intensity $[I] = [ML^0T^{-3}]$

(C) Pressure gradient $= \frac{P}{\text{length}} = [ML^{-2}T^{-2}]$

(D) Compressibility $[K] = [M^{-1}LT^{-2}]$

22. **(d)** Young's modulus,

$$[Y] = \frac{[F]}{[A]\left[\frac{\Delta\ell}{\ell}\right]} = \frac{MLT^{-2}}{L^2} = ML^{-1}T^{-2}$$

$[\text{Torque}] = [r_\perp F] = L \times MLT^{-2} = ML^2T^{-2}$

Coefficient of viscosity,

$$[\eta] = \left[\frac{F}{A}\right][t] = \frac{MLT^{-2}}{L^2} \times T = ML^{-1}T^{-1}$$

Gravitational constant,

$$[G] = \frac{[F.r^2]}{[m_1m_2]} = \frac{MLT^{-2} \times L^2}{M^2} = M^{-1}L^3T^{-2}$$

23. **(b)** $[\text{Surface Tension}] = \frac{[F]}{[L]} = \frac{[MLT^{-2}]}{[L]} = [MT^{-2}]$

$[\text{Impulse}] = [F][t] = [MLT^{-2}][T] = [MLT^{-1}]$

$\therefore$ Surface tension and impulse have different dimensions.

24. **(a)** By principle of homogeneity,

$[At^2] = [v] \qquad \therefore [A] = \left[\frac{v}{t^2}\right]$

$[c] = [t]$

$[B] = [v]$

$$\therefore \quad [ABC] = \left[\frac{v}{t^2}\right][v][t] = \frac{[v^2]}{[t]} = \frac{[L^2T^{-2}]}{[T]} = [L^2T^{-3}]$$

25. **(a)** Angular impulse $= [ML^2T^{-1}]$

Latent Heat $= [M^0L^2T^{-2}]$

Electrical resistivity $= [ML^3T^{-3}A^{-2}]$

Electromotive force $= [ML^2T^{-3}A^{-1}]$

26. **(0)** Dimension of modulus of elasticity $= ML^{-1}T^{-2}$

Torque $= ML^2T^{-2}$

$$\frac{\text{modulus of elasticity}}{\text{Torque}} = \frac{\text{Stress}}{\text{strain} \times \text{torque}}$$

$$=\frac{[\text{Force}]}{[\text{Area}]\times[\text{Force}\times\text{length}]}$$

$$=\frac{1}{[\text{Area}\times\text{length}]}=\left[L^{-3}\right]=\left[M^aL^bT^c\right]\ \therefore\ c=0$$

27. (d) Magnetic field is related to the force experienced by a moving charge as F = qvB

$$B=\left[\frac{MLT^{-2}}{AT\,L/T}\right]=\left[MT^{-2}A^{-1}\right]$$

$$B=\frac{\mu_0 I}{2\pi r}$$

$$\Rightarrow[\mu_0]=\left[\frac{B\times r}{I}\right]=\left[\frac{MT^{-2}A^{-1}\times L}{A}\right]=\left[MLT^{-2}A^{-2}\right]$$

$$[M]=[L^2A]$$

$$[\tau]=[c\theta]\Rightarrow[c]=\left[\frac{\tau}{\theta}\right]=\left[ML^2T^{-2}\right]$$

28. (c) $x(t)=A\sin(t)+B\cos^2(t)+Ct^2+D$

We can say that $x(t)$ represents displacement, its dimension is $[x(t)]=[L]$

The sine function is dimensionless

Similarly, the cosine function is also dimensionless.

∴ By principle of homogeneity,

$[A]=[L]$

$[B]=[L]$

$[C]=[LT^{-2}]$

$[D]=[L]$

$$\left[\frac{ABC}{D}\right]=\left[\frac{L\times L\times LT^{-2}}{L}\right]=[L^2T^{-2}]$$

29. (c) Linear charge density, $\lambda=\frac{Q}{L}$

The expression of the electric flux is

$\phi=\alpha\sigma+\beta\lambda$

$[\phi]=[\alpha\sigma]=[\beta\lambda]$

Surface charge density, $\sigma=\frac{Q}{A}$

$$\left[\frac{\alpha}{\beta}\right]=\frac{[\lambda]}{[\sigma]}=\frac{\left[\frac{Q}{L}\right]}{\left[\frac{Q}{\text{Area}}\right]}=\left[\frac{\text{Area}}{\text{Length}}\right]=L$$

30. (b) Relation between magnetic field and permeability is $B=\mu_0 ni$

$$\Rightarrow\left[\frac{B}{\mu_0}\right]=[ni]=L^{-1}A^1$$

31. (d) The capacitance is

$$[F]=\frac{[q]}{[V]}=\frac{[q.q]}{[V.q]}=\frac{[q^2]}{[W]}=\frac{[C^2]}{[ML^2T^{-2}]}=\left[C^2M^{-1}L^{-2}T^2\right]$$

32. (c) Heat, Q = mL where, L = latent heat

$$\therefore\ L=\frac{Q}{m}=\frac{ML^2T^{-2}}{M}=M^0L^2T^{-2}$$

33. (b) Energy, E = hν

$\Rightarrow[ML^2T^{-2}]=h[T^{-1}]$

$$\Rightarrow h=\frac{\left[ML^2T^{-2}\right]}{\left[T^{-1}\right]}\Rightarrow h=[ML^2T^{-1}]$$

34. (b) Electric field is given as,

$$E=\frac{Q}{4\pi\varepsilon_0R^2}\Rightarrow\varepsilon_0=\frac{Q}{4\pi R^2E}$$

$$\varepsilon_0E^2=\frac{Q}{4\pi R^2}.E$$

$$[\varepsilon_0E^2]=\left[\frac{QE}{R^2}\right]=\frac{[Q]}{[R^2]}\frac{[W]}{[Q][R]}$$

$$=\frac{[W]}{[R^3]}=\frac{ML^2T^{-2}}{L^3}=ML^{-1}T^{-2}$$

35. (b) $\left[\vec{\tau}\right]=\vec{r}\times\vec{F}=[L.MLT^{-2}]=\left[ML^2T^{-2}\right]$

Magnetic field is given by;

$$B=\frac{F}{qV}=\left[\frac{MLT^{-2}}{ATLT^{-1}}\right]=\left[MA^{-1}T^{-2}\right]$$

Magnetic moment is $[M]=[I\times\text{Area}]=\left[AL^2\right]$

Permeability of free space

$$[\mu]=\left[\frac{Br^2}{Idl}\right]=\left[\frac{MT^{-2}A^{-1}\times L^2}{AL}\right]$$

$=[MLT^{-2}A^{-2}]$

36. (c) Specific heat,

$$C=\frac{\Delta Q}{m\Delta T}$$

$$[C]=\left[\frac{ML^2T^{-2}}{MK}\right]=[L^2T^{-2}K^{-1}]$$

∴ Statement I is correct.

Ideal gas equation is,

$$PV=nRT\Rightarrow R=\frac{PV}{nT}$$

$$[R]=\frac{[ML^{-1}T^{-2}][L^3]}{[mol][K]}=[ML^2T^{-2}mol^{-1}K^{-1}]$$

∴ Statement II is incorrect.

37. (b) $[uG]=[ML^{-1}T^{-2}][M^{-1}L^3T^{-2}]$

$[uG]=[M^0L^2T^{-4}]$

$\therefore[uG]=[L^1T^{-2}]$

38. (b) As, $[b]=[V]$

Also, $[P]=\left[\frac{a}{V^2}\right]\Rightarrow[a]=\left[PV^2\right]$

$$\therefore\left[ab^{-1}\right]=\left[\frac{a}{b}\right]=\left[\frac{PV^2}{V}\right]=[PV]=[\text{work}]=[ML^2T^{-2}]$$

39. **(c)** When comparing the given equation with standard equation of stationary wave $\frac{2\pi n}{\lambda} = \omega$ and $\frac{2\pi}{\lambda} = k$

$$\left[\frac{n}{\lambda}\right] = [\omega] = T^{-1}; [nt] = [\lambda] = L$$

$[n] = [\lambda\omega] = LT^{-1}$; $[x] = [\lambda] = L$

40. **(c)** According to principle of homogeneity, dimensions of LHS should be equal to dimensions of RHS

Dimensions of $[G] = \left[M^{-1}L^3T^{-2}\right]; [M] = M$

$[r] = L, \left[T^2\right] = T^2$

$$T^2 = \frac{4\pi^2 r^3}{GM} = \frac{L^3}{M^{-1}L^3T^{-2}M} = T^2$$

41. **(d)** Angular impulse = change in angular momentum

i.e., $I = \Delta L$

$[I] = [mvr] = M.LT^{-1}.L = [M^1L^2T^{-1}]$

42. **(b)** Given

$F = ax^2 + bt^{1/2}$

From the principle of homogeneity of dimension

$[F] = [ax^2]$

Also $[F] = [bt^{1/2}]$

$$\therefore [a] = \frac{[F]}{\left[x^2\right]} = [M^1L^{-1}T^{-2}] \text{ and } [b] = \frac{[F]}{\left[t^{1/2}\right]} = [M^1L^1T^{-5/2}]$$

$$\text{Dimension of } \left[\frac{b^2}{a}\right] = \frac{[M^2L^2T^{-5}]}{[M^1L^{-1}T^{-2}]} = [M^1L^3T^{-3}]$$

43. **(b)** According to principle of homogeneity, only quantity with same dimension can be subtracted

$\therefore \quad [B] = L^2$

$$E = \frac{B}{At} - \frac{x^2}{At}$$

$$[E] = \left[\frac{B}{At}\right] = \left[\frac{x^2}{At}\right] \quad \therefore A = \frac{x^2}{tE} = \frac{L^2}{TML^2T^{-2}} = \frac{1}{MT^{-1}}$$

$\Rightarrow A = M^{-1}T \quad \therefore [AB] = [L^2M^{-1}T^1]$

44. **(c)** Viscous force, $F = \eta A\frac{dv}{dy}$

$[MLT^{-2}] = \eta[L^2][T^{-1}]$

Coefficient of viscosity $\eta = [ML^{-1}T^{-1}]$

Surface tension, $S.T = \frac{F}{\ell} = \frac{[MLT^{-2}]}{[L]} = [ML^0T^{-2}]$

Angular momentum, $L = mvr = [ML^2T^{-1}]$

$K.E = \frac{1}{2}I\omega^2 = [ML^2T^{-2}]$

45. **(a)** $m = k\, c^P G^{-1/2} h^{1/2}$

$M^1L^0T^0 = [LT^{-1}]^P[M^{-1}L^3T^{-2}][ML^2T^{-1}]^{1/2}$

$$[M^1L^0T^0] = [M^{\frac{1}{2}}M^{\frac{1}{2}}][L^PL^{\frac{-3}{2}}L^1][T^{-P}T^1T^{\frac{-1}{2}}]$$

$$[M^1L^0T^0] = [m^1L^{P-\frac{1}{2}}][T^{-P+\frac{1}{2}}]$$

By comparing, $P = \frac{1}{2}$

46. **(a)** Planck's constant, $h = E/\nu \Rightarrow [h] = ML^2T^{-1}$

Angular Momentum, $L = mvr \Rightarrow [L] = ML^2T^{-1}$

momentum, $P = mv \Rightarrow [P] = MLT^{-1}$

Moment of force, $\tau = F \times r_{\perp} \Rightarrow [\tau] = ML^2T^{-2}$

(Here h is Planck's constant, L is angular momentum, P is linear momentum and τ is moment of force)

47. **(b)** Dimensions of power

$$[P] = \left[\frac{a}{V^2}\right] \Rightarrow [a] = [PV^2]$$

And volume, $[V] = [b]$

$$\therefore \text{Dimensions of } \frac{[a]}{[b^2]} = \frac{[PV^2]}{[V^2]} = [P]$$

48. **(a)** $v = \lambda^a g^b \rho^C$

using dimension formula

$\Rightarrow [M^0L^1T^{-1}] = [L^1]^a[L^1T^{-2}]^b[M^1L^{-3}]^c$

$\Rightarrow [M^0L^1T^{-1}] = [M^cL^{a+b-3c}T^{-2b}]$

$\therefore c = 0, a + b - 3c = 1, -2b = -1 \Rightarrow b = \frac{1}{2}$

Now $a + b - 3c = 1 \Rightarrow a + \frac{1}{2} - 0 = 1 \Rightarrow a = \frac{1}{2}$

$\therefore a = \frac{1}{2}, b = \frac{1}{2}, c = 0$

49. **(a)** From the given equation it is clear,

X and $\frac{a}{Y^2}$ have same dimensions

Y and b have same dimensions

$\therefore [a] = [ML^5T^{-2}]$ and $[b] = [L^3]$

$\therefore \frac{[a]}{[b]} = \left[ML^2T^{-2}\right]$ The dimensions of energy.

50. **(a)** Angular momentum,

$L = mvr = [M][LT^{-1}][L] = [ML^2T^{-1}]$

Moment of inertia,

$I = mr^2 = [M][L^2] = [ML^2]$

Spring constant

$$k = \frac{F}{x} = \frac{[MLT^{-2}]}{[L]} = MT^{-2}$$

Angular speed

$$\omega = \frac{\theta}{t} = \frac{1}{T} = T^{-1}$$

51. **(b)** Given

density, $\rho \propto F^aV^bT^c$

$[ML^{-3}] = [MLT^{-2}]^a[LT^{-1}]^b[T]^c = [M^aL^{a+b}T^{-2a-b+c}]$

Comparing powers

a = 1,
a + b = –3, ⇒ b = –4,
– 2a – b + c = 0 ⇒ –2 + 4 + c = 0
⇒ c = –2

52. **(c)** **A.** Torque ⇒ $\vec{\tau} = \vec{r} \times \vec{F}$
$= [L][MLT^{-2}] = ML^2T^{-2}$

B. Stress $= \frac{F}{A} \Rightarrow \frac{MLT^{-2}}{L^2}$
stress $= ML^{-1}T^{-2}$

C. Pressure gradient $= \frac{\Delta P}{\Delta X}$

$\Rightarrow \frac{[F/A]}{[L]} \Rightarrow \frac{MLT^{-2}}{L^3} \Rightarrow ML^{-2}T^{-2}$

D. Coefficient of viscosity ⇒ $F = 6\pi\eta rv$
$MLT^{-2} = \eta[L][LT^{-1}]$

$\eta = \frac{MLT^{-2}}{L^2T^{-1}}$

$\eta = ML^{-1}T^{-1}$

53. **(d)** $[b] = [V], [P] = \left[\frac{a}{V^2}\right] = \left[\frac{a}{b^2}\right]$

$\therefore \left[\frac{b^2}{a}\right] = \frac{1}{[P]} = \frac{1}{[B]} = [K]$

54. **(a)** Let $m = Kh^a c^b G^c$
$\Rightarrow [m] = [h]^a [c]^b [G]^c$
$\Rightarrow M^1 = (ML^2T^{-1})^a (LT^{-1})^b (M^{-1}L^3T^{-2})^c$
⇒ 1 = a – c, 2a + b + 3c = 0, –a – b – 2c = 0
Solving above, we get

$a = \frac{1}{2}, b = \frac{1}{2}$ and $c = -\frac{1}{2}$

55. **(b)** Unit of $\left(\frac{R}{\sqrt{x_L x_C}}\right) = \frac{Ohm}{\sqrt{Ohm^2}} = \frac{Ohm}{Ohm}$ i.e. it is unit less

56. **(b)** We have
[Angular Momentum] $= ML^2T^{-1}$
[Torque] $= ML^2T^{-2}$
[Stress] $= ML^{-1}T^{-2}$

[Pressure Gradient] $= \frac{dP}{dx} = ML^{-2}T^{-2}$

57. **(c)** Dimension of pressure gradient

$\frac{dp}{dx} = \frac{\left(\frac{F}{A}\right)}{L} = \frac{[ML^{-1}T^{-2}]}{[L]} = [M^1L^{-2}T^{-2}]$

Energy density (u) $= \frac{(E)}{(V)} = \frac{[ML^2T^{-2}]}{[L^3]}$

$= [M^1L^{-1}T^{-2}]$

Electric field (E) $= \frac{(F)}{(q)} = \frac{[MLT^{-2}]}{[A.T]} = [M^1L^1T^{-3}A^{-1}]$

Latent heat (L) $= \frac{(Q)}{(m)} = \frac{[ML^2T^{-2}]}{[M]} = [M^0L^2T^{-2}]$

58. **(b)** $(x - At)^2 + \left(y - \frac{t}{B}\right)^2 = a^2$

$[x] = [At] \Rightarrow [A] = \frac{[x]}{[t]} = \frac{L}{T^{-1}} = LT$

and, $[y] = \left[\frac{t}{B}\right] \Rightarrow [B] = \frac{[t]}{[y]} = \frac{T^{-1}}{L} = T^{-1}L^{-1}$.

59. **(b)** $Y = \frac{Stress}{Strain} = \frac{Fl}{A\Delta l}$

So, $[Y] = \frac{[F][l]}{[A\Delta l]} = \frac{MLT^{-2}L}{L^2L} = ML^{-1}T^{-2}$

$F = 6\pi\eta rv \Rightarrow [\eta] = \frac{[F]}{[6\pi rv]} \Rightarrow [\eta] = \frac{MLT^{-2}}{LLT^{-1}} = ML^{-1}T^{-1}$

$E = h\nu \Rightarrow h = \frac{E}{\nu} \Rightarrow [h] = \frac{[E]}{[\nu]} \Rightarrow [h] = \frac{Ml^2T^{-2}}{T^{-1}} = ML^2T^{-1}$
Work function has same dimension as that of energy
So, $[\phi] = ML^2T^{-2}$

60. **(b)** $[h] = ML^2T^{-1}$

$[V_S] = \frac{[W]}{[Q]} = \frac{ML^2T^{-2}}{AT} = ML^2T^{-3}A^{-1}$

$[\phi] = ML^2T^{-2}$ and $[P] = MLT^{-1}$

61. **(a)** $\nu = r^a \rho^b s^c \Rightarrow [\nu] = [r]^a [\rho]^b [s]^c$

$\Rightarrow \quad [T^{-1}] = [L^1]^a [M^1L^{-3}]^b \left[\frac{MLT^{-2}}{L}\right]^c$

$\Rightarrow \quad T^{-1} = M^{b+c} . L^{a-3b} . T^{-2c}$
$\Rightarrow \quad b + c = 0, a - 3b = 0, -2c = -1$

$\Rightarrow \quad c = \frac{1}{2}, b = -\frac{1}{2}, a - 3b = 0; \; a + \frac{3}{2} = 0 \; \Rightarrow \; a = -\frac{3}{2}$

62. **(d)** Given, $T = k\sqrt{\frac{\rho r^3}{S^{\frac{3}{2}}}}$

Dimension of LHS $= M^0L^0T$

Dimension of RHS $= \frac{\left[M^{\frac{1}{2}}L^{-\frac{3}{2}}\right]\left[L^{\frac{3}{2}}\right]}{[MT^{-2}]^{\frac{3}{4}}} = M^{-\frac{1}{4}}L^0T^{\frac{3}{2}}$

Dimension of L.H.S ≠ Dimension of R.H.S
So, assertion is false and reason is true.

63. **(c)** Energy per unit volume, $u = \frac{B^2}{2\mu_0}$

$\left[\frac{B^2}{\mu_0}\right] = [u] = \frac{\left[ML^2T^{-2}\right]}{\left[L^3\right]} = \left[ML^{-1}T^{-2}\right]$

64. **(d)** $[\alpha\beta]=[\eta]=$ Dimensionless

So, $[\alpha]=\dfrac{1}{[\beta]} \Rightarrow [\alpha]\neq[\beta]$

$[\eta^{-1}\sin\theta]=\dfrac{1}{[\alpha\beta]}=$ Dimensionless

65. **(d)** $\sin\left(\dfrac{\alpha x}{kt}\right)=$ Dimensionless

$\therefore \dfrac{\alpha[L]}{[ML^2T^{-2}]}=[M^0L^0T^0] \Rightarrow \alpha=[ML^1T^{-2}]$

$\dfrac{\alpha}{\beta}=\dfrac{\text{Energy}}{\text{Volume}}=\dfrac{[ML^2T^{-2}]}{[L^3]} \Rightarrow \beta=\dfrac{[ML^1T^{-2}][L^3]}{ML^2T^{-2}}$
$=[M^0L^2T^0]$

66. **(a)** $F=\eta A\dfrac{dV}{dx} \Rightarrow \eta=\dfrac{F\,dx}{A\,dV}$

$\Rightarrow[\eta]=\dfrac{[F][dx]}{[A][dV]}=\dfrac{[F][dt][dx]}{[A][dt][dV]}=\dfrac{[P][dx]}{[A][dx]}=[PA^{-1}T^{\circ}]$

67. **(a)** As current and electric field have different units. So options (b) is incorrect. As current density $=\dfrac{I}{A}$ and surface charge density $=\dfrac{Q}{A}$. So both have different unit.

As electric potential $=\dfrac{W}{Q}$ and energy has unit of 'W'. So both have different unit. Now, only option left is (A), which must be correct

Also, $[D]=[\epsilon_o E]=\left[\dfrac{Q}{A}\right]=[ATL^{-2}]$

68. **(c)** We have $[P]=\left[\dfrac{a}{V^2}\right] \Rightarrow [a]=[P][V^2]$

And, $[V]=[b]$. So, $\dfrac{[a]}{[b]}=[P][V]=[PV]$

69. **(c)** As $F_{drag}=6\pi\eta rv \Rightarrow \eta=\dfrac{F}{6\pi rv}$

So, unit of η is $\dfrac{NS}{m^2}$

Now, unit of $P\times t$ is $\dfrac{Ns}{m^2}$

So, Assertion is correct

Also, $F=\eta A\dfrac{dv}{dx} \Rightarrow \eta=\dfrac{F}{A\dfrac{dv}{dx}}$ So, reason is incorrect.

70. **(d)** RC and $\dfrac{L}{R}$ are time constants.

Where $\sqrt{LC}$ is reciprocal of angular frequency and has dimension of time. So $\left(\dfrac{L}{C}\right)$ is only option left.

71. **(a)** Pascal second $=\dfrac{NS}{m^2}=\dfrac{[MLT^{-2}][T]}{[L^2]}=ML^{-1}T^{-1}$

72. **(c)** As $\left[\dfrac{kt}{\beta x}\right]=1$ $[\beta]=\left[\dfrac{kt}{x}\right]=\dfrac{ML^2T^{-2}}{L}=MLT^{-2}$

$[\because [E]=[k_BT]]$

Now, $[P]=\dfrac{[\alpha]}{[\beta]}$ $\quad\therefore\ [\alpha]=[P][\beta]=MLT^{-2}$

73. **(c)** Mutual indnctance, $M=-\dfrac{e_2}{\dfrac{di_1}{dt}}$

$[M]=\dfrac{[e_2]}{\left[\dfrac{di_1}{dt}\right]}=\dfrac{\left[\dfrac{W}{q}\right]}{\left[\dfrac{q}{t^2}\right]}=\dfrac{ML^2T^{-2}}{\dfrac{A^2T^2}{T^2}}=ML^2A^{-2}T^{-2}$

74. **(d)** Wave number and Rydberg constant have same unit m^{-1}. Stress and coefficient of elasticity have same unit Pascal (P).
Coercivity and magnetisation have same unit A/m.

Whereas, specific heat capacity, $S=\dfrac{\Delta Q}{m\Delta T}$ has unit $\dfrac{J}{Kg\,K}$ and latent heat, $L=\dfrac{\Delta Q}{m}$ has unit $\dfrac{J}{Kg}$.

75. **(a)** Velocity gradient $=\dfrac{dv}{dx}$

So, its unit will be $\dfrac{m/s}{m}=s^{-1}$

As $N=N_0e^{-\lambda t}$, $\lambda=$ decay constant
As power of e is dimensionless

So, $[\lambda t]=1$ or, $[\lambda]=\left(\dfrac{1}{t}\right)$

So, unit of λ is s^{-1}

76. **(a)** Torque τ = force × perpendicular distance
$=[MLT^{-2}][L]=[ML^2T^{-2}]$
Impulse I = force × time $=[MLT^{-2}][T]=[MLT^{-1}]$
Tension force $=MLT^{-2}$

Surface tension $S=\dfrac{F}{\ell}=MT^{-2}$

77. **(a)** (i) $\dfrac{\pi pa^4}{8\eta L}=\dfrac{dv}{dt}=$ Volumetric flow rate (Poiseuille's law)

(ii) $\because h\rho g=\dfrac{2s}{r}\cos\theta \quad \therefore h=\dfrac{2s\cos\theta}{\rho rg}$

(iii) $RHS \Rightarrow \varepsilon\times\dfrac{1}{4\pi\varepsilon_0}\dfrac{q}{r^2}\times\dfrac{1}{t}=\dfrac{1}{t}\times\dfrac{q}{r^2}=\dfrac{I}{L^2}=IL^{-2}$

LHS $\Rightarrow J=\dfrac{I}{A}=IL^{-2}$

(iv) $W=\tau\theta$.

78. **(b)** $[M]=k[F]^a[T]^b[V]^c$

$[M^1]=[M^1L^1T^{-2}]^a[T^1]^b[L^1T^{-1}]^c=M^aL^{a+c}T^{-2+b-c}$
$a=1, b=1, c=-1$
$\therefore [M]=[FTV^{-1}]$

79. (c) As $B=\frac{\mu_0 i}{2\pi r}$

$$[\mu_0]=\left[\frac{B\times 2\pi r}{i}\right]=\left[\frac{N}{Am}\times\frac{m}{A}\right]=\left[\frac{N}{A^2}\right]=MLT^{-2}A^{-2}$$

Clearly, μ_0 is not a dimensionless quantity.

80. (a) As, density $=[F]^a[L^b][T^c]$

$[ML^{-3}]=[MLT^{-2}]^a[L]^b[T]^c$

$[ML^{-3}]=[M^a L^a T^{-2a} L^b T^c]$

$[M^1L^{-3}]=[M^aL^{a+b}T^{-2a+c}]$

On comparing

$a=1, a+b=-3, 1+b=-3, b=-4$

$-2a+c=0 \Rightarrow c=2a$

$c=2$ ∴ Density $=[F^1L^{-4}T^2]$

81. (d) $[E]=ML^2T^{-2}$

$[L]=ML^2T^{-1}$

$[G]=M^{-1}L^{+3}T^{-2}$

$$P=\frac{EL^2}{M^5G^2}\Rightarrow[P]=\frac{(ML^2T^{-2})(M^2L^4T^{-2})}{M^5(M^{-2}L^6T^{-4})}=M^0L^0T^0$$

82. (d) (a) Dimension of magnetic Induction

F = BIL

$$\Rightarrow B=\frac{F}{IL}=\frac{[MLT^{-2}]}{[A][L]}\Rightarrow B=MT^{-2}A^{-1}$$

(b) Dimension of magnetic flux = $BA\cos\theta$

$=[MT^{-2}A^{-1}][L^2]=ML^2T^{-2}A^{-1}$

(c) Dimension of magnetic Permeability = $MLT^{-2}A^{-2}$

(d) Dimension of magnetization = $M^0L^{-1}A$

83. (a) Charge Q = CV

$$\Rightarrow C=\frac{Q}{V}=\frac{(A\times T)^2}{ML^2T^{-2}}=M^{-1}L^{-2}T^4A^2$$

Electric Field, $E=\frac{F}{q}=\frac{MLT^{-2}}{AT}=MLT^{-3}A^{-1}$

Force, $F=\frac{q_1 q_2}{4\pi\epsilon_0 r^2}\Rightarrow \epsilon_0=M^{-1}L^{-3}T^4A^2$

Speed of light, $c=\frac{1}{\sqrt{\mu_0\epsilon_0}}$

$$\Rightarrow \mu_0=\frac{1}{\epsilon_0 c^2}=\frac{1}{[M^{-1}L^{-3}T^4A^2][LT^{-1}]^2}$$

$=[M^1L^1T^{-2}A^{-2}]$

84. (b) Dimension of $A=MLT^{-2}$, $D=T^{-1}$, $B=L^{-1}$

Dimension $=\frac{AD}{B}=\frac{MLT^{-2}T^{-1}}{L^{-1}}=ML^2T^{-3}$

85. (b) Planck's constant,

$$h=\frac{E}{\nu}=[ET]=ML^2T^{-2}\times T=[ML^2T^{-1}]$$

Kinetic energy,

$$[E]=\frac{1}{2}mv^2=[ML^2T^{-2}]$$

Electric potential,

$$[V]=\left[\frac{U}{q}\right]=\frac{ML^2T^{-2}}{AT}=[ML^2T^{-3}A^{-1}]$$

Linear momentum,

$$[P]=mv=M\times LT^{-1}=[MLT^{-1}]$$

86. (d) $[E]=\frac{hc}{\lambda}$ also $[E]=\frac{e^2}{4\pi\varepsilon_0 r}$

or, $\frac{[E]}{[E]}=\frac{e^2\lambda}{4\pi\varepsilon_0 r\cdot hc}$

or, $[M^0L^0T^0]=\left[\frac{e^2}{4\pi\varepsilon_0 r}\frac{\lambda}{hc}\right]=\left[\frac{1}{4}\frac{|e|^2}{\pi\varepsilon_0 hc}\right]$.

87. (b) $W=\alpha^2\beta e^{-\frac{\beta x^2}{kT}}$

As exponents are dimensionless, so, $\frac{\beta x^2}{kT}$ should be dimensionless.

$$[\beta]=\left[\frac{kT}{x^2}\right]=\frac{ML^2T^{-2}}{L^2}=MT^{-2}$$

From the dimensional homogeneity, $\alpha^2\beta$ should have dimension of work.

∴ $[\alpha^2\beta]=ML^2T^{-2}$

$$\Rightarrow [\alpha^2]=\frac{ML^2T^{-2}}{MT^{-2}}\Rightarrow[\alpha]=M^0LT^0$$

88. (a) We know that

Speed of light, $c=\frac{1}{\sqrt{\mu_0\varepsilon_0}}=x$

Also, $c=\frac{E}{B}=y$

Time constant, $\tau=Rc=t$ ∴ $z=\frac{l}{Rc}=\frac{l}{t}=$ Speed

Thus, x, y, z will have the same dimension of speed.

89. (d) From formula, $\frac{dQ}{dt}=kA\frac{dT}{dx}$

$$\Rightarrow k=\frac{\left(\frac{dQ}{dt}\right)}{A\left(\frac{dT}{dx}\right)}\Rightarrow[k]=\frac{[ML^2T^{-3}]}{[L^2][KL^{-1}]}=[MLT^{-3}K^{-1}]$$

90. (d) Dimension of Force $F=M^1L^1T^{-2}$

Dimension of velocity $V=L^1T^{-1}$

Dimension of work $=M^1L^2T^{-2}$

Dimension of length = L

Moment of inertia = ML^2

$$\therefore x=\frac{IFv^2}{WL^4}=\frac{(M^1L^2)(M^1L^1T^{-2})(L^1T^{-1})^2}{(M^1L^2T^{-2})(L^4)}$$

$$=\frac{M^1L^{-2}T^{-2}}{L^3}=M^1L^{-1}T^{-2}=\text{Energy density}$$

91. **(b)** Solar constant $= \dfrac{\text{Energy}}{\text{Time Area}}$

Dimension of Energy, $E = ML^2T^{-2}$

Dimension of Time = T

Dimension of Area = L^2

$\therefore$ Dimension of Solar constant $= \dfrac{M^1L^2T^{-2}}{TL^2} = M^1L^0T^{-3}$.

92. **(d)** Young's modulus, $Y = \dfrac{\text{stress}}{\text{strain}}$

$\Rightarrow Y = \dfrac{F}{A}\Big/\dfrac{\Delta\ell}{\ell_0} = FA^{-1}V^0$

93. **(c)** $E = F \times s$

$[E] = [F] \times [s]$

$\Rightarrow [E] = \left[\dfrac{P}{t}\right] \times [A]^{1/2} \Rightarrow [E] = PT^{-1}A^{1/2}$

94. **(a)** $\sqrt{\dfrac{\mu_0}{\varepsilon_0}} = \sqrt{\dfrac{\mu_0^2}{\varepsilon_0\mu_0}} = \mu_0 c \quad \left(\dfrac{1}{\sqrt{\mu_0\varepsilon_0}} = c\right)$

$\mu_0 c \rightarrow MLT^{-2}A^{-2} \times LT^{-1}$

$ML^2T^{-3}A^{-2}$

Dimensions of resistance, $[R] = \left[\dfrac{V}{I}\right] = \left[\dfrac{W}{qI}\right] = \left[\dfrac{W}{I^2t}\right]$

$= \dfrac{ML^2T^{-2}}{A^2T} = ML^2T^{-3}A^{-2}$

95. **(a)** $X = 5YZ^2 \Rightarrow Y \propto \dfrac{X}{Z^2}$ ...(i)

$X = \text{Capacitance} = \dfrac{Q}{V} = \dfrac{Q^2}{W} = \dfrac{[A^2T^2]}{[ML^2T^{-2}]}$

$X = [M^{-1}L^{-2}T^4A^2]$

$Z = B = \dfrac{F}{IL} \qquad [\because F = ILB]$

$Z = [MT^{-2}A^{-1}]$

$Y = \dfrac{[M^{-1}L^{-2}T^4A^2]}{[MT^{-2}A^{-1}]^2} \Rightarrow Y = [M^{-3}L^{-2}T^8A^4]$ (Using (i))

96. **(b)** As we know,

$\left[\dfrac{\ell}{r}\right] = [T]$ and $[cv] = [AT]$

$\therefore \left[\dfrac{\ell}{rcv}\right] = \left[\dfrac{T}{AT}\right] = [A^{-1}]$

97. **(b)** Dimension of $[h] = [ML^2T^{-1}]$

$[C] = [LT^{-1}]$

$[G] = [M^{-1}L^3T^{-2}]$

Hence dimension of

$\left[\sqrt{\dfrac{hC^5}{G}}\right] = \dfrac{[ML^2T^{-1}]^{\frac{1}{2}} \cdot [L^5T^{-5}]^{\frac{1}{2}}}{[M^{-1}L^3T^{-2}]^{\frac{1}{2}}} = [ML^2T^{-2}] = \text{energy}$

98. **(None)** Stopping potential $(V_0) \propto h^x I^y G^Z C^r$

Here, h = Planck's constant $= [ML^2T^{-1}]$

I = current = $[A]$

G = Gravitational constant $= [M^{-1}L^3T^{-2}]$

and c = speed of light $= [LT^{-1}]$

V_0 = potential $= [ML^2T^{-3}A^{-1}]$

$\therefore [ML^2T^{-3}A^{-1}] = [ML^2T^{-1}]^x [A]^y [M^{-1}L^3T^{-2}]^z [LT^{-1}]^r$

$M^{x-z};\ L^{2x+3z+r};\ T^{-x-2z-r};\ A^y$

Comparing dimension of M, L, T, A, we get

$y = -1, x = 0, z = -1, r = 5 \ \therefore\ V_0 \propto h^0 I^{-1} G^{-1} C^5$

99. **(d)** The quantity $\dfrac{B^2}{2\mu_0}$ is the energy density of magnetic field.

$\Rightarrow \left[\dfrac{B^2}{2\mu_0}\right] = \dfrac{\text{Energy}}{\textit{Volume}} = \dfrac{\text{Force} \times \text{displacement}}{(\text{displacement})^3}$

$= \left[\dfrac{ML^2T^{-2}}{L^3}\right] = ML^{-1}T^{-2}$

100. **(b)** Plank length is a unit of length, $l_p = 1.616229 \times 10^{-35}$ m

$l_p = \sqrt{\dfrac{hG}{c^3}}$

101. **(a)** $M = \dfrac{L}{Vr}$

$[M] = \dfrac{L}{C.Ct} = \dfrac{[L]}{[C^2][t]} = hT^{-1}C^{-2}$

102. **(d)** Dimension of A $\neq$ dimension of (C)

Hence A – C is not possible.

103. **(b)** $\sigma = \dfrac{ne^2\tau}{m}$

$[\sigma] = \dfrac{L^{-3}I^2T^2T}{M} = M^{-1}L^{-3}I^2T^3$

104. **(c)** Option (a), (b) and (d) is dimensionally independent of mass. But, $[\mu]$ has mass dependence dimensionally so, option (c) should be correct.

105. **(d)** Let unit 'u' related with e, a_0, h and c as follows.

$[u] = [e]^a [a_0]^b [h]^c [C]^d$

Using dimensional method,

$[M^{-1}L^{-2}T^{+4}A^{+2}] = [A^1T^1]^a[L]^b[ML2T^{-1}]^c[LT^{-1}]^d$

$[M^{-1}L^{-2}T^{+4}A^{+2}] = [M^c L^{b+2c+d} T^{a-c-d} A^a]$

$a = 2,\ b = 1,\ c = -1,\ d = -1$

$\therefore \quad u = \dfrac{e^2a_0}{hc}$

106. **(b)** The dimensional formulae of

$e = [M^0L^0T^1A^1]$

$\varepsilon_0 = [M^{-1}L^{-3}T^4A^2]$

$G = [M^{-1}L^3T^{-2}]$ and $m_e = [M^1L^0T^0]$

Now, $\dfrac{e^2}{2\pi\varepsilon_0 Gm_e^2}$

$= \dfrac{[M^0L^0T^1A^1]^2}{2\pi[M^{-1}L^{-3}T^4A^2][M^{-1}L^3T^{-2}][M^1L^0T^0]^2} = \dfrac{1}{2\pi}$

$\because \frac{1}{2\pi}$ is dimensionless thus the combination $\frac{e^2}{2\pi\varepsilon_0 Gm_e^2}$ would have the same value in different systems of units.

107. (b) As we know, $F = \frac{1}{4\pi\varepsilon_0}\frac{q_1q_2}{R^2} \Rightarrow \varepsilon_0 = \frac{q_1q_2}{4\pi FR^2}$

Hence, $\varepsilon_0 = \frac{C^2}{N.m^2} = \frac{[AT]^2}{[MLT^{-2}][L^2]} = [M^{-1}L^{-3}T^4A^2]$

108. (c) $[t] = [r]^b[s]^{c/2}[d]^{a/4}$

$T = [L]^b[MT^{-2}]^{c/2}[ML^{-3}]^{a/4}$

So, $\frac{c}{2} + \frac{a}{4} = 0$, $c = -1$ and $b - \frac{3a}{4} = 0$

Solving above, we get $b = \frac{3}{2}$

109. (d) Angular momentum $= m \times v \times r = ML^2T^{-1}$

Latent heat $L = \frac{Q}{m} = \frac{ML^2T^{-2}}{M} = L^2T^{-2}$

Capacitance $C = \frac{\text{Charge}}{\text{Potential Difference}} = M^{-1}L^{-2}T^4A^2$

110. (a) Surface tension, $T = \frac{F}{\ell} = \frac{F}{\ell}.\frac{\ell}{\ell}.\frac{T^2}{T^2}$

(As, $F.\ell = K$ (energy); $\frac{T^2}{\ell^2} = V^{-2}$)

Therefore, surface tension $= [KV^{-2}T^{-2}]$

111. (c) Magnitude of Lorentz formula $F = qvB\sin\theta$

$$B = \frac{F}{qv} = \frac{MLT^{-2}}{C \times LT^{-1}} = [MT^{-1}C^{-1}]$$

112. (b) Mutual inductance $= \frac{\phi}{I} = \frac{BA}{I}$

$$[\text{Henry}] = \frac{[MT^{-1}Q^{-1}L^2]}{[QT^{-1}]} = ML^2Q^{-2}$$

113. (d) Moment of Inertia, $I = MR^2$

$[I] = [ML^2]$

Moment of force, $\vec{\tau} = \vec{r} \times \vec{F}$

$\vec{\tau} = [L][MLT^{-2}] = [ML^2T^{-2}]$

114. (a) According to Stokes law,

$$F = 6\pi\eta rv \Rightarrow \eta = \frac{F}{6\pi rv}$$

$$\eta = \frac{[MLT^{-2}]}{[L][LT^{-1}]} \Rightarrow \eta = [ML^{-1}T^{-1}]$$

115. (c) As we know, the velocity of light in free space is given by

$$c = \frac{1}{\sqrt{\mu_o\varepsilon_o}}$$

$$\therefore \frac{1}{\mu_o\varepsilon_o} = c^2 = [LT^{-1}]^2 = [M^0L^2T^{-2}]$$

116. (b) Momentum, $= mv = [MLT^{-1}]$

Planck's constant,

$$h = \frac{E}{v} = \frac{[ML^2T^{-2}]}{[T^{-1}]} = [ML^2T^{-1}]$$

117. (a) Work $W = \vec{F}\cdot\vec{s} = [MLT^{-2}][L] = [ML^2T^{-2}]$;

Torque, $\vec{\tau} = \vec{r} \times \vec{F} = [L][MLT^{-2}] = [ML^2T^{-2}]$

118. (c) Given, quantity $Q = X^{-2}Y^{+\frac{3}{2}}Z^{-\frac{2}{5}}$

Fractional error $= 2\frac{\Delta X}{X} + \frac{3}{2}\frac{\Delta Y}{Y} + \frac{2}{5}\frac{\Delta Z}{Z}$

$= 2(0.1) + \frac{3}{2}(0.2) + \frac{2}{5}(0.5) = 0.2 + 0.3 + 0.2 = 0.7$

119. (15) $C = P^1 q^2 r^{-3} s^{1/2}$

$$\frac{dC}{C} = \frac{dP}{P} + \frac{2dq}{q} + \frac{3dr}{r} + \frac{1}{2}\frac{ds}{s}$$

Here, $\frac{dp}{p} = 1\%$, $\frac{dq}{q} = 2\%$, $\frac{dr}{r} = 3\%$ and $\frac{ds}{s} = 2\%$

$$\therefore \frac{dC}{C} = (1 + 2 + 2 + 3 \times 3 + \frac{1}{2} \times 2)\% = 15\%$$

120. (c) The sum of three masses is

$m = m_1 + m_2 + m_3 = 435.42 + 226.3 + 0.125 = 661.845$ g

According to least significant digits, m = 661.8 g

121. (77) Given $Q = \frac{ab^4}{cd}$

$\therefore$ % error in Q is

$$\left(\frac{\Delta Q}{Q} \times 100\right)\% = \left[\frac{\Delta a}{a} + 4\frac{\Delta b}{b} + \frac{\Delta c}{c} + \frac{\Delta d}{d}\right] \times 100\%$$

$$\Rightarrow \frac{x}{1000} = \left[\frac{3}{60} + 4\left(\frac{0.1}{20}\right) + \left(\frac{0.2}{40}\right) + \frac{0.1}{50}\right] \times 100\%$$

$\Rightarrow x = 77$

122. (d) $y = \frac{32.3 \times 1125}{27.4} = 1326.186$

By rounding off the result to 3 significant digits, we get y = 1330

123. (c) Given, $E = \alpha^3 e^{-\beta t}$

$\ln E = 3\ln\alpha - \beta t$

$$\left(\frac{dE}{E}\right)_{max} = \frac{3d\alpha}{\alpha} + \beta\frac{dt}{t} \times t$$

$= 3 \times 1.2\% + (0.3 \times 1.6 \times 5)\% = 6\%$

124. (b) Density of a wire, $\rho = \frac{m}{\pi R^2\ell}$

$$\Rightarrow \frac{d\rho}{\rho} = \frac{dm}{m} + \frac{2dR}{R} + \frac{d\ell}{\ell}$$

$\because$ % error is $\rho = \left(\frac{d\rho}{\rho} \times 100\%\right)$

$$= \left(\frac{0.003}{0.6} + \frac{2 \times 0.01}{0.5} + \frac{0.05}{10}\right) \times 100 = 5\%$$

125. (b) In an expression $a \times 10^b$
if $a \le 5$ then order for the given expression is b
$a > 5$ then order for the given expression is $b + 1$

126. (d) Time period of spring is $T = 2\pi\sqrt{\frac{m}{k}}$

Squaring both sides,

$$T^2 \propto \frac{m}{k}$$

on differentiation $\frac{2\Delta T}{T} = \frac{\Delta m}{m} - \frac{\Delta k}{k}$

$$\frac{\Delta k}{k} = \frac{\Delta m}{m} - \frac{2\Delta T}{T} = \frac{\Delta k}{k} \times 100 = \left(\frac{\Delta m}{m} - \frac{2\Delta T}{T}\right) \times 100$$

$$\frac{\Delta k}{k} \times 100 = (-1)\% - 2(2)\% = 5\%$$

127. (b) Resistivity is given by,

$$\rho = R\frac{A}{\ell} = R\frac{\pi r^2}{\ell}$$

Taking log and differentiate

$$\frac{\Delta\rho}{\rho} = \frac{\Delta R}{R} + 2\frac{\Delta r}{r} + \frac{\Delta \ell}{\ell}$$

$$\frac{\Delta\rho}{\rho} \times 100 = \left(\frac{10}{100} + 2 \times \frac{0.05}{0.35} + \frac{0.2}{15}\right) \times 100$$

$$= \left(\frac{1}{10} + \frac{2}{7} + \frac{1}{75}\right) \times 100 = 39.93\%$$

128. (a) Digits between non-zero integers are significant. Hence in 1001, total significant figure is 4.
Zeroes before non-zero integers are not significant. Thus significant figures in 0.0010010 is 5.

129. (c) Resistance, $R = \frac{\rho l}{A} \Rightarrow R = \frac{\rho l}{\pi \frac{d^2}{4}} = \frac{4\rho \ell}{\pi d^2}$

$$\therefore \frac{\Delta R}{R} = \frac{\Delta l}{l} + \frac{2\Delta d}{d} = 0.1\% + 2(0.1\%) \quad \Rightarrow \frac{\Delta R}{R} = 0.3\%$$

130. (c) Given,
Time taken for 50 oscillations, $T = 40$ s
$\Delta l = 2\text{ mm} = 0.2\text{ cm}$
$\Delta T = 1$ second

$$T = 2\pi\sqrt{\frac{\ell}{g}} \Rightarrow g = \frac{4\pi^2 \ell}{T^2}$$

$$\frac{\Delta g}{g} = \frac{\Delta \ell}{\ell} + \frac{2\Delta T}{T} = \frac{0.2}{20} + 2\left(\frac{1}{40}\right) = \frac{1.2}{20}$$

$$\frac{\Delta g}{g} \times 100 = \frac{1.2}{20} \times 100 = 6\%$$

131. (a) $V = 200\text{ V}, \Delta V = 5\text{V}$

$I = 20\text{ A}, \Delta I = 0.2\text{ A}$

$$R = \frac{V}{I}$$

Using error analysis

$$\frac{\Delta R}{R} = \frac{\Delta V}{V} + \frac{\Delta I}{I} \Rightarrow \frac{\Delta R}{R} = \frac{5}{200} + \frac{0.2}{20} = \frac{7}{200}$$

$$\therefore \%\text{ error} = \frac{\Delta R}{R} \times 100 = \frac{7}{200} \times 100 = 3.5\%$$

132. (c) $Q = \frac{a^4 b^3}{c^2}$

By error analysis,

$$\frac{\Delta Q}{Q} = 4\left(\frac{\Delta a}{a}\right) + 3\left(\frac{\Delta b}{b}\right) + 2\left(\frac{\Delta c}{c}\right)$$

$$\frac{\Delta Q}{Q} \times 100 = 4\left(\frac{\Delta a}{a} \times 100\right) + 3\left(\frac{\Delta b}{b} \times 100\right) + 2\left(\frac{\Delta c}{c} \times 100\right)$$

$$\%\text{ error in } Q = 4 \times 3\% + 3 \times 4\% + 2 \times 5\%$$
$$= 12\% + 12\% + 10\%$$
$$= 34\%$$

133. (a) Kinetic energy, $k = \frac{1}{2}mv^2 \quad \therefore \frac{\Delta k}{k} = \frac{\Delta m}{m} + \frac{2\Delta v}{v}$

$$k = \frac{1}{2} \times 5 \times 400 = 5 \times 200 = 1000\text{ J}$$

$$\therefore \frac{\Delta k}{k} = \frac{0.5}{5} + \frac{2 \times 0.4}{20}$$

$$\therefore \Delta k = 1000\left(\frac{1}{10} + \frac{4}{100}\right) = 1000\left(\frac{10+4}{100}\right) = 140\text{ J}$$

134. (c) Density, $\rho = \frac{m}{V}$

$$\Rightarrow \left|\frac{\Delta\rho}{\rho}\right|_{max} = \frac{m}{\pi r^2 l} = \left|\frac{\Delta m}{m}\right| + 2\left|\frac{\Delta r}{r}\right| + \left|\frac{\Delta l}{l}\right|$$

$$= \frac{0.01}{0.4} + \frac{2(0.03)}{6} + \frac{0.04}{8}$$

$$\%\text{ error in density} = \left(\frac{\Delta\rho}{\rho}\right) \times 100\% = \left(\frac{1}{0.4} + \frac{6}{6} + \frac{4}{8}\right)\%$$

$$= (2.5 + 1 + 0.5)\% = 4\%$$

135. (c) Equivalent resistance in parallel combination is given by

$$\frac{1}{R_{eq}} = \frac{1}{R_1} + \frac{1}{R_2} \Rightarrow R_{eq} = \frac{R_1 R_2}{R_1 + R_2} = \frac{10 \times 15}{10 + 15} = 6$$

Differentiating both sides, we get

$$\frac{\Delta R_{eq}}{R^2_{eq}} = \frac{\Delta R_1}{R_1^2} + \frac{\Delta R_2}{R_2^2}$$

$$\Rightarrow \frac{\Delta R_{eq}}{R_{eq}} = \left(\frac{\Delta R_1}{R_1^2} + \frac{\Delta R_2}{R_2^2}\right) R_{eq}$$

$$= \left(\frac{0.5}{100} + \frac{0.5}{225}\right) 6 = \left(\frac{6 \times 0.5}{25}\right)\left(\frac{1}{4} + \frac{1}{9}\right) = \frac{13}{300}$$

$$\frac{\Delta R_{eq}}{R_{eq}} \times 100 = \frac{13}{3} = 4.33\%$$

136. (5) Time period of simple pendulum is given as

$$T = 2\pi\sqrt{\frac{\ell}{g}}$$

$$\Rightarrow g = \frac{1}{4\pi^2}\frac{T^2}{\ell} \Rightarrow \frac{\Delta g}{g} = \frac{2\Delta T}{T} + \frac{\Delta \ell}{\ell}$$

$$\Rightarrow \frac{\Delta g}{g} = 2\cdot\frac{1}{100\times 0.5} + \frac{1\,mm}{10\,cm} \Rightarrow \frac{\Delta g}{g} = \frac{5}{100} = 5\%$$

137. (b) As $[\tau] = ML^2T^{-2}$

$$\text{So, } \frac{\Delta\tau}{\tau} = \frac{\Delta M}{M} + \frac{2\Delta L}{L} + \frac{2\Delta T}{T}$$

$$= 0.05 + 2\times 0.05 + 2\times 0.05 = 0.25$$

$$\text{So, } \frac{\Delta\tau}{\tau}\times 100 = 0.25\times 100 = 25\%$$

138. (d) Dissipated heat is given as,

$E_H = I^2R\times T$

$$\frac{\Delta E}{E}\times 100 = \frac{2\Delta I}{I}\times 100 + \frac{\Delta R}{R}\times 100 + \frac{\Delta T}{T}\times 100$$

$= 2\times 2 + 1 + 3 = 8\%$

139. (a) We have

$$\rho = \frac{m}{V} = \frac{m}{\pi r^2 h} \Rightarrow \frac{\Delta\rho}{\rho} = \frac{\Delta m}{m} + \frac{2\Delta r}{r} + \frac{\Delta h}{h}$$

$$= \frac{0.006}{0.6} + \frac{2\times 0.005}{0.5} + \frac{0.04}{4}$$

$$= 0.01 + 2\times 0.01 + 0.01 = 0.01 + 0.02 + 0.01 = 0.04$$

$$\text{So, } \frac{\Delta\rho}{\rho}\times 100 = 4\%$$

140. (18) $z = a^2x^3y^{1/2}$

$$\frac{\Delta z}{z} = 2\frac{\Delta a}{a} + \frac{3\Delta x}{x} + \frac{1}{2}\frac{\Delta y}{y}$$

a is constant

$$\text{So, } \frac{\Delta z}{z}\times 100\% = \left(3\frac{\Delta x}{x} + \frac{1}{2}\frac{\Delta y}{y}\right)\times 100\%$$

$$= 3\times 4 + \frac{1}{2}\times 12 = 18\%$$

141. (b) $Y = \dfrac{MgL^3}{4bd^3\delta}$

Fraction error

$$\therefore \frac{\Delta Y}{Y} = \frac{\Delta M}{M} + \frac{3\Delta L}{L} + \frac{\Delta b}{b} + \frac{3\Delta d}{d} + \frac{\Delta\delta}{\delta}$$

If error is not specified, then least count is only error.

$$\Rightarrow \frac{\Delta Y}{Y} = \frac{1}{2000} + \frac{3}{1000} + \frac{1}{400} + \frac{3}{400} + \frac{1}{500} = \frac{31}{2000} = 0.0155$$

142. (c) The equivalent resistance in parallel combination,

$$\frac{1}{R_{eq}} = \frac{1}{R_1} + \frac{1}{R_2} = \frac{1}{4} + \frac{1}{4} = \frac{1}{2}$$

$\therefore R_{eq} = 2\Omega.$

Fractional error

$$\text{Also } \frac{\Delta R_{eq}}{R_{eq}^2} = \frac{\Delta R_1}{R_1^2} + \frac{\Delta R_2}{R_2^2}$$

$$\Rightarrow \frac{\Delta R_{eq}}{4} = \frac{0.8}{16} + \frac{0.4}{16} = \frac{1.2}{16} \Rightarrow \Delta R_{eq} = \frac{4.8}{16} = 0.3$$

$$\therefore R_{eq} = (2\pm 0.3)\Omega$$

143. (c) Time period, $T = 2\pi\sqrt{\dfrac{\ell}{g}}$

$$\therefore \frac{\Delta T}{T} = \frac{1}{2}\frac{\Delta\ell}{\ell}$$

Error in time period in one day

$$\Rightarrow \Delta T = \frac{1}{2}\times\frac{0.1}{100}\times 24\times 3600 = \frac{1}{2}\times\frac{0.1}{100}\times 86400$$

$$\Rightarrow \Delta T = 43.2\,s$$

144. (1) Time period,

$$T = 2\pi\sqrt{\frac{\ell}{g}} \Rightarrow g = \frac{4\pi^2\ell}{T^2} \Rightarrow \frac{\Delta g}{g} = \frac{\Delta\ell}{\ell} + \frac{2\Delta T}{T}$$

$$\Delta T = \frac{\text{least count of time }(\Delta T_0)}{\text{number of oscillations }(n)}$$

$$\therefore \frac{\Delta g}{g} = \frac{\Delta\ell}{\ell} + \frac{2\Delta T_0}{nT_0},\ T_0 = \text{Time for 1 oscillation}$$

$\dfrac{\Delta g}{g}$ is minimum for highest value of $n\,T_0$ and ℓ

Hence minimum percentage error in g is obtained by student no. 1

145. (34) Volume of sphere, $V = \dfrac{4}{3}\pi R^3$

Percentage error in volume

$$\frac{\Delta V}{V}\times 100 = (3)\times\frac{\Delta R}{R}\times 100 = (3)\times\frac{0.85}{7.5}\times 100 = 34\ \%$$

Hence $x = 34$

146. (5) Resistance, $R = \frac{V}{I}$

Percentage error in resistance,

$$\Rightarrow \frac{\Delta R}{R}\times 100 = \frac{\Delta V}{V}\times 100 + \frac{\Delta I}{I}\times 100$$

$$= \frac{2}{50}\times 100 + \frac{0.2}{20}\times 100 = 5\%.$$

147. (a) Given, $T = 2\pi\sqrt{\frac{l}{g}}$

Squaring on both sides and rearranging, we get

$$g = \frac{4\pi^2 l}{T^2} \Rightarrow \frac{\Delta g}{g} = \frac{\Delta l}{l} + \frac{2\Delta T}{T}$$

(Here, $\Delta l = 1$ mm, $l = 1$ m, $\Delta T = 0.01$ s, $T = 1.95$ s)

$$\therefore \frac{\Delta g}{g} = \frac{1\times 10^{-3}}{1} + 2\times\frac{0.01}{1.95} \Rightarrow \frac{\Delta g}{g} = 0.0113$$

Percentage error in g

$$\frac{\Delta g}{g}\times 100 = 1.13\%$$

148. (1050)

Density, $\rho = \frac{M}{V} = \frac{M}{\frac{4}{3}\pi\left(\frac{D}{2}\right)^3} \Rightarrow \rho = \frac{6}{\pi}MD^{-3}$

$$\therefore \%\left(\frac{\Delta\rho}{\rho}\right) = \frac{\Delta m}{m} + 3\left(\frac{\Delta D}{D}\right) = 6 + 3\times 1.5 = 10.5\%$$

$$\%\left(\frac{\Delta\rho}{\rho}\right) = \frac{1050}{100}\% = \left(\frac{x}{100}\right)\% \quad \therefore x = 1050.00$$

149. (d) Given : $Z = \frac{a^2 b^{2/3}}{\sqrt{c}d^3}$

Percentage error in z,

$$\frac{\Delta z}{z} = \frac{2\Delta a}{a} + \frac{2}{3}\frac{\Delta b}{b} + \frac{1}{2}\frac{\Delta c}{c} + \frac{3\Delta d}{d}$$

$$= 2\times 2 + \frac{2}{3}\times 1.5 + \frac{1}{2}\times 4 + 3\times 2.5 = 14.5\%.$$

150. (None)

$D_1 = A_1 + B_1 + C_1 = 24.36 + 0.0724 + 256.2 = 280.6$

$D_2 = A_2 + B_2 + C_2 = 24.44 + 16.082 + 240.2 = 280.7$

$D_3 = A_3 + B_3 + C_3 = 25.2 + 19.2812 + 236.183 = 280.7$

$D_4 = A_4 + B_4 + C_4 = 25 + 236.191 + 19.5 = 281$

None of the option matches.

151. (c) Given, Length of simple pendulum, $l = 25.0$ cm

Time of 40 oscillation, $T = 50s$

Time period of pendulum

$$T = 2\pi\sqrt{\frac{\ell}{g}} \Rightarrow T^2 = \frac{4\pi^2\ell}{g} \Rightarrow g = \frac{4\pi^2\ell}{T^2}$$

$$\Rightarrow \text{Fractional error in g} = \frac{\Delta g}{g} = \frac{\Delta l}{l} + \frac{2\Delta T}{T}$$

$$\Rightarrow \frac{\Delta g}{g} = \left(\frac{0.1}{25.0}\right) + 2\left(\frac{1}{50}\right) = 0.044$$

$$\therefore \text{Percentage error in } g = \frac{\Delta g}{g}\times 100 = 4.4\%$$

152. (None) $\rho = \frac{M}{V} = \frac{M}{l^3} = Ml^{-3}$

$$\frac{\Delta\rho}{\rho} = \frac{\Delta M}{M} + 3\frac{\Delta l}{l} = \frac{0.10}{10.00} + 3\left(\frac{0.01}{0.10}\right) = 0.31\ \text{kg/m}^3$$

153. (d) $A = 7\times 5.29 = 37.03\ \text{cm}^2$

The result should have three significant figures, so $A = 37.0\ \text{cm}^2$

154. (c) $\frac{\Delta d}{d}\times 100 = 1.5\ \% + 3\ (1\%) = 4.5\%$

155. (c) From Kepler's law, time period of a satellite,

$$\text{T} = 2\pi\sqrt{\frac{\text{r}^3}{\text{Gm}}} \qquad \text{T}^2 = \frac{4\pi^2}{\text{GM}}\text{r}^3$$

Relative uncertainty in the mass of the earth

$$\left|\frac{\Delta \text{M}}{\text{M}}\right| = 2\frac{\Delta \text{T}}{\text{T}} = 2\times 10^{-2}$$ ($\because$ 4π & G constant and relative uncertainty in radius $\frac{\Delta \text{r}}{\text{r}}$ negligible)

156. (c) Relative error in Surface area, $\frac{\Delta \text{s}}{\text{s}} = 2\times\frac{\Delta \text{r}}{\text{r}} = \alpha$ and relative error in volume, $\frac{\Delta \text{v}}{\text{v}} = 3\times\frac{\Delta \text{r}}{\text{r}}$

$\therefore$ Relative error in volume w.r.t. relative error in area,

$$\frac{\Delta \text{v}}{\text{v}} = \frac{3}{2}\alpha$$

157. (c) Sum of all observations = 90 + 91 + 95 + 92 = 368s

Average $= \frac{368}{4} = 92$ sec

$\Delta T_1 = 92 - 90 = 2,\ \Delta T_2 = 92 - 91 = 1,$

$\Delta T_3 = 95 - 92 = 3,\ \Delta T_4 = 92 - 92 = 0$

$$\Delta T = \frac{|\Delta T_1| + |\Delta T_2| + |\Delta T_3| + |\Delta T_4|}{4} = \frac{2+1+3+0}{4} = 1.5s$$

As the resolution of measuring clock is 1s, therefore the mean time should be given by rounding off 1.5s to 2s which is 92 ± 2.

158. (d) As, $g = 4\pi^2 \frac{L}{T^2}$

So, $\frac{\Delta g}{g} \times 100 = \frac{\Delta L}{L} \times 100 + 2\frac{\Delta T}{T} \times 100$

$= \frac{0.1}{20} \times 100 + 2 \times \frac{1}{90} \times 100 = 2.72 \simeq 3\%$

159. (a) The current voltage relation of diode is

$I = (e^{1000\,V/T} - 1)$ mA (given)

When, $I = 5mA,\ e^{1000\,V/T} = 6mA$

Also, $dI = (e^{1000\,V/T}) \times \frac{1000}{T}.\, dV$

Error $= \pm 0.01V$ (By exponential function)

$= (6\,mA) \times \frac{1000}{300} \times (0.01) = 0.2$ mA

160. (a) Rotation period of earth is about 24 hrs $\simeq 10^5$ s
Revolution period of earth is about 365 days $\simeq 10^7$ s
Speed of light wave $c = 3 \times 10^8$ m/s
Wavelength of visible light of spectrum
$\lambda = 4000 - 7800$ Å

$c = f\lambda \left(\text{and } T = \frac{1}{f}\right)$

Therefore period of light wave is 10^{-15} s (approx)

161. (b) According to the question.

$t = (90 \pm 1)$ or, $\frac{\Delta t}{t} = \frac{1}{90}$

$l = (20 \pm 0.1)$ or, $\frac{\Delta l}{l} = \frac{0.1}{20}$ $\qquad \frac{\Delta g}{g}\% = ?$

As we know,

$t = 2\pi\sqrt{\frac{l}{g}} \Rightarrow g = \frac{4\pi^2 l}{t^2}$

or, $\frac{\Delta g}{g} = \pm\left(\frac{\Delta l}{l} + 2\frac{\Delta t}{t}\right) = \left(\frac{0.1}{20} + 2 \times \frac{1}{90}\right) = 0.027$

$\Rightarrow \frac{\Delta g}{g}\% = 2.7\%$

162. (a) According to ohm's law, V = IR

$R = \frac{V}{I}$

$\therefore$ Percentage error $= \frac{\text{Absolute error}}{\text{Measurement}} \times 10^2$

where, $\frac{\Delta V}{V} \times 100 = \frac{\Delta I}{I} \times 100 = 3\%$

then, $\frac{\Delta R}{R} \times 100 = \frac{\Delta V}{V} \times 10^2 + \frac{\Delta I}{I} \times 10^2$

$= 3\% + 3\% = 6\%$

163. (a) Number of significant figures in 23.023 = 5
Number of significant figures in 0.0003 = 1
Number of significant figures in $2.1 \times 10^{-3} = 2$

164. (a) Momentum, $p = m \times v$
Given, mass of a body = 3.513 kg speed of body = 5.00m/s
$\therefore p = (3.513) \times (5.00) = 17.565$ kg m/s
= 17.6 (Rounding off to get three significant figures)

Motion in a Straight Line

2

1. **(a)** For option (A)
$\phi = kt + c$
For SHM (1D)
$x = A\sin\phi = A\sin(kt + c)$
For option (B)
$v^2 + x^2 = \text{constant}$
$\Rightarrow 2v\dfrac{dv}{dx} + 2x = 0$
$\therefore a = -x$
So, it is 1D motion
For option (C)
Time can't be negative, so motion is not possible.
For option (D)
Distance, s = f(t), so it can be 1D motion.

2. **(d)** Total round $= 1 + \dfrac{1}{2}$
$\therefore$ Displacement = 2r
Distance $= 2\pi r + \dfrac{2\pi r}{2} = 3\pi r$

3. **(10)** Average velocity,
$$v_{avg} = \frac{x_1 + x_2}{t_1 + t_2} = \frac{x_1 + x_2}{\frac{x_1}{v_1} + \frac{x_2}{v_2}}$$
$$\Rightarrow \frac{50}{7} = \frac{x + \frac{3x}{2}}{\frac{x}{5} + \frac{3x}{2v_2}} \Rightarrow \frac{50}{7} = \frac{5/2}{\frac{1}{5} + \frac{3}{2v_2}}$$
$$\Rightarrow \frac{1}{5} + \frac{3}{2v_2} = \frac{7}{20} \Rightarrow \frac{3}{2v_2} = \frac{7}{20} - \frac{1}{5} = \frac{7-4}{20}$$
or, $\dfrac{3}{2v_2} = \dfrac{3}{20}$ $\therefore v_2 = 10$ m/s

4. **(d)**

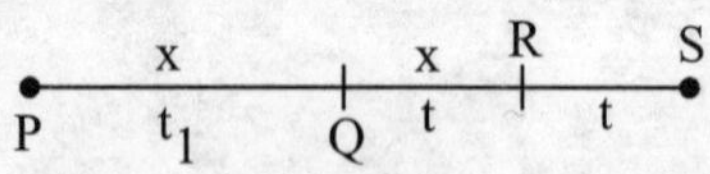

$QS = x = 9t + 15t = 24t$
$PQ = x = 6t_1 = 24t \Rightarrow t_1 = 4t$
$$\langle speed \rangle = \frac{\text{Total distance}}{\text{Total time}} = \frac{48t}{2t + t_1}$$
$$= \frac{48t}{2t + 4t} = \frac{48t}{6t} = 8\,\text{m/s}$$

5. **(b)** Here, $S_1 = \dfrac{1}{2}at^2 = \dfrac{1}{2}\left(\dfrac{v}{t}\right)t^2 = \dfrac{1}{2}vt$
$S_2 = v \times 3t$
$$v_{avg} = \frac{S_1 + S_2}{T} = \frac{\frac{80 \times t}{2} + 80 \times 3t}{4t} = 70\,\text{km/hr}$$

6. **(b)**

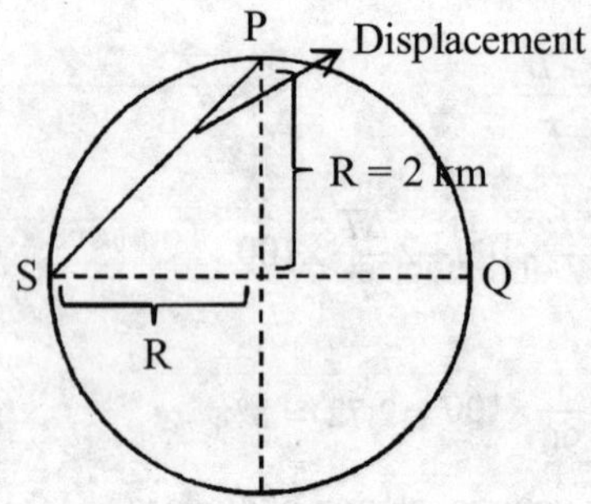

Displacement of cyclist = PS
$\therefore$ Displacement $PS = \sqrt{R^2 + R^2} = R\sqrt{2} = 2\sqrt{2} = \sqrt{8}$ km

7. **(c)** Average velocity, $v = \dfrac{\text{Total Displacement}}{\text{Total time taken}} = \dfrac{x + x}{t_1 + t_2}$
$$\Rightarrow v = \frac{2x}{\frac{x}{v_1} + \frac{x}{v_2}} \quad \left(\because t_1 = \frac{x}{v_1} \text{ and } t_2 = \frac{x}{v_2}\right)$$
$$\Rightarrow v = \frac{2v_1v_2}{v_1 + v_2} \Rightarrow \frac{1}{v_1} + \frac{1}{v_2} = \frac{2}{v}$$

8. **(d)**

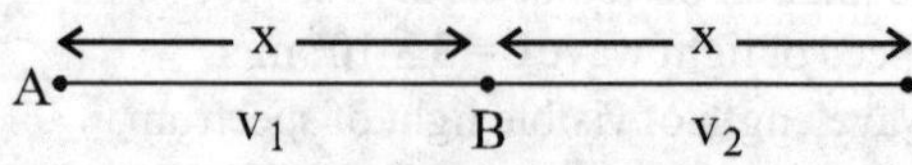

$$\text{Average velocity} = \frac{\text{Total displacement}}{\text{Total time}} = \frac{2x}{t_1 + t_2}$$
$$= \frac{2x}{\frac{x}{v_1} + \frac{x}{v_2}} = \frac{2v_1v_2}{v_1 + v_2}$$

9. **(a)** $PQ = R\sqrt{2}$
Let v be instantaneous velocity

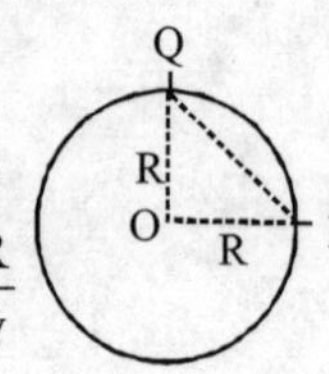

$$t = \frac{\text{Arc length}(PQ)}{v} = \frac{2\pi\frac{R}{4}}{v} = \frac{\pi R}{2v}$$
Average velocity,
$$\langle v \rangle = \frac{\text{Displacement}}{\text{Time}} = \frac{PQ}{t} = \frac{R\sqrt{2} \times 2v}{\pi R} = \frac{2\sqrt{2}v}{\pi}$$
$$\Rightarrow \frac{\pi}{x\sqrt{2}} = \frac{\pi}{2\sqrt{2}}$$
$\therefore x = 2$

10. **(d)** $|\langle\vec{v}\rangle| = \dfrac{|\vec{r}_f - \vec{r}_i|}{\Delta t}$

$$= \frac{2R\cos\left[\frac{\pi-\theta}{2}\right]}{\frac{2\pi R}{3v}} = \frac{3\times 2R\cos\left[\frac{180-120}{2}\right]}{2\pi R} = 3\cos 30^\circ$$

$= 1.5\sqrt{3}\,\text{m/s}$

11. **(b)** Consider,

$AB = x$

$BC = x$

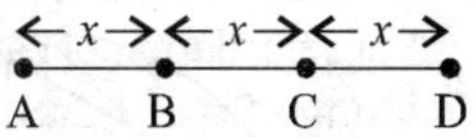

$2x + CD = 3x \Rightarrow CD = 3x - 2x = x$

Average speed of the object $\langle v\rangle = \dfrac{\text{Total distance}}{\text{Total time}}$

$$\langle v\rangle = \frac{3x}{\frac{x}{v_1}+\frac{x}{v_2}+\frac{x}{v_3}} = \frac{3v_1v_2v_3}{v_2v_3+v_1v_3+v_1v_2}$$

12. **(d)**

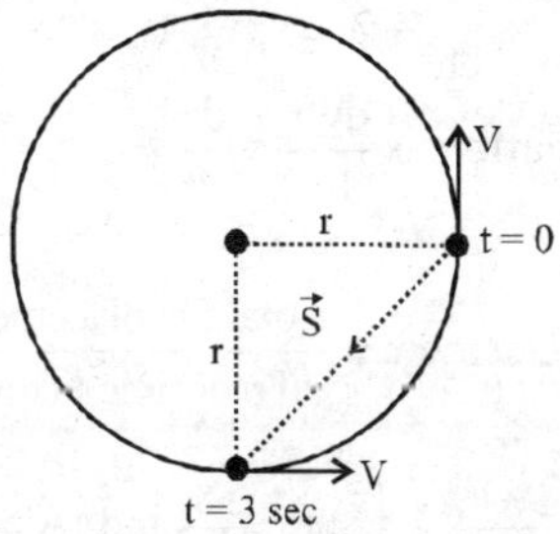

$|\vec{S}| = \sqrt{r^2+r^2} = \sqrt{2}\,r = \sqrt{2}\times 10 = 10\sqrt{2}\,\text{m}$

13. **(50)**

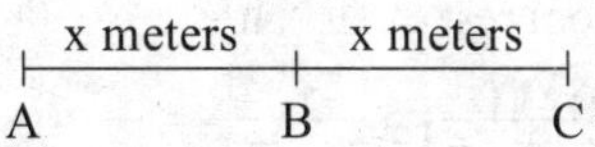

In motion AB,

$t_{AB} = \dfrac{x}{5}$

In motion BC

$x = d_1 + d_2$

where d_1 & d_2 we the distance travelled with 10 m/s and 15 m/s respectively in equal time intervals $\dfrac{'t'}{2}$ each

$d_1 = \dfrac{10t}{2}$, $d_2 = \dfrac{15t}{2}$

$d_1 + d_2 = x = \dfrac{t}{2}(10+15) = \dfrac{25t}{2}$

$\therefore t_{BC} = \dfrac{2x}{25}$

$$\langle v\rangle = \frac{2x}{\frac{x}{5}+\frac{2x}{25}} = \frac{2\times 25}{5+2} = \frac{50}{7}\,\text{m/s}$$

14. **(d)** Given, First distance travelled, $S_1 = 4$ km

Second distance travelled, $S_2 = 4$ km

$$V_{av} = \frac{\text{Distance}}{\text{time}} = \frac{S_1+S_2}{\frac{S_1}{V_1}+\frac{S_1}{V_2}}$$

Here, V_1 and V_2 are the velocities during first and second distance travelled

$$\Rightarrow V_{av} = \frac{4+4}{\frac{4}{3}+\frac{4}{5}} \Rightarrow V_{av} = \frac{15}{4} = 3.75\,\text{km/h}$$

15. **(b)** From ΔAOB

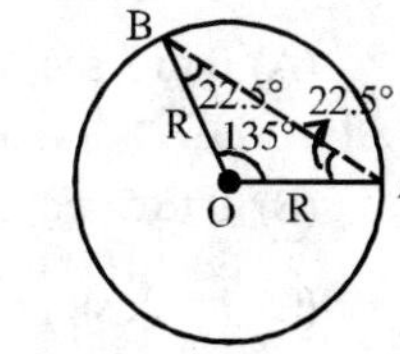

$\dfrac{AB}{\sin 135} = \dfrac{OB}{\sin 22.5}$

$AB = \dfrac{\sin 135}{\sin 22.5} OB$

$$= \frac{\sin(135)}{\sin(22.5)}\frac{\text{arc}(AB)}{\frac{3\pi}{4}} = \frac{\sin(45)}{\sin(22.5)}\times\frac{60\times 4}{3\pi} = 47\,\text{m}$$

16. **(3)** $V_{av} = \dfrac{S_{total}}{T_{total}} = \dfrac{200+200+\frac{2\pi\times 200}{4}}{238}$

$$= \frac{400+100\pi}{238} = \frac{714}{238} = 3\,m/s$$

17. **(18)** $v_{avg} = \dfrac{l_{total}}{t_{total}} = \dfrac{l/3 + l/3 + l/3}{\frac{l}{3v_1}+\frac{l}{3v_2}+\frac{l}{3v_3}}$

$$= \frac{l}{\frac{l}{3}\left(\frac{1}{v_1}+\frac{1}{v_2}+\frac{1}{v_3}\right)} = \frac{1}{\frac{1}{3}\left(\frac{1}{11}+\frac{1}{22}+\frac{1}{33}\right)}$$

$$= \frac{33}{1+\frac{1}{2}+\frac{1}{3}} = \frac{198}{6+3+2} = \frac{198}{11} = 18\,\text{m/s}.$$

18. **(b)** Given, $v = b\sqrt{x}$ or $\dfrac{dx}{dt} = b\,x^{1/2}$

or $\displaystyle\int_0^x x^{-1/2}dx = \int_0^t b\,dt \Rightarrow \frac{x^{1/2}}{1/2} = bt \Rightarrow x = \frac{b^2t^2}{4}$

Differentiating w. r. t. time, we get

$\dfrac{dx}{dt} = \dfrac{b^2\times 2t}{4}$ or $v = \dfrac{b^2\tau}{2}$ $(t=\tau)$

19. **(b)** Graphs in option (c) position-time and option (a) velocity-position are corresponding to velocity-time graph option (d) and its distance-time graph is as given below. Hence distance-time graph option (b) is incorrect.

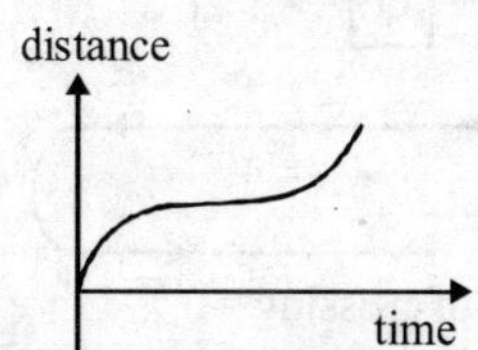

20. **(c)** Average speed $= \dfrac{\text{Total distance travelled}}{\text{Total time taken}} = \dfrac{x}{T}$

$$= \frac{x}{\frac{x}{2\times 40}+\frac{x}{2\times 60}} = 48\,\text{km/h}$$

21. **(d)** The slope of displacement-time (*x* - *t*) graph gives velocity of object.

$$<\vec{v}> = \frac{\Delta\vec{s}}{\Delta t} = \frac{S_f - S_i}{t_f - t_i}$$

$$\vec{v} = \frac{ds}{dt} = \text{slope}$$

verifying each and every option

(A) 0 to 3 sec; $<\vec{v}> = \frac{5-0}{3} = \frac{5}{3}$ m/s

(B) 0 to 5 sec; $<\vec{v}> = \frac{5-5}{2} = 0$

(C) $t = 2$; slope $= \vec{v} = 5$ m/s

(D) $t = 5$ to 7 sec; $<\vec{v}> = \frac{0-5}{2} = -2.5$ m/s

At $t = 6.5$ sec; $\vec{v} = 10$

(E) $t = 0$ to $t = 9$; $<\vec{v}> = 0$.

22. **(d)** For particle moving along x-axis,
$x = C_0(t^2 - 2) + c(t-2)^2$

$\therefore$ Velocity, $v = \frac{dx}{dt} = 2tc_0 + 2c(t-2)$

$\therefore$ Acceleration, $a = \frac{dv}{dt} = 2c_0 + 2c = 2(c_0 + c)$

23. **(3)** For two cars P and Q,
$a_P = kt$, k is constant
$a_Q = a$, a is constant
$\therefore$ $a_{Q/P} = a_Q - a_P = a - kt$

Case-I :
If $u_{Q/P}$ and $a_{Q/P}$ in same direction,

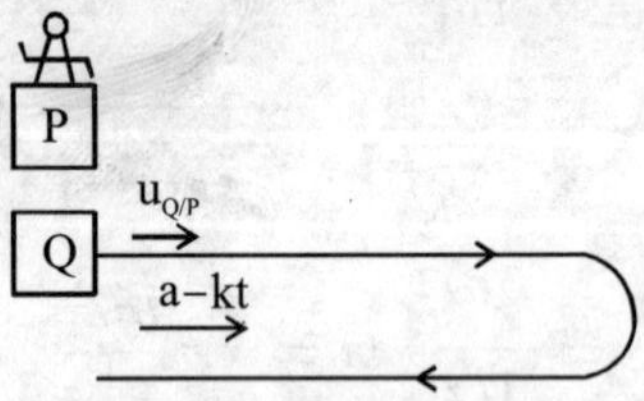

Total number of crossing = 2

Case-II :
If $u_{Q/P}$ and $a_{Q/P}$ in opposite direction,

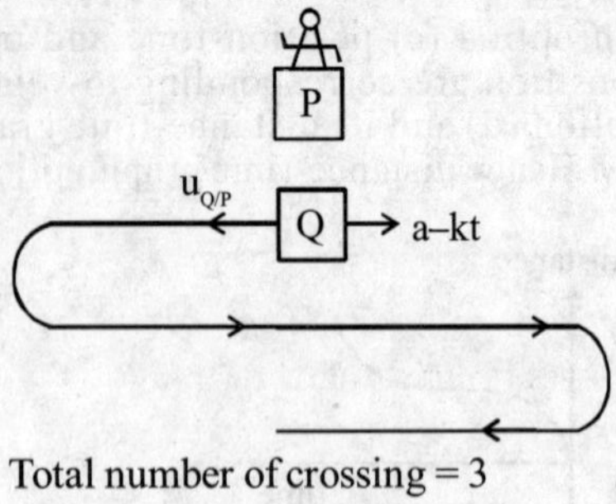

Total number of crossing = 3

$\therefore$ Maximum possible number of crossings = 3

24. **(a)** The area under the velocity-time graph gives distance.

Distance $= \frac{1}{2} \times 2 \times 10 + 2 \times 10 = 30$m

25. **(d)** The area under velocity-time graph gives displacement

$\therefore S = \frac{1}{2} \times (200 + 400) \times 2 + 28.5 \times 400 = 12000\text{m} = 12\text{km}$

26. **(c)** Displacement, $x = t^3 - 6t^2 + 20t + 15$

$\therefore$ Velocity, $v = \frac{dx}{dt} = 3t^2 - 12t + 20$

$\therefore$ Acceleration, $a = \frac{dv}{dt} = 6t - 12$

When a = 0
$\Rightarrow 6t - 12 = 0 \Rightarrow t = 2$ s
At t = 2s, $v = 3(2)^2 - 12(2) + 20 = 8$ m/s

27. **(c)** Given, speed of car, $u = 20\ ms^{-1}$
For retardation of car, $a = -2\ ms^{-2}$
Using $v^2 = u^2 + 2aS$
$\Rightarrow (0)^2 = (20)^2 + 2(-2)(S)$
S = 100 m
Distance travelled by two car = 2 × 100 = 200 m
Remaining distance = 300 – 200 = 100 m

28. **(3)** Displacement at time t is,
$x^2 = 1 + t^2$

On differentiation, $2x\frac{dx}{dt} = 2t \Rightarrow xv = t$

Again on differentiation, $x\frac{dv}{dt} + v\frac{dx}{dt} = 1$

$x.a + v^2 = 1$

$$a = \frac{1-v^2}{x} = \frac{1 - t^2/x^2}{x} = x^{-3}$$

29. **(19)** Acceleration is constant $\Rightarrow S_n = u + \frac{a}{2}(2n-1)$

$\because$ Body starts from rest $\quad \therefore u = 0$

$\therefore \frac{S_{n-1}}{S_n} = \frac{[2(n-1)-1]}{[2n-1]} = 1 - \frac{2}{2n-1} \quad \therefore x = 2n - 1$

For n = 10 $\Rightarrow$ x = 19

30. **(b)** Since $S_n = u + \frac{a}{2}(2n-1)$

Given, $102.5 = u + \frac{a}{2}(2n-1)$ and $115 = u + \frac{a}{2}(2n+3)$

$\Rightarrow 102.5 = u + an - \frac{a}{2}$ & $115 = u + an + \frac{3a}{2}$

$\therefore 12.5 = 2a \Rightarrow a = 6.25\ m/s^2$

31. **(40)** Given initial velocity u = 72 km/h = 20 m/s, v = 0, t = 4s
$v = u + at \Rightarrow 0 = 20 + a \times 4$
$a = -5\ m/s^2$
Using, $v^2 - u^2 = 2as$
$\Rightarrow 0^2 - 20^2 = 2(-5).s$
$\therefore$ s = 40 m

32. **(52)** Given, $x = -3t^3 + 18t^2 + 16t$
On differentiation

$$\upsilon = \frac{dx}{dt} = -9t^2 + 36t + 16$$

Again on differentiation

$$a = \frac{dv}{dt} = -18t + 36$$

If $a=0, t=2s$

Velocity at $t=2s$

$u=-9(2)^2+36\times 2+16=52$ m/s.

33. (8) Relation between acceleration and velocity is:

$$a=v\frac{dv}{dx}=4\sqrt{x}\times 4\times\frac{1}{2\sqrt{x}}=8m/s^2$$

34. (a) Given

$t=\alpha x^2+\beta x$

$$\Rightarrow \frac{dt}{dx}=2\alpha x+\beta \Rightarrow v=\frac{dx}{dt}=\frac{1}{2\alpha x+\beta}$$

$$a=v\frac{dv}{dx}=\frac{-1}{2\alpha x+\beta}\left[\frac{1}{2\alpha x+\beta}\right]^2 2\alpha$$

$$\Rightarrow a=\frac{-2\alpha}{(2\alpha x+\beta)^3} \Rightarrow a=-2\alpha v^3$$

35. (175) Considering acceleration is constant

$v=u+at$

$u+50=u+a\Rightarrow a=50\ m/s^2$

$$\therefore S_{n^{th}}=u+\frac{a}{2}[2n-1]$$

Displacement in $(t+1)^{th}$ second

$$125=u+\frac{a}{2}[2t+1]$$

$$125=u+\frac{50}{2}[2t+1]$$

$$125=u+50t+25$$

$$100=u+50t \quad \text{..... (i)}$$

$$S_{t+2}=u+\frac{a}{2}[2(t+2)-1]$$

$$S_{t+2}=u+25(2t+3)$$

Using equation (i)

$$\therefore S_{t+2}=u+50t+75=175\ m$$

36. (a) Using $s=ut+\frac{1}{2}at^2$

$$s_1=\frac{1}{2}a(p-1)^2$$

$$s_2=\frac{1}{2}ap^2 \quad [\because u=0, t_1=(p-1), t_2=(p)]$$

$$\text{Displacement}=s_1+s_2=\frac{1}{2}a[(p-1)^2+p^2]=\frac{1}{2}at^2$$

$$\therefore t=\sqrt{(2p^2-2p+1)}\ s$$

37. (None) Using $v^2-u^2=2aS$

After travelling 4 cm bullet losses one third of its velocity

$$\left(\frac{2u}{3}\right)^2=u^2+2(-a)(4\times 10^{-2})$$

$$\Rightarrow \frac{4u^2}{9}=u^2-2a(4\times 10^{-2})$$

$$\Rightarrow -\frac{5u^2}{9}=-2a(4\times 10^{-2}) \quad \text{...(i)}$$

When bullet comes to rest,

$$0=\left(\frac{2u}{3}\right)^2+2(-a)(x)$$

$$-\frac{4u^2}{9}=-2ax \quad \text{...(ii)}$$

Dividing eq. (i) by (ii)

$$\frac{5}{4}=\frac{4\times 10^{-2}}{x}\Rightarrow x=\frac{16}{5}\times 10^{-2}$$

$\therefore x=3.2\times 10^{-2}$ m $=32\times 10^{-3}$ m $=D\times 10^{-3}$ m (given)

$\therefore D=32$

38. (c) We know that, $v=\frac{dx}{dt}\Rightarrow dx=v\,dt$

Integrating, $\int_0^x dx=\int_0^t v\,dt$

or $x=\int_0^t (v_0+gt+ft^2)dt=\left[v_0t+\frac{gt^2}{2}+\frac{ft^3}{3}\right]_0^t$

$$\Rightarrow x=v_0t+\frac{gt^2}{2}+\frac{ft^3}{3}$$

At $t=1$, $x=v_0+\frac{g}{2}+\frac{f}{3}$.

39. (a) $v-\alpha\sqrt{x},\Rightarrow \frac{dx}{dt}=\alpha\sqrt{x}\Rightarrow \frac{dx}{\sqrt{x}}=\alpha\,dt$

Integrating both sides,

$$\int_0^x\frac{dx}{\sqrt{x}}=\alpha\int_0^t dt\Rightarrow\left[\frac{2\sqrt{x}}{1}\right]_0^x=\alpha[t]_0^t$$

$$\Rightarrow 2\sqrt{x}=\alpha t\Rightarrow x=\frac{\alpha^2}{4}t^2\Rightarrow x\ \alpha\ t^2$$

40. (c) $x=5t^2-4t+5$

$v=10t-4$; At $t=2s\Rightarrow v=16$ m/s

41. (d) Speed $(v)=\frac{ds}{dt}=\frac{d}{dt}\{(2.5)t^2\}$ $[\because s=(2.5)t^2$ given$]$

$v=5t$

At $t=5$, $v=5t=5\times 5=25$ m/s.

42. (c) Distance = area under v – t graph with positive value

$$\text{Distance}=\frac{1}{2}\times 10\times 5+5\times 10+5\times 10+\frac{1}{2}\times 5\times 10+\frac{1}{2}\times 5\times 20+\frac{1}{2}\times 5\times 20=250\ m$$

Displacement = area of graph with sign $=(200-50)$m $=150$ m

$$\therefore\frac{\text{Distance}}{\text{Displacement}}=\frac{250}{150}=\frac{5}{3}$$

43. (a) Area under velocity time graph gives distance of body in given time.

$$v=\frac{dx}{dt}\Rightarrow x=\int v\,dt=\text{Area}$$

Area under acceleration time graph gives change in velocity in the given time.

$$a=\frac{dv}{dt}\Rightarrow v=\int a\,dt=\text{Area}$$

44. **(d)** Initial velocity, $u = 10\ ms^{-1}$
Final velocity, $v = 60\ ms^{-1}$
Acceleration, $a = 2\ ms^{-2}$
From the equation of motion,
$v = u + at$
$\Rightarrow 60 = 10 + 2t$
$\Rightarrow t = 25$ sec.

45. (200) We have $v^2 = u^2 + 2as$
If $v = 0$, then $s = 500$ m
So, $0 = 20^2 + 2a \times 500$
$\Rightarrow 1000a = -400 \Rightarrow a = -0.4\ m/s^2$
If $s = 250$ m, then
$v^2 = 20^2 + 2 \times -0.4 \times 250$
$v^2 = 400 - 200 \Rightarrow v^2 = 200$
$v = \sqrt{200}$ m/s

46. **(a)** (A) $\frac{dx}{dt}$ = slope ≥ 0 always increasing
(B) $\frac{dx}{dt} < 0$; and at $t \to \infty\ \frac{dx}{dt} \to 0$
(C) $\frac{dx}{dt} > 0$ for first half $\frac{dx}{dt} < 0$ for second half.
(D) $\frac{dx}{dt}$ = constant

47. **(a)** Given, $x = 4t^2$
$v = \frac{dx}{dt} = 8t$
The velocity of the particle at $t = 5$s is
$v = 8 \times 5 = 40$ m/s.

48. **(d)** Displacement $= \sum$area $= 16 - 8 + 16 - 8 = 16$ m
Distance $= \sum|\text{area}| = 48$ m
So, $\frac{\text{displacement}}{\text{Distance}} = \frac{16m}{48m} = \frac{1}{3}$

49. **(b)** Given, $t = \sqrt{x} + 4$
$\Rightarrow x = (t-4)^2 = t^2 - 8t + 16$
$\Rightarrow \frac{dx}{dt} = 2t - 8 \Rightarrow \left.\frac{dx}{dt}\right|_{t=4} = 2 \times 4 - 8 = 0$

50. **(c)** Let initial velocity of bullet = u
Final velocity of bullet $= \frac{u}{3}$
Distance travelled by bullet, s = 4 cm
Using, $v^2 - u^2 = 2as$
$\Rightarrow \left(\frac{u}{3}\right)^2 = u^2 - 2a(4) \Rightarrow a = \frac{8u^2}{9(8)} = \frac{u^2}{9}$
Using $v^2 - u^2 = 2as$ again
$0 = u^2 - 2a(4 + x)$
$\Rightarrow u^2 = 2\left(\frac{u^2}{9}\right)(4 + x) \Rightarrow 4.5 = 4 + x \Rightarrow x = 0.5$

51. **(3)** If speed becomes 'n' times by keeping 'a' constant, then stopping distance becomes n^2 times
Here, $V \to V/3$
So, $S \to \frac{S}{9}$ i.e $\frac{27}{9}$ m i.e. 3 m

52. **(100)** $\frac{dv}{ds} = 5\ ms^{-1}$
So, $a = v\frac{dv}{ds} = 20 \times 5 = 100\ m/sec^2$

53. **(c)** In 't' sec, $S = 10$ m
So, $10 = \frac{1}{2}at^2$ ...(i)
In next 't' sec, suppose body travel S' distance
Then, $10 + S' = \frac{1}{2}a(2t)^2$ ...(ii)
Dividing (ii) by (i), we get
$\Rightarrow \frac{10 + S'}{10} = 4 \Rightarrow 10 + S' = 40 \Rightarrow S' = 30m$

54. **(d)** **For bus P**
$x_p(t) = \alpha t + \beta t^2$
$V_p(t) = \alpha + 2\beta t\ \ [\therefore V_p = \frac{dx_p}{dt}]$
For bus Q
$x_q(t) = ft - t^2$
$V_q(t) = f - 2t\ [\because V_q = \frac{dx_q}{dt}]$
As, $V_p(t) = V_q(t) \Rightarrow \alpha + 2\beta t = f - 2t$
$\Rightarrow \alpha - f = -2\beta t - 2t \Rightarrow f - \alpha = 2\beta t + 2t$
$\Rightarrow t = \frac{f - \alpha}{(2\beta + 2)}$

55. **(1)** As $v^2 = u^2 + 2ax$
At point 'A'
$40 = u^2 + 2a.10$ ..... (i)
At point 'B'
$60 = u^2 + 2a.20$ ..... (ii)
Substracting (ii) from (i), we get
$a = 1\ m/s^2$

56. **(12)** Acceleration equals to $a = v\frac{dv}{dx}$
$a = \frac{vdv}{dx}$ & $v = \sqrt{5000 + 24x}$
$\Rightarrow \frac{dv}{dx} = \frac{1}{2\sqrt{5000 + 24x}} \times 24 = \frac{12}{\sqrt{5000 + 24x}}$
$\Rightarrow a = v\frac{dv}{dx} = \sqrt{5000 + 24x} \times \frac{12}{\sqrt{5000 + 24x}} = 12\ m/s^2$

57. **(b)** We have given,
$v = \alpha t + \beta t^2$
$\Rightarrow \frac{ds}{dt} = \alpha t + \beta t^2 \Rightarrow \int_{s_1}^{s_2} ds = \int_1^2 \left(\alpha t + \beta t^2\right)dt$
$\Rightarrow s_2 - s_1 = \left[\frac{\alpha t^2}{2} + \frac{\beta t^3}{3}\right]_1^2$
As particle is moving in a straight line,
$\therefore$ Distance = Displacement
$\therefore$ Distance $= \left[\frac{\alpha[4-1]}{2} + \frac{\beta[8-1]}{3}\right] = \frac{3\alpha}{2} + \frac{7\beta}{3}$

58. **(c)** For time interval t_1
$v_1 = u_1 + a_1 t_1$
$\Rightarrow v_1 = a_1 t_1$ $[\because u_1 = 0]$

For time interval t_2
$v_2 = u_2 + (-a_2 t_2)$ or, $0 = v_1 - a_2 t_2 \Rightarrow v_1 = a_2 t_2$

$\therefore \ a_1 t_1 = a_2 t_2$ or, $\frac{t_1}{t_2} = \frac{a_2}{a_1}$

59. **(a)**

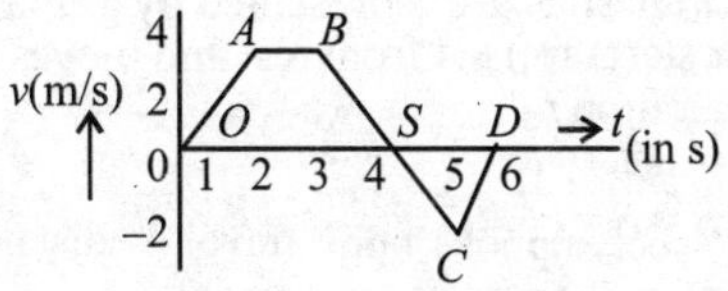

$OS = 4 + \frac{1}{3} = \frac{13}{3}$s; $SD = 2 - \frac{1}{3} = \frac{5}{3}$s

Distance covered by the body = area of v-t graph
= area ($OABS$) + area (SCD)

$$= \frac{1}{2}\left(\frac{13}{3} + 1\right) \times 4 + \frac{1}{2} \times \frac{5}{3} \times 2 = \frac{32}{3} + \frac{5}{3} = \frac{37}{3} \text{ m}$$

60. **(20)**

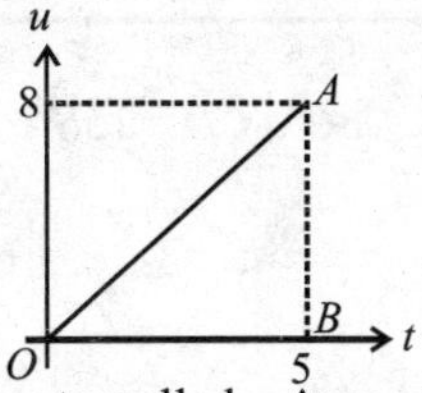

Distance travelled = Area of speed-time graph

$$= \frac{1}{2} \times 5 \times 8 = 20 \text{ m}$$

61. **(3)** Distance X varies with time t as $x^2 = at^2 + 2bt + c$

$$\Rightarrow 2x\frac{dx}{dt} = 2at + 2b$$

$$\Rightarrow x\frac{dx}{dt} = at + b \Rightarrow \frac{dx}{dt} = \frac{(at+b)}{x} \Rightarrow x\frac{d^2x}{dt^2} + \left(\frac{dx}{dt}\right)^2 = a$$

$$\Rightarrow \frac{d^2x}{dt^2} = \frac{a - \left(\frac{dx}{dt}\right)^2}{x} = \frac{a - \left(\frac{at+b}{x}\right)^2}{x}$$

$$= \frac{ax^2 - (at+b)^2}{x^3} = \frac{ac - b^2}{x^3}$$

$\Rightarrow \ a \propto x^{-3}$ Hence, n = 3

62. **(b)** From the third equation of motion
$v^2 - u^2 = 2aS$

But, $a = \frac{F}{m}$

$$\therefore v^2 = u^2 - 2\left(\frac{F}{m}\right)S \Rightarrow v^2 = (1)^2 - (2)\left[\frac{2.5\times10^{-2}}{20\times10^{-3}}\right]\frac{20}{100}$$

$$\Rightarrow v^2 = 1 - \frac{1}{2} \Rightarrow v = \frac{1}{\sqrt{2}} \text{m/s} = 0.7 \text{m/s}$$

63. **(b)** $x = at + bt^2 - ct^3$

Velocity, $v = \frac{dx}{dt} = \frac{d}{dt}(at + bt^2 - ct^3)$
$= a + 2bt - 3ct^2$

Acceleration, $\frac{dv}{dt} = \frac{d}{dt}(a + 2bt - 3ct^2)$

or $0 = 2b - 3c \times 2t \quad \therefore t = \left(\frac{b}{3c}\right)$

and $v = a + 2b\left(\frac{b}{3c}\right) - 3c\left(\frac{b}{3c}\right)^2 = \left(a + \frac{b^2}{3c}\right)$

64. **(d)** For constant acceleration, there is straight line parallel to t-axis on $a-t$ graph.

Inclined straight line on $v-t$ graph, and parabola on $x-t$ graph.

65. **(d)** Position of the particle,
S = area under graph (time t = 0 to 5s)

$$= \frac{1}{2} \times 2 \times 2 + 2 \times 2 + 3 \times 1 = 9 \text{m}$$

66. **(c)** Let time taken by car A to reach finishing point is t_0
$\therefore$ Time taken by car B to reach finishing point $= t_0 + t$

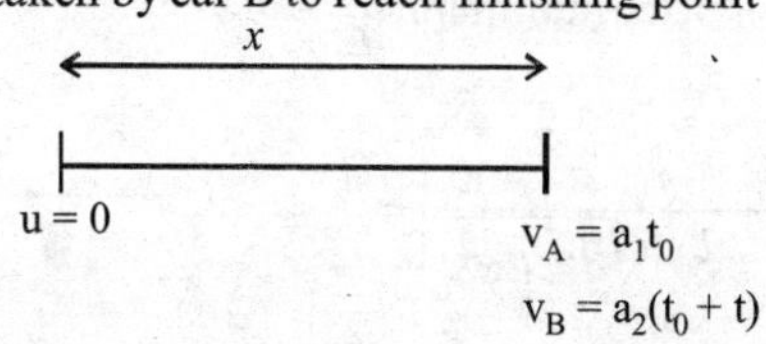

$v_A - v_B = v$
$\Rightarrow v = a_1 t_0 - a_2(t_0 + t) = (a_1 - a_2)t_0 - a_2 t$ ...(i)

$$x_B = x_A = \frac{1}{2}a_1 t_0^2 = \frac{1}{2}a_2(t_0 + t)^2$$

$$\Rightarrow \sqrt{a_1}t_0 = \sqrt{a_2}(t_0 + t) \Rightarrow \left(\sqrt{a_1} - \sqrt{a_2}\right)t_0 = \sqrt{a_2}t$$

$$\Rightarrow t_0 = \frac{\sqrt{a_2}t}{\sqrt{a_1} - \sqrt{a_2}}$$

Putting this value of t_0 in equation (i)

$$v = (a_1 - a_2)\frac{\sqrt{a_2}t}{\sqrt{a_1} - \sqrt{a_2}} - a_2 t$$

$$= \left(\sqrt{a_1} + \sqrt{a_2}\right)\sqrt{a_2}t - a_2 t = \sqrt{a_1 a_2}t + a_2 t - a_2 t$$

or, $v = \sqrt{a_1 a_2}t$

67. **(b)** Here, $s = \frac{v^2}{2a} \Rightarrow s \propto v^2$

$$\Rightarrow \quad s_2 = s_1\left(\frac{v_2}{v_1}\right)^2 \Rightarrow s_2 = 40\left(\frac{80}{40}\right)^2 \Rightarrow s_2 = 160 \text{ m}$$

68. **(c)** Using equation, $a = \frac{v-u}{t}$ and $S = ut + \frac{1}{2}at^2$

Distance travelled by car in 15 sec $= \frac{1}{2}\frac{(45)}{15}(15)^2$

$$= \frac{675}{2} \text{m}$$

Distance travelled by scooter in 15 seconds = 30 × 15 = 450 m
($\because$ distance = speed × time)
Difference between distance travelled by car and scooter in 15 sec, 450 – 337.5 = 112.5 m
Let car catches scooter in time t;

$$\frac{675}{2} + 45(t - 15) = 30t \Rightarrow 337.5 + 45t - 675 = 30t \Rightarrow 15t = 337.5$$

$\Rightarrow \quad$ t = 22.5 sec

69. **(a)** Let the car turn of the highway at a distance 'x' from the point M. So, RM = x
And if speed of car in field is v, then time taken by the car to cover the distance $QR = QM - x$ on the highway,

$$t_1 = \frac{QM - x}{2v} \quad .....(i)$$

Time taken to travel the distance 'RP' in the field

$$t_2 = \frac{\sqrt{d^2+x^2}}{v} \quad \text{..... (ii)}$$

Total time elapsed to move the car from Q to P

$$t = t_1 + t_2 = \frac{QM - x}{2v} + \frac{\sqrt{d^2+x^2}}{v}$$

For 't' to be minimum $\frac{dt}{dx} = 0$

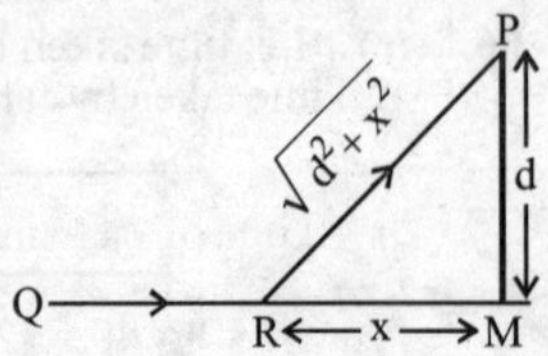

$$\frac{1}{v}\left[-\frac{1}{2} + \frac{x}{\sqrt{d^2+x^2}}\right] = 0$$

or $x = \frac{d}{\sqrt{2^2-1}} = \frac{d}{\sqrt{3}}$

70. (b) Distance along a line i.e., displacement (s) $= t^3$ ($\because s \propto t^3$ given)

By double differentiation of displacement, we get acceleration.

$$v = \frac{ds}{dt} = \frac{dt^3}{dt} = 3t^2 \text{ and } a = \frac{dv}{dt} = \frac{d(3t^2)}{dt} = 6t$$

$a = 6t$ or $a \propto t$

Hence graph (b) is correct.

71. (a) Instantaneous velocity $v = \frac{\Delta x}{\Delta t}$

From graph, $v_A = \frac{\Delta x_A}{\Delta t_A} = \frac{4m}{8s} = 0.5$ m/s

and $v_B = \frac{\Delta x_B}{\Delta t_B} = \frac{8m}{16s} = 0.5$ m/s

i.e., $v_A = v_B = 0.5$ m/s

72. (a) Given, $\frac{dv}{dt} = -2.5\sqrt{v} \Rightarrow \frac{dv}{\sqrt{v}} = -2.5\,dt$

Integrating,

$$\int_{6.25}^{0} v^{-1/2} dv = -2.5\int_0^t dt$$

$$\Rightarrow \left[\frac{v^{+1/2}}{(1/2)}\right]_{6.25}^{0} = -2.5[t]_0^t \Rightarrow -2(6.25)^{1/2} = -2.5t$$

$\Rightarrow -2 \times 2.5 = -2.5t \Rightarrow t = 2$s

73. (b) For the body starting from rest, distance travelled (x_1) is given by

$$x_1 = 0 + \frac{1}{2}at^2$$

$$\Rightarrow x_1 = \frac{1}{2}at^2$$

For the body moving with constant speed

$x_2 = vt$

$$\therefore x_1 - x_2 = \frac{1}{2}at^2 - vt$$

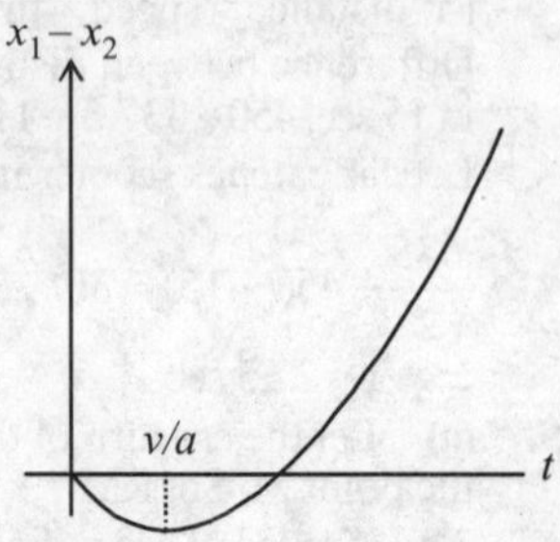

at $t = 0$, $x_1 - x_2 = 0$

This equation is of parabola.

For $t < \frac{v}{a}$; the slope is negative

For $t = \frac{v}{a}$; the slope is zero

For $t > \frac{v}{a}$; the slope is positive

These characteristics are represented by graph (b).

74. (d) Let car starts from A from rest and moves up to point B with acceleration f.

Distance, $AB = S = \frac{1}{2}ft_1^2$

Distance, $BC = (ft_1)t$

Distance, $CD = \frac{u^2}{2a} = \frac{(ft_1)^2}{2(f/2)} = ft_1^2 = 2S$

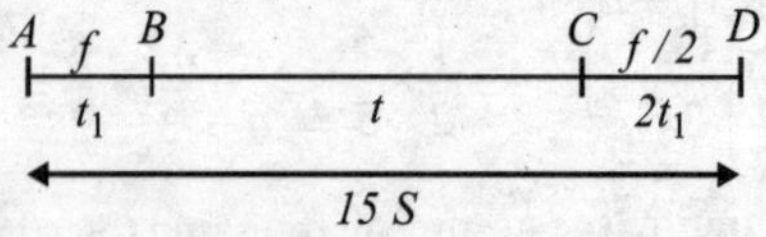

Total distance, $AD = AB + BC + CD = 15S$

$\Rightarrow S + f\,t_1 t + 2S = 15S$

$\Rightarrow f\,t_1 t = 12S$ .... (i)

$\frac{1}{2}f\,t_1^2 = S$ .... (ii)

Dividing (i) by (ii), we get $t_1 = \frac{t}{6}$

$$\Rightarrow S = \frac{1}{2}f\left(\frac{t}{6}\right)^2 = \frac{ft^2}{72}$$

75. (c)

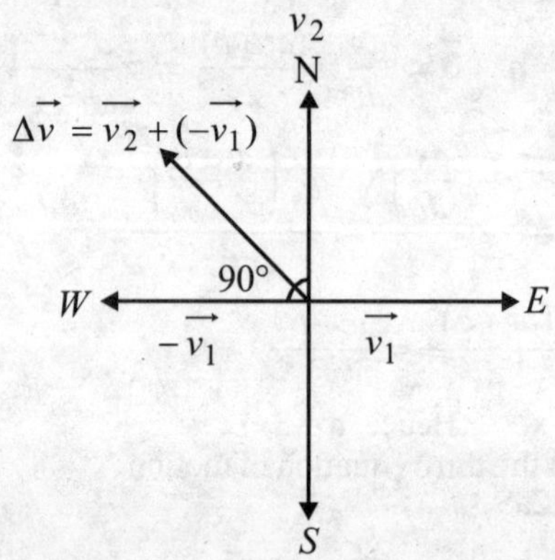

Initial velocity, $\vec{v_1} = 5\hat{i}$, Final velocity, $\vec{v_2} = 5\hat{j}$,

Change in velocity $\Delta\vec{v} = (\vec{v}_2 - \vec{v}_1)$

$$= \sqrt{v_1^2 + v_2^2 + 2v_1v_2\cos 90} = \sqrt{5^2+5^2+0} = 5\sqrt{2}\text{m/s}$$

[As $|v_1| = |v_2| = 5$ m/s]

Avg. acceleration $= \frac{\Delta\vec{v}}{t} = \frac{5\sqrt{2}}{10} = \frac{1}{\sqrt{2}}$ m/s^2

Also $\tan\theta = \frac{5}{-5} = -1$

which means θ is in the second quadrant. (towards north-west)

76. (d) Given, $t = ax^2 + bx$;

Diff. with respect to time (t)

$$\frac{d}{dt}(t) = a\frac{d}{dt}(x^2) + b\frac{dx}{dt} = a.2x\frac{dx}{dt} + \text{b.v.}$$

$\Rightarrow 1 = 2axv + bv = v(2ax + b)$ ($\because v$ = velocity)

$2ax + b = \frac{1}{v}$.

Again differentiating, we get

$$2a\frac{dx}{dt} + 0 = -\frac{1}{v^2}\frac{dv}{dt}$$

$$\Rightarrow \text{Acceleration} = \frac{dv}{dt} = -2av^3 \quad \left(\because \frac{dx}{dt} = v\right)$$

77. (a) **In first case:**

$u_1 = u\,;\, v_1 = \frac{u}{2}\,,\, s_1 = 3 \text{ cm},\, a_1 = ?$

Using, $v_1^2 - u_1^2 = 2a_1 s_1$ ...(i)

$$\Rightarrow \left(\frac{u}{2}\right)^2 - u^2 = 2 \times a \times 3 \Rightarrow a = \frac{-u^2}{8}$$

In second case: Assuming the same retardation

$u_2 = u/2\,;\, v_2 = 0\,;\, s_2 = ?;\, a_2 = \frac{-u^2}{8}$

$v_2^2 - u_2^2 = 2a_2 \times s_2$ ...(ii)

$$\therefore\ 0 - \frac{u^2}{4} = 2\left(\frac{-u^2}{8}\right) \times s_2 \Rightarrow s_2 = 1 \text{ cm}$$

78. (a) Height of body when projected vertically upwards,

$-\text{H} = \text{ut}_1 - \frac{1}{2}\text{gt}_1^2$ ...(i)

Height of body when projected vertically downwards,

$\text{H} = \text{ut}_2 + \frac{1}{2}\text{gt}_2^2$ ...(ii)

On comparing equation (i) and (ii),

$$\text{ut}_1 - \frac{1}{2}\text{gt}_1^2 = -\left(\text{ut}_2 + \frac{1}{2}\text{gt}_2^2\right)$$

$$\text{u}(\text{t}_1 + \text{t}_2) = \frac{1}{2}\text{g}\left(\text{t}_1^2 - \text{t}_2^2\right) = \frac{1}{2}\text{g}(\text{t}_1 + \text{t}_2)(\text{t}_1 - \text{t}_2)$$

$$\text{u} = \frac{1}{2}\text{g}(\text{t}_1 - \text{t}_2)$$

Height of body when it is dropped from the tower,

$$\text{H} = \frac{1}{2}\text{gt}^2$$

Substitute "u" and "H" in equation (i)

$$-\frac{1}{2}\text{gt}^2 = \frac{1}{2}\text{g}(\text{t}_1 - \text{t}_2)\text{t}_1 - \frac{1}{2}\text{gt}_1^2$$

$$\text{t}^2 = \text{t}_1\text{t}_2 \Rightarrow \text{t} = \sqrt{\text{t}_1\text{t}_2}$$

79. (b) $\text{V}_\text{A} = 20$ m/s and $\text{V}_\text{B} = -30$ m/s

Velocity of train B wrt train A,

$\text{V}_{\text{BA}} = \text{V}_\text{B} - \text{V}_\text{A} = -30 - 20 = -50$ m/s

Velocity of ground wrt train B.

$\text{V}_{\text{GB}} = \text{V}_\text{G} - \text{V}_\text{B} = 0 - (-30) = 30$ m/s

80. (c) Time, $\text{t} = \frac{\text{distance}}{\text{velocity}}$

$$\therefore \frac{60\ell + 4\ell}{20} - \frac{61\ell}{30} = 35 \Rightarrow \ell = \frac{1050}{35}$$

Total length of tunnel given,

$$\text{L} = 60\ell = \frac{1050}{35} \times 60 = 1800 \text{ m}$$

81. (d) Velocity of train B w.r.t. train A $= \vec{V}_B - \vec{V}_A$

$= 54 - (-90) = 144$ km/h

$= \frac{144 \times 5}{18} = 40$ m/s

$$\text{Time of crossing} = \frac{\text{length of train}}{\text{relative velocity}} \Rightarrow (8) = \frac{\ell}{40}$$

$\therefore \ell = 8 \times 40 = 320$ m

82. (a) As slope of B > Slope of A

$\therefore$ Speed of B (V_B) > Speed of A (V_A)

Also, $\text{x}_\text{B} > \text{x}_\text{A}$ at a given time

83. (c) Let L be the length of escalator.

Speed of man w.r.t escalator $= \frac{\text{L}}{\text{t}_1} = \text{v}_1$

Speed of escalator $= \frac{\text{L}}{\text{t}_2} = \text{v}_2$

Time taken when escalator is moving and man is also walking on it

$$= \text{t} = \frac{\text{L}}{\text{v}_1 + \text{v}_2} \Rightarrow \frac{1}{\text{t}} = \frac{\text{v}_1 + \text{v}_2}{\text{L}} = \frac{1}{\text{t}_1} + \frac{1}{\text{t}_2} \Rightarrow \text{t} = \frac{\text{t}_1\text{t}_2}{\text{t}_1 + \text{t}_2}$$

84. (a) According to question, train *A* and *B* are running on parallel tracks in the opposite direction.

$V_A = 36$ km/h $= 10$ m/s

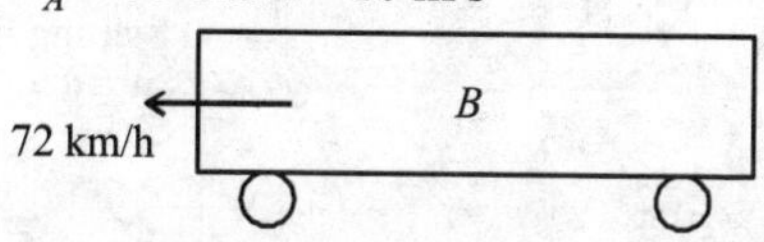

$V_B = -72$ km/h $= -20$ m/s ; $V_{MA} = -1.8$ km/h $= -0.5$ m/s

$V_{\text{man},\,B} = V_{\text{man},\,A} + V_{A,\,B}$

$= V_{\text{man},\,A} + V_A - V_B = -0.5 + 10 - (-20)$

$= -0.5 + 30 = 29.5$ m/s.

85. (a) $\text{t}_1 = \frac{(60 + 120)/1000 \text{ km}}{80 - 30}$

$$\text{t}_2 = \frac{(60 + 120)/1000 \text{ km}}{80 + 30} \quad \therefore \frac{\text{t}_1}{\text{t}_2} = \frac{11}{5}$$

86. (d)

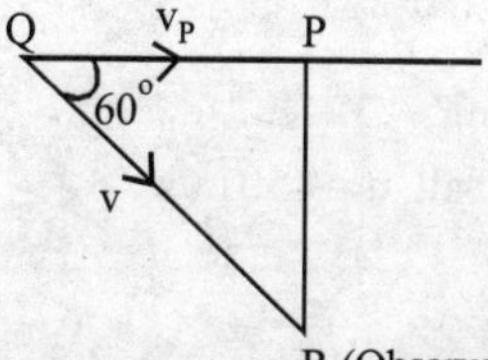

Distance, $\text{PQ} = \text{v}_\text{p} \times \text{t}$ ($\because$ Distance = speed × time)

Distance, QR = v.t

$$\cos 60° = \frac{\text{PQ}}{\text{QR}} \Rightarrow \frac{1}{2} = \frac{\text{v}_\text{p} \times \text{t}}{\text{v.t}} \Rightarrow \text{v}_\text{p} = \frac{\text{v}}{2}$$

87. (c) $\longrightarrow 4\ m/sec^2$ $\longrightarrow 2\ m/sec^2$

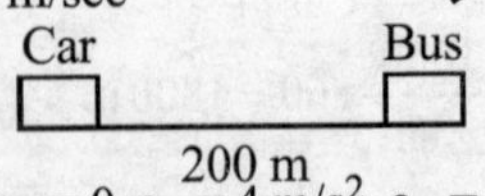

Given, $u_C = u_B = 0$, $a_C = 4\ m/s^2$, $a_B = 2\ m/s^2$
hence relative acceleration, $a_{CB} = 2\ m/sec^2$

Now, we know, $s = ut + \frac{1}{2}at^2 \Rightarrow 200 = \frac{1}{2} \times 2t^2$ $[\because u = 0]$

Hence, the car will catch up with the bus after time $t = 10\sqrt{2}$ second

88. (c) Person's speed walking only is $\frac{l}{60}$ (l is length of escalator)

Standing the escalator without walking the speed is $\frac{l}{40}$

Walking with the escalator going.

The person's speed is $\frac{l}{60} + \frac{l}{40} = \frac{5l}{120}$

So, the time to go up the escalator $t = \frac{120}{5} = 24$ second.

89. (d) Let 'S' be the distance between two ends 'a' be the constant acceleration

As we know $v^2 - u^2 = 2aS$ or, $aS = \frac{v^2 - u^2}{2}$

Let v_c be velocity at mid point.

Therefore, $v_c^2 - u^2 = 2a\frac{S}{2}$

$$v_c^2 = u^2 + aS \Rightarrow v_c^2 = u^2 + \frac{v^2 - u^2}{2} \Rightarrow v_c = \sqrt{\frac{u^2 + v^2}{2}}$$

90. (45) Using $S = ut + \frac{1}{2}gt^2$

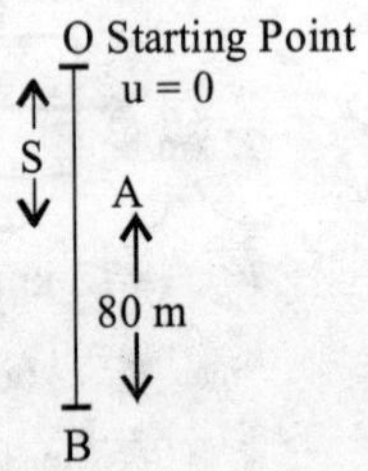

From $A \rightarrow B$

$-80 = -v_1 t - \frac{1}{2} \times 10t^2$

$-80 = -2v_1 - \frac{1}{2} \times 10 \times 2^2$

$-80 = -2v_1 - 20$

$-60 = -2v_1 \quad \therefore v_1 = 30\ m/s$

From O to A

$v^2 = u^2 + 2gS$

$30^2 = 0 + 2 \times (10)(S) \Rightarrow 900 = 20\,S$

$\therefore S = 45\ m$

91. (b) Given

Initial velocity of ball, $u = 150\ m/s$

$g = 10\ m/s^2$

Using

$\therefore v = u - gt$

$\therefore v(t = 3s) = 150 - 10 \times 3 = 120$

$\therefore v(t = 5s) = 150 - 10 \times 5 = 100$

$\therefore \frac{V(t = 3s)}{V(t = 5s)} = \frac{120}{100} = \frac{x+1}{x} = \frac{6}{5} \Rightarrow x = 5$

92. (120) Given height from ball is dropped, $h = 9.8\ m$

$v_{initial} = \sqrt{2gh}$

$= \sqrt{2 \times 10 \times 9.8} = 14\ m/s$ (downward direction)

Height to which ball rebounds, $h = 5m$

$v_f = \sqrt{2gh} = \sqrt{2 \times 10 \times 5}$

$= 10\ m/s$ (upward direction)

$$\left|\vec{a}_{avg}\right| = \left|\frac{\Delta\vec{v}}{\Delta t}\right| = \frac{|10m/s + 14m/s|}{0.2} = \frac{24}{0.2} = 120\ m/s^2.$$

93. (b) Maximum height, $h = \frac{u^2}{2g}$ [$\because$ At maximum height $v = 0$. so, $0^2 = u^2 - 2gh$]

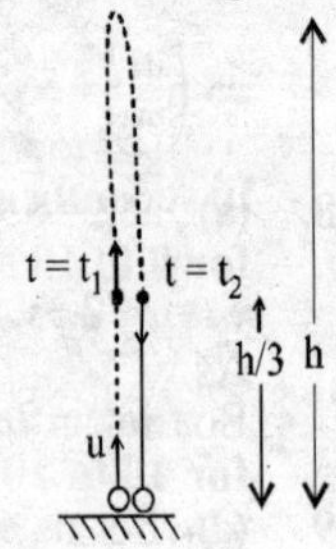

$\Rightarrow u = \sqrt{2gh}$

As, $s = ut + \frac{1}{2}at^2$

$\Rightarrow \frac{h}{3} = t\sqrt{2gh} + \frac{1}{2}(-g)t^2$

$\Rightarrow \frac{gt^2}{2} - t\sqrt{2gh} + \frac{h}{3} = 0$

Let t_1 and t_2 be two roots of above equation.

Then, $$\frac{t_2}{t_1} = \frac{\sqrt{2gh} + \sqrt{2gh - 4 \times \frac{g}{2} \times \frac{h}{3}}}{\sqrt{2gh} - \sqrt{2gh - 4 \times \frac{g}{2} \times \frac{h}{3}}}$$

$$= \frac{\sqrt{2gh} + \sqrt{\frac{4gh}{3}}}{\sqrt{2gh} - \sqrt{\frac{4gh}{3}}} = \frac{\sqrt{3} + \sqrt{2}}{\sqrt{3} - \sqrt{2}}$$

94. (d) From equation of motion, $s = ut + \frac{1}{2}gt^2$

For first $\frac{h}{2}$, $\frac{h}{2} = \frac{1}{2}gt_1^2$ ...(i)

For total height h,

$h = \frac{1}{2}g(t_1 + t_2)^2$ ...(ii)

Divide equation (ii) by (i) we have $\frac{1}{2} = \frac{t_1^2}{(t_1 + t_2)^2}$

$\frac{1}{\sqrt{2}} = \frac{t_1}{t_1 + t_2}$; $1 + \frac{t_2}{t_1} = \sqrt{2}$

$\frac{t_1}{t_2} = \frac{1}{\sqrt{2} - 1} \Rightarrow t_2 = (\sqrt{2} - 1)t_1$

95. (d) Time taken by ball to reach highest point, $t = \frac{v}{g}$

Frequency of throw, $n = \frac{1}{t} = \frac{g}{v}$

$\Rightarrow v = \frac{g}{n}$ ...(i)

The maximum height the balls can reach,

$$H_{max} = \frac{v^2}{2g} = \frac{\left(\frac{g}{n}\right)^2}{2g} = \frac{g}{2n^2} \text{ from equation (i)}$$

96. **(a)** From $h = ut + \frac{1}{2}gt^2 \Rightarrow t = \sqrt{\frac{2h}{g}}$

Time taken by mango, $t = \sqrt{\frac{2h}{g}} = \sqrt{\frac{2 \times 19.6}{9.8}} = 2\,s$

Distance $x = vt = \frac{5}{2} \times 2 = 5$ m

97. **(392)** $t_a = \frac{u}{g} = \frac{19.6}{9.8} = 2s$ $\left[\begin{array}{l} \because v = u - gt \\ 0 = u - gt \\ \Rightarrow t = u/g \end{array}\right]$

$t_d = (6-2)s = \sqrt{\frac{2h_{max}}{g}}$

$\Rightarrow h_{max} = \frac{16 \times 9.8}{2} = \frac{392}{5} m = \left(\frac{k}{5}\right) m$

$\therefore$ k = 392

98. **(a)** at $t = 0$, $u = 100$ m/s downwards
for 0 to 10 sec
$v = u - gt = -100 - 10 \times 10 = -200$ m/s
As $v = -100 - gt$ is straight line equation between v and t.
So the curve will be straight line with –100 as intercept
for 10 to 20 sec, v = 0, so option (A) is correct.

99. **(d)** Let us suppose both ball meet at $t = t$ sec at high h below.

For 'A', $h = \frac{1}{2}gt^2 \Rightarrow -80 = \frac{-1}{2} \times 10 \times t^2$

$\Rightarrow t^2 = 16 \Rightarrow t = 4$ sec.
For ball 'B',

$h = u(t-2) - \frac{1}{2}g(t-2)^2 \Rightarrow -80 = -2u - \frac{1}{2}g \times 2^2$

$\Rightarrow -80 = -2u - 20 \Rightarrow 2u = 60 \quad \therefore u = 30$ m/s

100. **(5)** $v^2 = u^2 + 2gh'$
$g^2 = 0^2 + 2gh'$

$h' = \frac{g}{2}$

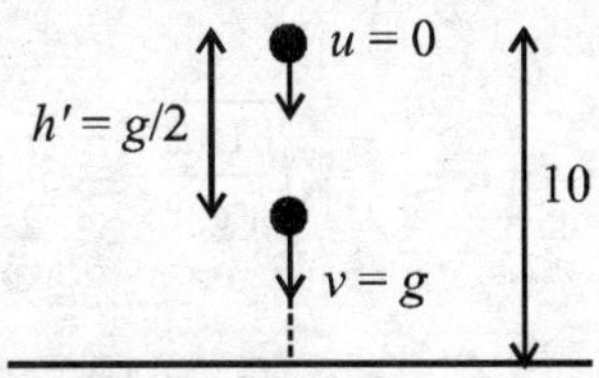

So, req. height

$= 10 - h' = 10 - \frac{g}{2}$

$= 10 - \frac{10}{2} = 5$ m

101. **(6)** Let the two ball meet at $t = t$ sec.
Then, first ball get 't' sec and second ball get $(t-2)$ sec and they will meet at same height.
So, $h_1 = h_2$

$\Rightarrow 50t - \frac{1}{2}gt^2 = 50(t-2) - \frac{1}{2}g(t-2)^2$

$\Rightarrow 0 = -100 + \frac{1}{2}g[t^2 - (t-2)^2] \Rightarrow 20 = 4t - 4$

$\Rightarrow 24 = 4t \Rightarrow t = 6$ sec.

102. **(c)** For 1st droplet, $h_1 = \frac{1}{2}gt^2 = \frac{1}{2} \times 9.8 \times (4)^2 = 78.4$m

For 2nd droplet, $h_2 = (78.4 - 34.3)$ m = 44.1 m

and $t_2 = \sqrt{\frac{2h_2}{g}} = \sqrt{\frac{2 \times 44.1}{9.8}} = 3s$

$\therefore$ Time intered, $\Delta t = (4-3)s = 1s$

$\therefore$ Rate of droplets fall, $R = \frac{1}{\Delta t} = 1$ droplet/s

103. **(3)** Taking upward direction as $+ve$, we get

$h = u \times 6 - \frac{1}{2}g \times 6^2 \rightarrow$ Case I

$h = -u \times 1.5 - \frac{1}{2}g \times 1.5^2 \rightarrow$ Case II

$h = -\frac{1}{2}gt^2 \rightarrow$ Case III

So, $6u - \frac{1}{2}g.6^2 = -1.5u - \frac{1}{2}g \times 1.5^2$ [from I and II]

$7.5u = \frac{1}{2}g(6^2 - 1.5^2)$; $u = \frac{g(33.75)}{2 \times 7.5} \Rightarrow u = \frac{45}{2}$ m/s.

Again, $-\frac{1}{2}gt^2 = 6u - \frac{1}{2}g.6^2$ [from I and III]

$t^2 = \frac{-2}{g}\left(6u - \frac{1}{2}g \times 36\right)$

$t^2 = \frac{-2}{10}\left(6 \times \frac{45}{2} - \frac{1}{2} \times 10 \times 36\right)$

$\Rightarrow t^2 = -\frac{2}{10}(135 - 180) \Rightarrow t^2 = \frac{2 \times 45}{10} \Rightarrow t^2 = 9 \Rightarrow t = 3$s.

104. **(d)** Using $H = \frac{1}{2}gt^2 \Rightarrow 2H/g = t^2$

$\Rightarrow \frac{9.8 \times 2}{9.8} = t^2 \Rightarrow t = \sqrt{2}\,s$

For first drop, $t = 0$; For second drop, $t = \Delta t$
For third drop, $t = 2\Delta t$
Let h be the distance travelled by second drop when third drop begins to fall

$\therefore 2\Delta t = \sqrt{2} \Rightarrow \Delta t = \frac{1}{\sqrt{2}}s$

$\therefore h = \frac{1}{2}g\left(\sqrt{2} - \frac{1}{\sqrt{2}}\right)^2 = \frac{1}{2} \times 9.8 \times \frac{1}{2} = \frac{9.8}{4} = 2.45$m

Height of second drop from floor,
$H - h = 9.8 - 2.45 = 7.35$ m

105. **(c)**

106. **(c)** Given,
Initial velocity of hot air balloon, u = 10 m/s
For stone
Using,

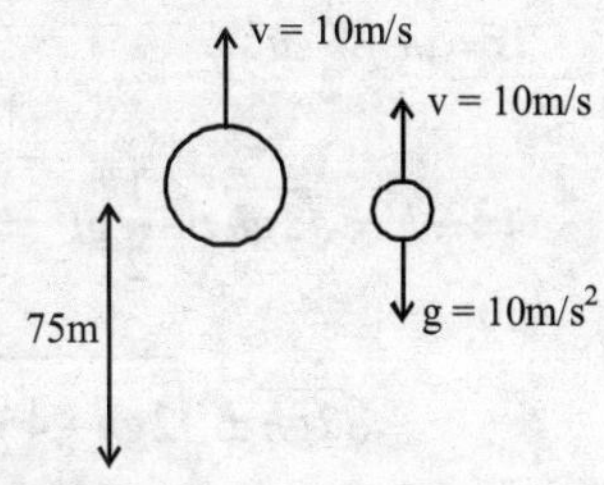

$h = ut + \frac{1}{2}gt^2$

$\Rightarrow 75 = -10t + \frac{1}{2}gt^2$

$\Rightarrow 75 = -10t + 5t^2$
$\Rightarrow t^2 - 2t - 15 = 0 \Rightarrow t = 5$ sec.
Height of balloon
$H = vt + 75 \Rightarrow H = 10 \times 5 + 75 = 125$ m.

107. **(a)** As $v = u + at$ and a = constant
So, v-t graph will be st. line.
a-t graph will be parallel to t-axis

and as $x = x_0 + ut + \frac{1}{2}at^2$ and a = constant

So, x-t graph will be parabolic.

108. (b)

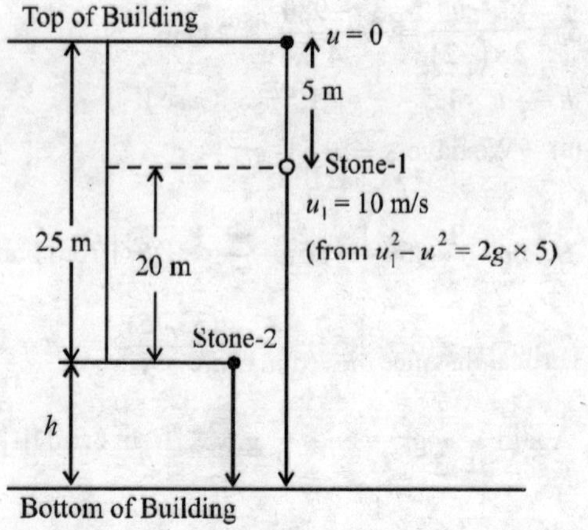

Height of the building = $h + 25$

For stone-1, $20 + h = 10t + \frac{1}{2}gt^2$ ...(i)

And for stone-2, $h = \frac{1}{2}gt^2$ ...(ii)

Putting value of h from eq. (ii) in eq. (i)

$20 + \frac{1}{2}gt^2 = 10t + \frac{1}{2}gt^2 \quad \therefore t = 2s$

Therefore, $h = \frac{1}{2}gt^2 = \frac{1}{2} \times 10 \times 2^2 = 20$ m

$\therefore$ Height of the building = $h + 25 = 20 + 25 = 45$ m.

109. (c) For upward motion of helicopter,

$v^2 = u^2 + 2gh \Rightarrow v^2 = 0 + 2gh \Rightarrow v = \sqrt{2gh}$

Now, packet will start moving under gravity.

Let 't' be the time taken by the food packet to reach the ground.

$s = ut + \frac{1}{2}at^2$

$\Rightarrow -h = \sqrt{2gh}\,t - \frac{1}{2}gt^2 \Rightarrow \frac{1}{2}gt^2 - \sqrt{2gh}\,t - h = 0$

or, $t = \dfrac{\sqrt{2gh} \pm \sqrt{2gh + 4 \times \frac{g}{2} \times h}}{2 \times \frac{g}{2}} \Rightarrow t = \sqrt{\dfrac{2gh}{g}}(1 + \sqrt{2})$

$\Rightarrow t = \sqrt{\dfrac{2h}{g}}(1 + \sqrt{2}) \Rightarrow t = 3.4\sqrt{\dfrac{h}{g}}$

110. (c) For uniformly accelerated/ deaccelerated motion :

$v^2 = u^2 \pm 2gh$

As equation is quadratic, so, v-h graph will be a parabola

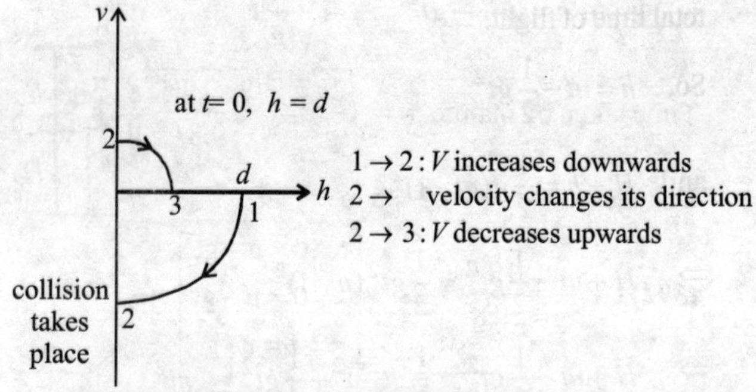

Initially velocity is downwards (–ve) and then after collision it reverses its direction with lesser magnitude, i.e. velocity is upwards (+ve).

Note that time $t = 0$ corresponds to the point on the graph where $h = d$.

111. (08.00) Let the ball takes time t to reach the ground

Using, $S = ut + \frac{1}{2}gt^2 \Rightarrow S = 0 \times t + \frac{1}{2}gt^2$

$\Rightarrow \quad 200 = gt^2 \qquad [\because 2S = 200m]$

$\Rightarrow t = \sqrt{\dfrac{200}{g}}$ ...(i)

In last $\frac{1}{2}s$, body travels a distance of 19 m, so in $\left(t - \frac{1}{2}\right)$ distance travelled = 81

Now, $\frac{1}{2}g\left(t - \frac{1}{2}\right)^2 = 81 \Rightarrow g\left(t - \frac{1}{2}\right)^2 = 81 \times 2$

$\Rightarrow \left(t - \frac{1}{2}\right) = \sqrt{\dfrac{81 \times 2}{g}} \Rightarrow \frac{1}{2} = \frac{1}{\sqrt{g}}(\sqrt{200} - \sqrt{81 \times 2})$

using (i)

$\Rightarrow \sqrt{g} = 2(10\sqrt{2} - 9\sqrt{2}) \Rightarrow \sqrt{g} = 2\sqrt{2}$

$\therefore$ g = 8 m/s^2

112. (a) For a body thrown vertically upwards acceleration remains constant ($a = -g$) and velocity at anytime t is given by $V = u - gt$

During rise velocity decreases linearly and during fall velocity increases linearly and direction is opposite to each other.

Hence graph (a) correctly depicts velocity versus time.

113. (a) For 1st stone, $-240 = 10t - \frac{1}{2}gt^2$

$\Rightarrow 5t^2 - 10t - 240 = 0 \Rightarrow t = 8s$

For 2nd stone, $-240 = 40t - \frac{1}{2}gt^2$

$\Rightarrow 5t^2 - 40t - 240 \Rightarrow t = 12s$

$\therefore$ For $t \leq 85$, $y_2 - y_1 = (40 - 10)t = 30t$ (Straight line)

For $t > 8s$, $y_2 - y_1 = 240 + 40t - 5t^2$ (Parabola)

114. (c) Let t be time to reach maximum height. Then 'nt' is total time of flight.

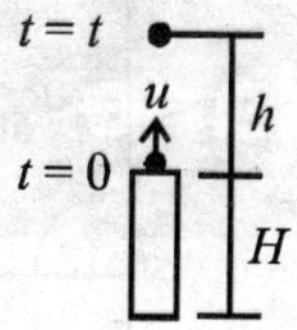

So, $h = ut - \frac{1}{2}gt^2$

and $H + h = \frac{1}{2}g(nt - t)^2$

$\Rightarrow \quad H + ut - \frac{1}{2}gt^2 = \frac{1}{2}gt^2(n-1)^2$

$\Rightarrow \quad H + ut - \frac{1}{2}gt^2 = \frac{1}{2}gt^2 n^2 + \frac{1}{2}gt^2 - gnt^2$

$\Rightarrow \quad H + ut - gt^2 = gt^2\left(\frac{n^2}{2} - n\right)$

$\Rightarrow \quad H + \frac{u^2}{g} - \frac{gu^2}{g^2} = \frac{gu^2}{2g^2}(n^2 - 2n) \quad \left[\because t = \frac{u}{g}\right]$

$\Rightarrow \quad H = \frac{u^2}{2g}n(n-2)$

$\Rightarrow \quad 2gH = u^2 n(n-2)$

115. (d) Initial velocity of parachute after bailing out,

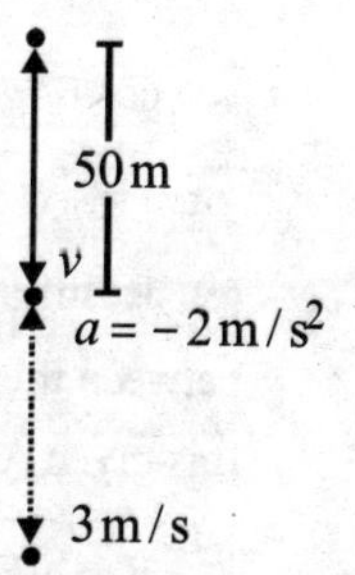

$u = \sqrt{2gh} \Rightarrow u = \sqrt{2 \times 9.8 \times 50} = 14\sqrt{5}$

The velocity at ground,

$v = 3\text{m/s}$

$S = \frac{v^2 - u^2}{2 \times (-2)} = \frac{3^2 - 980}{4} \approx 243 \text{ m}$

116. (a) We have $s = ut + \frac{1}{2}gt^2$,

$\Rightarrow \; h = 0 \times T + \frac{1}{2}gT^2 \Rightarrow h = \frac{1}{2}gT^2$

Vertical distance moved in time $\frac{T}{3}$ is

$h' = \frac{1}{2}g\left(\frac{T}{3}\right)^2 \Rightarrow h' = \frac{1}{2} \times \frac{gT^2}{9} = \frac{h}{9}$

$\therefore$ Position of ball from ground $= h - \frac{h}{9} = \frac{8h}{9}$

117. (b) Ball A is thrown upwards with velocity u from the building. During its downward journey when it comes back to the point of throw, its speed is equal to the speed of throw (u). So, for the journey of both the balls from point A to B. We can apply $v^2 - u^2 = 2gh$. As u, g, h are same for both the balls, hence, $v_A = v_B$

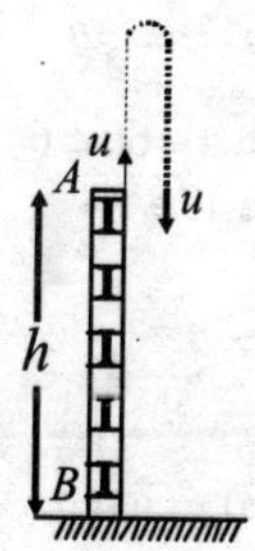

Motion in a Plane

3

1. **(3)** For two vectors $\vec{A}$ and $\vec{B}$ to be perpendicular,
$\vec{A}\cdot\vec{B}=0$
$\Rightarrow (2\hat{i}+3n\hat{j}+2\hat{k})\cdot(2\hat{i}-2\hat{j}+4p\hat{k})=0$
$4-6n+8p=0 \qquad \text{...(i)}$
Also, $|\vec{A}|=|\vec{B}|$
$\Rightarrow 4+9n^2+4=4+4+16p^2$
$\Rightarrow 9n^2=16p^2$
$\therefore P=\pm\frac{3}{4}n$
$\Rightarrow 4-6n\pm6n=0 \qquad$ [From eq(i)]
$\Rightarrow 12n=4$
$\Rightarrow n=\frac{1}{3} \qquad \therefore n^{-1}=3$

2. **(3)** Given, Angle between two vectors, $\theta=\cos^{-1}\left(\frac{5}{9}\right)$
$\therefore \quad \cos\theta=\frac{5}{9}$
Given, $|\vec{a}+\vec{b}|=\sqrt{2}\,|\vec{a}-\vec{b}|$
$\Rightarrow \quad a^2+b^2+2\vec{a}\cdot\vec{b}=2a^2+2b^2-4\vec{a}\cdot\vec{b}$
$\Rightarrow \quad 6\vec{a}\cdot\vec{b}=a^2+b^2 \Rightarrow 6ab\cos\theta=a^2+b^2$
$\Rightarrow \quad 6\times\frac{5}{9}ab=a^2+b^2 \Rightarrow \frac{10}{3}ab=a^2+b^2$
$\Rightarrow \quad \frac{10}{3}nb^2=n^2b^2+b^2 \qquad (\because \ a=nb \text{ given})$
$\Rightarrow \quad 3n^2-10n+3=0$
On Solving $n=\frac{1}{3}$ and $n=3$
Integer value $n=3$

3. **(150)**

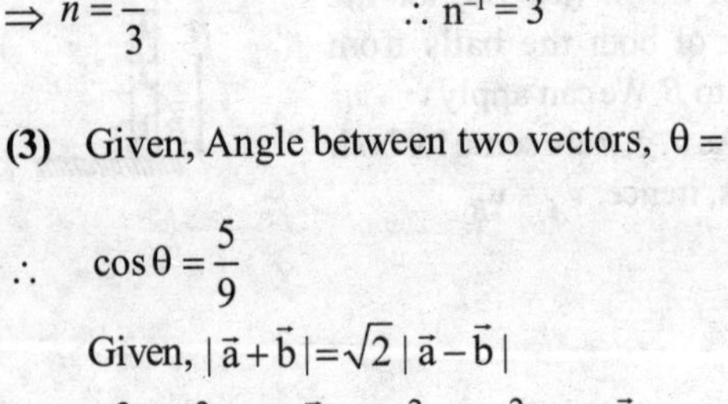

From the figure
$B\cos\theta=\frac{B}{2}$
$\Rightarrow \theta=60°$
So, angle between two vectors is $90°+60°=150°$

4. **(4)** $\vec{B}\times\vec{C}=\begin{vmatrix}\hat{i} & \hat{j} & \hat{k}\\ -1 & 4 & 3\\ -8 & -1 & 3\end{vmatrix}$
$\Rightarrow \hat{i}(12+3)-\hat{j}(-3+24)+\hat{k}(1+32)=15\hat{i}-21\hat{j}+33\hat{k}$
$\Rightarrow \vec{A}\cdot(\vec{B}\times\vec{C})=(-x\hat{i}-6\hat{j}-2\hat{k})\cdot(15\hat{i}-21\hat{j}+33\hat{k})$
$\Rightarrow 0=-15x+126-66 \Rightarrow x=4$

5. **(6)** As per question if $|\vec{F}_1|=F$ then $|\vec{F}_R|=|\vec{F}_2|=3F$
$F_R^2=F_1^2+F_2^2+2F_1F_2\cos\theta$
$\therefore 9F^2=F^2+9F^2+6F^2\cos\theta \Rightarrow 9=10+6\cos\theta$
$\Rightarrow \cos\theta=-\frac{1}{6} \Rightarrow \theta=\cos^{-1}\left(\frac{1}{-6}\right)$
or, $n=-6 \qquad \therefore |n|=6$

6. **(c)** Resultant of two vector $\vec{A}$ and $\vec{B}$ having angle θ between them $R'=\sqrt{A^2+B^2+2AB\cos\theta}$
It is given $A=B=R$
$\therefore \quad R'=\sqrt{R^2+R^2+2R^2\cos\theta}$
$=R\sqrt{2}\sqrt{1+\cos\theta}=\sqrt{2}R\sqrt{2\cos^2\frac{\theta}{2}}=2R\cos\frac{\theta}{2}$

7. **(4)** $\vec{N}=|\vec{A}|\hat{B}=|\vec{A}|\frac{\vec{B}}{|\vec{B}|}=\frac{\left(\sqrt{3^2+4^2}\right)(4\hat{i}+3\hat{j})}{\sqrt{4^2+3^2}}$
$=\frac{5(4\hat{i}+3\hat{j})}{5}=4\hat{i}+3\hat{j} \quad \therefore x=4$

8. **(d)** Here $A_y=2\sqrt{3}$
$A_y=A\cos30°=2\sqrt{3}$
$\Rightarrow A\frac{\sqrt{3}}{2}=2\sqrt{3} \Rightarrow A=4$
Now $A_x=A\,sn30°=4\times\frac{1}{2}=2$

9. **(None)** Given,
$\vec{B}-\vec{A}=2\hat{j}$
$\Rightarrow \vec{B}=\vec{A}+2\hat{j} \Rightarrow \vec{B}=2\hat{i}+3\hat{j}+2\hat{k}+2\hat{j}$
$\Rightarrow \vec{B}=2\hat{i}+5\hat{j}+2\hat{k}$
$|\vec{B}|=\sqrt{(2)^2+(5)^2+(2)^2}=\sqrt{4+25+4}=\sqrt{33}$

10. **(4)** The unit vector in the direction of $\vec{P} \times \vec{Q}$,

$$\hat{n} = \frac{\vec{P} \times \vec{Q}}{|\vec{P} \times \vec{Q}|}$$

$$\vec{P} \times \vec{Q} = \begin{vmatrix} \hat{i} & \hat{j} & \hat{k} \\ 3 & \sqrt{3} & 2 \\ 4 & \sqrt{3} & 2.5 \end{vmatrix} = \frac{\sqrt{3}}{2}\hat{i} + \frac{\hat{j}}{2} - \sqrt{3}\hat{k}$$

$$|\vec{P} \times \vec{Q}| = \sqrt{\left(\frac{\sqrt{3}}{2}\right)^2 + \left(\frac{1}{2}\right)^2 + (-\sqrt{3})^2}$$

$$= \sqrt{\frac{3}{4} + \frac{1}{4} + 3} = \sqrt{\frac{3+1+12}{4}} = \sqrt{\frac{16}{4}} = 2$$

$$\Rightarrow \frac{\vec{P} \times \vec{Q}}{|\vec{P} \times \vec{Q}|} = \frac{1}{2}\left(\sqrt{3}\frac{\hat{i}}{2} + \frac{\hat{j}}{2} - \sqrt{3}\hat{k}\right)$$

$$= \frac{1}{4}(\sqrt{3}\,\hat{i} + \hat{j} - 2\sqrt{3}\hat{k})$$

$x = 4$

11. **(1)** Given $\frac{a}{b} = \frac{x}{2}$ and $3a + 2b = 7$

If $\vec{a}$ and $\vec{b}$ are perpendicular

$\vec{a}.\vec{b} = 0$

$\Rightarrow 2a - 3b + 4 = 0$

$\Rightarrow 2a - 3b = -4$ ...(i)

Also, $3a + 2b = 7$ ...(ii)

solving (i) and (ii), we get

$a = 1, b = 2$

So, $\frac{a}{b} = \frac{1}{2} \Rightarrow x = 1$

12. **(d)** We have, $\vec{P}.\vec{Q} = 0$

$\Rightarrow \left(\hat{i} + 2m\hat{j} + m\hat{k}\right).\left(4\hat{i} - 2\hat{j} + m\hat{k}\right) = 0$

$\Rightarrow 4 - 4m + m^2 = 0 \Rightarrow (m-2)^2 = 0 \Rightarrow m = 2$

13. **(5)** Here, $\vec{a}.\vec{b} = 0$

$\therefore \Rightarrow (2\hat{i} + 4\hat{j} - 2\hat{k}).(\hat{i} + 2\hat{j} + \alpha\hat{k}) = 0$

$\therefore 2 \times 1 + 4 \times 2 - 2 \times \alpha = 0 \quad \therefore \alpha = 5$

14. **(2)** Given that

$\vec{A} = \left(2\hat{i} + 3\hat{j} - \hat{k}\right)$m and $\vec{B} = \left(\hat{i} + 2\hat{j} + 2\hat{k}\right)$m

Component of vector $\vec{A}$ along vector $\vec{B} = \vec{A} \cdot \hat{B}$

$$= \frac{\vec{A}.\vec{B}}{|\vec{B}|} = \frac{2+6-2}{\sqrt{1^2+2^2+2^2}} \qquad \left(\because \hat{B} = \frac{\vec{B}}{|\vec{B}|}\right)$$

$$= \frac{6}{3} = 2\text{m}$$

15. **(c)** Given $|\vec{A} + \vec{B}| = |\vec{A} - \vec{B}| \times 2$

$\Rightarrow |\vec{A} + \vec{B}|^2 = |\vec{A} - \vec{B}|^2 \times 4$

$\Rightarrow 0 = 3|\vec{A}|^2 + 3|\vec{B}|^2 - 10|\vec{A}||\vec{B}|\cos\theta$

$\Rightarrow 0 = 3|\vec{A}|^2 + 3|\vec{A}|^2 - 10|\vec{A}|^2\cos\theta$

$\Rightarrow 10|\vec{A}|^2\cos\theta = 6|\vec{A}|^2$

$\Rightarrow \cos\theta = \frac{3}{5}$

16. **(b)** $|\hat{A} + \hat{B}|^2 = |\hat{A}|^2 + |\hat{B}|^2 + 2|\hat{A}||\hat{B}|\cos\theta$

$= 1^2 + 1^2 + 2.1.1.\cos\theta = 2(1 + \cos\theta)$

$= 2 \times 2\cos^2\left(\frac{\theta}{2}\right)$; So, $|\hat{A} + \hat{B}| = 2\cos\left(\theta/2\right)$

and, $|\hat{A} - \hat{B}|^2 = |\hat{A}|^2 + |\hat{B}|^2 - 2|\hat{A}||\hat{B}|\cos\theta$

$= 1^2 + 1^2 - 2.1.1\cos\theta = 2(1 - \cos\theta) = 2 \times 2\sin^2\left(\frac{\theta}{2}\right)$

So, $|\hat{A} - \hat{B}| = 2\sin\left(\theta/2\right)$

Therefore, $\frac{|\hat{A} + \hat{B}|}{|\hat{A} - \hat{B}|} = \frac{2\cos\left(\theta/2\right)}{2\sin\left(\theta/2\right)} = \cot\left(\theta/2\right)$

$|\hat{A} - \hat{B}| = |\hat{A} + \hat{B}|\tan\left(\theta/2\right)$

17. **(c)** $|\vec{A}| \neq 0$

$\vec{A} \times \vec{A} = |\vec{A}||\vec{A}|\sin 0°\hat{n} = 0$

18. **(b)**

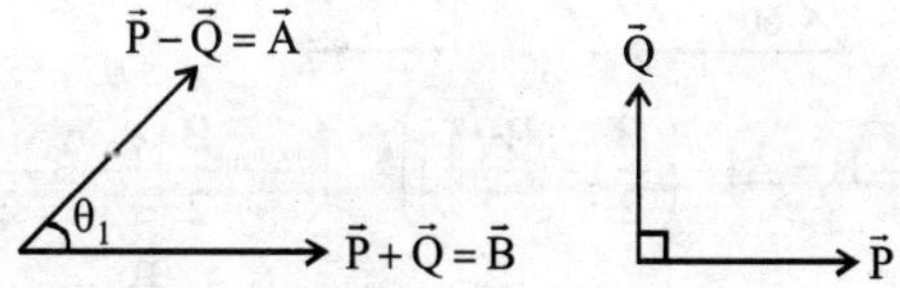

$\left|\vec{P} + \vec{Q}\right|^2 = P^2 + Q^2$

$\left|\vec{P} - \vec{Q}\right|^2 = P^2 + Q^2,$

Let $\vec{A} = \vec{P} - \vec{Q}$

$\vec{B} = \vec{P} + \vec{Q}$

$\left|\vec{A} + \vec{B}\right|^2 = |\vec{A}|^2 + |\vec{B}|^2 + 2\left|\vec{A}\right|\left|\vec{B}\right|\cos\theta_1$

$\Rightarrow 3P^2 + 3Q^2 = P^2 + Q^2 + P^2 + Q^2 + 2(P^2 + Q^2)\cos\theta_1$

$\Rightarrow P^2 + Q^2 = 2(P^2 + Q^2)\cos\theta_1$

$\Rightarrow \cos\theta_1 = \frac{1}{2} \Rightarrow \theta_1 = 60°$

Similarly,

$2P^2 + 2Q^2 = P^2 + Q^2 + P^2 + Q^2 + 2(P^2 + Q^2)\cos\theta_2$

$\Rightarrow 0 = 2(P^2 + Q^2)\cos\theta_2$

$\Rightarrow \cos\theta_2 = 0 \Rightarrow \theta_2 = 90°$

Clearly $\theta_1 < \theta_2$ and $\theta_1 = 60°$ and $\theta_2 = 90°$

So, option (b) is correct.

19. **(a)**

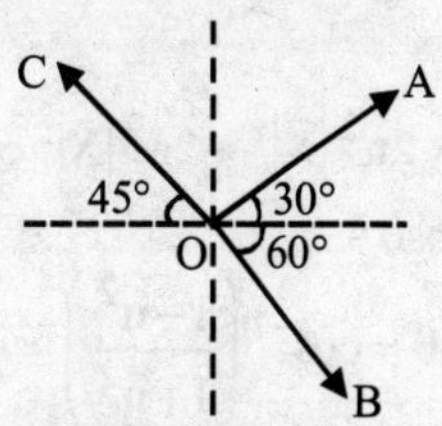

Say, magnitudes of vectors is r.

$\overrightarrow{OA} = r\left[\cos 30°\hat{i} + \sin 30\hat{j}\right] = r\left[\frac{\sqrt{3}}{2}\hat{i} + \frac{1}{2}\hat{j}\right]$

$\overrightarrow{OC} = r\left[\cos 45°(-\hat{i}) + \sin 45\hat{j}\right] = r\left[-\frac{1}{\sqrt{2}}\hat{i} + \frac{1}{\sqrt{2}}\hat{j}\right]$

$\overrightarrow{OB} = r\left[\cos 60°\hat{i} - \sin 60\hat{j}\right] = r\left[\frac{1}{2}\hat{i} - \frac{\sqrt{3}}{2}\hat{j}\right]$

$\therefore\ \overrightarrow{OA} + \overrightarrow{OB} - \overrightarrow{OC}$

$= r\left[\left(\frac{\sqrt{3}+1}{2} + \frac{1}{\sqrt{2}}\right)\hat{i} + \left(\frac{1}{2} - \frac{\sqrt{3}}{2} - \frac{1}{\sqrt{2}}\right)\hat{j}\right]$

$\therefore$ Angle made by vector with x-axis

$\tan^{-1}\left[\dfrac{\frac{1}{2} - \frac{\sqrt{3}}{2} - \frac{1}{\sqrt{2}}}{\frac{\sqrt{3}}{2} + \frac{1}{2} + \frac{1}{\sqrt{2}}}\right] = \tan^{-1}\left[\frac{1-\sqrt{3}-\sqrt{2}}{\sqrt{3}+1+\sqrt{2}}\right]$

$= \tan^{-1}\left[\frac{1-\sqrt{3}-\sqrt{2}}{1+\sqrt{3}+\sqrt{2}}\right]$

20. (c)

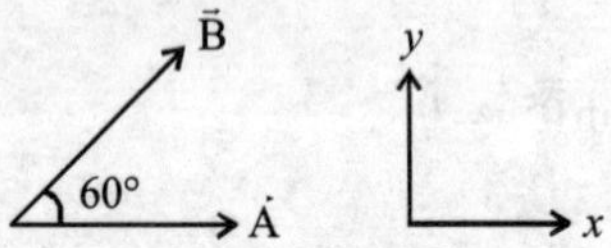

$\vec{A} - \vec{B} = A\hat{i} - \left(\frac{B}{2}\hat{i} + \frac{B\sqrt{3}}{2}\hat{j}\right) = \left(A - \frac{B}{2}\right)\hat{i} - B\frac{\sqrt{3}}{2}\hat{j}$

So, $\cos\beta = \dfrac{\vec{A}.(\vec{A}-\vec{B})}{|\vec{A}||\vec{A}-\vec{B}|} = \dfrac{A\left(A - \frac{B}{2}\right)}{A\sqrt{\left(A-\frac{B}{2}\right)^2 + \left(\frac{B\sqrt{3}}{2}\right)^2}}$

$= \dfrac{A - \frac{B}{2}}{\sqrt{A^2 + B^2 - AB}}$

So, $\tan\beta = \dfrac{\frac{\sqrt{3}B}{2}}{\frac{2A-B}{2}} = \dfrac{\sqrt{3}B}{2A-B}$

21. (a) We have given,

$|\vec{X} - \vec{Y}| = n|\vec{X} + \vec{Y}|$

As $|\vec{A}+\vec{B}|^2 = |\vec{A}|^2 + |\vec{B}|^2 + 2|\vec{A}||\vec{B}|\cos\theta$

and $|\vec{A}-\vec{B}|^2 = |\vec{A}|^2 + |\vec{B}|^2 - 2|\vec{A}||\vec{B}|\cos\theta$

$\therefore\ |\vec{X}|^2 + |\vec{Y}|^2 - 2|\vec{X}||\vec{Y}|\cos\theta$

$= n^2\left[|\vec{X}|^2 + |\vec{Y}|^2 + 2|\vec{X}||\vec{Y}|\cos\theta\right]$

Given, $|\vec{X}| = |\vec{Y}|$

$\Rightarrow\ 2|\vec{X}|^2 - 2|\vec{X}|^2\cos\theta = 2n^2|\vec{X}|^2 + 2n^2|\vec{X}|^2\cos\theta$

$\Rightarrow\ 1 - \cos\theta = n^2 + n^2\cos\theta$

$\Rightarrow\ \cos\theta = \left(\frac{1-n^2}{1+n^2}\right) \Rightarrow \theta = \cos^{-1}\left(\frac{1-n^2}{1+n^2}\right)$

22. (d) Projection of vector $\vec{A}$ on vector $\vec{B}$ is given by

$(A\cos\theta)\hat{B} = A\left(\frac{\vec{A}.\vec{B}}{AB}\right)\hat{B} = \frac{\vec{A}.\vec{B}}{B}\hat{B}$

$= \frac{2}{\sqrt{2}}\left(\frac{\hat{i}+\hat{j}}{\sqrt{2}}\right) = (\hat{i}+\hat{j})$

23. (3) Direction of particle P,

$\hat{v}_1 = \frac{\vec{A}\times\vec{B}}{|\vec{A}\times\vec{B}|} = \frac{\hat{i}-\hat{j}+\hat{k}}{\sqrt{3}}$

Direction of particle Q,

$\hat{v}_2 = \frac{\vec{A}\times\vec{C}}{|\vec{A}\times\vec{C}|} = \frac{2\hat{k}}{2} = \hat{k}$

Angle between $\hat{v}_1$ and $\hat{v}_2$

$\cos\theta = \dfrac{\hat{v}_1.\hat{v}_2}{|\hat{v}_1||\hat{v}_2|} = \dfrac{\frac{1}{\sqrt{3}}}{(1)(1)} = \dfrac{1}{\sqrt{3}}$

Hence the angle between the direction of motion of P and Q is $\theta = \cos^{-1}\left(\frac{1}{\sqrt{3}}\right)$

$\therefore$ value of x = 3

24. (c)

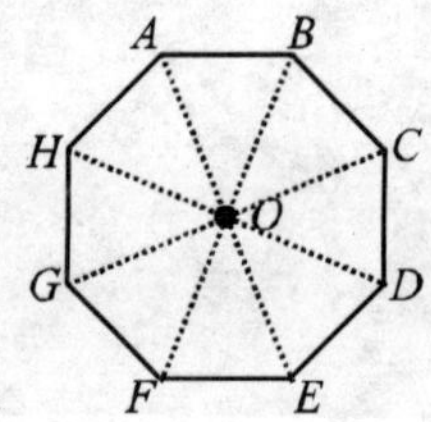

From figure,

$\overrightarrow{AO} + \overrightarrow{OB} = \overrightarrow{AB},\ \overrightarrow{AO} + \overrightarrow{OC} = \overrightarrow{AC},\ \overrightarrow{AO} + \overrightarrow{OD} = \overrightarrow{AD}$

$\overrightarrow{AO} + \overrightarrow{OE} = \overrightarrow{AE},\ \overrightarrow{AO} + \overrightarrow{OF} = \overrightarrow{AF},\ \overrightarrow{AO} + \overrightarrow{OG} = \overrightarrow{AG}$

$\overrightarrow{AO} + \overrightarrow{OH} = \overrightarrow{AH}$

Now, adding we get

$8\overrightarrow{AO} = (\overrightarrow{AB} + \overrightarrow{AC} + \overrightarrow{AD} + \overrightarrow{AE} + \overrightarrow{AF} + \overrightarrow{AG} + \overrightarrow{AH})$

$\Rightarrow 8(2\hat{i} + 3\hat{j} - 4\hat{k})$

$= \overrightarrow{AB} + \overrightarrow{AC} + \overrightarrow{AD} + \overrightarrow{AE} + \overrightarrow{AF} + \overrightarrow{AG} + \overrightarrow{AH}$

$\therefore\ \overrightarrow{AB} + \overrightarrow{AC} + \overrightarrow{AD} + \overrightarrow{AE} + \overrightarrow{AF} + \overrightarrow{AG} + \overrightarrow{AH}$

$= 16\hat{i} + 24\hat{j} - 32\hat{k}$

25. (180) $\vec{P}\times\vec{Q} = \vec{Q}\times\vec{P}$ only if $\vec{P} = 0$ or $\vec{Q} = 0$

$\Rightarrow \vec{P}\times\vec{Q} = 0$

So, $\theta = 0°$ or $180°$ and $0° < \theta < 360°$. $\therefore \theta = 180°$

26. (195) Given : $\vec{F} = (\hat{i} + 2\hat{j} + 3\hat{k})$ N

And, $\vec{r} = [(4\hat{i} + 3\hat{j} - \hat{k}) - (\hat{i} + 2\hat{j} + \hat{k})] = 3\hat{i} + \hat{j} - 2\hat{k}$

Torque, $\tau = \vec{r}\times\vec{F} = (3\hat{i} + \hat{j} - 2\hat{k})\times(\hat{i} + 2\hat{j} + 3\hat{k})$

$\tau = \begin{vmatrix} \hat{i} & \hat{j} & \hat{k} \\ 3 & 1 & -2 \\ 1 & 2 & 3 \end{vmatrix} = 7\hat{i} - 11\hat{j} + 5\hat{k}$

Magnitude of torque, $|\vec{\tau}| = \sqrt{195}$ N–m.

$\therefore$ x = 195

27. **(90)** Given,

$|\vec{R}| = |\vec{P}| \Rightarrow |\vec{P} + \vec{Q}| = |\vec{P}|$

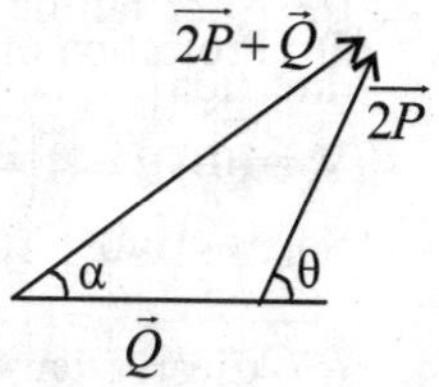

$P^2 + Q^2 + 2PQ.\cos\theta = P^2$

$\Rightarrow Q + 2P\cos\theta = 0$

$\tan\alpha = \dfrac{2P\sin\theta}{Q + 2P\cos\theta} = \infty \; (\because 2P\cos\theta + Q = 0)$

$\Rightarrow \alpha = 90°$

28. **(d)** Using,

$R^2 = A_1{}^2 + A_2{}^2 + 2A_1A_2\cos\theta$

$5^2 = 3^2 + 5^2 + 2 \times 3 \times 5\cos\theta$ or $\cos\theta = -0.3$

$\left(2\vec{A_1} + 3\vec{A_2}\right) \cdot \left(3\vec{A_1} - 2\vec{A_2}\right) = 2A_1 \times 3A_1$

$+ (3A_2)(3A_1)\cos\theta - (2A_1)(2A_2)\cos\theta - 3A_2 \times 2A_2$

$= 6A_1{}^2 + 9A_1A_2\cos\theta - 4A_1A_2\cos\theta - 6A_2{}^2$

$= 6A_1{}^2 - 6A_2{}^2 + 5A_1A_2\cos\theta$

$= 6 \times 3^2 - 6 \times 5^2 + 5 \times 3 \times 5(-0.3) = -118.5$

29. **(c)** From figure, $\vec{r}_G = \frac{a}{2}\hat{i} + \frac{a}{2}\hat{k}$

$\vec{r}_H = \frac{a}{2}\hat{j} + \frac{a}{2}\hat{k}$

$\therefore \vec{r}_H - \vec{r}_G = \left(\frac{a}{2}\hat{j} + \frac{a}{2}\hat{k}\right) - \left(\frac{a}{2}\hat{i} + \frac{a}{2}\hat{k}\right) = \frac{a}{2}\left(\hat{j} - \hat{i}\right)$

30. **(a)** Using, $R^2 = P^2 + Q^2 + 2PQ\cos\theta$

$4F^2 + 9F^2 + 12F^2\cos\theta = R^2$

When forces Q is doubled,

$4F^2 + 36F^2 + 24F^2\cos\theta = 4R^2$

$4F^2 + 36F^2 + 24F^2\cos\theta$

$= 4(13F^2 + 12F^2\cos\theta) = 52F^2 + 48F^2\cos\theta$

$\therefore \cos\theta = -\dfrac{12F^2}{24F^2} = -\dfrac{1}{2} \Rightarrow \theta = 120°$

31. **(a)** If $\vec{C} = a\hat{i} + b\hat{j}$ then $\vec{A}.\vec{C} = \vec{A}.\vec{B}$

$a + b = 1$ ..... (i)

$\vec{B}.\vec{C} = \vec{A}.\vec{B}$

$2a - b = 1$ ..... (ii)

Solving equation (i) and (ii) we get

$a = \frac{1}{3}, b = \frac{2}{3}$

$\therefore$ Magnitude of coplanar vector, $|\vec{C}| = \sqrt{\frac{1}{9} + \frac{4}{9}} = \sqrt{\frac{5}{9}}$

32. **(a)** Arc length = radius × angle

So, $|\vec{B} - \vec{A}| = |\vec{A}|\Delta\theta \; [\because |\vec{A}| = |\vec{B}|]$

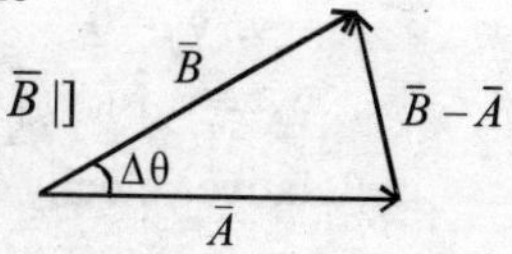

33. **(c)** $\vec{A} \times \vec{B} - \vec{B} \times \vec{A} = 0$

$\Rightarrow \vec{A} \times \vec{B} + \vec{A} \times \vec{B} = 0$

$\therefore \vec{A} \times \vec{B} = 0$

Angle between them is $0, \pi$, or 2π

from the given options, $\theta = \pi$

34. **(d)** Helicopter's horizontal speed,

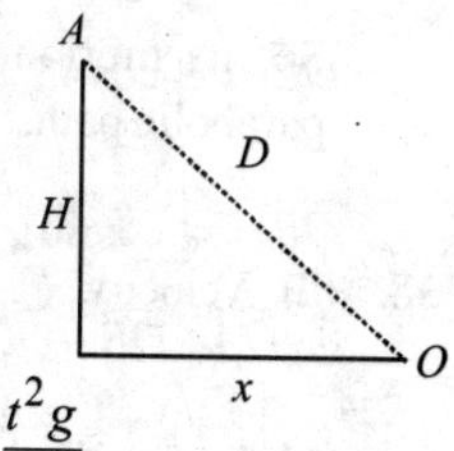

$u = 360 \times \frac{5}{18} = 100$ m/s

Horizontal displacement,

$x = u \times t = 2 \times 10^3$ m

Time of fall, $t = \sqrt{\dfrac{2H}{g}} \Rightarrow H = \dfrac{t^2 g}{2}$

Altitude, $H = 2$ km $\Rightarrow H = 2000$ m

$D = \sqrt{x^2 + H^2} \Rightarrow D = 2\sqrt{2}$ km

35. **(c)** For the moving body,

$\vec{r} = 5t^2\hat{i} - 5t\hat{j} \;\; \therefore \vec{v} = \dfrac{d\vec{r}}{dt} = 10t\hat{i} - 5\hat{j}$

$\therefore$ At $t = 2$s, $\vec{v} = 20\hat{i} - 5\hat{j}$

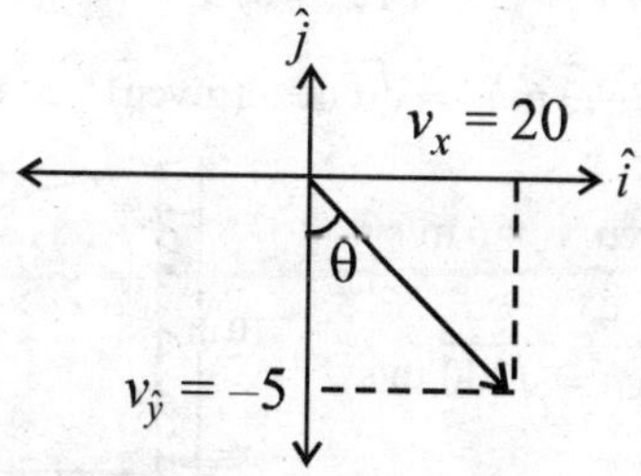

$\therefore v = \sqrt{(-5)^2 + (20)^2} = 5\sqrt{17}\, m/s$

$\therefore \tan\theta = \dfrac{20}{5} = 4; \theta = \tan^{-1}4$ with -ve y-axis

36. **(100)** Horizontal range from height H is,

$R = v\sqrt{\dfrac{2H}{g}}; R = vT$

Horizontal range for height 4H is, $R' = \dfrac{v}{2}\sqrt{\dfrac{2(4H)}{g}}$

$R' = v\sqrt{\dfrac{2H}{g}} = R = 100$ m

37. **(d)** Given $x = 2 + 4t$

$\therefore v_x = \dfrac{dx}{dt} = \dfrac{d(2+4t)}{dt} = 4$

$\therefore a_x = \dfrac{dv_x}{dt} = 0$

Since, $y = 3t + 8t^2$ [given]

$\therefore v_y = \dfrac{dy}{dt} = \dfrac{d}{dt}(3t + 8t^2) = 3 + 16t$

$\therefore a_y = \frac{dy_y}{dt} = \frac{d}{dt}(3+16t) = 16$

So, its motion should be uniformly accelerated and parabolic path.

38. (d) Velocity. $\vec{v} = \frac{\overrightarrow{ds}}{dt} = \frac{d\left(2t^2\hat{j}+5\hat{k}\right)}{dt} = 4t\hat{j}$

So, at t = 1 sec $\vec{v} = 4\hat{j}$

$\Rightarrow v = 4\text{m/s}$ in y-direction

39. (673) Using $v^2 - u^2 = 2as$

$u_x = 5$ m/s, $a_x = 3$ m/s^2 and $x = 84$ m

$v_x^2 - u_x^2 = 2ax \Rightarrow v_x^2 - 25 = 2(3)(84)$

$\therefore V_x = 23$ m/s

$v_x - u_x = a_x t \Rightarrow t = \frac{23-5}{3} = 6s$

$v_y = 0 + a_y t = 0 + 2 \times (6) = 12$ m/s

$v^2 = v_x^2 + v_y^2 = 23^2 + 12^2 = 673$

$\Rightarrow v = \sqrt{673}$ m/s $= \sqrt{\alpha}$ m/s (given) $\therefore \alpha = 673$

40. (a) Given, $v_x = 5$ m/s

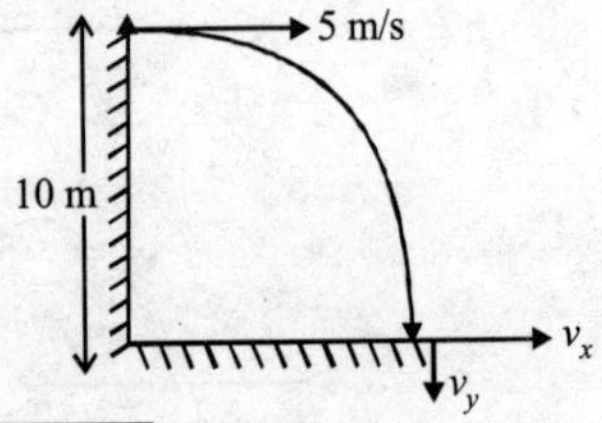

$v_y = \sqrt{2gh} = \sqrt{200}$ m/s

$v_{net} = \sqrt{v_x^2 + v_y^2} = \sqrt{(5)^2 + (\sqrt{200})^2} = \sqrt{25+200} = 15$ m/s

41. (13)

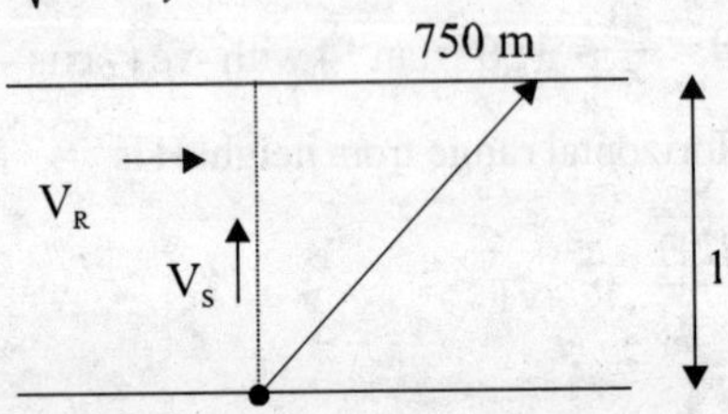

We have,
1 km = $V_s \times t$

and, 750 m = $V_R \times t$ So, $\frac{1000 \text{ m}}{750 \text{ m}} = \frac{V_S}{V_R}$

$\Rightarrow \frac{4}{3} = \frac{4}{V_R} \Rightarrow V_R = 3$ km/hr

42. (b) We have, $\vec{r} = 3t\hat{i} + 5t^3\hat{j} + 7k$

So, Acceleration, $\vec{a} = \frac{d^2\vec{r}}{dt^2} = 30t\hat{j}$

At t = 1s, $\vec{a} = 30\hat{j}$

43. (a) At t = 20 sec

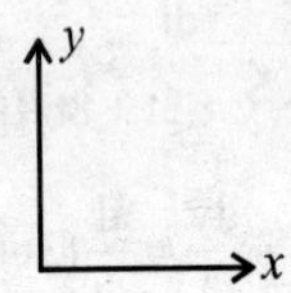

$V_x = at = 5 \times 20 = 100$m/s

At t = 1 sec

[$\because$ After falling there is no acceleration in horizontal direction]

$V_y = (0 + g \times 1) = 10 \times 1 = 10$ m/s

So, $\vec{V} = 100\hat{i} - 10\hat{j}$

44. (c) Clearly, $\tan 45° = \frac{|V_{rm}|}{|V_m|}$

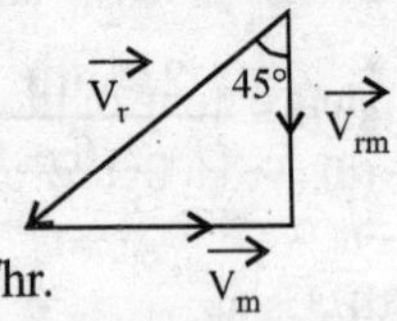

$\Rightarrow V_{rm} = V_m$

$\Rightarrow V_{rm} = 15\sqrt{2} \times \frac{\sqrt{2}}{\sqrt{2}} = \frac{30}{\sqrt{2}}$ km/hr.

45. (60)

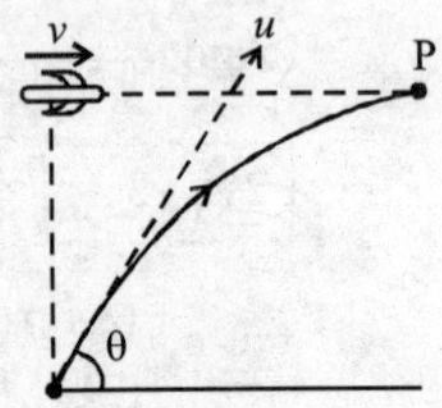

If fighter jet and bullet have same horizontal component of velocity, then it will strike after some time as shown above because horizontal displacement of both of them in given time interval will always be same.

i.e. $v = u\cos\theta \Rightarrow 200 = 400\cos\theta$

$\Rightarrow \cos\theta = \frac{1}{2} \Rightarrow \theta = 60°$

46. (c) Time of fall of packet

$t = \sqrt{\frac{2h}{g}}$, $x = \sqrt{\frac{2h}{g}}v$

Horizontal range (x) = time × horizontal component of velocity (v)

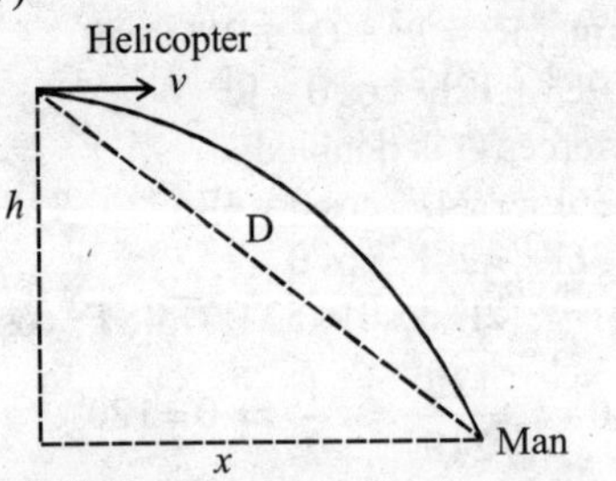

$\therefore$ Required distance

$$D = \sqrt{x^2 + h^2} = \sqrt{\left(\sqrt{\frac{2h}{g}}v\right)^2 + h^2}$$

or, $D = \sqrt{\frac{2hv^2}{g} + h^2}$

47. (c) For bomb

$v_x = v$, $v_y = -gt$

$\therefore \vec{v}_B = v_x\hat{i} + v_y\hat{j} = v\hat{i} - gt\hat{j}$

For observer

$v_x = v$, $v_y = 0$,

$\therefore \vec{v}_A = v_x\hat{i} + v_y\hat{j} = v\hat{i}$

Relative velocity of bomb w.r.t. observer = $\vec{v}_B - \vec{v}_A$

$= v\hat{i} - gt\hat{j} - v\hat{i} = -gt\hat{j}$

So, bomb will fall vertically down only.

So, trajectory of bomb is a straight line.

48. (None) Given, $\vec{v} = (0.5t^2\hat{i} + 3t\hat{j} + 9\hat{k})$ m/s

After $t = 2$ s

$\vec{v} = (2\hat{i} + 6\hat{j} + 9\hat{k})$

From y-axis $\theta = \tan^{-1}\frac{\sqrt{2^2+9^2}}{6} \Rightarrow \theta = \tan^{-1}\frac{\sqrt{85}}{6}$

From x-axis $\theta = \tan^{-1}\frac{\sqrt{6^2+9^2}}{2}$ or, $\theta = \tan^{-1}\frac{\sqrt{117}}{2}$

Hence none of the option matches.

49. **(d)** From figure/ trigonometry,

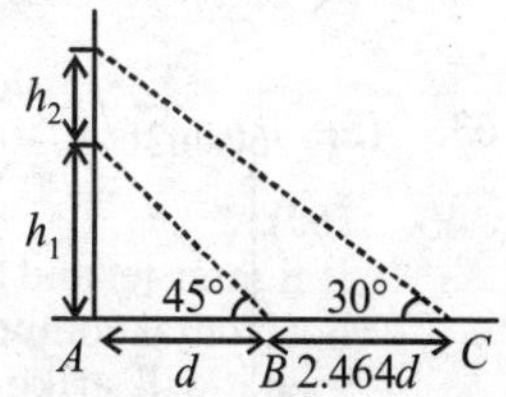

$\frac{h_1}{d} = \tan 45° \quad \therefore h_1 = d$

And, $\frac{h_1 + h_2}{d + 2.464d} = \tan 30°$

$\Rightarrow (h_1 + h_2) \times \sqrt{3} = 3.46d$

$\Rightarrow (h_1 + h_2) = \frac{3.46d}{\sqrt{3}} \Rightarrow d + h_2 = \frac{3.46d}{\sqrt{3}}$

$\therefore h_2 = d$

50. **(b)** Given : $\vec{u} = 5\hat{j}$ m/s

Acceleration, $\vec{a} = 10\hat{i} + 4\hat{j}$ and final coordinate $(20, y_0)$ in time t.

$S_x - u_x t + \frac{1}{2}a_x t^2 \qquad [\because u_x = 0]$

$\Rightarrow 20 = 0 + \frac{1}{2} \times 10 \times t^2 \Rightarrow t = 2$ s

$S_y = u_y \times t + \frac{1}{2}a_y t^2$

$y_0 = 5 \times 2 + \frac{1}{2} \times 4 \times 2^2 = 18$ m

51. **(d)** $\vec{r} = 15t^2\hat{i} + (4 - 20t^2)\hat{j}$

$\vec{v} = \frac{d\vec{r}}{dt} = 30t\hat{i} - 40t\hat{j}$

Acceleration, $\vec{a} = \frac{d\vec{v}}{dt} = 30\hat{i} - 40\hat{j}$

$\therefore a = \sqrt{30^2 + 40^2} = 50$ units

52. **(b)** As $\vec{s} = \vec{u}t + \frac{1}{2}\vec{a}t^2$

$\vec{s} = (5\hat{i} + 4\hat{j})2 + \frac{1}{2}(4\hat{i} + 4\hat{j})4 = 10\hat{i} + 8\hat{j} + 8\hat{i} + 8\hat{j}$

$\vec{r}_f - \vec{r}_i = 18\hat{i} + 16\hat{j}$ [as $\vec{s}$ = change in position $= \vec{r}_f - \vec{r}_i$]

$\vec{r}_f = 20\hat{i} + 20\hat{j}$

$|\vec{r}_f| = 20\sqrt{2}$ m

53. **(c)** From given equation,

$\vec{V} = K(y\hat{i} + x\hat{j})$

$\frac{dx}{dt} = Ky$ and $\frac{dy}{dt} = Kx$

Now $\frac{\frac{dy}{dt}}{\frac{dx}{dt}} = \frac{dy}{dx} = \frac{x}{y}, \Rightarrow ydy = xdx$

Integrating both side

$y^2 = x^2 + c$

54. **(580)** For pariticle 'A' For particle 'B'

$X_A = -3t^2 + 8t + 10 \qquad Y_B = 5 - 8t^3$

$\vec{v}_A = (8 - 6t)\hat{i} \qquad \vec{v}_B = -24t^2\hat{j}$

$\vec{a}_A = -6\hat{i} \qquad \vec{a}_B = -48t\hat{j}$

At $t = 1$ sec

$\vec{v}_A = (8 - 6t)\hat{i} = 2\hat{i}$ and $\vec{v}_B = -24\hat{j}$

$\therefore \quad \vec{v}_{B/A} = -\vec{v}_A + \vec{v}_B = -2\hat{i} - 24\hat{j}$

$\therefore \quad$ Speed of B w.r.t. A, $\sqrt{v} = \sqrt{2^2 + 24^2}$

$= \sqrt{4 + 576} = \sqrt{580}$ m/s

$\therefore \quad v = 580$

55. **(d)** Given, Position vector,

$\vec{r} = \cos\omega t\hat{i} + \sin\omega t\,\hat{j}$

Velocity, $\vec{v} = \frac{d\vec{r}}{dt} = \omega(-\sin\omega t\hat{i} + \cos\omega t\,\hat{j})$

Acceleration,

$\vec{a} = \frac{d\vec{v}}{dt} = -\omega^2(\cos\omega t\hat{i} + \sin\omega t\,\hat{j})$

$\vec{a} = -\omega^2\vec{r}$

$\therefore \vec{a}$ is antiparallel to $\vec{r}$ and directed towards origin.

Also $\vec{v}.\vec{r} = 0 \quad \therefore \vec{v} \perp \vec{r}$

56. **(a)** Given $\vec{u} = 3\hat{i} + 4\hat{j}$, $\vec{a} = 0.4\hat{i} + 0.3\hat{j}$, t=10s

From 1st equatoin of motion.

$a = \frac{v - u}{t} \qquad \therefore v = u + at$

$\Rightarrow v = (0.4\hat{i} + 0.3\hat{j}) \times 10 + (3\hat{i} + 4\hat{j})$

$= 4\hat{i} + 3\hat{j} + 3\hat{i} + 4\hat{j} = 7\hat{i} + 7\hat{j}$

$\Rightarrow \quad |\vec{v}| = \sqrt{7^2 + 7^2} = 7\sqrt{2}$ unit.

57. **(b)** Coordinates of moving particle at time 't' are $x = \alpha t^3$ and $y = \beta t^3$

$v_x = \frac{dx}{dt} = 3\alpha t^2$ and $v_y = \frac{dy}{dt} = 3\beta t^2$

$\therefore \; v = \sqrt{v_x^2 + v_y^2} = \sqrt{9\alpha^2 t^4 + 9\beta^2 t^4} = 3t^2\sqrt{\alpha^2 + \beta^2}$

58. (a) Maximum height $H_{max} = \dfrac{u^2 \sin^2 \theta}{2g}$

and time of flight, $T = \dfrac{2u \sin\theta}{g}$

Given, $(H_{max})_1 = 8 \times (H_{max})_2$

$$\Rightarrow \frac{u^2 \sin^2 \theta_1}{2g} = 8 \times \frac{u^2 \sin^2 \theta_2}{2g}$$

$$\Rightarrow \sin\theta_1 = 2\sqrt{2}\sin\theta_2$$

$$\therefore \frac{T_1}{T_2} = \frac{\frac{2u\sin\theta_1}{g}}{\frac{2u\sin\theta_2}{g}} = \frac{\sin\theta_1}{\sin\theta_2} = \frac{2\sqrt{2}\sin\theta_2}{\sin\theta_2} = 2\sqrt{2}$$

59. (d) For two projectiles,

$\theta_1 = 45 + \alpha;\ \theta_2 = 45 - \alpha$

Time of flight, $T = \dfrac{2u\sin\theta}{g} \Rightarrow T \propto \sin\theta$

$$\therefore \frac{T_1}{T_2} = \frac{\sin(45+\alpha)}{\sin(45-\alpha)} = \frac{\frac{1}{\sqrt{2}}\cos\alpha + \frac{1}{\sqrt{2}}\sin\alpha}{\frac{1}{\sqrt{2}}\cos\alpha - \frac{1}{\sqrt{2}}\sin\alpha}$$

$$= \frac{\cos\alpha + \sin\alpha}{\cos\alpha - \sin\alpha} = \frac{1+\tan\alpha}{1-\tan\alpha}.$$

60. (d) For the projectile,

Range, $R = 3H_{max}$

$$\Rightarrow \frac{u^2 \sin 2\theta}{g} = \frac{3u^2 \sin^2\theta}{2g}$$

$$\Rightarrow 2\sin\theta\cos\theta = \frac{3}{2}\sin^2\theta$$

$$\Rightarrow \tan\theta = \frac{4}{3} \Rightarrow \theta = 53°$$

$$\therefore \quad R = \frac{u^2(2\sin\theta.\cos\theta)}{g} = \frac{u^2\left(2\times\frac{4}{5}\times\frac{3}{5}\right)}{g} = \frac{24\mu^2}{25g}$$

$= \dfrac{nu^2}{25g}$ (given)

$\therefore\ n = 24$

61. (c)

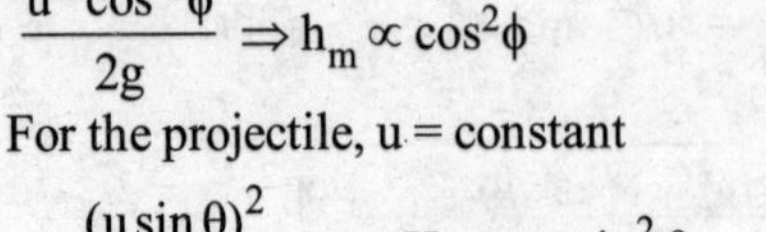

The maximum height is

$$h_m = \frac{u^2\cos^2\phi}{2g} \Rightarrow h_m \propto \cos^2\phi$$

62. (c) For the projectile, u = constant

$$H_{max} = \frac{(u\sin\theta)^2}{2g} \Rightarrow H_{max} \propto \sin^2\theta$$

$$\therefore \frac{(H_{max})_1}{(H_{max})_2} = \frac{\sin^2(45°-\alpha)}{\sin^2(45°+\alpha)}$$

$$= \frac{\left(\frac{1}{\sqrt{2}}\cos\alpha - \frac{1}{\sqrt{2}}\sin\alpha\right)^2}{\left(\frac{1}{\sqrt{2}}\cos\alpha + \frac{1}{\sqrt{2}}\sin\alpha\right)^2} = \frac{1-\sin 2\alpha}{1+\sin 2\alpha}$$

63. (5) 60sin30° = 30 m/s

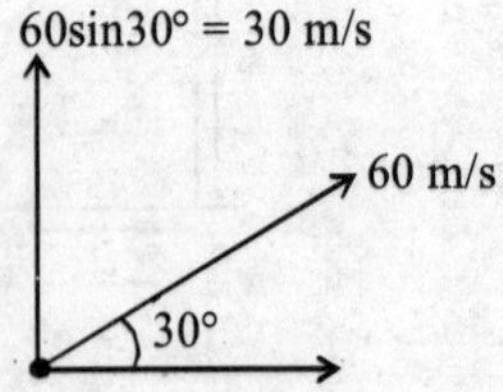

Time to reach maximum height, $t = \dfrac{U_y}{g} = \dfrac{30}{10} = 3s$

Using, $S = u_y + \dfrac{1}{2}a_y t^2$

$$h_0 = 30 \times 1 - \frac{1}{2} \times 10 \times 1 = 25m$$

$$h_1 = 30 + \left(\frac{-10}{2}\right) \times (2\times 3 - 1) = 5m$$

$$\therefore \frac{h_0}{h_1} = \frac{25}{5} = 5$$

64. (b) Given maximum height = horizontal range.

$$\frac{u^2\sin^2\theta}{2g} = \frac{u^2\sin 2\theta}{g}$$

$4\sin\theta\cos\theta = \sin^2\theta \Rightarrow \tan\theta = 4$

65. (16) For a projectile,

$$H_{max} = \frac{u^2\sin^2\theta}{2g} \Rightarrow H_{max} \propto u^2$$

$$\Rightarrow \frac{H_{1max}}{H_{2max}} = \frac{u_1^2}{u_2^2} \Rightarrow \frac{64}{H_{2max}} = \frac{u^2}{(u/2)^2}$$

$\therefore H_{2max} = 16m$

66. (d) For projectile A,

$u_y = u_A\cos 45° = \dfrac{u_A}{\sqrt{2}}$, H = –400m

$$\therefore H = u_y t - \frac{1}{2}gt^2 \Rightarrow -400 = \left(\frac{u_A}{\sqrt{2}}\right)t - \frac{1}{2}\times 10\,t^2$$

$$\Rightarrow 5t^2 - \left(\frac{u_A}{\sqrt{2}}\right)t - 400 = 0$$

$\therefore$ Time of flight $= t_1 + t_2 = \dfrac{-\left(-\dfrac{u_A}{\sqrt{2}}\right)}{2 \times 5} = \dfrac{u_A}{10\sqrt{2}}$ ...(i)

For projetile B

$u_y = u_B \cos 60° = \dfrac{u_B}{2}$, $H = -400$ m

$\therefore H = u_y t - \dfrac{1}{2} g t^2 \Rightarrow -400 = \left(\dfrac{u_B}{2}\right) - \dfrac{1}{2} \times 10 \times t^2$

$\Rightarrow 5t^2 - \left(\dfrac{u_B}{2}\right) t - 400 = 0$

$\therefore$ Time of flight $= t_1 + t_2 = \dfrac{-\left(-\dfrac{u_A}{2}\right)}{2 \times 5} = \dfrac{u_B}{20}$ ...(ii)

From equation (i) and (ii)

$\dfrac{u_A}{10\sqrt{2}} = \dfrac{u_B}{20}$ $\quad \therefore u_A : u_B = 1 : \sqrt{2}$

67. (2)

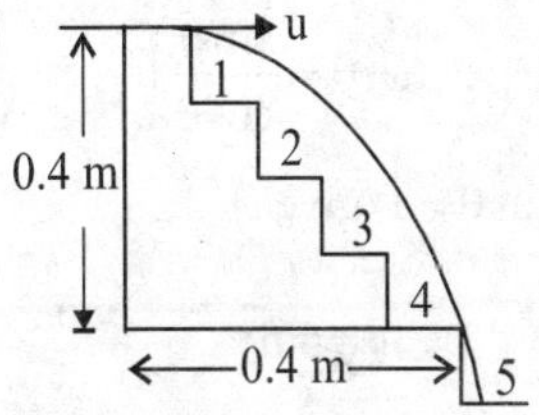

In x-direction

$t = \dfrac{x}{u} = \dfrac{0.4}{u}$

In y-direction

$H = \dfrac{1}{2} g t^2 \Rightarrow 0.4 = \dfrac{1}{2} \times 10 \times \left(\dfrac{0.4}{u}\right)^2$

$\Rightarrow u = \sqrt{2}$ m/s $= \sqrt{x}$ m/s (given)

$\therefore x = 2$

68. (80) Time of flight, T = 3 + 5 = 8 sec

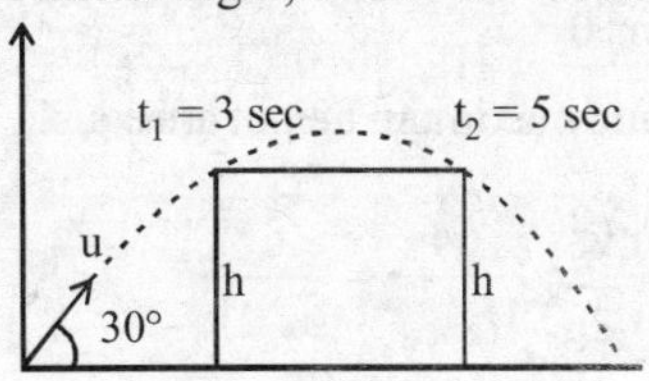

$T = \dfrac{2u \sin\theta}{g}$

$\Rightarrow 8 = \dfrac{2u \sin(30°)}{10}$ $\quad [\because \theta = 30° \text{ given}]$

$\Rightarrow u = 80$ m/s

69. (a) Given,

Initial velocity of projectile, u = 40 m/s

Angle, θ = 30°

Time of flight

$T = \dfrac{2u\sin\theta}{g} = \dfrac{2 \times 40 \times 1}{10 \times 2} = 4\text{s}$ $\quad (\because g = 10 \text{ m/s}^2)$

It means projectile is at maximum height at t = 2 s. At maximum height vertical component of velocity is zero.

Velocity at $t = 2s = V_x = u\cos\theta = 40 \cos 30°$

$= 20\sqrt{3}\text{ms}^{-1}$.

70. (c) Range of projectile

$R = \dfrac{v^2 \sin 2\theta}{g}$ $\quad (\because R \propto \sin(2\theta))$

$\dfrac{R_1}{R_2} = \dfrac{\sin(2\theta_1)}{\sin(2\theta_2)} = \dfrac{\sin(2 \times 15)}{\sin(2 \times 45)} = \dfrac{\sin 30°}{\sin 90°}$

$\Rightarrow \dfrac{50}{R_2} = \dfrac{1}{2} \Rightarrow R_2 = 100\text{m}$

71. (c) Maximum height in projectile motion

$H_{max} = \dfrac{u^2 \sin^2\theta}{2g}$

For two projectiles $\theta_1 = 30°$ and $\theta_2 = 60°$

$\therefore \quad \dfrac{H_1}{H_2} = \dfrac{\dfrac{u^2 \sin^2 30°}{2g}}{\dfrac{u^2 \sin^2 60°}{2g}} = \dfrac{1}{3}$

72. (d) The formula of range of projectile,

$R = \dfrac{u^2 \sin 2\theta}{g}$

$R_1 = \dfrac{u_1^2 \sin 2\theta_1}{g}; R_2 = \dfrac{u_2^2 \sin 2\theta_2}{g}$

$\dfrac{R_1}{R_2} = \dfrac{u_1^2 \sin 2\theta_1}{u_2^2 \sin 2\theta_2} = \dfrac{40^2 \sin(2 \times 30°)}{60^2 \sin(2 \times 60°)} = \dfrac{4}{9}$

73. (a) The trajectory of projectile projected. From the ground is given by

$y = x - \dfrac{x^2}{20}$

For maximum height,

$\dfrac{dy}{dx} = 0 \Rightarrow 1 - \dfrac{2x}{20} = 0$ $\quad \therefore x = 10\text{m}$

Hence, the maximum height attained by the projectile,

$y_{max} = 10 - \dfrac{100}{20} = 5\text{m}$

74. (b) Range, $R = \dfrac{u^2}{g} \sin 2\theta$

For maximum range, $\sin(2\theta) = 1 \Rightarrow \theta = 45°$

75. (b) At high .t point, only horizontal component of velocity exists

So, $u\cos\theta = \dfrac{\sqrt{3}}{2} u \Rightarrow \cos\theta = \dfrac{\sqrt{3}}{2} \Rightarrow \theta = 30°$

$T = \dfrac{2u\sin\theta}{g} = \dfrac{2 \times u \times \sin 30°}{g} = \dfrac{u}{g}$

76. **(80)** Given, $v_1 = v_2 = 40$ m/s
$R_1 = R_2$
$\theta_1 = 60°$. So, $\theta_2 = (90 - 60°) = 30°$
[$\because$ For same range, angle of projections are complementary]
We have,

$$R_1 = \frac{40^2 \sin 2\times 60°}{g} = \frac{1600}{10}\times\frac{\sqrt{3}}{2} = 80\sqrt{3}\text{ m}$$

As, $R_2 = R_1 \Rightarrow R_2 = 80\sqrt{3}$ m

$$\text{So, } H_1 + H_2 = \frac{R_1}{4}\tan\theta_1 + \frac{R_2}{4}\tan\theta_2 \quad \left[\because H = \frac{R}{4}\tan\theta\right]$$

$= 20\sqrt{3}\tan 60° + 20\sqrt{3}\tan 30° = 60 + 20 = 80$ m

77. **(d)** Let u be the speed of projectile at initial point.

$\therefore \quad KE = \frac{1}{2}mu^2$

At maximum height speed = u cos 30°

$\therefore \quad KE = \frac{1}{2}m\,(u\cos 30°)^2$

$$\frac{KE_{initial}}{KE_{top}} = \frac{\frac{1}{2}m(u)^2}{\frac{1}{2}m(u\cos 30°)^2} = \frac{4}{3}$$

78. **(d)** Range for projection angle α,

$$R_1 = \frac{u^2\sin 2\alpha}{g}$$

Range for projection angle β,

$$R_2 = \frac{u^2\sin 2\beta}{g}$$

$\alpha + \beta = 90° \Rightarrow \beta = 90° - \alpha$

$$R_2 = \frac{u^2\sin 2(90° - \alpha)}{g} = \frac{u^2\sin(180° - 2\alpha)}{g}$$

$$= \frac{u^2\sin 2\alpha}{g} \Rightarrow \frac{R_1}{R_2} = \frac{\left(\frac{u^2\sin 2\alpha}{g}\right)}{\left(\frac{u^2\sin 2\alpha}{g}\right)} = \frac{1}{1}$$

79. **(c)** As $v^2 = u^2 + 2as$

$\Rightarrow 0 = u^2 - 2gH \Rightarrow u^2 = 2gH \Rightarrow u = \sqrt{2\times 10\times 136}$

$$R_{max} = \frac{u^2}{g} = \frac{2\times 10\times 136}{10} = 272\text{ m}$$

80. **(15 or 75)** We know that range $(R) = \frac{u^2\sin 2\theta}{g}$

For 1st object,

$$\text{So, } R_{max} = \frac{u^2\sin 2(45°)}{g} = \frac{u^2}{g}$$

For 2nd object,

$$\frac{R_{max}}{2} = \frac{u^2}{2g} = \frac{u^2\sin 2\theta}{g}$$

$\Rightarrow \sin 2\theta = \frac{1}{2} \Rightarrow 2\theta = 30°, 150° \Rightarrow \theta = 15°, 75°$

81. **(1)** Here, time of flight is same.
So, vertical component of velocity is same.
Therefore H_{max} is same.

82. **(d)** Equation of trajectory of projectile

$$y = x\tan\theta - \frac{1}{2}\frac{gx^2}{u^2\cos^2\theta} = x\tan 45° - \frac{1}{2}\frac{10x^2}{u^2\cos^2 45°}$$

$$\Rightarrow y = x - \frac{10x^2}{2u^2\left(\frac{1}{2}\right)} \Rightarrow 10 = 20 - \frac{(10)(400)}{u^2} \Rightarrow u = 20\text{m/s}$$

$$\text{Time period, } T = \frac{2u\sin\theta}{g} = \frac{(2)(20)}{\sqrt{2}(10)} = 2\sqrt{2}\text{ s}$$

$\vec{v} = u\cos\theta\hat{i} + (u\sin\theta - gt)\hat{j} = [10\sqrt{2}\hat{i} + (10\sqrt{2} - 10(2)\hat{j}]$

Momentum $\vec{P} = M\vec{v} = 100\sqrt{2}\hat{i} + (100\sqrt{2} - 200)\hat{j}$

83. **(5)** Equation of trajectory of projectile,
$y = 5x(1 - x) = 5x - 5x^2$
Comparing the above equation with standard equation of projectile $y = (\tan\theta)x - \frac{g}{2u^2\cos^2\theta}x^2$

we get $\tan\theta = 5$

But $\tan\theta = \frac{u\sin\theta}{u\cos\theta} = 5 \quad \left(\because \tan\theta = \frac{\sin\theta}{\cos\theta}\right)$

$\Rightarrow \frac{u\sin\theta}{1} = 5 \; (\because u\cos\theta = 1 \text{ given})$

$\therefore \quad u_y = u\sin\theta = 5 \qquad \therefore \vec{u}_y = 5\hat{j}$

84. **(c)** As $\frac{T_1}{2} = \frac{T_2}{2}$

$$\Rightarrow T_1 = T_2 \Rightarrow \frac{2u_1\sin\theta_1}{g} = \frac{2u_2\sin\theta_2}{g}$$

$$\Rightarrow \frac{u_1}{u_2} = \frac{\sin\theta_2}{\sin\theta_1} \Rightarrow \frac{u_1}{u_2} = \frac{\sin 45°}{\sin 30°} \Rightarrow \frac{u_1}{u_2} = \sqrt{2}$$

85. **(c)** Projection speed is v.

$$R_1 = \frac{v^2\sin(90°)}{g}; R_2 = \frac{v^2\sin(60°)}{g}$$

$$\therefore \frac{R_1}{R_2} = \frac{2}{\sqrt{3}}$$

86. **(d)** As range and maximum height are equal,
R = H

$$\Rightarrow \frac{2v_x - v_y}{g} = \frac{v_y^2}{2g}$$

$$\Rightarrow v_x = \frac{v_y}{4} \Rightarrow u\cos\theta = \frac{u\sin\theta}{4} \Rightarrow \tan\theta = 4$$

87. **(c)** As, $H_{max} = \frac{u^2\sin^2\theta}{2g} \Rightarrow u^2 \propto \frac{1}{\sin^2\theta}$

$$\Rightarrow u \propto \frac{1}{\sin\theta} \Rightarrow \frac{\sin\theta_2}{\sin\theta_1} = \frac{u_1}{u_2}$$

$$\Rightarrow \sin\theta_1 = \frac{u_2}{u_1}\times\sin 60° = \frac{\sqrt{2}}{\sqrt{3}}\times\frac{\sqrt{3}}{2} = \frac{1}{\sqrt{2}}$$

$$\text{So, } \theta_1 = \sin^{-1}\left(\frac{1}{\sqrt{2}}\right) = 45°$$

88. **(a)** $R=\dfrac{u^2 \sin 2\theta}{g}$

For, $R = R_{max}$

$\sin 2\theta = 1 \Rightarrow 2\theta = \dfrac{\pi}{2}$

$\theta = \dfrac{\pi}{4}$. So, $R_{\max} = \dfrac{u^2}{g} = 100m$

So, $H = \dfrac{u^2 \sin^2\left(\dfrac{\pi}{4}\right)}{2g} = \dfrac{u^2 \times \dfrac{1}{2}}{2g} = \dfrac{u^2}{4g} = \dfrac{100}{4} = 25$ m

89. **(b)** $v_x = 20\cos\alpha$ $[\because a_x = 0]$

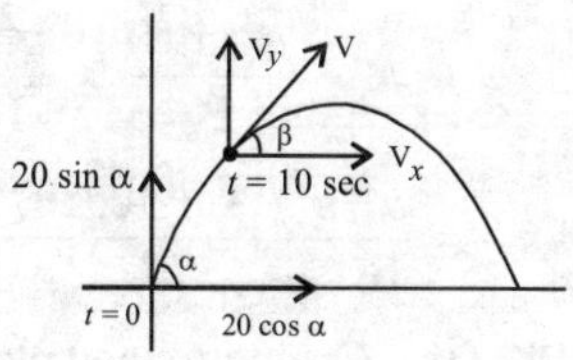

$v_y = 20 \sin \alpha - g \times t$

$[\because a_y = -g]$

$= 20 \sin \alpha - 10 \times 10$

$= 20 \sin \alpha - 100$

So, $\tan\beta = \dfrac{v_y}{v_x} = \dfrac{20\sin\alpha - 100}{20\cos\alpha}$

$= \tan\alpha - 5\sec\alpha$

90. **(60)**

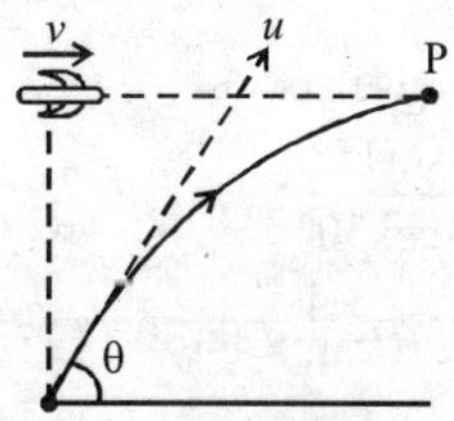

If fighter jet and bullet have same horizontal component of velocity, then it will strike after some time as shown above because horizontal displacement of both of them in given time interval will always be same.

i.e. $v = u\cos\theta \Rightarrow 200 = 400\cos\theta$

$\Rightarrow \cos\theta = \dfrac{1}{2} \Rightarrow \theta = 60°$

91. **(a)** We know that if range is same for two angle of projection, then these angle must be complementary.

Let first angle of projection be 'θ' then second will be $(90-\theta)$

$\therefore h_1 = \dfrac{u^2 \sin^2\theta}{2g}$ and $h_2 = \dfrac{u^2 \sin^2(90-\theta)}{2g} = \dfrac{u^2\cos^2\theta}{2g}$

$\therefore h_1 h_2 = \dfrac{u^2\sin^2\theta}{2g} \cdot \dfrac{u^2\cos^2\theta}{2g}$. So, reason is correct

$\Rightarrow \sqrt{h_1 h_2} = \dfrac{u^2 \sin\theta\cos\theta}{2g} \quad \Rightarrow 4\sqrt{h_1 h_2} = \dfrac{4u^2\sin\theta\cos\theta}{2g}$

$\Rightarrow 4\sqrt{h_1 h_2} = \dfrac{u^2(2\sin\theta\cos\theta)}{g} \quad \Rightarrow 4\sqrt{h_1 h_2} = \dfrac{u^2 \sin 2\theta}{g} = R$

So, assertion is correct and reason is correct explanation of assertion.

92. **(d)** At maximum height inclination of projectile with horizontal becomes zero

So, $t = \dfrac{u\sin\theta}{g} \Rightarrow u = \dfrac{gt}{\sin\theta}$ ...(i)

Now, $R = u\cos\theta \times T \Rightarrow R = u\cos\theta \times 2t$

$\Rightarrow \cos\theta = \dfrac{R}{2ut} \Rightarrow \cos\theta = \dfrac{R\sin\theta}{2t.gt}$ [From (i)]

$\Rightarrow \tan\theta = \dfrac{2gt^2}{R} \Rightarrow \tan\theta = \dfrac{20t^2}{R} \Rightarrow \cot\theta = \dfrac{R}{20t^2}$

$\therefore \theta = \cot^{-1}\left(\dfrac{R}{20t^2}\right)$

93. **(20)** Let $\vec{v}_1 = a\hat{i} + a\hat{j}$ $[\because \theta = 45° \Rightarrow u_x = u_y]$

Then, $\vec{V}_f = a\hat{i} + (a - 2\times 10)\hat{j} = a\hat{i} + (a-20)\hat{j}$

As, $|\vec{V}_f| = 20$

$\Rightarrow a^2 + (a-20)^2 = 400$

$\Rightarrow a^2 + a^2 + 400 - 40a = 400 \Rightarrow 2a^2 - 40a = 0$

$\Rightarrow a^2 - 20a = 0 \Rightarrow a(a-20) = 0$

$\Rightarrow a = 20$. So, $V_i = \sqrt{20^2 + 20^2} = 20\sqrt{2}$ m/s

$\therefore h_{\max} = \dfrac{V_i^2 \sin^2 45°}{2g} = \dfrac{800 \times \dfrac{1}{2}}{2\times 10} = 20$m

94. **(b)** For projectile motion,

Range $R = \dfrac{u^2 \sin 2\theta}{g}$

As range is same for angle of projection θ and 90– θ. So, range is same for 42° and 48°.

$\therefore R_1 = R_2$

Height in projectile motion, $H = \dfrac{u^2 \sin^2\theta}{2g} \Rightarrow H \propto \sin^2\theta$

So H is higher for 48° than H for 42°.

$\therefore H_1 < H_2$

95. **(c)** Given,

Initial speed of football, $u = 25\ ms^{-1}$.

Maximum height in projectile motion

$H_{\max} = \dfrac{u^2 \sin^2\theta}{2g} = \dfrac{(25)^2.(\sin 45)^2}{2\times 10} = \dfrac{625 \times \dfrac{1}{2}}{2\times 10}$

$= 15.625$ m

Time period,

$T = \dfrac{u\sin\theta}{g} = \dfrac{25\times\sin 45°}{10} = \dfrac{25\times\dfrac{1}{\sqrt{2}}}{10} = 1.77s$

96. **(5)** Change in momentum,

$\Delta\vec{P} = \vec{P}_B - \vec{P}_A$

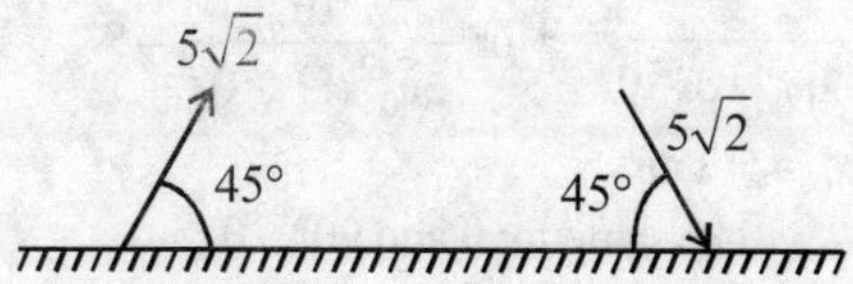

$\Delta\vec{P} = -mV\cos\theta\,\hat{j} - mV\sin\theta\,\hat{j}$

$= -m\left(5\sqrt{2}\cos 45°\,\hat{j} + 5\sqrt{2}\sin 45°\,\hat{j}\right)$

$= -m\left(5\sqrt{2}\times\frac{1}{\sqrt{2}}\hat{j}+5\sqrt{2}\times\frac{1}{\sqrt{2}}\hat{j}\right) = -2m\times 5\sqrt{2}\times\frac{1}{\sqrt{2}}\hat{j}$

$|\Delta\vec{P}| = 2\times(5\times10^{-3})\times(5) = 5\times10^{-2}\ \text{kg ms}^{-1}$

$\therefore\ x = 5$

97. **(d)** Given : $y = \alpha x - \beta x^2$

Comparing it with $y = x\tan\theta - \frac{1}{2}\frac{gx^2}{u^2\cos^2\theta}$, we get

$\tan\theta = \alpha$

i.e., $\theta = \tan^{-1}\alpha$

For maximum height, maximum value of y

$\frac{dy}{dx} = 0 \Rightarrow \frac{d(\alpha x - \beta x^2)}{dx} = 0 \Rightarrow \alpha - 2\beta x = 0$

$\therefore\ x = \frac{\alpha}{2\beta}$

$\therefore\ H_{max} = \alpha\left(\frac{\alpha}{2\beta}\right) - \beta\left(\frac{\alpha^2}{4\beta^2}\right) = \left(\frac{\alpha^2}{2\beta}\right) - \left(\frac{\alpha^2}{4\beta}\right) = \frac{\alpha^2}{4\beta}$

So, $H_{max} = \frac{\alpha^2}{4\beta}$

Hence option (d) is correct answer.

98. **(a)** Using principle of conservation of linear momentum for horizontal motion, we have

$2mv_x = mu + mu\cos 60°$

$v_x = \frac{3u}{4}$

For vertical motion

$h = 0 + \frac{1}{2}gT^2 \Rightarrow T = \sqrt{\frac{2h}{g}}$

Let R is the horizontal distance travelled by the body.

$R = v_xT + \frac{1}{2}(0)(T)^2$ (For horizontal motion)

Also, $h = \frac{u^2\sin^2 60°}{2g} = \frac{3u^2}{8g}$

$R = v_xT = \frac{3u}{4}\times\sqrt{\frac{2h}{g}} \Rightarrow R = \frac{3\sqrt{3}u^2}{8g}$

99. **(c)** Given, $y = 2x - 9x^2$

On comparing with,

$y = x\tan\theta_0 - \frac{gx^2}{2u_0^2\cos^2\theta_0}$,

We have, $\tan\theta_0 = 2$ or $\cos\theta_0 = \frac{1}{\sqrt{5}}$

$\Rightarrow \theta_0 = \cos^{-1}\left(\frac{1}{\sqrt{5}}\right)$

and $\frac{g}{2u_0^2\cos^2\theta} = 9$ or $\frac{10}{2u_0^2(1/\sqrt{5})^2} = 9$

$\therefore\ u_0 = 5/3$ m/s

100. **(d)** R will be same for θ and $90° - \theta$.

Time of flights:

$t_1 = \frac{2u\sin\theta}{g}$ and

$t_2 = \frac{2u\sin(90° - \theta)}{g} = \frac{2u\cos\theta}{g}$

Now, $t_1t_2 = \left(\frac{2u\sin\theta}{g}\right)\left(\frac{2u\cos\theta}{g}\right) = \frac{2}{g}\left(\frac{u^2\sin 2\theta}{g}\right) = \frac{2R}{g}$

101. **(b)** For same range, the angle of projections are : θ and $90° - \theta$. So,

$h_1 = \frac{u^2\sin^2\theta}{2g}$ and $h_2 = \frac{u^2\sin^2(90° - \theta)}{2g} = \frac{u^2\cos^2\theta}{2g}$

Also, $R = \frac{u^2\sin 2\theta}{g}$

$h_1h_2 = \frac{u^2\sin^2\theta}{2g}\times\frac{u^2\cos^2\theta}{2g}$

$= \frac{u^2}{16}\frac{u^2(2\sin\theta\cos\theta)^2}{g^2} = \frac{R^2}{16}$ or $R^2 = 16\,h_1h_2$

102. **(a)** On an inclined plane, time of flight (T) is given by

$T = \frac{2u\sin\theta}{g\cos\alpha}$

Substituting the values, we get

$T = \frac{(2)(2\sin 15°)}{g\cos 30°} = \frac{4\sin 15°}{10\cos 30°}$

Distance, $S = (2\cos 15°)T - \frac{1}{2}g\sin 30°(T)^2$

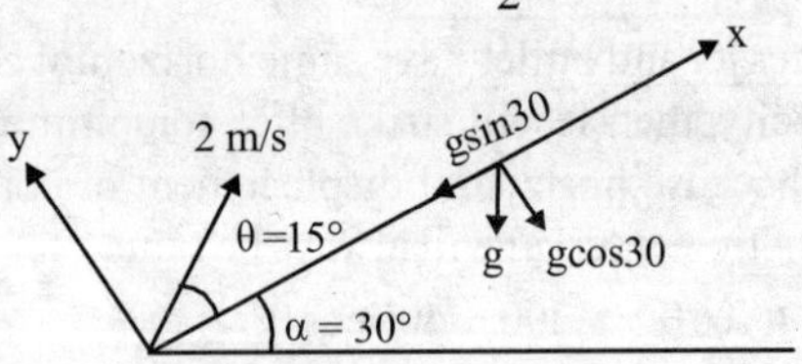

$= (2\cos 15°)\frac{4}{10}\frac{\sin 15°}{10\cos 30°} - \left(\frac{1}{2}\times 10\sin 30°\right)\frac{16\sin^2 15°}{100\cos^2 30°}$

$= \frac{16\sqrt{3} - 16}{60} \simeq 0.1952\text{m} \simeq 20\text{cm}$

103. **(b)**

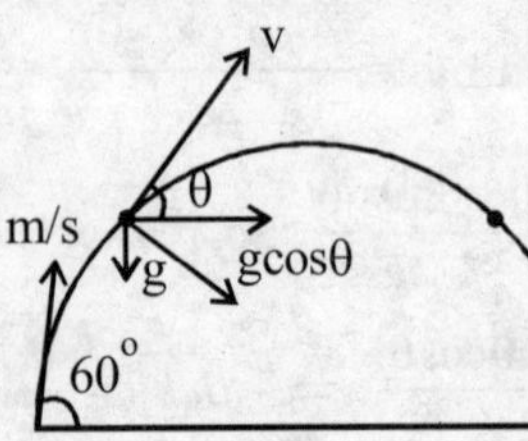

Horizontal component of velocity

$v_x = 10\cos 60° = 5$ m/s

vertical component of velocity

$v_y = 10\cos 30° = 5\sqrt{3}$ m/s

After t = 1 sec.

Horizontal component of velocity $v_x = 5$ m/s

Vertical component of velocity

$v_y = \left(5\sqrt{3} - 10\right)$ m / s

Centripetal, acceleration $a_n = \frac{v^2}{R}$

$\Rightarrow \quad R = \frac{v_x^2 + v_y^2}{a_n} = \frac{25+100+75-100\sqrt{3}}{10\cos\theta}$ ...(i)

From figure (using (i))

$\tan\theta = \frac{v_y}{v_x} = \frac{10-5\sqrt{3}}{5} = 2-\sqrt{3} \Rightarrow \theta = 15°$

$R = \frac{100\left(2-\sqrt{3}\right)}{10\cos 15} = 2.8\,\text{m}$

104. (a) As we know, range $R = \frac{u^2 \sin 2\theta}{g}$

and, area $A = \pi R^2$

$\therefore \quad A \propto R^2$ or, $A \propto u^4$

$\therefore \quad \frac{A_1}{A_2} = \frac{u_1^4}{u_2^4} = \left[\frac{1}{2}\right]^4 = \frac{1}{16}$

105. (c) Let 't' be the time taken by the bullet to hit the target.
For horizontal motion,
$\therefore \quad 700\,\text{m} = 630\,\text{ms}^{-1}\, t$

$\Rightarrow t = \frac{700\text{m}}{630\text{ms}^{-1}} = \frac{10}{9}\,\text{sec}$

For vertical motion,
Here, $u = 0$

$\therefore \quad h = \frac{1}{2}gt^2 = \frac{1}{2} \times 10 \times \left(\frac{10}{9}\right)^2 = \frac{500}{81}\,\text{m} = 6.1\,\text{m}$

Therefore, the rifle must be aimed 6.1 m above the centre of the target to hit the target.

106. (c) From question,
Horizontal velocity (initial),

$u_x = \frac{40}{2} = 20\,\text{m/s}$

Vertical velocity (initial), $50 = u_y t - \frac{1}{2}gt^2$

$= \quad u_y \times 2 - \frac{1}{2}(10) \times 4$

or, $50 = 2u_y - 20$ or, $u_y = \frac{70}{2} = 35\,\text{m/s}$

$\therefore \quad \tan\theta = \frac{u_y}{u_x} = \frac{35}{20} = \frac{7}{4}$

$\Rightarrow$ Angle $\theta = \tan^{-1}\frac{7}{4}$

107. (b) From equation, $\vec{u} = \hat{i} + 2\hat{j} \Rightarrow u_x = 1\,m/s, V_y = 2\,m/s$

$\Rightarrow x = u_x t = 1 \times t = t$ ...(i)

$y = v_y t - \frac{1}{2}gt^2 = 2t - \frac{1}{2}(10t^2)$ ...(ii)

From (i) and (ii), $y = 2x - 5x^2$

108. (b) $R = 10\,\text{m}, \theta = 45°$

$h = \frac{R}{4}\tan\theta = \frac{10}{4}\tan 45° = 2.5\,\text{m}$

So, the nearest possible answer is 2.4 m.

109. (d) $H = \frac{u^2\sin^2\theta}{2g}$

$\therefore H_{max} = \frac{u^2}{2g} \Rightarrow \frac{u^2}{2g} = 10 \Rightarrow u^2 = 10g \times 2$

$R = \frac{u^2 \sin 2\theta}{g} \Rightarrow R_{max} = \frac{u^2}{g}$

$R_{max} = \frac{10 \times g \times 2}{g} = 20$ metre

110. (a) Let, total area around fountain

$A = \pi R_{max}^2$ ...(i)

Where $R_{max} = \frac{v^2 \sin 2\theta}{g} = \frac{v^2 \sin 90°}{g} = \frac{v^2}{g}$ ...(ii)

From equation (i) and (ii)

$A = \pi\frac{v^4}{g^2}$

111. (a) A projectile have same range for two angle
Let one angle be θ, then other is 90° – θ

$T_1 = \frac{2u\sin\theta}{g}, \; T_2 = \frac{2u\cos\theta}{g}$

then, $T_1 T_2 = \frac{4u^2 \sin\theta\cos\theta}{g} = 2R \quad (\because R = \frac{u^2\sin 2\theta}{g})$

Thus, it is proportional to R. (Range)

112. (c) Yes, Man will catch the ball, if the horizontal component of velocity becomes equal to the constant speed of man.

$\frac{v_o}{2} = v_o\cos\theta$ or $\theta = 60°$

113. (d) Horizontal range is required

$R = \frac{u^2 \sin 2\theta}{g} = \frac{(10)^2 \sin(2 \times 30°)}{10} = 5\sqrt{3} = 8.66\,\text{m}$

114. (b) Let s be the width of the river

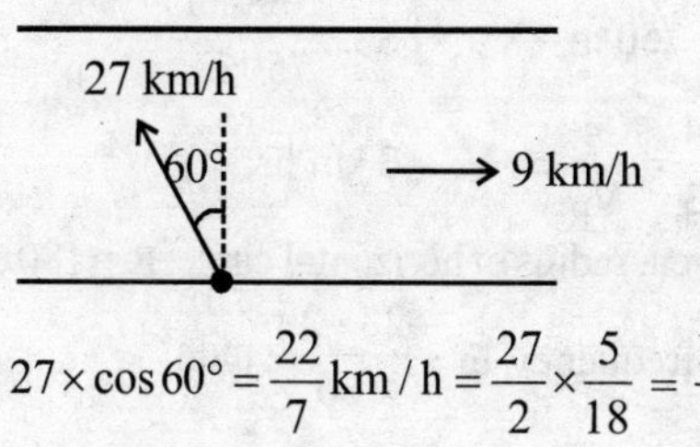

$V = 27 \times \cos 60° = \frac{22}{7}\,\text{km/h} = \frac{27}{2} \times \frac{5}{18} = \frac{15}{4}\,\text{m/s}$

Time taken = 30 sec.

$\therefore S = Vt = \frac{15}{4} \times 30 = 112.5\text{m}$

115. (2000)

$v_{br} = 27\text{ km/h}, v_r = 9\text{ km/h}$

For boat moving downstream to river,

$v_b = v_{br} + v_r = 27 + 9 = 36\text{ km/h} = 10\text{ m/s}$

$\therefore$ Time of flight,

$$T = \frac{2u_y}{g} = \frac{2 \times 10}{10} = 2\text{s}$$

$\therefore$ Range of the ball is

$R = u_x T = u_b T = 10 \times 2 = 20\text{ m} = 2000\text{ cm}$

116. (a) Maximum height is given by:

$$H_{max} = \frac{u^2 \sin^2 \theta}{2g}$$

substitute $H_{max} = 4R$ and $u = \frac{2\pi R}{T}$

$$\Rightarrow 4R = \frac{4\pi^2 R^2}{T^2 2g} \sin^2 \theta \Rightarrow \theta = \sin^{-1}\left[\frac{2gT^2}{\pi^2 R}\right]^{1/2}$$

117. (8) As, $|\vec{a}_c| = |\vec{a}_t|$

$$\frac{v^2}{r} = \frac{dv}{dt} \Rightarrow \int_4^v \frac{dv}{v^2} = \int_0^t \frac{dt}{r}$$

$$\Rightarrow \left[\frac{-1}{v}\right]_4^v = \frac{t}{r} \Rightarrow \frac{-1}{v} + \frac{1}{4} = 2t \Rightarrow v = \frac{4}{1-8t} = \frac{ds}{dt}$$

$$\Rightarrow 4\int_0^t \frac{dt}{1-8t} = \int_0^\pi ds \quad (\because = 0.5\text{ m} \Rightarrow s = 2\pi r = \pi)$$

$$\Rightarrow 4 \times \frac{[\ell n(1-8t)]_0^t}{-8} = \pi$$

$$\Rightarrow \ell n(1-8t) = -2\pi \Rightarrow 1 - 8t = e^{-2\pi}$$

$$\Rightarrow t = \left(1 - e^{-2\pi}\right)\frac{1}{8} s = \frac{1}{\alpha}(1 - e^{-2\pi})\text{ s (given)}$$

$\therefore \alpha = 8$

118. (3)

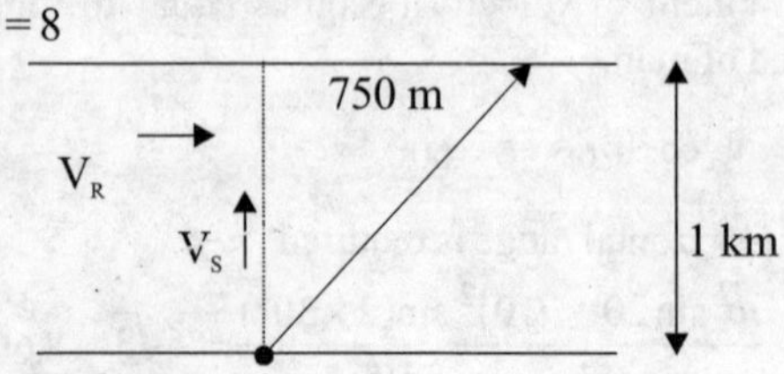

We have,

$1\text{ km} = V_s \times t$

and, $750\text{ m} = V_R \times t$ So, $\frac{1000\text{ m}}{750\text{ m}} = \frac{V_S}{V_R}$

$$\Rightarrow \frac{4}{3} = \frac{4}{V_R} \Rightarrow V_R = 3\text{ km/hr}$$

119. (125) Given, radius of horizontal circle, $R = 180\text{ cm} = 1.8\text{ m}$

Angular frequency, $\omega = \frac{28 \times 2\pi}{60}\text{ rad s}^{-1}$

Acceleration, $a = \omega^2 R = \left(\frac{28 \times 2\pi}{60}\right)^2 \times 1.8$

$$= \left(\frac{56}{60} \times \frac{22}{7}\right)^2 \times 1.8 = \frac{(44)^2}{225} \times 1.8 = \frac{1936 \times 1.8}{225} = \frac{1936}{x}$$

$\Rightarrow x = 125$

120. (b)

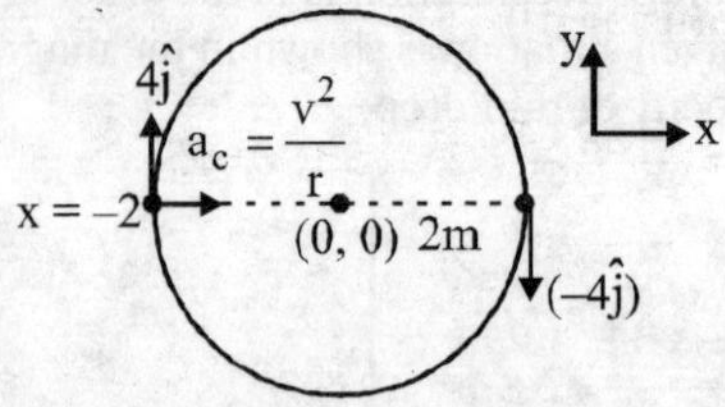

Centripetal acceleration is given as

$$a_c = \frac{V^2}{r} = \frac{4^2}{2} = \frac{16}{2} = 8\text{m/s}^2$$

At $x = -2$, from figure we can clearly see that $\vec{V} = 4\hat{j}$ and $\vec{a}_c = 8\hat{i}$

121. (c) Clearly, $\tan 45° = \frac{|-\vec{V}_m|}{|\vec{V}_{rm}|}$

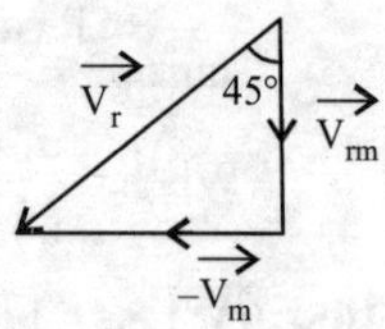

$\Rightarrow V_{rm} = V_m$

$\Rightarrow V_{rm} = 15\sqrt{2} \times \frac{\sqrt{2}}{\sqrt{2}} = \frac{30}{\sqrt{2}}\text{ km/hr.}$

122. (30) As per question, magnitude of velocity of the swimmer is same as that of the river.

So resultant would pass exactly midway through them along AB

$\therefore$ Angle $\theta = 30°$

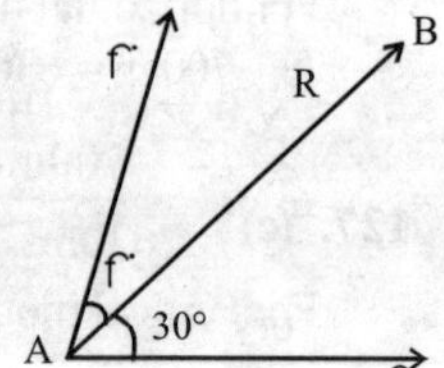

123. (5)

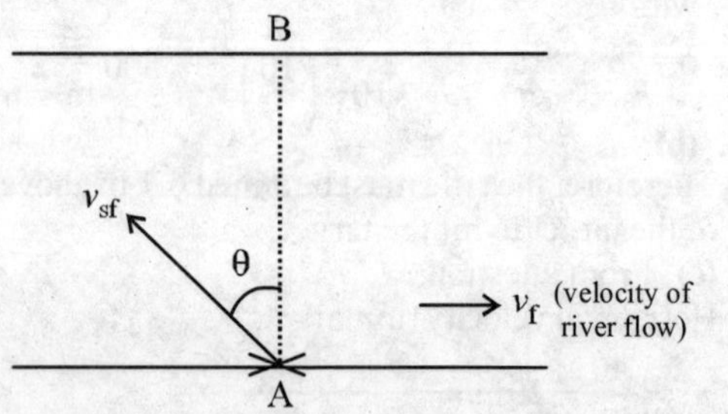

Velocity of swimmer w.r.t. river flow = v_{sf} = 10 m/s

Person is swimming at an angle of 120º.

$\therefore \quad \theta = 120º - 90º = 30º$

From figure, $\sin\theta = \frac{v_f}{v_{sf}}$

$$\Rightarrow \sin 30º = \frac{x}{10} \Rightarrow \frac{1}{2} = \frac{x}{10} \Rightarrow x = \frac{10}{2} = 5\text{ m/s}$$

124. (120) To reach directly opposite point swimmer should swim with an angle $90° + \theta$.

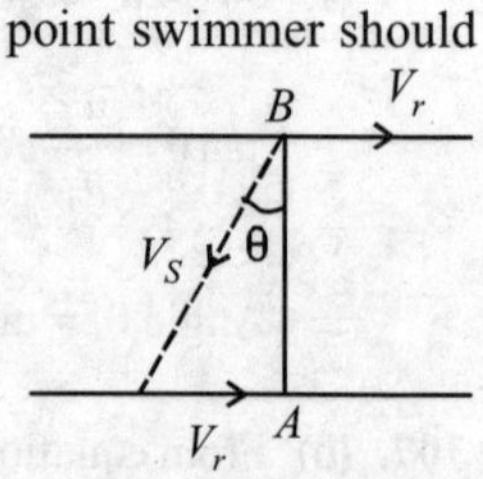

From figure

$\sin\theta = \frac{v_r}{v_s} \Rightarrow \sin\theta = \frac{6}{12} = \frac{1}{2}$

$\therefore \theta = 30°$

$\therefore$ Angle w.r.t. flow $= 90° + \theta = 90° + 30° = 120°$.

125. **(a)** Here, $R = 0.1$ m

$\omega = \frac{2\pi}{T} = \frac{2\pi}{60} = 0.105$ rad/s

Acceleration of the tip of the clock second's hand,

$a = \omega^2 R = (0.105)^2 (0.1) = 0.0011 = 1.1 \times 10^{-3}$ m/s^2

Hence, average acceleration is of the order of 10^{-3}.

126. **(a)** The given situation is shown in the diagram. Here v_r be the velocity of rain drop.

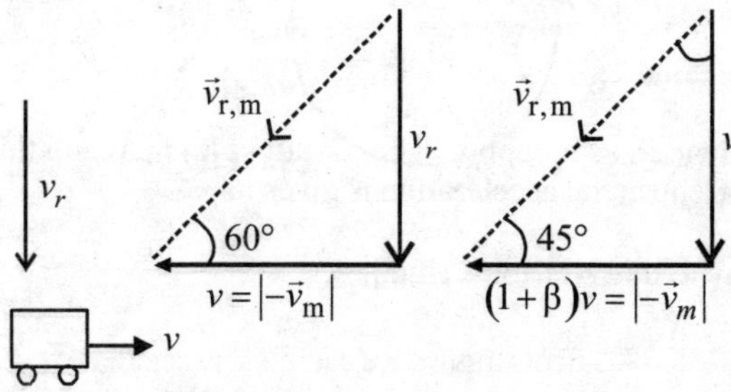

When car is moving with speed v,

$\tan 60° = \frac{v_r}{v}$ ...(i)

When car is moving with speed $(1+\beta)v$,

$\tan 45° = \frac{v_r}{(\beta+1)v}$ ...(ii)

Dividing (i) by (ii) we get,

$\sqrt{3}v = (\beta+1)v \Rightarrow \beta = \sqrt{3} - 1 = 0.732.$

127. **(c)** $\sin\theta = \frac{u}{v} = \frac{2}{4} = \frac{1}{2}$

or $\theta = 30°$

with respect to flow,

$\phi = 90° + 30° = 120°$

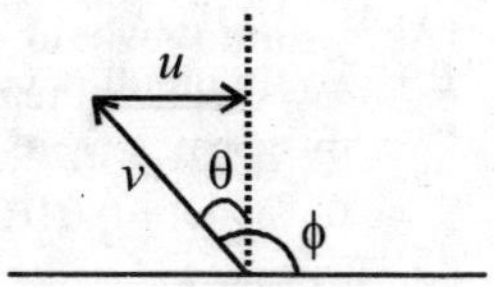

128. **(b)**

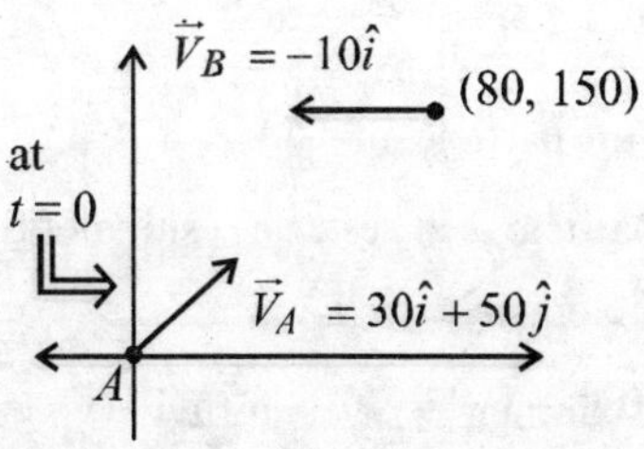

After time 't'

$\vec{r}_A = 30t\hat{i} + 50t\hat{j}$

$\vec{r}_B = (80 - 10t)\hat{i} + 150\hat{j}$

$|\vec{r}| = |\vec{r}_B - \vec{r}_A| = (80 - 40t)\hat{i} + (150 - 50t)\hat{j}$

$|r|^2 = (80 - 40t)^2 + (150 - 50t)^2$

For 'r' min, $\frac{d}{dt}(|r|^2) = 0$

$\Rightarrow 2(80 - 40t) \times -40 + 2(150 - 5t) \times -50 = 0$

Solving above we get $t = 2.6$ hr.

129. **(c)** From, $\theta = \omega t = \omega \frac{\pi}{2\omega} = \frac{\pi}{2}$

So, both have completed quater circle

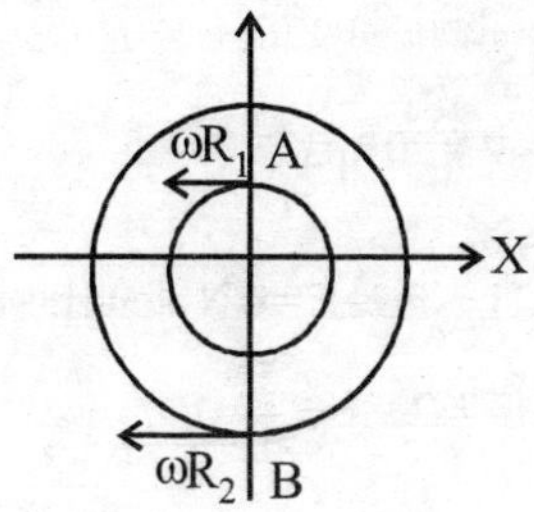

Relative velocity,

$\vec{v}_A - \vec{v}_B = \omega R_1(-\hat{i}) - \omega R_2(-i) = \omega(R_2 - R_1)i$

130. **(d)**

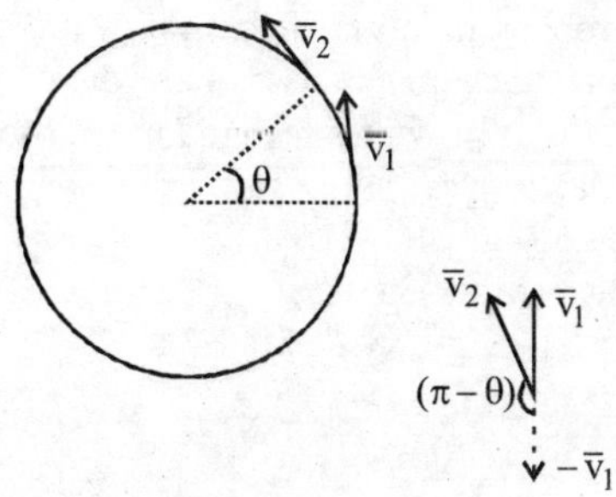

Change in velocity,

$|\Delta\vec{v}| = \sqrt{v_1^2 + v_2^2 + 2v_1 v_2 \cos(\pi - \theta)}$

$= 2v\sin\frac{\theta}{2}$ $\quad (\because |\vec{v}_1| = |\vec{v}_2| = v)$

$= (2 \times 10) \times \sin(30°) = 2 \times 10 \times \frac{1}{2} = 10$ m/s

131. **(c)** Speed, V = constant (from question)

Centripetal acceleration, $a = \frac{V^2}{r} \Rightarrow a \propto \frac{1}{r}$

Hence graph (c) correctly describes relation between acceleration and radius.

Laws of Motion

4

1. **(d)** $F = M \times a$ (Force acting on body)

Given : $v = 4\sqrt{x}$

Squaring on both side, $v^2 = 16x$

Differentiating on both side w.r.t. x,

$$2v\frac{dv}{dx} = 16 \Rightarrow a = \frac{vdv}{dx} = \frac{16}{2} = 8 \text{ m/s}^2$$

$\therefore F = 0.5 \times 8 = 4$ N.

2. **(b)** Given mass $m = 2$ kg, force $F = 6$ N along positive z-axis.

$\therefore$ Acceleration $\vec{a} = 3\hat{k},\ t = \frac{5}{3} s$

$\vec{u} = 3\hat{i} + 4\hat{j}$

$\therefore \vec{v} = \vec{u} + \vec{a}t = 3\hat{i} + 4\hat{j} + 3\hat{k} \times \frac{5}{3} = 3\hat{i} + 4\hat{j} + 5\hat{k}$

3. **(c)** The problem involves the buoyancy force (F) acting on a balloon and change in its motion when a small mass is released. From Newton's second Law of Motion.

$F - Mg = Ma$

$\Rightarrow F = Ma + Mg$ ...(i)

$F - (M - x)g = (M - x)3a$

$\Rightarrow Ma + Mg - Mg + xg = 3Ma - 3xa$ [using (i)]

$\Rightarrow x = \frac{2Ma}{g + 3a}$

4. **(c)** Given,

Mass of cricket ball, m = 150 g

Speed of cricket ball, V = 20 m/s

Force, $F = \frac{\Delta P}{\Delta t} = \frac{(mv - 0)}{0.1} = \frac{150 \times 20}{1000 \times 0.1} = 30$ N

5. **(c)**

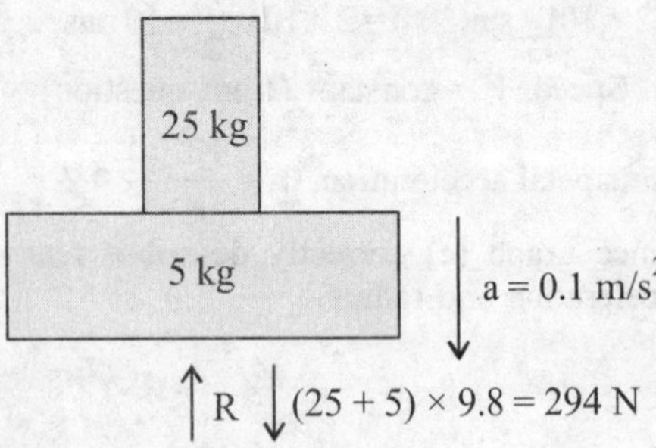

By Newton's 2nd law, $F_{net} = ma$

$\Rightarrow 294 - R = 30\,(0.1)\ \therefore R = 291$ N

6. **(a)** $\vec{P} = \cos(kt)\hat{i} - \sin(kt)\hat{j}$

$\therefore \vec{F} = \frac{d\vec{P}}{dt} = -k\sin(kt)\hat{i} - k\cos(kt)\hat{j}$

$$\cos\theta = \frac{\vec{F} \cdot \vec{P}}{|\vec{F}| \cdot |\vec{P}|} = \frac{-k\cos(kt) \cdot \sin(kt) + k\sin(kt) \cdot \cos(kt)}{1 \times k} = 0$$

$\therefore \theta = \frac{\pi}{2}$

7. **(b)** Here force applied on the block increases linearly with time, so

$F = ma \Rightarrow a = \frac{F}{m} = \frac{kt}{m}$ or, $a \propto t$

Hence a vs t graph will be a straight line passing through origin.

8. **(a)** Using Newton's IInd law, $F = \frac{\Delta p}{\Delta t} = \frac{mv}{t}$

$F = \frac{0.12 \times 25}{0.1} = 30$N

9. **(d)** Impulse, $I = \Delta P = P_f - P_i = m(v_f - v_i)$

$(\because v = \sqrt{2gh})$

Mass of body, $m = 0.1$ kg

$I = \Delta P = 0.1\left(\sqrt{2 \times 9.8 \times 5} - \left(-\sqrt{2. \times 9.8 \times 10}\right)\right)$

$= 0.1\left(14 + 7\sqrt{2}\right) \approx 2.39$ kg ms^{-1}

10. **(d)** From principle of momentum conservation,

$m_1v_1 = m_2v_2$

or, $1000 \times 6 = 1200 \times v \Rightarrow v = 5$m/s

11. **(b)** By momentum conservation,

$m_1u_1 + m_2u_2 = m_1v_1 + m_2v_2$

$0 = 3(-v) + 0.01(600 - v)$

$v \simeq 2$m/s

Impulse on gun = 3 × 2 = 6 Ns

12. **(a)** Given,

Mass of the particle, m = 5 kg

As the particle is at rest, So resultant of $\overrightarrow{F_2}$ and $\overrightarrow{F_3}$ should be opposite to $\overrightarrow{F_1}$

$F_{net} = \sqrt{F_2^2 + F_3^2} = \sqrt{6^2 + 8^2} = 10$ N

$\therefore$ Acceleration, $a = \frac{F_{net}}{m} = \frac{10}{5} = 2\text{ m/s}^2$

13. **(b)** Given, mass of each bullets, m = 10g

Speed of each bullets, v = 250 m/s

Force F = n m v

Here n = number of bullets fired per second

$\therefore n = \frac{F}{mv} = \frac{125}{10 \times 10^{-3} \times 250} = 50$

14. **(b)** Given,

mass of body, $m = \frac{500}{1000} = 0.5$kg

velocity, $v = 10\sqrt{x} \Rightarrow v^2 = 100x$

$\therefore\ 2v\dfrac{dv}{dx} = 100 \Rightarrow a = \dfrac{v\,dv}{dx} = 50\ m/s^2$

$\therefore$ Force, F = ma = 0.5 × 50 = 25 N

15. **(a)** Mass of particle,

m = 500 g = 0.5 kg

velocity of a particle,

$\vec{v} = 2t\hat{i} + 3t^2\hat{j}$

$\vec{a} = \dfrac{d\vec{v}}{dt} = 2\hat{i} + 6t\hat{j}$

at t = 1s, $\vec{a} = 2\hat{i} + 6\hat{j}$

Force acting on the particle,

$\vec{F} = m\vec{a} = 0.5(2\hat{i} + 6\hat{j}) = \hat{i} + 3\hat{j}$

$\vec{F} = \hat{i} + x\hat{j}$ Hence x = 3

16. **(b)** We have, Impulse = Area under F-t curve, for fig (D), area is highest. So, impulse is maximum.

17. **(c)** $\left|\dfrac{d\vec{p}}{dt}\right| = |\vec{F}|$ = Slope of curve

As, shown in figure maximum slope represent (c) and minimum slope represent (b).

18. **(d)** Given, mass of machine gun, M = 10 kg

mass of bullet, m = 20 g = 20×10^{-3} kg

velocity of bullet, V = 100 ms^{-1}, let V be the recoil velocity of gun, using conservation of momentum

nmv = MV

$\Rightarrow 20\times10^{-3} \times \dfrac{180}{60} \times 100 = 10V$

$\because$ n = 180 bullets per minute

$\Rightarrow$ v = 0.6 m/s

19. **(b)**

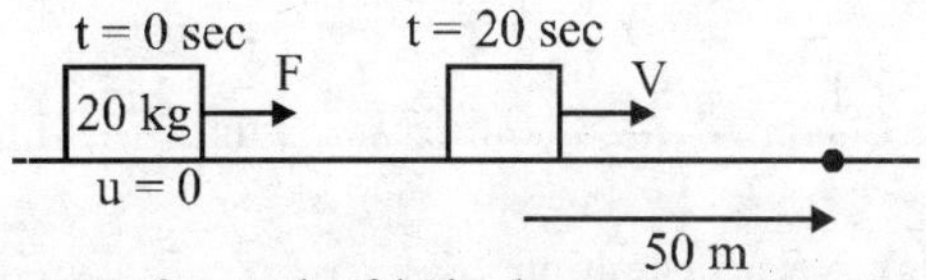

Assume surface to be frictionless

Then, 50 = V × 10 $\Rightarrow$ V = 5 m/s

As v = u + at $\Rightarrow$ V = 0 + a × 20

$\Rightarrow 5 = a \times 20 \Rightarrow a = \dfrac{1}{4}\ m/s^2$

So, F = ma = $20 \times \dfrac{1}{4}$ = 5N

20. **(b)** Impulse, I = change in momentum, ΔP

$F_{avg} = \dfrac{\Delta P}{\Delta t} \quad \because \Delta P_1 = \Delta P_2 \ \therefore I_1 = I_2$

Given $\Delta t_1 = 3s$ and $\Delta t_2 = 5$ s

Hence, F_{avg} in case (i), when Δt_2 = 3s is more than (ii) when Δt_2 = 5s

21. **(b)** Force, $F = \dfrac{dm}{dt}v = \dfrac{10}{5} \times 4.5 = 9$ dyne

22. **(b)** Initial momentum $\vec{P}_i = 0.15\times12(\hat{i})$

Final momentum $\vec{P}_f = 0.15\times12(-\hat{i})$

$|\Delta\vec{P}| = 3.6$ kg m/s or, $3.6 = F\Delta t$

$3.6 = 100\,\Delta t \qquad \therefore \Delta t = 0.036$ sec

23. **(c)** $F = \dfrac{dp}{dt} = v\dfrac{dm}{dt} = 10 \times 1 = 10\,N$

$a = \dfrac{F}{m} = \dfrac{10}{2} = 5\ m/s^2$

24. **(6)** By law of conservation of linear momentum $\vec{P_i} = \vec{P_f}$

$\Rightarrow 60 \times V = (120 + 60) \times 2 \Rightarrow 60V = 360 \Rightarrow V = 6$ m/s

25. **(c)** $a = \dfrac{vdv}{dx}$; $adx = vdx$; $\displaystyle\int_{0.5}^{1.5} -\frac{kx}{m}dx = \int_4^v vdv$

$\Rightarrow -\dfrac{k}{2m}\left[1.5^2 - 0.5^2\right] = \dfrac{v^2 - 4^2}{2} \Rightarrow -\dfrac{12}{2\times2}[2] = \dfrac{v^2 - 16}{2}$

$\Rightarrow -3 \times 4 = v^2 - 16 \Rightarrow v^2 = 4 \Rightarrow v = 2$ m/s

26. **(12)** Impulse = ΔP

$= P_f - P_i = mv - (-mv) = 2mv = 2 \times 0.4 \times 15 = 12$ Ns

27. **(2)** $\vec{F} = 10\hat{i} + 5\hat{j}$; $m = 0.1$ kg

$\vec{a} = \dfrac{\vec{F}}{m} = \dfrac{10\hat{i} + 5\hat{j}}{0.1} = 100\hat{i} + 50\hat{j}$

$\vec{S} = \vec{u}t + \dfrac{1}{2}\vec{a}t^2 = 0 + \dfrac{1}{2}(100\hat{i} + 50\hat{j}) \times 2^2 = 200\hat{i} + 100\hat{j}$

So, a = 200 and b = 100; Then, $\dfrac{a}{b} = \dfrac{200}{100} = 2$

28. **(b)** While going upward, $a_1 = -(g + a)$

$= -\left(10 + \dfrac{10}{5}\right) = -12\ m/s^2$

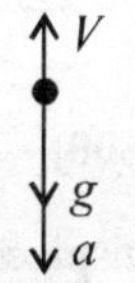

Now, $V = u + a_1t_1 \Rightarrow u = -a_1t_1$

and, $S = ut_1 + \dfrac{1}{2}a_1t_1^2$

$S = ut_1 - \dfrac{1}{2} \times 12 \times t_1^2$

$= -a_1t_1^2 - 6t_1^2 = 12t_1^2 - 6t_1^2 = 6t_1^2$

While coming downward,

$a_2 = g - a = 10 - \dfrac{10}{5} = 8\ m/s^2$

Here, $u = 0$

So, $S = \dfrac{1}{2}a_2t_2^2 = \dfrac{1}{2} \times 8 \times t_2^2 = 4t_2^2$

So, $S = 6t_1^2 = 4t_2^2 \Rightarrow \dfrac{t_1}{t_2} = \sqrt{\dfrac{4}{6}} = \sqrt{\dfrac{2}{3}}$

29. **(d)** Thrust force on rocket is given by

$F_{thrust} = \left(V_{rel} \cdot \dfrac{dm}{dt}\right)$

$$\Rightarrow \left(V_{rel}\cdot\frac{dm}{dt}-mg\right)=ma$$

$$\Rightarrow 500\left(\frac{dm}{dt}\right)-10^3\times10=10^3\times20$$

$$\Rightarrow \frac{dm}{dt}=60\text{ kg/s}$$

30. (c) From the Newton's second law of motion,

F = ma

$$\Rightarrow a=\frac{F}{M}\Rightarrow a=\frac{F_0}{M}\left[1-\left(\frac{T-t}{T}\right)^2\right]$$

$$\Rightarrow a=\frac{dv}{dt}=\frac{F_0}{M}\left[1-\left(\frac{T-t}{T}\right)^2\right]$$

$$\Rightarrow \int_0^v dv=\frac{F_0}{M}\int_0^{2T}\left[1-\left(\frac{T-t}{T}\right)^2\right]dt$$

$$\Rightarrow v=\frac{F_0}{M}\left[t+\frac{1}{3T^2}(T-t)^3\right]_0^{2T}$$

$$\Rightarrow v=\frac{F_0}{M}\left\{\left[2T+\frac{1}{3T^2}(T-2T)^3\right]-\left[0+\frac{T^3}{3T^2}\right]\right\}$$

$$\Rightarrow v=\frac{F_0}{M}\left[\frac{4T}{3}\right]=\frac{4F_0T}{3M}$$

31. (c) Acceleration, $\vec{a}=\frac{\vec{F}}{m}=\frac{40\hat{i}+10\hat{j}}{5}=8\hat{i}+2\hat{j}\text{ m/s}^2$

Using,

$$\Rightarrow \vec{s}=\vec{u}t+\frac{1}{2}\vec{a}t^2\Rightarrow \vec{s}=\frac{1}{2}\left(8\hat{i}+2\hat{j}\right)\times100$$

$$\Rightarrow \vec{s}=\left(400\hat{i}+100\hat{j}\right)\text{m}$$

32. (500) $\vec{a}=\frac{\vec{F}}{m}=10\hat{i}+5\hat{j}$

Displacement of the box along x-axis,

$$x=\frac{1}{2}a_xt^2=\frac{1}{2}\times10\times100=500\text{ m}$$

33. (a) Given

$$\vec{F}=k(v_y\hat{i}+v_x\hat{j})$$

$$\therefore F_x=kv_y,\ F_y=kv_x$$

$$\frac{mdv_x}{dt}=kv_y\Rightarrow\frac{dv_x}{dt}=\frac{k}{m}v_y$$

Similarly, $\frac{dv_y}{dt}=\frac{k}{m}v_x$

$$\frac{dv_y}{dv_x}=\frac{v_x}{v_y}\Rightarrow\int v_ydv_y=\int v_xdv_x$$

$$v_y^2=v_x^2+C$$

$$\Rightarrow v_x^2-v_y^2=\text{constant}=C_0\ (\text{let})$$

$$\vec{v}\times\vec{a}=(v_x\hat{i}+v_y\hat{j})\times\frac{k}{m}(v_y\hat{i}+v_x\hat{j})$$

$$=(v_x^2\hat{k}-v_y^2\hat{k})\frac{k}{m}=(v_x^2-v_y^2)\frac{k}{m}\hat{k}=\frac{C_0k}{m}\hat{k}=\text{constant}$$

34. (b) From the Newton's second law,

$$F=\frac{dp}{dt}=\frac{d(Mv)}{dt}=v\left(\frac{dM}{dt}\right)\quad\text{...(i)}$$

We have given, $\frac{dM(t)}{dt}=bv^2(t)$ ...(ii)

Thrust on the satellite,

$$F=v_{rel}\left(\frac{dM}{dt}\right)=(0-v)\left(\frac{dM}{dt}\right)=-v(bv^2)=-bv^3$$

[Using (i) and (ii)]

$$\Rightarrow F=M(t)a=-bv^3\Rightarrow a=\frac{-bv^3}{M(t)}$$

35. (d)

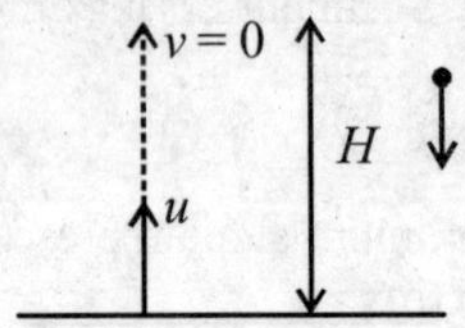

$(g+kv^2)=a$ (acceleration)

$$F=-mkv^2-mg$$

($\because mg$ and mkv^2 act in downward direction)

$$a=\frac{F}{m}=-[kv^2+g]$$

$$\Rightarrow v\cdot\frac{dv}{dh}=-[kv^2+g]\qquad\left(\because a=v\frac{dv}{dh}\right)$$

$$\Rightarrow\int_u^0\frac{v\cdot dv}{kv^2+g}=-\int_0^h dh\Rightarrow\frac{1}{2k}\ln\left[kv^2+g\right]_u^0=-h$$

$$\Rightarrow\frac{1}{2k}\ln\left[\frac{ku^2+g}{g}\right]=h\Rightarrow h=\frac{1}{2k}\ln\left[\frac{ku^2}{g}+1\right]$$

36. (a) Net acceleration

$$\frac{dv}{dt}=a=-(g+\gamma v^2)$$

Let time t required to rise to its zenith (v = 0) so,

$$\int_{v_0}^0\frac{-dv}{g+\gamma v^2}=\int_0^t dt\qquad[\text{for }H_{max},\ v=0]$$

$$\Rightarrow\frac{1}{\gamma}\int_{v_0}^0\frac{-1}{\left(\sqrt{\frac{g}{\gamma}}\right)^2+v^2}dv=t$$

$$\Rightarrow\frac{1}{\gamma}\cdot\frac{1}{\sqrt{\frac{g}{\gamma}}}\left[\tan^{-1}\left(\frac{v}{\sqrt{\frac{g}{\gamma}}}\right)\right]_{v_0}^0=-t$$

$$\Rightarrow\frac{-1}{\sqrt{\gamma g}}\tan^{-1}\left(\frac{\sqrt{\gamma}}{\sqrt{g}}v_0\right)=-t\ \therefore\ t=\frac{1}{\sqrt{\gamma g}}\tan^{-1}\left(\frac{\sqrt{\gamma}v_0}{\sqrt{g}}\right)$$

37. **(d)** $v^2 = u^2 - 2gh$ or $v = \sqrt{u^2 - 2gh}$

Momentum, $P = mv = m\sqrt{u^2 - 2gh}$

At $h = 0, P = \pm mu$ and at $h = \frac{u^2}{2g}, P = 0$

upward direction is positive and downward direction is negative.

38. **(b)** From Newton's second law

$\frac{dp}{dt} = F = kt$

Integrating both sides we get,

$\int_p^{3p} dp = \int_0^T kt\,dt \Rightarrow [p]_p^{3p} = k\left[\frac{t^2}{2}\right]_0^T$

$\Rightarrow 2p = \frac{kT^2}{2} \Rightarrow T = 2\sqrt{\frac{p}{k}}$

39. **(a)** From $F = \frac{R}{t^2}v(t) \Rightarrow m\frac{dv}{dt} = \frac{R}{t^2}v(t)$

Integrating both sides $\int \frac{dv}{v(t)} = \int \frac{Rdt}{mt^2}$

$\ln v(t) = -\frac{R}{mt}$

$\therefore \ln v(t) \propto \frac{1}{t}$

40. **(a)** From question,

Mass of body, $m = 5$ kg

Velocity at $t = 0$,

$\vec{u} = (6\hat{i} - 2\hat{j})$ m/s

Velocity at $t = 10$ s,

$\vec{v} = +6\hat{j}$ m/s

Force, $F = ?$

Acceleration, $\vec{a} = \frac{\vec{v} - \vec{u}}{t} = \frac{6\hat{j} - (6\hat{i} - 2\hat{j})}{10}$

$= \frac{-3\hat{i} + 4\hat{j}}{5}$ m/s^2

Force, $\vec{F} = m\vec{a} = 5 \times \frac{(-3\hat{i} + 4\hat{j})}{5} = (-3\hat{i} + 4\hat{j})N$

41. **(c)** Given that $F(t) = F_0 e^{-bt} \Rightarrow m\frac{dv}{dt} = F_0 e^{-bt}$

$\frac{dv}{dt} = \frac{F_0}{m}e^{-bt}$

$\int_0^v dv = \frac{F_0}{m}\int_0^t e^{-bt}dt$

$v = \frac{F_0}{m}\left[\frac{e^{-bt}}{-b}\right]_0^t = \frac{F_0}{mb}\left[-\left(e^{-bt} - e^{-0}\right)\right]$

$\Rightarrow v = \frac{F_0}{mb}\left[1 - e^{-bt}\right]$

42. **(c)** If you push a cart with some force then according to Newton third law the cart will exert an equal and opposite force on you. So the cart push you with same amount of force in opposite direction. So Statement 1 is correct. The action reaction pairs mentioned in statement one cannot cancel each other because action and reaction forces act on two different bodies. So they cannot cancel each other. So Statement 2 is wrong.

43. **(d)** $mg\sin\theta = ma$

$\therefore a = g\sin\theta$

$\therefore$ Vertical component of acceleration $= a\sin\theta$

$= g\sin^2\theta$

$\therefore$ Relative vertical acceleration of A with respect to B is

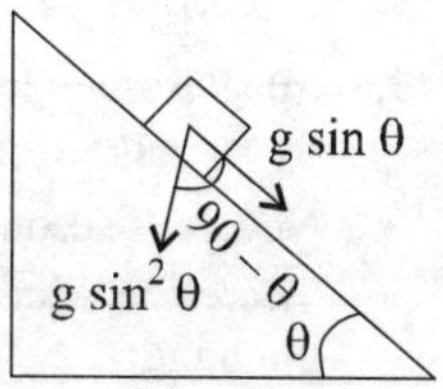

$g(\sin^2 60 - \sin^2 30) = g\left(\frac{3}{4} - \frac{1}{4}\right) = \frac{g}{2} = 4.9$ m/s^2

in vertical direction

44. **(d)** For the motion of ball, just after the throwing

$v = 0, s = 2\text{m}, a = -g = -10\text{ms}^{-2}$

$v^2 - u^2 = 2as$ for upward journey

$\Rightarrow -u^2 = 2(-10) \times 2 \Rightarrow u^2 = 40$

When the ball is in the hands of the thrower

$u = 0, v = \sqrt{40}$ ms^{-1}

$s = 0.2$ m

$v^2 - u^2 = 2as$

$\Rightarrow 40 - 0 = 2\,(a)\,0.2 \Rightarrow a = 100$ m/s^2

$\therefore F = ma = 0.2 \times 100 = 20\ N$

$\Rightarrow N - mg = 20 \Rightarrow N = 20 + 2 = 22N$

Note :

$W_{hand} + W_{gravity} = \Delta K$

$\Rightarrow F(0.2) - (0.2)(10)(2.2) = 0 \Rightarrow F = 22\ N$

45. **(c)** Given, mass of cricket ball, $m = 150\,g = 0.15$ kg

Initial velocity, $u = 20$ m/s

Force,

$F = \frac{m(u - v)}{t} = \frac{0.15(20 - 0)}{0.1} = 30\,N$

46. **(c)** Mass $(m) = 0.3$ kg

Force, $F = m.a = -kx$

$\Rightarrow ma = -15x \Rightarrow 0.3a = -15x$

$\Rightarrow a = -\frac{15}{0.3}x = \frac{-150}{3}x = -50x$

At $x = 0.2$m, $a = -50 \times 0.2 = -10$m/s^2 (towards origin)

$\therefore$ $a = 10$m/s^2 (away from origin)

47. **(c)** When the incline is given an acceleration a towards the right, the block experience a pseudo force *ma* towards left.

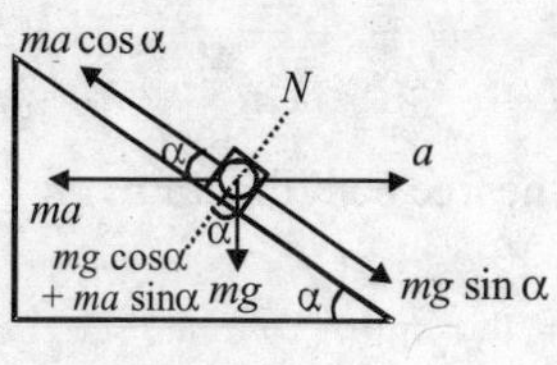

For block to remain stationary, Net force along the incline should be zero.

$mg \sin \alpha = ma \cos \alpha \Rightarrow a = g \tan \alpha$

48. **(b)** In the absence of air resistance, if the rocket moves up with an acceleration a, then thrust

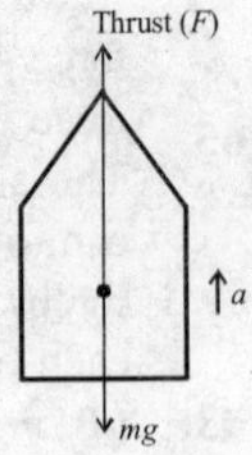

$F = mg + ma$

$\therefore\ F = m(g+a) = 3.5 \times 10^4\ (10+10)$

$= 7 \times 10^5 N$

49. **(d)** Resultant force is zero, as three forces are represented by the sides of a triangle taken in the same order. From Newton's second law, $\vec{F}_{net} = m\vec{a}$.
Therefore, acceleration is also zero *i.e.*, velocity remains unchanged.

50. **(d)** This is a case of sliding (if plane is frictionless) and therefore the acceleration of all the bodies is same.

51. **(b)** Given : $T_1 = \sqrt{3}\, T_2$

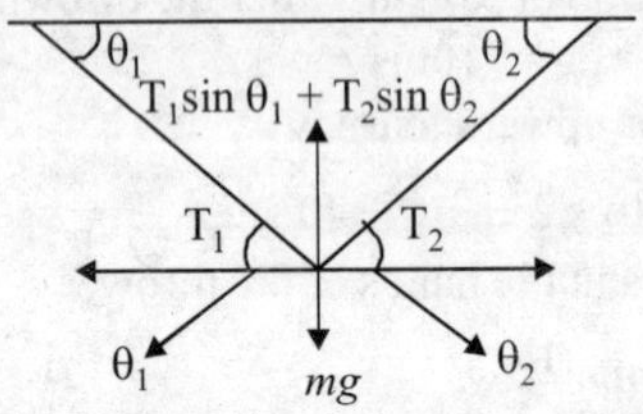

$T_1 \sin\theta_1 + T_2 \sin\theta_2 = mg$ and $T_1 = \sqrt{3}\, T_2$

$\Rightarrow T_2\left[\sqrt{3}\sin\theta_1 + \sin\theta_2\right] = mg$

From the option,

For $\theta_1 = 60°$ and $\theta_2 = 30°$

$\therefore T_2 = \dfrac{mg}{2}$.

52. **(b)** Using free body diagram

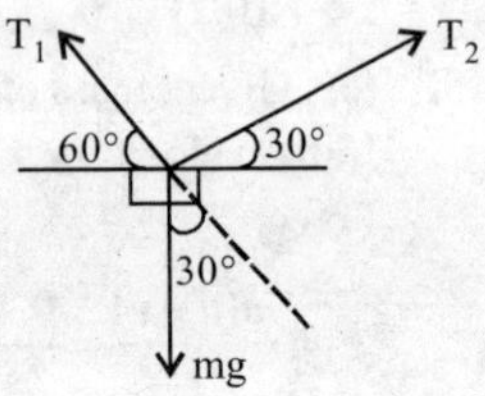

$T_1 = mg\cos 30°$

$= 1 \times 10 \times \dfrac{\sqrt{3}}{2} = 5\sqrt{3}$

$T_2 = mg\sin 30° = 1 \times 10 \times \dfrac{1}{2} = 5$

53. **(c)** For massless spring, $F = kx$

$\therefore kx_1 = 5N$

$\therefore kx_2 = 7N$

$\therefore k(5x_1 - 2x_2) = 5kx_1 - 2kx_2$

$= 5 \times 5 - 2 \times 7 = 11$ N

54. **(a)** Acceleration of the system is given by

$a = \left(\dfrac{m_2 - m_1}{m_1 + m_2}\right) g = \dfrac{g}{8}$

$\Rightarrow 8(m_2 - m_1) = m_1 + m_2 \Rightarrow 7m_2 = 9m_1$

$\Rightarrow \dfrac{m_2}{m_1} = \dfrac{9}{7}$

55. **(d)** From the free body diagram

$T_1 \sin 45° = F$

$T_1 \cos 45° = T_2 = mg$

$\therefore \tan 45° = \dfrac{F}{mg} \Rightarrow F = mg$

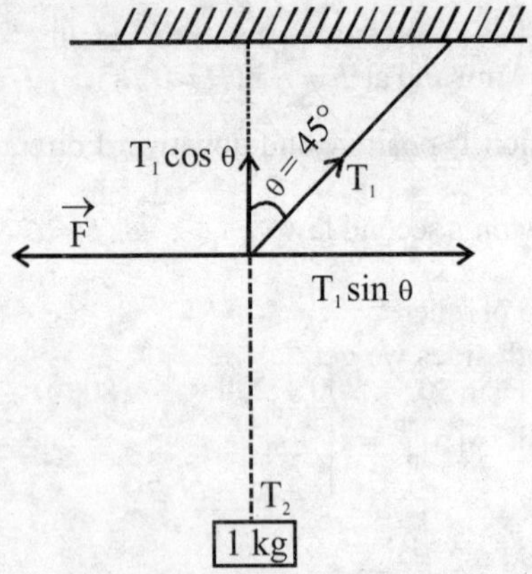

$\Rightarrow F = 1 \times 10 = 10$ N $\quad (\because m = 1$ kg$)$

56. **(a)** Acceleration is given as:

$a = \left(\dfrac{m_2 - m_1}{m_1 + m_2}\right) g$

$\Rightarrow \dfrac{g}{\sqrt{2}} = \left(\dfrac{m_2 - m_1}{m_1 + m_2}\right) g$

$\Rightarrow \sqrt{2}(m_2 - m_1) = m_1 + m_2$

$\Rightarrow \dfrac{m_1}{m_2} = \left(\dfrac{\sqrt{2}-1}{\sqrt{2}+1}\right)$

57. **(240)** Taking masses M_1, M_2 and M_3 as system, $\therefore F_{net} = ma$

$\Rightarrow T_1 - (4+6+10) \times 10 = (4+6+10) \times (2)$

$\therefore\ T_1 = 20\,(10+2) = 240$ N

58. **(d)** For 2 kg block

$T - 2g \sin 30 = 2a$ ..(i)

$\because\ T = g + 2a$

For 4 kg block

$4g - 2T = 4a$

$\Rightarrow 4g - 2(g + 2a) = 4a$

$\Rightarrow 4g - 2g - 4a = 4a \Rightarrow 2g = 8a$

$\therefore a = \dfrac{g}{4}$

59. **(a)** Horizontal force, $F = 80$ N

$a_A = a_B = a_C = \dfrac{F}{5+3+2} = \dfrac{80}{10} = 8 m/s^2$

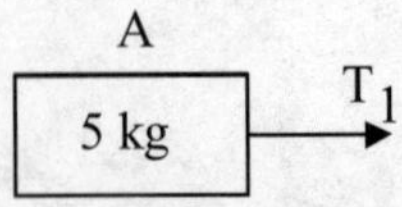

$T_1 = 5 \times 8 = 40$ N

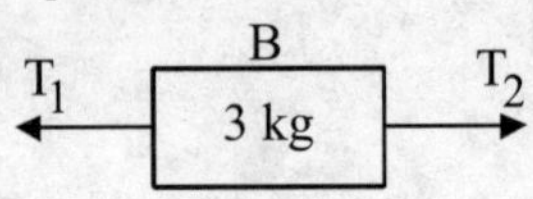

$T_2 - T_1 = 3 \times 8 \Rightarrow T_2 = 64$ N

60. **(c)**

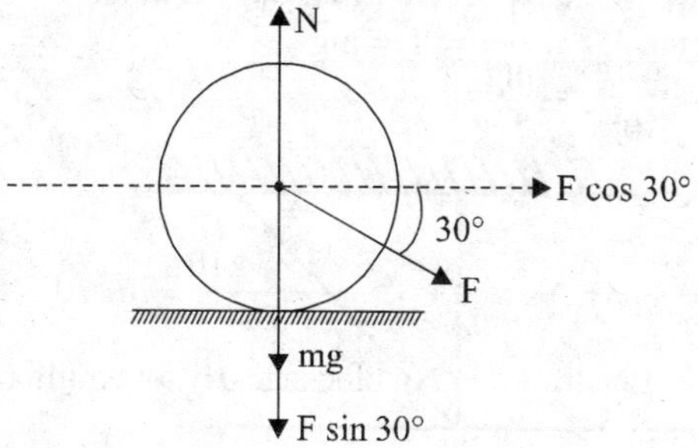

From FBD of roller

$N = mg + F\sin 30° = 700 + 200 \times \frac{1}{2} = 800\,N$

61. **(a)**

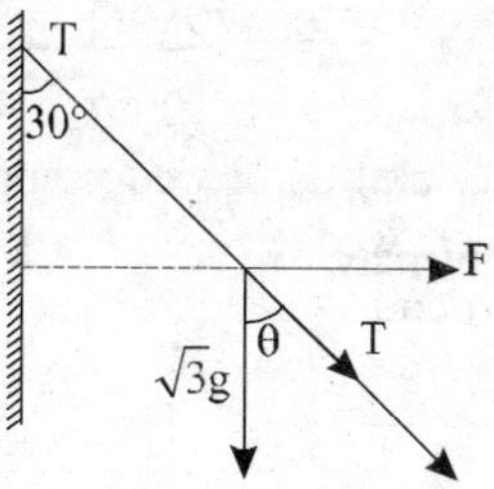

From the free body diagram shown above,

$\cos\theta = \frac{\sqrt{3}g}{T} \quad \because \theta = 30° \quad \therefore \frac{\sqrt{3}}{2} = \frac{\sqrt{3}g}{T}$

$\Rightarrow T = 20\,N$

62. **(b)** From FBD of lift,

$T_1 = T_2 + mg$, m = mass of lift

$\Rightarrow T_1 - T_2 = mg$

$\Rightarrow T_{net} = mg$.

So, (I) is true

From FBD of person,

$N + ma = mg$

$N = mg - ma \Rightarrow N < mg$

So, (II) is false.

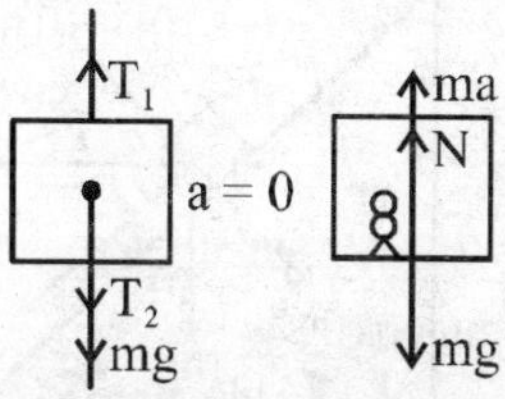

63. **(b)** For 4 kg block

$4g\sin 30° - T = 4a$ ...(i)

For 1 kg block

$T - 1g\sin 30° = 1a$

$4(T - g\sin 30°) = 4a$ ...(ii)

From (i) from (ii), we get

$4(T - g\sin 30°) = 4g\sin 60° - T$

$5T = 20\sqrt{3} + 20$

$T = 4(\sqrt{3} + 1)N$

64. **(b)** For equilibrium condition, $m_2g = m_1g\sin\theta$

$\sin\theta = \frac{m_2}{m_1} = \frac{3}{5}$

$\cos\theta = \frac{4}{5}$

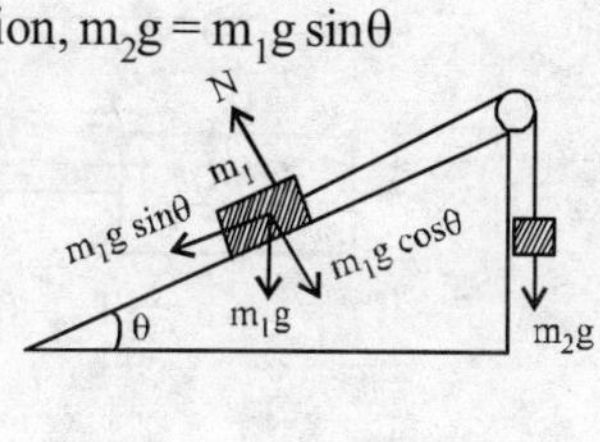

Normal force (N) on $m_1 = 5g\cos\theta$

$= 5 \times 10 \times \frac{4}{5} = 40\,N$

65. **(c)** Acceleration of block on smooth inclined plane,

$a = g\sin\theta$

Using, $s = ut + \frac{1}{2}at^2$

$s = \frac{1}{2}g\sin 30°(2)^2$

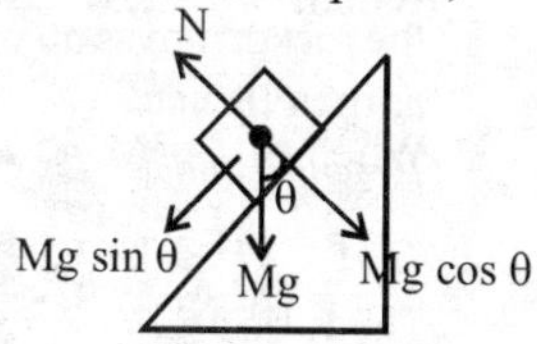

When the incline is changed to 45°

$s = \frac{1}{2}g\sin 45°\,t^2$

As distance travelled is same

$\therefore \left(\frac{1}{2}\right)(4) = \frac{1}{\sqrt{2}}t^2 \quad \Rightarrow t = \sqrt{2\sqrt{2}} \simeq 1.68$

66. **(d)** Acceleration in such system is given as

$a = \frac{(m_2 - m_1)}{(m_2 + m_1)}g$

$\Rightarrow \frac{g}{2} = \frac{(\lambda(L-\ell) - \lambda\ell)g}{\lambda L} \Rightarrow \ell = \frac{L}{4} = \frac{L}{x}$

So, x = 4

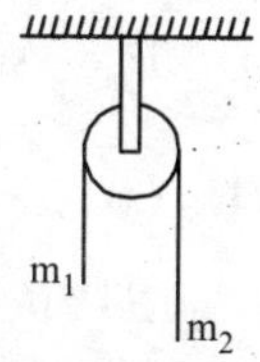

67. **(c)** Given that mass of monkey, m = 50 kg

Acceleration due to gravity, $g = 10\,m/s^2$

Tension (T) = 350 N

Given monkey climbs downward, acceleration of monkey, $a = 4\,m/s^2$

When monkey climbs upward, acceleration of monkey, $a = 5\,m/s^2$

(For upward)

$T - mg = ma \Rightarrow T = mg + ma = 50(10 + 5) = 750\,N$

Rope will break while climbing upward

(For downward)

$T = m(g - a) = 50(10 - 4) = 300\,N$

Rope will not break while climbing downward

68. **(a)** Let a_1 be the acceleration of 100 kg block

FBD of 100 kg block w.r.t. ground

$F - T - N_1 = 100\,a_1$ ... (i)

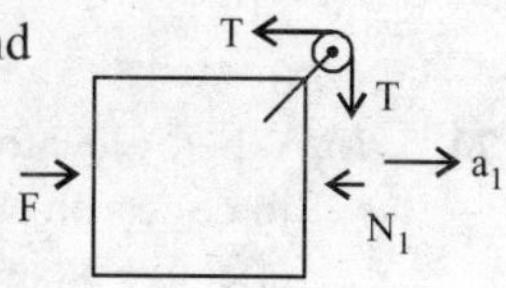

FBD of 20 kg block w.r.t. 100 kg

$T - 20g = 20(2) \Rightarrow T = 40 + 200$

$\Rightarrow T = 240$ ... (ii)

$N_1 = 20a_1$ ... (iii)

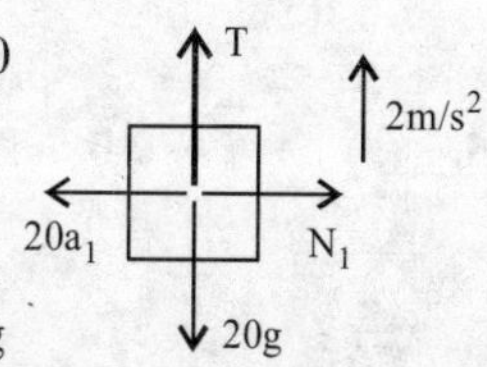

FBD of 10 kg block w.r.t. 100 kg

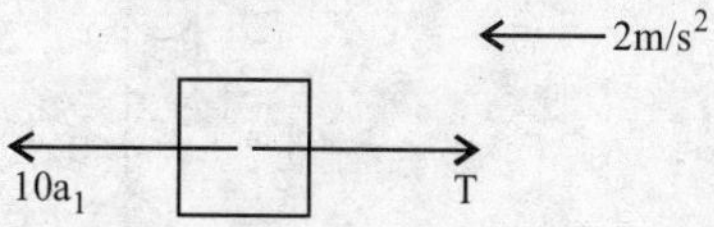

$10a_1 - 240 = 10(2) \Rightarrow a_1 = 26\,m/s^2$

$F = 240 + 20(26) + 100 \times 26 \Rightarrow F = 3360\,N$

69. **(b)** $a = \dfrac{M_2 g - M_1 g}{M_1 + M_2}$

When, $M_2 = 2M_1 \Rightarrow a_1 = \dfrac{2M_1 g - M_1 g}{3M_1} = \dfrac{g}{3}$

When $M_2 = 3M_1 \Rightarrow a_2 = \dfrac{3M_1 g - M_1 g}{4M_1} = \dfrac{g}{2}$

The ratio $\dfrac{a_1}{a_2} = \dfrac{\frac{g}{3}}{\frac{g}{2}} = \dfrac{2}{3}$

70. **(3)** Along horizontal

$F_1 + 1\cos 45° = 2\sin 45°$

$F_1 = \dfrac{2}{\sqrt{2}} - \dfrac{1}{\sqrt{2}} = \dfrac{1}{\sqrt{2}}$

Along vertical

$F_2 = 1\sin 45 + 2\sin 45$

$F_2 = 3\sin 45 = \dfrac{3}{\sqrt{2}}$ so, $\dfrac{F_1}{F_2} = \dfrac{1}{3} = \dfrac{1}{x}$ so, $x = 3$

71. **(a)** Let addtion force required be $= \vec{F}$

$\vec{F} + 5\hat{i} - 6\hat{i} + 7\hat{j} - 8\hat{j} = 0 \Rightarrow \vec{F} = \hat{i} + \hat{j}, |\vec{F}| = \sqrt{1^2 + 1^2} = \sqrt{2}\text{N}$

Angle with x-axis: $\tan\theta = \dfrac{\text{y component}}{\text{x component}} = \dfrac{1}{1}$

So, $\theta = \tan^{-1}(1) = 45°$

72. **(c)** For observer in box

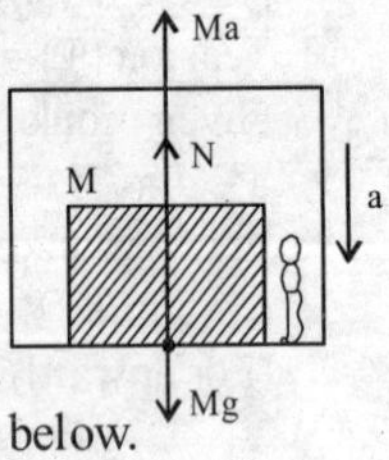

$N + Ma = Mg$

$\Rightarrow N = M(g - a)$

$\Rightarrow \dfrac{Mg}{4} = M(g - a) \Rightarrow a = \dfrac{3g}{4}$

73. **(3)** FBD of middle point is as shown below.

From FBD

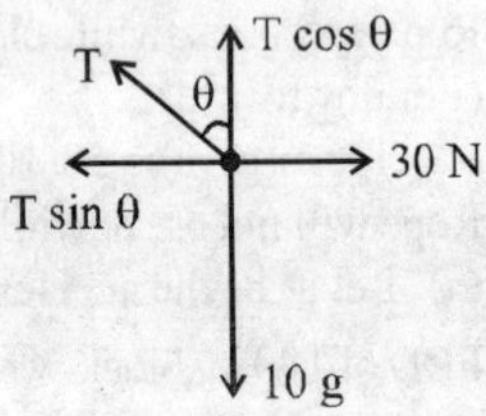

$T\sin\theta = 30$

$T\cos\theta = 100$

So, $\tan\theta = 0.3$

$= 3 \times 10^{-1}$

So, $\theta = \tan^{-1}(3 \times 10^{-1})$

74. **(b)** When elevator moves downward, then a pseudo force acts on object in upward direction, due to which effective weight of object decreases.

75. **(a)** $\sum \vec{T} \cdot \vec{a} = 0$

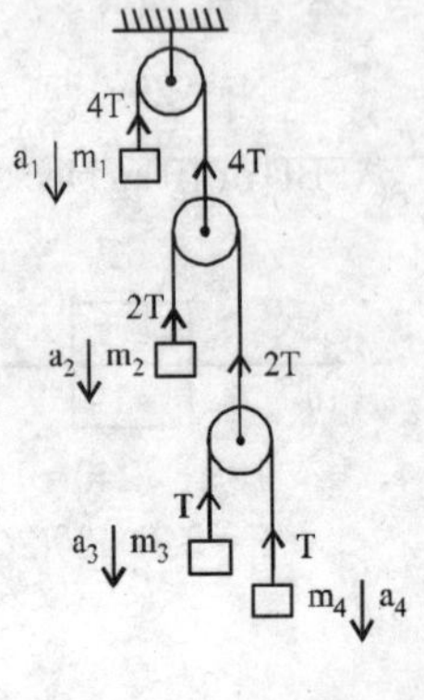

$\Rightarrow -4Ta_1 - 2Ta_2 - Ta_3 - Ta_4$

$= 0$

$\Rightarrow 4a_1 + 2a_2 + a_3 + a_4$

$= 0$

76. **(36)** We have, acceleration of system as

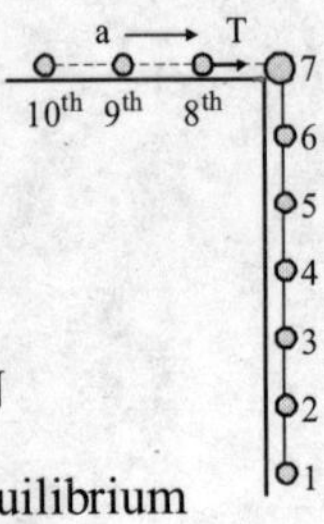

$a = \dfrac{6mg}{10m} = \dfrac{3g}{5}$

taking 8, 9, 10 together

$T = 3ma = 3m \times \dfrac{3g}{5} = \dfrac{3 \times 2 \times 3 \times 10}{5} = 36\text{N}$

77. **(12)** Let draw *FBD* of block clearly for equilibrium

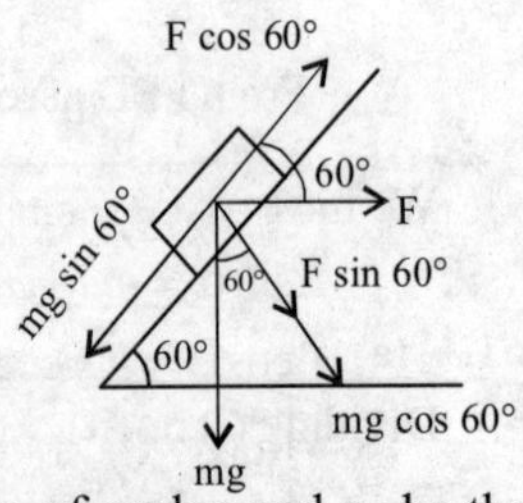

$F\cos 60° = mg\sin 60°$

$\Rightarrow \dfrac{F}{mg} = \tan 60°$

$\Rightarrow \dfrac{\sqrt{x}}{0.2 \times 10} = \sqrt{3}$

$\Rightarrow \sqrt{x} = 2\sqrt{3} \Rightarrow x = 12$

78. **(d)** Let a_w be the acceleration of wedge and a_b be the acceleration of block w.r.t wedge

For the wedge w.r.t. ground

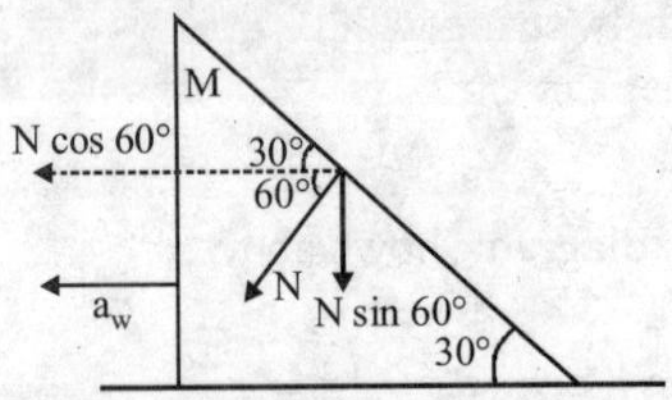

$N\cos 60° = Ma_w \Rightarrow N \times \dfrac{1}{2} = 16\,a_w$

$\Rightarrow N = 32a_w$ ...(i)

For block w.r.t. wedge

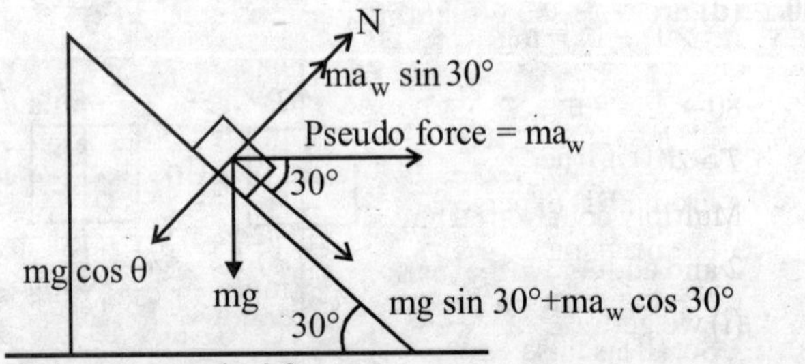

Balancing vertical forces

$N + ma_w\sin 30° = mg\cos 30°$

$\Rightarrow N = 8g\cos 30° - 8a_w\sin 30°$

$\Rightarrow 32a_w = 4\sqrt{3}g - 4a_w$ (Using (i))

$\Rightarrow a_w = \dfrac{\sqrt{3}}{9}g$

Along incline plane

$\Rightarrow m\,g\sin 30° + ma_w\cos 30° = ma_b = 8a_b$

$\Rightarrow a_b = \dfrac{8 \times g \times \frac{1}{2} + 8 \times \frac{\sqrt{3}}{9} \times g.\frac{\sqrt{3}}{2}}{8}$

$\Rightarrow a_b = g \times \dfrac{1}{2} + \dfrac{\sqrt{3}}{9}g.\dfrac{\sqrt{3}}{2} = \dfrac{2g}{3}$

79. (30) Let draw FBD of Bob

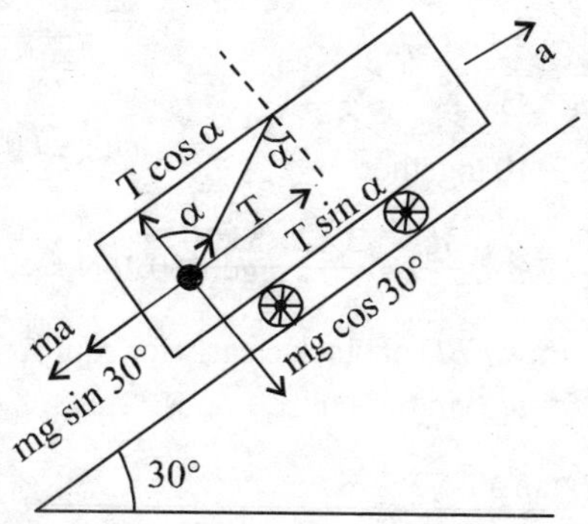

We have

$T\sin\alpha = ma + mg\sin 30°$ ..... (i)

$T\cos\alpha = mg\cos 30°$ ..... (ii)

Dividing (i) by (ii), we get

$$\tan\alpha = \frac{a + g\sin 30°}{g\cos 30°} = \frac{10+5}{5\sqrt{3}} = \sqrt{3}$$

So, $\alpha = 60°$

Let 'θ' be angle made by string with vertical

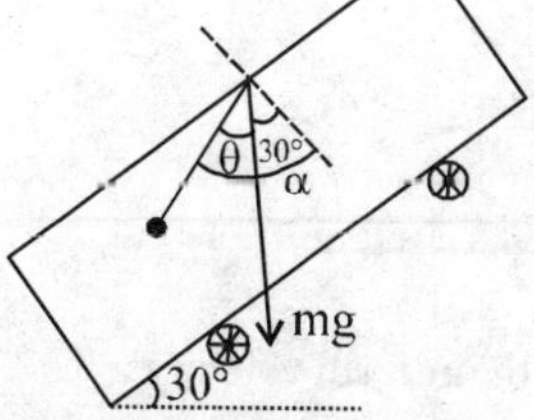

From diagram, $\alpha = \theta + 30°$

$\Rightarrow \quad 60° = \theta + 30° \Rightarrow \theta = 30°$

80. **(d)** From free body diagram

$80 - 2T = 8a$ ...(i)

$T - 20 = 4a$ ...(ii)

Multiply equation (ii) by 2 and adding with equation (i) we get

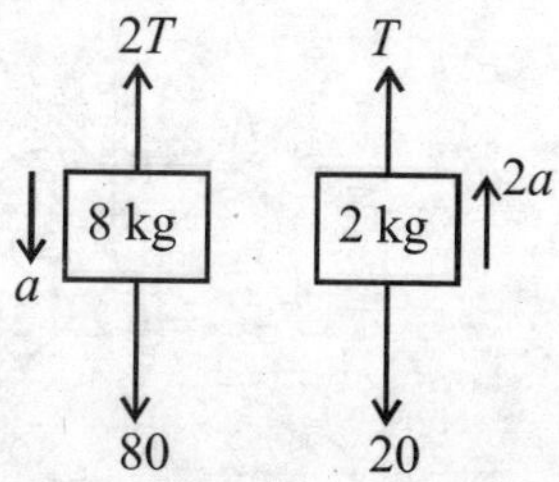

$$(8+8)a = 40 \Rightarrow a = \frac{40}{16} = \frac{10}{4} m/s^2$$

Using $S = \frac{1}{2}at^2 \Rightarrow t^2 = \frac{2S}{a}$

$$\Rightarrow \frac{0.2 \times 2 \times 4}{10} = t^2 \Rightarrow t = 0.4 \text{ sec}$$

81. **(b)** Total weight acting downward,

$W = (20 \times 3 + 10)g = 70\,g$

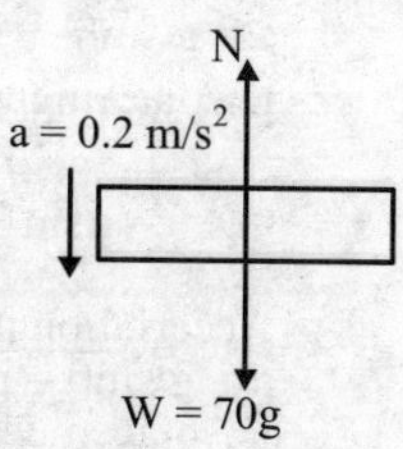

Now from FBD,

$W - N = 70a = 70 \times 0.2$

$\therefore \quad N = 70 \times 10 - 70 \times 0.2 = 686\text{ N}$

82. **(82)** $\vec{P}$ makes angle of 35° with AC

So, component along AC = 100 cos 35 = 81.9 N ≃ 82 N

83. **(492)** Mass of the person, $M = 60$ kg

Tension in the rope of the lift when it moves downward with acceleration a,

$T = M(g - a) = 60\,(10 - 1.8) = 492$ N

84. **(a)** The free body diagram is

By Lami's theorem

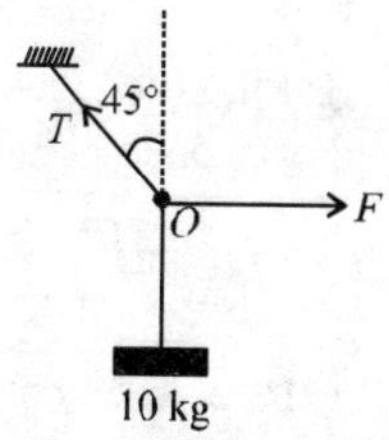

$$\Rightarrow \frac{mg}{\sin 135°} = \frac{F}{\sin 135°}$$

$\Rightarrow F = mg = 100$ N

85. **(d)** $a = \dfrac{480}{20+12+8} = 12 \text{ m/s}^2$

$480 - T = 24a$

$480 - T = 24 \times 12$

$480 - T = 288$

$T = 192$ N

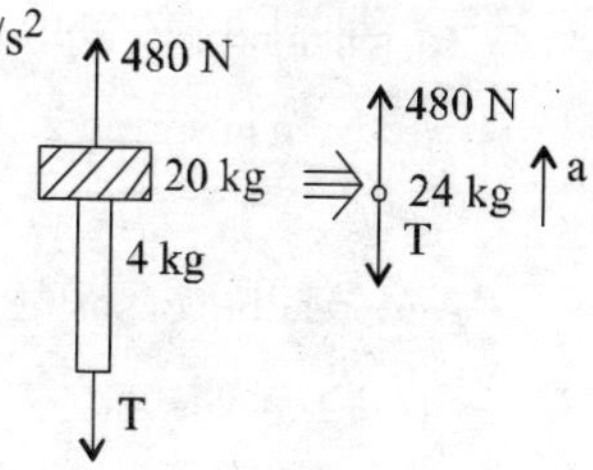

86. **(d)** Writing free body-diagrams for m & M,

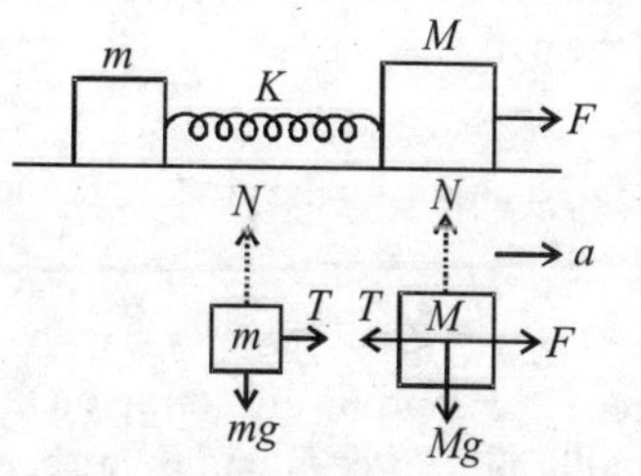

we get $T = ma$ and $F - T = Ma$

where T is force due to spring

$\Rightarrow F - ma = Ma$ or, $F = Ma + ma$

$\therefore$ Acceleration of the system

$$a = \frac{F}{M+m}.$$

Now, force acting on the block of mass m is

$$ma = m\left(\frac{F}{M+m}\right) = \frac{mF}{m+M}.$$

87. **(c)** Here, $m_1 = 5$kg and $m_2 = 4.8$ kg.

If a is the acceleration of the masses,

$m_1 a = m_1 g - T$ ...(i)

$m_2 a = T - m_2 g$ ...(ii)

Solving (i) and (ii) we get

$$a = \left(\frac{m_1 - m_2}{m_1 + m_2}\right)g \Rightarrow a = \frac{(5-4.8)\times 9.8}{(5+4.8)} m/s^2 = 0.2 \text{ m/s}^2$$

88. **(a)** When lift is stationary, $W_1 = mg$ ...(i)

When the lift descends with acceleration, a

$W_2 = m(g - a)$

$$W_2 = \frac{49}{9.8}(9.8 - 5) = 24\,N$$

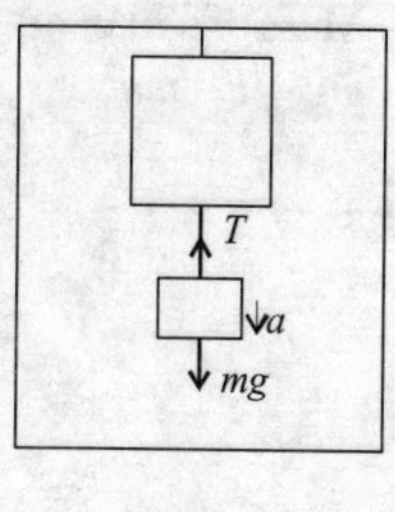

89. **(d)** Taking the rope and the block as a system we get $P = (m + M)a$

$\therefore$ Acceleration produced,

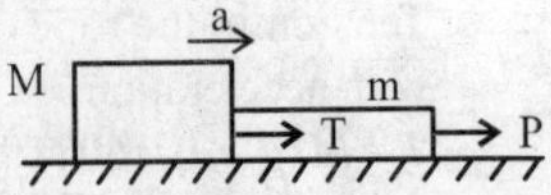

$$a = \frac{P}{m+M}$$

Taking the block as a system,
Force on the block, $T = Ma$

$$\therefore T = \frac{MP}{m+M}$$

90. **(a)** As springs are massless
So, tension in spring 1
= Tension in spring 2
= Mg
And, reading of spring balance is equal to tension of spring so, both will show same reading.

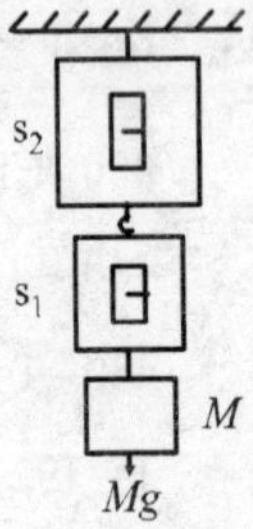

91. **(c)** **Case - I:** For the man standing in the lift, the acceleration of the ball

$a_{bm} = a_b - a_m \Rightarrow a_{bm} = g - a$

Case - II: The man standing on the ground, the acceleration of the ball

$a_{bm} = a_b - a_m \Rightarrow a_{bm} = g - 0 = g$

92. **(a)** When forces F_1, F_2 and F_3 are acting on the particle, it remains in equilibrium. Force F_2 and F_3 are perpendicular to each other,

$\therefore\ F_1 = \sqrt{F_2^2 + F_3^2}$

The force F_1 is now removed, so, resultant of F_2 and F_3 will now make the particle move with force equal to F_1.

Then, acceleration, $a = \frac{F_1}{m}$

93. **(b)** Force = mass × acceleration

$\therefore\ F = (m + m + m) \times a$

$F = 3m \times a$

$\Rightarrow a = \frac{F}{3m} \Rightarrow a = \frac{10.2}{6} m/s^2$

$\therefore T_2 = ma = 2 \times \frac{10.2}{6} = 3.4N$

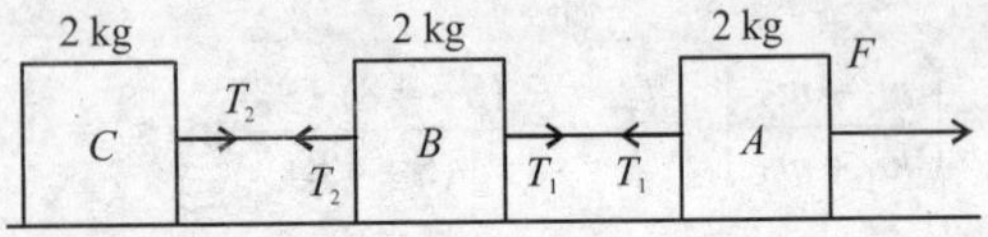

94. **(c)** Tension, $T = 360$ N
Mass of a man m = 60 kg

$mg - T = ma$

$\Rightarrow a = g - \frac{T}{m} = 10 - \frac{360}{60} = 4\ m/s^2$

95. **(a)** The FBD of the block is

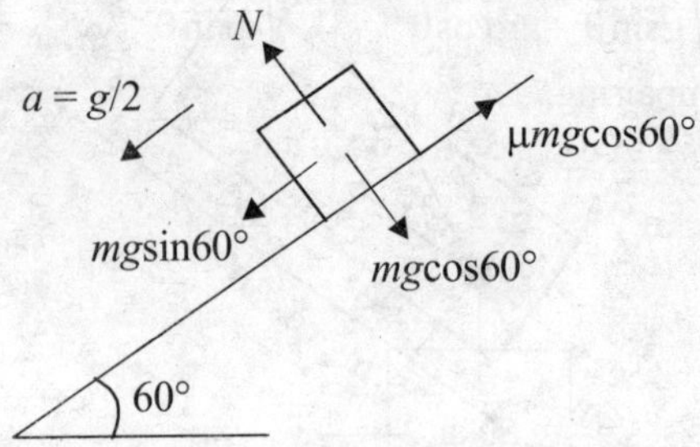

$\therefore F_{net} = ma$

$\Rightarrow mg \sin 60° - \mu mg \cos 60° = ma$

$\Rightarrow g \sin 60° - \mu g \cos 60° = \frac{g}{2}$ $\left[\because a = \frac{g}{2}\right]$

$\Rightarrow \frac{\sqrt{3}}{2} - \frac{\mu}{2} = \frac{1}{2}$

$\therefore \mu = \sqrt{3} - 1.$

96. **(c)** Given : $m = 25$ Kg, $\mu = 0.25$, $S = 5$ m

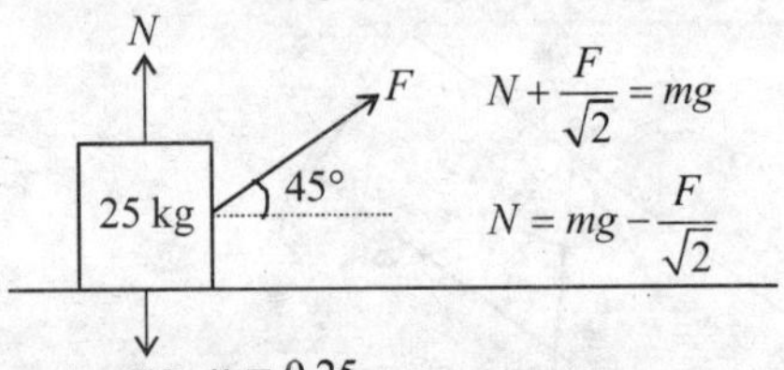

$N + \frac{F}{\sqrt{2}} = mg$

$N = mg - \frac{F}{\sqrt{2}}$

Block moves with uniform velocity.

$F - f = ma$

So, $a = 0 \Rightarrow F \cos 45° = f$

$f = \mu \times N$

$\Rightarrow \frac{F}{\sqrt{2}} = \mu\left[mg - \frac{F}{\sqrt{2}}\right]$

$\Rightarrow \frac{F}{\sqrt{2}} = 0.25\left[25 \times 9.8 - \frac{F}{\sqrt{2}}\right]$

$\Rightarrow 1.25 \frac{F}{\sqrt{2}} = 61.25$

$\Rightarrow F = \frac{61.25 \times \sqrt{2}}{1.25} = 49\sqrt{2}$

$W_{ext} = FS \cos 45° = 49\sqrt{2} \times 5 \times \frac{1}{\sqrt{2}} = 245$ J.

97. **(a)** Accerlation if plane is smooth,
a = g sinθ
Displacement of object

$\ell = \frac{1}{2}(g \sin\theta)t_1^2$ $\quad \therefore t_1 = \sqrt{\frac{2\ell}{g \sin\theta}}$

Accerlation if plane is rough,
a = g sinθ – μg cosθ
Displacement of object

$\ell = \frac{1}{2}(g \sin\theta - \mu g \cos\theta)t_2^2$

$$t_2 = \sqrt{\frac{2\ell}{g\sin\theta - \mu g\cos\theta}} = nt_1 n\sqrt{\frac{2\ell}{g\sin\theta}}$$

on compairing,

$$\mu = 1 - \frac{1}{n^2}$$

98. (b)

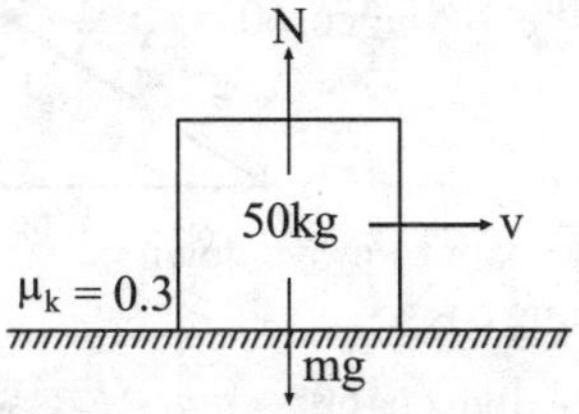

$F_k = \mu_k N = 0.3 \times 50 \times 9.8 = 147$ N $[\because N = mg]$

99. (a) From FBD,

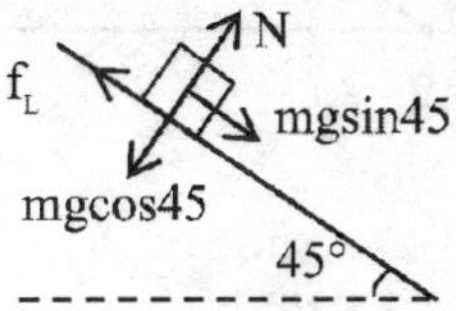

For brick begins to slide,

Frictional force $f_L = mg \sin 45°$

Normal reaction $N = mg \cos 45° = N$

$\therefore$ Coefficient of static friction

$$\mu_s = \frac{f_L}{N} = mg \sin 45°/mg \cos 45° = \tan 45° = 1$$

100. (c) FBD for 6 kg block

$6a = 6g - T$ ...(i)

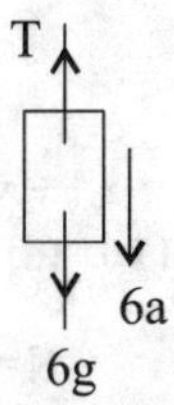

FBD for 20 kg block

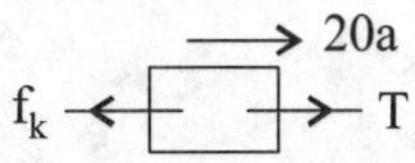

$T - f_k = 20$ a

$T - \mu(20)g = 20a$ ...(ii)

comparing (i) and (ii)

$\Rightarrow 6g - 6a = 20a + 20\,\mu g$

$\Rightarrow 6g - 20 \times 0.04g = 26a$

$$a = \frac{10(6-0.8)}{26} = 2\text{m/s}^2$$

101. (b) $N = mg$

$F = m\omega^2 r$

For no slipping $F \le \mu N$

$$\Rightarrow m\omega^2 r \le \mu mg \Rightarrow \omega^2 = \frac{\mu g}{r} \quad \Rightarrow \omega = \sqrt{\frac{\mu g}{r}}$$

102. (b)

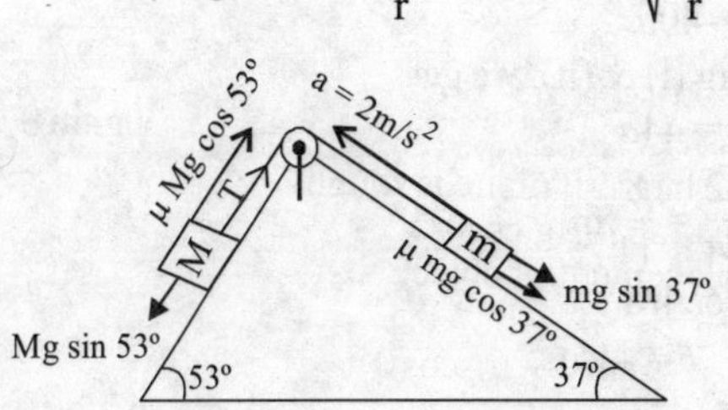

For M block

$Mg \sin 53° - \mu (Mg) \cos 53° - T = Ma$

$\Rightarrow T = 80 - 15 - 20 \Rightarrow T = 45$ N

For m block

$T - mg \sin 37° - \mu mg \cos 37° = 2m$

$\Rightarrow 45 = 10\,m \Rightarrow m = 4.5$ kg

103. (b) Maximum static friction

$$(f_s)_{max} = \mu mg \cos 30° = 0.1 \times \frac{50 \times \sqrt{3}}{2} = 2.5\sqrt{3}\ \text{N}$$

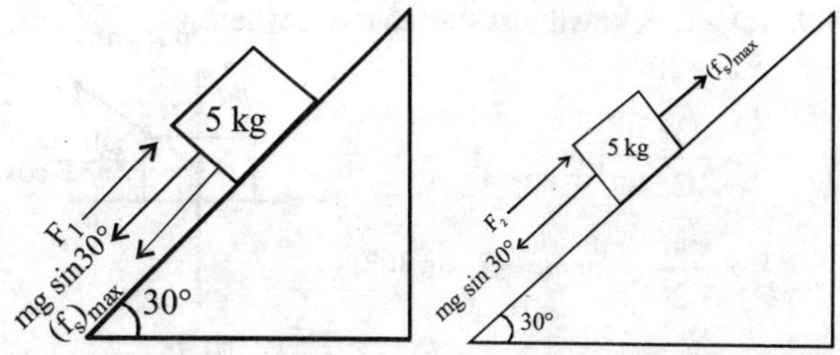

$F_1 = mg \sin 30° + (f_s)_{max} = (25 + 2.5\sqrt{3})$N

$F_2 = mg \sin 30° - (f_s)_{max} = (25 - 2.5\sqrt{3})$N

$\therefore\ F_1 - F_2 = 25 + 2.5\sqrt{3} - 25 + 2.5\sqrt{3} = 5\sqrt{3}$ N

104. (a) Coefficient of friction, $\mu = 0.5$

$$\tan\theta = \mu = \frac{1}{2} = \frac{dy}{dx} = \frac{x}{2}$$

$$\text{So } x = 1\text{m},\ y = \frac{x^2}{4} = \frac{(1)^2}{4} = \frac{1}{4}\text{m}$$

105. (b) Co-efficient of friction (μ) is independent of the area of contact and depends on nature of surface in contact, so, it depends on material of object.

106. (d) Initially,

Friction, $f = \mu\, mg = m\,\omega^2 R$ ...(i)

Finally,

$$\mu\, mg = m\left(\frac{\omega^2}{4}\right)R' \quad ...(ii)$$

From (i) and (ii)

$$1 = \frac{4R}{R'}$$

$\Rightarrow R' = 4R = 4 \times 1 = 4$ cm $(\because R = 10$ m given$)$

So distance from the center will be 4 cm.

107. (d)

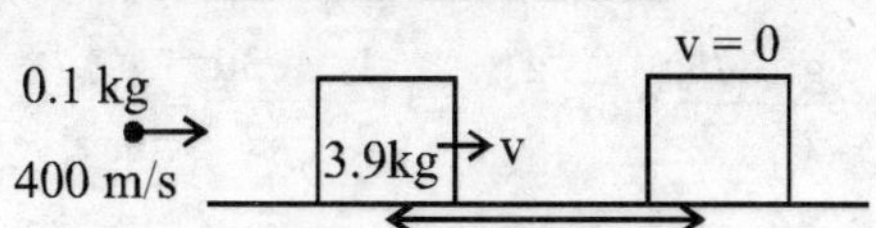

Apply the conservation of momentum

$P_i = P_f$ (Collision)

$$\Rightarrow (0.1)(400) = (0.1 + 3.9)\,v \Rightarrow v = \frac{0.1 \times 400}{4} = 10\text{m/s}$$

Apply equation of motion,

$$v^2 = u^2 + 2as \quad \because a = \frac{\mu mg}{m} = \mu g$$

$\Rightarrow 0 = (10)^2 - 2\mu g \times 20$

$$\Rightarrow \mu = \frac{100}{2 \times 10 \times 20} = \frac{1}{4} = 0.25$$

108. (a) From equation of motion

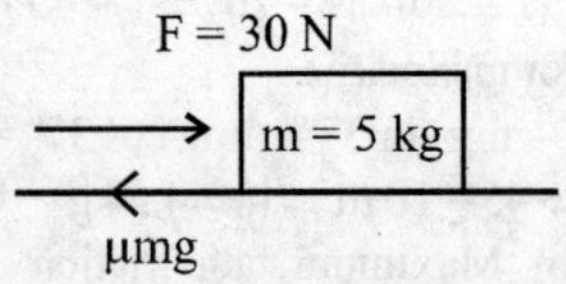

$S = ut + \frac{1}{2}at^2$

$\Rightarrow 50 = 0 + \frac{1}{2} \times a \times 100$

$a = 1\ m/s^2;\ F - \mu mg = ma$

$\Rightarrow 30 - \mu \times 50 = 5 \times 1 \Rightarrow 50\mu = 25;\ \mu = \frac{1}{2} = 0.50$

109. (c) Block will just start to move when

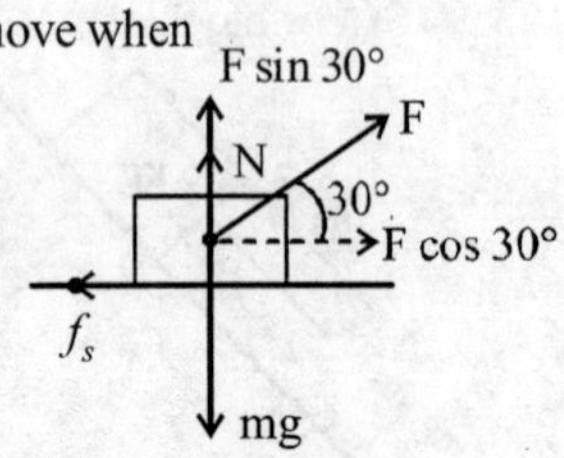

$F\cos 30° = f_s$

$\Rightarrow \frac{\sqrt{3}F}{2} = \mu N$

$\Rightarrow \frac{\sqrt{3}F}{2} = \mu(mg - F\sin 30°)$

$\Rightarrow \frac{\sqrt{3}F}{2} = 0.25(100 - \frac{F}{2}) \Rightarrow \frac{\sqrt{3}F}{2} = \frac{0.25}{2}(200 - F)$

$\Rightarrow \sqrt{3}\,F = 50 - 0.25F \Rightarrow F(\sqrt{3} + 0.25) = 50$

$\Rightarrow F = \frac{50}{0.25 + \sqrt{3}} = 25.2\ N$

110. (d) We have

$a = \frac{f}{m} = \frac{\mu mg}{m} = \mu g$

Also, $|a| = \frac{|V - u|}{t} = \frac{|20 - 0|}{5} = 4\ m/s^2$

So, $\mu g = 4 \Rightarrow \mu = \frac{4}{10} = 0.4$

111. (b) From the free body diagram shown.

mg sin 30° - μmg cos 30° = ma

$\Rightarrow a = g\left(\frac{1}{2} - \frac{\sqrt{3}\mu}{2}\right)$

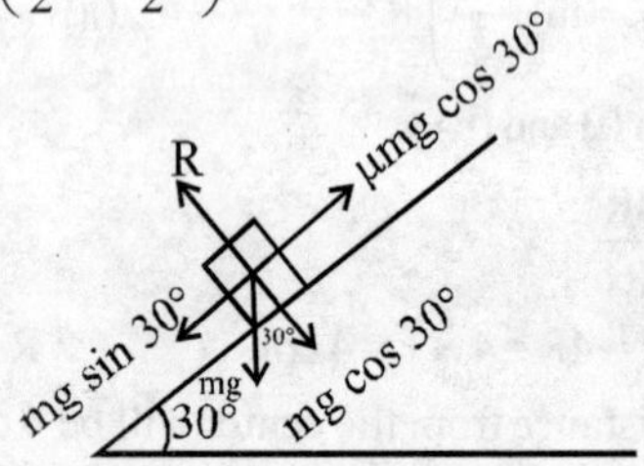

$\Rightarrow \frac{g}{2} - \frac{\sqrt{3}}{2}.\mu g = \frac{g}{4} \Rightarrow \frac{1}{2} - \frac{\sqrt{3}\mu}{2} = \frac{1}{4}$

$\Rightarrow \frac{\sqrt{3}}{2}\mu = \frac{1}{4} \Rightarrow \mu = \frac{1}{2\sqrt{3}}$

112. (d) Let a_1 be the acceleration when it slide down smooth incline plane.

Then, $a_1 = g\sin 45° = g/\sqrt{2}$

Let a_2 be the acceleration when it slide down rough inclined plane

Then, $a_2 = g\sin 45° - \mu_k g\cos 45° = \frac{g}{\sqrt{2}} - \frac{\mu_k g}{\sqrt{2}}$

Let 't_1' be the time taken when it slide down smooth surface and 't_2' be the time taken when it slide down rough surface.

$t_2 = nt_1\ \&\ \frac{1}{2}a_1t_1^2 = \frac{1}{2}a_2t_2^2$

$\Rightarrow \frac{1}{2}\frac{g}{\sqrt{2}}t_1^2 = \frac{1}{2}\left(\frac{g}{\sqrt{2}} - \frac{\mu_k g}{\sqrt{2}}\right)n^2t_1^2 \Rightarrow \mu_k = 1 - \frac{1}{n^2}$

113. (a) When block is just about to move up

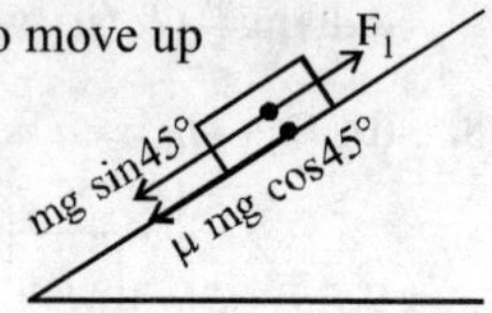

When block

$F_1 = mg\sin 45° + \mu mg\cos 45°$

$= \frac{mg}{\sqrt{2}}(1 + \mu)$

When block is just about to move down

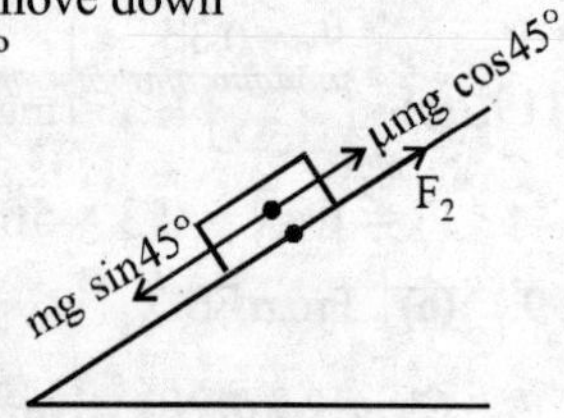

$F_2 = mg\sin 45° - \mu mg\cos 45°$

$= \frac{mg}{\sqrt{2}}(1 - \mu)$

As, $F_1 = 2F_2$

$\Rightarrow \frac{mg}{\sqrt{2}}(1 + \mu) = 2\frac{mg}{\sqrt{2}}(1 - \mu)$

$\Rightarrow 1 + \mu = 2(1 - \mu) \Rightarrow 1 + \mu = 2 - 2\mu$

$\Rightarrow 3\mu = 1 \Rightarrow \mu = \frac{1}{3} = 0.33$

114. (b) Retardation due to friction $= \frac{\mu Mg}{M} = \mu g \Rightarrow a = -\mu g$

Now, $s = \frac{v^2 - u^2}{2a} = \frac{0^2 - 2^2}{-2\mu g} = \frac{2}{0.4 \times 10} = 0.5m$

115. (a) From the free diagram shown

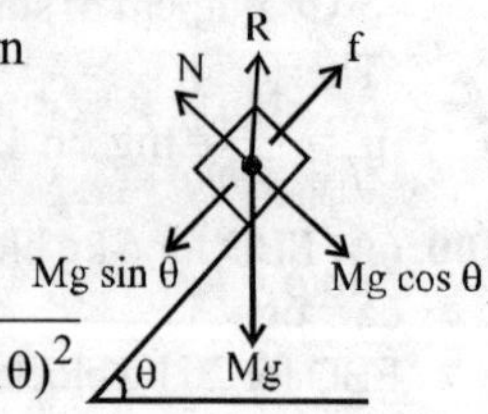

$N = Mg\cos\theta$

$f = Mg\sin\theta$

Contact force, $R = \sqrt{N^2 + f^2}$

$\Rightarrow R = \sqrt{(Mg\cos\theta)^2 + (Mg\sin\theta)^2}$

$= \sqrt{(Mg)^2(\cos^2\theta + \sin^2\theta)} \Rightarrow R = Mg$

116. (d) By FBD of block, we have

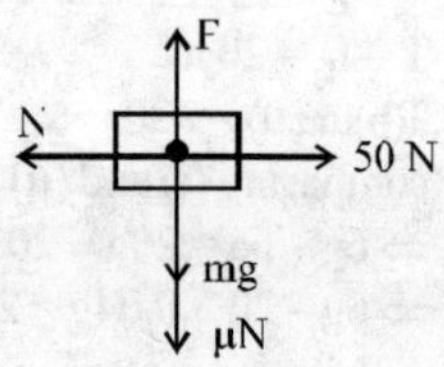

N = 50 N

And $F \le mg + \mu N$

i.e $F \le 2g + 0.5 \times 50$

$\le 20 + 25 \le 45N$

So, Maximum force that can be applied is 45N.

117. (d) For 4 kg block

$4g - T = 4a$ ...(i)

For 40 kg block

$T - 40g \times 0.02 = 40a\ [\because f_k = \mu mg]$

$T - 8 = 40a$ ...(ii)

Adding (i) & (ii), we get

$40 - 8 = 44a$

$a = \frac{32}{44} = \frac{8}{11}\ m/s^2.$

118. (6) For 4M

$2Mg - T = 4Ma$ ...(i)

For M

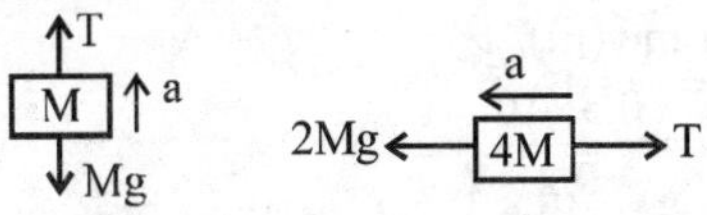

$T - Mg = Ma$ ...(ii)

Adding (i) & (ii), we get

$Mg = 5\,Ma \Rightarrow a = \dfrac{g}{5}$

So, $T = Ma + Mg = \dfrac{Mg}{5} + Mg = \dfrac{6}{5}Mg$

119. **(c)** 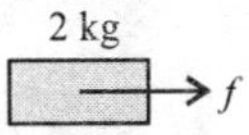 $f = 2a$ ...(i)

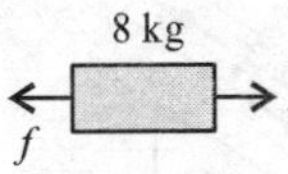 $F - f = 8a$ ...(ii)

Clearly, from (i) 'a' will be maximum when $f = f_{lim}$

So, $a_{max} = \dfrac{f_{lim}}{2} = \dfrac{\mu \times 2g}{2} = \mu g$

From (ii), $F_{max} = 8a_{max} + f_{lim} = 8\mu g + 2\mu g = 10\mu g = 49\text{N}$

120. **(b)** For beaker to move with disc

$f_s = m\omega^2 R$

So, R will be maximum, when $f_s = f_{lim}$

Therfore,

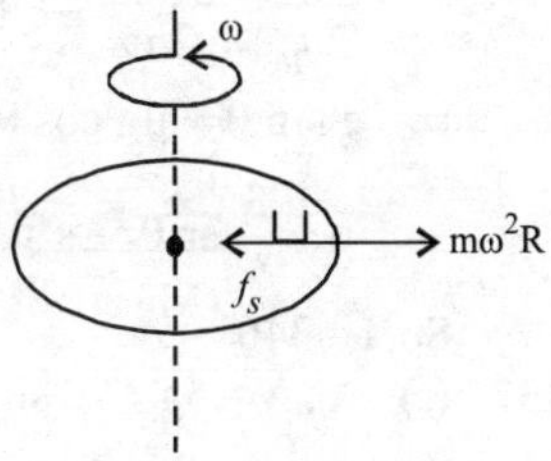

$f_{lim} = m\omega^2 R_{max}$

$\mu mg = m\omega^2 R_{max}$

$R_{max} = \dfrac{\mu g}{\omega^2}$

So, $R \le \dfrac{\mu g}{\omega^2}$

121. **(b)** $a = \dfrac{f}{m} = \dfrac{\mu mg}{m} = \mu g$

So, $S = \dfrac{u^2}{2a}$ $[\because v = 0]$

$S = \dfrac{(9.8)^2}{2 \times \mu g} = \dfrac{(9.8)^2}{2 \times 0.5 \times 9.8} = 9.8\text{ m}$

122. **(3)** Acceleration on smooth inclined plane

$a = g \sin 30° = g/2$

Using $S = ut + \dfrac{1}{2}at^2$

$\Rightarrow S = \dfrac{1}{2}\dfrac{g}{2}T^2 = \dfrac{g}{4}T^2$ ...(i) $(\because u = 0)$

Acceleration on rough inclined plane

$a = g\sin 30° - \mu g \cos 30° = \dfrac{g}{2} - \dfrac{\mu g\sqrt{3}}{2}$

$\Rightarrow a = \dfrac{g}{2}\left(1 - \mu\sqrt{3}\right)$

Using again $S = ut + \dfrac{1}{2}at^2$

$\Rightarrow S = \dfrac{1}{4}g\left(1 - \sqrt{3}\mu\right)(\alpha T)^2$ ...(ii)

By (i) and (ii)

$= \dfrac{1}{4}gT^2 = \dfrac{1}{4}g\left(1 - \sqrt{3}\mu\right)\alpha^2 T^2 \Rightarrow 1 - \sqrt{3}\mu = \dfrac{1}{\alpha^2}$

$\Rightarrow \mu = \left(\dfrac{\alpha^2 - 1}{\alpha^2}\right)\dfrac{1}{\sqrt{3}} \Rightarrow x = 3.00$

123. **(15)** For block of mass 1 kg

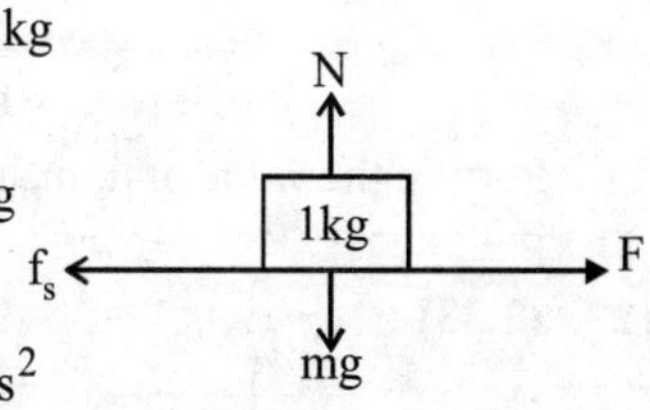

$F = f_{s,max}$

$\Rightarrow 1\text{kg} \times a = \mu N = \mu \times 1 \times g$

$(\because N = mg)$

$\Rightarrow a = 0.5 \times 1 \times 10 = 5\text{ m/s}^2$

$\therefore$ Maximum horizontal force, $F_{max} = ma = (1+2) \times 5 = 15\text{N}$

124. **(3)** From question,

$t_a = \dfrac{1}{2}t_d$

$\sqrt{\dfrac{2s}{a_a}} = \dfrac{1}{2}\sqrt{\dfrac{2s}{a_d}}$

or, $a_a = 4a_d$ ...(i)

$g \sin\theta + \mu g \cos\theta - 4(g \sin\theta - \mu g \cos\theta)$

$\Rightarrow 5\,\mu g \cos\theta = 3g \sin\theta \Rightarrow \mu = \dfrac{3\tan\theta}{5}$

$\Rightarrow \mu = \dfrac{\sqrt{3}}{5}$ $[\because \theta = 30°]$

So, $x = 3$

125. **(30)** From FBD shown, $T = mg - N$

$R = Mg + N = (M + m)g - T$

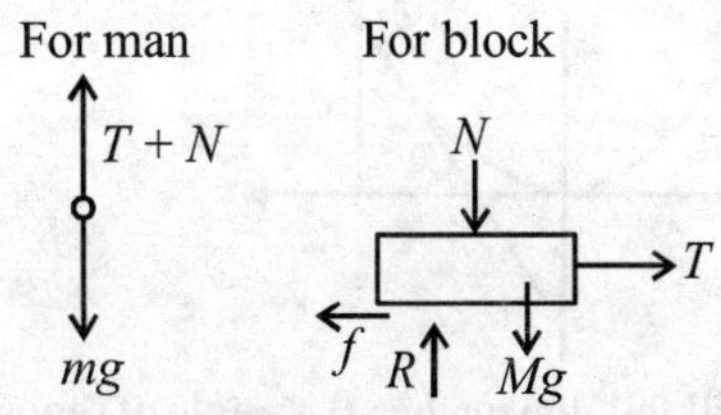

For no movement of block,

$T \le \mu R \Rightarrow T \le \mu[(M + m)g - T]$

$\Rightarrow T \le \dfrac{\mu(M + m)g}{1 + \mu} \Rightarrow T \le \dfrac{(0.5)(5 + 4) \times 10}{1 + 0.5}$

$\Rightarrow T \le 30\text{N}$

$\therefore T_{max} = 30\text{ N}$

126. **(5)** As block is at rest

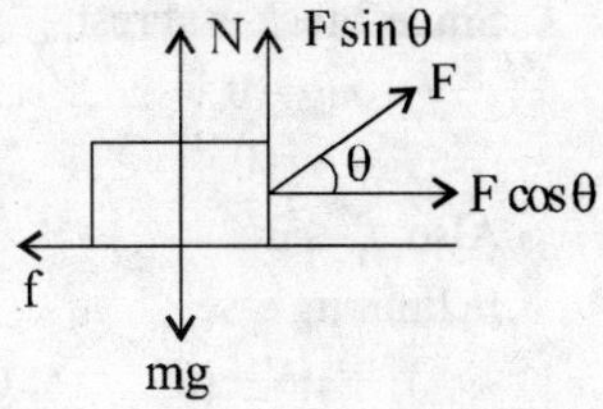

So, $F \cos\theta = f = \mu N$

$F \cos\theta = \mu\,(mg - F \sin\theta)$

$F(\cos\theta + \mu \sin\theta) = \mu mg$

$F = \dfrac{\mu mg}{\cos\theta + \mu \sin\theta}$

For F_{min}

$(\cos\theta + \mu \sin\theta)$ should be maximum

i.e. $\frac{d}{d\theta}(\cos\theta + \mu\sin\theta) = 0 \Rightarrow \tan\theta = \mu$

So, $\sin\theta = \frac{\mu}{\sqrt{1+\mu^2}}$ and $\cos\theta = \frac{1}{\sqrt{1+\mu^2}}$

Thus, $F_{min} = \frac{\mu mg}{\frac{1}{\sqrt{1+\mu^2}} + \frac{\mu.\mu}{\sqrt{1+\mu^2}}} = \frac{\mu mg}{\sqrt{1+\mu^2}}$

Putting the value of μ, m and g

$F_{min} = 5N$

127. **(3.33)**

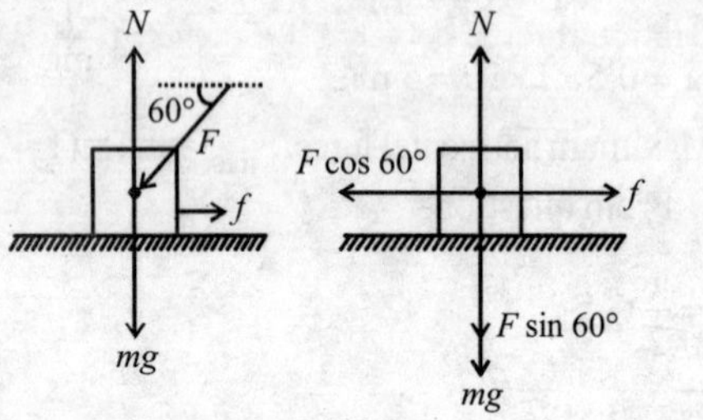

From the condition of equilibrium

$$N = mg + F\sin 60^\circ$$

For no movement of the block

$$F\cos 60^\circ \le f$$

$\Rightarrow F\cos 60^\circ \le \mu(mg + F\sin 60^\circ)$ $\quad (\because f = \mu N)$

$\Rightarrow F \le \frac{\mu mg}{\cos 60^\circ - \mu\sin 60^\circ} \Rightarrow F_{critical} = 10\text{ N}$

As $F_{critical} = 3x$

So, $3x = 10 \Rightarrow x = 3.33$

128. **(25)**

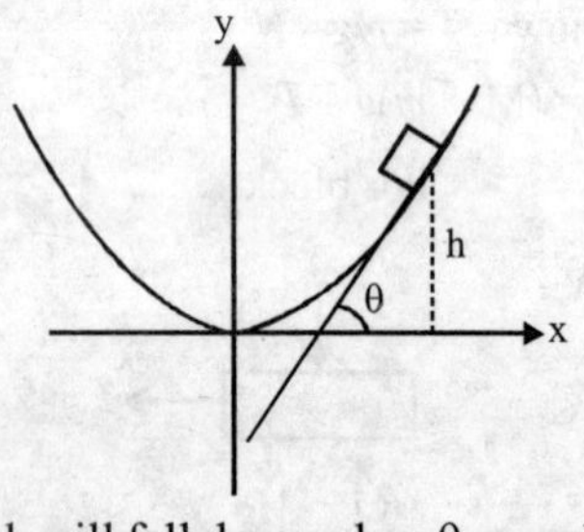

Block will fall down when θ = angle of repose i.e.,

$\tan\theta = \mu$

$\therefore \tan\theta = \frac{dy}{dx} = \frac{d}{dx}\left(\frac{x^2}{4}\right) = \frac{x}{2}$ and at time of maximum height $\tan\theta = \mu = 0.5$

$\Rightarrow x = 1$ and therefore $y = \frac{x^2}{4} = \frac{(1)^2}{4} = 0.25\text{ m} = 25\text{ cm}$

129. **(25)** F.B.D. of the block is shown in the diagram

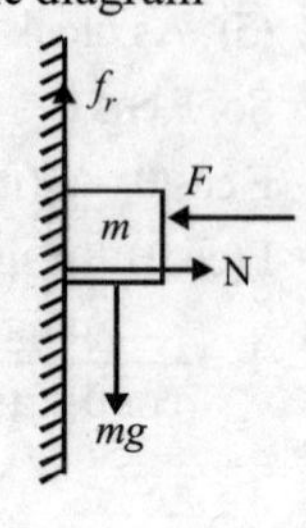

Since, block is at rest,

$\therefore f_r - mg = 0$ ... (i)

$F - N = 0$ ... (ii)

Also, $f_r \le \mu N$

In limiting case,

$f_r = \mu N = \mu F$ ... (iii)

Using equation (i) and (iii),

$$F = \frac{mg}{\mu} \Rightarrow F = \frac{0.5\times 10}{0.2} = 25\,N$$

130. **(a)** For balancing, $mg\sin\theta = f = \mu mg\cos\theta$

$\Rightarrow \tan\theta = \mu = \frac{3}{4} = 0.75$

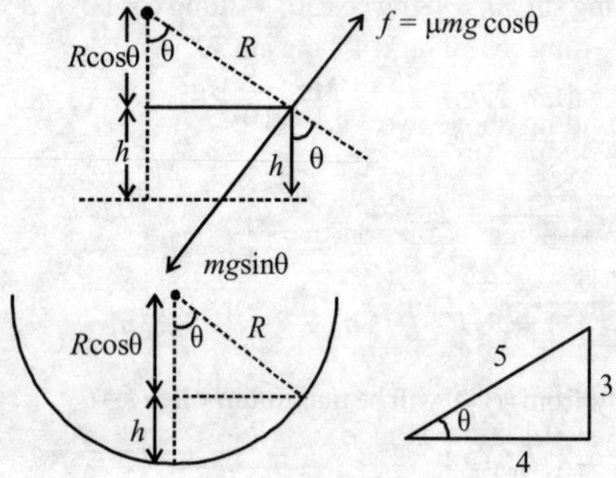

$h = R - R\cos\theta = R - R\left(\frac{4}{5}\right) = \frac{R}{5}$

$\therefore h = \frac{R}{5} = 0.2\text{ m}$ $\quad [\because \text{radius}, R = 1\text{m}]$

131. **(346)** $S_v = S_d$

$\frac{v_0^2}{2a_u} = \frac{v_0^2/4}{2a_d} \Rightarrow a_u = 4a_d$

$\Rightarrow g\sin\theta + \mu g\cos\theta = 4(g\sin\theta - \mu g\cos\theta)$

$\Rightarrow \mu = \frac{3}{5}\tan\theta \Rightarrow \mu = \frac{3}{5}.\frac{1}{\sqrt{3}} = \frac{\sqrt{3}}{5} = 0.346 = \frac{346}{1000}$

So, I = 346

132. **(c)** A : $N = 5g + 20\sin 30^\circ$

$= 50 + 20 \times \frac{1}{2} = 60\text{ N}$

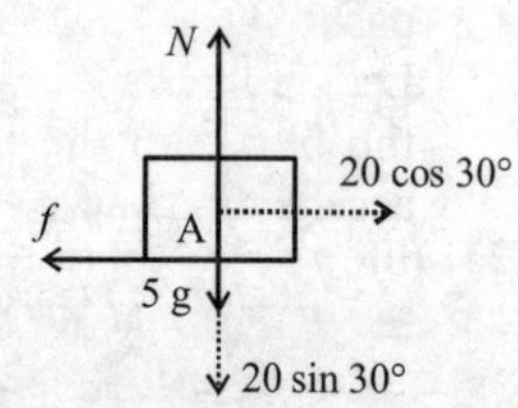

Accelaration, $a_1 = \frac{F-f}{m}$

$= \frac{20\cos 30^\circ - \mu N}{5}$

$= \left[\frac{20\times\frac{\sqrt{3}}{2} - 0.2\times 60}{5}\right]$

$= 1.06\text{ m/s}^2$

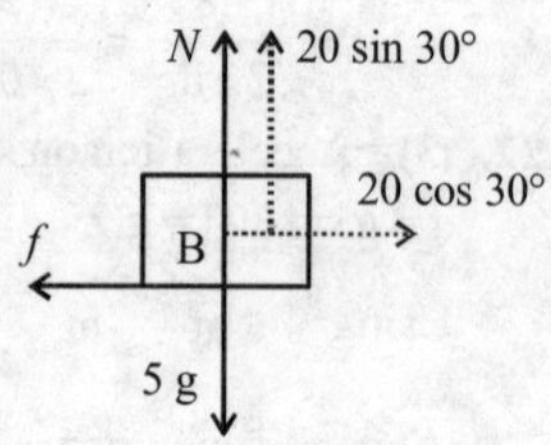

B : $N = 5g - 20\sin 30^\circ$

$= 50 - 20 \times \frac{1}{2} = 40\text{ N}$

$a_2 = \frac{F-f}{m} = \left[\frac{20\cos 30^\circ - 0.2\times 40}{5}\right] = 1.86\text{ m/s}^2$

Now $a_2 - a_1 = 1.86 - 1.06 = 0.8\text{ m/s}^2$

133. **(b)** Taking (A + B) as system

$F - \mu(M+m)g = (M+m)a$

$\Rightarrow a = \dfrac{F - \mu(M+m)g}{(M+m)} \Rightarrow a = \dfrac{F-(0.2)4\times10}{4} = \left(\dfrac{F-8}{4}\right)$

But, $a_{max} = \mu g = 0.2 \times 10 = 2 \therefore \dfrac{F-8}{4} = 2 \Rightarrow F = 16\text{ N}$

134. (a) From figure, $2 + mg\sin 30° = \mu mg\cos 30°$ and $10 = mg\sin 30° + \mu\, mg\cos 30° = 2\mu mg\cos 30° - 2$
$\Rightarrow 6 = \mu mg\cos 30°$ and $4 = mg\sin 30°$

By dividing above two $\Rightarrow \dfrac{3}{2} = \mu \times \sqrt{3}$

$\therefore$ Coefficient of friction, $\mu = \dfrac{\sqrt{3}}{2}$

135. (b) Given : $m_1 = 5\text{kg}; m_2 = 10\text{kg}; \mu = 0.15$

FBD for m_1, $m_1 g - T = m_1 a$

$\Rightarrow 50 - T = 5 \times a$

and, $T - 0.15(m + 10)g = (10 + m)a$

For rest $a = 0$

or, $50 = 0.15(m + 10)\, 10$

$\Rightarrow 5 = \dfrac{3}{20}(m + 10)$

$\dfrac{100}{3} = m + 10 \therefore m = 23.3\text{kg};$ close to option (b)

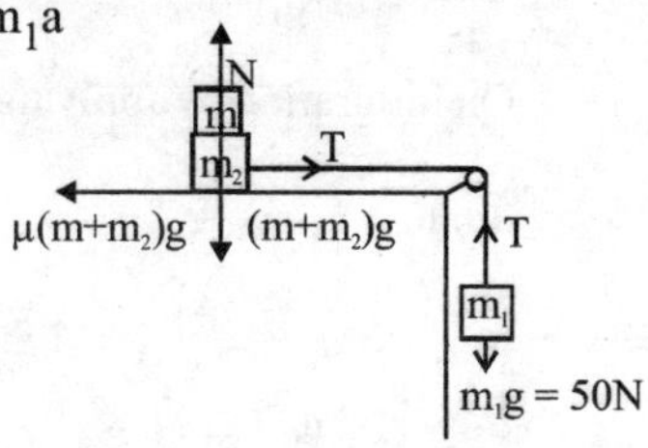

136. (d) Equation of motion when the mass slides down

$Mg\sin\theta - f = Ma$

$\Rightarrow 10 - f = 6$ ($M = 2$ kg, $a = 3$ m/s^2, $\theta = 30°$ given)

$\therefore f = 4\text{N}$

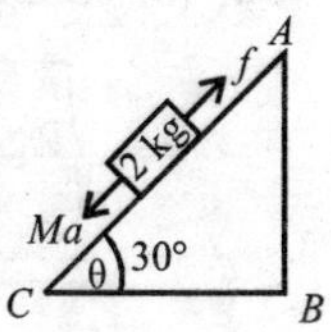

Equation of motion when the block is pushed up

Let the external force required to take the block up the plane with same acceleration be F

$F - Mg\sin\theta - f = Ma$

$\Rightarrow F - 10 - 4 = 6$

$F = 20\text{ N}$

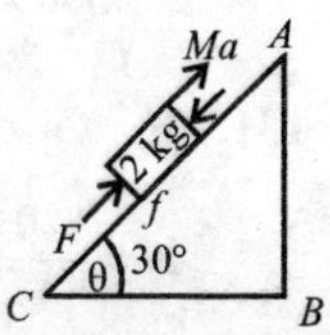

137. (b) Let μ be the minimum coefficient of friction

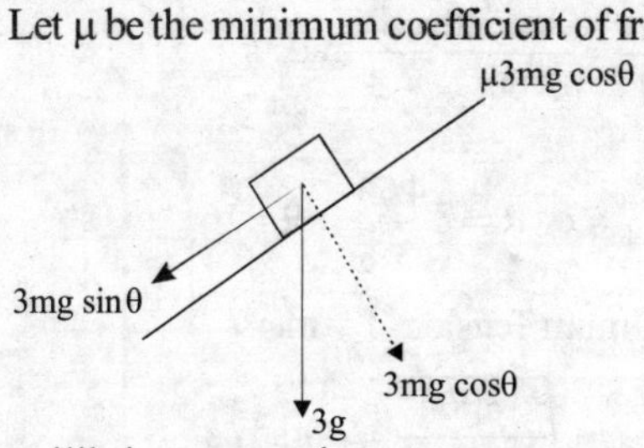

At equilibrium, mass does not move so,

$3mg\sin\theta = \mu 3\, mg\cos\theta \therefore \mu_{min} = \tan\theta$

138. (a)

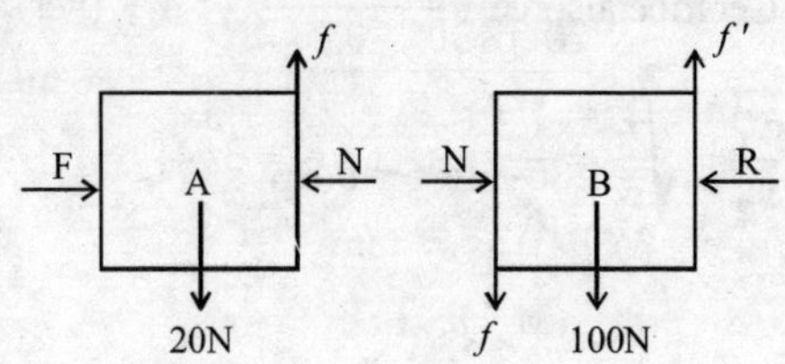

Along vertical direction

$A \rightarrow f = 20\text{ N}$

$B \rightarrow f' = f + 100 = 20 + 100 = 120\text{ N}$

139. (b) Initial speed at point A, $u = v_0$

Speed at point B, $v = ?$

$v^2 - u^2 = 2gh$

$v^2 = v_0^2 + 2gh$

Let ball travels distance 'S' before coming to rest

$S = \dfrac{v^2}{2\mu g} = \dfrac{v_0^2 + 2gh}{2\mu g} = \dfrac{v_0^2}{2\mu g} + \dfrac{2gh}{2\mu g} = \dfrac{h}{\mu} + \dfrac{v_0^2}{2\mu g}$

140. (c) Minimum force on A

= frictional force between the surfaces = 12 N

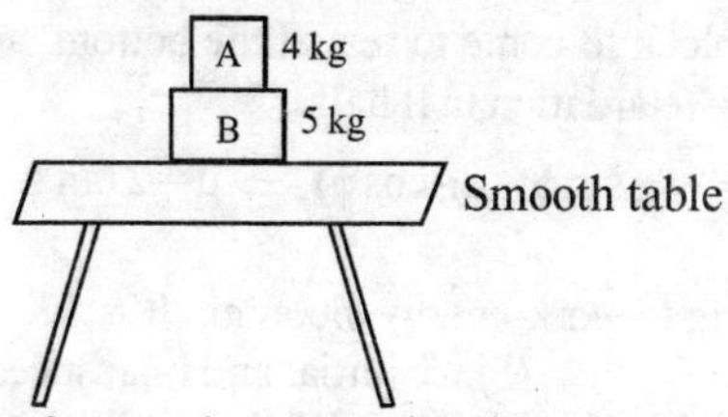

Therefore maximum acceleration

$a_{max} = \dfrac{12\text{N}}{4\text{kg}} = 3\text{m/s}^2$

Hence maximum force,

F_{max} = total mass $\times a_{max}$

$= 9 \times 3 = 27\text{ N}$

141. (b)

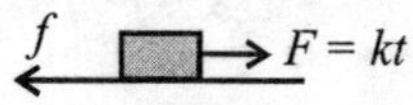

$a = \dfrac{kt - \mu mg}{m}$

$a = \dfrac{kt}{m} - \mu g$

So, a-t graph will be as shown

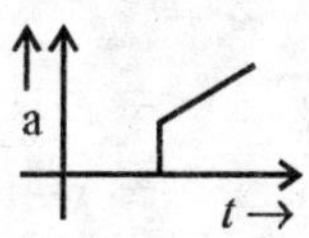

142. (d) $a = g\sin 45° - \mu g\cos 45°$

$a = g\sin 45° - 0.3\, xg\cos 45°$

$= \dfrac{g}{\sqrt{2}} - \dfrac{0.3gx}{\sqrt{2}} = 5\sqrt{2} - 0.3(5\sqrt{2})x = 5\sqrt{2} - 1.5\sqrt{2}x$

Velocity will increase until $a = 0$ and when $v = v_{max}$, then $a = 0$

$0 = 5\sqrt{2} - 1.5\sqrt{2}x$

$x = \dfrac{5\sqrt{2}}{1.5\sqrt{2}}$

$x = 3.33\text{ m}$

143. (c)

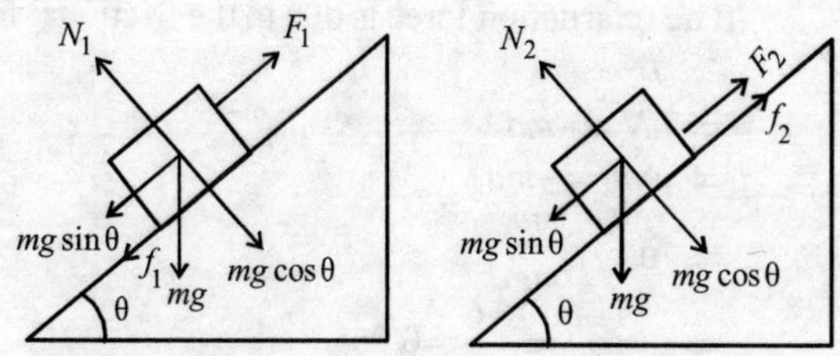

When the body slides up the inclined plane, then

$mg\sin\theta + f_1 = F_1$

or, $F_1 = mg\sin\theta + \mu mg\cos\theta$
When the body slides down the inclined plane, then
$mg\sin\theta - f_2 = F_2$
or $F_2 = mg\sin\theta - \mu mg\cos\theta$

$$\therefore \frac{F_1}{F_2} = \frac{\sin\theta + \mu\cos\theta}{\sin\theta - \mu\cos\theta}$$

$$\Rightarrow \frac{F_1}{F_2} = \frac{\tan\theta + \mu}{\tan\theta - \mu} = \frac{2\mu + \mu}{2\mu - \mu} = \frac{3\mu}{\mu} = 3$$

144. (d) For first half
acceleration $= g\sin\phi$;
For second half
acceleration $= -(g\sin\phi - \mu g\cos\phi)$
For the block to come to rest at the bottom, acceleration in I half = retardation in II half.

$g\sin\phi = -(g\sin\phi - \mu g\cos\phi) \Rightarrow \mu = 2\tan\phi$

NOTE :
According to work-energy theorem, $W = \Delta K = 0$
(Since initial and final speeds are zero)
$\therefore$ Workdone by friction + Work done by gravity = 0

i.e., $-(\mu\, mg\cos\phi)\frac{\ell}{2} + mg\ell\sin\phi = 0$

or $\frac{\mu}{2}\cos\phi = \sin\phi$ or $\mu = 2\tan\phi$

145. (a) Given, initial velocity, $u = 100$m/s.
Final velocity, $v = 0$.
Acceleration, $a = \mu_k g = 0.5 \times 10$
$v^2 - u^2 = 2as$ or
$\Rightarrow 0^2 - u^2 = 2(-\mu_k g)s$
$\Rightarrow -(100)^2 = 2 \times -\frac{1}{2} \times 10 \times s$
$\Rightarrow s = 1000$ m

146. (c) Since the body is at rest on the inclined plane,
$mg\sin 30° =$ Force of friction
$\Rightarrow m \times 10 \times \sin 30° = 10$
$\Rightarrow m \times 5 = 10 \Rightarrow m = 2.0$ kg

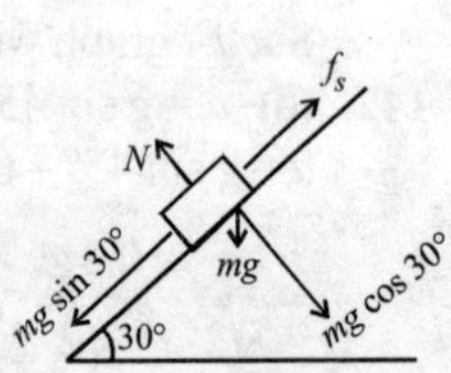

147. (d) $u = 6$ m/s, $v = 0$, $t = 10s$, $a = ?$

Acceleration $a = \frac{v-u}{t}$

$\Rightarrow a = \frac{0-6}{10}$

$\Rightarrow a = \frac{-6}{10} = -0.6\text{m/s}^2$

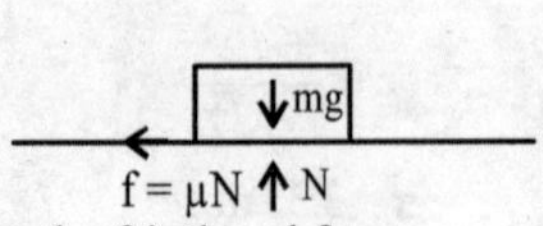

The retardation force is due to the frictional force
$\therefore f = -ma$
$\Rightarrow \mu N = -ma$
$\Rightarrow \mu mg = -ma$
$\Rightarrow \mu = -\frac{ma}{mg}$
$\Rightarrow \mu = \frac{-a}{g} = \frac{0.6}{10} = 0.06$

148. (c)

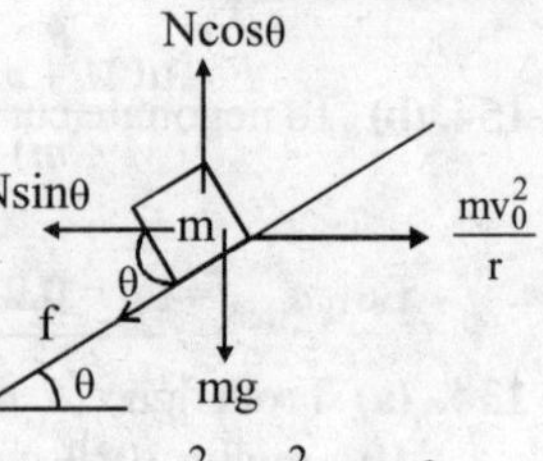

$N\sin\theta + f\cos\theta = \frac{mv_0^2}{r}$ ...(i)
$N\cos\theta - f\sin\theta = mg$ ...(ii)
From eq. (i) and (ii),

$$\frac{\sin\theta + \mu\cos\theta}{\cos\theta - \mu\sin\theta} = \frac{v_0^2}{rg}; \; rg\tan\theta + \mu rg = v_0^2 - v_0^2\mu\tan\theta$$

$$\mu = \frac{v_0^2 - rg\tan\theta}{rg + v_0^2\tan\theta}$$

149. (2) Centripetal accerlation is,
$a_c = \omega^2 x$

$$\frac{vdv}{dx} = \omega^2 x \qquad \left[\because a_c = v\frac{dv}{dx}\right]$$

On intigration by applying limits,

$$\int_0^v vdv = \int_1^3 \omega^2 x dx$$

$$\left[\frac{v^2}{2}\right]_0^v = \omega^2\left[\frac{x^2}{2}\right]_1^3$$

$$\frac{v^2}{2} = \frac{\omega^2}{2}\left[3^2 - 1^1\right] \Rightarrow v = 2\sqrt{2}\omega \quad \therefore x = 2$$

150. (b) Maximum velocity is given by :

$$V = \sqrt{Rg\left[\frac{\tan\theta + \mu}{1 - \mu\tan\theta}\right]} = \sqrt{300 \times g \times \left[\frac{\tan 30° + 0.2}{1 - 0.2 \times \tan 30°}\right]}$$

$$= \sqrt{300 \times 10 \times \left[\frac{0.57 + 0.2}{1 - 0.2 \times 0.57}\right]} = 51.4 \text{ m/s}$$

151. (b) Given : R = 9m,

$$N = \frac{120}{3}\text{rpm} = 40\text{rpm}$$

$$\omega = \frac{2\pi N}{60} = \frac{2\pi \times 40}{60} = \frac{4\pi}{3}\text{rad/s}$$

$$a_{centripetal} = \omega^2 R = \left(\frac{4\pi}{3}\right)^2 \times 9 = 16\pi^2 \text{m/s}^2$$

152. (b) Maximum Tension, $T = m\omega^2\ell$

$$\omega = \sqrt{\frac{T}{m\ell}} = \sqrt{\frac{400}{0.5 \times 0.5}} = 40\text{rad/s}$$

153. (c) Centripetal force, $F = \frac{mv^2}{r} \Rightarrow v^2 \propto r$ (For same m & F)

$$\Rightarrow \frac{v_1}{v_2} = \sqrt{\frac{r_1}{r_2}} = \frac{\sqrt{3}}{2} \quad [\because r_1 : r_2 = 3 : 4]$$

154. (b) To negotiate curve, $\tan\theta = \dfrac{v^2}{Rg} = \dfrac{12\times12}{10\times400}$

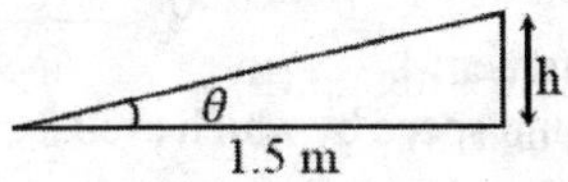

$\tan\theta = \dfrac{h}{1.5} \Rightarrow \dfrac{h}{1.5} = \dfrac{144}{4000}$

Therefore, outer rail raised with height, $h = 5.4$ cm

155. (d) Given that

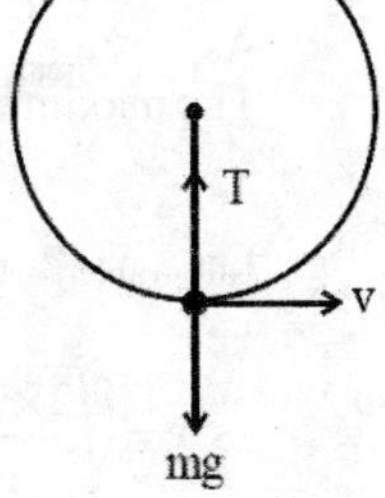

$m = 900\text{ g} = \dfrac{900}{1000}\text{kg} = \dfrac{9}{10}\text{kg}$

$r = 1\text{ m}, N = 10\text{ rpm}$

$\omega = \dfrac{2\pi N}{60} = \dfrac{2\pi(10)}{60} = \dfrac{\pi}{3}\text{rad/s}$

For circular motion, at lowest point

$F_{net} = mr\omega^2$

$\Rightarrow T - mg = mr\omega^2$

$\Rightarrow T = mg + mr\omega^2$

$= \dfrac{9}{10}\times9.8 + \dfrac{9}{10}\times1\left(\dfrac{\pi}{3}\right)^2 = 8.82 + 0.98 = 9.80\text{ N}$

156. (a) Centripetal force $F_c = \dfrac{mv^2}{r}$ and $v = r\times\omega$

$\therefore F_c = m\omega^2 r = 200\times(0.2)^2\times70 = 560\text{ N}$

157. (d) Given, mass of child, $m = 5$ kg

Radius of merry-go-round, $R = 2$ m

Angular velocity, $\omega = \dfrac{2\pi}{3.14} = 2\text{rad/s}$

The centrifugal force on the child will be

$F = m\omega^2 R = 5\times2^2\times2 = 40\text{ N}$

158. (c) Centripetal force, $f_c = \dfrac{mv^2}{r}$

Frictional force = μmg

Here, centripetal force for motion is being provided by the friction.

$\therefore \dfrac{mV^2}{r} = \mu mg$

$V_{max} = \sqrt{\mu rg} = \sqrt{0.34\times50\times10} \approx 13\text{m/s}$

159. (40) $v\dfrac{dv}{dx} = a \Rightarrow v\dfrac{dv}{dx} = \dfrac{v^2}{R} \Rightarrow \displaystyle\int_{15}^{v}\dfrac{dv}{v} = \dfrac{1}{R}\int_0^x dx$

$\Rightarrow \ln\dfrac{v}{15} = \dfrac{x}{R} \Rightarrow e^{x/R} = \dfrac{v}{15} \Rightarrow v = 15e^{x/R}$

$\Rightarrow \dfrac{dx}{dt} = 15e^{x/R} \Rightarrow \displaystyle\int_0^{\frac{\pi R}{2}} e^{-x/R}dx = 15\int_0^{t_0} dt$

$\Rightarrow t_0 = 40(1 - e^{-\pi/2})$

160. (c) Given that speed, $v = 20$ m/s

Radius of circular track, $R = 40$ m

$T\cos\theta = mg$ ...(i)

$T\sin\theta = \dfrac{mv^2}{R}$ ...(ii)

Divide equation (ii) by (i), we have

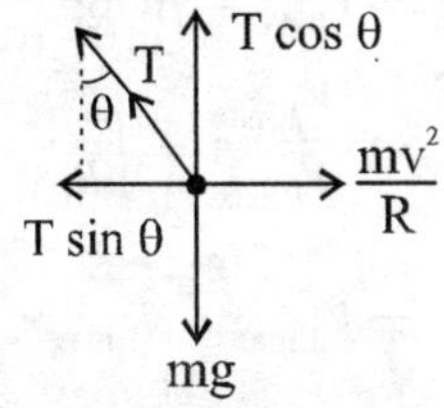

$\tan\theta = \dfrac{v^2}{Rg}$

$\tan\theta = \dfrac{20^2}{40\times10}$

$\tan\theta = 1 \Rightarrow \theta = \dfrac{\pi}{4}$

161. (a) For observer on block,

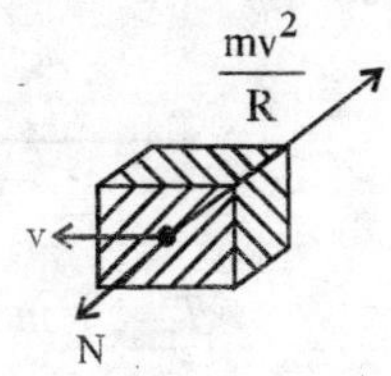

$N = \dfrac{mv^2}{r} \Rightarrow N \propto v^2$

So, curve is parabola, symmetric about N – axis

162. (b) $\Delta\vec{V} = |\vec{V} - \vec{u}| = \sqrt{V^2 + u^2}$ $[\because \vec{V}\perp\vec{u}]$

Now, $K_1 + P_1 = K_2 + P_2$

$\Rightarrow \dfrac{1}{2}mu^2 + 0 = \dfrac{1}{2}mV^2 + mg\,l \Rightarrow V^2 = u^2 - 2g\,l$

$\Rightarrow V = \sqrt{u^2 - 2gl}$

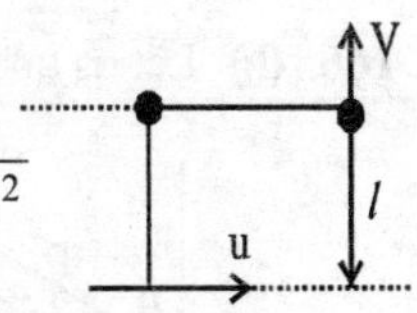

So, $\Delta V = \sqrt{V^2 + u^2} = \sqrt{u^2 - 2gl + u^2}$

$= \sqrt{2(u^2 - gl)} \Rightarrow x = 2$

163. (c) By law of conservation of mechanical energy,

$K_P + P_P = K_Q + P_Q$

$\Rightarrow 0 + mgR\sin\alpha = \dfrac{1}{2}mv^2 \Rightarrow v^2 = 2Rg\sin\alpha$

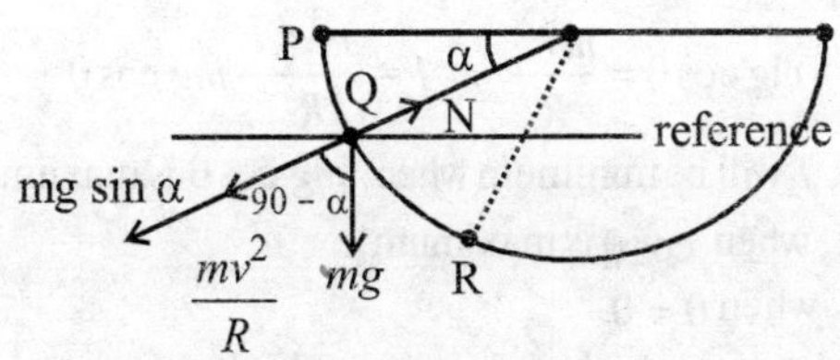

At point 'Q', $N = \dfrac{mv^2}{R} + mg\sin\alpha$

$\Rightarrow \dfrac{N}{\left(\dfrac{mv^2}{R}\right)} = 1 + \dfrac{Rg}{v^2}\sin\alpha = 1 + \dfrac{Rg\sin\alpha}{2Rg\sin\alpha} = \dfrac{3}{2}$

$\Rightarrow A = \text{constant}$

So, graph between A and α will be as shown below.

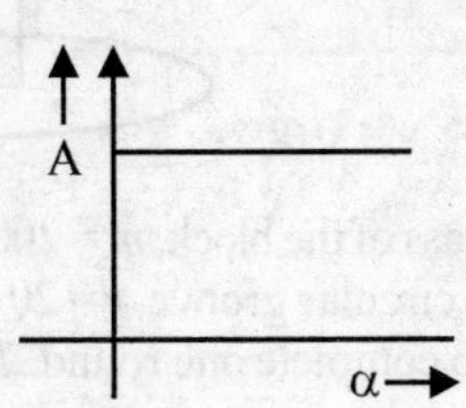

164. (24) $f_s, \max = \dfrac{mV_{\max}^2}{R}$

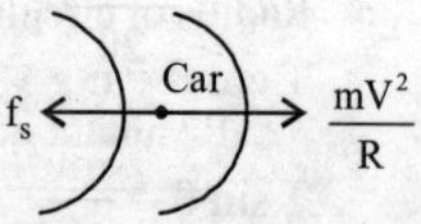

$\Rightarrow \mu mg = \dfrac{mV_{\max}^2}{R}$

$\Rightarrow V_{\max} = \sqrt{\mu Rg}$

So, $\dfrac{V_{2,\max}}{V_{1,\max}} = \sqrt{\dfrac{R_2}{R_1}}$

$\Rightarrow V_{2,\max} = V_{1,\max} \times \sqrt{\dfrac{R_2}{R_1}} = 30 \times \sqrt{\dfrac{48}{75}} = 30 \times \dfrac{4}{5} = 24\,\text{m/s}.$

165. (c)

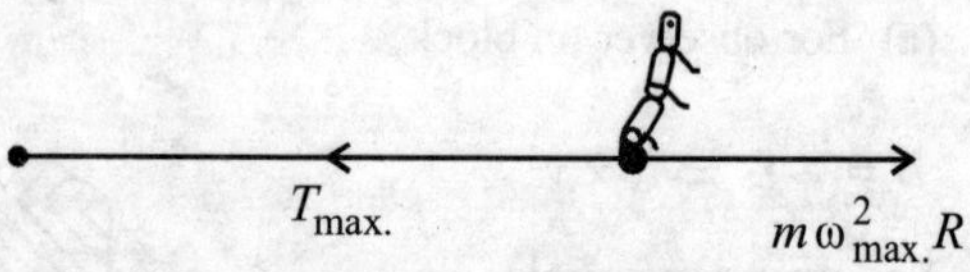

$T_{\max} = m\omega_{\max}^2 R$

$80 = 0.1 \times \left(\dfrac{k}{30}\right)^2 \times 2 \qquad \left[\because\ \omega = \dfrac{k}{\pi} \times \dfrac{2\pi}{60} = \dfrac{k}{30}\right]$

$\Rightarrow k^2 = \dfrac{30^2 \times 80}{2 \times 0.1} \Rightarrow k^2 = 360000 \Rightarrow k = 600$

166. (b) Let us take a general point '*P*'

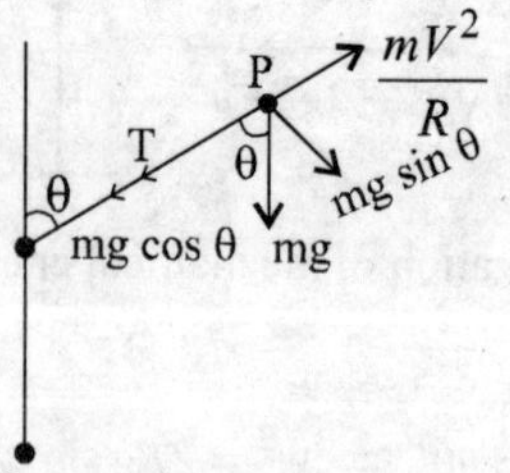

At *P*

$T + mg\cos\theta = \dfrac{mv^2}{R} \Rightarrow T = \dfrac{mv^2}{R} - mg\cos\theta$

So, *T* will be minimum when, $mg\cos\theta$ is maximum

i.e., when $\cos\theta$ is maximum

i.e. when $\theta = 0$

and θ is zero when string is at highest point.

167. (a) From figure

$\sin\theta = \dfrac{r}{L} = \dfrac{L}{L\sqrt{2}} = \dfrac{1}{\sqrt{2}} \Rightarrow \theta = 45°$

$T\sin\theta = \dfrac{mv^2}{r}$ ...(i)

$T\cos\theta = mg$ ... (ii)

From (i) and (ii)

$\tan\theta = \dfrac{v^2}{rg} \Rightarrow v = \sqrt{rg}$

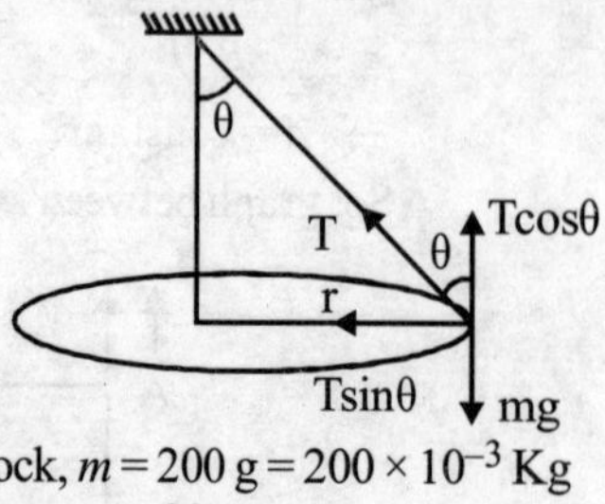

168. (b) Given, mass of the block, $m = 200\ \text{g} = 200 \times 10^{-3}$ Kg

Radius of the circular groove, $r = 20$ cm

Time taken to complete one round, $T = 40$ s

Here, normal force will provide the necessary centripetal force.

$N = m\omega^2 r = 200 \times 10^{-3} \times \left(\dfrac{2\pi}{40}\right)^2 \times 0.2 = 9.859 \times 10^{-4}\ \text{N}.$

169. (c) For statement-I

The maximum speed by which cyclist can take a turn on a circular path

$\Rightarrow v \le \sqrt{\mu rg} \le \sqrt{0.2 \times 2 \times 9.8}$

$\Rightarrow v \le \sqrt{3.92} \Rightarrow v_{\max} = 1.97\,m/s$

Speed of cyclist, $v = 7\ \text{kmh}^{-1} = 7 \times \dfrac{5}{18} = 1.94$ m/s

As, $v < v_{\max} \Rightarrow$ cyclist will not slip

The maximum safe speed on a banked frictional road

$v_{\text{allowable}} = \sqrt{rg\dfrac{(\mu + \tan\theta)}{1 - \mu\tan\theta}}$

$\Rightarrow v = \sqrt{\dfrac{2 \times 9.8(0.2 + \tan 45°)}{1 - 0.2 \times \tan 45°}} = \sqrt{\dfrac{2 \times 9.8 \times 1.2}{0.8}} = 5.42$ m/s

Speed of cyclist, $v = 18.5\ \text{kmh}^{-1} = 5.13$ m/s

As, $v < v_{\text{allowable}} \Rightarrow$ cyclist cross the curve without slipping

So, both the statements are true.

170. (5) At the highest position,

$T_{\min} = \dfrac{mv_{\min}^2}{l} - mg$

At the lowest position,

$T_{\max} = \dfrac{mv^2}{l} + mg$ (Given)

And we know,

$T_{\max} - T_{\min} = 6mg \Rightarrow 5T_{\min} - T_{\min} = 6mg$

$\therefore T_{\min} = \dfrac{3}{2}mg = \dfrac{mv_{\min}^2}{l} - mg$

$\therefore v_{\min} = \sqrt{\dfrac{5}{2}gl} = \sqrt{\dfrac{5}{2} \times 10 \times 1} = 5$ m/s.

171. (d) Using, $\mu mg = \dfrac{mv^2}{r} = mr\omega^2$

$\omega = 2\pi n = 2\pi \times 3.5 = 7\pi$ rad/sec

Radius, $r = 1.25\ \text{cm} = 1.25 \times 10^{-2}$ m

Coefficient of friction, $\mu = ?$

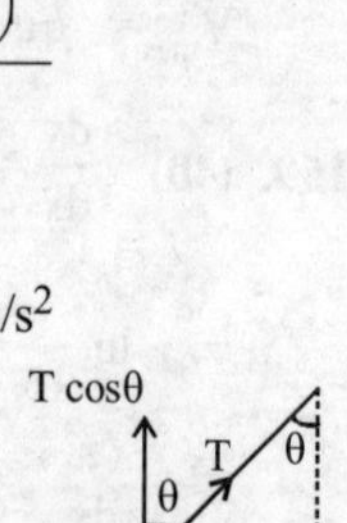

$\Rightarrow \quad \mu = \dfrac{r\omega^2}{g} = \dfrac{1.25 \times 10^{-2} \times \left(7 \times \dfrac{22}{7}\right)^2}{10}$

$= \dfrac{1.25 \times 10^{-2} \times 22^2}{10} = 0.6$

172. (d) Given, $\theta = 45°$, $r = 0.4$ m, $g = 10\ \text{m/s}^2$

$T\sin\theta = \dfrac{mv^2}{r}$ ...... (i)

$T\cos\theta = mg$ ...... (ii)

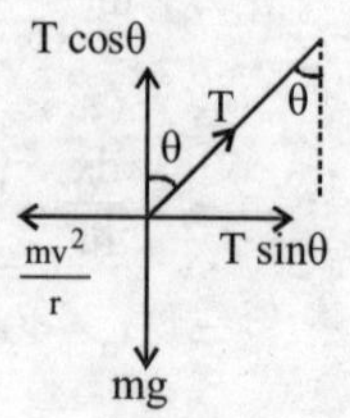

From equation (i) & (ii) we have,

$\tan\theta = \dfrac{v^2}{rg}$

$v^2 = rg \quad \because \theta = 45°$

Hence, speed of the pendulum in its circular path,

$v = \sqrt{rg} = \sqrt{0.4 \times 10} = 2$ m/s

173. (c)

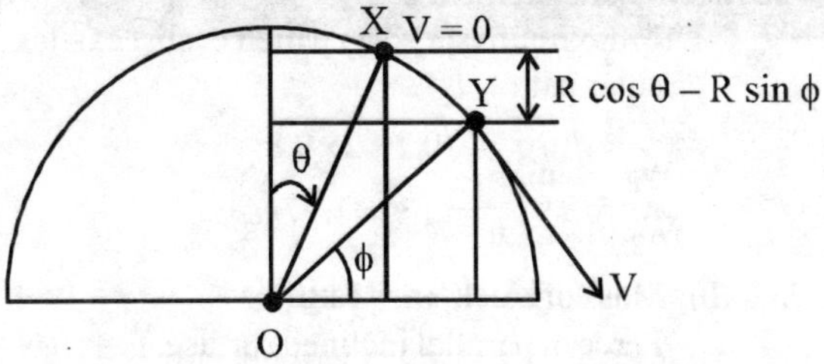

$K.E_X + P.E_X = K.E_Y + P.E_Y$

$\Rightarrow \quad mgR(\cos\theta - \sin\phi) = \frac{1}{2}mV^2$

$\Rightarrow \quad V^2 = 2gR(\cos\theta - \sin\phi)$

at Y, $\frac{mV^2}{R} = mg\sin\phi \quad [\because N_Y = 0]$

$\Rightarrow \quad 2\,mg(\cos\theta - \sin\phi) = mg\sin\phi$

$\Rightarrow \quad \sin\phi = \frac{2}{3}\cos\theta$

174. (a) Initially, $k \times 1 = m\omega^2 R$ ... (i)

Finally, $k \times 5 = m(2\omega)^2 (R+5)$ ... (ii)

From (i) & (ii), we get : $R = 15$ cm

So, original length, R = 15 cm.

175. (d) $s = t^3 + 5$

$\Rightarrow$ velocity, $v = \frac{ds}{dt} = 3t^2$

Tangential acceleration, $a_t = \frac{dv}{dt} = 6t$

Radial acceleration, $a_c = \frac{v^2}{R} = \frac{9t^4}{R}$

At $t = 2s$, $a_t = 6 \times 2 = 12$ m/s^2

$a_c = \frac{9 \times 16}{20} = 7.2$ m/s^2

$\therefore$ Resultant acceleration $= \sqrt{a_t^2 + a_c^2}$

$= \sqrt{(12)^2 + (7.2)^2} = \sqrt{144 + 51.84} = \sqrt{195.84} = 14$ m/s^2

176. (c) Clearly $\vec{a} = a_c\cos\theta(-\hat{i}) + a_c\sin\theta(-\hat{j})$

$= \frac{-v^2}{R}\cos\theta\,\hat{i} - \frac{v^2}{R}\sin\theta\,\hat{j}$

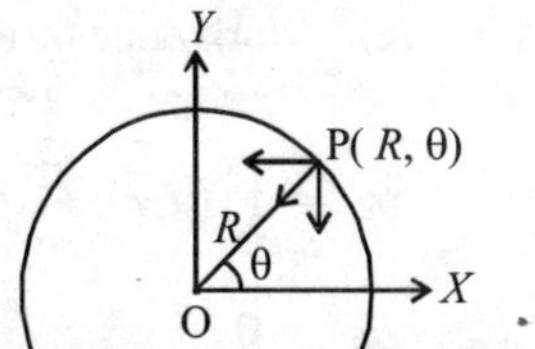

177. (c) Let m is the mass of each particle and ω is the angular speed of the annular ring.

$a_1 = \frac{v_1^2}{R_1} = \frac{\omega^2 R_1^2}{R_1} = \omega^2 R_1$

$a_2 = \frac{v_2^2}{R_2} = \omega^2 R_2$

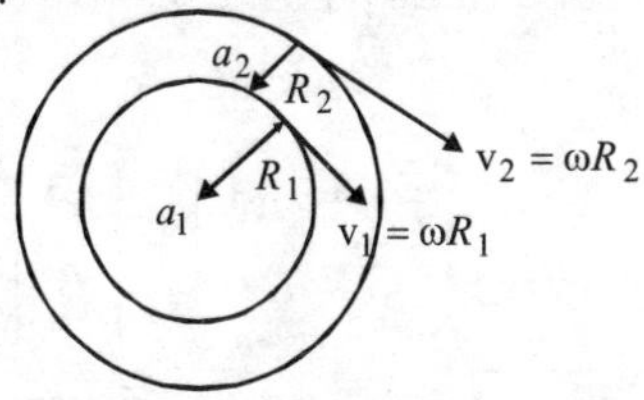

$\frac{F_1}{F_2} = \frac{ma_1}{ma_2} = \frac{mR_1\omega^2}{mR_2\omega^2} = \frac{R_1}{R_2}$

NOTE :

The force experienced by any particle is only along radial direction.

Force experienced by the particle, $F = m\omega^2 R$

$\therefore \frac{F_1}{F_2} = \frac{R_1}{R_2}$

178. (b) Only option (b) is false since acceleration vector is always radial (i.e. towards the center) for uniform circular motion.

5 Work, Energy and Power

1. **(c)** Work done by a variable force is $W = \int_0^x F dx$

$\Rightarrow 5 = \int_0^x (\alpha + \beta x^2) dx \Rightarrow 5 = \alpha x + \left.\frac{\beta x^3}{3}\right|_0^1 \Rightarrow 5 = \alpha + \frac{\beta}{3}$

$5 = 1 + \frac{\beta}{3} \quad [\because \alpha = 1\text{ N}] \Rightarrow 4 = \frac{\beta}{3} \Rightarrow \beta = 12\text{ N/m}^2$

2. **(152)** The expression to calcualte work done is

$$W = \int_{x_i}^{x_F} F dx = \int_0^4 x^2(10-x)dx + \int_0^2 y^2 dy$$

$$= \left[\frac{10x^3}{3} - \frac{x^4}{4}\right]_0^4 + \left[\frac{y^3}{3}\right]_0^2 = \frac{640}{3} - 64 + \frac{8}{3} = 152 \textit{ joule}$$

3. **(b)** Given, Force, $\vec{F} = (2\hat{i} + b\hat{j} + \hat{k})$,

Displacement, $\vec{S} = (\hat{i} - 2\hat{j} - \hat{k})$

$\therefore$ Work done, $W = \vec{F}\cdot\vec{S} = (2\hat{i} + b\hat{j} + \hat{k})\cdot(\hat{i} - 2\hat{j} - \hat{k})$

$= 2 - 2b - 1 = 0 \quad \therefore b = \frac{1}{2}$

4. **(c)** Given, $v = \alpha\sqrt{x}$
at $x = 0$; $v = 0$

at $x = d$; $v = \alpha\sqrt{d}$
Using work energy theorem
Work done = Change in kinetic energy

$= \frac{1}{2}m(\alpha\sqrt{d})^2 - \frac{1}{2}m(0)^2$

$= \frac{m\alpha^2 d}{2}$

5. **(58)** Given, Force, $F = (3x^2 + 2x - 5)$ N

Work done $W = \int_{x_1}^{x_2} F dx$

$W = \int_2^4 (3x^2 + 2x - 5) dx$

$\Rightarrow W = \left[x^3 + x^2 - 5x\right]_2^4 = [(64 + 16 - 20) - (8 + 4 - 10)]$

$\Rightarrow W = [60 - 2]\text{ J} = 58\text{ J}$

6. **(a)** As, force of gravity is a conservative force. So, work done by it depends on initial and final positions, and independent of path followed by the body.

$\therefore \frac{W_1}{W_2} = \frac{-mgh}{-mgh} = 1:1$

7. **(b)** Mass of black, m = 1 kg
Force of parallel inclined surface, F = 10 N
Work done against frictional force
$= \mu_s N \times 10 = \mu_s Mg \times 10 = 0.1 \times 5 \times 10 = 5\text{ J}$

8. **(a)** Using $v^2 - u^2 = 2as$
$\Rightarrow u^2 = 2 \times \mu g s\ [\because v = 0\ \&\ a = \mu g]$
$\Rightarrow u^2 = 2 \times (.4) \times 10 \times 10 \Rightarrow u^2 = 80$

$\therefore W_f = \Delta k = 0 - \frac{1}{2} \times 100 \times 80$

$\therefore W_f = -4000\text{ J}$
$\therefore$ WD against friction $= -W_f = 4000\text{ J}$

9. **(30)** Work done $W = \int_{x_1}^{x_2} F dx$

$= \int_2^4 5x dx = 5\left[\frac{x^2}{2}\right]_2^4 = \frac{5}{2}[16 - 4] = 30\text{J}$

10. **(784)** Given,
Mass of bus, m = 500 kg.
Distance Travelled, x = 4 km
Coefficient of friction $\mu = 0.04$
Work done by engine = –Work done by friction
$= -[-\mu mgx]$
$= 0.04 \times 500 \times 9.8 \times 4 \times 10^3 = 784\text{ kJ}$

11. **(40)** $W = \vec{F}.(\vec{r}_f - \vec{r}_1)$

$= (5\hat{i} + 2\hat{j} + 7\hat{k}).((5\hat{i} - 2\hat{j} + \hat{k}) - (2\hat{i} + 3\hat{j} - 4\hat{k}))$

W = 40 J

12. **(132)** We know that

$W = \int_{y_1}^{y_2} F dy = \int_2^5 (5 + 3y^2) dy$

$= [5y + y^3]_2^5 = [(25 + 125) - (10 + 8)] = 150 - 18 = 132\text{ J}$

13. **(a)** Work done by a force is positive if displacement is in direction of force and work done by a force is negative if displacement is in direction opposite to that of force.

14. **(b)** By work-energy thorem
$W = \Delta K$

$\Rightarrow W = \frac{1}{2}m(v_f^2 - v_i^2) \Rightarrow W = \frac{1}{2} \times 0.5 \times (16^2 - 4^2)$

$\Rightarrow W = \frac{1}{4} \times 240 \Rightarrow W = 60\text{ J}$

15. **(d)** By work – energy theorem

$$W = \Delta K = \frac{1}{2}mv_f^2 - \frac{1}{2}mv_i^2 = \frac{1}{2}\times 0.5\times\left(b^2.4^5\right)$$

$$= \frac{1}{2}\times\frac{1}{2}\times\frac{1}{4^2}\times 4^5 = 16\text{ J}$$

16. **(2)** Work done by A = Work done by B

$F_A \text{ d}\cos 45° = F_B \text{ d}\cos 60°$

$$\Rightarrow F_A\times\frac{1}{\sqrt{2}} = F_B\times\frac{1}{2} \Rightarrow \frac{F_A}{F_B} = \frac{\sqrt{2}}{2} = \frac{1}{\sqrt{2}} \Rightarrow x = 2$$

17. **(450)** Given,

Force, $F = (5y+20)\hat{j}$

Work done, $W = \int_{y_1}^{y_2} F.dy$

$$\Rightarrow W = \int_0^{10}(5y+20)dy = \left[\frac{5y^2}{2}+20y\right]_0^{10}$$

$$= \frac{5}{2}\times 100 + 20\times 10 = 450\text{J}$$

18. **(b)** From work-energy theorem,

$W_{Porter} + W_{mg} = \Delta K.E. = 0$ ($\because$ velocity constant)

or, $W_{Porter} = -W_{mg} = -mgh$

$$\therefore \quad W_{Porter} = -80\times 9.8\times\frac{80}{100} = -627.2\text{J}$$

19. **(d)** The given situation can be drawn graphically as shown in figure.

Work done = Area under *F-x* graph

= Area of rectangle *ABCD* + Area of trapezium *BCFE*

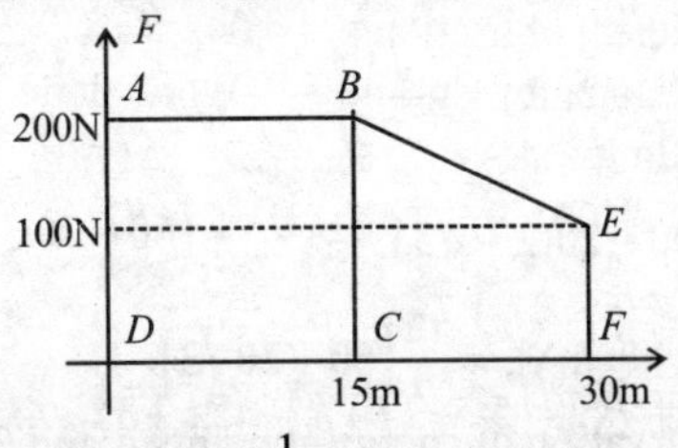

$$W = (200\times 15) + \frac{1}{2}(100+200)\times 15 = 3000 + 2250$$

$$\Rightarrow W = 5250\text{ J}$$

20. **(3)** If $AC = l$ then according to question, $BC = 2l$ and $AB = 3l$.

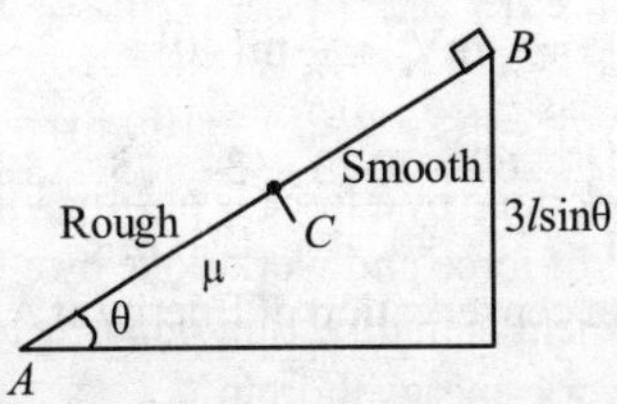

Here, work done by all the forces is zero.

$W_{friction} + W_{mg} = 0$

$mg(3l)\sin\theta - \mu mg\cos\theta(l) = 0$

$\Rightarrow \mu mg\cos\theta l = 3mgl\sin\theta \Rightarrow \mu = 3\tan\theta = k\tan\theta$

$\therefore k = 3$

21. **(c)** Work done, $W = \int \vec{F}\cdot\vec{ds} = \left(-x\hat{i}+y\hat{j}\right)\cdot\left(dx\hat{i}+dy\hat{j}\right)$

$$\Rightarrow W = -\int_1^0 xdx + \int_0^1 ydy = \left(0+\frac{1}{2}\right)+\frac{1}{2} = 1J$$

22. **(d)** Here, $N - mg = ma = \frac{mg}{2} \Rightarrow N = \frac{3mg}{2}$

N = normal reaction

Now, work done by normal reaction 'N' on block in time t, $W = \vec{N}.\vec{S} = \left(\frac{3mg}{2}\right)\left(\frac{1}{2}\frac{g}{2}t^2\right)$

or, $W = \frac{3mg^2t^2}{8}$

23. **(b)** Mass of hanging part of the chain

$$m' = \frac{4}{2}\times(0.6)\text{kg} = 1.2\text{ kg}$$

Weight of hanging part of the chain = 1.2 × 10 = 12 N

C.M. of hanging part = 0.3 m below the table

Workdone in putting the entire chain on the table = 12 × 0.30 = 3.6 J.

24. **(b)** Spring constant, $k = 5\times 10^3$ N/m

Let x_1 and x_2 be the initial and final stretched position of the spring, then

Work done, $W = \frac{1}{2}k\left(x_2^2 - x_1^2\right)$

$$= \frac{1}{2}\times 5\times 10^3\left[(0.1)^2 - (0.05)^2\right]$$

$$= \frac{5000}{2}\times 0.15\times 0.05 = 18.75\text{ Nm}$$

25. **(c)** Potential energy is defined only for conservative forces.

While friction forces are non-conservative.

26. **(d)** The acceleration of the block is

$$a = \frac{F_r}{m} = \frac{-kx}{1} = -10x$$

$$\Rightarrow \quad v\frac{dv}{dx} = -10x$$

$$\Rightarrow \quad \int_{10}^{v} vdv = -10\int_{0.1}^{1.9} xdx$$

$$\Rightarrow \quad \frac{v^2-100}{2} = -10\left(\frac{1.9^2-0.1^2}{2}\right)$$

$\therefore \quad v = 8$ m/s

27. **(d)** At height x from the earth surface

$V^2 = 0 + 2g(S-x)$

$V^2 = 2g(S-x)$ ... (i)

Potential energy, U = mgx

Here, K = 3U

$$\Rightarrow \quad \frac{1}{2}mv^2 = 3mgx$$

$\therefore \quad v^2 = 6gx$ ....(ii)

From eq (i) and (ii), we get

$$x = \frac{S}{4}$$

$$\therefore\ V = \sqrt{2g \times \frac{3S}{4}} = \sqrt{\frac{3gS}{2}}$$

28. **(d)** In vertical circular motion,

$$v_B^2 = v_T^2 + 2g(2R)$$

$$\therefore\ \frac{(KE)_B}{(KE)_T} = \frac{v_T^2 + 4gR}{v_T^2} = \frac{n^2 gR + 4gR}{n^2 gR} = \frac{n^2+4}{n^2}$$

29. **(d)**

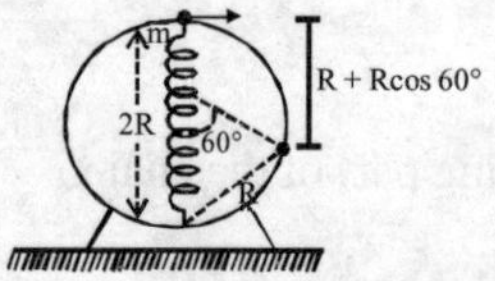

From work energy theorem

$$mg(R + R\cos 60°) + \frac{1}{2}k\left(R^2 - 0^2\right) = \frac{1}{2}mv^2$$

$$mg\frac{3R}{2} + \frac{kR^2}{2} = \frac{1}{2}mv^2\ ;\quad V = \sqrt{3gR + \frac{kR^2}{m}}$$

30. **(b)** Since central forces are conservative in nature, work done is independent of path. Friction is nonconservative. Both statement are correct but Reason is not the correct explanation of Assertion.

31. **(36)**

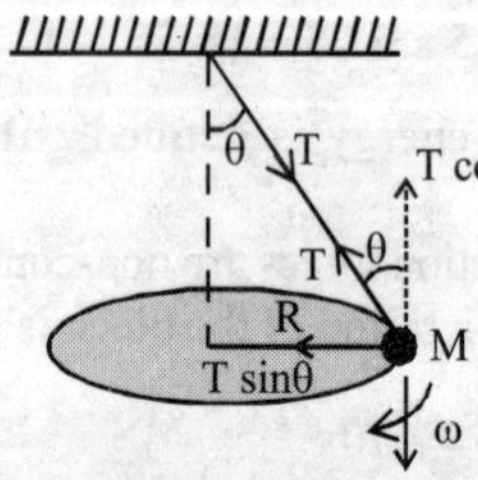

$$T \sin\theta = M\omega^2 R$$

$$\Rightarrow T = M\omega^2\left(\frac{R}{\sin\theta}\right) = M\omega^2 L = M\left(\frac{3}{\pi}\times 2\pi\right)^2 L$$

$$\therefore\ T = 36\,ML.$$

32. **(b)** When a projectile is launched at an intital velocity V_0 at an angle θ, its velocity components are

$V_{0x} = V_0 \cos\theta$

$V_{0y} = V_0 \text{Sin}\theta$

At the highest point, the vertical velocity component becomes zero.

Initial K.E, $(KE)_1 = \frac{1}{2}mu^2$

Speed at heighest point

$$V = u\cos 60° = \frac{u}{2}$$

$$\therefore\ (KE)_2 = \frac{1}{2}m\left(\frac{u}{2}\right)^2 = \frac{1}{4}\times\frac{1}{2}mu^2$$

$$= \frac{(KE)_1}{4} = \frac{KE}{4}$$

33. **(b)**

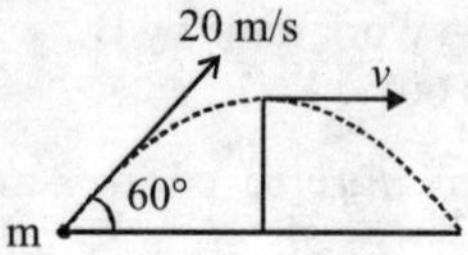

Given,

Initial velocity, u = 20m/s,

v = 20 cos 60° = 10 m/s

Initially, $K_1 = \frac{1}{2}mu^2$

At highest point,

$$K_2 = \frac{1}{2}mv^2 = \frac{1}{2}m\left(u\cos 60°\right)^2 = \frac{1}{8}mu^2$$

$$\Delta K = K_2 - K_1 = \frac{3}{8}mu^2 = \frac{3}{8}\times 0.1\times(20)^2 = 15J$$

34. **(c)**

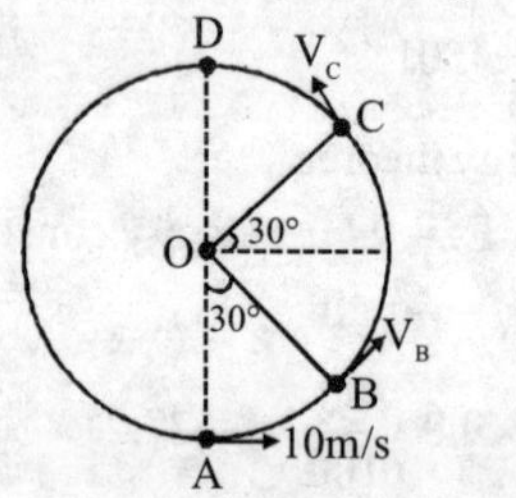

From conservation of energy at points A and B,

$$\frac{1}{2}m\times 100 + 0 = \frac{1}{2}mV_B^2 + mg\left(R - \frac{R\sqrt{3}}{2}\right)$$

$$\Rightarrow 100 = V_B^2 + 2gR\left(1 - \frac{\sqrt{3}}{2}\right)$$

$$\Rightarrow V_B = 100 - 20(2-\sqrt{3})\ \Rightarrow V_B^2 = \left(60 + 20\sqrt{3}\right)$$

$$\therefore\ (K.E.)_B = \frac{1}{2}mV_B^2 = \frac{m}{2}\left(60 + 20\sqrt{3}\right)$$

From conservation of energy at points A and C,

$$\Rightarrow \frac{1}{2}m(100) = \frac{1}{2}mV_C^2 + mg\left(\frac{3R}{2}\right)\ \Rightarrow 100 = V_C^2 + 60$$

$$\therefore\ V_C^2 = 40$$

$$\therefore\ (K.E.)_C = \frac{1}{2}mV_C^2 = \frac{1}{2}m(40)$$

$$\therefore\ \frac{(K.E)_B}{(K.E)_C} = \frac{60 + 20\sqrt{3}}{40} = \frac{3}{2} + \frac{\sqrt{3}}{2} = \frac{3+\sqrt{3}}{2}$$

35. **(a)** From conservation of Energy at A and B

$$\frac{1}{2}mv_0^2 = KE_B + mgh$$

$$\Rightarrow\ \frac{1}{2}m(5g\ell) = KE_B + mg\frac{\ell}{2}$$

$$\Rightarrow\ \frac{5mg\ell}{2} - \frac{mg\ell}{2} = KE_B \Rightarrow KE_B = 2mg\,\ell$$

Using conservation of energy again at C and D

$$KE_C = \frac{1}{2}mv_D^2 + mg\frac{\ell}{2}$$

$$\Rightarrow \quad KE_C = \frac{1}{2}mg\ell + mg\frac{\ell}{2} = mg\ell$$

$$\therefore \frac{KE_B}{KE_C} = 2$$

36. **(a)** Relation between kinetic energy and linear momentum is

$P = \sqrt{2m\,KE}$

$\therefore$ $P \propto \sqrt{m}$ ($\because$ KE is same)

$\therefore P_A : P_B : P_C = \sqrt{400} : \sqrt{1200} : \sqrt{1600} = 1 : \sqrt{3} : 2$

37. **(a)** From the conservation of momentum

$|P_A| = |P_B|$

$\therefore \quad m_A v_A = m_B v_B$

$$\therefore \quad \frac{m_B}{m_A} = \frac{v_A}{v_B} \quad ...(i)$$

$$\frac{K_A}{K_B} = \frac{\frac{1}{2}m_A v_A^2}{\frac{1}{2}m_B v_B^2} = \frac{m_A v_A^2}{m_B v_B^2} = \frac{v_A}{v_B} \text{ (using (i))}$$

$$\Rightarrow \frac{K_B}{K_A} = \frac{v_B}{v_A}$$

38. **(b)** Using work-energy theorem,

$W_g + W_{Fr} + W_s = \Delta KE$

$mgh \sin 30° + \mu mgd - \frac{1}{2}Kx^2 = 0$

$5 \times 10 \times 10 \times \frac{1}{2} - 0.5 \times 5 \times 10 \times 2 - \frac{1}{2}Kx^2 = 0$

$\Rightarrow 250 - 50 = \frac{1}{2} \times 100\,x^2 \Rightarrow x = 2$

39. **(d)** Kinetic energy is $KE = \frac{1}{2}mv^2$

Initial kinetic energy $KE_i = \frac{1}{2}(0.05)(100)^2$

$K_f = \frac{1}{2}(0.05)(40)^2$

Percentage loss of kinetic energy is

$$\% \Delta KE = \frac{|KE_f - KE_i|}{KE_i} \times 100$$

$$= \frac{\left|\frac{1}{2}0.05\,(40)^2 - \frac{1}{2}0.05\,(100)^2\right|}{\frac{1}{2}0.05\,(100)^2} \times 100$$

$$= \frac{|1600 - 100 \times 100|}{100} = 84\%$$

40. **(c)** Kinetic energy is

$$K = \frac{p^2}{2m} \Rightarrow K \propto \frac{1}{\sqrt{m}}$$

Particle of lowest mass $\frac{m}{2}$ will have maximum kinetic energy.

41. **(a)** Kinetic energy is given as, $KE = \frac{P^2}{2m}$

Given $\therefore KE_f = 36\,KE_i$

$$\frac{P_f^2}{2m} = \frac{P_i^2}{2m}(36) \Rightarrow P_f = 6P_i$$

$$\text{Percentage increase in momentum} = \frac{P_f - P_i}{P_i} \times 100\%$$

$$= \frac{6P_i - P_i}{P_i} \times 100\% = 500\%$$

42. **(b)** From work energy theorem

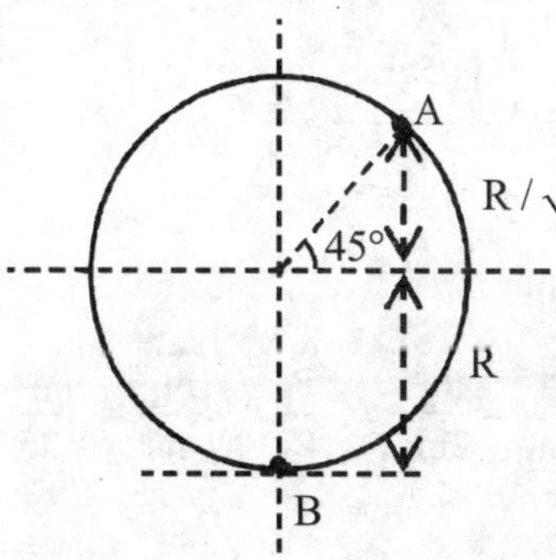

$W_{mg} = K_B - K_A$

$$\Rightarrow mg \times \left(\frac{R}{\sqrt{2}} + R\right) = \frac{1}{2}mv_B^2 - 0 \quad (\because v_A = 0)$$

$$\Rightarrow mgR\frac{(\sqrt{2}+1)}{\sqrt{2}} = \frac{1}{2}mv_B^2 \Rightarrow v_B = \sqrt{gR\frac{2(\sqrt{2}+1)}{\sqrt{2}}}$$

$$\Rightarrow v_B = \sqrt{\frac{10 \times 14 \times 2(2.4)}{1.4}}$$

$\therefore v_B = 21.9$ m/s

43. **(b)** Using conservation of linear momentum:

$mu = (M + m)v$

$$v = \frac{mu}{M+m} = \frac{0.01 \times 200}{1.01} = 1.98\text{m/s.}$$

Using law of conservation of energy

$$h = \frac{v^2}{2g} = \frac{(1.98)^2}{2 \times 9.8} = 0.20\text{m}$$

44. **(d)** By conservation of linear momentum

$|\vec{P}_1| = |\vec{P}_2|$

Kinetic energy, $KE = \frac{P^2}{2M}$

$$\therefore \frac{KE_1}{KE_2} = \frac{P^2/2M_1}{P^2/2M_2} = \frac{M_2}{M_1} \ (\because \text{P is same})$$

45. **(b)** From the conservation of energy

$$KE_A + U_A = KE_B + U_B$$

$$0 + mg(1) = \frac{1}{2}mv^2 + mg \times 0.5$$

$$v = \sqrt{g} = \sqrt{10} \text{ m/s}$$

46. **(d)** Potential Energy, $U = 2x^2 + 3y^3 + 2z$

The x-component of force is

$$F_x = -\frac{\partial U}{\partial x} = -4x$$

At $x = 1$, $|F_x| = |-4 \times 1| = 4N$

47. **(b)** Applying energy conservation between points A & B

$$\frac{1}{2}mV_L^2 = \frac{1}{2}mV_H^2 + mg(2L)$$

For complete circle, $V_L = \sqrt{5gL}$

$\therefore V_H = \sqrt{gL}$

$$\frac{(K.E)_A}{(K.E)_B} = \frac{\frac{1}{2}mV_L^2}{\frac{1}{2}mV_H^2} = \frac{\frac{1}{m}\left(\sqrt{5gL}\right)^2}{\frac{1}{2}m\left(\sqrt{gL}\right)^2} = \frac{5}{1}$$

48. **(c)** Using, $P = \sqrt{2mE}$

Here $E_1 = E_2 \Rightarrow \frac{P_1^2}{2m_1} = \frac{P_2^2}{2m_2} \Rightarrow \frac{P_1}{P_2} = \sqrt{\frac{m_1}{m_2}} = \sqrt{\frac{4}{25}} = \frac{2}{5}$

49. **(b)** Here,

loss in kinetic energy = gain in potential energy

$$\Rightarrow \frac{1}{2}mv^2 = mg\ell(1 - \cos\theta)$$

$$\Rightarrow \frac{v^2}{\ell} = 2g(1 - \cos\theta)$$

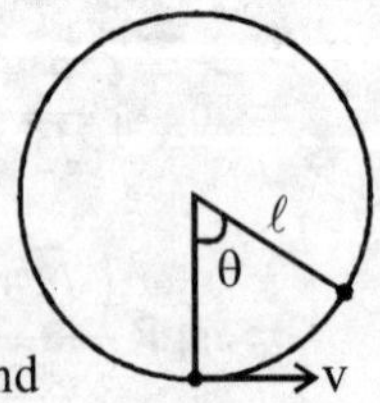

Acceleration at lowest point $= \frac{v^2}{\ell}$ and

at extreme point $= g \sin\theta$

Hence, $\frac{v^2}{\ell} = g\sin\theta \therefore \sin\theta = 2(1 - \cos\theta)$

$$\Rightarrow \tan\frac{\theta}{2} = \frac{1}{2} \Rightarrow \theta = 2\tan^{-1}\left(\frac{1}{2}\right)$$

50. **(d)** $K.E. = \frac{P^2}{2m}$

$$\therefore \frac{K_1}{K_2} = \frac{p_1^2}{2m_1} \times \frac{2m_2}{p_2^2} = \frac{m_2}{m_1} = \frac{16}{9} \quad (\because P_1 = P_2)$$

$$\therefore \frac{m_1}{m_2} = \frac{9}{16}$$

51. **(a)** Using, $v^2 - u^2 = 2aS$

$$\Rightarrow 0 - u^2 = 2(-a)S \Rightarrow S = \frac{u^2}{2a} = \frac{mu^2}{2F} = \frac{K \cdot E}{F}$$

As truck and car have same kinetic energy and retarding force. So, both will come to rest in equal distance.

Statement 1 is correct

As direction of motion is changing, velocity is also changing.

So, acceleration $a \neq 0$

52. **(125)** Given, $p_f = p_i + 0.5\,p_i = 1.5\,p_i$

Initial kinetic energy $= \frac{p_i^2}{2m}$

Final kinetic energy $= \frac{p_f^2}{2m} = \frac{(1.5p_i)^2}{2m} = \frac{2.25p_i^2}{2m}$

$$\% \text{ increase in KE} = \frac{2.25\frac{p_i^2}{2m} - \frac{p_i^2}{2m}}{\frac{p_i^2}{2m}} \times 100 = 125\%$$

53. **(30)** Given,

Mass of body M = 5 kg

Initial momentum $P_i = 10$ kg m/s

Impulse $= F\Delta t = \Delta P = P_f - P_i$

$\Rightarrow 2 \times 5 = P_f - 10$

final momentum $P_f = 20$ kg m/s

Increase in KE $= KE_f - KE_i \left(\because KE = \frac{P^2}{2m}\right)$

$$= \frac{P_f^2}{2m} - \frac{P_i^2}{2m} = \frac{400}{2 \times 5} - \frac{100}{2 \times 5} = 40 - 10 = 30J$$

54. **(b)**

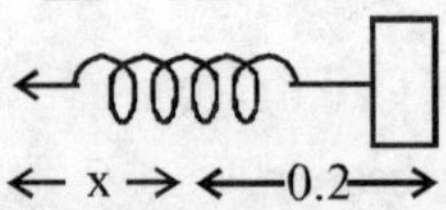

Given,

Mass of block, m = 100 g = 0.1 kg

Spring constant, k = 7.5 N/m

Let x be the extension in length of spring.

Radius of circle, r = 0.2 + x

Tension, $T = kx = m\omega^2 r$

$\Rightarrow 7.5x = 0.1\,(25)\,(0.2 + x)$

$$\Rightarrow \frac{15}{2}x = \frac{5}{2}\left(x + \frac{1}{5}\right) \Rightarrow 3x = \frac{5x + 1}{5}$$

$$\Rightarrow x = \frac{1}{10} = 0.1m$$

$\therefore$ Tension in spring $= kx = 7.5 \times 0.1 = 0.75N$

55. **(2)** Given,

Mass of particle, m = 10 g

Retardation, a = 2x

Loss of K.E = work done against retarding force.

$$\therefore \Delta KE = \int_0^x madx = \int_0^x m2xdx = mx^2 = \frac{10}{1000} \times x^2 J$$

$$= \left(\frac{10}{x}\right)^{-2} J$$

$\therefore n = 2$

56. **(375)** Let V_1 be the velocity before hitting the floor and V_2 be the velocity just after hitting the floor.

$$\frac{V_1}{V_2} = 4 \Rightarrow V_1 = 4V_2$$

$KE_{before} = \frac{1}{2}mV_1^2$

$KE_{after} = \frac{1}{2}mV_2^2 = \frac{1}{2}\frac{m.V_1^2}{16}$

$\Delta KE = KE_{after} - KE_{before} = \frac{1}{2}\frac{mV_1^2}{16} - \frac{1}{2}mV_1^2$

$\Delta KE = \frac{1}{2}mV_1^2\left(\frac{1}{16} - 1\right) = \frac{-15}{32}mV_1^2$

The percentage loss in kinetic energy is given by

$\frac{-(\Delta KE)}{KE_{before}} \times 100\% = \frac{15}{16} \times 100 = \frac{375}{4}\%$

The value of x is 375.

57. **(c)**

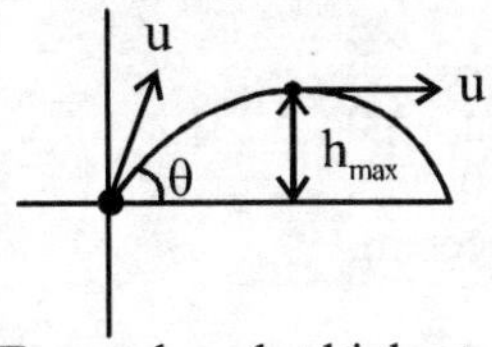

As P.E. = mgh and at highest point, 'h' is maximum, So, P.E. is maximum.

58. **(50)** We have $M.E_i = M.E_f$

$\Rightarrow K_i + P_i = M.E_f \Rightarrow 0 + \frac{1}{2}kA^2 = 0.25$

$\Rightarrow \frac{1}{2} \times k \times 0.1^2 = 0.25 \Rightarrow k \times 0.01 = 0.5$

$\Rightarrow k = \frac{0.50}{0.01} = 50\ N/m$

59. **(7)** $V^2 = u^2 + 2as$

$= 2^2 + 2 \times 2 \times 6 = 28$

So, $K.E = \frac{1}{2}mV^2 = \frac{1}{2} \times 500 \times 28 = 7000J = 7\ kJ$

60. **(300)** Given, mass of the body, m = 0.4 kg

Displacement in n^{th} sec.

$D_n = u + \frac{1}{2}g(2n - 1)$

$\Rightarrow D_8 = 0 + \frac{1}{2} \times 10 \times (2 \times 8 - 1) \Rightarrow D_8 = 5 \times 15$

$\Delta U = mg D_8 = 0.4 \times 10 \times 5 \times 15 \Rightarrow \Delta U = 20 \times 15 = 300J$

61. **(2)** From the figure, draw the free body diagram

$F\cos\theta = ma$ (Here, action of force F = 2N)

$2\cos(kx) = \frac{mvdv}{dx}$

$m\int_0^v vdv = 2\int_0^x \cos(kx)dx$

$\frac{mv^2}{2} = \frac{2}{k}\sin kx$

$K.E. = \frac{2}{k}\sin\theta$ $\left(\because K.E. = \frac{1}{2}mv^2\right)$

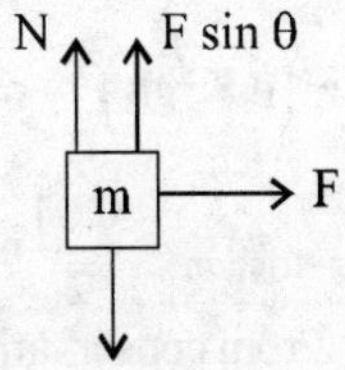

Compare with $E = \frac{n}{k}\sin\theta$

Hence, n = 2

62. **(40)** We have, $\frac{1}{2}mv_f^2 = 10000$

$\Rightarrow \frac{1}{2} \times 2 \times v_f^2 = 10000 \Rightarrow v_f^2 = 10000$

$\Rightarrow v_f = 100\ m/s$

So, $a = \frac{v_f - v_i}{t} = \frac{100 - 0}{5} = 20\ m/s^2$

$F = ma = 20 \times 2 = 40\ N$

63. **(c)**

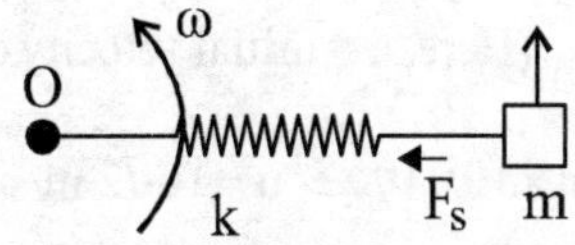

Natural length = L_0

Extension = x

$F_{spring} = F_{centripetal}$

$\Rightarrow kx = m(L_0 + x)\omega^2 \Rightarrow 12.5x = \frac{1}{5}(L_0 + x)25$

$\Rightarrow 1.5x = L_0 \Rightarrow \frac{x}{L_0} = \frac{2}{3}$

64. **(c)**

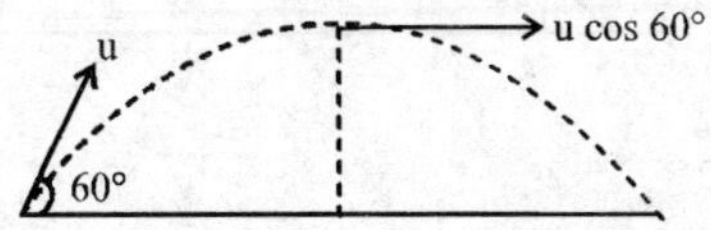

At maximum height, we only have horizontal component of velocity. So, Velocity $v = u\cos 60° = \frac{u}{2}$

$\therefore$ K.E at top most point $= \frac{1}{2}m\left(\frac{u}{2}\right)^2 = \frac{E}{4}$

65. **(c)** Momentum of a body is increased by

$P' = P + \frac{20}{100}P = 1.2\ P$

Percentage change in KE $= \left(\frac{K' - K}{K} \times 100\right)\%$

$= \left(\frac{\frac{P'^2}{2m} - \frac{P^2}{2m}}{\frac{P^2}{2m}}\right) \times 100 = [(1.2)^2 - 1] \times 100 = 44\%$

66. **(24)** Using work-energy theorem, $W_s = (K_f - K_i)$

$\Rightarrow -\frac{1}{2}Kx^2 = \frac{1}{2}m\left(\frac{v}{2}\right)^2 - \frac{1}{2}mv^2 = \frac{E}{4} - E$

$\Rightarrow \frac{1}{2}Kx^2 = \frac{3E}{4} \Rightarrow K = \frac{3E}{2x^2}$

$\Rightarrow K = \frac{3E}{2 \times \left(\frac{1}{4}\right)} = 24E = nE$

$\therefore$ n = 24

67. **(a)** Using $mv = \sqrt{2mk} \Rightarrow v = \frac{1}{m}\sqrt{2mk}$

So, $u = \frac{1}{0.2}\sqrt{2 \times 0.2 \times 90} = 30$ m/s

$v = \frac{1}{0.2}\sqrt{2 \times 0.2 \times 40} = 20$ m/s

$a = \frac{20-30}{1} = -10$ m/s^2

So, $s = -\frac{u^2}{2a} = 45$ m

68. **(2)** Using energy conservation for plane AB

$\frac{1}{2}mu^2 = mgh$ (Here, u = initial velocity of block)

$\Rightarrow \frac{1}{2} \times m \times u^2 = m \times 10 \times 10 \Rightarrow u = 10\sqrt{2}$ m/s

At point B

Acceleration, $a = -g\sin 45° = \frac{-10}{\sqrt{2}}$ m/s^2

Using $v = u + at_1$

$\Rightarrow 0 = 10\sqrt{2} - \frac{10}{\sqrt{2}}t_1$

$\Rightarrow t_1 = 2$ sec

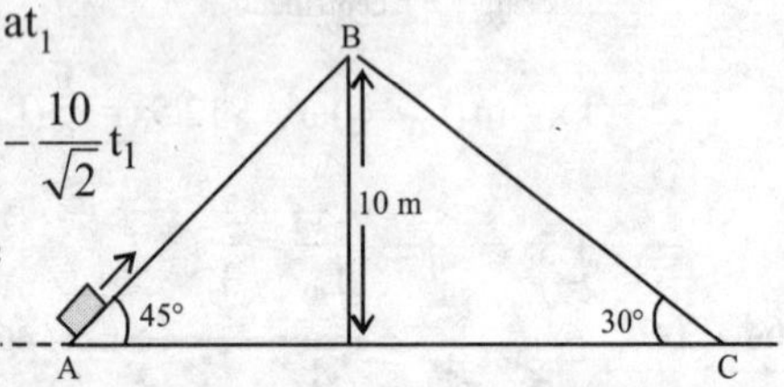

For plane BC

Using $s = ut_2 + \frac{1}{2}at_2^2$

$\Rightarrow \frac{10}{\sin 30°} = \frac{1}{2}(10\sin 30°)t_2^2 \quad \left(\because s = \frac{10}{\sin 30°}\right)$

$\Rightarrow t_2 = 2\sqrt{2}$ sec

So total time $T = t_1 + t_2 = 2\sqrt{2} + 2 = t(\sqrt{2}+1)$sec

$\therefore t = 2$

69. **(b)** Given, spring constant of spring, $K = 2\text{Nm}^{-1}$

Mass of block, $m = 250 \text{ g} = \frac{250}{1000}\text{g} = \frac{1}{4}$kg

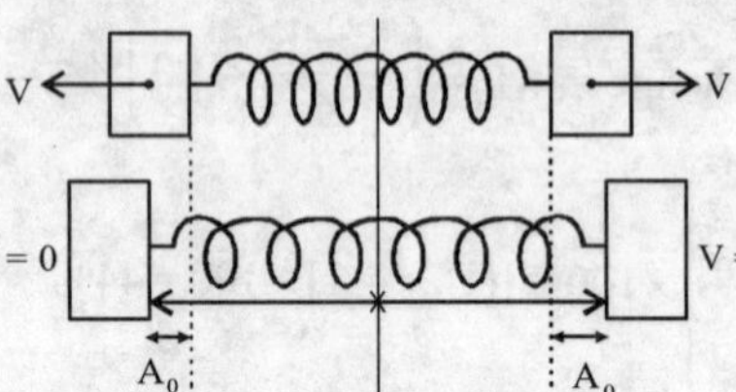

Using energy conservation

$\frac{1}{2}mv^2 \times 2 = \frac{1}{2}kx^2 \Rightarrow \frac{1}{4}v^2 = \frac{1}{2} \times 2 \times x^2$

$\therefore x = \frac{v}{2}$

70. **(b)** K.E. and momentum P relation, $K.E = \frac{P^2}{2m}$

$K_1 = \frac{P_1^2}{2(8)}; K_2 = \frac{P_2^2}{2(2)}$

Given, $K_1 = K_2$

So, $\frac{P_1^2}{16} = \frac{P_2^2}{4} \Rightarrow P_1^2 = 4P_2^2 \Rightarrow \frac{P_1}{P_2} = 2$

71. **(b)** Kinetic energy, $K.E = \frac{p^2}{2m}$

$$\frac{K.E_1}{K.E_2} = \left(\frac{P_1}{P_2}\right)^2 \times \left(\frac{m_2}{m_1}\right) = \left(\frac{1}{2}\right)^2 \times \frac{8}{5} = \frac{2}{5}$$

72. **(d)** By law of conservation of mechanical energy

$\Delta K = -\Delta U$

$\Rightarrow K_f - K_i = U_i - U_f \Rightarrow K_f = mgy - mg[y - y_0]$

$[\because K_i = 0, U_i = mgy \text{ and } U_f = mg(y - y_0)]$

$\Rightarrow K_f = mgy_0$

73. **(a)** At maximum height, $v = 0$

$\Rightarrow mv = 0 \Rightarrow P = 0$

74. **(c)** By work-energy theorem,

$$\Delta K = W_{\text{all forces}} = \int \vec{F}.d\vec{r}$$

$$= \int (4x\hat{i} + 3y^2\hat{j}).(dx\hat{i} + dy\hat{j}) = \int_1^2 4x\,dx + \int_2^3 3y^2 dy$$

$$= 4\left[\frac{x^2}{2}\right]_1^2 + 3\left[\frac{y^3}{3}\right]_2^3 = 2[2^2 - 1^2] + [3^3 - 2^3]$$

$= 6 + 19 = 25$ J.

75. **(16)** Given,

Mass of engine - wagon system, m = 40,000 kg

Velocity, $v = 72 \times 5/18 = 20$m/s

$K.E = \frac{1}{2}mv^2 = \frac{1}{2} \times (40{,}000) \times (20)^2 = 8000000$ J

As 90% of K.E of system lost in friction, only 10% is transfered to spring.

$\therefore \frac{1}{2}Kx^2 = \frac{10}{100} \times 8000000 \Rightarrow \frac{1}{2} \times K \times 1 \times 1 = 800000$

$\Rightarrow K = 16 \times 10^5$ N/m

76. **(a)** $W_{\text{air-friction}} = \Delta K + \Delta u = \frac{1}{2}mv^2 - mgh$

$= \frac{1}{2}m\left(0.8\sqrt{gh}\right)^2 - mgh$

$W_{\text{air-friction}} = \frac{0.64}{2}mgh - mgh = -0.68$ mgh

77. **(1)** From conservation of linear momentum

$P_{\text{initial}} = P_{\text{final}}$ (and $P = mv$)

$m \times 40 = \frac{m}{2} \times v + \frac{m}{2} \times 60 \Rightarrow 40 = \frac{v}{2} + 30$

$\therefore v = 20$

Initial kinetic energy, $E_1 = \frac{1}{2}m \times (40)^2 = 800m$

Final kinetic energy, E_2

$= \frac{1}{2}\frac{m}{2}.(20)^2 + \frac{1}{2}.\frac{m}{2}(60)^2 = 1000m$

$\Delta E = E_2 - E_1 = 1000m - 800m = 200m$

$\frac{\Delta E}{E_1} = \frac{200m}{800m} = \frac{1}{4} = \frac{x}{4} \Rightarrow x = 1$

78. **(40)**

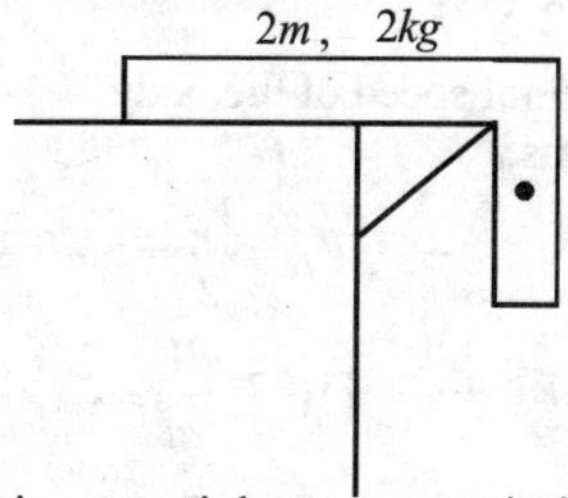

Loss in potential energy = gain in kinetic energy
Take zero potential energy at table, initial potential energy

$= -1 \times 10 \times \frac{1}{2} = -5J$

Final potential energy $= -3 \times 10 \times \frac{3}{2} = -45J$

Change in kinetic energy $= u_1 - u_2 = -5-(-45)J = 40J$

$\therefore \quad k = 40$

79. **(a)** Relation between kinetic energy and linear momentum is given as

$K.E. = K = \frac{P^2}{2m} \Rightarrow P \propto \sqrt{K}$

$\frac{P_2}{P_1} = \sqrt{\frac{K_2}{K_1}} \Rightarrow \frac{P_2}{P_1} = \sqrt{\frac{4K}{K}} \quad \Rightarrow \frac{P_2}{P_1} = 2$

$\therefore$ % change in KE

$= \frac{P_2 - P_1}{P_1} \times 100\% = \left(\frac{P_2}{P_1} - 1\right) \times 100 = (2-1) \times 100 = 100\%$

80. **(10)** By mechanical energy conservation,

$T.E_A = T.E_B$

$PE_A + KE_A = PE_B + KE_B$

$mg(10) + 0 = mg(5) + \frac{1}{2}mv^2 \Rightarrow v = \sqrt{2 \times g \times 5} = 10$ m/s

$\therefore \quad x = 10$

81. **(6)** Here kinetic energy of ball is equal to P.E. stored in spring i.e., $\frac{1}{2}mv^2 = \frac{1}{2}k(\Delta x)^2$

$\Rightarrow \quad \frac{1}{2} \times 4 \times (10)^2 = \frac{1}{2} \times 100 \times (\Delta x)^2 \Rightarrow \Delta x = 2$ m

Therefore length of the compressed spring

$x = 8 - 2 = 6$ m

82. **(1)** From $P = \sqrt{2mk} \Rightarrow P \propto \sqrt{m}$ as k is equal for two particles.

$\therefore \frac{P_1}{P_2} = \sqrt{\frac{m_1}{m_2}} = \sqrt{\frac{4}{16}} = \frac{1}{2} \qquad \therefore n = 1.$

83. **(2)** Kinetic energy of a body, $K = \frac{P^2}{2m}$

$\therefore K \propto \frac{1}{m} \quad (\because P_A = P_B)$

Now, $\frac{K_A}{K_B} = \frac{m_B}{m_A} = \frac{2}{1} = \frac{A}{1}.$

$\therefore A = 2$

84. **(150)** From work energy theorem,

$W = F \cdot s = \Delta KE = \frac{1}{2}mv^2$

Here $V^2 = 2gh$

$\therefore F \cdot s = F \times \frac{2}{10} = \frac{1}{2} \times \frac{15}{100} \times 2 \times 10 \times 20 \therefore F = 150$ N.

85. **(10)** From energy conservation, $\Delta K = -\Delta U$

$\Rightarrow KE - 0 = mg(h_1 - h_2)$

Given, $m = 1$ kg,

$h_1 - h_2 = 2 - 1 = 1m$

$\therefore KE = 1 \times 10 \times 1 = 10J$

86. **(c)** We know area under F-x graph gives the work done by the body

$\therefore \; W = \frac{1}{2} \times (3+2) \times (3-2) + 2 \times 2 = 2.5 + 4 = 6.5$ J

Using work energy theorem,

Δ K.E = work done

$\Rightarrow K_2 - 0 = 6.5$

$\therefore K_2 = 6.5$ J

87. **(c)** $l_1 + l_2 = l$ and $l_1 = nl_2$

$\therefore \quad l_1 = \frac{nl}{n+1}$ and $l_2 = \frac{l}{n+1}$

As $k \propto \frac{1}{l}$, $\therefore \frac{k_1}{k_2} = \frac{l/(n+1)}{(nl)/(n+1)} = \frac{1}{n}$

88. **(None)** Velocity of 1 kg block just before it collides with 3 kg block $= \sqrt{2gh} = \sqrt{2000}$ m/s

Using principle of conservation of linear momentum just before and just after collision, we get

$1 \times \sqrt{2000} = 4v \Rightarrow v = \frac{\sqrt{2000}}{4}$ m/s

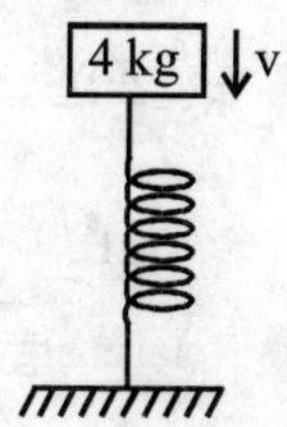

Initial compression of spring

$1.25 \times 10^6 x_0 = 30 \Rightarrow x_0 \approx 0$

using work energy theorem,

$W_g + W_{sp} = \Delta KE$

$\Rightarrow \; 40 \times x + \frac{1}{2} \times 1.25 \times 10^6 (0^2 - x^2)$

$= 0 - \frac{1}{2} \times 4 \times v^2$

solving $x \approx 2$ cm

89. (a) $W = U_f - U_i = 0 - \left(-\frac{Mg}{n} \times \frac{L}{2n}\right) = \frac{MgL}{2n^2}$.

90. (c) $mv = (m+M)V'$

or $V' = \frac{mv}{m+M} = \frac{mv}{m+4m} = \frac{v}{5}$

Using conservation of ME, we have

$$\frac{1}{2}mv^2 = \frac{1}{2}(m+4m)\left(\frac{v}{5}\right)^2 + mgh \text{ or } h = \frac{2}{5}\frac{v^2}{g}$$

91. (d) Work done $= \vec{F}.\vec{d} = (3\hat{i} - 12\hat{j}).(4\hat{i}) = 12J$

From work energy theorem,

$W_{net} = \Delta K.E. = K_f - K_i \Rightarrow 12 = K_f - 3$

$\therefore K_f = 15J$

92. (d) When force 'F' is applied, initially $F > F_s$.

As F_s will be increase, suppose after x distance $F = F_s$ and there is equilibrium. At this moment block has maximum velocity.

$F = kx \Rightarrow x = \frac{F}{k}$

From work-energy theorem,

$W_F + W_{sp} = \Delta KE$

$F(x) - \frac{1}{2}kx^2 = \frac{1}{2}mv^2_{max} - 0$

$\Rightarrow F\left(\frac{F}{k}\right) - \frac{1}{2}k\left(\frac{F}{k}\right)^2 = \frac{1}{2}mv^2_{max}$

$\Rightarrow \frac{1}{2}\frac{F^2}{K} = \frac{1}{2}mv^2_{max}$ or, $v_{max} = \frac{F}{\sqrt{mk}}$

93. (d) Position, $x = 3t^2 + 5$

$\therefore$ Velocity, $v = \frac{dx}{dt} \Rightarrow v = \frac{d(3t^2+5)}{dt} = 6t$

At $t = 0, v = 0$

And, at $t = 5$ sec, $v = 30$ m/s

According to work-energy theorem, $W = \Delta KE$

or, $W = \frac{1}{2}mv^2 - 0 = \frac{1}{2}(2)(30)^2 = 900J$

94. (c) $F = -\frac{\partial U}{\partial r} = \frac{K}{r^3}$

Since particle is moving in circular path

$F = \frac{mv^2}{r} = \frac{K}{r^3} \Rightarrow mv^2 = \frac{K}{r^2}$

$\therefore K.E. = \frac{1}{2}mv^2 = \frac{K}{2r^2}$

Total energy = P.E. + K.E.

$= -\frac{K}{2r^2} + \frac{K}{2r^2} = \text{Zero}$ $\quad (\because P.E. = -\frac{K}{2r^2}$ given)

95. (b) As the particles moving in circular orbits, So

$\frac{mv^2}{r} = -\frac{16}{r} - r^3$

Kinetic energy, $K = \frac{1}{2}mv^2 = -\frac{1}{2}[16 + r^4]$

For first particle, $r = 1$, $K_1 = -\frac{1}{2}m(16+1)$

Similarly, for second particle, $r = 4$, $K_2 = -\frac{1}{2}m(16+256)$

$$\therefore \frac{K_1}{K_2} = \frac{-\left(\frac{16+1}{2}\right)}{-\left(\frac{16+256}{2}\right)} = \frac{17}{272} \simeq 6 \times 10^{-2}$$

96. (a) Let V_f is the final speed of the body.

From questions,

$\frac{1}{2}mV_f^2 = \frac{1}{8}mV_0^2 \Rightarrow V_f = \frac{V_0}{2} = 5\,m/s$

$F = m\left(\frac{dV}{dt}\right) = -kV^2 \quad \therefore \quad (10^{-2})\frac{dV}{dt} = -kV^2$

$\int_{10}^{5}\frac{dV}{V^2} = -100k\int_0^{10} dt$

$\frac{1}{5} - \frac{1}{10} = 100k(10)$ or, $k = 10^{-4}\,kgm^{-1}$

97. (a) $h \propto \frac{V^2}{2g}$

$h \propto k.E.$

As k.E. becomes half after every collision. So height will also become half.

So, total distance $= h + 2\left(\frac{h}{2} + \frac{h}{4} + ...\right)$

$= h + 2h\left(\frac{\frac{1}{2}}{1-\frac{1}{2}}\right) = 3h$

98. (c) Using, $F = ma = m\frac{dV}{dt}$

$6t = 1.\frac{dV}{dt}$ $\quad [\because m = 1$ kg given$]$

$\int_0^v dV = \int_0^1 6t\,dt \Rightarrow V = 6\left[\frac{t^2}{2}\right]_0^1 = 3\,ms^{-1}$

$[\because t = 1$ sec given$]$

From work-energy theorem,

$W = \Delta KE = \frac{1}{2}m\left(V^2 - u^2\right) = \frac{1}{2} \times 1 \times 9 = 4.5$ J

99. (a) Work done by friction at QR = μmgx

In triangle, $\sin 30° = \frac{1}{2} = \frac{2}{PQ} \Rightarrow PQ = 4m$

Work done by friction at PQ = $\mu mg \times \cos 30° \times 4$

$= \mu mg \times \frac{\sqrt{3}}{2} \times 4 = 2\sqrt{3}\,\mu mg$

Since work done by friction on parts *PQ* and *QR* are equal,

$\mu mg\,x = 2\sqrt{3}\,\mu mg \Rightarrow x = 2\sqrt{3} \cong 3.5m$

Using work energy theorem, $mg\sin 30° \times 4 = 2\sqrt{3}\,\mu mg + \mu mgx$

$\Rightarrow 2 = 4\sqrt{3}\,\mu \Rightarrow \mu = 0.29$

100. (b) $\eta = \frac{W}{\text{Input}} = \frac{mgh \times 1000}{\text{Input}} = \frac{10 \times 9.8 \times 1 \times 1000}{\text{Input}}$

$\text{Input} = \frac{98000}{0.2} = 49 \times 10^4 J$

$\text{Fat used} = \frac{49 \times 10^4}{3.8 \times 10^7} = 12.89 \times 10^{-3} kg.$

101. (b) As we know, $dU = -\vec{F}.d\vec{r}$

$U = -\int_0^r \alpha r^2 dr \cos 180° = \frac{\alpha r^3}{3}$ ...(i)

As, $F = \frac{mv^2}{r} = \alpha r^2 \Rightarrow mv^2 = \alpha r^3$

$\therefore K.E. = \frac{1}{2}mv^2 = \frac{\alpha r^3}{2}$ ...(ii)

Total energy = Potential energy + kinetic energy
Now, from eqn (i) and (ii)
Total energy = K.E. + P.E.

$= \frac{\alpha r^3}{3} + \frac{\alpha r^3}{2} = \frac{5}{6}\alpha r^3$

102. (a) Let u be the initial velocity of the bullet of mass m. After passing through a plank of width x, its velocity decreases to v.

$\therefore \quad u - v = \frac{u}{n}$ or, $v = u - \frac{u}{n} = \frac{u(n-1)}{n}$

If F be the retarding force applied by each plank, then using work – energy theorem,

$Fx = \frac{1}{2}mu^2 - \frac{1}{2}mv^2 = \frac{1}{2}mu^2 - \frac{1}{2}mu^2\frac{(n-1)^2}{n^2}$

$= \frac{1}{2}mu^2\left[\frac{n^2-(n-1)^2}{n^2}\right]$

$Fx = \frac{1}{2}mu^2\left(\frac{2n-1}{n^2}\right)$

Let P be the number of planks required to stop the bullet. Total distance travelled by the bullet before coming to rest = Px
Using work-energy theorem again,

$F(Px) = \frac{1}{2}mu^2 - 0$

or, $P(Fx) = P\left[\frac{1}{2}mu^2\frac{(2n-1)}{n^2}\right] = \frac{1}{2}mu^2 \quad \therefore P = \frac{n^2}{2n-1}$

103. (a) Given : $k_A = 300\,N/m$, $k_B = 400\,N/m$

Let when the combination of springs is compressed by force F. Spring A is compressed by x. Therefore compression in spring B

$x_B = (8.75 - x)$ cm. In series force is same across both spring

So, $F = 300 \times x = 400(8.75 - x)$

Solving we get, $x = 5$ cm

$x_B = 8.75 - 5 = 3.75$ cm

$\frac{E_A}{E_B} = \frac{\frac{1}{2}k_A(x_A)^2}{\frac{1}{2}k_B(x_B)^2} = \frac{300 \times (5)^2}{400 \times (3.75)^2} = \frac{4}{3}$

104. (c) $K.E. \propto t$

$K.E. = ct$ [Here, c = constant]

$\Rightarrow \frac{1}{2}mv^2 = ct \Rightarrow \frac{(mv)^2}{2m} = ct$

$\Rightarrow \frac{p^2}{2m} = ct \; (\because p = mv)$

$\Rightarrow p = \sqrt{2ctm}$

$\Rightarrow F = \frac{dp}{dt} = \frac{d(\sqrt{2ctm})}{dt} \Rightarrow F = \sqrt{2\,cm} \times \frac{1}{2\sqrt{t}}$

$\Rightarrow F \propto \frac{1}{\sqrt{t}}$

105. (c) At equilibrium : $F = \frac{-dU(x)}{dx} = 0$

$\Rightarrow F = \frac{-d}{dx}\left[\frac{a}{x^{12}} - \frac{b}{x^6}\right] = -\left[-\frac{12a}{x^{13}} + \frac{6b}{x^7}\right] = 0$

$\Rightarrow \frac{12a}{x^{13}} = \frac{6b}{x^7} \Rightarrow x = \left(\frac{2a}{b}\right)^{\frac{1}{6}}$

$\therefore U_{\text{at equilibrium}} = \frac{a}{\left(\frac{2a}{b}\right)^2} - \frac{b}{\left(\frac{2a}{b}\right)} = -\frac{b^2}{4a}$ and $U_{(x=\infty)} = 0$

$\therefore D = 0 - \left(-\frac{b^2}{4a}\right) = \frac{b^2}{4a}$

106. (d) The average speed of the athlete

$v = \frac{S}{t} = \frac{100}{10} = 10\,m/s \quad \therefore \; K.E. = \frac{1}{2}mv^2$

Assuming the mass of athlete to 40 kg his average K.E would be

$K.E = \frac{1}{2} \times 40 \times (10)^2 = 2000 J$

Assuming mass to 100 kg average kinetic energy

$K.E. = \frac{1}{2} \times 100 \times (10)^2 = 5000 J$

107. (b) Suppose the spring gets compressed by x before stopping.
kinetic energy of the block = P.E. stored in the spring + work done against friction.

$\Rightarrow \frac{1}{2} \times 2 \times (4)^2 = \frac{1}{2} \times 10{,}000 \times x^2 + 15 \times x$

$\Rightarrow 10{,}000x^2 + 30x - 32 = 0 \Rightarrow 5000x^2 + 15x - 16 = 0$

$\therefore \; x = \frac{-15 \pm \sqrt{(15)^2 - 4 \times (5000)(-16)}}{2 \times 5000}$

$= 0.055 m = 5.5 cm.$

108. (d) Let u be the velocity with which the particle is thrown and m be the mass of the particle. Then

$K = \frac{1}{2}mu^2.$ ...(1)

At the highest point the velocity is u cos 60° (only the horizontal component remains, the vertical component being zero at the top-most point). Therefore kinetic energy at the highest point.

$$K' = \frac{1}{2}m(u\cos 60°)^2 = \frac{1}{2}mu^2\cos^2 60° = \frac{K}{4} \quad \text{[From (1)]}$$

109. (b) Given, Mass of the particle, $m = 100$g
Initial speed of the particle, u = 5 m/s
Final speed of the particle, $v = 0$
Work done by the force of gravity
= Loss in kinetic energy of the body.

$$= \frac{1}{2}m(v^2 - u^2) = \frac{1}{2}\times\frac{100}{1000}(0^2 - 5^2) = -1.25 \text{ J}$$

110. (a) Potential energy

$$V(x) = \left(\frac{x^4}{4} - \frac{x^2}{2}\right) \text{ joule}$$

For minimum P.E,

$$\frac{dV}{dx} = 0 \Rightarrow x^3 - x = 0 \Rightarrow x = \pm 1$$

$$\Rightarrow \text{Min. P.E.} = \frac{1}{4} - \frac{1}{2} = -\frac{1}{4} \text{ J}$$

$$\text{K.E.}_{(max.)} + \text{P.E.}_{(min.)} = 2 \text{ (Given)}$$

$$\therefore \text{K.E.}_{(max.)} = 2 + \frac{1}{4} = \frac{9}{4}$$

$$\text{K.E.}_{max.} = \frac{1}{2}mv^2_{max.}$$

$$\Rightarrow \frac{1}{2}\times 1\times v^2_{max.} = \frac{9}{4} \Rightarrow v_{max.} = \frac{3}{\sqrt{2}} \text{ m/s}$$

111. (d) Work done by tension + Work done by force (applied) + Work done by gravitational force = change in kinetic energy
Work done by tension is zero

$$\Rightarrow 0 + F\times AB - Mg\times AC = 0$$

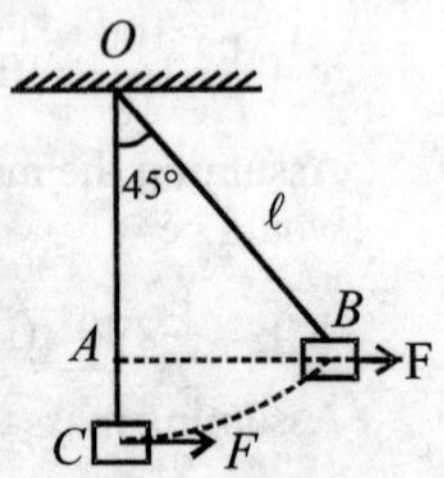

$$\Rightarrow F = Mg\left(\frac{AC}{AB}\right) = Mg\left[\frac{1-\frac{1}{\sqrt{2}}}{\frac{1}{\sqrt{2}}}\right]$$

$[\because AB = \ell\sin 45° = \frac{\ell}{\sqrt{2}}$ and

$AC = OC - OA = \ell - \ell\cos 45° = \ell\left(1 - \frac{1}{\sqrt{2}}\right)$

where ℓ = length of the string.]

$$\Rightarrow F = Mg(\sqrt{2} - 1)$$

112. (b)

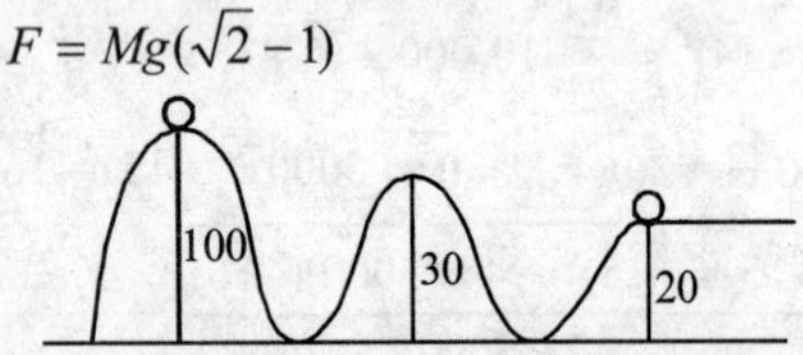

Using conservation of energy,
Total energy at 100 m height
= Total energy at 20m height

$$m(10\times 100) = m\left(\frac{1}{2}v^2 + 10\times 20\right)$$

or $\frac{1}{2}v^2 = 800$ or $v = \sqrt{1600} = 40$ m/s

113. (c) Given : retardation ∝ displacement
i.e., $a = -kx$ [Here, k = constant]

But $a = v\frac{dv}{dx}$

$$\therefore \frac{vdv}{dx} = -kx \Rightarrow \int_{v_1}^{v_2} v\,dv = -\int_0^x kx dx$$

$$\Rightarrow \left[\frac{v^2}{2}\right]_{v_1}^{v_2} = -k\left[\frac{x^2}{2}\right]_0^x \Rightarrow \left(v_2^2 - v_1^2\right) = -kx^2$$

$$\therefore \text{Loss in } K.E, \Delta K = \frac{1}{2}m\left(v_2^2 - v_1^2\right) = \frac{1}{2}m\left(-kx^2\right)$$

$$\therefore \Delta K \propto x^2$$

114. (a) Work done by such force is always zero since force is acting in a direction perpendicular to velocity.
∴ From work-energy theorem = ΔK = 0
K remains constant.

115. (d) The elastic potential energy

$$= \frac{1}{2}\times \text{Force} \times \text{extension}$$

$$\left[\because U = \frac{1}{2}Kx^2 = \frac{1}{2}\frac{F}{x}.x^2 = \frac{1}{2}\times F\times x\right]$$

$$= \frac{1}{2}\times F\times x = \frac{1}{2}\times 200\times 0.001 = 0.1 \text{ J}$$

116. (c) Let u be the speed with which the ball of mass m is projected. Then the kinetic energy (E) at the point of projection is

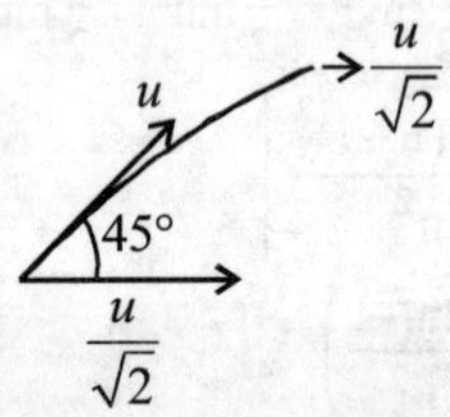

$$E = \frac{1}{2}mu^2 \quad \text{...(i)}$$

When the ball is at the highest point of its flight, the speed of the ball is $\frac{u}{\sqrt{2}}$ (Remember that the horizontal component of velocity does not change during a projectile motion).
∴ The kinetic energy at the highest point

$$= \frac{1}{2}m\left(\frac{u}{\sqrt{2}}\right)^2 = \frac{1}{2}\frac{mu^2}{2} = \frac{E}{2} \quad \text{[From (i)]}$$

117. (d) $\vec{F} = (2t\hat{i} + 3t\hat{j})N$, $m = 1000$ gm = 1 kg

$$\therefore \text{Acceleration } \vec{a} = \frac{\vec{F}}{m} = 2t\hat{i} + 3t^2\hat{j}$$

$$\Rightarrow \frac{d\vec{v}}{dt} = 2t\hat{i} + 3t^2\hat{j}$$

$$\vec{V} = \int(2t\hat{i} + 3t^2\hat{j})\,dt = t^2\hat{i} + t^3\hat{j}$$

$\therefore$ Power, $P = \vec{F}\cdot\vec{v}$

$\Rightarrow P = (2t\hat{i} + 3t^2\hat{j})\cdot(t^2\hat{i} + t^3\hat{j}) = (2t^3 + 3t^5)$ W

118. (d) Power, P = $\vec{F}.\vec{V}$

Also, $F = \frac{dp}{dt} = \left(\frac{dm}{dt}\right)v = C\left(\sqrt{v}\right)v = Cv^{\frac{3}{2}}$

$\therefore \quad P = C\left(v^{3/2}\right)v = Cv^{5/2} \Rightarrow p^2 \propto v^5$

119. (b) The expression of average power is

$<P> = \vec{F}.\Delta\vec{v}$

$= \frac{(2\hat{i}+3\hat{j}).(3\hat{i}+6\hat{j})}{4} = 6\text{watt}$

Now, $\vec{a} = \frac{\vec{F}}{m} = \frac{1}{2}\hat{i} + \frac{3}{4}\hat{j}$

$\vec{v}_f = \left(\frac{1}{2}\hat{i} + \frac{3}{4}\hat{j}\right)\times 4 = (2\hat{i} + 3\hat{j})$

$\therefore P_{ins} = \vec{F}.\vec{v}_f = \left(2\hat{i}+3\hat{j}\right).\left(2\hat{i}+3\hat{j}\right) = 13\text{watt}$

$\therefore \frac{<P>}{P_{ins}} = \frac{6}{13}$

120. (c) Given, P = costant $\Rightarrow$ FV = constant = K

$\Rightarrow m\frac{dV}{dt}V = \text{constant} = K$

$\Rightarrow m\int_0^V VdV = (K)\int_0^t dt \Rightarrow \left(\frac{mV^2}{2}\right) = Kt$

$\Rightarrow V \propto t^{1/2}$

$\Rightarrow \frac{ds}{dt} \propto t^{1/2} \Rightarrow \int_0^S ds = C\int_0^t t^{1/2}dt \Rightarrow S = C\times\frac{2}{3}t^{3/2}$

$\therefore s \propto t^{3/2}$

121. (d) Given,

Mass of the body, m = 2 kg

Force, $\vec{F} = m\vec{a} = (6t\hat{i} + 6t^2\hat{j})$ N

$\vec{a} = \frac{\vec{F}}{m} = 3t\hat{i} + 3t^2\hat{j} \Rightarrow \frac{d\vec{v}}{dt} = 3t\hat{i} + 3t^2\hat{j}$

$\Rightarrow \vec{v} = \frac{3t^2}{2}\hat{i} + t^3\hat{j}$

$\therefore$ Power, $P = \vec{F}.\vec{v} = (6t\hat{i} + 6t^2\hat{j}).(\frac{3t^2}{2}\hat{i} + t^3\hat{j})$

$= (9t^3 + 6t^5)$W

122. (d) Power, $P = \frac{W}{t} = \frac{mgh}{t}$

$\therefore \frac{P_1}{P_2} = \frac{\frac{m_1gh}{t_1}}{\frac{m_2gh}{t_2}} = \frac{m_1}{t_1}\frac{t_2}{m_2}$

$\therefore \frac{P_1}{P_2} = \frac{300\times 2}{5\times 50} = \frac{12}{5} = \frac{3\sqrt{x}}{\sqrt{x}+1}$

So, $12\sqrt{x} + 12 = 15\sqrt{x} \Rightarrow 3\sqrt{x} = 12$

$\therefore x = 16$

123. (300) Given, acceleration of block, $a = 1\text{ms}^{-2}$

Mass of block, m = 5 kg

$\therefore \; v = u + at = 0 + 1(10) = 10$ m/s

$F - mg\sin\theta = ma \Rightarrow F - 5g\sin 30° = 5a$

$\Rightarrow F = 5 + 25 = 30$N

Power, P = Fv = 300 W

124. (48) Given,

Mass of an elevator, m = 1400 kg

Friction = 2000 N

Speed of elevator, $v = 3\text{ms}^{-1}$

Maximum force, F_{max} = mg + friction

= 14000 + 2000 = 16000 N

Power, $P_{max} = F_{max} \times v$

$\Rightarrow P_{max} = 16000 \times 3 = 48000$ W = 48 kW

125. (4) Power, $P = \frac{\text{Total work done}}{\text{time}} = \frac{\frac{1}{2}mv^2}{t}$

$\Rightarrow \frac{1}{2}mV^2 = Pt$

$\Rightarrow V = \sqrt{\frac{2Pt}{m}} \Rightarrow \frac{dx}{dt} = \sqrt{\frac{2Pt}{m}} \Rightarrow x = \sqrt{\frac{2P}{m}}\frac{2}{3}[t^{3/2}]_0^4$

$\Rightarrow x = \frac{16\sqrt{P}}{3} = \frac{1}{3}\alpha^2\sqrt{P}$ ($\because$ mass m = 2 kg, time t = 4s)

$\Rightarrow \alpha = 4$

126. (100) We have $\vec{F} = t\hat{i} + 3t^2\hat{j}$

$\Rightarrow \frac{md\vec{v}}{dt} = t\hat{i} + 3t^2\hat{j} \Rightarrow 1.d\vec{v} = \left(t\hat{i} + 3t^2\hat{j}\right)dt$

$\Rightarrow \int_0^{\vec{v}} d\vec{v} = \int_0^t tdt\hat{i} + \int_0^t 3t^2dt\hat{j} \Rightarrow \vec{v} = \frac{t^2}{2}\hat{i} + t^3\hat{j}$

And, Power = $\vec{F}.\vec{V} = \frac{t^3}{2} + 3t^5$

At t = 2sec, power = $\frac{8}{2} + 3\times 32 = 100$ W

127. (d) $F_{thrust} = V_{rel}\frac{dm}{dt} = 5\times 0.5 = 2.5$ N

So, Power = Force × Velocity = 2.5 × 5 = 12.5 watt.

128. (c) $a_c = k^2rt^2 = \frac{v^2}{r} \Rightarrow v = krt$

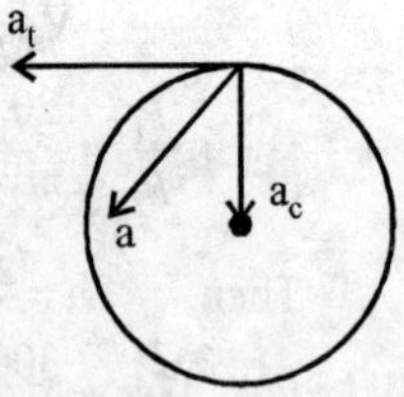

$a_t = \frac{dv}{dt} = kr$

Now, $P = (\vec{F_r} + \vec{F_t}).\vec{v}$

$= 0 + F_t v = ma_t v = mkr\,(krt) = mk^2r^2t$

129. (18) Given, Mass of the body, $m = 2$ kg
Power delivered by engine, $P = 1$ J/s
Time, $t = 9$ seconds
Power, $P = Fv$

$\Rightarrow P = mav \qquad [\because F = ma]$

$\Rightarrow m\frac{dv}{dt}v = P \qquad \left(\because a = \frac{dv}{dt}\right)$

$\Rightarrow v\,dv = \frac{P}{m}dt$

Integrating both sides we get

$\Rightarrow \int_0^v v\,dv = \frac{P}{m}\int_0^t dt \Rightarrow \frac{v^2}{2} = \frac{Pt}{m} \Rightarrow v = \left(\frac{2Pt}{m}\right)^{1/2}$

$\Rightarrow \frac{dx}{dt} = \sqrt{\frac{2P}{m}}t^{1/2} \qquad \left(\because v = \frac{dx}{dt}\right)$

$\Rightarrow \int_0^x dx = \sqrt{\frac{2P}{m}}\int_0^t t^{1/2}\,dt$

$\therefore$ Distance, $x = \sqrt{\frac{2P}{m}}\frac{t^{3/2}}{3/2} = \sqrt{\frac{2P}{m}} \times \frac{2}{3}t^{3/2}$

$\Rightarrow x = \sqrt{\frac{2\times1}{2}} \times \frac{2}{3} \times 9^{3/2} = \frac{2}{3} \times 27 = 18\,m$

130. (b) We know that
Power, $P = Fv$

But $F = mav = m\frac{dv}{dt}v$

$\therefore P = mv\frac{dv}{dt} \Rightarrow P\,dt = mv\,dv$

Integrating both sides $\int_0^t P\,dt = m\int_0^v v\,dv$

$P.\,t = \frac{1}{2}mv^2 \Rightarrow v = \left(\sqrt{\frac{2P}{m}}\right)t^{1/2}$

Distance, $s = \int_0^t v\,dt = \sqrt{\frac{2P}{m}}\int_0^t t^{1/2}\,dt = \sqrt{\frac{2P}{m}} \cdot \frac{t^{3/2}}{3/2}$

$\Rightarrow s = \sqrt{\frac{8P}{9m}} \cdot t^{3/2} \Rightarrow s \propto t^{3/2}$

So, graph (b) is correct.

131. (b) Total force required to lift maximum load capacity against frictional force = 4000 N

$F_{total} = Mg + \text{friction}$

$= 2000 \times 10 + 4000 = 20{,}000 + 4000 = 24000\,N$

Using power, $P = F \times v$

$60 \times 746 = 24000 \times v \Rightarrow v = 1.86$ m/s $\approx 1.9\,m/s$

Hence speed of the elevator at full load is close to 1.9 ms^{-1}

132. (b) Centripetel acceleration, $a_c = n^2Rt^2$

$a_c = \frac{v^2}{R} = n^2Rt^2$

$v^2 = n^2R^2t^2$

$v = nRt$

Here power is delivered by tangential force only because power by centripetal force is zero. $\left[\text{Since } \vec{F}_c \perp \vec{V}\right]$

$a_t = \frac{dv}{dt} = nR$

Power $= Ma_tv = M\,nR\,nRt = Mn^2R^2t$.

133. (c) Force exerted by wind, $F = m\frac{dv}{dt} + v\frac{dm}{dt}$

as, v = cons.

$F = v\frac{dm}{dt} = v\rho\frac{d}{dt}(Al) = v\rho A\frac{dl}{dt} = v\rho Av = \rho Av^2$

So, power = Fv

$= \rho Av^3$

i.e. power $\propto v^3$

134. (b) Let a be the acceleration of body
Using, $v = u + at$

$v_1 = 0 + at_1 \Rightarrow a = \frac{v_1}{t_1}$

Velocity of the body at instant t,

$v = at \Rightarrow v = \frac{v_1t}{t_1}$

Instantaneous powr, $P = \vec{F}\cdot\vec{v} = (m\vec{a})\cdot\vec{v}$

$= \left(\frac{mv_1}{t_1}\right)\left(\frac{v_1t}{t_1}\right) = m\left(\frac{v_1}{t_1}\right)^2 t$

135. (d) In elastic collision for same mass, velocities gets interchanged.
For elastic collision between A & B,

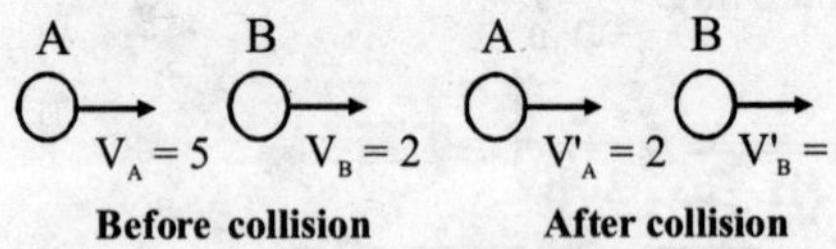

Before collision **After collision**

For elastic collision between B & C,

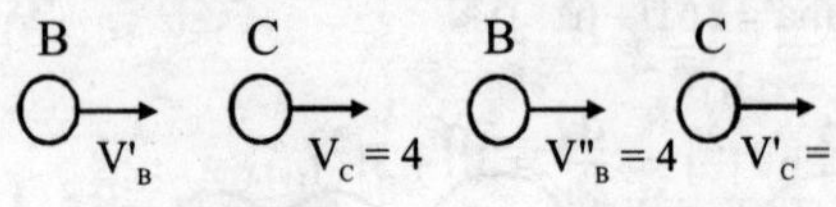

Before collision **After collision**

$\therefore \quad V'_A = 2\text{m/s} \quad V''_B = 4\text{m/s} \quad V'_C = 5\text{m/s}$

136. (a)

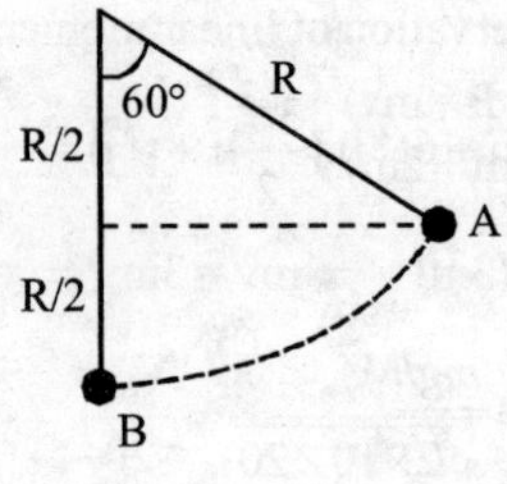

Velocity of the bob A just before hitting is,

$$u=\sqrt{2g\frac{R}{2}}=\sqrt{gR}$$

Just after collision, let velocity of two bobs A and B are v_1 and v_2, respectively, then

The velocity of bob A after elastic collision is

$$v_1=\left(\frac{m_1-m_2}{m_1+m_2}\right)u=\left(\frac{m-\frac{m}{2}}{m+\frac{m}{2}}\right)\sqrt{gR}=\frac{1}{3}\sqrt{gR}$$

137. (d) Using $v^2-u^2=2gh$

Velocity just before collision $=\sqrt{2gh}$

Velocity just after collision $=\sqrt{2g\left(\frac{h}{2}\right)}$

$\therefore \Delta KE=\frac{1}{2}m(2gh)-\frac{1}{2}mgh=\frac{1}{2}mgh$

$\therefore$ % Loss in kinetic energy

$$=\frac{\Delta KE}{KE_i}\times100=\frac{\frac{1}{2}mgh}{\frac{1}{2}mg2h}\times100=50\%$$

138. (2)

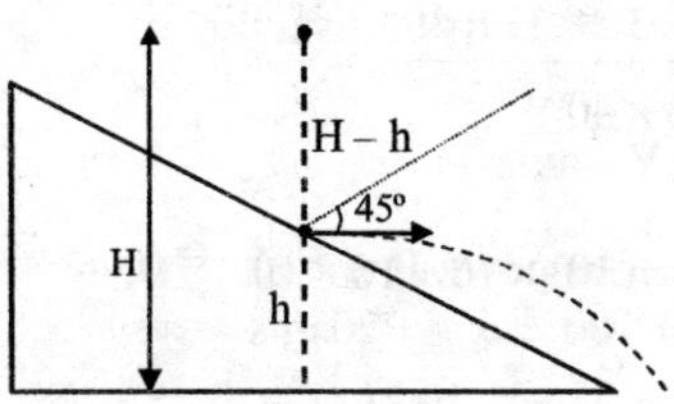

T = Total time of flight

$$T=\sqrt{\frac{2h}{g}}+\sqrt{\frac{2(H-h)}{g}}$$

For maximum time, $\frac{dT}{dh}=0$

$$\therefore \sqrt{\frac{2}{g}}\left(\frac{-1}{2\sqrt{H-h}}+\frac{1}{2\sqrt{h}}\right)=0$$

$$\Rightarrow 2\sqrt{h}=2\sqrt{H-h}\Rightarrow\sqrt{H-h}=\sqrt{h}\Rightarrow H-h=h$$

$$\Rightarrow h=\frac{H}{2}\Rightarrow\frac{H}{h}=2$$

139. (c)

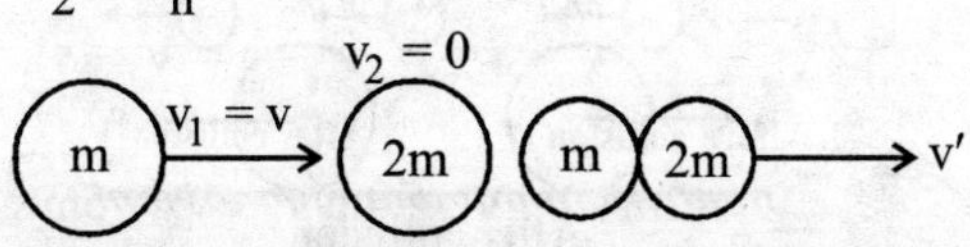

Applying conservation of linear momentum

$\Rightarrow \vec{P}_i=\vec{P}_f$ ($\because$ P = mv)

$mv_1+2mv_2=(m+2m)v'$

$mv+2m\times0=(3m)v'\Rightarrow mv=3mv'\Rightarrow v'=\frac{v}{3}$

140. (5) $e=\frac{V_{separation}}{V_{approach}}=\frac{V_{sep}}{\sqrt{2\times10\times20}}=\frac{V_{sep}}{20}$

$\Rightarrow 0.5=\frac{V_{sep}}{20}\Rightarrow V_{sep}=10\text{ m/s}$

Let 'h' be the maximum height reached

Then, $h=\frac{V_{sep}^2}{2g}=\frac{10^2}{2\times10}=5m$

141. (c) The velocities will be interchanged after collision.

Speed of P just before collision = $\sqrt{2gh}$

$=\sqrt{2\times10\times0.2}=2\text{m/s}$ ($\because$ h = 20 cm = 0.2 m)

$\therefore$ Speed of Q just after collision = 2m/s

142. (d) From equation of motion $h=ut+\frac{1}{2}gt^2$

$h=\frac{1}{2}gt^2$

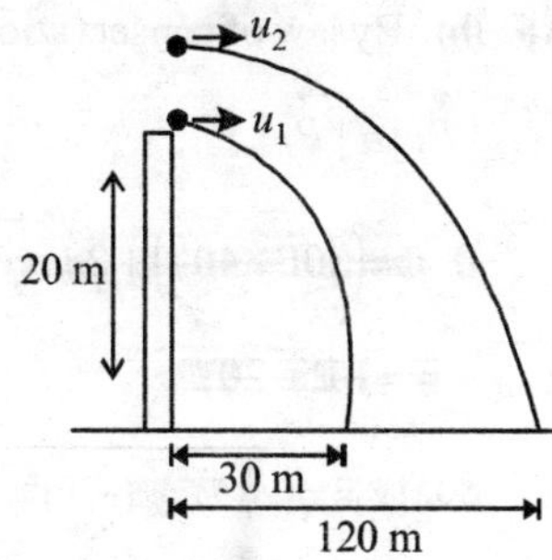

$t=\sqrt{\frac{2h}{g}}$; $v_1=\frac{30}{\sqrt{\frac{2h}{g}}}$,

$v_2=\frac{120}{\sqrt{\frac{2h}{g}}}$

apply the conservation of momentum $m_2u=m_1v_1+m_2v_2$

$(0.01)u=(0.2)\frac{30\sqrt{g}}{\sqrt{2h}}+(0.01)\frac{120\sqrt{g}}{\sqrt{2h}}$

$u=300+60=360\text{ ms}^{-1}$

143. (4) Before Collision

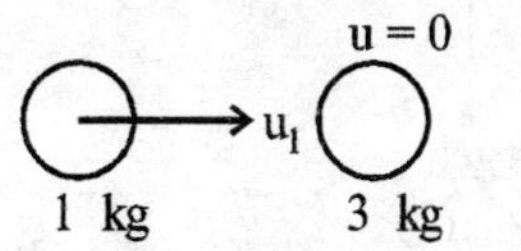

After Collision

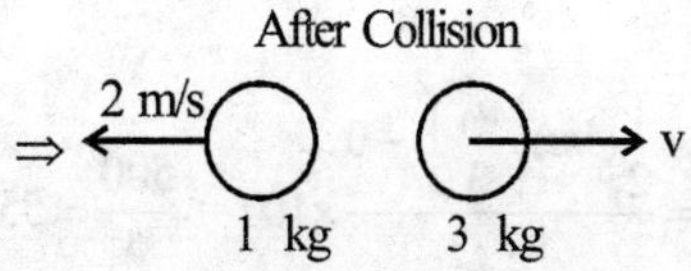

Applying conservation of momentum,

$1\times u_1+3\times0=-2+3v \quad\Rightarrow u_1=-2+3v$ ...(1)

Also, $e=1=\frac{v+2}{u_1} \quad\Rightarrow v+2=u_1$

Using equation (1)

$\Rightarrow v+2=-2+3v$

$\therefore$ v = 2 m/s

Put this in equation (1)

$\Rightarrow u_1=-2+3v=-2+3\times2=4\text{ m/s}$

144. (b) As no external force is acting on system so, $P_i=P_f$

$0.2\times10=10\times v\Rightarrow v=0.2\text{ m/sec}$

Loss in K.E. $=\frac{1}{2}\times(0.2)\times10^2-\frac{1}{2}\times10(0.2)^2$

$=\frac{1}{2}\times10\times(0.2)[10-0.2]=9.8\text{ J}$

145. (b)

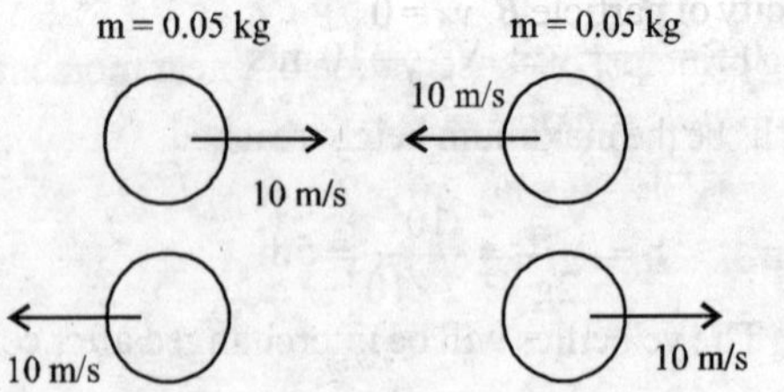

Chnage in momentum of any one ball

$|\Delta\vec{P}| = 2\times0.05\times10 = 1$

$|\vec{F}_{av}| = \frac{|\Delta\vec{P}|}{\Delta t} = \frac{1}{0.005} = \frac{1000}{5} = 200\text{ N}$

146. (b) By law of conservation of momentum

$|\vec{P}_i| = |\vec{P}_f|$

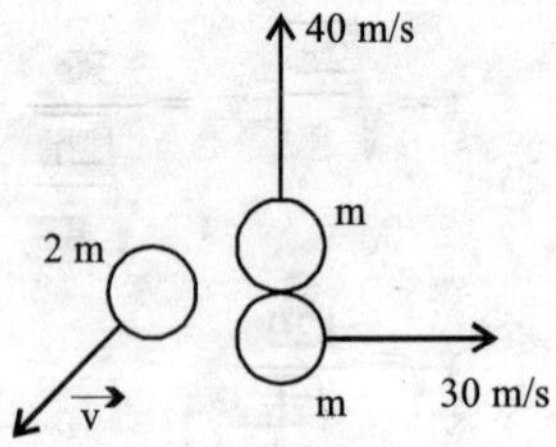

$0 = m(30\hat{i} + 40\hat{j}) + 2m\vec{v}$

$\vec{v} = -15\hat{i} - 20\hat{j}$

So, $|\vec{v}| = \sqrt{(-15)^2 + (-20)^2}$

$= \sqrt{625} = 25$ m/s.

147. (c) Let the velocity of striking particle be u_0

Then, $mu_0 = mv_1 + 5mv_2$

$u_0 = v_1 + 5v_2$ ...(i)

as, collision is elastic

So, $e = 1$

$\Rightarrow \frac{v_2 - v_1}{u_0} = 1$

$\Rightarrow v_2 - v_1 = u_0$ ...(ii)

Adding (i) and (ii), we get

$2u_0 = 6v_2 \Rightarrow v_2 = u_0/3$

So, $\%\ \Delta K.E_2 = \frac{\frac{1}{2}(5m)\left(\frac{u_0}{3}\right)^2 - 0}{\frac{1}{2}mu_0^2}\times100 = \frac{500}{9} \simeq 55.6\%$

148. (c)

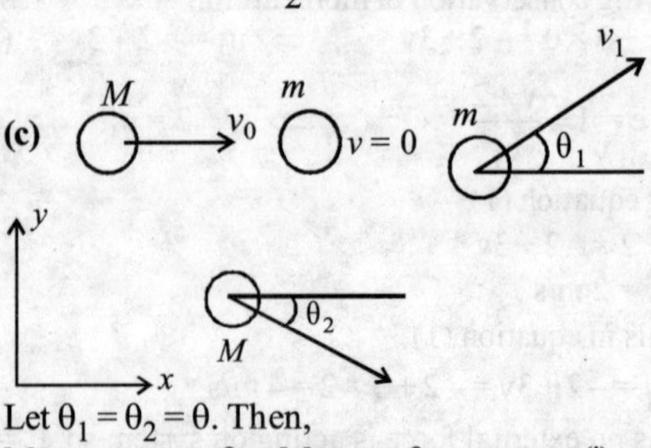

Let $\theta_1 = \theta_2 = \theta$. Then,

$Mv_0 = mv_1\cos\theta + Mv_2\cos\theta$ ...(i)

and, $0 = mv_1\sin\theta - Mv_2\sin\theta$

$\Rightarrow v_2 = \frac{mv_1}{M}$ ...(ii)

from (i) & (ii), we get

$Mv_0 = mv_1\cos\theta + M\left(\frac{mv_1}{M}\right)\cos\theta$

$Mv_0 = 2mv_1\cos\theta$ ....(iii)

By conservation of K.E.

$\frac{1}{2}Mv_0^2 = \frac{1}{2}mv_1^2 + \frac{1}{2}Mv_2^2$

$\Rightarrow \frac{1}{2}M\left(\frac{2mv_1\cos\theta}{M}\right)^2 = \frac{1}{2}mv_1^2 + \frac{1}{2}M.\frac{m^2v_1^2}{M^2}$

$\Rightarrow \frac{1}{2}\left(\frac{4m^2v_1^2\cos^2\theta}{M}\right)^2 = \frac{1}{2}mv_1^2 + \frac{m^2v_1^2}{2M}$

$\Rightarrow \frac{4m\cos^2\theta}{M} = 1 + \frac{m}{M} \Rightarrow 4\cos^2\theta = \frac{M}{m} + 1$

For largest $\frac{M}{m}$, $\cos\theta = 1$ So, $\frac{M}{m} = 3$

149. (c)

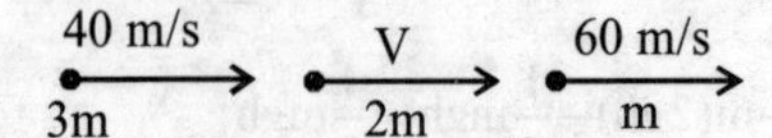

Using momentum conservation

$3m\times40 = 2mv + m60 \Rightarrow v = 30$ m/s

Fractional change in K.E is given by

$\frac{\Delta KE}{K.E} = \frac{\text{Final K.E} - \text{Initial K.E}}{\text{Initial K.E}} = \frac{\frac{1}{2}m60^2 + \frac{1}{2}2m\times30^2 - \frac{1}{2}3m\,40^2}{\frac{1}{2}3m\,40^2}$

$= \frac{60^2 + 2\times30^2 - 3\times40^2}{3\times40^2} = \frac{1}{8}$

150. (20) Along Y – axis

$P_i = P_f$

$\Rightarrow 0 = -(x\sin30)\times10 + 10\times10$

$\Rightarrow 0 = -5x + 100 \Rightarrow x = +20$ m/s

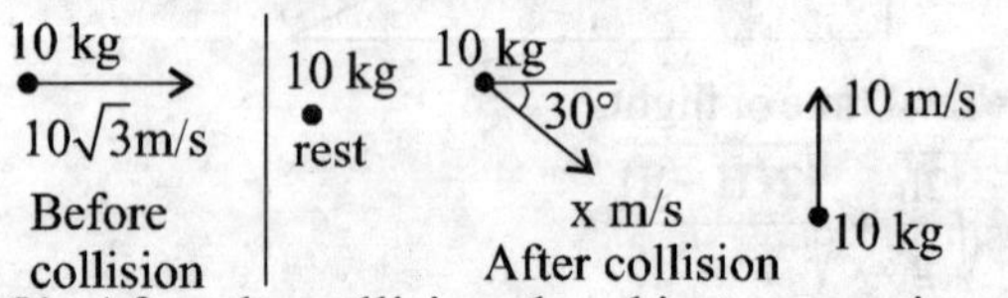

151. (b) After the collision the objects move in opposite direction, let with velocity v_1, then from law of conservation of momentum, $P_i = P_f$

$m_1v = (m_2 - m_1)v_1$

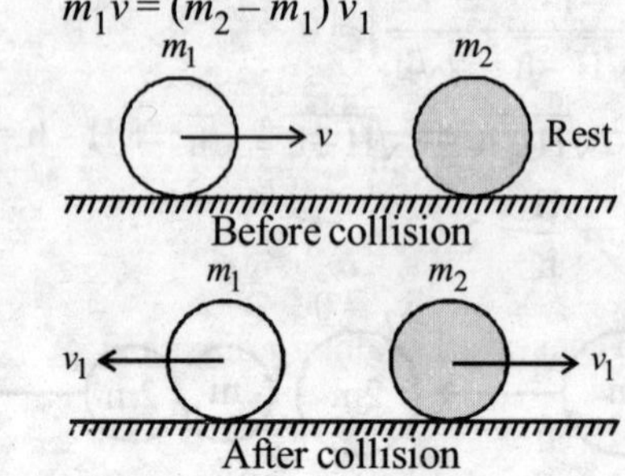

$\Rightarrow v_1 = \frac{m_1v}{(m_2 - m_1)}$ ...(i)

If collision is elastic then, $e = 1 = \frac{v_1 - (-v_1)}{v - 0}$

also, $2v_1 = v \Rightarrow v_1 = \frac{v}{2}$ ...(ii)

From equation (i) & (ii)

$\frac{1}{2} = \frac{m_1}{m_2 - m_1} \quad \therefore \quad \frac{m_2}{m_1} = \frac{3}{1}$

152. (b) From energy conservation

$(M+m)gh = \frac{1}{2}(M+m)V_1^2 \Rightarrow V_1 = \sqrt{2gh}$

Now applying momentum conservation,
momentum before collision = momentum after collision

$mV = (M+m)\,V_1$

$\Rightarrow V = \left(\frac{M+m}{m}\right)V_1$

$\Rightarrow V = \frac{5.99+0.01}{0.01} \times \sqrt{2\times 9.8 \times 9.8 \times 10^{-2}}$

$= 831.4$ m/s.

153. (a) As the body Q was at rest and the collision is perfectly elastic.

$e = \frac{v_{sep}}{v_{app}} = 1$

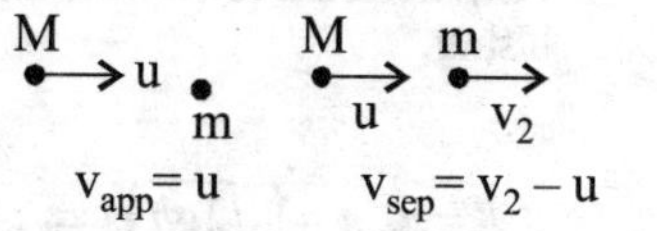

So, $\frac{v_2 - u}{u} = 1 \Rightarrow v_2 = 2u$

During elastic collision, the momentum and kinetic energy both are conserved.

154. (1)

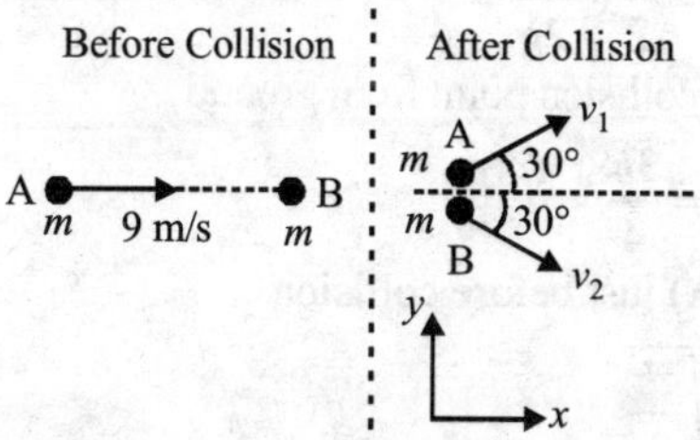

From conservation of momentum along y-axis.

$\vec{P}_{iy} = \vec{P}_{fy}$

$0 + 0 = mv_1 \sin 30°\ \hat{j} + mv_2 \sin 30°(-\hat{j})$

$mv_2 \sin 30° = mv_1 \sin 30°$

$v_2 = v_1$ or $\frac{v_1}{v_2} = 1$

155. (120)

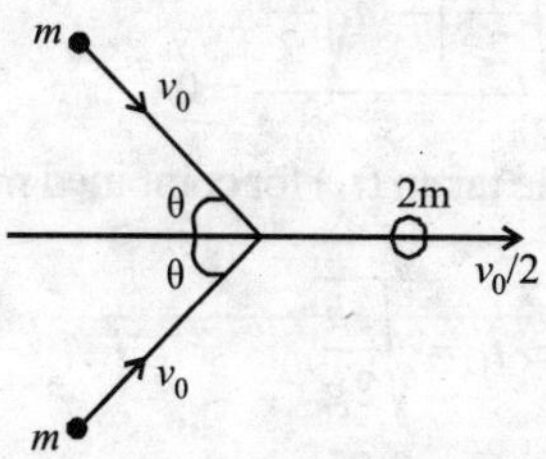

Momentum conservation along x direction,

$2mv_0 \cos\theta = 2m\frac{v_0}{2} \Rightarrow \cos\theta = \frac{1}{2}$ or $\theta = 60°$

Hence angle between the initial velocities of the two bodies

$= \theta + \theta = 60° + 60° = 120°$.

156. (d) Before collision,

Velocity of particle A, $u_1 = (\sqrt{3}\hat{i} + \hat{j})$ m/s

Velocity of particle B, $u_2 = 0$

After collision,

Velocity of particle A, $v_1 = (\hat{i} + \sqrt{3}\hat{j})$

Velocity of particle B, $v_2 = 0$

Using principal of conservation of linear momentum

$m_1\vec{u}_1 + m_2\vec{u}_2 = m_1\vec{v}_1 + m_2\vec{v}_2$

$\Rightarrow 2m_2(\sqrt{3}\hat{i} + \hat{j}) + m_2 \times 0 = 2m_2(\hat{i} + \sqrt{3}\hat{j}) + m_2 \times \vec{v}_2$

$\Rightarrow 2\sqrt{3}\hat{i} + 2\hat{j} = 2\hat{i} + 2\sqrt{3}\hat{j} + \vec{v}_2$

$\Rightarrow \vec{v}_2 = \left[(\sqrt{3}-1)\hat{i} - (\sqrt{3}-1)\hat{j}\right] \times 2$

$\Rightarrow \vec{v}_1 = \hat{i} + \sqrt{3}\hat{j}$

For angle between $\vec{v}_1$ and $\vec{v}_2$,

$\cos\theta = \frac{\vec{v}_1 \cdot \vec{v}_2}{v_1 v_2} = \frac{2(\sqrt{3}-1)(1-\sqrt{3})}{2\times 2\sqrt{2}(\sqrt{3}-1)} = \frac{1-\sqrt{3}}{2\sqrt{2}} \Rightarrow \theta = 105°$

Angle between $\vec{v}_1$ and $\vec{v}_2$ is 105°

157. (b) According to the question, all collisions are perfectly inelastic, so after the final collision, all blocks are moving together.

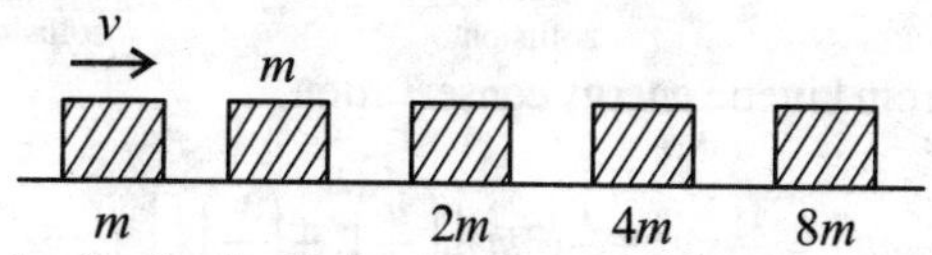

Let the final velocity be v', using momentum conservation

$mv = 16mv' \Rightarrow v' = \frac{v}{16}$

Now initial energy $E_i = \frac{1}{2}mv^2$

Final energy : $E_f = \frac{1}{2} \times 16m \times \left(\frac{v}{16}\right)^2 = \frac{1}{2}\frac{mv^2}{16}$

Energy loss : $E_i - E_f = \frac{1}{2}mv^2 - \frac{1}{2}m\frac{v^2}{16}$

$\Rightarrow \frac{1}{2}mv^2\left[1 - \frac{1}{16}\right] \Rightarrow \frac{1}{2}mv^2\left[\frac{15}{16}\right]$

The total energy loss is P% of the original energy.

$\therefore\ \%P = \frac{\text{Energy loss}}{\text{Original energy}} \times 100$

$= \frac{\frac{1}{2}mv^2\left[\frac{15}{16}\right]}{\frac{1}{2}mv^2} \times 100 = 93.75\%$

Hence, value of P is close to 94.

158. (b) Given,

Mass of block, $m_1 = 1.9$ kg

Mass of bullet, $m_2 = 0.1$ kg

Velocity of bullet, $v_2 = 20$ m/s

Let v be the velocity of the combined system. It is an inelastic collision.

Using conservation of linear momentum

$m_1 \times 0 + m_2 \times v_2 = (m_1 + m_2)v$

$\Rightarrow 0.1 \times 20 = (0.1 + 1.9) \times v \Rightarrow v = 1$ m/s

Using work – energy theorem,

Work done = Change in Kinetic energy

Let K be the Kinetic energy of combined system.

$(m_1 + m_2)gh$

$= K - \frac{1}{2}(m_1 + m_2)v^2$

$\Rightarrow 2 \times g \times 1 = K - \frac{1}{2} \times 2 \times 1^2 \Rightarrow K = 21\text{ J}$

159. (c) From conservation of linear momentum,

$mu\hat{i} + 0 = mv\hat{j} + 3m\vec{v'}$

$\vec{v'} = \frac{u}{3}\hat{i} - \frac{v}{3}\hat{j}$

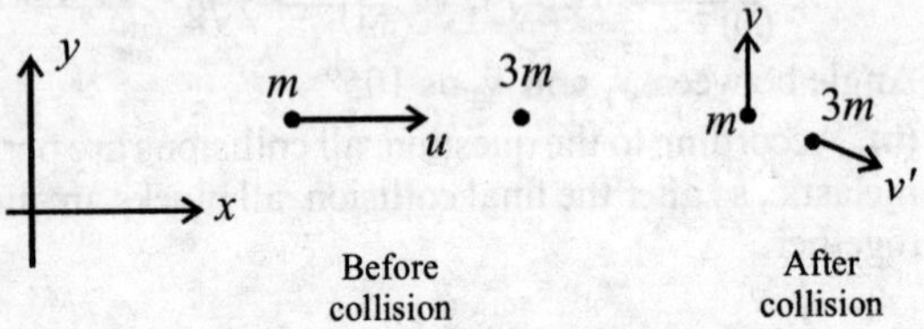

From kinetic energy conservation,

$$\frac{1}{2}mu^2 = \frac{1}{2}mv^2 + \frac{1}{2}(3m)\left(\left(\frac{u}{3}\right)^2 + \left(\frac{v}{3}\right)^2\right)$$

or, $mu^2 = mv^2 + \frac{mu^2}{3} + \frac{mv^2}{3}$ $\quad \therefore v = \frac{u}{\sqrt{2}}$

160. (10.00) From momentum conservation in perpendicular direction of initial motion.

$mu_1 \sin\theta_1 = 10mv_1 \sin\theta_2$ ...(i)

It is given that energy of m reduced by half. If u_1 be velocity of m after collision, then

$$\left(\frac{1}{2}mu^2\right)\frac{1}{2} = \frac{1}{2}mu_1^2 \Rightarrow u_1 = \frac{u}{\sqrt{2}}$$

If v_1 be the velocity of mass 10 m after collision, then

$$\frac{1}{2} \times 10m \times v_1^2 = \frac{1}{2}m\frac{u^2}{2} \Rightarrow v_1 = \frac{u}{\sqrt{20}}$$

From equation (i), we have

$\sin\theta_1 = \sqrt{10}\sin\theta_2$

161. (b)

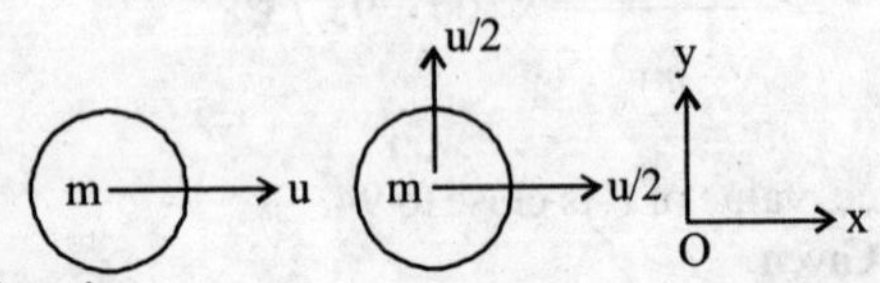

x-direction

$mu + \frac{mu}{2} = 2mv'_x \Rightarrow v'_x = \frac{3u}{4}$

y-direction $0 + \frac{mu}{2} = 2mv'_y \Rightarrow v'_y = \frac{u}{4}$

$$K.E._i = \frac{1}{2}m\,u^2 + \frac{1}{2}m\left[\left(\frac{u}{2}\right)^2 + \left(\frac{u}{2}\right)^2\right]$$

$$= \frac{1}{2}mu^2 + \frac{mu^2}{4} = \frac{3mu^2}{4}$$

$$K.E._f = \frac{1}{2}(2m)\left(v'_x\right)^2 + \frac{1}{2}(2m)\left(v'_y\right)^2$$

$$= \frac{1}{2}2m\left[\left(\frac{3u}{4}\right)^2 + \left(\frac{u}{4}\right)^2\right] = \frac{5}{8}mu^2$$

$$\therefore \text{Loss in } KE = KE_{\hat{i}} - KE_{\hat{f}} = mu^2\left(\frac{6}{8} - \frac{5}{8}\right) = \frac{mu^2}{8}$$

162. (1) For elastic collision $KE_i = KE_f$

$$\frac{1}{2}m \times 25 + \frac{1}{2} \times m \times 9 = \frac{1}{2}m \times 32 + \frac{1}{2}mv_B^2$$

$34 = 32 + v_B^2 \Rightarrow v_B = \sqrt{2}\text{ m/s}$

$KE_B = \frac{1}{2}mv_B^2 = \frac{1}{2} \times 0.1 \times 2 = 0.1J = \frac{x}{10}J \quad \therefore x = 1$

163. (d) Let t be the time taken by the particle to dropped from height h to collide with particle thrown upward. Using,

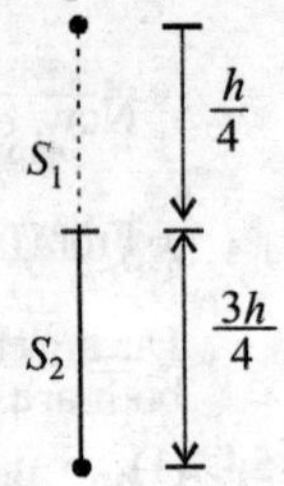

$\therefore h = u_{rel}t = \left(\sqrt{2gh}\right).t \Rightarrow t = \sqrt{\frac{h}{2g}}$

Downward distance travelled

$S_1 = \frac{1}{2}gt^2 = \frac{1}{2}g.\frac{h}{2g} = \frac{h}{4}$

Distance of collision point from ground

$s_2 = h - \frac{h}{4} = \frac{3h}{4}$

Speed of (A) just before collision

$v_1 = gt = \sqrt{\frac{gh}{2}}$

And speed of (B) just before collision

$v_2 = \sqrt{2gh} - \sqrt{\frac{gh}{2}}$

Using principle of conservation of linear momentum

$mv_1 + mv_2 = 2mv_f$

$$v_f = \frac{m\left(\sqrt{2gh} - \sqrt{\frac{gh}{2}}\right) - m\sqrt{\frac{gh}{2}}}{2m} = 0$$

After collision, time taken (t_1) for combined mass to reach the ground is

$\Rightarrow \frac{3h}{4} = \frac{1}{2}gt_1^2 \Rightarrow t_1 = \sqrt{\frac{3h}{2g}}$

164. (b) $v_{s,m} = v_s - v_m \Rightarrow 0.7 = v_s - v_m$

Let $v_m = -v$

$\therefore v_s = 0.7 - v$

$P_i = P_f$

or $\quad 0 = 20(0.7 - v) - 50v$

or $\quad v = 0.2\text{ m/s}$

165. (a) Apply concervation of linear momentum in X and Y direction for the system then

$M(10\cos30°) + 2M(5\cos45°) = 2M(v_1\cos30°)$
$+ M(v_2\cos45°)$

$$5\sqrt{3}+5\sqrt{2}=\sqrt{3}\,v_1+\frac{v_2}{\sqrt{2}} \quad ....(1)$$

Also

$$2M(5\sin 45°)-M(10\sin 30°)=2Mv_1\sin 30°-Mv_2\sin 45°$$

$$5\sqrt{2}-5=v_1-\frac{v_2}{\sqrt{2}} \quad ....(2)$$

Solving equation (1) and (2)

$$\left(\sqrt{3}+1\right)v_1=5\sqrt{3}+10\sqrt{2}-5 \Rightarrow v_1=6.5\text{m/s}$$

$v_2 = 6.3$ m/s

166. (d) By law of conservation of momentum

$$2u=2\frac{u}{4}+mv \Rightarrow \frac{3u}{2}=mv \quad ...(i)$$

$$\text{Now, } e=\frac{v-\frac{u}{4}}{u}=1 \Rightarrow u=v-\frac{u}{4} \quad \Rightarrow \frac{5u}{4}=v \quad ...(ii)$$

From (i) and (ii),

$$\Rightarrow \frac{5mu}{4}=\frac{3u}{2} \quad \Rightarrow m=\frac{6}{5}=1.2\text{kg}$$

167. (b) $m(2v)+2mv=0+2mv'\cos 45°$ or $v'=2\sqrt{2}v$

168. (b) $m_1v_1+m_2v_2=m_1v_3+m_2v_4$

or $m_1v_1+(0.5m_1)v_2=m_1 0.5v_1+(0.5m_1)v_4$

On solving, $v_1=v_4-v_2$

169. (d) Using conservation of momentum,

$mv_0 = Mv_2 - mv_1$

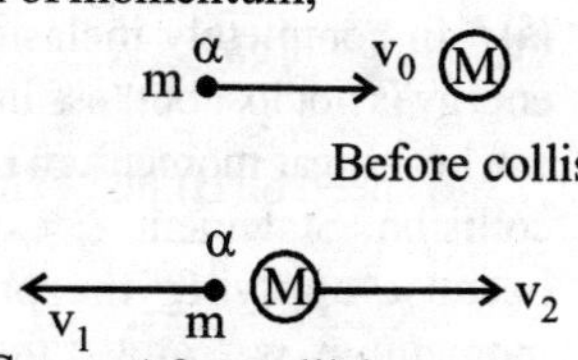

Before collision

After collision

$$\frac{1}{2}mv_1^2=0.36\times\frac{1}{2}mv_0^2$$

$$\Rightarrow v_1=0.6v_0$$

The collision is elastic. So,

$$\frac{1}{2}Mv_2^2=0.64\times\frac{1}{2}mv_0^2 \; [\therefore M=\text{mass of nucleus}]$$

$$\Rightarrow v_2=\sqrt{\frac{m}{M}}\times 0.8v_0$$

$$mv_0=\sqrt{mM}\times 0.8v_0-m\times 0.6v_0$$

$$\Rightarrow 1.6m=0.8\sqrt{mM} \Rightarrow 4m^2=mM \qquad \therefore M=4m$$

170. (c) Kinetic energy of block A

$$k_1=\frac{1}{2}mv_0^2$$

$\therefore$ From principle of linear momentum conservation

$$mv_0=(2m+M)v_f \Rightarrow v_f=\frac{mv_0}{2m+M}$$

According to question, $\frac{5}{6}$th of the initial kinetic energy is lost in whole process.

$$\therefore \frac{k_i}{k_f}=6 \Rightarrow \frac{\frac{1}{2}mv_0^2}{\frac{1}{2}(2m+M)\left(\frac{mv_0}{2m+M}\right)^2}=6$$

$$\Rightarrow \frac{2m+M}{m}=6 \therefore \frac{M}{m}=4$$

171. (b) Before Collision — After Collision

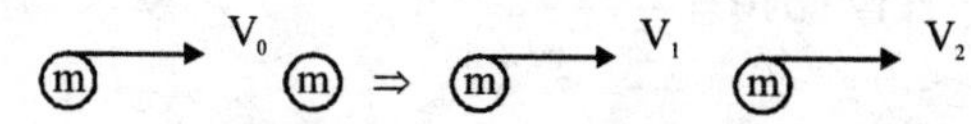

Stationary

Here, $K_f = 1.5K_i$

$$\Rightarrow \frac{1}{2}mv_1^2+\frac{1}{2}mv_2^2=\frac{3}{2}\left(\frac{1}{2}mv_0^2\right)$$

$$\Rightarrow v_1^2+v_2^2=\frac{3}{2}v_0^2 \quad ....(i)$$

From momentum conservation

$mv_0 = m(v_1 + v_2)$

Squarring both sides,

$(v_1+v_2)^2=v_0^2 \Rightarrow v_1^2+v_2^2+2v_1v_2=v_0^2$

$$\Rightarrow 2v_1v_2=-\frac{v_0^2}{2} \quad \text{[using (i)]}$$

$$(v_1-v_2)^2=v_1^2+v_2^2-2v_1v_2=\frac{3}{2}v_0^2+\frac{v_0^2}{2}$$

Solving we get relative velocity between the two particles

$v_1-v_2=\sqrt{2}v_0$

172. (a) Change in momentum

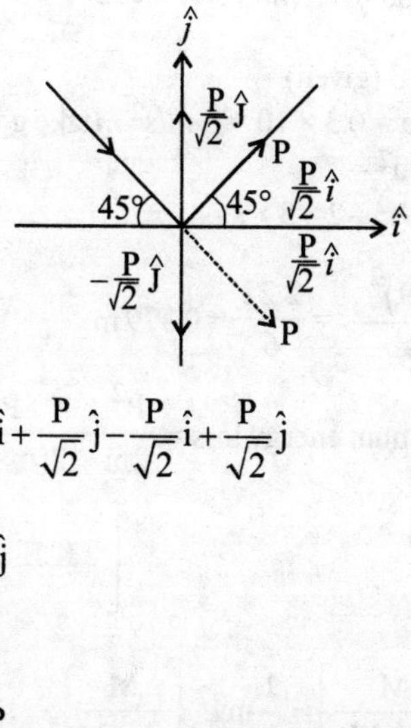

$$\Delta\vec{P}=\frac{P}{\sqrt{2}}\hat{i}+\frac{P}{\sqrt{2}}\hat{j}-\frac{P}{\sqrt{2}}\hat{i}+\frac{P}{\sqrt{2}}\hat{j}$$

$$\Delta\vec{P}=\frac{2P}{\sqrt{2}}\hat{j}$$

Pressure,

$$P=\frac{F}{A}=\frac{\sqrt{2}P}{A}n \quad (\because n=\text{no. of particles striking per second})$$

$$=\frac{\sqrt{2}\times 3.32\times 10^{-27}\times 10^{3}\times 10^{23}}{2\times 10^{-4}}=2.35\times 10^{3}\text{N/m}^2$$

173. (a) For particle C,
According to law of conservation of linear momentum, verticle component,
$2\,mv'\sin\theta = mv\sin 45° + mv\sin 60°$

$$2mv'\sin\theta = \frac{mv}{\sqrt{2}} + \frac{mv\sqrt{3}}{2} \quad \text{...... (i)}$$

Horizontal component,
$2\,mv'\cos\theta = mv\cos 60° - mv\cos 45°$

$$2mv'\cos\theta = \frac{mv}{2} - \frac{mv}{\sqrt{2}} \quad \text{...... (ii)}$$

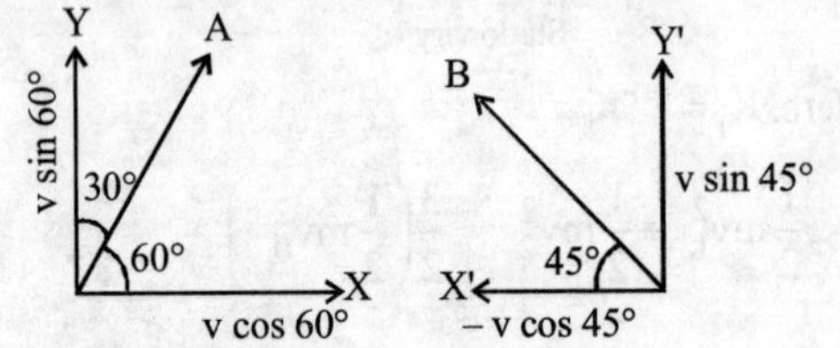

For particle A **For particle B**

Dividing eq^n (i) by eq^n (ii),

$$\tan\theta = \frac{\frac{1}{\sqrt{2}} + \frac{\sqrt{3}}{2}}{\frac{1}{2} - \frac{1}{\sqrt{2}}} = \frac{\sqrt{2}+\sqrt{3}}{1-\sqrt{2}}$$

174. (b) Given, $m_1 = 4g$, $u_1 = 300$ m/s
$m_2 = 0.8$ kg $= 800$ g, $u_2 = 0$ m/s
From law of conservation of momentum,
$m_1u_1 + m_2u_2 = m_1v_1 + m_2v_2$
Let the velocity of combined system = v m/s then,

$$4 \times 300 + 800 \times 0 = (800+4) \times v \Rightarrow v = \frac{1200}{804} = 1.49 \text{ m/s}$$

Now, $\mu = 0.3$ (given)
$a = \mu g \Rightarrow a = 0.3 \times 10 = 3$ m/s^2 (take g = 10 m/s^2)
using $v^2 = u^2 = 2as$
$\Rightarrow 0 = (1.49)^2 - 2 \times 3 \times s$

$$\therefore s = \frac{(1.49)^2}{6} = \frac{2.22}{6} = 0.379 \text{ m}$$

175. (d) Maximum energy loss $= \frac{P^2}{2m} - \frac{P^2}{2(m+M)}$

$$\left[\because K.E. = \frac{P^2}{2m} = \frac{1}{2}mv^2\right]$$

$$= \frac{P^2}{2m}\left[\frac{M}{(m+M)}\right] = \frac{1}{2}mv^2\left\{\frac{M}{m+M}\right\}$$

Statement II is a case of perfectly inelastic collision.
By comparing the equation given in statement I with above equation, we get

$$f = \left(\frac{M}{m+M}\right) \text{ instead of } \left(\frac{m}{M+m}\right)$$

Hence statement I is wrong and statement II is correct.

176. (c)

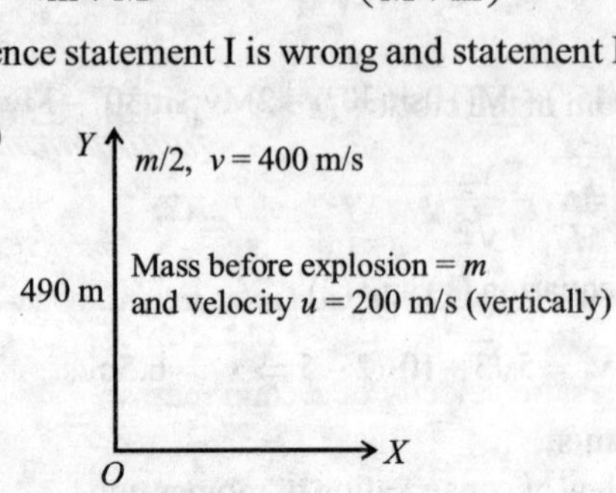

Momentum before explosion = Momentum after explosion

$$m \times 200\hat{j} = \frac{m}{2} \times 400\hat{j} + \frac{m}{2}\vec{u} = \frac{m}{2}\left(400\hat{j} + \vec{u}\right)$$

$\Rightarrow \quad 400\hat{j} - 400\hat{j} = \vec{u} \qquad \therefore \quad \vec{u} = 0$

i.e., the velocity of the other part of the mass, $v = 0$
Let time taken to reach the earth by this part be t

Applying formula, $h = ut + \frac{1}{2}gt^2$

$$490 = 0 + \frac{1}{2} \times 9.8 \times t^2$$

$$\Rightarrow \quad t^2 = \frac{980}{9.8} = 100 \qquad \therefore \quad t = \sqrt{100} = 10 \sec$$

177. (a) In completely inelastic collision, all initial kinetic energy is not lost but loss in kinetic energy is as large as it can be. Linear momentum remain conserved in all types of collision. Statement -2 explains statement -1 correctly because applying the principle of conservation of momentum, we can get the common velocity and hence the kinetic energy of the combined body.

178. (b) Let the velocity and mass of 4 kg piece be v_1 and m_1 and that of 12 kg piece be v_2 and m_2.

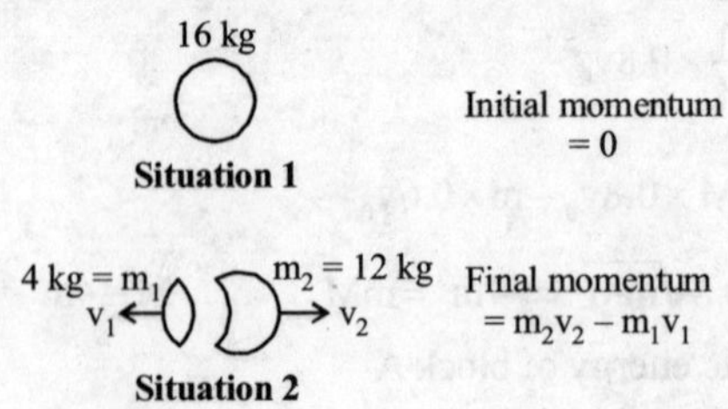

Applying conservation of linear momentum

$$16 \times 0 = -4 \times v_1 + 12 \times 4 \Rightarrow v_1 = \frac{12 \times 4}{4} = 12 \; ms^{-1}$$

Kinetic energy of 4 kg mass

$$\therefore K.E. = \frac{1}{2}m_1v_1^2 = \frac{1}{2} \times 4 \times 144 = 288 \; J$$

179. (b) When the spring gets compressed by length L.
K.E. lost by mass M = P.E. stored in the compressed spring.

$$\frac{1}{2}Mv^2 = \frac{1}{2}kL^2 \Rightarrow \quad v = \sqrt{\frac{k}{M}} \cdot L$$

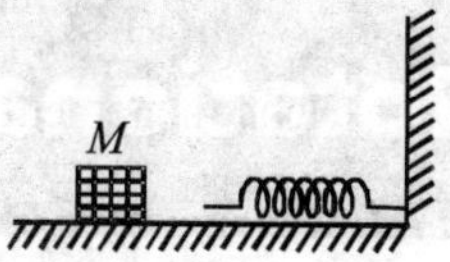

Momentum of the block $= M \times v$

$$= M \times \sqrt{\frac{k}{M}} \cdot L = \sqrt{kM} \cdot L$$

180. **(d)** Considering conservation of momentum along *x*-direction,

$mv = mv_1 \cos\theta$ ...(1)

where v_1 is the velocity of second mass

In *y*-direction,

$$0 = \frac{mv}{\sqrt{3}} - mv_1 \sin\theta$$

or $mv_1 \sin\theta = \dfrac{mv}{\sqrt{3}}$ ...(2)

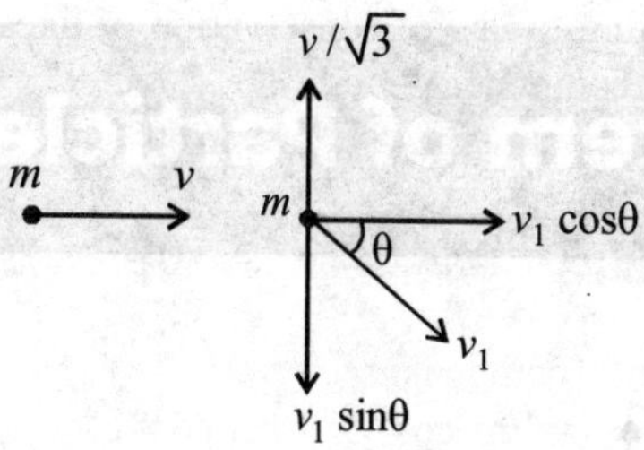

Squaring and adding eqns. (1) and (2) we get

$$v_1^2 = v^2 + \frac{v^2}{3} \Rightarrow v_1 = \frac{2}{\sqrt{3}}v$$

181. **(c)** Kinetic energy of a system of particle is zero only when the speed of each particles is zero. This implies momentum of each particle is zero, thus linear momentum of the system of particle has to be zero.

Also if linear momentum of the system is zero it does not mean linear momentum of each particle is zero. This is because linear momentum is a vector quantity. In this case the kinetic energy of the system of particles will not be zero.

∴ A does not imply B but B implies A.

6 System of Particles and Rotational Motion

1. **(d)**

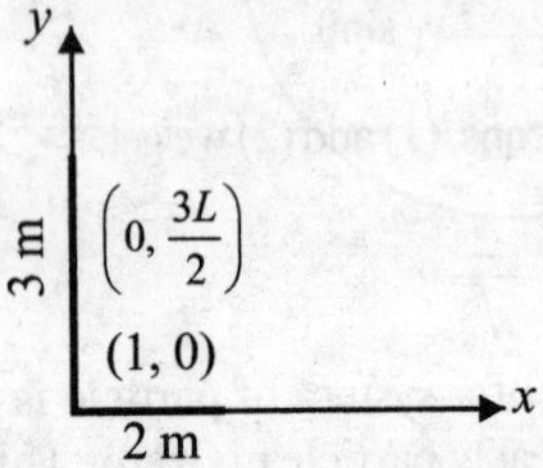

Here, $m_1 = 2$ m, $m_2 = 3$ m, $x_1 = L = 10$ cm, $x_2 = 0, y_1 = 0$,

$$y_2 = \frac{3L}{2} = \frac{3\times10}{2} = 15 \text{ cm}$$

$$x_{com} = \frac{m_1x_1 + m_2x_2}{m_1+m_2} = \frac{2\text{ m}(10)+3\text{ m}(0)}{5\text{ m}} = 4 \text{ cm}$$

$$y_{com} = \frac{m_1y_1 + m_2y_2}{m_1+m_2} = \frac{2\text{ m}(0)+3\text{ m}(15)}{5\text{ m}} = 9 \text{ cm}$$

$$\vec{r}_{com} = x_{com}\hat{i} + y_{com}\hat{j} = 4\hat{i} + 9\hat{j}.$$

2. **(a)** The centre of mass of a 2D plate is given by

$$x_{cm} = \frac{\int x dm}{\int dm}; \quad y_{cm} = \frac{\int y dm}{\int dm}$$

where dm is the mass of a small elemental area.
$dm = \sigma dA$
σ is constant in y-direction. So, $y_{cm} = b/2$

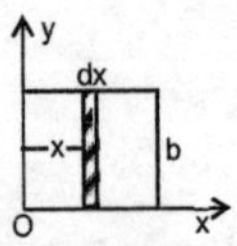

$$x_{cm} = \frac{\int_0^a x dm}{\int_0^a dm} = \frac{\int_0^a x\sigma_x dA}{\int_0^a \sigma_x dA} = \frac{\int_0^a x\frac{\sigma_o x}{ab} bdx}{\int_0^a \frac{\sigma_o x}{ab} bdx}$$

$$x_{cm} = \frac{\int_0^a x^2 dx}{\int_0^a x dx} = \frac{\left(\frac{x^3}{3}\right)_0^a}{\left(\frac{x^2}{2}\right)_0^a} = \frac{a^3/3}{a^2/2} = \frac{2a}{3}$$

3. **(d)**

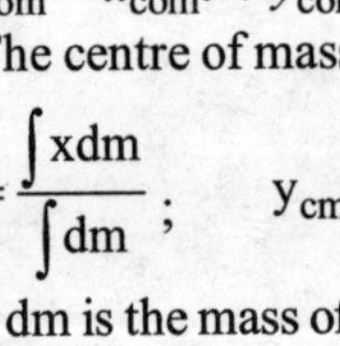

The expression for calculating center of mass of a system of particles

$$X_{cm} = \frac{m_1x_1 + m_2x_2 + ... + m_nx_n}{m_1 + m_2 + ... + m_n}$$

mass of disc = m

mass of cut part = $\frac{m}{16}$

$$X_{cm} = \frac{m\times0 - \frac{m}{16}\times15}{m - \frac{m}{16}} = 1 \text{ cm.}$$

4. **(15)** Consider metal plate is made up of three equal parts each of mass 2 kg

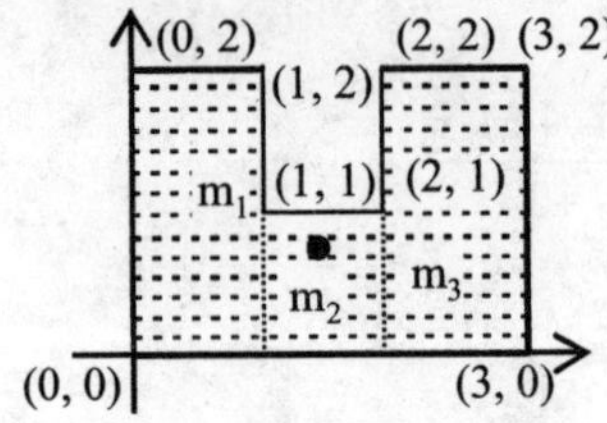

$$\Rightarrow x_{cm} = \frac{4\times0.5 + 2\times1.5 + 4\times2.5}{10} = 1.5$$

$$\Rightarrow y_{cm} = \frac{4\times1 + 2\times0.5 + 4\times1}{10} = 0.9$$

$$\frac{x_{cm}}{y_{cm}} = \frac{1.5}{0.9} = \frac{15}{9} = \frac{n}{9} \quad \therefore n = 15$$

5. **(3)** Using $m_1\Delta x_1 = m_2\Delta x_2$

or, $3\times2 = 2\,\Delta x_2$

$\therefore \Delta x_2 = 3$ cm

6. **(2)** Here, $\vec{r}_{cm} = \frac{m_1\vec{r}_1 + m_2\vec{r}_2}{m_1 + m_2}$

$$= \frac{2M(4\hat{i}) + 2M(4\hat{j})}{2M + 2M} = 2\hat{i} + 2\hat{j}$$

$$|\vec{r}_{cm}| = \sqrt{4+4} = 2\sqrt{2} = \frac{4\sqrt{2}}{x} \text{ (given)}$$

$\therefore$ x = 2

7. **(a)** Position of COM of the mass system is given as,

$$\vec{r}_{com} = \frac{m_1\vec{r}_1 + m_2\vec{r}_2}{m_1 + m_2} = \frac{1(\hat{i} + 2\hat{j} + \hat{k}) + 3(-3\hat{i} - 2\hat{j} + \hat{k})}{1+3}$$

$$= -2\hat{i} - \hat{j} + \hat{k}$$

$$|\vec{r}_{cm}| = |-2\hat{i} - \hat{j} + \hat{k}| = \sqrt{(2)^2 + (1)^2 + (1)^2} = \sqrt{6}$$

Only option (A) magnitude is $\sqrt{1^2 + (-2)^2 + 1^2} = \sqrt{6}$
So option(A) is correct.

8. **(8)** Let us consider a elemental length 'dx' at 'x' distance from one end as shown.

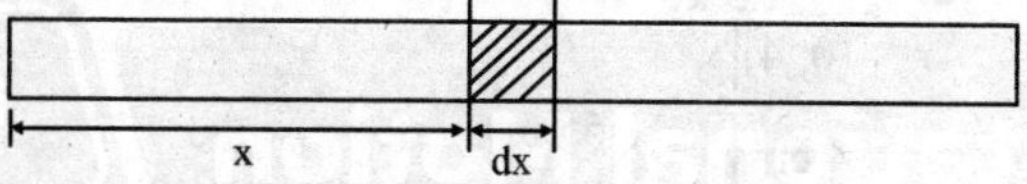

Let 'dm' be the mass of elemental length

$$dm = \lambda . dx = \lambda_0 \left(1 - \frac{x^2}{L^2}\right) dx$$

$$x_{cm} = \frac{\int x dm}{\int dm} = \frac{\lambda_0 \int_0^L \left(x - \frac{x^3}{L^2}\right) dx}{\int_0^L \lambda_0 \left(1 - \frac{x^2}{L^2}\right) dx} = \frac{\frac{L^2}{2} - \frac{L^4}{4L^2}}{L - \frac{L^3}{3L^2}} = \frac{3L}{8} m$$

Comparing it with $x_{cm} = \frac{3L}{\alpha} m$, we get $\alpha = 8$

9. **(2)** $\vec{r}_{com} = \frac{M_1\vec{r}_1 + M_2\vec{r}_2 + M_3\vec{r}_3}{M_1 + M_2 + M_3}$

$$\Rightarrow \vec{r}_{com} = \frac{M(0\hat{i} + 0\hat{j}) + M(3\hat{i}) + M(3\hat{j})}{3M}$$

$$\Rightarrow \vec{r}_{com} = \hat{i} + \hat{j}$$

$$|\vec{r}_{com}| = \sqrt{1^2 + 1^2} = \sqrt{2} m = \sqrt{x} m$$

so, $x = 2$

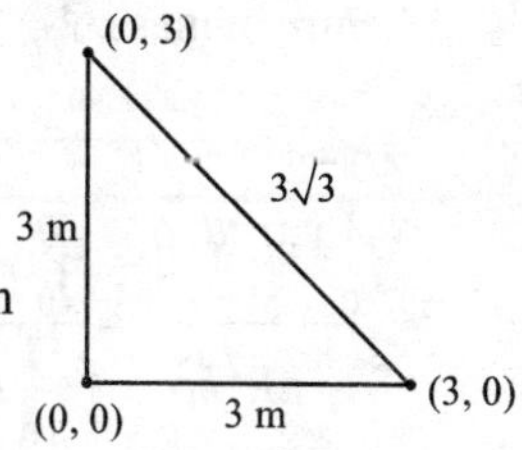

10. **(c)** Now, $d\vec{x}_{cm} = \frac{m_1 d\vec{x}_1 + m_2 d\vec{x}_2}{m_1 + m_2}$

$$\Rightarrow 0 = \frac{m_1 d\vec{x}_1 + m_2 d\vec{x}_2}{m_1 + m_2}$$

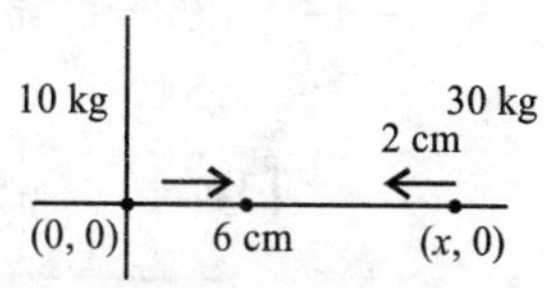

Here, $d\vec{x}_1 = +6 cm\ \hat{i}$

So, $10 \times 6\ \hat{i} + 30 \times d\vec{x}_2 = 0$

$\Rightarrow 30\, d\vec{x}_2 = -60\ \hat{i} \Rightarrow d\vec{x}_2 = -2cm\ \hat{i} \Rightarrow d\vec{x}_2 = 2$ cm towards 10 kg block

11. **(2)** COM of ring is at centre of ring and of semi-circular ring is at $\frac{2R}{\pi}$ as shown in figure.

Hence value of $x = 2$

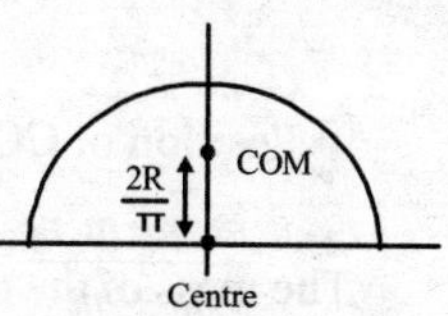

12. **(4)** Mass of element,

$$dm = \sigma \frac{1}{2} a \times a d\theta \Rightarrow dm = \frac{\sigma a^2 d\theta}{2}$$

$$x_{cm} = \frac{\int x\, dm}{\int dm} = \frac{\int_0^{\pi/2} \frac{\sigma a^2}{2} d\theta \left(\frac{2a}{3}\cos\theta\right)}{\int_0^{\pi/2} \frac{\sigma a^2}{2} d\theta}$$

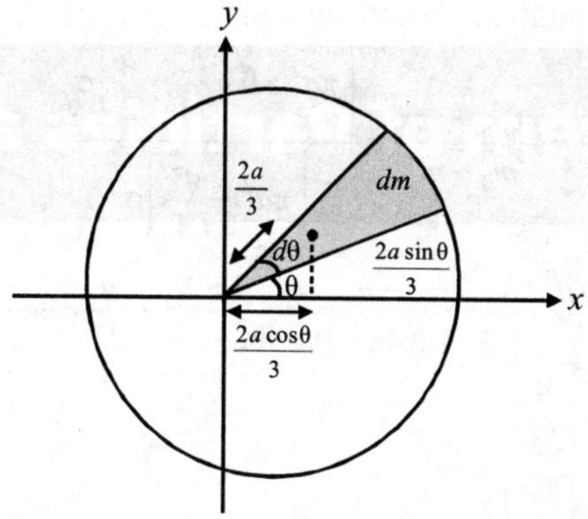

$$= \frac{2a}{3} \frac{\int_0^{\pi/2} \cos\theta\, d\theta}{\int_0^{\pi/2} d\theta} = \frac{2a}{3}\left(\frac{2}{\pi}\right) = \frac{4a}{3\pi} = \frac{xa}{3\pi} \qquad \therefore x = 4.$$

13. **(6)** By law of conservation of linear momentum $\vec{P}_i = \vec{P}_f$

$\Rightarrow 60 \times V = (120 + 60) \times 2 \Rightarrow 60V = 360 \Rightarrow V = 6$ m/s

14. **(c)** Let M be the mass of circular disc.

The density of disc; $\sigma = \frac{M}{\pi a^2}$

Now let M' be the mass of cut out disc.

$$\therefore M' = \sigma \times \pi\left(\frac{a}{2}\right)^2 = \frac{M}{\pi a^2} \times \pi \frac{a^2}{4} = \frac{M}{4}$$

$$X_{com} = \frac{M \times a - \frac{M}{4} \times \frac{3}{2} a}{M - \frac{M}{4}} = \frac{5}{6} a.$$

15. **(3)** Centre of mass of solid hemisphere of radius R lies at a distance $\frac{3R}{8}$ above the centre of flat side of hemisphere.

$$\therefore h_{cm} = \frac{3R}{8} = \frac{3 \times 8}{8} = 3 \text{ cm}$$

16. **(23)** Let σ be the mass density of circular disc.

Original mass of the disc, $m_0 = \pi a^2 \sigma$

Removed mass, $m = \frac{a^2}{4}\sigma$

Remaining, mass, $m' = \left(\pi a^2 - \frac{a^2}{4}\right)\sigma = a^2\left(\frac{4\pi - 1}{4}\right)\sigma$

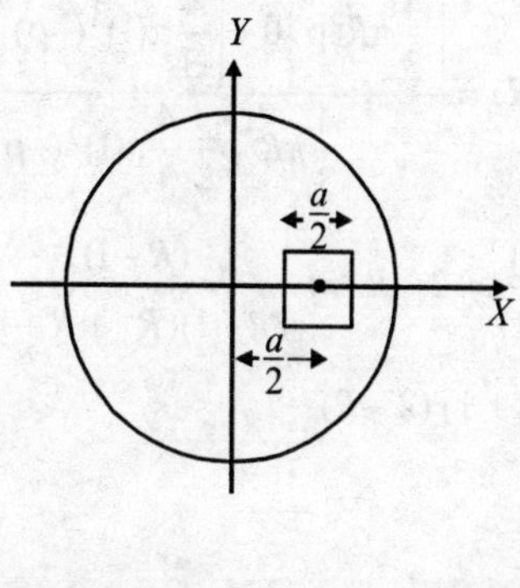

New position of centre of mass

$$X_{CM} = \frac{m_0 x_0 - mx}{m_0 - m} = \frac{\left[\pi a^2 \times 0 - \frac{a^2}{4} \times \frac{a}{2}\right]\sigma}{\left[\pi a^2 - \frac{a^2}{4}\right]\sigma}$$

$$= \frac{-a^3/8}{\left(\pi - \frac{1}{4}\right)a^2} = \frac{-a}{2(4\pi - 1)} = \frac{-a}{8\pi - 2} = -\frac{a}{23} = -\frac{a}{x} \text{ (given)}$$

$\therefore x = 23$

17. (b)

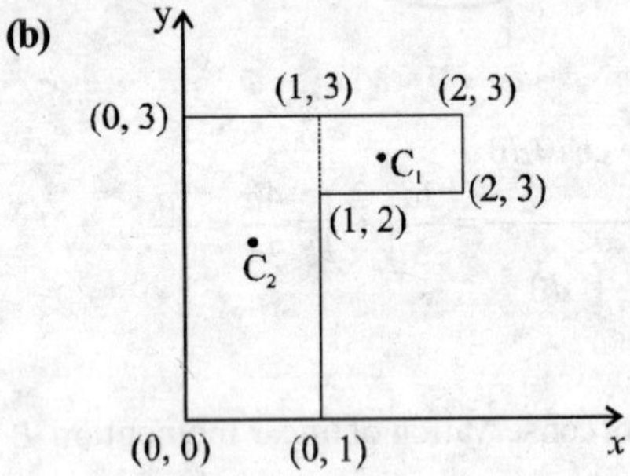

For given Lamina

$x \quad y$

$m_1 = 1, C_1 = (1.5, 2.5)$

$m_2 = 3, C_2 = (0.5, 1.5)$

$$X_{cm} = \frac{m_1 x_1 + m_2 x_2}{m_1 + m_2} = \frac{1.5 + 1.5}{4} = 0.75m$$

$$Y_{cm} = \frac{m_1 y_1 + m_2 y_2}{m_1 + m_2} = \frac{2.5 + 4.5}{4} = 1.75m$$

$\therefore$ Coordinate of centre of mass of flag shaped lamina $(0.75m, 1.75m)$

18. (a) Mass of sphere = volume of sphere x density of sphere

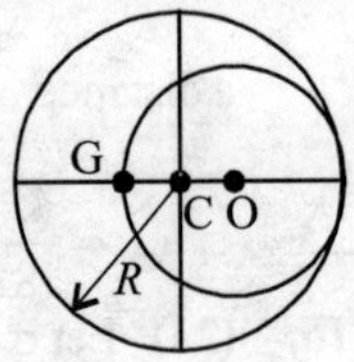

$= \frac{4}{3}\pi R^3 \rho$

Mass of cavity, $M_{\text{cavity}} = \frac{4}{3}\pi(1)^3 \rho$

Mass of remaining,

$$M_{(\text{Remaining})} = \frac{4}{3}\pi R^3 \rho - \frac{4}{3}\pi(1)^3 \rho$$

Centre of mass of remaining part,

$$X_{COM} = \frac{M_1 r_1 + M_2 r_2}{M_1 + M_2}$$

$$\Rightarrow -(2 - R) = \frac{\left[\frac{4}{3}\pi R^3 \rho\right]0 + \left[\frac{4}{3}\pi(1)^3(-\rho)\right][R - 1]}{\frac{4}{3}\pi R^3 \rho + \frac{4}{3}\pi(1)^3(-\rho)}$$

$$\Rightarrow \frac{(R-1)}{(R^3 - 1)} = 2 - R \Rightarrow \frac{(R-1)}{(R-1)(R^2 + R + 1)} = 2 - R$$

$$\Rightarrow (R^2 + R + 1)(2 - R) = 1$$

19. (d)

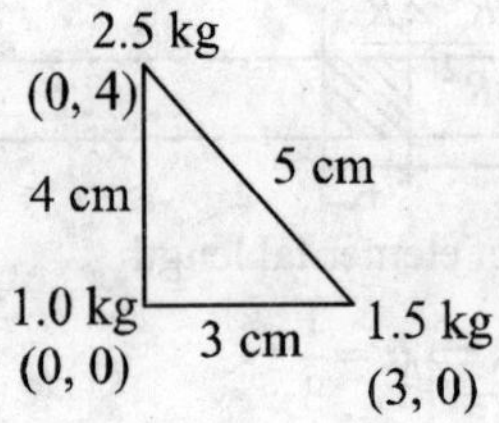

$$X_{cm} = \frac{m_1 x_1 + m_2 x_2 + m_3 x_3}{m_1 + m_2 + m_3}$$

$$X_{cm} = \frac{1 \times 0 + 1.5 \times 3 + 2.5 \times 0}{1 + 1.5 + 2.5} = \frac{1.5 \times 3}{5} = 0.9cm$$

$$Y_{cm} = \frac{m_1 y_1 + m_2 y_2 + m_3 y_3}{m_1 + m_2 + m_3}$$

$$Y_{cm} = \frac{1 \times 0 + 1.5 \times 0 + 2.5 \times 4}{1 + 1.5 + 2.5} = \frac{2.5 \times 4}{5} = 2cm$$

Hence, centre of mass of system is at point (0.9, 2)

20. (d) Let density of cone = ρ.

Centre of mass, $y_{cm} = \frac{\int y dm}{\int dm}$

$$= \frac{\int_0^h y \pi r^2 dy \rho}{\frac{1}{3}\pi R^2 h \rho} = \frac{\int_0^h r^2 y dy}{\frac{1}{3}R^2 h} \quad \text{...(i)}$$

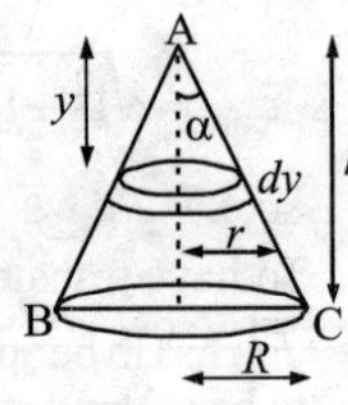

For a cone, we know that

$$\frac{r}{R} = \frac{y}{h} \quad \therefore r = \frac{y}{h}R$$

$$y_{cm} = \frac{\int_0^h 3y^3 dy}{h^3} = \frac{3\left[\frac{y^4}{4}\right]_0^h}{h^3} = \frac{3}{4}h$$

21. (b) Let σ be the mass per unit area of the disc. Then the mass of the complete disc $= \sigma(\pi(2R)^2)$

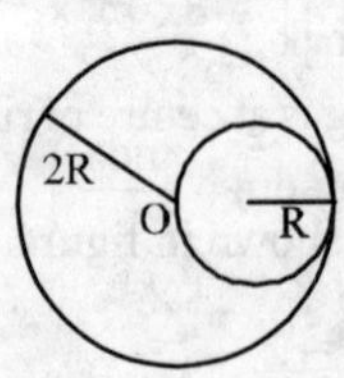

The mass of the removed disc $= \sigma(\pi R^2) = \pi\sigma R^2$

Let us consider the above situation to be a complete disc of radius 2R on which a disc of radius R of negative mass is superimposed. Let O be the origin. Then the above figure can be redrawn keeping in mind the concept of centre of mass as :

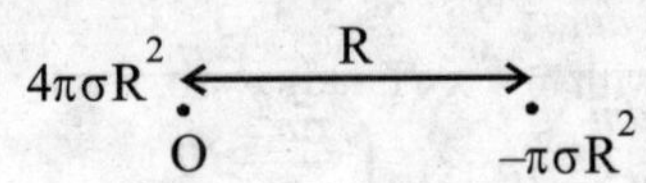

$$x_{c.m} = \frac{4\pi\sigma R^2 \times 0 - \pi\sigma R^2 \times R}{4\pi\sigma R^2 - \pi\sigma R^2}$$

$$\therefore\ x_{c.m} = \frac{-\pi\sigma R^2 \times R}{3\pi\sigma R^2}$$

$$\therefore\ x_{c.m} = -\frac{R}{3} = -\alpha R \Rightarrow \alpha = \frac{1}{3}$$

22. **(c)** $m_1 dx_1 + m_2 dx_2 = 0$

$\Rightarrow\ m_1 d + m_2 dx_2 = 0$

$\Rightarrow\ dx_2 = -\frac{m_1 d}{m_2}$

$|dx_2| = \frac{m_1}{m_2} d$

23. **(a)** The centre of mass of bodies B and C taken together does not shift as no external force acts. The centre of mass of the system continues its original path. It is only the internal forces which comes into play while breaking.

24. **(d)**

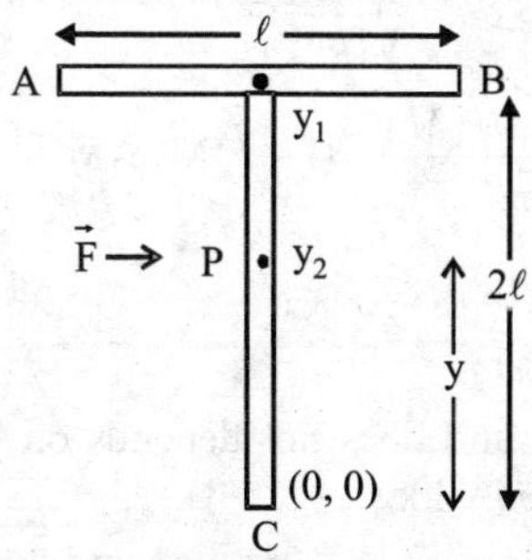

To have translational motion without rotation, the force $\vec{F}$ has to be applied at centre of mass. i.e. the point 'P' has to be at the centre of mass.

Taking point C at the origin position, positions of y_1 and y_2 are $y_1 = 2l$, $y_2 = l$ and $m_1 = m$ and $m_2 = 2m$

$$y = \frac{m_1 y_1 + m_2 y_2}{m_1 + m_2} = \frac{m \times 2\ell + 2m \times \ell}{3m} = \frac{4\ell}{3}$$

25. **(4)** The force exerted by the liuqid at the other end,

$$F = \int (dm)\omega^2 x = \int_0^L \left(\frac{m}{L}dx\right)\omega^2 x$$

$$\left(\text{for small element } dm = \frac{mdx}{L}\right)$$

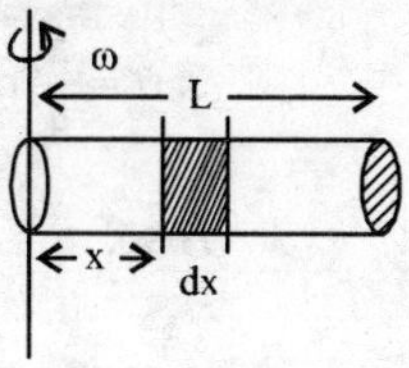

$$= \frac{m}{L}\omega^2 \frac{L^2}{2} = \frac{m\omega^2 L}{2}$$

The angular velocity will be

$$\omega = \sqrt{\frac{2}{mL}}\sqrt{F} = \sqrt{\frac{2F}{0.25 \times 0.5}} = \sqrt{16}\sqrt{F}$$

$\omega = 4\sqrt{F}$ rad s^{-1}

Compair with $\omega = x\sqrt{F}$ rad s^{-1} so, $x = 4$

26. **(b)** $\alpha = 6t^2 - 2t$

$$\int_{10}^{\omega} d\omega = \int_0^t (6t^2 - 2t)dt \Rightarrow \omega - 10 = \frac{6t^3}{3} - \frac{2t^2}{2}$$

$$\omega = 10 + 2t^3 - t^2 \text{ as, } \int_4^{\theta} d\theta = \int_0^t \omega\, dt$$

$$\Rightarrow \theta - 4 = \int_0^t (10 + 2t^3 - t^2)dt \Rightarrow \theta = 4 + 10t + \frac{t^4}{2} - \frac{t^3}{3}$$

27. **(b)** We have $\theta = \omega_0 t + \frac{1}{2}\alpha t^2$

Here, $\omega_0 = 0, \theta = 5rad$ and $t = 1$ sec

$\Rightarrow 5 = \frac{1}{2}\alpha \times 1^2 \Rightarrow \alpha = 10$ rad/s^2

Now, $\theta_n = \frac{\alpha}{2}(2n-1) + \omega_0$

$\theta_2 = \frac{10}{2}(2 \times 2 - 1) + 0 = 15$ rad/sec.

28. **(728)** From $\omega = \omega_0 + \alpha t$

$$\Rightarrow\ \alpha = \frac{\omega - \omega_0}{t} = \frac{(2460 - 900)}{26} \times \frac{2\pi}{60}$$

$\Rightarrow\ \alpha = 2\pi$ rad/s^2

$$\theta = \omega_0 t + \frac{1}{2}\alpha t^2$$

or, $2\pi n = 900 \times \frac{2\pi}{60} \times 26 + \frac{1}{2} \times 2\pi \times (26)^2\ \therefore n = 728$

29. **(a)** $y = 4Cx^2 \Rightarrow \frac{dy}{dx} = \tan\theta = 8Cx$

At P, $\tan\theta = 8Ca$

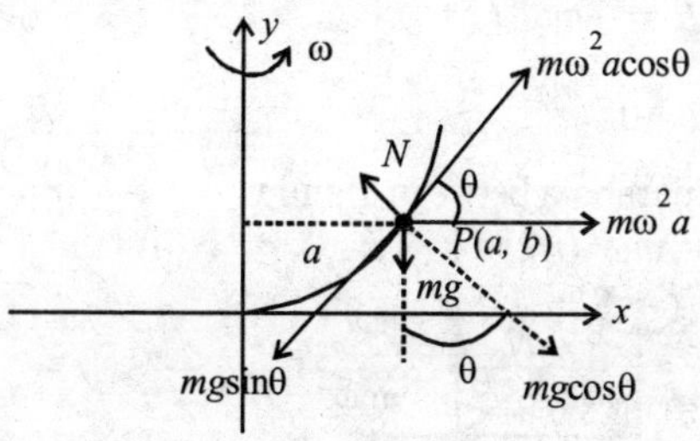

For steady circular motion

$$m\omega^2 a\cos\theta = mg\sin\theta \Rightarrow \omega = \sqrt{\frac{g\tan\theta}{a}}$$

$$\therefore \omega = \sqrt{\frac{g \times 8aC}{a}} = 2\sqrt{2gC}$$

30. **(c)** $P_0 - P_0 = \frac{1}{2}\rho\omega^2 R^2 - \rho gh$

$$h = \frac{R^2\omega^2}{2g}$$

$$h = \frac{25\omega^2}{2g} \quad [\because R = 5 \text{ cm}]$$

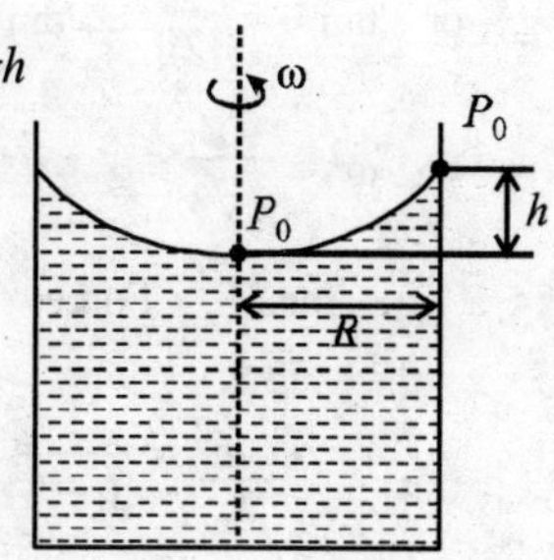

31. **(c)** Free body diagram in the frame of disc

$$\xleftarrow{kx} \boxed{m} \longrightarrow m\omega^2(\ell_0 + x)$$

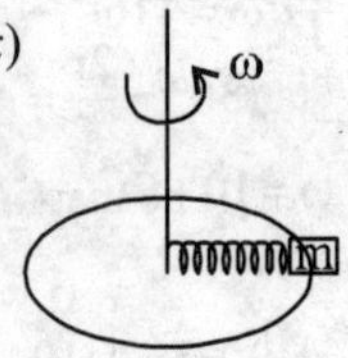

$\therefore \; m\omega^2(\ell_0 + x) = kx$

$\Rightarrow \; x = \dfrac{m\ell_0\omega^2}{k - m\omega^2}$

For $k >> m\omega^2$

$\Rightarrow \; \dfrac{x}{\ell_0} = \dfrac{m\omega^2}{k}$

32. **(b)** At elongated position (x),

$F_{\text{radial}} = \dfrac{mv^2}{r} = mr\omega^2$

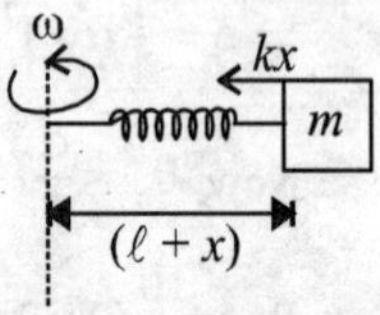

$\therefore kx = m(\ell + x)\omega^2$

$(\because r = \ell + x \text{ here})$

$kx = m\ell\omega^2 + mx\omega^2 \therefore x = \dfrac{m\ell\omega^2}{k - m\omega^2}$

33. **(b)** $\displaystyle\int_0^T dT = \int_x^l (dm)\omega^2 x$

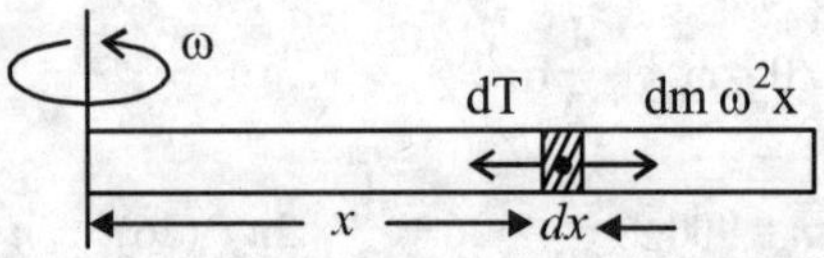

$T = \displaystyle\int_x^l \left(\frac{m}{l}dx\right)\omega^2 x \Rightarrow T = \frac{m\omega^2}{2l}(l^2 - x^2)$

when $x = 0$, $T = \dfrac{1}{2}m\omega^2 l$

when $x = l$, $T = 0$

Also, it is a parabola between T and x.

34. **(b)** $N\sin\theta = m\omega^2(r/2)$ ...(i)

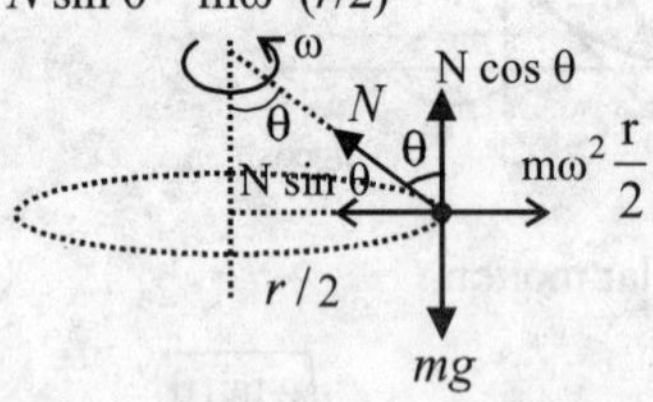

$\sin\theta = \dfrac{r/2}{r} = \dfrac{1}{2} \Rightarrow \theta = 30°$

and $N\cos\theta = mg$ ...(ii)

or $\tan\theta = \dfrac{\omega^2 r}{2g} \Rightarrow \tan 30° = \dfrac{\omega^2 r}{2g} \Rightarrow \dfrac{1}{\sqrt{3}} = \dfrac{\omega^2 r}{2g}$

$\Rightarrow \omega^2 = \dfrac{2g}{r\sqrt{3}}$.

35. **(c)** $m\omega^2 R = \text{Force} \propto \dfrac{1}{R^n}$

$\Rightarrow \omega^2 \propto \dfrac{1}{R^{n+1}} \Rightarrow \omega \propto \dfrac{1}{R^{\frac{n+1}{2}}}$

Time period $T = \dfrac{2\pi}{\omega}$

Time period, $T \propto R^{\frac{n+1}{2}}$

36. **(b)** Distance of COM from

$m_1 = \dfrac{(1-f)MR}{(1-f)M + fM} = R(1-f)$

Distance of COM from

$m_2 = \dfrac{(fM)R}{(1-f)M + fM} = Rf$

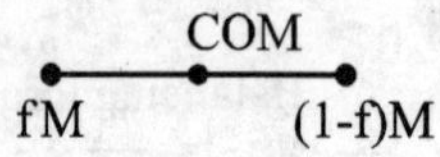

Now, both mass are rotating about COM. So,

For m_1, $\dfrac{G(1-f)M(fM)}{R^2} = (fM)\omega_1^2 R(1-f)$

$\Rightarrow \omega_1^2 = \dfrac{GM}{R^3} \Rightarrow \omega_1 = \sqrt{\dfrac{GM}{R^3}}$

For m_2, $\dfrac{G(1-f)Mf(M)}{R^2} = (1-f)M\omega_2^2 Rf$

$\Rightarrow \omega_2 = \sqrt{\dfrac{GM}{R^3}}$

Clearly, $\omega_1 = \omega_2$ and does not depends on 'f'.

37. **(40)** Given, mass $m = 1$ kg

$\omega_i = 1800 \text{ rpm} = 1800 \times \dfrac{2\pi}{60} = 60\pi \dfrac{\text{rad}}{\text{sec}}$

$\omega_f = 2100 \text{ rpm} = 2100 \times \dfrac{2\pi}{60} = 70\pi \dfrac{\text{rad}}{\text{sec}}$

$\tau_{\text{ext}} = 25\pi$ Nm and $t = 40$ sec

Using equation of motion, $\omega_f = \omega_i + \alpha t$

$\Rightarrow 70\pi = 60\pi + \alpha(40) \Rightarrow \alpha = \dfrac{10\pi}{40}$

$\therefore \alpha = \dfrac{\pi}{4}$ rad/sec^2

Also, $\tau = I\alpha = \dfrac{mR^2}{4}\alpha$

$\Rightarrow 25\pi = \dfrac{1 \times R^2}{4} \times \dfrac{\pi}{4} \Rightarrow R = 20$ m

Hence, diameter of disk $= 2R = 2 \times 20 = 40$ m

38. **(3)** Half volume of the unknown mass is inside the water

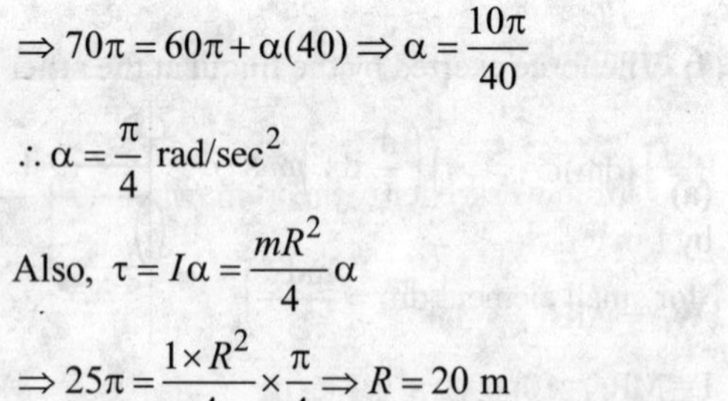

Given, volume of block $= (10 \times 10^{-2})^3 = 10^{-3}$ m^3

Let density of block = ρ kg/m^3

mass of block = $\rho \times 10^{-3}$ kg

Buoyant Force $(F_B) = \rho V g = 1000 \times \frac{10^{-3}}{2} \times 10 = 5$ N

F.B.D. of blocks

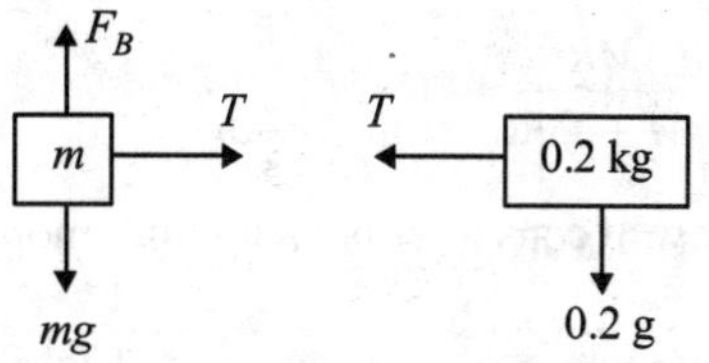

Balancing torque about point O, we get

$mg(2\times10^{-2}) - F_B(2\times10^{-2}) = 0.2g(25\times10^{-2})$

$\Rightarrow \rho\times10^{-3}\times10\times2-10=50$

$\Rightarrow \rho = 3000$ kg/m^3

Hence, mass of block $= \rho V = \rho\times10^{-3}$
$= 3000\times10^{-3} = 3$ kg

39. **(c)** $\vec{\tau} = \vec{r}\times\vec{L}$ is incorrect, torque is defined as $\vec{\tau} = \vec{r}\times\vec{F}$, remaining options are correct.

40. **(a)** Force $(F) = m\vec{a}$

$\vec{a} = -\omega^2\vec{r}$

$F = -m\omega^2\vec{r}$

$\therefore \vec{F}$ opposite to $\vec{r}$.

41. **(1)** Here FR = Iα

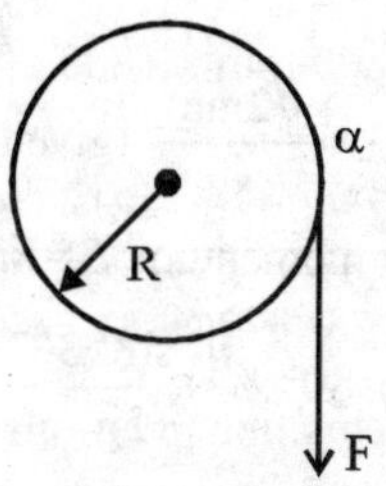

$\therefore$ Moment of inertia $I = \frac{FR}{\alpha} = \frac{10\times0.2}{2} = 1\,\text{kg m}^2$

42. **(a)** From work-energy theorem, when cord is unwound by 1 m

$W_F = \Delta KE = \frac{1}{2}I\omega^2$

$I = MR^2 = 10\times0.1^2 = 0.1\,\text{kg m}^2$

$\therefore\ 20 = \frac{1}{2}\times0.1\times\omega^2$

$\therefore\ \omega = 20$ rad/sec

43. **(c)**

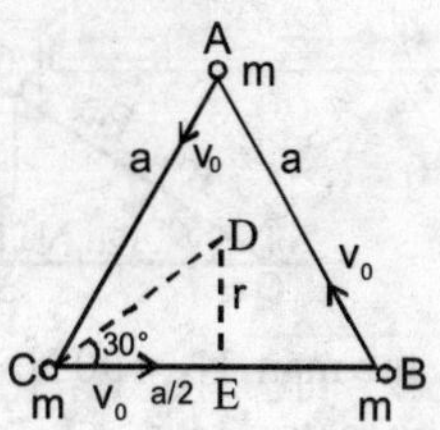

In ΔCDE,

$\tan 30° = \frac{2r}{a} = \frac{1}{\sqrt{3}} \Rightarrow r = \frac{a}{2\sqrt{3}}$

$\therefore$ The magnitude of angular momentum of the system about D is

$L = (mvr_\perp)\times3 = mv_0\frac{a}{2\sqrt{3}}\times3 = \frac{\sqrt{3}}{2}mv_0a$

44. **(c)** As lamina remains stationary so

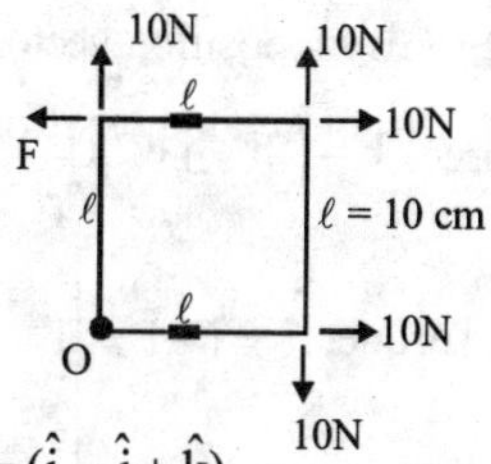

$F_{net} = 0$ and $\tau_{net} = 0$

$\therefore 0 = 10\ell - F\ell$

$\therefore F = 10$ N

45. **(2)** $\vec{r} = (\hat{i}+\hat{j}+\hat{k})$, $\vec{F} = (\hat{i}-\hat{j}+\hat{k})$

$\therefore$ Torque, $\vec{\tau} = \vec{r}\times\vec{F} = \begin{vmatrix}\hat{i} & \hat{j} & \hat{k}\\ 1 & 1 & 1\\ 1 & -1 & 1\end{vmatrix} = \hat{k}(-1-1) = -2\hat{k}$

$\therefore\ |\vec{\tau}| = 2$ Nm

46. **(b)** Use the concept of principle of moments sum of clockwise moments = Sum of counter clockwise moments

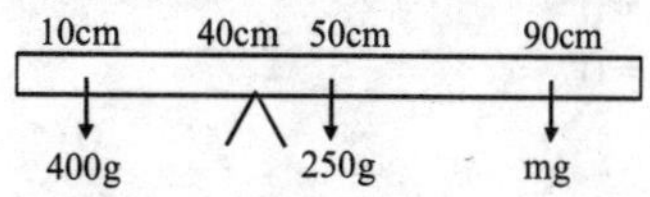

$\therefore\ \tau_{Net} = 0 \Rightarrow (400g\times30) = (250g\times10) + (mg\times50)$

$m = \frac{12000-2500}{50} = \frac{9500}{50} = 190$ g

47. **(d)**

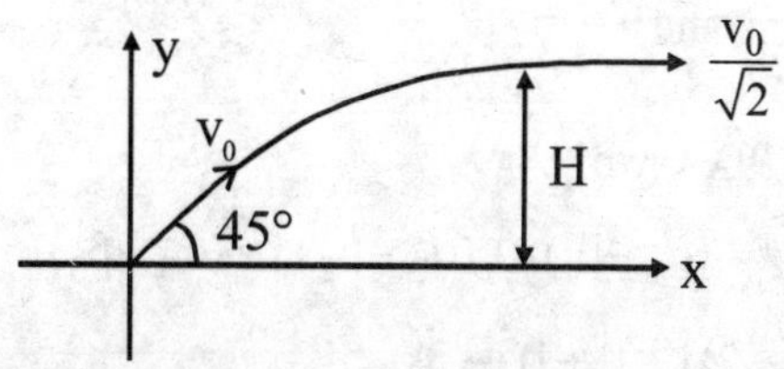

Maximum height,

$H = \frac{u^2}{2g} = \frac{(v_0\sin45°)^2}{2g} = \frac{\left(\frac{v_0}{\sqrt{2}}\right)^2}{2g} = \frac{v_0^2}{4g}$

Angular momentum, $L = mvr_\perp = mu_xH = m\frac{v_0}{\sqrt{2}}\frac{v_0^2}{4g}$

$= \frac{mv_0^3}{4\sqrt{2}g}$ along negative z-axis

48. **(a)** The given torque expression is

$\theta = 5t^2 - 8t$

$\omega = \frac{d\theta}{dt} = 10t - 8$

$\alpha = \frac{d\omega}{dt} = 10$

The power delivered by the applied torque is given by

$\therefore P = \tau\omega$

$= (I\alpha)\omega$

The moment of inertia for a circular disc rotating about an axis perpendicular to the disc $I = \frac{1}{2}MR^2$

$\therefore p = \left(\frac{MR^2}{2}\right)\alpha\,\omega = \left(\frac{MR^2}{2}\right)(10)(10t-8)$

At, t = 2, P = 60 mR^2

49. **(c)** Given, position vector, $\vec{r} = (\hat{i}+\hat{j}+\hat{k})$

Force, $\vec{F} = (2\hat{i}+\hat{j}+2\hat{k})$

$\therefore$ Torque, $\vec{\tau} = \vec{r}\times\vec{F} = \begin{vmatrix}\hat{i} & \hat{j} & \hat{k}\\ 1 & 1 & 1\\ 2 & 1 & 2\end{vmatrix} = \hat{i} - 0\hat{j} - \hat{k} = \hat{i} - \hat{k}$

50. **(90)** $\vec{V}_A = \frac{d\vec{r}_A}{dt} = (2t\hat{i} + 3n\hat{j} + 2\hat{k})$

$\vec{V}_B = \frac{d\vec{r}_B}{dt} = (2\hat{i} - 2t\hat{j} + 4p\hat{k})$

For $\vec{V}_A$ and $\vec{V}_B$ to be orthogonal,

At t = 1s, $\vec{V}_A \cdot \vec{V}_B = 0$

$\Rightarrow 4 - 6n + 8p = 0$

$\Rightarrow 2 - 3n + 4p = 0$

$\Rightarrow 3n = 2 + 4p$ ...(i)

Also, $|\vec{V}_A| = |\vec{V}_B|$

$4 + 9n^2 + 4 = 4 + 4 + 16p^2$ [Using eq (i)]

$p = \frac{-1}{4}$ and $n = \frac{1}{3}$

$\vec{L} = m_A(\vec{r}_{A/B}\times\vec{V}_A)$

$\vec{r}_{A/B} = (\alpha_1-\beta_1)\hat{i} + (\alpha_2-\beta_2)\hat{j} + (\alpha_3-\beta_3)\hat{k}$

$= (1-2)\hat{i} + (1+1)\hat{j} + 3\hat{k}$

$\overrightarrow{V_A} = (2\hat{i}+\hat{j}+2\hat{k})$

$\therefore \vec{L} = \begin{vmatrix}\hat{i} & \hat{j} & \hat{k}\\ -1 & 2 & 3\\ 2 & 1 & 2\end{vmatrix} = \hat{i} + 8\hat{j} - 5\hat{k}$

$\Rightarrow |\vec{L}| = \sqrt{1+64+25} = \sqrt{90}$

51. **(a)**

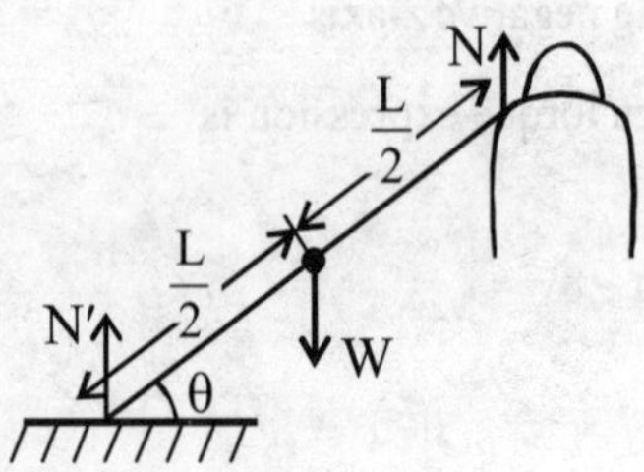

Let N be the force of reaction by shoulder

Balancing torque about point of contact on ground:

$W\left(\frac{L}{2}\cos\theta\right) = N(L\cos\theta) \Rightarrow N = \frac{W}{2}$

52. **(c)** Using law of conservation of angular momentum,

$I_1\omega = I_2\omega_2$

$\frac{MR^2}{2}\omega = \frac{3}{2}\left(\frac{MR^2}{2}\right)\omega_2 \Rightarrow \omega_2 = \frac{2}{3}\omega$

53. **(13)** Using law of conservation of angular momentum

$I_1\omega_1 = I_2\omega_2$

$\left(\frac{2}{5}MR^2\right)\frac{2\pi}{T_1} = \frac{2}{5}M\left(\frac{3}{4}R\right)^2\frac{2\pi}{T_2}$

$\Rightarrow \frac{1}{T_1} = \frac{9}{16T_2} \Rightarrow T_2 = \frac{16}{9T_1}$

$T_2 = \frac{16}{9\times 24} = 13$ hr 30 mins.

54. **(8)**

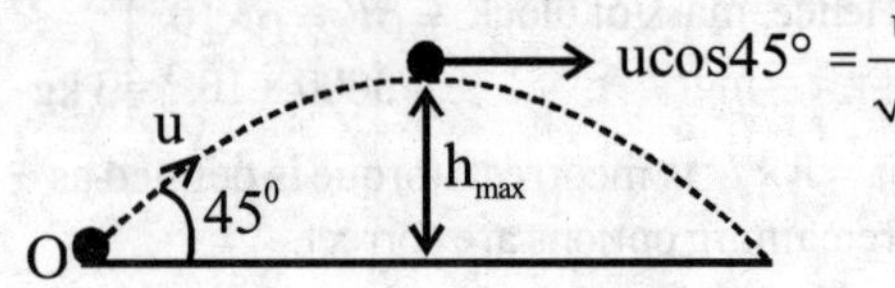

Maximum height, $h_{max} = \frac{u^2\sin^2\theta}{2g}$

Angular momentum, $L = mvh_{max} = mu\cos\theta\frac{u^2\sin^2\theta}{2g}$

$= mu^3\frac{1}{4\sqrt{2}g} = \frac{\sqrt{2}mu^3}{xg}$ (given) $\Rightarrow x = 8$

55. **(a)** Angular momentum, $L = mvH = mu\cos 30°.H$

$= mu\cos 30° \times \frac{u^2\sin^2 30°}{2g}$ $\left[\because H = \frac{u^2\sin^2\theta}{2g}\right]$

$= \frac{mu^3}{2g}\times\frac{\sqrt{3}}{2}\times\left(\frac{1}{2}\right)^2 = \frac{\sqrt{3}mu^3}{16g}$

56. **(60)**

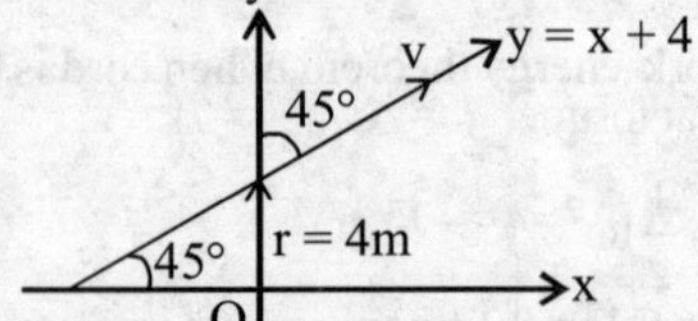

The angular momentum about origin is

$L = mvr\sin\theta = 5\times 3\sqrt{2}\times 4\sin 45° = 60$ kgm^2/s

57. **(c)** Torque (τ) about O = 0

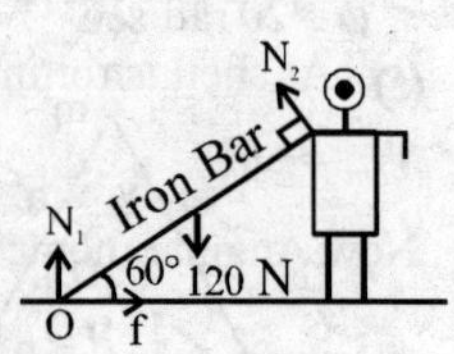

$120\left(\frac{L}{2}\cos 60°\right) - N_2L = 0$

$\Rightarrow 30L = N_2L$

$\therefore$ Weight experienced by the man, $N_2 = 30$ N

$\therefore$ Mass experienced by the man = 3 kg

58. **(15)** Torque, $\tau = I\alpha$ where α = angular acceleration

$\Rightarrow FR = mR^2\alpha$

$\therefore \alpha = \frac{F}{mR} = \frac{52.5}{5\times 0.7} = 15 \text{rad s}^{-2}$

59. **(3)** Given,

Force, $\vec{F} = -P\hat{k}$

Position vector, $\vec{r} = 2\hat{i} - 3\hat{j}$

Torque, $\vec{\tau} = \vec{r}\times\vec{F}$

$\Rightarrow \vec{\tau} = (2\hat{i} - 3\hat{j})\times(-P\hat{k})$

$= P(2\hat{j} + 3\hat{i})$

On comparing with $P(a\hat{i} + b\hat{j})$

We get a = 3, b = 2

$\therefore \frac{a}{b} = \frac{3}{2} = \frac{x}{2}$ (given)

$\therefore x = 3$

60. **(800)** Use $\Delta L = \int_0^t \tau dt$

at t = t, $\tau = mg(v_x t)$

$L_2 - 0 = \int_0^2 mg(v_x t)dt$

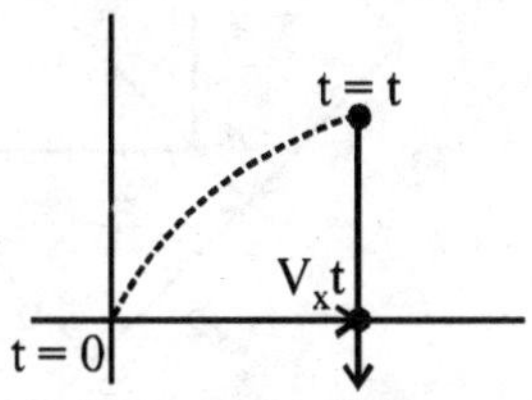

$\Rightarrow L_2 = mgv_x\frac{t^2}{2} = (0.1)(10)(10\sqrt{2})\frac{2^2}{2}$

$\Rightarrow L_2 = 20\sqrt{2} \Rightarrow L_2 = \sqrt{800}$ kgm²/s

61. **(c)** As shown in figure

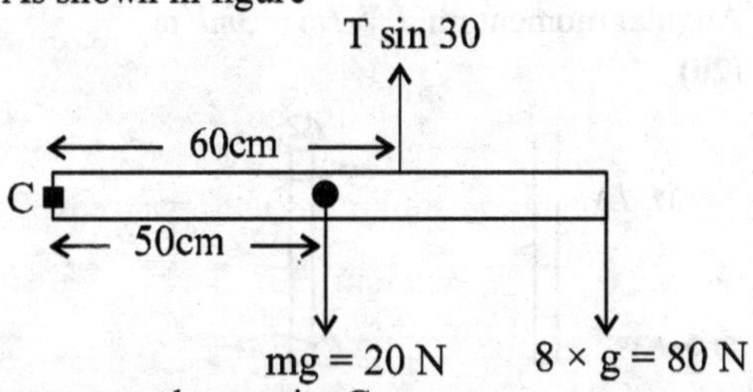

Taking torque about point C

$\frac{T}{2}\times 60 = 20\times 50 + 80\times 100 \Rightarrow 3T = 100 + 800$

T = 300 N

62. **(c)** Given that torque, $\vec{\tau} = 5\hat{i} + 3\hat{j} - 7\hat{k}$

Position vector, $\vec{r} = 2\hat{i} + 2\hat{j} + \hat{k}$

$$\vec{\tau} = \vec{r}\times\vec{F} = \begin{vmatrix} i & j & k \\ 2 & 2 & 1 \\ 5 & 3 & -7 \end{vmatrix}$$

$= \hat{i}(-14-3) - \hat{j}(-14-5) + \hat{k}(6-10) = -17\hat{i} + 19\hat{j} - 4\hat{k}$

63. **(c)** As shell is rolling

$v_{cm} = \omega R$

Now, Angular momentum (L) = $Mv_{cm}R + I\omega$

$= 1\times\omega R^2 + \frac{2}{3}\times 1\times R^2\omega = \frac{5}{3}\omega R^2 = \frac{a}{3}\omega R^2$ So, $a = 5$.

64. **(91)** Angular momentum, $\vec{L} = m(\vec{r}\times\vec{v})$

$= 1[(3\hat{i} - \hat{j})\times(3\hat{j} + \hat{k})] = (9\hat{k} - 3\hat{j} - \hat{i})$

$|\vec{L}| = \sqrt{9^2 + (-3)^2 + (-1)^2} = \sqrt{91} Ns$

65. **(c)** From *FBD* of ladder

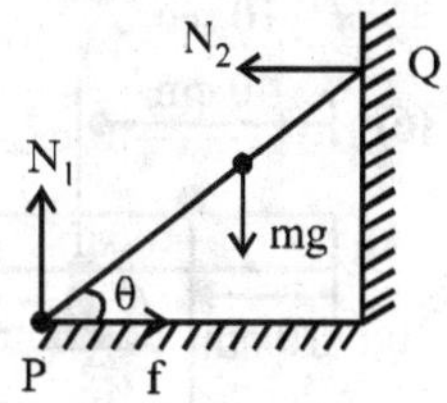

$f = N_2$ and $N_1 = mg$

about '*P*', $\vec{\tau}_{net} = 0$

$N_2 l\sin\theta = mg\frac{l}{2}\cos\theta$

$N_2 = \frac{mg}{2}\cot\theta$

So, $\frac{F_w}{F_f} = \frac{N_2}{\sqrt{N_1^2 + f^2}} = \frac{\frac{mg}{2}\cot\theta}{\sqrt{(mg)^2 + \left(\frac{mg}{2}\cot\theta\right)^2}}$

$= \frac{\frac{\cot\theta}{2}}{\sqrt{1 + \frac{\cot^2\theta}{4}}} = \frac{\cot\theta}{\sqrt{4 + \cot^2\theta}} = \frac{1}{\sqrt{\frac{4}{\cot^2\theta} + 1}}$

$= \frac{1}{\sqrt{\frac{4}{9}\times 25 + 1}} = \frac{3}{\sqrt{109}}$

66. **(10)** $2g - T = 2a$ ...(i)

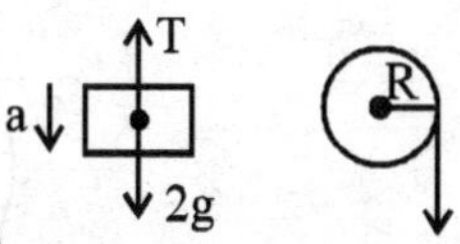

$TR = I\alpha \Rightarrow TR = \frac{1}{2}MR^2\alpha$

$\Rightarrow T = \frac{1}{2}MR(\alpha R)$ $\left(\because \alpha = \frac{a}{R}\right)$

$\Rightarrow T = \frac{1}{2}MRa$ ...(ii)

Adding (i) and (ii), we get

$2g - \frac{1}{2}.4a = 2a \Rightarrow 2g = 4a \Rightarrow a = \frac{g}{2}$

So, $T = \frac{1}{2}\times 4\times\frac{g}{2} = \frac{40}{4} = 10$ N

67. **(c)** We have, $L_i = MR^2\omega_i$

and, $L_f = (M + 2m)R^2\omega_f$

As no external torque act on the system

So, $\vec{L}$ = constant $\Rightarrow \vec{L}_i = \vec{L}_f$

$\Rightarrow MR^2\omega_i = (M + 2m)R^2\omega_f$

$\Rightarrow \omega_f = \frac{M}{M + 2m}\omega_i = \frac{2M}{(M + 2m)}$

68. **(10)** $\vec{r} = 10\alpha t^2\hat{i} + 5\beta(t - 5)\hat{j}$

So, $\vec{v} = \frac{d\vec{r}}{dt} = 20\alpha t\hat{i} + 5\beta\hat{j}$

As $\vec{L} = m(\vec{r}\times\vec{v}) = m[50\alpha\beta t^2\hat{k} - 100\alpha\beta t(t - 5)\hat{k}]$

So, at $t = 0$, we have $L = 0$
Let at $t = t$ again we have $L = 0$

So, $0 = m[50\alpha\beta t^2\hat{k} - 100\alpha\beta t(t-5)\hat{k}]$

$\Rightarrow 0 = 50t - 100(t-5)$
$\Rightarrow 0 = 50t - 100t + 500$
$\Rightarrow 50t = 500$
$\Rightarrow t = 10$ sec

69. (6)

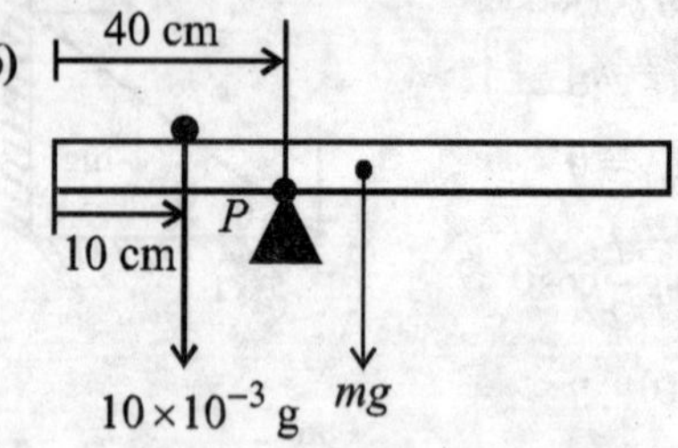

$= 10^{-2}$ g

Let mass of scale be 'm'. Then,

About 'P' $\hat{\tau}_{net} = 0$

$\Rightarrow 2 \times 10^{-2} g \times 30 - mg \times 10 = 0$

$\Rightarrow 6 \times 10^{-1} - 10m = 0 \Rightarrow 0.6 = 10m$

$\Rightarrow m = \frac{0.6}{10} \Rightarrow m = 0.06$ kg

70. (b) Angular momentum, $|\vec{L}| = mvr$
Particle in circular motion moving with constant speed.

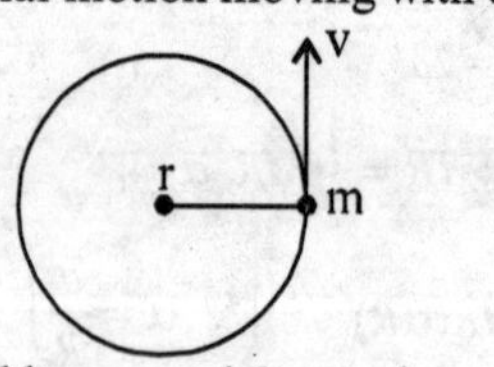

$\therefore$ Direction will be upward & speed remains constant.

71. (a)

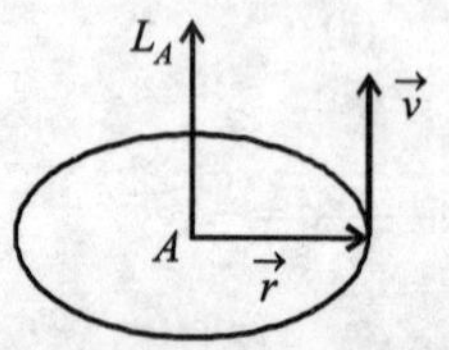

$\vec{L}_A = m(\vec{r} \times \vec{v})$ $\vec{r} \perp \vec{v}$, so $\vec{r} \perp \vec{p}$

Hence, $\vec{L}_A$ is constant in magnitude and direction.

But $\vec{L}_B$ is constant in magnitude with varying direction.

72. (20) Given, mass of the circular disc, $m = 20$ kg
Radius of the circular disc, $R = 0.2$ m
Torque required to attain angular speed of 50 rad s^{-1},

$\tau = I\alpha$

$\therefore \alpha = \frac{F \cdot R}{I}$ $\quad (\because \tau = F \cdot R)$

$= \frac{20 \times 0.2}{\frac{1}{2} \times 20 \times 0.2^2} = 10$ rad/s^2

Using, $\omega^2 = \omega_0^2 + 2\alpha\theta$

$\Rightarrow (50)^2 = 0^2 + 2(10)\theta \Rightarrow \theta = 125$ rad

$\therefore$ No. of revolutions, $n = \frac{125}{2\pi} \approx 20.$

73. (20) Torque, $\vec{\tau} = \vec{r} \times \vec{F}$

$= \left[(2-2)\hat{i} + (0-3)\hat{j} + (0-4)\hat{k}\right] \times (4\hat{i} + 3\hat{j} + 4\hat{k})$

$= (-3\hat{j} - 4\hat{k}) \times (4\hat{i} + 3\hat{j} + 4\hat{k}) = -16\hat{j} + 12\hat{k}$

$|\vec{\tau}| = \sqrt{(-16)^2 + 12^2} = 20$ units

74. (c) Angular momentum, $L = I\omega$

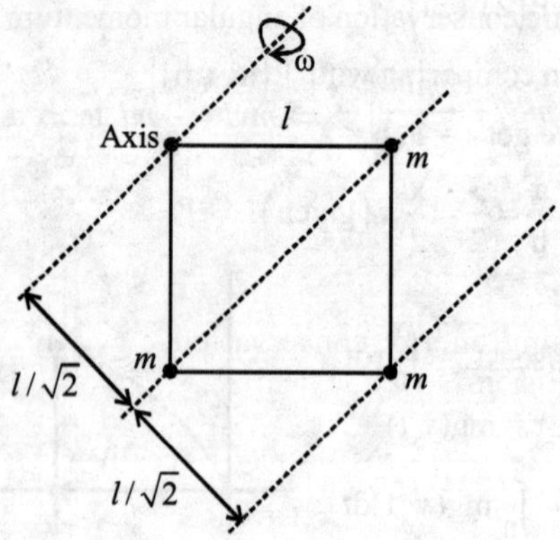

$I = m(0)^2 + m\left(\frac{l}{\sqrt{2}}\right)^2 \times 2 + m(\sqrt{2}l)^2$

$= \frac{2ml^2}{2} + 2ml^2 = 3ml^2$

Angular momentum $L = I\omega = 3ml^2\omega$

75. (20)

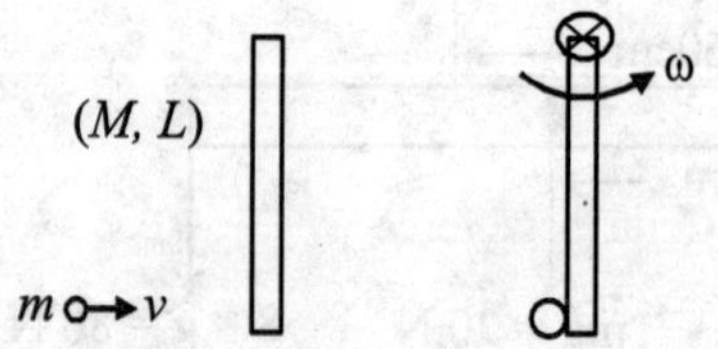

Before collision After collision

Using principal of conservation of angular momentum we have

$\vec{L}_i = \vec{L}_f \Rightarrow mvL = I\omega$

$\Rightarrow mvL = \left(\frac{ML^2}{3} + mL^2\right)\omega$

$\Rightarrow 0.1 \times 80 \times 1 = \left(\frac{0.9 \times 1^2}{3} + 0.1 \times 1^2\right)\omega$

$\Rightarrow 8 = \left(\frac{3}{10} + \frac{1}{10}\right)\omega \Rightarrow 8 = \frac{4}{10}\omega \Rightarrow \omega = 20$ rad/sec.

76. (9)
Here $M_0 = 200$ kg, $m = 80$ kg
Using conservation of angular momentum, $L_i = L_f$

$I_1\omega_1 = I_2\omega_2$

$$I_1 = (I_M + I_m) = \left(\frac{M_0R^2}{2} + mR^2\right)$$

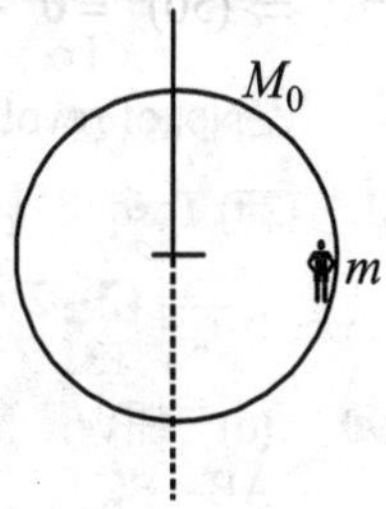

$$I_2 = \frac{1}{2}M_0R^2 \text{ and } \omega_1 = 5 \text{ rpm}$$

$$\therefore \omega_2 = \left(\frac{M_0R^2}{2} + mR^2\right) \times \frac{5}{\frac{M_0R^2}{2}}$$

$$= \frac{5R^2}{R^2} \times \frac{(80+100)}{100} = 9 \text{ rpm.}$$

77. **(a)** Using conservation of angular momentum

$$mvl = \left(ml^2 + \frac{2ml^2}{3}\right)\omega \Rightarrow mvl = \frac{5}{3}ml^2\omega \Rightarrow \omega = \frac{3v}{5l}$$

or, $\omega = \frac{3\times 6}{5\times 1} = \frac{18}{5}$ rad/s

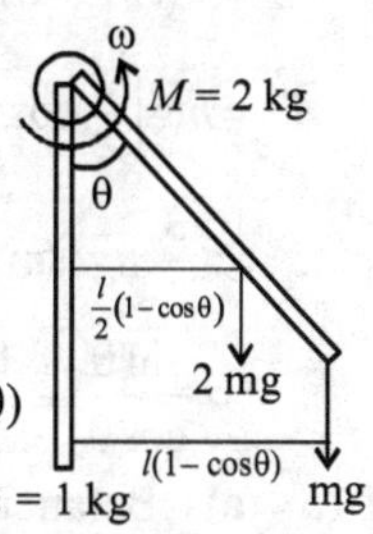

Now using energy conservation, after collision

$$\frac{1}{2}I\omega^2 = 2mg\frac{l}{2}(1-\cos\theta) + mgl(1-\cos\theta)$$

$$\Rightarrow \frac{1}{2}\left(\frac{5}{3}ml^2\right)\frac{9v^2}{25l^2} = 2mgl(1-\cos\theta)$$

$$\Rightarrow \frac{3}{5\times 2}mv^2 = 2mgl(1-\cos\theta)$$

$$\Rightarrow \frac{3}{10}\times\frac{36}{2\times 10} = 1-\cos\theta \Rightarrow 1-\frac{27}{50} = \cos\theta$$

or, $\cos\theta = \frac{23}{50}$ $\quad\therefore \theta = 63°$.

78. **(d)** Vertical force $= mg$

Horizontal force = Centripetal force

$$= m\omega^2\frac{l}{2}\sin\theta$$

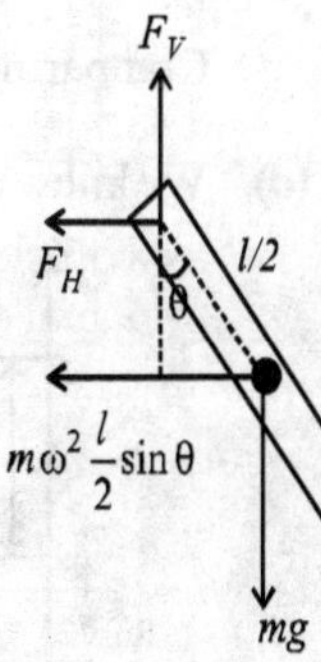

Torque due to vertical force

$$= mg\left(\frac{l}{2}\sin\theta\right)$$

Torque due to horizontal force

$$= m\omega^2\frac{l}{2}\sin\theta\left(\frac{l}{2}\cos\theta\right)$$

Net Torque about point of contact = Angular momentum

$$mg\frac{l}{2}\sin\theta - m\omega^2\frac{l}{2}\sin\theta\frac{l}{2}\cos\theta = \frac{ml^2}{12}\omega^2\sin\theta\cos\theta$$

$$\Rightarrow \cos\theta = \frac{3}{2}\frac{g}{\omega^2 l}$$

79. **(d)** Net torque, τ_{net} about B is zero at equilibrium

$$\therefore T_A\times 100 - mg\times 50 - 2mg\times 25 = 0$$

$$\Rightarrow T_A\times 100 = 100mg$$

$$\Rightarrow T_A = 1mg \text{ (Tension in the string at } A)$$

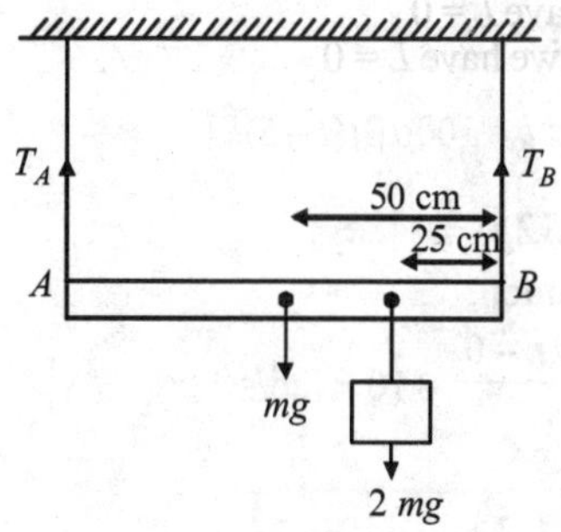

80. **(a)**

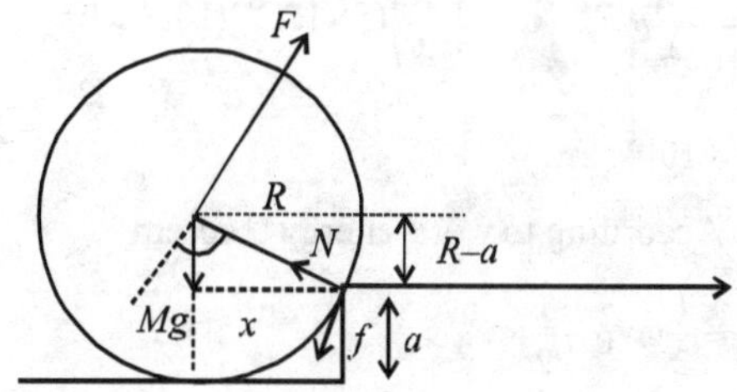

For step up

$$\tau_F \ge \tau_{Mg}$$

i.e., $F\times R \ge Mg\times x$

$x = \sqrt{R^2-(R-a)^2}$ from figure

$$F_{min} = \frac{Mg}{R}\times\sqrt{R^2-(R-a)^2} = Mg\sqrt{1-\left(\frac{R-a}{R}\right)^2}$$

81. **(c)**

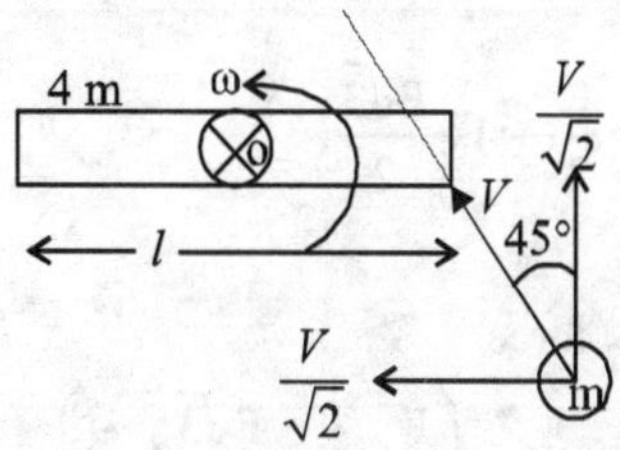

About point O angular momentum

$$L_{initial} = L_{final}$$

$$\Rightarrow \frac{mV}{\sqrt{2}}\times\frac{l}{2} = \left[\frac{4ml^2}{12} + \frac{ml^2}{4}\right]\times\omega \Rightarrow \omega = \frac{6V}{7\sqrt{2}l} = \frac{3\sqrt{2}V}{7l}$$

82. **(b)** Given that, $x = x_0 + a\cos\omega_1 t$
$y = y_0 + b\sin\omega_2 t$

$$\Rightarrow \frac{dx}{dt} = v_x = -a\omega_1\sin(\omega_1 t), \text{ and } \frac{dy}{dt} = v_y = b\omega_2\cos(\omega_2 t)$$

$$\frac{dv_x}{dt} = a_x = -a\omega_1^2\cos(\omega_1 t), \quad \frac{dv_y}{dt} = a_y = -b\omega_2^2\sin(\omega_2 t)$$

At $t = 0$, $x = x_0 + a$, $y = y_0$
$a_x = -a\omega_1^2$, $a_y = 0$

Now, $\vec{\tau} = \vec{r}\times\vec{F} = m(\vec{r}\times\vec{a})$

$$= [(x_0+a)\hat{i} + y_0\hat{j}]\times m(-a\omega_1^2\hat{i}) = +my_0 a\omega_1^2\hat{k}$$

83. **(d)** We have given $\vec{r} = 2t\hat{i} - 3t^2\hat{j}$

$$\vec{r}(\text{at } t=2) = 4\hat{i} - 12\hat{j}$$

Velocity, $\vec{v} = \frac{d\vec{r}}{dt} = 2\hat{i} - 6t\hat{j}$

$\vec{v}(\text{at } t=2) = 2\hat{i} - 12\hat{j}$

$\vec{L} = mvr\sin\theta\hat{n} = m(\vec{r}\times\vec{v})$

$= 2(4\hat{i} - 12\hat{j})\times(2\hat{i} - 12\hat{j}) = -48\hat{k}$

84. **(d)** Angular acceleration,

$$\alpha = \frac{\omega - \omega_0}{t} = \frac{25\times 2\pi - 0}{5} = 10\,\pi\ \text{rad/s}^2$$

$\tau = I\alpha$

$$\Rightarrow \tau = \left(\frac{5}{4}mR^2\right)\alpha \approx \left(\frac{5}{4}\right)(5\times10^{-3})(10^{-4})10\pi$$

$= 2.0\times 10^{-5}\,\text{Nm}$

85. **(c)** According to work-energy theorem

$$\text{mgh} = \frac{1}{2}\text{mv}_B^2 - \frac{1}{2}\text{mv}_A^2$$

$2\text{gh} = \text{v}_B^2 - \text{v}_A^2$

$2\times10\times10 = \text{v}_B^2 - 5^2$

$\Rightarrow v_B = 15\ \text{m/s}$

Angular momentum about O,

$L_O = mvr$

$= 20\times10^{-3}\times15\times20$

$L_O = 6\ \text{kg.m}^2/\text{s}$

86. **(a)** Given, $\vec{F}_2 = \frac{F}{2}(-\hat{i}) + \frac{F\sqrt{3}}{2}(-\hat{j})$

$\vec{r}_2 = 0\hat{i} + 6\hat{j}$

Torque due to F_2 force

$$\vec{\tau}_{F_2} = \vec{r}_2\times\vec{F}_2 = 6\hat{j}\times\left(\frac{F}{2}(-\hat{i}) + \frac{F\sqrt{3}}{2}(-\hat{j})\right) = 3F(\hat{k})$$

Torque due to F_1 force

$\vec{\tau}_{F_1} = (2\hat{i} + 3\hat{j})\times F\hat{k} = 3F\hat{i} + 2F(-\hat{j})$

$\vec{\tau}_{net} = \vec{\tau}_{F_1} + \vec{\tau}_{F_2} = 3F\hat{i} + 2F(-\hat{j}) + 3F(\hat{k})$

$= (3\hat{i} - 2\hat{j} + 3\hat{k})F$

87. **(a)** Torque about the origin

$= r\,F\sin\theta \Rightarrow 2.5 = 1\times5\sin\theta$

$\sin\theta = 0.5 = \frac{1}{2} \Rightarrow \theta = \frac{\pi}{6}$

88. **(d)** Consider a elemental ring of radius x and thickness dx,

Torque on ring of width dx, $d\tau = dF\times x$

$$= \frac{\mu F}{\pi R^2}\times(2\pi\,x dx)\times x$$

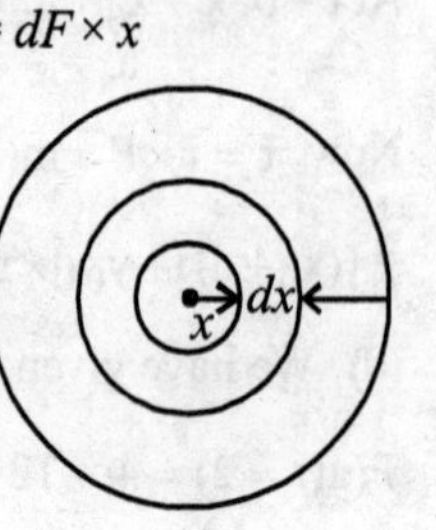

$$\int d\tau = \int_0^R \frac{\mu F.2\pi x^2 dx}{\pi R^2}$$

$$\Rightarrow \tau = \frac{2\mu F}{R^2}\cdot\frac{R^3}{3} = \frac{2\mu FR}{3}$$

89. **(a)** Applying torque equation about point P.

$\tau = I\,\alpha = [2M_0(2\ell)^2 + 5M_0\ell^2]\alpha$

$\Rightarrow 5M_0g\ell - 4\,M_0g\ell = [2M_0(2\ell)^2 + 5\,M_0\ell^2]\alpha$

$\Rightarrow M_0g\ell = (13M_0g\ell^2)\alpha$

$\therefore \alpha = \frac{g}{13\ell}$

90. **(d)** Given that, the rod is of uniform mass density and AB = BC

Let mass of one rod is m.

Balancing torque about hinge point.

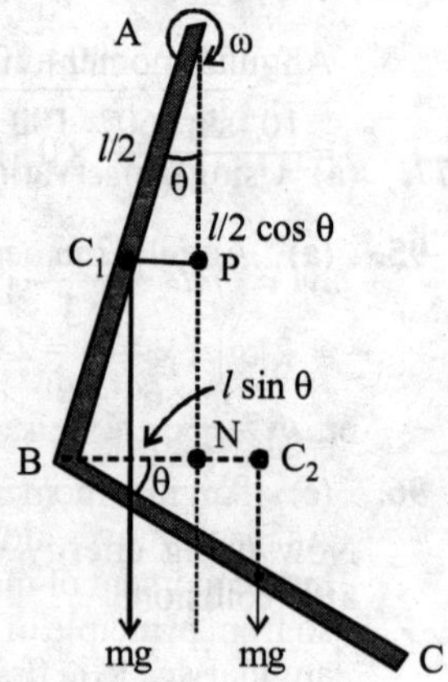

$mg\,(C_1P) = mg\,(C_2N)$

$$mg\left(\frac{l}{2}\sin\theta\right)$$

$$= mg\left(\frac{l}{2}\cos\theta - l\sin\theta\right)$$

$$\Rightarrow \frac{3}{2}mgl\sin\theta = \frac{mgl}{2}\cos\theta$$

$$\Rightarrow \frac{\sin\theta}{\cos\theta} = \frac{1}{3}\ \text{or,}\ \tan\theta = \frac{1}{3}$$

91. **(a)** Balancing torque w.r.t. point of suspension

$$mg\,x = Mg\left(\frac{\ell}{2} - x\right)$$

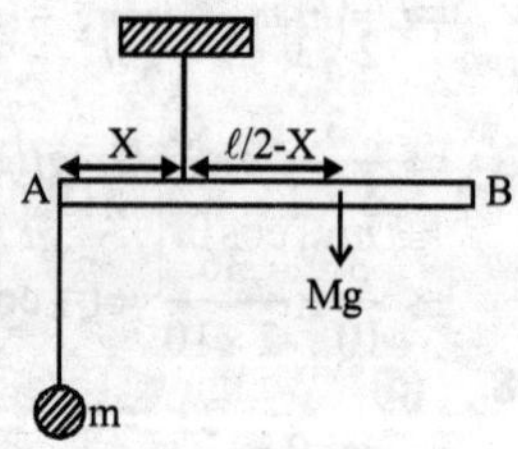

$$\Rightarrow mx = M\frac{\ell}{2} - Mx$$

$$m = \left(M\frac{\ell}{2}\right)\frac{1}{x} - M$$

$y = (\tan\theta)x + C - C$ Straight line equation.

Comparing, we get variable $\left(m, \frac{1}{x}\right)$

92. **(d)** We know that $|L| = mvr_{\perp}$

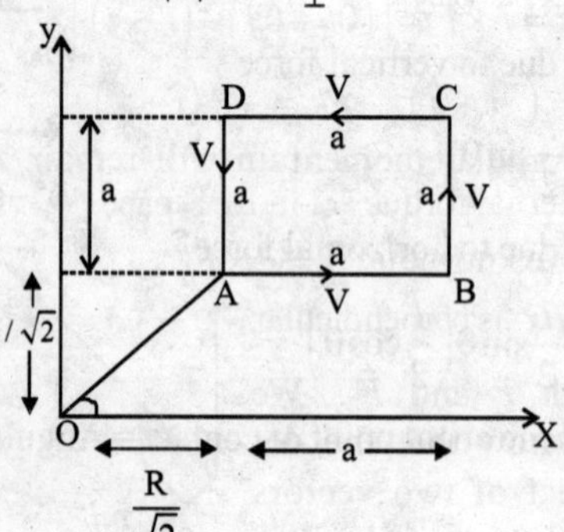

When particle moves from C to D,

we have $r_{\perp} = \left(\frac{R}{\sqrt{2}} + a\right)$

So, $\vec{L}_{C\to D} = mV\left(\frac{R}{\sqrt{2}} + a\right)\hat{k}$

93. (c) Torque working on the bob of mass m is, $\tau = mg \times \ell \sin\theta$. (Direction parallel to plane of rotation of particle)

As $\vec{\tau}$ is perpendicular to $\vec{L}$, direction of $\vec{L}$ changes but magnitude remains same.

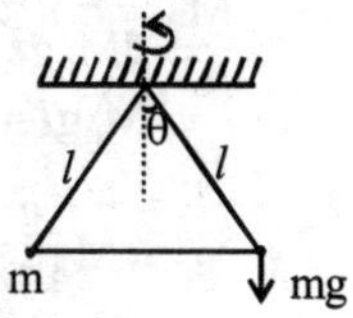

94. (b) Given : m = 0.160 kg

θ = 60°

v = 10 m/s

Angular momentum $L = \vec{r} \times m\vec{v} = H\,mv\cos\theta$

$$= \frac{10^2 \sin^2 60°}{2 \times 10} \times 0.16 \times 10 \cos 60° = 3 \text{ kg m}^2\text{/s}$$

95. (a) Angular momentum $\vec{L} = m(\vec{r} \times \vec{v})$

$$= 2\text{kg}\left(\vec{r} \times \frac{d\vec{r}}{dt}\right) = 2\text{ kg}(5\hat{i} - 2t^2\hat{j}) \times (-4t\hat{j})$$

$$= 2\text{ kg}(-20t\hat{k}) = 2 \times (-20) \times 2\hat{k} = -80\hat{k}$$

96. (c) Angular momentum, $L = I\omega \Rightarrow L = mr^2\omega$

As insect moves along a diameter, the effective mass and hence moment of inertia (I) first decreases then increases so from principle of conservation of angular momentum, angular speed ω first increases then decreases.

97. (c) $\vec{L} = m(\vec{r} \times \vec{v})$

$$\vec{L} = m\left[v_0\cos\theta t\hat{i} + (v_0\sin\theta t - \frac{1}{2}gt^2)\hat{j}\right] \times \left[v_0\cos\theta\hat{i} + (v_0\sin\theta - gt)\hat{j}\right]$$

$$= mv_0\cos\theta t\left[-\frac{1}{2}gt\right]\hat{k} = -\frac{1}{2}mgv_0t^2\cos\theta\hat{k}$$

98. (d) We know Torque $\vec{\tau_c} = \frac{d\vec{L_c}}{dt}$

where $\vec{L_c}$ = Angular momentum about the center of mass of the body. Central forces act along the center of mass. Therefore torque about center of mass is zero.

$$\therefore \tau = \frac{dL}{dt} = 0 \Rightarrow \vec{L_c} = \text{const.}$$

99. (b) $\vec{\tau} = \vec{r} \times \vec{F} = \left[(0\hat{i} + 0\hat{j}) - (\hat{i} - \hat{j})\right] \times -F\hat{k}$

$$= (-\hat{i} + \hat{j}) \times -F\hat{k} = -F(\hat{i} + \hat{j})$$

100. (b) Angular momentum will remain the same since no external torque act in free space.

101. (d) We know that $\vec{\tau} = \vec{r} \times \vec{F}$

Vector $\vec{\tau}$ is perpendicular to both $\vec{r}$ and $\vec{F}$. We also know that the dot product of two vectors which have an angle of 90° between them is zero.

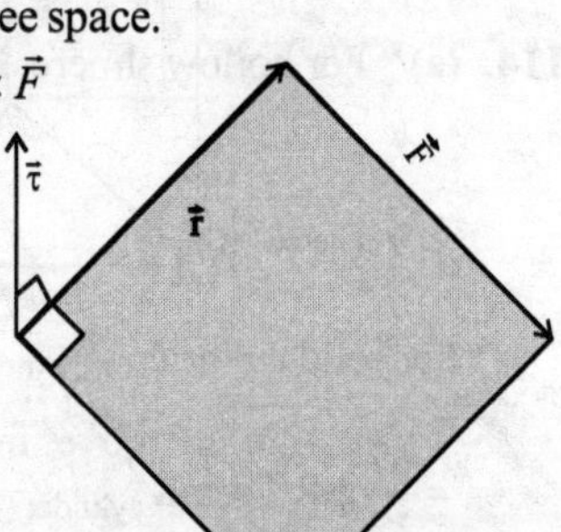

$\therefore \vec{r} \cdot \vec{T} = 0$ and $\vec{F} \cdot \vec{T} = 0$

102. (d) Angular momentum (L) = (linear momentum) × (perpendicular distance of the line of action of momentum from the axis of rotation)

As the particle moves with velocity V along line PC, the line of motion passes through P.

$\therefore L_P = mv \times r = mv \times 0 = 0$

103. (d) Linear mass density, $\lambda = \frac{M}{L}$

$$L = 2\pi R \Rightarrow R = \frac{L}{2\pi}$$

Moment of inertia, $I = \frac{MR^2}{2} = \frac{\lambda \times L}{2} \times \left(\frac{L}{2\pi}\right)^2 = \frac{\lambda L^3}{8\pi^2}$.

104. (9) MOI about its central perpendicular axis $(I_{big}) = \frac{MR^2}{2}$

Mass of small disc $= \frac{M}{\pi R^2} \times \left(\frac{R}{3}\right)^2 \pi = \left(\frac{M}{9}\right)$

M.I. of small disc $I_2 = \dfrac{\frac{M}{9}\left(\frac{R}{3}\right)^2}{2} + \frac{M}{9} \times \left(\frac{2R}{3}\right)^2 = \frac{MR^2}{18}$

$$I_{remaining} = I_{big} - I_{small} = \frac{MR^2}{2} - \frac{MR^2}{18} = \frac{4MR^2}{9}$$

∴ $(n = 9)$

105. (199)

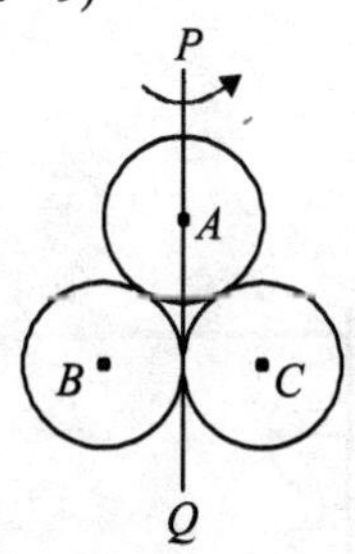

The MOI of disc about its diameter is $I = \frac{MR^2}{4}$

$$\therefore I_{PQ} = \frac{MR^2}{4} + \left(\frac{2}{5}MR^2 + MR^2\right) + \left(\frac{2}{3}MR^2 + MR^2\right)$$

$$= \frac{15MR^2 + 24MR^2 + 60MR^2 + 40MR^2 + 60MR^2}{60}$$

$$= \frac{199}{60}MR^2 = \frac{199}{15}\left(\frac{MR^2}{4}\right) = \frac{199}{15}I = \frac{x}{15}I \text{ (given)}$$

$\therefore x = 199.$

106. (32) Sphere is of radius R, its mass is M, when radius is reduced to $\frac{R}{2}$, mass will reduced to $\frac{M}{8}$

Since no external torque is acting on the sphere the angular momentum is conserved.

Moment of Intertia (I) for solid sphere $= \frac{2}{5}mR^2$

$$L_1 = L_2$$

$$I_1\omega_1 = I_2\omega_2$$

$$\Rightarrow \left(\frac{2}{5}MR^2\right)\omega_1 = \left(\frac{2}{5}\left(\frac{M}{8}\right)\left(\frac{R}{2}\right)^2\right)\omega_2$$

$\omega_2 = 32\omega_1$, value of x is 32. (Answer is 32.)

107. (b) Moment of inertia of a ring about a tangential axis lying in the plane of the ring, $= \frac{3}{2}MR^2$

Diameter is given as R.

$\therefore$ Radius = R/2

$$I_{tangent} = \frac{3}{2}M\left(\frac{R}{2}\right)^2 = \frac{3}{8}MR^2$$

108. (b) Moment of inertia of a rod, $\alpha = \frac{ML^2}{12}$

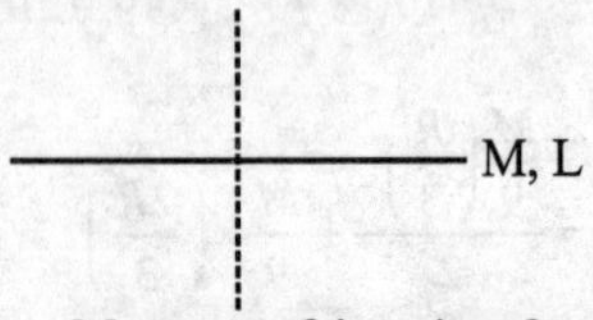

$\therefore$ Moment of inertia of cross about an axis passing through its centre and normal to plane containing cross.

$$\alpha' = 2\left[\frac{\frac{M}{2}\left(\frac{L}{2}\right)^2}{12}\right]$$

or, $\alpha' = \frac{ML^2}{48} = \frac{\alpha}{4}$

109. (4)

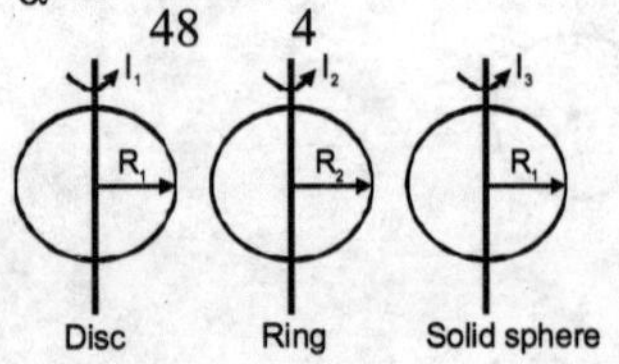

Given : $M_{disc} = M_{ring} = M_{sphere} = M$ and $R_{sphere} = R_{disc}$

$$I_1 = \frac{MR_1^2}{4}, I_2 = \frac{MR_2^2}{2}, I_3 = \frac{2MR_1^2}{5}$$

According to problem

$$\frac{I_1}{I_2} = 2.5 \Rightarrow \frac{\frac{MR_1^2}{4}}{\frac{MR_2^2}{2}} = \frac{5}{2} \Rightarrow \frac{R_1^2}{R_2^2} = 5 \quad \text{........ (i)}$$

Now we are provided with information that $\frac{I_3}{I_2} = n$

$$\Rightarrow \frac{\frac{2MR_1^2}{5}}{\frac{MR_2^2}{2}} = n \Rightarrow \frac{4R_1^2}{5R_2^2} = n \quad \text{.........(ii)}$$

From Eq, (i) and (ii)

$\Rightarrow$ n = 4

110. (16)

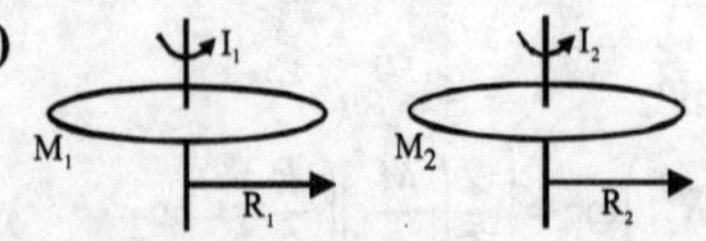

Given $R_2 = 2R_1$

For first disc, $I_1 = \frac{1}{2}M_1R_1^2$

For second disc, $I_2 = \frac{1}{2}M_2R_2^2$

$$M_1 = \sigma \times \pi R_1^2 = M_0$$

$$M_2 = \sigma \times \pi R_2^2 = \sigma \times \pi[2R_1]^2 = \sigma \times 4\pi R_1^2 = 4M_0$$

$$\frac{I_1}{I_2} = \frac{\frac{M_1R_1^2}{2}}{\frac{M_2R_2^2}{2}} = \frac{M_1R_1^2}{M_2R_2^2} = \frac{1}{4} \times \frac{1}{4} = \frac{1}{16}$$

111. (d) Moment of inertia of disc, $I = \frac{MR^2}{2}$

$$I_{remaining} = I - I_{removed}$$

$$= \frac{MR^2}{2} - \left[\frac{\frac{M}{4}\left(\frac{R}{2}\right)^2}{2} + \frac{M}{4}\left(\frac{R}{2}\right)^2\right]$$

$$I_{remaining} = \frac{13}{32}MR^2$$

112. (100) Given, Moment of inertia of wheel, $I = 0.40$ kg m^2

Radius of wheel, R = 10 cm

Force applied, F = 40 N

Torque $\tau = FR = I\alpha$

$\Rightarrow 40 \times 0.1 = 0.4\alpha$

$\Rightarrow \alpha = 10$ rad/s^2

$\omega_F = 10 \times 10 = 100$ rad/s

113. (4)

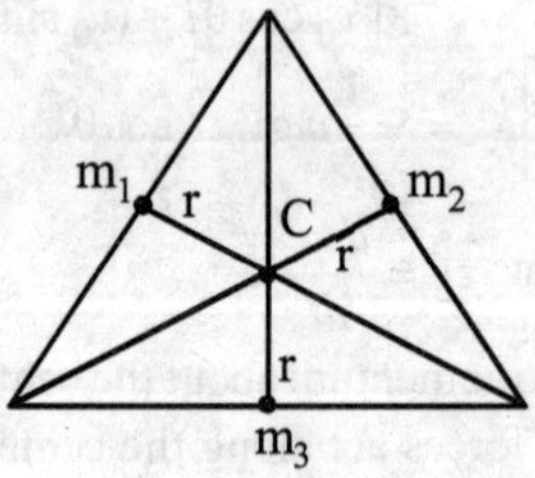

Moment of inertia about centroid and perpendicular to the plane of triangle

$$I = m_1r^2 + m_2r^2 + m_3r^2$$

$$I = r^2[2 + 4 + 6] = \frac{1}{3} \times 12 = 4 \text{ kg-m}^2 \quad \left[\because r = \frac{1}{\sqrt{3}}\right]$$

114. (a) For hollow shpere, $I_{sphere} \Rightarrow \frac{2}{3}mR^2 = mK_{sphere}^2$

$$\therefore K_{sphere} = \sqrt{\frac{2}{3}}R$$

For solid cylinder, $I_{cylinder} = \frac{1}{4}mR^2 + \frac{m}{3}(4R)^2$

$$= \frac{67}{12}mR^2 = mK_{cylinder}^2$$

$$\therefore K_{cylinder} = \sqrt{\frac{67}{12}}R \Rightarrow \frac{K_{sphere}}{K_{cylinder}} = \sqrt{\frac{2}{3} \times \frac{12}{67}}$$

$$= \sqrt{\frac{8}{67}} = \sqrt{\frac{8}{x}} \quad \text{(given)} \qquad \therefore x = 67$$

115. (4) Impulse is given by :

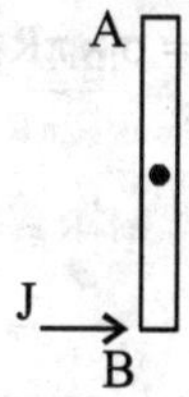

$J = \int F dt = 0.2 \text{ N-s}$

Angular impulse is

$Q = \int \tau dt$

$= \int \frac{FL}{2} dt = \int \frac{FL}{2} dt = \frac{L}{2} \int F dt = \frac{L}{2} \times J$

$= \frac{0.3}{2} \times 0.2 = 0.03$

MOI of rod,

$I_{cm} = \frac{ML^2}{12} = \frac{2 \times (0.3)^2}{12} = \frac{0.09}{6} = 0.015$

$Q = I_{cm}(\omega_2 - \omega_1)$

$0.03 = 0.015\, \omega_2 \Rightarrow \omega_2 = 2 \text{ rad/s}$

Also $\theta = \omega t \Rightarrow t = \frac{\theta}{\omega} = \frac{\pi}{2 \times 2} = \frac{\pi}{4} s = \frac{\pi}{x} s$ (given)

$\therefore x = 4$

116. (53) Radius of sphere, $R = 50 \text{ cm} = \frac{1}{2} \text{m}$

Mass of solid sphere, $m = 2$ kg

Moment of inertia of solid sphere, $I = \frac{2}{5} mR^2$

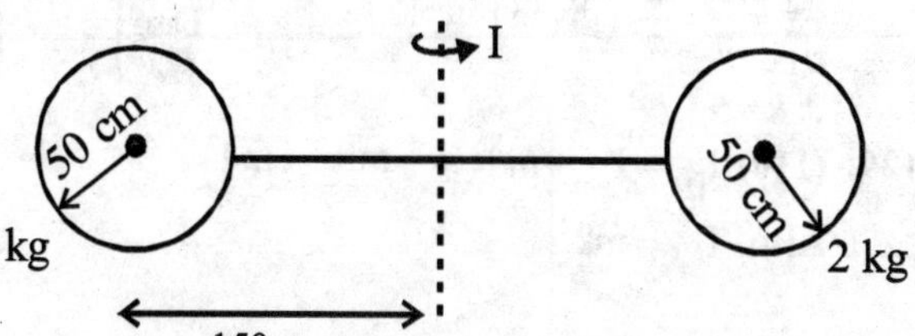

Using parallel axis theorem

$I_{net} = \left(\frac{2}{5} mR^2 + md^2\right) \times 2$

$\Rightarrow I_{net} = 2\left(\frac{2}{5} \times 2 \times \left(\frac{1}{2}\right)^2 + 2 \times \left(\frac{3}{4}\right)^2\right) = \frac{53}{20} \text{ kg-m}^2$

$= \frac{x}{20} \text{kg-m}^2$ (given) $\therefore x = 53$

117. (250) Angular momentum,

$\vec{L}_i = I\omega_i = \frac{MR^2}{2} . \omega = 10 \times 10 = 100 \text{ kgm}^2/\text{s}$

$E_i = \frac{1}{2} . \frac{MR^2}{2} . \omega^2 = 500 \text{ J}$

Apply the conservation of angular momentum

$\vec{L}_i = \vec{L}_f \Rightarrow 100 = 2I\omega_f$

$\omega_f = 5$ rad/s

$E_f = 2 \times \frac{1}{2} \frac{5(2)^2}{2} . (5)^2 = 250 \text{ J}$

$\therefore \quad \Delta E = E_i - E_f = 250 \text{ J}$

118. (24) Moment of intertia of disc, $I_1 = 4 \text{kg m}^2$ and $I_2 = 2 \text{kgm}^2$

Angular speed, $\omega_1 = 10$ rad/s and $\omega_2 = 4$ rad/s

Apply the angular momentum conservation,

$I_1\omega_1 + I_2\omega_2 = (I_1 + I_2)\omega_0$

$\omega_0 = \frac{I_1\omega_1 + I_2\omega_2}{I_1 + I_2} = \frac{4 \times 10 + 2 \times 4}{4 + 2} = 8 \text{ rad/s}$

$E_1 = \frac{1}{2} I_1\omega_1^2 + \frac{1}{2} I_2\omega_2^2 = \frac{1}{2} \times 4 \times 10^2 + \frac{1}{2} \times 2 \times 4^2 = 216 \text{ J}$

$E_2 = \frac{1}{2}(I_1 + I_2)\omega_0^2 = \frac{1}{2} \times (4 + 2) \times 8^2 = 192 \text{J}$

$\therefore$ Loss in kinetic energy, $\Delta E = 24$J

119. (16) Moment of inertia of system about an axis perpendicular to its plane and passing through one of its vertex.

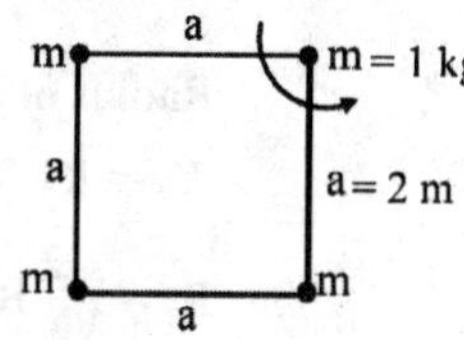

$I = ma^2 + ma^2 + m\left(\sqrt{2}a\right)^2 = 4\, ma^2$

$\therefore \quad I = 4 \times 1 \times (2)^2 = 16 \text{ kgm}^2$

120. (5) For slid sphere, $\frac{2}{5} mR^2 = mk_{sph}^2 \Rightarrow k_{sph} = \sqrt{\frac{2}{5}} R$

For solid cylinder, $\frac{mR^2}{2} = mk_{cyl}^2 \Rightarrow k_{cyl} = \frac{R}{\sqrt{2}}$

$\frac{k_{sph}}{k_{cyl}} = \frac{\sqrt{\frac{2}{5}} R}{\frac{R}{\sqrt{2}}} = \frac{2}{\sqrt{5}} = \frac{2}{\sqrt{x}} \qquad \therefore x = 5$

121. (35) Moment of inertia of sphere about diameter, $I = \frac{2}{5} mR^2$

Here, m = mass of sphere,

R = radius of sphere

Using theorem of parallel axis, moment of inertia about tangent, $I_T = \frac{2}{5} mR^2 + mR^2 = \frac{7}{5} mR^2$

$I_T = x \times 10^{-2}$ L (Here, L = angular momentum)

($\because L = I\omega$)

$\therefore \frac{7}{5} mR^2 = x \times 10^{-2} \times \frac{2}{5} mR^2 \omega$

$\Rightarrow \frac{7}{2} = x \times 10 \times 10^{-2}$

$\therefore x = \frac{70}{2} = 35$

122. (3) $K.E=\frac{L^2}{2I}\Rightarrow K.E\propto\frac{1}{I}$

$\therefore \frac{KE_{final}}{KE_{initial}}=\frac{I_{initial}}{I_{final}}\Rightarrow\frac{KE_{final}}{E}=\frac{1}{3}$

$\Rightarrow KE_{final}=\frac{E}{3}=\frac{E}{x}$ (given)

$\therefore x=3$

123. (1) The moment of inertia of semicircular ring about axis passing through centre of ring and perpendicular to plane of ring, $I=MR^2=\frac{1}{x}MR^2$ (given). So, $x=1$

124. (5) Moment of inertia for ring, $I=mR_1^2=mK_1^2$

$\therefore$ Radius of gyration $K_1=R_1$

Moment of inertia for solid sphere

$I'=\frac{2}{5}m'R_2^2$

$\therefore$ Radius of gyration $K_2=\sqrt{\frac{I'}{m'}}=\sqrt{\frac{2}{5}}R_2$

$\because K_1=K_2$ (given)

$\therefore R_1=\sqrt{\frac{2}{5}}R_2$

$\therefore \frac{R_1}{R_2}=\sqrt{\frac{2}{5}}=\sqrt{\frac{2}{x}}$ (given) $\therefore x=5$

125. (3) By parallel axes theorem

$I_P=I_C+MR^2$

$I_P=\frac{1}{2}MR^2+MR^2$

$I_P=\frac{3}{2}MR^2=\frac{x}{2}MR^2$ (given)

So, $x=3$

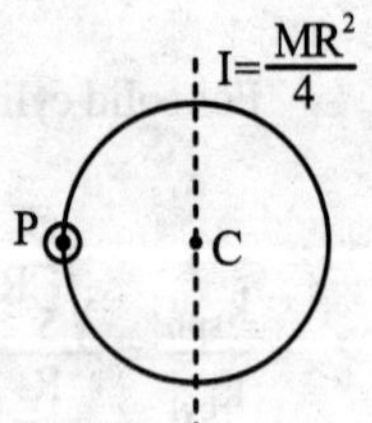

126. (5) We have

$m_1=m_2, t_1=1cm, t_2=0.5\text{ cm}, \frac{\rho_1}{\rho_2}=\frac{3}{5}$

$\frac{I_1}{I_2}=\frac{\frac{1}{4}m_1r_1^2}{\frac{1}{4}m_2r_2^2}=\left(\frac{r_1}{r_2}\right)^2$

As $m_1=m_2\Rightarrow A_1t_1\rho_1=A_2t_2\rho_2\Rightarrow r_1^2t_1\rho_1=r_2^2t_2\rho_2$

$\Rightarrow\left(\frac{r_1}{r_2}\right)^2=\frac{\rho_2}{\rho_1}\cdot\frac{t_2}{t_1}\Rightarrow\left(\frac{r_1}{r_2}\right)^2=\frac{5}{3}\times\frac{0.5}{1}=\frac{5}{6}$

So, $\frac{I_1}{I_2}=\frac{5}{6}$

127. (17) Moment of inertia of circular disc,

$I_{CM}=\frac{mR^2}{2}$

From parallel axis theorem

$I_{AB}=I_{CM}+md^2$ $\left(\text{Here } d=\frac{2R}{3}\right)$

$I_{AB}=\frac{mR^2}{2}+m\left(\frac{2R}{3}\right)^2=\frac{17}{18}mR^2$

$\frac{I_{AB}}{I_{CM}}=\frac{17}{9}=\frac{x}{9}$ (given) $\Rightarrow x=17$

128. (5) Moment of inertia of solid sphere, $I'_{cm}=\frac{2}{5}m_1R^2$

Apply parallel axis theorem,

$I_1=I'_{cm}+m_1R^2$

$I_1=\frac{2}{5}m_1R^2+m_1R^2$

$I_1=m_1R^2\left(\frac{7}{5}\right)$

$I_1=7R^2$

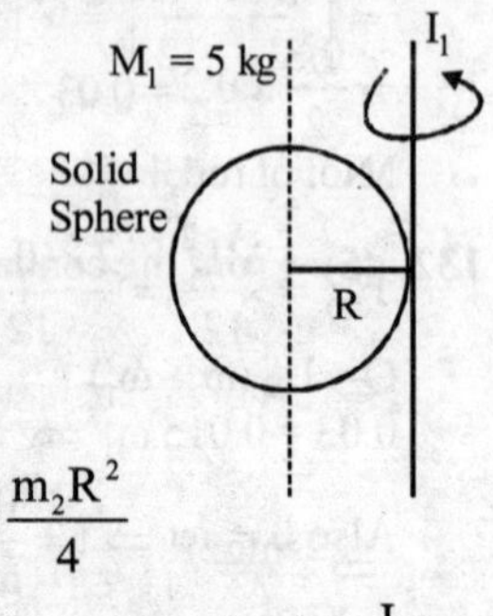

Moment of inertia of disc, $I''_{cm}=\frac{m_2R^2}{4}$

$I_2=\frac{m_2R^2}{4}+m_2R^2$

$I_2=\frac{5}{4}m_2R^2$

$I_2=5R^2$

$\frac{I_2}{I_1}=\frac{5}{7}=\frac{x}{7}$ (given)

$\therefore x=5$

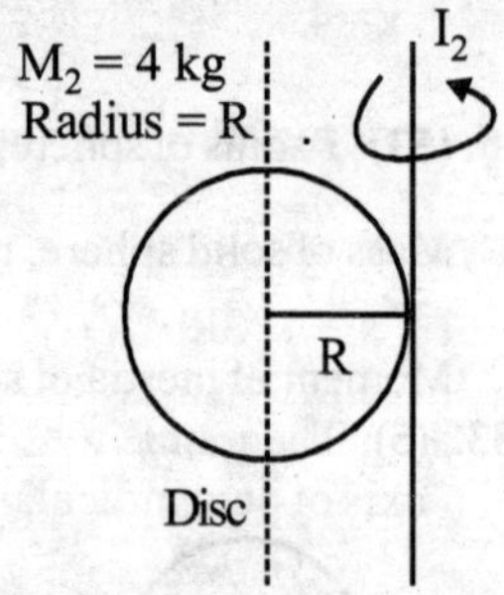

129. (110) $I_{PQ}=I_{cm}+md^2=\frac{2}{5}mR^2+md^2$

Also, $I_{PQ}=mk^2$

So, $k^2=\frac{2}{5}R^2+d^2=\frac{2}{5}\times5^2+10^2=110$

So, $k=\sqrt{110}$ cm

130. (3) From figure, moment of inertia,

$I_1=I_3=\frac{MR^2}{4}$

And $I_2=\frac{MR^2}{4}+MR^2=\frac{5}{4}MR^2=I_4$

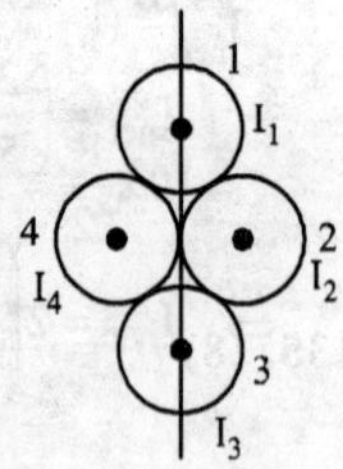

$\therefore$ Moment of inertia of system,

$I=I_1+I_2+I_3+I_4$

$=\frac{MR^2}{2}+\frac{5}{2}MR^2=3MR^2$

$\therefore I=\frac{3Ma^2}{4}=\frac{x}{4}Ma^2$ (given) $\left(\because R=\frac{a}{2}\right)$

$\therefore x=3$

131. (18) As $\tau=I\alpha\Rightarrow(12t-3t^2)1.5=4.5\alpha$

$\Rightarrow \alpha=4t-t^2$

$\Rightarrow \frac{d\omega}{dt}=4t-t^2 \Rightarrow \omega=\int_0^t(4t-t^2)dt$

$\Rightarrow \omega=2t^2-\frac{t^3}{3}$

For $\omega = 0 = 2t^2 - \frac{t^3}{3} \Rightarrow t^2\left(2 - \frac{t}{3}\right) = 0 \Rightarrow t = 0, 6$

$\frac{d\theta}{dt} = 2t^2 - \frac{t^3}{3} \Rightarrow \theta = \int_0^6 \left(2t^2 - \frac{t^3}{3}\right)dt$

$= \left[\frac{2t^3}{3} - \frac{t^4}{12}\right]_0^6 = 6^3\left(\frac{2}{3} - \frac{6}{12}\right) = 6^3\left(\frac{8-6}{12}\right) = \frac{6^3}{6} = 36$

No. of revolutions $= \frac{36}{2\pi} = \frac{18}{\pi} = \frac{K}{\pi}$ (given) So, $K = 18$

132. **(5)** From the conservation of mechanical energy

$\Rightarrow \text{mg}2\text{R} = \frac{1}{2}\text{I}_{\text{disc}}\omega^2 + \frac{1}{2}\text{I}_{\text{particle}}\omega^2$

$\Rightarrow \text{mg}2\text{R} = \frac{\omega^2}{2}\left[\frac{\text{mR}^2}{2} + \text{mR}^2\right] \left[\because \text{I}_{\text{disc}} = \frac{\text{mR}^2}{2}\right]$

$\Rightarrow \text{mg}2\text{R} = \frac{\omega^2}{2}\frac{3}{2}\text{mR}^2 \Rightarrow \frac{3}{4}\omega^2 = \frac{2\text{g}}{\text{R}}$

$\Rightarrow \omega^2 = \frac{8\text{g}}{3\text{R}} \Rightarrow \omega = \sqrt{\frac{80}{3\text{R}}} = 4\sqrt{\frac{\text{x}}{3\text{R}}}$ (given)

$\therefore$ x = 5

133. **(5)** The moment of inertia of a rod of mass m about an axis of perpendicular given by

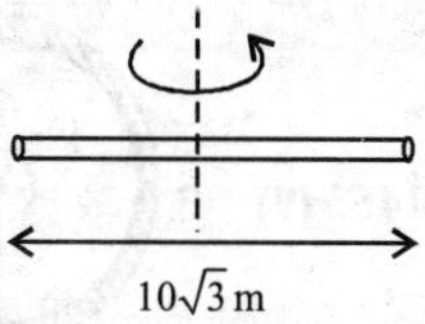

$\text{I} = \frac{\text{m}\ell^2}{12}$

The general equation of moment at inertia, $\text{I} = \text{mK}^2$

$\text{I} = \frac{\text{m}\ell^2}{12} = \text{mK}^2 \Rightarrow \text{K}^2 = \frac{\ell^2}{12} \Rightarrow \text{K} = \frac{\ell}{\sqrt{12}} = \frac{\ell}{2\sqrt{3}} = \frac{10\sqrt{3}}{2\sqrt{3}} = 5$

134. **(5)** $\text{I} = (\text{m}_A + \text{m}_B + \text{m}_C + \text{m}_D)\text{r}^2$

$\text{I}_{\text{net}} = (1+2+3+4)\cdot\left(\frac{a}{\sqrt{2}}\right)^2$, where a = side of square

$= 10 \times \frac{1^2}{2} = 5\,\text{kg m}^2$

135. **(8)** $I_1 = \frac{ml^2}{3}$

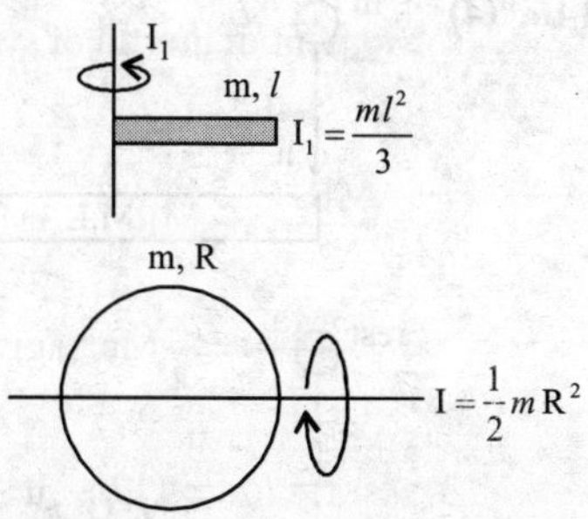

$I_2 = \frac{1}{2}mR^2$

$= \frac{1}{2}m\left(\frac{l}{2\pi}\right)^2 = \frac{ml^2}{8\pi^2}$

So, $\frac{I_1}{I_2} = \frac{8\pi^2}{3} = \frac{x\pi^3}{3}$ (given) $\Rightarrow x = 8$

136. **(a)** $I_{cm} = \frac{2}{5}mR^2$

$I = I_{cm} + MR^2$

$= \frac{2}{5}MR^2 + MR^2 = \frac{7}{5}MR^2$

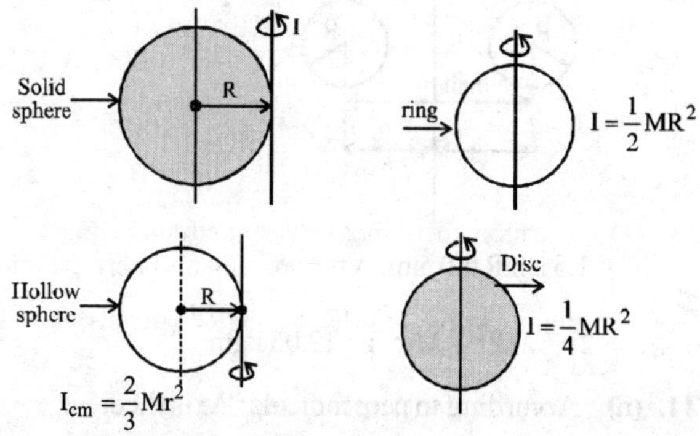

$I = I_{cm} + MR^2$

$= \frac{2}{3}MR^2 + MR^2 = \frac{5}{3}MR^2$

137. **(b)**

138. **(5)** We have

$I_1 = \frac{2}{5}M(2R)^2 \Rightarrow M(2R)^2 = \frac{5I_1}{2}$

$I_2 = \frac{1}{2}M(2R)^2$

$\Rightarrow I_2 = \frac{1}{2} \times \frac{5I_1}{2}$

$I_3 = \frac{1}{4}M(2R)^2 \Rightarrow I_3 = \frac{1}{4} \times \frac{5I_1}{2}$

$I_4 = \frac{1}{2}M(2R)^2 \Rightarrow I_4 = \frac{1}{2} \times \frac{5I_1}{2}$

So, $2(I_2 + I_3) + I_4$

$= 2\left(\frac{3}{4}\right) \times \frac{5I_1}{2} + \frac{1}{2} \times \frac{5I_1}{2}$

$= \frac{5I_1}{2}\left(\frac{3}{2} + \frac{1}{2}\right) = 5I_1$

139. **(4.24)** Given,

Length of the steel rod, $\ell = 0.6$m

Moment of inertia of rod, $\text{I} = \frac{\text{m}\ell^2}{3}$

Using conservation of energy

$\frac{1}{2}\text{I}\omega^2 = \frac{1}{2}\frac{\text{m}\ell^2\omega^2}{3} = \frac{\text{mg}\ell}{2}$

$\Rightarrow \omega^2 = \frac{3\,\text{mg}\ell}{\text{m}\ell^2} = \frac{3\text{g}}{\ell}$

$\Rightarrow \omega = \sqrt{\frac{3\text{g}}{\ell}}$

Speed $\text{v} = \omega\ell = \sqrt{3\text{g}\ell}$

$= \sqrt{\frac{3 \times 10 \times 6}{10}} = \sqrt{18}\,\text{m/s} = 4.24\,\text{m/s}$

140. (c)

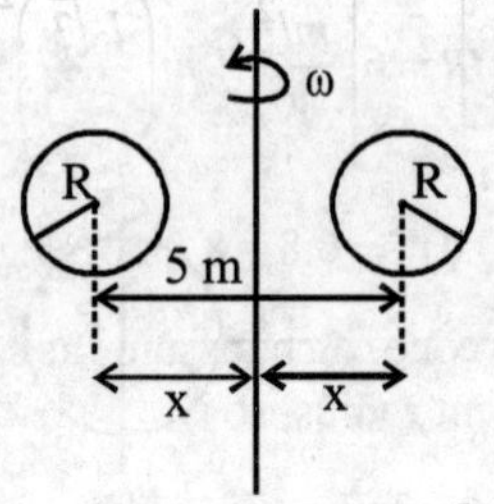

$m = 1.5$ kg, $R = 0.5$ m, $x = \frac{5}{2}m$

$I = 2\left(\frac{2}{5}MR^2 + Mx^2\right) = 19.05$ kgm²

141. (d) According to perpendicular Axis theorem.

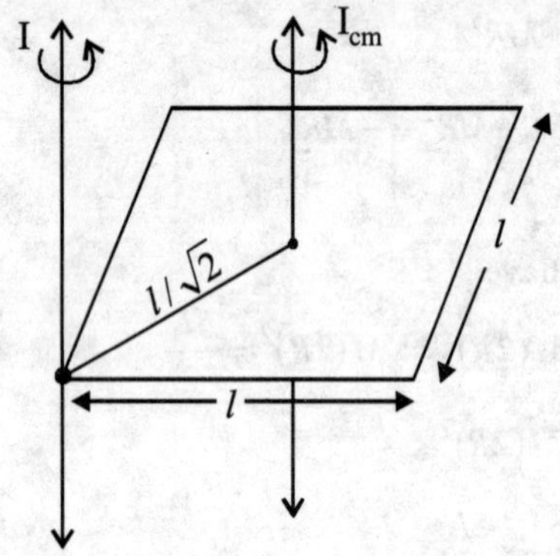

$I_{cm} = \frac{1}{6}ml^2$

$I = I_{cm} + md^2 = \frac{1}{6}ml^2 + m\left(\frac{l}{\sqrt{2}}\right)^2 = \frac{2}{3}ml^2 \left[\because d = \frac{2}{\sqrt{2}}\right]$

142. (c) Using conservation of angular momentum we get

$I_1\omega_1 + I_2\omega_2 = (I_1 + I_2)\omega \Rightarrow \omega = \frac{I_1\omega_1 + I_2 + \omega_2}{I_1 + I_2}$

Initial Kinetic energy

$K_i = \frac{1}{2}I_1\omega_1^2 + \frac{1}{2}I_2\omega_2^2$

Final Kinetic energy

$K_r = \frac{1}{2}(I_1 + I_2)\omega^2$

Loss in Kinetic energy

$= K_i - K_f = \frac{1}{2}\left[I_1\omega_1^2 + I_2\omega_2^2 - \frac{(I_1\omega_1 + I_2\omega_2)^2}{I_1 + I_2}\right]$

$\Rightarrow K_i - K_f = \frac{1}{2}\left(\frac{I_1 I_2}{I_1 + I_2}\right)(\omega_1 - \omega_2)^2$

143. (d) Given,

Length of cylinder, L = 80 cm

Radius of the cylinder, r = 20 cm

Moment of inertia of the cylinder I = 2.7 kgm²

From the parallel axis theorem

$I = I_{CM} + Md^2$

$\Rightarrow I = \frac{Mr^2}{2} + M\left(\frac{L}{2}\right)^2 \qquad (\because d = L/2)$

$\Rightarrow 2.7 = M\frac{(0.2)^2}{2} + M\left(\frac{0.8}{2}\right)^2 = M\left[\frac{2}{100} + \frac{16}{100}\right]$

$\Rightarrow M = 15$kg

$\Rightarrow \rho = \frac{M_{cylinder}}{V_{cylinder}} = \frac{M}{\pi r^2 L} = \frac{15}{\pi(0.2)^2 \times 0.8}$

$= 0.1492 \times 10^3$ kg/m³

144. (d) Let mass density be σ

Mass of disc of radius R,

$M_1 = \pi R^2\sigma$

Mass of disc of radius r,

$M_2 = \pi r^2\sigma$

Moment of inertia of disc of radius R,

$I_1 = \frac{1}{2}M_1R^2 = \frac{\pi\sigma R^4}{2}$ ... (i)

Moment of inertia of disc of radius r,

$I_2 = \frac{1}{4}M_2r^2 = \frac{\pi\sigma r^4}{4}$ ... (ii)

From (i) and (ii) we get $\Rightarrow \frac{I_1}{I_2} = \frac{2R^4}{r^4}$

145. (9)

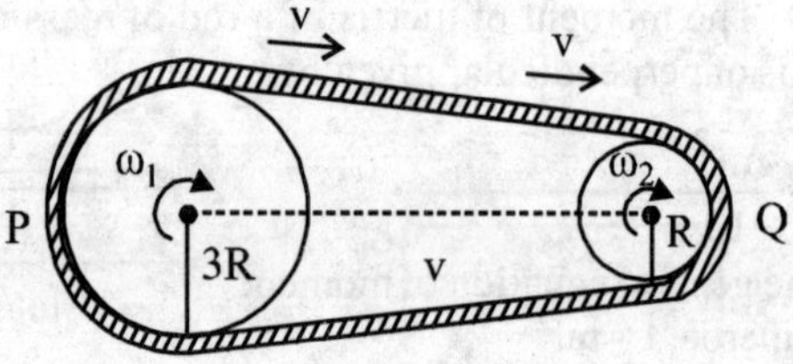

As per question, same rotational kinetic energy for wheel P and Q

$\therefore \quad KE_{rot\,1} = KE_{rot\,2}$ or, $\frac{1}{2}I_1(\omega_1)^2 = \frac{1}{2}I_2(\omega_2)^2$

$I_1\left(\frac{v}{3R}\right)^2 = I_2\left(\frac{v}{R}\right)^2 \qquad \left[\because \omega = \frac{v}{R}\right]$

$\therefore \quad \frac{I_1}{I_2} = \frac{9}{1}$

146. (4)

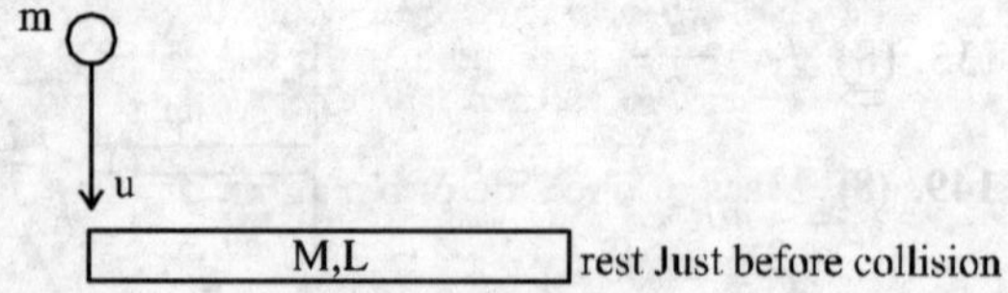

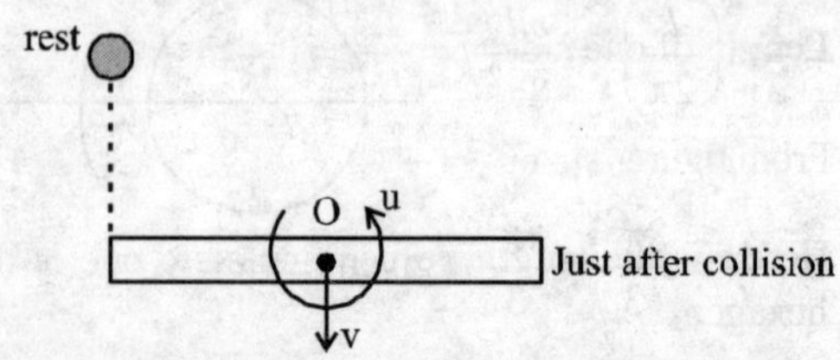

From linear momentum conservation, $P_i = P_f$

$mu = Mv \Rightarrow v = \frac{mu}{M}$

From angular momentum conservation about O, mid-point

$$mu.\frac{L}{2} = \frac{ML^2}{12}\omega$$

$$\therefore \quad \omega = \frac{6mu}{ML} \qquad ...(ii)$$

In elastic collision,

$$e = \frac{\text{relative velocity of separation}}{\text{relative velocity of approach}} = 1$$

$$\text{or, } 1 = \frac{V + \frac{\omega L}{2}}{u} \Rightarrow v + \frac{\omega L}{2} = u$$

$$\Rightarrow \quad V + \frac{6mu}{ML} \times \frac{L}{2} = u \Rightarrow v + \frac{3mu}{M} = u$$

$$\Rightarrow \quad \frac{mu}{M} + \frac{3mu}{M} = u \Rightarrow \frac{4mu}{M} = u$$

$$\text{or, } \frac{m}{M} = \frac{1}{4} = \frac{1}{x} \text{ (given)}$$

Hence value of x = 4

147. (d) $I + I = 2MR^2$

$2I = 2MR^2$

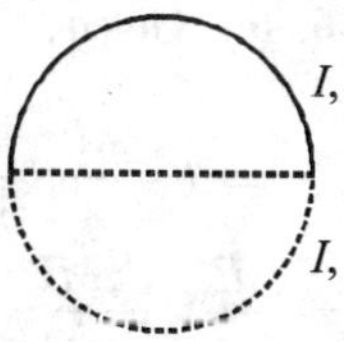

$$I = M\left(\frac{L}{\pi}\right)^2 \qquad [\because L = \pi R]$$

148. (c)

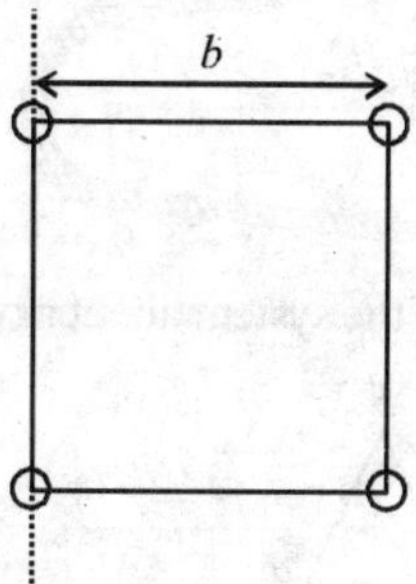

Moment of inertia of a sphere $= \frac{2}{5}ma^2$

Using the parallel axis theorem of moment of inertia, we have

$$I = 2 \times \frac{2}{5}ma^2 + 2\left(\frac{2}{5}ma^2 + mb^2\right)$$

$$\Rightarrow I = \frac{2}{5}ma^2 \times 4 + 2 \times mb^2 = \frac{8}{5}ma^2 + 2mb^2$$

149. (8) Mass of one side of hexagon,

$$m = \frac{6 \text{ Kg}}{6} = 1 \text{ Kg}$$

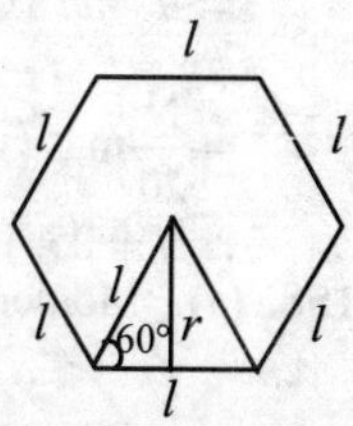

Length of one side, $l = \frac{2.4}{6} = 0.4$ m

From figure, $\sin 60° = \frac{r}{l}$

Here r is the length perpendicular to one of the side of hexagon.

$$r = l \sin 60° = \frac{l\sqrt{3}}{2}$$

Moment of inertia of hexagon,

$$I = 6\left[\frac{ml^2}{12} + mr^2\right] = 6\left[\frac{ml^2}{12} + m\left(\frac{l\sqrt{3}}{2}\right)^2\right]$$

$$= 5ml^2 = 5 \times 1 \times (0.4)^2 = 0.8 = 8 \times 10^{-1} \text{ kg m}^2$$

150. (a) Hollow ice-cream cone can be assume as several parts of discs having different radius, so

$$I = \int dI = \int dm(r^2) \qquad ...(i)$$

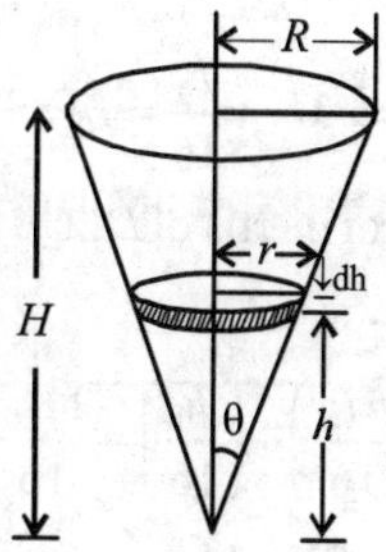

From diagram,

$$\frac{r}{h} = \tan\theta = \frac{R}{H} \text{ or } r = \frac{R}{H}h \qquad ...(ii)$$

Mass of element, $dm = \rho(\pi r^2)dh \qquad ...(iii)$

$$I = \int dI - \int_0^H dm(r^2) = \int_0^H \rho(\pi r^2)dh\left(\frac{R}{H}\cdot h\right)^2$$

$$= \int_0^H \rho\left(\pi\left(\frac{R}{H}\cdot h\right)^4\right)dh$$

Solving we get, $I = \frac{MR^2}{2}$

151. (d) By angular momentum conservation, $L_i = L_f$

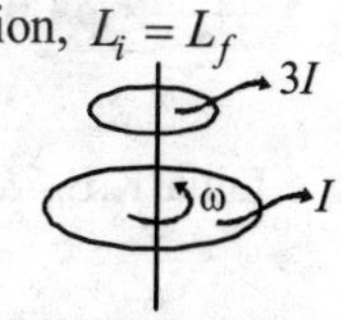

$$\omega I + 3I \times 0 = 4I\omega' \Rightarrow \omega' = \frac{\omega}{4}$$

$$(KE)_i = \frac{1}{2}I\omega^2$$

$$(KE)_f = \frac{1}{2}(3I + I)\omega'^2 = \frac{1}{2} \times (4I) \times \left(\frac{\omega}{4}\right)^2 = \frac{I\omega^2}{8}$$

$$\Delta KE = \frac{1}{2}I\omega^2 - \frac{1}{8}I\omega^2 = \frac{3}{8}I\omega^2$$

$$\therefore \text{Fractional loss in K.E.} = \frac{\Delta KE}{(KE)_i} = \frac{\frac{3}{8}I\omega^2}{\frac{1}{2}I\omega^2} = \frac{3}{4}.$$

152. (11) Let mass of triangular lamina = m, and length of side = l, then moment of inertia of lamina about an axis passing through centroid G perpendicular to the plane.

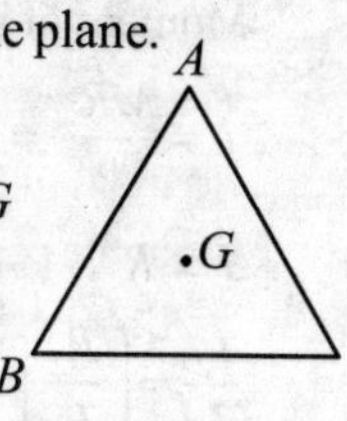

$I_0 \propto ml^2$

Let moment of inertia of $DEF = I_1$ about G

$$\text{So, } I_1 \propto \left(\frac{m}{4}\right)\left(\frac{l}{2}\right)^2 \propto \frac{ml^2}{16} \text{ or, } I_1 = \frac{I_0}{16}$$

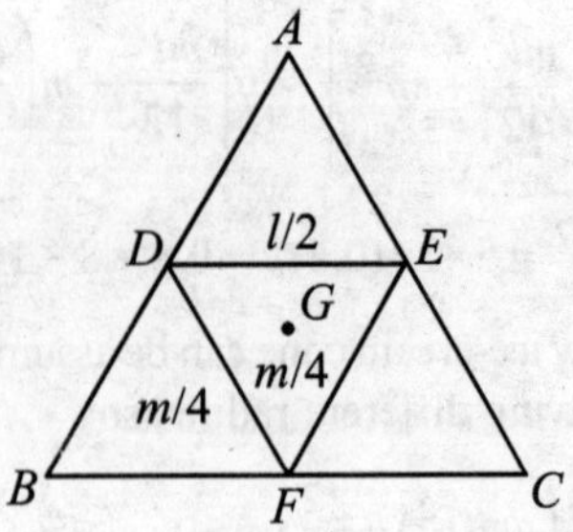

Let $I_{ADE} = I_{BDF} = I_{EFC} = I_2$

$\therefore 3I_2 + I_1 = I_0 \Rightarrow 3I_2 + \frac{I_0}{16} = I_0 \Rightarrow I_2 = \frac{5I_0}{16}$

Hence, moment of inertia of *DECB* i.e., after removal part *ADE*

$= 2I_2 + I_1 = 2\left(\frac{5I_0}{16}\right) + \left(\frac{I_0}{16}\right) = \frac{11I_0}{16} = \frac{NI_0}{16}$

Therefore value of $N = 11$.

153. (20) As we know moment of inertia disc, $I_{\text{disc}} = \frac{1}{2}MR^2$

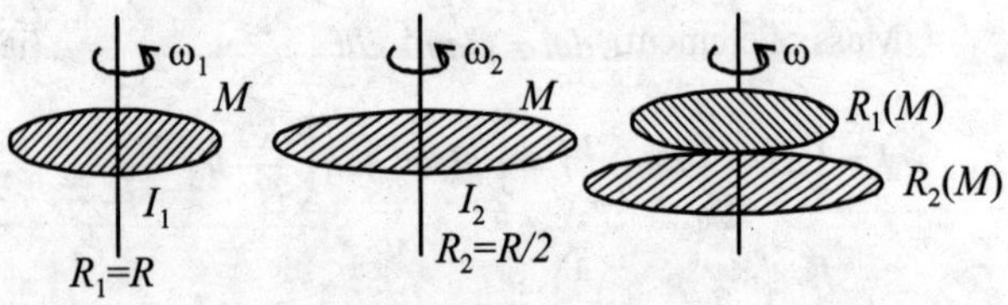

Using angular momentum conservation

$I_1\omega_1 + I_2\omega_2 = (I_1 + I_2)\times\omega_f$

$\frac{MR^2}{2}\times\omega + 0 = \left(\frac{MR^2}{2} + \frac{MR^2}{8}\right)\omega_f \Rightarrow \omega_f = \frac{4}{5}\omega$

Initial K.E., $K_i = \frac{1}{2}I\omega^2 = \frac{1}{2}\left(\frac{MR^2}{2}\right)\omega^2 = \frac{MR^2\omega^2}{4}$

Final K.E., $K_f = \frac{1}{2}\left(\frac{MR^2}{2} + \frac{MR^2}{8}\right)\frac{16}{25}\omega^2 = \frac{MR^2\omega^2}{5}$

Percentage loss in kinetic energy

$$= \frac{\frac{MR^2\omega^2}{4} - \frac{MR^2\omega^2}{5}}{\frac{MR^2\omega^2}{4}} \times 100 = 20\% = P\%$$

Hence, value of $P = 20$.

154. (c) Let ρ be the density of the discs and t is the thickness of discs.

Moment of inertia of disc is given by

$I = \frac{MR^2}{2} = \frac{[\rho(\pi R^2)t]R^2}{2}$

$I \propto R^4$ (As ρ and t are same)

$\frac{I_2}{I_1} = \left(\frac{R_2}{R_1}\right)^4 \Rightarrow \frac{16}{1} = \alpha^4 \Rightarrow \alpha = 2$

155. (b) Moment of inertia of rectangular sheet about an axis passing through O,

$I_O = \frac{M}{12}(a^2 + b^2) = \frac{M}{12}[(80)^2 + (60)^2]$

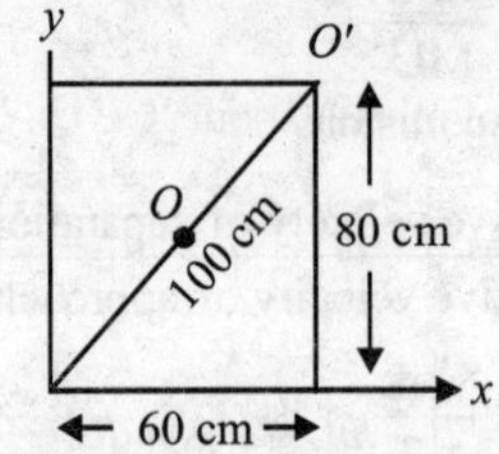

From the parallel axis theorem, moment of inertia about O',

$I_{O'} = I_O + M(50)^2$

$$\frac{I_O}{I_{O'}} = \frac{\frac{M}{12}(80^2 + 60^2)}{\frac{M}{12}(80^2 + 60^2) + M(50)^2} = \frac{1}{4}$$

156. (c) Given : $I = \frac{MR^2}{4} + \frac{ML^2}{12}$

$\Rightarrow I = \frac{M}{4}\times\frac{V}{\pi L} + \frac{ML^2}{12} \Rightarrow I = \frac{MV}{4\pi L} + \frac{ML^2}{12}$

$\frac{dI}{dL} = -\frac{MV}{4\pi L^2} + \frac{M\times 2L}{12} = 0$

$\Rightarrow V = \frac{2}{3}\pi L^3 \Rightarrow \pi R^2 L = \frac{2}{3}\pi L^3$

$\therefore \frac{L}{R} = \sqrt{\frac{3}{2}}$

157. (25) Moment of inertia of the system about axis *XE*.

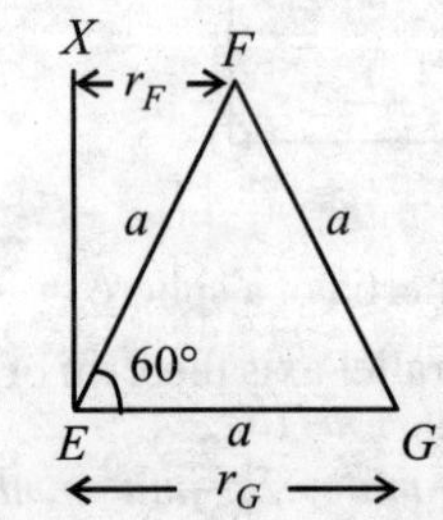

$I = I_E + I_F + I_G \Rightarrow I = m(r_E)^2 + m(r_F)^2 + m(r_G)^2$

$\Rightarrow I = m\times 0^2 + m\left(\frac{a}{2}\right)^2 + ma^2 = \frac{5}{4}ma^2 = \frac{25}{20}ma^2$

$= \frac{N}{20}ma^2$ (given)

$\therefore N = 25.$

158. (a) Moment of inertia,

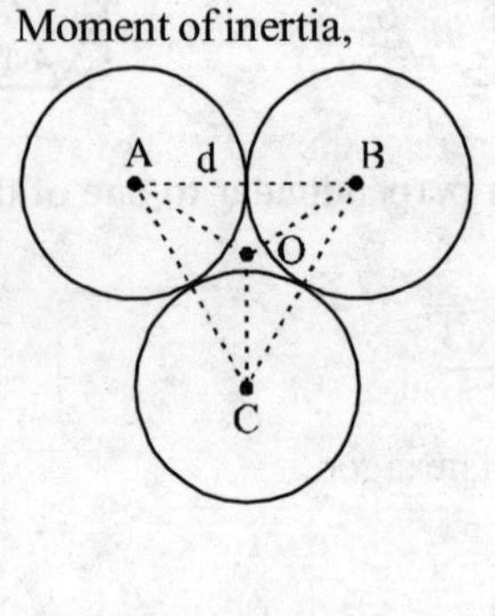

$$I_1 = \frac{2}{5}m\left(\frac{d}{2}\right)^2 + m(AO)^2$$

and $AO = \frac{d}{\sqrt{3}}$

Moment of inertia about 'O'

$$I_O = 3I_1 = 3\left[\frac{2}{5}m\left(\frac{d}{2}\right)^2 + m\left(\frac{d}{\sqrt{3}}\right)^2\right] = \frac{13}{10}md^2$$

$$\text{And } I_A = 2\left[\frac{2}{5}m\left(\frac{d}{2}\right)^2 + md^2\right] + \frac{2}{5}m\left(\frac{d}{2}\right)^2$$

$$\Rightarrow I_A = \frac{23}{10}md^2 \qquad \therefore \frac{I_O}{I_A} = \frac{\frac{13}{10}md^2}{\frac{23}{10}md^2} = \frac{13}{23}$$

159. (15) Here, length of bar, $l = 1$ m
angle, $\theta = 30°$

$$\Delta PE = \Delta KE \text{ or } mgh = \frac{1}{2}I\omega^2$$

$$\Rightarrow (\text{mg})\frac{l}{2}\sin 30° = \frac{1}{2}\left(\frac{ml^2}{3}\right)\omega^2$$

$$\Rightarrow \text{mg}\frac{l}{2}\times\frac{1}{2} = \frac{1}{2}\left(\frac{ml^2}{3}\right)\omega^2 \Rightarrow \omega = \sqrt{15}\text{ rad/s}$$

160. (a) Using principal of conservation of energy

$$(m_1 - m_2)gh = \frac{1}{2}(m_1 + m_2)v^2 + \frac{1}{2}I\omega^2$$

$$\Rightarrow (m_1 - m_2)gh = \frac{1}{2}(m_1 + m_2)(\omega R)^2 + \frac{1}{2}I\omega^2$$

$(\because v = \omega R)$

$$\Rightarrow (m_1 - m_2)gh = \frac{\omega^2}{2}\left[(m_1 + m_2)R^2 + I\right]$$

$$\Rightarrow \omega = \sqrt{\frac{2(m_1 - m_2)gh}{(m_1 + m_2)R^2 + I}}$$

161. (a) When the bob covered a distance 'h'

$$\text{Using } mgh = \frac{1}{2}mv^2 + \frac{1}{2}I\omega^2$$

$$= \frac{1}{2}\text{m}(\omega r)^2 + \frac{1}{2}\times\frac{mr^2}{2}\times\omega^2 \;(\because v = \omega r \textit{ no slipping})$$

$$\Rightarrow mgh = \frac{3}{4}m\omega^2 r^2 \Rightarrow \omega = \sqrt{\frac{4gh}{3r^2}} = \frac{1}{r}\sqrt{\frac{4gh}{3}}$$

162. (c) Moment inertia of the rod passing through a point $\frac{\ell}{4}$ away from the centre of the rod

$$I = \text{I}_\text{g} + \text{Md}^2$$

$$\Rightarrow I = \frac{Ml^2}{12} + M\times\left(\frac{l^2}{16}\right) = \frac{7Ml^2}{48}\left[\because d = \frac{l}{4}\right]$$

Using $I = MK^2 = \frac{7Ml^2}{48}$ (K = radius of gyration)

$$\Rightarrow K = \sqrt{\frac{7}{48}}l$$

163. (a) Given,
mass per unit area of circular disc, $\sigma = A + Br$
Area of the ring $= 2\pi r dr$
Mass of the ring, $dm = \sigma 2\pi r dr$
Moment of inertia,

$$I = \int dmr^2 = \int \sigma 2\pi r dr.r^2$$

$$\Rightarrow I = 2\pi\int_0^a (A + Br)r^3 dr = 2\pi\left[\frac{Aa^4}{4} + \frac{Ba^5}{5}\right]$$

$$\Rightarrow I = 2\pi a^4\left[\frac{A}{4} + \frac{Ba}{5}\right]$$

164. (c)

$$I = \int_a^b (dm)r^2$$

$$= \int_a^b \left(\frac{\sigma_0}{r}\times 2\pi r\,dr\right)r^2 = \frac{2\pi\sigma_0}{3}|r^3|_a^b = \frac{2\pi\sigma_0}{3}(b^3 - a^3)$$

Mass of the disc,

$$m = \int_a^b \frac{\sigma_0}{r}\times 2\pi r\,dr = 2\pi\sigma_0(b - a)$$

Radius of gyration,

$$k = \sqrt{\frac{I}{m}} = \sqrt{\frac{(2\pi\sigma_0/3)(b^3 - a^3)}{2\pi\sigma_0(b - a)}} = \sqrt{\frac{a^2 + b^2 + ab}{3}}$$

165. (b)

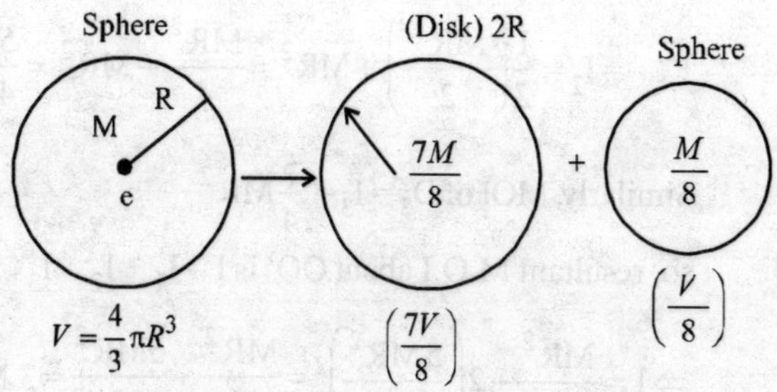

$$I_1 = \left(\frac{7M}{8}\right)(2R)^2\frac{1}{2} = \left(\frac{7}{16}\times 4\right)MR^2 = \frac{14}{8}\text{mR}^2$$

$$I_2 = \frac{2}{5}\left(\frac{M}{8}\right)r^2$$

$$\Rightarrow I_2 = \frac{2}{5}\left(\frac{M}{8}\right)\left(\frac{R^2}{4}\right) = \frac{MR^2}{80} \quad \left[\frac{4}{3}\pi r^3\rho = \frac{1}{8}\frac{4}{3}\pi R^3\times\rho \;\Rightarrow r = R/2\right]$$

$$\frac{I_1}{I_2} = \frac{14\times 80}{8} = 140$$

166. (d) $\frac{1}{2}I\omega^2 = k\theta^2$

or $\omega = \left(\sqrt{\frac{2k}{I}}\right)\theta$

or $\alpha = \frac{d\omega}{dt} = \sqrt{\frac{2K}{I}}\left(\frac{d\theta}{dt}\right) = \left(\sqrt{\frac{2k}{I}}\right)\omega$

$= \left(\sqrt{\frac{2k}{I}}\right)\left(\sqrt{\frac{2k}{I}}\right)\theta = \frac{2k\theta}{I}$

167. (b) $\omega = \alpha t = 20t$

Given, $\frac{1}{2}I\omega^2 = 1200$ or $\frac{1}{2}\times 1.5\times(20t)^2 = 1200$

or $t = 2$ s

168. (b) $I_{thin} = mk^2$

$I_{hollow} = \frac{m}{2}(10^2 + 20^2)$

As, $I_{thin} = I_{hollow} \Rightarrow k = \sqrt{\frac{10^2+20^2}{2}} = 16$ (approx.)

169. (d) According to parallel axes theorem

$I = \frac{2}{5}mR^2 + mx^2$

Hence graph (d) correctly depicts I vs x.

170. (b) Moment of inertia of disc D_1 about $OO' = I_1 = \frac{MR^2}{2}$

M.O.I of D_2 about OO'

$= I_2 = \frac{1}{2}\left(\frac{MR^2}{2}\right) + MR^2 = \frac{MR^2}{4} + MR^2 = \frac{5}{4}MR^2$

Similarly, MOI of $D_3 = I_3 = \frac{5}{4}MR^2$

so, resultant M.O.I about OO' is $I = I_1 + I_2 + I_3$

$\Rightarrow I = \frac{MR^2}{2} + 2\left(\frac{5MR^2}{4}\right) = \frac{MR^2}{2} + \frac{5MR^2}{2} = 3\ MR^2$

171. (a) Using parallel axes theorem, for ball moment of inertaia,

$I_{ball} = \frac{2}{5}MR^2 + M(2R)^2 = \frac{22}{5}MR^2$

For two balls $I_{balls} = 2\times\frac{22}{5}MR^2$

$I_{rod} = \frac{M(2R)^2}{12} = \frac{MR^2}{3}$

$I_{system} = I_{balls} + I_{rod} = \frac{44}{5}MR^2 + \frac{MR^2}{3} = \frac{137}{15}MR^2$

172. (c) $I = m\left(\frac{l}{3}\right)^2 + \frac{m}{2}\left(\frac{2l}{3}\right)^2$

$= \frac{ml^2}{9} + \frac{m}{2}\times\frac{4l^2}{9} = \frac{ml^2}{9} + \frac{2ml^2}{9}$

$= \frac{3ml^2}{9} = \frac{ml^2}{3}$

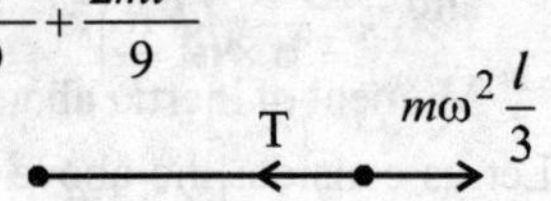

By energy conservation,

$\frac{1}{2}k\theta_0^2 = \frac{1}{2}\left(\frac{ml^2}{3}\right)\omega^2 \Rightarrow \omega^2 = \frac{3k\theta_0^2}{ml^2}$

$\therefore\quad T = \frac{m\omega^2 l}{3} = \frac{m\times 3k\theta_0^2}{ml^2}\times\frac{l}{3} = \frac{k\theta_0^2}{l}$

173. (b)

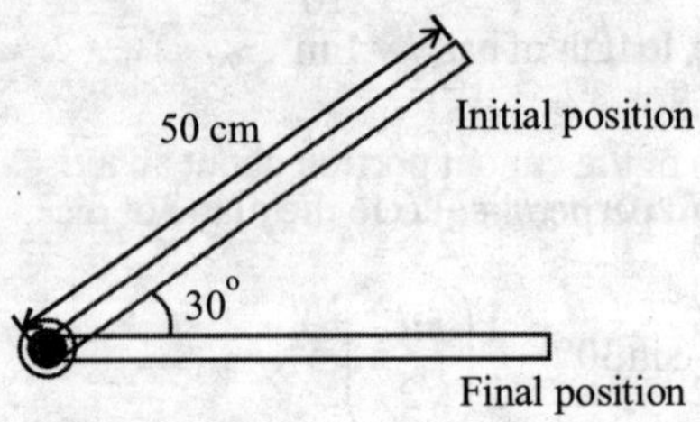

By the low of conservation of energy,

P.E. of rod = Rotational K.E.

$mg\frac{\ell}{2}\sin\theta = \frac{1}{2}I\omega^2$

$\Rightarrow mg\frac{\ell}{2}\sin 30° = \frac{1}{2}\frac{m\ell^2}{3}\omega^2 \quad\Rightarrow mg\frac{\ell}{2}\times\frac{1}{2} = \frac{1}{2}\frac{m\ell^2}{3}\omega^2$

For complete length of rod,

$\omega = \sqrt{3g/(2\ell)} = \sqrt{30}\ \text{rad s}^{-1}$

174. (d) Using parallel axes theorem, moment of inertia about 'O'

$I_o = I_{cm} + md^2$

$= \frac{7MR^2}{2} + 6(M\times(2R)^2) = \frac{55MR^2}{2}$

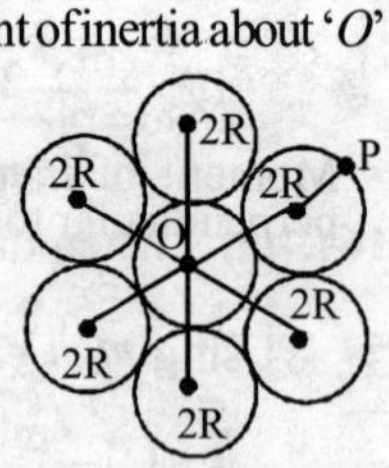

Again, moment of inertia about point P, $I_p = I_o + md^2$

$= \frac{55MR^2}{2} + 7M(3R)^2 = \frac{181}{2}MR^2$

175. (a) Let σ be the mass per unit area.

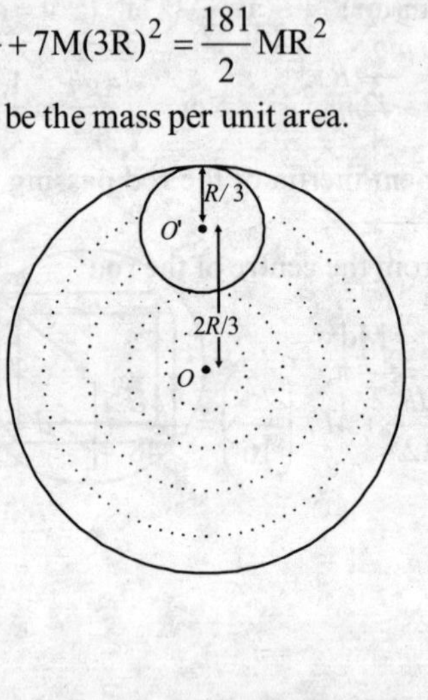

The total mass of the disc

$= \sigma \times \pi R^2 = 9M$

The mass of the circular disc cut

$$= \sigma \times \pi \left(\frac{R}{3}\right)^2 = \sigma \times \frac{\pi R^2}{9} = M$$

Let us consider the above system as a complete disc of mass $9M$ and a negative mass M super imposed on it.

Moment of inertia (I_1) of the complete disc $= \frac{1}{2} 9MR^2$ about an axis passing through O and perpendicular to the plane of the disc.

M.I. of the cut out portion about an axis passing through O' and perpendicular to the plane of disc

$$= \frac{1}{2} \times M \times \left(\frac{R}{3}\right)^2$$

$\therefore$ *M.I.* (I_2) of the cut out portion about an axis passing through O and perpendicular to the plane of disc

$$= \left[\frac{1}{2} \times M \times \left(\frac{R}{3}\right)^2 + M \times \left(\frac{2R}{3}\right)^2\right]$$

[Using parallel axis theorem]

$\therefore$ The total *M.I.* of the system about an axis passing through O and perpendicular to the plane of the disc is

$$I = I_1 - I_2 = \frac{1}{2} 9MR^2 - \left[\frac{1}{2} \times M \times \left(\frac{R}{3}\right)^2 + M \times \left(\frac{2R}{3}\right)^2\right]$$

$$= \frac{9MR^2}{2} - \frac{9MR^2}{18} = \frac{(9-1)MR^2}{2} = 4MR^2$$

176. (c) As we know, moment of inertia of a disc about an axis passing through C.G. and perpendicular to its plane,

$$I_z = \frac{mR^2}{2}$$

Moment of inertia of a disc about a tangential axis perpendicular to its own plane,

$$I_z' = \frac{3}{2} mR^2$$

$$\therefore \quad I_z / I_z' = \frac{mR^2}{2} \Big/ \frac{3mR^2}{2} = 1/3$$

177. (a) Here $a = \frac{2}{\sqrt{3}} R$

$$\text{Now, } \frac{M}{M'} = \frac{\frac{4}{3}\pi R^3}{a^3}$$

$$= \frac{\frac{4}{3}\pi R^3}{\left(\frac{2}{\sqrt{3}} R\right)^3} = \frac{\sqrt{3}}{2}\pi.$$

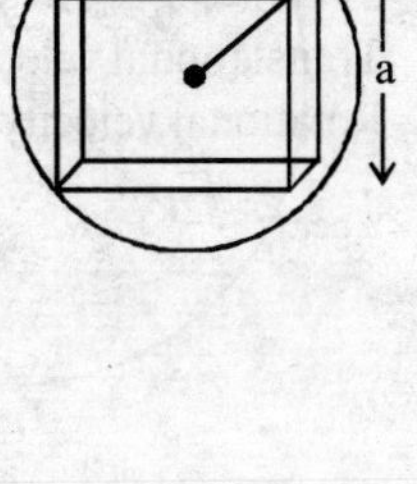

$$\Rightarrow M' = \frac{2M}{\sqrt{3}\pi}$$

Moment of inertia of the cube about the given axis,

$$I = \frac{M'a^2}{6} = \frac{\frac{2M}{\sqrt{3}\pi} \times \left(\frac{2}{\sqrt{3}} R\right)^2}{6} = \frac{4MR^2}{9\sqrt{3}\pi}$$

178. (d) For a thin uniform square sheet

$$I_1 = I_2 = I_3 = \frac{ma^2}{12}$$

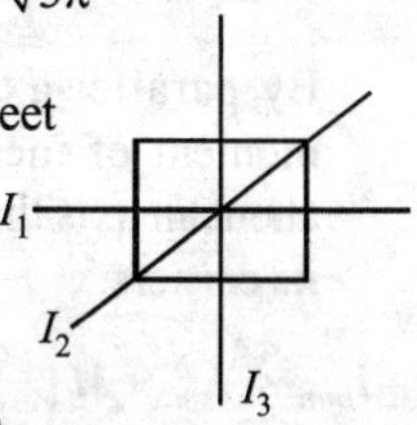

179. (b) For translational motion,

$mg - T = ma$ .....(i)

For rotational motion,

$T.R = I\alpha$

$$\Rightarrow T.R = \frac{1}{2} mR^2 \alpha$$

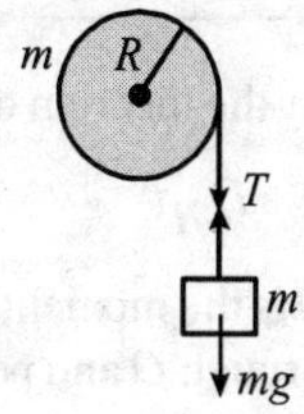

Also, acceleration, $a = R\alpha$

$$\therefore T = \frac{1}{2} mR\alpha = \frac{1}{2} ma$$

Substituting the value of T is equation (i) we get

$$mg - \frac{1}{2} ma = ma \Rightarrow a = \frac{2}{3} g$$

180. (a) Given,

Force, $F = (20t - 5t^2)$

Radius, $r = 2m$

Torque, $T = rf = I\alpha \Rightarrow 2(20t - 5t^2) = 10\alpha$

$\therefore \alpha = 4t - t^2$

$$\Rightarrow \frac{d\omega}{dt} = 4t - t^2 \Rightarrow \int_0^{\omega} d\omega = \int_0^{t} \left(4t - t^2\right) dt$$

$$\Rightarrow \omega = 2t^2 - \frac{t^3}{3}$$

$$\int_0^{\theta} d\theta = \int_0^{6} \left(2t^2 - \frac{t^3}{3}\right) dt \text{ (as } \omega = 0 \text{ at } t = 0, 6\text{s)}$$

$$\Rightarrow \theta = 36 \text{ rad} \Rightarrow 2\pi n = 36 \Rightarrow n = \frac{36}{2\pi} < 6$$

181. (c)

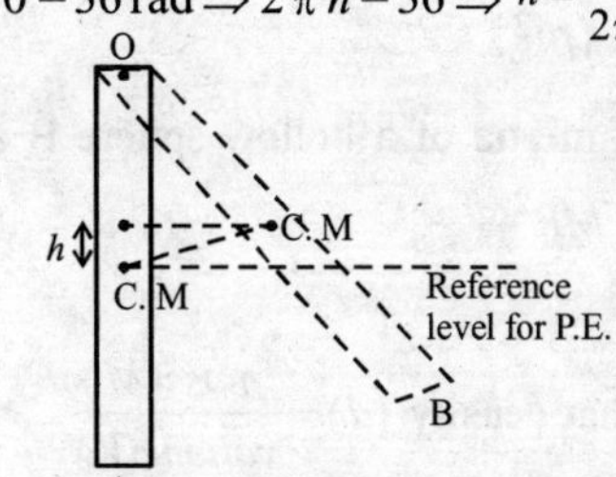

The moment of inertia of the rod about O is $\frac{1}{3} m\ell^2$. The kinetic energy of the rod at position A $= \frac{1}{2} I\omega^2$ where I is the moment of inertia of the rod about O. When the rod is at position B, its angular velocity is zero. In this case, the energy of the rod is mgh where h is the maximum height to which the centre of mass (C.M) rises.

Gain in potential energy = Loss in kinetic energy

$\therefore\ mgh = \frac{1}{2}I\omega^2 = \frac{1}{2}\left(\frac{1}{3}ml^2\right)\omega^2 \Rightarrow h = \frac{\ell^2\omega^2}{6g}$

182. (d) $I_{nn'} = \frac{1}{12}M(a^2 + a^2) = \frac{Ma^2}{6}$

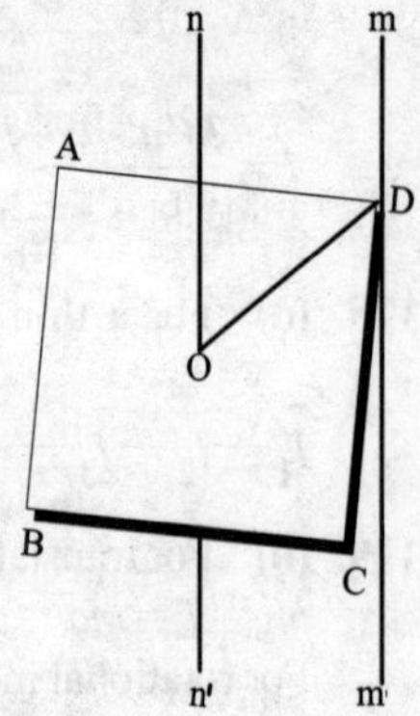

Also, $DO = \frac{DB}{2} = \frac{\sqrt{2}a}{2} = \frac{a}{\sqrt{2}}$

By parallel axis theorem, moment of inertia of plate about an axis through one of its corners.

$I_{mm'} = I_{nn'} + M\left(\frac{a}{\sqrt{2}}\right)^2 = \frac{Ma^2}{6} + \frac{Ma^2}{2}$

$= \frac{Ma^2 + 3Ma^2}{6} = \frac{2}{3}Ma^2$

183. (d) By the theorem of perpendicular axis,

$I = I_{EF} + I_{GH}$

Here, I is the moment of inertia of square lamina about an axis through O and perpendicular to its plane.

$\therefore\ I_{EF} = I_{GH}$ (By Symmetry of Figure)

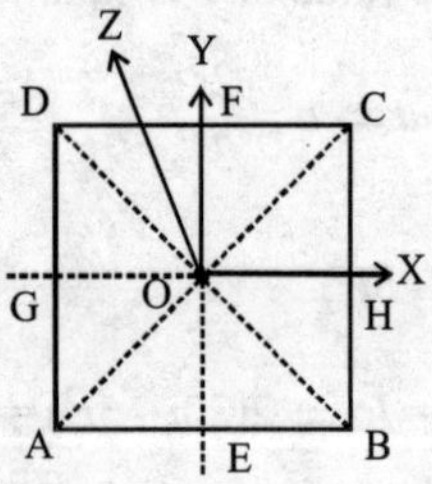

$\therefore\ I_{EF} = \frac{I}{2}$ ...(i)

Again, by the same theorem $I = I_{AC} + I_{BD} = 2I_{AC}$

($\therefore\ I_{AC} = I_{BD}$ by symmetry of the figure)

$\therefore\ I_{AC} = \frac{I}{2}$ ...(ii)

From (i) and (ii), we get, $I_{EF} = I_{AC}$.

184. (a) The moment of inertia of solid sphere A about its diameter $I_A = \frac{2}{5}MR^2$.

The moment of inertia of a hollow sphere B about its diameter $I_B = \frac{2}{3}MR^2$.

$\therefore I_A < I_B$

185. (d) We know that density $(d) = \frac{mass(M)}{volume(V)}$

$\therefore\ M = d \times V = d \times (\pi R^2 \times t)$.

The moment of inertia of a disc is given by $I = \frac{1}{2}MR^2$

$\therefore\ I = \frac{1}{2} \times d \times \pi R^2 \times t \times R^2$

$\Rightarrow I \propto tR^4$

$\therefore\ \frac{I_X}{I_Y} = \frac{t_X R_X^4}{t_Y R_Y^4} = \frac{t \times R^4}{\frac{t}{4} \times (4R)^4} = \frac{1}{64}$

$\therefore\ I_Y = 64 I_X$

186. (a) Rotational kinetic energy $= \frac{1}{2}I\omega^2$,

Angular momentum, $L = I\omega \Rightarrow I = \frac{L}{\omega}$

$\therefore\ K.E. = \frac{1}{2}\frac{L}{\omega} \times \omega^2 = \frac{1}{2}L\omega$

$L = \frac{2K.E}{\omega}$

When ω is doubled and K.E is halved.

New angular momentum,

$L' = \frac{2\frac{K.E}{2}}{2\omega} \Rightarrow L' = \frac{L}{4}$

187. (a) M. I of a circular wire about an axis nn' passing through the centre of the circle and perpendicular to the plane of the circle $= MR^2$

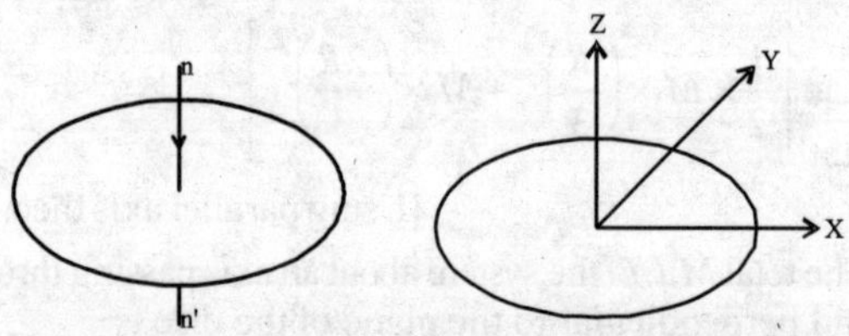

As shown in the figure, X-axis and Y-axis lie along diameter of the ring. Using perpendicular axis theorem

$I_X + I_Y = I_Z$

Here, I_X and I_Y are the moment of inertia about the diameter.

$\Rightarrow 2I_X = MR^2$ [$\because I_X = I_Y$ (by symmetry) and $I_Z = MR^2$]

$\therefore I_X = \frac{1}{2}MR^2$

188. (a)

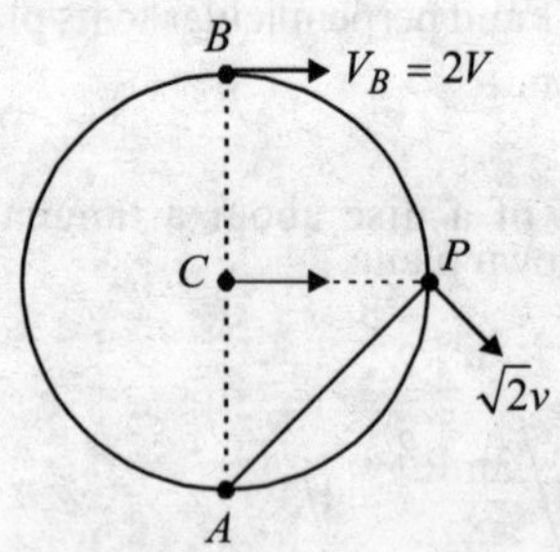

Translational velocity : V - rightward

Rotational velocity : V - rightward

$V_{top} = V + V = 2V$

If $V_B = 2V$

Given $V_B = 8$ m/s

$2V = 8 \Rightarrow V = 4$ m/s

Translational velocity : Rightward, V

Rotational velocity : Upward, V

$V_{net} = \sqrt{2}V$

$V_P=\sqrt{2}V \Rightarrow V_P=4\sqrt{2}$ m/s.

Correct (a).

189. (a) The FBD of solid sphere is

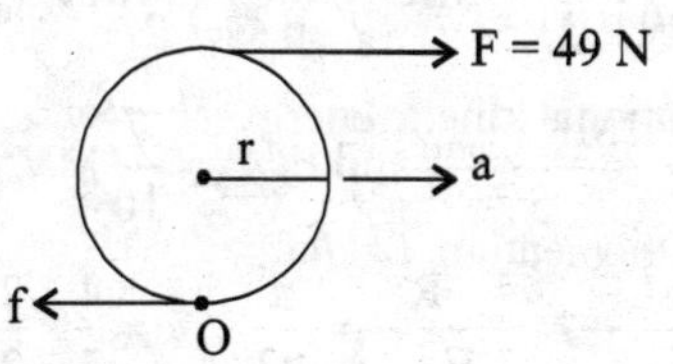

For rotational motion, $\tau_0 = I\alpha$

$\Rightarrow F\times 2r = I\alpha$

$\Rightarrow 49\times 2r = \left(\frac{2}{5}mr^2+mr^2\right)\alpha$

$\therefore \quad r\alpha=\frac{14}{4}$ ...(i)

As sphere rolls without slipping

$\therefore \quad a=r\alpha=\frac{14}{4}=\frac{7}{2}=3.5\text{ m/s}^2$ [Using (i)]

190. (b) In pure rolling,

$$\frac{\text{Linear KE}}{\text{Rotational K.E.}}=\frac{\frac{1}{2}mv_{cm}^2}{\frac{1}{2}I\omega^2}=\frac{mR^2\omega^2}{\frac{2}{5}mR^2\omega^2}=\frac{5}{2}$$

191. (a)

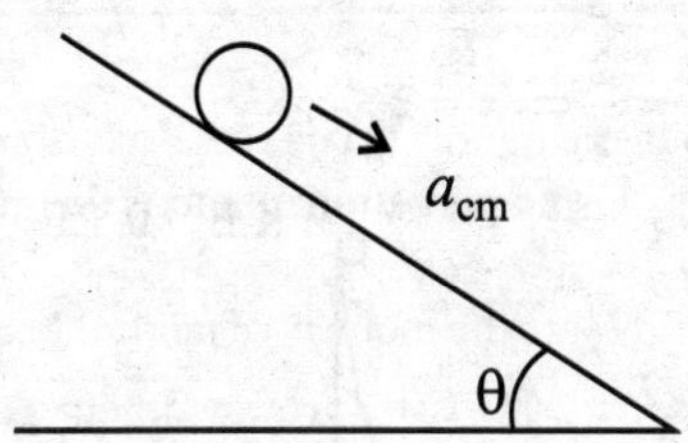

$$t=\sqrt{\frac{2\ell}{a_{cm}}}=\sqrt{\frac{2\ell\left(1+\frac{K^2}{R^2}\right)}{g\sin\theta}}\Rightarrow t\propto\left(\frac{K^2}{R^2}\right)$$

For solid sphere, $\frac{K^2}{R^2}=\frac{2}{5}$

For hollow sphere, $\frac{K^2}{R^2}=\frac{2}{3}\ \therefore \left(\frac{K^2}{R^2}\right)_{SS}<\left(\frac{K^2}{R^2}\right)_{HS}$

$\therefore t_1<t_2$

192. (c) In pure rolling,

$$\text{Acceleration, } a=\frac{g\sin\theta}{1+\frac{K^2}{R^2}}\Rightarrow a=\frac{g\sin 45^\circ}{1+\frac{1}{2}}=\frac{2\frac{g}{\sqrt{2}}}{3}=\frac{\sqrt{2}g}{3}$$

193. (a)

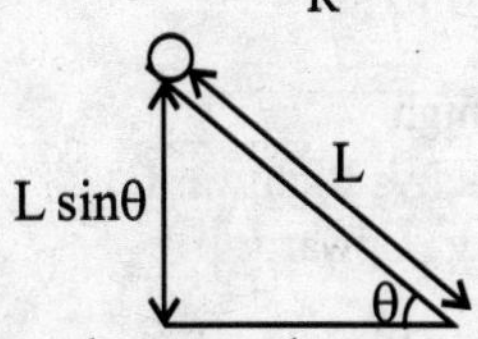

Using work-energy theorem,

$W_g = k_f - k_i$

$Mg\,L\sin\theta = k_f - k_i$ ...(i)

K.E. in pure rolling $=\frac{1}{2}mV_{cm}^2+\frac{1}{2}I_{cm}\omega^2$

$=\frac{1}{2}mV^2+\frac{1}{2}\times\frac{2}{5}mR^2\frac{V^2}{R^2}=\frac{7}{10}mV^2$

From equation (i)

$mgL\sin\theta=\frac{7}{10}mv^2-0$

$V^2\propto\sin\theta$

$\left(\frac{V_1}{V_2}\right)^2=\frac{\sin\theta_1}{\sin\theta_2}=\frac{\sin 30^\circ}{\sin 45^\circ}=\frac{1}{\sqrt{2}}$

194. (3) During slipping

Acceleration, $a_1 = g\sin\theta$

Distance travelled, $S=\frac{1}{2}a_1t^2\Rightarrow t=\sqrt{\frac{2S}{a_1}}=\sqrt{\frac{2S}{g\sin\theta}}$

For rolling motion

Acceleration, $a_2=\frac{g\sin\theta}{1+\frac{k^2}{R^2}}$

for disc, $k=\frac{R}{\sqrt{2}}$

$\therefore a_2=\frac{2g\sin\theta}{3}$

Distance travelled, $S=\frac{1}{2}a_2(t_2)^2$

$\Rightarrow t_2=\sqrt{\frac{2S}{a_2}}=\sqrt{\frac{6S}{2g\sin\theta}}=\sqrt{\frac{3}{2}}\sqrt{\frac{2S}{g\sin\theta}}=\sqrt{\frac{3}{2}}t_1$

$=\sqrt{\frac{a}{2}}t_1$ (given)

$\therefore \alpha=3$

195. (2) For hollow sphere,

$I=\frac{2}{3}mR^2=mK^2\Rightarrow\frac{R^2}{k^2}=\frac{3}{2}$

$\therefore\frac{(KE)_R}{(KE)_{Total}}=\frac{1}{\left(1+\frac{R^2}{k^2}\right)}=\frac{1}{1+\frac{3}{2}}=\frac{2}{5}=\frac{x}{5}$ (given)

$\therefore \quad x=2$

196. (7) For a body rolling without slipping, velocity

$$V=\sqrt{\frac{2gh}{1+\frac{K^2}{R^2}}}$$

Here $V_{\text{solid sphere}}=V_{\text{hollow cylinder}}$ and height of solid sphere $=h_1$ & of hollow cylinder $=h_2$

And $V_{\text{solid sphere}}=\sqrt{\frac{10}{7}gh_1}$

$V_{hollow\ cylinder} = \sqrt{gh_2}$ or, $\sqrt{\frac{10}{7}gh_1} = \sqrt{gh_2}$

$\therefore \frac{h_1}{h_2} = \frac{7}{10}$ or, $\frac{h_1}{h_2} = \frac{7}{10} = \frac{n}{10}$ (given) $\quad \therefore n = 7$

197. (d) Here translation kinetic energy is converted into potential energy.

$$\frac{1}{2}mv^2 = mgh_{max}$$

$$h_{max} = \frac{v^2}{2g}$$

198. (6) Given, Mass of disc, m = 50 kg
Speed of centre of mass, v = 0.4 m/s
From the work energy theorem

$$W = \text{Change in Kinetic energy} = 0 - \left(\frac{1}{2}mv^2 + \frac{1}{2}I\omega^2\right)$$

$$\Rightarrow W = 0 - \frac{1}{2}mv^2\left(1 + \frac{K^2}{R^2}\right) = \frac{-3}{4}mv^2$$

$$= -\frac{3}{4} \times 50 \times 0.4^2 = -6J$$

Absolute work, $|W| = 6J$

199. (10)

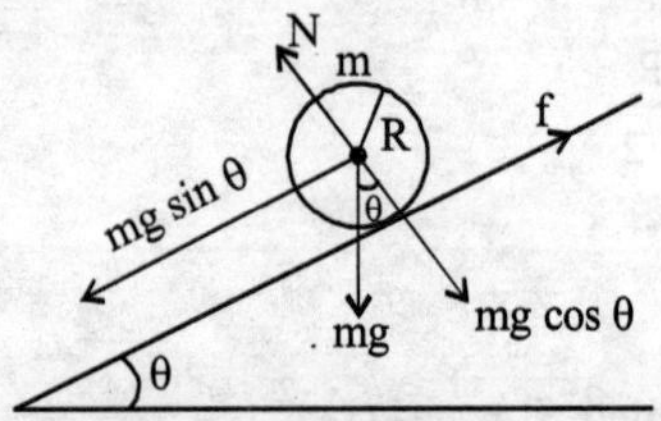

For rolling on inclined plane,

$$a = \frac{g\sin\theta}{1 + \frac{I}{MR^2}} = \frac{10\sin 60°}{1 + \frac{1}{2}} = \frac{10}{\sqrt{3}} m/s^2$$

$$= \frac{x}{\sqrt{3}} m/s^2 \text{ (given)} \therefore x = 10$$

200. (7) In pure rolling work done by friction, $W_{friction} = 0$. Therefore, PE = KE. Initially, $PE_{ring} = PE_{sphere}$ so finally, $KE_{ring} = KE_{sphere}$

$$\therefore \text{Ratio of kinetic energy} = \frac{KE_{ring}}{KE_{sphere}} = 1$$

$$\therefore \frac{7}{x} = 1 \quad \therefore \quad x = 7$$

201. (a) From figure,

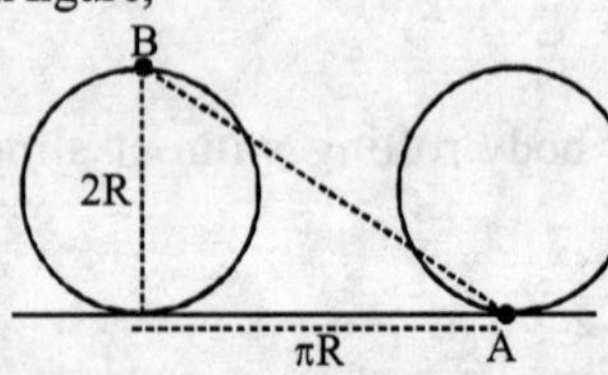

Displacement, $BA = \sqrt{(2R)^2 + (\pi R)^2} = R\sqrt{4 + \pi^2}$

202. (4) Angular momentum

$$L = (I_{com})(\omega) \text{ and } E_{total} = \frac{1}{2}(I_{com})(\omega^2) + \frac{1}{2}MV_{com}^2$$

$$L = \frac{2}{5}MR^2\frac{V_{com}}{R} \text{ and}$$

$$E_{total} = \frac{1}{2}\left(\frac{2}{5}MR^2\right)\frac{V_{com}^2}{R^2} + \frac{1}{2}MV_{com}^2$$

$$L = \frac{2MRV_{com}}{5} \text{ and } E_{total} = \frac{7}{10}MV_{com}^2$$

$$\therefore \frac{L}{E_{total}} = \frac{4}{7}\frac{R}{V_{com}} = \frac{\pi}{22} \quad \therefore \omega = \frac{4}{7} \times \frac{22}{22} \times 7 = 4$$

203. (2) Moment of inertia of spherical shell, $I = \frac{2}{3}mR^2$

$$\frac{K_{rotational}}{K_{Total}} = \frac{\frac{1}{2}I_{cm}\omega^2}{\frac{1}{2}mv_{cm}^2 + \frac{1}{2}I_{cm}\omega^2}$$

$$\frac{K_{rotational}}{K_{Total}} = \frac{\frac{1}{2}\left(\frac{2}{3}mR^2\right)\left(\frac{v}{R}\right)^2}{\frac{1}{2}mv^2 + \frac{1}{2}\left(\frac{2}{3}mR^2\right)\left(\frac{v}{R}\right)^2}$$

$$= \frac{\frac{2}{3}}{\frac{5}{3}} \quad (\because v = \omega R)$$

$$\therefore \frac{x}{5} = \frac{2}{5} \Rightarrow x = 2$$

204. (75)

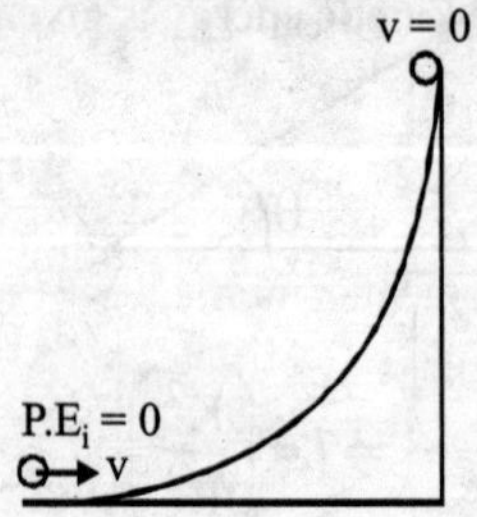

At highest point $KE_f = 0$

Initial KE_i = Translational KE + Rotational KE

$$= \frac{1}{2}mv^2 + \frac{1}{2}I\omega^2$$

In case of rolling $v = R\omega$

$$= \frac{1}{2}mv^2 + \frac{1}{2} \times \frac{2}{3}mR^2 \times \frac{v^2}{R^2} \left(\because I = \frac{2}{3}mR^2\right)$$

$$= \frac{5}{6}mv^2$$

Apply energy conservation

$$KE_i + PE_i = KE_f + PE_f$$

$$\Rightarrow \frac{5}{6}mv^2 + 0 = 0 + mgh$$

$$\Rightarrow \frac{5}{6}mv^2 = mgh$$

$$\Rightarrow h = \frac{5}{6 \times 10} \times 9m = \frac{15}{20}m = 75cm$$

205. (2) Using energy conservation

$$mgh = \frac{1}{2}mv^2_{cm} + \frac{1}{2}\frac{mR^2}{2}\frac{v^2_{cm}}{R^2};\ mg(0.3) = \frac{3}{4}mv^2_{cm}$$

Where $h = 60\sin 30° = 30\text{ cm} = 0.3\text{ m}$

$v_{cm} = 2\text{ m/sec}$

206. (10) As $(K.E)_{Rolling} = (K.E)_{Trans} + (K.E)_{Rot}$

$$\Rightarrow 7\times10^{-3} = \frac{1}{2}mV^2 + \frac{1}{2}\left(\frac{2}{5}mR^2\right)\frac{V^2}{R^2}$$

$$\Rightarrow 7\times10^{-3} = \frac{1}{2}mV^2 + \frac{1}{5}mV^2$$

$$\Rightarrow 7\times10^{-3} = \frac{7}{10}\times1\times V^2$$

$\Rightarrow V^2 = 10^{-2}$

$\Rightarrow V = 0.1\text{ m/s} = 10\text{ cm/s}$

207. (10) As $(K.E)_{Rolling} = (K.E)_{Trans} + (K.E)_{Rot}$

$$\Rightarrow 7\times10^{-3} = \frac{1}{2}mV^2 + \frac{1}{2}\left(\frac{2}{5}mR^2\right)\frac{V^2}{R^2}$$

$$\Rightarrow 7\times10^{-3} = \frac{1}{2}mV^2 + \frac{1}{5}mV^2$$

$$\Rightarrow 7\times10^{-3} = \frac{7}{10}\times1\times V^2$$

$\Rightarrow V^2 = 10^{-2}$

$\Rightarrow V = 0.1\text{ m/s} = 10\text{ cm/s}$

208. (3) The rotational kinetic energy is given by

$$E = \frac{1}{2}I\omega^2 \Rightarrow E = \frac{1}{2}\frac{ml^2}{12}\omega^2$$

(for uniform rod moment of inertia, $I = \frac{ml^2}{12}$)

$$E = \frac{1}{2}\frac{dAl^3}{12}\omega^2 \qquad \left(\because d = \frac{m}{v} = \frac{m}{A.l}\right)$$

$$E = \frac{dA(2)^3}{24}\omega^2 \Rightarrow \omega = \sqrt{\frac{3E}{dA}} = \sqrt{\frac{\alpha E}{dA}}\ \text{(given)}$$

$\Rightarrow \alpha = 3$

209. (54) Acceleration in rough horizontal surface, $a = -\mu_k g = -3$

Using $v = u + at$

$v = 18 - 3\times2$

$\Rightarrow v = 12\text{ m/s}$

$$KE = \frac{1}{2}mv^2 + \frac{1}{2}I\omega^2$$

Here, I = moment of inertia of disc $= \frac{1}{2}mr^2$

$$KE = \frac{1}{2}mv^2 + \frac{1}{2}\frac{mr^2}{2}\frac{v^2}{r^2} = \frac{3}{4}mv^2$$

$$\Rightarrow KE = \frac{3}{4}\times0.5\times(12)^2 = 54\text{ J}$$

210. (40) Given, mass of solid sphere, m = 2 kg

Kinetic energy, KE = 2240 J

$$\therefore\ KE = \frac{1}{2}mv^2 + \frac{1}{2}I\omega^2$$

$$\Rightarrow 2240 = \frac{1}{2}2(v)^2 + \frac{1}{2}\times\frac{2}{5}(2)R^2.\left(\frac{v}{R}\right)^2$$

($\because$ Moment of inertia of solid sphere, $I = \frac{2}{5}mR^2$)

$$\Rightarrow 2240 = v^2 + \frac{2}{5}v^2$$

$\Rightarrow v = 40\text{ m/s}.$

211. (120) From the energy conservation,

$$mgh = \frac{1}{2}mv^2 + \frac{1}{2}I\omega^2$$

$$\Rightarrow mgh = \frac{1}{2}mv^2 + \frac{1}{2}\frac{mR^2}{2}\omega^2 \qquad \left(\because I = \frac{mR^2}{2}\right)$$

$$\Rightarrow mgh = \frac{1}{2}mv^2 + \frac{1}{2}\frac{mR^2}{2}.\frac{v^2}{R^2}$$

$$\Rightarrow mgh = \frac{1}{2}mv^2 + \frac{1}{4}mv^2$$

$$\Rightarrow 10h = \frac{16}{2} + \frac{16}{4} \Rightarrow h = 1.2\text{ m} = 120\text{ cm}$$

212. (d) $V = \sqrt{\dfrac{2gh}{1+\dfrac{I}{mR^2}}}$

$$\frac{V_{cylinder}}{V_{sphere}} = \frac{\sqrt{1+\dfrac{I_{sphere}}{mR^2}}}{\sqrt{1+\dfrac{I_{cylinder}}{mR^2}}} = \frac{\sqrt{1+\frac{2}{5}}}{\sqrt{1+\frac{1}{2}}} = \frac{\sqrt{\frac{7}{5}}}{\sqrt{\frac{3}{2}}} = \sqrt{\frac{14}{15}}$$

213. (3) When wheel reaches bottom, block reaches the top. As they are connected by same string, so there velocity will be same at any time.

By law of conservation of mechanical energy.

$K_i + P_i = K_f + P_f$

$$\Rightarrow 0 + 12gh = \frac{1}{2}\times3\times V^2 + \frac{1}{2}\times12\times V^2 + \frac{1}{2}I\frac{V^2}{R^2} + 3gh$$

$$\Rightarrow 9gh = \frac{1}{2}\times15V^2 + \frac{1}{2}\times12R^2\times\frac{V^2}{R^2}\quad [\because I = 12R^2]$$

$$\Rightarrow 9gh = \frac{V^2}{2}\times27$$

$$\Rightarrow 9gh = \frac{V^2}{2}\times27 \Rightarrow V^2 = \frac{2}{3}gh \Rightarrow V^2 = \frac{1}{4}\times\frac{8}{3}gh$$

$$\Rightarrow V = \frac{1}{2}\sqrt{\frac{8}{3}gh} = \frac{1}{2}\sqrt{xgh}\ \text{(given)}.\ \text{So } x = \frac{8}{3}\approx3$$

214. (5) From the conservation of mechanical energy

$$\Rightarrow mg2R = \frac{1}{2}I_{disc}\omega^2 + \frac{1}{2}I_{particle}\omega^2$$

$$\Rightarrow mg2R = \frac{\omega^2}{2}\left[\frac{mR^2}{2} + mR^2\right] \qquad \left[\because I_{disc} = \frac{mR^2}{2}\right]$$

$$\Rightarrow mg2R = \frac{\omega^2}{2}\frac{3}{2}mR^2$$

$\Rightarrow \frac{3}{4}\omega^2 = \frac{2g}{R} \Rightarrow \omega^2 = \frac{8g}{3R}$

$\Rightarrow \omega = \sqrt{\frac{80}{3R}}$

215. (b) Here $v_{cm} = \omega R$

$K_{rot} = \frac{1}{2}I_{cm}\omega^2 = \frac{1}{2}\times\frac{2}{5}mR^2\times\omega^2 = \frac{1}{5}mv_{cm}^2$

$K_{trans} = \frac{1}{2}m\,v_{cm}^2 \;\therefore \frac{K_{rot}}{K_{total}} = \frac{\frac{1}{5}mv_{cm}^2}{\frac{1}{5}mv_{cm}^2 + \frac{1}{2}mv_{cm}^2} = \frac{\frac{1}{5}}{\frac{7}{10}} = \frac{2}{7}$

216. (2) For rolling without slipping $v_0 = \omega R$

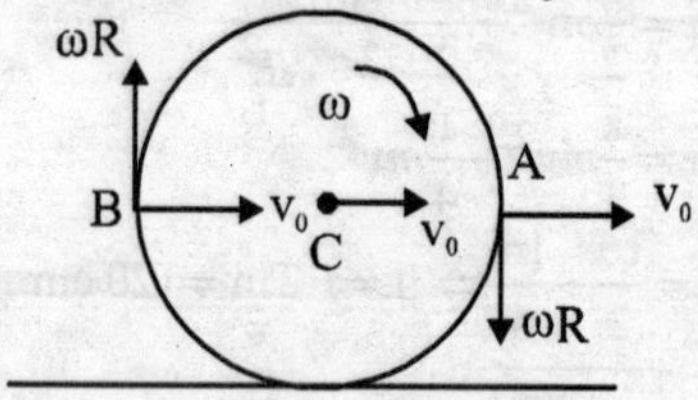

Now $v_A = v_B = \sqrt{v_0^2 + (\omega R)^2} = \sqrt{2}v_0 = \sqrt{x}v_0$ (given)

$\therefore$ value of x = 2

217. (2) When disc slides then acceleration, $a_1 = g\sin\theta$ and, distance travelled

$S = ut_1 + \frac{1}{2}a_1t_1^2 = \frac{1}{2}g\sin\theta.t_1^2$ ... (i)

Again, when disc do pure rolling

$a_2 = \frac{g\sin\theta}{1+\frac{I}{mr^2}} = \frac{2}{3}g\sin\theta \quad \left[\because I_{disc} = \frac{1}{2}mr^2\right]$

$\therefore S = ut_2 + \frac{1}{2}a_2t_2^2 = \frac{1}{2}.\frac{2}{3}g\sin\theta.t_2^2 = \frac{g\sin\theta}{3}t_2^2$ ... (ii)

From eq. (i) & (ii),

$\frac{S}{S} = 1 = \frac{g\sin\theta\, t_1^2}{2}\Big/\frac{g\sin\theta\, t_2^2}{3}$

$\therefore \quad \frac{t_2}{t_1} = \sqrt{\frac{3}{2}} = \sqrt{\frac{3}{x}}$ (given)

$\therefore \quad x = 2$

218. (3) Velocity at the lowest point of inclined plane

$V = \sqrt{\frac{2gh}{1+\frac{K^2}{R^2}}}$

$\left(\frac{K^2}{R^2}\right)_{Ring} = 1 \,\&\, \left(\frac{K^2}{R^2}\right)_{Solid\ cylinder} = \frac{1}{2}$

$\therefore \frac{V_R}{V_C} = \sqrt{\frac{2gh}{1+1}}\Big/\sqrt{\frac{2gh}{1+\frac{1}{2}}} = \sqrt{\frac{3}{2}\times\frac{1}{2}} = \frac{\sqrt{3}}{2} = \frac{\sqrt{x}}{2}$ (given)

$\therefore x = 3$

219. (b) Translational kinetic energy $= \frac{1}{2}mv^2$

Rotational kinetic energy $= \frac{1}{2}I\omega^2$

According to question

$\frac{1}{2}I\omega^2 = \frac{1}{2}\times\frac{1}{2}mv^2$

As $v = R\omega$ (pure rolling)

$\frac{1}{2}I\omega^2 = \frac{1}{4}mR^2\omega^2 \Rightarrow I = \frac{1}{2}mR^2$

Thus, body should be solid cylinder.

220. (4) The body having maximum acceleration (*a*) will reach the bottom first.

$a = \frac{g\sin\theta}{\left(1+\frac{I}{MR^2}\right)} = \frac{g\sin\theta}{1+\frac{K^2}{R^2}}$

$\left(\frac{K^2}{R^2}\right.$ for ring = 1, disc & solid cylinder $= \frac{1}{2}$, solid sphere $\left.= \frac{2}{5}\right)$ i.e., $\frac{K^2}{R^2}$ is least for solid sphere so solid sphere will reach first at the bottom of the inclined plane.

221. (3) Acceleration,

$a = \frac{g\sin\theta}{1+\frac{K^2}{R^2}} = \frac{2}{3}g\sin\theta = \frac{2}{b}g\sin\theta$ (given)

$\left[\because \frac{K^2}{R^2} = 0.5 \text{ for disc}\right]$

$\therefore b = 3$

222. (a) *K.E* of the sphere = translational *K.E* + rotational *K.E*

$= \frac{1}{2}mv^2 + \frac{1}{2}I\omega^2$

$\because$ Moment of inertia of sphere $I = \frac{2}{5}mR^2$

$\therefore \quad K.E = \frac{1}{2}mv^2 + \frac{1}{2}\times\frac{2}{5}mR^2\times\omega^2$

$\Rightarrow K.E = \frac{1}{2}mv^2 + \frac{1}{2}\times\frac{2}{5}mR^2\times\left(\frac{v}{R}\right)^2 \left(\because \omega = \frac{v}{R}\right)$

$\Rightarrow KE = \frac{1}{2}mv^2 + \frac{1}{5}mv^2$

$\Rightarrow KE = \frac{7}{10}mv^2 = \frac{7}{10}\times\frac{1}{2}\times\frac{25}{10^4}$

$\Rightarrow KE = \frac{35}{4}\times10^{-4}$ joule

$\Rightarrow KE = 8.75\times10^{-4}$ joule

223. (50) For the box to be slide

$F = f = \mu mg = 0.4\,mg$

For no toppling

$F\left(\frac{a}{2}+b\right) \le mg\frac{a}{2}$

$\Rightarrow \quad 0.4\,mg\left(\frac{a}{2}+b\right) \le mg\frac{a}{2}$

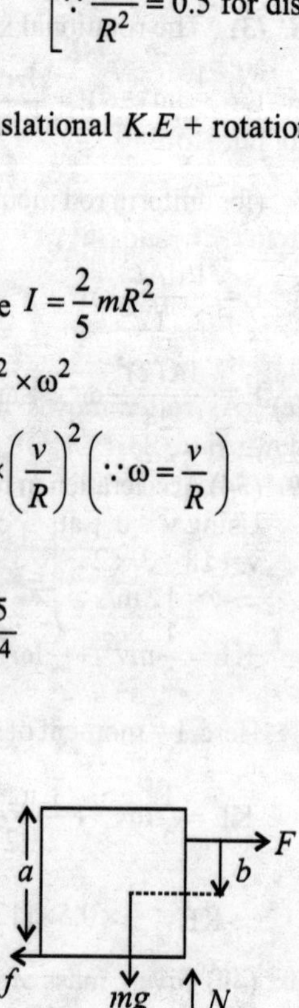

$\Rightarrow \quad 0.2a + 0.4b \le 0.5a \Rightarrow \dfrac{b}{a} \le \dfrac{3}{4}$ i.e. $b \le 0.75a$

As the maximum value of b can be equal to $0.5a$.

$\therefore \left(\dfrac{b}{a}\right)_{max} = \dfrac{1}{2}$

So, $\dfrac{100b}{a} = 50$

224. (c) For sphere,

$$\frac{1}{2}mv^2 + \frac{1}{2}I\omega^2 = mgh$$

or $\dfrac{1}{2}mv^2 + \dfrac{1}{2}\left(\dfrac{2}{5}mR^2\right)\dfrac{v^2}{R^2} = mgh$ or $h = \dfrac{7v^2}{10g}$

For cylinder

$$\frac{1}{2}mv^2 + \frac{1}{2}\left(\frac{mR^2}{2}\right) = mgh' \text{ or } h' = \frac{3v^2}{4g}$$

$$\therefore \frac{h}{h'} = \frac{7v^2/10g}{3v^2/4g} = \frac{14}{15}$$

225. (d)

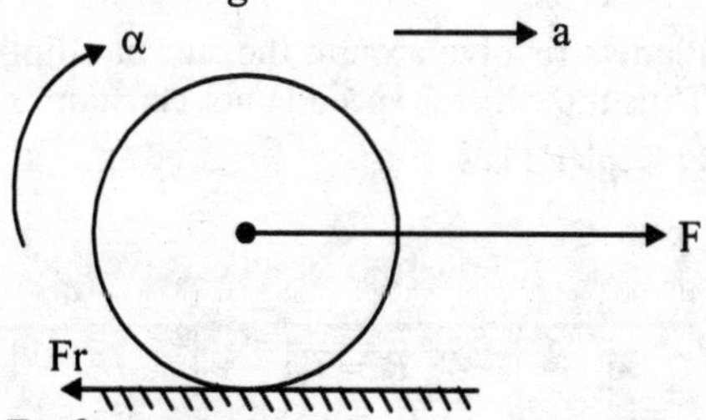

$F - f_r = ma$ ...(i)

$f_r R = I\alpha = \dfrac{mR^2}{2}\alpha$ ...(ii)

for pure rolling

$a = \alpha R$ ...(iii)

from (i) (ii) and (iii)

$$F - \frac{mR\alpha}{2} = m\alpha R \Rightarrow F = \frac{3}{2}mR\alpha \Rightarrow \alpha = \frac{2F}{3mR}$$

226. (c) As roller moves forward left part comes slightly downward. So, PL < QR.

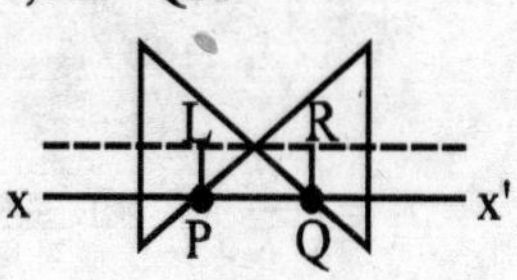

Line PQ is instantaneous axis of rotation as $\vec{V}_P = \vec{V}_Q = 0$. Since $V_L = \omega(PL)$ and $V_R = \omega(QR)$. So, $V_L < V_R$ [$\because$ PL < QR]. So roller will turn left.

227. (d) As we know,

$$\text{Acceleration, } a = \frac{g\sin\theta}{1 + \dfrac{I}{mr^2}}$$

Here, $a_c = a_s$

$$\Rightarrow \frac{g\sin\theta_c}{1+\dfrac{1}{2}} = \frac{g\sin\theta_s}{1+\dfrac{2}{5}} \Rightarrow \frac{\sin\theta_c}{\sin\theta_s} = \frac{\dfrac{3}{2}}{\dfrac{7}{5}} = \frac{15}{14}$$

228. (c)

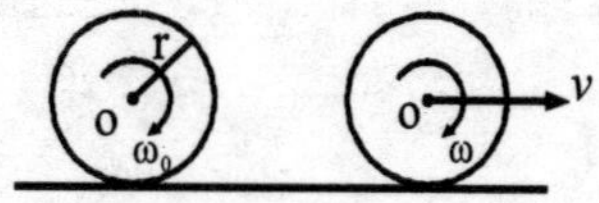

Let v be the speed of centre of mass and ω be the angular speed, when slipping stops.

From conservation of angular momentum about any fix point on the surface,

$$mr^2\omega_0 = mr^2\omega + mvr$$

$$\Rightarrow mr^2\omega_0 = mr^2\omega + mr^2\omega \Rightarrow mr^2\omega_0 = 2\,mr^2\omega$$

$$\Rightarrow \omega = \omega_0/2 \Rightarrow v = \frac{\omega_0 r}{2} \quad [\because v = r\omega]$$

229. (d) For solid sphere, $\dfrac{K^2}{R^2} = \dfrac{2}{5}$

$$\therefore \quad v = \sqrt{\frac{2gh}{1 + \dfrac{K^2}{R^2}}} = \sqrt{\frac{10}{7}gh}$$

230. (b) Acceleration of the body rolling down an inclined plane is given by.

$$a = \frac{g\sin\theta}{1 + \dfrac{I}{MR^2}}$$

7 Gravitation

1. **(a)** Areal velocity of the planet is constant, from Kepler's second law

$$\frac{dA}{dt} = \frac{L}{2m}$$

There is no external torque acting on it. As a result, its angular momentum remains constant.

2. **(b)** By Kepler's period law,

$T^2 \propto r^3$

$$\frac{T_1^2}{r_1^3} = \frac{T_2^2}{r_2^3}$$

$T_1 = 200$ days, $r_2 = \frac{r_1}{4}$

$$\frac{(200)^2}{r_1^3} = \frac{T_2^2}{\left(\frac{r_1}{4}\right)^3}$$

$$T_2 = \frac{200}{4 \times 2} = 25 \text{ days}$$

3. **(b)** Angular speed,

$\omega = \frac{2\pi}{T} \Rightarrow \omega \propto \frac{1}{T}$ [T = time period]

$T_{moon} = 27$ days and

$T_{earth} = 365$ days 4 hour

$T_{earth} > T_{moon}$ So, $\omega_{moon} > \omega_{earth}$

4. **(16)** From conservation of angular momentum, $(\because L = I\omega)$

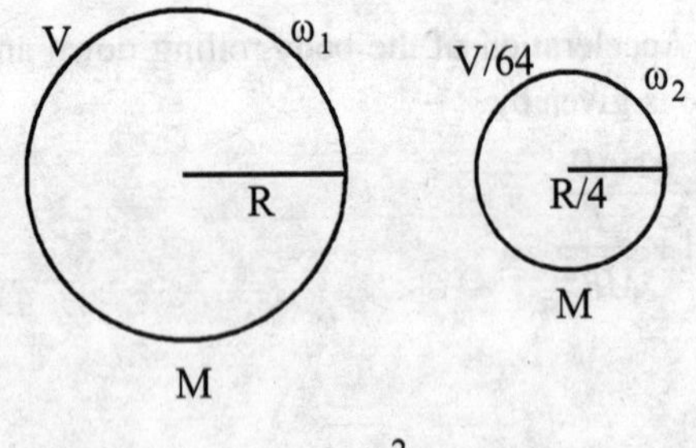

$$\frac{2}{5}MR^2\omega_1 = \frac{2}{5}M\left(\frac{R}{4}\right)^2 \omega_2$$

$$\left(\text{Here, } I = \frac{2}{5}MR^2\right) \Rightarrow MR^2\omega_1 = \frac{MR^2}{16}\omega_2$$

$$\Rightarrow \frac{\omega_1}{\omega_2} = \frac{1}{16} \Rightarrow \frac{T_2}{T_1} = \frac{1}{16} \quad \left(\because \omega = \frac{2\pi}{T}\right)$$

Here $T_1 = 24$

$$T_2 = \frac{T_1}{16} = \frac{24}{16} = \frac{24}{x}(\text{given})(\therefore x = 16)$$

5. **(d)** Angular momentum, $L = mvr$

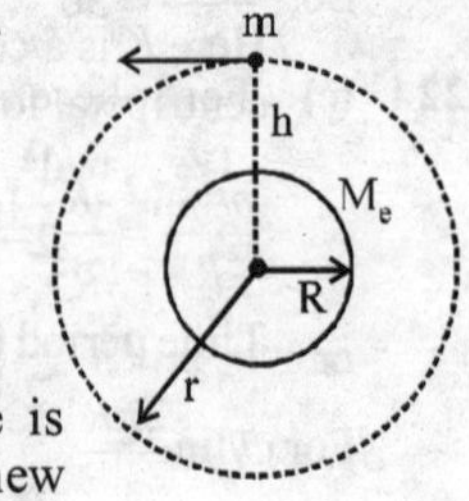

Here, $v = \sqrt{\frac{GM_e}{r}}$

$$L = m\sqrt{\frac{GM_e}{r}}.r \Rightarrow L \propto r^{\frac{1}{2}}$$

Now distance from centre is increased by 8 times. So new distance from centre $= r + 8r = 9r$

New angular momentum, $L' \propto (9r)^{1/2}$

$$\frac{L}{L'} = \frac{r^{1/2}}{(9r)^{1/2}} = \frac{1}{3} \Rightarrow L' = 3L$$

6. **(d)** Planets revolve around the sun in elliptical paths. Thus their linear speed is not constant.

7. **(b)** By Kepler's law

$T^2 \propto R^3$

$$T_2^2 = T_1^2 \times \left(\frac{R_2}{R_1}\right)^3 \Rightarrow T_2 = T_1\left(\frac{R_2}{R_1}\right)^{\frac{3}{2}}$$

$$= 7(3)^{\frac{3}{2}} = 7 \times 3\sqrt{3} = 21\sqrt{3} \text{ hours} \simeq 36 \text{ hours}$$

8. **(c)** By kepler's law $T^2 \propto r^3$

$$\Rightarrow \left(\frac{T_A}{T_B}\right)^2 = \left(\frac{r_A}{r_B}\right)^3 \Rightarrow 2^2 = \left(\frac{r_A}{r_B}\right)^3 \Rightarrow r_A^3 = 4r_B^3$$

9. **(d)** Basically 1 year is equal to time period of earth revolution around sun.

So, $T_i = 1$ year

Now, $T^2 \propto R^3 \therefore \left(\frac{T_2}{T_1}\right)^2 = \left(\frac{R_2}{R_1}\right)^3$

$$\Rightarrow T_2 = \left(\frac{R_2}{R_1}\right)^{3/2}.T_1 = \left(\frac{3R}{R}\right)^{3/2} \times 1 = 3\sqrt{3} \text{ years.}$$

10. **(d)** When distance of the planet from the sun is maximum i.e., x_2 at apogee so velocity is minimum and vice-versa.

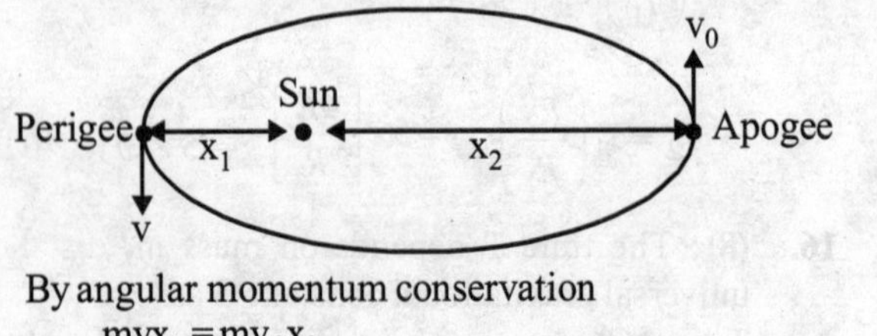

By angular momentum conservation

$mvx_1 = mv_0x_2$

$$\therefore \quad v = \frac{v_0 x_2}{x_1}$$

11. **(b)** According to Kepler's law, when a planet revolves around the sun, its areal velocity is constant.

$$\frac{dA}{dt} = \text{constant}$$

12. **(c)** We have given $F \propto \frac{1}{R^3} \Rightarrow F = \frac{K}{R^3}$

Here K is a constant. This force will provide the required centripetal force to the particle for revolution.

$$\frac{mv^2}{R} = \frac{K}{R^3} \Rightarrow v \propto \frac{1}{R}$$

Time period of revolution,

$$T = \frac{2\pi R}{v} \quad \therefore T \propto R^2$$

13. **(c)** Areal velocity; $= \frac{dA}{dt}$

$$dA = \frac{1}{2} r^2 d\theta$$

$$\Rightarrow \frac{dA}{dt} = \frac{1}{2} r^2 \frac{d\theta}{dt} = \frac{1}{2} r^2 \omega$$

Also, $L = mvr = mr^2\omega$

$$\therefore \frac{dA}{dt} = \frac{1}{2}\frac{L}{m}$$

14. **(c)** Let area of ellipse abcd = x

Area of SabcS $= \frac{x}{2} + \frac{x}{4}$ (i.e., ar of abca + SacS)

(Area of half ellipse + Area of triangle)

$$= \frac{3x}{4}$$

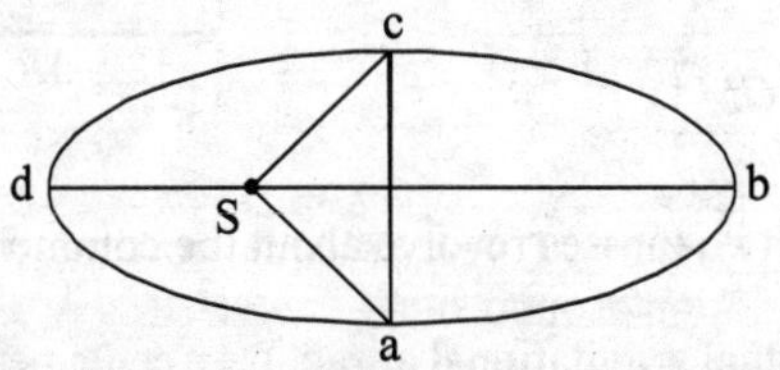

Area of SadcS $= x - \frac{3x}{4} = \frac{x}{4}$

$$\frac{\text{Area of SabcS}}{\text{Area of SadcS}} = \frac{3x/4}{x/4} = \frac{t_1}{t_2}$$

$$\frac{t_1}{t_2} = 3 \text{ or, } t_1 = 3t_2$$

15. **(c)** According to Kepler's law of periods $T^2 \propto R^3$

$$\therefore \left(\frac{T_2}{T_1}\right)^2 = \left(\frac{R_2}{R_1}\right)^3$$

$$\Rightarrow T_2 = T_1\left(\frac{R_2}{R_1}\right)^{3/2} = 5 \times \left[\frac{4R}{R}\right]^{3/2} = 5 \times 2^3 = 40 \text{ hours}$$

16. **(8)** The time T depends on mass m, distance a and universal gravitational constant G as

$T \propto m^x G^y a^z$

$T \propto M^x [M^{-1} L^3 T^{-2}]^y [L]^z$

$T \propto M^{x-y} L^{3y+z} T^{-2y}$

$x - y = 0 \Rightarrow x = y$

$-2y = 1 \Rightarrow y = -\frac{1}{2}, x = -\frac{1}{2}$

$\Rightarrow 3y + z = 0$

$z = -3y = \frac{3}{2}$

$\therefore \quad T \propto m^{-1/2} G^{-1/2} a^{3/2}$

$$\Rightarrow T \propto \left(\frac{a^3}{m}\right)^{1/2}$$

$$T = 4 \times \left(\frac{2^3}{2}\right)^{1/2} = 8s$$

17. **(c)** Initially,

Gravitational Force, $F_1 = \frac{GMm}{(2R)^2} = \frac{GMm}{4R^2}$

Finally,

$$F_2 = \frac{GMm}{(2R)^2} - \left(\frac{G\left(\frac{M}{27}\right)m}{\left(\frac{4R}{3}\right)^2}\right)$$

$$\Rightarrow F_2 = \frac{11}{48}\frac{GMm}{R^2}$$

$F_1 : F_2 = 12 : 11$

18. **(d)** From figure

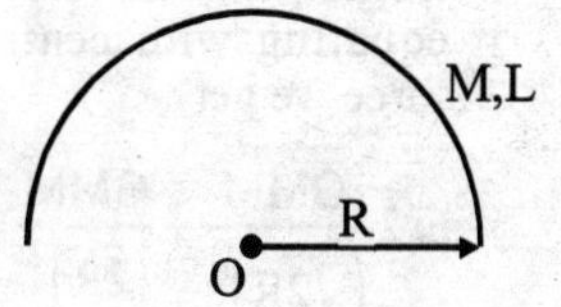

$$R = \frac{L}{\pi}$$

$$g_0 = \frac{2G\frac{M}{L}}{R} = \frac{2GM\pi}{L^2}$$

$\therefore$ Gravitational force $F_m = mg_0 = \frac{2GM\pi m}{L^2}$

19. **(b)** $F_{net} = 2F\cos 45° + F' = \sqrt{2}F + F'$

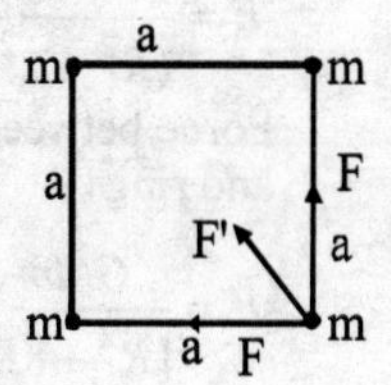

Here, $F = \frac{Gm^2}{a^2}$ and $F' = \frac{Gm^2}{(\sqrt{2}a)^2}$

$$\therefore F_{net} = \sqrt{2}\frac{Gm^2}{a^2} + \frac{Gm^2}{2a^2}$$

$$\left(\frac{2\sqrt{2}+1}{32}\right)\frac{Gm^2}{L^2} = \frac{Gm^2}{a^2}\left(\frac{2\sqrt{2}+1}{2}\right)$$

$\Rightarrow 2a^2 = 32L^2 \Rightarrow a = 4L$

20. **(c)** Radius of circle, r = a

Centrifugal force, $F = m\omega^2 r$

$$\Rightarrow \frac{Gmm}{(2a)^2} = m\omega^2 a \quad \left(\because F = \frac{GM_1M_2}{d}\right)$$

Here $M_1 = m$; $M_2 = m$ and $d = 2a$

$\Rightarrow$ angular speed, $\omega = \sqrt{\frac{Gm}{4a^3}}$

21. (b)

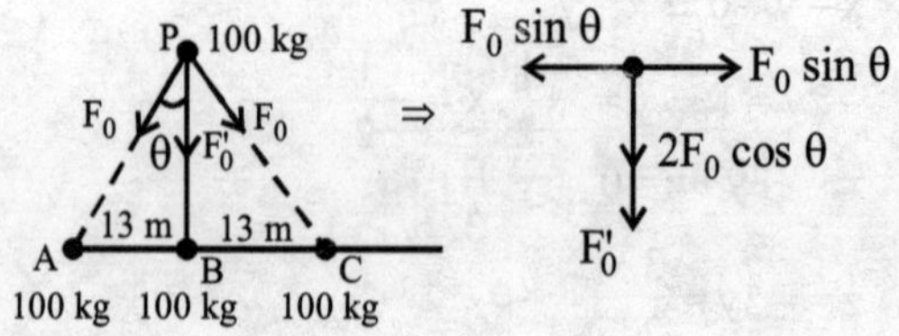

So, $F_{net} = F_o' + 2F_o \cos\theta$

$$= \frac{G\times 100^2}{(13)^2} + \frac{2G(100)^2}{(13\sqrt{2})^2}\cdot\frac{13}{13\sqrt{2}}$$

$$= \frac{G100^2}{13^2}\left(1+\frac{1}{\sqrt{2}}\right) \simeq 100\,G$$

22. (b) Centripetal Force $F_{net} = \frac{Mv^2}{R}$

Gravitational force between two masses $= \frac{GM_1M_2}{d^2}$

So, $F_{12} = F_{13} = \frac{GM^2}{\left(\sqrt{2}R\right)^2}$

$(\because M_1 = M_2 = M)$

Resultant of these two forces

$= \sqrt{2}\frac{GM^2}{2R^2}$

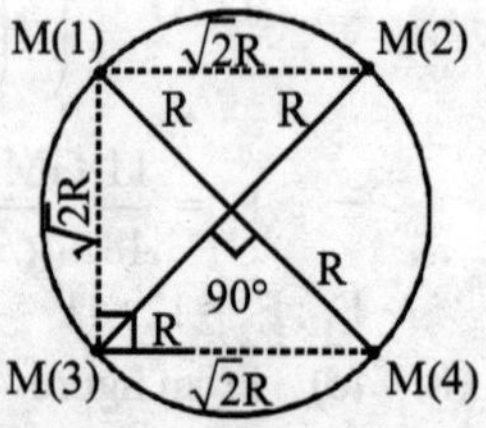

Combining all forces and equating with centripetal force we get

$$\sqrt{2}\frac{GMM}{\left(\sqrt{2}R\right)^2} + \frac{GMM}{(2R)^2} = \frac{Mv^2}{R} \Rightarrow \frac{GM}{R}\left(\frac{1}{\sqrt{2}}+\frac{1}{4}\right) = v^2$$

$$\Rightarrow \frac{GM}{R}\left(\frac{4+\sqrt{2}}{4\sqrt{2}}\right) = v^2 \Rightarrow v = \frac{1}{2}\sqrt{\frac{GM\left(2\sqrt{2}+1\right)}{R}}$$

23. (c) Gravitational field of ring

$$E = \frac{-Gmx}{(R^2+x^2)^{3/2}}$$

Force between sphere and ring

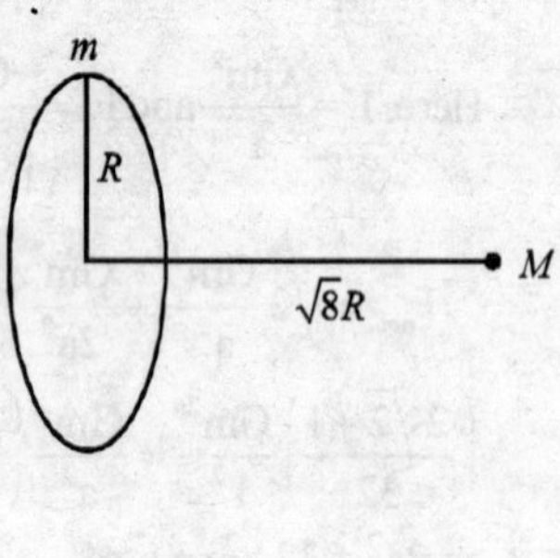

$$F = \frac{GMm\sqrt{8}R}{[R^2+8R^2]^{3/2}}$$

$$\Rightarrow F = \frac{\sqrt{8}GMm}{27R^2}$$

24. (d) Given $\lambda = (A + Bx^2)$,
Taking small element dm of length dx at a distance x from x = 0
so, $dm = \lambda\, dx$
$dm = (A + Bx^2)dx$

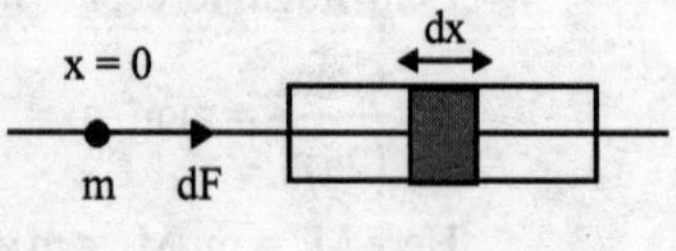

$$dF = \frac{Gmdm}{x^2}$$

$$\Rightarrow F = \int_a^{a+L}\frac{Gm}{x^2}(A+Bx^2)dx$$

$$= Gm\left[-\frac{A}{x}+Bx\right]_a^{a+L} = Gm\left[A\left(\frac{1}{a}-\frac{1}{a+L}\right)+BL\right]$$

25. (a) As we know, Gravitational force of attraction,

$$F = \frac{GMm}{R^2}$$

$$F_1 = \frac{GM_e m}{r_1^2} \text{ and } F_2 = \frac{GM_e M_s}{r_2^2}$$

$$\Delta F_1 = -2\frac{GM_e m}{r_1^3}\Delta r_1 \text{ and } \Delta F_2 = -2\frac{GM_e M_s}{r_2^3}\Delta r_2$$

$$\frac{\Delta F_1}{\Delta F_2} = \frac{m\Delta r_1}{r_1^3}\frac{r_2^3}{M_s \Delta r_2} = \left(\frac{m}{M_s}\right)\left(\frac{r_2^3}{r_1^3}\right)\left(\frac{\Delta r_1}{\Delta r_2}\right)$$

Using $\Delta r_1 = \Delta r_2 = 2\,R_{earth}$; $m = 8 \times 10^{22}$ kg;
$M_s = 2 \times 10^{30}$ kg
$r_1 = 0.4 \times 10^6$ km and $r_2 = 150 \times 10^6$ km

$$\frac{\Delta F_1}{\Delta F_2} = \left(\frac{8\times10^{22}}{2\times10^{30}}\right)\left(\frac{150\times10^6}{0.4\times10^6}\right)^3 \times 1 \cong 2$$

26. (a) Let *M'* be the mass of removed section

Then, $\frac{M}{\frac{4}{3}\pi R^3} = \frac{M'}{\frac{4}{3}\pi\left(\frac{R}{2}\right)^3} \Rightarrow M' = \frac{M}{8}$

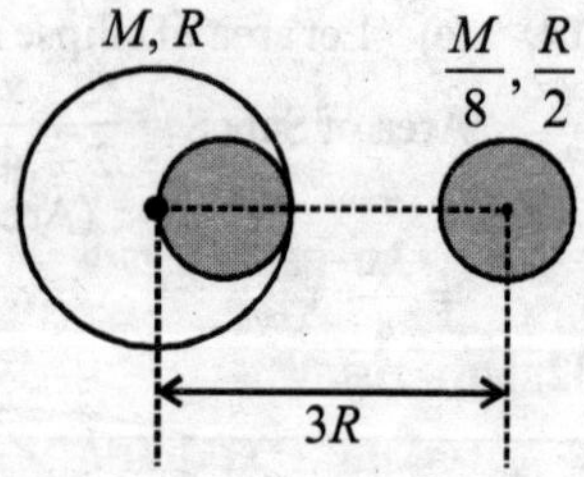

$$F = \frac{GM.\frac{M}{8}}{(3R)^2} - \frac{G\frac{M}{8}\frac{M}{8}}{\left(\frac{5R}{2}\right)^2}$$

$$= \frac{41}{3600}\frac{GM^2}{R^2}$$

27. (a) As two masses revolve about the common centre of mass *O*.
$\therefore$ Mutual gravitational attraction = centripetal force

$$\frac{Gm^2}{(2R)^2} = m\omega^2 R$$

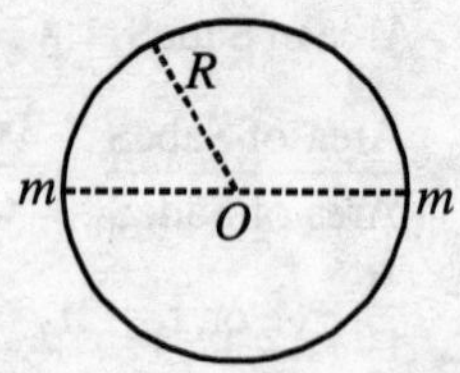

$$\Rightarrow \frac{Gm}{4R^3} = \omega^2$$

$$\Rightarrow \omega = \sqrt{\frac{Gm}{4R^3}}$$

If the velocity of the two particles with respect to the centre of gravity is v then
$v = \omega R$

$$v = \sqrt{\frac{Gm}{4R^3}} \times R = \sqrt{\frac{Gm}{4R}}$$

28. (c) We know that
Force = mass × acceleration.

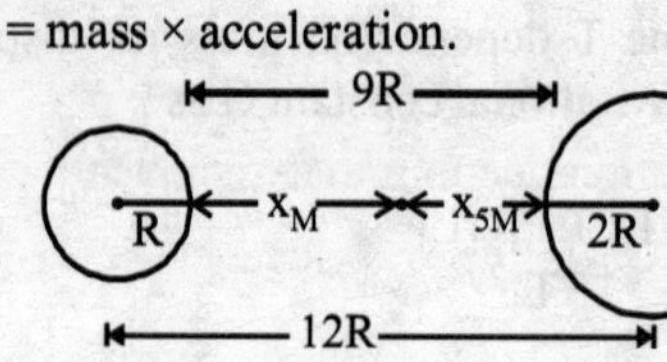

$x_M + x_{5M} = 9R$ ..... (i)

and, $Mx_M = 5Mx_{5M} \Rightarrow x_M = 5x_{5M}$ ..... (ii)

Solving (i) & (ii), we get $x_M = 7.5R$

29. (9) Acceleration due to gravity on the earth surface is

$$g = \frac{GM}{R_e^2} \Rightarrow g \propto \frac{1}{R_e^2}$$

$$\therefore \frac{g'}{g} = \left(\frac{R_e}{R_e'}\right)^2 \quad \therefore g' = g(3)^2 = 9g$$

30. (d) On surface of earth,

mg = 300 N

At depth $\frac{R}{4}$ acceleration due to gravity is,

$$g' = g\left[1 - \frac{d}{R}\right]$$

$$g' = g\left[1 - \frac{R}{4R}\right] = \frac{3g}{4}$$

Weight at depth $\frac{R}{4} = mg' = m \times \frac{3g}{4} = \frac{3}{4} \times 300 = 225$ N

31. (d) Gravitational pull = mg

Value of g varies with height 'h' as $g_s\left(1 + \frac{h}{R}\right)^{-2}$

$$= g_s(1+2)^{-2} = \frac{g_s}{9}$$

$\therefore$ Gravitational pull = mg = $90 \times \frac{g_s}{9} = 90 \times \frac{10}{9} = 100$N

32. (d) Using $W_{(R+x)} = W_{(R-x)}$

$$\frac{GMm}{(R+x)^2} = \frac{GMm(R-x)}{R^3}$$

$[\because W = mg]$

$\Rightarrow R^3 = (R+x)^2(R-x) \Rightarrow R^3 = (R^2 - x^2)(R+x)$

$\Rightarrow x^2 + Rx - R^2 = 0$

$$\therefore x = \frac{-R \pm \sqrt{R^2 + 4R^2}}{2}; \; x = \frac{\sqrt{5}R - R}{2}$$

33. (d) Acceleration due to gravity $g = \frac{GM}{R^2} \Rightarrow g \propto \frac{1}{R^2}$

$$\therefore \frac{g_2}{g_1} = \frac{R_1^2}{R_2^2} = \frac{R_1^2}{\left(\frac{R_1}{2}\right)^2} \quad \left(\because R_2 = \frac{R_1}{2}\right)$$

Therefore, $g_2 = 4g_1 = 4g$

34. (c) Acceleration due to gravity,

$$g = \frac{GM}{R^2} = \frac{4}{3}\pi G\rho R$$

$$\therefore \frac{g_2}{g_1} = \frac{\rho_2}{\rho_1} \times \frac{R_2}{R_1} = \frac{1}{2} \times 1.5 = \frac{3}{4}$$

35. (c) Acceleration due to gravity is given by

$$g = \frac{GM}{R^2} = \frac{G}{R^2} \times \rho \times \frac{4\pi}{3}R^3 = \left(\frac{4\pi}{3}G\right)\rho R$$

$$\therefore \frac{g_A}{g_B} = \frac{\frac{4}{3}\pi G\rho R}{\frac{4}{3}\pi G\left(\frac{\rho}{3}\right)4R}$$

$$\Rightarrow \frac{g_A}{g_B} = \frac{R \times \rho}{4R \times \frac{\rho}{3}} = \frac{3}{4}$$

36. (d) Weight on the surface of earth, W = 200N

$$m = \frac{W}{g} = \frac{200}{10} = 20\text{kg}$$

Acceleration due to gravity at a depth, $g' = g\left(1 - \frac{d}{R}\right)$

$d \rightarrow$ depth from surface; $d = \frac{R}{2}$

$$g' = g\left(1 - \frac{R/2}{R}\right) = \frac{g}{2} = \frac{10}{2} = 5\text{m/s}^2$$

Weight at depth R/2 = mg'

$= 20 \times 5 = 100$N

37. (d) Weight of a body, W = mg = 400 N

At depth d, acceleration due to gravity $g' = g\left(1 - \frac{d}{R}\right)$

For $d = \frac{R}{2}, g' = g\left(1 - \frac{R}{2R}\right) = \frac{g}{2}$

$$W' = mg' = \frac{mg}{2} = 200\text{N}$$

38. (a) For point outside the surface of earth

$$g = \frac{GM}{r^2}$$

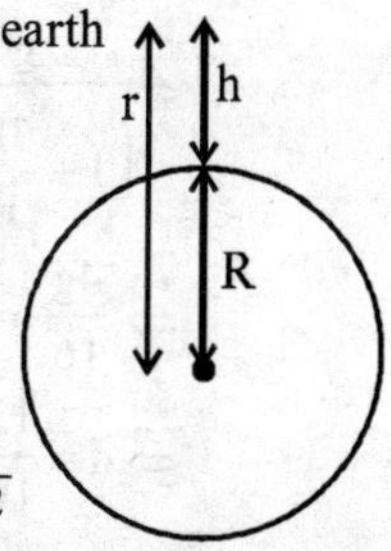

The acceleration due to gravity at height h above the earth is given by

$$\Rightarrow g' = \frac{GM}{(R+h)^2} = \frac{GM}{R^2\left(1 + \frac{h}{R}\right)^2}$$

$$\Rightarrow g' = \frac{GM}{R^2}\left(1 + \frac{h}{R}\right)^{-2}$$

If $h <<< R$, $\left(1 + \frac{h}{R}\right)^{-2} \approx 1 - \frac{2h}{R}$

$$\Rightarrow g' = \frac{GM}{R^2}\left(1 - \frac{2h}{R}\right) \Rightarrow g' = g\left(1 - \frac{2h}{R}\right)$$

39. (c) Let R_e and R_p be the radius of earth and planet respectively.

Density of earth = Density of planet

$$\frac{M}{\frac{4}{3}\pi R_e^3} = \frac{2M}{\frac{4}{3}\pi R_p^3} \Rightarrow R_p = 2^{\frac{1}{3}} R_e$$

Weight, $W = mg = \frac{mGM}{R^2} \; \therefore W \propto \frac{M}{R^2}$

$$\frac{W_e}{W_p}=\frac{M_e}{M_p}\times\left(\frac{R_p}{R_e}\right)^2$$

$$\Rightarrow\frac{W_e}{W_p}=\frac{M_e}{2M_e}\times\left(2^{\frac{1}{3}}\right)^2 \quad [\because M_p=2M_e \text{ given}]$$

$$\Rightarrow W_p=W_e\,2^{\frac{1}{3}}$$

In question, $W_e=W \therefore W_p=W2^{\frac{1}{3}}$

40. **(b)** Weight on the surface of earth, w = mg = 100 N
Variation in acceleration due to gravity with height,

$$g'=\frac{gR^2}{(r+h)^2}=\frac{gR^2}{\left(R+\frac{R}{4}\right)^2}=\frac{16g}{25} \quad \left[\because h=\frac{R}{4}\right]$$

$\therefore$ Weight on height $\frac{R}{4}=mg'=\frac{16}{25}mg=\frac{16}{25}\times100=64N$

41. **(d)** Acceleration due to gravity is different at different places
$g_{eff}=g-\omega^2R_e\cos^2\theta$, $\theta\to$ latitude angle
Acceleration due to gravity below the earth's surface

$$g_{eff}=g\left(1-\frac{d}{R_e}\right),$$

Hence, d is depth.

42. **(d)** We have
$4g_h=g_d$

$$4\frac{g}{\left(1+\frac{3R}{R}\right)^2}=g\left(1-\frac{d}{R}\right) \quad [\because h=3R]$$

$$\Rightarrow\frac{4g}{16}=g\left(1-\frac{d}{R}\right)\Rightarrow\frac{4}{16}=1-\frac{d}{R}\Rightarrow\frac{d}{R}=1-\frac{4}{16}$$

$$\Rightarrow d=\frac{12}{16}R=\frac{3}{4}\times6400=4800\text{ km}$$

43. **(b)** $W_{eff}=\frac{mg}{\left(1+\frac{h}{R}\right)^2}=\frac{mg}{\left(1+\frac{9R}{R}\right)^2}=\frac{mg}{100}=\frac{W}{100}$

44. **(2)** At 'h' height above the ground ($h\ll R$)

$$g_h=g\left(1-\frac{2h}{R}\right)$$

At depth 'd' below the surface of earth

$$g_d=g\left(1-\frac{d}{R}\right)$$

Given, $g\left(1-\frac{2h}{R}\right)=g\left(1-\frac{d}{R}\right)$

$$\Rightarrow\frac{2h}{R}=\frac{d}{R}\Rightarrow2h=\alpha h=d\Rightarrow\alpha=2$$

45. **(d)** Relation between acceleration due to gravity (g) and radius of the earth (R), $g=\frac{GM}{R^2}$

M = constant $\Rightarrow g\propto\frac{1}{R^2}$

$$100\times\frac{\Delta g}{g}=-2\frac{\Delta R}{R}\times100$$

% change in g = – 2(–2) = 4%
So acceleration due to gravity on the earth's surface will increase by 4%.

46. **(a)** Given, radius of earth = 6400 km

We have, $g'=g\left(1-\frac{2h}{R}\right)$

The percentage decrease in the weight,

$$\frac{g'-g}{g}\times100=\frac{2h}{R}\times100=\frac{2\times32\times100}{6400}=1\%$$

47. **(a)** Effective value of 'g' at 'h' height is given as

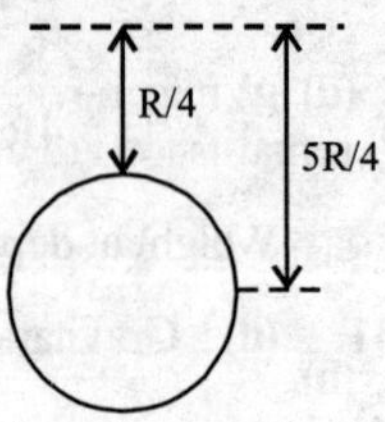

$$g_{eff}=\frac{g}{\left(1+\frac{h}{R}\right)^2}$$

$$\Rightarrow g_{eff}=\frac{g}{\left(1+\frac{1}{4}\right)^2}=\frac{16g}{25}$$

$$\text{Change in 'g'}=\frac{g_{eff}-g}{g}\times100=\frac{\frac{16}{25}-1}{1}\times100$$

$$=\frac{-9}{25}\times100=-36\%$$

Hence % decrease in the weight = 36%

48. **(d)** Since $g=\frac{GM}{R^2}$

$$\Rightarrow g=\frac{G}{R^2}\times\rho\times\frac{4}{3}\pi R^3$$

$$\Rightarrow g\propto\rho R$$

So, $\frac{g_1}{g_2}=\frac{\rho_1}{\rho_2}\cdot\frac{R_1}{R_2}=\frac{3}{5}\cdot\frac{2}{3}=\frac{2}{5}$

49. **(a)** Law of gravitation is universal law so it is valid for any two pair of bodies.
At centre, $d=R$

So, $g=\left(1-\frac{d}{R}\right)=\left(1-\frac{R}{R}\right)=0\Rightarrow W=mg=0$

50. **(a)** The expression for acceleration due to gravity is as shown,

$$\vec{g}=-\frac{GMr}{R^3}(\hat{r}),\quad r<R$$

$$=-\frac{GM}{r^2}\hat{r},\quad r>R$$

$$\Rightarrow|g|\propto r,\ r<R$$

$$\propto\frac{1}{r^2},\quad r>R$$

So, graph will be as shown below.

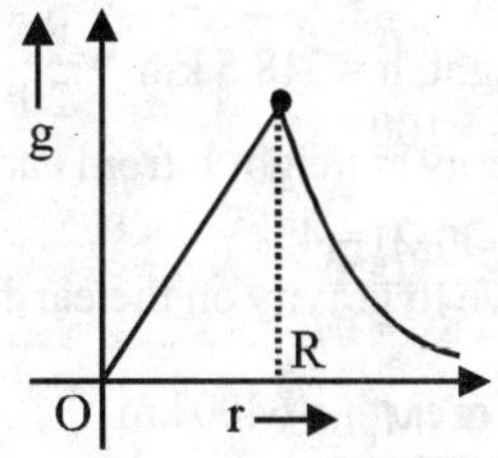

51. **(6)** Enlongation of wire due to its own weight is given by

$\Delta l = \frac{mgl}{2yA}$ or, $\Delta l \propto g$

$\Rightarrow \frac{\Delta l_e}{\Delta l_p} = \frac{g_e}{g_p} \Rightarrow g_p = g_e \frac{\Delta l_p}{\Delta l_e}$

$\Rightarrow g_p = \frac{10 \times 6 \times 10^{-5}}{10^{-4}} = 6 \text{ m/s}^2$

52. **(d)** While moving from pole to equator the direction of 'g' remains toward centre of earth but magnitude change as $g_{eff} = g - \omega^2 R\cos^2\theta$, where θ depends on latitude

53. **(b)** $W_h = \frac{W_{surface}}{3} \Rightarrow Mg' = \frac{Mg}{3}$

$\Rightarrow \frac{g}{\left(1+\frac{h}{R}\right)^2} = \frac{g}{3} \Rightarrow \left(1+\frac{h}{R}\right) = \sqrt{3}$

$\Rightarrow \frac{h}{R} = \sqrt{3} - 1 \Rightarrow h = (\sqrt{3}-1)R$

$\Rightarrow h = 0.732 \times 6400 \simeq 4685 \text{ km}$

54. **(c)** As density is same, $2M_E = M_P$

$\Rightarrow \quad 2\rho \times \frac{4}{3}R_E^3 = \rho \times \frac{4}{3}\pi R_P^3 \Rightarrow R_P = 2^{1/3} R_E$

Acceleration due to gravity on the surface of planet,

$g_P = \frac{GM_P}{R_P^2}$

$\Rightarrow g_P = \frac{G2M_E}{\left(2^{1/3}R_E\right)^2} = \frac{G2M_E}{2^{2/3}R_E^2}$

$\Rightarrow g_P = 2^{1/3} g_e$

Weight on planet = $2^{1/3}$ Weight on earth

$\Rightarrow \quad W_P = 2^{1/3} W$

55. **(4)** At the surface of earth, $g = \frac{GM}{R^2}$

At point C $g_c = \frac{GM}{\left(R+\frac{R}{2}\right)^2} = \frac{4}{9}g$

At point A $g_A = g\left(1-\frac{d}{R}\right)$ or, $g_A = g\left(1-\frac{AB}{R}\right)$

From question,

$g_A = g_C \Rightarrow \frac{4}{9}g = g\left(1-\frac{AB}{R}\right)$

$\Rightarrow \quad AB = \frac{5R}{9}$

$\therefore \quad OA = OB - AB = R - \frac{5}{9}R = \frac{4R}{9}$

$\therefore \quad \frac{OA}{AB} = \frac{x}{y} = \frac{\frac{4R}{9}}{\frac{5R}{9}} = \frac{4}{5} \quad \therefore \quad x = 4$

56. **(b)** Weight of body at pole = $mg = 49$ N

Weight of body at equator due to rotation,

$g_e = g - R\omega^2$

so $W_e = mg_e = m(g - R\omega^2)$

$\therefore \; W_P > W_e$

$W_P = 49 \text{ N} \Rightarrow W_e < 49$ N.

So, $W_e = 48.83$ N.

57. **(b)** According to question, $g_h = g_d = g_1$

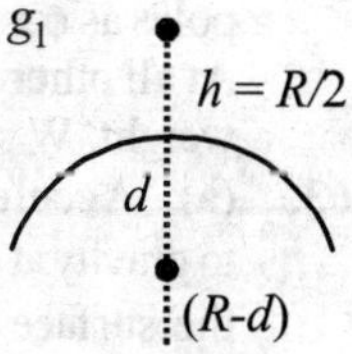

$g_h = \frac{GM}{\left(R+\frac{R}{2}\right)^2}$ and $g_d = \frac{GM(R-d)}{R^3}$

$\frac{GM}{\left(\frac{3R}{2}\right)^2} = \frac{GM(R-d)}{R^3} \Rightarrow \frac{4}{9} = \frac{(R-d)}{R}$

$\Rightarrow 4R = 9R - 9d \Rightarrow 5R = 9d$

$\therefore \frac{d}{R} = \frac{5}{9}$

58. **(a)** Value of g at equator, $g_A = g - R\omega^2$

Value of g at height h above the pole,

$g_B = g \cdot \left(1 - \frac{2h}{R}\right)$

As object is weighed equally at the equator and poles, it means g is same at these places.

$g_A = g_B$

$\Rightarrow g - R\omega^2 = g\left(1-\frac{2h}{R}\right)$

$\Rightarrow R\omega^2 = \frac{2gh}{R} \Rightarrow h = \frac{R^2\omega^2}{2g}$

59. **(d)** Weight at pole, $w = mg = 196\,N$

$\Rightarrow \quad m = 19.6$ kg

Weight at equator, $w' = mg' = m(g - \omega^2 R)$

$= 19.6\left[10 - \left(\frac{2\pi}{24\times 3600}\right)^2 \times 6400 \times 10^3\right]$ N

$\left(\because \; \omega = \frac{2\pi}{T}\right)$

$= 19.6\,[10 - 0.034] = 195.33\,N$

60. (d) $\frac{W_e}{W_p} = \frac{mg_e}{mg_p} = \frac{9}{4}$ or $\frac{g_e}{g_p} = \frac{9}{4}$

or $\frac{GM/R^2}{G(M/9)/R_p^2} = \frac{9}{4}$

$\therefore R_p = \frac{R}{2}$

61. (a) $g_{\text{eff}} = \frac{g}{\left(1+\frac{h}{R}\right)^2}$

$\Rightarrow \frac{g}{2} = \frac{g}{\left(1+\frac{h}{R}\right)^2} \Rightarrow \sqrt{2} = 1+\frac{h}{R} \Rightarrow \frac{h}{R} = \sqrt{2}-1$

$\Rightarrow h = (\sqrt{2}-1)\times 6400\times 10^3 \text{ m} = 2.6\times 10^6 \text{ m}$

62. (d) With rotation of earth or latitude, acceleration due to gravity vary as $g' = g - \omega^2 R\cos^2\phi$
Where ϕ is latitude, there will be no change in gravity at poles as $\phi = 90°$
At all other points as ω increases, g' will decreases hence, weight, W = mg decreases.

63. (a) Acceleration due to gravity at depth d from the surface of the earth or at a distance r from the centre '*O*' of the earth

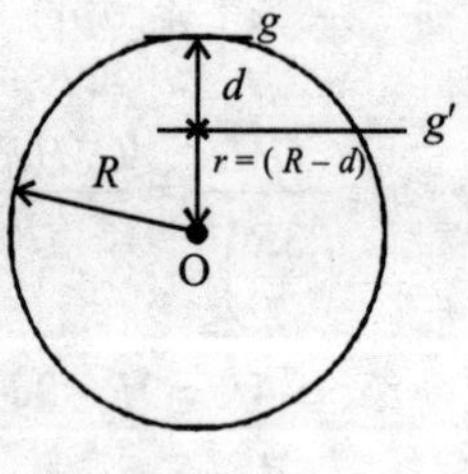

$g' = \frac{4}{3}\pi\rho G r$

Hence $g' \propto r$

64. (d) Value of *g* with altitude is,

$g_h = g\left[1-\frac{2h}{R}\right]$;

Value of *g* at depth d below earth's surface,

$g_d = g\left[1-\frac{d}{R}\right]$

Equating g_h and g_d, we get $d = 2h$

65. (d) Value of *g* on earth's surface,

$g = \frac{GM}{R^2} = \frac{G\rho\times V}{R^2} \Rightarrow g = \frac{G\times\rho\times\frac{4}{3}\pi R^3}{R^2}$

$g = \frac{4}{3}\rho\pi G.R$ where $\rho \to$ average density

$\rho = \left(\frac{3g}{4\pi GR}\right)$

$\Rightarrow \rho$ is directly proportional to *g*.

66. (a) Total energy at Earth's surface

$T.E_i = \frac{-GM_e m}{R_e}$

Radius of earth, $R_e = 6400$ km

$\therefore$ given height, $h = 318.5 \text{ km} \approx \frac{Re}{20}$

Total energy at height h from earth surface

$T.E_f = \frac{-GM_e m}{2(R_e+h)}$

$\Rightarrow T.E_f = \frac{-10GM_e m}{21R_e}$

Change in total mechanical energy
$= TE_f - TE_i$

$= \frac{GM_e m}{R_e}\left[1-\frac{10}{21}\right] = \frac{11GM_e m}{21R_e} = \frac{xGM_e m}{21R_e}$

$\therefore x = 11$

67. (a) $V_e = \frac{-GM}{R_E+h} = -5.12\times 10^{-2}$ ..... (i)

Acceleration due to gravity,

$g = \frac{GM}{(R_E+h)^2} = 6.4$ ..... (ii)

$\frac{5.12\times 10^{-7}}{R_E+h} = 6.4$

$h = (8000-6400)$ km $= 1600$ km

68. (a) Apply conservation of energy

$U_i + K_i = U_f + K_i$

$-\frac{GMm}{2R} + 0 = -\frac{GMm}{R} + \frac{1}{2}mv^2$

$-\frac{GMm}{2R} + \frac{GMm}{R} = \frac{1}{2}mv^2$

$\frac{GMm}{2R} = \frac{1}{2}mv^2 \Rightarrow v = \sqrt{\frac{GM}{R}} = \sqrt{gR}$

69. (a) Potential due to sphere

$V = -\frac{GM}{2R^3}\left(3R^2 - r^2\right)$

At surface $r = R \Rightarrow V = \left(-\frac{GM}{R}\right)$

At centre, r = 0

$\therefore V_0 = -\frac{3GM}{2R} = \left(\frac{3V}{2}\right)$

70. (a) Intial potential energy, $U_i = \frac{-GMm}{R}$

Final potential energy, $U_f = -\frac{GMm}{4R}$

The gain in potential energy of the object will be

$\Delta U = U_f - U_i = -\frac{GMm}{4R} + \frac{GMm}{R} = \frac{3GMm}{4R}$ $(\because GM = gR^2)$

$= \frac{3}{4}mgR = \frac{3}{4}\times 1\times 10\times 64\times 10^5 = 48$ MJ

71. **(a)**

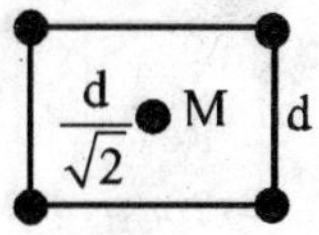

$$U_{net} = -\frac{Gmm}{d}\times 4 - \frac{GMm}{d}\times 4\sqrt{2} - \frac{Gmm}{\sqrt{2}d}\times 2$$

$$= -\frac{Gm}{d}\left[(4+\sqrt{2})m + 4\sqrt{2}M\right]$$

72. **(2)** $U = -G\left[\frac{(M-m)m}{a}\times 4 + \frac{m^2}{\sqrt{2}a} + \frac{(M-m)^2}{\sqrt{2}a}\right]$

$$= -\frac{G}{a}\left[4Mm - 4m^2 + \frac{m^2}{\sqrt{2}} + \frac{(M-m)^2}{\sqrt{2}}\right]$$

$$\frac{dU}{dm} = -\frac{G}{a}\left[4M - 8m + \frac{2m}{\sqrt{2}} - \frac{2(M-m)}{\sqrt{2}}\right]$$

For maximum 'U', $\frac{dU}{dm} = 0$

$\Rightarrow\ 0 = 4M - 8m + 2\sqrt{2}m - \sqrt{2}M$

$\Rightarrow\ 0 = M(4-\sqrt{2}) - 2m(4-\sqrt{2})$

$\Rightarrow\ M - 2m \Rightarrow \frac{M}{m} = 2$

73. **(d)** The gravitational potential at a point A due to mass from the centre is

$$V_1 = -\frac{GM}{r}$$

Gravitational potential at A due to shell is

$$V_2 = -\frac{GM}{R}$$

$$V_A = V_1 + V_2 \Rightarrow V_A = \left[-\frac{GM_1}{r} - \frac{GM_2}{R}\right]$$

$$= \left[-\frac{50}{25}G - \frac{100}{50}G\right] = -4G$$

74. **(a)** Inside uniform spherical shall, field is zero & hence potential is constant and same as on its surface.

75. **(3)** Binding energy of uniform sphere $= -\frac{3}{5}\frac{GM^2}{R}$

Energy given

$$E = U_f - U_i = 0 - \left(-\frac{3}{5}\frac{GM^2}{R}\right) = \frac{3}{5}\frac{GM^2}{R} = \frac{x}{5}\frac{GM^2}{R}$$

$\therefore x = 3$

76. **(d)**

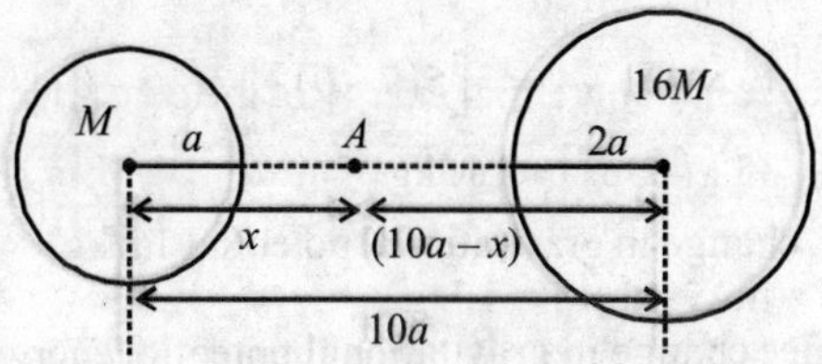

Let A be the point where gravitation field of both planets cancel each other i.e. zero. After this field due to small mass will dominate and 'm' will easily reach small mass surface.

$$\frac{GM}{x^2} = \frac{G(16M)}{(10a-x)^2}$$

$$\Rightarrow \frac{1}{x} = \frac{4}{(10a-x)} \Rightarrow 4x = 10a - x \Rightarrow x = 2a \quad \text{...(i)}$$

Using conservation of energy, we have

$$-\frac{GMm}{8a} - \frac{G(16M)m}{2a} + KE = -\frac{GMm}{2a} - \frac{G(16M)m}{8a}$$

$$KE = GMm\left[\frac{1}{8a} + \frac{16}{2a} - \frac{1}{2a} - \frac{16}{8a}\right]$$

$$\Rightarrow KE = GMm\left[\frac{1+64-4-16}{8a}\right]$$

$$\Rightarrow \frac{1}{2}mv^2 = GMm\left[\frac{45}{8a}\right] \Rightarrow v = \sqrt{\frac{90GM}{8a}} \Rightarrow v = \frac{3}{2}\sqrt{\frac{5GM}{a}}$$

77. **(a)** Given : Gravitational field,

$$E = \frac{Ax}{(x^2+a^2)^{3/2}},\ V_\infty = 0$$

$$\int_{V_\infty}^{V_x} dV = -\int_{\infty}^{x} \vec{E}\cdot\vec{d_x} \Rightarrow V_x - V_\infty = -\int_{\infty}^{x} \frac{Ax}{(x^2+a^2)^{3/2}}dx$$

$$\therefore V_x = \frac{A}{(x^2+a^2)^{1/2}} - 0 = \frac{A}{(x^2+a^2)^{1/2}}$$

78. **(d)** Mass of small element of planet of radius r and thickness dr.

$$dm = \rho\times 4\pi r^2 dr = \rho_0\left(1 - \frac{r^2}{R^2}\right)\times 4\pi r^2 dr$$

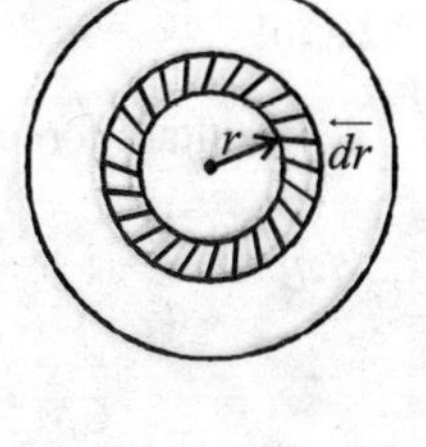

Mass of the planet

$$M = 4\pi\rho_0\int_0^r\left(r^2 - \frac{r^4}{R^2}\right)dr$$

$$\Rightarrow M = 4\pi\rho_0\left|\frac{r^3}{3} - \frac{r^5}{5R^2}\right|$$

Gravitational field,

$$E = \frac{GM}{r^2} = \frac{G}{r^2}\times 4\pi\rho_0\left(\frac{r^3}{3} - \frac{r^5}{5R^2}\right)$$

$$\Rightarrow E = 4\pi G\rho_0\left(\frac{r}{3} - \frac{r^3}{5R^2}\right)$$

E is maximum when $\frac{dE}{dr} = 0$

$$\Rightarrow \frac{dE}{dr} = 4\pi G\rho_0\left(\frac{1}{3} - \frac{3r^2}{5R^2}\right) = 0$$

$$\Rightarrow r = \sqrt{\frac{5}{9}}R$$

79. (b) Gravitation field at the surface

$$E = \frac{Gm}{r^2}$$

$$\therefore E_1 = \frac{Gm_1}{r_1^2} \text{ and } E_2 = \frac{Gm_2}{r_2^2}$$

From the diagram given in question,

$$\frac{E_1}{E_2} = \frac{2}{3} \ (r_1 = 1\text{m}, R_2 = 2m \text{ given})$$

$$\therefore \frac{E_1}{E_2} = \left(\frac{r_2}{r_1}\right)^2\left(\frac{m_1}{m_2}\right) \Rightarrow \frac{2}{3} = \left(\frac{2}{1}\right)^2\left(\frac{m_1}{m_2}\right)$$

$$\Rightarrow \left(\frac{m_1}{m_2}\right) = \frac{1}{6}$$

80. (16.00) Using law of conservation of energy
Total energy at height 10 R = total energy at earth

$$-\frac{GM_E m}{10R} + \frac{1}{2}mV_0^2 = -\frac{GM_E m}{R} + \frac{1}{2}mV^2$$

$$\left[\because \text{Gravitational potential energy} = -\frac{GMm}{r}\right]$$

$$\Rightarrow \frac{GM_E}{R}\left(1 - \frac{1}{10}\right) + \frac{V_0^2}{2} = \frac{V^2}{2} \Rightarrow V^2 = V_0^2 + \frac{9}{5}gR$$

$$\Rightarrow V = \sqrt{V_0^2 + \frac{9}{5}gR} \approx 16 \ km/s \quad [\because V_0 = 12 \text{ km/s given}]$$

81. (c) $E_g = \frac{GM}{(3a)^2} + \frac{G(2M)}{(3a)^2} = \frac{GM}{3a^2}$

82. (b) $AC = a\sqrt{2} \quad \because r = \frac{AC}{2} = \frac{a\sqrt{2}}{2} = \frac{a}{\sqrt{2}}$

Resultant force on the body

$$B = \frac{GM^2}{a^2}\hat{i} + \frac{GM^2}{a^2}\hat{j} + \frac{GM^2}{(a\sqrt{2})^2}(\cos 45°\hat{i} + \sin 45°\hat{j})$$

$$\Rightarrow |F| = \frac{GM^2}{a^2}(\sqrt{2}) + \frac{GM^2}{2a^2}$$

$$\frac{Mv^2}{r} = \text{Resultant force towards centre}$$

$$\therefore \frac{Mv^2}{\left(\frac{a}{\sqrt{2}}\right)} = \frac{GM^2}{a^2}\left(\sqrt{2} + \frac{1}{2}\right)$$

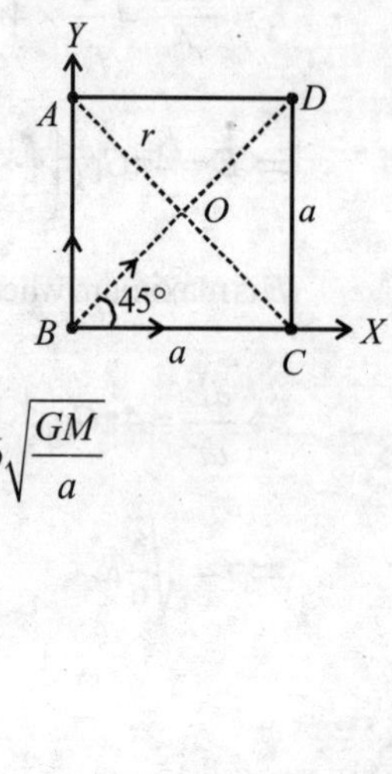

$$\Rightarrow v^2 = \frac{GM}{a}\left(1 + \frac{1}{2\sqrt{2}}\right)$$

$$\Rightarrow v = \sqrt{\frac{GM}{a}\left(1 + \frac{1}{2\sqrt{2}}\right)} = 1.16\sqrt{\frac{GM}{a}}$$

83. (a) $F = \int \frac{Gm}{r^2} dM = \int_0^R \frac{Gm}{r^2} \rho 4\pi r^2 dr$

$$= mG\int_0^R \frac{k}{r^2}\frac{4\pi r^2 dr}{r^2} = -4\pi kGm\left(\frac{1}{r}\right)_0^R = -\frac{4\pi kGm}{R}$$

Using Newton's second law, we have

$$\frac{mv_0^2}{R} = -\frac{4\pi kGm}{R}$$

or $v_0 = \text{C (const.)}$

Time period, $T = \frac{2\pi R}{v_0} = \frac{2\pi R}{C}$ or $\frac{T}{R} = \text{constant.}$

84. (c) Initial gravitational potential energy, $E_i = -\frac{GMm}{2R}$

Final gravitational potential energy,

$$E_f = -\frac{GMm/2}{2\left(\frac{R}{2}\right)} - \frac{GMm/2}{2\left(\frac{3R}{2}\right)} = -\frac{GMm}{2R} - \frac{GMm}{6R}$$

$$= -\frac{4GMm}{6R} = -\frac{2GMm}{3R}$$

$\therefore$ Difference between final and initial energy,

$$E_f - E_i = \frac{GMm}{R}\left(-\frac{2}{3} + \frac{1}{2}\right) = -\frac{GMm}{6R}$$

85. (d) Due to complete solid sphere, potential at point P

$$V_{\text{sphere}} = \frac{-GM}{2R^3}\left[3R^2 - \left(\frac{R}{2}\right)^2\right]$$

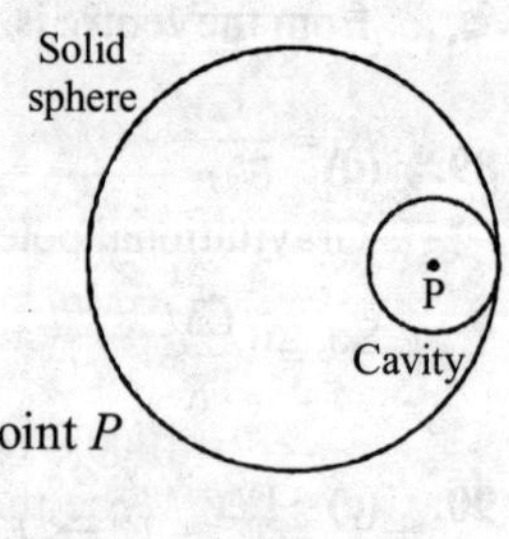

$$= \frac{-GM}{2R^3}\left(\frac{11R^2}{4}\right) = -11\frac{GM}{8R}$$

Due to cavity part potential at point P

$$V_{\text{cavity}} = -\frac{3}{2}\frac{\frac{GM}{8}}{\frac{R}{2}} = -\frac{3GM}{8R}$$

So potential at the centre of cavity

$$= V_{\text{sphere}} - V_{\text{cavity}} = -\frac{11GM}{8R} - \left(-\frac{3}{8}\frac{GM}{R}\right) = \frac{-GM}{R}$$

86. (c) As, $V = \frac{GM}{2R^3}(3R^2 - r^2)$

Graph (c) most closely depicts the correct variation of $v(r)$.

87. (d) Gravitational field, $I = \left(5\hat{i} + 12\hat{j}\right)$ N/kg

$$\Delta v = -\left[\int_0^x I_x dx + \int_0^y I_y dy\right]$$

$$= -\left[I_x . x + I_y . y\right] = -\left[5(7-0) + 12(-3-0)\right]$$

$$= -\left[35 + (-36)\right] = 1 \text{ J/kg}$$

i.e., change in gravitational potential 1 J/kg.
$\Delta U = m\Delta v = 1 \times 1 = 1$ J
Hence change in gravitational potential energy 1 J

88. (c) Let mass of smaller sphere (which has to be removed) is m

Radius $= \frac{R}{2}$ (from figure)

$$\frac{M}{\frac{4}{3}\pi R^3} = \frac{m}{\frac{4}{3}\pi\left(\frac{R}{2}\right)^3} \Rightarrow m = \frac{M}{8}$$

Mass of the left over part of the sphere

$$M' = M - \frac{M}{8} = \frac{7}{8}M$$

Therefore gravitational field due to the left over part of the sphere

$$\vec{E} = \left(-\frac{GM}{x^2} + \frac{GM}{8\left(x+\frac{R}{2}\right)^2}\right)\hat{i}$$

$$= +GM\left[\frac{1}{8\left(x+\frac{R}{2}\right)^2} - \frac{1}{x^2}\right]\hat{i}$$

$$= GM\left[\frac{x^2 - 8\left(x+\frac{R}{2}\right)^2}{8\left(x+\frac{R}{2}\right)^2 x^2}\right]\hat{i} = GM\left[\frac{x^2-8x^2}{8x^4}\right]\hat{i}$$

$$= \frac{-7GM}{8x^2}\hat{i} \therefore |\vec{E}| = \frac{7GM}{8x^2}$$

89. (d) $U_i = \frac{-GMm}{R}, U_f = 0$

So, $\Delta U = \frac{GMm}{R} = mgR \quad \left[\because g = \frac{GM}{R^2}\right]$

$= 1000 \times 10 \times 6400 \times 10^3 = 6.4 \times 10^{10}$ J

90. (c) Let 'M' be the mass of the particle

Now, $E_{initial} = E_{final}$

i.e. $-\frac{GMm}{\sqrt{2}r} + 0 = -\frac{GMm}{r} + \frac{1}{2}MV^2$

or, $\frac{1}{2}MV^2 = \frac{GMm}{r}\left[1 - \frac{1}{\sqrt{2}}\right] \Rightarrow \frac{1}{2}V^2 = \frac{Gm}{r}\left[1-\frac{1}{\sqrt{2}}\right]$

or, $V = \sqrt{\frac{2Gm}{r}\left(1-\frac{1}{\sqrt{2}}\right)}$

91. (c) Let P be the point where gravitational field is zero.

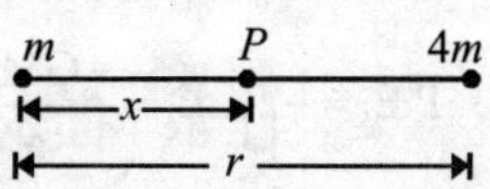

$\therefore \frac{Gm}{x^2} = \frac{4Gm}{(r-x)^2}$

$\Rightarrow \frac{1}{x} = \frac{2}{r-x}$

$\Rightarrow r - x = 2x \Rightarrow x = \frac{r}{3}$

Gravitational potential at P,

$$V = -\frac{Gm}{\frac{r}{3}} - \frac{4Gm}{\frac{2r}{3}} = -\frac{9Gm}{r}$$

92. (b) Gravitational field, $E = \frac{GM}{r^2}$. So, $E \propto \frac{1}{r^2}$

Here, $m \leftrightarrow q$ and $\frac{1}{\varepsilon_0} \leftrightarrow 4\pi G$

So, $\phi_{sides} = \oint \vec{E}_g . d\vec{A} = m \times 4\pi G = 4\pi Gm$

93. (c) Initial P.E. $U_i = -\frac{GMm}{R}$

When the particle is far away from the sphere, the P.E. of the system is zero.

$\therefore \quad U_f = 0$

$$W = \Delta U = U_f - U_i = 0 - \left[\frac{-GMm}{R}\right]$$

$$W = \frac{6.67\times10^{-11}\times100}{0.1}\times\frac{10}{1000} = 6.67\times10^{-10}\text{ J}$$

94. (b) On earth's surface potential energy,

$U = -\frac{GmM}{R}$

At a height R from the earth's surface, $P.E.$ of system $= -\frac{GmM}{2R}$

$\therefore \Delta U = \frac{-GmM}{2R} + \frac{GmM}{R} \Rightarrow \Delta U = \frac{GmM}{2R}$

Now $\frac{GM}{R^2} = g; \therefore \frac{GM}{R} = gR$

$\therefore \Delta U = \frac{1}{2}mgR$

95. (d) Gravitational potential energy of mass m in an orbit of radius R

$U = -\frac{GMm}{R}$

Energy required = potential energy at 3R − potential energy at 2R

$$= \frac{-GMm}{3R} - \left(\frac{-GMm}{2R}\right) = \frac{-GMm}{3R} + \frac{GMm}{2R}$$

$$= \frac{-2GMm + 3GMm}{6R} = \frac{GMm}{6R}$$

96. (a) Escape velocity, $V_{esc} = \sqrt{\frac{2GM}{r}}$

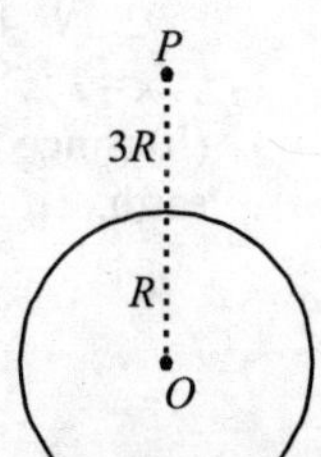

$\Rightarrow r = R + 3R = 4R$

$\therefore V_P = \sqrt{\frac{GM}{2R}} \Rightarrow V_{esc} = \sqrt{\frac{2GM}{4R}}.$

97. (3) Kinetic energy

$$KE = \frac{1}{2}mv^2 = \frac{1}{2}m\frac{GM_e}{r} = \frac{GM_eM}{2r} = \frac{GM_em}{2(R_E+h)}$$

$$= \frac{6.67\times10^{-11}\times6\times10^{24}\times6.4\times10^6}{2\left(6.4\times10^6+2.7\times10^5\right)} = 3\times10^{10}\text{ J}$$

98. **(b)** The expression of escape velocity is given as

$$V_{escape} = \sqrt{\frac{2GM}{R}}$$

$$\frac{(V_{escape})_{Planet}}{(V_{escape})_{Earth}} = \sqrt{\left(\frac{M_P}{M_E}\right) \times \left(\frac{R_E}{R_P}\right)} = \sqrt{\frac{1}{8} \times 2} = \frac{1}{2}$$

$$(V_{escape})_{Planet} = \frac{1}{2}(V_{escape})_{Earth} = 5.6\,\text{km/s}$$

99. **(b)** Time period of a satellite is

$$T = 2\pi\sqrt{\frac{r^3}{GM}} \Rightarrow T^2 \propto r^3$$

$$\therefore \frac{2\Delta T}{T} = \frac{3\Delta r}{r} \Rightarrow \frac{2\Delta T}{T} = \frac{3 \times 0.03R}{R}$$

$$\Rightarrow \frac{\Delta T}{T} = \frac{3 \times 0.03}{2} \times 100 = 4.5\%$$

100. **(a)** From Kepler's Third law,
$T^2 \propto R^3$

$$\left(\frac{T_m}{T_s}\right)^2 = \left(\frac{R}{R/9}\right)^3$$

$$\frac{T_m}{T_s} = (3)^3 \Rightarrow T_s = \frac{T_m}{27} = \left(\frac{27}{27}\right) = 1\,\text{day}$$

101. **(3)** (i) If earth is assumed to be stationary

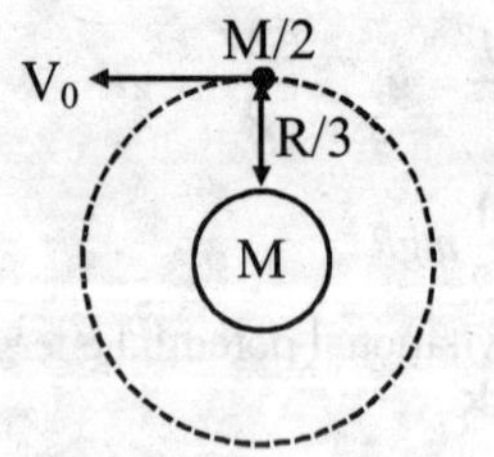

Orbital velocity of the satellite,

$$V_0 = \sqrt{\frac{GM}{4R/3}} = \sqrt{\frac{3GM}{4R}}$$

Angular momentum is given by $L = mvr$

Angular momentum of satellite $= \frac{M}{2} V_0 \frac{4R}{3}$

$$= \frac{M}{2} \cdot \sqrt{\frac{3GM}{4R}} \cdot \frac{4R}{3} = M\sqrt{\frac{GMR}{3}} = M\sqrt{\frac{GMR}{x}}$$

$\therefore x = 3$

(ii) Since mass of satellite is comparable to the mass of earth.

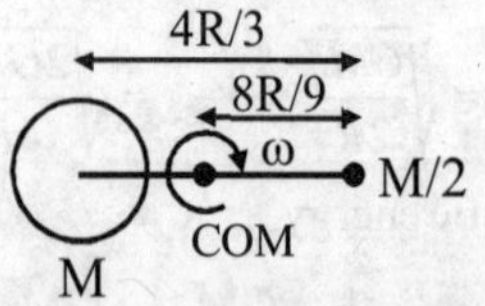

For a satellite in a stable circular orbit, the gravitational force provides the necessary centripetal force.

$$\frac{G.M.\frac{M}{2}}{\left(\frac{4R}{3}\right)^2} = \frac{M}{2}\omega^2 \cdot \frac{8R}{9}$$

$$\omega = \sqrt{\frac{81GM}{128R^3}}$$

Angular momentum of satellite about common centre of mass,

$$L = \frac{M}{2} \cdot \left(\frac{8R}{9}\right)^2 \cdot \omega$$

$$L = M\sqrt{GMR\left(\frac{8}{81}\right)} = M\sqrt{\frac{GMR}{x}}$$

$$\therefore x = \frac{81}{8} \simeq 10$$

102. **(d)** Initial total energy of satellite, $E_1 = \frac{-GMm}{2(2R)}$

When energy $\frac{10^4 R}{6}$ is added, then $E_2 = \frac{-GMm}{4R} + \frac{10^4 R}{6}$

$$E_2 = \frac{-GMm}{4R} + \frac{10^4 R}{6} = \frac{-GMm}{2r} \quad \text{...(i)}$$

Here r is radius of new circular orbit

Using $g = \frac{GM}{R^2}$ in (i) we get

$$-\frac{mgR}{4} + \frac{10^4 R}{6} = -\frac{mgR^2}{2r}$$

$$\Rightarrow -\frac{10^3 \times 10 \times R}{4} + \frac{10^4 R}{6} = -\frac{10^3 \times 10 \times R^2}{2r} \quad (\because m = 10^3\,\text{kg})$$

$$\Rightarrow -\frac{1}{4} + \frac{1}{6} = -\frac{R}{2r}$$

$\Rightarrow r = 6R$

103. **(b)** Using law of conservation of energy

$$KE = \frac{1}{2}mv_e^2 = \frac{GMm}{R_E} \quad \left[\text{Using } g = \frac{GM}{R_E^2}\right]$$

$$KE = mgR_E$$

104. **(a)** $K.E. = \frac{GMm}{2a}$

$$P.E. = -2K.E. = \frac{-GMm}{a}$$

$$M.E. = -K.E. = \frac{-GMm}{2a} \text{ and}$$

$$\text{Escape Energy} = \frac{Gm}{r}$$

105. (d) Time period of revolution of a satellite is

$$T = 2\pi\sqrt{\frac{r^3}{GM}}$$

$$\Rightarrow \frac{T_1}{T_2} = \left(\frac{r_1}{r_2}\right)^{3/2}\left(\frac{M_2}{M_1}\right)^{1/2}$$

Given, $T_2 = 24h$ and $r_2 = 4.2 \times 10^4$ km

$$\Rightarrow \frac{6}{24} = \frac{(r_1)^{3/2}}{(4.2\times10^4)^{3/2}}\left(\frac{M}{M/4}\right)^{1/2}$$

$\therefore r_1 = 1.05 \times 10^4$ km

106. (b) At height h satellite revolves so gravitational force = centripetal force

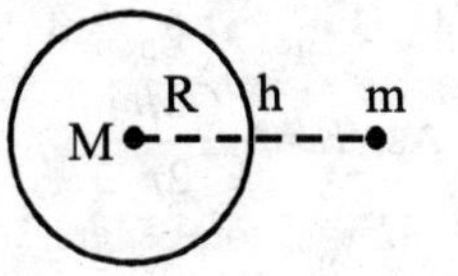

$$\frac{GMm}{(R+h)^2} = \frac{mv^2}{(R+h)}$$

$$\Rightarrow \frac{GM}{(R+h)} = v^2 \quad \text{...(i)}$$

$v = (R+h)\,\omega$

$$\Rightarrow v = (R+h)\frac{2\pi}{T} \quad \text{...(ii)}$$

Since, $\frac{GM}{R^2} = g$

$\Rightarrow GM = gR^2$ ... (iii)

From eq. (i), (ii) & (iii)

$$\frac{gR^2}{(R+h)} = (R+h)^2\left(\frac{2\pi}{T}\right)^2$$

$$\Rightarrow \frac{T^2R^2g}{(2\pi)^2} = (R+h)^3$$

$$\therefore h = \left[\frac{T^2R^2g}{4\pi^2}\right]^{\frac{1}{3}} - R$$

107. (c) Force between star and planet,

$$F = \frac{GMm}{R^{3/2}} \quad \text{...(i)}$$

Due to circular motion, $F = m\omega^2 R$ ... (ii)

Comparing equation (i) and (ii)

$$\omega^2 R \propto \frac{1}{R^{3/2}} \Rightarrow \omega^2 \propto \frac{1}{R^{5/2}} \qquad \left(\text{Also, } T = \frac{2\pi}{\omega}\right)$$

$\therefore T^2 \propto R^{5/2}$

108. (a) Escape speed is given by

$$v_{escape} = \sqrt{\frac{2GM}{R}}$$

For planet

$$v_{Planet} = \sqrt{\frac{2GM_p}{R_p}} = v$$

For moon

$$v_{Moon} = \sqrt{\frac{2GM_m}{R_m}} = \sqrt{\frac{2GM_p \times 16}{144R_p}} = \frac{1}{3}\sqrt{\frac{2GM_p}{R_p}}$$

$$\therefore v_{Moon} = \frac{v_{Planet}}{3} = \frac{v}{3}$$

109. (d) Escape velocity of body is given as $v = \sqrt{\frac{2GM}{R}}$

Given, $R_P = \frac{R_e}{3}, M_P = \frac{M_e}{6}$

$$v_e = \sqrt{\frac{2GM_e}{R_e}} \quad \text{...(i)}$$

$$v_P = \sqrt{\frac{2GM_P}{R_P}} = \sqrt{\frac{2GM_e \times 3}{6R_e}} = \sqrt{\frac{Gm_e}{R_e}} \quad \text{...(ii)}$$

From (i) and (ii),

$$\frac{v_e}{v_P} = \sqrt{2} \Rightarrow v_P = \frac{v_e}{\sqrt{2}} = \frac{11.2}{\sqrt{2}} = 7.9 \text{ km/s}$$

110. (b) Escape velocity $V = \sqrt{\frac{2GM}{R}}$

$$\therefore V_p = \sqrt{\frac{2GM_p}{R_p}} \text{ and } V_E = \sqrt{\frac{2GM_E}{R_E}}$$

$$\therefore \frac{V_P}{V_E} = \frac{\sqrt{\frac{2GM_P}{R_P}}}{\sqrt{\frac{2GM_E}{R_E}}} = \sqrt{\frac{R_E}{R_P} \times \frac{M_P}{M_E}}$$

$$\Rightarrow V_P = \sqrt{\frac{1}{4} \times 9 \times V_E} = \frac{3}{2}V_E$$

$$\therefore V_P = \frac{3}{2} \times 11.2 \text{ km/s} = 16.8 \text{ km/s}$$

111. (b) Escape velocity, $V_e = \sqrt{\frac{2GM}{R}} \Rightarrow V_e \propto \sqrt{\frac{M}{R}}$

As $\frac{M}{R}$ increases $\Rightarrow V_e$ increases

Also $V_e \propto \frac{1}{\sqrt{R}}$

Therefore statement (1) is correct and statement (2) is incorrect

112. (d) Let m = mass of satellite,
M = mass of earth
Potential energy of satellite

$$P.E = -\frac{GM\,m}{R}$$

Here, G = gravitational constant
Kinetic energy of satellite

$$K.E = +\frac{GM\,m}{2R}$$

Total energy of satellite

$$T.E = -\frac{GM\,m}{2R}$$

Speed of satellite $v = \sqrt{\frac{GM}{R}}$

It is clear that only speed of satellite does not depend on the mass of satellite. So, speed is same for both satellite.

113. (b) Escape speed of earth is

$$V_e = \sqrt{\frac{2GM_E}{R_E}}$$

Here M_E = Mass of earth

R_E = radius of earth

G = gravitational constant

Escape speed of planet

$$V_p = \sqrt{\frac{2G(16M_E)}{(4R_E)}} = 2\sqrt{\frac{2GM_E}{R_E}} \quad \therefore \frac{V_p}{V_e} = \frac{2}{1}$$

114. (c) Given, radius of earth, R = 6400 km

Acceleration due to gravity, $g = 10\ m/s^2$

$$V_{escape} = \sqrt{\frac{2GM}{R}} = \sqrt{2gR}$$

$$= \sqrt{\frac{2\times10\times6400}{1000}} = 8\sqrt{2}\ km/s$$

$$V_{orbit} = \sqrt{\frac{GM}{R}} = \sqrt{gR} = \sqrt{\frac{10\times6400}{1000}} = 8 km/s$$

Additional velocity required,

$$\Delta v = V_{esc} - V_{orb} = 8(\sqrt{2}-1)\ km/s$$

115. (c) Orbital speed, $v = \sqrt{\frac{GM}{r}}$

$$v \propto \frac{1}{\sqrt{r}} \Rightarrow \frac{v_1}{v_2} = \sqrt{\frac{r_2}{r_1}} = \sqrt{\frac{3r}{r}} = \sqrt{3}:1$$

116. (c) The centripetal force required for revolution of satellite is provided by the gravitational force of attraction between earth and satellite.

$$\therefore \frac{m_s v^2}{2R} = \frac{GM_E m_s}{(2R)^2}$$

$$\Rightarrow v = \sqrt{\frac{GM_E}{2R}} = \sqrt{\frac{Rg}{2}} \left(\because g = \frac{GM_E}{R^2}\right)$$

$$\therefore \text{Time period } T = \frac{2\pi(2R)}{v} = \frac{4\pi R\sqrt{2}}{\sqrt{Rg}} = \frac{4\pi R\sqrt{2}}{\sqrt{R\pi^2}} = \sqrt{32}\,R$$

117. (b) Energy of satellite in orbit, $E = \frac{-GMm}{2R}$

Potential energy of satellite in orbit, $PE = \frac{-GMm}{R}$

$\Rightarrow U = 2E$

Kinetic energy of satellite in orbit, KE = E – PE

$$KE = \frac{GMm}{2R} = (-E)$$

118. (c) Given, $M_E = 9M_P, R_E = 2R_P$

$$v_{(escape)planet} = \sqrt{\frac{2GM_P}{R_P}}$$

$$= \sqrt{\frac{2G\left(\frac{M_E}{9}\right)}{\left(\frac{R_E}{2}\right)}} = \frac{v_e\sqrt{2}}{3} = \frac{v_e}{3}\sqrt{x} \qquad \therefore x = 2$$

119. (c) We have

$$V_e = \sqrt{2Rg} \Rightarrow \frac{V_{eA}^2}{V_{eB}^2} = \frac{R_A}{R_B}\frac{g_A}{g_B}$$

$$\Rightarrow \frac{1}{4} = \frac{1}{3}\frac{g_A}{g_B} \Rightarrow \frac{g_A}{g_B} = \frac{3}{4}$$

120. (b) As, $E = -\frac{GMm}{2r}$

$$E \propto \frac{M}{r}$$

$$\frac{E_A}{E_B} = \frac{M_A}{M_B} \times \frac{r_B}{r_A} = \frac{4}{3} \times \frac{4r}{3r} = \frac{16}{9}$$

121. (b) Initial kinetic energy

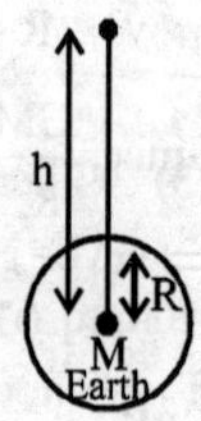

$$K_i = \frac{1}{2}mv_i^2 = \frac{1}{2}m(\lambda v_e)^2$$

Initial potential energy,

$$U_i = \frac{GMm}{R}$$

Final potential energy,

$$U_f = \frac{-GMm}{h}$$

Using law of conservation of energy

$$K_i + U_i = K_f + U_f$$

$$-\frac{GMm}{R} + \frac{1}{2}m\lambda^2 V_e^2 = -\frac{GMm}{h}$$

($\because$ Final kinetic energy = 0)

$$\Rightarrow -\frac{GMm}{R} + \frac{1}{2}\lambda^2 \frac{2GMm}{R} = -\frac{GMm}{h}$$

$$\Rightarrow \frac{\lambda^2}{R} - \frac{1}{R} = \frac{-1}{h} \Rightarrow \frac{1}{h} = \frac{1-\lambda^2}{R} \Rightarrow h = \frac{R}{1-\lambda^2}$$

122. (a) Escape velocity, $v_e = \sqrt{\frac{2Gm}{R}}$

from conservation of energy

$$\frac{-GMm}{R} + \frac{1}{2}m\frac{V_e^2}{9} = -\frac{GMm}{R+h}$$

$$\frac{GM}{R+h} = \frac{GM}{R} - \frac{V_e^2}{18}$$

$$\frac{GM}{R+h} = \frac{GM}{R} - \frac{GM}{9R}$$

$$\frac{GM}{R+h}=\frac{8GM}{9R} \Rightarrow \frac{1}{R+h}=\frac{8}{9R}$$

$9R=8R+8h$

The maximum height attained by the body,

$$h=\frac{R}{8} \Rightarrow \frac{6400}{8} \Rightarrow 800 \text{ km}$$

123. (a) We know that, escape velocity is given as

$$v_e=\sqrt{\frac{2GM}{R}}$$

$$=\sqrt{\frac{2G\times\rho\times\frac{4}{3}\pi R^3}{R}} \Rightarrow v_e \propto \sqrt{\rho R^2}$$

So, $v_{e_2}=v_{e_1}\sqrt{\left(\frac{\rho_2}{\rho_1}\right)^2\times\left(\frac{R_2}{R_1}\right)^2}=12\sqrt{4\times\left(\frac{1}{2}\right)^2}=12$ km/s.

124. (2) We know that orbital velocity is given as

$$V=\sqrt{\frac{GM}{r}}$$

$$\therefore V \propto \frac{1}{\sqrt{r}}$$

$$\frac{V_1}{V_2}=\sqrt{\frac{r_2}{r_1}}=\sqrt{\frac{800}{3200}}=\frac{1}{2}$$

125. (3) Given, $T_1=1$ hour

$\Rightarrow \omega_1=2\pi$ rad/hour

$T_2=8$ hours $\Rightarrow \omega_2=\frac{\pi}{4}$ rad/hour

$R_1=2\times10^3$ km

From the Kepler's law of orbit,

As $T^2 \propto R^3$

$$\therefore \left(\frac{R_2}{R_1}\right)^3=\left(\frac{T_2}{T_1}\right)^2 \Rightarrow \frac{R_2}{R_1}=\left(\frac{8}{1}\right)^{2/3}=4$$

$\Rightarrow R_2=4R_1=8\times10^3$ km

$V_1=\omega_1R_1=4\pi\times10^3$ km/h

$V_2=\omega_2R_2=2\pi\times10^3$ km/h

$$\omega=\frac{V_{rel}}{R_{rel}}=\frac{V_1-V_2}{R_2-R_1}=\frac{2\pi\times10^3}{6\times10^3}$$

$$=\frac{\pi}{3}\text{ rad/hour}=\frac{\pi}{x}\text{ rad/h}$$

$\therefore x=3$

126. (b) Total energy at middle point

$=$ K.E $+$ P.E of M_1 & m $+$ P.E of M_2 & m

To get escape velocity total energy should be zero.

$$\frac{1}{2}mV^2-\frac{GM_1m}{r/2}-\frac{GM_2m}{r/2}=0$$

$$\Rightarrow \frac{1}{2}mV^2=\frac{2Gm}{r}(M_1+M_2) \;\therefore V=\sqrt{\frac{4G(M_1+M_2)}{r}}$$

127. (64) Escape velocity,

$$v_e=\sqrt{\frac{2GM}{R}}$$

Let R' be the radius so that escape velocity is increased 10 times.

$$v'=10v_e=\sqrt{\frac{2GM}{R'}} \Rightarrow 10\times\sqrt{\frac{2GM}{R}}=\sqrt{\frac{2GM}{R'}}$$

$$\therefore \quad R'=\frac{R}{100}=\frac{6400}{100}=64 \text{ km}$$

128. (b) From angular momentum conservation

$\therefore \quad mv_1r_1=mv_2r_2$

or, $6\times10^4\times8\times10^{10}=v_2\times1.6\times10^{12}$

$$\Rightarrow v_2=\frac{6\times8}{1.6}\times10^2=3\times10^3 \text{ m/s}$$

129. (b) Speed of satellite, $v=\sqrt{\frac{GM}{r}}$

Time, $T=\frac{2\pi r}{v}=\frac{2\pi r}{\sqrt{\frac{GM}{r}}}=2\pi\sqrt{\frac{r^3}{GM}}$

$$\therefore T_B-T_A=\frac{2\pi}{\sqrt{GM}}\left[r_B^{3/2}-r_A^{3/2}\right]$$

$$=\frac{2\pi}{\sqrt{GM}}\left[(8\times10^6)^{3/2}-(7\times10^6)^{3/2}\right]$$

$$=\frac{2\pi}{\sqrt{6.67\times10^{-11}\times6\times10^{24}}}\times10^9[8^{3/2}-7^{3/2}]$$

$=1.33\times10^3$ s

130. (10). From energy conservation,

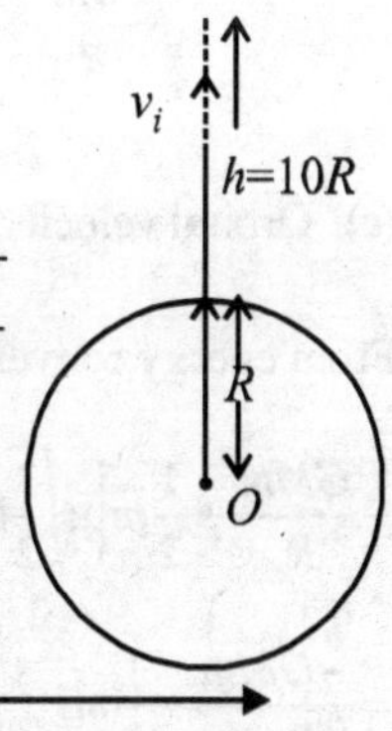

$$-\frac{GM_em}{R}+\frac{1}{2}mv_i^2=-\frac{GM_em}{11R}$$

$$v_i=\sqrt{\frac{20}{11}\frac{GM_e}{R}} \quad \because v_e=\sqrt{\frac{2GM_e}{R}}$$

$$\therefore v_i=\sqrt{\frac{10}{11}}V_e=\sqrt{\frac{x}{11}}V_e$$

$\therefore x=10.$

131. (b)

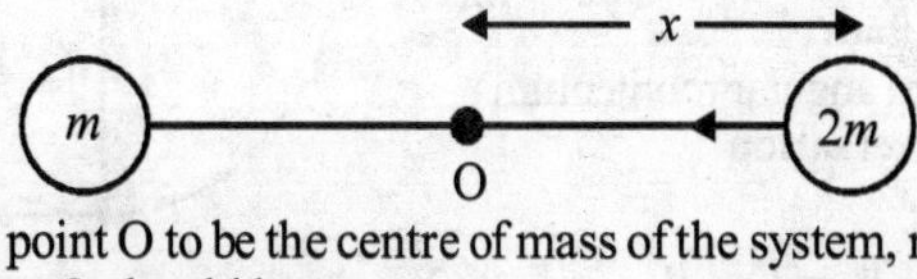

For point O to be the centre of mass of the system, moment about O should be zero.

$\therefore \; 2mx=m(d-x)$

$$\Rightarrow 3mx=md \Rightarrow x=\frac{d}{3}$$

For equilibrium,

$$F_{gravitational}=F_{centripetal}$$

$$\therefore\ F=\frac{G(2m)m}{d^2}=(2m)\omega^2\left(\frac{d}{3}\right)$$

$$\Rightarrow \frac{Gm}{d^2}=\omega^2\frac{d}{3}\Rightarrow \omega^2=\frac{3Gm}{d^3}\Rightarrow \omega=\sqrt{\frac{3Gm}{d^3}}$$

$\therefore$ Period of revolution, $T=\frac{2\pi}{\omega}=2\pi\sqrt{\frac{d^3}{3Gm}}$

132. (a) By angular momentum conservation

$r_{min}v_{max}=r_{max}v_{min}$

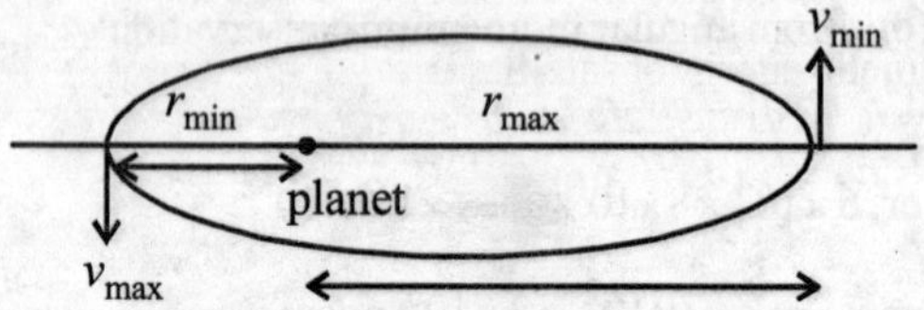

Given, $v_{min}=\frac{v_{max}}{6}$ $\quad\therefore \frac{r_{min}}{r_{max}}=\frac{v_{min}}{v_{max}}=\frac{1}{6}$

133. (a) Orbital speed of the body when it revolves very close to the surface of planet

$$V_0=\sqrt{\frac{GM}{R}} \quad ...(i)$$

Here, G = gravitational constant

Escape speed from the surface of planet

$$V_e=\sqrt{\frac{2GM}{R}} \quad ...(ii)$$

Dividing (i) by (ii), we have

$$\frac{V_0}{V_e}=\frac{\sqrt{\frac{GM}{R}}}{\sqrt{\frac{2GM}{R}}}=\frac{1}{\sqrt{2}}$$

134. (c) Orbital velocity, $V_0=\sqrt{\frac{GM}{R_e}}$

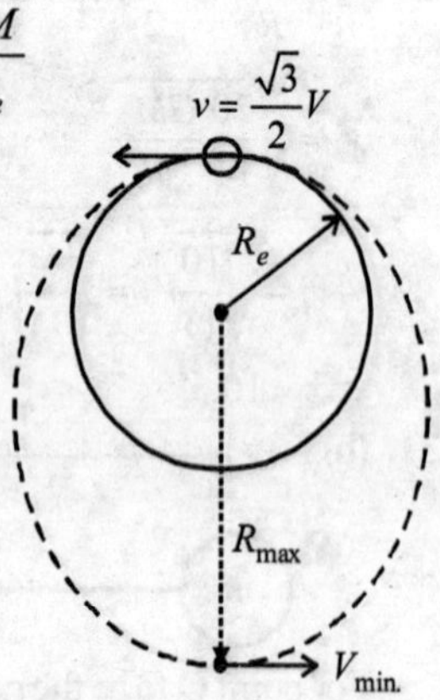

From energy conversation,

$$-\frac{GMm}{R_e}+\frac{1}{2}m\left(\sqrt{\frac{3}{2}}V\right)^2$$

$$=\frac{-GMm}{R_{max}}+\frac{1}{2}mV_{min}^2 \quad ...(1)$$

From angular momentum conversation

$$\sqrt{\frac{3}{2}}VR_e=V_{min}R_{max} \quad ...(2)$$

Solving equation (1) and (2) we get,

$R_{max}=3R_e$

135. (a) According to question, mass density of a spherical galaxy varies as $\frac{k}{r}$.

Mass, $M=\int\rho dV$

$$\Rightarrow M=\int_0^R\frac{k}{r}4\pi r^2\,dr\ \Rightarrow M=4\pi k\int_0^R r\,dr$$

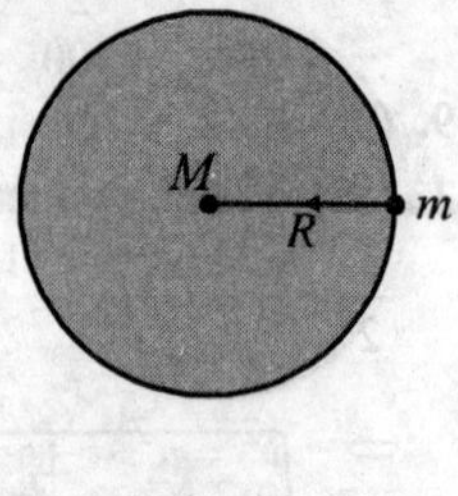

or, $M=\frac{4\pi kR^2}{2}=2\pi kR^2$

$$F_G=\frac{GMm}{R^2}=m\omega_0^2R\ (=F_C)$$

$$\Rightarrow \frac{G\left(2\pi kR^2\right)}{R^2}=\omega_0^2R\Rightarrow \omega_0=\sqrt{\frac{2\pi KG}{R}} \quad \left(\because \omega=\frac{2\pi}{T}\right)$$

$$\therefore T=\frac{2\pi}{\omega_0}=\frac{2\pi\sqrt{R}}{\sqrt{2\pi KG}}=\sqrt{\frac{2\pi R}{KG}}\Rightarrow T^2=\frac{2\pi R}{KG}$$

$\because 2\pi$, K and G are constants $\therefore T^2\propto R$.

136. (d) From law of conservation of momentum, $\vec{p}_i=\vec{p}_f$

$m_1u_1+m_2u_2=MV_f$

$$\Rightarrow v_f=\frac{\left(mv+\frac{mv}{4}\right)}{\frac{3m}{2}}=\frac{5v}{6}$$

Clearly, $v_f<v_i$ $\quad\therefore$ Path will be elliptical

137. (b) $U_{surface}+E_1=U_h$

[$\because$ E_1 is minimum, at height 'h', K.E. = 0]

$$-\frac{GM_em}{R_e}+E_1=-\frac{GM_em}{(R_e+h)}$$

$$\Rightarrow E_1=GM_em\left(\frac{1}{R_e}-\frac{1}{R_e+h}\right)\ \Rightarrow E_1=\frac{GM_em}{(R_e+h)}\times\frac{h}{R_e}$$

Gravitational attraction

$$F_G=ma_c=\frac{mv^2}{(R_e+h)}=\frac{GM_em}{(R_e+h)^2}$$

$$mv^2=\frac{GM_em}{(R_e+h)}\Rightarrow E_2=\frac{mv^2}{2}=\frac{GM_em}{2(R_e+h)}$$

As $E_1=E_2$

Clearly, $\frac{h}{R_e}=\frac{1}{2}\Rightarrow h=\frac{R_e}{2}=3200\text{km}$

138. (a) Escape velocity of the planet A is $V_A=\sqrt{\frac{2GM_A}{R_A}}$

where M_A and R_A be the mass and radius of the planet A.

According to given problem

$$M_B=\frac{M_A}{2},\ R_B=\frac{R_A}{2}$$

$$\therefore\ V_B=\sqrt{\frac{2G\frac{M_A}{2}}{\frac{R_A}{2}}}\qquad \therefore\ \frac{V_A}{V_B}=\sqrt{\frac{\frac{2GM_A}{R_A}}{\frac{2GM_A/2}{R_A/2}}}=1=\frac{n}{4}$$

$\therefore\quad n=4$

139. (b) Let v be the speed of satellite just before ejection of rocket then,

$$\frac{1}{2}mu^2-\frac{GMm}{R}=\frac{1}{2}mv^2-\frac{GMm}{2R}$$

$$\Rightarrow\quad v^2=u^2-\frac{GM}{R}\qquad .....(i)$$

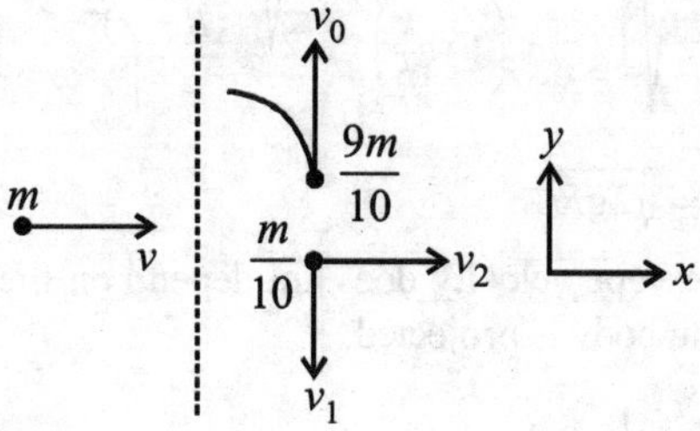

Along x:

$$(P_i)_x=(P_f)_x\Rightarrow mv=\frac{m}{10}v_2$$

$$\Rightarrow\quad v_2=10v\Rightarrow v_2^2=100\left(u^2-\frac{GM}{R}\right)\ \text{[From (i)]}$$

Along y:

$$(P_i)_y=(P_f)_y\Rightarrow 0=\frac{9m}{10}v_0-\frac{m}{10}v_1$$

$$\Rightarrow\quad v_1=9v_0\Rightarrow v_1^2=81\frac{GM}{2R}$$

$$\therefore\quad (\text{K.E.})_R=\frac{1}{2}\frac{m}{10}\left(v_1^2+v_2^2\right)=5m\left[u^2-\frac{119}{200}\frac{GM}{R}\right]$$

140. (d) Time period of revolution of satellite,

$$T=\frac{2\pi r}{v}$$

$$\therefore T=2\pi r\sqrt{\frac{r}{GM}}=2\pi\sqrt{\frac{r^3}{GM}}\qquad\left(\because v=\sqrt{\frac{GM}{r}}\right)$$

Substituting the values, we get

$$T=2\pi\sqrt{\frac{(202)^3\times10^{12}}{6.67\times10^{-11}\times8\times10^{22}}}\ \text{sec}$$

T = 7812.2 s

$$T\simeq2.17\ \text{hr}\Rightarrow n=\frac{24}{2.17}\simeq11\ \text{revolutions}$$

141. (d) $K_E=E=\frac{1}{2}m\times\frac{2GM_e}{R_e}=\frac{GMM_e}{R_e}$

as, $64\,V_m=V_e\Rightarrow 64\times\frac{4}{3}\pi R_m^3=\frac{4}{3}\pi R_e^3\Rightarrow 4R_m=R_e$

and, $M\propto V$

So, $M_e=64M_m$

$$K_{moon}=\frac{1}{2}m\times\frac{2GM_m}{R_m}=\frac{GMM_m}{R_m}$$

$$=\frac{GMM_e/64}{\frac{R_e}{4}}=\frac{GMM_e}{16R_m}=\frac{K_E}{16}=\frac{E}{16}$$

142. (b) Orbital, velocity, $v=\sqrt{\frac{GM}{r}}$

Kinetic energy of satellite A,

$$T_A=\frac{1}{2}m_AV_A^2$$

Kinetic energy of satellite B,

$$T_B=\frac{1}{2}m_BV_B^2\Rightarrow\frac{T_A}{T_B}=\frac{m\times\frac{GM}{R}}{2m\times\frac{GM}{2R}}=1$$

143. (d) For a satellite orbiting close to the earth, orbital velocity is given by

$$v_0=\sqrt{g(R+h)}\approx\sqrt{gR}$$

Escape velocity (v_e) is

$$v_e=\sqrt{2g(R+h)}\approx\sqrt{2gR}\quad[\because h<<R]$$

$$\Delta v=v_e-v_0=(\sqrt{2}-1)\sqrt{gR}$$

144. (b) At height r from center of earth, orbital velocity

$$v=\sqrt{\frac{GM}{r}}$$

By principle of energy conservation

$$\text{KE of 'm'}+\left(-\frac{GMm}{r}\right)=0+0\quad(\because\text{At infinity, PE}=\text{KE}=0)$$

or $\quad\text{KE of 'm'}=\frac{GMm}{r}=\left(\sqrt{\frac{GM}{r}}\right)^2m=mv^2$

145. (d) Let M is mass of star m is mass of meteroite

By energy convervation between O and ∞.

$$-\frac{GMm}{r}+\frac{-GMm}{r}+\frac{1}{2}mV_{esc}^2=0+0$$

$$\therefore\quad v=\sqrt{\frac{4GM}{r}}=\sqrt{\frac{4\times6.67\times10^{-11}\times3\times10^{31}}{10^{11}}}$$

$$\simeq2.8\times10^5\ \text{m/s}$$

146. (a) Due to infinite wire of mass 'm' at 'r' distance

$$E_g=\frac{2G\lambda}{r}\quad\text{so } F_g=mE_g$$

So force on star $=\frac{2Gm\lambda}{r}=\frac{mv^2}{r}\Rightarrow v=\sqrt{2G\lambda}$

as $T=\frac{2\pi r}{v}=\frac{2\pi r}{\sqrt{2G\lambda}}\Rightarrow T\propto r$

147. (a) As we know,

Gravitational potential energy $=\frac{-GMm}{r}$

and orbital velocity, $v_0=\sqrt{GM/R+h}$

$$E_f = \frac{1}{2}mv_0^2 - \frac{GMm}{3R} = \frac{1}{2}m\frac{GM}{3R} - \frac{GMm}{3R}$$

$$= \frac{GMm}{3R}\left(\frac{1}{2}-1\right) = \frac{-GMm}{6R}$$

$$E_i = \frac{-GMm}{R} + K$$

$E_i = E_f$

Therefore minimum required energy, $K = \frac{5GMm}{6R}$

148. (c) Escape velocity on earth,

$$v_e = \sqrt{\frac{2GM_e}{R_e}} = 11\ km\,s^{-1}$$

$$\therefore \frac{(v_e)_p}{(v_e)_e} = \frac{\sqrt{\frac{2GM_p}{R_p}}}{\sqrt{\frac{2GM_e}{R_e}}} = \sqrt{\frac{M_p}{M_e} \times \frac{R_e}{R_p}}$$

$$= \sqrt{\frac{10M_e}{M_e} \times \frac{R_e}{R_e/10}} = 10$$

$$\therefore (v_e)_p = 10 \times (v_e)_e = 10 \times 11 = 110\ km/s$$

149. (c) Gravitational force, $F = KR^{-n}$

This force provides the centripetal force $MR\omega^2$ to the planet at height R above earth's surface.

$\therefore \quad F = KR^{-n} = MR\omega^2$

$$\Rightarrow \omega^2 \propto KR^{-(n+1)} \Rightarrow \omega \propto KR^{\frac{-(n+1)}{2}}$$

$$\frac{2\pi}{T} \propto R^{\frac{-(n+1)}{2}} \qquad \therefore T \propto R^{\frac{+(n+1)}{2}}$$

150. (c) Time period of satellite is given by

$$T = 2\pi\sqrt{\frac{(R+h)^3}{GM}}$$

Where $R + h$ = radius of orbit of satellite

M = mass of earth.

Time period is independent of mass of satellite.

151. (d) $v_0 = \sqrt{(R+x)g_x}$

$$= \sqrt{(R+x)\frac{g}{\left(1+\frac{x}{R}\right)^2}} = \sqrt{\frac{(R+x)gR^2}{(R+x)^2}} = \sqrt{\frac{gR^2}{R+x}}$$

152. (c) $v_e = \sqrt{2gR}$

Clearly escape velocity does not depend on the angle at which the body is projected.

153. (c) $K.E = \frac{1}{2}mv_e^2$

Here v_e = escape velocity is independent of mass of the body

Escape velocity, $v_e = \sqrt{2gR}$

Substituting value of v_e in above equation we get

$$K.E = \frac{1}{2}m \times 2gR = mgR$$

154. (a) Escape velocity, $v_e = \sqrt{2gR} = \sqrt{\frac{2GM}{R}}$

$\Rightarrow v_e \propto m^0$

8 Mechanical Properties of Solids

1. **(a)** Young's modulus, $Y=\dfrac{\text{Stress}}{\text{Strain}}=\dfrac{F/A}{\Delta l/l}$

$$\Rightarrow \frac{50g}{\pi r^2}=Y\cdot\frac{\Delta l}{l}$$

$$\Rightarrow \frac{50\times 3\pi}{\pi\times(3\times10^{-3})^2}=P\times10^{11}\times\frac{0.1\times10^{-3}}{3}$$

$$\Rightarrow P=\frac{50\times3\times3}{3^2\times10^{-6}\times10^{11}\times0.1\times10^{-3}}=5$$

2. **(b)** The elongations in two wires A and B is

$$\Delta L_A=\frac{F_AL_A}{A_AY_A}\text{ and }\Delta L_B=\frac{F_BL_B}{A_BY_B}$$

Given, $F_A=F_B, Y_A=Y_B, \dfrac{L_A}{L_B}=\dfrac{1}{3}$ and $\dfrac{d_A}{d_B}=2$

$$\therefore \frac{\Delta L_A}{\Delta L_B}=\frac{\dfrac{F_AL_A}{A_AY_A}}{\dfrac{F_BL_B}{A_BY_B}}=\left(\frac{L_A}{L_B}\right)\left(\frac{A_B}{A_A}\right)$$

$$=\left(\frac{L_A}{L_B}\right)\left(\frac{\frac{\pi}{4}d_B^2}{\frac{\pi}{4}d_A^2}\right)=\left(\frac{L_A}{L_B}\right)\left(\frac{d_B}{d_A}\right)^2=\left(\frac{1}{3}\right)\left(\frac{1}{2}\right)^2=\frac{1}{12}.$$

3. **(1)** Tension $T=K(\ell-\ell_0)$

$5=K(1.4-\ell_0)$ ...(i)

$7=K(1.56-\ell_0)$ ...(ii)

From eq. (i) & (ii)

$$\frac{5}{1.4-\ell_0}=\frac{7}{1.56-\ell_0}\quad\therefore \ell_0=1\text{m}$$

4. **(20)** Given, Force on the wire, $F=200$ N

Young's modulus of material of wire, $Y=1\times10^{11}\text{ Nm}^{-2}$

Initial length of wire, $\ell=2$m

Area of cross section of wire, $A=2\text{ cm}^2$

$$Y=\frac{F/A}{\Delta\ell/\ell}\quad\Rightarrow\quad \Delta\ell=\frac{F\ell}{AY}$$

$$\Delta\ell=\frac{200\times2}{2\times10^{-4}\times10^{11}}=2\times10^{-5}=20\mu\text{m}$$

5. **(1)** Along y-axis,

$2T\sin\theta=2g$

For small angle $\sin\theta\approx\theta$

$2T\theta=2g$

$$T=\frac{g}{\theta}=\frac{10}{1}\times100=1000\text{ N}$$

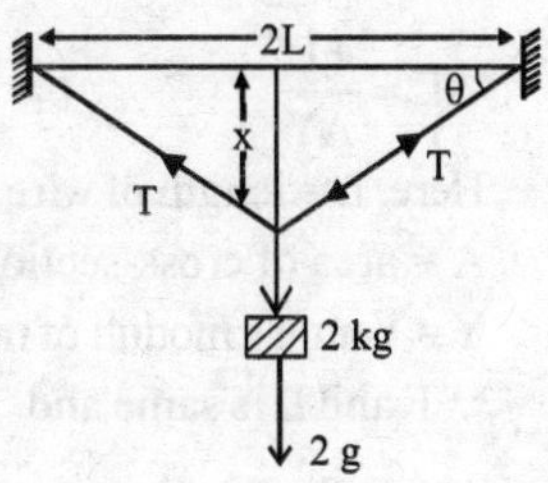

Change in length is given as,

$$\Delta L=2\sqrt{x^2+L^2}-2L$$

$$=2L\left[1+\frac{x^2}{2L^2}-1\right]=\frac{x^2}{L}$$

Hence, modulus of elasticity $=\dfrac{\text{stress}}{\text{strain}}$

$$2\times10^{11}=\frac{10^3}{A\times\frac{x^2}{L}}\times2L\Rightarrow A=1\times10^{-4}\text{ m}^2$$

6. **(600)** Breaking stress $=\dfrac{T}{A}=\dfrac{mg_p}{A}\left[\because m=\rho Al\ \&\ g_p=\dfrac{g}{3}\right]$

$$\Rightarrow 1.2\times10^8=\frac{\rho A(l)g}{3A}$$

$$\Rightarrow l=\frac{1.2\times10^8\times3}{6\times10^4\times10}=600\text{m}$$

7. **(b)** Young's modulus, $Y=\dfrac{\text{Stress}}{\text{Strain}}$

If the temperature increases, strain also increases

Hence young's modulus decreases

8. **(3)** Longitudinal strain is, $e=\dfrac{F}{AY}=\dfrac{T}{AY}$

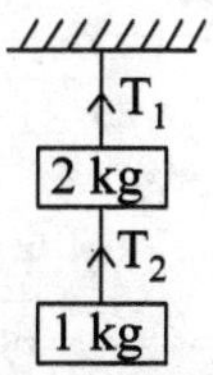

$$\therefore \frac{e_1}{e_2}=\frac{T_1}{T_2}=\frac{30\text{N}}{10\text{N}}=3$$

9. **(12)** Given,

Radius of wire, $r=4\times10^{-5}$ m

Young's modulus of wire, $Y=2\times10^{11}\text{ N/m}^2$

Mass of blocks

$m_1=2$ kg, $m_2=4$ kg

Tension in the string, $T=\left(\dfrac{2m_1m_2}{m_1+m_2}\right)g=\dfrac{80}{3}$N

Area of cross section of wire, $A=\pi r^2=16\pi\times10^{-10}\text{ m}^2$

Strain $=\dfrac{\Delta\ell}{\ell}=\dfrac{T}{AY}$

$$=\frac{80/3}{16\pi\times10^{-10}\times2\times10^{11}}=\frac{1}{12\pi}=\frac{1}{\alpha\pi}\text{ (given)}$$

$\therefore\ \alpha=12$

10. **(c)** Young's modulus depends on the material not depends on length and cross sectional area. So Young's modulus remains constant.

11. **(2)** From free body diagram,

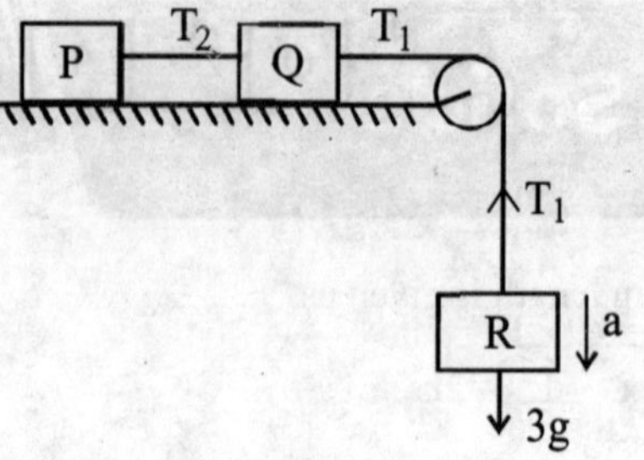

$$a=\frac{3g}{3+3+3}=\frac{g}{3}=\frac{10}{3}\text{m/s}^2$$

$$30-T_1=3\times a=3\times\frac{10}{3}$$

$$T_1=20\,\text{N}$$

$$strain=\frac{stress}{Y}=\frac{T_1}{YA}=\frac{20}{2\times10^{11}\times0.005\times10^{-4}}$$

$$=2\times10^{-4}$$

12. **(d)** The increase in length is

$$\delta\ell=\frac{FL}{AY}\Rightarrow\frac{\delta\ell_2}{\delta\ell_1}=\left(\frac{F_2}{F_1}\right)\left(\frac{A_1}{A_2}\right)$$

$$\because F_2=\frac{F_1}{2},\ r_2=\frac{r_1}{2}\Rightarrow\delta\ell_2=\ell\left(\frac{1}{2}\right)(2)^2=2\,\ell$$

13. **(16)** Extension, $\Delta\ell=\frac{F\ell}{AY}$

As, $V=A\ell\ \Rightarrow\ \ell=\frac{V}{A}$

$$\therefore\ \Delta\ell=\frac{FV}{A^2Y}$$

Y, $\Delta\ell$ & V are same for both the wires.

$$\Rightarrow\quad F\propto A^2\ \Rightarrow\ \frac{F_P}{F_Q}=\left(\frac{A_P}{A_Q}\right)^2=\left(\frac{4}{1}\right)^2$$

$$\therefore\quad\frac{F_1}{F_2}=16\qquad[\because A_P:A_Q=4:1]$$

14. **(d)** Elasticity is the property by virtue of which body tends to regain its original shape after removal of applied external force.

15. **(b)**

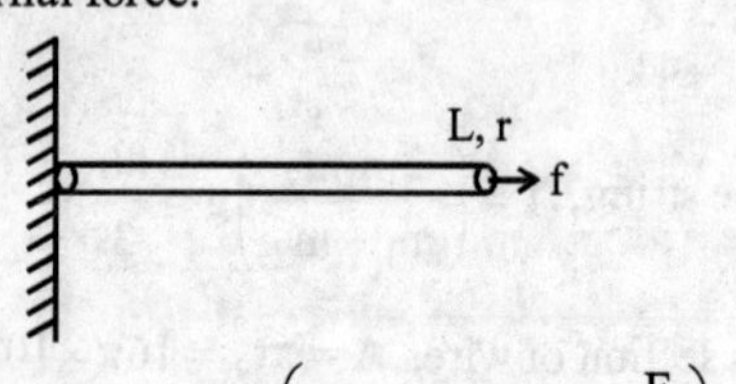

$$\frac{f}{\pi r^2}=Y\frac{\ell}{L}\quad\left(\because Y=\frac{\text{Stres}}{\text{Strain}}=\frac{\frac{F}{A}}{\frac{\Delta\ell}{\ell}}\right)$$

(Here $A=\pi r^2$)

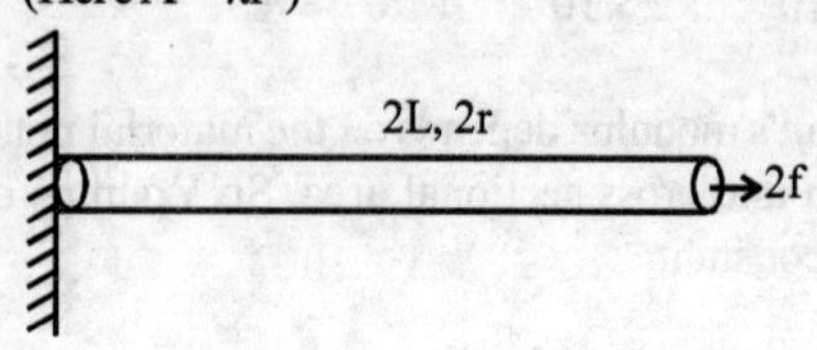

Similarly, for another wire

$$\frac{2f}{\pi(2r)^2}=Y\frac{\ell'}{2L}\Rightarrow\frac{2}{1}=\frac{2\ell}{\ell'}\Rightarrow\ell'=\ell$$

16. **(2)** Let ℓ_0 be the original length.

When $T_1=100$ N

$$100=K(\ell_1-\ell_0)\qquad\text{...(i)}$$

When $T_2=120$ N

$$120=K(\ell_2-\ell_0)\qquad\text{...(ii)}$$

Dividing (ii) by (i) we get

$$\frac{120}{100}=\frac{K(\ell_2-\ell_0)}{K(\ell_1-\ell_0)}$$

$$\Rightarrow5\,\ell_2-5\ell_0=6\ell_1-6\ell_0\Rightarrow\ell_0=6\ell_1-5\ell_2$$

$$\Rightarrow\ell_0=6\ell_1-5\left(\frac{11\ell_1}{10}\right)\ [\because 10\,\ell_2=11\,\ell_1\text{ given}]$$

$$\Rightarrow\ell_0=6\ell_1-\frac{11\ell_1}{2}\Rightarrow\ell_0=\frac{\ell_1}{2}=\frac{1}{x}\ell_1$$

$\therefore$ x = 2

17. **(20)**

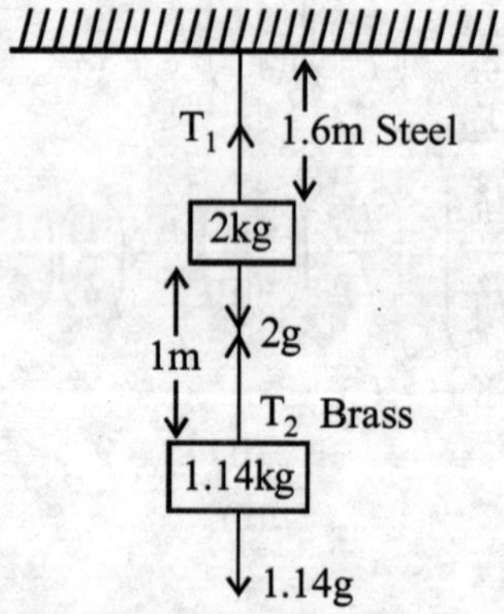

Tension in steel wire, $T_1=2g+T_2$

$$T_1=20+11.4=31.4\,\text{N}$$

$$\because T_2=1.14\,g=11.4\,\text{N}$$

Young modulus of steel, $Y=\dfrac{\frac{T_1}{A}}{\frac{\Delta L}{L}}$

Elongation in steel wire, $\Delta L=\frac{T_1L}{AY}$

$$\Delta L=\frac{31.4\times1.6}{\pi\left(0.2\times10^{-2}\right)^2\times2\times10^{11}}=\frac{16}{2\times4\times10^{-6}\times10^{11}}$$

$$=20\times10^{-6}\,\text{m}.$$

18. **(b)** Elongation produced in the wire is given by

$$\Delta L=\frac{FL}{AY}$$

Here, L = length of wire

A = area of cross-section of wire

Y = Young's moduli of material of wire.

$\because$ F and L is same and

$\frac{Y_A}{Y_B}=\frac{1}{4}$ (given)

$\frac{A_A}{A_B}=\frac{1}{3}$ (given)

$\therefore \frac{\Delta L_A}{\Delta L_B}=\frac{A_B}{A_A}\frac{Y_B}{Y_A}=\frac{3}{1}\times\frac{4}{1}=12$

19. (25) Given,

Radius of rod, r = 20 mm = 0.02m

Length of rod, ℓ = 2m

Force, F = 62.8 kN

Young's modulus of steel, $Y = 2\times10^{11}$ N/m^2

$$\text{Strain}=\frac{\text{stress}}{Y}=\frac{\frac{62.8\times10^3}{\pi\times(0.02)^2}}{2\times10^{11}}$$

$$=\frac{62.8\times10^3}{3.14\times4\times10^{-4}\times2\times10^{11}}$$

$= 2.5\times10^{-4} = 25\times10^{-5}$

20. (11) Given,
Diameter of wire, D = 14 mm = 14×10^{-3} m
Tensile stress, $\sigma = 7\times10^5$ Nm^{-2}

Tensile stress, $\sigma=\frac{F}{A}=\frac{mg}{\pi\frac{D^2}{4}}=\frac{4mg}{\pi D^2}$

$\Rightarrow m=\frac{\pi D^2\sigma}{4g}$

$=\frac{22}{7}\times\frac{(14\times10^{-3})^2\times7\times10^5}{4\times9.8}=11$ kg

21. (d) As $Y=\frac{F\ell}{A\Delta\ell}\Rightarrow\Delta\ell=\frac{F\ell}{YA}$

$\Rightarrow\Delta\ell=\frac{4\times g/4\times6}{2\times10^{11}\times3\times10^{-6}}\Rightarrow\Delta\ell=\frac{60}{6}\times10^{-5}$

$\Rightarrow\Delta\ell=10^{-4}$ m = 0.1mm

22. (b) We have
$\Delta l_A=\Delta l_B$

$\Rightarrow\frac{Fl_A}{Y_AA_1}=\frac{Fl_B}{Y_BA_2}\Rightarrow\frac{5}{Y_A\times2.5\times10^{-5}}=\frac{6}{Y_B\times3\times10^{-5}}$

$\Rightarrow\frac{Y_A}{Y_B}=\frac{15\times10^{-5}}{15\times10^{-5}}=\frac{1}{1}$

23. (d) Young's modulus,

$Y=\frac{F/A}{\frac{\Delta\ell}{\ell}}\Rightarrow F=\frac{YA}{\ell}\Delta\ell$

As F and Y is same for two wire.

$\therefore\frac{F}{Y}=\frac{A_1\Delta\ell_1}{\ell_1}=\frac{A_2\Delta\ell_2}{\ell_2}$

$\therefore\frac{\Delta\ell_2}{\Delta\ell_1}=\frac{A_1}{A_2}\times\frac{\ell_2}{\ell_1}$

Here, $\Delta\ell_1=0.2$ mm, $\ell_2=2\ell_1$, $A_2=(2.4)^2A_1$

$\Rightarrow\frac{\Delta\ell_2}{0.2}=\frac{1}{2.4\times2.4}\times\frac{2}{1}$

$\Rightarrow\Delta\ell_2=6.9\times10^{-2}$ mm

24. (5) From graph
F = ΔL
Youngs modulus,

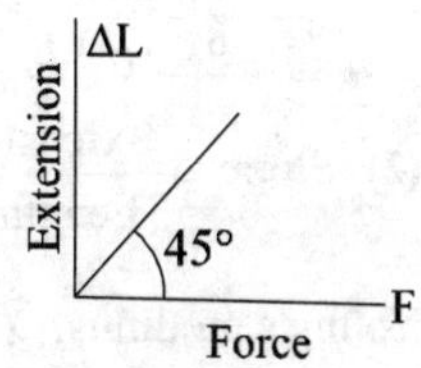

$Y=\frac{\text{stress}}{\text{strain}}=\frac{F/A}{\Delta L/L}$

$Y=\frac{FL}{A\Delta L}$

$Y=\frac{L}{A}=\frac{L}{\pi r^2}$

(Here length of wire, L = 62.8 cm = 62.8×10^{-2} m)

$Y=\frac{62.8\times10^{-2}}{\pi(2\times10^{-3})^2}=5\times10^4$ N/m^2

25. (d) $\Delta\ell=\frac{F\ell}{YA}=\frac{250\times100}{10^{10}\times6.25\times10^{-4}}=40\times10^{-4}$ m

$=4\times10^{-3}$ m

26. (c) Steel is used in construction because it has high elastic limit and it binds well to concrete.

27. (a) Young modulus depends on material of wire not on length and area of wire's cross section.

28. (15) The velocity of transverse wave, $v_w=\sqrt{\frac{T}{\mu}}$

$60=\sqrt{\frac{T}{10\times10^{-3}}\times0.5}$

$\Rightarrow T=\frac{(60)^2\times10^{-2}}{0.5}=72$ N

$Y=\frac{\text{stress}}{\text{strain}}$; $Y=\frac{F/A}{\frac{\Delta\ell}{\ell}}$

$\Rightarrow\Delta\ell=\frac{F\ell}{AY}=\frac{72\times0.5}{2\times10^{-6}\times1.2\times10^{11}}=15\times10^{-5}$ m

Compare with $x\times10^{-5}$ m, we have x = 15

29. (50) Given, length of metal wire, ℓ = 0.5 m
Cross-sectional area, A = 10^{-4} m^2
Breaking stress = 5×10^8 Nm^{-2}
Mass of block m = 10 kg
T_{max} = Breaking stress × Area

$\frac{mv^2}{\ell}=5\times10^8\times10^{-4}=5\times10^4$

$\frac{10v^2}{0.5}=5\times10^4\Rightarrow v=\sqrt{\frac{0.5\times5\times10^4}{10}}=50$ m/s

30. **(c)** Force $F = YA\frac{\Delta l}{l}$

$= 2\times10^{11}\times10^{-4}\left(\frac{2l-l}{l}\right) = 2\times10^{7}\,\text{N}$

31. **(30) Strain** $= \text{F/AY} = \dfrac{\text{mg}+\dfrac{\text{mv}^2}{\text{R}}}{\text{AY}}$

$= \dfrac{20+\dfrac{2(5)^2}{0.5}}{4\times10^{-6}\times10^{11}} = 30\times10^{-5}$

32. **(2)** Slope $= \dfrac{\text{Extension/Load}}{\text{Length of wire}} = \dfrac{\Delta l/\text{w}}{\text{L}} = \dfrac{\Delta l}{\text{wL}}$

Young's modulus, $\text{Y} = \dfrac{\text{w/A}}{\Delta\ell/\text{L}} = \dfrac{\text{wL}}{\Delta\ell\text{A}}$

$\therefore\ \text{Y} = \dfrac{1}{(\text{slope})\,\text{A}} \Rightarrow \text{Y} = \dfrac{1}{2\times10^{-6}(0.25\times10^{-5})}$

$\Rightarrow\ \text{Y} = 2\times10^{11}\,\text{N/m}^2$

33. **(d)** Young's modulus,

$\text{Y} = \dfrac{\text{F/A}}{\dfrac{\Delta\ell}{\ell}} = \dfrac{\text{F}\ell}{\text{A}\Delta\ell} \Rightarrow \Delta\ell = \dfrac{\text{F}\ell}{\text{AY}}$

Net elongation, $\Delta\ell = \Delta\ell_1 + \Delta\ell_2$

$\Rightarrow\ \Delta\ell = \dfrac{\text{F}\ell_1}{\text{A}_1\text{Y}_\text{s}} + \dfrac{\text{F}\ell_2}{\text{A}_2\text{Y}_\text{c}}$

$\Rightarrow \text{F} = \dfrac{\Delta\ell}{\dfrac{\ell_1}{\text{A}_1\text{Y}_\text{s}} + \dfrac{\ell_2}{\text{A}_2\text{Y}_\text{c}}} = \dfrac{\Delta\ell\text{A}}{\dfrac{\ell_1}{\text{Y}_\text{s}} + \dfrac{\ell_2}{\text{Y}_\text{c}}}\ (\because\ \text{A}_1 = \text{A}_2 = \text{A})$

$= \dfrac{1.4\times10^{-3}\times\pi\times(1.4\times10^{-3})^2}{\dfrac{3.2}{2\times10^{11}} + \dfrac{4.4}{1.1\times10^{11}}}$

$= 1.54\times10^{2} = 154\,\text{N}$

34. **(2)** Given,

Length of wire, L = 1 m

Extension in the length of wire, $l = 0.4\,\text{mm} = 0.4\times10^{-3}\,\text{m}$

Diameter of wire, d = 0.4 mm

Uncertainty in extension of the length, $\Delta l = \pm0.02$ mm

Uncertainty in diameter of wire, $\Delta\text{d} = \pm0.01$ mm

Young's modulus, $\text{Y} = \dfrac{\text{stress}}{\text{strain}} = \dfrac{\dfrac{\text{mg}}{\text{A}}}{\dfrac{l}{\text{L}}}$

$= \dfrac{\text{mg}\times\text{L}}{\text{A}\times l} = \dfrac{\text{mgL}}{\pi\left(\dfrac{\text{d}}{2}\right)^2 l} = \dfrac{4\text{mgL}}{\pi(\text{d})^2 l}$

$= \dfrac{4\times1\times10\times1}{\pi(0.4\times10^{-3})^2(0.4\times10^{-3})} = 0.199\times10^{+12}\,\text{N/m}^2$

$\dfrac{\Delta\text{Y}}{\text{Y}} = 2\dfrac{\Delta\text{d}}{\text{d}} + \dfrac{\Delta l}{l}$

$\Delta\text{Y} = \left(2\times\dfrac{0.01}{0.4} + \dfrac{0.02}{0.4}\right)\times0.199\times10^{+12}$

$= (0.05+0.05)\times0.199\times10^{+12} = 1.99\times10^{10}\,\text{Nm}^{-2}$

35. **(a)** Since breaking stress (Maximum lifting capacity) is the property of material so it will remain same.

$\text{breaking stress} = \dfrac{\text{Maximum lifting capacity}}{\text{Area of cross-section of rope}}$

$\Rightarrow \dfrac{10}{2.5\times10^{-4}} = \dfrac{25}{\text{A}}$

$\Rightarrow \text{A} = \dfrac{25\times2.5\times10^{-4}}{10} = 625\times10^{-6} = 6.25\times10^{-4}\,\text{m}^2$

36. **(25)** Given mass of rod m = 20 kg

Cross-section area, $\text{A} = 0.4\,\text{m}^2$

Length $\ell = 20$ m

Let extension is dy in length dx.

$\text{Y} = \dfrac{\text{stress}}{\text{strain}} = \dfrac{\dfrac{\text{T}}{\text{A}}}{\dfrac{\text{dy}}{\text{dx}}} = \dfrac{\text{T}}{\text{A}}\cdot\dfrac{\text{dx}}{\text{dy}} \Rightarrow \text{dy} = \dfrac{\text{Tdx}}{\text{AY}}$

Tension at a distance x from lower end $= \dfrac{\text{mg}}{\ell}\text{x}$

$\int_0^{\Delta\ell}\text{dy} = \int_0^{\ell}\dfrac{\text{mg}}{\ell}\text{x}\dfrac{\text{dx}}{\text{AY}}$

$\Delta\ell = \dfrac{\text{mg}}{\ell\text{AY}}\left[\dfrac{\text{x}^2}{2}\right]_0^{\ell}$

$\Delta\ell = \dfrac{\text{mg}\ell}{2\text{AY}} = \dfrac{20\times10\times20}{2\times0.4\times2\times10^{11}} = 25\times10^{-9}\,\text{m}$

Compare with $\text{x}\times10^{-9}$

we have, x = 25

37. **(5)** $\Delta\ell = \dfrac{\text{F}\ell}{\gamma\text{A}} \Rightarrow \Delta\ell \propto \dfrac{\text{F}\ell}{\text{r}^2}$

$\Rightarrow \Delta\ell_2 = \Delta\ell_1\left(\dfrac{\text{F}_2}{\text{F}_1}\right)\left(\dfrac{\ell_2}{\ell_1}\right)\left(\dfrac{\text{r}_1}{\text{r}_2}\right)^2$

$= 5\times4\times4\times\dfrac{1}{16} = 5\,\text{cm}$

38. **(c)** When length '*L*' is hanging from fixed support

$\Delta L = \dfrac{mgL}{YA} \Rightarrow \Delta L \propto m$

$\Rightarrow \dfrac{\Delta L_1}{\Delta L_2} = \dfrac{m_1}{m_2} = \dfrac{1}{2} \Rightarrow \dfrac{L_1 - L}{L_2 - L} = \dfrac{1}{2}$

$\Rightarrow 2L_1 - 2L = L_2 - L \Rightarrow 2L_1 - L_2 = \text{L}$

$\therefore\ \ L = 2L_1 - L_2$

39. **(b)** Given, Weight of uniform heavy rod. $W = 100$ N

Length of rod, $L = 20\,\text{cm} = 0.2\,\text{m}$

Consider an element *dx* at distance *x* from top.

Force acting on part below *dx*

$= \dfrac{(L-x)W}{L}$

Elongation for small element *dx*,

$dl = \left(\dfrac{L-x}{L}\right)\dfrac{Wdx}{AY}$

$\int_0^{\Delta l} dl = \int_0^{L}\left(\dfrac{L-x}{L}\right)W\dfrac{dx}{AY} \Rightarrow \Delta\ell = \dfrac{WL}{2AY} = 5\times10^{-9}\text{m}$

40. (b) Force exerted on each column $F = \dfrac{mg}{4}$

$$\because Y = \frac{F}{A} \bigg/ \frac{\Delta \ell}{\ell}$$

$$\therefore \quad \text{Strain} = \left(\frac{\Delta \ell}{\ell}\right) = \frac{F}{AY} = \frac{mg}{4AY}$$

$$= \frac{50 \times 10^3 \times 9.8}{4 \times \left[\pi \times 1^2 - \pi (.5)^2\right] \times 2 \times 10^{11}}$$

$$= 2.6 \times 10^{-7} \qquad \left[\because A = \pi r^2\right]$$

41. (40) Let m be mass that can be placed in the pan. For wire W_1

$$\text{Stress} = \frac{\text{Maximum weight}}{\text{Area}} = \frac{(m+30)g}{8 \times 10^{-7}}$$

$$\Rightarrow \quad m + 30 = 1.25 \times 10^9 \times 8 \times 10^{-7} \times 10^{-1}$$

$$\Rightarrow \quad m + 30 = 100 \Rightarrow m = 70 \text{ kg}$$

For wire W_2

$$\text{Stress} = \frac{(m+10)g}{4 \times 10^{-7}} = 1.25 \times 10^9$$

$$\Rightarrow \quad m + 10 = 50 \Rightarrow m = 40 \text{ kg}$$

Maximum mass that can be placed = 40 kg.

42. (c) Here, $m_1 = 3\text{kg}, m_2 = 5 \text{ kg}$

Tension in the metal wire.

$$T = \frac{2m_1 m_2 g}{m_1 + m_2} = \frac{2 \times 3 \times 5 \times 10}{8} = \frac{75}{2} \text{N}$$

$$\text{Stress} = \frac{T}{A} = \frac{T}{\pi r^2}$$

$$\Rightarrow \frac{24}{\pi} \times 10^2 = \frac{75}{2 \times \pi r^2} \Rightarrow r^2 = \frac{75}{2 \times 24 \times 100} = \frac{3}{8 \times 24}$$

$$\Rightarrow r = 0.125 \text{ m} = 12.5 \text{ cm}$$

43. (d) $T \propto \Delta l \quad [\because T = K\Delta l]$

$$\therefore \quad \frac{T_1}{T_2} = \frac{l_1 - l_0}{l_2 - l_0} \text{ or, } T_1 l_2 - T_1 l_0 = T_2 l_1 - T_2 l_0$$

$$\therefore \; l_0 = \frac{T_2 l_1 - T_1 l_2}{T_2 - T_1}$$

44. (2) As $\Delta l = \dfrac{F \cdot l}{Y \cdot \pi r^2} \Rightarrow \Delta l \propto \dfrac{l}{r^2}$

$$\Delta l_2 = \left(\frac{l_2}{l_1}\right)\left(\frac{r_1}{r_2}\right)^2 \Delta l_1 = (2)\left(\frac{1}{2}\right)^2 (0.04) \text{ m} = 2 \text{ cm.}$$

45. (4) $T = ml\omega^2$

$$\sigma = \frac{T}{A} = \frac{ml\omega^2}{A}$$

$$\frac{ml\omega^2}{A} \le 4.8 \times 10^7 \Rightarrow \omega^2 \le \frac{\left(4.8 \times 10^7\right)A}{ml}$$

$$\Rightarrow \omega^2 \le \frac{\left(4.8 \times 10^7\right)\left(10^{-6}\right)}{10 \times 0.3} = 16 \Rightarrow \omega_{max} = 4 \text{ rad/s}$$

46. (c) $\Delta \ell_{temp} = \Delta \ell_{force}$

or $L\alpha(\Delta T) = \dfrac{FL}{AY} \qquad \therefore \; \alpha = \dfrac{F}{AY\Delta T} = \dfrac{F}{\pi r^2 YT}$

Coefficient of volume expression

$$\gamma = 3\alpha = \frac{3F}{\pi r^2 YT}.$$

47. (b) $\text{Stress} = \dfrac{F}{A} = \dfrac{400 \times 4}{\pi d^2} = 379 \times 10^6 \text{ N/m}^2$

$$\Rightarrow d^2 = \frac{400 \times 4}{379 \times 10^6 \pi} \Rightarrow d = 1.16 \text{ mm}$$

48. (c) Given,

Radius of wire, $r = 2$ mm, Mass of the load $m = 4$ kg

$$\text{Stress} = \frac{F}{A} = \frac{mg}{\pi (r)^2} = \frac{4 \times 9.8}{\pi \times (2 \times 10^{-3})^2} = 3.1 \times 10^6 \text{ N/m}^2$$

49. (c) $\Delta \ell_1 = \Delta \ell_2$

$$\text{or} \quad \frac{Fl_1}{\pi r_1^2 Y_1} = \frac{Fl_2}{\pi r_2^2 Y_2} \text{ or } \frac{2}{R^2 \times 7} = \frac{1.5}{2^2 \times 4}$$

$$\therefore \; R = 1.7 \text{ mm}$$

50. (a) Young's modulus $Y = \dfrac{F}{A} / \dfrac{\Delta \ell}{\ell}$

$$Y = \frac{F\ell}{\pi r^2 \Delta \ell}$$

Given, radius r – 5mm, force F = 50πk N,

$$\frac{\Delta \ell}{\ell} = 0.01 \text{ mm} \quad \therefore \; Y = \frac{F}{\pi r^2} \frac{\ell}{\Delta \ell} = 2 \times 10^{14} \text{ N/m}^2.$$

51. (b) Tension in the wire, $T = \left(\dfrac{2mM}{m+M}\right)g$

$$\text{Stress} = \frac{\text{Force (Tension)}}{\text{Area}} = \frac{2mM}{A(m+M)} g$$

$$= \frac{2(m \times 2m)g}{A(m+2m)} = \frac{4m^2}{3mA} g = \frac{4mg}{3A} \; (\because M = 2\,m \text{ given})$$

52. (d) $Y_c \times (\Delta L_c / L_c) = Y_s \times (\Delta L_s / L_s)$

$$\Rightarrow \quad 1 \times 10^{11} \times \left(\frac{1 \times 10^{-3}}{1}\right) = 2 \times 10^{11} \times \left(\frac{\Delta L_s}{0.5}\right)$$

$$\therefore \quad \Delta L_s = \frac{0.5 \times 10^{-3}}{2} = 0.25 \text{ mm}$$

Therefore, total extension of the composite wire

$$= \Delta L_c + \Delta L_s = 1 \text{ mm} + 0.25 \text{ mm} = 1.25 \text{ mm}$$

53. (c) According to questions,

$$\frac{\ell_s}{\ell_b} = a, \frac{r_s}{r_b} = b, \frac{Y_s}{Y_b} = c, \frac{\Delta \ell s}{\Delta \ell_b} = ?$$

$$\text{As, } Y = \frac{F\ell}{A\Delta \ell} \Rightarrow \Delta \ell = \frac{F\ell}{AY}$$

$$\Delta \ell_s = \frac{3mg\ell_s}{\pi r_s^2 . Y_s} \qquad [\because F_s = (M + 2M)g]$$

$$\Delta\ell_b = \frac{2Mg\ell_b}{\pi r_b^2 . Y_b} \qquad [\because F_b = 2Mg]$$

$$\therefore \quad \frac{\Delta\ell_s}{\Delta\ell_b} = \frac{\dfrac{3Mg\ell_s}{\pi r_s^2 . Y_s}}{\dfrac{2Mg.\ell_b}{\pi r_b^2 . Y_b}} = \frac{3a}{2b^2c}$$

54. **(c)** Breaking force $\propto$ area of cross section of wire
Load hold by wire is independent of length of the wire.

55. **(a)** Let m = slope of the given curve

$$F = \frac{\Delta l}{l} YA \Rightarrow F = m\Delta l, m = \frac{YA}{l} \Rightarrow m \propto A$$

As slope of OA is minimum, so its area is also minimum.

56. **(c)** $\Delta l_1 = \Delta l_2$

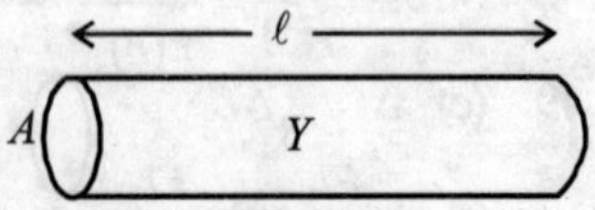

Wire (1)

$$\frac{F_1 l_1}{Y_1 A_1} = \frac{F_2 l_2}{Y_2 A_2}$$

$$\frac{F_1 V}{A_1^2} = \frac{F_2 V}{A_2^2} \quad \left[\because l = \frac{V}{A}\right]$$

$$F_2 = F_1 \times \left(\frac{A_2}{A_1}\right)^2 = F_1 \times 9 = 9F$$

57. **(a)** Case (i)

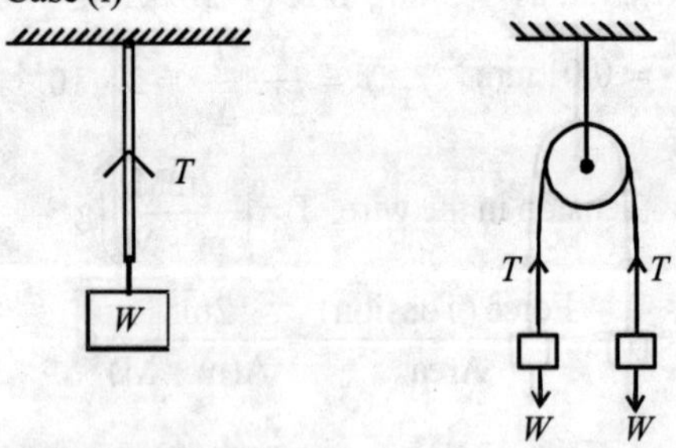

At equilibrium, $T = W$

Young's modules, $Y = \dfrac{W/A}{\ell/L}$ .....(1)

Elongation, $\ell = \dfrac{W}{A} \times \dfrac{L}{Y}$

Case (ii) At equilibrium $T = W$

$$\Delta l_{right} = \frac{W l/2}{YA}$$

$$\Delta l_{left} = \frac{W l/2}{YA}$$

$$\therefore \quad \Delta l_{net} = \Delta l_{right} + \Delta l_{left} = \frac{Wl}{YA}$$

So we have same elongation.

58. **(4)** Bulk modulus $B = \dfrac{-dP}{(dV/V)} \Rightarrow V = -B\left(\dfrac{dV}{dP}\right)$

Given, initial pressure of liquid $(P_i) = 1$ atm
Final pressure of liquid $(P_f) = 5$ atm
Change in pressure $(dP) = P_f - P_i = 4$ atm
$= 4 \times 10^5$ Pa
Change in volume $(dV) = -0.8$ cm^3
Bulk modulus $(B) = 2 \times 10^9$ Pa

Volume (V) = ?

$$V = -B\left(\frac{dV}{dP}\right) = -2\times10^9 \times \frac{(-0.8\times10^{-6})}{4\times10^5}$$

$$= 4 \times 10^{-3}\ m^3 = 4 \text{ litre}$$

59. **(a)** Given : $r = 4$ cm, $A = \pi\times16\times10^{-4}\ m^2$

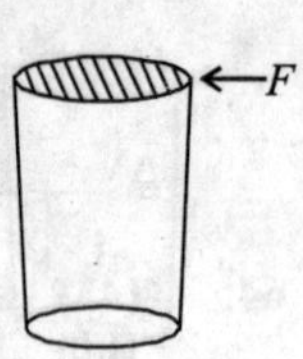

$$\text{Shear moduli} = \frac{\sigma_{shear}}{\theta}$$

$$\sigma_{shear} = \frac{F}{A} = \frac{F}{\pi r^2}$$

$$\Rightarrow 10^{10} = \frac{10^5}{\pi\times16\times10^{-4}} \times \frac{1}{\theta}$$

$$\Rightarrow \theta = \frac{1}{160\pi} \text{ Radian.}$$

60. **(1)** Deformation angle, $2\theta_1 = \theta_2$ (Given)
$n_1 = 4 \times 10^9$ N/m^2 (Given)

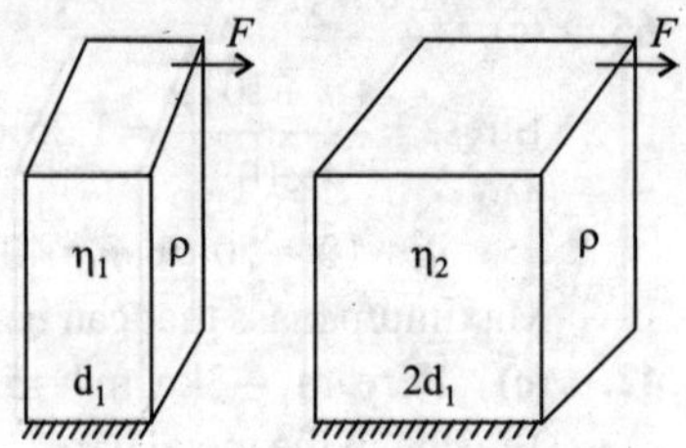

$$\Rightarrow 2\frac{\sigma_1}{\eta_1} = \frac{\sigma_2}{\eta_2}$$

$$\Rightarrow \sigma = \frac{F}{A}$$

$$\Rightarrow 2\left(\frac{F}{ld_1\eta_1}\right) = \frac{F}{ld_2\eta_2}$$

$$\therefore d_2 = 2d_1$$

$$\Rightarrow \eta_2 = \frac{\eta_1}{4} = 1\times10^9 \Rightarrow x = 1.$$

61. **(25)** Elastic potential energy density $u = \dfrac{1}{2} \times y \times \left(\dfrac{\Delta\ell}{\ell}\right)^2$

$$\mu = -\frac{\left(\dfrac{\Delta r}{r}\right)}{\left(\dfrac{\Delta\ell}{\ell}\right)} \Rightarrow \frac{\Delta\ell}{\ell} = \frac{1}{\mu} \times \left(\frac{\Delta r}{r}\right) = \frac{1}{0.2} \times \left(10^{-3}\right)$$

or, $\dfrac{\Delta\ell}{\ell} = 5\times10^{-3}$

$$\therefore u = \frac{1}{2} \times y \times \left(\frac{\Delta\ell}{\ell}\right)^2 = \frac{1}{2} \times 2\times10^{11} \times \left[5\times10^{-3}\right]^2 = 25$$

62. **(d)** Given
Bulk modulus of water, $B = 2 \times 10^9$ Nm^{-2}
Density of water, $\rho = 10^3$ kg m^{-3}
The bulk modulus of water at depth h is

$$B = \frac{\Delta P}{\left(\dfrac{\Delta v}{v}\right)} = \frac{\rho gh}{\left(\dfrac{\Delta v}{v}\right)}$$

$\therefore$ Fractional compression of water is

$$\left(\frac{\Delta v}{v} \times 100\right)\% = \left(\frac{\rho gh}{B} \times 100\right)\%$$

$$= \frac{1000\times10\times2.5\times10^3}{2\times10^9} \times 100\% = 1.25\%$$

63. (50) The expression of Bulk Modulus is given as

$$B = \frac{\Delta P}{\frac{\Delta V}{V}} \Rightarrow \Delta V = \left(\frac{\Delta P}{B}\right)V$$

$$\Delta V = \frac{7 \times 10^6}{1.4 \times 10^{11}} \times (10 \times 10^{-2})^3 = 50\,\text{mm}^3$$

64. (43) Given,
Bulk modules of water, $B = 2.15 \times 10^9\,\text{Nm}^{-2}$
Since bulk modulus of water sample is

$$B = \frac{-\Delta P}{\left(\frac{\Delta V}{V}\right)} \Rightarrow 2.15 \times 10^9 = \frac{-\Delta P}{-\left(\frac{0.2}{100}\right)}$$

$$\therefore \Delta P = 2.15 \times 10^9 \times 2 \times 10^{-3} = 43 \times 10^5\,\text{N/m}^2$$

65. (c) (A) stress $= \frac{F_{restoring}}{A}$

If A = 1 ⇒ Stress = $F_{restoring}$
(A) - (III)

(B) 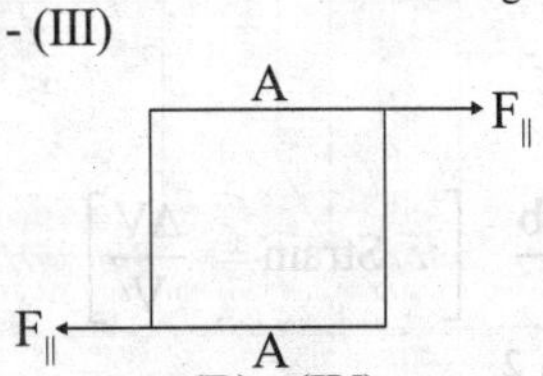
$\Rightarrow \frac{F_{\parallel}}{A}$ = Shear stress
(B) - (IV)

(C) 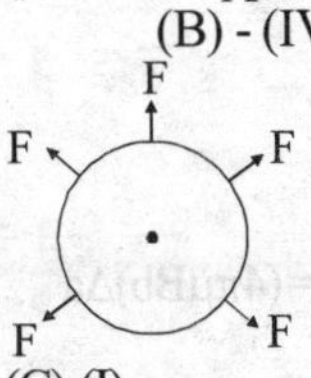
$\Rightarrow \frac{F}{A}$ = Volumetric stress
(C)-(I)

(D) 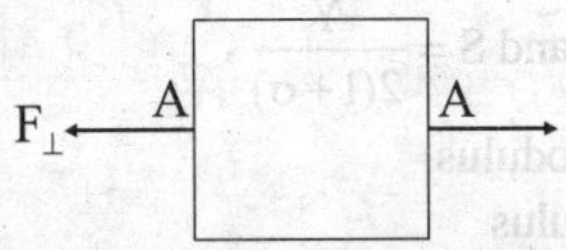
$\Rightarrow \frac{F_{\perp}}{A}$
= Longitudinal stress

66. (18) Given,
Bulk modulus of rubber
$B = 9 \times 10^8\,\text{Nm}^{-2}$
Density of sea water, $\rho = 10^3\,\text{kgm}^{-3}$
Bulk modulus is given by

$$B = \frac{-\Delta P}{\frac{\Delta V}{V}} \Rightarrow \Delta P = -B\frac{\Delta V}{V}$$

$$\Rightarrow \rho g h = -B\frac{\Delta V}{V} = -9 \times 10^8\left(-\frac{0.02}{100}\right)$$

$$\Rightarrow 10^3 \times 10 \times h = 18 \times 10^4 \Rightarrow h = 18\text{m}$$

67. (2) Bulk modulus, $B = -\frac{\Delta P}{\left(\frac{\Delta V}{V}\right)}$

$$\Rightarrow -\left(\frac{\Delta V}{V}\right) = \frac{\rho g h}{B} = \frac{1000 \times 10 \times 4000}{2 \times 10^9} \quad [\because P = \rho g h]$$

$$\text{or, } -\left(\frac{\Delta V}{V}\right) = 2 \times 10^{-2} = \alpha \times 10^{-2} \text{ (given)} \therefore \alpha = 2$$

68. (1) $B_{water} = \frac{-\Delta P}{\left(\frac{\Delta V_{water}}{V}\right)} = \frac{-\Delta P}{\frac{0.01}{100}}$

$$B_{liquid} = \frac{-\Delta P}{\left(\frac{\Delta V_{liquid}}{V}\right)} = \frac{-\Delta P}{\frac{0.03}{100}} \Rightarrow \frac{B_{water}}{B_{liquid}} = 3 = \frac{3}{x} \therefore x = 1$$

69. (48) $\eta = \frac{F}{A \tan\theta} = \frac{F}{A\left(\frac{x}{\ell}\right)}$

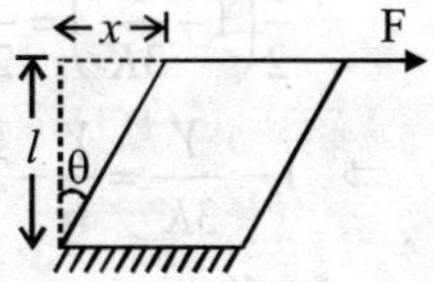

$$\frac{F}{A} = \eta\frac{x}{\ell} \Rightarrow \frac{F\ell}{A\eta} = x$$

$$\Rightarrow x = \frac{18 \times 10^4 \times 60 \times 10^{-2}}{60 \times 10^{-2} \times 15 \times 10^{-2} \times 25 \times 10^9}$$

$$\therefore \Delta P = -3 \times 10^{10} \times \left(-\frac{2}{10}\right)$$

$$= 6 \times 10^8\,\text{Nm}^{-2}$$

70. (c) Bulk modulus, $B = -\frac{\Delta P}{\frac{\Delta V}{V}} \Rightarrow \Delta P = -B\frac{\Delta V}{V}$

$$\therefore \quad \Delta P| = -3 \times 10^{10} \times \left(\frac{-2}{100}\right) = 6 \times 10^8\,\text{Nm}^{-2}$$

71. (500) Bulk modulus, $B = -\frac{\Delta P}{\left(\frac{\Delta V}{V}\right)}$

$$\Rightarrow B = -\frac{V}{\Delta V}(\rho g h) \Rightarrow h = -\frac{B\left(\frac{\Delta V}{V}\right)}{\rho g}$$

$$\therefore \quad h = \frac{9.8 \times 10^8 \times 0.5}{100 \times 10^3 \times 9.8} = 500\,\text{m}.$$

72. (d) $\frac{\text{Hydraulic stress}}{\text{Hydraulic strain}} = \frac{\Delta P}{\left(\frac{\Delta V}{V}\right)} = \frac{\rho g h}{\left(\frac{\Delta V}{V}\right)}$

$$= \frac{10^3 \times 9.8 \times 2 \times 10^3}{\left(\frac{1.36}{100}\right)} = 1.44 \times 10^9\,\text{N/m}^2$$

73. (a) Bulk modulus (K) is given by $K = \frac{P}{\left(-\frac{\Delta V}{V}\right)}$

Density, $\rho = \frac{m}{V}$

So, $\frac{\Delta\rho}{\rho} = \frac{-\Delta V}{V} \therefore K = \frac{P}{\left(\frac{\Delta\rho}{\rho}\right)} \Rightarrow \frac{\Delta\rho}{\rho} = \frac{P}{K} \Rightarrow \Delta\rho = \frac{\rho P}{K}$

74. **(d)** We know that

$$Y = 3K(1-2\sigma)$$

Here,

σ = Poisson's ratio

$$\sigma = \frac{1}{2}\left(1-\frac{Y}{3K}\right) \quad ...(i)$$

Also,

$$Y = 2\eta(1+\sigma)$$

$$\sigma = \frac{Y}{2\eta} - 1 \quad ...(ii)$$

From equation (i) and equation (ii),

$$\frac{1}{2}\left(1-\frac{Y}{3K}\right) = \frac{Y}{2\eta} - 1$$

$$\Rightarrow 1 - \frac{Y}{3K} = \frac{Y}{\eta} - 2 \Rightarrow \frac{Y}{3K} = 3 - \frac{Y}{\eta} \Rightarrow \frac{Y}{3K} = \frac{3\eta - Y}{\eta}$$

$$\Rightarrow \frac{\eta Y}{3K} = 3\eta - Y \Rightarrow K = \frac{\eta Y}{9\eta - 3Y}$$

75. **(a)** If force F acts along the length L of the wire of cross-section A, then energy stored in unit volume of wire is given by

$$\text{Energy density} = \frac{1}{2} \times \text{stress} \times \text{strain}$$

$$= \frac{1}{2} \times \frac{F}{A} \times \frac{F}{AY} \quad \left(\because \text{stress} = \frac{F}{A} \text{ and strain} = \frac{F}{AY}\right)$$

$$= \frac{1}{2}\frac{F^2}{A^2Y} = \frac{1}{2}\frac{F^2 \times 16}{(\pi d^2)^2 Y} = \frac{1}{2}\frac{F^2 \times 16}{\pi d^4 Y}$$

If u_1 and u_2 are the densities of two wires, then

$$\frac{u_1}{u_2} = \left(\frac{d_2}{d_1}\right)^4 \Rightarrow \frac{d_1}{d_2} = (4)^{1/4} \Rightarrow \frac{d_1}{d_2} = \sqrt{2}:1$$

76. **(a)** When a catapult is stretched up to length l, then the stored energy in it = Δk. E $\Rightarrow \frac{1}{2}\cdot\left(\frac{YA}{L}\right)(\Delta l)^2 = \frac{1}{2}mv^2$

$$\Rightarrow Y = \frac{mv^2 L}{A(\Delta l)^2}$$

$m = 0.02$ kg, $v = 20\ ms^{-1}$,

$L = 0.42$ m, $A = (\pi d^2)/(4)$

$d = 6 \times 10^{-3}$ m, $\Delta l = 0.2$ m

$$Y = \frac{0.02 \times 400 \times 0.42 \times 4}{\pi \times 36 \times 10^{-6} \times 0.04} = 2.97 \times 10^6\ N/m^2$$

So, order is 10^6.

77. **(c)** Bulk modulus, $K = \frac{\text{volumetric stress}}{\text{volumetric strain}}$

$$K = \frac{mg}{a\left(\frac{dV}{V}\right)} \quad \left[\because \text{Stress} = \frac{F}{A} = \frac{mg}{a}\right]$$

$$\Rightarrow \frac{dV}{V} = \frac{mg}{Ka} \quad ....(i)$$

volume of sphere, $V = \frac{4}{3}\pi r^3$

Fractional change in volume $\frac{dV}{V} = \frac{3dr}{r}$ ...(ii)

Using eq. (i) & (ii) $\frac{3dr}{r} = \frac{mg}{Ka}$

$\therefore \frac{dr}{r} = \frac{mg}{3Ka}$ (fractional decrement in radius)

78. **(d)** $\text{Stress} = \frac{\text{Normal force}}{\text{Area}} = \frac{N}{A} = \frac{N}{(2\pi a)b}$

$V = \pi a^2 b$

$\Delta V = 2\pi ab\,\Delta a$

Stress = B×strain

$$\frac{N}{(2\pi a)b} = B\frac{2\pi a\Delta a \times b}{\pi a^2 b} \quad \left[\because \text{Strain} = \frac{\Delta V}{V}\right]$$

$$\Rightarrow N = B\frac{(2\pi a)^2 \Delta ab^2}{\pi a^2 b}$$

Force needed to push the cork.

$$f = \mu N = \frac{\mu \times B \times 4\pi^2 a^2 b^2 \Delta a}{\pi a^2 b} = (4\pi\mu Bb)\Delta a$$

79. **(c)** We have

$$B = \frac{Y}{3(1-2\sigma)} \text{ and } S = \frac{Y}{2(1+\sigma)},$$

Y = Young's modulus

S = Shear modulus

B = Bulk modulus

σ = Poisson's ratio

For metal like aluminium and copper, the Poisson's ratio is nearly 0.35.

So, $B = \frac{Y}{3(1-2\sigma)} = \frac{Y}{3(1-2\times 0.35)} = \frac{Y}{0.9} \Rightarrow B > Y$ ... (i)

and $S = \frac{Y}{2(1+\sigma)} = \frac{Y}{2(1+0.35)} = \frac{Y}{2.7} \Rightarrow Y > S$ ... (ii)

From (i) & (ii), we get

S < Y < B

9 Mechanical Properties of Fluids

1. **(d)**

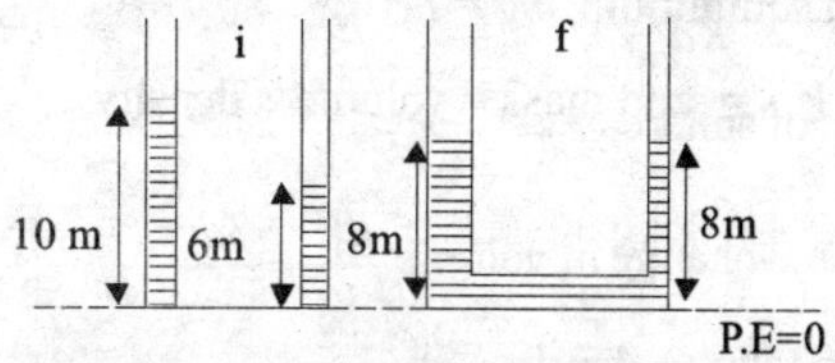

Initial potential energy is

$U_i = (\rho A \times 10)g \times 5 + (\rho A6)g \times 3 = 68\rho Ag$

Final potential energy is

$U_f = (\rho A \times 16)g \times 4 = (\rho Ag) \times 64$

$\therefore \quad WD = \Delta U = 4 \times \rho Ag = 4 \times 1000 \times 2 \times 10 = 8 \times 10^4$ J

2. **(150)**

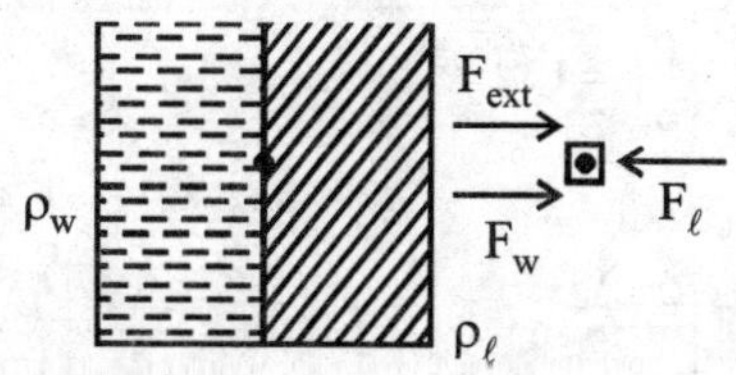

In equilibrium

$F_{ext} + F_w = F_\ell$

$\Rightarrow \quad F_{ext} = F_\ell - F_w$

$= (P_0 + \rho_\ell gh)A - (P_0 + \rho_w gh)A$

$= (\rho_\ell - \rho_w)ghA$

$= (1500 - 1000) \times 10 \times 3 \times (100 \times 10^{-4}) = 150$ m

3. **(5)**

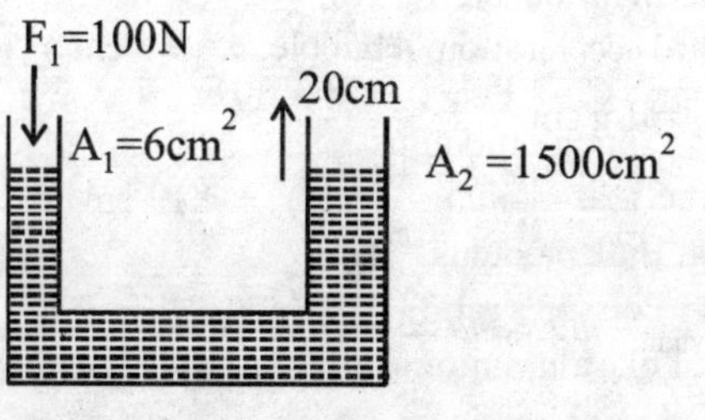

By principle of hydraulic machine,

$$\frac{F_1}{A_1} = \frac{F_2}{A_2} \Rightarrow \frac{100}{6} = \frac{F_2}{1500}$$

$\therefore \quad F_2 = 50 \times 500 = 25 \times 10^3$ N

$\therefore$ Work done, $W = F_2\ell_2 = 25 \times 10^3 \times \dfrac{20}{100}$

$= 5 \times 10^3 = 5$ kJ

4. **(d)** For floating solid cube,

$Mg = F_B$

$(400 \times 10^{-3}) = 10^3 \times V_i$

$V_i = 400 \times 10^{-6}\ m^3$

Volume of cube outside water can be calculated as

$(Vol.)_{outside} = (10 \times 10^{-2})^3 - 400 \times 10^{-6}$

$= 600 \times 10^{-6}\ m^3 = 600\ cm^3$

5. **(a)** From Bernoulli equation;

$$P + \frac{1}{2}\rho V^2 + \rho gh = \text{constant}$$

$$P_1 + \frac{1}{2}\rho V_1^2 = P_2 + \frac{1}{2}\rho v_2^2$$

$V_1 = 0$ (water is initially at rest)

$$P_1 + \frac{1}{2} \times \rho \times 0^2 = P_2 + \frac{1}{2}\rho v^2$$

$$v = \sqrt{\frac{2}{\rho}(P_1 - P_2)}$$

$$\therefore V \propto \sqrt{(P_1 - P_2)}$$

6. **(a)** Weight of the sphere,

$$W = mg = \frac{4}{3}\pi\left(\frac{D^3 - d^3}{8}\right)\sigma g$$

$$\text{Buoyant force } (F_b) = 1 \times \frac{4}{3}\pi\left(\frac{D^3}{8}\right).g$$

For floating

$W = F_b$

$$\frac{4}{3}\pi\frac{(D^3 - d^3)6}{8} = \frac{4}{3}\pi\frac{D^3}{8}$$

$\Rightarrow 6D^3 - 6d^3 = D^3$

$\Rightarrow (D^3 - d^3)\sigma = D^3$

$$\Rightarrow 1 - \frac{d^3}{D^3} = \frac{1}{\sigma} \Rightarrow 1 - \frac{1}{\sigma} = \left(\frac{d}{D}\right)^3$$

$$\Rightarrow \quad \left(\frac{\sigma}{\sigma - 1}\right)^{\frac{1}{3}} = \left(\frac{D}{d}\right)$$

7. **(d)** Let V_1 be, the volume immersed in water.

V_2 be the volume immersed in oil.

In equilibrium condition,

$v_1\rho_w g + v_2\rho_0 g = (v_1 + v_2)\rho_c g$

$$v_1 + \frac{v_2\rho_0}{\rho_w} = (v_1 + v_2)\frac{\rho_c}{\rho_w}$$

$\Rightarrow v_1 + 0.8v_2 = 0.9v_1 + 0.9v_2 \Rightarrow 0.1v_1 = 0.1v_2$

$\Rightarrow v_1 : v_2 = 1 : 1$

8. **(1000)** As, $P_1 = P_2$

$$\therefore \frac{F_1}{A_1} = \frac{F_2}{A_2} \Rightarrow \frac{F_1}{\pi(7)^2} = \frac{10}{\pi \times (0.7)^2}$$

$\therefore F_1 = 1000\text{N}$

9. **(177)** Here, pressure $(P_0 + \rho gh) = \frac{F}{A}$

$$\Rightarrow F = P_0 A + \rho_m ghA$$

$$= 10^5 \times \frac{22}{7} \times \left(2 \times 10^{-2}\right)^2$$

$$+1.36 \times 10^4 \times 10 \times \left(30 \times 10^{-2}\right)\left(\frac{22}{7} \times \left(2 \times 10^{-2}\right)^2\right)$$

$$F = 51.29 + 125.71 = 177\text{N}$$

10. **(c)** Area of cross section, $A = 250\text{ cm}^2 = 250 \times 10^{-4}\text{ m}^2$

$$\text{maximum pressure} = \frac{\text{Force}}{\text{area of cross section}} = \frac{mg}{A}$$

$$= \frac{5000g}{250 \times 10^{-4}} = \frac{20 \times g}{10^{-4}} = 2 \times 10^6\text{ Pa}$$

11. **(c)** Reason is the statement of Pascal's law.
When we press any part of the toothpaste, the pressure applied get transmitted equally to all part of toothpaste and it comes out from other end.

12. **(d)** F = Momentum transferred by water per sec $= \rho aV \times V$

$F = \rho Av^2 = 10^3 \times 10 \times 10^{-4} \times 20 \times 20$

$F = 400$ N

13. **(b)** Let 'x' be the rise (or fall) in water.
Final volume of both vessel will be same

So, $16 \times 10^{-4} \times (150 - x) = 16 \times 10^{-4} \times (100 + x)$

$\Rightarrow 50 = 2x$

$\Rightarrow x = 25$ cm

Work done = Potential Energy of extra water that enters in cylindrical vessel $= m_{extra} \times g \times x$

$= \rho_w A_w x \times g \times x$

$= \rho_w A_w g \times x^2$

$= 10^3 \times 16 \times 10^{-4} \times 10 \times (25 \times 10^{-2})^2$

= 1 joule

14. **(25600)** Using Pascal's law,

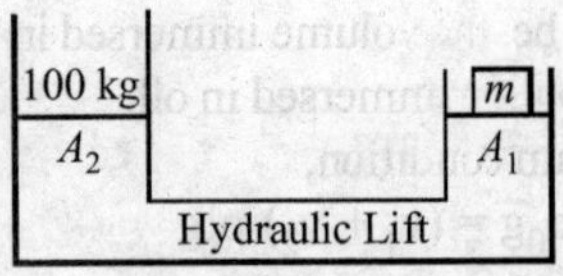

$$\frac{100 \times g}{A_2} = \frac{mg}{A_1} (A_1 < A_2) \quad \text{...(i)}$$

Let m mass can lift M_0 in second case then

$$\frac{M_0 g}{16A_2} = \frac{mg}{\frac{A_1}{16}} \quad \left(\because A = \frac{\pi d^2}{4}\right) \quad \text{...(ii)}$$

From equation (i) and (ii), we get

$$\frac{M_0}{16 \times 100} = 16 \Rightarrow M_0 = 25600\text{ kg}$$

15. **(a)** In equilibrium, $mg = F_e$

$F_B = V\rho_0 g$ and mass = volume × density

$$\frac{4}{3}\pi(R^3 - r^3)\rho_0 g = \frac{4}{3}\pi R^3 \rho_w g$$

Given, relative density, $\frac{\rho_0}{\rho_w} = \frac{27}{8}$

$$\left[1 - \left(\frac{r}{R}\right)^3\right]\rho_0 = \rho_w$$

$$\left[1 - \left(\frac{r}{R}\right)^3\right]\frac{\rho_0}{\rho_w} = 1$$

$$\left[1 - \left(\frac{r}{R}\right)^3\right] = \frac{8}{27}$$

$$\frac{19}{27} = \left(\frac{r}{R}\right)^3$$

$$r = 0.89R \Rightarrow r = \frac{8}{9}R$$

16. **(c)** Given :
Radius of air bubble = 1 cm,
Upward acceleration of bubble, $a = 9.8\text{ cm/s}^2$,

$\rho_{water} = 1\text{ g cm}^{-3}$

Volume $V = \frac{4\pi}{3}r^3 = \frac{4\pi}{3} \times (1)^3 = 4.19\text{ cm}^3$

$$F_{buoyant} - mg = ma \Rightarrow m = \frac{F_{buoyant}}{g + a}$$

$$\therefore m = \frac{(V\rho_\omega g)}{g + a} = \frac{V\rho_\omega}{1 + \frac{a}{g}} = \frac{(4.19) \times 1}{1 + \frac{9.8}{980}} = \frac{4.19}{1.01} = 4.15\text{g}$$

17. **(d)** Initial potential energy,

$$U_1 = (\rho Sx_1)g \cdot \frac{x_1}{2} + (\rho Sx_2)g \cdot \frac{x_2}{2}$$

Final potential energy,

$$U_f = (\rho Sx_f)g \cdot \frac{x_f}{2} \times 2$$

By volume conservation,

$$Sx_1 + Sx_2 = S(2x_f) \Rightarrow x_f = \frac{x_1 + x_2}{2}$$

When valve is opened loss in potentail energy occur till water level become same.

$\Delta U = U_i - U_f$

$$\Delta U = \rho Sg\left[\left(\frac{x_1^2}{2}+\frac{x_2^2}{2}\right)-x_f^2\right] = \rho Sg\left[\frac{x_1^2}{2}+\frac{x_2^2}{2}-\left(\frac{x_1+x_2}{2}\right)^2\right]$$

$$= \frac{\rho Sg}{2}\left[\frac{x_1^2}{2}+\frac{x_2^2}{2}-x_1x_2\right] = \frac{\rho Sg}{4}(x_1-x_2)^2$$

Here, $\rho = d$. So, $\Delta U = \frac{dSg}{4}(x_1-x_2)^2$

18. **(c)** When cylinder is floating in water at 0°C

Net thrust $= A(h_2-h_1)\rho_{0^\circ C}\,g = \text{weight (w)}$

$= A(100-20)\rho_{0^\circ C}\,g = \text{w}$ ...(i)

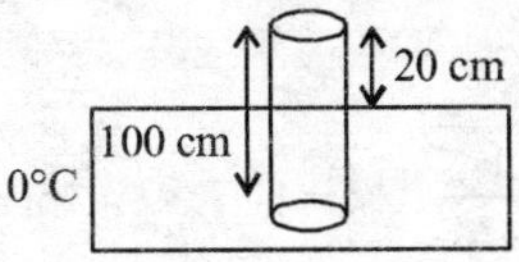

When cylinder is floating in water at 4° C

Net thrust $= A(h_2-h_1)\rho_{4^\circ C}\,g = A(100-21)\rho_{4^\circ C}\,g = \text{w}$ ...(ii)

Dividing (i) by (ii), we get

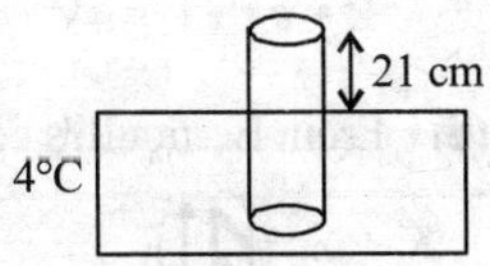

$\therefore \frac{\rho_{4^\circ C}}{\rho_{0^\circ C}} = \frac{80}{79} = 1.01$

19. **(c)** For minimum density of liquid, solid sphere has to float (completely immersed) in the liquid.

$mg = F_B$ (also $V_{\text{immersed}} = V_{\text{total}}$)

Now, $m = \int \rho dV$ and, $F_B = \frac{4}{3}\pi R^3 \rho_\ell g$

So, $\int_0^R \rho_0 4\pi\left(1-\frac{r^2}{R^2}\right).\, r^2 dr = \frac{4}{3}\pi R^3 \rho_\ell$

$\Rightarrow 4\pi\rho_0\left[\frac{r^3}{3}-\frac{r^5}{5R^2}\right]_0^R = \frac{4}{3}\pi R^3 \rho_\ell$

$\Rightarrow \frac{4\pi\rho_0 R^3}{3}\times\frac{2}{5} = \frac{4}{3}\pi R^3\rho_\ell \qquad \therefore \rho_\ell = \frac{2\rho_0}{5}$

20. **(d)** Let P_1, P_2 and P_3 be the pressure at points M, N and O respectively.

Pressure is given by $P = \rho gh$

Now, $P_1 = 0\ (\because h = 0)$

$P_2 = \rho g(5)$

$P_3 = \rho g(15) = 15\rho g$

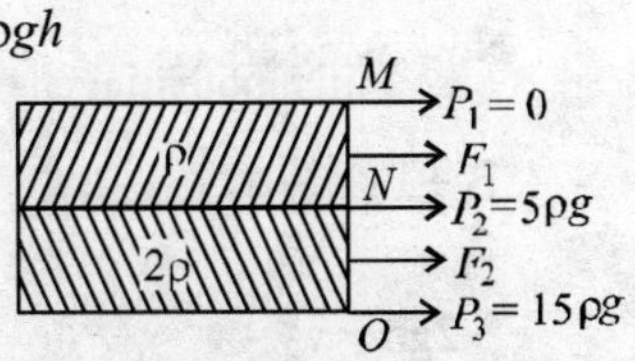

Force on upper part, $F_1 = \frac{(P_1+P_2)}{2}A$

Force on lower part, $F_2 = \frac{(P_2+P_3)}{2}A$

$\therefore \frac{F_1}{F_2} = \frac{5\rho g}{20\rho g} = \frac{5}{20} = \frac{1}{4}$

21. **(b)** When a body floats then the weight of the body = upthrust

$\therefore (50)^3 \times \frac{30}{100}\times(1)\times g = M_{\text{cube}}g$ ...(i)

Let m mass should be placed, then

$(50)^3\times(1)\times g = (M_{\text{cube}}+m)g$ ...(ii)

Subtracting equation (i) from equation (ii), we get

$\Rightarrow mg = (50)^3\times g(1-0.3) = 125\times 0.7\times 10^3 g \Rightarrow m = 87.5\,\text{kg}$

22. **(a)** $P_1 = P_0 + \rho g d_1$

$P_2 = P_0 + \rho g d_2$

$\Delta P = P_2 - P_1 = \rho g \Delta d$

$3.03\times 10^6 = 10^3\times 10\times\Delta d \Rightarrow \Delta d \simeq 300\,\text{m}$

23. **(c)** $Mg = \left(\frac{4V}{5}\right)\rho_\omega g$ or $\left(\frac{M}{V}\right) = \frac{4\rho_\omega}{5}$ or $\rho = \frac{4\rho_\omega}{5}$

When block floats fully in water and oil, then

$Mg = F_{b_1} + F_{b_2}$

$(\rho V)g = \left(\frac{V}{2}\right)\rho_{\text{oil}}\,g + \frac{V}{2}\rho_\omega g$ or $\rho_{\text{oil}} = \frac{3}{5}\rho_\omega = 0.6\rho_\omega$

24. **(a)** Using $\frac{F}{A} = Y\cdot\frac{\Delta\ell}{\ell}$

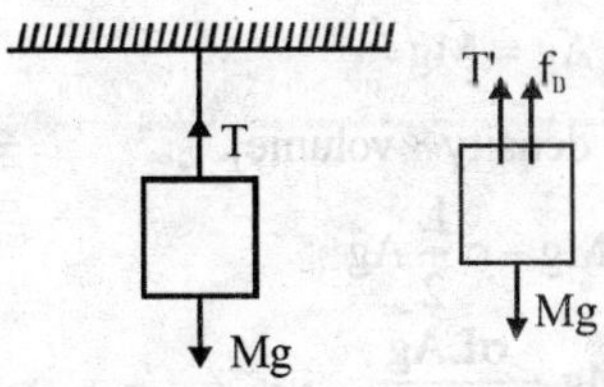

$\Rightarrow \Delta\ell \propto F$ ...(i)

$T = Mg$

$T' = Mg - f_B = Mg - \frac{M}{\rho_b}\cdot\rho_\ell\cdot g = \left(1-\frac{\rho_\ell}{\rho_b}\right)Mg = \left(1-\frac{2}{8}\right)Mg$

$T' = \frac{3}{4}Mg$

From eqn (i)

$\frac{\Delta\ell'}{\Delta\ell} = \frac{T'}{T} = \frac{3}{4}$ [Given: $\Delta\ell = 4$ mm]

$\therefore \Delta\ell' = \frac{3}{4}\cdot\Delta\ell = \frac{3}{4}\times 4 = 3\,\text{mm}$

25. **(d)** $V = kt$ or, $\frac{4}{3}\pi r^3 = kt \Rightarrow r = kt^{\frac{1}{3}}$

$P = P_0 + \frac{4T}{k\,t^{1/3}} \Rightarrow P = P_0 + c\left(\frac{1}{t^{1/3}}\right)$,

where $c = \frac{4T}{k}$ = constant

26. **(b)** Mass per unit time of the liquid $= \rho av$

Momentum per second carried by liquid $= \rho av \times v$

Net force due to bounced back liquid,

$F_1 = 2\times\left[\frac{1}{4}\rho av^2\right]$

Net force due to stopped liquid, $F_2 = \frac{1}{4}\rho av^2$

Total force,

$$F = F_1 + F_2 = \frac{1}{2}\rho a v^2 + \frac{1}{4}\rho a v^2 = \frac{3}{4}\rho a v^2$$

Net pressure $= \frac{3}{4}\rho v^2$

27. **(b)** Pressure at interface A must be same from both the sides to be in equilibrium.

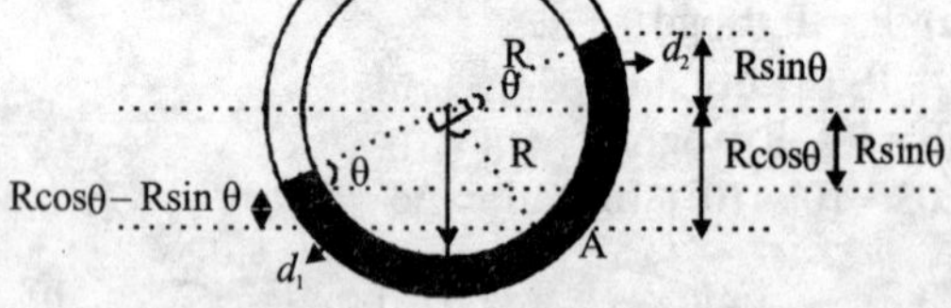

$$\therefore \quad (R\cos\theta + R\sin\theta)\rho_2 g = (R\cos\theta - R\sin\theta)\rho_1 g$$

$$\Rightarrow \frac{\rho_1}{\rho_2} = \frac{\cos\theta + \sin\theta}{\cos\theta - \sin\theta} = \frac{1+\tan\theta}{1-\tan\theta}$$

$$\Rightarrow \rho_1 - \rho_1 \tan\theta = \rho_2 + \rho_2 \tan\theta$$

$$\Rightarrow (\rho_1 + \rho_2)\tan\theta = \rho_1 - \rho_2$$

$$\therefore \quad \theta = \tan^{-1}\left(\frac{\rho_1 - \rho_2}{\rho_1 + \rho_2}\right)$$

28. **(c)** From figure, $kx_0 + F_B = Mg$

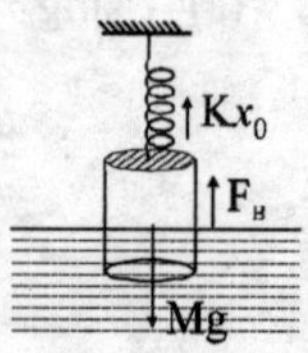

$$kx_0 + \sigma\frac{L}{2}Ag = Mg$$

[$\because$ mass = density × volume]

$$\Rightarrow kx_0 = Mg - \sigma\frac{L}{2}Ag$$

$$\Rightarrow x_0 = \frac{Mg - \frac{\sigma LAg}{2}}{k} = \frac{Mg}{k}\left(1 - \frac{LA\sigma}{2M}\right)$$

Hence, extension of the spring when it is in equilibrium is,

$$x_0 = \frac{Mg}{k}\left(1 - \frac{LA\sigma}{2M}\right)$$

29. **(b)** Oil will float on water so, (b) or (d) is the correct option. But density of ball is more than that of oil,, hence it will sink in oil.

30. **(d)** Given, $A = 0.5\ m^2$, $M = 50\ kg$, $h = (160 - 90) = 70\ cm$

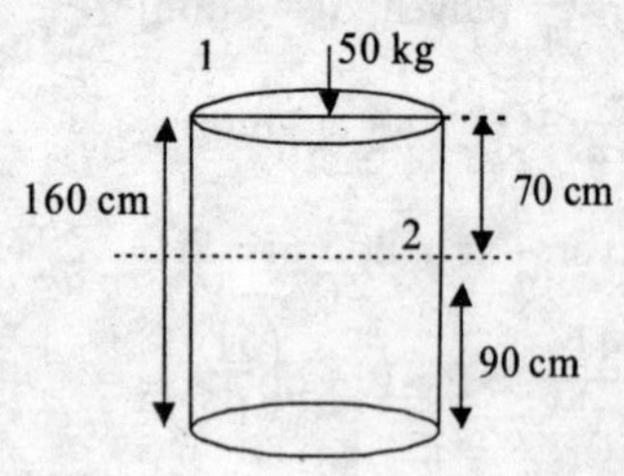

Apply Bernouli equation between points 1 & 2

$$P_1 + \frac{1}{2}\rho v_1^2 + \rho gh = P_2 + \frac{1}{2}\rho v_2^2 + 0$$

Here, $P_1 = P_0 + \frac{mg}{A}$, $P_2 = P_0$, $v_1 = 0$

$$\Rightarrow \quad P_0 + \frac{mg}{A} + \rho gh = P_0 + \frac{1}{2}\rho v_2^2$$

$$\Rightarrow \quad \frac{5000}{0.5} + 10^3 \times 10\frac{70}{100} = \frac{1}{2} \times 10^3 v_2^2$$

$$\Rightarrow \quad 10^3 + 10^3 \times 7 = \frac{10^3}{2} v_2^2$$

$$\therefore \quad v_2 = 4\ m/s$$

31. **(d)** Given, $r_1 = 2$ cm, $r_2 = 1$ cm, $v_1 = 2$m/s

$\therefore$ By equation of continuity,

$$A_1 v_1 = A_2 v_2 \Rightarrow \pi(2)^2 \times 2 = \pi(1)^2 \times v_2$$

$$\therefore v_2 = 8\ m/s$$

32. **(1)**

The force exerted by the liquid at the other and is

$$F = 2M\omega^2\frac{\ell}{2} = M\omega^2\ell$$

$$\therefore \omega = \sqrt{\frac{F}{M\ell}} = \sqrt{\frac{F}{M}} = \sqrt{\frac{F}{\propto M}} \therefore \propto = 1 \quad (\because \ell = 1m \text{ given})$$

33. **(b)** From the Bernoulli's theorem

$$P + \rho gh + \frac{1}{2}\rho V^2 = \text{constant}$$

34. **(d)** From Bernoulli's equation,

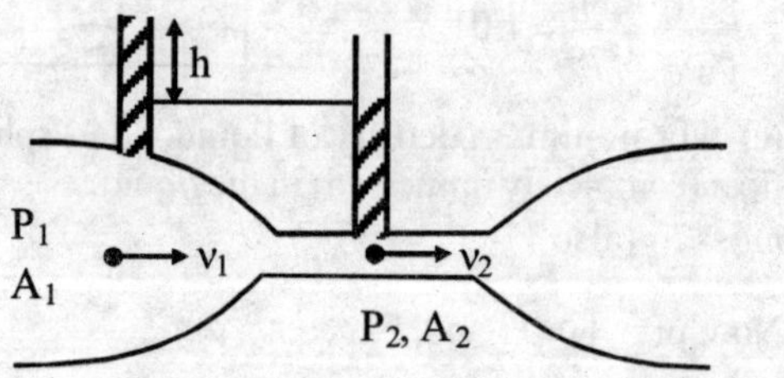

$$P_1 + \rho gh_1 + \frac{1}{2}\rho v_1^2 = P_2 + \rho gh_2 + \frac{1}{2}\rho v_2^2$$

Here, h_1 & h_2 are height of point from any reference level

If $V_1 = V_2 = 0$ (for statement-1)

$\therefore P_1 - P_2 = \rho g(h_2 - h_1)$ ...(i)

For statement-2

$$P_1 + \frac{1}{2}\rho v_1^2 = P_2 + \frac{1}{2}\rho v_2^2$$

Now using eq. (i)

$$P_1 - P_2 = \frac{1}{2}\rho v_2^2 - \frac{1}{2}\rho v_1^2 \Rightarrow \rho gh = \frac{1}{2}\rho v_2^2 - \frac{1}{2}\rho v_1^2$$

$$\therefore 2gh = v_2^2 - v_1^2$$

35. **(9600)** According to Bernoulli's equation

$$P_2 - P_1 = \frac{1}{2}\rho\left(V_1^2 - V_2^2\right)$$

$$\Rightarrow \quad mg = \frac{1}{2} \times \left(70^2 - 50^2\right) \times 80$$

$$\Rightarrow \quad mg = 96000 \Rightarrow m = 9600\ kg$$

36. **(810)** $v_1 = 70$ m/s, $v_2 = 65$ m/s, $A = 2m^2$
The lift force on the wing is

$$F = \frac{1}{2}\rho(v_1^2 - v_2^2)$$

$$F = \frac{1}{2} \times 1.2 \times (70^2 - 65^2) \times 2 = 810\text{N}$$

37. **(50)** Change in pressure, $\Delta P = \frac{1}{2}\rho v^2$

$$4.5 \times 10^4 - 2.0 \times 10^4 = \frac{1}{2} \times 10^3 \times v^2$$

$$\left[\because \rho_{water} = 1 \times 10^3 \text{kgm}^{-3}\right]$$

$$2.5 \times 10^4 = \frac{1}{2} \times 10^3 \times v^2 \Rightarrow v^2 = 50 \quad \therefore v = \sqrt{50} \text{ m/s}$$

Velocity of water $= \sqrt{V}$ m/s $= \sqrt{50}$ m/s
$\therefore V = 50$

38. **(a)** From theorem of continuity $A_A V_A = A_B V_B$
$1.5 \times V_A = 25 \times 10^{-2} \times 60$

$$\therefore V_A = \frac{25 \times 60 \times 10^{-2} \times 10}{1.5} = 10 \text{cm/s}$$

By Bernoulli's theorem, $P_A + \frac{1}{2}\rho V_A{}^2 = P_B + \frac{1}{2}\rho V_B{}^2$

$$P_A + \frac{1}{2} \times 1000 \times (0.1)^2 = P_B + \frac{1}{2} \times 1000 \times (0.6)^2$$

$$\Rightarrow P_A + 5 = P_B + \frac{1}{2} \times 1000 \times 36 \times 10^{-2}$$

$P_A + 5 = P_B + 180$
$\therefore P_A - P_B = 175$ Pa

39. **(4)** Given,
Density of Glycerine, $p = 1.25 \times 10^3$ kg m^{-3}

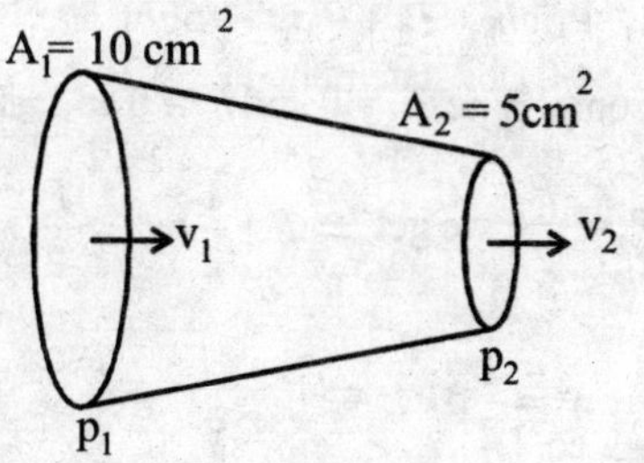

$\Delta P = P_1 - P_2 = 3$ N/m^2 (given)
From the equation of continuity
$A_1 v_1 = A_2 v_2$

$$\therefore v_1 = \frac{A_1}{A_2} v_2 \quad \text{...(i)}$$

From the Bernoulli's equation

$$P_1 + \frac{1}{2}\rho v_1{}^2 = P_2 + \frac{1}{2}\rho v_2{}^2 \Rightarrow P_1 - P_2 = \frac{1}{2}\rho\left(v_2^2 - v_1^2\right)$$

$$\Rightarrow \Delta P = \frac{1}{2}\rho\left(v_2^2 - \frac{A_2^2}{A_1^2}v_2^2\right) \Rightarrow \Delta P = \frac{1}{2}\rho\left[1 - \left(\frac{A_2}{A_1}\right)^2\right]v_2^2$$

$$\Rightarrow 3 = \frac{1}{2} \times 1.25 \times 10^3 \left[1 - \left(\frac{5}{10}\right)^2\right] v_2^2$$

$$\Rightarrow 3 = \frac{1}{2} \times 1.25 \times 10^3 \left[1 - \frac{1}{4}\right] v_2^2$$

$$\Rightarrow 3 = \frac{1}{2} \times 1.25 \times 10^3 \times \frac{3}{4} v_2^2$$

$\therefore v_2 = 8 \times 10^{-2}$ m/s
So rate of flow of glycerine
$= A_2 v_2 = 5 \times 10^{-4} \times 8 \times 10^{-2}$
$= 4 \times 10^{-5}$ m^3/s

40. **(2)**

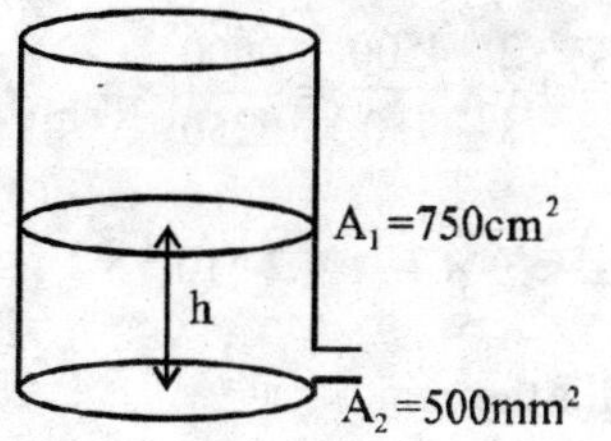

By continuity equation
$A_1 V_1 = A_2 V_2 =$ volume flow rate

Also, volume flow rate $= A_1 \frac{dh}{dt}$

So, volume flow rate $= A_2 V_2 = A_1 \frac{dh}{dt} \rightarrow A_2 V_2 = A_1 \frac{dh}{dt}$

$$\Rightarrow 500 \times 10^{-6} \times 0.3 = 750 \times 10^{-4} \times \frac{dh}{dt}$$

$$\Rightarrow \frac{dh}{dt} = 2 \times 10^{-3} \text{ m/s} \quad \because x = 2$$

41. **(d)** $P_2 A - P_1 A = 5.4 \times 10^5 \times g$

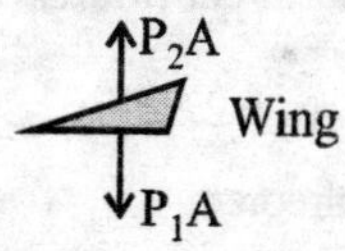

$$P_2 - P_1 = \frac{5.4 \times 10}{500} = 5.4 \times 2 \times 10^2 \times 10 = 10.8 \times 10^3$$

By Bernoulli's theorem

$$P_2 + 0 + \frac{1}{2}\rho V_2^2 = P_1 + 0 + \frac{1}{2}\rho V_1^2$$

$$\Rightarrow \quad P_2 - P_1 = \frac{1}{2}\rho(V_1^2 - V_2^2) = \frac{1}{2}\rho(V_1 - V_2)(V_1 + V_2)$$

$$\Rightarrow \quad 10.8 \times 10^3 = \frac{1}{2} \times 1.2(V_1 - V_2) \times 2 \times 3 \times 10^2$$

$$\Rightarrow \quad 10.8 \times 10 = 3.6(V_1 - V_2) \Rightarrow V_1 - V_2 = 30$$

$$\text{So, } \left(\frac{V_1 - V_2}{V}\right) \times 100 = \frac{30}{300} \times 100 = 10\%$$

42. **(c)** Apply Bernoulli's theorem between Piston and hole

$$P_A + \rho gh = P_0 + \frac{1}{2}\rho v_e^2$$

Assuming there is no atmospheric pressure on piston

$$\frac{5 \times 10^5}{\pi(1)^2} + 10^3 \times 10 \times 10 = 1.01 \times 10^5 + \frac{1}{2} \times 10^3 \times v_e^2$$

$\Rightarrow v_e = 17.8$ m/s

43. **(24)** Here, h = constant

So, $P+\frac{1}{2}\rho V^2=\text{constant}$

$\Rightarrow P_1+\frac{1}{2}\rho V_1^2=P_2+\frac{1}{2}\rho V_2^2 \Rightarrow P_1-P_2=\frac{1}{2}\rho(V_2^2-V_1^2)$

Now, by equation of continuty.

$A_1V_1=A_2V_2$

$\Rightarrow 2A_2V_1=A_2V_2 \Rightarrow V_2=2V_1$

So, $P_1-P_2=\frac{1}{2}\rho 3V_1^2$

$\Rightarrow V_1^2=\frac{2}{3}\frac{P_1-P_2}{\rho}=\frac{2}{3}\times\frac{4500}{750}=\frac{9000}{2250}=4$

So, $V_1=2$ m/s.

So, Volume flow rate $=A_1V_1=1.2\times10^{-2}\times 2$

$=24\times10^{-3}\ \text{m}^3/\text{s}$

44. **(300)** By Bernoull's theorem

$P+\rho gh+\frac{1}{2}\rho V^2=\text{constant}$

$\Rightarrow\left(P_0+\frac{mg}{A}\right)+\rho gh=P_0+\frac{1}{2}\rho V^2$

[$\because$ Speed of water in tank $=0$]

$\Rightarrow \frac{25\times10}{0.5}+1000\times10\times0.4=\frac{1}{2}\times1000\times V^2$

$\Rightarrow 500+4000=500V^2 \Rightarrow V^2=9$

$\Rightarrow V=3$ m/s $=300$ cm/sec

45. **(363)** By equation of continuity,

$av_1=\frac{a}{2}v_2 \Rightarrow v_2=2v_1$

From Bernoulli's theorem,

$P_1+\rho gh_1+\frac{1}{2}\rho v_1^2=P_2+\frac{1}{2}\rho v_2^2$

$P_1-P_2=\frac{1}{2}\rho(v_2^2-v_1^2)-\rho gh_1$

$4100=\frac{1}{2}\times800\,(4v_1^2-v_1^2)-800\times10$

$4100=400\times3v_1^2-8000$

$12100=1200\times v_1^2$

$v_1^2=\frac{121}{12}$

$v_1^2=\frac{121\times3}{12\times3}\Rightarrow v_1=\frac{\sqrt{363}}{6}$ m/s.

46. **(c)** Volume of liquid coming out of hole of area a with velocity v = av

Mass of liquid = density × volume = ρav

Force exerted by leaving water on the container,

$F=\left(\frac{dm}{dt}\right)v=(\rho av)v=\rho av^2$

Velocity of liquid coming out of hole, $v=\sqrt{2gh}$

For the container to be not moving, force exerted must be equal to the force of friction.

$\mu mg=\rho a(2gh)$

$\Rightarrow \mu\rho(Ah)g=\rho a2gh \Rightarrow \mu=\frac{2a}{A}$

47. **(6)** Given,

Height of the water, H = 12 cm

Velocity of water coming out of hole, $v=\sqrt{2gh}$

Range of water, R = vt

$\Rightarrow R=\sqrt{2gh}\times\sqrt{\frac{2(H-h)}{g}}$

$=2\sqrt{h(H-h)}$

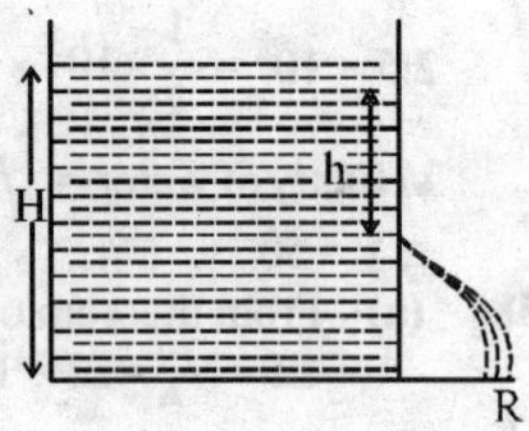

For maximum range $\frac{dR}{dh}=0$

$\therefore\ h=\frac{H}{2}$

Range is maximum when $h=\frac{12}{2}=6$m

48. **(3)** Applying Bernoulli's theorem at point A and B.

$P_0+\frac{mg}{A}+\rho gh+\frac{1}{2}\rho v^2=P_0+\frac{1}{2}\rho v^2$

$\Rightarrow \frac{mg}{A}+\rho gh=\frac{1}{2}\rho v^2$ [$\because$ At A, $v=0$]

$\Rightarrow \frac{240}{0.4}+1000\times10\times0.4=\frac{1}{2}\times1000\times v^2$

$\therefore\ v=3$ m/s

49. **(d)** Using Bernoulli's equation

$P_1+\frac{1}{2}\rho v_1^2+\rho gh_1=P_2+\frac{1}{2}\rho v_2^2+\rho gh_2$

For horizontal pipe, $h_1=0$ and $h_2=0$ and taking

$P_1=P$, $P_2=\frac{P}{2}$, we get $\Rightarrow P+\frac{1}{2}\rho v^2=\frac{P}{2}+\frac{1}{2}\rho V^2$

$\Rightarrow \frac{P}{2}+\frac{1}{2}\rho v^2=\frac{1}{2}\rho V^2 \Rightarrow V=\sqrt{v^2+\frac{P}{\rho}}$

50. **(b)** According to question, area of cross-section at A, a_A = 40 cm^2 and at B, a_B = 20 cm^2

Let velocity of liquid flow at A, $=V_A$ and at B, $=V_B$

Using equation of continuity $a_AV_A=a_BV_B$

$40V_A=20V_B \Rightarrow 2V_A=V_B$

Now, using Bernoulli's equation

$P_A+\frac{1}{2}\rho V_A^2=P_B+\frac{1}{2}\rho V_B^2 \Rightarrow P_A-P_B=\frac{1}{2}\rho\left(V_B^2-V_A^2\right)$

$\Rightarrow \Delta P=\frac{1}{2}1000\left(V_B^2-\frac{V_B^2}{4}\right)\Rightarrow \Delta P=500\times\frac{3V_B^2}{4}$

$\Rightarrow V_B = \sqrt{\frac{(\Delta P)\times 4}{1500}} = \sqrt{\frac{(700)\times 4}{1500}}\text{m/s} = 1.37\times 10^2\text{ cm/s}$

Volume flow rate $Q = a_B \times v_B$

$= 20 \times 100 \times V_B = 2732\text{ cm}^3/\text{s} \approx 2720\text{ cm}^3/\text{s}$

51. **(a)** From the equation of continuity

$A_1v_1 = A_2v_2$

Here, v_1 and v_2 are the velocities at two ends of pipe. A_1 and A_2 are the area of pipe at two ends

$$\Rightarrow \frac{v_1}{v_2} = \frac{A_2}{A_1} = \frac{\pi(4.8)^2}{\pi(6.4)^2} = \frac{9}{16}$$

52. **(b)** Using Bernoullie's equation

$$P + \frac{1}{2}(v_1^2 - v_2^2) + \rho gh = P$$

$$\Rightarrow v_2^2 = v_1^2 + 2gh \Rightarrow v_2 = \sqrt{v_1^2 + 2gh}$$

Equation of continuity

$A_1v_1 = A_2v_2$

$$\Rightarrow (1\text{ cm}^2)(1\text{ m/s}) = (A_2)\left(\sqrt{(1)^2 + 2\times 10\times\frac{15}{100}}\right)$$

$$\Rightarrow 10^{-4}\times 1 = A_2 \times 2 \Rightarrow A_2 = \frac{10^{-4}}{2} = 5\times 10^{-5}\text{ m}^2$$

53. **(a)** Since height of water column is constant therefore, water inflow rate (Q_{in}) = water outflow rate

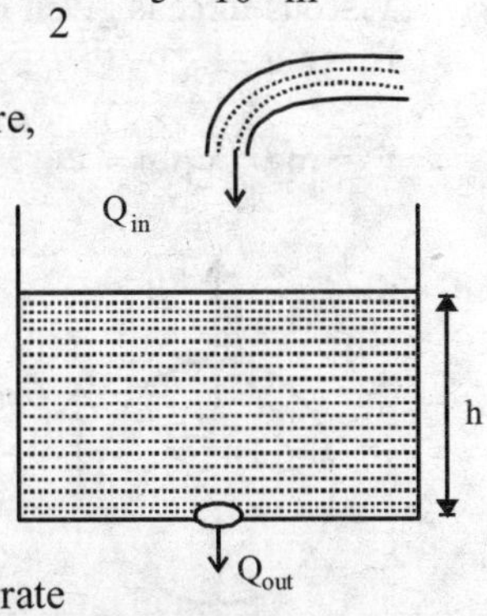

$Q_{in} = 10^{-4}\text{ m}^3\text{s}^{-1}$

$Q_{out} = Au = 10^{-4} \times \sqrt{2gh}$

$\therefore 10^{-4} = 10^{-4} \times \sqrt{20\times h}$

$\therefore h = \frac{1}{20}\text{m} = 5\text{ cm}$

54. **(b)** Here, volume tric flow rate

$$= \frac{0.74}{60} = \pi r^2 v = (\pi\times 4\times 10^{-4})\times\sqrt{2gh}$$

$$\Rightarrow \sqrt{2gh} = \frac{74\times 100}{240\pi} \Rightarrow \sqrt{2gh} = \frac{740}{24\pi}$$

$$\Rightarrow 2gh = \frac{740\times 740}{24\times 24\times 10}(\because \pi^2 = 10) \Rightarrow h = \frac{74\times 74}{2\times 24\times 24} \approx 4.8\text{m}$$

i.e., The depth of the centre of the opening from the level of water in the tank is close to 4.8 m

55. **(d)** The volume of liquid flowing through both the tubes i.e., rate of flow of liquid is same i.e., Q = cons.

$$Q = \frac{\Delta P\pi r^4}{8\eta L} \text{ [By Poiseuille equation]}$$

i.e., $\frac{\pi P_1 r_1^4}{8\eta l_1} = \frac{\pi P_2 r_2^4}{8\eta l_2} \Rightarrow \frac{P_1 r_1^4}{l_1} = \frac{P_2 r_2^4}{l_2}$

$\because P_2 = 4P_1$ and $l_2 = l_1/4$

$$\frac{P_1 r_1^4}{l_1} = \frac{4P_1 r_2^4}{l_1/4} \Rightarrow r_2^4 = \frac{r_1^4}{16}$$

$r_2 = r_1/2$

56. **(a)** According to Bernoulli's Principle,

$$\frac{1}{2}\rho v_1^2 + \rho gh = \frac{1}{2}\rho v_2^2 \Rightarrow v_1^2 + 2gh = v_2^2$$

$$\Rightarrow 2gH + 2gh = v_2^2 \quad ...(i)$$

$a_1v_1 = a_2v_2$

$$\pi r^2\sqrt{2gh} = \pi x^2 v_2 \Rightarrow v_2 = \frac{r^2}{x^2}\sqrt{2gh}$$

Substituting the value of v_2 in equation (i)

$$2gH + 2gh = \frac{r^4}{x^4}2gh \text{ or, } x = r\left[\frac{H}{H+h}\right]^{\frac{1}{4}}$$

57. **(b)** Volume flow rate = AV

Mass flow rate = ρAV

Momentum flow rate = ρAV^2

$\therefore F = \rho AV^2 = 1000\times 10^{-2}\times 1.5^2 = 22.5\text{ N}$

58. **(c)** Pressure difference

$$P_2 - P_1 = \frac{1}{2}\rho\left(v_2^2 - v_1^2\right) = \frac{1}{2}\times 1.2\left((150)^2 - (100)^2\right)$$

$$= \frac{1}{2}\times 1.2(22500 - 10000) \simeq 7500\text{ Nm}^{-2}$$

59. **(b)**

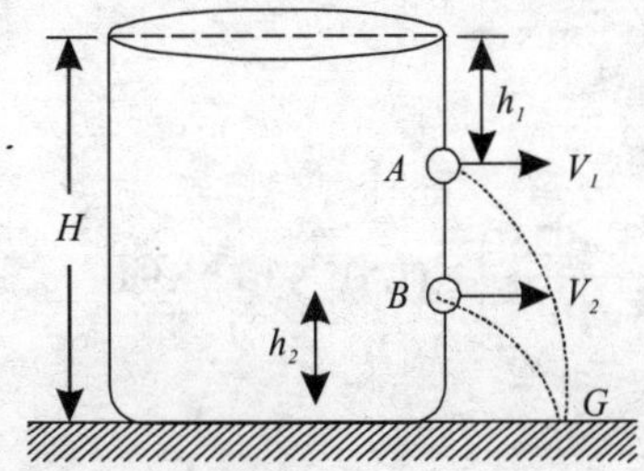

i.e. $R_1 = R_2 = R$

or, $v_1 t_1 = v_2 t_2$ ... (i)

Where v_1 = velocity of efflux at $A = \sqrt{(2gh_1)}$ and

v_2 = velocity of efflux at $B = \sqrt{(2g(H - h_2)}$

t_1 = time of fall water stream through $A = \sqrt{(\frac{2(H-h_1)}{g})}$

t_2 = time of fall of the water stream through $B = \sqrt{\frac{2h_2}{g}}$

Putting these values is eqn (i) we get

$(H - h_1)h_1 = (H - h_2)h_2$

or $[H - (h_1 + h_2)][h_1 - h_2] = 0$

Here, $H = h_1 + h_2$ is irrelevant because the holes are at two different heights. Hence $h_1 = h_2$ or, $\frac{h_1}{h_2} = 1$

60. **(a)** As $A_1v_1 = A_2v_2$ (Principle of continuity)

or, $\ell^2\sqrt{2gh} = \pi r^2\sqrt{2g \times 4h}$ (Efflux velocity= $\sqrt{2gh}$)

$\therefore \quad r^2 = \frac{\ell^2}{2\pi}$ or $r = \sqrt{\frac{\ell^2}{2\pi}} = \frac{\ell}{\sqrt{2\pi}}$

61. **(c)** Using Bernoulli's theorem, for horizontal flow

$$P_0 + \frac{1}{2}\rho v_1^2 + \rho gh = P_0 + \frac{1}{2}\rho v_2^2 + 0$$

$$v_2 = \sqrt{v_1^2 + 2gh} = \sqrt{0.16 + 2 \times 10 \times 0.2} = 2.03 \text{ m/s}$$

According to equation of continuity

$$A_2v_2 = A_1v_1$$

$$\pi\frac{D_2^2}{4} \times v_2 = \pi\frac{D_1^2}{4}v_1 \Rightarrow D_2 = D_1\sqrt{\sqrt{\frac{v_1}{v_2}}} = 3.55 \times 10^{-3} \text{ m}$$

62. **(b)** Given, Height of cylinder, h=20 cm Acceleration due to gravity, g=10 ms^{-2}

Velocity of efflux $v = \sqrt{2gh}$

Where h is the height of the free surface of liquid from th e hole

$\Rightarrow v = \sqrt{2 \times 10 \times 20} = 20 \text{ m/s}$

63. **(d)** The terminal velocity of ball falling through a viscous medium is

$$V_T = \frac{2}{9}\frac{(\rho_0 - \rho_\ell)r^2g}{\eta}$$

Here, $r = \frac{d}{2} = \frac{3.6}{2} \text{ mm} = 1.8 \text{ mm}$

$V_T = 2.45 \times 10^{-2}$ m/s

$\rho_0 = 7825$ kg/m^3

$\rho_\ell = 925$ kg/m^3

$$\therefore \eta = \frac{2}{9}\left(\frac{7825 - 925}{2.45 \times 10^{-2}}\right) \times (1.8)^2 \times 10^{-6} \times 9.8$$

≈ 1.99

64. **(b)**

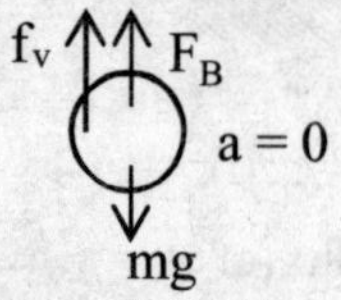

By Newton's 2nd law, $F_{net} = ma$

$mg - F_B - F_v = 0$

$\Rightarrow mg - \frac{mg}{2} - F_v = 0 \qquad \therefore F_v = \frac{mg}{2}$

65. **(c)** Given,

Radius of spherical ball, $r = 1 \times 10^{-4}$m

Density of spherical ball, $\rho = 10^5$ kg/m^3

Coefficient of viscosity of water, $\eta = 9.8 \times 10^{-6}$ N s/m^2

Density of water, $\sigma = 10^3$ kg/m^3

Terminal velocity is given by

$$v = \frac{2g}{9}\frac{r^2[\rho - \sigma]}{\eta}$$

$$\Rightarrow v = \frac{2}{9} \times \frac{10 \times \left(10^{-4}\right)^2}{9.8 \times 10^{-6}}\left[10^5 - 10^3\right] \Rightarrow v = 224.5$$

Velocity of ball falling from height (h)

$v = \sqrt{2gh}$

$$\Rightarrow h = \left(\frac{v^2}{2g}\right) = \frac{(224.5)^2}{2 \times 10} \approx 2518\text{m}$$

66. **(40)** Terminal velocity is, $V_t = \frac{2}{9}\frac{R^2(\rho - \sigma)g}{\eta} \Rightarrow V_t \propto R^2$

Since volume remains constant

$8V_s = V_B \quad 8 \times \frac{4}{3}\pi r^3 = \frac{4}{3}\pi R^3$

$8r^3 = R^3 \Rightarrow R = 2r$

$V_t = V_t \times R^2 = 4 \times 10 = 40$ cm/s

67. **(d)**

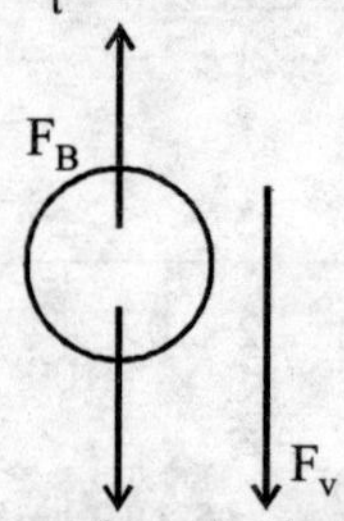

Viscous force is given by:

$F_v = mg - F_B$

$$F_v = mg - V\rho_0 g = mg - \frac{m}{\rho}\rho_0 g = mg\left(1 - \frac{\rho_0}{\rho}\right)$$

68. **(d)**

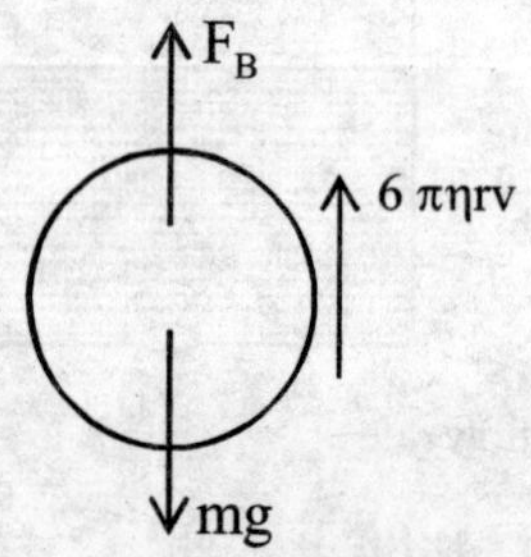

$F_{net} = mg - F_B - 6\pi nrv$

$$\Rightarrow m\frac{dv}{dt} = \left(\rho\frac{4}{3}\pi r^3\right)g - \left(\rho_L\frac{4}{3}\pi r^3\right)g - 6\pi\eta rv$$

$$\Rightarrow \text{Let } A = \frac{4}{3m}\pi R^3 g(\rho - \rho_L) \text{ and } B = \frac{6\pi\eta r}{m}$$

$$\frac{dv}{dt} = A - Bv$$

$$\Rightarrow \int_0^v \frac{dv}{A - Bv} = \int_0^t dt \qquad \Rightarrow -\frac{1}{B}\ln[A - Bv]_0^v = t$$

$$\Rightarrow \ln\left(\frac{A - B}{A}\right) = -Bt \Rightarrow A - Bv = Ae^{-Bt}$$

$$\Rightarrow v = \frac{A}{B}\left[1 - e^{-Bt}\right]$$

69. **(a)** For negligible density, buoyancy force will be zero
At terminal velocity.
weight of spherical ball = viscous force
$\Rightarrow$ $mg = 6\pi\eta rv$

$\therefore$ $v \propto \frac{1}{r}$ (m is constant)

According to question

$$\frac{v}{v'} = \frac{r'}{r}$$

$\because$ $r' = 2r$

$\therefore$ $v' = \frac{v}{2}$

70. **(d)** Terminal velocity of a spherical body in fluid

$$V_t = \frac{2}{9}\frac{r^2(\rho - \sigma)g}{\eta}$$

$\Rightarrow V_t \propto r^2$

$\therefore \frac{\Delta V_t}{V_t} = 2.\frac{\Delta r}{r}$

$\therefore$ Percentage error,

$$\frac{\Delta V_t}{V_t} \times 100\% = 2\frac{(0.1)}{5} \times 100 = 4\%$$

Hence A is true but R is false

71. **(b)** Let r be the radius of small drops of water.
R = radius of big drop formed
as volume remain same.

$\therefore 8.\frac{4}{3}\pi r^3 - \frac{4}{3}\pi R^3 \Rightarrow R = 2r$

Terminal velocity,

$$v_T = \frac{2}{9\eta}(\rho - \sigma)r^2 g$$

$\therefore v_T \propto r^2$

$$\therefore \frac{v_1}{v_2} = \left(\frac{r}{R}\right)^2$$

$$\Rightarrow \frac{10}{v_2} = \left(\frac{1}{2}\right)^2 \qquad (\because v_1 = 10 \text{ cm/s given})$$

$\Rightarrow v_2 = 40$ cm/s

72. **(10)** Diameter of air bubble, $D = 6 \times 10^{-3}$
Radius of air bubble, $R = 3 \times 10^{-3}$
Speed of air bubble, $v = 0.35 \times 10^{-2}$ m/s

Since the bubble is moving at constant speed. So
Buoyant force Fb = Viscous drag force Fv

$$\frac{4}{3}\pi R^3 \rho g = 6\pi\eta Rv$$

$$\eta = \frac{2R^2\rho g}{9v} = \frac{2\times(3\times10^{-3})^2 \times 1750 \times 10}{9 \times 0.35 \times 10^{-2}} = 10\text{Pas}$$

73. **(a)**

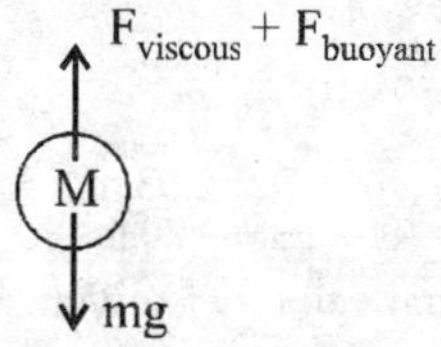

At terminal velocity, ball move with constant velocity

$\therefore F_{net} = 0$

$\Rightarrow F_{viscous} + F_{buoyant} = $ Weight of ball

$\Rightarrow F_{viscous} + \rho_0 Vg = \rho Vg$

$\Rightarrow F_{viscous} = \rho Vg - \rho_0 Vg = (\rho - \rho_0)\, Vg$

$$= \rho Vg\left(1 - \frac{\rho_0}{\rho}\right) = Mg\left(1 - \frac{\rho_0}{\rho}\right)$$

74. **(7)** When the ball attain terminal velocity, $mg = F_V + F_B$
So, $F_V = (mg - F_B)\,(\because a = 0) = V\sigma_b g - V\rho_\ell g$

$= Vg\,(\sigma_b\ \rho_\ell) - \frac{4}{3}\pi\,(10^{-3})^3 \times 9.8\,(10.5 - 1.5) \times 10^3$

$= 3696 \times 10^{-7}$ N So, $x = 7$

75. **(25)** $F_V + F_B = mg$ (V = constant)

$F_V = mg - F_B = \rho_B Vg - \rho_L Vg = (\rho_B - \rho_L)Vg$

$$= (8 - 1.3) \times 10^{+3} \times \frac{0.3\times10^{-3}}{8\times10^3} \times 10 = 25\times10^{-4}\text{ N}$$

Hence, the value of viscous force acting on ball will be 25×10^{-4} N.
Compair with $x \times 10^{-4}$, we get $x = 25$

76. **(11)** When bubble is rising steadily the net force acting on it will be zero.
So, buoyant force (B) = drag force (F)

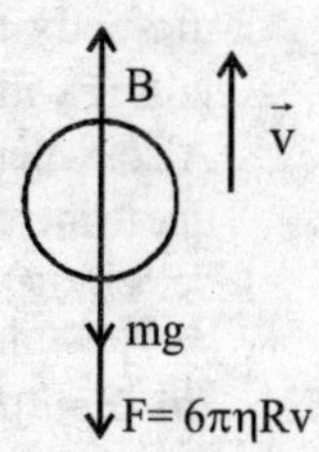

$\Rightarrow \frac{4\pi}{3}R^3\rho g = 6\pi\eta Rv$

$\therefore \quad \eta = \frac{4}{18}\frac{R^2\rho g}{v}$

or, $\eta = \frac{4}{18}\frac{(10^{-3})^2 \times 1.75\times10^3 \times 10}{0.35\times10^{-2}}$

$\therefore \quad \eta = \frac{10}{9} = 1.11 = 11$ poise

77. **(c)** We know that for air bubble in liquid

$$v_T = v = \frac{2r^2 g}{9\eta}\left(\rho_f - \rho\right) \Rightarrow \eta = \frac{2r^2 g}{9v}(\rho_f - \rho)$$

$$\Rightarrow \eta = \frac{2r^2 g}{9v}(\sigma) \quad [\because \rho_f = \sigma \text{ and } \rho \approx o]$$

78. **(20)** Speed after falling 'h' height = Terminal velocity

$$\Rightarrow \sqrt{2gh} = \frac{2r^2 g}{9\eta}(\rho - \rho_f)$$

$$\Rightarrow \sqrt{2\times 10\times h} = \frac{2\times(0.1\times 10^{-3})^2\times 10}{9\times 10^{-5}}\times(10^4 - 10^3)$$

$$\Rightarrow \sqrt{20h} = \frac{2\times 10^{-8}\times 10}{9\times 10^{-5}}\times 9\times 10^3$$

$$\Rightarrow \sqrt{20h} = 20 \;\Rightarrow 20h = 400 \;\Rightarrow h = 20\text{ m}$$

79. **(d)** We know that terminal velocity is given by

$$V_T = \frac{2gr^2}{9\eta}(\rho - \rho_\ell)$$

Here, we have no involvement of buoyant force. So remove ρ_ℓ.

$$\text{Then, } v_T = \frac{2gr^2\rho}{9\eta} = \frac{2\times 10\times 10^{-12}\times 10^3}{9\times 1.8\times 10^{-5}}$$

$= 123.4\times 10^{-6}$ m/s.

80. **(b)** We have, $F_v = mg - F_B$

$= mg - \rho_l V$

$$= mg - d_2.\frac{m}{d_1}g = mg\left(1 - \frac{d_2}{d_1}\right)$$

81. **(a)** $27\times\frac{4}{3}\pi r^3 = \frac{4}{3}\pi r^3$ or $r = \frac{R}{3}$.

Terminal velocity, $v \propto r^2$

$$\therefore \frac{v_1}{v_2} = \frac{r_1^2}{r_2^2} \text{ or } v_2 = \left(\frac{r_2}{r_1}\right)^2 v_1 = \left(\frac{R/3}{R}\right)^2 v_1 = \frac{1}{9}$$

or $\frac{v_1}{v_2} = 9$.

82. **(c)** When a point mass is falling vertically in a viscous medium, the medium or viscous fluid exerts drag force on the body to oppose its motion and at one stage body falling with constant terminal velocity.
Also, when body fall vertically "a_{net}" keep on decreasing due to increase in viscous force and when body attain terminal velocity net force on it becomes zero i.e., "a_{net}" become zero.

83. **(b)** $F = \eta A\frac{dv}{dx}$

$$\frac{F}{A} = \eta\frac{dv}{dx} = 10^{-2}\times\frac{5}{5} = 10^{-2}\text{ N/m}^2$$

84. **(c)** Initially

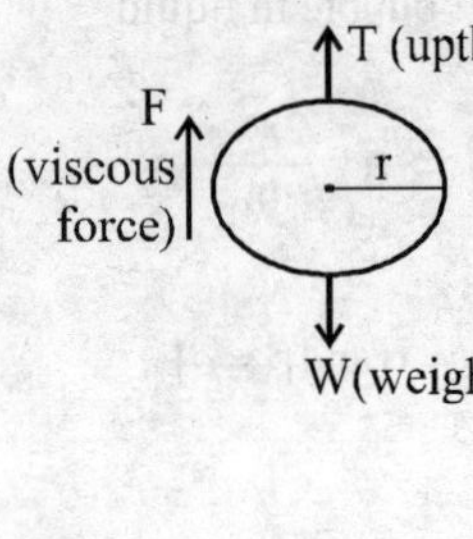

W = T + F

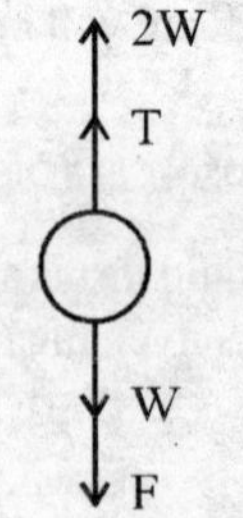

Finally
Here
2W + T = W + F
⇒ W = T + F

So in both case W = T + F
So, in both case state of body will be same.

85. **(a)** Terminal velocity in a viscous medium is given by:

$$V_T = \frac{2a^2(\rho - \sigma)g}{9\eta} \qquad \therefore V_T \propto a^2$$

86. **(a)** $V_T = \frac{2r^2}{9\eta}(\rho - \rho_l)g$

$$\frac{v'}{v} = \left(\frac{\rho - \rho'_l}{\rho - \rho_l}\right)\times\frac{\eta_w}{\eta'} \Rightarrow v' = \frac{(\rho - \rho'_\ell)}{(\rho - \rho_\ell)}\times\frac{v\eta_w}{\eta'}$$

$$\Rightarrow v' = \frac{(\rho - \rho'_\ell)}{(\rho - \rho_\ell)}\times\frac{v\eta_w}{\eta'} = \frac{(7.8 - 1.2)}{(7.8 - 1)}\times\frac{10\times 8.5\times 10^{-4}}{13.2}$$

$\therefore \quad v' = 6.25\times 10^{-4}$ cm/s

87. **(a)** When the ball attains terminal velocity
Weight of the ball = Buoyant force + Viscous force

$$\therefore V\rho_1 g = V\rho_2 g + kv_t^2$$

$$\Rightarrow Vg(\rho_1 - \rho_2)g = kv_t^2$$

$$\Rightarrow v_t = \sqrt{\frac{Vg(\rho_1 - \rho_2)}{k}}$$

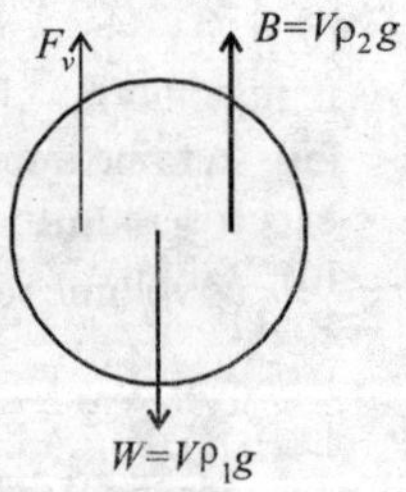

88. **(c)** Terminal velocity, $v_T = \frac{2r^2(\rho - \sigma)g}{9\eta}$

$$\therefore \frac{v_{T_2}}{0.2} = \frac{(10.5 - 1.5)}{(19.5 - 1.5)} \Rightarrow v_{T_2} = 0.2\times\frac{9}{18} = 0.1\text{ m/s}$$

89. **(a)**

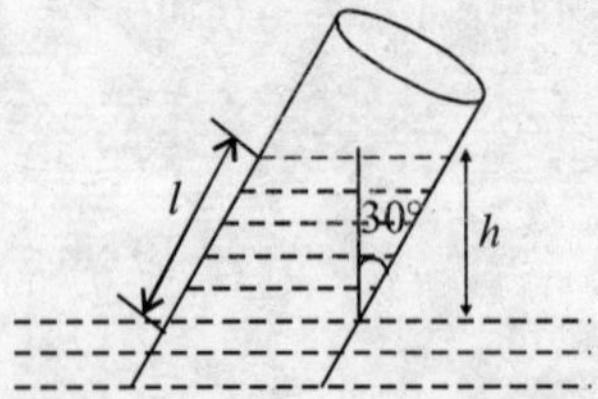

Given : $r = 0.1$ mm $= 1$ cm $\times 10^{-2}$, $l = ?$, $g = 980$ cm/s^2, $T = 70$ dyn/cm

$$h = \frac{2T\cos\theta}{\rho g r} = \frac{2\times 70\times 1}{1\times 980\times 10^{-2}}$$

$$\Rightarrow h = \frac{100}{7}\text{ cm}$$

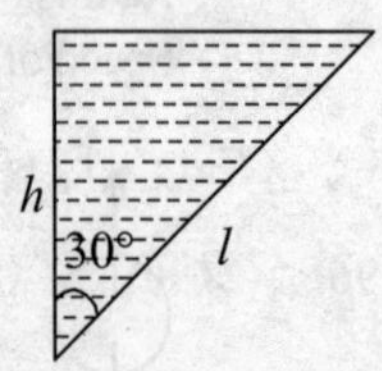

$$\cos 30° = \frac{h}{l} \Rightarrow l = \frac{h\times 2}{\sqrt{3}}$$

$$\therefore l = \frac{100}{7}\times\frac{2}{\sqrt{3}} = \frac{200}{7\times\sqrt{3}} = 16.49\text{ cm}$$

90. **(c)** Given, $K = \dfrac{\cos\theta_A}{\cos\theta_B}$

It is negative when $\cos\theta_A$ and $\cos\theta_B$ are of opposite sign.

Therefore, if K is negative, one liquid forms a concave meniscus and the other forms a convex meniscus.

91. **(8)** $(\Delta P)_A = \dfrac{4T}{R_A}$ and $(\Delta P)_B = \dfrac{4T}{R_B}$

$\Rightarrow \dfrac{(\Delta p)_B}{(\Delta P)_A} = \dfrac{R_A}{R_B} = 2$

$\Rightarrow \dfrac{V_A}{V_B} = \left(\dfrac{R_A}{R_B}\right)^3 = 8$

$\therefore \quad V_A = 8V_B$

92. **(a)** Since volume is conserved so

$2 \times \dfrac{4}{3}\pi R^3 = \dfrac{4}{3}\pi r^3 \Rightarrow r = 2^{1/3}R$

Surface energy

$U_i = 2 \times 4\pi R^2 T$

$U_f = 4\pi r^2 T = 4\pi R^2 T 2^{2/3}$

Surface energy released $= 4\pi R^2 T[2 - 2^{2/3}]$

93. **(b)** (i) Surface tension arises because molecules at the surface possess higher energy compared to those in the interior of a liquid.

(ii) The viscosity of a liquid decreases with increasing temperature.

(iii) A flow becomes turbulent when the Reynolds number exceeds 2000. In a steady flow, streamlines never intersect.

94. **(d)** Given,

Radius of air bubble, $R = 0.1$ cm

Depth, $h = 20$ cm

Density of liquid, $\rho = 1000$ kg/m^3

Pressure inside air bubble, $P_{in} = P_0 + \rho gh + \dfrac{2T}{R}$

$P_{in} - P_0 = \rho gh + \dfrac{2T}{R} \Rightarrow 2100 = \rho gh + \dfrac{2T}{R}$

$\therefore T = \dfrac{1}{20}(2100 - 2000) \times 10^{-2} = 0.05\text{Nm}^{-1}$

95. **(a)** The amount of work done to break a big water drop into n small drops is

$W = 4\pi R^2 T(n^{1/3} - 1)$

$\therefore \dfrac{W_2}{W_1} = \dfrac{(n_2^{1/3} - 1)}{(n_1^{1/3} - 1)}$

$\Rightarrow \dfrac{W_2}{10} = \dfrac{(64)^{1/3} - 1}{(27)^{1/3} - 1} \qquad (\because \text{given } W_1 = 10\text{ J})$

$\therefore W_2 = \dfrac{3}{2} \times 10 = 15\text{J}$

96. **(2190)**

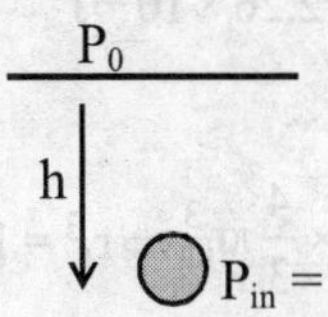

$P_{in} = P_0 + \rho gh + \dfrac{2T}{R}$

The difference between pressure inside the bubble and atmospheric pressure can be calculated as

$\Delta P = P_{in} - P_0 = \rho gh + \dfrac{2T}{R}$

$= \dfrac{1000 \times 10 \times 20}{100} + \dfrac{2 \times 0.095}{10^{-3}} = 2190$

97. **(b)** Hot water molecuels have higher kinetic energy.

Hot water is less viscous then cold water.

Soap reduces surface tension of water.

So, soap water has lower surface tension than fresh water.

98. **(4)** Given, radii of soap bubble, $r_1 = 2$ cm and $r_2 = 4$ cm

Radius of common surface, $r = \dfrac{r_1 r_2}{r_1 - r_2} = \dfrac{2 \times 4}{4 - 2} = 4$ cm

99. **(d)** Excess pressure for soap bubble having radius r,

$P_1 - P_0 = \dfrac{4T}{r_1}$

Excess pressure for soap bubble having radius r_2

$P_2 - P_0 = \dfrac{4T}{r_2}$

According to question

$P_1 - P_0 = 3(P_2 - P_0)$

$\Rightarrow \dfrac{4T}{r_1} = 3\dfrac{4T}{r_2} \Rightarrow r_1 = \dfrac{r_2}{3}$

$\therefore \quad \dfrac{V_1}{V_2} = \dfrac{\frac{4}{3}\pi r_1^3}{\frac{4}{3}\pi r_2^3} = \dfrac{1}{27}$

100. **(7)** Given,

Density of liquid, $\rho = 8 \times 10^3$ kg m^{-3}

Surface tension of soap solution,

$S = 0.28$ Nm^{-1}

Height of liquid column, $h = 0.04$ cm

Excess pressure due to soap bubble $= \dfrac{4S}{R}$

Pressure due to liquid column $= h\rho g$

$\therefore \rho gh = \dfrac{4S}{\frac{D}{2}} \qquad \left(\because R = \dfrac{D}{2}\right)$

$\Rightarrow \dfrac{D}{2} = \dfrac{4 \times 0.28}{8 \times 10^3 \times 10 \times 4 \times 10^{-4}}$

$\Rightarrow \dfrac{D}{2} = \dfrac{28}{8}$ cm

$\Rightarrow D = 7$ cm

101. **(1)** $1000\dfrac{4}{3}\pi r^3 = \dfrac{4}{3}\pi R^3$

$\Rightarrow R = 10r$

$$\frac{\text{Surface energy of 1000 drops}}{\text{Surface energy of Big drop}} = \frac{1000(4\pi r^2)T}{4\pi R^2 T}$$

$$= \frac{1000 \times r^2}{(10r)^2} = 10 = \frac{10}{x} \Rightarrow x = 1$$

102. (a) Change in pressure in soap bubble is,

$$\Delta P = 2\left(\frac{2S}{R}\right) = \frac{4S}{R}$$

103. (b) Capillary rise, $h = \frac{2T\cos\theta}{r\rho g}$

For $\theta = 0 \Rightarrow h \neq 0$

Also contact angle is the property of the materials in contact i.e., solid as well as liquid.

104. (7) Work done $\Delta W = T\Delta A = 8\pi(R_2^2 - R_1^2)T$

Where T = surface tension, R_1 = original radius, R_2 = final radius

$$\text{or}, 36960 = 8 \times \frac{22}{7} \times 40\left(R_2^2 - \frac{49}{4}\right) \quad \therefore R_2 = 7\text{ cm}$$

105. (c) Angle of contact between a solid and a liquid is a property of the material of the solid and liquid as well. Hence statement I is correct.

Statement II is incorrect because height of liquid or capillary rise $h = \frac{2T\cos\theta_c}{\rho g r}$ i.e., $h \propto \frac{1}{r}$ where r is radius of the tube.

106. (b) Since volume conserved, so $V_B = V_S$

$$\frac{4}{3}\pi R^3 = 1000\frac{4}{3}\pi r^3 \Rightarrow R = 10r$$

Here,

r = Radius of small drop

R = Radius of big drop

Initial energy $E_i = 1000\,(4\pi r^2 S)$

final energy $E_f = 4\pi R^2 S = 100\,(4\pi r^2 S)$

$$\therefore \frac{E_f}{E_i} = \frac{1}{10} \Rightarrow E_f = \frac{1}{10}E_i$$

107. (10) Volume of small drop × 1000 = Volume of big drop

$$\left(\frac{4}{3}\pi r^3\right)1000 = \frac{4}{3}\pi R^3 \Rightarrow R = 10r$$

$E_1 = 1000 \times 4\pi r^2 \times S$

$E_2 = 4\pi R^2 \times S = 4\pi(10r)^2 S$

$$\frac{E_1}{E_2} = \frac{10}{1} \therefore x = 1$$

108. (c) As, $h = \frac{2S\cos\theta}{\rho r g} \Rightarrow h \propto S$

Also, $S \propto \frac{1}{\text{Temperature}} \quad \therefore \; h \propto \frac{1}{\text{Temperature}}$

Height of capillary rise will be smaller in hot water and larger in cold water.

109. (b) Volume remains constant

$$\frac{4}{3}\pi R^3 = 27 \times \frac{4}{3} \times \pi r^3$$

$$R^3 = 27r^3 \Rightarrow r^2 = \frac{R^2}{9}$$

Work done = T.ΔA

$= T(27 \times 4\pi r^2 - 4\pi R^2)$

$= 4\pi T(27 \times \frac{R^2}{9} - R^2)$

$= 8\pi R^2 T$

110. (b) Viscosity of gases is smaller than that of liquids. Presence of insoluble impurities like detergent, surface tension decreases.

111. (1150) Radius of air bubble, $r = 1 \times 10^{-3}$ m

Surface tension, $T = 0.075\ \frac{N}{m}$

Density of liquid, $\rho = 1000\text{ kg/m}^3$

at depth, h = 10 cm = 0.10 m

P_0 = atmospheric pressure

Pressure inside the bubble,

$$P = P_0 + h\rho g + \frac{2T}{r}$$

$$P - P_0 = h\rho g + \frac{2T}{r} = 0.1 \times 1000 \times 10 + \frac{2 \times .075}{10^{-3}}$$

$= 1000 + (0.15)(1000) = 1150$ Pa

112. (264) Initial radius, $r_1 = 10$ cm

Final radius, $r_2 = 20$ cm

$$W = T.\,(\Delta A) \Rightarrow W = 2T\left(4\pi\left(r_2^2 - r_1^2\right)\right)$$

$$\Rightarrow \quad W = 8 \times \frac{22}{7} \times 3.5 \times 10^{-2}((20 \times 10^{-2})^2 - (10 \times 10^{-2})^2)$$

$$= 8 \times \frac{22}{7} \times 3.5 \times 10^{-2} \times 300 \times 10^{-4}$$

$\Rightarrow \quad W = 264 \times 10^{-4}$ J

113. (b) Given, radius of mercury drop, $r = 10^{-3}$ m

Initial surface energy E_1 = surface tension × Area

$E_1 = 0.45 \times 4\pi r^2 = 0.45 \times 4\pi(10^{-3})^2$

Initial volume = n × final volume

$$\frac{4}{3}\pi(10^{-3})^3 = 125 \times \frac{4\pi}{3}R_{new}^3$$

$$\therefore \quad 10^{-3} = 5R_{new} \therefore R_{new} = \frac{10^{-3}}{5}\text{ m}$$

So, final surface energy, $E_2 = 0.45 \times 125 \times 4\pi\left(\frac{10^{-3}}{5}\right)^2$

Increase in energy $= E_2 - E_1$

Increase in energy $= 0.45 \times 4\pi \times (10^{-3})^2\left[\frac{125}{25} - 1\right]$

$= 4 \times 0.45 \times 4\pi \times 10^{-6} = 2.26 \times 10^{-5}$ J

114. (b) We have

$V_f = V_i$

$$\Rightarrow \frac{4}{3}\pi r_f^3 = 1000 \times \frac{4}{3}\pi r_i^3 \Rightarrow r_f^3 = 1000\,r_i^3$$

$\Rightarrow r_f = 10\,r_i$

So, released energy
$= \text{Initial surface energy} - \text{final surface energy}$
$= 1000 \times T \times 4\pi r_i^2 - T \times 4\pi r_f^2$
$= 4\pi T(1000 r_i^2 - r_f^2)$
$= 4\pi \times 0.07\,(1000\, r_i^2 - 100\, r_i^2)$
$= 4\pi \times 0.07 \times 900\, r_i^2$
$= 4\pi \times 63 \times 10^{-6} = 7.92 \times 10^{-4}\ \text{J}$

115. (c) The height of liquid column in a capillary tube,

$$h = \frac{2S\cos\theta}{r\rho g}$$

$$\therefore \frac{h_1}{h_2} = \frac{S_1}{S_2}\frac{\rho_2}{\rho_1} \qquad (\because S_2 = 2S_1\ \rho_2 = 2\rho_1)$$

$$\frac{5}{h_2} = \left[\frac{1}{2}\right]\left[\frac{2}{1}\right] \Rightarrow h_2 = 5\ \text{cm} = 0.05\ \text{m}$$

116. (c) Given, surface tension of soap bubble, $T = 2 \times 10^{-2}\ \text{Nm}^{-1}$
Surface area of soap bubble $= 2 \times 4\pi R^2$
Work done = change in surface area of soap bubble × T

$= T \times 8\pi \times \left(R_2^2 - R_1^2\right)$
$= 2 \times 10^{-2} \times 8\pi(7^2 - (3.5)^2) \times 10^{-4}$
$= 2 \times 10^{-2} \times 8 \times \frac{22}{7} \times 49 \times \frac{3}{4} \times 10^{-4}$
$= 18.48 \times 10^{-4}\ \text{J}$

117. (a)

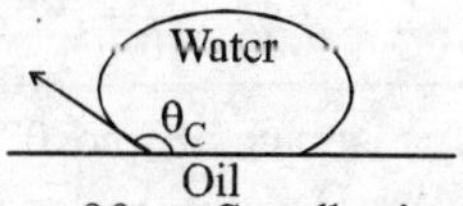

As $\theta_C > 90 \Rightarrow$ So adhesive force will dominate between water and oil. And water molecule will stick to oil rather than removing it.

118. (c) Given that radius of water drop, $R = 1\ \text{cm} = 10^{-2}\ \text{m}$

Surface tension of water $T = 75\ \text{dyne/cm} = \frac{75 \times 10^{-5}}{10^{-2}}$

$= 75 \times 10^{-3}\ \text{N/m}$

From volume conservation

$$\frac{4}{3}\pi R^3 = 729 \times \frac{4}{3}\pi r^3$$

$$\Rightarrow R = 9r \Rightarrow r = \frac{R}{9} = \frac{10^{-2}}{9}\ \text{m}$$

Initial surface energy, $E_1 = T \times 4\pi R^2$
Final surface energy, $E_2 = T \times [4\pi r^2 \times 729]$
Change in surface energy $= E_2 - E_1 = T[4\pi r^2 \times 729 - 4\pi R^2]$

$$= 4\pi \times 75 \times 10^{-3}\left[\left(\frac{10^{-2}}{9}\right)^2 \times 729 - (9)^2 \times \left(\frac{10^{-2}}{9}\right)^2\right]$$

$$= 4\pi \times 75 \times 10^{-3}\left[(9-1) \times 10^{-4}\right]$$

$$= 4\pi \times 75 \times 10^{-3} \times 8 \times 10^{-4} = 7.5 \times 10^{-4}\ \text{J}$$

119. (500) $P = P_0 + \frac{2T}{R} \Rightarrow P - P_0 = \frac{2T}{R}$

$$500 = \frac{2 \times T}{2 \times 10^{-3}}$$

$T = 500 \times 10^{-3}$
So, $x = 500$

120. (a) $E_i = T \times 4\pi r_i^2$

We have, $V_i = V_f \Rightarrow \frac{4}{3}\pi r_i^3 = 64 \times \frac{4}{3}\pi r_f^3 \Rightarrow r_i^3 = 64 r_f^3$

$\Rightarrow r_i = 4r_f$

So, $E_f = T \times 64 \times 4\pi r_f^2 = 256T \times \pi\frac{r_i^2}{16} = 16\pi T r_i^2$

So, $\Delta E = E_f - E_i = 12\pi T r_i^2 = 12\pi \times 0.075 \times 0.1^2$
$= 2.82 \times 10^{-4}\ \text{J} \simeq 2.8 \times 10^{-4}\ \text{J}$

121. (b) Pressure at same horizontal level is equal

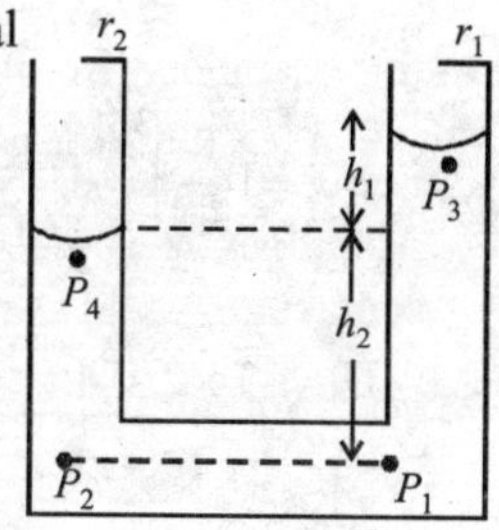

So, $P_1 = P_2$

$$P_3 = P_{atm} - \frac{2T}{r_1}$$

$$\& \ P_4 = P_{atm} - \frac{2T}{r_2}$$

$$P_2 = P_4 + \rho g h_2 = P_{atm} - \frac{2T}{r_2} + \rho g h_2$$

$$\& \ P_1 = P_3 + \rho g(h_1 + h_2) = P_{atm} - \frac{2T}{r_1} + \rho g(h_1 + h_2)$$

$$\therefore P_{atm} - \frac{2T}{r_1} + \rho g(h_1 + h_2) = P_{atm} - \frac{2T}{r_2} + \rho g h_2$$

$$\therefore \rho g h_1 = 2T\left[\frac{1}{r_1} - \frac{1}{r_2}\right]$$

$$= 2 \times 7.3 \times 10^{-2}\left[\frac{1}{2.5 \times 10^{-3}} - \frac{1}{4 \times 10^{-3}}\right]$$

$\Rightarrow h_1 = 2.19 \times 10^{-3}\ \text{m} = 2.19\ \text{mm}$

122. (2)

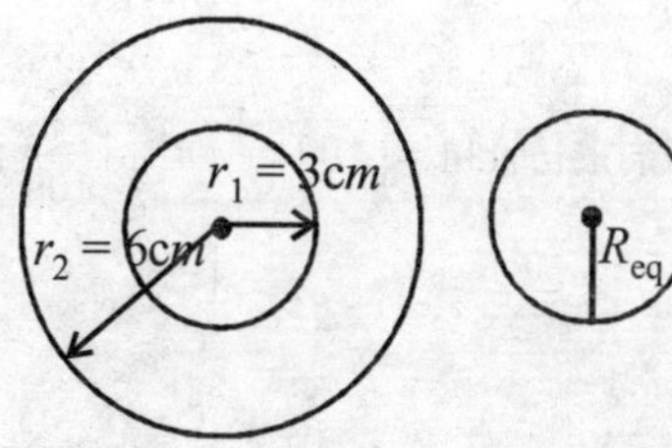

Excess pressure inside bigger bubble $= \frac{4T}{r_2}$

So, Excess pressure inside the smaller soap bubble

$$\Delta P_1 = \frac{4T}{r_1} + \frac{4T}{r_2}$$

The excess pressure inside the equivalent soap bubble.

$$\Delta P_2 = \frac{4T}{R_{eq}}$$

$$\Delta P_1 = \Delta P_2$$

$$\Rightarrow \frac{4T}{R_{eq}} = \frac{4T}{r_1} + \frac{4T}{r_2} \Rightarrow R_{eq} = \frac{r_1 r_2}{r_1 + r_2}$$

Putting $r_1 = 3\text{cm}, r_2 = 6\text{ cm}$

$\Rightarrow R_{eq} = 2\text{ cm}$

123. (a) As volume remain same i.e volume of two smaller drops will be equal to volume of one big drop.

$$2 \times \frac{4}{3}\pi r^3 = \frac{4}{3}\pi R^3 \Rightarrow 2r^3 = R^3$$

$$\Rightarrow 2 = \left(\frac{R}{r}\right)^3 \Rightarrow \frac{R}{r} = 2^{\frac{1}{3}}$$

$$\frac{U_i}{U_f} = \frac{T \times 2 \times (4\pi r^2)}{T \times 4\pi R^2} = \frac{2r^2}{2^{\frac{2}{3}} R^2} = 2^{\frac{1}{3}}$$

124. (a) Let R be the radius of curvature of common surface

$P_1 = P_0 + \frac{4T}{a}$ and

$P_2 = P_0 + \frac{4T}{b}$

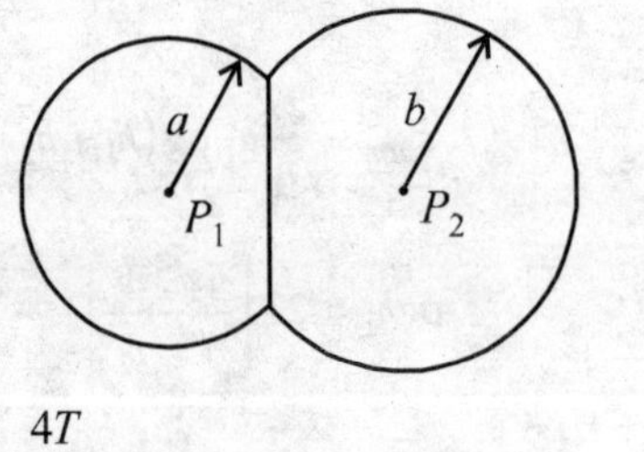

And $P_1 - P_2 = \frac{4T}{R}$

$$\left(P_0 + \frac{4T}{a}\right) - \left(P_0 + \frac{4T}{b}\right) = \frac{4T}{R}$$

$$\Rightarrow \frac{1}{a} - \frac{1}{b} = \frac{1}{R} \qquad \therefore \quad R = \frac{ab}{(b-a)}$$

125. (25) Volume of oleic acid $= \left(100 \times \frac{4}{3}\pi r^3\right) \times \frac{1}{100} = A \times t$

$[\because V_{\text{oleic acid}} = V_{\text{thin film}}]$

$$r = \left(\frac{3}{40\pi}\right)^{\frac{1}{3}} \times 10^{-3} \text{ cm} = \left(\frac{3}{40\pi}\right)^{\frac{1}{3}} \times 10^{-5} \text{ m}$$

and $A = 4\text{ cm}^2 = 4 \times 10^{-4}\text{ m}^2$

$$\frac{4}{3}\pi r^3 = A \times t$$

$$t = \frac{4}{3}\frac{\pi r^3}{A} = \frac{4}{3}\pi\left[\left(\frac{3}{40\pi}\right)^{\frac{1}{3}}\right]^3 \times (10^{-5})^3 \times \frac{1}{4 \times 10^{-4}}$$

$\Rightarrow t = 25 \times 10^{-14}$ m

$\therefore x = 25.$

126. (c) As volume remains unchanged

$$n \times \frac{4}{3}\pi r^3 = \frac{4}{3}\pi R^3 \Rightarrow nr^3 = R^3$$

Work done $= T \times$ Increment in surface area

$$W = T[4\pi nr^2 - 4\pi R^2] \quad ...(i)$$

By conservation of energy

Loss in thermal energy = work done against surface tension

$$JQ = W$$

Where J = mechanical equivalent of heat

$$Q = \frac{W}{J} = \frac{4\pi T(nr^2 - R^2)}{J} \qquad \text{[Using (i)]}$$

Heat energy per unit volume

$$Q = \frac{4\pi T}{Jn \cdot \frac{4}{3}\pi r^3}[nr^2 - R^2] = \frac{4\pi T}{J \times \frac{4}{3}\pi}\left[\frac{1}{r} - \frac{R^2}{nR^3}\right]$$

$$= \frac{3T}{J}\left[\frac{1}{r} - R^2\right] = \frac{3T}{J}\left[\frac{1}{r} - \frac{1}{R}\right]$$

127. (101) Given : Radius of capillary tube,

$r = 0.015\text{ cm} = 15 \times 10^{-5}$ mm

$h = 15\text{ cm} = 15 \times 10^{-2}$ mm

Using, $h = \frac{2T\cos\theta}{\rho g r}$ $\qquad [\cos\theta = \cos 0° = 1]$

Surface tension,

$$T = \frac{rh\rho g}{2} = \frac{15 \times 10^{-5} \times 15 \times 10^{-2} \times 900 \times 10}{2} = 101$$

milli newton m^{-1}

128. (c) According to question, pressure inside, 1st soap bubble,

$$\Delta P_1 = P_1 - P_0 = 0.01 = \frac{4T}{R_1} \quad ...(i)$$

$$\text{And } \Delta P_2 = P_2 - P_0 = 0.02 = \frac{4T}{R_2} \quad ...(ii)$$

Dividing, equation (ii) by (i),

$$\frac{1}{2} = \frac{R_2}{R_1} \Rightarrow R_1 = 2R_2$$

$$\text{Volume } V = \frac{4}{3}\pi R^3 \Rightarrow \frac{V_1}{V_2} = \frac{R_1^3}{R_2^3} = \frac{8R_2^3}{R_2^3} = \frac{8}{1}$$

129. (d) For the drops to be in equilibrium upward force on drop = downward force on drop

$$T.2\pi R = \frac{4}{3}\pi R^3 dg - \frac{2}{3}\pi R^3 \rho g$$

$$\Rightarrow T(2\pi R) = \frac{2}{3}\pi R^3(2d - \rho)g$$

$$\Rightarrow T = \frac{R^2}{3}(2d - \rho)g \Rightarrow R = \sqrt{\frac{3T}{(2d - \rho)g}}$$

130. (b) As we know that

$$\frac{2T\cos\theta}{r\rho g}=h.\quad \text{Given, } \frac{T_{Hg}}{T_{Water}}=7.5$$

$$\frac{\rho_{Hg}}{\rho_W}=13.6\ \&\ \frac{\cos\theta_{Hg}}{\cos\theta_W}=\frac{\cos 135°}{\cos 0°}=\frac{1}{\sqrt{2}}$$

$$\frac{r_{Hg}}{r_{Water}}=\left(\frac{T_{Hg}}{T_W}\right)\left(\frac{\rho_W}{\rho_{Hg}}\right)\left(\frac{\cos\theta_{Hg}}{\cos\theta_W}\right)$$

$$=7.5\times\frac{1}{13.6}\times\frac{1}{\sqrt{2}}=0.4=\frac{2}{5}$$

131. (d) We have, $h=\dfrac{2T\cos\theta}{r\rho g}$

Mass of the water in the capillary

$$m=\rho V=\rho\times\pi r^2 h=\rho\times\pi r^2\times\frac{2T\cos\theta}{r\rho g}\Rightarrow m\propto r$$

$$\therefore \frac{m_1}{m_2}=\frac{r}{2r}\quad \text{or } m_2=2m_1=2\,M.$$

132. (a) Inside a drop or an air bubble, towards concave side

$$P=P_0+\frac{2T}{R}\Rightarrow \Delta P=\frac{2T}{R}$$

Here, the figure shows that the water between the plates formed a curved surface, similar to that formed by air bubble inside the water. So, pressure in water between the plates is lowered by $\dfrac{2T}{R}$.

133. (a) When the bubble gets detached,

Buoyant force = force due to surface tension from diagram

$$\sin\theta=\frac{r}{R}$$

$$\rho_w Vg=(T\sin\theta)\times(2\pi r)$$

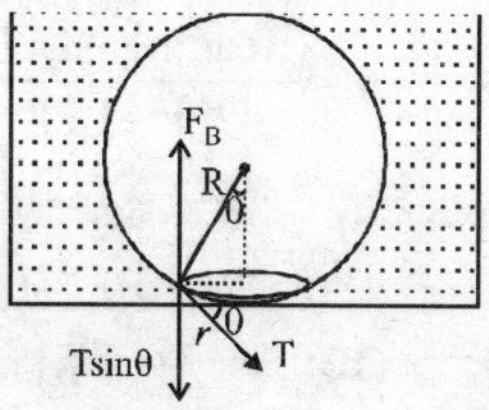

$$\Rightarrow \rho_w\times\frac{4}{3}\pi R^3 g=T\cdot\frac{r}{R}\times 2\pi r$$

$$\Rightarrow \rho_w\frac{4}{3}R^3 g=\frac{2T}{R}r^2\Rightarrow r=R^2\sqrt{\frac{2\rho_w g}{3T}}$$

134. (d) When drops combine to form a single drop of radius R.

$$\Delta E=T[4\pi R^2-n\times 4\pi r^2]$$

$$=4\pi T\left[R^2-\frac{R^3}{r^3}r^2\right]\qquad [\because R^3=nr^3]$$

$$=4\pi TR^3\left[\frac{1}{R}-\frac{1}{r}\right]$$

So energy released, $|E|=4\pi TR^3\left[\dfrac{1}{r}-\dfrac{1}{R}\right]$

If this energy is converted into kinetic energy then

$$\frac{1}{2}mv^2=4\pi R^3 T\left[\frac{1}{r}-\frac{1}{R}\right]$$

$$\frac{1}{2}\times\left[\frac{4}{3}\pi R^3\rho\right]v^2=4\pi R^3 T\left[\frac{1}{r}-\frac{1}{R}\right]$$

$$\Rightarrow v^2=\frac{6T}{\rho}\left[\frac{1}{r}-\frac{1}{R}\right]\Rightarrow v=\sqrt{\frac{6T}{\rho}\left[\frac{1}{r}-\frac{1}{R}\right]}$$

135. (a) As we know,

$$\rho_{Excess}=h\rho g+\frac{2s}{r}$$

$$\Rightarrow 1100=h\times 10^3\times 9.8+\frac{2\times 6\times 10^{-2}}{10^{-3}}$$

$$\Rightarrow 1100=9800h+120\Rightarrow 9800h=1100-120$$

$$\Rightarrow h=\frac{980}{9800}=0.1\text{ m}$$

136. (a) Acceleration due to gravity changes with the depth,

$$g'=g\left(1-\frac{d}{R}\right)$$

Pressure, $P=\rho gh$, hence ratio, $\dfrac{x}{y}$ is $\left(1-\dfrac{d}{R}\right)$

137. (c) Due to waxing, adhesion decreases so, contact angle increases i.e. $\theta\uparrow\uparrow$. As $h=\dfrac{2T\cos\theta}{r\rho g}$. So, $h\downarrow\downarrow$.

138. (b) Surface tension of a liquid decreases with the rise in temperture. At the boiling point of liquid, surface tension is zero.

Capillary rise $h=\dfrac{2T\cos\theta}{rdg}$

As surface tension T decreases with rise in temperature hence capillary rise also decreases.

139. (d) Let T is the force due to surface tension per unit length, then

$F=2lT$

l = length of the slider.

At equilibrium, $F=W$

$\therefore 2Tl=mg$

$$\Rightarrow T=\frac{mg}{2l}=\frac{1.5\times 10^{-2}}{2\times 30\times 10^{-2}}=\frac{1.5}{60}=0.025\text{ Nm}^{-1}$$

140. (c) Work done = increase in surface area × surface tension

$$\Rightarrow W=2T4\pi[(5^2)-(3)^2]\times 10^{-4}$$

$$=2\times 0.03\times 4\pi[25-9]\times 10^{-4}\text{ J}$$

$$=0.4\pi\times 10^{-3}\text{ J}=0.4\pi\text{ mJ}$$

141. (c) Water fills the tube entirely in gravity less condition i.e., 20 cm.

142. (a) Let pressure outside be P_0 and r and R be the radius of smaller bubble and bigger bubble respectively.

$\therefore$ Pressure P_1 For smaller bubble $=P_0+\dfrac{2T}{r}$

P_2 For bigger bubble $=P_0+\dfrac{2T}{R}\quad (R>r)$

$\therefore P_1>P_2$

hence air moves from smaller bubble to bigger bubble.

Thermal Properties of Matter

10

1. **(b)** The equation for Celcius and Fahrenheit scale temperature is

$$\frac{C}{5}=\frac{F-32}{9} \Rightarrow C=\frac{5F}{9}-\frac{160}{9}$$

$\therefore$ C – F graph is

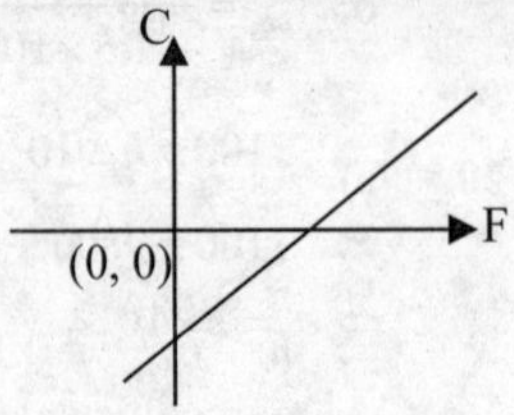

2. **(c)** Since $\frac{F-32}{9}=\frac{C}{5}$

$$\Rightarrow \Delta C=\frac{5}{9}\Delta F$$

$$\Rightarrow 40=\frac{5}{9}\Delta F \Rightarrow \Delta F=72°F$$

3. **(d)** Given,
Boiling point of water, = UFP 65°
Freezing point of water, = LFP – 15°

$$\frac{X-LFP}{UFP-LFP}=\frac{T_F-32}{212-32}$$

$$\Rightarrow \frac{-95-(-15)}{65-(-15)}=\frac{T_F-32}{180}$$

$$\Rightarrow \frac{-80}{80}=\frac{T_F-32}{180} \Rightarrow T_F=-180+32$$

$$\Rightarrow T_F=-148°\,F$$

4. **(4)** Stress $=$ Y $\times$ strain $= Y\times\frac{\Delta\ell}{\ell}$

$$= Y\times\frac{\ell\alpha\Delta T}{\ell}=Y\alpha\Delta T \quad (\because \Delta\ell=\ell\alpha\Delta T)$$

Compressive Tension = Stress × Area of cross section
$= YA\alpha\Delta T = 2\times10^{11}\times10^{-4}\times10^{-5}\times(200-0)=4\times10^{4}N$

5. **(60)** We have
$|\Delta l_i| = |\Delta l_f|$

$$\Rightarrow l\alpha\Delta T=\frac{Mgl}{YA} \Rightarrow M=\frac{YA\alpha\Delta T}{g}$$

$$\Rightarrow M=\frac{2\times10^{11}\times3\times10^{-6}\times2\times10^{-5}\times50}{10}$$

$$\Rightarrow M=60\text{ kg}$$

6. **(d)** The volume expansion coefficient of the gas,

$$\gamma=\frac{dv}{vdT} \quad ...(i)$$

PT^2 = constant, using PV = nRT

$$P=\frac{nRT}{V}$$

$$PT^2=\frac{nRT}{V}\times T^2=\text{constant}$$

$\Rightarrow T^3 = KV$ ...(ii)

So, $\frac{d}{dT}(KV)=3T^2 \Rightarrow \frac{KdV}{dT}=3T^2$

$$\frac{dV}{dT}=\frac{3T^2}{K}$$

Put this value in equation (i)

$$\Rightarrow \gamma=\frac{3T^2}{KV}=\frac{3T^2}{T^3}=\frac{3}{T}$$

7. **(313)** We know that

$$\frac{\text{Reading} - \text{Lower fixed point}}{\text{Upper fixed point} - \text{Lower fixed point}}=\text{Constant}$$

$$\frac{41°-5°}{95°-5°}=\frac{C-0°}{100°-0°}$$

$$\Rightarrow C=\frac{36}{90}\times100=40°C=313\text{ K}$$

8. **(b)**

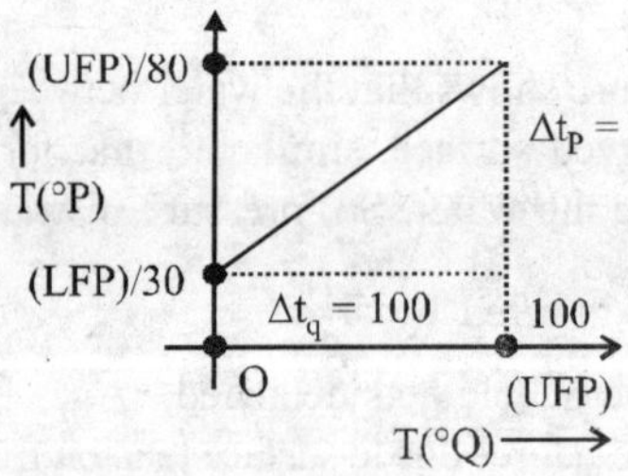

As, $\frac{\text{Reading}-\text{LFP}}{\text{UFP}-\text{LFP}}=\text{Constant}$

$$\frac{t_P-30}{180-30}=\frac{t_Q-0}{100-0} \Rightarrow \frac{t_P-30}{150}=\frac{t_Q}{100}$$

9. **(12)** As $\Delta D=\alpha D\Delta T \Rightarrow \Delta D=1.6\times10^{-5}\times5\times150$
$=1200\times10^{-5}=12\times10^{-3}$ cm

10. **(60)** For no change in length
$\Rightarrow \Delta\ell_{Iron}=\Delta\ell_{Brass}$
$\Rightarrow \ell_I\alpha_I\Delta T=\ell_B\alpha_B\Delta T$
$\Rightarrow \ell_I\times1.2\times10^{-5}=40\times1.8\times10^{-5}$
$\Rightarrow \ell_I=\frac{40\times1.8}{1.2}=\frac{40\times3}{2}=60\text{ cm}$

11. **(d)** We know that
$\Delta l = l_0\alpha\Delta T$
$\Rightarrow (6.241-6.230)=6.230\times1.4\times10^{-5}\times\Delta T$
$\Rightarrow 0.011=6.230\times1.4\times10^{-5}\times\Delta T$
$\Rightarrow \Delta T=126.1$
$\Rightarrow T_f-27=126.1$
$\Rightarrow T_f=153.11$ °C.
So nearest option is (d).

12. **(b)** We have $\Delta V = V_0 \gamma \Delta T$

$\Delta V = a^3.(3\alpha)\,\Delta T$

Now, $6a^2 = 24$ [$\because$ Total surface area of cube $= 6a^2$]

$\Rightarrow a^2 = 4 \Rightarrow a = 2$

So, $\Delta V = 2^3(3 \times 5 \times 10^{-4}) \times 10 = 1200 \times 10^{-4}\,m^3$

$= 1200 \times 10^2\,cm^3 = 1.2 \times 10^5\,cm^3$

13. **(8)** Original length $= l_0(1 + \alpha \Delta T)$

Compression $= l_0 \alpha \Delta T$

$\therefore$ stress $= \gamma$(strain) $= \dfrac{\gamma l_0 \alpha \Delta T}{l_0(1+\alpha \Delta T)} = \gamma \alpha \Delta T$

[$\because$ α is very small]

So, F = Stress × Area $= \gamma \alpha A \Delta T$

$\Rightarrow F = (10 \times 10^{-4})(2 \times 10^{11})(10^{-5})(400)$

$\Rightarrow F = 8 \times 10^5\,N = x \times 10^5\,N \Rightarrow x = 8$

14. **(c)** PT^3 = constant

$\left(\dfrac{nRT}{V}\right)T^3$ = constant

$T^4 V^{-1}$ = constant

[using $PV = nRT$]

Taking log both sides

$4 \ln T - 1 \ln V = \ln$ constant

differention w.r.t v

$$4\frac{dT}{T} = \frac{dV}{V} \quad ...(i)$$

$$\Rightarrow V\left(\frac{4}{T}\right)dT = dV$$

$$dV = V\gamma \Delta T \quad ...(ii)$$

from (i) & (ii)

$$\gamma = \frac{4}{T}$$

15. **(5)** Elastic energy stored, $= \dfrac{Y}{2}(\text{strain})^2 \times \text{area} \times \text{length}$

$\therefore$ Elastic energy stored per unit length

$= \dfrac{Y}{2}(\text{strain})^2 \times \text{area}$

$= \dfrac{Y}{2}(\alpha \Delta T)^2 \times A \left[\because \text{strain} = \dfrac{\Delta \ell}{\ell} = \alpha \Delta T\right]$

$= \dfrac{10^{11}}{2} \times (10^{-5} \times 10)^2 \times 10^{-2} = \dfrac{10^{11}}{2} \times (10^{-4})^2 \times 10^{-2} = 5\,J/m$

16. **(c)** As rod is lying freely so in free expansion, thermal stress = 0, but length of the rod increases on heating. Statement R is also true but it is not correct reason for A.

17. **(a)** $\Delta V = V\gamma \Delta T$

Since, $V = a^3$ and $\gamma = 3\alpha$

$\therefore \Delta V = 3a^3 \alpha \Delta T$

18. **(a)** $\Delta l_{net} = L_1 \alpha_1 \Delta T + L_2 \alpha_2 \Delta T$

$(L_1 + L_2)\alpha \Delta T = (L_1 \alpha_1 + L_2 \alpha_2)\Delta T$

$$\alpha = \frac{L_1 \alpha_1 + L_2 \alpha_2}{L_1 + L_2}$$

19. **(20.00)** $\Delta V_{beaker} = \Delta V_{mercury}$

$\Rightarrow V_0 \gamma_{beaker} = V_m \gamma_M$

$\therefore V_m = \dfrac{V_0 \gamma_{beaker}}{\gamma_M}$

$\Rightarrow V_m = \dfrac{500 \times 6 \times 10^{-6}}{1.5 \times 10^{-4}} = 20$ cc.

20. **(a)** As $\alpha = \dfrac{\Delta l}{l \Delta T} = \dfrac{0.02}{100 \times 10}$

$\Rightarrow \alpha = 2 \times 10^{-5}$

Volume coefficient of expansion, $\gamma = 3\alpha = 6 \times 10^{-5}$

$\because \rho = \dfrac{M}{V}$

$\dfrac{\Delta V}{V} \times 100 = \gamma \Delta T$

$= (6 \times 10^{-5} \times 10 \times 100) = 6 \times 10^{-2}$

Volume increase by 0.06% therefore density decrease by 0.06%.

21. **(60.00)** Volume, $V = lbh$

$\Rightarrow \dfrac{\Delta V}{V} = \dfrac{\Delta l}{l} + \dfrac{\Delta b}{b} + \dfrac{\Delta h}{h}$

$\therefore \gamma = \dfrac{\Delta V}{V \Delta T} = \dfrac{\Delta \ell}{\ell \Delta T} + \dfrac{\Delta b}{b \Delta T} + \dfrac{\Delta h}{h \Delta T}$

$\Rightarrow \gamma = 5 \times 10^{-5} + 5 \times 10^{-6} + 5 \times 10^{-6}$

$= 60 \times 10^{-6}/°C$

$\therefore$ Value of $C = 60.00$

22. **(None)** $\Delta_{temp} = \Delta_{load}$

and $A = \pi r^2$

$= \pi(10^{-3})^2 = \pi \times 10^{-6}$

$L \alpha \Delta T = \dfrac{FL}{AY}$

$\Rightarrow 0.2 \times 10^{-5} \times 20 = \dfrac{F \times 0.2}{(\pi \times 10^{-6}) \times 10^{11}}$

$\therefore F = 20\pi N$

$\therefore m = \dfrac{F}{g} = 2\pi = 6.28$ kg

23. **(a)** Change in length in both rods are same i.e.

$\Delta \ell_1 = \Delta \ell_2$

$\ell \alpha_1 \Delta \theta_1 = \ell \alpha_2 \Delta \theta_2$

$\dfrac{\alpha_1}{\alpha_2} = \dfrac{\Delta \theta_2}{\Delta \theta_1}$ $\left[\because \dfrac{\alpha_1}{\alpha_2} = \dfrac{4}{3}\right]$

$\dfrac{4}{3} = \dfrac{\theta - 30}{180 - 30} \Rightarrow \theta = 230°C$

24. (a) Let required temperature = T°C

$\Rightarrow T°C = \frac{x_0}{2} - \frac{x_0}{3} = \frac{x_0}{6}$

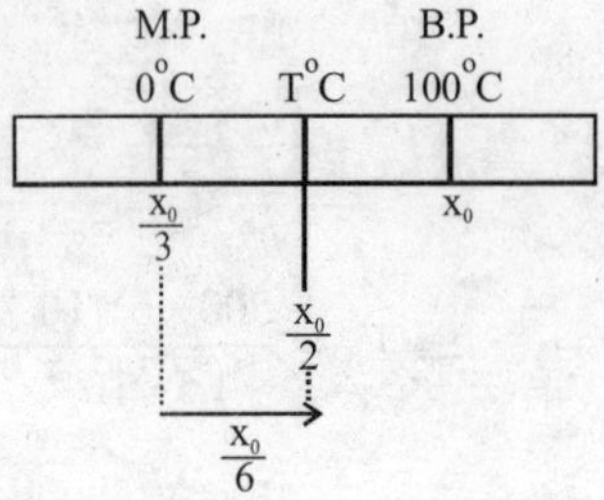

$\& \left(x_0 - \frac{x_0}{3}\right) = (100 - 0°C)$

$\Rightarrow \frac{2x_0}{3} = 100 \Rightarrow x_0 = \frac{300}{2}$

$\Rightarrow T°C = \frac{x_0}{6} = \frac{150}{6} = 25°C$

25. (a) Young's modulus $Y = \frac{\text{stress}}{\text{strain}}$

$= \frac{F/A}{A(\Delta\ell/\ell)}$

Using, coefficient of linear expansion,

$\alpha = \frac{\Delta\ell}{\ell\Delta T} \Rightarrow \frac{\Delta\ell}{\ell} = \alpha\Delta T$

$\therefore Y = \frac{F}{A(\alpha\Delta T)}$

26. (b) $\Delta V_c = \Delta V_l$

$\Rightarrow V_c\gamma_c\Delta T = V_l\Delta T\gamma_l$

$\Rightarrow \frac{V_l}{V_c} = \frac{\gamma_c}{\gamma_l} = \frac{1}{4}$

27. (c) $\frac{\text{Reading on any scale} - \text{LFP}}{\text{UFP} - \text{LFP}}$

= constant for all scales

$\frac{340-273}{373-273} = \frac{°Y-(-160)}{-50-(-160)}$

$\Rightarrow \frac{67}{100} = \frac{Y+160}{110}$

$\therefore Y = -86.3° Y$

28. (d) $\Delta L_{\text{ring}} = (\Delta L)_{\text{heat}} = L\alpha\Delta T$

Also, $(\Delta L) = \frac{FL}{YS}$.

So, $L\alpha\Delta T = \frac{FL}{Y} \Rightarrow F = YS\alpha\Delta T$

So, $F_{\text{req}} = 2F = 2YS\alpha\Delta T$

29. (d) Energy is conserved, so $mgh = ms\Delta T$

$\therefore \Delta T = \frac{gh}{s} = \frac{10\times 200}{4200} K = \frac{10}{21} K = 0.48 K.$

30. (360)

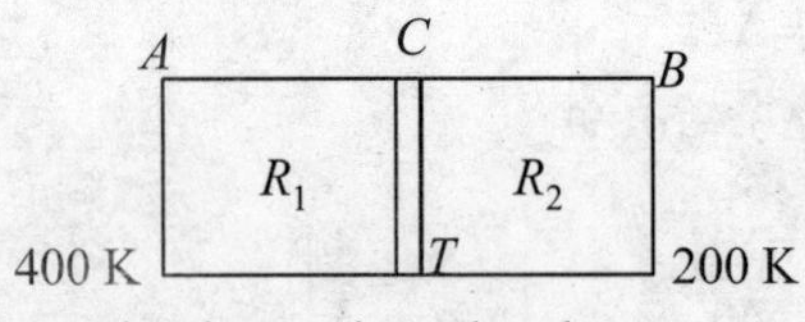

Thermal resistance in each rod

$R_1 = \frac{l_1}{K_1A_1},\ R_2 = \frac{l_2}{K_2A_2}$

Rate of change of heat, $\frac{dQ}{dt} = \frac{\Delta T}{R}$

$\Rightarrow \left(\frac{dQ}{dt}\right)_1 = \left(\frac{dQ}{dt}\right)_2 \Rightarrow \frac{400-T_C}{R_1} = \frac{T_C-200}{R_2}$

$\Rightarrow \frac{400-T_C}{T_C-200} = \frac{R_1}{R_2} = \left(\frac{l_1}{l_2}\right)\left(\frac{r_2}{r_1}\right)^2 \times \frac{K_2}{K_1}$

$= \frac{1}{2} \times \left(\frac{1}{2}\right)^2 \times 2 \Rightarrow \frac{400-T_C}{T_C-200} = \left(\frac{1}{4}\right)$

$\frac{400-T_C}{T_C-200} = \frac{1}{4}$

$\Rightarrow 1600 - 4T_C = T_C - 200$

$\Rightarrow 5T_C = 18000 \Rightarrow T_C = 360\ K$

31. (a) Given : $l = 9$ cm, $d = 4$ am, $\alpha = 3.1 \times 10^{-5} K^{-1}$, $m = 0.1$ kg, $C_v = 900\ J/Kg^{-1} K^{-1}$, $\Delta Q = 8.1 \times 10^2$ J

$\Delta Q = ms\Delta T$

$\Rightarrow 8.1\times 10^2 = 0.1\times 900\times \Delta T \Rightarrow \Delta T = 9\ K$

$\Delta A = A_0 2\alpha\Delta T = 2.0\times 10^{-6}\ m^2\ (\because \beta = 2\alpha.)$

32. (c) The heat required to raise the temperature of bullet, $Q_1 = mC\Delta T$

Heat required to melt the bullet, $Q_2 = mL$

$Q = Q_1 + Q_2$

$625 = ms\Delta T + mL$

$\Rightarrow 625 = m[125 \times 300 + 2.5 \times 10^4]$

$\Rightarrow 625 = m[62500]$

$\therefore m = \frac{1}{100}$ kg = 10 grams

33. (b) Given, Radius of bigger body, $R_1 = 0.8$m

Radius of smaller body, $R_2 = 0.2$m

Temperature of smaller body, $T_1 = 800$ K

Temperature of bigger body, $T_2 = 400$ K

$E = \sigma eAT^4 \Rightarrow E \propto AT^4$

Here, σ = Stefan's constant

$\therefore \frac{E_{\text{smaller}}}{E_{\text{larger}}} = \frac{4\pi(0.2)^2 \times 800^4}{4\pi(0.8)^2 \times 400^4} = 1 \quad \therefore E_{\text{larger}} = E_{\text{smaller}} = E$

34. (b) Given, mass of ice, $m = 10^{-3}$ kg

Specific heat of ice, $S_i = 2100\ Jkg^{-1} K^{-1}$

Specific heat of water, $S_w = 4180\ Jkg^{-1} K^{-1}$

Specific heat of steam, $S_v = 1920\ Jkg^{-1} K^{-1}$

Latent heat of ice, $L_f = 3.35 \times 10^5\ Jkg^{-1}$ and

Latent heat of steam, $L_v = 2.25 \times 10^6\ Jkg^{-1}$

$\Delta Q_1 = m \times S_1 \times \Delta T = 10^{-3} \times 2100 \times 10 = 21$ J

$\Delta Q_2 = m \times L_f = 10^{-3} \times 3.35 \times 10^5 = 335$ J

$\Delta Q_3 = m \times S_w \times \Delta T = 10^{-3} \times 4180 \times 100 = 418$ J

$\Delta Q_4 = m \times L_v = 10^{-3} \times 2.25 \times 10^6 = 2250$ J

$\Delta Q_5 = m \times S_v \times \Delta T = 10^{-3} \times 1920 \times 10 = 19.2$ J

$\Delta Q_{\text{net}} = \Delta Q_1 + \Delta Q_2 + \Delta Q_3 + \Delta Q_4 + \Delta Q_5 = 3043.2$ J

35. **(40)** The equivalent thermal circuit diagram is

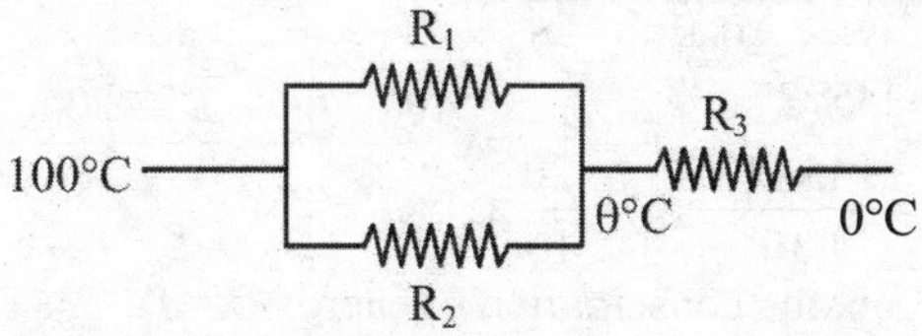

The thermal resistance across the rods are

$$R_1 = \frac{2L}{K_1A} \text{ and } R_2 = \frac{2L}{K_2A}$$

$$\text{Also, } R_3 = \frac{L}{K_3A}$$

$$\Rightarrow \frac{\theta - 100}{\frac{R_1R_2}{R_1 + R_2}} + \frac{\theta - 0}{R_3} = 0 \Rightarrow \theta = 40°C$$

36. **(d)** When heat is supplied the temperature of ice increase from –10°C to 0°C. It represent a straight line. At 0°C the heat is used in converting into water. Here it represents horizontal straight line. After that temperature of water rises from 0°C to 100°C. It represent again straight line. At 100°C the heat is used in converting water into steam. This stage is represented by horizontal straight line.

37. **(a)** Given $\Delta Q = 184 \times 10^3$
m = 0.600 kg at –12°C
s = 2222.3 J/kg/°C
and, $L = 336 \times 10^3$ J/kg
Heat required for conversion of ice at –12°C to 0°C is given as
$Q_1 = 0.600 \times 2222.3 \times 12 = 16000.56$ J
Remaining heat $\Delta Q_1 = 184000 - 16000.56$
= 167999.44 J
For melting at 0°C
$\Delta Q_2 = 0.600 \times 336000 = 201600$ J needed
∴ 100% ice is not melted. So, final temperature of system is 0°C.
Amount of ice melted
$167999.44 = m \times 336000 = 0.4999$ kg
∴ mass of water = 0.4999 kg
and, mass of ice = 0.6 – 0.4999 = 0.1001 kg

$$\therefore \text{Ratio} = \frac{0.1001}{0.4999} \approx 1:5$$

38. **(c)** In series $\frac{dQ}{dt}$ = constant

$$\Rightarrow \frac{T_1 - T_2}{R} = \text{constant}$$

$$\Rightarrow \frac{450 - T}{\frac{l_2}{K_2A_2}} = \frac{T - 0}{\frac{l_1}{K_1A_1}} \Rightarrow \frac{450 - T}{\frac{l_1}{l_2} \cdot \frac{K_2}{K_1} \cdot \frac{A_2}{A_1}} = T$$

$$\Rightarrow \frac{450 - T}{2.9.\frac{1}{2}} = T \Rightarrow 450 - T = 9T$$

$\Rightarrow 450 = 10T \Rightarrow T = 45$ °C

39. **(b)** Given that latent heat of fusion of ice $L = 3.4 \times 10^5$ J kg^{-1}
Thermal conductivity of insulation wall, $K = 0.05$ Wm^{-1} C^{-1}
Change in temperature, $\Delta T = 40°C$
Area of ice cube, $A = 2(0.6 \times 0.5 + 0.5 \times 0.2 + 0.2 \times 0.6)$
$= 2(0.3 + 0.1 + 0.12) = 1.04$ m^2
We have

$$\frac{dQ}{dt} = \frac{KA\Delta T}{\ell} = \frac{0.05 \times 1.04 \times 40}{10^{-2}} = 2.08 \times 10^2 \text{ J/s}$$

$$\frac{dQ}{dt} = mL$$

$2.08 \times 10^2 = m \times 3.4 \times 10^5$

$$m = \frac{2.08}{3.4 \times 10^3} \Rightarrow 0.61 \times 10^{-3} \text{ kg/s}$$

$= 61 \times 10^{-5}$ Kg/s

40. **(90)** Energy released by water = M(ΔT)S
$= 0.3 \times 25 \times 4200 = 31500$ J
Let m kg ice metls = mL
$m \times 3.5 \times 10^5 = 31500$ [∵ Heat lost = Heat absorbed]

$$\Rightarrow m = \frac{31500 \times 10^{-5}}{3.5} = 9000 \times 10^{-5}$$

m = 0.09 kg = 90 g ⇒ x = 90

41. **(21)**

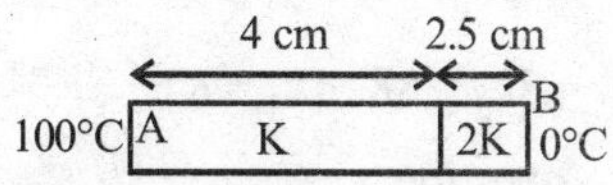

In series, $R_{eq} = R_1 + R_2$

$$\Rightarrow \frac{L}{KA} = \frac{L_1}{K_1A} + \frac{L_2}{K_2A}$$

$$\Rightarrow \frac{L}{K} = \frac{L_1}{K_1} + \frac{L_2}{K_2} \Rightarrow \frac{6.5}{K_{eq}} = \frac{4}{K} + \frac{2.5}{2K}$$

$$\Rightarrow \frac{6.5}{K_{eq}} = \frac{1}{K}\left(\frac{8 + 2.5}{2}\right) \Rightarrow \frac{6.5}{K_{eq}} = \frac{1}{K}\left(\frac{10.5}{2}\right)$$

$$\therefore K_{eq} = \frac{13K}{10.5} = \frac{26}{21}K = \left(1 + \frac{5}{21}\right)K$$

So, α = 21.

42. **(c)** Number of moles of gas, $n = \frac{44.8}{22.4} = 2$

$\Delta Q = nC_V \Delta T$ [$\because V = \text{constant} \Rightarrow C = C_V$]

$= 2.\frac{3}{2}R \times 20 = 60R = 60 \times 8.3 = 498$ J.

43. **(b)** 40% of K.E $= ms\,\Delta T + mL_f$

$$\Rightarrow \frac{40}{100} \times \frac{1}{2}mV^2 = ms(327 - 127) + mL_f$$

$$\Rightarrow \frac{20}{100} \times V^2 = 200s + L_f$$

$$\Rightarrow \frac{1}{5}V^2 = 200 \times 125 + 2.5 \times 10^4$$

$\Rightarrow V^2 = 1000 \times 125 + 2.5 \times 10^4 \times 5$
$\Rightarrow V^2 = 250000$
$\Rightarrow V = 500$ m/s

44. **(42)** Heat absorbed by water per min $= 2 \times 4200 \times 40$
= 336000J

Now, 8×10^3J of heat is produced by 1 gm.
So, 336000 J of heat is produced by 1 gm

$$= \frac{1}{8\times10^3}\times 336000 gm$$

So, rate of combustion = 42 gm / min.

45. **(c)** Here, heat lost by Cu block = Heat gain by ice

$$\Rightarrow M_{cu} \cdot S_{cu}(T_{500} - T_0) = M_{ice} L_f$$

$$\Rightarrow M_{ice} = \frac{M_{cu}.S_{cu}(T_{500}-T_0)}{L_f}$$

$$= \frac{5\times390\times500}{335\times10^3} = 2.9 \text{ kg}$$

46. **(b)** As Q = constant in series
So, $Q_{M1} = Q_{M2}$

$$\Rightarrow \frac{(100-80)}{\frac{16}{k_1A}} = \frac{80-0}{\frac{8}{k_2A}} \left[\because Q = \frac{T_1-T_2}{l/kA}\right]$$

$$\Rightarrow \frac{20}{16}\times k_1A = \frac{80}{8}k_2A \Rightarrow k_2 = \frac{20\times8}{16\times80}k_1$$

$$\Rightarrow k_2 = \frac{k_1}{8} \Rightarrow k_1 = 8k_2 = 8K \quad [k_2 = K]$$

47. **(c)** According to question

$$\frac{1}{4}\times(\text{K.E. of Hammer}) = \text{Heat energy received by nail}$$

$$\Rightarrow \frac{1}{4}\times\frac{1}{2}m_hv_h^2 = m_ns_n\Delta T$$

$$\Rightarrow \frac{1}{8}\times1.5\times60^2 = 0.1\times420\times\Delta T$$

$$\Rightarrow \Delta T = \frac{1.5\times60^2}{8\times0.1\times420} = 16.07\,°\text{C}$$

48. **(a)** Resistance of spherical shell of thickness dr and radius r is

$$dR = \frac{dr}{K(4\pi r^2)} \quad \left[\text{using } R = \frac{1}{K}\frac{\ell}{A}\right]$$

Total resistance

$$R = \int_{r_1}^{r_2} \frac{dr}{K(4\pi r^2)}$$

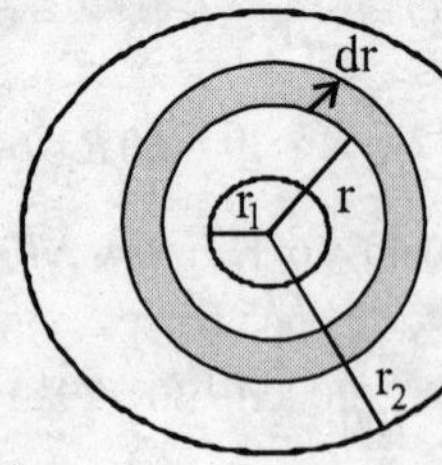

$$\Rightarrow \quad R = \frac{1}{4\pi K}\left(\frac{1}{r_1}-\frac{1}{r_2}\right)$$

$$\text{current} = \frac{\theta_2-\theta_1}{R} = \frac{4\pi K r_1 r_2}{r_2-r_1}(\theta_2-\theta_1)$$

49. **(2)** [As AB & CD identical]
$R_{AB} = R_{CD} = 10\ KW^{-1}$
$R_{AC} = R_{CB} = 5\ KW^{-1}$
[As AC = BC]
$R_{AC} = R_{CB} = 5\ KW^{-1}$
at point C
$Q = Q_1 + Q_2$

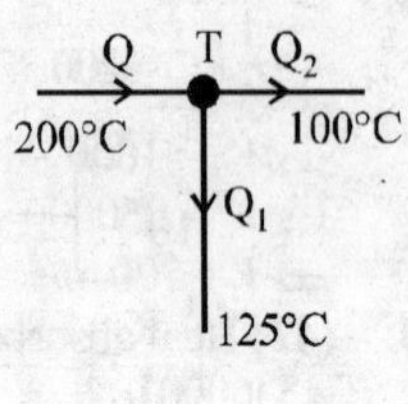

$$\frac{200-T}{5} = \frac{T-125}{10} + \frac{T-100}{5}$$

$$\Rightarrow T = 145°\text{C}$$

$$\Rightarrow P = \frac{145-125}{10}\text{W} = \frac{20}{10}w \Rightarrow P = 2\text{W}$$

50. **(a)** From the conservation of energy
Change in potential energy = Heat energy

$$\Rightarrow \quad mgh = mc\Delta T$$

$$\Rightarrow \Delta T = \frac{gh}{c} = \frac{10\times63}{4200 J/kgC} = 0.147°\text{C}$$

51. **(c)** Let m be the mass of each liquid, s_1, s_2 and s_3 be the specific heat of x, y and z.
When x and y are mixed
Heat gained by x = Heat lost by y
$ms_1(16-10) = ms_2(20-16)$

$$\Rightarrow s_1 = \frac{2}{3}s_2 \quad \text{...(i)}$$

When y and z are mixed
Heat gained by y = Heat lost by z
$\Rightarrow ms_2(26-20) = ms_3(30-26)$

$$\Rightarrow s_2\times6 = s_3\times4 \Rightarrow s_3 = \frac{3}{2}s_2 \quad \text{...(ii)}$$

From (i) and (ii)

$$s_3 = \frac{3}{2}\times\frac{3}{2}s_1 = \frac{9}{4}s_1$$

When x and z are mixed, let the final temperature be θ
$\Rightarrow ms_1(\theta-10) = ms_3(30-\theta)$

$$\theta - 10 = \frac{9}{4}(30-\theta) \quad \left(\because \frac{s_3}{s_1} = \frac{9}{4}\right)$$

$$\Rightarrow \theta = 23.84°\text{C}$$

52. **(a)** Equivalent thermal resistance,

$$R_{eq} = R_1 + R_2$$

$$\Rightarrow \quad \frac{2\ell}{K_{eq}\times A} = \frac{\ell}{K_1A} + \frac{\ell}{K_2A} \quad \left[\because R = \frac{1}{K}\frac{\ell}{A}\right]$$

$$\Rightarrow \quad \frac{2}{K_{eq}} = \frac{K_1+K_2}{K_1K_2} \Rightarrow K_{eq} = \frac{2K_1K_2}{K_1+K_2}$$

53. **(d)** Heat current is same in the two conductors

$$\therefore \quad \frac{\theta_2-\theta}{R_2} = \frac{\theta-\theta_1}{R_1}$$

$$\Rightarrow \quad \theta_2R_1 - \theta\times R_1 = \theta R_2 - \theta_1R_2$$

$$\Rightarrow \quad \theta = \frac{\theta_1R_2+\theta_2R_1}{R_1+R_2}$$

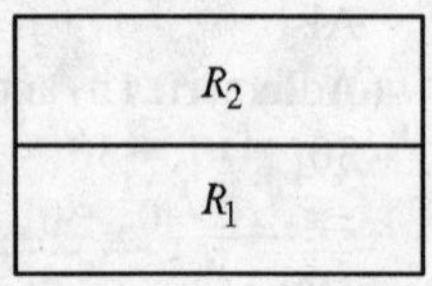

54. **(a)** As the rods are identical, so they have same length (l) and area of cross-section (A). They are connected in series. So, heat current will be same for all rods.

$$\text{Heat current} = \left(\frac{\Delta Q}{\Delta t}\right)_{AB} = \left(\frac{\Delta Q}{\Delta t}\right)_{BC} = \left(\frac{\Delta Q}{\Delta t}\right)_{CD}$$

$$\Rightarrow \frac{(100-70)K_1A}{l} = \frac{(70-20)K_2A}{l} = \frac{(20-0)K_3A}{l}$$

$\Rightarrow K_1(100-70) = K_2(70-20) = K_3(20-0)$

$\Rightarrow K_1(30) = K_2(50) = K_3(20)$

$\Rightarrow \frac{K_1}{10} = \frac{K_2}{6} = \frac{K_3}{15} \Rightarrow K_1 : K_2 : K_3 = 10 : 6 : 15$

$\Rightarrow K_1 : K_3 = 2 : 3$, and $K_2 : K_3 = 2 : 5$

55. (a) According to question, one half of its kinetic energy is converted into heat in the wood.

$\frac{1}{2}mv^2 \times \frac{1}{2} = ms\Delta T$

$\Rightarrow \Delta T = \frac{v^2}{4 \times s} = \frac{210 \times 210}{4 \times 4.2 \times 0.03 \times 1000} = 87.5°C$

56. (a) Here ice melts due to water.
Let the amount of ice melts $= m_{ice}$

$m_w s_w \Delta\theta = m_{ice} L_{ice}$

$\therefore m_{ice} = \frac{m_w s_w \Delta\theta}{L_{ice}}$

$= \frac{0.2 \times 4200 \times 25}{3.4 \times 10^5} = 0.0617 \text{ kg} = 61.7 \text{ g}$

57. (a) Heat given by water $= m_w C_w (T_{mix} - T_w)$

$= 200 \times 1 \times (31 - 25)$

Heat taken by steam $= m L_{stem} + m C_w (T_s - T_{mix})$

$= m \times 540 + m(1) \times (100 - 31) = m \times 540 + m(1) \times (69)$

From the principal of calorimeter,
Heat lost = Heat gained

$\therefore (200)(31-25) = m \times 540 + m(1)(69)$

$\Rightarrow 1200 = m(609) \Rightarrow m \approx 2.$

58. (50.00) Let Q_1, Q_2, Q_3 be the temperatures of container C_1, C_2 and C_3 respectively.
Using principle of calorimetry in container C_1, we have

ms $(\theta_1 - 60) = 2$ ms$(60 - \theta_2)$

$\Rightarrow \theta_1 - 60 = 120 - 2\theta_2$

$\Rightarrow \theta_1 = 180 - 2\theta_2$ ...(i)

For container C_2

ms $(\theta_2 - 30) = 2$ms $(30 - \theta_3)$

$\Rightarrow \theta_2 = 90 - 2\theta_3$ ...(ii)

For container C_3

2ms $(\theta_1 - 60) =$ ms $(60 - \theta_3)$

$\Rightarrow 2\theta_1 - 120 = 60 - \theta_3$

$\Rightarrow 2\theta_1 + \theta_3 = 180$ ...(iii)

Also, $\theta_1 + \theta_2 + \theta_3 = 3\theta$ ...(iv)

Adding (i), (ii) and (iii)

$3\theta_1 + 3\theta_2 + 3\theta_3 = 450$

$\Rightarrow \theta_1 + \theta_2 + \theta_3 = 150 \Rightarrow 3\theta = 150 \Rightarrow \theta = 50$ °C

59. (40) Using the principal of calorimetry

$M_{ice} L_f + m_{ice}(40-0) C_w = m_{stream} L_v + m_{stream}(100-40) C_w$

$\Rightarrow M(540) + M \times 1 \times (100-40) = 200 \times 80 + 200 \times 1 \times 40$

$\Rightarrow 600M = 24000 \Rightarrow M = 40g$

60. (a) $M_1 C_{ice} \times (10) + M_1 L = M_2 C_w (50)$

or $M_1 \times 0.5 \times 10 + M_1 L = M_2 \times 1 \times 50$

$\Rightarrow L = \frac{50M_2}{M_1} - 5$

61. (b) $\frac{1}{2}.kx^2 = m_m C_m (\Delta T) + m_w C_w \Delta T$

or $\frac{1}{2} \times 800 \times 0.02^2 = 0.5 \times 400 \times \Delta T + 1 \times 4184 \times \Delta T$

$\therefore \Delta T \simeq 1 \times 10^{-5} K$

62. (a) $H_1 = H_2$ θ_2 | $3k$ (d) | k ($3d$) | θ_1 (interface temperature θ)

$\Rightarrow (3k)A\left(\frac{\theta_2 - \theta}{d}\right) = kA\left(\frac{\theta - \theta_1}{3d}\right) \Rightarrow \theta = \left(\frac{\theta_1 + 9\theta_2}{10}\right)$

63. (d) Inner and outer cylinder are in parallel.
So, $R = R_1 + R_2$

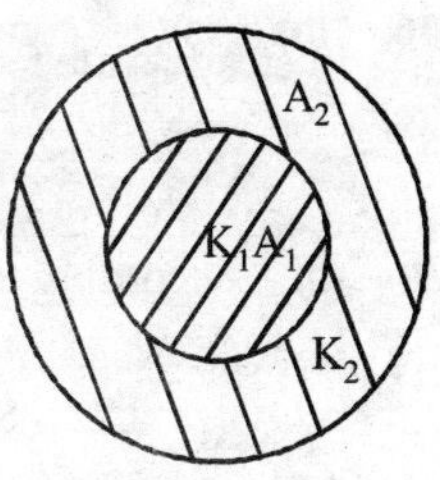

$\frac{KA}{l} = \frac{K_1A_1}{l} + \frac{K_2A_2}{l}$

$K = \frac{K_1A_1 + K_2A_2}{A}$

$= \frac{K_1\pi R^2 + K_2[\pi(2R)^2 - \pi R^2]}{\pi(2R)^2}$

$= \frac{K_1(\pi R^2) + K_2(3\pi R^2)}{4\pi R^2} = \frac{K_1 + 3K_2}{4}$

64. (d) Let m gram of ice is added.
From principal of calorimeter
heat gained (by ice) = heat lost (by water)

$\therefore 20 \times 2.1 \times m + (m - 20) \times 334 = 50 \times 4.2 \times 40$

$\Rightarrow 376m = 8400 + 6680 \Rightarrow m = 40.1$

65. (c) Heat loss = Heat gain = $mS\Delta\theta$

So, $m_A S_A \Delta\theta_A = m_B S_B \Delta\theta_B$

$\Rightarrow 100 \times S_A \times (100 - 90) = 50 \times S_B \times (90 - 75)$

$2S_A = 1.5S_B \Rightarrow S_A = \frac{3}{4} S_B$

Now, $100 \times S_A \times (100 - \theta) = 50 \times S_B \times (\theta - 50)$

$2 \times \left(\frac{3}{4}\right) \times (100 - \theta) = (\theta - 50)$

$\Rightarrow 300 - 3\theta = 2\theta - 100 \Rightarrow 400 = 5\theta \Rightarrow \theta = 80°C$

66. (d) Assume final temperature = T°C
Heat loss = Heat gain = $ms\Delta T$

$\Rightarrow m_B s_B \Delta T_B = m_w s_w \Delta T_w$

$0.1 \times 400 \times (500 - T) = 0.5 \times 4200 \times (T - 30) + 800(T - 30)$

$\Rightarrow 40(500 - T) = (T - 30)(2100 + 800)$

$\Rightarrow 20000 - 40T = 2900T - 30 \times 2900$

$\Rightarrow 20000 + 30 \times 2900 = T(2940)$

$T = 36.4°C \Rightarrow \frac{\Delta T}{T} \times 100 = \frac{6.4}{30} \times 100 = 21\%$,

so the closest answer is 20%.

67. (a) Temp. of heat source   Temp. of heat reservoir

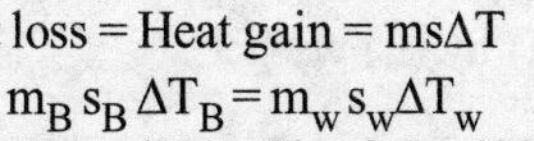

$\left(\frac{dQ}{dt}\right) = \frac{kA\Delta T}{\ell}$

Energy flux, $\frac{1}{A}\left(\frac{dQ}{dt}\right) = \frac{k\Delta T}{\ell}$

$= \frac{(0.1)(900)}{1} = 90\ W/m^2$

68. **(a)** $\frac{\Delta T_{AB}}{R_{AB}} = \frac{120}{\frac{8}{5}R} = \frac{120 \times 5}{8R}$

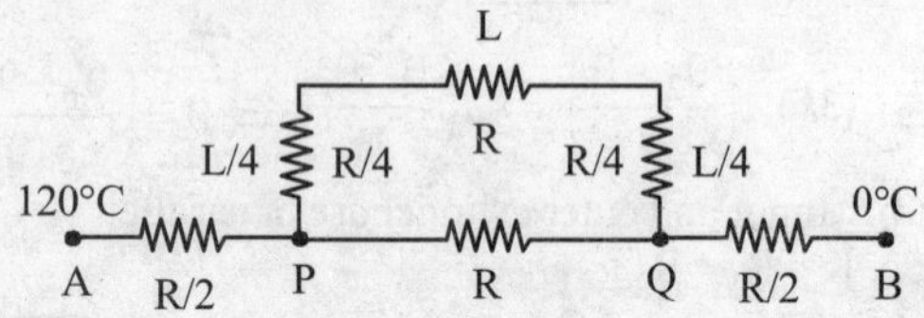

In steady state temperature difference between P and Q,

$$\Delta T_{PQ} = \frac{120 \times 5}{8R} \times \frac{3}{5}R = \frac{360}{8} = 45°C$$

69. **(d)** According to principle of calorimetry,

$Q_{given} = Q_{used}$

$\Delta Q_{used} = 0.22 \times Sa \times 110$

$\Delta Q_{used} = 22 \times Sa$ ...(i)

$\Delta Q_{given} = (m_w + W)s_w DT$ ...(ii)

$\Delta T = T_w - T = 40 - 27 = 13°C$

$m_w = V \times \rho$

$m_w = 150 \times 10^{-6} \times 10^3$ $\therefore \rho_w = 10^3\ kg/m^3$

$m_w = 150 \times 10^{-3}\ kg$

$w = 0.025$ kg (given)

$S_w = 4.2 \times 10^3 J/kg.°C$

$\Delta Q_{given} = (0.15 + 0.025) \times 4.2 \times 10^3$

$\Delta Q_{given} = 9555J$

$\Delta Q_{given} = \Delta Q_{used}$

$9555 = 22 \times Sa$

$$Sa = \frac{9555}{22}$$

$Sa = 434.3 J/kg.°C$

70. **(b)** As $Pt = mC\Delta T$

So, $P \times 10 \times 60 = mC\ 100$ ...(i)

and $P \times 55 \times 60 = mL$ ...(ii)

Dividing equation (i) by (ii) we get

$\frac{10}{55} = \frac{C \times 100}{L} \Rightarrow L = 550$ cal./g.

71. **(c)** Rate of heat flow is given by,

$$Q = \frac{KA(\theta_1 - \theta_2)}{l}$$

Where, K = coefficient of thermal conductivity

l = length of rod and A = area of cross-section of rod

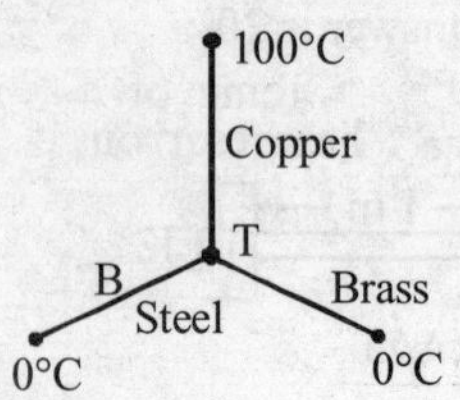

If the junction temperature is T, then

$Q_{Copper} = Q_{Brass} + Q_{Steel}$

$$\frac{0.92 \times 4(100 - T)}{46} = \frac{0.26 \times 4 \times (T - 0)}{13} + \frac{0.12 \times 4 \times (T - 0)}{12}$$

$\Rightarrow\ 200 - 2T = 2T + T$

$\Rightarrow T = 40°C$

$$\therefore\ Q_{Copper} = \frac{0.92 \times 4 \times 60}{46} = 4.8\ cal/s$$

72. **(a)** From question,

In 1 sec heat gained by water

= 1 KW – 160 J/s = 1000 J/s – 160 J/s = 840 J/s

Total heat required to raise the temperature of water (volume 2L) from 27°c to 77°c

$= m_{water} \times sp.\ ht \times \Delta\theta$

$= 2 \times 10^3 \times 4.2 \times 50$ [$\because$ mass = density × volume]

And, $840 \times t = 2 \times 10^3 \times 4.2 \times 50$

$$\text{or, } t = \frac{2 \times 10^3 \times 4.2 \times 50}{840}$$

= 500 s = 8 min 20s

73. **(d)** When radius is decreased by ΔR,

Decrease in surface energy = Heat lost as latent heat

$$(4\pi R^2 \Delta R \rho) L = 4\pi[R^2 - (R - \Delta R)^2]T$$

$$\Rightarrow \rho R^2 \Delta R L = T[R^2 - R^2 + 2R\Delta R - \Delta R^2]$$

$$\Rightarrow \rho R^2 \Delta R L = T(2R)\Delta R \quad [\Delta R \text{ is very small}]$$

$$\Rightarrow R = \frac{2T}{\rho L}$$

74. **(d)** As the surrounding is identical, vessel is identical time taken to cool both water and liquid (from 30°C to 25°C) is same 2 minutes, therefore

$$\left(\frac{dQ}{dt}\right)_{water} = \left(\frac{dQ}{dt}\right)_{liquid}$$

$$\text{or, } \frac{(m_w C_w + W)\Delta T}{t} = \frac{(m_\ell C_\ell + W)\Delta T}{t}$$

(W = water equivalent of the vessel)

or, $m_w C_w = m_\ell C_\ell$

$\therefore$ Specific heat of liquid, $C_\ell = \frac{m_w C_w}{m_\ell} = \frac{50 \times 1}{100}$

= 0.5 kcal/kg

75. **(b)** As 1g of steam at 100°C melts 8g of ice at 0°C.

10 g of steam will melt 8× 10 g of ice at 0°C

Water in calorimeter = 500 + 80 + 10g = 590g

76. **(a)** According to Stefan's law

$E = \sigma T^4$

Heat radiated per unit area in 1 hour (3600s) is

$= 5.7 \times 10^{-8} \times (3000)^4 \times 3600 = 1.7 \times 10^{10}$ J

77. **(a)** $\Delta U = \Delta Q = mc\Delta T$

$$= \frac{100}{1000} \times \frac{4184}{1000}(50 - 30) \approx 8.4\ kJ$$

78. (d) Required work = energy released

Here, $Q=\int mc\,dT$

$$=\int_{20}^{4}0.1\times32\times\left(\frac{T^3}{400^3}\right)dT=\int_{20}^{4}\frac{3.2}{64\times10^6}T^3dT$$

$$=5\times10^{-8}\int_{20}^{4}T^3dT=0.002kJ$$

Therefore, required work = 0.002 kJ

79. (a) Let θ be the temperature at a distance x from hot end of bar. Let θ_1 is the temperature of hot end.

The heat flow rate is given by

$$\frac{dQ}{dt}=\frac{kA(\theta_1-\theta)}{x}$$

$$\Rightarrow\theta_1-\theta=\frac{x}{kA}\frac{dQ}{dt}\Rightarrow\theta=\theta_1-\frac{x}{kA}\frac{dQ}{dt}$$

Thus, the graph of θ versus x is a straight line with a positive intercept and a negative slope.

The above equation can be graphically represented by option (a).

80. (d) Let T be the temperature of the interface. In the steady state, $Q_1=Q_2$

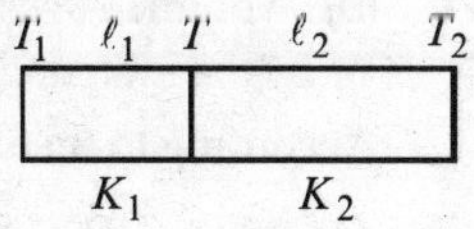

$$\therefore\ \frac{K_1A(T_1-T)}{\ell_1}=\frac{K_2A(T-T_2)}{\ell_2},$$

where A is the area of cross-section.

$$\Rightarrow\ K_1A(T_1-T)\ell_2=K_2A(T-T_2)\ell_1$$

$$\Rightarrow\ K_1T_1\ell_2-K_1T\ell_2=K_2T\ell_1-K_2T_2\ell_1$$

$$\Rightarrow\ (K_2\ell_1+K_1\ell_2)T=K_1T_1\ell_2+K_2T_2\ell_1$$

$$\Rightarrow\ T=\frac{K_1T_1\ell_2+K_2T_2\ell_1}{K_2\ell_1+K_1\ell_2}=\frac{K_1\ell_2T_1+K_2\ell_1T_2}{K_1\ell_2+K_2\ell_1}.$$

81. (b) From Stefan's law, total power radiated by Sun, $E=\sigma T^4\times4\pi R^2$

The intensity of power Per unit area incident on earth's surface $=\frac{\sigma T^4\times4\pi R^2}{4\pi r^2}$

Total power received by Earth

$$E'=\frac{E}{4\pi r^2}\times\text{Cross – Section area of earth facing the sun}$$

$$=\frac{\sigma T^4R^2}{r^2}(\pi r_0^2)=\frac{\pi r_0^2R^2\sigma T^4}{r^2}$$

82. (d) From stefan's law, energy radiated by sun per second

$E=\sigma AT^4$; $\therefore A\propto R^2$

$\therefore E\propto R^2T^4$

$$\therefore\ \frac{E_2}{E_1}=\frac{R_2^2T_2^4}{R_1^2T_1^4}$$

put $R_2=2R, R_1=R$; $T_2=2T, T_1=T$

$$\Rightarrow\frac{E_2}{E_1}=\frac{(2R)^2(2T)^4}{R^2T^4}=64$$

83. (d) The thermal resistance is given by

$$\frac{x}{KA}+\frac{4x}{2KA}=\frac{x}{KA}+\frac{2x}{KA}=\frac{3x}{KA}$$

Amount of heat flow per second,

$$\frac{dQ}{dt}=\frac{\Delta T}{\frac{3x}{KA}}=\frac{(T_2-T_1)KA}{3x}$$

$$=\frac{1}{3}\left\{\frac{A(T_2-T_1)K}{x}\right\}\qquad\therefore f=\frac{1}{3}$$

84. (b) Heat required for raising the temperature of a body through 1°C is called its thermal capacity.

85. (a) Black body is one which absorb all incident radiation. Black board paint is quite approximately equal to black bodies.

86. (c) When water is cooled at 0°C to form ice, energy is released from water in the form of heat. As energy is equivalent to mass, therefore, when water is cooled to ice, its mass decreases.

87. (a) From stefan's law, the energy radiated per second is given by $E=e\sigma T^4A$

Here, T = temperature of the body

A = surface area of the body

For same material e is same. σ is stefan's constant

Let T_1 and T_2 be the temperature of two spheres. A_1 and A_2 be the area of two spheres.

$$\therefore\ \frac{E_1}{E_2}=\frac{T_1^4A_1}{T_2^4A_2}=\frac{T_1^4\,4\pi r_1^2}{T_2^4\,4\pi r_2^2}=\frac{(4000)^4\times1^2}{(2000)^4\times4^2}=\frac{1}{1}$$

11 Thermodynamics

1. **(d)** During ice melting at 273 K, $\Delta V = -$ ve, Q = +ve
$\Rightarrow \Delta W = -$ve (by ice-water system)
$\therefore$ $\Delta W = +$ve (on ice-water system)
$\therefore$ Internal energy, $\Delta U = \Delta Q - \Delta W = \Delta Q - P\Delta V$
$= (+ve) - (-ve) = +ve$

2. **(c)** Extensive properties are dependent on size or mass of system. For example, internal energy, volume, mass.

3. **(a)** Given, Mass of water, m = 1kg
Latent heat of vaporisation $L_v = 2257$ kJ/kg.
Atmospheric pressure, $P = 1 \times 10^5$ Pa
Change in volume, $\Delta V = 1.671\ m^3 - 1 \times 10^{-3}$
$= 1.67\ m^3$
$\Delta Q = \Delta U + \Delta W$
$\Rightarrow \Delta U = \Delta Q - \Delta W = mL_v - P\Delta V$
$= (1kg)(2257 \times 10^3\ J/kg) - (1 \times 10^5\ Pa)(1.67)$
$= 2257 \times 10^3\ J - 167 \times 10^3\ J = 2090$ kJ

4. **(d)** According to 1st law of thermodynamics
$\Delta Q = \Delta U + W$
If $\Delta Q > 0$, $\Delta U < 0$ and $W > 0$ is also possible.
Hence $\Delta T < 0$, so T decreases.
Statement I is false
$W > 0$; $\int PdV > 0$
Therefore volume of system must increase during positive work done by the system.
Statement II is true.

5. **(d)** Given,
Rate of heat supplied, $\frac{dQ}{dt} = 1000$W
Rate of work performed, $\frac{dw}{dt} = 200$W
Using first law of thermodynamics
$dQ = dU + dw$
$\Rightarrow \frac{dU}{dt} = \frac{dQ}{dt} - \frac{dw}{dt} \Rightarrow \frac{dU}{dt} = 1000 - 200 = 800$ W

6. **(d)** $|\Delta W|$ = Area under P-V diagram
$= \frac{1}{2} \times (50 + 10) \times 10^3 \times 150 \times 10^{-6}$
$= 30 \times 150 \times 10^{-3} = 4500 \times 10^{-3}$
$= 4.5$ J. As volume is decreasing. So $\Delta W = -4.5$ J
By Ist law of thermodynamics
$\Delta Q = \Delta U + \Delta W$
$\Rightarrow 0 = \Delta U - 4.5$
$\Rightarrow \Delta U = +4.5$ J

7. **(c)** First law of thermodynamics is based on law of conservation of energy. From the first law of thermodynamics, dQ = dU – dW.
Here, dW is negative as work is done on the system.

8. **(b)** In complete cycle
$\Delta Q = \Delta W \qquad [\because \Delta U_{cylce} = 0]$
$\Rightarrow 40 + 0 - 60 = 0 - 50 + W_{CA}$
$\Rightarrow -20 = -50 + W_{CA}$
$\Rightarrow W_{CA} = 30$ J

9. **(c)** Heat and work depends on the path taken to reach a specific value. Hence, heat and work are path functions.

10. **(a)** ΔU remains same for both paths ACB and ADB
$\Delta Q_{ACB} = \Delta W_{ACB} + \Delta U_{ACB}$
$\Rightarrow 60\ J = 30\ J + \Delta U_{ACB} \Rightarrow U_{ACB} = 30$ J

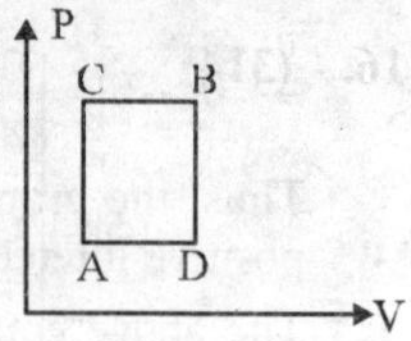

As change in internal energy depends only on initial and final point
$\therefore \Delta U_{ADB} = \Delta U_{ACB} = 30$ J
$\Delta Q_{ADB} = \Delta U_{ADB} + \Delta W_{ADB} = 30\ J + 10\ J = 40$ J

11. **(d)** Volume of water does not change, no work is done on or by the system (W = 0)
According to first law of thermodynamics
$Q = \Delta U + W$
Boiling of water is an isochoric process. So W = 0 i.e.,
$Q = \Delta U$
$\Delta U = \mu c dT = 2 \times \frac{4184}{1000} \times 20 = 16.7$ kJ.

12. **(a)** Here $Q = 0$ and $W = 0$. Therefore from first law of thermodynamics $\Delta U = Q + W = 0$
Internal energy of first vessle + Internal energy of second vessel = Internal energy of combined vessel
$n_1 C_v T_1 + n_2 C_v T_2 = (n_1 + n_2) C_v T$
$\Rightarrow T = \frac{n_1 T_1 + n_2 T_2}{n_1 + n_2}$
For first vessel $n_1 = \frac{P_1 V_1}{RT_1}$, for second vessel $n_2 = \frac{P_2 V_2}{RT_2}$
$\therefore T = \frac{\frac{P_1 V_1}{RT_1} \times T_1 + \frac{P_2 V_2}{RT_2} \times T_2}{\frac{P_1 V_1}{RT_1} + \frac{P_2 V_2}{RT_2}}$
$= \frac{T_1 T_2 (P_1 V_1 + P_2 V_2)}{P_1 V_1 T_2 + P_2 V_2 T_1}$

13. **(b)** First law is applicable to a cyclic process. For a cyclic process, $\Delta U = 0$. So, $\Delta Q = \Delta W$

14. (c) $PV^\gamma = \text{Constant}$

$$P_i V_i^\gamma = P_f V_f^\gamma$$

$$\Rightarrow \frac{P_f}{P_i} = \left(\frac{V_i}{V_f}\right)^\gamma = \left(\frac{V}{V/8}\right)^\gamma = (8)^{5/3}$$

$$\therefore \frac{P_f}{P_i} = 32$$

15. (c)

(A) Isothermal $\rightarrow \Delta T = 0 \rightarrow \Delta U = 0$ (IV)

$\therefore (\Delta U = nC_v \Delta T)$

(B) Adiabatic $\rightarrow \Delta Q = 0$ (II) $\quad \because \Delta U = -W$

(C) Isobaric $\rightarrow \Delta P = 0 \rightarrow \Delta U \neq 0$ (III)

(D) Isochoric $\rightarrow \Delta V = 0 \rightarrow \Delta W = 0$ (I) $\therefore W = P\Delta V$

16. (314)

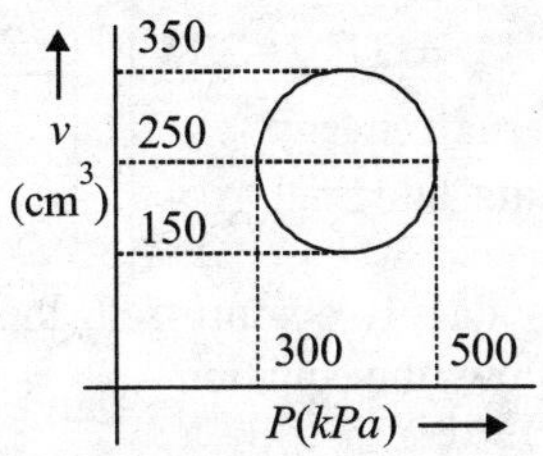

Work done, W = Area of P-V diagram $= \frac{\pi}{4} d_1 d_2$

Here, $d_1 = (500 - 300)\text{ kPa} = 200 \times 10^3\text{ Pa}$

$d_2 = (350 - 150)\text{ cm}^3 = 200 \times 10^{-6}\text{ m}^3$

$$= \frac{\pi}{4} \times 200 \times 10^3 \times 200 \times 10^{-6}$$

$$= 31.4\text{ J} = 314 \times 10^{-4}\text{ J}.$$

17. (c)

(A) Isobaric : Pressure will remain constant

$\boxed{\Delta Q = \Delta U + P\Delta V}$

(B) Isochoric : Volume will remain constant

$\boxed{\Delta Q = \Delta U}$ $\quad \therefore \Delta W = 0\ (\Delta V = 0)$

(C) Adiabatic : No heat exchange

$\boxed{\Delta Q = 0}$

(D) Isothermal ($\Delta U = 0$) : Temperature remain constant.

$\boxed{\Delta Q = \Delta W}$

18. (a) The P-V diagram for the process is

Here, $P_A = P_0$, $V_A = V_0$

$P_B = \frac{P_0}{4}$, $V_B = 4V_0$

$P_C = \frac{P_0}{4}$, $V_C = V_0$

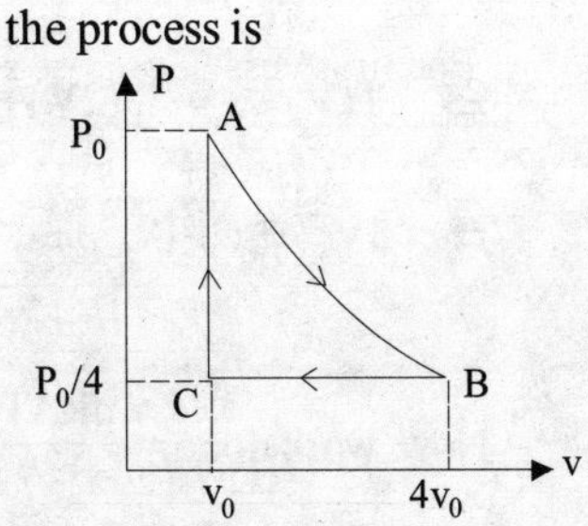

$$W_{AB} = P_A V_A \ell n\left(\frac{V_B}{V_A}\right) = P_0 V_0 \ell n4 = 2P_0 V_0 \ell n2$$

$$W_{BC} = P_B (V_C - V_B) = \frac{P_0}{4}(V_0 - 4V_0) = -\frac{3P_0 V_0}{4}$$

$W_{CA} = 0$

In cyclic process,

$$Q_{net} = W_{net} = W_{AB} + W_{BC} + W_{CA}$$

$$= P_0 V_0 \left(\ell n4 - \frac{3}{4}\right)$$

$$= P_0 V_0 (2\ell n2 - 0.75)$$

19. (d) Given, $V_1 = 800\text{ cm}^3$, $V_2 = 200\text{ cm}^3$, $T_1 = 300\text{ K}$

For adiabatic process,

$TV^{\gamma-1} = \text{const.}$

$\Rightarrow (300)(800)^{1.5-1} = T_2(200)^{1.5-1}$

$$\Rightarrow T_2 = 300\left[\frac{800}{200}\right]^{0.5} = 300 \times (2^2)^{1/2}$$

$\therefore T_2 = 600\text{ K}$

$\therefore$ Change in temperature is

$\Delta T = 600 - 300 = 300\text{ K}$

20. (c) From 1st law of thermodynamics $Q = \Delta U + W = 0$

$\Rightarrow -\Delta U = W$

$W = -nC_v\Delta T \Rightarrow |W| = nC_v\Delta T \propto T_2 - T_1.$

$\therefore$ B and E are correct.

21. (c) In an adiabatic process, $dQ = 0$

$\therefore$ Molar heat capacity $= 0$

From 1st law of thermodynamics

$dQ = du + dW$

$\because dQ = 0 \Rightarrow dU = -dW$

Also $dU = \frac{f}{2} nRdT$

22. (d) During isothermal process,

$PV = \text{constant}$

$P\Delta V + V\Delta P = 0$

$$\therefore \left(\frac{\Delta P}{\Delta V}\right)_{\text{isothermal}} = -\frac{P}{V}$$

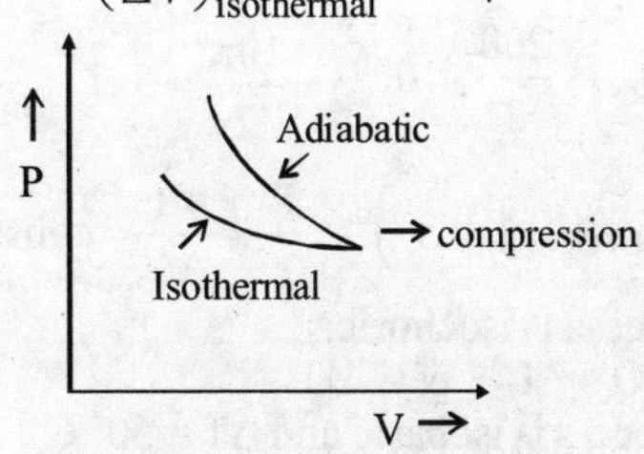

Similarly, for adiabatic process, $PV^\gamma = c$

$$\Rightarrow \left(\frac{\Delta P}{\Delta V}\right)_{\text{adiabatic}} = -\gamma\frac{P}{V}$$

$$\therefore \left(\frac{\Delta P}{\Delta V}\right)_{\text{Adiabatic}} > \left(\frac{\Delta P}{\Delta V}\right)_{\text{Isothermal}}$$

$\therefore$ Volume falls off more rapidly in an adiabatic process than that of an isothermal process.

23. (b) In adiabatic process, $\Delta Q = 0$

$\therefore \Delta Q_{AB} = 0$

From 1st law of thermodynamics

$\Delta Q = \Delta U + W$

In isothermal process BC, $\Delta U = 0$

$\therefore \quad \Delta Q_{BC} = \Delta W_{BC}$

$$= nRT\ell n\left(\frac{V_C}{V_B}\right) = 450R\ell n\left(\frac{8\times10^{-6}}{6\times10^{-6}}\right)$$

$$= 450\ell n\left(\frac{4}{3}\right) = 450R\ell n(\ell n4 - \ell n3)$$

$$\therefore \quad (\Delta Q)_{net} = \Delta Q_{AB} + \Delta Q_{BC} = 450R(\ell n4 - \ell n3)$$

24. (d) In adiabatic process,

$\Delta W = -\Delta U = -nC_v\Delta T$

$\therefore$ Work done in adiabatic process depends only on change in temperature.

25. (c) For adiabatic process,

$$T_1V_1^{\gamma-1} = T_2V_2^{\gamma-1} \Rightarrow T \propto \frac{1}{V}$$

$\therefore$ V decreases, then temperature increases

Free expansion is assumed fast process, so it is an adiabatic processes.

26. (b)

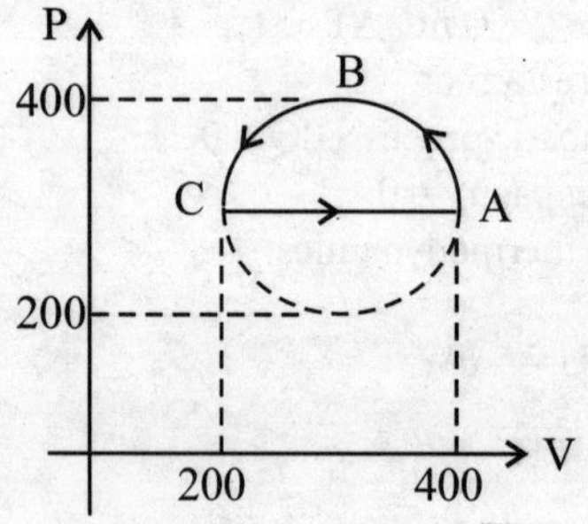

For cyclic process,

Q = W = Area of P-V graph

$$= \frac{1}{2}\pi\left(\frac{P_2 - P_1}{2}\right)\left(\frac{V_2 - V_1}{2}\right)$$

$$= \frac{1}{2}\times\pi\times\left(\frac{200}{2}\times10^3\right)\times\frac{200}{2}\times10^{-6} = \frac{10\pi}{2} = 5\pi J$$

27. (b) For the given process, $P \propto T \therefore \frac{P}{T} = \frac{nR}{V} = \text{constant}$

$\therefore V = c \Rightarrow$ the process is isochoric.

$\therefore W = P\Delta V = 0 \therefore Q = \Delta U + W = \Delta U$

28. (15) Given that process is isobaric and $\Delta T = 50°$ C

$\therefore Q = nC_p\Delta T = E_1$

$\Delta U = nC_v\Delta T = E_2$

For monoatomic gas, f = 3

$$\therefore \frac{E_1}{E_2} = \frac{C_P}{C_V} = \gamma = 1 + \frac{2}{f} = 1 + \frac{2}{3} = \frac{5}{3} = \frac{x}{9} \quad \therefore x = 15$$

29. (d) For a substance with specific heat capacity s, the heat added when changing temperature by dT is

dQ = msdT

$$dS = \frac{dQ}{T} = \frac{msdT}{T}$$

$$\Delta S = \int_{T_1}^{T_2}\frac{msdT}{T} = ms\,ln\frac{T_2}{T_1} \qquad [\because S = 1Jkg^{-1}K^{-1}]$$

30. (d) The work done by a gas during a process is given by area enclosed under P-V curve

$\therefore W_{ABCD} = W_{AB} + W_{BC} + W_{CD}$

$= P_0V_0 + 0 + (-2P_0 \times 2V_0)$

$= P_0V_0 - 4P_0V_0 = -3P_0V_0$

31. (d) For isothermal process T = c

$$\therefore P \propto \frac{1}{V}$$

In isobaric, pressure is constant

$W \neq 0, \Delta U \neq 0, \Delta Q \neq 0$ [only isobaric]

In adiabatic, no heat is transferred into or out of the system.

$\Delta Q = 0$

In isochoric, volume is constant, $W = P\Delta V = 0$

32. (273) For adiabatic compression of an ideal gas

Initial temprature, $T_1 = 0°C = 273$ K

Final volume, $V_2 = \frac{V_1}{4}$

Ratio of specific heat, $\gamma = \frac{C_P}{C_V} = \frac{3}{2}$

Since, the gas is suddenly compressed, the process is adiabatic which follows the equation

$TV^{\gamma-1} = \text{constant}$

$$273(V_1)^{0.5} = T\left(\frac{V_1}{4}\right)^{0.5}$$

$T = 273 \times 2 = 546$

$\Delta T = 273$ K

33. (c) Given, Initial volume of an ideal gas,

$V_i = 5L$

Final volume of an ideal gas, $V_f = 4L$

For Adiabatic process, $PV^\gamma = \text{constant}$

$P_iV_i^\gamma = P_fV_f^\gamma$

$\Rightarrow P_i(5)^{1.5} = P_f(4)^{1.5}$

$$\Rightarrow \frac{P_i}{P_f} = \left(\frac{4}{5}\right)^{\frac{3}{2}} = \frac{4}{5}\sqrt{\frac{4}{5}} = \frac{8}{5\sqrt{5}}$$

34. (a) For the adiabatic process

$TV^{\lambda-1} = \text{constant}$

$$\Rightarrow T(V)^{\frac{3}{2}-1} = T_f(2V)^{\frac{3}{2}-1}$$

$$\Rightarrow TV^{\frac{1}{2}} = T_f(2)^{\frac{1}{2}}(V)^{\frac{1}{2}} \Rightarrow T_f = \left(\frac{T}{\sqrt{2}}\right)$$

$$\text{Now, work done} = \frac{nR\Delta T}{1-\gamma} = \frac{1.R\left[\frac{T}{\sqrt{2}} - T\right]}{1 - \frac{3}{2}}$$

$$= 2RT\left[1 - \frac{1}{\sqrt{2}}\right] = RT\left[2 - \sqrt{2}\right]$$

35. (b) AC is an isochoric process
$\therefore W_{AC} = 0$
BC is an isobaric process
$\therefore W_{BC} = P\Delta V = 10(2-4) = -20$ J
Work done during process AB

$$W_{AB} = \int PdV = \int_2^4 \frac{RT}{V^3} dV = \frac{-RT}{2}\left[\frac{1}{V^2}\right]_2^4$$

$$\frac{-RT}{2}\left[\frac{1}{11} - \frac{1}{4}\right] = 225 \text{ J}$$

$\therefore$ Total work done = –20 J + 225 J = 205 J

36. (c) For adiabatic process
$TV^{\gamma-1}$ = constant
For adiabatic curve ad

$$\Rightarrow T_a \cdot V_a^{\gamma-1} = T_d \cdot V_d^{\gamma-1}$$

$$\Rightarrow \left(\frac{V_a}{V_d}\right)^{\gamma-1} = \frac{T_d}{T_a} \quad \text{...(i)}$$

For adiabatic curve bc

$$T_b \cdot V_b^{\gamma-1} = T_c \cdot V_c^{\gamma-1}$$

$$\Rightarrow \left(\frac{V_b}{V_c}\right)^{\gamma-1} = \frac{T_c}{T_b} \quad \text{...(ii)}$$

As the curves ab and cd are isothermal expansion and isothermal compression respectively.

$\therefore \quad T_d = T_c$ and $T_a = T_b$

$$\therefore \quad \frac{V_a}{V_d} = \frac{V_b}{V_c} \quad \text{[using (i) and (ii)]}$$

37. (a) For Isobaric process work done is given by:
$W = P\Delta V = nR\Delta T = 100$ J
According to 1st law of thermodynamics $\Delta Q = \Delta U + W$

$$\Delta Q = \frac{f}{2}nR\Delta T + nR\Delta T$$

$$= \left(\frac{f}{2}+1\right)nR\Delta T \quad \left(\frac{5}{2}+1\right)100 = 350 \text{ J}$$

38. (d) According to Ist law of thermodynamics :
$dQ = dU + dW$
$CdT = CvdT + PdV \quad$ ...(i)
Now $PV^2 = RT$
On differentiation, $P(2VdV) = RdT$

$$\Rightarrow PdV = \frac{RdT}{2V}$$

Substitute in equation (i)

$$C_P = C_V + \frac{R}{2V}$$

39. (d) According to first law of thermodynamics
$dQ = dU + W$
$\Rightarrow 48 = nC_v\Delta T + W$

$$\Rightarrow 48 = (1)\left(\frac{3R}{2}\right)(2) + W \quad \left[\because C_v = \frac{3R}{2}\right]$$

$\Rightarrow W = 48 - 3 \times R = 48 - 3 \times (8.3)$
$\Rightarrow W = 23.1$ Joule

40. (a) By 1st law of thermodynamics,
$\Delta Q = \Delta U + W = W$
[$\because$ For cyclic process, $\Delta U = 0$ & W = Area of P-V diagram]

$$\therefore \Delta Q = \frac{\pi}{4} \times (280 \times 10^3 \text{ Pa}) \times (280 \times 10^{-6} \text{ m}^3) = 61.6 \text{ J}$$

41. (c) Given, $P \propto T^3$ ...(i)
in an adiabatic process

$$T^{\gamma}P^{1-\gamma} = \text{constant} \quad \left[\text{as } \gamma = \frac{C_P}{C_V}\right]$$

$$T \propto \frac{1}{P^{\left(\frac{1-\gamma}{\gamma}\right)}} \Rightarrow T^{\left(\frac{\gamma}{\gamma-1}\right)} \propto P \quad \text{...(ii)}$$

comparing equation (i) and (ii), we get

$$\left(\frac{\gamma}{\gamma-1}\right) = 3$$

$3\gamma - 3 = \gamma$ or $2\gamma = 3$

$$\gamma = \frac{C_P}{C_P} = \frac{3}{2}$$

42. (d) Work done, $W = \frac{nR\Delta T}{1-\gamma}$

In adiabatic process, $TV^{\gamma-1} = \text{constant} = T_f(2V)^{\gamma-1}$

$$T_f = T\left(\frac{1}{2}\right)^{1/2} = \frac{T}{\sqrt{2}}$$

$$\therefore W = \frac{R\left(\frac{T}{\sqrt{2}} - T\right)}{1 - \frac{3}{2}} = 2RT\frac{(\sqrt{2}-1)}{\sqrt{2}}$$

or $W = RT(2-\sqrt{2})$

43. (a) Work done $W = \frac{P_1V_1 - P_2V_2}{\gamma - 1}$

$$= \frac{P_1V_1 - P_2V_2}{\left(\frac{3}{2}-1\right)} = 2(P_1V_1 - P_2V_2) \quad \left(\because \gamma = \frac{3}{2} \text{ given}\right)$$

44. (a) From the ideal gas equation
$PV = nRT$

$$\Rightarrow \frac{V}{T} = \frac{nR}{P} \Rightarrow V = \left(\frac{nR}{P}\right)T \quad \because \text{Slope of graph} \propto \frac{1}{P}$$

$\because (\text{Slope})_2 > (\text{Slope})_1 \quad \therefore P_2 < P_1$

45. (a) Curve (B) represents adiabatic

$$\text{Adiabatic} \Rightarrow PV^{\gamma} = \text{constant or } P\left(\frac{T}{P}\right)^{\gamma} = \text{constant}$$

$$\frac{T^{\gamma}}{P^{\gamma-1}} = \text{constant}$$

Curve (A) represents isothermal
T = constant
PV = constant

46. (d)

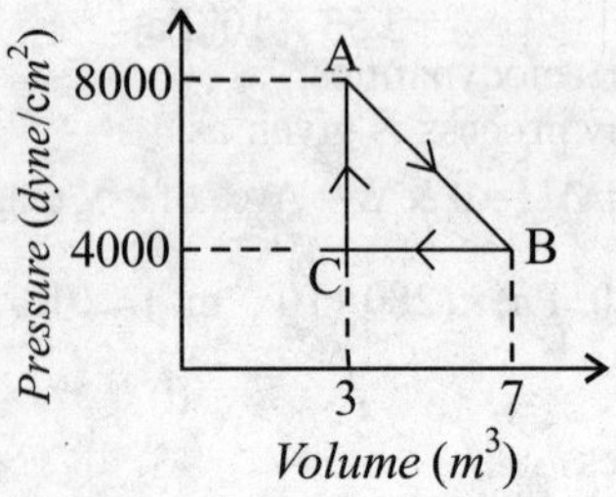

For process ABC,
$W_{ABC} = W_{AB} + W_{BC} + W_{CA}$
$[\because W_{CA} = 0]$

$$\therefore W_{ABC} = W_{ABCA} = \frac{1}{2} \times \frac{4000}{10} \times 4 = 800\,J$$

$$\left[\because 1\,dyne/cm^2 = \frac{1}{10} N/m^2\right]$$

47. (c) Since, $Q = \Delta U + W$
$\Rightarrow Q = \Delta U$ as work done is zero $[\because$ volume constant]
$\therefore \Delta U = ms\Delta T = 0.08 \times (170 \times 4.18) \times 5 \simeq 284J$

48. (b) Pressure, $P \propto T^3 \Rightarrow PT^{-3} =$ constant

$$PV^{\gamma} = \text{constant} \Rightarrow P\left(\frac{nRT}{P}\right)^{\gamma} = \text{constant}$$

$$P^{1-\gamma}T^{\gamma} = \text{constant} \Rightarrow PT^{\frac{\gamma}{1-\gamma}} = \text{constant}$$

$$\frac{\gamma}{1-\gamma} = -3 \Rightarrow \gamma = -3 + 3\gamma \Rightarrow 3 = 2\gamma \therefore \gamma = \frac{3}{2}$$

49. (b) Work done = Area under the curve

$$\Rightarrow W = \frac{1}{2} \times (4-2) \times (400-100) = \frac{1}{2}(2) \times 300$$

$W = 300J$

50. (b) Given pressure under isothermal condition, $Pv^3 = a$
Differentiating w.r.t to pressure

$$V^3 + P3V^2 \frac{dv}{dP} = 0$$

$$\Rightarrow V = -3\frac{PdV}{dP} = 0 \Rightarrow V = -3\frac{PdV}{dP}$$

$$\frac{dP.V}{dV} = -3P\left[\because \text{Bulk modulus } B = -\frac{dP}{dV/V}\right]$$

$$\therefore B = -\left(\frac{dPV}{dV}\right) = -(-3P) = 3P$$

51. (b) If gas is compressed suddenly, the processes is adiabatic.
And equation for adiabatic process $PV^{\gamma} =$ constant or $P_1 V_1^{\gamma} = P_2 V_2^{\gamma}$

$$\Rightarrow P_0 V_0^{\gamma} = P_2\left(\frac{V_0}{4}\right)^{\gamma} \quad \therefore P_2 = P_0 (4)^{\gamma}$$

52. (b) In isothermal process
Temperature, T = constant
$\therefore$ Internal energy = constant

53. (c) Consider initial Volume $V_1 = V$ and final volume $V_2 = 2V$
Let intial temperature $T_1 = T$

$$T_1V_1^{\gamma-1} = T_2V_2^{\gamma-1};\ TV^{1/2} = T_2(2V)^{1/2};\ T_2 = \frac{T}{\sqrt{2}}$$

The workdone by the gas in a process is

$$W = \frac{R(T_1 - T_2)}{\gamma - 1} = \frac{R\left(T - \frac{T}{\sqrt{2}}\right)}{\left(\frac{3}{2} - 1\right)} = RT\left(2 - \sqrt{2}\right)$$

54. (a) In adiabatic process
$PV^{\gamma} =$ constant

$$\Rightarrow P_1V_1^{\gamma} = P_2V_2^{\gamma}$$

$$\Rightarrow \frac{P_1}{P_2} = \left(\frac{V_2}{V_1}\right)^{\gamma} \Rightarrow \frac{81}{16} = \left(\frac{27}{8}\right)^{\gamma}$$

$$\Rightarrow \left(\frac{9}{4}\right)^2 = \left(\frac{9\times3}{4\times2}\right)^{\gamma} \Rightarrow \left(\frac{9}{4}\right)^{2-\gamma} = \left(\frac{3}{2}\right)^{\gamma}$$

$$\Rightarrow \left(\frac{9}{4}\right)^{2-\gamma} = \left(\frac{9}{4}\right)^{\gamma/2} \Rightarrow 2 - \gamma = \frac{\gamma}{2}$$

$$\Rightarrow 2 = \frac{3}{2}\gamma \Rightarrow \gamma = \frac{4}{3}$$

55. (d) From first law of thermodynamics
$dQ = dU + dW \Rightarrow dU = nC_v\,dT$
$dU = 0$ (for isothermal, $dT = 0$)
$\therefore U =$ constant
Hence, internal energy of the gas will not change.
Also $dQ > 0$ (Supplied)
Hence, $dW > 0$

56. (b) $\Delta U = nC_v\Delta T$
For isothermal process T is constant
So $\Delta U = 0$
$A \longrightarrow II$
Adiabatic process
$\Delta Q = 0$
$\Delta Q = \Delta U + \Delta W$
$\Delta U = -\Delta W$
Work done by gas is positive
So, ΔU is negative
$B \longrightarrow I$
For Isochoric process $\Delta W = 0$
$C \longrightarrow IV$
For Isobaric process
$\Delta W = P\Delta V \neq 0$
$\Delta U = nC_v\Delta T \neq 0$
Heat absorbed goes partly to increase internal energy and partly do work.

57. (d) For isothermal process P–V graph is rectangular hyperbola

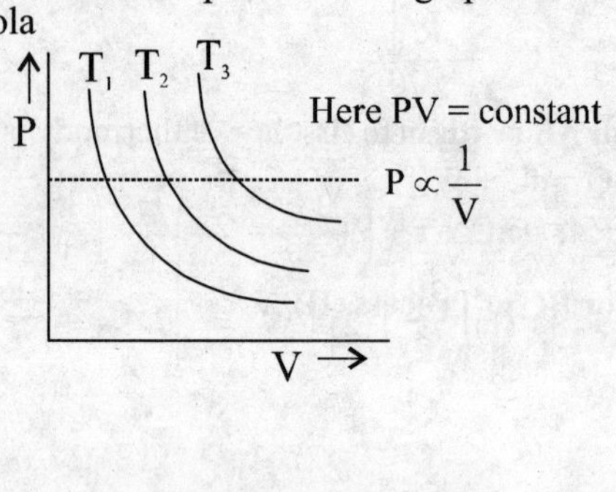

$\because PV \propto T$
$\therefore$ Curve having large area of PV graph will have more temperature.

58. **(4)** For adiabatic process, PV^γ = constant

$$\Rightarrow p\left(\frac{m}{d}\right)^\gamma = \text{constant} \quad \left[\because d = \frac{m}{V}\right]$$

$$\Rightarrow \frac{p}{d^\gamma} = \text{constant}$$

$$\Rightarrow \frac{p_1}{p_2} = \left(\frac{d_1}{d_2}\right)^\gamma = \left(\frac{1}{32}\right)^{7/5} = \frac{1}{128} \quad \left[\because \frac{d_2}{d_1} = 32\right]$$

as $PV = nRT \Rightarrow T \propto PV \Rightarrow \frac{T_1}{T_2} = \frac{P_1V_1}{P_2V_2} = \frac{1}{128}32 = \frac{1}{4}$

$\Rightarrow T_2 = 4T_1$

59. **(b)** Work done by the gas = Area under the curve
The work done by the gas from D to E,

$$W_{DE} = \frac{1}{2}(600+300)\,3\,J = 1350\,J$$

The work done by the gas from E to F,
$W_{EF} = -300 \times 3 = -900\,J$
The total work done by the gas from D to E to F will be
$W_{DEF} = 1350\,J - 900\,J = 450\,J$

60. **(b)** Given that change in temperature $\Delta T = 40\,K$
Number of moles n = 7 mol
$R = 8.3\,JK^{-1}\,mol^{-1}$

For monoatomic gas $C_V = \frac{3}{2}R$

The increase in the internal energy of the gas will be,

$$\Delta U = nC_V\Delta T = n\frac{3}{2}R\Delta T = 7 \times \frac{3}{2} \times 8.3 \times 40$$

$\Rightarrow \Delta U = 3486\,J$

61. **(c)** For monoatomic gas, $\gamma = \frac{5}{3}$
For an adiabatic process, PV^γ = constant
For an adiabatic process, $PV^\gamma = P'(V')^\gamma$

$$\Rightarrow PV^{5/3} = P'\left(\frac{V}{8}\right)^{5/3} \quad \left(\text{Given, } V' = \frac{V}{8}\right)$$

$$\Rightarrow P' = (8)^{5/3}P = (2)^5P = 32\,P$$

62. **(b)**

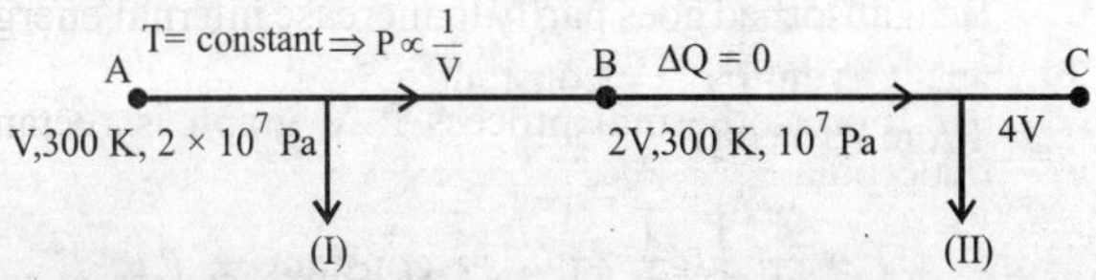

For AB or Process (I)

$$P \propto \frac{1}{V} \Rightarrow P_2 = P_1\left(\frac{V_1}{V_2}\right) = 2\times10^7 \times \left(\frac{1}{2}\right) = 10^7\,Pa$$

For BC or Process (II)
PV^γ = Constant

$$P_2 = P_1\left(\frac{V_1}{V_2}\right)^\gamma = 10^7\left(\frac{1}{2}\right)^{1.5} = 3.53 \times 10^6\,Pa$$

63. **(b)** Work done in any process is given as

$$W = \frac{P_2V_2 - P_1V_1}{1-x}$$

In adiabatic process,
$P_2V_2^\gamma = P_1V_1^\gamma$

$$P_2 = P_1\left(\frac{V_1}{V_2}\right)^\gamma = 75\left(\frac{1200}{150}\right)^{\frac{5}{3}} = 75(8)^{\frac{5}{3}}$$

$= 75 \times 32 = 2400\,kPa]$
Here, $x = \gamma$

$$\text{So, } W = \frac{P_2V_2 - P_1V_1}{1-\gamma} = \left(\frac{2400\times150 - 75\times1200}{1-\frac{5}{3}}\right)\times10^{-3}$$

$= 405000 \times 10^{-3} = 405\,J$

64. **(d)** Let draw the P–V diagram for all the process.

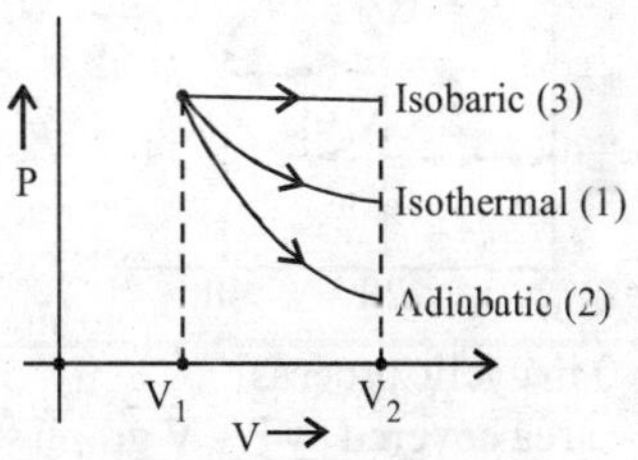

As area under P–V diagram is work
So, $W_2 < W_1 < W_3$

65. **(a)** Work alone in a polytropic process is given by

$$W = \frac{\mu R(T_2 - T_1)}{1-x}$$

For adiabatic process, $x = \gamma$

So, $W = \frac{\mu R(T_2 - T_1)}{1-\gamma}$ So statement I is true

For adiabatic process, $\Delta Q = 0$
So, $\Delta U = -\Delta W$
$\Delta W = -\Delta U$
If work is done on gas, then ΔW is –ve. So, ΔU will be +ve as ΔU is + ve ⇒ Temperature will rise.
So, statement II is also true.

66. **(1400)** Work done = $P\Delta V$
$\Rightarrow 400 = P\Delta V$
$\Rightarrow 400 = nR\Delta T \quad [\because P\Delta V = nR\Delta T$ at constant pressure]
Now, $Q = nC_P\Delta T$

$$= n\frac{R\gamma}{\gamma-1}\Delta T = 400\times\frac{\gamma}{\gamma-1} = 400\times\frac{1.4}{0.4} = 1400\,J$$

67. **(2)** By Ist law of thermodynamics, $\Delta Q = \Delta U + \Delta W$
$\Delta U = \Delta Q - \Delta W = Q - Q/4 = 3Q/4$

Now, $\Delta U = \frac{3}{2}nR\Delta T \quad [\because \Delta U = \frac{f}{2}nR\Delta T]$

$\Rightarrow \frac{3Q}{4} = \frac{3}{2} Rn\Delta T \Rightarrow n\Delta T = \frac{Q}{2R}$

As, $Q = nC\Delta T$

$\Rightarrow Q = \frac{Q}{2R} \times C \; [\because n\Delta T = \frac{Q}{2R}] \Rightarrow C = 2R$

68. **(480)** In adiabatic process

$p_1 V_1^{\gamma} = p_2 V_2^{\gamma}$

$\Rightarrow (200)(1200)^{1.5} = p_2 (300)^{1.5}$

$p_2 = 200\,(4)^{3/2} = 1600$ kPa

Work done $= \frac{p_2 V_2 - p_1 V_1}{1-\gamma} = \left(\frac{480-240}{-0.5}\right) = -480 J$

Absolute value = 480

69. **(100)** From first law of thermodynamics

$\Delta Q = \Delta U + W$ or, $\Delta Q = W$

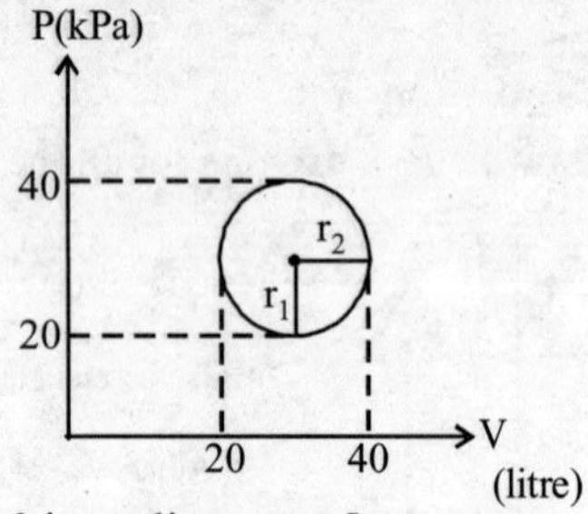

$[\because \Delta U = 0$ in cyclic process$]$

$\Delta Q = W =$ area covered by P – V graph $= \pi r_1 r_2$

or, $\Delta Q = \pi(10 \times 10^3) \times (10 \times 10^{-3}) = 100\pi J$

70. **(d)** Given,

Change in temperature, $\Delta T = 50°C$

Specific heat at constant volume for diatomic gas,

$C_V = \frac{5R}{2}$

As work done = 0. It means the process is isochoric

Using first law of thermodynamics, $Q = \Delta U + W$

$\Rightarrow Q = \Delta U$

$= nC_V\Delta T = 4 \times \frac{5R}{2} \times 50 = 500R$

71. **(17258)** In isothermal process, work done

$W = nRT \ell n\left(\frac{V_2}{V_1}\right) = 1 \times 8.3 \times 300 \; \ell n\left(\frac{4}{2}\right)$

$= 1 \times 8.3 \times 300 \times \ell n2 = 2490 \times 0.6931$

or $W = 17258 \times 10^{-1}$ J

[**Note :** There is anamoly in problem as PV = nRT]

72. **(a)** Work done during adiabatic process *CD*,

$W_{CD} = \frac{P_C V_C - P_D V_D}{\gamma - 1}$

$= \frac{100 \times 4 - 200 \times 3}{1.4 - 1} = \frac{400 - 600}{\frac{2}{5}}$ J = – 500 J

73. **(d)** Gas is identical, so after removal of the piston, the total entropy of the system, $S = S_1 + S_2$

74. **(b)** In adiabatic process PV^{γ} = constant

Differentiating both sides, we get

$\Rightarrow (P)(\gamma)V^{\gamma-1}dV + V^{\gamma}dP = 0$

$\Rightarrow dP = \frac{-\gamma P}{V}dV \Rightarrow \frac{dP}{P} = -\frac{\gamma dV}{V}$

75. **(a)** For adiabatic process, $V \propto T^{\frac{1}{1-\gamma}}$

And for isothermal process, temperature is constant.

Let $\gamma = 1.5$

then, $V \propto \frac{1}{T^2}$

Hence, (3) is the correct *V-T* graph.

Also for adiabatic process, $P \propto T^{\frac{\gamma}{\gamma-1}}$

Let $\gamma = 1.5$

Then, $P \propto T^3$

Hence, (4) is the correct *P-T* graph.

76. **(25.00)** From first law of thermodynamics

$Q = \Delta U + \Delta W = \Delta U + \frac{Q}{5}$

$\Rightarrow \Delta U = \frac{4Q}{5} \Rightarrow \frac{5R}{2}\Delta T = \frac{4}{5}Q \quad \left[\because \Delta U = \frac{f}{2} nR\Delta T\right]$

or, $\frac{Q}{\Delta T} = \frac{\frac{5R}{2}}{\frac{4}{5}} = \frac{25R}{8}$

Therefore molar heat capacity of the gas during the process $C = \frac{Q}{\Delta T} = \frac{25}{8} R$

$\therefore \; x = 25$

77. **(60.00)** $V = KT^{2/3}$

$TV^{-\frac{3}{2}} =$ cons.

Comparing it with $TV^{m-1} =$ cons.

$m = -\frac{1}{2}$

So, $W = \frac{nR\Delta T}{1-m} = \frac{1R \times 90}{1-\left(-\frac{1}{2}\right)} = 60R \; \therefore x = 60$

78. **(50)** Given : $P = KV^3$

$\Rightarrow PV^{-3} =$ Constant (K) $\quad \therefore \gamma = -3.$

Work done, $W = \frac{nR(T_1 - T_2)}{\gamma - 1} = \frac{nR(100-300)}{-3-1}$

or, $W = \frac{200nR}{4} = 50\; nR$

79. **(a)** Given, $PV^{1/2}$ = constant

From $PV = nRT$

$P \propto \frac{T}{V} \; \therefore \; \frac{T}{V} \times V^{1/2} = \text{constant} \Rightarrow T \propto \sqrt{V}$

$\therefore \frac{T_2}{T_1} = \sqrt{\frac{V_2}{V_1}} = \sqrt{\frac{2}{1}} = \sqrt{2} \quad [\because V_2 = 2V_1 \text{ Given}]$

80. **(d)** Work done during isothermal process,

$$W_{AB} = nRT\ln\left(\frac{V_2}{V_1}\right) = nRT\ln\left(\frac{2V}{V}\right) = nRT\ln 2$$

$$W_{BC} = -P_2V_1 = -\frac{nRT}{2V_1}.V_1 = \frac{-nRT}{2}$$

$$W_{CA} = P.dV = 0$$

$$\Rightarrow W_{net} = W_{AB} + W_{BC} + W_{CA}$$

$$\Rightarrow W_{net} = nRT\ln 2 - \frac{nRT}{2} = nRT\left(\ln 2 - \frac{1}{2}\right)$$

81. **(c)** In adiabatic process

PV^γ = constant

$\therefore P\left(\frac{m}{\rho}\right)^\gamma$ = constant $\quad\left(\because V = \frac{m}{\rho}\right)$

As mass is constant

$\therefore P \propto \rho^\gamma$

If P_i and P_f be the initial and final pressure of the gas and ρ_i and ρ_f be the initial and final density of the gas. Then

$$\frac{P_f}{P_i} = \left(\frac{\rho_f}{\rho_i}\right)^\gamma = (32)^{7/5} \Rightarrow \frac{nP_i}{P_i} = (2^5)^{7/5} = 2^7$$

$\Rightarrow n = 2^7 = 128.$

82. **(d)**

(I) **Adiabatic process:** No exchange of heat takes place with surroundings. $\therefore \Delta Q = 0$

(II) **Isothermal process:** Temperature remains constant.

$$\therefore \Delta T = 0 \Rightarrow \Delta U = \frac{f}{2}nR\Delta T \Rightarrow \Delta U = 0$$

No change in internal energy [$\Delta U = 0$].

(III) **Isochoric process:** Volume remains constant

$$\Delta V = 0 \Rightarrow W = \int P\cdot dV = 0$$

Hence work done is zero.

(IV) **In isobaric process:** Pressure remains constant.

$W = P\cdot\Delta V \neq 0$

$$\Delta U = \frac{f}{2}nR\Delta T = \frac{f}{2}[P\Delta V] \neq 0$$

$$\therefore \Delta Q = nC_p\Delta T \neq 0$$

83. **(b)** Bursting of helium balloon is irreversible and in this process $\Delta Q = 0$, so adiabatic.

84. **(46)** For adiabatic process, $TV^{\gamma-1}$ = constant

or, $T_1V_1^{\gamma-1} = T_2V_2^{\gamma-1}$

$T_1 = 20°\text{C} + 273 = 293$ K, $V_2 = \frac{V_1}{10}$ and $\gamma = \frac{7}{5}$

$$\Rightarrow T_1(V_1)^{\gamma-1} = T_2\left(\frac{V_1}{10}\right)^{\gamma-1}$$

$$\Rightarrow 293 = T_2\left(\frac{1}{10}\right)^{2/5} \Rightarrow T_2 = 293(10)^{2/5} \simeq 736 \text{ K}$$

$\Delta T = 736 - 293 = 443$ K

During the process, change in internal energy

$$\Delta U = NC_V\Delta T = 5 \times \frac{5}{2} \times 8.3 \times 443 \simeq 46 \times 10^3 \text{ J} = X \text{ kJ}$$

$\therefore X = 46.$

85. **(c)** For process $3 \rightarrow 1$ volume is constant

$\therefore$ Graph given in option (d) is wrong.

And process $1 \rightarrow 2$ is adiabatic $\therefore$ graph in option (a) is wrong

$\because$ v = constant

$P\uparrow, T\uparrow$

For Process $2 \rightarrow 3$ Pressure constant

i.e., P = constant

$\therefore$ $V\downarrow T\downarrow$

Hence graph (c) is the correct $V-T$ graph of given $P-V$ graph

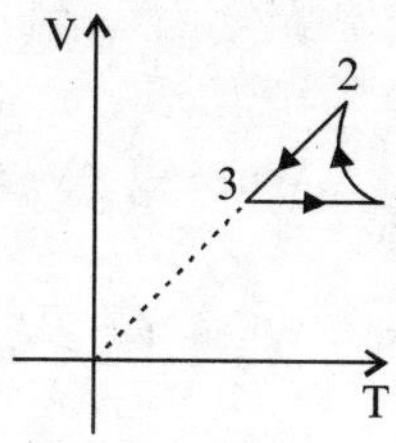

86. **(1818)** For an adiabatic process,

$TV^{\gamma-1}$ = constant

$$\therefore T_1V_1^{\gamma-1} = T_2V_2^{\gamma-1}$$

$$\Rightarrow T_2 = (300) \times \left(\frac{V_1}{\frac{V_1}{16}}\right)^{1.4-1} \Rightarrow T_2 = 300 \times (16)^{0.4}$$

Ideal gas equation, $PV = nRT$

$$\therefore V = \frac{nRT}{P}$$

$\Rightarrow V = kT$ (since pressure is constant for isobaric process)

So, during isobaric process

$V_2 = kT_2$ ...(i)

$2V_2 = kT_f$ ...(ii)

Dividing (i) by (ii)

$$\frac{1}{2} = \frac{T_2}{T_f} \Rightarrow T_f = 2T_2 = 300 \times 2 \times (16)^{0.4} = 1818 K$$

87. **(b)** Given, $V_1 = 1$ litre, $P_1 = 1$ atm

$V_2 = 3$ litre, $\gamma = 1.40$,

Using, PV^γ = constant $\Rightarrow P_1V_1^\gamma = P_2V_2^\gamma$

$$\Rightarrow P_2 = P_1 \times \left(\frac{1}{3}\right)^{1.4} = \frac{1}{4.6555} atm$$

$$\therefore \text{Work done, } W = \frac{P_1V_1 - P_2V_2}{\gamma - 1}$$

$$= \frac{\left(1\times1 - \frac{1}{4.6555}\times3\right)1.01325\times10^5\times10^{-3}}{0.4} = 90.1 J$$

Closest value of $W = 90.5 J$

88. **(b)** $\Delta U_{CA} = -\Delta U_{AC}$

$-180 = -(Q_{AC} - \Delta W_{AC})$

$180 = (310 - \Delta W_{AC}) \Rightarrow \Delta W_{AC} = 130$ J

89. (c) As the process is isochoric so,

$$Q = nC_V\Delta T = \frac{67.2}{22.4} \times \frac{3R}{2} \times 20 = 90R$$

$= 90 \times 8.31 \simeq 748$ Joule.

90. (a) At constant volume

Work done (W) = $nR\Delta T$

Heat given $Q = C_v\Delta T + nR\Delta T$

$$\text{So, } \therefore \frac{W}{Q} = \frac{nR\Delta T}{C_v\Delta T + nR\Delta T} = \frac{nR}{C_V + nR}$$

91. (b) We have given,

$$P = P_0\left[1 - \frac{1}{2}\left(\frac{V_0}{V}\right)^2\right]$$

When $V_1 = V_0 \Rightarrow P_1 = P_0\left[1 - \frac{1}{2}\right] = \frac{P_0}{2}$

When $V_2 = 2V_0 \Rightarrow P_2 = P_0\left[1 - \frac{1}{2}\left(\frac{1}{4}\right)\right] = \left(\frac{7P_0}{8}\right)$

$$\Delta T = T_2 - T_1 = \left|\frac{P_1V_1}{nR} - \frac{P_2V_2}{nR}\right| \left[\because T = \frac{PV}{nR}\right]$$

$$\Delta T = \left|\left(\frac{1}{nR}\right)(P_1V_1 - P_2V_2)\right| = \left(\frac{1}{nR}\right)\left|\left(\frac{P_0V_0}{2} - \frac{7P_0V_0}{4}\right)\right|$$

$$= \frac{5P_0V_0}{4nR} = \frac{5P_0V_0}{4R} \qquad (\because n = 1)$$

92. (c) Internal energy depends only on initial and final state

So, $\Delta U_A = \Delta U_B$

Also $\Delta Q = \Delta U + W$

As $W_A > W_B \Rightarrow \Delta Q_A > \Delta Q_B$

93. (b) Suppose amount of water evaporated be M gram. Then (150 – M) gram water converted into ice.

so, heat consumed in evoporation = Heat released in fusion

$M \times L_v = (150 - M) \times L_s$

$M \times 2.1 \times 10^6 = (150 - M) \times 3.36 \times 10^5$

$\Rightarrow M = 20$ g

94. (d) a → Isobaric, b → Isothermal, c → Adiabatic, d → Isochoric. So correct option will be dabc.

95. (b) Total work done by the gas during the cycle is equal to area of triangle ABC.

As cyclic process is clockwise. So sign will be +ve.

$$\therefore \Delta W = \frac{1}{2} \times 4 \times 5 = 10 \text{ J}$$

96. (b) Equation of adiabatic change is

$TV^{\gamma-1}$ = constant

Put $\gamma = \frac{7}{5}$, we get: $\gamma - 1 = \frac{7}{5} - 1 \quad \therefore x = \frac{2}{5}$

97. (b) Work done,

$$W = P\Delta V = nR\Delta T = \frac{1}{2} \times 8.31 \times 70 \simeq 291\text{J}$$

98. (a) Equation of the BC is given by two point straight line equation

$$\Rightarrow \frac{P - P_0}{V - 2V_0} = \frac{3P_0 - P_0}{V_0 - 2V_0} \Rightarrow P = P_0 - \frac{2P_0}{V_0}(V - 2V_0)$$

using PV = nRT

$$\text{Temperature, } T = \frac{P_0V - \dfrac{2P_0V^2}{V_0} + 4P_0V}{1 \times R}$$

$(\because$ n = 1 mole given)

$$T = \frac{P_0}{R}\left[5V - \frac{2V^2}{V_0}\right], \ \frac{dT}{dV} = 0 \Rightarrow 5 - \frac{4V}{V_0} = 0 \Rightarrow V = \frac{5}{4}V_0$$

$$T = \frac{P_0}{R}\left[5 \times \frac{5V_0}{4} - \frac{2}{V_0} \times \frac{25}{16}V_0^2\right] = \frac{25}{8}\frac{P_0V_0}{R}$$

99. (d) Work done on gas = nRT $\ell n\left(\frac{P_f}{P_i}\right)$ = R(300) $\ell n(2)$

$$= 300 \text{ Rln2} \left(\because \frac{P_f}{P_i} = 2\right)$$

100. (a) $W = P\Delta V$

$$Q = nC_P\Delta T = \frac{nR\gamma\Delta T}{\gamma - 1} = nR\Delta T\frac{\gamma}{\gamma - 1} = P\Delta V\left(\frac{\gamma}{\gamma - 1}\right)$$

$$\therefore \frac{W}{Q} = \frac{\gamma - 1}{\gamma} = \frac{\frac{5}{3} - 1}{\frac{5}{3}} = \frac{2}{5}$$

101. (a) $\frac{U}{V} = KT^4$

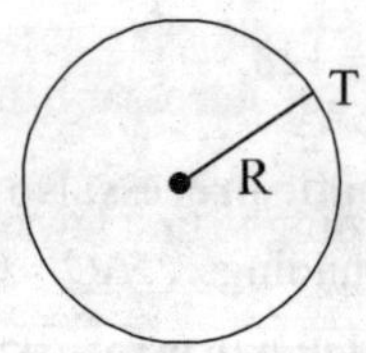

$P = \frac{1}{3}KT^4$

For adiabatic process,

$dU = -dW$

$\Rightarrow d(KVT^4) = -PdV$

$\Rightarrow K[4VT^3dT + T^4dV] = -\frac{1}{3}KT^4dV$

$\Rightarrow 4KVT^3dT = -\frac{4}{3}KT^4dV \Rightarrow \frac{dT}{T} = -\frac{1}{3}\frac{dV}{V}$

$\Rightarrow \ln T = -\frac{1}{3}\ln V \Rightarrow \ln T + \ln V^{1/3}$ = cons.

$\Rightarrow TV^{1/3}$ = cons. $\Rightarrow T \times \left(\frac{4}{3}\pi R^3\right)^{1/3}$ = constant

$\Rightarrow$ TR = constant $\Rightarrow T \propto \frac{1}{R}$

102. (b) In *VT* graph

ab-process : Isobaric, temperature increases.

bc process : Adiabatic, pressure decreases.

cd process : Isobaric, volume decreases.

da process : Adiabatic, pressure increases.

The above processes correctly represented in *P-V* diagram (b).

103. (d) In cyclic process, change in total internal energy is zero.

$\Delta U_{cyclic} = 0$

$$\Delta U_{BC} = nC_v\Delta T = 1 \times \frac{5R}{2}\Delta T$$

Where, C_v = molar specific heat at constant volume.
For BC, $\Delta T = -200$ K
$\therefore \quad \Delta U_{BC} = -500R$

104. (c) Given : work done, $W = 830$ J
No. of moles of gas, $\mu = 2$
For diatomic gas $\gamma = 1.4$
Work done during an adiabatic change

$$W = \frac{\mu R(T_1 - T_2)}{\gamma - 1} \Rightarrow 830 = \frac{2 \times 8.3(\Delta T)}{1.4 - 1} = \frac{2 \times 8.3(\Delta T)}{0.4}$$

$$\Rightarrow \Delta T = \frac{830 \times 0.4}{2 \times 8.3} = 20\,\text{K}$$

105. (b) Volume of the gas

$$v = \frac{m}{d} \text{ and}$$

Using PV^γ = constant

$$\frac{P'}{P} = \frac{V}{V'} = \left(\frac{d'}{d}\right)^\gamma \text{ or } 128 = (32)^\gamma$$

$$\therefore \gamma = \frac{7}{5} = 1.4$$

106. (d) Given $P = \alpha V$

$$\text{Work done, } w = \int_V^{mV} P dV = \int_V^{mV} \alpha V dV = \frac{\alpha V^2}{2}(m^2 - 1).$$

107. (b) In an adiabatic process, $\delta H = 0$
But, it is not necessary that temperature will remain constant.
And according to first law of thermodynamics
$\delta H = \delta U + W \therefore W = -\delta U$

108. (d) It is the free expansion
$\therefore$ So, T remains constant

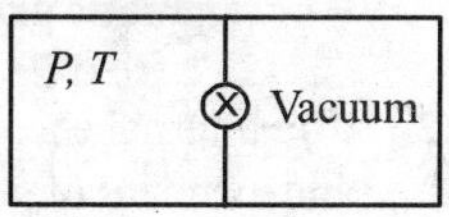

$$\Rightarrow P_1V_1 = P_2V_2 \Rightarrow P\frac{V}{2} = P_2(V)$$

$$P_2 = \left(\frac{P}{2}\right)$$

109. (b) The process $A \to B$ is isobaric.
$\therefore$ work done $W_{AB} = nR(T_2 - T_1)$
$= 2R(500 - 300) = 400R$

110. (d) $W = nRT\ln\left(\frac{P_1}{P_2}\right) = 2R \times 300 \ln\left(\frac{10^5}{2 \times 10^5}\right) = -414\,R$

111. (a) The net work in the cycle $ABCDA$ is

$$W = W_{AB} + W_{BC} + W_{CD} + W_{DA}$$

$$= 400R + 2.303nRT\log\frac{P_B}{P_C} + (-400R) - 414R$$

$$= 2.303 \times 2R \times 500\log\frac{2 \times 10^5}{1 \times 10^5} - 414\,R$$

$$= 693.2\,R - 414\,R = 279.2\,R$$

112. (a) Work done in adiabatic compression is given by

$$W = \frac{nR\Delta T}{1 - \gamma} \Rightarrow -146000 = \frac{1000 \times 8.3 \times 7}{1 - \gamma}$$

$$\text{or } 1 - \gamma = -\frac{58.1}{146} \Rightarrow \gamma = 1 + \frac{58.1}{146} = 1.4$$

Hence the gas is diatomic.

113. (c) Work is not a state function. The remaining three parameters are state function.

114. (c) Efficiency of Carnot engine

$$n = 1 - \frac{T_{freezing}}{T_{boiling}} \Rightarrow \eta = \left(1 - \frac{273}{373}\right) \times 100 = 26.8\%$$

115. (102) We have, $\eta = 1 - \frac{T_2}{T_1} = 1 - \frac{Q_2}{Q_1}$

$$\Rightarrow \frac{Q_2}{Q_1} = \frac{T_2}{T_1} \Rightarrow T_2 = \frac{Q_2}{Q_1} \times T_1 = \frac{225}{300} \times 500 = 375K = 102°C$$

116. (a) Efficiency of heat engine

$$\eta = 1 - \frac{T_s}{T} \Rightarrow \frac{1}{4} = 1 - \frac{T_s}{T} \qquad [T_s \to \text{sink temperature}]$$

$$\Rightarrow 4T_s = 3T \text{ or, } T_s = \frac{3}{4}T \qquad ...(i)$$

$$\Rightarrow \frac{1}{2} = 1 - \frac{T_s - 58}{T} \qquad ...(ii)$$

From equation (i) & (ii)

$$\frac{3}{4} = \frac{58}{T} + \frac{1}{2} \Rightarrow \frac{1}{4} = \frac{58}{T} \Rightarrow T = 232$$

$$\therefore T_s = \frac{3}{4}T = 174°C$$

117. (500) Given,
Heat taken, $Q_1 = 300$ J
Heat delivered, $Q_2 = 240$ J
Work done by heat engine, $W = Q_1 - Q_2 = 300 - 240 = 60$J

$$\text{Efficiency} = \frac{W}{Q_{in}} = \frac{60}{300} = 1 - \frac{T_2}{T_1}$$

$$\Rightarrow \frac{1}{5} = 1 - \frac{400}{T_1} \Rightarrow \frac{400}{T_1} = \frac{4}{5} \Rightarrow T_1 = 500\,K.$$

118. (a) Coefficient of performance $= \frac{\theta_2}{W} = \frac{T_L}{T_H - T_L}$

$$\Rightarrow \frac{\frac{d\theta_2}{dt}}{\frac{dW}{dt}} = \frac{263}{298 - 263} = \frac{263}{35}$$

$$\Rightarrow \frac{d\theta_2}{dt} = \frac{263}{35} \times \frac{dW}{dt} = \frac{263}{35} \times 35 \Rightarrow \frac{d\theta_2}{dt} = 263\,\text{watt}$$

119. (19)

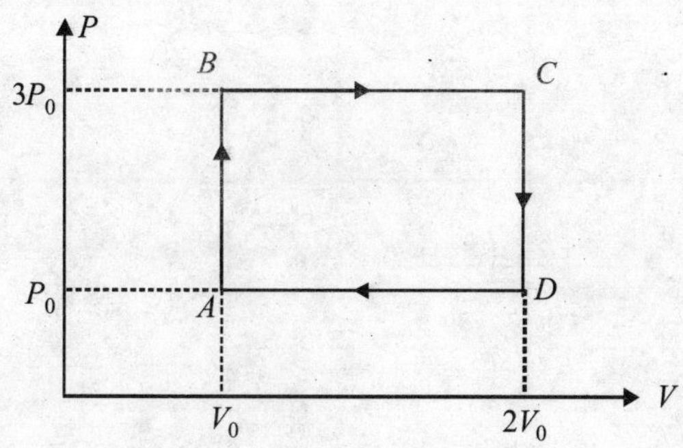

From the figure,

Work, $W = 2P_0V_0$

Heat given, $Q_{in} = Q_{AB} + Q_{BC} = n \cdot C_V \Delta T_{AB} + nC_P \Delta T_{BC}$

$= n\frac{3R}{2}(T_B - T_A) + \frac{n5R}{2}(T_C - T_B)$

$\left(\because C_V = \frac{3R}{2} \text{ and } C_P = \frac{5R}{2}\right)$

$= \frac{3}{2}(P_BV_B - P_AV_A) + \frac{5}{2}(P_CV_C - P_BV_B)$

$= \frac{3}{2} \times [3P_0V_0 - P_0V_0] + \frac{5}{2}[6P_0V_0 - 3P_0V_0]$

$= 3P_0V_0 + \frac{15}{2}P_0V_0 = \frac{21}{2}P_0V_0$

Efficiency, $\eta = \frac{W}{Q_{in}} = \frac{2P_0V_0}{\frac{21}{2}P_0V_0} = \frac{4}{21} \Rightarrow \eta\% = \frac{400}{21} \approx 19.$

120. **(8791)** Given,

Heat absorbed, $Q_2 = mL = 80 \times 100 = 8000$ Cal

Temperature of ice, $T_2 = 273$ K

Temperature of surrounding,

$T_1 = 273 + 27 = 300$ K

Efficiency $= \frac{w}{Q_2} = \frac{Q_1 - Q_2}{Q_2} = \frac{T_1 - T_2}{T_2} = \frac{300 - 273}{273}$

$\Rightarrow \frac{Q_1 - 8000}{8000} = \frac{27}{273} \Rightarrow Q_1 = 8791$ Cal

121. **(c)** Efficiency, $\eta = \frac{\text{Work done}}{\text{Heat absorbed}} = \frac{W}{\Sigma Q}$

$= \frac{Q_1 + Q_2 + Q_3 + Q_4}{Q_1 + Q_3} = 0.5$

Here, $Q_1 = 1915$ J, $Q_2 = -40$ J and $Q_3 = 125$ J

$\therefore \frac{1915 - 40 + 125 + Q_4}{1915 + 125} = 0.5$

$\Rightarrow 1915 - 40 + 125 + Q_4 = 1020 \Rightarrow Q_4 = 1020 - 2000$

$\Rightarrow Q_4 = -Q = -980 \text{ J} \Rightarrow Q = 980 \text{ J}$

122. **(d)** In a refrigerator, the heat dissipated in the atmosphere is more than that taken from the cooling chamber, therefore the room is heated. If the door of a refrigerator is kept open.

123. **(d)** Q_1 = area under BC = $T_0S_0 + \frac{1}{2}T_0S_0$

Q_2 = area under AC = $T_0(2S_0 - S_0) = T_0S_0$ and $Q_3 = 0$

Efficiency, $\eta = \frac{W}{Q_1} = \frac{Q_1 - Q_2}{Q_1}$

Now, Q_1 is heat absorbed because entropy increases in going from B → C

Q_2 is heat lost as entropy decreases in going from C → A

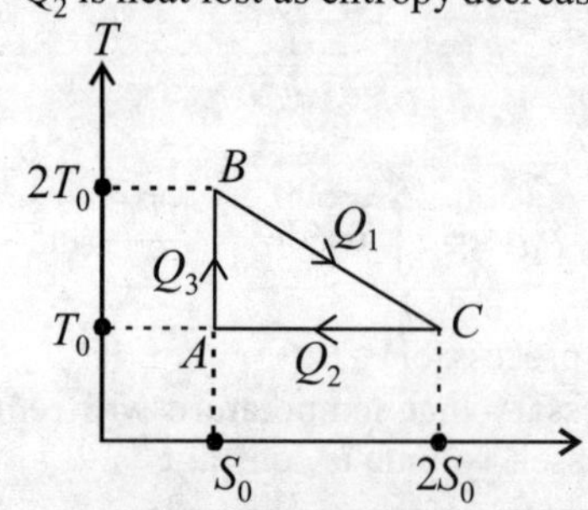

$= 1 - \frac{Q_2}{Q_1} = 1 - \frac{T_0S_0}{\frac{3}{2}T_0S_0} = \frac{1}{3}$

124. **(b)** Internal energy and entropy are state function, they are independent of path taken.

Kinetic Theory 12

1. **(b)** Given : $P_1 = 8 \times 10^3$ Pa, $T_1 = 1000$ K, $P_2 = 7 \times 10^3$ Pa, $T_2 = 500$ K, $T_f = 600$ K
Using idle gas equation, $Pv = nRT$

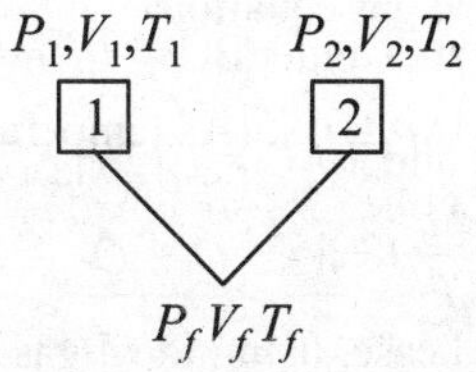

Number of masses will remain constant
$n_1 + n_2 = n_f$
$$\Rightarrow \frac{P_1V_1}{RT_1} + \frac{P_2V_2}{RT_2} = \frac{P_fV_f}{RT_f}$$
$$\Rightarrow \frac{8\times 2V}{R\times 1000} + \frac{7\times V}{R\times 500} = \frac{P_f(3V)}{R\times 600}$$
$$\Rightarrow \frac{16}{1000} + \frac{14}{1000} = \frac{P_f}{R\times 600} \Rightarrow \frac{30}{1000} = \frac{P_f}{200}$$
$\therefore P_f = 6$ kPa.

2. **(c)** For closed vessel, volume is constant
$\therefore \quad P \propto T$
$$\frac{\Delta P}{P} = \frac{\Delta T}{T} \Rightarrow \frac{0.4}{100} = \frac{1}{T} \qquad \therefore \quad T = 250\,K$$

3. **(327)** For a gas container of fixed volume,
$$\frac{P_1}{T_1} = \frac{P_2}{T_2} \Rightarrow \frac{P}{300} = \frac{2P}{T_2}$$
$\therefore T_2 = 600\,K = 327°C$

4. **(b)** For ideal gas, equation
$$PV = nRT \Rightarrow PV = \frac{m}{M}RT \Rightarrow P = \left(\frac{m}{V}\right)\frac{RT}{M} = \frac{\rho RT}{M}$$
(Where m is mass of gas and M is molecular mass of gas)

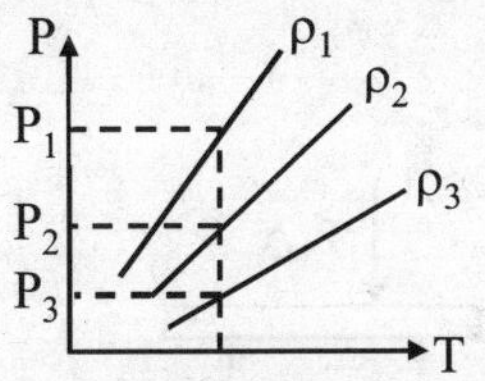

For same temperature $P_1 > P_2 > P_3$. So $\rho_1 > \rho_2 > \rho_3$

5. **(b)** $n_A = \frac{1}{2}$ mol, $n_B = \frac{1}{32}$ mol
By ideal gas equation,
$$PV = nRT \Rightarrow P \propto n \quad \therefore \frac{P_A}{P_B} = \frac{n_A}{n_B} = \frac{32}{2} = 16$$

6. **(b)** By ideal gas equation
$$PV = nRT \Rightarrow PV = \frac{N}{N_A}RT$$
N = Total no. of molecules
$$P = \frac{N}{V}kT \qquad \left[\because k = \frac{R}{N_A}\right]$$
$$\Rightarrow 1.38 \times 1.01 \times 10^5 = 2 \times 10^{25} \times 1.38 \times 10^{-23} \times T$$
$$\Rightarrow 1.01 \times 10^5 = 2 \times 10^2 \times T \qquad \therefore T = \frac{1.01\times 10^3}{2} \approx 500\,K$$

7. **(a)** By ideal gas equation
$$PV = nRT \Rightarrow P = \left(\frac{nR}{V}\right)T$$
So, when $T = 0\,K \Rightarrow P = 0$
i.e. for zero pressure, temperature should be 0K or –273°C irrespective of volume.

8. **(c)** From the ideal gas equation PV = nRT
Here, volume is constant $\therefore \frac{P_1}{T_1} = \frac{P_2}{T_2}$
Here, $T_1 = 27 + 273 = 300\,K$
$P_1 = 270$ kPa
$T_2 = 36 + 273 = 309\,K$
$$\Rightarrow \quad P_2 = \frac{P_1}{T_1} \times T_2 = \frac{270\times(309)}{300} = 278\text{ kPa.}$$

9. **(d)** At constant volume
$P \propto T$
$$\text{So, } T_2 = T_1 \times \frac{P_2}{P_1} = 273 \times \frac{180}{100} = 273 \times \frac{9}{5} \approx 491K$$

10. **(c)** Number of moles $(n) = n_{H_2} + n_{O_2}$
$$= \frac{16}{2} + \frac{128}{32} = 8 + 4 = 12$$
At STP, 1 moles occupy 22.4 l
So, 12 moles occupy $(22.4 \times 12)l$
i.e. $268.8l$
i.e. $268.8 \times 10^3 cm^3$
i.e. $27 \times 10^4 cm^3$

11. **(b)**

12. **(a)** We have $PV = nRT \Rightarrow V = \frac{nR}{P}T$
$V = mT$, where m = slope of $v - T_{curve} = \frac{nR}{P}$
Now, as $m_2 > m_1 \Rightarrow \frac{nR}{P_2} > \frac{nR}{P_1} \Rightarrow P_2 < P_1$

13. **(b)** We have, $\lambda = \dfrac{RT}{\sqrt{2}\pi P N_A d^2}$

If P = constant $\Rightarrow \lambda \propto T$. So statement (C) is correct.

and, increasing density, increases collision frequency so mean free path decreases. So option (B) is correct.

14. **(a)** Given $\dfrac{dp}{dv} = -ap$

Integrating both sides we get

$$\Rightarrow \int_{p_0}^{p} \frac{dp}{P} = -a\int_0^v dv \Rightarrow \ell n\left(\frac{p}{p_0}\right) = -av$$

$$\Rightarrow p = p_0 e^{-av}$$

Using ideal gas equation pv = nRT

$$T = \frac{pv}{nR} = \frac{p_0 v e^{-av}}{R} \quad (\because n = 1)$$

$$\frac{dT}{dv} = 0 \Rightarrow \frac{p_0}{R}\{e^{-av} + ve^{-av}(-a)\} = 0$$

$$\Rightarrow \frac{p_0 e^{-av}}{R}\{1 - av\} = 0 \Rightarrow v = \frac{1}{a}$$

$$\therefore T_{\max} = \frac{p_0 1}{Rae} = \frac{p_0}{Rae}$$

15. **(c)** From, ideal gas equation

$PV = nRT$

$$\Rightarrow 400 \times 10^3 \times 500 \times 10^{-6} = n\left(\frac{25}{3}\right)(300) \Rightarrow n = \frac{2}{25}$$

Let n_1 & n_2 be no. of moles of hydrogen & oxygen respectively

Also, $n_1 + n_2 = n$

Using $n = \dfrac{m}{M}$

$$\frac{m_1}{2} + \frac{m_2}{32} = \frac{2}{25}$$

and $m_1 + m_2 = 0.76$ gm

$$\frac{m_2}{m_1} = \frac{16}{3}$$

16. **(c)** Given,

Volume of cylinder, $V = 4 \times 10^{-3}\,m^3$

n = 1 + 2 = 3 moles

Temperature of the mixture, T = 400K

From ideal gas equation

$$\Rightarrow PV = nRT \Rightarrow P = \frac{nRT}{V}$$

$$\Rightarrow P = \frac{3 \times 8.3 \times 400}{4 \times 10^{-3}} \Rightarrow P = 24.9 \times 10^5 \text{ Pa}$$

17. **(c)** From the ideal gas equation,

PV = nRT

$\Rightarrow$ PV = CT

Therefore, PV v/s T graph is straight line.

18. **(5)** Using ideal gas equation, $PV = nRT$

$$\Rightarrow P_1V_1 = nR \times 250 \qquad [\because T_1 = 250 \text{ K}] \qquad ...(i)$$

$$P_2(2V_1) = \frac{5n}{4}R \times 2000 \qquad [\because T_2 = 2000 \text{ K}] \qquad ...(ii)$$

Dividing eq. (i) by (ii),

$$\frac{P_1}{2P_2} = \frac{4 \times 250}{5 \times 2000} \Rightarrow \frac{P_1}{P_2} = \frac{1}{5}$$

$$\therefore \frac{P_2}{P_1} = 5.$$

19. **(150)** In first case,

From ideal gas equation

$PV = nRT$

$P\Delta V + V\Delta P = 0$ (As temperature is constant)

$$\Delta V = -\frac{\Delta P}{P}V \qquad ...(i)$$

In second case, using ideal gas equation again

$P\Delta V = -nR\Delta T$

$$\Delta V = -\frac{nR\Delta T}{P} \qquad ...(ii)$$

Equating (i) and (ii), we get

$$\frac{nR\Delta T}{P} = -\frac{\Delta P}{P}V \Rightarrow \Delta T = \Delta P\frac{V}{nR}$$

Comparing the above equation with $|\Delta T| = C|\Delta P|$, we have

$$C = \frac{V}{nR} = \frac{\Delta T}{\Delta P} = \frac{300 \text{ K}}{2 \text{ atm}} = 150 \text{ K/atm}$$

20. **(a)** $N = \int \rho(dv)$

$$= \int_0^\infty n_0 e^{-\alpha r^4} \times 4\pi r^2 dr = 4\pi n_0 \int_0^\infty r^2(e^{-\alpha r^4})dr$$

$$= 4\pi n_0 \cdot \frac{2!}{\alpha^{3/4}} \quad \left[\because \int_0^\infty r^n e^{-r^m}\, dr = \frac{n!}{\alpha^{n+1/m}}\right]$$

[**Note:** For more details refer to gamma integration]

So, $N \propto n_0 . \alpha^{-3/4}$

21. **(d)** Clearly from figure,

$P_2A = P_1A + mg$

$$\Rightarrow \frac{nRT \cdot A}{A\ell_2} = \frac{nRT \cdot A}{A\ell_1} + mg$$

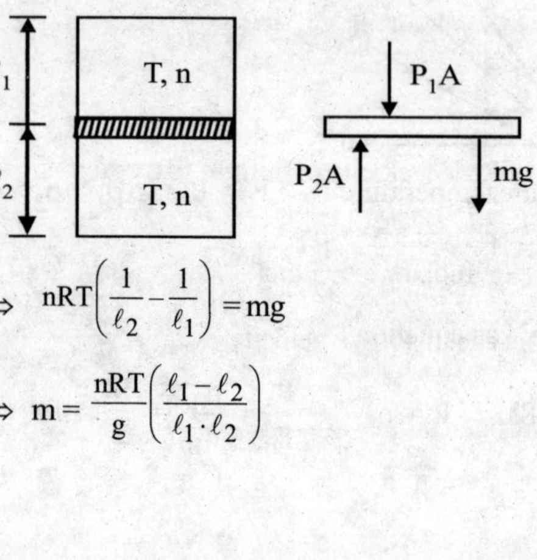

$$\Rightarrow nRT\left(\frac{1}{\ell_2} - \frac{1}{\ell_1}\right) = mg$$

$$\Rightarrow m = \frac{nRT}{g}\left(\frac{\ell_1 - \ell_2}{\ell_1 \cdot \ell_2}\right)$$

22. (b) Given: Temperature $T_i = 17 + 273 = 290\,K$
Temperature $T_f = 27 + 273 = 300\,K$
Atmospheric pressure, $P_0 = 1 \times 10^5\,Pa$
Volume of room, $V_0 = 30\,m^3$
Difference in number of molecules, $n_f - n_i = ?$
Using ideal gas equation, $PV = nRT(N_0)$,
N_0 = Avogadro's number

$$\Rightarrow n = \frac{PV}{RT}(N_0) \therefore n_f - n_i = \frac{P_0V_0}{R}\left(\frac{1}{T_f} - \frac{1}{T_i}\right) N_0$$

$$= \frac{1 \times 10^5 \times 30}{8.314} \times 6.023 \times 10^{23}\left(\frac{1}{300} - \frac{1}{290}\right) = -2.5 \times 10^{25}$$

23. (b) In free expansion of ideal gas, temperature do not change as there is no change in internal energy. But, in free expansion of real gas potential energy increases, so kinetic energy decreases and therefore temperature decreases.

24. (c) The centre of mass of gas molecules also moves with lorry with uniform speed. As there is no relative motion of gas molecule. So, kinetic energy and hence temperature remain same.

25. (d) Using formula for kinetic theory of gas

$$K.E. = \frac{f}{2}KT$$

f = Degree of freedom, K = Boltzman constant
For He and Ar, Degree of freedom, $f - 3$

$$\frac{K.E._{He}}{K.E._{Ar}} = \frac{1}{1}$$

26. (c) Given,

$$\frac{\rho_1}{\rho_2} = \frac{4}{25}$$

RMS speed $\propto \frac{1}{\sqrt{\rho}}$

$\therefore$ Ratio of rms velocities $= \sqrt{\frac{\rho_2}{\rho_1}} = \frac{5}{2}$

27. (a) The expression of mean squared velocity is given by

$$V_{rms} = \sqrt{\frac{3RT}{M}}, \quad V_{rms}^2 = 3RT/M$$

Hence, we can conclude that V_{rms}^2 is directly proportional to temperature.
$y = mx \Rightarrow$ Graph will be straight

28. (b) RMS velocity is $V_{rms} = \sqrt{\frac{3RT}{M}}$

$$V_{rms} \propto \frac{1}{\sqrt{M}}$$

Ratio of RMS velocity of helium to oxygen is:

$$\frac{V_H}{V_O} = \sqrt{\frac{32}{4}} = \frac{2\sqrt{2}}{1}$$

29. (d) Root mean square velocity, $V_{rms} = \sqrt{\frac{3RT}{M}}$

For hydrogen $V_H = \sqrt{\frac{3RT}{M_H}} = \sqrt{\frac{3RT}{2}}$ ... (i)

For oxygen $V_O = \sqrt{\frac{3RT}{M_O}} = \sqrt{\frac{3RT}{32}}$ ... (ii)

Divide (i) by (ii)

Now $\frac{V_H}{V_O} = \sqrt{\frac{32}{2}} \Rightarrow \frac{2}{V_O} = 4 \quad \therefore V_O = 0.5\,km/s$

30. (d) $T_{O_2} = 47 + 273 = 320\,K$

$$v_{H_2} = v_{O_2}$$

$$\sqrt{\frac{3RT_{H_2}}{m_{H_2}}} = \sqrt{\frac{3RT_{O_2}}{m_{O_2}}} \quad \because v_{rms} = \sqrt{\frac{3RT}{m}}$$

$$\Rightarrow \sqrt{\frac{T_{H_2}}{2}} = \sqrt{\frac{320}{32}} \Rightarrow T_{H_2} = \frac{320}{16} = 20K$$

31. (b) Average kinetic energy, $K_{avg} = \frac{f}{2}K_BT$
For monoatomic molecule, degree of freedom f = 3.

$$\therefore K_{avg} = \frac{3}{2}K_BT \Rightarrow T = \frac{0.414 \times 1.6 \times 10^{-19} \times 2}{3 \times 1.38 \times 10^{-23}} = 3200\,K$$

32. (c) Kinetic energy, $K = \frac{f}{2}nRT$
And for diatomic molecule, degree of freedom, f = 5

or, $K = \frac{5}{2} \times 1 \times 8.31 \times 300\,J = 6232.5\,J$

33. (a) Average KE per molecule, $K.E = \frac{3}{2}kT$
Average Kinetic energy per molecule of the gas is independent of mass.

$$\frac{K.E_{Ar}}{K.E_H} = \frac{1}{1}$$

34. (a) $V_{rms} = \sqrt{\frac{3RT}{M}}$ and $V_{av} = \sqrt{\frac{8RT}{\pi M}}$

$$\therefore \sqrt{\frac{3RT}{M}} = \left(1 + \frac{5}{x}\right)^{\frac{1}{2}} \sqrt{\frac{8RT}{\pi M}} \Rightarrow \frac{3 \times 22}{7 \times 8} = 1 + \frac{5}{x}$$

$\therefore x = 28$

35. (c) Given,
R.M.S. speed of chlorine molecule,
$V_{Cl} = 490\,m/s$
R.M.S. Speed,

$$v_{rms} = \sqrt{\frac{3RT}{M}} \Rightarrow v_{rms} \propto \sqrt{\frac{1}{M}}$$

$$\therefore \frac{v_{Ar}}{v_{Cl}} = \sqrt{\frac{M_{Cl}}{M_{Ar}}} \Rightarrow \frac{v_{Ar}}{490} = \sqrt{\frac{70.9}{39.9}} = 1.33$$

$\Rightarrow v_{Ar} = 1.33 \times 490 = 651.7\,m/s$

36. (b) R. M. S. speed, $v_{rms} = \sqrt{\frac{\gamma RT}{M}}$

$\because M_{Ne} > M_{Cl} > M_{\text{uranium hexafloride}}$
$\therefore v_{rms}(\text{mono}) > v_{rms}(\text{dia}) > v_{rms}(\text{poly})$

37. **(a)** Given,
Mass of nitrogen molecule, $m = 4.6 \times 10^{-26}$ kg
Temperature, T = 27°C = 300K
RMS speed (v_{rms}) is given by

$$v_{rms} = \sqrt{\frac{3k_BT}{m}} = \sqrt{\frac{3\times1.4\times10^{-23}\times300}{4.6\times10^{-26}}} = 523 \text{ m/s}$$

38. **(c)** If K_1 is kinetic energy at $T_1 = 27°C = 27 + 273 = 300K$
K_2 is kinetic energy at new temperature T.

$$\text{Kinetic energy} = \frac{f}{2}kT$$

$$\frac{K_1}{K_2} = \frac{T_1}{T_2} \Rightarrow \frac{1}{2} = \frac{300}{T} \quad (\because K_2 = 2K_1)$$

$\Rightarrow$ T = 600 K $\Rightarrow$ T = 327°C

39. **(d)** Moon has less escape velocity due to its small radius and low value of acceleration due to gravity. Also the rms velocity of molecules at moon is greater than its escape velocity. Hence molecules of air escape and there is no atmosphere at Moon.

40. **(d)** RMS speed,

$$V_{rms} = \sqrt{\frac{3RT}{M}}$$

$\Rightarrow V_{rms} \propto \sqrt{T}$
Here, $T_{initial} = 200$ K
$T_{final} = 800$ K
Initial RMS speed = v_0

$$\therefore \frac{v_0}{v_{rms}} = \sqrt{\frac{200}{800}} \Rightarrow v_{rms} = 2v_0$$

41. **(a)** Translational *K.E* on average of a molecule is $\frac{3}{2}$KT. Which is independent of nature of gas, pressure and volume.

42. **(b)** Average kinetic energy for diatomic gases

$$K_{av} = \frac{5}{2}kT$$

$$\therefore \frac{(K_{av})_H}{(K_{av})_O} = \frac{(27+273)}{(27+273)} = 1$$

43. **(b)** We have, $V_{rms} = \sqrt{\frac{\alpha+5}{\alpha}}V_{av}$

$$\sqrt{\frac{3RT}{M}} = \sqrt{\frac{\alpha+5}{\alpha}}\sqrt{\frac{8}{\pi}\frac{RT}{M}} \Rightarrow 3 = \frac{\alpha+5}{\alpha}\frac{8}{\pi} \Rightarrow \alpha = 28$$

44. **(c)** The rms speed of a gas molecule is

$$V_{rms} = \sqrt{\frac{3RT}{M}} \quad \therefore V_{rms} \propto \sqrt{T}$$

45. **(b)** $v_{rms} = \sqrt{\frac{3KT}{m}} \quad V_{rms} \propto \sqrt{T}$

$$\frac{(V_{rms})_f}{(V_{rms})_i} = \left(\frac{T_2}{T_1}\right)^{\frac{1}{2}} = \left(\frac{800}{200}\right)^{\frac{1}{2}} = 2 \Rightarrow (V_{rms})_f = 2(V_{rms})_i$$

$[\because T_1 = -73°C = 200K; T_2 = 527°C = 800K]$

So, (I) is correct.
As, PV = nRT for ideal gas and translational kinetic energy $= \frac{3}{2}nRT$

46. **(c)** The root mean square speed of smoke particles is given by

$$v_{rms} = \sqrt{\frac{3kT}{m}} = \sqrt{\frac{3\times1.38\times10^{-23}\times293}{5\times10^{-17}}} \approx 15 \text{ mm/s}$$

47. **(d)** Due to random motion of gas molecules, average momentum $P_{avg} = 0$, so P_{avg} do not depend on temperature.

$$v_{rms} = \sqrt{\frac{3RT}{M}}$$

When temperature is doubled, $T_{new} = 2T$

$$M_{new} = \frac{M}{2}$$

$$\frac{v_{new}}{v} = \frac{\sqrt{\frac{2T}{M/2}}}{\sqrt{\frac{T}{M}}} = 2$$

So, $v_{new} = 2v$

48. **(c)** $v_{rms} = \sqrt{\frac{3RT}{M_0}}$ and, $v_{rms_f} = 2v_{rms_i} \Rightarrow \sqrt{\frac{T_f}{T_i}} = 2$

$\Rightarrow T_f = 4T_i = 1200$ k

Now, $T_f = 1200K, T_i = 300K, n = \frac{14}{28} = \frac{1}{2}$

$$\text{So, } Q = nC_v\Delta T = \frac{1}{2}\times\frac{5R}{2}\times900$$

Q = 9360 J

49. **(c)** $v_{rms} = \sqrt{\frac{3RT}{M_0}}$

$v_{rms} \propto \frac{\sqrt{T}}{\sqrt{M_0}}$. As T and M_0 are same for both gas. So, V_{rms} will be same and point (A) is correct.
As, $PV = NKT$
as, V and T is constant $\Rightarrow P \propto N$

$$\Rightarrow \frac{P_1}{P_2} = \frac{N_1}{N_2} \Rightarrow \frac{P_1}{P_2} = \frac{1}{4}$$

So, point (B) is correct
As point (A) and (B) is correct. So (C) and (D) must be wrong.

50. **(a)** $k_{av} \propto T$, so, average kinetic energy of gas molecules depends an only temperature.
$k_{av}\uparrow\uparrow$, Þ $T_{av}\uparrow\uparrow$
k_{av} does not depend on the pressure and volume.
So, the only statement (A) is correct option (a) is right answer.

51. **(b)** Speed of sound is given as

$$v_S = \sqrt{\frac{\gamma RT}{M}}$$

and, $v_{rms} = \sqrt{\frac{3RT}{M}}$

So, $\frac{v_S}{v_{rms}} = \sqrt{\frac{\gamma}{3}} = \frac{1}{\sqrt{2}} \Rightarrow \frac{\gamma}{3} = \frac{1}{2} \Rightarrow \gamma = \frac{3}{2}$

As we know

$$\gamma = 1 + \frac{2}{f_{mix}}$$

and $f_{mix.} = \frac{2\times 3 + n\times 5}{n+2} = \frac{6+n\times 5}{(n+2)}$

$$\gamma = 1 + \frac{2(n+2)}{6+n\times 5} = \frac{6+5n+2n+4}{6+5n} \Rightarrow \gamma = \frac{7n+10}{6+5n} = \frac{3}{2}$$

$\Rightarrow 14n + 20 = 18 + 15n \Rightarrow n = 2$

52. **(b)** $v_{rms} = \sqrt{\frac{3RT}{M_0}}$; $v_{rms} \propto \sqrt{\frac{T}{M_0}}$

as, T becomes double and M_0 becomes half

So, v_{rms} will become $\sqrt{4}$ times i.e. 2 times.

53. **(25)** In isobaric process, pressure does not change.

$\therefore$ Change in internal energy, $\Delta U = nC_v\Delta T = \frac{5nR\Delta T}{2}$

Workdone, $W = nR\Delta T$

$$\therefore \frac{\Delta U}{W} = \frac{5nR\Delta T}{2nR\Delta T} = \frac{5}{2} = \frac{x}{10} \Rightarrow x = 25.00$$

54. **(a)** Root mean square speed is given by

$$v_{rms} = \sqrt{\frac{3KT}{M}} \Rightarrow v_{rms} \propto \frac{1}{\sqrt{M}}$$

$$\therefore \frac{(v_{rms})_{O_2}}{(v_{rms})_{H_2}} = \sqrt{\frac{M_{H_2}}{M_{O_2}}} \Rightarrow \frac{(v_{rms})_{O_2}}{(v_{rms})_{H_2}} = \sqrt{\frac{2}{32}} = \frac{1}{4}$$

$$\Rightarrow (v_{rms})_{H_2} = 4\times(v_{rms})_{O_2} \quad \left(\because (v_{rms})_{O_2} = 160\, m/s\right)$$

$= 4 \times 160 = 640$ m/s

55. **(a)** Root mean square velocity is given by

$$V_{rms} = \sqrt{\frac{3RT}{M}} \Rightarrow V_{rms} \propto \frac{1}{\sqrt{M}} \therefore\ V_H > V_O > V_{CO_2}$$

56. **(c)** Total kinetic energy $= \frac{f}{2}PV = \frac{3}{2}PV$

According the question $\frac{3}{2}PV = N \times 2 \times 10^{-9}J$

$$\Rightarrow \frac{3}{2}\times 2\times 10^5 \times 1000\times 10^{-6} = N\times 2\times 10^{-9}$$

$$\Rightarrow N = \frac{3}{4}\times 2\times \frac{10^5\times 10^3\times 10^{-6}}{10^{-9}} = \frac{3}{2}\times 10^{11} = 1.5\times 10^{11}$$

57. **(d)** RMS speed of gas molecule at NTP

$$V_{rms} = \sqrt{\frac{3RT}{M}}$$

Average speed of gas molecule at NTP

$$V_{avg} = \sqrt{\frac{8RT}{\pi M}}$$

The required ratio

$$\Rightarrow \frac{V_{rms}}{V_{avg}} = \sqrt{\frac{3\pi}{8}}$$

58. **(400)** rms speed of molecule is $v_{rms} = \sqrt{\frac{3RT}{M}}$

$\therefore v_{rms} \propto \sqrt{T}$

Now,

$$\therefore \frac{(v_{rms})_2}{(v_{rms})_1} = \sqrt{\frac{T_2}{T_1}}$$

$T_1 = 300K$ and $T_2 = 400K$

$$\therefore \frac{(v_{rms})_2}{(v_{rms})_1} = \sqrt{\frac{400}{300}} = \frac{2}{\sqrt{3}}$$

$$\Rightarrow (v_{rms})_2 = \frac{2}{\sqrt{3}}(v_{rms})_1 = \frac{2}{\sqrt{3}}\times 200 = \frac{400}{\sqrt{3}} \text{ m/s}$$

59. **(3600)** Here, $nC_v\Delta T = \frac{1}{2}mv^2$

For monoatomic gas, $C_v = \frac{3}{2}R$

$$\Rightarrow \Delta T = \frac{1}{2}\left(\frac{m}{n}\right)\cdot\frac{2}{3R}(30)^2$$

Lets assume mass per mole = 4 kg/mol

$$= \frac{1}{2}\times 4\times \frac{2}{3R}\times 30\times 30 = \frac{3600}{3R}$$

$\therefore x = 3600.$

60. **(41)** Room mean square speed is given by

$$v_{rms} = \sqrt{\frac{3RT}{M}}$$

Here, M = Molar mass of gas molecule

T = temperature of the gas molecule

We have given $v_{N_2} = v_{H_2}$

$$\therefore \sqrt{\frac{3RT_{N_2}}{M_{N_2}}} = \sqrt{\frac{3RT_{H_2}}{M_{H_2}}} \Rightarrow \frac{T_{H_2}}{2} = \frac{573}{28} \Rightarrow T_{H_2} = 41\text{ K}$$

61. **(c)** $V_{rms} = \sqrt{\frac{3RT}{M}}$

$$\frac{v_1}{v_2} = \sqrt{\frac{T_1}{T_2}} = \frac{(273+127)}{(273+227)} = \sqrt{\frac{400}{500}} = \sqrt{\frac{4}{5}} = \frac{2}{\sqrt{5}}$$

$$\therefore v_2 = \frac{\sqrt{5}}{2}v_1 = \frac{\sqrt{5}}{2}\times 200 = 100\sqrt{5}\text{ m/s.}$$

62. **(None)** Rate of change of momentum during collision

$$= \frac{mv-(-mv)}{\Delta t} = \frac{2mv}{\Delta t}N$$

so pressure $P = \frac{N\times(2mv)}{\Delta t\times A}$

$$= \frac{10^{22}\times 2\times 10^{-26}\times 10^4}{1\times 1} = 2\,N/m^2$$

63. **(c)** $v_{rms} = v_e$

$$\sqrt{\frac{3RT}{M}} = 11.2\times 10^3 \Rightarrow \sqrt{\frac{3\times 8.314\times T}{2\times 10^{-3}}} = 11200$$

$\Rightarrow T = 10058\text{ K} \simeq 10^4\text{K}$

64. **(a)** Using $\frac{V_{1rms}}{V_{2rms}} = \sqrt{\frac{M_2}{M_1}}$

$$\frac{V_{rms}(He)}{V_{rms}(Ar)} = \sqrt{\frac{M_{Ar}}{M_{He}}} = \sqrt{\frac{40}{4}} = 3.16$$

65. **(c)** Pressure, $P = \frac{1}{3}\frac{mN}{V}V_{rms}^2$

or, $P \propto \frac{(mN)T}{V}$ $\quad \because V_{rms} = \sqrt{\frac{3RT}{M}}$

If the gas mass and temperature are constant then

$P \propto (V_{rms})^2 \propto T$

So, force $\propto (V_{rms})^2 \propto T$

i.e., Value of $q = 1$

66. **(a)** $\because C = \sqrt{\frac{3RT}{M}}$

$$\Rightarrow (1930)^2 = \frac{3\times 8.314\times 300}{M}$$

$$M = \frac{3\times 8.314\times 300}{1930\times 1930} \approx 2\times 10^{-3}\text{kg}$$

The gas is H_2.

67. **(a)** Given, $V_1 = V$

$V_2 = 2V$

$T_1 = 27° + 273 = 300\text{ K}$

$T_2 = ?$

From charle's law

$$\frac{V_1}{T_1} = \frac{V_2}{T_2} \quad (\because \text{Pressure is constant})$$

or, $\frac{V}{300} = \frac{2V}{T_2}$ $\quad \therefore T_2 = 600\text{ K} = 600 - 273 = 327°\text{C}$

68. **(a)** As DOF is not stated, we will assume C_V is same for all gases

$U_i = U_f$

$\Rightarrow n_1 C_V T_1 + n_2 C_V T_2 + n_3 C_V T_3 = (n_1 + n_2 + n_3) C_V T$

$$\Rightarrow T = \frac{n_1T_1 + n_2T_2 + n_3T_3}{n_1 + n_2 + n_3}$$

69. **(a)** Given, mass = 1 kg

Density = 4 kg m^{-3}

$$\text{Volume} = \frac{\text{mass}}{\text{density}} = \frac{1}{4}\text{m}^3$$

Internal energy of the diatomic gas

$$= \frac{5}{2}PV = \frac{5}{2}\times 8\times 10^4\times\frac{1}{4} = 5\times 10^4 J$$

70. **(a)** The ratio of specific heat capacities γ is

$$\gamma = \frac{C_p}{C_v} = 1 + \frac{2}{f}$$

For triatomic rigid gas, $f = 6$

$$\gamma = 1 + \frac{2}{6} = \frac{4}{3}$$

For diatomic non-rigid gas, $f = 7$

$$\gamma = 1 + \frac{2}{7} = \frac{9}{7}$$

For diatomic rigid gas, $f = 5$

$$\gamma = 1 + \frac{2}{5} = \frac{7}{5}$$

For monoatomic gas, $f = 3$

$$\gamma = 1 + \frac{2}{3} = \frac{5}{3}.$$

71. **(c)** Collision frequency $= \frac{\text{average speed}}{\text{mean free path}}$

$$\text{frequency} = \frac{1}{T} = \frac{V_{avg}}{\lambda} = \frac{600}{3\times 10^{-7}} = 2\times 10^9\text{ sec}^{-1}.$$

72. **(12)** To find the internal energy of gas in the room

$$U = nC_VT = n\frac{5RT}{2} = \frac{5}{2}Pv\left[v = 4\times 4\times 3 = 48\text{ m}^3\right]$$

or, $u = \frac{5}{2}\times 10^5\times 48 = 12\times 10^6\text{ J}$

73. **(3)** Here $\frac{\gamma_A}{\gamma_B} = \frac{f_A+2}{f_A}\times\frac{f_B}{f_B+2}$

$$= \frac{3+2}{3}\times\frac{(6+2)}{(6+2)+2} = \frac{5}{3}\times\frac{8}{10} = \frac{40}{30}$$

$$\because \frac{40}{30} = 1 + \frac{1}{n} \Rightarrow \frac{40}{30} - 1 = \frac{1}{n}$$

$\therefore n = 3$

74. **(b)** Kinetic energy of translation $= \frac{3}{2}nRT$

$$= \frac{3}{2}\times\frac{50}{44}\times 8.31\times 290$$

$= 4108.644\text{ J}$

75. **(d)** $\gamma = \frac{C_p}{C_v} = \frac{2}{f} + 1$

For diatomic gas,

Without vibration, $f = 5 \therefore \gamma_1 = 1.4$

With vibration, $f = 7 \therefore \gamma_2 = 1.14 \quad \therefore \gamma_2 < \gamma_1$

76. **(b)** Initial temperature, $T_i = 273 + [-78°C] = 195K$
According to kinetic theory of gases;

$$K.E = \frac{nf_1RT}{2}$$

$\therefore$ $K.E \propto T$

$\therefore$ To double the K.E temperature should be $T_f = 2 \times T_i$

$\Rightarrow T_f = 2 \times 195 = 390\,K = 390 - 273 = 117°C$

77. **(c)** For monoatomic gas, $C_v = \frac{3}{2}R$

For diatomic gas, $C'_v = \frac{5}{2}R$

$$\therefore \text{Ratio} = \frac{C_V}{C'_V} = \frac{\frac{3}{2}R}{\frac{5}{2}R} = \frac{3}{5}$$

78. **(d)** Mean free path is given by,

$$\lambda = \frac{RT}{\sqrt{2}\pi d^2 N_A P} \Rightarrow \lambda \propto \frac{1}{d^2}$$

Kinetic energy is given by, $K = \frac{f}{2}nRT \Rightarrow K \propto T$

79. **(c)** Collision frequecny, $\upsilon = \sqrt{2}\pi d^2.n.\upsilon_{rms}$

$$= \sqrt{2}\pi d^2 \frac{N}{V}\sqrt{\frac{3RT}{m}} \Rightarrow \nu \propto \sqrt{T}$$

$$\therefore \frac{\nu_2}{\nu_1} = \sqrt{\frac{T_2}{T_1}} = \sqrt{\frac{(127+273)}{(27+273)}}$$

$$\therefore \nu_2 = \frac{2}{\sqrt{3}}Z \qquad [\because \nu_1 = Z]$$

80. **(c)** The average distance between two successive collisions of molecule is

$$\lambda = \frac{1}{\sqrt{2}\pi d^2 n}$$

81. **(b)** For polyatomic, non-linear molecules translational degree of freedom (f_t) = rotational degree of freedom (f_r). Since CH_4 is polyatomic non-linear

$\therefore f_t = f_r = 3$

82. **(d)** Degree of freedom, $f = 5 + 2(2n - 5)$
$f = 5 + 2(3 \times 2 - 5) = 7$

Energy of each molecule $= \frac{f}{2}K_BT$

$[\because n = 2 \text{ for diatomic molecule}]$

Energy of 10 molecules

$$E = 10\left(\frac{f}{2}K_BT\right) = 10\left(\frac{7}{2}K_BT\right) = 35\,K_BT$$

83. **(a)** Molar specific heat of mixture

$$C_V = \left(\frac{n_1C_1 + n_2C_2}{n_1 + n_2}\right) = \left(\frac{2\times\frac{3}{2} + 6\times\frac{5}{2}}{2+6}\right)R = \frac{9}{4}R$$

84. **(c)** Heat supplied is given by

$$Q = nC_p\Delta T = \left(\frac{\gamma}{\gamma-1}\right)nR\Delta T$$

$$Q = \left(\frac{\gamma}{\gamma-1}\right)W = \left(\frac{1.4}{1.4-1}\right)\times 200 = 700J$$

85. **(c)** Internal energy of the system is given by $U = nC_vT$
Argon is a monoatomic gas

$$\therefore C_{V_1} = \frac{3R}{2}$$

For diatomic gas $C_{V_2} = \frac{5R}{2}$ (for oxygen)

$$U_{Total} = n_1C_{V_1}T + n_2C_{V_2}T$$
$$= 8 \times \frac{3R}{2} \times T + 6 \times \frac{5R}{2} \times T = 27RT$$

86. **(c)**

Monoatomic	Diatomic
$f_1 = 3$,	$f_2 = 5$
$n_1 = 3$,	$n_2 = 2$

$$f_{mixture} = \frac{n_1f_1 + n_2f_2}{n_1 + n_2} = \frac{9+10}{f} = \frac{19}{5}$$

$$\gamma_{mixture} = 1 + \frac{2}{f_{mixture}} \Rightarrow \gamma_{mixture} = 1 + \frac{2\times 2}{19} = \frac{29}{19} = 1.52$$

87. **(a)** When two gases of degree of freedom f_1 and f_2 are mixed, then

$$f_{eq} = \frac{n_1f_1 + n_2f_2}{n_1 + n_2}$$

For diatomic gas, $f_{eq} = 5$
$n_1 = N, n_2 = 2$ moles
$f_1 = 6, f_2 = 3$

$$\therefore 5 = \frac{(N)(6)+(2)(3)}{N+2} \Rightarrow 5N + 10 = 6N + 6$$

$\therefore N = 4$

88. **(b)** Mean free path $\lambda = \frac{RT}{\sqrt{2}\pi d^2 N_A P}$ $\therefore \lambda \propto T$

$$\frac{1500d}{\lambda} = \frac{273}{373} \therefore \lambda = 2049d$$

89. **(c)**

Type of gases	No. of degrees of freedom
Monoatomic gas	3T
Diatomic + rigid	3T + 2R
Diatomic + non-rigid	3T + 2R + 1V
Polyatomic	3T + 3R + More than 1V

T = Translational degree of freedom
R = Rotational degree of freedom
V = Vibrational degree of freedom

90. **(d)** Given, $n_1 = 2, n_2 = 4$
Specific heat of mixture at constant volume

$$(C_v)_{mix} = \frac{n_1Cv_1 + n_2Cv_2}{n_1 + n_2} = \frac{2\times\frac{5}{2}R + 4\times\frac{3}{2}R}{2+4} = \frac{11R}{6}$$

Total internal energy of mixture

$$\Delta U = n(C_v)_{mix}T = 6\frac{11R}{6}\times T = 11RT$$

91. (b) Given,
Initial number of molecules per cm^3, $n_1 = 3\times10^{19}$
Final number of molecules per cm^3, $n_2 = 12\times10^{19}$

Mean free path, $\lambda = \frac{1}{\sqrt{2}\pi d^2 n}$
Collision frequency (f) is given by

$$f = \frac{v}{\lambda} = \frac{v}{\left(\frac{1}{\sqrt{2}\pi d^2 n_v}\right)} = \sqrt{2}\pi d^2 vn$$

$\because f \propto n$

$$\therefore \frac{f_1}{f_2} = \frac{n_1}{n_2} = \frac{3\times10^{19}}{12\times10^{-19}} = 0.25$$

92. (b) As $\gamma = 1 + \frac{2}{f}$
and at low temperature 'f' is independent of temperature.
So, γ is also independent of temperature.

93. (a) By Ist law of thermodynamics
$\Delta Q = \Delta U + \Delta W$

$$\Rightarrow 735 = \frac{f}{2}P\Delta V + P\Delta V \Rightarrow 735 = \left(\frac{5}{2}+1\right)P\Delta V$$

$$\Rightarrow 735 = \frac{7}{2}P\Delta V \Rightarrow 210 = P\Delta V$$

So, $\Delta U = \frac{f}{2}P\Delta V = \frac{5}{2}\times 210 = 525\text{ J}$

94. (d) We know that $C_v = \frac{f}{2}R$
Here, $f = 5 + 2$ [$\because$ 1 Vib.mode = 2DOF]
$= 7$

So, $C_v = \frac{7}{2}R$

95. (c) For monoatomic gas $\gamma_1 = \frac{5}{3}$
For diatomic gas at low temperatures

$\gamma_2 = \frac{7}{5}$

So, required ratio $= \frac{\gamma_1}{\gamma_2} = \frac{\frac{5}{3}}{\frac{7}{5}} = \frac{25}{21}$

96. (3) $(C_V)_{mix}$ of two mixed is given as,

$$(C_V)_{mix} = \frac{n_1C_{V_1} + n_2C_{V_2}}{n_1 + n_2}$$

$$= \frac{1.\frac{3R}{2} + 3.\frac{5R}{2}}{1+3} = \frac{9R}{4} = \frac{\alpha^2}{4}R \Rightarrow \alpha = 3$$

97. (750) $W = nR\Delta T = 150\text{ J}$ [$\because P\Delta V = nR\Delta T$]

$$Q = \Delta U + W = \frac{f}{2}nR\Delta T + nR\Delta T$$

$$\Rightarrow Q = \left(\frac{f}{2}+1\right)nR\Delta T = \left(\frac{8}{2}+1\right)150 = 750\text{J}$$

98. (b) According to law of equipartition of energy associated with each degree of freedom is $\frac{1}{2}RT$ per mole.
CH_4 is tetrahedron molecule.
Degree of freedom due to rotation = 3
Degree of freedom due to translation = 3

99. (a) $C_v = \frac{nR}{2}, C_p = \frac{(n+2)R}{2}$ ($\because C_p - C_v = R$)

$$\frac{C_v}{C_p} = \frac{n}{n+2}$$

100. (b) Here, the entire external kinetic energy of gases will be converted to internal energy.

i.e. $\frac{1}{2}(nM)V^2 = nC_V\Delta T$

$$\frac{1}{2}nMV^2 = \frac{nR}{\gamma-1}\Delta T$$

$$\Delta T = \frac{MV^2(\gamma-1)}{2R} = \frac{MV^2\times0.4}{2R} = \frac{MV^2}{5R}$$

101. (None) Average kinetic energy per molecule $= \frac{f}{2}KT$

$$\frac{Av.K.E_{Ar}}{Av.K.E_{O_2}} = \frac{\frac{3}{2}KT}{\frac{5}{2}KT} = \frac{3}{5}$$

102. (b) We have $C_v = \frac{f}{2}R$. and $C_p = \left(\frac{f}{2}+1\right)R$

So, $$\gamma = \frac{C_p}{C_v} = \frac{1+\frac{2}{f}}{\frac{f}{2}}\cdot\frac{}{} = \frac{2}{f}+1 = 1+\frac{2}{f}$$

103. (b) Total degree of freedom,
$f = 3 + 3 + (4\times2)$ [$\because$ 1 vibrational mode is equal to 2DOF]
$\Rightarrow f = 14$
Work done in adiabatic process is given by

$$W = \frac{P_1V_1 - P_2V_2}{\gamma-1} = \frac{nR(T_1-T_2)}{\gamma-1}$$

$$= \frac{1\times8.314\times(-10)}{\left(\frac{8}{7}-1\right)} = -582\text{ J} \qquad \left(\because \gamma = 1+\frac{2}{f} = \frac{8}{7}\right)$$

104. (c) According to Equi-partition law each degree of freedom contributes $\frac{1}{2}k_BT$

Average energy $= f\left(\frac{1}{2}K_BT\right)$

$\therefore$ Average energy $= 3\times\frac{1}{2}k_BT = \frac{3}{2}k_BT$

$[\because f_{monoatomic} = 3]$

105. (d) Here, initial internal energy = Final internal energy

$$\frac{F_1}{2}n_1RT_1 + \frac{F_2}{2}n_2RT_2 = \frac{F_1}{2}n_1RT + \frac{F_2}{2}n_2RT$$

or $T = \dfrac{n_1F_1T_1 + n_2F_2T_2}{n_1F_1 + n_2F_2}$

106. (d) Given, temperature of oxygen molecules, $T = 27°C$
Pressure of oxygen

$$l_{mean} = \frac{RT}{\sqrt{2}\pi d^2 N_A P} = \frac{1.38\times300\times10^{-23}}{\sqrt{2}\times3.14\times(0.3\times10^{-9})^2\times1.01\times10^5}$$

$= 102\times10^{-9}$ m $= 102$ nm

107. (25) As this is a case of free expansion.

So, $\Delta V = 0 \Rightarrow \Delta T = 0$

So, $T_i = T_f = T$

Now, $(n_1 + n_2)_{initial} = (n_1 + n_2)_{final}$

$$\Rightarrow \frac{P_1V_1}{RT} + \frac{P_2V_2}{RT} = \frac{PV_1}{RT} + \frac{PV_2}{RT}$$

$$\Rightarrow P = \frac{P_1V_1 + P_2V_2}{V_1 + V_2} = 2.55 \text{ atm}$$

108. (a) At constant pressure,

$$dU = nC_VdT = n\frac{5}{2}RdT$$

$$dQ = nC_PdT = n\left(\frac{7R}{2}\right)dT$$

and $dW = nRdT$

$$\therefore dU : dQ : dW = \frac{5}{2} : \frac{7}{2} : 1 = 5 : 7 : 2.$$

109. (c) Given, $U = 3PV + 4$

$$\because U = \frac{f}{2}nRT = \frac{f}{2}PV \therefore \frac{f}{2}PV = 3PV + 4$$

or, $f = 6 + \dfrac{8}{PV}$

So $f_{min} = 6$

Hence the gas is polyatomic.

110. (b) Mean free path, $\lambda = \dfrac{1}{\sqrt{2}\pi nd^2}$

where, d = diameter of the molecule
n = number of molecules per unit volume

But, mean time of collision, $\tau = \dfrac{\lambda}{v_{rms}}$

But $v_{rms} = \sqrt{\dfrac{3kT}{R}}$

$$\therefore \tau = \frac{\lambda}{\sqrt{\frac{3kT}{m}}} \Rightarrow \tau \propto \frac{1}{\sqrt{T}}$$

111. (c) As we know,

$\gamma = \dfrac{C_p}{C_v} = 1 + \dfrac{2}{f}$, where f = degree of freedom

(A) Monatomic, $f = 3$

$\therefore \gamma = 1 + \frac{2}{3} = \frac{5}{3}$

(B) Diatomic rigid molecules, $f = 5$

$\therefore \gamma = 1 + \frac{2}{5} = \frac{7}{5}$

(C) Diatomic non-rigid molecules, $f = 7$

$\therefore \gamma = 1 + \frac{2}{7} = \frac{9}{7}$

(D) Triatomic rigid molecules, $f = 6$

$\therefore \gamma = 1 + \frac{2}{6} = \frac{4}{3}$

112. (266.67) Heat gain by 0.1 mole = Heat lost by 0.05 mole

$\Rightarrow 0.1\,C_V(T - 200) = 0.05\,C_V(400 - T)$

$\Rightarrow 0.1T - 20 = 20 - 0.05\,T \Rightarrow 0.15T = 40 \Rightarrow T = 266.67$

113. (d) Here degree of freedom, $f = 3 + 3 = 6$ for triatomic non-linear molecule.

Internal energy of a mole of the gas at temperature T,

$$U = \frac{f}{2}nRT = \frac{6}{2}RT = 3RT$$

114. (b) Let S_p and S_v be the specific heat per unit mass of the gas at constant pressure and volume.

At constant pressure, heat required

$$\Delta Q_1 = mS_p \times \Delta T$$

$$\Rightarrow 160 = mS_p \times 50 \quad ...(i)$$

At constant volume, heat required

$$\Delta Q_2 = mS_v\Delta T$$

$$\Rightarrow 240 = mS_v \times 100 \quad ...(ii)$$

Dividing (i) by (ii), we get

$$\frac{160}{240} = \frac{S_p}{S_v} \Rightarrow \frac{S_p}{S_v} = \frac{4}{3} \Rightarrow \frac{C_p}{C_v} = \frac{4}{3} \quad \left[\because S = \frac{C}{M_0} \; S \propto C\right]$$

$$\gamma = \frac{C_p}{C_v} = \frac{4}{3} = 1 + \frac{2}{f} \quad \text{(Here, } f = \text{degree of freedom)}$$

$\Rightarrow f = 6.$

115. (a)

116. (d) Specific heat of gas at constant volume

$C_v = \frac{1}{2}fR$; f = degree of freedom

For gas A (diatomic)

f = 5 (3 translational + 2 rotational)

$\therefore C_v^A = \frac{5}{2}R$

For gas B (diatomic) in addition to (3 translational + 2 rotational) and 2 vibrational degree of freedom.

[Note → 1 vibration mode has 2DOF]

$$\therefore C_v^B = \frac{7}{2}R \text{ Hence } \frac{C_v^A}{C_v^B} = \frac{\frac{5}{2}R}{\frac{7}{2}R} = \frac{5}{7}$$

117. (None) Mean free path of a gas molecule is given by

$$\lambda = \frac{1}{\sqrt{2}\pi d^2 n}$$

Here, n = number of collisions per unit volume
d = diameter of the molecule
If average speed of molecule is v then

Mean free time, $\tau = \frac{\lambda}{v}$

$$\Rightarrow \tau = \frac{1}{\sqrt{2}\pi n d^2 v} = \frac{1}{\sqrt{2}\pi n d^2}\sqrt{\frac{M}{3RT}} \quad \left(\because v = \sqrt{\frac{3RT}{M}}\right)$$

$$\therefore \tau \propto \frac{\sqrt{M}}{d^2}$$

$$\therefore \frac{\tau_1}{\tau_2} = \frac{\sqrt{M_1}}{d_1^2} \times \frac{d_2^2}{\sqrt{M_2}} = \sqrt{\frac{40}{140}} \times \left(\frac{0.1}{0.07}\right)^2 = 1.09$$

118. (c) Relaxation time $(\tau) \propto \frac{\text{mean free path}}{\text{speed}} \Rightarrow \tau \propto \frac{1}{\text{v}}$

and, $v \propto \sqrt{T} \Rightarrow \tau \propto \frac{1}{\sqrt{T}}$

Hence graph between τ v/s $\frac{1}{\sqrt{T}}$ is a straight line which is correctly depicted by graph shown in option (c).

119. (a) Helium is a monoatomic gas and Oxygen is a diatomic gas.

For helium, $C_{V_1} = \frac{3}{2}R$ and $C_{P_1} = \frac{5}{2}R$

For oxygen, $C_{V_2} = \frac{5}{2}R$ and $C_{P_2} = \frac{7}{2}R$

$$\gamma = \frac{N_1 C_{P_1} + N_2 C_{P_2}}{N_1 C_{V_1} + N_2 C_{V_2}} \Rightarrow \gamma = \frac{n.\frac{5}{2}R + 2n.\frac{7}{2}R}{n.\frac{3}{2}R + 2n.\frac{5}{2}R} = \frac{19nR \times 2}{2(13nR)}$$

$$\therefore \left(\frac{C_P}{C_V}\right)_{mixture} = \frac{19}{13}$$

120. (d) Using, $\gamma_{\text{mixture}} = \frac{n_1 C_{p_1} + n_2 C_{p_2}}{n_1 C_{v_1} + n_2 C_{v_2}}$

As $C_p = \frac{R\gamma}{\gamma - 1}$ and $C_v = \frac{R}{\gamma - 1}$

So, $\frac{n_1}{\gamma_1 - 1} + \frac{n_2}{\gamma_2 - 1} = \frac{n_1 + n_2}{\gamma_m - 1} \Rightarrow \frac{3}{\frac{4}{3} - 1} + \frac{2}{\frac{5}{3} - 1} = \frac{5}{\gamma_m - 1}$

$$\Rightarrow \frac{9}{1} + \frac{2 \times 3}{2} = \frac{5}{\gamma_m - 1} \Rightarrow \gamma_m - 1 = \frac{5}{12}$$

$$\Rightarrow \gamma_m = \frac{17}{12} = 1.42$$

121. (b) $Q = nC_p \Delta T$

$$= n\frac{\gamma R}{\gamma - 1}\Delta T = (nR\Delta T)\frac{\gamma}{\gamma - 1} = P\Delta T\frac{\gamma}{\gamma - 1} = 10 \times \frac{1.4}{0.4} = 35\text{J}$$

122. (c) $V = 25 \times 10^{-3}\ \text{m}^3$, N = 1 mole of O_2
T = 300 K
$V_{rms} = 200$ m/s

$$\therefore \lambda = \frac{1}{\sqrt{2}N\pi r^2}$$

Average time $\frac{1}{\tau} = \frac{<V>}{\lambda} = 200.N\pi r^2.\sqrt{2}$

$$= \frac{\sqrt{2} \times 200 \times 6.023 \times 10^{23}}{25 \times 10^{-3}} \times \pi \times 10^{-18} \times 0.09$$

The closest value in the given option is = 10^{10}

123. (c) Amount of heat required (Q) to raise the temperature at constant volume
$Q = nC_v\Delta T$ ...(i)
Amount of heat required (Q_1) at constant pressure
$Q_1 = nC_P\Delta T$ ...(ii)
Dividing equation (ii) by (i), we get

$$\therefore \frac{Q_1}{Q} = \frac{C_P}{C_v} \Rightarrow Q_1 = (Q)\left(\frac{7}{5}\right) \quad \left(\because \gamma = \frac{C_P}{C_v} = \frac{7}{5}\right)$$

124. (b) $V_{rms}^2 = \frac{3k_B T}{m}$

$$T = \frac{mV_{rms}^2}{3k_B} = \frac{m\bar{v}^2}{3k_B}$$

125. (b) $\gamma_A = \frac{C_P}{C_v} = \frac{29}{22} = 1.32 < 1.4$ (diatomic)

and $\gamma_B = \frac{30}{21} = \frac{10}{7} = 1.43 > 1.4$

Gas A has more than 5-degrees of freedom and $\gamma = 1 + \frac{2}{f}$

So A must have vibrational DOF.

126. (a) Energy of the gas, E

$$= \frac{f}{2}nRT = \frac{f}{2}PV = \frac{f}{2}(3 \times 10^6)(2) = f \times 3 \times 10^6$$

Considering gas is monoatomic i.e., f = 3
Energy, E = 9×10^6 J

127. (b) Using, $\tau = \frac{1}{2n\pi d^2 V_{avg}}$, where $V_{av} = \sqrt{\frac{8RT}{\pi M_0}}$

$$\therefore \tau \propto \frac{\sqrt{T}}{P}\left[\because n = \frac{\text{no. of molecules}}{\text{Volume}}\right]$$

or, $\frac{\tau_1}{6 \times 10^{-8}} = \frac{\sqrt{500}}{2P} \times \frac{P}{\sqrt{300}} \approx 4 \times 10^{-8}$

128. **(a)** $U = \frac{f_1}{2} n_1 RT + \frac{f_2}{2} n_2 RT$

Considering translational and rotational modes, degrees of freedom $f_1 = 5$ and $f_2 = 3$

$$\therefore u = \frac{5}{2}(3RT) + \frac{3}{2} \times 5RT \Rightarrow U = 15RT$$

129. **(a)** According to question VT = K

and, $TV^{x-1} = \text{cons.}$

$\therefore x - 1 = 1 \Rightarrow 1 - x = -1$

$\because C = \frac{R}{1-x} + C_V$ (For polytropic process)

$$C = \frac{R}{-1} + \frac{3R}{2} = \frac{R}{2}$$

$\therefore \Delta Q = nC\,\Delta T$

$= \frac{R}{2} \times \Delta T$ [here, n = 1 mole]

130. **(c)** Thermal energy of N molecule

$$= N\left(\frac{3}{2}kT\right) = \frac{N}{N_A}\frac{3}{2}RT = \frac{3}{2}(nRT) = \frac{3}{2}PV$$

$$= \frac{3}{2}P\left(\frac{m}{\rho}\right) = \frac{3}{2}P\left(\frac{2}{8}\right) = \frac{3}{2} \times 4 \times 10^4 \times \frac{2}{8} = 1.5 \times 10^4 \text{ J}$$

therefore, order = 10^4 J

131. **(c)** Heat transferred,

$Q = nC_V \Delta T$ as gas in closed vessel

To double the rms speed, temperature should be 4 times i.e., $T' = 4T$ as $v_{rms} = \sqrt{3RT/M}$

$$\therefore Q = \frac{15}{28} \times \frac{5 \times R}{2} \times (4T - T)$$

$$\left[\therefore \frac{C_p}{C_v} = \gamma_{diatomic} = \frac{7}{5} \& C_p - C_v = R\right]$$

or, Q = 10000 J = 10 kJ

132. **(c)** In an adiabatic process

$TV^{\gamma-1} = \text{Constant}$ or, $T_1 V_1^{\gamma-1} = T_2 V_2^{\gamma-1}$

For monoatomic gas $\gamma = \frac{5}{3}$

$(300)V^{2/3} = T_2(2V)^{2/3} \Rightarrow T_2 = \frac{300}{(2)^{2/3}}$

$T_2 = 189$ K (final temperature)

Change in internal energy $\Delta U = n\frac{f}{2}R\,\Delta T$

$$= 2\left(\frac{3}{2}\right)\left(\frac{25}{3}\right)(-111) = -2.7 \text{ kJ}$$

133. **(a)** $S_p - S_v = a \Rightarrow C_p - C_v = Ma$

So, for H $\rightarrow C_p - C_v = 2a$ ...(i)

for $N_2 \rightarrow C_p - C_v = 28b$ ...(ii)

from (i) and (ii)

$2a = 28b = R \Rightarrow a = 14b$

134. **(d)** The ratio of specific heats at constant pressure (C_p) and constant volume (C_v)

$$\frac{C_P}{C_V} = \gamma = \left(1 + \frac{2}{f}\right)$$

where f is degree of freedom

$$\frac{C_P}{C_V} = \left(1 + \frac{2}{5}\right) = \frac{7}{5}$$

135. **(d)** For a polytropic process

$$C = C_V + \frac{R}{1-n} \qquad \therefore C - C_V = \frac{R}{1-n}$$

$$\therefore 1 - n = \frac{R}{C - C_V} \qquad \therefore 1 - \frac{R}{C - C_V} = n$$

$$\therefore n = \frac{C - C_V - R}{C - C_V} = \frac{C - C_V - C_P + C_V}{C - C_V}$$

$$\Rightarrow n = \frac{C - C_P}{C - C_V} \quad (\because C_P - C_V = R)$$

136. **(d)** In solid, we don't have translational and rotational DOF. But at high temp, we have 6 vibrational DOF.

So, $U = \frac{f}{2}RT = \frac{6}{2}RT = 3RT$

$$C = \frac{dU}{dT} = 3R$$

$$S = \frac{C}{M_0} = \frac{3R}{M_0} = \frac{3 \times 8.314}{27 \times 10^{-3}} = 924 \text{J/kg K}$$

137. **(c)** $TV^{\gamma-1} = \text{constant}$

$T_1 V_1^{\gamma-1} = T_2 V_2^{\gamma-1}$

$$\Rightarrow T(V)^{1/2} = T_2\left(\frac{V}{4}\right)^{1/2}$$

$$\left[\because \gamma = 1.5, T_1 = T, V_1 = V \text{ and } V_2 = \frac{V}{4}\right]$$

$$\therefore T_2 = \left(\frac{4V}{V}\right)^{1/2} T = 2T$$

138. **(b)** $S_p - S_v$

$$= \frac{C_P}{M_0} - \frac{C_V}{M_0} = \frac{C_P - C_V}{M_0} = \frac{R}{28}$$

139. **(d)** $P \propto T^3 \Rightarrow PT^{-3} = \text{constant}$ ....(i)

But for an adiabatic process, the pressure temperature relationship is given by

$P^{1-\gamma}\, T^{\gamma} = \text{constant}$

$\Rightarrow PT^{\frac{\gamma}{1-\gamma}} = \text{constt.}$ ....(ii)

From (i) and (ii) $\frac{\gamma}{1-\gamma} = -3 \Rightarrow \gamma = -3 + 3\gamma \Rightarrow \gamma = \frac{3}{2}$

13 Oscillations

1. **(a)** Given $x_1 = \sqrt{7}\sin 5t$ and $x_2 = 2\sqrt{7}\sin\left(5t+\frac{\pi}{3}\right)$

From phasor,

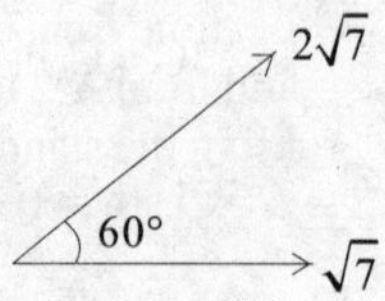

$\therefore$ Amplitude of resultant SHM = 7

$$\phi = \tan^{-1}\frac{2\sqrt{7}\times\sqrt{3}/2}{\sqrt{7}+2\sqrt{7}\times\frac{1}{2}} = \tan^{-1}\frac{\sqrt{21}}{2\sqrt{7}} = \tan^{-1}\frac{\sqrt{3}}{2}$$

$\therefore X_R = 7\sin(5t+\phi)$

$a_R = -7\times 25\sin(5t+\phi)$

$\therefore a_{max} = 175\text{ cm/sec} = 175\times 10^{-2}\text{ m/sec}$

2. **(d)** The general expression for SHM is given as

$x = A\sin(\omega t+\phi)$

$x_0 = A\sin\phi$ ...(i)

Both position and linear momentum of a particle can be expressed as a function of time.

$p = mA\omega\cos(\omega t+\phi)$

$p_0 = mA\omega\cos\phi$ ...(ii)

Dividing equation (ii) by (i), we get

$$\tan\phi = \left(\frac{x_0}{p_0}\right)m\omega$$

$$\text{From (i), } A = \frac{x_0}{\sin\phi} = \frac{\sqrt{(m\omega x_0)^2+p_0^2}}{m\omega}$$

This means we can explain assertion with the given reason.

3. **(b)** In SHM,

A = 1 cm, T = 2S

Number of cycles, $n = \frac{12.5}{2} = 6.25$ cycles

$\therefore D = 4\times 6+1 = 25$ cm and d = 1 cm $\therefore \frac{D}{d} = 25$

4. **(17)** Acceleration in simple harmonic motion is given by

$|a| = \omega^2 x$

$\Rightarrow 16 = \omega^2(4) \Rightarrow \omega = 2$ rad/s

Velocity, $v = \omega\sqrt{A^2-x^2}$

$$\Rightarrow A = \sqrt{\frac{v^2}{\omega^2}+x^2} \Rightarrow A = \sqrt{\frac{4}{4}+16} \Rightarrow A = \sqrt{17}\text{ m}$$

5. **(12)** Maximum velocity is,

$v_{max} = \omega A$

$$\Rightarrow v_{max} = \frac{2\pi}{T}A = \frac{2\pi}{\pi}\times 0.06 = 0.12\text{ m/sec} = 12\text{ cm/sec}$$

6. **(10)** Time period

Since, $T = 3.14 = \frac{2\pi}{\omega} \therefore \omega = 2$ rad / s

Given $x = 10\sin\left(\omega t+\frac{\pi}{3}\right)$

Velocity of the particle, $v = \frac{dx}{dt} = 10\omega\cos\left(\omega t+\frac{\pi}{3}\right)$

Therefore at t = 0

$$v = 10\,\omega\cos\left(\frac{\pi}{3}\right) = 10\times 2\times\frac{1}{2} \qquad [\because \omega = 2\text{ rad/s}]$$

$\therefore v = 10$ m/s

7. **(7)** Velocity of a particle in SHM is given by

$V = \omega\sqrt{A^2-x^2}$

Here, A = amplitude of motion

x = displacement of particle

ω = angular velocity

$$\text{At } x = \frac{2A}{3} \qquad v = \omega\sqrt{A^2-\left(\frac{2A}{3}\right)^2} = \frac{\sqrt{5}A\omega}{3}$$

Let A' be the new amplitude of particle when speed is increased to three times.

$$\therefore 3v = \sqrt{5}A\omega = \omega\sqrt{(A')^2-\left(\frac{2A}{3}\right)^2}$$

$$\Rightarrow 5A^2 = (A')^2 - \frac{4A^2}{9}$$

$$\Rightarrow (A')^2 = \frac{49A^2}{9} \Rightarrow A' = \frac{7A}{3} = \frac{nA}{3}\text{ (given)}$$

$\therefore n = 7$

8. **(2)** The phasor diagram of SHM is

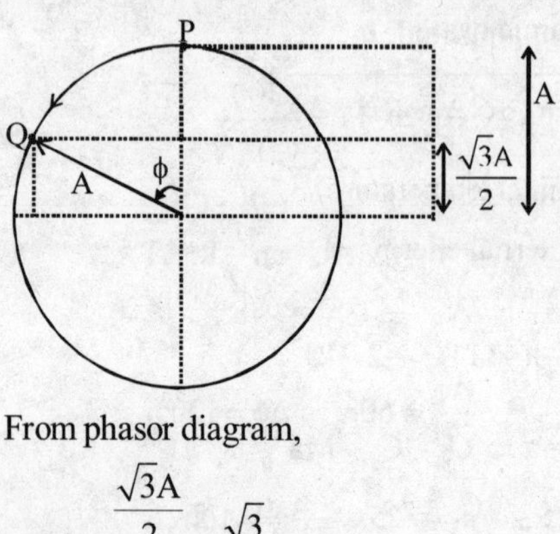

From phasor diagram,

$$\cos\phi = \frac{\frac{\sqrt{3}A}{2}}{A} = \frac{\sqrt{3}}{2}$$

$$\therefore \phi = \omega t = \frac{\pi}{6} \Rightarrow \frac{2\pi}{T}t = \frac{\pi}{6}$$

$\therefore t = \frac{\pi}{2}s = \frac{\pi}{x}s$ (given) $\quad [\because T = 6\pi\ s]$

$\therefore x = 2$

9. **(12)** Velocity, $V_{\text{at mean position}} = A\omega \Rightarrow 10 = 4\omega$

$\therefore \omega = \frac{5}{2}$

$v = \omega\sqrt{A^2 - x^2}$ or, $5 = \frac{5}{2}\sqrt{4^2 - x^2} \Rightarrow x^2 = 16 - 4$

$\therefore x = \sqrt{12}$ cm $= \sqrt{\alpha}$ cm (given) $\therefore \alpha = 12$

10. **(a)** $X = A\sin(\omega t + \delta)$

$\frac{A}{2} = A\sin(\omega t + \delta)$ At $t = 0$

$\sin\delta = \frac{1}{2} \Rightarrow \delta = \frac{\pi}{6}, \frac{5\pi}{6} \therefore \delta = \frac{\pi}{6}$

11. **(a)**

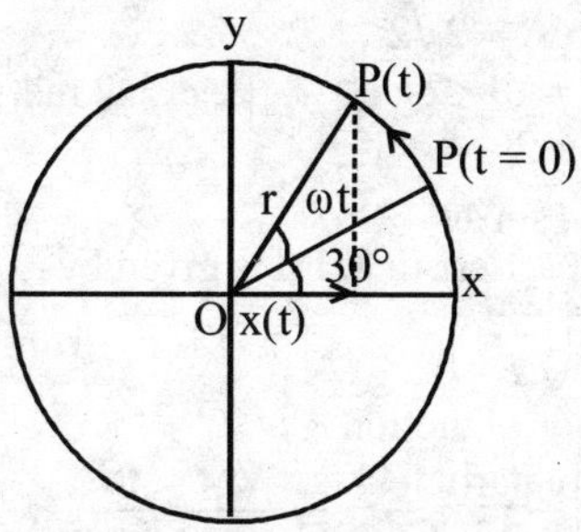

$x(t) = r\cos(\omega t + 30°) = r\cos(\omega t + \pi/6)$

12. **(88)** Given, $4v^2 = 50 - x^2$

$\Rightarrow v^2 = \frac{1}{4}(50 - x^2) \Rightarrow v = \frac{1}{2}\sqrt{50 - x^2}$

Comparing with velocity in SHM

$v = \omega\sqrt{A^2 - x^2}$

Angular, frequency

$\omega = \frac{1}{2}$

Time period, $T = \frac{2\pi}{\omega} = 4\pi = \frac{88}{7}$

$x = 88$

13. **(120)** Resultant amplitude

$A_R = \sqrt{A_0^2 + A_0^2 + 2A_0A_0\cos\Delta\phi}$

$A_R = \sqrt{4A_0^2\cos^2\frac{\Delta\phi}{2}}$

$\Rightarrow 2A_0\cos\left(\frac{\Delta\phi}{2}\right) = A_R \Rightarrow 2A_0\cos\left(\frac{\Delta\phi}{2}\right) = A_0$

$\Rightarrow \cos\frac{\Delta\phi}{2} = \frac{1}{2} \Rightarrow \frac{\Delta\phi}{2} = 60°, \quad \Delta\phi = 120°$

14. **(40)** $F = -25x \Rightarrow \frac{1}{4}a = -25x \Rightarrow a = -100x$

Comparing it with $a = -\omega^2 x$

we get $\omega^2 = 100 \Rightarrow \omega = 10$,

Now, $V_{max} = 4$ m/s

$\Rightarrow \omega A = 4\text{m/s} \Rightarrow 10A = 4$

$\Rightarrow A = 0.4\,\text{m} = 40$ cm.

15. **(d)**

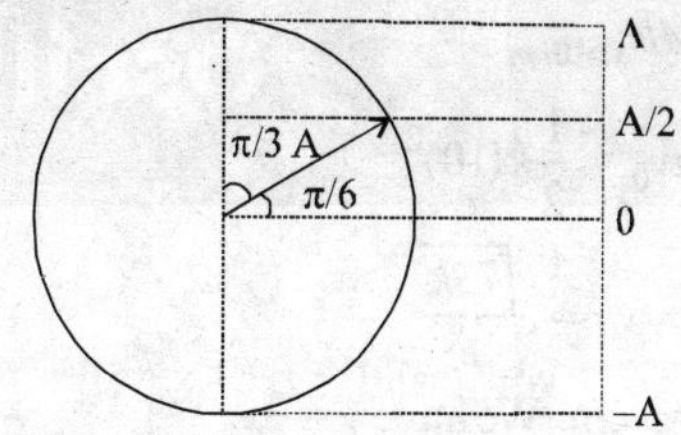

Let time from 0 to A/2 is t_1

and from A/2 to A is t_2

From the standard equation of SHM,

$x = A_0\sin(\omega t)$

$\Rightarrow \frac{A}{2} = A\sin(\omega t_1)$

$\Rightarrow \omega t_1 = \sin^{-1}\left(\frac{1}{2}\right) = \frac{\pi}{6}$ ...(i)

then $\omega t_1 = \pi/6$

Using $x = A_0\sin\omega t$ again

$A = A\sin\omega(t_1 + t_2)$

$\omega(t_1 + t_2) = \sin^{-1}(1) = \frac{\pi}{2}$

Using (i)

$\omega t_2 = \frac{\pi}{2} - \frac{\pi}{6} = \frac{\pi}{3}$ ...(ii)

Dividing equation (i) by (ii) we get

$\frac{t_1}{t_2} = \frac{1}{2}$

$\Rightarrow t_2 = 2t_1 = 2 \times 2 = 4$ sec

16. **(16)** $\frac{1}{2}kA^2 = \frac{1}{2}mv_{max}^2$

$\Rightarrow \frac{1}{2}kA^2 = \frac{(mv_{max})^2}{2m} \Rightarrow \frac{1}{2}kA^2 = \frac{p^2}{2m}$

$\Rightarrow \left(\frac{A_1}{A_2}\right)^2 = \frac{m_2}{m_1} = \frac{1024}{900}$

$\Rightarrow \frac{A_1}{A_2} = \frac{32}{30} = \frac{16}{15} = \frac{16}{16-1} \quad \therefore \alpha = 16$

17. **(b)** For simple harmonic motion, $x = A\sin\omega t$

$\Rightarrow v = \frac{dx}{dt} = A\omega\cos\omega t$

$\frac{x}{A} = \sin\omega t$ ..... (i)

$\frac{v}{A\omega} = \cos\omega t$ ..... (ii)

From (i) and (ii) we have

$\left(\frac{x}{A}\right)^2 + \left(\frac{v}{A\omega}\right)^2 = \sin^2\omega t + \cos^2\omega t$

$\left(\frac{x}{A}\right)^2 + \left(\frac{v}{A\omega}\right)^2 = 1$

The nature of the graph will be elliptical.

18. **(700)** Initially, at $x = 5$ cm, Let $v = v_0$
by applying law of conservation of mechanical energy

$$M.E_{x=5m} = ME_{x=10cm}$$

$$\frac{1}{2}k(5)^2 + \frac{1}{2}mv_0^2 = \frac{1}{2}k(10)^2$$

$$mv_0^2 = 75k \Rightarrow v_0 = \sqrt{\frac{75k}{m}}$$

Finally, at $x = 5$ cm. We have $v = 3v_0$
So amplitude will increase but mechanical energy is still conserved.

$$\text{So, } \frac{1}{2}k(5)^2 + \frac{1}{2}m\left(3\sqrt{\frac{75k}{m}}\right)^2 = \frac{1}{2}kA^2$$

$$\Rightarrow 25k + 675k = kA^2 \Rightarrow 700 = A^2 \quad \therefore A = \sqrt{700} \text{ cm}$$

19. **(a)** $x = 4\sin\left(\frac{\pi}{2} - \omega t\right) = 4\cos\omega t$

$Y = 4\sin\omega t$

So, $x^2 + y^2 = 4^2$, which is equation of circle.

20. **(1)** We have
$A = 8$ cm, $T = 6$ sec
Let us suppose at $t = 0$, particle is at maximum amplitude and at $t = t$, $x = \frac{A}{2}$

$$\text{So, } \frac{A}{2} = A\cos\omega t \Rightarrow \cos\omega t = \frac{1}{2}$$

$$\text{So, } \omega t = \frac{\pi}{3}$$

$$\frac{2\pi}{T} \times t = \frac{\pi}{3}$$

$$t = \frac{T}{6} = \frac{6}{6} = 1 \text{ sec}$$

Note: We have use $x = A\cos\omega t$ as equation of SHM because at $t = 0$, particle is at $x = A$.

21. **(b)** At $t = 1$ sec

$$x = \sin\pi\left(1 + \frac{1}{3}\right) = \sin\frac{4\pi}{3} = -\frac{\sqrt{3}}{2}$$

$$\text{So, } V = \omega\sqrt{A^2 - x^2} = \pi\sqrt{1 - \frac{3}{4}} = \frac{\pi}{2} m/s = 157 \text{ cm/s}$$

22. **(1)** Given,

$$y_1 = 10\sin\left(3\pi t + \frac{\pi}{3}\right)$$

Amplitude, $A_1 = 10$

$$y_2 = 5\left(\sin 3\pi t + \sqrt{3}\cos 3\pi t\right)$$

$$\Rightarrow y_2 = 10\left(\frac{1}{2}\sin 3\pi t + \frac{\sqrt{3}}{2}\cos 3\pi t\right)$$

$$\Rightarrow y_2 = 10\left(\cos\frac{\pi}{3}\sin 3\pi t + \sin\frac{\pi}{3}\cos 3\pi t\right)$$

$$\Rightarrow y_2 = 10\sin\left(3\pi t + \frac{\pi}{3}\right)$$

$\therefore$ Amplitude, $A_2 = 10$

$$\therefore \text{Ratio of amplitudes } = \frac{A_1}{A_2} = \frac{10}{10} = 1$$

23. **(2)** Given:

$$x(t) = A\sin(\omega t + \phi)$$

$$v(t) = \frac{dx}{dt}$$

$$\therefore \quad v(t) = A\omega\cos(\omega t + \phi)$$

$2 = A\sin\phi$ ...(i)

$2\omega = A\omega\cos\phi$ ...(ii)

From eq (i) and (ii)
$\tan\phi = 1$
$\phi = 45°$
Putting value of ϕ in equation (i)

$$2 = A\left\{\frac{1}{\sqrt{2}}\right\} \Rightarrow A = 2\sqrt{2}$$

$\therefore$ Value of x = 2

24. **(6)** From $Y = A\sin\omega t$

$$\Rightarrow \frac{A}{2} = A\sin\left(\frac{2\pi t}{T}\right) \qquad \left[\because Y = \frac{A}{2} \text{ from question}\right]$$

$$\Rightarrow \sin\left(\frac{2\pi t}{T}\right) = \frac{A}{2\times A} = \frac{1}{2} \Rightarrow \frac{2\pi t}{T} = \frac{\pi}{6}$$

$$\Rightarrow t = \frac{T}{12} = \frac{2}{12} = \frac{1}{6} \text{ seconds}$$

$\therefore \quad a = 6$

25. **(03.00)** Maximum speed = $a\omega$

$$V = \omega\sqrt{a^2 - y^2}$$

$$\Rightarrow \frac{a\omega}{2} = \omega\sqrt{a^2 - y^2} \Rightarrow \frac{a^2}{4} = a^2 - y^2$$

$$\Rightarrow y^2 = \frac{3a^2}{4} \Rightarrow y = \frac{a\sqrt{3}}{2}$$

$\therefore \quad x = 3$

26. **(a)** Restoration force, $F = m\omega^2 A$

$$\therefore \frac{F}{m} = \omega^2 A = \left(\frac{2\pi}{T}\right)^2 A$$

Point A covers $30° = \frac{\pi}{6}$ in 0.1 s

$$\therefore 2\pi \text{ covered in } \frac{2\pi \times 0.1}{\pi/6} = 1.2 \text{ s or } T = 1.2 \text{ s}$$

$\therefore$ Restoration force per unit mass,

$$\frac{F}{m} = \frac{4\pi^2}{(1.2)^2} \times 0.36 = \pi^2 = (3.14)^2 \simeq 9.87 \text{ N}$$

27. **(a)** Here, $v_x = -a\omega \sin \omega t$, $v_y = a\omega \cos \omega t$ and $v_z = a\omega$

$\Rightarrow v = \sqrt{v_x^2 + v_y^2 + v_z^2}$

$\Rightarrow v = \sqrt{(-a\omega \sin \omega t)^2 + (a\omega \cos \omega t)^2 + (a\omega)^2}$

$\Rightarrow v = \sqrt{2}a\omega$

28. **(c)** From the two mutually perpendicular S.H.M.'s, the general equation of Lissajous figure is given as

$$\frac{x^2}{A^2} + \frac{y^2}{B^2} - \frac{2xy}{AB}\cos\delta = \sin^2\delta$$

$x = A \sin (at + \delta)$

$y = B \sin (bt + r)$

Clearly $A \neq B$ hence ellipse.

29. **(b)** Maximum velocity in SHM, $v_{max} = a\omega$

Maximum acceleration in SHM, $A_{max} = a\omega^2$

where a and ω are maximum amplitude and angular frequency.

Given that, $\frac{A_{max}}{v_{max}} = 10$

i.e., $\omega = 10\ s^{-1}$

Displacement is given by

$x = a \sin (\omega t + \pi/4)$

at $t = 0$, $x = 5$

$5 = a \sin \pi/4$

$5 = a \sin 45° \Rightarrow a = 5\sqrt{2}$

Maximum acceleration $A_{max} = a\omega^2 = 500\sqrt{2}\ m/s^2$

30. **(b)** We know that $V = \omega\sqrt{A^2 - x^2}$

Initially $V = \omega\sqrt{A^2 - \left(\frac{2A}{3}\right)^2}$

Finally $3V = \omega\sqrt{A'^2 - \left(\frac{2A}{3}\right)^2}$

Where A'= final amplitude (Given at $x = \frac{2A}{3}$, velocity to trebled)

On dividing we get

$$\frac{3}{1} = \frac{\sqrt{A'^2 - \left(\frac{2A}{3}\right)^2}}{\sqrt{A^2 - \left(\frac{2A}{3}\right)^2}} \Rightarrow 9\left[A^2 - \frac{4A^2}{9}\right] = A'^2 - \frac{4A^2}{9}$$

$$\Rightarrow A' = \frac{7A}{3}$$

31. **(d)** For particle at $x = A$, equation of SHM is given as

$x_1 = A \cos \omega t$

For particle at $x = -\frac{A}{2}$, equation of SHM is given as

$$x_2 = A\sin\left(\omega t - \frac{\pi}{6}\right) \qquad \left[\because \frac{-A}{2} = A\sin(\omega.0 + \phi) \Rightarrow -\frac{1}{2} = \sin\phi \Rightarrow \phi = \frac{-\pi}{6}\right]$$

When they will meet

$x_1 = x_2$

$\Rightarrow A\cos\omega t = A\sin\left(\omega t - \frac{\pi}{6}\right)$

$\Rightarrow \cos\omega t = \sin\omega t \cos\frac{\pi}{6} - \cos\omega t \sin\frac{\pi}{6}$

$\Rightarrow \cos\omega t = \sin\omega t.\frac{\sqrt{3}}{2} - \cos\omega t.\frac{1}{2}$

$\Rightarrow \frac{3}{2}\cos\omega t = \sin\omega t.\frac{\sqrt{3}}{2}$

$\Rightarrow \tan\omega t = \sqrt{3} \Rightarrow \omega t = \frac{\pi}{3} \Rightarrow t = \frac{\pi}{3\omega}$

$\Rightarrow t = \frac{\pi \times T}{3 \times 2\pi} \Rightarrow t = \frac{T}{6}$

32. **(c)** At $t = 0$, $x(t) = 0$; $y(t) = 0$

$x(t)$ is a sinusoidal function

At $t = \frac{\pi}{2\omega}$; $x(t) = a$ and $y(t) = 0$

Hence trajectory of particle will look like as (c).

33. **(d)** In linear S.H.M., the restoring force acting on particle should always be proportional to the displacement of the particle and directed towards the equilibrium position.

i.e., $F \propto x$

or $F = -bx$ where b is a positive constant.

34. **(a)** Displacement $y(t) = A \sin (wt + \phi)$[Given]

For $\phi = \frac{2\pi}{3}$

at $t = 0$; $y = A \sin \phi = A \sin \frac{2\pi}{3}$

$= A \sin 120° = 0.87 A$ [$\because \sin 120° \approx 0.866$]

Graph (a) depicts $y = 0.87A$ at $t = 0$

35. **(a)** Let, $x_1 = A \sin \omega t$ and $x_2 = A \sin (\omega t + \phi)$

$x_2 - x_1 = 2A\cos\left(\omega t + \frac{\phi}{2}\right)\sin\frac{\phi}{2}$

$\Rightarrow (x_2 - x_1)_{max} = 2A\sin\left(\frac{\phi}{2}\right)$

$\Rightarrow (x_0 + x_2 - x_1)_{max} = x_0 + A$ (Given)

$\Rightarrow (x_2 - x_1)_{max} = A$

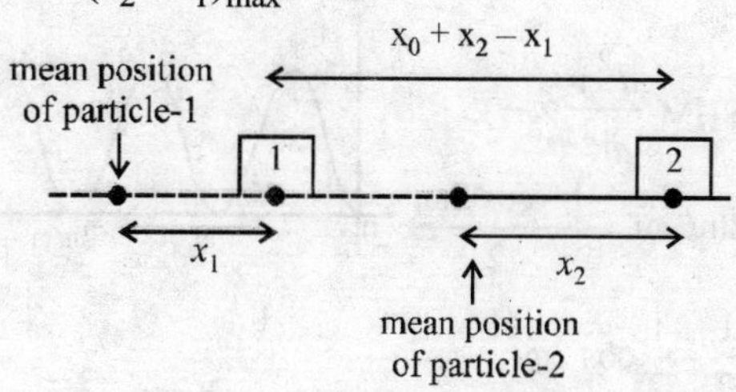

$\therefore 2A\sin\frac{\phi}{2} = A \Rightarrow \sin\frac{\phi}{2} = \frac{1}{2} \Rightarrow \phi = \frac{\pi}{3}$

36. **(c)** At mean position, $F\text{net} = 0$

Therefore, by principal of conservation of linear momentum.

$\therefore Mv_1 = (M + m)v_2$

$$MA_1\sqrt{\frac{k}{M}} = (M + m)A_2\sqrt{\frac{k}{m + M}} \qquad \left[\because V_0 = \omega A = \sqrt{\frac{k}{m}}A\right]$$

$\Rightarrow A_1\sqrt{M} = A_2\sqrt{M + m} \Rightarrow \frac{A_1}{A_2} = \sqrt{\frac{m + M}{M}}$

37. (a) Given,
Displacement, $x = x_0 \cos(\omega t - \pi/4)$

$\therefore$ Velocity, $v = \frac{dx}{dt} = -x_0\omega \sin\left(\omega t - \frac{\pi}{4}\right)$

Acceleration,

$$a = \frac{dv}{dt} = -x_0\omega^2 \cos\left(\omega t - \frac{\pi}{4}\right) = x_0\omega^2 \cos\left[\pi + \left(\omega t - \frac{\pi}{4}\right)\right]$$

$$= x_0\omega^2 \cos\left(\omega t + \frac{3\pi}{4}\right) \quad ...(1)$$

Acceleration, $a = A\cos(\omega t + \delta)$ ...(2)
Comparing the two equations, we get

$A = x_0\omega^2$ and $\delta = \frac{3\pi}{4}$.

38. (b) For block A to move in SHM.

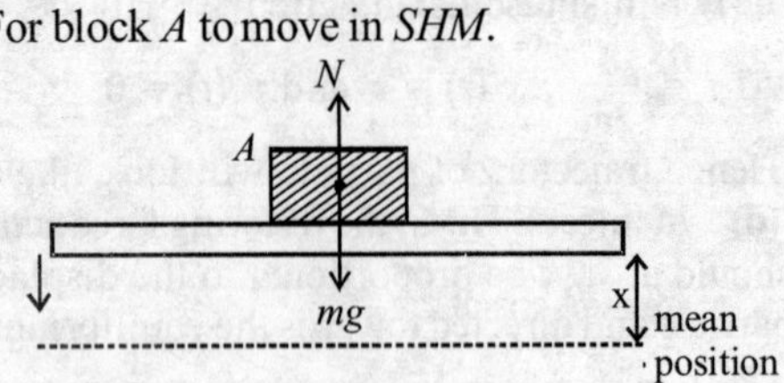

$mg - N = m\omega^2 x$
where x is the distance from mean position
For block to leave contact $N = 0$

$\Rightarrow mg = m\omega^2 x \Rightarrow x = \frac{g}{\omega^2}$

39. (a) Maximum velocity,
$v_{max} = a\omega$
Here, a = amplitude of SHM
ω = angular velocity of SHM

$v_{max} = a \times \frac{2\pi}{T} \quad \left(\because \omega = \frac{2\pi}{T}\right)$

$\Rightarrow T = \frac{2\pi a}{v_{max}} = \frac{2 \times 3.14 \times 7 \times 10^{-3}}{4.4} \approx 0.01\ s$

40. (a) Clearly $\sin 2\omega t$ is a periodic function with period $\frac{\pi}{\omega}$

For SHM $\frac{d^2y}{dt^2} \propto -y$

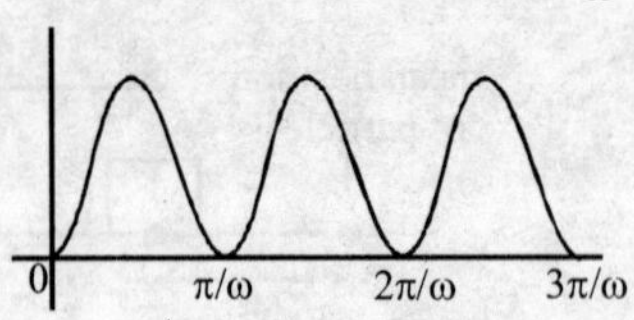

$y = \sin^2 \omega t = \frac{1 - \cos 2\omega t}{2}$

$= \frac{1}{2} - \frac{1}{2}\cos 2\omega t$

$v = \frac{dy}{dt} = \frac{1}{2} \times 2\omega \sin 2\omega t = 2\omega \sin \omega t \cos \omega t$

$= \omega \sin 2\omega t$

Acceleration, $a = \frac{d^2y}{dt^2} = 2\omega^2 \cos 2\omega t$ which is not proportional to $-y$. Hence, it is not in SHM.

41. (b) Velocity of particle 1,

$v_1 = \frac{dy_1}{dt} = 0.1 \times 100\pi \cos\left(100\pi t + \frac{\pi}{3}\right)$

Velocity of particle 2,

$v_2 = \frac{dy_2}{dt} = -0.1\pi \sin \pi t = 0.1\pi \cos\left(\pi t + \frac{\pi}{2}\right)$

$\therefore$ Phase difference of velocity of particle 1 with respect to the velocity of particle 2 is

$= \phi_1 - \phi_2 = \frac{\pi}{3} - \frac{\pi}{2} = \frac{2\pi - 3\pi}{6} = -\frac{\pi}{6}$

42. (c) Maximum velocity during SHM, $V_{max} = A\omega$
But $k = m\omega^2$

$\therefore \omega = \sqrt{\frac{k}{m}}$

$\therefore$ Maximum velocity of the body in SHM

$= A\sqrt{\frac{k}{m}}$

As maximum velocities are equal

$\therefore A_1\sqrt{\frac{k_1}{m}} = A_2\sqrt{\frac{k_2}{m}} \Rightarrow A_1\sqrt{k_1} = A_2\sqrt{k_2}$

$\Rightarrow \frac{A_1}{A_2} = \sqrt{\frac{k_2}{k_1}}$

43. (a) For oscillation of particle,

$x = x_o \sin^2\frac{t}{2} = x_0\left(\frac{1 - \cos t}{2}\right) \quad \therefore \cos t = \left(1 - \frac{2x}{x_o}\right)$

$\therefore K = \frac{1}{2}mv^2 = \frac{1}{2}m\left(\frac{x_o}{2}\sin t\right)^2 = \frac{x_o^2 m}{8}\left[1 - \left(1 - \frac{2x}{x_o}\right)^2\right]$

At $x_o = 1$m, $K = \frac{m}{2}(x - x^2)$

When, $K = 0 \Rightarrow x = 0, 1$ and $K - x$ graph is inverted parabola.

44. (6) Total energy, $E = K + U$
$E = 0.5 + 0.4 = 0.9$ J
$E = m\omega^2 A^2 = 0.9$ J

$= \frac{1}{2} \times 0.2\left(2\pi \times \frac{25}{\pi}\right)^2 \times A^2 = 0.9 \quad \Rightarrow A = 6$ cm

45. (b) Energy $E = \frac{1}{2}kA^2$

Since energy E independent of mass so new energy of the system for same amplitude will remain unchanged i.e., E

46. (9) For SHM oscillator,

Total energy $= E = \frac{1}{2}KA^2$

$PE = \frac{1}{2}kx^2 = \frac{1}{2}K\left(\frac{A}{3}\right)^2 = \frac{KA^2}{2 \times 9} = \frac{E}{9} \left[\because x = \frac{A}{3}\right]$

$KE = TE - PE = E - \frac{E}{9} = \frac{8E}{9}$

Ratio $= \frac{TE}{KE} = \frac{E}{\frac{8E}{9}} = \frac{9}{8} = \frac{x}{8}$ (given)

$\therefore x = 9$

47. **(4)** General equation for displacement is $x = A\sin(\omega t + \phi)$
Comparing with given equation, $y = A\cos(30°)$
$y = A\sin 60°$
$\omega t + \phi = 60°$
$\therefore x = 40 \times \frac{\sqrt{3}}{2} \Rightarrow 20\sqrt{3}\ cm$ and $A = 40\ cm$
Kinetic energy, $K.E = \frac{1}{2}k\left(A^2 - x^2\right) = 200$
$200 = \frac{1}{2}k\left(\frac{1600 - 1200}{100 \times 100}\right)$
$\Rightarrow 400 \times 100 \times 100 = k \times 400 \Rightarrow k = 10^4 \therefore x = 4$

48. **(c)** Let the distance from the mean position is X.
Given KE = PE
So, $\frac{1}{2}M\omega^2\left(A^2 - x^2\right) = \frac{1}{2}M\omega^2 x^2$
$A^2 - x^2 = x^2 \Rightarrow A^2 = 2 \times 2 \ \therefore x = \pm\frac{A}{\sqrt{2}}$

49. **(d)** Total energy in SHM $= \frac{1}{2}kA^2$
Here, k = force constant
A = amplitude of S.H.M.
Potential energy, $= \frac{1}{2}kx^2$
x = displacement from mean position
Kinetic energy $= \frac{1}{2}kA^2 - \frac{1}{2}kx^2$
$$\therefore \frac{P.E}{K.E} = \frac{x^2}{A^2 - x^2} = \frac{\left(\frac{A}{2}\right)^2}{A^2 - \left(\frac{A}{2}\right)^2} = \frac{A^2}{4\left(\frac{3A^2}{4}\right)} = \frac{1}{3}$$

50. **(d)** T.E = K.E + P.E
$\Rightarrow K.E = T.E - K.E = \frac{1}{2}kA^2 - \frac{1}{2}kx^2$
$\Rightarrow KE = \frac{1}{2}m\omega^2\ (A^2 - x^2)\ \ [\because k = m\omega^2]$
KE is maximum at x = 0 & K.E is zero when x = A
∴ K.E vs x graph is parabola.

51. **(2)** Given amplitude A = 3 cm
K.E = P.E + 25% of PE; $K.E = PE + \frac{PE}{4}$
$$KE = \frac{5}{4}PE \qquad \left(\begin{array}{l}\because K.E = \frac{1}{2}m\omega^2(A^2 - x^2) \\ P.E = \frac{1}{2}m\omega^2 x^2\end{array}\right)$$
$\frac{1}{2}m\omega^2(A^2 - x^2) = \frac{5}{4} \times \frac{1}{2}m\omega^2 x^2$
$\Rightarrow A^2 - x^2 = \frac{5}{4}x^2 \Rightarrow \frac{9x^2}{4} = A^2$
$\Rightarrow \frac{9x^2}{4} = 9 \Rightarrow x^2 = 4 \Rightarrow x = 2$ cm

52. **(c)** $(P \cdot E)_{max} = \frac{1}{2}m\omega^2 A^2 = 25j$
$K \cdot E$ at $\frac{A}{2} = \frac{1}{2}m\omega^2\left(A^2 - x^2\right) = \frac{1}{2}m\omega^2\left(A^2 - \frac{A^2}{4}\right)$
$K \cdot E = \frac{1}{2}m\omega^2\frac{3A^2}{4} = \frac{3}{4}\left(\frac{1}{2}m\omega^2 A^2\right)$
$K \cdot E = \frac{3}{4} \times 25 = 18.75j$
Hence option (c) is correct.

53. **(8)** The displacement of a simple harmonic oscillator is given by
$x = A\sin(\omega t)$
The potential energy of simple harmonic oscillator
$U = \frac{1}{2}kx^2,$
$\frac{dU}{dt} = \frac{1}{2}k2x\frac{dx}{dt} \qquad \left(\because \frac{dx}{dt} = A\omega\cos\omega t\right)$
$= kA^2\ \omega\sin\omega t\cos\omega t$
$\left(\frac{dU}{dt}\right)_{max} = \frac{kA^2\omega}{2}(\sin 2\omega t)_{max}$
$2\omega t = \frac{\pi}{2} \Rightarrow t = \frac{\pi}{4\omega} = \frac{T}{8} \Rightarrow \beta = 8$

54. **(8)** K.E. = P.E.
$\Rightarrow \frac{1}{2}m(V_0^2 - \omega^2 x^2) = \frac{1}{2}m\omega^2 x^2$
$\Rightarrow V_0^2 = 2\omega^2 x^2 \Rightarrow \omega^2 A^2 = 2\omega^2 x^2 \Rightarrow x = \frac{A}{\sqrt{2}}$
$\Rightarrow A\sin\omega t = \frac{A}{\sqrt{2}} \Rightarrow \omega t = \frac{\pi}{4} \Rightarrow \frac{2\pi}{T}t = \frac{\pi}{4} \Rightarrow t = \frac{T}{8}$
So, $x = 8$

55. **(a)** Spring force is conservative so mechanical energy will be constant. Also average value of kinetic energy & potential energy is constant in a time period.
$\langle K.E\rangle = \langle P.E\rangle = \frac{1}{2}kA^2$

56. **(a)** $T = 2\pi\sqrt{\frac{m}{K}}, 0.2 = 2\pi\sqrt{\frac{0.5}{K}}$
$K = 50\pi^2 = 500$
$x = A\sin(\omega t + \phi)$
$= 5 \times \sin\left(\frac{\omega T}{4} + 0\right)$
$x = 5 \times \sin\frac{\pi}{2} = 5cm$
$PE = \frac{1}{2}Kx^2 = \frac{1}{2} \times 500 \times \left(\frac{5}{100}\right)^2$
$\boxed{PE = 0.625j}$

57. **(c)** From graph equation of SHM
$X = A\cos\omega t$
(1) At $\frac{3T}{4}$ particle is at mean position.
∴ Acceleration = 0, Force = 0

(2) At T particle again at extreme position so acceleration is maximum.

(3) At $t = \frac{T}{4}$, particle is at mean position so velocity is maximum.
Acceleration = 0

(4) When KE = PE $\Rightarrow \frac{1}{2}k(A^2 - x^2) = \frac{1}{2}kx^2$
Here, A = amplitude of SHM
x = displacement from mean position

$$\Rightarrow A^2 = 2x^2 \Rightarrow x = \frac{+A}{\sqrt{2}} \Rightarrow \frac{A}{\sqrt{2}} = A\cos\omega t \Rightarrow t = \frac{T}{8}$$

$\therefore$ 1, 2 and 3 are correct.

58. **(d)** Kinetic energy, $k = \frac{1}{2}m\omega^2 A^2 \cos^2 \omega t$

Potential energy, $U = \frac{1}{2}m\omega^2 A^2 \sin^2 \omega t$

$$\frac{k}{U} = \cot^2 \omega t = \cot^2 \frac{\pi}{90}(210) = \frac{1}{3}$$

59. **(a)** $K_1 = \frac{1}{2}m\omega^2 A^2$

as $\omega^2 \propto l \Rightarrow \omega_2^2 = \omega_1^2 \times \frac{l_2}{l_1} \Rightarrow \omega_2^2 = 2\omega^2$

So, $K_2 = \frac{1}{2}m\omega_2^2 A^2 = 2 \times \frac{1}{2}m\omega^2 A^2 = 2K_1$

60. **(b)** For a particle executing SHM
At mean position; $t = 0, \omega t = 0, y = 0, V = V_{max} = a\omega$

$\therefore$ $K.E. = KE_{max} = \frac{1}{2}m\omega^2 a^2$

At extreme position : $t = \frac{T}{4}, \omega t = \frac{\pi}{2}, y = A, V = V_{min} = 0$

$\therefore$ $K.E. = KE_{min} = 0$

Kinetic energy in SHM, $KE = \frac{1}{2}m\omega^2(a^2 - y^2)$

$= \frac{1}{2}m\omega^2 a^2 \cos^2 \omega t$

Hence graph (b) correctly depicts kinetic energy time graph.

61. **(d)** $K.E = \frac{1}{2}k(A^2 - d^2)$ and $P.E. = \frac{1}{2}kd^2$
As both KE and PE are proportional to square of distance, so graph must be parabolic in nature. At mean position, d = 0. So, PE will be zero and KE will be maximum. Only graph in option (d) satisfies this.

62. **(b)** Work done, $w = \frac{1}{2}kx^2$

Work done by spring S_1, $w_1 = \frac{1}{2}k_1 x^2$

Work done by spring S_2, $w_2 = \frac{1}{2}k_2 x^2$

Since $w_1 > w_2$ Thus $(k_1 > k_2)$

63. **(b)** $K = \frac{1}{2}m\omega^2(A^2 - x^2)$

$= \frac{1}{2}m\omega^2 A^2 - \frac{1}{2}m\omega^2 x^2$

$= \frac{1}{2}m\omega^2 A^2 - \frac{1}{2}m\omega^2 A^2 \sin^2 \omega t$

$< K > = \frac{1}{2}m\omega^2 A^2 - \frac{1}{2}m\omega^2 A^2 \frac{1}{2}$

$= \frac{1}{4}m\omega^2 A^2 = mA^2\pi^2\nu^2$ $[\because \omega = 2\pi\nu]$

64. **(a)** K.E. of a body undergoing SHM is given by,

$K.E. = \frac{1}{2}ma^2\omega^2 \cos^2 \omega t$

Total energy in S.H.M $= \frac{1}{2}ma^2\omega^2$

Given K.E. = 75% T.E.

$$\frac{1}{2}ma^2\omega^2 \cos^2 \omega t = \frac{75}{100} \times \frac{1}{2}ma^2\omega^2$$

$$\Rightarrow 0.75 = \cos^2 \omega t \Rightarrow \omega t = \frac{\pi}{6}$$

$$\Rightarrow t = \frac{\pi}{6 \times \omega} \Rightarrow t = \frac{\pi \times 2}{6 \times 2\pi} \Rightarrow t = \frac{1}{6}s$$

65. **(a)** At any instant the total energy in SHM is

$\frac{1}{2}kA_0^2$ = constant,
where A_0 = amplitude
k = spring constant
hence total energy is independent of x.

66. **(a)** K.E. of simple harmonic motion

$= \frac{1}{2}m\omega^2(a^2 - x^2)$
When x = 0,

$K.E = \frac{1}{2}m\omega^2(a^2 - 0) = \frac{1}{2}m\omega^2 a^2$ = maximum

67. **(c)** The kinetic energy (K. E.) of particle in SHM is given by,

$K.E = \frac{1}{2}k(A^2 - x^2)$;

Potential energy of particle in SHM is $U = \frac{1}{2}kx^2$

Where A = amplitude and $k = m\omega^2$
x = displacement from the mean position
At the mean position $x = 0$

$\therefore$ $K.E. = \frac{1}{2}kA^2$ = Maximum, and U = 0

68. **(b)** Given, Natural length of spring = 2 m
Initial compression in spring (x_i) = 1 m
Final compression in spring $(x_f) = (2 - x)$ m

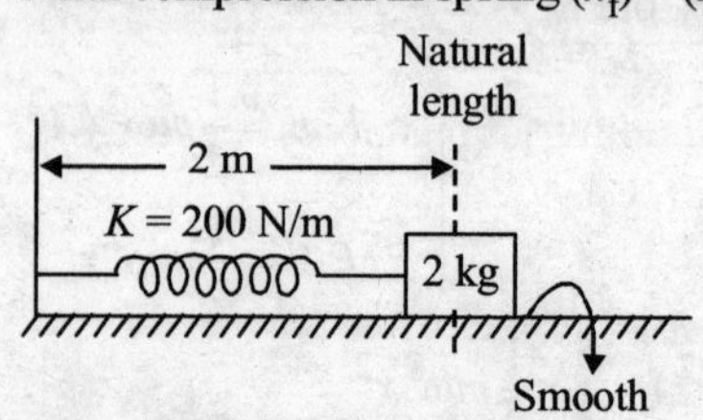

Using energy conservation

$$K_i + U_i = K_f + U_f$$

$$\Rightarrow 0 + \frac{1}{2}Kx_i^2 = \frac{1}{2}mv^2 + \frac{1}{2}Kx_f^2$$

$$\Rightarrow \frac{1}{2}mv^2 = \frac{1}{2}K(x_i^2 - x_f^2)$$

$$\Rightarrow \frac{1}{2}\times 2\times v^2 = \frac{1}{2}\times 200\times(1^2-(2-x)^2)$$

$$\Rightarrow v^2 = 100[1-(2-x)^2]$$

$$\therefore v = 10[1-(2-x)^2]^{1/2} \text{ m/s}$$

69. (d) $T = 2\pi\sqrt{\frac{l}{g}},\ \omega = \frac{2\pi}{T} \Rightarrow \omega = \sqrt{\frac{g}{l}}$

$\alpha = -\omega^2\theta \quad \therefore \frac{g}{l_1}\theta_1 = \frac{g}{l_2}\theta_2 \Rightarrow \theta_1 l_2 = \theta_2 l_1.$

70. (c) By conservation of linear momentum, $P_i = P_f$

$m_1v_1 + m_2v_2 = (m_1+m_2)v_{cm}$

Here, $m_1 = 10$ kg, $m_2 = 5$ kg, $v_1 = 3$m/s, $v_2 = 0$

$$\therefore v_{cm} = \frac{10\times 3}{10+5} \Rightarrow \frac{30}{15} = 2\text{m/s}$$

By WET,

$$W_{sp} = \frac{1}{2}(m_1+m_2)V_{cm}^2 - \frac{1}{2}m_1v_1^2$$

$$\Rightarrow -\frac{1}{2}kx^2 = \frac{1}{2}(15)(2)^2 - \frac{1}{2}(10)(3)^2$$

$$\Rightarrow x^2 = \frac{30}{3000} = \frac{1}{100} \quad \therefore x = \frac{1}{10}\text{m} = 0.1\text{ m}$$

71. (a) (A) As both blocks moving together so

Time period, $T = 2\pi\sqrt{\frac{M+m}{K}}$

(B) When blocks are displaced by x, then restoring force on them is $F = -Kx$

$\Rightarrow (M+m)a = -Kx$

$\therefore$ Acceleration, $a = -\frac{Kx}{(M+m)}$

(C) The magnitude of frictional force on upper block is

$f = m|a| = \frac{mK|x|}{(M+m)}$

(D) For maximum amplitude,

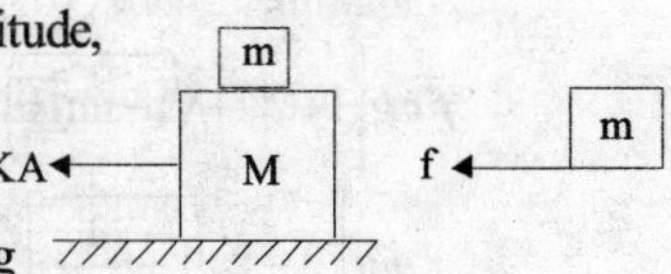

$f = \mu mg$

$\Rightarrow ma = \mu mg$

$\Rightarrow m\left(\frac{kA}{M+m}\right) = \mu mg$

$\therefore A = \frac{\mu(M+m)g}{K}$

(E) The maximum frictional force between the blocks is $f_{max} = \mu mg$

72. (c) Using WET

$W_{gravity} + W_{spring} + W_{gas} + W_{ext} = 0$

$$-Mg(L_1-L_0) - \int_{L_0}^{L_1} kx^3dx - nRT\ell n\left[\frac{L_1A}{L_0A}\right] + W_{ext} = 0$$

$$-Mg(L_1-L_0) - \frac{K}{4}\left[x^4\right]_{L_0}^{L_1} - nRT\ell n\left[\frac{L_1}{L_0}\right] + W_{ext} = 0$$

$$-Mg(L_1-L_0) - \frac{k}{4}\left(L_1^4-L_0^4\right) - nRT\ell n\left[\frac{L_1}{L_0}\right] + W_{ext} = 0$$

$$\therefore \quad W_{ext} = \frac{k}{4}\left(L_1^4-L_0^4\right) + Mg(L_1-L_0) + nRT\ell n\left[\frac{L_1}{L_0}\right]$$

73. (a) In SHM, maximum velocity is $V = A\omega$

For body A, $V_1 = A\omega_1$

For body B, $V_2 = A\omega_2$

$$\therefore \quad \frac{V_1}{V_2} = \frac{\omega_1}{\omega_2} = \frac{\sqrt{\frac{k_1}{m}}}{\sqrt{\frac{k_2}{m}}} = \sqrt{\frac{k_1}{k_2}} \quad \left(\because \omega = \sqrt{\frac{k}{m}}\right)$$

74. (b) The time period of a simple pendulum at height h above the earth surface is

$T = T_{earth}\left(1+\frac{h}{R}\right) \quad \therefore T \propto h$ and $T \propto \frac{1}{g_{earth}}$

75. (a) In simple harmonic motion, $F_{restoring} = -kx$

Buoyant force, $F_B = \rho Vg$

$F_{restoring} = -a^2x\,\rho g$ (Using archimedes principle)

$\Rightarrow k = a^2\rho g$

The time period of oscillation is given by

$$T = 2\pi\sqrt{\frac{m}{k}} = 2\pi\sqrt{\frac{m}{a^2\rho g}}$$

Given, $m = 10$ g, $a = 10$ cm, $\rho = 1000$ kg/m³

$$T = 2\pi\sqrt{\frac{m}{a^2\rho g}} = 2\pi\sqrt{\frac{10\times10^{-3}}{10^{-2}\times10^3\times10}}$$

$= 2\pi\times10^{-2} = \pi y\times10^{-2}$

$\therefore y = 2$

76. (a) For earth, $g = \frac{GM}{R^2}$ and $T = 2\pi\sqrt{\frac{l}{g}}$

For planet, $g' = \frac{G(4M)}{(2R)^2} = g \quad \therefore T' = 2\pi\sqrt{\frac{l}{g'}} = T$

77. (22) Given, mass of particle, $m = 0.5$ kg

Force, $F = -50(x) = ma$

$\Rightarrow 0.5a = -50x \Rightarrow a = (-100x)$

Comparing with $a = \omega^2 x$ we get

$\omega^2 = 100$

$$\therefore \text{Time period } T = \frac{2\pi}{10} = \left(\frac{\pi}{5}\right) = \frac{22}{7\times5} = \left(\frac{22}{35}\right) = \frac{x}{35}$$

$\Rightarrow x = 22$

78. (c) Time period of simple pendulum,

$T = 2\pi\sqrt{\frac{l}{g}}$ & $g = \frac{GM}{(R+h)^2}$

$\therefore T \propto (R+h)$

$$\Rightarrow \frac{T_1}{T_2} = \frac{(R+h_1)}{(R+h_2)} = \frac{(R+R)}{(R+2R)} = \frac{2}{3} \Rightarrow 3T_1 = 2T_2$$

79. (b) Acceleration due to gravity on earth surface, $g = \dfrac{GM}{R^2}$

At height, $h = 2R$

Acceleration due to gravity, $g' = \dfrac{GM}{(3R)^2} = \dfrac{1}{9}g$

Time period of pendulum,

$$T = 2\pi\sqrt{\frac{\ell}{g'}} \Rightarrow \ell = \left(\frac{T}{2\pi}\right)^2 g' \Rightarrow \ell = \frac{4}{4\pi^2}\frac{g}{9} = \frac{1}{9}\text{m}$$

80. (3) Frequency, $f_1 = \dfrac{1}{2\pi}\sqrt{\dfrac{k}{m}}$

If mass 9m is suspended from spring, frequency is

$$f_2 = \frac{1}{2\pi}\sqrt{\frac{k}{9m}} \qquad \therefore \frac{f_1}{f_2} = \sqrt{\frac{m_2}{m_1}} = \sqrt{\frac{9}{1}} = 3$$

81. (12) Equivalent spring constant of system

$$k_{eq} = \frac{2k.k}{3k} + k = \frac{5k}{3}$$

Time period of oscillation

$$T = 2\pi\sqrt{\frac{M}{k_{eq}}} = 2\pi\sqrt{\frac{M(3)}{5k}}$$

$$= \pi\sqrt{\frac{12M}{5k}} = \pi\sqrt{\frac{\alpha M}{5k}} \quad \text{(given)}$$

$\therefore \quad \alpha = 12$

82. (8) Length of the string, $L = 4$ m

Acceleration due to gravity, $g' = \dfrac{g}{4}$

$$T = 2\pi\sqrt{\frac{L}{g'}}$$

$$T = 2\pi\sqrt{\frac{4L}{g}} = 2\pi\sqrt{\frac{4\times 4}{g}} = 2\pi\frac{4}{\pi} = 8\text{s}$$

83. (a) By energy conservation at points A and B,

$$mg\ell = \frac{1}{2}mv_B^2 + \frac{1}{10}mg\ell$$

$$\Rightarrow \frac{9}{10}mg\ell = \frac{1}{2}mv_B^2$$

$$\Rightarrow \frac{9}{10}\times 10\times 10 = \frac{1}{2}v_B^2$$

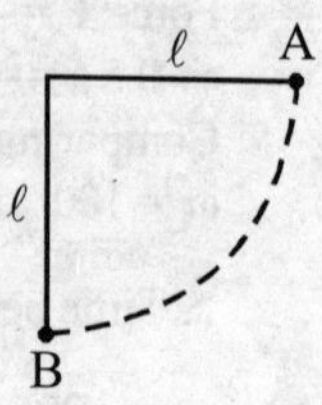

$v_B = \sqrt{180} = 6\sqrt{5}\,\text{m/s}$

84. (a) Restoring force, $F = -kx$; $F \propto x$ A true

acceleration, $a = -\omega^2 x$ B true

Velocity is maximum at mean position C true

$v = \omega\sqrt{A^2 - x^2}$ at main position, $x = 0$ $v_{max} = A\omega$

Acceleration is maximum at extreme points, $|a| = \omega^2 x$ D false

85. (b) Given,

$y = \sin\omega t + \cos\omega t$

$\Rightarrow y = \sin\omega t + \sin\left(\omega t + \dfrac{\pi}{2}\right)$

$\therefore$ Phase difference, $\Delta\phi = \dfrac{\pi}{2}$

$$A_{net} = \sqrt{A_1^2 + A_2^2 + 2A_1A_2\cos(\Delta\phi)}$$

$$\Rightarrow A_{net} = \sqrt{1^2 + 1^2 + 2\times 1\times 1\times\cos\left(\frac{\pi}{2}\right)} \Rightarrow A_{net} = \sqrt{2}$$

86. (20) Given,

Mass of block, $m = 5$ kg

Amplitude of SHM, $A = 1$ m

Time period, $T = 3.14 = \pi$

$\Rightarrow T = \dfrac{2\pi}{\omega} \Rightarrow \omega = 2$

Maximum force,

$F_{max} = m\,a_{max} = m(A\omega^2) = mA(2)^2 = 5\times 1\times 4 = 20$ N

87. (c) As spring is connected in parallel

$\therefore K_{eq} = K_1 + K_2$ $\therefore$ Angular frequency

$$\omega = \sqrt{\frac{K_{eq}}{m}} \qquad \therefore \text{Time period, } T = \frac{2\pi}{\omega} = 2\pi\sqrt{\frac{m}{k_1 + k_2}}$$

88. (99) Mass of bob, $m = 250$ g

Using conservation of energy

$\dfrac{1}{2}mv^2 = mgl(1 - \cos\theta)$

Maximum tension occurs at mean position.

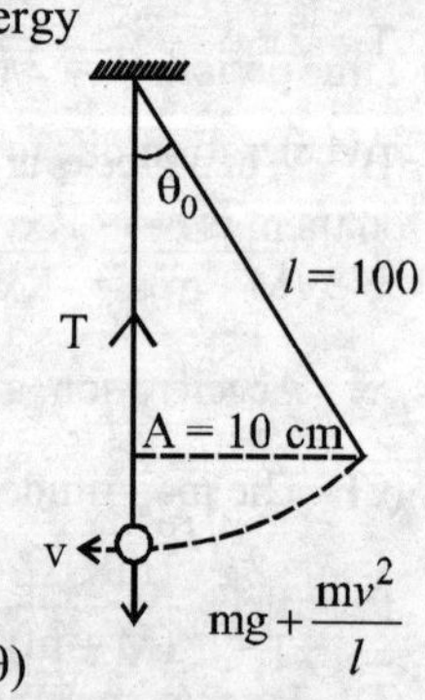

$\therefore T - mg = \dfrac{mv^2}{l}$

$\Rightarrow T = mg + \dfrac{mv^2}{l}$

$\therefore T = mg + 2mg(1 - \cos\theta)$

$= mg\left[1 + 2\left(1 - \sqrt{1 - \sin^2\theta}\right)\right] \quad \left(\because \sin\theta = \dfrac{A}{\ell} = \dfrac{10}{100} = \dfrac{1}{10}\right)$

$= mg\left[3 - 2\sqrt{1 - \dfrac{1}{100}}\right]$

$= \dfrac{250}{1000}\times 9.8\left[3 - 2\left(1 - \dfrac{1}{200}\right)\right] = \dfrac{99}{40}$

$\therefore x = 99$

89. (d) We have $T = 2\pi\sqrt{\dfrac{L}{g}} \Rightarrow T^2 = \dfrac{4\pi^2 L}{g}$

$\Rightarrow T^2 = mL \Rightarrow$ curve between T^2 and L will be straight line passing through origin having slope $= \dfrac{4\pi^2}{g}$.

90. (20) 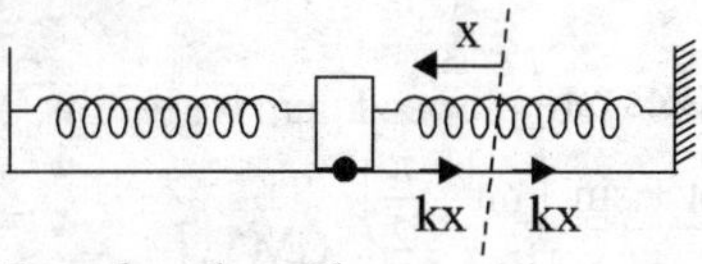

$F_{res} = kx + kx + 2kx$

Also, $F_{res} = m\omega^2 x$

So, $2kx = m\omega^2 x$

$$\omega = \sqrt{\frac{2k}{m}} \Rightarrow f = \frac{1}{2\pi}\sqrt{\frac{2k}{m}} = \frac{1}{2\pi}\sqrt{\frac{2\times 2}{0.49}} = \frac{1}{2\pi}\sqrt{\frac{400}{49}}$$

So, no. of oscillation in 14π second $= f \times 14\pi$

$$= \frac{1}{2\pi}\sqrt{\frac{400}{49}} \times 14\pi = 7 \times \frac{20}{7} = 20$$

91. (b) Angular frequency in spring mass system is given by

$$\omega = \sqrt{\frac{k}{m}}$$

When m = 1 kg

$$\omega_1 = \sqrt{\frac{k}{m}} = \sqrt{\frac{k}{1}}$$

When m = 2 kg, $\omega_2 = \sqrt{\frac{k}{2}}$

$$\frac{\omega_2}{\omega_1} = \sqrt{\frac{k/2}{k}} = \sqrt{\frac{1}{2}}$$

92. (b) At surface of earth time period,

$$T = 2\pi\sqrt{\frac{\ell}{g}} \qquad ...(i)$$

Acceleration due to gravity at height R

$$g' = \frac{g}{\left(1+\frac{h}{R}\right)^2} = \frac{g}{\left(1+\frac{R}{R}\right)^2} = \frac{g}{2^2} = \frac{g}{4}$$

$$xT = 2\pi\sqrt{\frac{\ell}{(g/4)}} \qquad (\because h = R)$$

$$\Rightarrow xT = 2\times 2\pi\sqrt{\frac{\ell}{g}} \qquad \text{From equation (i)}$$

$\Rightarrow xT = 2T \Rightarrow x = 2$

93. (b) Consider at some time particle is at a distance x from centre of Earth then acceleration of particle

$$\vec{a} = -\frac{GM}{R^3}\vec{x}$$

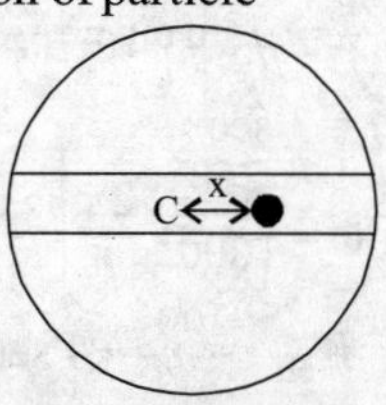

$$-\omega^2 x = -\frac{GM}{R^3}x$$

$$\Rightarrow \omega = \sqrt{\frac{GM}{R^3}} = \sqrt{\frac{g}{R}} \qquad (\because GM = gR^2)$$

The time period of the motion of, particle will be

$$T = \frac{2\pi}{\omega} = 2\pi\sqrt{\frac{R}{g}} \Rightarrow T = 2\times 3.14\times\sqrt{\frac{6400\times 10^3}{10}}$$

$= 2\times 3.14\times 800$ sec $\approx$ 1 hour 24 minutes

94. (5) 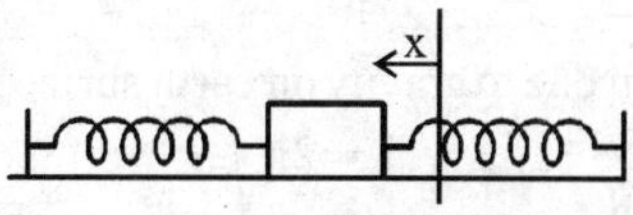

Equilibrium position

If block is shift 'x_0' distance towards left then, $\vec{F}_{res} = -k\vec{x}_0 - k\vec{x}_0 = -2k\vec{x}_0$

In SHM, $\vec{F}_{res} = -m\omega^2\vec{x}_0$

So, $m\omega^2 = 2K$

$$\omega = \sqrt{\frac{2K}{m}}$$

$$T = 2\pi\sqrt{\frac{m}{2K}} = 2\pi\sqrt{\frac{2}{2\times 20}} = \frac{\pi}{\sqrt{5}}$$

95. (c) As $g_{eff} = \dfrac{g}{\left(1+\frac{h}{R}\right)^2}$. So if h $\uparrow\uparrow \Rightarrow g_{eff} \downarrow\downarrow$

So, at mount everest value of g will be less

as, $T = 2\pi\sqrt{\frac{l}{g}}$. Here g $\downarrow\downarrow \Rightarrow T\uparrow\uparrow$

$\Downarrow$

Pendulum will become slow

96. (1) Time period of spring mass system is given as

$$T = 2\pi\sqrt{\frac{m}{k}}$$

$$\text{So,} \quad T = 2\pi\sqrt{\frac{m}{k}} = 1 \qquad ...(i)$$

$$\text{and,} \quad T' = 2\pi\sqrt{\frac{m+3}{k}} = 2 \qquad ...(ii)$$

Dividing (ii) by (i), we get

$$\frac{T}{T'} = \sqrt{\frac{m}{m+3}} = \frac{1}{2} \Rightarrow \frac{m}{m+3} = \frac{1}{4} \Rightarrow m = 1\text{ kg}$$

97. (a) Here, $g_{eff} = g\cos\alpha$

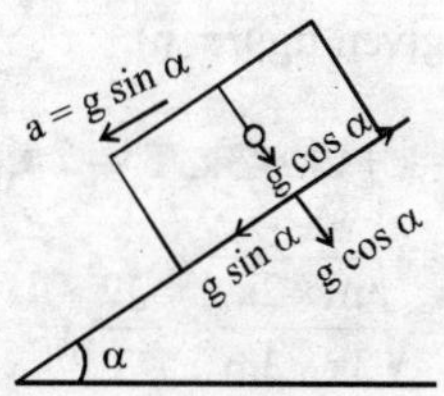

$$T = 2\pi\sqrt{\frac{\ell}{g_{eff}}}$$

$$\Rightarrow T = 2\pi\sqrt{\frac{\ell}{g\cos\alpha}}$$

98. (5) $mg' = mg - F_B$

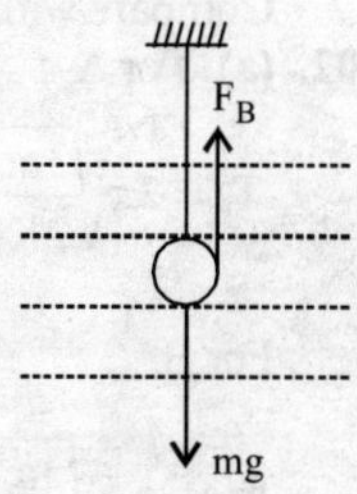

$$g' = \frac{mg - F_B}{m} = \frac{mg - m_w g}{m}$$

$$= \frac{\rho_B Vg - \rho_w Vg}{\rho_B V} \quad (\because m = \rho v)$$

$$= \left(\frac{\rho_B - \rho_w}{\rho_B}\right)g$$

$$= \frac{5-1}{5}\times g = \frac{4}{5}g$$

$$\frac{T'}{T}=\sqrt{\frac{g}{g'}}=\sqrt{\frac{\frac{g}{4}}{\frac{4}{5}g}}=\sqrt{\frac{5}{4}}\quad T=2\pi\sqrt{\frac{\ell}{g}}$$

$$\Rightarrow\ T'=T\sqrt{\frac{5}{4}}=\frac{10}{2}\sqrt{5}=5\sqrt{5}$$

Compair with $5\sqrt{x}$, we have $x=5$

99. **(2)** $U=4(1-\cos 4x)$

$$F=-\frac{dU}{dx}=-4(+\sin 4x)\,4=-16\sin(4x)$$

For small θ

$\sin\theta\approx\theta$

$\Rightarrow \sin 4x\approx 4x$

So, $F=-64x$

$a=-64\,x/m=-16x$

Comparing with $a=-\omega^2 x$

we get, $\omega^2=16$

$\Rightarrow \omega=4$

So, $T=\frac{2\pi}{\omega}=\frac{\pi}{2}$. So, $K=2$

100. **(c)** $T\propto\frac{1}{\sqrt{g}}$ and $g\propto\frac{1}{(R+h)^2}$

$$\frac{T_1}{T_2}=\sqrt{\frac{g'}{g}}=\sqrt{\frac{R^2}{(R+h)^2}}$$

$$\frac{T_1}{T_2}=\frac{4}{6}=\frac{R}{(R+h)}\Rightarrow h=3200\text{ km}$$

101. **(2)** For given figure (a):

$$k_{eq}=\frac{k\times 2k}{k+2k}=\frac{2k}{3}$$

$$T=2\pi\sqrt{\frac{m}{k_{eq}}}=2\pi\sqrt{\frac{m}{2k/3}}=2\pi\sqrt{\frac{3m}{2k}}$$

For given figure (b):

$$k_{eq}=k+2k=3k,\ T'=2\pi\sqrt{\frac{m}{3K}}$$

$$\frac{T'}{T}=\sqrt{\frac{m\times 2k}{3k\times 3m}}=\frac{\sqrt{2}}{3}\quad(\because T=3s)$$

$T'=\sqrt{2}s$

Compare with $\sqrt{x}$, we have $\Rightarrow x=2$

102. **(a)** For A

$$T_1=2\pi\sqrt{\frac{3m}{2K}}\quad[\because \text{A and B are in parallel. So, } K_{eq}=K+K=2K \text{ and, } m_{net}=2m+m=3m]$$

For B

$$T_2=2\pi\sqrt{\frac{m}{3K}}\quad \text{So, } \frac{T_1}{T_2}=\frac{\sqrt{\frac{3}{2}}}{\sqrt{\frac{1}{3}}}=\frac{3}{\sqrt{2}}$$

[$\because$ A and B are again in parallel So, $K_{eq}=K+2K=3K$ and, $m_{net}=m$]

103. **(d)** Time period of simple pendulum is given as, $T=2\pi\sqrt{\frac{L}{g}}$,

$$g'=\frac{GM}{9R^2}=\frac{g}{9}=\frac{\pi^2}{9}\quad\left[\because g'=\frac{GM}{(R+h)^2}\right]$$

and $2=2\pi\sqrt{\frac{L}{\pi^2}\times 9}$

[$\because$ Time period of second pendulum = 2 sec]

$$\Rightarrow 1=\pi\sqrt{L}\times\frac{3}{\pi}\Rightarrow L=\frac{1}{9}\text{ m}$$

104. **(c)** As $T=2\pi\sqrt{\frac{l}{g}}$

$$\Rightarrow \omega=\sqrt{\frac{g}{l}}\Rightarrow \pi^2=\frac{g}{l}\Rightarrow l=\frac{g}{\pi^2}=\frac{9.8}{\pi^2}m$$

$\simeq 99.4$ cm

105. **(5)** At mean position, pendulum will have maximum velocity

So, By conservation of energy

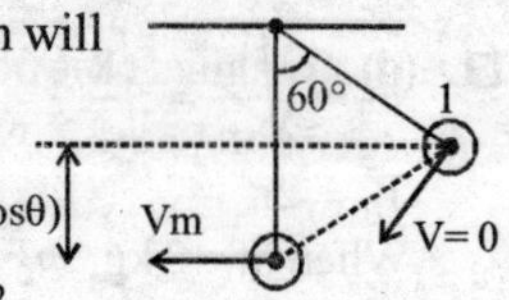

$$P_i+K_1=P_2+K_2$$

$$mgl(1-\cos\theta)+0=0+\frac{1}{2}mV_m^2$$

$$V_m=\sqrt{2gl(1-\cos\theta)}$$

$$=\sqrt{2\times 10\times 2.5\times(1-\cos 60°)}=\sqrt{25}=5\text{ m/s.}$$

106. **(10)** To complete the entire vertical circle, the minimum speed of bob should be $\sqrt{5Rg}$

By law of conservation of momentum $\vec{P_i}=\vec{P_f}$

$$75V+0=50\sqrt{5Rg}+75V/3$$

$$\Rightarrow 75\times\frac{2V}{3}=50\sqrt{5Rg}$$

$$\Rightarrow 50V=50\sqrt{5Rg}$$

$$\Rightarrow V=\sqrt{5Rg}=\sqrt{5\times 2\times 10}=10\text{ m/s}$$

107. **(d)** We have

$x=A\sin\omega t$

$$\Rightarrow\frac{A}{2}=A\sin\omega t\Rightarrow\frac{1}{2}=\sin\omega t\Rightarrow\omega t=\pi/6\Rightarrow t=\frac{\pi}{6\omega}$$

As $t=3$ sec

$$\text{So, }\omega=\frac{\pi}{18}\Rightarrow\frac{2\pi}{T}=\frac{\pi}{18}\Rightarrow T=36\text{ sec}$$

108. **(c)** $W_{eff.}=mg+\text{Pseudo force}$

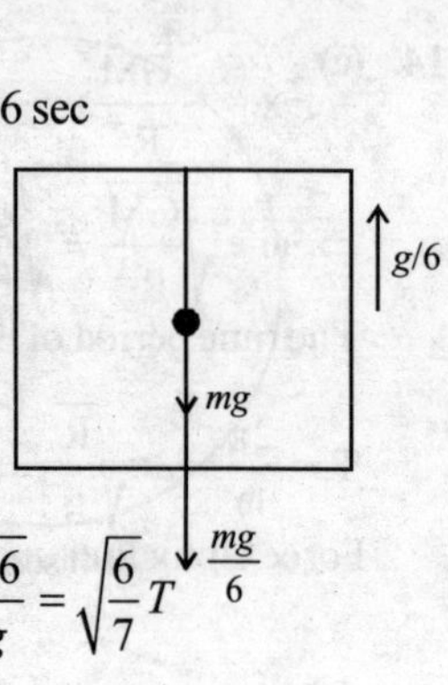

$$\Rightarrow mg_{eff}=mg+\frac{mg}{6}$$

$$\Rightarrow g_{eff}=\frac{7g}{6}$$

So, $T=$

$$2\pi\sqrt{\frac{l}{g_{eff}}}=2\pi\sqrt{\frac{l\times 6}{7g}}=\sqrt{\frac{6}{7}}T$$

109. (b) In SHM, $V_{max} = \omega A$

ωA = constant

$$\frac{A_1}{A_2} = \frac{\omega_2}{\omega_1} = \sqrt{\frac{k_2}{m_2} \times \frac{m_1}{k_1}} = \sqrt{\frac{9k}{100} \times \frac{50}{2k}} = \frac{3}{2}$$

110. (d) Time period of a pendulum is given by $T = 2\pi\sqrt{\frac{\ell}{g_{eff}}}$

$T = T_1 = 2\pi\sqrt{\ell_1 / g}$ [T_1 is time period outside liquid]

$$g_{eff} = g\left(1 - \frac{\rho_\ell}{\rho}\right) = g\left(1 - \frac{1}{4}\right) = \frac{3}{4}g$$

$T_2 = 2\pi\sqrt{\frac{\ell_2}{g_{eff}}}$ [T_2 is time period inside liquid]

$$\ell_2 = \ell_1 + \frac{\ell_1}{3} = \frac{4\ell_1}{3},\ T_2 = 2\pi\sqrt{\frac{4\ell_1}{3 \times \frac{3g}{4}}}$$

$$T_2 = \frac{4}{3}T_1 = \frac{4}{3}T$$

111. (d) Original time period,

$$T_0 = 2\pi\sqrt{\frac{\ell}{g}}$$

New time period $T = 2\pi\sqrt{\frac{\ell/16}{g}} = \frac{2\pi}{4}\sqrt{\frac{\ell}{g}}$ $\therefore T = \frac{T_0}{4}$

112. (2) In series, fig. (2) equivalent spring constant $k' = \frac{k}{2}$

$$T_a = 2\pi\sqrt{\frac{m}{k}} \text{ and } T_b = 2\pi\sqrt{\frac{m}{k/2}}$$

$$\therefore \quad \frac{T_b}{T_a} = \frac{2\pi\sqrt{\frac{m}{k/2}}}{2\pi\sqrt{\frac{m}{k}}} = \sqrt{2}. \text{ So } x = 2.$$

113. (a) In stationary lift, $T = 2\pi\sqrt{\frac{l}{g}}$

In upward moving lift, $T' = 2\pi\sqrt{\frac{l}{g+a}}$

(Here, a = acceleration of lift)

$$\frac{T'}{T} = \sqrt{\frac{g}{g+a}} = \sqrt{\frac{g}{g + \frac{g}{2}}} \Rightarrow T' = \sqrt{\frac{2}{3}}T$$

114. (c)

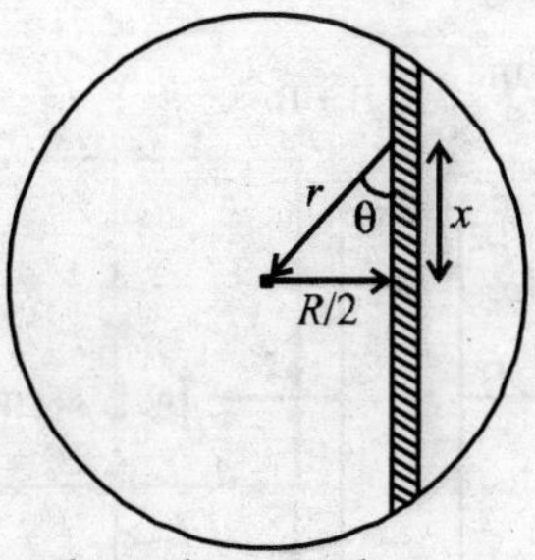

Force along the tunnel

$$F = \left(\frac{GMmr}{R^3}\right)\cos\theta \Rightarrow F = \frac{GMr}{R^3} \times m \times \left(\frac{x}{r}\right)$$

$$\Rightarrow a = \frac{d^2x}{dt^2} = -\frac{GM}{R^3}x = \frac{-g}{R}x$$

Comparing with $a = -\omega^2 x$, we get $\omega^2 = \frac{g}{R}$

$$\therefore \quad T = 2\pi\sqrt{\frac{R^3}{GM}} = 2\pi\sqrt{\frac{R}{g}} \qquad \left[\because g = \frac{GM}{R^2}\right]$$

115. (c) In series combination, equal forces acts on spring but extension in springs are different. Spring constant of combination,

$$\frac{1}{K_S} = \frac{1}{K_1} + \frac{1}{K_1} \Rightarrow K_S = \frac{K_1}{2}$$

Initial time period, $T = 2\pi\sqrt{\frac{m}{K_1}}$

New time period, $T' = 2\pi\sqrt{\frac{m}{K_S}}$

$$\Rightarrow T' = 2\pi\sqrt{\frac{m}{K_1}}\sqrt{2} \Rightarrow T' = \sqrt{2}T$$

116. (7) $\frac{5}{8}$ oscillation $= \frac{1}{2}$ oscillation $+ \frac{1}{8}$ oscillation

Displacement in $\frac{1}{2}$ oscillation $= 4A \times \frac{1}{2} = 2A$

Displacement in $\frac{1}{8}$ oscillation $= 4A \times \frac{1}{8} = \frac{A}{2}$

Time for $\frac{1}{2}$ oscillation $= \frac{T}{2}$

Time for $\frac{1}{8}$ oscillation (or $\frac{A}{2}$ displacement)

$$\frac{A}{2} = A\sin\omega t$$

$$\Rightarrow \frac{1}{2} = \sin\omega t \Rightarrow \frac{\pi}{6} = \left(\frac{2\pi}{T}\right)t \Rightarrow t = \frac{T}{12}$$

$$\therefore T_{net} = \frac{T}{2} + \frac{T}{12} = \frac{7T}{12}$$

117. (c) The two springs are in parallel, so the effective spring constant is

$$K_{eq} = K_1 + K_2 = K + K = 2K$$

The period of oscillation,

$$T = 2\pi\sqrt{\frac{m}{K_{eq}}} = 2\pi\sqrt{\frac{m}{2K}}$$

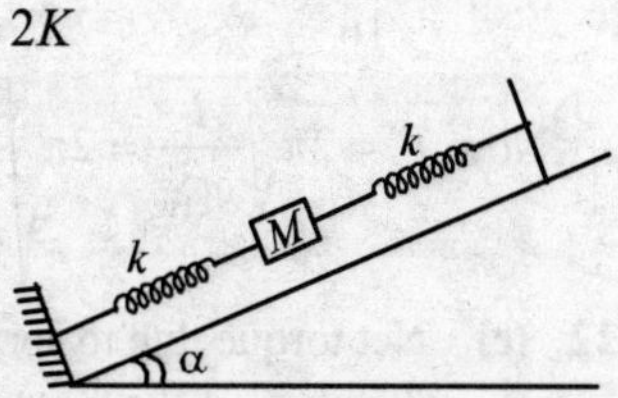

Frequency of oscillation,

$$f = \frac{1}{T} = \frac{1}{2\pi}\sqrt{\frac{2K}{m}}$$

118. (b) An elastic wire can be treated as a spring and its spring constant.

$$k = \frac{YA}{L} \qquad \left[\because Y = \frac{F}{A}\Big/\frac{\Delta l}{l_0}\right]$$

Frequency of oscillation,

$$f = \frac{1}{2\pi}\sqrt{\frac{k}{m}} = \frac{1}{2\pi}\sqrt{\frac{YA}{mL}}$$

119. (c) $y = y_0 \sin^2 \omega t$

$$\Rightarrow y = \frac{y_0}{2}(1-\cos 2\omega t) \qquad \left(\because \sin^2 \omega t = \frac{1-\cos 2\omega t}{2}\right)$$

$\Rightarrow y - \frac{y_0}{2} = \frac{-y_0}{2}\cos 2\omega t$. So from this equation we can say mean position is shifted by $\frac{y_0}{2}$ distance and frequency of this SHM is 2ω.

So, at equilibrium $\frac{ky_0}{2} = mg \Rightarrow \frac{k}{m} = \frac{2g}{y_0}$

Also, spring constant $k = m(2\omega)^2$

$$\Rightarrow 2\omega = \sqrt{\frac{k}{m}} = \sqrt{\frac{2g}{y_0}} \Rightarrow \omega = \frac{1}{2}\sqrt{\frac{2g}{y_0}} = \sqrt{\frac{g}{2y_0}}$$

120. (d) $P_i = P_f$

$$\frac{\omega'}{\omega}\times\frac{A'}{A} = 2$$

$$\because \omega \times \frac{1}{\sqrt{m}} \Rightarrow \frac{\sqrt{m}}{\sqrt{m'}}\times\frac{A'}{A} = 2$$

where $m' = m/2 \Rightarrow \sqrt{2}\frac{A'}{A} = 2$

$$\Rightarrow \frac{A'}{A} = \frac{2}{\sqrt{2}}\times\frac{\sqrt{2}}{\sqrt{2}}$$

$$\boxed{A' = \sqrt{2}A}$$

121. (c) $T = 2\pi\sqrt{\frac{l}{g}}$

When immersed non viscous liquid

$$\rho v g_{eff} = \rho vg - \rho_l vg = vg(\rho - \rho_l) = vg\left(\rho - \frac{\rho}{16}\right) = \frac{15}{16}\rho vg$$

$$g_{eff} = \frac{15g}{16}$$

$$\text{Now } T' = 2\pi\sqrt{\frac{l}{g_{eff}}} = 2\pi\sqrt{\frac{l}{\frac{15g}{16}}} = \frac{4}{\sqrt{15}}T$$

122. (c) Net torque due to spring force:

Suppose the rod is tilted by angle $d\theta$ then,

$$F_{res} = Kx = K\frac{l}{2}d\theta$$

$$\tau_{res} = F_{res} \times \text{arm of couple} = K\cdot\frac{l}{2}d\theta \times l = \frac{Kl^2}{2}d\theta$$

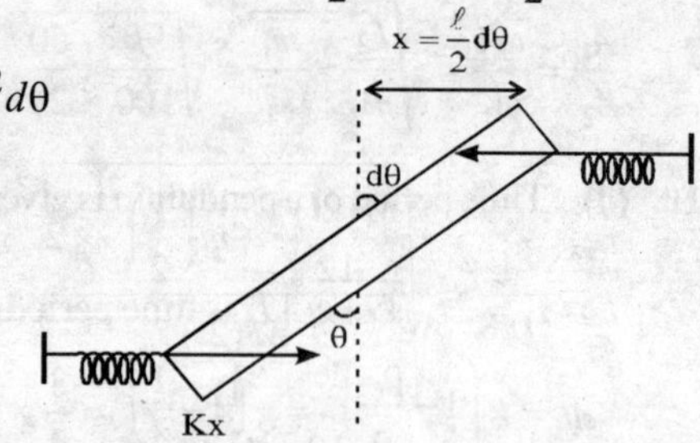

As, $\tau_{res} = I\omega^2 d\theta$

$$\frac{Kl^2}{2} = I\omega^2$$

$$\frac{Kl^2}{2} = \frac{ml^2}{12}\omega^2$$

$$\omega = \sqrt{\frac{6K}{m}}$$

$$\text{So, } f = \frac{\omega}{2\pi} = \frac{1}{2\pi}\sqrt{\frac{6K}{m}}$$

123. (a) Angular frequency of pendulum $\omega = \sqrt{\frac{g}{\ell}}$

$\therefore$ relative change in angular frequency

$\frac{\Delta\omega}{\omega} = \frac{1}{2}\frac{\Delta g}{g}$ [as length remains constant]

Now, $g_{min} = g - \omega_s^2 A$

$g_{max} = g + \omega_s^2 A$

So, $\Delta g = 2A\omega_s^2$ [ω_s = angular frequency of support and, A = amplitude]

$$\frac{\Delta\omega}{\omega} = \frac{1}{2}\times\frac{2A\omega_s^2}{g} \Rightarrow \Delta\omega = \frac{1}{2}\times\frac{2\times1^2\times10^{-2}}{10}$$

$= 10^{-3}$ rad/sec.

124. (d) Acceleration due to gravity $g = \frac{GM}{R^2}$

$$\frac{g_p}{g_e} = \frac{M_p}{M_e}\left(\frac{R_e}{R_p}\right)^2 = 3\left(\frac{1}{3}\right)^2 = \frac{1}{3}$$

Also $T \propto \frac{1}{\sqrt{g}} \Rightarrow \frac{T_p}{T_e} = \sqrt{\frac{g_e}{g_p}} = \sqrt{3} \Rightarrow T_p = 2\sqrt{3}$ s

125. (c) Velocity, $v = \omega\sqrt{A^2 - x^2}$ ...(i)

acceleration, $a = -\omega^2 x$ ...(ii)

and according to question,

$$|v| = |a| \Rightarrow \omega\sqrt{A^2 - x^2} = \omega^2 x \Rightarrow A^2 - x^2 = \omega^2 x^2$$

$$\Rightarrow 5^2 - 4^2 = \omega^2(4^2) \Rightarrow 3 = \omega \times 4 \Rightarrow \omega = \frac{3}{4}$$

$$\therefore T = 2\pi/\omega = \frac{2\pi}{3/4} = \frac{8\pi}{3}$$

126. (None) $B_0 + B = mg + ma$

$\therefore B = ma = \rho Axg$

$= (\pi r^2 \rho g)x$

$$a = \frac{(\pi r^2 \rho g)x}{m}$$

using, $a = \omega^2 x$

$$\Rightarrow \omega = \sqrt{\frac{\pi r^2 \rho g}{m}}$$

$W \simeq 7.95$ rads^{-1}

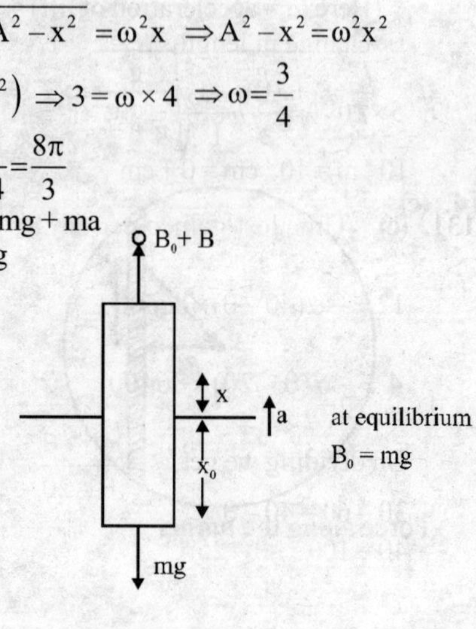

127. (c) $f = \frac{1}{2\pi}\sqrt{\frac{K}{I}} \Rightarrow f \propto \frac{1}{\sqrt{I}}$

So, $\frac{f_1}{f_2} = \sqrt{\frac{I_2}{I_1}}$

$$\Rightarrow \frac{1}{0.8} = \sqrt{\frac{\frac{M(2l)^2}{12} + m\left(\frac{l}{2}\right)^2 \times 2}{\frac{M(2l)^2}{12}}} \Rightarrow \frac{5}{4} = \sqrt{1 + \frac{3m}{2M}}$$

Solving, we get $\frac{m}{M} = 0.375$

128. (b) As we know, frequency in SHM

$$f = \frac{1}{2\pi}\sqrt{\frac{k}{m}} = 10^{12}$$

where m = mass of one atom

Mass of one atom of silver, $= \frac{108}{\left(6.02 \times 10^{23}\right)} \times 10^{-3}$ kg

$$\frac{1}{2\pi}\sqrt{\frac{k}{108 \times 10^{-3}} \times 6.02 \times 10^{23}} = 10^{12}$$

Solving we get, spring constant,

K = 7.1N/m

129. (d) Using y = A sin ωt

$a = A \sin \omega t_0$

$b = A \sin 2\omega t_0$

$c = A \sin 3\omega t_0$

$a + c = A[\sin \omega t_0 + A \sin 3\omega t_0] = 2A \sin 2\omega t_0 \cos \omega t_0$

$$\frac{a+c}{b} = 2\cos \omega t_0$$

$$\Rightarrow \omega = \frac{1}{t_0}\cos^{-1}\left(\frac{a+c}{2b}\right) \Rightarrow f = \frac{1}{2\pi t_0}\cos^{-1}\left(\frac{a+c}{2b}\right)$$

130. (b) As we know, Time-period of simple pendulum, $T \propto \sqrt{l}$

differentiating both side, $\frac{\Delta T}{T} = \frac{1}{2}\frac{\Delta l}{l}$

∵ change in length $\Delta l = r_1 - r_2$

$5 \times 10^{-4} = \frac{1}{2}\frac{r_1 - r_2}{1} \Rightarrow r_1 - r_2 = 10 \times 10^{-4}$

10^{-3} m $= 10^{-1}$ cm $= 0.1$ cm

131. (c) Time lost/gained per day $= \frac{1}{2}\alpha\Delta\theta \times 86400$ second

$$12 = \frac{1}{2}\alpha(40 - \theta) \times 86400 \quad \text{.... (i)}$$

$$4 = \frac{1}{2}\alpha(\theta - 20) \times 86400 \quad \text{....(ii)}$$

On dividing we get, $3 = \frac{40 - \theta}{\theta - 20}$

$3\theta - 60 = 40 - \theta$

$4\theta = 100 \Rightarrow \theta = 25°C$

Putting θ = 25°C in equation (i), we get

$\alpha = 1.85 \times 10^{-5}/°C$

132. (b) Washer contact with piston is lost when, N = 0

$N - mg = ma$

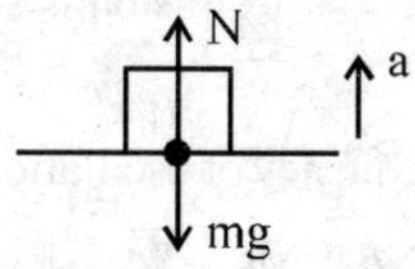

when $N = 0$

$|a_{max}| = g$

$a_{max} = g = \omega^2 A$

The frequency of piston

$$f = \frac{\omega}{2\pi} = \sqrt{\frac{g}{A}}\frac{1}{2\pi} = \sqrt{\frac{1000}{7}}\frac{1}{2\pi} = 1.9 \text{ Hz.}$$

133. (c) As we know, time period, $T = 2\pi\sqrt{\frac{\ell}{g}}$

When additional mass M is added then

$$T_M = 2\pi\sqrt{\frac{\ell + \Delta\ell}{g}}$$

$$\frac{T_M}{T} = \sqrt{\frac{\ell + \Delta\ell}{\ell}} \Rightarrow \left(\frac{T_M}{T}\right)^2 = \frac{\ell + \Delta\ell}{\ell}$$

$$\Rightarrow \left(\frac{T_M}{T}\right)^2 = 1 + \frac{Mg}{AY} \quad \left[\because \Delta\ell = \frac{Mg\ell}{AY}\right]$$

$$\therefore \frac{1}{Y} = \left[\left(\frac{T_M}{T}\right)^2 - 1\right]\frac{A}{Mg}$$

134. (d) In simple harmonic motion, starting from rest,

At $t = 0$, $x = A$

$x = A\cos\omega t$ ..... (i)

When $t = \tau$, $x = A - a$

When $t = 2\tau$, $x = A - 3a$

From equation (i)

$A - a = A\cos\omega\tau$ ......(ii)

$A - 3a = A\cos 2\omega\tau$ ....(iii)

As $\cos 2\omega\tau = 2\cos^2\omega\tau - 1$...(iv)

From equation (ii), (iii) and (iv)

$$\frac{A - 3a}{A} = 2\left(\frac{A - a}{A}\right)^2 - 1$$

$$\Rightarrow \frac{A - 3a}{A} = \frac{2A^2 + 2a^2 - 4Aa - A^2}{A^2}$$

$$\Rightarrow A^2 - 3aA = A^2 + 2a^2 - 4Aa \Rightarrow 2a^2 = aA$$

$$\Rightarrow A = 2a \Rightarrow \frac{a}{A} = \frac{1}{2}$$

Now, $A - a = A\cos\omega\tau$

$$\Rightarrow \cos\omega\tau = \frac{A - a}{A} \Rightarrow \cos\omega\tau = \frac{1}{2}$$

$$\Rightarrow \frac{2\pi}{T}\tau = \frac{\pi}{3} \Rightarrow T = 6\tau$$

135. (c) Initially, $\frac{Mg}{A} = P_0$

$\Rightarrow Mg = P_0 A$

Let piston is displaced by distance x

$P_0V_0^{\gamma} = PV^{\gamma}$

$P_0(Ax_0)^{\gamma} = P[A(x_0 - x)]^{\gamma}$ [∵ The displacement of piston is so fast that the process is adiabatic]

$\Rightarrow \quad P = \frac{P_0 x_0^{\gamma}}{(x_0 - x)^{\gamma}}$

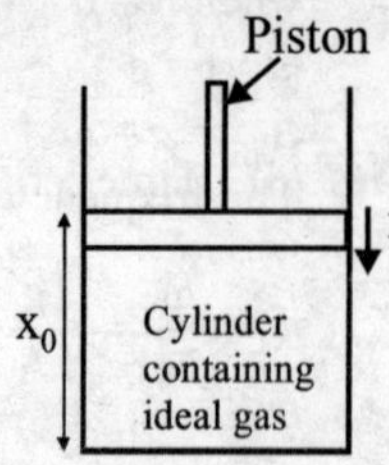

Now, Mg – PA = $F_{restoring}$

$$Mg - \left(\frac{P_0 x_0^{\gamma}}{(x_0 - x)^{\gamma}}\right)A = F_{restoring}$$

$$\Rightarrow \quad P_0 A\left(1 - \frac{x_0^{\gamma}}{(x_0 - x)^{\gamma}}\right) = F_{restoring}$$

$$\Rightarrow F_{restoring} = P_0 A\left(1 - \frac{1}{\left(1 - \frac{x}{x_0}\right)^{\gamma}}\right) = P_0 A\left[1 - \left(1 - \frac{x}{x_0}\right)^{-\gamma}\right]$$

$$= P_0 A\left[1 - \left(1 + \frac{\gamma x}{x_0}\right)\right] \quad \left[\because \frac{x}{x_0} << 1\right]$$

$$F_{restoring} = -\frac{\gamma P_0 A x}{x_0} \Rightarrow M\omega^2 x = \frac{\gamma P_0 A x}{x_0} \Rightarrow \omega = \sqrt{\frac{P_0 A \gamma}{x_0 M}}$$

∴ Frequency with which piston executes SHM.

$$f = \frac{1}{2\pi}\sqrt{\frac{\gamma P_0 A}{x_0 M}} = \frac{1}{2\pi}\sqrt{\frac{\gamma P_0 A^2}{M V_0}}$$

136. (a) $\omega_1 = \sqrt{\frac{g}{1}} = \omega_0$ (let)

$\omega_2 = \sqrt{\frac{g}{4}} = \frac{\omega_0}{2}$

$|\omega_{rel}| = |\omega_1 - \omega_2| = \frac{\omega_0}{2}$

So, to come in same phase $\theta_{rel} = 2\pi$

So, $t = \frac{\theta_{rel}}{\omega_{rel}} = \frac{2\pi}{\frac{\omega_0}{2}} = 2T$

So, no. of oscillation = 2

137. (a) Let the cube be at a depth x from the equilibrium position.

Force acting on the cube = up thrust on the portion of length x.

$F = -\rho\ell^2 xg$ [∴ mass = density × volume] ....(i)

Clearly $F \propto -x$, Hence it is a SHM.

Equation of SHM is $F = -kx$ ....(ii)

Comparing equation (i) and (ii) we have

$k = \rho\ell^2 g$

Now, Time period, $T = 2\pi\sqrt{\frac{m}{k}}$

$$\Rightarrow T = 2\pi\sqrt{\frac{\ell^3 d}{\rho\ell^2 g}} = 2\pi\sqrt{\frac{\ell d}{\rho g}}$$

138. (a) For an SHM, the acceleration

$a = -\omega^2 x$ where ω^2 is a constant.

$$a = \frac{-4\pi^2 x}{T^2} \Rightarrow \frac{aT}{x} = \frac{-4\pi^2}{T}$$

The time period T is also constant. Therefore, $\frac{aT}{x}$ is a constant.

139. (a) The two springs are in parallel.

∴ Effective spring constant,

$k = k_1 + k_2$

Initial frequency of oscillation is given by

$$f = \frac{1}{2\pi}\sqrt{\frac{k_1 + k_2}{m}} \quad ....(i)$$

When both k_1 and k_2 are made four times their original values, the new frequency is given by

$$f' = \frac{1}{2\pi}\sqrt{\frac{4k_1 + 4k_2}{m}}$$

$$= \frac{1}{2\pi}\sqrt{\frac{4(k_1 + 4k_2)}{m}} = 2\left(\frac{1}{2\pi}\sqrt{\frac{k_1 + k_2}{m}}\right) = 2f$$

140. (b) Here, Displacement, $x = 2 \times 10^{-2} \cos \pi t$

Velocity is given by

$$v = \frac{dx}{dt} = 2 \times 10^{-2}\, \pi \sin \pi t$$

For the first time, the when velocity becomes maximum, $\sin \pi t = 1$

$\Rightarrow \sin \pi t = \sin \frac{\pi}{2}$

$\Rightarrow \pi t = \frac{\pi}{2}$ or, $t = \frac{1}{2} = 0.5$ sec.

141. (b) When plugged hole near the bottom of the oscillating bob gets suddenly unplugged, centre of mass of combination of liquid and hollow portion (at position ℓ), first goes down (to $\ell + \Delta\ell$) and when total water is drained out, centre of mass regain its original position (to ℓ),

Time period, $T = 2\pi\sqrt{\frac{\ell}{g}}$

∴ 'T' first increases and then decreases to original value.

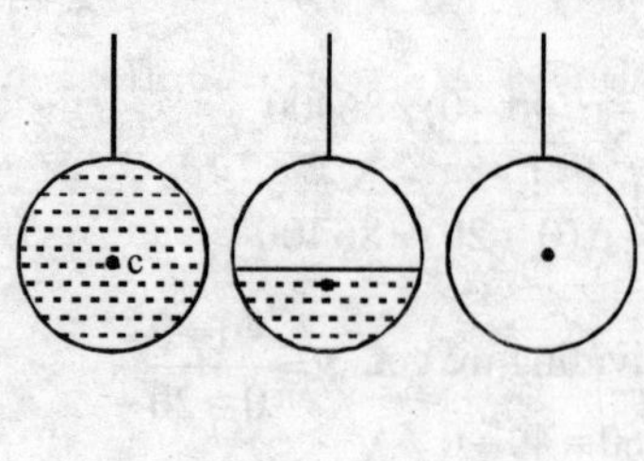

142. (a) The equation of SHM, $\frac{d^2x}{dt^2}+\alpha x=0$

$$\frac{d^2x}{dt^2} = -\alpha x = -\omega^2 x$$

$$\Rightarrow \omega=\sqrt{\alpha} \Rightarrow T=\frac{2\pi}{\omega}=\frac{2\pi}{\sqrt{\alpha}}$$

143. (a) Time period, $t=2\pi\sqrt{\frac{\ell}{g_{\text{eff}}}}$;

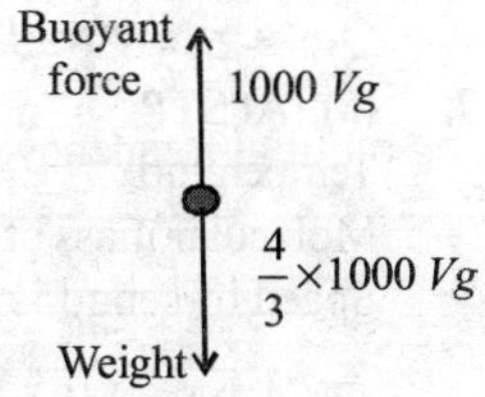

In air, $t_0=2\pi\sqrt{\frac{\ell}{g}}$

Net force $=\left(\frac{4}{3}-1\right)\times 1000\ Vg$

$$=\frac{1000}{3}Vg$$

$$g_{\text{eff}}=\frac{1000\ Vg}{3\times\frac{4}{3}\times 1000V}=\frac{g}{4}$$

$$\therefore t=2\pi\sqrt{\frac{\ell}{g/4}}=2\times 2\pi\frac{\ell}{g}$$

$$t=2t_0$$

144. (b) Time period for first spring, $t_1=2\pi\sqrt{\frac{m}{k_1}}$,

Time period for second spring, $t_2=2\pi\sqrt{\frac{m}{k_2}}$

Force constant of the series combination $k_{\text{eff}}=\frac{k_1k_2}{k_l+k_2}$

$\therefore$ Time period of oscillation for series combination

$$T=2\pi\sqrt{\frac{m(k_l+k_2)}{k_lk_2}}$$

$$\therefore T=2\pi\sqrt{\frac{m}{k_2}+\frac{m}{k_1}}=2\pi\sqrt{\frac{t_2^2}{(2\pi)^2}+\frac{t_1^2}{(2\pi)^2}}\Rightarrow T^2=t_1^2+t_2^2$$

where x is the displacement from the mean position

145. (c) With mass M, the time period of the spring.

$$T=2\pi\sqrt{\frac{M}{k}}$$

With mass $M+m$, the time period becomes,

$$T'=2\pi\sqrt{\frac{M+m}{k}}=\frac{5T}{3}$$

$$\therefore\ 2\pi\sqrt{\frac{M+m}{k}}=\frac{5}{3}\times 2\pi\sqrt{\frac{M}{k}}$$

$$\Rightarrow\quad M+m=\frac{25}{9}\times M \Rightarrow 1+\frac{m}{M}=\frac{25}{9}$$

$$\Rightarrow\frac{m}{M}=\frac{25}{9}-1=\frac{16}{9}$$

146. (d) Time period, $T=2\pi\sqrt{\frac{\ell}{g}}$

New length, $\ell'=\ell+21\%$ of ℓ

$\ell'=\ell+0.21\,\ell\Rightarrow\ell'=1.21\,\ell$

$$T'=2\pi\sqrt{\frac{1.21\ell}{g}}$$

$$\%\text{ increase in length}=\frac{T'-T}{T}\times 100=\frac{\sqrt{1.21\ell}-\sqrt{\ell}}{\sqrt{\ell}}\times 100$$

$$=\left(\sqrt{1.21}-\sqrt{1}\right)\times 100\ =(1.1-1)\times 100=10\%$$

147. (b) Let k be the spring constant of the original spring.

Time period T $=2\pi\sqrt{\frac{m}{k}}$ where m is the mas s of oscillating body and $K\propto\frac{1}{l}\ \left[\because K=\frac{\gamma A}{l}\right]$

So, when the spring is cut into n equal parts, the spring constant of one part becomes nk. Therefore the new time period,

$$T'=2\pi\sqrt{\frac{m}{nk}}=\frac{T}{\sqrt{n}}$$

148. (b) The time period $T=2\pi\sqrt{\frac{\ell}{g}}$ where ℓ = distance between the point of suspension and the centre of mass of the child.

As the child stands up, her centre of mass is raised. The distance between point of suspension and centre of mass decreases ie length ℓ decreases.

$\therefore\ \ell'<\ell$

$\therefore\ T'<T$ *i.e.*, the period decreases.

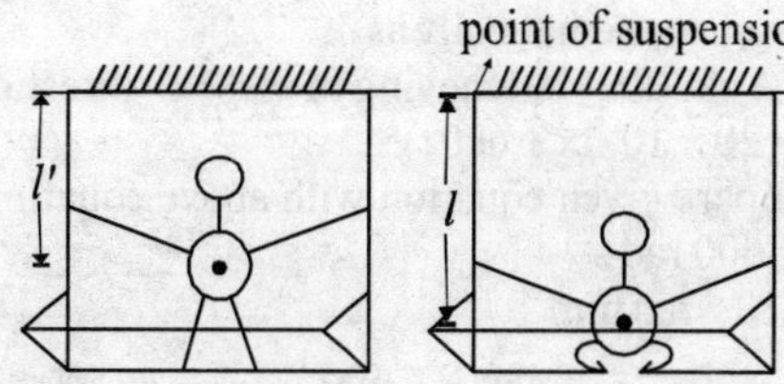

Case (ii) child standing Case (i) child sitting

Waves

14

1. **(a)** $y(x, t) = A\sin(kx \pm \omega t + \phi)$

Using general equation for a wave travelling on string.

Minimum distance between 2 points is $\frac{\lambda}{2}$, when the speed is same on comparing, $K = 20\pi$

$\Rightarrow \lambda = \frac{2\pi}{k} = \frac{1}{10}$ m = 10 cm

$\therefore$ Distance $= \frac{\lambda}{2} = 5$ cm

2. **(b)** Given : $\omega = 628$

Velocity $(v_{\omega}) = 300$ m/s

$x(t) = 5\cos\left[628t + \frac{\pi}{2}\right]m$

$v_{\omega} = \frac{\omega}{K}$

$\Rightarrow 300 = \frac{628}{K} \Rightarrow K = \frac{628}{300}$

$\Rightarrow K = \frac{2\pi}{\lambda} = \frac{628}{300} \Rightarrow \lambda = \frac{2\times3.14\times300}{628} \Rightarrow \lambda = 3$ m.

3. **(a)** Wave equation, $y = A\cos(Kx - \omega t)$

$v = \frac{\text{distance}}{\text{time}}$

$v = \frac{12}{0.3} = 4$ cm/s

$k = \frac{2\pi}{\lambda} = \frac{2\pi}{7.5} = \frac{4\pi}{15} = 0.83$

$v = \frac{\omega}{k} \Rightarrow \omega = vk = 4\times\frac{4\pi}{15} = 3.35$

So $y = 2\cos(0.83x - 3.35t)$

4. **(d)** In gases, the speed of sound depends on the bulk modulus

$V_{gas} = \sqrt{\frac{B}{\rho}}$, solids have higher value of bulk modulus than gases. So, $V_{solid} > V_{gas}$

5. **(c)** A transverse wave travelling along a string can be represented mathematically as

$y(x, t) = A\sin(kx - \omega t)$ moving in the + x - direction

$y = 4\sin(20\times10^{-3}x + 600t)$

Now compare given equation with above equation.

Here $\omega = 600$ rad s^{-1}

$k = 20\times10^{-3}$ rad mm^{-1}

$\therefore$ Wave velocity, $v = \frac{\omega}{k} = \frac{600}{20\times10^{-3}}$

$= 30\times10^{3}$ mm/s = 30 m/s

& direction of wave is towards –ve x axis

$\therefore v = -30$ m/s

6. **(b)** Equation of wave is, $y = 2\cos 2\pi(330t - x)$ m

$y = A\cos(\omega t - kx)$

On comparing $\omega = 2\pi\times330$ $[\because \omega = 2\pi f]$

$2\pi f = 2\pi\times330 \Rightarrow f = 330$ Hz

7. **(a)** At STP

Temperature, T = 273K

Molecular mass of oxygen, $M = 32\times10^{-3}$kg

Speed of sound is given by

$v = \sqrt{\frac{\gamma RT}{M}} = \sqrt{\frac{1.4\times8.3\times273}{32\times10^{-3}}}$

$= 314.8541 \simeq 315$m/s

8. **(1152)** From the equation $\omega = 160$ and $k = 0.5$

Speed of wave

$V = \frac{\omega}{k}$

$= \frac{160}{0.5}\times\frac{18}{5}$ km/h = 1152 km

9. **(20)** $y(x, t) = 5\sin(6t + 0.003x)$

Compair with $y(x, t) = y.\sin(\omega t + kx)$

$k = 0.003$ cm^{-1}, $\omega = 6$ rad/s,

wave velocity, $v = \frac{\omega}{k} \Rightarrow \frac{6}{0.003\times10^{2}} = 20$ ms^{-1}

10. **(a)** Given $M_{H_2} = 2$; $M_{O_2} = 32$

Speed of sound, $v = \sqrt{\frac{\gamma RT}{M}}$

$\Rightarrow v \propto \frac{1}{\sqrt{M}}$

$\therefore \frac{v_{H_2}}{v_{O_2}} = \sqrt{\frac{M_{O_2}}{M_{H_2}}} = \sqrt{\frac{32}{2}} = 4:1$

11. **(a)** The speed of transverse wave in the wire is

$v = \sqrt{\frac{T}{\mu}} = \sqrt{\frac{70}{7\times10^{-3}}} = 100$ m/s

12. **(d)** In solid, speed of sound is given as

$V = \sqrt{\frac{Y}{\rho}} = \sqrt{\frac{3.2\times10^{11}}{8\times10^{3}}} = \sqrt{0.4\times10^{8}}$

$= \sqrt{40\times10^{6}} = 6.32\times10^{3}$ ms^{-1}

13. **(18)** Phase difference $\Delta\phi = 60\times\frac{\pi}{180} = \frac{\pi}{3}$

The phase difference is given by

$\Delta\phi = \frac{2\pi}{\lambda}\Delta x$

$\frac{\pi}{3}=\frac{2\pi}{\lambda}(6\,m)$

$\Rightarrow \lambda = 36\,m$ $\quad$ ($\because$ frequency, f = 500 Hz)

$v = f\lambda = (500\,Hz)(36\,m)$

$= 18000\,m/s = 18\,km/s$

14. **(c)** Velocity of wave is

$V=\frac{\omega}{K}=\frac{4}{8}=\frac{1}{2}\,m/s$

15. **(c)** Comparing the given $y = 0.5\sin\left(\frac{2\pi}{\lambda}400t - \frac{2\pi}{\lambda}x\right)$ with standard equation

$\omega = \frac{2\pi}{\lambda}400$ and $k=\frac{2\pi}{\lambda}$

$\therefore$ velocity of the wave $v=\frac{\omega}{k}=\frac{\frac{2\pi}{\lambda}400}{\frac{2\pi}{\lambda}}$

$\therefore v = 400\,m/s$

16. **(a)** $y = 2\sin(\omega t - kx)$

Maximum particle velocity, $v_m = A\omega$

Wave velocity, $v_p=\frac{\omega}{k}$

$v_p = v_m$

$\frac{\omega}{k}=A\omega \Rightarrow k=\frac{1}{A}=\frac{2\pi}{\lambda}$

$\lambda = 2\pi A = 2\pi\times 2 = 4\pi$

17. **(d)** We have

$k=\frac{2\pi}{\lambda}=\frac{2\pi}{4}=0.5\pi\,cm^{-1}$

$\omega = 2\pi f = 2\pi\times 100 = 200\,\pi s^{-1}$

18. **(b)** Given,

(V_{max}) particle $= 4\,V_{wave}$

$\Rightarrow \omega A=\frac{4\omega}{K} \Rightarrow KA = 4$

$\Rightarrow \frac{2\pi}{\lambda}\times 10 = 4 \Rightarrow \lambda=\frac{20\pi}{4}=5\pi$

19. **(7)** $y_1 = A_1\sin k(x - vt)$

$y_1 = 12\sin 6.28\,(x - vt)$

$y_2 = 5\sin 6.28\,(x - vt + 3.5)$

Phase difference, $\Delta\phi = k(\Delta x) = 6.28\times 3.5 = 7\pi$

$A^2 = A_1^2 + A_2^2 + 2A_1A_2\cos\phi$

$\Rightarrow A=\sqrt{(12)^2+(5)^2+2(12)(5)\cos(7\pi)}$

$=\sqrt{144+25-120}=\sqrt{49}=7\,mm$

20. **(2)** At $t = 0$; $y=\frac{1}{1+x^2}$

We know that displacement of wave is a function of $(x - vt)$.

$\therefore\ y=\frac{1}{1+(x-vt)^2}$ at $t = t$; $y=\frac{1}{1+(x-vt)^2}$

We have given, at $t = 1$, $y=\frac{1}{1+(x-2)^2}$

By comparing

$v = 2\,m/s$

21. **(b)** Travelling wave $Y(x,t) = A\sin(Kx-\omega t)$

Here,

$A=\frac{0.06}{2}=0.03,\ \omega=2\pi f=2\pi\times 245=1.5\times 10^3$

$K=\frac{\omega}{v}=\frac{1.5\times 10^3}{300}\simeq 5.1$

$\therefore Y(x,t)=0.03\sin[5.1x-(1.5\times 10^3)t]$

22. **(a)** As we know,

Pressure amplitude, $\Delta P_0 = aKB = SKB = S\times\frac{\omega}{V}\times\rho V^2$

$\left[\because K=\frac{\omega}{V}, V=\sqrt{\frac{B}{\rho}}\right]$

$\Rightarrow S=\frac{\Delta P_0}{\rho V\omega}\approx\frac{10}{1\times 300\times 1000}\,m=\frac{1}{30}\,mm\approx\frac{3}{100}\,mm$

[**Note:** - '2π' is ignored here. So as to match the answer]

23. **(b)** Given : Distance between one crest and one trough $= 1.5\,m$

$=(2n_1+1)\frac{\lambda}{2}$

Distance between two crests = 5 m $= n_2\lambda$

$\frac{1.5}{5}=\frac{(2n_1+1)}{2n_2}\Rightarrow 3n_2 = 10n_1 + 5$

Here n_1 and n_2 are integer.

If $n_1 = 1, n_2 = 5$ $\quad\therefore \lambda = 1$

$n_1 = 4, n_2 = 15$ $\quad\therefore \lambda = 1/3$

$n_1 = 7, n_2 = 25$ $\quad\therefore \lambda = 1/5$

Hence possible wavelengths $\frac{1}{1},\frac{1}{3},\frac{1}{5}$ metre.

24. **(b)** At $t = 0, x = 0, y = 0$

$\phi = \pi\,rad$

25. **(a)** Loudness (dB) $= 10\log_{10}\left(\frac{I_2}{I_{ref}}\right)$

or $120 = 10\log_{10}\left(\frac{I}{10^{-12}}\right)\Rightarrow I = 1$

Also $I=\frac{P}{4\pi r^2}=\frac{2}{4\pi r^2}\Rightarrow 1=\frac{2}{4\pi r^2}\Rightarrow r=\sqrt{\frac{1}{2\pi}}$

$\Rightarrow r = 40\,cm$

26. **(a)** On comparing with $P = P_0\sin(\omega t - kx)$, we have

$\omega = 1000\,rad/s$, $K = 3\,m^{-1}$

$\therefore v_0=\frac{\omega}{k}=\frac{1000}{3}=333.3\,m/s$

$\frac{v_1}{v_2}=\sqrt{\frac{T_1}{T_2}}$ $\quad\left[\because V=\sqrt{\frac{\gamma RT}{M_o}}\right]$

or $\frac{333.3}{336}=\sqrt{\frac{273+0}{273+T_2}}\Rightarrow T_2 = 277\,K$

$\therefore T_2 = 4°C$

27. (a) Comparing the given equation
$y = 10^{-3}\sin(50t + 2x)$ with standard equation,
$y = a\sin(\omega t - kx)$
$\Rightarrow$ wave is moving along –ve x-axis with speed

$$v = \frac{\omega}{K} \Rightarrow v = \frac{50}{2} = 25 \text{ m/sec.}$$

28. (a) $V = 2(V_P)_{max}$
$V = f\lambda$
$f\lambda = 2\omega A \quad \because [\omega = 2\pi f]$
$\lambda = 4\pi A$

$$= 4\cancel{\pi} \times \frac{10}{\cancel{\pi}}$$

$\boxed{\lambda = 40\text{cm}}$

29. (b) $V = \frac{8}{0.4} = 20$ cm/s and $\lambda = 4 \times 2 = 8$ cm

and, $\omega = VK = 20 \times \frac{2\pi}{8} = 5\pi$ rad/sec

So, $(V_{max})_{particle} = \omega A = 5\pi \times \frac{1\text{ cm}}{2} = \frac{5\pi}{2}$ cm/s

30. (c) The equation of wave at any time is obtained by putting $X = x - vt$

$$y = \frac{1}{1+x^2} = \frac{1}{1+(x-vt)^2} \quad ...(i)$$

We know at $t = 2$ sec,

$$y = \frac{1}{1+(x-1)^2} \quad ...(ii)$$

On comparing (i) and (ii) we get

$$vt = 1 \Rightarrow v = \frac{1}{t}$$

As $t = 2$ sec $\quad \therefore v = \frac{1}{2} = 0.5$ m/s.

31. (a) Given

$$y(x,t) = e^{\left(-ax^2+bt^2+2\sqrt{ab}\,xt\right)}$$

$$= e^{-[(\sqrt{a}x)^2+(\sqrt{b}t)^2+2\sqrt{a}x.\sqrt{b}t]}$$

$$= e^{-(\sqrt{a}x+\sqrt{b}t)^2} \qquad \therefore V = \frac{\text{coeff of } t}{\text{coeff of } x} = \frac{\sqrt{b}}{\sqrt{a}}$$

and, coeff of x and coeff of y has same sign. So, wave is travelling along –ve x-direction.

32. (a) Given,
Wavelength, $l = 0.08$m
Time period, T = 2.0S
$y(x,t) = 0.005\cos(\alpha x - \beta t)$ (Given)
Comparing it with the standard equation of wave
$y(x,t) = a\cos(kx - \omega t)$ we get

$$k = \alpha = \frac{2\pi}{\lambda} \text{ and } \omega = \beta = \frac{2\pi}{T}$$

$$\therefore \alpha = \frac{2\pi}{0.08} = 25\pi \text{ and } \beta = \frac{2\pi}{2} = \pi$$

33. (b) Given, $y = 10^{-6}\sin\left(100t + 20x + \frac{\pi}{4}\right)m$
Comparing it with standard equation, we get
$\omega = 100$ and $k = 20$

$$v = \frac{\omega}{k} = \frac{100}{20} = 5\text{m/s}$$

34. (b) Velocity of wave in string, $v = \sqrt{\frac{T}{\mu}} = \sqrt{\frac{T}{f\pi R^2}}$

i.e., $V \propto \frac{1}{R}$

$$\therefore \frac{v_2}{v_1} = \frac{R_1}{R_2} = \frac{R}{R/2} = 2$$

35. (a) $\lambda_1 = 4l,\ f_1 = \frac{v}{4l}$

l

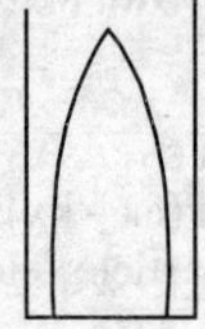

when a closed organ pipe is filled with water up to $\frac{1}{5}$ of its volume, the effective length of the air column decreases to $\frac{4}{5}$ of its original length.

$\frac{4}{5}l$

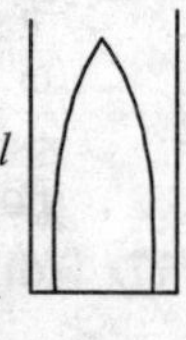

$$\lambda_2 = \frac{16l}{5},\ f_2 = \frac{5V}{16l}$$

$$\Delta f = f_2 - f_1 \Rightarrow \Delta F = \frac{5V}{16l} - \frac{4V}{16l} \Rightarrow \Delta F = \frac{V}{16l}$$

$$\therefore \frac{\Delta F}{f} = \frac{\frac{V}{l}\left(\frac{1}{16}\right)}{\frac{V}{4l}} \times 100 = 25\%.$$

36. (c) Given, length of closed tube, $l_1 = 10$cm

Frequency of nth harmonic in closed pipe $= \frac{nv}{4\ell}$

For 9th harmonic, frequency $= \frac{9v_1}{4\ell_1}$

For open pipe, frequency of nth harmonic $= \frac{nv}{2L}$

For 4th harmonic, frequency $= \frac{4 \times v_2}{2 \times L} = \frac{2v_2}{\ell_2}$

According to question,

$$\therefore \frac{9v_1}{4\ell_1} = \frac{2v_2}{\ell_2} \qquad \because v = \sqrt{\frac{B}{\rho}}$$

$$\therefore \frac{9}{4\ell_1}\sqrt{\frac{B}{\rho_1}} = \frac{2}{\ell_2}\sqrt{\frac{B}{\rho_2}} \Rightarrow \frac{\ell_2}{\ell_1} = \frac{8}{9}\sqrt{\frac{\rho_1}{\rho_2}}$$

$$\Rightarrow \ell_2 = \ell_1 \times \frac{8}{9} \times \frac{1}{4} = \frac{20}{9} \text{ cm}$$

37. **(16)** Frequency of closed organ pipe

$$f_c = (2n+1)\frac{v}{4\ell}$$

For seventh overtones, n = 7

$$\therefore f_c = \frac{(2\times 7+1)v}{4\ell} = \frac{15v}{4\ell}$$

Frequincy of open organ pipe $= (n+1)\frac{v}{2\ell}$

For n = 7 $\quad f_0 = \frac{8v}{2\ell} \Rightarrow \frac{f_c}{f_0} = \frac{15}{16} = \frac{a-1}{a} \Rightarrow a = 16$

38. **(740)** The difference in frequency in open organ pipe is given by:

$$f = \frac{nv}{2L}$$

Change in frequency,

$$\Delta f = f_1 - f_2 = \frac{n_1 v}{2L_1} - \frac{n_2 v}{2L_2} \Rightarrow \Delta f = \frac{6v}{2\times 0.6} - \frac{5v}{2\times 0.9}$$

$$\Delta f = \frac{v}{2}\left[\frac{6}{0.6} - \frac{5}{0.9}\right] \Rightarrow \Delta f = \frac{333}{2}\left(\frac{6}{0.6} - \frac{5}{0.9}\right) = 740 \text{ Hz}$$

39. **(60)** Fundamental frequency of vibration of sonometer is

$$v_0 = \frac{1}{2\ell}\sqrt{\frac{T}{m}}$$

For same T and m, $v_0 \propto \frac{1}{\ell}$

$$\Rightarrow \frac{(v_0)_2}{(v_0)_1} = \frac{\ell_1}{\ell_2} \Rightarrow \ell_2 = \frac{(v_0)_1}{(v_0)_2} \times \ell_1$$

$$\therefore \ell_2 = \frac{90}{600} \times 400 = 60\text{cm}$$

40. **(d)** Given,

Length of open organ pipe, $\ell_1 = 60$cm

Let ℓ_2 be the length of closed organ pipe.

Fundamental frequency of closed organ pipe, $f_1 = \frac{v}{4\ell_2}$

Frequency of first overtone of open organ pipe, $f_2 = 2\frac{v}{2\ell_1}$

According to question

$$\frac{v}{4\ell_2} = 2\left(\frac{v}{2\ell_1}\right) \Rightarrow \ell_2 = \frac{\ell_1}{4} \Rightarrow \ell_2 = \frac{60}{4} \Rightarrow \ell_2 = 15 \text{ cm}$$

41. **(400)** Area of cross-section, $A = 2 \text{ cm}^2$

Speed of sound, $v = 330$ m/s

$$\frac{v}{4\ell_1} = \frac{330}{4\ell_1} = 30 \Rightarrow \ell_1 = \frac{11}{4}\text{ m}$$

$$\frac{v}{4\ell_2} = \frac{330}{4\ell_2} = 110 \Rightarrow \ell_2 = \frac{3}{4}\text{ m}$$

$\Delta\ell = \ell_1 - \ell_2 = 2$ m

Change in volume $= A\Delta\ell = 400 \text{ cm}^3$

Mass of water $= \rho A(\ell_1 - \ell_2) \quad (\because \rho = 1\text{g}/\text{cm}^3)$

$= 1 \times 4 \times 10^{-4} = 0.4 \text{ kg} = 400$ g

42. **(90)**

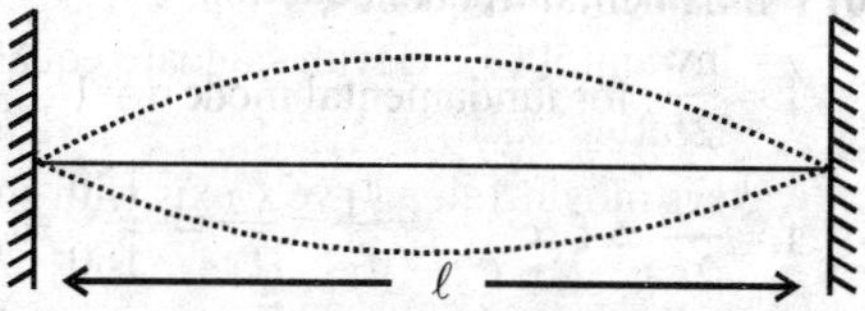

Fundamental frequency, f = 50 Hz

linear mass density μ = mass/ length = 20 g/m

mass of string, m = 18g

length of string, $\ell = \frac{m}{\mu} = \frac{18}{20}\text{m} = \frac{9}{10}\text{m}$

from diagram $\frac{\lambda}{2} = \ell \Rightarrow \lambda = 2\ell = \frac{9}{5}\text{m}$

The speed of the transverse waves in the string

$$v = f\lambda = 50 \times \frac{9}{5} = 90 \text{ m/s}$$

43. **(500)** Frequency, $f = \frac{1}{2\ell}\sqrt{\frac{T}{\mu}}$ (T = Tension)

$$\frac{f_2}{f_1} = \sqrt{\frac{T_2}{T_1}} \text{ or, } \left(\frac{f_2}{f_1}\right)^2 = \frac{T_2}{T_1}$$

$$\left(\frac{50}{30}\right)^2 = \frac{mg}{(180g)g} \Rightarrow \frac{25}{9} = \frac{m}{180g}$$

m = 500g

44. **(1)** In organ pipe, for 5th harmonic

frequency $f_5 = \frac{5v}{4\ell}$

Here,

v = Speed of sound

ℓ = length of organ pipe

On substituting respective values,

we get $405 = \frac{5\times 324}{4\ell} \Rightarrow \ell = 1\text{m}$

45. **(80)** Given,

Density of wire $\rho = 8 \times 10^3 \text{ kg/m}^3$

Extension developed in the wire, $\Delta L = 3.2 \times 10^{-4}$ m

Length of wire, L = 0.5 m

Young modulus of wire $Y = 8 \times 10^{10} \text{ N/m}^2$

Fundamental frequency,

$$f = \frac{1}{2L}\sqrt{\frac{T}{\mu}} = \frac{1}{2L}\sqrt{\frac{YA\Delta L}{\rho L}} \quad [\because \mu = \text{density} \times A]$$

$$= \frac{1}{2\times 0.5}\sqrt{\frac{8\times 10^{10} \times 3.2\times 10^{-4}}{8\times 10^3 \times 0.5}} \Rightarrow f = 80 \text{ Hz}$$

46. **(900)** For second harmonic of open organ pipe,

the length given as $L = \frac{n\lambda}{2}; n = 2$

So frequency of vibration is $f = \frac{v}{\lambda}$

$$f = \frac{v}{\lambda} = \frac{v}{L} = \frac{360}{\frac{40}{100}} = 900\text{Hz}$$

47. **(60)** Fundamental frequency,

$$f = \frac{nv}{2\ell}, \text{ for fundamental mode } n = 1$$

$$\Rightarrow \quad f = \frac{v}{2\ell} \Rightarrow f \propto \frac{1}{\ell} \quad \therefore \frac{f_1}{f_2} = \frac{\ell_2}{\ell_1} \Rightarrow \frac{120}{180} = \frac{\ell_2}{90}$$

Hence, $\ell_2 = 60$ cm

48. **(3)** For n^{th} harmonic,

Frequency, $f_1 = \frac{nv}{0.6} = 400$

For $(n+1)^{th}$ harmonic,

$$f_2 = \frac{(n+1)v}{0.6} = 450 \Rightarrow \left[\frac{0.6 \times 400}{v} + 1\right]\frac{v}{0.6} = 450$$

$\Rightarrow v = 30$

$\Rightarrow$ Speed of wave on a string $v = \sqrt{\frac{T}{\mu}} = 30$

$$\Rightarrow \quad \frac{2700}{\mu} = 900 \Rightarrow \mu = 3$$

49. **(50)** If we don't ignore the word fundamental mode, you cannot observe resonance once again whatever be the height of water in column.

$$\lambda = \frac{V}{f} = \frac{340}{340} = 1\text{m}.$$

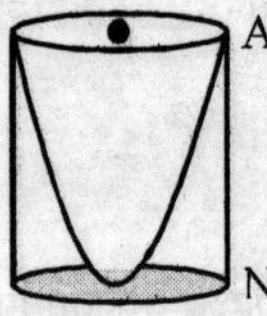

In first resonance

$\lambda = 4L_1$

$L_1 = \frac{\lambda}{4} = 25$ cm

In second resonance $\lambda = \frac{4L}{3}$

$L = \frac{3\lambda}{4} = 75$ cm

In third resonance

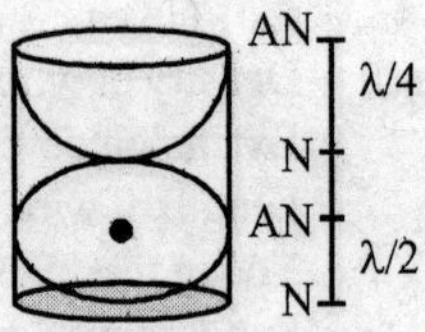

$\lambda = \frac{4L}{5} \qquad L = \frac{\lambda}{2} + \frac{\lambda}{4} = \frac{4\lambda}{3}$

$L = \frac{5\lambda}{4} = 125$ cm, So mode of vibration is $L = \frac{5\lambda}{4}$

Now, if 50 cm of water is added, it will vibrate in second resonance mode.
So, height of water required = (125 – 75) cm = 50 cm

50. **(10)** Linear mass density $\mu = 9.0 \times 10^{-4}\ \frac{kg}{m}$

Tension, $T = 900\ N$
Velocity of wire,

$$V = \sqrt{\frac{T}{\mu}} = \sqrt{\frac{900}{9 \times 10^{-4}}} = 10^3\ m/s$$

$$f_n = \frac{nV}{2\ell} = 500 \qquad ...(i)$$

$$\text{And } \frac{(n+1)v}{2\ell} = 550 \qquad ...(ii)$$

From (i) & (ii), $f_{n+1} - f_n = \frac{v}{2\ell} = 50 \therefore \ell = \frac{10^3}{2 \times 50} = 10m$

51. **(34)** The resonant frequency of a closed organ pipe of length L is

$$f = \frac{nv}{4L}$$

Here, n = odd positive integer
v = speed of sound in air
For L to be minimum, $n = 1$

$$\therefore 250 = \frac{v}{4L} \Rightarrow 250 = \frac{340}{4L} \Rightarrow L = \frac{34}{4 \times 25} = 0.34\ m$$

$\Rightarrow \ L = 34$ cm

52. **(4)** Velocity $v = \sqrt{\frac{B}{\rho}} = \sqrt{\frac{1}{k\rho}}$

[$\because$ Compressibility, $K = \frac{1}{B \text{ (Bulk modulus)}}$]

And both closed and open organ pipes vibrating in their first overtone with same frequency

$\therefore f_{\text{closed}} = f_{\text{open}}$

$$\frac{3v_1}{4L} = \frac{2v_2}{2L'} \Rightarrow L' = \frac{4}{3}L\left(\frac{v_2}{v_1}\right)$$

$$L' = \frac{4}{3}L\sqrt{\frac{\rho_1}{\rho_2}} \qquad [\because \text{compressibility are equal}]$$

$\therefore x = 4$

53. **(c)** As tension is different at every point of rope. So velocity of wave will be different at different point.

$$f = \text{constant} \Rightarrow \frac{v_1}{\lambda_1} = \frac{v_2}{\lambda_2} \Rightarrow \lambda_2 = \frac{v_2}{v_1}\lambda_1$$

$$\lambda_2 = \sqrt{\frac{T_2}{T_1}}\ \lambda_1 \quad [\because v \propto \sqrt{T}]$$

$$= \sqrt{\frac{8g}{2g}}.\lambda_1 = 2 \times 6 \text{ cm}$$

= 12 cm

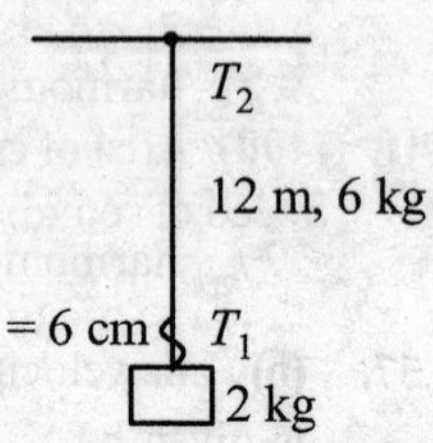

54. **(a)** Using $f = \frac{1}{2\ell}\sqrt{\frac{T}{\mu}}$,

where, T = tension and $\mu = \frac{\text{mass}}{\text{length}}$

$$f_x = \frac{1}{2\ell}\sqrt{\frac{T_x}{\mu}} \text{ and } f_z = \frac{1}{2\ell}\sqrt{\frac{T_z}{\mu}}$$

$$\frac{f_x}{f_z} = \frac{450}{300} = \sqrt{\frac{T_x}{T_z}}$$

$$\therefore \frac{T_x}{T_z} = \frac{9}{4} = 2.25.$$

55. **(35.00)** Given,

Density of wire, $\sigma = 9 \times 10^{-3}$ kg cm^{-3}

Young's modulus of wire, $Y = 9 \times 10^{10}$ Nm^{-2}

Strain $= 4.9 \times 10^{-4}$

$$Y = \frac{\text{Stress}}{\text{Strain}} = \frac{T/A}{\text{Strain}}$$

$$\therefore \frac{T}{A} = Y \times \text{Strain} = 9 \times 10^9 \times 4.9 \times 10^{-4}$$

Also, mass of wire, $m = Al\sigma$

Mass per unit length, $\mu = \frac{m}{l} = A\sigma$

Fundamental frequency in the string

$$f = \frac{1}{2l}\sqrt{\frac{T}{\mu}} = \frac{1}{2l}\sqrt{\frac{T}{\sigma A}} = \frac{1}{2\times 1}\sqrt{\frac{9\times 10^9 \times 4.9\times 10^{-4}}{9\times 10^3}}$$

$$= \frac{1}{2}\sqrt{49\times 10^{9-4-3}} = \frac{1}{2}\times 70 = 35 \text{ Hz}$$

56. **(106)** Given : $v_{air} = 300$ m/s, $\rho_{gas} = 2\rho_{air}$

$$\text{Using, } v = \sqrt{\frac{B}{\rho}};\ \frac{v_{gas}}{v_{air}} = \frac{\sqrt{\frac{B}{2\rho_{air}}}}{\sqrt{\frac{B}{\rho_{air}}}}$$

$$\Rightarrow v_{gas} = \frac{v_{air}}{\sqrt{2}} = \frac{300}{\sqrt{2}} = 150\sqrt{2}\text{m/s}$$

And f_{nth} harmonic $= \frac{nv}{2L}$ (in open organ pipe)

(L = 1 metre given)

$$\therefore f_{2nd} \text{ harmonic} - f_{fundamental} = \frac{2v}{2\times 1} - \frac{v}{2\times 1} = \frac{v}{2}$$

$$\therefore f_{2nd} \text{ harmonic} - f_{fundamental} = \frac{150\sqrt{2}}{2} = \frac{150}{\sqrt{2}} \approx 106\,Hz$$

57. **(b)** The velocity of a transverse wave in a stretched wire is given by

$$v = \sqrt{\frac{T}{\mu}}$$

Where,

T = Tension in the wire

μ = linear density of wire $(\because V \propto T)$

$$\therefore \frac{v_1}{v_2} = \sqrt{\frac{T_1}{T_2}} \Rightarrow \frac{v}{v}\times 2 = \sqrt{\frac{2.06\times 10^4}{T_2}}$$

$$\Rightarrow T_2 = \frac{2.06\times 10^4}{4} = 0.515\times 10^4\,N \Rightarrow T_2 = 5.15\times 10^3\,N$$

58. **(a)** Given, $l = 60$ cm, $m = 6$ g, $A = 1$ mm^2, $v = 90$ m/s and $Y = 16 \times 10^{11}$ Nm^{-2}

$$\text{Using, } v = \sqrt{\frac{T}{m}\times l} \Rightarrow T = \frac{mv^2}{l}$$

$$\text{Again from, } Y = \frac{T}{A}\Delta L / L_0$$

$$\Delta L = \frac{Tl}{YA} = \frac{mv^2\times l}{l(\text{YA})} = \frac{6\times 10^{-3}\times 90^2}{16\times 10^{11}\times 10^{-6}} = 3\times 10^{-4}\,m$$

= 0.03 mm

59. **(c)** We have given,

y = 0.03 sin(450 t − 9x)

Comparing it with standard equation of wave, we get $\omega = 450$, k = 9

$$\therefore \text{v} = \frac{\omega}{\text{k}} = \frac{450}{9} = 50\text{m/s}$$

Velocity of travelling wave on a stretched string is given by

$$\text{v} = \sqrt{\frac{\text{T}}{\mu}} \Rightarrow \frac{\text{T}}{\mu} = 2500$$

μ = linear mass density

$\Rightarrow$ T = 2500 × 5 × 10^{-3} $\Rightarrow$ 12.5 N

60. **(b)** Wave speed $\text{v} = \sqrt{\frac{\text{T}}{\mu}}$

when car is at rest a = 0

$$\therefore 60 = \sqrt{\frac{\text{Mg}}{\mu}}$$

Similarly when the car is moving with acceleration a,

$$60.5 = \sqrt{\frac{\text{M}\left(\text{g}^2+\text{a}^2\right)^{1/2}}{\mu}} \Rightarrow \frac{60.5}{60} = \sqrt{\sqrt{\frac{\text{g}^2+\text{a}^2}{\text{g}^2}}}$$

$$\Rightarrow \left(1+\frac{0.5}{60}\right)^4 = \frac{\text{g}^2+\text{a}^2}{\text{g}^2} = 1+\frac{2}{60}$$

$$\Rightarrow \text{g}^2+\text{a}^2 = \text{g}^2+\text{g}^2\times\frac{2}{60}$$

$$\text{a} = \text{g}\sqrt{\frac{2}{60}} = \frac{\text{g}}{\sqrt{30}}$$ [which is closest to g/5]

61. **(a)** Fundamental frequency, $f = (490 - 420)$ Hz $= 70$ Hz.
The fundamental frequency of wire vibrating under tension T is given by

$$f = \frac{1}{2L}\sqrt{\frac{T}{\mu}}.$$

Here, μ = mass per unit length of the wire
L = length of wire

$$70 = \frac{1}{2L}\sqrt{\frac{540}{6\times10^{-3}}} \Rightarrow L \approx 2.14\,\text{m}$$

62. **(b)** $\frac{3\lambda}{2} = 2$ or $\lambda = \frac{4}{3}m$

Velocity, $v = f = 240 \times \frac{4}{3} = 320\,\text{m/sec}$

Also $f_1 = \frac{240}{3} = 80\,\text{Hz}$

63. **(b)** Given, $y = 0.3 \sin(0.157x)\cos(200\pi t)$
So $k = 0.157$ and $\omega = 200\pi$

or $f = 100$ Hz, $v = \frac{\omega}{k} = \frac{200\pi}{0.157} = 4000\,\text{m/s}$

Now, using $f = \frac{nv}{2l} = \frac{4v}{2l} = \frac{2v}{l}$

$$\therefore l = \frac{2v}{f} = \frac{2\times4000}{100} = 80m$$

64. **(c)** $f = \frac{n}{2l}\sqrt{\frac{T}{\mu}}$, where n = nth harmonic and is equal to number of antinodes.

$$f_A = \frac{p}{2l}\sqrt{\frac{T}{\rho A_o}} \Rightarrow f_B = \frac{q}{2l}\sqrt{\frac{T}{4\rho A_o}}$$

$$\therefore \frac{f_A}{f_B} = \frac{2p}{q} \Rightarrow \frac{p}{q} = \frac{1}{2}$$

65. **(a)** If a closed pipe vibration in N[th] mode then frequency of vibration $n = \frac{(2N-1)v}{4l} = (2N-1)n_1$
(where n_1 = fundamental frequency of vibration)
Hence $20{,}000 = (2N - 1) \times 1500$
$\Rightarrow N = 7.1 \approx 7$
$\therefore$ Number of over tones = (No. of mode of vibration) – 1
$= 7 - 1 = 6$

66. **(d)** Velocity of wave on string

$$v = \sqrt{\frac{T}{\mu}} = \sqrt{\frac{8}{5}\times1000} = 40\text{m/s}$$

Here, T = tension and μ = mass/length

Wavelength of wave $\lambda = \frac{v}{n} = \frac{40}{100}$ m

Separation b/w successive nodes,

$$\frac{\lambda}{2} = \frac{40}{2\times100} = \frac{20}{100}\text{m} = 20\text{ cm}$$

67. **(a)** In solids, Velocity of wave

$$v = \sqrt{\frac{Y}{\rho}} = \sqrt{\frac{9.27\times10^{10}}{2.7\times10^{3}}} \Rightarrow v = 5.85\times10^{3}\text{ m/sec}$$

Since rod is clamped at middle fundamental wave shape is as follow

$\frac{\lambda}{2} = L \Rightarrow \lambda = 2L$

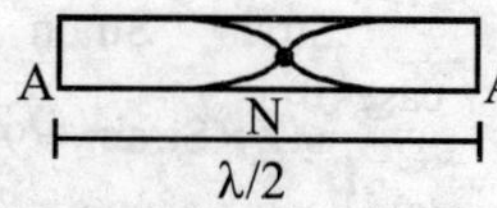

$\lambda = 1.2\text{m}$ ($\because$ L = 60 cm = 0.6m (given))
Using $v = f\lambda$

$$\Rightarrow f = \frac{v}{\lambda} = \frac{5.85\times10^{3}}{1.2} = 4.88\times10^{3}\text{ Hz} \simeq 5\text{ KHz}$$

68. **(a)** For first resonance, $\frac{\lambda}{4} = \ell_1 + e = 11$ cm

($\because$ end correction e = 1 cm given)

For second resonance, $\frac{3\lambda}{4} = \ell_2 + e$

$\Rightarrow \ell_2 = 3\times11 - 1 = 32$ cm

69. **(b)** $n_1 = n_2$
T $\rightarrow$ Same
r $\rightarrow$ Same
$l \rightarrow$ Same

Frequency of vibration $n = \frac{p}{2l}\sqrt{\frac{T}{\pi r^2 \rho}}$

As T, r, and l are same for both the wires
$n_1 = n_2$

$$\frac{p_1}{\sqrt{\rho_1}} = \frac{p_2}{\sqrt{\rho_2}} \Rightarrow \frac{p_1}{p_2} = \frac{1}{2} \qquad [\because \rho_2 = 4\rho_1]$$

70. **(a)** We know that velocity in string is given by

$$v = \sqrt{\frac{T}{\mu}} \qquad \text{...(i)}$$

where $\mu = \frac{m}{l} = \frac{\text{mass of string}}{\text{length of string}}$

The tension $T = \frac{m}{\ell}\times x\times g$ ..(ii)

From (i) and (ii)

$$\frac{dx}{dt}=\sqrt{gx} \Rightarrow x^{-1/2}dx=\sqrt{g}\,dt$$

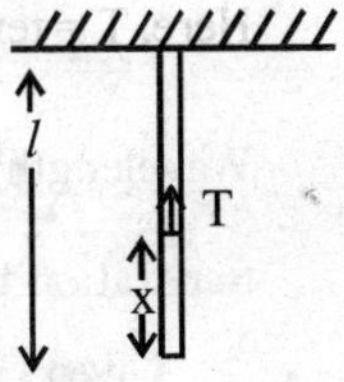

$$\therefore \int_0^{\ell} x^{-1/2}dx=\sqrt{g}\int_0^{\ell} dt$$

$$\Rightarrow 2\sqrt{\ell}=\sqrt{g}t$$

$$2\sqrt{\ell}=\sqrt{g}\times t \quad \therefore\ t=2\sqrt{\frac{\ell}{g}}=2\sqrt{\frac{20}{10}}=2\sqrt{2}$$

71. **(b)** The fundamental frequency

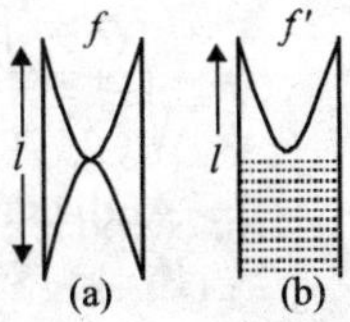

in case (a) is $f=\frac{v}{2\ell}$

The fundamental frequency in

case (b) is $f'=\frac{v}{4(\ell/2)}=\frac{v}{2\ell}=f$

72. **(c)** Length of pipe = 85 cm = 0.85m

Frequency of oscillations of air column in closed organ pipe is given by,

$$f=\frac{(2n-1)v}{4L}$$

$$f=\frac{(2n-1)v}{4L}\le 1250$$

$$\Rightarrow \frac{(2n-1)\times 340}{0.85\times 4}\le 1250$$

$$\Rightarrow\ 2n-1\le 12.5\approx 6$$

73. **(b)** Total length of sonometer wire, l = 110 cm = 1.1 m

Length of wire is in ratio, 6 : 3 : 2 i.e. 60 cm, 30 cm, 20 cm.

Tension in the wire, T = 400 N

Mass per unit length, μ = 0.01 kg

Minimum common frequency = ?

As we know,

$$\text{Frequency, } \nu=\frac{1}{2l}\sqrt{\frac{T}{\mu}}=\frac{1000}{11}\text{Hz}$$

$$\text{Similarly, } \nu_1=\frac{1000}{6}\text{Hz}$$

$$\nu_2=\frac{1000}{3}\text{Hz}$$

$$\nu_3=\frac{1000}{2}\text{Hz}$$

Hence common frequency = 1000 Hz

74. **(b)** Fundamental frequency,

$$f=\frac{v}{2\ell}=\frac{1}{2\ell}\sqrt{\frac{T}{\mu}}=\frac{1}{2\ell}\sqrt{\frac{T}{A\rho}}\quad\left[\because\ v=\sqrt{\frac{T}{\mu}}\text{ and }\mu=\frac{m}{\ell}\right]$$

$$\text{Also, } Y=\frac{T\ell}{A\Delta\ell}\Rightarrow\frac{T}{A}=\frac{Y\Delta\ell}{\ell}$$

$$\Rightarrow f=\frac{1}{2\ell}\sqrt{\frac{\gamma\Delta\ell}{\ell\rho}} \qquad ....(i)$$

$$\ell=1.5\text{ m},\ \frac{\Delta\ell}{\ell}=0.01,$$

$\rho=7.7\times10^3\ kg/m^3$ (given)

$\gamma=2.2\times10^{11}\ N/m^2$ (given)

Putting the value of $\ell, \frac{\Delta\ell}{\ell}, \rho$ and γ in eq^n. (i) we get,

$$f=\sqrt{\frac{2}{7}}\times\frac{10^3}{3}\quad\text{or}\quad f\approx 178.2\text{ Hz}$$

75. **(d)** Total length of the wire, L = 114 cm

$n_1 : n_2 : n_3 = 1 : 3 : 4$

Let L_1, L_2 and L_3 be the lengths of the three parts

As $n\propto\frac{1}{L}$

$$\therefore\ L_1:L_2:L_3=\frac{1}{1}:\frac{1}{3}:\frac{1}{4}=12:4:3$$

$$\therefore\ L_1=\left(\frac{12}{12+4+3}\times114\right)=72\text{cm}$$

$$L_2=\left(\frac{4}{19}\times114\right)=24\text{ cm}$$

$$\text{and } L_3=\left(\frac{3}{19}\times114\right)=18\text{ cm}$$

Hence the bridges should be placed at 72 cm and 72 + 24 = 96 cm from one end.

76. **(a)** Initially for open organ pipe, fundamental frequency

$$\nu_0=\frac{v}{2l_0} \qquad ...(i)$$

where l_0 is the length of the tube

v = speed of sound

But when it is half dipped in water, it becomes closed organ pipe of length $\frac{\ell_0}{2}$.

Fundamental frequency of closed organ pipe

$$\nu_c=\frac{v}{4l_c} \qquad ...(ii)$$

$$\text{New length, } l_c=\frac{l_0}{2}$$

$$\text{Thus } \nu_c=\frac{v}{4l_0/2}\Rightarrow\nu_c=\frac{v}{2l_0} \qquad ...(iii)$$

From equations (i) and (iii)

$$\nu_0=\nu_c$$

Thus, $\nu_c=f$ ($\because\ \nu_0=f$ is given)

77. **(b)** Given : Frequency of tuning fork, $n = 264$ Hz
Length of column $L = ?$
For closed organ pipe

$$n = \frac{v}{4l} \Rightarrow l = \frac{v}{4n} = \frac{330}{4 \times 264} = 0.3125$$

or, $l = 0.3125 \times 100 = 31.25$ cm

In case of closed organ pipe only odd harmonics are possible.

Therefore value of l will be $(2n-1)\,l$

Hence option (b) i.e. $3 \times 31.25 = 93.75$ cm is correct.

78. **(d)** Two lowest frequencies to which tube will resonates are 272 Hz and 544 Hz.

79. **(a)** It is given that 315 Hz and 420 Hz are two resonant frequencies, let these be n[th] and $(n+1)^{th}$ harmonies, then

we have $\frac{nv}{2\ell} = 315$

and $(n+1)\frac{v}{2\ell} = 420 \Rightarrow \frac{n+1}{n} = \frac{420}{315} \Rightarrow n = 3$

Hence $3 \times \frac{v}{2\ell} = 315 \Rightarrow \frac{v}{2\ell} = 105\ Hz$

The lowest resonant frequency is when
n = 1
Therefore lowest resonant frequency
= 105 Hz.

80. **(c)** The fundamental frequency for tube B closed at one end is given by

$$\upsilon_B = \frac{v}{4\ell} \qquad \left[\because \ell = \frac{\lambda}{4}\right]$$

Where ℓ = length of the tube and v is the velocity of sound in air.

The fundamental frequency for tube A open with both ends is given by

$$\upsilon_A = \frac{v}{2\ell} \qquad \left[\because \ell = \frac{\lambda}{2}\right]$$

$$\therefore \frac{\upsilon_A}{\upsilon_B} = \frac{v}{2\ell} \times \frac{4\ell}{v} = \frac{2}{1}$$

81. **(b)** To form a node there should be superposition of this wave with the reflected wave. The reflected wave should travel in opposite direction with a phase change of π. The equation of the reflected wave will be

$y = a\sin(\omega t + kx + \pi)$

$\Rightarrow y = -a\sin(\omega t + kx)$

82. **(d)** Resultant amplitude, $A = \sqrt{A_1^2 + A_2^2 + 2A_1A_2\cos\theta}$

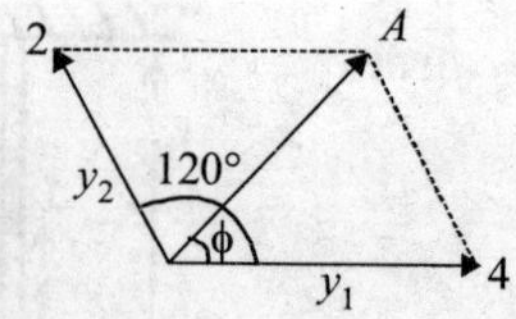

Given, $y_1(x, t) = 4\sin(Kx - \omega t)$

and $y_2(x, t) = 2\sin\left(Kx - \omega t + \frac{2\pi}{3}\right)$

$$\therefore A = \sqrt{2^2 + 4^2 + 2 \times 2 \times 4 \times \cos 120^\circ} = \sqrt{12} = 2\sqrt{3}$$

And $\tan\phi = \frac{2\sin 120^\circ}{4 + 2\cos 120^\circ} = \frac{\sqrt{3}}{3} = \frac{1}{\sqrt{3}}$

$\therefore$ Phase $\phi = \frac{\pi}{6}$

83. **(d)** The equation of resultant wave is

$x = a\cos(1.5t)\cos(50.5t)$

$= \frac{a}{2}\cos[1.5 + 50.5]t + \frac{a}{2}\cos[50.5 - 1.5]$

$= \frac{a}{2}\cos[52t] + \frac{a}{2}\cos[49t]$

$\Rightarrow 2\pi f_1 = 52$ and $2\pi f_2 = 49$

$\therefore f_1 = \frac{52}{2\pi},\ f_2 = \frac{49}{2\pi}$

$\therefore f_{Beat} = f_1 - f_2 = \frac{52}{2\pi} - \frac{49}{2\pi} = \frac{3}{2\pi}$ Hz

$\therefore T_{Beat} = \frac{1}{f_{Beat}} = \frac{2\pi}{3}$ sec $= 2.09$ sec ≈ 2 sec.

84. **(d)** Fundamental frequency of two closed air columns are

$f_1 = \frac{v}{4\ell_1}$ & $f_2 = \frac{v}{4\ell_2}$

Beat $= (f_1 - f_2) = \frac{v}{4}\left(\frac{1}{\ell_1} - \frac{1}{\ell_2}\right)$

Here, Beat = 15 beats,
$\ell_1 = 100$ cm = 1m
$\ell_2 = 120$ cm = 1.2m

$15 = \frac{v}{4}\left(\frac{1}{1} - \frac{1}{1.2}\right)$

$v = \left(\frac{15 \times 4 \times 1.2}{0.2}\right) = 60 \times 6 = 360$ m/s

85. **(6)** Frequency of tuning fork $f = \frac{1}{2L}\sqrt{\frac{T}{\mu}}$

For T = 6N and L = 1m

$f_1 = \frac{1}{2}\sqrt{\frac{6}{\mu}}$ ...(i)

For T = 54N and L = 1m

$$f_2 = \frac{1}{2}\sqrt{\frac{54}{\mu}} \quad \text{...(ii)}$$

$\frac{f_1}{f_2} = \frac{1}{3}$, Given $f_2 - f_1 = 12$

$\therefore f_1 = 6$ Hz

86. **(13)** Intensity for coherent wave

$I_2 = I_A + I_B + 2\sqrt{I_A I_B} \cos 60°$

$\Rightarrow I_2 = I_0 + 9I_0 + 2\sqrt{9I_0^2}.\frac{1}{2} = 13I_0$

Intensity for incoherent wave

$I_1 = I_A + I_B = I_0 + 9I_0 \Rightarrow I_1 = 10I_0$

$\therefore \frac{I_1}{I_2} = \frac{10}{13} = \frac{10}{x}$ (given) $\therefore x = 13$

87. **(294)** Frequency for closed organ pipe $f_c = \frac{v}{4\ell_1}$

and for open organ pipe, $f_0 = \frac{v}{2\ell_2}$

Closed organ pipe of length, $\ell_1 = 150$ cm

Open organ pipe of length, $\ell_2 = 350$ cm

Given $|f_c - f_0| = 7$

$$\frac{v}{4\times150} - \frac{v}{2\times350} = 7$$

$$\Rightarrow \frac{v}{600\text{ cm}} - \frac{v}{700\text{ cm}} = 7 \Rightarrow \frac{v}{6\text{ m}} - \frac{v}{7\text{ m}} = 7$$

$\Rightarrow v\left(\frac{1}{42}\right) = 7 \quad \therefore v = 42 \times 7 = 294$ m/s

88. **(20)** We have $A_1 = 10$ cm

$Y = 5\sin\omega t + 5\sqrt{3}\cos\omega t$

$= A_2\cos\theta\sin\omega t + A_2\sin\theta\cos\omega t = A_2\sin(\omega t + \theta)$

Here, $A_2\cos\theta = 5$ and $A_2\sin\theta = 5\sqrt{3}$

So, $\frac{A_2\cos\theta}{A_2\sin\theta} = \frac{5}{5\sqrt{3}} \Rightarrow \cot\theta = \frac{1}{\sqrt{3}} \Rightarrow \theta = \frac{\pi}{3}$

Squaring and adding, we get

$A_2^2 = 5^2 + (5\sqrt{3})^2$

$A_2 = 10$ cm

As, $A_R^2 = A_1^2 + A_2^2 + 2A_1A_2\cos\Delta\phi$

$= 10^2 + 10^2 + 2.10.10\cos\left(\frac{\pi}{3} - \frac{\pi}{3}\right)$

$= 100 + 100 + 200\cos 0° = 400$

So, $A_R = 20$ cm

89. **(60)** The resultant amplitude is given as,

$A_{resultant} = \sqrt{A_1^2 + A_2^2 + 2A_1A_2\cos\phi}$

$\Rightarrow \sqrt{3}A = \sqrt{A^2 + A^2 + 2A^2\cos\phi}$

$\Rightarrow 3A^2 = 2A^2 + 2A^2\cos\phi \Rightarrow \cos\phi = \frac{1}{2}$

$\therefore \phi = 60°$ $\therefore$ Phase difference = 60 degree

90. **(d)** Beat frequency $= f_1 - f_2$

$\frac{40}{12} = \frac{v}{4.08} - \frac{v}{4.16} \Rightarrow \frac{10}{3} = v\left(\frac{0.08}{4.08\times4.16}\right) \Rightarrow v = 707.2$ m/s

91. **(152)** Let $f_1 = f_0$. Then

$f_2 = f_0 + 4$

$f_3 = f_0 + 2\times4$

$f_4 = f_0 + 3\times4$

$\vdots \quad \vdots$

$f_{20} = f_0 + 19\times4$

Now, $f_{20} = 2f_1 \Rightarrow f_0 + 19\times4 = 2f_0 \Rightarrow f_0 = 76$ Hz

So, $f_{20} = 76 + 19\times4 = 152$ Hz.

92. **(a)** $A = \sqrt{A_1^2 + A_2^2 + 2A_1A_2\cos\phi}$

where ϕ = phase difference

$= [(x - vt + 1.5) - (x - vt)] \times 2\pi$

$= 2\pi \times 1.5 = 3\pi$

So, $A = \sqrt{5^2 + 3^2 + 2.5.3\cos 3\pi} = \sqrt{4} = 2$ cm.

93. **(5)** The standing wave equation is

$y = 10\cos(\pi x)\sin\left(\frac{2\pi t}{T}\right)$

Denotes Amplitude ($10\cos(\pi x)$)

$\therefore A = 10\cos\pi x$

$A = 10\cos\pi\times\frac{4}{3}$

Clearly, $|A| = \left|10\cos\frac{4\pi}{3}\right| = 5cm$

94. **(a)** $f_A = 340 \pm 5 = 335$ or 345 Hz

When fork A is filled, then beat frequency decreases to 2 beats/s.

It is possible only when

$f_A = 335$ Hz

95. **(a)** $A_{res} = \sqrt{2}A + A$

$= (\sqrt{2}+1)A$ as $I \propto A^2$

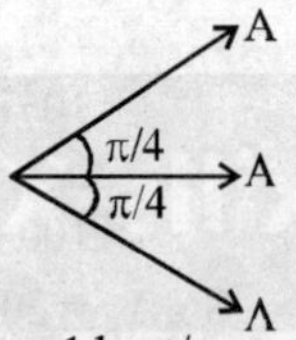

$I_{res} = (\sqrt{2}+1)^2 I_o = 5.8 I_o$

96. **(d)** According to question, tuning fork gives 1 beat/second with (N) 3rd normal mode. Therefore, organ pipe will have frequency (256 ± 1) Hz. In open organ pipe, frequency

$$n = \frac{NV}{2\ell} \Rightarrow 255 = \frac{3 \times 340}{2 \times \ell} \Rightarrow \ell = 2\,m = 200\text{ cm}$$

97. **(a)** Probable frequencies of tuning fork be $n \pm 5$

Frequency of sonometer wire, $n \propto \frac{1}{l}$

$$\therefore \quad \frac{n+5}{n-5} = \frac{100}{95} \Rightarrow 95(n+5) = 100(n-5)$$

or, $95n + 475 = 100n - 500$

or, $5n = 975$ or, $n = \frac{975}{5} = 195$ Hz

98. **(a)** Given, $y(x, t) = 0.5 \sin\left(\frac{5\pi}{4}x\right) \cos(200\pi t)$,

comparing with equation $- y(x, t) = 2a \sin kx \cos \omega t$

$$\omega = 200\pi, k = \frac{5\pi}{4}$$

speed of travelling wave $v = \frac{\omega}{k} = \frac{200\pi}{5\pi/4} = 160$ m/s

99. **(b)** Since the point $x = 0$ is a node and reflection is taking place from point $x = 0$. This means that reflection must be taking place from the fixed end and hence the reflected ray must suffer an additional phase change of π or a path change of $\frac{\lambda}{2}$.

So, if $y_{incident} = a\cos(kx - \omega t)$

$\Rightarrow y_{incident} = a\cos(-kx - \omega t + \pi) = -a\cos(\omega t + kx)$

Hence equation for the other wave

$$y = a\cos(kx + \omega t + \pi)$$

100. **(d)** In case of destructive interference

Phase difference $\phi = 180°$ or π

So wave pair (i) and (ii) will produce destructive interference.

Stationary or standing waves will be produce by equations (iii) & (iv) as two waves travelling along the same line but in opposite direction.

$n' = n + x$

101. **(d)** $y = A\sin(\omega t - kx) + A\sin(\omega t + kx)$

$y = 2A \sin \omega t \cos kx$

This is an equation of standing wave. For position of nodes $\cos kx = 0$

$$\Rightarrow \quad \frac{2\pi}{\lambda}.x = (2n+1)\frac{\pi}{2}$$

$$\Rightarrow x = \frac{(2n+1)\lambda}{4}, \; n = 0, 1, 2, 3, \ldots\ldots\ldots$$

102. **(a)** Intensity of a wave

$$I = \frac{1}{2} p\omega^2 A^2 v$$

Since, $I \propto A^2\omega^2 \quad \therefore I_1 \propto (2a)^2\omega^2$

and $I_2 \propto a^2(2\omega)^2$

$I_1 = I_2$

In the same medium, p and v are same.

Intensity depends on amplitude and frequency.

103. **(b)** Maximum number of beats

= Maximum frequency – Minimum frequency

$= (\nu + 1) - (\nu - 1) = 2$ Beats per second

104. **(d)** Frequency of fork 1, no = 200 Hz

No. of beats heard when fork 2 is sounded with fork 1 = $\Delta n = 4$

Now on loading (attaching tape) on unknown fork, the mass of tuning fork increases, So the beat frequency increases (from 4 to 6 in this case) then the frequency of the unknown fork 2 is given by,

$n = n_0 - \Delta n = 200 - 4 = 196$ Hz

105. **(c)** It is given that tuning fork of frequency 256 Hz makes 5 beats/second with the vibrating string of a piano. Therefore, possible frequency of the piano are (256 ± 5) Hz. i.e., either 261Hz or 251 Hz. When the tension in the piano string increases, its frequency will increases. As the original frequency was 261Hz, the beat frequency should decreases, we can conclude that the frequency of piano string is 251Hz

106. **(b)** Frequency of unknown fork = known frequency ± Beat frequency = 288 + 4 cps or 288 – 4 cps i.e. 292 cps or 284 cps. When a little wax is placed on the unknown fork, it produces 2 beats/sec. When a little wax is placed on the unknown fork, its frequency decreases and simultaneously the beat frequency decreases confirming that the frequency of the unknown fork is 292 cps.

Note : Had the frequency of unknown fork been 284 cps, then on placing wax its frequency would have decreased thereby increasing the gap between its frequency and the frequency of known fork. This would produce high beat frequency.

Experimental Skills-XI

15

1. (3) Since least count of screw gauge

$$\text{Least count} = \frac{\text{pitch length}}{\text{No. of division on circular scale}}$$

$$= \frac{0.75}{15} = 0.05\text{mm}$$

Given L = 5 mm & W = 2.5 mm

∴ We know that A = L.W

Fractional error is given by

$$\frac{dA}{A} = \frac{dL}{L} + \frac{dW}{W}$$

$$\Rightarrow \frac{dA}{A} = \frac{0.05}{5} + \frac{0.05}{2.5} \quad (\text{Here, } dL = dW = 0.05\text{ mm})$$

$$\Rightarrow \frac{dA}{A} = \frac{1}{100} + \frac{2}{100} = \frac{3}{100} = \frac{x}{100}(\text{given})$$

So, x = 3

2. (35) $(LC)_I = \frac{\text{Pitch}}{\text{No. of circular turn}} = \frac{P}{N} = 0.01\text{mm}$

$$(LC)_F = \frac{P(1+0.75)}{N(1-0.5)} = \frac{P}{N}\left[\frac{1.75}{0.5}\right] = 3.5(LC)_I$$

$= (0.01)\, 3.5 = 0.035\text{ mm} = 35 \times 10^{-3}\text{ mm}$

3. (a) In general one vernier scale division is smaller than one main scale division but it is not correct in all the cases.

Also, $LC = \frac{\text{One main scale division}}{\text{Number of vernier scale division}}$

4. (b) Given 1 Main scale division = m

n MSD = (n + 1) VSD

$$\Rightarrow 1\text{ VSD} = \frac{n}{n+1}\text{MSD}$$

Least count of vernier caliper = 1 MSD – 1 VSD

$$\Rightarrow \text{L.C} = m - m\left(\frac{n}{n+1}\right) = \left(1 - \frac{n}{n+1}\right)m$$

$$= m\left(\frac{n+1-n}{n+1}\right) = \left(\frac{1}{n+1}\right)m \Rightarrow \text{L.C} = \left(\frac{m}{n+1}\right)$$

5. (d) Given,

Mass of sphere, m = 8.635 g

Given 9MSD = 10VSD

Least count of vernier caliper = 1 MSD – 1 VSD

$$\therefore \text{LC} = 1\text{ MSD} - \frac{9}{10}\text{MSD} = \frac{1}{10}\text{MSD}$$

$$\Rightarrow \text{LC} = \frac{1}{10}\text{mm} = 0.01\text{ cm}$$

Diameter Reading = MSR + LC × VSR

= 2 cm + 2(0.01) = 2.02 cm

$$\therefore \text{Radius, } r = \frac{2.02}{2} = 1.01$$

$$\text{Volume of sphere, } V = \frac{4}{3}\pi(r)^3 = \frac{4}{3}\pi(1.01)^3 = 4.32\text{ cm}^3$$

$$\text{Density of sphere} = \frac{m}{V} = \frac{8.635}{4.32} = 1.998 \sim 2.00\text{ g/cm}^3$$

6. (b) Least of screw guage,

$$\text{L.C.} = \frac{1}{100}\text{mm} = 0.01\text{ mm}$$

zero error of scale is 0.05 mm

Diameter, D = 4 × 1 mm + 60 × 0.01 mm – 0.05 mm = 4.55 mm

7. (b) Least count is given as L.C. = 1 MSR – 1 VSR

Let main scale M division coincide with N vernier scale division

$$1\text{VSR} = \frac{M \times 1\text{mm}}{N} \Rightarrow \frac{1}{20N}\text{cm} = 1\text{mm} - \frac{M \times 1\text{mm}}{N}$$

$$\frac{1}{2N}\text{mm} = 1\text{mm} - \frac{M}{N}\text{mm}$$

$$M = \left(1 - \frac{1}{2N}\right)N \Rightarrow M = \frac{2N-1}{2}$$

8. (b) Least count is given by

$$\text{LC} = \frac{\text{pitch}}{\text{CSR}} = \left(\frac{1}{100}\right) = 0.01\text{mm}$$

Diameter = MSD + LC × CSR

⇒ 1 + (0.01) × 42 mm = 1.42 mm

On compairing with $\frac{x}{50}$ mm

∴ x = 71

9. (a) Least count is given as,

$$\text{LC} = 1\text{MSR} - 1\text{VSR} = \frac{1\text{ cm}}{20} - \frac{49}{50}(\text{MSR})$$

$$= 0.05\text{ cm} - \frac{49}{50} \times 0.05 = 0.001\text{ cm}$$

For mark on paper thickness is t_1 = MSR + VSR × LC

= 8.45 cm + 26 × 0.001 cm

= 84.76 mm

For mark on paper thickness is t_2 = MSR + VSR × LC

t_2 = 7.12 cm + 41 × 0.001 cm = 71.61 mm

For powder particle, T = 4.05 cm + 1 × 0.001 cm = 40.51 mm

$\ell_1 = 84.76 - 40.51 = 44.25$ mm

$\ell_2 = 71.61 - 40.51 = 31.10$ mm

Refractive index is given as,

$$\mu=\frac{\ell_1}{\ell_2}=\frac{44.25}{31.10}=1.42$$

10. **(c)** Least count is given by,

$$LC=1\,MSR-1\,VSR=1\,MSR-\left(\frac{49}{50}\right)MSR$$

$$0.01=\frac{MSR}{50}\Rightarrow MSR=0.5\,mm$$

0.05 cm have 1 division thus 1 cm have 20 divisions.

11. **(a)** $\langle T\rangle=\frac{\Sigma T}{n}=\frac{4.62+4.632+4.6+4.64}{4}=\frac{18.492}{4}$

$=4.623s$

12. **(c)** 20 VSD = 19 MSD

$$\Rightarrow 1VSD=\frac{19}{20}MSD$$

Now, L.C. = 1 MSD – 1 VSD

$$\Rightarrow 0.1mm=1MSD-\frac{19}{20}MSD$$

$$\therefore 1MSD=\frac{0.1\times 20}{20-19}=2mm$$

13. **(d)** Least count is given as,

LC = 1MSR – 1VSR

Given, 10 MSR = 11 VSR

$$\therefore LC=1MSR-\frac{10}{11}MSR=\frac{1MSR}{11}=\frac{5}{11}\text{units}$$

14. **(d)** Here, 49 MSD = 50 VSD

$$1\,VSD=\frac{49}{50}MSD$$

$$LC=1\,MSD-1\,VSD=\left(1-\frac{49}{50}\right)MSD$$

The vernier constant of travelling microscope

$$=\frac{1}{50}\times 0.5=0.01\,mm$$

15. **(b)** Spherometer is used to measure diameter of sphere or curvature of surface.

16. **(b)** In vernier calliper, reading taken > actual reading, if +(ve) zero error exists.

17. **(a)** Given

Least count of vernier callipers LC = 0.1 mm

$\therefore$ Zero Error = 6 × LC = 0.6 mm

Positive zero error means error will have to subtracted

Diameter of spherical bob = MSR + VSR × LC – Zero Error

= [32mm + 0.1 × 4mm] – 0.6 mm = 31.8 mm = 3.18 cm

18. **(22)** Least count $=\frac{\text{Pitch}}{\text{No. of circular divisions}}=\frac{0.5\,mm}{100}$

Least count $=5\times10^{-3}$ mm

Positive Error = MSR + CSR (LC) = 0 mm + 6 (5×10^{-3} mm)

Reading of Diameter = MSR + CSR (LC) – Positive zero error

$=4\times0.5\,mm+(46(5\times10^{-3}))-6(5\times10^{-3})\,mm$

$=2\,mm+40\times5\times10^{-3}\,mm=2.2\,mm=22\times10^{-2}\,cm$

19. **(41)** We have $u=\frac{h}{h'}=\frac{5.25}{5.00}$

$$\text{Least count}=\frac{1}{20}cm-\frac{49}{50}\cdot\frac{1}{20}cm=\frac{1}{20}\left(1-\frac{49}{50}\right)cm$$

$$=\frac{1}{20}\times\frac{1}{50}cm=0.01\,mm$$

As $u=\frac{h}{h'}\Rightarrow \ln u=\ln h-\ln h$

$$\Rightarrow\quad \frac{du}{u}=\frac{dh}{h}-\frac{dh'}{h'}\Rightarrow du=\left[\frac{dh}{h}-\frac{dh'}{h}\right]u$$

$$\Rightarrow\quad du=\left[\frac{0.01}{5.25}+\frac{0.01}{5.00}\right]\frac{5.25}{5.00}$$

$$=0.01\left[\frac{5+5.25}{5.25\times5}\right]\times\frac{5.25}{5}=\frac{0.1025}{25}=4.1\times10^{-3}$$

$$=\frac{41}{10}\times10^{-3}\quad\text{So, }x=41$$

20. **(c)** $1\,MSD=\frac{1}{20}cm$

$$1\,VSD=\frac{24}{25}MSD=\frac{24}{25}\times\frac{1}{20}cm$$

$$\therefore\ \text{Least count}=\frac{1}{20}\left(1-\frac{24}{25}\right)cm$$

[$\because$ Least count = 1MSD – 1VSD]

$$=\frac{1}{20}\times\frac{1}{25}=\frac{1}{500}cm=0.002\,cm$$

21. **(c)** MSR = 2.5 mm

$$CSR=45\times\frac{0.5}{50}mm=0.45\,mm$$

Diameter reading = Reading of screw gauge

= MSR + CSR – zero error

= 2.5 + 0.45 – (–0.03) = 2.98 mm

22. **(412)** Since 10 VSD = 9 MSD

$$\Rightarrow\ 1\,VSD=\frac{9}{10}MSD$$

So, least count = 1 MSD – 1 VSD

$$=1\,MSD-\frac{9}{10}MSD=\frac{1}{10}MSD=0.1\,mm=0.01\,cm$$

Error = 4 × L.C = 4 × 0.01 = 0.04 cm

So, Diameter = MSR + CSR × L.C – error

= 4.1 + 0.01 × 6 – 0.04 = 4.12 cm

$=412\times10^{-2}$ cm

23. **(b)** $L.C. = \frac{P}{N} = \frac{0.5 \text{ mm}}{50} = 0.01 \text{ mm}$

Length of wire = 6.8 cm

Diameter of wire = 1.5 mm + 7 × L.C.

$= 1.5 \text{ mm} + 7 \times 0.01 = 1.57 \text{ mm}$

Curved surface area = $\pi D l$

$= 3.14 \times 6.8 \times 1.57 \times 10^{-1} \text{ cm}^2 = 3.4 \text{ cm}^2$

24. **(c)** Given that 1 M.S.D = 1 mm

9 M.S.D = 10 V.S.D

1 V.S.D = 0.9 M.S.D = 0.9 mm

L.C of vernier caliper = 1 – 0.9 = 0.1 mm = 0.01 cm zero error

$= -(10-4) \times 0.1 \text{ mm} = -0.6 \text{ mm}$

The diameter of the spherical body = M.S.R + V.S.R – Zero error

$= 3 \text{ cm} + 6 \times 0.01 - [-0.06] = 3 + 0.06 + 0.06 = 3.12 \text{ cm}$

25. **(a)** n MSD = (n + 1) VSD

$1 \text{ VSD} = \left(\frac{n}{n+1}\right) \text{MSD}$

$= \left(\frac{n}{n+1}\right) \times \frac{1 \text{cm}}{5} = \frac{0.2n}{n+1} \text{cm} = \frac{2n}{n+1} \text{mm}$

26. **(180)** Measured value of diameter

$= MSR + V.S.R \times LC$ – zero error

$= 1.7 + 5 \times 0.01 - (-0.05)$

$= 1.8 \text{ cm} = 180 \times 10^{-2} \text{ cm}$

27. **(150)** We have

$\bar{x} = \frac{1.22 + 1.23 + 1.19 + 1.20}{4} \Rightarrow \bar{x} = 1.21$

So, $|\bar{x} - x_1| = 0.01$

$|\bar{x} - x_2| = 0.02$

$|\bar{x} - x_3| = 0.02$

$|\bar{x} - x_4| = 0.01$

Then, $\Delta x = \frac{0.01 + 0.02 + 0.02 + 0.01}{4} = \frac{0.06}{4} = 0.015$

So, % error $= \frac{0.015}{1.21} \times 100 = \frac{150}{121}$

28. **(5)** $1 \text{ MSD} = \frac{1}{20} \text{ cm}$

Now, 10 VSD = 9 MSD

$1 \text{ VSD} = \frac{9}{10} \text{ MSD}$

So, vernier constant = 1 MSD – 1 VSD

$= \frac{1}{10} \text{MSD} = \frac{1}{10} \times \frac{1}{20} \text{ cm} = \frac{1}{200} \text{ cm}$

$= \frac{1}{20} \text{ mm} = 5 \times 10^{-2} \text{ mm}$

29. **(5)** We have

49 MSD = 50 VSD

$1 \text{ VSD} = \frac{49}{50} \text{MSD}$

So, L.C = 1 MSD – 1 VSD = 1 MSD – $\frac{49}{50}$ MSD

$= \frac{1}{50} \text{ MSD} = \frac{1}{50} \times \frac{1}{40} \text{ cm} = \frac{1}{2000} \text{ cm}$

$= \frac{10000}{2000 \times 10^4} = 5 \times 10^{-4} \text{ cm} = 5 \times 10^{-6} \text{ m}$

30. **(13)** For screw gauge, (A)

Reading = MSR + CSR + error

or, 0.322 = 0.300 + CSR + 5 × LC

0.322 = 0.300 + CSR + 0.005

∴ CSR = 0.017

For screw gauge B

$LC = \frac{\text{Pitch}}{\text{no. of parts on circular scale}}$

Reading = MSR + CSR + error

0.322 = 0.200 + CSR + 0.092

CSR = 0.030

Difference = 0.030 – 0.017 = 0.013 cm

Absolute value of difference,

$= \frac{0.013}{0.001} = 13$

31. **(52)** 9 MSD = 10 VSD ⇒ 1 VSD = $\frac{9}{10}$ MSD

L.C. = 1 MSD – 1 VSD

$= 1 \text{ MSD} - \frac{9}{10} \text{MSD} = \frac{1}{10} \text{MSD} = \frac{1}{10} \times 1 \text{ mm} = 0.1 \text{ mm}$

diameter = M.S.R + L.C × V.S.R – Error

= 10 + 0.8 – 0.4 = 10.4 mm

So, radius = 5.2 mm = 5.2×10^{-1} cm

$= 52 \times 10^{-2}$ cm.

32. **(c)** Least count (L.C)

$= \frac{\text{Pitch}}{\text{no. of division on circular scale}} = \frac{0.5}{50} = .01 \text{ mm}$

Zero error = 5 × .01 mm

True reading = 5 + .01 mm × 20 – .01mm × 5 = 5.15 mm

33. **(a)** Least count $= \frac{\text{Pitch}}{\text{total division on circular scale}}$

For 5 revolution, distance travel 5mm,

For one revolution, distance travel 1mm.

∴ LC = 1mm/50 = 0.02 mm

Assertion is wrong. Reason is right

34. **(a)** Correct measured value

= MSR + (VSD × LC) – Zero error

$= 8.5 + (6 \times 0.01) - 0.2 \times 10^{-1} = 8.54 \text{ cm}$

35. (d) One main scale division, 1 MSD = a cm

One vernier scale division,

$$1\text{ VSD} = \frac{(n-1)\times(a\text{ cm})}{n}$$

$\therefore$ Least count = 1 MSD – 1 VSD

$$= a - \frac{(n-1)a}{n} = \frac{a(n-n+1)}{n}$$

$$= \frac{a}{n}\text{ cm} = \frac{10a}{n}\text{ mm} \quad (\because 1\text{ cm} = 10\text{ mm})$$

36. (b) Given : No. of division on circular scale of screw gauge = 50

Pitch = 0.5 mm

Least count of screw gauge

$$= \frac{\text{Pitch}}{\text{No. of division on circular scale}} = \frac{0.5}{50}\text{mm} = 10\ \mu\text{m}$$

And nature of zero error is positive.

37. (d) Average diameter, $d_{av} = 5.5375$ mm

Deviation of data, $\Delta d = 0.07395$ mm

As the measured data have two digits after decimal, therefore answer should be in two digits after decimal.

$\therefore d = (5.54 \pm 0.07)$ mm

38. (b) Thickness = M.S. Reading + Circular Scale Reading (L.C.)

$$\text{Here LC} = \frac{\text{Pitch}}{\text{Circular scale division}} = \frac{0.1}{50} = 0.002\text{ cm per division}$$

So, correct measurement is measurement of integral multiple of L.C.

39. (c) L.C. of vernier callipers = 1 MSD – 1 VSD

$$= \left(1 - \frac{9}{10}\right)\times 1 = 0.1\text{ mm} = 0.01\text{ cm}$$

Here 7th division of vernier scale coincides with a division of main scale and the zero of vernier scale is lying right side of the zero of main scale.

Zero error = $7 \times 0.1 = 0.7$ mm = 0.07 cm.

Length of the cylinder = measured value – zero error

$= (3.1 + 4 \times 0.01) - 0.07 = 3.07$ cm.

40. (a) When screw on a screw-gauge is given six rotations, it moves by 3mm on the main scale

$$\therefore \quad \text{Pitch} = \frac{3}{6} = 0.5\text{ mm}$$

$$\therefore \quad \text{Least count L.C.} = \frac{\text{Pitch}}{CSD} = \frac{0.5\text{ mm}}{50}$$

$$= \frac{1}{100}\text{ mm} = 0.01\,mm = 0.001\text{ cm}$$

41. (b) Least count of main scale of M.S.D. is equal to pitch. So pitch = 1 mm

Now, least count of screw gauge (L.C)

$$= \frac{\text{Pitch}}{\text{Number of division on circular scale}}$$

$$\Rightarrow \text{ no. of div } (n) = \frac{\text{Pitch}}{\text{L.C}} \Rightarrow n = \frac{1\text{ mm}}{5\mu m} \Rightarrow n = 200$$

42. (c) Least count of screw gauge,

$$\text{LC} = \frac{\text{Pitch}}{\text{No. of division}} = \frac{0.5}{100}\text{ mm} = 0.5 \times 10^{-2}\text{ mm}$$

+ve error = $3 \times 0.5 \times 10^{-2}$ mm = 1.5×10^{-2} mm = 0.015 mm

Reading = MSR + CSR – (+ve error)

$= 5.5\text{ mm} + (48 \times 0.5 \times 10^{-2}) - 0.015$

$= 5.5 + 0.24 - 0.015 = 5.725$ mm

43. (d) $\text{Least count} = \dfrac{\text{Value of 1 part on main scale}}{\text{Number of parts on vernier scale}}$

$$= \frac{0.25}{5\times 100}\text{ cm} = 5 \times 10^{-4}\text{ cm}$$

Reading = 4×0.05 cm + $30 \times 5 \times 10^{-4}$ cm

= (0.2 + 0.0150) cm = 0.2150 cm (Thickness of wire)

44. (d) $\text{L.C.} = \dfrac{0.5}{50} = 0.01$ mm

Zero error = (50 – 45) = 5 × 0.01 = 0.05 mm (Negative)

Reading = (0.5 + 25 × 0.01) + 0.05 = 0.80 mm

45. (b) $\text{Least count} = \dfrac{0.1}{10} = 0.01$ cm

$d_1 = 0.5 + 8 \times 0.01 + 0.03 = 0.61$ cm

$d_2 = 0.5 + 4 \times 0.01 + 0.03 = 0.57$ cm

$d_3 = 0.5 + 6 \times 0.01 + 0.03 = 0.59$ cm

$$\text{Mean diameter} = \frac{0.61 + 0.57 + 0.59}{3} = 0.59\text{ cm}$$

46. (b) Measured length of rod = 3.50 cm

For Vernier Scale with 1 Main Scale Division = 1 mm

9 Main Scale Division = 10 Vernier Scale Division,

Least count = 1 MSD – 1 VSD = 0.1 mm

47. (d) No. of divisions on main scale = N

No. of divisions on vernier scale = $N + 1$

size of main scale division = a

Let size of vernier scale division be b

then we have

$$aN = b\,(N+1) \Rightarrow b = \frac{aN}{N+1}$$

$$\text{Least count is a} - \text{b} = \text{a} - \frac{aN}{N+1}$$

$$= a\left[\frac{N+1-N}{N+1}\right] = \frac{a}{N+1}$$

48. **(d)** The least count (L.C.) of a screw guage is the smallest length which can be measured accurately with it.

As least count is 0.001 cm $= \frac{1}{1000}$ cm

Hence measured value should be recorded upto 3 decimal places i.e., 5.320 cm

49. **(c)** $\because$ Reading of Vernier = Main scale reading + Vernier scale reading × least count.

Main scale reading = 58.5

Vernier scale reading = 09 division

least count of Vernier = 0.5°/30

Thus, $R = 58.5° + 9 \times \frac{0.5°}{30}$

R = 58.65°

50. **(a)** Least count, L.C. $= \frac{1}{100}$ mm

Diameter of wire = *MSR* + *CSR* × *L.C.*

$\because$ 1 mm = 0.1 cm

$= 0 + \frac{1}{100} \times 52 = 0.52\ \text{mm} = 0.052\ \text{cm}$

51. **(d)** Least count of screw gauge = 0.01 mm

$$\left[\because LC = \frac{\text{Pitch}}{\text{no. of division}}\right]$$

Reading = [M.S.R. + C.S.R. × L.C.] – (zero error)
= [3 + 35 × 0.01] – (–0.03) = 3.38 mm

52. **(b)** The expression for the terminal velocity of ball of radius R inside a liquid.

$$V_T = \frac{2}{9} R^2 \frac{g}{\eta}(\sigma - \rho)$$

V_T is directly proportional to R^2, meaning the graph of V_T vs R is a parabola.

53. **(2)** Slope $= \frac{\text{Extension/Load}}{\text{Length of wire}} = \frac{\Delta l / w}{wL} \frac{\Delta l}{wL}$

Young's modulus, $Y = \frac{mg/A}{\Delta\ell/L} = \frac{wL}{\Delta\ell A}$

$\therefore\quad Y = \frac{1}{(\text{slope})A}$

$\Rightarrow\ Y = \frac{1}{2 \times 10^{-6}(0.25 \times 10^{-5})}$

$\Rightarrow\ Y = 2 \times 10^{11}\ \text{N/m}^2$

54. **(2)** Given,
Length of wire, L = 1 m
Extension in the length of wire, $l = 0.4\ \text{mm} = 0.4 \times 10^{-3}$ m
Diameter of wire, d = 0.4 mm
Uncertainty in extension of the length, $\Delta l = \pm 0.02$ mm
Uncertainty in diameter of wire, $\Delta d = \pm 0.01$ mm

Young's modulus, $Y = \frac{\text{stress}}{\text{strain}} = \frac{\frac{mg}{A}}{\frac{\Delta l}{L}}$

$$= \frac{mg \times L}{A \times \Delta l} = \frac{mgL}{\pi\left(\frac{d}{2}\right)^2 \Delta l} = \frac{4mgL}{\pi(d)^2 \Delta l}$$

$$= \frac{4 \times 1 \times 10 \times 1}{\pi(0.4 \times 10^{-3})^2(0.4 \times 10^{-3})} = 0.199 \times 10^{+12}\ \text{N/m}^2$$

$$\frac{\Delta Y}{Y} = 2\frac{\Delta d}{d} + \frac{\Delta l}{l}$$

$$\Delta Y = \left(2 \times \frac{0.01}{0.4} + \frac{0.02}{0.4}\right) \times 0.199 \times 10^{+12}$$

$$= (0.05 + 0.05) \times 0.199 \times 10^{+12} = 1.99 \times 10^{10}\ \text{Nm}^{-2}$$

55. **(104)** For Ist resonance

$l_1 + e = \frac{\lambda}{4}$

$20 + e = \frac{336}{400} \times 100\text{cm} \times \frac{1}{4}$

$20 + e = \frac{84}{4}$

$e = 21 - 20 = 1$ cm

For IIIrd resonance

$l_2 + e = \frac{5\lambda}{4} \Rightarrow l_2 + 1 = 5 \times 21 \Rightarrow l_2 = 105 - 1 = 104$ cm

56. **(6)**

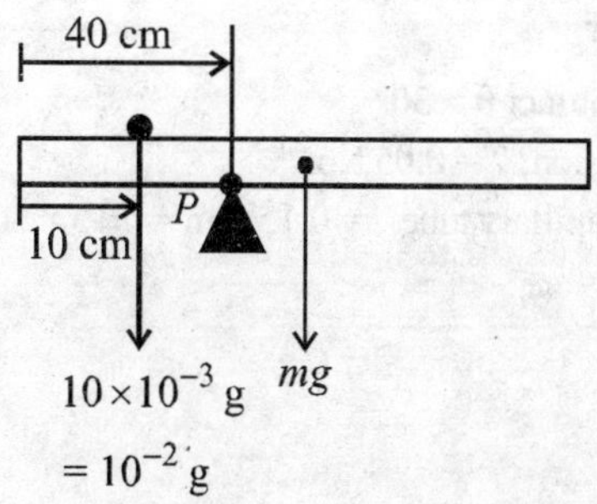

Let mass of scale be '*m*'. Then,

About '*P*' $\hat{\tau}_{net} = 0$

$\Rightarrow 2 \times 10^{-2}\,g \times 30 - mg \times 10 = 0$

$\Rightarrow 6 \times 10^{-1} - 10\,m = 0 \Rightarrow 0.6 = 10\,m$

$\Rightarrow m = \frac{0.6}{10} \Rightarrow m = 0.06\ \text{kg} = 6 \times 10^{-2}\ \text{kg}$

57. **(c)** Let '*l*' be the length of water level when the first resonance occurs

$\lambda = \frac{v}{\nu} = \frac{336}{504}\text{m} = \frac{2}{3}\text{m}$

$\therefore \frac{\lambda}{4} = l + e = l + 0.3d \quad (d = \text{diameter of column})$

$\Rightarrow \frac{\frac{2}{3} \times 100}{4} = l + .3 \times 6 \Rightarrow 16.66 = l + 1.8$

$\therefore l = 14.8$ cm

58. **(a)** Using, $v^2 - u^2 = 2gh \Rightarrow v^2 - 0^2 = 2gh \Rightarrow v = \sqrt{2gh}$

Terminal velocity, $V_T = \frac{2}{9}\frac{r^2(\rho - \sigma)g}{\eta}$

After falling through h the velocity should be equal to terminal velocity

$$\therefore \sqrt{2gh} = \frac{2}{9}\frac{r^2(\rho-\sigma)g}{\eta} \Rightarrow 2gh = \frac{4}{81}\frac{r^4 g^2(\rho-\sigma)^2}{\eta^2}$$

$$\Rightarrow h = \frac{2r^4 g(\rho-\sigma)^2}{81\eta^2} \Rightarrow h \propto r^4$$

59. **(a)** Here, l_1 = 17 cm and l_2 = 24.5 cm, V = 330 m/s, f = ?

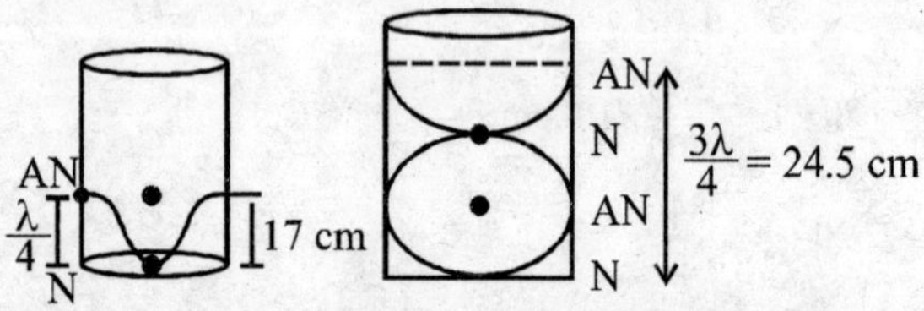

$\lambda = 2(l_2 - l_1) = 2 \times (24.5 - 17) = 15$ cm

Now, from $v = f\lambda \Rightarrow 330 = f \times 15 \times 10^{-2}$

$\therefore f = \frac{330}{15} \times 100 = \frac{1100 \times 100}{5} = 2200$ Hz

60. **(b)** Given,

Angle of contact $\theta = 30°$

Surface tension, $T = 0.05$ Nm^{-1}

Radius of capillary tube, $r = 0.15$ mm $= 0.15 \times 10^{-3}$m

Density of methylene iodide, $\rho = 667$ kg m^{-3}

Capillary rise, $h = \frac{2T\cos\theta}{\rho g r}$

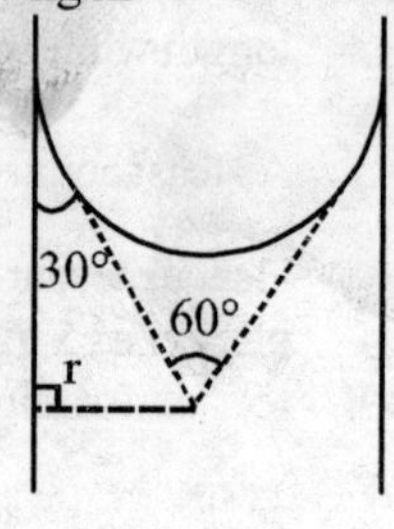

$$= \frac{2 \times 0.05 \times \frac{\sqrt{3}}{2}}{667 \times 10 \times 0.15 \times 10^{-3}} = 0.087 \text{ m}$$

61. **(b)** $v = f\lambda = f \times 2(\ell_2 - \ell_1)$

$= 480 \times 2(0.70 - 0.30)$

$= 384$ m/s

62. **(d)** Using $\lambda = \frac{v}{4f}$, $\lambda_1 = 11$cm. So $\frac{v}{512 \times 4} = 11$cm ...(i)

$\lambda_2 = 27$cm. So, $\frac{v}{256 \times 4} = 27$cm ...(ii)

Substracting (ii) from (i), we get $\frac{v}{256 \times 4} \times 0.5 = 0.16$m

$v = 328$ m/s

63. **(b)** Fundamental frequency for first resonant length

$\nu = \frac{v}{4\ell_1} = \frac{v}{4 \times 18}$ (in winter)

Fundamental frequency for second resonant length

$\nu' = \frac{3v'}{4\ell_2} = \frac{3v'}{4x}$ (in summer)

According to questions,

$\therefore \frac{v}{4 \times 18} = \frac{3v'}{4 \times x} \Rightarrow x = 3 \times 18 \times \frac{v'}{v}$

$\therefore x = 54 \times \frac{v'}{v}$cm

$v' > v$ because velocity of light is greater in summer as compared to winter $(v \propto \sqrt{T})$

$\therefore x > 54$cm

CLASS XII

1 Electric Charges and Fields

TREND ANALYSIS
JEE MAIN (2025-2021)

	2025				2024				2023				2022				2021			
	Session II		Session I		Session II		Session I		Session II		Session I		Session II		Session I		Session II		Session I	
	MCQs	NVQs	MCQs	NVQs	MCQs	NVQs	MCQs	NVQs	MCQs	NVQs	MCQs	NVQs	MCQs	NVQs	MCQs	NVQs	MCQs	NVQs	MCQs	NVQs
Topic 1	2	–	1	1	1	–	1	4	1	2	1	2	3	–	4	–	3	–	–	2
Topic 2	7	–	2	–	–	1	1	–	–	1	3	2	2	–	4	–	2	1	2	–
Topic 3	2	1	6	1	4	5	4	2	4	–	1	3	–	4	4	–	3	–	–	3
Topic 4																				
Total	11	1	9	2	5	6	6	6	5	3	5	7	5	4	12	–	8	1	2	5

How to Code Qns for Better & Faster Revision

Type 1: Able to solve (No code Req)

Type 2: Solved with struggle or by seeing solution **(Use code NR means Needs Revision or orange stickers)**

Type 3: Tricky/ Could not solve on my own. Need help from Teacher/ Friend. **(Use code TQ means Tricky Question or red stickers)**

You can either hand write the code in boxes or else paste coloured stickers given at end of book.

Topic 1 - Electric Charges and Coulomb's Law

1. Electric charge is transferred to an irregular metallic disk as shown in figure. If σ_1, σ_2, σ_3 and σ_4 are charge densities at given points then, choose the correct answer from the options given below:

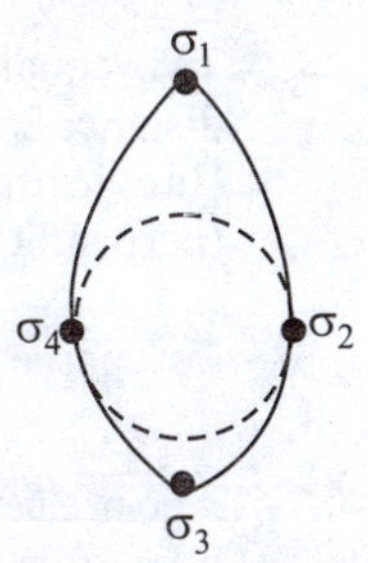

(A) $\sigma_1 > \sigma_3;\ \sigma_2 = \sigma_4$

(B) $\sigma_1 > \sigma_2;\ \sigma_3 > \sigma_4$

(C) $\sigma_1 > \sigma_3 > \sigma_2 = \sigma_4$

(D) $\sigma_1 < \sigma_3 < \sigma_2 = \sigma_4$

(E) $\sigma_1 = \sigma_2 = \sigma_3 = \sigma_4$

[NCERT : P L-28 | April 8, 2025 (II)]

(a) A, B and C Only (b) A and C only
(c) D and E only (d) B and C only

2. A small bob of mass 100 mg and charge +10 μC is connected to an insulating string of length 1 m. It is brought near to an infinitely long non-conducting sheet of charge density 'σ' as shown in figure. If string subtends an angle of 45° with the sheet at equilibrium the charge density of sheet will be :

(Given $\varepsilon_0 = 8.85 \times 10^{-12}\ \frac{F}{m}$ add acceleration due to gravity, $g = 10\ m/s^2$) **[NCERT : P L-7 | April 2, 2025 (I)]**

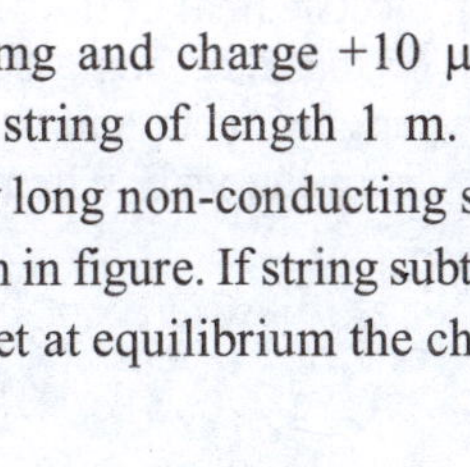

(a) 0.885 nC/m^2

(b) 17.7 nC/m^2

(c) 885 nC/m^2

(d) 1.77 nC/m^2

3. A small uncharged conducting sphere is placed in contact with an identical sphere but having 4×10^{-8} C charge and then removed to a distance such that the force of repulsion between them is 9×10^{-3} N. The distance between them is

(Take $\frac{1}{4\pi\varepsilon_0}$ as 9×10^9 in SI units)

[NCERT : P L-7 | Jan. 24, 2025 (II)]

(a) 2 cm (b) 3 cm (c) 4 cm (d) 1 cm

4. A positive ion A and a negative ion B has charges 6.67×10^{-19} C and 9.6×10^{-10} C, and masses 19.2×10^{-27} kg and 9×10^{-27} kg respectively. At an instant, the ions are separated by a certain distance r. At that instant the ratio of the magnitudes of electrostatic force to gravitational force is $P \times 10^{44}$, where the value of P is ____.

(Take $\frac{1}{4\pi\varepsilon_0} = 9 \times 10^9$ Nm^{2}C^{-1} and universal gravitational constant as 6.67×10^{-11} Nm^2 kg^{-2})

[NCERT : P L-7 | Jan. 23, 2025 (I)]

5. Two identical conducting spheres P and S with charge Q on each, repel each other with a force 16N. A third identical uncharged conducting sphere R is successively brought in contact with the two spheres. The new force of repulsion between P and S is :

[NCERT : P L-7 | April 6, 2024 (II)]

(a) 4N (b) 6N (c) 1N (d) 12N

6. Two identical charged spheres are suspended by strings of equal lengths. The strings make an angle θ with each other. When suspended in water the angle remains the same. If density of the material of the sphere is 1.5 g/cc, the dielectric constant of water will be ______ (Take density of water = 1 g/cc)

[NCERT : P L-7 | Feb. 1, 2024 (I)]

7. Suppose a uniformly charged wall provides a uniform electric field of 2×10^4 N/C normally. A charged particle of mass 2g being suspended through a silk thread of length 20 cm and remain stayed at a distance of 10 cm from the wall. Then the charge on the particle will be $\frac{1}{\sqrt{x}}\mu C$ where x = _________ . [use g = 10 m/s^2]

[NCERT : P L-14 | Feb. 1, 2024 (II)]

8. Force between two point charges q_1 and q_2 placed in vacuum at 'r' cm apart is F. Force between them when placed in a medium having dielectric K = 5 at 'r/5' cm apart will be: **[NCERT : P L-7| Jan. 31, 2024 (II)]**

(a) F/25 (b) 5F (c) F/5 (d) 25F

9. Two identical charged spheres are suspended by strings of equal lengths. The strings make an angle of 37° with each other. When suspended in a liquid of density 0.7 *g/cm*^{3}, the angle remains same. If density of material of the sphere is 1.4 *g/cm*^{3}, the dielectric constant of the liquid is _________ $\left(\tan 37° = \frac{3}{4}\right)$

[NCERT : P L-7 | Jan. 30, 2024 (II)]

10. A thin metallic wire having cross sectional area of 10^{-4} m^2 is used to make a ring of radius 30 cm. A positive charge of 2π C is uniformly distributed over the ring, while another positive charge of 30 pC is kept at the centre of the ring. The tension in the ring is ______ N ; provided that the ring does not get deformed (neglect the influence of gravity). (given, $\frac{1}{4\pi\epsilon_0} = 9 \times 10^9$ SI units)

[NCERT : P L-7 | Jan. 27, 2024 (I)]

11. A 10 μC charge is divided into two parts and placed at 1 cm distance so that the repulsive force between them is maximum. The charges of the two parts are :

[NCERT : P L-7 | April 13, 2023 (II)]

(a) 9 μC, 1 μC (b) 5 μC, 5 μC
(c) 7 μC, 3 μC (d) 8 μC, 2 μC

12. Three point charges q, –2q and 2q are placed on x-axis at a distance x = 0, $x = \frac{3}{4}R$ and x = R respectively from origin as shown. If $q = 2 \times 10^{-6}$ C and R = 2 cm, the magnitude of net force experienced by the charge –2q is ______ N.

[NCERT : P L-12 | April 13, 2023 (II)]

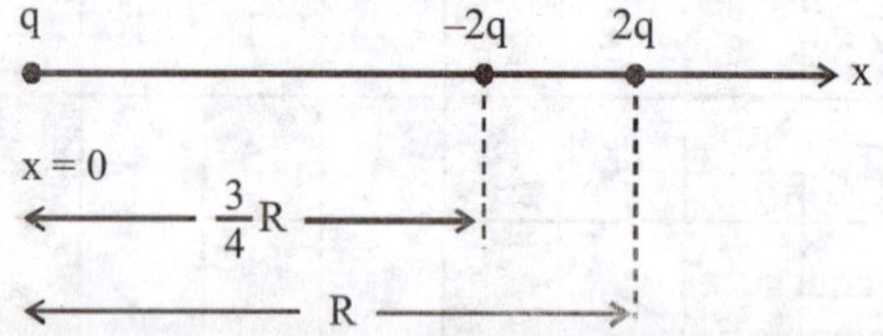

13. As shown in the figure, a configuration of two equal point charges ($q_0 = +2\mu$ C) is placed on an inclined plane. Mass of each point charge is 20 g. Assume that there is no friction between charge and plane. For the system of two point charges to be in equilibrium (at rest) the height $h = x \times 10^{-3}$ m. The value of x is ____.

[NCERT : P L-7 | April 11, 2023 (I)]

(Take $\frac{1}{4\pi\varepsilon_0} = 9 \times 10^9$ N m^{2}C^{-2}, g = 10 ms^{-1})

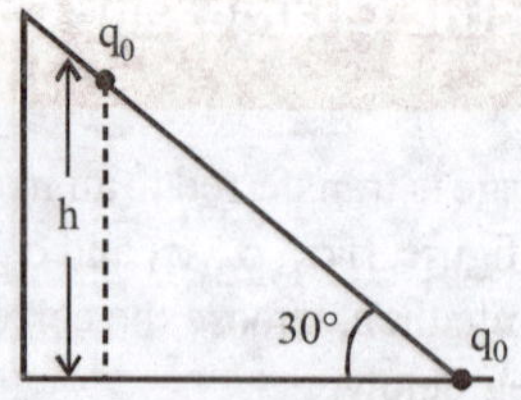

14. Two equal positive point charges are separated by a distance 2a. The distance of a point from the centre of the line joining two charges on the equatorial line (perpendicular bisector) at which force experienced by a test charge q_0 becomes maximum is $\frac{a}{\sqrt{x}}$. The value of x is ____.

[NCERT : P L-7 | Feb. 1, 2023 (I)]

15. A point charge $q_1 = 4q_0$ is placed at origin. Another point charge $q_1 = -q_0$ is placed at x = 12 cm. Charge of proton is q_0. The proton is placed on x-axis so that the electrostatic force on the proton is zero. In this situation, the position of the proton from the origin is _____ cm.

[NCERT : P L-7 | Jan. 29, 2023 (I)]

16. If two charges q_1 and q_2 are separated with distance 'd' and placed in a medium of dielectric constant K. What will be the equivalent distance between charges in air for the same electrostatic force ? **[Jan. 24, 2023 (I)]**

(a) $d\sqrt{K}$ (b) $K\sqrt{d}$ (c) $1.5d\sqrt{K}$ (d) $2d\sqrt{K}$

17. Two identical metallic spheres A and B when placed at certain distance in air repel each other with a force of F. Another identical uncharged sphere C is first placed in contact with A and then in contact with B and finally placed at midpoint between spheres A and B. The force experienced by sphere C will be :

[NCERT : P L-7 | July 29, 2022 (II)]

(a) 3F/2 (b) 3F/4 (c) F (d) 2F

18. Two identical positive charges Q each are fixed at a distance of '2a' apart from each other. Another point charge q_0 with mass 'm' is placed at midpoint between two fixed charges. For a small displacement along the line joining the fixed charges, the charge q_0 executes SHM. The time period of oscillation of charge q_0 will be :

[NCERT : P L-7 | July 27, 2022 (I)]

(a) $\sqrt{\dfrac{4\pi^3\varepsilon_0 ma^3}{q_0 Q}}$ (b) $\sqrt{\dfrac{q_0 Q}{4\pi^3\varepsilon_0 ma^3}}$

(c) $\sqrt{\dfrac{2\pi^2\varepsilon_0 ma^3}{q_0 Q}}$ (d) $\sqrt{\dfrac{8\pi^3\varepsilon_0 ma^3}{q_0 Q}}$

19. A charge of 4 μC is to be divided into two parts. The distance between the two divided charges is constant. The magnitude of the divided charges so that the force between them is maximum, will be:

[NCERT : P L-7 | July 27, 2022 (II)]

(a) 1 μC and 3 μC (b) 2 μC and 2 μC

(c) 0 and 4 μC (d) 1.5 μC and 2.5 μC

20. Two point charges Q each are placed at a distance d apart. A third point charge q is placed at a distance x from mid-point on the perpendicular bisector. The value of x at which charge q will experience the maximum Coulomb's force is :

[NCERT : P L-7 | June 29, 2022 (II)]

(a) $x = d$ (b) $x = \dfrac{d}{2}$ (c) $x = \dfrac{d}{\sqrt{2}}$ (d) $x = \dfrac{d}{2\sqrt{2}}$

21. Three identical charged balls each of charge 2C are suspended from a common point P by silk threads of 2m each (as shown in figure). They form an equilateral triangle of side 1m.

The ratio of net force on a charged ball to the force between any two charged balls will be :

[NCERT : P L-12 | June 27, 2022 (II)]

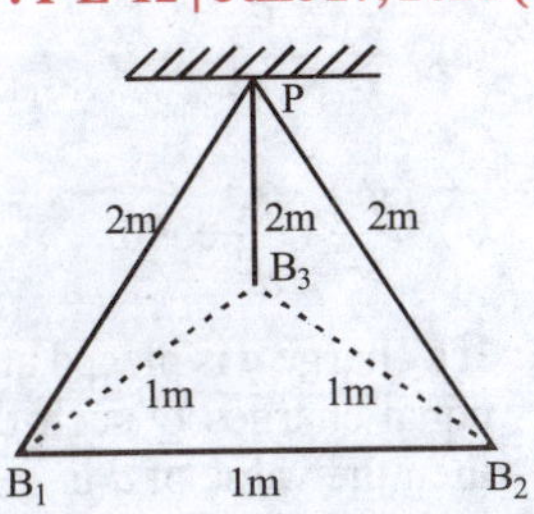

(a) 1 : 1

(b) 1 : 4

(c) $\sqrt{3} : 2$

(d) $\sqrt{3} : 1$

22. Sixty four conducting drops each of radius 0.02 m and each carrying a charge of 5 μC are combined to form a bigger drop. The ratio of surface density of bigger drop to the smaller drop will be :

[NCERT : P L-28 | June 26, 2022 (II)]

(a) 1 : 4 (b) 4 : 1 (c) 1 : 8 (d) 8 : 1

23. Two identical charged particles each having a mass 10 g and charge 2.0×10^{-7} C placed on a horizontal table with a separation of L between then such that they stay in limited equilibrium. If the coefficient of friction between each particle and the table is 0.25, find the value of L. [Use $g = 10\ ms^{-2}$] **[NCERT : P L-7 | June 24, 2022 (II)]**

(a) 12 cm (b) 10 cm (c) 8 cm (d) 5 cm

24. Two particles A and B having charges 20 μC and –5 μC respectively are held fixed with a separation of 5 cm. At what position a third charged particle should be placed so that it does not experience a net electric force?

[NCERT : P L-7 | Aug. 31, 2021 (I)]

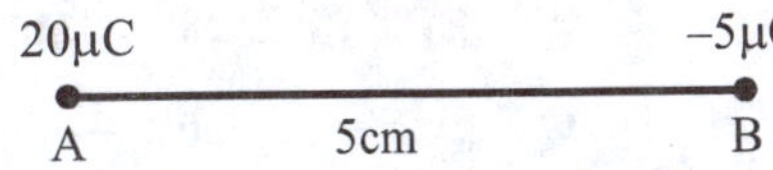

(a) At 5 cm from 20 μC on the left side of system

(b) At 5 cm from – 5 μC on the right side

(c) At 1.25 cm from – 5 μC between two charges

(d) At midpoint between two charges

25. Two identical tennis balls each having mass 'm' and charge 'q' are suspended from a fixed point by threads of length 'l'. What is the equilibrium separation when each thread makes a small angle 'θ' with the vertical ?

[NCERT : P L-7 | July 27, 2021 (I)]

(a) $d = \left(\dfrac{q^2 l}{2\pi\varepsilon_0 mg}\right)^{\frac{1}{2}}$ (b) $d = \left(\dfrac{q^2 l}{2\pi\varepsilon_0 mg}\right)^{\frac{1}{3}}$

(c) $d = \left(\dfrac{q^2 l^2}{2\pi\varepsilon_0 m^2 g}\right)^{\frac{1}{3}}$ (d) $d = \left(\dfrac{q^2 l^2}{2\pi\varepsilon_0 m^2 g^2}\right)^{\frac{1}{3}}$

26. A certain charge Q is divided into two parts q and (Q–q). How should the charges Q and q be divided so that q and (Q–q) placed at a certain distance apart experience maximum electrostatic repulsion ?

[NCERT : P L-7 | July 20, 2021 (I)]

(a) $Q = \dfrac{q}{2}$ (b) $Q = 2q$ (c) $Q = 4q$ (d) $Q = 3q$

27. An infinite number of point charges, each carrying 1 μC charge, are placed along the y-axis at y = 1 m, 2 m, 4 m, 8 m_______.

The total force on a 1 C point charge, placed at the origin, is $x \times 10^3$ N.

The value of x, to the nearest integer, is ________.

[Take $\dfrac{1}{4\pi\varepsilon_0} = 9 \times 10^9\ Nm^2/C^2$]

[NCERT : P L-7 | March 18, 2021 (II)]

28. Two identical conducting spheres with negligible volume have 2.1 nC and –0.1 nC charges, respectively. They are brought into contact and then separated by a distance of 0.5 m. The electrostatic force acting between the spheres is ________ $\times 10^{-9}$ N.

[Given : $4\pi\varepsilon_0 = \dfrac{1}{9\times10^9}$ SI unit]

[NCERT : P L-7 | Feb. 25, 2021 (II)]

29. Three charges + Q, q, + Q are placed respectively, at origin, d/2 and d from the origin, on the x-axis. If the net force experienced by + Q, placed at $x = 0$, is zero, then value of q is: **[NCERT : P L-12 | 9 Jan. 2019 (I)]**

(a) – Q/4 (b) + Q/2 (c) + Q/4 (d) – Q/2

30. Charge is distributed within a sphere of radius R with a volume charge density $p(r) = \dfrac{A}{r^2}e^{-2r/a}$ where A and a are constants. If Q is the total charge of this charge distribution, the radius R is:

[NCERT : P L-28 | 9 Jan. 2019, (II)]

(a) $a\log\left(1-\dfrac{Q}{2\pi aA}\right)$ (b) $\dfrac{a}{2}\log\left(\dfrac{1}{1-\dfrac{Q}{2\pi aA}}\right)$

(c) $a\log\left(\dfrac{1}{1-\dfrac{Q}{2\pi aA}}\right)$ (d) $\dfrac{a}{2}\log\left(1-\dfrac{Q}{2\pi aA}\right)$

31. Two identical conducting spheres A and B, carry equal charge. They are separated by a distance much larger than their diameter, and the force between them is F. A third identical conducting sphere, C, is uncharged. Sphere C is first touched to A, then to B, and then removed. As a result, the force between A and B would be equal to

[NCERT : P L-7 | Online April 16, 2018]

(a) $\dfrac{3F}{4}$ (b) $\dfrac{F}{2}$ (c) F (d) $\dfrac{3F}{8}$

32. Shown in the figure are two point charges +Q and –Q inside the cavity of a spherical shell. The charges are kept near the surface of the cavity on opposite sides of the centre of the shell.

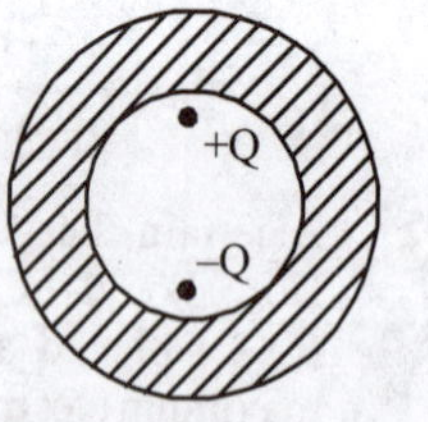

If σ_1 is the surface charge on the inner surface and Q_1 net charge on it and σ_2 the surface charge on the outer surface and Q_2 net charge on it then:

[NCERT : P L-16 | Online April 10, 2015]

(a) $\sigma_1 \neq 0, Q_1 = 0$; $\sigma_2 = 0, Q_2 = 0$
(b) $\sigma_1 \neq 0, Q_1 = 0$; $\sigma_2 \neq 0, Q_2 = 0$
(c) $\sigma_1 = 0, Q_1 = 0$; $\sigma_2 = 0, Q_2 = 0$
(d) $\sigma_1 \neq 0, Q_1 \neq 0$; $\sigma_2 \neq 0, Q_2 \neq 0$

33. Two charges, each equal to q, are kept at x = – a and x = a on the x-axis. A particle of mass m and charge $q_0 = \dfrac{q}{2}$ is placed at the origin. If charge q_0 is given a small displacement (y<<a) along the y-axis, the net force acting on the particle is proportional to

[NCERT : P L-12 | 2013]

(a) y (b) –y (c) $\dfrac{1}{y}$ (d) $-\dfrac{1}{y}$

34. Two identical charged spheres suspended from a common point by two massless strings of length l are initially a distance $d(d << l)$ apart because of their mutual repulsion. The charge begins to leak from both the spheres at a constant rate. As a result charges approach each other with a velocity v. Then as a function of distance x between them, **[NCERT : P L-7 | 2011]**

(a) $v \propto x^{-1}$ (b) $v \propto x^{1/2}$ (c) $v \propto x$ (d) $v \propto x^{-1/2}$

35. A charge Q is placed at each of the opposite corners of a square. A charge q is placed at each of the other two corners. If the net electrical force on Q is zero, then Q/q equals: **[NCERT : P L-12 | 2009]**

(a) –1 (b) 1 (c) $-\dfrac{1}{\sqrt{2}}$ (d) $-2\sqrt{2}$

36. If g_E and g_M are the accelerations due to gravity on the surfaces of the earth and the moon respectively and if Millikan's oil drop experiment could be performed on the two surfaces, one will find the ratio

[NCERT : P L-4 | 2007]

$\dfrac{\text{electronic charge on the moon}}{\text{electronic charge on the earth}}$ to be

(a) g_M / g_E (b) 1 (c) 0 (d) g_E / g_M

37. Two spherical conductors B and C having equal radii and carrying equal charges on them repel each other with a force F when kept apart at some distance. A third spherical conductor having same radius as that B but uncharged is brought in contact with B, then brought in contact with C and finally removed away from both. The new force of repulsion between B and C is **[NCERT : P L-7 | 2004]**

(a) $F/8$ (b) $3F/4$ (c) $F/4$ (d) $3F/8$

38. Three charges $-q_1$, $+q_2$ and $-q_3$ are place as shown in the figure. The x - component of the force on $-q_1$ is proportional to **[NCERT : P L-11 | 2003]**

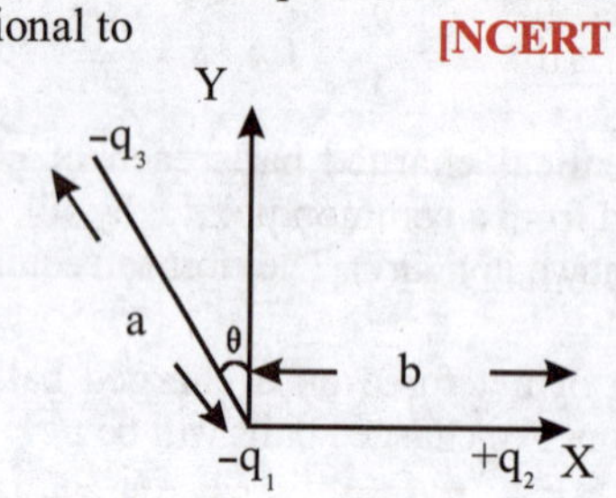

(a) $\dfrac{q_2}{b^2} - \dfrac{q_3}{a^2}\cos\theta$ (b) $\dfrac{q_2}{b^2} + \dfrac{q_3}{a^2}\sin\theta$

(c) $\dfrac{q_2}{b^2} + \dfrac{q_3}{a^2}\cos\theta$ (d) $\dfrac{q_2}{b^2} - \dfrac{q_3}{a^2}\sin\theta$

39. If a charge q is placed at the centre of the line joining two equal charges Q such that the system is in equilibrium then the value of q is **[NCERT : P L-7 | 2002]**

(a) $Q/2$ (b) $-Q/2$ (c) $Q/4$ (d) $-Q/4$

Topic 2 - Electric Field and Electric Field Lines

40. Given below are two statements : one is labelled as **Assertion (A)** and the other is labelled as **Reason (R)**.
Assertion (A) : The outer body of an air craft is made of metal which protects persons sitting inside from lightning-strikes.
Reason (R) : The electric field inside the cavity enclosed by a conductor is zero.
In the light of the above statements, choose the **most appropriate answer** from the options given below :
[NCERT : P L-16 | April 7, 2025 (II)]
(a) Both **(A)** and **(R)** are correct and **(R)** is the correct explanation of **(A)**
(b) **(A)** is correct but **(R)** is not correct
(c) Both **(A)** and **(R)** are correct but **(R)** is not correct explanation of **(A)**
(d) **(A)** is not correct but **(R)** is correct

41. A metallic ring is uniformly charged as shown in figure. *AC* and *BD* are two mutually perpendicular diameters. Electric field due to arc *AB* to '*O*' is '*E*' is magnitude. What would be the magnitude of electric field at '*O*' due to arc *ABC* ?
[NCERT : P L-16 | April 4, 2025 (II)]
(a) $2E$
(b) $\sqrt{2}E$
(c) $\frac{E}{2}$
(d) Zero

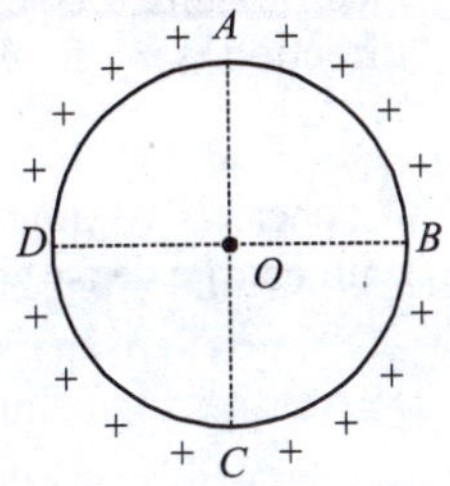

42. Two infinite identical charged sheets and a charged spherical body of charge density 'ρ' are arranged as shown in figure. Then the correct relation between the electrical fields at *A, B, C* and *D* points is :
[NCERT : P L-16 | April 4, 2025 (I)]

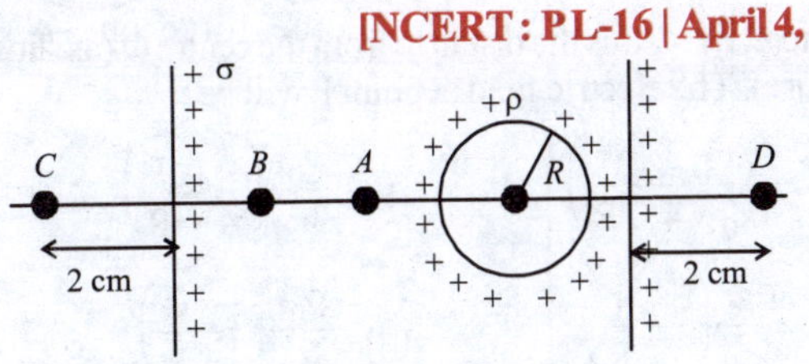

(a) $\vec{E}_A = \vec{E}_B; \vec{E}_C = \vec{E}_D$ (b) $\vec{E}_A > \vec{E}_B; \vec{E}_C = \vec{E}_D$
(c) $\vec{E}_C \neq \vec{E}_D; \vec{E}_A > \vec{E}_B$ (d) $|\vec{E}_A| = |\vec{E}_B|; \vec{E}_C > \vec{E}_D$

43. Two small spherical balls of mass 10g each with charges $-2\mu C$ and $2\mu C$, are attached to two ends of very light rigid rod of length 20 cm. The arrangement is now placed near an infinite non-conducting charge sheet with uniform charge density of 100 μC/m² such that length of rod makes an angle of 30° with electric field generated by charge sheet. Net torque acting on the rod is:
(Take ε_0 : 8.85×10^{-12} C²/Nm²)
[NCERT : P L-27 | April 4, 2025 (I)]
(a) 112Nm (b) 1.12Nm
(c) 2.24Nm (d) 11.2Nm

44. Given below are two statements : one is labelled as **Assertion (A)** and the other is labelled as **Reason (R)**.
Assertion (A) : Net dipole moment of a polar linear isotropic dielectric substance is not zero even in the absence of an external electric field.
Reason (R) : In absence of an external electric field, the different permanent dipoles of a polar dielectric substance are oriented in random directions.
In the light of the above statements, choose the **most appropriate answer** from the options given below :
[NCERT : P L-25 | April 2, 2025 (II)]
(a) **(A)** is correct but **(R)** is not correct
(b) Both **(A)** and **(R)** are correct but **(R)** is **not** the correct explanation of **(A)**
(c) Both **(A)** and **(R)** are correct and **(R)** is the correct explanation of **(A)**
(d) **(A)** is not correct but **(R)** is correct

45. Consider a circular loop that is uniformly charged and has a radius $a\sqrt{2}$. Find the position along the positive z–axis of the cartesian coordinate system where the electric field is maximum if the ring was assumed to be placed in xy–plane at the origin :
[NCERT : P L-16 | April 2, 2025 (II)]
(a) $\frac{a}{\sqrt{2}}$ (b) $\frac{a}{2}$ (c) a (d) 0

46. A point charge +q is placed at the origin. A second point charge +9q is placed at (d, 0, 0) in Cartesian coordinate system. The point in between them where the electric field vanishes is : **[NCERT : P L-15, 16 | April 2, 2025 (I)]**
(a) (4d/3, 0, 0) (b) (d/4, 0, 0)
(c) (3d/4, 0, 0) (d) (d/3, 0, 0)

47. A particle of mass '*m*' and charge '*q*' is fastened to one end 'A' of a massless string having equilibrium length 'ℓ' whose other end is fixed at point 'O'. The whole system is placed on a frictionless horizontal plane and is initially at rest. If uniform electric field is switched on along the direction as shown in figure, then the speed of the particle when it crosses the *x*-axis is
[NCERT : P L-14, 15, 16 | Jan. 28, 2025 (I)]

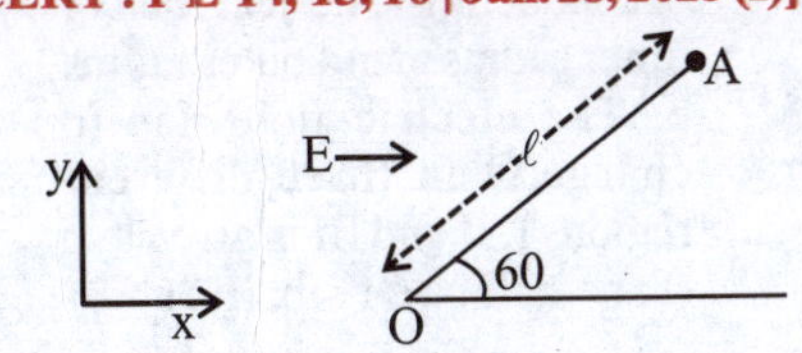

(a) $\sqrt{\frac{2qE\ell}{m}}$
(b) $\sqrt{\frac{qE\ell}{4m}}$ (c) $\sqrt{\frac{qE\ell}{m}}$ (d) $\sqrt{\frac{qE\ell}{2m}}$

48. An electron is made to enters symmetrically between two parallel and equally but oppositely charged metal plates, each of 10 cm length. The electron emerges out of the field region with a horizontal component of velocity 10^6 m/s. If the magnitude of the electric field between the plates is 9.1 V/cm, then the vertical component of velocity of electron is (mass of electron = 9.1×10^{-31} kg and charge of electron = 1.6×10^{-19} C) **[NCERT : P L-14, 15, 16 | Jan. 22, 2025 (I)]**

(a) 1×10^6 m/s (b) 0
(c) 16×10^6 m/s (d) 16×10^4 m/s

49. If the net electric field at point P (as shown in figure) along Y axis is zero, then the ratio of $\left|\frac{q_2}{q_3}\right|$ is $\frac{8}{5\sqrt{x}}$, where x= ....

[NCERT : P L-14 | April 8, 2024 (II)]

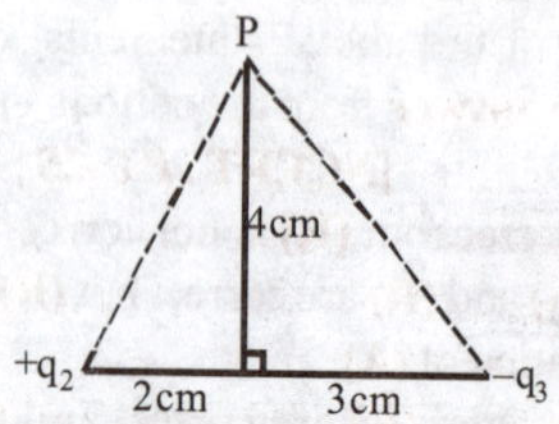

50. Two charges q and 3q are separated by a distance 'r' in air. At a distance x from charge q, the resultant electric field is zero. The value of x is :

[NCERT : P L-7 | Jan. 31, 2024 (I)]

(a) $\frac{r}{(1+\sqrt{3})}$ (b) $\frac{(1+\sqrt{3})}{r}$
(c) $r(1+\sqrt{3})$ (d) $\frac{r}{3(1+\sqrt{3})}$

51. A thin infinite sheet charge and an infinite line charge of respective charge densities $+\sigma$ and $+\lambda$ are placed parallel at 5 m distance from each other. Points 'P' and 'Q' are at $\frac{3}{\pi}$ m and $\frac{4}{\pi}$ m perpendicular distance from line charge towards sheet charge, respectively. 'E_P' and 'E_Q' are the magnitudes of resultant electric field intensities at point 'P' and 'Q', respectively. If $\frac{E_P}{E_Q} = \frac{4}{a}$ for $2|\sigma| = |\lambda|$. Then the value of a is ________.

[NCERT : P L-34 | April 13, 2023 (I)]

52. As shown in the figure, a point charge Q is placed at the centre of conducting spherical shell of inner radius a and outer radius b. The electric field due to charge Q in three different regions I, II and III is given by: (I : r < a, II : a < r < b, III : r > b)

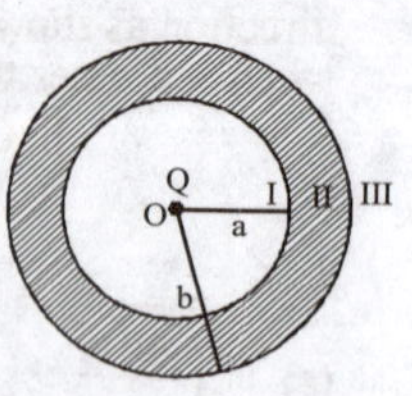

[NCERT : P L-35 | Jan. 30, 2023 (II)]

(a) $E_I = 0, E_{II} = 0, E_{III} \neq 0$ (b) $E_I \neq 0, E_{II} = 0, E_{III} \neq 0$
(c) $E_I \neq 0, E_{II} = 0, E_{III} = 0$ (d) $E_I = 0, E_{II} = 0, E_{III} = 0$

53. A point charge 2×10^{-2} C is moved from P to S in a uniform electric field of 30 NC^{-1} directed along positive x-axis. If coordinates of P and S are (1, 2, 0) m and (0, 0, 0) m respectively, the work done by electric field will be

[NCERT : P L-14 | Jan. 29, 2023 (II)]

(a) 1200mJ (b) 600mJ (c) –600mJ (d) –1200mJ

54. A uniform electric field of 10 N/C is created between two parallel charged plates (as shown in figure). An electron enters the field symmetrically between the plates with a kinetic energy 0.5 eV. The length of each plate is 10 cm. The angle (θ) of deviation of the path of electron as it comes out of the field is ___ (in degree). **[NCERT : P L-17 | Jan. 25, 2023 (I)]**

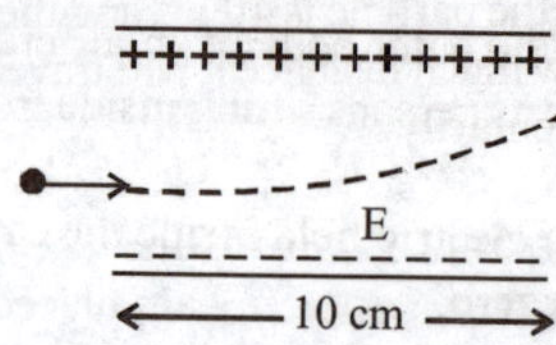

55. Point charge of 10 μC is placed at the origin. At what location on the X-axis should a point charge of 40μC be placed so that the net electric field is zero at x = 2 cm on the X-axis ? **[NCERT : P L-16 | Jan. 25, 2023 (II)]**

(a) x = 6 cm (b) x = 4 cm
(c) x = 8 cm (d) x = –4 cm

56. A stream of a positively charged particles having $\frac{q}{m} = 2\times10^{11}\frac{C}{kg}$ and velocity $\vec{v}_0 = 3\times10^7\hat{i}\,m/s$ is deflected by an electric field $1.8\hat{j}\,KV/m$. The electric field exists in a region of 10 cm along x direction. Due to the electric field, the deflection of the charge particles in the y direction is ______ *mm.*

[NCERT : P L-17 | Jan. 24, 2023 (I)]

57. A spherically symmetric charge distribution is considered with charge density varying as

[NCERT : P L-29 | July 29, 2022 (I)]

$$\rho(r) = \begin{cases} \rho_0\left(\frac{3}{4} - \frac{r}{R}\right) & \text{for } r \le R \\ \text{Zero} & \text{for } r > R \end{cases}$$

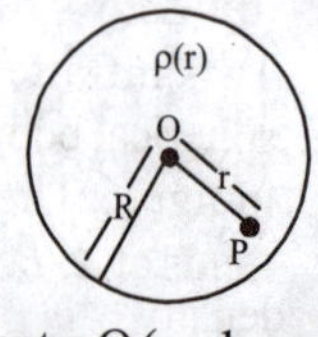

Where, r(r < R) is the distance from the centre O (as shown in figure). The electric field at point P will be :

(a) $\frac{\rho_0 r}{4\varepsilon_0}\left(\frac{3}{4} - \frac{r}{R}\right)$ (b) $\frac{\rho_0 r}{3\varepsilon_0}\left(\frac{3}{4} - \frac{r}{R}\right)$
(c) $\frac{\rho_0 r}{4\varepsilon_0}\left(1 - \frac{r}{R}\right)$ (d) $\frac{\rho_0 r}{5\varepsilon_0}\left(1 - \frac{r}{R}\right)$

58. A uniform electric field E = (8m/e) V/m is created between two parallel plates of length 1 m as shown in figure, (where m = mass of electron and e = charge of electron). An electron enters the field symmetrically between the plates with a speed of 2 m/s. The angle of the deviation (θ) of the path of the electron as it comes out of the field will be ______. **[NCERT : P L-17 | July 28, 2022 (II)]**

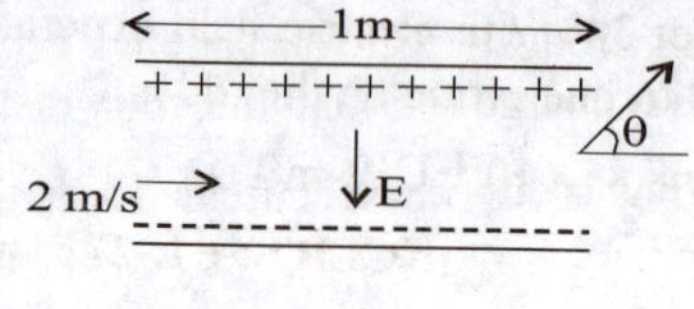

(a) $\tan^{-1}(4)$ (b) $\tan^{-1}(2)$

(c) $\tan^{-1}\left(\frac{1}{3}\right)$ (d) $\tan^{-1}(3)$

59. A positive charge particle of 100 μg is thrown in opposite direction to a uniform electric field of strength 1×10^5 NC^{-1}. If the charge on the particle is 40 μC and the initial velocity is 200 ms^{-1}, how much distance it will travel before coming to the rest momentarily :

[NCERT : P L-17 | June 29, 2022 (I)]

(a) 1m (b) 5m (c) 10m (d) 0.5m

60. The three charges $q/2$, q and $q/2$ are placed at the corners A, B and C of a square of side 'a' as shown in figure. The magnitude of electric field (E) at the corner D of the square, is : **[NCERT : P L-16 | June 28, 2022 (I)]**

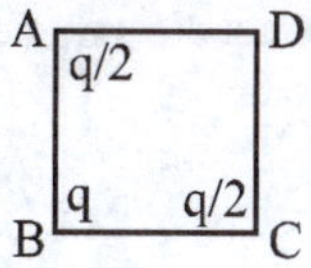

(a) $\frac{q}{4\pi \epsilon_0 a^2}\left(\frac{1}{\sqrt{2}}+\frac{1}{2}\right)$ (b) $\frac{q}{4\pi \epsilon_0 a^2}\left(1+\frac{1}{\sqrt{2}}\right)$

(c) $\frac{q}{4\pi \epsilon_0 a^2}\left(1-\frac{1}{\sqrt{2}}\right)$ (d) $\frac{q}{4\pi \epsilon_0 a^2}\left(\frac{1}{\sqrt{2}}-\frac{1}{2}\right)$

61. Two point charges A and B of magnitude $+8 \times 10^{-6}$ C and -8×10^{-6} C respectively are placed at a distance d apart. The electric field at the middle point O between the charges is 6.4×10^4 NC^{-1}. The distance 'd' between the point charges A and B is:

[NCERT : P L-14 | June 28, 2022 (II)]

(a) 2.0m (b) 3.0m (c) 1.0m (d) 4.0m

62. A vertical electric field of magnitude 4.9×10^5 N/C just prevents a water droplet of a mass 0.1 g from falling. The value of charge on the droplet will be :

(Given g = 9.8 m/s^2) **[NCERT : P L-14 | June 24, 2022 (I)]**

(a) 1.6×10^{-9} C (b) 2.0×10^{-9} C

(c) 3.2×10^{-9} C (d) 0.5×10^{-9} C

63. A uniformly charged disc of radius R having surface charge density σ is placed in the xy plane with its center at the origin. Find the electric field intensity along the z-axis at a distance Z from origin :-

[NCERT : P L-14 | Aug. 27, 2021 (I)]

(a) $E = \frac{\sigma}{2\varepsilon_0}\left(1 - \frac{Z}{(Z^2+R^2)^{1/2}}\right)$

(b) $E = \frac{\sigma}{2\varepsilon_0}\left(1 + \frac{Z}{(Z^2+R^2)^{1/2}}\right)$

(c) $E = \frac{2\varepsilon_0}{\sigma}\left(\frac{1}{(Z^2+R^2)^{1/2}} + Z\right)$

(d) $E = \frac{\sigma}{2\varepsilon_0}\left(\frac{1}{(Z^2+R^2)} + \frac{1}{Z^2}\right)$

64. Figure shows a rod AB, which is bent in a 120° circular arc of radius R. A charge (–Q) is uniformly distributed over rod AB. What is the electric field $\vec{E}$ at the centre of curvature O ? **[NCERT : P L-14 | Aug. 27, 2021 (II)]**

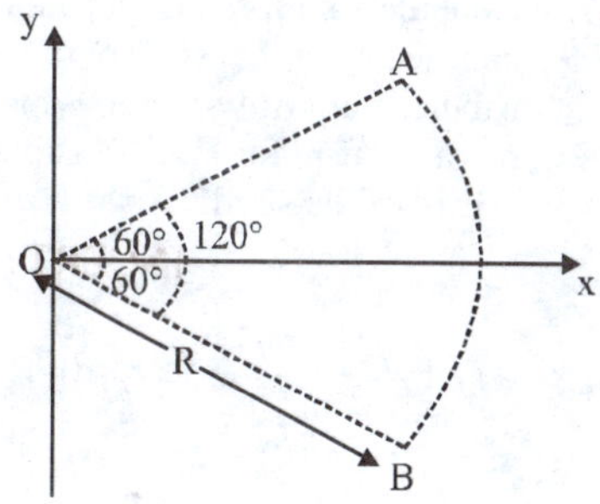

(a) $\frac{3\sqrt{3}Q}{8\pi\varepsilon_0 R^2}(\hat{i})$

(b) $\frac{3\sqrt{3}Q}{8\pi^2\varepsilon_0 R^2}(\hat{i})$

(c) $\frac{3\sqrt{3}Q}{16\pi^2\varepsilon_0 R^2}(\hat{i})$

(d) $\frac{3\sqrt{3}Q}{8\pi^2\varepsilon_0 R^2}(-\hat{i})$

65. A body having specific charge 8 μC/g is resting on a frictionless plane at a distance 10 cm from the wall (as shown in the figure). It starts moving towards the wall when a uniform electric field of 100 V/m is applied horizontally towards the wall. If the collision of the body with the wall is perfectly elastic, then the time period of the motion will be ______ s.

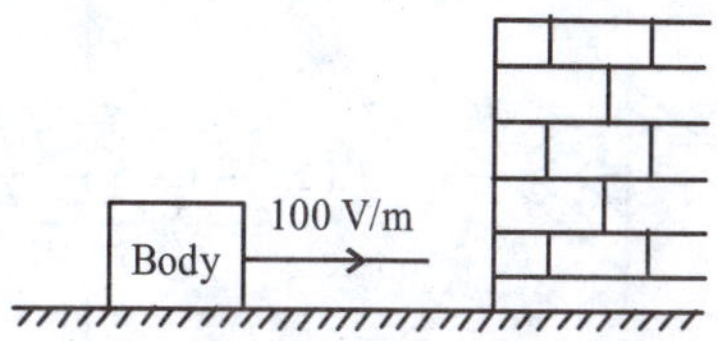

[NCERT : P L-17 | NA, July 20, 2021 (I)]

66. An oil drop of radius 2 mm with a density 3 g cm–3 is held stationary under a constant electric field 3.55×10^5 V m^{-1} in the Millikan's oil drop experiment. What is the number of excess electrons that the oil drop will process? Consider g = 9.81 m/s^2

[NCERT : P L-14 | March 18, 2021 (I)]

(a) 1.73×10^{12} (b) 1.73×10^{10}

(c) 48.8×10^{11} (d) 17.3×10^{10}

67. Find the electric field at point P (as shown in figure) on the perpendicular bisector of a uniformly charged thin wire of length L carrying a charge Q. The distance of the point P from the centre of the rod is $a = \frac{\sqrt{3}}{2}L$.

[NCERT : P L-34 | Feb. 26, 2021 (I)]

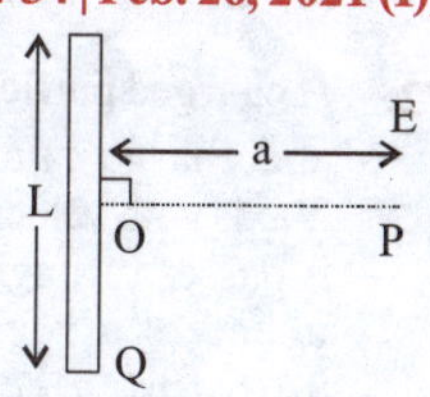

(a) $\frac{Q}{3\pi\varepsilon_0 L^2}$ (b) $\frac{Q}{4\pi\varepsilon_0 L^2}$

(c) $\frac{\sqrt{3}Q}{4\pi\varepsilon_0 L^2}$ (d) $\frac{Q}{2\sqrt{3}\pi\varepsilon_0 L^2}$

68. Charges Q_1 and Q_2 are at points A and B of a right angle triangle OAB (see figure). The resultant electric field at point O is perpendicular to the hypotenuse, then Q_1/Q_2 is proportional to :

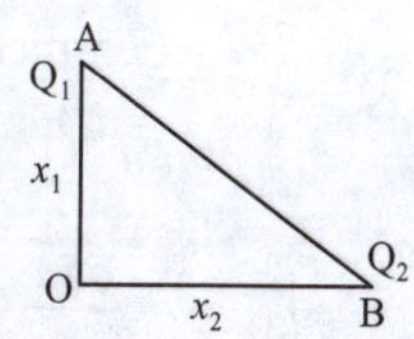

[NCERT : P L-14 | Sep. 06, 2020 (I)]

(a) $\frac{x_1^3}{x_2^3}$ (b) $\frac{x_2}{x_1}$ (c) $\frac{x_1}{x_2}$ (d) $\frac{x_2^2}{x_1^2}$

69. Consider the force F on a charge 'q' due to a uniformly charged spherical shell of radius R carrying charge Q distributed uniformly over it. Which one of the following statements is true for F, if 'q' is placed at distance r from the centre of the shell?

[NCERT : P L-35 | Sep. 06, 2020 (II)]

(a) $F = \frac{1}{4\pi\varepsilon_0}\frac{Qq}{R^2}$ for $r<R$ (b) $\frac{1}{4\pi\varepsilon_0}\frac{Qq}{R^2} > F > 0$ for $r<R$

(c) $F = \frac{1}{4\pi\varepsilon_0}\frac{Qq}{R^2}$ for $r>R$ (d) $F = \frac{1}{4\pi\varepsilon_0}\frac{Qq}{R^2}$ for all r

70. Two charged thin infinite plane sheets of uniform surface charge density σ_+ and σ_-, where $|\sigma_+|>|\sigma_-|$, intersect at right angle. Which of the following best represents the electric field lines for this system ?

[NCERT : P L-34 | Sep. 04, 2020 (I)]

(a) 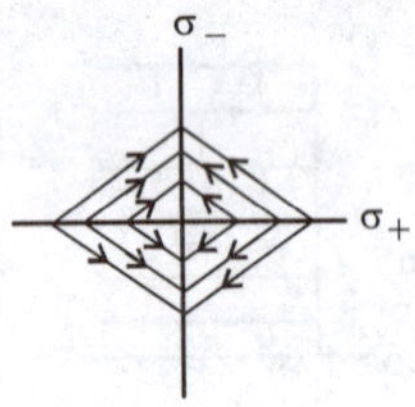

(b) 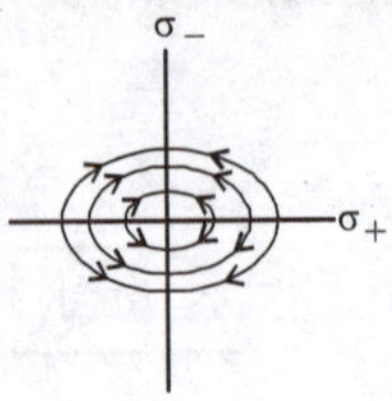

(c) 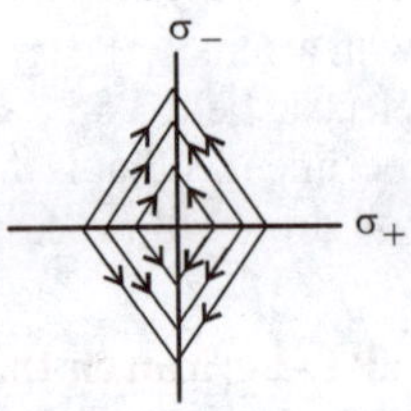

(d) 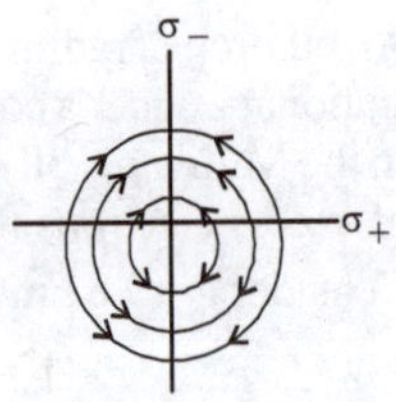

71. A particle of charge q and mass m is subjected to an electric field $E = E_0(1 - ax^2)$ in the x-direction, where a and E_0 are constants. Initially the particle was at rest at $x = 0$. Other than the initial position the kinetic energy of the particle becomes zero when the distance of the particle from the origin is : **[NCERT : P L-17 | Sep. 04, 2020 (II)]**

(a) a (b) $\sqrt{\frac{2}{a}}$ (c) $\sqrt{\frac{3}{a}}$ (d) $\sqrt{\frac{1}{a}}$

72. A charged particle (mass m and charge q) moves along X axis with velocity V_0. When it passes through the origin it enters a region having uniform electric field $\vec{E} = -E\hat{j}$ which extends upto $x = d$. Equation of path of electron in the region $x > d$ is : **[NCERT : P L-17 | Sep. 02, 2020 (I)]**

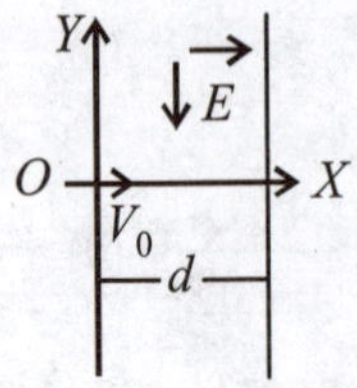

(a) $y = \frac{qEd}{mV_0^2}(x-d)$ (b) $y = \frac{qEd}{mV_0^2}\left(\frac{d}{2} - x\right)$

(c) $y = \frac{qEd}{mV_0^2}x$ (d) $y = \frac{qEd^2}{mV_0^2}x$

73. A small point mass carrying some positive charge on it, is released from the edge of a table. There is a uniform electric field in this region in the horizontal direction.

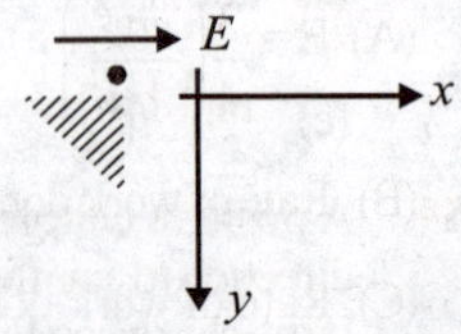

Which of the following options then correctly describe the trajectory of the mass ? (Curves are drawn schematically and are not to scale).

[NCERT : P L-17 | Sep. 02, 2020 (II)]

(a) 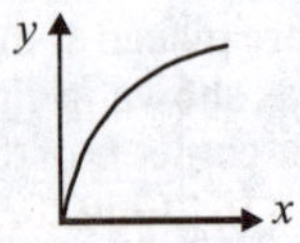

(b) 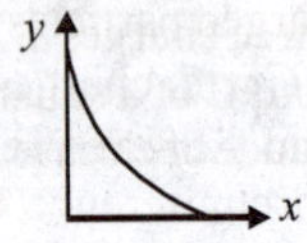

(c) 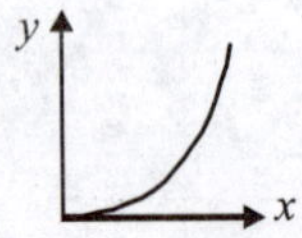

(d) 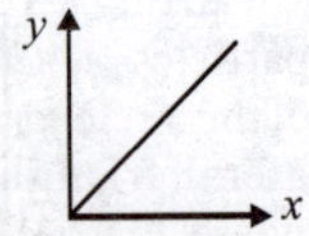

74. Consider a sphere of radius R which carries a uniform charge density ρ. If a sphere of radius $\frac{R}{2}$ is carved out of it, as shown, the ratio $\frac{|\vec{E}_A|}{|\vec{E}_B|}$ of magnitude of electric field $\vec{E}_A$ and $\vec{E}_B$, respectively, at points A and B due to the remaining portion is:

[NCERT : P L-29 | 9 Jan. 2020, (I)]

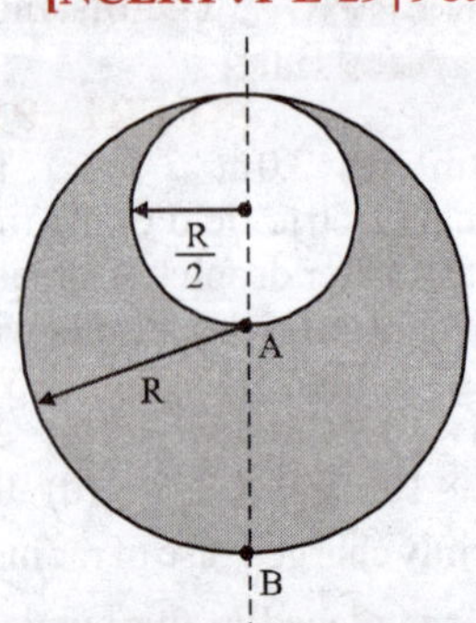

(a) $\frac{21}{34}$

(b) $\frac{18}{34}$

(c) $\frac{17}{54}$

(d) $\frac{18}{54}$

75. An electric dipole of moment $\vec{p} = (\hat{i} - 3\hat{j} + 2\hat{k}) \times 10^{-29}$ C.m is at the origin (0, 0, 0). The electric field due to this dipole at $\vec{r} = +\hat{i} + 3\hat{j} + 5\hat{k}$ (note that $\vec{r}.\vec{p} = 0$) is parallel to: **[9 Jan. 2020, (I)]**

(a) $(+\hat{i} - 3\hat{j} - 2\hat{k})$ (b) $(-\hat{i} + 3\hat{j} - 2\hat{k})$

(c) $(+\hat{i} + 3\hat{j} - 2\hat{k})$ (d) $(-\hat{i} - 3\hat{j} + 2\hat{k})$

76. A charged particle of mass 'm' and and charge 'q' moving under the influence of uniform electric field $E\hat{i}$ and a uniform magnetic field $B\hat{k}$ follows a trajectory from point P to Q as shown in figure.

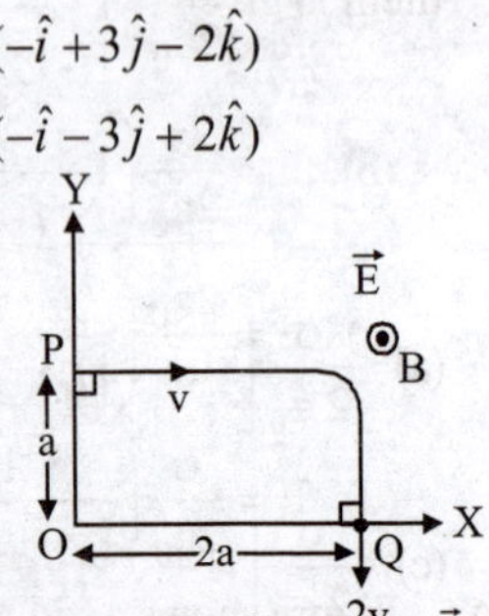

The velocities at P and Q are respectively, $v\vec{i}$ and $-2v\vec{j}$. Then which of the following statements (A, B, C, D) are the correct? (Trajectory shown is schematic and not to scale) **[NCERT : P L-17 | 9 Jan. 2020, (I)]**

(A) $E = \frac{3}{4}\left(\frac{mv^2}{qa}\right)$

(B) Rate of work done by the electric field at P is $\frac{3}{4}\left(\frac{mv^3}{a}\right)$

(C) Rate of work done by both the fields at Q is zero

(D) The difference between the magnitude of angular momentum of the particle at P and Q is 2 ma$v$.

(a) (A), (C), (D) (b) (B), (C), (D)

(c) (A), (B), (C) (d) (A), (B), (C), (D)

77. Three charged particles A, B and C with charges $-4q$, $2q$ and $-2q$ are present on the circumference of a circle of radius d. The charged particles A, C and centre O of the circle formed an equilateral triangle as shown in figure. Electric field at O along x-direction is:

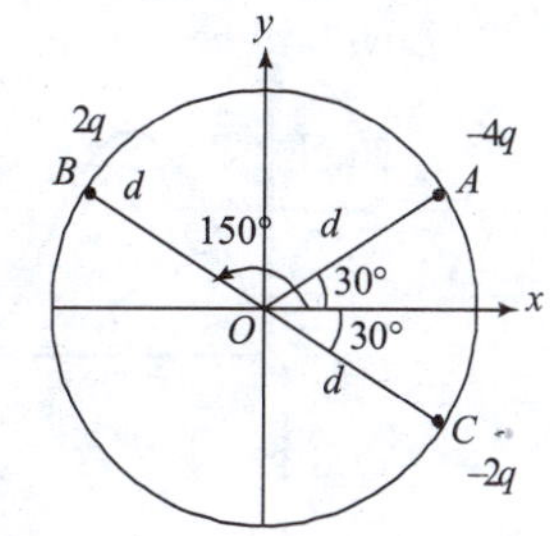

[NCERT : P L-16 | 8 Jan. 2020, (I)]

(a) $\frac{\sqrt{3}q}{\pi \epsilon_0 d^2}$ (b) $\frac{2\sqrt{3}q}{\pi \epsilon_0 d^2}$ (c) $\frac{\sqrt{3}q}{4\pi \epsilon_0 d^2}$ (d) $\frac{3\sqrt{3}q}{4\pi \epsilon_0 d^2}$

78. A particle of mass m and charge q is released from rest in a uniform electric field. If there is no other force on the particle, the dependence of its speed v on the distance x travelled by it is correctly given by (graphs are schematic and not drawn to scale)

[NCERT : P L-17 | 8 Jan. 2020, (II)]

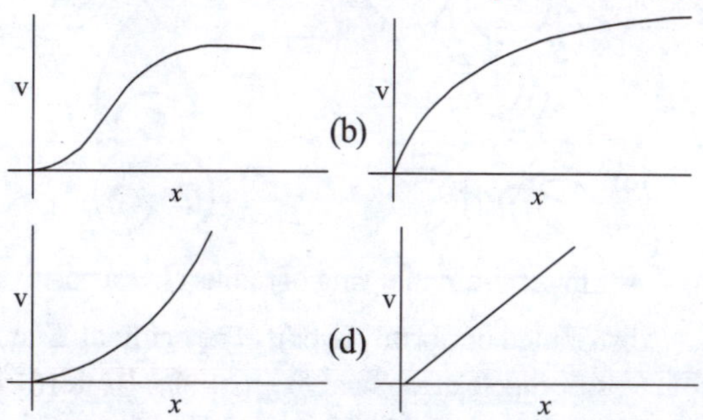

79. Two infinite planes each with uniform surface charge density $+\sigma$ are kept in such a way that the angle between them is 30°. The electric field in the region shown between them is given by: **[NCERT : P L-35 | 7 Jan. 2020, (I)]**

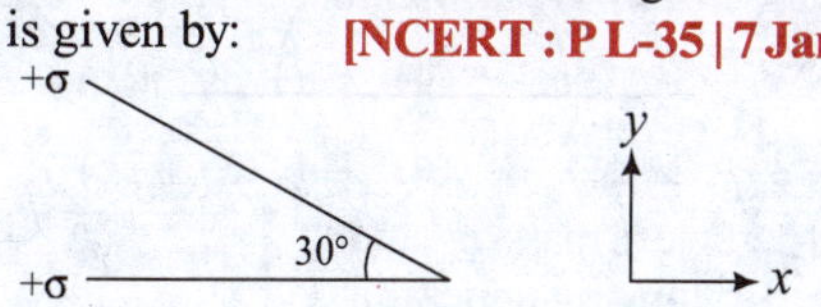

(a) $\frac{\sigma}{2\epsilon_0}\left[(1+\sqrt{3})\hat{y} - \frac{\hat{x}}{2}\right]$ (b) $\frac{\sigma}{\epsilon_0}\left[\left(1+\frac{\sqrt{3}}{2}\right)\hat{y} + \frac{\hat{x}}{2}\right]$

(c) $\frac{\sigma}{2\epsilon_0}\left[\left(1+\sqrt{3}\right)\hat{y} + \frac{\hat{x}}{2}\right]$ (d) $\frac{\sigma}{2\epsilon_0}\left[\left(1-\frac{\sqrt{3}}{2}\right)\hat{y} - \frac{\hat{x}}{2}\right]$

80. A particle of mass m and charge q has an initial velocity $\vec{v} = v_0\hat{j}$. If an electric field $\vec{E} = E_0\hat{i}$ and magnetic field $\vec{B} = B_0\hat{i}$ act on the particle, its speed will double after a time:

[NCERT : P L-17 | 7 Jan 2020, (II)]

(a) $\frac{2mv_0}{qE_0}$ (b) $\frac{3mv_0}{qE_0}$ (c) $\frac{\sqrt{3}mv_0}{qE_0}$ (d) $\frac{\sqrt{2}mv_0}{qE_0}$

81. A simple pendulum of length L is placed between the plates of a parallel plate capacitor having electric field E, as shown in figure. Its bob has mass m and charge q. The time period of the pendulum is given by :

[NCERT : P L-14 | 10 April 2019, (II)]

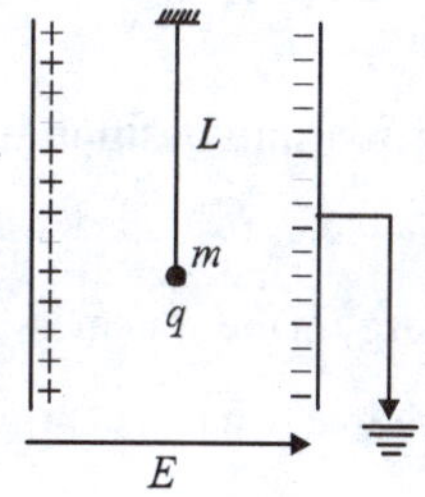

(a) $2\pi\sqrt{\frac{L}{\left(g+\frac{qE}{m}\right)}}$ (b) $2\pi\sqrt{\frac{L}{\sqrt{g^2-\frac{q^2E^2}{m^2}}}}$

(c) $2\pi\sqrt{\frac{L}{\left(g-\frac{qE}{m}\right)}}$ (d) $2\pi\sqrt{\frac{L}{\sqrt{g^2+\left(\frac{qE}{m}\right)^2}}}$

82. Four point charges $-q$, $+q$, $+q$ and $-q$ are placed on y-axis at $y = -2d$, $y = -d$, $y = +d$ and $y = +2d$, respectively. The magnitude of the electric field E at a point on the x-axis at $x = D$, with $D >> d$, will behave as:

[NCERT : P L-16 | 9 April 2019, (II)]

(a) $E \propto \frac{1}{D^3}$ (b) $E \propto \frac{1}{D}$ (c) $E \propto \frac{1}{D^4}$ (d) $E \propto \frac{1}{D^2}$

83. The bob of a simple pendulum has mass 2 g and a charge of 5.0 μC. It is at rest in a uniform horizontal electric field of intensity 2000 V/m. At equilibrium, the angle that the pendulum makes with the vertical is :

[NCERT : P L-14 | 8 April 2019, (I)]

(take $g = 10$ m/s²)

(a) $\tan^{-1}(2.0)$ (b) $\tan^{-1}(0.2)$

(c) $\tan^{-1}(5.0)$ (d) $\tan^{-1}(0.5)$

84. For a uniformly charged ring of radius R, the electric field on its axis has the largest magnitude at a distance h from its centre. Then value of h is:

[NCERT : P L-29 | 9 Jan. 2019, (I)]

(a) $\frac{R}{\sqrt{5}}$ (b) $\frac{R}{\sqrt{2}}$ (c) R (d) $R\sqrt{2}$

85. Two point charges q_1 ($\sqrt{10}$ μC) and q_2 (– 25 μC) are placed on the x-axis at $x = 1$ m and $x = 4$ m respectively. The electric field (in V/m) at a point $y = 3$ m on y-axis is, $\left[\text{take } \frac{1}{4\pi\epsilon_0} = 9\times10^9 \text{ Nm}^2\text{C}^{-2}\right]$

[NCERT : P L-16 | 9 Jan 2019, (II)]

(a) $(63\hat{i} - 27\hat{j}) \times 10^2$ (b) $(-63\hat{i} + 27\hat{j}) \times 10^2$

(c) $(81\hat{i} - 81\hat{j}) \times 10^2$ (d) $(-81\hat{i} + 81\hat{j}) \times 10^2$

86. A body of mass M and charge q is connected to a spring of spring constant k. It is oscillating along x-direction about its equilibrium position, taken to be at $x = 0$, with an amplitude A. An electric field E is applied along the x-direction. Which of the following statements is correct? **[NCERT : P L-14 | Online April 15, 2018]**

(a) The total energy of the system is $\frac{1}{2}m\omega^2 A^2 + \frac{1}{2}\frac{q^2E^2}{k}$

(b) The new equilibrium position is at a distance: $\frac{2qE}{k}$ from $x = 0$

(c) The new equilibrium position is at a distance: $\frac{qE}{2k}$ from $x = 0$

(d) The total energy of the system is $\frac{1}{2}m\omega^2 A^2 - \frac{1}{2}\frac{q^2E^2}{k}$

87. A solid ball of radius R has a charge density ρ given by $\rho = \rho_0\left(1 - \frac{r}{R}\right)$ for $0 \le r \le R$. The electric field outside the ball is: **[NCERT : P L-29 | Online April 15, 2018]**

(a) $\frac{\rho_0 R^3}{\varepsilon_0 r^2}$ (b) $\frac{4\rho_0 R^3}{3\varepsilon_0 r^2}$ (c) $\frac{3\rho_0 R^3}{4\varepsilon_0 r^2}$ (d) $\frac{\rho_0 R^3}{12\varepsilon_0 r^2}$

88. A long cylindrical shell carries positive surface charge σ in the upper half and negative surface charge - σ in the lower half. The electric field lines around the cylinder will look like figure given in : (*figures are schematic and not drawn to scale*) **[NCERT : P L-23 | 2015]**

(a) 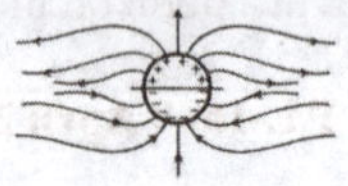 (b) 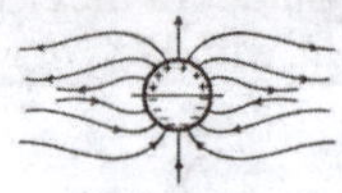

(c) 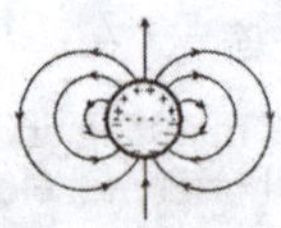 (d) 

89. A wire of length L (=20 cm), is bent into a semicircular arc. If the two equal halves of the arc were each to be uniformly charged with charges ± Q, $[|Q| = 10^3\varepsilon_0$ Coulomb where ε_0 is the permittivity (in SI units) of free space] the net electric field at the centre O of the semicircular arc would be : **[NCERT : P L-23 | Online April 11, 2015]**

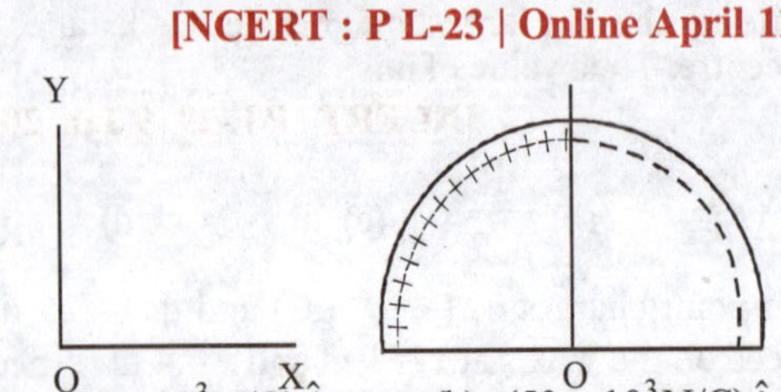

(a) $(50 \times 10^3 N/C)\hat{j}$ (b) $(50 \times 10^3 N/C)\hat{i}$

(c) $(25 \times 10^3 N/C)\hat{j}$ (d) $(25 \times 10^3 N/C)\hat{i}$

90. The magnitude of the average electric field normally present in the atmosphere just above the surface of the Earth is about 150 N/C, directed inward towards the center of the Earth. This gives the total net surface charge carried by the Earth to be: **[NCERT : P L-29 | Online April 9, 2014]**

[Given $\varepsilon_0 = 8.85 \times 10^{-12}$ C²/N-m², $R_E = 6.37 \times 10^6$ m]

(a) +670 kC (b) – 670 kC (c) –680 kC (d) +680 kC

91. A liquid drop having 6 excess electrons is kept stationary under a uniform electric field of 25.5 kVm^{-1}. The density of liquid is 1.26×10^3 kg m^{-3}. The radius of the drop is (neglect buoyancy). **[NCERT : P L-14 | Online April 23, 2013]**

(a) 4.3×10^{-7} m (b) 7.8×10^{-7} m

(c) 0.078×10^{-7} m (d) 3.4×10^{-7} m

92. In a uniformly charged sphere of total charge Q and radius R, the electric field E is plotted as function of distance from the centre, The graph which would correspond to the above will be: **[NCERT : P L-29 | 2012]**

(a) 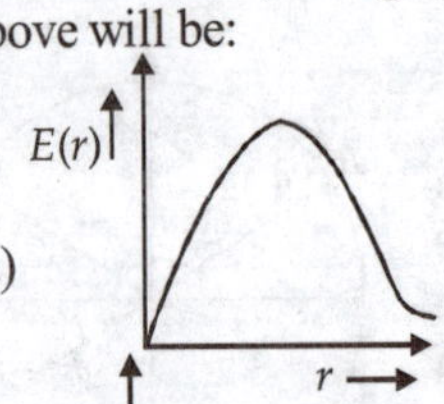

(b) 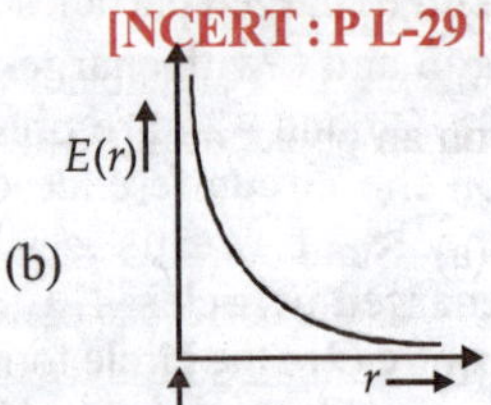

(c) 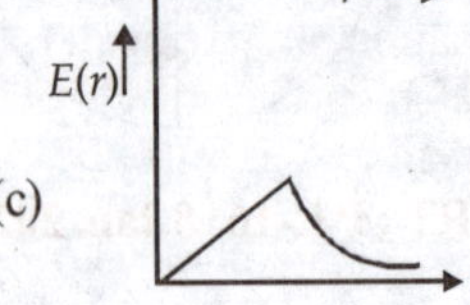

(d) 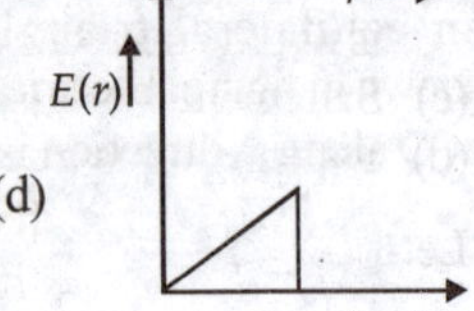

93. Three positive charges of equal value q are placed at vertices of an equilateral triangle. The resulting lines of force should be sketched as in **[NCERT : P L-21 | Online May 26, 2012]**

(a) 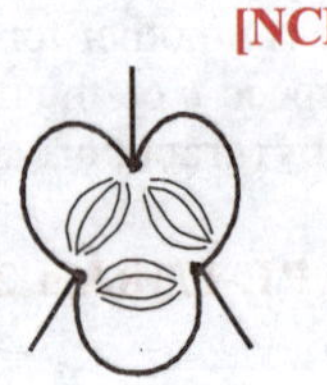 (b) 

(c)  (d) 

94. A thin semi-circular ring of radius r has a positive charge q distributed uniformly over it. The net field $\vec{E}$ at the centre O is **[NCERT : P L-14 | 2010]**

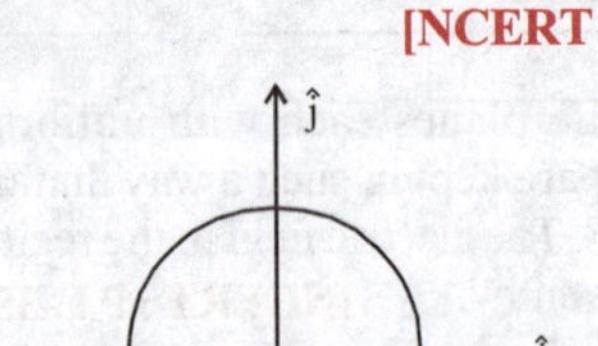

(a) $\frac{q}{4\pi^2\varepsilon_0 r^2}\hat{j}$ (b) $-\frac{q}{4\pi^2\varepsilon_0 r^2}\hat{j}$

(c) $-\frac{q}{2\pi^2\varepsilon_0 r^2}\hat{j}$ (d) $\frac{q}{2\pi^2\varepsilon_0 r^2}\hat{j}$

95. Let there be a spherically symmetric charge distribution with charge density varying as $\rho(r) = \rho_0\left(\frac{5}{4} - \frac{r}{R}\right)$ upto $r = R$, and $\rho(r) = 0$ for $r > R$, where r is the distance from the origin. The electric field at a distance $r (r < R)$ from the origin is given by **[NCERT : P L-35 | 2010]**

(a) $\frac{\rho_0 r}{4\varepsilon_0}\left(\frac{5}{3}-\frac{r}{R}\right)$ (b) $\frac{4\pi\rho_0 r}{3\varepsilon_0}\left(\frac{5}{3}-\frac{r}{R}\right)$

(c) $\frac{\rho_0 r}{4\varepsilon_0}\left(\frac{5}{4}-\frac{r}{R}\right)$ (d) $\frac{\rho_0 r}{3\varepsilon_0}\left(\frac{5}{4}-\frac{r}{R}\right)$

96. This question contains Statement-1 and Statement-2. Of the four choices given after the statements, choose the one that best describes the two statements.

Statement-1 : For a charged particle moving from point P to point Q, the net work done by an electrostatic field on the particle is independent of the path connecting point P to point Q.

Statement-2 : The net work done by a conservative force on an object moving along a closed loop is zero.

[NCERT : P L-21 | 2009]

(a) Statement-1 is true, Statement-2 is true; Statement-2 is the correct explanation of Statement-1.
(b) Statement-1 is true, Statement-2 is true; Statement-2 is not the correct explanation of Statement-1.
(c) Statement-1 is false, Statement-2 is true.
(d) Statement-1 is true, Statement-2 is false.

97. Let $\rho(r)=\frac{Q}{\pi R^4}r$ be the charge density distribution for a solid sphere of radius R and total charge Q. For a point 'P' inside the sphere at distance r_1 from the centre of the sphere, the magnitude of electric field is :

[NCERT : P L-30 | 2009]

(a) $\frac{Q}{4\pi\epsilon_0 r_1^2}$ (b) $\frac{Qr_1^2}{4\pi\epsilon_0 R^4}$

(c) $\frac{Qr_1^2}{3\pi\epsilon_0 R^4}$ (d) 0

98. A thin spherical shell of radus R has charge Q spread uniformly over its surface. Which of the following graphs most closely represents the electric field $E(r)$ produced by the shell in the range $0 \le r < \infty$, where r is the distance from the centre of the shell? **[NCERT : P L-35 | 2008]**

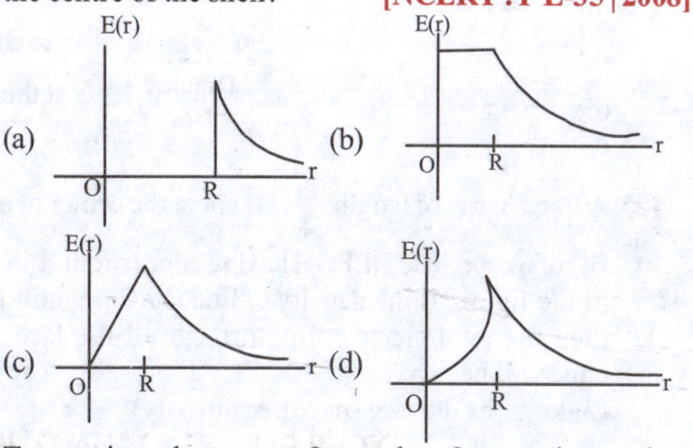

99. Two point charges + $8q$ and – $2q$ are located at $x = 0$ and $x = L$ respectively. The location of a point on the x axis at which the net electric field due to these two point charges is zero is **[NCERT : P L-16 | 2005]**

(a) $\frac{L}{4}$ (b) $2L$ (c) $4L$ (d) $8L$

100. A charged ball B hangs from a silk thread S, which makes an angle θ with a large charged conducting sheet P, as shown in the figure. The surface charge density σ of the sheet is proportional to **[NCERT : P L-28 | 2005]**

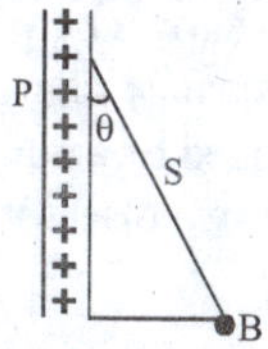

(a) cot θ
(b) cos θ
(c) tan θ
(d) sin θ

101. Four charges equal to $-Q$ are placed at the four corners of a square and a charge q is at its centre. If the system is in equilibrium the value of q is **[NCERT : P L-12 | 2004]**

(a) $-\frac{Q}{2}(1+2\sqrt{2})$ (b) $\frac{Q}{4}(1+2\sqrt{2})$

(c) $-\frac{Q}{4}(1+2\sqrt{2})$ (d) $\frac{Q}{2}(1+2\sqrt{2})$

102. A charged oil drop is suspended in a uniform field of 3×10^4 v/m so that it neither falls nor rises. The charge on the drop will be (Take the mass of the charge = 9.9×10^{-15} kg and $g = 10$ m/s^2) **[NCERT : P L-14 | 2004]**

(a) 1.6×10^{-18} C (b) 3.2×10^{-18} C
(c) 3.3×10^{-18} C (d) 4.8×10^{-18} C

Topic 3 - Electric Dipole, Electric Flux and Gauss's Law

103. An infinitely long wire has uniform linear charge density λ = 2 nC/m. The net flux through a Gaussian cube of side length $\sqrt{3}$ cm, if the wire passes through any two corners of the cube, that are maximally displaced from each other, would be xNm^{2}C^{-1}, where x is :

[Neglect any edge effects and use $\frac{1}{4\pi\varepsilon_0} = 9 \times 10^9$ SI units] **[NCERT : P L-33, 34 | April 8, 2025 (II)]**

(a) 0.72π (b) 1.44π (c) 6.48π (d) 2.16π

104. The electric field in a region is given by $\vec{E} = (2\hat{i} + 4\hat{j} + 6\hat{k}) \times 10^3$ N/C. The flux of the field through a rectangular surface parallel to x-z plane is 6.0 Nm^{2}C^{-1}. The area of the surface is ______ cm^2.

[NCERT : P L-21, 22 | April 7, 2025 (II)]

105. An electric dipole is placed at a distance of 2 cm from an infinite plane sheet having positive charge density σ_0. Choose the correct option from the following.

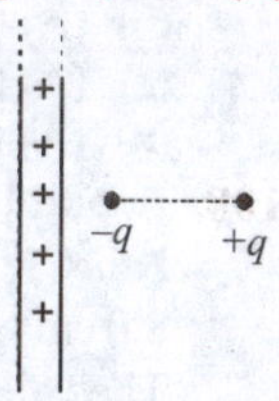

[NCERT : P L-27 | Jan. 29, 2025 (II)]

(a) Torque on dipole is zero and net force is directed away from the sheet.
(b) Torque on dipole is zero and net force acts towards the sheet.
(c) Potential energy of dipole is minimum and torque is zero.
(d) Potential energy and torque both are maximum

106. Consider two infinitely large plane parallel conducting plates as shown below. The plates are uniformly charged with a surface charge density +σ and –2σ. The force experienced by a point charge +q placed at the mid point between two plates will be :

[NCERT : P L-34 | April 2, 2025 (I)]

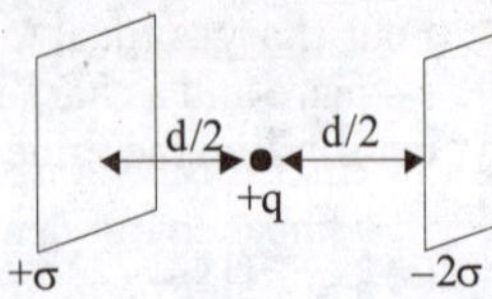

(a) $\frac{\sigma q}{4\epsilon_0}$ (b) $\frac{3\sigma q}{2\epsilon_0}$

(c) $\frac{3\sigma q}{4\epsilon_0}$ (d) $\frac{\sigma q}{2\epsilon_0}$

107. Match List-I with List-II.

[NCERT : P L-33, 34, 35, 36 | Jan. 29, 2025 (I)]

List-I	List-II
(A) Electric field inside (distance $r > 0$ from center) of a uniformly charged spherical shell with surface charge density σ, and radius R.	(I) $\frac{\sigma}{\varepsilon_0}$
(B) Electric field at distance r > 0 from a uniformly charged infinite plane sheet with surface charge density σ	(II) $\frac{\sigma}{2\varepsilon_0}$
(C) Electric field outside (distance $r > 0$ from center) of a uniformly charged spherical shell with surface charge density σ, and radius R	(III) 0
(D) Electric field between 2 oppositely charged infinite plane parallel sheets with uniform surface charge density σ.	(IV) $\frac{\sigma}{\varepsilon_0 r^2}$

Choose the correct answer from the options given below :

(a) (A)-(IV), (B)-(I), (C)-(III), (D)-(II)
(b) (A)-(IV), (B)-(II), (C)-(III), (D)-(I)
(c) (A)-(II), (B)-(I), (C)-(IV), (D)-(III)
(d) (A)-(III), (B)-(II), (C)-(IV), (D)-(I)

108. A point charge causes an electric flux of -2×10^4 Nm^2C^{-1} to pass through a spherical Gaussian surface of 8.0 cm radius, centred on the charge.
The value of the point charge is :
(Given, $\epsilon_0 = 8.85 \times 10^{-12} C^2N^{-1}m^{-2}$)

[NCERT : P L-30 | Jan. 29, 2025 (II)]

(a) -17.7×10^{-8} C (b) -15.7×10^{-8} C
(c) 17.7×10^{-8} C (d) 15.7×10^{-8} C

109. An electric dipole of mass m, charge q, and length l is placed in a uniform electric field $\vec{E} = E_0\hat{i}$. When the dipole is rotated slightly from its equilibrium position and released, the time period of its oscillations will be :

[NCERT : P L-27 | Jan. 29, 2025 (I)]

(a) $\frac{1}{2\pi}\sqrt{\frac{2ml}{qE_0}}$ (b) $2\pi\sqrt{\frac{ml}{qE_0}}$

(c) $\frac{1}{2\pi}\sqrt{\frac{ml}{2qE_0}}$ (d) $2\pi\sqrt{\frac{ml}{2qE_0}}$

110. A square loop of sides a = 1 m is held normally in front of a point charge q = 1C. The flux of the electric field through the shaded region is $\frac{5}{P} \times \frac{1}{\varepsilon_0} \frac{Nm^2}{C}$, where the value of p is_______.

[NCERT : P L-22 | Jan. 24, 2025 (I)]

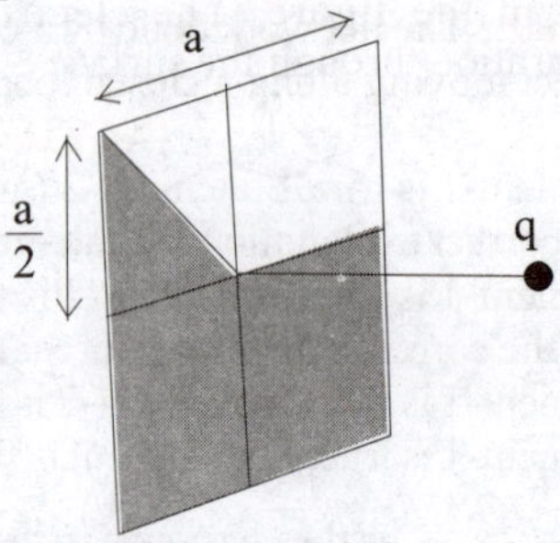

111. A point particle of charge Q is located at P along the axis of an electric dipole 1 at a distance r as shown in the figure. The point P is also on the equatorial plane of a second electric dipole 2 at a distance r. The dipoles are made of opposite charge q separated by a distance 2a. For the charge particle at P not to experience any net force, which of the following correctly describes the situation?

[NCERT : P L-27 | Jan. 23, 2025 (I)]

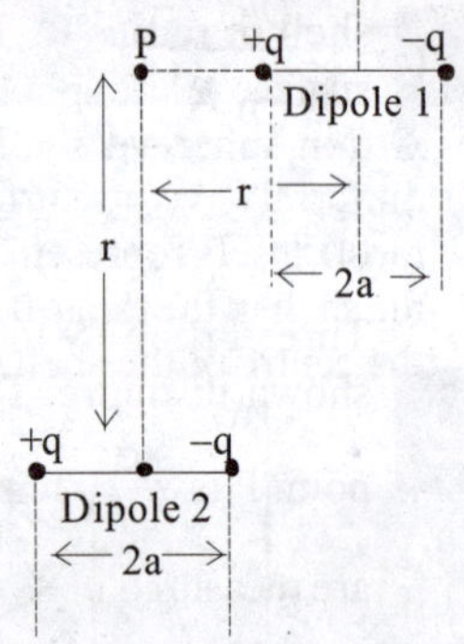

(a) $\frac{a}{r} \approx 20$

(b) $\frac{a}{r} \approx 10$

(c) $\frac{a}{r} \approx 0.5$

(d) $\frac{a}{r} \approx 3$

112. A line charge of length $\frac{'a'}{2}$ is kept at the center of an edge BC of a cube ABCDEFGH having edge length 'a' as shown in the figure. If the density of line is λC per unit length, then the total electric flux through all the faces of the cube will be _____.
(Take ϵ_0 as the free space permittivity)

[NCERT : P L-22, 23 | Jan. 22, 2025 (I)]

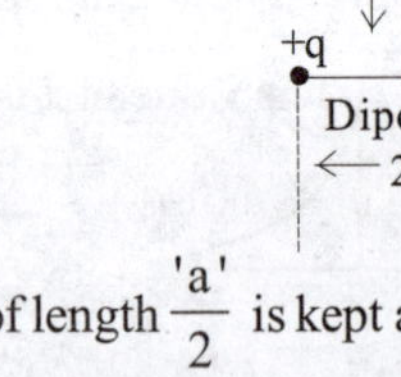

(a) $\frac{\lambda a}{8\epsilon_0}$ (b) $\frac{\lambda a}{16\epsilon_0}$ (c) $\frac{\lambda a}{2\epsilon_0}$ (d) $\frac{\lambda a}{4\epsilon_0}$

113. An electric field $\vec{E} = (2x\hat{i})\ NC^{-1}$ exists in space. A cube of side 2m is placed in the space as per figure given below. The electric flux through the cube is .... Nm^2/C.
[NCERT : P L-22 | April 9, 2024 (II)]

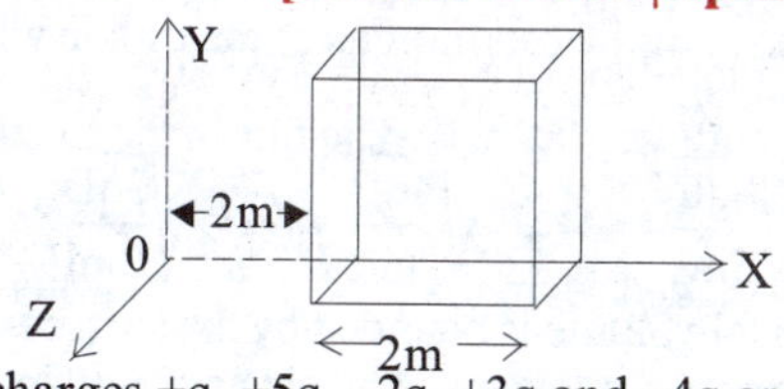

114. Five charges +q, +5q, –2q, +3q and –4q are situated as shown in the figure. The electric flux due to this configuration through the surface S is :
[NCERT : P L-22 | April 9, 2024 (II)]

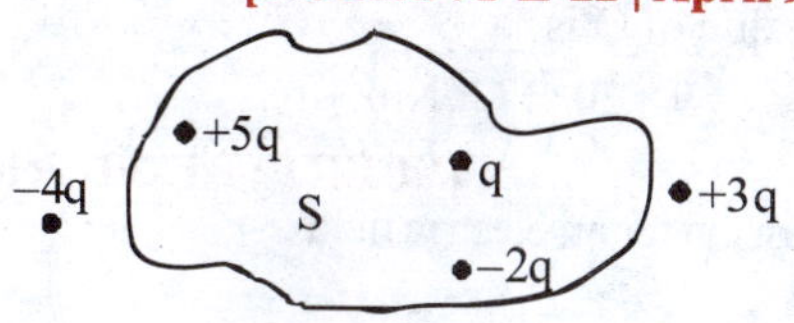

(a) $\frac{5q}{\epsilon_0}$ (b) $\frac{4q}{\epsilon_0}$ (c) $\frac{3q}{\epsilon_0}$ (d) $\frac{q}{\epsilon_0}$

115. An electric field, $\vec{E} = \frac{2\hat{i} + 6\hat{j} + 8\hat{k}}{\sqrt{6}}$ passes through the surface of 4 m^2 area having unit vector $\hat{n} = \left(\frac{2\hat{i} + \hat{j} + \hat{k}}{\sqrt{6}}\right)$. The electric flux for that surface is ________ V m.
[NCERT : P L-22 | April 8, 2024 (I)]

116. σ is the uniform surface charge density of a thin spherical shell of radius R. The electric field at any point on the surface of the spherical shell is :
[NCERT : P L-35 | April 6, 2024 (I)]

(a) $\frac{\sigma}{\epsilon_0 R}$ (b) $\frac{\sigma}{2\epsilon_0}$ (c) $\frac{\sigma}{\epsilon_0}$ (d) $\frac{\sigma}{4\epsilon_0}$

117. Three infinitely long charged thin sheets are placed as shown in figure. The magnitude of electric field at the point P is $\frac{x\sigma}{\epsilon_0}$. The value of x is ________ (all quantities are measured in SI units).
[NCERT : P L-35 | April 6, 2024 (I)]

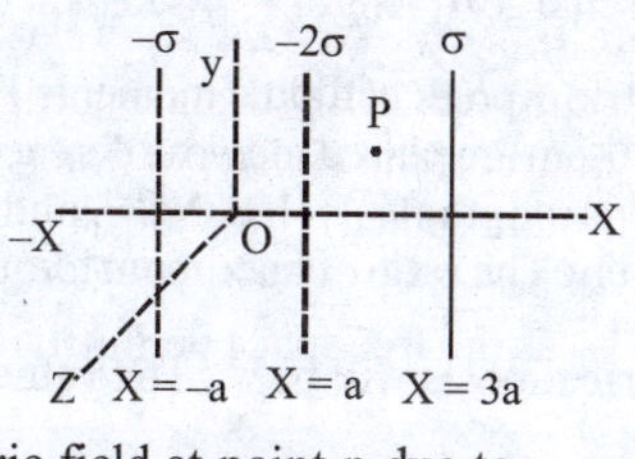

118. The electric field at point p due to an electric dipole is E. The electric field at point R on equitorial line will be $\frac{E}{x}$. The value of x:
[NCERT : PL-24 | April 5, 2024 (II)]

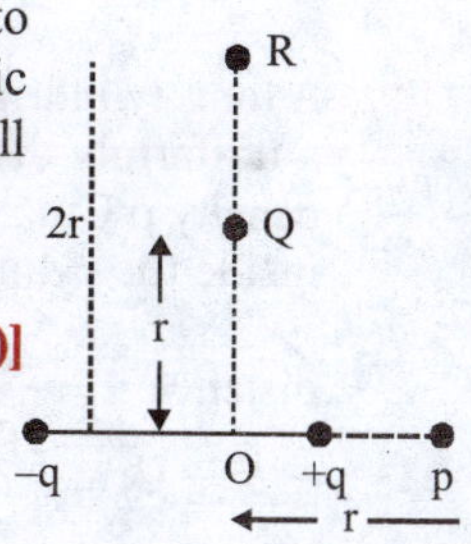

119. An infinite plane sheet of charge having uniform surface charge density $+\sigma_s$ C/m^2 is placed on x-y plane. Another infinitely long line charge having uniform linear charge density $+\lambda_e$ C/m is placed at z = 4m plane and parallel to y-axis. If the magnitude values $|\sigma_s| = 2|\lambda_e|$ then at point (0, 0, 2), the ratio of magnitudes of electric field values due to sheet charge to that of line charge is $\pi\sqrt{n} : 1$. The value of n is ______.
[NCERT : P L-35 | April 4, 2024 (I)]

120. An infinitely long positively charged straight thread has a linear charge density λ Cm^{-1}. An electron revolves along a circular path having axis along the length of the wire. The graph that correctly represents the variation of the kinetic energy of electron as a function of radius of circular path from the wire is :
[NCERT : P L-34 | April 4, 2024 (I)]

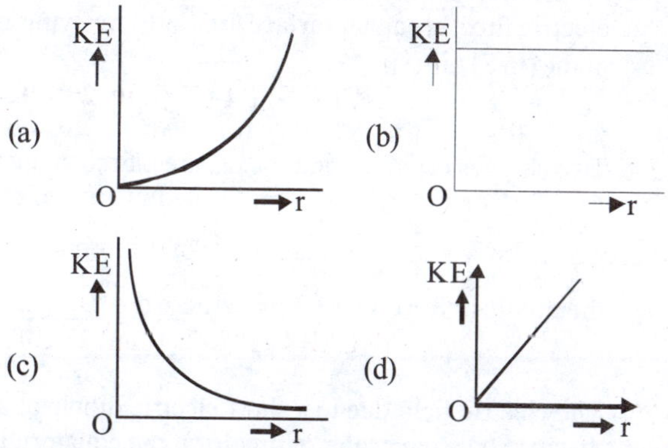

121. A charge q is placed at the center of one of the surface of a cube. The flux linked with the cube is :-
[NCERT : P L-30 | April 4, 2024 (II)]

(a) $\frac{q}{4\epsilon_0}$ (b) $\frac{q}{2\epsilon_0}$ (c) $\frac{q}{8\epsilon_0}$ (d) Zero

122. C_1 and C_2 are two hollow concentric cubes enclosing charges 2Q and 3Q respectively as shown in figure . The ratio of elecetric flux passing through C_1 and C_2 is :
[NCERT : P L-30 | Feb. 1, 2024 (II)]

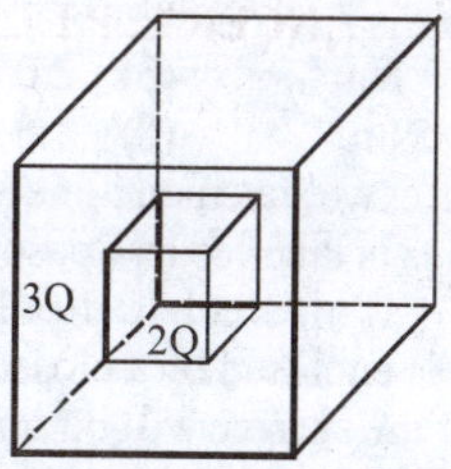

(a) 3:2 b) 5:2 (c) 2:5 (d) 2:3

123. A particle of charge '$-q$' and mass 'm' moves in a circle of radius 'r' around an infinitely long line charge of linear charge density '$+\lambda$'. Then time period will be given as: (Consider k as Coulomb's constant)
[NCERT : P L-14, 34 | Jan. 30, 2024 (II)]

(a) $T^2 = \frac{4\pi^2 m}{2k\lambda q} r^3$ (b) $T = 2\pi r\sqrt{\frac{m}{2k\lambda q}}$

(c) $T = \frac{1}{2\pi r}\sqrt{\frac{m}{2k\lambda q}}$ (d) $T = \frac{1}{2\pi}\sqrt{\frac{2k\lambda q}{m}}$

124. Two charges of 5Q and –2Q are situated at the points (3a, 0) and (–5a, 0) respectively. The electric flux through a sphere of radius '4a' having center at origin is:

[NCERT : P L-30 | Jan. 29, 2024 (I)]

(a) $\frac{3Q}{\varepsilon_0}$ (b) $\frac{5Q}{\varepsilon_0}$ (c) $\frac{7Q}{\varepsilon_0}$ (d) $\frac{2Q}{\varepsilon_0}$

125. An electron is moving under the influence of the electric field of a uniformly charged infinite plane sheet S having surface charge density $+\sigma$. The electron at t = 0 is at a distance of 1 m from S and has a speed of 1 m/s. The maximum value of σ if the electron strikes S at t = 1 s is $\alpha\left[\frac{m\epsilon_0}{e}\right]\frac{C}{m^2}$, the value of α is ______.

[NCERT : P L-30 | Jan. 29, 2024 (I)]

126. An electric field is given by $(6\hat{i}+5\hat{j}+3\hat{k})N/C$. The electric flux through a surface area $30\hat{i}\ m^2$ lying in YZ-plane (in SI unit) is ;

[NCERT : P L-22 | Jan. 29, 2024 (II)]

(a) 180 (b) 90 (c) 150 (d) 60

127. Two charges of – 4 μC and + 4 μC are placed at the points A(1, 0, 4)m and B(2,–1, 5) m located in an electric field $\vec{E} = 0.20\hat{i}\ V/cm$. The magnitude of the torque acting on the dipole is $8\sqrt{\alpha}\times 10^{-5}$ Nm , Where α = ______.

[NCERT : P L-27 | Jan. 27, 2024 (II)]

128. The electric field due to a short electric dipole at a large distance (r) from center of dipole on the equatorial plane varies with distance as :

[NCERT : P L-24 | April 15, 2023 (I)]

(a) r (b) $\frac{1}{r}$ (c) $\frac{1}{r^3}$ (d) $\frac{1}{r^2}$

129. Two charges each of magnitude 0.01 C and separated by a distance of 0.4 mm constitute an electric dipole. If the dipole is placed in an uniform electric field '$\vec{E}$' of 10 dyne/C making 30° angle with $\vec{E}$, the magnitude of torque acting on dipole is : **[NCERT : P L-29 | April 13, 2023 (I)]**

(a) 4.0×10^{-10} Nm (b) 2.0×10^{-10} Nm
(c) 1.0×10^{-8} Nm (d) 1.5×10^{-9} Nm

130. Given below are two statement: one is labelled as Assertion A and the other is labelled as Reason R.

Assertion A : If an electric dipole of dipole moment 30×10^{-5} cm is enclosed by a closed surface, the net flux coming out of the surface will be zero.

Reason R : Electric dipole consists of two equal and opposite charges. **[NCERT : P L-29 | April 12, 2023 (I)]**

In the light of above, statements, choose the correct answer from the options given below:

(a) Both A and R are true and R is the correct explanation of A
(b) A is true but R is false
(c) Both A and R true but R is NOT the correct explanation of A
(d) A is false but R is true

131. A dipole comprises of two charged particles of identical magnitude q and opposite in nature. The mass 'm' of the positive charged particle is half of the mass of the negative charged particle. The two charges are separated by a distance 'ℓ'. If the dipole is placed in a uniform electric field '$\vec{E}$'; in such a way that dipole axis makes a very small angle with the electric field, '$\vec{E}$'. The angular frequency of the oscillations of the dipole when released is given by: **[NCERT : P L-29 | April 6, 2023 (II)]**

(a) $\sqrt{\frac{8qE}{3ml}}$ (b) $\sqrt{\frac{4qE}{ml}}$ (c) $\sqrt{\frac{3qE}{2ml}}$ (d) $\sqrt{\frac{8qE}{ml}}$

132. A cubical volume is bounded by the surfaces x = 0, x = a, y = 0, y = a, z = 0, z = a. The electric field in the region is given by $\vec{E} = E_0 x\hat{i}$. Where $E_0 = 4\times 10^4\ NC^{-1}m^{-1}$. If a = 2 cm, the charge contained in the cubical volume is $Q\times 10^{-14}$ C. The value of Q is ______.

Take $\epsilon_0 = 9\times 10^{-12}\ C^2/Nm^2$)

[NCERT : P L-30 | Feb. 1, 2023 (II)]

133. Expression for an electric field is given by $\vec{E} = 4000x^2\hat{i}\frac{V}{m}$. The electric flux through the cube of side 20 cm when placed in electric field (as shown in the figure) is ______ V cm.

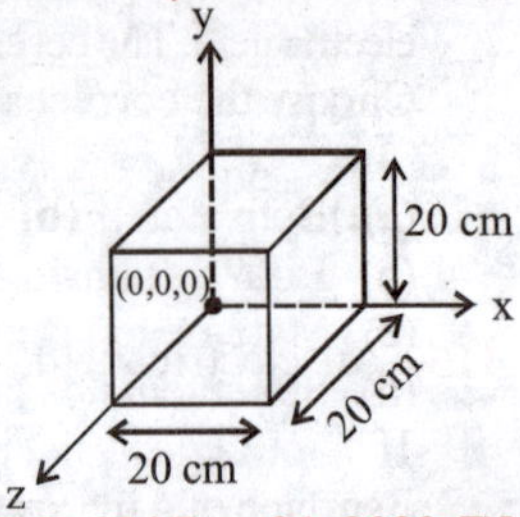

[NCERT : P L-22 | Jan. 31, 2023 (I)]

134. As shown in figure, a cuboid lies in a region with electric field $E = 2x^2\hat{i} - 4y\hat{j} + 6\hat{k}$. The magnitude of charge within the cuboid is $n\epsilon_0$ C N/C. The value of n is ______ (if dimension of cuboid is $1\times 2\times 3\ m^3$)

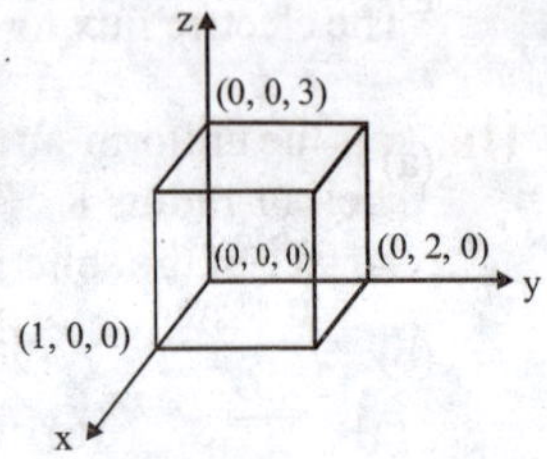

[NCERT : P L-30 | Jan. 30, 2023 (II)]

135. In a cuboid of dimension $2L\times 2L\times L$, a charge q is placed at the centre of the surface 'S' having area of $4L^2$. The flux through the opposite surface to 'S' is given by

[NCERT : P L-30 | Jan. 29, 2023 (I)]

(a) $\frac{q}{12\varepsilon_0}$ (b) $\frac{q}{3\varepsilon_0}$ (c) $\frac{q}{2\varepsilon_0}$ (d) $\frac{q}{6\varepsilon_0}$

136. Two electric dipoles of dipole moments 1.2×10^{-30} cm and 2.4×10^{-30} cm are placed in two difference uniform electric fields of strengths $5\times 10^4\ NC^{-1}$ and $15\times 10^4\ NC^{-1}$ respectively. The ratio of maximum torque experienced by the electric dipoles will be $\frac{1}{x}$. The value of x is ______.

[NCERT : P L-27 | July 28, 2022 (I)]

137. A long cylindrical volume contains a uniformly distributed charge of density $\rho\ Cm^{-3}$. The electric field inside the cylindrical volume at a distance $x = \frac{2\varepsilon_0}{\rho}$ m from its axis is ______ Vm^{-1}

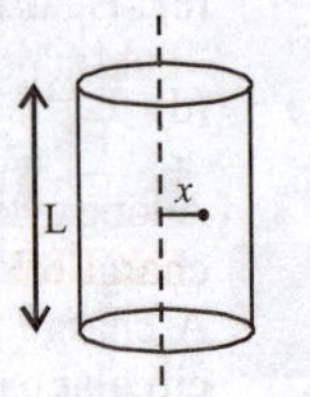

[NCERT : P L-30 | July 27, 2022 (I)]

138. Three point charges of magnitude 5μC, 0.16μC and 0.3μC are located at the vertices A, B, C of a right angled triangle whose sides are AB = 3cm, BC = $3\sqrt{2}$ cm and CA = 3 cm and point A is the right angle corner. Charge at point A experiences ______ N of electrostatic force due to the other two charges. **[NCERT : P L-12 | July 26, 2022 (II)]**

139. The volume charge density of a sphere of radius 6 m is 2 μC cm^{-3}. The number of lines of force per unit surface area coming out from the surface of the sphere is ______ $\times 10^{10}$ NC^{-1}. **[NCERT : P L-30 | July 25, 2022 (I)]**
(Given : Permittivity of vacuum $\varepsilon_0 = 8.85 \times 10^{-12} C^2N^{-1} - m^{-2}$)

140. Given below are two statements :
Statement-I : A point charge is brought in an electric field. The value of electric field at a point near to the charge may increase if the charge is positive.
Statement-II : An electric dipole is placed in a non-uniform electric field. The net electric force on the dipole will not be zero.
Choose the correct answer from the options given below: **[NCERT : P L-27, 14 | June 28, 2022 (I)]**
(a) Both statement-I and statement-II are true.
(b) Both statement-I and statement-I are false.
(c) Statement-I is true but statement-II is false.
(d) Statement-I is false but statement-II is true.

141. If a charge q is placed at the centre of a closed hemispherical non-conducting surface, the total flux passing through the flat surface would be : **[NCERT : P L-30 | June 27, 2022 (II)]**
(a) $\frac{q}{\varepsilon_0}$
(b) $\frac{q}{2\varepsilon_0}$
(c) $\frac{q}{4\varepsilon_0}$
(d) $\frac{q}{2\pi\varepsilon_0}$

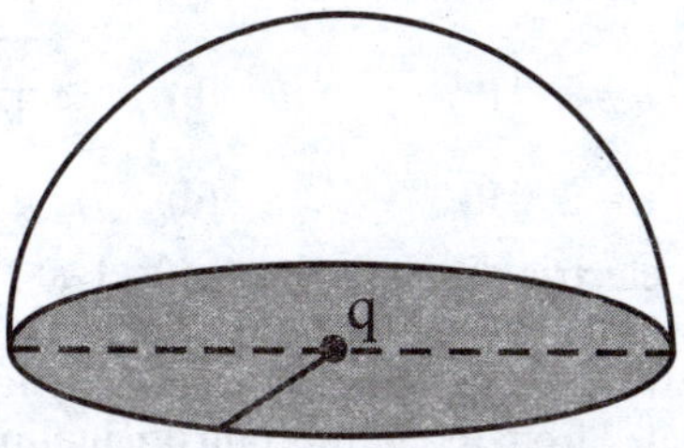

142. In the figure, a very large plane sheet of positive charge is shown. P_1 and P_2 are two points at distance l and $2l$ from the charge distribution. If σ is the surface charge density, then the magnitude of electric fields E_1 and E_2 at P_1 and P_2 respectively are : **[NCERT : P L-34 | June 25, 2022 (I)]**
(a) $E_1 = \sigma/\varepsilon_0$, $E_2 = \sigma/2\varepsilon_0$
(b) $E_1 = 2\sigma/\varepsilon_0$, $E_2 = \sigma/\varepsilon_0$
(c) $E_1 = E_2 = \sigma/2\varepsilon_0$
(d) $E_1 = E_2 = \sigma/\varepsilon_0$

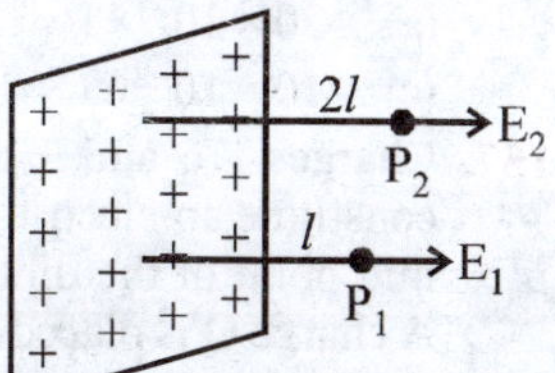

143. A long cylindrical volume contains a uniformly distributed charge of density ρ. The radius of cylindrical volume is R. A charge particle (q) revolves around the cylinder in a circular path. The kinetic energy of the particle is : **[NCERT : P L-30 | June 24, 2022 (II)]**
(a) $\frac{\rho q R^2}{4\varepsilon_0}$ (b) $\frac{\rho q R^2}{2\varepsilon_0}$
(c) $\frac{q\rho}{4\varepsilon_0 R^2}$ (d) $\frac{4\varepsilon_0 R^2}{q\rho}$

144. A cube is placed inside an electric field, $\vec{E} = 150y^2\hat{j}$. The side of the cube is 0.5 m and is placed in the field as shown in the given figure. The charge inside the cube is : **[NCERT : P L-30 | Sep. 1, 2021 (II)]**
(a) 3.8×10^{-11} C
(b) 8.3×10^{-11} C
(c) 3.8×10^{-12} C
(d) 8.3×10^{-12} C

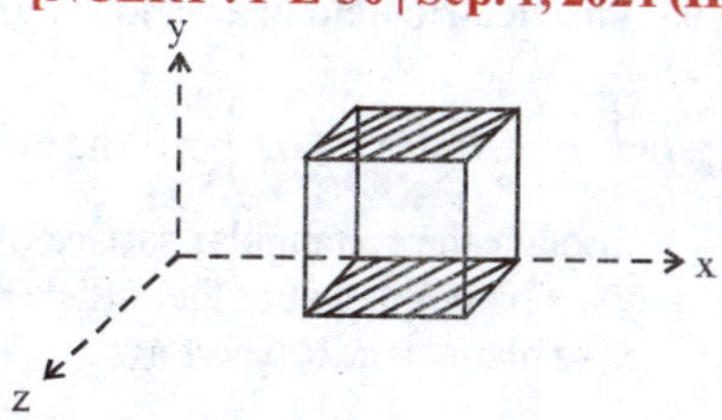

145. Choose the **incorrect** statement :
(A) The electric lines of force entering into a Gaussian surface provide negative flux.
(B) A charge 'q' is placed at the centre of a cube. The flux through all the faces will be the same.
(C) In a uniform electric field net flux through a closed Gaussian surface containing no net charge, is zero.
(D) When electric field is parallel to a Gaussian surface, it provides a finite non-zero flux.
Choose the most appropriate answer from the options given below **[NCERT : P L-29 | Aug. 31, 2021 (II)]**
(a) (C) and (D) only (b) (B) and (D) only
(c) (D) only (d) (A) and (C) only

146. Two ideal electric dipoles A and B, having their dipole moment p_1 and p_2 respectively are placed on a plane with their centres at O as shown in the figure.

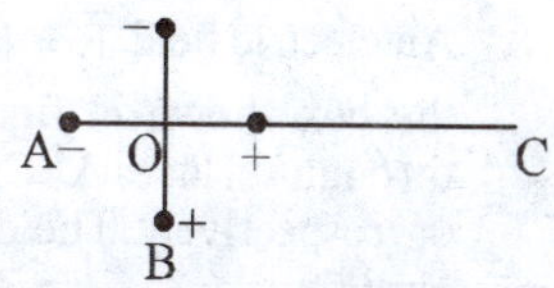

At point C on the axis of dipole A, the resultant electric field is making an angle of 37° with the axis. The ratio of the dipole moment of A and B, $\frac{p_1}{p_2}$ is : (take $\sin 37° = \frac{3}{5}$) **[NCERT : P L-23, 24 | July 25, 2021 (II)]**
(a) $\frac{3}{8}$ (b) $\frac{3}{2}$ (c) $\frac{2}{3}$ (d) $\frac{4}{3}$

147. The electric field in a region is given by $\vec{E} = \frac{2}{5}E_0\hat{i} + \frac{3}{5}E_0\hat{j}$ with $E_0 = 4.0 \times 10^3 \frac{N}{C}$. The flux of this field through a rectangular surface area 0.4 m^2 parallel to the Y – Z plane is ______ $Nm^2 C^{-1}$ **[NCERT : P L-22 | March 17, 2021 (II)]**

148. A point charge of $+12\ \mu C$ is at a distance 6 cm vertically above the centre of side of a square of side 12 cm as shown in figure. The magnitude of the electric flux through the square will be ______ $\times 10^3\ Nm^2/C$.

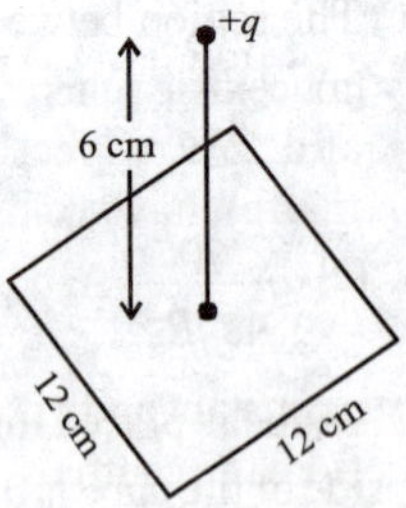

[NCERT : P L-30 | Feb. 24, 2021 (II)]

149. The electric field in a region is given by $\vec{E} = \left(\frac{3}{5}E_0\hat{i} + \frac{4}{5}E_0\hat{j}\right)\frac{N}{C}$. The ratio of flux of reported field through the rectangular surface of area $0.2\ m^2$ (parallel to y – z plane) to that of the surface of area $0.3\ m^2$ (parallel to x – z plane) is a : b, where a = ______.

[NCERT : P L-29 | Feb. 25, 2021 (I)]

150. Two identical electric point dipoles have dipole moments $\vec{P_1} = P\hat{i}$ and $\vec{P_2} = -P\hat{i}$ and are held on the x axis at distance 'a' from each other. When released, they move along x-axis with the direction of their dipole moments remaining unchanged. If the mass of each dipole is 'm', their speed when they are infinitely far apart is :

[Sep. 06, 2020 (II)]

(a) $\frac{P}{a}\sqrt{\frac{1}{\pi\varepsilon_0 ma}}$ (b) $\frac{P}{a}\sqrt{\frac{1}{2\pi\varepsilon_0 ma}}$

(c) $\frac{P}{a}\sqrt{\frac{2}{\pi\varepsilon_0 ma}}$ (d) $\frac{P}{a}\sqrt{\frac{2}{2\pi\varepsilon_0 ma}}$

151. An electric field $\vec{E} = 4x\hat{i} - (y^2+1)\hat{j}$ N/C passes through the box shown in figure. The flux of the electric field through surfaces ABCD and BCGF are marked as ϕ_I and ϕ_{II} respectively. The difference between $(\phi_I - \phi_{II})$ is (in Nm^2/C) ______. [NCERT : P L-29 | 9 Jan 2020, (II)]

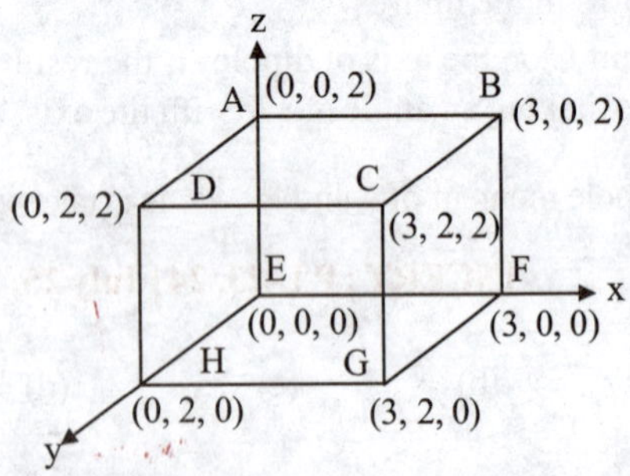

152. In finding the electric field using Gauss law the formula $|\vec{E}| = \frac{q_{enc}}{\epsilon_0 |A|}$ is applicable. In the formula ϵ_0 is permittivity of free space, A is the area of Gaussian surface and q_{enc} is charge enclosed by the Gaussian surface. This equation can be used in which of the following situation?

[NCERT : P L-29 | 8 Jan 2020, (I)]

(a) Only when the Gaussian surface is an equipotential surface.

(b) Only when the Gaussian surface is an equipotential surface and $|\vec{E}|$ is constant on the surface.

(c) Only when $|\vec{E}|$ = constant on the surface.

(d) For any choice of Gaussian surface.

153. Shown in the figure is a shell made of a conductor. It has inner radius a and outer radius b, and carries charge Q. At its centre is a dipole $\vec{p}$ as shown. In this case :

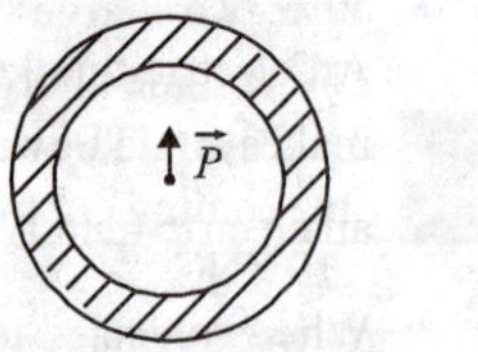

[NCERT : P L-35 | 12 April 2019, (I)]

(a) surface change density on the inner surface is uniform and equal to $\frac{Q/2}{4\pi a^2}$

(b) electric field outside the shell is the same as that of a point charge at the centre of the shell.

(c) surface charge density on the outer surface depends on $|\vec{p}|$

(d) surface charge density on the inner surface of the shell is zero everywhere.-

154. Let a total charge 2 Q be distributed in a sphere of radius R, with the charge density given by $\rho(r) = kr$, where r is the distance from the centre. Two charges A and B, of – Q each, are placed on diametrically opposite points, at equal distance, a, from the centre. If A and B do not experience any force, then. [NCERT : P L-29 | 12 April 2019, (II)]

(a) $a = 8^{-1/4}R$ (b) $a = \frac{3R}{2^{1/4}}$

(c) $a = 2^{-1/4}R$ (d) $a = R/\sqrt{3}$

155. An electric dipole is formed by two equal and opposite charges q with separation d. The charges have same mass m. It is kept in a uniform electric field E. If it is slightly rotated from its equilibrium orientation, then its angular frequency ω is : [NCERT : P L-27 | 8 April 2019, (II)]

(a) $\sqrt{\frac{qE}{md}}$ (b) $\sqrt{\frac{2qE}{md}}$ (c) $2\sqrt{\frac{qE}{md}}$ (d) $\sqrt{\frac{qE}{2md}}$

156. An electric field of 1000 V/m is applied to an electric dipole at angle of 45°. The value of electric dipole moment is 10^{-29} C.m. What is the potential energy of the electric dipole? [11 Jan 2019, (II)]

(a) -20×10^{-18} J (b) -7×10^{-27} J

(c) -10×10^{-29} J (d) -9×10^{-20} J

157. Charges – q and + q located at A and B, respectively, constitute an electric dipole. Distance AB = 2a, O is the mid point of the dipole and OP is perpendicular to AB. A charge Q is placed at P where OP = y and $y >> 2a$. The charge Q experiences an electrostatic force F. If Q is now moved along the equatorial line to P′ such that OP′ $= \left(\frac{y}{3}\right)$, the force on Q will be close to: $\left(\frac{y}{3} >> 2a\right)$

[NCERT : P L-24 | 10 Jan 2019, (II)]

(a) 3 F

(b) $\frac{F}{3}$

(c) 9 F

(d) 27 F

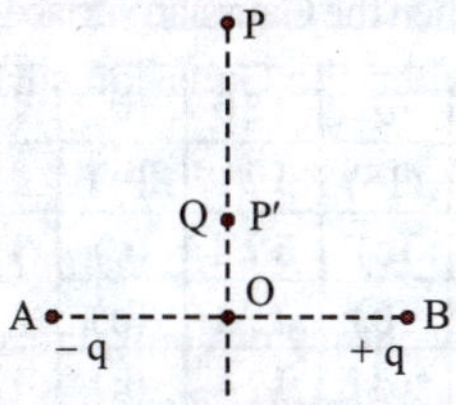

158. An electric dipole has a fixed dipole moment $\vec{p}$, which makes angle θ with respect to x-axis. When subjected to an electric field $\vec{E_1} = E\hat{i}$, it experiences a torque $\vec{T_1} = \tau\hat{k}$. When subjected to another electric field $\vec{E_2} = \sqrt{3}E_1\hat{j}$ it experiences torque $\vec{T_2} = -\vec{T_1}$. The angle θ is :

[NCERT : P L-27 | 2017]

(a) 60° (b) 90°

(c) 30° (d) 45°

159. Four closed surfaces and corresponding charge distributions are shown below.

[NCERT : P L-31 | Online April 9, 2017]

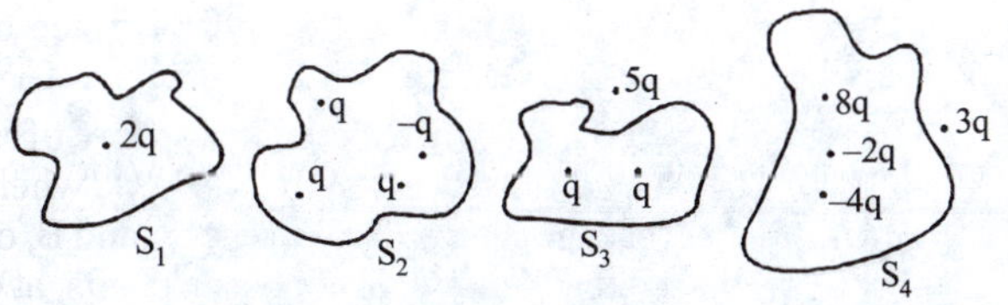

Let the respective electric fluxes through the surfaces be ϕ_1, ϕ_2, ϕ_3, and ϕ_4. Then :

(a) $\phi_1 < \phi_2 = \phi_3 > \phi_4$ (b) $\phi_1 > \phi_2 > \phi_3 > \phi_4$

(c) $\phi_1 = \phi_2 = \phi_3 = \phi_4$ (d) $\phi_1 > \phi_3 ; \phi_2 < \phi_4$

160. The region between two concentric spheres of radii 'a' and 'b', respectively (see figure), have volume charge density $\rho = \frac{A}{r}$, where A is a constant and r is the istance from the centre.

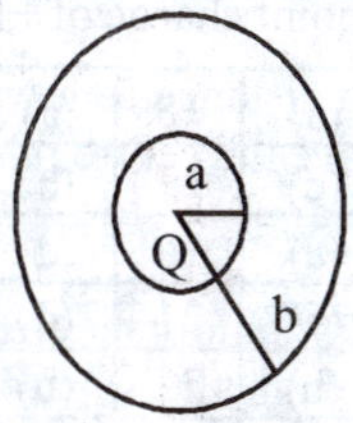

At the centre of the spheres is a point charge Q. The value of A such that the electric field in the region between the spheres will be constant, is:

[NCERT : P L-29 | 2016]

(a) $\frac{2Q}{\pi(a^2 - b^2)}$ (b) $\frac{2Q}{\pi a^2}$

(c) $\frac{Q}{2\pi a^2}$ (d) $\frac{Q}{2\pi(b^2 - a^2)}$

161. An electric dipole is placed at an angle of 30° to a non-uniform electric field. The dipole will experience

[NCERT : P L-27 | 2006]

(a) a translational force only in the direction of the field

(b) a translational force only in a direction normal to the direction of the field

(c) a torque as well as a translational force

(d) a torque only

162. If the electric flux entering and leaving an enclosed surface respectively is ϕ_1 and ϕ_2, the electric charge inside the surface will be **[NCERT : P L-30 | 2003]**

(a) $(\phi_2 - \phi_1)\varepsilon_o$ (b) $(\phi_1 - \phi_2)/\varepsilon_o$

(c) $(\phi_2 - \phi_1)/\varepsilon_o$ (d) $(\phi_1 - \phi_2)\varepsilon_o$

ANSWER KEYS

1	(a)	**18**	(a)	**35**	(d)	**52**	(b)	**69**	(c)	**86**	(a)	**103**	(d)	**120**	(b)	**137**	(1)	**154**	(a)
2	(d)	**19**	(b)	**36**	(b)	**53**	(c)	**70**	(c)	**87**	(d)	**104**	(15)	**121**	(b)	**138**	(17)	**155**	(b)
3	(a)	**20**	(d)	**37**	(d)	**54**	(45)	**71**	(c)	**88**	(c)	**105**	(c)	**122**	(c)	**139**	(45)	**156**	(b)
4	(5)	**21**	(d)	**38**	(b)	**55**	(a)	**72**	(b)	**89**	(d)	**106**	(b)	**123**	(b)	**140**	(a)	**157**	(d)
5	(b)	**22**	(b)	**39**	(d)	**56**	(2)	**73**	(d)	**90**	(c)	**107**	(d)	**124**	(b)	**141**	(b)	**158**	(a)
6	(3)	**23**	(a)	**40**	(a)	**57**	(c)	**74**	(b)	**91**	(b)	**108**	(a)	**125**	(8)	**142**	(c)	**159**	(c)
7	(3)	**24**	(b)	**41**	(b)	**58**	(b)	**75**	(c)	**92**	(c)	**109**	(d)	**126**	(a)	**143**	(a)	**160**	(c)
8	(b)	**25**	(b)	**42**	(c)	**59**	(d)	**76**	(c)	**93**	(c)	**110**	(48)	**127**	(2)	**144**	(b)	**161**	(c)
9	(2)	**26**	(b)	**43**	(b)	**60**	(a)	**77**	(a)	**94**	(c)	**111**	(d)	**128**	(c)	**145**	(c)	**162**	(a)
10	(3)	**27**	(12)	**44**	(d)	**61**	(b)	**78**	(b)	**95**	(a)	**112**	(a)	**129**	(b)	**146**	(c)		
11	(b)	**28**	(36)	**45**	(c)	**62**	(b)	**79**	(d)	**96**	(a)	**113**	(16)	**130**	(a)	**147**	(640)		
12	(5440)	**29**	(a)	**46**	(b)	**63**	(a)	**80**	(c)	**97**	(b)	**114**	(b)	**131**	(c)	**148**	(226)		
13	(300)	**30**	(b)	**47**	(c)	**64**	(b)	**81**	(d)	**98**	(a)	**115**	(12)	**132**	(288)	**149**	(1)		
14	(2)	**31**	(d)	**48**	(c)	**65**	(1)	**82**	(d)	**99**	(b)	**116**	(c)	**133**	(640)	**150**	(b)		
15	(24)	**32**	(c)	**49**	(5)	**66**	(b)	**83**	(d)	**100**	(c)	**117**	(2)	**134**	(12)	**151**	(–48)		
16	(a)	**33**	(a)	**50**	(a)	**67**	(d)	**84**	(b)	**101**	(b)	**118**	(16)	**135**	(d)	**152**	(b)		
17	(b)	**34**	(d)	**51**	(6)	**68**	(c)	**85**	(a)	**102**	(c)	**119**	(16)	**136**	(6)	**153**	(b)		

Electrostatic Potential and Capacitance

2

TREND ANALYSIS
JEE MAIN (2025-2021)

	2025				2024				2023				2022				2021			
	Session II		Session I		Session II		Session I		Session II		Session I		Session II		Session I		Session II		Session I	
	MCQs	NVQs	MCQs	NVQs	MCQs	NVQs	MCQs	NVQs	MCQs	NVQs	MCQs	NVQs	MCQs	NVQs	MCQs	NVQs	MCQs	NVQs	MCQs	NVQs
Topic 1	4	–	2	–	2	1	3	2	2	2	4	1	2	–	1	–	1	–	–	2
Topic 2	1	–	3	1	–	–	–	–	–	1	–	–	–	–	–	1	–	–	–	–
Topic 3	4	3	6	1	2	2	1	2	3	5	2	4	6	3	10	1	4	2	3	6
Topic 4																				
Total	9	3	11	2	4	3	4	4	5	8	6	5	8	3	11	2	5	2	3	8

How to Code Qns for Better & Faster Revision

Type 1: Able to solve (No code Req)
Type 2: Solved with struggle or by seeing solution **(Use code NR means Needs Revision or orange stickers)**
Type 3: Tricky/ Could not solve on my own. Need help from Teacher/ Friend. **(Use code TQ means Tricky Question or red stickers)**

You can either hand write the code in boxes or else paste coloured stickers given at end of book.

Topic 1 - Electrostatic Potential and Equipotential Surfaces

1. Given below are two statements : one is labelled as **Assertion A** and the other is labelled as **Reason R**
Assertion A : Work done in moving a test charge between two points inside a uniformly charged spherical shell is zero, no matter which path is choosen.
Reason R : Electrostatic potential inside a uniformly charged spherical shell is constant and is same as that on the surface of the shell.
[NCERT : PL-52 | April 8, 2025 (II)]
In the light of the above statements, choose the correct answer from the options given below
(a) A is true but R is false
(b) Both A and R are true and R is the correct explanation of A
(c) A is false but R is true
(d) Both A and R are true but R is NOT the correct explanation of A

2. Two metal spheres of radius R and 3R have same surface charge density σ. If they are brought in contact and then separated, the surface charge density on smaller and bigger sphere becomes σ_1 and σ_2, respectively. The ratio $\frac{\sigma_1}{\sigma_2}$ is:
[NCERT : PL-52 | April 8, 2025 (II)]
(a) $\frac{1}{9}$ (b) 9 (c) $\frac{1}{3}$ (d) 3

3. The electrostatic potential on the surface of uniformly charged spherical shell of radius R = 10 cm is 120 V. The potential at the centre of shell, at a distance r = 5 cm from centre, and at a distance r = 15 cm from the centre of the shell respectively, are : **[NCERT : PL-52 | April 3, 2025 (I)]**
(a) 120V, 120V, 80V (b) 40V, 40V, 80V
(c) 0V, 0V, 80V (d) 0V, 120V, 40V

4. Two large plane parallel conducting plates are kept 10 cm apart as shown in figure. The potential difference between them is V. The potential difference between the points A and B (shown in the figure) is :
[NCERT : PL-48, 51, 52 | April 2, 2025 (II)]

(a) $\frac{1}{4}$V

(b) $\frac{2}{5}$V

(c) $\frac{3}{4}$V

(d) 1V

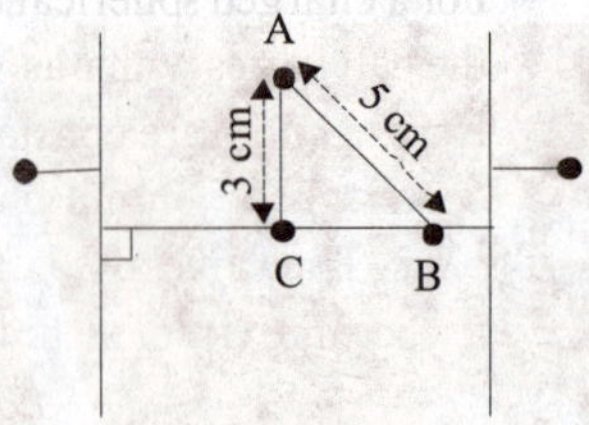

5. Three infinitely long wires with linear charge density λ are placed along the x-axis, y-axis and z-axis respectively. Which of the following denotes an equipotential surface ?

[NCERT : P L-54 | Jan. 28, 2025 (I)]

(a) $xy + yz + zx$ = constant
(b) $(x + y)(y + z)(z + x)$ = constant
(c) $(x^2 + y^2)(y^2 + z^2)(z^2 + x^2)$ = constant
(d) xyz = constant

6. For a short dipole placed at origin O, the dipole moment P is along x-axis, as shown in the figure.
If the electric potential and electric field at A are V_0 and E_0, respectively, then the correct combination of the electric potential and electric field, respectively, at point B on the y-axis is given by **[NCERT : P L-48 | Jan. 22, 2025 (II)]**

(a) $\frac{V_0}{2}$ and $\frac{E_0}{16}$

(b) zero and $\frac{E_0}{8}$

(c) zero and $\frac{E_0}{16}$

(d) V_0 and $\frac{E_0}{4}$

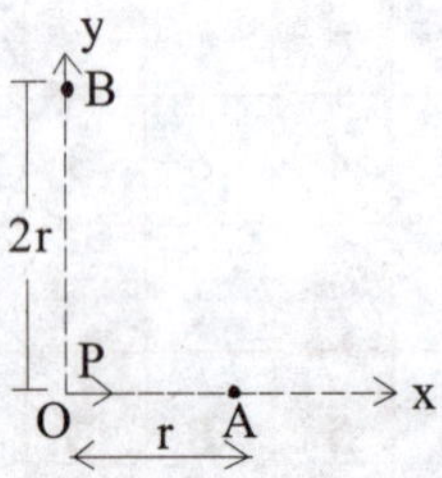

7. At the centre of a half ring of radius R = 10 cm and linear charge density 4n C m^{-1}, the potential is x π V. The value of x is ________. **[NCERT : P L-48 | April 9, 2024 (I)]**

8. Two charged conducting spheres of radii a and b are connected to each other by a conducting wire. The ratio of charges of the two spheres respectively is:

[NCERT : P L-48 | April 8, 2024 (I)]

(a) $\sqrt{ab}$ (b) ab (c) $\frac{a}{b}$ (d) $\frac{b}{a}$

9. The vehicles carrying inflammable fluids usually have metallic chains touching the ground :

[NCERT : P L-63 | April 8, 2024 (II)]

(a) To conduct excess charge due to air friction to ground and prevent sparking.
(b) To alert other vehicles.
(c) To protect tyres from catching dirt from ground.
(d) It is a custom.

10. The distance between charges +q and –q is $2l$ and between +2 q and –2 q is $4l$. The electrostatic potential at point P at a distance r from centre O is $-\alpha\left[\frac{ql}{r^2}\right] \times 10^9$ V, where the value of α is ________.

[NCERT : P L-52 | Jan. 31, 2024 (II)]

(Use $\frac{1}{4\pi\varepsilon_0} = 9 \times 10^9$ Nm^2 C^{-2})

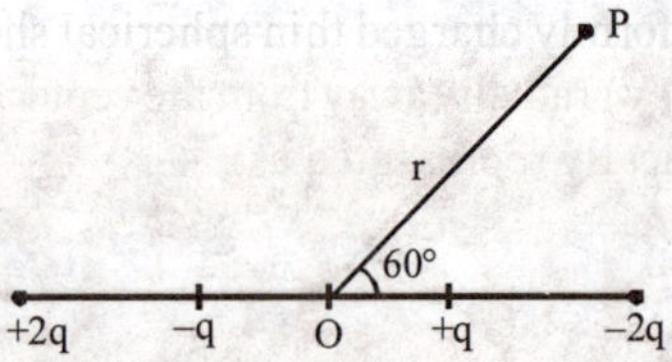

11. The electrostatic potential due to an electric dipole at a distance 'r' varies as: **[NCERT : P L-51 | Jan. 30, 2024 (I)]**

(a) r (b) $\frac{1}{r^2}$ (c) $\frac{1}{r^3}$ (d) $\frac{1}{r}$

12. An electric charge 10^{-6} μC is placed at origin (0, 0) m of X – Y co-ordinate system. Two points P and Q are situated at $(\sqrt{3}, \sqrt{3})$ m and $(\sqrt{6}, 0)$ m respectively. The potential difference between the points P and Q will be :

[NCERT : P L-48 | Jan. 27, 2024 (I)]

(a) $\sqrt{3}$V (b) $\sqrt{6}$V (c) 0 V (d) 3V

13. The electric potential at the surface of an atomic nucleus (Z = 50) of radius 9×10^{-13} cm is ________ $\times 10^6$ V.

[NCERT : P L-48 | Jan. 27, 2024 (II)]

14. Given below are two statements : one is labelled a Assertion (A) and the other is labelled as Reason(R)

Assertion (A) : Work done by electric field on moving a positive charge on an equipotential surface is always zero.

Reason (R) : Electric lines of forces are always perpendicular to equipotential surfaces.

In the light of the above statements, choose the most appropriate answer from the options given below :

[NCERT : P L-54 | Jan. 27, 2024 (II)]

(a) Both (A) and (R) are correct but (R) is not the correct explanation of (A)
(b) (A) is correct but (R) is not correct
(c) (A) is not correct but (R) is correct
(d) Both (A) and (R) are correct and (R) is the correct explanation of (A).

15. 64 identical drops each charged upto potential or 10 mV are combined to form a bigger drop. The potential of the bigger drop will be ________ mV.

[NCERT : P L-52 | April 12, 2023 (I)]

16. Three concentric spherical metallic shells X, Y and Z of radius a, b and c respectively [a < b < c] have surface charge densities σ, –σ and σ, respectively. The shells X and Z are at same potential. If the radii of X & Y are 2 cm and 3 cm, respectively. The radius of shell Z is ______ cm.

[NCERT : P L-48 | April 10, 2023 (I)]

17. Electric potential at a point 'P' due to a point charge of 5×10^{-9} C is 50 V. The distance of 'P' from the point charge is: (Assume, $\frac{1}{4\pi\varepsilon_0} = 9 \times 10^{+9}$ Nm^{2}C^{-2})

[NCERT : P L-48 | April 8, 2023 (II)]

(a) 3 cm (b) 9 cm (c) 90 cm (d) 0.9 cm

18. For a uniformly charged thin spherical shell, the electric potential (V) radially away from the center (O) of shell can be graphically represented as **[NCERT : P L-52 | April 6, 2023 (I)]**

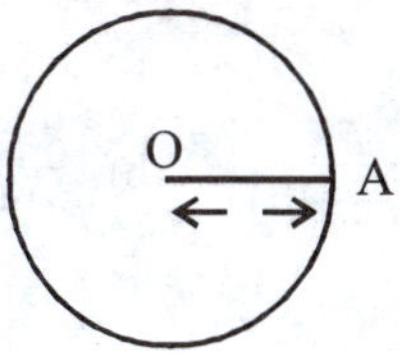

(a) 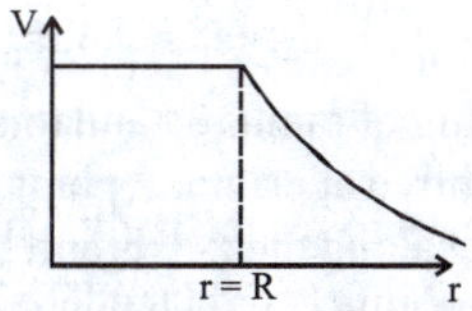

(b) 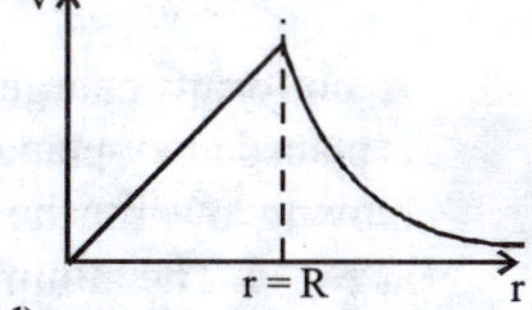

(c) 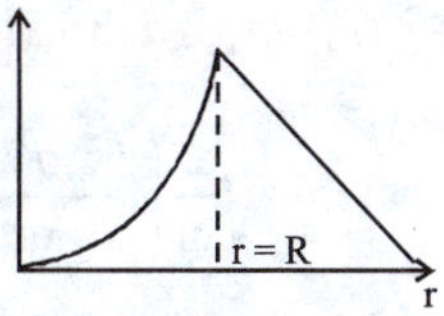

(d) 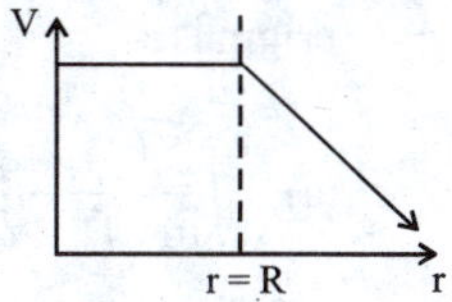

19. Which of the following correctly represents the variation of electric potential (V) of a charged spherical conductor of radius (R) with radial distance (r) from the centre? **[NCERT : P L-52 | Jan. 31, 2023 (I)]**

(a) 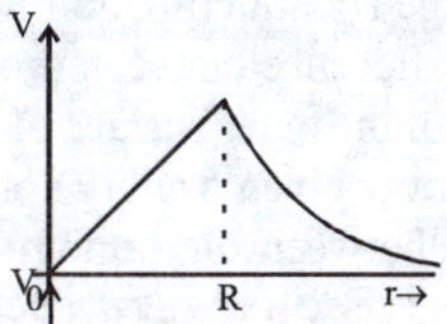

(b) 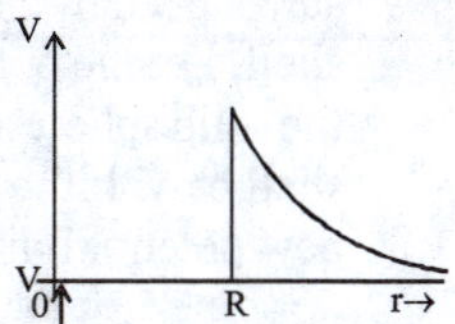

(c) 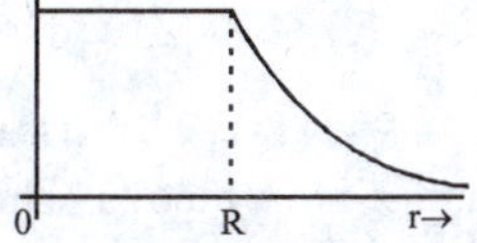

(d) 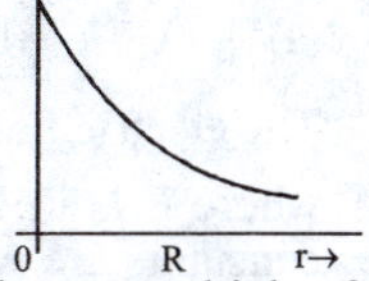

20. Considering a group of positive charges, which of the following statements is correct? **[NCERT : P L-51, 52 | Jan. 31, 2023 (II)]**

(a) Net potential of the system cannot be zero at a point but net electric field can be zero at that point.

(b) Net potential of the system at a point can be zero but net electric field can't be zero at that point.

(c) Both the net potential and the net field can be zero at a point.

(d) Both the net potential and the net electric field cannot be zero at a point.

21. Two isolated metallic solid spheres of radii R and 2R are charged such that both have same charge density σ. The spheres are then connected by a thin conducting wire. If the new charge density of the bigger sphere is σ'. The ratio $\frac{\sigma'}{\sigma}$ is : **[NCERT : P L-52 | Jan. 30, 2023 (I)]**

(a) $\frac{9}{4}$ (b) $\frac{4}{3}$ (c) $\frac{5}{3}$ (d) $\frac{5}{6}$

22. For a charged spherical ball, electrostatic potential inside the ball varies with r as $V = 2ar^2 + b$.
Here, a and b are constant and r is the distance from the center. The volume charge density inside the ball is $-\lambda a\varepsilon$. The value of λ is ________.
ε = permittivity of medium.
[NCERT : P L-55 | Jan. 29, 2023 (II)]

23. The electric potential at the centre of two concentric half rings of radii R_1 and R_2, having same linear charge density λ is **[NCERT : P L-48 | Jan. 24, 2023 (II)]**

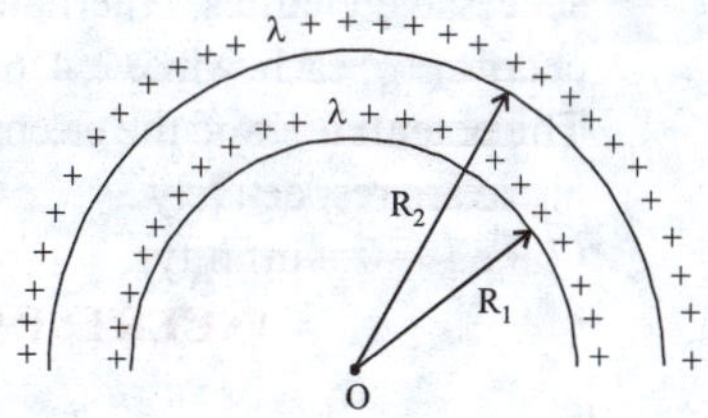

(a) $\frac{2\lambda}{\varepsilon_0}$ (b) $\frac{\lambda}{2\varepsilon_0}$

(c) $\frac{\lambda}{4\varepsilon_0}$ (d) $\frac{\lambda}{\varepsilon_0}$

24. Given below are two statements.

Statement I : Electric potential is constant within and at the surface of each conductor.

Statement II : Electrict field just outside a charged conductor is perpendicular to the surface of the conductor at every point.

In the light of the above statements, choose the most appropriate answer from the options given below. **[NCERT : P L-52 | July 29, 2022 (I)]**

(a) Both statement I and statement II are correct

(b) Both statement I and statement II are incorrect

(c) Statement I is correct but statement II is incorrect

(d) Statement I is incorrect but and statement II is correct

25. Two uniformly charged spherical conductors A and B of radii 5 mm and 10 mm are separated by a distance of 2 cm. If the spheres are connected by a conducting wire, then in equilibrium condition, the ratio of the magnitudes of the electric fields at the surface of the sphere A and B will be : **[NCERT : P L-52 | July 26, 2022 (II)]**

(a) 1 : 2 (b) 2 : 1 (c) 1 : 1 (d) 1 : 4

26. If the electric potential at any point (*x*, *y*, *z*) m in space is given by $V = 3x^2$ volt. The electric field at the point (1, 0, 3) m will be : **[NCERT : P L-55 | June 29, 2022 (II)]**

(a) 3 Vm^{-1}, directed along positive *x*-axis.

(b) 3 Vm^{-1}, directed along negative *x*-axis.

(c) 6 Vm^{-1}, directed along positive *x*-axis.

(d) 6 Vm^{-1}, directed along negative x-axis.

27. The two thin coaxial rings, each of radius 'a' and having charges +Q and –Q respectively are separated by a distance of 's'. The potential difference between the centres of the two rings is : **[NCERT : P L-48 | Aug. 26, 2021 (II)]**

(a) $\frac{Q}{2\pi\varepsilon_0}\left[\frac{1}{a}+\frac{1}{\sqrt{s^2+a^2}}\right]$ (b) $\frac{Q}{4\pi\varepsilon_0}\left[\frac{1}{a}+\frac{1}{\sqrt{s^2+a^2}}\right]$

(c) $\frac{Q}{4\pi\varepsilon_0}\left[\frac{1}{a}-\frac{1}{\sqrt{s^2+a^2}}\right]$ (d) $\frac{Q}{2\pi\varepsilon_0}\left[\frac{1}{a}-\frac{1}{\sqrt{s^2+a^2}}\right]$

28. 27 similar drops of mercury are maintained at 10 V each. All these spherical drops combine into a single big drop. The potential energy of the bigger drop is _____ times that of a smaller drop. **[NCERT : P L-56 | Feb. 26, 2021 (II)]**

29. 512 identical drops of mercury are charged to a potential of 2 V each. The drops are joined to form a single drop. The potential of this drop is _____ V.

[NCERT : P L-52 | Feb. 25, 2021 (I)]

30. Ten charges are placed on the circumference of a circle of radius R with constant angular separation between successive charges. Alternate charges 1, 3, 5, 7, 9 have charge $(+q)$ each, while 2, 4, 6, 8, 10 have charge $(-q)$ each. The potential V and the electric field E at the centre of the circle are respectively :
(Take $V = 0$ at infinity)

[NCERT : P L-51, 52 | Sep. 05, 2020 (II)]

(a) $V = \frac{10q}{4\pi\varepsilon_0 R}$; $E = 0$ (b) $V = 0$; $E = \frac{10q}{4\pi\varepsilon_0 R^2}$

(c) $V = 0$; $E = 0$ (d) $V = \frac{10q}{4\pi\varepsilon_0 R}$; $E = \frac{10q}{4\pi\varepsilon_0 R^2}$

31. Two isolated conducting spheres S_1 and S_2 of radius $\frac{2}{3}R$ and $\frac{1}{3}R$ have 12 μC and –3 μC charges, respectively, and are at a large distance from each other. They are now connected by a conducting wire. A long time after this is done the charges on S_1 and S_2 are respectively :

[NCERT : P L-52 | Sep. 03, 2020 (I)]

(a) 4.5 μC on both (b) +4.5 μC and –4.5 μC
(c) 3 μC and 6 μC (d) 6 μC and 3 μC

32. Concentric metallic hollow spheres of radii R and $4R$ hold charges Q_1 and Q_2 respectively. Given that surface charge densities of the concentric spheres are equal, the potential difference $V(R) - V(4R)$ is :

[NCERT : P L-52 | Sep. 03, 2020 (II)]

(a) $\frac{3Q_1}{16\pi\varepsilon_0 R}$ (b) $\frac{3Q_2}{4\pi\varepsilon_0 R}$ (c) $\frac{Q_2}{4\pi\varepsilon_0 R}$ (d) $\frac{3Q_1}{4\pi\varepsilon_0 R}$

33. A charge Q is distributed over two concentric conducting thin spherical shells radii r and R $(R > r)$. If the surface charge densities on the two shells are equal, the electric potential at the common centre is :

[NCERT : P L-52 | Sep. 02, 2020 (II)]

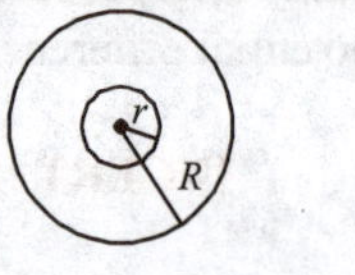

(a) $\frac{1}{4\pi\varepsilon_0}\frac{(R+r)}{2(R^2+r^2)}Q$ (b) $\frac{1}{4\pi\varepsilon_0}\frac{(2R+r)}{(R^2+r^2)}Q$

(c) $\frac{1}{4\pi\varepsilon_0}\frac{(R+2r)Q}{2(R^2+r^2)}$ (d) $\frac{1}{4\pi\varepsilon_0}\frac{(R+r)}{(R^2+r^2)}Q$

34. A point dipole $\vec{P} = P_0\hat{x}$ kept at the origin. The potential and electric field due to this dipole on the y-axis at a distance d are, respectively : (Take V = 0 at infinity)

[NCERT : P L-51 | 12 April 2019 (I)]

(a) $\frac{|\vec{p}|}{4\pi\varepsilon_0 d^2}, \frac{\vec{p}}{4\pi\varepsilon_0 d^3}$ (b) $0, \frac{-\vec{p}}{4\pi\varepsilon_0 d^3}$

(c) $0, \frac{|\vec{p}|}{4\pi\varepsilon_0 d^3}$ (d) $\frac{|\vec{p}|}{4\pi\varepsilon_0 d^2}, \frac{-\vec{p}}{4\pi\varepsilon_0 d^3}$

35. A uniformly charged ring of radius 3a and total charge q is placed in xy-plane centred at origin. A point charge q is moving towards the ring along the z-axis and has speed v at z = 4a. The minimum value of v such that it crosses the origin is : **[NCERT : P L-48 | 10 April 2019 (I)]**

(a) $\sqrt{\frac{2}{m}}\left(\frac{4}{15}\frac{q^2}{4\pi\varepsilon_0 a}\right)^{1/2}$ (b) $\sqrt{\frac{2}{m}}\left(\frac{1}{5}\frac{q^2}{4\pi\varepsilon_0 a}\right)^{1/2}$

(c) $\sqrt{\frac{2}{m}}\left(\frac{2}{15}\frac{q^2}{4\pi\varepsilon_0 a}\right)^{1/2}$ (d) $\sqrt{\frac{2}{m}}\left(\frac{1}{15}\frac{q^2}{4\pi\varepsilon_0 a}\right)^{1/2}$

36. A solid conducting sphere, having a charge Q, is surrounded by an uncharged conducting hollow spherical shell. Let the potential difference between the surface of the solid sphere and that of the outer surface of the hollow shell be V. If the shell is now given a charge of – 4 Q, the new potential difference between the same two surfaces is : **[NCERT : P L-55 | 8 April 2019 (I)]**

(a) –2V (b) 2V (c) 4V (d) V

37. The electric field in a region is given by $\vec{E} = (Ax + B)\hat{i}$, where E is in NC^{-1} and x is in metres. The values of constants are A = 20 SI unit and B = 10 SI unit. If the potential at $x = 1$ is V_1 and that at $x = -5$ is V_2, then $V_1 - V_2$ is : **[NCERT : P L-55 | 8 Jan. 2019 (II)]**

(a) 320 V (b) –48V (c) 180V (d) –520 V

38. The given graph shows variation (with distance r from centre) of : **[NCERT : P L-52 | 11 Jan. 2019 (I)]**

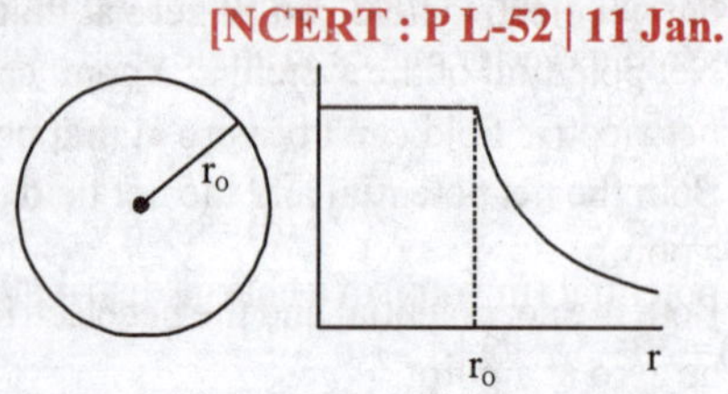

(a) Electric field of a uniformly charged sphere
(b) Potential of a uniformly charged spherical shell
(c) Potential of a uniformly charged sphere
(d) Electric field of a uniformly charged spherical shell

39. A charge Q is distributed over three concentric spherical shells of radii a, b, c (a < b < c) such that their surface charge densities are equal to one another.
The total potential at a point at distance r from their common centre, where r < a, would be:

[NCERT : P L-52 | 10 Jan. 2019 (I)]

(a) $\frac{Q}{12\pi\epsilon_0}\frac{ab+bc+ca}{abc}$ (b) $\frac{Q(a^2+b^2+c^2)}{4\pi\epsilon_0(a^3+b^3+c^3)}$

(c) $\frac{Q}{4\pi\epsilon_0(a+b+c)}$ (d) $\frac{Q(a+b+c)}{4\pi\epsilon_0(a^2+b^2+c^2)}$

40. Two electric dipoles, A, B with respective dipole moments $\vec{d}_A = -4\,qa\,\hat{i}$ and $\vec{d}_B = -2\,qa\,\hat{i}$ are placed on the *x*-axis with a separation R, as shown in the figure

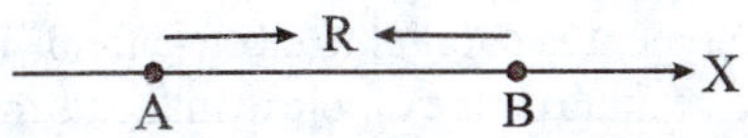

The distance from A at which both of them produce the same potential is: **[NCERT : P L-48 | 10 Jan. 2019 (I)]**

(a) $\frac{R}{\sqrt{2}+1}$ (b) $\frac{\sqrt{2}R}{\sqrt{2}+1}$ (c) $\frac{R}{\sqrt{2}-1}$ (d) $\frac{\sqrt{2}R}{\sqrt{2}-1}$

41. Consider two charged metallic spheres S_1 and S_2 of radii R_1 and R_2, respectively. The electric fields E_1 (on S_1) and E_2 (on S_2) on their surfaces are such that $E_1/E_2 = R_1/R_2$. Then the ratio V_1(on S_1)/V_2(on S_2) of the electrostatic potentials on each sphere is:

[NCERT : P L-52 | 8 Jan. 2019 (II)]

(a) R_1/R_2 (b) $(R_1/R_2)^2$

(c) (R_2/R_1) (d) $\left(\frac{R_1}{R_2}\right)^3$

42. Three concentric metal shells A, B and C of respective radii *a*, *b* and *c* ($a < b < c$) have surface charge densities $+\sigma, -\sigma$ and $+\sigma$ respectively. The potential of shell B is:

[NCERT : P L-52 | 2018]

(a) $\frac{\sigma}{\epsilon_0}\left[\frac{a^2-b^2}{a}+c\right]$ (b) $\frac{\sigma}{\epsilon_0}\left[\frac{a^2-b^2}{b}+c\right]$

(c) $\frac{\sigma}{\epsilon_0}\left[\frac{b^2-c^2}{b}+a\right]$ (d) $\frac{\sigma}{\epsilon_0}\left[\frac{b^2-c^2}{c}+a\right]$

43. There is a uniform electrostatic field in a region. The potential at various points on a small sphere centred at P, in the region, is found to vary between in the limits 589.0 V to 589.8 V. What is the potential at a point on the sphere whose radius vector makes an angle of 60° with the direction of the field? **[NCERT : P L-55 | Online April 8, 2017]**

(a) 589.5 V (b) 589.2 V

(c) 589.4 V (d) 589.6 V

44. The potential (in volts) of a charge distribution is given by

$V(z) = 30 - 5z^2$ for $|z| \le 1$m

$V(z) = 35 - 10|z|$ for $|z| \ge 1$ m.

V(z) does not depend on x and y. If this potential is generated by a constant charge per unit volume ρ_0 (in units of ε_0) which is spread over a certain region, then choose the correct statement.

[NCERT : P L-55 | Online April 9, 2016]

(a) $\rho_0 = 20\,\varepsilon_0$ in the entire region

(b) $\rho_0 = 10\,\varepsilon_0$ for $|z| \le 1$ m and $\rho_0 = 0$ elsewhere

(c) $\rho_0 = 20\,\varepsilon_0$ for $|z| \le 1$ m and $\rho_0 = 0$ elsewhere

(d) $\rho_0 = 40\,\varepsilon_0$ in the entire region

45. A uniformly charged solid sphere of radius R has potential V_0 (measured with respect to ∞) on its surface. For this sphere the equipotential surfaces with potentials $\frac{3V_0}{2}, \frac{5V_0}{4}, \frac{3V_0}{4}$ and $\frac{V_0}{4}$ have radius R_1, R_2, R_3 and R_4 respectively. Then **[NCERT : P L-54 | 2015]**

(a) $R_1 = 0$ and $R_2 < (R_4 - R_3)$

(b) $2R = R_4$

(c) $R_1 = 0$ and $R_2 > (R_4 - R_3)$

(d) $R_1 \ne 0$ and $(R_2 - R_1) > (R_4 - R_3)$

46. An electric field $\vec{E} = (25\hat{i} + 30\hat{j})NC^{-1}$ exists in a region of space. If the potential at the origin is taken to be zero then the potential at $x = 2$ *m*, $y = 2$ *m* is :

[NCERT : P L-55 | Online April 11, 2015]

(a) –110 V (b) –140 V (c) –120 V (d) –130 V

47. Assume that an electric field $\vec{E} = 30x^2\hat{i}$ exists in space. Then the potential difference $V_A - V_O$, where V_O is the potential at the origin and V_A the potential at x = 2 m is:

[NCERT : P L-55 | 2014]

(a) 120 J/C (b) –120 J/C (c) –80 J/C (d) 80 J/C

48. Consider a finite insulated, uncharged conductor placed near a finite positively charged conductor. The uncharged body must have a potential :

[NCERT : P L-47, 48 | Online April 23, 2013]

(a) less than the charged conductor and more than at infinity.

(b) more than the charged conductor and less than at infinity.

(c) more than the charged conductor and more than at infinity.

(d) less than the charged conductor and less than at infinity.

49. A point charge of magnitude + 1 μC is fixed at (0, 0, 0). An isolated uncharged spherical conductor, is fixed with its center at (4, 0, 0). The potential and the induced electric field at the centre of the sphere is :

[NCERT : P L-48 | Online April 22, 2013]

(a) 1.8×10^5 V and -5.625×10^6 V/m

(b) 0 V and 0 V/m

(c) 2.25×10^5 V and -5.625×10^6 V/m

(d) 2.25×10^5 V and 0 V/m

50. The electric potential $V(x)$ in a region around the origin is given by $V(x) = 4x^2$ volts. The electric charge enclosed in a cube of 1 m side with its centre at the origin is (in coulomb)

[NCERT : P L-55 | Online May 7, 2012]

(a) $8\varepsilon_0$ (b) $-4\varepsilon_0$ (c) 0 (d) $-8\varepsilon_0$

51. The electrostatic potential inside a charged spherical ball is given by $\phi = ar^2 + b$ where *r* is the distance from the centre and *a*, *b* are constants. Then the charge density inside the ball is: **[NCERT : P L-55 | 2011]**

(a) $-6a\varepsilon_0 r$ (b) $-24\pi a\varepsilon_0$ (c) $-6a\varepsilon_0$ (d) $-24\pi a\varepsilon_0 r$

52. An electric charge $10^{-3}\,\mu$C is placed at the origin (0, 0) of *X – Y* co-ordinate system. Two points *A* and *B* are situated at $(\sqrt{2}, \sqrt{2})$ and (2, 0) respectively. The potential difference between the points *A* and *B* will be

[NCERT : P L-48 | 2007]

(a) 4.5 volts (b) 9 volts (c) Zero (d) 2 volt

53. Charges are placed on the vertices of a square as shown. Let $\vec{E}$ be the electric field and V the potential at the centre. If the charges on A and B are interchanged with those on D and C respectively, then **[NCERT : P L-47, 48 | 2007]**

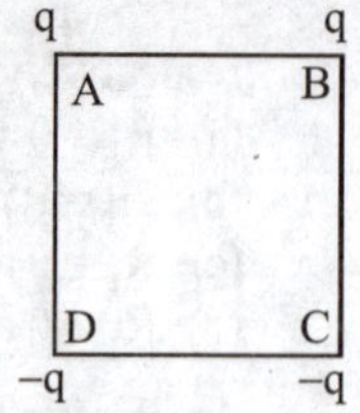

(a) $\vec{E}$ changes, V remains unchanged
(b) $\vec{E}$ remains unchanged, V changes
(c) both $\vec{E}$ and V change
(d) $\vec{E}$ and V remain unchanged

54. The potential at a point x (measured in μ m) due to some charges situated on the x-axis is given by $V(x) = 20/(x^2 - 4)$ volt. The electric field E at $x = 4$ μ m is given by **[NCERT : P L-55 | 2007]**
(a) (10/9) volt/ μ m and in the +ve x direction
(b) (5/3) volt/ μ m and in the −ve x direction
(c) (5/3) volt/ μ m and in the +ve x direction
(d) (10/9) volt/ μ m and in the −ve x direction

55. Two thin wire rings each having a radius R are placed at a distance d apart with their axes coinciding. The charges on the two rings are $+q$ and $-q$. The potential difference between the centres of the two rings is **[NCERT : P L-48 | 2005]**
(a) $\frac{q}{2\pi \epsilon_0}\left[\frac{1}{R} - \frac{1}{\sqrt{R^2 + d^2}}\right]$ (b) $\frac{qR}{4\pi \epsilon_0 d^2}$
(c) $\frac{q}{4\pi \epsilon_0}\left[\frac{1}{R} - \frac{1}{\sqrt{R^2 + d^2}}\right]$ (d) zero

56. A thin spherical conducting shell of radius R has a charge q. Another charge Q is placed at the centre of the shell. The electrostatic potential at a point P, a distance $\frac{R}{2}$ from the centre of the shell is **[NCERT : P L-52 | 2003]**
(a) $\frac{2Q}{4\pi\varepsilon_o R}$ (b) $\frac{2Q}{4\pi\varepsilon_o R} - \frac{2q}{4\pi\varepsilon_o R}$
(c) $\frac{2Q}{4\pi\varepsilon_o R} + \frac{q}{4\pi\varepsilon_o R}$ (d) $\frac{(q+Q)2}{4\pi\varepsilon_o R}$

Topic 2 - Electric Potential Energy and Work Done in Carrying a Charge

57. Two charges q_1 and q_2 are separated by a distance of 30 cm. A third charge q_3 initially at 'C' as shown in the figure, is moved along the circular path of radius 40 cm from C to D. If the difference in potential energy due to movement of q_3 from C to D is given by $\frac{q_3 K}{4\pi \epsilon_0}$, the value of K is : **[NCERT : P L-55, 56 | April 7, 2025 (I)]**

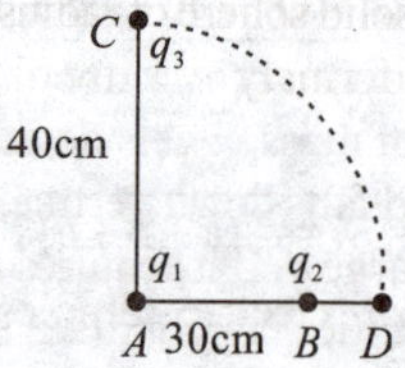

(a) $8q_2$
(b) $6q_2$
(c) $8q_1$
(d) $6q_1$

58. An electric dipole of dipole moment 6×10^{-6} Cm is placed in uniform electric field of magnitude 10^6 V/m. Initially, the dipole moment is parallel to electric field. The work that needs to be done on the dipole to make its dipole moment opposite to the field, will be ____ J. **[NCERT : P L-60 | Jan. 28, 2025 (II)]**

59.

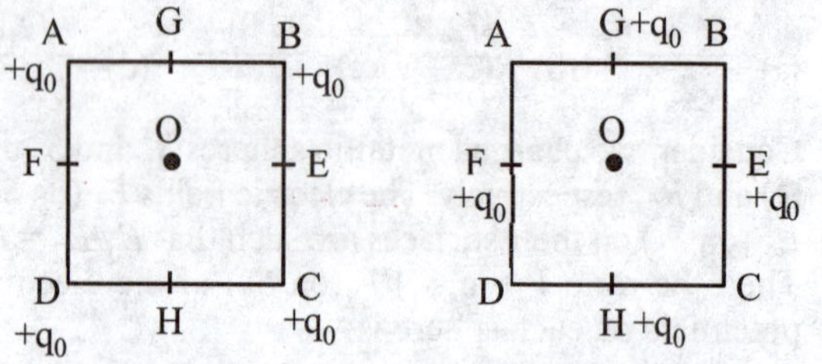

Configuration (1) Configuration (2)

In the first configuration (1) as shown in the figure, four identical charges (q_0) are kept at the corners A,B,C and D of square of side length 'a'. In the second configuration (2), the same charges are shifted to mid points G,E,H and F, of the square, If $K = \frac{1}{4\pi\varepsilon_0}$, the difference between the potential energies of configuration (2) and (1) is given by **[NCERT : P L-55, 56 | Jan. 24, 2025 (II)]**
(a) $\frac{Kq_0^2}{a}(4\sqrt{2} - 2)$ (b) $\frac{Kq_0^2}{a}(3 - \sqrt{2})$
(c) $\frac{Kq_0^2}{a}(4 - 2\sqrt{2})$ (d) $\frac{Kq_0^2}{a}(3\sqrt{2} - 2)$

60. Two charges 7 μc and −4 μc are placed at (−7 cm, 0, 0) and (7 cm, 0, 0) respectively. Given, $\epsilon_0 = 8.85 \times 10^{-12}$ $C^2 N^{-1} m^{-2}$, the electrostatic potential energy of the charge configuration is **[NCERT : P L-55, 56 | Jan. 23, 2025 (II)]**
(a) −1.5 J (b) −2.0 J (c) −1.2 J (d) −1.8 J

61. Two point charges – 4 μc and 4 μc, constituting an electric dipole, are placed at (– 9, 0, 0) cm and (9, 0, 0) cm in a uniform electric field of strength 10^4 NC^{-1}. The work done on the dipole in rotating it from the equilibrium through 180° is **[NCERT : P L-60 | Jan. 23, 2025 (II)]**
(a) 14.4 mJ (b) 18.4 mJ (c) 12.4 mJ (d) 16.4 mJ

62. An electric dipole of dipole moment is 6.0×10^{-6} Cm placed in a uniform electric field of 1.5×10^3 NC^{-1} in such a way that dipole moment is along electric field. The work done in rotating dipole by 180° in this field will be ____ mJ. **[NCERT : P L-60 | April 8, 2023 (I)]**

63. Eight similar drops of mercury are maintained at 12 V each. All these spherical drops combine into a single big drop. The potential energy of bigger drop will be ____ E. Where E is the potential energy of a single smaller drop. **[NCERT : P L-74 | June 30, 2022 (I)]**

64. A solid sphere of radius R carries a charge $Q+q$ distributed uniformaly over its volume. A very small point like piece of it of mass m gets detached from the bottom of the sphere and falls down vertically under gravity. This piece carries charge q. If it acquires a speed v when it has fallen through a vertical height y (see figure), then : (assume the remaining portion to be spherical). **[NCERT : P L-56 | Sep. 05, 2020 (I)]**

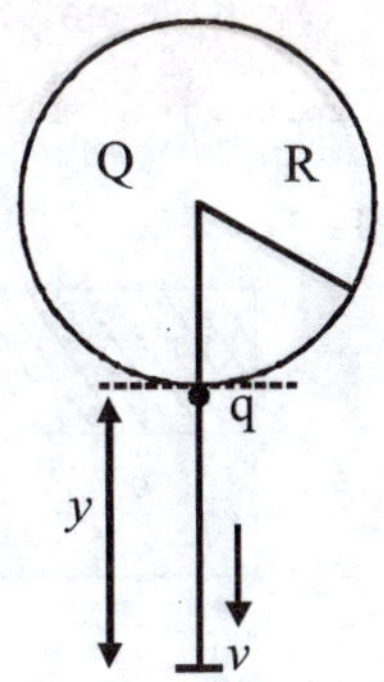

(a) $v^2 = y\left[\frac{qQ}{4\pi\varepsilon_0 R^2 ym}+g\right]$

(b) $v^2 = y\left[\frac{qQ}{4\pi\varepsilon_0 R(R+y)m}+g\right]$

(c) $v^2 = 2y\left[\frac{Qq\,R}{4\pi\varepsilon_0 (R+y)^3 m}+g\right]$

(d) $v^2 = 2y\left[\frac{qQ}{4\pi\varepsilon_0 R(R+y)m}+g\right]$

65. A two point charges $4q$ and $-q$ are fixed on the x-axis at $x=-\frac{d}{2}$ and $x=\frac{d}{2}$, respectively. If a third point charge 'q' is taken from the origin to $x=d$ along the semicircle as shown in the figure, the energy of the charge will : **[NCERT : P L-56 | Sep. 04, 2020 (I)]**

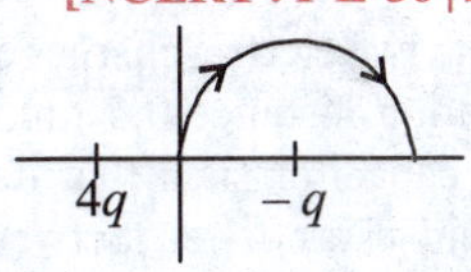

(a) increase by $\frac{3q^2}{4\pi\varepsilon_0 d}$ (b) increase by $\frac{2q^2}{3\pi\varepsilon_0 d}$

(c) decrease by $\frac{q^2}{4\pi\varepsilon_0 d}$ (d) decrease by $\frac{4q^2}{3\pi\varepsilon_0 d}$.

66. Hydrogen ion and singly ionized helium atom are accelerated, from rest, through the same potential difference. The ratio of final speeds of hydrogen and helium ions is close to : **[NCERT : P L-56 | Sep. 03, 2020 (II)]**

(a) 1 : 2 (b) 10 : 7

(c) 2 : 1 (d) 5 : 7

67. In free space, a particle A of charge 1 mC is held fixed at a point P. Another particle B of the same charge and mass 4 mg is kept at a distance of 1 mm from P. If B is released, then its velocity at a distance of 9 mm from P is :

$\left[Take\ \frac{1}{4\pi\varepsilon_0} = 9\times10^9\ Nm^2C^{-2}\right]$

[NCERT : P L-56 | 10 April 2019 (II)]

(a) 1.0m/s (b) 3.0×10^4 m/s

(c) 2.0×10^3 m/s (d) 1.5×10^2 m/s

68. A system of three charges are placed as shown in the figure:

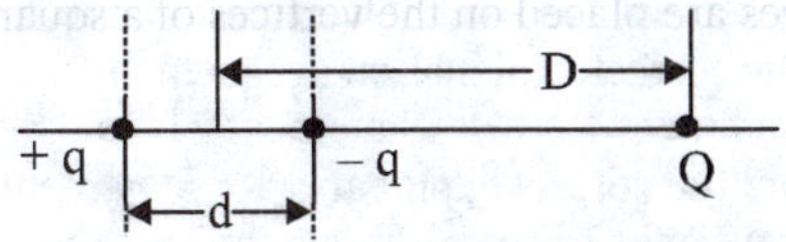

If D >> d, the potential energy of the system is best given by **[NCERT : P L-56 | 9 April 2019 (I)]**

(a) $\frac{1}{4\pi\epsilon_0}\left[\frac{-q^2}{d}-\frac{qQd}{2D^2}\right]$ (b) $\frac{1}{4\pi\epsilon_0}\left[\frac{-q^2}{d}+\frac{2qQd}{D^2}\right]$

(c) $\frac{1}{4\pi\epsilon_0}\left[+\frac{q^2}{d}+\frac{qQd}{D^2}\right]$ (d) $\frac{1}{4\pi\epsilon_0}\left[-\frac{q^2}{d}-\frac{qQd}{D^2}\right]$

69. A positive point charge is released from rest at a distance r_0 from a positive line charge with uniform density. The speed (v) of the point charge, as a function of instantaneous distance r from line charge, is proportional to: **[NCERT : P L-59 | 8 April 2019 (II)]**

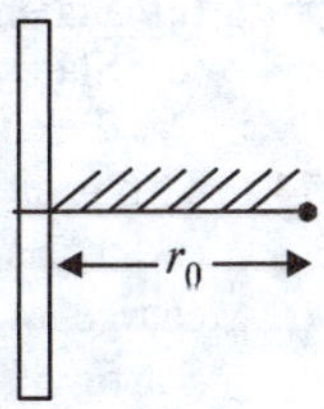

(a) $v \propto e^{+r/r_0}$ (b) $v \propto \sqrt{ln\left(\frac{r}{r_0}\right)}$

(c) $v \propto ln\left(\frac{r}{r_0}\right)$ (d) $v \propto \left(\frac{r}{r_0}\right)$

70. Three charges Q, + q and + q are placed at the vertices of a right-angle isosceles triangle as shown below. The net electrostatic energy of the configuration is zero, if the value of Q is: **[NCERT : P L-56 | 11 Jan. 2019 (I)]**

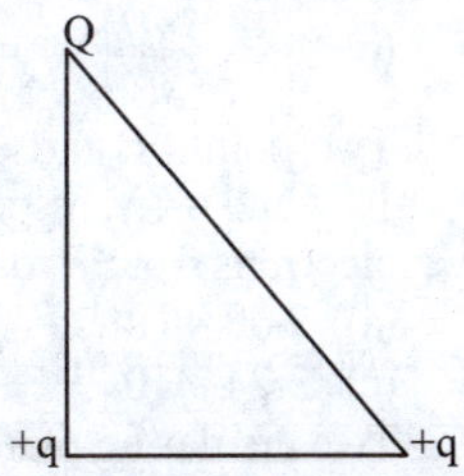

(a) + q

(b) $\frac{-\sqrt{2}q}{\sqrt{2}+1}$

(c) $\frac{-q}{1+\sqrt{2}}$

(d) –2q

71. Four equal point charges Q each are placed in the xy plane at (0, 2), (4, 2), (4, – 2) and (0, – 2). The work required to put a fifth charge Q at the origin of the coordinate system will be: **[NCERT : P L-59 | 10 Jan. 2019 (II)]**

(a) $\frac{Q^2}{4\pi\varepsilon_0}\left(1+\frac{1}{\sqrt{3}}\right)$ (b) $\frac{Q^2}{4\pi\varepsilon_0}\left(1+\frac{1}{\sqrt{5}}\right)$

(c) $\frac{Q^2}{2\sqrt{2}\,\pi\varepsilon_0}$ (d) $\frac{Q^2}{4\pi\varepsilon_0}$

72. **Statement 1 :** No work is required to be done to move a test charge between any two points on an equipotential surface.

Statement 2 : Electric lines of force at the equipotential surfaces are mutually perpendicular to each other.

[NCERT : P L-54 | Online April 25, 2013]

(a) Statement 1 is true, Statement 2 is true, Statement 2 is the correct explanation of Statement 1.
(b) Statement 1 is true, Statement 2 is true, Statement 2 is not the correct explanation of Statement 1.
(c) Statement 1 is true, Statement 2 is false.
(d) Statement 1 is false, Statement 2 is true.

73. An insulating solid sphere of radius R has a uniformly positive charge density ρ. As a result of this uniform charge distribution there is a finite value of electric potential at the centre of the sphere, at the surface of the sphere and also at a point outside the sphere. The electric potential at infinite is zero. **[NCERT : P L-59 | 2012]**

Statement -1 When a charge q is taken from the centre to the surface of the sphere its potential energy changes by $\frac{q\rho}{3\varepsilon_0}$.

Statement -2 The electric field at a distance r ($r<R$) from the centre of the sphere is $\frac{\rho r}{3\varepsilon_0}$.

(a) Statement 1 is true, Statement 2 is true; Statement 2 is not the correct explanation of statement 1.
(b) Statement 1 is true Statement 2 is false.
(c) Statement 1 is false Statement 2 is true.
(d) Statement 1 is true, Statement 2 is true, Statement 2 is the correct explanation of Statement 1

74. Two positive charges of magnitude 'q' are placed, at the ends of a side (side 1) of a square of side '$2a$'. Two negative charges of the same magnitude are kept at the other corners. Starting from rest, if a charge Q moves from the middle of side 1 to the centre of square, its kinetic energy at the centre of square is **[NCERT : P L-59 | 2011 RS]**

(a) zero (b) $\frac{1}{4\pi\varepsilon_0}\frac{2qQ}{a}\left(1+\frac{1}{\sqrt{5}}\right)$
(c) $\frac{1}{4\pi\varepsilon_0}\frac{2qQ}{a}\left(1-\frac{2}{\sqrt{5}}\right)$ (d) $\frac{1}{4\pi\varepsilon_0}\frac{2qQ}{a}\left(1-\frac{1}{\sqrt{5}}\right)$

75. Two points P and Q are maintained at the potentials of 10 V and – 4 V, respectively. The work done in moving 100 electrons from P to Q is: **[NCERT : P L-59 | 2009]**
(a) 9.60×10^{-17} J (b) -2.24×10^{-16} J
(c) 2.24×10^{-16} J (d) -9.60×10^{-17} J

76. Two insulating plates are both uniformly charged in such a way that the potential difference between them is $V_2 - V_1 = 20$ V. (i.e., plate 2 is at a higher potential). The plates are separated by $d = 0.1$ m and can be treated as infinitely large. An electron is released from rest on the inner surface of plate 1. What is its speed when it hits plate 2? ($e = 1.6 \times 10^{-19}$ C, $m_e = 9.11 \times 10^{-31}$ kg) **[NCERT : P L-59 | 2006]**

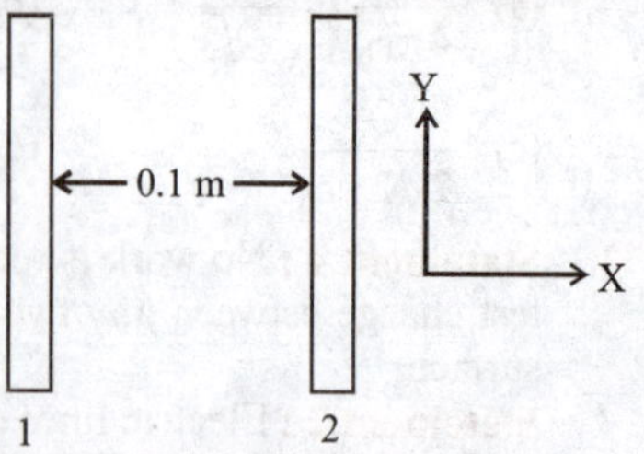

(a) 2.65×10^6 m/s
(b) 7.02×10^{12} m/s
(c) 1.87×10^6 m/s
(d) 32×10^{-19} m/s

77. A charged particle 'q' is shot towards another charged particle 'Q' which is fixed, with a speed 'v'. It approaches 'Q' upto a closest distance r and then returns. If q were given a speed of '$2v$' the closest distances of approach would be **[NCERT : P L-59 | 2004]**
(a) $r/2$ (b) $2r$ (c) r (d) $r/4$

78. On moving a charge of 20 coulomb by 2 cm, 2 J of work is done, then the potential difference between the points is **[NCERT : P L-59 | 2002]**
(a) 0.1 V (b) 8V (c) 2V (d) 0.5V

Topic 3 - Capacitors, Grouping of Capacitors and Energy Stored in a Capacitor

79.

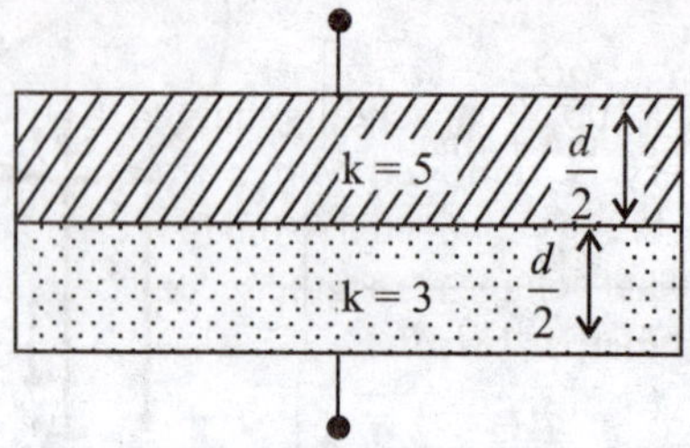

Space between the plates of a parallel plate capacitor of plate area 4 cm^2 and separation of (d) 1.77 mm, is filled with uniform dielectric materials with dielectric constants (3 and 5) as shown in figure. Another capacitor of capacitance 7.5 pF is connected in parallel with it. The effective capacitance of this combination is ______ pF.
(Given $\varepsilon_0 = 8.85 \times 10^{-12}$ F/m)
[NCERT : P L-70, 72 | April 8, 2025 (II)]

80. A dipole with two electric charges of 2 μC magnitude each, with separation distance 0.5 μm, is placed between the plates of a capacitor such that its axis is parallel to an electric field established between the plates when a potential difference of 5 V is applied. Separation between the plates is 0.5 mm. If the dipole is rotated by 30^0 from the axis, it tends to realign in the direction due to a torque. The value of torque is : **[NCERT : P L-55, 67 | April 7, 2025 (II)]**
(a) 5×10^{-9} Nm (b) 5×10^{-3} Nm
(c) 2.5×10^{-12} Nm (d) 2.5×10^{-9} Nm

81. A parallel plate capacitor has charge 5×10^{-6} C. A dielectric slab is inserted between the plates and almost fills the space between the plates. If the induced charge on one face of the slab is 4×10^{-6} C then the dielectric constant of the slab is ______. **[NCERT : P L-70 | April 7, 2025 (II)]**

82. Three parallel plate capacitors C_1, C_2 and C_3 each of capacitance 5 μF are connected as shown in figure. The effective capacitance between points A and B, when the space between the parallel plates of C_1 capacitor is filled with a dielectric medium having dielectric constant of 4, is: **[NCERT : P L-70, 72 | April 4, 2025 (II)]**

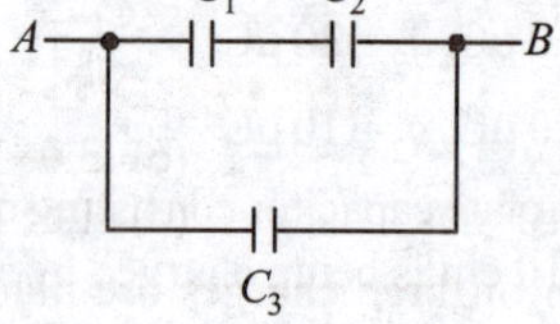

(a) 22.5 μF (b) 7.5 μF
(c) 9 μF (d) 30 μF

83. Four capacitor each of capacitance 16 μF are connected as shown in the figure. The capacitance between points A and B is : _____ (in μF).

[NCERT : P L-72 | April 4, 2025 (I)]

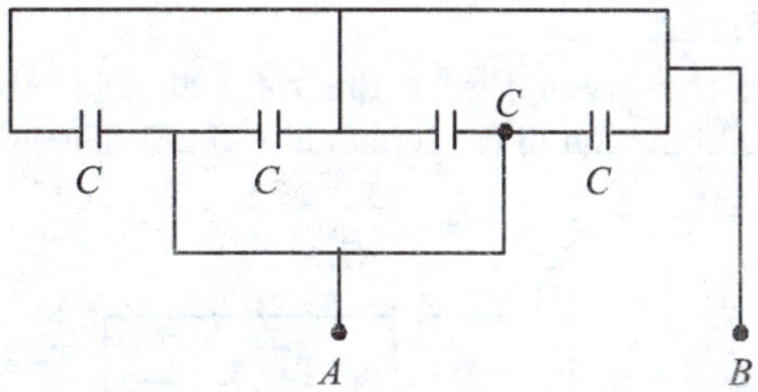

84. Using a battery, a 100 pF capacitor is charged to 60 V and then the battery is removed. After that, a second uncharged capacitor is connected to the first capacitor in parallel. If the final voltage across the second capacitor is 20 V, its capacitance is : (in pF)

[NCERT : P L-72 | April 3, 2025 (II)]

(a) 600 (b) 200 (c) 400 (d) 100

85. A parallel plate capacitor is filled equally (half) with two dielectrics of dielectric constant ε_1 and ε_2, as shown in figures. The distance between the plates is d and area of each plate is A. If capacitance in first configuration and second configuration are C_1 and C_2 respectively, then $\frac{C_1}{C_2}$ is : **[NCERT : P L-70 | April 3, 2025 (I)]**

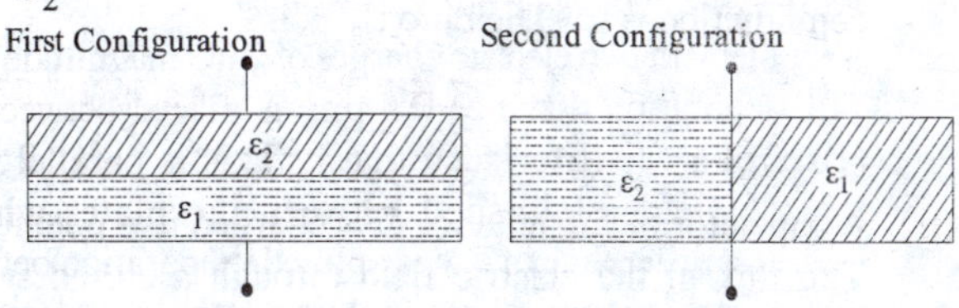

(a) $\frac{\varepsilon_1\varepsilon_2^2}{(\varepsilon_1+\varepsilon_2)^2}$ (b) $\frac{4\varepsilon_1\varepsilon_2}{(\varepsilon_1+\varepsilon_2)^2}$

(c) $\frac{\varepsilon_1\varepsilon_2}{\varepsilon_1+\varepsilon_2}$ (d) $\frac{\varepsilon_0(\varepsilon_1+\varepsilon_2)}{2}$

86. A capacitor, $C_1 = 6\mu F$ is charged to a potential difference of $V_0 = 5V$ using a 5V battery. The battery is removed and another capacitor, $C_2 = 12\ \mu F$ is inserted in place of the battery. When the switch 'S' is closed, the charge flows between the capacitors for some time until equilibrium condition is reached. What are the charges (q_1 and q_2) on the capacitors C_1 and C_2 when equilibrium condition is reached. **[NCERT : P L-72 | Jan. 29, 2025 (II)]**

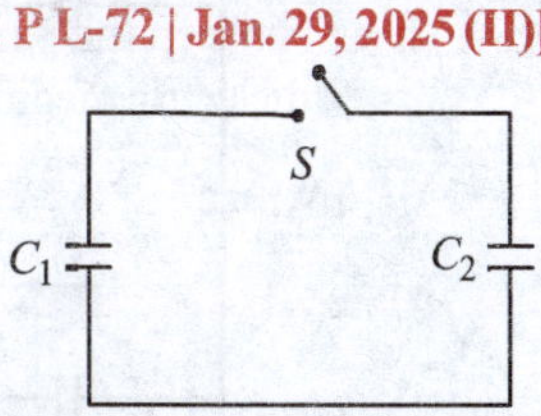

(a) $q_1 = 15\ \mu C, q_2 = 30\ \mu C$

(b) $q_1 = 30\ \mu C, q_2 = 15\ \mu C$

(c) $q_1 = 10\ \mu C, q_2 = 20\ \mu C$

(d) $q_1 = 20\ \mu C, q_2 = 10\ \mu C$

87. A parallel plate capacitor consisting of two circular plates of radius 10 cm is being charged by a constant current of 0.15 A. If the rate of change of potential difference between the plates is 7×10^8 V/s, then the integer value of the distance between the parallel plates is –

$\left(\text{Take, } \epsilon_0 = 9\times 10^{-12}\frac{F}{m}, \pi = \frac{22}{7}\right)$ ________ μm.

[NCERT : P L-69 | Jan. 29, 2025 (II)]

88. A parallel plate capacitor of capacitance 1 μF is charged to a potential difference of 20 V. The distance between plates is 1 μm. The energy density between plates of capacitor is : **[NCERT : P L-74 | Jan. 28, 2025 (II)]**

(a) $1.8\times 10^3\ J/m^3$ (b) $2\times 10^{-4}\ J/m^3$

(c) $2\times 10^2\ J/m^3$ (d) $1.8\times 10^5\ J/m^3$

89. Two capacitors C_1 and C_2 are connected in parallel to a battery. Charge-time graph is shown below for the two capacitors. The energy stored with them are U_1 and U_2, respectively. Which of the given statements is true ?

[NCERT : P L-72 | Jan. 28, 2025 (I)]

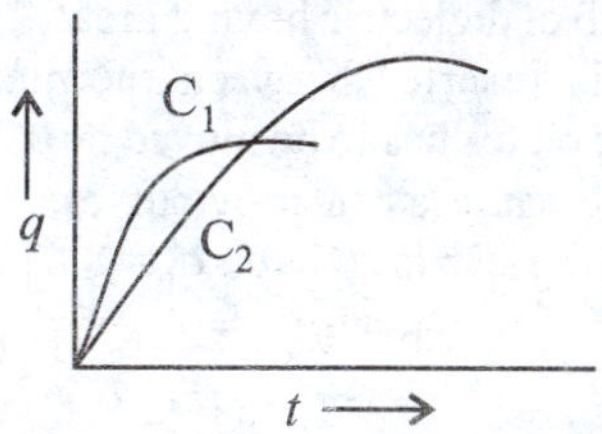

(a) $C_1 > C_2, U_1 > U_2$ (b) $C_2 > C_1, U_2 < U_1$

(c) $C_1 > C_2, U_1 < U_2$ (d) $C_2 > C_1, U_2 > U_1$

90. A parallel plate capacitor was made with two rectangular plates, each with a length of l = 3 cm and breadth of b = 1 cm. The distance between the plates is 3 μm. Out of the following, which are the ways to increase the capacitance by a factor of 10?

[NCERT : P L-69 | Jan. 24, 2025 (I)]

A. l = 30 cm, b = 1 cm, d = 1 μm

B. l = 3 cm, b = 1 cm, d = 30 μm

C. l = 6 cm, b = 5 cm, d = 3 μm

D. l = 1 cm, b = 12 cm, d = 10 μm

E. l = 5 cm, b = 2 cm, d = 1 μm

Choose the correct answer from the options given below

(a) C and E only (b) B and D only

(c) A only (d) C only

91. Consider a parallel plate capacitor of area A (of each plate) and separation 'd' between the plates. If E is the electric field and ε_0 is the permittivity of free space between the plates, then potential energy stored in the capacitor is

[NCERT : P L-74 | Jan. 24, 2025 (I)]

(a) $\frac{1}{2}\varepsilon_0 E^2 Ad$ (b) $\frac{3}{4}\varepsilon_0 E^2 Ad$

(c) $\frac{1}{4}\varepsilon_0 E^2 Ad$ (d) $\varepsilon_0 E^2 Ad$

92. A parallel-plate capacitor of capacitance 40μF is connected to a 100 V power supply. Now the intermediate space between the plates is filled with a dielectric material of dielectric constant K = 2. Due to the introduction of dielectric material, the extra charge and the change in the electrostatic energy in the capacitor, respectively, are

[NCERT : P L-70, 74 | Jan. 22, 2025 (I)]

(a) 2 mC and 0.2 J (b) 8 mC and 2.0 J

(c) 4 mC and 0.2 J (d) 2 mC and 0.4 J

93. A capacitor is made of a flat plate of area A and a second plate having a stair-like structure as shown in figure. If the area of each stair is $\frac{A}{3}$ and the height is d, the capacitance of the arrangement is:

[NCERT : P L-69, 72 | April 9, 2024 (I)]

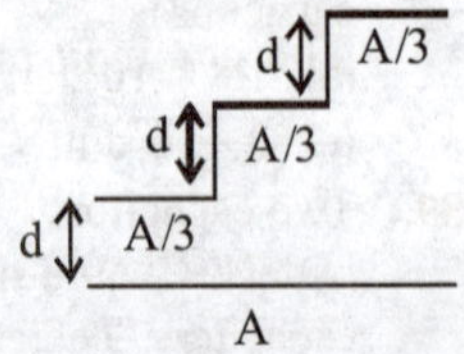

(a) $\frac{11\varepsilon_0 A}{18d}$ (b) $\frac{13\varepsilon_0 A}{17d}$

(c) $\frac{11\varepsilon_0 A}{20d}$ (d) $\frac{18\varepsilon_0 A}{11d}$

94. A capacitor has air as dielectric medium and two conducting plates of area 12 cm^2 and they are 0.6 cm apart. When a slab of dielectric having area 12 cm^2 and 0.6 cm thickness is inserted between the plates, one of the conducting plates has to be moved by 0.2 cm to keep the capacitance same as in previous case. The dielectric constant of the slab is : (Given $\epsilon_0 = 8.834 \times 10^{-12}$ F/m)

[NCERT : P L-70 | April 8, 2024 (II)]

(a) 1.50 (b) 1.33 (c) 0.66 (d) 1

95. A capacitor of 10 μF capacitance whose plates are separated by 10 mm through air and each plate has area 4 cm^2 is now filled equally with two dielectric media of $K_1 = 2$, $K_2 = 3$ respectively as shown in figure. If new force between the plates is 8 N. The supply voltage is ______ V.

[NCERT : P L-70 | April 6, 2024 (II)]

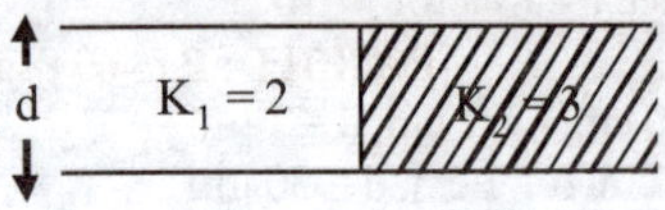

96. Three capacitors of capacitances 25 μF, 30 μF and 45 μF are connected in parallel to a supply of 100 V. Energy stored in the above combination is E. When these capacitors are connected in series to the same supply, the stored energy is $\frac{9}{x}$E. The value of x is ....

[NCERT : P L-72, 74 | April 5, 2024 (I)]

97. Two identical capacitors have same capacitance C. One of them is charged to the potential V and other to the potential 2V. The negative ends of both are connected together. When the positive ends are also joined together, the decrease in energy of the combined system is:

[NCERT : P L-74 | Feb. 1, 2024 (I)]

(a) $\frac{1}{4}CV^2$ (b) $2CV^2$ (c) $\frac{1}{2}CV^2$ (d) $\frac{3}{4}CV^2$

98. A parallel plate capacitor with plate separation 5 mm is charged up by a battery. It is found that on introducing a dielectric sheet of thickness 2 mm, while keeping the battery connections intact, the capacitor draws 25% more charge from the battery than before. The dielectric constant of the sheet is _____.

[NCERT : P L-70 | Jan. 31, 2024 (I)]

99. A capacitor of capacitance C and potential V has energy E. It is connected to another capacitor of capacitance 2C and potential 2V. Then the loss of energy is $\frac{x}{3}$E, where x is_______.

[NCERT : P L-74 | Jan. 30, 2024 (I)]

100. In the given figure the total charge stored in the combination of capacitors is 100 μC. The value of 'x' is ________.

[NCERT : P L-67 | April 15, 2023 (I)]

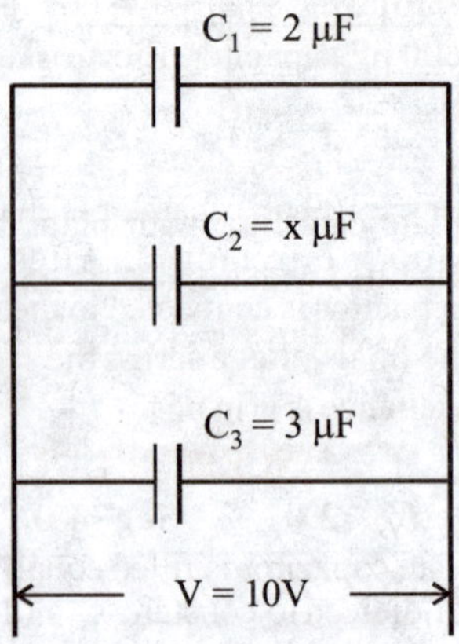

101. A parallel plate capacitor of capacitance 2 F is charged to a potential V. The energy stored in the capacitor is E_1. The capacitor is now connected to another uncharged identical capacitor in parallel combination. The energy stored in the combination is E_2. The ratio E_2/E_1 is

[NCERT : P L-74 | April 11, 2023 (I)]

(a) 2 : 1 (b) 1 : 2 (c) 1 : 4 (d) 2 : 3

102. A capacitor of capacitance C is charged to a potential V. The flux of the electric field through a closed surface enclosing the positive plate of the capacitor is:

[NCERT : P L-67 | April 11, 2023 (II)]

(a) $\frac{CV}{2\varepsilon_0}$ (b) $\frac{2CV}{\varepsilon_0}$ (c) $\frac{CV}{\varepsilon_0}$ (d) Zero

103. In the given circuit,

$C_1 = 2$ μF, $C_2 = 0.2$ μF, $C_3 = 2$ μF, $C_4 = 4$ μF, $C_5 = 2$ μF, $C_6 = 2$ μF, the charge stored on capacitor C_4 is ____ μC.

[NCERT : P L-72 | April 11, 2023 (I)]

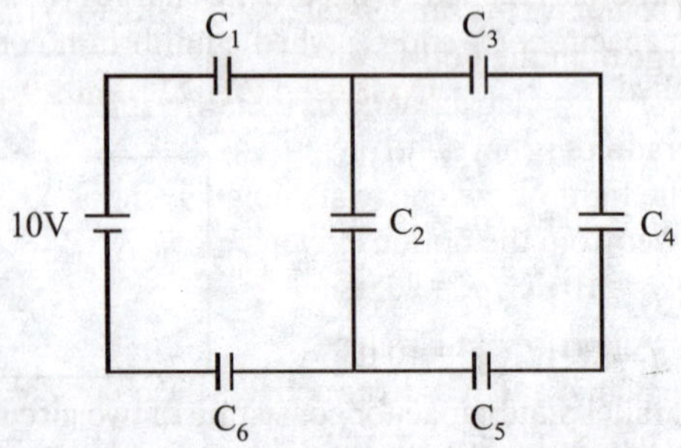

104. The equivalent capacitance of the combination shown is

[NCERT : P L-72 | April 10, 2023 (I)]

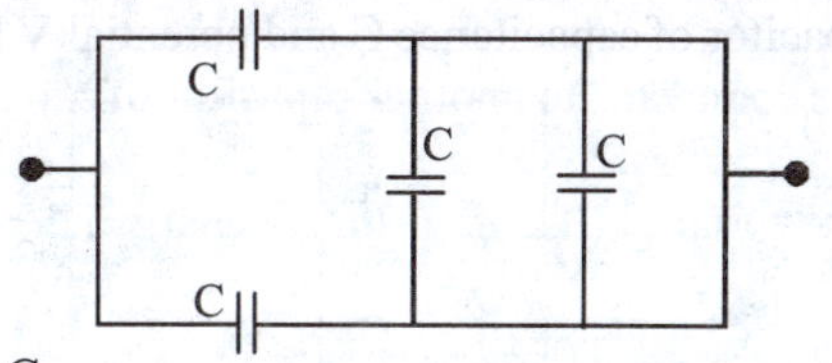

(a) $\frac{C}{2}$ (b) 4C (c) 2C (d) $\frac{5}{3}C$

105. A 600 pF capacitor is charged by 200V supply. It is then disconnected from the supply and is connected to another uncharged 600 pF capacitor. Electrostatic energy lost in the process is ______ μJ.

[NCERT : P L-74 | April 8, 2023 (II)]

106. A parallel plate capacitor with plate area A and plate separation d is filled with a dielectric material of dielectric constant K = 4. The thickness of the dielectric material is x, where x < d. **[NCERT : P L-70 | April 6, 2023 (I)]**

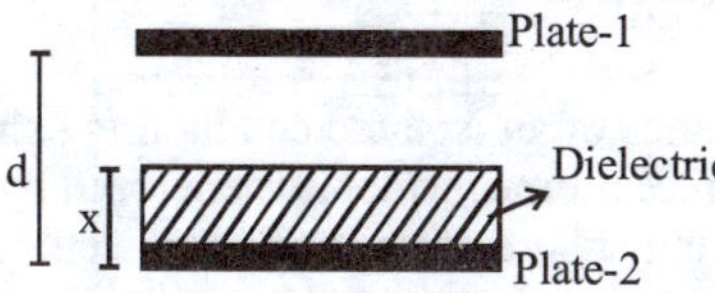

Let C_1 and C_2 be the capacitance of the system for $x = \frac{1}{3}d$ and $x - \frac{2d}{3}$, respectively. If $C_1 = 2\mu F$ the value of C_2 is ______ μF

107. As shown in the figure, two parallel plate capacitors having equal plate area of 200 cm^2 are joined in such a way that a ≠ b. The equivalent capacitance of the combination is x ϵ_0 F. The value of x is ______.

[NCERT : P L-70, 72 | April 6, 2023 (II)]

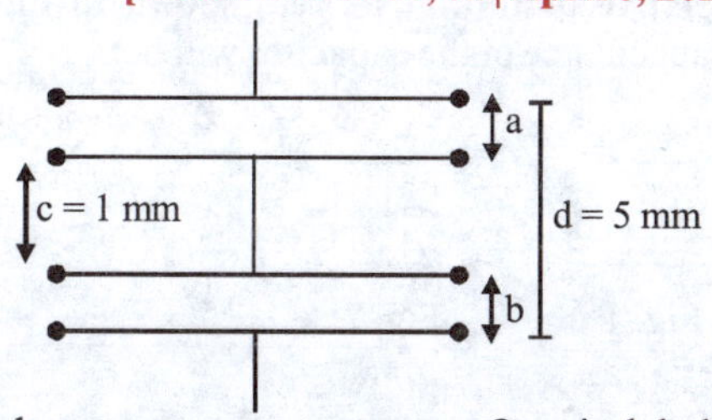

108. Given below are two statements : One is labelled as **Assertion A** and the other is labelled as **Reason R**.

Assertion A : Two metallic spheres are charged to the same potential. One of them is hollow and another is solid, and both have the same radii. Solid sphere will have lower charge than the hollow one.

Reason R : Capacitance of metallic spheres depend on the radii of spheres.

In the light of the above statements, choose the **correct** answer from the options given below.

[NCERT : P L-67 | Feb. 1, 2023 (II)]

(a) A is false but R is true
(b) Both A and R are true and R is the correct explanation of A
(c) A is true but R is false
(d) Both A and R are true but R is not the correct explanation of A

109. Two parallel plate capacitors C_1 and C_2 each having capacitance of 10 μF are individually charged by a 100 V D.C. source. Capacitor C_1 is kept connected to the source and a dielectric slab is inserted between its plates. Capacitor C_2 is disconnected from the source and then a dielectric slab is inserted in it. Afterwards the capacitor C_1 is also disconnected from the source and the two capacitors are finally connected in parallel combination. The common potential of the combination will be ______ V. (Assuming Dielectric constant = 10)

[NCERT : P L-70, 72 | Jan. 31, 2023 (II)]

110. A capacitor of capacitance 900 μF is charged by a 100 V battery. The capacitor is disconnected from the battery and connected to another uncharged identical capacitor such that one plate of uncharged capacitor connected to positive plate and another plate of uncharged capacitor connected to negative plate of the charged capacitor. The loss of energy in this process is measured as x × 10^{-2} J. The value of x is ______.

[NCERT : P L-74 | Jan. 30, 2023 (I)]

111. A parallel plate capacitor has plate area 40cm^2 and plates separation 2 mm. The space between the plates is filled with a dielectric medium of a thickness 1 mm and dielectric constant 5. The capacitance of the system is :

[NCERT : P L-70, 72 | Jan. 25, 2023 (I)]

(a) $24\epsilon_0 5$ (b) $\frac{3}{10}\epsilon_0 F$ (c) $\frac{10}{3}\epsilon_0 F$ (d) $10\epsilon_0 F$

112. A capacitor has capacitance 5μF when it's parallel plates are separated by air medium of thickness d. A slab of material of dielectric constant 1.5 having area equal to that of plates but thickness $\frac{d}{2}$ is inserted between the plates. Capacitance of the capacitor in the presence of slab will be ______ μF.

[NCERT : P L-70 | Jan. 25, 2023 (II)]

113. A parallel plate capacitor with air between the plate has a capacitance of 15pF. The separation between the plate becomes twice and the space between them is filled with a medium of dielectric constant 3.5. Then the capacitance becomes $\frac{x}{4}$ pF. The value of x is ______.

[NCERT : P L-70 | Jan. 24, 2023 (II)]

114. Two identical thin metal plates has charge q_1 and q_2 respectively such that $q_1 > q_2$. The plates were brought close to each other to form a parallel plate capacitor of capacitance C. The potential difference between them is :

[NCERT : P L-69 | July 29, 2022 (II)]

(a) $\frac{(q_1+q_2)}{C}$ (b) $\frac{(q_1-q_2)}{C}$

(c) $\frac{(q_1-q_2)}{2C}$ (d) $\frac{2(q_1-q_2)}{C}$

115. Two capacitors, each having capacitance 40 μF are connected in series. The space between one of the capacitors is filled with dielectric material of dielectric constant K such that the equivalence capacitance of the system became 24 μF . The value of K will be :

[NCERT : P L-70, 72 | July 28, 2022 (I)]

(a) 1.5 (b) 2.5 (c) 1.2 (d) 3

116. A slab of dielectric constant K has the same cross sectional area as the plate of a parallel plate capacitor and thickness $\frac{3}{4}$ d, where d is the separation of plates.

The capacitance of the capacitor when the slab is inserted between the plates will be :

(Given C_0 = capacitance of capacitor with air as medium between plates.) NCERT : P L-70 | July 28, 2022 (II)]

(a) $\frac{4KC_0}{3+K}$ (b) $\frac{3KC_0}{3+K}$

(c) $\frac{3+K}{4KC_0}$ (d) $\frac{K}{4+K}$

117. A parallel plate capacitor with width 4 cm, length 8 cm and separation between the plates of 4mm is connected to a battery of 20 V. A dielectric slab of dielectric constant 5 having length 1cm, width 4 cm and thickness 4 mm is inserted between the plates of parallel plate capacitor. The electrostatic energy of this system will be ______ ϵ_0 J. (Where ϵ_0 is the permittivity of free space)

[NCERT : P L-74 | July 27, 2022 (II)]

118. A composite parallel plate capacitor is made up of two different dielectric materials with different thickness (t_1 and t_2) as shown in figure. The two different dielectric materials are separated by a conducting foil F. The voltage of the conducting foil is ______ V.

[NCERT : P L-70, 72 | July 26, 2022 (I)]

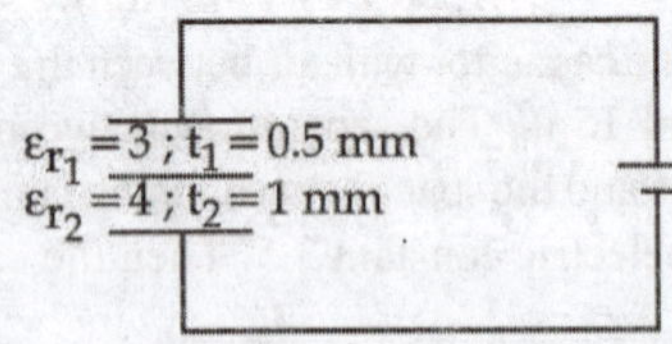

119. The total charge on the system of capacitors C_1 = 1 μF, C_2 = 2 μF, C_3 = 4 μF and C_4 = 3 μF connected in parallel is: (Assume a battery of 20 V is connected to the combination) [NCERT : P L-72 | July 26, 2022 (I)]

(a) 200 μC (b) 200 C (c) 10 μC (d) 10 C

120. A source of potential difference V is connected to the combination of two identical capacitors as shown in the figure. When key 'K' is closed, the total energy stored across the combination is E_1. Now key 'K' is opened and dielectric of dielectric constant 5 is introduced between the plates of the capacitors. The total energy stored across the combination is now E_2. The ratio E_1/E_2 will be :

[NCERT : P L-74 | July 26, 2022 (II)]

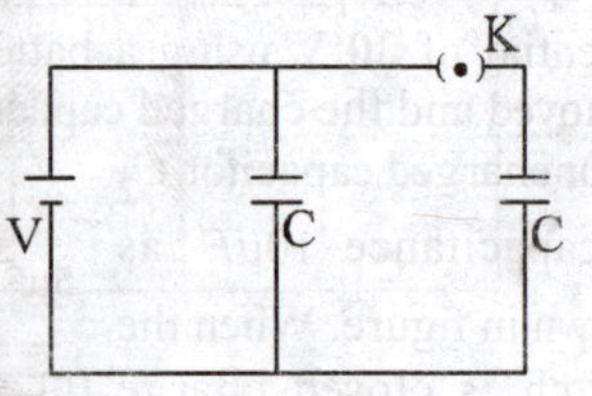

(a) $\frac{1}{10}$

(b) $\frac{2}{5}$

(c) $\frac{5}{13}$

(d) $\frac{5}{26}$

121. Two parallel plate capacitors of capacity C and 3C are connected in parallel combination and charged to a potential difference 18 V. The battery is then disconnected and the space beteen the plates of the capacitor of capacity C is completely filled with a material of dielectric constant 9. The final potential difference across the combination of capacitors will be ______ V

[NCERT : P L-69, 72 | July 25, 2022 (II)]

122. Capacitance of an isolated conducting sphere of radius R_1 becomes n times when it is enclosed by a concentric conducting sphere of radius R_2 connected to earth. The ratio of their radii $\left(\frac{R_2}{R_1}\right)$ is:

[NCERT : P L-67 | July 25, 2022 (II)]

(a) $\frac{n}{n-1}$ (b) $\frac{2n}{2n+1}$ (c) $\frac{n+1}{n}$ (d) $\frac{2n+1}{n}$

123. C_o is the capacitance of a parallel plate capacitor with air as a medium between the plates (as shown in Fig. 1). If half space between the plates is filled with a dielectric of relative permittivity ε_r (as shown in Fig. 2), the new capacitance of the capacitor will be:

[NCERT : P L-69, 72 | June 30, 2022 (I)]

Fig. 1 Fig. 2

(a) $\frac{C_o}{2}(1+\varepsilon_r)$ (b) $C_o + \varepsilon_r$

(c) $\frac{C_o\varepsilon_r}{2}$ (d) $C_o(1+\varepsilon_r)$

124. A parallel plate capacitor filled with a medium of dielectric constant 10, is connected across a battery and is charged. The dielectric slab is replaced by another slab of dielectric constant 15. Then the energy of capacitor will :

[NCERT : P L-74 | June 29, 2022 (I)]

(a) increase by 50% (b) decrease by 15%
(c) increase by 25% (d) increase by 33%

125. A capacitor is discharging through a resistor R. Consider in time t_1, the energy stored in the capacitor reduces to half of its initial value and in time t_2, the charge stored reduces to one eighth of its initial value. The ratio t_1/t_2 will be : [NCERT : P L-74 | June 29, 2022 (II)]

(a) 1/2 (b) 1/3 (c) 1/4 (d) 1/6

126. A capacitor C_1 of capacitance $5\mu F$ is charged to a potential of 30 V using a battery. The battery is then removed and the charged capacitor is connected to an uncharged capacitor C_2 of capacitance $10\mu F$ as shown in figure. When the switch is closed charge flows between the capacitors. At equilibrium, the charge on the capacitor C_2 is ______ μC.

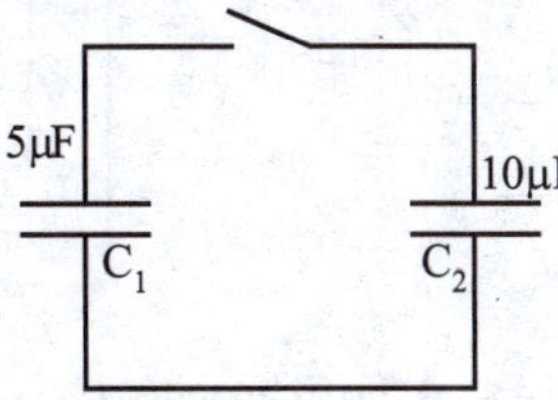

[NCERT : P L-69, 72 | June 28, 2022 (II)]

127. A force of 10N acts on a charged particle placed between two plates of a charged capacitor. If one plate of capacitor is removed, then the force acting on that particle will be :

[NCERT : P L-69 | June 27, 2022 (I)]

(a) 5 N (b) 10 N (c) 20 N (d) Zero

128. Two capacitors having capacitance C_1 and C_2 respectively are connected as shown in figure. Initially, capacitor C_1 is charged to a potential difference V volt by a battery. The battery is then removed and the charged capacitor C_1 is now connected to uncharged capacitor C_2 by closing the switch S. The amount of charge on the capacitor C_2, after equilibrium is : **[NCERT : P L-69, 72 | June 26, 2022 (I)]**

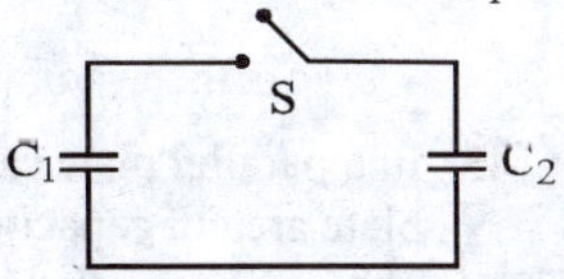

(a) $\frac{C_1 C_2}{(C_1+C_2)}V$ (b) $\frac{(C_1+C_2)}{C_1 C_2}V$

(c) $(C_1+C_2)V$ (d) $(C_1-C_2)V$

129. The charge on capacitor of capacitance 15μF in the figure given below is :

[NCERT : P L-69, 72 | June 26, 2022 (II)]

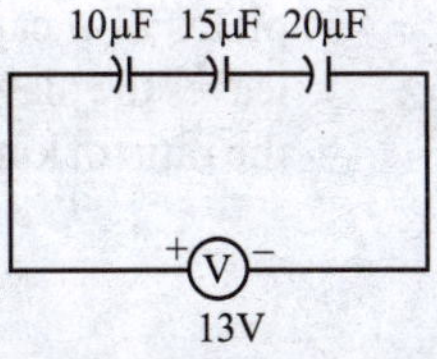

(a) 60 μC

(b) 130 μC

(c) 260 μC

(d) 585 μC

130. A parallel plate capacitor with plate area A and plate separation $d = 2$ m has a capacitance of 4 μF. The new capacitance of the system if half of the space between them is filled with a dielectric material of dielectric constant K= 3 (as shown in figure) will be:

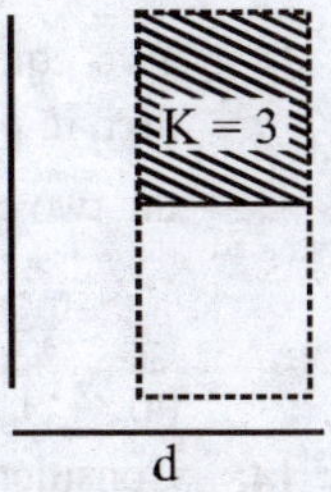

[NCERT : P L-70 | June 26, 2022 (II)]

(a) 2μF (b) 32μF

(c) 6μF (d) 8μF

131. Two metallic plates form a parallel plate capacitor. The distance between the plates is 'd'. A metal sheet of thickness $\frac{d}{2}$ and of area equal to area of each plate is introduced between the plates. What will be the ratio of the new capacitance to the original capacitance of the capacitor ?

[NCERT : P L-70 | June 25, 2022 (II)]

(a) 2 : 1 (b) 1 : 2 (c) 1 : 4 (d) 4 : 1

132. A parallel plate capacitor is formed by two plates each of area 30π cm^2 separated by 1 mm. A material of dielectric strength 3.6×10^7 Vm^{-1} is filled between the plates. If the maximum charge that can be stored on the capacitor without causing any dielectric breakdown is 7×10^{-6} C, the value of dielectric constant of the material is :

[NCERT : P L-69, 70 | June 24, 2022 (I)]

$$\left\{\text{use: } \frac{1}{4\pi\varepsilon_o} = 9\times10^9\ \text{Nm}^2\text{C}^{-2}\right\}$$

(a) 1.66 (b) 1.75 (c) 2.25 (d) 2.33

133. If the charge on a capacitor is increased by 2 C, the energy stored in it increases by 44%. The original charge on the capacitor is (in C) : **[NCERT : P L-74 | June 24, 2022 (II)]**

(a) 10 (b) 20 (c) 30 (d) 40

134. A capacitor is connected to a 20 V battery through a resistance of 10 Ω. It is found that the potential difference across the capacitor rises to 2 V in 1 μs. The capacitance of the capacitor is ______ μF. Given : $\ln\left(\frac{10}{9}\right) = 0.105$

[Sep. 1, 2021 (II)]

(a) 9.52 (b) 0.95 (c) 0.105 (d) 1.85

135. A capacitor of 50 μF is connected in a circuit as shown in figure. The charge on the upper plate of the capacitor is ______ μC. **[NCERT : P L-69 | NA, Aug. 31, 2021 (I)]**

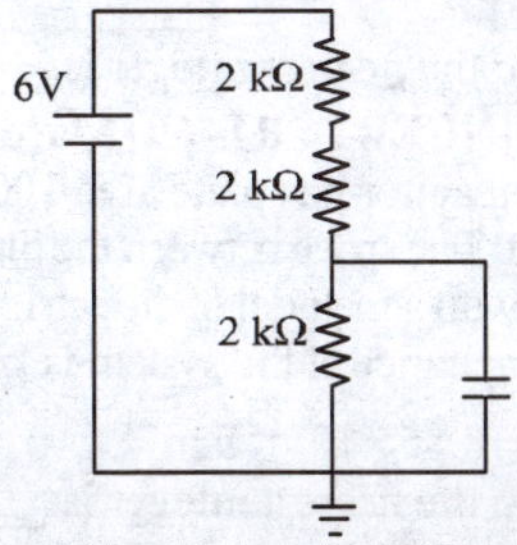

136. A parallel plate capacitor of capacitance 200 μF is connected to a battery of 200 V. A dielectric slab of dielectric constant 2 is now inserted into the space between plates of capacitor while the battery remain connected. The change in the electrostatic energy in the capacitor will be ______ J.

[NCERT : P L-74 | NA, Aug. 31, 2021 (II)]

137. Calculate the amount of charge on capacitor of 4 μF. The internal resistance of battery is 1Ω :

[NCERT : P L-69, 72 | Aug. 27, 2021 (I)]

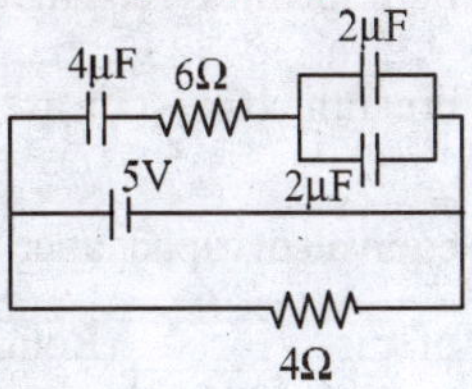

(a) 8 μC

(b) zero

(c) 16 μC

(d) 4 μC

138. The material filled between the plates of a parallel plate capacitor has resistivity 200 Ωm. The value of capacitance of the capacitor is 2 pF. If a potential difference of 40 V is applied across the plates of the capacitor, then the value of leakage current flowing out of the capacitor is : (given the value of relative permitivity of material is 50)

[NCERT : P L-69 | Aug. 26, 2021 (I)]

(a) 9.0 μA (b) 9.0 mA (c) 0.9 mA (d) 0.9 μA

139. A parallel - plate capacitor with plate area A has separation d between the plates. Two dielectric slabs of dielectric constant K_1 and K_2 of same area A/2 and thickness d/2 are inserted in the space between the plates. The capacitance of the capacitor will be given by :

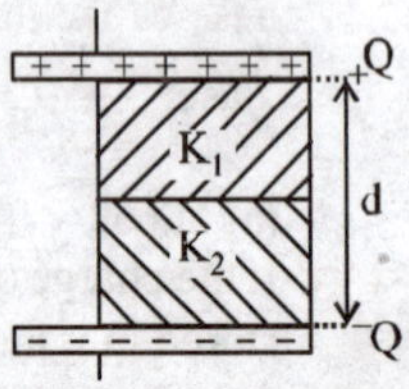

[NCERT : P L-70, 72 | Aug. 26, 2021 (II)]

(a) $\frac{\varepsilon_0 A}{d}\left(\frac{1}{2}+\frac{K_1K_2}{K_1+K_2}\right)$ (b) $\frac{\varepsilon_0 A}{d}\left(\frac{1}{2}+\frac{K_1K_2}{2(K_1+K_2)}\right)$

(c) $\frac{\varepsilon_0 A}{d}\left(\frac{1}{2}+\frac{K_1+K_2}{K_1K_2}\right)$ (d) $\frac{\varepsilon_0 A}{d}\left(\frac{1}{2}+\frac{2(K_1+K_2)}{K_1K_2}\right)$

140. The circuit shown in the figure consists of a charged capacitor of capacity 3 μF and a charge of 30 μC. At time t = 0, when the key is closed, the value of current flowing through the 5 MΩ resistor is 'x' μA. The value of 'x' to the nearest integer is ______.

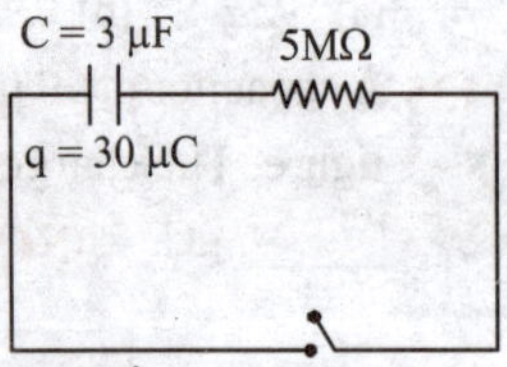

[NCERT : P L-69 | March 18, 2021 (I)]

141. A parallel plate capacitor has plate area 100 m^2 and plate separation of 10 m. The space between the plates is filled up to a thickness 5 m with a material of dielectric constant of 10. The resultant capacitance of the system is 'x' pF.

The value of $\varepsilon_0 = 8.85\times10^{-12}\,F.m^{-1}$

The value of 'x' to the nearest integer is ___.

[NCERT : P L-70 | March 18, 2021 (I)]

142. A parallel plate capacitor whose capacitance C is 14 pF is charged by a battery to a potential difference V = 12 V between its plates. The charging battery is now disconnected and a porcelin plate with k = 7 is inserted between the plates, then the plate would oscillate back and forth between the plates with a constant mechanical energy of ________ pJ. (Assume no friction)

[NCERT : P L-74 | March 17, 2021 (I)]

143. Four identical rectangular plates with length, l = 2 cm and breadth, $b=\frac{3}{2}$ cm are arranged as shown in figure. The equivalent capacitance between A and C is $\frac{x\varepsilon_0}{d}$. The value of x is ________. (Round off to the Nearest Integer)

[NCERT : P L-72 | March 17, 2021 (I)]

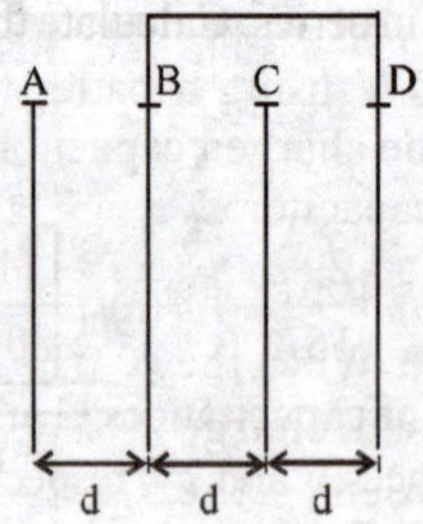

144. A 2 μF capacitor C_1 is first charged to a potential difference of 10 V using a battery. Then the battery is removed and the capacitor is connected to an uncharged capacitor C_2 of 8 μF. The charge in C_2 on equilibrium condition is ______ μC. (Round off to the nearest integer)

[NCERT : P L-69 | March 17, 2021 (II)]

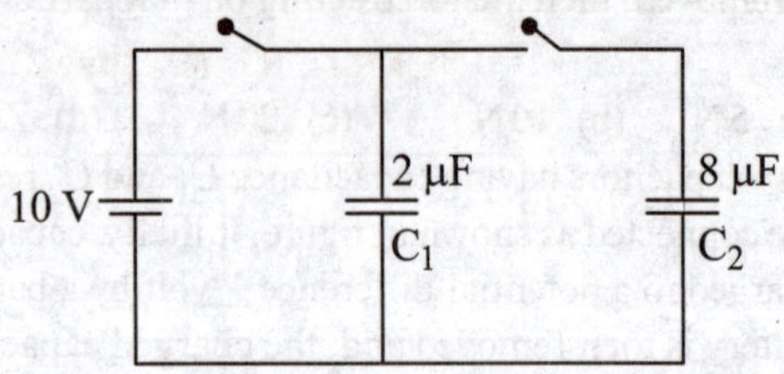

145. In a parallel plate capacitor set up, the plate area of capacitor is 2 m^2 and the plates are separated by 1 m. If the space between the plates are filled with a dielectric material of thickness 0.5 m and area 2 m^2 (see fig) the capacitance of the set-up will be ______ ε_0. (Dielectric constant of the material = 3.2) (Round off to the Nearest Integer).

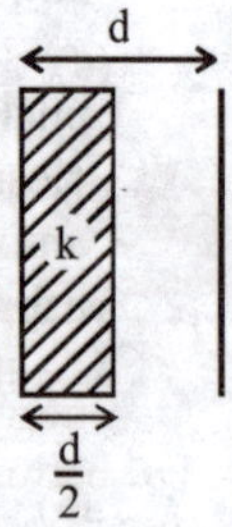

[NCERT : P L-70 | March 16, 2021 (II)]

146. An electron with kinetic energy K_1 enters between parallel plates of a capacitor at an angle 'α' with the plates. It leaves the plates at angle 'β' with kinetic energy K_2. Then the ratio of kinetic energies $K_1 : K_2$ will be :

[NCERT : P L-69 | Feb. 25, 2021 (II)]

(a) $\frac{\cos\beta}{\sin\alpha}$ (b) $\frac{\cos\beta}{\cos\alpha}$ (c) $\frac{\cos^2\beta}{\cos^2\alpha}$ (d) $\frac{\sin^2\beta}{\cos^2\alpha}$

147. Two equal capacitors are first connected in series and then in parallel. The ratio of the equivalent capacities in the two cases will be :

[NCERT : P L-72 | Feb. 24, 2021 (I)]

(a) 4 : 1 (b) 2 : 1 (c) 1 : 4 (d) 1 : 2

148. Consider the combination of 2 capacitors C_1 and C_2, with $C_2 > C_1$, when connected in parallel, the equivalent capacitance is $\frac{15}{4}$ times the equivalent capacitance of the

same connected in series. Calculate the ratio of capacitors, $\frac{C_2}{C_1}$. **[NCERT : P L-72 | Feb. 26, 2021 (I)]**

(a) $\frac{15}{4}$ (b) $\frac{29}{15}$ (c) $\frac{111}{80}$ (d) $\frac{15}{11}$

149. Two capacitors of capacitances C and 2C are charged to potential differences V and 2V, respectively. These are then connected in parallel in such a manner that the positive terminal of one is connected to the negative terminal of the other. The final energy of this configuration is : **[NCERT : P L-69, 74 | Sep. 05, 2020 (I)]**

(a) $\frac{25}{6}CV^2$ (b) $\frac{3}{2}CV^2$ (c) zero (d) $\frac{9}{2}CV^2$

150. In the circuit shown, charge on the 5 μF capacitor is : **[NCERT : P L-69 | Sep. 05, 2020 (II)]**

(a) 18.00 μC
(b) 10.90 μC
(c) 16.36 μC
(d) 5.45 μC

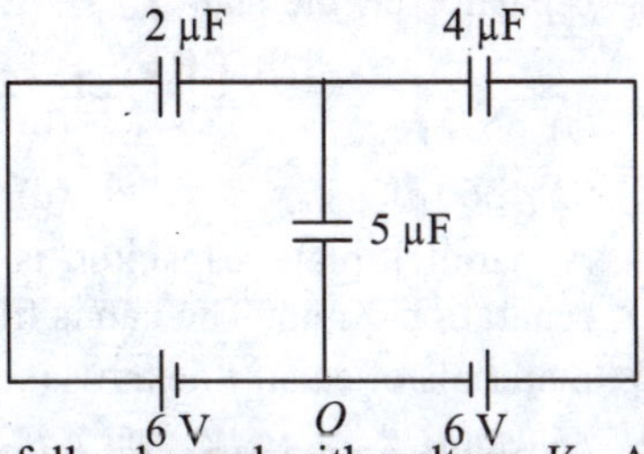

151. A capacitor C is fully charged with voltage V_0. After disconnecting the voltage source, it is connected in parallel with another uncharged capacitor of capacitance $\frac{C}{2}$. The energy loss in the process after the charge is distributed between the two capacitors is : **[NCERT : P L-74 | Sep. 04, 2020 (II)]**

(a) $\frac{1}{2}CV_0^2$ (b) $\frac{1}{3}CV_0^2$

(c) $\frac{1}{4}CV_0^2$ (d) $\frac{1}{6}CV_0^2$

152. In the circuit shown in the figure, the total charge is 750 μC and the voltage across capacitor C_2 is 20 V. Then the charge on capacitor C_2 is : **[NCERT : P L-69, 72 | Sep. 03, 2020 (I)]**

(a) 450 μC
(b) 590 μC
(c) 160 μC
(d) 650 μC

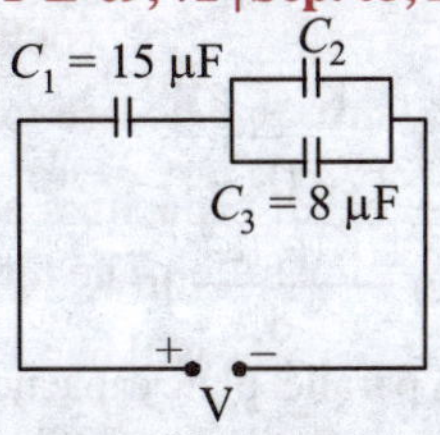

153. A 5 μF capacitor is charged fully by a 220 V supply. It is then disconnected from the supply and is connected in series to another uncharged 2.5 μF capacitor. If the energy change during the charge redistribution is $\frac{X}{100}$ J then value of X to the nearest integer is ______. **[NCERT : P L-74 | NA Sep. 02, 2020 (I)]**

154. A 10 μF capacitor is fully charged to a potential difference of 50 V. After removing the source voltage it is connected to an uncharged capacitor in parallel. Now the potential difference across them becomes 20 V. The capacitance of the second capacitor is : **[NCERT : P L-69, 74 | Sep. 02, 2020 (II)]**

(a) 15 μF (b) 30 μF (c) 20 μF (d) 10 μF

155. Effective capacitance of parallel combination of two capacitors C_1 and C_2 is 10 μF. When these capacitors are individually connected to a voltage source of 1 V, the energy stored in the capacitor C_2 is 4 times that of C_1. If these capacitors are connected in series, their effective capacitance will be: **[NCERT : P L-72, 74 | 8 Jan. 2020 (I)]**

(a) 4.2 μF (b) 3.2 μF (c) 1.6 μF (d) 8.4 μF

156. A capacitor is made of two square plates each of side 'a' making a very small angle α between them, as shown in figure. The capacitance will be close to: **[NCERT : P L-69 | 8 Jan. 2020 (II)]**

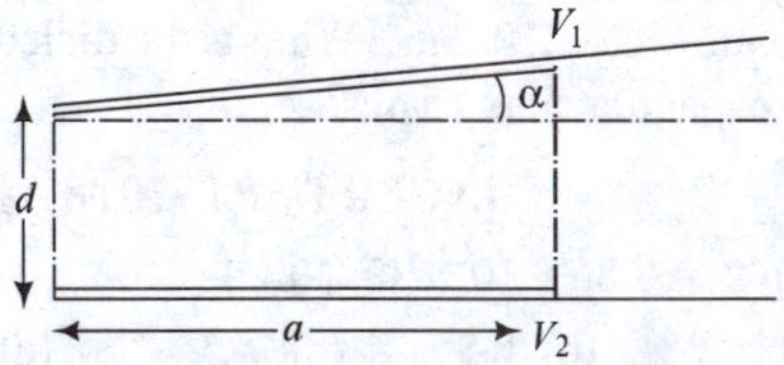

(a) $\frac{\epsilon_0 a^2}{d}\left(1-\frac{\alpha a}{2d}\right)$ (b) $\frac{\epsilon_0 a^2}{d}\left(1-\frac{\alpha a}{4d}\right)$

(c) $\frac{\epsilon_0 a^2}{d}\left(1+\frac{\alpha a}{d}\right)$ (d) $\frac{\epsilon_0 a^2}{d}\left(1-\frac{3\alpha a}{2d}\right)$

157. A 60 pF capacitor is fully charged by a 20 V supply. It is then disconnected from the supply and is connected to another uncharged 60 pF capacitor in parallel. The electrostatic energy that is lost in this process by the time the charge is redistributed between them is (in nJ) **[NCERT : P L-74 | NA 7 Jan. 2020 (II)]**

158. The parallel combination of two air filled parallel plate capacitors of capacitance C and nC is connected to a battery of voltage, V. When the capacitors are fully charged, the battery is removed and after that a dielectric material of dielectric constant K is placed between the two plates of the first capacitor. The new potential difference of the combined system is: **[NCERT : P L-70, 72 | 9 April 2020 (II)]**

(a) $\frac{nV}{K+n}$ (b) V (c) $\frac{V}{K+n}$ (d) $\frac{(n+1)V}{(K+n)}$

159. In the given circuit, the charge on 4 μF capacitor will be : **[NCERT : P L-69, 72 | 12 April 2019 (II)]**

(a) 5.4 μC
(b) 9.6 μC
(c) 13.4 μC
(d) 24 μC

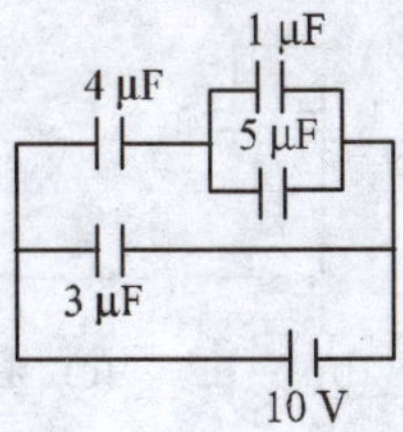

160. Figure shows charge (q) versus voltage (V) graph for series and parallel combination of two given capacitors. The capacitances are :

[NCERT : P L-69, 72 | 10 April 2019 (I)]

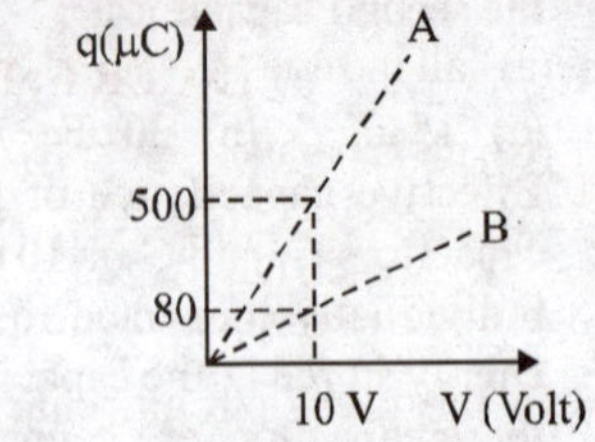

(a) 40 μF and 10 μF
(b) 60 μF and 40 μF
(c) 50 μF and 30 μF
(d) 20 μF and 30 μF

161. A capacitor with capacitance 5μF is charged to 5 μC. If the plates are pulled apart to reduce the capacitance to 2 μF, how much work is done?

[NCERT : P L-74 | 9 April 2019 (I)]

(a) 6.25×10^{-6} J (b) 3.75×10^{-6} J
(c) 2.16×10^{-6} J (d) 2.55×10^{-6} J

162. Voltage rating of a parallel plate capacitor is 500 V. Its dielectric can withstand a maximum electric field of 10^6 V/m. The plate area is 10^{-4} m^2. What is the dielectric constant if the capacitance is 15 pF ?

[NCERT : P L-70 | 8 April 2019 (I)]

(given $\varepsilon_0 = 8.86 \times 10^{-12}$ C^2/Nm^2)

(a) 3.8 (b) 8.5 (c) 4.5 (d) 6.2

163. A parallel plate capacitor has 1μF capacitance. One of its two plates is given + 2μC charge and the other plate, +4μC charge. The potential difference developed across the capacitor is : **[NCERT : P L-69 | 8 April 2019 (II)]**

(a) 3V (b) 1V
(c) 5V (d) 2V

164. A parallel plate capacitor with plates of area 1 m^2 each, are at a separation of 0.1 m. If the electric field between the plates is 100 N/C, the magnitude of charge on each plate is :

(Take $\epsilon_0 = 8.85 \times 10^{-12} \dfrac{C^2}{N-m^2}$)

[NCERT : P L-69 | 12 Jan. 2019 (II)]

(a) 7.85×10^{-10} C (b) 6.85×10^{-10} C
(c) 8.85×10^{-10} C (d) 9.85×10^{-10} C

165. In the circuit shown, find C if the effective capacitance of the whole circuit is to be 0.5 μF. All values in the circuit are in μF. **[NCERT : P L-72 | 12 Jan. 2019 (II)]**

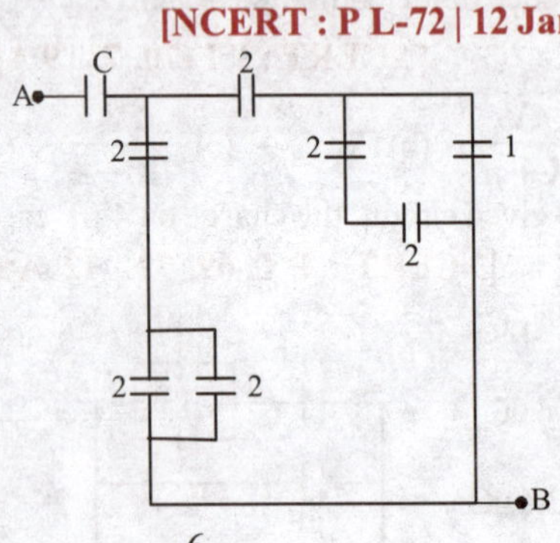

(a) $\dfrac{7}{11}$ μF (b) $\dfrac{6}{5}$ μF (c) 4 μF (d) $\dfrac{7}{10}$ μF

166. In the figure shown below, the charge on the left plate of the 10 μF capacitor is –30μC. The charge on the right plate of the 6μF capacitor is :

[NCERT : P L-69, 72 | 11 Jan. 2019 (I)]

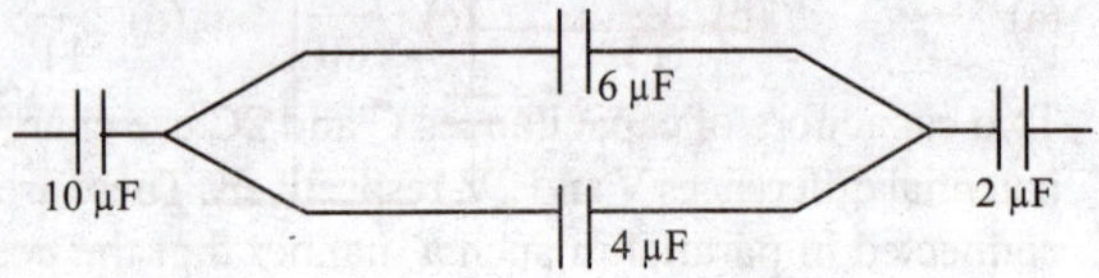

(a) –12 μC (b) +12 μC
(c) –18 μC (d) +18 μC

167. A parallel plate capacitor having capacitance 12 pF is charged by a battery to a potential difference of 10 V between its plates. The charging battery is now disconnected and a porcelain slab of dielectric constant 6.5 is slipped between the plates. The work done by the capacitor on the slab is:

[NCERT : P L-74 | 10 Jan. 2019 (II)]

(a) 692 pJ (b) 508 pJ
(c) 560 pJ (d) 600 pJ

168. A parallel plate capacitor is of area 6 cm^2 and a separation 3 mm. The gap is filled with three dielectric materials of equal thickness (see figure) with dielectric constants $K_1 = 10$, $K_2 = 12$ and $K_3 = 14$. The dielectric constant of a material which when fully inserted in above capacitor, gives same capacitance would be:

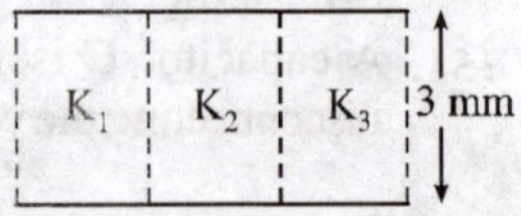

[NCERT : P L-70, 72 | 10 Jan. 2019 (I)]

(a) 4 (b) 14
(c) 12 (d) 36

169. A parallel plate capacitor is made of two square plates of side 'a', separated by a distance d (d<<a). The lower triangular portion is filled with a dielectric of dielectric constant K, as shown in the figure. Capacitance of this capacitor is:

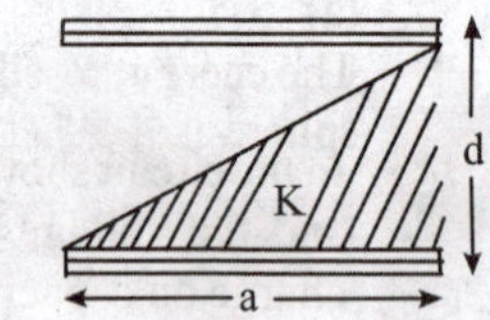

[NCERT : P L-70 | 9 Jan. 2019 (I)]

(a) $\dfrac{K \epsilon_0 a^2}{2d(K+1)}$ (b) $\dfrac{K \epsilon_0 a^2}{d(K-1)} \ln K$

(c) $\dfrac{K \epsilon_0 a^2}{d} \ln K$ (d) $\dfrac{1}{2}\dfrac{K \epsilon_0 a^2}{d}$

170. A parallel plate capacitor of capacitance 90 pF is connected to a battery of emf 20V. If a dielectric material of dielectric constant $k = \dfrac{5}{3}$ is inserted between the plates, the magnitude of the induced charge will be:

[NCERT : P L-70 | 2018]

(a) 1.2 n C (b) 0.3 n C
(c) 2.4 n C (d) 0.9 n C

171. The equivalent capacitance between A and B in the circuit given below is:

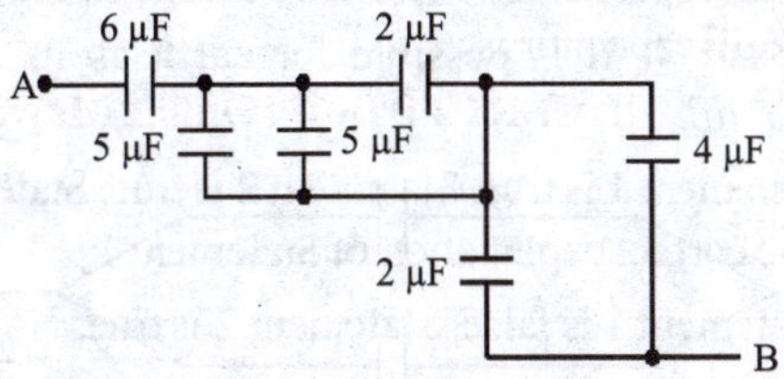

[NCERT : P L-72 | Online April 15, 2018]

(a) $4.9\,\mu F$ (b) $3.6\,\mu F$ (c) $5.4\,\mu F$ (d) $2.4\,\mu F$

172. A capacitor $C_1 = 1\mu F$ is charged up to a voltage $V = 60V$ by connecting it to battery B through switch (1), Now C_1 is disconnected from battery and connected to a circuit consisting of two uncharged capacitors $C_2 = 3.0\mu F$ and $C_3 = 6.0\mu F$ through a switch (2) as shown in the figure. The sum of final charges on C_2 and C_3 is:

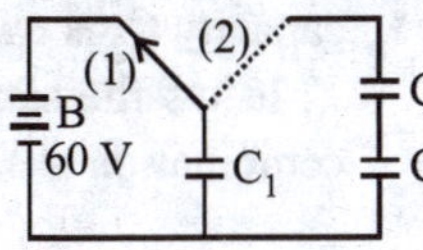

[NCERT : P L-69, 72 | Online April 15, 2018]

(a) $36\mu C$ (b) $20\mu C$ (c) $54\mu C$ (d) $40\mu C$

173. A capacitance of $2\mu F$ is required in an electrical circuit across a potential difference of 1.0 kV. A large number of $1\mu F$ capacitors are available which can withstand a potential difference of not more than 300 V. The minimum number of capacitors required to achieve this is

[NCERT : P L-72 | 2017]

(a) 24 (b) 32 (c) 2 (d) 16

174. A parallel plate capacitors is maintained at a certain potential difference. When a 3 mm thick slab is introduced between the plates, in order to maintain the same potential difference, the distance between the plates is increased by 2.4 mm.

Find the dielectric constant of the slab.

[NCERT : P L-70 | Online April 9, 2017]

(a) 3 (b) 4 (c) 5 (d) 6

175. The energy stored in the electric field produced by a metal sphere is 4.5 J. If the sphere contains 4 μC charge, its radius will be : [Take : $\frac{1}{4\pi\varepsilon_0} = 9\times10^9\,N-m^2/C^2$]

[NCERT : P L-74 | Online April 8, 2017]

(a) 20mm (b) 32mm (c) 28mm (d) 16mm

176. A combination of capacitors is set up as shown in the figure. The magnitude of the electric field, due to a point charge Q (having a charge equal to the sum of the charges on the 4 μ F and 9 μ F capacitors), at a point distance 30 m from it, would equal : **[NCERT : P L-69, 72 | 2016]**

(a) 420 N/C

(b) 480 N/C

(c) 240 N/C

(d) 360 N/C

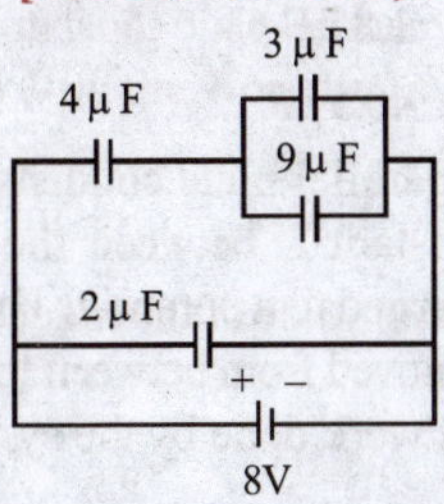

177. Three capacitors each of 4 μF are to be connected in such a way that the effective capacitance is 6μF. This can be done by connecting them :

[NCERT : P L-72 | Online April 9, 2016]

(a) all in series

(b) all in parallel

(c) two in parallel and one in series

(d) two in series and one in parallel

178. In the given circuit, charge Q_2 on the 2μF capacitor changes as C is varied from 1μF to 3μF. Q_2 as a function of 'C' is given properly by: (*figures are drawn schematically and are not to scale*) **[NCERT : P L-72 | 2015]**

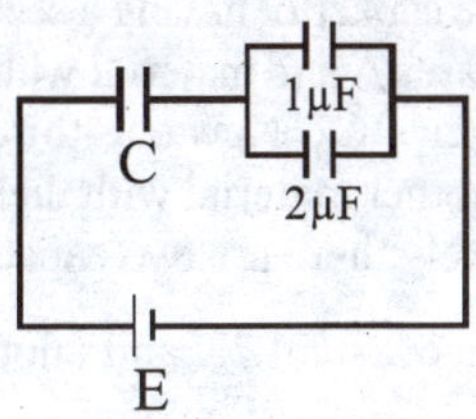

(a)

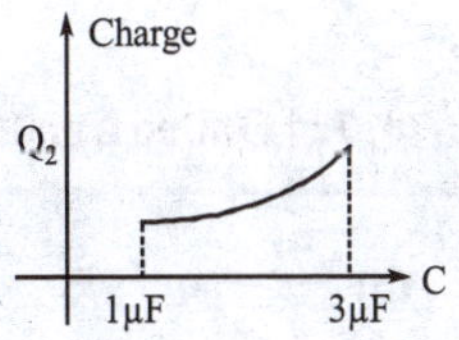

(b)

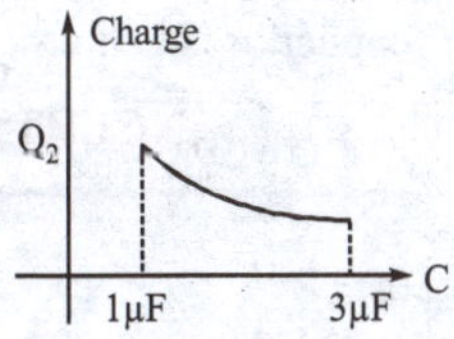

(c)

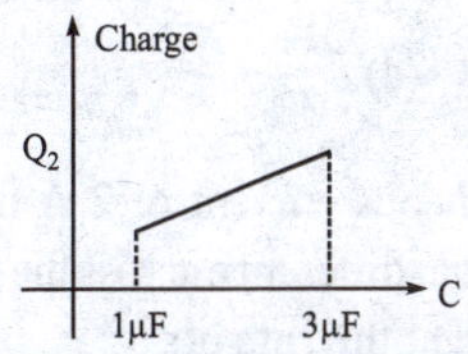

(d)

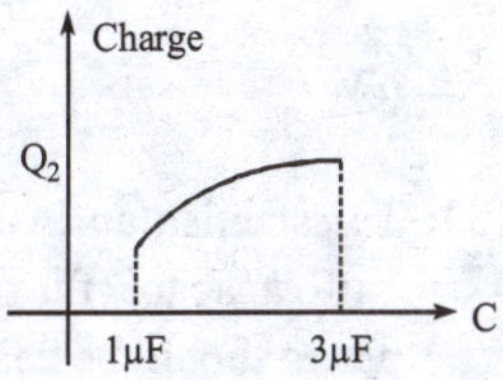

179. In figure a system of four capacitors connected across a 10 V battery is shown.

[NCERT : P L-69 | Online April 11, 2015]

Charge that will flow from switch S when it is closed is :

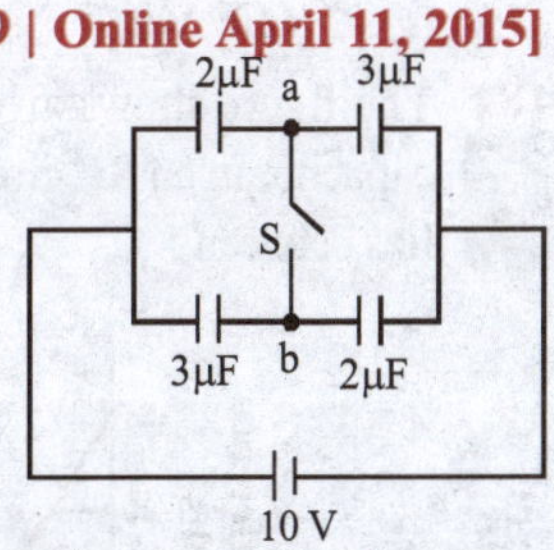

(a) 5 μC from b to a

(b) 20 μC from a to b

(c) zero

(d) 5 μC from a to b

180. A parallel plate capacitor is made of two circular plates separated by a distance 5 mm and with a dielectric of dialectric constant 2.2 between them. When the electric field in the dielectric is 3×10^4 V/m the charge density of the positive plate will be close to:

[NCERT : P L-70 | 2014]

(a) $6\times10^{-7}\,C/m^2$ (b) $3\times10^{-7}\,C/m^2$

(c) $3\times10^{4}\,C/m^2$ (d) $6\times10^{4}\,C/m^2$

181. The space between the plates of a parallel plate capacitor is filled with a 'dielectric' whose 'dielectric constant' varies with distance as per the relation:

$K(x) = K_o + \lambda x$ (λ = a constant)

The capacitance C, of the capacitor, would be related to its vacuum capacitance C_o for the relation :

[NCERT : P L-70 | Online April 12, 2014]

(a) $C = \frac{\lambda d}{ln(1+K_o\lambda d)}C_o$ (b) $C = \frac{\lambda}{d.ln(1+K_o\lambda d)}C_o$

(c) $C = \frac{\lambda d}{ln(1+\lambda d/K_o)}C_o$ (d) $C = \frac{\lambda}{d.ln(1+K_o/\lambda d)}C_o$

182. A parallel plate capacitor having a separation between the plates d, plate area A and material with dielectric constant K has capacitance C_0. Now one-third of the material is replaced by another material with dielectric constant 2K, so that effectively there are two capacitors one with area $\frac{1}{3}$A, dielectric constant 2K and another with area $\frac{2}{3}$A and dielectric constant K. If the capacitance of this new capacitor is C then $\frac{C}{C_0}$ is

[NCERT : P L-70, 72 | Online April 25, 2013]

(a) 1 (b) $\frac{4}{3}$

(c) $\frac{2}{3}$ (d) $\frac{1}{3}$

183. To establish an instantaneous current of 2 A through a 1 μF capacitor ; the potential difference across the capacitor plates should be changed at the rate of :

[NCERT : P L-67 | Online April 22, 2013]

(a) 2×10^4 V/s (b) 4×10^6 V/s

(c) 2×10^6 V/s (d) 4×10^4 V/s

184. The figure shows an experimental plot discharging of a capacitor in an RC circuit. The time constant τ of this circuit lies between : **[2012]**

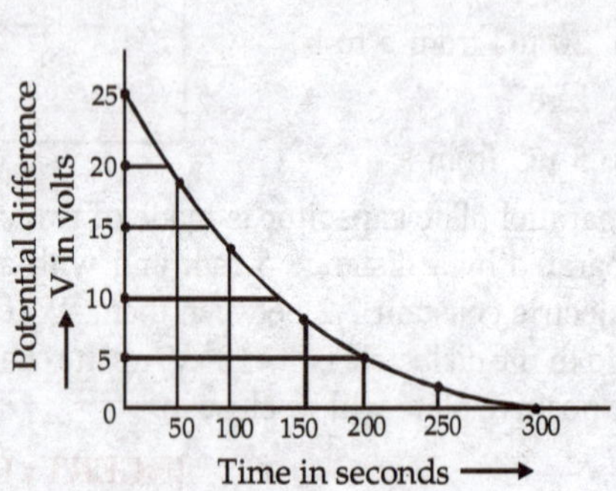

(a) 150 sec and 200 sec (b) 0 sec and 50 sec

(c) 50 sec and 100 sec (d) 100 sec and 150 sec

185. **Statement 1:** It is not possible to make a sphere of capacity 1 farad using a conducting material.

Statement 2: It is possible for earth as its radius is 6.4×10^6 m. **[NCERT : P L-67 | Online May 26, 2012]**

(a) Statement 1 is true, Statement 2 is true, Statement 2 is the correct explanation of Statement 1.

(b) Statement 1 is false, Statement 2 is true.

(c) Statement 1 is true, Statement 2 is true, Statement 2 is not the correct explanation of Statement 1.

(d) Statement 1 is true, Statement 2 is false.

186. A series combination of n_1 capacitors, each of capacity C_1 is charged by source of potential difference 4 V. When another parallel combination of n_2 capacitors each of capacity C_2 is charged by a source of potential difference V, it has the same total energy stored in it as the first combination has. The value of C_2 in terms of C_1 is then

[NCERT : P L-72, 74 | Online May 12, 2012]

(a) $16\frac{n_2}{n_1}C_1$ (b) $\frac{2C_1}{n_1 n_2}$

(c) $2\frac{n_2}{n_1}C_1$ (d) $\frac{16C_1}{n_1 n_2}$

187. Let C be the capacitance of a capacitor discharging through a resistor R. Suppose t_1 is the time taken for the energy stored in the capacitor to reduce to half its initial value and t_2 is the time taken for the charge to reduce to one-fourth its initial value. Then the ratio t_1/t_2 will be

[NCERT : P L-74 | 2010]

(a) 1 (b) $\frac{1}{2}$

(c) $\frac{1}{4}$ (d) 2

188. A parallel plate capacitor with air between the plates has capacitance of 9 pF. The separation between its plates is 'd'. The space between the plates is now filled with two dielectrics. One of the dielectrics has dielectric constant $k_1 = 3$ and thickness $\frac{d}{3}$ while the other one has dielectric constant $k_2 = 6$ and thickness $\frac{2d}{3}$. Capacitance of the capacitor is now **[NCERT : P L-70, 72 | 2008]**

(a) 1.8 pF (b) 45 pF

(c) 40.5 pF (d) 20.25 pF

189. A parallel plate condenser with a dielectric of dielectric constant K between the plates has a capacity C and is charged to a potential V volt. The dielectric slab is slowly removed from between the plates and then reinserted. The net work done by the system in this process is

[NCERT : P L-70, 74 | 2007]

(a) zero (b) $\frac{1}{2}(K-1)\,CV^2$

(c) $\frac{CV^2(K-1)}{K}$ (d) $(K-1)\,CV^2$

190. A parallel plate capacitor is made by stacking n equally spaced plates connected alternatively. If the capacitance between any two adjacent plates is 'C' then the resultant capacitance is **[NCERT : P L-72 | 2005]**

(a) $(n+1)\,C$ (b) $(n-1)\,C$

(c) nC (d) C

191. A fully charged capacitor has a capacitance 'C'. It is discharged through a small coil of resistance wire embedded in a thermally insulated block of specific heat capacity 's' and mass 'm'. If the temperature of the block is raised by 'ΔT', the potential difference 'V' across the capacitance is **[NCERT : P L-74 | 2005]**

(a) $\frac{mC\Delta T}{s}$ (b) $\sqrt{\frac{2mC\Delta T}{s}}$

(c) $\sqrt{\frac{2ms\Delta T}{C}}$ (d) $\frac{ms\Delta T}{C}$

192. A sheet of aluminium foil of negligible thickness is introduced between the plates of a capacitor. The capacitance of the capacitor **[NCERT : P L-70 | 2003]**

(a) decreases (b) remains unchanged

(c) becomes infinite (d) increases

193. The work done in placing a charge of 8×10^{-18} coulomb on a condenser of capacity 100 micro-farad is **[NCERT : P L-74 | 2003]**

(a) 16×10^{-32} joule (b) 3.1×10^{-26} joule

(c) 4×10^{-10} joule (d) 32×10^{-32} joule

194. If there are n capacitors in parallel connected to V volt source, then the energy stored is equal to **[NCERT : P L-72, 74 | 2002]**

(a) CV (b) $\frac{1}{2}nCV^2$

(c) CV^2 (d) $\frac{1}{2n}CV^2$

195. Capacitance (in F) of a spherical conductor with radius 1 m is **[NCERT : P L-67 | 2002]**

(a) 1.1×10^{-10} (b) 10^{-6}

(c) 9×10^{-9} (d) 10^{-3}

ANSWER KEYS																			
1	(b)	**21**	(d)	**41**	(b)	**61**	(a)	**81**	(5)	**101**	(b)	**121**	(6)	**141**	(161)	**161**	(b)	**181**	(c)
2	(d)	**22**	(12)	**42**	(b)	**62**	(18)	**82**	(c)	**102**	(c)	**122**	(a)	**142**	(864)	**162**	(b)	**182**	(b)
3	(a)	**23**	(b)	**43**	(b)	**63**	(32)	**83**	(64)	**103**	(4)	**123**	(a)	**143**	(2)	**163**	(b)	**183**	(c)
4	(b)	**24**	(a)	**44**	(b)	**64**	(d)	**84**	(b)	**104**	(c)	**124**	(a)	**144**	(16)	**164**	(c)	**184**	(d)
5	(c)	**25**	(b)	**45**	(a)	**65**	(d)	**85**	(b)	**105**	(6)	**125**	(d)	**145**	(3)	**165**	(a)	**185**	(d)
6	(c)	**26**	(d)	**46**	(a)	**66**	(c)	**86**	(c)	**106**	(3)	**126**	(100)	**146**	(c)	**166**	(d)	**186**	(d)
7	(36)	**27**	(d)	**47**	(c)	**67**	(c)	**87**	(1320)	**107**	(5)	**127**	(a)	**147**	(c)	**167**	(b)	**187**	(c)
8	(c)	**28**	(243)	**48**	(a)	**68**	(d)	**88**	(a)	**108**	(a)	**128**	(a)	**148**	(None)	**168**	(c)	**188**	(c)
9	(a)	**29**	(128)	**49**	(c)	**69**	(b)	**89**	(d)	**109**	(55)	**129**	(a)	**149**	(b)	**169**	(b)	**189**	(a)
10	(27)	**30**	(c)	**50**	(d)	**70**	(b)	**90**	(a)	**110**	(225)	**130**	(c)	**150**	(c)	**170**	(a)	**190**	(b)
11	(b)	**31**	(d)	**51**	(c)	**71**	(b)	**91**	(a)	**111**	(c)	**131**	(a)	**151**	(d)	**171**	(d)	**191**	(c)
12	(c)	**32**	(a)	**52**	(c)	**72**	(c)	**92**	(c)	**112**	(6)	**132**	(d)	**152**	(b)	**172**	(d)	**192**	(b)
13	(8)	**33**	(d)	**53**	(a)	**73**	(c)	**93**	(a)	**113**	(105)	**133**	(a)	**153**	(4)	**173**	(b)	**193**	(d)
14	(d)	**34**	(b)	**54**	(a)	**74**	(d)	**94**	(a)	**114**	(c)	**134**	(b)	**154**	(a)	**174**	(c)	**194**	(b)
15	(160)	**35**	(c)	**55**	(a)	**75**	(c)	**95**	(0.0095)	**115**	(a)	**135**	(100)	**155**	(c)	**175**	(d)	**195**	(a)
16	(5)	**36**	(d)	**56**	(c)	**76**	(a)	**96**	(86)	**116**	(a)	**136**	(4)	**156**	(a)	**176**	(a)		
17	(c)	**37**	(c)	**57**	(a)	**77**	(d)	**97**	(a)	**117**	(240)	**137**	(a)	**157**	(6)	**177**	(d)		
18	(a)	**38**	(b)	**58**	(12)	**78**	(a)	**98**	(2)	**118**	(60)	**138**	(c)	**158**	(d)	**178**	(d)		
19	(c)	**39**	(d)	**59**	(d)	**79**	(15)	**99**	(2)	**119**	(a)	**139**	(a)	**159**	(d)	**179**	(a)		
20	(a)	**40**	(d)	**60**	(d)	**80**	(a)	**100**	(5)	**120**	(c)	**140**	(2)	**160**	(a)	**180**	(a)		

3 Current Electricity

TREND ANALYSIS
JEE MAIN (2025-2021)

	2025				2024				2023				2022				2021			
	Session II		Session I		Session II		Session I		Session II		Session I		Session II		Session I		Session II		Session I	
	MCQs	NVQs	MCQs	NVQs	MCQs	NVQs	MCQs	NVQs	MCQs	NVQs	MCQs	NVQs	MCQs	NVQs	MCQs	NVQs	MCQs	NVQs	MCQs	NVQs
Topic 1	2	–	2	–	1	4	2	1	4	5	6	1	2	4	3	–	2	1	2	4
Topic 2	2	1	2	3	3	4	2	4	2	3	5	–	3	2	3	4	3	3	1	4
Topic 3	–	1	1	1	1	2	2	3	2	4	–	5	1	1	3	–	1	–	–	1
Topic 4	3	1	1	–	4	1	2	1	2	1	2	–	2	1	1	2	4	2	–	1
Total	7	3	6	4	9	11	8	9	10	13	13	6	8	8	10	6	10	6	3	10

How to Code Qns for Better & Faster Revision →

Type 1: Able to solve (No code Req)
Type 2: Solved with struggle or by seeing solution **(Use code NR means Needs Revision or orange stickers)**
Type 3: Tricky/ Could not solve on my own. Need help from Teacher/ Friend. **(Use code TQ means Tricky Question or red stickers)**

You can either hand write the code in boxes or else paste coloured stickers given at end of book.

Topic 1 - Electric Current, Drift of Electrons, Ohm's Law, Resistance and Resistivity

1. Current passing through a wire as function of time is given as $I(t) = 0.02\,t + 0.01$ A. The charge that will flow through the wire from $t = 1$s to $t = 2$s is :
[NCERT : P L-82 | April 4, 2025 (I)]
(a) 0.06 C (b) 0.02 C
(c) 0.07 C (d) 0.04 C

2. A wire of length 25 m and cross-sectional area 5 mm^2 having resistivity of $2 \times 10^{-6}\,\Omega$ m is bent into a complete circle. The resistance between diametrically opposite points will be **[NCERT : P L-84 | April 3, 2025 (I)]**
(a) 12.5 Ω (b) 50 Ω
(c) 100 Ω (d) 2.5 Ω

3. Identify the valid statements relevant to the given circuit at the instant when the key is closed.
[NCERT : P L-83 | Jan. 23, 2025 (I)]

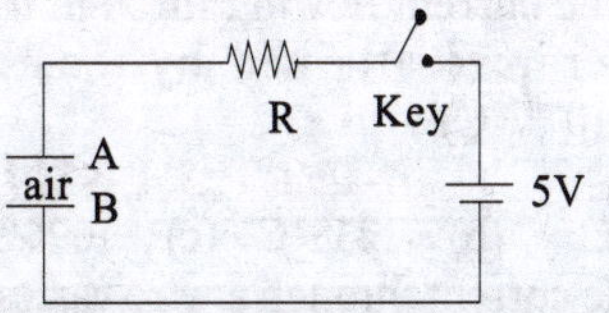

A. There will be no current through resistor R.
B. There will be maximum current in the connecting wires.
C. Potential difference between the capacitor plates A and B is minimum.
D. Charge on the capacitor plates is minimum.
Choose the correct answer from the options given below:
(a) C, D only (b) B, C, D only
(c) A, C only (d) A, B, D only

4. Which of the following resistivity (ρ) v/s temperature (T) curves is most suitable to be used in wire bound standard resistors? **[NCERT : P L-91 | Jan. 22, 2025 (I)]**

(a) 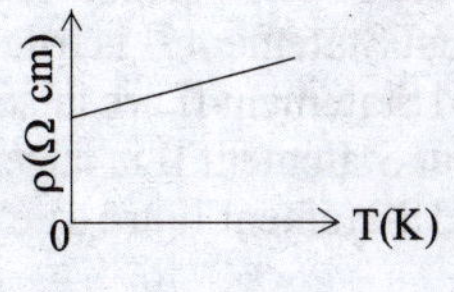

(b) 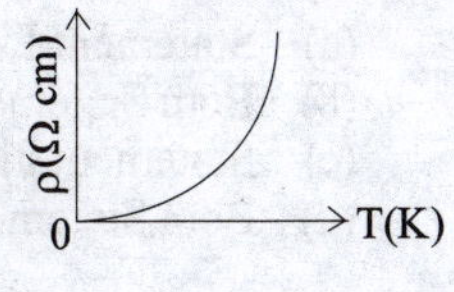

(c) 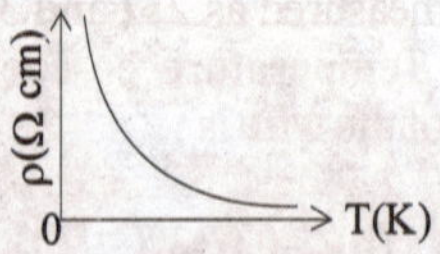 (d) 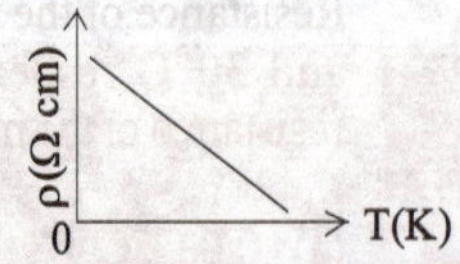

5. At room temperature (27°C), the resistance of a heating element is 50Ω. The temperature coefficient of the material is $2.4 \times 10^{-4}\ ^\circ C^{-1}$. The temperature of the element, when its resistance is 62 Ω, is .... °C.
[NCERT : P L-91 | April 9, 2024 (II)]

6. Resistance of a wire at 0 °C, 100 °C and t °C is found to be 10 Ω, 10.2 Ω and 10.95 Ω respectively. The temperature t in Kelvin scale is ______.
[NCERT : P L-91 | April 8, 2024 (I)]

7. A wire of resistance R and radius r is stretched till its radius became $\frac{r}{2}$. If new resistance of the stretched wire is x R, then value of x is ______.
[NCERT : P L-84 | April 6, 2024 (I)]

8. The resistances of the platinum wire of a platinum resistance thermometer at the ice point and steam point are 8 Ω and 10 Ω respectively. After inserting in a hot bath of temperature 400°C, the resistance of platinum wire is :
[NCERT : P L-91 | April 4, 2024 (I)]
(a) 2 Ω (b) 16 Ω (c) 8 Ω (d) 10 Ω

9. Two wires A and B are made up of the same material and have the same mass. Wire A has radius of 2.0 mm and wire B has radius of 4.0 mm. The resistance of wire B is 2Ω. The resistance of wire A is _____Ω.
[NCERT : P L-84 | April 4, 2024 (II)]

10. The current in a conductor is expressed as $I = 3t^2 + 4t^3$, where I is Ampere and t is second. The amount of electric charge that flows through a section of the conductor during t = 1s to to t = 2s is __________ C.
[NCERT : P L-82 | Feb. 1, 2024 (I)]

11. An electric toaster has resistance of 60 Ω at room temperature (27 °C). The toaster is connected to a 220 V supply. If the current flowing through it reaches 2.75 A, the temperature attained by toaster is around : (if $\alpha = 2 \times 10^{-4}/^\circ C$)
[NCERT : P L-91 | Jan. 30, 2024 (I)]
(a) 694 °C (b) 1235°C (c) 1694 °C (d) 1667 °C

12. The electric current through a wire varies with time as $I = I_0 + \beta t$, where $I_0 = 20$ A and $\beta = 3$ A/s. The amount of electric charge crossed through a section of the wire in 20 s is: **[NCERT : P L-82 | Jan. 29, 2024 (I)]**
(a) 1600 C (b) 800 C (c) 80 C (d) 1000 C

13. Given below are two statements:
Statement I : The equivalent resistance of resistors in a series combination is smaller than least resistance used in the combination.
Statement II : The resistivity of the material is independent of temperature.
In the light of the above statements, choose the correct answer from the options given below:
[NCERT : P L-90 | April 15, 2023 (I)]
(a) Statement I is false but Statement II is true
(b) Both Statement I and Statement II are false
(c) Statement I is true but Statement II is false
(d) Both Statement I and Statement II are true

14. A wire of resistance 160 Ω is melted and drawn in wire of one-fourth of its length. The new resistance of the wire will be **[NCERT : P L-84 | April 12, 2023 (I)]**
(a) 10 Ω (b) 640 Ω (c) 40 Ω (d) 16 Ω

15. The current flowing through a conductor connected across a source is 2A and 1.2 A at 0°C and 100°C respectively. The current flowing through the conductor at 50°C will be ________ $\times 10^2$ mA.
[NCERT : P L-91 | April 12, 2023 (I)]

16. In a metallic conductor, under the effect of applied electric field, the free electrons of the conductor
[NCERT : P L-86 | April 10, 2023 (II)]
(a) drift from higher potential to lower potential
(b) move in the curved paths from lower potential to higher potential
(c) move with the uniform velocity throughout from lower potential to higher potential
(d) move in the straight line paths in the same direction

17. A rectangular parallelopiped is measured as 1 cm × 1 cm × 100 cm. If its specific resistance is 3×10^{-7} Ωm, then the resistance between its two opposite rectangular faces will be________$\times^{-7}$ Ω.
[NCERT : P L-84 | April 10, 2023 (II)]

18. A current of 2 A flows through a wire of cross-sectional area 25.0 mm². The number of free electrons in a cubic meter are 2.0×10^{28}. The drift velocity of the electrons is ______ $\times 10^{-6}$ ms^{-1} **[NCERT : P L-86 | April 8, 2023 (I)]**
(given, charge on electron = 1.6×10^{-19} C)

19. The number density of free electrons in copper is nearly 8×10^{28} m^{-3}. A copper wire has its area of cross-section = 2×10^{-6} m² and is carrying a current of 3.2 A. The drift speed of the electrons is _______ $\times 10^{-6}$ ms^{-1}.
[NCERT : P L-86 | April 8, 2023 (II)]

20. The length of a metallic wire is increased by 20% and its area of cross section is reduced by 4%. The percentage change in resistance of the metallic wire is _____%.
[NCERT : P L-84 | April 6, 2023 (I)]

21. The resistivity (ρ) of semiconductor varies with temperature. Which of the following curve represents the correct behaviour **[NCERT : P L-91 | April 6, 2023 (I)]**

(a) 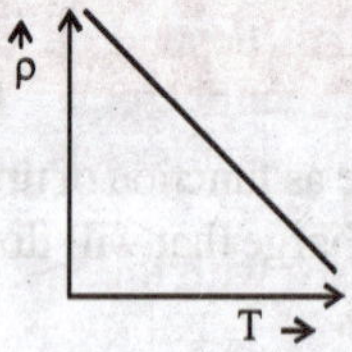 (b) 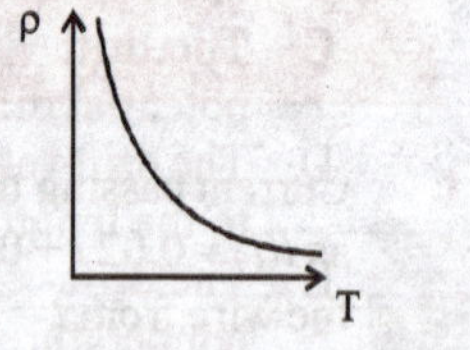

(c) 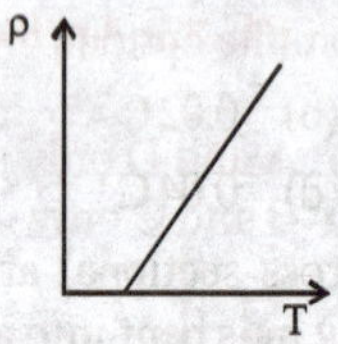 (d) 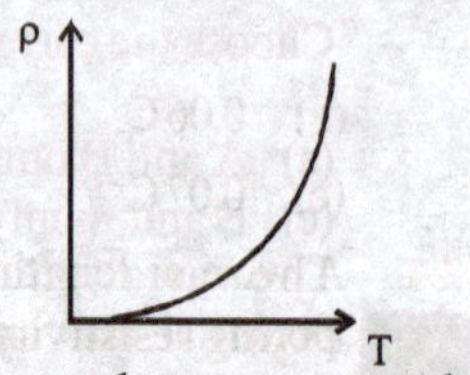

22. The drift velocity of electrons for a conductor connected in an electrical circuit is V_d. The conductor in now replaced by another conductor with same material and same length but double the area of cross section. The applied voltage remains same. The new drift velocity of electrons will be
[NCERT : P L-86 | Jan. 31, 2023 (I)]
(a) V_d (b) $\frac{V_d}{2}$ (c) $\frac{V_d}{4}$ (d) $2V_d$

23. The charge flowing in a conductor changes with time as $Q(t) = \alpha t - \beta t^2 + \gamma t^3$. Where α, β and γ are constants. Minimum value of current is :
[NCERT : P L-82 | Jan. 30, 2023 (I)]
(a) $\alpha - \frac{3\beta^2}{\gamma}$ (b) $\alpha - \frac{\gamma^2}{3\beta}$ (c) $\beta - \frac{\alpha^2}{3\gamma}$ (d) $\alpha - \frac{\beta^2}{3\gamma}$

24. A uniform metallic wire carries a current 2 A. when 3.4 V battery is connected across it. The mass of uniform metallic wire is 8.92×10^{-3} kg, density is 8.92×10^3 kg/m³ and resistivity is 1.7×10^{-8} Ω–m. The length of wire is :
[NCERT : P L-84 | Jan. 25, 2023 (I)]
(a) $l = 6.8$ m (b) $l = 10$ m (c) $l = 5$ m (d) $l = 100$ m

25. The resistance of a wire is 5 Ω. It's new resistance in ohm if stretched to 5 times of its original length will be :
[NCERT : P L-84 | Jan. 25, 2023 (II)]
(a) 625 (b) 5 (c) 125 (d) 25

26. A hollow cylindrical conductor has length of 3.14 m, while its inner and outer diameters are 4 mm and 8 mm respectively. The resistance of the conductor is $n \times 10^{-3}$ Ω.
If the resistivity of the material is 2.4×10^{-8} Ωm. The value of n is ______.
[NCERT : P L-84 | Jan. 24, 2023 (I)]

27. If a copper wire is stretched to increase its length by 20%. The percentage increase in resistance of the wire is ______%.
[NCERT : P L-84 | Jan. 24, 2023 (II)]

28. A 1 m long wire is broken into two unequal parts X and Y The X part of the wire is streched into another wire W. Length of W is twice the length of X and the resistance of W is twice that of Y. Find the ratio of length of X and Y.
[NCERT : P L-84 | July 29, 2022 (II)]
(a) 1 : 4 (b) 1 : 2 (c) 4 : 1 (d) 2 : 1

29. A wire of resistance R_1 is drawn out so that its length is increased by thrice of its original length. The ratio of new resistance to original resistance is:
[NCERT : P L-84 | July 28, 2022 (I)]
(a) 9 : 1 (b) 1 : 9 (c) 4 : 1 (d) 3 : 1

30. A. The drift velocity of electrons decreases with the increase in the temperature of conductor.
B. The drift velocity is inversely proportional to the area of cross-section of given conductor.
C. The drift velocity does not depend on the applied potential difference to the conductor.
D. The drift velocity of electron is inversely proportional to the length of the conductor.
E. The drift velocity increases with the increase in the temperature of conductor.
Choose the correct answer from the options given below:
[NCERT : P L-86 | July 27, 2022 (II)]
(a) A and B only (b) A and D only
(c) B and E only (d) B and C only

31. The variation of applied potential and current flowing through a given wire is shown in figure. The length of wire is 31.4 cm. The diameter of wire is measured as 2.4 cm. The resistivity of the given wire is measured as $x \times 10^{-3}$ Ω cm. The value of x is ______.

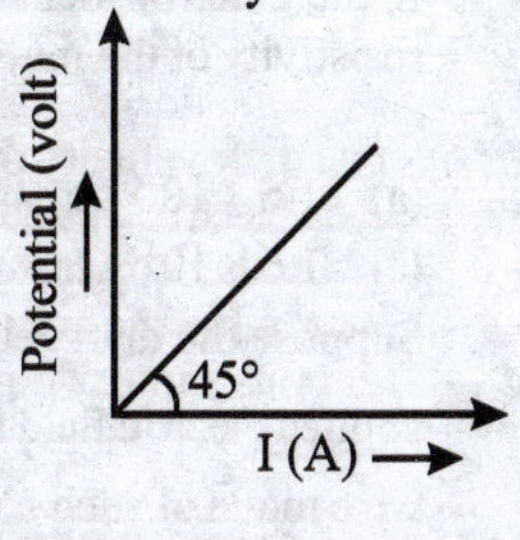

[NCERT : P L-84 | June 29, 2022 (I)]

32. Resistance of the wire is measured as 2Ω and 3Ω at 10°C and 30°C respectively. Temperature co-coefficient of resistance of the material of the wire is :
[NCERT : P L-91 | June 28, 2022 (II)]
(a) 0.033°C⁻¹ (b) –0.033°C⁻¹
(c) 0.011°C⁻¹ (d) 0.055°C⁻¹

33. The current density in a cylindrical wire of radius 4 mm is 4×10^6 Am⁻². The current through the outer portion of the wire between radial distance $\frac{R}{2}$ and R is ____ π A.
[NCERT : P L-86 | June 27, 2022 (I)]

34. The current density in a cylindrical wire of radius r = 4.0 mm is 1.0×10^6 A/m². The current through the outer portion of the wire between radial distances r/2 and r is $x\pi$ A; where x is ______.
[NCERT : P L-86 | June 27, 2022 (II)]

35. An aluminium wire is stretched to make its length, 0.4% larger. Then percentage change in resistance is:
[NCERT : P L-84 | June 26, 2022 (I)]
(a) 0.4% (b) 0.2% (c) 0.8% (d) 0.6%

36. The length of a given cylindrical wire is increased to double of its original length. The percentage increase in the resistance of the wire will be ______%.
[NCERT : P L-84 | June 25, 2022 (II)]

37. In the given figure, a battery of emf E is connected across a conductor PQ of length 'l' and different area of cross-sections having radii r_1 and r_2 ($r_2 < r_1$).
[NCERT : P L-86 | July 27, 2021 (I)]
Choose the correct option as one moves from P to Q :
(a) Drift velocity of electron increases.
(b) Electric field decreases.
(c) Electron current decreases.
(d) All of these

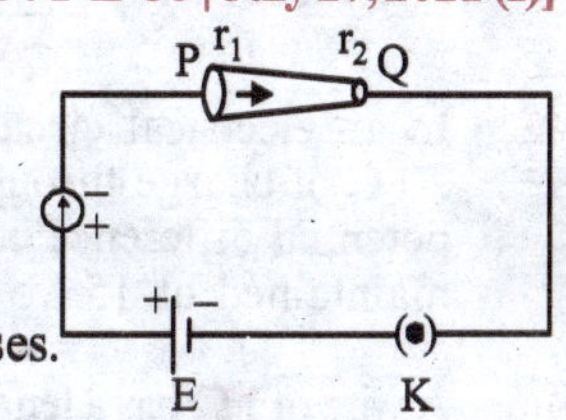

38. For the circuit shown, the value of current at time t = 3.2 s will be ______ A.

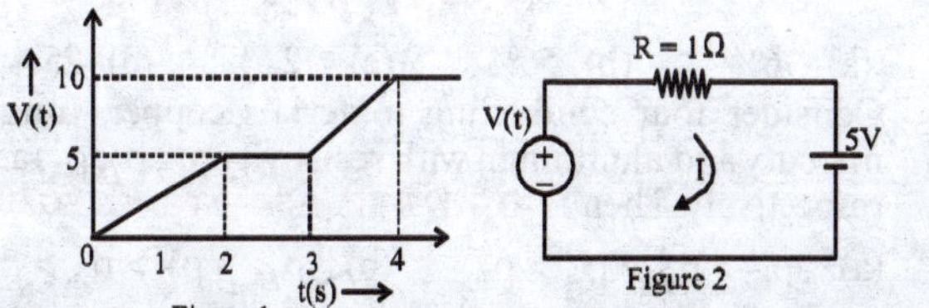

Figure 1 Figure 2

[Voltage distribution V(t) is shown by Fig. (1) and the circuit is shown in Fig. (2)]
[NCERT : P L-83 | NA, July 27, 2021 (II)]

39. A Copper (Cu) rod of length 25 cm and cross-sectional area 3 mm² is joined with a similar Aluminium (Al) rod as shown in figure. Find the resistance of the combination between the ends A and B.
(Take Resistivity of Copper = 1.7×10^{-8} Ωm Resistivity of Aluminium = 2.6×10^{-8} Ωm)
[NCERT : P L-84 | July 22, 2021 (II)]
(a) 2.170 mΩ
(b) 1.420 mΩ
(c) 0.0858 mΩ
(d) 0.858 mΩ

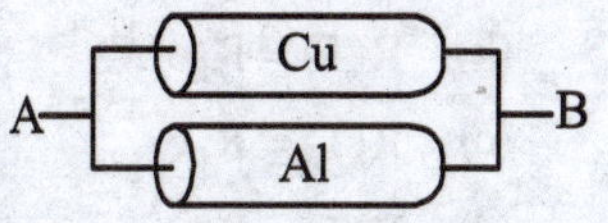

40. Two wires of same length and thickness having specific resistances 6 Ω cm and 3 Ω cm respectively are connected in parallel. The effective resistivity is ρΩ cm. The value of ρ, to the nearest integer, is ______.
[NCERT : P L-84 | March 18, 2021 (II)]

41. A current of 10 A exists in a wire of cross-sectional area of 5 mm^2 with a drift velocity of 2×10^{-3} ms^{-1}. The number of free electrons in each cubic meter of the wire is ______
[NCERT : P L-86 | March 17, 2021 (I)]
(a) 2×10^{25} (b) 2×10^{23} (c) 625×10^{25} (d) 2×10^{6}

42. A conducting wire of length 'l', area of cross-section A and electric resistivity ρ is connected between the terminals of a battery. A potential difference V is developed between its ends, causing an electric current.
If the length of the wire of the same material is doubled and the area of cross-section is halved, the resultant current would be:
[NCERT : P L-84 | March 16, 2021 (I)]
(a) $4\frac{VA}{\rho l}$ (b) $\frac{3}{4}\frac{VA}{\rho l}$ (c) $\frac{1}{4}\frac{\rho l}{VA}$ (d) $\frac{1}{4}\frac{VA}{\rho l}$

43. A current through a wire depends on time as $i = \alpha_0 t + \beta t^2$ where $\alpha_0 = 20$ A/s and $\beta = 8$ As^{-2}. Find the charge crossed through a section of the wire in 15 s.
[NCERT : P L-82 | Feb. 24, 2021 (I)]
(a) 2250 C (b) 11250 C (c) 2100 C (d) 260 C

44. A cylindrical wire of radius 0.5 mm and conductivity 5×10^7 S/m is subjected to an electric field of 10 mV/m. The expected value of current in the wire will be $x^3\pi$ mA. The value of x is ______
[NCERT : P L-86 | Feb. 24, 2021 (II)]

45. In an electrical circuit, a battery is connected to pass 20 C of charge through it in a certain given time. The potential difference between two plates of the battery is maintained at 15 V. The work done by the battery is ______ J. **[NCERT : P L-92 | Feb. 26, 2021 (I)]**

46. A wire of 1 Ω has a length of 1 m. It is stretched till its length increases by 25%. The percentage change in resistance to the nearest integer is
[NCERT : P L-84 | Feb. 26, 2021 (II)]
(a) 76% (b) 56% (c) 12.5% (d) 25%

47. Consider four conducting materials copper, tungsten, mercury and aluminium with resistivity ρ_C, ρ_T, ρ_M and ρ_A respectively. Then : **[NCERT : P L-84 | Sep. 02, 2020 (I)]**
(a) $\rho_C > \rho_m > \rho_A > \rho_T$ (b) $\rho_M > \rho_T > \rho_A > \rho_C$
(c) $\rho_A > \rho_m > \rho_T > \rho_C$ (d) $\rho_A > \rho_T > \rho_M > \rho_C$

48. A current of 5 A passes through a copper conductor (resistivity) = 1.7×10^{-8} Ωm) of radius of cross-section 5 mm. Find the mobility of the charges if their drift velocity is 1.1×10^{-3} m/s. **[NCERT : P L-88 | 10 Apr. 2019 I]**
(a) 1.8 m^2/Vs (b) 1.5 m^2/Vs (c) 1.3 m^2/Vs (d) 1.0 m^2/Vs

49. In an experiment, the resistance of a material is plotted as a function of temperature (in some range). As shown in the figure, it is a straight line. **[10 Apr. 2019 I]**
One may canclude that:

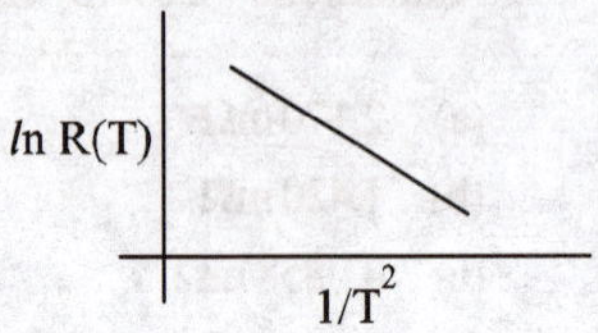

(a) $R(T) = \frac{R_0}{T^2}$
(b) $R(T) = R_0 e^{-T_0^2/T^2}$
(c) $R(T) = R_0 e^{-T^2/T_0^2}$
(d) $R(T) = R_0 e^{T^2/T_0^2}$

50. Space between two concentric conducting spheres of radii a and b ($b > a$) is filled with a medium of resistivity ρ. The resistance between the two spheres will be :
[NCERT : P L-84 | 10 Apr. 2019 II]
(a) $\frac{\rho}{4\pi}\left(\frac{1}{a}-\frac{1}{b}\right)$ (b) $\frac{\rho}{2\pi}\left(\frac{1}{a}-\frac{1}{b}\right)$
(c) $\frac{\rho}{2\pi}\left(\frac{1}{a}+\frac{1}{b}\right)$ (d) $\frac{\rho}{4\pi}\left(\frac{1}{a}+\frac{1}{b}\right)$

51. In a conductor, if the number of conduction electrons per unit volume is 8.5×10^{28} m^{-3} and mean free time is 25 *fs* (femto second), it's approximate resistivity is: ($m_e = 9.1 \times 10^{-31}$ kg)
[NCERT : P L-91 | 9 Apr. 2019 II]
(a) 10^{-6} Ωm (b) 10^{-7} Ωm (c) 10^{-8} Ωm (d) 10^{-5} Ωm

52. The charge on a capacitor plate in a circuit, as a function of time, is shown in the figure:
[NCERT : P L-82 | 12 Jan. 2019 II]

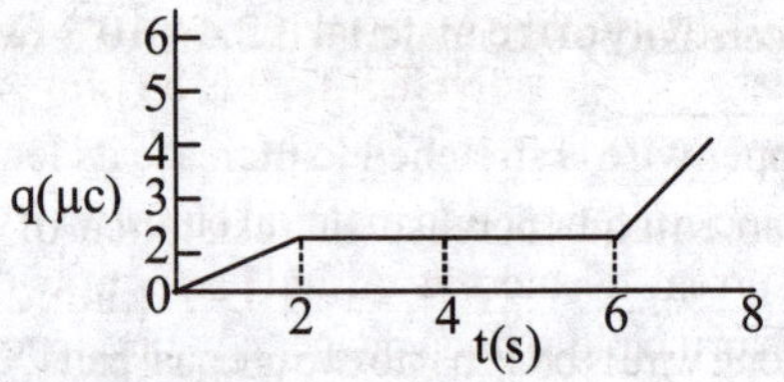

What is the value of current at t = 4 s ?
(a) Zero (b) 3 μA (c) 2 μA (d) 1.5 μA

53. Drift speed of electrons, when 1.5 A of current flows in a copper wire of cross section 5 mm^2, is v. If the electron density in copper is 9×10^{28}/m^3 the value of v in mm/s close to (Take charge of electron to be = 1.6×10^{-19}C)
[NCERT : P L-86 | 9 Jan. 2019 I]
(a) 0.02 (b) 3 (c) 2 (d) 0.2

54. A copper wire is stretched to make it 0.5% longer. The percentage change in its electrical resistance if its volume remains unchanged is:
[NCERT : P L-84 | 9 Jan. 2019 I]
(a) 2.0% (b) 2.5% (c) 1.0% (d) 0.5%

55. A uniform wire of length l and radius r has a resistance of 100 Ω. It is recast into a wire of radius $\frac{r}{2}$. The resistance of new wire will be : **[NCERT : P L-84 | Online April 9, 2017]**
(a) 1600 Ω (b) 400 Ω (c) 200 Ω (d) 100 Ω

56. When 5V potential difference is applied across a wire of length 0.1 m, the drift speed of electrons is 2.5×10^{-4} ms^{-1}. If the electron density in the wire is 8×10^{28} m^{-3}, the resistivity of the material is close to :
[NCERT : P L-86 | 2015]
(a) 1.6×10^{-6} Ωm (b) 1.6×10^{-5} Ωm
(c) 1.6×10^{-8} Ωm (d) 1.6×10^{-7} Ωm

57. Suppose the drift velocity v_d in a material varied with the applied electric field E as $v_d \propto \sqrt{E}$. Then V – I graph for a wire made of such a material is best given by :
[Online April 10, 2015]

(a) 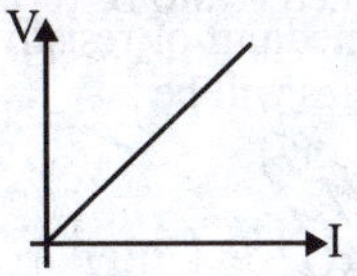

(b) 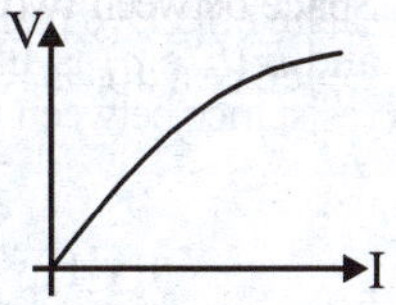

(c) 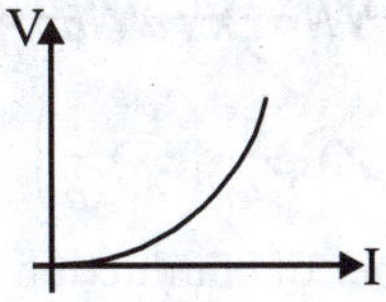

(d) 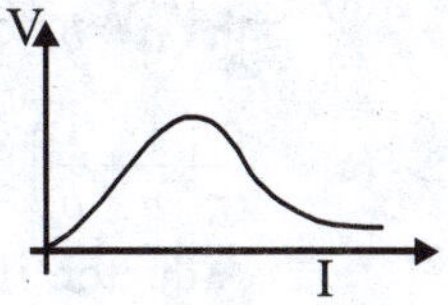

58. The resistance of a wire is R. It is bent at the middle by 180° and both the ends are twisted together to make a shorter wire. The resistance of the new wire is **[NCERT : P L-84 | Online May 26, 2012]**

(a) $2R$ (b) $R/2$ (c) $R/4$ (d) $R/8$

59. If a wire is stretched to make it 0.1% longer, its resistance will : **[NCERT : P L-84 | 2011]**

(a) increase by 0.2% (b) decrease by 0.2%
(c) decrease by 0.05% (d) increase by 0.05%

Directions : Question No. 60 and 61 are based on the following paragraph.

Consider a block of conducting material of resistivity 'ρ' shown in the figure. Current 'I' enters at 'A' and leaves from 'D'. We apply superposition principle to find voltage 'ΔV' developed between 'B' and 'C'. The calculation is done in the following steps:

(i) Take current 'I' entering from 'A' and assume it to spread over a hemispherical surface in the block.
(ii) Calculate field E(r) at distance 'r' from A by using Ohm's law E = ρ j, where j is the current per unit area at 'r'.
(iii) From the 'r' dependence of E(r), obtain the potential V(r) at r.
(iv) Repeat (i), (ii) and (iii) for current 'I' leaving 'D' and superpose results for 'A' and 'D'.

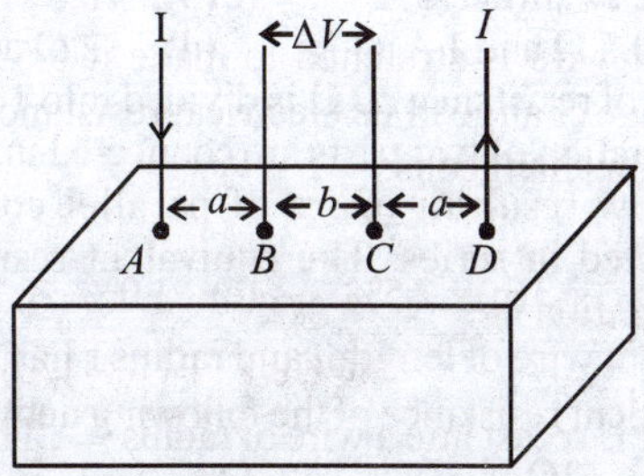

60. ΔV measured between B and C is **[NCERT : P L-86 | 2008]**

(a) $\frac{\rho I}{\pi a}-\frac{\rho I}{\pi(a+b)}$ (b) $\frac{\rho I}{a}-\frac{\rho I}{(a+b)}$

(c) $\frac{\rho I}{2\pi a}-\frac{\rho I}{2\pi(a+b)}$ (d) $\frac{\rho I}{2\pi(a-b)}$

61. For current entering at A, the electric field at a distance 'r' from A is **[NCERT : P L-86 | 2008]**

(a) $\frac{\rho I}{8\pi r^2}$ (b) $\frac{\rho I}{r^2}$ (c) $\frac{\rho I}{2\pi r^2}$ (d) $\frac{\rho I}{4\pi r^2}$

62. The resistance of a wire is 5 ohm at 50°C and 6 ohm at 100°C. The resistance of the wire at 0°C will be **[NCERT : P L-91 | 2007]**

(a) 3 ohm (b) 2 ohm (c) 1 ohm (d) 4 ohm

63. A material 'B' has twice the specific resistance of 'A'. A circular wire made of 'B' has twice the diameter of a wire made of 'A'. then for the two wires to have the same resistance, the ratio l_B/l_A of their respective lengths must be **[NCERT : P L-84 | 2006]**

(a) 1 (b) $\frac{1}{2}$ (c) $\frac{1}{4}$ (d) 2

64. An electric current is passed through a circuit containing two wires of the same material, connected in parallel. If the lengths and radii are in the ratio of $\frac{4}{3}$ and $\frac{2}{3}$, then the ratio of the current passing through the wires will be **[NCERT : P L-84 | 2004]**

(a) 8/9 (b) 1/3 (c) 3 (d) 2

Topic 2 - Combination of Resistances

65. A wire of resistance R is bent into a triangular pyramid as shown in figure with each segment having same length. The resistance between points A and B is R/n. The value of n is : **[April 7, 2025 (I)]**

(a) 16 (b) 14 (c) 10 (d) 12

66. From the combination of resistors with resistance values $R_1 = R_2 = R_3 = 5\ \Omega$ and $R_4 = 10\ \Omega$, which of the following combination is the best circuit to get an equivalent resistance of 6Ω ? **[April 4, 2025 (II)]**

(a) 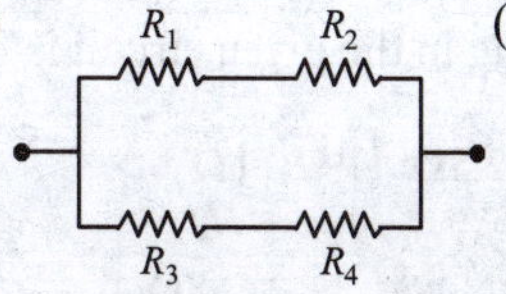

(b) 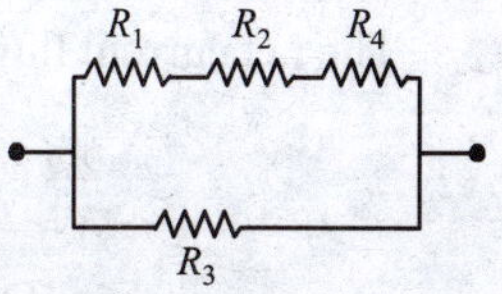

(c) 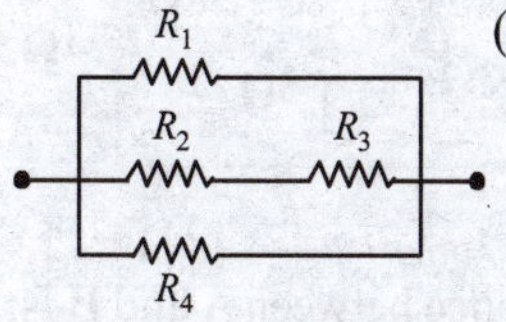

(d) 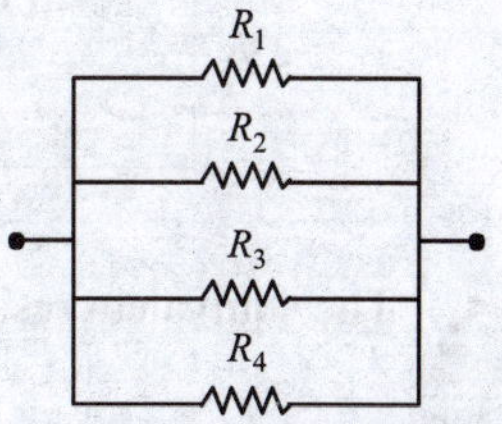

67. In the figure shown below, a resistance of 150.4 Ω is connected in series to an ammeter A of resistance 240 Ω. A shunt resistance of 10 Ω is connected in parallel with the ammeter. The reading of the ammeter is ______ mA. **[April 3, 2025 (I)]**

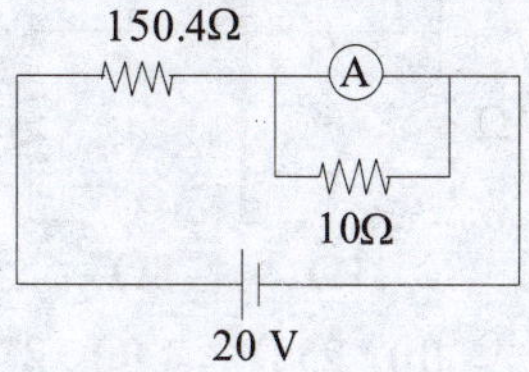

68. Find the equivalent resistance between two ends of the following circuit. **[Jan. 28, 2025 (I)]**

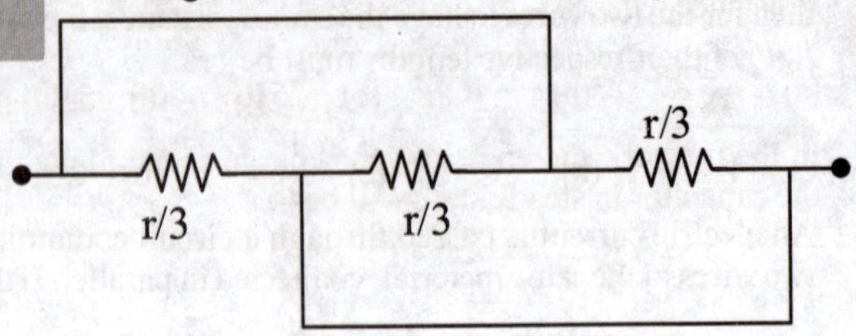

(a) r (b) $\frac{r}{6}$ (c) $\frac{r}{9}$ (d) $\frac{r}{3}$

69. A wire of resistance R is bent into an equilateral triangle and an identical wire is bent into a square. The ratio of resistance between the two end points of an edge of the triangle to that of the square is **[Jan. 28, 2025 (I)]**

(a) $\frac{9}{8}$ (b) $\frac{8}{9}$ (c) $\frac{27}{32}$ (d) $\frac{32}{27}$

70. A wire of resistance 9 Ω is bent to form an equilateral triangle. Then the equivalent resistance across any two vertices will be __ ohm. **[Jan. 24, 2025 (I)]**

71. At steady state the charge on the capacitor, as shown in the circuit below, is ____ μC.

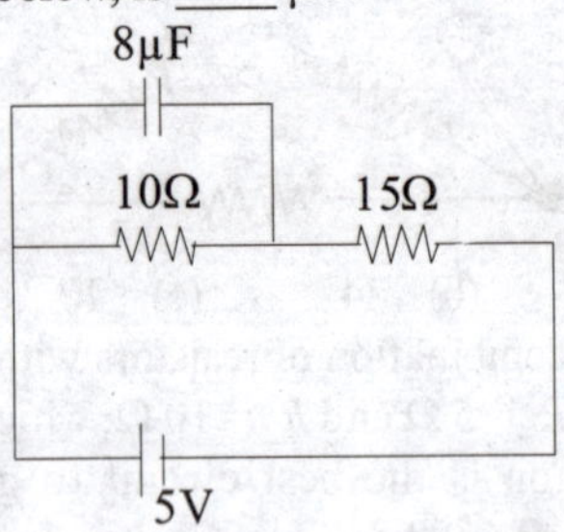

[Jan. 23, 2025 (II)]

72. The net current flowing in the given circuit is ____ A.

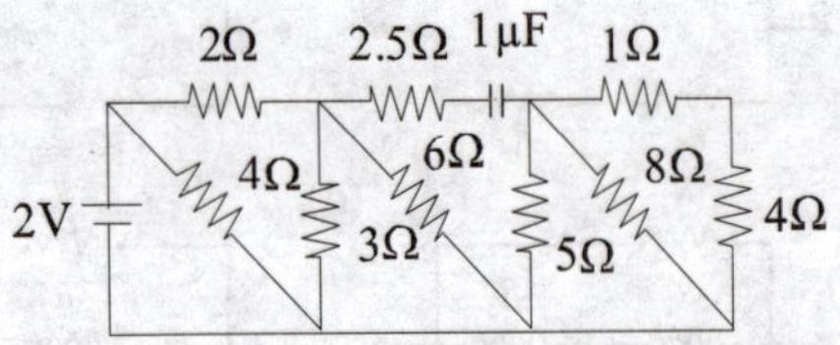

[Jan. 22, 2025 (II)]

73. The equivalent resistance between A and B is: **[April 9, 2024 (I)]**

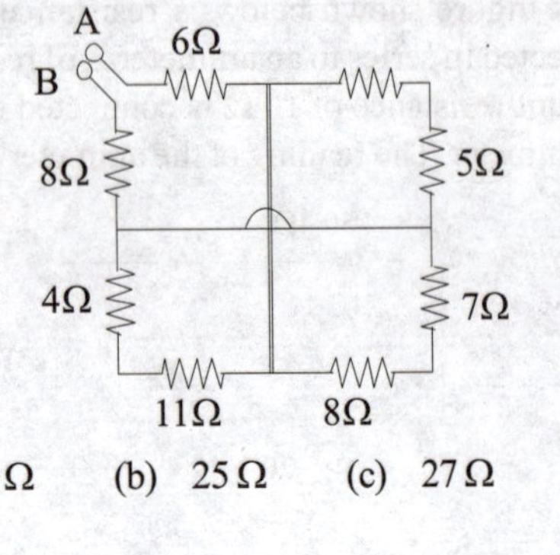

(a) 18 Ω (b) 25 Ω (c) 27 Ω (d) 19 Ω

74. The effective resistance between A and B, if resistance of each resistor is R, will be **[April 9, 2024 (II)]**

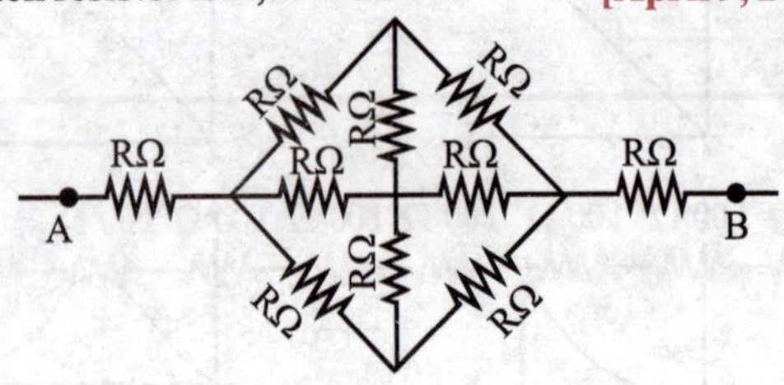

(a) $\frac{2}{3}R$ (b) $\frac{8R}{3}$ (c) $\frac{5R}{3}$ (d) $\frac{4R}{3}$

75. In the given figure an ammeter A consists of a 240Ω coil connected in parallel to a 10 Ω shunt. The reading of the ammeter is ______ mA.

[NCERT : P L-83 | April 6, 2024 (II)]

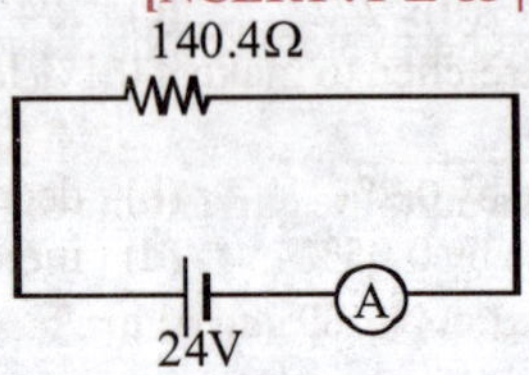

76. In the given figure $R_1 = 10\Omega$, $R_2 = 8\Omega$, $R_3 = 4\Omega$ and $R_4 = 8\Omega$. Battery is ideal with emf 12V. Equivalent resistant of the circuit and current supplied by battery are respectively.

[NCERT : P L-83 | April 5, 2024 (I)]

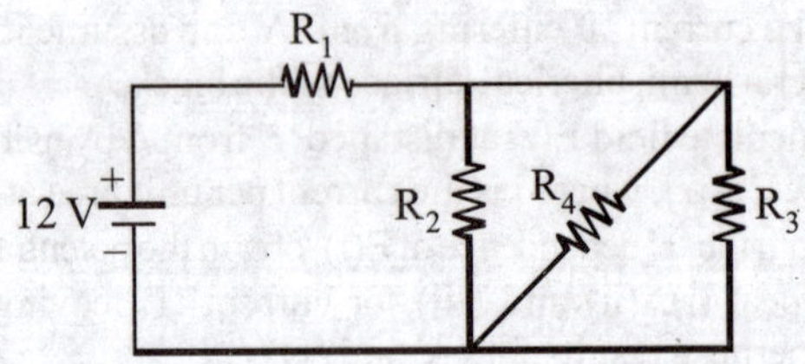

(a) 12 Ω and 11.4 A (b) 10.5 Ω and 1.14 A
(c) 10.5 Ω and 1 A (d) 12 Ω and 1 A

77. A wire of resistance 20 Ω is divided into 10 equal parts. A combination of two parts are connected in parallel and so on. Now resulting pairs of parallel combination are connected in series. The equivalent resistance of final combination is ______ Ω.

[NCERT : P L-84 | April 5, 2024 (II)]

78. Equivalent resistance of the following network is _____ Ω.

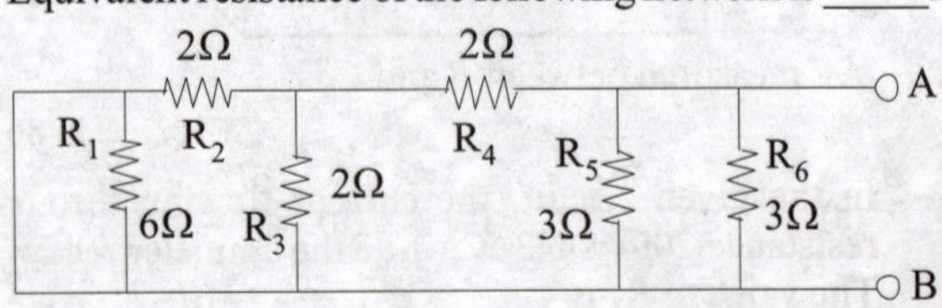

[Jan. 31, 2024 (I)]

79. Two conductors have the same resistances at 0°C but their temperature coefficients of resistance are α_1 and α_2. The respective temperature coefficients for their series and parallel combinations are :

[NCERT : P L-91 | Jan. 31, 2024 (I)]

(a) $\alpha_1 + \alpha_2, \frac{\alpha_1 + \alpha_2}{2}$ (b) $\frac{\alpha_1 + \alpha_2}{2}, \frac{\alpha_1 + \alpha_2}{2}$

(c) $\frac{\alpha_1 + \alpha_2}{2}, \alpha_1 + \alpha_2$ (d) $\alpha_1 + \alpha_2, \frac{\alpha_1 \alpha_2}{\alpha_1 + \alpha_2}$

80. A potential divider circuit is shown in figure. The output voltage V_0 is: **[NCERT : P L-83 | Jan. 30, 2024 (I)]**

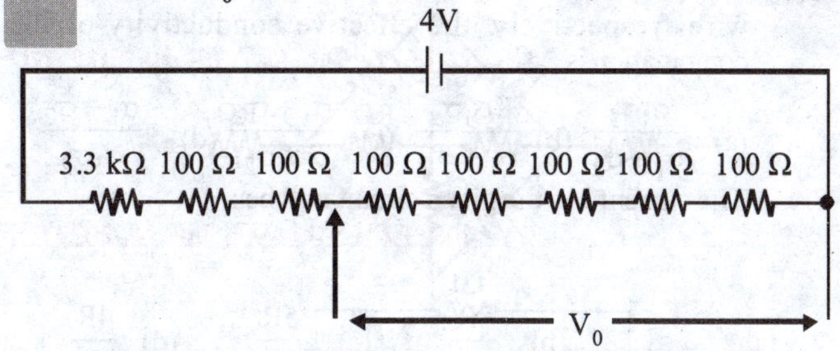

(a) 4V (b) 2mV (c) 0.5 V (d) 12mV

81. A 16Ω wire is bend to form a square loop. A 9V battery with internal resistance 1Ω is connected across one of its sides. If a 4μF capacitor is connected across one of its diagonals, the energy stored by the capacitor will be $\frac{x}{2}$ μJ, where x = ____. **[NCERT : P L-83 | Jan. 29, 2024 (I)]**

82. In the given circuit, the current in resistance R_3 is : **[NCERT : P L-95 | Jan. 29, 2024 (II)]**

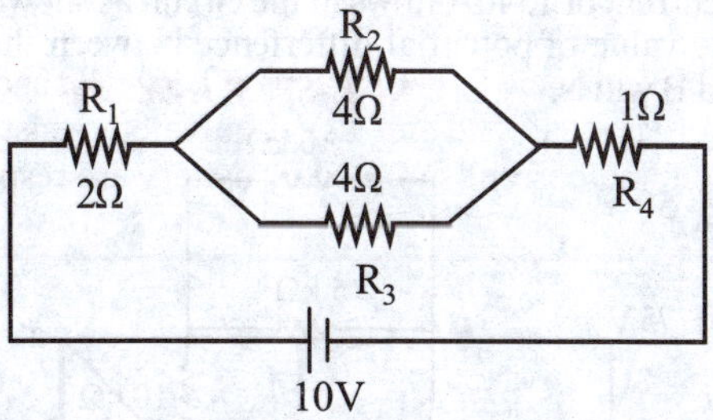

(a) 2.5 A (b) 1 A (c) 1.5 A (d) 2 A

83. In the given figure, the charge stored in 6μF capacitor, when points A and B are joined by a connecting wire is ____ μC. **[NCERT : P L-96, 97 | Jan. 29, 2024 (II)]**

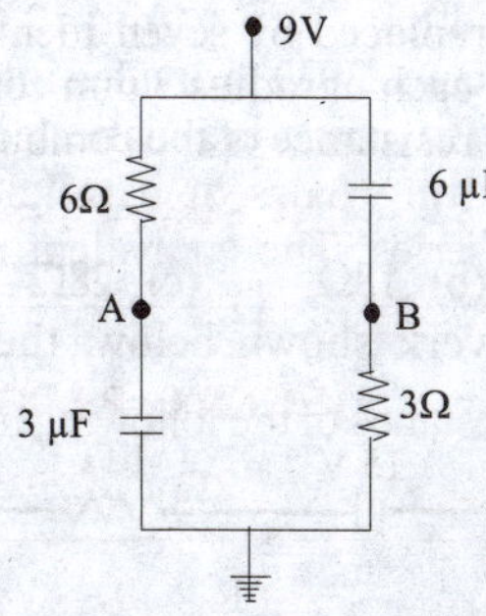

84. In the given circuit, the current flowing through the resistance 20Ω is 0.3 A, while the ammeter reads 0.9 A. The value of R_1 is ____ Ω. **[NCERT : P L-97 | Jan. 29, 2024 (II)]**

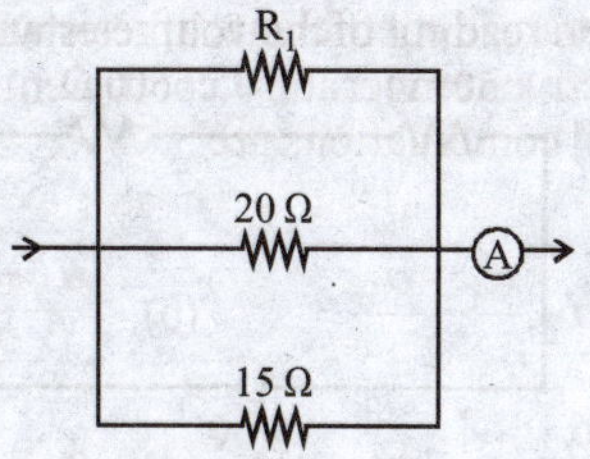

85. A wire of resistance R and length L is cut into 5 equal parts. If these parts are joined parallely, then resultant resistance will be : **[NCERT : P L-84 | Jan. 27, 2024 (I)]**

(a) $\frac{1}{25}$R (b) $\frac{1}{5}$R (c) 25 R (d) 5 R

86. In the network shown below, the charge accumulated in the capacitor in steady state will be: **[NCERT : P L-83 | April 13, 2023 (II)]**

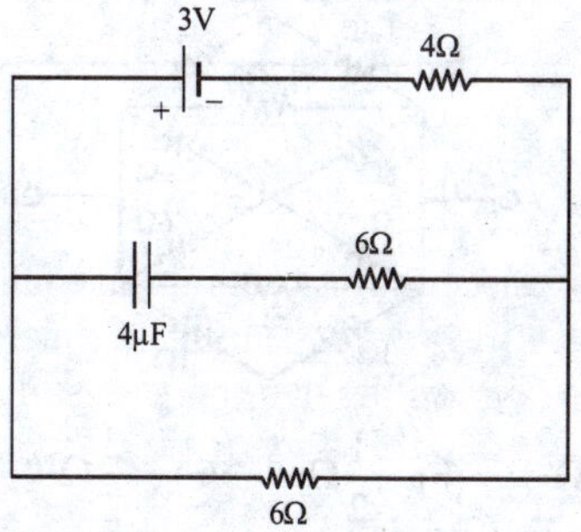

(a) 7.2 μC (b) 4.8 μC (c) 10.3 μC (d) 12 μC

87. The equivalent resistance of the circuit shown below between points a and b is : **[April 10, 2023 (I)]**

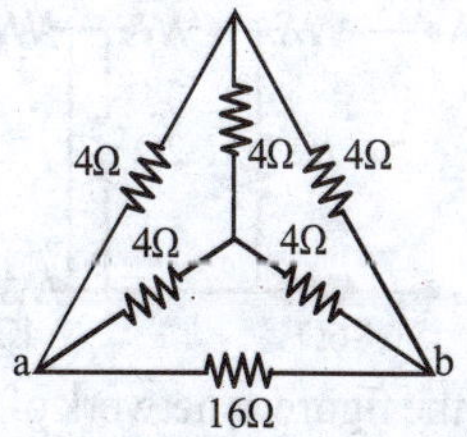

(a) 24Ω (b) 3.2Ω (c) 20Ω (d) 16Ω

88. 10 resistors each of resistance 10Ω can be connected in such as to get maximum and minimum equivalent resistance. The ratio of maximum and minimum equivalent resistance will be ____. **[April 10, 2023 (I)]**

89. The equivalent resistance between A and B as shown in figure is: **[April 8, 2023 (II)]**

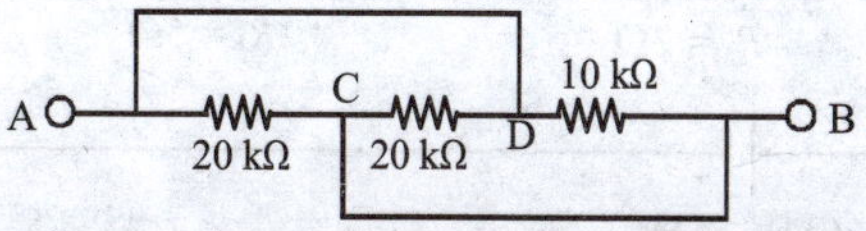

(a) 5 kΩ (b) 30 kΩ (c) 10 kΩ (d) 20 kΩ

90. As shown in the figure, the voltmeter reads 2V across 5 Ω resistor. The resistance of the voltmeter is ____ Ω. **[NCERT : P L-98 | April 6, 2023 (II)]**

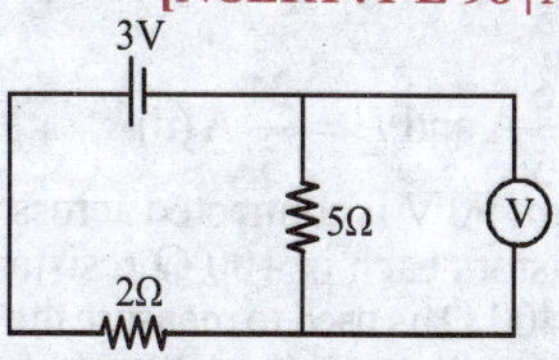

91. The equivalent resistance between *A* and *B* of the network shown in figure: **[NCERT : P L-101 | Feb. 1, 2023 (I)]**

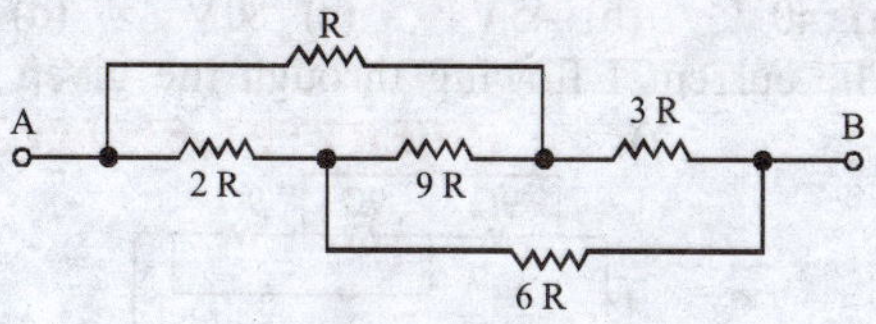

(a) 14 R (b) $\frac{2}{3}$R (c) $\frac{8}{3}$R (d) 21 R

92. Equivalent resistance between the adjacent corners of a regular n-sided polygon of uniform wire of resistance R would be: [Feb. 1, 2023 (II)]

(a) $\frac{(n-1)R}{(2n-1)}$ (b) $\frac{n^2R}{n-1}$ (c) $\frac{(n-1)R}{n^2}$ (d) $\frac{(n-1)R}{n}$

93. The equivalent resistance between A and B is ...... [Jan. 30, 2023 (II)]

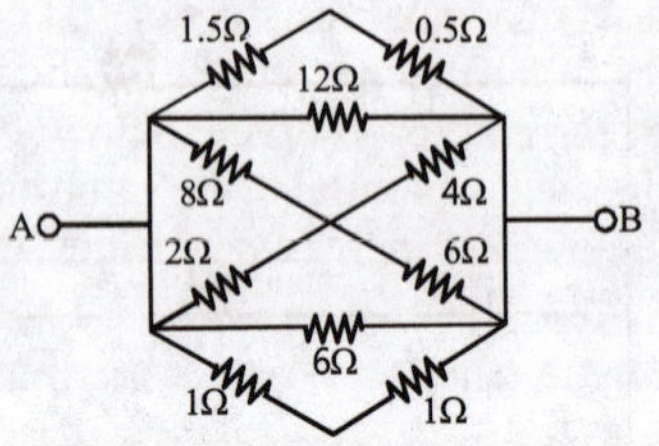

(a) $\frac{2}{3}\Omega$ (b) $\frac{1}{2}\Omega$ (c) $\frac{3}{2}\Omega$ (d) $\frac{1}{3}\Omega$

94. In the given circuit, the equivalent resistance between the terminal A and B is_____Ω. [Jan. 25, 2023 (I)]

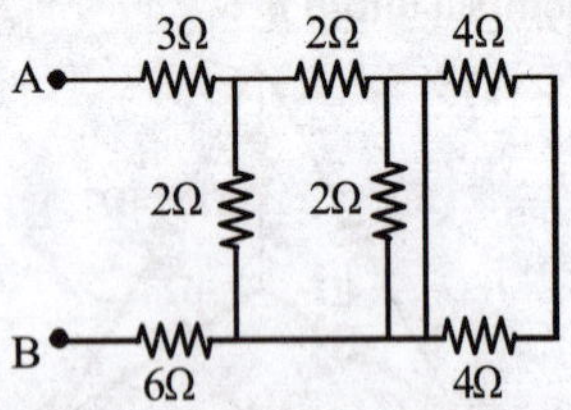

95. As shown in the figure, a network of resistors is connected to a battery of 24 V with an internal resistance of 3Ω. The currents through the resistors R_4 and R_5 are I_4 and I_5 respectively. The values of I_4 and I_5 are :

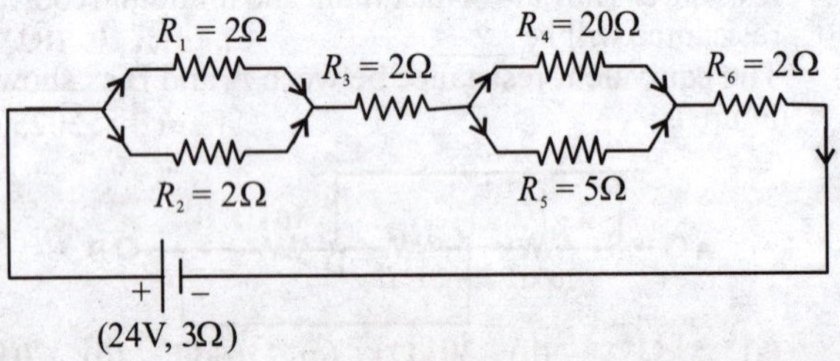

[NCERT : P L-83 | Jan. 24, 2023 (I)]

(a) $I_4 = \frac{8}{5}$A and $I_5 = \frac{2}{5}$A (b) $I_4 = \frac{24}{5}$A and $I_5 = \frac{6}{5}$A

(c) $I_4 = \frac{6}{5}$A and $I_5 = \frac{24}{5}$A (d) $I_4 = \frac{2}{5}$A and $I_5 = \frac{8}{5}$A

96. A cell of emf 90 V is connected across series combination of two resistors each of 100 Ω resistance. A voltmeter of resistance 400 Ω is used to measure the potential difference across each resistor. The reading of the voltmeter will be: [NCERT : P L-83 | Jan. 24, 2023 (II)]

(a) 40 V (b) 45 V (c) 80 V (d) 90 V

97. The current I flowing through the given circuit will be_______ A. [NCERT : P L-83 | July 29, 2022 (I)]

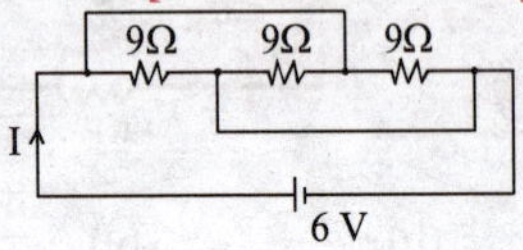

98. Two metallic wires of identical dimensions are connected is series. If σ_1 and σ_2 are the conductivities of the these wires respectively, the effective conductivity of the combination is : [NCERT : P L-84 | July 29, 2022 (I)]

(a) $\frac{\sigma_1\sigma_2}{\sigma_1+\sigma_2}$ (b) $\frac{2\sigma_1\sigma_2}{\sigma_1+\sigma_2}$ (c) $\frac{\sigma_1+\sigma_2}{2\sigma_1\sigma_2}$ (d) $\frac{\sigma_1+\sigma_2}{\sigma_1\sigma_2}$

99. The current I in the given circuit will be : [NCERT : P L-98 | July 26, 2022 (I)]

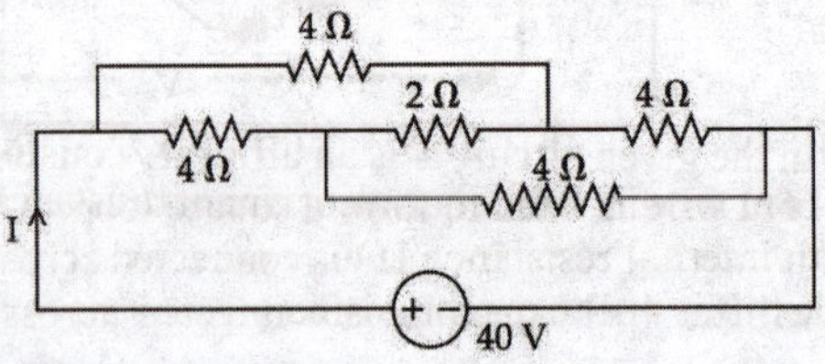

(a) 10 A (b) 20 A (c) 4 A (d) 40 A

100. Eight copper wire of length l and diameter d are joined in parallel to form a single composite conductor of resistance R. If a single copper wire of length $2l$ have the same resistance (R) then its diameter will be ______ d. [NCERT : P L-84 | July 25, 2022 (I)]

101. A current of 15 mA flows in the circuit as shown in figure. The value of potential difference between the points A and B will be [NCERT : P L-98 | July 25, 2022 (II)]

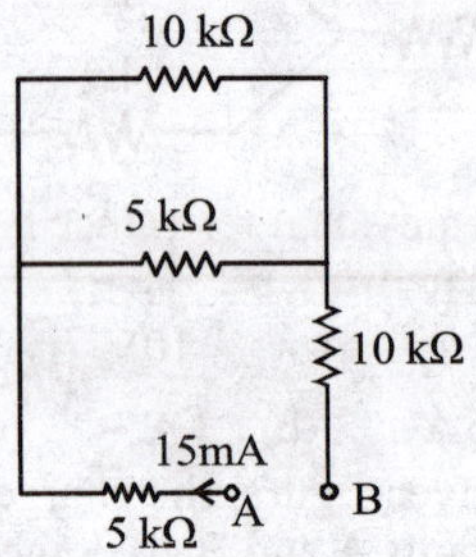

(a) 50 V

(b) 75 V

(c) 150 V

(d) 275 V

102. An electric cable of copper has just one wire of radius 9 mm. Its resistance is 14Ω. If this single copper wire of the cable is replaced by seven identical well insulated copper wires each of radius 3 mm connected in parallel, then the new resistance of the combination will be: [NCERT : P L-84 | June 30, 2022 (I)]

(a) 9Ω (b) 18Ω (c) 28Ω (d) 126Ω

103. For the network shown below, the value $V_B - V_A$ is _______ V. [NCERT : P L-83 | June 29, 2022 (I)]

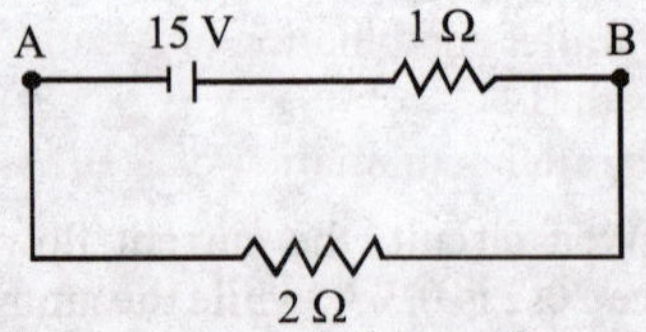

104. Two resistors are connected in series across a battery as shown in figure. If a voltmeter of resistance 2000 Ω is used to measure the potential difference across 500 Ω resister, the reading of the voltmeter will be ____ V.

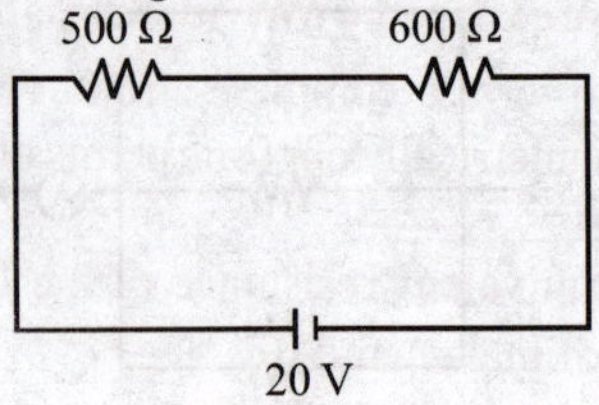

[NCERT : P L-83 | June 29, 2022 (II)]

105. All resistances in figure are 1Ω each. The value of current 'I' is $\frac{a}{5}A$. The value of a is ______.

[NCERT : P L-83 | June 28, 2022 (II)]

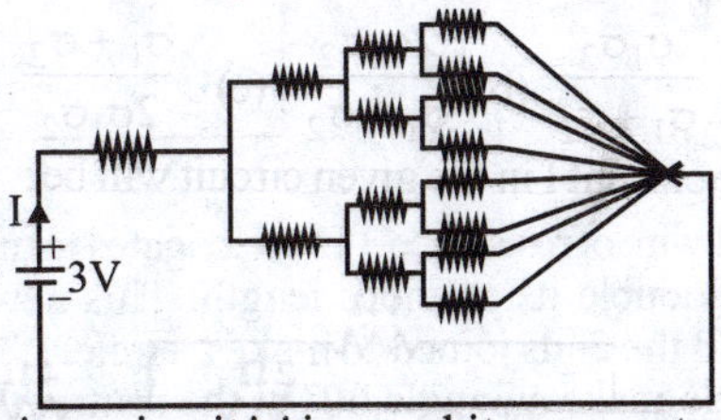

106. In the given circuit 'a' is an arbitrary constant. The value of m for which the equivalent circuit resistance is minimum, will be $\sqrt{\frac{x}{2}}$. The value of x is ______.

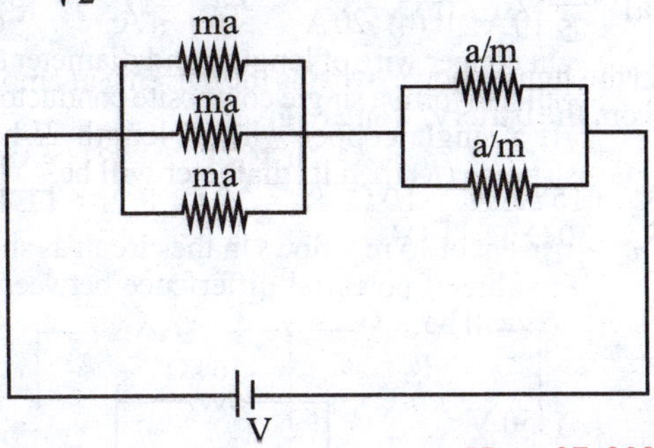

[June 27, 2022 (II)]

107. The equivalent resistance between points A and B in the given network is : **[June 26, 2022 (II)]**

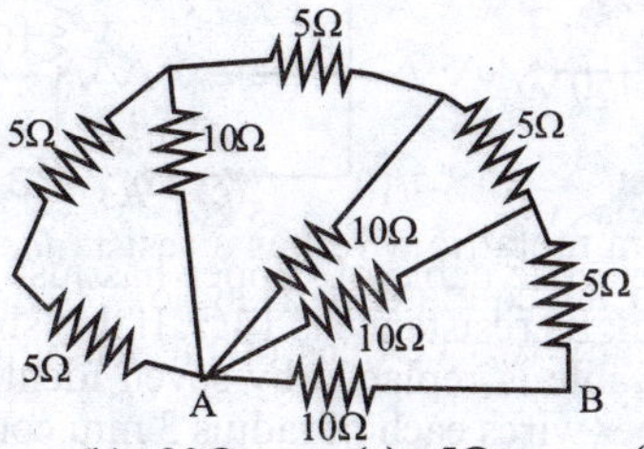

(a) 65Ω (b) 20Ω (c) 5Ω (d) 2Ω

108. What will be the most suitable combination of three resistors $A = 2\Omega$, $B = 4\Omega$, $C = 6\Omega$ so that $\left(\frac{22}{3}\right)\Omega$ is equivalent resistance of combination? **[June 24, 2022 (II)]**

(a) Parallel combination of A and C connected in series with B.
(b) Parallel combination of A and B connected in series with C.
(c) Series combination of A and C connected in parallel with B.
(d) Series combination of B and C connected in parallel with A.

109. A square shaped wire with resistance of each side 3Ω is bent to form a complete circle. The resistance between two diametrically opposite points of the circle in unit of Ω will be ______. **[NA, Aug. 31, 2021 (I)]**

110. The equivalent resistance of the given circuit between the terminals A and B is : **[Aug. 31, 2021 (II)]**

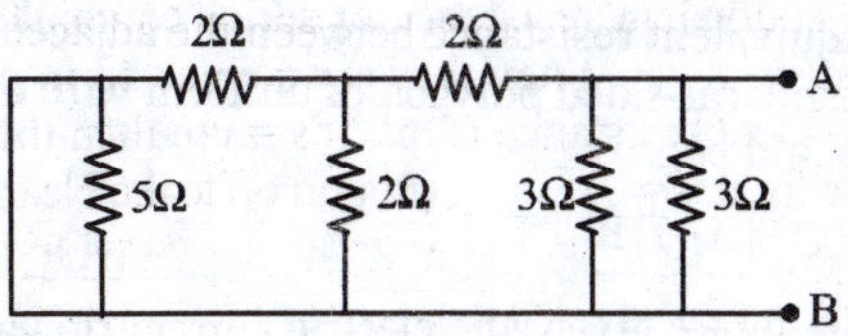

(a) $0\,\Omega$ (b) $3\,\Omega$ (c) $\frac{9}{2}\Omega$ (d) $1\,\Omega$

111. First, a set of n equal resistors of $10\,\Omega$ each are connected in series to a battery of emf 20V and internal resistance $10\,\Omega$. A current I is observed to flow. Then, the n resistors are connected in parallel to the same battery. It is observed that the current is increased 20 times, then the value of n is ______. **[NCERT : P L-83 | NA, Aug. 27, 2021 (I)]**

112. The ratio of the equivalent resistance of the network (shown in figure) between the points a and b when switch is open and switch is closed is x : 8. The value of x is ______. **[NA, Aug. 27, 2021 (II)]**

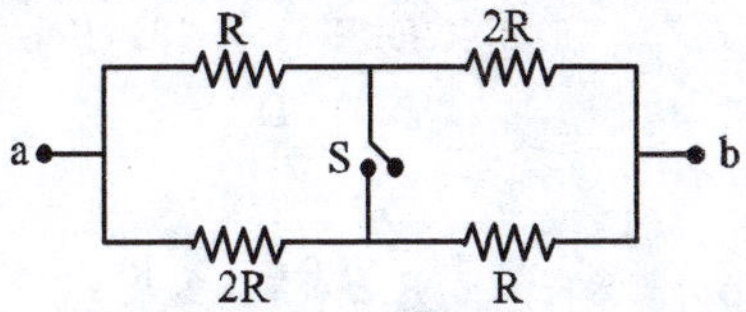

113. What equal length of an iron wire and a copper-nickel alloy wire, each of 2 mm diameter connected parallel to give an equivalent resistance of 3Ω?
(Given resistivities of iron and copper-nickel alloy wire are $12\ \mu\Omega$ cm and $51\ \mu\Omega$ cm respectively)

[NCERT : P L-84 | Aug. 26, 2021 (I)]

(a) 82m (b) 97m (c) 110m (d) 90m

114. If you are provided a set of resistances 2Ω, 4Ω, 6Ω and 8Ω. Connect these resistances so as to obtain an equivalent resistance of $\frac{46}{3}\Omega$. **[Aug. 26, 2021 (II)]**

(a) 4Ω and 6Ω are in parallel with 2Ω and 8Ω in series
(b) 6Ω and 8Ω are in parallel with 2Ω and 4Ω in series
(c) 2Ω and 6Ω are in parallel with 4Ω and 8Ω in series
(d) 2Ω and 4Ω are in parallel with 6Ω and 8Ω in series

115. In the given figure switches S_1 and S_2 are in open condition. The resistance across ab when the switches S_1 and S_2 are closed is ______ Ω. **[NA, July 20, 2021 (II)]**

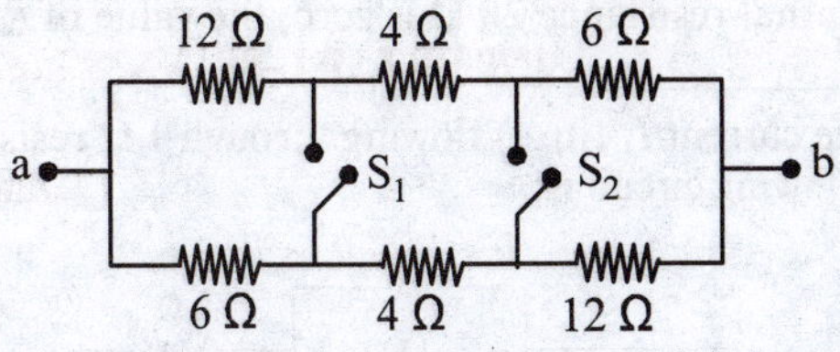

116. The voltage across the $10\,\Omega$ resistor in the given circuit is x volt.

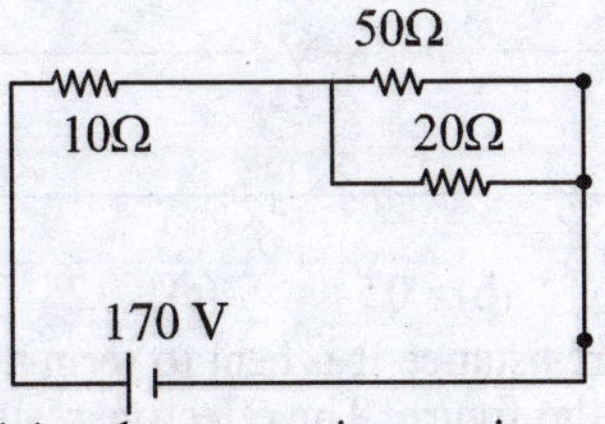

The value 'x' to the nearest integer is ______.

[March 18, 2021 (I)]

117. The equivalent resistance of series combination of two resistors is 's'. When they are connected in parallel, the equivalent resistance is 'p'. If s = np, then the minimum value for n is ________. (Round off to the Nearest Integer) **[March 17, 2021 (I)]**

118. In the figure given, the electric current flowing through the 5 kΩ resistor is x mA. **[NCERT : P L-83 | March 16, 2021 (I)]**

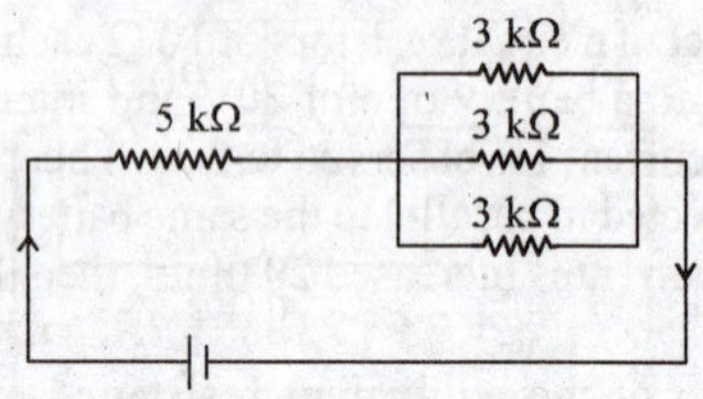

The value of x to the nearest integer is ________.

119. Five equal resistances are connected in a network as shown in figure. The net resistance between the points A and B is: **[NCERT : P L-101 | Feb. 26, 2021 (I)]**

(a) $\frac{R}{2}$

(b) $\frac{3R}{2}$

(c) R

(d) 2R

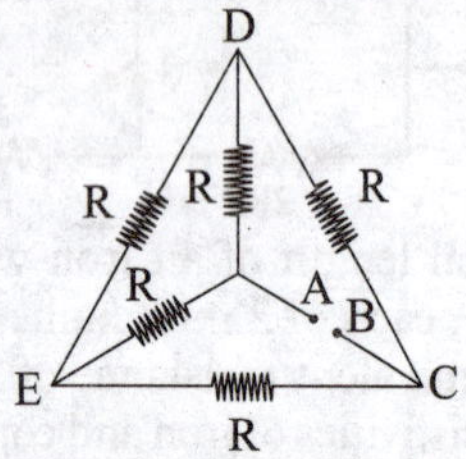

120. In the given circuit diagram, a wire is joining points B and D. The current in this wire is: **[NCERT : P L-97 | 9 Jan. 2020 I]**

(a) 0.4A

(b) 2A

(c) 4A

(d) zero

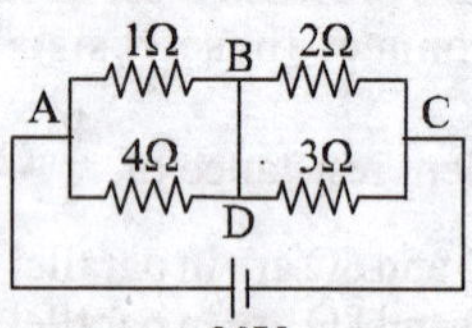

121. The series combination of two batteries, both of the same emf 10 V, but different internal resistance of 20 Ω and 5 Ω, is connected to the parallel combination of two resistors 30 Ω and R Ω. The voltage difference across the battery of internal resistance 20 Ω is zero, the value of *R* (in Ω) is ________. **[NCERT : P L-98 | NA. 8 Jan. 2020 II]**

122. The current I_1 (in *A*) flowing through 1 Ω resistor in the following circuit is: **[7 Jan. 2020 I]**

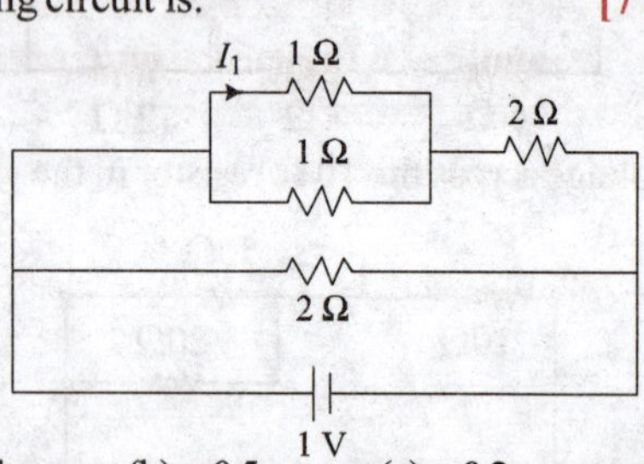

(a) 0.4 (b) 0.5 (c) 0.2 (d) 0.25

123. A wire of resistance R is bent to form a square ABCD as shown in the figure. The effective resistance between E and C is: (E is mid-point of arm CD) **[9 April 2019 I]**

(a) R

(b) $\frac{7}{64}R$

(c) $\frac{3}{4}R$

(d) $\frac{1}{16}R$

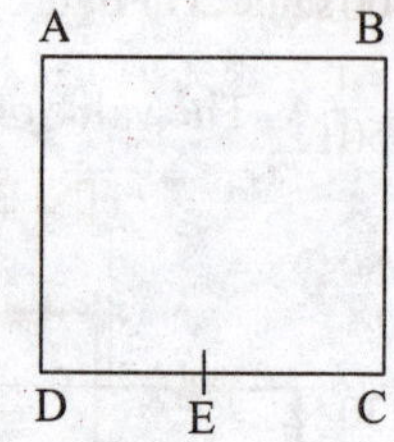

124. A metal wire of resistance 3 Ω is elongated to make a uniform wire of double its previous length. This new wire is now bent and the ends joined to make a circle. If two points on the circle make an angle 60° at the centre, the equivalent resistance between these two points will be: **[NCERT : P L-84 | 9 Apr. 2019 II]**

(a) $\frac{12}{5}\Omega$ (b) $\frac{5}{2}\Omega$ (c) $\frac{5}{3}\Omega$ (d) $\frac{7}{2}\Omega$

125. In the figure shown, what is the current (in ampere) drawn from the battery? You are given : **[NCERT : P L-83 | 8 Apr. 2019 II]**

$R_1 = 15\ \Omega$, $R_2 = 10\ \Omega$, $R_3 = 20\ \Omega$, $R_4 = 5\ \Omega$, $R_5 = 25\ \Omega$, $R_6 = 30\ \Omega$, $E = 15$ V

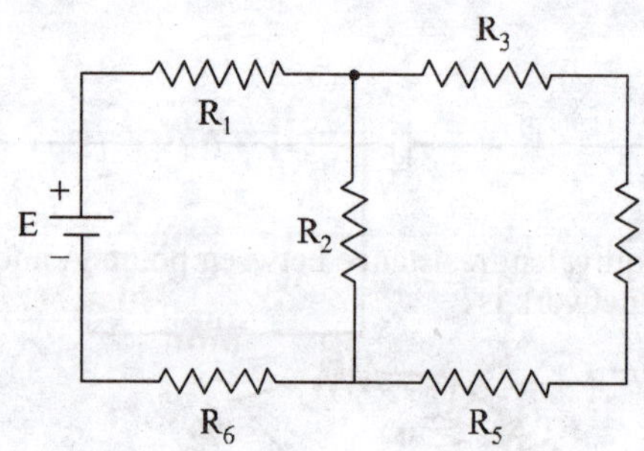

(a) 13/24 (b) 7/18 (c) 9/32 (d) 20/3

126. A uniform metallic wire has a resistance of 18 Ω and is bent into an equilateral triangle. Then, the resistance between any two vertices of the triangle is: **[10 Jan. 2019 I]**

(a) 4 Ω (b) 8 Ω (c) 12 Ω (d) 2 Ω

127. The actual value of resistance R, shown in the figure is 30 Ω. This is measured in an experiment as shown using the standard formula $R = \frac{V}{I}$, where V and I are the reading of the voltmeter and ammeter, respectively. If the measured value of R is 5% less, then the internal resistance of the voltmeter is: **[10 Jan. 2019 II]**

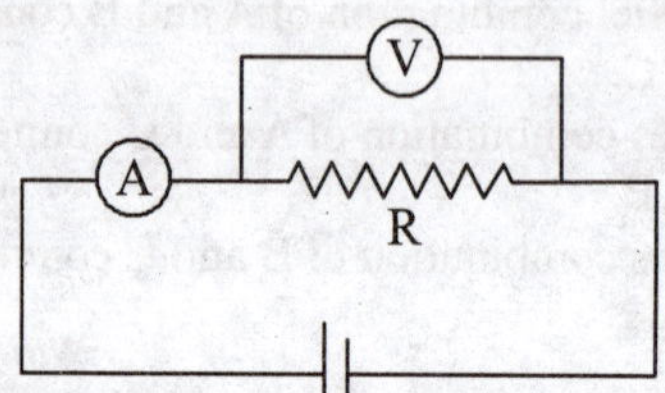

(a) 600 Ω (b) 570 Ω (c) 35 Ω (d) 350 Ω

128. In the given circuit the internal resistance of the 18 V cell is negligible. If $R_1 = 400\Omega$, $R_3 = 100\ \Omega$ and $R_4 = 500\ \Omega$ and the reading of an ideal voltmeter across R_4 is 5 V, then the value of R_2 will be: **[NCERT : P L-97 | 9 Jan. 2019 II]**

(a) $300\,\Omega$

(b) $450\,\Omega$

(c) $550\,\Omega$

(d) $230\,\Omega$

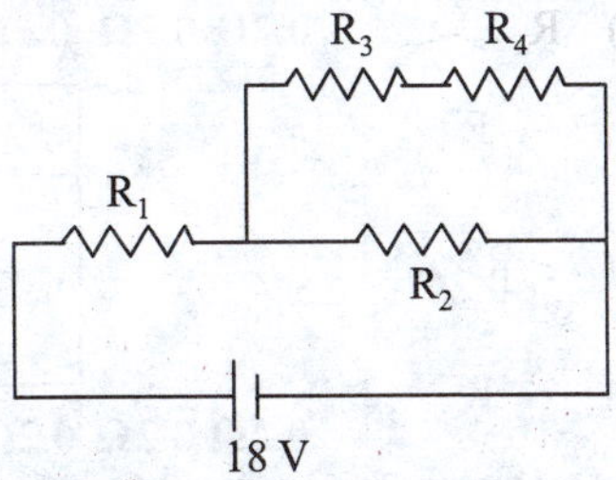

129. In the given circuit all resistances are of value R ohm each. The equivalent resistance between A and B is : **[Online April 15, 2018]**

(a) $2R$

(b) $\frac{5R}{2}$

(c) $\frac{5R}{3}$

(d) $3R$

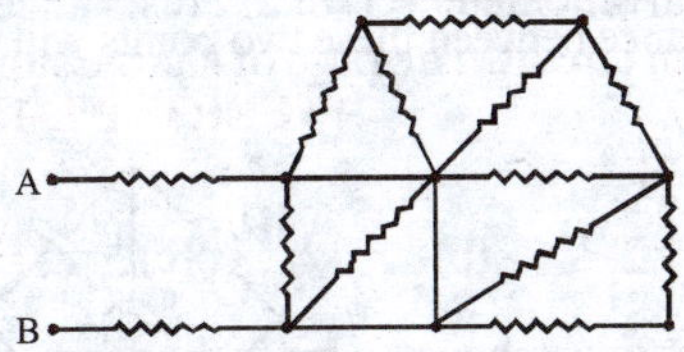

130. In the given circuit diagram when the current reaches steady state in the circuit, the charge on the capacitor of capacitance C will be : **[NCERT : P L-97 | 2017]**

(a) $CE\frac{r_2}{(r+r_2)}$

(b) $CE\frac{r_1}{(r_1+r)}$

(c) CE

(d) $CE\frac{r_1}{(r_2+r)}$

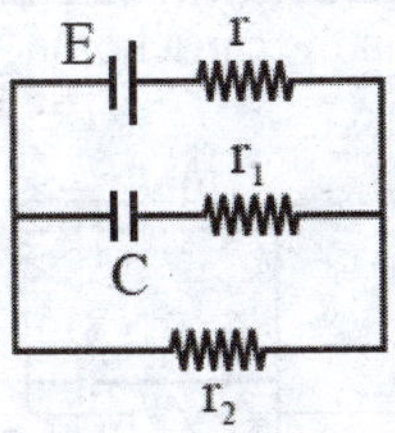

131. In the above circuit the current in each resistance is **[NCERT : P L-97 | 2017]**

(a) 0.5A

(b) 0 A

(c) 1 A

(d) 0.25 A

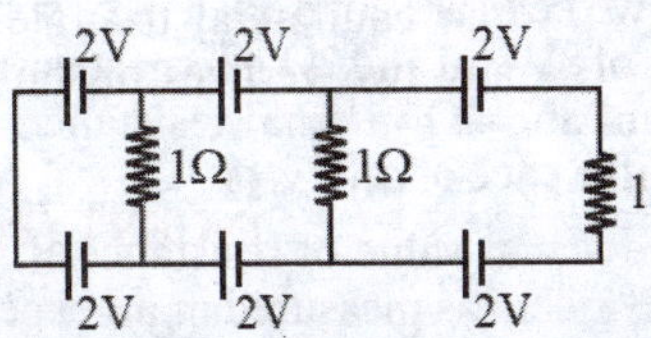

132. Six equal resistances are connected between points P, Q and R as shown in figure. Then net resistance will be maximum between : **[Online April 25, 2013]**

(a) P and R

(b) P and Q

(c) Q and R

(d) Any two points

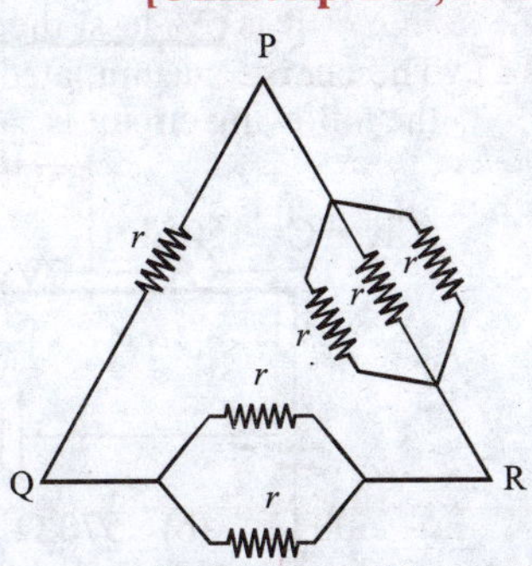

133. Two conductors have the same resistance at 0°C but their temperature coefficients of resistance are α_1 and α_2. The respective temperature coefficients of their series and parallel combinations are nearly **[NCERT : P L-91 | 2010]**

(a) $\frac{\alpha_1+\alpha_2}{2}, \alpha_1+\alpha_2$

(b) $\alpha_1+\alpha_2, \frac{\alpha_1+\alpha_2}{2}$

(c) $\alpha_1+\alpha_2, \frac{\alpha_1\alpha_2}{\alpha_1+\alpha_2}$

(d) $\frac{\alpha_1+\alpha_2}{2}, \frac{\alpha_1+\alpha_2}{2}$

134. The current I drawn from the 5 volt source will be **[NCERT : P L-101 | 2006]**

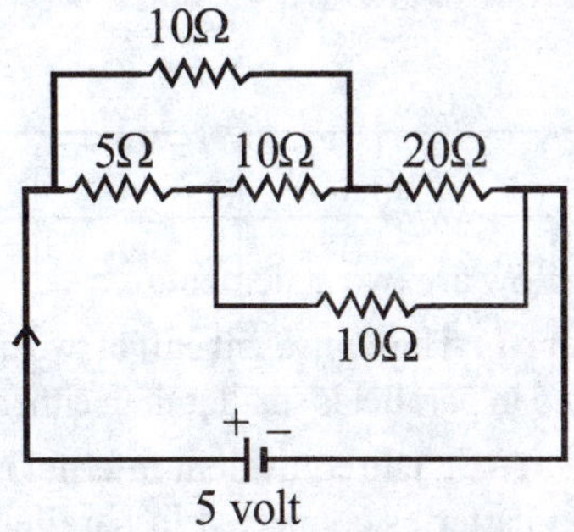

(a) 0.33 A (b) 0.5 A (c) 0.67 A (d) 0.17 A

135. The total current supplied to the circuit by the battery is **[NCERT : P L-83 | 2004]**

(a) 4 A

(b) 2 A

(c) 1 A

(d) 6 A

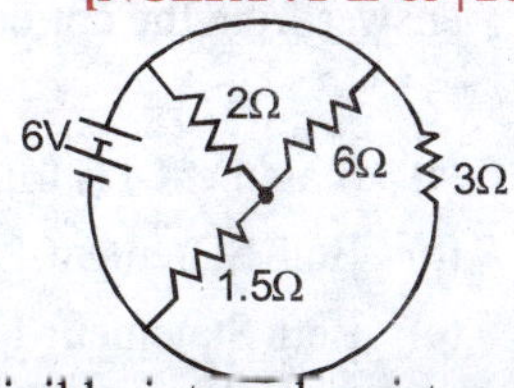

136. A 3 volt battery with negligible internal resistance is connected in a circuit as shown in the figure. The current I, in the circuit will be **[NCERT : P L-83 | 2003]**

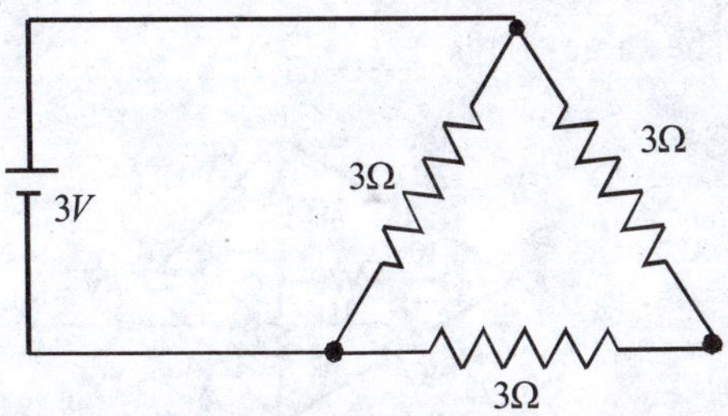

(a) 1 A (b) 1.5 A (c) 2 A (d) 1/3 A

Topic 3 - Kirchhoff's Laws and Cells

137. Two cells of emf 1V and 2V and internal resistance $2\,\Omega$ and $1\,\Omega$, respectively, are connected in series with an external resistance of $6\,\Omega$. The total current in the circuit is I_1. Now the same two cells in parallel configuration are connected to same external resistance. In this case, the total current drawn is I_2. The value of $\left(\frac{I_1}{I_2}\right)$ is $\frac{x}{3}$. The value of x is ____. **[NCERT : P L-95, 96 | April 3, 2025 (II)]**

138. The value of current I in the electrical circuit as given below, when potential at A is equal to the potential at B, will be ________ A. **[NCERT : P L-97, 98 | Jan. 28, 2025 (II)]**

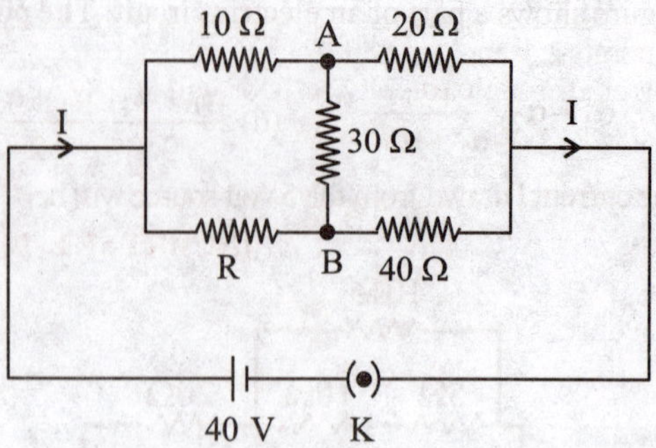

139. Given below are two statements :

Statement-I : The equivalent emf of two nonideal batteries connected in parallel is smaller than either of the two emfs.

Statement-II : The equivalent internal resistance of two nonideal batteries connected in parallel is smaller than the internal resistance of either of the two batteries.

In the light of the above statements, choose the correct answer from the options given below.

[NCERT : P L-95, 96 | Jan. 22, 2025 (I)]

(a) Statement-I is true but Statement-II is false

(b) Both Statement-I and Statement-II are false

(c) Both Statement-I and Statement-II are true

(d) Statement-I is false but Statement-II is true

140. The current flowing through the 1 Ω resistor is $\frac{n}{10}$ A. The value of n is ______.

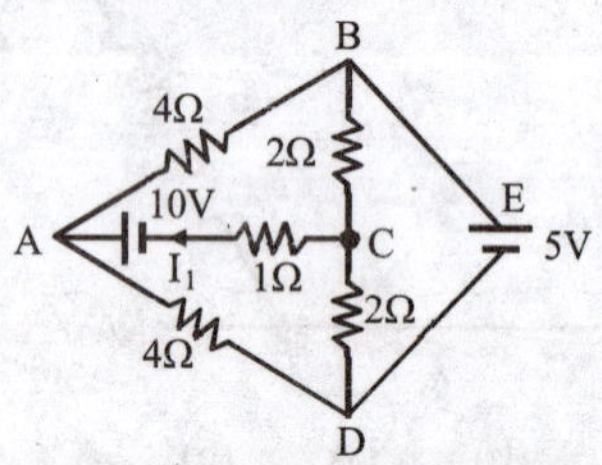

[NCERT : P L-83 | April 9, 2024 (I)]

141. In the given circuit, the terminal potential difference of the cell is : **[NCERT : P L-95 | April 8, 2024 (I)]**

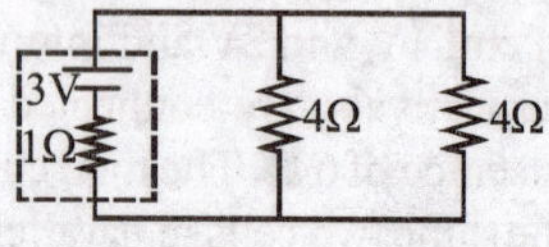

(a) 2V (b) 4V (c) 1.5V (d) 3V

142. The electric field between the two parallel plates of a capacitor of 1.5 μF capacitance drops to one third of its initial value in 6.6 μs when the plates are connected by a thin wire. The resistance of this wire is .... Ω. (Given, log 3 = 1.1) **[April 5, 2024 (I)]**

143. The reading in the ideal voltmeter (V) shown in the given circuit diagram is : **[NCERT : P L-98 | Feb. 1, 2024 (I)]**

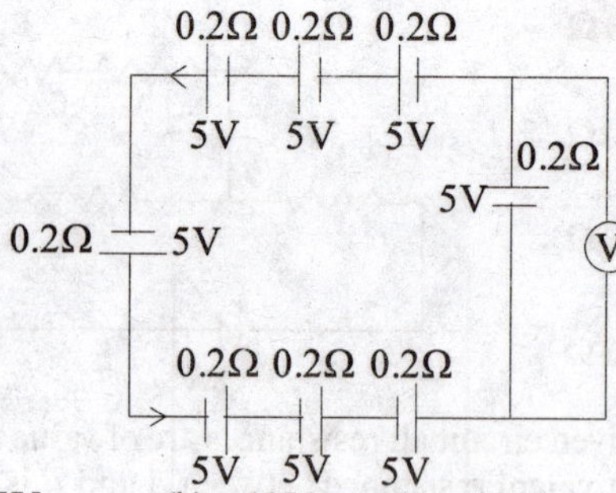

(a) 5V (b) 10V (c) 0V (d) 3V

144. A galvanometer (G) of 2Ω resistance is connected in the given circuit. The ratio of charge stored in C_1 and C_2 is: **[NCERT : P L-100 | Feb. 1, 2024 (II)]**

(a) 1

(b) $\frac{2}{3}$

(c) $\frac{3}{2}$

(d) $\frac{1}{2}$

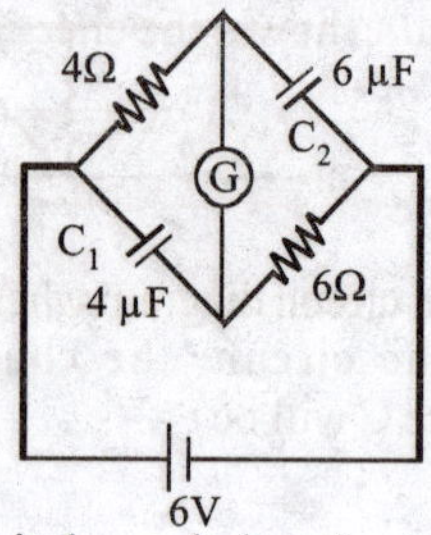

145. In an electrical circuit drawn below the amount of charge stored in the capacitor is ______ μC. **[NCERT : P L-97 | Feb. 1, 2024 (II)]**

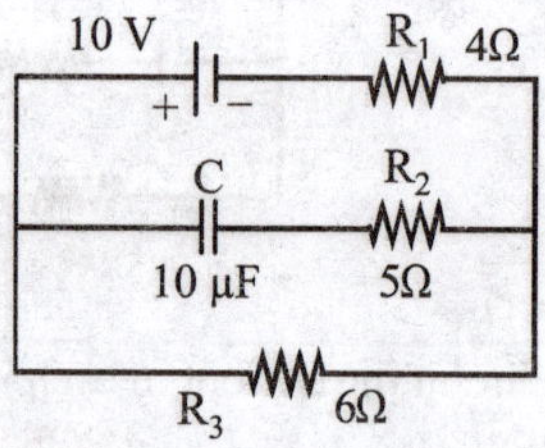

146. Two cells are connected in opposition as shown. Cell E_1 is of 8V emf and 2Ω internal resistance; the cell E_2 is of 2V emf and 4Ω internal resistance. The terminal potential difference of cell E_2 is ______ V. **[NCERT : P L-95 | Jan. 30, 2024 (I)]**

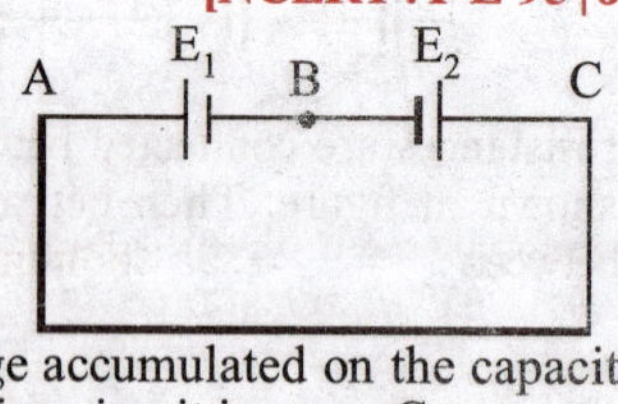

147. The charge accumulated on the capacitor connected in the following circuit is ______ μC **[NCERT : P L-97 | Jan. 27, 2024 (I)]**

(Given C = 150 μF)

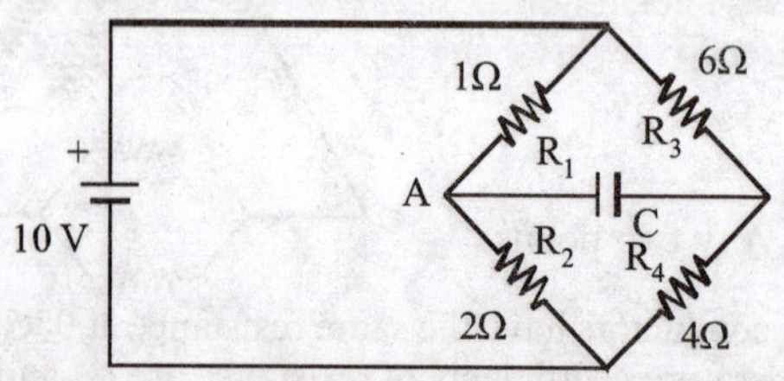

148. A network of four resistances is connected to 9 V battery, as shown in figure. The magnitude of voltage difference between the points A and B is ______ V. **[NCERT : P L-98 | April 15, 2023 (I)]**

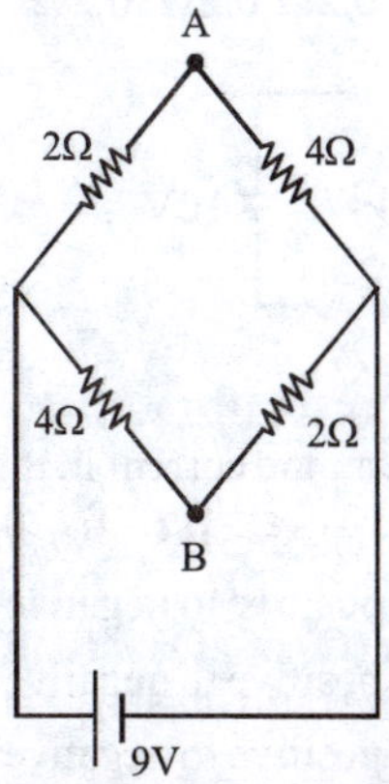

149. In the circuit shown, the energy stored in the capacitor is n μJ. The value of n is ______.

[NCERT : P L-98 | April 13, 2023 (II)]

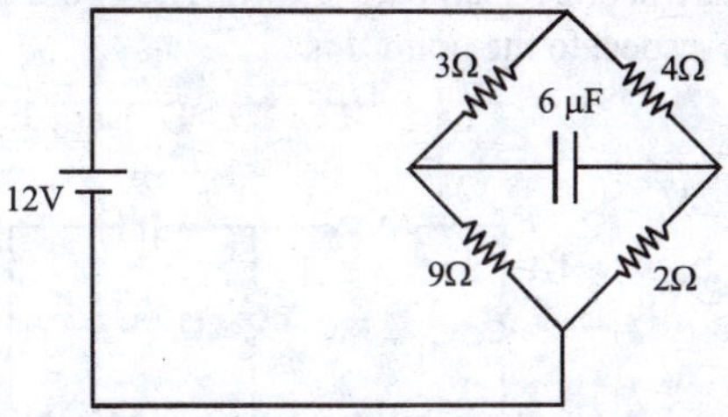

150. In the circuit diagram shown in figure given below, the current flowing through resistance 3 Ω is $\frac{x}{3}$ A. The value of x is ______.

[NCERT : P L-97 | April 11, 2023 (I)]

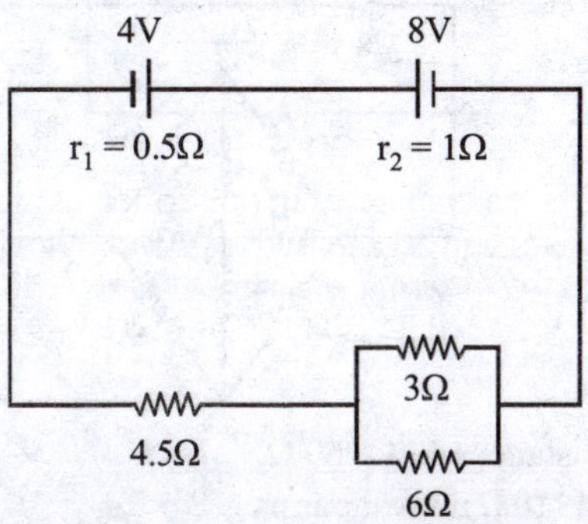

151. Two identical cells each of emf 1.5 V are connected in series across a 10 Ω resistance. An ideal voltmeter connected across 10 Ω resistance reads 1.5 V. The internal resistance of each cell is ______ Ω.

[NCERT : P L-95 | April 11, 2023 (II)]

152. In this figure the resistance of the coil of galvanometer G is 2 Ω. The emf of the cell is 4 V. The ratio of potential difference across C_1 and C_2 is:

[NCERT : P L-98 | April 11, 2023 (I)]

(a) 1

(b) $\frac{4}{5}$

(c) $\frac{3}{4}$

(d) $\frac{5}{4}$

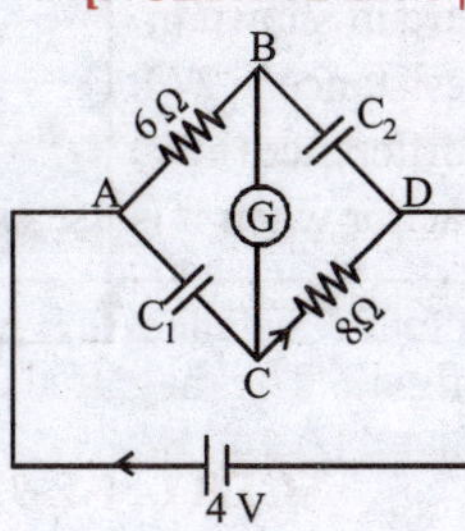

153. Figure shows a part of an electric circuit. The potentials at points a, b and c are 30 V, 12 V and 2V respectively. The current through the 20Ω resistor will be.

[NCERT : P L-97 | April 6, 2023 (II)]

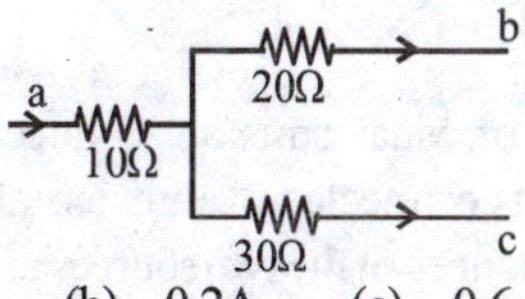

(a) 0.4 A (b) 0.2 A (c) 0.6 A (d) 1.0 A

154. In the given circuit, the value of $\left|\frac{I_1 + I_3}{I_2}\right|$ is ______

[NCERT : P L-98 | Feb. 1, 2023 (II)]

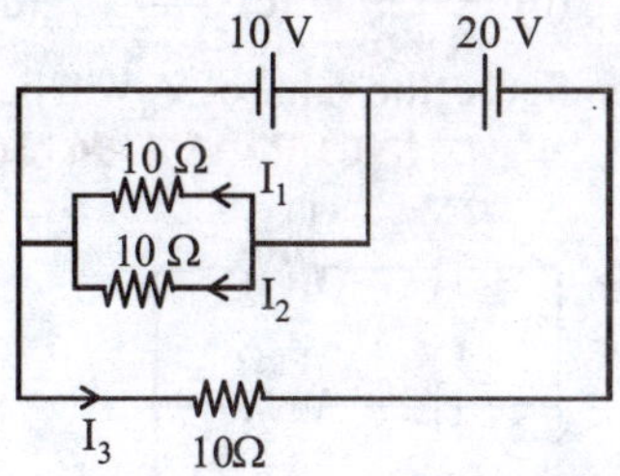

155. Two identical cells, when connected either in parallel or in series gives same current in an external resistance 5Ω. The internal resistance of each cell will be ______ Ω.

[NCERT : P L-95 | Jan. 31, 2023 (I)]

156. For the given circuit, in the steady state, $|V_B - V_D|$ = ______ V. **[NCERT : P L-98 | Jan. 31, 2023 (II)]**

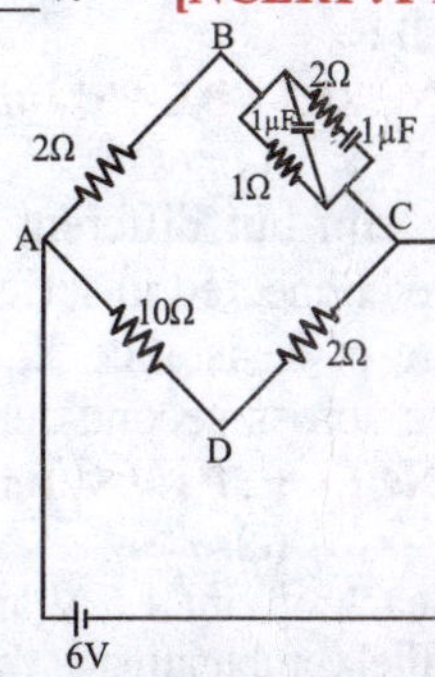

157. In the following circuit, the magnitude of current I_1, is ______ A. **[NCERT : P L-97 | Jan. 30, 2023 (I)]**

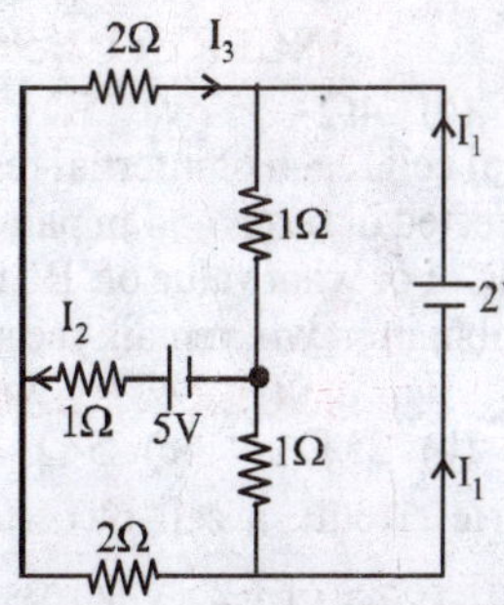

158. Two cells are connected between points A and B as shown. Cell 1 has emf of 12 V and internal resistance of 3Ω. Cell 2 has emf of 6V and internal resistance of 6Ω. An external resistor R of 4Ω is connected across A and B. The current flowing through R will be ______ A.

[NCERT : P L-96 | Jan. 25, 2023 (II)]

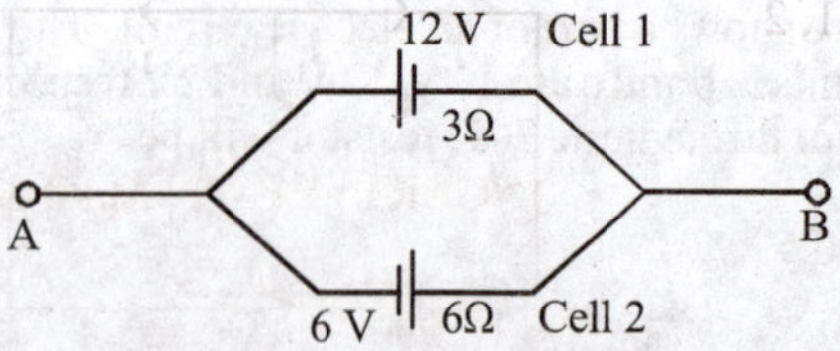

159. Two sources of equal emfs are connected in series. This combination is connected to an external resistance R. The internal resistances of the two sources are r_1 and r_2 ($r_1 > r_2$). If the potential difference across the source of internal resistance r_1 is zero then the value of R will be

[NCERT : P L-95 | July 27, 2022 (I)]

(a) $r_1 - r_2$ (b) $\frac{r_1 r_2}{r_1 + r_2}$ (c) $\frac{r_1 + r_2}{2}$ (d) $r_2 - r_1$

160. In the given figure, the value of V_0 will be ________ V.

[NCERT : P L-96 | July 25, 2022 (I)]

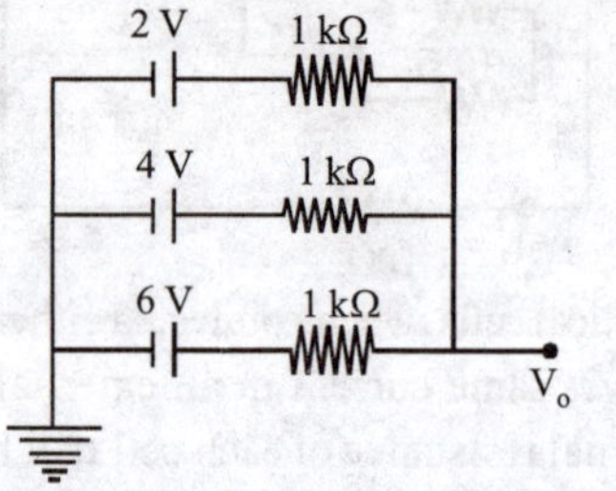

161. The combination of two identical cells, whether connected in series or parallel combination provides the same current through an external resistance of 2Ω. The value of internal resistance of each cell is :

[NCERT : P L-95 | June 29, 2022 (II)]

(a) 2Ω (b) 4Ω (c) 6Ω (d) 8Ω

162. Two cells of same emf but different internal resistances r_1 and r_2 are connected in series with a resistance R. The value of resistance R, for which the potential difference across second cell is zero, is

[NCERT : P L-95 | June 25, 2022 (II)]

(a) $r_2 - r_1$ (b) $r_1 - r_2$ (c) r_1 (d) r_2

163. Two identical cells each of emf 1.5 V are connected in parallel across a parallel combination of two resistors each of resistance 20Ω. A voltmeter connected in the circuit measures 1.2 V. The internal resistance of each cell is

[NCERT : P L-96 | June 24, 2022 (I)]

(a) 2.5Ω (b) 4Ω (c) 5Ω (d) 10Ω

164. Five identical cells each of internal resistance 1Ω and emf 5V are connected in series and in parallel with an external resistance 'R'. For what value of 'R', current in series and parallel combination will remain the same?

[NCERT : P L-96 | Aug. 27, 2021 (I)]

(a) 1 Ω (b) 25 Ω (c) 5 Ω (d) 10 Ω

165. In an electric circuit, a cell of certain emf provides a potential difference of 1.25 V across a load resistance of 5 Ω. However, it provides a potential difference of 1 V across a load resistance of 2 Ω. The emf of the cell is given by $\frac{x}{10}$ V. Then the value of x is ________.

[NCERT : P L-95 | NA, July 22, 2021 (II)]

166.

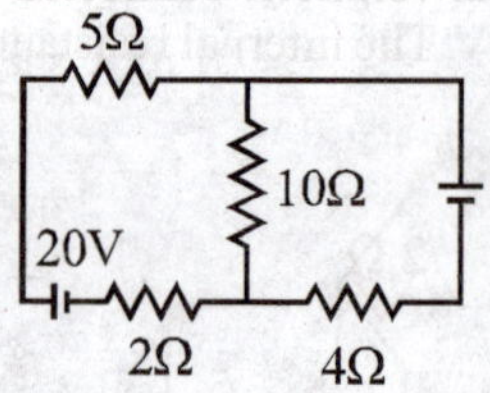

In the figure shown, the current in the 10 V battery is close to :

[NCERT : P L-98 | Sep. 06, 2020 (II)]

(a) 0.71 A from positive to negative terminal

(b) 0.42 A from positive to negative terminal

(c) 0.21 A from positive to negative terminal

(d) 0.36 A from negative to positive terminal

167. In the circuit, given in the figure currents in different branches and value of one resistor are shown. Then potential at point B with respect to the point A is :

[NCERT : P L-97 | Sep. 05, 2020 (II)]

(a) +2 V

(b) −2 V

(c) −1 V

(d) +1 V

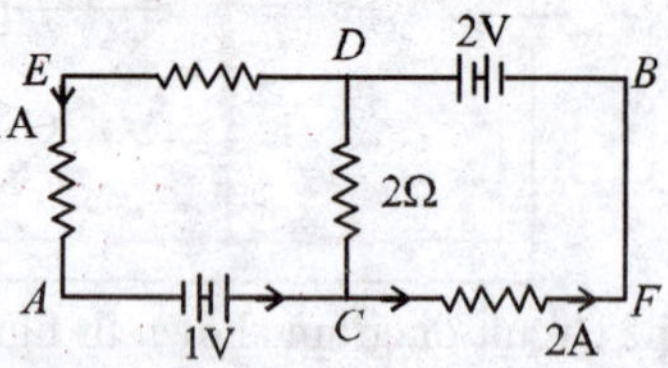

168. The value of current i_1 flowing from A to C in the circuit diagram is :

[NCERT : P L-97 | Sep. 04, 2020 (II)]

(a) 2 A

(b) 4 A

(c) 1 A

(d) 5 A

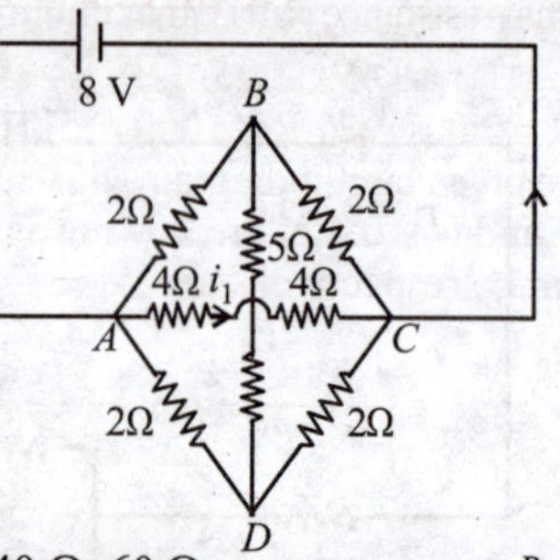

169. Four resistances 40 Ω, 60 Ω, 90 Ω and 110 Ω make the arms of a quadrilateral $ABCD$. Across AC is a battery of emf 40 V and internal resistance negligible. The potential difference across BD in V is ________.

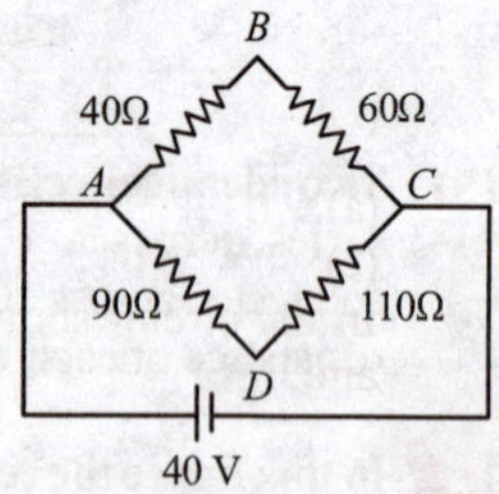

[NCERT : P L-97 | NA. Sep. 04, 2020 (II)]

170. An ideal cell of emf 10 V is connected in circuit shown in figure. Each resistance is 2 Ω. The potential difference (in V) across the capacitor when it is fully charged is ________.

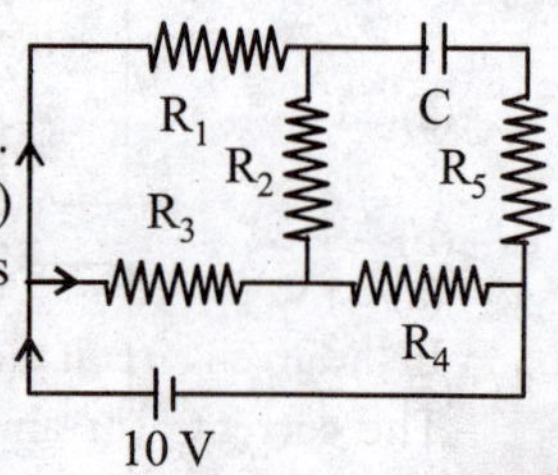

[NCERT : P L-97 | Sep. 02, 2020 (II)]

171. In the given circuit, an ideal voltmeter connected across the 10 Ω resistance reads 2V. The internal resistance r, of each cell is : **[NCERT : P L-97 | 10 Apr. 2019 I]**

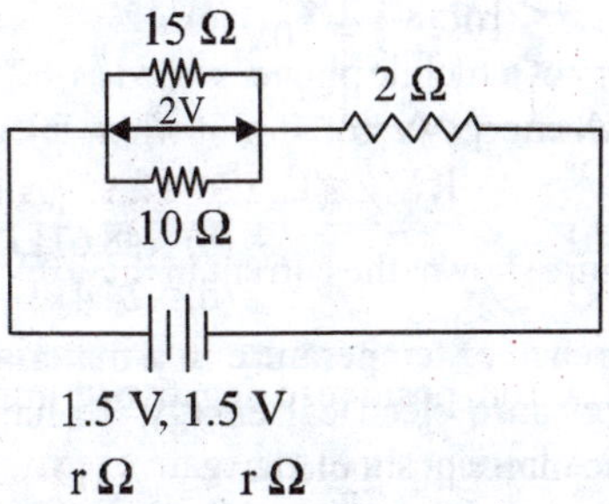

(a) 1 Ω (b) 0.5 Ω (c) 1.5 Ω (d) 0 Ω

172. For the circuit shown, with $R_1 = 1.0\,\Omega$, $R_2 = 2.0\,\Omega$, $E_1 = 2$ V and $E_2 = E_3 = 4$ V, the potential difference between the points '*a*' and '*b*' is approximately (in V) : **[NCERT : P L-95 | 8 April 2019 I]**

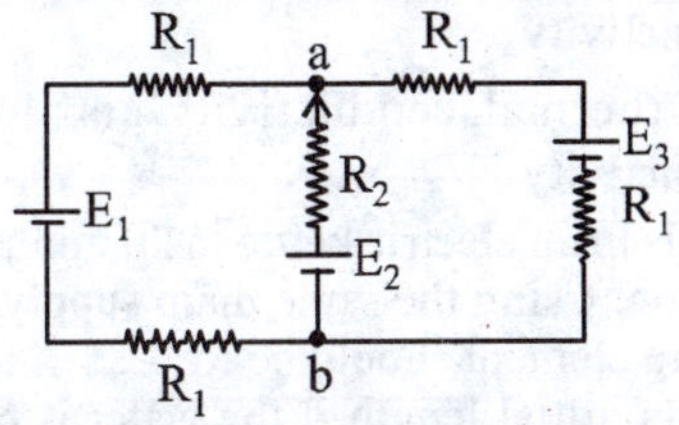

(a) 2.7 (b) 2.3 (c) 3.7 (d) 3.3

173. A cell of internal resistance *r* drives current through an external resistance R. The power delivered by the cell to the external resistance will be maximum when : **[NCERT : P L-95 | 8 Apr. 2019 II]**

(a) $R = 0.001r$ (b) $R = 1000r$ (c) $R = 2r$ (d) $R = r$

174. In the given circuit diagram, the currents, $I_1 = -0.3$ A, $I_4 = 0.8$ A and $I_5 = 0.4$ A, are flowing as shown. The currents I_2, I_3 and I_6, respectively, are : **[NCERT : P L-97 | 12 Jan. 2019 II]**

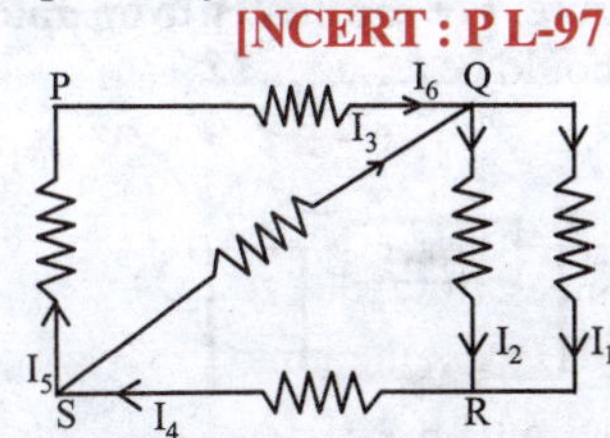

(a) 1.1 A, –0.4 A, 0.4 A (b) 1.1 A, 0.4 A, 0.4 A
(c) 0.4 A, 1.1 A, 0.4 A (d) –0.4 A, 0.4 A, 1.1 A

175. In the circuit shown, the potential difference between A and B is : **[NCERT : P L-95 | 11 Jan. 2019 II]**

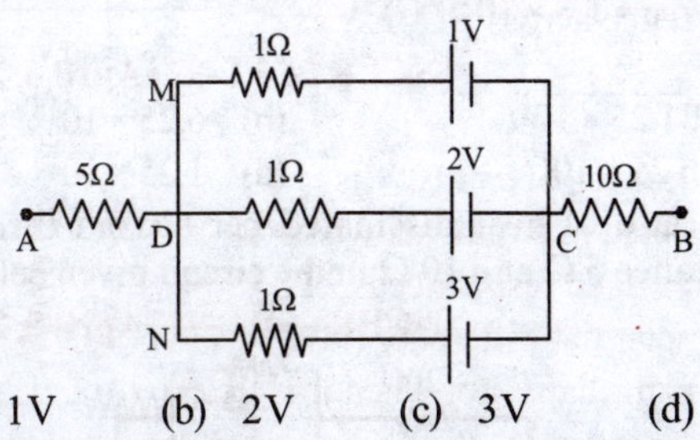

(a) 1V (b) 2V (c) 3V (d) 6V

176. In the given circuit the cells have zero internal resistance. The currents (in amperes) passing through resistance R_1 and R_2 respectively, are: **[NCERT : P L-97 | 10 Jan. 2019 I]**

(a) 1, 2
(b) 2, 2
(c) 0.5, 0
(d) 0, 1

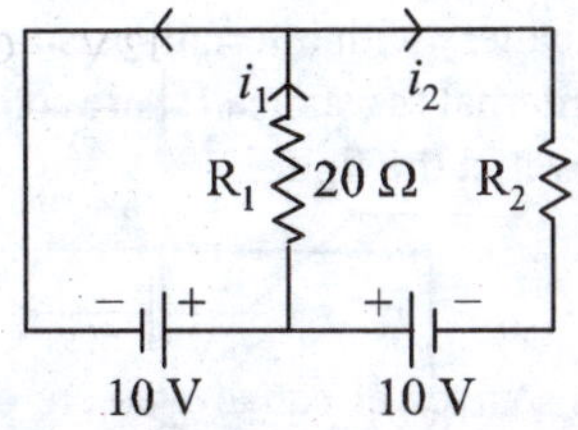

177. When the switch S, in the circuit shown, is closed then the valued of current *i* will be: **[NCERT : P L-97 | 9 Jan. 2019 I]**

(a) 3A
(b) 5A
(c) 4A
(d) 2A

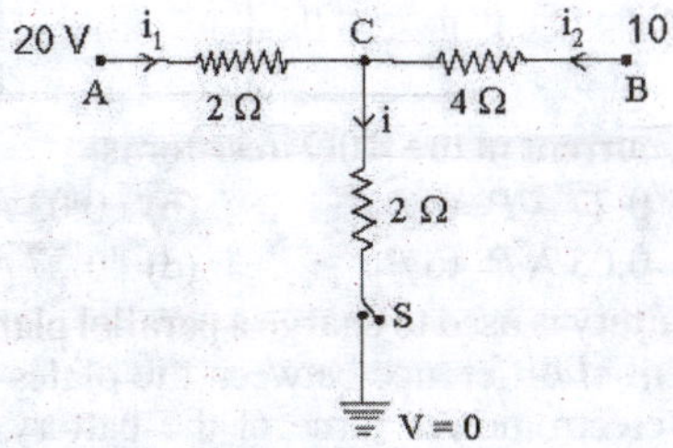

178. Two batteries with e.m.f. 12 V and 13 V are connected in parallel across a load resistor of 10Ω. The internal resistances of the two batteries are 1Ω and 2Ω respectively. The voltage across the load lies between: **[NCERT : P L-97 | 2018]**

(a) 11.6 V and 11.7 V (b) 11.5 V and 11.6 V
(c) 11.4 V and 11.5 V (d) 11.7 V and 11.8 V

179. In the circuit shown, the current in the 1Ω resistor is: **[NCERT : P L-97 | 2015]**

(a) 0.13 A, from Q to P
(b) 0.13 A, from P to Q
(c) 1.3A from P to Q
(d) 0A

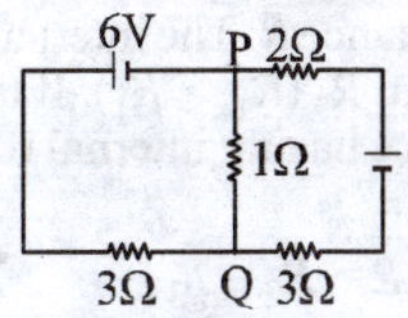

180. A d.c. main supply of e.m.f. 220 V is connected across a storage battery of e.m.f. 200 V through a resistance of 1Ω. The battery terminals are connected to an external resistance 'R'. The minimum value of 'R', so that a current passes through the battery to charge it is: **[NCERT : P L-97 | Online April 9, 2014]**

(a) 7 Ω (b) 9 Ω (c) 11 Ω (d) Zero

181. A dc source of emf $E_1 = 100$ V and internal resistance $r = 0.5\,\Omega$, a storage battery of emf $E_2 = 90$ V and an external resistance R are connected as shown in figure. For what value of R no current will pass through the battery? **[NCERT : P L-96 | Online April 22, 2013]**

(a) 5.5 Ω
(b) 3.5 Ω
(c) 4.5 Ω
(d) 2.5 Ω

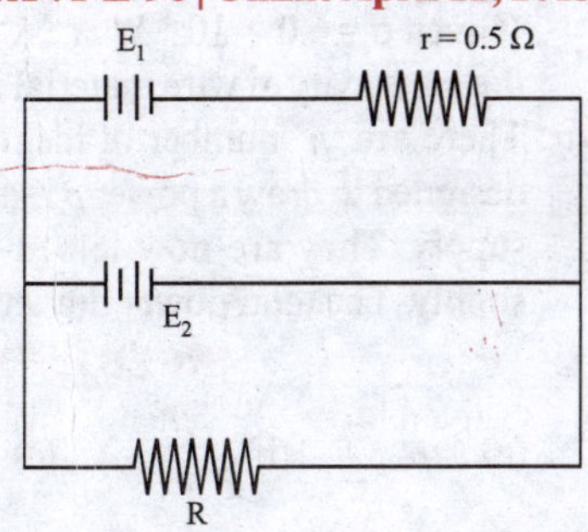

182. A 5V battery with internal resistance 2Ω and a 2V battery with internal resistance 1Ω are connected to a 10Ω resistor as shown in the figure. **[NCERT : P L-98 | 2008]**

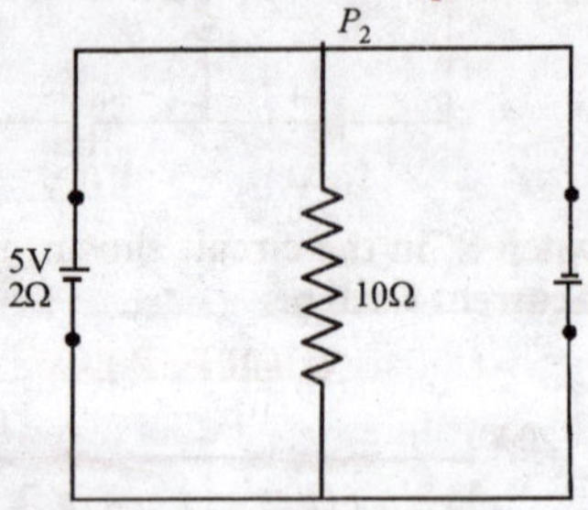

The current in the 10Ω resistor is

(a) 0.27 A P_2 to P_1 (b) 0.03 A P_1 to P_2
(c) 0.03 A P_2 to P_1 (d) 0.27 A P_1 to P_2

183. A battery is used to charge a parallel plate capacitor till the potential difference between the plates becomes equal to the electromotive force of the battery. The ratio of the energy stored in the capacitor and the work done by the battery will be **[2007]**

(a) 1/2 (b) 1 (c) 2 (d) 1/4

184. The Kirchhoff's first law ($\Sigma i = 0$) and second law ($\Sigma iR = \Sigma E$), where the symbols have their usual meanings, are respectively based on **[NCERT : P L-97 | 2006]**

(a) conservation of charge, conservation of momentum
(b) conservation of energy, conservation of charge
(c) conservation of momentum, conservation of charge
(d) conservation of charge, conservatrion of energy

185. Two sources of equal emf are connected to an external resistance R. The internal resistance of the two sources are R_1 and R_2 ($R_1 > R_1$). If the potential difference across the source having internal resistance R_2 is zero, then **[NCERT : P L-95 | 2005]**

(a) $R = R_2 - R_1$
(b) $R = R_2 \times (R_1 + R_2)/(R_2 - R_1)$
(c) $R = R_1 R_2 / (R_2 - R_1)$
(d) $R = R_1 R_2 / (R_1 - R_2)$

Topic 4 - Heating Effect of Current

186. A wire of length 10 cm and diameter 0.5 mm is used in a bulb. The temperature of the wire is 1727°C and power radiated by the wire is 94.2 W. Its emissivity is $\frac{x}{8}$ where x = ____ **[NCERT : P L-92, 93 | April 7, 2025 (I)]**
(Given $\sigma = 6.0 \times 10^{-8}\ W m^{-2} K^{-4}$, $\pi = 3.14$ and assume that the emissivity of wire material is same at all wavelength.)

187. There are 'n' number of identical electric bulbs, each is designed to draw a power p independently from the mains supply. They are now joined in series across the main supply. The total power drawn by the combination is : **[NCERT : PL-92, 93 | April 4, 2025 (II)]**

(a) np (b) $\frac{p}{n^2}$ (c) $\frac{p}{n}$ (d) p

188. A motor operating on 100 V draws a current of 1 A. If the efficiency of the motor is 91.6%, then the loss of power in units of cal/s is **[NCERT : P L-92, 93 | April 3, 2025 (II)]**

(a) 4 (b) 8.4 (c) 2 (d) 6.2

189. The battery of a mobile phone is rated as 4.2 V, 5800 mAh. How much energy is stored in it when fully charged ? **[NCERT : P L-92, 93 | April 2, 2025 (I)]**

(a) 43.8 kJ (b) 48.67 kJ
(c) 87.7 kJ (d) 24.4 kJ

190. The difference of temperature in a material can convert heat energy into electrical energy. To harvest the heat energy, the material should have **[Jan. 29, 2025 (II)]**

(a) low thermal conductivity and low electrical conductivity
(b) high thermal conductivity and high electrical conductivity
(c) low thermal conductivity and high electrical conductivity
(d) high thermal conductivity and low electrical conductivity

191. Water boils in an electric kettle in 20 minutes after being switched on. Using the same main supply, the length of the heating element should be .... to .... times of its initial length if the water is to be boiled in 15 minutes. **[NCERT : P L-92 | April 8, 2024 (II)]**

(a) increased, $\frac{3}{4}$ (b) increased, $\frac{4}{3}$
(c) decreased, $\frac{3}{4}$ (d) decreased, $\frac{4}{3}$

192. A heater is designed to operate with a power of 1000 W in a 100 V line. It is connected in combination with a resistance of 10 Ω and a resistance R, to a 100 V mains as shown in figure. For the heater to operate at 62.5 W, the value of R should be .... Ω. **[NCERT : P L-93 | April 8, 2024 (II)]**

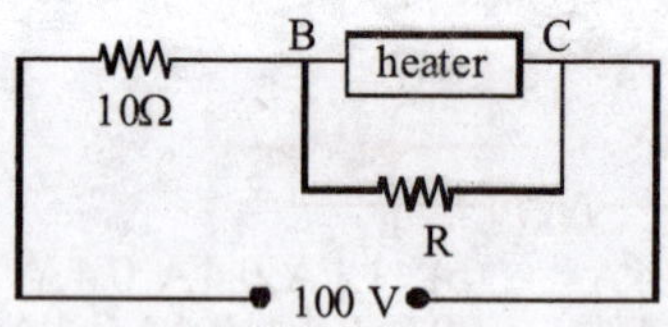

193. The number of electrons flowing per second in the filament of a 110 W bulb operating at 220 V is :
(Given $e = 1.6 \times 10^{-19}$ C) **[NCERT : P L-92 | April 6, 2024 (II)]**

(a) 31.25×10^{17} (b) 6.25×10^{18}
(c) 6.25×10^{17} (d) 1.25×10^{19}

194. The ratio of heat dissipated per second through the resistance 5 Ω and 10 Ω in the circuit given below is : **[NCERT : P L-92 | April 5, 2024 (II)]**

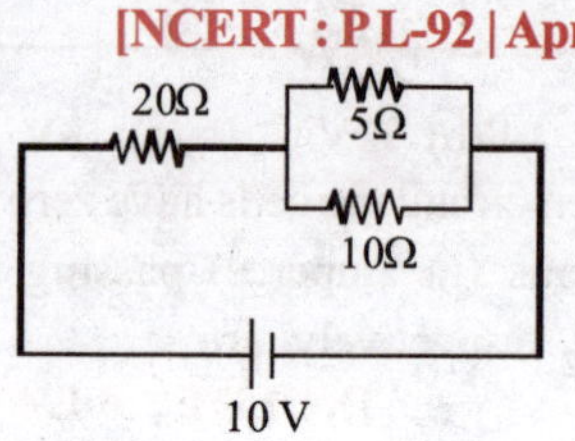

(a) 1 : 2
(b) 2 : 1
(c) 4 : 1
(d) 1 : 1

195. An electric bulb rated 50 W – 200 V is connected across a 100 V supply. The power dissipation of the bulb is :
[NCERT : P L-92 | April 4, 2024 (II)]
(a) 12.5 W (b) 25 W (c) 50 W (d) 100 W

196. By what percentage will the illumination of the lamp decrease if the current drops by 20%?
[NCERT : P L-93 | Jan. 31, 2024 (II)]
(a) 46% (b) 26% (c) 36% (d) 56%

197. In the following circuit, the battery has an emf of 2 V and an internal resistance of $\frac{2}{3}\Omega$. The power consumption in the entire circuit is ______ W.
[NCERT : P L-93 | Jan. 31, 2024 (II)]

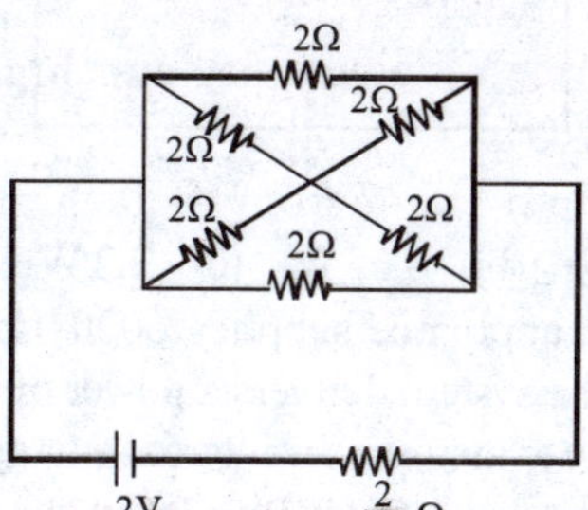

198. When a potential difference V is applied across a wire of resistance R, it dissipates energy at a rate W. If the wire is cut into two halves and these halves are connected mutually parallel across the same supply, the energy dissipation rate will become:
[NCERT : P L-93 | Jan. 30, 2024 (II)]
(a) $1/4W$ (b) $1/2W$ (c) $2W$ (d) $4W$

199. Different combination of 3 resistors of equal resistance R are shown in the figures.
The increasing order for power dissipation is:
[NCERT : P L-93 | April 13, 2023 (I)]

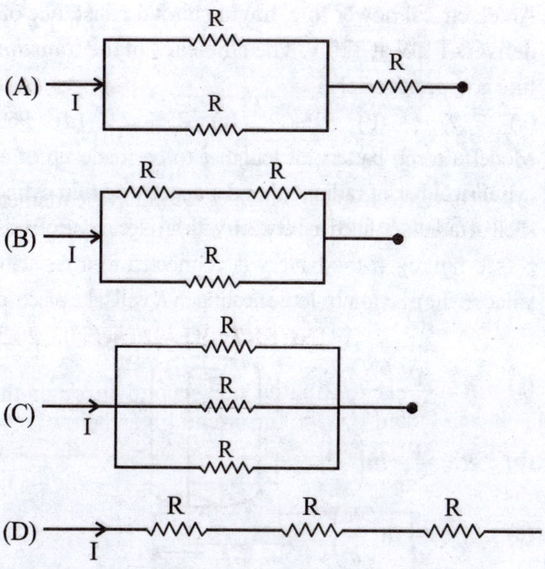

(a) $P_A < P_B < P_C < P_D$ (b) $P_C < P_D < P_A < P_B$
(c) $P_B < P_C < P_D < P_A$ (d) $P_C < P_B < P_A < P_D$

200. A potential V_0 is applied across a uniform wire of resistance R. The power dissipation is P_1. The wire is then cut into two equal halves and a potential of V_0 is applied across the length of each half. The total power dissipation across two wires is P_2. The ratio $P_2 : P_1$ is $\sqrt{x} : 1$. The value of x is ______.
[NCERT : P L-93 | April 13, 2023 (I)]

201. Two identical heater filaments are connected first in parallel and then in series. At the same applied voltage, the ratio of heat produced in same time for parallel to series will be:
[NCERT : P L-92 | April 11, 2023 (I)]
(a) 4 : 1 (b) 2 : 1 (c) 1 : 2 (d) 1 : 4

202. The H amount of thermal energy is developed by a resistor in 10 s when a current of 4A is passed through it. If the current is increased to 16A, the thermal energy developed by the resistor in 10 s will be:
[NCERT : P L-92 | Jan. 31, 2023 (II)]
(a) H (b) 16H (c) $\frac{H}{4}$ (d) 4H

203. Ratio of thermal energy released in two resistor R and 3R connected in parallel in an electric circuit is :
[NCERT : P L-92 | Jan. 29, 2023 (I)]
(a) 3 : 1 (b) 1 : 1 (c) 1 : 3 (d) 1 : 27

204. Given below are two statements: One is labelled as **Assertion A** and the other is labelled as **Reason R**.
Assertion A: Alloys such as constantan and manganin are used in making standard resistance coils.
Reason R: Constantan and manganin have very small value of temperature coefficient of resistance.
In the light of the above statements, choose the correct answer from the options given below.
[NCERT : P L-90 | July 29, 2022 (II)]
(a) Both A and R are true and R is the correct explanation of A.
(b) Both A and R are true but R is NOT the correct explanation of A.
(c) A is true but R is false.
(d) A is false but R is true.

205. An electrical bulb rated 220 V, 100 W, is connected in series with another bulb rated 220 V, 60 W. If the voltage across combination is 220 V, the power consumed by the 100 W bulb will be about ______ W.
[NCERT : P L-93 | July 28, 2022 (II)]

206. Given below are two statements :
Statement I : A unifrom wire of resistance 80 Ω is cut into four equal parts. These parts are now connected in parallel. The equivalent resistance of the combination will be 5 Ω.
Statement II : Two resistances 2R and 3R are connected in parallel in a electric circuit. The value of thermal energy developed in 3R and 2R will be in the ratio 3 : 2.

In the light of the above statments, choose the most appropriate answer from the option given below

[NCERT : P L-93 | July 28, 2022 (II)]

(a) Both statement I and statement II are correct
(b) Both statement I and statement II are incorrect
(c) Statement I is correct but statement II is incorrect
(d) Statement I is incorrect but statement II is correct

207. Two coils require 20 minutes and 60 minutes respectively to produce same amount of heat energy when connected separately to the same source. If they are connected in parallel arrangement to the same source; the time required to produce same amount of heat by the combination of coils, will be ______ min.

[NCERT : P L-92 | June 29, 2022 (I)]

208. Water fall from a 40 m high dam at the rate of 9×10^4 kg per hour. Fifty percentage of gravitational potential energy can be converted into electrical energy. Using this hydroelectric energy number of 100W lamps, that can be lit, is: (Take $g = 10\ ms^{-2}$)

[NCERT : P L-92 | June 28, 2022 (II)]

(a) 25 (b) 50 (c) 100 (d) 18

209. A 220 V, 50 Hz AC source is connected to a 25 V, 5 W lamp and an additional resistance R in series (as shown in figure) to run the lamp at its peak brightness, then the value of R (in ohm) will be ______.

[NCERT : P L-93 | June 27, 2022 (I)]

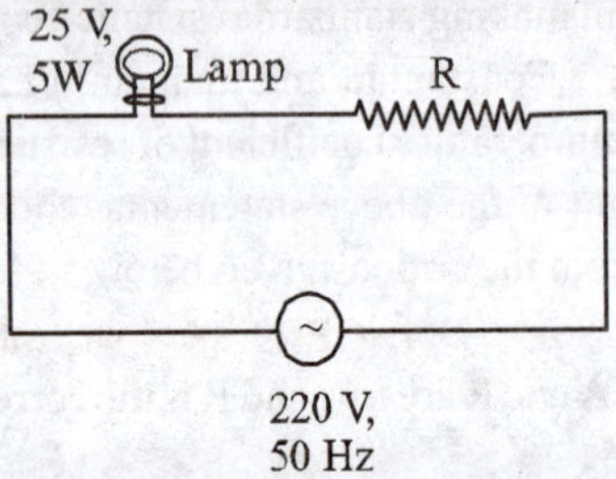

210. A uniform heating wire of resistance 36 Ω is connected across a potential difference of 240 V. The wire is then cut into half and potential difference of 240 V is applied across each half separately. The ratio of power dissipation in first case to the total power dissipation in the second case would be 1 : x, where x is ______.

[NCERT : P L-93 | NA, Sep. 1, 2021 (II)]

211. Due to cold weather a 1 m water pipe of cross–sectional area 1 cm² is filled with ice at –10°C. Resistive heating is used to melt the ice. Current of 0.5 A is passed through 4 kΩ resistance. Assuming that all the heat produced is used for melting, what is the minimum time required? (Given latent heat of fusion for water/ice = $3.33 \times 10^5\ J\ kg^{-1}$, specific heat of ice = $2 \times 10^3\ J\ kg^{-1}$ and density of ice = $10^3\ kg/m^3$ **[NCERT : P L-92 | Sep. 1, 2021 (II)]**

(a) 0.353 s (b) 35.3 s (c) 3.53 s (d) 70.6 s

212. A resistor dissipates 192 J of energy in 1 s when a current of 4A is passed through it. Now, when the current is doubled, the amount of thermal energy dissipated in 5 s in ______ J. **[NCERT : P L-93 | NA, Aug. 31, 2021 (II)]**

213. In the given figure, the emf of the cell is 2.2 V and if internal resistance is 0.6Ω. Calculate the power dissipated in the whole circuit : **[NCERT : P L-93 | Aug. 26, 2021 (I)]**

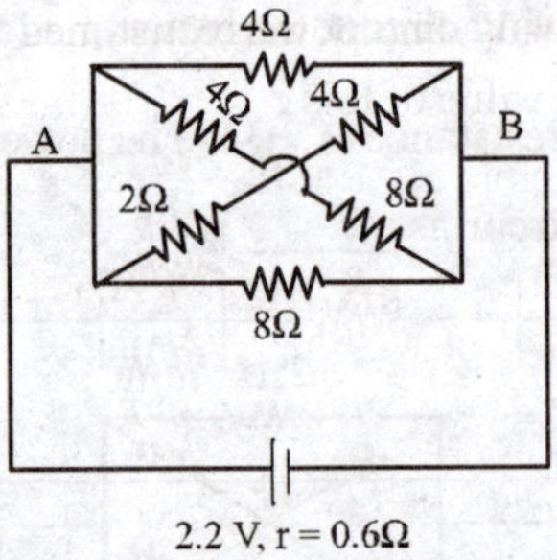

(a) 1.32 W (b) 0.65 W (c) 2.2 W (d) 4.4 W

214. An electric appliance supplies 6000 J/min heat to the system. If the system delivers a power of 90W. How long it would take to increase the internal energy by 2.5×10^3 J?

[NCERT : P L-92 | Aug. 26, 2021 (I)]

(a) 2.5×10^2 s (b) 4.1×10^1 s
(c) 2.4×10^3 s (d) 2.5×10^1 s

215. An electric bulb of 500 watt at 100 volt is used in a circuit having a 200 V supply. Calculate the resistance R to be connected in series with the bulb so that the power delivered by the bulb is 500 W.

[NCERT : P L-92 | Aug. 26, 2021 (II)]

(a) 20 Ω (b) 30 Ω (c) 5 Ω (d) 10 Ω

216. The energy dissipated by a resistor is 10 mJ in 1 s when an electric current of 2 mA flows through it. The resistance is ______Ω. (Round off to the Nearest Integer).

[NCERT : P L-93 | March 16, 2021 (II)]

217. An electrical power line, having a total resistance of 2 Ω, delivers 1 kW at 220 V. The efficiency of the transmission line is approximately : **[Sep. 05, 2020 (I)]**

(a) 72% (b) 91% (c) 85% (d) 96%

218. Model a torch battery of length l to be made up of a thin cylindrical bar of radius 'a' and a concentric thin cylindrical shell of radius 'b' filled in between with an electrolyte of resistivity ρ (see figure). If the battery is connected to a resistance of value R, the maximum Joule heating in R will take place for :

[NCERT : P L-92 | Sep. 03, 2020 (I)]

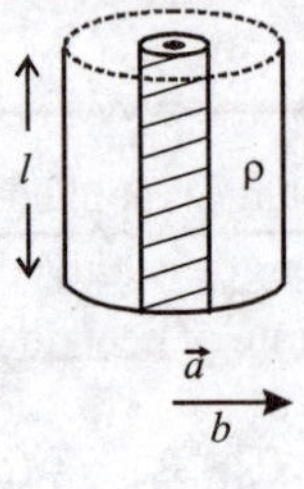

(a) $R = \frac{\rho}{2\pi l}\left(\frac{b}{a}\right)$

(b) $R = \frac{\rho}{2\pi l} \ln\left(\frac{b}{a}\right)$

(c) $R = \frac{\rho}{\pi l} \ln\left(\frac{b}{a}\right)$

(d) $R = \frac{2\rho}{\pi l} \ln\left(\frac{b}{a}\right)$

219. In a building there are 15 bulbs of 45 W, 15 bulbs of 100 W, 15 small fans of 10 W and 2 heaters of 1 kW. The voltage of electric main is 220 V. The minimum fuse capacity (rated value) of the building will be: **[NCERT : P L-92 | 7 Jan. 2020 II]**

(a) 10 A (b) 25 A (c) 15 A (d) 20 A

220. The resistive network shown below is connected to a D.C. source of 16 V. The power consumed by the network is 4 watt. The value of R is : **[NCERT : P L-93 | 12 Apr. 2019 I]**

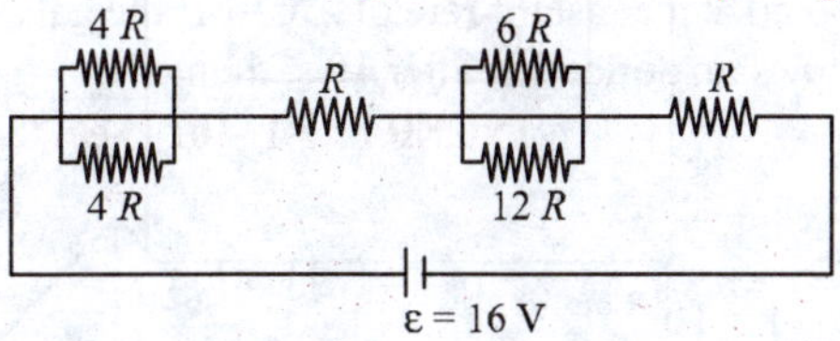

(a) 6Ω (b) 8Ω (c) 1Ω (d) 16Ω

221. One kg of water, at 20°C, is heated in an electric kettle whose heating element has a mean (temperature averaged) resistance of 20 Ω. The rms voltage in the mains is 200 V. Ignoring heat loss from the kettle, time taken for water to evaporate fully, is close to :

[Specific heat of water = 4200 J/(kg°C), Latent heat of water = 2260 kJ/kg] **[NCERT : P L-93 | 12 Apr. 2019 II]**

(a) 16 minutes (b) 22 minutes
(c) 3 minutes (d) 3 minutes

222. Two electric bulbs, rated at (25 W, 220 V) and (100 W, 220 V), are connected in series across a 220 V voltage source. If the 25 W and 100 W bulbs draw powers P_1 and P_2 respectively, then: **[NCERT : P L-93 | 12 Jan. 2019 I]**

(a) P_1=16 W, P_2=4 W (b) P_1=16 W, P_2=9 W
(c) P_1=9 W, P_2=16 W (d) P_1=4 W, P_2=16 W

223. Two equal resistances when connected in series to a battery, consume electric power of 60 W. If these resistance are now connected in parallel combination to the same battery, the electric power consumed will be : **[NCERT : P L-93 | 11 Jan. 2019 I]**

(a) 60 W (b) 240 W (c) 120 W (d) 30 W

224. A current of 2 mA was passed through an unknown resistor which dissipated a power of 4.4 W. Dissipated power when an ideal power supply of 11 V is connected across it is: **[NCERT : P L-93 | 10 Jan. 2019 II]**

(a) 11×10^{-5} W (b) 11×10^{-3} W
(c) 11×10^{-4} W (d) 11×10^{5} W

225. A constant voltage is applied between two ends of a metallic wire. If the length is halved and the radius of the wire is doubled, the rate of heat developed in the wire will be: **[NCERT : P L-93 | Online April 15, 2018]**

(a) Increased 8 times (b) Doubled
(c) Halved (d) Unchanged

226. In the circuit shown, the resistance r is a variable resistance. If for r = fR, the heat generation in r is maximum then the value of f is : **[NCERT : P L-93 | Online April 9, 2016]**

(a) $\frac{1}{2}$

(b) 1

(c) $\frac{1}{4}$

(d) $\frac{3}{4}$

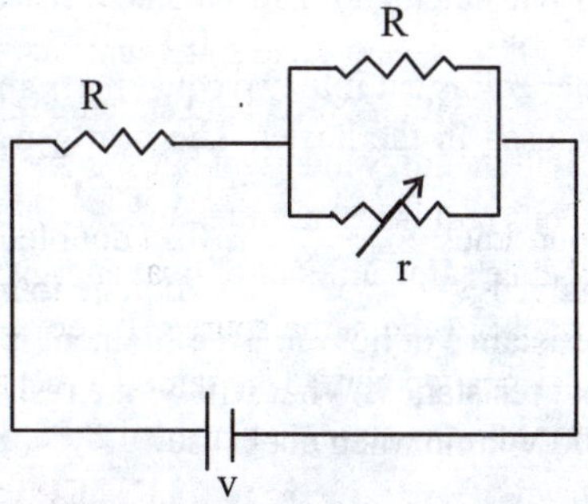

227. In a large building, there are 15 bulbs of 40 W, 5 bulbs of 100 W, 5 fans of 80 W and 1 heater of 1 kW. The voltage of electric mains is 220 V. The minimum capacity of the main fuse of the building will be: **[NCERT : P L-93 | 2014]**

(a) 8 A (b) 10 A (c) 12 A (d) 14 A

228. Four bulbs B_1, B_2, B_3 and B_4 of 100 W each are connected to 220 V main as shown in the figure. **[NCERT : P L-93 | Online April 19, 2014]**

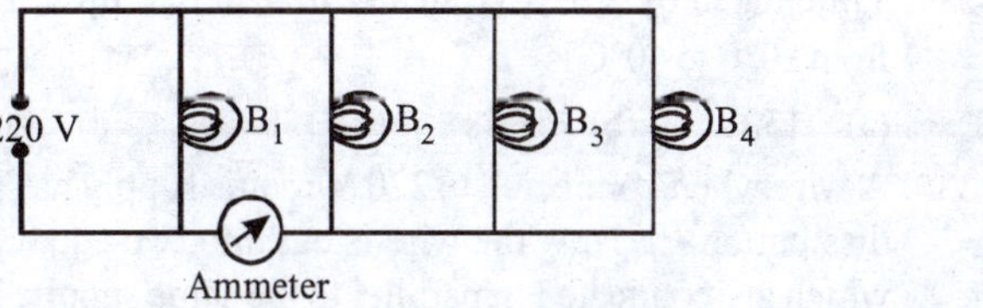

The reading in an ideal ammeter will be:

(a) 0.45 A (b) 0.90 A (c) 1.35 A (d) 1.80 A

229. The supply voltage to room is 120V. The resistance of the lead wires is 6Ω. A 60 W bulb is already switched on. What is the decrease of voltage across the bulb, when a 240 W heater is switched on in parallel to the bulb? **[NCERT : P L-93 | 2013]**

(a) zero (b) 2.9 volt
(c) 13.3 volt (d) 10.04 volt

230. Two electric bulbs rated 25W – 220 V and 100W – 220V are connected in series to a 440 V supply. Which of the bulbs will fuse? **[NCERT : P L-93 | 2012]**

(a) Both (b) 100 W (c) 25 W (d) Neither

231. Three resistors of 4 Ω, 6 Ω and 12 Ω are connected in parallel and the combination is connected in series with a 1.5 V battery of 1 Ω internal resistance. The rate of Joule heating in the 4 Ω resistor is **[NCERT : P L-93 | Online May 12, 2012]**

(a) 0.55 W (b) 0.33 W (c) 0.25 W (d) 0.86 W

232. The resistance of a bulb filmanet is 100Ω at a temperature of 100°C. If its temperature coefficient of resistance be 0.005 per °C, its resistance will become 200 Ω at a temperature of **[NCERT : P L-91 | 2006]**

(a) 300°C (b) 400°C (c) 500°C (d) 200°C

233. An electric bulb is rated 220 volt - 100 watt. The power consumed by it when operated on 110 volt will be **[NCERT : P L-93 | 2006]**

(a) 75 watt (b) 40 watt
(c) 25 watt (d) 50 watt

234. A heater coil is cut into two equal parts and only one part is now used in the heater. The heat generated will now be **[NCERT : P L-92 | 2005]**

(a) four times (b) doubled
(c) halved (d) one fourth

235. The resistance of hot tungsten filament is about 10 times the cold resistance. What will be the resistance of 100 W and 200 V lamp when not in use ? **[NCERT : P L-93 | 2005]**

(a) 20Ω (b) 40Ω (c) 200Ω (d) 400Ω

236. The thermistors are usually made of **[NCERT : P L-90 | 2004]**

(a) metal oxides with high temperature coefficient of resistivity
(b) metals with high temperature coefficient of resistivity
(c) metals with low temperature coefficient of resistivity
(d) semiconducting materials having low temperature coefficient of resistivity

237. Time taken by a 836 W heater to heat one litre of water from 10°C to 40°C is **[NCERT : P L-92 | 2004]**

(a) 150 s (b) 100 s (c) 50 s (d) 200 s

238. A wire when connected to 220 V mains supply has power dissipation P_1. Now the wire is cut into two equal pieces which are connected in parallel to the same supply. Power dissipation in this case is P_2. Then $P_2 : P_1$ is **[NCERT : P L-93 | 2002]**

(a) 1 (b) 4 (c) 2 (d) 3

239. If in the circuit, power dissipation is 150 W, then R is **[NCERT : P L-93 | 2002]**

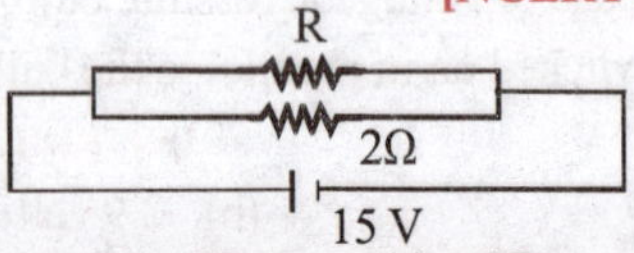

(a) 2Ω (b) 6Ω (c) 5Ω (d) 4Ω

Topic 5 - Wheatstone Bridge and Different Measuring Instruments

240. The value of unknown resistance (x) for which the potential difference between B and D will be zero in the arrangement shown, is : **[NCERT : P L-101 | April 6, 2024 (I)]**

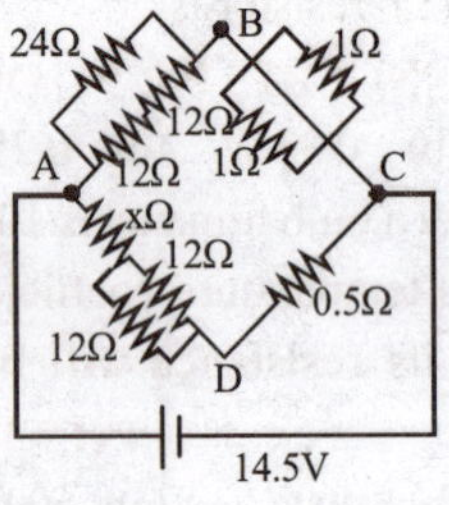

(a) 3 Ω (b) 9 Ω (c) 6 Ω (d) 42 Ω

241. In a metre - bridge when a resistance in the left gap is 2Ω and unknown resistance in the right gap, the balance length is found to be 40 cm. On shunting the unknown resistance with 2Ω, the balance length changes by: **[Feb. 1, 2024 (II)]**

(a) 62.5 cm (b) 22.5 cm (c) 20 cm (d) 65 cm

242. To measure the temperature coefficient of resistivity α of a semiconductor, an electrical arrangement shown in the figure is prepared. The arm BC is made up of the semiconductor. The experiment is being conducted at 25°C and resistance of the semiconductor arm is 3mΩ. Arm BC is cooled at a constant rate of 2°C/s. If the galvanometer G shows no deflection after 10 s, then α is: **[NCERT : P L-101 | Feb. 1, 2024 (II)]**

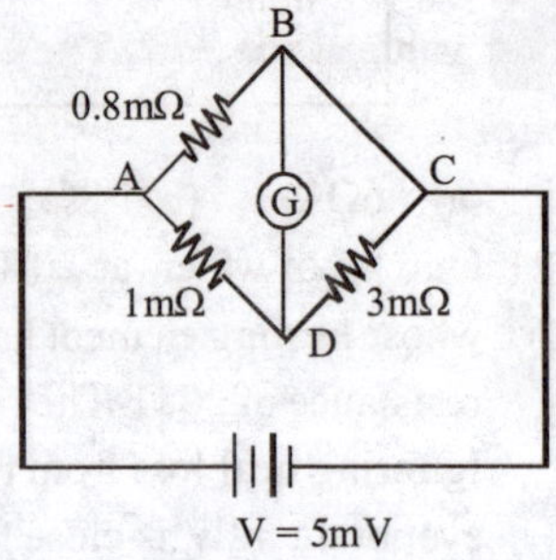

(a) $-1 \times 10^{-2}\,°C^{-1}$

(b) $-2 \times 10^{-2}\,°C^{-1}$

(c) $-2.5 \times 10^{-2}\,°C^{-1}$

(d) $-1.5 \times 10^{-2}\,°C^{-1}$

243. The resistance per centimetreof a meter bridge wire is r, with XΩ resistance in left gap. Balancing length from left end is at 40 cm with 25 Ω resistance in right gap. Now the wire is replaced by another wire of 2r resistance per centimeter. The new balancing length for same settings will be at **[Jan. 31, 2024 (II)]**

(a) 20 cm (b) 10 cm (c) 80 cm (d) 40 cm

244. Two resistance of 100 Ω and 200 Ω are connected in series with a battery of 4 V and negligible internal resistance. A voltmeter is used to measure voltage across 100Ω resistance, which gives reading as 1 V. The resistance of voltmeter must be _______ Ω. **[NCERT : P L-97 | Jan. 30, 2024 (II)]**

245. A wire of length 10 cm and radius $\sqrt{7} \times 10^{-4}$m connected across the right gap of a meter bridge. When a resistance of 4.5 Ω is connected on the left gap by using a resistance box, the balance length is found to be at 60 cm from the left end. If the resistivity of the wire is R × 10^{-7}Ωm, then value of R is : **[NCERT : P L-101 | Jan. 27, 2024 (I)]**

(a) 63 (b) 70 (c) 66 (d) 35

246. Wheatstone bridge principle is used to measure the specific resistance (S_1) of given wire, having length L, radius r. If X is the resistance of wire, then specific resistance is :

$S_1 = X\left(\frac{\pi r^2}{L}\right)$. If the length of the wire gets doubled then the value of specific resistance will be : **[NCERT : P L-84 | Jan. 27, 2024 (II)]**

(a) $\frac{S_1}{4}$ (b) $2S_1$ (c) $\frac{S_1}{2}$ (d) S_1

247. Three voltmeters, all having different internal resistances are joined as shown in figure. When some potential difference is applied across A and B, their readings are V_1, V_2 and V_3. Choose the correct option.

[Jan. 27, 2024 (II)]

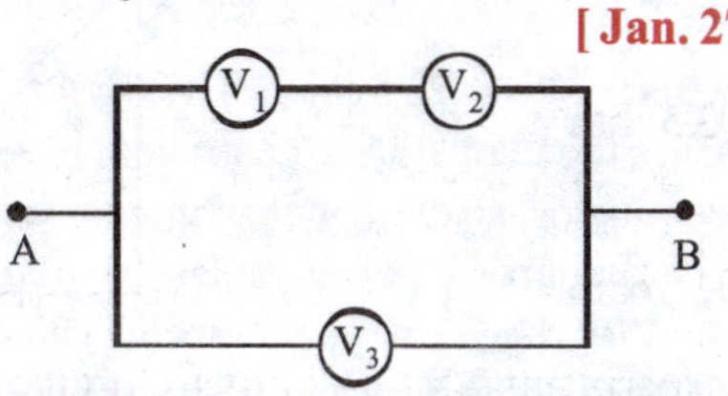

(a) $V_1 = V_2$ (b) $V_1 \neq V_3 - V_2$
(c) $V_1 + V_2 > V_3$ (d) $V_1 + V_2 = V_3$

248. If the potential difference between B and D is zero, the value of x is $\frac{1}{n}\Omega$. The value of n is ....

[NCERT : P L-101 | Jan. 30, 2023 (II)]

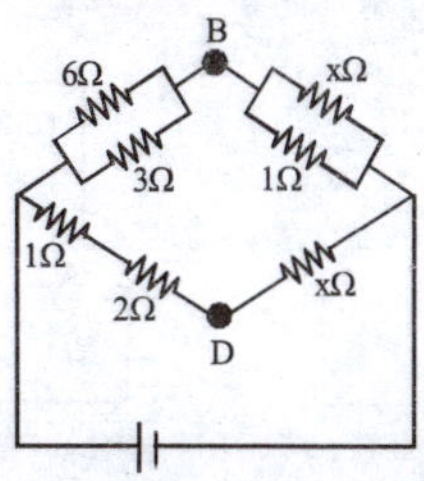

249. When two resistance R_1 and R_2 connected in series and introduced into the left gap of a meter bridge and a resistance of 10 Ω is introduced into the right gap, a null point is found at 60 cm from left side. When R_1 and R_2 are connected in parallel and introduced into the left gap, a resistance of 3 Ω is introduced into the right-gap to get null point at 40 cm from left end. The product of R_1 R_2 is ______ Ω^2

[Jan. 29, 2023 (II)]

250. Resistances are connected in a meter bridge circuit as shown in the figure. The balancing length l_1 is 40 cm. Now an unknown resistance x is connected in series with P and new balancing length is found to be 80 cm measured from the same end. Then the value of x will be ______ Ω.

[July 26, 2022 (I)]

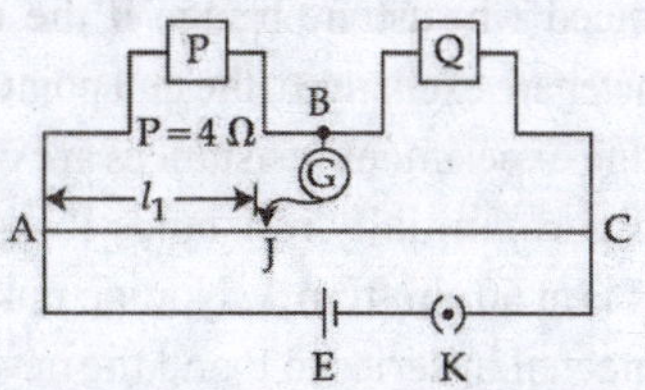

251. A battery of 6 V is connected to the circuit as shown below. The current I drawn from the battery is :

[NCERT : P L-101 | July 26, 2022 (II)]

(a) 1A
(b) 2A
(c) $\frac{6}{11}$A
(d) $\frac{4}{3}$A

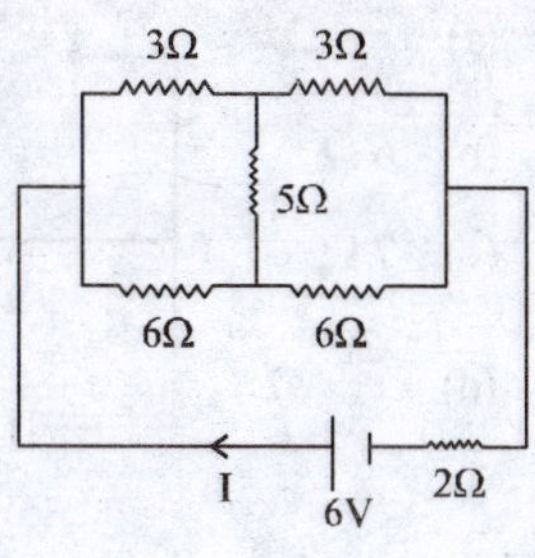

252. A meter bridge setup is shown in the figure. It is used to determine an unknown resistance R using a given resistor of 15 Ω. The galvanometer (G) shows null deflection when tapping key is at 43 cm mark from end A. If the end correction for end A is 2 cm then the determined value of R will be ______ Ω.

[NCERT : P L-101 | June 28, 2022 (I)]

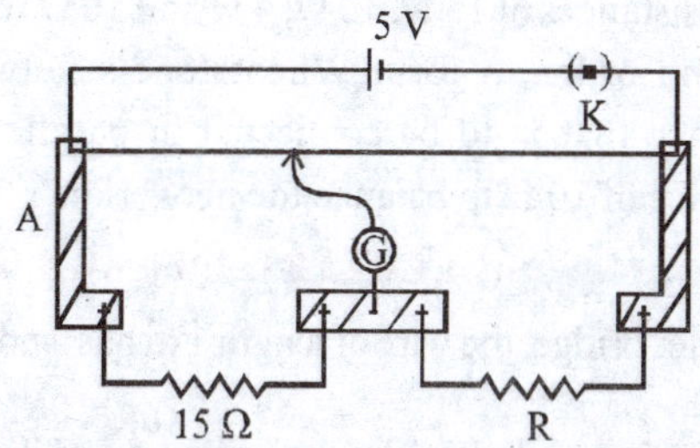

253. Current measured by the ammeter (A) in the reported circuit when no current flows through 10 Ω resistance will be ______ A.

[NCERT : P L-101 | June 28, 2022 (I)]

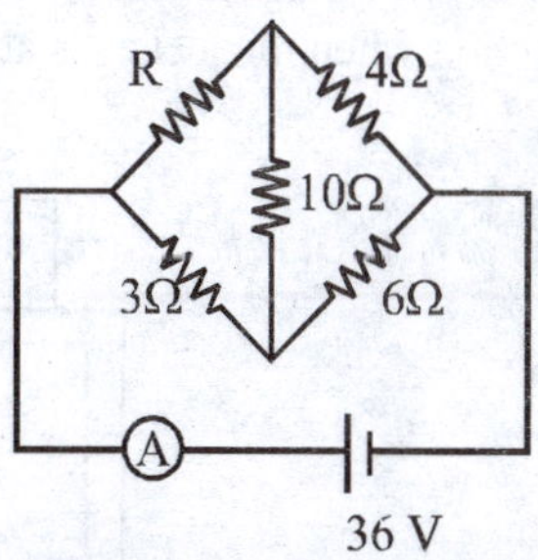

254. Consider a 72 cm long wire AB as shown in the figure. The galvanometer jockey is placed at P on AB at a distance x cm from A. The galvanometer shows zero deflection.

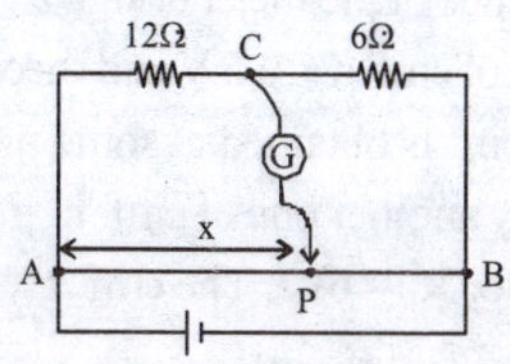

The value of x, to the nearest integer, is ______.

[NCERT : P L-101 | March 18, 2021 (II)]

255. The four arms of a Wheatstone bridge have resistances as shown in the figure. A galvanometer of 15Ω resistance is connected across BD. Calculate the current through the galvanometer when a potential difference of 10 V is maintained across AC.

[NCERT : P L-101 | March 17, 2021 (II)]

(a) 2.44 μA
(b) 4.87 μA
(c) 2.44 mA
(d) 4.87 mA

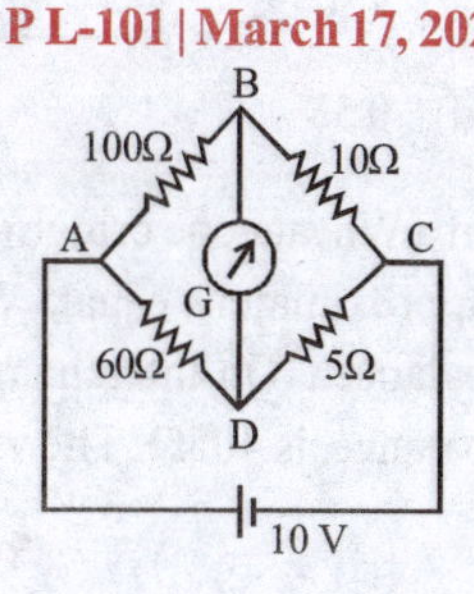

256. Two resistors 400Ω and 800Ω are connected in series across a 6 V battery. The potential difference measured by a voltmeter of 10 kΩ across 400Ω resistor is close to:

[NCERT : P L-83 | Sep. 03, 2020 (II)]

(a) 2V (b) 1.8V
(c) 2.05V (d) 1.95V

257. Four resistances of 15 Ω, 12 Ω, 4 Ω and 10 Ω respectively in cyclic order to form Wheatstone's network. The resistance that is to be connected in parallel with the resistance of 10 Ω to balance the network is ______ Ω.

[NCERT : P L-101 | NA. 8 Jan. 2020 I]

258. In a meter bridge, the wire of length 1 m has a non-uniform cross-section such that, the variation $\frac{dR}{dl}$ of its resistance R with length l is $\frac{dR}{dl} \propto \frac{1}{\sqrt{l}}$. Two equal resistances are connected as shown in the figure. The galvanometer has zero deflection when the jockey is at point P. What is the length AP? **[12 Jan. 2019 I]**

(a) 0.2m
(b) 0.3m
(c) 0.25m
(d) 0.35m

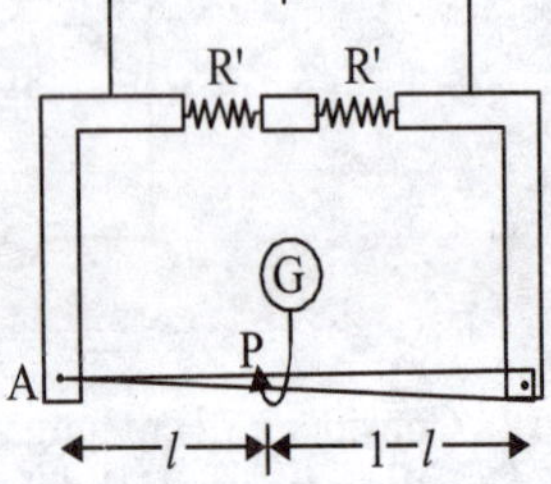

259. The resistance of the meter bridge AB in given figure is 4Ω. With a cell of emf ε = 0.5 V and rheostat resistance $R_h = 2\Omega$ the null point is obtained at some point J. When the cell is replaced by another one of emf $\varepsilon = \varepsilon_2$ the same null point J is found for $R_h = 6\Omega$. The emf ε_2 is: **[11 Jan. 2019 I]**

(a) 0.4 V
(b) 0.3 V
(c) 0.6 V
(d) 0.5 V

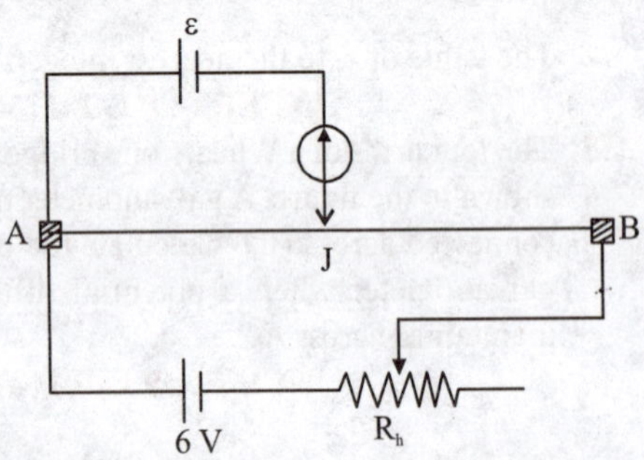

260. In a Wheatstone bridge (see fig.), Resistances P and Q are approximately equal. When R = 400 Ω, the bridge is balanced. On interchanging P and Q, the value of R, for balance, is 405Ω. The value of X is close to :

[NCERT : P L-101 | 11 Jan. 2019 I]

(a) 401.5 ohm
(b) 404.5 ohm
(c) 403.5 ohm
(d) 402.5 ohm

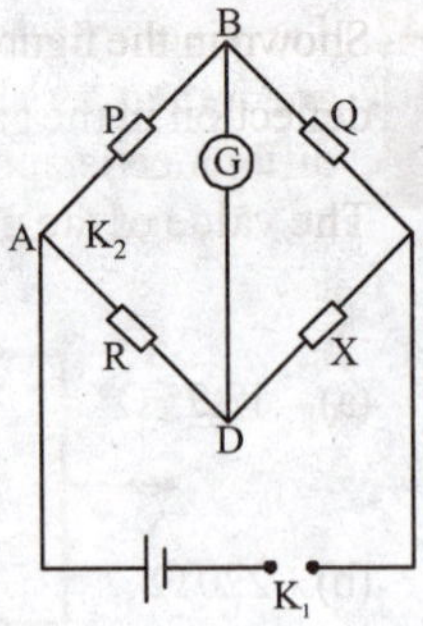

261. In the experimental set up of metre bridge shown in the figure, the null point is obtained at a distance of 40 cm from A. If a 10 Ω resistor is connected in series with R_1, the null point shifts by 10 cm. The resistance that should be connected in parallel with $(R_1 + 10)$ Ω such that the null point shifts back to its initial position is : **[11 Jan. 2019 II]**

(a) 20 Ω
(b) 40 Ω
(c) 60 Ω
(d) 30 Ω

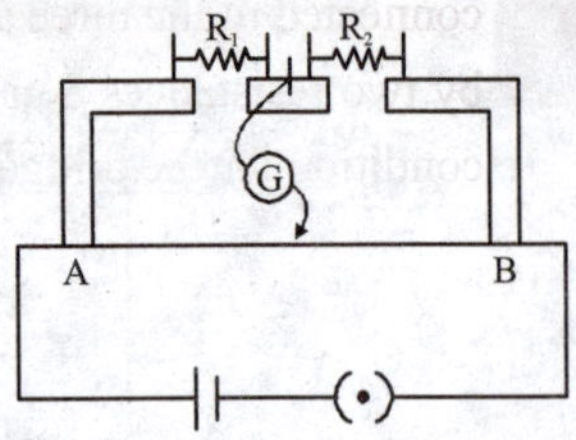

262. On interchanging the resistances, the balance point of a meter bridge shifts to the left by 10 cm. The resistance of their series combination is 1kΩ. How much was the resistance on the left slot before interchanging the resistances? **[2018]**

(a) 990 Ω (b) 505 Ω
(c) 550 Ω (d) 910 Ω

263. Which of the following statements is false ?

[NCERT : P L-101 | 2017]

(a) A rheostat can be used as a potential divider
(b) Kirchhoff's second law represents energy conservation
(c) Wheatstone bridge is the most sensitive when all the four resistances are of the same order of magnitude
(d) In a balanced wheatstone bridge if the cell and the galvanometer are exchanged, the null point is disturbed.

264. In a meter bridge experiment resistances are connected as shown in the figure. Initially resistance P = 4 Ω and the neutral point N is at 60 cm from A. Now an unknown resistance R is connected in series to P and the new position of the neutral point is at 80 cm from A. The value of unknown resistance R is : **[Online April 9, 2017]**

(a) $\frac{33}{5}$ Ω
(b) 6 Ω
(c) 7 Ω
(d) $\frac{20}{3}$ Ω

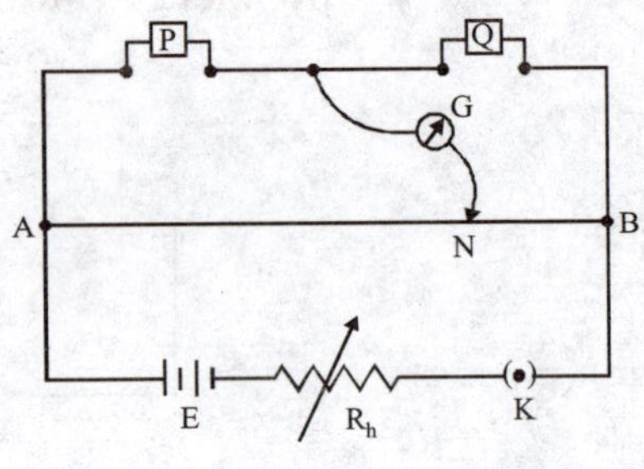

265. Shown in the figure below is a meter-bridge set up with null deflection in the galvanometer. **[2008]**

The value of the unknown resistor R is

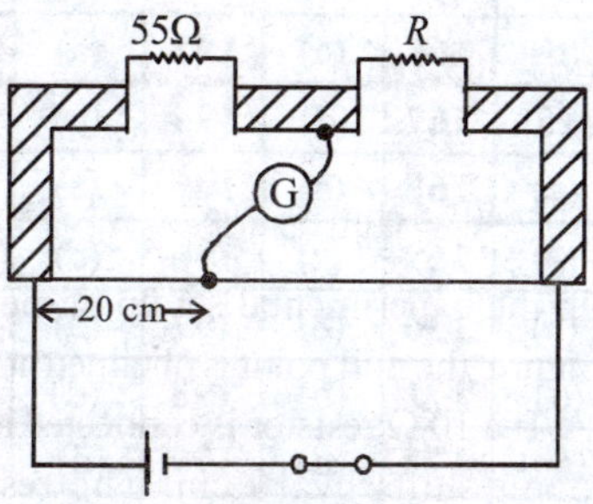

(a) 13.75 Ω

(b) 220 Ω

(c) 110 Ω

(d) 55 Ω

266. In a Wheatstone's bridge, three resistances P, Q and R connected in the three arms and the fourth arm is formed by two resistances S_1 and S_2 connected in parallel. The condition for the bridge to be balanced will be

[NCERT : P L-101 | 2006]

(a) $\frac{P}{Q} = \frac{2R}{S_1 + S_2}$ (b) $\frac{P}{Q} = \frac{R(S_1 + S_2)}{S_1 S_2}$

(c) $\frac{P}{Q} = \frac{R(S_1 + S_2)}{2S_1 S_2}$ (d) $\frac{P}{Q} = \frac{R}{S_1 + S_2}$

267. In a meter bridge experiment null point is obtained at 20 cm. from one end of the wire when resistance X is balanced against another resistance Y. If $X < Y$, then where will be the new position of the null point from the same end, if one decides to balance a resistance of $4X$ against Y **[2004]**

(a) 40 cm (b) 80 cm

(c) 50 cm (d) 70 cm

268. An ammeter reads upto 1 ampere. Its internal resistance is 0.81ohm. To increase the range to 10 A the value of the required shunt is **[2003]**

(a) 0.03 Ω (b) 0.3 Ω

(c) 0.9 Ω (d) 0.09 Ω

ANSWER KEYS																			
1	(d)	**28**	(b)	**55**	(a)	**82**	(b)	**109**	(3)	**136**	(b)	**163**	(c)	**190**	(c)	**217**	(d)	**244**	(200)
2	(d)	**29**	(a)	**56**	(b)	**83**	(36)	**110**	(d)	**137**	(4)	**164**	(a)	**191**	(c)	**218**	(b)	**245**	(c)
3	(b)	**30**	(b)	**57**	(c)	**84**	(30)	**111**	(20)	**138**	(2)	**165**	(15)	**192**	(5)	**219**	(d)	**246**	(d)
4	(a)	**31**	(144)	**58**	(c)	**85**	(a)	**112**	(9)	**139**	(d)	**166**	(c)	**193**	(a)	**220**	(b)	**247**	(d)
5	(1027)	**32**	(a)	**59**	(a)	**86**	(a)	**113**	(b)	**140**	(25)	**167**	(d)	**194**	(b)	**221**	(b)	**248**	(2)
6	(748)	**33**	(48)	**60**	(a)	**87**	(b)	**114**	(d)	**141**	(a)	**168**	(c)	**195**	(a)	**222**	(a)	**249**	(30)
7	(16)	**34**	(12)	**61**	(c)	**88**	(100)	**115**	(10)	**142**	(4)	**169**	(2)	**196**	(c)	**223**	(b)	**250**	(20)
8	(b)	**35**	(c)	**62**	(d)	**89**	(a)	**116**	(70)	**143**	(c)	**170**	(8)	**197**	(3)	**224**	(a)	**251**	(a)
9	(32)	**36**	(300)	**63**	(d)	**90**	(20)	**117**	(4)	**144**	(d)	**171**	(b)	**198**	(d)	**225**	(a)	**252**	(19)
10	(22)	**37**	(a)	**64**	(b)	**91**	(c)	**118**	(3)	**145**	(60)	**172**	(d)	**199**	(d)	**226**	(a)	**253**	(10)
11	(c)	**38**	(1)	**65**	(d)	**92**	(c)	**119**	(c)	**146**	(6)	**173**	(d)	**200**	(16)	**227**	(c)	**254**	(48)
12	(d)	**39**	(d)	**66**	(a)	**93**	(a)	**120**	(b)	**147**	(400)	**174**	(b)	**201**	(a)	**228**	(c)	**255**	(d)
13	(b)	**40**	(4)	**67**	(5)	**94**	(10)	**121**	(30)	**148**	(3)	**175**	(b)	**202**	(b)	**229**	(d)	**256**	(d)
14	(a)	**41**	(c)	**68**	(c)	**95**	(d)	**122**	(c)	**149**	(75)	**176**	(c)	**203**	(a)	**230**	(c)	**257**	(10)
15	(15)	**42**	(d)	**69**	(d)	**96**	(a)	**123**	(b)	**150**	(1)	**177**	(b)	**204**	(a)	**231**	(c)	**258**	(c)
16	(b)	**43**	(b)	**70**	(2)	**97**	(2)	**124**	(c)	**151**	(5)	**178**	(b)	**205**	(14)	**232**	(b)	**259**	(b)
17	(3)	**44**	(5)	**71**	(16)	**98**	(b)	**125**	(c)	**152**	(b)	**179**	(a)	**206**	(c)	**233**	(c)	**260**	(d)
18	(25)	**45**	(300)	**72**	(1)	**99**	(a)	**126**	(a)	**153**	(a)	**180**	(c)	**207**	(15)	**234**	(b)	**261**	(c)
19	(125)	**46**	(b)	**73**	(d)	**100**	(4)	**127**	(b)	**154**	(2)	**181**	(c)	**208**	(b)	**235**	(b)	**262**	(c)
20	(25)	**47**	(b)	**74**	(b)	**101**	(d)	**128**	(a)	**155**	(5)	**182**	(c)	**209**	(975)	**236**	(a)	**263**	(d)
21	(b)	**48**	(d)	**75**	(160)	**102**	(b)	**129**	(a)	**156**	(1)	**183**	(a)	**210**	(4)	**237**	(a)	**264**	(d)
22	(a)	**49**	(b)	**76**	(d)	**103**	(10)	**130**	(a)	**157**	(1.5)	**184**	(d)	**211**	(b)	**238**	(b)	**265**	(b)
23	(d)	**50**	(a)	**77**	(5)	**104**	(8)	**131**	(b)	**158**	(1)	**185**	(a)	**212**	(3840)	**239**	(b)	**266**	(b)
24	(b)	**51**	(c)	**78**	(1)	**105**	(8)	**132**	(b)	**159**	(a)	**186**	(5)	**213**	(c)	**240**	(c)	**267**	(c)
25	(c)	**52**	(a)	**79**	(b)	**106**	(3)	**133**	(d)	**160**	(4)	**187**	(c)	**214**	(a)	**241**	(b)	**268**	(d)
26	(2)	**53**	(a)	**80**	(c)	**107**	(c)	**134**	(b)	**161**	(a)	**188**	(c)	**215**	(a)	**242**	(a)		
27	(44)	**54**	(c)	**81**	(81)	**108**	(b)	**135**	(a)	**162**	(a)	**189**	(c)	**216**	(2500)	**243**	(d)		

4 Moving Charges and Magnetism

TREND ANALYSIS
JEE MAIN (2025-2021)

	2025				2024				2023				2022				2021			
	Session II		Session I		Session II		Session I		Session II		Session I		Session II		Session I		Session II		Session I	
	MCQs	NVQs	MCQs	NVQs	MCQs	NVQs	MCQs	NVQs	MCQs	NVQs	MCQs	NVQs	MCQs	NVQs	MCQs	NVQs	MCQs	NVQs	MCQs	NVQs
Topic 1	3	1	1	1	3	1	2	2	3	2	–	1	4	–	4	2	2	–	2	–
Topic 2	2	1	4	3	1	2	1	4	1	2	9	–	4	2	5	–	5	1	1	–
Topic 3	1	1	1	1	–	2	2	2	–	1	3	2	3	–	3	1	–	–	–	–
Topic 4	1	–	2	–	2	–	5	1	4	1	3	–	1	–	2	–	2	–	–	–
Total	7	3	8	5	6	5	10	9	8	6	15	3	12	2	14	3	9	1	3	–

How to Code Qns for Better & Faster Revision

Type 1: Able to solve (No code Req)
Type 2: Solved with struggle or by seeing solution **(Use code NR means Needs Revision or orange stickers)**
Type 3: Tricky/ Could not solve on my own. Need help from Teacher/ Friend. **(Use code TQ means Tricky Question or red stickers)**

You can either hand write the code in boxes or else paste coloured stickers given at end of book.

Topic 1 - Motion of Charged Particle in Magnetic Field

1. Uniform magnetic fields of different strengths (B_1 and B_2), both normal to the plane of the paper exist as shown in the figure. *A* charged particle of mass m and charge *q*, at the interface at an instant, moves into the region 2 with velocity *v* and returns to the interface. It continues to move into region 1 and finally reaches the interface. What is the displacement of the particle during this movement along the interface ? **[NCERT : P L-112, 113 | April 7, 2025 (I)]**

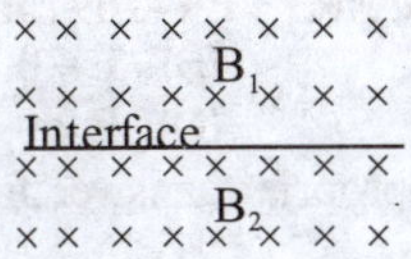

(Consider the velocity of the particle to be normal to the magnetic field and $B_2 > B_1$)

(a) $\frac{mv}{qB_1}\left(1-\frac{B_2}{B_1}\right)\times 2$ (b) $\frac{mv}{qB_1}\left(1-\frac{B_1}{B_2}\right)$

(c) $\frac{mv}{qB_1}\left(1-\frac{B_2}{B_2}\right)$ (d) $\frac{mv}{qB_1}\left(1-\frac{B_1}{B_2}\right)\times 2$

2. *A* particle of charge q, mass m and kinetic energy *E* enters in magnetic field perpendicular to its velocity and undergoes a circular arc of radius(r). Which of the following curves represents the variation of r with *E*? **[NCERT : P L-112, 113 | April 7, 2025 (I)]**

(a) 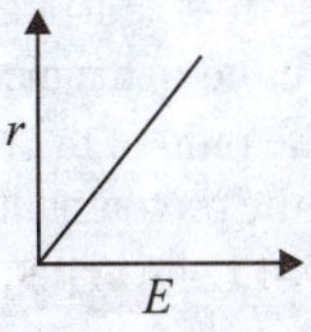

(b) 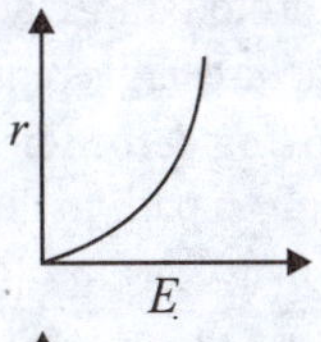

(c) 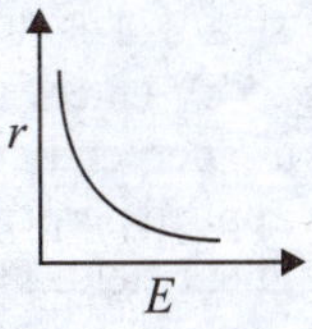

(d) 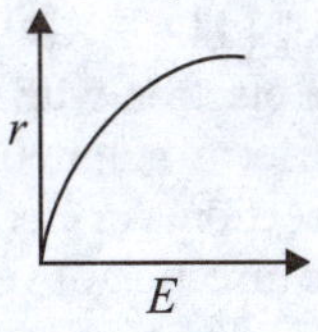

3. A particle of charge 1.6 μC and mass 16 μg is present in a strong magnetic field of 6.28 T. The particle is then fired perpendicular to magnetic field. The time required for the particle to return to original location for the first time is ________ s. ($\pi = 3.14$)

[NCERT : P L-112, 113 | April 4, 2025 (II)]

4. Given below are two statements : one is labelled as **Assertion A** and the other is labelled as **Reason R**.

Assertion A : If oxygen ion (O^{-2}) and Hydrogen ion (H^+) enter normal to the magnetic field with equal momentum, then the path of O^{-2} ion has a smaller curvature than that of H^+.

Reason R : A proton with same linear momentum as an electron will form a path of smaller radius of curvature on entering a uniform magnetic field perpendicularly.

In the light of the above statement, choose the **correct** answer from the options given below

[NCERT : P L-112, 113 | April 3, 2025 (II)]

(a) **A** is true but **R** is false.
(b) Both **A** and **R** are true but **R** is **NOT** the correct explanation of **A**.
(c) **A** is false but **R** is true.
(d) Both **A** and **R** are true and **R** is the correct explanation of **A**.

5. Given below are two statements. One is labelled as **Assertion (A)** and the other is labelled as **Reason(R)**.

Assertion (A) : A electron in a certain region of uniform magnetic field is moving with constant velocity in a straight line path.

Reason (R) : The magnetic field in that region is along the direction of velocity of the electron.

In the light of the above statements, choose the correct answer from the options given below :

[NCERT : P L-112, 113 | Jan. 24, 2025 (II)]

(a) (A) is false but (R) is true
(b) Both (A) and (R) are true and (R) is the correct explanation of (A)
(c) Both (A) and (R) are true but (R) is NOT the correct explanation of (A)
(d) (A) is true but (R) is false

6. A proton is moving undeflected in a region of crossed electric and magnetic fields at a constant speed of 2×10^5 ms^{-1}. When the electric field is switched off, the proton moves along a circular path of radius 2 cm. The magnitude of electric field is $x \times 10^4$ N/C. The value of x is _____. (Take the mass of the proton $= 1.6 \times 10^{-27}$ kg.)

[NCERT : P L-112 | Jan. 22, 2025 (II)]

7. A proton and a deutron (q= +e, m = 2.0u) having same kinetic energies enter a region of uniform magnetic field $\vec{B}$, moving perpendicular to $\vec{B}$. The ratio of the radius r_d of deutron path to the radius r_p of the proton path is :

[NCERT : P L-112 | April 9, 2024 (II)]

(a) 1 : 1 (b) $1:\sqrt{2}$ (c) $\sqrt{2}:1$ (d) 1 : 2

8. An electron with kinetic energy 5 eV enters a region of uniform magnetic field of 3 μT perpendicular to its direction. An electric field E is applied perpendicular to the direction of velocity and magnetic field. The value of E, so that electron moves along the same path, is ______ NC^{-1}.

(Given, mass of electron $= 9 \times 10^{-31}$ kg, electric charge $= 1.6 \times 10^{-19}$C) **[NCERT : P L-112 | April 8, 2024 (I)]**

9. The electrostatic force $(\vec{F}_1)$ and magnetic force $(\vec{F}_2)$ acting on a charge q moving with velocity v can be written:

[NCERT : P L-109 | April 5, 2024 (II)]

(a) $\vec{F}_1 = q\vec{V}.\vec{E}, \vec{F}_2 = q(\vec{B}.\vec{V})$ (b) $\vec{F}_1 = q\vec{B}, \vec{F}_2 = q(\vec{B}\times\vec{V})$

(c) $\vec{F}_1 = q\vec{E}, \vec{F}_2 = q(\vec{V}\times\vec{B})$ (d) $\vec{F}_1 = q\vec{E}, \vec{F}_2 = q(\vec{B}\times\vec{V})$

10. An electron is projected with uniform velocity along the axis inside a current carrying long solenoid. Then :

[NCERT : P L-112 | April 4, 2024 (I)]

(a) the electron will be accelerated along the axis.
(b) the electron will continue to move with uniform velocity along the axis of the solenoid.
(c) the electron path will be circular about the axis.
(d) the electron will experience a force at 45° to the axis and execute a helical path.

11. An electron moves through a uniform magnetic field $\vec{B} = (B_0\hat{i} + 2B_0\hat{j})T$. At a particular instant of time, the velocity of electron is $\vec{u} = (3\hat{i} + 5\hat{j})$ m/s. If the magnetic force acting on electron is $\vec{F} = 5e\hat{k}$ N, where e is the charge of electron, then the value of B_0 is _____ T.

[NCERT : P L-109 | Jan. 31, 2024 (I)]

12. Two particles X and Y having equal charges are being accelerated through the same potential difference. There after they enter normally in a region of uniform magnetic field and describes circular paths of radii R_1 and R_2 respectively. The mass ratio of X and Y is :

[NCERT : P L-112 | Jan. 29, 2024 (II)]

(a) $\left(\frac{R_1}{R_2}\right)$ (b) $\left(\frac{R_2}{R_1}\right)$ (c) $\left(\frac{R_2}{R_1}\right)^2$ (d) $\left(\frac{R_1}{R_2}\right)^2$

13. A charge of 4.0 μC is moving with a velocity of 4.0×10^6 ms^{-1} along the positive y-axis under a magnetic field $\vec{B}$ of strength $(2\hat{k})$ T. The force acting on the charge is $x\,\hat{i}$ N. The value of x is _____.

[NCERT : P L-109 | Jan. 29, 2024 (II)]

14. A proton moving with a constant velocity passes through a region of space without any change in its velocity. If $\vec{E}$ and $\vec{B}$ represent the electric and magnetic fields respectively, then the region of space may have :

[NCERT : P L-109 | Jan. 27, 2024 (I)]

(A) $E=0, B=0$ (B) $E=0, B\neq 0$
(C) $E\neq 0, B=0$ (D) $E\neq 0, B\neq 0$

Choose the most appropriate answer from the options given below :

(a) (A), (B) and (C) only (b) (A), (C) and (D) only
(c) (A), (B) and (D) only (d) (B), (C) and (D) only

15. An electron is moving along the positive x-axis. If the uniform magnetic field is applied parallel to the negative z-axis. Then

A. The electron will experience magnetic force along positive y-axis
B. The electron will experience magnetic force along negative y-axis
C. The electron will not experience any force in magnetic field
D. The electron will continue to move along the positive x-axis
E. The electron will move along circular path in magnetic field

Choose the correct answer from the options given below: **[NCERT : P L-109 | April 13, 2023 (II)]**

(a) B and E only (b) A and E only
(c) C and D only (d) B and D only

16. An electron is allowed to move with constant velocity along the axis of current carrying straight solenoid. **[NCERT : P L-109 | April 11, 2023 (II)]**

A. The electron will experience magnetic force along the axis of the solenoid.
B. The electron will not experience magnetic force.
C. The electron will continue to move along the axis of the solenoid.
D. The electron will be accelerated along the axis of the solenoid.
E. The electron will follow parabolic path–inside the solenoid.

Choose the correct answer from the options given below:

(a) B, C and D only (b) B and C only
(c) A and D only (d) B and E only

17. A charge particle moving in magnetic field B, has the components of velocity along B as well as perpendicular to B. The path of the charge particle will be **[NCERT : P L-112 | April 8, 2023 (I)]**

(a) helical path with the axis perpendicular to the direction of magnetic field B
(b) straight along the direction of magnetic field B
(c) helical path with the axis along magnetic field B
(d) circular path

18. A proton with a kinetic energy of 2.0 eV moves into a region of uniform magnetic field of magnitude $\frac{\pi}{2}\times10^{-3}$ T. The angle between the direction of magnetic field and velocity of proton is 60°. The pitch of the helical path taken by the proton is_______cm. (Take, mass of proton = 1.6×10^{-27} kg and Charge on proton = 1.6×10^{-19} C). **[NCERT : P L-113 | April 6, 2023 (II)]**

19. A charge particle of 2 μC accelerated by a potential difference of 100V enters a region of uniform magnetic field of magnitude 4 mT at right angle to the direction of field. The charge particle completes semicircle of radius 3 cm inside magnetic field. The mass of the charge particle is_____ $\times10^{-18}$ kg. **[NCERT : P L-112 | Feb. 1, 2023 (I)]**

20. A charge particle is moving in a uniform magnetic field $(2\hat{i}+3\hat{j})$ T. If it has an acceleration of $(\alpha\hat{i}-4\hat{j})$ m/s², then the value of α will be : **[NCERT : P L-109 | July 26, 2022 (I)]**

(a) 3 (b) 6 (c) 12 (d) 2

21. A velocity selector consists of electric field $\vec{E}=E\hat{k}$ and magnetic field $\vec{B}=B\hat{j}$ with B = 12 mT. The value E required for an electron of energy 728 eV moving along the positive x-axis to pass undeflected is : (Given, mass of electron = 9.1×10^{-31}kg) **[NCERT : P L-109 | July 26, 2022 (II)]**

(a) 192 kVm^{-1} (b) 192 mVm^{-1}
(c) 9600 kVm^{-1} (d) 16 kVm^{-1}

22. Two charged particles, having same kinetic energy, are allowed to pass through a uniform magnetic field perpendicular to the direction of motion. If the ratio of radii of their circular paths is 6 : 5 and their respective masses ratio is 9 : 4. Then, the ratio of their charges will be: **[NCERT : P L-113 | July 25, 2022 (I)]**

(a) 8 : 5 (b) 5 : 4 (c) 5 : 3 (d) 8 : 7

23. An electron with energy 0.1 keV moves at right angle to the magnetic field of 1×10^{-4} Wbm^{-2}. The frequency of revolution of the electron will be (Take mass of electron = 9.0×10^{-31} kg) **[NCERT : P L-112 | July 25, 2022 (II)]**

(a) 1.6×10^5 Hz (b) 5.6×10^5 Hz
(c) 2.8×10^6 Hz (d) 1.8×10^6 Hz

24. Given below are two statements :

Statement I: The electric force changes the speed of the charged particle and hence changes its kinetic energy: whereas the magnetic force does not change the kinetic energy of the charged particle.

Statement II: The electric force accelerates the positively charged particle perpendicular to the direction of electric field. The magnetic force accelerates the moving charged particle along the direction of magnetic field. In the light of the above statements, choose the most appropriate answer from the options given below: **[NCERT : P L-109 | June 29, 2022 (II)]**

(a) Both Statement I and Statement II are correct.
(b) Both Statement I and Statement II are incorrect.
(c) Statement I is correct but Statement II is incorrect.
(d) Statement I is incorrect but Statement II is correct.

25. A singly ionized magnesium atom (A = 24) ion is accelerated to kinetic energy 5 *keV* and is projected perpendicularly into a magnetic field *B* of the magnitude 0.5 *T*. The radius of path formed will be_______cm. **[NCERT : P L-112 | June 28, 2022 (I)]**

26. A deuteron and a proton moving with equal kinetic energy enter into to a uniform magnetic field at right angle to the field. If r_d and r_p are the radii of their circular paths respectively, then the ratio $\frac{r_d}{r_p}$ will be $\sqrt{x}:1$ where x is _______. **[NCERT : P L-112 | June 27, 2022 (II)]**

27. A proton and an alpha particle of the same velocity enter in a uniform magnetic field which is acting perpendicular to their direction of motion. The ratio of the circular paths described by the alpha particle and proton is:

[NCERT : P L-112 | June 26, 2022 (I)]

(a) 1 : 4 (b) 4 : 1 (c) 2 : 1 (d) 1 : 2

28. Given below are two statements : One is labelled as **Assertion (A)** and the other is labelled as **Reason (R)**.

Assertion (A) : In an uniform magnetic field, speed and energy remains the same for a moving charged particle.

Reason (R) : Moving charged particle experiences magnetic force perpendicular to its direction of motion.

[NCERT : P L-112 | June 24, 2022 (I)]

(a) Both (A) and (R) are true and (R) is the correct explanation of (A)
(b) Both (A) and (R) are true but (R) is NOT the correct explanation of (A)
(c) (A) is true but (R) is false
(d) (A) is false but (R) is true.

29. A proton, a deuteron and an α-particle with same kinetic energy enter into a uniform magnetic field at right angle to magnetic field. The ratio of the radii of their respective circular paths is : **[NCERT : P L-112 | June 24, 2022 (II)]**

(a) $1:\sqrt{2}:\sqrt{2}$ (b) $1:1:\sqrt{2}$
(c) $\sqrt{2}:1:1$ (d) $1:\sqrt{2}:1$

30. Two ions having same mass have charges in the ratio 1 : 2. They are projected normally in a uniform magnetic field with their speeds in the ratio 2 : 3. The ratio of the radii of their circular trajectories is :

[NCERT : P L-112 | July 25, 2021 (II)]

(a) 1 : 4 (b) 4 : 3 (c) 3 : 1 (d) 2 : 3

31. A deuteron and an alpha particle having equal kinetic energy enter perpendicular into a magnetic field. Let r_d and r_α be their respective radii of circular path. The value of $\frac{r_d}{r_\alpha}$ is equal to : **[NCERT : P L-112 | July 20, 2021 (I)]**

(a) $\frac{1}{\sqrt{2}}$ (b) $\sqrt{2}$ (c) 1 (d) 2

32. A proton and an α-particle, having kinetic energies K_p and K_α respectively, enter into a magnetic field at right angles. The ratio of the radii of trajectory of proton to that of α-particle is 2 : 1. The ratio of $K_p : K_\alpha$ is:

[NCERT : P L-113 | March 18, 2021 (II)]

(a) 1 : 8 (b) 8 : 1 (c) 1 : 4 (d) 4 : 1

33. A charge Q is moving $\overrightarrow{dl}$ distance in the magnetic field $\vec{B}$. Find the value of work done by $\vec{B}$.

[NCERT : P L-113 | March 16, 2021 (II)]

(a) –1 (b) Zero (c) 1 (d) Infinite

34. An electron is moving along + *x* direction with a velocity of 6×10^6 ms^{-1}. It enters a region of uniform electric field of 300 V/cm pointing along + *y* direction. The magnitude and direction of the magnetic field set up in this region such that the electron keeps moving along the *x* direction will be : **[NCERT : P L-112 | Sep. 06, 2020 (I)]**

(a) 3×10^{-4} T, along + *z* direction
(b) 5×10^{-3} T, along – *z* direction
(c) 5×10^{-3} T, along + *z* direction
(d) 3×10^{-4} T, along – *z* direction

35. A particle of charge *q* and mass *m* is moving with a velocity $-v\,\hat{i}(v \neq 0)$ towards a large screen placed in the Y-Z plane at a distance *d*. If there is a magnetic field $\vec{B} = B_0\hat{k}$, the minimum value of *v* for which the particle will not hit the screen is:

[NCERT : P L-109 | Sep. 06, 2020 (I)]

(a) $\frac{qdB_0}{3m}$ (b) $\frac{2qdB_0}{m}$ (c) $\frac{qdB_0}{m}$ (d) $\frac{qdB_0}{2m}$

36. A charged particle carrying charge 1 μC is moving with velocity $(2\hat{i}+3\hat{j}+4\hat{k})$ ms^{-1}. If an external magnetic field of $(5\hat{i}+3\hat{j}-6\hat{k})\times 10^{-3}$ T exists in the region where the particle is moving then the force on the particle is $\vec{F}\times 10^{-9}$ N. The vector $\vec{F}$ is :

[NCERT : P L-109 | Online May 12, 2012 S, Sep. 03, 2020 (I)]

(a) $-0.30\hat{i}+0.32\hat{j}-0.09\hat{k}$ (b) $-30\hat{i}+32\hat{j}-9\hat{k}$
(c) $-300\hat{i}+320\hat{j}-90\hat{k}$ (d) $-3.0\hat{i}+3.2\hat{j}-0.9\hat{k}$

37. A beam of protons with speed 4×10^5 ms^{-1} enters a uniform magnetic field of 0.3 T at an angle of 60° to the magnetic field. The pitch of the resulting helical path of protons is close to : (Mass of the proton = 1.67×10^{-27} kg, charge of the proton = 1.69×10^{-19} C) **[NCERT : P L-113 | Sep. 02, 2020 (I)]**

(a) 2 cm (b) 5 cm (c) 12 cm (d) 4 cm

38. The figure shows a region of length '*l*' with a uniform magnetic field of 0.3 T in it and a proton entering the region with velocity 4×10^5 ms^{-1} making an angle 60° with the field. If the proton completes 10 revolution by the time it cross the region shown, '*l*' is close to (mass of proton = 1.67×10^{-27} kg, charge of the proton = 1.6×10^{-19} C)

[NCERT : P L-112, 113 | Sep. 02, 2020 (II)]

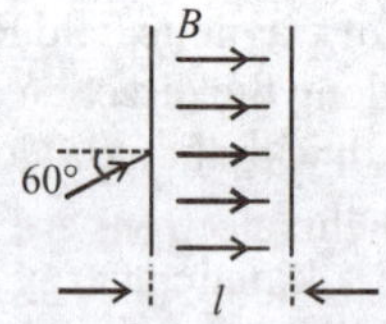

(a) 0.11 m
(b) 0.88 m
(c) 0.44 m
(d) 0.22 m

39. Proton with kinetic energy of 1 MeV moves from south to north. It gets an acceleration of 10^{12} m/s^2 by an applied magnetic field (west to east). The value of magnetic field: (Rest mass of proton is 1.6×10^{-27} kg)

[NCERT : P L-113 | 8 Jan 2020, (I); Online 23 April 2013]

(a) 0.71 mT (b) 7.1 mT (c) 0.071 mT (d) 71 mT

40. A particle having the same charge as of electron moves in a circular path of radius 0.5 cm under the influence of a magnetic field of 0.5T. If an electric field of 100V/m makes it to move in a straight path then the mass of the particle is (Given charge of electron = 1.6×10^{-19}C)

[NCERT : P L-109 | 12 April 2019, I]

(a) 9.1×10^{-31} kg (b) 1.6×10^{-27} kg
(c) 1.6×10^{-19} kg (d) 2.0×10^{-24} kg

41. An electron, moving along the x-axis with an initial energy of 100 eV, enters a region of magnetic field $\vec{B} = (1.5\times10^{-3}\text{T})\hat{k}$ at S (see figure). The field extends between $x = 0$ and $x = 2$ cm. The electron is detected at the point Q on a screen placed 8 cm away from the point S. The distance d between P and Q (on the screen) is : (Electron's charge = 1.6×10^{-19} C, mass of electron = 9.1×10^{-31} kg)

[NCERT : P L-112, 113 | 12 April 2019, II]

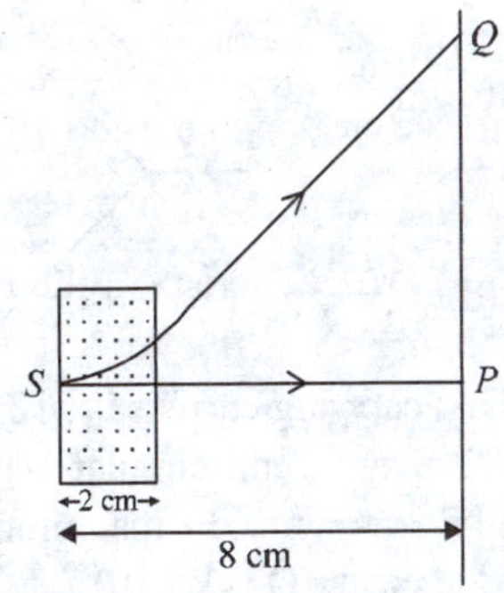

(a) 11.65 cm (b) 12.87 cm
(c) 1.22 cm (d) 2.25 cm

42. A proton, an electron, and a Helium nucleus, have the same energy. They are in circular orbits in a plane due to magnetic field perpendicular to the plane. Let r_p, r_e and r_{He} be their respective radii, then,

[NCERT : P L-112 | 10 April 2019, I; 2018, 2012]

(a) $r_e > r_p = r_{He}$ (b) $r_e < r_p = r_{He}$
(c) $r_e < r_p < r_{He}$ (d) $r_e > r_p > r_{He}$

43. A proton and an α-particle (with their masses in the ratio of 1 : 4 and charges in the ratio 1 : 2) are accelerated from rest through a potential difference V. If a uniform magnetic field (B) is set up perpendicular to their velocities, the ratio of the radii $r_p : r_\alpha$ of the circular paths descrfibed by them will be: **[NCERT : P L-112, 113 | 12 Jan 2019, I]**

(a) $1:\sqrt{2}$ (b) 1 : 2
(c) 1 : 3 (d) $1:\sqrt{3}$

44. In an experiment, electrons are accelerated, from rest, by applying a voltage of 500 V. Calculate the radius of the path if a magnetic field 100 mT is then applied.
[Charge of the electron = 1.6×10^{-19} C
Mass of the electron = 9.1×10^{-31} kg]

[NCERT : P L-112, 113 | 11 Jan 2019, I]

(a) 7.5×10^{-3} m (b) 7.5×10^{-2} m
(c) 7.5 m (d) 7.5×10^{-4} m

45. The region between $y = 0$ and $y = d$ contains a magnetic field $\vec{B} = B\hat{z}$. A particle of mass m and charge q enters the region with a velocity $\vec{v} = v\hat{i}$. if $d = \frac{mv}{2qB}$, the acceleration of the charged particle at the point of its emergence at the other side is : **[NCERT : P L-112 | 11 Jan 2019, II]**

(a) $\frac{qvB}{m}\left(\frac{1}{2}\hat{i} - \frac{\sqrt{3}}{2}\hat{j}\right)$ (b) $\frac{qvB}{m}\left(\frac{\sqrt{3}}{2}\hat{i} + \frac{1}{2}\hat{j}\right)$

(c) $\frac{qvB}{m}\left(\frac{-\hat{j}+\hat{i}}{\sqrt{2}}\right)$ (d) $\frac{qvB}{m}\left(\frac{\hat{i}+\hat{j}}{\sqrt{2}}\right)$

46. A negative test charge is moving near a long straight wire carrying a current. The force acting on the test charge is parallel to the direction of the current. The motion of the charge is : **[NCERT : P L-109 | Online April 9, 2017]**

(a) away from the wire
(b) towards the wire
(c) parallel to the wire along the current
(d) parallel to the wire opposite to the current

47. A proton (mass m) accelerated by a potential difference V flies through a uniform transverse magnetic field B. The field occupies a region of space by width 'd'. If α be the angle of deviation of proton from initial direction of motion (see figure), the value of $\sin\alpha$ will be :

[NCERT : P L-112 | Online April 10, 2015]

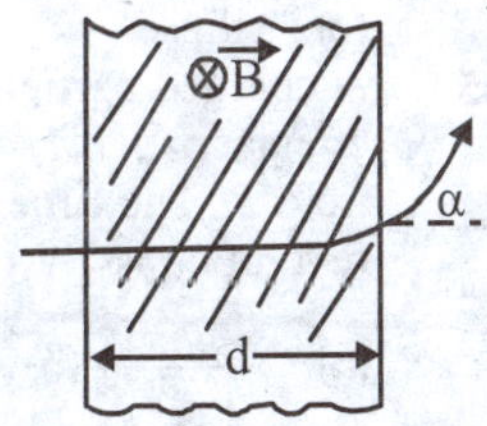

(a) $qV\sqrt{\frac{Bd}{2m}}$

(b) $\frac{B}{2}\sqrt{\frac{qd}{mV}}$

(c) $\frac{B}{d}\sqrt{\frac{q}{2mV}}$

(d) $Bd\sqrt{\frac{q}{2mV}}$

48. This question has **Statement 1** and **Statement 2**. Of the four choices given after the Statements, choose the one that best describes the two Statements.

Statement 1: A charged particle is moving at right angle to a static magnetic field. During the motion the kinetic energy of the charge remains unchanged.

Statement 2: Static magnetic field exert force on a moving charge in the direction perpendicular to the magnetic field.

[NCERT : P L-109 | Online May 26, 2012]

(a) Statement 1 is false, Statement 2 is true.
(b) Statement 1 is true, Statement 2 is true, Statement 2 is not the correct explanation of Statement 1.
(c) Statement 1 is true, Statement 2 is false.
(d) Statement 1 is true, Statement 2 is true, Statement 2 is the correct explanation of Statement 1.

49. The velocity of certain ions that pass undeflected through crossed electric field $E = 7.7$ k V/m and magnetic field $B = 0.14$ T is **[Online May 7, 2012]**

(a) 18 km/s (b) 77 km/s
(c) 55 km/s (d) 1078 km/s

50. An electric charge $+q$ moves with velocity $\vec{v} = 3\hat{i} + 4\hat{j} + \hat{k}$ in an electromagnetic field given by $\vec{E} = 3\hat{i} + \hat{j} + 2\hat{k}$ and $\vec{B} = \hat{i} + \hat{j} - 3\hat{k}$ The y - component of the force experienced by $+q$ is : **[NCERT : P L-109 | 2011 RS]**

(a) $11q$ (b) $5q$ (c) $3q$ (d) $2q$

51. A charged particle with charge q enters a region of constant, uniform and mutually orthogonal fields $\vec{E}$ and $\vec{B}$ with a velocity $\vec{v}$ perpendicular to both $\vec{E}$ and $\vec{B}$, and comes out without any change in magnitude or direction of $\vec{v}$. Then **[NCERT : P L-109 | 2007]**

(a) $\vec{v}=\vec{B}\times\vec{E}/E^2$ (b) $\vec{v}=\vec{E}\times\vec{B}/B^2$

(c) $\vec{v}=\vec{B}\times\vec{E}/B^2$ (d) $\vec{v}=\vec{E}\times\vec{B}/E^2$

52. A charged particle moves through a magnetic field perpendicular to its direction. Then **[NCERT : P L-112 | 2007]**

(a) kinetic energy changes but the momentum is constant
(b) the momentum changes but the kinetic energy is constant
(c) both momentum and kinetic energy of the particle are not constant
(d) both momentum and kinetic energy of the particle are constant

53. In a region, steady and uniform electric and magnetic fields are present. These two fields are parallel to each other. A charged particle is released from rest in this region. The path of the particle will be a **[NCERT : P L-109 | 2006]**

(a) helix (b) straight line
(c) ellipse (d) circle

54. A charged particle of mass m and charge q travels on a circular path of radius r that is perpendicular to a magnetic field B. The time taken by the particle to complete one revolution is **[NCERT : P L-112 | 2005]**

(a) $\frac{2\pi q^2 B}{m}$ (b) $\frac{2\pi mq}{B}$ (c) $\frac{2\pi m}{qB}$ (d) $\frac{2\pi qB}{m}$

55. A uniform electric field and a uniform magnetic field are acting along the same direction in a certain region. If an electron is projected along the direction of the fields with a certain velocity then **[NCERT : P L-109 | 2005]**

(a) its velocity will increase
(b) Its velocity will decrease
(c) it will turn towards left of direction of motion
(d) it will turn towards right of direction of motion

56. A particle of mass M and charge Q moving with velocity $\vec{v}$ describe a circular path of radius R when subjected to a uniform transverse magnetic field of induction B. The work done by the field when the particle completes one full circle is **[NCERT : P L-112, 113 | 2003]**

(a) $\left(\frac{Mv^2}{R}\right)2\pi R$ (b) zero

(c) $BQ2\pi R$ (d) $BQv2\pi R$

57. If an electron and a proton having same momenta enter perpendicular to a magnetic field, then **[NCERT : P L-112 | 2002]**

(a) curved path of electron and proton will be same (ignoring the sense of revolution)
(b) they will move undeflected
(c) curved path of electron is more curved than that of the proton
(d) path of proton is more curved.

58. The time period of a charged particle undergoing a circular motion in a uniform magnetic field is independent of its **[NCERT : P L-112 | 2002]**

(a) speed (b) mass
(c) charge (d) magnetic induction

Topic 2 - Magnetic Field Lines, Biot-Savart's Law and Ampere's Circuital Law

59. Figure shows a current carrying square loop ABCD of edge length is 'a' lying in a plane. If the resistance of the ABC part is r and that of ADC part is $2r$, then the magnitude of the resultant magnetic field at centre of the square loop is **[NCERT : P L-114, 116 | April 8, 2025 (II)]**

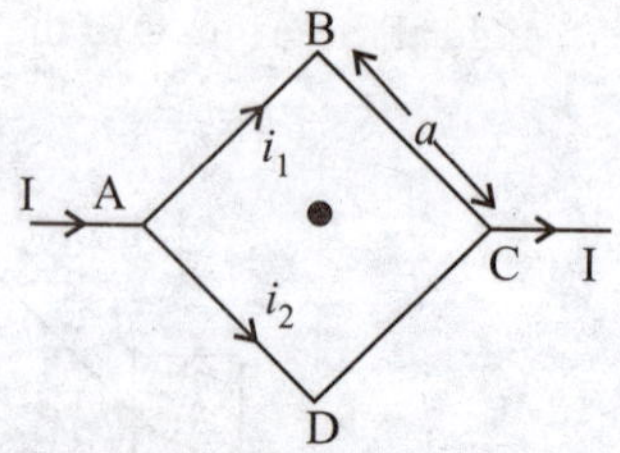

(a) $\frac{3\pi\mu_o I}{\sqrt{2}a}$ (b) $\frac{\mu_0 I}{2\pi a}$

(c) $\frac{\sqrt{2}\mu_0 I}{3\pi a}$ (d) $\frac{2\mu_0 I}{3\pi a}$

60. A loop ABCDA, carrying current I = 12 A, is placed in a plane, consists of two semi-circular segments of radius $R_1 = 6\pi$ m and $R_2 = 4\pi$ m. The magnitude of the resultant magnetic field at center O is $k \times 10^{-7}$ T. The value of k is ______. **[NCERT : P L-114, 116 | April 3, 2025 (I)]**
(Given $\mu_0 = 4\pi\times10^{-7}$ TmA^{-1})

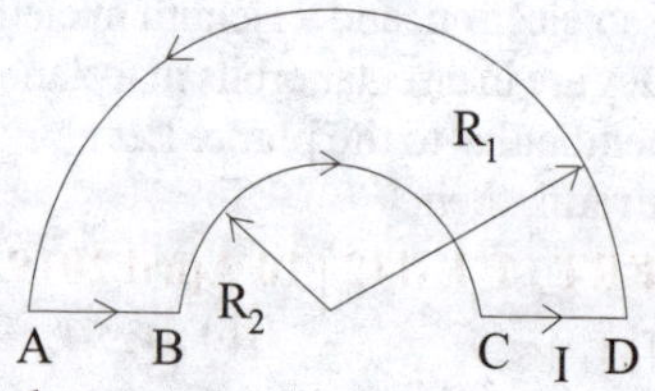

61. Let B_1 be the magnitude of magnetic field at center of a circular coil of radius R carrying current I. Let B_2 be the magnitude of magnetic field at an axial distance 'x' from the center. For x : R = 3 : 4, $\frac{B_2}{B_1}$ is : **[NCERT : P L-116 | April 2, 2025 (I)]**

(a) 4 : 5 (b) 16 : 25 (c) 64 : 125 (d) 25 : 16

62. The magnetic field inside a 200 turns solenoid of radius 10 cm is 2.9×10^{-4}. If the solenoid carries a current of 0.29 A, then the length of the solenoid is ______ π cm. **[NCERT : P L-122 | Jan. 29, 2025 (II)]**

63. Consider a long straight wire of a circular cross-section (radius a) carrying a steady current I. The current is uniformly distributed across this cross-section. The distances from the centre of the wire's cross-section at which the magnetic field [inside the wire, outside the wire] is half of the maximum possible magnetic field, any where due to the wire, will be **[NCERT : P L-118 | Jan. 29, 2025 (I)]**

(a) $\left[\frac{a}{4},\frac{3a}{2}\right]$ (b) $\left[\frac{a}{2},2a\right]$

(c) $\left[\frac{a}{2},3a\right]$ (d) $\left[\frac{a}{4},2a\right]$

64. 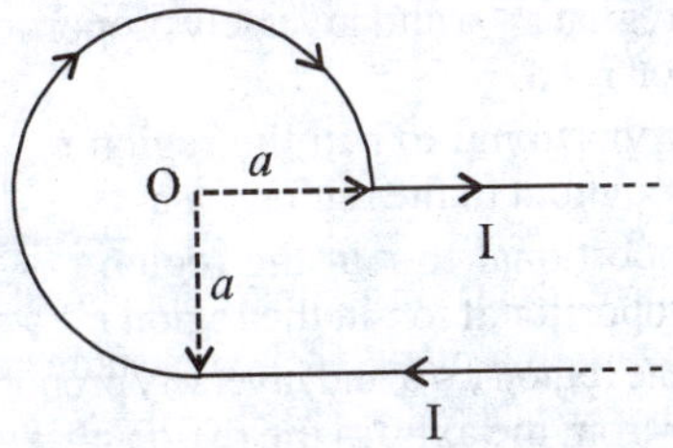

An infinite wire has a circular bend of radius a, and carrying a current I as shown in figure. The magnitude of magnetic field at the origin O of the arc is given by :

[NCERT : P L-114, 116 | Jan. 28, 2025 (II)]

(a) $\frac{\mu_0}{4\pi}\frac{I}{a}\left[\frac{\pi}{2}+1\right]$ (b) $\frac{\mu_0}{4\pi}\frac{I}{a}\left[\frac{3\pi}{2}+1\right]$

(c) $\frac{\mu_0}{2\pi}\frac{I}{a}\left[\frac{\pi}{2}+2\right]$ (d) $\frac{\mu_0}{4\pi}\frac{I}{a}\left[\frac{3\pi}{2}+2\right]$

65. 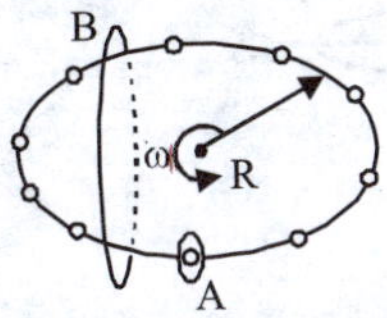

N equally spaced charges each of value q, are placed on a circle of radius R. The circle rotates about its axis with an angular velocity ω as shown in the figure. A bigger Amperian loop B encloses the whole circle where as a smaller Amperian loop A encloses a small segment. The difference between enclosed currents, $I_A - I_B$, for the given Amperian loops is

[NCERT : P L-118 | Jan. 24, 2025 (II)]

(a) $\frac{N^2}{2\pi}q\omega$ (b) $\frac{2\pi}{N}q\omega$ (c) $\frac{N}{2\pi}q\omega$ (d) $\frac{N}{\pi}q\omega$

66. A long straight wire of a circular cross-section with radius 'a' carries a steady current I. The current I is a uniformly distributed across this cross-section. The plot of magnitude of magnetic field B with distance r from the centre of the wire is given by **[NCERT : P L-118 | Jan. 24, 2025 (II)]**

(a) 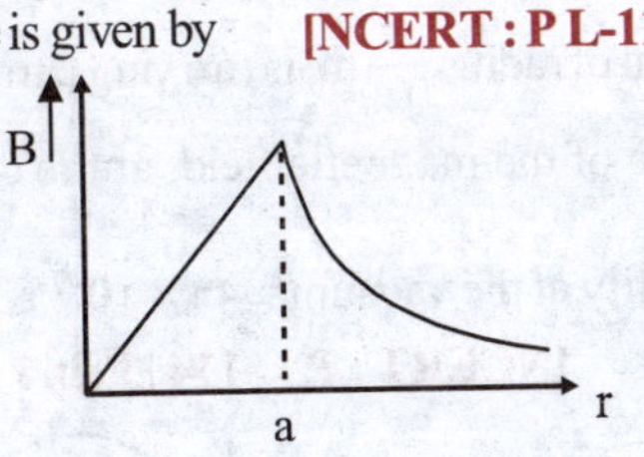

(b) 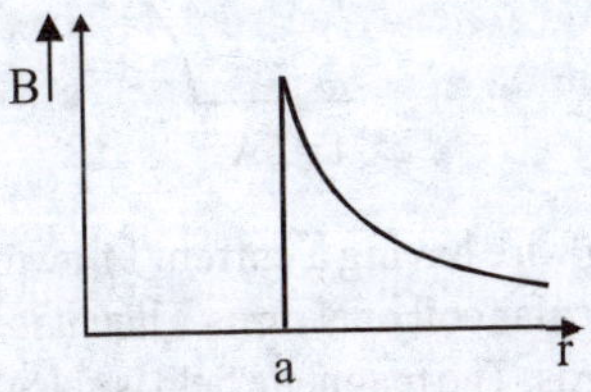

(c) 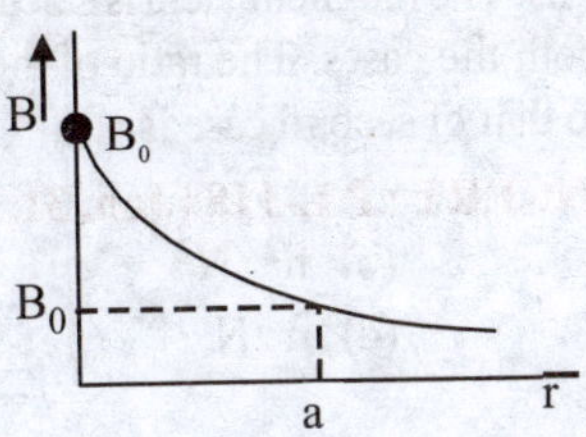

(d) 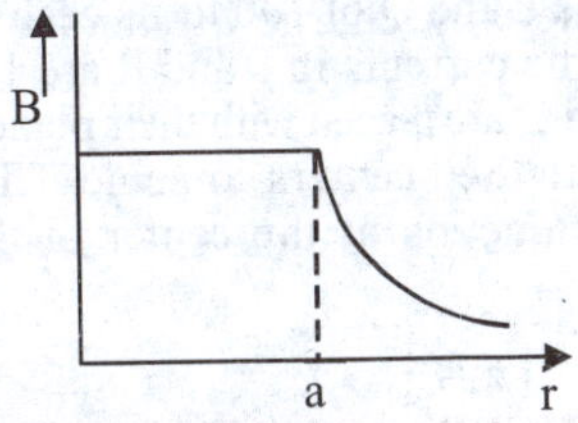

67. A tightly wound long solenoid carries a current of 1.5 A. An electron is executing uniform circular motion inside the solenoid with a time period of 75ns. The number of turns per metre in the solenoid is ______.
[Take mass of electron $m_e = 9 \times 10^{-31}$ kg, charge of electron $|q_e| = 1.6 \times 10^{-19}$ C, $\mu_0 = 4\pi\times10^{-7}\ \frac{N}{A^2}$, 1 ns $= 10^{-9}$ s]

[NCERT : P L-122 | Jan. 24, 2025 (II)]

68. A current of 5A exists in a square loop of side $\frac{1}{\sqrt{2}}$ m. Then the magnitude of the magnetic field B at the centre of the square loop will be $p \times 10^{-6}$ T. where, value of p is ______.
[Take $\mu_0 = 4\pi \times 10^{-7}$ T mA^{-1}].

[NCERT : P L-114, 116 | Jan. 24, 2025 (I)]

69. A long straight wire of radius a carries a steady current I. The current is uniformly distributed across its cross section. The ratio of the magnetic field at $\frac{a}{2}$ and 2a from axis of the wire is :

[NCERT : P L-118 | April 8, 2024 (II)]

(a) 1 : 4 (b) 4 : 1 (c) 1 : 1 (d) 3 : 4

70. A solenoid of length 0.5 m has a radius of 1 cm and is made up of 'm' number of turns. It carries a current of 5A. If the magnitude of the magnetic field inside the solenoid is 6.28×10^{-3} T, then the value of m is :

[NCERT : P L-122 | April 5, 2024 (II)]

71. Two parallel long current carrying wire separated by a distance 2r are shown in the figure. The ratio of magnetic field at A to the magnetic field produced at C is $\frac{x}{7}$. The value of x is ____.

[NCERT : P L-118 | April 4, 2024 (II)]

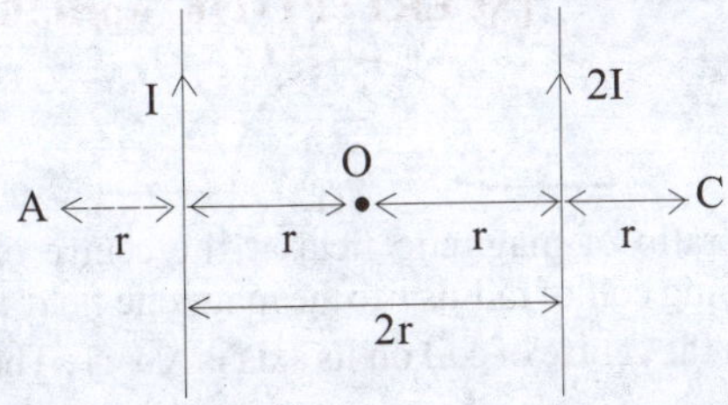

72. A regular polygon of 6 sides is formed by bending a wire of length 4π meter. If an electric current of $4\pi\sqrt{3}$ A is flowing through the sides of the polygon, the magnetic field at the centre of the polygon would be $x \times 10^{-7}$ T. The value of x is ____. **[NCERT : P L-114 | Feb. 1, 2024 (I)]**

73. Two circular coils P and Q of 100 turns each have same radius of π cm. The currents in P and R are 1 A and 2 A respectively. P and Q are placed with their planes mutually perpendicular with their centers coincide. The resultant magnetic field induction at the center of the coils is $\sqrt{x}$ mT, where x = ________.

[Use $\mu_0 = 4\pi \times 10^{-7}$ TmA^{-1}]

[NCERT : P L-116 | Jan. 31, 2024 (II)]

74. Two insulated circular loop A and B of radius 'a' carrying a current of 'I' in the anti clockwise direction as shown in the figure. The magnitude of the magnetic induction at the centre will be: **[NCERT : P L-116 | Jan. 30, 2024 (I)]**

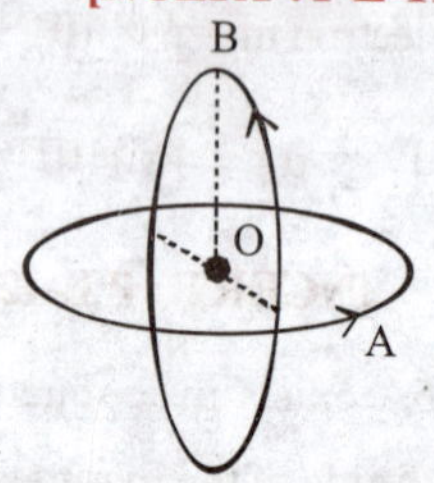

(a) $\frac{\sqrt{2}\mu_0 I}{a}$ (b) $\frac{\mu_0 I}{2a}$ (c) $\frac{\mu_0 I}{\sqrt{2}a}$ (d) $\frac{2\mu_0 I}{a}$

75. The current of 5A flows in a square loop of sides 1 *m* is placed in air. The magnetic field at the centre of the loop is $X\sqrt{2} \times 10^{-7}$ T. The value of *X* is ________.

[Jan. 30, 2024 (II)]

76. The magnetic field at the centre of a wire loop formed by two semicircular wires of radii $R_1 = 2\pi$m, and $R_2 = 4\pi$m carrying current I = 4A as per figure given below is $\alpha \times 10^{-7}$ T. The value of α is ______. (Centre O is common for all segments) **[NCERT : P L-116 | Jan. 27, 2024 (II)]**

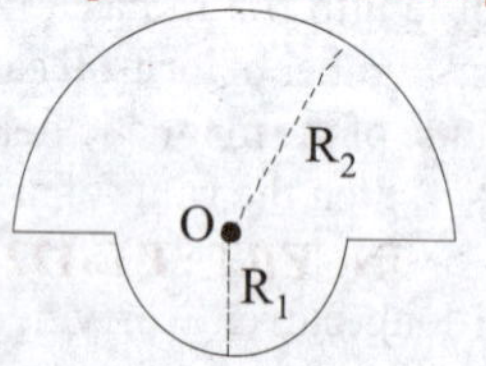

77. A straight wire carrying a current of 14 A is bent into a semicircular arc of radius 2.2 cm as shown in the figure. The magnetic field produced by the current at the centre (O) of the arc. is ______ $\times 10^{-4}$T

[NCERT : P L-116 | April 10, 2023 (II)]

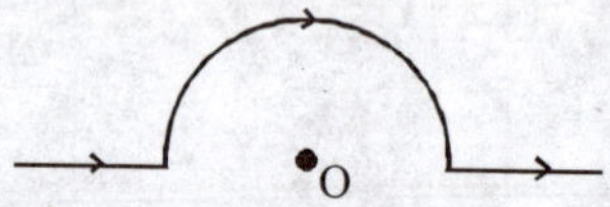

78. The ratio of magnetic field at the centre of a current carrying coil of radius *r* to the magnetic field at distance *r* from the centre of coil on its axis is $\sqrt{x}:1$. The value of x is ________

[NCERT : P L-115, 116 | April 8, 2023 (II)]

79. A long straight wire of circular cross–section (radius a) is carrying steady current I. The current I is uniformly distributed across this cross–section. The magnetic field is **[NCERT : P L-118 | April 6, 2023 (I)]**

(a) Zero in the region r < a and inversely proportional to r in the region r > a

(b) Inversely proportional to r in the region r < a and uniform throughout in the region r > a

(c) Directly proportional to r in the region r < a and inversely proportional to r in the region r > a

(d) Uniform in the region r < a and inversely proportional to distance r from the axis, in the region r > a

80. Two identical circular wires of radius 20 cm and carrying current $\sqrt{2}$ A are placed in perpendicular planes as shown in figure. The net magnetic field at the centre of the circular wire is ______ $\times 10^{-8}$ T. (Take $\pi = 3.14$)

[NCERT : P L-116 | April 6, 2023 (I)]

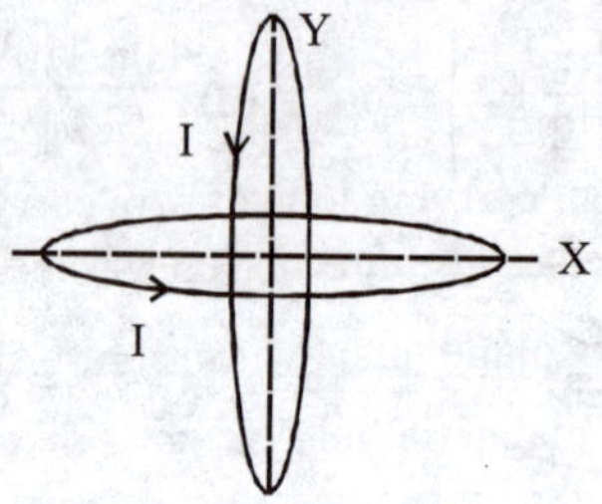

81. Find the magnetic field at the point P in figure. The curved portion is a semicircle connected to two long straight wires. **[NCERT : P L-114 | Feb. 1, 2023 (I)]**

(a) $\frac{\mu_0 i}{2r}\left(\frac{1}{2}+\frac{1}{2\pi}\right)$

(b) $\frac{\mu_0 i}{2r}\left(1+\frac{1}{\pi}\right)$

(c) $\frac{\mu_0 i}{2r}\left(1+\frac{2}{\pi}\right)$

(d) $\frac{\mu_0 i}{2r}\left(\frac{1}{2}+\frac{1}{\pi}\right)$

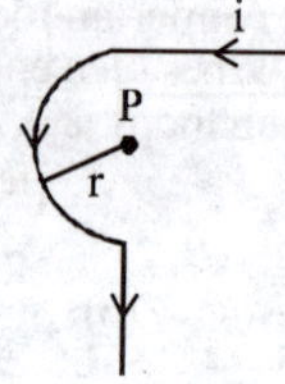

82. As shown in the figure, a long straight conductor with semicircular arc of radius $\frac{\pi}{10}$m is carrying current I = 3A. The magnitude of the magnetic field, at the center O of the arc is:

(The permeability of the vacuum = $4\pi \times 10^{-7}$ NA^{-2})

[NCERT : P L-114 | Feb. 1, 2023 (II)]

(a) 6μT
(b) 1μT
(c) 3μT
(d) 4μT

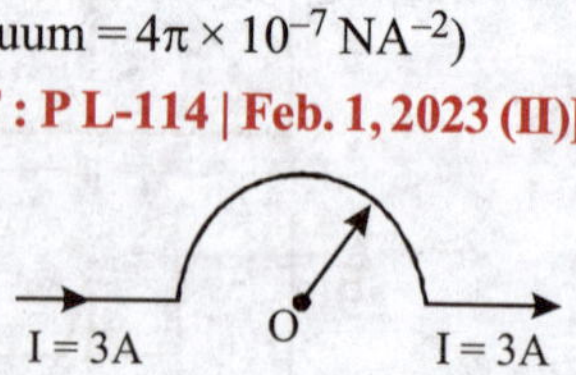

83. A long conducting wire having a current I flowing through it, is bent into a circular coil of N turns. Then it is bent into a circular coil of n turns. The magnetic field is calculated at the centre of coils in both the cases. The ratio of the magnetic field in first case to that of second case is:

[NCERT : P L-118 | Jan. 31, 2023 (II)]

(a) N : n (b) $n^2 : N^2$
(c) $N^2 : n^2$ (d) n : N

84. The magnetic moments associated with two closely wound circular coils A and B of radius $r_A = 10$ cm and $r_B = 20$ cm respectively are equal if: (Where N_A, I_A and N_B, I_B are number of turn and current of A and B respectively) **[Jan. 30, 2023 (I)]**

(a) $2N_AI_A = N_BI_B$ (b) $N_A = 2N_B$
(c) $N_AI_A = 4N_BI_B$ (d) $4N_AI_A = N_BI_B$

85. As shown in the figure, a current of 2A flowing in an equilateral triangle of side $4\sqrt{3}$cm. The magnetic field at the centroid O of the triangle is:
(Neglect the effect of earth's magnetic field.)

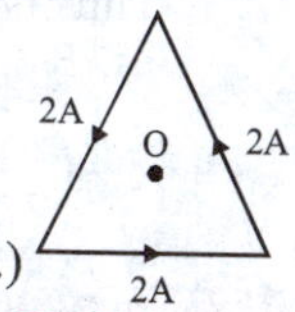

[Jan. 30, 2023 (II)]

(a) $4\sqrt{3}\times10^{-4}$ T (b) $4\sqrt{3}\times10^{-5}$ T
(c) $\sqrt{3}\times10^{-4}$ T (d) $3\sqrt{3}\times10^{-5}$ T

86. A single current carrying loop of wire carrying current I flowing in anticlockwise direction seen from +ve z direction and lying in xy plane in shown in figure. The plot of $\hat{j}$ component of magnetic field (B_y) at a distance 'a' (less than radius of the coil) and on yz plane vs z coordinate look like **[NCERT : P L-115 | Jan. 29, 2023 (I)]**

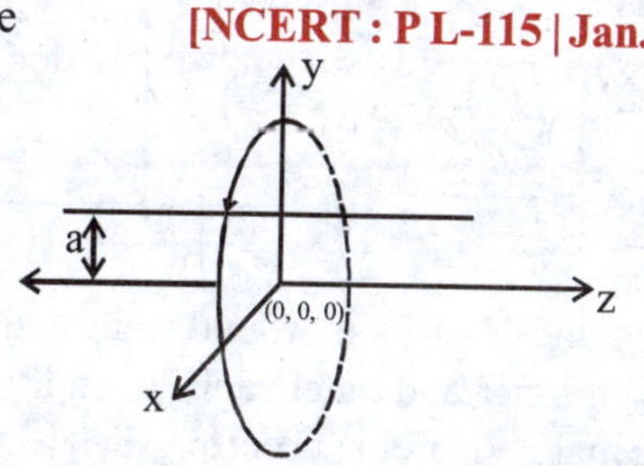

(a)
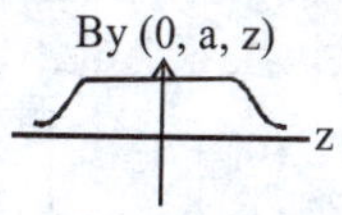

(b)
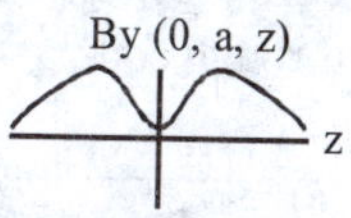

(c)
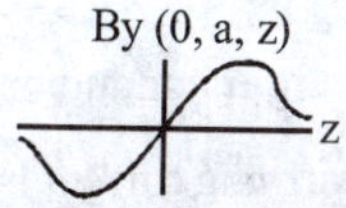

(d)
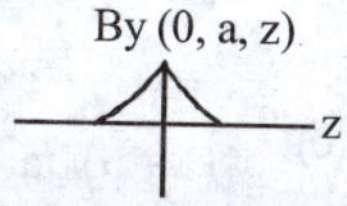

87. The magnitude of magnetic induction at mid-point O due to current arrangement as shown in Fig will be : **[NCERT : P L-114 | Jan. 29, 2023 (I)]**

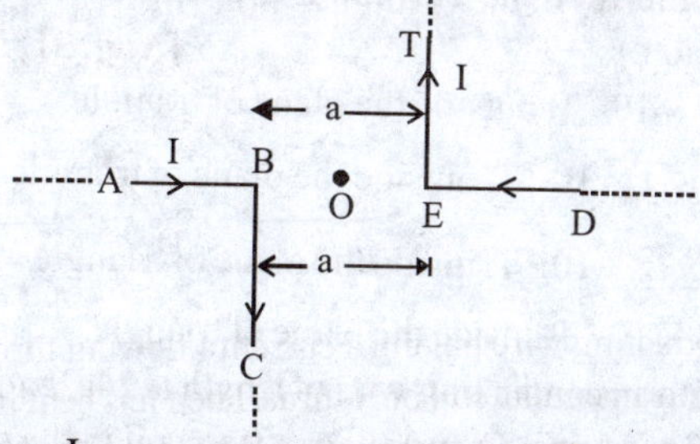

(a) $\frac{\mu_0 I}{2\pi a}$ (b) 0 (c) $\frac{\mu_0 I}{4\pi a}$ (d) $\frac{\mu_0 I}{\pi a}$

88. The electric current in a circular coil of four turns produces a magnetic induction 32 T at its centre. The coil is unwound and is rewound into a circular coil of single turn, the magnetic induction at the centre of the coil by the same current will be : **[NCERT : P L-116 | Jan. 29, 2023 (II)]**

(a) 8T (b) 4T (c) 2T (d) 16T

89. Math List I with List II

[NCERT : P L-114 | Jan. 25, 2023 (I)]

List – I (Current configuration)	List – II (Magnetic field at point O)
A. 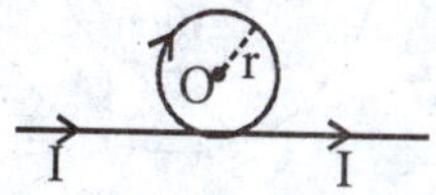	I. $B_0 = \frac{\mu_0 I}{4\pi r}[\pi + 2]$
B. 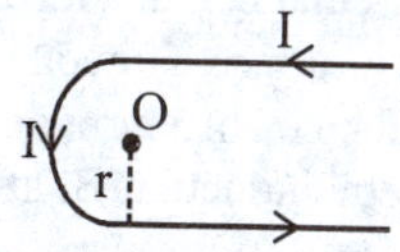	II. $B_0 = \frac{\mu_0}{4}\frac{I}{r}$
C. 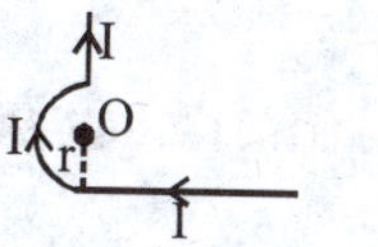	III. $B_0 = \frac{\mu_0 I}{2\pi r}[\pi - 1]$
D. 	IV. $B_0 = \frac{\mu_0 I}{4\pi r}[\pi + 1]$

Choose the correct answer from the option given below:

(a) A-III, B-IV, C-I, D-II (b) A-I, B-III, C-IV, D-II
(c) A-III, B-I, C-IV, D-II (d) A-II, B-I, C-IV, D-III

90. A long solenoid is formed by winding 70 turns cm^{-1}. If 2.0 A current flows, then the magnetic field produced inside the solenoid is ______ ($\mu_0 = 4\pi \times 10^{-7}$ TmA^{-1}) **[NCERT : P L-122 | Jan. 24, 2023 (II)]**

(a) 1232×10^{-4} T (b) 176×10^{-4} T
(c) 352×10^{-4} T (d) 88×10^{-4} T

91. A wire of length 314 cm carrying current of 14 A is bent to form a circle. The magnetic moment of the coil is ______ A-m^2. [Given $\pi = 3.14$] **[NCERT : P L-116, 118 | July 29, 2022 (I)]**

92. A closely wounded circular coil of radius 5 cm produces a magnetic field of 37.68×10^{-4} T at its center. The current through the coil is ______ A.
[Given, number of turns in the coil is 100 and $\pi = 3.14$] **[NCERT : P L-116 | July 29, 2022 (I)]**

93. The magnetic field at the center of current carrying circular loop is B_1. The magnetic field at a distance of $\sqrt{3}$ times radius of the given circular loop from the center on its axis is B_2. The value of B_1/B_2 will be **[NCERT : P L-116 | July 28, 2022 (II)]**

(a) 9 : 4 (b) $12:\sqrt{5}$ (c) 8 : 1 (d) $5:\sqrt{3}$

94. B_X and B_Y are the magnetic fields at the centre of two coils X and Y respectively each carrying equal current. If coil X has 200 turns and 20 cm radius and coil Y has 400 turns and 20 cm radius, the ratio of B_X and B_Y is : **[NCERT : P L-116 | July 26, 2022 (I)]**

(a) 1 : 1 (b) 1 : 2 (c) 2 : 1 (d) 4 : 1

95. Two concentric circular loops of radii $r_1 = 30$ cm and $r_2 = 50$ cm are placed in X-Y plane as shown in the figure. A current I = 7A is flowing through them in the direction as shown in figure. The net magnetic moment of this system of two circular loops is approximately: **[NCERT : P L-116 | July 26, 2022 (II)]**

(a) $\frac{7}{2}\hat{k}\,Am^2$

(b) $-\frac{7}{2}\hat{k}\,Am^2$

(c) $7\hat{k}\,Am^2$

(d) $-7\hat{k}\,Am^2$

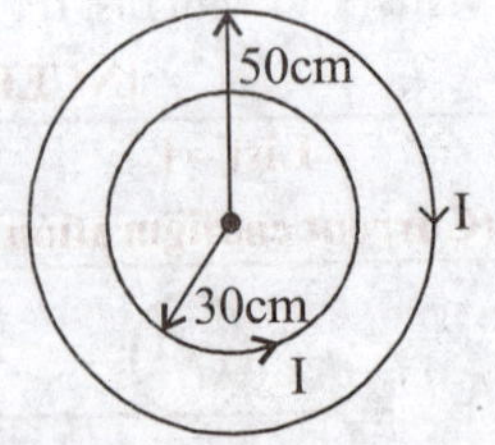

96. The electric current in a circular coil of 2 turns produces a magnetic induction B_1 at its centre. The coil is unwound and is rewound into a circular coil of 5 turns and the same current produces a magnetic induction B_2 at its centre. The ratio of $\frac{B_2}{B_1}$ is:

[NCERT : P L-116 | July 25, 2022 (II)]

(a) $\frac{5}{2}$ (b) $\frac{25}{4}$ (c) $\frac{5}{4}$ (d) $\frac{25}{2}$

97. A coil of n number of turns wound tightly in the form of a spiral with inner and outer radii r_1 and r_2 respectively. When a current of strength I is passed through the coil, the magnetic field at its centre will be:

[NCERT : P L-116 | June 30, 2022 (I)]

(a) $\frac{\mu_0 nI}{2(r_2-r_1)}$ (b) $\frac{\mu_0 nI}{r_2}$

(c) $\frac{\mu_0 nI}{r_2-r_1}\log_e\frac{r_1}{r_2}$ (d) $\frac{\mu_0 nI}{2(r_2-r_1)}\log_e\frac{r_2}{r_1}$

98. An infinitely long hollow conducting cylinder with radius R carries a uniform current along its surface. Choose the correct representation of magnetic field (B) as a function of radial distance (r) from the axis of cylinder.

[NCERT : P L-116 | June 28, 2022 (I)]

(a)
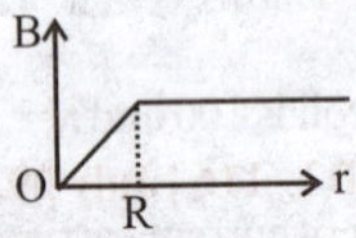

(b)
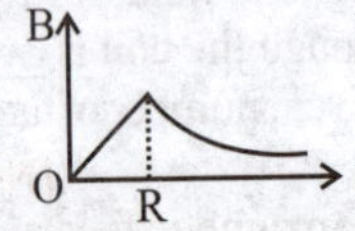

(c)
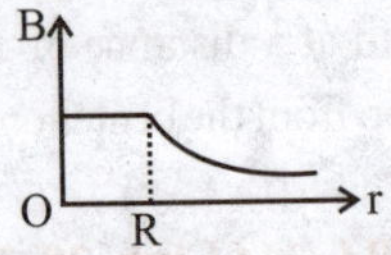

(d)
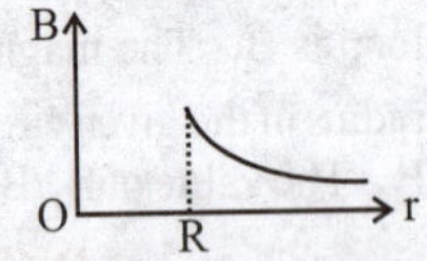

99. A long straight wire with a circular cross-section having radius R, is carrying a steady current I. The current I is uniformly distributed across this cross-section. Then the variation of magnetic field due to current I with distance r ($r < R$) from its centre will be :-

[NCERT : P L-114 | June 25, 2022 (I)]

(a) $B \propto r^2$ (b) $B \propto r$

(c) $B \propto \frac{1}{r^2}$ (d) $B \propto \frac{1}{r}$

100. A long solenoid carrying a current produces a magnetic field B along its axis. If the current is doubled and the number of turns per cm is halved, the new value of magnetic field will be equal to

[NCERT : P L-122 | June 25, 2022 (II)]

(a) B (b) $2B$

(c) $4B$ (d) $\frac{B}{2}$

101. The magnetic field at the centre of a circular coil of radius r, due to current I flowing through it, is B. The magnetic field at a point along the axis at a distance $\frac{r}{2}$ from the centre is: **[NCERT : P L-115, 116 | June 24, 2022 (I)]**

(a) $B/2$ (b) $2B$

(c) $\left(\frac{2}{\sqrt{5}}\right)^3 B$ (d) $\left(\frac{2}{\sqrt{3}}\right)^3 B$

102. There are two infinitely long straight current carrying conductors and they are held at right angles to each other so that their common ends meet at the origin as shown in the figure given below. The ratio of current in both conductor is 1 : 1. The magnetic field at point P is ____.

[Sep. 1, 2021 (II)]

(a) $\frac{\mu_0 I}{4\pi xy}\left[\sqrt{x^2+y^2}+(x+y)\right]$

(b) $\frac{\mu_0 I}{4\pi xy}\left[\sqrt{x^2+y^2}-(x+y)\right]$

(c) $\frac{\mu_0 Ixy}{4\pi}\left[\sqrt{x^2+y^2}-(x+y)\right]$

(d) $\frac{\mu_0 Ixy}{4\pi}\left[\sqrt{x^2+y^2}+(x+y)\right]$

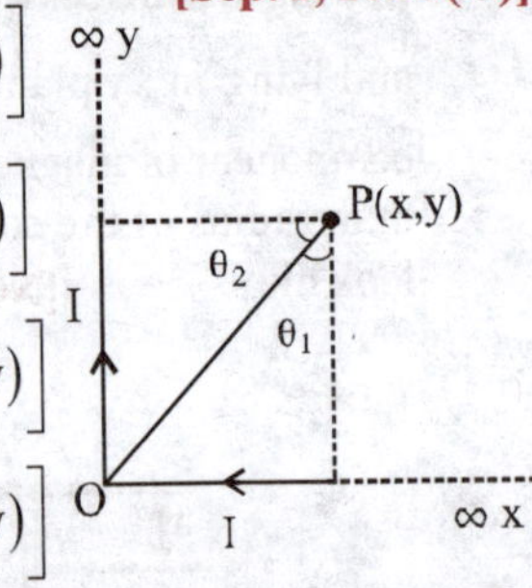

103. A coil having N turns is wound tightly in the form of a spiral with inner and outer radii 'a' and 'b' respectively. Find the magnetic field at centre, when a current I passes through coil : **[NCERT : P L-116 | Aug. 31, 2021 (I)]**

(a) $\frac{\mu_0 IN}{2(b-a)}\log_e\left(\frac{b}{a}\right)$ (b) $\frac{\mu_0 I}{8}\left[\frac{a+b}{a-b}\right]$

(c) $\frac{\mu_0 I}{4(a-b)}\left[\frac{1}{a}-\frac{1}{b}\right]$ (d) $\frac{\mu_0 I}{8}\left(\frac{a-b}{a+b}\right)$

104. A current of 1.5 A is flowing through a triangle, of side 9 cm each. The magnetic field at the centroid of the triangle is : (Assume that the current is flowing in the clockwise direction.) **[Aug. 31, 2021 (II)]**

(a) 3×10^{-7} T, outside the plane of triangle.

(b) $2\sqrt{3} \times 10^{-7}$ T, outside the plane of triangle.

(c) $2\sqrt{3} \times 10^{-5}$ T, inside the plane of triangle.

(d) 3×10^{-5} T, inside the plane of triangle.

105. A uniform conducting wire of length is 24a, and resistance R is wound up as a current carrying coil in the shape of an equilateral triangle of side 'a' and then in the form of a square of side 'a'. The coil is connected to a voltage source V_0. The ratio of magnetic moment of the coils in case of equilateral triangle to that for square is $1:\sqrt{y}$ where y is ________. **[NA, Aug. 27, 2021 (I)]**

106. A coaxial cable consists of an inner wire of radius 'a' surrounded by an outer shell of inner and outer radii 'b' and 'c' respectively. The inner wire carries an electric current i_0, which is distributed uniformly across cross-sectional area. The outer shell carries an equal current in opposite direction and distributed uniformly. What will be the ratio of the magnetic field at a distance x from the axis when (i) $x < a$ and (ii) $a < x < b$?

[NCERT : P L-118 | Aug. 27, 2021 (II)]

(a) $\frac{x^2}{a^2}$ (b) $\frac{a^2}{x^2}$ (c) $\frac{x^2}{b^2 - a^2}$ (d) $\frac{b^2 - a^2}{x^2}$

107. The fractional change in the magnetic field intensity at a distance 'r' from centre on the axis of current carrying coil of radius 'a' to the magnetic field intensity at the centre of the same coil is : (Take $r < a$)

[NCERT : P L-115, 116 | Aug. 26, 2021 (I)]

(a) $\frac{3}{2}\frac{a^2}{r^2}$ (b) $\frac{2}{3}\frac{a^2}{r^2}$ (c) $\frac{2}{3}\frac{r^2}{a^2}$ (d) $\frac{3}{2}\frac{r^2}{a^2}$

108. Figure A and B shown two long straight wires of circular cross-section (a and b with $a < b$), carrying current I which is uniformly distributed across the cross-section. The magnitude of magnetic field B varies with radius r and can be represented as:

[NCERT : P L-118 | July 27, 2021 (II)]

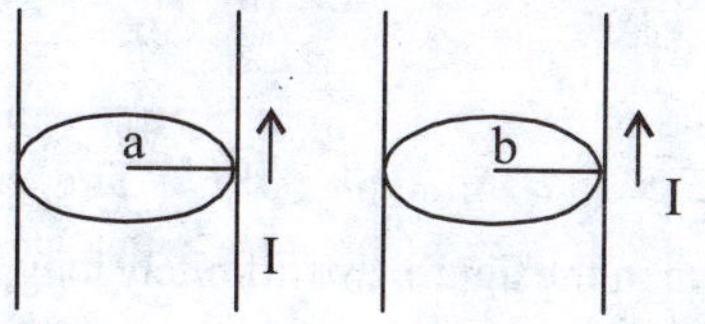

Fig. A Fig. B

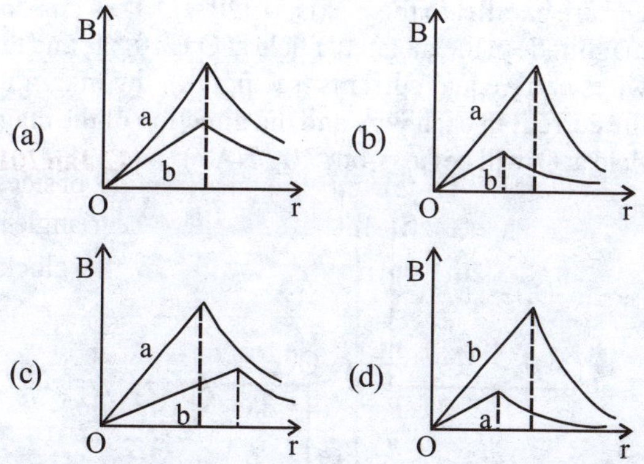

109. A hairpin like shape as shown in figure is made by bending a long current carrying wire. What is the magnitude of a magnetic field at point P which lies on the centre of the semicircle?

[NCERT : P L-114, 117 | March 17, 2021 (II)]

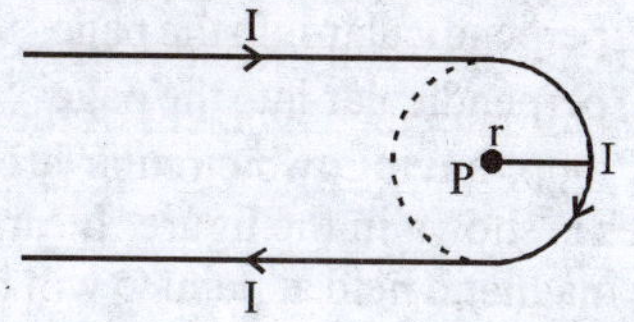

(a) $\frac{\mu_0 I}{4\pi r}(2+\pi)$ (b) $\frac{\mu_0 I}{4\pi r}(2-\pi)$

(c) $\frac{\mu_0 I}{2\pi r}(2-\pi)$ (d) $\frac{\mu_0 I}{2\pi r}(2+\pi)$

110. A charged particle going around in a circle can be considered to be a current loop. A particle of mass m carrying charge q is moving in a plane with speed v under the influence of magnetic field $\vec{B}$. The magnetic moment of this moving particle : **[NCERT : P L-116 | Sep. 06, 2020 (II)]**

(a) $\frac{mv^2 \vec{B}}{2B^2}$ (b) $-\frac{mv^2 \vec{B}}{2\pi B^2}$ (c) $-\frac{mv^2 \vec{B}}{B^2}$ (d) $-\frac{mv^2 \vec{B}}{2B^2}$

111. A wire A, bent in the shape of an arc of a circle, carrying a current of 2 A and having radius 2 cm and another wire B, also bent in the shape of arc of a circle, carrying a current of 3 A and having radius of 4 cm, are placed as shown in the figure. The ratio of the magnetic fields due to the wires A and B at the common centre O is :

[NCERT : P L-116 | Sep. 04, 2020 (I)]

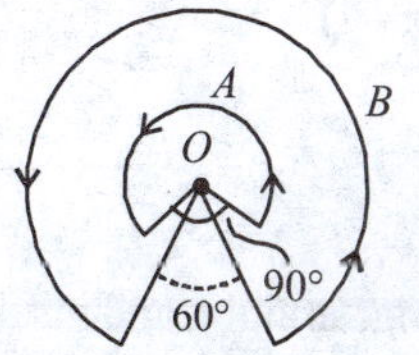

(a) 4 : 6 (b) 6 : 4 (c) 2 : 5 (d) 6 : 5

112. Magnitude of magnetic field (in SI units) at the centre of a hexagonal shape coil of side 10 cm, 50 turns and carrying current I (Ampere) in units of $\frac{\mu_0 I}{\pi}$ is: **[Sep. 03, 2020 (I)]**

(a) $250\sqrt{3}$ (b) $50\sqrt{3}$ (c) $500\sqrt{3}$ (d) $5\sqrt{3}$

113. A long, straight wire of radius a carries a current distributed uniformly over its cross-section. The ratio of the magnetic fields due to the wire at distance $\frac{a}{3}$ and $2a$, respectively from the axis of the wire is:

[NCERT : P L-118 | 2007 S, 9 Jan 2020, I]

(a) $\frac{2}{3}$ (b) 2 (c) $\frac{1}{2}$ (d) $\frac{3}{2}$

114. An electron gun is placed inside a long solenoid of radius R on its axis. The solenoid has n turns/length and carries a current I. The electron gun shoots an electron along the radius of the solenoid with speed v. If the electron does not hit the surface of the solenoid, maximum possible value of v is (all symbols have their standard meaning):

[NCERT : P L-122 | 9 Jan 2020, II]

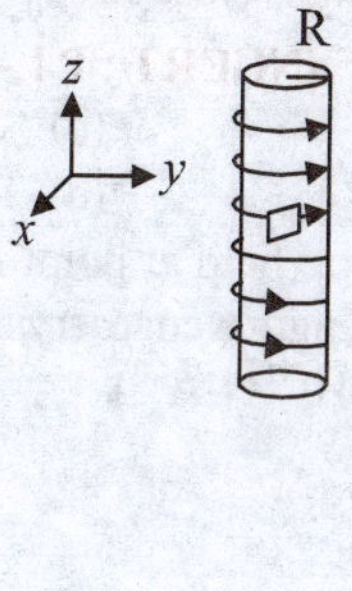

(a) $\frac{e\mu_0 nIR}{m}$ (b) $\frac{e\mu_0 nIR}{2m}$

(c) $\frac{e\mu_0 nIR}{4m}$ (d) $\frac{2e\mu_0 nIR}{m}$

115. A very long wire $ABDMNDC$ is shown in figure carrying current I. AB and BC parts are straight, long and at right angle. At D wire forms a circular turn $DMND$ of radius R.
AB, BC parts are tangential to circular turn at N and D. Magnetic field at the centre of circle is:

[NCERT : P L-116, 118 | 8 Jan 2020, II]

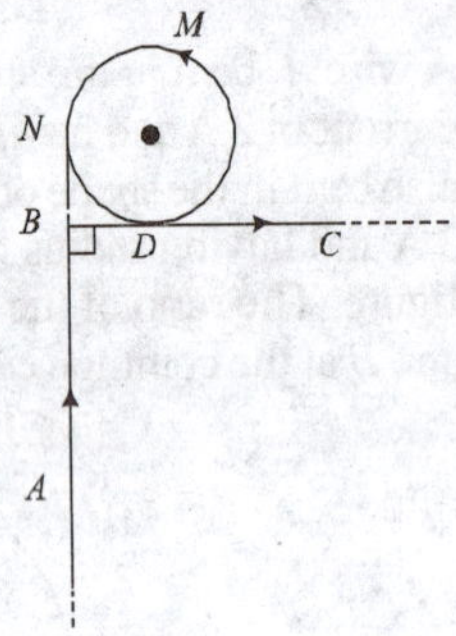

(a) $\frac{\mu_0 I}{2\pi R}\left(\pi+\frac{1}{\sqrt{2}}\right)$

(b) $\frac{\mu_0 I}{2\pi R}\left(\pi-\frac{1}{\sqrt{2}}\right)$

(c) $\frac{\mu_0 I}{2\pi R}(\pi+1)$

(d) $\frac{\mu_0 I}{2R}$

116. Two very long, straight, and insulated wires are kept at 90° angle from each other in *xy*-plane as shown in the figure.

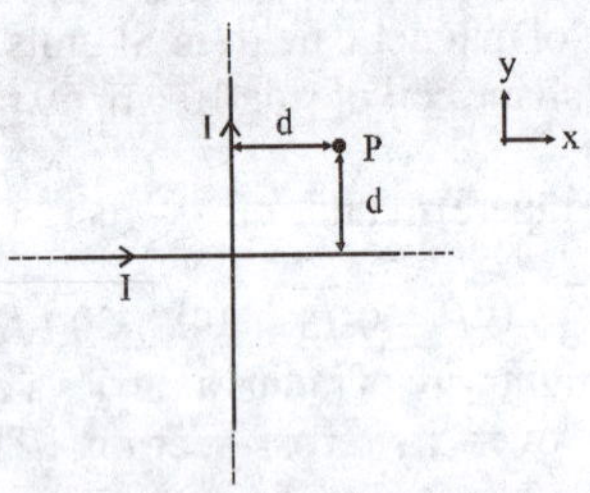

These wires carry currents of equal magnitude I, whose directions are shown in the figure. The net magnetic field at point P will be : **[NCERT : P L-118 | 12 April 2019, I]**

(a) Zero (b) $-\frac{\mu_0 I}{2\pi d}(\hat{x}+\hat{y})$

(c) $\frac{+\mu_0 I}{\pi d}(\hat{z})$ (d) $\frac{\mu_0 I}{2\pi d}(\hat{x}+\hat{y})$

117. A thin ring of 10 cm radius carries a uniformly distributed charge. The ring rotates at a constant angular speed of 40π rad s^{-1} about its axis, perpendicular to its plane. If the magnetic field at its centre is 3.8×10^{-9} T, then the charge carried by the ring is close to ($\mu_0 = 4\pi \times 10^{-7}$ N/A^2).

[NCERT : P L-116 | 12 April 2019, I]

(a) 2×10^{-6}C (b) 3×10^{-5}C
(c) 4×10^{-5}C (d) 7×10^{-6}C

118. Find the magnetic field at point P due to a straight line segment AB of length 6 cm carrying a current of 5 A. (See figure) ($\mu_o=4\pi\times10^{-7}$ N-A^{-2}) **[12 April 2019, II]**

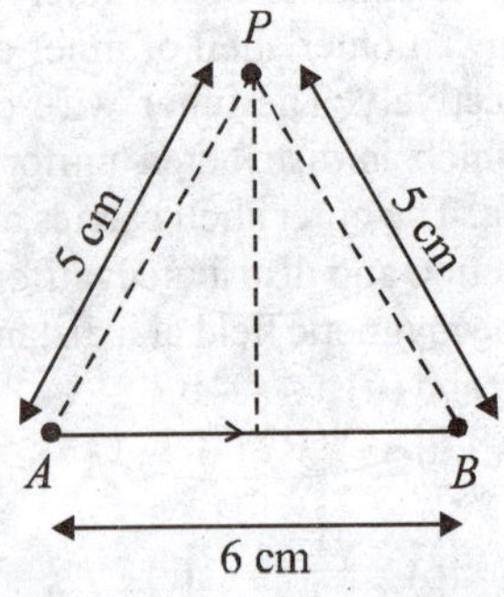

(a) 2.0×10^{-5}T (b) 1.5×10^{-5}T
(c) 3.0×10^{-5}T (d) 2.5×10^{-5}T

119. The magnitude of the magnetic field at the center of an equilateral triangular loop of side 1 m which is carrying a current of 10 A is : [Take $\mu_o = 4\pi\times10^{-7}$ NA^{-2}]

[Online April 15, 2018 S, 10 April 2019, II]

(a) 18 μT (b) 9 μT (c) 3 μT (d) 1 μT

120. A square loop is carrying a steady current I and the magnitude of its magnetic dipole moment is m. If this square loop is changed to a circular loop and it carries the same current, the magnitude of the magnetic dipole moment of circular loop will be :

[NCERT : P L-126 | 10 April 2019, II]

(a) $\frac{m}{\pi}$ (b) $\frac{3m}{\pi}$

(c) $\frac{2m}{\pi}$ (d) $\frac{4m}{\pi}$

121. As shown in the figure, two infinitely long, identical wires are bent by 90° and placed in such a way that the segments LP and QM are along the *x*-axis, while segments PS and QN are parallel to the *y*-axis. If OP = OQ = 4 cm, and the magnitude of the magneticf field at O is 10^{-4} T, and the two wires carry equal currents (see figure), the magnitude of the current in each wire and the direction of the magnetic field at O will be ($\mu_0 = 4\pi \times 10^{-7}$ NA^{-2}): **[12 Jan 2019, I]**

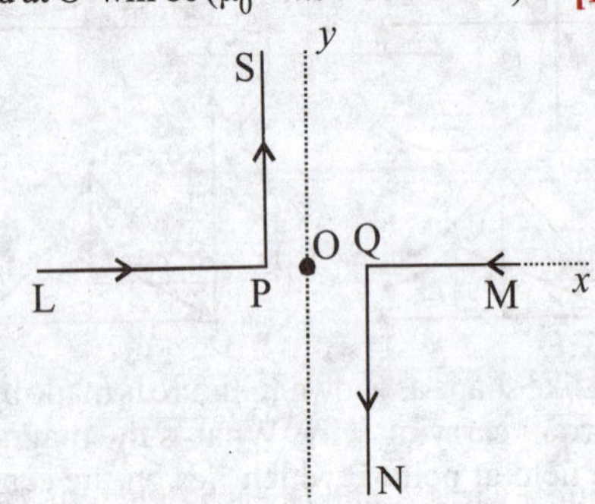

(a) 20 A, perpendicular out of the page
(b) 40 A, perpendicular out of the page
(c) 20 A, perpendicular into the page
(d) 40 A, perpendicular into the page

122. A current loop, having two circular arcs joined by two radial lines is shown in the figure. It carries a current of 10 A. The magnetic field at point O will be close to:

[9 Jan. 2019 I]

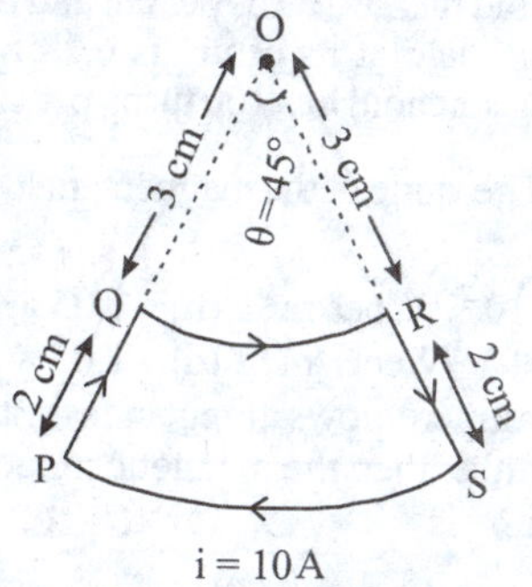

(a) 1.0×10^{-7}T (b) 1.5×10^{-7}T
(c) 1.5×10^{-5}T (d) 1.0×10^{-5}T

123. One of the two identical conducting wires of length L is bent in the form of a circular loop and the other one into a circular coil of N identical turns. If the same current is passed in both, the ratio of the magnetic field at the centre of the loop (B_L) to that at the centre of the coil (B_C), *i.e.*, $\frac{B_L}{B_C}$ will be: **[NCERT : P L-116 | 9 Jan 2019, II]**

(a) N (b) $\frac{1}{N}$ (c) N^2 (d) $\frac{1}{N^2}$

124. The dipole moment of a circular loop carrying a current I, is m and the magnetic field at the centre of the loop is B_1. When the dipole moment is doubled by keeping the current constant, the magnetic field at the centre of the loop is B_2. The ratio $\frac{B_1}{B_2}$ is: **[NCERT : P L-126 | 2018]**

(a) 2 (b) $\sqrt{3}$ (c) $\sqrt{2}$ (d) $\frac{1}{\sqrt{2}}$

125. A Helmholtz coil has pair of loops, each with *N* turns and radius *R*. They are placed coaxially at distance *R* and the same current *I* flows through the loops in the same direction. The magnitude of magnetic field at *P*, midway between the centres *A* and *C*, is given by (Refer to figure): **[Online April 15, 2018]**

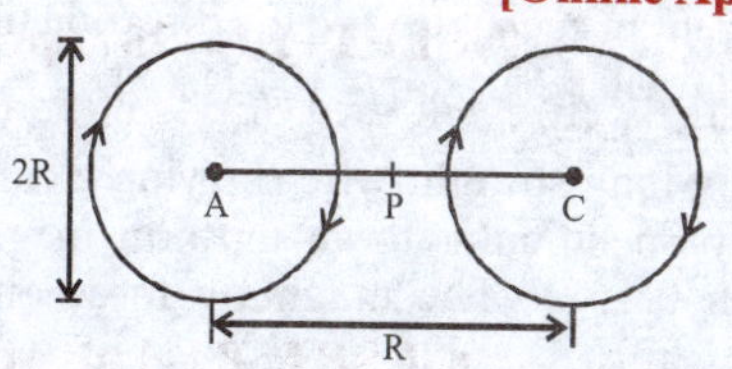

(a) $\frac{4N\mu_0 I}{5^{3/2} R}$ (b) $\frac{8N\mu_0 I}{5^{3/2} R}$

(c) $\frac{4N\mu_0 I}{5^{1/2} R}$ (d) $\frac{8N\mu_0 I}{5^{1/2} R}$

126. Two identical wires A and B, each of length '*l*', carry the same current I. Wire A is bent into a circle of radius R and wire B is bent to form a square of side 'a'. If B_A and B_B are the values of magnetic field at the centres of the circle and square respectively, then the ratio $\frac{B_A}{B_B}$ is: **[2016]**

(a) $\frac{\pi^2}{16}$ (b) $\frac{\pi^2}{8\sqrt{2}}$ (c) $\frac{\pi^2}{8}$ (d) $\frac{\pi^2}{16\sqrt{2}}$

127. Consider two thin identical conducting wires covered with very thin insulating material. One of the wires is bent into a loop and produces magnetic field B_1, at its centre when a current I passes through it. The second wire is bent into a coil with three identical loops adjacent to each other producer magnetic field B_2 at the centre of loops when current I/3 passer through it. The ratio $B_1 : B_2$ is: **[NCERT : P L-115, 116 | Online April 12, 2014]**

(a) 1 : 1 (b) 1 : 3 (c) 1 : 9 (d) 9 : 1

128. A current *i* is flowing in a straight conductor of length *L*. The magnetic induction at a point on its axis at a distance $\frac{L}{4}$ from its centre will be: **[NCERT : P L-118 | Online April 22, 2013]**

(a) Zero (b) $\frac{\mu_0 i}{2\pi L}$ (c) $\frac{\mu_0 i}{\sqrt{2}L}$ (d) $\frac{4\mu_0 i}{\sqrt{5}\pi L}$

129. An electric current is flowing through a circular coil of radius R. The ratio of the magnetic field at the centre of the coil and that at a distance $2\sqrt{2}R$ from the centre of the coil and on its axis is: **[NCERT : P L-115, 116 | Online April 9, 2013]**

(a) $2\sqrt{2}$ (b) 27
(c) 36 (d) 8

130. A charge *Q* is uniformly distributed over the surface of non-conducting disc of radius *R*. The disc rotates about an axis perpendicular to its plane and passing through its centre with an angular velocity ω. As a result of this rotation a magnetic field of induction B is obtained at the centre of the disc. If we keep both the amount of charge placed on the disc and its angular velocity to be constant and vary the radius of the disc then the variation of the magnetic induction at the centre of the disc will be represented by the figure: **[2012]**

(a) 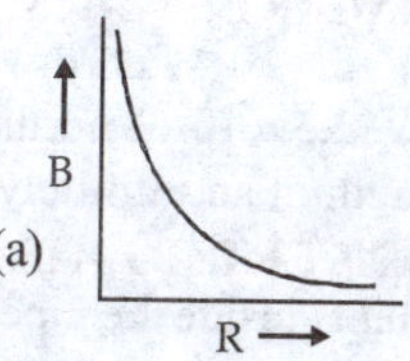

(b) 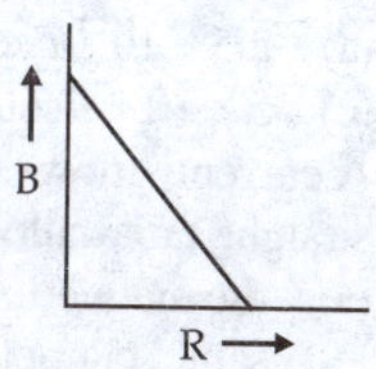

(c) 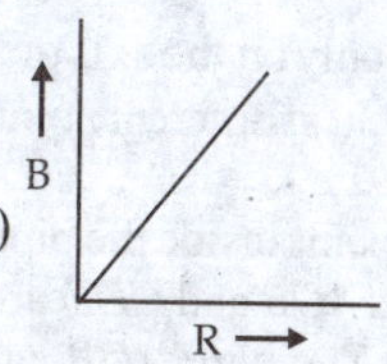

(d) 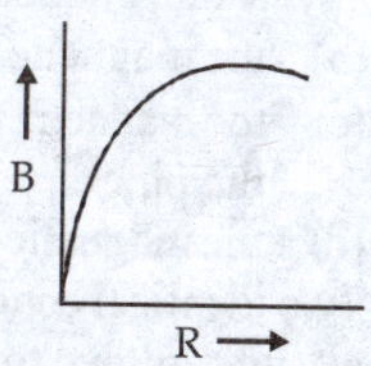

131. A current *I* flows in an infinitely long wire with cross section in the form of a semi-circular ring of radius *R*. The magnitude of the magnetic induction along its axis is: **[NCERT : P L-118 | 2011]**

(a) $\frac{\mu_0 I}{2\pi^2 R}$ (b) $\frac{\mu_0 I}{2\pi R}$

(c) $\frac{\mu_0 I}{4\pi R}$ (d) $\frac{\mu_0 I}{\pi^2 R}$

132. Two long parallel wires are at a distance 2d apart. They carry steady equal currents flowing out of the plane of the paper as shown. The variation of the magnetic field B along the line XX' is given by **[NCERT : P L-118 | 2010]**

(a) 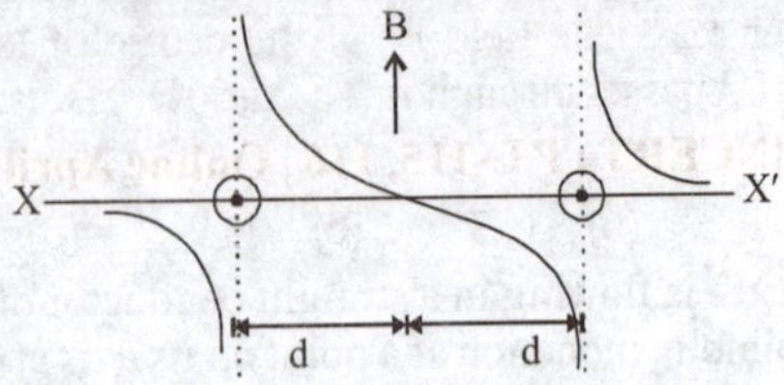

(b) 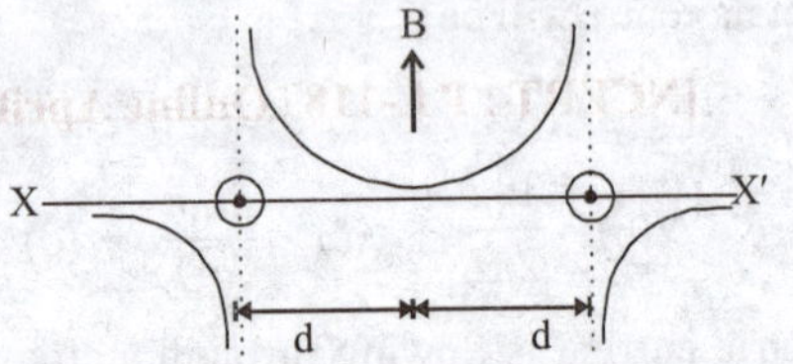

(c) 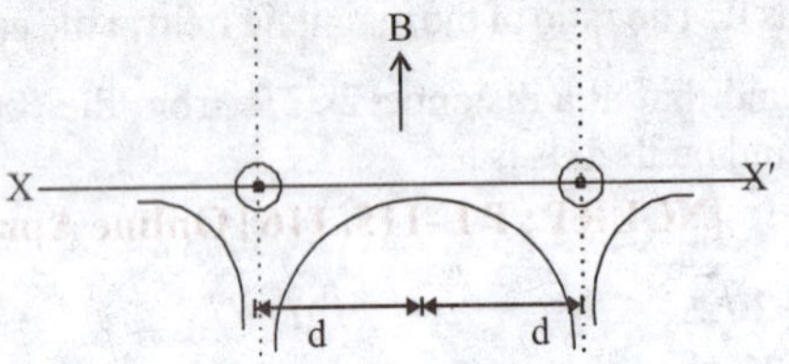

(d) 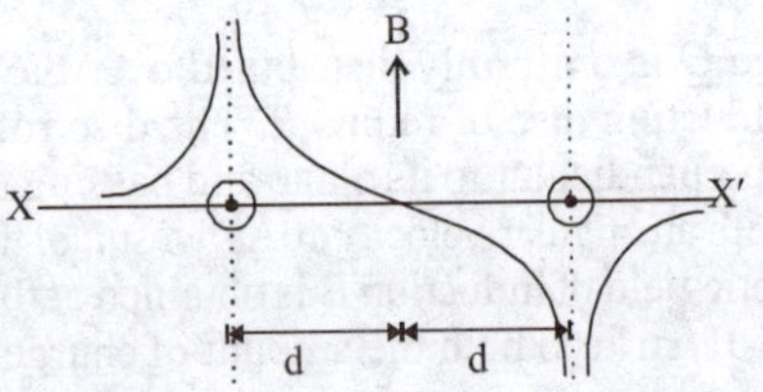

133. A horizontal overhead powerline is at height of 4m from the ground and carries a current of 100A from east to west. The magnetic field directly below it on the ground is ($\mu_0 = 4\pi \times 10^{-7}$ Tm A^{-1}) **[NCERT : P L-118 | 2008]**

(a) 2.5×10^{-7} *T* southward (b) 5×10^{-6} *T* northward
(c) 5×10^{-6} *T* southward (d) 2.5×10^{-7} *T* northward

134. A current *I* flows along the length of an infinitely long, straight, thin walled pipe. Then **[NCERT : P L-118 | 2007]**

(a) the magnetic field at all points inside the pipe is the same, but not zero
(b) the magnetic field is zero only on the axis of the pipe
(c) the magnetic field is different at different points inside the pipe
(d) the magnetic field at any point inside the pipe is zero

135. Two identical conducting wires *AOB* and *COD* are placed at right angles to each other. The wire *AOB* carries an electric current I_1 and *COD* carries a current I_2. The magnetic field on a point lying at a distance d from *O*, in a direction perpendicular to the plane of the wires *AOB* and *COD*, will be given by **[NCERT : P L-118 | 2007]**

(a) $\frac{\mu_0}{2\pi d}(I_1^2 + I_2^2)$ (b) $\frac{\mu_0}{2\pi}\left(\frac{I_1 + I_2}{d}\right)^{\frac{1}{2}}$

(c) $\frac{\mu_0}{2\pi d}\left(I_1^2 + I_2^2\right)^{\frac{1}{2}}$ (d) $\frac{\mu_0}{2\pi d}(I_1 + I_2)$

136. A long solenoid has 200 turns per cm and carries a current *i*. The magnetic field at its centre is 6.28×10^{-2} Weber/m^2. Another long solenoid has 100 turns per cm and it carries a current $\frac{i}{3}$. The value of the magnetic field at its centre is **[NCERT : P L-122 | 2006]**

(a) 1.05×10^{-2} Weber/m^2 (b) 1.05×10^{-5} Weber/m^2
(c) 1.05×10^{-3} Weber/m^2 (d) 1.05×10^{-4} Weber/m^2

137. A current *i* ampere flows along an infinitely long straight thin walled tube, then the magnetic induction at any point inside the tube is **[NCERT : P L-118 | 2004]**

(a) $\frac{\mu_0}{4\pi} \cdot \frac{2i}{r}$ tesla (b) zero
(c) infinite (d) $\frac{2i}{r}$ tesla

138. A long wire carries a steady current. It is bent into a circle of one turn and the magnetic field at the centre of the coil is *B*. It is then bent into a circular loop of n turns. The magnetic field at the centre of the coil will be **[NCERT : P L-116 | 2004]**

(a) $2n\,B$ (b) $n^2\,B$ (c) nB (d) $2\,n^2\,B$

139. The magnetic field due to a current carrying circular loop of radius 3 cm at a point on the axis at a distance of 4 cm from the centre is 54 μT. What will be its value at the centre of loop? **[NCERT : P L-115, 116 | 2004]**

(a) 125 μT (b) 150 μT (c) 250 μT (d) 75 μT

140. If in a circular coil *A* of radius *R*, current *I* is flowing and in another coil *B* of radius 2*R* a current 2*I* is flowing, then the ratio of the magnetic fields B_A and B_B, produced by them at their centres will be **[NCERT : P L-116 | 2002]**

(a) 1 (b) 2 (c) 1/2 (d) 4

Topic 3 - Force and Torque on Current Carrying Conductor

141. A magnetic dipole experiences a torque of $80\sqrt{3}$ N m when placed in uniform magnetic field in such a way that dipole moment makes angle of 60° with magnetic field. The potential energy of the dipole is : **[NCERT : P L-125 | April 3, 2025 (II)]**

(a) 80 J (b) $-40\sqrt{3}$ J (c) – 60 J (d) – 80 J

142. A 4.0 cm long straight wire carrying a current of 8A is placed perpendicular to an uniform magnetic field of strength 0.15 T. The magnetic force on the wire is ________ mN. **[NCERT : P L-110 | April 3, 2025 (I)]**

143. Consider a long thin conducting wire carrying a uniform current I. A particle having mass "m" and charge "*q*" is released at a distance "*a*" from the wire with a speed v_o along the direction of current in the wire. The particle gets attracted to the wire due to magnetic force. The particle turns round when it is at distance *x* from the wire. The value of *x* is [μ_0 is vacuum permeability] **[NCERT : P L-123 | Jan. 28, 2025 (I)]**

(a) $a\left[1 - \frac{mv_0}{2q\mu_0 I}\right]$ (b) $\frac{a}{2}$

(c) $a\left[1 - \frac{mv_0}{q\mu_0 I}\right]$ (d) $ae^{\frac{-4\pi mv_0}{q\mu_0 I}}$

144. Two long parallel wires X and Y, separated by a distance of 6 cm, carry currents of 5A and 4A, respectively, in opposite directions as shown in the figure. Magnitude of the resultant magnetic field at point P at a distance of 4 cm from wire Y is $x \times 10^{-5}$ T. The value of x is ______. (Take permeability of free space as $\mu_0 = 4\pi \times 10^{-7}$ SI units.)

[NCERT : P L-123 | Jan. 22, 2025 (II)]

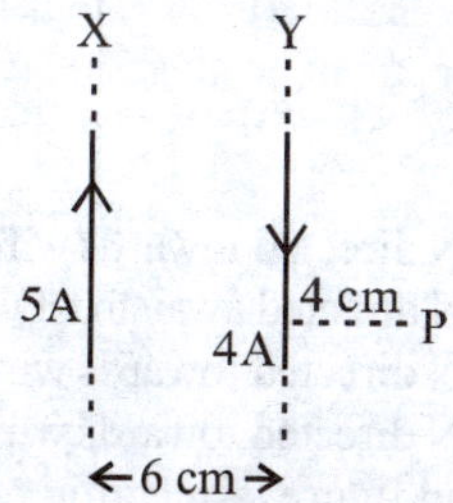

145. A 2A current carrying straight metal wire of resistance 1 Ω, resistivity 2×10^{-6} Ωm, area of cross-section 10 mm^2 and mass 500 g is suspended horizontally in mid air by applying a uniform magnetic field $\vec{B}$. The magnitude of B is .... $\times 10^{-1}$ T (given, $g = 10$ m/s^2)

[NCERT : P L-110 | April 5, 2024 (I)]

146. The magnetic field existing in a region is given by $\vec{B} - 0.2(1 + 2x)\hat{k}T$ A square loop of edge 50 cm carrying 0.5 A current is placed in x-y plane with its edges parallel to the x-y axes, as shown in figure. The magnitude of the net magnetic force experienced by the loop is ______ mN.

[NCERT : P L-110 | April 4, 2024 (I)]

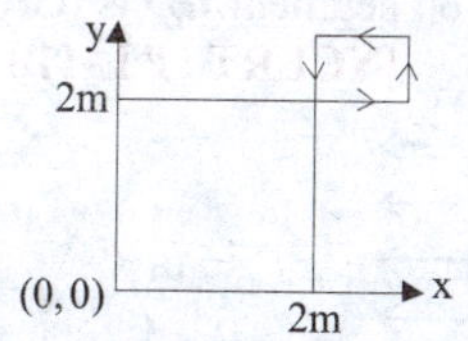

147. A rigid wire consists of a semicircular portion of radius R and two straight sections. The wire is partially immerged in a perpendicular magnetic field $B = B_0\ \hat{k}$ as shown in figure. The magnetic force on the wire if it has a current i is :

[NCERT : P L-110 | Jan. 31, 2024 (I)]

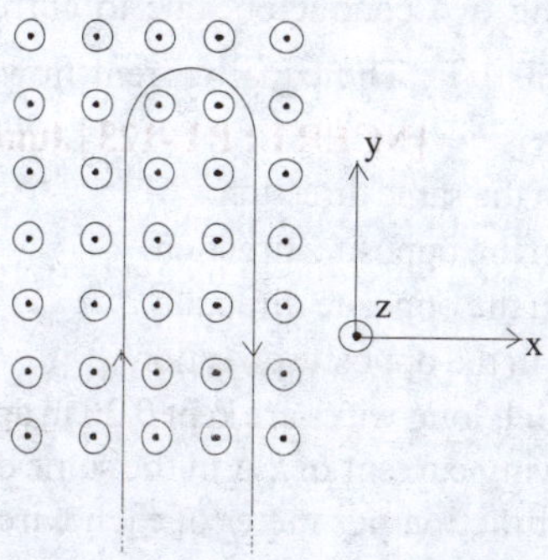

(a) $-2iBR\ \hat{j}$
(b) $2iBR\ \hat{j}$
(c) $-iBR\ \hat{j}$
(d) $iBR\ \hat{j}$

148. A uniform magnetic field of 2×10^{-3}T acts along positive Y-direction. A rectangular loop of sides 20 cm and 10 cm with current of 5 A is Y-Z plane. The current is in anticlockwise sense with reference to negative X axis. Magnitude and direction of the torque is :

[NCERT : P L-126 | Jan. 31, 2024 (II)]

(a) 2×10^{-4} N – m along positive Z -direction
(b) 2×10^{-4} N – m along negative Z -direction
(c) 2×10^{-4} N – m along positive X -direction
(d) 2×10^{-4} N – m along positive Y -direction

149. The horizontal component of earth's magnetic field at a place is 3.5×10^{-5} T. A very long straight conductor carrying current of $\sqrt{2}$ A in the direction from South east to North West is placed. The force per unit length experienced by the conductor is ________ $\times 10^{-6}$ N/m.

[NCERT : P L-110 | Jan. 30, 2024 (I)]

150. Two long, straight wires carry equal currents in opposite directions as shown in figure. The separation between the wires is 5.0 cm. The magnitude of the magnetic field at a point P midway between the wires is _____ μT (Given : $\mu_0 = 4\pi \times 10^{-7}$ TmA^{-1})

[NCERT : P L-123 | Jan. 27, 2024 (I)]

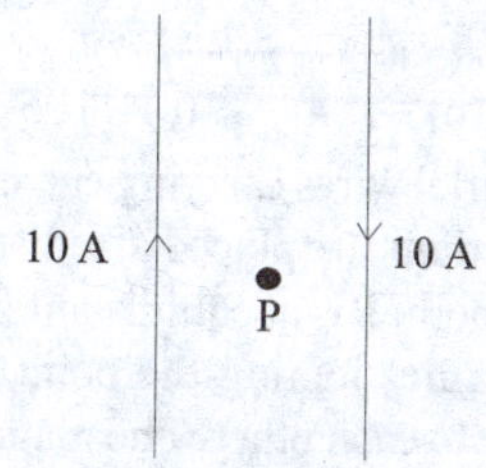

151. A straight wire AB of mass 40 g and length 50 cm is suspended by a pair of flexible leads in uniform magnetic field of magnitude 0.40 T as shown in the figure. The magnitude of the current required in the wire to remove the tension in the supporting leads is ______ A. (Take $g = 10\ ms^{-2}$).

[NCERT : P L-110 | April 13, 2023 (II)]

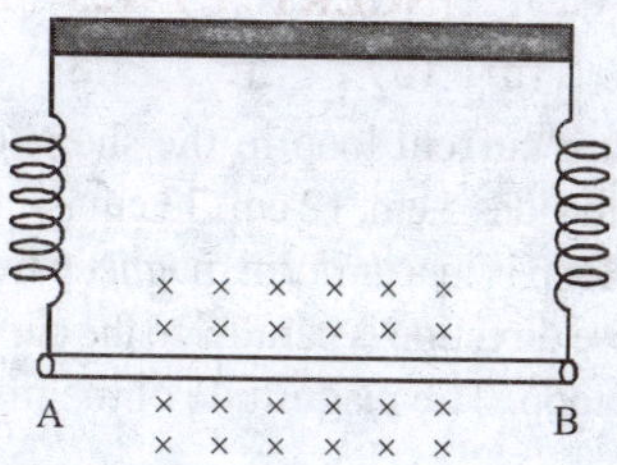

152. A massless square loop, of wire of resistance 10Ω. supporting a mass of 1 g. hangs vertically with one of its sides in a uniform magnetic field of 10^3 G, directed outwards in the shaded region. A dc voltage V is applied to the loop. For what value of V. the magnetic force will exactly balance the weight of the supporting mass of 1g? (If sides of the loop = 10 cm, $g = 10\ ms^{-2}$)

[NCERT : P L-110 | Jan. 30, 2023 (I)]

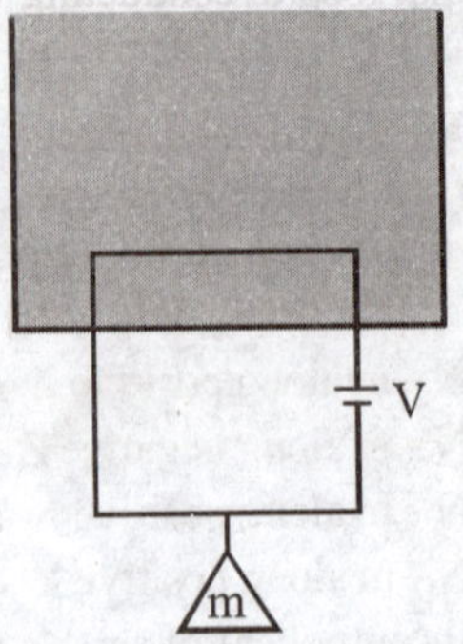

(a) $\frac{1}{10}$V (b) 100V
(c) 1V (d) 10V

153. A current carrying rectangular loop PQRS is made of uniform wire. The length PR = QS = 5 cm and PQ = RS = 100 cm. If ammeter current reading changes from I to 2I, the ratio of magnetic forces per unit length on the wire PQ due to wire RS in the two cases respectively $f_{PQ}^{I} : f_{PQ}^{2I}$ is:

[NCERT : P L-123 | Jan. 30, 2023 (II)]

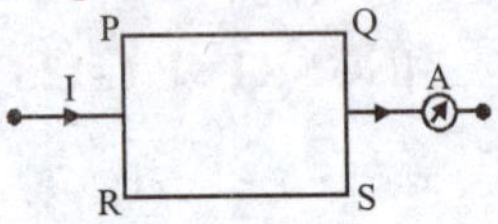

(a) 1 : 2 (b) 1 : 4 (c) 1 : 5 (d) 1 : 3

154. Two long parallel wires carrying currents 8A and 15 A in opposite directions are placed at a distance of 7 cm from each other. A point P is at equidistant from both the wires such that the lines joining the point P to the wires are perpendicular to each other. The magnitude of magnetic field at P is ____ $\times 10^{-6}$ T. (Given : $\sqrt{2} = 1.4$)

[NCERT : P L-123 | Jan. 25, 2023 (II)]

155. Two long straight wires *P* and *Q* carrying equal current 10*A* each were kept parallel to each other at 5 cm distance. Magnitude of magnetic force experienced by 10 cm length of wire *P* is F_1. If distance between wires is halved and currents on them are doubled, force F_2 on 10 cm length of wire P will be : **[NCERT : P L-123 | Jan. 24, 2023 (I)]**

(a) $8F_1$ (b) $10F_1$ (c) $F_1/8$ (d) $F_1/10$

156. A single turn current loop in the shape of a right angle triangle with sides 5 cm, 12 cm, 13 cm is carrying a current of 2A. The loop is in a uniform magnetic field of magnitude 0.75 T whose direction is parallel to the current in the 13 cm side of the loop. The magnitude of the magnetic force on the 5 cm side will be $\frac{x}{130}$ N. The value of x is ______.

[NCERT : P L-110 | Jan. 24, 2023 (II)]

157. A wire X of length 50 cm carrying a current of 2 A is placed parallel to a long wire Y of length 5 m. The wire Y carries a current of 3 A. The distance between two wires is 5 cm and currents flow in the same direction. The force acting on the wire Y is: **[NCERT : P L-123 | July 29, 2022 (II)]**

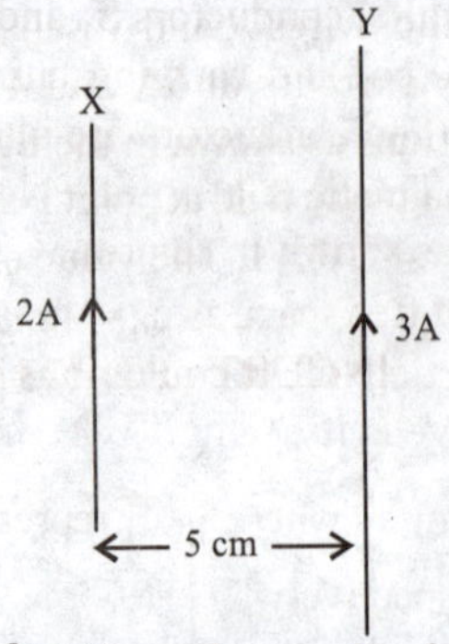

(a) 1.2×10^{-5} N directed towards wire X.
(b) 1.2×10^{-4} N directed away from wire X.
(c) 1.2×10^{-4} N directed towards wire X.
(d) 2.4×10^{-5} N directed towards wire X.

158. As shown in the figure, a metallic rod of linear density 0.45 kg m^{-1} is lying horizontally on a smooth incline plane which makes an angle of 45° with the horizontal. The minimum current flowing in the rod required to keep it stationary, when 0.15 T magnetic field is acting on it in the vertical upward direction, will be : {Use g = 10 m/s^2}

[NCERT : P L-110 | July 28, 2022 (I)]

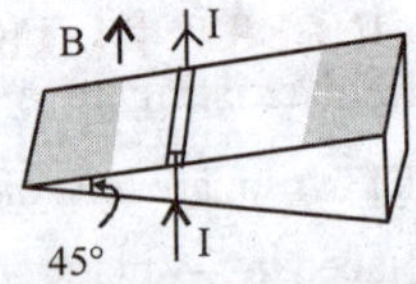

(a) 30 A (b) 15 A (c) 10 A (d) 3 A

159. A triangular shaped wire carrying 10 A current is placed in a uniform magnetic field of 0.5 T, as shown in figure. The magnetic force on segment CD is (Given BC = CD = BD = 5 cm.) **[NCERT : P L-110 | July 28, 2022 (II)]**

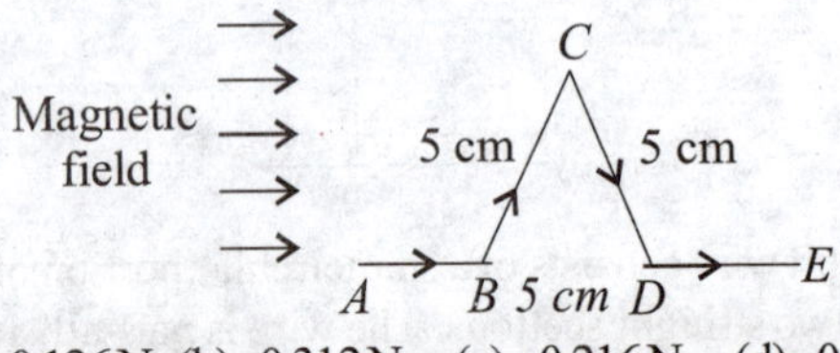

(a) 0.126N (b) 0.312N (c) 0.216N (d) 0.245N

160. Two long current carrying conductors are placed parallel to each other at a distance of 8 cm between them. The magnitude of magnetic field produced at mid-point between the two conductors due to current flowing in them is 300 μT. The equal current flowing in the two conductors is : **[NCERT : P L-123 | June 29, 2022 (II)]**

(a) 30A in the same direction.
(b) 30A in the opposite direction.
(c) 60A in the opposite direction.
(d) 300A in the opposite direction

161. Two parallel, long wires are kept 0.20 m apart in vacuum, each carrying current of x A in the same direction. If the force of attraction per meter of each wire is 2×10^{-6} N, then the value of x is approximately:

[NCERT : P L-123 | June 28, 2022 (II)]

(a) 1 (b) 2.4 (c) 1.4 (d) 2

162. Two long parallel conductors S_1 and S_2 are separated by a distance 10 cm and carrying currents of 4A and 2A respectively. The conductors are placed along x-axis in X-Y plane. There is a point P located between the conductors (as shown in figure).

[NCERT : P L-123 | June 27, 2022 (II)]

A charge particle of 3π coulomb is passing through the point P with velocity.

$\vec{v} = (2\hat{i} + 3\hat{j})$ m/s ; where $\hat{i}$ & $\hat{j}$ represent unit vector along x & y axis respectively.

The force acting on the charge particle is $4\pi \times 10^{-5}(-x\hat{i} + 2\hat{j})$ N. The value of x is:

(a) 2

(b) 1

(c) 3

(d) –3

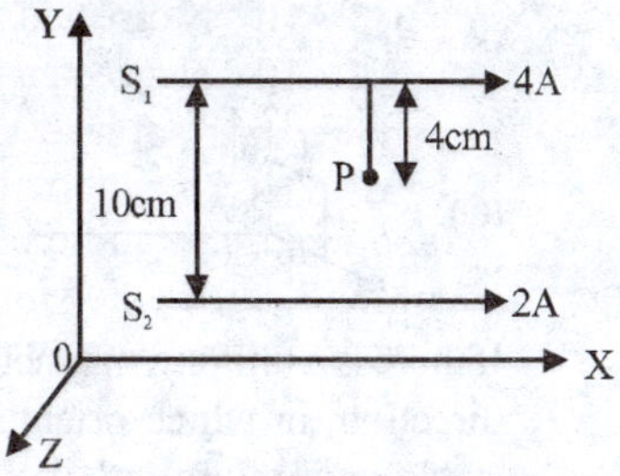

163. Two 10 cm long, straight wires, each carrying a current of 5A are kept parallel to each other. If each wire experienced a force of 10^{-5} N, then separation between the wires is ________cm. **[NCERT : P L-123 | June 26, 2022 (II)]**

164. A square loop of side $2a$ and carrying current I is kept is xz plane with its centre at origin. A long wire carrying the same current I is placed parallel to z-axis and passing through point $(0, b, 0)$, $(b >> a)$. The magnitude of torque on the loop about z-axis will be :

[NCERT : P L-125 | Sep. 06, 2020 (II)]

(a) $\dfrac{2\mu_0 I^2 a^2}{\pi b}$ (b) $\dfrac{2\mu_0 I^2 a^2 b}{\pi(a^2+b^2)}$

(c) $\dfrac{\mu_0 I^2 a^2 b}{2\pi(a^2+b^2)}$ (d) $\dfrac{\mu_0 I^2 a^2}{2\pi b}$

165. A wire carrying current I is bent in the shape $ABCDEFA$ as shown, where rectangle $ABCDA$ and $ADEFA$ are perpendicular to each other. If the sides of the rectangles are of lengths a and b, then the magnitude and direction of magnetic moment of the loop $ABCDEFA$ is :

[NCERT : P L-126 | Sep. 02, 2020 (II)]

(a) abI, along $\left(\dfrac{\hat{j}}{\sqrt{2}} + \dfrac{\hat{k}}{\sqrt{2}}\right)$

(b) $\sqrt{2}abI$, along $\left(\dfrac{\hat{j}}{\sqrt{2}} + \dfrac{\hat{k}}{\sqrt{2}}\right)$

(c) $\sqrt{2}abI$, along $\left(\dfrac{\hat{j}}{\sqrt{5}} + \dfrac{2\hat{k}}{\sqrt{5}}\right)$

(d) abI, along $\left(\dfrac{\hat{j}}{\sqrt{5}} + \dfrac{2\hat{k}}{\sqrt{5}}\right)$

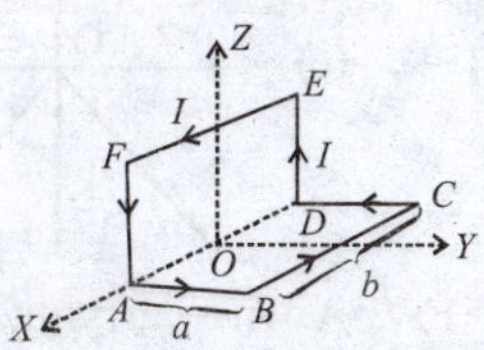

166. A small circular loop of conducting wire has radius a and carries current I. It is placed in a uniform magnetic field B perpendicular to its plane such that when rotated slightly about its diameter and released, it starts performing simple harmonic motion of time period T. If the mass of the loop is m then :

[NCERT : P L-125 | 9 Jan 2020, II]

(a) $T = \sqrt{\dfrac{2m}{IB}}$ (b) $T = \sqrt{\dfrac{\pi m}{2IB}}$

(c) $T = \sqrt{\dfrac{2\pi m}{IB}}$ (c) $T = \sqrt{\dfrac{\pi m}{IB}}$

167. Two wires A & B are carrying currents I_1 and I_2 as shown in the figure. The separation between them is d. A third wire C carrying a current I is to be kept parallel to them at a distance x from A such that the net force acting on it is zero. The possible values of x are :

[NCERT : P L-123 | 10 April 2019, I]

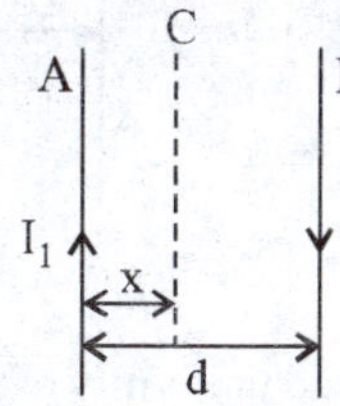

(a) $x = \left(\dfrac{I_1}{I_1 - I_2}\right)d$ and $x = \dfrac{I_2}{(I_1 + I_2)}d$

(b) $x = \left(\dfrac{I_2}{(I_1 + I_2)}\right)d$ and $x = \left(\dfrac{I_2}{(I_1 - I_2)}\right)d$

(c) $x = \left(\dfrac{I_1}{(I_1 + I_2)}\right)d$ and $x = \left(\dfrac{I_2}{(I_1 - I_2)}\right)d$

(d) $x = \pm\dfrac{I_1 d}{(I_1 - I_2)}$

168. A rectangular coil (Dimension 5 cm × 2.5 cm) with 100 turns, carrying a current of 3 A in the clock-wise direction, is kept centered at the origin and in the X-Z plane. A magnetic field of 1 T is applied along X-axis. If the coil is tilted through 45° about Z-axis, then the torque on the coil is: **[NCERT : P L-125 | 9 April 2019 I]**

(a) 0.38Nm (b) 0.55Nm (c) 0.42Nm (d) 0.27Nm

169. A rigid square of loop of side 'a' and carrying current I_2 is lying on a horizontal surface near a long current I_1 carrying wire in the same plane as shown in figure. The net force on the loop due to the wire will be:

[NCERT : P L-123 | 9 April 2019 I]

(a) Repulsive and equal to $\dfrac{\mu_o I_1 I_2}{2\pi}$

(b) Attractive and equal to $\dfrac{\mu_o I_1 I_2}{3\pi}$

(c) Repulsive and equal to $\dfrac{\mu_o I_1 I_2}{4\pi}$

(d) Zero

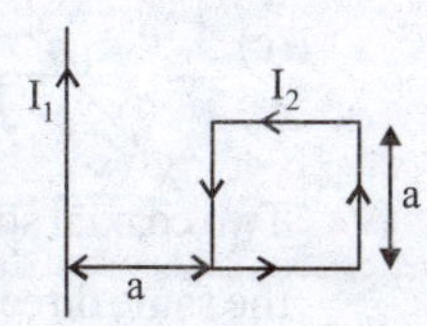

170. A circular coil having N turns and radius r carries a current I. It is held in the XZ plane in a magnetic field B. The torque on the coil due to the magnetic field is : **[NCERT : P L-125 | 8 April 2019 I]**

(a) $\frac{Br^2 I}{\pi N}$ (b) $B\pi r^2 IN$

(c) $\frac{B\pi r^2 I}{N}$ (d) Zero

171. An infinitely long current carrying wire and a small current carrying loop are in the plane of the paper as shown. The redius of the loop is a and distance of its centre from the wire is d (d>>a). If the loop applies a force F on the wire then: **[NCERT : P L-126 | 9 Jan. 2019 I]**

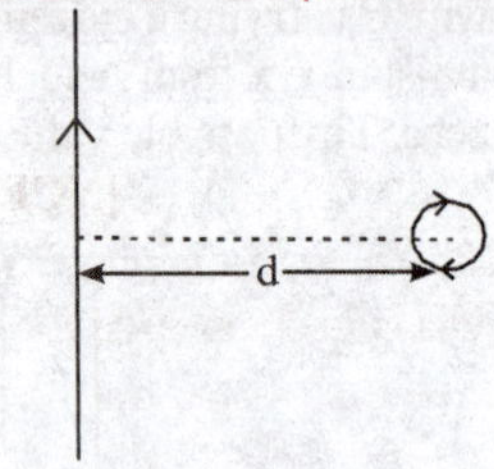

(a) F = 0

(b) $F \propto \left(\frac{a}{d}\right)$

(c) $F \propto \left(\frac{a^2}{d^3}\right)$

(d) $F \propto \left(\frac{a}{d}\right)^2$

172. A charge q is spread uniformly over an insulated loop of radius r . If it is rotated with an angular velocity ω with respect to normal axis then the magnetic moment of the loop is **[NCERT : P L-126 | Online April 16, 2018]**

(a) $\frac{1}{2}q\omega r^2$ (b) $\frac{4}{3}q\omega r^2$

(c) $\frac{3}{2}q\omega r^2$ (d) $q\omega r^2$

173. A uniform magnetic field B of 0.3 T is along the positive Z-direction. A rectangular loop (abcd) of sides 10 cm × 5 cm carries a current I of 12 A. Out of the following different orientations which one corresponds to stable equilibrium? **[NCERT : P L-126 | Online April 9, 2017]**

(a) 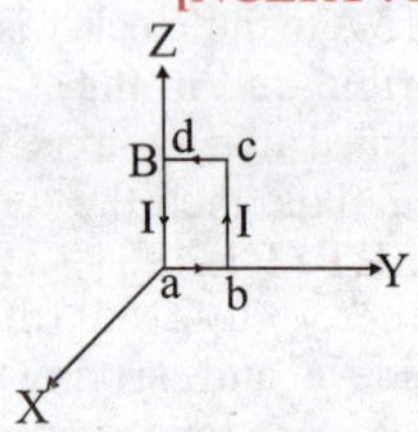

(b) 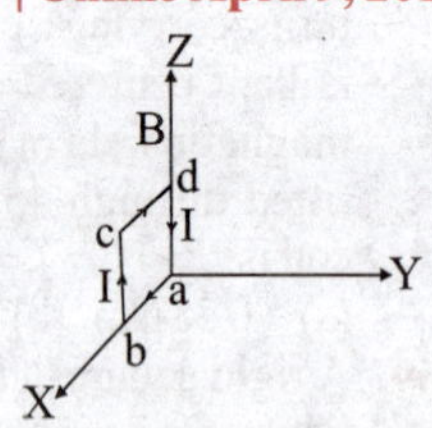

(c) 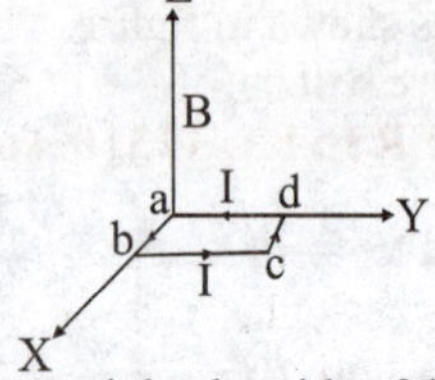

(d) 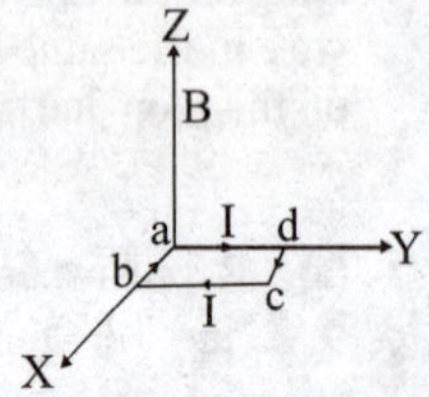

174. Two coaxial solenoids of different radius carry current I in the same direction. $\vec{F_1}$ be the magnetic force on the inner solenoid due to the outer one and $\vec{F_2}$ be the magnetic force on the outer solenoid due to the inner one. Then : **[NCERT : P L-122 | 2015]**

(a) $\vec{F_1}$ is radially inwards and $\vec{F_2} = 0$

(b) $\vec{F_1}$ is radially outwards and $\vec{F_2} = 0$

(c) $\vec{F_1} = \vec{F_2} = 0$

(d) $\vec{F_1}$ is radially inwards and $\vec{F_2}$ is radially outwards

175. A rectangular loop of sides 10 cm and 5 cm carrying a current I of 12 A is placed in different orientations as shown in the figures below :

(A) 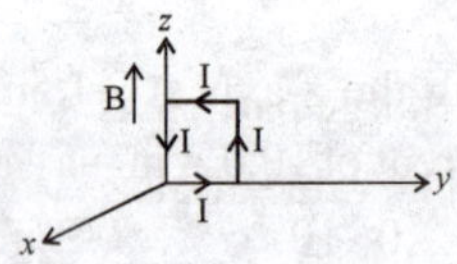

(B) 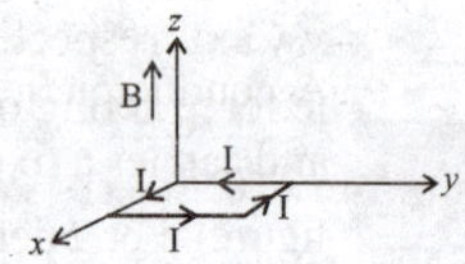

(C) 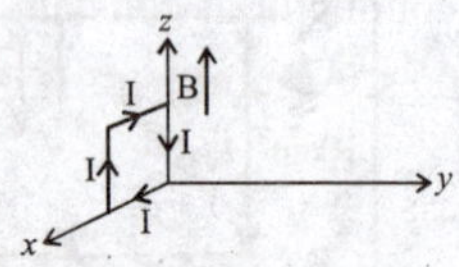

(D) 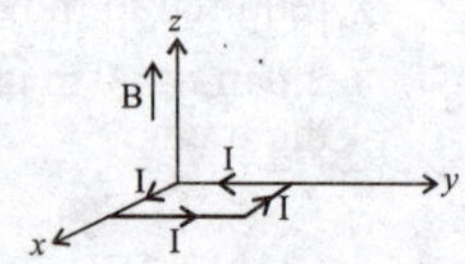

If there is a uniform magnetic field of 0.3 T in the positive z direction, in which orientations the loop would be in (i) stable equilibrium and (ii) unstable equilibrium ?

(a) (B) and (D), respectively **[NCERT : P L-126 | 2015]**

(b) (B) and (C), respectively

(c) (A) and (B), respectively

(d) (A) and (C), respectively

176. Two long straight parallel wires, carrying (adjustable) current I_1 and I_2, are kept at a distance d apart. If the force 'F' between the two wires is taken as 'positive' when the wires repel each other and 'negative' when the wires attract each other, the graph showing the dependence of 'F', on the product $I_1 I_2$, would be : **[NCERT : P L-123 | Online April 11, 2015]**

(a) 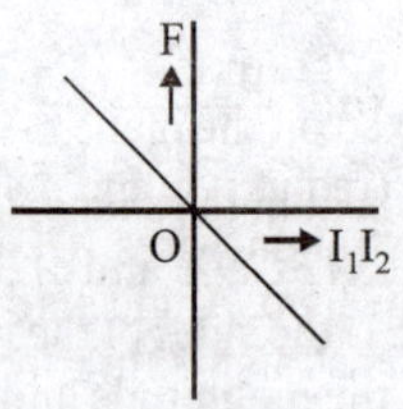

(b) 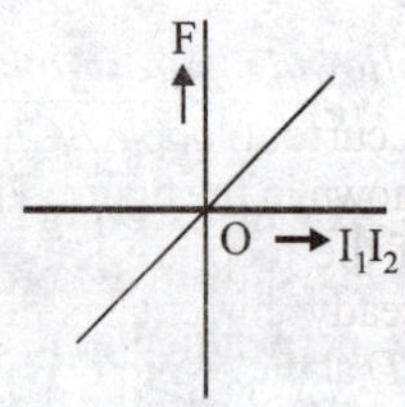

(c) 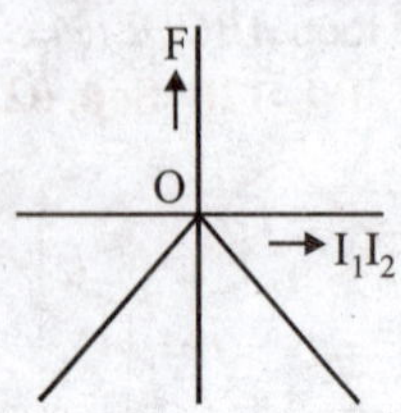

(d) 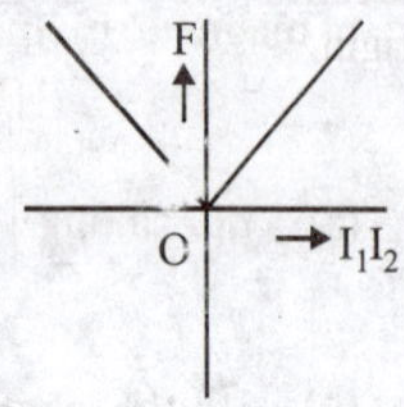

177. A wire carrying current I is tied between points P and Q and is in the shape of a circular arc of radius R due to a uniform magnetic field B (perpendicular to the plane of the paper, shown by xxx) in the vicinity of the wire. If the wire subtends an angle $2\theta_0$ at the centre of the circle (of which it forms an arc) then the tension in the wire is : **[NCERT : P L-110 | Online April 11, 2015]**

(a) $\dfrac{IBR}{2\sin\theta_0}$

(b) $\dfrac{IBR\theta_0}{\sin\theta_0}$

(c) IBR

(d) $\dfrac{IBR}{\sin\theta_0}$

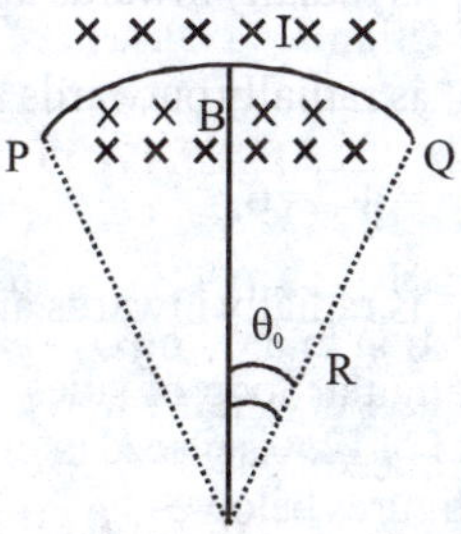

178. A conductor lies along the z-axis at $-1.5\text{m} \le z < 1.5$ m and carries a fixed current of 10.0 A in $-\hat{a}_z$ direction (see figure). For a field $\vec{B} = 3.0\times10^{-4} e^{-0.2x}\ \hat{a}_y$ T, find the power required to move the conductor at constant speed to x = 2.0 m, y = 0 m in 5×10^{-3} s. Assume parallel motion along the x-axis. **[2014]**

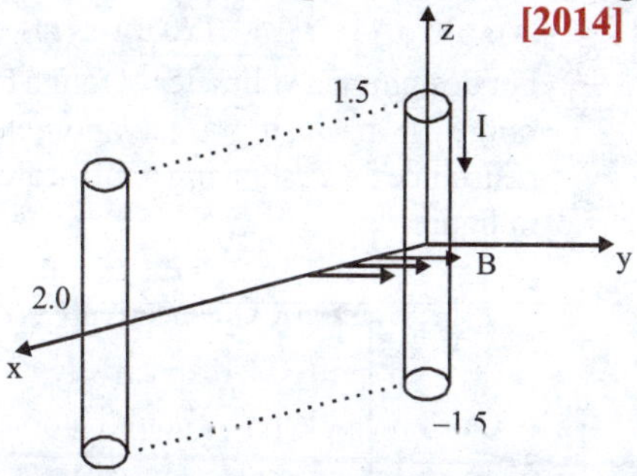

(a) 1.57 W

(b) 2.97 W

(c) 14.85 W

(d) 29.7 W

179. Currents of a 10 ampere and 2 ampere are passed through two parallel thin wires *A* and *B* respectively in opposite directions. Wire *A* is infinitely long and the length of the wire *B* is 2 m. The force acting on the conductor *B*, which is situated at 10 cm distance from *A* will be

[NCERT : P L-123 | Online May 26, 2012]

(a) 8×10^{-5} N (b) 5×10^{-5} N

(c) $8\pi\times10^{-7}$ N (d) $4\pi\times10^{-7}$ N

Directions : *Question numbers 180 and 181 are based on the following paragraph.*

A current loop *ABCD* is held fixed on the plane of the paper as shown in the figure. The arcs *BC* (radius = *b*) and *DA* (radius = *a*) of the loop are joined by two straight wires *AB* and *CD*. A steady current *I* is flowing in the loop. Angle made by *AB* and *CD* at the origin *O* is 30°. Another straight thin wire with steady current I_1 flowing out of the plane of the paper is kept at the origin. **[2009]**

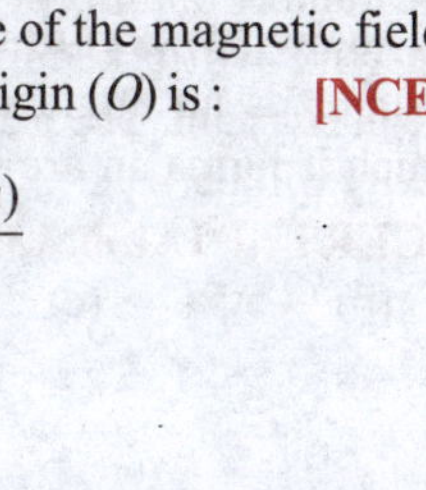

180. The magnitude of the magnetic field (*B*) due to the loop *ABCD* at the origin (*O*) is : **[NCERT : P L-116 | 2009]**

(a) $\dfrac{\mu_0 I(b-a)}{24ab}$

(b) $\dfrac{\mu_0 I}{4\pi}\left[\dfrac{b-a}{ab}\right]$

(c) $\dfrac{\mu_0 I}{4\pi}[2(b-a)+\pi/3(a+b)]$

(d) zero

181. Due to the presence of the current I_1 at the origin: **[NCERT : P L-110 | 2009]**

(a) The forces on *AD* and *BC* are zero.

(b) The magnitude of the net force on the loop is given by $\dfrac{I_1 I}{4\pi}\mu_0[2(b-a)+\pi/3(a+b)]$.

(c) The magnitude of the net force on the loop is given by $\dfrac{\mu_0 I I_1}{24ab}(b-a)$.

(d) The forces on *AB* and *DC* are zero.

182. Two long conductors, separated by a distance d carry current I_1 and I_2 in the same direction. They exert a force *F* on each other. Now the current in one of them is increased to two times and its direction is reversed. The distance is also increased to 3*d*. The new value of the force between them is **[NCERT : P L-123 | 2004]**

(a) $-\dfrac{2F}{3}$ (b) $\dfrac{F}{3}$ (c) $-2F$ (d) $-\dfrac{F}{3}$

183. If a current is passed through a spring then the spring will **[2002]**

(a) expand (b) compress

(c) remains same (d) none of these

184. Wires 1 and 2 carrying currents i_1 and i_2 respectively are inclined at an angle θ to each other. What is the force on a small element *dl* of wire 2 at a distance of *r* from wire 1 (as shown in figure) due to the magnetic field of wire 1? **[NCERT : P L-110 | 2002]**

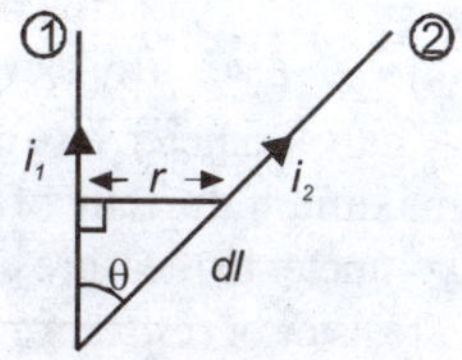

(a) $\dfrac{\mu_0}{2\pi r} i_1 i_2\, dl \tan\theta$

(b) $\dfrac{\mu_0}{2\pi r} i_1 i_2\, dl \sin\theta$

(c) $\dfrac{\mu_0}{2\pi r} i_1 i_2\, dl \cos\theta$

(d) $\dfrac{\mu_0}{4\pi r} i_1 i_2\, dl \sin\theta$

Topic 4 - Galvanometer and its Conversion into Ammeter and Voltmeter

185. In a moving coil galvanometer, two moving coils M_1 and M_2 have the following particulars :

$R_1 = 5\,\Omega, N_1 = 15, A_1 = 3.6\times10^{-3}\,m^2, B_1 = 0.25$ T

$R_2 = 7\,\Omega, N_2 = 21, A_2 = 1.8\times10^{-3}\,m^2, B_2 = 0.50$ T

Assuming that torsional constant of the springs are same for both coils, what will be the ratio of voltage sensitivity of M_1 and M_2 ?

[NCERT : P L-130, 131 | April 2, 2025 (II)]

(a) 1 : 1 (b) 1 : 4 (c) 1 : 3 (d) 1 : 2

186. A galvanometer having a coil of resistance 30 Ω need 20 mA of current for full-scale deflection. If a maximum current of 3 A is to be measured using this galvanometer, the resistance of the shunt to be added to the galvanometer should be $\frac{30}{X}\Omega$, where X is

[NCERT : P L-131 | Jan. 23, 2025 (II)]

(a) 447 (b) 298 (c) 149 (d) 596

187. Consider a moving coil galvanometer (MCG) :

[NCERT : P L-130, 131 | Jan. 23, 2025 (I)]

A: The torsional constant in moving coil galvanometer has dimensions $[ML^2T^{-2}]$

B: Increasing the current sensitivity may not necessarily increase the voltage sensitivity.

C: If we increase number of turns (N) to its double (2N), then the voltage sensitivity doubles.

D: MCG can be converted into an ammeter by introducing a shunt resistance of large value in parallel with galvanometer.

E: Current sensitivity of MCG depends inversely on number of turns of coil.

Choose the correct answer from options given below:

(a) A, B only (b) A, D, only
(c) B, D, E only (d) A, B, E only

188. A galvanometer has a coil of resistance 200 Ω with a full scale deflection at 20 μA. The value of resistance to be added to use it as an ammeter of range (0 – 20) mA is:

[NCERT : P L-131 | April 9, 2024 (I)]

(a) 0.40 Ω (b) 0.20 Ω
(c) 0.50 Ω (d) 0.10 Ω

189. A galvanometer of resistance 100 Ω when connected in series with 400 Ω measures a voltage of upto 10 V. The value of resistance required to convert the galvanometer into ammeter to read upto 10 A is $x \times 10^{-2}$ Ω. The value of x is : **[NCERT : P L-131 | April 5, 2024 (II)]**

(a) 2 (b) 800 (c) 20 (d) 200

190. A galvanometer has a resistance of 50Ω and it allows maximum current of 5 mA. It can be converted into voltmeter to measure upto to 100V by connected in series a resistor of resistance:

[NCERT : P L-131 | Feb. 1, 2024 (I)]

(a) 5975Ω (b) 20050Ω (c) 19950 Ω (d) 19500Ω

191. In an ammeter, 5% of the main current passes through the galvanometer. If resistance of the galvanometer is G, the resistance of ammeter will be:

[NCERT : P L-131 | Feb. 1, 2024 (II)]

(a) 199G (b) 200G (c) $\frac{G}{200}$ (d) $\frac{G}{19}$

192. A moving coil galvanometer has 100 turns and each turn has an area of 2.0 cm^2. The magnetic field produced by the magnet is 0.01 T and the deflection in the coil is 0.05 radian when a current of 10mA is passed through it. The torsional constant of the suspension wire is $x \times 10^{-5}$ N-m /rad. The value of x is ______.

[NCERT : P L-130 | Feb. 1, 2024 (II)]

193. A galvanometer having coil resistance 10Ω shows a full scale deflection for a current of 3mA. For it to measure a current of 8A, the value of the shunt should be:

[NCERT : P L-131 | Jan. 29, 2024 (I)]

(a) 4.85×10^{-3}Ω (b) 3×10^{-3}Ω
(c) 3.75×10^{-3}Ω (d) 2.75×10^{-3}Ω

194. The deflection in moving coil galvanometer falls from 25 divisions to 5 division when a shunt of 24Ω is applied. The resistance of galvanometer coil will be :

[NCERT : P L-131 | Jan. 29, 2024 (I)]

(a) 100 Ω (b) 48 Ω (c) 12 Ω (d) 96 Ω

195. A current of 200 μA deflects the coil of a moving coil galvanometer through 60^0. The current to cause deflection through $\frac{\pi}{10}$ radian is :

[NCERT : P L-131 | Jan. 27, 2024 (II)]

(a) 30 μA (b) 120 μA (c) 60 μA (d) 180 μA

196. For designing a voltmeter of range 50 V and an ammeter of range 10 mA using a galvanometer which has a coil of resistance 54 Ω showing a full scale deflection for 1 mA as in figure.

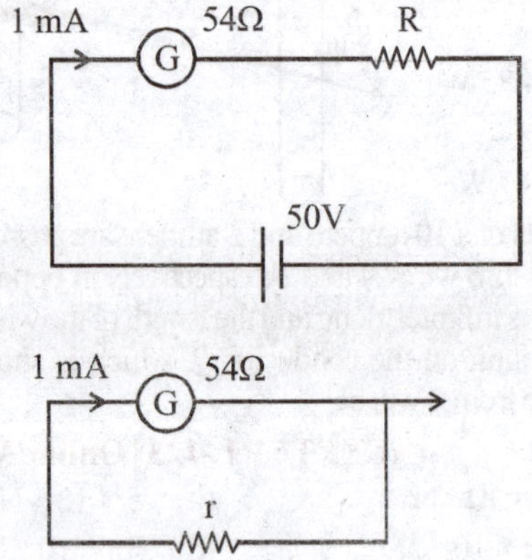

(A) for voltmeter R ≈ 50 kΩ
(B) for ammeter r ≈ 0.2 Ω
(C) for ammeter r ≈ 6 Ω
(D) for voltmeter R ≈ 5 kΩ
(E) for voltmeter R ≈ 500 kΩ

Choose the correct answer from the options given below:

[NCERT : P L-131 | April 15, 2023 (I)]

(a) (C) and (E) (b) (C) and (D)
(c) (A) and (C) (d) (A) and (B)

197. When a resistance of 5 Ω is shunted with a moving coil galvanometer, it shows a full scale deflection for a current of 250 mA, however when 1050 Ω resistance is connected with it in series, it gives full scale deflection for 25 volt. The resistance of galvanometer is ________ Ω.

[NCERT : P L-131 | April 13, 2023 (I)]

198. The current sensitivity of moving coil galvanometer is increased by 25%. This increase is achieved only by changing in the number of turns of coils and area of cross section of the wire while keeping the resistance of galvanometer coil constant. The percentage change in the voltage sensitivity will be:

[NCERT : P L-130, 131 | April 11, 2023 (I)]

(a) +25% (b) –50% (c) Zero (d) –25%

199. Given below are two statement :
Statement I : If the number of turns in the coil of a moving coil galvanometer is doubled then the current sensitivity becomes double.
Statement II : Increasing current sensitivity of a moving coil galvanometer by only increasing the number of turns in the coil will also increase its voltage sensitivity in the same ratio:
In the light of the above statement, choose the correct answer from the options given below :
[NCERT : P L-130, 131 | April 10, 2023 (I)]
(a) Statement I is false but Statement II is true
(b) Both Statement I and Statement II are true
(c) Both Statement I and Statement II are false
(d) Statement I is true but Statement II is false

200. Certain galvanometers have a fixed core made of non magnetic metallic material. The function of this metallic material is **[NCERT : P L-130 | April 8, 2023 (I)]**
(a) to oscillate the coil in magnetic field for longer period of time
(b) to bring the coil to rest quickly
(c) to produce large deflecting torque on the coil
(d) to make the magnetic field radial

201. Given below are two statements: one is labelled as **Assertion A** and the other is labelled as **Reason R**
[NCERT : P L-131 | Feb. 1, 2023 (II)]
Assertion A : For measuring the potential difference across a resistance of 600 Ω, the voltmeter with resistance 1000 Ω will be preferred over voltmeter with resistance 4000 Ω.
Reason R: Voltmeter with higher resistance will draw smaller current than voltmeter with lower resistance.
In the light of the above statements, choose the **most appropriate** answer from the options given below
(a) A is not correct but R is correct
(b) Both A and R are correct and R is the correct explanation of A
(c) Both A and R are correct but R is not the correct explanation of A
(d) A is correct R is not correct

202. The number of turns of the coil of a moving coil galvanometer is increased in order to increase current sensitivity by 50%. The percentage change in voltage sensitivity of the galvanometer will be :
[NCERT : P L-130, 131 | Jan. 31, 2023 (II)]
(a) 100% (b) 50% (c) 75% (d) 0%

203. For a moving coil galvanometer, the deflection in the coil is 0.05 rad when a current of 10 mA is passed through it. If the torsional constant of suspension wire is 4.0×10^{-5} Nm rad^{-1}, the magnetic field is 0.01 T and the number of turns in the coil is 200, the area of each turn (in cm^2) is :
[NCERT : P L-130 | Jan. 25, 2023 (II)]
(a) 2.0 (b) 1.0 (c) 1.5 (d) 0.5

204. The current sensitivity of a galvanometer can be increased by :
(A) decreasing the number of turns
(B) increasing the magnetic field
(C) decreasing the area of the coil
(D) decreasing the torsional constant of the spring
Choose the most appropriate answer from the options given below : **[NCERT : P L-130 | July 28, 2022 (I)]**
(a) (B) and (C) only (b) (C) and (D) only
(c) (A) and (C) only (d) (B) and (D) only

205. A 72 Ω galvanometer is shunted by a resistance of 8 Ω. The percentage of the total current which passes through the galvanometer is :
[NCERT : P L-131 | June 27, 2022 (I)]
(a) 0.1% (b) 10% (c) 25% (d) 0.25%

206. If n represents the actual number of deflections in a converted galvanometer of resistance G and shunt resistance S. Then the total current I when its figure of merit is K will be : **[NCERT : P L-131 | June 25, 2022 (II)]**
(a) $\dfrac{KS}{(S+G)}$ (b) $\dfrac{(G+S)}{nKS}$
(c) $\dfrac{nKS}{(G+S)}$ (d) $\dfrac{nK(G+S)}{S}$

207. Consider a galvanometer shunted with 5Ω resistance and 2% of current passes through it. What is the resistance of the given galvanometer ?
[NCERT : P L-131 | Aug. 31, 2021 (I)]
(a) 300 Ω (b) 344 Ω (c) 245 Ω (d) 226 Ω

208. For full scale deflection of total 50 divisions, 50 mV voltage is required in galvanometer. The resistance of galvanometer if its current sensitivity is 2 div/mA will be:
[NCERT : P L-130 | Aug. 27, 2021 (II)]
(a) 1 Ω (b) 5 Ω (c) 4 Ω (d) 2 Ω

209. A galvanometer of resistance G is converted into a voltmeter of ragne 0 – 1V by connecting a resistance R_1 in series with it. The additional resistance that should be connected in series with R_1 to increase the range of the voltmeter to 0 – 2V will be :
[NCERT : P L-131 | Sep. 05, 2020 (I)]
(a) G (b) R_1
(c) $R_1 - G$ (d) $R_1 + G$

210. A galvanometer is used in laboratory for detecting the null point in electrical experiments. If, on passing a current of 6 mA it produces a deflection of 2°, its figure of merit is close to : **[NCERT : P L-130 | Sep. 05, 2020 (II)]**
(a) 333° A/div. (b) 6×10^{-3} A/div.
(c) 666° A/div. (d) 3×10^{-3} A/div.

211. A galvanometer coil has 500 turns and each turn has an average area of 3×10^{-4} m^2. If a torque of 1.5 Nm is required to keep this coil parallel to a magnetic field when a current of 0.5 A is flowing through it, the strength of the field (in T) is ______. **[NCERT : P L-130 | NA Sep. 03, 2020 (II)]**

212. A galvanometer of resistance 100 Ω has 50 divisions on its scale and has sensitivity of 20 μA/division. It is to be converted to a voltmeter with three ranges, of 0–2V, 0–10 V and 0–20 V. The appropriate circuit to do so is :
[NCERT : P L-131 | 12 April 2019, I]

(a) 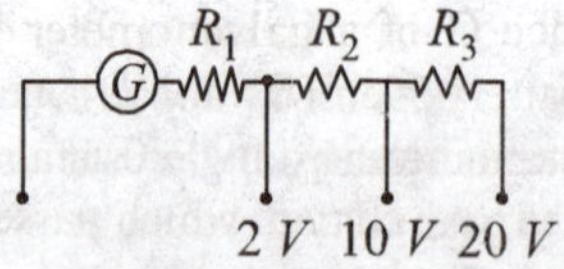

$R_1 = 2000\ \Omega$
$R_2 = 8000\ \Omega$
$R_3 = 10000\ \Omega$

(b) 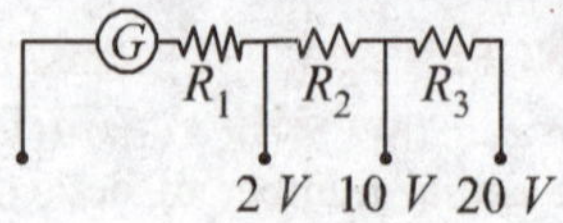

$R_1 = 1900\ \Omega$
$R_2 = 9900\ \Omega$
$R_3 = 19900\ \Omega$

(c) 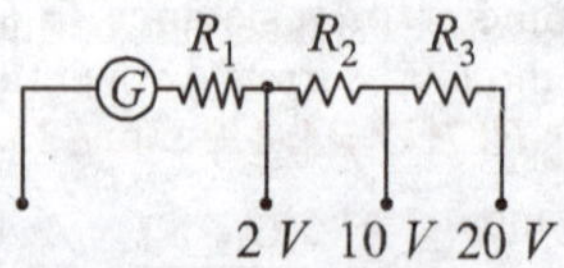

$R_1 = 1900\ \Omega$
$R_2 = 8000\ \Omega$
$R_3 = 10000\ \Omega$

(d) 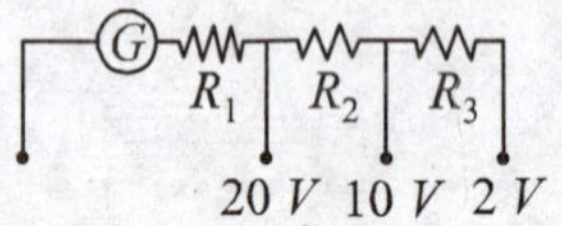

$R_1 = 19900\ \Omega$
$R_2 = 9900\ \Omega$
$R_3 = 1900\ \Omega$

213. A moving coil galvanometer, having a resistance G, produces full scale deflection when a current I_g flows through it. This galvanometer can be converted into (i) an ammeter of range 0 to I_0 ($I_0 > I_g$) by connecting a shunt resistance R_A to it and (ii) into a voltmeter of range 0 to V ($V = GI_0$) by connecting a series resistance R_V to it. Then,

[NCERT : P L-131 | 12 April 2019, II]

(a) $R_A R_V = G^2\left(\frac{I_0 - I_g}{I_g}\right)$ and $\frac{R_A}{R_V} = \left(\frac{I_g}{I_0 - I_g}\right)^2$

(b) $R_A R_V = G^2$ and $\frac{R_A}{R_V} = \left(\frac{I_g}{I_0 - I_g}\right)^2$

(c) $R_A R_V = G^2\left(\frac{I_g}{I_0 - I_g}\right)$ and $\frac{R_A}{R_V} = \left(\frac{I_0 - I_g}{I_g}\right)^2$

(d) $R_A R_V = G^2$ and $\frac{R_A}{R_V} = \frac{I_g}{(I_0 - I_g)}$

214. A moving coil galvanometer allows a full scale current of 10^{-4} A. A series resistance of 2 MΩ is required to convert the above galvanometer into a voltmeter of range 0 – 5 V. Therefore the value of shunt resistance required to convert the above galvanometer into an ammeter of range 0 – 10 mA is: **[NCERT : P L-131 | 10 April 2019, I]**

(a) 500 Ω (b) 100 Ω (c) 200 Ω (d) 10 Ω

215. A moving coil galvanometer has resistance 50 Ω and it indicates full deflection at 4 mA current. A voltmeter is made using this galvanometer and a 5 kΩ resistance. The maximum voltage, that can be measured using this voltmeter, will be close to:

[NCERT : P L-131 | 9 April 2019 I]

(a) 40 V (b) 15 V (c) 20 V (d) 10 V

216. A moving coil galvanometer has a coil with 175 turns and area 1 cm^2. It uses a torsion band of torsion constant 10^{-6} N-m/rad. The coil is placed in a magnetic field B parallel to its plane. The coil deflects by 1° for a current of 1 mA. The value of B (in tesla) is approximately:

[NCERT : P L-130 | 9 April 2019, II]

(a) 10^{-4} (b) 10^{-2} (c) 10^{-1} (d) 10^{-3}

217. The resistance of a galvanometer is 50 ohm and the maximum current which can be passed through it is 0.002 A. What resistance must be connected to it order to convert it into an ammeter of range 0 – 0.5 A?

[NCERT : P L-131 | 9 April 2019, II]

(a) 0.5 ohm (b) 0.002 ohm
(c) 0.02 ohm (d) 0.2 ohm

218. The galvanometer deflection, when key K_1 is closed but K_2 is open, equals θ_0 (see figure). On closing K_2 also and adjusting R_2 to 5Ω, the deflection in galvanometer becomes $\frac{\theta_0}{5}$. The resistance of the galvanometer is, then, given by [Neglect the internal resistance of battery]:

[NCERT : P L-131 | 12 Jan 2019, I]

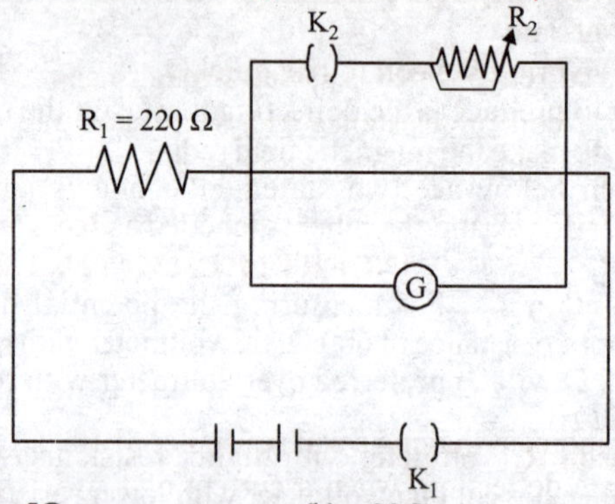

(a) 5Ω (b) 22Ω
(c) 25Ω (d) 12Ω

219. A galvanometer, whose resistance is 50 ohm, has 25 divisions in it. When a current of 4×10^{-4} A passes through it, its needle (pointer) deflects by one division. To use this galvanometer as a voltmeter of range 2.5 V, it should be connected to a resistance of :

[NCERT : P L-131 | 12 Jan 2019, II]

(a) 250 ohm (b) 200 ohm
(c) 6200 ohm (d) 6250 ohm

220. A galvanometer having a resistance of 20 Ω and 30 division on both sides has figure of merit 0.005 ampere/division. The resistance that should be connected in series such that it can be used as a voltmeter upto 15 volt, is:

[NCERT : P L-131 | 11 Jan 2019, II]

(a) 100 Ω (b) 120 Ω
(c) 80 Ω (d) 125 Ω

221. A galvanometer having a coil resistance 100 Ω gives a full scale deflection when a current of 1 mA is passed through it. What is the value of the resistance which can convert this galvanometer into a voltmeter giving full scale deflection for a potential difference of 10 V?

[NCERT : P L-131 | 8 Jan 2019, II]

(a) 10 kΩ (b) 8.9 kΩ
(c) 7.9 kΩ (d) 9.9 kΩ

222. In a circuit for finding the resistance of a galvanometer by half deflection method, a 6 V battery and a high resistance of 11kΩ are used. The figure of merit of the galvanometer 60μA/division. In the absence of shunt resistance, the galvanometer produces a deflection of $\theta = 9$ divisions when current flows in the circuit. The value of the shunt resistance that can cause the deflection of $\theta/2$, is closest to **[NCERT : P L-131 | Online April 16, 2018]**

(a) 55Ω (b) 110Ω (c) 220Ω (d) 550Ω

223. A galvanometer with its coil resistance 25Ω requires a current of 1mA for its full deflection. In order to construct an ammeter to read up to a current of 2A, the approximate value of the shunt resistance should be

[NCERT : P L-131 | Online April 16, 2018]

(a) $2.5 \times 10^{-2}\Omega$ (b) $1.25 \times 10^{-3}\Omega$
(c) $2.5 \times 10^{-3}\Omega$ (d) $1.25 \times 10^{-2}\Omega$

224. When a current of 5 mA is passed through a galvanometer having a coil of resistance 15 Ω, it shows full scale deflection. The value of the resistance to be put in series with the galvanometer to convert it into to voltmeter of range 0 – 10 V is **[NCERT : P L-131 | 2017]**

(a) $2.535\times10^3\Omega$ (b) $4.005\times10^3\Omega$
(c) $1.985\times10^3\Omega$ (d) $2.045\times10^3\Omega$

225. A galvanometer having a coil resistance of 100 Ω gives a full scale deflection, when a currect of 1 mA is passed through it. The value of the resistance, which can convert this galvanometer into ammeter giving a full scale deflection for a current of 10 A, is : **[NCERT : P L-131 | 2016]**

(a) 0.1 Ω (b) 3Ω (c) 0.01Ω (d) 2Ω

226. A 50 Ω resistance is connected to a battery of 5V. A galvanometer of resistance 100 Ω is to be used as an ammeter to measure current through the resistance, for this a resistance r_s is connected to the galvanometer. Which of the following connections should be employed if the measured current is within 1% of the current without the ammeter in the circuit?

[NCERT : P L-131 | Online April 9, 2016]

(a) $r_s = 0.5\ \Omega$ in series with the galvanometer
(b) $r_s = 1\ \Omega$ in series with galvanometer
(c) $r_s = 1\Omega$ in parallel with galvanometer
(d) $r_s = 0.5\ \Omega$ in parallel with the galvanometer.

227. To know the resistance G of a galvanometer by half deflection method, a battery of emf V_E and resistance R is used to deflect the galvanometer by angle θ. If a shunt of resistance S is needed to get half deflection then G, R and S related by the equation:

[NCERT : P L-131 | Online April 9, 2016]

(a) $S(R+G)=RG$ (b) $2S(R+G)=RG$
(c) $2G=S$ (d) $2S=G$

228. The AC voltage across a resistance can be measured using a : **[NCERT : P L-130 | Online April 11, 2015]**

(a) hot wire voltmeter
(b) moving coil galvanometer
(c) potential coil galvanometer
(d) moving magnet galvanometer

229. This questions has Statement I and Statement II. Of the four choices given after the Statements, choose the one that best describes into two Statements.

Statement-I : Higher the range, greater is the resistance of ammeter.

Statement-II : To increase the range of ammeter, additional shunt needs to be used across it.

[NCERT : P L-131 | 2013]

(a) Statement-I is true, Statement-II is true, Statement-II is the correct explanation of Statement-I.
(b) Statement-I is true, Statement-II is true, Statement-II is not the correct explanation of Statement-I.
(c) Statement-I is true, Statement-II is false.
(d) Statement-I is false, Statement-II is true.

230. A shunt of resistance 1 Ω is connected across a galvanometer of 120 Ω resistance. A current of 5.5 ampere gives full scale deflection in the galvanometer. The current that will give full scale deflection in the absence of the shunt is nearly :

[NCERT : P L-131 | Online April 9, 2013]

(a) 5.5 ampere (b) 0.5 ampere
(c) 0.004 ampere (d) 0.045 ampere

231. A moving coil galvanometer has 150 equal divisions. Its current sensitivity is 10-divisions per milliampere and voltage sensitivity is 2 divisions per millivolt. In order that each division reads 1 volt, the resistance in ohms needed to be connected in series with the coil will be -

[NCERT : P L-130, 131 | 2005]

(a) 10^5 (b) 10^3 (c) 9995 (d) 99995

ANSWER KEYS																			
1	(d)	25	(10)	49	(c)	73	(20)	97	(d)	121	(c)	145	(5)	169	(c)	193	(c)	217	(d)
2	(d)	26	(2)	50	(a)	74	(c)	98	(d)	122	(d)	146	(50)	170	(b)	194	(d)	218	(b)
3	(0 or 10)	27	(c)	51	(b)	75	(40)	99	(b)	123	(d)	147	(a)	171	(d)	195	(c)	219	(b)
4	(a)	28	(a)	52	(b)	76	(3)	100	(a)	124	(c)	148	(b)	172	(a)	196	(c)	220	(c)
5	(b)	29	(d)	53	(b)	77	(2)	101	(c)	125	(b)	149	(35)	173	(c)	197	(50)	221	(d)
6	(2)	30	(b)	54	(c)	78	(8)	102	(a)	126	(b)	150	(160)	174	(c)	198	(a)	222	(b)
7	(c)	31	(b)	55	(b)	79	(c)	103	(a)	127	(b)	151	(2)	175	(a)	199	(d)	223	(d)
8	(4)	32	(d)	56	(b)	80	(628)	104	(d)	128	(a)	152	(d)	176	(a)	200	(b)	224	(c)
9	(c)	33	(b)	57	(a)	81	(a)	105	(3)	129	(b)	153	(b)	177	(c)	201	(a)	225	(c)
10	(b)	34	(c)	58	(a)	82	(c)	106	(a)	130	(a)	154	(68)	178	(b)	202	(d)	226	(d)
11	(5)	35	(c)	59	(c)	83	(c)	107	(d)	131	(d)	155	(a)	179	(a)	203	(b)	227	(a)
12	(d)	36	(b)	60	(1)	84	(c)	108	(c)	132	(a)	156	(9)	180	(a)	204	(d)	228	(b)
13	(32)	37	(d)	61	(c)	85	(d)	109	(a)	133	(c)	157	(a)	181	(d)	205	(b)	229	(d)
14	(c)	38	(c)	62	(8)	86	(c)	110	(d)	134	(d)	158	(a)	182	(a)	206	(d)	230	(d)
15	(a)	39	(a)	63	(b)	87	(d)	111	(d)	135	(c)	159	(c)	183	(b)	207	(c)	231	(d)
16	(b)	40	(d)	64	(b)	88	(c)	112	(c)	136	(a)	160	(b)	184	(c)	208	(d)		
17	(c)	41	(b)	65	(c)	89	(c)	113	(a)	137	(b)	161	(c)	185	(a)	209	(d)		
18	(40)	42	(b)	66	(a)	90	(b)	114	(b)	138	(b)	162	(c)	186	(c)	210	(d)		
19	(144)	43	(a)	67	(250)	91	(11)	115	(a)	139	(c)	163	(5)	187	(a)	211	(20)		
20	(b)	44	(d)	68	(8)	92	(3)	116	(a)	140	(a)	164	(a)	188	(b)	212	(c)		
21	(a)	45	(None)	69	(c)	93	(c)	117	(b)	141	(d)	165	(b)	189	(c)	213	(b)		
22	(b)	46	(b)	70	(500)	94	(b)	118	(b)	142	(48)	166	(c)	190	(c)	214	(None)		
23	(c)	47	(d)	71	(5)	95	(b)	119	(a)	143	(d)	167	(d)	191	(d)	215	(c)		
24	(c)	48	(d)	72	(72)	96	(b)	120	(d)	144	(1)	168	(d)	192	(4)	216	(d)		

Magnetism and Matter

5

TREND ANALYSIS
JEE MAIN (2025-2021)

	2025				2024				2023				2022				2021			
	Session II		Session I		Session II		Session I		Session II		Session I		Session II		Session I		Session II		Session I	
	MCQs	NVQs	MCQs	NVQs	MCQs	NVQs	MCQs	NVQs	MCQs	NVQs	MCQs	NVQs	MCQs	NVQs	MCQs	NVQs	MCQs	NVQs	MCQs	NVQs
Topic 1	1	–	1	–	1	3	–	1	–	–	–	1	1	–	1	–	–	2	–	1
Topic 2	2	–	–	–	2	1	–	–	2	2	2	–	3	–	–	–	4	–	2	–
Topic 3	–	–	–	–	–	–	–	–	–	–	–	–	1	–	–	–	–	–	–	–
Topic 4																				
Total	3	-	1	-	3	4	-	1	2	2	2	1	5	-	1	-	4	2	2	1

How to Code Qns for Better & Faster Revision

Type 1: Able to solve (No code Req)

Type 2: Solved with struggle or by seeing solution **(Use code NR means Needs Revision or orange stickers)**

Type 3: Tricky/ Could not solve on my own. Need help from Teacher/ Friend. **(Use code TQ means Tricky Question or red stickers)**

You can either hand write the code in boxes or else paste coloured stickers given at end of book.

Topic 1 - Magnetism, Magnetic Moment, Properties of Magnet

1. Given below are two statements : one is labelled as **Assertion (A)** and the other is labelled as **Reason(R)**.

Assertion (A) : Magnetic monopoles do not exist.

Reason (R) : Magnetic field lines are continuous and form closed loops.

In the light of the above statements, choose the **most appropriate answer** from the options given below :

[NCERT : P L-138 | April 7, 2025 (II)]

(a) Both **(A)** and **(R)** are correct but **(R)** is **not** the correct explanation of **(A)**

(b) **(A)** is correct but **(R)** is not correct

(c) Both **(A)** and **(R)** are correct and **(R)** is the correct explanation of **(A)**

(d) **(A)** is not correct but **(R)** is correct

2.

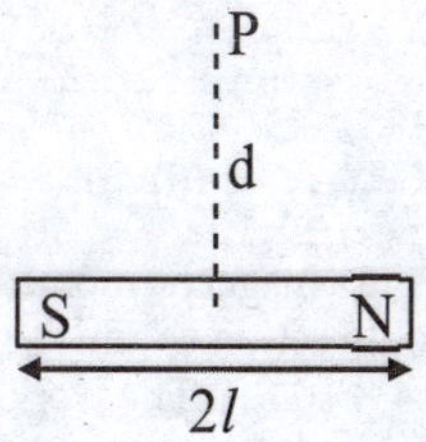

A bar magnet has total length $2l = 20$ units and the field point P is at a distance $d = 10$ units from the centre of the magnet. If the relative uncertainty of length measurement is 1%, then uncertainty of the magnetic field at point P is :

[NCERT : P L-140 | Jan. 28, 2025 (II)]

(a) 10% (b) 4% (c) 3% (d) 5%

3. A straight magnetic strip has a magnetic moment of 44 Am^2. If the strip is bent in a semicircular shape, its magnetic moment will be .... Am^2. (Given $\pi = \frac{22}{7}$)

[NCERT : P L-139 | April 9, 2024 (II)]

4. A circular coil having 200 turns, 2.5×10^{-4} m^2 area and carrying 100 μA current is placed in a uniform magnetic field of 1 T. Initially the magnetic dipole moment $(\vec{M})$ was directed along $\vec{B}$. Amount of work, required to rotate the coil through 90° from its initial orientation such that $\vec{M}$ becomes perpendicular to $\vec{B}$, is ______ μJ.
[NCERT : P L-139 | April 6, 2024 (I)]

5. A coil having 100 turns, area of 5×10^{-3}m^2, carrying current of 1 mA is placed in uniform magnetic field of 0.20 T such a way that plane of coil is perpendicular to the magnetic field. The work done in turning the coil through 90° is ______ μJ. **[NCERT : P L-139 | April 6, 2024 (II)]**

6. The magnetic moment of a bar magnet is 0.5 Am^2. It is suspended in a uniform magnetic field of 8×10^{-2} T. The work done in rotating it from its most stable to most unstable position is :
[NCERT : P L-139 | April 4, 2024 (II)]
(a) 16×10^{-2} J (b) 8×10^{-2} J
(c) 4×10^{-2} J (d) Zero

7. The magnetic potential due to a magnetic dipole at a point on its axis situated at a distance of 20 cm from its center is 1.5×10^{-5} T m. The magnetic moment of the dipole is ______ A m^2. (Given : $\frac{\mu_0}{4\pi} = 10^{-7}$ T m A^{-1})
[NCERT : P L-141 | Jan. 29, 2024 (I)]

8. A bar magnet with a magnetic moment 5.0 Am^2 is placed in parallel position relative to a magnetic field of 0.4 T. The amount of required work done in turning the magnet form parallel to antiparallel position relative to the field direction is ________.
[NCERT : P L-139 | Jan. 31, 2023 (I)]
(a) 4 J (b) 1 J (c) 2 J (d) Zero

9. The magnetic moment of an electron (e) revolving in an orbit around nucleus with an orbital angular momentum is given by: **[July 25, 2022 (I)]**
(a) $\vec{\mu}_L = \frac{e\vec{L}}{2m}$ (b) $\vec{\mu}_L = -\frac{e\vec{L}}{2m}$
(c) $\vec{\mu}_l = -\frac{e\vec{L}}{m}$ (d) $\vec{\mu}_l = \frac{2e\vec{L}}{m}$

10. A bar magnet having a magnetic moment of 2.0×10^5 JT^{-1}, is placed along the direction of uniform magnetic field of magnitude $B = 14 \times 10^{-5}$ T. The work done in rotating the magnet slowly through 60° from the direction of field is :
[NCERT : P L-139 | June 26, 2022 (II)]
(a) 14 J (b) 8.4 J
(c) 4 J (d) 1.4 J

11. Two short magnetic dipoles m_1 and m_2 each having magnetic moment of 1 Am^2 are placed at point O and P respectively. The distance between OP is 1 meter. The torque experienced by the magnetic dipole m_2 due to the presence of m_1 is ________ $\times 10^{-7}$ Nm.

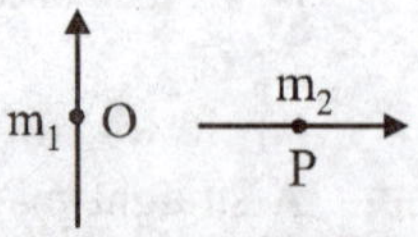

[NCERT : P L-139 | NA, Aug. 26, 2021 (I)]

12. A coil in the shape of an equilateral triangle of side 10 cm lies in a vertical plane between the pole pieces of permanent magnet producing a horizontal magnetic field 20 mT. The torque acting on the coil when a current of 0.2 A is passed through it and its plane becomes parallel to the magnetic field will be $\sqrt{x} \times 10^{-5}$ Nm. The value of x is ______
[NCERT : P L-139 | NA, Aug. 26, 2021 (II)]

13. In a uniform magnetic field, the magnetic needle has a magnetic moment 9.85×10^{-2} A/m^2 and moment of inertia 5×10^{-6} kg m^2. If it performs 10 complete oscillations in 5 seconds then the magnitude of the magnetic field is ________ mT. [Take π^2 as 9.85] **[NA, July 27, 2021 (I)]**

14. A small bar magnet placed with its axis at 30° with an external field of 0.06 T experiences a torque of 0.018 Nm. The minimum work required to rotate it from its stable to unstable equilibrium position is :
[NCERT : P L-139 | Sep. 04, 2020 (I)]
(a) 6.4×10^{-2} J (b) 9.2×10^{-3} J
(c) 7.2×10^{-2} J (d) 11.7×10^{-3} J

15. A circular coil has moment of inertia 0.8 kg m^2 around any diameter and is carrying current to produce a magnetic moment of 20 Am^2. The coil is kept initially in a vertical position and it can rotate freely around a horizontal diameter. When a uniform magnetic field of 4 T is applied along the vertical, it starts rotating around its horizontal diameter. The angular speed the coil acquires after rotating by 60° will be : **[NCERT : P L-139 | Sep. 04, 2020 (II)]**
(a) 10 rad s^{-1} (b) 10π rad s^{-1}
(c) 20π rad s^{-1} (d) 20 rad s^{-1}

16. Two magnetic dipoles X and Y are placed at a separation *d*, with their axes perpendicular to each other. The dipole moment of Y is twice that of X. A particle of charge *q* is passing through their midpoint P, at angle θ = 45° with the horizontal line, as shown in figure. What would be the magnitude of force on the particle at that instant? (*d* is much larger than the dimensions of the dipole)
[NCERT : P L-140 | 8 April 2019 II]

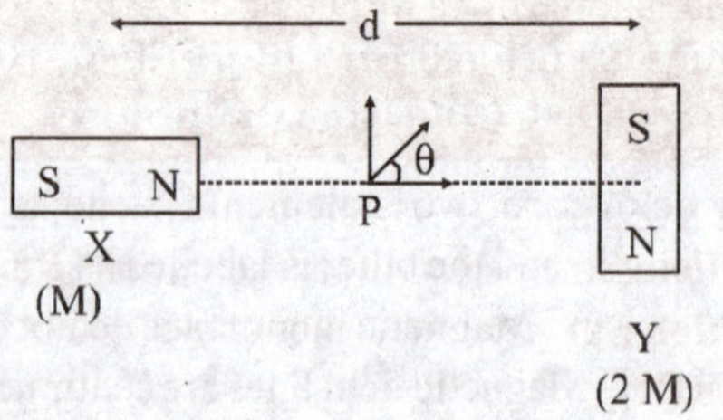

(a) $\left(\frac{\mu_0}{4\pi}\right)\frac{M}{\left(d/2\right)^3} \times qv$ (b) 0
(c) $\sqrt{2}\left(\frac{\mu_0}{4\pi}\right)\frac{M}{\left(d/2\right)^3} \times qv$ (d) $\left(\frac{\mu_0}{4\pi}\right)\frac{2M}{\left(d/2\right)^3} \times qv$

17. A magnet of total magnetic moment $10^{-2}\hat{i}$ A-m^2 is placed in a time varying magnetic field, $B\hat{i}$ (cosωt)where B = 1 tesla and ω = 0.125 rad/s. The work done for reversing the direction of the magnetic moment at t = 1 second, is: **[NCERT : P L-139 | 10 Jan. 2019 I]**
(a) 0.01 J (b) 0.007 J (c) 0.028 J (d) 0.014 J

18. A magnetic dipole in a constant magnetic field has : **[NCERT : P L-139 | Online April 8, 2017]**
(a) maximum potential energy when the torque is maximum
(b) zero potential energy when the torque is minimum.
(c) zero potential energy when the torque is maximum.
(d) minimum potential energy when the torque is maximum.

19. A magnetic dipole is acted upon by two magnetic fields which are inclined to each other at an angle of 75°. One of the fields has a magnitude of 15 mT. The dipole attains stable equilibrium at an angle of 30° with this field. The magntidue of the other field (in mT) is close to : **[Online April 9, 2016]**
(a) 1 (b) 11 (c) 36 (d) 1060

20. A 25 cm long solenoid has radius 2 cm and 500 total number of turns. It carries a current of 15 A. If it is equivalent to a magnet of the same size and magnetization $\vec{M}$ (magnetic moment/volume), then $|\vec{M}|$ is : **[Online April 10, 2015]**
(a) $30000\pi\ Am^{-1}$ (b) $3\pi\ Am^{-1}$
(c) $30000\ Am^{-1}$ (d) $300\ Am^{-1}$

21. A bar magnet of length 6 cm has a magnetic moment of $4\ J\ T^{-1}$. Find the strength of magnetic field at a distance of 200 cm from the centre of the magnet along its equatorial line. **[NCERT : P L-140 | Online May 7, 2012]**
(a) 4×10^{-8} tesla (b) 3.5×10^{-8} tesla
(c) 5×10^{-8} tesla (d) 3×10^{-8} tesla

22. A thin circular disc of radius R is uniformly charged with density $\sigma > 0$ per unit area. The disc rotates about its axis with a uniform angular speed ω. The magnetic moment of the disc is **[2011 RS]**
(a) $\pi R^4 \sigma\omega$ (b) $\frac{\pi R^4}{2}\sigma\omega$
(c) $\frac{\pi R^4}{4}\sigma\omega$ (d) $2\pi R^4 \sigma\omega$

23. A magnetic needle is kept in a non-uniform magnetic field. It experiences **[NCERT : P L-139 | 2005]**
(a) neither a force nor a torque
(b) a torque but not a force
(c) a force but not a torque
(d) a force and a torque

24. The length of a magnet is large compared to its width and breadth. The time period of its oscillation in a vibration magnetometer is 2s. The magnet is cut along its length into three equal parts and these parts are then placed on each other with their like poles together. The time period of this combination will be **[2004]**
(a) $2\sqrt{3}$ s (b) $\frac{2}{3}$ s (c) 2 s (d) $\frac{2}{\sqrt{3}}$ s

25. A magnetic needle lying parallel to a magnetic field requires W units of work to turn it through 60°. The torque needed to maintain the needle in this position will be **[NCERT : P L-139 | 2003]**
(a) $\sqrt{3}$W (b) W
(c) $\frac{\sqrt{3}}{2}$W (d) 2W

26. The magnetic lines of force inside a bar magnet **[NCERT : P L-138 | 2003]**
(a) are from north-pole to south-pole of the magnet
(b) do not exist
(c) depend upon the area of cross-section of the bar magnet
(d) are from south-pole to north-pole of the Magnet

Topic 2 - Magnetic Materials and their Properties

27. The percentage increase in magnetic field (B) when space within a current carrying solenoid is filled with magnesium (magnetic susceptibility $\chi_{mg} = 1.2 \times 10^{-5}$) is : **[NCERT : P L-146 | April 7, 2025 (I)]**
(a) $\frac{6}{5}\times10^{-3}\%$ (b) $\frac{5}{6}\times10^{-5}\%$
(c) $\frac{5}{6}\times10^{-4}\%$ (d) $\frac{5}{3}\times10^{-5}\%$

28. The relationship between the magnetic susceptibility (χ) and the magnetic permeability (μ) is given by : (μ_0 is the permeability of free space and μ_r is relative permeability) **[NCERT : P L-148 | April 2, 2025 (I)]**
(a) $\chi = \frac{\mu}{\mu_0} - 1$ (b) $\chi = \frac{\mu_r}{\mu_0} + 1$
(c) $\chi = \mu_r + 1$ (d) $\chi = 1 - \frac{\mu}{\mu_0}$

29. Paramagnetic substances:
A. align themselves along the directions of external magnetic field.
B. attract strongly towards external magnetic field.
C. has susceptibility little more than zero.
D. move from a region of strong magnetic field to weak magnetic field.
Choose the most appropriate answer from the options given below: **[NCERT : P L-148 | April 8, 2024 (I)]**
(a) A, B, C, D (b) B, D only
(c) A, B, C only (d) A, C only

30. The coercivity of a magnet is 5×10^3 A/m. The amount of current required to be passed in a solenoid of length 30 cm and the number of turns 150, so that the magnet gets demagnetised when inside the solenoid is ....A. **[NCERT : P L-146 | April 8, 2024 (II)]**

31. Match List-I with List-II **[NCERT : P L-146 | April 6, 2024 (II)]**

List - I (Y vs X)	List - II (Shape of Graph)
(A) Y = magnetic susceptibility X = magnetising field	(I) 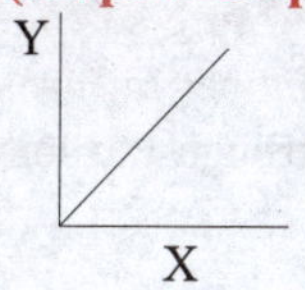

(B) Y = magnetic field X = distance from centre of a current carrying wire for x < a (where a = radius of wire) (II)

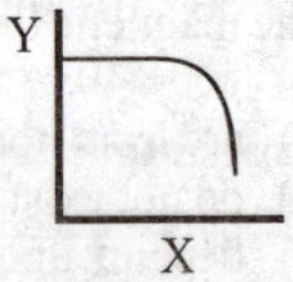

(C) Y = magnetic field X = distance from centre of a current carrying wire for x > a (where a = radius of wire) (III)

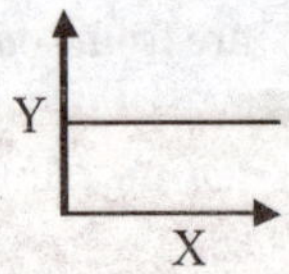

(D) Y= magnetic field inside solenoid X = distance from center (IV)

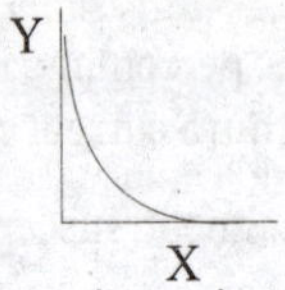

Choose the correct answer from the options given below:

(a) (A)-(III), (B)-(I), (C)-(IV), (D)-(II)
(b) (A)-(I), (B)-(III), (C)-(II), (D)-(IV)
(c) (A)-(IV), (B)-(I), (C)-(III), (D)-(II)
(d) (A)-(III), (B)-(IV), (C)-(I), (D)-(II)

32. Given below are two statements:

Statement I : The diamagnetic property depends on temperature.

Statement II : The included magnetic dipole moment in a diamagnetic sample is always opposite to the magnetizing field.

In the light of given statement, choose the correct answer from the options given below:

[NCERT : P L-147, 148 | April 12, 2023 (I)]

(a) Statement I is incorrect but Statement II is true
(b) Both Statement I and Statement II are true
(c) Both Statement I and Statement II are false
(d) Statement I is correct but Statement II is false

33. The current required to be passed through a solenoid of 15 cm length and 60 turns in order to demagnetise a bar magnet of magnetic intensity 2.4×10^3 Am^{-1} is____A.

[NCERT : P L-146 | April 10, 2023 (I)]

34. Given below are two statements:

Statement I: For diamagnetic substance $-1 \le \chi < 0$, where χ is the magnetic susceptibility.

Statement II: Diamagnetic substances when placed in an external magnetic field, tend to move from stronger to weaker part of the field.

[NCERT : P L-147, 148 | April 10, 2023 (II)]

In the light of the above statements, choose the correct answer from the options given below.

(a) Both Statement I and Statement II are false.
(b) Both Statement I and Statement II are true.
(c) Statement I is incorrect but Statement II is true.
(d) Statement I is correct but Statement II is false.

35. The magnetic intensity at the centre of a long current carrying solenoid is found to be $1.6 \times 10^3 Am^{-1}$. If the number of turns is 8 per cm, then the current flowing through the solenoid is____A.

[NCERT : P L-146 | April 8, 2023 (I)]

36. A solenoid of 1200 turns is wound uniformly in a single layer on a glass tube 2 m long and 0.2 m in diameter. The magnetic intensity at the center of the solenoid when a current of 2 A flows through it is:

[NCERT : P L-146 | Jan. 25, 2023 (I)]

(a) 2.4×10^3 A m^{-1} (b) 1.2×10^3 A m^{-1}
(c) 1 A m^{-1} (d) 2.4×10^{-3} A m^{-1}

37. The space inside a straight current carrying solenoid is filled with a magnetic material having magnetic susceptibility equal to 1.2×10^{-5}. What is fractional increase in the magnetic field inside solenoid with respect to air as medium inside the solenoid?

[NCERT : P L-146 | June 28, 2022 (II)]

(a) 1.2×10^{-5} (b) 1.2×10^{-3}
(c) 1.8×10^{-3} (d) 2.4×10^{-5}

38. The susceptibility of a paramagnetic material is 99. The permeability of the material in *Wb*/A-m is :

[Permeability of free space $\mu_0 = 4\pi \times 10^{-7}$ $Wb/A-m$]

[NCERT : P L-148 | June 27, 2022 (I)]

(a) $4\pi \times 10^{-7}$ (b) $4\pi \times 10^{-4}$
(c) $4\pi \times 10^{-5}$ (d) $4\pi \times 10^{-6}$

39. Given below are two statements:

Statement-I: Susceptibilities of paramagnetic and ferromagnetic substances increase with decrease in temperature.

Statement-II: Diamagnetism is a result of orbital motions of electrons developing magnetic moments opposite to the applied magnetic field.

Choose the **CORRECT** answer from the options given below : -

[NCERT : P L-147, 148, 149 | June 25, 2022 (II)]

(a) Both statement – I and statement -II are true.
(b) Both statement – I and Statement – II are false.
(c) Statement – I is true but statement – II is false.
(d) Statement-I is false but Statement-II is true.

40. **Statement I :** The ferromagnetic property depends on temperature. At high temperature, ferromagnet becomes paramagnet.

Statement II : At high temperature, the domain wall area of a ferromagnetic substance increases.

In the light of the above statements, choose the **most appropriate** answer from the options given below :

[July 22, 2021 (II)]

(a) **Statement I** is true but **Statement II** is false
(b) Both **Statement I** and **Statement II** are true
(c) Both **Statement I** and **Statement II** are false
(d) **Statement I** is false but **Statement II** is true

41. The magnetic susceptibility of a material of a rod is 499. Permeability in vacuum is $4\pi \times 10^{-7}$ H/m. Absolute permeability of the material of the rod is :

[July 20, 2021 (II)]

(a) $4\pi \times 10^{-4}$ H/m (b) $2\pi \times 10^{-4}$ H/m
(c) $3\pi \times 10^{-4}$ H/m (d) $\pi \times 10^{-4}$ H/m

42. Following plots show Magnetization (M) vs Magnetising field (H) and Magnetic susceptibility (χ) vs temperature (T) graph :

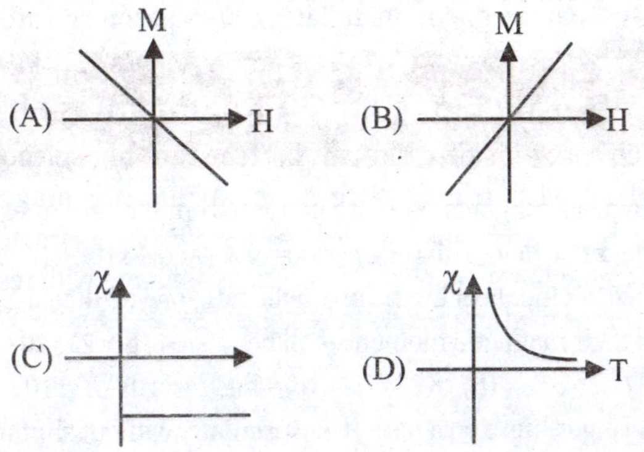

Which of the following combination will be represented by a diamagnetic material?

[NCERT : P L-146 | Sep. 1, 2021 (II)]

(a) (A), (C)
(b) (A), (D)
(c) (B), (D)
(d) (B), (C)

43. A long solenoid with 1000 turns/m has a core material with relative permeability 500 and volume 10^3 cm^3. If the core material is replaced by another material having relative permeability of 750 with same volume maintaining same current of 0.75 A in the solenoid, the fractional change in the magnetic moment of the core would be approximately $\left(\frac{x}{499}\right)$. Find the value of x.

[NCERT : PL-146 | NA, Aug. 31, 2021 (II)]

44. A soft ferromagnetic material is placed in an external magnetic field. The magnetic domains :

[NCERT : P L-149 | Feb. 24, 2021 (II)]

(a) increase in size but no change in orientation
(b) have no relation with external magnetic field
(c) decrease in size and changes orientation
(d) may increase or decrease in size and change its orientation

45. In a ferromagnetic material, below the curie temperature, a domain is defined as :

[NCERT : P L-149 | Feb. 25, 2021 (II)]

(a) a macroscopic region with randomly oriented magnetic dipoles.
(b) a macroscopic region with consecutive magnetic dipoles oriented in opposite direction.
(c) a macroscopic region with saturation magnetization.
(d) a macroscopic region with zero magnetization.

46. An iron rod of volume 10^{-3} m^3 and relative permeability 1000 is placed as core in a solenoid with 10 turns/cm. If a current of 0.5 A is passed through the solenoid, then the magnetic moment of the rod will be :

[NCERT : P L-146 | Sep. 05, 2020 (II)]

(a) 50×10^2 Am^2
(b) 5×10^2 Am^2
(c) 500×10^2 Am^2
(d) 0.5×10^2 Am^2

47. A paramagnetic sample shows a net magnetisation of 6 A/m when it is placed in an external magnetic field of 0.4 T at a temperature of 4 K. When the sample is placed in an external magnetic field of 0.3 T at a temperature of 24 K, then the magnetisation will be :

[Sep. 04, 2020 (II)]

(a) 1 A/m
(b) 4 A/m
(c) 2.25 A/m
(d) 0.75 A/m

48. A perfectly diamagnetic sphere has a small spherical cavity at its centre, which is filled with a paramagnetic substance. The whole system is placed in a uniform magnetic field $\vec{B}$. Then the field inside the paramagnetic substance is :

[NCERT : P L-147, 148 | Sep. 03, 2020 (II)]

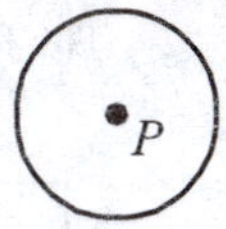

(a) $\vec{B}$
(b) zero
(c) much large than $|\vec{B}|$ and parallel to $\vec{B}$
(d) much large than $|\vec{B}|$ but opposite to $\vec{B}$

49. A paramagnetic material has 10^{28} atoms/m^3. Its magnetic susceptibility at temperature 350 K is 2.8×10^{-4}. Its susceptibility at 300 K is:

[12 Jan. 2019 II]

(a) 3.267×10^{-4}
(b) 3.672×10^{-4}
(c) 3.726×10^{-4}
(d) 2.672×10^{-4}

50. A paramagnetic substance in the form of a cube with sides 1 cm has a magnetic dipole moment of 20×10^{-6} J/T when a magnetic intensity of 60×10^3 A/m is applied. Its magnetic susceptibility is:

[NCERT : P L-145 | 11 Jan. 2019 II]

(a) 3.3×10^{-2}
(b) 4.3×10^{-2}
(c) 2.3×10^{-2}
(d) 3.3×10^{-4}

51. A bar magnet is demagnetized by inserting it inside a solenoid of length 0.2 m, 100 turns, and carrying a current of 5.2 A. The coercivity of the bar magnet is:

[NCERT : P L-146 | 9 Jan. 2019 I]

(a) 285 A/m
(b) 2600 A/m
(c) 520 A/m
(d) 1200 A/m

52. An example of a perfect diamagnet is a superconductor. This implies that when a superconductor is put in a magnetic field of intensity B, the magnetic field B_s inside the superconductor will be such that:

[NCERT : P L-146 | Online April 19, 2014]

(a) $B_s = -B$
(b) $B_s = 0$
(c) $B_s = B$
(d) $B_s < B$ but $Bs \neq 0$

53. Three identical bars A, B and C are made of different magnetic materials. When kept in a uniform magnetic field, the field lines around them look as follows:

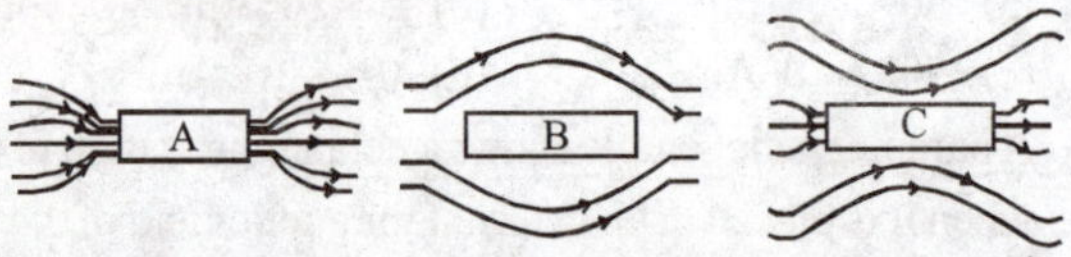

Make the correspondence of these bars with their material being diamagnetic (D), ferromagnetic (F) and paramagnetic (P):

[NCERT : P L-147, 148, 149 | Online April 11, 2014]

(a) $A \leftrightarrow D, B \leftrightarrow P, C \leftrightarrow F$

(b) $A \leftrightarrow F, B \leftrightarrow D, C \leftrightarrow P$

(c) $A \leftrightarrow P, B \leftrightarrow F, C \leftrightarrow D$

(d) $A \leftrightarrow F, B \leftrightarrow P, C \leftrightarrow D$

54. The mid points of two small magnetic dipoles of length d in end-on positions, are separated by a distance x, (x >> d). The force between them is proportional to x^{-n} where n is:

[NCERT : P L-139 | Online April 9, 2014]

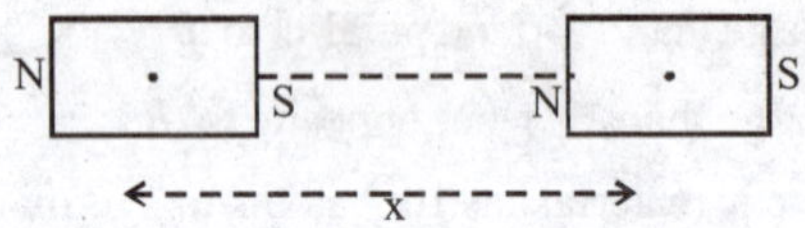

(a) 1 (b) 2

(c) 3 (d) 4

55. Relative permittivity and permeability of a material ε_r and μ_r, respectively. Which of the following values of these quantities are allowed for a diamagnetic material?

[NCERT : P L-147 | 2008]

(a) $\varepsilon_r = 0.5, \mu_r = 1.5$ (b) $\varepsilon_r = 1.5, \mu_r = 0.5$

(c) $\varepsilon_r = 0.5, \mu_r = 0.5$ (d) $\varepsilon_r = 1.5, \mu_r = 1.5$

56. Needles N_1, N_2 and N_3 are made of a ferromagnetic, a paramagnetic and a diamagnetic substance respectively. A magnet when brought close to them will

[NCERT : P L-147, 148, 149 | 2006]

(a) attract N_1 and N_2 strongly but repel N_3

(b) attract N_1 strongly, N_2 weakly and repel N_3 weakly

(c) attract N_1 strongly, but repel N_2 and N_3 weakly

(d) attract all three of them

57. Curie temperature is the temperature above which **[2003]**

(a) a ferromagnetic material becomes paramagnetic

(b) a paramagnetic material becomes diamagnetic

(c) a ferromagnetic material becomes diamagnetic

(d) a paramagnetic material becomes ferromagnetic

Topic 3 - Magnetic Equipment

58. Two bar magnets oscillate in a horizontal plane in earth's magnetic field with time periods of 3 s and 4 s respectively. If their moments of inertia are in the ratio of 3 : 2 then the ratio of their magnetic moments will be : **[July 27, 2022 (I)]**

(a) 2 : 1 (b) 8 : 3 (c) 1 : 3 (d) 27 : 16

59. A ring is hung on a nail. It can oscillate, without slipping or sliding (i) in its plane with a time period T_1 and, (ii) back and forth in a direction perpendicular to its plane, with a period T_2. The ratio $\frac{T_1}{T_2}$ will be : **[Sep. 05, 2020 (II)]**

(a) $\frac{2}{\sqrt{3}}$ (b) $\frac{2}{3}$ (c) $\frac{3}{\sqrt{2}}$ (d) $\frac{\sqrt{2}}{3}$

60. A hoop and a solid cylinder of same mass and radius are made of a permanent magnetic material with their magnetic moment parallel to their respective axes. But the magnetic moment of hoop is twice of solid cylinder. They are placed in a uniform magnetic field in such a manner that their magnetic moments make a small angle with the field. If the oscillation periods of hoop and cylinder are T_h and T_c respectively, then: **[10 Jan. 2019 II]**

(a) $T_h = T_c$ (b) $T_h = 2\,T_c$

(c) $T_h = 1.5\,T_c$ (d) $T_h = 0.5\,T_c$

61. A magnetic needle of magnetic moment 6.7×10^{-2} Am² and moment of inertia 7.5×10^{-6} kg m² is performing simple harmonic oscillations in a magnetic field of 0.01 T. Time taken for 10 complete oscillations is : **[2017]**

(a) 6.98 s (b) 8.76 s

(c) 6.65 s (d) 8.89 s

ANSWER KEYS

1	(c)	8	(a)	15	(a)	22	(b)	29	(d)	36	(b)	43	(250)	50	(d)	57	(a)
2	(b)	9	(b)	16	(b)	23	(d)	30	(10)	37	(a)	44	(d)	51	(b)	58	(b)
3	(28)	10	(a)	17	(c)	24	(b)	31	(d)	38	(c)	45	(c)	52	(b)	59	(a)
4	(5)	11	(1)	18	(c)	25	(a)	32	(a)	39	(a)	46	(b)	53	(b)	60	(a)
5	(100)	12	(3)	19	(b)	26	(d)	33	(6)	40	(a)	47	(d)	54	(d)	61	(c)
6	(b)	13	(8)	20	(c)	27	(a)	34	(b)	41	(b)	48	(b)	55	(b)		
7	(6)	14	(c)	21	(c)	28	(a)	35	(2)	42	(a)	49	(a)	56	(b)		

6 Electromagnetic Induction

TREND ANALYSIS
JEE MAIN (2025-2021)

	2025				2024				2023				2022				2021			
	Session II		Session I		Session II		Session I		Session II		Session I		Session II		Session I		Session II		Session I	
	MCQs	NVQs	MCQs	NVQs	MCQs	NVQs	MCQs	NVQs	MCQs	NVQs	MCQs	NVQs	MCQs	NVQs	MCQs	NVQs	MCQs	NVQs	MCQs	NVQs
Topic 1	–	–	1	–	–	–	2	3	1	3	3	2	–	3	2	1	1	2	1	–
Topic 2	1	2	4	1	2	3	–	5	3	5	5	2	1	–	2	3	2	1	3	–
Topic 3																				
Topic 4																				
Total	1	2	5	1	2	3	2	8	4	8	8	4	1	3	4	4	3	3	4	–

How to Code Qns for Better & Faster Revision

Type 1: Able to solve (No code Req)
Type 2: Solved with struggle or by seeing solution **(Use code NR means Needs Revision or orange stickers)**
Type 3: Tricky/ Could not solve on my own. Need help from Teacher/ Friend. **(Use code TQ means Tricky Question or red stickers)**

You can either hand write the code in boxes or else paste coloured stickers given at end of book.

Topic 1 – Magnetic Flux, Faraday's and Lenz's Law

1. A coil of area A and N turns is rotating with angular velocity ω in a uniform magnetic field $\vec{B}$ about an axis perpendicular to $\vec{B}$. Magnetic flux ϕ and induced emf ε across it, at an instant when $\vec{B}$ is parallel to the plane of coil, are : **[NCERT : P L-158 | Jan. 29, 2025 (I)]**

(a) $\phi = AB, \varepsilon = 0$
(b) $\phi = 0 \;\; \varepsilon = NAB\omega$
(c) $\phi = 0, \varepsilon = 0$
(d) $\phi = AB, \varepsilon = NAB\omega$

2. A coil of 200 turns and area 0.20 m^2 is rotated at half a revolution per second and is placed in uniform magnetic field of 0.01 T perpendicular to axis of rotation of the coil. The maximum voltage generated in the coil is $\frac{2\pi}{\beta}$ volt. The value of β is ______. **[NCERT : P L-171 | Feb. 1, 2024 (II)]**

3. A coil is placed perpendicular to a magnetic field of 5000 T. When the field is changed to 3000 T in 2 s, an induced emf of 22 V is produced in the coil. If the diameter of the coil is 0.02 m, then the number of turns in the coil is : **[NCERT : P L-158 | Jan. 31, 2024 (I)]**

(a) 70
(b) 140
(c) 35
(d) 7

4. The magnetic flux ϕ (in weber) linked with a closed circuit of resistance 8 Ω varies with time (in seconds) as $\phi = 5t^2 - 36t + 1$. The induced current in the circuit at t = 2s is ______ A. **[NCERT : P L-168 | Jan. 31, 2024 (II)]**

5. A square loop of side 10 cm and resistance 0.7Ω is placed vertically in east-west plane. A uniform magnetic field of 0.20 T is set up across the plane in north east direction. The magnetic field is decreased to zero in 1 s at a steady rate. Then, magnitude of induced emf is $\sqrt{x} \times 10^{-3}$V. The value of x is ______. **[NCERT : P L-158 | Jan. 29, 2024 (I)]**

6. A rectangular loop of length 2.5 m and width 2 m is placed at 60° to a magnetic field of 4 T. The loop is removed from the field in 10 sec. The average emf induced in the loop during this time is **[NCERT : P L-158 | Jan. 27, 2024 (I)]**
(a) $-2V$ (b) $+2V$ (c) $+1V$ (d) $-1V$

7. An insulated copper wire of 100 turns is wrapped around a wooden cylindrical core of the cross-sectional area 24 cm^2. The two ends of the wire are connected to a resistor. The total resistance in the circuit is 12 Ω. If an externally applied uniform magnetic field in the core along its axis changes from 1.5 T in one direction to 1.5 T in the opposite direction, the charge flowing through a point in the circuit during the change of magnetic field will be _____ mC.
[NCERT : P L-157 | April 13, 2023 (II)]

8. A conducting circular loop is placed in a uniform magnetic field of 0.4 T with its plane perpendicular to the field. Somehow, the radius of the loop starts expanding at a constant rate of 1 mm/s. The magnitude of induced emf in the loop at an instant when the radius of the loop is 2 cm will be ______ μV. **[NCERT : PL-158 | April 12, 2023 (I)]**

9. The magnetic field B crossing normally a square metallic plate of area 4 m^2 is changing with time as shown in figure. The magnitude of induced emf in the plate during t = 2 s to t = 4s, is _____ mV
[NCERT : P L-158 | April 10, 2023 (I)]

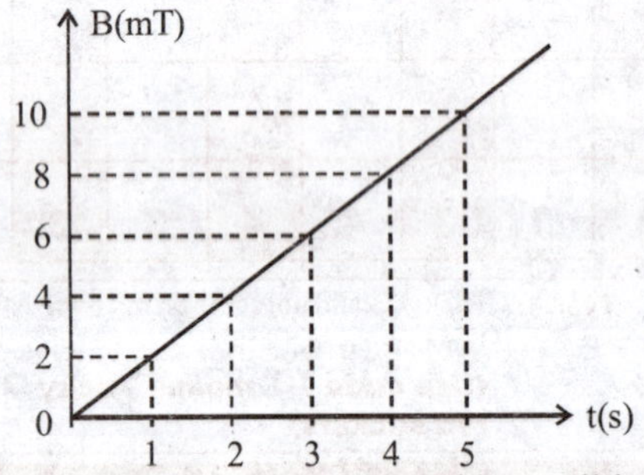

10. The induced emf can be produced in a coil by
A. moving the coil with uniform speed inside magnetic field
B. moving the coil with non–uniform speed inside uniform magnetic field
C. rotating the coil inside the uniform magnetic field
D. changing the area of the coil inside the uniform magnetic field

Choose the correct answer from the options given below:
[NCERT : P L-157 | April 6, 2023 (I)]
(a) B and D only (b) B and C only
(c) A and C only (d) C and D only

11. A coil is placed in magnetic field such that plane of coil is perpendicular to the direction of magnetic field. The magnetic flux through a coil can be changed:
(A) By changing the magnitude of the magnetic field within the coil.
(B) By changing the area of coil within the magnetic field.
(C) By changing the angle between the direction of magnetic field and the plane of the coil.
(D) By reversing the magnetic field direction abruptly without changing its magnitude.

Choose the most appropriate answer from the options given below: **[NCERT : P L-157 | Feb. 1, 2023 (II)]**
(a) A and C only (b) A, B and D only
(c) A and B only (d) A, B and C only

12. A square shaped coil of area 70 cm^2 having 600 turns rotates in a magnetic field of 0.4 wbm^{-2}, about an axis which is parallel to one of the side of the coil and perpendicular to the direction of field. If the coil completes 500 revolution in a minute, the instantaneous emf when the plane of the coil is inclined at 60° with the field, will be ______ V. $\left(\text{Take } \pi = \frac{22}{7}\right)$
[NCERT : P L-158 | Feb. 1, 2023 (II)]

13. A rod with circular cross-section area 2 cm^2 and length 40 cm is wound uniformly with 400 turns of an insulated wire. If a current of 0.4 A flows in the wire windings, the total magnetic flux produced inside windings is $4\pi \times 10^{-6}$ Wb. The relative permeability of the rod is (Given : Permeability of vacuum $\mu_0 = 4\pi \times 10^{-7}\ NA^{-2}$)
[NCERT : P L-157 | Jan. 31, 2023 (I)]
(a) 12.5 (b) $\frac{32}{5}$ (c) 125 (d) $\frac{5}{16}$

14. A certain elastic conducting material is stretched into a circular loop. It is placed with its plane perpendicular to a uniform magnetic field B = 0.8 T. When released the radius of the loop starts shrinking at a constant rate of 2 cm^{-1}. The induced emf in the loop at an instant when the radius of the loop is 10 cm will be ____ mV.
[NCERT : P L-158 | Jan. 29, 2023 (I)]

15. A conducting loop of radius $\frac{10}{\sqrt{\pi}}$ cm is placed perpendicular to a uniform magnetic field of 0.5*T*. The magnetic field is decreased to zero in 0.5 *s* at a steady rate. The induced emf in the circular loop at 0.25 *s* is:
[NCERT : P L-158 | Jan. 24, 2023 (I)]
(a) emf = 1 mV (b) emf = 10 mV
(c) emf = 100 mV (d) emf = 5 mV

16. A conducting circular loop is placed in X – Y plane in presence of magnetic field $\vec{B} = \left(3t^3\hat{j} + 3t^2\hat{k}\right)$ in SI unit. If the radius of the loop is 1 m, the induced emf in the loop, at time, t = 2s is nπV. The value of n is ________.
[NCERT : P L-157, 158 | July 27, 2022 (II)]

17. In a coil of resistance 8Ω, the magnetic flux due to an external magnetic field varies with time as $\phi = \frac{2}{3}\left(9 - t^2\right)$. The value of total heat produced in the coil, till the flux becomes zero, will be _______ J.
[NCERT : P L-158 | July 26, 2022 (II)]

18. Magnetic flux (in weber) in a closed circuit of resistance 20 Ω varies with time t(s) as $\phi = 8t^2 - 9t + 5$. The magnitude of the induced current at t = 0.25 s will be _____ mA
[NCERT : P L-158 | July 25, 2022 (II)]

19. A coil is placed in a time varying magnetic field. If the number of turns in the coil were to be halved and the radius of wire doubled, the electrical power dissipated due to the current induced in the coil would be:
(Assume the coil to be short circuited.)
[NCERT : P L-158 | June 28, 2022 (II)]
(a) Halved (b) Quadrupled
(c) The same (d) Doubled

20. The magnetic flux through a coil perpendicular to its plane is varying according to the relation $\phi = (5t^3 + 4t^2 + 2t - 5)$ Weber. If the resistance of the coil is 5 ohm, then the induced current through the coil at $t = 2$ sec will be:
[NCERT : P L-158, 159 | June 26, 2022 (I)]

(a) 15.6A (b) 16.6A (c) 17.6A (d) 18.6A

21. A circular coil of 1000 turns each with area $1 m^2$ is rotated about its vertical diameter at the rate of one revolution per second in a uniform horizontal magnetic field of 0.07T. The maximum voltage generation will be ________ V.
[NCERT : P L-171 | June 24, 2022 (II)]

22. A coil is placed in a magnetic field $\vec{B}$ as shown below:
[NCERT : P L-160 | Aug. 31, 2021 (II)]

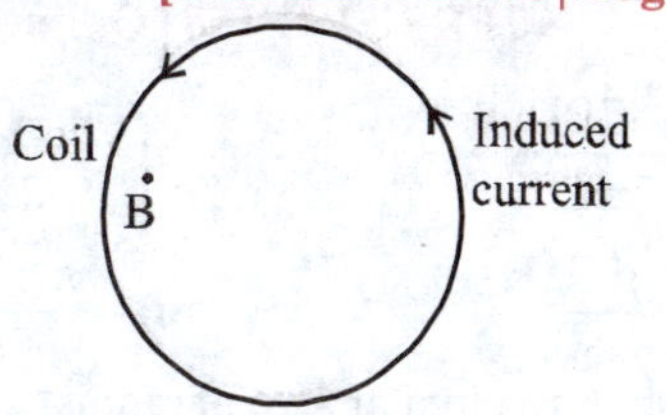

A current is induced in the coil because $\vec{B}$ is :

(a) Outward and decreasing with time
(b) Parallel to the plane of coil and decreasing with time
(c) Outward and increasing with time
(d) Parallel to the plane of coil and increasing with time

23. A circular coil of radius 8.0 cm and 20 turns is rotated about its vertical diameter with an angular speed of 50 rad s^{-1} in a uniform horizontal magnetic field of 3.0×10^{-2} T. The maximum emf induced the coil will be ________ $\times 10^{-2}$ volt (rounded off to the nearest integer)
[NCERT : P L-171 | NA, Aug. 26, 2021 (II)]

24. In the given figure the magnetic flux through the loop increases according to the relation $\phi_B(t) = 10t^2 + 20t$, where ϕ_B is in milliwebers and t is in seconds.
The magnitude of current through $R = 2\Omega$ resistor at $t = 5$ s is ________ mA.
[NCERT : P L-157, 158 | NA, July 27, 2021 (II)]

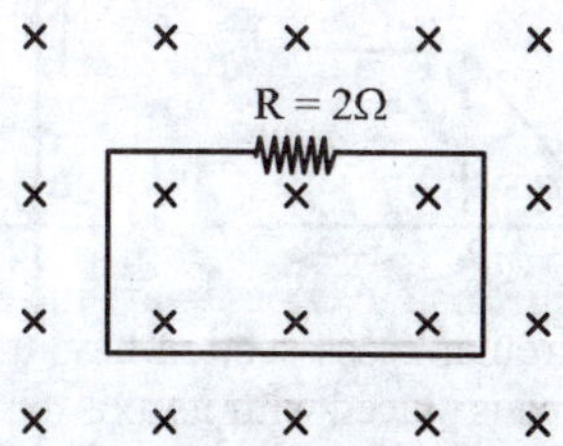

25. Four identical long solenoids A, B, C and D are connected to each other as shown in the figure. If the magnetic field at the center of A is 3 T, the field at the centre of C would be: (Assume that the magnetic field is confined within the volume of respective solenoid). **[March 18, 2021 (I)]**

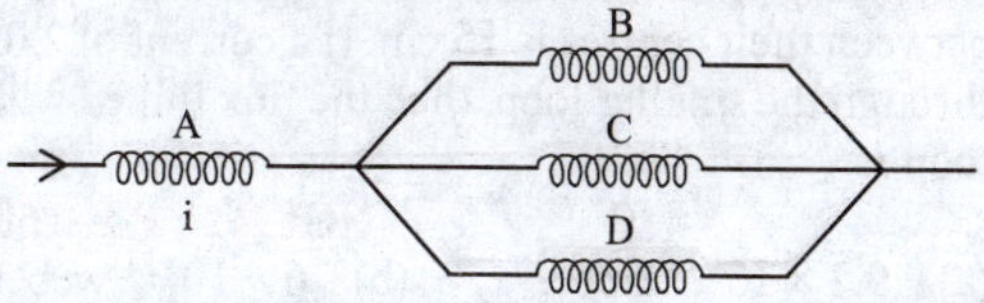

(a) 12T (b) 9T (c) 1T (d) 6T

26. Two concentric circular coils, C_1 and C_2, are placed in the XY plane. C_1 has 500 turns, and a radius of 1 cm. C_2 has 200 turns and radius current 20 cm. C_2 carries a time dependent current $I(t) = (5t^2 - 2t + 3)$ A Where t is in s. The emf induced in C_1 (in mV), at the instant $t = 1$ s is $\frac{4}{x}$. The value of x is ______.
[NCERT : P L-158 | NA Sep. 05, 2020 (I)]

27. An elliptical loop having resistance R, of semi major axis a, and semi minor axis b is placed in a magnetic field as shown in the figure. If the loop is rotated about the x-axis with angular frequency ω, the average power loss in the loop due to Joule heating is :
[NCERT : P L-158 | Sep. 03, 2020 (I)]

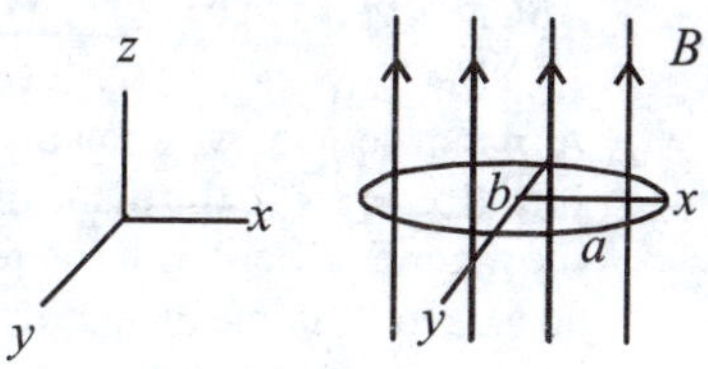

(a) $\frac{\pi^2 a^2 b^2 B^2 \omega^2}{2R}$

(b) zero

(c) $\frac{\pi ab B\omega}{R}$

(d) $\frac{\pi^2 a^2 b^2 B^2 \omega^2}{R}$

28. A uniform magnetic field B exists in a direction perpendicular to the plane of a square loop made of a metal wire. The wire has a diameter of 4 mm and a total length of 30 cm. The magnetic field changes with time at a steady rate $dB/dt = 0.032\ Ts^{-1}$. The induced current in the loop is close to (Resistivity of the metal wire is $1.23 \times 10^{-8}\ \Omega m$)
[NCERT : P L-157, 158 | Sep. 03, 2020 (II)]

(a) 0.43 A (b) 0.61 A (c) 0.34 A (d) 0.53 A

29. A circular coil of radius 10 cm is placed in a uniform magnetic field of 3.0×10^{-5} T with its plane perpendicular to the field initially. It is rotated at constant angular speed about an axis along the diameter of coil and perpendicular to magnetic field so that it undergoes half of rotation in 0.2 s. The maximum value of EMF induced (in μV) in the coil will be close to the integer ______.
[NCERT : P L-158 | NA Sep. 02, 2020 (I)]

30. At time $t = 0$ magnetic field of 1000 Gauss is passing perpendicularly through the area defined by the closed loop shown in the figure. If the magnetic field reduces linearly to 500 Gauss, in the next 5 s, then induced EMF in the loop is:
[NCERT : P L-158 | NA 8 Jan. 2020 I]

(a) 56μV
(b) 28μV
(c) 48μV
(d) 36μV

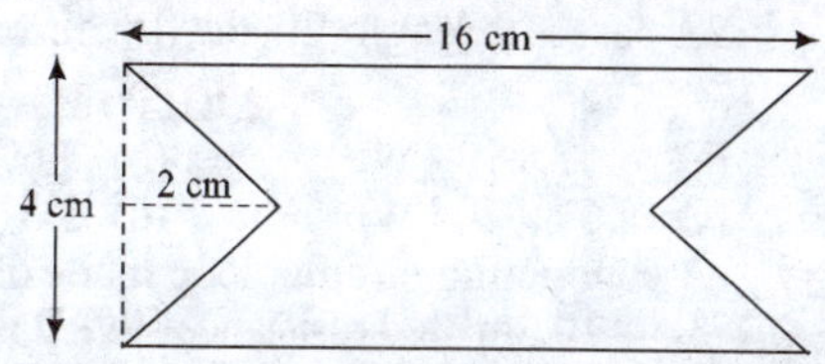

31. Consider a circular coil of wire carrying constant current I, forming a magnetic dipole. The magnetic flux through an infinite plane that contains the circular coil and excluding the circular coil area is given by ϕ_i. The magnetic flux through the area of the circular coil area is given by ϕ_0. Which of the following option is correct?
[NCERT : P L-160 | 7 Jan. 2020 I]

(a) $\phi_i = \phi_0$ (b) $\phi_i > \phi_0$ (c) $\phi_i < \phi_0$ (d) $\phi_i = -\phi_0$

32. A long solenoid of radius R carries a time (t) - dependent current $I(t) = I_0 t(1 - t)$. A ring of radius $2R$ is placed coaxially near its middle. During the time interval $0 \leq t \leq 1$, the induced current (I_R) and the induced $EMF(V_R)$ in the ring change as: **[NCERT : P L-158 | 7 Jan. 2020 I]**

(a) Direction of I_R remains unchanged and V_R is maximum at $t = 0.5$
(b) At $t = 0.25$ direction of I_R reverses and V_R is maximum
(c) Direction of I_R remains unchanged and V_R is zero at $t = 0.25$
(d) At $t = 0.5$ direction of I_R reverses and V_R is zero

33. A loop ABCDEFA of straight edges has six corner points $A(0, 0, 0)$, $B\{5, 0, 0)$, $C(5, 5, 0)$, $D(0, 5, 0)$, $E(0, 5, 5)$ and $F(0, 0, 5)$. The magnetic field in this region is $\vec{B} = (3\hat{i} + 4\hat{k})$T. The quantity of flux through the loop $ABCDEFA$ (in Wb) is ________.
[NCERT : P L-157 | NA 7 Jan. 2020 I]

34. A planar loop of wire rotates in a uniform magnetic field. Initially, at $t = 0$, the plane of the loop is perpendicular to the magnetic field. If it rotates with a period of 10 s about an axis in its plane then the magnitude of induced emf will be maximum and minimum, respectively at:
[NCERT : P L-158 | 7 Jan. 2020 II]

(a) 2.5 s and 7.5 s (b) 2.5 s and 5.0 s
(c) 5.0 s and 7.5 s (d) 5.0 s and 10.0 s

35. A very long solenoid of radius R is carrying current $I(t) = kte^{-\alpha t}$ ($k > 0$), as a function of time ($t \geq 0$). Counter clockwise current is taken to be positive. A circular conducting coil of radius 2R is placed in the equatorial plane of the solenoid and concentric with the solenoid. The current induced in the outer coil is correctly depicted, as a function of time, by:
[NCERT : P L-157, 158 | 9 Apr. 2019 II]

(a)
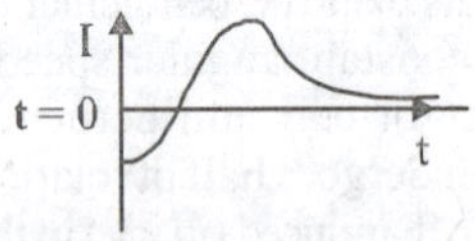

(b)
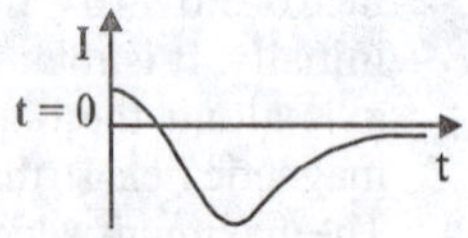

(c)
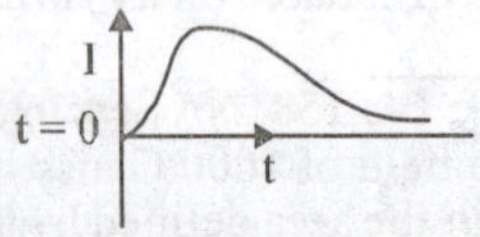

(d)
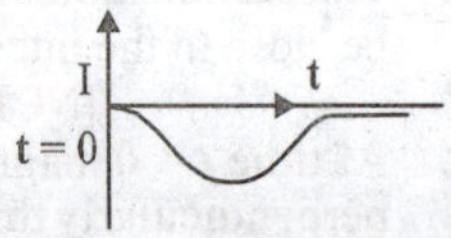

36. Two coils 'P' and 'Q' are separated by some distance. When a current of 3A flows through coil 'P', a magnetic flux of 10^{-3} Wb passes through 'Q'. No current is passed through 'Q'. When no current passes through 'P' and a current of 2A passes through 'Q', the flux through 'P' is:
[NCERT : P L-157 | 9 Apr. 2019 II]

(a) 6.67×10^{-4} Wb (b) 3.67×10^{-3} Wb
(c) 6.67×10^{-3} Wb (d) 3.67×10^{-4} Wb

37. A conducting circular loop made of a thin wire, has area 3.5×10^{-3}m^2 and resistance 10Ω. It is placed perpendicular to a time dependent magnetic field $B(t) = (0.4T) \sin(50\pi t)$. The the net charge flowing through the loop during t = 0 s and t = 10 ms is close to: **[9 Jan. 2019 I]**

(a) 14mC (b) 7mC (c) 21mC (d) 6mC

38. In a coil of resistance 100 Ω, a current is induced by changing the magnetic flux through it as shown in the figure. The magnitude of change in flux through the coil is
[NCERT : P L-158 | 2017, Online May 12, 2012]

(a) 250 Wb
(b) 275 Wb
(c) 200 Wb
(d) 225 Wb

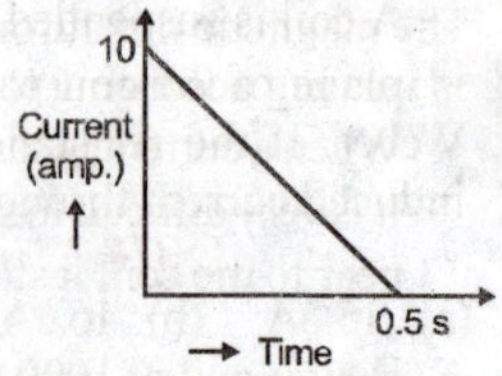

39. A conducting metal circular–wire–loop of radius r is placed perpendicular to a magnetic field which varies with time as $B = B_0 e^{-t/\tau}$, where B_0 and τ are constants, at time t = 0. If the resistance of the loop is R then the heat generated in the loop after a long time ($t \to \infty$) is ;
[NCERT : P L-158 | Online April 10, 2016]

(a) $\frac{\pi^2 r^4 B_0^4}{2\tau R}$ (b) $\frac{\pi^2 r^4 B_0^2}{2\tau R}$
(c) $\frac{\pi^2 r^4 B_0^2 R}{\tau}$ (d) $\frac{\pi^2 r^4 B_0^2}{\tau R}$

40. Figure shows a circular area of radius R where a uniform magnetic field $\vec{B}$ is going into the plane of paper and increasing in magnitude at a constant rate.

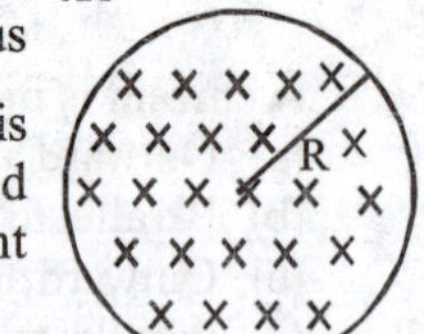

In that case, which of the following graphs, drawn schematically, correctly shows the variation of the induced electric field E(r)?
[NCERT : P L-157 | Online April 19, 2014]

(a)
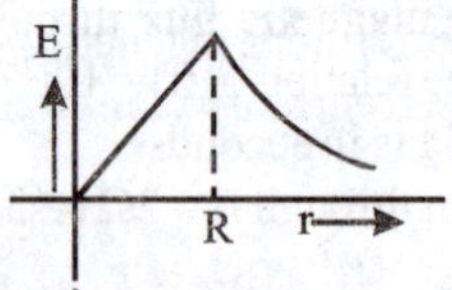

(b)
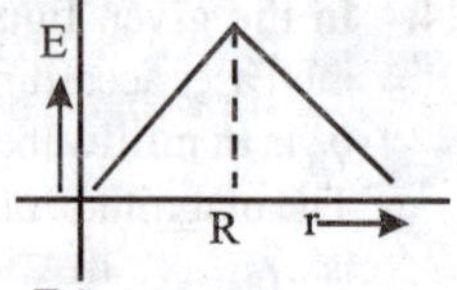

(c)
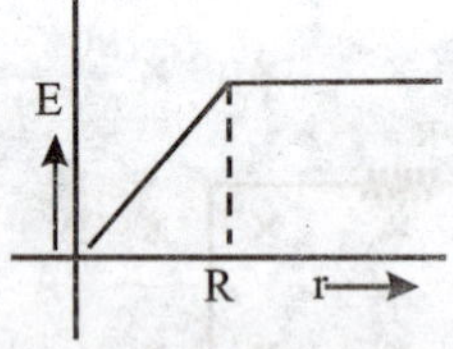

(d)
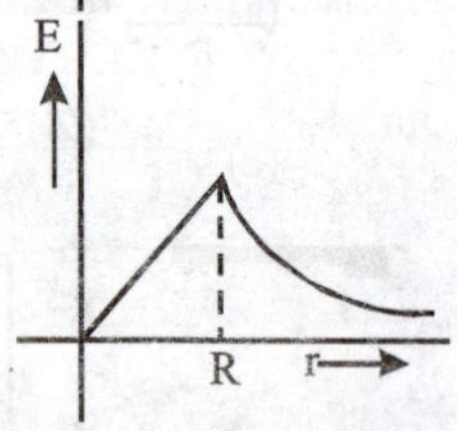

41. A coil of circular cross-section having 1000 turns and 4 cm^2 face area is placed with its axis parallel to a magnetic field which decreases by 10^{-2} Wb m^{-2} in 0.01 s. The e.m.f. induced in the coil is:
[NCERT : P L-158 | Online April 11, 2014]

(a) 400mV (b) 200mV (c) 4mV (d) 0.4mV

42. A circular loop of radius 0.3 cm lies parallel to a much bigger circular loop of radius 20 cm. The centre of the small loop is on the axis of the bigger loop. The distance between their centres is 15 cm. If a current of 2.0 A flows through the smaller loop, then the flux linked with bigger loop is
[NCERT : P L-158 | 2013]

(a) 9.1×10^{-11} weber (b) 6×10^{-11} weber
(c) 3.3×10^{-11} weber (d) 6.6×10^{-9} weber

43. A coil is suspended in a uniform magnetic field, with the plane of the coil parallel to the magnetic lines of force. When a current is passed through the coil it starts oscillating; It is very difficult to stop. But if an aluminium plate is placed near to the coil, it stops. This is due to :

[NCERT : P L-160 | 2012]

(a) developement of air current when the plate is placed
(b) induction of electrical charge on the plate
(c) shielding of magnetic lines of force as aluminium is a paramagnetic material.
(d) electromagnetic induction in the aluminium plate giving rise to electromagnetic damping.

44. Magnetic flux through a coil of resistance 10 Ω is changed by $\Delta\phi$ in 0.1 s. The resulting current in the coil varies with time as shown in the figure. Then $|\Delta\phi|$ is equal to (in weber)

[NCERT : P L-158 | Online May 12, 2012]

(a) 6
(b) 4
(c) 2
(d) 8

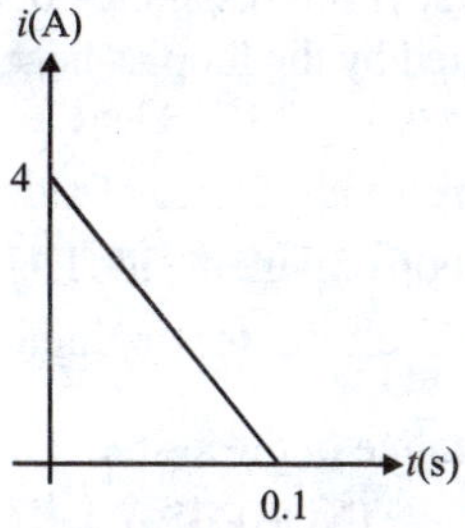

45. The flux linked with a coil at any instant 't' is given by $\phi = 10t^2 - 50t + 250$. The induced emf at $t = 3s$ is

[NCERT : P L-158 | 2006]

(a) –190 V (b) –10 V (c) 10 V (d) 190 V

46. A coil having n turns and resistance $R\Omega$ is connected with a galvanometer of resistance $4R\Omega$. This combination is moved in time t seconds from a magnetic field W_1 weber to W_2 weber. The induced current in the circuit is

[NCERT : P L-158 | 2004]

(a) $-\dfrac{(W_2 - W_1)}{Rnt}$ (b) $-\dfrac{n(W_2 - W_1)}{5\,Rt}$

(c) $-\dfrac{(W_2 - W_1)}{5\,Rnt}$ (d) $-\dfrac{n(W_2 - W_1)}{Rt}$

Topic 2 - Motional & Static EMI and Application of EMI

47. An inductor of self inductance 1 H connected in series with a resistor of 100 π ohm and an ac supply of 100 π volt, 50 Hz. Maximum current flowing in the circuit is ______ A.

[NCERT : P L-168 | April 4, 2025 (II)]

48. Conductor wire $ABCDE$ with each arm 10 cm in length is placed in magnetic field of $\dfrac{1}{\sqrt{2}}$ Tesla, perpendicular to its plane. When conductor is pulled towards right with constant velocity of 10 cm/s, induced emf between points A and E is ________ mV.

[NCERT : P L-163 | April 4, 2025 (I)]

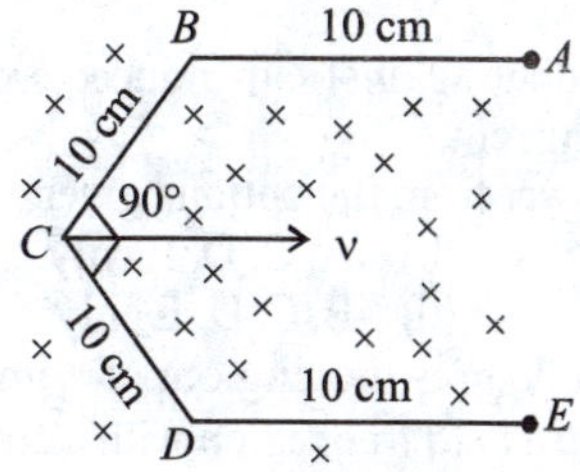

$B = \dfrac{1}{\sqrt{2}}$ Tesla

49. A solenoid having area A and length 'l' is filled with a material having relative permeability 2. The magnetic energy stored in the solenoid is :

[NCERT : P L-168 | April 2, 2025 (II)]

(a) $\dfrac{B^2Al}{\mu_0}$ (b) $\dfrac{B^2Al}{2\mu_0}$ (c) B^2Al (d) $\dfrac{B^2Al}{4\mu_0}$

50. Consider I_1 and I_2 are the currents flowing simultaneously in two nearby coils 1 & 2, respectively. If L_1 = self inductance of coil 1, M_{12} = mutual inductance of coil 1 with respect to coil 2, then the value of induced emf in coil 1 will be **[NCERT : P L-166 | Jan. 29, 2025 (I)]**

(a) $\varepsilon_1 = -L_1\dfrac{dI_1}{dt} + M_{12}\dfrac{dI_2}{dt}$ (b) $\varepsilon_1 = -L_1\dfrac{dI_1}{dt} - M_{12}\dfrac{dI_1}{dt}$

(c) $\varepsilon_1 = -L_1\dfrac{dI_1}{dt} - M_{12}\dfrac{dI_2}{dt}$ (d) $\varepsilon_1 = -L_1\dfrac{dI_2}{dt} - M_{12}\dfrac{dI_1}{dt}$

51. A uniform magnetic field of 0.4 T acts perpendicular to a circular copper disc 20 cm in radius. The disc is having a uniform angular velocity of 10 π rad s^{-1} about an axis through its centre and perpendicular to the disc. What is the protential difference developed between the axis of the disc and the rim ? (π = 3.14)

[NCERT : P L-165 | Jan. 28, 2025 (II)]

(a) 0.0628 V (b) 0.5024 V
(c) 0.2512 V (d) 0.1256 V

52.

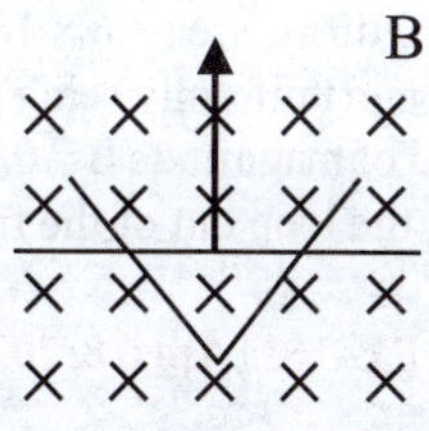

A conducting bar moves on two conducting rails as shown in the figure. A constant magnetic field B exists into the page. The bar starts to move from the vertex at time t = 0 with a constant velocity. If the induced EMF is $E \propto t^n$, then value of n is ___. **[NCERT : P L-163 | Jan. 28, 2025 (II)]**

53. Regarding self-inductance :

[NCERT : P L-168 | Jan. 23, 2025 (I)]

A: The self-inductance of the coil depends on its geometry.
B: Self-inductance does not depend on the permeability of the medium.
C: Self-induced e.m.f. opposes any change in the current in a circuit.

D: Self-inductance is electromagnetic analogue of mass in mechanics.

E: Work needs to be done against self-induced e.m.f. in establishing the current.

Choose the correct answer from the options given below:

(a) A, B, C, D only (b) A, C, D, E only

(c) A, B, C, E only (d) B, C, D, E only

54. A rectangular metallic loop is moving out of a uniform magnetic field region to a field free region with a constant speed. When the loop is partially inside the magnate field, the plot of magnitude of induced emf (ε) with time (t) is given by **[NCERT : P L-163 | Jan. 22, 2025 (II)]**

(a) 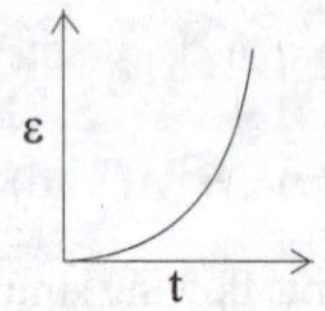

(b) 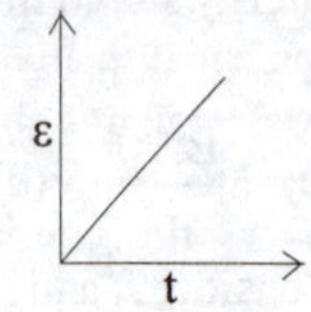

(c) 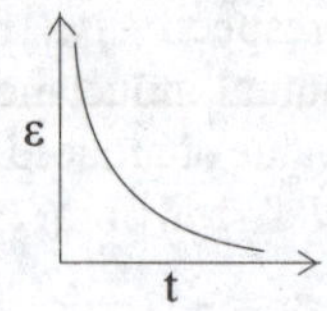

(d) 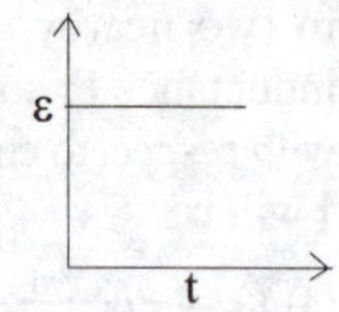

55. A square loop of side 15 cm being moved towards right at a constant speed of 2 cm/s as shown in figure. The front edge enters the 50 cm wide magnetic field at t = 0. The value of induced emf in the loop at t = 10 s will be : **[NCERT : P L-163 | April 9, 2024 (II)]**

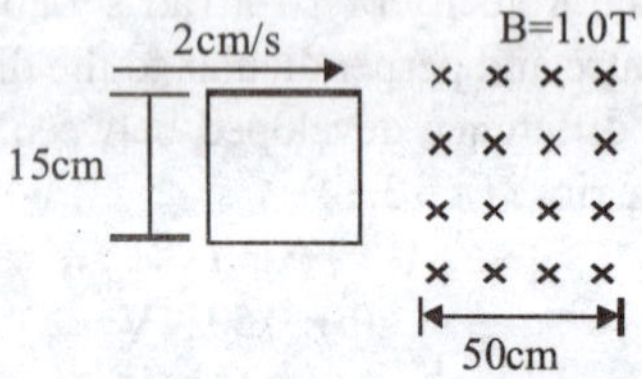

(a) 0.3 mV (b) 4.5 mV (c) zero (d) 3 mV

56. A square loop PQRS having 10 turns, area 3.6×10^{-3} m^2 and resistance 100 Ω is slowly and uniformly being pulled out of a uniform magnetic field of magnitude B = 0.5 T as shown. Work done in pulling the loop out of the field in 1.0 s is ________ $\times 10^{-6}$ J. **[NCERT : P L-163 | April 8, 2024 (I)]**

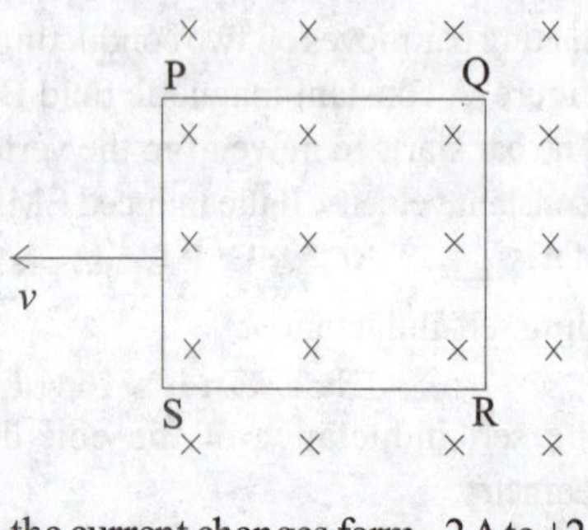

57. In a coil, the current changes form –2 A to +2 A in 0.2 s and induces an emf of 0.1 V. The selfinductance of the coil is : **[NCERT : P L-168 | April 6, 2024 (II)]**

(a) 5 mH (b) 1 mH (c) 2.5 mH (d) 4 mH

58. The current in an inductor is given by I = (3t + 8) where t is in second. The magnitude of induced emf produced in the inductor is 12 mV. The self inductance of the inductor ________ mH. **[NCERT : P L-168 | April 5, 2024 (II)]**

59. A rod of length 60 cm rotates with a uniform angular velocity 20 rad s^{-1} about its perpendicular bisector, in a uniform magnetic field 0.5 T. The direction of magnetic field is parallel to the axis of rotation. The potential difference between the two ends of the rod is _____ V. **[NCERT : P L-163 | April 4, 2024 (II)]**

60. A rectangular loop of sides 12cm and 5 cm, width its sides parallel to the x - axis and y - axis respectively, moves with a velocity of 5cm/s in the positive x axis direction, in a space containing a variable magnetic field in the positive z direction. The field has a gradient of 10^{-3}T/cm along the negative x direction and it is decreasing with time at the rate of 10^{-3}T/s. If the resistance of the loop is 6mΩ, the power dissipated by the loop as heat is ________ $\times 10^{-9}$ W. **[NCERT : P L-163 | Feb. 1, 2024 (I)]**

61. A small square loop of wire of side ℓ is placed inside a large square loop of wire of side L ($L = \ell^2$). The loops are coplanar and their centers coincide. The value of the mutual inductance of the system is $\sqrt{x} \times 10^{-7}$ H, where x = ______. **[NCERT : P L-166 | Jan. 31, 2024 (I)]**

62. A ceiling fan having 3 blades of length 80 cm each is rotating with an angular velocity of 1200 rpm. The Magnetic field of earth in that region is 0.5 G and angle of dip is 30°. The emf induced across the blades is $N\pi \times 10^{-5}$V. The value of N is ________. **[NCERT : P L-165 | Jan. 30, 2024 (I)]**

63. A horizontal straight wire 5 m long extending from east to west falling freely at right angle to horizontal component of earths magnetic field 0.60×10^{-4} Wbm^{-2}. The instantaneous value of emf induced in the wire when its velocity is 10 ms^{-1} is ______ $\times 10^{-3}$ V. **[NCERT : P L-163 | Jan. 29, 2024 (II)]**

64. Two coils have mutual inductance 0.002 H. The current changes in the first coil according to the relation $i = i_0 \sin \omega t$, where i_0 = 5A and $\omega = 50\pi$ rad/s. The maximum value of emf in the second coil is $\frac{\pi}{\alpha}$ V. The value of α is _____. **[NCERT : P L-166 | Jan. 27, 2024 (I)]**

65. A 12 V battery connected to a coil of resistance 6 Ω through a switch, drives a constant current in the circuit. The switch is opened in 1 ms. The emf induced across the coil is 20 V. The inductance of the coil is : **[NCERT : P L-168 | April 15, 2023 (I)]**

(a) 5 mH (b) 12 mH (c) 8 mH (d) 10 mH

66. A 20 cm long metallic rod is rotated with 210 rpm about an axis normal to the rod passing through its one end. The order end of the rod is in contact with a circular metallic ring. A constant and uniform magnetic field 0.2 T parallel to the axis exists everywhere. The emf developed between the centre and the ring is ________ mV.

Take $\pi = \frac{22}{7}$ **[NCERT : P L-165 | April 15, 2023 (I)]**

67. Given below are two statements : one is labelled as **Assertion A** and the other is labelled as **Reason R**.

Assertion A: A bar magnet dropped through a metallic cylindrical pipe takes more time to come down compared to a non-magnetic bar with same geometry and mass.

Reason R: For the magnetic bar, Eddy currents are produced in the metallic pipe which oppose the motion of the magnetic bar. **[April 11, 2023 (II)]**

In the light of the above statements, choose the correct answer from the options given below

(a) Both A and R are true but R is NOT the correct explanation of A
(b) A is true but R is false
(c) Both A and R are true but R is the correct explanation of A
(d) A is false but R is true

68. A metallic cube of side 15 cm moving along y-axis at a uniform velocity of 2 ms^{-1}. In a region of uniform magnetic field of magnitude 0.5 T directed along z-axis. In equilibrium the potential difference between the faces of higher and lower potential developed because of the motion through the field will be _____ mV.

[NCERT : P L-163 | April 11, 2023 (II)]

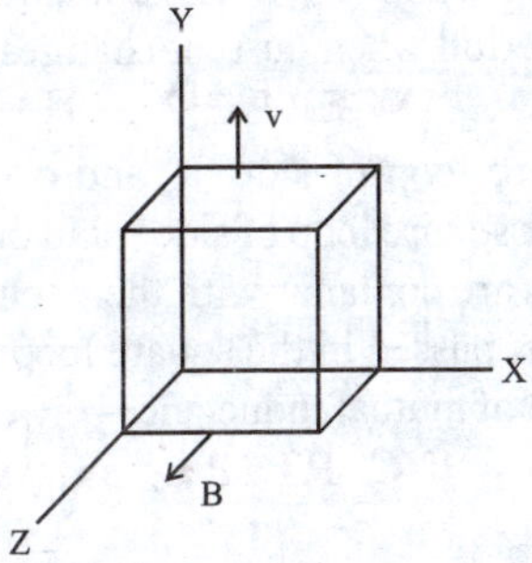

69. A 1m long metal rod XY completes the circuit as shown in figure. The plane of the circuit is perpendicular to the magnetic field of flux density 0.15 T. If the resistance of the circuit is 5Ω , the force needed to move the rod in direction, as indicated, with a constant speed of 4 m/s will be_______ 10^{-3} N.

[NCERT : P L-163 | April 10, 2023 (I)]

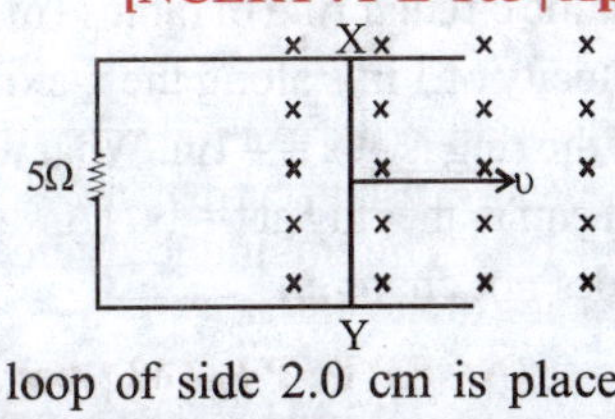

70. A square loop of side 2.0 cm is placed inside a long solenoid that has 50 turns per centimetre and carries a sinusoidally varying current of amplitude 2.5 A and angular frequency 700 rad s^{-1}. The central axes of the loop and solenoid coincide. The amplitude of the emf induced in the loop is x × 10^{-4} V. The value of x is_________

$\left(\text{Take, } \pi = \frac{22}{7}\right)$ **[NCERT : P L-168 | April 10, 2023 (II)]**

71. An emf of 0.08 V is induced in a metal rod of length 10 cm held normal to a uniform magnetic field of 0.4 T, when moves with a velocity of:

[NCERT : P L-163 | April 8, 2023 (II)]

(a) 2 ms^{-1} (b) 3.2 ms^{-1}
(c) 0.5 ms^{-1} (d) 20 ms^{-1}

72. Two concentric circular coils with radii 1 cm and 1000 cm, and number of turns 10 and 200 respectively are placed coaxially with centers coinciding. The mutual inductance of this arrangement will be ________ × 10^{-8} H. (Take, π^2=10)

[NCERT : P L-166 | April 6, 2023 (II)]

73. Spherical insulating ball and a spherical metallic ball of same size and mass are dropped from the same height. Choose the correct statement out of the following {Assume negligible air friction} **[Jan. 31, 2023 (I)]**

(a) Time taken by them to reach the earth's surface will be independent of the properties of their materials
(b) Insulating ball will reach the earth's surface earlier than the metal ball
(c) Both will reach the earth's surface simultaneously
(d) Metal ball will reach the earth's surface earlier than the insulating ball.

74. As per the given figure, if $\frac{dI}{dt} = -1$ A/s then the value of V_{AB} at this instant will be_____ V.

[Jan. 30, 2023 (I)]

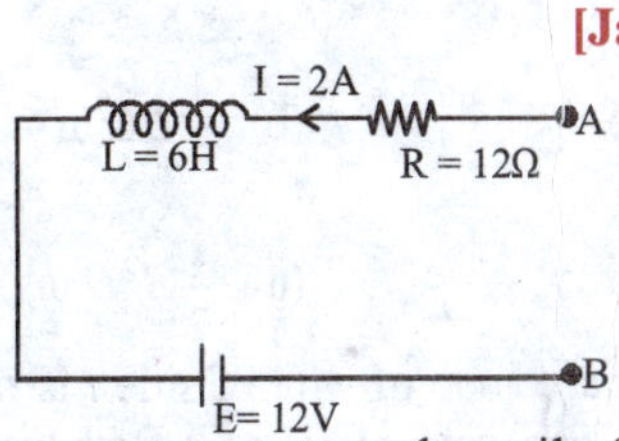

75. In an ac generator, a rectangular coil of 100 turns each having area 14 × 10^{-2} m^2 is rotated at 360 rev/min about an axis perpendicular to a uniform magnetic field of magnitude 3.0 T. The maximum value of the emf produced will be _______ V. $\left(\text{Take } \pi = \frac{22}{7}\right)$

[NCERT : P L-171 | Jan. 30, 2023 (II)]

76. Find the mutual inductance in the arrangement, when a small circular loop of wire of radius 'R' is placed inside a large square loop of wire of side L (L >> R). The loops are coplanar and their centres coincide :

[NCERT : P L-166 | Jan. 29, 2023 (I)]

(a) $M = \frac{\sqrt{2}\mu_0 R^2}{L}$

(b) $M = \frac{\sqrt{2}\mu_0 R}{L^2}$

(c) $M = \frac{2\sqrt{2}\mu_0 R^2}{L}$

(d) $M = \frac{\sqrt{2}\mu_0 R}{L^2}$

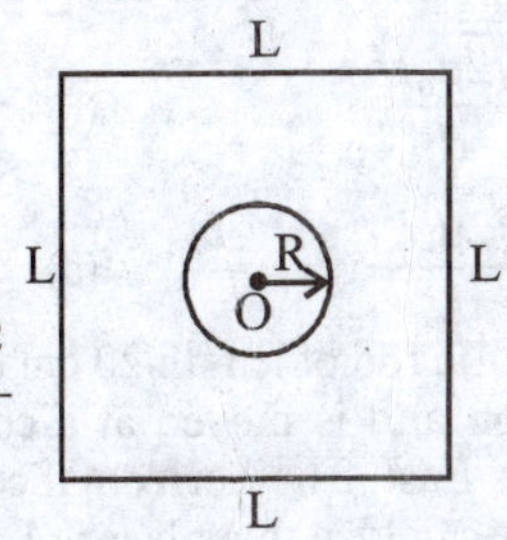

77. A square loop of area $25 cm^2$ has a resistance of 10Ω. The loop is placed in uniform magnetic field of magnitude 40.0 T. The plane of loop is perpendicular to the magnetic field. The work done in pulling the loop out of the magnetic field slowly and uniformly in 1.0 sec, will be

[NCERT : P L-163 | Jan. 29, 2023 (II)]

(a) 2.5×10^{-3} J (b) 1.0×10^{-3} J
(c) 1.0×10^{-4} J (d) 5×10^{-3} J

78. A wire of length 1 m moving with velocity 8 m/s at right angles to a magnetic field of 2T. The magnitude of induced emf, between the ends of wire will be ______:

[NCERT : P L-163 | Jan. 25, 2023 (II)]

(a) 20V (b) 8V (c) 12V (d) 16V

79. A metallic rod of length 'L' is rotated with an angular speed of 'ω' normal to a uniform magnetic field 'B' about an axis passing through one end of rod as shwon in figure. The induced emf will be:

[NCERT : P L-165 | Jan. 24, 2023 (II)]

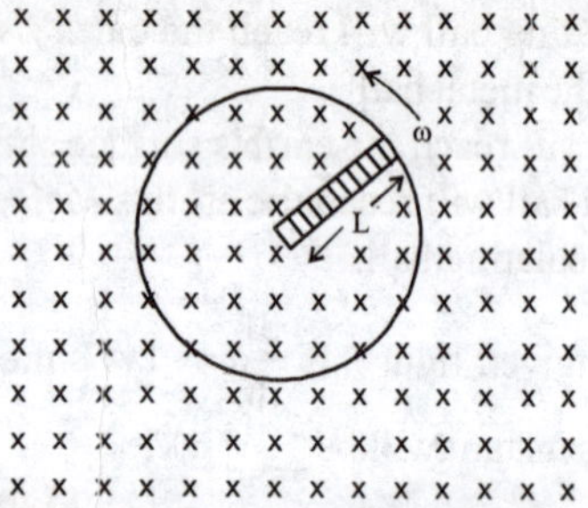

(a) $\frac{1}{4}B^2L\omega$ (b) $\frac{1}{4}BL^2\omega$

(c) $\frac{1}{2}BL^2\omega$ (d) $\frac{1}{2}B^2L^2\omega$

80. A small square loop of wire of side l is placed inside a large square loop of wire L (L >> l). Both loops are coplanar and their centres coincide at point O as shown in figure. The mutual inductance of the system is:

[NCERT : P L-166 | July 25, 2022 (I)]

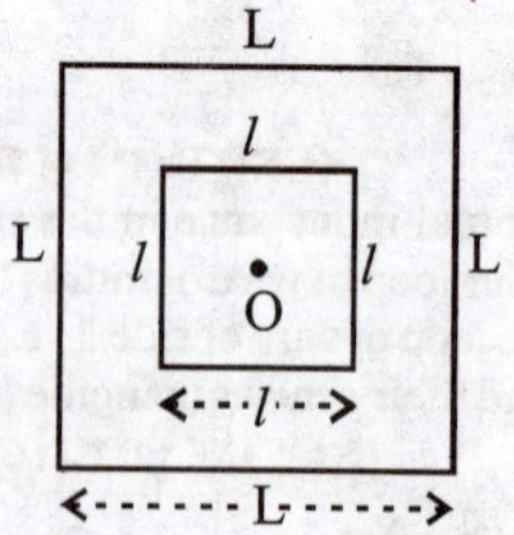

(a) $\frac{2\sqrt{2}\mu_0 L^2}{\pi l}$ (b) $\frac{\mu_0 l^2}{2\sqrt{2}\pi L}$

(c) $\frac{2\sqrt{2}\mu_0 l^2}{\pi L}$ (d) $\frac{\mu_0 L^2}{2\sqrt{2}\pi l}$

81. A metallic rod of length 20 cm is placed in North-South direction and is moved at a constant speed of 20 m/s towards East. The horizontal component of the Earth's magnetic field at that place is 4×10^{-3} T and the angle of dip is 45°. The emf induced in the rod is ______ mV.

[NCERT : P L-163 | June 27, 2022 (II)]

82. A 10Ω, 20 mH coil carrying constant current is connected to a battery of 20 V through a switch is opened current becomes zero in 100μs. The average emf induced in the coil is .... V. **[NCERT : P L-168 | June 26, 2022 (I)]**

83. Two coils of self inductance L_1 and L_2 are connected in series combination having mutual inductance of the coils as M. The equivalent self inductance of the combination will be : **[NCERT : P L-166, 168 | June 26, 2022 (II)]**

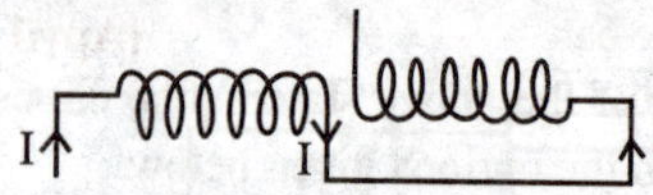

(a) $\frac{1}{L_1}+\frac{1}{L_2}+\frac{1}{M}$ (b) L_1+L_2+M

(c) L_1+L_2+2M (d) L_1+L_2-2M

84. A metallic conductor of length 1m rotates in a vertical plane parallel to east-west direction about one of its end with angular velocity 5 rad/s. If the horizontal component of earth's magnetic field is 0.2×10^{-4} T, then emf induced between the two ends of the conductor is :

[NCERT : P L-165 | June 26, 2022 (II)]

(a) 5μV (b) 50μV
(c) 5mV (d) 50mV

85. The current in a coil of self inductance 2.0 H is increasing according to $I = 2\sin(t^2)$A. The amount of energy spent during the period when current changes from 0 to 2A is ______ J. **[NCERT : P L-169 | June 25, 2022 (I)]**

86. A small square loop of side 'a' and one turn is placed inside a larger square loop of side b and one turn (b >> a). The two loops are coplanar with their centres coinciding. If a current I is passed in the square loop of side 'b', then the coefficient of mutual inductance between the two loops is : **[NCERT : P L-166 | Aug. 31, 2021 (I)]**

(a) $\frac{\mu_0}{4\pi}8\sqrt{2}\frac{a^2}{b}$ (b) $\frac{\mu_0}{4\pi}\frac{8\sqrt{2}}{a}$

(c) $\frac{\mu_0}{4\pi}8\sqrt{2}\frac{b^2}{a}$ (d) $\frac{\mu_0}{4\pi}\frac{8\sqrt{2}}{b}$

87. A constant magnetic field of 1 T is applied in the x > 0 region. A metallic circular ring of radius 1m is moving with a constant velocity of 1 m/s along the x-axis. At t = 0s, the centre of O of the ring is at x = – 1m. What will be the value of the induced emf in the ring at t = 1s? (Assume the velocity of the ring does not change.)

[NCERT : P L-163 | Aug. 27, 2021 (II)]

(a) 1V

(b) 2πV

(c) 2V

(d) 0V

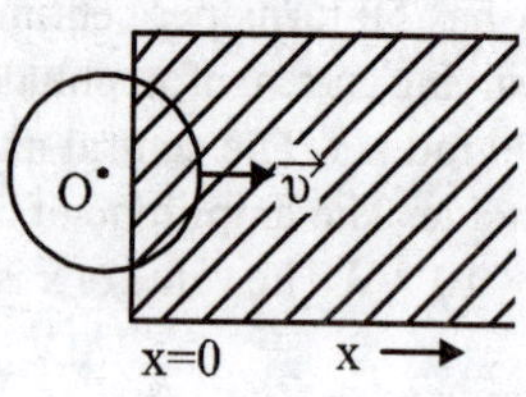

88. Consider an electrical circuit containing a two way switch 'S'. Initially S is open and then T_1 is connected to T_2. As the current in R = 6 Ω attains a maximum value of steady state level, T_1 is disconnected from T_2 and immediately connected to T_3. Potential drop across r = 3 Ω resistor immediately after T_1 is connected to T_3 is______V. (Round off to the Nearest Integer) **[NA, July 27, 2021 (I)]**

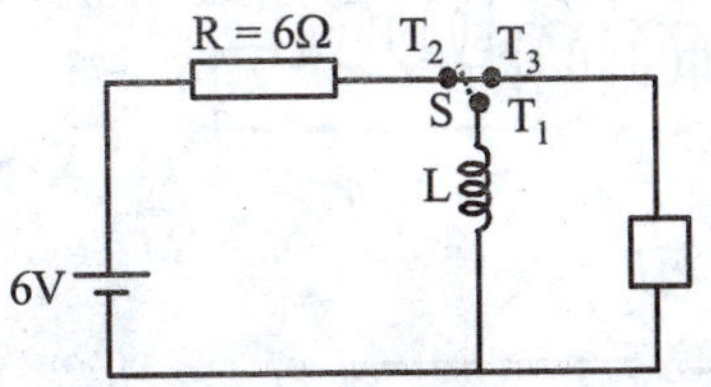

89. The time taken for the magnetic enegy to reach 25% of its maximum value, when a solenoid of resistance R, inductance L is connected to a battery, is: **[NCERT : P L-169 | March 18, 2021 (II)]**

(a) $\frac{L}{R}\ln 5$ (b) Infinite

(c) $\frac{L}{R}\ln 10$ (d) $\frac{L}{R}\ln 2$

90. The magnetic field in a region is given by $\vec{B} = B_0\left(\frac{x}{a}\right)\hat{k}$. A square loop of side d is placed with its edges along the x and y-axis. The loop is moved with a constant velocity $\vec{v} = v_0\hat{i}$. The emf induced in the loop is: **[NCERT : P L-163 | March 16, 2021 (II)]**

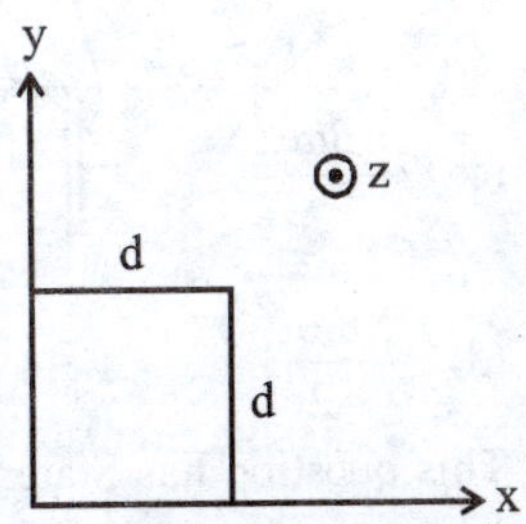

(a) $\frac{B_0 v_0 d^2}{2a}$

(b) $\frac{B_0 v_0 d^2}{a}$

(c) $\frac{B_0 v_0 d}{2a}$

(d) $\frac{B_0 v_0^2 d}{2a}$

91. An aeroplane, with its wings spread 10 m, is flying at a speed of 180 km/h in a horizontal direction. The total intensity of earth's field at that part is 2.5×10^{-4} Wb/m^2 and the angle of dip is 60°. The emf induced between the tips of the plane wings will be ______. **[NCERT : P L-163 | Feb. 26, 2021 (II)]**

(a) 108.25 mV
(b) 88.37 mV
(c) 62.50 mV
(d) 54.125 mV

92. An infinitely long straight wire carrying current I, one side opened rectangular loop and a conductor C with a sliding connector are located in the same plane, as shown in the figure. The connector has length l and resistance R. It slides to the right with a velocity v. The resistance of the conductor and the self inductance of the loop are negligible. The induced current in the loop, as a function of separation r, between the connector and the straight wire is : **[NCERT : P L-163 | Sep. 05, 2020 (II)]**

(a) $\frac{\mu_0}{4\pi}\frac{Ivl}{Rr}$

(b) $\frac{\mu_0}{\pi}\frac{Ivl}{Rr}$

(c) $\frac{2\mu_0}{\pi}\frac{Ivl}{Rr}$

(d) $\frac{\mu_0}{2\pi}\frac{Ivl}{Rr}$

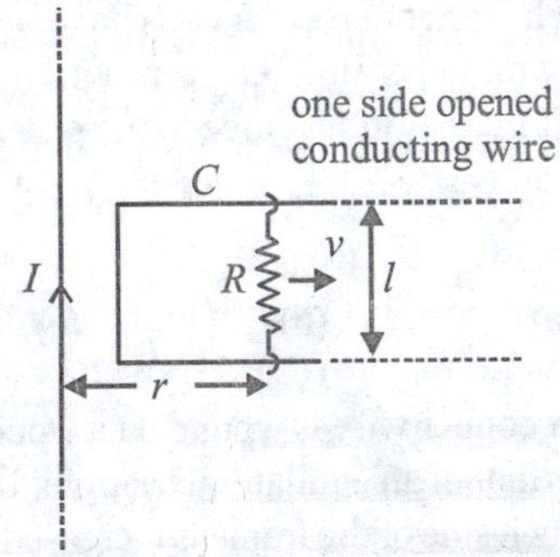

93. In a fluorescent lamp choke (a small transformer) 100 V of reverse voltage is produced when the choke current changes uniformly from 0.25 A to 0 in a duration of 0.025 ms. The self-inductance of the choke (in mH) is estimated to be______. **[NCERT : P L-168 | NA 9 Jan. 2020 I]**

94. The figure shows a square loop L of side 5 cm which is connected to a network of resistances. The whole setup is moving towards right with a constant speed of 1 cm s^{-1}. At some instant, a part of L is in a uniform magnetic field of 1 T, perpendicular to the plane of the loop. If the resistance of L is 1.7Ω the current in the loop at that instant will be close to : **[NCERT : P L-163 | 12 Apr. 2019 I]**

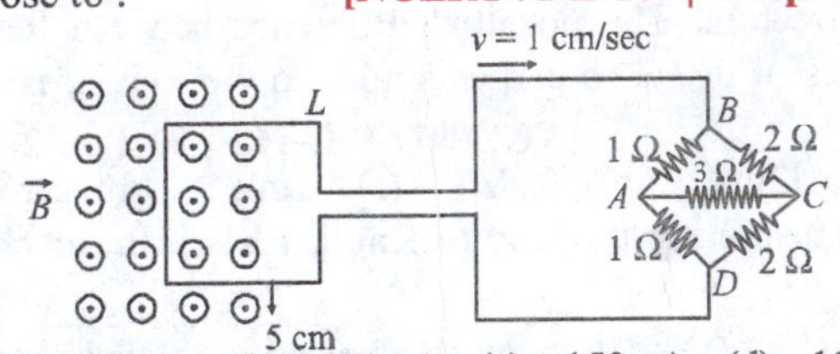

(a) 60μA (b) 170 μA (c) 150 μA (d) 115 μA

95. The total number of turns and cross-section area in a solenoid is fixed. However, its length L is varied by adjusting the separation between windings. The inductance of solenoid will be proportional to: **[NCERT : P L-168 | 9 April 2019 I]**

(a) L (b) L^2 (c) 1/L^2 (d) 1/L

96. A thin strip 10 cm long is on a U shaped wire of negligible resistance and it is connected to a spring of spring constant 0.5 Nm^{-1} (see figure). The assembly is kept in a uniform magnetic field of 0.1 T. If the strip is pulled from its equilibrium position and released, the number of oscillations it performs before its amplitude decreases by a factor of e is N. If the mass of strip is 50 grams, its resistance 10Ω and air drag negligible, N will be close to : **[8 April 2019 I]**

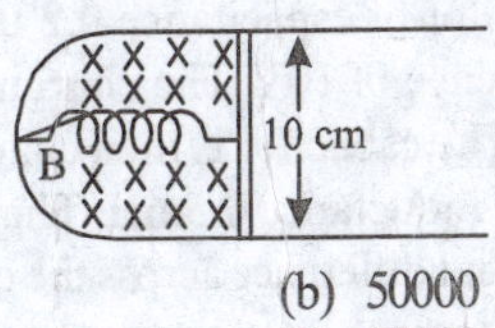

(a) 1000 (b) 50000
(c) 5000 (d) 10000

97. A 10 m long horizontal wire extends from North East to South West. It is falling with a speed of $5.0\ ms^{-1}$, at right angles to the horizontal component of the earth's magnetic field, of $0.3 \times 10^{-4}\ Wb/m^2$. The value of the induced emf in wire is : **[NCERT : P L-163 | 12 Jan. 2019 II, 2011 RS]**
(a) 1.5×10^{-3} V (b) 1.1×10^{-3} V
(c) 2.5×10^{-3}V (d) 0.3×10^{-3} V

98. There are two long co-axial solenoids of same length l. The inner and outer coils have radii r_1 and r_2 and number of turns per unit length n_1 and n_2, respectively. The ratio of mutual inductance to the self-inductance of the inner-coil is : **[NCERT : P L-166 | 11 Jan. 2019 I]**
(a) $\frac{n_1}{n_2}$ (b) $\frac{n_2}{n_1} \cdot \frac{r_1}{r_2}$ (c) $\frac{n_2}{n_1} \cdot \frac{r_2^2}{r_1^2}$ (d) $\frac{n_2}{n_1}$

99. A copper wire is wound on a wooden frame, whose shape is that of an equilateral triangle. If the linear dimension of each side of the frame is increased by a factor of 3, keeping the number of turns of the coil per unit length of the frame same, then the self inductance of the coil: **[NCERT : PL-168 | 11 Jan. 2019 II]**
(a) decreases by a factor of 9
(b) increases by a factor of 27
(c) increases by a factor of 3
(d) decreases by a factor of $9\sqrt{3}$

100. A solid metal cube of edge length 2 cm is moving in a positive y-direction at a constant speed of 6 m/s. There is a uniform magnetic field of 0.1 T in the positive z-direction. The potential difference between the two faces of the cube perpendicular to the x-axis, is: **[NCERT : P L-163 | 10 Jan. 2019 I]**
(a) 12mV (b) 6mV (c) 1mV (d) 2mV

101. An insulating thin rod of length l has a linear charge density $\rho(x) = \rho_0 \frac{x}{l}$ on it. The rod is rotated about an axis passing through the origin ($x = 0$) and perpendicular to the rod. If the rod makes n rotations per second, then the time averaged magnetic moment of the rod is: **[10 Jan. 2019 I]**
(a) $\pi n \rho_0 l^3$ (b) $\frac{\pi}{3} n \rho_0 l^3$ (c) $\frac{\pi}{4} n \rho_0 l^3$ (d) $n \rho_0 l^3$

102. The self induced emf of a coil is 25 volts. When the current in it is changed at uniiform rate from 10 A to 25 A in 1s, the change in the energy of the inductance is: **[NCERT : P L-168 | 9 Jan. 2019 II]**
(a) 740 J (b) 437.5 J (c) 540 J (d) 637.5 J

103. An ideal capacitor of capacitance 0.2 μF is charged to a potential difference of 10V. The charging battery is then disconnected. The capacitor is then connected to an ideal inductor of self inductance 0.5mH. The current at a time when the potential difference across the capacitor is 5V, is: **[NCERT : PL-169 | Online April 15, 2018]**
(a) 0.17A (b) 0.15A (c) 0.34A (d) 0.25A

104. When current in a coil changes from 5 A to 2 A in 0.1 s, average voltage of 50 V is produced. The self - inductance of the coil is : **[NCERT : P L-168 | 2003, Online April 10, 2015]**
(a) 2 H (b) 2.3 H (c) 3 H (d) 1.67 H

105. A square frame of side 10 cm and a long straight wire carrying current 1 A are in the plane of the paper. Starting from close to the wire, the frame moves towards the right with a constant speed of $10\ ms^{-1}$ (see figure).

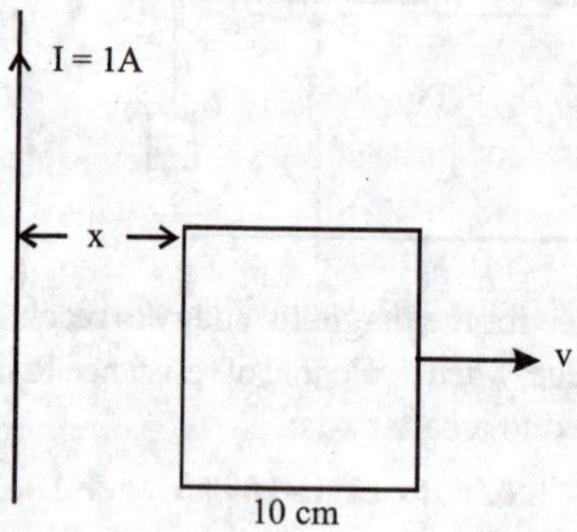

The e.m.f induced at the time the left arm of the frame is at x = 10 cm from the wire is: **[NCERT : P L-163 | Online April 19, 2014]**
(a) 2 μV (b) 1 μV (c) 0.75 μV (d) 0.5 μV

106. A metallic rod of length 'ℓ' is tied to a string of length 2ℓ and made to rotate with angular speed ω on a horizontal table with one end of the string fixed. If there is a vertical magnetic field 'B' in the region, the e.m.f. induced across the ends of the rod is **[NCERT : P L-165 | 2013]**

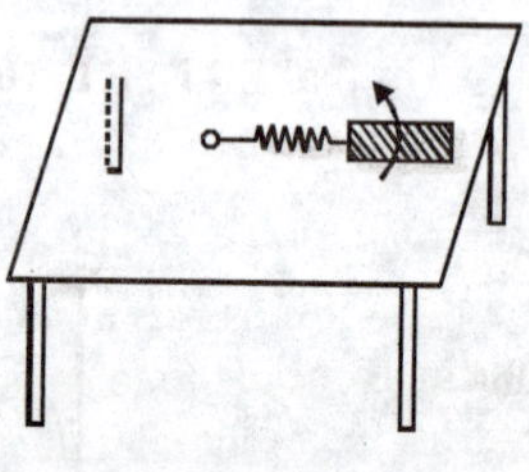

(a) $\frac{2B\omega\ell^2}{2}$
(b) $\frac{3B\omega\ell^2}{2}$
(c) $\frac{4B\omega\ell^2}{2}$
(d) $\frac{5B\omega\ell^2}{2}$

107. This question has Statement 1 and Statement 2. Of the four choices given after the Statements, choose the one that best describes the two Statements.
Statement 1: Self inductance of a long solenoid of length L, total number of turns N and radius r is less than $\frac{\pi\mu_0 N^2 r^2}{L}$.
Statement 2: The magnetic induction in the solenoid in Statement 1 carrying current I is $\frac{\mu_0 NI}{L}$ in the middle of the solenoid but becomes less as we move towards its ends. **[NCERT : P L-168 | Online May 19, 2012]**

(a) Statement 1 is true, Statement 2 is false.
(b) Statement 1 is true, Statement 2 is true, Statement 2 is the correct explanation of Statement 1.
(c) Statement 1 is false, Statement 2 is true.
(d) Statement 1 is true, Statement 2 is true, Statement 2 is not the correct explanation of Statement 1.

108. A boat is moving due east in a region where the earth's magnetic field is 5.0×10^{-5} NA^{-1} m^{-1} due north and horizontal. The boat carries a vertical aerial 2 m long. If the speed of the boat is 1.50 ms^{-1}, the magnitude of the induced emf in the wire of aerial is: **[NCERT : P L-163 | 2011]**

(a) 0.75 mV (b) 0.50 mV (c) 0.15 mV (d) 1 mV

109. A rectangular loop has a sliding connector PQ of length l and resistance R Ω and it is moving with a speed v as shown. The set-up is placed in a uniform magnetic field going into the plane of the paper. The three currents I_1, I_2 and I are **[NCERT : P L-163 | 2010]**

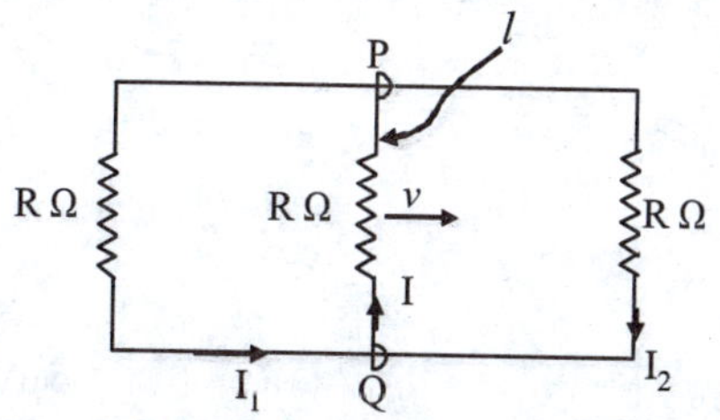

(a) $I_1 = -I_2 = \frac{Blv}{6R},\ I = \frac{2Blv}{6R}$

(b) $I_1 = I_2 = \frac{Blv}{3R}, I = \frac{2Blv}{3R}$

(c) $I_1 = I_2 = I = \frac{Blv}{R}$

(d) $I_1 = I_2 = \frac{Blv}{6R},\ I = \frac{Blv}{3R}$

110. Two coaxial solenoids are made by winding thin insulated wire over a pipe of cross-sectional area $A = 10$ cm^2 and length = 20 cm. If one of the solenoid has 300 turns and the other 400 turns, their mutual inductance is **[NCERT : P L-166 | 2008]**

$(\mu_0 = 4\pi \times 10^{-7}$ Tm A$^{-1})$

(a) $2.4\pi \times 10^{-5}$ H (b) $4.8\pi \times 10^{-4}$ H
(c) $4.8\pi \times 10^{-5}$ H (d) $2.4\pi \times 10^{-4}$ H

111. One conducting U tube can slide inside another as shown in figure, maintaining electrical contacts between the tubes. The magnetic field B is perpendicular to the plane of the figure. If each tube moves towards the other at a constant speed v, then the emf induced in the circuit in terms of B, l and v where l is the width of each tube, will be **[NCERT : P L-163 | 2005]**

(a) $-Blv$
(b) Blv
(c) $2\ Blv$
(d) zero

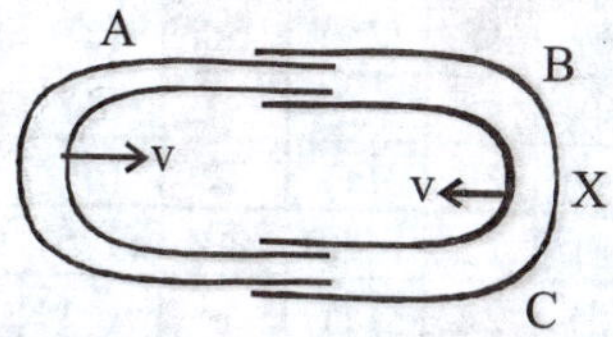

112. A metal conductor of length 1 m rotates vertically about one of its ends at angular velocity 5 radians per second. If the horizontal component of earth's magnetic field is 0.2×10^{-4}T, then the e.m.f. developed between the two ends of the conductor is **[NCERT : P L-165 | 2004]**

(a) 5mV (b) 50μV (c) 5μV (d) 50mV

113. Two coils are placed close to each other. The mutual inductance of the pair of coils depends upon **[NCERT : P L-166 | 2003]**

(a) the rates at which currents are changing in the two coils
(b) relative position and orientation of the two coils
(c) the materials of the wires of the coils
(d) the currents in the two coils

114. A conducting square loop of side L and resistance R moves in its plane with a uniform velocity v perpendicular to one of its sides. A magnetic induction B constant in time and space, pointing perpendicular and into the plane at the loop exists everywhere with half the loop outside the field, as shown in figure. The induced emf is **[NCERT : P L-163 | 2002]**

(a) zero
(b) RvB
(c) vBL/R
(d) vBL

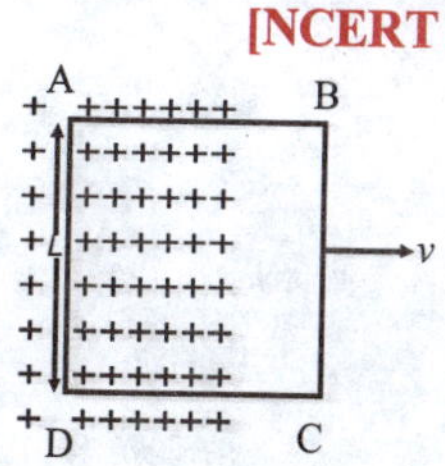

ANSWER KEYS

1	(b)	13	(c)	25	(c)	37	(None)	49	(d)	61	(128)	73	(b)	85	(4)	97	(a)	109	(b)
2	(5)	14	(10)	26	(5)	38	(a)	50	(c)	62	(32)	74	(30)	86	(a)	98	(d)	110	(d)
3	(a)	15	(b)	27	(a)	39	(b)	51	(c)	63	(3)	75	(1584)	87	(c)	99	(c)	111	(c)
4	(2)	16	(12)	28	(b)	40	(a)	52	(1)	64	(2)	76	(c)	88	(3)	100	(a)	112	(b)
5	(2)	17	(2)	29	(15)	41	(a)	53	(a)	65	(d)	77	(b)	89	(d)	101	(c)	113	(b)
6	(c)	18	(250)	30	(a)	42	(a)	54	(d)	66	(88)	78	(d)	90	(b)	102	(b)	114	(d)
7	(60)	19	(d)	31	(d)	43	(d)	55	(c)	67	(c)	79	(c)	91	(a)	103	(a)		
8	(50)	20	(a)	32	(d)	44	(c)	56	(3)	68	(150)	80	(c)	92	(d)	104	(d)		
9	(8)	21	(440)	33	(175)	45	(b)	57	(a)	69	(18)	81	(16)	93	(10)	105	(b)		
10	(d)	22	(a)	34	(b)	46	(b)	58	(4)	70	(44)	82	(400)	94	(b)	106	(d)		
11	(d)	23	(60)	35	(a)	47	(1)	59	(0)	71	(a)	83	(d)	95	(d)	107	(d)		
12	(44)	24	(60)	36	(a)	48	(10)	60	(216)	72	(4)	84	(b)	96	(c)	108	(c)		

7 Alternating Current

TREND ANALYSIS
JEE MAIN (2025-2021)

	2025				2024				2023				2022				2021			
	Session II		Session I		Session II		Session I		Session II		Session I		Session II		Session I		Session II		Session I	
	MCQs	NVQs	MCQs	NVQs	MCQs	NVQs	MCQs	NVQs	MCQs	NVQs	MCQs	NVQs	MCQs	NVQs	MCQs	NVQs	MCQs	NVQs	MCQs	NVQs
Topic 1	3	–	1	–	–	2	2	–	–	–	–	–	3	–	2	1	–	1	1	–
Topic 2	–	2	2	2	7	5	3	1	5	2	4	5	6	6	4	7	4	5	2	3
Topic 3	–	–	–	–	–	–	3	1	–	2	1	–	1	–	1	–	–	–	1	1
Topic 4																				
Total	3	2	3	2	7	7	8	2	5	4	5	5	10	6	7	8	4	6	4	4

How to Code Qns for Better & Faster Revision

Type 1: Able to solve (No code Req)

Type 2: Solved with struggle or by seeing solution **(Use code NR means Needs Revision or orange stickers)**

Type 3: Tricky/ Could not solve on my own. Need help from Teacher/ Friend. **(Use code TQ means Tricky Question or red stickers)**

You can either hand write the code in boxes or else paste coloured stickers given at end of book.

Topic 1 - Alternating Current, Voltage and Power

1. An ac current is represented as $i = 5\sqrt{2} + 10 \cos\left(650\pi t + \frac{\pi}{6}\right)$ Amp

[NCERT : P L-180 | April 7, 2025 (I)]

The r.m.s value of the current is

(a) 50 Amp (b) 100 Amp

(c) 10 Amp (d) $5\sqrt{2}$ Amp

2. An alternating current is represented by the equation, $i = 100\sqrt{2}\sin(100\pi t)$ ampere. The RMS value of current and the frequency of the given alternating current are

[NCERT : P L-180 | April 4, 2025 (I)]

(a) $100\sqrt{2}$ A, 100 Hz (b) $\frac{100}{\sqrt{2}}$ A, 100 Hz

(c) 100 A, 50 Hz (d) $50\sqrt{2}$ A, 50 Hz

3. An electric bulb rated as 100 W-220 V is connected to an ac source of rms voltage 220 V. The peak value of current through the bulb is :

[NCERT : P L-180 | April 3, 2025 (II)]

(a) 0.64 A (b) 0.45 A (c) 2.2 A (d) 0.32 A

4. An alternating current is given by $I = I_A \sin \omega t + I_B \cos \omega t$.
The r.m.s. current will be

[NCERT : P L-180 | Jan. 24, 2025 (I)]

(a) $\sqrt{I_A^2 + I_B^2}$ (b) $\frac{\sqrt{I_A^2 + I_B^2}}{2}$

(c) $\sqrt{\frac{I_A^2 + I_B^2}{2}}$ (d) $\frac{|I_A + I_B|}{\sqrt{2}}$

5. An alternating emf $E = 110\sqrt{2} \sin 100t$ volt is applied to a capacitor of $2\mu F$, the rms value of current in the circuit is ______ mA.

[NCERT : PL-184 | April 8, 2024 (II)]

6. A alternating current at any instant is given by $i = \left[6 + \sqrt{56}\sin\left(100\pi t + \frac{\pi}{3}\right)\right]$A. The rms value of the current is______ A.
[NCERT : P L-180 | April 4, 2024 (I)]

7. An alternating voltage $V(t) = 220 \sin 100\,\pi t$ volt is applied to a purely resistive load of 50 Ω. The time taken for the current to rise from half of the peak value to the peak value is: **[NCERT : P L-178 | Jan. 30, 2024 (II)]**
(a) 5 *ms* (b) 3.3 *ms* (c) 7.2 *ms* (d) 2.2 *ms*

8. In an a.c. circuit, voltage and current are given by:
$V = 100 \sin(100\,t)$ V and
$I = 100 \sin(100\,t + \frac{\pi}{3})$ mA respectively.
The average power dissipated in one cycle is:
[NCERT : P L-191 | Jan. 29, 2024 (II)]
(a) 10 W (b) 2.5 W (c) 25 W (d) 5 W

9. An alternating emf $E = 440 \sin 100\,\pi t$ is applied to a circuit containing an inductance of $\frac{\sqrt{2}}{\pi}$ H. If an a.c. ammeter is connected in the circuit, its reading will be :
[NCERT : P L-182 | July 29, 2022 (I)]
(a) 4.4A (b) 1.55A (c) 2.2A (d) 3.11A

10. The equation of current in a purely inductive circuit is $5 \sin(49\pi t - 30°)$. If the inductance is 30 mH then the equation for the voltage across the inductor, will be : $\left(\text{Let } \pi = \frac{22}{7}\right)$ **[NCERT : P L-182 | July 28, 2022 (I)]**
(a) $1.47 \sin(49\,\pi t - 30°)$ (b) $1.47 \sin(49\,\pi t + 60°)$
(c) $23.1 \sin(49\,\pi t - 30°)$ (d) $23.1 \sin(49\,\pi t + 60°)$

11. A direct current of 4 A and an alternating current of peak value 4 A flow through resistance of 3 Ω and 2 Ω respectively. The ratio of heat produced in the two resistances in same interval of time will be :
[NCERT : P L-191 | July 27, 2022 (I)]
(a) 3 : 2 (b) 3 : 1 (c) 3 : 4 (d) 4 : 3

12. The current flowing through an ac circuit is given by
$I = 5\sin(120\pi t)\,A$
How long will the current take to reach the peak value starting from zero?
[NCERT : P L-178 | June 27, 2022 (I)]
(a) $\frac{1}{60}s$ (b) 60*s* (c) $\frac{1}{120}s$ (d) $\frac{1}{240}s$

13. As shown in the figure an inductor of inductance 200 mH is connected to an AC source of emf 220 V and frequency 50 Hz. The instantaneous voltage of the source is 0 V when the peak value of current is $\frac{\sqrt{a}}{\pi}A$. The value of *a* is ______
[NCERT : P L-182 | June 24, 2022 (I)]

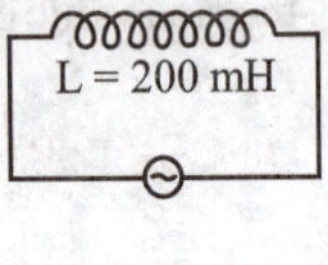

14. A resistance of 40Ω is connected to a source of alternating current rated 220 V, 50 Hz. Find the time taken by the current to change from its maximum value to rms value : **[NCERT : P L-180 | June 24, 2022 (I)]**
(a) 2.5 ms (b) 1.25 ms (c) 2.5 s (d) 0.25 s

15. An AC current is given by $I = I_1 \sin\omega t + I_2 \cos\omega t$. A hot wire ammeter will give a reading.
[NCERT : P L-180 | March 17, 2021 (I), Feb 26, 2021 (I) S]
(a) $\sqrt{\frac{I_1^2 + I_2^2}{2}}$ (b) $\frac{I_1 + I_2}{\sqrt{2}}$
(c) $\frac{I_1 + I_2}{2\sqrt{2}}$ (d) $\sqrt{\frac{I_1^2 - I_2^2}{2}}$

16. The alternating current is given by
$i = \left\{\sqrt{42}\sin\left(\frac{2\pi}{T}t\right) + 10\right\}$ A
The r.m.s. value of this current is ______ A.
[NCERT : P L-180 | NA, Aug. 27, 2021 (I)]

17. An alternating voltage $V(t) = 220 \sin 100\pi t$ volt is applied to a purely resistive load of 50Ω. The time taken for the current to rise from half of the peak value to the peak value is : **[NCERT : P L-178 | 8 April 2019 I]**
(a) 5 ms (b) 2.2 ms
(c) 7.2 ms (d) 3.3 ms

18. A sinusoidal voltage $V(t) = 100 \sin(500t)$ is applied across a pure inductance of L = 0.02 H. The current through the coil is: **[NCERT : P L-182 | Online April 12, 2014]**
(a) $10 \cos(500\,t)$ (b) $-10 \cos(500t)$
(c) $10 \sin(500t)$ (d) $-10 \sin(500t)$

19. In an a.c. circuit the voltage applied is $E = E_0 \sin\omega t$. The resulting current in the circuit is $I = I_0 \sin\left(\omega t - \frac{\pi}{2}\right)$. The power consumption in the circuit is given by
[NCERT : P L-191 | 2007]
(a) $P = \sqrt{2}E_0 I_0$ (b) $P = \frac{E_0 I_0}{\sqrt{2}}$
(c) P = zero (d) $P = \frac{E_0 I_0}{2}$

20. In a uniform magnetic field of induction *B* a wire in the form of a semicircle of radius r rotates about the diameter of the circle with an angular frequency ω. The axis of rotation is perpendicular to the field. If the total resistance of the circuit is R, the mean power generated per period of rotation is **[NCERT : P L-191 | 2004]**
(a) $\frac{(B\pi r\omega)^2}{2R}$ (b) $\frac{(B\pi r^2\omega)^2}{8R}$
(c) $\frac{B\pi r^2\omega}{2R}$ (d) $\frac{(B\pi r\omega^2)^2}{8R}$

21. Alternating current can not be measured by D.C. ammeter because **[NCERT : P L-177, 178 | 2004]**
(a) Average value of current for complete cycle is zero
(b) A.C. Changes direction
(c) A.C. can not pass through D.C. Ammeter
(d) D.C. Ammeter will get damaged.

Topic 2 - AC Circuit, LCR Circuit and Power Factor

22. An inductor of reactance 100 Ω, a capacitor of reactance 50 Ω, and a resistor of resistance 50 Ω are connected in series with an AC source of 10 V, 50 Hz. Average power dissipated by the circuit is _______ W.

[NCERT : P L-191 | April 7, 2025 (II)]

23. For ac circuit shown in figure, $R = 100\,k\Omega$ and $C = 100\,pF$ and the phase difference between V_{in} and $(V_B - V_A)$ is 90°. The input signal frequency is 10^x rad/sec, where 'x' is _____

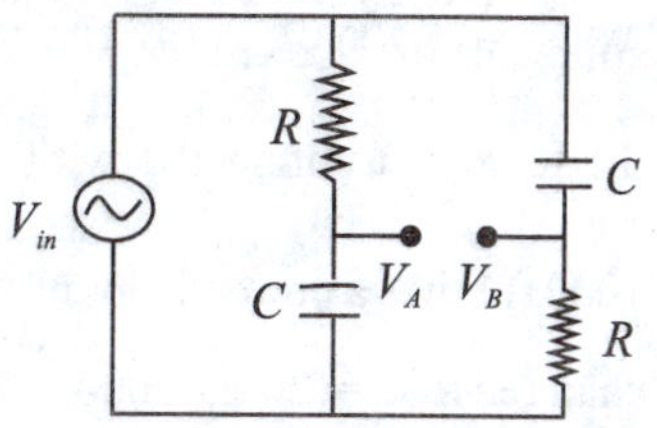

[NCERT : P L-189, 191 | April 6, 2025 (I)]

24. Given below are two statements : one is labelled as **Assertion (A)** and the other is labelled as **Reason (R)**.

Assertion (A) : Choke coil is simply a coil having a large inductance but a small resistance. Choke coils are used with fluorescent mercury-tube fittings. If household electric power is directly connected to a mercury tube, the tube will be damaged.

Reason (R) : By using the choke coil, the voltage across the tube is reduced by a factor $\left(\frac{R}{\sqrt{R^2 + \omega^2 L^2}}\right)$, where ω is frequency of the supply across resistor R and inductor L. If the choke coil were not used, the voltage across the resistor would be the same as the applied voltage.

In the light of the above statements, choose the most appropriate answer from the options given below:

[NCERT : P L-189 | Jan. 29, 2025 (I)]

(a) Both (A) and (R) are true but (R) is not the correct explanation of (A).
(b) (A) is false but (R) is true.
(c) Both (A) and (R) are true and (R) is the correct explanation of (A).
(d) (A) is true but (R) is false.

25. In a series LCR circuit, a resistor of 300 Ω, a capacitor of 25 nF and an inductor of 100 mH are used. For maximum current in the circuit, the angular frequency of the ac source is _____ $\times 10^4$ radians s^{-1}.

[NCERT : P L-191 | Jan. 23, 2025 (II)]

26.

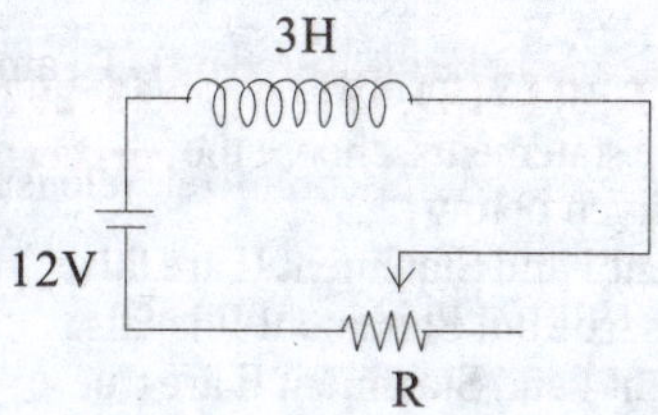

In the given circuit the sliding contact is pulled outwards such that electric current in the circuit changes at the rate of 8 A/s. At an instant when R is 12 Ω, the value of the current in the circuit will be ____A.

[NCERT : P L-189 | Jan. 23, 2025 (I)]

27. A series LCR circuit is connected to an alternating source of emf E. The current amplitude at resonant frequency is I_0. If the value of resistance R becomes twice of its initial value then amplitude of current at resonance will be

[NCERT : P L-191 | Jan. 22, 2025 (II)]

(a) I_0 (b) $\frac{I_0}{2}$ (c) $\frac{I_0}{\sqrt{2}}$ (d) $2I_0$

28. A bulb and a capacitor are connected in series across an ac supply. A dielectric is then placed between the plates of the capacitor. The glow of the bulb:

[NCERT : P L-187 | April 9, 2024 (I)]

(a) increases (b) remains same
(c) becomes zero (d) decreases

29. When a coil is connected across a 20 V dc supply, it draws a current of 5 A. When it is connected across 20 V, 50 Hz ac supply, it draws a current of 4 A. The self inductance of the coil is ______ mH. (Take π = 3)

[NCERT : P L-187 | April 9, 2024 (I)]

30. A capacitor of reactance $4\sqrt{3}\Omega$ and a resistor of resistance 4Ω are connected in series with an ac source of peak value $8\sqrt{2}$V. The power dissipation in the circuit is ______ W.

[NCERT : P L-191 | April 9, 2024 (II)]

31. A LCR circuit is at resonance for a capacitor C, inductance L and resistance R. Now the value of resistance is halved keeping all other parameters same. The current amplitude at resonance will be now:

[NCERT : P L-189 | April 8, 2024 (I)]

(a) zero (b) double
(c) same (d) halved

32. A coil of negligible resistance is connected in series with 90 Ω resistor across 120 V, 60 Hz supply. A voltmeter reads 36 V across resistance. Inductance of the coil is :

[NCERT : P L-187 | April 8, 2024 (II)]

(a) 0.76 H (b) 2.86 H (c) 0.286 H (d) 0.91 H

33. Given below are two statements :

Statement I : In an LCR series circuit, current is maximum at resonance.

Statement II : Current in a purely resistive circuit can never be less than that in a series LCR circuit when connected to same voltage source.

In the light of the above statements, choose the **correct** from the options given below :

[NCERT : P L-187, 189 | April 6, 2024 (I)]

(a) Statement I is true but Statement II is false
(b) Statement I is false but Statement II is true
(c) Both Statement I and Statement II are true
(d) Both Statement I and Statement II are false

34. When a dc voltage of 100V is applied to an inductor, a dc current of 5A flows through it. When an ac voltage of 200V peak value is connected to inductor, its inductive reactance is found to be $20\sqrt{3}\,\Omega$. The power dissipated in the circuit is _______W.

[NCERT : P L-191 | April 6, 2024 (I)]

35. For a given series LCR circuit it is found that maximum current is drawn when value of variable capacitance is 2.5 nF. If resistance of 200Ω and 100 mH inductor is being used in the given circuit. The frequency of ac source is _____ $\times 10^3$ Hz. (given $\pi^2 = 10$)

[NCERT : P L-189 | April 6, 2024 (II)]

36. An ac source is connected in given series LCR circuit. The rms potential difference across the capacitor of 20 μF is .... V.

[NCERT : P L-187 | April 5, 2024 (I)]

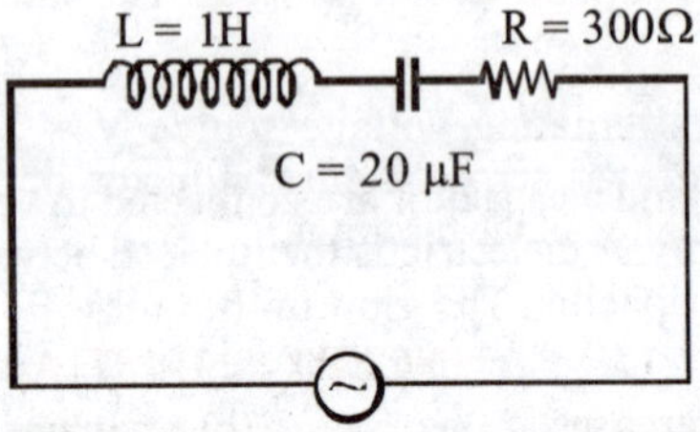

$V = 50\sqrt{2}\sin 100t$ volt

37. A series LCR circuit is subjected to an AC signal of 200 V, 50 Hz. If the voltage across the inductor (L = 10 mH) is 31.4 V, then the current in this circuit is _____ :

[NCERT : P L-182 | April 5, 2024 (II)]

(a) 68 A (b) 63 A (c) 10 A (d) 10 mA

38. In an ac circuit, the instantaneous current is zero, when the instantaneous voltage is maximum. In this case, the source may be connected to :

[NCERT : P L-187 | April 4, 2024 (I)]

A. pure inductor.
B. pure capacitor.
C. pure resistor.
D. combination of an inductor and capacitor.

Choose the correct answer from the options given below:

(a) A, B and C only (b) B, C and D only
(c) A and B only (d) A, B and D only

39. Match List I with List II

List - I	List - II
A. Purely capacitive circuit	I. 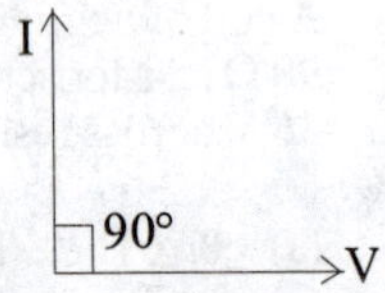
B. Purely inductive circuit	II. 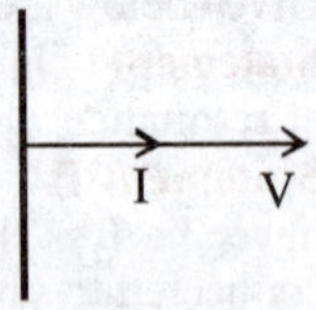
C. LCR series at resonance	III. 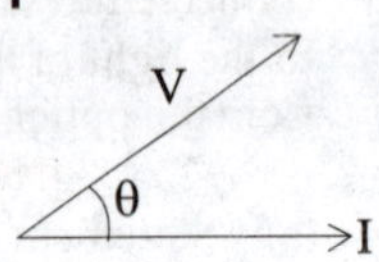
D. LCR series circuit	IV. 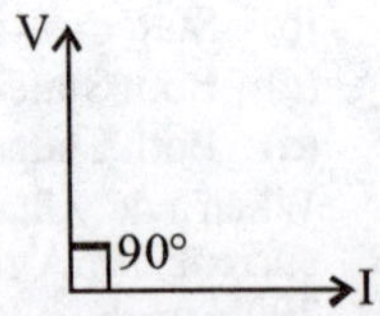

Choose the correct answer from the options given below:

[NCERT : P L-187, 189 | April 4, 2024 (II)]

(a) A-I, B-IV, C-III, D-II (b) A-IV, B-I, C-III, D-II
(c) A-IV, B-I, C-II, D-III (d) A-I, B-IV, C-II, D-III

40. In series LCR circuit, the capacitance is changed from C to 4C. To keep the resonance frequency unchanged, the new inductance should be:

[NCERT : P L-189 | Feb. 1, 2024 (I)]

(a) reduced by $\frac{1}{4}L$ (b) increased by 2L
(c) reduced by $\frac{3}{4}L$ (d) increased by 4L

41. An AC voltage $V = 20\sin 200\pi t$ is applied to a series LCR circuit which drives a current $I = 10\sin\left(200\pi t + \frac{\pi}{3}\right)$. The average power dissipated is:

[NCERT : P L-191 | Jan. 31, 2024 (II)]

(a) 21.6 W (b) 200 W (c) 173.2 W (d) 50 W

42. A series L. R circuit connected with an ac source E = (25 sin 1000 t) V has a power factor of $\frac{1}{\sqrt{2}}$. If the source of emf is changed to E = (20 sin 2000 t) V, the new power factor of the circuit will be:

[NCERT : P L-191 | Jan. 30, 2024 (I)]

(a) $\frac{1}{\sqrt{2}}$ (b) $\frac{1}{\sqrt{3}}$ (c) $\frac{1}{\sqrt{5}}$ (d) $\frac{1}{\sqrt{7}}$

43. A series LCR circuit with $L = \frac{100}{\pi}$ mH, $C = \frac{10^{-3}}{\pi}$ F and R = 10 Ω, is connected across an ac source of 220 V, 50 Hz supply. The power factor of the circuit would be _____.

[NCERT : P L-191 | Jan. 27, 2024 (II)]

44. In the given figure, an inductor and a resistor are connected in series with a battery of emf E volt. $\frac{E^a}{2b}$ J/s represents the maximum rate at which the energy is stored in the magnetic field (inductor). The numerical value of $\frac{b}{a}$ will be _____.

[April 13, 2023 (I)]

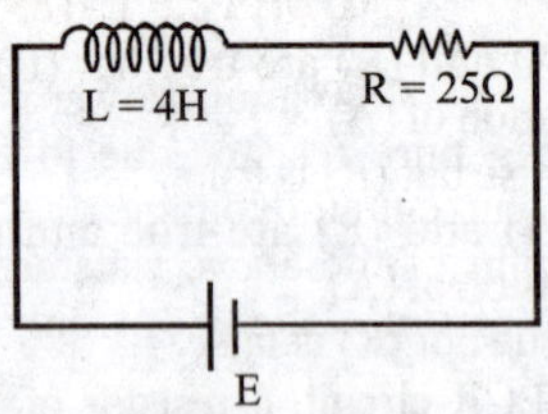

45. Given below are two statements :

Statement I : An AC circuit undergoes electrical resonance if it contains either a capacitor or an inductor.

Statement II : An AC circuit containing a pure capacitor or a pure inductor consumes high power due to its non-zero power factor.

[NCERT : P L-189, 191 | April 13, 2023 (II)]

In the light of above statements, choose the correct answer from the options given below :

(a) Both Statement I and Statement II are false
(b) Statement I is true but Statement II is false
(c) Both Statement I and Statement II are true
(d) Statement I is false but Statement II is true

46. Given below are two statements :
Statement I : When the frequency of an a.c. source in a series LCR circuit increases, the current in the circuit first increases, attains a maximum value and then decreases.
Statement II : In a series LCR circuit, the value of power factor at resonance is one.
[NCERT : P L-187, 191 | April 12, 2023 (I)]
In the light of given statements, choose the most appropriate answer from the options given below:
(a) Statement I is incorrect but Statement II is true
(b) Both Statement I and Statement II are false
(c) Statement I is correct but Statement II is false
(d) Both Statement I and Statement II are true

47. As per the given graph choose the correct representation for curve A and curve B.
[NCERT : P L-187 | April 11, 2023 (I)]
{Where X_C = reactance of pure capacitive circuit connected with A.C. source
X_L = reactance of pure inductive circuit connected with A.C. source
R = impedance of pure resistive circuit connected with A.C. source
Z = Impedance of the LCR series circuit}

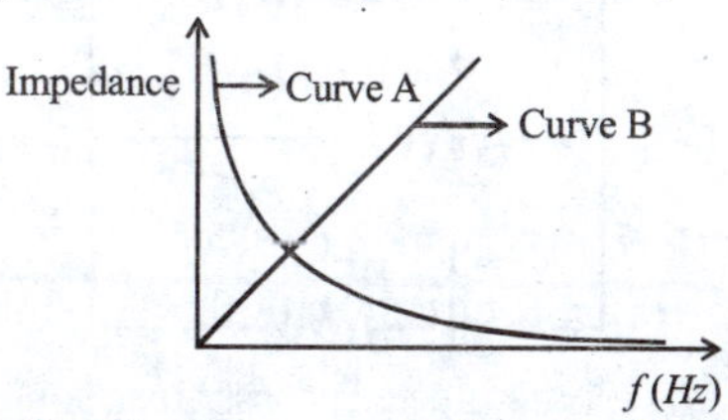

(a) $A = X_C, B = R$ (b) $A = X_L, B = Z$
(c) $A = X_C, B = X_L$ (d) $A = X_L, B = R$

48. A coil has an inductance of 2 H and resistance of 2 Ω. A 10 V is applied across the coil. The energy stored in the magnetic field after the current has built up to its equilibrium value will be ______ $\times 10^{-2}$ J.
[NCERT : P L-182 | April 11, 2023 (II)]

49. Given below are two statements:
Statement I : Maximum power is dissipated in a circuit containing an inductor, a capacitor and a resistor connected in series with an AC source, when resonance occurs.
Statement II : Maximum power is dissipated in a circuit containing pure resistor due to zero phase difference between current and voltage.
In the light of the above statements, choose the correct answer from the options given below :
[NCERT : P L-189, 191 | April 10, 2023 (I)]
(a) Statement I is false but Statement II is true
(b) Statement I is true but Statement II is false
(c) Both Statement I and Statement II are true
(d) Both Statement I and Statement II are false

50. A capacitor of capacitance 150.0 μF is connected to an alternating source of emf given by $E = 36 \sin(120\pi t)$ V. The maximum value of current in the circuit is approximately equal to: **[NCERT : P L-184 | April 6, 2023 (II)]**
(a) 2A (b) $\frac{1}{\sqrt{2}}$A (c) $\sqrt{2}$A (d) $2\sqrt{2}$A

51. A series LCR circuit is connected to an ac source of 220V, 50Hz. The circuit contain a resistance R = 100Ω and an inductor of inductive reactance X_L = 79.6Ω. The capacitance of the capacitor needed to maximize the average rate at which energy is supplied will be ______ μF.
[NCERT : P L-189 | Feb. 1, 2023 (I)]

52. An inductor of 0.5 mH, a capacitor of 20 μF and resistance of 20 Ω are connected in series with a 220 V ac source. If the current is in phase with the emf, the amplitude of current of the circuit is $\sqrt{x}$ A. The value of x is -
[NCERT : P L-189 | Jan. 31, 2023 (I)]

53. An alternating voltage source V = 260 sin (628t) is connected across a pure inductor of 5 mH. Inductive reactance in the circuit is:
[NCERT : P L-182 | Jan. 31, 2023 (II)]
(a) 3.14Ω (b) 6.28Ω (c) 0.5Ω (d) 0.318Ω

54. A series LCR circuit consists of R = 80Ω. X_L = 100Ω, and X_C = 40Ω. The input voltage is 2500 cos(100 πt) V. The amplitude of current, in the circuit, is ________ A.
[NCERT : P L-187 | Jan. 31, 2023 (II)]

55. In a series LR circuit with X_L = R. power factor is P_1. If a capacitor of capacitance C with $X_C = X_L$ is added to the circuit the power factor becomes P_2. The ratio of P_1 to P_2 will be : **[NCERT : P L-191 | Jan. 30, 2023 (I)]**
(a) 1 : 3 (b) $1 : \sqrt{2}$ (c) 1 : 1 (d) 1 : 2

56. In the given circuit, rms value of current (Irms) through the resistor R is : **[Jan. 30, 2023 (II)]**

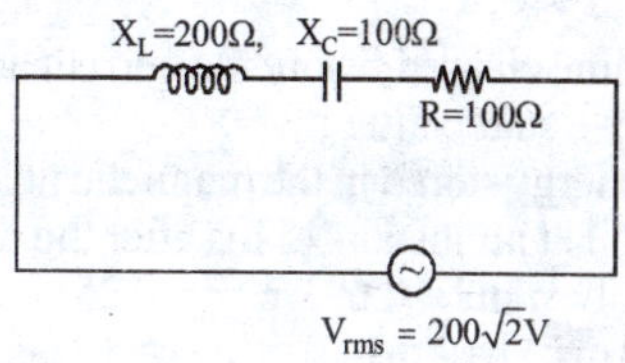

(a) 2A (b) $\frac{1}{\sqrt{2}}$A (c) 20 A (d) $2\sqrt{2}$A

57. For the given figures, choose the correct options:
[NCERT : P L-187 | Jan. 29, 2023 (II)]

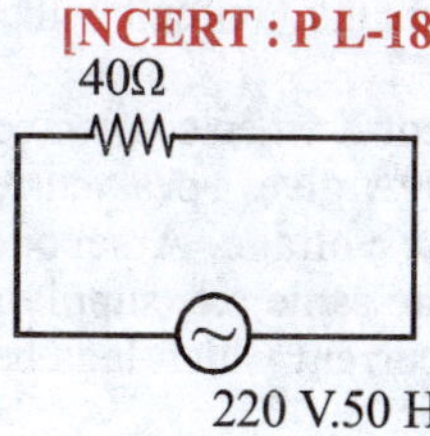

(a)

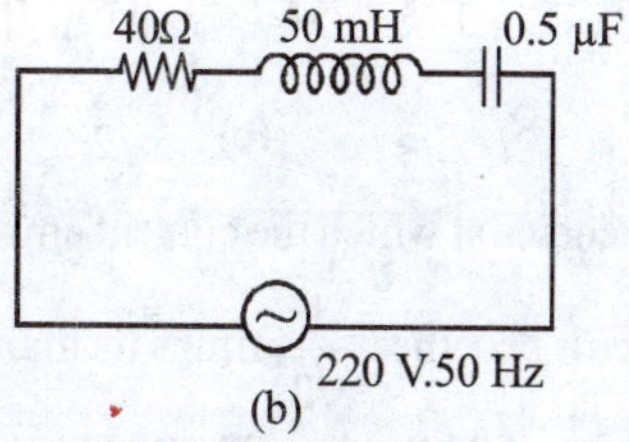

(b)

(a) The rms current in circuit (b) can never be larger than that in (a)
(b) The rms current in figure (a) is always equal to that in figure (b)
(c) The rms current in circuit (b) can be larger than that in (a)
(d) At resonance, current in (b) is less than that in (a)

58. An inductor of inductance 2 μH is connected in series with a resistance, a variable capacitor and an AC source of frequency 7 kHz. The value of capacitance for which maximum current is drawn into the circuit is $\frac{1}{x}$ F where the value of x is______.
(Take $\pi = \frac{22}{7}$) **[NCERT : P L-189 | Jan. 29, 2023 (II)]**

59. An LCR series circuit of capacitance 62.5 nF and resistance of 50 Ω. is connected to an A.C. source of frequency 2.0 kHz. For maximum value of amplitude of current in circuit, the value of inductance is______mH. (take $\pi^2 = 10$)
[NCERT : P L-189 | Jan. 25, 2023 (I)]

60. A series LCR circuit is connected to an AC source of 220 V, 50 Hz. The circuit contains a resistance R = 80Ω, an inductor of inductive reactance X_L = 70Ω, and a capacitor of capacitive reactance X_C = 130Ω. The power factor of circuit is $\frac{x}{10}$. The value of x is:
[NCERT : P L-191 | Jan. 25, 2023 (II)]

61. A coil of inductance 1 H and resistance 100 Ω is connected to a battery of 6V. Determine approximately.
[July 29, 2022 (I)]
(a) The time elapsed before the current acquires half of its steady - state value
(b) The energy stored in the magnetic field associated with the coil at an instant 15 ms after the circuit is switched on. (Given ln2 = 0.693, $e^{-3/2}$ = 0.25)
(a) t = 10 ms; U = 2 mJ (b) t = 10 ms; U = 1 mJ
(c) t = 7 ms; U = 1 mJ (d) t = 7 ms; U = 2 mJ

62. A capacitor of capacitance 500 μF is charged completely using a dc supply of 100 V. It is now connected to an inductor of inductance 50 mH to form an LC circuit. The maximum current in LC circuit will be______A.
[July 29, 2022 (II)]

63. A circuit element X when connected to an a.c. supply of peak voltage 100 V gives a peak current of 5 A which is in phase with the voltage. A second element Y when connected to the same a.c. supply also gives the same value of peak current which lags behind the voltage by $\frac{\pi}{2}$. If X and Y are connected in series to the same supply, what will be the rms value of the current in ampere?
[NCERT : P L-187 | July 29, 2022 (II)]
(a) $\frac{10}{\sqrt{2}}$ (b) $\frac{5}{\sqrt{2}}$ (c) $5\sqrt{2}$ (d) $\frac{5}{2}$

64. The frequencies at which the current amplitude in an LCR series circuit becomes $\frac{1}{\sqrt{2}}$ times its maximum value, are 212 rad s^{-1} and 232 rads^{-1}. The value of resistance in the circuit is R = 5 Ω. The self inductance in the circuit is ________ mH. **[NCERT : P L-189 | July 28, 2022 (I)]**

65. For the given circuit the current through battery of 6 V just after closing the switch 'S' will be______A.
[NCERT : P L-182 | July 28, 2022 (II)]

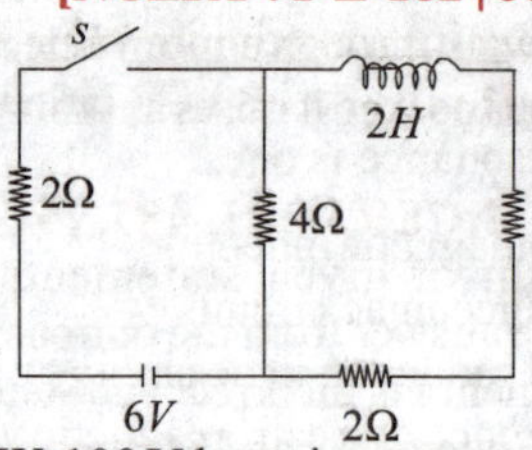

66. To light, a 50 W, 100 V lamp is connected, in series with a capacitor of capacitance $\frac{50}{\pi\sqrt{x}}\mu F$ with 200 V, 50Hz AC source. The value of x will be______.
[NCERT : P L-187, 191 | July 27, 2022 (I)]

67. As show in the figure, in steady state, the charge stored in the capacitor is______ $\times 10^{-6}$C.
[NCERT : P L-187 | July 27, 2022 (II)]

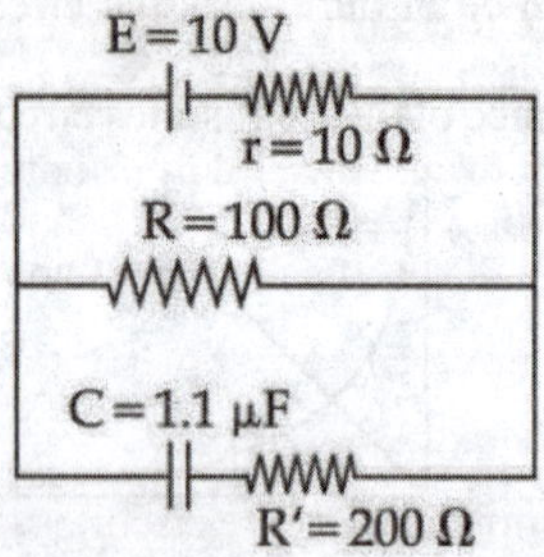

68. A series LCR circuit has L = 0.01 H, R = 10 Ω and C = 1 μF and it is connected to ac voltage of amplitude (V_m) 50 V. At frequency 60% lower than resonant frequency, the amplitude of current will be approximately:
[NCERT : P L-189 | July 27, 2022 (II)]
(a) 466 mA (b) 312 mA (c) 238 mA (d) 196 mA

69. The effective current I in the given circuit at very high frequencies will be______A.
[NCERT : P L-187 | July 26, 2022 (I)]

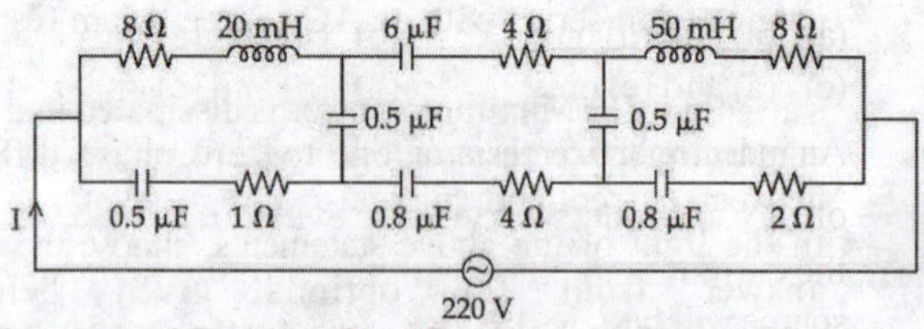

70. In a series LR circuit X_L = R and power factor of the circuit is P_1. When capacitor with capacitance C such that $X_L = X_C$ is put in series, the power factor becomes P_2. The ratio $\frac{P_1}{P_2}$ is : **[NCERT : P L-191 | July 26, 2022 (I)]**
(a) $\frac{1}{2}$ (b) $\frac{1}{\sqrt{2}}$ (c) $\frac{\sqrt{3}}{\sqrt{2}}$ (d) 2 : 1

71. To increase the resonant frequency in series LCR circuit,
[NCERT : P L-189 | July 25, 2022 (I)]
(a) source frequency should be increased
(b) another resistance should be added in series with the first resistance.

(c) another capacitor should be added in series with the first capacitor.

(d) the source frequency should be decreased.

72. When you walk through a metal detector carrying a metal object in your pocket, it raises an alarm. This phenomenon works on **[NCERT : P L-189 | July 25, 2022 (II)]**

(a) Electromagnetic induction

(b) Resonance in ac circuits

(c) Mutual induction in ac circuits

(d) interference of electromagnetic waves

73. A series LCR circuit with $R = \frac{250}{11}\,\Omega$ and $X_L = \frac{70}{11}\,\Omega$ is connected across a 220 V, 50 Hz supply. The value of capacitance needed to maximize the average power of the circuit will be ______ μF. (Take: $\pi = \frac{22}{7}$)

[NCERT : P L-191 | June 30, 2022 (I)]

74. For a series LCR circuit, I *vs* ω curve is shown :

[NCERT : P L-189 | June 29, 2022 (I)]

(a) To the left of ω_r, the circuit is mainly capacitive.

(b) To the left of ω_r, the circuit is mainly inductive.

(c) At ω_r, impedance of the circuit is equal to the resistance of the circuit.

(d) At ω_r, impedance of the circuit is 0.

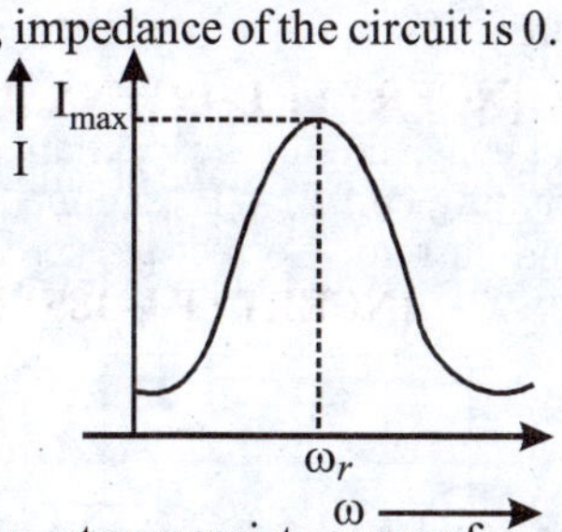

Choose the most appropriate answer from the options given below :

(a) (a) and (d) only (b) (b) and (d) only

(c) (a) and (c) only (d) (b) and (c) only

75. An inductor of 0.5 mH, a capacitor of 200 μF and a resistor of 2Ω are connected in series with a 220 V ac source. If the current is in phase with the emf, the frequency of ac source will be___ × 10^2 Hz.

[NCERT : P L-189 | June 29, 2022 (II)]

76. A telegraph line of length 100 km has a capacity of 0.01 μF/km and it carries an alternating current at 0.5 kilo cycle per second. If minimum impedance is required, then the value of the inductance that needs to be introduced in series is ________ mH. (if $\pi = \sqrt{10}$)

[NCERT : P L-189 | June 28, 2022 (I)]

77. An AC source is connected to an inductance of 100 mH, a capacitance of 100 μ F and a resistance of 120Ω as shown in figure. The time in which the resistance having a thermal capacity 2 J°/C will get heated by 16°C is _______ s.

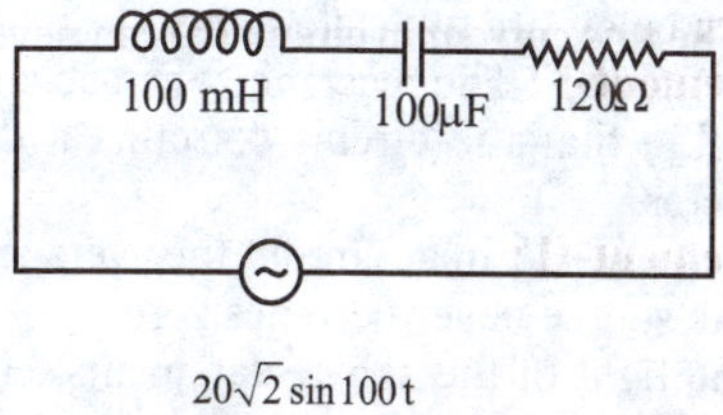

[NCERT : P L-187 | June 28, 2022 (I)]

78. In the given circuit, the magnitude of V_L and V_C are twice that of V_R. Given that $f = 50$ Hz, the inductance of the coil is $\frac{1}{K\pi}$ mH . The value of K is ________.

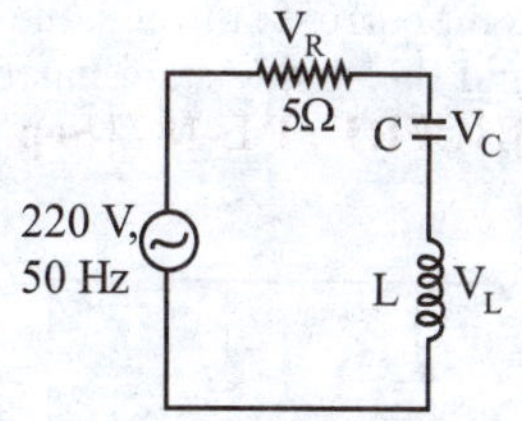

[NCERT : P L-187 | June 28, 2022 (II)]

79. A 110 V, 50 Hz, AC source is connected in the circuit (as shown in figure). The current through the resistance 55 Ω, at resonance in the circuit, will be .... A.

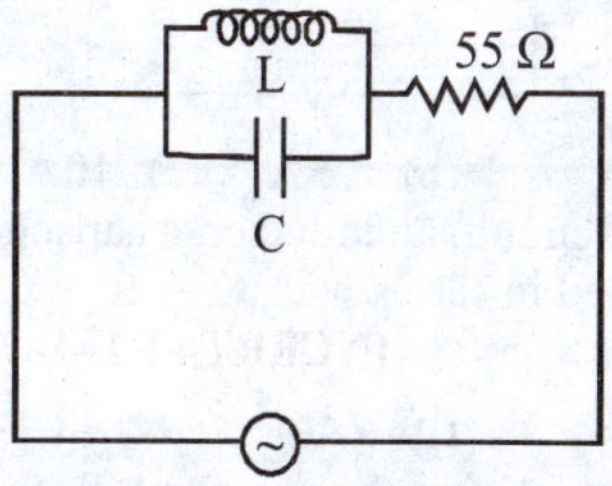

AC Source

[NCERT : P L-189 | June 26, 2022 (I)]

80. If wattless current flows in the AC circuit, then the circuit is **[NCERT : P L-191 | June 25, 2022 (I)]**

(a) Purely Resistive circuit (b) Purely Inductive circuit

(c) LCR series circuit (d) RC series circuit only

81. In a series LCR circuit, the inductance, capacitance and resistance are $L = 100$ mH, $C = 100$ μF and $R = 10\Omega$ respectively. They are connected to an AC source of voltage 220V and frequency of 50 Hz. The approximate value of current in the circuit will be________A.

[NCERT : P L-187 | June 25, 2022 (II)]

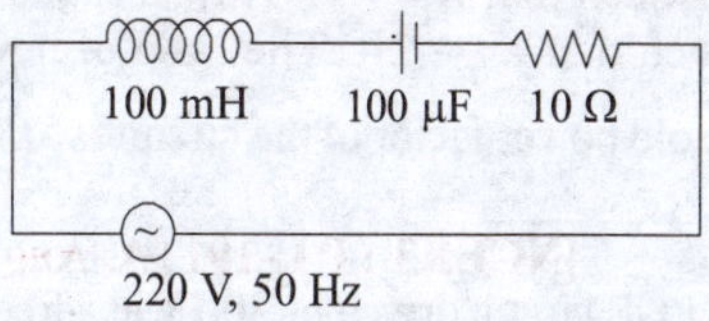

82. A sinusoidal voltage $V(t) = 210 \sin 3000\,t$ volt is applied to a series LCR circuit in which $L = 10$ mH, $C = 25$ μF and $R = 100\ \Omega$. The phase difference (Φ) between the applied voltage and resultant current will be :

[NCERT : P L-188 | June 25, 2022 (II)]

(a) $\tan^{-1}(0.17)$ (b) $\tan^{-1}(9.46)$

(c) $\tan^{-1}(0.30)$ (d) $\tan^{-1}(13.33)$

83. Given below are two statements :

Statement-I : The reactance of an ac circuit is zero. It is possible that the circuit contains a capacitor and an inductor.

Statement-II : In ac circuit, the average power delivered by the source never becomes zero.

In the light of the above statements, choose the correct answer from the options given below :

[NCERT : P L-191 | June 24, 2022 (II)]

(a) Both Statement I and Statement II are true.
(b) Both Statement I and Statement II are false.
(c) Statement I is true but Statement II in false.
(d) Statement I is false but Statement II is true.

84. For the given circuit the current i through the battery when the key in closed and the steady state has been reached is________. **[NCERT : P L-187 | Sep. 1, 2021 (II)]**

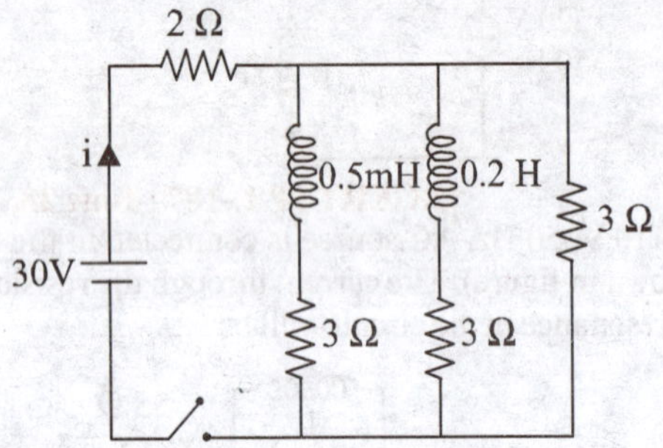

(a) 6 A (b) 25 A (c) 10 A (d) 0 A

85. In an ac circuit, an inductor, a capacitor and a resistor are connected in series with $X_L = R = X_C$. Impedance of this circuit is : **[NCERT : P L-187 | Aug. 31, 2021 (I)]**

(a) $2R^2$ (b) Zero (c) R (d) $R\sqrt{2}$

86. At very high frequencies, the effective impedance of the given circuit will be______ Ω.

[NCERT : P L-187 | NA, Aug. 31, 2021 (II)]

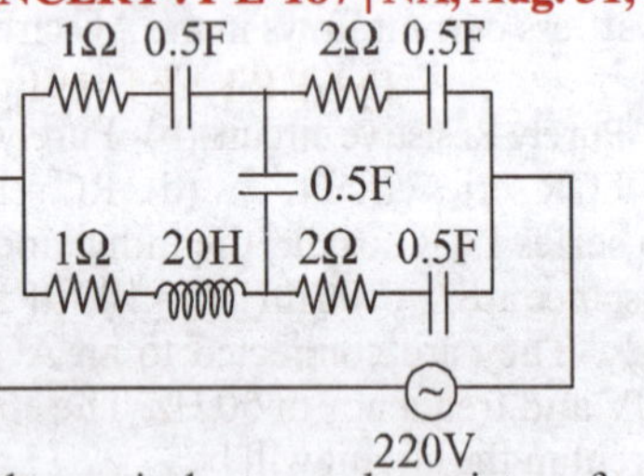

87. An ac circuit has an inductor and a resistor of resistance R in series, such that $X_L = 3R$. Now, a capacitor is added in series such that $X_C = 2R$. The ratio of new power factor with the old power factor of the circuit is $\sqrt{5} : x$. The value of x is __________

[NCERT : P L-191 | A, Aug. 27, 2021 (II)]

88. A series LCR circuit driven by 300 V at a frequency of 50 Hz contains a resistance R = 3 kΩ, an inductor of inductive reactance $X_L = 250\ \pi\Omega$ and an unknown capacitor. The value of capacitance to maximize the average power should be : (Take $\pi^2 = 10$) **[NCERT : P L-189, 191 | Aug. 26, 2021 (I)]**

(a) 4 μF (b) 25 μF (c) 400 μF (d) 40 μF

89. In the given circuit the AC source has ω = 100 rad s^{-1}. Considering the inductor and capacitor to be ideal, what will be the current I flowing through the circuit?

[NCERT : P L-187 | Aug. 26, 2021 (II)]

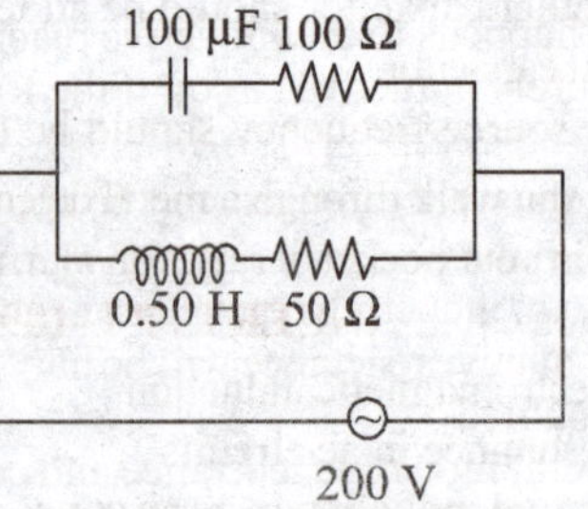

(a) 5.9 A (b) 4.24 A (c) 0.94 A (d) 6 A

90. An inductor of 10 mH is connected to a 20 V battery through a resistor of 10 kΩ and a switch. After a long time, when maximum current is set up in the circuit, the current is switched off. The current in the circuit after 1 μs is $\frac{x}{100}$ mA. Then x is equal to______. (Take $e^{-1} = 0.37$)

[NA, July 25, 2021 (I)]

91. Two circuits are shown in the figure (a) & (b). At a frequency of ______ rad/s the average power dissipated in one cycle will be same in both the circuits.

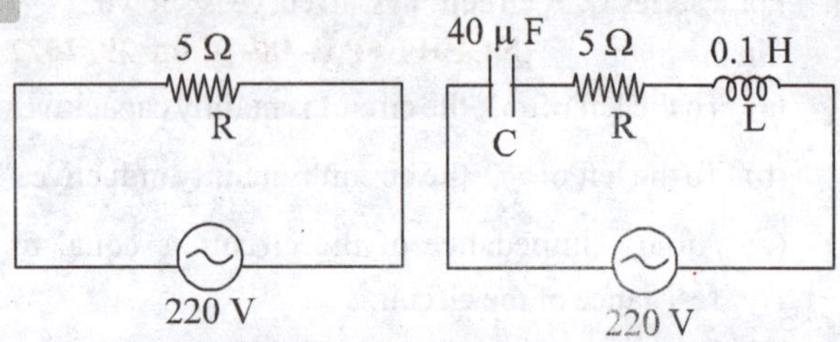

figure (a) figure (b)

[NCERT : P L-191 | NA, July 25, 2021 (II)]

92. In a circuit consisting of a capacitance and a generator with alternating emf $E_g = E_{g0} \sin \omega t$, V_C and I_C are the voltage and current. Correct phasor diagram for such circuit is : **[NCERT : P L-185 | July 22, 2021 (II)]**

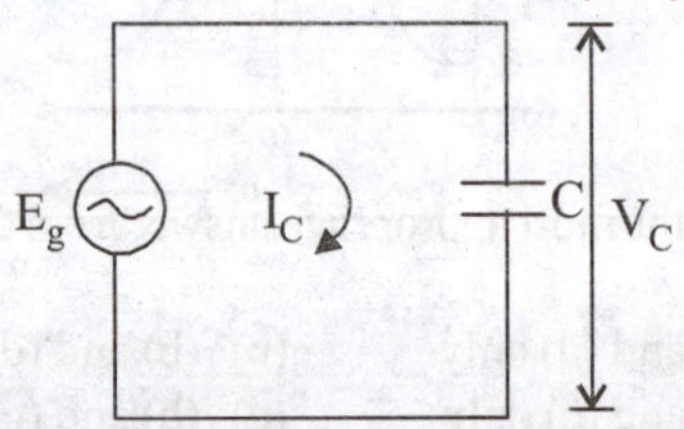

(a) 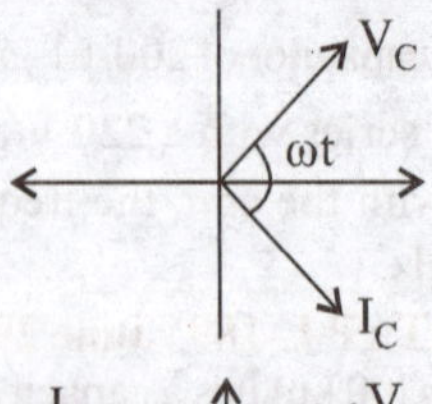

(b) 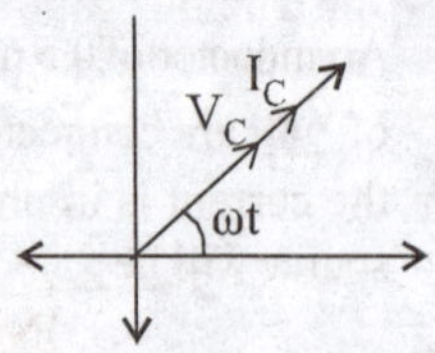

(c) 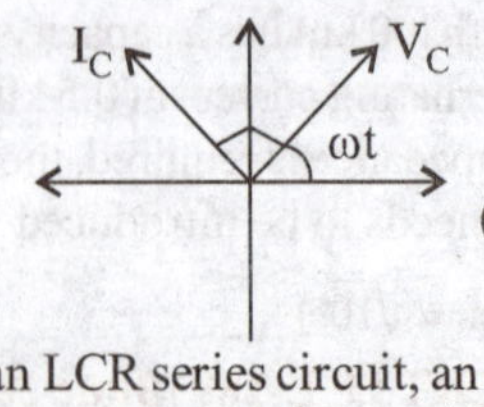

(d) 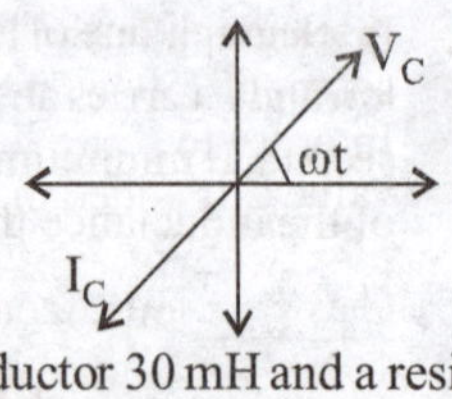

93. In an LCR series circuit, an inductor 30 mH and a resistor 1 Ω are connected to an AC source of angular frequency 300 rad/s. The value of capacitance for which, the current leads the voltage by 45° is $\frac{1}{x} \times 10^{-3}$ F. Then the value of x is ____. **[NCERT : P L-188 | NA, July 20, 2021 (I)]**

94. What happens to the inductive reactance and the current in a purely inductive circuit if the frequency is halved?
[NCERT : P L-182, 184 | March 17, 2021 (II)]
(a) Inductive reactance will be doubled and current will be halved
(b) Both, inducting reactance and current will be doubled
(c) Inductive reactance will be halved and current will be doubled
(d) Both, inductive reactance and current will be halved

95. A sinusoidal voltage of peak value 250 V is applied to a series LCR circuit, in which R = 8Ω, L = 24 mH and C = 60 μF. The value of power dissipated at resonant conditions is x kW.
The value of x to the nearest integer is ______.
[NCERT : P L-189, 191 | March 16, 2021 (I)]

96. A series LCR circuit is designed to resonate at an angular frequency $\omega_0 = 10^5$ rad/s. The circuit draws 16 W power from 120 V source at resonance. The value of resistance 'R' in the circuit is ______ Ω.
[NCERT : P L-189, 191 | Feb. 24, 2021 (II)]

97. The angular frequency of alternating current in a L-C-R circuit is 100 rad/s. The components connected are shown in the figure. Find the value of inductance of the coil ______ H and capacity of condenser is ______.

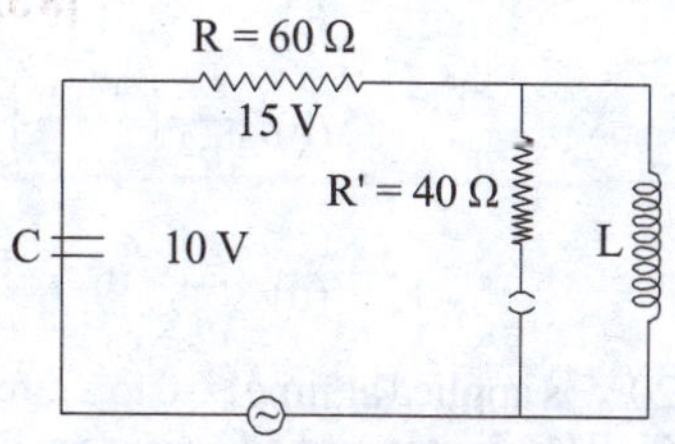

[NCERT : P L-187 | Feb. 25, 2021 (I)]

98. An LCR circuit contains resistance of 110 Ω and a supply of 220 V at 300 rad/s angular frequency. If only capacitance is removed from the circuit, current lags behind the voltage by 45°. If on the other hand, only inductor is removed the current leads by 45° with the applied voltage. The rms current flowing in the circuit will be
[NCERT : P L-189 | Feb. 25, 2021 (II) 2010 S]
(a) 1.5 A (b) 1 A (c) 2 A (d) 2.5 A

99. A part of a complete circuit is shown in the figure. At some instant, the value of current I is 1 A and it is decreasing at a rate of 10^2 A s^{-1}. The value of the potential difference $V_P - V_Q$, (in volts) at that instant, is ______.
[NA Sep. 06, 2020 (I)]

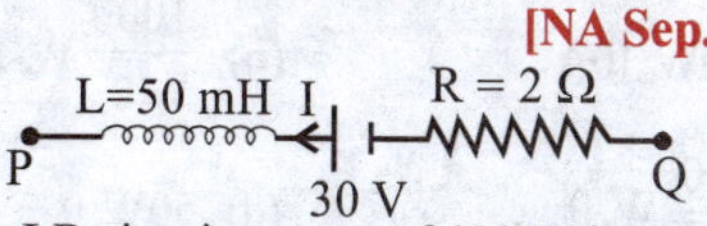

100. In a series LR circuit, power of 400 W is dissipated from a source of 250 V, 50 Hz. The power factor of the circuit is 0.8. In order to bring the power factor to unity, a capacitor of value C is added in series to the L and R. Taking the value C as $\left(\frac{n}{3\pi}\right)$ μF, then value of n is ______.
[NCERT : P L-191 | NA Sep. 06, 2020 (II)]

101. A series L-R circuit is connected to a battery of emf V. If the circuit is switched on at $t = 0$, then the time at which the energy stored in the inductor reaches $\left(\frac{1}{n}\right)$ times of its maximum value, is :
[Sep. 04, 2020 (II)]
(a) $\frac{L}{R}\ln\left(\frac{\sqrt{n}}{\sqrt{n}-1}\right)$ (b) $\frac{L}{R}\ln\left(\frac{\sqrt{n}+1}{\sqrt{n}-1}\right)$
(c) $\frac{L}{R}\ln\left(\frac{\sqrt{n}}{\sqrt{n}+1}\right)$ (d) $\frac{L}{R}\ln\left(\frac{\sqrt{n}-1}{\sqrt{n}}\right)$

102. A 750 Hz, 20 V (rms) source is connected to a resistance of 100 Ω, an inductance of 0.1803 H and a capacitance of 10 μF all in series. The time in which the resistance (heat capacity 2 J/°C) will get heated by 10°C. (assume no loss of heat to the surroundings) is close to:
[NCERT : P L-187, 191 | Sep. 03, 2020 (I)]
(a) 418 s (b) 245 s (c) 365 s (d) 348 s

103. An inductance coil has a reactance of 100 Ω. When an AC signal of frequency 1000 Hz is applied to the coil, the applied voltage leads the current by 45°. The self-inductance of the coil is : **[NCERT : P L-187, 188 | Sep. 02, 2020 (II)]**
(a) 1.1×10^{-2} H (b) 1.1×10^{-1} H
(c) 5.5×10^{-5} H (d) 6.7×10^{-7} H

104. Consider the LR circuit shown in the figure. If the switch S is closed at t = 0 then the amount of charge that passes through the battery between t = 0 and $t = \frac{L}{R}$ is :
[12 April 2019 (II)]
(a) $\frac{2.7EL}{R^2}$ (b) $\frac{EL}{2.7R^2}$
(c) $\frac{7.3EL}{R^2}$ (d) $\frac{EL}{7.3R^2}$

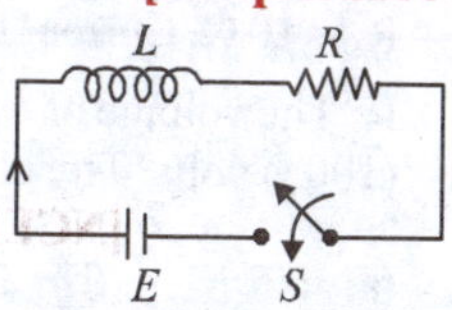

105. A coil of self inductance 10 mH and resistance 0.1 Ω is connected through a switch to a battery of internal resistance 0.9 Ω. After the switch is closed, the time taken for the current to attain 80% of the saturation value is [take $ln\,5 = 1.6$] **[10 April 2019 (II)]**
(a) 0.324 s (b) 0.103 s (c) 0.002 s (d) 0.016 s

106. A 20 Henry inductor coil is connected to a 10 ohm resistance in series as shown in figure. The time at which rate of dissipation of energy (Joule's heat) across resistance is equal to the rate at which magnetic energy is stored in the inductor, is : **[8 April 2019 (I)]**
(a) $\frac{2}{ln\,2}$ (b) $\frac{1}{2}ln\,2$
(c) $2\,ln\,2$ (d) $ln\,2$

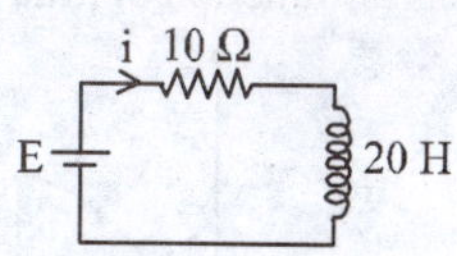

107. A circuit connected to an ac source of emf $e = e_0\sin(100t)$ with t in seconds, gives a phase difference of $\frac{\pi}{4}$ between the emf e and current i. Which of the following circuits will exhibit this ?
[NCERT : P L-187, 188 | 8 April 2019 (II)]
(a) RL circuit with R = 1 kΩ and L = 10 mH
(b) RL circuit with R = 1 kΩ and L = 1 mH
(c) RC circuit with R = 1 kΩ and C = 1 μF
(d) RC circuit with R = 1 kΩ and C = 10 μF.

108. In the figure shown, a circuit contains two identical resistors with resistance R = 5 Ω and an inductance with L = 2 mH. An ideal battery of 15 V is connected in the circuit. What will be the current through the battery long after the switch is closed?

[NCERT : P L-182 | 12 Jan. 2019 (I)]

(a) 5.5 A

(b) 7.5 A

(c) 3 A

(d) 6 A

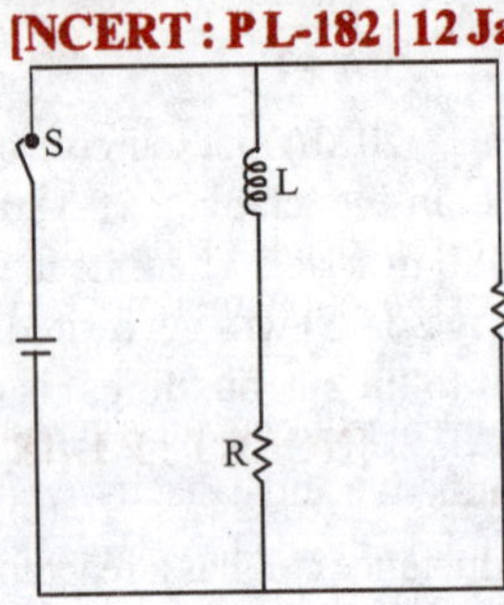

109.

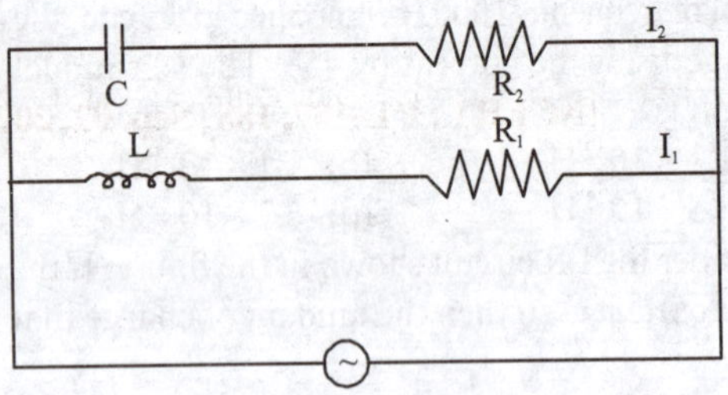

In the above circuit, $C = \frac{\sqrt{3}}{2}\mu F$, $R_2 = 20\,\Omega$, $L = \frac{\sqrt{3}}{10}$ H and $R_1 = 10\,\Omega$. Current in L-R_1 path is I_1 and in C-R_2 path it is I_2. The voltage of A.C source is given by, $V = 200\sqrt{2}$ sin (100 t) volts. The phase difference between I_1 and I_2 is :

[NCERT : P L-187, 188 | 12 Jan. 2019 (II)]

(a) 60° (b) 30° (c) 90° (d) 0

110. In the circuit shown,

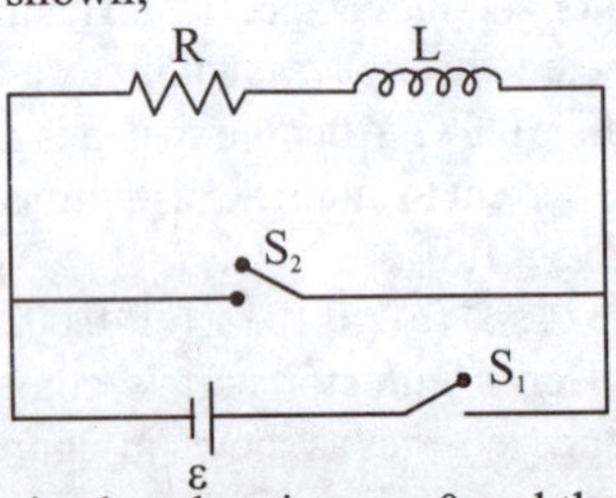

the switch S_1 is closed at time t = 0 and the switch S_2 is kept open. At some later time (t_0), the switch S_1 is opened and S_2 is closed. the behaviour of the current I as a function of time 't' is given by: **[11 Jan. 2019 (II)]**

(a) 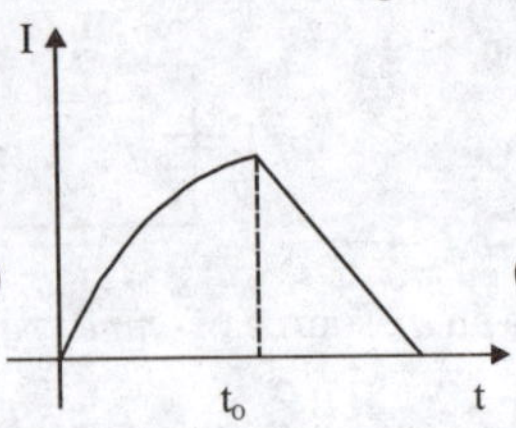

(b) 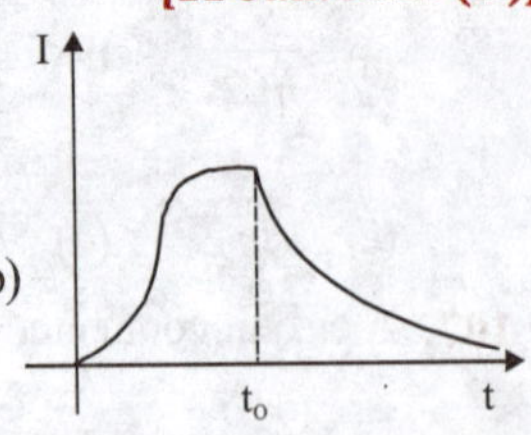

(c) 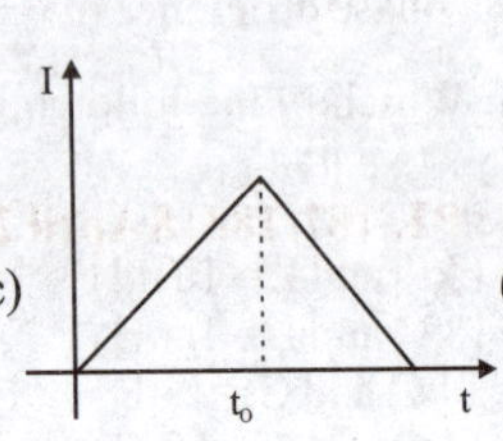

(d) 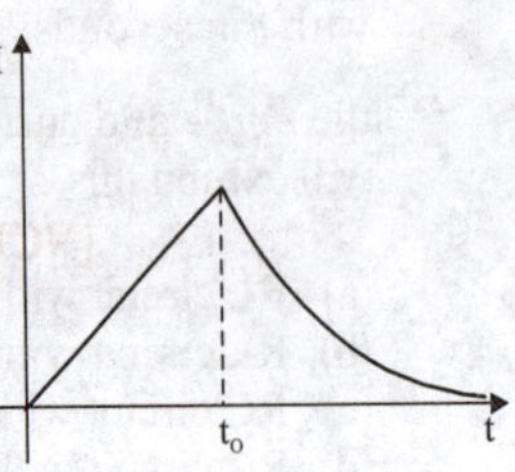

111. A series AC circuit containing an inductor (20 mH), a capacitor (120 μF) and a resistor (60 Ω) is driven by an AC source of 24 V/50 Hz. The energy dissipated in the circuit in 60 s is:

[NCERT : P L-191 | 9 Jan. 2019 (I)]

(a) 5.65×10^2 J (b) 2.26×10^3 J

(c) 5.17×10^2 J (d) 3.39×10^3 J

112. In LC circuit the inductance L = 40 mH and capacitance C = 100 μF. If a voltage $V(t) = 10 \sin(314\,t)$ is applied to the circuit, the current in the circuit is given as:

[NCERT : P L-187, 188 | 9 Jan. 2019 (II)]

(a) 0.52 cos 314 t (b) 10 cos 314 t

(c) 5.2 cos 314 t (d) 0.52 sin 314 t

113.

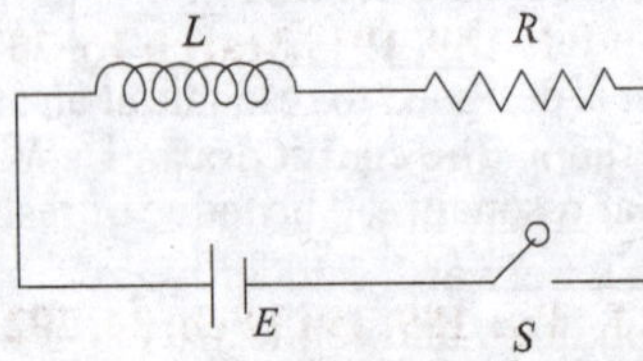

As shown in the figure, a battery of emf E is connected to an inductor L and resistance R in series. The switch is closed at t = 0. The total charge that flows from the battery, between t = 0 and $t = t_c$ (t_c is the time constant of the circuit) is:

[8 Jan. 2020 (II)]

(a) $\frac{ER}{R^2 e}$ (b) $\frac{EL}{R^2}\left(1-\frac{1}{e}\right)$

(c) $\frac{EL}{R^2}$ (d) $\frac{ER}{eL^2}$

114. An emf of 20 V is applied at time t = 0 to a circuit containing in series 10 mH inductor and 5 Ω resistor. The ratio of the currents at time t = ∞ and at t = 40 s is close to: (Take $e^2 = 7.389$) **[7 Jan. 2020 (II)]**

(a) 1.06 (b) 1.15 (c) 1.46 (d) 0.84

115. In an a.c. circuit, the instantaneous e.m.f. and current are given by

$e = 100 \sin 30\,t$

$i = 20 \sin\left(30t - \frac{\pi}{4}\right)$

In one cycle of a.c., the average power consumed by the circuit and the wattless current are, respectively:

NCERT : P L-191 | [2018]

(a) 50W, 10A (b) $\frac{1000}{\sqrt{2}}$ W, 10A

(c) $\frac{50}{\sqrt{2}}$ W, 0 (d) 50W, 0

116. A sinusoidal voltage of peak value 283 V and angular frequency 320rad/s is applied to a series LCR circuit. Given that R = 5 Ω, L = 25 mH and C = 1000 μF. The total impedance, and phase difference between the voltage across the source and the current will respectively be:

[NCERT : P L-187, 188 | Online April 9, 2017]

(a) 10 Ω and $\tan^{-1}\left(\frac{5}{3}\right)$ (b) 7 Ω and 45°

(c) 10 Ω and $\tan^{-1}\left(\frac{8}{3}\right)$ (d) 7 Ω and $\tan^{-1}\left(\frac{5}{3}\right)$

117. An arc lamp requires a direct current of 10 A at 80 V to function. If it is connected to a 220 V (rms), 50 Hz AC supply, the series inductor needed for it to work is close to : **[NCERT : P L-187 | 2016]**

(a) 0.044 H (b) 0.065 H (c) 80 H (d) 0.08 H

118. An inductor (L = 0.03 H) and a resistor (R = 0.15 kΩ) are connected in series to a battery of 15V emf in a circuit shown below. The key K_1 has been kept closed for a long time. Then at t = 0, K_1 is opened and key K_2 is closed simultaneously. At t = 1 ms, the current in the circuit will be : $(e^5 \cong 150)$ **[2015]**

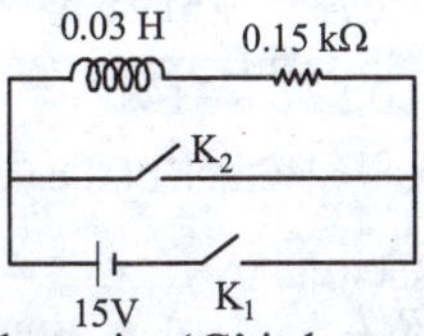

(a) 6.7 mA
(b) 0.67 mA
(c) 100 mA
(d) 67 mA

119. In the circuit shown here, the point 'C' is kept connected to point 'A' till the current flowing through the circuit becomes constant. Afterward, suddenly, point 'C' is disconnected from point 'A' and connected to point 'B' at time t = 0. Ratio of the voltage across resistance and the inductor at t = L/R will be equal to: **[NCERT : P L-187 | 2014]**

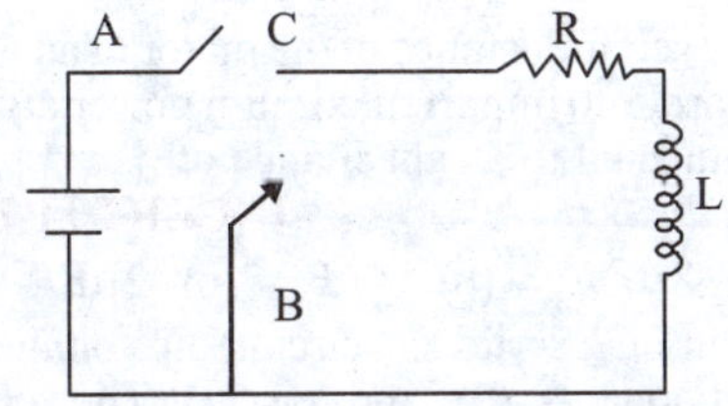

(a) $\frac{e}{1-e}$

(b) 1

(c) –1

(d) $\frac{1-e}{e}$

120. When the rms voltages V_L, V_C and V_R are measured respectively across the inductor L, the capacitor C and the resistor R in a series LCR circuit connected to an AC source, it is found that the ratio $V_L : V_C : V_R = 1 : 2 : 3$. If the rms voltage of the AC sources is 100 V, the V_R is close to: **NCERT : P L-187 | Online April 9, 2014]**

(a) 50 V (b) 70 V (c) 90 V (d) 100 V

121. In an LCR circuit as shown below both switches are open initially. Now switch S_1 is closed, S_2 kept open. (q is charge on the capacitor and τ = RC is Capacitive time constant). Which of the following statement is correct ? **[2013]**

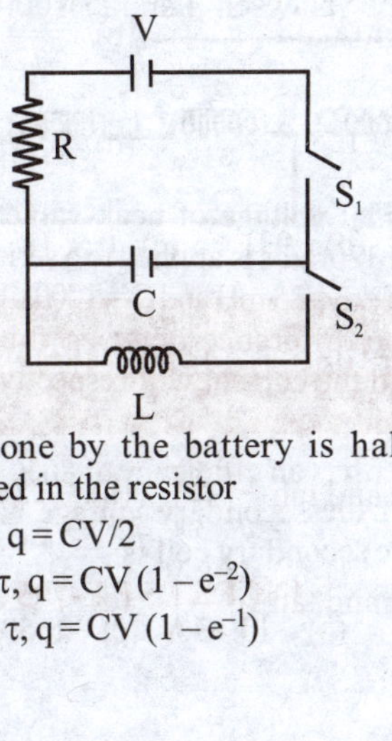

(a) Work done by the battery is half of the energy dissipated in the resistor
(b) At, t = τ, q = CV/2
(c) At, t = 2τ, $q = CV(1 - e^{-2})$
(d) At, t = 2 τ, $q = CV(1 - e^{-1})$

122. A series LR circuit is connected to an ac source of frequency ω and the inductive reactance is equal to 2R. A capacitance of capacitive reactance equal to R is added in series with L and R. The ratio of the new power factor to the old one is : **[NCERT : P L-191 | Online April 25, 2013]**

(a) $\sqrt{\frac{2}{3}}$ (b) $\sqrt{\frac{2}{5}}$ (c) $\sqrt{\frac{3}{2}}$ (d) $\sqrt{\frac{5}{2}}$

123. When resonance is produced in a series LCR circuit, then which of the following is not correct? **[NCERT : P L-189 | Online April 25, 2013]**

(a) Current in the circuit is in phase with the applied voltage.
(b) Inductive and capacitive reactances are equal.
(c) If R is reduced, the voltage across capacitor will increase.
(d) Impedance of the circuit is maximum.

124. In the circuit shown here, the voltage across L and C are respectively 300 V and 400 V. The voltage E of the ac source is : **[NCERT : P L-187 | Online April 9, 2013]**

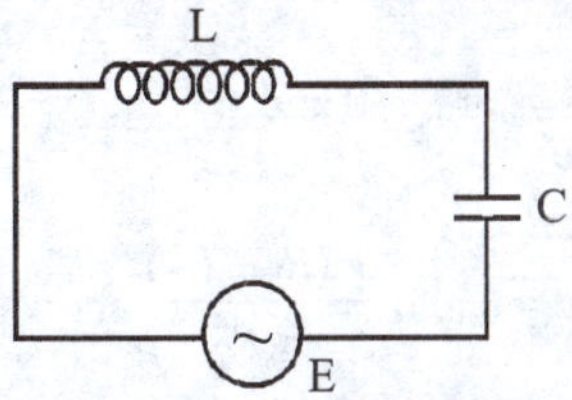

(a) 400 Volt (b) 500 Volt (c) 100 Volt (d) 700 Volt

125. In an *LCR* circuit shown in the following figure, what will be the readings of the voltmeter across the resistor and ammeter if an *a.c.* source of 220V and 100 Hz is connected to it as shown? **[NCERT : P L-187 | Online May 7, 2012]**

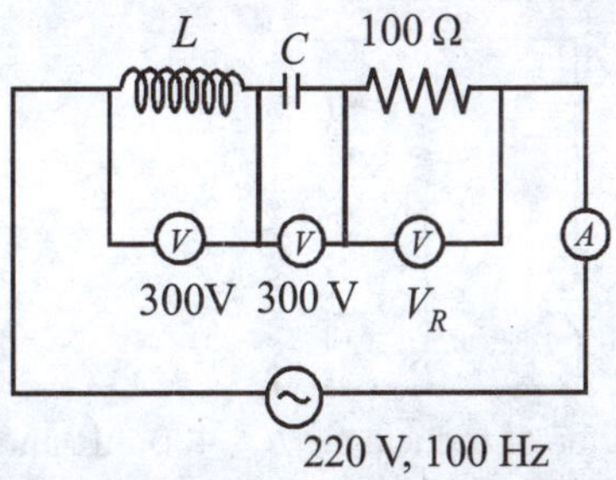

(a) 800 V, 8 A (b) 110 V, 1.1 A
(c) 300 V, 3 A (d) 220V, 2.2 A

126. A fully charged capacitor *C* with initial charge q_0 is connected to a coil of self inductance *L* at *t* = 0. The time at which the energy is stored equally between the electric and the magnetic fields is: **[2011]**

(a) $\frac{\pi}{4}\sqrt{LC}$ (b) $2\pi\sqrt{LC}$
(c) $\sqrt{LC}$ (d) $\pi\sqrt{LC}$

127. A resistor '*R*' and 2μ*F* capacitor in series is connected through a switch to 200 V direct supply. Across the capacitor is a neon bulb that lights up at 120 V. Calculate the value of *R* to make the bulb light up 5 s after the switch has been closed. ($\log_{10} 2.5 = 0.4$) **[2011]**

(a) $1.7 \times 10^5 \Omega$ (b) $2.7 \times 10^6 \Omega$
(c) $3.3 \times 10^7 \Omega$ (d) $1.3 \times 10^4 \Omega$

128. Combination of two identical capacitors, *a* resistor *R* and *a dc* voltage source of voltage 6V is used in an experiment on a (*C-R*) circuit. It is found that for a parallel combination of the capacitor the time in which the voltage of the fully charged combination reduces to half its original voltage is 10 second. For series combination the time needed for reducing the voltage of the fully charged series combination by half is **[2011 RS]**

(a) 10 second (b) 5 second
(c) 2.5 second (d) 20 second

129. In the circuit shown below, the key *K* is closed at $t = 0$. The current through the battery is **[2010]**

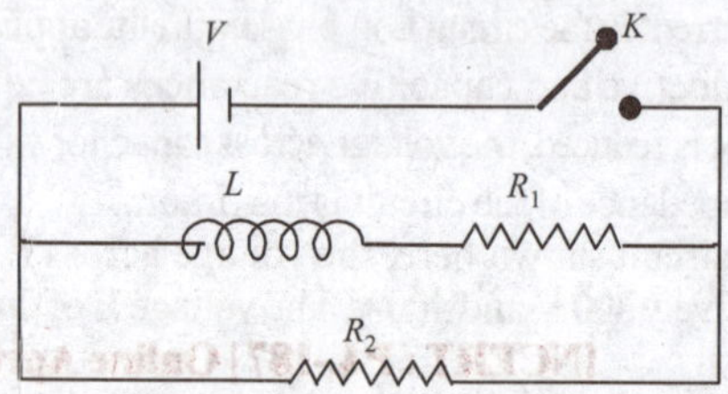

(a) $\dfrac{VR_1R_2}{\sqrt{R_1^2+R_2^2}}$ at $t=0$ and $\dfrac{V}{R_2}$ at $t=\infty$

(b) $\dfrac{V}{R_2}$ at $t=0$ and $\dfrac{V(R_1+R_2)}{R_1R_2}$ at $t=\infty$

(c) $\dfrac{V}{R_2}$ at $t=0$ and $\dfrac{VR_1R_2}{\sqrt{R_1^2+R_2^2}}$ at $t=\infty$

(d) $\dfrac{V(R_1+R_2)}{R_1R_2}$ at $t=0$ and $\dfrac{V}{R_2}$ at $t=\infty$

130.

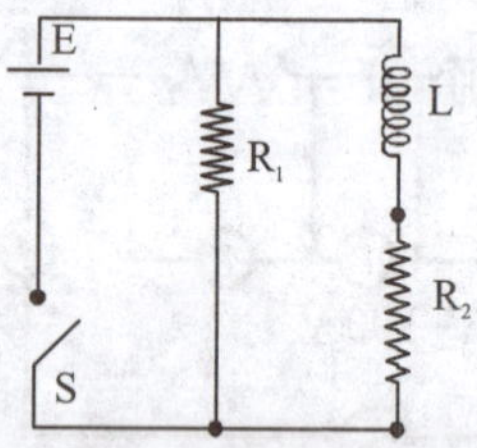

An inductor of inductance L = 400 mH and resistors of resistance $R_1 = 2\Omega$ and $R_2 = 2\Omega$ are connected to a battery of emf 12 V as shown in the figure. The internal resistance of the battery is negligible. The switch *S* is closed at $t = 0$. The potential drop across *L* as a function of time is **[2009]**

(a) $\dfrac{12}{t}e^{-3t}$V (b) $6\left(1-e^{-t/0.2}\right)$V
(c) $12e^{-5t}$V (d) $6e^{-5t}$V

131. In a series resonant LCR circuit, the voltage across *R* is 100 volts and *R* = 1 kΩ with C = 2μF. The resonant frequency ω is 200 rad/s. At resonance the voltage across *L* is **[NCERT : P L-189 | 2006]**

(a) 2.5×10^{-2} V (b) 40 V
(c) 250 V (d) 4×10^{-3} V

132. An inductor (*L* = 100 mH), a resistor (*R* = 100 Ω) and a battery (*E* = 100 V) are initially connected in series as shown in the figure. After a long time the battery is disconnected after short circuiting the points A and B. The current in the circuit 1 ms after the short circuit is **[2006]**

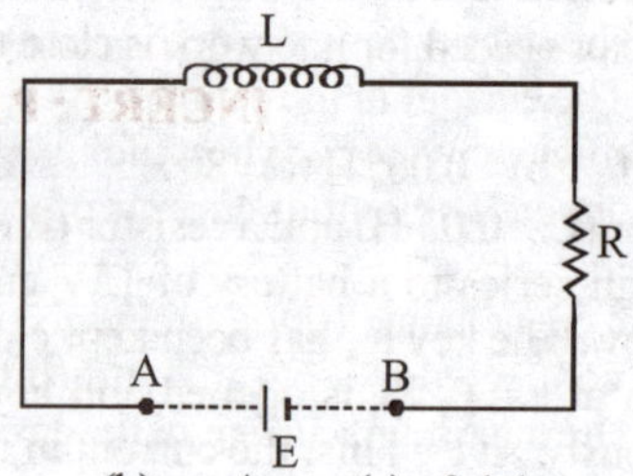

(a) 1/e A (b) e A (c) 0.1 A (d) 1 A

133. The phase difference between the alternating current and emf is $\dfrac{\pi}{2}$. Which of the following cannot be the constituent of the circuit? **[NCERT : P L-187 | 2005]**

(a) *R, L* (b) *C* alone (c) *L* alone (d) *L, C*

134. A circuit has a resistance of 12 ohm and an impedance of 15 ohm. The power factor of the circuit will be **[NCERT : P L-191 | 2005]**

(a) 0.4 (b) 0.8 (c) 0.125 (d) 1.25

135. A coil of inductance 300 mH and resistance 2 Ω is connected to a source of voltage 2V. The current reaches half of its steady state value in **[2005]**

(a) 0.1 s (b) 0.05 s (c) 0.3 s (d) 0.15 s

136. The self inductance of the motor of an electric fan is 10 H. In order to impart maximum power at 50 Hz, it should be connected to a capacitance of **[NCERT : P L-189, 191 | 2005]**

(a) 8μF (b) 4μF (c) 2μF (d) 1μF

137. In an *LCR* series a.c. circuit, the voltage across each of the components, *L*, *C* and *R* is 50V. The voltage across the *LC* combination will be **[NCERT : P L-187 | 2004]**

(a) 100 V (b) $50\sqrt{2}$ V
(c) 50 V (d) 0 V (zero)

138. In a *LCR* circuit capacitance is changed from *C* to 2 *C*. For the resonant frequency to remain unchanged, the inductance should be changed from *L* to **[NCERT : P L-189 | 2004]**

(a) *L*/2 (b) 2*L* (c) 4*L* (d) *L*/4

139. The power factor of an AC circuit having resistance (*R*) and inductance (*L*) connected in series and an angular velocity ω is **[NCERT : P L-191 | 2002]**

(a) $R/\omega L$ (b) $R/(R^2+\omega^2L^2)^{1/2}$
(c) $\omega L/R$ (d) $R/(R^2-\omega^2L^2)^{1/2}$

140. The inductance between *A* and *D* is **[2002]**

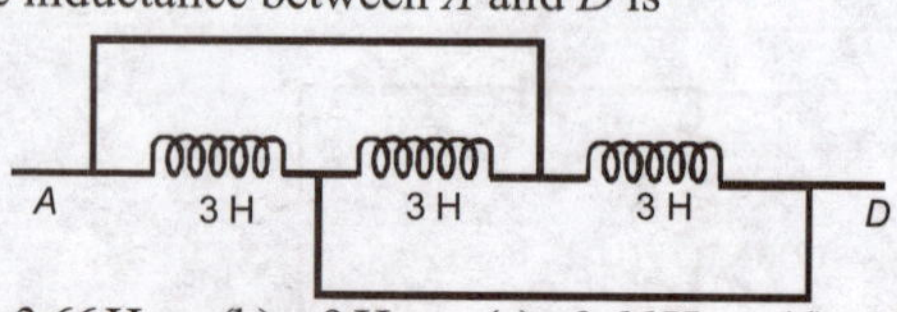

(a) 3.66 H (b) 9 H (c) 0.66 H (d) 1 H

Topic 3 - Transformers and LC Oscillations

141. A transformer has an efficiency of 80% and works at 10V and 4 kW. If the secondary voltage is 240 V, then the current in the secondary coil is: **[NCERT : P L-195 | Feb. 1, 2024 (II)]**

(a) 1.33A (b) 13.33A (c) 1.59 A (d) 15.1 A

142. A power transmission line feeds input power at $2.3kV$ to a step down transformer with its primary winding having 3000 turns. The output power is delivered at $230V$ by the transformer. The current in the primary of the transformer is $5A$ and its efficiency is 90%. The winding of transformer is made of copper. The output current of transformer is ________ A. **[NCERT : P L-195 | Jan. 30, 2024 (II)]**

143. A capacitor of capacitance 100μF is charged to a potential of 12 V and connected to a 6.4 mH inductor to produce oscillations. The maximum current in the circuit would be: **[Jan. 29, 2024 (I)]**

(a) 1.2 A (b) 1.5 A (c) 3.2 A (d) 2.0 A

144. Primary side of a transformer is connected to 230 V, 50 Hz supply. Turns ratio of primary to secondary winding is 10 : 1. Load resistance connected to secondary side is 46Ω. The power consumed in it is : **[NCERT : P L-195 | Jan. 27, 2024 (II)]**

(a) 12.5 W (b) 10.0 W (c) 11.5 W (d) 12.0 W

145. An oscillating LC circuit consists of a 75 mH inductor and a 1.2μF capacitor. If the maximum charge to the capacitor is 2.7μC. The maximum current in the circuit will be ____mA. **[April 8, 2023 (I)]**

146. An ideal transformer with purely resistive load operates at 12 kV on the primary side. It supplies electrical energy to a number of nearby houses at 120 V. The average rate of energy consumption in the houses served by the transformer is 60 kW. The value of resistive load (Rs) required in the secondary circuit will be _____mΩ. **[NCERT : P L-195 | April 6, 2023 (I)]**

147. In an LC oscillator, if values of inductance and capacitance become twice and eight times, respectively, then the resonant frequency of oscillator becomes x times its initial resonant frequency ω_0. The value of x is: **[Jan. 25, 2023 (I)]**

(a) 1/4 (b) 16 (c) 1/16 (d) 4

148. A transformer operating at primary voltage 8 kV and secondary voltage 160 V serves a load of 80 kW. Assuming the transformer to be ideal with purely resistive load and working on unity power factor, the loads in the primary and secondary circuit would be **[NCERT : P L-195 | July 28, 2022 (II)]**

(a) 800 Ω and 1.06 Ω (b) 10Ω and 500 Ω
(c) 800 Ω and 0.32 Ω (d) 1.06 Ω and 500 Ω

149. Match **List-I** with **List-II** **[NCERT : P L-194 | June 25, 2022 (I)]**

List-I	List-II
(A) AC generator	(I) Detects the presence of current in the circuit
(B) Galvanometer	(II) Converts mechanicalenergy into electrical energy
(C) Transformer	(III) Works on the principle of resonance in AC circuit
(D) Metal detector	(IV) Changes an alternating voltage for smaller or greater value

Choose the **correct answer** from the options given below :-

(a) (A) – (II), B – (I), (C) – (IV), (D) – (III)
(b) (A) – (II), B – (I), (C) – (III), (D) – (IV)
(c) (A) – (III), B – (IV), (C) – (II), (D) – (I)
(d) (A) – (III), B – (I), (C) – (II), (D) – (IV)

150. For the given circuit, comment on the type of transformer used. **[NCERT : P L-194, 195 | March 16, 2021 (II)]**

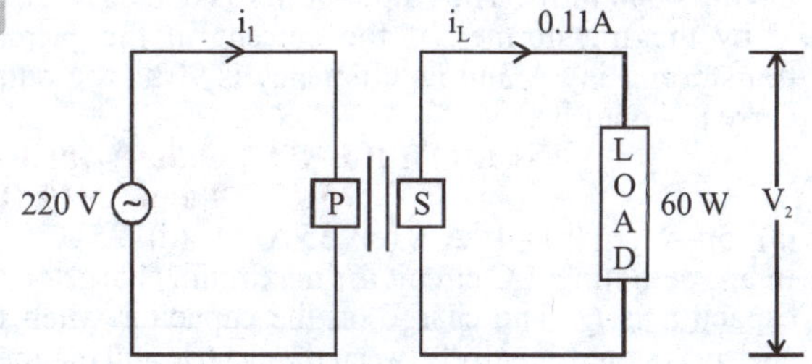

(a) Step-up transformer
(b) Auto transformer
(c) Step down transformer
(d) Auxilliary transformer

151. A radio set requires 12 V (D.C.) for its operation. The D.C. source is constructed by using a transformer and a rectifier circuit, which are operated at 220 V (A.C.) on standard domestic A.C. supply. The number of turns of secondary coil are 24, then the number of turns of primary are ________. **[NCERT : P L-195 | Feb. 24, 2021 (I)]**

152. For the given input voltage waveform $V_{in}(t)$, the output voltage waveform $V_o(t)$, across the capacitor is correctly depicted by : **[Sep. 06, 2020 (I)]**

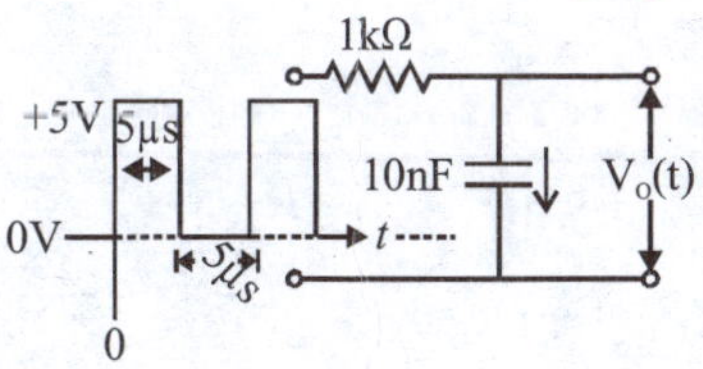

(a)
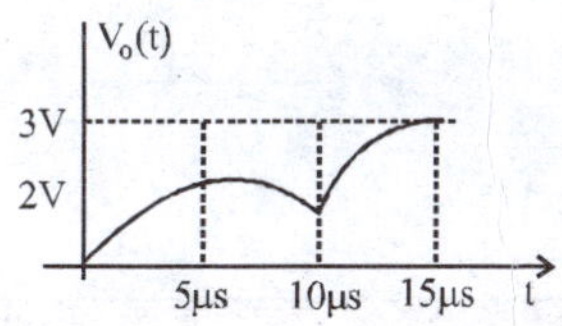

(b)
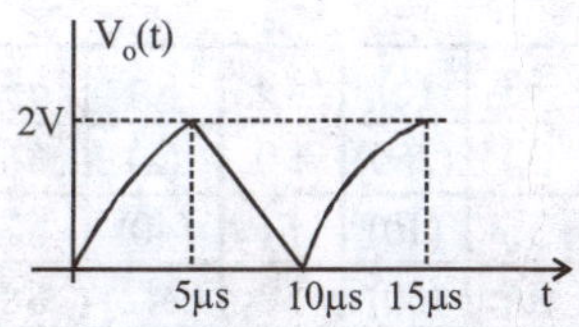

(c)
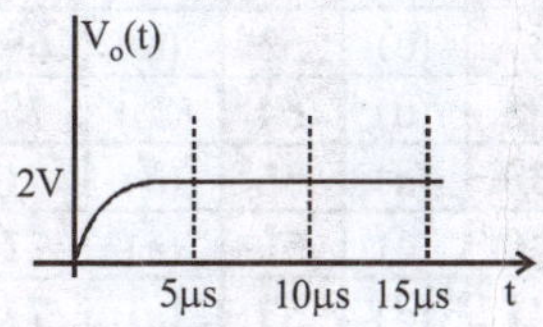

(d)
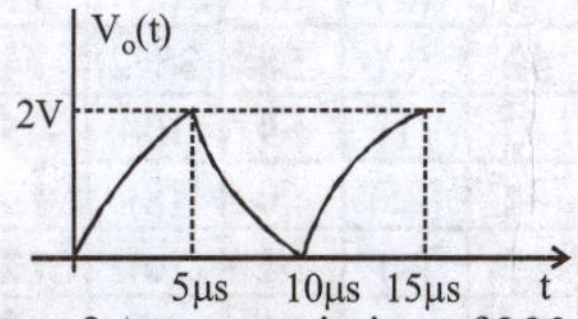

153. A transformer consisting of 300 turns in the primary and 150 turns in the secondary gives output power of 2.2kW. If the current in the secondary coil is 10 A, then the input voltage and current in the primary coil are : **[NCERT : P L-195 | 10 April 2019 (I)]**

(a) 220 V and 20 A (b) 440 V and 20 A
(c) 440 V and 5 A (d) 220 V and 10 A

154. A power transmission line feeds input power at 2300 V to a step down transformer with its primary windings having 4000 turns. The output power is delivered at 230 V by the transformer. If the current in the primary transformer is 5A and its efficiency is 90%, the output current would be:

[NCERT : P L-195 | Online April 16, 2018 S, 9 Jan. 2019 (II)]

(a) 50 A (b) 45 A (c) 35 A (d) 25 A

155. In an oscillating LC circuit the maximum charge on the capacitor is Q. The charge on the capacitor when the energy is stored equally between the electric and magnetic field is **[2003]**

(a) $\frac{Q}{2}$ (b) $\frac{Q}{\sqrt{3}}$ (c) $\frac{Q}{\sqrt{2}}$ (d) Q

156. The core of any transformer is laminated so as to

[NCERT : P L-195, 196 | 2003]

(a) reduce the energy loss due to eddy currents
(b) make it light weight
(c) make it robust and strong
(d) increase the secondary voltage

157. In a transformer, number of turns in the primary coil are 140 and that in the secondary coil are 280. If current in primary coil is 4 A, then that in the secondary coil is

[NCERT : P L-195 | 2002]

(a) 4 A (b) 2 A
(c) 6 A (d) 10 A

ANSWER KEYS

1	(c)	17	(d)	33	(c)	49	(c)	65	(1)	81	(22)	97	(0.8, 250)	113	(a)	129	(b)	145	(9)
2	(c)	18	(b)	34	(250)	50	(a)	66	(3)	82	(a)	98	(c)	114	(a)	130	(c)	146	(240)
3	(a)	19	(c)	35	(10)	51	(40)	67	(10)	83	(c)	99	(33)	115	(b)	131	(c)	147	(a)
4	(c)	20	(b)	36	(50)	52	(242)	68	(c)	84	(c)	100	(400)	116	(b)	132	(a)	148	(c)
5	(22)	21	(a)	37	(c)	53	(a)	69	(44)	85	(c)	101	(a)	117	(b)	133	(a)	149	(a)
6	(8)	22	(1)	38	(d)	54	(25)	70	(b)	86	(2)	102	(d)	118	(b)	134	(b)	150	(a)
7	(b)	23	(5)	39	(d)	55	(b)	71	(c)	87	(1)	103	(a)	119	(c)	135	(a)	151	(440)
8	(b)	24	(c)	40	(c)	56	(a)	72	(b)	88	(a)	104	(b)	120	(c)	136	(d)	152	(a)
9	(c)	25	(2)	41	(d)	57	(a)	73	(500)	89	(None)	105	(d)	121	(c)	137	(d)	153	(c)
10	(d)	26	(3)	42	(c)	58	(3872)	74	(c)	90	(74)	106	(c)	122	(d)	138	(a)	154	(b)
11	(b)	27	(b)	43	(1)	59	(100)	75	(5)	91	(500)	107	(d)	123	(d)	139	(b)	155	(c)
12	(d)	28	(a)	44	(25)	60	(8)	76	(100)	92	(c)	108	(d)	124	(c)	140	(d)	156	(a)
13	(242)	29	(10)	45	(a)	61	(c)	77	(15)	93	(3)	109	(None)	125	(d)	141	(b)	157	(b)
14	(a)	30	(4)	46	(d)	62	(10)	78	(0)	94	(c)	110	(b)	126	(a)	142	(45)		
15	(a)	31	(b)	47	(c)	63	(d)	79	(0)	95	(3.906)	111	(c)	127	(b)	143	(b)		
16	(11)	32	(a)	48	(625)	64	(250)	80	(b)	96	(900)	112	(a)	128	(c)	144	(c)		

8 Electromagnetic Waves

TREND ANALYSIS
JEE MAIN (2025-2021)

	2025				2024				2023				2022				2021			
	Session II		Session I		Session II		Session I		Session II		Session I		Session II		Session I		Session II		Session I	
	MCQs	NVQs	MCQs	NVQs	MCQs	NVQs	MCQs	NVQs	MCQs	NVQs	MCQs	NVQs	MCQs	NVQs	MCQs	NVQs	MCQs	NVQs	MCQs	NVQs
Topic 1	1	–	5	2	6	–	10	–	9	–	8	2	7	–	10	1	6	3	3	7
Topic 2	–	–	1	–	2	–	–	–	3	–	2	–	1	–	2	–	–	–	–	–
Topic 3																				
Topic 4																				
Total	1	–	6	2	8	–	10	–	12	–	10	2	8	–	12	1	6	3	3	7

How to Code Qns for Better & Faster Revision

Type 1: Able to solve (No code Req)

Type 2: Solved with struggle or by seeing solution **(Use code NR means Needs Revision or orange stickers)**

Type 3: Tricky/ Could not solve on my own. Need help from Teacher/ Friend. **(Use code TQ means Tricky Question or red stickers)**

You can either hand write the code in boxes or else paste coloured stickers given at end of book.

Topic 1 - Electromagnetic Waves, Conduction and Displacement Current

1. The radiation pressure exerted by a 450 W light source on a perfectly reflecting surface placed at 2m away from it, is : **[April 3, 2025 (I)]**

(a) 1.5×10^{-8} pascals (b) 0
(c) 6×10^{-8} pascals (d) 3×10^{-8} pascals

2. A plane electromagnetic wave propagates along the + x direction in free space. The components of the electric field, $\vec{E}$ and magnetic field, $\vec{B}$ vectors associated with the wave in Cartesian frame are : **[NCERT : P L-206 | Jan. 29, 2025 (II)]**

(a) E_y, B_x (b) E_y, B_z (c) E_x, B_y (d) E_z, B_y

3. The magnetic field of an E.M. wave is given by

$$\vec{B} = \left(\frac{\sqrt{3}}{2}\hat{i} + \frac{1}{2}\hat{j}\right) 30 \sin\left[\omega\left(t - \frac{z}{c}\right)\right] \text{ (S.I. Units)}$$

The corresponding electric field in S.I. units is : **[NCERT : P L-207 | Jan. 28, 2025 (II)]**

(a) $\vec{E} = \left(\frac{1}{2}\hat{i} - \frac{\sqrt{3}}{2}\hat{j}\right) 30c \sin\left[\omega\left(t - \frac{z}{c}\right)\right]$

(b) $\vec{E} = \left(\frac{3}{4}\hat{i} + \frac{1}{4}\hat{j}\right) 30c \cos\left[\omega\left(t - \frac{z}{c}\right)\right]$

(c) $\vec{E} = \left(\frac{1}{2}\hat{i} + \frac{\sqrt{3}}{2}\hat{j}\right) 30c \sin\left[\omega\left(t + \frac{z}{c}\right)\right]$

(d) $\vec{E} = \left(\frac{\sqrt{3}}{2}\hat{i} - \frac{1}{2}\hat{j}\right) 30c \sin\left[\omega\left(t + \frac{z}{c}\right)\right]$

4. Due to presence of an EM wave whose electric component is given by $E = 100 \sin(\omega t - kx)\ NC^{-1}$, a cylinder of length 200 cm holds certain amount of EM energy inside it. If another cylinder of same length but half diameter than previous one holds same amount of EM energy, the magnitude of the electric field of the corresponding EM wave should be modified as **[Jan. 28, 2025 (I)]**

(a) $25 \sin(\omega t - kx)\ NC^{-1}$ (b) $200 \sin(\omega t - kx)\ NC^{-1}$
(c) $400 \sin(\omega t - kx)\ NC^{-1}$ (d) $50 \sin(\omega t - kx)\ NC^{-1}$

5. A plane electromagnetic wave of frequency 20 MHz travels in free space along the +x direction. At a particular point in space and time, the electric field vector of the wave is $E_y = 9.3\ Vm^{-1}$. Then, the magnetic field vector of the wave at that point is
[NCERT : P L-206, 207 | Jan. 23, 2025 (II)]
(a) $B_z = 9.3 \times 10^{-8}$ T (b) $B_z = 1.55 \times 10^{-8}$ T
(c) $B_z = 6.2 \times 10^{-8}$ T (d) $B_z = 3.1 \times 10^{-8}$ T

6. A time varying potential difference is applied between the plates of a parallel plate capacitor of capacitance 2.5 μF. The dielectric constant of the medium between the capacitor plates is 1. It produces an instantaneous displacement current of 0.25 mA in the intervening space between the capacitor plates, the magnitude of the rate of change of the potential difference will be ______
[NCERT : P L-203 | Jan. 23, 2025 (II)]

7. The electric field of an electromagnetic wave in free space is $\vec{E} = 57\cos[7.5\times10^6 t - 5\times10^{-3}(3x+4y)]\left(4\hat{i}-3\hat{j}\right)$ N/C
The associated magnetic field in Tesla is
[NCERT : P L-206, 207 | Jan. 23, 2025 (I)]
(a) $\vec{B} = \frac{57}{3\times10^8}\cos\left[7.5\times10^6 t - 5\times10^{-3}(3x+4y)\right]\left(5\hat{k}\right)$
(b) $\vec{B} = \frac{57}{3\times10^8}\cos\left[7.5\times10^6 t - 5\times10^{-3}(3x+4y)\right]\left(\hat{k}\right)$
(c) $\vec{B} = -\frac{57}{3\times10^8}\cos\left[7.5\times10^6 t - 5\times10^{-3}(3x+4y)\right]\left(5\hat{k}\right)$
(d) $\vec{B} = -\frac{57}{3\times10^8}\cos\left[7.5\times10^6 t - 5\times10^{-3}(3x+4y)\right]\left(\hat{k}\right)$

8. A parallel plate capacitor of area $A = 16\ cm^2$ and separation between the plates 10 cm, is charged by a DC current. Consider a hypothetical plane surface of area $A_0 = 3.2\ cm^2$ inside the capacitor and parallel to the plates. At an instant, the current through the circuit is 6A. At the same instant the displacement current through A_0 is ______ mA.
[NCERT : P L-203 | Jan. 22, 2025 (II)]

9. A plane EM wave is propagating along x direction. It has a wavelength of 4 mm. If electric field is in y-direction with the maximum magnitude of 60 Vm^{-1}, the equation for magnetic field is: **[NCERT : PL-206, 207 | April 9, 2024 (I)]**
(a) $B_z = 60\sin\left[\frac{\pi}{2}\left(x - 3\times10^8 t\right)\right]\hat{k}T$
(b) $B_z = 2\times10^{-7}\sin\left[\frac{\pi}{2}\times10^3(x - 3\times10^8 t)\right]\hat{k}T$
(c) $B_x = 60\sin\left[\frac{\pi}{2}\left(x - 3\times10^8 t\right)\right]\hat{i}T$
(d) $B_z = 2\times10^{-7}\sin\left[\frac{\pi}{2}\left(x - 3\times10^8 t\right)\hat{k}T\right]$

10. Electromagnetic waves travel in a medium with speed of $1.5\times10^8\ ms^{-1}$. The relative permeability of the medium is 2.0. The relative permittivity will be :
[NCERT : PL-207 | April 6, 2024 (I)]
(a) 5 (b) 1 (c) 4 (d) 2

11. The magnetic field in a plane electromagnetic wave is $B_y = (3.5\times10^{-7})\sin(1.5\times10^3 x + 0.5\times10^{11} t)$T. The corresponding electric field will be
[NCERT : P L-206, 207 | April 9, 2024 (II)]
(a) $E_y = 1.17\sin(1.5\times10^3 x + 0.5\times10^{11} t)Vm^{-1}$
(b) $E_z = 105\sin(1.5\times10^3 x + 0.5\times10^{11} t)Vm^{-1}$
(c) $E_z = 1.17\sin(1.5\times10^3 x + 0.5\times10^{11} t)Vm^{-1}$
(d) $E_y = 10.5\sin(1.5\times10^3 x + 0.5\times10^{11} t)Vm^{-1}$

12. In the given electromagnetic wave $E_y = 600\sin(\omega t - kx)\ Vm^{-1}$, intensity of the associated light beam is (in W/m^2); (Given $\epsilon_0 = 9\times10^{-12}C^2N^{-1}m^{-2}$)
[NCERT : P L-206, 207 | April 6, 2024 (II)]
(a) 486 (b) 243 (c) 729 (d) 972

13. An alternating voltage of amplitude 40 V and frequency 4 kHz is applied directly across the capacitor of 12 μF. The maximum displacement current between the plates of the capacitor is nearly:
[NCERT : P L-203 | April 5, 2024 (I)]
(a) 13 A (b) 8 A (c) 10 A (d) 12 A

14. The electric field in an electromagnetic wave is given by $\vec{E} = \hat{i}40\cos\omega\left(t - \frac{z}{c}\right)NC^{-1}$. The magnetic field induction of this wave is (in SI unit):
[NCERT : P L-206, 207 | April 4, 2024 (I)]
(a) $\vec{B} = \hat{i}\frac{40}{c}\cos\omega\left(t - \frac{z}{c}\right)$ (b) $\vec{B} = \hat{j}40\cos\omega\left(t - \frac{z}{c}\right)$
(c) $\vec{B} = \hat{k}\frac{40}{c}\cos\omega\left(t - \frac{z}{c}\right)$ (d) $\vec{B} = \hat{j}\frac{40}{c}\cos\omega\left(t - \frac{z}{c}\right)$

15. A parallel plate capacitor has a capacitance C = 200 pF. It is connected to 230 V ac supply with an angular frequency 300 rad/s. The rms value of conduction current in the current and displacement current in the capacitor respectively are: **[NCERT : P L-203 | Feb. 1, 2024 (I)]**
(a) 1.38 μA and 1.38 μA (b) 14.3 μA and 143 μA
(c) 13.8 μA and 138 μA (d) 13.8 μA and 13.8 μA

16. If frequency of electromagnetic wave is 60 MHz and it travels in air along z direction then the corresponding electric and magnetic field vectors will be mutually perpendicular to each other and the wavelength of the wave (in m) is: **[NCERT : P L-207 | Feb. 1, 2024 (II)]**
(a) 2.5 (b) 5 (c) 10 (d) 2

17. In a plane EM wave, the electric field oscillates sinusoidally at a frequency of 5×10^{10} Hz and an amplitude of 50 Vm^{-1}. The total average energy density of the electromagnetic field of the wave is : [Use $\varepsilon_0 = 8.85\times10^{-12}C^2/Nm^2$] **[Jan. 31, 2024 (I)]**
(a) $1.106\times10^{-8}\ Jm^{-3}$ (b) $2.212\times10^{-10}\ Jm^{-3}$
(c) $4.425\times10^{-8}\ Jm^{-3}$ (d) $2.212\times10^{-8}\ Jm^{-3}$

18. The electric field of an electromagnetic wave in free space is represented as $\vec{E} = E_0\cos(\omega t - kz)\hat{i}$.
The corresponding magnetic induction vector will be:
[NCERT : P L-206, 207 | Jan. 30, 2024 (I)]
(a) $\vec{B} = E_0 C\cos(\omega t - kz)\hat{j}$
(b) $\vec{B} = \frac{E_0}{C}\cos(\omega t - kz)\hat{j}$

(c) $\vec{B} = E_0 C \cos(\omega t + kz)\hat{j}$

(d) $\vec{B} = \frac{E_0}{C}\cos(\omega t + kz)\hat{j}$

19. If the total energy transferred to a surface in time t is $6.48 \times 10^5 J$, then the magnitude of the total momentum delivered to this surface for complete absorption will be: **[Jan. 30, 2024 (II)]**

(a) 2.46×10^{-3} kg m/s (b) 2.16×10^{-3} kg m/s
(c) 1.58×10^{-3} kg m/s (d) 4.32×10^{-3} kg m/s

20. Match List I with List II **[NCERT : P L-205 | Jan. 30, 2024 (II)]**

List I	List II
A. Gauss's law of magnetostatics	I. $\oint \vec{E}\cdot d\vec{a} = \frac{1}{\varepsilon_0}\int \rho dV$
B. Faraday's law of electro magnetic induction	II. $\oint \vec{B}\cdot d\vec{a} = 0$
C. Ampere's law	III. $\int \vec{E}\cdot d\vec{l} = \frac{-d}{dt}\int \vec{B}\cdot d\vec{a}$
D. Gauss's law of electrostatics	IV. $\oint \vec{B}\cdot d\vec{l} = \mu_0 I$

Choose the correct answer from the options given below:
(a) A - I, B - III, C - IV, D - II
(b) A - III, B - IV, C - I, D - II
(c) A - IV, B - II, C - III, D - I
(d) A - II, B - III, C - IV, D - I

21. Match List I with List II **[NCERT : P L-205 | Jan. 29, 2024 (I)]**

LIST I	LIST II
A. $\oint \vec{B}.d\vec{l} = \mu_o i_c + \mu_o \varepsilon_o \frac{d\phi_E}{dt}$	I. Gauss' law for electricity
B. $\oint \vec{E}.d\vec{l} = \frac{d\phi_B}{dt}$	II. Gauss' law for magnetism
C. $\oint \vec{E}.d\vec{A} = \frac{Q}{\varepsilon_o}$	III. Faraday law
D. $\oint \vec{B}.d\vec{A} = 0$	IV. Ampere - Maxwell law

Choose the correct answer from the options given below:
(a) A-II, B-III, C-I, D-IV (b) A-IV, B-I, C-III, D-II
(c) A-IV, B-III, C-I, D-II (d) A-I, B-II, C-III, D-IV

22. A plane electromagnetic wave of frequency 35 Mhz travels in free space along the X-direction. At a particular point (in space and time) $\vec{E} = 9.6\hat{j}\,V/m$. The value of magnetic field at this point is : **[NCERT : P L-207 | Jan. 29, 2024 (II)]**

(a) $9.6\hat{j}\,T$ (b) $3.2 \times 10^{-8}\,\hat{i}\,T$
(c) $3.2 \times 10^{-8}\,\hat{k}\,T$ (d) $9.6 \times 10^{-8}\,\hat{k}\,T$

23. A plane electromagnetic wave propagating in x-direction is described by **[NCERT : P L-206, 207 | Jan. 27, 2024 (I)]**

$E_y = (200\,Vm^{-1})\sin[1.5 \times 10^7 t - 0.05\,x]$;
The intensity of the wave is :
(Use $\epsilon_0 = 8.85 \times 10^{-12}\,C^2N^{-1}m^{-2}$)

(a) $35.4\,Wm^{-2}$ (b) $53.1\,Wm^{-2}$
(c) $26.6\,Wm^{-2}$ (d) $106.2\,Wm^{-2}$

24. An object is placed in a medium of refractive index 3. An electromagnetic wave of intensity $6 \times 10^8\,W/m^2$ falls normally on the object and it is absorbed completely. The radiation pressure on the object would be (speed of light in free space = 3×10^8 m/s) **[Jan. 27, 2024 (II)]**

(a) $36\,Nm^{-2}$ (b) $18\,Nm^{-2}$ (c) $6\,Nm^{-2}$ (d) $2\,Nm^{-2}$

25. Which of the following Maxwell's equations is valid for time varying conditions but not valid for static conditions : **[NCERT : P L-205 | April 13, 2023 (I)]**

(a) $\oint \vec{B}.\vec{dl} = \mu_0 I$ (b) $\oint \vec{E}.\vec{dl} = 0$
(c) $\oint \vec{E}.\vec{dl} = -\frac{\partial \phi_B}{\partial t}$ (d) $\oint \vec{D}.\vec{dA} = Q$

26. The source of time varying magnetic field may be
(A) a permanent magnet
(B) an electric field changing linearly with time
(C) direct current
(D) a decelerating charge particle
(E) an antenna fed with a digital signal

Choose the correct answer from the options given below: **[NCERT : P L-205 | April 13, 2023 (I)]**

(a) (D) only (b) (C) and (E) only
(c) (A) only (d) (B) and (D) only

27. In an electromagnetic wave, at an instant and at a particular position, the electric field is along the negative z-axis and magnetic field is along the positive x-axis. Then the direction of propagation of electromagnetic wave is : **[NCERT : P L-206 | April 13, 2023 (II)]**

(a) at 45° angle from positive y-axis
(b) negative y-axis
(c) positive z-axis
(d) positive y-axis

28. The electric field in an electromagnetic wave is given as
$$\vec{E} = 20\sin\omega\left(t - \frac{x}{c}\right)\vec{j}\,NC^{-1}.$$
Where ω and c are angular frequency and velocity of electromagnetic wave respectively. The energy contained in a volume of $5 \times 10^{-4}\,m^3$ will be
(Given $\varepsilon_0 = 8.85 \times 10^{-12}\,C^2/Nm^2$) **[April 11, 2023 (I)]**

(a) 28.5×10^{-13} J (b) 17.7×10^{-13} J
(c) 8.85×10^{-13} J (d) 88.5×10^{-13} J

29. A plane electromagnetic wave of frequency 20 MHz propagates in free space along x-direction. At a particular space and time, $\vec{E} = 6.6\hat{j}$ V/m. What is $\vec{B}$ at this point ? **[NCERT : P L-206, 207 | April 11, 2023 (II)]**

(a) $-2.2 \times 10^{-8}\,\hat{i}T$ (b) $2.2 \times 10^{-8}\,\hat{k}T$
(c) $-2.2 \times 10^{-8}\,\hat{k}T$ (d) $2.2 \times 10^{-8}\,\hat{i}T$

30. The energy of an electromagnetic wave contained in a small volume oscillates with **[April 10, 2023 (I)]**
(a) zero frequency
(b) half the frequency of the wave
(c) double the frequency of the wave
(d) the frequency of the wave

31. The amplitude of magnetic field in an electromagnetic wave propagating along y-axis is 6.0×10^{-7} T. The maximum value of electric field in the electromagnetic wave is: **[NCERT : P L-207 | April 10, 2023 (II)]**
(a) $5 \times 10^{14}\ Vm^{-1}$ (b) $180\ Vm^{-1}$
(c) $2 \times 10^{15}\ Vm^{-1}$ (d) $6.0 \times 10^{-7}\ Vm^{-1}$

32. For the plane electromagnetic wave given by $E = E_0 \sin(\omega t - kx)$ and $B = B_0 \sin(\omega t - kx)$, the ratio of average electric energy density to average magnetic energy density is **[April 6, 2023 (I)]**
(a) 1 (b) 1/2 (c) 2 (d) 4

33. The energy density associated with electric field $\vec{E}$ and magnetic field $\vec{B}$ of an electromagnetic wave in free space is given by (ϵ_0 – permittivity of free space, μ_0 - permeability of free space) **[April 6, 2023 (II)]**
(a) $U_E = \frac{E^2}{2\epsilon_0}, U_B = \frac{B^2}{2\mu_0}$
(b) $U_E = \frac{E^2}{2\epsilon_0}, U_B = \frac{\mu_0 B^2}{2}$
(c) $U_E = \frac{\epsilon_0 E^2}{2}, U_B = \frac{\mu_0 B^2}{2}$
(d) $U_E = \frac{\epsilon_0 E^2}{2}, U_B = \frac{B^2}{2\mu_0}$

34. The ratio of average electric energy density and total average energy density of electromagnetic wave is : **[Feb. 1, 2023 (II)]**
(a) 2 (b) $\frac{1}{2}$ (c) 1 (d) 3

35. In a medium the speed of light wave decreases to 0.2 times to its speed in free space The ratio of relative permittivity to the refractive index of the medium is x : 1. The value of x is ______.
(Given speed of light in free space = $3 \times 10^8\ m\ s^{-1}$ and for the given medium $\mu_r = 1$) **[NCERT : P L-207 | Jan. 31, 2023 (I)]**

36. A small object at rest, absorbs a light pulse of power 20 mW and duration 300 ns. Assuming speed of light as 3×10^8 m/s. the momentum of the object becomes equal to: **[Jan. 30, 2023 (I)]**
(a) 0.5×10^{-17} kg m/s (b) 2×10^{-17} kg m/s
(c) 3×10^{-17} kg m/s (d) 1×10^{-17} kg m/s

37. A point source of light is placed at the centre of curvature of a hemispherical surface. The source emits a power of 24 W The radius of curvature of hemisphere is 10 cm and the inner surface is completely reflecting. The force on the hemisphere due to the light falling on it is ______ $\times 10^{-8}$ N. **[Jan. 30, 2023 (I)]**

38. A point source of 100 W emits light with 5% efficiency. At a distance of 5 m from the source, the intensity produced by the electric field component is : **[Jan. 30, 2023 (II)]**
(a) $\frac{1}{2\pi}\frac{W}{m^2}$ (b) $\frac{1}{40\pi}\frac{W}{m^2}$ (c) $\frac{1}{10\pi}\frac{W}{m^2}$ (d) $\frac{1}{20\pi}\frac{W}{m^2}$

39. Given below are two statements: **[NCERT : P L-207 | Jan. 29, 2023 (II)]**
Statement I : Electromagnetic waves are not deflected by electric and magnetic field.
Statement II : The amplitude of electric field and the magnetic field in electromagnetic waves are related to each other as $E_0 = \sqrt{\frac{\mu_0}{\varepsilon_0}} B_0$
In the light of the above statements, choose the correct answer from the options given below:
(a) Statement I is true but statement II is false
(b) Both Statement I and Statement II are true
(c) Statement I is false but statement II is true
(d) Both Statement I and Statement II are false

40. All electromagnetic wave is transporting energy in the negative z direction. At a certain point and certain time the direction of electric field of the wave is along positive y direction. What will be the direction of the magnetic field of the wave at that point and instant? **[Jan. 25, 2023 (I)]**
(a) Positive direction of x (b) Positive direction of z
(c) Negative direction of x (d) Negative direction of y

41. Match List I with List II : **[NCERT : P L-205 | Jan. 25, 2023 (II)]**

A.	Gauss's Law in Electrostatics	I.	$\oint \vec{E}.d\vec{l} = -\frac{d\phi_B}{dt}$
B.	Faraday's Law	II.	$\oint \vec{B}.d\vec{A} = 0$
C.	Gauss's Law in Magnetism	III.	$\oint \vec{B}.d\vec{l} = \mu_0 i_C + \mu_0 \epsilon_0 \frac{d\phi_E}{dt}$
D.	Ampere-Maxwell Law	IV.	$\oint \vec{E}.d\vec{s} = \frac{q}{\epsilon_0}$

Choose the correct answer from the options given below:
(a) A-IV, B-I, C-II, D-III (b) A-I, B-II, C-III, D-IV
(c) A-III, B-IV, C-I, D-II (d) A-II, B-III, C-IV, D-I

42. If $\vec{E}$ and $\vec{K}$ represent electric field and propagation vectors of the EM waves in vacuum, then magnetic field vector is given by : (ω - angular frequency) : **[NCERT : P L-206 | Jan. 24, 2023 (I)]**
(a) $\frac{1}{\omega}(\vec{K} \times \vec{E})$ (b) $\omega(\vec{E} \times \vec{K})$
(c) $\omega(\vec{K} \times \vec{E})$ (d) $\vec{K} \times \vec{E}$

43. The electric field and magnetic field components of an electromagnetic wave going through vacuum is described by **[NCERT : P L-207 | Jan. 24, 2023 (II)]**
$E_x = E_0 \sin(kz - \omega t)$
$B_y = B_0 \sin(kz - \omega t)$

Then the correct relation between E_0 and B_0 is given by

(a) $kE_0 = \omega B_0$ (b) $E_0 B_0 = \omega k$

(c) $\omega E_0 = kB_0$ (d) $E_0 = kB_0$

44. Nearly 10% of the power of a 110 W light bulb is converted to visible radiation. The change in average intensities of visible radiation, at a distance of 1 m from the bulb to a distance of 5 m is $a \times 10^{-2}$ W/m². The value of 'a' will be **[July 29, 2022 (II)]**

45. Sun light falls normally on a surface of area 36 cm² and exerts an average force of 7.2×10^{-9} N within a time period of 20 minutes. Considering a case of complete absorption, the energy flux of incident light is **[July 28, 2022 (II)]**

(a) 25.92×10^2 W/cm² (b) 8.64×10^{-6} W/cm²

(c) 6.0 W/cm² (d) 0.06 W/cm²

46. Identify the correct statements from the following descriptions of various properties of electromagnetic waves. **[NCERT : P L-207 | July 27, 2022 (II)]**

A. In a plane electromagnetic wave electric field and magnetic field must be perpendicular to each other and direction of propagation of wave should be along electric field or magnetic field.

B. The energy in electromagnetic wave is divided equally between electric and magnetic fields.

C. Both electric field and magnetic field are parallel to each other and perpendicular to the direction of propagation of wave.

D. The electric field, magnetic field and direction of propagation of wave must be perpendicular to each other.

E. The ratio of amplitude of magnetic field to the amplitude of electric field is equal to speed of light.

Choose the most appropriate answer from the options given below:

(a) D only (b) B and D only

(c) B, C and E only (d) A, B and E only

47. The magnetic field of a plane electromagnetic wave is given by: **[NCERT : P L-206, 207 | July 26, 2022 (I)]**

$\vec{B} = 2\times10^{-8}\sin(0.5\times10^3 x + 1.5\times10^{11} t)\hat{j}$ T.

The amplitude of the electric field would be:

(a) 6 Vm⁻¹ along x-axis (b) 3 Vm⁻¹ along z-axis

(c) 6 Vm⁻¹ along z-axis (d) 2×10^{-8} Vm⁻¹ along z-axis

48. The oscillating magnetic field in a plane electromagnetic wave is given by $B_y = 5\times10^{-6}\sin 1000\pi(5x - 4\times10^8 t)$T. The amplitude of electric field will be : **[NCERT : P L-207 | July 26, 2022 (II)]**

(a) 15×10^2 Vm⁻¹ (b) 5×10^{-6} Vm⁻¹

(c) 16×10^{12} Vm⁻¹ (d) 4×10^2 Vm⁻¹

49. The rms value of conduction current in a parallel plate capacitor is 6.9 μA. The capacity of this capacitor, if it is connected to 230 V ac supply with an angular frequency of 600 rad/s, will be : **[July 25, 2022 (I)]**

(a) 5 pF (b) 50 pF (c) 100 pF (d) 200 pF

50. Light wave traveling in air along x-direction is given by $E_y = 540\sin\pi\times10^4(x - ct)$ Vm⁻¹. Then, the peak value of magnetic field of wave will be (Given $c = 3\times10^8$ ms⁻¹) **[NCERT : P L-207 | July 25, 2022 (II)]**

(a) 18×10^{-7} T (b) 54×10^{-7} T

(c) 54×10^{-8} T (d) 18×10^{-9} T

51. An expression for oscillating electric field in a plane electromagnetic wave is given as

$E_z = 300\sin(5\pi\times10^3 x - 3\pi\times10^{11} t)$Vm⁻¹

Then, the value of magnetic field amplitude will be: (Given: speed of light in Vacuum $c = 3\times10^8$ ms⁻¹) **[NCERT : P L-207 | June 30, 2022 (I)]**

(a) 1×10^{-6} T (b) 5×10^{-6} T

(c) 18×10^9 T (d) 21×10^9 T

52. The intensity of the light from a bulb incident on a surface is 0.22 W/m². The amplitude of the magnetic field in this light-wave is______$\times10^{-9}$ T.

(Given : Permittivity of vacuum $\varepsilon_0 = 8.85\times10^{-12}$ C²N⁻¹m⁻², speed of light in vacuum $c = 3\times108$ ms⁻¹) **[June 29, 2022 (I)]**

53. The displacement current of 4.425 μA is developed in the space between the plates of parallel plate capacitor when voltage is changing at a rate of 10^6 Vs⁻¹. The area of each plate of the capacitor is 40 cm². The distance between each plate of the capacitor is $x\times10^{-3}$m. The value of x is, (Permittivity of free space, $E_0 = 8.85\times10^{-12}$ C² N⁻¹ m⁻²) **[NCERT : P L-203 | June 29, 2022 (II)]**

54. A radar sends an electromagnetic signal of electric field $(E_0) = 2.25$ V/m and magnetic field $(B_0) = 1.5\times10^{-8}$ T which strikes a target on line of sight at a distance of 3 km in a medium. After that, a pail of signal (echo) reflects back towards the radar with same velocity and by same path. If the signal was transmitted at time t_0 from radar then after how much time echo will reach to the radar? **[NCERT : P L-207 | June 28, 2022 (I)]**

(a) 2.0×10^{-5} s (b) 4.0×10^{-5} s

(c) 1.0×10^{-5} s (d) 8.0×10^{-5} s

55. An EM wave propagating in x-direction has a wavelength of 8 mm. The electric field vibrating y-direction has maximum magnitude of 60 Vm⁻¹. Choose the correct equations for electric and magnetic fields if the EM wave is propagating in vacuum : **[NCERT : P L-206, 207 | June 28, 2022 (II)]**

(a) $E_y = 60\sin\left[\frac{\pi}{4}\times10^3(x-3\times10^8 t)\right]\hat{j}$Vm⁻¹

$B_z = 2\sin\left[\frac{\pi}{4}\times10^3(x-3\times10^8 t)\right]\hat{k}$T

(b) $E_y = 60\sin\left[\frac{\pi}{4}\times10^3(x-3\times10^8 t)\right]\hat{j}$Vm⁻¹

$B_z = 2\times10^{-7}\sin\left[\frac{\pi}{4}\times10^3(x-3\times10^8 t)\right]\hat{k}$T

(c) $E_y = 2\times10^{-7}\sin\left[\frac{\pi}{4}\times10^3(x-3\times10^8 t)\right]\hat{j}Vm^{-1}$

$B_z = 60\sin\left[\frac{\pi}{4}\times10^3(x-3\times10^8 t)\right]\hat{k}T$

(d) $E_y = 2\times10^{-7}\sin\left[\frac{\pi}{4}\times10^4(x-4\times10^8 t)\right]\hat{j}Vm^{-1}$

$B_z = 60\sin\left[\frac{\pi}{4}\times10^4(x-4\times10^8 t)\right]\hat{k}T$

56. Given below are two statements:

Statement I : A time varying electric field is a source of changing magnetic field and vice-versa. Thus a disturbance in electric or magnetic field creates EM waves.

Statement II : In a material medium. The EM wave travels with speed $v = \frac{1}{\sqrt{\mu_0 \varepsilon_0}}$.

In the light of the above statements, choose the correct answer from the options given below:

[NCERT : P L-205, 207 | June 27, 2022 (II)]

(a) Both statement I and statement II are true.
(b) Both statement I and statement II are false.
(c) Statement I is correct but statement II is false.
(d) Statement I is incorrect but statement II is true.

57. If electric field intensity of a uniform plane electromagnetic wave is given as

$E = -301.6 \sin(kz - \omega t)\, \hat{a}_x + 452.4 \sin(kz - \omega t)\, \hat{a}_y \frac{V}{m}$

Then, magnetic intensity H of this wave in Am^{-1} will be:'

[Given: Speed of light in vacuum $c = 3 \times 10^8\ ms^{-1}$, permeability of vacuum $\mu_0 = 4\pi \times 10^{-7}\ NA^{-2}$]

[NCERT : P L-206, 207 | June 26, 2022 (I)]

(a) $+0.8 \sin(kz - \omega t)\, \hat{a}_y + 0.8 \sin(kz - \omega t)\, \hat{a}_x$
(b) $+1.0 \times 10^{-6} \sin(kz - \omega t)\, \hat{a}_y + 1.5 \times 10^{-6} (kz - \omega t)\, \hat{a}_x$
(c) $-0.8 \sin(kz - \omega t)\, \hat{a}_y - 1.2 \sin(kz - \omega t)\, \hat{a}_x$
(d) $-1.0 \times 10^{-6} \sin(kz - \omega t)\, \hat{a}_y - 1.5 \times 10^{-6} \sin(kz - \omega t)\, \hat{a}_x$

58. The electric field in an electromagnetic wave is given by $E = 56.5 \sin \omega(t - x/c) NC^{-1}$. Find the intensity of the wave if it is propagating along x-axis in the free space. (Given $\varepsilon_0 = 8.85 \times 10^{-12} C^2 N^{-1} m^{-2}$) **[June 25, 2022 (I)]**

(a) $5.65\ Wm^{-2}$ (b) $4.24\ Wm^{-2}$
(c) $1.9 \times 10^{-7}\ Wm^{-2}$ (d) $56.5\ Wm^{-2}$

59. The electromagnetic waves travel in a medium at a speed of 2.0×10^8 m/s. The relative permeability of the medium is 1.0. The relative permittivity of the medium will be:

[NCERT : P L-207 | June 25, 2022 (II)]

(a) 2.25 (b) 4.25 (c) 6.25 (d) 8.25

60. A plane electromagnetic wave travels in a medium of relative permeability 1.61 and relative permittivity 6.44. If magnitude of magnetic intensity is $4.5 \times 10^{-2}\ Am^{-1}$ at a point, what will be the approximate magnitude of electric field intensity at that point ?

[NCERT : P L-207 | June 24, 2022 (I)]

(Given: permeability of free space $\mu_0 = 4\pi \times 10^{-7}\ NA^{-2}$, speed of light in vacuum $c = 3 \times 10^8\ ms^{-1}$)

(a) $16.96\ Vm^{-1}$ (b) $2.25 \times 10^{-2}\ Vm^{-1}$
(c) $8.48\ Vm^{-1}$ (d) $6.75 \times 10^6\ Vm^{-1}$

61. An electric bulb is rated as 200 W. What will be the peak magnetic field at 4 m distance produced by the radiations coming from this bulb? Consider this bulb as a point source with 3.5% efficiency. **[June 24, 2022 (II)]**

(a) 1.19×10^{-8} T (b) 1.71×10^{-8} T
(c) 0.84×10^{-8} T (d) 3.36×10^{-8} T

62. Electric field of plane electromagnetic wave propagating through a non–magnetic medium is given by E = 20cos $(2 \times 10^{10} t - 200x)$ V/m. The dielectric constant of the medium is equal to : (Take $\mu_r = 1$)

[NCERT : P L-207 | Sep. 1, 2021 (II)]

(a) 9 (b) 2 (c) $\frac{1}{3}$ (d) 3

63. The electric field in an electromagnetic wave is given by $E = (50\ NC^{-1}) \sin\omega(t - x/c)$

The energy contained in a cylinder of volume V is 5.5×10^{-12} J. The value of V is ________ cm^3. (given $\epsilon_0 = 8.8 \times 10^{-12} C^2 N^{-1} m^{-2}$) **[NA, Aug. 31, 2021 (I)]**

64. The magnetic field vector of an electromagnetic wave is given by $B = B_0 \frac{\hat{i} + \hat{j}}{\sqrt{2}} \cos(kz - \omega t)$; where $\hat{i}, \hat{j}$ represents unit vector along x and y-axis respectively. At t = 0 s, two electric charges q_1 of 4π coulomb and q_2 of 2π coulomb located at $\left(0, 0, \frac{\pi}{k}\right)$ and $\left(0, 0, \frac{3\pi}{k}\right)$ respectively, have the same velocity of $0.5\, c\, \hat{i}$, (where c is the velocity of light). The ratio of the force acting on charge q_1 to q_2 is

[Aug. 31, 2021 (II)]

(a) $2\sqrt{2} : 1$ (b) $1 : \sqrt{2}$ (c) $2 : 1$ (d) $\sqrt{2} : 1$

65. Electric field in a plane electromagnetic wave is given by $E = 50 \sin(500x - 10 \times 10^{10} t)$ V/m. The velocity of electromagnetic wave in this medium is : (Given C = speed of light in vacuum)

[NCERT : P L-207 | Aug. 27, 2021 (I)]

(a) $\frac{3}{2}$C (b) C (c) $\frac{2}{3}$C (d) $\frac{C}{2}$

66. A plane electromagnetic wave with frequency of 30 MHz travels in free space. At particular point in space and time, electric field is 6 V/m. The magnetic field at this point will be $x \times 10^{-8}$ T. The value of x is ____.

[NCERT : P L-207 | NA, Aug. 27, 2021 (II)]

67. The electric field in a plane electromagnetic wave is given by $\vec{E} = 200 \cos\left[\left(\frac{0.5 \times 10^3}{m}\right)x - \left(1.5 \times 10^{11} \frac{rad}{s} \times t\right)\right] \frac{V}{m} \hat{j}$

If this wave falls normally on a perfectly reflecting surface having an area of 100 cm^2. If the radiation pressure exerted by the E.M. wave on the surface during a 10 minute exposure is $\frac{x}{10^9} \frac{N}{m^2}$. Find the value of x.

[NA, Aug. 26, 2021 (I)]

68. The relative permittivity of distilled water is 81. The velocity of light in it will be : (Given $\mu_r = 1$)

[NCERT : P L-207 | July 27, 2021 (I)]

(a) 4.33×10^7 m/s (b) 2.33×10^7 m/s
(c) 3.33×10^7 m/s (d) 5.33×10^7 m/s

69. Intensity of sunlight is observed as $0.092\ Wm^{-2}$ at a point in free space. What will be the peak value of magnetic field at that point ? **[July 22, 2021 (II)]**

($\varepsilon_0 = 8.85 \times 10^{-12} C^2 N^{-1} m^{-2}$)

(a) 2.77×10^{-8}T (b) 1.96×10^{-8}T
(c) 8.31 T (d) 5.88 T

70. In an electromagnetic wave the electric field vector and magnetic field vector are given as $\vec{E} = E_0\hat{i}$ and $\vec{B} = B_0\hat{k}$ respectively. The direction of propagation of electromagnetic wave is along : **[July 20, 2021 (II)]**

(a) $(\hat{k})$ (b) $\hat{j}$ (c) $(-\hat{k})$ (d) $(-\hat{j})$

71. A plane electromagnetic wave of frequency 100 MHz is travelling in vacuum along the x-direction. At a particular point in space and time, $\vec{B} = 2.0 \times 10^{-8}\hat{k}$T. (where, $\hat{k}$ is unit vector along z-direction) What is $\vec{E}$ at this point? (speed of light $c = 3 \times 10^8$ m/s)
[NCERT : P L-207 | March 18, 2021 (I)]

(a) $6.0\hat{j}$ V/m (b) $0.6\hat{j}$ V/m
(c) $0.6\hat{k}$ V/m (d) $6.0\hat{k}$ V/m

72. If 2.5×10^{-6} N average force is exerted by a light wave on a non-reflecting surface of 30 cm^2 area during 40 minutes of time span, the energy flux of light just before it falls on the surface is ____ W/cm^2. (Round off to the Nearest Integer) (Assume complete adsorption and normal incidence conditions are there) **[March 17, 2021 (I)]**

73. The electric field intensity produced by the radiation coming from a 100 W bulb at a distance of 3 m is E. The electric field intensity produced by the radiation coming from 60 W at the same distance is $\sqrt{\frac{x}{5}}$ E. Where the value of x = ____. **[March 17, 2021 (II)]**

74. Seawater at a frequency $f = 9 \times 10^2$ Hz, has permittivity $\varepsilon = 80\varepsilon_0$ and resistivity $\rho = 0.25\ \Omega$m. Imagine a parallel plate capacitor is immersed in seawater and is driven by an alternating voltage source $V(t) = V_0 \sin(2\pi ft)$. Then the conduction current density becomes 10^x time the displacment current density after time $t = \frac{1}{800}$ s. The value of x is ____. $\left(\text{Given: } \frac{1}{4\pi\varepsilon_0} = 9 \times 10^9\ \text{Nm}^2\text{C}^{-2}\right)$
[NCERT : P L-203 | March 17, 2021 (II)]

75. For an electromagnetic wave travelling in free space, the relation between average energy densities due to electric (U_e) and magnetic (U_m) fields is:
[Jan. 9, 2019 (II); March 16, 2021 (I)]

(a) $U_e > U_m$ (b) $U_e = U_m$
(c) $U_e \neq U_m$ (d) $U_e < U_m$

76. A plane electromagnetic wave of frequency 500 MHz is travelling in vacuum along y-direction. At a particular point in space and time, $\vec{B} = 8.0 \times 10^{-8}\hat{z}$T. The value of electric field at this point is :
[NCERT : P L-207 | March 16, 2021 (I)]
(speed of light $= 3 \times 10^8$ ms^{-1})
$\hat{x}, \hat{y}, \hat{z}$ are unit vectors along x, y and z directions.

(a) $-24\hat{x}$ V/m (b) $2.6\hat{x}$ V/m
(c) $24\hat{x}$ V/m (d) $-2.6\hat{y}$ V/m

77. An electromagnetic wave of frequency 5 GHz, is travelling in a medium whose relative electric permittivity and relative magnetic permeability both are 2. Its velocity in this medium is ____ $\times 10^7$ m/s.
[NCERT : P L-207 | Feb. 24, 2021 (I)]

78. An electromagnetic wave of frequency 3 GHz enters a dielectric medium of relative electric permittivity 2.25 from vacuum. The wavelength of this wave in that medium will be ____ $\times 10^{-2}$ cm.
[NCERT : P L-207 | Feb. 24, 2021 (II)]

79. The peak electric field produced by the radiation coming from the 8 W bulb at a distance of 10 m is $\frac{x}{10}\sqrt{\frac{\mu_0 c}{\pi}}\frac{V}{m}$. The efficiency of the bulb is 10% and it is a point source. The value of x is ____. **[Feb. 25, 2021 (II)]**

80. A radiation is emitted by 1000 W bulb and it generates an electric field and magnetic field at P, placed at a distance of 2 m. The efficiency of the bulb is 1.25%. The value of peak electric field at P is $x \times 10^{-1}$ V/m. Value of x is ____. (Rounded-off to the nearest integers)
[Take $\varepsilon_0 = 8.85 \times 10^{-12}$ C^2 N^{-1} m^{-2}, $c = 3 \times 10^8$ ms^{-1}]
[Feb. 26, 2021 (I)]

81. For a plane electromagnetic wave, the magnetic field at a point x and time t is

$$\vec{B}(x,t) = [1.2 \times 10^{-7} \sin(0.5 \times 10^3 x + 1.5 \times 10^{11} t)\hat{k}]\text{T}$$

The instantaneous electric field $\vec{E}$ corresponding to $\vec{B}$ is: (speed of light $c = 3 \times 10^8$ ms^{-1})
[NCERT : P L-206, 207 | Sep. 06, 2020 (II)]

(a) $\vec{E}(x,t) = [-36\sin(0.5 \times 10^3 x + 1.5 \times 10^{11} t)\hat{j}]\frac{V}{m}$

(b) $\vec{E}(x,t) = [36\sin(1 \times 10^3 x + 0.5 \times 10^{11} t)\hat{j}]\frac{V}{m}$

(c) $\vec{E}(x,t) = [36\sin(0.5 \times 10^3 x + 1.5 \times 10^{11} t)\hat{k}]\frac{V}{m}$

(d) $\vec{E}(x,t) = [36\sin(1 \times 10^3 x + 1.5 \times 10^{11} t)\hat{i}]\frac{V}{m}$

82. An electron is constrained to move along the y-axis with a speed of 0.1 c (c is the speed of light) in the presence of electromagnetic wave, whose electric field is $\vec{E} = 30\hat{j}$ $\sin(1.5 \times 10^7 t - 5 \times 10^{-2} x)$ V/m. The maximum magnetic force experienced by the electron will be :
(given $c = 3 \times 10^8$ ms^{-1} & electron charge $= 1.6 \times 10^{-19}$C)
[NCERT : P L-207 | Sep. 05, 2020 (I)]

(a) 3.2×10^{-18}N (b) 2.4×10^{-18}N
(c) 4.8×10^{-19}N (d) 1.6×10^{-19}N

83. The electric field of a plane electromagnetic wave is given by $\vec{E} = E_0(\hat{x} + \hat{y})\sin(kz - \omega t)$
Its magnetic field will be given by:
[NCERT : P L-206 | Sep. 04, 2020 (II)]

(a) $\frac{E_0}{c}(-\hat{x}+\hat{y})\sin(kz-\omega t)$

(b) $\frac{E_0}{c}(\hat{x}+\hat{y})\sin(kz-\omega t)$

(c) $\frac{E_0}{c}(\hat{x}-\hat{y})\sin(kz-\omega t)$

(d) $\frac{E_0}{c}(\hat{x}-\hat{y})\cos(kz-\omega t)$

84. The magnetic field of a plane electromagnetic wave is $\vec{B}=3\times10^{-8}\sin[200\pi(y+ct)]\hat{i}T$ where $c = 3\times10^8\ ms^{-1}$ is the speed of light. The corresponding electric field is : **[NCERT : P L-206, 207 | Sep. 03, 2020 (I)]**

(a) $\vec{E}=9\sin[200\pi(y+ct)]\hat{k}$ V/m

(b) $\vec{E}=-10^{-6}\sin[200\pi(y+ct)]\hat{k}$ V/m

(c) $\vec{E}=3\times10^{-8}\sin[200\pi(y+ct)]\hat{k}$ V/m

(d) $\vec{E}=-9\sin[200\pi(y+ct)]\hat{k}$ V/m

85. The electric field of a plane electromagnetic wave propagating along the x direction in vacuum is $\vec{E}=E_0\hat{j}\cos(\omega t-kx)$. The magnetic field $\vec{B}$, at the moment $t=0$ is : **[NCERT : P L-207 | Sep. 03, 2020 (II)]**

(a) $\vec{B}=\frac{E_0}{\sqrt{\mu_0\varepsilon_0}}\cos(kx)\hat{k}$

(b) $\vec{B}=E_0\sqrt{\mu_0\varepsilon_0}\cos(kx)\hat{j}$

(c) $\vec{B}=E_0\sqrt{\mu_0\varepsilon_0}\cos(kx)\hat{k}$

(d) $\vec{B}=\frac{E_0}{\sqrt{\mu_0\varepsilon_0}}\cos(kx)\hat{j}$

86. A plane electromagnetic wave, has frequency of 2.0×10^{10} Hz and its energy density is $1.02\times10^{-8}\ J/m^3$ in vacuum. The amplitude of the magnetic field of the wave is close to ($\frac{1}{4\pi\varepsilon_0}=9\times10^9\frac{Nm^2}{C^2}$ and speed of light $=3\times10^8\ ms^{-1}$) : **[Sep. 02, 2020 (I)]**

(a) 150 nT (b) 160 nT (c) 180 nT (d) 190 nT

87. In a plane electromagnetic wave, the directions of electric field and magnetic field are represented by $\hat{k}$ and $2\hat{i}-2\hat{j}$, respectively. What is the unit vector along direction of propagation of the wave. **[NCERT : P L-206, 207 | Sep. 02, 2020 (II)]**

(a) $\frac{1}{\sqrt{2}}(\hat{i}+\hat{j})$ (b) $\frac{1}{\sqrt{2}}(\hat{j}+\hat{k})$

(c) $\frac{1}{\sqrt{5}}(\hat{i}+2\hat{j})$ (d) $\frac{1}{\sqrt{5}}(2\hat{i}+\hat{j})$

88. The electric fields of two plane electromagnetic plane waves in vacuum are given by $\vec{E}_1=E_0\hat{j}\cos(\omega t-kx)$ and $\vec{E}_2=E_0\hat{k}\cos(\omega t-ky)$.

At $t=0$, a particle of charge q is at origin with a velocity $\vec{v}=0.8c\hat{j}$ (c is the speed of light in vacuum). The instantaneous force experienced by the particle is: **[NCERT : P L-206, 207 | 9 Jan 2020, (I)]**

(a) $E_0q(0.8\hat{i}-\hat{j}+0.4\hat{k})$ (b) $E_0q(0.4\hat{i}-3\hat{j}+0.8\hat{k})$

(c) $E_0q(-0.8\hat{i}+\hat{j}+\hat{k})$ (d) $E_0q(0.8\hat{i}+\hat{j}+0.2\hat{k})$

89. A plane electromagnetic wave of frequency 25 GHz is propagating in vacuum along the z-direction. At a particular point in space and time, the magnetic field is given by $\vec{B}=5\times10^{-8}\hat{j}\,T$. The corresponding electric field $\vec{E}$ is (speed of light $c=3\times10^8\ ms^{-1}$) **[NCERT : P L-207 | 8 Jan 2020, (II)]**

(a) $1.66\times10^{-16}\ \hat{i}$ V/m (b) $-1.66\times10^{-16}\ \hat{i}$ V/m

(c) $-15\ \hat{i}$ V/m (d) $15\ \hat{i}$ V/m

90. If the magnetic field in a plane electromagnetic wave is given by $\vec{B}=3\times10^{-8}\sin(1.6\times10^3x+48\times10^{10}t)\ \hat{j}$ T, then what will be expression for electric field? **[NCERT : P L-207 | 7 Jan 2020, (I)]**

(a) $\vec{E}=(60\sin(1.6\times10^3x+48\times10^{10}t)\ \hat{k}$ v/m)

(b) $\vec{E}=(9\sin(1.6\times10^3x+48\times10^{10}t)\ \hat{k}$ v/m)

(c) $\vec{E}=(3\times10^{-8}\sin(1.6\times10^3x+48\times10^{10}t)\ \hat{k}$ v/m)

(d) $\vec{E}=(3\times10^{-8}\sin(1.6\times10^3x+48\times10^{10}t)\ \hat{k}$ v/m)

91. The electric field of a plane electromagnetic wave is given by $\vec{E}=E_0\frac{\hat{i}+\hat{j}}{\sqrt{2}}\cos(kz+\omega t)$

At $t=0$, a positively charged particle is at the point $(x,y,z)=\left(0,0,\frac{\pi}{k}\right)$. If its instantaneous velocity at $(t=0)$ is $v_0\hat{k}$, the force acting on it due to the wave is: **[NCERT : P L-206, 207 | 7 Jan 2020, (II)]**

(a) parallel to $\frac{\hat{i}+\hat{j}}{\sqrt{2}}$ (b) zero

(c) antiparallel to $\frac{\hat{i}+\hat{j}}{\sqrt{2}}$ (d) parallel to $\hat{k}$

92. An electromagnetic wave is represented by the electric field $\vec{E}=E_0\hat{n}\sin[\omega t+(6y-8z)]$. Taking unit vectors in x, y and z directions to be $\hat{i},\hat{j},\hat{k}$, the direction of propogation $\hat{s}$ is : **NCERT : P L-206, 207 | 12 April 2019, (I)]**

(a) $\hat{s}=\frac{3\hat{i}-4\hat{j}}{5}$ (b) $\hat{s}=\frac{-4\hat{k}+3\hat{j}}{5}$

(c) $\hat{s}=\left(\frac{-3\hat{j}+4\hat{k}}{5}\right)$ (d) $\hat{s}=\frac{3\hat{j}-3\hat{k}}{5}$

93. A plane electromagnetic wave having a frequency $v = 23.9$ GHz propagates along the positive z-direction in free space. The peak value of the Electric Field is 60 V/m. Which among the following is the acceptable magnetic field component in the electromagnetic wave? **[NCERT : P L-207 | 12 April 2019, (I)]**

(a) $\vec{B} = 2\times10^{7}\sin(0.5\times10^{3}z + 1.5\times10^{11}t)\hat{i}$

(b) $\vec{B} = 2\times10^{-7}\sin(0.5\times10^{3}z - 1.5\times10^{11}t)\hat{i}$

(c) $\vec{B} = 60\sin(0.5\times10^{3}x + 1.5\times10^{11}t)\hat{k}$

(d) $\vec{B} = 2\times10^{-7}\sin(1.5\times10^{2}x + 0.5\times10^{11}t)\hat{j}$

94. The electric field of a plane electromagnetic wave is given by $\vec{E} = E_0\hat{i}\cos(kz)\cos(\omega t)$

The corresponding magnetic field is then given by : **[NCERT : P L-206, 207 | 10 April 2019, (I)]**

(a) $\vec{B} = \frac{E_0}{C}\hat{j}\sin(kz)\sin(\omega t)$

(b) $\vec{B} = \frac{E_0}{C}\hat{j}\sin(kz)\cos(\omega t)$

(c) $\vec{B} = \frac{E_0}{C}\hat{j}\cos(kz)\sin(\omega t)$

(d) $\vec{B} = \frac{E_0}{C}\hat{k}\sin(kz)\cos(\omega t)$

95. Light is incident normally on a completely absorbing surface with an energy flux of 25 Wcm^{-2}. If the surface has an area of 25 cm^2, the momentum transferred to the surface in 40 min time duration will be: **[10 April 2019, (II)]**

(a) 6.3×10^{-4} Ns (b) 1.4×10^{-6} Ns
(c) 5.0×10^{-3} Ns (d) 3.5×10^{-6} Ns

96. The magnetic field of a plane electromagnetic wave is given by: $\vec{B} = B_0\hat{i}[\cos(kz-\omega t)] + B_1\hat{j}\cos(kz+\omega t)$

Where $B_0 = 3\times10^{-5}$ T and $B_1 = 2\times10^{-6}$ T.

The rms value of the force experienced by a stationary charge Q = 10^{-4} C at z = 0 is closest to: **[NCERT : P L-207 | 9 April 2019, (I)]**

(a) 0.6 N (b) 0.1 N (c) 0.9 N (d) 3×10^{-2} N

97. A plane electromagnetic wave of frequency 50 MHz travels in free space along the positive x-direction. At a particular point in space and time, $\vec{E} = 6.3\,\hat{j}$ V/m. The corresponding magnetic field $\vec{B}$, at that point will be: **[NCERT : P L-207 | 9 April 2019, (I)]**

(a) $18.9\times10^{-8}\,\hat{k}$T (b) $2.1\times10^{-8}\,\hat{k}$T

(c) $6.3\times10^{-8}\,\hat{k}$T (d) $18.9\times10^{8}\,\hat{k}$T

98. 50 W/m^2 energy density of sunlight is normally incident on the surface of a solar panel. Some part of incident energy (25%) is reflected from the surface and the rest is absorbed. The force exerted on 1m^2 surface area will be close to (c = 3×10^8 m/s): **[9 April 2019, (II)]**

(a) 15×10^{-8} N (b) 20×10^{-8} N
(c) 10×10^{-8} N (d) 35×10^{-8} N

99. A plane electromagnetic wave travels in free space along the x-direction. The electric field component of the wave at a particular point of space and time is E = 6 Vm^{-1} along y-direction. Its corresponding magnetic field component, B would be: **[NCERT : P L-207 | 8 April 2019, (I)]**

(a) 2×10^{-8} T along z-direction
(b) 6×10^{-8} T along x-direction
(c) 6×10^{-8} T along z-direction
(d) 2×10^{-8} T along y-direction

100. The mean intensity of radiation on the surface of the Sun is about 10^8 W/m^2. The rms value of the corresponding magnetic field is closest to : **[12 Jan 2019, (II)]**

(a) 1 T (b) 10^2 T (c) 10^{-2} T (d) 10^{-4} T

101. An electromagnetic wave of intensity 50 Wm^{-2} enters in a medium of refractive index 'n' without any loss. The ratio of the magnitudes of electric fields, and the ratio of the magnitudes of magnetic fields of the wave before and after entering into the medium are respectively, given by: **[11 Jan 2019, (I)]**

(a) $\left(\frac{1}{\sqrt{n}}, \frac{1}{\sqrt{n}}\right)$ (b) $\left(\sqrt{n}, \sqrt{n}\right)$

(c) $\left(\sqrt{n}, \frac{1}{\sqrt{n}}\right)$ (d) $\left(\frac{1}{\sqrt{n}}, \sqrt{n}\right)$

102. A 27 mW laser beam has a cross-sectional area of 10 mm^2. The magnitude of the maximum electric field in this electromagnetic wave is given by :
[Given permittivity of space $\epsilon_0 = 9\times10^{-12}$ SI units, Speed of light c = 3×10^8 m/s] **[11 Jan 2019, (II)]**

(a) 2kV/m (c) 0.7kV/m
(b) 1kV/m (d) 1.4kV/m

103. If the magnetic field of a plane electromagnetic wave is given by (The speed of light = 3×10^8 m/s)

$$B = 100\times10^{-6}\sin\left[2\pi\times2\times10^{15}\left(t - \frac{x}{c}\right)\right]$$

then the maximum electric field associated with it is: **[NCERT : P L-207 | 10 Jan. 2019, (I)]**

(a) 6×10^4 N/C (b) 3×10^4 N/C
(c) 4×10^4 N/C (d) 4.5 10^4 N/C

104. An EM wave from air enters a medium. The electric fields are $\vec{E}_1 = E_{01}\hat{x}\cos\left[2\pi v\left(\frac{z}{c} - t\right)\right]$ in air and

$\vec{E}_2 = E_{02}\hat{x}\cos[k(2z - ct)]$ in medium, where the wave number k and frequency v refer to their values in air. The medium is nonmagnetic. If ϵ_{r_1} and ϵ_{r_2} refer to relative permittivities of air and medium respectively, which of the following options is correct? **[NCERT : P L-207 | 9 Jan 2019, (I)]**

(a) $\frac{\epsilon_{r_1}}{\epsilon_{r_2}} = 4$ (b) $\frac{\epsilon_{r_1}}{\epsilon_{r_2}} = 2$

(c) $\frac{\epsilon_{r_1}}{\epsilon_{r_2}} = \frac{1}{4}$ (d) $\frac{\epsilon_{r_1}}{\epsilon_{r_2}} = \frac{1}{2}$

105. Magnetic field in a plane electromagnetic wave is given by

$\vec{B} = B_0\sin(kx + \omega t)\hat{j}$T

Expression for corresponding electric field will be :
Where c is speed of light. **[NCERT : P L-206, 207 | Online April 8, 2017]**

(a) $\vec{E} = B_0 c\sin(kx + \omega t)\hat{k}$ V/m

(b) $\vec{E} = \frac{B_0}{c}\sin(kx+\omega t)\hat{k}\,V/m$

(c) $\vec{E} = -B_0 c\sin(kx+\omega t)\hat{k}\,V/m$

(d) $\vec{E} = B_0 c\sin(kx-\omega t)\hat{k}\,V/m$

106. For plane electromagnetic waves propagating in the z-direction, which one of the following combination gives the correct possible direction for $\vec{E}$ and $\vec{B}$ field respectively? **[NCERT : P L-206 | Online April 11, 2015]**

(a) $(2\hat{i}+3\hat{j})$ and $(\hat{i}+2\hat{j})$ (b) $(-2\hat{i}-3\hat{j})$ and $(3\hat{i}-2\hat{j})$

(c) $(3\hat{i}+4\hat{j})$ and $(4\hat{i}-3\hat{j})$ (d) $(\hat{i}+2\hat{j})$ and $(2\hat{i}-\hat{j})$

107. During the propagation of electromagnetic waves in a medium: **[2014]**

(a) Electric energy density is double of the magnetic energy density.

(b) Electric energy density is half of the magnetic energy density.

(c) Electric energy density is equal to the magnetic energy density.

(d) Both electric and magnetic energy densities are zero.

108. An electromagnetic wave of frequency 1×10^{14} hertz is propagating along z-axis. The amplitude of electric field is 4 V/m. If $\varepsilon_0 = 8.8 \times 10^{-12}\ C^2/N\text{-}m^2$, then average energy density of electric field will be: **[Online April 11, 2014]**

(a) $35.2 \times 10^{-10}\ J/m^3$ (b) $35.2 \times 10^{-11}\ J/m^3$

(c) $35.2 \times 10^{-12}\ J/m^3$ (d) $35.2 \times 10^{-13}\ J/m^3$

109. The magnetic field in a travelling electromagnetic wave has a peak value of 20 nT. The peak value of electric field strength is : **[NCERT : P L-207 | 2013]**

(a) 3 V/m (b) 6 V/m (c) 9 V/m (d) 12 V/m

110. Select the correct statement from the following : **[NCERT : P L-205 | Online April 9, 2013]**

(a) Electromagnetic waves cannot travel in vacuum.

(b) Electromagnetic waves are longitudinal waves.

(c) Electromagnetic waves are produced by charges moving with uniform velocity.

(d) Electromagnetic waves carry both energy and momentum as they propagate through space.

111. An electromagnetic wave in vacuum has the electric and magnetic field $\vec{E}$ and $\vec{B}$, which are always perpendicular to each other. The direction of polarization is given by $\vec{X}$ and that of wave propagation by $\vec{k}$. Then **[NCERT : P L-206 | 2012]**

(a) $\vec{X} \parallel \vec{B}$ and $\vec{k} \parallel \vec{B}\times\vec{E}$ (b) $\vec{X} \parallel \vec{E}$ and $\vec{k} \parallel \vec{E}\times\vec{B}$

(c) $\vec{X} \parallel \vec{B}$ and $\vec{k} \parallel \vec{E}\times\vec{B}$ (d) $\vec{X} \parallel \vec{E}$ and $\vec{k} \parallel \vec{B}\times\vec{E}$

112. An electromagnetic wave with frequency ω and wavelength λ travels in the $+y$ direction. Its magnetic field is along $+x$-axis. The vector equation for the associated electric field (of amplitude E_0) is **[NCERT : P L-206 | Online May 19, 2012]**

(a) $\vec{E} = -E_0\cos\left(\omega t + \frac{2\pi}{\lambda}y\right)\hat{x}$

(b) $\vec{E} = E_0\cos\left(\omega t - \frac{2\pi}{\lambda}y\right)\hat{x}$

(c) $\vec{E} = E_0\cos\left(\omega t - \frac{2\pi}{\lambda}y\right)\hat{z}$

(d) $\vec{E} = -E_0\cos\left(\omega t + \frac{2\pi}{\lambda}y\right)\hat{z}$

113. An electromagnetic wave of frequency $\nu = 3.0$ MHz passes from vacuum into a dielectric medium with permittivity $\epsilon = 4.0$. Then **[NCERT : P L-207 | 2004]**

(a) wave length is halved and frequency remains unchanged

(b) wave length is doubled and frequency becomes half

(c) wave length is doubled and the frequency remains unchanged

(d) wave length and frequency both remain unchanged.

114. Electromagnetic waves are transverse in nature is evident by **[NCERT : P L-205 | 2002]**

(a) polarization (b) interference

(c) reflection (d) diffraction

Topic 2 - Electromagnetic Spectrum

115. Arrange the following in the ascending order of wavelength (λ) **[NCERT : P L-211 | Jan. 24, 2025 (II)]**

(A) Microwaves (λ_1) (B) Ultraviolet rays (λ_2)

(C) Infrared rays (λ_3) (D) X-rays (λ_4)

Choose the most appropriate answer from the options given below

(a) $\lambda_4<\lambda_3<\lambda_2<\lambda_1$ (b) $\lambda_3<\lambda_4<\lambda_2<\lambda_1$

(c) $\lambda_4<\lambda_2<\lambda_3<\lambda_1$ (d) $\lambda_4<\lambda_3<\lambda_1<\lambda_2$

116. Match **List-I** with **List-II**

List-I EM-Wave	List-II Wavelength Range
(A) Infra-red	(I) $<10^{-3}$ nm
(B) Ultraviolet	(II) 400 nm to 1 nm
(C) X-rays	(III) 1 mm to 700 nm
(D) Gamma rays	(IV) 1 nm to 10^{-3} nm

Choose the correct answer from the options given below: **[NCERT : P L-211 | April 5, 2024 (II)]**

(a) (A)-(II), (B)-(I), (C)-(IV), (D)-(III)

(b) (A)-(III), (B)-(II), (C)-(IV), (D)-(I)

(c) (A)-(IV), (B)-(III), (C)-(II), (D)-(I)

(d) (A)-(I), (B)-(III), (C)-(II), (D)-(IV)

117. Arrange the following in the ascending order of wavelength :

(A) Gamma rays (λ_1) (B) x-ray (λ_2)

(C) Infrared waves (λ_3) (D) Microwaves (λ_4)

Choose the most appropriate answer from the options given below : **[NCERT : P L-211 | April 4, 2024 (II)]**

(a) $\lambda_4<\lambda_3<\lambda_1<\lambda_2$ (b) $\lambda_4<\lambda_3<\lambda_2<\lambda_1$

(c) $\lambda_1<\lambda_2<\lambda_3<\lambda_4$ (d) $\lambda_2<\lambda_1<\lambda_4<\lambda_3$

118. Match List I with List II of Electromagnetic waves with corresponding wavelength range :

List I	List II
(A) Microwave	(I) 400 nm to 1 nm
(B) Ultraviolet	(II) 1 nm to 10^{-3} nm
(C) X-Ray	(III) 1 mn to 700 nm
(D) Infra-red	(IV) 0.1 m to 1 mm

Choose the correct answer from the options given below: **[NCERT : P L-211 | April 15, 2023 (I)]**

(a) (A)-(I), (B)-(IV), (C)-(II), (D)-(III)

(b) (A)-(IV), (B)-(I), (C)-(II), (D)-(III)

(c) (A)-(IV), (B)-(II), (C)-(I), (D)-(III)

(d) (A)-(IV), (B)-(I), (C)-(III), (D)-(II)

119. Given below are two statement : one is labelled as Assertion A and the other is labelled as Reason R.

Assertion A : EM waves used for optical communication have longer wavelengths than that of microwave, employed in Radar technology.

Reason R : Infrared EM waves are more energetic than microwaves, (used in Radar)

[NCERT : P L-209 | April 12, 2023 (I)]

In the light of given statements, choose the correct answer from the options given below:

(a) A is false but R is true
(b) A is true but R is false
(c) Both A and R true but R is NOT the correct explanation of A
(d) Both A and R true and r is the correct explanation of A

120. The waves emitted when a metal target is bombarded with high energy electrons are

[NCERT : P L-211 | April 8, 2023 (II)]

(a) Microwaves (b) X-rays
(c) Infrared rays (d) Radio waves

121. Match List I with List II:

[NCERT : P L-211 | Feb. 1, 2023 (I)]

List I	List II
(A) Microwaves	(I) Radio active decay of the nucleus
(B) Gamma rays	(II) Rapid acceleration and deccleration of electron in aerials
(C) Radio waves	(III) Inner shell electrons
(D) X-rays	(IV) Klystron valve

Choose the correct answer from the options given below:

(a) (A) – (I), (B) – (III), (C) – (IV), (D) – (II)
(b) (A) – (IV), (B) – (III), (C) – (II), (D) – (I)
(c) (A) – (IV), (B) – (I), (C) – (II), (D) – (III)
(d) (A) – (I), (B) – (II), (C) – (III), (D) – (IV)

122. Match List-I with List-II.

[NCERT : P L-209, 210, 211 | Jan. 31, 2023 (II)]

List-I	List-II
A. Microwaves	I. Physiotherapy
B. UV rays	II. Treatment of cancer
C. Infra-red rays	III. Lasik eye surgery
D. X-rays	IV. Aircraft navigation

Choose the correct answer from the option given below:

(a) A-II, B-IV, C-III, D-I (b) A-IV, B-I, C-II, D-III
(c) A-IV, B-III, C-I, D-II (d) A-III, B-II, C-I, D-IV

123. Match List - I with List - II

[NCERT : P L-209, 210, 211 | July 29, 2022 (I)]

	List - I		List - II
(a)	UV rays	(i)	Diagnostic tool in medicine
(b)	X-rays	(ii)	Water purification
(c)	Microwave	(iii)	Communication, Radar
(d)	Infrared wave	(iv)	Improving visibility in foggy days

Choose the correct answer from the options given below :

(a) (a)-(iii), (b)-(ii), (c)-(i), (d)-(iv)
(b) (a)-(ii), (b)-(i), (c)-(iii), (d)-(iv)
(c) (a)-(ii), (b)-(iv), (c)-(iii), (d)-(i)
(d) (a)-(iii), (b)-(i), (c)-(ii), (d)-(iv)

124. Match List–I with List – II :

List – I	List – II
(A) Ultraviolet rays	(i) Study crystal structure
(B) Microwaves	(ii) Greenhouse effect
(C) Infrared waves	(iii) Sterilizing surgical instrument
(D) X-rays	(iv) Radar system

Choose the correct answer from the options given below:

[NCERT : P L-209, 210, 211 | June 27, 2022 (I)]

(a) (A) – (iii), (B) – (iv), (C) – (ii), (D) – (i)
(b) (A) – (iii), (B) – (i), (C) – (ii), (D) – (iv)
(c) (A) – (iv), (B) – (iii), (C) – (ii), (D) – (i)
(d) (A) – (iii), (B) – (iv), (C) – (i), (D) – (ii)

125. Which is the correct ascending order of wavelengths?

[NCERT : P L-211 | June 26, 2022 (II)]

(a) $\lambda_{visible} < \lambda_{X-ray} < \lambda_{gamma-ray} < \lambda_{microwave}$
(b) $\lambda_{gamma-ray} < \lambda_{X-ray} < \lambda_{visible} < \lambda_{microwave}$
(c) $\lambda_{X-ray} < \lambda_{gamma-ray} < \lambda_{visible} < \lambda_{microwave}$
(d) $\lambda_{microwave} < \lambda_{visible} < \lambda_{gamma-ray} < \lambda_{X-ray}$

126. The correct match between the entries in Column I and Column II are : **[NCERT : P L-211 | Sep. 05, 2020 (II)]**

Column-I Radiation	Column-II Wavelength
(A) Microwave	(i) 100 m
(B) Gamma rays	(ii) 10^{-15} m
(C) A.M. radio waves	(iii) 10^{10} m
(D) X-rays	(iv) 10^{-3} m

(a) (A)-(ii), (B)-(i), (C)-(iv), (D)-(iii)
(b) (A)-(i), (B)-(iii), (C)-(iv), (D)-(ii)
(c) (A)-(iii), (B)-(ii), (C)-(i), (D)-(iv)
(d) (A)-(iv), (B)-(ii), (C)-(i), (D)-(iii)

127. Chosse the correct option relating wavelengths of different parts of electromagnetic wave spectrum :

[NCERT : P L-211 | Sep. 04, 2020 (I)]

(a) $\lambda_{visible} < \lambda_{micro\ waves} < \lambda_{radio\ waves} < \lambda_{X-rays}$
(b) $\lambda_{radio\ waves} > \lambda_{micro\ waves} > \lambda_{visible} > \lambda_{x-rays}$
(c) $\lambda_{x-rays} < \lambda_{micro\ waves} < \lambda_{radio\ waves} < \lambda_{visible}$
(d) $\lambda_{visible} > \lambda_{x-rays} > \lambda_{radio\ waves} > \lambda_{micro\ waves}$

128. Given below in the left column are different modes of communication using the kinds of waves given in the right column.

[NCERT : P L-209, 210, 211 | 10 April 2019, (I)]

A. Optical Fibre Communication	P. Ultrasound
B. Radar	Q. Infrared Light
C. Sonar	R. Microwaves
D. Mobile Phones	S. Radio Waves

From the options given below, find the most appropriate match between entries in the left and the right column.

(a) A – Q, B – S, C – R, D – P
(b) A – S, B – Q, C – R, D – P
(c) A – Q, B – S, C – P, D – R
(d) A – R, B – P, C – S, D – Q

129. Arrange the following electromagnetic radiations per quantum in the order of increasing energy : **[NCERT : P L-209 | 2016]**

A : Blue light B : Yellow light
C : X-ray D : Radiowave.

(a) C, A, B, D (b) B, A, D, C
(c) D, B, A, C (d) A, B, D, C

130. Microwave oven acts on the principle of : **[NCERT : P L-209 | Online April 9, 2016]**

(a) giving rotational energy to water molecules
(b) giving translational energy to water molecules
(c) giving vibrational energy to water molecules
(d) transferring electrons from lower to higher energy levels in water molecule

131. Match List - I (Electromagnetic wave type) with List - II (Its association/application) and select the correct option from the choices given below the lists: **[NCERT : P L-209, 210, 211 | 2014]**

	List-I		List-II
1.	Infrared waves	(i)	To treat muscular strain
2.	Radio waves	(ii)	For broadcasting
3.	X-rays	(iii)	To detect fracture of bones
4.	Ultraviolet rays	(iv)	Absorbed by the ozone layer of the atmosphere

	1	2	3	4
(a)	(iv)	(iii)	(ii)	(i)
(b)	(i)	(ii)	(iv)	(iii)
(c)	(iii)	(ii)	(i)	(iv)
(d)	(i)	(ii)	(iii)	(iv)

132. If microwaves, X rays, infrared, gamma rays, ultra-violet, radio waves and visible parts of the electromagnetic spectrum are denoted by M, X, I, G, U, R and V then which of the following is the arrangement in ascending order of wavelength ? **[NCERT : P L-211 Online April 19, 2014]**

(a) R, M, I, V, U, X and G (b) M, R, V, X, U, G and I
(c) G, X, U, V, I, M and R (d) I, M, R, U, V, X and G

133. Match List I (Wavelength range of electromagnetic spectrum) with List II (Method of production of these waves) and select the correct option from the options given below the lists. **[NCERT : P L-211 Online April 9, 2014]**

	List I		List II
(1)	700 nm to 1 mm	(i)	Vibration of atoms and molecules.
(2)	1 nm to 400 nm	(ii)	Inner shell electrons in atoms moving from one energy level to a lower level.
(3)	$< 10^{-3}$ nm	(iii)	Radioactive decay of the nucleus.
(4)	1 mm to 0.1 m	(iv)	Magnetron valve.

(a) (1)-(iv), (2)-(iii), (3)-(ii), (4)-(i)
(b) (1)-(iii), (2)-(iv), (3)-(i), (4)-(ii)
(c) (1)-(ii), (2)-(iii), (3)-(iv), (4)-(i)
(d) (1)-(i), (2)-(ii), (3)-(iii), (4)-(iv)

134. Photons of an electromagnetic radiation has an energy 11 keV each. To which region of electromagnetic spectrum does it belong ? **[NCERT : P L-209, 210, 211 | Online April 9, 2013]**

(a) X-ray region
(b) Ultra violet region
(c) Infrared region
(d) Visible region

135. The frequency of *X*-rays; γ-rays and ultraviolet rays are respectively *a*, *b* and *c* then **[NCERT : P L-209 Online May 26, 2012]**

(a) $a < b; b > c$ (b) $a > b ; b > c$
(c) $a < b < c$ (d) $a = b = c$

ANSWER KEYS

1	(c)	15	(d)	29	(b)	43	(a)	57	(c)	71	(a)	85	(c)	99	(a)	113	(a)	127	(b)
2	(b)	16	(b)	30	(c)	44	(84)	58	(b)	72	(25)	86	(b)	100	(d)	114	(a)	128	(c)
3	(a)	17	(a)	31	(b)	45	(d)	59	(a)	73	(3)	87	(a)	101	(c)	115	(c)	129	(c)
4	(b)	18	(b)	32	(a)	46	(b)	60	(None)	74	(6)	88	(d)	102	(d)	116	(b)	130	(c)
5	(d)	19	(b)	33	(d)	47	(a)	61	(b)	75	(b)	89	(d)	103	(b)	117	(c)	131	(d)
6	(100)	20	(d)	34	(b)	48	(d)	62	(a)	76	(a)	90	(b)	104	(c)	118	(b)	132	(c)
7	(d)	21	(c)	35	(5)	49	(b)	63	(500)	77	(155)	91	(c)	105	(a)	119	(a)	133	(d)
8	(1200)	22	(c)	36	(b)	50	(a)	64	(c)	78	(667)	92	(c)	106	(b)	120	(b)	134	(a)
9	(b)	23	(b)	37	(4)	51	(b)	65	(c)	79	(2)	93	(b)	107	(c)	121	(c)	135	(a)
10	(d)	24	(c)	38	(b)	52	(43)	66	(2)	80	(137)	94	(a)	108	(c)	122	(c)		
11	(b)	25	(c)	39	(a)	53	(8)	67	(354)	81	(a)	95	(c)	109	(b)	123	(b)		
12	(a)	26	(a)	40	(a)	54	(b)	68	(c)	82	(c)	96	(a)	110	(d)	124	(a)		
13	(d)	27	(b)	41	(a)	55	(b)	69	(a)	83	(a)	97	(b)	111	(b)	125	(b)		
14	(d)	28	(c)	42	(a)	56	(c)	70	(d)	84	(d)	98	(b)	112	(c)	126	(d)		

9 Ray Optics and Optical Instruments

TREND ANALYSIS
JEE MAIN (2025-2021)

	2025				2024				2023				2022				2021			
	Session II		Session I		Session II		Session I		Session II		Session I		Session II		Session I		Session II		Session I	
	MCQs	NVQs	MCQs	NVQs	MCQs	NVQs	MCQs	NVQs	MCQs	NVQs	MCQs	NVQs	MCQs	NVQs	MCQs	NVQs	MCQs	NVQs	MCQs	NVQs
Topic 1	3	2	1	1	–	–	2	–	3	–	1	2	–	1	–	1	3	–	2	1
Topic 2	3	2	4	1	1	1	–	1	3	3	1	1	4	1	4	2	2	–	–	–
Topic 3	8	1	13	–	5	–	1	3	–	5	2	3	2	1	1	2	2	–	–	–
Topic 4	1	1	2	–	1	1	2	–	–	1	–	–	–	–	–	2	1	2	–	–
Total	15	6	20	2	7	2	5	4	6	9	4	6	6	3	5	7	8	2	2	1

How to Code Qns for Better & Faster Revision →

Type 1: Able to solve (No code Req)

Type 2: Solved with struggle or by seeing solution **(Use code NR means Needs Revision or orange stickers)**

Type 3: Tricky/ Could not solve on my own. Need help from Teacher/ Friend. **(Use code TQ means Tricky Question or red stickers)**

You can either hand write the code in boxes or else paste coloured stickers given at end of book.

Topic 1 - Plane Mirror, Spherical Mirror and Reflection of Light

1. A mirror is used to produce an image with magnification of $\frac{1}{4}$. If the distance between object and its image is 40 cm, then the focal length of the mirror is ____.

[NCERT : P L-225 | April 7, 2025 (II)]

(a) 10 cm (b) 12.7 cm (c) 10.7 cm (d) 15 cm

2. A finite size object is placed normal to the principal axis at a distance of 30 cm from a convex mirror of focal length 30 cm. A plane mirror is now placed in such a way that the image produced by both the mirrors coincide with each other. The distance between the two mirrors is :

[NCERT : P L-225 | April 4, 2025 (II)]

(a) 45 cm (b) 7.5 cm (c) 22.5 cm (d) 15 cm

3. When an object is placed 40 cm away from a spherical mirror an image of magnification $\frac{1}{2}$ is produced. To obtain an image with magnification of $\frac{1}{3}$, the object is to be moved:

[NCERT : P L-225 | April 4, 2025 (I)]

(a) 40 cm away from the mirror.
(b) 80 cm away from the mirror.
(c) 20 cm towards the mirror.
(d) 20 cm away from the mirror.

4. Distance between object and its image (magnified by $-\frac{1}{3}$) is 30 cm. The focal length of the mirror used is $\left(\frac{x}{4}\right)$ cm, where magnitude of value of x is ____.

[NCERT : P L-225 | April 4, 2025 (I)]

5. Two identical objects are placed in front of convex mirror and concave mirror having same radii of curvature of 12 cm, at same distance of 18 cm from the respective mirrors. The ratio of sizes of the images formed by convex mirror and by concave mirror is :

[NCERT : P L-225 | April 2, 2025 (II)]

(a) 1/2 (b) 2 (c) 3 (d) 1/3

6. A concave mirror produces an image of an object such that the distance between the object and image is 20 cm. If the magnification of the image is '–3', then the magnitude of the radius of curvature of the mirror is :

[NCERT : P L-225 | Jan. 28, 2025 (II)]

(a) 3.75 cm (b) 30 cm
(c) 7.5 cm (d) 15 cm

7. The driver sitting inside a parked car is watching vehicles approaching from behind with the help of his side view mirror, which is a convex mirror with radius of curvature R = 2 m. Another car approaches him from behind with a uniform speed of 90 km/hr. When the car is at a distance of 24 m from him, the magnitude of the acceleration of the image of the side view mirror is 'a'. The value of 100a is ______ m/s^2. [NCERT : P L-225 | Jan. 22, 2025 (I)]

8. A convex mirror of radius of curvature 30 cm forms an image that is half the size of the object. The object distance is: [NCERT : P L-225, 226 | Jan. 29, 2024 (I)]

(a) – 15 cm (b) – 45 cm (c) 45 cm (d) 15 cm

9. If the distance between object and its two times magnified virtual image produced by a curved mirror is 15 cm, the focal length of the mirror must be:

[NCERT : P L-225, 226 | Jan. 29, 2024 (II)]

(a) $\frac{10}{3}$ cm (b) – 12 cm (c) – 10 cm (d) 15 cm

10. When one light ray is reflected from a plane mirror with 30° angle of reflection, the angle of deviation of the ray after reflection is: [NCERT : P L-222 | April 11, 2023 (II)]

(a) 140° (b) 120° (c) 110° (d) 130°

11. An object is placed at a distance of 12 cm in front of a plane mirror. The virtual and erect image is formed by the mirror. Now the mirror is moved by 4 cm towards the stationary object. The distance by which the position of image would be shifted, will be:

[NCERT : P L-222 | April 10, 2023 (I)]

(a) 4 cm towards mirror (b) 8 cm towards mirror
(c) 8 cm away from mirror (d) 2 cm towards mirror

12. Two vertical parallel mirrors A and B are separated by 10 cm. A point object O is placed at a distance of 2 cm from mirror A. The distance of the second nearest image behind mirror A from the mirror A is ____ cm.

[NCERT : P L-222 | April 8, 2023 (I)]

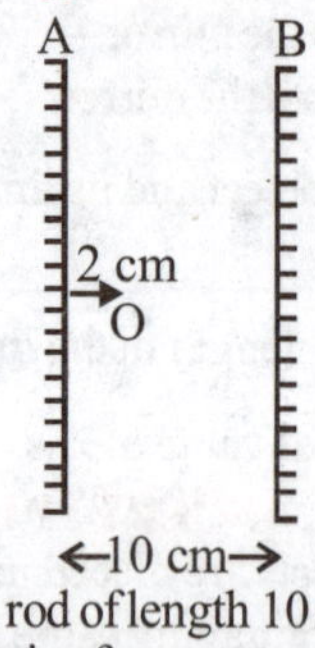

13. A thin cylindrical rod of length 10 cm is placed horizontally on the principal axis of a concave mirror of focal length 20 cm. The rod is placed in a such a way that mid point of the rod is at 40 cm from the pole of mirror. The length of the image formed by the mirror will be $\frac{x}{3}$ cm . The value of x is ______. [NCERT : P L-225 | Feb. 1, 2023 (I)]

14. Two objects A and B are placed at 15 cm and 25 cm from the pole in front of a concave mirror having radius of curvature 40 cm. The distance between images formed by the mirror is ______.

[NCERT : P L-225 | Feb. 1, 2023 (II)]

(a) 60 cm (b) 40 cm (c) 160 cm (d) 100 cm

15. In an experiment for estimating the value of focal length of converging mirror, image of an object placed at 40 cm from the pole of the mirror is formed at distance 120 cm from the pole of the mirror. These distances are measured with a modified scale in which there are 20 small divisions in 1 cm. The value of error in measurement of focal length of the mirror is 1/K cm. The value of K is ______ .

[NCERT : P L-225 | Jan. 30, 2023 (I)]

16. An object 'o' is placed at a distance of 100 cm in front of a concave mirror of radius of curvature 200 cm as shown in the figure. The object starts moving towards the mirror at a speed 2 cm/s. The position of the image from the mirror after 10 s will be at ______ cm.

[NCERT : P L-225 | July 28, 2022 (II)]

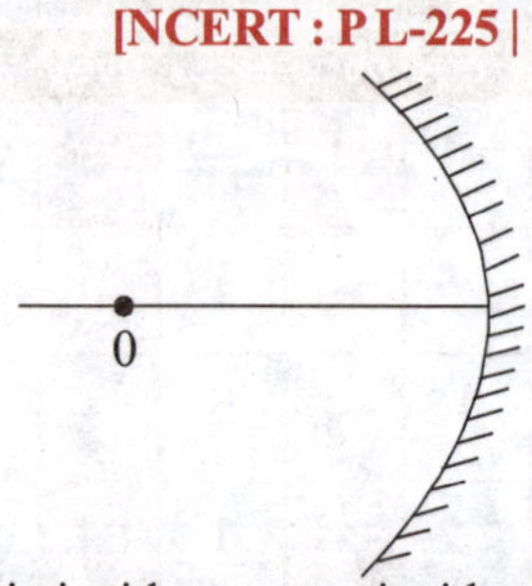

17. A light ray is incident, at an incident angle θ_1, on the system of two plane mirrors M_1 and M_2 having an inclination angle 75° between them (as shown in figure). After reflecting from mirror M_1 it gets reflected back by the mirror M_2 with an angle of reflection 30°. The total deviation of the ray will be .... degree.

[NCERT : P L-222 | June 26, 2022 (I)]

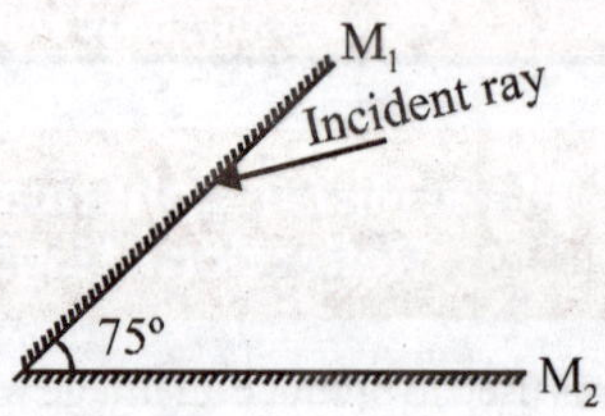

18. Two plane mirrors M_1 and M_2 are at right angle to each other shown. A point source 'P' is placed at 'a' and '2a' meter away from M_1 and M_2 respectively. The shortest distance between the images thus formed is :

(Take $\sqrt{5} = 2.3$) [NCERT : P L-222 | Aug. 31, 2021 (I)]

(a) 3a
(b) 4.6 a
(c) 2.3 a
(d) $2\sqrt{10}$a

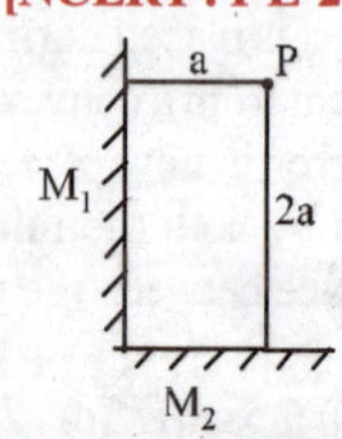

19. An object is placed beyond the centre of curvature C of the given concave mirror. If the distance of the object is d_1 from C and the distance of the image formed is d_2 from C, the radius of curvature of this mirror is : [Aug. 27, 2021 (I)]

(a) $\frac{2d_1d_2}{d_1-d_2}$ (b) $\frac{2d_1d_2}{d_1+d_2}$ (c) $\frac{d_1d_2}{d_1+d_2}$ (d) $\frac{d_1d_2}{d_1-d_2}$

20. Car B overtakes another car A at a relative speed of 40 ms^{-1}. How fast will the image of car B appear to move in the mirror of focal length 10 cm fitted in car A, when the car B is 1.9 m away from the car A? **[NCERT : P L-226 | Aug. 26, 2021 (I)]**

(a) 4 ms^{-1} (b) 0.2 ms^{-1} (c) 40 ms^{-1} (d) 0.1 ms^{-1}

21. The focal length f is related to the radius of curvature r of the spherical convex mirror by : **[NCERT : P L-224 | Feb. 24, 2021 (I)]**

(a) $f = +\frac{1}{2}r$ (b) $f = -r$

(c) $f = -\frac{1}{2}r$ (d) $f = r$

22. A short straight object of height 100 cm lies before the central axis of a spherical mirror whose focal length has absolute value $|f| = 40$ cm. The image of object produced by the mirror is of height 25 cm and has the same orientation of the object. One may conclude from the information: **[NCERT : P L-226 | Feb. 26, 2021 (I)]**

(a) Image is virtual, opposite side of convex mirror
(b) Image is virtual, opposite side of concave mirror
(c) Image is real, same side of convex mirror
(d) Image is real, same side of concave mirror

23. A point source of light S, placed at a distance 60 cm in front of the centre of a plane mirror of width 50 cm, hangs vertically on a wall. A man walks in front of the mirror along a line parallel to the mirror at a distance 1.2 m from it (see in the figure). The distance between the extreme points where he can see the image of the light source in the mirror is ______ cm.
[NCERT : P L-222 | 12 June 2019 (I); Feb. 26, 2021 (II)]

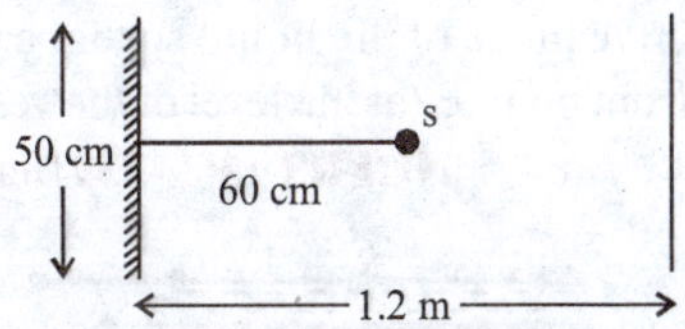

24. When an object is kept at a distance of 30 cm from a concave mirror, the image is formed at a distance of 10 cm from the mirror. If the object is moved with a speed of 9 cms^{-1}, the speed (in cms^{-1}) with which image moves at that instant is ________. **[NCERT : P L-226 | NA Sep. 03, 2020 (II)]**

25.

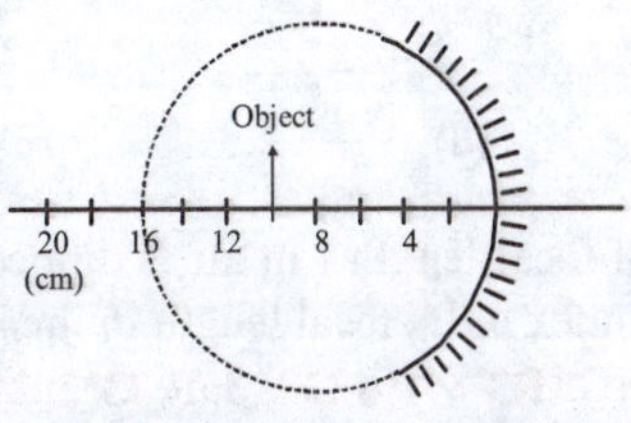

A spherical mirror is obtained as shown in the figure from a hollow glass sphere. If an object is positioned in front of the mirror, what will be the nature and magnification of the image of the object? (Figure drawn as schematic and not to scale) **[NCERT : P L-224 | Sep. 02, 2020 (I)]**

(a) Inverted, real and magnified
(b) Erect, virtual and magnified
(c) Erect, virtual and unmagnified
(d) Inverted, real and unmagnified

26. A concave mirror for face viewing has focal length of 0.4 m. The distance at which you hold the mirror from your face in order to see your image upright with a magnification of 5 is: **[NCERT : P L-225 | 9 April 2019 I]**

(a) 0.24m (b) 1.60m (c) 0.32m (d) 0.16m

27. Two plane mirrors are inclined to each other such that a ray of light incident on the first mirror (M_1) and parallel to the second mirror (M_2) is finally reflected from the second mirror (M_2) parallel to the first mirror (M_1). The angle between the two mirrors will be: **[NCERT : P L-222 | 9 Jan. 2019 II]**

(a) 45° (b) 60° (c) 75° (d) 90°

28. An object is gradually moving away from the focal point of a concave mirror along the axis of the mirror. The graphical representation of the magnitude of linear magnification (m) versus distance of the object from the mirror (x) is correctly given by
(Graphs are drawn schematically and are not to scale)
[NCERT : P L-226 | 8 Jan. 2020 II]

(a)
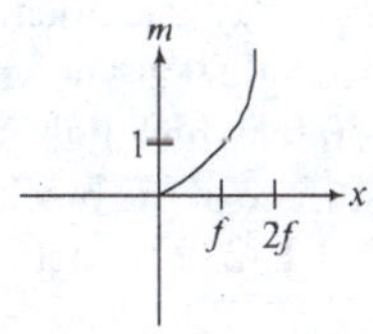

(b)
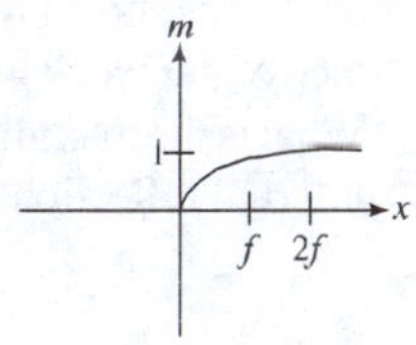

(c)
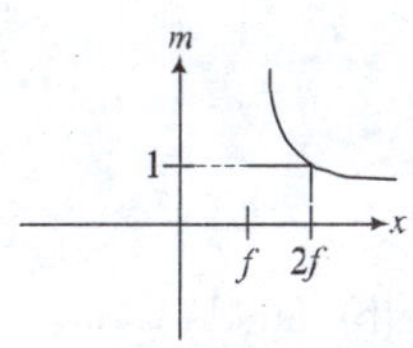

(d)
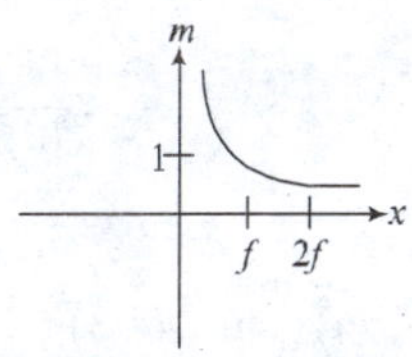

29. You are asked to design a shaving mirror assuming that a person keeps it 10 cm from his face and views the magnified image of the face at the closest comfortable distance of 25 cm. The radius of curvature of the mirror would then be : **[NCERT : P L-225 | Online April 10, 2015]**

(a) 60 cm (b) –24 cm (c) – 60 cm (d) 24 cm

30. A car is fitted with a convex side-view mirror of focal length 20 cm. A second car 2.8 m behind the first car is overtaking the first car at a relative speed of 15 m/s. The speed of the image of the second car as seen in the mirror of the first one is : **[NCERT : P L-225 | 2011]**

(a) $\frac{1}{15}$ m/s (b) 10 m/s (c) 15 m/s (d) $\frac{1}{10}$ m/s

31. To get three images of a single object, one should have two plane mirrors at an angle of **[NCERT : P L-222 | 2003]**

(a) 60° (b) 90° (c) 120° (d) 30°

32. If two plane mirrors are kept at 60° to each other, then the number of images formed by them is **[NCERT : P L-222 | 2002]**

(a) 5 (b) 6 (c) 7 (d) 8

Topic 2 - Refraction of Light at Plane Surface and Total Internal Reflection

33. Given below are two statements : one is labelled as **Assertion (A)** and the other is labelled as **Reason (R)**

Assertion (A) : Refractive index of glass is higher than that of air.

Reason (R) : Optical density of a medium is directly proportionate to its mass density which results in a proportionate refractive index.

In the light of the above statements, choose the **most appropriate answer** from the options given below :

[NCERT : P L-228 | April 7, 2025 (II)]

(a) **(A)** is not correct but **(R)** is correct
(b) Both **(A)** and **(R)** are correct and **(R)** is the correct explanation of **(A)**
(c) **(A)** is correct but **(R)** is not correct
(d) Both **(A)** and **(R)** are correct but **(R)** is **not** the correct explanation of **(A)**

34. A transparent block A having refractive index $\mu = 1.25$ is surrounded by another medium of refractive index $\mu = 1.0$ as shown in figure. A light ray is incident on the flat face of the block with incident angle θ as shown in figure. What is the maximum value of θ for which light suffers total internal reflection at the top surface of the block ?

[NCERT : P L-230 | April 7, 2025 (II)]

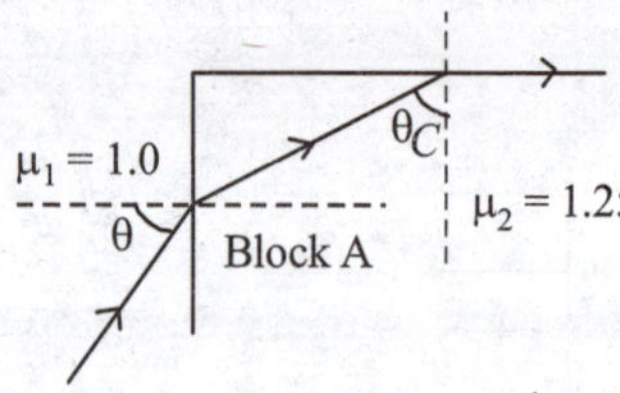

(a) $\tan^{-1}(4/3)$ (b) $\tan^{-1}(3/4)$
(c) $\sin^{-1}(3/4)$ (d) $\cos^{-1}(3/4)$

35. A container contains a liquid with refractive index of 1.2 up to a height of 60 cm and another liquid having refractive index 1.6 is added to height H above first liquid. If viewed from above, the apparent shift in the position of bottom of container is 40 cm. The value of H is___cm. (Consider liquids are immisible) **[NCERT : P L-229 | April 7, 2025 (I)]**

36. If an optical medium possesses a relative permeability of $\frac{10}{\pi}$ and relative permittivity of $\frac{1}{0.0885}$, then the velocity of light is greater in vacuum than that in this medium by ______ times.

$(\mu_0 = 4\pi \times 10^{-7}$ H/m, $\epsilon_0 = 8.85 \times 10^{-12}$ F/m, $c = 3 \times 10^8$ m/s) **[NCERT : P L-228 | April 4, 2025 (II)]**

37. A monochromatic light of frequency 5×10^{14} Hz travelling through air, is incident on a medium of refractive index '2'. Wavelength of the refracted light will be :

[NCERT : P L-228 | April 3, 2025 (II)]

(a) 300 nm (b) 600 nm (c) 400 nm (d) 500 nm

38. At the interface between two materials having refractive indices n_1 and n_2, the critical angle for reflection of an em wave is θ_{1C}. The n_2 material is replaced by another material having refractive index n_3, such that the critical angle at the interface between n_1 and n_3 materials is θ_{2C}. If $n_3 > n_2 > n_1$; $\frac{n_2}{n_3} = \frac{2}{5}$ and $\sin\theta_{2C} - \sin\theta_{1C} = \frac{1}{2}$, then θ_{1C} is

[NCERT : P L-230 | Jan. 29, 2025 (I)]

(a) $\sin^{-1}\left(\frac{1}{6n_1}\right)$ (b) $\sin^{-1}\left(\frac{2}{3n_1}\right)$
(c) $\sin^{-1}\left(\frac{5}{6n_1}\right)$ (d) $\sin^{-1}\left(\frac{1}{3n_1}\right)$

39. Two light beams fall on a transparent material block at point 1 and 2 with angle θ_1 and θ_2, respectively, as shown in figure. After refraction, the beams intersect at point 3 which is exactly on the interface at other end of the block. Given : the distance between 1 and 2, $d = 4\sqrt{3}$ cm and $\theta_1 = \theta_2 = \cos^{-1}\left(\frac{n_2}{2n_1}\right)$, where refractive index of the block $n_2 >$ refractive index of the outside medium n_1, then the thickness of the block is ______ cm.

[NCERT : P L-228, 229, 230 | Jan. 29, 2025 (I)]

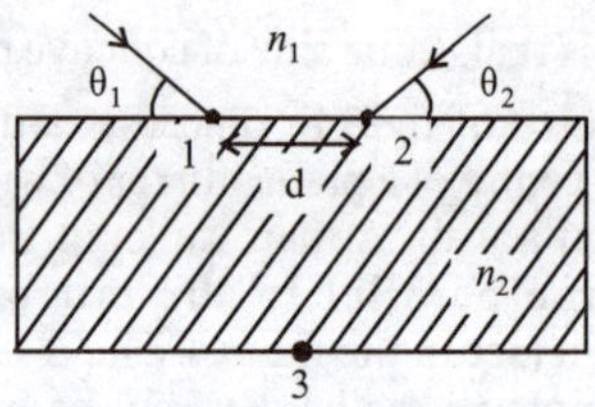

40. A hemispherical vessel is completely filled with a liquid of refractive index μ. A small coin is kept at the lowest point (O) of the vessel as shown in figure. The minimum value of the refractive index of the liquid so that a person can see the coin from point E (at the level of the vessel) is______.

[NCERT : P L-229 | Jan. 28, 2025 (I)]

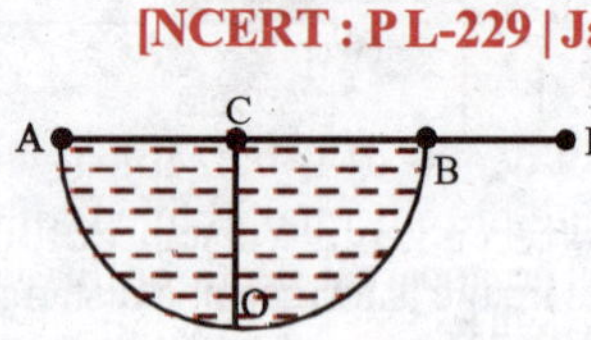

(a) $\sqrt{3}$ (b) $\frac{3}{2}$
(c) $\sqrt{2}$ (d) $\frac{\sqrt{3}}{2}$

41. A concave mirror of focal length f in air is dipped in a liquid of refractive index μ. Its focal length in the liquid will be **[NCERT : P L-228 | Jan. 23, 2025 (II)]**

(a) $\frac{f}{\mu}$ (b) $\frac{f}{(\mu - 1)}$
(c) μf (d) f

42. What is the lateral shift of a ray refracted through a parallel-sided glass slab of thickness 'h' in terms of the angle of incidence 'i' and angle of refraction 'r', if the glass slab is placed in air medium? **[NCERT : P L-229 | Jan. 23, 2025 (I)]**

(a) $\frac{h\tan(i-r)}{\tan r}$ (b) $\frac{h\cos(i-r)}{\sin r}$

(c) h (d) $\frac{h\sin(i-r)}{\cos r}$

43. Critical angle of incidence for a pair of optical media is 45°. The refractive indices of first and second media are in the ratio: **[NCERT : P L-230 | April 8, 2024 (I)]**

(a) $\sqrt{2}:1$ (b) $1:2$ (c) $1:\sqrt{2}$ (d) $2:1$

44. A light ray is incident on a glass slab of thickness $4\sqrt{3}$ cm and refractive index $\sqrt{2}$. The angle of incidence is equal to the critical angle for the glass slab with air. The lateral displacement of ray after passing through glass slab is ______ cm. **[NCERT : P L-229 | April 4, 2024 (II)]**

45. Two immiscible liquids of refractive indices $\frac{8}{5}$ and $\frac{3}{2}$ respectively are put in a beaker as shown in the figure. The height of each column is 6 cm. A coin is placed at the bottom of the beaker. For near normal vision, the apparent depth of the coin is $\frac{\alpha}{4}$ cm. The value of α is ______. **[NCERT : P L-229 | Jan. 27, 2024 (I)]**

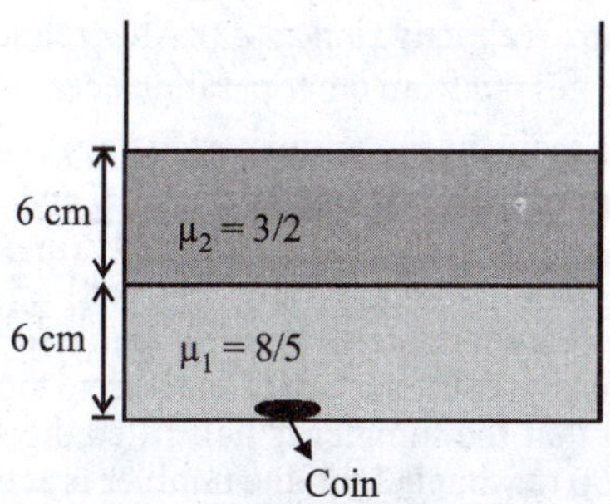

46. A vessel of depth 'd' is half filled with oil of refractive index n_1 and the other half is filled with water of refractive index n_2. The apparent depth of this vessel when viewed from above will be- **[NCERT : PL-229 | April 13, 2023 (I)]**

(a) $\frac{dn_1n_2}{(n_1+n_2)}$ (b) $\frac{d(n_1+n_2)}{2n_1n_2}$

(c) $\frac{dn_1n_2}{2(n_1+n_2)}$ (d) $\frac{2d(n_1+n_2)}{n_1n_2}$

47. A fish rising vertically upward with a uniform velocity of 8 ms^{-1}, observes that a bird is diving vertically downward towards the fish with the velocity of 12 ms^{-1}. If the refractive index of water is $\frac{4}{3}$, then the actual velocity of the diving bird to pick the fish, will be ______ ms^{-1}. **[NCERT : P L-229 | April 13, 2023 (I)]**

48. An ice cube has a bubble inside. When viewed from one side the apparent distance of the bubble is 12 cm. When viewed from the opposite side, the apparent distance of the bubble is observed as 4 cm. If the side of the ice cube is 24 cm, the refractive index of the ice cube is **[NCERT : P L-229 | April 12, 2023 (I)]**

(a) $\frac{4}{3}$ (b) $\frac{3}{2}$ (c) $\frac{2}{3}$ (d) $\frac{6}{5}$

49. As shown in the figure, a plane mirror is fixed at a height of 50 cm from the bottom of tank containing water $\left(\mu = \frac{4}{3}\right)$. The height of water in the tank is 8 cm. A small bulb is placed at the bottom of the water tank. The distance of image of the bulb formed by mirror from the bottom of the tank is ______ cm. **[NCERT : P L-229 | April 11, 2023 (II)]**

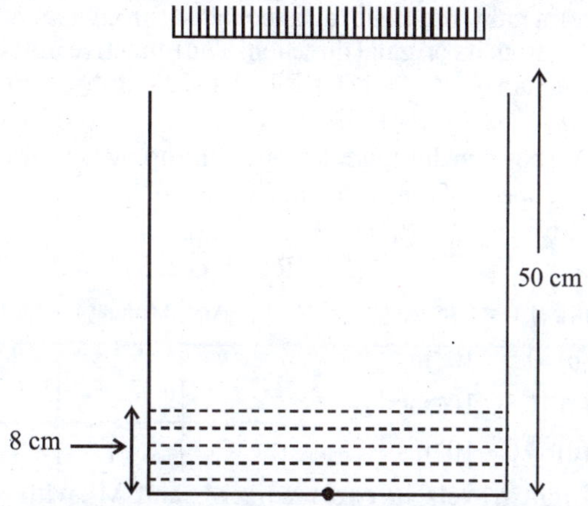

50. A monochromatic light wave with wavelength λ_1 and frequency v_1 in air enters another medium. If the angle of incidence and angle of refraction at the interface are 45° and 30° respectively, then the wavelength λ_2 and frequency v_2 of the refracted wave are: **[NCERT : P L-228 | April 6, 2023 (I)]**

(a) $\lambda_2 = \lambda_1, v_2 = \sqrt{2v_1}$ (b) $\lambda_2 = \frac{1}{\sqrt{2}}\lambda_1, v_2 = v_1$

(c) $\lambda_2 = \sqrt{2}\lambda_1, v_2 = v_1$ (d) $\lambda_2 = \lambda_1, v_2 = \frac{1}{\sqrt{2}}v_1$

51. A pole is vertically submerged in swimming pool, such that it gives a length of shadow 2.15 m within water when sunlight is incident at an angle of 30° with the surface of water. If swimming pool is filled to a height of 1.5 m, then the height of the pole above the water surface in centimetres is ($n_w = 4/3$) ______. **[NCERT : P L-228 | April 6, 2023 (I)]**

52. A microscope is focused on an object at the bottom of a bucket. If liquid with refractive index $\frac{5}{3}$ is poured inside the bucket, then microscope have to be raised by 30 cm to focus the object again. The height of the liquid in the bucket is : **[NCERT : P L-229 | Jan. 31, 2023 (II)]**

(a) 75 cm (b) 50 cm (c) 18 cm (d) 12 cm

53. A ray of light is incident from air on a glass plate having thickness $\sqrt{3}$ cm and refractive index $\sqrt{2}$. The angle of incidence of a ray is equal to the critical angle for glass-air interface. The lateral displacement of the ray when it passes through the plate is___ $\times 10^{-2}$cm. (given sin 15° = 0.26)
[NCERT : P L-229 | Jan. 25, 2023 (I)]

54. The X-Y plane be taken as the boundary between two transparent media M_1 and M_2. M_1 in $Z \geq 0$ has a refractive index of $\sqrt{2}$ and M_2 with $Z < 0$ has a refractive index of $\sqrt{3}$. A ray of light travelling in M_1 along the direction given by the vector $\vec{A} = 4\sqrt{3}\hat{i} - 3\sqrt{3}\hat{j} - 5\hat{k}$, is incident on the plane of separation. The value of difference between the angle of incident in M_1 and the angle of refraction in M_2 will be _____ degree. **[NCERT : P L-228 | July 29, 2022 (I)]**

55. Light enters from air into a given medium at an angle of 45° with interface of the air-medium surface. After refraction, the light ray is deviated through an angle of 15° from its original direction. The refractive index of the medium is : **[NCERT : P L-228 | July 29, 2022 (II)]**
(a) 1.732 (b) 1.333 (c) 1.414 (d) 2.732

56. As shown in the figure, after passing through the medium 1. The speed of light v_2 in medium 2 will be : (Given $c = 3 \times 10^8$ ms^{-1})
[NCERT : P L-228 | July 28, 2022 (I)]
(a) 1.0×10^8 ms^{-1}
(b) 0.5×10^8 ms^{-1}
(c) 1.5×10^8 ms^{-1}
(d) 3.0×10^8 ms^{-1}

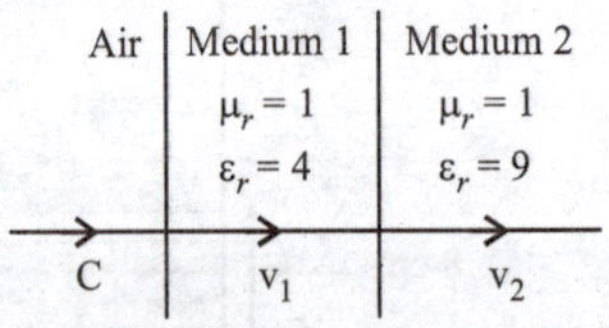

57. Light travels in two media M_1 and M_2 with speeds 1.5×10^8 ms^{-1} and 2.0×10^8 ms^{-1} respectively. The critical angle between them is:
[NCERT : P L-230 | July 26, 2022 (II)]
(a) $\tan^{-1}\left(\frac{3}{\sqrt{7}}\right)$ (b) $\tan^{-1}\left(\frac{2}{3}\right)$
(c) $\cos^{-1}\left(\frac{3}{4}\right)$ (d) $\sin^{-1}\left(\frac{2}{3}\right)$

58. Time taken by light to travel in two different materials A and B of refractive indices μ_A and μ_B of same thickness is t_1 and t_2 respectively. If $t_2 - t_1 = 5 \times 10^{-10}$ s and the ratio of μ_A to μ_B is 1 : 2. Then, the thickness of material, in meter is: (Given υ_A and υ_B are velocities of light in A and B materials respectively.) **[NCERT : P L-229 | July 25, 2022 (I)]**
(a) $5 \times 10^{-10} \upsilon_A$ m (b) 5×10^{-10} m
(c) 1.5×10^{-10} m (d) $5 \times 10^{-10} \upsilon_B$ m

59. The speed of light in media 'A' and 'B' are 2.0×10^{10} cm/s and 1.5×10^{10} cm/s respectively. A ray of light enters from the medium B to A at an incident angle 'θ'. If the ray suffers total internal reflection, then
[NCERT : P L-230 | June 29, 2022 (II)]
(a) $\theta = \sin^{-1}\left(\frac{3}{4}\right)$ (b) $\theta > \sin^{-1}\left(\frac{2}{3}\right)$
(c) $\theta < \sin^{-1}\left(\frac{3}{4}\right)$ (d) $\theta > \sin^{-1}\left(\frac{3}{4}\right)$

60. Consider a light ray travelling in air is incident into a medium of refractive index $\sqrt{2n}$. The incident angle is twice that of refracting angle. Then, the angle of incidence will be : **[NCERT : P L-228 | June 27, 2022 (I)]**
(a) $\sin^{-1}(\sqrt{n})$ (b) $\cos^{-1}\left(\sqrt{\frac{n}{2}}\right)$
(c) $\sin^{-1}(\sqrt{2n})$ (d) $2\cos^{-1}\left(\sqrt{\frac{n}{2}}\right)$

61. If a wave gets refracted into a denser medium, then which of the following is true?
[NCERT : P L-228 | June 27, 2022 (II)]
(a) wavelength speed and frequency decreases.
(b) wavelength increases, speed decreases and frequency remains constant.
(c) wavelength and speed decreases but frequency remains constant.
(d) wavelength, speed and frequency increases.

62. A small bulb is placed at the bottom of a tank containing water to a depth of $\sqrt{7}$ m. The refractive index of water is $\frac{4}{3}$. The area of the surface of water through which light from the bulb can emerge out is $x\pi$ m^2. The value of x is______. **[NCERT : P L-230 | June 26, 2022 (II)]**

63. The difference of speed of light in the two media A and B $(v_A - v_B)$ is 2.6×10^7 m/s. If the refractive index of medium B is 1.47, then the ratio of refractive index of medium B to medium A is : (Given : speed of light in vacuum $c = 3 \times 10^8$ ms^{-1}) **[NCERT : P L-228 | June 25, 2022 (I)]**
(a) 1.303 (b) 1.318 (c) 1.13 (d) 0.12

64. A ray of light is incident at an angle of incidence 60° on the glass slab of refractive index $\sqrt{3}$. After refraction, the light ray emerges out from other parallel faces and lateral shift between incident ray and emergent ray is $4\sqrt{3}$ cm. The thickness of the glass slab is _______ cm.
[NCERT : P L-229 | June 24, 2022 (II)]

65. A glass tumbler having inner depth of 17.5 cm is kept on a table. A student starts pouring water ($\mu = 4/3$) into it while looking at the surface of water from the above. When he feels that the tumbler is half filled, he stops pouring water. Up to what height, the tumbler is actually filled?
[NCERT : P L-229 | Sep. 1, 2021 (II)]
(a) 11.7 cm (b) 10 cm (c) 7.5 cm (d) 8.75 cm

66. A ray of light entering from air into a denser medium of refractive index $\frac{4}{3}$, as shown in figure. The light ray suffers total internal reflection at the adjacent surface as shown. The maximum value of angle θ should be equal to:
[NCERT : P L-230 | July 25, 2021 (II)]
(a) $\sin^{-1}\frac{\sqrt{7}}{3}$
(b) $\sin^{-1}\frac{\sqrt{5}}{4}$
(c) $\sin^{-1}\frac{\sqrt{7}}{4}$
(d) $\sin^{-1}\frac{\sqrt{5}}{3}$

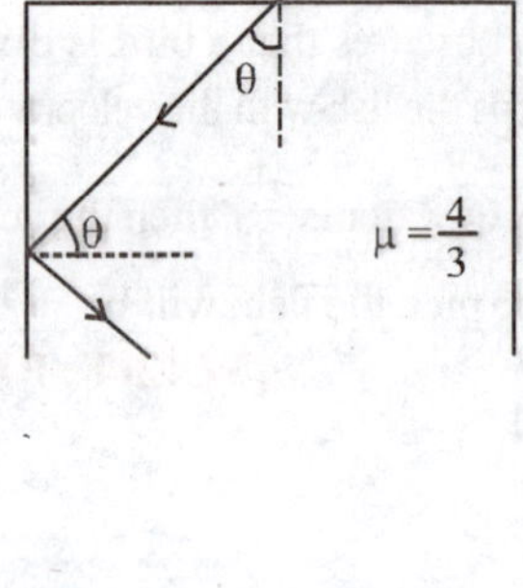

67. An observer can see through a small hole on the side of a jar (radius 15 cm) at a point at height of 15 cm from the bottom (see figure). The hole is at a height of 45 cm. When the jar is filled with a liquid up to a height of 30 cm the same observer can see the edge at the bottom of the jar. If the refractive index of the liquid is $N/100$, where N is an integer, the value of N is ________.

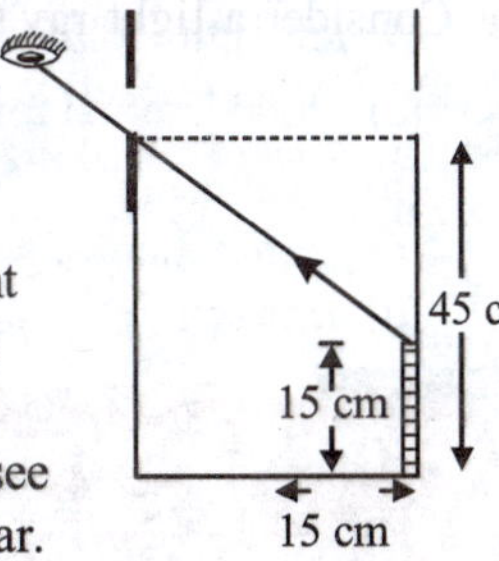

[NCERT : P L-229 | NA Sep. 03, 2020 (I)]

68. A light ray enters a solid glass sphere of refractive index $\mu = \sqrt{3}$ at an angle of incidence 60°. The ray is both reflected and refracted at the farther surface of the sphere. The angle (in degrees) between the reflected and refracted rays at this surface is ________. **[NCERT : P L-228 | NA Sep. 02, 2020 (II)]**

69. A vessel of depth 2h is half filled with a liquid of refractive index $2\sqrt{2}$ and the upper half with another liquid of refractive index $\sqrt{2}$. The liquids are immiscible. The apparent depth of the inner surface of the bottom of vessel will be: **[NCERT : P L-229 | 9 Jan. 2020 I]**

(a) $\frac{h}{\sqrt{2}}$ (b) $\frac{h}{2(\sqrt{2}+1)}$

(c) $\frac{h}{3\sqrt{2}}$ (d) $\frac{3}{4}h\sqrt{2}$

70. There is a small source of light at some depth below the surface of water (refractive index = $\frac{4}{3}$) in a tank of large cross sectional surface area. Neglecting any reflection from the bottom and absorption by water, percentage of light that emerges out of surface is (nearly):
[Use the fact that surface area of a spherical cap of height h and radius of curvature r is $2\pi rh$] **[NCERT : P L-229 | 9 Jan. 2020 II]**

(a) 21% (b) 34% (c) 17% (d) 50%

71. The critical angle of a medium for a specific wavelength, if the medium has relative permittivity 3 and relative permeability $\frac{4}{3}$ for this wavelength, will be: **[NCERT : P L-230 | 8 Jan. 2020 I]**

(a) 15° (b) 30° (c) 45° (d) 60°

72. A concave mirror has radius of curvature of 40 cm. It is at the bottom of a glass that has water filled up to 5 cm (see figure). If a small particle is floating on the surface of water, its image as seen, from directly above the glass, is at a distance d from the surface of water. The value of d is close to : **[NCERT : P L-229 | 12 Apr. 2019 I]**
(Refractive index of water = 1.33)

(a) 6.7 cm

(b) 13.4 cm

(c) 8.8 cm

(d) 11.7 cm

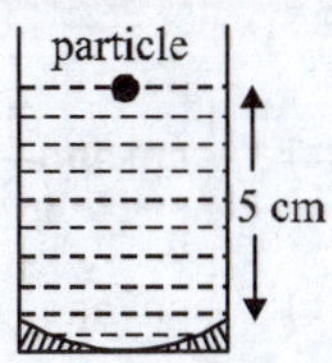

73. A ray of light AO in vacuum is incident on a glass slab at angle 60° and refracted at angle 30° along OB as shown in the figure. The optical path length of light ray from A to B is : **[NCERT : P L-228 | 10 Apr. 2019 I]**

(a) $\frac{2\sqrt{3}}{a} + 2b$

(b) $2a + \frac{2b}{3}$

(c) $2a + \frac{2b}{\sqrt{3}}$

(d) $2a + 2b$

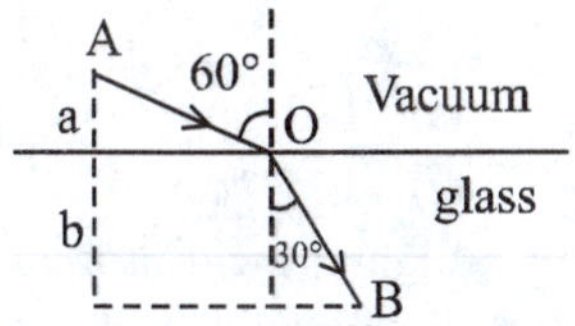

74. In figure, the optical fiber is $l = 2$ m long and has a diameter of $d = 20$ μm. If a ray of light is incident on one end of the fiber at angle $\theta_1 = 40°$, the number of reflections it makes before emerging from the other end is close to :
(refractive index of fiber is 1.31 and sin 40° = 0.64) **[NCERT : P L-231 | 8 April 2019 I]**

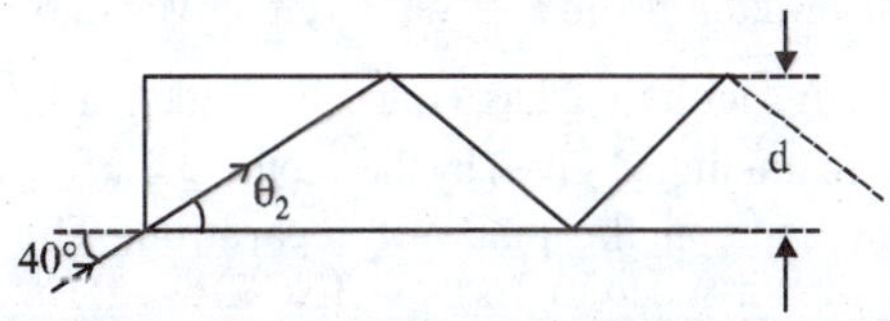

(a) 55000 (b) 66000 (c) 45000 (d) 57000

75. A light wave is incident normally on a glass slab of refractive index 1.5. If 4% of light gets reflected and the amplitude of the electric field of the incident light is 30 V/m, then the amplitude of the electric field for the wave propogating in the glass medium will be: **[12 Jan. 2019 I]**

(a) 30 V/m (b) 10 V/m (c) 24 V/m (d) 6 V/m

76. Let the refractive index of a denser medium with respect to a rarer medium be n_{12} and its critical angle be θ_C. At an angle of incidence A when light is travelling from denser medium to rarer medium, a part of the light is reflected and the rest is refracted and the angle between reflected and refracted rays is 90°. Angle A is given by : **[NCERT : P L-230 | Online April 8, 2017]**

(a) $\frac{1}{\cos^{-1}(\sin\theta_C)}$ (b) $\frac{1}{\tan^{-1}(\sin\theta_C)}$

(c) $\cos^{-1}(\sin\theta_C)$ (d) $\tan^{-1}(\sin\theta_C)$

77. A diver looking up through the water sees the outside world contained in a circular horizon. The refractive index of water is $\frac{4}{3}$, and the diver's eyes are 15 cm below the surface of water. Then the radius of the circle is: **[NCERT : P L-229 | 2005 (S), Online April 9, 2014]**

(a) $15 \times 3 \times \sqrt{5}$ cm (b) $15 \times 3\sqrt{7}$ cm

(c) $\frac{15 \times \sqrt{7}}{3}$ cm (d) $\frac{15 \times 3}{\sqrt{7}}$ cm

78. A printed page is pressed by a glass of water. The refractive index of the glass and water is 1.5 and 1.33, respectively. If the thickness of the bottom of glass is 1 cm and depth of water is 5 cm, how much the page will appear to be shifted if viewed from the top ? **[NCERT : P L-229 | Online April 25, 2013]**

(a) 1.033 cm (b) 3.581 cm
(c) 1.3533 cm (d) 1.90 cm

79. Light is incident from a medium into air at two possible angles of incidence (A) 20° and (B) 40°. In the medium light travels 3.0 cm in 0.2 ns. The ray will : **[NCERT : P L-230 | Online April 9, 2013]**

(a) suffer total internal reflection in both cases (A) and (B)
(b) suffer total internal reflection in case (B) only
(c) have partial reflection and partial transmission in case (B)
(d) have 100% transmission in case (A)

80. Let the x-y plane be the boundary between two transparent media. Medium 1 in $z \geq 0$ has a refractive index of $\sqrt{2}$ and medium 2 with $z < 0$ has a refractive index of $\sqrt{3}$. A ray of light in medium 1 given by the vector $\vec{A} = 6\sqrt{3}\hat{i} + 8\sqrt{3}\hat{j} - 10\hat{k}$ is incident on the plane of separation. The angle of refraction in medium 2 is: **[NCERT : P L-228 | 2011]**

(a) 45° (b) 60° (c) 75° (d) 30°

81. A beaker contains water up to a height h_1 and kerosene of height h_2 above water so that the total height of (water + kerosene) is $(h_1 + h_2)$. Refractive index of water is μ_1 and that of kerosene is μ_2. The apparent shift in the position of the bottom of the beaker when viewed from above is **[NCERT : P L-229 | 2011 RS]**

(a) $\left(1+\frac{1}{\mu_1}\right)h_1 - \left(1+\frac{1}{\mu_2}\right)h_2$ (b) $\left(1-\frac{1}{\mu_1}\right)h_1 + \left(1-\frac{1}{\mu_2}\right)h_2$

(c) $\left(1+\frac{1}{\mu_1}\right)h_2 - \left(1+\frac{1}{\mu_2}\right)h_1$ (d) $\left(1-\frac{1}{\mu_1}\right)h_2 + \left(1-\frac{1}{\mu_2}\right)h_1$

82. A transparent solid cylindrical rod has a refractive index of $\frac{2}{\sqrt{3}}$. It is surrounded by air. A light ray is incident at the mid-point of one end of the rod as shown in the figure.

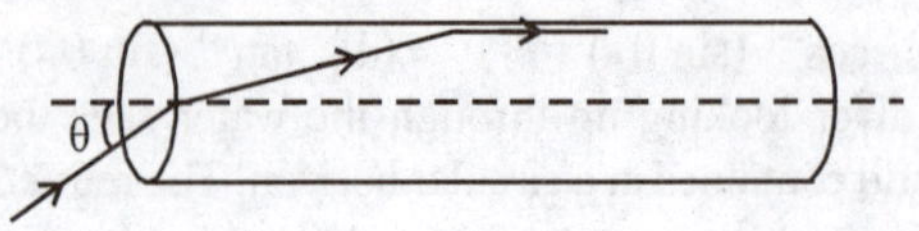

The incident angle θ for which the light ray grazes along the wall of the rod is : **[NCERT : P L-228, 230 | 2009]**

(a) $\sin^{-1}(\sqrt{3}/2)$ (b) $\sin^{-1}\left(\frac{2}{\sqrt{3}}\right)$

(c) $\sin^{-1}\left(\frac{1}{\sqrt{3}}\right)$ (d) $\sin^{-1}(1/2)$

83. Which of the following is used in optical fibres? **[2002]**

(a) total internal reflection
(b) scattering
(c) diffraction
(d) refraction

Topic 3 - Refraction at Curved Surface, Lenses and Power of Lens

84. A concave-convex lens of refractive index 1.5 and the radii of curvature of its surfaces are 30 cm and 20 cm, respectively. The concave surface is upwards and is filled with a liquid of refractive index 1.3. The focal length of the liquid-glass combination will be **[NCERT : P L-234 | April 8, 2025 (II)]**

(a) $\frac{500}{11}$ cm (b) $\frac{800}{11}$ cm (c) $\frac{700}{11}$ cm (d) $\frac{600}{11}$ cm

85. A convex lens of focal length 30 cm is placed in contact with a concave lens of focal length 20 cm. An object is placed at 20 cm to the left of this lens system. The distance of the image from the lens in cm is ______ **[NCERT : P L-235, 236 | April 8, 2025 (II)]**

(a) 30 (b) 45 (c) $\frac{60}{7}$ (d) 15

86. *A* lens having refractive index 1.6 has focal length of 12 cm, when it is in air. Find the focal length of the lens when it is placed in water. **[NCERT : P L-234 | April 7, 2025 (I)]**
(Take refractive index of water as 1.28)

(a) 355 mm (b) 288 mm (c) 555 mm (d) 655 mm

87. Two thin convex lenses of focal lengths 30 cm and 10 cm are placed coaxially, 10 cm apart. The power of this combination is : **[NCERT : P L-238 | April 7, 2025 (I)]**

(a) 5 D (b) 1 D (c) 20 D (d) 10 D

88. Light from a point source in air falls on a spherical glass surface (refractive index, $\mu = 1.5$ and radius of curvature = 50 cm). The image is formed at a distance of 200 cm from the glass surface inside the glass. The magnitude of distance of the light source from the glass surface is ______ m. **[NCERT : P L-233 | April 3, 2025 (II)]**

89. The radii of curvature for a thin convex lens are 10 cm and 15 cm respectively. The focal length of the lens is 12 cm. The refractive index of the lens material is **[NCERT : P L-234 | April 3, 2025 (I)]**

(a) 1.2 (b) 1.4 (c) 1.5 (d) 1.8

90. A bi – convex lens has radius of curvature of both the surfaces same as 1/6 cm. If this lens is required to be replaced by another convex lens having different radii of curvatures on both sides ($R_1 \neq R_2$), without any change in lens power then possible combination of R_1 and R_2 is : **[NCERT : P L-238 | April 2, 2025 (II)]**

(a) $\frac{1}{3}$ cm and $\frac{1}{3}$ cm (b) $\frac{1}{5}$ cm and $\frac{1}{7}$ cm

(c) $\frac{1}{3}$ cm and $\frac{1}{7}$ cm (d) $\frac{1}{6}$ cm and $\frac{1}{9}$ cm

91. A slanted object AB is placed on one side of convex lens as shown in the diagram. The image is formed on the opposite side. Angle made by the image with principal axis is : **[NCERT : P L-235, 236 | April 2, 2025 (I)]**

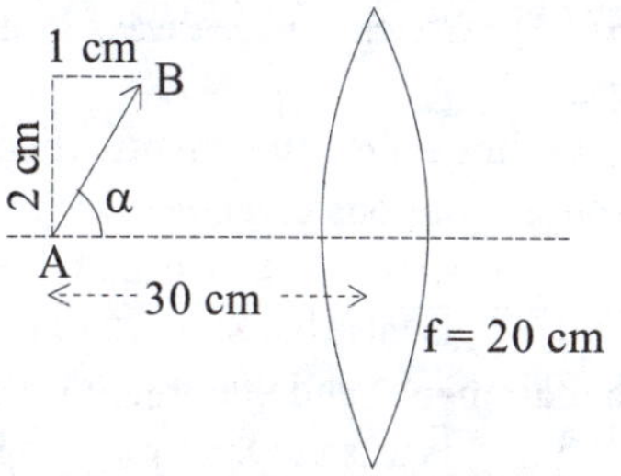

(a) $-\frac{\alpha}{2}$ (b) $-45°$ (c) $+45°$ (d) $-\alpha$

92.

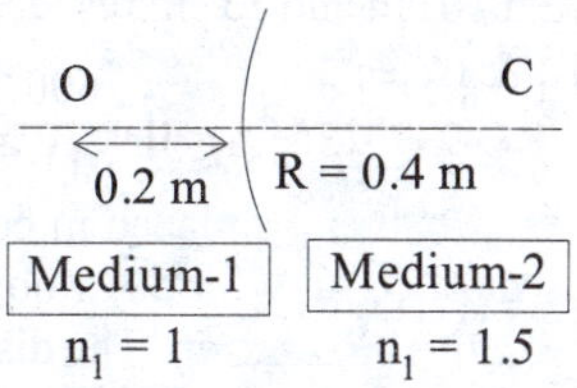

A spherical surface separates two media of refractive indices 1 and 1.5 as shown in figure. Distance of the image of an object 'O', is :
(C is the center of curvature of the spherical surface and R is the radius of curvature)
[NCERT : P L-233 | April 2, 2025 (I)]

(a) 0.24 m right to the spherical surface
(b) 0.4 m left to the spherical surface
(c) 0.24 m left to the spherical surface
(d) 0.4 m right to the spherical surface

93. Two identical symmetric double convex lenses of focal length f are cut into two equal parts L_1, L_2 by AB plane and L_3, L_4 by XY plane as shown in figure respectively. The ratio of focal lengths of lenses L_1 and L_3 is **[NCERT : P L-236 | Jan. 29, 2025 (II)]**

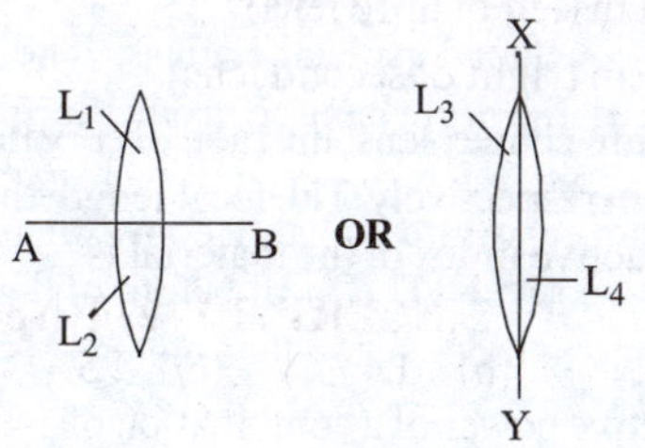

(a) 1 : 4 (b) 1 : 1
(c) 2 : 1 (d) 1 : 2

94.

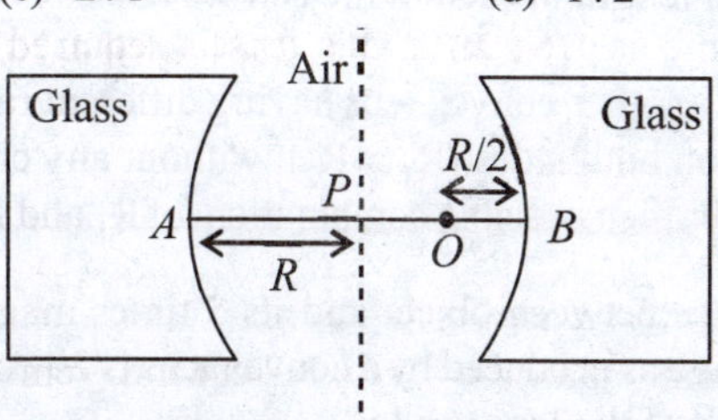

Two concave refracting surfaces of equal radii of curvature and refractive index 1.5 face each other in air as shown in figure. A point object O is placed midway, between P and B. The separation between the images of O, formed by each refracting surface is : **[NCERT : P L-233 | Jan. 29, 2025 (II)]**

(a) 0.214R (b) 0.114R (c) 0.411R (d) 0.124R

95. A convex lens mode of glass (refractive index = 1.5) has focal length 24 cm in air. When it is totally immersed in water (refractive index = 1.33), its focal length changes to **[NCERT : P L-234 | Jan. 29, 2025 (II)]**

(a) 72 cm (b) 96 cm (c) 24 cm (d) 48 cm

96. In a long glass tube, mixture of two liquids A and B with refractive indices 1.3 and 1.4 respectively, forms a convex refractive meniscus towards A. If an object placed at 13 cm from the vertex of the meniscus in A forms an image with a magnification of '–2' then the radius of curvature of meniscus is : **[NCERT : P L-233, 234 | Jan. 28, 2025 (II)]**

(a) 1 cm (b) $\frac{1}{3}$ cm (c) $\frac{2}{3}$ cm (d) $\frac{4}{3}$ cm

97. A photograph of a landscape is captured by a drone camera at a height of 18 km. The size of the camera film is 2 cm × 2 cm and the area of the landscape photographed is 400 km^2. The focal length of the lens in the drone camera is **[NCERT : P L-235, 236 | Jan. 24, 2025 (II)]**

(a) 1.8 cm (b) 2.8 cm
(c) 2.5 cm (d) 0.9 cm

98. What is the relative decrease in focal length of a lens for an increase in optical power by 0.1 D from 2.5 D? ['D' stands for dioptre] **[NCERT : P L-238 | Jan. 24, 2025 (I)]**

(a) 0.04 (b) 0.40 (c) 0.1 (d) 0.01

99. A thin plano convex lens made of glass of refractive index 1.5 is immersed in a liquid of refractive index 1.2. When the plane side of the lens is silver coated for complete reflection, the lens immersed in the liquid behaves like a concave mirror of focal length 0.2 m. The radius of curvature of the curved surface of the lens is **[NCERT : P L-234, 235, 236 | Jan. 24, 2025 (I)]**

(a) 0.15m (b) 0.10m (c) 0.20m (d) 0.25m

100. A plano-convex lens having radius of curvature of first surface 2 cm exhibits focal length of f_1 in air. Another plano-convex lens with first surface radius of curvature 3 cm has focal length of f_2, when it is immersed in a liquid of refractive index 1.2. If both the lenses are made of same glass of refractive index 1.5, the ratio of f_1 and f_2 will be **[NCERT : P L-234 | Jan. 24, 2025 (I)]**

(a) 3 : 5 (b) 1 : 3 (c) 1 : 2 (d) 2 : 3

101. A spherical surface of radius of curvature R, separates air from glass (refractive index = 1.5). The centre of curvature is in the glass medium, A point object 'O' placed in air on the optic axis of the surface, so that its real image is formed at 'I' inside glass. The line OI intersects the spherical surface at P and PO = PI. The distance PO equals to **[NCERT : P L-233 | Jan. 23, 2025 (I)]**

(a) 5R (b) 3R (c) 2R (d) 1.5R

102. Given a thin convex lens (refractive index μ_2), kept in a liquid (refractive index μ_1, $\mu_1 < \mu_2$) having radii of curvature $|R_1|$ and $|R_2|$. Its second surface is silver polished. Where should an object be placed on the optic axis so that a real and inverted image is formed at the same place?

[NCERT : P L-234 | Jan. 23, 2025 (I)]

(a) $\frac{\mu_1|R_1|.|R_2|}{\mu_2(|R_1|+|R_2|)-\mu_1|R_1|}$

(b) $\frac{\mu_1|R_1|.|R_2|}{\mu_2(|R_1|+|R_2|)-\mu_1|R_2|}$

(c) $\frac{\mu_1|R_1|.|R_2|}{\mu_2(2|R_1|+|R_2|)-\mu_1\sqrt{|R_1|.|R_2|}}$

(d) $\frac{(\mu_2+\mu_1)|R_1|}{(\mu_2-\mu_1)}$

103. A symmetric thin biconvex lens is cut into four equal parts by two planes AB and CD as shown in figure. If the power of original lens is 4D then the power of a part of the divided lens is **[NCERT : P L-238 | Jan. 22, 2025 (II)]**

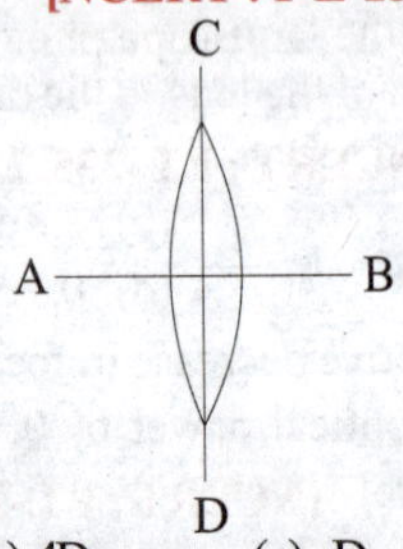

(a) 8D (b) 4D (c) D (d) 2D

104. In the diagram given below, there are three lenses formed. Considering negligible thickness of each of them as compared to $|R_1|$ and $|R_2|$, i.e., the radii of curvature for upper and lower surfaces of the glass lens, the power of the combination is **[NCERT : P L-234, 238 | Jan. 22, 2025 (I)]**

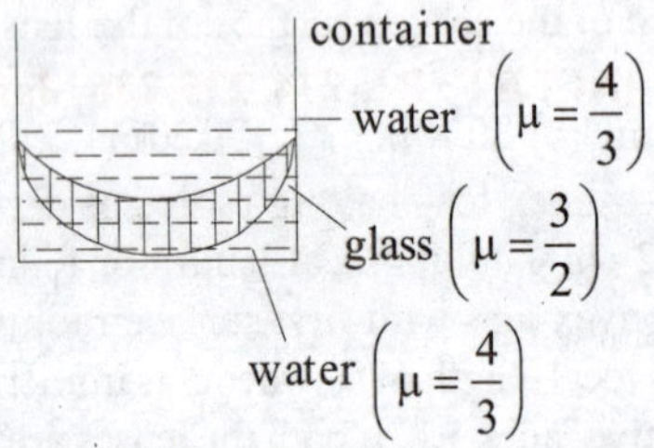

(a) $-\frac{1}{6}\left(\frac{1}{|R_1|}+\frac{1}{|R_2|}\right)$ (b) $-\frac{1}{6}\left(\frac{1}{|R_1|}-\frac{1}{|R_2|}\right)$

(c) $\frac{1}{6}\left(\frac{1}{|R_1|}+\frac{1}{|R_2|}\right)$ (d) $\frac{1}{6}\left(\frac{1}{|R_1|}-\frac{1}{|R_2|}\right)$

105. Given is a thin convex lens of glass (refractive index μ) and each side having radius of curvature R. One side is polished for complete reflection. At what distance from the lens, an object be placed on the optic axis so that the image gets formed on the object itself.

[NCERT : P L-234, 235, 236 | Jan. 22, 2025 (I)]

(a) R/μ (b) R/(2μ–3)
(c) μR (d) R/(2μ–1)

106. Given below are two statements:

Statement (I) : When an object is placed at the centre of curvature of a concave lens, image is formed at the centre of curvature of the lens on the other side.

Statement (II) : Concave lens always forms a virtual and erect image.

In the light of the above statements, choose the **correct** answer from the options given below:

[NCERT : P L-235 | April 9, 2024 (I)]

(a) **Statement I** is false but **Statement II** is true.
(b) Both **Statement I** and **Statement II** are false.
(c) **Statement I** is true but **Statement II** is false.
(d) Both **Statement I** and **Statement II** are true.

107. The following figure represents two biconvex lenses L_1 and L_2 having focal length 10 cm and 15 cm respectively. The distance between L_1 & L_2 is :

[NCERT : P L-237, 238 | April 9, 2024 (II)]

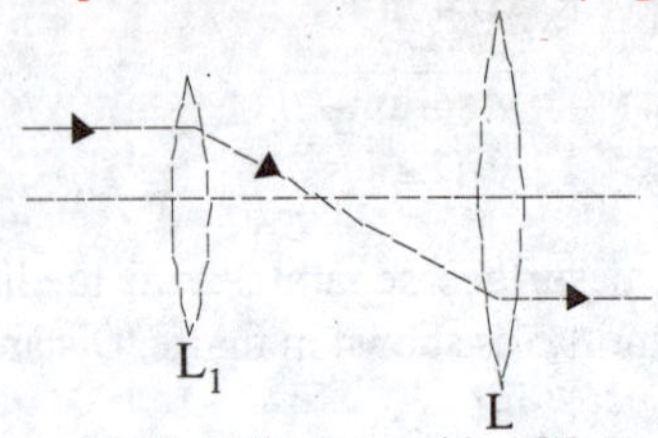

(a) 10 cm (b) 15 cm (c) 25 cm (d) 35 cm

108. The position of the image formed by the combination of lenses is : **[NCERT : P L-237, 238 | April 8, 2024 (II)]**

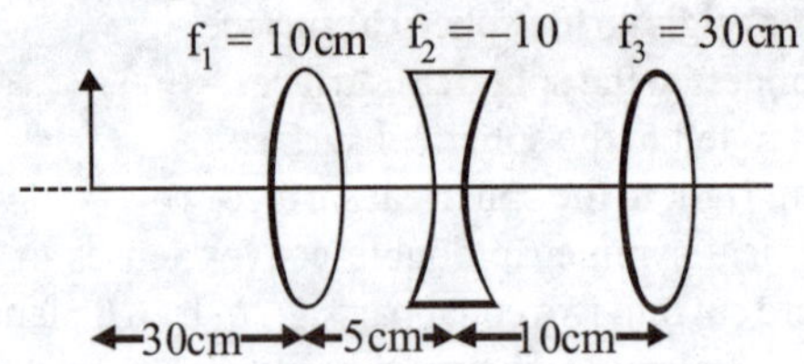

(a) 30 cm (right of third lens)
(b) 15 cm (left of second lens)
(c) 30 cm (left of third lens)
(d) 15 cm (right of second lens)

109. For the thin convex lens, the radii of curvature are at 15 cm and 30 cm respectively. The focal length the lens is 20 cm. The refractive index of the material is :

[NCERT : P L- 234 | April 6, 2024 (II)]

(a) 1.2 (b) 1.4 (c) 1.5 (d) 1.8

110. An effective power of a combination of 5 identical convex lenses which are kept in contact along the principal axis is 25 D. Focal length of each of the convex lens is :

[NCERT : P L-238 | April 4, 2024 (I)]

(a) 20 cm
(b) 50 cm
(c) 500 cm
(d) 25 cm

111. The distance between object and its 3 times magnified virtual image as produced by a convex lens is 20 cm. The focal length of the lens used is ______ cm.

[NCERT : P L-235, 236 | Feb. 1, 2024 (I)]

112. Light from a point source in air falls on a convex curved surface of radius 20 cm and refractive index 1.5. If the source is located at 100 cm from the convex surface, the image will be formed at____ cm from the object.

[NCERT : P L-233 | Jan. 31, 2024 (II)]

113. The distance between object and its two times magnified real image as produced by a convex lens is 45cm. The focal length of the lens used is _________ cm.

[NCERT : P L-235, 236 | Jan. 30, 2024 (I)]

114. A biconvex lens of refractive index 1.5 has a focal length of 20 cm in air. Its focal length when immersed in a liquid of refractive index 1.6 will be:

[NCERT : P L-234 | Jan. 29, 2024 (I)]

(a) – 16 cm (b) + 16 cm (c) + 160 cm (d) –160 cm

115. A bi convex lens of focal length 10 cm is cut in two identical parts along a plane perpendicular to the principal axis. The power of each lens after cut is ________ D.

[NCERT : P L-238 | April 13, 2023 (II)]

116. Two convex lenses of focal length 20 cm each are placed coaxially with a separation of 60 cm between them. The image of the distant object formed by the combination is at______ cm from the first lens.

[NCERT : P L-237, 238 | April 12, 2023 (I)]

117. The radius of curvature of each surface of a convex lens having refractive index 1.8 is 20 cm. The lens is now immersed in a liquid of refractive index 1.5. The ratio of power of lens in air to its power in the liquid will be x : 1. The value of x is _____.

[NCERT : P L-234 |April 11, 2023 (I)]

118. A point object, 'O' is placed in front of two thin symmetrical coaxial convex lenses L_1 and L_2 with focal length 24 cm and 9 cm respectively. The distance between two lenses is 10 cm and the object is placed 6 cm away from lens L_1 as shown in the figure. The distance between the object and the image formed by the system of two lenses is______cm. **[NCERT : P L-237, 238 |April 10, 2023 (II)]**

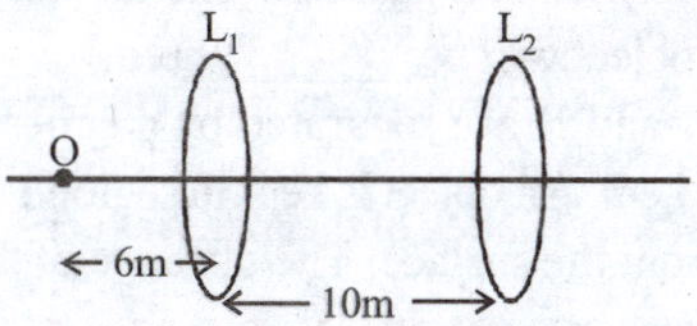

119. Two transparent media having refractive indices 1.0 and 1.5 are separated by a spherical refracting surface of radius of curvature 30 cm. The centre of curvature of surface is towards denser medium and a point object is placed on the principle axis in rarer medium at a distance of 15 cm from the pole of the surface. The distance of image from the pole of the surface is _______ cm.

[NCERT : P L-233 | April 8, 2023 (II)]

120. A person has been using spectacles of power-1.0 diopter for distant vision and a separate reading glass of power 2.0 diopters. What is the least distance of distinct vision for this person: **[NCERT : P L-235 | Jan. 30, 2023 (I)]**

(a) 10 cm (b) 40 cm
(c) 30 cm (d) 50 cm

121. An object is placed on the principal axis of convex lens of focal length 10 cm as shown. A plane mirror is placed on the other side of lens at a distance of 20 cm. The image produced by the plane mirror is 5 cm inside the mirror. The distance of the object from the lens is ____ cm.

[Jan. 25, 2023 (II)]

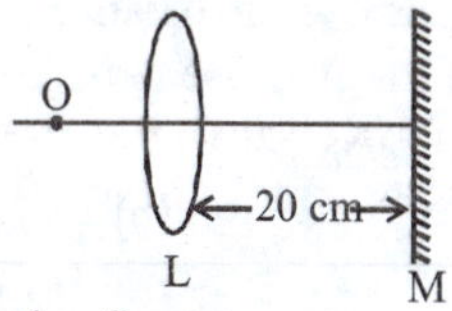

122. As shown in the figure, a combination of a thin plano concave lens and a thin plano convex lens is used to image an object placed at infinity. The radius of curvature of both the lenses is 30 cm and refraction index of the material for both the lenses is 1.75. Both the lenses are placed at distance of 40 cm from each other. Due to the combination, the image of the object is formed at distance x = _____ cm, from concave lens.

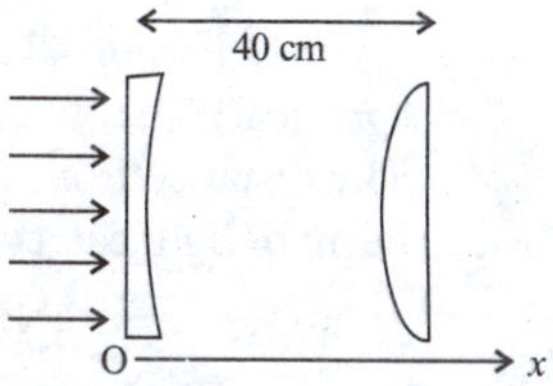

[NCERT : P L-234 | Jan. 24, 2023 (I)]

123. When a beam of white light is allowed to pass through convex lens parallel to principal axis, the different colours of light converge at different point on the principle axis after refraction. This is called : **[Jan. 24, 2023 (II)]**

(a) Scattering (b) Chromatic aberration
(c) Spherical aberration (d) Polarisation

124. A convex lens of refractive index 1.5 and focal length 18 cm in air is immersed in water. The change in focal length of the lens will be ________ cm.

(Given refractive index of water = $\frac{4}{3}$)

[NCERT : P L-234 | Jan. 24, 2023 (II)]

125. The power of a lens (biconvex) is 1.25 m^{-1} in particular medium. Refractive index of the lens is 1.5 and radii of curvature are 20 cm and 40 cm respectively. The refractive index of surrounding medium :

[NCERT : P L-234 | July 28, 2022 (II)]

(a) 1.0 (b) $\frac{9}{7}$ (c) $\frac{3}{2}$ (d) $\frac{4}{3}$

126. The graph between $\frac{1}{u}$ and $\frac{1}{v}$ for a thin convex lens in order to determine its focal length is plotted as shown in the figure. The refractive index of lens is 1.5 and its both the surfaces have same radius of curvature R. The value of R will be________ cm. **[NCERT : P L-235 | July 26, 2022 (I)]**
(where u = object distance, v = image distance)

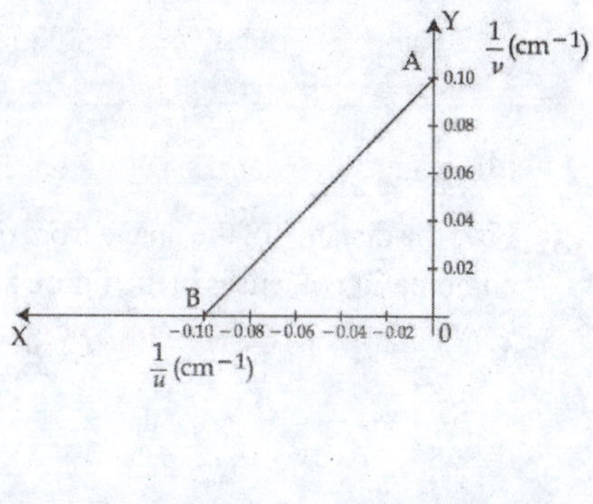

127. For an object placed at a distance 2.4 m from a lens, a sharp focused image is observed on a screen placed at a distance 12 cm from the lens. A glass plate of refractive index 1.5 and thickness 1 cm is introduced between lens and screen such that the glass plate plane faces parallel to the screen. By what distance should the object be shifted so that sharp focused image is observed again on the screen? **[NCERT : P L-235 | July 25, 2022 (II)]**

(a) 0.8m (b) 3.2m

(c) 1.2m (d) 5.6m

128. A parallel beam of light is allowed to fall on a transparent spherical globe of diameter 30 cm and refractive index 1.5. The distance from the centre of the globe at which the beam of light can converge is ______ mm. **[NCERT : P L-233 | June 29, 2022 (I)]**

129. A convex lens has power P. It is cut into two halves along its principal axis. Further one piece (out of the two halves) is cut into two halves perpendicular to the principal axis (as shown in figure). Choose the incorrect option for the reported pieces. **[NCERT : P L-236 | June 27, 2022 (II)]**

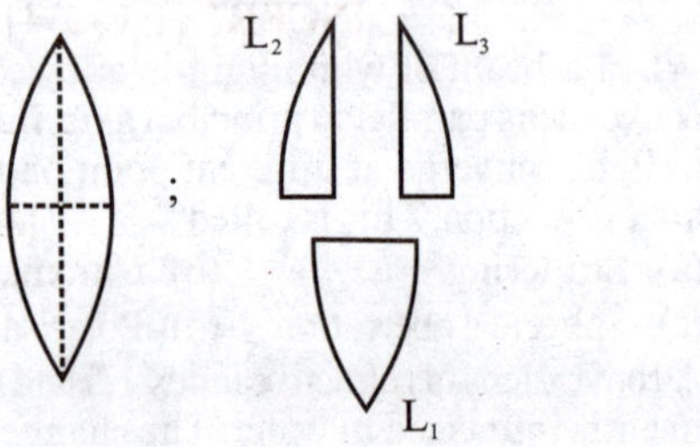

(a) Power of $L_1 = \frac{P}{2}$ (b) Power of $L_2 = \frac{P}{2}$

(c) Power of $L_3 = \frac{P}{2}$ (d) Power of $L_1 = P$

130. Two identical thin biconvex lenses of focal length 15 cm and refractive index 1.5 are in contact with each other. The space between the lenses is filled with a liquid of refractive index 1.25. The focal length of the combination is ______ cm. **[NCERT : P L-234 | June 24, 2022 (I)]**

131. An object is placed at the focus of concave lens having focal length *f*. What is the magnification and distance of the image from the optical centre of the lens? **[NCERT : P L-235, 236 | Aug. 31, 2021 (I)]**

(a) $1, \infty$

(b) Very high, ∞

(c) $\frac{1}{2}, \frac{f}{2}$

(d) $\frac{1}{4}, \frac{f}{4}$

132. Find the distance of the image from object O, formed by the combination of lenses in the figure : **[NCERT : P L-237, 238 | Aug. 27, 2021 (I)]**

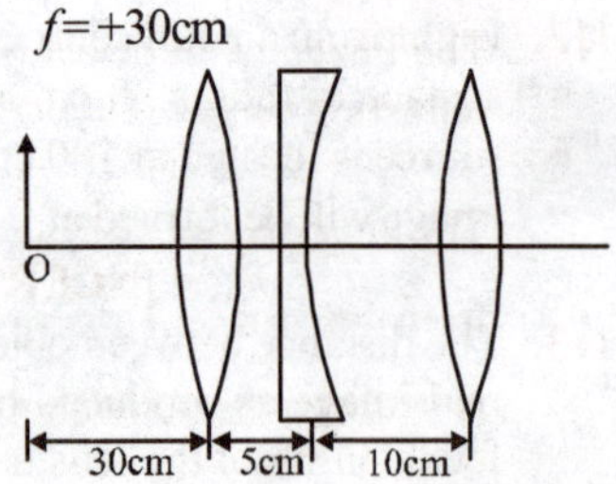

(a) 75 cm

(b) 10 cm

(c) 20 cm

(d) infinity

133. Curved surfaces of a plano-convex lens of refractive index μ_1 and a plano-concave lens of refractive index μ_2 have equal radius of curvature as shown in figure. Find the ratio of radius of curvature to the focal length of the combined lenses. **[NCERT : P L-234 | 10 April 2019 (I), 12 Jan 2019 (II); Aug. 27, 2021 (II)]**

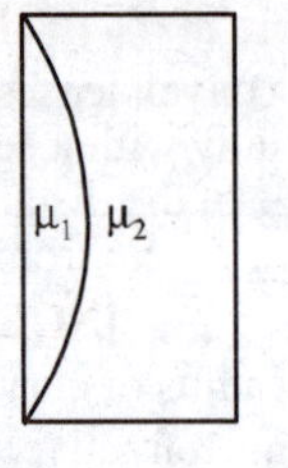

(a) $\frac{1}{\mu_2 - \mu_1}$

(b) $\mu_1 - \mu_2$

(c) $\frac{1}{\mu_1 - \mu_2}$

(d) $\mu_2 - \mu_1$

134. An object is placed at a distance of 12 cm from a convex lens. A convex mirror of focal length 15 cm is placed on other side of lens at 8 cm as shown in the figure. Image of object coincides with the object. **[NA, Aug. 26, 2021 (II)]**

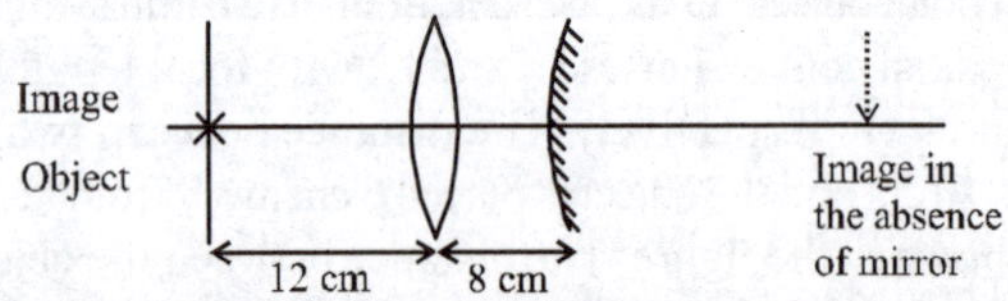

When the convex mirror is removed, a real and inverted image is formed at a position. The distance of the image from the object will be ______ (cm)

135. Region I and II are separated by a spherical surface of radius 25 cm. An object is kept in region I at a distance of 40 cm from the surface. The distance of the image from the surface is : **[NCERT : P L-233 | July 20, 2021 (I)]**

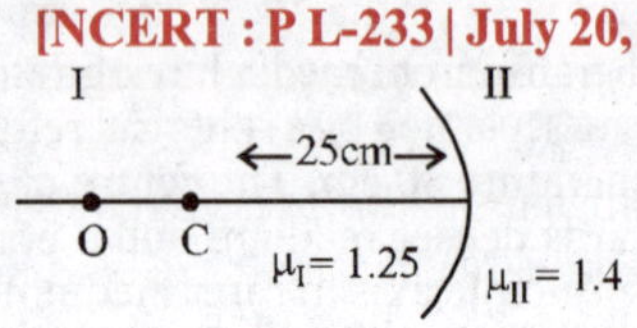

(a) 55.44 cm

(b) 9.52 cm

(c) 18.23 cm

(d) 37.58 cm

136. The thickness at the centre of a plano convex lens is 3 mm and the diameter is 6 cm. If the speed of light in the material of the lens is 2×10^8 ms^{-1}. The focal length of the lens is ______. **[NCERT : P L-234 | 2013 (S), March 17, 2021 (I)]**

(a) 1.5 cm (b) 0.30 cm

(c) 15 cm (d) 30 cm

137. The image of an object placed in air formed by a convex refracting surface is at a distance of 10 m behind the surface. The image is real and is at $\frac{2}{3}^{rd}$ of the distance of the object from the surface. The wavelength of light inside the surface is $\frac{2}{3}$ times the wavelength in air. The radius of the curved surface is $\frac{x}{13}$ m. The value of 'x' is ________.

[NCERT : P L-233 | March 17, 2021 (II)]

138. The same size images are formed by a convex lens when the object is placed at 20 cm or at 10 cm from the lens. The focal length of convex lens is ________ cm.

[NCERT : P L-235 | 9 April 2019 (II) (S); Feb. 25, 2021 (I)]

139. A point like object is placed at a distance of 1 m in front of a convex lens of focal length 0.5 m. A plane mirror is placed at a distance of 2 m behind the lens. The position and nature of the final image formed by the system is : **[Sep. 06, 2020 (I)]**

(a) 2.6 m from the mirror, real
(b) 1 m from the mirror, virtual
(c) 1 m from the mirror, real
(d) 2.6 m from the mirror, virtual

140. A double convex lens has power P and same radii of curvature R of both the surfaces. The radius of curvature of a surface of a plano-convex lens made of the same material with power $1.5\,P$ is : **[NCERT : P L-234 | Sep. 06, 2020 (II)]**

(a) $2R$ (b) $\frac{R}{2}$
(c) $\frac{3R}{2}$ (d) $\frac{R}{3}$

141. The distance between an object and a screen is 100 cm. A lens can produce real image of the object on the screen for two different positions between the screen and the object. The distance between these two positions is 40 cm. If the power of the lens is close to $\left(\frac{N}{100}\right)D$ where N is an integer, the value of N is ________.

[NCERT : P L-236 | NA Sep. 04, 2020 (II)]

142. A point object in air is in front of the curved surface of a *plano-convex* lens. The radius of curvature of the curved surface is 30 cm and the refractive index of the lens material is 1.5, then the focal length of the lens (in cm) is________. **[NCERT : P L-234 | NA 8 Jan. 2020 I]**

143. A thin lens made of glass (refractive index = 1.5) of focal length f = 16 cm is immersed in a liquid of refractive index 1.42. If its focal length in liquid is f_l, then the ratio f_l/f is closest to the integer: **[NCERT : P L-234 | 7 Jan. 2020 II]**

(a) 1 (b) 9
(c) 5 (d) 17

144. The graph shows how the magnification m produced by a thin lens varies with image distance v. What is the focal length of the lens used ?

[NCERT : P L-235 | 10 Apr. 2019 II]

(a) $\frac{b^2}{ac}$ (b) $\frac{b^2c}{a}$
(c) $\frac{a}{c}$ (d) $\frac{b}{c}$

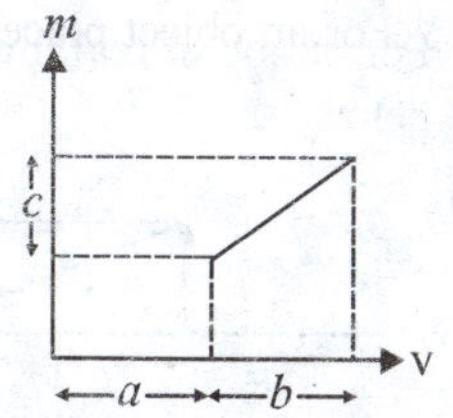

145. A thin convex lens L (refractive index = 1.5) is placed on a plane mirror M. When a pin is placed at A, such that OA = 18 cm, its real inverted image is formed at A itself, as shown in figure. When a liquid of refractive index μ_l is put between the lens and the mirror, the pin has to be moved to A', such that OA' = 27 cm, to get its inverted real image at A' itself. The value of μ_l will be: **[NCERT : P L-234 | 9 Apr. 2019 II]**

(a) $\frac{4}{3}$ (b) $\frac{3}{2}$
(c) $\sqrt{3}$ (d) $\sqrt{2}$

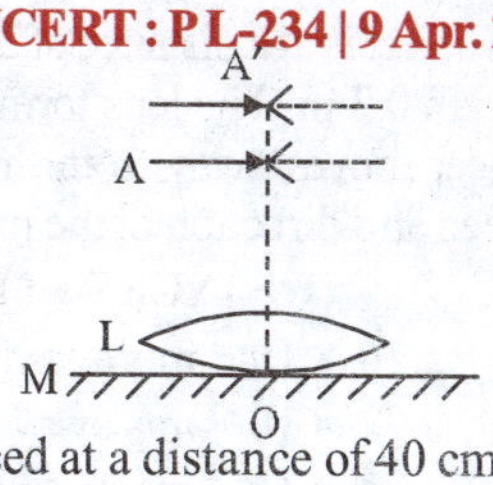

146. An upright object is placed at a distance of 40 cm in front of a convergent lens of focal length 20 cm. A convergent mirror of focal length 10 cm is placed at a distance of 60 cm on the other side of the lens. The position and size of the final image will be : **[8 April 2019 I]**

(a) 20 cm from the convergent mirror, same size as the object
(b) 40 cm from the convergent mirror, same size as the object
(c) 40 cm from the convergent lens, twice the size of the object
(d) 20 cm from the convergent mirror, twice the size of the object

147. A convex lens (of focal length 20 cm) and a concave mirror, having their principal axes along the same lines, are kept 80 cm apart from each other. The concave mirror is to the right of the convex lens. When an object is kept at a distance of 30 cm to the left of the convex lens, its image remains at the same position even if the concave mirror is removed. The maximum distance of the object for which this concave mirror, by itself would produce a virtual image would be : **[8 Apr. 2019 II]**

(a) 30 cm (b) 25 cm (c) 10 cm (d) 20 cm

148. What is the position and nature of image formed by lens combination shown in figure? (f_1, f_2 are focal lengths)

[NCERT : P L-237, 238 | 12 Jan. 2019 I]

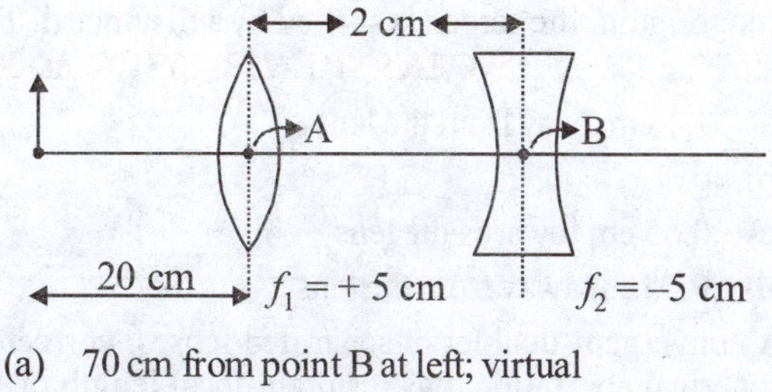

(a) 70 cm from point B at left; virtual
(b) 40 cm from point B at right; real
(c) $\frac{20}{3}$ cm from point B at right, real
(d) 70 cm from point B at right; real

149. Formation of real image using a biconvex lens is shown below : **[NCERT : P L-234 | 12 Jan. 2019 II]**

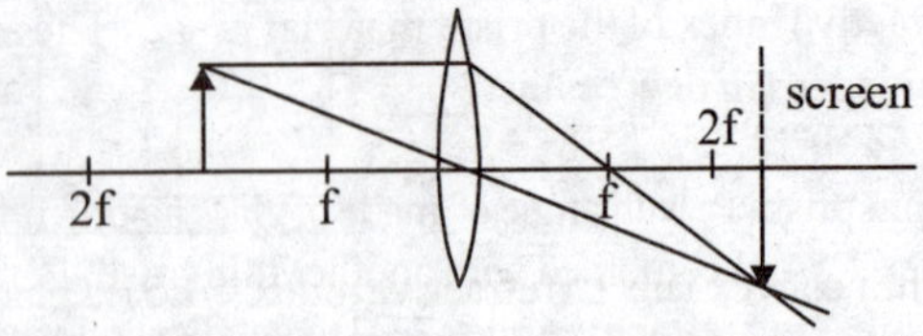

If the whole set up is immersed in water without disturbing the object and the screen positions, what will one observe on the screen ?

(a) Image disappears (b) Magnified image
(c) Erect real image (d) No change

150. An object is at a distance of 20 m from a convex lens of focal length 0.3 m. The lens forms an image of the object. If the object moves away from the lens at a speed of 5m/s, the speed and direction of the image will be : **[NCERT : P L-235, 236 | 11 Jan. 2019 I]**

(a) 2.26×10^{-3} m/s away from the lens
(b) 0.92×10^{-3} m/s away from the lens
(c) 3.22×10^{-3} m/s towards the lens
(d) 1.16×10^{-3} m/s towards the lens

151. A plano convex lens of refractive index μ_1 and focal length f_1 is kept in contact with another plano concave lens of refractive index μ_2 and focal length f_2 If the radius of curvature of their spherical faces is R each and $f_1 = 2f_2$, then μ_1 and μ_2 are related as: **[NCERT : P L-234 | 10 Jan. 2019 I]**

(a) $\mu_1 + \mu_2 = 3$ (b) $2\mu_1 - \mu_2 = 1$
(c) $3\mu_2 - 2\mu_1 = 1$ (d) $2\mu_2 - \mu_1 = 1$

152. The eye can be regarded as a single refracting surface. The radius of curvature of this surface is equal to that of cornea (7.8 mm). This surface separates two media of refractive indices 1 and 1.34. Calculate the distance from the refracting surface at which a parallel beam of light will come to focus. **[NCERT : P L-233 | 10 Jan. 2019 II]**

(a) 1 cm (b) 2 cm
(c) 4.0 cm (d) 3.1 cm

153. A convex lens is put 10 cm from a light source and it makes a sharp image on a screen, kept 10 cm from the lens. Now a glass block (refractive index 1.5) of 1.5 cm thickness is placed in contact with the light source. To get the sharp image again, the screen is shifted by a distance d. Then d is: **[NCERT : P L-229, 235 | 9 Jan. 2019 I]**

(a) 1.1 cm away from the lens
(b) 0
(c) 0.55 cm towards the lens
(d) 0.55 cm away from the lens

154. A convergent doublet of separated lenses, corrected for spherical aberration, has resultant focal length of 10cm. The separation between the two lenses is 2cm. The focal lengths of the component lenses **[Online April 15, 2018]**

(a) 18cm, 20cm (b) 10cm, 12cm
(c) 12cm, 14cm (d) 16cm, 18cm

155. In an experiment a convex lens of focal length 15 cm is placed coaxially on an optical bench in front of a convex mirror at a distance of 5 cm from it. It is found that an object and its image coincide, if the object is placed at a distance of 20 cm from the lens. The focal length of the convex mirror is : **[Online April 9, 2017]**

(a) 27.5 cm (b) 20.0 cm (c) 25.0 cm (d) 30.5 cm

156. A hemispherical glass body of radius 10 cm and refractive index 1.5 is silvered on its curved surface. A small air bubble is 6 cm below the flat surface inside it along the axis. The position of the image of the air bubble made by the mirror is seen : **[Online April 10, 2016]**

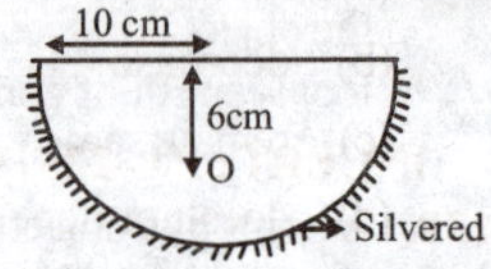

(a) 14 cm below flat surface
(b) 20 cm below flat surface
(c) 16 cm below flat surface
(d) 30 cm below flat surface

157. A thin convex lens of focal length 'f' is put on a plane mirror as shown in the figure. When an object is kept at a distance 'a' from the lens - mirror combination, its image is formed at a distance $\frac{a}{3}$ in front of combination. The value of 'a' is : **[Online April 11, 2015]**

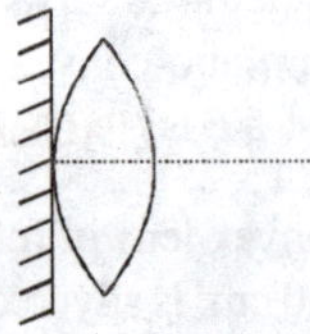

(a) 3f
(b) $\frac{3}{2}f$
(c) f
(d) 2f

158. A thin convex lens made from crown glass $\left(\mu = \frac{3}{2}\right)$ has focal length f. When it is measured in two different liquids having refractive indices $\frac{4}{3}$ and $\frac{5}{3}$, it has the focal lengths f_1 and f_2 respectively. The correct relation between the focal lengths is: **[NCERT : P L-234 | 2014]**

(a) $f_1 = f_2 < f$
(b) $f_1 > f$ and f_2 becomes negative
(c) $f_2 > f$ and f_1 becomes negative
(d) f_1 and f_2 both become negative

159. An object is located in a fixed position in front of a screen. Sharp image is obtained on the screen for two positions of a thin lens separated by 10 cm. The size of the images in two situations are in the ratio 3 : 2. What is the distance between the screen and the object? **[NCERT : P L-236 | Online April 11, 2014]**

(a) 124.5 cm (b) 144.5 cm (c) 65.0 cm (d) 99.0 cm

160. The image of an illuminated square is obtained on a screen with the help of a converging lens. The distance of the square from the lens is 40 cm. The area of the image is 9 times that of the square. The focal length of the lens is : **[NCERT : P L-235 | Online April 22, 2013]**

(a) 36 cm (b) 27 cm
(c) 60 cm (d) 30 cm

161. An object at 2.4 m in front of a lens forms a sharp image on a film 12 cm behind the lens. A glass plate 1 cm thick, of refractive index 1.50 is interposed between lens and film with its plane faces parallel to film. At what distance (from lens) should object shifted to be in sharp focus of film? **[NCERT : P L-229, 235 | 2012]**

(a) 7.2 m (b) 2.4 m
(c) 3.2 m (d) 5.6 m

162. When monochromatic red light is used instead of blue light in a convex lens, its focal length will

(a) increase **[NCERT : P L-234 | 2011 RS]**
(b) decrease
(c) remain same
(d) does not depend on colour of light

163. Two lenses of power $-15\ D$ and $+5\ D$ are in contact with each other. The focal length of the combination is **[NCERT : P L-238 | 2007]**

(a) +10 cm (b) –20 cm
(c) –10 cm (d) +20 cm

164. A thin glass (refractive index 1.5) lens has optical power of $-5\ D$ in air. Its optical power in a liquid medium with refractive index 1.6 will be **[NCERT : P L-234 | 2005]**

(a) $-1D$ (b) $1D$
(c) $-25D$ (d) $25D$

165. A plano convex lens of refractive index 1.5 and radius of curvature 30 cm. Is silvered at the curved surface. Now this lens has been used to form the image of an object. At what distance from this lens an object be placed in order to have a real image of size of the object **[NCERT : P L-234 | 2004]**

(a) 60 cm (b) 30 cm
(c) 20 cm (d) 80 cm

Topic 4 - Refraction of Light through a Prism

166. Consider following statements for refraction of light through prism, when angle of deviation is minimum. **[NCERT : P L-239, 240 | April 3, 2025 (I)]**

(A) The refracted ray inside prism becomes parallel to the base.
(B) Larger angle prisms provide smaller angle of minimum deviation.
(C) Angle of incidence and angle of emergence becomes equal.
(D) There are always two sets of angle of incidence for which deviation will be same except at minimum deviation setting.
(E) Angle of refraction becomes double of prism angle.

Choose the **correct** answer from the options given below.

(a) A, C and D Only (b) B, C and D Only
(c) A, B and E Only (d) B, D and E Only

167. A ray of light suffers minimum deviation when incident on a prism having angle of the prism equal to 60°. The refractive index of the prism material is $\sqrt{2}$. The angle of incidence (in degrees) is ______. **[NCERT : P L-239, 240 | April 2, 2025 (II)]**

168. A thin prism P_1 with angle 4° made of glass having refractive index 1.54, is combined with another thin prism P_2 made of glass having refractive index 1.72 to get dispersion without deviation. The angle of the prism P_2 in degrees is **[NCERT : P L-239, 240 | Jan. 28, 2025 (I)]**

(a) 4 (b) 3
(c) $\frac{16}{3}$ (d) 1.5

169. The refractive index of the material of a glass prism is $\sqrt{3}$. The angle of minimum deviation is equal to the angle of the prism. What is the angle of the prism? **[NCERT : P L-239, 240 | Jan. 23, 2025 (II)]**

(a) 50° (b) 60°
(c) 58° (d) 48°

170. The refractive index of prism is $\mu = \sqrt{3}$ and the ratio of the angle of minimum deviation to the angle of prism is one. The value of angle of prism is ______°. **[NCERT : P L-240 | April 6, 2024 (I)]**

171. Given below are two statements :

Statement I : When the white light passed through a prism, the red light bends lesser than yellow and violet.

Statement II : The refractive indices are different for different wavelengths in dispersive medium.

In the light of the above statements, choose the correct answer from the options given below : **[NCERT : P L-240 | April 5, 2024 (II)]**

(a) Both Statement I and Statement II are true.
(b) Statement I is true but Statement II is false.
(c) Both Statement I and Statement II are false
(d) Statement I is false but Statement II is true.

172. The refractive index of a prism with apex angle A is cot A/2. The angle of minimum deviation is : **[NCERT : P L-240 | Jan. 31, 2024 (I)]**

(a) $\delta_m = 180° - 4A$ (b) $\delta_m = 180° - 3A$
(c) $\delta_m = 180° - 2A$ (d) $\delta_m = 180° - A$

173. If the refractive index of the material of a prism is $\cot\left(\frac{A}{2}\right)$, where A is the angle of prism then the angle of minimum deviation will be **[NCERT : P L-240 | Jan. 27, 2024 (I)]**

(a) $\pi - 2A$ (b) $\frac{\pi}{2} - 2A$
(c) $\pi - A$ (d) $\frac{\pi}{2} - A$

174. The refractive index of a transparent liquid filled in an equilateral hollow prism is $\sqrt{2}$. The angle of minimum deviation for the liquid will be ______°.

[NCERT : P L-240 | April 15, 2023 (I)]

175. The refractive index of an equilateral prism is $\sqrt{2}$. The angle of emergence under minimum deviation position of prism, in degree, is _____.

[NCERT : P L-240 | June 30, 2022 (I)]

176. The refracting angle of a prism is A and refractive index of the material of the prism is cot (A/2). Then the angle of minimum deviation will be -

[NCERT : P L-240 | June 28, 2022 (I)]

(a) $180-2A$ (b) $90-A$
(c) $180+2A$ (d) $180-3A$

177. Cross–section view of a prism is the equilateral triangle ABC in the figure. The minimum deviation is observed using this prism when the angle of incidence is equal to the prism angle. The time taken by light to travel from P (midpoint of BC) to A is______ $\times 10^{-10}$ s.

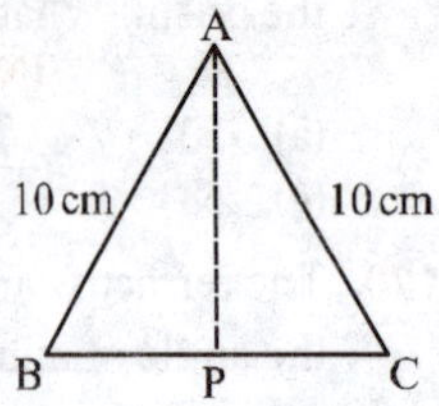

(Given, speed of light in vacuum = 3×10^8 m/s and $\cos 30° = \frac{\sqrt{3}}{2}$) **[NCERT : PL-240 | NA, Aug. 31, 2021 (II)]**

178. A prism of refractive index μ and angle of prism A is placed in the position of minimum angle of deviation. If minimum angle of deviation is also A, then in terms of refractive index, A = __________.

[NCERT : P L-240 | July 25, 2021 (II)]

(a) $2\cos^{-1}\left(\frac{\mu}{2}\right)$ (b) $\sin^{-1}\left(\frac{\mu}{2}\right)$
(c) $\sin^{-1}\left(\sqrt{\frac{\mu-1}{2}}\right)$ (d) $\cos^{-1}\left(\frac{\mu}{2}\right)$

179. A ray of light passing through a prism $(\mu=\sqrt{3})$ suffers minimum deviation. It is found that the angle of incidence is double the angle of refraction within the prism. Then, the angle of prism is ________ (in degrees).

[NCERT : P L-240 | NA, July 22, 2021 (II)]

180. A monochromatic light is incident at a certain angle on an equilateral triangular prism and suffers minimum deviation. If the refractive index of the material of the prism is $\sqrt{3}$, then the angle of incidence is :

[NCERT : P L-239 | 11 Jan. 2019 II]

(a) 90° (b) 30°
(c) 60° (d) 45°

181. A ray of light is incident at an angle of 60° on one face of a prism of angle 30°. The emergent ray of light makes an angle of 30° with incident ray. The angle made by the emergent ray with second face of prism will be:

[NCERT : P L-239 | Online April 16, 2018]

(a) 30° (b) 90°
(c) 0° (d) 45°

182. Monochromatic light is incident on a glass prism of angle A. If the refractive index of the material of the prism is μ, a ray, incident at an angle θ, on the face AB would get transmitted through the face AC of the prism provided :

[NCERT : P L-240 | 2015]

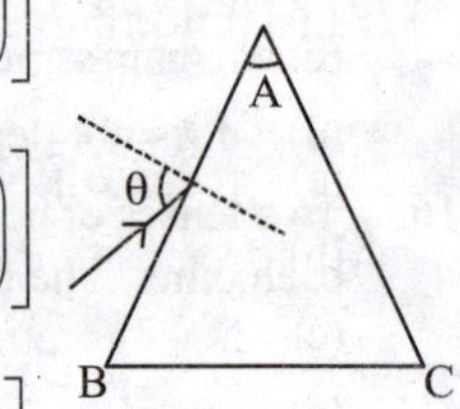

(a) $\theta > \cos^{-1}\left[\mu \sin\left(A + \sin^{-1}\left(\frac{1}{\mu}\right)\right)\right]$

(b) $\theta < \cos^{-1}\left[\mu \sin\left(A + \sin^{-1}\left(\frac{1}{\mu}\right)\right)\right]$

(c) $\theta > \sin^{-1}\left[\mu \sin\left(A - \sin^{-1}\left(\frac{1}{\mu}\right)\right)\right]$

(d) $\theta < \sin^{-1}\left[\mu \sin\left(A - \sin^{-1}\left(\frac{1}{\mu}\right)\right)\right]$

183. A beam of light consisting of red, green and blue colours is incident on a right-angled prism on face *AB*. The refractive indices of the material for the above red, green and blue colours are 1.39, 1.44 and 1.47 respectively. A person looking on surface *AC* of the prism will see

[NCERT : P L-232 | Online May 26, 2012]

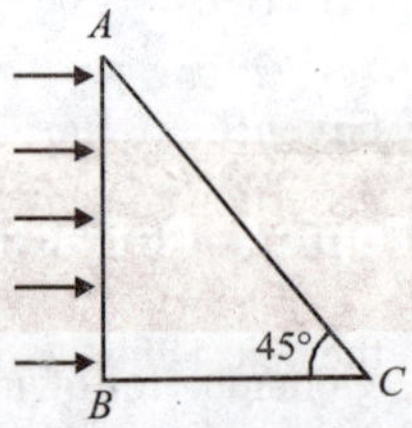

(a) no light
(b) green and blue colours
(c) red and green colours
(d) red colour only

184. The refractive index of a glass is 1.520 for red light and 1.525 for blue light. Let D_1 and D_2 be angles of minimum deviation for red and blue light respectively in a prism of this glass. Then, **[NCERT : P L-240 | 2006]**

(a) $D_1 < D_2$
(b) $D_1 = D_2$
(c) D_1 can be less than or greater than D_2 depending upon the angle of prism
(d) $D_1 > D_2$

185. A light ray is incident perpendicularly to one face of a 90° prism and is totally internally reflected at the glass-air interface. If the angle of reflection is 45°, we conclude that the refractive index *n* **[NCERT : P L-232 | 2004]**

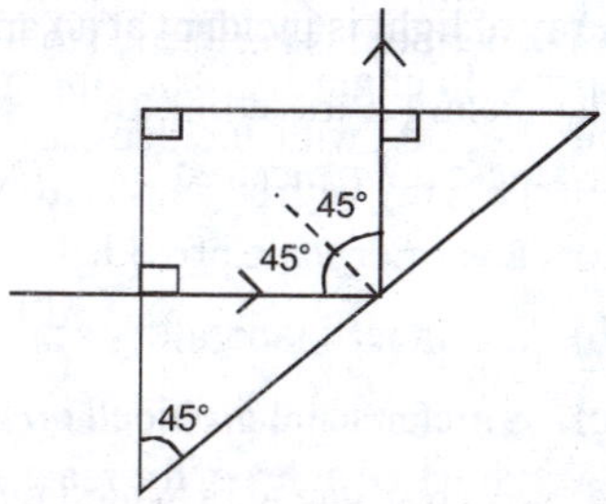

(a) $n > \frac{1}{\sqrt{2}}$

(b) $n > \sqrt{2}$

(c) $n < \frac{1}{\sqrt{2}}$

(d) $n < \sqrt{2}$

Topic 5 - Optical Instruments

186. In a reflecting telescope, a secondary mirror is used to:

[NCERT : P L-246 | April 8, 2023 (I)]

(a) reduce the problem of mechanical support

(b) remove spherical aberration

(c) make chromatic aberration zero

(d) move the eyepiece outside the telescopic tube

187. In normal adjustment, for a refracting telescope, the distance between objective and eye piece is 30 cm. The focal length of the objective, when the angular magnification of the telescope is 2, will be:

[NCERT : P L-245 | July 28, 2022 (I)]

(a) 20 cm (b) 30 cm (c) 10 cm (d) 15 cm

188. An object viewed from a near point distance of 25 cm, using a microscopic lens with magnification '6', gives an unresolved image. A resolved image is observed at infinite distance with a total magnification double the earlier using an eyepiece along with the given lens and a tube of length 0.6 m, if the focal length of the eyepiece is equal to ______ cm.

[NCERT : P L-242 | NA, July 20, 2021 (I)]

189. Given below are two statements: one is labelled as Assertion (A) and the other is labelled as Reason (R).

Assertion A : For a simple microscope, the angular size of the object equals the angular size of the image.

Reason R : Magnification is achieved as the small object can be kept much closer to the eye than 25 cm and hence it subtends a large angle.

In the light of the above statements, choose the most appropriate answer from the options given below :

[NCERT : P L-242 | Feb. 26, 2021 (II)]

(a) A is false but R is true

(b) A is true but R is false

(c) Both A and R are true but R is NOT the correct explanation of A

(d) Both A and R are true and R is the correct explanation of A

190. A compound microscope consists of an objective lens of focal length 1 cm and an eye piece of focal length 5 cm with a separation of 10 cm.

The distance between an object and the objective lens, at which the strain on the eye is minimum is $\frac{n}{40}$ cm.

The value of n is ______.

[NCERT : P L-244 | NA Sep. 05, 2020 (I)]

191. In a compound microscope, the magnified virtual image is formed at a distance of 25 cm from the eye-piece. The focal length of its objective lens is 1 cm. If the magnification is 100 and the tube length of the microscope is 20 cm, then the focal length of the eye-piece lens (in cm) is __________.

[NCERT : P L-244 | NA Sep. 04, 2020 (I)]

192. The magnifying power of a telescope with tube length 60 cm is 5. What is the focal length of its eye piece?

[NCERT : P L-245 | 8 Jan. 2020 I]

(a) 20 cm (b) 40 cm (c) 30 cm (d) 10 cm

193. If we need a magnification of 375 from a compound microscope of tube length 150 mm and an objective of focal length 5 mm, the focal length of the eye-piece, should be close to: **[NCERT : P L-244 | 7 Jan. 2020 I]**

(a) 22mm (b) 12mm (c) 2mm (d) 33mm

194. An observer looks at a distant tree of height 10 m with a telescope of magnifying power of 20. To the observer the tree appears : **[NCERT : P L-245 | 2016]**

(a) 20 times taller (b) 20 times nearer

(c) 10 times taller (d) 10 times nearer

195. A telescope has an objective lens of focal length 150 cm and an eyepiece of focal length 5 cm. If a 50 m tall tower at a distance of 1 km is observed through this telescope in normal setting, the angle formed by the image of the tower is θ, then θ is close to :

[NCERT : P L-245 | Online April 10, 2015]

(a) 30° (b) 15° (c) 60° (d) 1°

196. The focal lengths of objective lens and eye lens of a Galilean telescope are respectively 30 cm and 3.0 cm. telescope produces virtual, erect image of an object situated far away from it at least distance of distinct vision from the eye lens. In this condition, the magnifying power of the Galilean telescope should be:

[NCERT : P L-245 | Online April 9, 2014]

(a) +11.2 (b) –11.2

(c) –8.8 (d) +8.8

197. This question has Statement-1 and Statement-2. Of the four choices given after the Statements, choose the one that best describes the two Statements.

Statement 1: Very large size telescopes are reflecting telescopes instead of refracting telescopes.

Statement 2: It is easier to provide mechanical support to large size mirrors than large size lenses.

[NCERT : P L-246 | Online April 23, 2013]

(a) Statement-1 is true and Statement-2 is false.

(b) Statement-1 is false and Statement-2 is true.

(c) Statement-1 and statement-2 are true and Statement-2 is correct explanation for statement-1.

(d) Statements-1 and statement-2 are true and Statement-2 is not the correct explanation for statement-1.

198. An experment is performed to find the refractive index of glass using a travelling microscope. In this experiment distances are measured by **[NCERT : P L-244 | 2008]**

(a) a vernier scale provided on the microscope

(b) a standard laboratory scale

(c) a meter scale provided on the microscope

(d) a screw gauge provided on the microscope

199. The image formed by an objective of a compound microscope is **[NCERT : P L-244 | 2003]**

(a) virtual and diminished

(b) real and diminished

(c) real and enlarged

(d) virtual and enlarged

ANSWER KEYS

1	(c)	21	(a)	41	(d)	61	(c)	81	(b)	101	(a)	121	(30)	141	(476.19)	161	(d)	181	(c)
2	(b)	22	(a)	42	(d)	62	(9)	82	(c)	102	(b)	122	(120)	142	(60)	162	(a)	182	(c)
3	(a)	23	(150.00)	43	(a)	63	(c)	83	(a)	103	(d)	123	(b)	143	(b)	163	(c)	183	(d)
4	(45)	24	(1)	44	(2)	64	(12)	84	(d)	104	(b)	124	(54)	144	(d)	164	(b)	184	(a)
5	(a)	25	(d)	45	(31)	65	(a)	85	(d)	105	(d)	125	(b)	145	(a)	165	(c)	185	(b)
6	(d)	26	(c)	46	(b)	66	(a)	86	(b)	106	(a)	126	(10)	146	(b)	166	(a)	186	(d)
7	(8)	27	(b)	47	(3)	67	(158)	87	(d)	107	(c)	127	(b)	147	(c)	167	(45)	187	(a)
8	(a)	28	(c)	48	(b)	68	(90.00)	88	(4)	108	(a)	128	(225)	148	(d)	168	(b)	188	(25)
9	(c)	29	(c)	49	(98)	69	(d)	89	(c)	109	(c)	129	(a)	149	(a)	169	(b)	189	(d)
10	(b)	30	(a)	50	(b)	70	(c)	90	(b)	110	(a)	130	(10)	150	(d)	170	(60)	190	(50)
11	(b)	31	(b)	51	(50)	71	(b)	91	(b)	111	(15)	131	(c)	151	(b)	171	(a)	191	(4.48)
12	(18)	32	(a)	52	(a)	72	(c)	92	(b)	112	(200)	132	(a)	152	(d)	172	(c)	192	(d)
13	(32)	33	(c)	53	(52)	73	(d)	93	(d)	113	(10)	133	(b)	153	(d)	173	(a)	193	(a)
14	(c)	34	(c)	54	(15)	74	(d)	94	(b)	114	(d)	134	(50)	154	(a)	174	(30)	194	(b)
15	(32)	35	(80)	55	(c)	75	(c)	95	(b)	115	(5)	135	(d)	155	(a)	175	(45)	195	(c)
16	(400)	36	(6)	56	(a)	76	(d)	96	(c)	116	(100)	136	(d)	156	(b)	176	(a)	196	(d)
17	(210)	37	(a)	57	(a)	77	(d)	97	(a)	117	(4)	137	(30)	157	(d)	177	(5)	197	(c)
18	(b)	38	(d)	58	(a)	78	(c)	98	(a)	118	(34)	138	(15)	158	(b)	178	(a)	198	(a)
19	(a)	39	(6)	59	(d)	79	(b)	99	(b)	119	(30)	139	(a)	159	(d)	179	(60)	199	(c)
20	(d)	40	(c)	60	(d)	80	(a)	100	(b)	120	(d)	140	(d)	160	(d)	180	(c)		

Wave Optics

10

TREND ANALYSIS
JEE MAIN (2025-2021)

	2025				2024				2023				2022				2021			
	Session II		Session I		Session II		Session I		Session II		Session I		Session II		Session I		Session II		Session I	
	MCQs	NVQs	MCQs	NVQs	MCQs	NVQs	MCQs	NVQs	MCQs	NVQs	MCQs	NVQs	MCQs	NVQs	MCQs	NVQs	MCQs	NVQs	MCQs	NVQs
Topic 1	2	1	1	1	1	1	–	–	1	–	–	–	1	2	3	–	1	1	1	–
Topic 2	3	1	5	1	–	4	1	2	2	1	2	4	1	1	3	2	2	2	3	1
Topic 3	2	1	1	–	–	3	6	1	1	1	3	2	1	–	1	1	–	1	–	1
Topic 4																				
Total	7	3	7	2	1	8	7	3	4	2	5	6	3	3	7	3	3	4	4	2

How to Code Qns for Better & Faster Revision

Type 1: Able to solve (No code Req)

Type 2: Solved with struggle or by seeing solution **(Use code NR means Needs Revision or orange stickers)**

Type 3: Tricky/ Could not solve on my own. Need help from Teacher/ Friend. **(Use code TQ means Tricky Question or red stickers)**

You can either hand write the code in boxes or else paste coloured stickers given at end of book.

Topic 1 – Wavefront, Interference of Light, Coherent and Incoherent Sources

1. Two monochromatic light beams have intensities in the ratio 1:9. An interference pattern is obtained by these beams. The ratio of the intensities of maximum to minimum is **[NCERT : PL-264 | April 3, 2025 (II)]**

(a) 8 : 1 (b) 9 : 1
(c) 3 : 1 (d) 4 : 1

2. Two coherent monochromatic light beams of intensities 4I and 9I are superimposed. The difference between the maximum and minimum intensities in the resulting interference pattern is xI. The value of x is ________. **[NCERT : PL-264 | April 3, 2025 (I)]**

3. A light wave is propagating with plane wave fronts of the type $x + y + z = \text{constant}$. The angle made by the direction of wave propagation with the x-axis is : **[NCERT : PL-261 | April 2, 2025 (I)]**

(a) $\cos^{-1}\left(\frac{1}{\sqrt{3}}\right)$ (b) $\cos^{-1}\left(\frac{2}{3}\right)$
(c) $\cos^{-1}\left(\frac{1}{3}\right)$ (d) $\cos^{-1}\left(\sqrt{\frac{2}{3}}\right)$

4. A thin transparent film with refractive index 1.4, is held on circular ring of radius 1.8 cm. The fluid in the film evaporates such that transmission through the film at wavelength 560 nm goes to a minimum every 12 seconds. Assuming that the film is flat on its two sides, the rate of evaporation is ________ $\pi \times 10^{-13}\ m^3/s$. **[NCERT : P L-264 | Jan. 28, 2025 (II)]**

5. A transparent film of refractive index, 2.0 is coated on a glass slab of refractive index, 1.45. What is the minimum thickness of transparent film to be coated for the maximum transmission of Green light of wavelength 550 nm. [Assume that the light is incident nearly perpendicular to the glass surface.] **[NCERT : P L-264 | Jan. 22, 2025 (II)]**

(a) 94.8 nm (b) 68.7 nm
(c) 137.5 nm (d) 275 nm

6. Two coherent monochromatic light beams of intensities I and 4I are superimposed. The difference between maximum and minimum possible intensities in the resulting beam is x I. The value of x is______.
[NCERT : P L-264 | April 6, 2024 (II)]

7. Light emerges out of a convex lens when a source of light kept at its focus. The shape of wavefront of the light is:
[NCERT : P L-261 | April 5, 2024 (I)]
(a) Both spherical and cylindrical
(b) Cylindrical
(c) Spherical
(d) Plane

8. Given below are two statements: one is labelled as **Assertion A** and the other is labelled as **Reason R**
Assertion A: The phase difference of two light waves change if they travel through different media having same thickness, but different indices of refraction.
Reason R: The wavelengths of waves are different in different media.
In the light of the above statements, choose the most appropriate answer from the options given below
[NCERT : P L-264 | April 6, 2023 (II)]
(a) Both **A** and **R** are correct but **R** is **NOT** the correct explanation of **A**
(b) **A** is correct but **R** is not correct
(c) Both **A** and **R** are correct and **R** is the correct explanation of **A**
(d) **A** is not correct but **R** is correct

9. Two light beams of intensities 4I and 9I interfere on a screen. The phase difference between these beams on the screen at point A is zero and at point B is π. The difference of resultant intensities, at the point A and B, will be _____ I.
[NCERT : P L-264 | July 29, 2022 (I)]

10. Two beams of light having intensities I and 4I interfere to produce a fringe pattern on a screen. The phase difference between the two beams are $\pi/2$ and $\pi/3$ at points A and B respectively. The difference between the resultant intensities at the two points is xI. The value of x will be _____.
[NCERT : P L-264 | July 27, 2022 (I)]

11. Two coherent sources of light interfere. The intensity ratio of two sources is 1 : 4. For this interference pattern if the value of $\frac{I_{max}+I_{min}}{I_{max}-I_{min}}$ is equal to $\frac{2\alpha+1}{\beta+3}$, then $\frac{\alpha}{\beta}$ will be:
[NCERT : P L-264 | July 27, 2022 (II)], [April 13, 2023 (II)]
(a) 1.5 (b) 2 (c) 0.5 (d) 1

12. The two light beams having intensities I and 9I interfere to produce a fringe pattern on a screen. The phase difference between the beams is $\frac{\pi}{2}$ at point P and π at point Q. Then the difference between the resultant intensities at P and Q will be :
[NCERT : P L-264 | June 25, 2022 (I)]
(a) 2I (b) 6I (c) 5I (d) 7I

13. The interference pattern is obtained with two coherent light sources of intensity ratio 4 : 1. And the ratio $\frac{I_{max}+I_{min}}{I_{max}-I_{min}}$ is $\frac{5}{x}$. Then, the value of x will be equal to :
[NCERT : P L-264 | June 25, 2022 (II)]
(a) 3 (b) 4 (c) 2 (d) 1

14. Two light beams of intensities in the ratio of 9 : 4 are allowed to interfere. The ratio of the intensity of maxima and minima will be :
[NCERT : P L-264 | June 24, 2022 (II)]
(a) 2 : 3 (b) 16 : 81 (c) 25 : 169 (d) 25 : 1

15. The light waves from two coherent sources have same intensity $I_1 = I_2 = I_0$. In interference pattern the intensity of light at minima is zero. What will be the intensity of light at maxima ?
[NCERT : P L-264 | Aug. 27, 2021 (II)]
(a) I_0 (b) $2I_0$ (c) $5I_0$ (d) $4I_0$

16. The difference in the number of waves when yellow light propagates through air and vacuum columns of the same thickness is one. The thickness of the air column is _____ mm. [Refractive index of air = 1.0003, wavelength of yellow light in vacuum = 6000 Å]
[NCERT : P L-264 | NA, July 27, 2021 (II)]

17. The coherent light sources having intensity in the ratio 2x produce an interference pattern.
The ratio $\frac{I_{max}-I_{min}}{I_{max}+I_{min}}$ will be
[NCERT : P L-264 | Feb. 25, 2021 (I)]
(a) $\frac{\sqrt{2x}}{2x+1}$ (b) $\frac{\sqrt{2x}}{x+1}$ (c) $\frac{2\sqrt{2x}}{x+1}$ (d) $\frac{2\sqrt{2x}}{2x+1}$

18. In the figure below, P and Q are two equally intense coherent sources emitting radiation of wavelength 20 m. The separation between P and Q is 5 m and the phase of P is ahead of that of Q by 90°. A, B and C are three distinct points of observation, each equidistant from the midpoint of PQ. The intensities of radiation at A, B, C will be in the ratio :
[NCERT : P L-264 | Sep. 06, 2020 (I)]

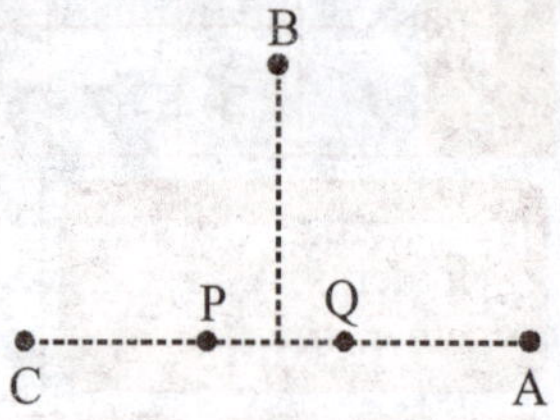

(a) 0 : 1 : 4
(b) 2 : 1 : 0
(c) 0 : 1 : 2
(d) 4 : 1 : 0

19. Two coherent sources of sound, S_1 and S_2, produce sound waves of the same wavelength, $\lambda = 1$ m, in phase. S_1 and S_2 are placed 1.5 m apart (see fig.). A listener, located at L, directly in front of S_2 finds that the intensity is at a minimum when he is 2 m away from S_2. The listener moves away from S_1, keeping his distance from S_2 fixed. The adjacent maximum of intensity is observed when the listener is at a distance d from S_1. Then, d is :

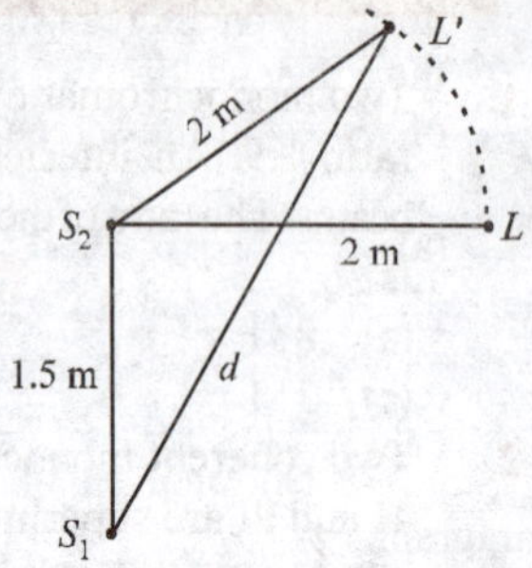

[NCERT : P L-264 | Sep. 05, 2020 (II)]
(a) 12m (b) 5m (c) 2m (d) 3m

20. Two light waves having the same wavelength λ in vacuum are in phase initially. Then the first wave travels a path L_1 through a medium of refractive index n_1 while the second wave travels a path of length L_2 through a medium of refractive index n_2. After this the phase difference between the two waves is : **[NCERT : P L-264 | Sep. 03, 2020 (II)]**

(a) $\frac{2\pi}{\lambda}\left(\frac{L_2}{n_1}-\frac{L_1}{n_2}\right)$ (b) $\frac{2\pi}{\lambda}\left(\frac{L_1}{n_1}-\frac{L_2}{n_2}\right)$

(c) $\frac{2\pi}{\lambda}(n_1L_1-n_2L_2)$ (d) $\frac{2\pi}{\lambda}(n_2L_1-n_1L_2)$

21. In an interference experiment the ratio of amplitudes of coherent waves is $\frac{a_1}{a_2}=\frac{1}{3}$. The ratio of maximum and minimum intensities of fringes will be : **[NCERT : P L-264 | 8 April 2019 I]**

(a) 2 (b) 18 (c) 4 (d) 9

22. Two coherent sources produce waves of different intensities which interfere. After interference, the ratio of the maximum intensity to the minimum intensity is 16. The intensity of the waves are in the ratio: **[NCERT : P L-264 | 9 Jan. 2019 I]**

(a) 16 : 9 (b) 25 : 9 (c) 4 : 1 (d) 5 : 3

23. On a hot summer night, the refractive index of air is smallest near the ground and increases with height from the ground. When a light beam is directed horizontally, the Huygens' principle leads us to conclude that as it travels, the light beam : **[NCERT : P L-257 | 2015]**

(a) bends downwards
(b) bends upwards
(c) becomes narrower
(d) goes horizontally without any deflection

24. Two monochromatic light beams of intensity 16 and 9 units are interfering. The ratio of intensities of bright and dark parts of the resultant pattern is: **[NCERT : P L-264 | Online April 11, 2014]**

(a) $\frac{16}{9}$ (b) $\frac{4}{3}$ (c) $\frac{7}{1}$ (d) $\frac{49}{1}$

25. n identical waves each of intensity I_0 interfere with each other. The ratio of maximum intensities if the interference is (i) coherent and (ii) incoherent is : **[NCERT : P L-263, 264 | Online April 23, 2013]**

(a) n^2 (b) $\frac{1}{n}$ (c) $\frac{1}{n^2}$ (d) n

26. Two coherent plane light waves of equal amplitude makes a small angle α (< < 1) with each other. They fall almost normally on a screen. If λ is the wavelength of light waves, the fringe width Δx of interference patterns of the two sets of waves on the screen is **[Online May 19, 2012]**

(a) $\frac{2\lambda}{\alpha}$ (b) $\frac{\lambda}{\alpha}$

(c) $\frac{\lambda}{(2\alpha)}$ (d) $\frac{\lambda}{\sqrt{\alpha}}$

Directions : Questions number 27-28 are based on the following paragraph.

An initially parallel cylindrical beam travels in a medium of refractive index $\mu(I)=\mu_0+\mu_2 I$, where μ_0 and μ_2 are positive constants and I is the intensity of the light beam. The intensity of the beam is decreasing with increasing radius

27. As the beam enters the medium , it will **[NCERT : P L-259, 260 | 2010]**

(a) diverge
(b) converge
(c) diverge near the axis and converge near the periphery
(d) travel as a cylindrical beam

28. The initial shape of the wavefront of the beam is **[NCERT : P L-261 | 2010]**

(a) convex
(b) concave
(c) convex near the axis and concave near the periphery
(d) planar

29. The speed of light in the medium is **[NCERT : P L-259, 260 | 2010]**

(a) minimum on the axis of the beam
(b) the same everywhere in the beam
(c) directly proportional to the intensity I
(d) maximum on the axis of the beam

30. To demonstrate the phenomenon of interference, we require two sources which emit radiation **[NCERT : P L-262, 263 | 2003]**

(a) of nearly the same frequency
(b) of the same frequency
(c) of different wavelengths
(d) of the same frequency and having a definite phase relationship

Topic 2 - Young's Double Slit Experiment

31. In a Young's double slit experiment, the source is white light. One of the slits is covered by red filter and another by a green filter. In this case **[NCERT : P L-266 | April 8, 2025 (II)]**

(a) There shall be an interference pattern for red distinct from that for green.
(b) There shall be no interference fringes.
(c) There shall be alternate interference fringes of red and green.
(d) There shall be an interference pattern, where each fringe's pattern center is green and outer edges is red.

32. In a Young's double slit experiment, two slits are located 1.5 mm apart. The distance of screen from slits is 2 m and the wavelength of the source is 400 nm. If the 20 maxima of the double slit pattern are contained within the centre maximum of the single slit diffraction pattern, then the width of each slit is $x \times 10^{-3}$ cm, where x-value is _____. **[NCERT : P L-266 | April 4, 2025 (II)]**

33. In a Young's double slit experiment, the slits are separated by 0.2 mm. If the slits separation is increased to 0.4 mm, the percentage change of the fringe width is: **[NCERT : P L-266 | April 4, 2025 (I)]**

(a) 0% (b) 100% (c) 50% (d) 25%

34. Width of one of the two slits in a Young's double slit interference experiment is half of the other slit. The ratio of the maximum to the minimum intensity in the interference pattern is : **[NCERT : P L-266 | April 3, 2025 (II)]**

(a) $(2\sqrt{2}+1):(2\sqrt{2}-1)$ (b) $(3+2\sqrt{2}):(3-2\sqrt{2})$

(c) 9 : 1 (d) 3 : 1

35. A double slit interference experiment performed with a light of wavelength 600 nm forms an interference fringe pattern on a screen with 10th bright fringe having its centre at a distance of 10 mm from the central maximum. Distance of the centre of the same 10th bright fringe from the central maximum when the source of light is replaced by another source of wavelength 660 nm would be_____mm.
[NCERT : P L-266 | Jan. 28, 2025 (I)]

36. Young's double slit interference apparatus is immersed in a liquid of refractive index 1.44. It has slit separation of 1.5mm. The slits are illuminated by a parallel beam of light whose wavelength in air is 690 nm. The fringe-width on a screen placed behind the plane of slits at a distance of 0.72m, will be **[NCERT : P L-266 | Jan. 24, 2025 (II)]**
(a) 0.23 mm (b) 0.33 mm
(c) 0.63 mm (d) 0.46 mm

37. The Young's double slit interference experiment is performed using light consisting of 480 nm and 600 nm wavelengths to form interference patterns. The least number of the bright fringes of 480 nm light that are required for the first coincidence with the bright fringes formed by 600 nm light is
[NCERT : P L-266 | Jan. 24, 2025 (I)]
(a) 4 (b) 8 (c) 6 (d) 5

38. The width of one of the two slits in Young's double slit experiment is d while that of the other slit is xd. If the ratio of the maximum to the minimum intensity in the interference pattern on the screen is 9 : 4, then what is the value of x? (Assume that the field strength varies according to the slit width.) **[NCERT : P L-266 | Jan. 23, 2025 (II)]**
(a) 2 (b) 3 (c) 5 (d) 4

39. Given below are two statements. One is labelled as **Assertion (A)** and the other is labelled as **Reason (R)**.
Assertion (A): In Young's double slit experiment, the fringes produced by red light are closer as compared to those produced by blue light.
Reason (R): The fringe width is directly proportional to the wavelength of light.
In the light of above statements, choose the correct answer from the options given below:
[NCERT : P L-266 | Jan. 22, 2025 (II)]
(a) Both (A) and (R) are true and (R) is the correct explanation of (A).
(b) (A) is false but (R) is true.
(c) Both (A) and (R) are true but (R) is NOT the correct explanation of (A).
(d) (A) is true but (R) is false.

40. Given below are two statements: one is labelled as **Assertion (A)** and the other is labelled as **Reason (R)**.
Assertion (A) : If Young's double slit experiment is performed in an optically denser medium than air, then the consecutive fringes come closer.
Reason (R) : The speed of light reduces in an optically denser medium than air while its frequency does not change.
In the light of the above statements, choose the most appropriate answer from the options given below :
[NCERT : P L-266 | Jan. 22, 2025 (I)]
(a) Both (A) and (R) are true and (R) is the correct explanation of (A).
(b) (A) is false but (R) is true.
(c) Both (A) and (R) are true but (R) is not the correct explanation of (A)
(d) (A) is true but (R) is false.

41. In a Young's double slit experiment, the intensity at a point is $\left(\frac{1}{4}\right)^{th}$ of the maximum intensity, the minimum distance of the point from the central maximum is ________ μm.
[NCERT : P L-264 | April 9, 2024 (I)]
(Given : $\lambda = 600$ nm, $d = 1.0$ mm, $D = 1.0$ m)

42. Monochromatic light of wavelenlgth 500 nm is used in Young's double slit experiment. An interference pattern is obtained on a screen When one of the slits is covered with a very thin glass plate (refractive index = 1.5), the central maximum is shifted to a position previously occupied by the 4th bright fringe. The thickness of the glass-plate is ....μm.
[NCERT : P L-266 | April 9, 2024 (II)]

43. In Young's double slit experiment, carried out with light of wavelength 5000Å, the distance between the slits is 0.3 mm and the screen is at 200 cm from the slits. The central maximum is at x = 0 cm. The value of x for third maxima is .... mm. **[NCERT : P L-266 | April 5, 2024 (I)]**

44. Two wavelengths λ_1 and λ_2 are used in Young's double slit experiment $\lambda_1 = 450$ nm and $\lambda_2 = 650$ nm. The minimum order of fringe produced by λ_2 which overlaps with the fringe produced by λ_1 is n. The value of n is ______.
[NCERT : P L-266 | April 4, 2024 (I)]

45. In Young's double slit experiment, monochromatic light of wavelength 5000 Å is used. The slits are 1.0 mm apart and screen is placed at 10. m away from slits. The distance from the centre of the screen where intensity becomes half of the maximum intensity for the first time is ______ $\times 10^{-6}$ m. **[NCERT : P L-264 | Feb. 1, 2024 (II)]**

46. In a double slit experiment shown in figure, when light of wavelength 400 nm is used, dark fringe is observed at P. If D = 0.2 m, the minimum distance between the slits S_1 and S_2 is ______ mm. **[NCERT : P L-266 | Jan. 29, 2024 (I)]**

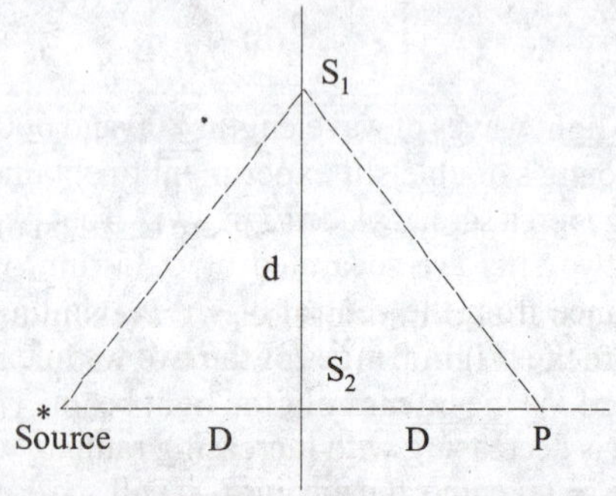

47. In Young's double slit experiment, light from two identical sources are superimposing on a screen. The path difference between the two lights reaching at a point on

the screen is $\frac{7\lambda}{4}$. The ratio of intensity of fringe at this point with respect to the maximum intensity of the fringe is: **[NCERT : P L-264 | Jan. 29, 2024 (II)]**

(a) $\frac{1}{4}$ (b) $\frac{1}{3}$ (c) $\frac{3}{4}$ (d) $\frac{1}{2}$

48. The ratio of intensities at two points P and Q on the screen in a Young's double slit experiment where phase difference between two wave of same amplitude are $\frac{\pi}{3}$ and $\frac{\pi}{2}$, respectively are

[NCERT : P L-264 | April 10, 2023 (II)]

(a) 1 : 3 (b) 3 : 1 (c) 3 : 2 (d) 2 : 3

49. The width of fringe is 2 mm on the screen in a double slits experiment for the light of wavelength of 400 nm. The width of the fringe for the light of wavelength 600 nm will be: **[NCERT : P L-266 | April 8, 2023 (II)]**

(a) 4mm (b) 1.33 mm (c) 3mm (d) 2mm

50. A beam of light consisting of two wavelengths 7000 Å and 5500 Å is used to obtain interference pattern in Young's double slit experiment. The distance between the slits is 2.5 mm and the distance between the plane of slits and the screen is 150 cm. The least distance from the central fringe, where the bright fringes due to both the wavelengths coincide, is n × 10^{-5} m. The value of n is________. **[NCERT : P L-266 | April 6, 2023 (II)]**

51. As shown in the figure, in Young's double slit experiment, a thin plate of thickness t = 10μm and refractive index μ = 1.2 is inserted infront of slit S_1. The experiment is conducted in air (μ = 1) and uses a monochromatic light of wavelength λ = 500 nm. Due to the insertion of the plate, central maxima is shifted by a distance of $x\beta_0$. β_0 is the fringe-width before the insertion of the plate. The value of the x is _______. **[NCERT : P L-266 | Feb. 1, 2023 (II)]**

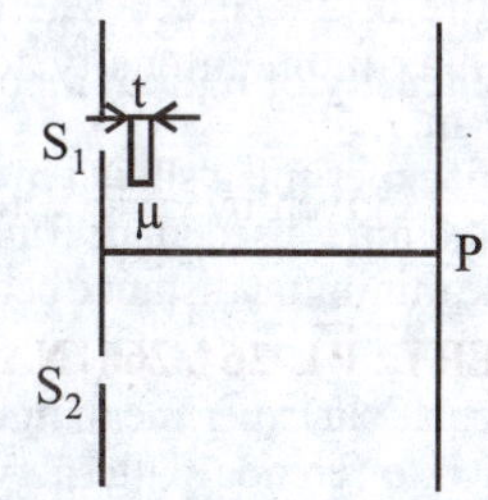

52. Two light waves of wavelengths 800 and 600 nm are used in Young's double slit experiment to obtain interference fringes on a screen placed 7 m away from plane of slits. If the two slits are separated by 0.35 mm, then shortest distance from the central bright maximum to the point where the bright fringes of the two wavelength coincide will be ________ mm.

[NCERT : P L-266 | Jan. 31, 2023 (II)]

53. In Young's double slit experiment, two slits S_1 and S_2 are 'd' distance apart and the separation from slits to screen is D (as shown in figure). Now if two transparent slabs of equal thickness 0.1 mm but refractive index 1.51 and 1.55 are introduced in the path of beam (λ = 4000Å) from S_1 and S_2 respectively. The central bright fringe spot will shift by ________ number of fringes.

[NCERT : P L-266 | Jan. 30, 2023 (I)]

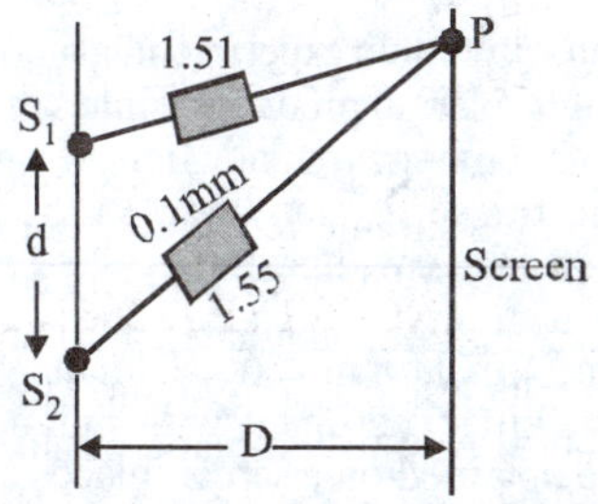

54. In a Young's double slit experiment, the intensities at two points, for the path difference $\frac{\lambda}{4}$ and $\frac{\lambda}{3}$ (λ being the wavelength of light used) are I_1 and I_2 respectively. If I_0 denotes the intensity produced by each one of the individual slits, then $\frac{I_1 + I_2}{I_0}$ = ...

[NCERT : P L-264 | Jan. 30, 2023 (II)]

55. In a Young's double slit experiment, two slits are illuminated with a light of wavelength 800 nm. The line joining A_1P is perpendicular to A_1A_2 as shown in the figure. If the first minimum is detected at P, the value of slits separation 'a' will be : **[NCERT : P L-266 | Jan. 29, 2023 (I)]**

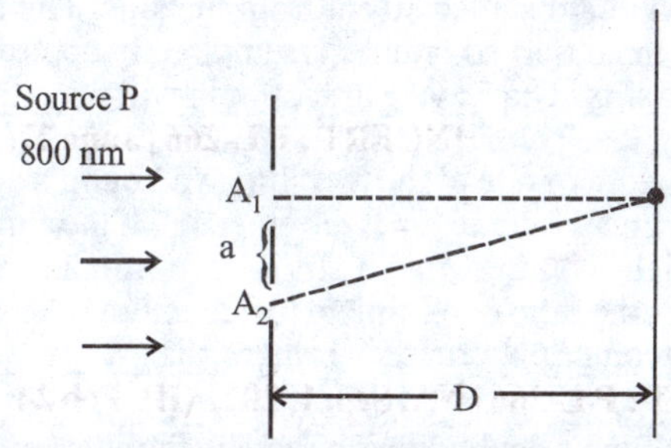

The distance of screen from slits D = 5 cm

(a) 0.4 mm (b) 0.5 mm (c) 0.2 mm (d) 0.1 mm

56. In Young's double slits experiment, the position of 5th bright fringe from the central maximum is 5 cm. The distance between slits and screen is 1 m and wavelength of used monochromatic light is 600 nm. The separation between the slits is: **[NCERT : P L-266 | Jan. 25, 2023 (I)]**

(a) 60 μm (b) 48 μm
(c) 12 μm (d) 36 μm

57. In a Young's double slit experiment, a laser light of 560 nm produces an interference pattern with consecutive bright fringes' separation of 7.2 mm. Now another light is used to produce an interference pattern with consecutive bright fringes' separation of 8.1 mm. The wavelength of second light is ______ nm. **[NCERT : P L-266 | July 28, 2022 (I)]**

58. In Young's double slit experiment, the fringe width is 12 mm. If the entire arrangement is placed in water of refractive index $\frac{4}{3}$, then the fringe width becomes (in mm) : **[NCERT : P L-266 | July 26, 2022 (I)]**

(a) 16 (b) 9 (c) 48 (d) 12

59. Find the ratio of maximum intensity to the minimum intensity in the interference pattern if the widths of the two slits in Young's experiment are in the ratio of 9 : 16. (Assuming intensity of light is directly proportional to the width of slits) **[NCERT : P L-264 | June 30, 2022 (I)]**
(a) 3 : 4 (b) 4 : 3 (c) 7 : 1 (d) 49 : 1

60. Using Young's double slit experiment, a monochromatic light of wavelength 5000 Å produces fringes of fringe width 0.5 mm. If another monochromatic light of wavelength 6000Å is used and the separation between the slits is doubled, then the new fringe width will be : **[NCERT : P L-266 | June 29, 2022 (I)]**
(a) 0.5 mm (b) 1.0 mm (c) 0.6 mm (d) 0.3 mm

61. In a double slit experiment with monochromatic light, fringes are obtained on a screen placed at some distance from the plane of slits. If the screen is moved by 5×10^{-2} m towards the slits, the change in fringe width is 3×10^{-3} cm. If the distance between the slits is 1 mm, then the wavelength of the light will be ______ nm. **[NCERT : P L-266 | June 29, 2022 (II)]**

62. In young's double slit experiment performed using a monochromatic light of wavelength λ, when a glass plate ($\mu = 1.5$) of thickness $x\lambda$ is introduced in the path of the one of the interfering beams, the intensity at the position where the central maximum occurred previously remains unchanged. The value of x will be: **[NCERT : P L-266 | June 28, 2022 (II)]**
(a) 3 (b) 2 (c) 1.5 (d) 0.5

63. In Young's double slit experiment the two slits are 0.6 mm distance apart. Interference pattern is observed on a screen at a distance 80 cm from the slits. The first dark fringe is observed on the screen directly opposite to one of the slits. The wavelength of light will be ______ nm. **[NCERT : P L-266 | June 27, 2022 (I)]**

64. The width of one of the two slits in a Young's double slit experiment is three times the other slit. If the amplitude of the light coming from a slit is proportional to the slit-width, the ratio of minimum to maximum intensity in the interference pattern is x : 4 where x is ______. **[NCERT : P L-266 | NA, Sep. 1, 2021 (II), Feb 24, 2021 (I)]**

65. In a Young's double slit experiment, the slits are separated by 0.3 mm and the screen is 1.5 m away from the plane of slits. Distance between fourth bright fringes on both sides of central bright is 2.4 cm. The frequency of light used is ______ $\times 10^{14}$ Hz. **[NCERT : PL-266 | NA, Aug. 31, 2021 (II)]**

66. White light is passed through a double slit and interference is observed on a screen 1.5 m away. The separation between the slits is 0.3 mm. The first violet and red fringes are formed 2.0 mm and 3.5 mm away from the central white fringes. The difference in wavelengths of red and violet light is ______ nm. **[NCERT : PL-266 | NA, Aug. 26, 2021 (I)]**

67. In Young's double slit experiment, if the source of light changes from orange to blue then : **[NCERT : P L-266 | July 27, 2021 (I)]**
(a) the central bright fringe will become a dark fringe.
(b) the distance between consecutive fringes will decrease.
(c) the distance between consecutive fringes will increase.
(d) the intensity of the minima will increase.

68. In the Young's double slit experiment, the distance between the slits varies in time as $d(t) = d_0 + a_0 \sin\omega t$; where d_0, ω and a_0 are constants. The difference between the largest fringe width and the smallest fringe width obtained over time is given as : **[NCERT : P L-266 | July 25, 2021 (I)]**
(a) $\frac{2\lambda D(d_0)}{(d_0^2 - a_0^2)}$ (b) $\frac{2\lambda D a_0}{(d_0^2 - a_0^2)}$
(c) $\frac{\lambda D}{d_0^2} a_0$ (d) $\frac{\lambda D}{d_0 + a_0}$

69. In Young's double slit arrangement, slits are separated by a gap of 0.5 mm, and the screen is placed at a distance of 0.5 m from them. The distance between the first and the third bright fringe formed when the slits are illuminated by a monochromatic light of 5890 Å is: **[NCERT : P L-266 | March 18, 2021 (I)]**
(a) 1178×10^{-9} m (b) 1178×10^{-6} m
(c) 1178×10^{-12} m (d) 5890×10^{-7} m

70. A fringe width of 6 mm was produced for two slits separated by 1 mm apart. The screen is placed 10 m away. The wavelength of light used is 'x' nm.
The value of 'x' to the nearest integer is ______. **[NCERT : P L-266 | March 16, 2021 (I)]**

71. If the source of light used in a Young's double slit experiment is changed from red to violet : **[NCERT : P L-266 | Feb. 24, 2021 (II)]**
(a) consecutive fringe lines will come closer
(b) the central bright fringe will become a dark fringe
(c) the fringes will become brighter
(d) the intensity of minima will increase

72. In a Young's double slit experiment two slits are separated by 2 mm and the screen is placed one meter away. When a light of wavelength 500 nm is used, the fringe separation will be : **[NCERT : P L-266 | Feb. 26, 2021 (I)]**
(a) 0.25 mm (b) 0.75 mm (c) 0.50 mm (d) 1 mm

73. A young's double-slit experiment is performed using monocromatic light of wavelength λ. The intensity of light at a point on the screen, where the path difference is λ, is K units. The intensity of light at a point where the path difference is $\frac{\lambda}{6}$ is given by $\frac{nK}{12}$, where n is an integer. The value of n is ______. **[NCERT : P L-264, 266 | NA Sep. 06, 2020 (II)]**

74. In a Young's double slit experiment, light of 500 nm is used to produce an interference pattern. When the distance between the slits is 0.05 mm, the angular width (in degree) of the fringes formed on the distance screen is close to : **[Sep. 03, 2020 (I)]**
(a) 0.17° (b) 0.57° (c) 1.7° (d) 0.07°

75. Interference fringes are observed on a screen by illuminating two thin slits 1 mm apart with a light source (λ = 632.8 nm). The distance between the screen and the slits is 100 cm. If a bright fringe is observed on a screen at a distance of 1.27 mm from the central bright fringe, then the path difference between the waves, which are reaching this point from the slits is close to : **[NCERT : P L-266 | Sep. 02, 2020 (I)]**
(a) 1.27 μm (b) 2.87 nm (c) 2 nm (d) 2.05 μm

76. In a Young's double slit experiment, 16 fringes are observed in a certain segment of the screen when light of wavelength 700 nm is used. If the wavelength of light is changed to 400 nm, the number of fringes observed in the same segment of the screen would be :
[NCERT : P L-266 | Sep. 02, 2020 (II), NA 09, Jan 2020 (II)]

(a) 24 (b) 30 (c) 18 (d) 28

77. In a Young's double slit experiment 15 fringes are observed on a small portion of the screen when light of wavelength 500 nm is used. Ten fringes are observed on the same section of the screen when another light source of wavelength λ is used. Then the value of λ is (in nm) ______. **[NCERT : P L-266 | NA 9 Jan 2020 II]**

78. In a double-slit experiment, at a certain point on the screen the path difference between the two interfering waves is $\frac{1}{8}$ th of a wavelength. The ratio of the intensity of light at that point to that at the centre of a bright fringe is:
[NCERT : P L-264 | 8 Jan 2020 II, 11 Jan 2019 I]

(a) 0.853 (b) 0.672 (c) 0.568 (d) 0.760

79. In a Young's double slit experiment, the separation between the slits is 0.15 mm. In the experiment, a source of light of wavelength 589 nm is used and the interference pattern is observed on a screen kept 1.5 m away. The separation between the successive bright fringes on the screen is: **[NCERT : P L-266 | 7 Jan 2020 II]**

(a) 6.9 mm (b) 3.9 mm (c) 5.9 mm (d) 4.9 mm

80. In a double slit experiment, when a thin film of thickness t having refractive index μ is introduced in front of one of the slits, the maximum at the centre of the fringe pattern shifts by one fringe width. The value of t is ($\approx$ is the wavelength of the light used) :
[NCERT : P L-266 | 12 April 2020 I]

(a) $\frac{2\lambda}{(\mu-1)}$ (b) $\frac{\lambda}{2(\mu-1)}$

(c) $\frac{\lambda}{(\mu-1)}$ (d) $\frac{\lambda}{(2\mu-1)}$

81. The figure shows a Young's double slit experimental setup. It is observed that when a thin transparent sheet of thickness t and refractive index μ is put in front of one of the slits, the central maximum gets shifted by a distance equal to n fringe widths. If the wavelength of light used is λ, t will be: **[NCERT : P L-266 | 9 April 2019 I]**

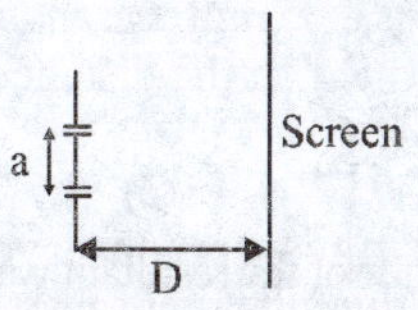

(a) $\frac{2nD\lambda}{a(\mu-1)}$ (b) $\frac{nD\lambda}{a(\mu-1)}$ (c) $\frac{D\lambda}{a(\mu-1)}$ (d) $\frac{2D\lambda}{a(\mu-1)}$

82. In a Young's double slit experiment with slit separation 0.1 mm, one observes a bright fringe at angle $\frac{1}{40}$ rad by using light of wavelength λ_1 When the light of wavelength λ_2 is used a bright fringe is seen at the same angle in the same set up. Given that λ_1 and λ_2 are in visible range (380 nm to 740 nm), their values are:
[NCERT : P L-266 | 10 Jan. 2019 I]

(a) 625 nm, 500 nm (b) 380 nm, 525 nm
(c) 380 nm, 500 nm (d) 400 nm, 500 nm

83. Consider a Young's double slit experiment as shown in figure. What should be the slit separation d in terms of wavelength λ such that the first minima occurs directly in front of the slit (S_1)?
[NCERT : P L-266 | 10 Jan 2019 II]

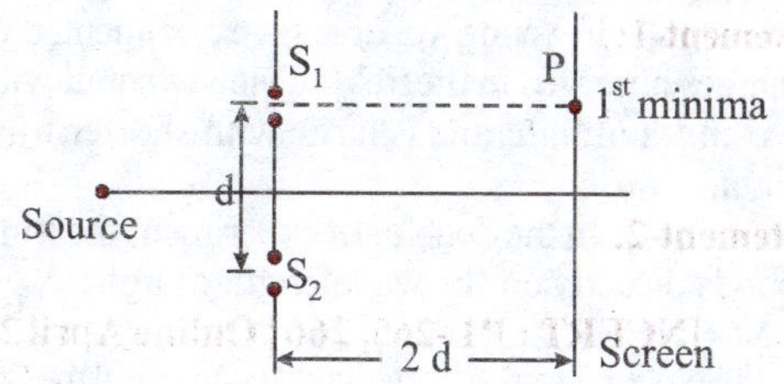

(a) $\frac{\lambda}{2(\sqrt{5}-2)}$ (b) $\frac{\lambda}{(\sqrt{5}-2)}$

(c) $\frac{\lambda}{2(5-\sqrt{2})}$ (d) $\frac{\lambda}{(5-\sqrt{2})}$

84. In a Young's double slit experiment, the slits are placed 0.320 mm apart. Light of wavelength λ = 500 nm is incident on the slits. The total number of bright fringes that are observed in the angular range $-30° \le \theta \le 30°$ is **[NCERT : P L-266 | 9 Jan 2019 II]**

(a) 640 (b) 320 (c) 321 (d) 641

85. In a Young's double slit experiment, slits are separated by 0.5 mm, and the screen is placed 150 cm away. A beam of light consisting of two wavelengths, 650 nm and 520 nm, is used to obtain interference fringes on the screen. The least distance from the common central maximum to the point where the bright fringes due to both the wavelengths coincide is : **[NCERT : P L-266 | 2017]**

(a) 9.75 mm (b) 15.6 mm (c) 1.56 mm (d) 7.8 mm

86. In a Young's double slit experiment with light of wavelength λ the separation of slits is d and distance of screen is D such that D >> d >> λ. If the fringe width is β, the distance from point of maximum intensity to the point where intensity falls to half of maximum intensity on either side is:
[NCERT : P L-264, 266 | Online April 11, 2015]

(a) $\frac{\beta}{6}$ (b) $\frac{\beta}{3}$ (c) $\frac{\beta}{4}$ (d) $\frac{\beta}{2}$

87. In a Young's double slit experiment, the distance between the two identical slits is 6.1 times larger than the slit width. Then the number of intensity maxima observed within the central maximum of the single slit diffraction pattern is:
[Online April 19, 2014]

(a) 3 (b) 6 (c) 12 (d) 24

88. Two coherent point sources S_1 and S_2 are separated by a small distance 'd' as shown. The fringes obtained on the screen will be **[NCERT : P L-265, 266 | 2013]**

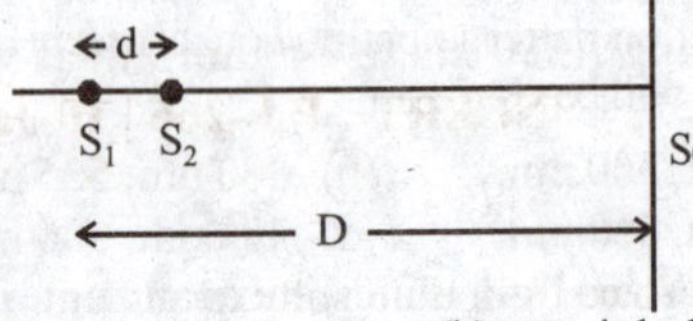

(a) points (b) straight lines
(c) semi-circles (d) concentric circles

89. This question has Statement-1 and Statement-2. Of the four choices given after the Statements, choose the one that best describes the two Statements.

Statement-1: In Young's double slit experiment, the number of fringes observed in the field of view is small with longer wavelength of light and is large with shorter wavelength of light.

Statement-2: In the double slit experiment the fringe width depends directly on the wavelength of light.

[NCERT : P L-265, 266 | Online April 22, 2013]

(a) Statement-1 is true, Statement-2 is true and the Statement-2 is correct explanation of the Statement-1.
(b) Statement-1 is false and the Statement-2 is true.
(c) Statement-1 is true Statement-2 is true and the Statement-2 is not correct explanation of the Statement-1.
(d) Statement-1 is true and the Statement-2 is false.

90. In Young's double slit experiment, one of the slit is wider than other, so that amplitude of the light from one slit is double of that other slit. If I_m be the maximum intensity, the resultant intensity I when they interfere at phase difference ϕ is given by : **[NCERT : P L-264 | 2012]**

(a) $\frac{I_m}{9}(4+5\cos\phi)$ (b) $\frac{I_m}{3}\left(1+2\cos^2\frac{\phi}{2}\right)$
(c) $\frac{I_m}{5}\left(1+4\cos^2\frac{\phi}{2}\right)$ (d) $\frac{I_m}{9}\left(1+8\cos^2\frac{\phi}{2}\right)$

91. In Young's double slit interference experiment, the slit widths are in the ratio 1 : 25. Then the ratio of intensity at the maxima and minima in the interference pattern is **[NCERT : P L-264 | Online May 26, 2012]**

(a) 3 : 2 (b) 1 : 25 (c) 9 : 4 (d) 1 : 5

92. The maximum number of possible interference maxima for slit separation equal to 1.8λ, where λ is the wavelength of light used, in a Young's double slit experiment is **[NCERT : P L-266 | Online May 12, 2012]**

(a) zero (b) 3 (c) infinite (d) 5

93. In a Young's double slit experiment with light of wavelength λ, fringe pattern on the screen has fringe width β. When two thin transparent glass (refractive index μ) plates of thickness t_1 and t_2 ($t_1 > t_2$) are placed in the path of the two beams respectively, the fringe pattern will shift by a distance **[NCERT : P L-266 | Online May 7, 2012]**

(a) $\frac{\beta(\mu-1)}{\lambda}\left(\frac{t_1}{t_2}\right)$ (b) $\frac{\mu\beta}{\lambda}\frac{t_1}{t_2}$
(c) $\frac{\beta(\mu-1)}{\lambda}(t_1-t_2)$ (d) $(\mu-1)\frac{\lambda}{\beta}(t_1+t_2)$

94. At two points P and Q on screen in Young's double slit experiment, waves from slits S_1 and S_2 have a path difference of 0 and $\frac{\lambda}{4}$, respectively. The ratio of intensities at P and Q will be : **[NCERT : P L-264 | 2011 RS]**

(a) 2 : 1 (b) $\sqrt{2}$: 1
(c) 4 : 1 (d) 3 : 2

95. In a Young's double slit experiment, the two slits act as coherent sources of wave of equal amplitude A and wavelength λ. In another experiment with the same arrangement the two slits are made to act as incoherent sources of waves of same amplitude and wavelength. If the intensity at the middle point of the screen in the first case is I_1 and in the second case is I_2, then the ratio $\frac{I_1}{I_2}$ is **[NCERT : P L-263, 264 | 2011 RS]**

(a) 2 (b) 1 (c) 0.5 (d) 4

96. A mixture of light, consisting of wavelength 590 nm and an unknown wavelength, illuminates Young's double slit and gives rise to two overlapping interference patterns on the screen. The central maximum of both lights coincide. Further, it is observed that the third bright fringe of known light coincides with the 4th bright fringe of the unknown light. From this data, the wavelength of the unknown light is: **[NCERT : P L-266 | 2009]**

(a) 885.0 nm (b) 442.5 nm
(c) 776.8 nm (d) 393.4 nm

97. In a Young's double slit experiment the intensity at a point where the path difference is $\frac{\lambda}{6}$ (λ being the wavelength of light used) is I. If I_0 denotes the maximum intensity, $\frac{I}{I_0}$ is equal to **[NCERT : P L-264 | 2007]**

(a) $\frac{3}{4}$ (b) $\frac{1}{\sqrt{2}}$ (c) $\frac{\sqrt{3}}{2}$ (d) $\frac{1}{2}$

98. A Young's double slit experiment uses a monochromatic source. The shape of the interference fringes formed on a screen is **[NCERT : P L-265, 266 | 2005]**

(a) circle (b) hyperbola
(c) parabola (d) straight line

99. The maximum number of possible interference maxima for slit-separation equal to twice the wavelength in Young's double-slit experiment is **[NCERT : P L-266 | 2004]**

(a) three (b) five (c) infinite (d) zero

Topic 3 - Diffraction, Polarisation of Light

100. Two plane polarized light waves combine at a certain point whose electric field components are

$E_1 = E_0 \sin\omega t$

$E_2 = E_0 \sin(\omega t + \frac{\pi}{3})$

Find the amplitude of the resultant wave. **[April 7, 2025 (I)]**

(a) 0.9 E (b) E_0
(c) 1.7 E_0 (d) 3.4 E_0

101. Two polarisers P_1 and P_2 are placed in such a way that the intensity of the transmitted light will be zero. A third polariser P_3 is inserted in between P_1 and P_2, at the particular angle between P_2 and P_3. The transmitted

intensity of the light passing the through all three polarisers is maximum. The angle between the polarisers P_2 and P_3 is: **[NCERT : P L-270 | April 4, 2025 (II)]**

(a) $\frac{\pi}{4}$ (b) $\frac{\pi}{6}$ (c) $\frac{\pi}{8}$ (d) $\frac{\pi}{3}$

102. If the measured angular separation between the second minimum to the left of the central maximum and the third minimum to the right of the central maximum is 30° in a single slit diffraction pattern recorded using 628 nm light, then the width of the slit is _____ μm.

[NCERT : P L-267 | April 2, 2025 (I)]

103. In a Young's double slit experiment, three polarizers are kept as shown in the figure. The transmission axes of P_1 and P_2 are orthogonal to each other. The polarizer P_3 covers both the slits with its transmission axis at 45° to those of P_1 and P_2. An unpolarized light of wavelength λ and intensity I_0 is incident on P_1 and P_2. The intensity at a point after P_3 where the path difference between the light waves from s_1 and s_2 is $\frac{\lambda}{3}$, is

[NCERT : P L-270 | Jan. 24, 2025 (II)]

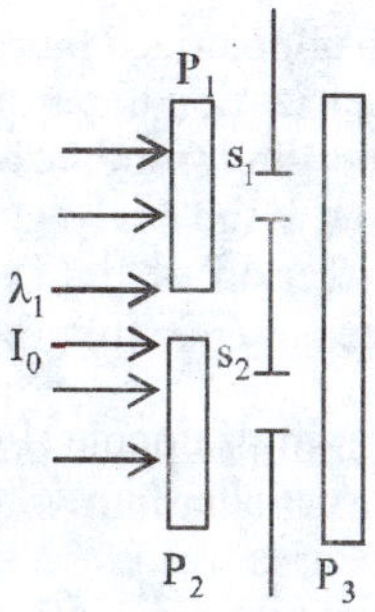

(a) $\frac{I_0}{2}$

(b) $\frac{I_0}{4}$

(c) I_0

(d) $2I_0$

104. A parallel beam of monochromatic light of wavelength 600 nm passes through single slit of 0.4 mm width. Angular divergence corresponding to second order minima would be_____ $\times 10^{-3}$ rad.

[NCERT : P L-267 | April 8, 2024 (I)]

105. Two slits are 1 mm apart and the screen is located 1 m away from the slits. A light wavelength 500 nm is used. The width of each slit to obtain 10 maxima of the double slit pattern within the central maximum of the single slit pattern is.......... $\times 10^{-4}$ m.

[NCERT : P L-267 | April 8, 2024 (II)]

106. In a single slit experiment, a parallel beam of green light of wavelength 550 nm passes through a slit of width 0.20 mm. The transmitted light is collected on a screen 100 cm away. The distance of first order minima from the central maximum will be x $\times 10^{-5}$ m. The value of x is :

[NCERT : P L-267 | April 5, 2024 (II)]

107. A monochromatic light of wavelength 6000 Å is incident on the single slit of width 0.01 mm. If the diffraction pattern is formed at the focus of the convex lens of focal length 20 cm, the linear width of the central maximum is:

[NCERT : P L-267 | Feb. 1, 2024 (I)]

(a) 60mm (b) 24mm (c) 120mm (d) 12mm

108. A microwave of wavelength 2.0 cm falls normally on a slit of width 4.0 cm. The angular spread of the central maxima of the diffraction pattern obtained on a screen 1.5 m away from the slit, will be:

[NCERT : P L-267 | Feb. 1, 2024 (II)]

(a) 60° (b) 45° (c) 15° (d) 30°

109. When unpolarized light is incident at an angle of 60° on a transparent medium from air. The reflected ray is completely polarized. The angle of refraction in the medium is

[Jan. 31, 2024 (II)]

(a) 30° (b) 60° (c) 90° (d) 45°

110. The diffraction pattern of a light of wavelength 400 nm diffracting from a slit of width 0.2 mm is focused on the focal plane of a convex lens of focal length 100 cm. The width of the 1st secondary maxima will be:

[NCERT : P L-267 | Jan. 30, 2024 (I)]

(a) 2mm (b) 2 cm (c) 0.02 mm (d) 0.2 mm

111. A beam of unpolarised light of intensity I_o is passed through a polaroid *A* and then through another polaroid *B* which is oriented so that its principal plane makes an angle of 45° relative to that of *A*. The intensity of emergent light is: **[NCERT : P L-270 | Jan. 30, 2024 (II)]**

(a) $I_o/4$ (b) I_o (c) $I_o/2$ (d) $I_o/8$

112. In a single slit diffraction pattern, a light of wavelength 6000 Å is used. The distance between the first and third minima in the diffraction pattern is found to be 3 mm when the screen in placed 50 cm away from slits. The width of the slit is _____ $\times 10^{-4}$ m.

[NCERT : P L-267 | Jan. 29, 2024 (II)]

113. When a polaroid sheet is rotated between two crossed polaroids then the transmitted light intensity will be maximum for a rotation of:

[NCERT : P L-270 | Jan. 27, 2024 (II)]

(a) 60° (b) 30° (c) 90° (d) 45°

114. A parallel beam of monochromatic light of wavelength 5000 Å is incident normally on a single narrow slit of width 0.001 mm. The light is focused by convex lens on screen, placed on its focal plane. The first minima will be formed for the angle of diffraction of ________ (degree).

[NCERT : P L-267 | Jan. 27, 2024 (II)]

115. A single slit of a width *a* is illuminated by a monochromatic light of wavelength 600 nm. The value of 'a' for which first minimum appears at θ = 30° on the screen will be :

[NCERT : P L-267 | April 15, 2023 (I)]

(a) 0.6 μm (b) 1.2 μm (c) 1.8 μm (d) 3 μm

116. Unpolarised light of intensity 32 Wm^{-2} passes through the combination of three polaroids such that the pass axis of the last polaroid is perpendicular to that of the pass axis of first polaroid. If intensity of emerging light is 3 Wm^{-2}, then the angle between pass axis of first two polaroids is ________°.

[NCERT : PL-270 | April 10, 2023 (I)]

117. '*n*' polarizing sheets are arranged such that each makes an angle 45° with the preceding sheet. An unpolarized light of intensity I is incident into this arrangement. The output intensity is found to be I/64. The value of *n* will be:

[NCERT : P L-270 | Feb. 1, 2023 (I)]

(a) 6 (b) 3 (c) 5 (d) 4

118. Two polaroid A and B are placed in such a way that the pass-axis of polaroids are perpendicular to each other. Now, another polaroid C is placed between A and B bisecting angle between them. If intensity of unpolarised light is I_0 then intensity of transmitted light after passing through polaroid B will be :

[NCERT : P L-270 | Jan. 31, 2023 (I)]

(a) $\frac{I_0}{4}$ (b) $\frac{I_0}{2}$ (c) $\frac{I_0}{8}$ (d) Zero

119. As shown in figures, three identical polaroids P_1, P_2 and P_3 are placed one after another. The pass axis of P_2 and P_3 are inclined at angle of 60° and 90° with respect to axis of P_1. The source S has an intensity of 256 $\frac{W}{m^2}$. The intensity of light at point O is --------- $\frac{W}{m^2}$.

[NCERT : P L-270 | Jan. 29, 2023 (I)]

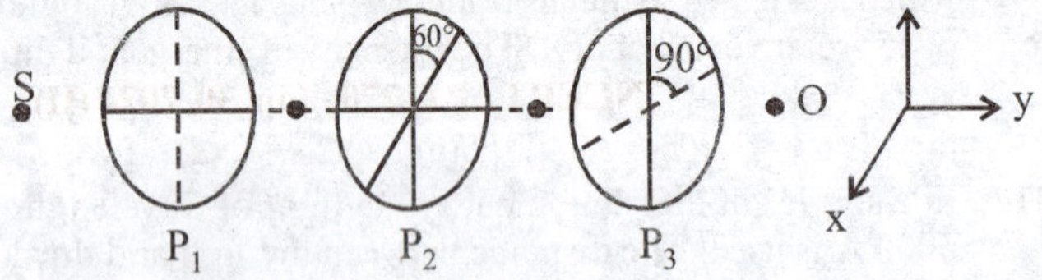

120. Unpolarised light is incident on the boundary between two dielectric media, whose dielectric constants are 2.8 (medium – 1) and 6.8 (medium – 2), respectively. To satisfy the condition, so that the reflected and refracted rays are perpendicular to each other, the angle of incidence should be $\tan^{-1}\left(1+\frac{10}{\theta}\right)^{\frac{1}{2}}$ the value of θ is _____.

(Given for dielectric media, $\mu_r = 1$) **[Jan. 29, 2023 (II)]**

121. Given below are two statements :

Statement I : If the Brewster's angle for the light propagating from air to glass is θ_B, then Brewster's angle for the light propagating from glass to air is $\frac{\pi}{2}-\theta_B$

Statement II : The Brewster's angle for the light propagating from glass to air is $\tan^{-1}(\mu_g)$ where μ_g is the refractive index of glass.

In the light of the above statements, choose the correct answer from the options given below : **[Jan. 24, 2023 (I)]**

(a) Both Statements I and Statement II are true.
(b) Statement I is true but Statement II is false.
(c) Both Statement I and Statement II are false.
(d) Statement I is false but Statement II is true.

122. An unpolarised light beam of intensity $2I_0$ is passed through a polaroid P and then through another polaroid Q which is oriented in such a way that its passing axis makes an angle of 30° relative to that of P. The intensity of the emergent light is

[NCERT : P L-270 | July 29, 2022 (II)]

(a) $\frac{I_0}{4}$ (b) $\frac{I_0}{2}$ (c) $\frac{3I_0}{4}$ (d) $\frac{3I_0}{2}$

123. A light whose electric field vectors are completely removed by using a good Polaroid, allowed to incident on the surface of the prism at Brewster's angle. Choose the most suitable option for the phenomenon related to the prism.

[June 25, 2022 (II)]

(a) Reflected and refracted rays will be perpendicular to each other
(b) Wave will propagate along the surface of prism
(c) No refraction, and there will be total reflection of light.
(d) No reflection and there will be total transmission of light.

124. Sodium light of wavelengths 650 nm and 655 nm is used to study diffraction at a single slit of aperture 0.5 mm. The distance between the slit and the screen is 2.0 m. The separation between the positions of the first maxima of diffraction pattern obtained in the two cases is ____ $\times 10^{-5}$ m. **[NCERT : P L-267 | June 24, 2022 (I)]**

125. A source of light is placed in front of a screen. Intensity of light on the screen is I. Two Polaroids P_1 and P_2 are so placed in between the source of light and screen that the intensity of light on screen is I/2. P_2 should be rotated by an angle of _________ (degrees) so that the intensity of light on the screen becomes $\frac{3I}{8}$.

[NCERT : PL-270 | NA, Aug. 26, 2021 (II)]

126. An unpolarized light beam is incident on the polarizer of a polarization experiment and the intensity of light beam emerging from the analyzer is measured as 100 Lumens. Now, if the analyzer is rotated around the horizontal axis (direction of light) by 30° in clockwise direction, the intensity of emerging light will be ______ Lumens.

[NCERT : P L-270 | Feb. 24, 2021 (I)]

127. A beam of plane polarised light of large cross-sectional area and uniform intensity of 3.3 Wm^{-2} falls normally on a polariser (cross sectional area 3×10^{-4} m^2) which rotates about its axis with an angular speed of 31.4 rad/s. The energy of light passing through the polariser per revolution, is close to :

[NCERT : P L-270 | Sep. 04, 2020 (I)]

(a) 1.0×10^{-5} J (b) 1.0×10^{-4} J
(c) 1.5×10^{-4} J (d) 5.0×10^{-4} J

128. Orange light of wavelength 6000×10^{-10} m illuminates a single slit of width 0.6×10^{-4} m. The maximum possible number of diffraction minima produced on both sides of the central maximum is ________.

[NCERT : P L-267 | NA Sep. 04, 2020 (II)]

129. A polarizer - analyser set is adjusted such that the intensity of light coming out of the analyser is just 10% of the original intensity. Assuming that the polarizer - analyser set does not absorb any light, the angle by which the analyser need to be rotated further to reduce the output intensity to be zero, is:

[NCERT : P L-270 | 7 Jan. 2020 I]

(a) 71.6° (b) 18.4° (c) 90° (d) 45°

130. A system of three polarizers P_1, P_2, P_3 is set up such that the pass axis of P_3 is crossed with respect to that of P_1. The pass axis of P_2 is inclined at 60° to the pass axis of P_3. When a beam of unpolarized light of intensity I_o is incident on P_1, the intensity of light transmitted by the three polarizers is I. The ratio (I_o/I) equals (nearly) :

[NCERT : P L-270 | 12 April 2019 II]

(a) 5.33 (b) 16.00 (c) 10.67 (d) 1.80

131. In a double-slit experiment, green light (5303Å) falls on a double slit having a separation of 19.44 μm and a width of 4.05 μm. The number of bright fringes between the first and the second diffraction minima is :

[NCERT : P L-267 | 11 Jan 2019 II]

(a) 10 (b) 05 (c) 04 (d) 09

132. The angular width of the central maximum in a single slit diffraction pattern is 60°. The width of the slit is 1 μm. The slit is illuminated by monochromatic plane waves. If another slit of same width is made near it, Young's fringes can be observed on a screen placed at a distance 50 cm from the slits. If the observed fringe width is 1 cm, what is slit separation distance?
(i.e. distance between the centres of each slit.)

[NCERT : P L-267 | 2018]

(a) 25 μm (b) 50 μm
(c) 75 μm (d) 100 μm

133. Unpolarized light of intensity I passes through an ideal polarizer A. Another indentical polarizer B is placed behind A. The intensity of light beyond B is found to be $\frac{I}{2}$. Now another identical polarizer C is placed between A and B. The intensity beyond B is now found to be $\frac{I}{8}$. The angle between polarizer A and C is:

[NCERT : P L-270 | 2018]

(a) 0° (b) 30°
(c) 45° (d) 60°

134. Unpolarized light of intensity I is incident on a system of two polarizers, A followed by B. The intensity of emergent light is I/2. If a third polarizer C is placed between A and B, the intensity of emergent light is reduced to I/3. The angle between the polarizers A and C is θ. Then

[NCERT : P L-270 | Online April 16, 2018]

(a) $\cos\theta = \left(\frac{2}{3}\right)^{1/4}$ (b) $\cos\theta = \left(\frac{1}{3}\right)^{1/4}$

(c) $\cos\theta = \left(\frac{1}{3}\right)^{1/2}$ (d) $\cos\theta = \left(\frac{2}{3}\right)^{1/2}$

135. A single slit of width 0.1 mm is illuminated by a parallel beam of light of wavelength 6000 Å and diffraction bands are observed on a screen 0.5 m from the slit. The distance of the third dark band from the central bright band is :

[NCERT : P L-267 | Online April 9, 2017]

(a) 3mm (b) 9mm
(c) 4.5mm (d) 1.5mm

136. A single slit of width b is illuminated by a coherent monochromatic light of wavelength λ. If the second and fourth minima in the diffraction pattern at a distance 1 m from the slit are at 3 cm and 6 cm respectively from the central maximum, what is the width of the central maximum? (i.e. distance between first minimum on either side of the central maximum)

[NCERT : P L-267 | Online April 8, 2017]

(a) 1.5 cm (b) 3.0 cm (c) 4.5 cm (d) 6.0 cm

137. The box of a pin hole camera, of length L, has a hole of radius a. It is assumed that when the hole is illuminated by a parallel beam of light of wavelength λ the spread of the spot (obtained on the opposite wall of the camera) is the sum of its geometrical spread and the spread due to diffraction. The spot would then have its minimum size (say b_{min}) when :

[NCERT : P L-267 | 2016]

(a) $a = \sqrt{\lambda L}$ and $b_{min} = \sqrt{4\lambda L}$

(b) $a = \frac{\lambda^2}{L}$ and $b_{min} = \sqrt{4\lambda L}$

(c) $a = \frac{\lambda^2}{L}$ and $b_{min} = \left(\frac{2\lambda^2}{L}\right)$

(d) $a = \sqrt{\lambda l}$ and $b_{min} = \left(\frac{2\lambda^2}{L}\right)$

138. Two beams, A and B, of plane polarized light with mutually perpendicular planes of polarization are seen through a polaroid. From the position when the beam A has maximum intensity (and beam B has zero intensity), a rotation of polaroid through 30° makes the two beams appear equally bright. If the initial intensities of the two beams are I_A and I_B respectively, then $\frac{I_A}{I_B}$ equals:

[NCERT : P L-270 | 2014]

(a) 3 (b) $\frac{3}{2}$ (c) 1 (d) $\frac{1}{3}$

139. A ray of light is incident from a denser to a rarer medium. The critical angle for total internal reflection is θ_{iC} and Brewster's angle of incidence is θ_{iB}, such that $\sin\theta_{iC}/\sin\theta_{iB} = \eta = 1.28$. The relative refractive index of the two media is:

[Online April 19, 2014]

(a) 0.2 (b) 0.4 (c) 0.8 (d) 0.9

140. In an experiment of single slit diffraction pattern, first minimum for red light coincides with first maximum of some other wavelength. If wavelength of red light is 6600 Å, then wavelength of first maximum will be:

[NCERT : P L-267 | Online April 12, 2014]

(a) 3300Å (b) 4400Å (c) 5500Å (d) 6600Å

141. Abeam of unpolarised light of intensity I_0 is passed through a polaroidAand then through another polaroid B which is oriented so that its principal plane makes an angle of 45° relative to that of A. The intensity of the emergent light is

[NCERT : P L-270 | 2013]

(a) I_0 (b) $I_0/2$ (c) $I_0/4$ (d) $I_0/8$

142. This question has Statement-1 and Statements-2. Of the four choices given after the Statements, choose the one that best describes the two Statements.

Statement-1 : Out of radio waves and microwaves, the radio waves undergo more diffraction.

Statement-2 : Radio waves have greater frequency compared to microwaves.

[NCERT : P L-267, 268 | Online April 25, 2013]

(a) Statement-1 is true, Statement-2 is true and Statement-2 is the correct explanation of Statement-1
(b) Statement-1 is false, Statement-2 is true.
(c) Statement-1 is true, Statement-2 is false.
(d) Statement-1 is true, Statement-2 is true but Statement-2 is not the correct explanation of Statement-1

143. The first diffraction minimum due to the single slit diffraction is seen at $\theta = 30°$ for a light of wavelength 5000 Å falling perpendicularly on the slit. The width of the slit is **[NCERT : P L-267, 268 | Online May 12, 2012]**

(a) 2.5×10^{-5} cm (b) 1.25×10^{-5} cm
(c) 10×10^{-5} cm (d) 5×10^{-5} cm

144. Two polaroids have their polarizing directions parallel so that the intensity of a transmitted light is maximum. The angle through which either polaroid must be turned if the intensity is to drop by one-half is **[NCERT : P L-270 | Online May 7, 2012]**

(a) 135° (b) 90° (c) 120° (d) 180°

145. Statement - 1: On viewing the clear blue portion of the sky through a Calcite Crystal, the intensity of transmitted light varies as the crystal is rotated.

Statement - 2: The light coming from the sky is polarized due to scattering of sun light by particles in the atmosphere. The scattering is largest for blue light. **[NCERT : P L-269, 270 | 2011 RS]**

(a) Statement -1 is true, statement-2 is false.
(b) Statement-1 is true, statement-2 is true, statement-2 is the correct explanation of statement-1
(c) Statement-1 is true, statement-2 is true, statement-2 is not the correct explanation of statement-1
(d) Statement-1 is false, statement-2 is true.

146. In an experiment, electrons are made to pass through a narrow slit of width '*d*' comparable to their de Broglie wavelength. They are detected on a screen at a distance '*D*' from the slit (see figure).

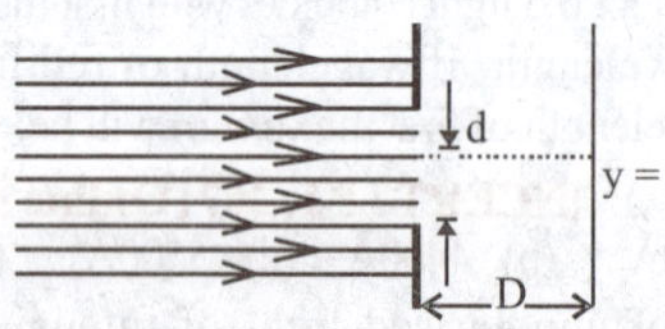

Which of the following graphs can be expected to represent the number of electrons '*N*' detected as a function of the detector position '*y*'(*y* = 0 corresponds to the middle of the slit) **[NCERT : P L-267 | 2008]**

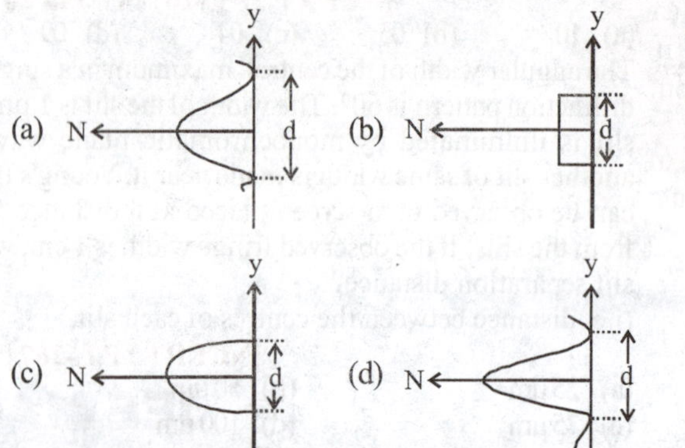

147. If I_0 is the intensity of the principal maximum in the single slit diffraction pattern, then what will be its intensity when the slit width is doubled? **[NCERT : P L-267 | 2005]**

(a) $4I_0$ (b) $2I_0$ (c) $\frac{I_0}{2}$ (d) I_0

148. When an unpolarized light of intensity I_0 is incident on a polarizing sheet, the intensity of the light which does not get transmitted is **[NCERT : P L-270 | 2005]**

(a) $\frac{1}{4}I_0$ (b) $\frac{1}{2}I_0$
(c) I_0 (d) zero

149. The angle of incidence at which reflected light is totally polarized for reflection from air to glass (refractive index *n*), is **[2004]**

(a) $\tan^{-1}(1/n)$ (b) $\sin^{-1}(1/n)$
(c) $\sin^{-1}(n)$ (d) $\tan^{-1}(n)$

ANSWER KEYS

1	(d)	16	(2)	31	(b)	46	(0.20)	61	(600)	76	(d)	91	(c)	106	(275)	121	(b)	136	(b)
2	(24)	17	(d)	32	(15)	47	(d)	62	(b)	77	(750)	92	(b)	107	(b)	122	(c)	137	(a)
3	(a)	18	(b)	33	(c)	48	(c)	63	(450)	78	(a)	93	(c)	108	(a)	123	(d)	138	(d)
4	(54)	19	(d)	34	(b)	49	(c)	64	(1)	79	(c)	94	(a)	109	(a)	124	(3)	139	(c)
5	(c)	20	(c)	35	(11)	50	(462)	65	(5)	80	(c)	95	(a)	110	(a)	125	(30)	140	(b)
6	(8)	21	(c)	36	(a)	51	(4)	66	(300)	81	(None)	96	(b)	111	(a)	126	(75)	141	(c)
7	(d)	22	(b)	37	(d)	52	(48)	67	(b)	82	(a)	97	(a)	112	(2)	127	(b)	142	(c)
8	(c)	23	(b)	38	(c)	53	(10)	68	(b)	83	(a)	98	(d)	113	(d)	128	(198)	143	(c)
9	(24)	24	(d)	39	(b)	54	(3)	69	(b)	84	(d)	99	(b)	114	(30)	129	(b)	144	(a)
10	(2)	25	(d)	40	(a)	55	(c)	70	(600)	85	(d)	100	(c)	115	(b)	130	(c)	145	(b)
11	(b)	26	(b)	41	(200)	56	(a)	71	(a)	86	(c)	101	(a)	116	(30)	131	(b)	146	(d)
12	(b)	27	(b)	42	(4)	57	(630)	72	(a)	87	(c)	102	(6)	117	(a)	132	(a)	147	(a)
13	(b)	28	(d)	43	(10)	58	(b)	73	(9)	88	(d)	103	(b)	118	(c)	133	(c)	148	(b)
14	(d)	29	(a)	44	(9)	59	(d)	74	(b)	89	(a)	104	(6)	119	(24)	134	(a)	149	(d)
15	(d)	30	(d)	45	(125)	60	(d)	75	(a)	90	(d)	105	(2)	120	(7)	135	(b)		

11

Dual Nature of Radiation and Matter

TREND ANALYSIS
JEE MAIN (2025-2021)

	2025				2024				2023				2022				2021			
	Session II		Session I		Session II		Session I		Session II		Session I		Session II		Session I		Session II		Session I	
	MCQs	NVQs	MCQs	NVQs	MCQs	NVQs	MCQs	NVQs	MCQs	NVQs	MCQs	NVQs	MCQs	NVQs	MCQs	NVQs	MCQs	NVQs	MCQs	NVQs
Topic 1	2	–	5	–	3	–	2	–	5	–	5	–	5	–	4	–	5	–	4	–
Topic 2	4	–	7	1	8	–	8	–	6	1	6	–	5	–	6	2	5	3	3	1
Topic 3																				
Topic 4																				
Total	6	-	12	1	11	-	10	-	11	1	11	-	10	-	10	2	10	3	7	1

How to Code Qns for Better & Faster Revision →

Type 1: Able to solve (No code Req)
Type 2: Solved with struggle or by seeing solution
(Use code NR means Needs Revision or orange stickers)
Type 3: Tricky/ Could not solve on my own. Need help from Teacher/ Friend.
(Use code TQ means Tricky Question or red stickers)

You can either hand write the code in boxes or else paste coloured stickers given at end of book.

Topic 1 - Matter Waves, Cathode and Positive Rays

1. An electron is released from rest near an infinite non-conducting sheet of uniform charge density '$-\sigma$'. The rate of change of de-Broglie wave length associated with the electron varies inversely as n^{th} power of time. The numerical value of n is ______.

[NCERT : P L-285 | April 8, 2025 (II)]

2. An electron with mass 'm' with an initial velocity (t = 0) $\vec{v} = v_0\hat{i}$ $(v_0 > 0)$ enters a magnetic field $\vec{B} = B_0\hat{j}$. If the initial de-Broglie wavelength at t = 0 is λ_0 then its value after time 't' would be : **[NCERT : PL-285 | April 2, 2025 (II)]**

(a) $\dfrac{\lambda_0}{\sqrt{1-\dfrac{e^2B_0^2t^2}{m^2}}}$ (b) $\dfrac{\lambda_0}{\sqrt{1+\dfrac{e^2B_0^2t^2}{m^2}}}$

(c) $\lambda_0\sqrt{1+\dfrac{e^2B_0^2t^2}{m^2}}$ (d) λ_0

3. If λ and K are de-Broglie wavelength and kinetic energy, respectively, of a particle with constant mass. The correct graphical representation for the particle will be :-

[NCERT : P L-285 | Jan. 29, 2025 (I)]

(a) 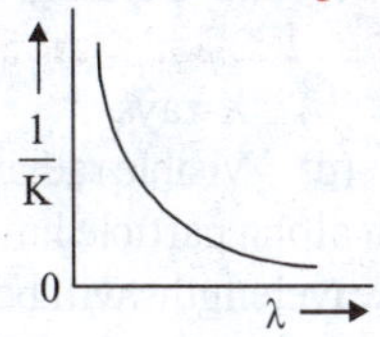

(b) 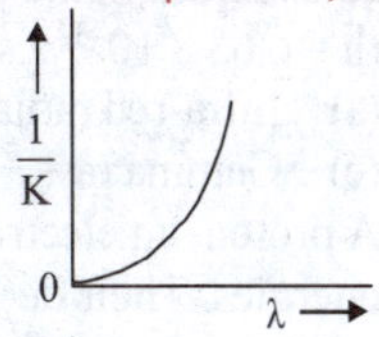

(c) 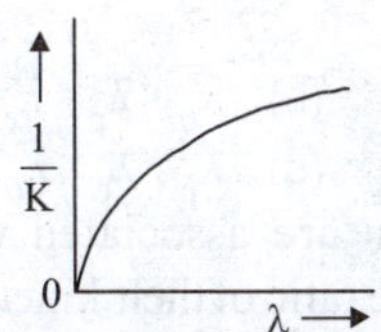

(d) 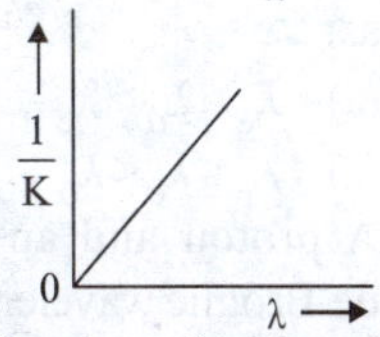

4. A photo-emissive substance is illuminated with a radiation of wavelength λ_i so that it releases electrons with de-Broglie wavelength λ_e. The longest wavelength of radiation that can emit photoelectron is λ_0. Expression for de-Broglie wavelength is given by : (m : mass of the electron, h : Planck's constant and c : speed of light)

[NCERT : P L-281, 282 | April 7, 2025 (II)]

(a) $\lambda_e = \dfrac{h}{\sqrt{2mc\left(\dfrac{1}{\lambda_i} - \dfrac{1}{\lambda_0}\right)}}$ (b) $\lambda_e = \sqrt{\dfrac{h\lambda_0}{2mc}}$

(c) $\lambda_e = \dfrac{h}{\sqrt{2mc\left(\dfrac{1}{\lambda_i} - \dfrac{1}{\lambda_0}\right)}}$ (d) $\lambda_e = \sqrt{\dfrac{h\lambda_i}{2mc}}$

5. A proton of mass 'm_p' has same energy as that of a photon of wavelength 'λ'. If the proton is moving at non-relativistic speed, then ratio of its de-Broglie wavelength to the wavelength of photon is. **[NCERT : P L-285 | Jan. 28, 2025 (I)]**

(a) $\dfrac{1}{c}\sqrt{\dfrac{2E}{m_p}}$ (b) $\dfrac{1}{c}\sqrt{\dfrac{E}{m_p}}$

(c) $\dfrac{1}{c}\sqrt{\dfrac{E}{2m_p}}$ (d) $\dfrac{1}{2c}\sqrt{\dfrac{E}{m_p}}$

6. The energy E and momentum p of a moving body of mass m are related by some equation. Given that c represents the speed of light, identify the correct equation. **[NCERT : P L-285 | Jan. 24, 2025 (II)]**

(a) $E^2 = pc^2 + m^2c^4$ (b) $E^2 = pc^2 + m^2c^2$
(c) $E^2 = p^2c^2 + m^2c^2$ (d) $E^2 = p^2c^2 + m^2c^4$

7. An electron of mass 'm' with an initial velocity $\vec{v} = v_0\hat{i}\,(v_0 > 0)$ enters an electric field $\vec{E} = -E_0\hat{k}$. If the initial de-Broglie wavelength is λ_0, then its value after time t would be **[NCERT : P L-285 | Jan. 24, 2025 (I)]**

(a) $\dfrac{\lambda_0}{\sqrt{1+\dfrac{e^2E_0^2t^2}{m^2v_0^2}}}$ (b) $\dfrac{\lambda_0}{\sqrt{1-\dfrac{e^2E_0^2t^2}{m^2v_0^2}}}$

(c) λ_0 (d) $\lambda_0\sqrt{1+\dfrac{e^2E_0^2t^2}{m^2v_0^2}}$

8. A sub-atomic particle of mass 10^{-30} kg is moving with a velocity 2.21×10^6 m/s. Under the matter wave consideration, the particle will behave closely like ______. ($h = 6.63 \times 10^{-34}$ J.s) **[NCERT : P L-285 | Jan. 23, 2025 (I)]**

(a) Infra-red radiation (b) X-rays
(c) Gamma rays (d) Visible radiation

9. A proton, an electron and an alpha particle have the same energies. Their de-Broglie wavelengths will be compared as: **[NCERT : P L-285 | April 9, 2024 (I)]**

(a) $\lambda_e > \lambda_\alpha > \lambda_p$ (b) $\lambda_\alpha < \lambda_p < \lambda_e$
(c) $\lambda_p < \lambda_e < \lambda_\alpha$ (d) $\lambda_p > \lambda_e > \lambda_\alpha$

10. A proton and an electron are associated with same de-Broglie wavelength. The ratio of their kinetic energies is:
(Assume $h = 6.63 \times 10^{-34}$ J s, $m_e = 9.0 \times 10^{-31}$ kg and $m_p = 1836$ times m_e) **[NCERT : P L-285 | April 8, 2024 (I)]**

(a) 1 : 1836 (b) $1 : \dfrac{1}{1836}$

(c) $1 : \dfrac{1}{\sqrt{1836}}$ (d) $1 : \sqrt{1836}$

11. A proton and an electron have the same de Broglie wavelength. If K_p and K_e be the kinetic energies of proton and electron respectively. Then choose the correct relation: **[NCERT : P L-285 | April 8, 2024 (II)]**

(a) $K_p > K_e$ (b) $K_p = K_e$ (c) $K_p = K_e^2$ (d) $K_p < K_e$

12. The de Brogile wavelength of a proton and an α particle are λ and 2λ respectively. The ratio of the velocities of proton and α particle will be: **[NCERT : P L-285 | Feb. 1, 2024 (I)]**

(a) 1 : 8 (b) 1 : 2 (c) 4 : 1 (d) 8 : 1

13. The de-Broglie wavelength of an electron is the same as that of a photon. If velocity of electron is 25% of the velocity of light, then the ratio of K.E. of electron and K.E. of photon will be: **[NCERT : P L-285 | Jan. 29, 2024 (I)]**

(a) $\dfrac{1}{4}$ (b) $\dfrac{1}{1}$ (c) $\dfrac{8}{1}$ (d) $\dfrac{1}{8}$

14. The de Broglie wavelength of an electron having kinetic energy E is λ. If the kinetic energy of electron becomes $\dfrac{E}{4}$, then its de-Broglie wavelength will be : **[NCERT : P L-285 | April 15, 2023 (I)]**

(a) $\dfrac{\lambda}{\sqrt{2}}$ (b) $\dfrac{\lambda}{2}$ (c) 2λ (d) $\sqrt{2}\lambda$

15. A proton and an α-particle are accelerated from rest by 2V and 4V potentials, respectively. The ratio of their de-Broglie wavelength is: **[NCERT : P L-285 | April 12, 2023 (I)]**

(a) 4 : 1 (b) 2 : 1 (c) 8 : 1 (d) 16 : 1

16. The ratio of the de-Broglie wavelengths of proton and electron having same kinetic energy: **[NCERT : P L-285 | April 11, 2023 (II)]**
(Assume $m_p = m_e \times 1849$)

(a) 1 : 43 (b) 1 : 30 (c) 1 : 62 (d) 2 : 43

17. Proton (P) and electron (e) will have same de-Broglie wavelength when the ratio of their momentum is (assume, $m_p = 1849m_e$) **[NCERT : P L-285 | April 8, 2023 (I)]**

(a) 1 : 43 (b) 43 : 1 (c) 1 : 1849 (d) 1 : 1

18. The kinetic energy of an electron, α–particle and a proton are given as 4K, 2K and K respectively. The de–Broglie wavelength associated with electron (λ_e) α–particle (λ_α) and the proton (λ_p) are as follows: **[NCERT : P L-285 | April 6, 2023 (I)]**

(a) $\lambda_\alpha = \lambda_p < \lambda_e$ (b) $\lambda_\alpha > \lambda_p > \lambda_e$
(c) $\lambda_\alpha < \lambda_p < \lambda_e$ (d) $\lambda_\alpha = \lambda p > \lambda_e$

19. A proton moving with one tenth of velocity of light has a certain de Broglie wavelength of λ. An alpha particle having certain kinetic energy has the same de-Broglie wavelength λ. The ratio of kinetic energy of proton and that of alpha particle is: **[NCERT : P L-285 | Feb. 1, 2023 (I)]**

(a) 4 : 1 (b) 1 : 4 (c) 1 : 2 (d) 2 : 1

20. An electron accelerated through a potential difference V_1 has a de-Broglie wavelength of λ. When the potential is changed to V_2, its de-Broglie wavelength increases by 50%. The value of $\left(\frac{V_1}{V_2}\right)$ is equal to:

[NCERT : P L-285 | Jan. 30, 2023 (II)]

(a) 3 (b) $\frac{9}{4}$ (c) $\frac{3}{2}$ (d) 4

21. The ratio of de-Broglie wavelength of an α-particle and a proton accelerated from rest by the same potential is $\frac{1}{\sqrt{m}}$, the value of m is [NCERT : P L-285 | Jan. 29, 2023 (II)]

(a) 4 (b) 16 (c) 8 (d) 2

22. Electron beam used in an electron microscope, when accelerated by a voltage of 20 kV. has a de-Broglie wavelength of λ_o. If the voltage is increased to 40 kV. then the de-Broglie wavelength associated with the electron beam would be: [NCERT : P L-285 | Jan. 25, 2023 (I)]

(a) $3\lambda_o$ (b) $9\lambda_o$ (c) $\frac{\lambda_o}{2}$ (d) $\frac{\lambda_o}{\sqrt{2}}$

23. An α–particle, a proton and an electron have the same kinetic energy. Which one of the following is correct in case of their De-Brogile wavelength

[NCERT : P L-285 | Jan. 24, 2023 (II)]

(a) $\lambda_\alpha > \lambda_p > \lambda_e$ (b) $\lambda_\alpha < \lambda_p < \lambda_e$
(c) $\lambda_\alpha = \lambda_p = \lambda_e$ (d) $\lambda_\alpha > \lambda_p < \lambda_e$

24. An α particle and a proton are accelerated from rest through the same potential difference. The ratio of linear momenta acquired by above two particals will be :

[NCERT : P L-285 | July 29, 2022 (II)]

(a) $\sqrt{2}:1$ (b) $2\sqrt{2}:1$ (c) $4\sqrt{2}:1$ (d) 8 : 1

25. The equation $\lambda = \frac{1.227}{x}$ nm can be used to find the de-Broglie wavelength of an electron. In this equation x stands for : [NCERT : P L-285 | July 28, 2022 (I)]

Where,
m = mass of electron
P = momentum of electron
K = Kinetic energy of electron
V = Accelerating potential in volts for electron

(a) $\sqrt{mK}$ (b) $\sqrt{P}$ (c) $\sqrt{K}$ (d) $\sqrt{V}$

26. An electron (mass m) with an initial velocity $\vec{v} = v_0\hat{i}$ $(v_0 > 0)$ is moving in an electric field $\vec{E} = -E_0\hat{i}$ $(E_0 > 0)$ where E_0 is constant. If at $t = 0$ de Broglie wavelength is $\lambda_0 = \frac{h}{mv_0}$, then its de Broglie wavelength after time t is given by [NCERT : P L-285 | July 27, 2022 (I)]

(a) λ_0 (b) $\lambda_0\left(1+\frac{eE_0t}{mv_0}\right)$

(c) $\lambda_0 t$ (d) $\frac{\lambda_0}{\left(1+\frac{eE_0t}{mv_0}\right)}$

27. A nucleus of mass M at rest splits into two parts having masses $\frac{M'}{3}$ and $\frac{2M'}{3}$ $(M' < M)$. The ratio of de Broglie wavelength of two parts will be :

[NCERT : P L-285 | July 26, 2022 (II)]

(a) 1 : 2 (b) 2 : 1 (c) 1 : 1 (d) 2 : 3

28. The ratio of wavelengths of proton and deuteron accelerated by potential V_p and V_d is $1:\sqrt{2}$. Then, the ratio of V_p to V_d will be [NCERT : P L-285 | July 25, 2022 (II)]

(a) 1 : 1 (b) $\sqrt{2}:1$ (c) 3 : 1 (d) 4 : 1

29. The de Broglie wavelengths for an electron and a photon are λ_e and λ_p respectively. For the same kinetic energy of electron and photon. Which of the following presents the correct relation between the de Broglie wavelengths of two ? [NCERT : P L-285 | June 28, 2022 (I)]

(a) $\lambda_p \propto \lambda_e^2$ (b) $\lambda_p \propto \lambda_e$
(c) $\lambda_p \propto \sqrt{\lambda_e}$ (d) $\lambda_p \propto \sqrt{\frac{1}{\lambda_e}}$

30. An α particle and a carbon-12 atom has same kinetic energy K. The ratio of their de-Broglie wavelength $(\lambda_a : \lambda_{C_{12}})$ is:

[NCERT : P L-285 | June 27, 2022 (I)]

(a) $1:\sqrt{3}$ (b) $\sqrt{3}:1$ (c) 3 : 1 (d) $2:\sqrt{3}$

31. An electron with speed v and a photon with speed c have the same de-Broglie wavelength. If the kinetic energy and momentum of electron are E_e and p_e and that of photon are E_{ph} and p_{ph} respectively. Which of the following is correct?

[NCERT : P L-285 | June 26, 2022 (I)]

(a) $\frac{E_e}{E_{ph}} = \frac{2c}{v}$ (b) $\frac{E_e}{E_{ph}} = \frac{v}{2c}$

(c) $\frac{p_e}{p_{ph}} = \frac{2c}{v}$ (d) $\frac{p_e}{p_{ph}} = \frac{v}{2c}$

32. A proton, a neutron, an electron and an α-particle have same energy. If λ_p, λ_n, λ_e and λ_α are the de Broglie's wavelengths of proton, neutron, electron and α particle respectively, then choose the correct relation from the following : [NCERT : P L-285 | June 25, 2022 (II)]

(a) $\lambda_p = \lambda_n > \lambda_e > \lambda_\alpha$ (b) $\lambda_a < \lambda_n < \lambda_p < \lambda_e$
(c) $\lambda_e < \lambda_p = \lambda_n > \lambda_\alpha$ (d) $\lambda_e = \lambda_p = \lambda_n = \lambda_a$

33. The temperature of an ideal gas in 3-dimensions is 300 K. The corresponding de-Broglie wavelength of the electron approximately at 300 K, is :

[NCERT : P L-285 | Sep. 1, 2021 (II)]

[me = mass of electron = 9×10^{-31} kg
h = Planck constant = 6.6×10^{-34} Js
k_B = Boltzmann constant = 1.38×10^{-23} JK–1]

(a) 6.26 nm (b) 8.46 nm (c) 2.26 nm (d) 3.25 nm

34. A moving proton and electron have the same de-Broglie wavelength. If K and P denote the K.E. and momentum respectively. Then choose the correct option :

[NCERT : P L-285 | Aug. 31, 2021 (I)]

(a) Kp < Ke and Pp = Pe (b) Kp = Ke and Pp = Pe
(c) Kp < Ke and Pp < Pe (d) Kp > Ke and Pp = Pe

35. Consider two separate ideal gases of electrons and protons having same number of particles. The temperature of both the gases are same. The ratio of the uncertainty in determining the position of an electron to that of a proton is proportional to :- **[Aug. 31, 2021 (II)]**

(a) $\left(\frac{m_p}{m_e}\right)^{3/2}$ (b) $\sqrt{\frac{m_e}{m_p}}$ (c) $\sqrt{\frac{m_p}{m_e}}$ (d) $\frac{m_p}{m_e}$

36. The de-Broglie wavelength of a particle having kinetic energy E is λ. How much extra energy must be given to this particle so that the de-Broglie wavelength reduces to 75% of the initial value? **[NCERT : P L-285 | Aug. 26, 2021 (II)]**

(a) $\frac{1}{9}$E (b) $\frac{7}{9}$E (c) E (d) $\frac{16}{9}$E

37. A particle of mass 4M at rest disintegrates into two particles of mass M and 3M respectively having non zero velocities. The ratio of de-Broglie wavelength of particle of mass M to that of mass 3M will be: **[NCERT : P L-285 | July 25, 2021 (I)]**

(a) 1 : 3 (b) 3 : 1 (c) $1 : \sqrt{3}$ (d) 1 : 1

38. A particle is travelling 4 times as fast as an electron. Assuming the ratio of de-Broglie wavelength of a particle to that of electron is 2 : 1, the mass of the particle is **[NCERT : P L-285 | March 18, 2021 (I)]**

(a) 16 times the mass of e^- (b) 8 times the mass of e^-
(c) $\frac{1}{16}$ times the mass of e^- (d) $\frac{1}{8}$ times the mass of e^-

39. An α particle and a proton are accelerated from rest by a potential difference of 200 V. After this, de Broglie wavelengths are λ_α and λ_p respectively. The ratio $\frac{\lambda_p}{\lambda_\alpha}$ is : **[NCERT : P L-285 | Feb. 25, 2021 (I)]**

(a) 8 (b) 2.8 (c) 3.8 (d) 7.8

40. An electron of mass m_e and a proton of mass $m_p = 1836\, m_e$ are moving with the same speed. The ratio of their de Broglie wavelength $\frac{\lambda_{electron}}{\lambda_{proton}}$ will be **[NCERT : P L-285 | Feb. 25, 2021 (II)]**

(a) $\frac{1}{1836}$ (b) 918 (c) 1 (d) 1836

41. The deBroglie wavelength of a proton and α-particle are equal. The ratio of their velocities is : **[NCERT : P L-285 | Feb. 24, 2021 (II)]**

(a) 4 : 3 (b) 4 : 1 (c) 4 : 2 (d) 1 : 4

42. An electron, a doubly ionized helium ion (He^{++}) and a proton are having the same kinetic energy. The relation between their respective de-Broglie wavelengths λ_e, $\lambda_{He^{++}}$ and λ_p is : **[NCERT : P L-285 | Sep. 06, 2020 (I)]**

(a) $\lambda_e > \lambda_{He^{++}} > \lambda_p$ (b) $\lambda_e < \lambda_{He^{++}} = \lambda_p$
(c) $\lambda_e > \lambda_p > \lambda_{He^{++}}$ (d) $\lambda_e < \lambda_p < \lambda_{He^{++}}$

43. Assuming the nitrogen molecule is moving with r.m.s. velocity at 400 K, the de-Broglie wavelength of nitrogen molecule is close to :
(Given : nitrogen molecule weight : 4.64×10^{-26} kg, Boltzman constant : 1.38×10^{-23} J/K, Planck constant : 6.63×10^{-34} J.s) **[NCERT : P L-285 | Sep. 06, 2020 (II)]**

(a) 0.24 Å (b) 0.20 Å (c) 0.34 Å (d) 0.44 Å

44. Particle A of mass $m_A = \frac{m}{2}$ moving along the x-axis with velocity v_0 collides elastically with another particle B at rest having mass $m_B = \frac{m}{3}$. If both particles move along the x-axis after the collision, the change $\Delta\lambda$ in de-Broglie wavelength of particle A, in terms of its de-Broglie wavelength (λ_0) before collision is : **[NCERT : P L-285 | Sep. 04, 2020 (I)]**

(a) $\Delta\lambda = \frac{3}{2}\lambda_0$ (b) $\Delta\lambda = \frac{5}{2}\lambda_0$
(c) $\Delta\lambda = 2\lambda_0$ (d) $\Delta\lambda = 4\lambda_0$

45. A particle is moving 5 times as fast as an electron. The ratio of the de-Broglie wavelength of the particle to that of the electron is 1.878×10^{-4}. The mass of the particle is close to : **[NCERT : P L-285 | Sep. 02, 2020 (II)]**

(a) 4.8×10^{-27} kg (b) 9.1×10^{-31} kg
(c) 1.2×10^{-28} kg (d) 9.7×10^{-28} kg

46. A particle moving with kinetic energy E has de Broglie wavelength λ. If energy ΔE is added to its energy, the wavelength become $\frac{\lambda}{2}$. Value of ΔE, is: **[NCERT : P L-285 | 9 Jan. 2020 I]**

(a) E (b) 4E (c) 3E (d) 2E

47. An electron of mass m and magnitude of charge |e| initially at rest gets accelerated by a constant electric field E. The rate of change of de-Broglie wavelength of this electron at time t ignoring relativistic effects is: **[NCERT : P L-285 | 9 Jan. 2020 II]**

(a) $-\frac{h}{|e|E\sqrt{t}}$ (b) $\frac{|e|Et}{h}$
(c) $-\frac{h}{|e|Et}$ (d) $\frac{-h}{|e|Et^2}$

48. An electron (mass m) with initial velocity $\vec{v} = v_0\hat{i} + v_0\hat{j}$ is in an electric field $\vec{E} = -E_0\hat{k}$. If λ_0 is initial de-Broglie wavelength of electron, its de-Broglie wave length at time t is given by: **[NCERT : P L-285 | 8 Jan. 2020 II]**

(a) $\frac{\lambda_0\sqrt{2}}{\sqrt{1+\frac{e^2E^2t^2}{m^2v_0^2}}}$ (b) $\frac{\lambda_0}{\sqrt{1+\frac{e^2E_0^2t^2}{m^2v_0^2}}}$
(c) $\frac{\lambda_0}{\sqrt{1+\frac{e^2E_0^2t^2}{2m^2v_0^2}}}$ (d) $\frac{\lambda_0}{\sqrt{2+\frac{e^2E^2t^2}{m^2v_0^2}}}$

49. A particle 'P' is formed due to a completely inelastic collision of particles 'x' and 'y' having de-Broglie wavelengths 'λ_x' and 'λ_y' respectively. If x and y were moving in opposite directions, then the de-Broglie wavelength of 'P' is:
[NCERT : P L-285 | 9 Apr. 2019 II]

(a) $\frac{\lambda_x \lambda_y}{\lambda_x + \lambda_y}$ (b) $\frac{\lambda_x \lambda_y}{|\lambda_x - \lambda_y|}$

(c) $\lambda_x - \lambda_y$ (d) $\lambda_x + \lambda_y$

50. Two particles move at right angle to each other. Their de Broglie wavelengths are λ_1 and λ_2 respectively. The particles suffer perfectly inelastic collision. The de Broglie wavelength λ, of the final particle, is given by :
[NCERT : P L-285 | 8 April 2019 I]

(a) $\frac{1}{\lambda^2} = \frac{1}{\lambda_1^2} + \frac{1}{\lambda_2^2}$ (b) $\lambda = \sqrt{\lambda_1 \lambda_2}$

(c) $\lambda = \frac{\lambda_2 + \lambda_2}{2}$ (d) $\frac{2}{\lambda} = \frac{1}{\lambda_1} + \frac{1}{\lambda_2}$

51. A particle A of mass 'm' and charge 'q' is accelerated by a potential difference of 50v Another particle B of mass '4m' and charge'q' is accelerated by a potential diffrence of 2500V. The ratio of de-Broglie wavelength $\frac{\lambda_A}{\lambda_B}$ is
[NCERT : P L-285 | 12 Jan. 2019 I]

(a) 10.00 (b) 0.07 (c) 14.14 (d) 4.47

52. If the deBroglie wavelength of an electron is equal to 10^{-3} times the wavelength of a photon of frequency 6×10^{14} Hz, then the speed of electron is equal to :
(Speed of light = 3×10^8 m/s)
Planck's constant = 6.63×10^{-34}J.s
Mass of electron = 9.1×10^{-31} kg)
[NCERT : P L-285 | 11 Jan. 2019 I]

(a) 1.1×10^6 m/s (b) 1.7×10^6 m/s
(c) 1.8×10^6 m/s (d) 1.45×10^6 m/s

53. In an electron microscope, the resolution that can be achieved is of the order of the wavelength of electrons used. To resolve a width of 7.5×10^{-12} m, the minimum electron energy required is close to:
[NCERT : P L-285 | 10 Jan. 2019 I]

(a) 500 keV (b) 100 keV (c) 1 keV (d) 25 keV

54. If the de Broglie wavelengths associated with a proton and an α-particle are equal, then the ratio of velocities of the proton and the α-particle will be:
[NCERT : P L-285 | Online April 15, 2018]

(a) 1 : 4 (b) 1 : 2 (c) 4 : 1 (d) 2 : 1

55. A particle A of mass m and initial velocity v collides with a particle B of mass $\frac{m}{2}$ which is at rest. The collision is head on, and elastic. The ratio of the de-Broglie wavelengths λ_A to λ_B after the collision is
[NCERT : P L-285 | 2017]

(a) $\frac{\lambda_A}{\lambda_B} = \frac{2}{3}$ (b) $\frac{\lambda_A}{\lambda_B} = \frac{1}{2}$

(c) $\frac{\lambda_A}{\lambda_B} = \frac{1}{3}$ (d) $\frac{\lambda_A}{\lambda_B} = 2$

56. de-Broglie wavelength of an electron accelerated by a voltage of 50 V is close to ($|e| = 1.6 \times 10^{-19}$ C, $m_e = 9.1 \times 10^{-31}$ kg, $h = 6.6 \times 10^{-34}$ Js) :
[NCERT : P L-285 | Online April 10, 2015]

(a) 2.4 Å (b) 0.5 Å (c) 1.7 Å (d) 1.2 Å

57. For which of the following particles will it be most difficult to experimentally verify the de-Broglie relationship?
[NCERT : P L-285 | Online April 9, 2014]

(a) an electron (b) a proton
(c) an α-particle (d) a dust particle

58. Electrons are accelerated through a potential difference V and protons are accelerated through a potential difference 4 V. The de-Broglie wavelengths are λ_e and λ_p for electrons and protons respectively. The ratio of $\frac{\lambda_e}{\lambda_p}$ is given by :
(given m_e is mass of electron and m_p is mass of proton).
[NCERT : P L-285 | Online April 23, 2013]

(a) $\frac{\lambda_e}{\lambda_p} = \sqrt{\frac{m_p}{m_e}}$ (b) $\frac{\lambda_e}{\lambda_p} = \sqrt{\frac{m_e}{m_p}}$

(c) $\frac{\lambda_e}{\lambda_p} = \frac{1}{2}\sqrt{\frac{m_e}{m_p}}$ (d) $\frac{\lambda_e}{\lambda_p} = 2\sqrt{\frac{m_p}{m_e}}$

59. If the kinetic energy of a free electron doubles, it's de-Broglie wavelength changes by the factor
[NCERT : P L-285 | 2005]

(a) 2 (b) $\frac{\lambda}{2}$ (c) $\sqrt{2}$ (d) $\frac{\lambda}{\sqrt{2}}$

Topic 2 - Photon, Photoelectric Effect

60. A small mirror of mass m is suspended by a massless thread of length l. Then the small angle through which the thread will be deflected when a short pulse of laser of energy E falls normal on the mirror
[NCERT : P L-283 | April 4, 2025 (I)]
(c = speed of light in vacuum and g = acceleration due to gravity)

(a) $\theta = \frac{3E}{4mc\sqrt{gl}}$ (b) $\theta = \frac{E}{mc\sqrt{gl}}$

(c) $\theta = \frac{E}{2mc\sqrt{gl}}$ (d) $\theta = \frac{2E}{mc\sqrt{gl}}$

61. Given below are two statements, one is labelled as **Assertion A** and the other is labelled as **Reason R.**
Assertion A: In photoelectric effect, on increasing the intensity of incident light the stopping potential increases.
Reason R : Increase in intensity of light increases the rate of photoelectrons emitted, provided the frequency of incident light is greater than threshold frequency.
In the light of the above statements, choose the **correct** answer from the options given below
[NCERT : P L-282 | April 4, 2025 (I)]

(a) Both **A** and **R** are true but **R** is **NOT** the correct explanation of **A**
(b) **A** is false but **R** is true
(c) **A** is true but **R** is false
(d) Both **A** and **R** are true and **R** is the correct explanation of **A**

62. The work function of a metal is 3 eV. The color of the visible light that is required to cause emission of photoelectrons is **[NCERT : P L-282 | April 3, 2025 (I)]**
(a) Green (b) Blue (c) Red (d) Yellow

63. A monochromatic light is incident on a metallic plate having work function ϕ. An electron, emitted normally to the plate from a point A with maximum kinetic energy, enters a constant magnetic field, perpendicular to the initial velocity of electron. The electron passes through a curve and hits back the plate at a point B. The distance between A and B is :
(Given : The magnitude of charge of an electron is e and mass is m, h is Planck's constant and c is velocity of light. Take the magnetic field exists throughout the path of electron) **[NCERT : P L-281, 282 | April 2, 2025 (I)]**
(a) $\sqrt{2m\left(\frac{hc}{\lambda}-\phi\right)}/eB$ (b) $\sqrt{m\left(\frac{hc}{\lambda}-\phi\right)}/eB$
(c) $\sqrt{8m\left(\frac{hc}{\lambda}-\phi\right)}/eB$ (d) $2\sqrt{m\left(\frac{hc}{\lambda}-\phi\right)}/eB$

64. In an experiment with photoelectric effect, the stopping potential. **[NCERT : P L-282 | Jan. 29, 2025 (II)]**
(a) increases with increase in the wavelength of the incident light
(b) increases with increase in the intensity of the incident light
(c) is $\left(\frac{1}{e}\right)$ times the maximum kinetic energy of the emitted photoelectrons
(d) decreases with increase in the intensity of the incident light

65. Given below are two statements : one is labelled as **Assertion (A)** and the other is labelled as **Reason (R)**.
Assertion (A): Emission of electrons in photoelectric effect can be suppressed by applying a sufficiently negative electrode potential to the photoemissive substance.
Reason (R): A negative electric potential, which stops the emission of electrons from the surface of a photoemissive substance, varies linearly with frequency of incident radiation. **[NCERT : P L-282 | Jan. 29, 2025 (I)]**
In the light of the above statements, choose the most appropriate answer from the options given below:
(a) (A) is false but (R) is true.
(b) (A) is true but (R) is false.
(c) Both (A) and (R) are true and (R) is the correct explanation of (A).
(d) Both (A) and (R) are true but (R) is not the correct explanation of (A).

66. Given below are two statements : one is labelled as **Assertion (A)** and other is labelled as **Reason (R).**
Assertion (A) : Electromagnetic waves carry energy but not momentum.
Reason (R) : Mass of a photon is zero.
In the light of the above statements, choose the most appropriate answer from the options given below : **[NCERT : P L-283 | Jan. 29, 2025 (I)]**
(a) (A) is true but (R) is false.
(b) (A) is false but (R) is true.
(c) Both (A) and (R) are true but (R) is not the correct explanation of (A).
(d) Both (A) and (R) are true and (R) is the correct explanation of (A).

67. In photoelectric effect, the stopping potential (V_0) v/s frequency (ν) curve is plotted.
(h is the Planck's constant and ϕ_0 is work function of metal)
(A) V_0 v/s ν is linear
(B) The slope of V_0 v/s ν curve $= \frac{\phi_0}{h}$
(C) h constant is related to the slope of V_0 v/s ν line
(D) The value of electric charge of electron is not required to determine h using the V_0 v/s ν curve.
(E) The work function can be estimated without knowing the value of h.
Choose the correct answer from the options given below : **[NCERT : P L-282 | Jan. 24, 2025 (II)]**
(a) (A),(B) and (C) only (b) (C) and (D) only
(c) (A),(C) and (E) only (d) (D) and (E) only

68. The ratio of the power of a light source S_1 to that the light source S_2 is 2. S_1 is emitting 2×10^{15} photons per second at 600 nm. If the wavelength of the source S_2 is 300 nm, then the number of photons per second emitted by S_2 is_____ $\times 10^{14}$. **[NCERT : P L-283 | Jan. 24, 2025 (II)]**

69. In photoelectric effect an em-wave is incident on a metal surface and electrons are ejected from the surface. If the work function of the metal is 2.14 eV and stopping potential is 2V, what is the wavelength of the em-wave?
(Given hc = 1242 eV-nm where h is the Planck's constant and c is the speed of light in vaccum.) **[NCERT : P L-281, 282 | Jan. 23, 2025 (II)]**
(a) 400 nm (b) 600 nm
(c) 200 nm (d) 300 nm

70. A light source of wavelength λ illuminates a metal surface and electrons are ejected with maximum kinetic energy of 2 eV. If the same surface is illuminated by a light source of wavelength $\frac{\lambda}{2}$, then the maximum kinetic energy of ejected electrons will be (The work function of metal is 1 eV) **[NCERT : P L-281, 282 | Jan. 22, 2025 (II)]**
(a) 2 eV (b) 6 eV (c) 5 eV (d) 3 eV

71. The work functions of cesium (Cs) and lithium (Li) metals are 1.9 eV and 2.5 eV, respectively. If we incident a light of wavelength 550 nm on these two metal surface, then photo-electric effect is possible for the case of **[NCERT : P L-281, 282 | Jan. 22, 2025 (I)]**
(a) Li only (b) Cs only
(c) Neither Cs nor Li (d) Both Cs and Li

72. UV light of 4.13 eV is incident on a photosensitive metal surface having work function 3.13 eV. The maximum kinetic energy of ejected photoelectrons will be :
[NCERT : P L-281 | April 9, 2024 (II)]
(a) 4.13 eV (b) 1 eV (c) 3.13 eV (d) 7.26 eV

73. Average force exerted on a non-reflecting surface at normal incidence is 2.4×10^{-4}N. If 360 W/cm^2 is the light energy flux during span of 1 hour 30 minutes. Then the area of the surface is: **[NCERT : P L-283 | April 8, 2024 (I)]**
(a) 0.2 m^2 (b) 0.02 m^2 (c) 20 m^2 (d) 0.1 m^2

74. Which of the following phenomena does not explain by wave nature of light.
[NCERT : P L-280 | April 6, 2024 (I)]
(A) reflection (B) diffraction
(C) photoelectric effect (D) interference
(E) polarization
Choose the **most appropriate** answer from the options given below :
(a) E only (b) C only
(c) B, D only (d) A, C only

75. When UV light of wavelength 300 nm is incident on the metal surface having work function 2.13 eV, electron emission takes place. The stopping potential is : (Given hc = 1240 eV nm) **[NCERT : P L-285 | April 6, 2024 (II)]**
(a) 4V (b) 4.1 V (c) 2V (d) 1.5 V

76. Given below are two statements:
Statement-I: Figure shows the variation of stopping potential with frequency (ν) for the two photosensitive materials M_1 and M_2. The slope gives value of $\frac{h}{e}$, where h is Planck's constant, e is the charge of electron.

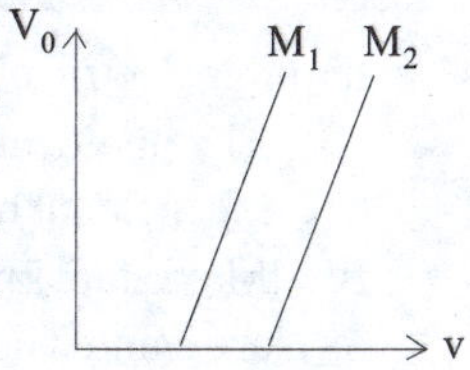

Statement-II: M_2 will emit photoelectrons of greater kinetic energy for the incident radiation having same frequency.
In the light of the above statements, choose the most appropriate answer from the options given below.
[NCERT : P L-282 | April 5, 2024 (I)]
(a) **Statement-I** is correct and **Statement-II** is incorrect.
(b) **Statement-I** is incorrect but **Statement-II** is correct.
(c) Both **Statement-I** and **Statement-II** are incorrect.
(d) Both **Statement-I** and **Statement-II** are correct.

77. Which of the following statement is not true about stopping potential (V_0)?
[NCERT : P L-282 | April 5, 2024 (II)]
(a) It depends on the nature of emitter material.
(b) It depends upon frequency of the incident light.
(c) It increases with increase in intensity of the incident light.
(d) It is 1/e times the maximum kinetic energy of electrons emitted.

78. Which figure shows the correct variation of applied potential difference (V) with photoelectric current (I) at two different intensities of light ($I_1 < I_2$) of same wavelengths : **[NCERT : P L-278 | April 4, 2024 (I)]**

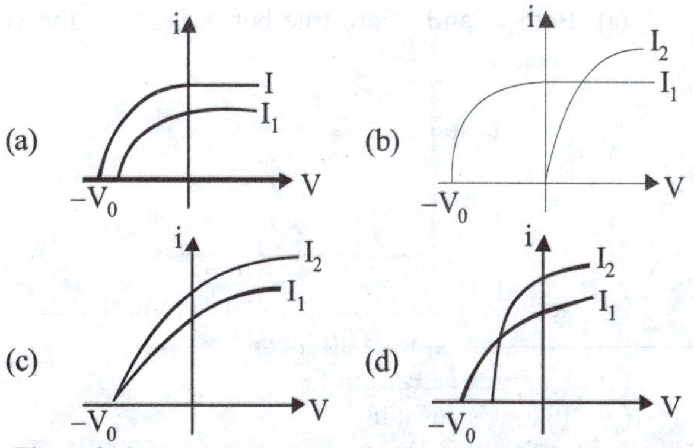

79. Given below are two statements : one is labelled as **Assertion A** and the other is labelled as **Reason R.**
Assertion A : Number of photons increases with increase in frequency of light.
Reason R : Maximum kinetic energy of emitted electrons increases with the frequency of incident radiation.
In the light of the above statements, choose the **most appropriate** answer from the options given below :
[NCERT : P L-283 | April 4, 2024 (II)]
(a) Both **A** and **R** are correct and **R** is NOT the correct explanation of A.
(b) **A** is correct but **R** is not correct.
(c) Both **A** and **R** are correct and **R** is the correct explanation of **A**.
(d) **A** is not correct but **R** is correct.

80. Monochromatic light of frequency 6×10^{14} Hz is produced by a laser. The power emitted is 2×10^{-3} W. How many photons per second on an average, are emitted by the source? **[NCERT : P L-283 | Feb. 1, 2024 (II)]**
(Given h = 6.63×10^{-34}Js)
(a) 5×10^{15} (b) 7×10^{16}
(c) 6×10^{15} (d) 9×10^{18}

81. When a metal surface is illuminated by light of wavelength λ, the stopping potential is 8V. When the same surface is illuminated by light of wavelength 3λ, stopping potential is 2V. The threshold wavelength for this surface is :
[NCERT : P L-281 | Jan. 31, 2024 (I)]
(a) 9λ (b) 3λ (c) 4.5λ (d) 5λ

82. In a photoelectric effect experiment a light of frequency 1.5 times the threshold frequency is made to fall on the surface of photosensitive material. Now if the frequency is halved and intensity is doubled, the number of photo electrons emitted will be:
[NCERT : P L-282 | Jan. 31, 2024 (II)]
(a) Doubled (b) Quadrupled
(c) Zero (d) Halved

83. The work function of a substance is 3.0 eV. The longest wavelength of light that can cause the emission of photoelectrons from this substance is approximately;
[NCERT : P L-281 | Jan. 30, 2024 (I)]
(a) 215nm (b) 414 nm (c) 400 nm (d) 200 nm

84. For the photoelectric effect, the maximum kinetic energy (E_k) of the photoelectrons is plotted against the frequency (ν) of the incident photons as shown in figure. The slope of the graph gives **[NCERT : P L-281 | Jan. 30, 2024 (II)]**

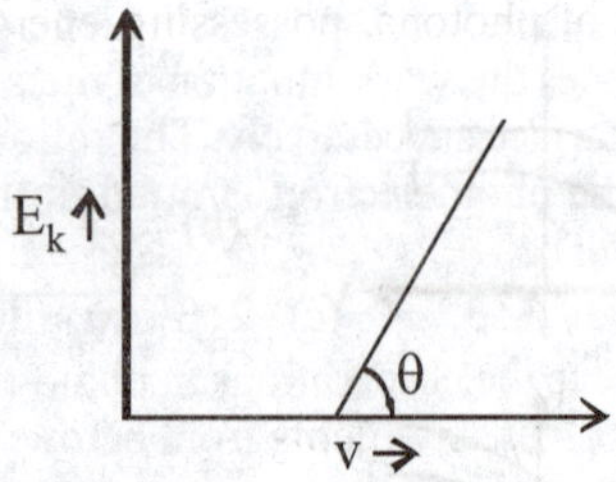

(a) Ratio of Planck's constant to electric charge
(b) Work function of the metal
(c) Charge of electron
(d) Planck's constant

85. Two sources of light emit with a power of 200W. The ratio of number of photons of visible light emitted by each source having wavelengths 300nm and 500 nm respectively, will be : **[NCERT : P L-283 | Jan. 29, 2024 (II)]**
(a) 3 : 5 (b) 1 : 5 (c) 5 : 3 (d) 1 : 3

86. A convex lens of focal length 40 cm forms an image of an extended source of light on a photoelectric cell. A current I is produced. The lens is replaced by another convex lens having the same diameter but focal length 20 cm. The photoelectric current now is : **[NCERT : P L-278| Jan. 27, 2024 (I)]**
(a) $\frac{I}{2}$ (b) 4I (c) 2I (d) I

87. The threshold frequency of a metal with work function 6.63 eV is : **[NCERT : P L-282 | Jan. 27, 2024 (II)]**
(a) 16×10^{15} Hz (b) 16×10^{12} Hz
(c) 1.6×10^{12} Hz (d) 1.6×10^{15} Hz

88. The difference between threshold wavelengths for two metal surfaces A and B having work function $\phi_A = 9$ eV and $\phi_B = 4.5$ eV in nm is:
(Given, hc = 1242 eV nm) **[NCERT : P L-283 | April 13, 2023 (I)]**
(a) 264 (b) 138 (c) 276 (d) 540

89. Given below are two statements :
Statement I : Out of microwaves, infrared rays and ultraviolet rays, ultraviolet rays are the most effective for the emission of electrons from a metallic surface.
Statement II : Above the threshold frequency, the maximum kinetic energy of photoelectrons is inversely proportional to the frequency of the incident light.
In the light of above statements, choose the correct answer from the options given below **[NCERT : P L-281 | April 13, 2023 (II)]**
(a) Statement I is true but Statement II is false
(b) Both Statement I and Statement II are true
(c) Statement I is false but Statement II is true
(d) Both Statement I and Statement II are false

90. An atom absorbs a photon of wavelength 500 nm and emits another photon of wavelength 600 nm. The net energy absorbed by the atom in this process is n × 10^{-4} eV. The value of n is ______.
[Assume the atom to be stationary during the absorption and emission process]
(Take h = 6.6×10^{-34} Js and c = 3×10^8 m/s). **[NCERT : P L-281 | April 13, 2023 (ii)]**

91. A metallic surface is illuminated with radiation of wavelength λ, the stopping potential is V_0. If the same surface is illuminated with radiation of wavelength 2λ, the stopping potential becomes $\frac{V_0}{4}$. The threshold wavelength for this metallic surface will be - **[NCERT : P L-282 | April 11, 2023 (I)]**
(a) $\frac{\lambda}{4}$ (b) 4λ (c) $\frac{3}{2}\lambda$ (d) 3λ

92. The variation of stopping potential (V_0) as a function of the frequency (ν) of the incident light for a metal is shown in figure. The work function of the surface is

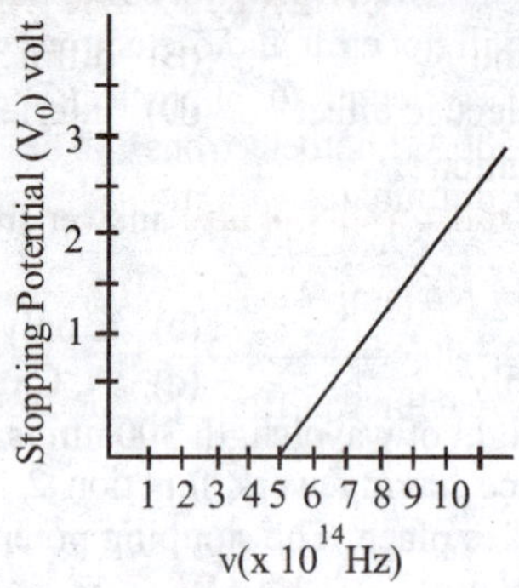

[NCERT : P L-282 | April 10, 2023 (II)]
(a) 18.6 eV (b) 2.98 eV (c) 2.07 eV (d) 1.36 eV

93. In photoelectric effect
A. The photocurrent is proportional to the intensity of the incident radiation.
B. Maximum kinetic energy with which photoelectrons are emitted depends on the intensity of incident light.
C. Max. K.E with which photoelectrons are emitted depends on the frequency of incident light.
D. The emission of photoelectrons require a minimum threshold intensity of incident radiation.
E. Max. K.E of the photoelectrons is independent of the frequency of the incident light.
Choose the correct answer from the options given below: **[NCERT : P L-281 | April 8, 2023 (II)]**
(a) A and C only (b) A and E only
(c) B and C only (d) A and B only

94. The work functions of Aluminium and Gold are 4.1 eV and 5.1 eV respectively. The ratio of the slope of the stopping potential versus frequency plot for Gold to that of Aluminium is **[NCERT : P L-282 | April 6, 2023 (II)]**
(a) 1.24 (b) 2 (c) 1 (d) 1.5

95. The threshold frequency of a metal is f_0. When the light of frequency $2f_0$ is incident on the metal plate, the maximum velocity of photoelectrons is v_1. When the frequency of incident radiation is increased to $5f_0$, the maximum velocity of photoelectrons emitted is v_2. The ratio of v_1 to v_2 is : **[NCERT : P L-281 | Feb. 1, 2023 (II)]**
(a) $\frac{v_1}{v_2} = \frac{1}{2}$ (b) $\frac{v_1}{v_2} = \frac{1}{8}$ (c) $\frac{v_1}{v_2} = \frac{1}{4}$ (d) $\frac{v_1}{v_2} = \frac{1}{16}$

96. If a source of electromagnetic radiation having power 15 kW produces 10^{16} photons per second, the radiation belongs to a part of spectrum is.
(Take Planck constant $h = 6 \times 10^{-34}$ Js)
[NCERT : P L-283 | Jan. 31, 2023 (I)]
(a) Micro waves (b) Ultraviolet rays
(c) Gamma rays (d) Radio waves

97. If the two metals A and B are exposed to radiation of wavelength 350 nm. The work functions of metals A and B are 4.8 eV and 2.2 eV. Then choose the correct option
[NCERT : P L-282 | Jan. 31, 2023 (II)]
(a) Metal B will not emit photo-electrons
(b) Both metals A and B will emit photo-electrons
(c) Both metals A and B will not emit photo-electrons
(d) Metal A will not emit photo-electrons

98. The threshold wavelength for photoelectric emission from a material is 5500Å. Photoelectrons will be emitted, when this material is illuminated with monochromatic radiation from a
A. 75 W infra-red lamp B. 10 W infra-red lamp
C. 75 W ultra-violet lamp D. 10 W ultra-violet lamp
Choose the correct answer from the options given below:
[NCERT : P L-283 | Jan. 29, 2023 (I)]
(a) B and C only (b) A and D only
(c) C only (d) C and D only

99. Given below are two statements :
Statement I : Stopping potential in photoelectric effect does not depend on the power of the light source.
Statement II : For a given metal, the maximum kinetic energy of the photoelectron depends on the wavelength of the incident light.
In the light of above statements, choose the most appropriate answer from the options given below.
Options : **[NCERT : P L-282 | Jan. 25, 2023 (II)]**
(a) Statement I is incorrect but statement II is correct
(b) Both Statement I and Statement II are incorrect
(c) Statement I is correct but statement II is incorrect
(d) Both statement I and statement II are correct

100. From the photoelectric effect experiment, following observations are made. Identify which of these are correct
A. The stopping potential depends only on the work function of the metal.
B. The saturation current increases as the intensity of incident light increases.
C. The maximum kinetic energy of a photo electron depends on the intensity of the incident light.
D. Photoelectric effect can be explained using wave theory of light.
Choose the correct answer from the options given below:
[NCERT : P L-282 | Jan. 24, 2023 (I)]
(a) B, C only (b) A, C, D only
(c) B only (d) A, B, D only

101. The kinetic energy of emitted electron is E when the light incident on the metal has wavelength λ. To double the kinetic energy, the incident light must have wavelength :
[NCERT : P L-281 | July 29, 2022 (I)]
(a) $\frac{hc}{E\lambda - hc}$ (b) $\frac{hc\lambda}{E\lambda + hc}$ (c) $\frac{h\lambda}{E\lambda + hc}$ (d) $\frac{hc\lambda}{E\lambda - hc}$

102. Two streams of photons, possessing energies equal to five and ten times the work function of metal are incident on the metal surface successively. The ratio of maximum velocities of the photoelectron emitted, in the two cases respectively, will be **[NCERT : P L-281 | July 28, 2022 (II)]**
(a) 1 : 2 (b) 1 : 3 (c) 2 : 3 (d) 3 : 2

103. With reference to the observations in photo-electric effect, identify the correct statements from below:
[NCERT : P L-281 | July 27, 2022 (II)]
A. The square of maximum velocity of photoelectrons varies linearly with frequency of incident light.
B. The value of saturation current increases on moving the source of light away from the metal surface.
C. The maximum kinetic energy of photo-electrons decreases on decreasing the power of LED (light emitting diode) source of light.
D. The immediate emission of photo-electrons out of metal surface can not be explained by particle nature of light/electromagnetic waves.
E. Existence of threshold wavelength cannot be explained by wave nature of light/electromagnetic waves.
Choose the correct answer from the options given below:
(a) A and B only (b) A and E only
(c) C and E only (d) D and E only

104. A parallel beam of light of wavelength 900 nm and intensity 100 Wm^{-2} is incident on a surface perpendicular to the beam. The number of photons crossing 1 cm^2 area perpendicular to the beam in one second is :
[NCERT : P L-283 | July 26, 2022 (I)]
(a) 3×10^{16} (b) 4.5×10^{16}
(c) 4.5×10^{17} (d) 4.5×10^{20}

105. A metal exposed to light of wavelength 800 nm and emits photoelectrons with a certain kinetic energy. The maximum kinetic energy of photo-electron doubles when light of wavelength 500 nm is used. The workfunction of the metal is : (Take hc = 1230 eV-nm).
[NCERT : P L-281 | July 25, 2022 (I)]
(a) 1.537 eV (b) 2.46 eV (c) 0.615 eV (d) 1.23 eV

106. A source of monochromatic light liberates 9×10^{20} photon per second with wavelength 600 nm when operated at 400 W. The number of photons emitted per second with wavelength of 800 nm by the source of monochromatic light operating at same power will be:
[NCERT : P L-283 | June 30, 2022 (I)]
(a) 12×10^{20} (b) 6×10^{20}
(c) 9×10^{20} (d) 24×10^{20}

107. Given below are two statements : one is labelled as Assertion A and the other is labelled as Reason R.
Assertion A : The photoelectric effect does not take place, if the energy of the incident radiation is less than the work function of a metal.
Reason R : Kinetic energy of the photoelectrons is zero, if the energy of the incident radiation is equal to the work function of a metal.
In the light of the above statements, choose the most appropriate answer from the options given below.
[NCERT : P L-281 | June 29, 2022 (I)]

(a) Both A and R are correct and R is the correct explanation of A
(b) Both A and R are correct but R is not the correct explanation of A
(c) A is correct but R is not correct
(d) A is not correct but R is correct

108. The electric field at the point associated with a light wave is given by
$E = 200\,[\sin(6 \times 10^{15})t + \sin(9 \times 10^{15})t]\ \text{Vm}^{-1}$
Given : $h = 4.14 \times 10^{-15}\ eVs$
If this light falls on a metal surface having a work function of 2.50 eV, the maximum kinetic energy of the photoelectrons will be :
[NCERT : P L-281 | June 29, 2022 (II)]
(a) 1.90 eV (b) 3.27 eV (c) 3.60 eV (d) 3.42 eV

109. Let K_1 and K_2 be the maximum kinetic energies of photo–electrons emitted when two monochromatic beams of wavelength λ_1 and λ_2, respectively are incident on a metallic surface. If $\lambda_1 = 3\lambda_2$ then:
[NCERT : P L-281 | June 28, 2022 (II)]
(a) $K_1 > \frac{K_2}{3}$ (b) $K_1 < \frac{K_2}{3}$
(c) $K_1 = \frac{K_2}{3}$ (d) $K_2 = \frac{K_1}{3}$

110. A metal surface is illuminated by a radiation of wavelength 4500 Å. The ejected photo-electron enters a constant magnetic field of 2 mT making an angle of 90° with the magnetic field. If it starts revolving in a circular path of radius 2 mm, the work function of the metal is approximately: **[NCERT : P L-281 | June 26, 2022 (II)]**
(a) 1.36 eV (b) 1.69 eV (c) 2.78 eV (d) 2.23 eV

111. The stopping potential for photoelectrons emitted from a surface illuminated by light of wavelength 6630 Å is 0.42 eV. If the threshold frequency is $x \times 10^{13}$/s, where x is ______ (nearest integer).
(Given, speed light = 3×10^8 m/s, Planck's constant = 6.63×10^{-34} Js)**[NCERT : P L-282 | June 26, 2022 (II)]**

112. When light of frequency twice the threshold frequency is incident on the metal plate, the maximum velocity of emitted election is v_1. When the frequency of incident radiation is increased to five times the threshold value, the maximum velocity of emitted electron becomes v_2. If $v_2 = x\,v_1$, the value of x will be____.
[NCERT : P L-281 | June 24, 2022 (I)]

113. The light of two different frequencies whose photons have energies 3.8 eV and 1.4 eV respectively, illuminate a metallic surface whose work function is 0.6 eV successively. The ratio of maximum speeds of emitted electrons for the two frequencies respectivly will be :
[NCERT : P L-281 | June 24, 2022 (II)]
(a) 1 : 1 (b) 2 : 1 (c) 4 : 1 (d) 1 : 4

114. In a photoelectric experiment, increasing the intensity of incident light : **[NCERT : P L-282 | Aug. 27, 2021 (I)]**
(a) increases the number of photons incident and also increases the K.E. of the ejected electrons
(b) increases the frequency of photons incident and increases the K.E. of the ejected electrons.
(c) increases the frequency of photons incident and the K.E. of the ejected electrons remains unchanged
(d) increases the number of photons incident and the K.E. of the ejected electrons remains unchanged

115. A monochromatic neon lamp with wavelength of 670.5 nm illuminates a photo-sensitive material which has a stopping voltage of 0.48 V. What will be the stopping voltage if the source light is changed with another source of wavelength of 474.6 nm? **[NCERT : P L-282 | Aug. 27, 2021 (II); Jan. 11, 2019 II S]**
(a) 0.96 V (b) 1.25 V (c) 0.24 V (d) 1.5 V

116. In a photoelectric experiment ultraviolet light of wavelength 280 nm is used with lithium cathode having work function $\phi = 2.5$ eV. If the wavelength of incident light is switched to 400 nm, find out the change in the stopping potential. (h = 6.63×10^{-34} Js, c = 3×10^8 ms^{-1})
[NCERT : P L-282 | Aug. 26, 2021 (I)]
(a) 1.3 V (b) 1.1 V (c) 1.9 V (d) 0.6 V

117. A particle of mass 9.1×10^{-31} kg travels in a medium with a speed of 10^6 m/s and a photon of a radiation of linear momentum 10^{-27} kg m/s travels in vacuum. The wavelength of photon is _______ times the wavelength of the particle.**[NCERT : P L-285 | NA, July 27, 2021 (I)]**

118. The K_α X-ray of molybdenum has wavelength 0.071 nm. If the energy of a molybdenum atoms with a K electron knocked out is 27.5 keV, the energy of this atom when an L electron is knocked out will be_____ keV. (Round off to the nearest integer) [h = 4.14×10^{-15} eVs, c = 3×10^8 ms^{-1}]
[NCERT : P L-283 | NA, July 27, 2021 (II)]

119. A light beam of wavelength 500 nm is incident on a metal having work function of 1.25 eV, placed in a magnetic field of intensity B. The electrons emitted perpendicular to the magnetic field B, with maximum kinetic energy are bent into circular arc of radius 30 cm. The value of B is_______ $\times 10^{-7}$ T. Given hc = 20×10^{-26} J-m, mass of electron = 9×10^{-31} kg
[NCERT : P L-281 | NA, July 25, 2021 (II)]

120. When radiation of wavelength λ_0 is incident on a metallic surface, the stopping potential of ejected photoelectrons is 4.8 V. If the same surface is illuminated by radiation of double the previous wavelength, then the stopping potential becomes 1.6 V. The threshold wavelength of the metal is : **[NCERT : P L-282 | July 25, 2021 (II), NA July 20, 2021 (II) S]**
(a) $2\lambda_0$ (b) $4\lambda_0$ (c) $8\lambda_0$ (d) $6\lambda_0$

121. An electron moving with speed v and a photon moving with speed c, have same D-Broglie wavelength. The ratio of kinetic energy of electron to that of photon is :
[NCERT : P L-285 | July 25, 2021 (II)]
(a) $\frac{3c}{v}$ (b) $\frac{v}{3c}$ (c) $\frac{v}{2c}$ (d) $\frac{2c}{v}$

122. The radiation corresponding to 3 to 2 transition of a hydrogen atom falls on a gold surface to generate photoelectrons. These electrons are passed through a magnetic field of 5×10^{-4}T. Assume that the radius of the largest circular path followed by these electrons is 7 mm, the work function of the metal is: (Mass of electron = 9.1×10^{-31} kg)

[NCERT : P L-281 | July 20, 2021 (I)]

(a) 1.36 eV (b) 1.88 eV (c) 0.16 eV (d) 0.82 eV

123. The stopping potential in the context of photoelectric effect depends on the following property of incident electromagnetic radiation:

[NCERT : P L-282 | March 16, 2021 (I)]

(a) Intensity (b) Amplitude

(c) Frequency (d) Phase

124. Two stream of photons, possessing energies equal to twice and ten times the work function of metal are incident on the metal surface successively. The value of ratio of maximum velocities of the photoelectrons emitted in the two respective cases is x : y. The value of x is ________.

[NCERT : P L-282 | Feb. 26, 2021 (II)]

125. The stopping potential for electrons emitted from a photosensitive surface illuminated by light of wavelength 491 nm is 0.710 V. When the incident wavelength is changed to a new value, the stopping potential is 1.43 V. The new wavelength is : **[NCERT : P L-282 | Feb. 25, 2021 (II)]**

(a) 329 nm (b) 400 nm (c) 382 nm (d) 309 nm

126. An *X*-ray tube is operated at 1.24 million volt. The shortest wavelength of the produced photon will be :

[NCERT : P L-283 | Feb. 24, 2021 (II)]

(a) 10^{-3} nm (b) 10^{-1} nm (c) 10^{-2} nm (d) 10^{-4} nm

127. A beam of electrons of energy E scatters from a target having atomic spacing of 1Å. The first maximum intensity occurs at $\theta = 60°$. Then E (in eV) is ______.

(Plank constant $h = 6.64 \times 10^{-34}$ Js, 1 eV = 1.6×10^{-19} J, electron mass $m = 9.1 \times 10^{-31}$ kg) **[NA Sep. 05, 2020 (I)]**

128. The surface of a metal is illuminated alternately with photons of energies $E_1 = 4$ eV and $E_2 = 2.5$ eV respectively. The ratio of maximum speeds of the photoelectrons emitted in the two cases is 2. The work function of the metal in (eV) is ________. **[NCERT : P L-282 | NA Sep. 05, 2020 (II)]**

129. Given figure shows few data points in a photo electric effect experiment for a certain metal. The minimum energy for ejection of electron from its surface is : (Plancks constant $h = 6.62 \times 10^{-34}$ J.s)

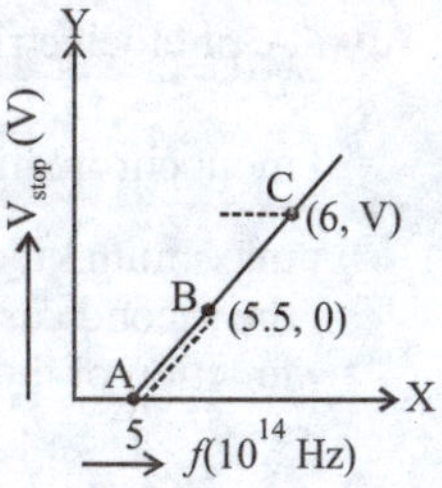

[NCERT : P L-282 | Sep. 04, 2020 (I)]

(a) 2.27 eV (b) 2.59 eV (c) 1.93 eV (d) 2.10 eV

130. In a photoelectric effect experiment, the graph of stopping potential *V* versus reciprocal of wavelength obtained is shown in the figure. As the intensity of incident radiation is increased : **[NCERT : P L-282 | Sep. 04, 2020 (II)]**

(a) Straight line shifts to right

(b) Slope of the straight line get more steep

(c) Straight line shifts to left

(d) Graph does not change

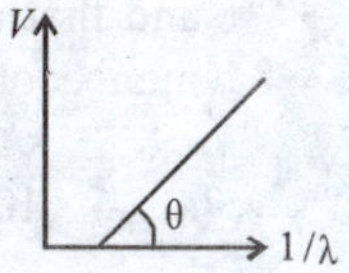

131. When the wavelength of radiation falling on a metal is changed from 500 nm to 200 nm, the maximum kinetic energy of the photoelectrons becomes three times larger. The work function of the metal is close to :

[NCERT : P L-281 | Sep. 03, 2020 (I)]

(a) 0.81 eV (b) 1.02 eV (c) 0.52 eV (d) 0.61 eV

132. Two sources of light emit X-rays of wavelength 1 nm and visible light of wavelength 500 nm, respectively. Both the sources emit light of the same power 200 W. The ratio of the number density of photons of X-rays to the number density of photons of the visible light of the given wavelengths is : **[NCERT : P L-284 | Sep. 03, 2020 (II)]**

(a) $\frac{1}{500}$ (b) 250 (c) $\frac{1}{250}$ (d) 500

133. When radiation of wavelength λ is used to illuminate a metallic surface, the stopping potential is *V*. When the same surface is illuminated with radiation of wavelength 3λ, the stopping potential is $\frac{V}{4}$. If the theshold wavelength for the metallic surface is *n*λ then value of *n* will be ________. **[NCERT : P L-282 | NA Sep. 02, 2020 (I)]**

134. Radiation, with wavelength 6561 Å falls on a metal surface to produce photoelectrons. The electrons are made to enter a uniform magnetic field of 3×10^{-4} T. If the radius of the largest circular path followed by the electrons is 10 mm, the work function of the metal is close to:

[NCERT : P L-281 | 9 Jan. 2020 I]

(a) 1.1 eV (b) 0.8 eV (c) 1.6 eV (d) 1.8 eV

135. When photon of energy 4.0 eV strikes the surface of a metal *A*, the ejected photoelectrons have maximum kinetic energy T_A eV and de-Broglie wavelength λ_A. The maximum kinetic energy of photoelectrons liberated from another metal *B* by photon of energy 4.50 eV is $T_B = (T_A - 1.5)$eV. If the de-Broglie wavelength of these photoelectrons $\lambda_B = 2\lambda_A$, then the work function of metal B is:

[NCERT : P L-285 | 8 Jan. 2020 I]

(a) 4 eV (b) 2 eV (c) 1.5 eV (d) 3 eV

136. A beam of electromagnetic radiation of intensity 6.4×10^{-5} W/cm² is comprised of wavelength, λ = 310 nm. It falls normally on a metal (work function φ = 2eV) of surface area of 1 cm². If one in 10^3 photons ejects an electron, total number of electrons ejected in 1 s is 10^x. (*hc* = 1240 eVnm, 1 eV = 1.6×10^{-19} J), then *x* is ______.

[NCERT : P L-283 | NA 7 Jan. 2020 I]

137. The stopping potential V_0 (in volt) as a function of frequency (*v*) for a sodium emitter, is shown in the figure. The work function of sodium, from the data plotted in the figure, will be :

(Given : Planck's constant (*h*) = 6.63×10^{-34} Js, electron charge $e = 1.6 \times 10^{-19}$ C) **[NCERT : P L-282 | 12 Apr. 2019 I]**

(a) 1.82 eV

(b) 1.66 eV

(c) 1.95 eV

(d) 2.12 eV

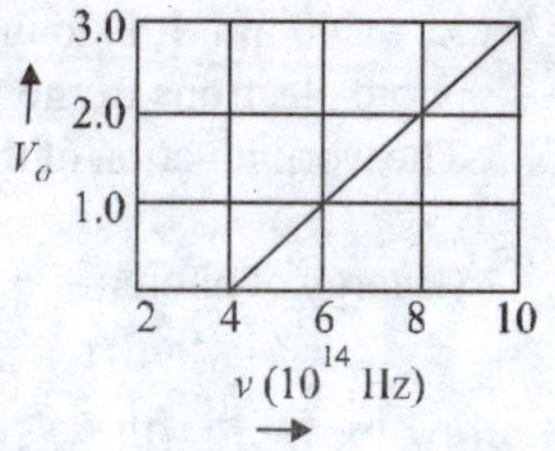

138. In a photoelectric effect experiment the threshold wavelength of light is 380 nm. If the wavelength of incident light is 260 nm, the maximum kinetic energy of emitted electrons will be: Given E (in eV) =

[NCERT : P L-281 | 10 Apr. 2019 I]

(a) 1.5 eV (b) 3.0 eV (c) 4.5 eV (d) 15.1 eV

139. A 2 mW laser operates at a wavelength of 500 nm. The number of photons that will be emitted per second is : [Given Planck's constant h= 6.6×10^{-34} Js, speed of light $c = 3.0\times10^8$ m/s] [NCERT : P L-283 | 10 Apr. 2019 II]

(a) 5×10^{15} (b) 1.5×10^{16} (c) 2×10^{16} (d) 1×10^{16}

140. The electric field of light wave is given as

$$\vec{E} = 10^3 \cos\left(\frac{2\pi x}{5\times10^{-7}} - 2\pi\times6\times10^{14}t\right)\hat{x}\frac{N}{C}$$

This light falls on a metal plate of work function 2eV. The stopping potential of the photo-electrons is:

Given, $E\,(\text{in eV}) = \dfrac{12375}{\lambda\,(\text{in Å})}$

[NCERT : P L-101 | 9 April 2019 I]

(a) 2.0 V (b) 0.72 V (c) 0.48 V (d) 2.48 V

141. When a certain photosensitive surface is illuminated with monochromatic light of frequency v, the stopping potential for the photo current is $-V_0/2$. When the surface is illuminated by monochromatic light of frequency $v/2$, the stoppoing potential is $-V_0$. The threshold frequency for photoelectric emission is : [NCERT : P L-282 | 12 Jan. 2019 II]

(a) $\frac{5v}{3}$ (b) $\frac{4}{3}v$ (c) $2v$ (d) $\frac{3v}{2}$

142. In a Frank-Hertz experiment, an electron of energy 5.6 eV passes through mercury vapour and emerges with an energy 0.7 eV. The minimum wavelength of photons emitted by mercury atoms is close to : [NCERT : P L-283 | 12 Jan. 2019 II]

(a) 1700 nm (b) 2020 nm (c) 220 nm (d) 250 nm

143. A metal plate of area 1×10^{-4} m^2 is illuminated by a radiation of intensity 16 mW/m^2. The work function of the metal is 5 eV. The energy of the incident photons is 10 eV and only 10% of it produces photo electrons. The number of emitted photo electrons per second and their maximum energy, respectively, will be: [1 eV = 1.6×10^{-19} J] [NCERT : P L-283 | 10 Jan. 2019 II]

(a) 10^{14} and 10 eV (b) 10^{12} and 5 eV
(c) 10^{11} and 5 eV (d) 10^{10} and 5 eV

144. Surface of certain metal is first illuminated with light of wavelength $\lambda_1 = 350$ nm and then, by light of wavelength $\lambda_2 = 540$ nm. It is found that the maximum speed of the photo electrons in the two cases differ by a factor of (2) The work function of the metal (in eV) is close to:

(Energy of photon $= \dfrac{1240}{\lambda\,(\text{in nm})}$ eV) [NCERT : P L-281 | 9 Jan. 2019 I]

(a) 1.8 (b) 2.5 (c) 5.6 (d) 1.4

145. The magnetic field associated with a light wave is given at the origin by
$B = B_0\,[\sin(3.14\times10^7)ct + \sin(6.28\times10^7)ct]$.
If this light falls on a silver plate having a work function of 4.7 eV, what will be the maximum kinetic energy of the photoelectrons? [NCERT : P L-281 | 9 Jan. 2019 II]
($c = 3\times10^8$ ms^{-1}, $h = 6.6\times10^{-34}$J-s)

(a) 6.82 eV (b) 12.5 eV (c) 8.52 eV (d) 7.72 eV

146. An electron beam is accelerated by a potential difference V to hit a metallic target to produce X-rays. It produces continuous as well as characteristic X-rays. If λ_{min} is the smallest possible wavelength of X-ray in the spectrum, the variation of log λ_{min} with log V is correctly represented in : [NCERT : P L-282 | 2017]

(a) 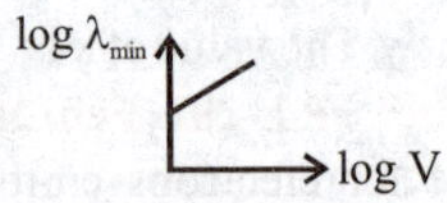

(b) 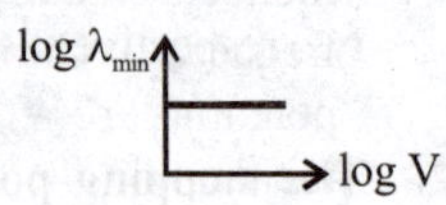

(c) 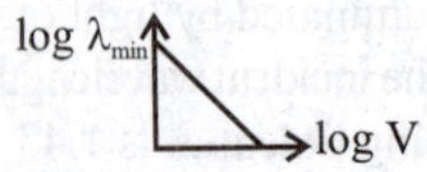

(d) 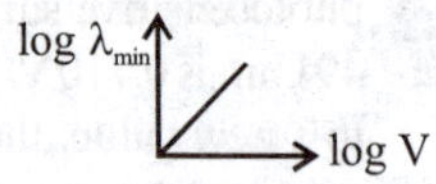

147. A Laser light of wavelength 660 nm is used to weld Retina detachment. If a Laser pulse of width 60 ms and power 0.5 kW is used the approximate number of photons in the pulse are : [Take Planck's constant h = 6.62×10^{-34} Js] [NCERT : P L-284 | Online April 9, 2017]

(a) 10^{20} (b) 10^{18} (c) 10^{22} (d) 10^{19}

148. Radiation of wavelength λ, is incident on a photocell. The fastest emitted electron has speed v. If the wavelength is changed to $\frac{3\lambda}{4}$, the speed of the fastest emitted electron will be: [NCERT : P L-281 | 2016]

(a) $= v\left(\frac{4}{3}\right)^{\frac{1}{2}}$ (b) $= v\left(\frac{3}{4}\right)^{\frac{1}{2}}$

(c) $> v\left(\frac{4}{3}\right)^{\frac{1}{2}}$ (d) $< v\left(\frac{4}{3}\right)^{\frac{1}{2}}$

149. A photoelectric surface is illuminated successively by monochromatic light of wavelengths λ and $\frac{\lambda}{2}$. If the maximum kinetic energy of the emitted photoelectrons in the second case is 3 times that in the first case, the work function of the surface is : [NCERT : P L-281 | Online April 10, 2016]

(a) $\frac{hc}{2\lambda}$ (b) $\frac{hc}{\lambda}$ (c) $\frac{hc}{3\lambda}$ (d) $\frac{3hc}{\lambda}$

150. A beam of light has two wavelengths of 4972Å and 6216Å with a total intensity of 3.6×10^{-3} Wm^{-2} equally distributed among the two wavelengths. The beam falls normally on an area of 1 cm^2 of a clean metallic surface of work function 2.3 eV. Assume that there is no loss of light by reflection and that each capable photon ejects one electron. The number of photoelectrons liberated in 2s is approximately: [NCERT : P L-282 | Online April 12, 2014]

(a) 6×10^{11} (b) 9×10^{11} (c) 11×10^{11} (d) 15×10^{11}

151. Photoelectrons are ejected from a metal when light of frequency υ falls on it. Pick out the wrong statement from the following. [NCERT : P L-281 | Online May 26, 2012]
(a) No electrons are emitted if υ is less than W/h, where W is the work function of the metal
(b) The ejection of the photoelectrons is instantaneous.
(c) The maximum energy of the photoelectrons is hυ.
(d) The maximum energy of the photoelectrons is independent of the intensity of the light.

152. This question has Statement – 1 and Statement – 2. Of the four choices given after the statements, choose the one that best describes the two statements. [NCERT : P L-282 | 2011]
Statement – 1: A metallic surface is irradiated by a monochromatic light of frequency $v > v_0$ (the threshold frequency). The maximum kinetic energy and the stopping potential are K_{max} and V_0 respectively. If the frequency incident on the surface is doubled, both the K_{max} and V_0 are also doubled.
Statement – 2 : The maximum kinetic energy and the stopping potential of photoelectrons emitted from a surface are linearly dependent on the frequency of incident light.
(a) Statement–1 is true, Statement–2 is true, Statement – 2 is the correct explanation of Statement – 1.
(b) Statement–1 is true, Statement–2 is true, Statement – 2 is not the correct explanation of Statement – 1.
(c) Statement – 1 is false, Statement – 2 is true.
(d) Statement – 1 is true, Statement – 2 is false.

153. **Statement -1 :** When ultraviolet light is incident on a photocell, its stopping potential is V_0 and the maximum kinetic energy of the photoelectrons is K_{max}. When the ultraviolet light is replaced by X-rays, both V_0 and K_{max} increase.
Statement -2 : Photoelectrons are emitted with speeds ranging from zero to a maximum value because of the range of frequencies present in the incident light. [NCERT : P L-282 | 2010]
(a) Statement -1 is true, Statement -2 is true ; Statement - 2 is the correct explanation of Statement -1.
(b) Statement -1 is true, Statement -2 is true; Statement - 2 is **not** the correct explanation of Statement -1
(c) Statement -1 is false, Statement -2 is true.
(d) Statement -1 is true, Statement -2 is false.

154. The surface of a metal is illuminted with the light of 400 nm. The kinetic energy of the ejected photoelectrons was found to be 1.68 eV. The work function of the metal is : (hc = 1240 eV.nm) [NCERT : P L-281 | 2009]
(a) 1.41 eV (b) 1.51 eV (c) 1.68 eV (d) 3.09 eV

155. Photon of frequency ν has a momentum associated with it. If c is the velocity of light, the momentum is [NCERT : P L-283 | 2007]
(a) hν / c (b) ν /c (c) h ν c (d) $h\nu / c^2$

156. The threshold frequency for a metallic surface corresponds to an energy of 6.2 eV and the stopping potential for a radiation incident on this surface is 5 V. The incident radiation lies in [NCERT : P L-282 | 2006]
(a) ultra-violet region (b) infra-red region
(c) visible region (d) X-ray region

157. The time taken by a photoelectron to come out after the photon strikes is approximately [NCERT : P L-280 | 2006]
(a) 10^{-4} s (b) 10^{-10} s (c) 10^{-16} s (d) 10^{-1} s

158. The anode voltage of a photocell is kept fixed. The wavelength λ of the light falling on the cathode is gradually changed. The plate current I of the photocell varies as follows [NCERT : P L-277 | 2006]

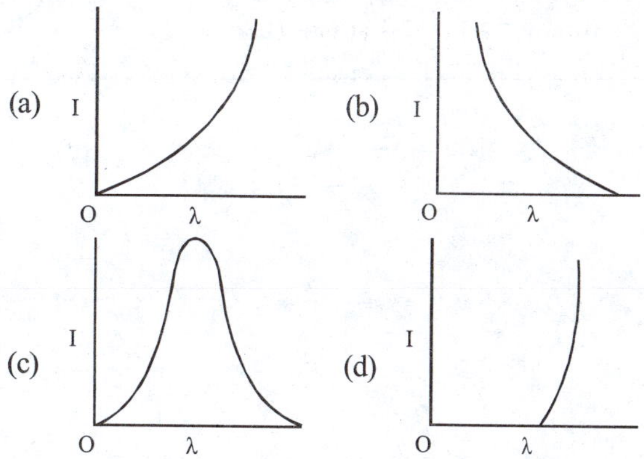

159. A photocell is illuminated by a small bright source placed 1 m away. When the same source of light is placed $\frac{1}{2}$ m away, the number of electrons emitted by photocathode would [NCERT : P L-278 | 2005]
(a) increase by a factor of 4
(b) decrease by a factor of 4
(c) increase by a factor of 2
(d) decrease by a factor of 2

160. A radiation of energy E falls normally on a perfectly reflecting surface. The momentum transferred to the surface is [NCERT : P L-283 | 2004]
(a) Ec (b) $2E/c$ (c) E/c (d) E/c^2

161. According to Einstein's photoelectric equation, the plot of the kinetic energy of the emitted photo electrons from a metal Versus the frequency, of the incident radiation gives a straight line whose slope [NCERT : P L-281 | 2004]
(a) depends both on the intensity of the radiation and the metal used
(b) depends on the intensity of the radiation
(c) depends on the nature of the metal used
(d) is the same for the all metals and independent of the intensity of the radiation

162. The work function of a substance is 4.0 eV. The longest wavelength of light that can cause photoelectron emission from this substance is approximately [NCERT : P L-281 | 2004]
(a) 310 nm (b) 400 nm (c) 540 nm (d) 220 nm

163. Two identical photocathodes receive light of frequencies f_1 and f_2. If the velocities of the photo electrons (of mass m) coming out are respectively v_1 and v_2, then [NCERT : P L-281 | 2003]

(a) $v_1^2 - v_2^2 = \frac{2h}{m}(f_1 - f_2)$

(b) $v_1 + v_2 = \left[\frac{2h}{m}(f_1 + f_2)\right]^{1/2}$

(c) $v_1^2 + v_2^2 = \frac{2h}{m}(f_1 + f_2)$

(d) $v_1 - v_2 = \left[\frac{2h}{m}(f_1 - f_2)\right]^{1/2}$

164. Sodium and copper have work functions 2.3 eV and 4.5 eV respectively. Then the ratio of their cutoff wavelengths is nearest to **[NCERT : P L-281 | 2002]**

(a) 1 : 2 (b) 4 : 1

(c) 2 : 1 (d) 1 : 4

ANSWER KEYS

1	(2)	18	(c)	35	(c)	52	(d)	69	(d)	86	(d)	103	(b)	120	(b)	137	(b)	154	(a)
2	(d)	19	(a)	36	(b)	53	(d)	70	(c)	87	(d)	104	(b)	121	(c)	138	(a)	155	(a)
3	(b)	20	(b)	37	(d)	54	(c)	71	(b)	88	(b)	105	(c)	122	(d)	139	(a)	156	(a)
4	(a)	21	(c)	38	(d)	55	(d)	72	(b)	89	(a)	106	(a)	123	(c)	140	(c)	157	(b)
5	(c)	22	(d)	39	(b)	56	(c)	73	(b)	90	(4125)	107	(b)	124	(1)	141	(d)	158	(b)
6	(d)	23	(b)	40	(d)	57	(d)	74	(b)	91	(d)	108	(d)	125	(c)	142	(d)	159	(a)
7	(a)	24	(b)	41	(b)	58	(d)	75	(c)	92	(c)	109	(b)	126	(a)	143	(c)	160	(b)
8	(b)	25	(d)	42	(c)	59	(d)	76	(a)	93	(a)	110	(a)	127	(50)	144	(a)	161	(d)
9	(b)	26	(d)	43	(a)	60	(d)	77	(c)	94	(c)	111	(35)	128	(2)	145	(d)	162	(a)
10	(a)	27	(c)	44	(d)	61	(b)	78	(c)	95	(a)	112	(2)	129	(a)	146	(c)	163	(a)
11	(d)	28	(d)	45	(d)	62	(b)	79	(d)	96	(c)	113	(b)	130	(d)	147	(a)	164	(c)
12	(d)	29	(a)	46	(c)	63	(c)	80	(a)	97	(d)	114	(d)	131	(d)	148	(c)		
13	(d)	30	(b)	47	(d)	64	(c)	81	(a)	98	(d)	115	(b)	132	(a)	149	(a)		
14	(c)	31	(b)	48	(c)	65	(d)	82	(c)	99	(d)	116	(a)	133	(9)	150	(b)		
15	(a)	32	(b)	49	(b)	66	(b)	83	(b)	100	(c)	117	(910)	134	(a)	151	(c)		
16	(a)	33	(a)	50	(a)	67	(c)	84	(d)	101	(b)	118	(10)	135	(a)	152	(c)		
17	(d)	34	(a)	51	(c)	68	(5)	85	(a)	102	(c)	119	(122)	136	(11)	153	(d)		

12

Atoms

TREND ANALYSIS
JEE MAIN (2025-2021)

	2025				2024				2023				2022				2021			
	Session II		Session I		Session II		Session I		Session II		Session I		Session II		Session I		Session II		Session I	
	MCQs	NVQs	MCQs	NVQs	MCQs	NVQs	MCQs	NVQs	MCQs	NVQs	MCQs	NVQs	MCQs	NVQs	MCQs	NVQs	MCQs	NVQs	MCQs	NVQs
Topic 1	–	–	–	–	–	1	1	–	–	–	–	–	–	–	1	1	–	–	–	–
Topic 2	6	2	5	–	6	4	5	5	4	6	5	4	3	2	5	2	2	2	6	1
Topic 3																				
Topic 4																				
Total	6	2	5	–	6	5	6	5	4	6	5	4	3	2	6	3	2	2	6	1

How to Code Qns for Better & Faster Revision

Type 1: Able to solve (No code Req)
Type 2: Solved with struggle or by seeing solution **(Use code NR means Needs Revision or orange stickers)**
Type 3: Tricky/ Could not solve on my own. Need help from Teacher/ Friend. **(Use code TQ means Tricky Question or red stickers)**

You can either hand write the code in boxes or else paste coloured stickers given at end of book.

Topic 1 - Atomic Structure and Rutherford's Nuclear Model

1. In an alpha particle scattering experiment distance of closest approach for the α particle is 4.5×10^{-14} m. If target nucleus has atomic number 80, then maximum velocity of α-particle is ______ $\times 10^5$ m/s approximately.

$\left(\frac{1}{4\pi \epsilon_0} = 9 \times 10^9\right.$ SI unit, mass of α particle $= 6.72 \times 10^{-27}$ kg)

[NCERT : P L-295 | April 8, 2024 (I)]

2. Given below are two statements:

Statement I : Most of the mass of the atom and all its positive charge are concentrated in a tiny nucleus and the electrons revolve around it, is Rutherford's model.

Statement II : An atom is a spherical cloud of positive charges with electrons embedded in it, is a special case of Rutherford's model.

In the light of the above statements, choose the **most appropriate** from the options given below

[NCERT : P L-292, 293, 294 | Jan. 29, 2024 (II)]

(a) Statement I is true but Statement II is false
(b) Both Statement I and Statement II are true
(c) Both Statement I and Statement II are false
(d) Statement I is false but Statement II is true

3. $\sqrt{d_1}$ and $\sqrt{d_2}$ are the impact parameters corresponding to scattering angles 60° and 90° respectively, when an α particle is approaching a gold nucleus. For $d_1 = x\, d_2$, the value of x will be______.

[NCERT : P L-294 | June 29, 2022 (I)]

4. Choose the correct option from the following options given below : **[NCERT : P L-291, 292, 293 | June 24, 2022 (I)]**

(a) In the ground state of Rutherford's model electrons are in stable equilibrium. While in Thomson's model electrons always experience a net-force.
(b) An atom has a nearly continuous mass distribution in a Rutherford's model but has a highly non-uniform mass distribution in Thomson's model
(c) A classical atom based on Rutherford's model is doomed to collapse.
(d) The positively charged part of the atom possesses most of the mass in Rutherford's model but not in Thomson's model.

5. The graph which depicts the results of Rutherford gold foil experiment with α-particles is: [NCERT : P L-293 | 8 Jan. 2020 I]
θ: Scattering angle
Y: Number of scattered α-particles detected
(Plots are schematic and not to scale)

(a) 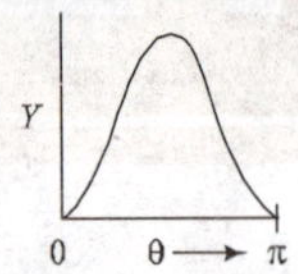

(b) 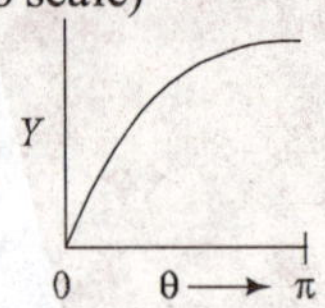

(c) 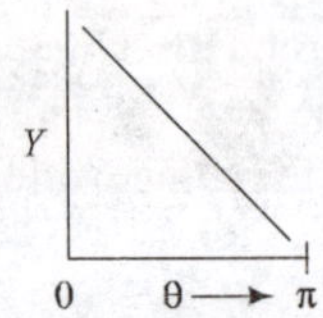

(d) 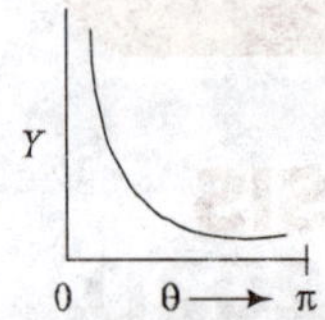

6. In the Rutherford experiment, α-particles are scattered from a nucleus as shown. Out of the four paths, which path is not possible? NCERT : P L-292 | [Online May 7, 2012]

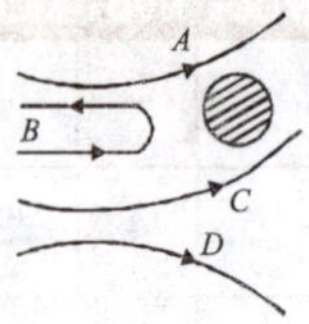

(a) *D*
(b) *B*
(c) *C*
(d) *A*

7. An alpha nucleus of energy $\frac{1}{2}mv^2$ bombards a heavy nuclear target of charge Ze. Then the distance of closest approach for the alpha nucleus will be proportional to [NCERT : P L-295 | 2006]

(a) v^2 (b) $\frac{1}{m}$ (c) $\frac{1}{v^2}$ (d) $\frac{1}{Ze}$

8. An α-particle of energy 5 MeV is scattered through 180° by a fixed uranium nucleus. The distance of closest approach is of the order of [NCERT : P L-295 | 2004]

(a) 10^{-12} cm (b) 10^{-10} cm
(c) 10^{-14} cm (d) 10^{-15} cm

Topic 2 - Bohr's Model and the Spectra of the Hydrogen Atom

9. In a hydrogen like ion, the energy difference between the 2nd excitation energy state and ground is 108.8 *eV*. The atomic number of the ion is [NCERT : P L-299 | April 7, 2025 (I)]

(a) 4 (b) 2 (c) 1 (d) 3

10. For a hydrogen atom, the ratio of the largest wavelength of Lyman series to that of the Balmer series is. [April 7, 2025 (I)]

(a) 5 : 36 (b) 5 : 27 (c) 3 : 4 (d) 27 : 5

11. Given below are two statements :
Statement (I) : The dimensions of Planck's constant and angular momentum are same.
Statement (II) : In Bohr's model electron revolve around the nucleus only in those orbits for which angular momentum is integral multiple of Planck's constant. [NCERT : P L-299 | April 4, 2025 (II)]

In the light of the above statements, choose the **most appropriate answer** from the options given below :
(a) Both Statement I and Statement II are correct
(b) Statement I is incorrect but Statement II is correct
(c) Statement I is correct but Statement II is incorrect
(d) Both Statement I and Statement II are incorrect

12. Considering the Bohr model of hydrogen like atoms, the ratio of the ratio of the radius 5th orbit of the electron in Li^{2+} and He^{+} is [NCERT : P L-299 | April 4, 2025 (I)]

(a) $\frac{3}{2}$ (b) $\frac{4}{9}$ (c) $\frac{9}{4}$ (d) $\frac{2}{3}$

13. Given below are two statements : one is labelled as **Assertion A** and the other is labelled as **Reason R**.
Assertion A : The Bohr model is applicable to hydrogen and hydrogen-like atoms only.
Reason R : The formulation of Bohr model does not include repulsive force between electrons.
In the light of the above statements, choose the **correct** answer from the options given below : [NCERT : P L-299 | April 3, 2025 (II)]
(a) Both **A** and **R** are true but **R** is **NOT** the correct explanation of **A**.
(b) **A** is false but **R** is true.
(c) Both **A** and **R** are true and **R** is the correct explanation of **A**.
(d) **A** is true but **R** is false.

14. An electron in the hydrogen atom initially in the fourth excited state makes a transition to n^{th} energy state by emitting a photon of energy 2.86 eV. The integer value of n will be ________. [NCERT : P L-299 | April 3, 2025 (II)]

15. Assuming the validity of Bohr's atomic model for hydrogen like ions the radius of Li^{++} ion in its ground state is given by $\frac{1}{X}a_0$, where X = ______.
(Where a_0 is the first Bohr's radius.) [NCERT : P L-299 | April 2, 2025 (II)]

(a) 2 (b) 1 (c) 3 (d) 9

16. Considering Bohr's atomic model for hydrogen atom : [NCERT : P L-299 | April 2, 2025 (I)]
(A) the energy of H atom in ground state is same as energy of He^{+} ion in its first excited state.
(B) the energy of H atom in ground state is same as that for Li^{++} ion in its second excited state.
(C) the energy of H atom in its ground state is same as that of He^{+} ion for its ground state.
(D) the energy of He^{+} ion in its first excited state is same as that for Li^{++} ion in its ground state
Choose the **correct** answer from the options given below :
(a) (B), (D) only (b) (A), (B) only
(c) (A), (D) only (d) (B), (C) only

17. The number of spectral lines emitted by atomic hydrogen that is in the 4th energy level, is [NCERT : P L-300 | Jan. 29, 2025 (II)]

(a) 6 (b) 0 (c) 3 (d) 1

18. The frequency of revolution of the electron in Bohr's orbit varies with n, the principal quantum number as [NCERT : P L-299 | Jan. 28, 2025 (II)]

(a) $\frac{1}{n}$ (b) $\frac{1}{n^3}$ (c) $\frac{1}{n^4}$ (d) $\frac{1}{n^2}$

19. During the transition of electron from state A to state C of a Bohr atom, the wavelength of emitted radiation is 2000 Å and it becomes 6000 Å when the electron jumps from state B to state C. Then the wavelength of the radiation emitted during the transition of electrons from state A to state B is [Jan. 24, 2025 (I)]

(a) 3000 Å (b) 6000 Å
(c) 4000 Å (d) 2000 Å

20. An electron projected perpendicular to a uniform magnetic field B moves in a circle. If Bohr's quantization is applicable, then the radius of the electronic orbit in the first excited state is : [NCERT : P L-299 | Jan. 22, 2025 (II)]

(a) $\sqrt{\frac{2h}{\pi eB}}$ (b) $\sqrt{\frac{4h}{\pi eB}}$
(c) $\sqrt{\frac{h}{2\pi eB}}$ (d) $\sqrt{\frac{h}{\pi eB}}$

21. An electron in the ground state of the hydrogen atom has the orbital radius of 5.3×10^{-11} m while that for the electron in third excited state is 8.48×10^{-10} m. The ratio of the de-Broglie wavelengths of electron in the ground state to that in excited state is [NCERT : P L-299 | Jan. 22, 2025 (I)]

(a) $\frac{1}{4}$ (b) $\frac{1}{9}$ (c) $\frac{1}{3}$ (d) $\frac{1}{16}$

22. A hydrogen atom in ground state is given an energy of 10.2 eV. How many spectral lines will be emitted due to transition of electrons ? [NCERT : P L-300 | April 9, 2024 (II)]

(a) 6 (b) 3 (c) 10 (d) 1

23. The ratio of the shortest wavelength of Balmer series to the shortest wavelength of Lyman series for hydrogen atom is : [April 6, 2024 (I)]

(a) 4 : 1 (b) 1 : 2 (c) 1 : 4 (d) 2 : 1

24. Radius of a certain orbit of hydrogen atom is 8.48 Å. If energy of electron in this orbit is E/x, then x = ________. [NCERT : P L-299 | April 6, 2024 (I)]
(Given $a_0 = 0.529$Å, E = energy of electron in ground state)

25. The longest wavelength associated with Paschen series is : (Given $R_H = 1.097 \times 10^7$ SI unit) [April 6, 2024 (II)]

(a) 1.094×10^{-6}m (b) 2.973×10^{-6}m
(c) 3.646×10^{-6}m (d) 1.876×10^{-6}m

26. In Franck-Hertz experiment, the first dip in the current-voltage graph for hydrogen is observed at 10.2 V. The wavelength of light emitted by hydrogen atom when excited to the first excitation level is ______ nm.
(Given hc = 1245 eV nm, e = 1.6×10^{-19}C). [NCERT : P L-300 | April 6, 2024 (II)]

27. An electron rotates in a circle around a nucleus having positive charge Ze. Correct relation between total energy (E) of electron to its potential energy (U) is: [NCERT : P L-296 | April 5, 2024 (I)]

(a) E = 2U (b) 2E = 3U (c) E = U (d) 2E = U

28. The angular momentum of an electron in a hydrogen atom is proportional to : (Where r is the radius of orbit of electron) [NCERT : P L-299 | April 5, 2024 (II)]

(a) $\sqrt{r}$ (b) $\frac{1}{r}$ (c) r (d) $\frac{1}{\sqrt{r}}$

29. The shortest wavelength of the spectral lines in the Lyman series of hydrogen spectrum is 915 Å. The longest wavelength of spectral lines in the Balmer series will be ______ Å. [April 5, 2024 (II)]

30. A hydrogen atom changes its state from n = 3 to n = 2. Due to recoil, the percentage change in the wavelength of emitted light is approximately 1×10^{-n}. The value of n is__________.
[Given Rhc = 13.6 eV, hc = 1242 eV nm, h = 6.6×10^{-34} J s, mass of the hydrogen atom = 1.6×10^{-27} kg] [NCERT : P L-299 | April 4, 2024 (I)]

31. According to Bohr's theory, the moment of momentum of an electron revolving in 4th orbit of hydrogen atom is : [NCERT : P L-299 | April 4, 2024 (II)]

(a) $8\frac{h}{\pi}$ (b) $\frac{h}{\pi}$ (c) $2\frac{h}{\pi}$ (d) $\frac{h}{2\pi}$

32. The minimum energy required by a hydrogen atom in ground state to emit radiation in Balmer series is nearly: [Feb. 1, 2024 (I)]

(a) 1.5 eV (b) 13.6 eV
(c) 1.9 eV (d) 12.1 eV

33. A particular hydrogen - like ion emits the radiation of frequency 3×10^{15} Hz when it makes transition from n = 2 to n = 1. The frequency of radiation emitted in transition from n = 3 to n = 1 is $\frac{x}{9} \times 10^{15}$ Hz, when x = ________. [NCERT : P L-299 | Feb. 1, 2024 (II)]

34. If the wavelength of the first member of Lyman series of hydrogen is λ. The wavelength of the second member will be [Jan. 31, 2024 (I)]

(a) $\frac{5}{27}\lambda$ (b) $\frac{27}{32}\lambda$ (c) $\frac{32}{27}\lambda$ (d) $\frac{27}{5}\lambda$

35. The ratio of the magnitude of the kinetic energy to the potential energy of an electron in the 5th excited state of a hydrogen atom is: [NCERT : P L-296 | Jan. 30, 2024 (I)]

(a) 4 (b) $\frac{1}{4}$ (c) $\frac{1}{2}$ (d) 1

36. A electron of hydrogen atom on an excited state is having energy $E_n = -0.85$ eV. The maximum number of allowed transitions to lower energy level is ________. [NCERT : P L-299 | Jan. 30, 2024 (I)]

37. An electron revolving in n^{th} Bohr orbit has magnetic moment μ_n. If $\mu_n \propto n^x$, the value of x is **[Jan. 30, 2024 (II)]**
(a) 2 (b) 1 (c) 3 (d) 0

38. When a hydrogen atom going from n = 2 to n = 1 emits a photon, its recoil speed is $\frac{x}{5}$ m/s. Where x = ______. (Use, mass of hydrogen atom = 1.6×10^{-27} kg) **[NCERT : P L-299 | Jan. 29, 2024 (I)]**

39. Hydrogen atom is bombarded with electrons accelerated through a potential difference of V, which causes excitation of hydrogen atoms. If the experiment is being performed at T = 0 K, the minimum potential difference needed to observe any Balmer series lines in the emission spectra will be $\frac{\alpha}{10}$ V, where α = ______. **[Jan. 29, 2024 (II)]**

40. The radius of third stationary orbit of electron for Bohr's atom is R. The radius of fourth stationary orbit will be: **[NCERT : P L-299 | Jan. 27, 2024 (I)]**
(a) $\frac{4}{3}R$ (b) $\frac{16}{9}R$ (c) $\frac{3}{4}R$ (d) $\frac{9}{16}R$

41. If Rydberg's constant is R, the longest wavelength of radiation in Paschen series will be $\frac{\alpha}{7R}$, where α = __________. **[Jan. 27, 2024 (II)]**

42. As per given figure A, B and C are the first, second and third excited energy level of hydrogen atom respectively. If the ratio of the two wavelengths $\left(\text{i.e. } \frac{\lambda_1}{\lambda_2}\right)$ is $\frac{7}{4n}$, then the value of n will be __________. **[April 15, 2023 (I)]**

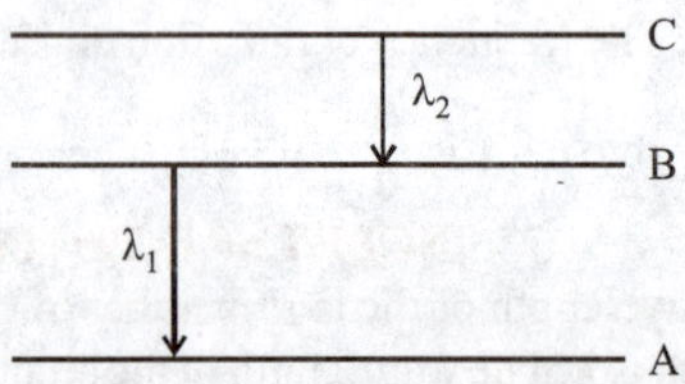

43. The radius of 2nd orbit of He^+ of Bohr's model is r_1 and that of fourth orbit of Be^{3+} is represented as r_2. Now the ratio $\frac{r_2}{r_1}$ is x : 1. The value of x is ______. **[NCERT : P L-299 | April 13, 2023 (I)]**

44. A 12.5 eV electron beam is used to bombard gaseous hydrogen at room temperature. The number of spectral lines emitted will be: **[NCERT : P L-300 | April 12, 2023 (I)]**
(a) 2 (b) 1 (c) 3 (d) 4

45. A monochromatic light is incident on a hydrogen sample in ground state. Hydrogen atoms absorb a fraction of light and subsequently emit radiation of six different wavelengths. The frequency of incident light is $x \times 10^{15}$ Hz. The value of x is ______. (Given h = 4.25×10^{-15} eVs) **[NCERT : P L-299 | April 11, 2023 (I)]**

46. The energy of He^+ ion in its first excited state is. (The ground state energy for the Hydrogen atom is –13.6 eV): **[NCERT : P L-299 | April 11, 2023 (II)]**
(a) –3.4 eV (b) –54.4 eV (c) –13.6 eV (d) –27.2 eV

47. The angular momentum for the electron in Bohr's orbit is L. If the electron is assumed to revolve in second orbit of hydrogen atom, then the change in angular momentum will be : **[NCERT : P L-299 | April 10, 2023 (I)]**
(a) $\frac{L}{2}$ (b) zero (c) L (d) 2L

48. The ratio of wavelength of spectral lines H_α and H_β in the Balmer series is $\frac{x}{20}$. The value of x is ______ **[April 8, 2023 (II)]**

49. The energy levels of an hydrogen atom are shown below. The transition corresponding to emission of shortest wavelength is **[NCERT : P L-299, 300 | April 6, 2023 (I)]**
(a) C
(b) D
(c) B
(d) A

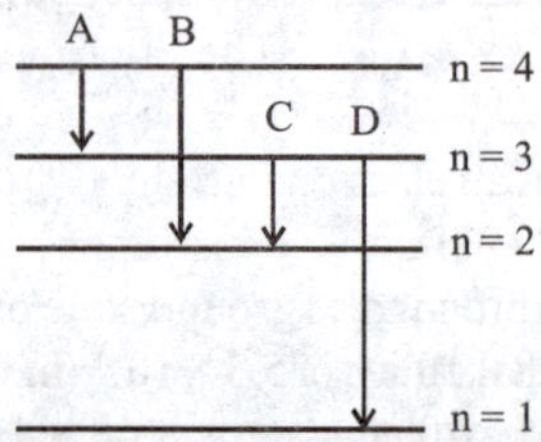

50. The radius of fifth orbit of the Li^{2+} is ______ $\times 10^{-12}$ m. Take : radius of hydrogen atom = 0.51Å **[NCERT : P L-299 | April 6, 2023 (I)]**

51. Experimentally it is found that 12.8 eV energy is required to separate a hydrogen atom into a proton and an electron. So, the orbital radius of the electron in a hydrogen atom is $\frac{9}{x} \times 10^{-10}$ m. The value of the x is ______. ($1 eV = 1.6 \times 10^{-19}$ J, $\frac{1}{4\pi\epsilon_0} = 9 \times 10^9 Nm^2/C^2$ and electronic charge = 1.6×10^{-19} C) **[April 6, 2023 (II)]**

52. A light of energy 12.75 eV is incident on a hydrogen atom in its ground state. The atom absorbs the radiation and reaches to one of its excited states. The angular momentum of the atom in the excited state is $\frac{x}{\pi} \times 10^{-17}$ eVs. The value of x is ___ (use h = 4.14×10^{-15} e Vs, c = 3×10^8 ms^{-1}). **[NCERT : P L-299, 300 | Feb. 1, 2023 (I)]**

53. An electron of a hydrogen like atom, having Z = 4, jumps from 4th energy state to 2nd energy state. The energy released in this process, will be: **[NCERT : P L-299 | Feb. 1, 2023 (II)]**
(Given Rch = 13.6 eV)
Where R = Rydberg constant, c = Speed of light in vacuum, h = Planck's constant
(a) 10.5 eV (b) 13.6 eV (c) 3.4 eV (d) 40.8 eV

54. For hydrogen atom, λ_1 and λ_2 are the wavelengths corresponding to the transitions 1 and 2 respectively as shown in figure. The ratio of λ_1 and λ_2 is $\frac{x}{32}$. The value of x is ______. **[Jan. 31, 2023 (I)]**

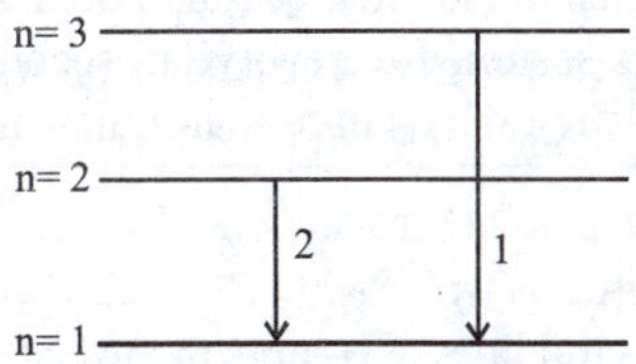

55. The radius of electron's second stationary orbit in Bohr's atom is R. The radius of 3rd orbit will be **[NCERT : P L-299 | Jan. 31, 2023 (II)]**

(a) $\frac{R}{3}$ (b) 2.25R (c) 3R (d) 9R

56. If the binding energy of ground state electron in a hydrogen atom is 13.6 eV, then, the energy required to remove the electron from the second excited state of Li^{2+} will be: $x \times 10^{-1}$ eV. The value of x is ______. **[NCERT : P L-299 | Jan. 31, 2023 (II)]**

57. Speed of an electron in Bohr's 7th orbit for Hydrogen atom is 3.6×10^6 m/s. The corresponding speed of the electron in 3rd orbit, in m/s is : **[Jan. 30, 2023 (I)]**

(a) (1.8×10^6) (b) (7.5×10^6)
(c) (3.6×10^6) (d) (8.4×10^6)

58. The wavelength of the radiation emitted is λ_0 when an electron jumps from the second excited state to the first excited state of hydrogen atom. If the electron jumps from the third excited state to the second orbit of the hydrogen atom, the wavelength of the radiation emitted will be $\frac{20}{x}\lambda_0$. The value of x is ______. **[Jan. 25, 2023 (I)]**

59. The energy levels of an atom is shown is figure. Which one of these transitions will result in the emission of a photon of wavelength 124.1 nm ? **[NCERT : P L-299 | Jan. 25, 2023 (II)]**

Given ($h = 6.62 \times 10^{-34}$ Js)

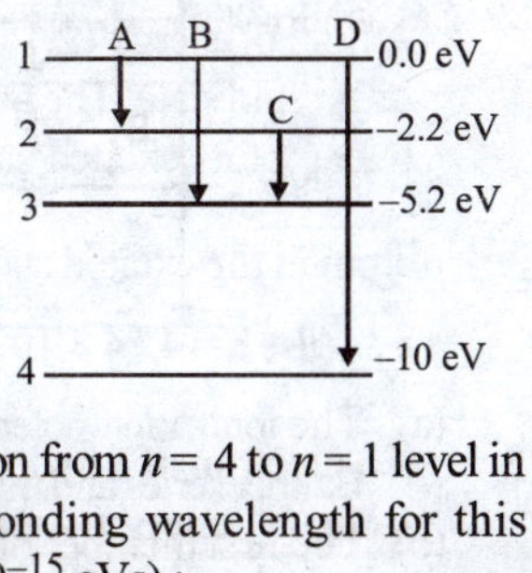

(a) B
(b) A
(c) C
(d) D

60. A photon is emitted in transition from $n = 4$ to $n = 1$ level in hydrogen atom. The corresponding wavelength for this transition is (given, $h = 4 \times 10^{-15}$ eVs) : **[NCERT : P L-299 | Jan. 24, 2023 (II)]**

(a) 94.1 nm (b) 941 nm
(c) 97.4 nm (d) 99.3 nm

61. Find the ratio of energies of photons produced due to transition of an election of hydrogen atom from its (i) second permitted energy level to the first level. and (ii) the highest permitted energy level to the first permitted level. **[NCERT : P L-299 | July 29, 2022 (I)]**

(a) 3:4 (b) 4:3 (c) 1: 4 (d) 4:1

62. In the hydrogen spectrum, λ the wavelength of first transition line of Lyman series. The wavelength difference will be "$a\lambda$" between the wavelength of 3rd transition line of Paschen series and that of 2nd transition line of Balmer series where a = ________. **[July 26, 2022 (I)]**

63. The momentum of an electron revolving in nth orbit is given by: **[NCERT : P L-299 | July 25, 2022 (I)]**

(a) $\frac{nh}{2\pi r}$ (b) $\frac{nh}{2r}$ (c) $\frac{nh}{2\pi}$ (d) $\frac{2\pi r}{nh}$

64. $\frac{x}{x+4}$ is the ratio of energies of photons produced due to transition of an electron of hydrogen atom from its (i) third permitted energy level to the second level and (ii) the highest permitted energy level to the second permitted level. **[NCERT : P L-299 | July 25, 2022 (II)]** The value of x will be

65. Hydrogen atom from excited state comes to the ground by emitting a photon of wavelength λ. The value of principal quantum number 'n' of the excited state will be: **[NCERT : P L-299 | July 25, 2022 (II)]**

(a) $\sqrt{\frac{\lambda R}{\lambda - 1}}$ (b) $\sqrt{\frac{\lambda R}{\lambda R - 1}}$

(c) $\sqrt{\frac{\lambda}{\lambda R - 1}}$ (d) $\sqrt{\frac{\lambda R^2}{\lambda R - 1}}$

66. A hydrogen atom in its first excited state absorbs a photon of energy $x \times 10^{-2}$ eV and exited to a higher energy state where the potential energy of electron is –1.08 eV. The value of x is ________. **[NCERT : P L-299 | June 30, 2022 (I)]**

67. A hydrogen atom in ground state absorbs 12.09eV of energy. The orbital angular momentum of the electron is increased by: **[NCERT : P L-299 | June 30, 2022 (I)]**

(a) 1.05×10^{-34} Js (b) 2.11×10^{-34} Js
(c) 3.16×10^{-34} Js (d) 4.22×10^{-34} Js

68. A beam of monochromatic light is used to excite the electron in Li^{++} from the first orbit to the third orbit. The wavelength of monochromatic light is found to be $x \times 10^{-10} m$. The value of x is ______. [Given hc = 1242 eV nm] **[NCERT : P L-299 | June 27, 2022 (I)]**

69. A hydrogen atom in is ground state absorbs 10.2 eV of energy. The angular momentum of electron of the hydrogen atom will increase by the value of: **[NCERT : P L-299 | June 27, 2022 (I)]** (Given, Planck's constant = 6.6×10^{-34} Js)

(a) 2.10×10^{-34} Js (b) 1.05×10^{-34} Js
(c) 3.15×10^{-34} Js (d) 4.2×10^{-34} Js

70. Given below are two statements:
Statement I : In hydrogen atom, the frequency of radiation emitted when an electron jumps from lower energy orbit (E_1) to higher energy orbit (E_2), is given as $hf = E_1 - E_2$.
Statement II : The jumping of electron from higher energy orbit (E_2) to lower energy orbit (E_1) is associated with frequency of radiation given as $f = (E_2 - E_1)/h$
This condition is Bohr's frequency condition.
In the light of the above statements, choose the correct answer from the options given below:
[NCERT : P L-299 | June 27, 2022 (II)]
(a) Both statement I and statement II are true.
(b) Both statement I and statement II are false
(c) Statement I is correct but statement II is false
(d) Statement I is incorrect but statement II is true.

71. The ratio for the speed of the electron in the 3rd orbit of He^+ to the speed of the electron in the 3rd orbit of hydrogen atom will be :- **[June 25, 2022 (I)]**
(a) 1 : 1 (b) 1 : 2 (c) 4 : 1 (d) 2 : 1

72. In Bohr's atomic model of hydrogen, let *K, P* and *E* are the kinetic energy, potential energy and total energy of the electron respectively. Choose the correct option when the electron undergoes transitions to a higher level :
[NCERT : P L-296, 299 | June 24, 2022 (II)]
(a) All K, P and E increase.
(b) K decreases, P and E increase.
(c) P decreases, K and E increase.
(d) K increases, P and E decrease.

73. A free electron of 2.6 eV energy collides with a H^+ ion. This results in the formation of a hydrogen atom in the first excited state and a photon is released. Find the frequency of the emitted photon.
($h = 6.6 \times 10^{-34}$ Js)
[NCERT : P L-296, 299 | Aug. 31, 2021 (II)]
(a) 1.45×10^{16} MHz (b) 0.19×10^{15} MHz
(c) 1.45×10^{9} MHz (d) 9.0×10^{27} MHz

74. X different wavelengths may be observed in the spectrum from a hydrogen sample if the atoms are exited to states with principal quantum number n = 6 ? The value of X is________. **[NA, Aug. 27, 2021 (II)]**

75. A particular hydrogen like ion emits radiation of frequency 2.92×10^{15} Hz when it makes transition from n = 3 to n = 1. The frequency in Hz of radiation emitted in transition from n = 2 to n = 1 will be :
[NCERT : P L-299 | Aug. 26, 2021 (I)]
(a) 0.44×10^{15} (b) 6.57×10^{15}
(c) 4.38×10^{15} (d) 2.46×10^{15}

76. In Bohr's atomic model, the electron is assumed to revolve in a circular orbit of radius 0.5 Å. If the speed of electron is 2.2×16^{6} m/s, then the current associated with the electron will be________$\times 10^{-2}$ mA. [Take π as $\frac{22}{7}$]
[NCERT : P L-299 | NA, July 27, 2021 (I)]

77. Imagine that the electron in a hydrogen atom is replaced by a muon (μ). The mass of muon particle is 207 times that of an electron and charge is equal to the charge of an electron. The ionization potential of this hydrogen atom will be : **[March 18, 2021 (I)]**
(a) 331.2 eV (b) 2815.2 eV
(c) 13.6 eV (d) 27.2 eV

78. A particle of mass m moves in a circular orbit in a central potential field $U(r) = U_0 r^4$. If Bohr's quantization conditions are applied, radii of possible orbitals r_n vary with $n^{1/\alpha}$, where α is________. **[NCERT : P L-299 | March 17, 2021 (II)]**

79. The first three spectral lines of H-atom in the Balmer series are given $\lambda_1, \lambda_2, \lambda_3$ considering the Bohr atomic model, the ratio of wavelengths of first and third spectral lines $\left(\frac{\lambda_1}{\lambda_3}\right)$ are ratio of related by a factor of approximately $x \times 10^{-1}$. The value of x, to the nearest integer, is
[March 16, 2021 (I)]

80. If λ_1 and λ_2 are the wavelengths of the third member of Lyman and first member of the Paschen series respectively, then the value of $\lambda_1 : \lambda_2$ is **[Feb. 26, 2021 (I)]**
(a) 1 : 3 (b) 7 : 108 (c) 7 : 135 (d) 1 : 9

81. The recoil speed of a hydrogen atom after it emits a photon in going from n = 5 state to n = 1 state will be
[NCERT : P L-299 | Feb. 26, 2021 (II)]
(a) 4.34 m/s (b) 2.19 m/s (c) 4.17 m/s (d) 3.25 m/s

82. The wavelength of the photon emitted by a hydrogen atom when an electron makes a transition from n = 2 to n = 1 state is **[Feb. 25, 2021 (II)]**
(a) 490.7 nm (b) 121.8 nm (c) 913.3 nm (d) 194.8 nm

83. In the given figure, the energy levels of hydrogen atom have been shown along with some transitions marked A, B, C, D and E. The transitions A, B and C respectively represent : **[Feb. 24, 2021 (I)]**

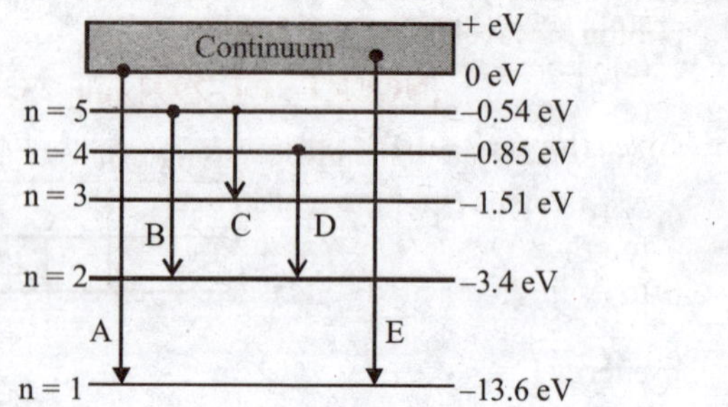

(a) The ionization potential of hydrogen, second member of Balmer series and third member of Paschen series.
(b) The first member of the Lyman series, third member of Balmer series and second member of Paschen series.
(c) The series limit of Lyman series, third member of Balmer series and second member of Paschen series.
(d) The series limit of Lyman series, second member of Balmer series and second member of Paschen series.

84. According to Bohr atom model, in which of the following transitions will the frequency be maximum? **[NCERT : P L-299 | Feb. 24, 2021 (II)]**
(a) $n=4$ to $n=3$ (b) $n=2$ to $n=1$
(c) $n=5$ to $n=4$ (d) $n=3$ to $n=2$

85. A particle of mass 200 MeV/c^2 collides with a hydrogen atom at rest. Soon after the collision the particle comes to rest, and the atom recoils and goes to its first excited state. The initial kinetic energy of the particle (in eV) is $\frac{N}{4}$. The value of N is : **[NA Sep. 05, 2020 (I)]**
(Given the mass of the hydrogen atom to be 1 GeV/c^2)

86. In the line spectra of hydrogen atom, difference between the largest and the shortest wavelengths of the Lyman series is 304 Å. The corresponding difference for the Paschan series in Å is : ________. **[NA Sep. 04, 2020 (I)]**

87. In a hydrogen atom the electron makes a transition from $(n+1)^{th}$ level to the nth level. If $n >> 1$, the frequency of radiation emitted is proportional to : **[NCERT : P L-299 | Sep. 02, 2020 (II)]**
(a) $\frac{1}{n}$ (b) $\frac{1}{n^3}$ (c) $\frac{1}{n^2}$ (d) $\frac{1}{n^4}$

88. The energy required to ionise a hydrogen like ion in its ground state is 9 Rydbergs. What is the wavelength of the radiation emitted when the electron in this ion jumps from the second excited state to the ground state? **[9 Jan. 2020 II]**
(a) 24.2 nm (b) 11.4 nm (c) 35.8 nm (d) 8.6 nm

89. The first member of the Balmer series of hydrogen atom has a wavelength of 6561 Å. The wavelength of the second member of the Balmer series (in nm) is ______. **[NA 8 Jan. 2020 II]**

90. The time period of revolution of electron in its ground state orbit in a hydrogen atom is 1.6×10^{-16} s. The frequency of revolution of the electron in its first excited state (in s^{-1}) is: **[7 Jan. 2020 I]**
(a) 1.6×10^{14} (b) 7.8×10^{14}
(c) 6.2×10^{15} (d) 5.6×10^{12}

91. An excited He^+ ion emits two photons in succession, with wavelengths 108.5 nm and 30.4 nm, in making a transition to ground state. The quantum number n, corresponding to its initial excited state is (for photon of wavelength λ, energy $E = \frac{1240 eV}{\lambda(in\ nm)}$ **[NCERT : P L-299 | 12 April 2019 II]**
(a) $n=4$ (b) $n=5$ (c) $n=7$ (d) $n=6$

92. The electron in a hydrogen atom first jumps from the third excited state to the second excited state and subsequently to the first excited state. The ratio of the respective wavelengths, λ_1/λ_2, of the photons emitted in this process is : **[12 April 2019 II]**
(a) 20/7 (b) 27/5 (c) 7/5 (d) 9/7

93. Consider an electron in a hydrogen atom, revolving in its second excited state (having radius 4.65 Å). The de-Broglie wavelength of this electron is : **[NCERT : P L-301, 302 | 12 April 2019 II]**
(a) 3.5 Å (b) 6.6 Å (c) 12.9 Å (d) 9.7 Å

94. In Li^{++}, electron in first Bohr orbit is excited to a level by a radiation of wavelength λ. When the ion gets deexcited to the ground state in all possible ways (including intermediate emissions), a total of six spectral lines are observed. What is the value of l ? **[10 April 2019 II]**
(Given : $h = 6.63 \times 10^{-34}$ Js; $c = 3 \times 10^8 ms^{-1}$)
(a) 11.4 nm (b) 9.4 nm (c) 12.3 nm (d) 10.8 nm

95. Taking the wavelength of first Balmer line in hydrogen spectrum (n = 3 to n = 2) as 660 nm, the wavelength of the 2nd Balmer line (n = 4 to n = 2) will be; **[9 April 2019 I]**
(a) 889.2 nm (b) 488.9 nm
(c) 642.7 nm (d) 388.9 nm

96. A He^+ ion is in its first excited state. Its ionization energy is: **[NCERT : P L-299 | 9 April 2019 II]**
(a) 48.36 eV (b) 54.40 eV
(c) 13.60 eV (d) 6.04 eV

97. Radiation coming from transitions $n=2$ to $n=1$ of hydrogen atoms fall on He^+ ions in $n=1$ and $n=2$ states. The possible transition of helium ions as they absorb energy from the radiation is : **[NCERT : P L-299 | 8 April 2019 I]**
(a) $n=2 \rightarrow n=3$ (b) $n=1 \rightarrow n=4$
(c) $n=2 \rightarrow n=5$ (d) $n=2 \rightarrow n=4$

98. A hydrogen atom, initially in the ground state is excited by absorbing a photon of wavelength 980Å. The radius of the atom in the excited state, in terms of Bohr radius a_0, will be: **[NCERT : P L-299 | 11 Jan 2019 I]**
(a) $25a_0$ (b) $9a_0$ (c) $16a_0$ (d) $4a_0$

99. In a hydrogen like atom, when an electron jumps from the M-shell to the L-shell, the wavelength of emitted radiation is λ. If an electron jumps from N-shell to the L-shell, the wavelength of emitted radiation will be: **[11 Jan 2019 II]**
(a) $\frac{27}{20}\lambda$ (b) $\frac{16}{25}\lambda$ (c) $\frac{25}{16}\lambda$ (d) $\frac{20}{27}\lambda$

100. An electron from various excited states of hydrogen atom emit radiation to come to the ground state. Let λ_n, λ_g be the de-Broglie wavelength of the electron in the nth state and the ground state respectively. Let Λ_n be the wavelength of the emitted photon in the transition from the nth state to the ground state. For large n, (A, B are constants) Λ_n is **[NCERT : P L-302 | 2018]**
(a) $\Lambda_n \approx A + \frac{B}{\lambda_n^2}$ (b) $\Lambda_n \approx A + B\lambda_n$
(c) $\Lambda_n^2 \approx A + B\lambda_n^2$ (d) $\Lambda_n^2 \approx \lambda$

101. If the series limit frequency of the Lyman series is ν_L, then the series limit frequency of the P-fund series is : **[2018]**
(a) $25\,\nu_L$ (b) $16\,\nu_L$ (c) $\nu_L/16$ (d) $\nu_L/25$

102. The energy required to remove the electron from a singly ionized Helium atom is 2.2 times the energy required to remove an electron from Helium atom. The total energy required to ionize the Helium atom completely is:
[NCERT : P L-299 | Online April 15, 2018]
(a) 20 eV (b) 79 eV
(c) 109 eV (d) 34 eV

103. Some energy levels of a molecule are shown in the figure. The ratio of the wavelengths $r = \lambda_1/\lambda_2$, is given by
[NCERT : P L-299, 300 | 2017]

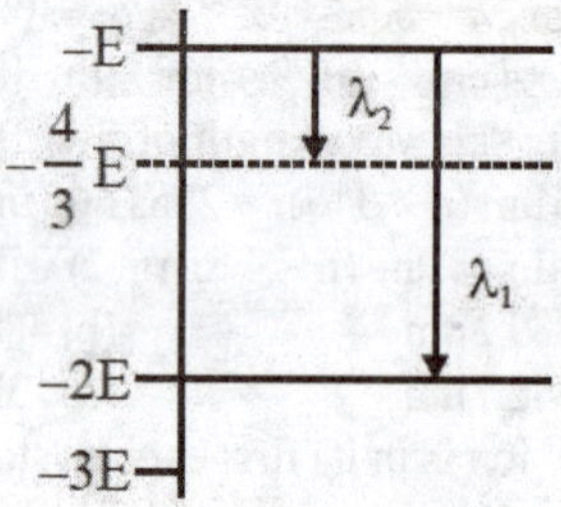

(a) $r = \frac{3}{4}$
(b) $r = \frac{1}{3}$
(c) $r = \frac{4}{3}$
(d) $r = \frac{2}{3}$

104. The acceleration of an electron in the first orbit of the hydrogen atom (z = 1) is : **[Online April 9, 2017]**
(a) $\frac{h^2}{\pi^2 m^2 r^3}$ (b) $\frac{h^2}{8\pi^2 m^2 r^3}$
(c) $\frac{h^2}{4\pi^2 m^2 r^3}$ (d) $\frac{h^2}{4\pi m^2 r^3}$

105. According to Bohr's theory, the time averaged magnetic field at the centre (i.e. nucleus) of a hydrogen atom due to the motion of electrons in the nth orbit is proportional to : (n = principal quantum number)
[NCERT : P L-299 | Online April 8, 2017]
(a) n^{-4} (b) n^{-5} (c) n^{-3} (d) n^{-2}

106. A hydrogen atom makes a transition from n = 2 to n = 1 and emits a photon. This photon strikes a doubly ionized lithium atom (z = 3) in excited state and completely removes the orbiting electron. The least quantum number for the excited state of the ion for the process is : **[Online April 9, 2016]**
(a) 2 (b) 4 (c) 5 (d) 3

107. As an electron makes a transition from an excited state to the ground state of a hydrogen - like atom/ion :
[NCERT : P L-296, 300 | 2015]
(a) kinetic energy decreases, potential energy increases but total energy remains same
(b) kinetic energy and total energy decrease but potential energy increases
(c) its kinetic energy increases but potential energy and total energy decrease
(d) kinetic energy, potential energy and total energy decrease

108. The de–Broglie wavelength associated with the electron in the n = 4 level is :
[NCERT : P L-301, 302 | Online April 11, 2015]
(a) $\frac{1}{4}$ th of the de–Broglie wavelength of the electron in the ground state.
(b) four times the de–Broglie wavelength of the electron in the ground state
(c) two times the de–Broglie wavelength of the electron in the ground state
(d) half of the de–Broglie wavelength of the electron in the ground state

109. The radiation corresponding to 3 → 2 transition of hydrogen atom falls on a metal surface to produce photoelectrons. These electrons are made to enter a magnetic field of 3×10^{-4} T. If the radius of the largest circular path followed by these electrons is 10.0 mm, the work function of the metal is close to:
[NCERT : P L-299 | 2014]
(a) 1.8 eV (b) 1.1 eV (c) 0.8 eV (d) 1.6 eV

110. Hydrogen ($_1H^1$), Deuterium ($_1H^2$), singly ionised Helium $\left({}_2He^4\right)^+$ and doubly ionised lithium $\left({}_3Li^6\right)^{++}$ all have one electron around the nucleus. Consider an electron transition from n = 2 to n = 1. If the wavelengths of emitted radiation are $\lambda_1, \lambda_2, \lambda_3$ and λ_4 respectively then approximately which one of the following is correct? **[2014]**
(a) $4\lambda_1 = 2\lambda_2 = 2\lambda_3 = \lambda_4$ (b) $\lambda_1 = 2\lambda_2 = 2\lambda_3 = \lambda_4$
(c) $\lambda_1 = \lambda_2 = 4\lambda_3 = 9\lambda_4$ (d) $\lambda_1 = 2\lambda_2 = 3\lambda_3 = 4\lambda_4$

111. The binding energy of the electron in a hydrogen atom is 13.6 eV, the energy required to remove the electron from the first excited state of Li^{++} is:
[NCERT : P L-299 | Online April 9, 2014]
(a) 122.4 eV (b) 30.6 eV (c) 13.6 eV (d) 3.4 eV

112. In a hydrogen like atom electron make transition from an energy level with quantum number n to another with quantum number (n – 1). If n>>1, the frequency of radiation emitted is proportional to : **[NCERT : P L-299 | 2013]**
(a) $\frac{1}{n}$ (b) $\frac{1}{n^2}$ (c) $\frac{1}{n^{3/2}}$ (d) $\frac{1}{n^3}$

113. A 12.5 eV electron beam is used to bombard gaseous hydrogen at room temperature. It will emit :
[NCERT : P L-299 | Online April 25, 2013]
(a) 2 lines in the Lyman series and 1 line in the Balmar series
(b) 3 lines in the Lyman series
(c) 1 line in the Lyman series and 2 lines in the Balmar series
(d) 3 lines in the Balmer series

114. Orbits of a particle moving in a circle are such that the perimeter of the orbit equals an integer number of de-Broglie wavelengths of the particle. For a charged particle moving in a plane perpendicular to a magnetic field, the radius of the nth orbital will therefore be proportional to :
[NCERT : P L-302 | Online April 22, 2013]
(a) n^2 (b) n (c) $n^{1/2}$ (d) $n^{1/4}$

115. Hydrogen atom is excited from ground state to another state with principal quantum number equal to 4. Then the number of spectral lines in the emission spectra will be : **[NCERT : P L-300 | 2012]**

(a) 2 (b) 3 (c) 5 (d) 6

116. A diatomic molecule is made of two masses m_1 and m_2 which are separated by a distance r. If we calculate its rotational energy by applying Bohr's rule of angular momentum quantization, its energy will be given by : (n is an integer) **[NCERT : P L-299 | 2012]**

(a) $\frac{(m_1+m_2)^2 n^2 h^2}{2m_1^2 m_2^2 r^2}$ (b) $\frac{n^2 h^2}{2(m_1+m_2)r^2}$

(c) $\frac{2n^2 h^2}{(m_1+m_2)r^2}$ (d) $\frac{(m_1+m_2)n^2 \hbar^2}{2m_1 m_2 r^2}$

117. Which of the plots shown in the figure represents speed (v_n) of the electron in a hydrogen atom as a function of the principal quantum number (n)? **[Online May 26, 2012]**

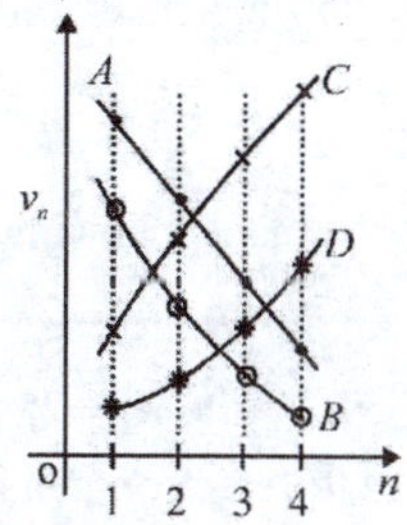

(a) B (b) D (c) C (d) A

118. The electron of a hydrogen atom makes a transition from the $(n+1)^{th}$ orbit to the nth orbit. For large n the wavelength of the emitted radiation is proportional to **[NCERT : P L-299 | Online May 7, 2012]**

(a) n (b) n^3 (c) n^4 (d) n^2

119. Energy required for the electron excitation in Li^{++} from the first to the third Bohr orbit is : **[NCERT : P L-299 | 2011]**

(a) 36.3 eV (b) 108.8 eV
(c) 122.4 eV (d) 12.1 eV

120. The transition from the state $n = 4$ to $n = 3$ in a hydrogen like atom results in ultraviolet radiation. Infrared radiation will be obtained in the transition from : **[NCERT : P L-299 | 2009]**

(a) $3 \to 2$ (b) $4 \to 2$ (c) $5 \to 4$ (d) $2 \to 1$

121. Suppose an electron is attracted towards the origin by a force $\frac{k}{r}$ where 'k' is a constant and 'r' is the distance of the electron from the origin. By applying Bohr model to this system, the radius of the nth orbital of the electron is found to be 'r_n' and the kinetic energy of the electron to be 'T_n'. Then which of the following is true? **[NCERT : P L-299 | 2008]**

(a) $T_n \propto \frac{1}{n^2}, r_n \propto n^2$ (b) T_n independent of n, $r_n \propto n$

(c) $T_n \propto \frac{1}{n}, r_n \propto n$ (d) $T_n \propto \frac{1}{n^3}, r_n \propto n^2$

122. Which of the following transitions in hydrogen atoms emit photons of highest frequency? **[NCERT : P L-299 | 2007]**

(a) $n=1$ to $n=2$ (b) $n=2$ to $n=6$
(c) $n=6$ to $n=2$ (d) $n=2$ to $n=1$

123. The diagram shows the energy levels for an electron in a certain atom. Which transition shown represents the emission of a photon with the most energy? **[NCERT : P L-299 | 2005]**

(a) IV
(b) III
(c) II
(d) I

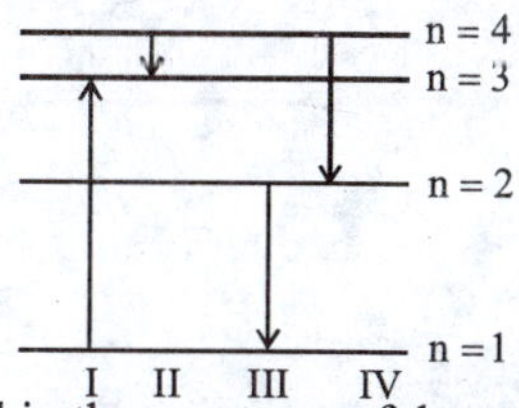

124. The wavelengths involved in the spectrum of deuterium $\left({}^2_1D\right)$ are slightly different from that of hydrogen spectrum, because **[2003]**

(a) the size of the two nuclei are different
(b) the nuclear forces are different in the two cases
(c) the masses of the two nuclei are different
(d) the attraction between the electron and the nucleus is differernt in the two cases

125. If 13.6 eV energy is required to ionize the hydrogen atom, then the energy required to remove an electron from $n=2$ is **[NCERT : P L-299 | 2002]**

(a) 10.2 eV (b) 0 eV (c) 3.4 eV (d) 6.8 eV

ANSWER KEYS

1	(156)	14	(2)	27	(d)	40	(b)	53	(d)	66	(286)	79	(15)	92	(a)	105	(d)	118	(b)
2	(a)	15	(c)	28	(a)	41	(144)	54	(27)	67	(b)	80	(c)	93	(d)	106	(b)	119	(b)
3	(3)	16	(b)	29	(6588)	42	(5)	55	(b)	68	(114)	81	(c)	94	(d)	107	(c)	120	(c)
4	(c)	17	(a)	30	(7)	43	(2)	56	(136)	69	(b)	82	(b)	95	(b)	108	(b)	121	(b)
5	(d)	18	(b)	31	(c)	44	(c)	57	(d)	70	(d)	83	(c)	96	(c)	109	(b)	122	(d)
6	(c)	19	(a)	32	(c)	45	(3)	58	(27)	71	(d)	84	(b)	97	(d)	110	(c)	123	(b)
7	(b, c)	20	(d)	33	(32)	46	(c)	59	(d)	72	(b)	85	(51)	98	(c)	111	(b)	124	(c)
8	(a)	21	(a)	34	(b)	47	(c)	60	(a)	73	(c)	86	(10553.14)	99	(d)	112	(d)	125	(c)
9	(d)	22	(d)	35	(c)	48	(27)	61	(a)	74	(15)	87	(b)	100	(a)	113	(a)		
10	(b)	23	(a)	36	(6)	49	(b)	62	(5)	75	(d)	88	(b)	101	(d)	114	(c)		
11	(c)	24	(16)	37	(b)	50	(425)	63	(a)	76	(112)	89	(486)	102	(b)	115	(d)		
12	(d)	25	(d)	38	(17)	51	(16)	64	(5)	77	(b)	90	(b)	103	(b)	116	(d)		
13	(c)	26	(122)	39	(121)	52	(828)	65	(b)	78	(3)	91	(b)	104	(c)	117	(a)		

13 Nuclei

TREND ANALYSIS JEE MAIN (2025-2021)

	2025				2024				2023				2022				2021			
	Session II		Session I		Session II		Session I		Session II		Session I		Session II		Session I		Session II		Session I	
	MCQs	NVQs	MCQs	NVQs	MCQs	NVQs	MCQs	NVQs	MCQs	NVQs	MCQs	NVQs	MCQs	NVQs	MCQs	NVQs	MCQs	NVQs	MCQs	NVQs
Topic 1	2	–	–	–	–	–	1	2	–	1	1	1	2	–	1	–	–	–	–	–
Topic 2	2	–	2	–	7	2	5	1	4	1	–	3	–	1	1	–	1	1	–	–
Topic 3																				
Topic 4																				
Total	4	-	2	-	7	2	6	3	4	2	1	4	2	1	2	-	1	1	-	-

How to Code Qns for Better & Faster Revision

Type 1: Able to solve (No code Req)

Type 2: Solved with struggle or by seeing solution **(Use code NR means Needs Revision or orange stickers)**

Type 3: Tricky/ Could not solve on my own. Need help from Teacher/ Friend. **(Use code TQ means Tricky Question or red stickers)**

You can either hand write the code in boxes or else paste coloured stickers given at end of book.

Topic 1 - Composition and Size of the Nuclei

1. For a nucleus of mass number A and radius R, the mass density of nucleus can be represented as

[NCERT : P L-309 | April 8, 2025 (II)]

(a) A^3 (b) $A^{\frac{1}{3}}$

(c) $A^{\frac{2}{3}}$ (d) Independent of A

2. Given below are two statements : one is labelled as **Assertion (A)** and the other is labelled as **Reason (R)**.

Assertion (A) : The density of the copper ($^{64}_{29}Cu$) nucleus is greater than that of the carbon $\left(^{12}_{6}C\right)$ nucleus.

Reason (R) : The nucleus of mass number A has a radius proportional to $A^{1/3}$.

In the light of the above statements, choose the **most appropriate answer** from the options given below :

[NCERT : P L-309 | April 7, 2025 (II)]

(a) **(A)** is correct but **(R)** is not correct

(b) **(A)** is not correct but **(R)** is correct

(c) Both **(A)** and **(R)** are correct and **(R)** is the correct explanation of **(A)**

(d) Both **(A)** and **(R)** are correct but **(R)** is not the correct explanation of **(A)**

3. The radius of a nucleus of mass number 64 is 4.8 fermi. Then the mass number of another nucleus having raidus of 4 fermi is $\frac{1000}{x}$, where x is ______.

[NCERT : P L-309 | Feb. 1, 2024 (I)]

4. The mass number of nucleus having radius equal to half of the radius of nucleus with mass number 192 is:

[NCERT : P L-309 | Jan. 31, 2024 (II)]

(a) 24 (b) 32 (c) 40 (d) 20

5. A nucleus has mass number A_1 and volume V_1. Another nucleus has mass number A_2 and volume V_2. If relation between mass number is $A_2 = 4A_1$, then $\frac{V_2}{V_1}$ = ______.

[NCERT : P L-309 | Jan. 31, 2024 (II)]

6. A nucleus disintegrates into two nuclear parts, in such a way that ratio of their nuclear sizes is $1 : 2^{1/3}$. Their respective speed have a ratio of n : 1. The value of n is ______ **[NCERT : P L-309 | April 11, 2023 (II)]**

7. Given below are two statements : one is labelled as Assertion A and the other is labelled as Reason R.

Assertion A : The nuclear density of nuclides ${}^{10}_{5}B$, ${}^{6}_{3}Li$, ${}^{56}_{26}Fe$, ${}^{20}_{10}Ne$ and ${}^{209}_{83}Bi$ can be arranged as

$\rho^N_{Bi} > \rho^N_{Fe} > \rho^N_{Ne} > \rho^N_{B} > \rho^N_{Li}$.

Reason R : The radius R of nucleus is related to its mass number A as $R = R_0 A^{1/3}$, where R_0 is a constant.

In the light of the above statement, choose the correct answer from the options given below :

[NCERT : P L-309 | Jan. 30, 2023 (II)]

(a) Both A and R are true and R is the correct explanation of A
(b) A is false but R is true
(c) A is true but R is false
(d) Both A and R are true but R is NOT the correct explanation of A

8. The ratio of the density of oxygen nucleus $\left({}^{16}_{8}O\right)$ and helium nucleus $\left({}^{4}_{2}He\right)$ is

[NCERT : P L-309 | Jan. 25, 2023 (I)]

(a) 4:1 (b) 8:1
(c) 1:1 (d) 2:1

9. A nucleus disintegrates into two smaller parts, which have their velocities in the ratio 3 : 2. The ratio of their nuclear sizes will be $\left(\frac{x}{3}\right)^{\frac{1}{3}}$. The value of 'x' is:

[NCERT : P L-309 | Jan. 25, 2023 (II)]

10. Assume that protons and neutrons have equal masses. Mass of a nucleon is 1.6×10^{-27} kg and radius of nucleus is $1.5 \times 10^{-15} A^{1/3}$ m. The approximate ratio of the nuclear density and water density is $n \times 10^{13}$. The value of n is ______. **[NCERT : P L-309 | Jan. 24, 2023 (I)]**

11. Read the following statements:

[NCERT : P L-309 | July 29, 2022 (II)]

(A) Volume of the nucleus is directly proportional to the mass number.
(B) Volume of the nucleus is independent of mass number.
(C) Density of the nucleus is directly proportional to the mass number.
(D) Density of the nucleus is directly proportional to the cube root of the mass number.
(E) Density of the nucleus is independent of the mass number.

Choose the correct option from the following options.
(a) (A) and (D) only (b) (A) and (E) only.
(c) (B) and (E) only (d) (A) and (C) only

12. Mass numbers of two nuclei are in the ratio of 4:3. Their nuclear densities will be in the ratio of

[NCERT : P L-309 | July 26, 2022 (II)]

(a) 4:3 (b) $\left(\frac{3}{4}\right)^{\frac{1}{3}}$ (c) 1 : 1 (d) $\left(\frac{4}{3}\right)^{\frac{1}{3}}$

13. Which of the following figure represents the variation of $\ln\left(\frac{R}{R_0}\right)$ with $\ln A$ (If R = radius of a nucleus and A = its mass number)? **[NCERT : P L-309 | June 25, 2022 (II)]**

(a) 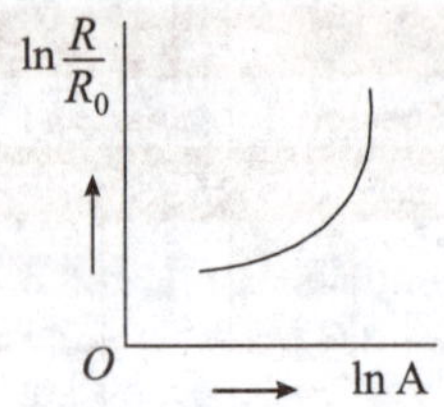

(b) 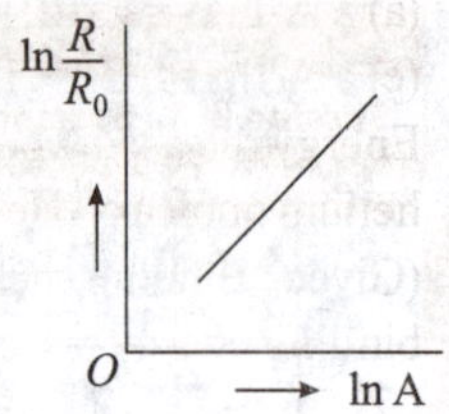

(c) 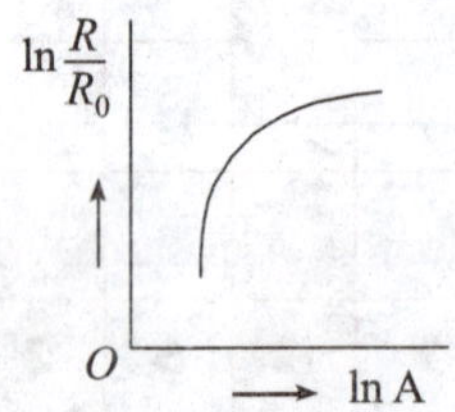

(d) 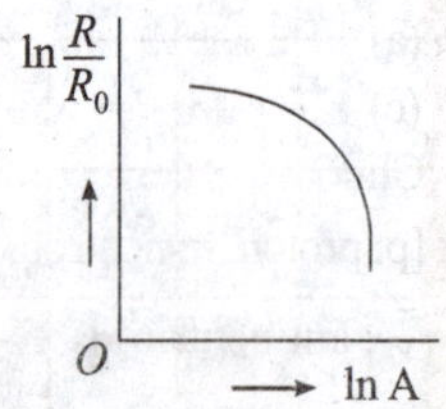

14. The radius R of a nucleus of mass number A can be estimated by the formula $R = (1.3 \times 10^{-15})A^{1/3}$ m. It follows that the mass density of a nucleus is of the order of :
$(M_{prot.} \cong M_{neut.} \simeq 1.67 \times 10^{-27}$ kg)

[NCERT : P L-309 | Sep. 03, 2020 (II)]

(a) 10^3 kg m^{-3} (b) 10^{10} kg m^{-3}
(c) 10^{24} kg m^{-3} (d) 10^{17} kg m^{-3}

15. The ratio of the mass densities of nuclei of ${}^{40}Ca$ and ${}^{16}O$ is close to : **[NCERT : P L-309 | 8 April 2019 II]**

(a) 1 (b) 0.1 (c) 5 (d) 2

16. An unstable heavy nucleus at rest breaks into two nuclei which move away with velocities in the ratio of 8:27. The ratio of the radii of the nuclei (assumed to be spherical) is:

[NCERT : P L-309 | Online April 15, 2018]

(a) 8 : 27 (b) 2 : 3 (c) 3 : 2 (d) 4 : 9

17. Which of the following are the constituents of the nucleus?

[NCERT : P L-308 | 2007]

(a) Electrons and protons (b) Neutrons and protons
(c) Electrons and neutrons (d) Neutrons and positrons

18. If radius of the ${}^{27}_{13}Al$ nucleus is estimated to be 3.6 fermi then the radius of ${}^{125}_{52}Te$ nucleus be nearly

[NCERT : P L-309 | 2005]

(a) 8 fermi (b) 6 fermi (c) 5 fermi (d) 4 fermi

Topic 2 - Mass-Energy Equivalence and Nuclear Reactions

19. Match the **LIST-I** with **LIST-II**

[NCERT : P L-315 | April 3, 2025 (I)]

LIST-I	LIST-II
A. $^1_0n + ^{235}_{92}U \rightarrow ^{140}_{50}Xe + ^{94}_{38}Sr + 2^1_0n$	I. Chemical reaction
B. $2H_2 + O_2 \rightarrow 2H_2O$	II. Fusion with +ve Q value
C. $^2_1H + ^2_1H \rightarrow ^3_2He + ^1_0n$	III. Fission
D. $^1_1H + ^3_1H \rightarrow ^2_1H + ^2_1H$	IV. Fusion with -ve Q value

Choose the **correct** answer from the options given below:

(a) A-II, B-I, C-III, D-IV (b) A-III, B-I, C-II, D-IV
(c) A-II, B-I, C-IV, D-III (d) A-III, B-I, C-IV, D-II

20. Energy released when two deuterons ($_1H^2$) fuse to form a helium nucleus ($_2He^4$) is :
(Given : Binding energy per nucleon of $_1H^2 = 1.1$ MeV and binding energy per nucleon of $_2He^4 = 7.0$ MeV)

[NCERT : P L-311 | April 2, 2025 (II)]

(a) 8.1 MeV (b) 5.9 MeV
(c) 23.6 MeV (d) 26.8 MeV

21. Choose the correct nuclear process from the below options [p: proton, n: neutron, e^- : electron, e^+ : positron, ν : neutrino, $\bar{\nu}$: antineutrino] **[NCERT : P L-315 | Jan. 28, 2025 (I)]**

(a) $n \rightarrow p + e^- + \bar{\nu}$ (b) $n \rightarrow p + e^- + \nu$
(c) $n \rightarrow p + e^+ + \bar{\nu}$ (d) $n \rightarrow p + e^+ + \nu$

22. Given below are two statements. One is labelled as **Assertion (A)** and the other is labelled as **Reason (R)**.
Assertion (A) : The binding energy per nucleon is found to be practically independent of the atomic number A, for nuclei with mass numbers between 30 and 170.
Reason (R) : Nuclear force is long range.
In the light of the above statements, choose the correct answer from the options given below:

[NCERT : P L-311 | Jan. 23, 2025 (II)]

(a) (A) is false but (R) is true
(b) (A) is true but (R) is false
(c) Both (A) and (R) are true and (R) is the correct explanation of (A)
(d) Both (A) and (R) are true but (R) is NOT the correct explanation of (A)

23. A star has 100% helium composition. It starts to convert three ^4He into one ^{12}C via triple alpha process as $^4He + {^4He} + {^4He} \rightarrow {^{12}C} + Q$. The mass of the star is 2.0×10^{32} kg and it generates energy at the rate of 5.808×10^{30} W. The rate of converting these ^4He to ^{12}C is $n \times 10^{42}\ s^{-1}$, where n is _______. **[NCERT : P L-316 | April 9, 2024 (I)]**
[Take, mass of $^4He = 4.0026$ u, mass of $^{12}C = 12$ u]

24. The energy released in the fusion of 2 kg of hydrogen deep in the sun is E_H and the energy released in the fission of 2 kg of ^{235}U is E_U. The ratio $\frac{E_H}{E_U}$ is approximately:

(Consider the fusion reaction as $4^1_1H + 2e^- \rightarrow ^4_2He + 2\nu + 6\gamma + 26.7$ MeV, energy released in the fission reaction of ^{235}U is 200 MeV per fission nucleus and $N_A = 6.023 \times 10^{23}$)

[NCERT : P L-315 | April 9, 2024 (II)]

(a) 9.13 (b) 15.04 (c) 7.62 (d) 25.6

25. A nucleus at rest disintegrates into two smaller nuclei with their masses in the ratio of 2:1. After disintegration they will move :- **[April 9, 2024 (II)]**

(a) In opposite directions with speed in the ratio of 1:2 respectively
(b) In opposite directions with speed in the ratio of 2:1 respectively
(c) In the same direction with same speed.
(d) In opposite directions with the same speed.

26. Binding energy of a certain nucleus is 18×10^8 J. How much is the difference between total mass of all the nucleons and nuclear mass of the given nucleus:

[NCERT : P L-311 | April 8, 2024 (I)]

(a) 0.2 μg (b) 20 μg (c) 2 μg (d) 10 μg

27. In a hypothetical fission reaction
$_{92}X^{236} \rightarrow {_{56}Y^{141}} + {_{36}Z^{92}} + 3R$

[NCERT : P L-315 | April 8, 2024 (II)]

The identity of emitted particles (R) is :

(a) Proton (b) Electron
(c) Neutron (d) γ-radiations

28. If M_o is the mass of isotope $^{12}_5B$, M_p and M_n are the masses of proton and neutron, then nuclear binding energy of isotope is :

[NCERT : P L-311 | April 8, 2024 (II)]

(a) $(5M_p + 7M_n - M_o)C^2$ (b) $(M_o - 5M_p)C^2$
(c) $(M_o - 12M_n)C^2$ (d) $(M_o - 5M_p - 7M_n)C^2$

29. If three helium nuclei combine to form a carbon nucleus then the energy released in this reaction is .... $\times 10^{-2}$ MeV. (Given 1 u = 931 MeV/c^2, atomic mass of helium = 4.002603 u) **[NCERT : P L-311 | April 5, 2024 (I)]**

30. Which of the following nuclear fragments corresponding to nuclear fission between neutron $\left(^1_0n\right)$ and uranium isotope $\left(^{235}_{92}U\right)$ is correct.

[NCERT : P L-315 | April 4, 2024 (I)]

(a) $^{144}_{56}Ba + ^{89}_{36}Kr + 4^1_0n$ (b) $^{140}_{56}Xe + ^{94}_{38}Sr + 3^1_0n$
(c) $^{153}_{51}Sb + ^{99}_{41}Nb + 3^1_0n$ (d) $^{144}_{56}Ba + ^{89}_{36}Kr + 3^1_0n$

31. From the statements given below:
(A) The angular momentum of an electron in n^{th} orbit is an integral multiple of $\hbar$.
(B) Nuclear forces do not obey inverse square law.
(C) Nuclear forces are spin dependent.
(D) Nuclear forces are central and charge independent.
(E) Stability of nucleus is inversely proportional to the value of packing fraction.
[NCERT : P L-313 | Feb. 1, 2024 (II)]
Choose the correct answer from the options given below:
(a) (B), (C), (D), (E) only (b) (A), (C), (D), (E) only
(c) (A), (B), (C), (E) only (d) (A), (B), (D), (E) only

32. In a nuclear fission reaction of an isotope of mass M, three similar daughter nuclei of same mass are formed. The speed of a daughter nuclei in terms of mass defect ΔM will be: **[NCERT : P L-315 | Jan. 30, 2024 (II)]**
(a) $\sqrt{\frac{2c\Delta M}{M}}$ (b) $\frac{\Delta Mc^2}{3}$ (c) $c\sqrt{\frac{2\Delta M}{M}}$ (d) $c\sqrt{\frac{3\Delta M}{M}}$

33. The explosive in a Hydrogen bomb is a mixture of ${}_1H^2$, ${}_1H^3$ and ${}_3Li^6$ in some condensed from. The chain reaction is given by **[NCERT : P L-316 | Jan. 29, 2024 (I)]**
${}_3Li^6 + {}_0n^1 \rightarrow {}_2He^4 + {}_1H^3$
${}_1H^2 + {}_1H^3 \rightarrow {}_2He^4 + {}_0n^1$
During the explosion the energy released is approximately
[Given : M (Li) = 6.01690 amu, M(${}_1H^2$) = 2.01471 amu, M (${}_2He^4$) = 4.00388 amu, and 1 amu = 931.5 MeV]
(a) 28.12 MeV (b) 16.48 MeV
(c) 12.64 MeV (d) 22.22 MeV

34. In a nuclear fission process, a high mass nuclide ($A \approx 236$) with binding energy 7.6 MeV/Nucleon dissociated into middle mass nuclides ($A \approx 118$), having binding energy of 8.6 MeV/Nucleon. The energy released in the process would be ____ MeV.
[NCERT : P L-311 | Jan. 27, 2024 (I)]

35. The atomic mass of ${}_6C^{12}$ is 12.000000 u and that of ${}_6C^{13}$ is 13.003354 u. The required energy to remove a neutron from ${}_6C^{13}$, if mass of neutron is 1.008665 u, will be :
[NCERT : P L-311 | Jan. 27, 2024 (II)]
(a) 62. 5 MeV (b) 6.25 MeV
(c) 4.95 MeV (d) 49.5 MeV

36. ${}^{238}_{92}A \rightarrow {}^{234}_{90}B + {}^{4}_{2}D + Q$
In the given nuclear reaction, the approximate amount of energy released will be:
[Given, mass of ${}^{238}_{92}A = 238.05079 \times 931.5$ MeV/c^2,
mass of ${}^{234}_{90}B = 234.04363 \times 931.5$ MeV/c^2,
mass of ${}^{4}_{2}D = 4.00260 \times 931.5$ MeV/c^2]
[NCERT : P L-311 | April 13, 2023 (I)]
(a) 3.82 MeV (b) 5.9 MeV
(c) 2.12 MeV (d) 4.25 MeV

37. Given below are two statements : one is labelled as **Assertion A** and the other is labelled as **Reason R**
Assertion A : The binding energy per nucleon is practically independent of the atomic number for nuclei of mass number in the range 30 to 170.
Reason R : Nuclear force is short ranged.
In the light of the above statements, choose the **correct** answer from the options given below:
[NCERT : P L-312 | April 13, 2023 (II)]
(a) Both **A** and **R** are true but **R** is **NOT** the correct explanation of **A**
(b) **A** is true but **R** is false
(c) **A** is false but **R** is true
(d) Both **A** and **R** are true and **R** is the correct explanation of **A**

38. For a nucleus ${}^{A}_{Z}X$ having mass number A and atomic number Z **[April 8, 2023 (I)]**
A. The surface energy per nucleon (b_s) = $a_1A^{2/3}$
B. The Coulomb contribution to the binding energy $b_c = -a_2\frac{Z(Z-1)}{A^{4/3}}$
C. The volume energy $b_v = a_3A$
D. Decrease in the binding energy is proportional to surface area.
E. While estimating the surface energy, it is assumed that each nucleon interacts with 12 nucleons, (a_1, a_2 and a_3 are constants)
Choose the **most appropriate** answer from the options given below:
(a) C, D only (b) B, C, E only
(c) A, B, C, D only (d) B, C only

39. A nucleus with mass number 242 and binding energy per nucleon as 7.6 MeV breaks into fragment each with mass number 121. If each fragment nucleus has binding energy per nucleon as 8.1 MeV, the total gain in binding energy is ____ MeV. **[NCERT : P L-311 | April 8, 2023 (I)]**

40. The mass of proton, neutron and helium nucleus are respectively 1.0073 u, 1.0087 u and 4.0015 u. The binding energy of helium nucleus is:
[NCERT : P L-311 | Feb. 1, 2023 (I)]
(a) 14.2 MeV (b) 56.8 MeV (c) 28.4 MeV (d) 7.1 MeV

41. Nucleus A having Z = 17 and equal number of protons and neutrons has 1.2 MeV binding energy per nucleon. Another nucleus B of Z = 12 has total 26 nucleons and 1.8 MeV binding energy per nucleons.
The difference of binding energy of B and A will be ______ MeV. **[NCERT : P L-311 | Feb. 1, 2023 (II)]**

42. The energy released per fission of nucleus of ${}^{240}X$ is 200 MeV. The energy released if all the atoms in 120 g of pure ${}^{240}X$ undergo fission is ________ $\times 10^{25}$ MeV. (Given $N_A = 6 \times 10^{23}$)
[NCERT : P L-311 | Jan. 24, 2023 (II)]

43. Two lighter nuclei combine to form a comparatively heavier nucleus by the relation given below:

$_1^2X + _1^2X = _2^4Y$

The binding energies per nucleon $_1^2X$ and $_2^4Y$ are 1.1 MeV and 7.6 MeV respectively. The energy released in this process is_______ MeV.

[NCERT : P L-311 | July 26, 2022 (II)]

44. Nucleus A is having mass number 220 and its binding energy per nucleon is 5.6 MeV. It splits in two fragments 'B' and 'C' of mass numbers 105 and 115. The binding energy of nucleons in 'B' and 'C' is 6.4 MeV per nucleon. The energy Q released per fission will be :

[NCERT : P L-311 | June 24, 2022 (I)]

(a) 0.8 MeV (b) 275 MeV
(c) 220 MeV (d) 176 MeV

45. From the given data, the amount of energy required to break the nucleus of aluminium $_{13}^{27}Al$ is_____ $x \times 10^{-3}$ J.

Mass of neutron = 1.00866 u
Mass of proton = 1.00726 u
Mass of Aluminium nucleus = 27.18846 u
(Assume 1 u corresponds to x J of energy)
(Round off to the nearest integer)

[NCERT : P L-311 | NA, July 25, 2021 (II)]

46. A nucleus of mass M emits γ-ray photon of frequency 'ν'. The loss of internal energy by the nucleus is :
[Take 'c' as the speed of electromagnetic wave]

[July 20, 2021 (I)]

(a) hν (b) 0
(c) $h\nu\left[1-\frac{h\nu}{2Mc^2}\right]$ (d) $h\nu\left[1+\frac{h\nu}{2Mc^2}\right]$

47. You are given that mass of $_3^7Li$ = 7.0160 u,

Mass of $_2^4He$ = 4.0026 u

and Mass of $_1^1H$ = 1.0079 u.

When 20 g of $_3^7Li$ is converted into $_2^4He$ by proton capture, the energy liberated, (in kWh), is :
[Mass of nucleon = 1 GeV/c^2]

[NCERT : P L-310 | Sep. 06, 2020 (I)]

(a) 4.5×10^5 (b) 8×10^6
(c) 6.82×10^5 (d) 1.33×10^6

48. Given the masses of various atomic particles m_p = 1.0072 u, m_n = 1.0087 u, m_e = 0.000548 u, $m_{\bar{\nu}}$ = 0, m_d = 2.0141 u, where p ≡ proton, n ≡ neutron, e ≡ electron, $\bar{\nu}$ ≡ antineutrino and d ≡ deuteron. Which of the following process is allowed by momentum and energy conservation?

[NCERT : P L-311 | Sep. 06, 2020 (II)]

(a) $n + n \rightarrow$ deuterium atom (electron bound to the nucleus)
(b) $p \rightarrow n + e^+ + \bar{\nu}$
(c) $n + p \rightarrow d + \gamma$
(d) $e^+ + e^- \rightarrow \gamma$

49. Find the Binding energy per neucleon for $_{50}^{120}Sn$. Mass of proton m_p = 1.00783 U, mass of neutron m_n = 1.00867 U and mass of tin nucleus m_{Sn} = 119.902199 U.
(take 1U = 931 MeV)

[NCERT : P L-311 | Sep. 04, 2020 (II)]

(a) 7.5 MeV (b) 9.0 MeV (c) 8.0 MeV (d) 8.5 MeV

50. In a reactor, 2 kg of $_{92}U^{235}$ fuel is fully used up in 30 days. The energy released per fission is 200 MeV. Given that the Avogadro number, N = 6.023×10^{26} per kilo mole and 1 eV = 1.6×10^{-19} J. The power output of the reactor is close to: **[Sep. 02, 2020 (I)]**

(a) 35 MW (b) 60 MW (c) 125 MW (d) 54 MW

51. Consider the nuclear fission
$Ne^{20} \rightarrow 2He^4 + C^{12}$
Given that the binding energy/nucleon of Ne^{20}, He^4 and C^{12} are, respectively, 8.03 MeV, 7.07 MeV and 7.86 MeV, identify the correct statement:

[NCERT : P L-311 | 10 Jan. 2019 II]

(a) energy of 12.4 MeV will be supplied
(b) 8.3 MeV energy will be released
(c) energy of 3.6 MeV will be released
(d) energy of 11.9 MeV has to be supplied

52. Imagine that a reactor converts all given mass into energy and that it operates at a power level of 10^9 watt. The mass of the fuel consumed per hour in the reactor will be :
(velocity of light, c is 3×10^8 m/s)

[NCERT : P L-310 | Online April 9, 2017]

(a) 0.96 gm (b) 0.8 gm (c) 4×10^{-2} gm (d) 6.6×10^{-5} gm

53. Two deuterons undergo nuclear fusion to form a Helium nucleus. Energy released in this process is : (given binding energy per nucleon for deuteron = 1.1 MeV and for helium = 7.0 MeV)

[NCERT : P L-311 | 2004, Online April 8, 2017]

(a) 30.2 MeV (b) 32.4 MeV
(c) 23.6 MeV (d) 25.8 MeV

54. Assume that a neutron breaks into a proton and an electron. The energy released during this process is : (mass of neutron = 1.6725×10^{-27} kg, mass of proton = 1.6725×10^{-27} kg, mass of electron = 9×10^{-31} kg).

[NCERT : P L-311 | 2012]

(a) 0.51 MeV (b) 7.10 MeV (c) 6.30 MeV (d) 5.4 MeV

55. After absorbing a slowly moving neutron of mass m_N (momentum ≈ 0) a nucleus of mass M breaks into two nuclei of masses m_1 and $5m_1$ ($6m_1 = M + m_N$) respectively. If the de-Broglie wavelength of the nucleus with mass m_1 is λ, the de-Broglie wavelength of the nucleus will be **[2011]**

(a) 5λ (b) λ/5 (c) λ (d) 25λ

56. The binding energy per nucleon for the parent nucleus is E_1 and that for the daughter nuclei is E_2. Then

[NCERT : P L-315 | 2010]

(a) $E_2 = 2E_1$ (b) $E_1 > E_2$ (c) $E_2 > E_1$ (d) $E_1 = 2E_2$

57. The speed of daughter nuclei is **[NCERT : P L-311 | 2010]**

(a) $c\dfrac{\Delta m}{M+\Delta m}$ (b) $c\sqrt{\dfrac{2\Delta m}{M}}$

(c) $c\sqrt{\dfrac{\Delta m}{M}}$ (d) $c\sqrt{\dfrac{\Delta m}{M+\Delta m}}$

58. **Statement-1:** Energy is released when heavy nuclei undergo fission or light nuclei undergo fusion.

Statement-2 : For heavy nuclei, binding energy per nucleon increases with increasing *Z* while for light nuclei it decreases with increasing *Z*. **[NCERT : P L-315 | 2008]**

(a) Statement-1 is false, Statement-2 is true

(b) Statement-1 is true, Statement-2 is true; Statement-2 is a correct explanation for Statement-1

(c) Statement-1 is true, Statement-2 is true; Statement-2 is not a correct explanation for Statement-1

(d) Statement-1 is true, Statement-2 is false

59. If M_O is the mass of an oxygen isotope $_8O^{17}$, M_P and M_N are the masses of a proton and a neutron respectively, the nuclear binding energy of the isotope is **[NCERT : P L-311 | 2007]**

(a) $(M_O - 17M_N)c^2$ (b) $(M_O - 8M_P)c^2$

(c) $(8M_P + 9M_N - M_O)c^2$ (d) M_Oc^2

60. When $_3Li^7$ nuclei are bombarded by protons, and the resultant nuclei are $_4Be^8$, the emitted particles will be **[NCERT : P L-314 | 2006]**

(a) alpha particles (b) beta particles

(c) gamma photons (d) neutrons

61. If the binding energy per nucleon in ^7_3Li and ^4_2He nuclei are 5.60 MeV and 7.06 MeV respectively, then in the reaction

$$p + {}^7_3Li \longrightarrow 2\,{}^4_2He$$

energy of proton must be **[NCERT : P L-311 | 2006]**

(a) 28.24 MeV (b) 17.28 MeV

(c) 1.46 MeV (d) 39.2 MeV

62. A nucleus disintegrated into two nuclear parts which have their velocities in the ratio of 2 : 1. The ratio of their nuclear sizes will be **[NCERT : P L-309 | 2004]**

(a) $3^{1/2} : 1$ (b) $1 : 2^{1/3}$

(c) $2^{1/3} : 1$ (d) $1 : 3^{1/2}$

63. When a U^{238} nucleus originally at rest, decays by emitting an alpha particle having a speed '*u*', the recoil speed of the residual nucleus is **[NCERT : P L-315 | 2003]**

(a) $\dfrac{4u}{238}$ (b) $-\dfrac{4u}{234}$ (c) $\dfrac{4u}{234}$ (d) $-\dfrac{4u}{238}$

64. In the nuclear fusion reaction

$$^2_1H + {}^3_1H \rightarrow {}^4_2He + n$$

given that the repulsive potential energy between the two nuclei is 7.7×10^{-14} J, the temperature at which the gases must be heated to initiate the reaction is nearly

[Boltzmann's Constant $k = 1.38 \times 10^{-23}$ J/K] **[NCERT : P L-311 | 2003]**

(a) 10^7 K (b) 10^5 K

(c) 10^3 K (d) 10^9 K

ANSWER KEYS

1	(d)	8	(c)	15	(a)	22	(b)	29	(737)	36	(d)	43	(26)	50	(b)	57	(b)	64	(d)
2	(b)	9	(2)	16	(c)	23	(5)	30	(d)	37	(d)	44	(d)	51	(None)	58	(d)		
3	(27)	10	(11)	17	(b)	24	(c)	31	(c)	38	(a)	45	(27.16)	52	(c)	59	(c)		
4	(a)	11	(b)	18	(b)	25	(a)	32	(c)	39	(121)	46	(d)	53	(c)	60	(c)		
5	(4)	12	(c)	19	(b)	26	(b)	33	(d)	40	(c)	47	(d)	54	(a)	61	(b)		
6	(2)	13	(b)	20	(c)	27	(c)	34	(236)	41	(6)	48	(c)	55	(c)	62	(b)		
7	(b)	14	(d)	21	(a)	28	(a)	35	(c)	42	(6)	49	(d)	56	(c)	63	(c)		

Semiconductor Electronics : Materials, Devices and Simple Circuits

14

TREND ANALYSIS
JEE MAIN (2025-2021)

	2025				2024				2023				2022				2021			
	Session II		Session I		Session II		Session I		Session II		Session I		Session II		Session I		Session II		Session I	
	MCQs	NVQs	MCQs	NVQs	MCQs	NVQs	MCQs	NVQs	MCQs	NVQs	MCQs	NVQs	MCQs	NVQs	MCQs	NVQs	MCQs	NVQs	MCQs	NVQs
Topic 1	4	–	5	–	5	1	6	–	4	–	7	–	1	5	–	–	4	5	2	4
Topic 2	5	–	5	–	6	–	4	–	5	–	4	–	1	–	3	–	4	–	3	1
Topic 3																				
Topic 4																				
Total	9	–	10	–	11	1	10	–	9	–	11	–	2	5	3	–	8	5	5	5

How to Code Qns for Better & Faster Revision →

Type 1: Able to solve (No code Req)
Type 2: Solved with struggle or by seeing solution **(Use code NR means Needs Revision or orange stickers)**
Type 3: Tricky/ Could not solve on my own. Need help from Teacher/ Friend. **(Use code TQ means Tricky Question or red stickers)**

You can either hand write the code in boxes or else paste coloured stickers given at end of book.

Topic 1 - Semiconductors and P-N Junction Diode

1. The output voltage in the following circuit is (Consider ideal diode case)

[NCERT : P L-335, 336 | April 8, 2025 (II)]

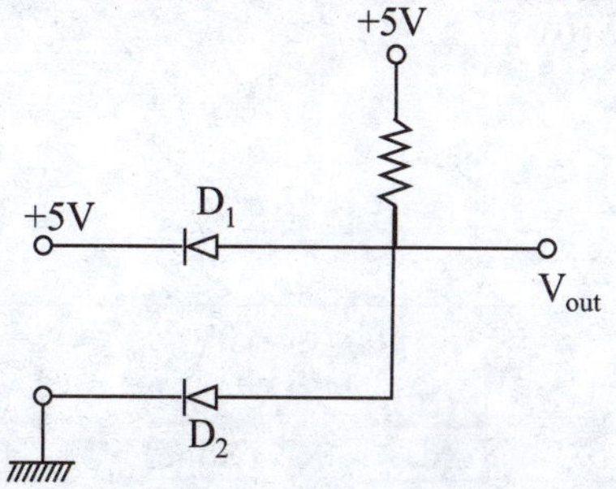

(a) 10V (b) 0V (c) + 5 V (d) –5 V

2. In the following circuit, the reading of the ammeter will be (Take Zener breakdown voltage = 4 V)

[April 7, 2025 (I)]

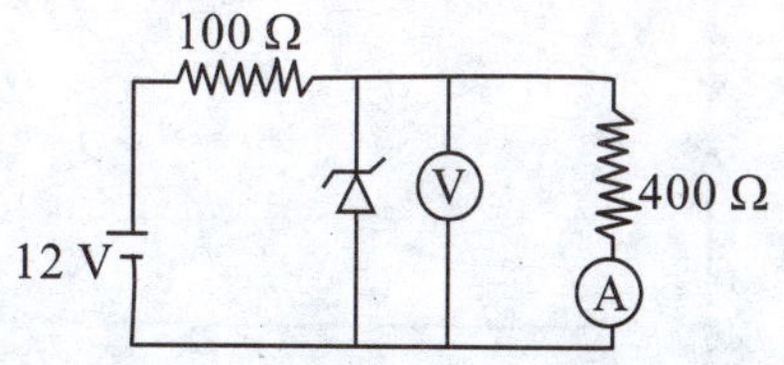

(a) 24 mA (b) 80 mA (c) 10 mA (d) 60 mA

3. Consider a *n*-type semiconductor in which n_e and n_h are number of electrons and holes, respectively.

(A) Holes are minority carriers

(B) The dopant is a pentavalent atom

(C) $n_e n_h \neq n_i^2$

(where n_i is number of electrons or holes in semiconductor when it is intrinsic form)

(D) $n_e n_h \geq n_i^2$

(E) The holes are not generated due to the donors

Choose the **correct** answer from the options given below:

[NCERT : P L-332 | April 4, 2025 (II)]

(a) (A), (C), (D) only (b) (A), (C), (E) only
(c) (A), (B), (E) only (d) (A), (B), (C) only

4. A zener diode with 5V zener voltage is used to regulate an unregulated dc voltage input of 25 V. For a 400 Ω resistor connected in series, the zener current is found to be 4 times load current. The load current (I_L) and load resistance (R_L) are : **[April 2, 2025 (I)]**

(a) $I_L = 20\text{ mA}; R_L = 250\,\Omega$
(b) $I_L = 10\text{ A}; R_L = 0.5\,\Omega$
(c) $I_L = 0.02\text{ mA}; R_L = 25\,\Omega$
(d) $I_L = 10\text{ mA}; R_L = 500\,\Omega$

5.

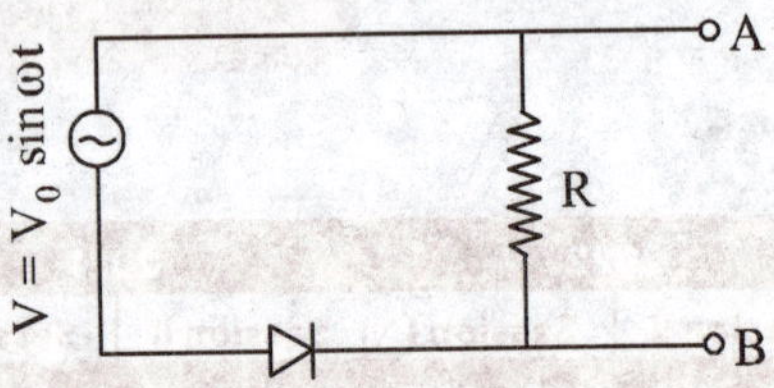

In the circuit shown here, assuming threshold voltage of diode is negligibly small, then voltage V_{AB} is correctly represented by :
[NCERT : P L-335, 336 | Jan. 28, 2025 (II)]

(a) V_{AB} would be zero at all times

(b)
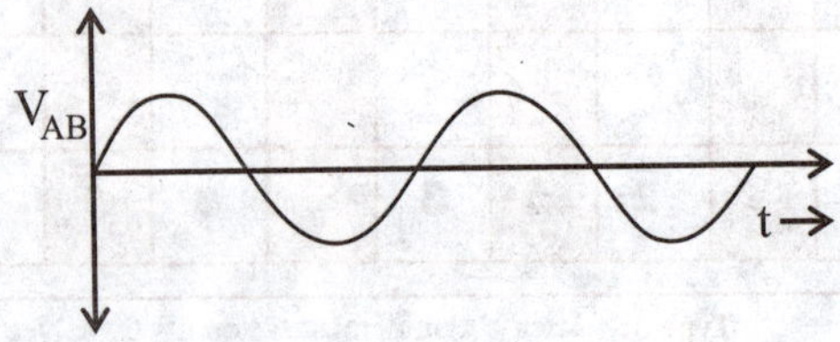

(c)
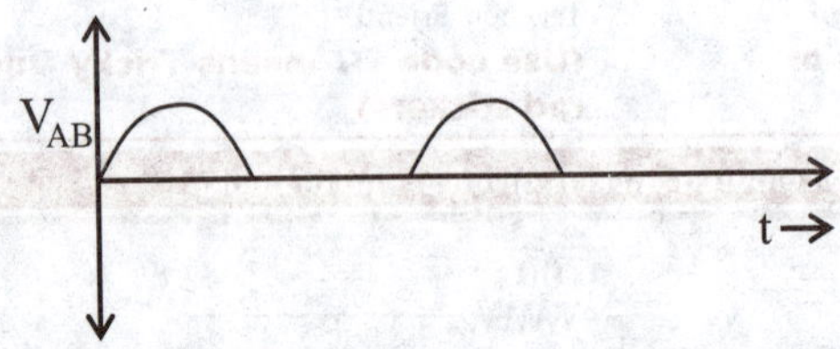

(d)
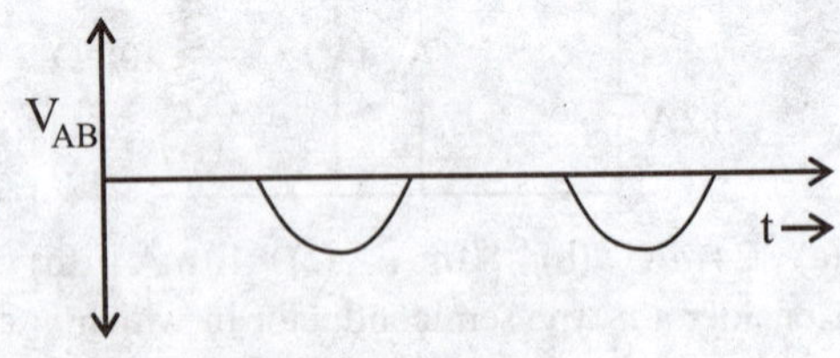

6. Consider the following statements:

A. The junction area of solar cell is made very narrow compared to a photodiode.
B. Solar cells are not connected with any external bias.
C. LED is made of lightly doped p-n junction.
D. Increase of forward current results in continuous increase of LED light intensity.
E. LEDs have to be connected in forward bias for emission of light.

[Jan. 24, 2025 (I)]

(a) B, D, E Only (b) A, C Only
(c) A, C, E Only (d) B, E Only

7. What is the current through the battery in the circuit shown below?
[NCERT : P L-335, 336 | Jan. 23, 2025 (II)]

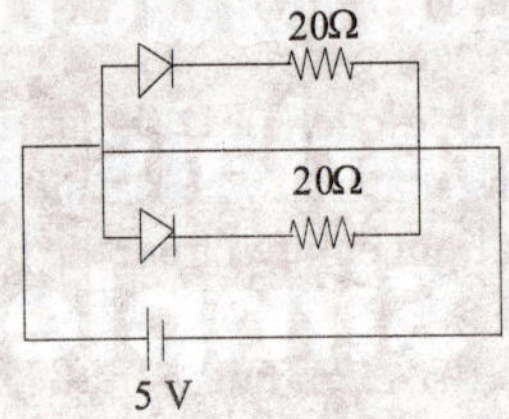

(a) 1.0A (b) 1.5A
(c) 0.5A (d) 0.25A

8. Refer to the circuit diagram given in the figure, which of the following observation are correct?
[NCERT : P L-335, 336 | Jan. 23, 2025 (I)]

A. Total resistance of circuit is 6 Ω.
B. Current in Ammeter is 1A.
C. Potential across AB is 4 volts.
D. Potential across CD is 4 volts.
E. Total resistance of the circuit is 8Ω.

Choose the correct answer from the options given below:

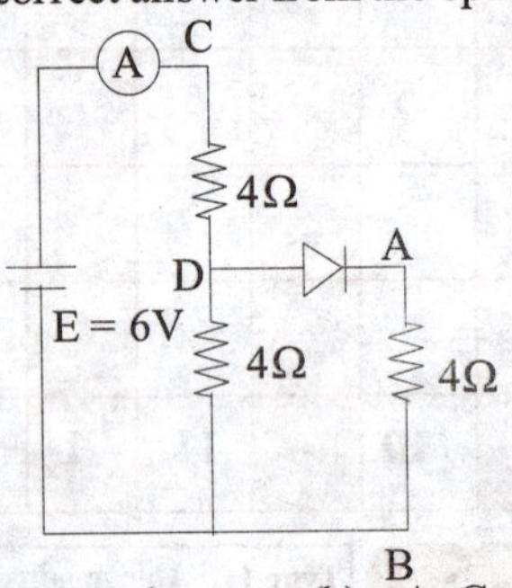

(a) A, B and D only (b) A, C and D only
(c) B, C and E only (d) A, B and C only

9. Which of the following circuits represents a forward biased diode? **[NCERT : P L-335, 336 | Jan. 22, 2025 (I)]**

(A)
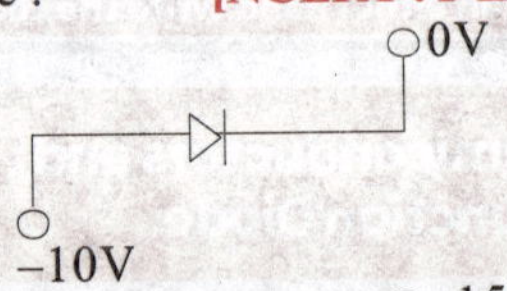

(B)
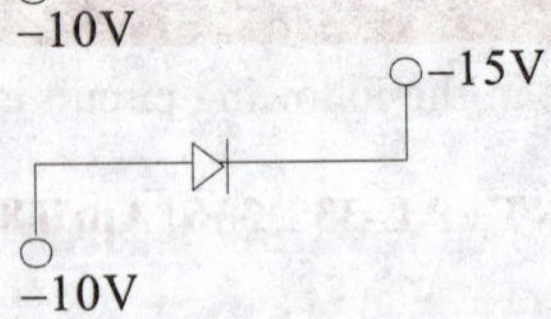

(C)
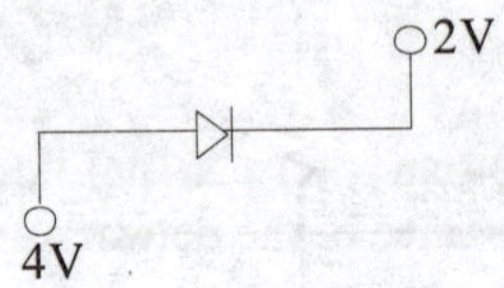

(D)
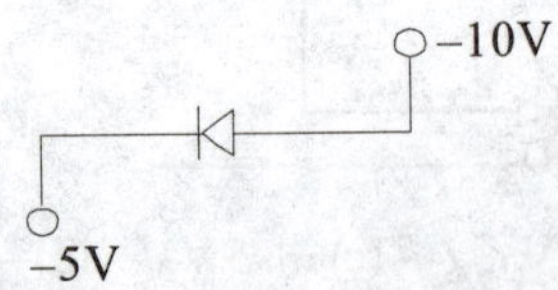

(E)
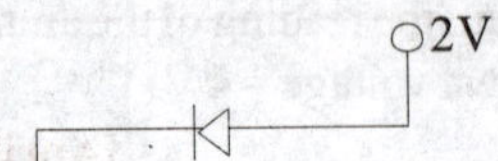

Choose the correct answer from the options given below:

(a) (B), (D) and (E) only
(b) (A) and (D) only
(c) (B), (C) and (E) only
(d) (C) and (E) only

10. A light emitting diode (LED) is fabricated using GaAs semiconducting material whose band gap is 1.42 eV. The wavelength of light emitted from the LED is: **[NCERT : P L-326 | April 9, 2024 (I)]**

(a) 650 nm (b) 1243 nm (c) 875 nm (d) 1400 nm

11. The I-V characteristics of an electronic device shown in the figure. The device is : **[April 9, 2024 (II)]**

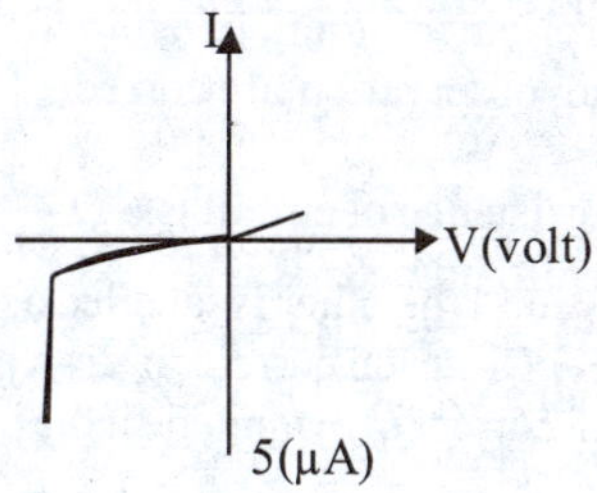

(a) a solar cell
(b) a transistor which can be used as an amplifier
(c) a zener diode which can be used as voltage regulator
(d) a diode which can be used as a rectifier

12. A potential divider circuit is connected with a dc source of 20 V, a light emitting diode of glow in voltage 1.8 V and a zener diode of breakdown voltage of 3.2 V. The length (PR) of the resistive wire is 20 cm. The minimum length of PQ to justg low the LED is .... cm. **[April 8, 2024 (II)]**

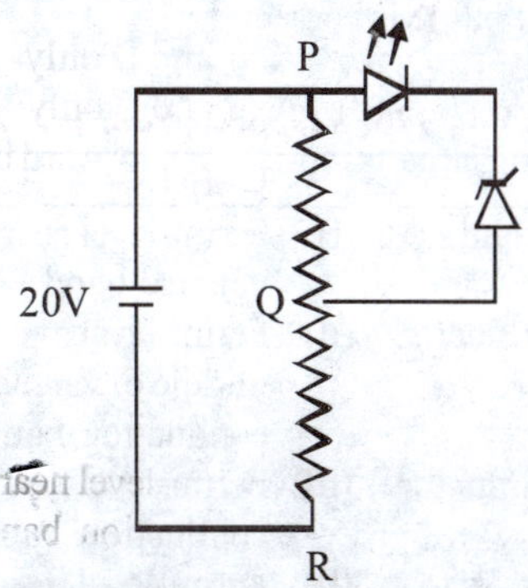

13. The acceptor level of a p-type semiconductor is 6eV. The maximum wavelength of light which can create a hole would be : Given hc = 1242 eV nm. **[NCERT : P L-331 | April 6, 2024 (II)]**

(a) 407 nm (b) 414 nm (c) 207 nm (d) 103.5 nm

14. The value of net resistance of the network as shown in the given figure is : **[NCERT : P L-335, 336 | April 4, 2024 (I)]**

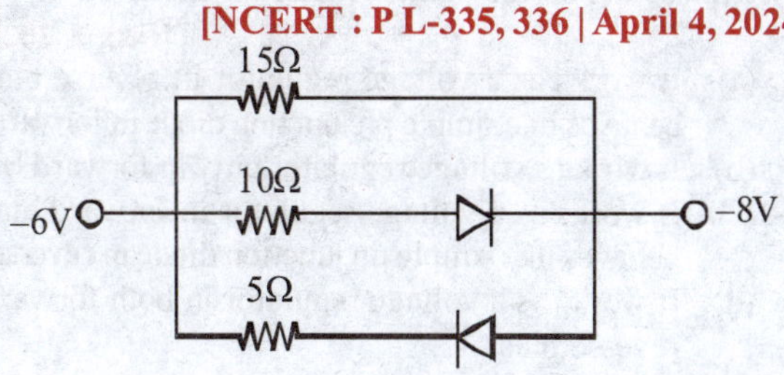

(a) $\left(\frac{5}{2}\right)\Omega$ (b) $\left(\frac{15}{4}\right)\Omega$ (c) 6Ω (d) $\left(\frac{30}{11}\right)\Omega$

15. Which of the diode circuit shows correct biasing used for the measurement of dynamic resistance of p-n junction diode : **[NCERT : P L-335 | April 4, 2024 (II)]**

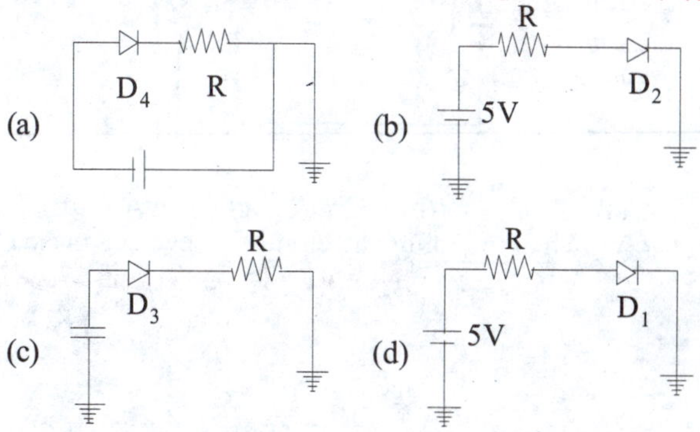

16. In the given circuit if the power rating of Zener diode is 10mW, the value of series resistance R_s to regulate the input unregulated supply is: **[Feb. 1, 2024 (I)]**

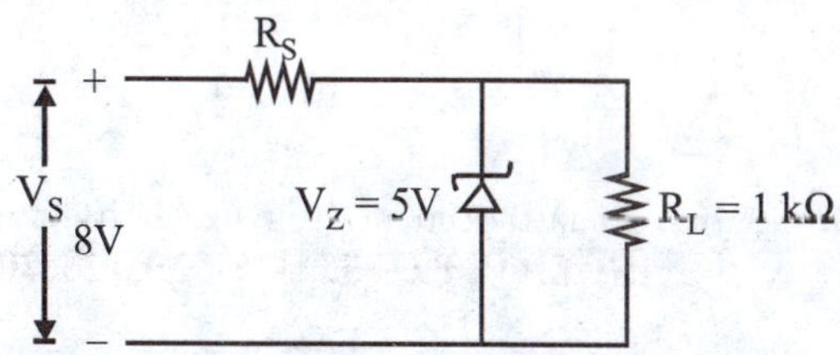

(a) 5 kΩ (b) 10 Ω (c) 1 kΩ (d) 10 kΩ

17. Conductivity of a photodiode starts changing only if the wavelength of incident light is less than 660 nm. The band gap of photodiode is found to be $\left(\frac{X}{8}\right)$ eV. The value of X is : (Given , $h = 6.6 \times 10^{-34}$ Js, $e = 1.6 \times 10^{-19}$C) **[NCERT : P L-326 | Feb. 1, 2024 (II)]**

(a) 11 (b) 13 (c) 15 (d) 21

18. A Zener diode of breakdown voltage 10V is used as a voltage regulator as shown in the figure. The current through the Zener diode is: **[Jan. 30, 2024 (II)]**

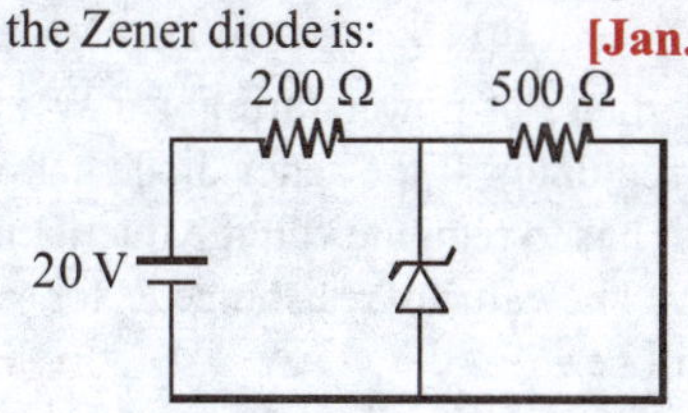

(a) 50mA (b) 0 (c) 30 mA (d) 20 mA

19.

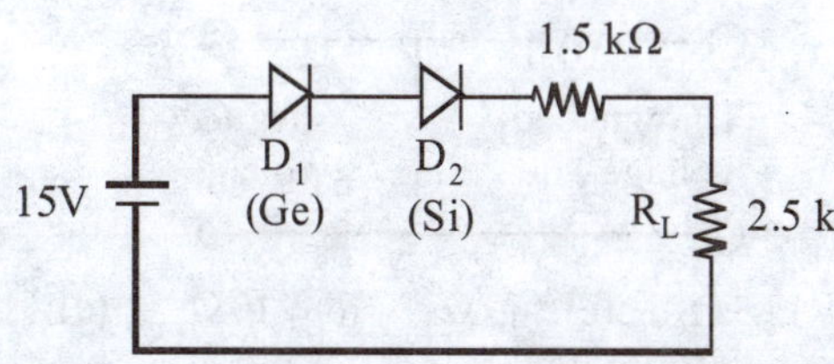

In the given circuit, the voltage across load resistance (R_L) is: **[NCERT : P L-335, 336 | Jan. 30, 2024 (II)]**

(a) 8.75 V (b) 9.00 V (c) 8.50 V (d) 14.00 V

20. In the given circuit, the breakdown voltage of the Zener diode is 3.0 V. What is the value of I_z? **[Jan. 29, 2024 (I)]**

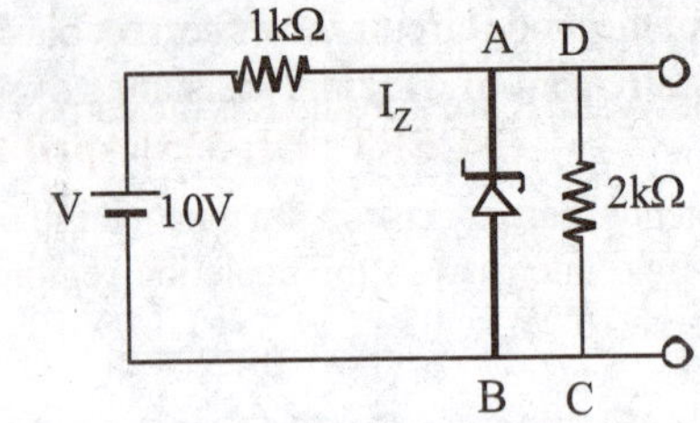

(a) 7 mA (b) 5.5 mA (c) 10 mA (d) 3.3 mA

21. Which of the following circuits is reverse - biased?
[NCERT : P L-336 | Jan. 27, 2024 (I)]

(a) 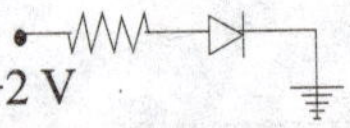

(b) 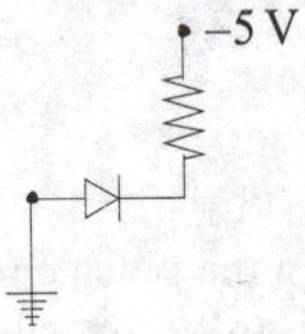

(c) 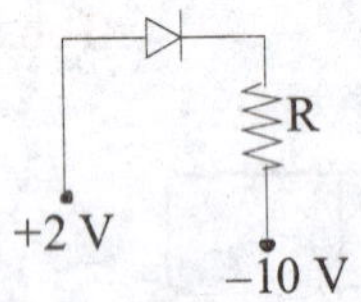

(d) 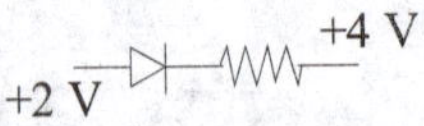

22. In the given circuit, the current (I) through the battery will be:
[NCERT : P L-335, 336 | April 15, 2023 (I)]

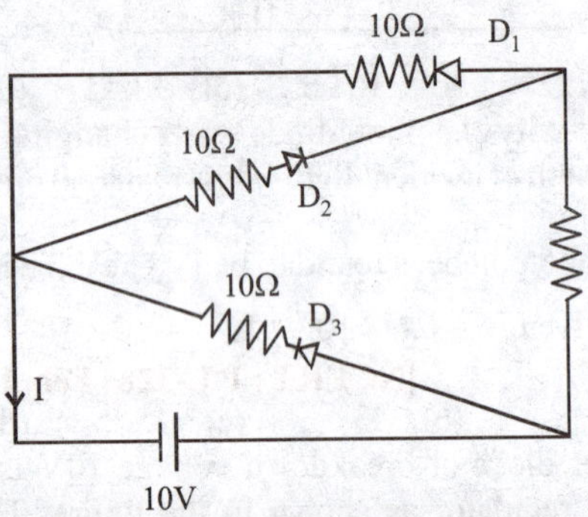

(a) 1.5 A (b) 1 A (c) 2.5 A (d) 2 A

23. A zener diode of power rating 1.6 W is to be used as voltage regulator. If the zener diode has a breakdown of 8V and it has to regulate voltage fluctuating between 3V and 10 V. The value of resistance R_s for safe operation of diode will be: **[April 10, 2023 (I)]**

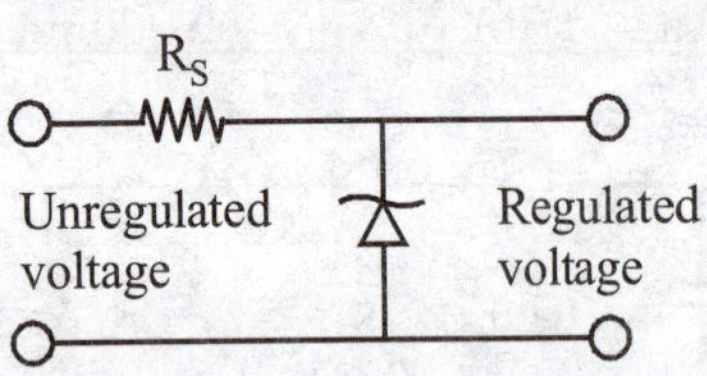

(a) 13.3 Ω (b) 12 Ω (c) 10 Ω (d) 13 Ω

24. If each diode has a forward bias resistance of 25 Ω in the below circuit,
[NCERT : P L-335, 336 | April 10, 2023 (II)]

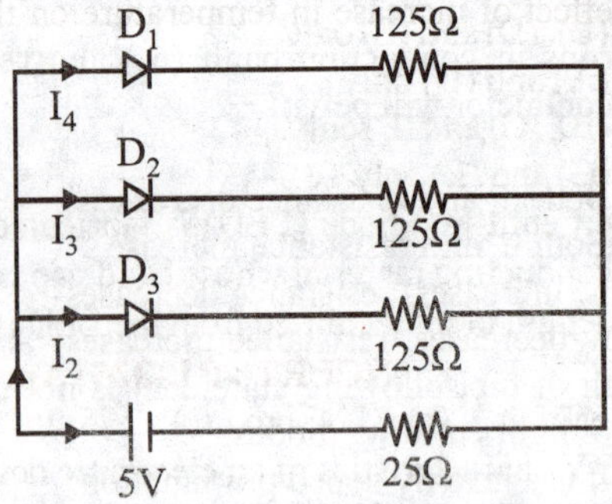

Which of the following options is correct:

(a) $\frac{I_3}{I_4}=1$ (b) $\frac{I_2}{I_3}=1$ (c) $\frac{I_1}{I_2}=1$ (d) $\frac{I_1}{I_2}=2$

25. Given below are two statements: one is labelled as **Assertion A** and the other is labelled as **Reason R**

Assertion A: Diffusion current in a p-n junction is greater than the drift current in magnitude if the junction is forward biased.

Reason R: Diffusion current in a p-n junction is from the n-side to the p-side if the junction is forward biased.

In the light of the above statements, choose the most appropriate answer from the options given below
[NCERT : P L-333, 335 | April 6, 2023 (II)]

(a) Both **A** and **R** are correct and **R** is the correct explanation of **A**
(b) Both **A** and **R** are correct but **R** is **NOT** the correct explanation of **A**
(c) **A** is correct but **R** is not correct
(d) **A** is not correct but **R** is correct

26. Match List I with List II:
[NCERT : P L-326, 329, 332 | Feb. 1, 2023 (I)]

List I	List II
(A) Intrinsic semiconductor	(I) Fermi-level near the valence band
(B) n-type semiconductor	(II) Fermi-level in the middle of valence and conduction band.
(C) p-type semiconductor	(III) Fermi-level near the conduction band
(D) Metals	(IV) Fermi-level inside the conduction band

Choose the **correct** answer from the options given below:

(a) (A) – (III), (B) – (I), (C) – (II), (D) – (IV)
(b) (A) – (I), (B) – (II), (C) – (III), (D) – (IV)
(c) (A) – (II), (B) – (III), (C) – (I), (D) – (IV)
(d) (A) – (II), (B) – (I), (C) – (III), (D) – (IV)

27. Choose the correct statement about Zener diode:
[Feb. 1, 2023 (II)]

(a) It works as a voltage regulator in reverse bias and behaves like simple pn junction diode in forward bias.
(b) It works a voltage regulator only in forward bias.
(c) It works as a voltage regulator in forward bias and behaves like simple pn junction diode in reverse bias.
(d) It works as a voltage regulator in both forward and reverse bias.

28. The effect of increase in temperature on the number of electrons in conduction band (n_e) and resistance of a semiconductor will be as:

[NCERT : P L-329 | Jan. 31, 2023 (I)]

(a) Both n_e and resistance decrease
(b) Both n_e and resistance increase
(c) n_e increases, resistance decreases
(d) n_e decreases, resistance increases

29. Which of the following statement is not correct in the case of light emitting diodes?

A. It is a heavily doped p-n junction.
B. It emits light only when it is forward biased.
C. It emits light only when it is reverse biased.
D. The energy of the light emitted is equal to or slightly less than the energy gap of the semiconductor used.

Choose the correct answer from the options given below:

[Jan. 29, 2023 (I)]

(a) C and D (b) A
(c) C (d) B

30. Given below are two statements : one is labeled as Assertion A and the other is labeled as Reason R

Assertion A: Photodiodes are used in forward bias usually for measuring the light intensity.

Reason R: For a p-n junction diode, at applied voltage V the current in the forward bias is more than the current in the reverse bias for $|V_z| \geq V \geq |V_0|$ where V_o is the threshold voltage and V_z is the breakdown voltage.

In the light of the above statements, choose the correct answer from the options given below

[NCERT : P L-336 | Jan. 25, 2023 (I)]

(a) Both A and R are true and R is correct explanation A
(b) Both A and R are true but R is NOT the correct explanation A
(c) A is false but R is true
(d) A is true but R is false

31. **Statement I :** When a Si sample is doped with Boron, it becomes P type and when doped by Arsenic it becomes N-type semi conductor such that P-type has excess holes and N-type has excess electrons.

Statement II : When such P-type and N-type semi-conductors, are fused to make a junction, a current will automatically flow which can be detected with an externally connected ammeter. In the light of above statements, choose the most appropriate answer from the options given below.

Options: **[NCERT : P L-333 | Jan. 25, 2023 (II)]**

(a) Both Statement I and statement II are incorrect
(b) Statement I is incorrect but statement II is correct
(c) Both Statement I and statement II are correct
(d) Statement I is correct but statement II is incorrect

32. Given below are two statements : one is labelled as **Assertion A** and the other is labelled as **Reason R**

Assertion A : Photodiodes are preferably operated in reverse bias condition for light intensity measurement.

Reason R : The current in the forward bias is more than the current in the reverse bias for a p – n junction diode.

In the light of the above statement, choose the correct answer from the options given below :

[NCERT : P L-335, 336 | Jan. 24, 2023 (I)]

(a) A is false but R is true
(b) Both A and R are true but R is NOT the correct explanation of A
(c) A is true but R is false
(d) Both A and R are true and R is the correct explanation of A

33. If the potential barrier across a p-n junction is 0.6 V. Then the electric field intensity, in the depletion region having the width of 6×10^{-6}m, will be ________ $\times 10^5$ N/C.

[NCERT : P L-333 | July 29, 2022 (I)]

34. A 8 V Zener diode along with a series resistance R is connected across a 20 V supply (as shown in the figure). If the maximum Zener current is 25 mA, then the minimum value of R will be _____Ω. **[July 29, 2022 (II)]**

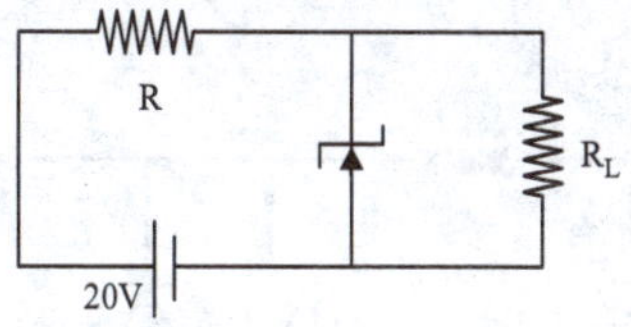

35. Identify the solar cell characteristics from the following options : **[July 28, 2022 (I)]**

(a)
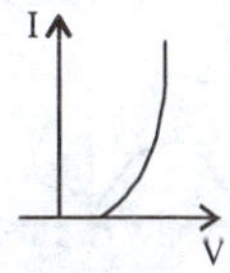

(b)
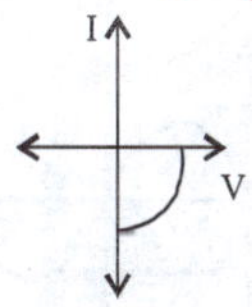

(c)
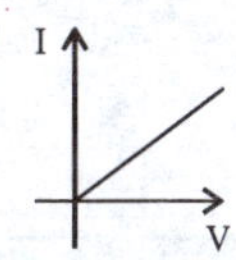

(d)
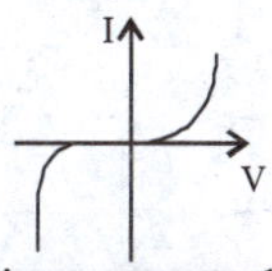

36. In the circuit shown below, maximum zener diode current will be ________ mA. **[July 26, 2022 (I)]**

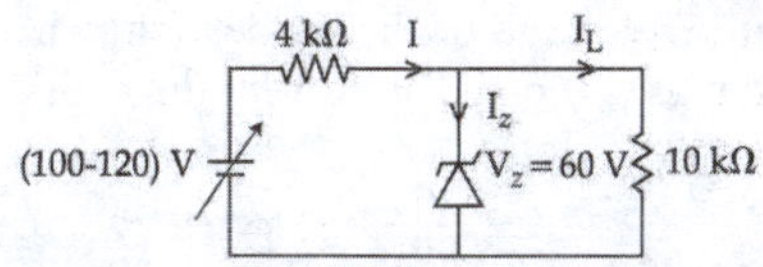

37. The energy band gap of semiconducting material to produce violet (wavelength = 4000 Å). LED is ________ eV. (Round off to the nearest integer)

[NCERT : P L-326 | July 25, 2022 (I)]

38. Two ideal diodes are connected in the network as shown in figure. The equivalent resistance between A and B is ____ Ω.

[NCERT : P L-335, 336 | July 25, 2022 (II)]

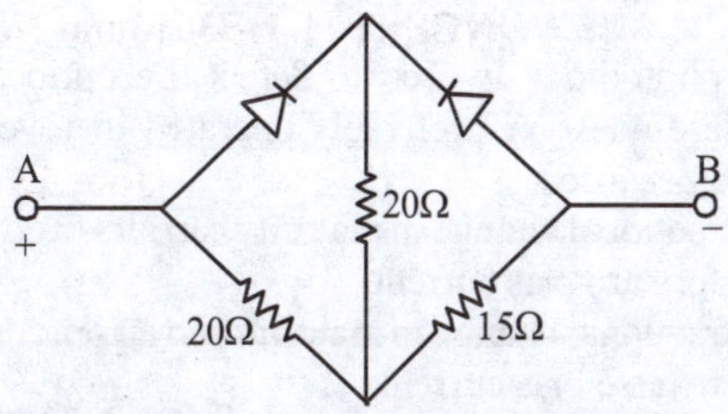

39. A Zener of breakdown voltage $V_Z = 8V$ and maximum zener current, $I_{ZM} = 10$ mA is subjected to an input voltage $V_i = 10$V with series resistance $R = 100\Omega$. In the given circuit R_L represents the variable load resistance. The ratio of maximum and minimum value of R_L is

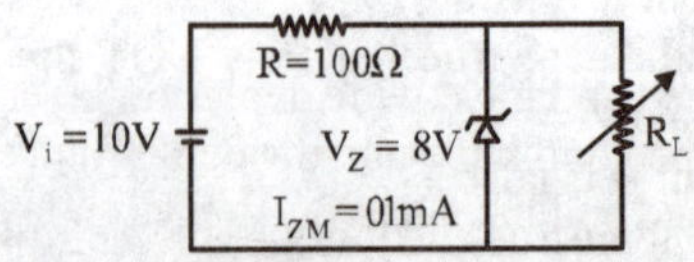

[June 28, 2022 (II)]

40. In the given circuit the input voltage V_{in} is shown in figure. The cut–in voltage of $p–n$ junction diode (D_1 or D_2) is 0.6 V. Which of the following output voltage (V_0) waveform across the diode is correct? **[NCERT : P L-335, 336 | June 28, 2022 (II)]**

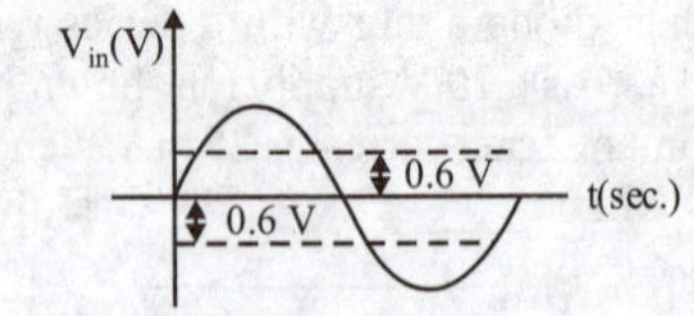

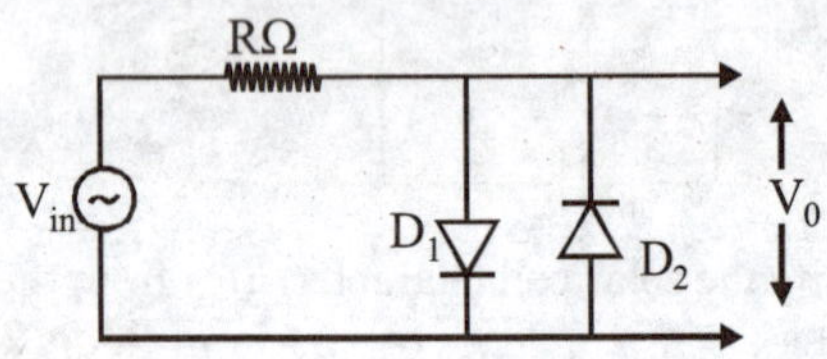

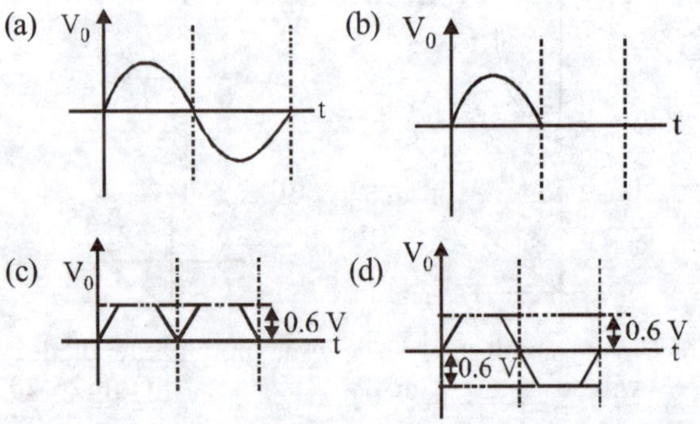

41. The cut-off voltage of the diodes (shown in figure) in forward bias is 0.6 V. The current through the resistor of 40 Ω is ______ mA.

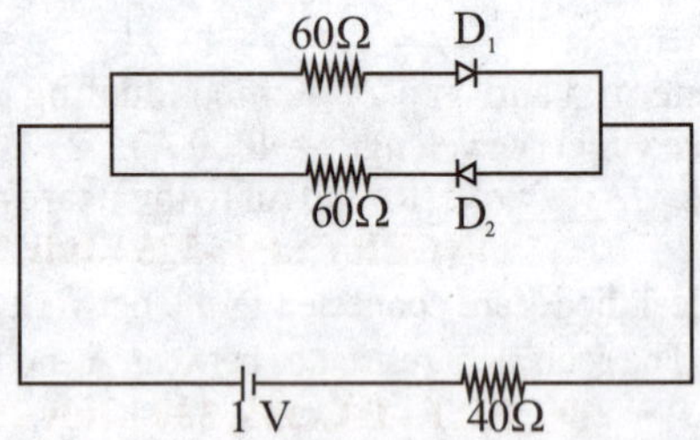

[NCERT : P L-336 | June 27, 2022 (II)]

42. The photodiode is used to detect the optical signals. These diodes are preferably operated in reverse biased mode because. **[June 25, 2022 (I)]**
(a) fractional change in majority carriers produce higher forward bias current
(b) fractional change in majority carriers produce higher reverse bias current
(c) fractional change in minority carriers produce higher forward bias current
(d) fractional change in minority carriers produce higher reverse bias current

43. In the given circuit- the value of current I_L will be ______ mA. (When $R_L = 1k\Omega$) **[June 24, 2022 (II)]**

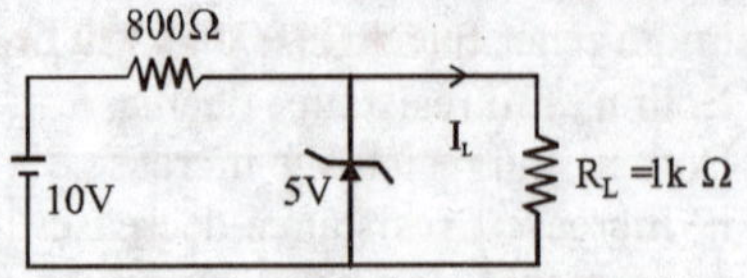

44. In the given figure, each diode has a forward bias resistance of 30 Ω and infinite resistance in reverse bias. The current I_1 will be : **[NCERT : P L-335, 336 | Sep. 1, 2021 (II)]**

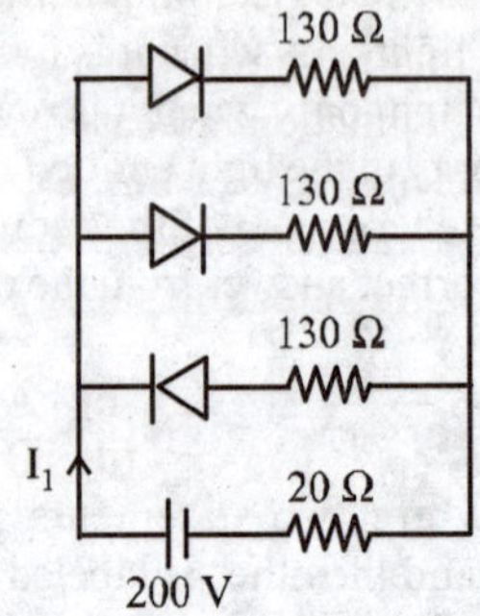

(a) 3.75 A
(b) 2.35 A
(c) 2 A
(d) 2.73 A

45. **Statement-I :** By doping silicon semiconductor with pentavalent material, the electrons density increases.
Statement-II : The n-type semiconductor has net negative charge. **[NCERT : P L-330 | Aug. 26, 2021 (I)]**
In the light of the above statements, choose the most appropriate answer from the options given below :
(a) Statement-I is true but Statement-II is false.
(b) Statement-I is false but Statement-II is true.
(c) Both Statement-I and Statement-II are true.
(d) Both Statement-I and Statement-II are false.

46. A zener diode of power rating 2W is to be used as a voltage regulator. If the zener diode has a breakdown of 10 V and it has to regulate voltage fluctuated between 6 V and 14 V, the value of R_s for safe operation should be ______ Ω. **[NA, Aug. 27, 2021 (II)]**

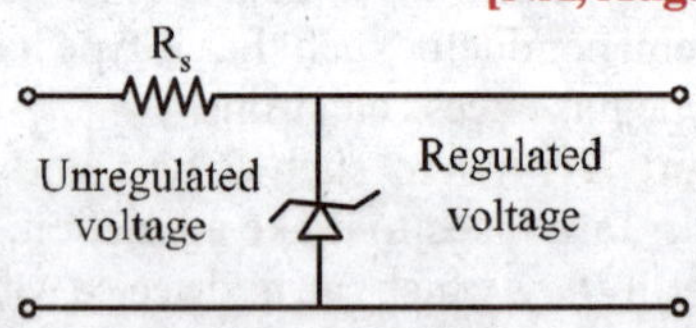

47. For the given circuit, the power across zener diode is ______ mW. **[NA, Aug. 26, 2021 (II)]**

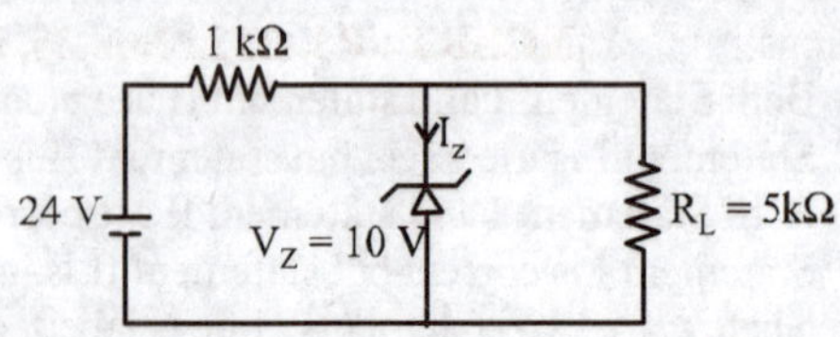

48. In a semiconductor, the number density of intrinsic charge carriers at 27°C is $1.5 \times 10^{16}/m^3$. If the semiconductor is doped with impurity atom, the hole density increases to 4.5×10^{22} / m^3. The electron density in the doped semiconductor is ______ $\times 10^9/m^3$.
[NCERT : P L-332 | NA, July 25, 2021 (II)]

49. In a given circuit diagram, a 5 V zener diode along with a series resistance is connected across a 50 V power supply. The minimum value of the resistance required, if the maximum zener current is 90 mA will be ______ Ω.

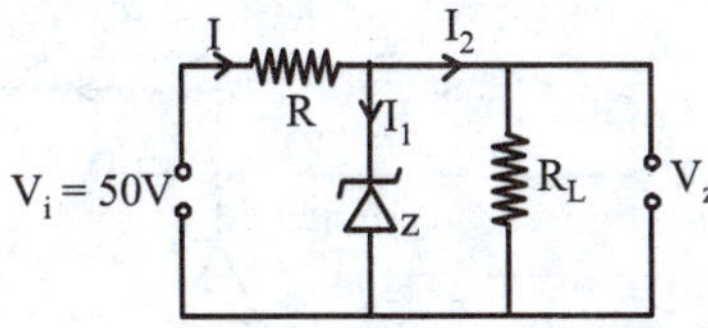

[NA, July 22, 2021 (II)]

50. Consider a situation in which reverse biased current of a particular P-N junction increases when it is exposed to a light of wavelength ≤ 621 nm. During this process, enhancement in carrier concentration takes place due to generation of hole-electron pairs. The value of band gap is nearly. **[July 22, 2021 (II)]**

(a) 2 eV (b) 4 eV (c) 1 eV (d) 0.5 eV

51. For the circuit shown below, calculate the value of I_z: **[July 20, 2021 (I)]**

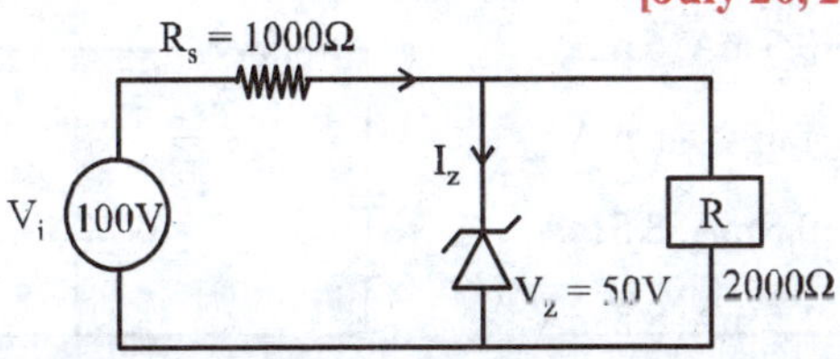

(a) 25 mA (b) 0.15 A (c) 0.1 A (d) 0.05 A

52. A zener diode having zener voltage 8 V and power dissipation rating of 0.5 W is connected across a potential divider arranged with maximum potential drop across zener diode is as shown in the diagram. The value of protective resistance R_p is ....Ω. **[NA, July 20, 2021 (II)]**

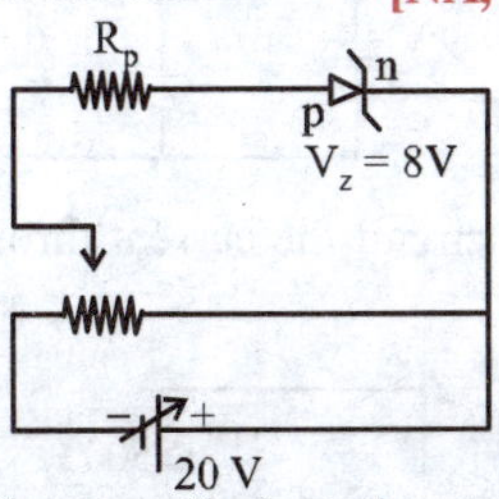

53. The value of power dissipated across the zener diode ($V_z = 15V$) connected in the circuit as shown in the figure is x × 10^{-1} watt. **[March 16, 2021 (I)]**

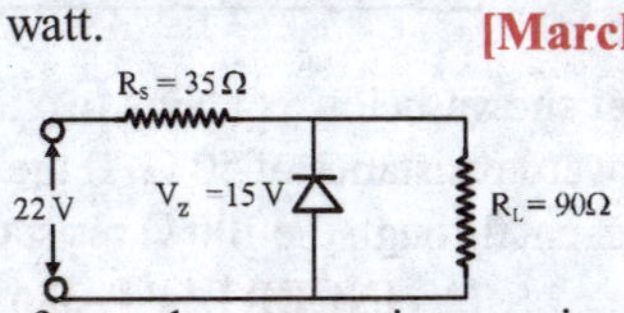

The value of x, to the nearest integer, is _____.

54. In connection with the circuit drawn below, the value of current flowing through 2 kΩ resistor is _______ × 10^{-4} A.

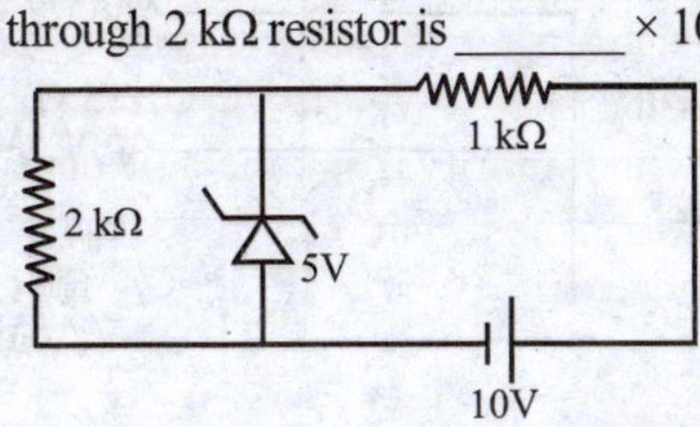

[Feb. 24, 2021 (I)]

55. Zener breakdown occurs in a *p-n* junction having *p* and *n* both : **[Feb. 24, 2021 (II)]**

(a) lightly doped and have wide depletion layer
(b) heavily doped and have narrow depletion layer
(c) lightly doped and have narrow depletion layer
(d) heavily doped and have wide depletion layer

56. LED is constructed from Ga-As-P semiconducting material. The energy gap of this LED is 1.9 eV. Calculate the wavelength of light emitted and its colour.
[$h = 6.63 \times 10^{-34}$ Js and $c = 3 \times 10^8$ ms^{-1}]

(a) 654 nm and red colour

[NCERT : P L-332 | Feb. 26, 2021 (I)]

(b) 1046 nm and blue colour
(c) 1046 nm and red colour
(d) 654 nm and orange colour

57. The circuit contains two diodes each with a forward resistance of 50 Ω and with infinite reverse resistance. If the battery voltage is 6 V, the current through the 120 Ω resistance is ______ mA. **[Feb. 26, 2021 (I)]**

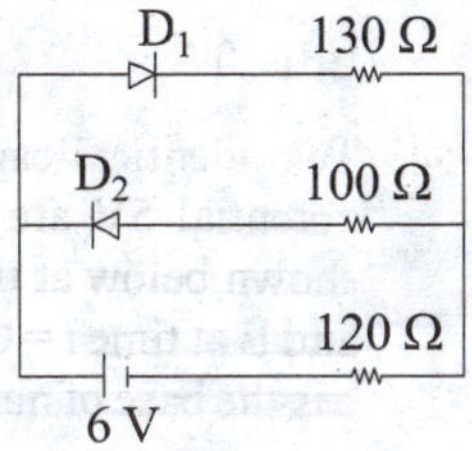

58. The zener diode has a $V_z = 30$ V. The current passing through the diode for the following circuit is _______ mA.

[NCERT : P L-336 | Feb. 26, 2021 (II)]

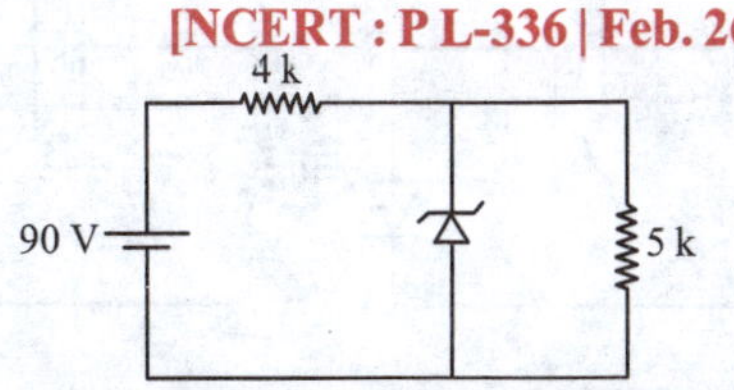

59. With increasing biasing voltage of a photodiode, the photocurrent magnitude :

[NCERT : P L-336 | Sep. 05, 2020 (I)]

(a) remains constant
(b) increases initially and after attaining certain value, it decreases
(c) Increases linearly
(d) increases initially and saturates finally

60. When a diode is forward biased, it has a voltage drop of 0.5 V. The safe limit of current through the diode is 10 mA. If a battery of emf 1.5 V is used in the circuit, the value of minimum resistance to be connected in series with the diode so that the current does not exceed the safe limit is :

[NCERT : P L-335 | Sep. 03, 2020 (I)]

(a) 300 Ω (b) 50 Ω (c) 100 Ω (d) 200 Ω

61. If a semiconductor photodiode can detect a photon with a maximum wavelength of 400 nm, then its band gap energy is:
Planck's constant, $h = 6.63 \times 10^{-34}$ J.s.
Speed of light, $c = 3 \times 10^8$ m/s

[NCERT : P L-326 | Sep. 03, 2020 (II)]

(a) 1.1 eV (b) 2.0 eV (c) 1.5 eV (d) 3.1 eV

62. Both the diodes used in the circuit shown are assumed to be ideal and have negligible resistance when these are forward biased. Built in potential in each diode is 0.7 V. For the input voltages shown in the figure, the voltage (in Volts) at point A is _______.

[NCERT : P L-335, 336 | NA 9 Jan. 2020 I]

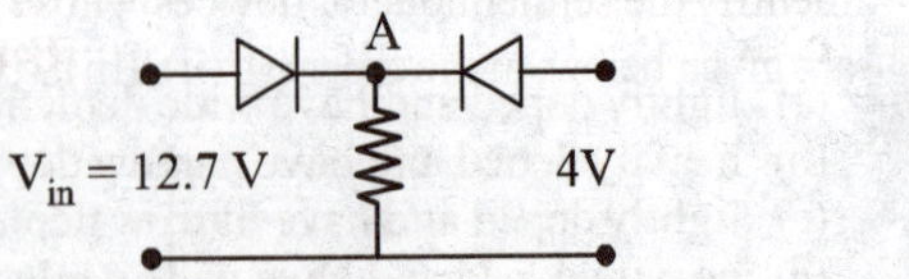

63. The current *i* in the network is: **[NCERT : P L-336 | 9 Jan. 2020 II]**

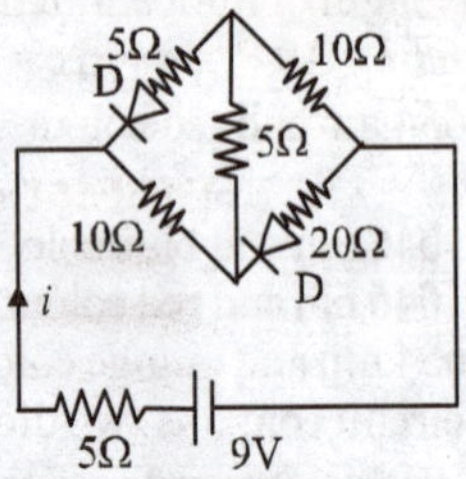

(a) 0.2 A
(b) 0.6 A
(c) 0.3 A
(d) 0 A

64. Two identical capacitors A and B, charged to the same potential 5V are connected in two different circuits as shown below at time $t = 0$. If the charge on capacitors A and B at time $t = CR$ is Q_A and Q_B respectively, then (Here e is the base of natural logarithm) **[NCERT : P L-336 | 9 Jan. 2020 II]**

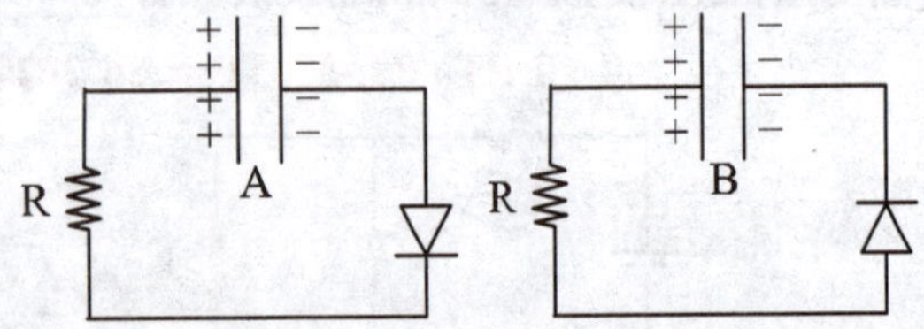

(a) $Q_A = \frac{VC}{e}, Q_B = \frac{CV}{2}$ (b) $Q_A = VC, Q_B = CV$

(c) $Q_A = VC, Q_B = \frac{VC}{e}$ (d) $Q_A = \frac{CV}{2}, Q_B = \frac{VC}{e}$

65. The circuit shown below is working as a 8 V dc regulated voltage source. When 12 V is used as input, the power dissipated (in mW) in each diode is; (considering both zener diodes are identical) ________ **[NA 9 Jan. 2020 II]**

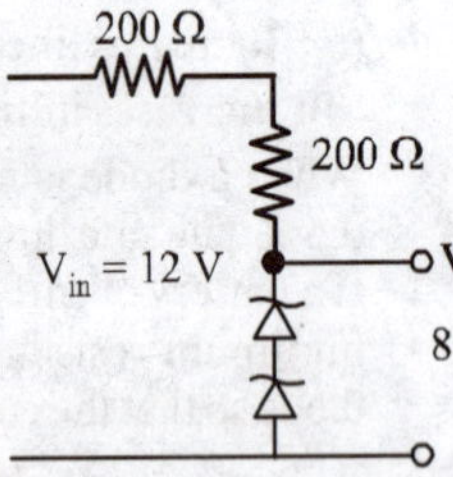

66. In the figure, potential difference between *A* and *B* is: **[NCERT : P L-335 | 7 Jan. 2020 II]**

(a) 10V
(b) 5V
(c) 15V
(d) zero

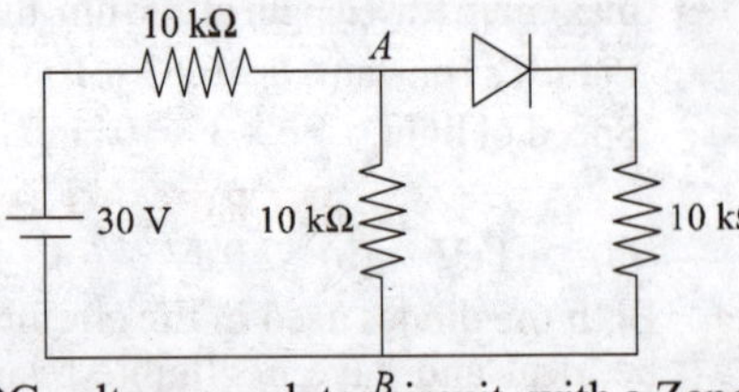

67. Figure shows a DC voltage regulator circuit, with a Zener diode of breakdown voltage = 6V. If the unregulated input voltage varies between 10 V to 16 V, then what is the maximum Zener current? **[12 Apr. 2019 II]**

(a) 2.5 mA
(b) 1.5 mA
(c) 7.5 mA
(d) 3.5 mA

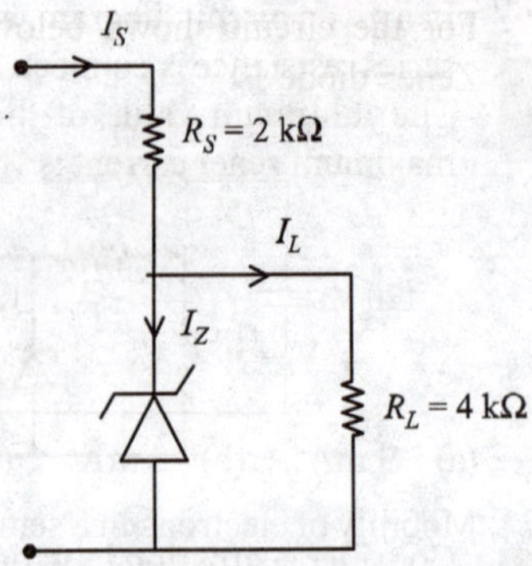

68. The figure represents a voltage regulator circuit using a Zener diode. The breakdown voltage of the Zener diode is 6 V and the load resistance is $R_L = 4\,k\Omega$. The series resistance of the circuit is $R_i = 1k\Omega$. If the battery voltage V_B varies from 8 V to 16 V, what are the minimum and maximum values of the current through Zener diode? **[10 Apr. 2019 II]**

(a) 0.5 mA; 6 mA
(b) 1 mA; 8.5 mA
(c) 0.5 mA; 8.5 mA
(d) 1.5 mA; 8.5 mA

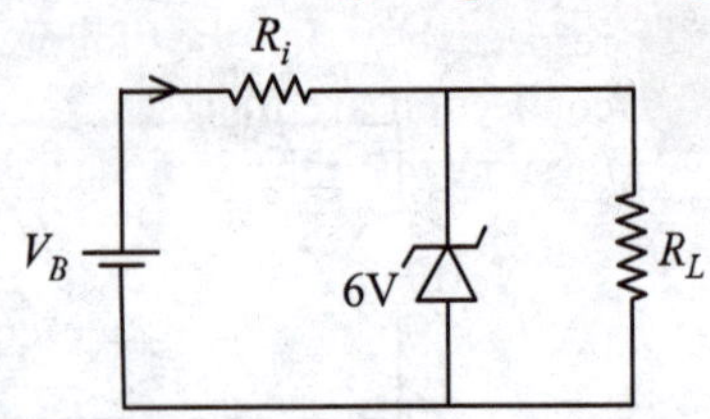

69. The reverse breakdown voltage of a Zener diode is 5.6 V in the given circuit. The current I_z through the Zener is: **[8 April 2019 I]**

(a) 10 mA
(b) 17 mA
(c) 15 mA
(d) 7 mA

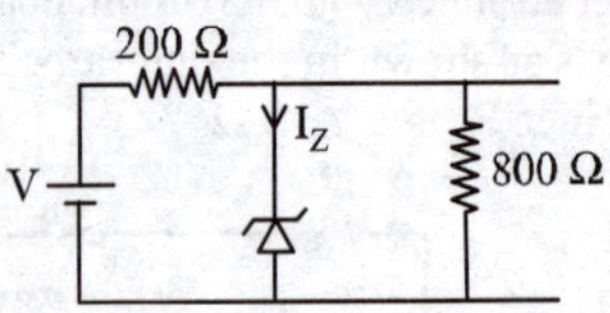

70. In the given circuit the current through Zener Diode is close to: **[11 Jan. 2019 I]**

(a) 0.0 mA
(b) 6.7 mA
(c) 4.0 mA
(d) 6.0 mA

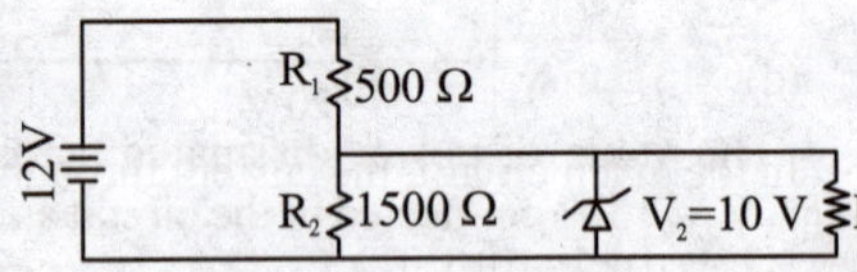

71. The circuit shown below contains two ideal diodes, each with a forward resistance of 50 Ω. If the battery voltage is 6V, the current through the 100 Ω resistance (in Amperes) is: **[NCERT : P L-336 | 11 Jan. 2019 II]**

(a) 0.036
(b) 0.020
(c) 0.027
(d) 0.030

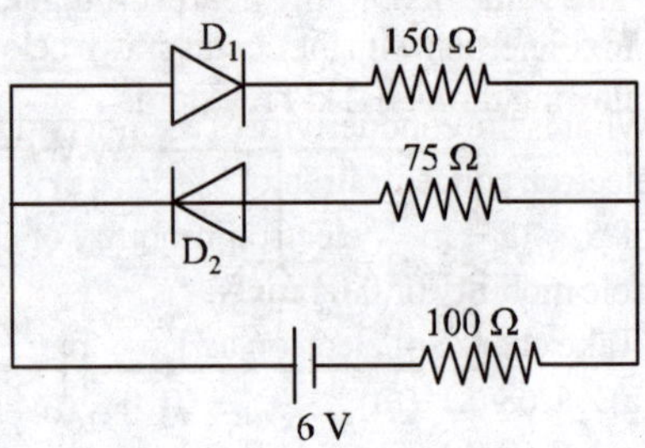

72. For the circuit shown below, the current through the Zener diode is: **[10 Jan. 2019 II]**

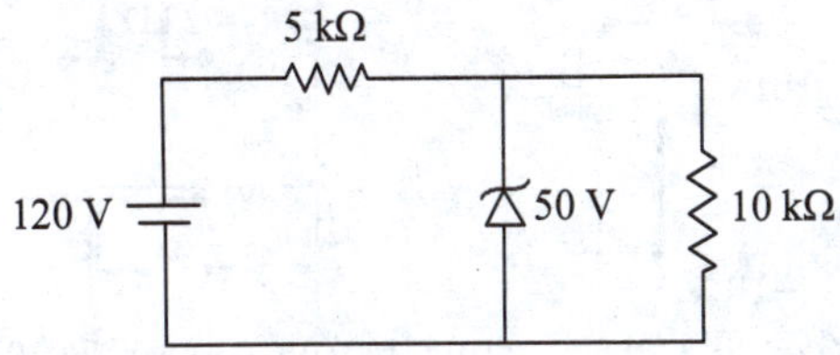

(a) 9 mA (b) 5 mA (c) Zero (d) 14 mA

73. Mobility of electrons in a semiconductor is defined as the ratio of their drift velocity to the applied electric field. If, for an n-type semiconductor, the density of electrons is $10^{19} m^{-3}$ and their mobility is $1.6 m^2/(V.s)$ then the resistivity of the semiconductor (since it is an n-type semiconductor contribution of holes is ignored) is close to: **[9 Jan. 2019 I]**

(a) $2 \Omega m$ (b) $4 \Omega m$ (c) $0.4 \Omega m$ (d) $0.2 \Omega m$

74. Ge and Si diodes start conducting at 0.3 V and 0.7 V respectively. In the following figure if Ge diode connection are reversed, the value of V_o changes by : (assume that the Ge diode has large breakdown voltage) **[NCERT : P L-332 | 9 Jan. 2019 II]**

(a) 0.8 V

(b) 0.6 V

(c) 0.2 V

(d) 0.4 V

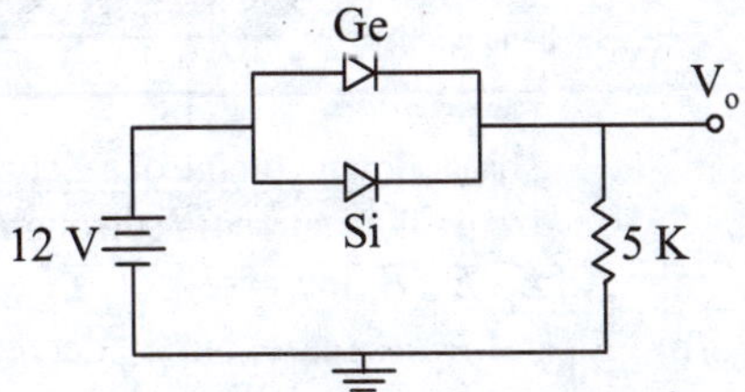

75. The reading of the ammeter for a silicon diode in the given circuit is : **[NCERT : P L-335 | 2018]**

(a) 0

(b) 15 mA

(c) 11.5 mA

(d) 13.5 mA

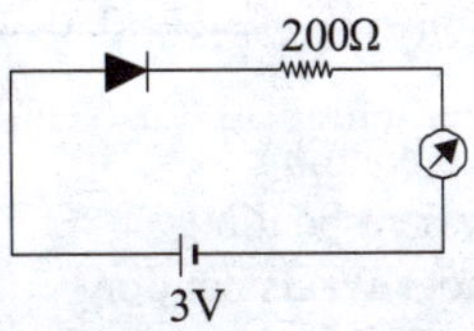

76. In the given circuit, the current through zener diode is: **[Online April 16, 2018]**

(a) 2.5mA

(b) 3.3mA

(c) 5.5mA

(d) 6.7mA

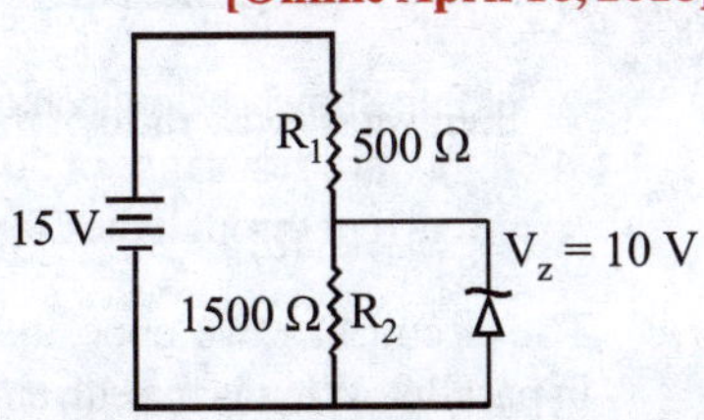

77. What is the conductivity of a semiconductor sample having electron concentration of $5 \times 10^{18} m^{-3}$, hole concentration of $5 \times 10^{19} m^{-3}$, electron mobility of $2.0 m^2 V^{-1} s^{-1}$ and hole mobility of $0.01 m^2 V^{-1} s^{-1}$? **[Online April 8, 2017]**
(Take charge of electron as 1.6×10^{-19} C)

(a) $1.68 (\Omega - m)^{-1}$ (b) $1.83 (\Omega - m)^{-1}$
(c) $0.59 (\Omega - m)^{-1}$ (d) $1.20 (\Omega - m)^{-1}$

78. Identify the semiconductor devices whose characteristics are given below, in the order (i), (ii), (iii), (iv) : **[NCERT : P L-335, 336 | 2016]**

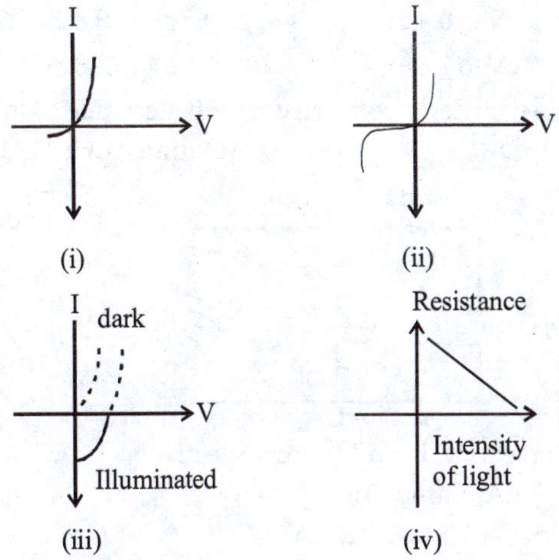

(a) Solar cell, Light dependent resistance, Zener diode, simple diode
(b) Zener diode, Solar cell, simple diode, Light dependent resistance
(c) Simple diode, Zener diode, Solar cell, Light dependent resistance
(d) Zener diode, Simple diode, Light dependent resistance, Solar cell

79. The temperature dependence of resistances of Cu and undoped Si in the temperature range 300-400 K, is best described by : **[2016]**

(a) Linear increase for Cu, exponential decrease of Si.
(b) Linear decrease for Cu, linear decrease for Si.
(c) Linear increase for Cu, linear increase for Si.
(d) Linear increase for Cu, exponential increase for Si.

80. In an unbiased n-p junction electrons diffuse from n-region to p-region because : **[NCERT : P L-333 | Online April 10, 2015]**

(a) holes in p-region attract them
(b) electrons travel across the junction due to potential difference
(c) only electrons move from n to p region and not the vice-versa
(d) electron concentration in n-region is more compared to that in p-region

81. The forward biased diode connection is: **[NCERT : P L-335 | 2014]**

(a) 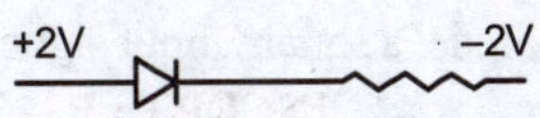

(b) 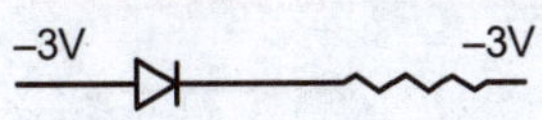

(c) 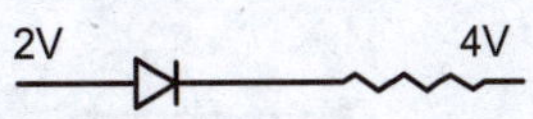

(d) 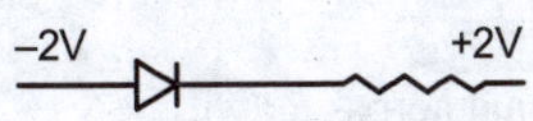

82. For LED's to emit light in visible region of electromagnetic light, it should have energy band gap in the range of: **[NCERT : P L-332 | Online April 12, 2014]**

(a) 0.1 eV to 0.4 eV (b) 0.5 eV to 0.8 eV
(c) 0.9 eV to 1.6 eV (d) 1.7 eV to 3.0 eV

83. A Zener diode is connected to a battery and a load as show below: **[Online April 11, 2014]**

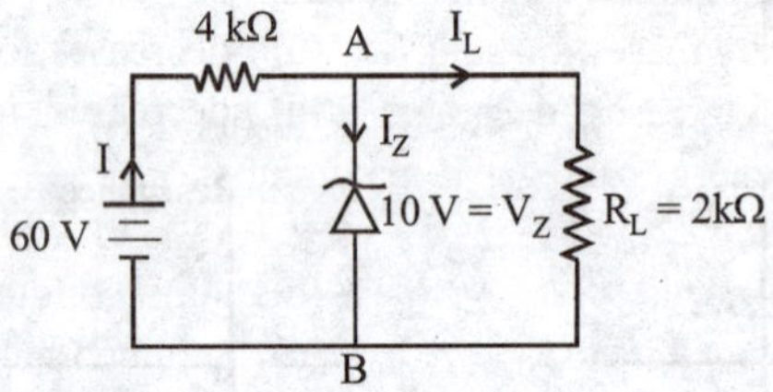

The currents, I, I_Z and I_L are respectively.

(a) 15 mA, 5 mA, 10 mA
(b) 15 mA, 7.5 mA, 7.5 mA
(c) 12.5 mA, 5 mA, 7.5 mA
(d) 12.5 mA, 7.5 mA, 5 mA

84. The I-V characteristic of an LED is **[2013]**

(a) 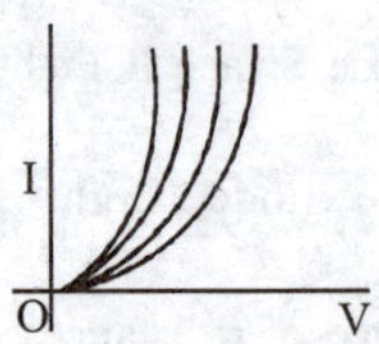

(b) 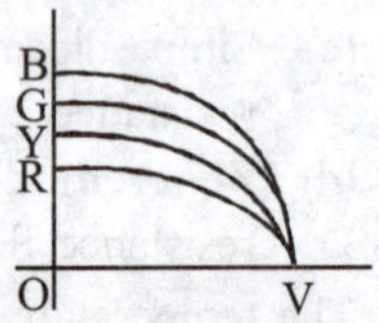

(c) 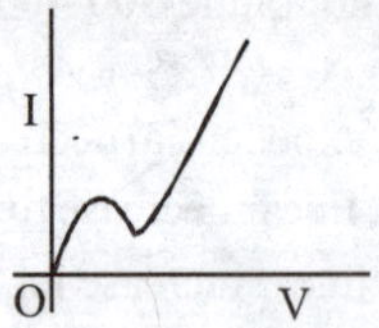

(d) 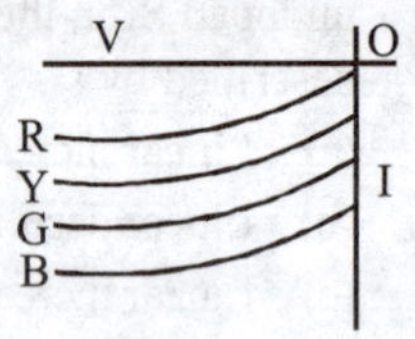

85. This question has Statement 1 and Statement 2. Of the four choices given after the Statements, choose the one that best describes the two Statements.

Statement 1: A pure semiconductor has negative temperature coefficient of resistance.

Statement 2: On raising the temperature, more charge carriers are released into the conduction band. **[Online May 12, 2012]**

(a) Statement 1 is false, Statement 2 is true.
(b) Statement 1 is true, Statement 2 is false.
(c) Statement 1 is true, Statement 2 is true, Statement 2 is not a correct explanation of Statement 1.
(d) Statement 1 is true, Statement 2 is true, Statement 2 is the correct explanation of Statement 1.

86. If in a *p-n* junction diode, a square input signal of 10 V is applied as shown **[NCERT : P L-333 | 2007]**

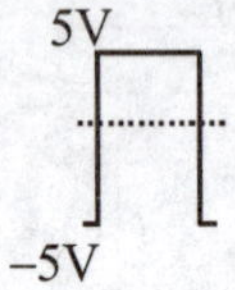

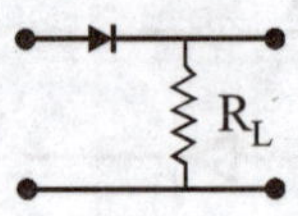

Then the output signal across R_L will be

(a) 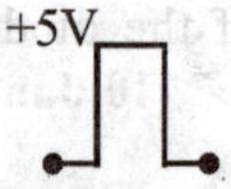

(b) 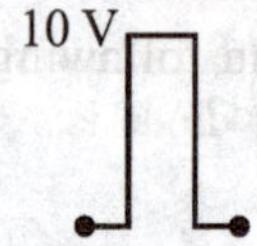

(c) 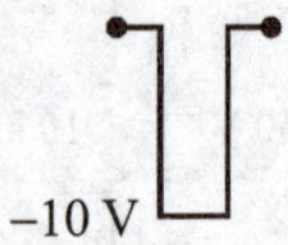

(d) 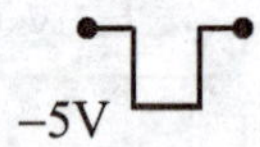

87. Carbon, silicon and germanium have four valence electrons each. At room temperature which one of the following statements is most appropriate ? **[NCERT : P L-332 | 2007]**

(a) The number of free electrons for conduction is significant only in Si and Ge but small in C.
(b) The number of free conduction electrons is significant in C but small in Si and Ge.
(c) The number of free conduction electrons is negligibly small in all the three.
(d) The number of free electrons for conduction is significant in all the three.

88. If the lattice constant of this semiconductor is decreased, then which of the following is correct? **[NCERT : P L-332 | 2006]**

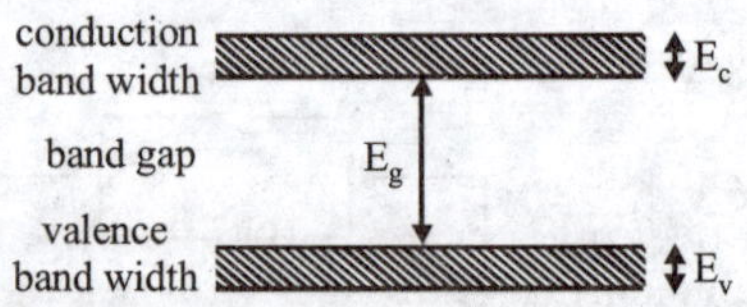

(a) All E_c, E_g, E_v increase
(b) E_c and E_v increase, but E_g decreases
(c) E_c and E_v decrease, but E_g increases
(d) All E_c, E_g, E_v decrease

89. A solid which is not transparent to visible light and whose conductivity increases with temperature is formed by **[NCERT : P L-333 | 2006]**

(a) Ionic bonding
(b) Covalent bonding
(c) Vander Waals bonding
(d) Metallic bonding

90. If the ratio of the concentration of electrons to that of holes in a semiconductor is $\frac{7}{5}$ and the ratio of currents is $\frac{7}{4}$, then what is the ratio of their drift velocities? **[2006]**

(a) $\frac{5}{8}$ (b) $\frac{4}{5}$ (c) $\frac{5}{4}$ (d) $\frac{4}{7}$

91. The circuit has two oppositively connected ideal diodes in parallel. What is the current flowing in the circuit? **[NCERT : P L-335, 336 | 2006]**

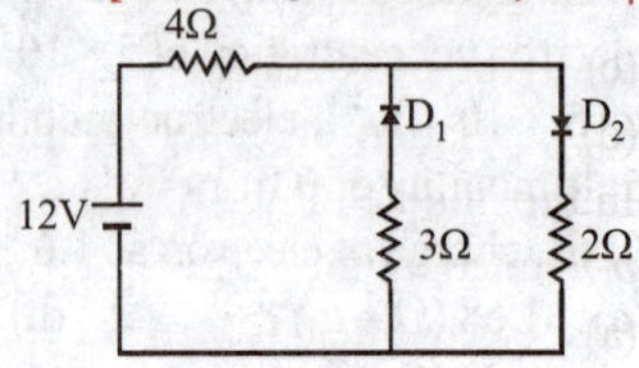

(a) 1.71 A
(b) 2.00 A
(c) 2.31 A
(d) 1.33 A

92. In the following, which one of the diodes reverse biased? **[NCERT : P L-336 | 2006]**

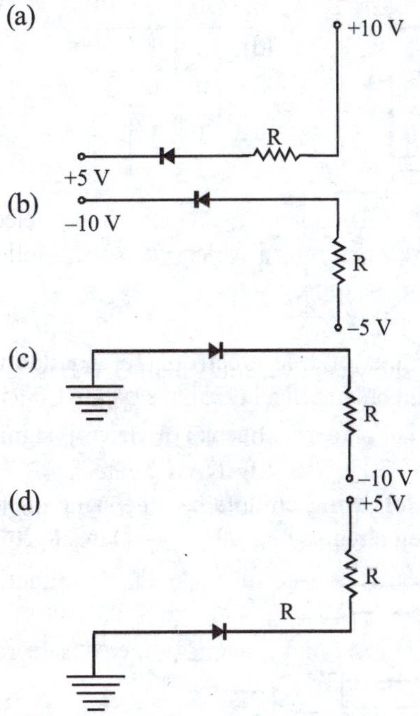

93. The electrical conductivity of a semiconductor increases when electromagnetic radiation of wavelength shorter than 2480 nm is incident on it. The band gap in (eV) for the semiconductor is **[NCERT : P L-332 | 2005]**
(a) 2.5 eV (b) 1.1 eV (c) 0.7 eV (d) 0.5 eV

94. When p-n junction diode is forward biased then **[NCERT : P L-335 | 2004]**
(a) both the depletion region and barrier height are reduced
(b) the depletion region is widened and barrier height is reduced
(c) the depletion region is reduced and barrier height is increased
(d) Both the depletion region and barrier height are increased

95. A strip of copper and another of germanium are cooled from room temperature to 80K. The resistance of **[NCERT : P L-332 | 2003]**
(a) each of these decreases
(b) copper strip increases and that of germanium decreases
(c) copper strip decreases and that of germanium increases
(d) each of these increases

96. The difference in the variation of resistance with temeperature in a metal and a semiconductor arises essentially due to the difference in the **[2003]**
(a) crystal sturcture
(b) variation of the number of charge carriers with temperature
(c) type of bonding
(d) variation of scattering mechanism with temperature

97. In the middle of the depletion layer of a reverse- biased *p-n* junction, the **[NCERT : P L-336 | 2003]**
(a) electric field is zero
(b) potential is maximum
(c) electric field is maximum
(d) potential is zero

98. At absolute zero, Si acts as **[NCERT : P L-332 | 2002]**
(a) non-metal (b) metal
(c) insulator (d) none of these

99. By increasing the temperature, the specific resistance of a conductor and a semiconductor **[2002]**
(a) increases for both (b) decreases for both
(c) increases, decreases (d) decreases, increases

100. The energy band gap is maximum in **[NCERT : P L-332 | 2002]**
(a) metals (b) superconductors
(c) insulators (d) semiconductors

Topic 2 - Digital Electronics and Logic Gates

101. Consider the following logic circuit.

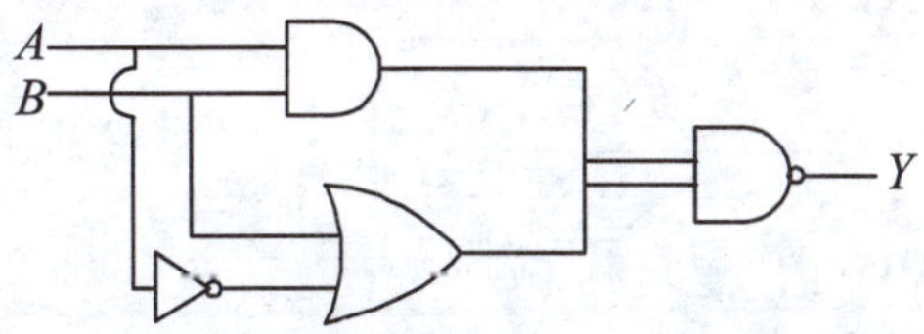

The output is $Y=0$ when : **[April 7, 2025 (II)]**
(a) $A=1$ and $B=1$ (b) $A=0$ and $B=1$
(c) $A=1$ and $B=0$ (d) $A=0$ and $B=0$

102. The Boolean expression $Y = A\overline{B}C + \overline{A}\,\overline{C}$ can be realised with which of the following gate configurations.
A. One 3-input AND gate, 3 NOT gates and one 2-input OR gate, One 2-input AND gate,
B. One 3-input AND gate, 1 NOT gate, One 2-input NOR gate and one 2-input OR gate
C. 3-input OR gate, 3 NOT gates and one 2-input AND gate
Choose the **correct** answer from the options given below **[April 4, 2025 (I)]**
(a) B, C Only (b) A,B Only
(c) A, B, C Only (d) A, C Only

103. The truth table corresponding to the circuit given below is **[April 3, 2025 (II)]**

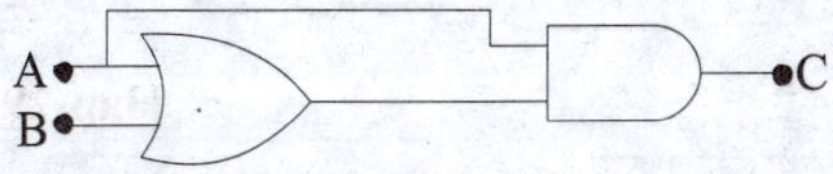

(a)

A	B	C
0	0	0
1	0	0
0	1	0
1	1	1

(b)

A	B	C
0	0	0
0	1	0
1	0	1
1	1	1

(c)

A	B	C
0	0	1
1	0	0
0	1	0
1	1	0

(d)

A	B	C
0	0	1
0	1	0
1	0	0
1	1	0

104. Choose the correct logic circuit for the given truth table having inputs A and B. **[April 3, 2025 (I)]**

Inputs		Output
A	B	Y
0	0	0
0	1	0
1	0	1
1	1	1

(a) 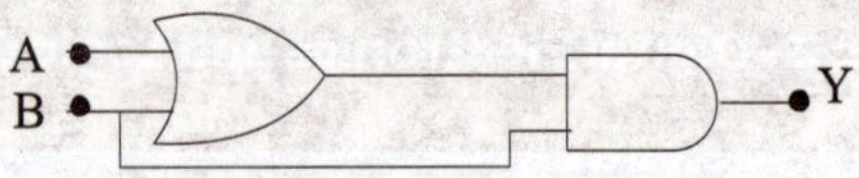

(b) 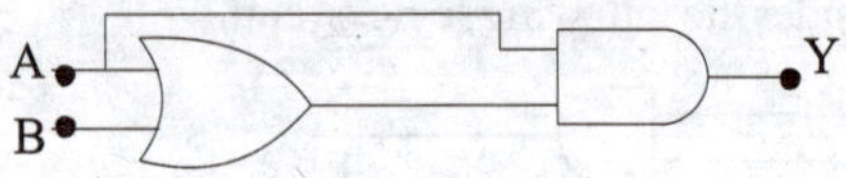

(c) 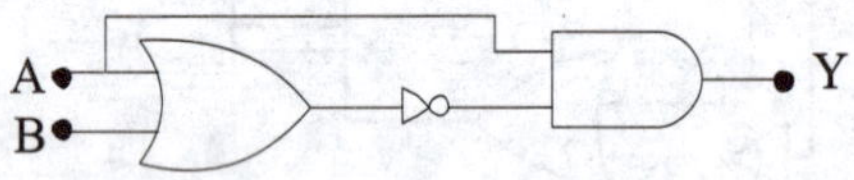

(d) 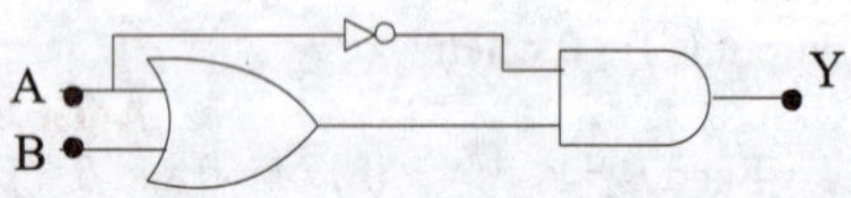

105. In the digital circuit shown in the figure, for the given inputs the P and Q values are : **[April 2, 2025 (II)]**

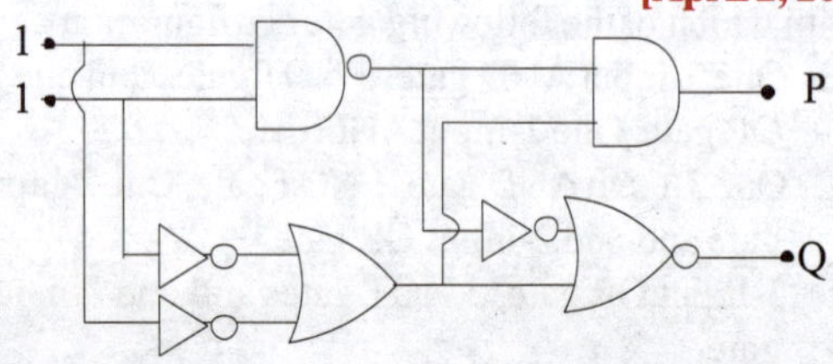

(a) P = 1, Q = 1 (b) P = 0, Q = 0
(c) P = 0, Q = 1 (d) P = 1, Q = 0

106. The truth table for the circuit given below is :

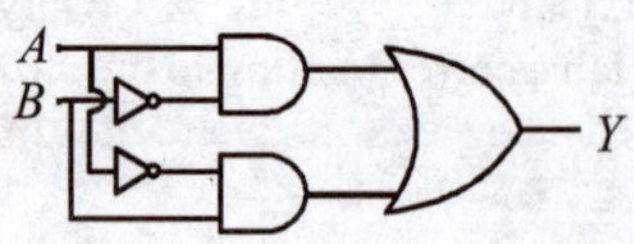

[Jan. 29, 2025 (II)]

(a)

A	B	Y
0	0	0
0	1	1
1	0	1
1	1	0

(b)

A	B	Y
0	0	0
1	0	0
1	1	0
0	1	1

(c)

A	B	Y
0	0	0
1	0	1
0	1	0
1	1	0

(d)

A	B	Y
0	0	0
1	1	1
1	0	1
0	1	1

107. 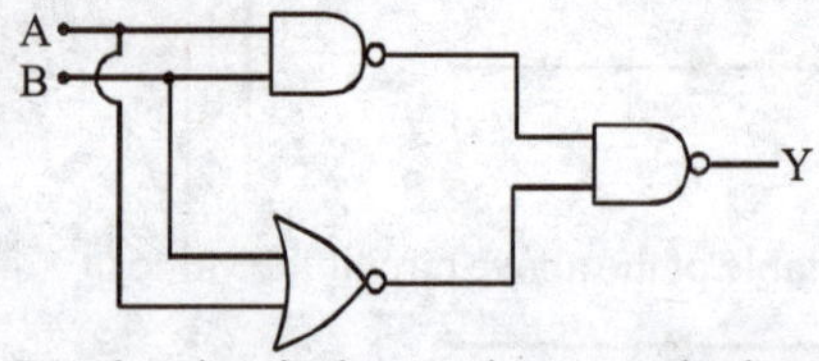

For the circuit shown above, equivalent GATE is: **[Jan. 29, 2025 (I)]**

(a) OR gate (b) NOT gate
(c) AND gate (d) NAND gate

108. Which of the following circuits has the same output as that of the given circuit? **[Jan. 28, 2025 (I)]**

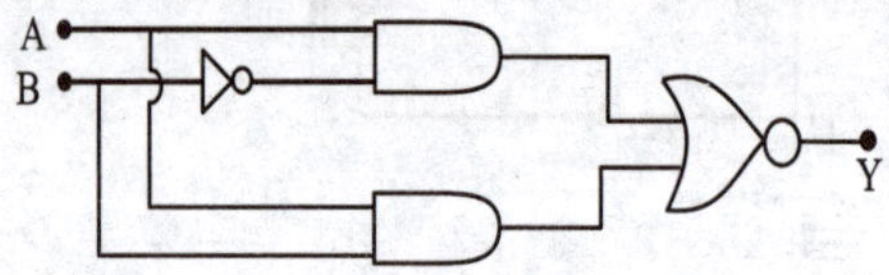

(a) 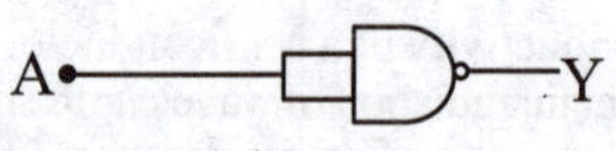

(b) 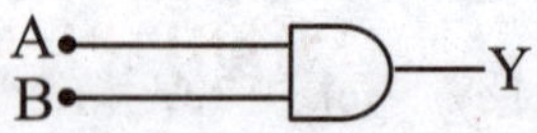

(c) 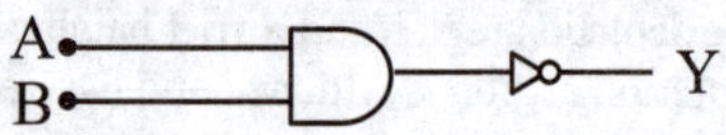

(d) 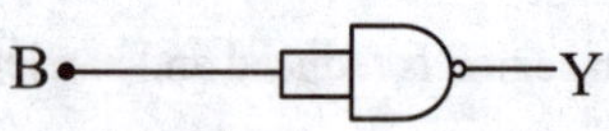

109. The output of the circuit is low (zero) for

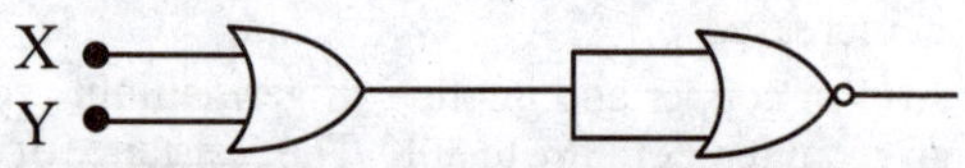

(A) X = 0, Y = 0 (B) X = 0, Y = 1
(C) X = 1, Y = 0 (D) X = 1, Y = 1

Choose the correct answer from the options given below **[Jan. 24, 2025 (II)]**

(a) (A), (C) and (D) only (b) (A), (B) and (C) only
(c) (B), (C) and (D) only (d) (A), (B) and (D) only

110. 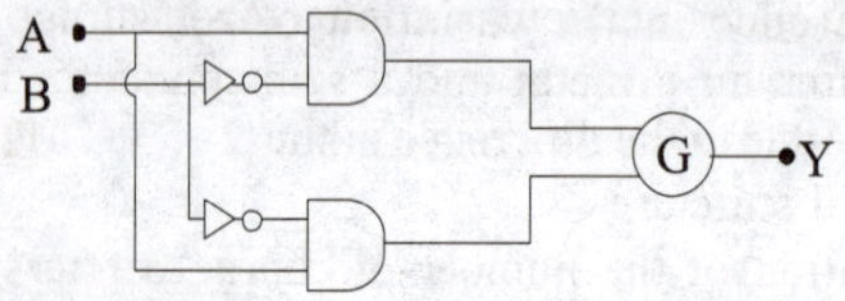

A	B	Y
0	0	1
0	1	0
1	0	0
1	1	1

To obtain the given truth table, following logic gate should be placed at G **[Jan. 22, 2025 (II)]**

(a) NOR Gate (b) AND Gate
(c) NAND Gate (d) OR Gate

111. 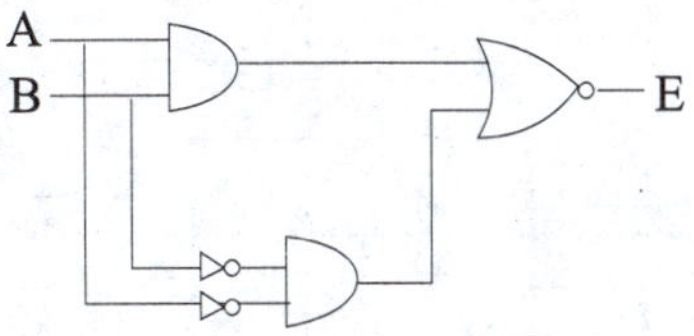

A	B	E
0	0	0
0	1	X
1	0	Y
1	1	0

In the truth table of the above circuit the value of X and Y are : **[April 9, 2024 (II)]**

(a) 1, 1 (b) 1, 0 (c) 0, 1 (d) 0, 0

112. The output Y of following circuit for given inputs is : **[April 8, 2024 (I)]**

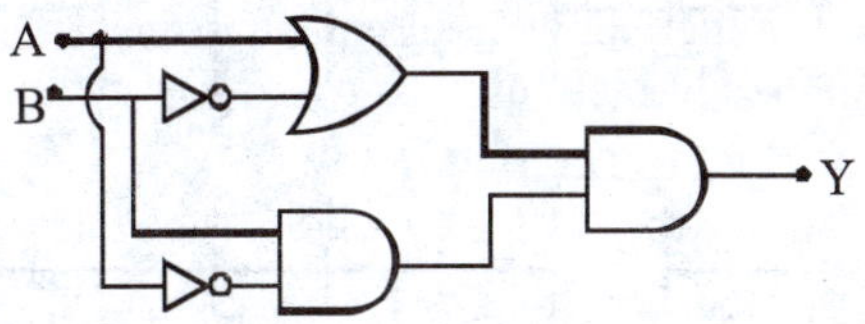

(a) A•B(A + B) (b) A • B
(c) 0 (d) $\bar{A}$•B

113. The correct truth table for the following logic circuit is : **[April 6, 2024 (I)]**

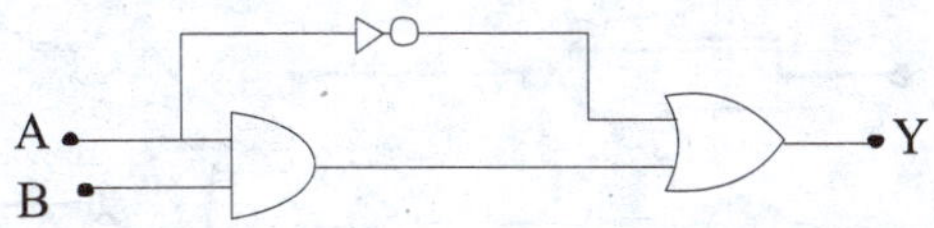

(a)

A	B	Y
0	0	0
0	1	1
1	0	0
1	1	1

(b)

A	B	Y
0	0	1
0	1	1
1	0	0
1	1	1

(c)

A	B	Y
0	0	1
0	1	1
1	0	0
1	1	0

(d)

A	B	Y
0	0	0
0	1	0
1	0	0
1	1	1

114. Following gates section is connected in a complete suitable circuit.

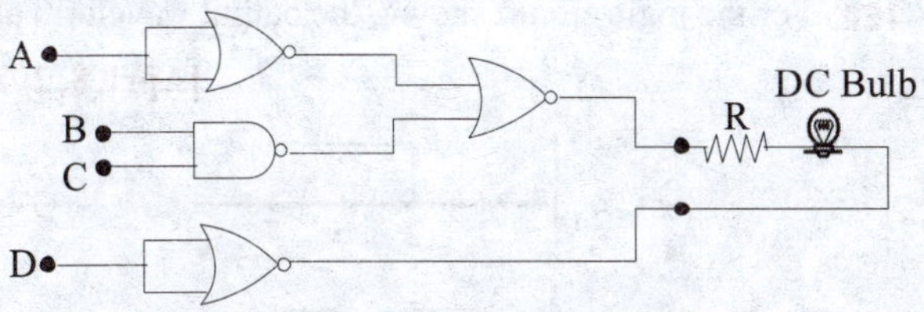

For which of the following combination, bulb will glow (ON): **[April 5, 2024 (I)]**

(a) A = 0, B = 1, C = 1, D = 1 (b) A = 1, B = 0, C = 0, D = 0
(c) A = 0, B = 0, C = 0, D = 1 (d) A = 1, B = 1, C = 1, D = 0

115. The output (Y) of logic circuit given below is 0 only when: **[April 5, 2024 (II)]**

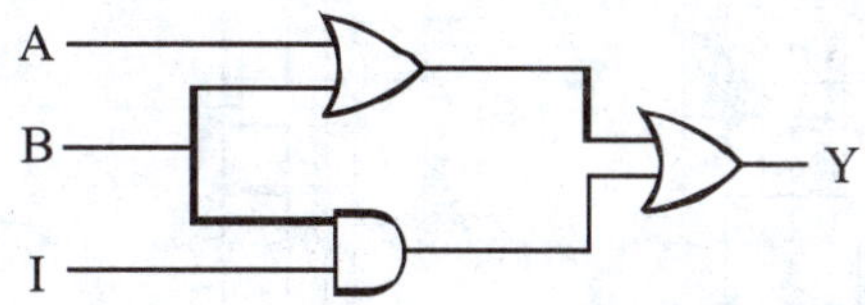

(a) A = 1, B = 0 (b) A = 0, B = 0
(c) A = 1, B = 1 (d) A = 0, B = 1

116. Identify the logic gate given in the circuit : **[April 4, 2024 (II)]**

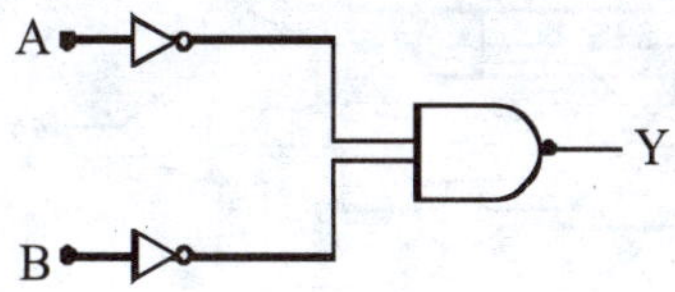

(a) NAND - gate (b) OR - gate
(c) AND gate (d) NOR gate

117. Identify the logic operation performed by the given circuit. **[Jan. 31, 2024 (I)]**

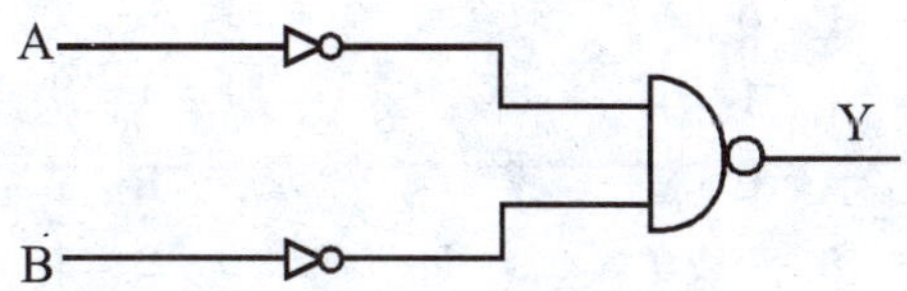

(a) AND (b) NOR
(c) NAND (d) OR

118. 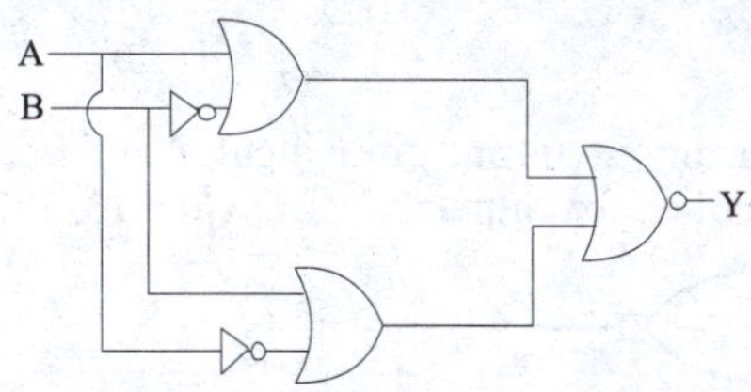

[Jan. 31, 2024 (II)]

The output of the given circuit diagram is

(a)

A	B	Y
0	0	0
1	0	0
0	1	0
1	1	1

(b)

A	B	Y
0	0	0
1	0	1
0	1	1
1	1	0

(c)

A	B	Y
0	0	0
1	0	0
0	1	0
1	1	0

(d)

A	B	Y
0	0	0
1	0	0
0	1	1
1	1	0

119. The truth table for this given circuit is : **[Jan. 29, 2024 (II)]**

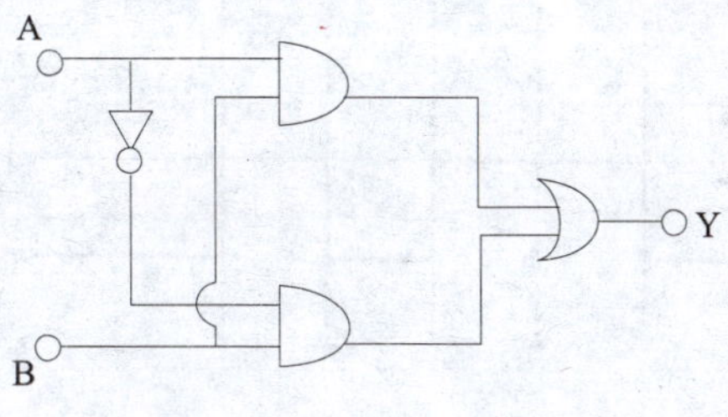

(a)

A	B	Y
0	0	1
0	1	1
1	0	1
1	1	0

(b)

A	B	Y
0	0	1
0	1	0
1	0	1
1	1	0

(c)

A	B	Y
0	0	0
0	1	1
1	0	0
1	1	1

(d)

A	B	Y
0	0	0
0	1	0
1	0	0
1	1	1

120. The truth table of the given circuit diagram is : **[Jan. 27, 2024 (II)]**

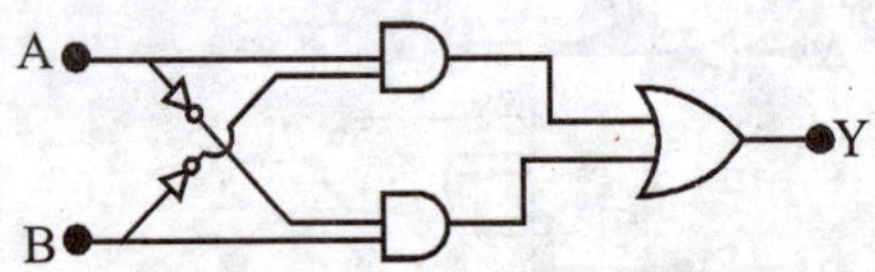

(a)

A	B	Y
0	0	1
0	1	0
1	0	0
1	1	1

(b)

A	B	Y
0	0	0
0	1	1
1	0	1
1	1	0

(c)

A	B	Y
0	0	0
0	1	0
1	0	0
1	1	1

(d)

A	B	Y
0	0	1
0	1	1
1	0	1
1	1	0

121. For the following circuit and given inputs A and B, choose the correct option for output 'Y' **[April 13, 2023 (I)]**

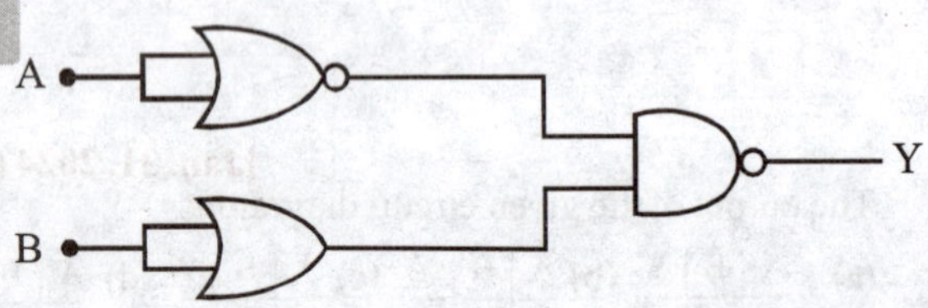

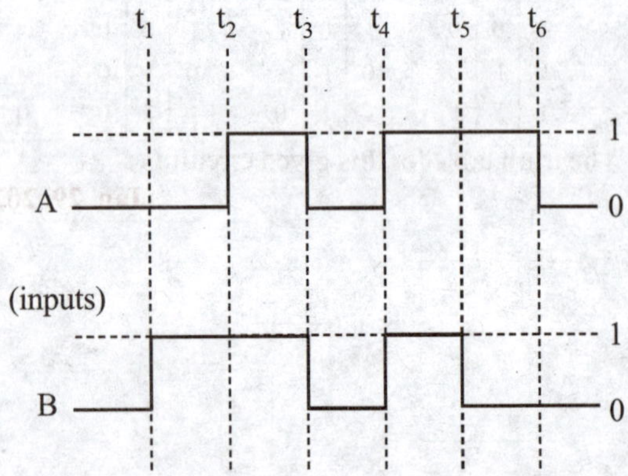

(a)

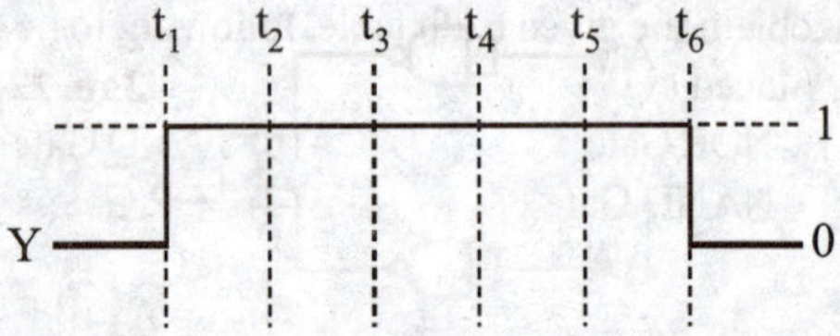

(b)

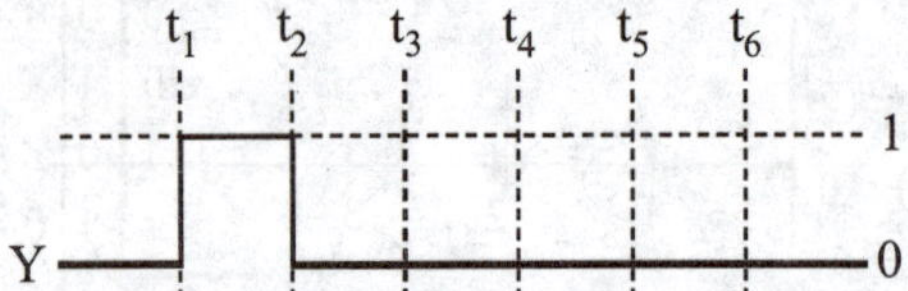

(c)

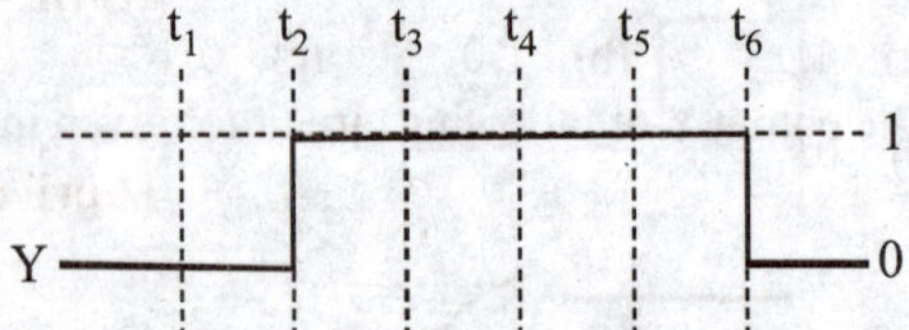

(d)

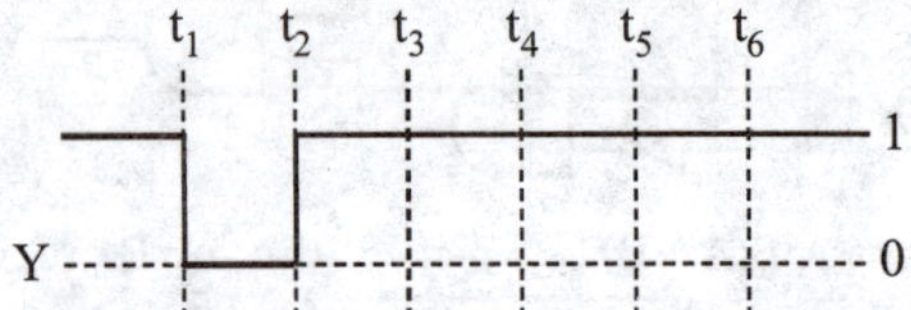

122. The logic performed by the circuit shown in figure is equivalent to: **[April 11, 2023 (I)]**

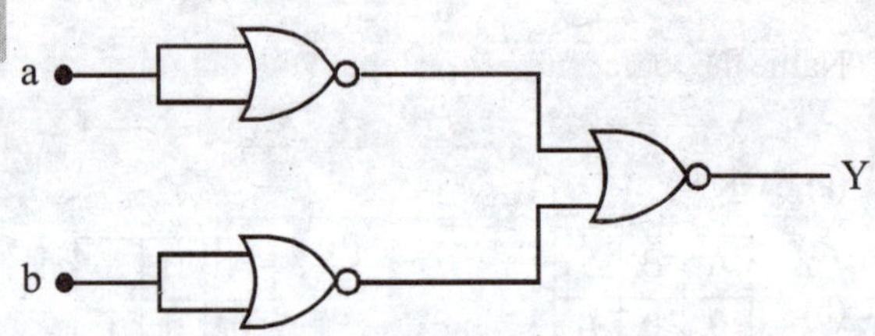

(a) AND (b) NAND (c) OR (d) NOR

123. The logic operations performed by the given digital circuit is equivalent to: **[April 11, 2023 (II)]**

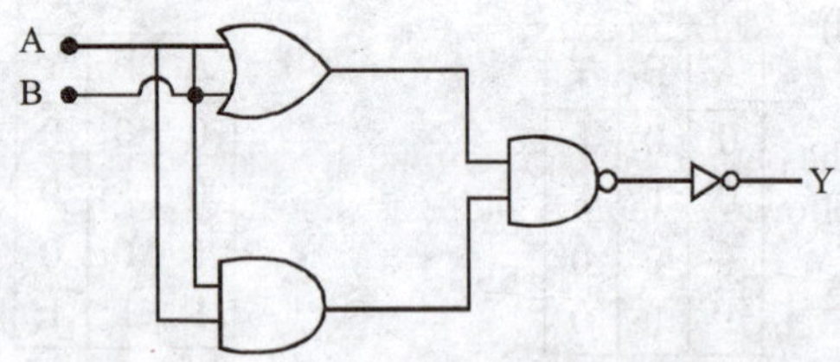

(a) AND (b) NOR (c) OR (d) NAND

124. For the logic circuit shown, the output waveform at Y is: **[April 8, 2023 (I)]**

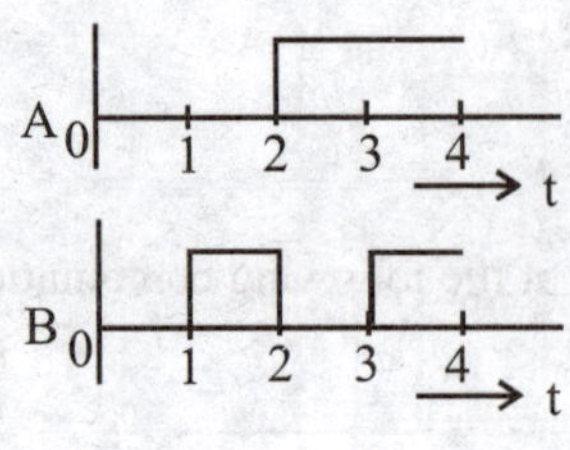

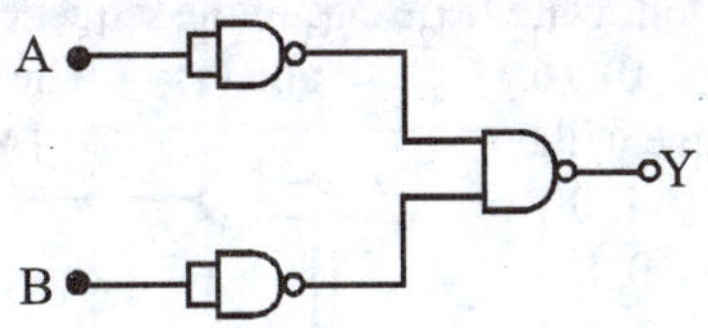

(a) 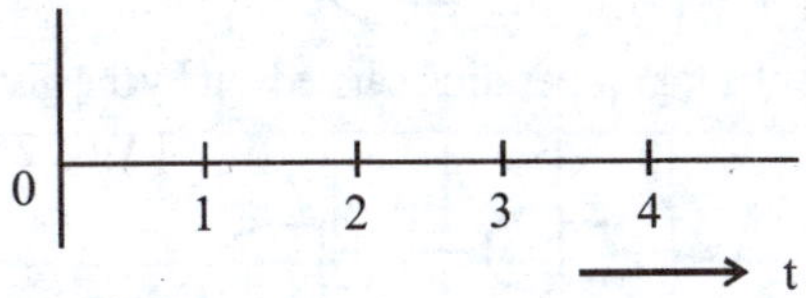

(b) 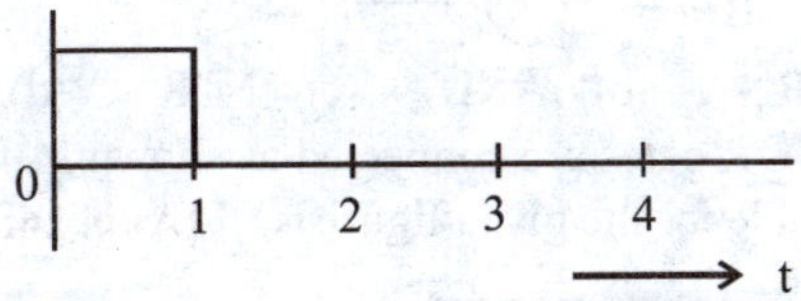

(c) 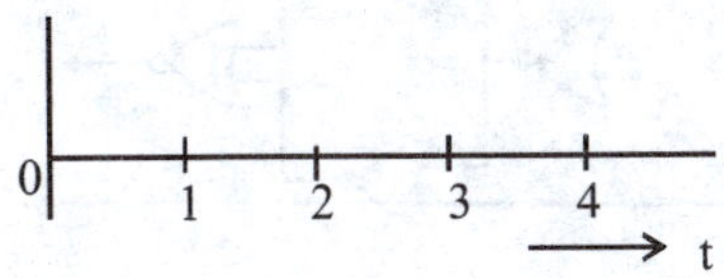

(d) 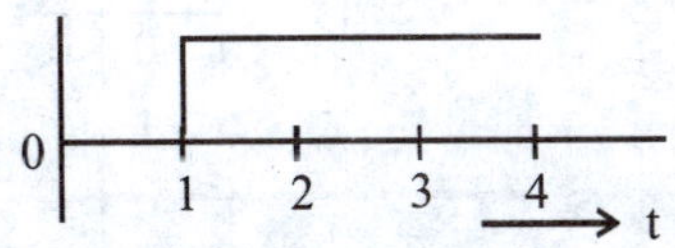

125. Name the logic gate equivalent to the diagram attached **[April 6, 2023 (I)]**

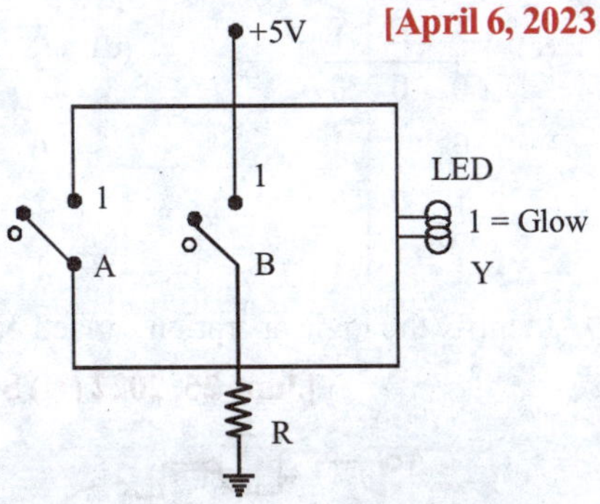

(a) OR

(b) NOR

(c) NAND

(d) AND

126. The output waveform of the given logical circuit for the following inputs A and B as shown below, is : **[Jan. 30, 2023 (I)]**

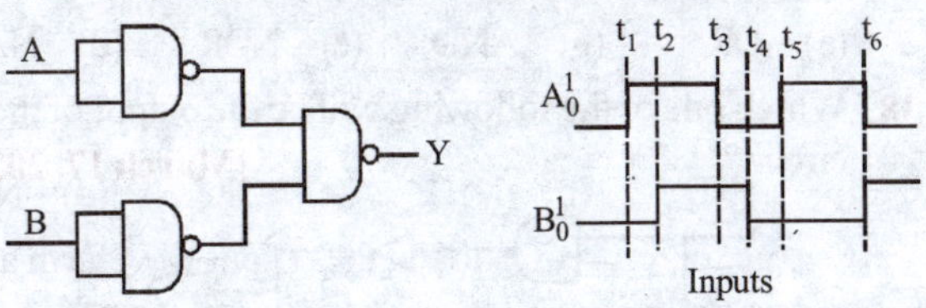

(a) 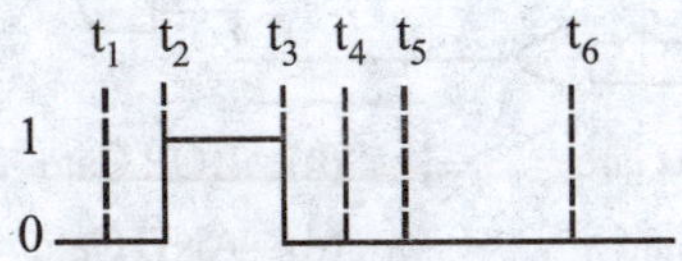

(b) 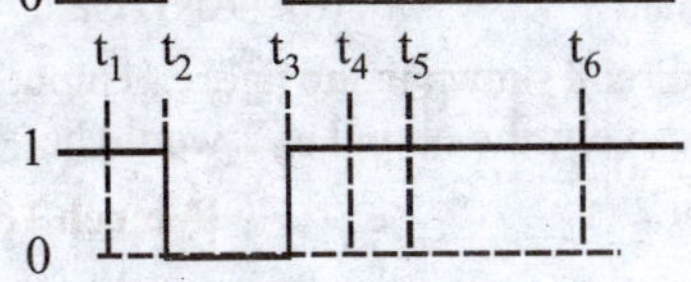

(c) 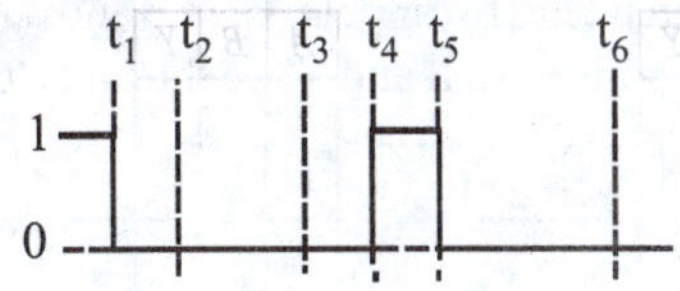

(d) 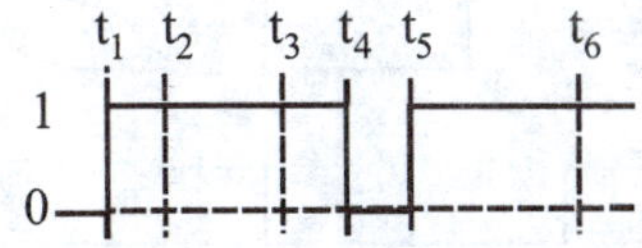

127. The output Y for the inputs A and B of circuit is given by

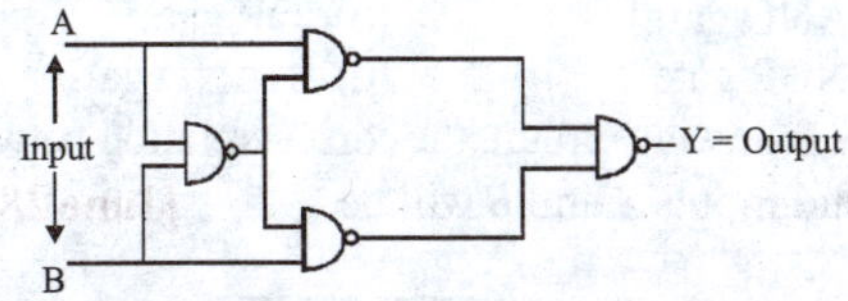

Truth table of the shown circuit is : **[Jan. 30, 2023 (II)]**

(a)

A	B	Y
0	0	1
0	1	1
1	0	1
1	1	0

(b)

A	B	Y
0	0	1
0	1	0
1	0	0
1	1	1

(c)

A	B	Y
0	0	0
0	1	1
1	0	1
1	1	1

(d)

A	B	Y
0	0	0
0	1	1
1	0	1
1	1	0

128. For the given logic gates combination, the correct truth table will be **[Jan. 29, 2023 (II)]**

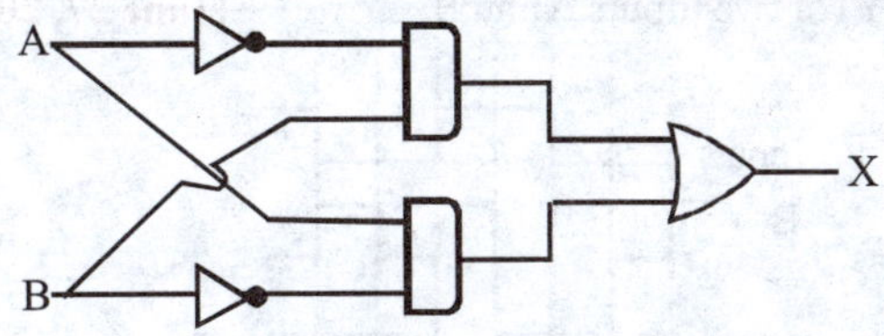

(a)

A	B	X
0	0	1
0	1	0
1	0	0
1	1	0

(b)

A	B	X
0	0	0
0	1	1
1	0	1
1	1	0

(c)

A	B	X
0	0	1
0	1	0
1	0	1
1	1	0

(d)

A	B	X
0	0	0
0	1	1
1	0	1
1	1	1

129. The logic gate equivalent to the given circuit diagram is : **[Jan. 24, 2023 (II)]**

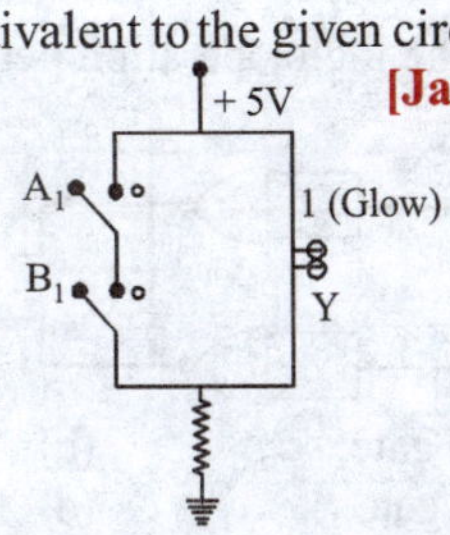

(a) OR

(b) NAND

(c) NOR

(d) AND

130. A logic gate circuit has two inputs A and B and output Y. The voltage waveforms of A, B and Y are shown below

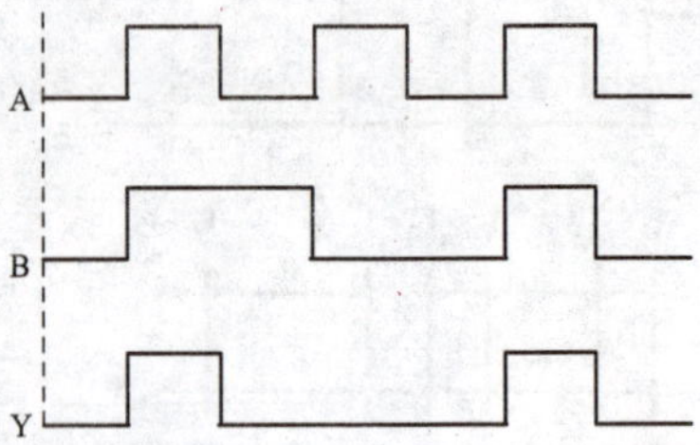

The logic gate circuit is **[July 27, 2022 (I)]**

(a) AND gate (b) OR gate
(c) NOR gate (d) NAND gate

131. In the following circuit, the correct relation between output (Y) and inputs A and B will be : **[June 28, 2022 (I)]**

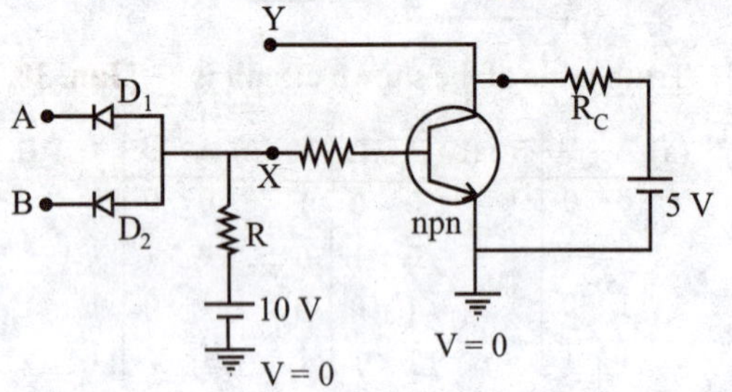

(a) $Y = AB$ (b) $Y = A + B$ (c) $Y = \overline{AB}$ (d) $Y = \overline{A + B}$

132. Identify the correct Logic Gate for the following output (Y) of two inputs A and B. **[June 27, 2022 (I)]**

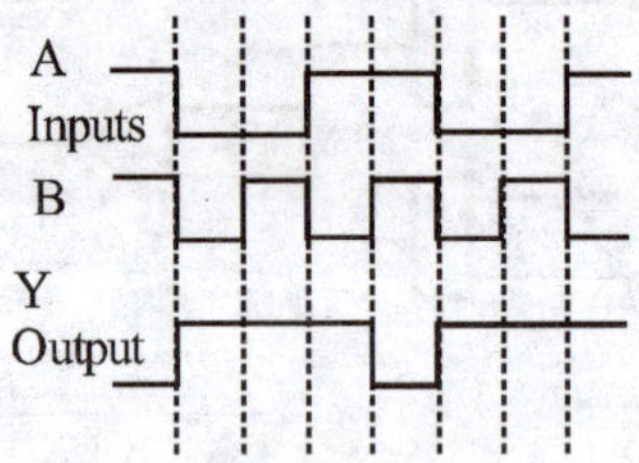

(a) 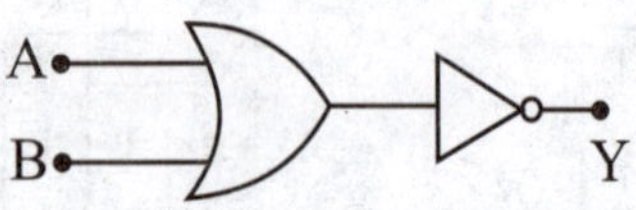

(b) 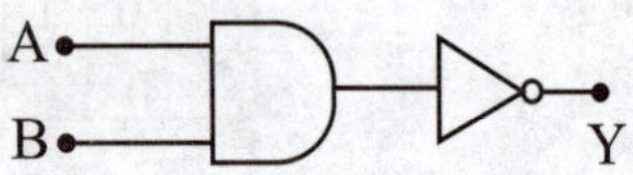

(c) 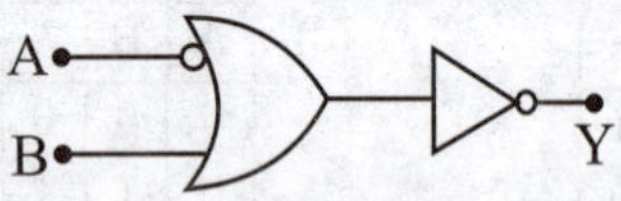

(d) 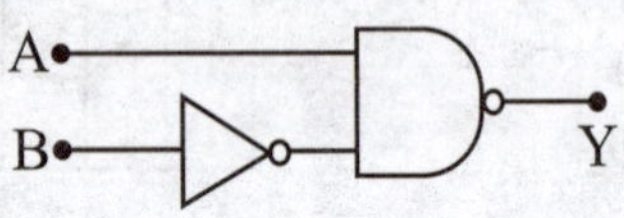

133. Identify the logic operation performed by the given circuit : **[June 25, 2022 (II)]**

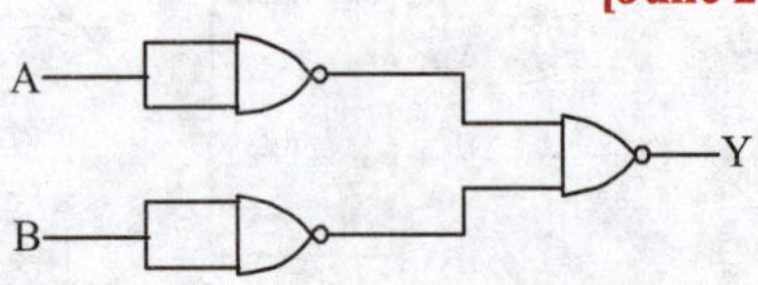

(a) AND gate (b) OR gate
(c) NOR gate (d) NAND gate

134. In the following logic circuit the sequence of the inputs A, B are (0, 0), (0,1), (1, 0) and (1, 1). The output Y for this sequence will be : **[Aug. 31, 2021 (I)]**

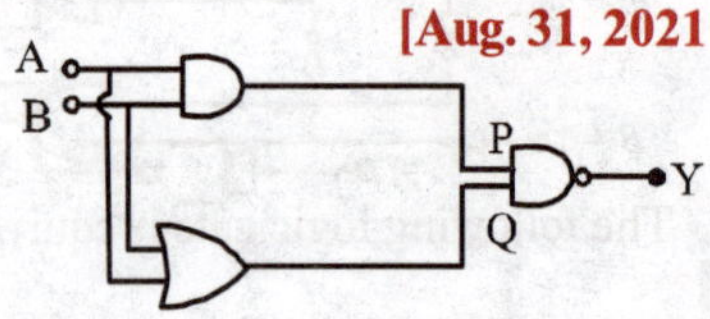

(a) 1, 0, 1, 0
(b) 0, 1, 0, 1
(c) 1, 1, 1, 0
(d) 0, 0, 1, 1

135. Identify the logic operation carried out by the given circuit :- **[Aug. 26, 2021 (I)]**

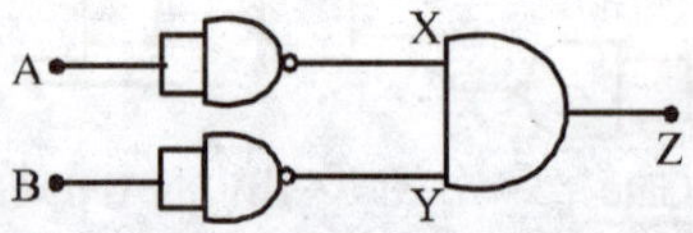

(a) OR (b) AND (c) NOR (d) NAND

136. Four NOR gates are connected as shown in figure. The truth table for the given figure is : **[Aug. 26, 2021 (II)]**

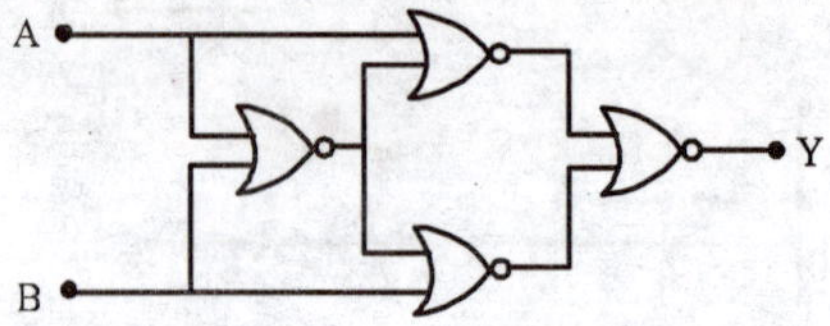

(a)

A	B	Y
0	0	1
0	1	0
1	0	1
1	1	0

(b)

A	B	Y
0	0	0
0	1	1
1	0	1
1	1	0

(c)

A	B	Y
0	0	0
0	1	1
1	0	0
1	1	1

(d)

A	B	Y
0	0	1
0	1	0
1	0	0
1	1	1

137. Identify the logic operation carried out. **[June 25, 2022 (II) S; July 25, 2021 (I)]**

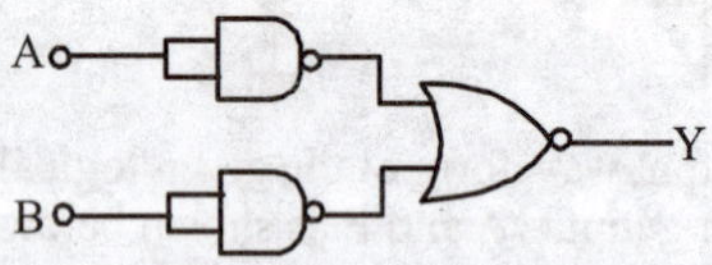

(a) OR (b) AND (c) NOR (d) NAND

138. Which one of the following will be the output of the given circuit? **[March 17, 2021 (II)]**

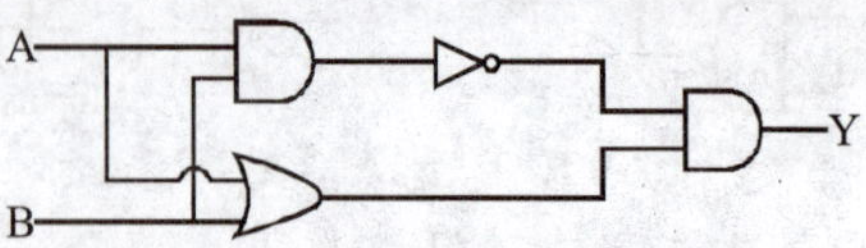

(a) NAND Gate (b) XOR Gate
(c) NOR Gate (d) AND Gate

139. In the logic circuit shown in the figure, if input A and B are 0 to 1 respectively, the output at Y would be 'x'. The value of x is ______. **[March 16, 2021 (I)]**

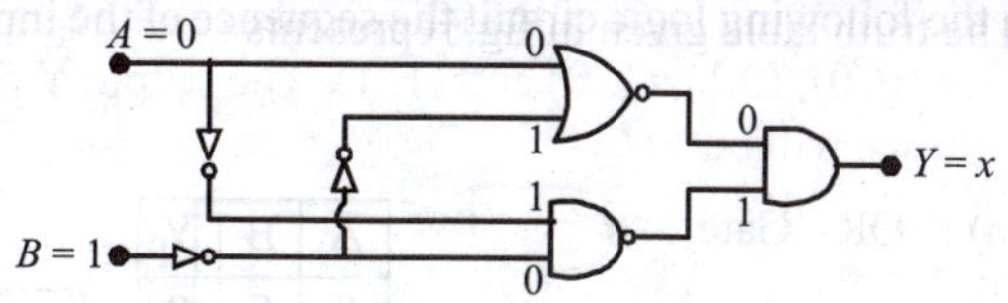

140. The following logic gate is equivalent to: **[March 16, 2021 (II)]**

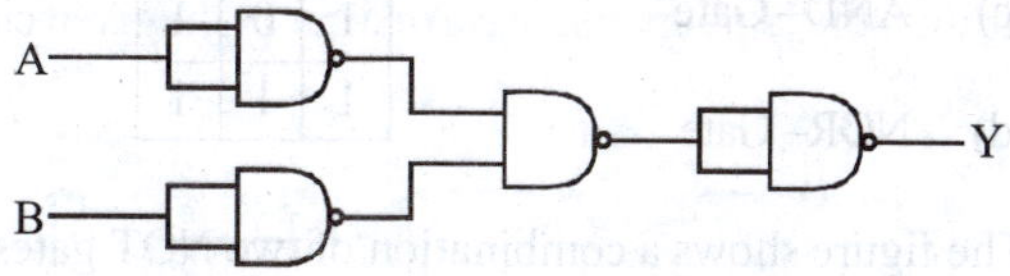

(a) OR Gate (b) NAND Gate
(c) NOR Gate (d) AND Gate

141. The truth table for the following logic circuit is

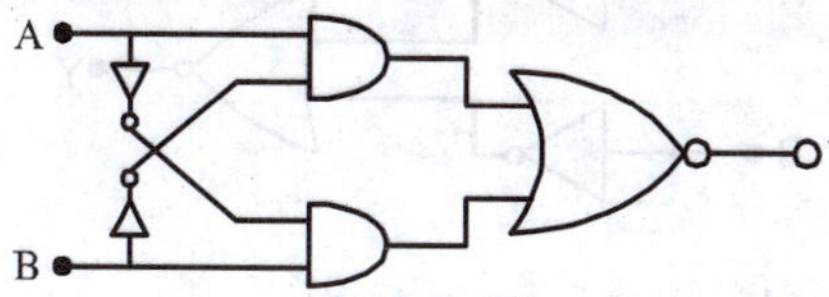

[Feb. 25, 2021 (II)]

(a)

A	B	Y
0	0	0
0	1	1
1	0	1
1	1	0

(b)

A	B	Y
0	0	1
0	1	0
1	0	1
1	1	0

(c)

A	B	Y
0	0	1
0	1	0
1	0	0
1	1	1

(d)

A	B	Y
0	0	0
0	1	1
1	0	0
1	1	1

142. Identify the operation performed by the circuit given below: **[Sep. 04, 2020 (II)]**

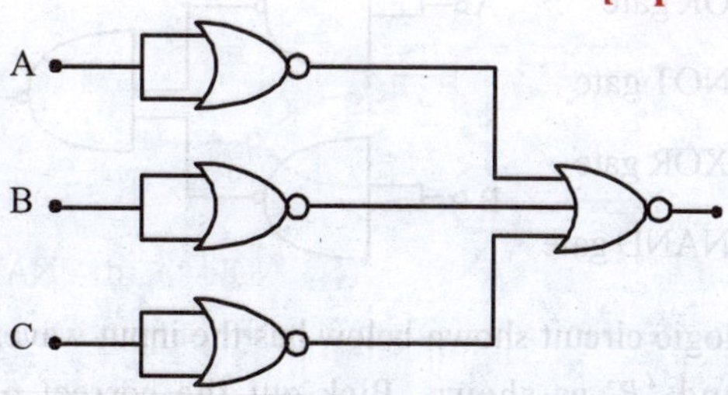

(a) NAND (b) OR (c) AND (d) NOT

143. In the following digitial circuit, what will be the output at 'Z', when the input (A, B) are (1, 0), (0, 0), (1, 1), (0, 1) : **[Sep. 02, 2020 (II)]**

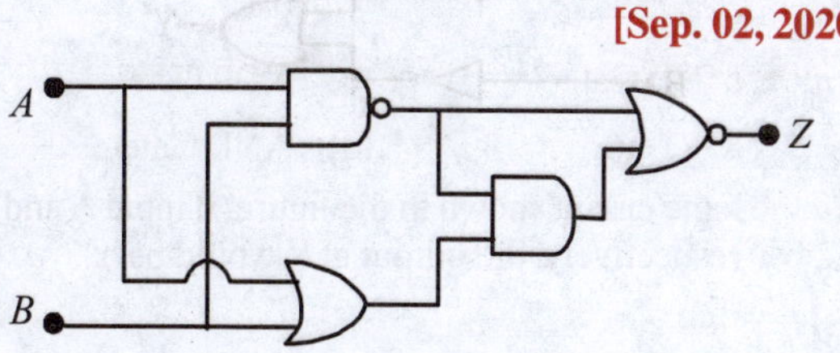

(a) 0, 0, 1, 0 (b) 1, 0, 1, 1
(c) 1, 1, 0, 1 (d) 0, 1, 0, 0

144. Boolean relation at the output stage-Y for the following circuit is: **[8 Jan. 2020 I]**

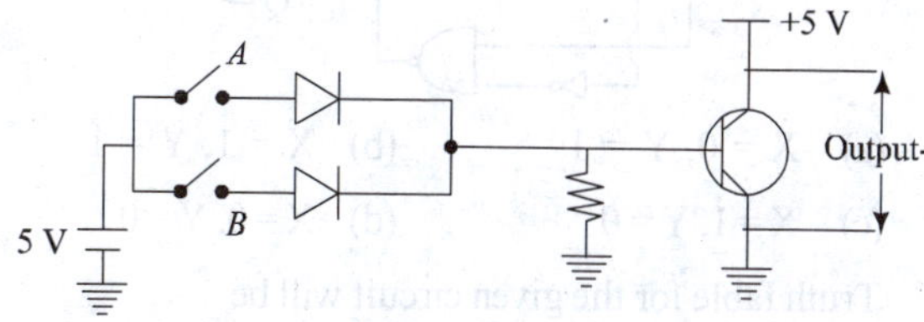

(a) $\overline{A}+\overline{B}$ (b) $A+B$
(c) $A.B$ (d) $\overline{A.B}$

145. In the given circuit, value of Y is: **[8 Jan. 2020 II]**

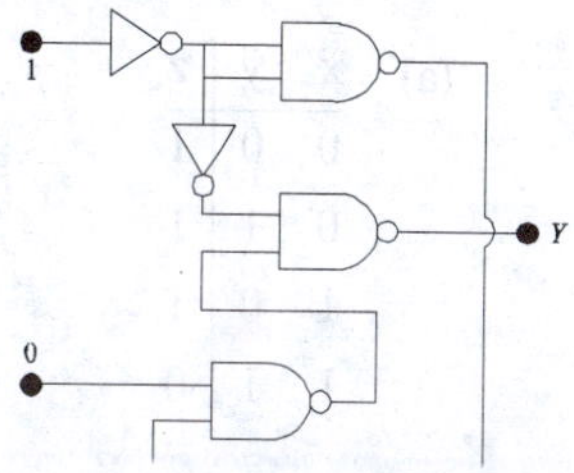

(a) 0
(b) toggles between 0 and 1
(c) will not execute
(d) 1

146. Which of the following gives a reversible operation? **[7 Jan. 2020 I]**

(a) 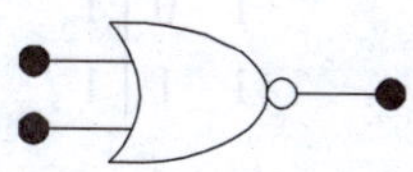 (b) 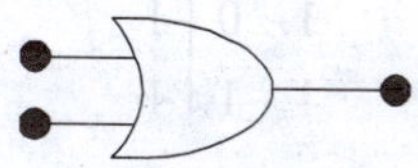

(c) 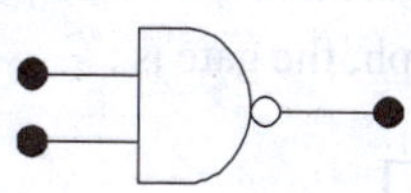 (d) 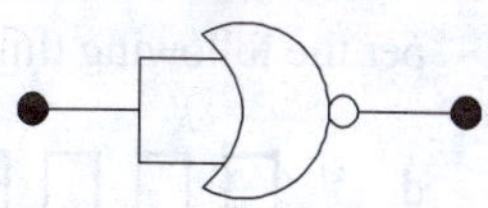

147. The logic gate equivalent to the given logic circuit is: **[9 Apr. 2019 II]**

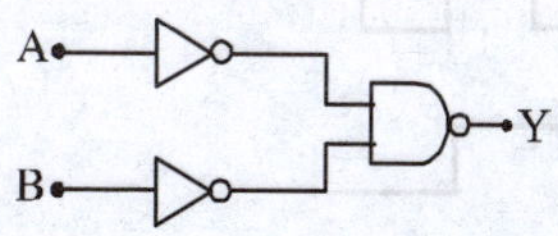

(a) NAND (b) OR
(c) NOR (d) AND

148. The ouput of the given logic cirfcuit is: **[12 Jan. 2019 I]**

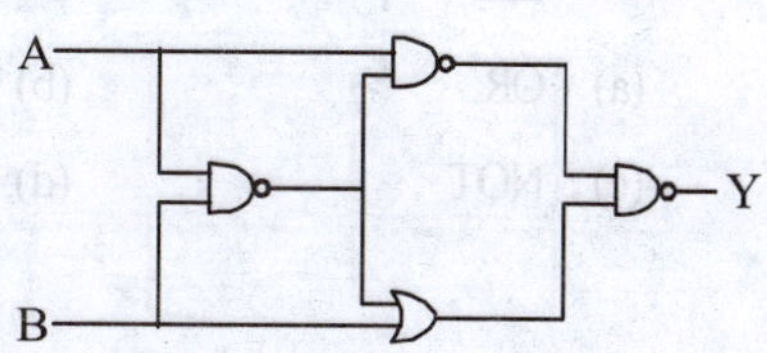

(a) $A\overline{B}+\overline{A}B$
(b) $AB+\overline{A}B$
(c) $A\overline{B}$
(d) $\overline{A}B$

149. To get output '1' at R, for the given logic gate circuit the input values must be: **[10 Jan. 2019 I]**

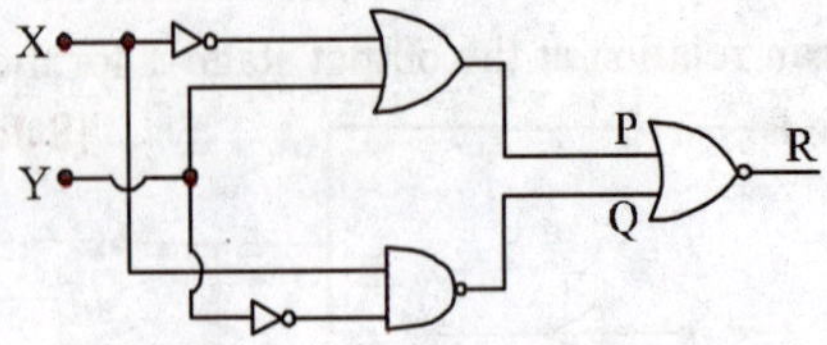

(a) X = 0, Y = 1 (b) X = 1, Y = 1
(c) X = 1, Y = 0 (d) X = 0, Y = 0

150. Truth table for the given circuit will be **[Online April 15, 2018]**

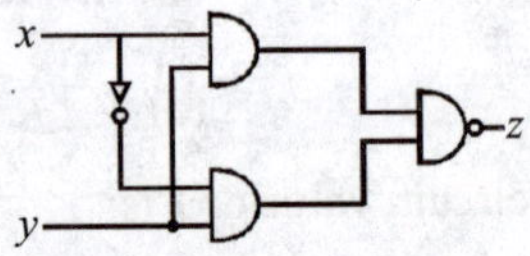

(a)

x	y	z
0	0	1
0	1	1
1	0	1
1	1	0

(b)

x	y	z
0	0	0
0	1	0
1	0	0
1	1	1

(c)

x	y	z
0	0	1
0	1	1
1	0	1
1	1	1

(d)

x	y	z
0	0	0
0	1	1
1	0	1
1	1	1

151. If a, b, c, d are inputs to a gate and x is its output, then, as per the following time graph, the gate is : **[2016]**

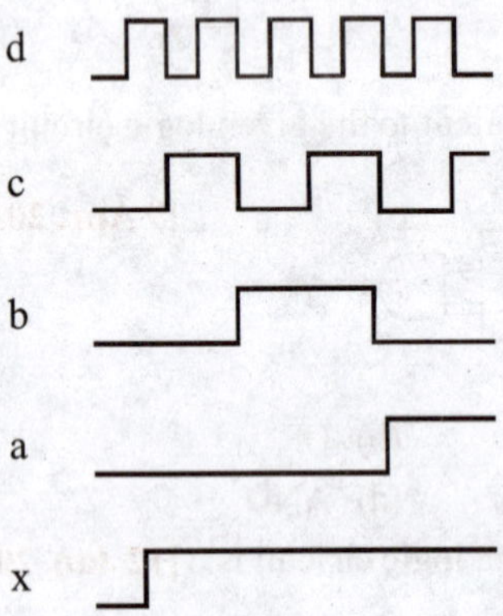

(a) OR (b) NAND
(c) NOT (d) AND

152. The truth table given in fig. represents : **[Online April 9, 2016]**

(a) OR – Gate
(b) NAND– Gate
(c) AND– Gate
(d) NOR– Gate

A	B	Y
0	0	0
0	1	1
1	0	1
1	1	1

153. The figure shows a combination of two NOT gates and a NOR gate. **[Online May 26, 2012]**

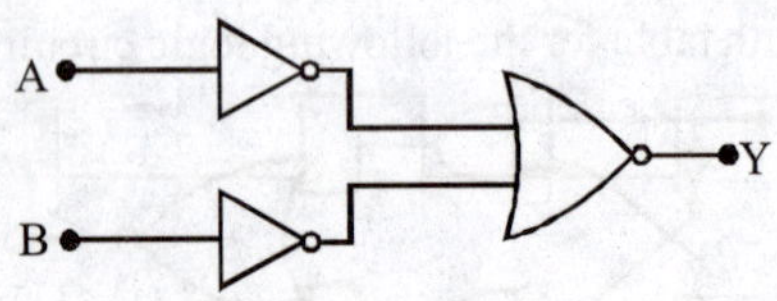

The combination is equivalent to a
(a) NAND gate (b) NOR gate
(c) AND gate (d) OR gate

154. Which one of the following is the Boolean expression for NOR gate? **[Online May 19, 2012]**

(a) $Y = \overline{A + B}$ (b) $Y = \overline{A.B}$
(c) $Y = A.B$ (d) $Y = \overline{A}$

155. The output of an OR gate is connected to both the inputs of a NAND gate. The combination will serve as a: **[2011 RS]**
(a) NOT gate (b) NOR gate
(c) AND gate (d) OR gate

156. The combination of gates shown below yields **[2010]**

(a) OR gate
(b) NOT gate
(c) XOR gate
(d) NAND gate

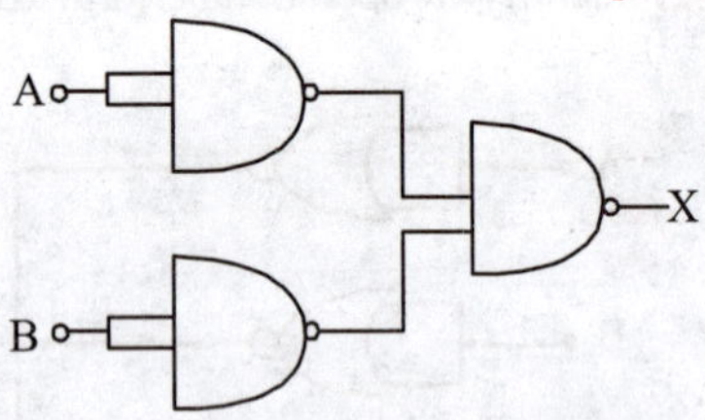

157. The logic circuit shown below has the input waveforms '*A*' and '*B*' as shown. Pick out the correct output waveform. **[2009]**

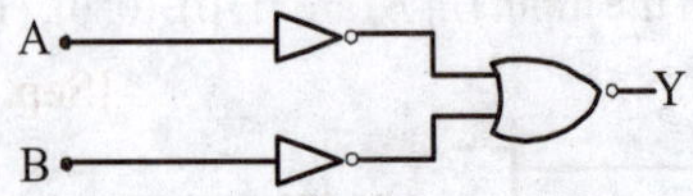

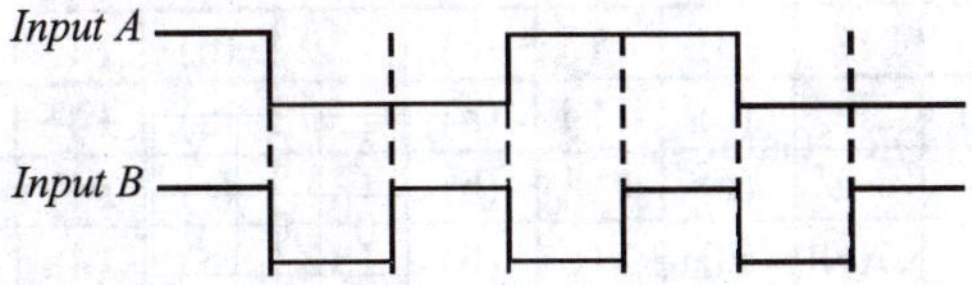

Output is

(a) 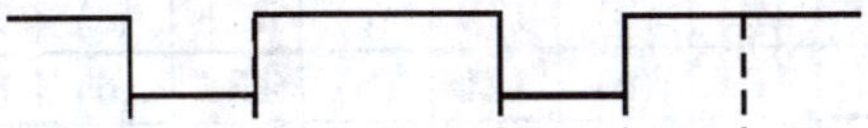

(b) 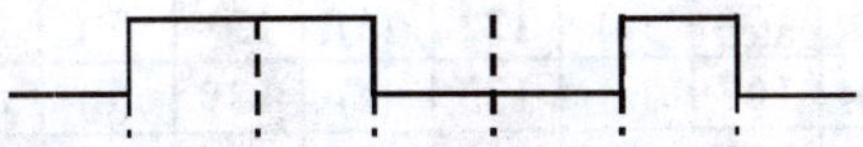

(c) 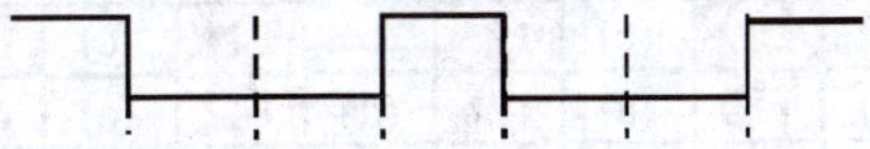

(d) 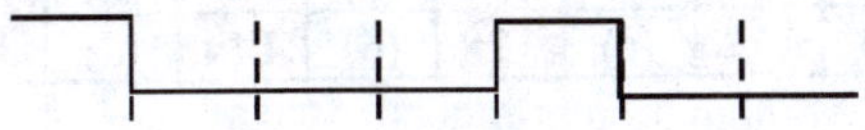

158. In the circuit below, *A* and *B* represent two inputs and *C* represents the output. **[2008]**

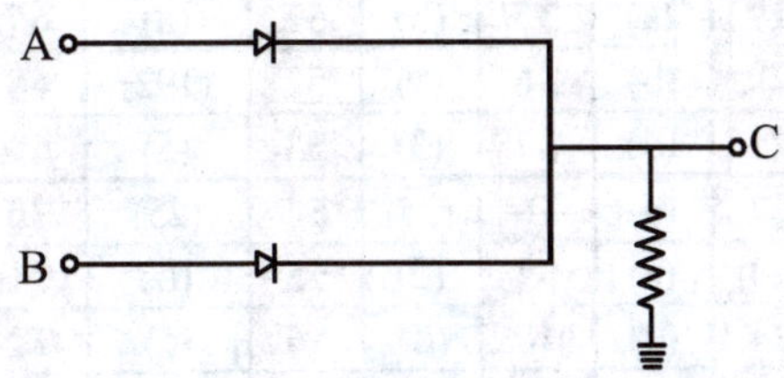

The circuit represents

(a) NOR gate
(b) AND gate
(c) NAND gate
(d) OR gate

ANSWER KEYS

1	(b)	17	(c)	33	(1)	49	(500)	65	(40)	81	(a)	97	(a)	113	(b)	129	(b)	145	(a)
2	(c)	18	(c)	34	(480)	50	(a)	66	(a)	82	(d)	98	(c)	114	(b)	130	(a)	146	(d)
3	(c)	19	(a)	35	(b)	51	(a)	67	(d)	83	(d)	99	(c)	115	(b)	131	(c)	147	(b)
4	(a)	20	(b)	36	(9)	52	(192)	68	(c)	84	(a)	100	(c)	116	(b)	132	(b)	148	(c)
5	(d)	21	(d)	37	(3)	53	(5)	69	(a)	85	(d)	101	(a)	117	(d)	133	(a)	149	(c)
6	(d)	22	(a)	38	(25)	54	(25)	70	(a)	86	(a)	102	(b)	118	(c)	134	(c)	150	(c)
7	(c)	23	(c)	39	(2)	55	(b)	71	(b)	87	(a)	103	(b)	119	(c)	135	(c)	151	(a)
8	(a)	24	(d)	40	(d)	56	(a)	72	(a)	88	(c)	104	(b)	120	(b)	136	(d)	152	(a)
9	(c)	25	(c)	41	(4)	57	(20)	73	(c)	89	(b)	105	(b)	121	(d)	137	(b)	153	(c)
10	(c)	26	(c)	42	(d)	58	(09.00)	74	(d)	90	(c)	106	(a)	122	(a)	138	(b)	154	(a)
11	(c)	27	(a)	43	(5)	59	(d)	75	(c)	91	(b)	107	(a)	123	(a)	139	(0)	155	(b)
12	(5)	28	(c)	44	(c)	60	(c)	76	(b)	92	(d)	108	(a)	124	(d)	140	(c)	156	(a)
13	(c)	29	(c)	45	(a)	61	(d)	77	(a)	93	(d)	109	(c)	125	(b)	141	(c)	157	(d)
14	(c)	30	(c)	46	(20)	62	(12)	78	(c)	94	(a)	110	(a)	126	(d)	142	(c)	158	(d)
15	(b)	31	(d)	47	(120)	63	(c)	79	(a)	95	(c)	111	(a)	127	(d)	143	(a)		
16	(c)	32	(b)	48	(5)	64	(c)	80	(d)	96	(b)	112	(c)	128	(b)	144	(d)		

Experimental Skills-XII

15

TREND ANALYSIS
JEE MAIN (2025-2021)

	2025				2024				2023				2022				2021			
	Session II		Session I		Session II		Session I		Session II		Session I		Session II		Session I		Session II		Session I	
	MCQs	NVQs	MCQs	NVQs	MCQs	NVQs	MCQs	NVQs	MCQs	NVQs	MCQs	NVQs	MCQs	NVQs	MCQs	NVQs	MCQs	NVQs	MCQs	NVQs
Topic 1	–	–	–	–	–	1	–	–	1	–	–	1	–	1	1	–	–	–	–	–
Topic 2	1	–	1	–	1	–	–	1	1	–	–	–	–	1	–	–	–	–	–	–
Topic 3																				
Topic 4																				
Total	1	–	1	–	1	1	–	1	2	–	–	1	–	2	1	–	–	–	–	–

How to Code Qns for Better & Faster Revision

Type 1: Able to solve (No code Req)

Type 2: Solved with struggle or by seeing solution **(Use code NR means Needs Revision or orange stickers)**

Type 3: Tricky/ Could not solve on my own. Need help from Teacher/ Friend. **(Use code TQ means Tricky Question or red stickers)**

You can either hand write the code in boxes or else paste coloured stickers given at end of book.

Topic 1 - Determination of resistivity of the material of a given wire using a Meter bridge To determine the resistance of a given wire using Ohm's law To determine the resistance and figure of merit of a galvanometer by half deflection method

1. In the experiment to determine the galvanometer resistance by half-deflection method, the plot of $\frac{1}{\theta}$ vs the resistance (R) of the resistance box is shown in the figure. The figure of merit of the galvanometer is .... $\times 10^{-1}$ A/division. [The source has emf 2V] **[April 5, 2024 (I)]**

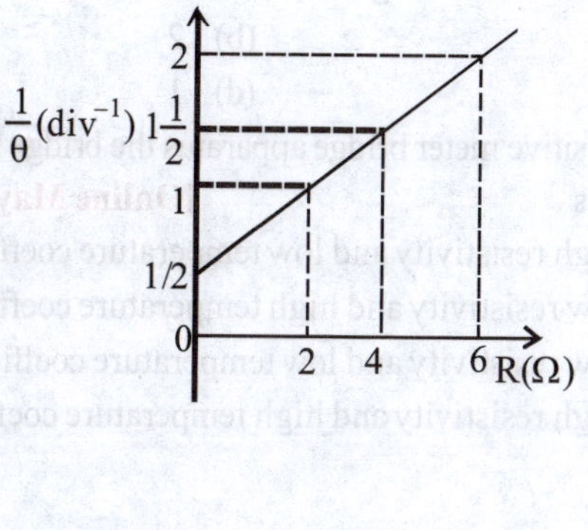

2. A student is provided with a variable voltage source V, a test resistor $R_T = 10\Omega$, two identical galvanometers G_1 and G_2 and two additional resistors, $R_1 = 10M\Omega$ and $R_2 = 0.001\Omega$. For conducting an experiment to verify ohm's, law, the most suitable circuit is: **[April 6, 2023 (II)]**

(a)
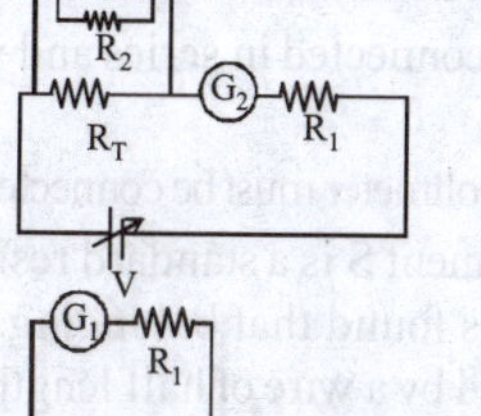

(b)
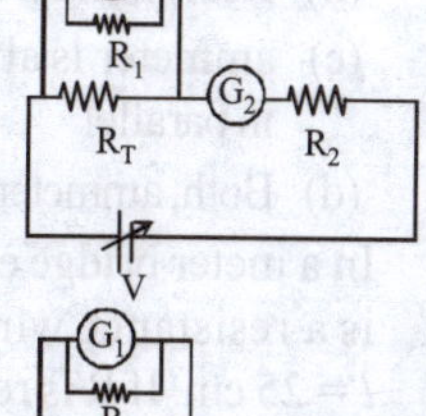

(c)
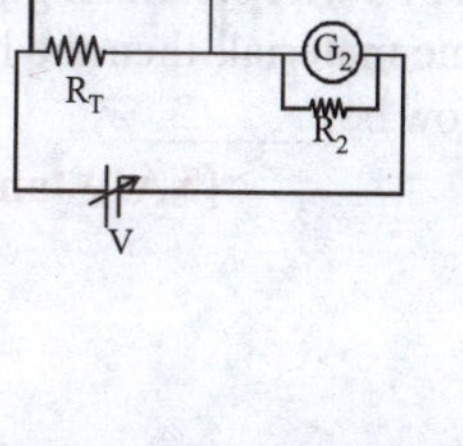

(d)
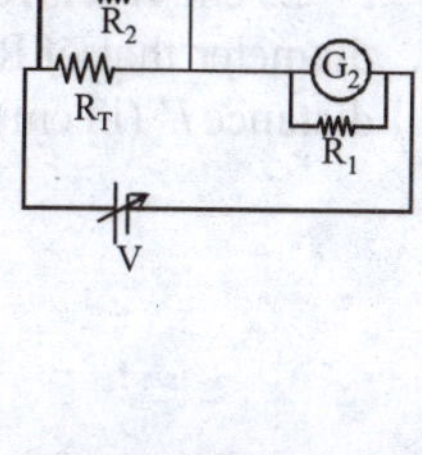

3. In a metre bridge experiment the balance point is obtained if the gaps are closed by 2Ω and 3Ω. A shunt of XΩ is added to 3Ω resistor to shift the balancing point by 22.5 cm. The value of X is ____ [Jan. 29, 2023 (I)]

4. In meter bridge experiment for measuring unknown resistance 'S', the null point is obtained at a distance 30 cm from the left side as shown at point D. If R is 5.6 k Ω, then the value of unknown resistance 'S' will be ______ Ω. [July 27, 2022 (I)]

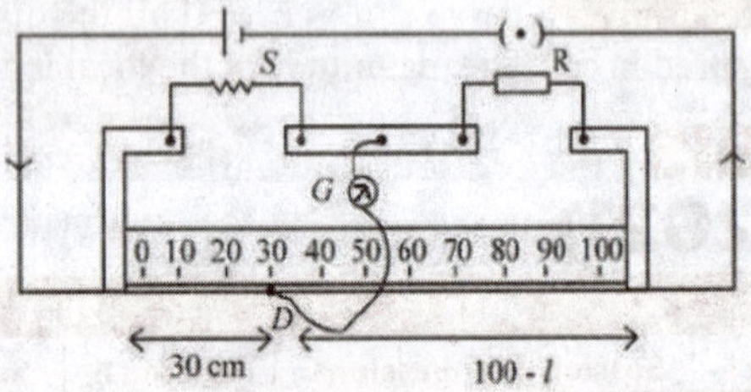

5. A teacher in his physics laboratory allotted an experiment to determine the resistance (G) of a galvanometer. Students took the observations for $\frac{1}{3}$ deflection in the galvanometer. Which of the below is **true** for measuring value of G? [June 25, 2022 (I)]

(a) $\frac{1}{3}$ deflection method cannot be used for determining the resistance of the galvanometer.

(b) $\frac{1}{3}$ deflection method can be used and in this case the G equals to twice the value of shunt resistance(s).

(c) $\frac{1}{3}$ deflection method can be used and in this case, the G equals to three times the value of shunt resistance(s)

(d) $\frac{1}{3}$ deflection method can be used and in this case the G value equals to the shunt resistance(s).

6. A circuit to verify Ohm's law uses ammeter and voltmeter in series or parallel connected correctly to the resistor. In the circuit : [Sep. 06, 2020 (II)]

(a) ammeter is always used in parallel and voltmeter is series

(b) Both ammeter and voltmeter must be connected in parallel

(c) ammeter is always connected in series and voltmeter in parallel

(d) Both, ammeter and voltmeter must be connected in series

7. In a meter bridge experiment S is a standard resistance. R is a resistance wire. It is found that balancing length is $l = 25$ cm. If R is replaced by a wire of half length and half diameter that of R of same material, then the balancing distance l' (in cm) will now be ________. [NA. 9 Jan. 2020 II]

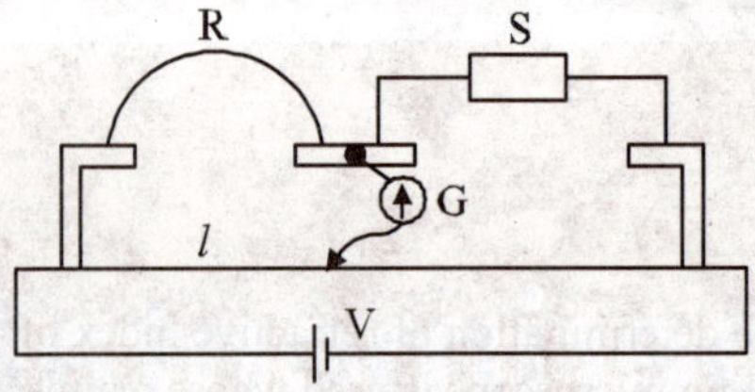

8. To verify Ohm's law, a student connects the voltmeter across the battery as, shown in the figure. The measured voltage is plotted as a function of the current, and the following graph is obtained : [12 Apr. 2019 I]

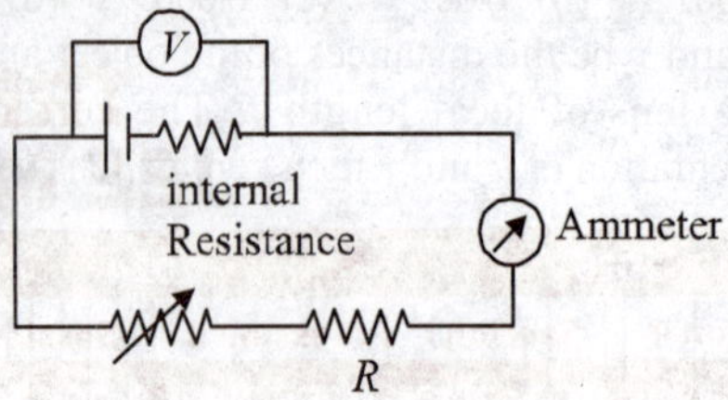

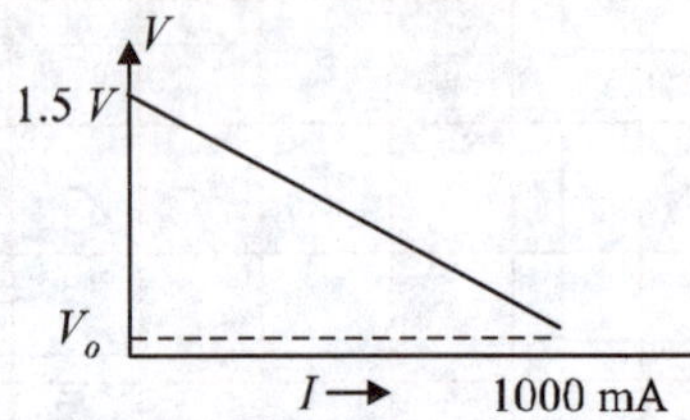

If V_0 is almost zero, identify the correct statement:

(a) The emf of the battery is 1.5 V and its internal resistance is 1.5 Ω

(b) The value of the resistance R is 1.5 Ω

(c) The potential difference across the battery is 1.5 V when it sends a current of 1000 mA

(d) The emf of the battery is 1.5 V and the value of R is 1.5 Ω

9. In a meter bridge experiment, the circuit diagram and the corresponding observation table are shown in figure.

Sl.No.	RΩ	l (cm)
1.	1000	60
2.	100	13
3.	10	1.5
4.	1	1.0

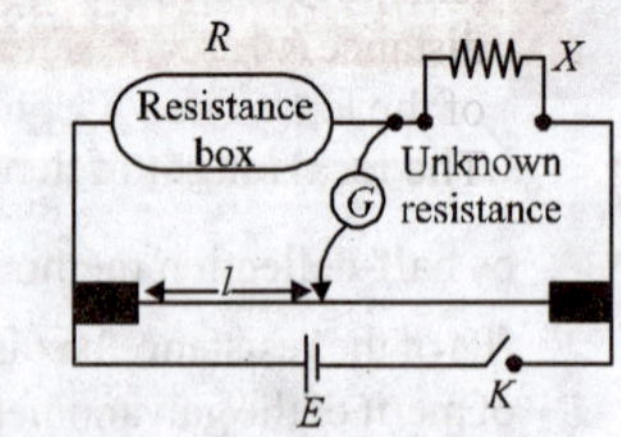

Which of the reading is inconsistent ? [10 Apr. 2019 I]

(a) 3 (b) 2
(c) 4 (d) 1

10. In a sensitive meter bridge apparatus the bridge wire should possess [Online May 12, 2012]

(a) high resistivity and low temperature coefficient.

(b) low resistivity and high temperature coefficient.

(c) low resistivity and low temperature coefficient.

(d) high resistivity and high temperature coefficient.

Topic 2 - Determination of focal length of a (i) convex mirror (ii) concave mirror and (iii) convex lens using the parallax method to plot angle of deviation versus angle of incidence for a triangular prism using parallax method to determine the refractive index of a glass slab using a travelling microscope

11. For the determination of refractive index of glass slab, a travelling microscope is used whose main scale contains 300 equal divisions equals to 15 cm. The vernier scale attached to the microscope has 25 divisions equals to 24 divisions of main scale. The least count (LC) of the travelling microscope is (in cm) : **[April 4, 2025 (II)]**

(a) 0.001 (b) 0.002 (c) 0.0005 (d) 0.0025

12. Let u and v be the distances of the object and the image from a lens of focal length f. The correct graphical representation of u and v for a convex lens when $|u| > f$, is **[Jan. 29, 2025 (I)]**

(a) 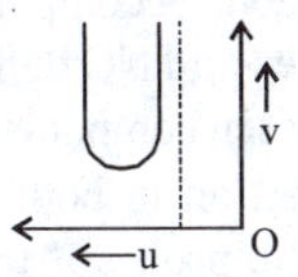

(b) 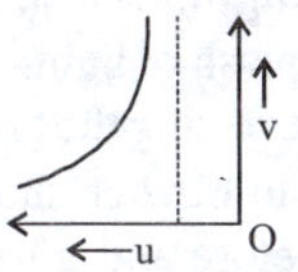

(c) 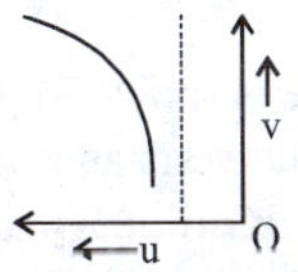

(d) 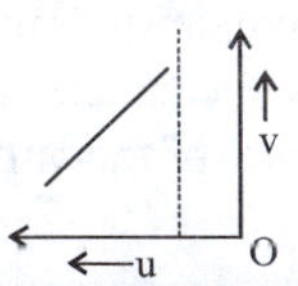

13. In an experiment to measure focal length (f) of convex lens, the least counts of the measuring scales for the position of object (u) and for the position of image (v) are Δu and Δv, respectively. The error in the measurement of the focal length of the convex lens will be : **[April 4, 2024 (I)]**

(a) $\frac{\Delta u}{u}+\frac{\Delta v}{v}$ (b) $f^2\left[\frac{\Delta u}{u^2}+\frac{\Delta v}{v^2}\right]$

(c) $2f\left[\frac{\Delta u}{u}+\frac{\Delta v}{v}\right]$ (d) $f\left[\frac{\Delta u}{u}+\frac{\Delta v}{v}\right]$

14. In an experiment to measure the focal length (f) of a convex lens, the magnitude of object distance (x) and the image distance (y) are measured with reference to the focal point of the lens. The y - x plot is shown in figure. The focal length of the lens is _____ cm. **[Jan. 30, 2024 (II)]**

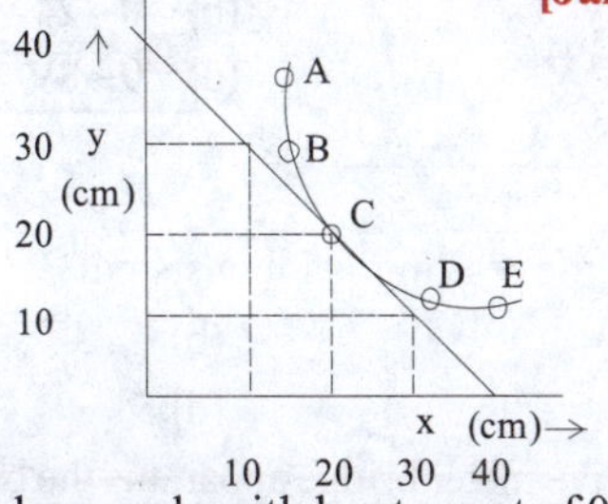

15. A 2 meter long scale with least count of 0.2 cm is used to measure the locations of objects on an optical bench. While measuring the focal length of a convex lens, the object pin and the convex lens are placed at 80 cm mark and lm mark, respectively. The image of the object pin on the other side of lens coincides with image pin that is kept at 180 cm mark. The % error in the estimation of focal length is: **[April 6, 2023 (II)]**

(a) 1.02 (b) 0.85 (c) 1.70 (d) 0.51

16. In an experiment with a convex lens, the plot of the image distance (v') against the object distance (u') measured from the focus gives a curve $v'u' = 225$. If all the distance are measured in cm. The magnitude of the focal length of the lens is _____ cm. **[July 28, 2022 (II)]**

17. In an experiment for determination of refractive index of glass of a prism by $i - \delta$ plot, it was found that a ray incident at angle 35°, suffers a deviation of 40° and that it emerges at angle 79°. In that case which of the following is closest to the maximum possible value of the refractive index? **[2016]**

(a) 1.7 (b) 1.8 (c) 1.5 (d) 1.6

18. The graph between angle of deviation (δ) and angle of incidence (i) for a triangular prism is represented by **[2013]**

(a) 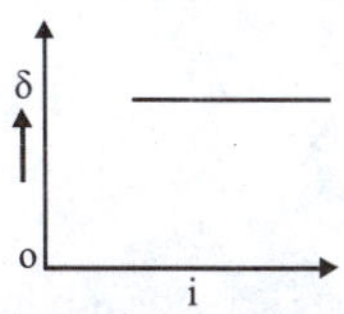

(b) 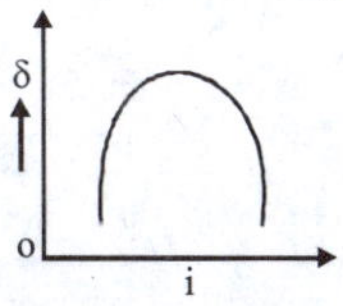

(c) 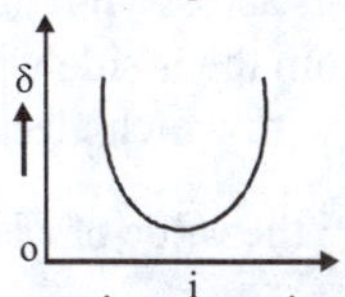

(d) 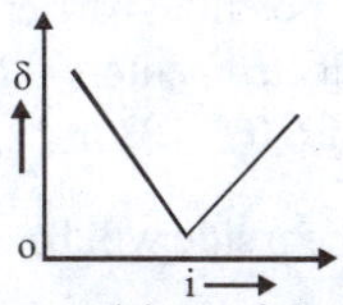

19. In an optics experiment, with the position of the object fixed, a student varies the position of a convex lens and for each position, the screen is adjusted to get a clear image of the object. A graph between the object distance u and the image distance v, from the lens, is plotted using the same scale for the two axes. A straight line passing through the origin and making an angle of 45° with the x-axis meets the experimental curve at P. The coordinates of P will be **[2009]**

(a) $\left(\frac{f}{2},\frac{f}{2}\right)$ (b) (f, f) (c) $(4f, 4f)$ (d) $(2f, 2f)$

20. A student measures the focal length of a convex lens by putting an object pin at a distance 'u' from the lens and measuring the distance 'v' of the image pin. The graph between 'u' and 'v' plotted by the student should look like **[2008]**

(a) 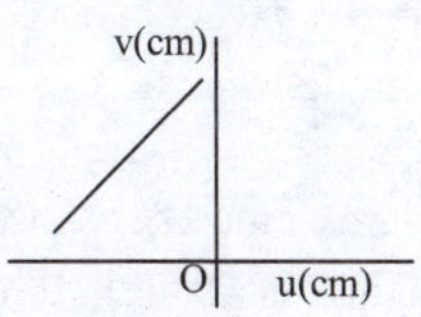

(b) 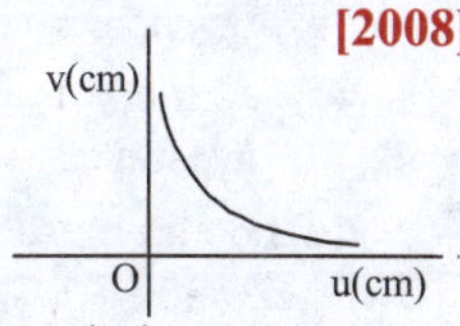

(c) 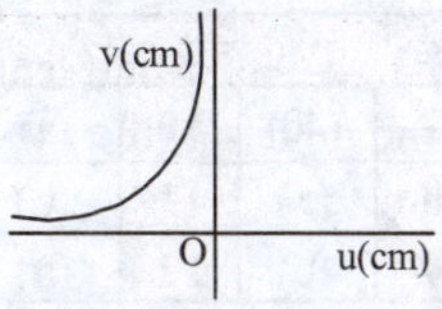

(d) 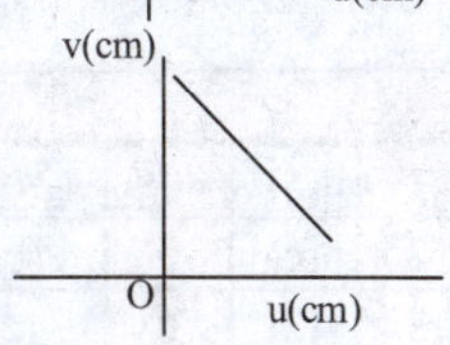

Topic 3 - To draw the characteristic curves of a p-n junction diode in forward and reverse bias. To draw the characteristic curves of a Zener diode and finding reverse break down voltage. Identification of diode, LED, resistor, capacitor from mixed collection of such items

21. The circuit diagram used to study the characteristic curve of a zener diode is connected to variable power supply (0–15V) as shown in figure. A zener diode with maximum potential $V_z = 10$ V and maximum power dissipation of 0.4 W is connected across a potential divider arrangement. The value of resistance R_P connected in series with the zener diode to protect it from the damage is ____ Ω **[June 30, 2022 (I)]**

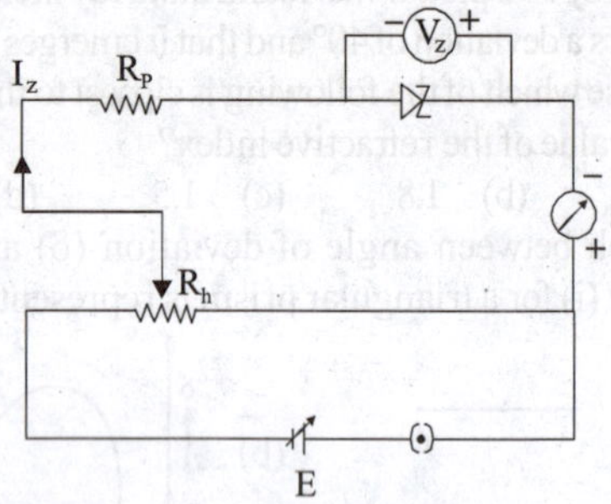

22. A potential barrier of 0.4 V exists across a p-n junction. An electron enters the junction from the n-side with a speed of 6.0×10^5 ms^{-1}. The speed with which electron enters the p side will be $\frac{x}{3} \times 10^5$ ms^{-1} the value of x is ____.
(Given mass of electron $= 9 \times 10^{-31}$ kg, charge on electron $= 1.6 \times 10^{-19}$ C.) **[June 29, 2022 (II)]**

23. The I-V characteristics of a p-n junction diode in forward bias is shown in the figure. The ratio of dynamic resistance, corresponding to forward bias voltages of 2V and 4V respectively, is : **[June 26, 2022 (I)]**

(a) 1 : 2
(b) 5 : 1
(c) 1 : 40
(d) 20 : 1

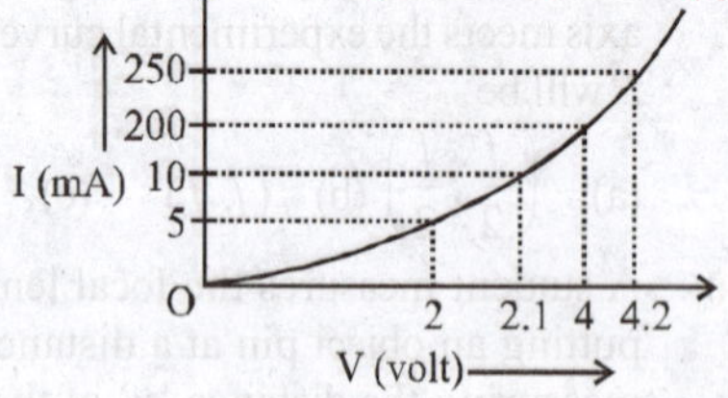

24. Which of the following will NOT be observed when a multimeter (operating in resistance measuring mode) probes connected across a component, are just reversed? **[Sep. 03, 2020 (II)]**

(a) Multimeter shows an equal deflection in both cases i.e. before and after reversing the probes if the chosen component is resistor.
(b) Multimeter shows NO deflection in both cases i.e. before and after reversing the probes if the chosen component is capacitor.
(c) Multimeter shows a deflection, accompanied by a splash of light out of connected and NO deflection on reversing the probes if the chosen component is LED.
(d) Multimeter shows NO deflection in both cases i.e. before and after reversing the probes if the chosen component is metal wire.

25. The V–I characteristic of a diode is shown in the figure. The ratio of forward to reverse bias resistance is : **[Online April 8, 2017]**

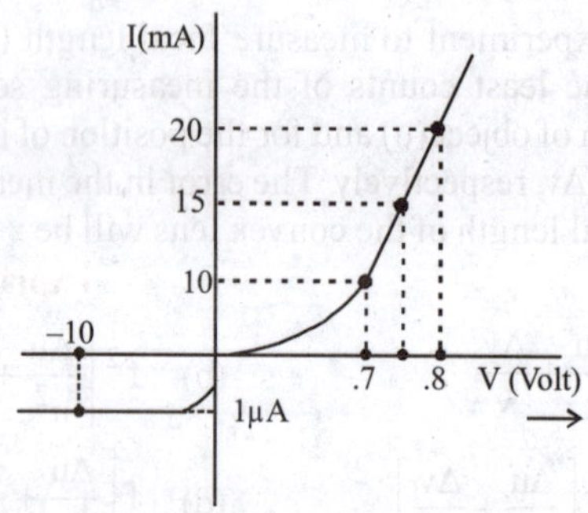

(a) 10 (b) 10^{-6} (c) 10^6 (d) 100

26. An experiment is performed to determine the I-V characteristics of a Zener diode, which has a protective resistance of R = 100 Ω, and a maximum power of dissipation rating of 1W. The minimum voltage range of the DC source in the circuit is : **[Online April 9, 2016]**

(a) 0 – 5V (b) 0 – 24 V
(c) 0 – 12 V (d) 0 – 8V

ANSWER KEYS

1	(5)	4	(2400)	7	(40)	10	(a)	13	(b)	16	(15)	19	(d)	22	(14)	25	(b)		
2	(c)	5	(b)	8	(a)	11	(b)	14	(20)	17	(c)	20	(c)	23	(b)	26	(b)		
3	(2)	6	(c)	9	(c)	12	(b)	15	(c)	18	(c)	21	(125)	24	(b)				

HINTS & SOLUTIONS

1 Electric Charges and Fields

1. **(a)** Charge accumulated maximum at sharp end.

So, $\sigma_1 > \sigma_3 > \sigma_2 = \sigma_4$.

2. **(d)** At equilibrium

$qE = mg$

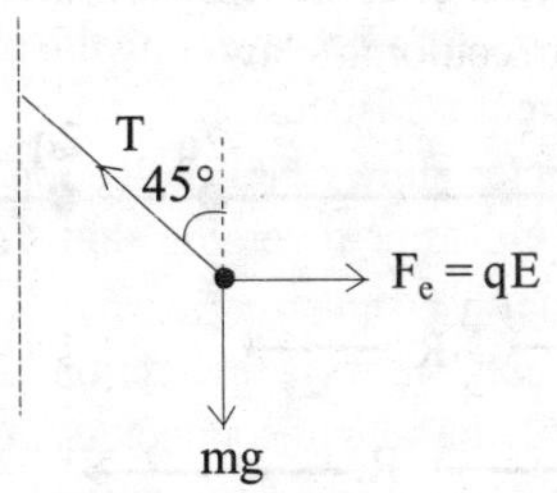

$$q\left[\frac{\sigma}{2\varepsilon_0}\right] = mg$$

$$\sigma = \frac{2\varepsilon_0 mg}{q} = \frac{2\times 8.85\times 10^{-12}\times 100\times 10^{-3}\times 10}{10\times 10^{-6}}$$

$\therefore \sigma = 1.77\ nC/m^2$

3. **(a)**

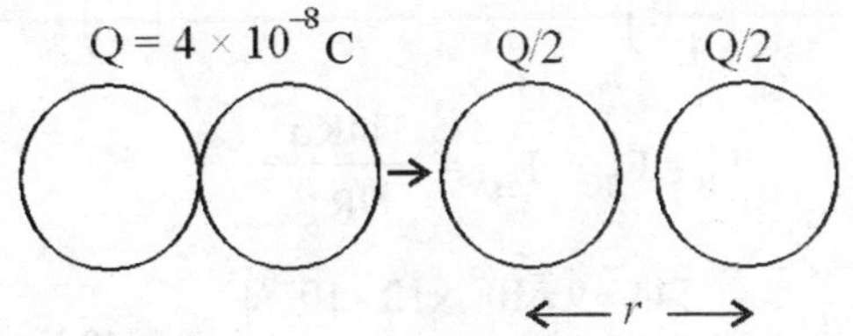

$$F = \frac{k\left(\frac{Q}{2}\right)\left(\frac{Q}{2}\right)}{r^2}$$

$$9\times 10^{-3} = \frac{9\times 10^9\times (4\times 10^{-8})\times (4\times 10^{-8})}{4\times r^2}$$

$\therefore r = 2\times 10^{-2} m = 2$ cm

4. **(5)** The expression for electrostatic force

$$F_e = \frac{kq_1q_2}{r_2}$$

The expression for gravitational force $F_g = \dfrac{Gm_1m_2}{r^2}$

$$\frac{Fe}{F_g} = \frac{9\times 10^9\times 6.67\times 10^{-19}\times 9.6\times 10^{-10}}{6.67\times 10^{-11}\times 19.2\times 10^{-27}\times 9\times 10^{-27}} = 5\times 10^{44}$$

5. **(b)** Using Coulomb's law,

$$F = K\frac{Q^2}{R^2}$$

$F_{PS} \propto Q^2$

$F_{PS} = 16$ N

If P & R are brought in contact then charges will equally divide

$$Q_P = Q_R = \frac{Q}{2}$$

Now, If S and R are brought in contact then

$$Q_S = \frac{3Q}{4}, Q_R = \frac{3Q}{4}$$

Force between P & S is,

$$F_{PS} \propto \frac{Q}{2}\times\frac{3Q}{4} = \frac{3}{8}\times 16 = 6\ N$$

6. **(3)**

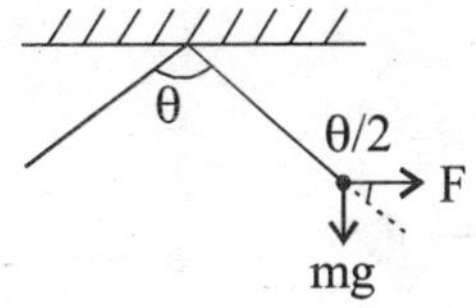

If spheres are in air,

$$\tan\frac{\theta}{2} = \frac{F}{mg} = \frac{q^2}{4\pi\epsilon_0 r^2 mg}$$

If spheres are in water,

$$\tan\frac{\theta}{2} = \frac{F'}{mg'} = \frac{q^2}{4\pi k\epsilon_0 r^2 mg'}$$

Since angle remains same.

$$\Rightarrow \frac{q^2}{4\pi\epsilon_0 r^2 mg} = \frac{q^2}{4\pi k\epsilon_0 r^2 mg'}$$

$$\Rightarrow g = kg' = g\left(1 - \frac{1}{1.5}\right)k$$

$\therefore k = 3$

7. **(3)**

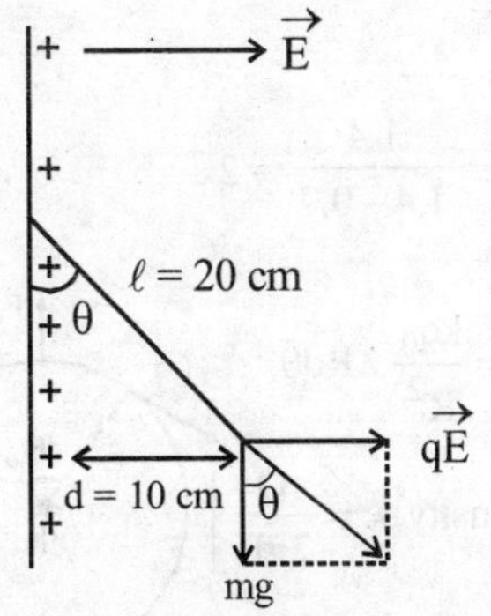

Here, $\sin\theta = \dfrac{d}{\ell} = \dfrac{10}{20} = \dfrac{1}{2}$ $\quad\therefore \theta = 30°$

$$\tan\theta = \frac{F_e}{F_g} = \frac{qE}{mg}$$

Substituting values,

$$\tan 30° = \frac{q \times 2 \times 10^4}{2 \times 10^{-3} \times 10} \Rightarrow q = \frac{1}{\sqrt{3}}\mu C = \frac{1}{\sqrt{x}}\mu C \text{ (given)}$$

$\therefore x = 3$

8. **(b)** Electrostatic force acting between two point charges in vacuum

$$F = \frac{1}{4\pi \epsilon_0} \frac{q_1 q_2}{r^2}$$

Electrostatic force acting between two point charges in a medium of dielectric constant K is

$$F' = \frac{1}{4\pi(K \epsilon_0)} \frac{q_1 q_2}{(r')^2} = \frac{25}{4\pi(5\epsilon_0)} \frac{q_1 q_2}{(r)^2} \Rightarrow F' = 5F$$

9. **(2)** Density of material of the sphere, $\rho_B = 1.4 \text{ g/cm}^3$

Density of liquid, $\rho_L = 0.7 \text{ g/cm}^3$

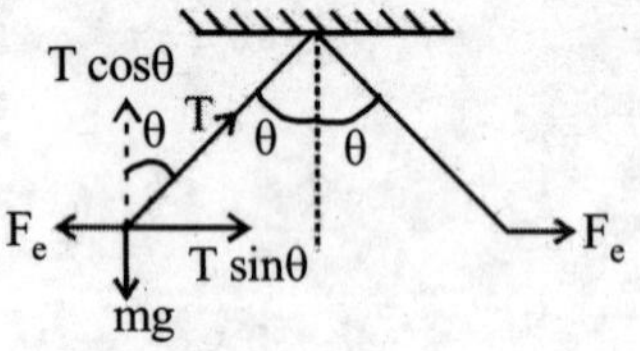

From the free body diagram

$T \cos \theta = mg$

$T \sin \theta = F_e$

$$\tan \theta = \frac{F_e}{mg}$$

If spheres are in air,

$$\tan \theta = \frac{F_e}{\rho_B V g} \quad \text{...(i)}$$

If spheres are in liquid,

$$\tan \theta = \frac{F_e}{k(\rho_B - \rho_L) V g} \quad \text{...(ii)}$$

From Eq. (i) & (ii),

$\rho_B V g = (\rho_B - \rho_L) k V g$

$\rho_B = (\rho_B - \rho_L) k$

$$\text{Dielectric constant, } k = \frac{1.4}{1.4 - 0.7} = 2$$

10. **(3)** Here, $2T \sin \frac{d\theta}{2} = \frac{kq_0}{R^2} . \lambda R d\theta$

$$\left[\because \text{Linear charge density, } \lambda = \frac{Q}{2\pi R}\right]$$

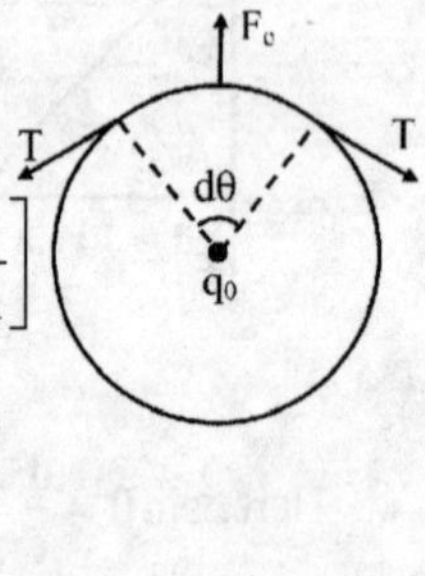

$$\Rightarrow T = \frac{Kq_0 Q}{(R^2) \times 2\pi}$$

$$= \frac{(9 \times 10^9)(2\pi \times 30 \times 10^{-12})}{(0.30)^2 \times 2\pi} = \frac{9 \times 10^{-3} \times 30}{9 \times 10^{-2}} = 3N$$

11. **(b)** Let charge $q = 10\ \mu C$ is divided into two parts x and $q - x$.

From coulomb's law, $F = k\frac{q_1 q_2}{r^2}$

$$F = \frac{Kx(q - x)}{r^2}$$

for F to be maximum

$$\frac{dF}{dx} = \frac{K}{r^2}(q - 2x) = 0 \qquad \therefore x = \frac{q}{2}$$

Hence charge of the two parts are 5 μC and 5μC

12. **(5440)** From coulomb's law

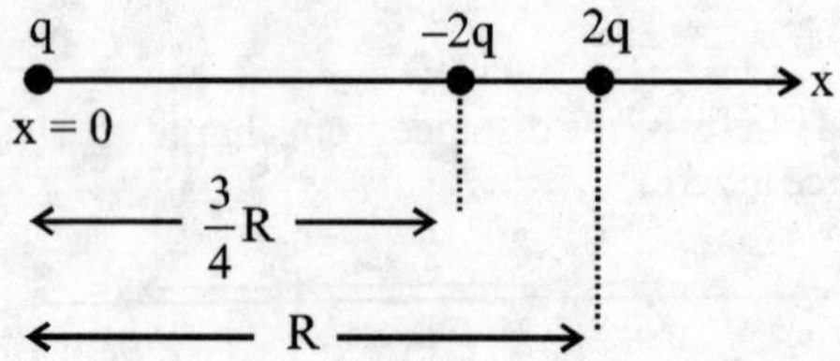

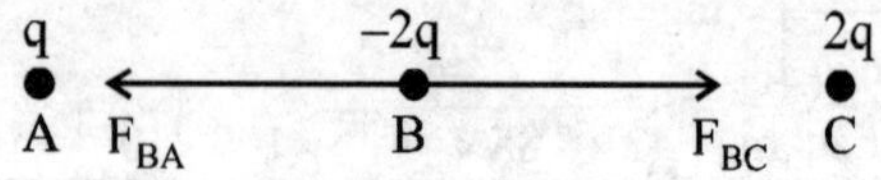

$$F_{BA} = \frac{Kq(2q)}{\left(\frac{3}{4}R\right)^2} = \frac{32Kq^2}{9R^2} \text{ and } F_{BC} = \frac{K(2q)(2q)}{\left(\frac{R}{4}\right)^2} = \frac{64Kq^2}{R^2}$$

$$F_B = F_{BC} - F_{BA} = \frac{544Kq^2}{9R^2}$$

$$= \frac{544 \times 9 \times 10^9 \times (2 \times 10^{-6})^2}{9 \times (2 \times 10^{-2})^2} = 5440 \text{ N}$$

13. **(300)** Given, Mass of charge, $m = 20 \text{ g} = 0.02 \text{ kg}$

Charge, $q_0 = +2\mu C$

For equilibrium

$mg \sin\theta = F_e$

$$\Rightarrow mg \sin \theta = \frac{1}{4\pi \epsilon_0} \times \frac{q_0^2}{(h \operatorname{cosec} 30°)^2}$$

$$\therefore h^2 = \frac{1}{4\pi \epsilon_0} \times \frac{q_0^2}{mg \operatorname{cosec} 30°}$$

$$= 9 \times 10^9 \times \frac{(2 \times 10^{-6})^2}{0.02 \times 10 \times 2}$$

$$\Rightarrow h = 3 \times 10^4 \times \frac{2 \times 10^{-6}}{0.2} = 0.3\text{m} = 300 \text{ mm}$$

14. **(2)** Force due to both charge at point P

$F' = 2F \cos \theta$

$$\text{Here, } F = \frac{Kqq_0}{a^2 + x^2}; \; F' = \frac{2Kqq_0}{a^2 + x^2} \frac{x}{(a^2 + x^2)^{1/2}}$$

$$F' = \frac{2Kqq_0 x}{(a^2 + x^2)^{3/2}}$$

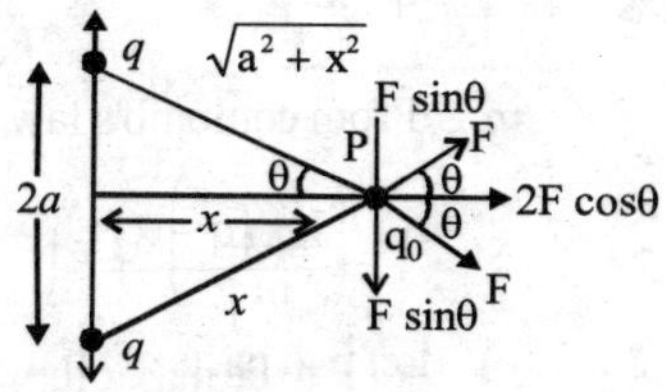

For F′ to be maximum

$$\frac{dF'}{dx} = 0$$

$$2Kqq_0\left[\frac{\frac{3x}{2}(x^2 + a^2)^{1/2} \times 2x - (x^2 + a^2)^{3/2}}{(x^2 + a^2)^3}\right] = 0$$

$$\therefore\ x = \frac{a}{\sqrt{2}}$$

15. **(24)** Let x be the distance from charge $-q_0$ where electric force is zero.

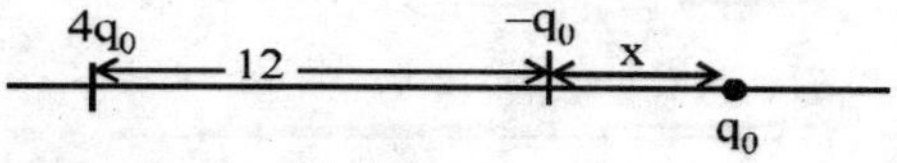

Using the Coulomb's law

$$\frac{1}{4\pi E_0}\cdot\frac{q_0^2}{x^2} = \frac{1}{4\pi E_0}\cdot\frac{4q_0^2}{(x+12)^2}$$

$\Rightarrow\ x + 12 = 2x$

$x = 12$

Distance from origin $= x + 12 = 24$ cm.

16. **(a)** As, $F_{air} = F_{med}$

$$\Rightarrow \frac{1}{4\pi\epsilon_0}\frac{q_1 q_2}{d_{air}^2} = \frac{1}{K}\frac{1}{4\pi\epsilon_0}\frac{q_1 q_2}{d^2} \Rightarrow d_{air} = d\sqrt{K}$$

17. **(b)** Consider two metallic spheres A and B both have charge q.

The force between two sphere will be

$$F = \frac{kq^2}{r^2}$$

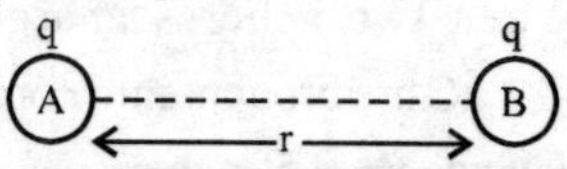

When sphere C is placed in contact with sphere A, charge on sphere A and sphere C will be $= \frac{q}{2}$

Now sphere C is placed in contact with sphere B, charge on sphere B and sphere C will be $= \frac{q + \frac{q}{2}}{2} = \frac{3q}{4}$

Now,

$$F_1 = \frac{k\frac{q}{2}\cdot\frac{3q}{4}}{\frac{r^2}{4}}$$

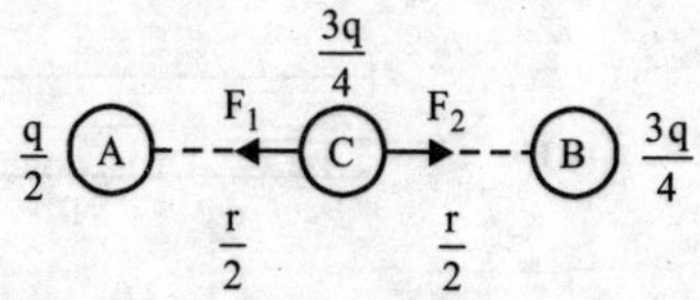

$$F_2 = \frac{k\frac{3q}{4}\cdot\frac{3q}{4}}{\frac{r^2}{4}}$$

The force experienced by sphere C,

$$F' = F_1 - F_2 = \frac{\left(k\frac{q}{2} - k\frac{3q}{4}\right)}{\frac{r^2}{4}}\cdot\frac{3q}{4} = \frac{3kq^2}{4r^2} = \frac{3F}{4}$$

18. **(a)** Force on charge 'Q'

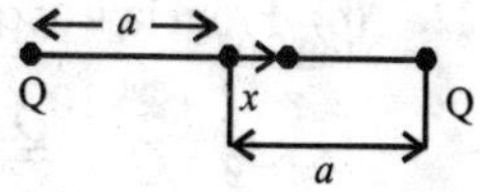

$$F = \frac{KQq_0}{(a-x)^2} - \frac{KQq_0}{(a+x)^2}$$

$$= KQq_0\left(\frac{4ax}{(a^2 - x^2)^2}\right) \text{ as } a << x$$

$$\text{So, } F = \frac{KQq_0 \times 4ax}{a^4} = \frac{4Kxq_0 Q}{a^3}$$

As, $F = m\omega^2 x$

$$\text{So, } m\omega^2 = \frac{4KQq_0}{a^3}$$

$$\omega = \sqrt{\frac{4KQq_0}{ma^3}} \Rightarrow T = \sqrt{\frac{4\pi^3\epsilon_0 ma^3}{q_0 Q}}$$

19. **(b)** Let distance between the two divided charges be r. From Coulomb's law, force between two charge,

$$F = \frac{Kq(4-q)}{r^2}$$

For F to be maximum,

$$\frac{dF}{dq} = \frac{K}{r^2}[4 - 2q] = 0 \Rightarrow q = 2\,\mu C$$

So, other charge will $(4-2)\mu C$ i.e., $2\,\mu C$

20. **(d)** We have, from the given figure

$$F_{net} = 2F\cos\theta$$

$$F_{net} = 2\frac{KQq}{x^2 + \frac{d^2}{4}}\cdot\frac{x}{\left(x^2 + \frac{d^2}{4}\right)^{1/2}}$$

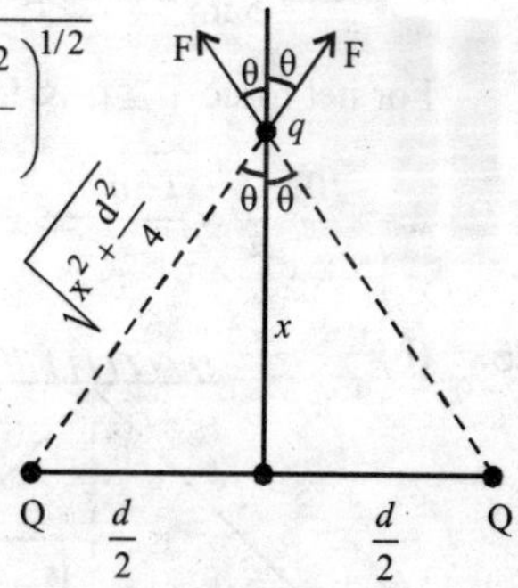

$$F_{net} = \frac{2KQqx}{\left(x^2 + \frac{d^2}{4}\right)^{3/2}}$$

For maximum F_{net}

$$\frac{dF_{net}}{dx} = 0$$

$$\Rightarrow x \times -\frac{3}{2}\left(x^2 + \frac{d^2}{4}\right)^{-5/2}.2x + \left(x^2 + \frac{d^2}{4}\right)^{-3/2} = 0$$

$$\Rightarrow \left(x^2 + \frac{d^2}{4}\right)^{-5/2}\left[-3x^2 + x^2 + \frac{d^2}{4}\right] = 0$$

$$\Rightarrow 2x^2 = \frac{d^2}{4} \Rightarrow x^2 = \frac{d^2}{8} \Rightarrow x = \frac{d}{2\sqrt{2}}$$

21. **(d)** Let force between two charge be F. Then

$$F_{net} = \sqrt{F^2 + F^2 + 2.F.F\cos 60°}$$

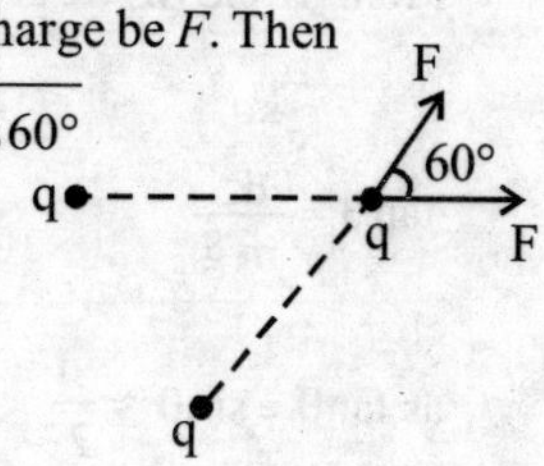

$= \sqrt{3}F$

so, required ratio

$$= \frac{F_{net}}{F} = \frac{\sqrt{3}F}{F} = \sqrt{3}$$

22. (b) From law of conservation of charge $q_i = q_f$

$$\Rightarrow 64q = Q \Rightarrow \frac{Q}{q} = 64 \text{ and, } \frac{4}{3}\pi R^3 = 64 \times \frac{4}{3}\pi r^3$$

$$\Rightarrow R = 4r \Rightarrow \frac{r}{R} = \frac{1}{4}$$

$$\text{So, } \frac{\sigma_{bigger}}{\sigma_{smaller}} = \frac{\frac{Q}{4\pi R^2}}{\frac{q}{4\pi r^2}} = \frac{Q}{q} \times \left(\frac{r}{R}\right)^2 = 64 \times \frac{1}{16} = \frac{4}{1}$$

23. (a) 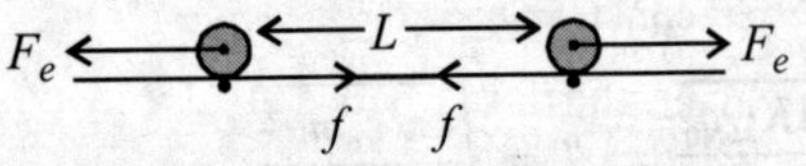

For charge to stay in equilibrium $F_e = f$

$$\frac{kq^2}{L^2} = \mu mg$$

$$L = \sqrt{\frac{kq^2}{\mu mg}} = \sqrt{\frac{k}{\mu mg}} . q = \sqrt{\frac{9 \times 10^9}{0.25 \times 10^{-2} \times 10}} \times 2 \times 10^{-7}$$

$= 12$ cm

24. (b) Let, charge q be placed at P.
At point P forces due to 20 μC & $-5\mu C$ should be in opposite direction

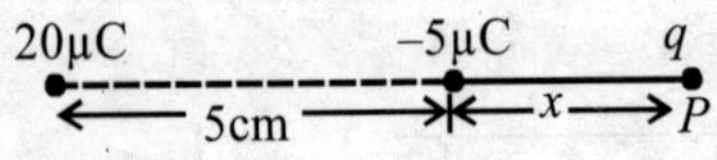

For net force $\vec{F} = 0$ & from coulomb's law force

$$k\frac{20q}{(5+x)^2} = \frac{k5q}{x^2} \Rightarrow x = 5 \text{ cm}$$

25. (b)

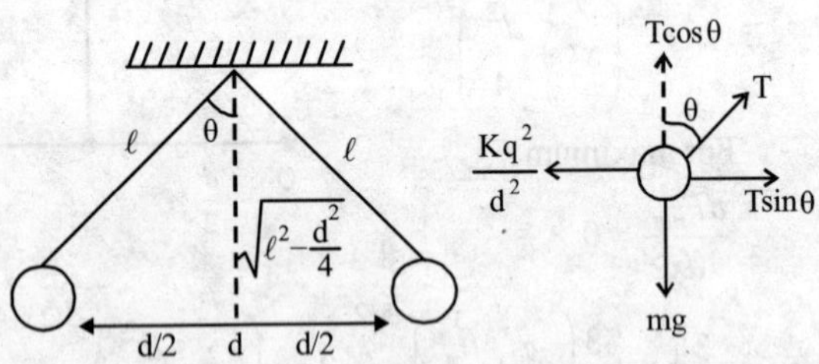

$T\cos\theta = mg$ .... (i)

Force due to charges $= \dfrac{kq^2}{d^2}$

$T\sin\theta = \dfrac{kq^2}{d^2}$ .... (ii)

From (i) and (ii) we get

$$\tan\theta = \frac{\frac{kq^2}{d^2}}{mg}$$

$$\text{as } \tan\theta \approx \sin\theta \approx \frac{d}{2\ell}$$

$$\frac{kq^2}{mgd^2} = \frac{d}{2\ell} \Rightarrow d^3 = \frac{2kq^2\ell}{mg}$$

$$\Rightarrow d = \left(\frac{2kq^2\ell}{mg}\right)^{1/3} = \left(\frac{q^2\ell}{2\pi\varepsilon_0 mg}\right)^{1/3}$$

26. (b) Let two parts of charge Q be divided into q and Q − q

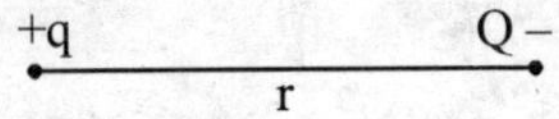

$$F = \frac{Kq_1q_2}{r^2} = \frac{K(q)(Q-q)}{r^2}$$

For maximum repulsion

$$\frac{dF}{dq} = 0 \text{ or, } \frac{d}{dq}\left[\frac{Kq(Q-q)}{r^2}\right] = 0 \Rightarrow Q - 2q = 0 \Rightarrow Q = 2q$$

27. (12) 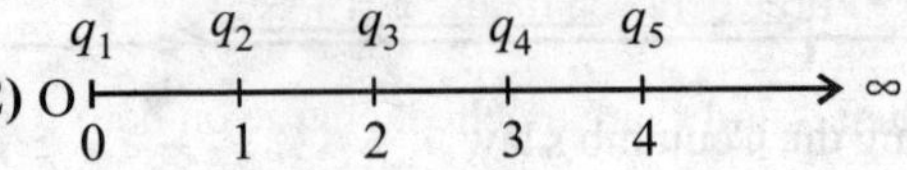

$$F_{\text{total}} = \left[\frac{kq_1q_2}{r_1^2} + \frac{kq_1q_3}{r_2^2} + \frac{kq_1q_4}{r_3^2} + ...\right]$$

$$\Rightarrow \quad F = (k)(10^{-6})\left[\frac{1}{1^2} + \frac{1}{2^2} + \frac{1}{4^2} + ...\right]$$

$$= \frac{(k)10^{-6}}{1 - \frac{1}{4}} = \frac{(9 \times 10^9) \times 4 \times 10^{-6} N}{3} = 12 \times 10^3 N$$

$\therefore \quad x = 12$

28. (36) When two spheres charges $q'_1 = 2.1$ nC and $q'_2 = -0.1$ nC are brought into contact and then separated by a distance $r = 0.5$ m then,

$$q'_1 = q'_2 = \frac{Q_1 + Q_2}{2} = 1 \text{ nC}$$

Electrostatic force between the two charged sphere,

$$F = \frac{1}{4\pi\varepsilon_0} \cdot \frac{q'_1 \, q'_2}{r^2} = 9 \times 10^9 \times \frac{10^{-9} \times 10^{-9}}{(0.5)^2} = 36 \times 10^{-9} \text{ N}$$

29. (a) 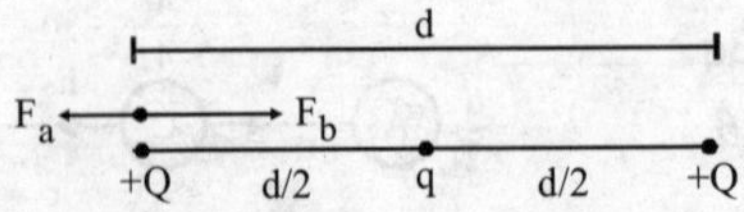

Force due to charge + Q, $F_a = \dfrac{KQQ}{d^2}$

Force due to charge q, $F_b = \dfrac{KQq}{\left(\frac{d}{2}\right)^2}$

For equilibrium, $\vec{F}_a + \vec{F}_b = 0$

$$\Rightarrow \frac{kQQ}{d^2} + \frac{kQq}{(d/2)^2} = 0 \qquad \therefore q = -\frac{Q}{4}$$

30. **(b)** $Q=\int \rho dv=\int_0^R \frac{A}{r^2}e^{-2r/a}\left(4\pi r^2 dr\right)$

$$\Rightarrow Q=4\pi A\int_0^R e^{-2r/a}dr=4\pi A\left(\frac{e^{-2r/a}}{-\frac{2}{a}}\right)_0^R$$

$$\Rightarrow Q=4\pi A\left(-\frac{a}{2}\right)\left(e^{-2R/a}-1\right)$$

$$\Rightarrow Q=2\pi aA(1-e^{-2R/a})$$

$$\Rightarrow R=\frac{a}{2}\log\left(\frac{1}{1-\frac{Q}{2\pi aA}}\right)$$

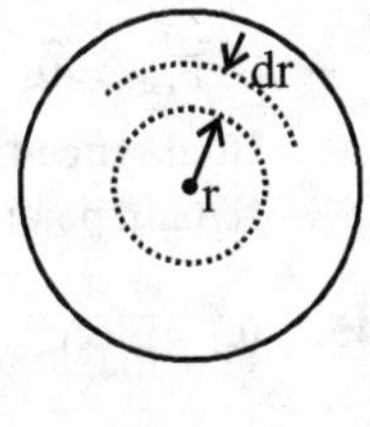

31. **(d)** Spheres A and B carry equal charge say 'q'

$\therefore$ Force between them, $F=\frac{kqq}{r^2}$

When A and C are touched, charge on both $q_A=q_C=\frac{q}{2}$

Then when B and C are touched, charge on B

$$q_B=\frac{\frac{q}{2}+q}{2}=\frac{3q}{4}$$

Now, the force between charge q_A and q_B

$$F'=\frac{kq_Aq_B}{r^2}=\frac{k\times\frac{q}{2}\times\frac{3q}{4}}{r^2}=\frac{3}{8}\frac{kq^2}{r^2}=\frac{3}{8}F$$

32. **(c)** Inside the cavity net charge is zero.

$\therefore$ $Q_1=0$ and $\sigma_1=0$

There is no effect of point charges $+Q$, $-Q$ and induced charge on inner surface on the outer surface.

$\therefore$ $Q_2=0$ and $\sigma_2=0$

33. **(a)** $F_{net}=2F\cos\theta$

$$F_{net}=\frac{2kq\left(\frac{q}{2}\right)}{\left(\sqrt{y^2+a^2}\right)^2}\cdot\frac{y}{\sqrt{y^2+a^2}}$$

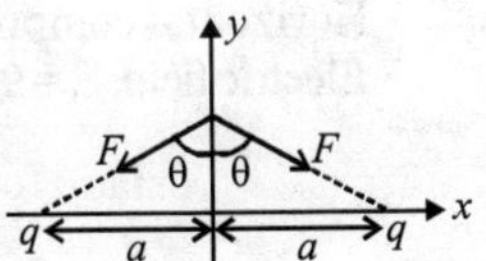

$$F_{net}=\frac{2kq\left(\frac{q}{2}\right)y}{(y^2+a^2)^{3/2}}\quad(\because y<<a)$$

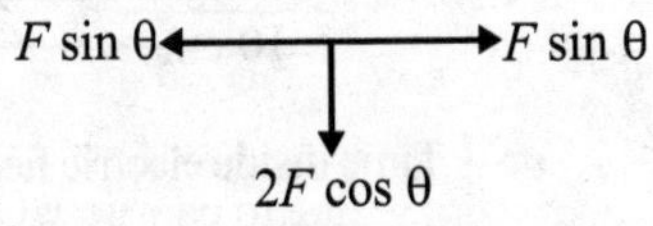

$\Rightarrow \frac{kq^2y}{a^3}$ So, $F\propto y$

34. **(d)** From figure

$T\cos\theta=mg$ ....(i)

$T\sin\theta=F_e$ ....(ii)

Dividing equation (ii) by (i), we get

$$\Rightarrow \frac{\sin\theta}{\cos\theta}=\frac{F_e}{mg},$$

$$\Rightarrow F_e=mg\tan\theta$$

$$\Rightarrow \frac{kq^2}{x^2}=mg\tan\theta$$

$$\Rightarrow q^2=\frac{x^2mg\tan\theta}{k}$$

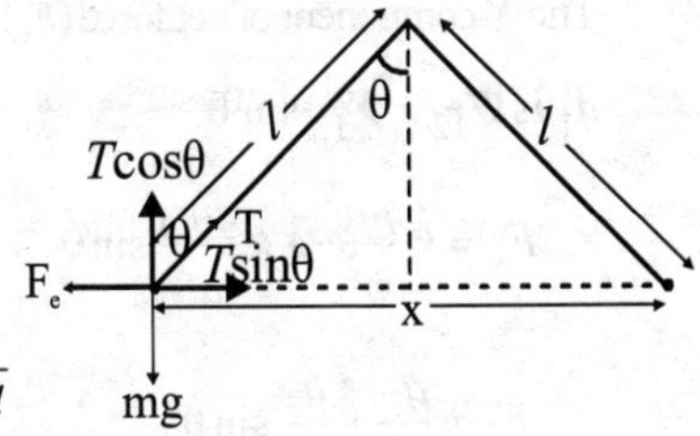

Since θ is small

$\therefore \tan\theta\approx\sin\theta=\frac{x}{2l}$

$$\therefore q^2=\frac{x^3mg}{2kl}\Rightarrow q^2\propto x^{3/2}$$

$$\Rightarrow \frac{dq}{dt}\propto\frac{3}{2}\sqrt{x}\frac{dx}{dt}=\frac{3}{2}\sqrt{x}V$$

Since $\frac{dq}{dt}=$ const. $\Rightarrow v\propto x^{-1/2}$

35. **(d)** Let F be the force between Q and Q. The force between q and Q should be attractive for net force on Q to be zero. Let F' be the force between Q and q. The resultant of F' and F' is R. For equilibrium

Net force on Q at C is zero.

$$\therefore \vec{R}+\vec{F}=0\Rightarrow\sqrt{2}F'=-F$$

$$\Rightarrow \sqrt{2}\times k\frac{Qq}{\ell^2}=-k\frac{Q^2}{(\sqrt{2}\,\ell)^2}$$

$$\Rightarrow \frac{Q}{q}=-2\sqrt{2}$$

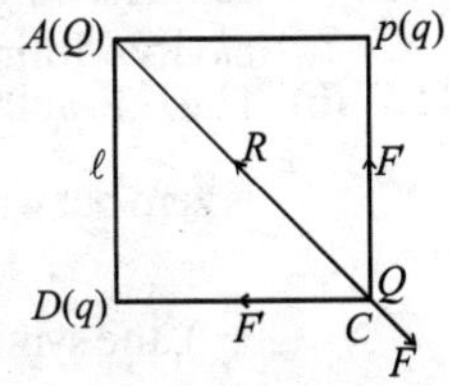

36. **(b)** It is obvious that by charge conservaiton law, electronic charge does not depend on acceleration due to gravity as it is a universal constant.

So, electronic charge on earth

= electronic charge on moon

$\therefore$ Required ratio = 1.

37. **(d)** Initial force, $F=K\frac{Q_BQ_C}{x^2}$

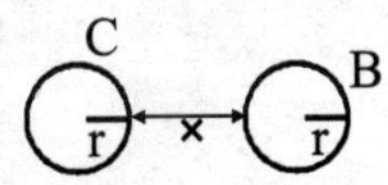

x is distance between the spheres. When third spherical conductor comes in contact with B charge on B is halved *i.e.*, $\frac{Q}{2}$ and charge on third sphere becomes $\frac{Q}{2}$. Now it is touched to C, charge then equally distributes themselves to make potential same, hence charge on C becomes $\left(Q+\frac{Q}{2}\right)\frac{1}{2}=\frac{3Q}{4}$.

$$\therefore F_{new}=k\frac{Q_C'Q_B'}{x^2}=k\frac{\left(\frac{3Q}{4}\right)\left(\frac{Q}{2}\right)}{x^2}=k\frac{3}{8}\frac{Q^2}{x^2}=\frac{3}{8}F$$

38. **(b)** Force applied by charge q_2 on q_1,

$$F_{12}=k\frac{q_1q_2}{b^2}$$

Force applied by charge q_3 on q_1

$$F_{13}=k\frac{q_1q_3}{a^2}$$

The X-component of net force (F_x) on

q_1 is $F_{12} + F_{13}\sin\theta$

$\therefore F_x = k\frac{q_1 q_2}{b^2} + k\frac{q_1 q_3}{a^2}\sin\theta$

$\therefore F_x \propto \frac{q_2}{b^2} + \frac{q_3}{a^2}\sin\theta$

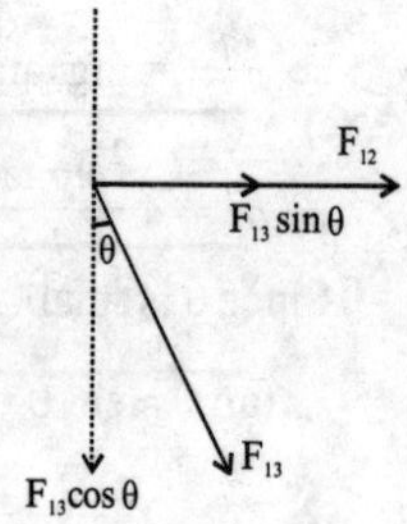

39. **(d)** At equilibrium net force is zero,

$\therefore k\frac{Q\times Q}{(2x)^2} + k\frac{Qq}{x^2} = 0$

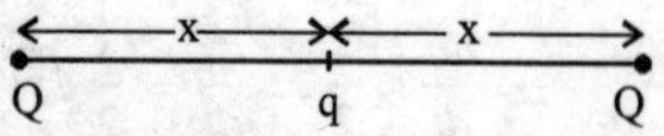

$\Rightarrow \frac{kQ}{x^2}\left(\frac{Q}{4} + q\right) = 0 \Rightarrow q = -\frac{Q}{4}$

40. **(a)** When lightning strikes, the electrical charge is distributed over the exterior surface, preventing it from penetrating the interior and thereby safeguarding passengers. In electrostatics, the electric field inside a cavity within a conductor is zero.

41. **(b)** The E.F. at the center of a full uniformly charged ring is zero but when only a $\frac{1}{4}$th arc remains (from $\theta = 0$ to $\frac{\pi}{2}$), the symmetry is broken.

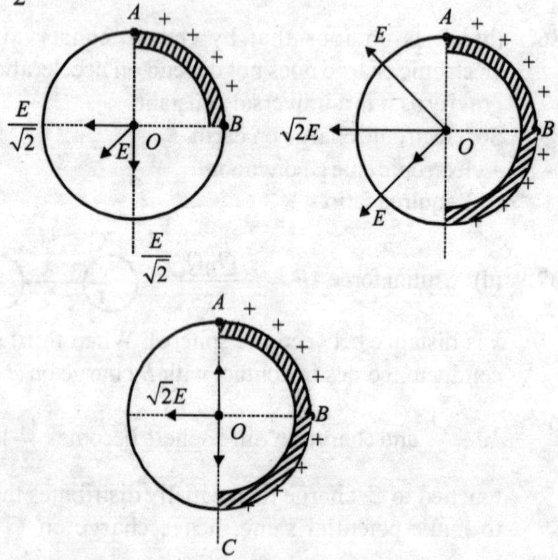

42. **(c)** $E = \frac{\sigma}{2\varepsilon_0}$ (due to sheet)

$\Rightarrow E = \frac{\rho R^3}{3\varepsilon_0 r^2}$ (due to sphere)

E (due to sheet) will be same at point C and D but E (due to sphere) will be maximum at point D.
Similarly for point A and B.

43. **(b)** $E = \frac{\sigma}{2\varepsilon_0}$

$\tau = \vec{P}\times\vec{E}$

$\tau = PE\sin\theta$

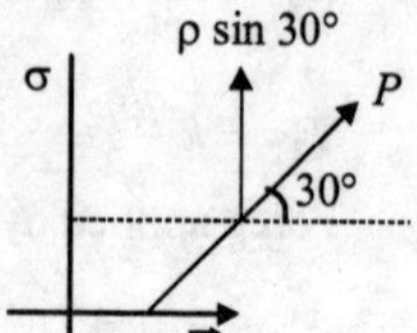

$= \left[(2\times10^{-6})\left(\frac{2}{10}\right)\right]\left[\frac{100\times10^{-6}}{2\times8.85\times10^{-12}}\right]\left(\frac{1}{2}\right)$

$= \frac{10}{8.85} = 1.12$ Nm.

44. **(d)** Since polar dielectrics are oriented in random directions

$\therefore \vec{P}_{net} = \vec{0}$.

In absence of an external electric field, polar dielectric remain polar and are randomly oriented.

45. **(c)** Electric field $E = \frac{KQr}{\left(x^2+R^2\right)^{3/2}}$

$\frac{dE}{dx} = 0 \quad \therefore x = \frac{R}{\sqrt{2}} = \frac{\sqrt{2}a}{\sqrt{2}} = a$

46. **(b)** Let at point P between the charge q and 9q electric field is zero i.e., $E_P = 0$

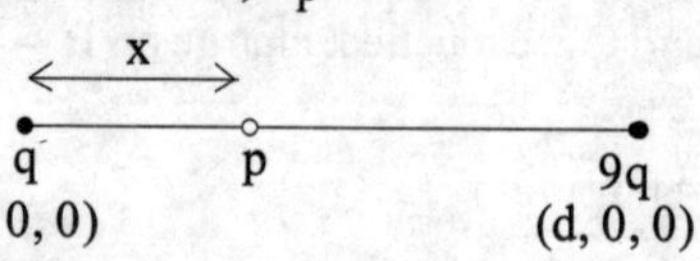

$\therefore \frac{kq}{x^2} = \frac{k9q}{(d-x)^2} \Rightarrow \frac{d-x}{x} = 3 \therefore x = \frac{d}{4}$

The point is between two charges q and 9q where the electric field vanishes is $\left(\frac{d}{4}, 0, 0\right)$

47. **(c)** Use work energy theorem

W = K.E$_f$ – K.E$_i$

$qE\frac{\ell}{2} = \frac{1}{2}mv^2 - 0 \Rightarrow v = \sqrt{\frac{qE\ell}{m}}$

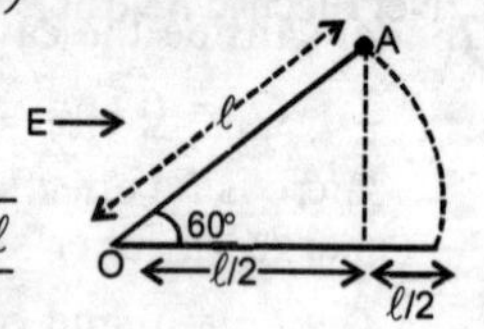

48. **(c)** Given, length of metal plates, $l = 10$ cm
Horizontal component of velocity, $v_x = 10^6$ m/s
Electric field, E = 9.1 V/cm

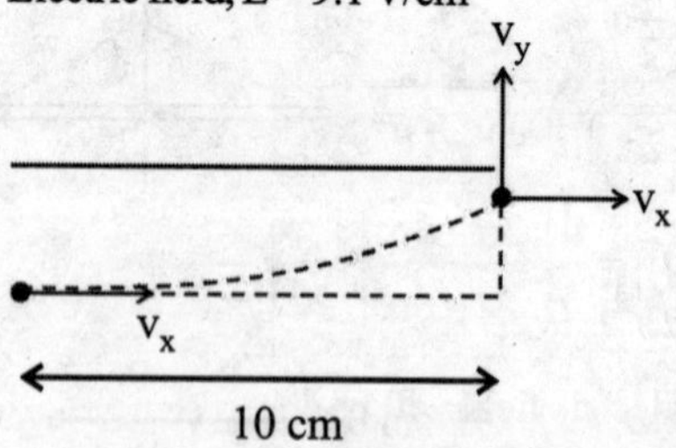

$\Rightarrow$ Time inside electric field, $t = \frac{\ell}{v_x} = \frac{10\times10^{-2}}{10^6} = 10^{-7}$

Acceleration along y-axis, $a = \frac{eE}{m}$

Using $v_y = u_y + a_y t$

$\Rightarrow v_y = 0 + \frac{eE}{m}\times10^{-7}$

$\Rightarrow v_y = \frac{1.6\times10^{-19}}{9.1\times10^{-31}}\times9.1\times10^2\times10^{-7} = 16\times10^6$ m/s

49. **(5)**

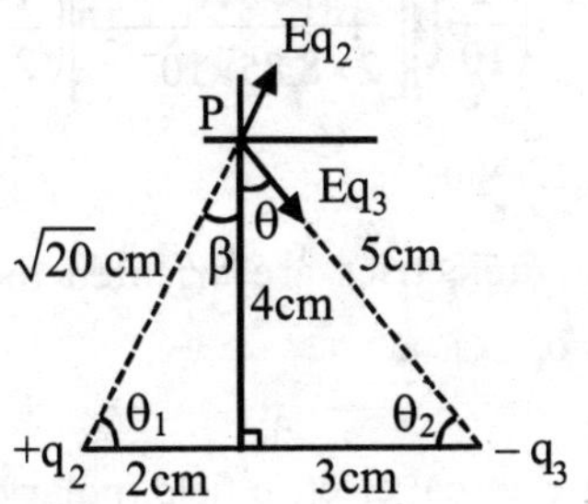

Since $E_P = 0$

$\therefore E_{q2} = Eq_3$

$$\frac{Kq_2}{20}\cos\beta = \frac{Kq_3}{25}\cos\theta$$

$$\frac{q_2}{20}\frac{4}{\sqrt{20}} = \frac{q_3}{25}\frac{4}{\sqrt{25}}$$

$$\frac{q_2}{q_3} = \frac{20}{25}\sqrt{\frac{20}{25}} = \frac{8}{5\sqrt{x}}$$

on comparing & rearranging.

$$\sqrt{x} = \frac{8\times 25\sqrt{25}}{5\times 20\sqrt{20}} \Rightarrow x = 5$$

50. **(a)**

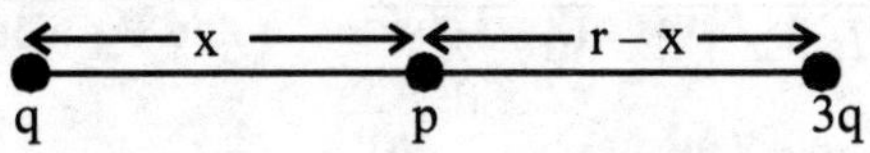

Net Electric field at P = 0

$$\therefore E_q = E_{3q} \Rightarrow \frac{kq}{x^2} = \frac{k\cdot 3q}{(r-x)^2}$$

$$\Rightarrow (r-x)^2 = 3x^2 \Rightarrow r - x = \sqrt{3}x \quad \Rightarrow x = \frac{r}{\sqrt{3}+1}$$

51. **(6)** Electric field intensity

$$E_P = \frac{\lambda}{2\pi\varepsilon_0 r_P} - \frac{\sigma}{2\varepsilon_0}\left\{r_P = \frac{3}{\pi}\right\} = \frac{1}{2\varepsilon_0}\left[\frac{\lambda}{3} - \sigma\right]$$

$$E_Q = \frac{\lambda}{2\pi\varepsilon_0 r_Q} - \frac{\sigma}{2\varepsilon_0}\left\{r_Q = \frac{4}{\pi}\right\} = \frac{1}{2\varepsilon_0}\left[\frac{\lambda}{4} - \sigma\right]$$

$$\frac{E_P}{E_Q} = \frac{4}{3}\left(\frac{\lambda - 3\sigma}{\lambda - 4\sigma}\right) = \frac{4}{3}\left[\frac{2\sigma - 3\sigma}{2\sigma - 4\sigma}\right] = \frac{4}{3}\left[\frac{-\sigma}{-2\sigma}\right] = \frac{4}{6}$$

$\therefore a = 6$

52. **(b)** In region I Electric field will be due to charge Q

$E_{II} = 0$. Since electric field inside conductor is zero.

$E_{III} \neq 0$, Electric field will be due to charge Q only or just due to free charge

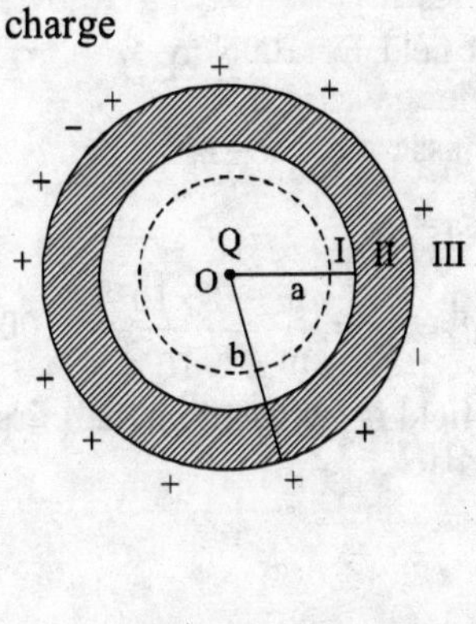

53. **(c)**

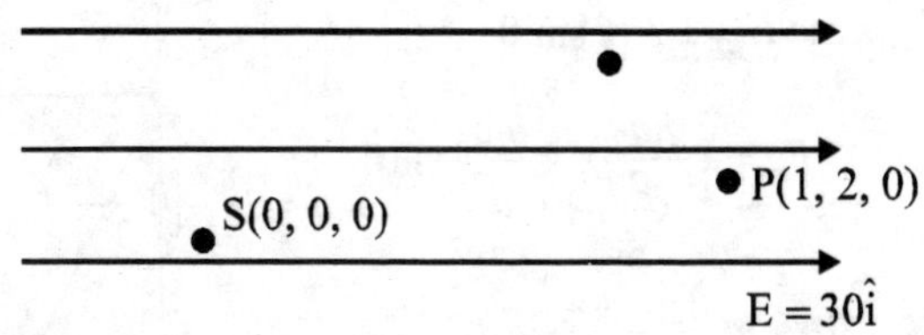

$$W = \vec{F}\cdot\vec{S} \Rightarrow W = q\vec{E}\cdot\vec{S}$$

$$W = 2\times 10^{-2}\left[30\hat{i}\cdot(-\hat{i})\right] = 2\times 10^{-2}(-30) = -600\text{ mJ}$$

54. **(45)** Kinetic theory $= \frac{1}{2}mv_x^2$

$$0.5\,e = \frac{1}{2}mv_x^2 \Rightarrow v_x = \sqrt{\frac{e}{m}} \quad \text{...(i)}$$

Along x, $L = v_x t = \sqrt{\frac{e}{m}}t$

Along y, $v_y = \frac{eE}{m}t$ ...(ii)

Divide equation (ii) by (i)

$$\frac{v_y}{L} = E\sqrt{\frac{e}{m}} = Ev_x$$

$$\Rightarrow \tan\theta = \frac{v_y}{v_x} = E\times L = 10\times 0.1 = 1$$

$\theta = 45°$

55. **(a)**

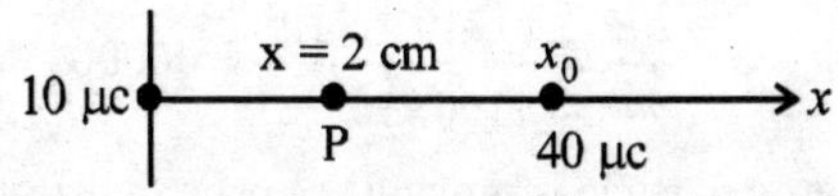

$$\text{As, } E_P = 0 \Rightarrow K\times\frac{10\times 10^{-6}}{2^2}\hat{i} - K\frac{(40\times 10^{-6})}{(x_0-2)^2}\hat{i} = 0$$

$$\Rightarrow \frac{10\times 10^{-6}}{40\times 10^{-6}} = \frac{2^2}{(x_0-2)^2}$$

$$\Rightarrow \frac{1}{2} = \frac{2}{x_0 - 2} \Rightarrow x_0 - 2 = 4 \Rightarrow x_0 = 6\text{ cm}$$

56. **(2)**

$$a = \frac{F}{m} = \frac{qE}{m} = 2\times 10^{11}\times 1.8\times 10^{3} = 3.6\times 10^{14}\text{ m/s}^2$$

$$\text{Time to cross the plates} = \frac{d}{v} = \frac{0.10}{3\times 10^7} = \frac{1}{3}\times 10^{-8}$$

$$Y = \frac{1}{2}at^2 = \frac{1}{2}\times 3.6\times 10^{14}\times\frac{1}{9}\times 10^{-16} = 2\times 10^{-3}\text{ m} = 2\text{mm}$$

57. **(c)** By Gauss law $= \oint \vec{E}.\vec{da} = \frac{q_{in}}{\varepsilon_0}$

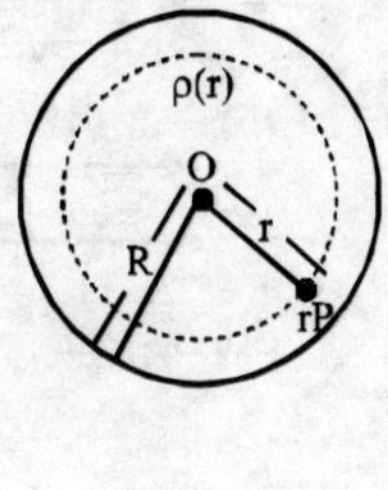

$$\text{Here, } E.4\pi r^2 = \frac{\int_0^r \rho_0\left(\frac{3}{4} - \frac{r}{R}\right)4\pi r^2 dr}{\varepsilon_0}$$

$$\Rightarrow E.4\pi r^2 = \frac{\rho_0 4\pi}{\varepsilon_0}\left(\frac{3}{4}\frac{r^3}{3} - \frac{r^4}{4R}\right)$$

$$\Rightarrow Er^2 = \frac{\rho_0 r^3}{4\varepsilon_0}\left(1-\frac{r}{R}\right) \quad \Rightarrow E = \frac{\rho_0 r}{4\varepsilon_0}\left(1-\frac{r}{R}\right)$$

58. (b)

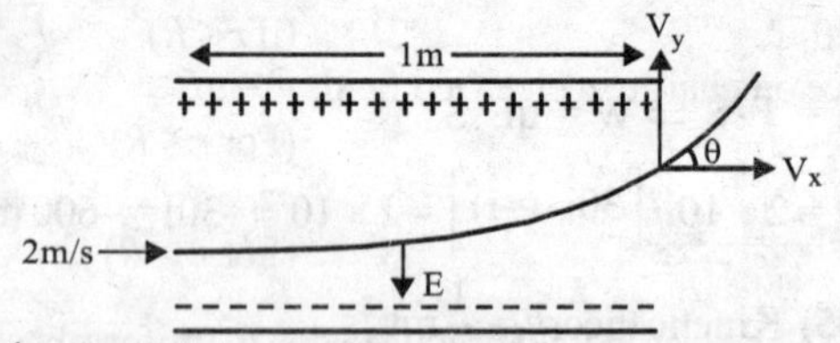

We have,

$$a_y = \frac{F_y}{m} = \frac{e(E)}{m} = \frac{e\left(\frac{8m}{e}\right)}{m} = 8 \text{ m/s}^2$$

and, $s_x = u_x t$

$$\Rightarrow 1 = 2 \times t \Rightarrow t = \frac{1}{2}\text{ sec}$$

and, $v_y = u_y + a_y t$

$$\Rightarrow v_y = 0 + 8 \times \frac{1}{2} \Rightarrow v_y = 4 \text{ m/s}$$

$$\tan\theta = \frac{v_y}{v_x} = \frac{4}{2} = 2 \Rightarrow \theta = \tan^{-1}(2)$$

59. (d) Here $a = \frac{qE}{m} = \frac{40\times10^{-6}\times10^5}{100\times10^{-6}} = 40000 \text{ m/s}^2$

As charge particle is moving opposite to direction of $\vec{E}$

So $\vec{E}$ will declerate the charge and charge will come to rest.

Now, $v^2 = u^2 + 2as$

$0 = 200^2 + 2\times(-40\times10^3)\times s$

$80\times10^3 s = 4\times10^4$

$$\therefore s = \frac{4\times10^4}{80\times10^3} = \frac{40}{80} = 0.5 \text{ m.}$$

60. (a) Here, sum of $\vec{E_C}$ and $\vec{E_A}$ will lie along $\vec{E_B}$.

$$\text{So, } \left|\vec{E}_{net}\right| = \left|\vec{E_B}\right| + \left|\vec{E_C} + \vec{E_B}\right|$$

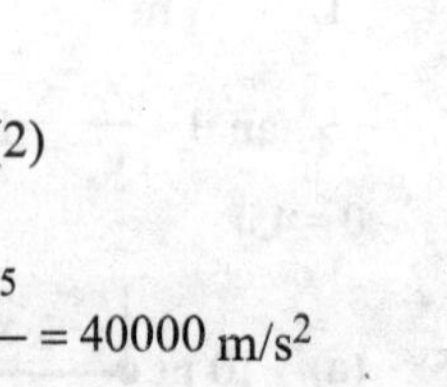

$$= \frac{Kq}{(\sqrt{2}a)^2} + \left[\sqrt{2}\frac{kq}{2a^2}\right]$$

$$= \frac{kq}{a^2}\left[\frac{1}{2} + \frac{1}{\sqrt{2}}\right]$$

61. (b)

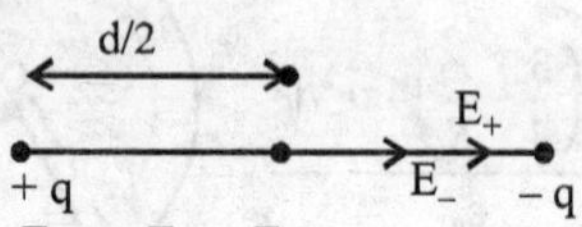

So, $E_{net} = E_+ + E_-$

$$= 2\frac{kq}{\left(\frac{d}{2}\right)^2} = \frac{8kq}{d^2} \Rightarrow 6.4\times10^4 = 8\times9\times10^9\times\frac{8\times10^{-6}}{d^2}$$

$$\Rightarrow d^2 = \frac{72\times8\times10^{+3}}{6.4\times10^4} \Rightarrow d^2 = 9 \Rightarrow d = 3 \text{ m}$$

62. (b) As water droplet is at rest

So, $\vec{F}_{net} = 0$

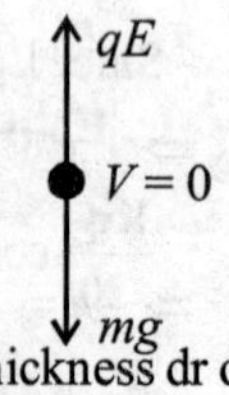

$$\Rightarrow mg = qE \Rightarrow q = \frac{mg}{E}$$

$$\Rightarrow q = \frac{0.1\times10^{-3}\times9.8}{4.9\times10^5}$$

$$\Rightarrow q = 2\times10^{-9} C$$

63. (a) Consider a small ring of radius r and thickness dr on disc.

$dq = \sigma 2\pi r\, dr$

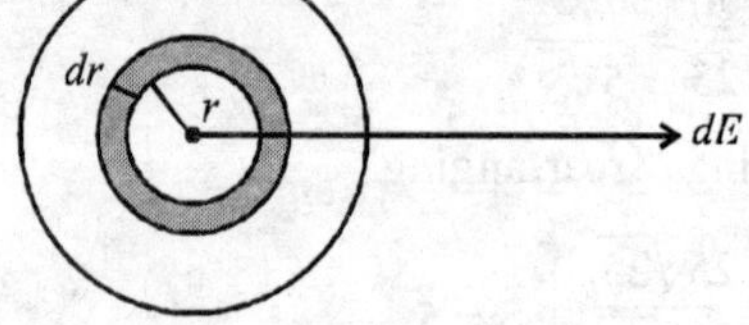

$$dE = \frac{Kdqz}{(r^2+z^2)^{3/2}} = \frac{K\sigma(2\pi r)\,drz}{(r^2+z^2)^{3/2}}$$

$$E = \int dE = \int dE$$

$$= \int_0^R \frac{K\sigma 2\pi r drz}{\left(z^2+r^2\right)^{3/2}} = \frac{\sigma}{2\epsilon_0}\left[1 - \frac{z}{\sqrt{R^2+z^2}}\right]$$

64. (b) Charge per unit length of the rod,

$$\lambda = \left(\frac{-Q}{R\theta}\right) = \left(\frac{-Q}{R\frac{2\pi}{3}}\right) = \frac{-3Q}{2\pi R}$$

$$E = \int dE\cos\theta \Rightarrow E = \int_{-\pi/3}^{\pi/3} \frac{K\times(Q)}{\frac{2\pi}{3}R}\times\frac{Rd\theta}{R^2}\cos\theta$$

$$\Rightarrow E = \frac{3}{2\pi}\frac{KQ}{R^2}[\sin\theta]_{-\pi/3}^{\pi/3} = \frac{3}{2\pi}\frac{KQ}{R^2}\times\frac{2\sqrt{3}}{2}$$

$$\Rightarrow E = \frac{3\sqrt{3}Q}{8\pi^2 \epsilon_0 R^2}(+\hat{i})$$

65. (1) Electric field, E = 100N/m

Charge to mass ratio, $\frac{q}{m} = 8\mu C/g$

Force, F = ma = qE

$$\Rightarrow ma = qE \Rightarrow a = \frac{qE}{m} = \frac{8\times10^{-6}}{10^{-3}}\times100 = 0.8 \text{m/s}^2$$

As electric field is switched on, ball first strikes to wall and returns back.

one oscillation.

Thus $s = ut + \frac{1}{2}at_1^2 \Rightarrow 0.1 = \frac{1}{2}\times 0.8t_1^2$

$(\because \text{initial velocity, } u = 0)$

$\Rightarrow \quad t_1 = \frac{1}{2}s$

Thus time period $T = 2\times\frac{1}{2} = 1\,\text{sec}$.

66. **(b)** The electrostatic field will balance the weight of oil drop,

$m_{oil} \times g_{oil} = qE$

$\Rightarrow \quad \frac{4}{3}\pi r^3 \times \rho \times g = neE$ (n = no. of excess electrons)

$\Rightarrow \quad n = \dfrac{\frac{4}{3}\pi r^3 \rho g}{eE} = \dfrac{\frac{4}{3}\times\pi\times(2\times10^{-3})^3\times(3\times10^3)\times 9.81}{1.6\times10^{-19}\times3.55\times10^5}$

$= 173.65\times10^8 \simeq 1.73\times10^{10}$

67. **(d)** Electric field at point P

$E = \frac{k\lambda}{a}(\sin\theta_1 + \sin\theta_2)$

$\Rightarrow \quad E = \frac{1}{4\pi\varepsilon_0}\times\frac{Q}{L}\times\frac{1}{\left(\frac{\sqrt{3}L}{2}\right)}\times 2\sin\theta \quad \left[\because \lambda = \frac{Q}{L}\right]$

From figure,

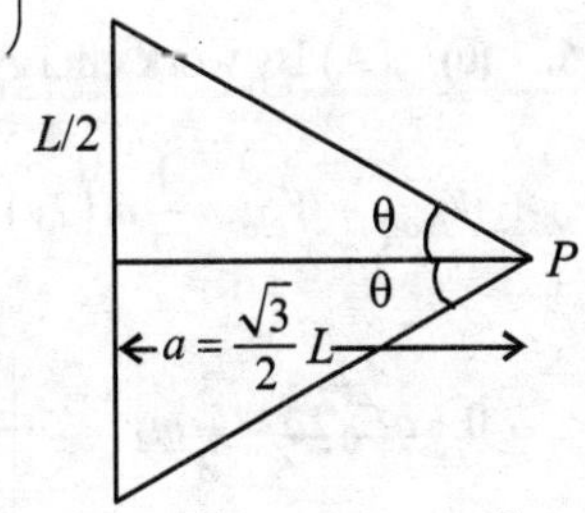

$\tan\theta = \dfrac{\frac{L}{2}}{\frac{\sqrt{3}L}{2}} = \frac{1}{\sqrt{3}}$

$\Rightarrow \quad \theta = \tan^{-1}\left(\frac{1}{\sqrt{3}}\right) = 30°$

$\Rightarrow \quad E = \frac{1}{4\pi\varepsilon_0}\times\frac{2Q}{\sqrt{3}L^2}\times\left(\frac{2\times1}{2}\right) \quad \left[\because \sin 30° = \frac{1}{2}\right]$

$\Rightarrow \quad E = \frac{Q}{2\sqrt{3}\pi\varepsilon_0 L^2}$

68. **(c)** Electric field due charge Q_2, $E_2 = \frac{kQ_2}{x_2^2}$

Electric field due charge Q_1, $E_1 = \frac{kQ_1}{x_1^2}$

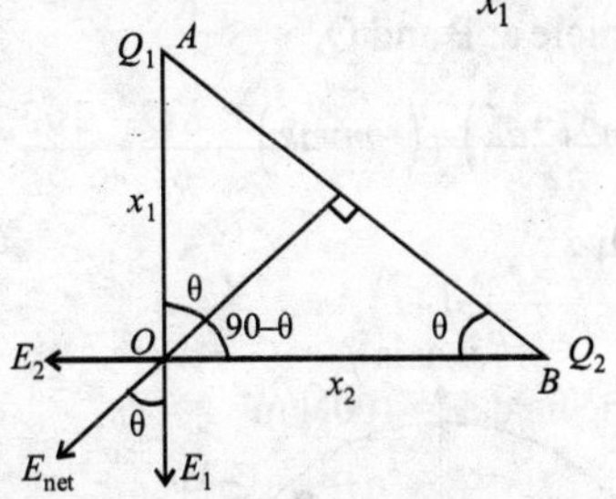

From figure,

$\tan\theta = \frac{E_2}{E_1} = \frac{x_1}{x_2} \Rightarrow \dfrac{kQ_2}{x_2^2\times\frac{kQ_1}{x_1^2}} = \frac{x_1}{x_2}$

$\Rightarrow \frac{Q_2x_1^2}{Q_1x_2^2} = \frac{x_1}{x_2} \Rightarrow \frac{Q_2}{Q_1} = \frac{x_2}{x_1}$ or, $\frac{Q_1}{Q_2} = \frac{x_1}{x_2}$.

69. **(c)** For spherical shell

$E = \frac{1}{4\pi\varepsilon_0}\frac{Q}{r^2}$ (if $r \geq R$)

$= 0$ (if $r < R$)

Force on charge in electried field, $F = qE$

$\therefore F = 0$ (For $r < R$)

$F = \frac{1}{4\pi\varepsilon_0}\frac{Qq}{r^2}$ (For $r > R$)

70. **(c)** The electric field produced due to uniformly charged infinite plane is uniform. So option (b) and (d) are wrong. And +ve charge density σ_+ is bigger in magnitude so its field along Y direction will be bigger than field of –ve charge density σ_- in X direction. Hence option (c) is correct.

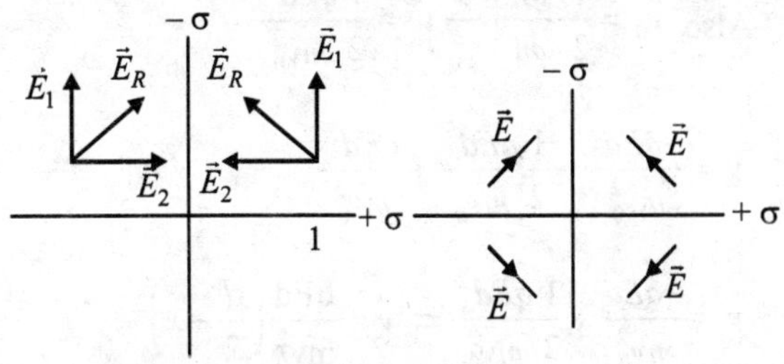

71. **(c)** Given, Electric field, $E = E_0(1 - ax^2)$

$\therefore$ Force, $F = qE = qE_0(1 - ax^2)$

Also, $F = ma = mv\frac{dv}{dx} \quad \left(\because a = v\frac{dv}{dx}\right)$

$\therefore mv\frac{dv}{dx} = qE_0(1 - ax^2) \Rightarrow v\,dv = \frac{qE_0(1 - ax^2)dx}{m}$

Integrating both sides we get,

$\Rightarrow \int_0^v v\,dv = \int_0^x \frac{qE_0(1 - ax^2)dx}{m} \Rightarrow \frac{v^2}{2} = \frac{qE_0}{m}\left(x - \frac{ax^3}{3}\right) = 0$

$\Rightarrow x = \sqrt{\frac{3}{a}}$

72. **(b)** $F_x = 0, a_x = 0, (v)_x = v_0 =$ constant

Time taken to reach at $'P' = \frac{d}{v_0} = t_0$ (let) ...(i)

(Along $-y$), $v_y = \frac{qE}{m}t_0$ ...(ii)

$\tan\theta = \frac{v_y}{v_x} = \frac{qEt_0}{m\cdot v_0}, \left(t_0 = \frac{d}{v_0}\right)$

$\tan\theta = \frac{qEd}{m\cdot v_0^2}$, Slope $= \frac{-qEd}{mv_0^2}$

No electric field $\Rightarrow F_{\text{net}} = 0, \vec{v} = $ const.

$$y = (\text{slope})x + c, \begin{cases} \text{slope} = -\dfrac{qEd}{mv_0^2} \\ (d, -y_0) \end{cases}$$

$$-y_0 = \frac{-qEd}{mv_0^2}, d + c$$

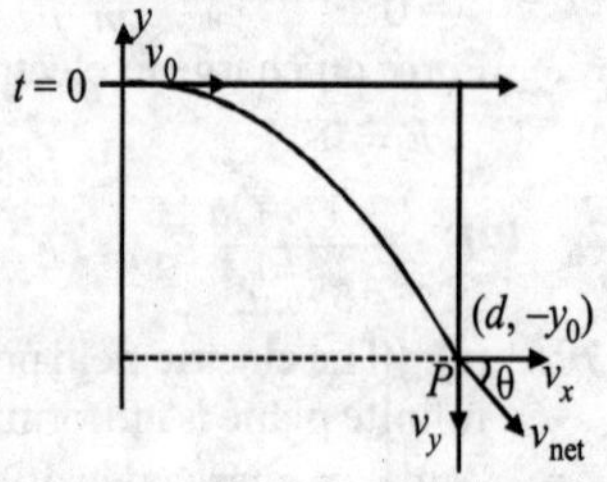

$$\Rightarrow c = -y_0 + \frac{qEd^2}{mv_0^2}$$

$$y = \frac{-qEd}{mv_0^2}x - y_0 + \frac{qEd^2}{mv_0^2}$$

Also $y_0 = \dfrac{1}{2} \cdot \dfrac{qE}{m}\left(\dfrac{d}{v_0}\right)^2 = \dfrac{1}{2}\dfrac{qEd^2}{mv_0^2}$

$$y = \frac{-qEdx}{mv_0^2} - \frac{1}{2}\frac{qEd^2}{mv_0^2} + \frac{qEd^2}{mv_0^2}$$

$$y = \frac{-qEd}{mv_0^2} + \frac{1}{2}\frac{qEd^2}{mv_0^2} \Rightarrow y = \frac{\text{qEd}}{\text{mv}_0^2}\left(\frac{d}{2} - x\right)$$

73. **(d)** Net force acting on the particle,

$\vec{F} = qE\hat{i} + mg\hat{j}$

Net acceleration of particle is constant, initial velocity is zero therefore path is straight line.

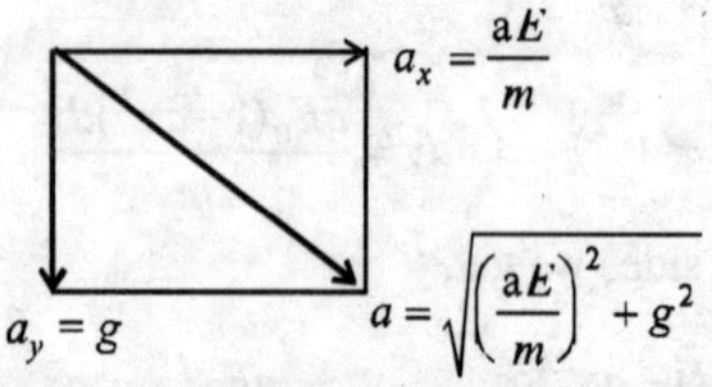

74. **(b)** Electric field at A $\left(R' = \dfrac{R}{2}\right)$

$$E_A.ds = \frac{q}{\varepsilon_0}$$

$$\Rightarrow E_A = \frac{\rho \times \frac{4}{3}\pi\left(\frac{R}{2}\right)^3}{\varepsilon_0 \cdot 4\pi\left(\frac{R}{2}\right)^2}$$

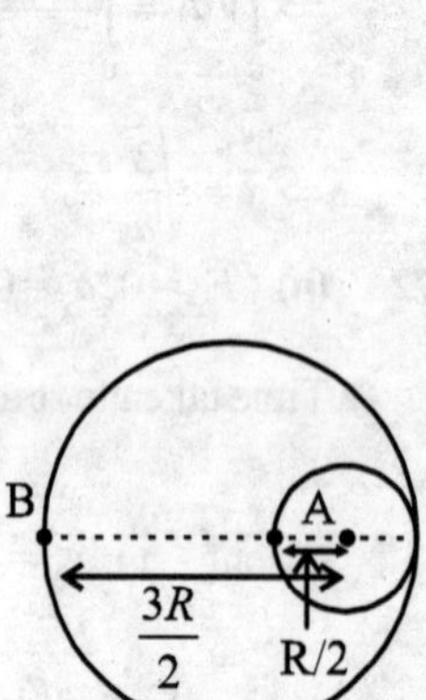

$$\Rightarrow E_A = \frac{\sigma(R/2)}{3\varepsilon_0} = \left(\frac{\sigma R}{6\varepsilon_0}\right)$$

Electric fields at 'B'

$$E_B = \frac{k \times \rho \times \frac{4}{3}\pi R^3}{R^2} - \frac{k \times \rho \times \frac{4}{3}\pi\left(\frac{R}{2}\right)^3}{\left(\frac{3R}{2}\right)^2}$$

$$\Rightarrow E_B = \frac{\sigma R}{3\varepsilon_0} - \left(\frac{1}{4\pi\varepsilon_0}\right)\frac{(\sigma)}{\left(\frac{3R}{2}\right)^2}\frac{4\pi}{3}\left(\frac{R}{2}\right)^3$$

$$\Rightarrow E_B = \frac{\sigma R}{3\varepsilon_0} - \frac{\sigma R}{54\varepsilon_0} \Rightarrow E_B = \frac{17}{54}\left(\frac{\sigma R}{\varepsilon_0}\right)$$

$$\left|\frac{E_A}{E_B}\right| = \frac{1 \times 54}{6 \times 17} = \left(\frac{9}{17}\right) = \frac{9}{17} \times \frac{2}{2} = \frac{18}{34}$$

75. **(c)** Since $\vec{r} \cdot \vec{p} = 0$

$\vec{E}$ must be antiparallel to $\vec{p}$

$\therefore\ \hat{E}$ is parallel to $\left(\hat{i} + 3\hat{j} - 2\hat{k}\right)$

76. **(c)** (A) By work energy theorem

$$W_{mag} + W_{ele} = \frac{1}{2}m(2v)^2 - \frac{1}{2}m(v)^2$$

$$0 + qE_0 2a = \frac{3}{2}mv^2 \quad \Rightarrow E_0 = \frac{3}{4}\frac{mv^2}{qa}$$

(B) Rate of work done at P = power of electric force

$$= qE_0 v = \frac{3}{4}\frac{mv^3}{a}$$

(C) At Q, $\dfrac{dw}{dt} = 0$ for both the fields

(D) The difference of magnitude of angular momentum of the particle at P and Q,

$$\Delta\vec{L} = \left(-m2v2a\hat{k}\right) - \left(-mva\hat{k}\right)$$

$$\left|\Delta\vec{L}\right| = 3mva$$

77. **(a)**

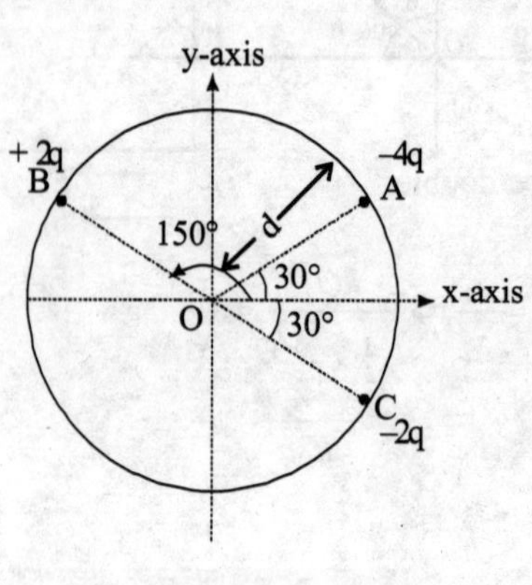

Electric field due to charge +2q at centre O

$$\vec{E}_1 = \frac{1}{4\pi\varepsilon_0} \times \frac{2q}{d^2}\left[\frac{\sqrt{3}\hat{i} - \hat{j}}{2}\right]$$

Electric field due to charge $-2q$ at centre O

$$\vec{E}_2 = \frac{1}{4\pi\varepsilon_0} \times \frac{2q}{d^2}\left[\frac{\sqrt{3}\hat{i} - \hat{j}}{2}\right]$$

Electric field due to charge $-4q$ at centre O

$$\vec{E}_3 = \frac{1}{4\pi\varepsilon_0} \times \frac{4q}{d^2}\left[\frac{\sqrt{3}\hat{i} + \hat{j}}{2}\right]$$

$\therefore$ Net electric field at point O

$$\vec{E}_0 = \vec{E}_1 + \vec{E}_2 + \vec{E}_3 = \frac{\sqrt{3}\,q}{\pi\varepsilon_0 d^2}\hat{i}$$

78. **(b)** Using

$v^2 - u^2 = 2aS$ ...(i)

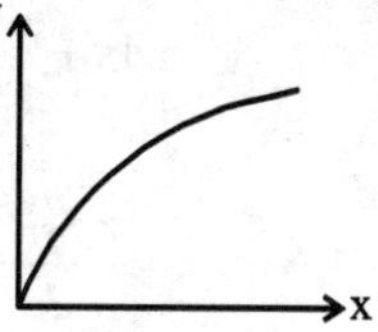

Here, $u = 0, s = x$

Also, $F_{electric} = ma$

$$\Rightarrow qE = ma \Rightarrow a = \frac{qE}{m} \Rightarrow a = \frac{qE}{m}$$

Substituting the values in (i) we get

$$v^2 = \frac{2qE}{m}.x$$

79. **(d)** From figure, $\vec{E}_1 = \frac{\sigma}{2\varepsilon_0}\hat{y}$ and

$$\vec{E}_2 = \frac{\sigma}{2\varepsilon_0}(-\cos 60°\hat{x} - \sin 60°\hat{y})$$

$$= \frac{\sigma}{2\varepsilon_0}\left(-\frac{1}{2}\hat{x} - \frac{\sqrt{3}}{2}\hat{y}\right)$$

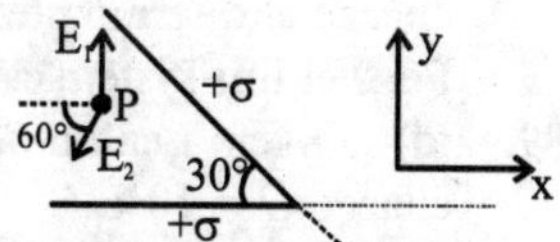

Electric field in the region shown in figure (P)

$$\vec{E}_P = \vec{E}_1 + \vec{E}_2 = \frac{\sigma}{2\varepsilon_0}\left[-\frac{1}{2}\hat{x} + \left(1 - \frac{\sqrt{3}}{2}\right)\hat{y}\right]$$

$$\text{or, } \vec{E}_P = \frac{\sigma}{2\varepsilon_0}\left[\left(1 - \frac{\sqrt{3}}{2}\right)\hat{y} - \frac{\hat{x}}{2}\right]$$

80. **(c)** In the x direction

$F_x = qE_0$

$\Rightarrow \; ma_x = qE_0$

$$\Rightarrow \; a_x = \frac{E_0 q}{m}$$

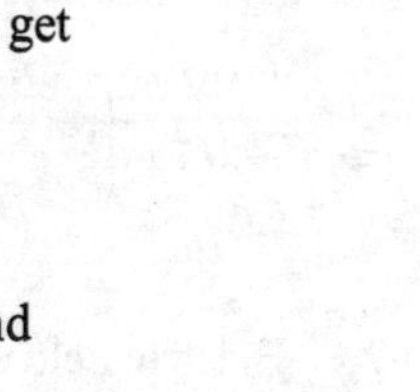

For speed to be double,

$$v_0^2 + v_x^2 = (2v_0)^2$$

$$\Rightarrow \; v_x = \sqrt{3}\, v_0 = a_x t$$

$$\Rightarrow \sqrt{3}v_0 = \frac{qE_0 t}{m} \;\Rightarrow\; t = \frac{\sqrt{3}v_0 m}{E_0 q}$$

81. **(d)** Time period of the pendulum (T) is given by

$$T = 2\pi\sqrt{\frac{L}{g_{eff}}} \Rightarrow g_{eff} = \frac{\sqrt{(mg)^2 + (qE)^2}}{m}$$

$$\Rightarrow g_{eff} = \sqrt{g^2 + \left(\frac{gE}{m}\right)^2} \Rightarrow T = 2\pi\sqrt{\frac{L}{\sqrt{g^2 + \left(\frac{qE}{m}\right)^2}}}$$

82. **(d)** $\vec{E} = (\vec{E}_1 + \vec{E}_2) + (\vec{E}_3 + \vec{E}_4)$ or $E = 2E\cos\alpha - 2E\cos\beta$

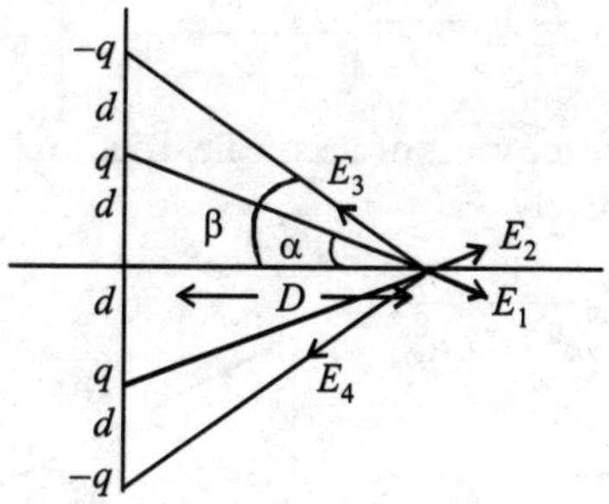

$$= \frac{2kq}{(D^2 + d^2)} \times \frac{D}{\sqrt{D^2 + d^2}} - \frac{2kq}{(D^2 + (2d)^2} \times \frac{D}{\sqrt{D^2 + (2d)^2}}$$

$$= \frac{2kqD}{(D^2 + d^2)^{3/2}} - \frac{2kqD}{[D^2 + (2d)^2]^{3/2}}$$

For d<<D

$$E \propto \frac{D}{D^3} \propto \frac{1}{D^2}$$

83. **(d)** At equilibrium resultant force on bob must be zero, so

$T\cos\theta = mg$ ..... (i)

$T\sin\theta = qE$ ..... (ii)

Solving (i) and (ii) we get

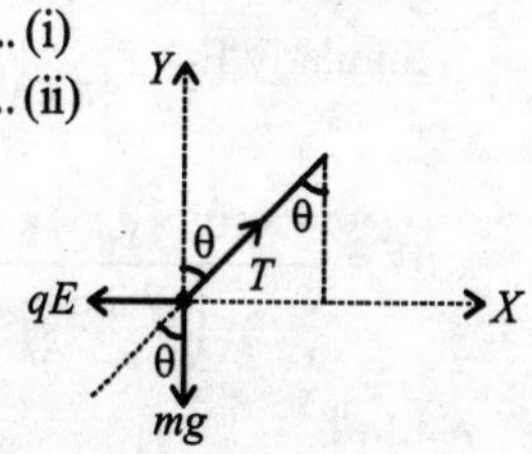

$$\tan\theta = \frac{qE}{mg}$$

$$\tan\theta = \frac{5\times10^{-6}\times2000}{2\times10^{-3}\times10} = \frac{1}{2}$$

[Here, $q = 5\times10^{-6}$C, E = 2000 v/m, m = 2×10^{-3} kg]

$$\Rightarrow \theta = \tan^{-1}\left(\frac{1}{2}\right)$$

84. **(b)** Electric field on the axis of a ring of radius R at a distance h from the centre,

$$E = \frac{kQh}{\left(h^2 + R^2\right)^{3/2}}$$

Condition: for maximum electric field $\frac{dE}{dh} = 0$

$$\Rightarrow \frac{d}{dh}\left[\frac{kQh}{\left(R^2 + h^2\right)^{3/2}}\right] = 0$$

By using the concept of maxima and minima we get,

$h = \frac{R}{\sqrt{2}}$

85. (a)

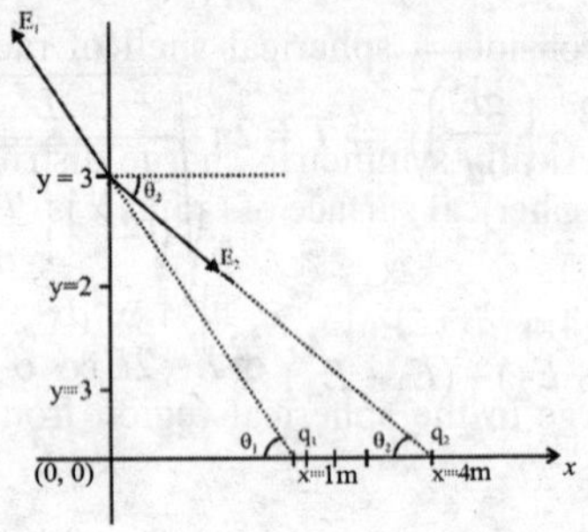

Let $\vec{E}_1$ and $\vec{E}_2$ are the vaues of electric field due to charge, q_1 and q_2 respectively

magnitude of $E_1 = \frac{1}{4\pi \epsilon_0} \frac{q_1}{r_1^2}$

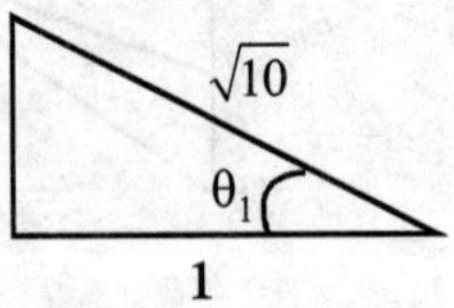

$= \frac{1}{4\pi \epsilon_0} \frac{\sqrt{10} \times 10^{-6}}{(1^2 + 3^2)}$

$= (9 \times 10^9) \times \sqrt{10} \times 10^{-7} \ = 9\sqrt{10} \times 10^2$

$\therefore \vec{E}_1 = 9\sqrt{10} \times 10^2 \left[\cos\theta_1 (-\vec{i}) + \sin\theta_1 \vec{j}\right]$

$\Rightarrow E_1 = 9 \times \sqrt{10} \times 10^2 \left[\frac{1}{\sqrt{10}}(-\hat{i}) + \frac{3}{\sqrt{10}}\hat{j}\right] A$

$\Rightarrow E_1 = 9 \times 10^2 \left[-\hat{i} + 3\hat{j}\right] = \left[-9\hat{i} + 27\hat{j}\right] 10^2$

Similarly, $E_2 = \frac{1}{4\pi \epsilon_0} \frac{q_2}{r^2}$

$E_2 = \frac{9 \times 10^9 \times (25) \times 10^{-6}}{(4^2 + 3^2)} \Rightarrow E_2 = 9 \times 10^3$ V/m

$\therefore \vec{E}_2 = 9 \times 10^3 (\cos\theta_2 \hat{i} - \sin\theta_2 \hat{j}) \ \because \tan\theta_2 = \frac{3}{4}$

$\therefore \vec{E}_2 = 9 \times 10^3 \left(\frac{4}{5}\hat{i} - \frac{3}{5}\hat{j}\right) = (72\hat{i} - 54\hat{j}) \times 10^2$

$\therefore \vec{E} = \vec{E}_1 + \vec{E}_2 = (63\hat{i} - 27\hat{j}) \times 10^2$ V/m

86. (a) Equilibrium position will shift to point where resultant force = 0

$kx_{eq} = qE \Rightarrow x_{eq} = \frac{qE}{k}$

Total energy $= \frac{1}{2} m\omega^2 A^2 + \frac{1}{2} kx_{eq}^2$

Total energy $= \frac{1}{2} m\omega^2 A^2 + \frac{1}{2} \frac{q^2 E^2}{k}$

87. (d) Charge density, $\rho = \rho_0 \left(1 - \frac{r}{R}\right)$

$dq = \rho dv$

$q_{in} = \int dq = \int \rho dv$

$= \int \rho_0 \left(1 - \frac{r}{R}\right) 4\pi r^2 dr \quad (\because dv = 4\pi r^2 dr)$

$= 4\pi\rho_0 \int_0^R \left(1 - \frac{r}{R}\right) r^2 dr = 4\pi\rho_0 \int_0^R r^2 dr - \frac{r^3}{R} dr$

$= 4\pi\rho_0 \left[\left[\frac{r^3}{3}\right]_0^R - \left[\frac{r^4}{4R}\right]_0^R\right] = 4\pi\rho_0 \left[\frac{R^3}{3} - \frac{R^4}{4R}\right]$

$= 4\pi\rho_0 \left[\frac{R^3}{3} - \frac{R^3}{4}\right] = 4\pi\rho_0 \left[\frac{R^3}{12}\right]$

$q_{in} = \frac{\pi\rho_0 R^3}{3}$

using Gauss law, $\oint \vec{E}.d\vec{A} = \frac{q_{in}}{\epsilon_0}$

$\Rightarrow E.4\pi r^2 = \left(\frac{\pi\rho_0 R^3}{3\epsilon_0}\right)$

$\therefore$ Electric field outside the ball, $E = \frac{\rho_0 R^3}{12 \epsilon_0 r^2}$

88. (c) Field lines originate perpendicular from positive charge and terminate perpendicular at negative charge. Further this system can be treated as an electric dipole.

89. (d) Given: Length of wire $L = 20$ cm

charge $Q = 10^3 \varepsilon_0$

We know, electric field at the centre of the semicircular arc

$E = \frac{2K\lambda}{r} \Rightarrow E = \frac{2K\left(\frac{2Q}{\pi r}\right)}{r} \left[As\lambda = \frac{2Q}{\pi r}\right]$

$= \frac{4KQ}{\pi r^2} = \frac{4KQ\pi^2}{\pi L^2} = \frac{4\pi KQ}{L^2} = 25 \times 10^3 N/C\hat{i}$

90. (c) Given,

Electric field E = 150 N/C

Total surface charge carried by earth q = ?

By Gauss law

$\frac{q_{in}}{\epsilon_o} = \int \vec{E}.\vec{da} \Rightarrow \frac{q_{in}}{\epsilon_o} = |E| \times 4\pi r^2$

$\Rightarrow q_{in} = 4\pi\epsilon_o |E| r^2$

$= 4\pi \times 8.85 \times 10^{-12} \times 150 \times (6.37 \times 10^6)^2 \simeq 680$ kC.

As electric field directed inward hence

q = – 680 kC

91. **(b)** $F = qE = mg$ $(q = 6e = 6 \times 1.6 \times 10^{-19})$

$$\text{Density (d)} = \frac{\text{mass}}{\text{volume}} = \frac{m}{\frac{4}{3}\pi r^3} \quad \text{or } r^3 = \frac{m}{\frac{4}{3}\pi d}$$

Putting the value of d and m $\left(= \frac{qE}{g}\right)$ and solving we get r $= 7.8 \times 10^{-7}$ m

92. **(c)** $|E| \propto r, r < 0$

$|E| \propto \frac{1}{r^2}, r > 0$

So the graph will be

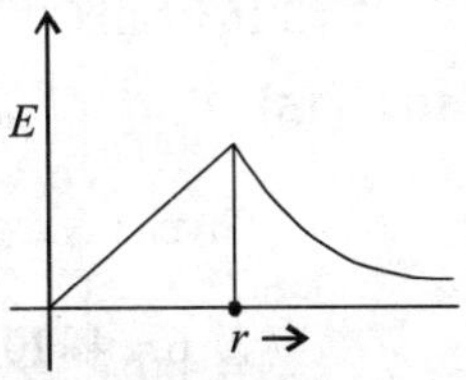

93. **(c)** Electric lines of force due to a positive charge is spherically symmetric.

All the charges are positive and equal in magnitude. So repulsion takes place. Due to which no lines of force are present inside the equilateral triangle and the resulting lines of force obtained as shown:

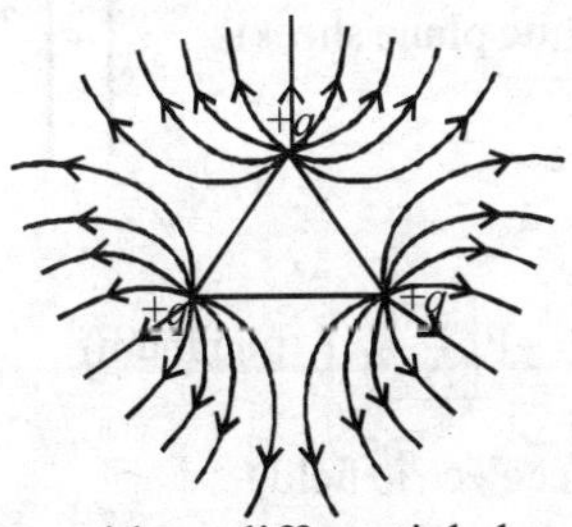

94. **(c)** Let us consider a differential element dl subtending at angle $d\theta$ at the centre Q as shown in the figure. Linear charge density

$$\lambda = \frac{q}{Qr}$$

Charge on the element,

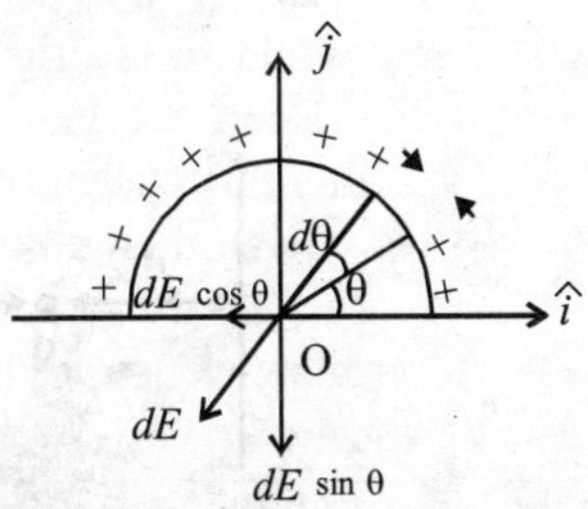

$$dq = \left(\frac{q}{\pi r}\right) dl$$

$$= \frac{q}{\pi r}(r d\theta) = \left(\frac{q}{\pi}\right) d\theta \quad (\because dl = r d\theta)$$

Electric field at the center O due to dq is

$$dE = \frac{1}{4\pi \epsilon_0} \cdot \frac{dq}{r^2} = \frac{1}{4\pi \epsilon_0} \cdot \frac{q}{\pi r^2} d\theta$$

Resolving dE into two rectangular component, we find the component $dE \cos\theta$ will be counter balanced by another element on left portion. Hence resultant field at O is the resultant of the component $dE \sin\theta$ only.

$$\therefore E = \int dE \sin\theta = \int_0^{\pi} \frac{q}{4\pi^2 r^2 \epsilon_0} \sin\theta d\theta$$

$$= \frac{q}{4\pi^2 r^2 \epsilon_0}[-\cos\theta]_0^{\pi} = \frac{q}{4\pi^2 r^2 \epsilon_0}(+1+1) = \frac{q}{2\pi^2 r^2 \epsilon_0}$$

The direction of E is towards negative y-axis.

$$\therefore \vec{E} = -\frac{q}{2\pi^2 r^2 \epsilon_0}\hat{j}$$

95. **(a)** Let us consider a spherical shell of radius x and thickness dx.

Due to shpherically symmetric charge distribution, the chrge on the spherical surface of radius x is

$$dq = \rho dv = \rho \cdot 4\pi x^2 dx = \rho_0\left(\frac{5}{4} - \frac{x}{R}\right) \cdot 4\pi x^2 dx$$

$\therefore$ Total charge in the spherical region from centre to r $(r < R)$ is

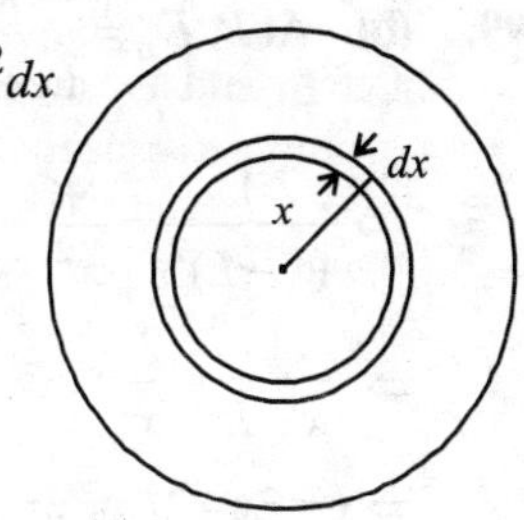

$$q_{in} = \int dq = 4\pi\rho_0 \int_0^r \left(\frac{5}{4} - \frac{x}{R}\right) x^2 dx$$

$$= 4\pi\rho_0\left[\frac{5}{4} \cdot \frac{r^3}{3} - \frac{1}{R} \cdot \frac{r^4}{4}\right]$$

$$= \pi\rho_0 r^3\left(\frac{5}{3} - \frac{r}{R}\right)$$

By Gauss law

$$\frac{q_{in}}{\epsilon_o} = \oint \vec{E} \cdot \vec{da}$$

$$\Rightarrow \frac{\pi\rho_o r^3}{\epsilon_o}\left(\frac{5}{3} - \frac{r}{R}\right) = |E| \times 4\pi r^2$$

$$\Rightarrow E = \frac{\rho_o r}{4\epsilon_o}\left(\frac{5}{3} - \frac{r}{R}\right)$$

96. **(a)** In conservative field work done is independent of path and work done by force over a closed loops is zero.

97. **(b)** Let us consider a spherical shell of thickness dr and radius r.

The area of this spherical shell $= 4\pi r^2$.

The volume of this spherical shell $= 4\pi r^2 dr$. The charge enclosed within shell

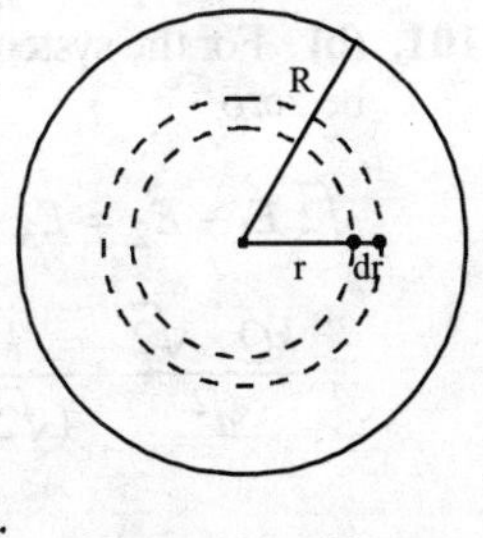

$$dq = \left[\frac{Q.r}{\pi R^4}\right][4\pi r^2 dr] = \frac{4Q}{R^4} r^3 dr$$

The charge enclosed in a sphere of radius r_1 can be calculated by

$$Q = \int dq = \frac{4Q}{R^4}\int_0^{r_1} r^3 dr = \frac{4Q}{R^4}\left[\frac{r^4}{4}\right]_0^{r_1} = \frac{Q}{R^4} r_1^4$$

By Gauss law,

$$\oint \vec{E} \cdot \vec{dA} = \frac{q_{in}}{\epsilon_o} \Rightarrow |E| \times 4\pi r_1^2 = \frac{1}{\epsilon_o}\left(\frac{Q r_1^4}{R^4}\right)$$

$$\Rightarrow |E| = \frac{Q r_1^2}{4\pi \epsilon_o R^4}$$

98. **(a)** The electric field inside a thin spherical shell of radius R has charge Q spread uniformly over its surface is zero.

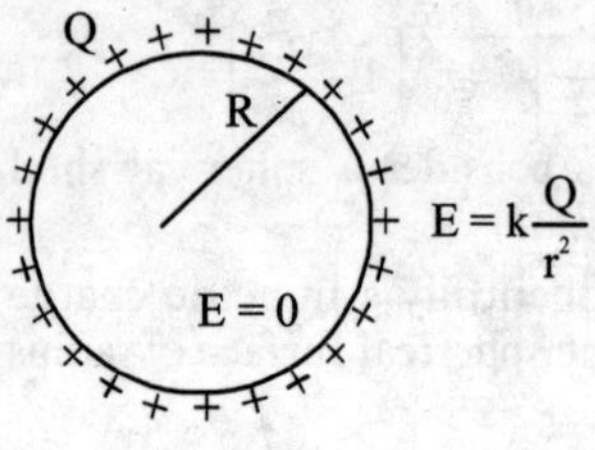

Outside the shell the electric field is $E = k\frac{Q}{r^2}$. These characteristics are represented by graph (a).

99. **(b)** At P, $E_p = \frac{-K2q}{(x-L)^2} + \frac{K8q}{x^2} = 0$

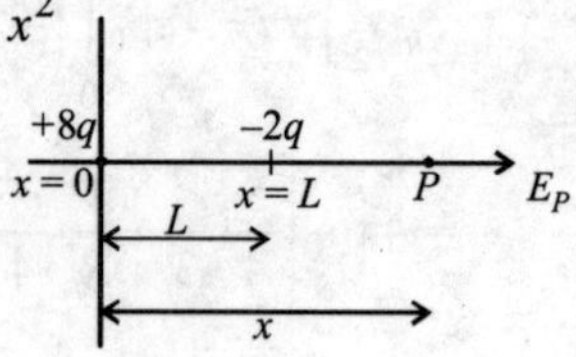

$\Rightarrow \frac{1}{(x-L)^2} = \frac{4}{x^2}$

$\Rightarrow \frac{1}{x-L} = \frac{2}{x}$

$\Rightarrow x = 2x - 2L$ or $x = 2L$

100. **(c)** $T\sin\theta = qE$ .... (i)

$T\cos\theta = mg$ .... (ii)

Dividing (i) by (ii),

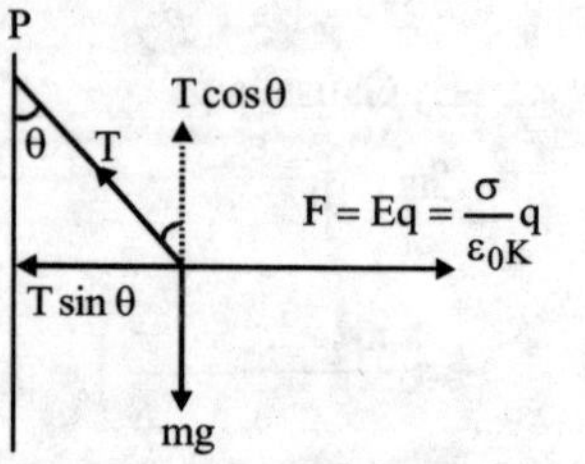

$\tan\theta = \frac{qE}{mg}$

$= \frac{q}{mg}\left(\frac{\sigma}{\varepsilon_0 K}\right) = \frac{\sigma q}{\varepsilon_0 K.mg}$

$\therefore \quad \sigma \propto \tan\theta$

101. **(b)** For the system to be equilibrium, net field at A should be zero

$\sqrt{2}\,E_1 + E_2 = E_3$

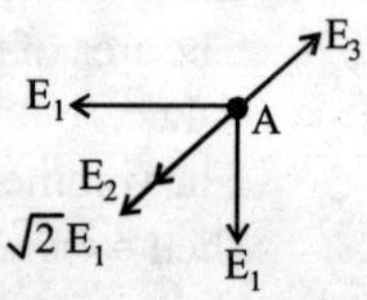

$\therefore \quad \frac{kQ\times\sqrt{2}}{a^2} + \frac{kQ}{(\sqrt{2}\,a)^2} = \frac{kq}{\left(\frac{a}{\sqrt{2}}\right)^2}$

$\Rightarrow \frac{Q\sqrt{2}}{1} + \frac{Q}{2} = 2q$

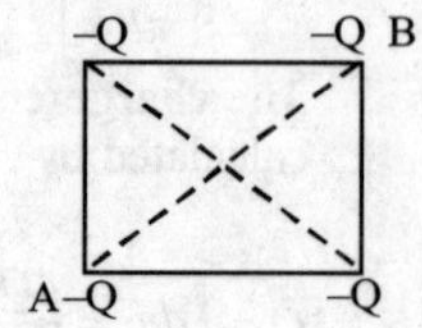

$\Rightarrow q = \frac{Q}{4}(2\sqrt{2}+1)$

102. **(c)** Given, Electric field, $E = 3\times10^4$
Mass of the drop, $m = 9.9\times10^{-15}$ kg
At equilibrium, coulomb force on drop balances weight of drop.
$qE = mg$

$\Rightarrow q = \frac{mg}{E} \Rightarrow q = \frac{9.9\times10^{-15}\times10}{3\times10^4} = 3.3\times10^{-18}$C

103. **(d)** Linear charge density,

$\lambda = \frac{q}{l}$

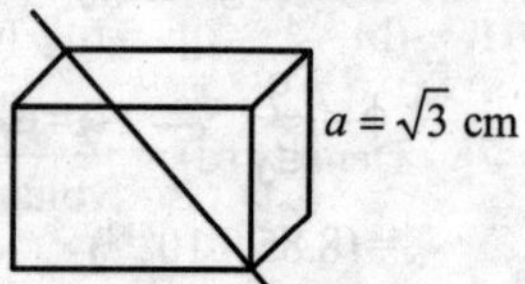

Net flux, $\phi = \frac{q_{enc}}{\varepsilon_0} = \frac{\lambda\cdot\sqrt{3}a}{\varepsilon_0}$

$= 2\times10^{-9}\times\sqrt{3}\times\sqrt{3}\times10^{-2}\times36\pi\times10^9$ Nm^{2}C^{-1}

$= 2.16\,\pi$ Nm^{2}C^{-1}

104. **(15)** $\phi = \vec{E}\cdot\vec{A} = (2\hat{i}+4\hat{j}+6\hat{k})\times10^3\cdot A(0\hat{i}+\hat{j}+0\hat{k})$

Given : $\phi = 6$ Nm^2/C

$\therefore\ 6 = 4\times10^3 A$

$\therefore\ A = 1.5\times10^{-3}$ m^2 = 15 cm^2

105. **(c)** $\tau = PE\sin\theta$

$\Rightarrow \quad \tau = 0 \qquad (\because\ \theta = 0°)$

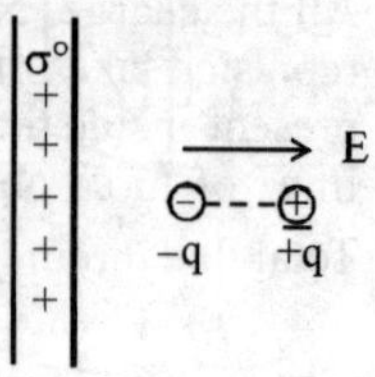

The expression of electric field of infinite plane sheet is

$E = \frac{\sigma_0}{2\epsilon_0}$

$U = -\vec{P}.\vec{E} = -P\left(\frac{\sigma_0}{2\epsilon_0}\right)$ (minimum)

106. **(b)** Here net electric field

$E = \frac{\sigma}{2\epsilon_0} + \frac{2\sigma}{2\epsilon_0} = \frac{3\sigma}{2\epsilon_0}$

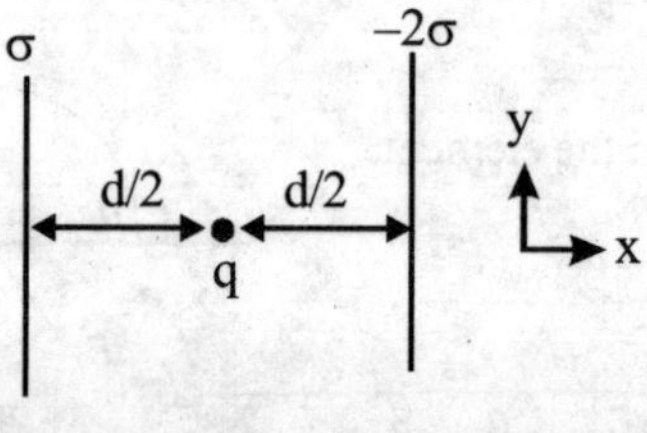

$\therefore F_{net} = qE = \frac{3\sigma}{2\epsilon_0}q$

107. **(d)** For spherical shell,

$E_{inside} = 0, E_{outside} = \frac{\sigma}{\epsilon_0}\left(\frac{R}{r}\right)^2$

For infinite plane sheet,

$E = \frac{\sigma}{2\epsilon_0}$

For two oppositely charged infinite parallel sheets,

$E = \frac{\sigma}{\epsilon_0}$

108. **(a)** Given, $\phi = -2\times10^4\ \frac{Nm^2}{C}$

Radius, r = 8.0 cm

Flux is given by

$$\phi = \frac{q}{\epsilon_0} \Rightarrow q = \epsilon_0 \phi$$

$= (8.85 \times 10^{-12}) \times (-2 \times 10^4)$

$q = -17.7 \times 10^{-8}\,C$

109. (d) The time period of an electric dipole is

$$T = 2\pi\sqrt{\frac{I}{PE}} = 2\pi\sqrt{\frac{m\left(\frac{l}{2}\right)^2 \times 2}{(ql).E_0}} = 2\pi\sqrt{\frac{ml}{2qE_0}}$$

110. (48)

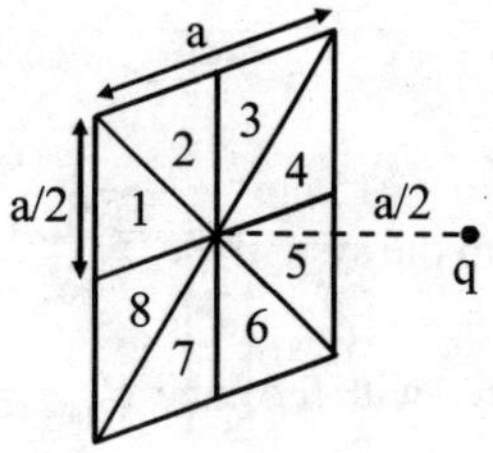

Total flux through square $= \phi_T = \frac{q}{\epsilon_0}\left(\frac{1}{6}\right)$

$\therefore$ Flux through shaded portion is $\phi_{shaded} = \frac{5}{8}\phi_T$

$= \frac{5}{8} \times \frac{q}{\epsilon_0}\frac{1}{6} = \frac{5}{48}\frac{1}{\epsilon_0} = \frac{5}{\rho} \times \frac{1}{\varepsilon_0}$ (given) $\therefore P = 48$

111. (d)

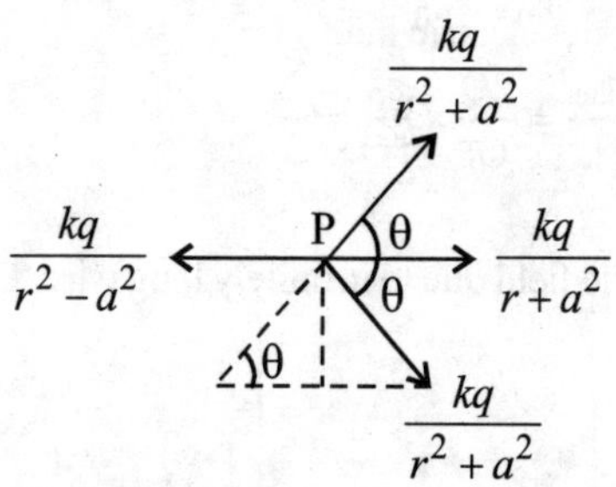

$$\frac{kq}{(r-a)^2} = \frac{kq}{(r+a)^2} + \frac{2kq}{(r^2+a^2)}\cos\theta$$

$$\frac{1}{(r-a)^2} = \frac{1}{(r+a)^2} + \frac{2a}{(r^2+a^2)^{\frac{3}{2}}}$$

$$\frac{1}{(r-a)^2} - \frac{1}{(r+a)^2} = \frac{2a}{(r^2+a^2)^{\frac{3}{2}}}$$

$$\frac{4ra}{(r^2-a^2)^2} = \frac{2a}{(r^2+a^2)^{\frac{3}{2}}}$$

$$\Rightarrow \frac{2r}{(r^2-a^2)^2} = \frac{1}{(r^2+a^2)^{\frac{3}{2}}}$$

$$\frac{4r^2}{(r^2-a^2)^4} = \frac{1}{(r^2+a^2)^3}$$

$$\Rightarrow 4r^2(r^2+a^2)^3 = (r^2-a^2)^4$$

$$4r^8\left(1+\frac{a^2}{r^2}\right)^3 = r^8\left(1-\frac{a^2}{r^2}\right)^4$$

$$4\left(1+\frac{a^2}{r^2}\right)^3 = \left(1-\frac{a^2}{r^2}\right)^4$$

$$\therefore \frac{a}{r} \approx 3$$

112. (a) Total charge inside the given cube

$$q_{in} = \frac{\lambda\frac{a}{2}}{4} = \frac{\lambda a}{8} \quad \therefore \text{Total flux } \phi = \frac{q_{in}}{\varepsilon_0} = \frac{\lambda a}{8\varepsilon_0}$$

113. (16)

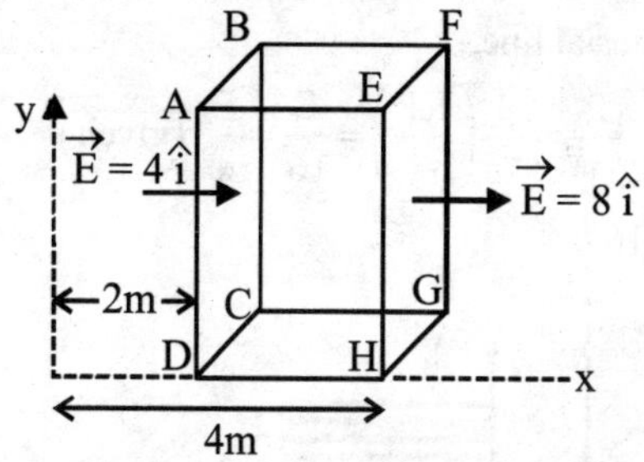

Electric field, $\vec{E} = 2x\hat{i}$

Electric flux,

$\phi = \vec{E}.\vec{A}$

Only those surfaces having area vector parallel to x axis will contribute to the flux.

Total flux entering

$\phi_1 = -2 \times 2 \times 4 = -16\,Nm^2/c$

Total flux leaving,

$\phi_2 = 8 \times 4 = 32\,Nm^2/c$

$\phi_{Total} = \phi_1 + \phi_2 = -16 + 32 = 16\,Nm^2/c$

114. (b) Using Gauss's law, $\phi = \frac{q}{\epsilon_0}$

Here, q = charge inside the closed surface

$$\therefore \phi = \frac{q + (-2q) + 5q}{\epsilon_0}$$

$$\Rightarrow \phi = \frac{4q}{\epsilon_0}$$

115. (12) Given,

Area of surface, $A = 4m^2$

Electric flux, $\phi = \vec{E}\cdot\vec{A} = \vec{E}.\hat{n}(A)$

$$= \left(\frac{2\hat{i}+6\hat{j}+8\hat{k}}{\sqrt{6}}\right)\cdot 4\left(\frac{2\hat{i}+\hat{j}+\hat{k}}{\sqrt{6}}\right)$$

$$= 4 \times \frac{(4+6+8)}{6} = 12\,Vm$$

116. (c) Using Gauss law

$$\int \vec{E}\cdot d\vec{A} = \frac{Q}{\varepsilon_0}$$

$$EdA = \frac{\sigma \times dA}{\varepsilon_0} \qquad [\because Q = \sigma dA]$$

$$\Rightarrow E = \frac{\sigma}{\varepsilon_0}$$

117. (2) Electric field at point P is,

$$\vec{E} = \left(\frac{\sigma}{2\varepsilon_0} + \frac{2\sigma}{2\varepsilon_0} + \frac{\sigma}{2\varepsilon_0}\right)(-\hat{i}) = -\frac{2\sigma}{\varepsilon_0}\hat{i}$$

$\therefore x = 2.$

118. (16) At axial line,

$$E_P = \frac{2KP}{r^3} = E$$

At equitorial line,

$$E_R = \frac{KP}{(2r)^3} = \frac{1}{16}\left(\frac{KP}{r^3}\right) = \frac{E}{16} = \frac{E}{x} \text{ (given)}$$

$\therefore x = 16$

119. (16)

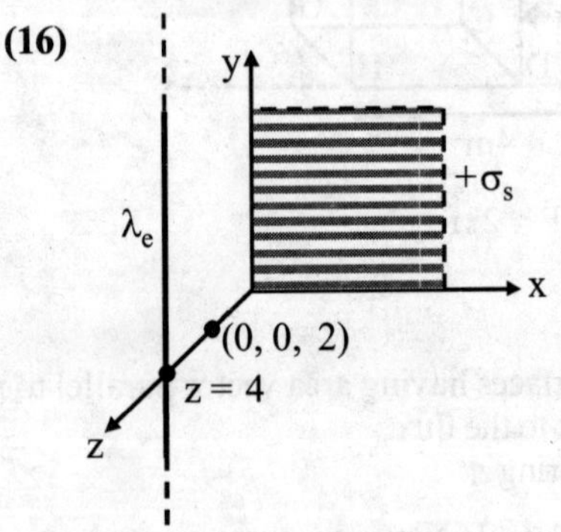

Here, $E_s = \frac{\sigma_s}{2\varepsilon_0}$ and $E_\ell = \frac{\lambda e}{2\pi r \varepsilon_0}$

$$\therefore \frac{E_s}{E_\ell} = \frac{\sigma}{2\varepsilon_0} \times \frac{2\pi\varepsilon_0 r}{\lambda}$$

$$= \frac{\pi \times \sigma r}{\lambda} = \frac{\pi \times 2\lambda \times 2}{\lambda} = \frac{4\pi}{1}$$

$\therefore 4\pi : 1 = \pi\sqrt{n} : 1 \qquad \therefore n = 16$

120. (b)

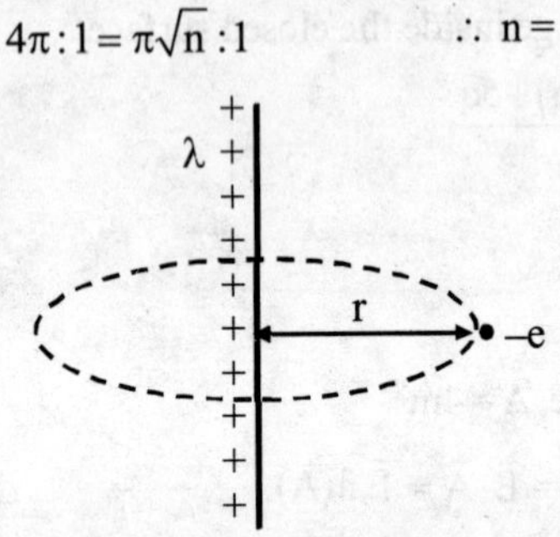

Electric field E at a distance r due to infinite long wire $E = \frac{2k\lambda}{r}$ where λ = linear charge density

Force on electron $F = eE$

$$F = e\left(\frac{2k\lambda}{r}\right) = \frac{2k\lambda e}{r}$$

This force will provide required centripetal force

$$F = \frac{mv^2}{r} = \frac{2k\lambda e}{r} \text{ or, } v = \sqrt{\frac{2k\lambda e}{m}}$$

$$\therefore KE = \frac{1}{2}mv^2 = \frac{1}{2}m\left(\frac{2k\lambda e}{m}\right) = k\lambda e \text{ (constant)}$$

So option (b) represents correct KE – r graph

121. (b)

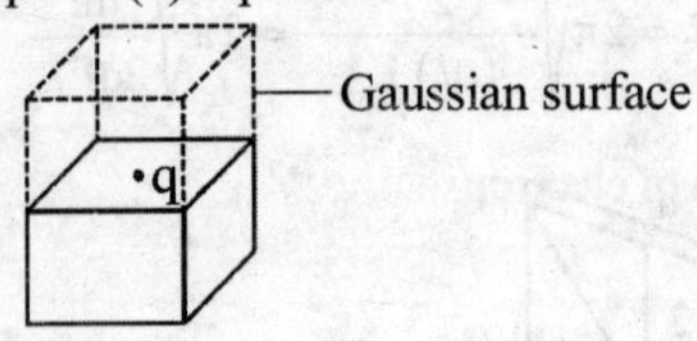

Total flux emerges from the system $\phi_T = \frac{q}{\varepsilon_0}$

Therefore the flux linked with the cube $\phi_{cube} = \frac{q}{2\varepsilon_0}$

122. (c) Flux is given by $\phi = \frac{q}{\varepsilon_0}$

$$\phi_{\text{Inner cube}} = \frac{q_1}{\varepsilon_0} = \frac{2Q}{\varepsilon_0}$$

$$\phi_{\text{Outer cube}} = \frac{q_2}{\varepsilon_0} = \frac{5Q}{\varepsilon_0}$$

$$\therefore \frac{\phi_{\text{Inner cube}}}{\phi_{\text{Outer cube}}} = \frac{2Q}{\varepsilon_0} \times \frac{\varepsilon_0}{5Q} = \frac{2}{5}$$

123. (b) Electric field due to infinitely long wire, $E = \frac{\lambda}{2\pi\varepsilon_0 r}$

$F = qE$

$$= \frac{\lambda . q}{2\pi\varepsilon_0 r} = m\omega^2 r \qquad \left(\because k = \frac{1}{4\pi\varepsilon_0}\right)$$

$$\frac{2k\lambda q}{r} = m\omega^2 r \Rightarrow \omega^2 = \frac{2k\lambda q}{mr^2}$$

$$\left(\frac{2\pi}{T}\right)^2 = \frac{2k\lambda q}{mr^2} \Rightarrow T = 2\pi r\sqrt{\frac{m}{2k\lambda q}}$$

124. (b)

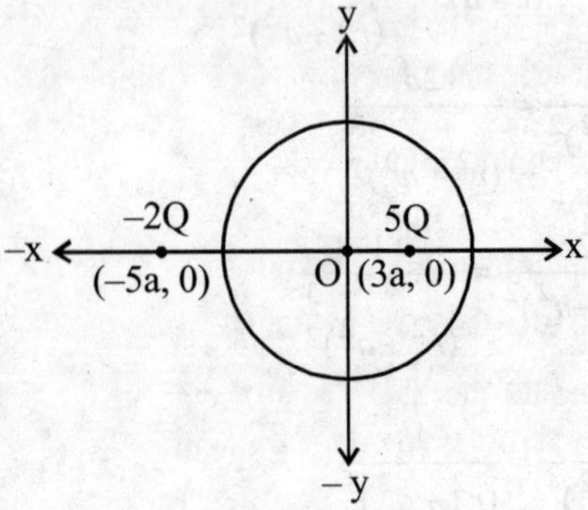

Using Gauss law,

Electric flux, $\phi = \frac{q_{enclosed}}{\varepsilon_0} = \frac{5Q}{\varepsilon_0}$

125. (8) For maximum value of σ, initially, electron must move away and strikes the plate due to retardation

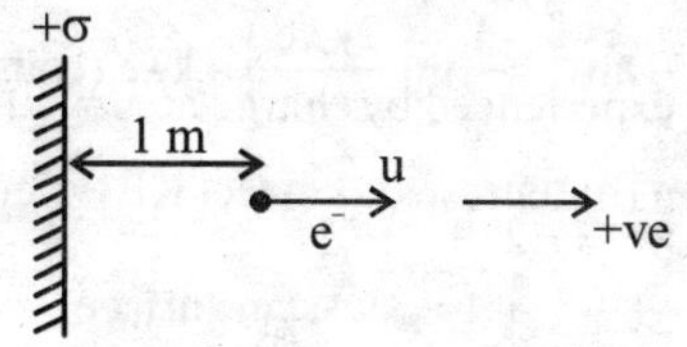

Acceleration of electron,

$$a = \frac{F}{m} = -\frac{eE}{m} = \frac{e\sigma}{2\varepsilon_0 m}$$

Now, $s = ut + \frac{1}{2}at^2$

$t = 1s, u = 1$ m/s, $s = -1$ m

$$-1 = 1 \times 1 - \frac{1}{2}\left(\frac{e\sigma}{2\varepsilon_0 m}\right) \times 1^2$$

$$\therefore \sigma_{max} = \left(\frac{8m\varepsilon_0}{e}\right)\frac{C}{m^2} = \alpha\left(\frac{m\varepsilon_0}{e}\right)\frac{C}{m^2} \text{ (given)}$$

$\therefore \alpha = 8$

126. (a) $\vec{E} = \left(6\hat{i} + 5\hat{j} + 3\hat{k}\right) N/C$

$\vec{A} = 30\hat{i}\, m^2$

Electric flux, $\phi = \vec{E}.\vec{A}$

$= \left(6\hat{i} + 5\hat{j} + 3\hat{k}\right).\left(30\hat{i}\right) = 6 \times 30 = 180$ V-m

127. (2) Torque acting on the dipole,

$\vec{\tau} = \vec{p} \times \vec{E}$

$\vec{p} = q\vec{\ell} = 4 \times \left(\hat{i} - \hat{j} + \hat{k}\right)$

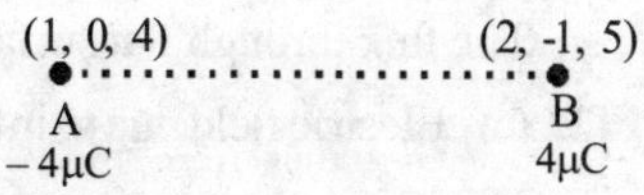

$= \left(4\hat{i} - 4\hat{j} + 4\hat{k}\right)\mu C - m$

$\therefore \vec{\tau} = \vec{P} \times \vec{E} = \left(4\hat{i} - 4\hat{j} + 4\hat{k}\right) \times \left(20\hat{i}\right) \times 10^{-6}$ Nm

$= \left(8\hat{k} + 8\hat{j}\right) \times 10^{-5} = 8\sqrt{2} \times 10^{-5}$ Nm

$= 8\sqrt{\alpha} \times 10^{-5}$ Nm (given) $\therefore \alpha = 2$

128. (c) Electric field due to an electric dipole at distance r on the equatorial plane, $E = \frac{kp}{r^3}$

129. (b) Dipole mement, $\left|\vec{P}\right| = qd$

$= 0.01 \times 0.4 \times 10^{-3} = 4 \times 10^{-6}$

And torque, $\left|\vec{\tau}\right| = PE \sin\theta$

$= 4 \times 10^{-6} \times 10 \times 10^{-5} \times \sin 30$

$\therefore \left|\vec{\tau}\right| = 2 \times 10^{-10}$

130. (a) Given,

Dipole moment,

$\vec{P} = 30 \times 10^{-5}$ Cm

An electric dipole consists of two equal and opposite charges.

∴ Net charge enclosed by a closed surface is zero.

Using Gauss law

$$\phi = \frac{Q_{in}}{\varepsilon_0} \Rightarrow \phi = 0$$

131. (c)

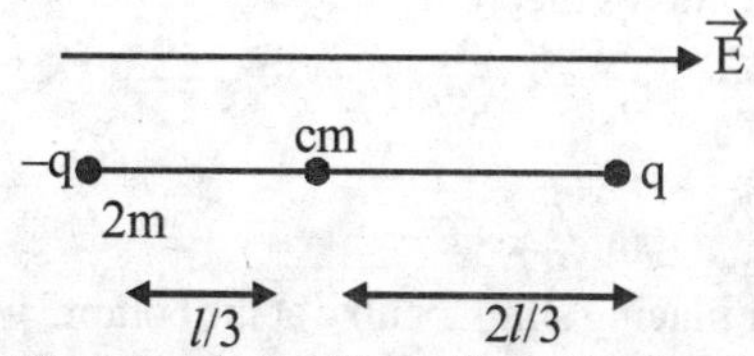

If released, it will oscillate about centre of mass.

$\tau_{com} = q E l \sin\theta$

For small 'θ'

$\tau_{com} = q E l \theta$

Using $T = I\alpha$

$$\Rightarrow qlE.\theta = \left[2m\frac{l^2}{9} + m\frac{4l^2}{9}\right]\alpha$$

$$\Rightarrow qlE.\theta = \frac{2ml^2}{3}\alpha \Rightarrow \alpha = \frac{3qE\theta}{2ml}$$

$$\Rightarrow \omega = \sqrt{\frac{3qE}{2ml}}$$

132. (288)

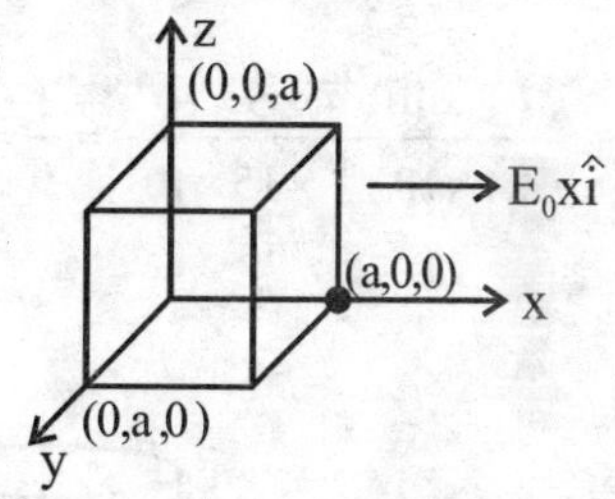

For plane, x = a

$$\phi(\text{flux}) = \iint E_0 a\hat{i}.dydz\hat{i} = \int_0^a\int_0^a E_0 a\,dydz$$

By Gauss law $\phi = \oint \vec{E}.\vec{dA} = \frac{q_{in}}{\epsilon_0}$

For plane x = 0, y = 0, y = a, z = 0 and z = a, we will have $\int \vec{E}.\vec{dA} = 0$ and there will be flux through x = a only

$$\Rightarrow \int_0^a\int_0^a E_0 a\,dydz = \frac{q_{in}}{\epsilon_0} \Rightarrow E_0 a^3 = \frac{q_{in}}{\epsilon_0} \Rightarrow q_{in} = \epsilon_0 E_0 a^3$$

$= 9 \times 10^{-12} \times 4 \times 10^4 \times (2 \times 10^{-2})^3 = 288 \times 10^{-14}$ C

133. (640) Flux through surface x = 0, y = 0, y = 20 cm, z = 0 and z = 20 cm)

will be zero as for these surface $\vec{E} \perp \vec{A}$

Flux through surface (x = 0.2 m) = $4000 \times 0.2^2 \times 0.2^2$

= 6.4 Vm = 640 Vcm

134. (12) Given, electric field

$\vec{E} = 2x^2\hat{i} - 4y\hat{j} + 6\hat{k}$

Electric flux for face, x = 1, $\phi = 2 \times (1)^2 \times 2 \times 3 = 12$

Electric flux for face, y = 2, $\phi = -4 \times 2 \times 3 \times 1 = -24$

Electric flux for face, z = 3, $\phi = 6 \times 1 \times 2 = 12$

Electric flux for faces x = 0, y = 0, is zero.

Electric flux for z = 0, $\phi = -12$

$\phi_{net} = -24 + 12 = -12$

From the Gauss's theorem

$-12 = \frac{q}{\epsilon_0}$

$|q| = 12\,\epsilon_0$

135. (d) After placing similar cubic at the bottom, we get cube of side 2L having charge q at its centre.

From the Gauss's law electric flux through whole cube, $\phi = \frac{q}{\varepsilon_0}$

Flux passing through shaded face

$\phi = \frac{q/\varepsilon_0}{6} = \frac{q}{6\varepsilon_0}$

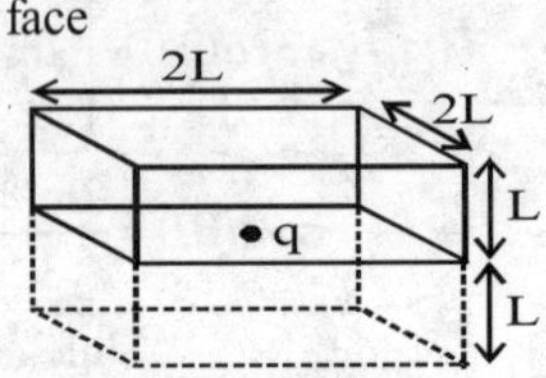

136. (6) Maximum torque,

$|\tau|_{max} = PE$ or, $\tau_1 = P_1E_1$ and $\tau_2 = P_2E_2$

$\therefore \frac{\tau_1}{\tau_2} = \frac{P_1E_1}{P_2E_2} = \frac{1.2 \times 10^{-30} \times 5 \times 10^4}{2.4 \times 10^{-30} \times 15 \times 10^4} = \frac{1}{6} = \frac{1}{x}$

$\therefore x = 6$

137. (1) By Gauss law

$\int Eds\cos 0 = \frac{q}{\varepsilon_0}$

$\Rightarrow E.2\pi xh = \frac{\rho \times \pi x^2 h}{\varepsilon_0}$

$\Rightarrow E = \frac{\rho x}{2\varepsilon_0}$

$\Rightarrow E = \frac{\rho}{2\varepsilon_0} \times \frac{2\varepsilon_0}{\rho} = 1\,Vm^{-1}$

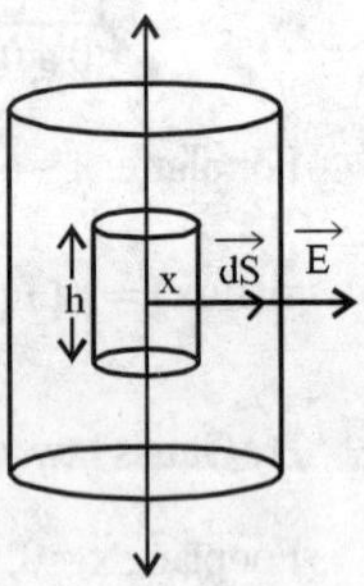

138. (17) The force between 5 μC and 0.3 μC given by

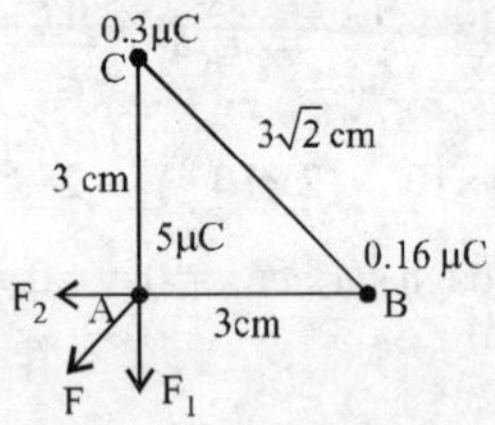

$F_1 = \frac{k \times 5 \times 0.3 \times 10^{-12}}{9 \times 10^{-4}}$ $(\because K = 9 \times 10^9)$

$= \frac{9 \times 10^9 \times 5 \times 0.3 \times 10^{-12}}{9 \times 10^{-4}} = 15\,N$

The force between 5 μC and 0.16 μC given by

$F_2 = \frac{9 \times 10^9 \times 5 \times 0.16 \times 10^{-12}}{9 \times 10^{-4}} = 8\,N$

Net force experienced by charge at A = $\sqrt{F_1^2 + F_2^2}$

$= \sqrt{15^2 + 8^2} = \sqrt{289} = 17\,N$

139. (45) Here, $\phi_e = \int \vec{E}.dA \Rightarrow \frac{\phi_e}{A} = E$

$\Rightarrow \frac{\phi_e}{A} = \frac{\rho R}{3\epsilon_o} = \frac{2 \times 10^0 \times 6}{3 \times 8.85 \times 10^{-12}} = 45 \times 10^{10}\,NC^{-1}$

140. (a)

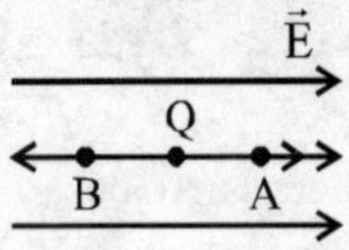

At point '*A*' electric field will increase because field due to charge 'Q' will add up in external electric field.

At point '*B*' electric field will decrease because field due to charge 'Q' will oppose the external electric field.

In non-uniform electric field, when dipole is placed at a point, then net force experienced by it is not zero.

141. (b) Total flux through complete spherical surface is $\frac{q}{\epsilon_0}$.

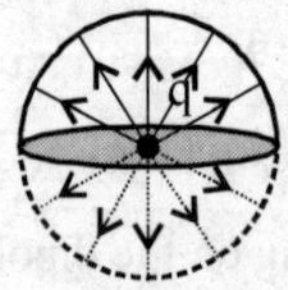

So flux, through curved surface will be $\frac{q}{2\,\epsilon_0}$.

The flux through flat surface will be zero.

142. (c) Electric field due to infinite sheet is given by $E = \frac{\sigma}{2\epsilon_0}$, clearly E is independent of distance

So, $E_1 = E_2 = \frac{\sigma}{2\,\epsilon_0}$

143. (a) As charge particle is moving in circular path. So,

$qE = \frac{mv^2}{r}$

$\Rightarrow mv^2 = qEr$

$\therefore K.E = \frac{mv^2}{2} = \frac{1}{2}qEr$ ...(i)

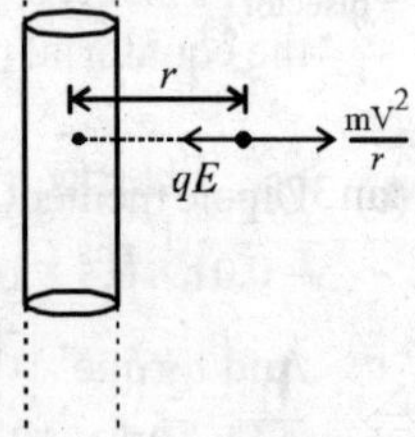

Now, by Gauss's law

$E \times 2\pi rl = \frac{q_{in}}{\epsilon_0}$

$E \times 2\pi rl = \frac{\rho \times \pi R^2 l}{\epsilon_0}$

$$E = \frac{\rho R^2}{2\epsilon_0 r} \quad ...(ii)$$

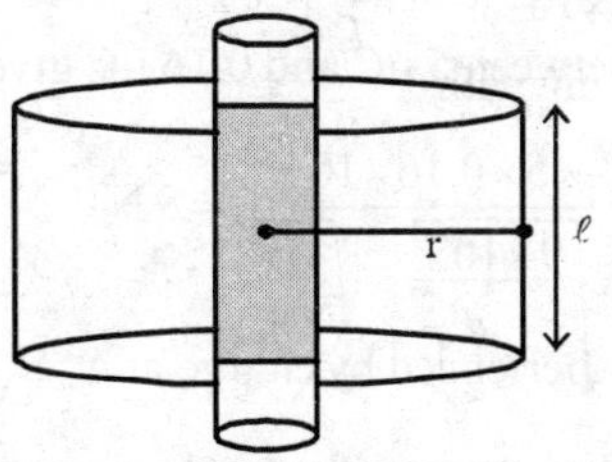

From (i) and (ii), we get

$$K \cdot E = \frac{1}{2}mv^2 = \frac{1}{2}q \times \frac{\rho R^2}{2\epsilon_0 r} \times r = \frac{\rho q R^2}{4\epsilon_0}$$

144. (b) At bottom surface, electric field is zero as $y = 0$

$\therefore$ Electric flux, $\phi_1 = 0$; At top surface, $y = 0.5$

$\therefore$ Electric flux, $\phi_2 = EA = (150y^2)(0.5)^2$

$= 150 \times (0.5)^2 \times (0.5)^2 = \frac{150}{16}$

And flux through all other surface is zero because $\vec{E} \perp \vec{A}$ for each of them

$\therefore \phi_{total} = \phi_1 + \phi_2 = \frac{150}{16}$

Using Gauss's law $\phi_{total} = \frac{Q_{in}}{\epsilon_0} \Rightarrow \frac{150}{16} - \frac{Q_{in}}{\epsilon_0}$

$\Rightarrow Q_{in} = \frac{150}{16} \times 8.85 \times 10^{-12} = 8.3 \times 10^{-11} C$

145. (c) When electric field is parallel to surface, it makes 90° angle with area vector so flux = 0

As $\phi = \vec{E}.\vec{A} = EA\cos 90° = 0$

146. (c)

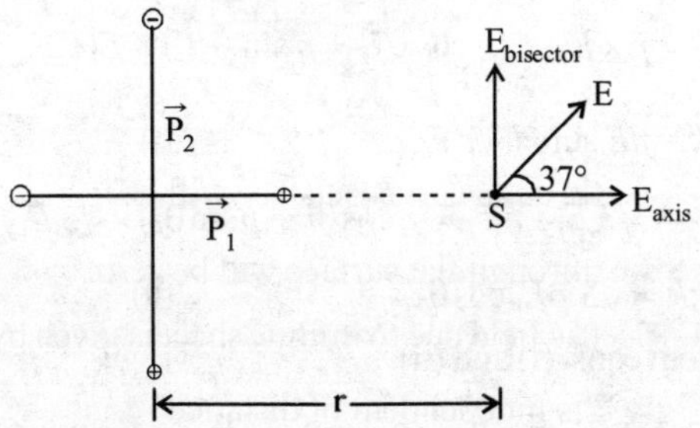

Electric field due to $\vec{P}_1$ at axis point S

$$E_{axis} = \frac{2kP_1}{r^3} \quad ...(i)$$

Electric field due to $\vec{P}_2$ at perpendicular bisector at point S.

$$E_{bisector} = \frac{kP_2}{r^3} \quad ...(ii)$$

$$\tan 37° = \frac{E_{bisector}}{E_{axis}} \Rightarrow \frac{3}{4} = \frac{\frac{KP_2}{r^3}}{K\frac{2P_1}{r^3}} \Rightarrow \frac{P_2}{P_1} = \frac{3}{2}$$

$$\Rightarrow \frac{P_1}{P_2} = \frac{2}{3}$$

147. (640) Flux, $\phi = \vec{E} \cdot \vec{A} = \left[\frac{2}{5}E_0\hat{i} + \frac{3}{5}E_0\hat{j}\right] \cdot \vec{A}$

$= \frac{E_0}{5}(2\hat{i} + 3\hat{j}) \cdot (0.4\hat{i}) = \frac{4000}{5}(2 \times 0.4) = 640 \text{ Nm}^2\text{C}^{-1}$

148. (226) $\phi_{Total} = \frac{q}{\varepsilon_0}$,

From symmetry electric flux

$\phi_{Sq} = \frac{1}{6}\left(\frac{q}{\varepsilon_0}\right) = \frac{12 \times 10^{-6}}{6 \times 8.85 \times 10^{-12}} = 225.98 \times 10^3$

$= 226 \times 10^3 \frac{Nm^2}{C}$

149. (1) Given : Electric field, $\vec{E} = \frac{3}{5}E_0\hat{i} + \frac{4}{5}E_0\hat{j}\ \frac{N}{C}$

Area (Parallel to *y-z* plane), $A_a = 0.2\hat{i}$ m^2

Area (Parallel to *x-z* plane), $A_b = 0.3\hat{j}$ m^2

Flux, $\phi_a = \vec{E} \cdot \vec{A}_a = \left(\frac{3}{5}E_0\hat{i} + \frac{4}{5}E_0\hat{j}\right) \cdot 0.2\hat{i} = \frac{6}{50}E_0$

$\phi_b = \vec{E} \cdot \vec{A}_b = \left(\frac{3}{5}E_0\hat{i} + \frac{4}{5}E_0\hat{j}\right) \cdot 0.3\hat{j} = \frac{12}{50}E_0$

$\therefore \frac{\phi_a}{\phi_b} = \frac{6/50 E_0}{12/50 E_0} = \frac{1}{2}$. So $a = 1$

150. (b) Let v be the speed of dipole.

Using energy conservation

$K_i + U_i = K_f + U_f$

$\Rightarrow 0 - \frac{2k \cdot p_1}{r^3} p_2 \cos(180°) = \frac{1}{2}mv^2 + \frac{1}{2}mv^2 + 0$

$\left(\because \text{Potential energy of interaction between dipole } -\vec{p_2} \cdot \vec{E_1} = \frac{-2p_1p_2\cos\theta}{4\pi\epsilon_0 r^3}\right)$

$\Rightarrow mv^2 = \frac{2kp_1p_2}{r^3} \Rightarrow v = \sqrt{\frac{2kp_1p_2}{mr^3}}$

When $p_1 = p_2 = p$ and $r = a$

$v = \frac{p}{a}\sqrt{\frac{1}{2\pi\epsilon_0 ma}}$

151. (–48) Flux of electric field $\vec{E}$ through any area $\vec{A}$ is defined as

$\phi = \int E.A\cos\theta$

Here, θ = angle between electric field and area vector of a surface

For surface *ABCD* Angle, $\theta = 90°$

$\therefore \phi_1 = \int E.A\cos 90° = 0$

For surface $BCGF$ $\phi_n = \int \vec{E}\,.\,\overrightarrow{dA}$

$\therefore\ \phi_{11} = \left[4x\hat{i} - (y^2+1)\hat{j}\right].4\hat{i} = 16x$

$\phi_{11} = 48\dfrac{Nm^2}{C}$; $\phi_1 - \phi_{11} = -48\dfrac{Nm^2}{C}$

152. (b) By Gauss law

$$\oint \vec{E}\cdot\overrightarrow{dA} = \frac{q_{in}}{\epsilon_o}$$

If E = cons. and $\vec{E} \parallel \vec{A}$, then $|E|\,|A| = \dfrac{q_{in}}{\epsilon_o}$

$\Rightarrow |E| = \dfrac{q_{in}}{\epsilon_o\, A}\cdot$ And $\vec{E}$ is always $\perp$ to equipotential surface so, gaussian surface is equipotential.

153. (b) By Gauss law,

$$\oint \vec{E}.d\vec{A} = \frac{Q_{in}}{\epsilon_o} = \frac{Q+q-q}{\epsilon_o} = \frac{Q}{\epsilon_o}$$

or $\quad E\times 4\pi r^2 = \dfrac{Q}{\varepsilon_0} \Rightarrow E = \dfrac{1}{4\pi\varepsilon_0}\dfrac{Q}{r^2},\ r \ge R$

So, electric field outside is same as point charge inside the shell.

154. (a) $\oint \vec{E}.d\vec{A} = \dfrac{q_{in}}{\varepsilon_0}$

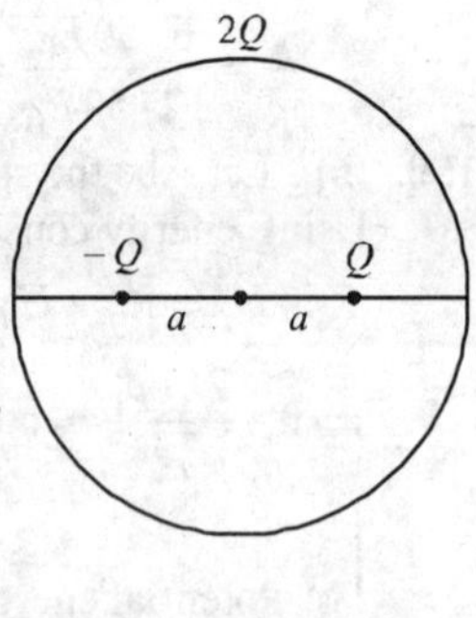

or $E\times 4\pi r^2 = \dfrac{1}{\varepsilon_0}\int \delta(4\pi r^2)dr$

$\Rightarrow E\times 4\pi r^2 = \dfrac{1}{\varepsilon_0}\int_0^r (kr)(4\pi r^2)dr$

or $\quad E\times 4\pi r^2 = \dfrac{4\pi k}{\varepsilon_0}\left(\dfrac{r^4}{4}\right)$

$\therefore\quad E = \dfrac{k}{4\varepsilon_0}r^2$ ...(i)

Also $2Q = \displaystyle\int_0^R (kr)\,(4\pi r^2)dr \ = 4\pi k\left|\frac{r^4}{4}\right|_0^R$

$Q = \dfrac{\pi k R^4}{2}$ ....(ii)

From above equations,

$E = \dfrac{Qr^2}{2\pi\varepsilon_0 R^4}$. For $r = a$, $E = \dfrac{Qa^2}{2\pi\in_o R^4}$ ....(iii)

According to given condition

$QE = \dfrac{1}{4\pi\in_o}\dfrac{Q^2}{(2a)^2}$

$= Q\cdot\dfrac{Qa^2}{2\pi\in_o R^4} = \dfrac{Q^2}{4\pi\in_o\cdot 4a^2} \Rightarrow R = a8^{1/4} \Rightarrow a = R8^{-1/4}$

155. (b) $\tau = -PE\sin\theta$

$I\alpha = -PE(\theta) \Rightarrow \alpha = \dfrac{PE}{I}(-\theta)$

On comparing with

$\alpha = -\omega^2\theta$

$$\omega = \sqrt{\frac{PE}{I}} = \sqrt{\frac{qdE}{2m\left(\frac{d}{2}\right)^2}} = \sqrt{\frac{2qE}{md}}$$

156. (b) Potential energy of a dipole is given by

$U = -\vec{P}.\vec{E} \ = -PE\cos\theta$

[Whereθ = angle between dipole and electric field]

$= -(10^{-29})(10^3)\cos 45°$

$= -0.707\times 10^{-26}\,J = -7\times 10^{-27}J$

157. (d) Electric field of equitorial plane of dipole $= -\dfrac{K\vec{P}}{r^3}$

$\therefore$ At point P, $F = +\dfrac{KP}{y^3}Q$

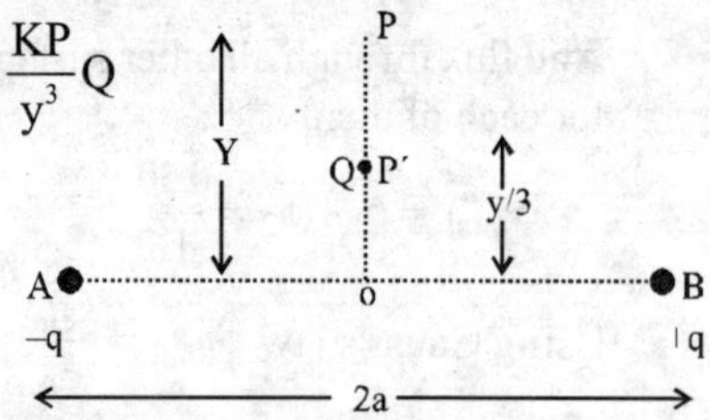

At Point P′,

$F' = +\dfrac{KPQ}{(y/3)^3}$

$= 27\,F.$

158. (a) $T = PE\sin\theta$ Torque experienced by the dipole in an electric field, $\vec{T} = \vec{P}\times\vec{E}$

$\vec{p} = p\cos\theta\ \hat{i} + p\sin\theta\ \hat{j}$; $\vec{E}_1 = E\vec{i}$

$\vec{T}_1 = \vec{p}\times\vec{E}_1 = (p\cos\theta\,\hat{i} + p\sin\theta\,\hat{j})\times E(\hat{i})$

$\tau\hat{k} = pE\sin\theta(-\hat{k})$ ...(i)

$\vec{E}_2 = \sqrt{3}E_1\hat{j}$; $\vec{T}_2 = p\cos\theta\hat{i} + p\sin\theta\hat{j})\times\sqrt{3}E_1\hat{j}$

$-\tau\hat{k} = \sqrt{3}\,pE_1\cos\theta\hat{k}$ ...(ii)

From eqns. (i) and (ii)

$pE\sin\theta = \sqrt{3}\,pE\cos\theta$; $\tan\theta = \sqrt{3}$ $\quad\therefore\quad \theta = 60°$

159. (c) The net flux linked with closed surfaces S_1, S_2, S_3 & S_4 are

For surface S_1, $\phi_1 = \dfrac{1}{\varepsilon_0}(2q)$

For surface S_2, $\phi_2 = \dfrac{1}{\varepsilon_0}(q+q+q-q) = \dfrac{1}{\varepsilon_0}2q$

For surface S_3, $\phi_3 = \dfrac{1}{\varepsilon_0}(q+q) = \dfrac{1}{\varepsilon_0}(2q)$

For surface S_4, $\phi_4 = \dfrac{1}{\varepsilon_0}(8q-2q-4q) = \dfrac{1}{\varepsilon_0}(2q)$

Hence, $\phi_1 = \phi_2 = \phi_3 = \phi_4$ i.e. net electric flux is same for all surfaces.

Keep in mind, the electric field due to a charge outside (S_3 and S_4), the Gaussian surface contributes zero net flux through the surface, because as the lines due to that charge enter the surface is same as leave it.

160. (c) Applying Gauss's law

$$\oint_S \vec{E}\cdot\vec{ds} = \frac{Q+q}{\epsilon_0};$$

$$q = \int_a^r \frac{A}{r} 4\pi r^2 dr = 2\pi A[r^2 - a^2]$$

$$\therefore E \times 4\pi r^2 = \frac{Q + 2\pi A r^2 - 2\pi A a^2}{\epsilon_0}$$

$$E \times 4\pi r^2 = (Q - 2\pi A a^2) + 2\pi A r^2$$

For E to be independent of 'r'

$$Q - 2\pi A a^2 = 0$$

$$\Rightarrow A = \frac{Q}{2\pi a^2}$$

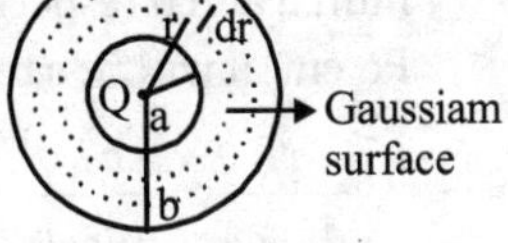

161. (c)

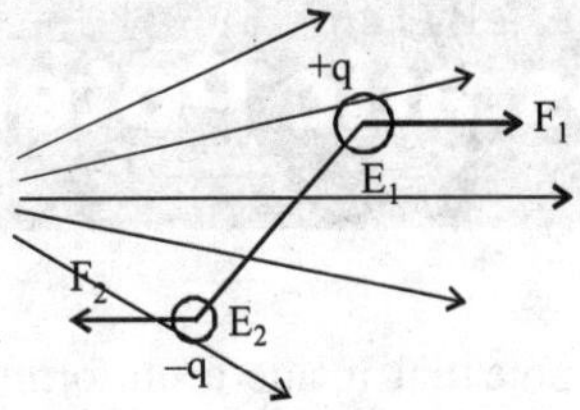

As the dipole is placed in non-uniform field, so the force acting on the dipole will not cancel each other. This will result in a translational force as well as a torque.

162. (a) The electric flux ϕ_1 entering an enclosed surface is taken as negative and the electric flux ϕ_2 leaving the surface is taken as positive, by convention. Therefore the net flux leaving the enclosed surface, $\phi = \phi_2 - \phi_1$

According to Gauss theorem

$$\phi = \frac{q}{\epsilon_0} \Rightarrow q = \epsilon_0\phi = \epsilon_0(\phi_2 - \phi_1)$$

2 Electrostatic Potential and Capacitance

1. **(b)** Electrostatic potential inside a uniformly charged spherical shell is constant and is same as that on the surface of the shell *i.e.*, V_{in} = constant = V_s
V_{in} constant *i.e.*, equipotential so work done in moving a test charge between two points inside a uniformly charged spherical shell is constant.

2. **(d)** For conducting sphere, $V = \frac{\sigma R}{\varepsilon_0}$
After contact, $V_1 = V_2$
$\therefore \sigma_1 R_1 = \sigma_2 R_2$ or, $\frac{\sigma_1}{\sigma_2} = \frac{R_2}{R_1}$
$\therefore \frac{\sigma_1}{\sigma_2} = \frac{3R}{R} = 3$

3. **(a)** Potential inside the spherical shell is equal to potential on its surface.
$\therefore$ At r = 5 cm, $V = V_{surface} = \frac{kQ}{R} = 120\text{ V}$
At r = 15 cm, $V = \frac{kQ}{r} = \frac{120 \times 10}{15} = 80\text{V}$

4. **(b)** Using $\Delta V = E(\Delta d)$
$V = E(10) \Rightarrow E = \frac{V}{10}$
Potential difference between points A and B
$V_{AB} = E.4 = \frac{V}{10} \times 4 = \frac{2V}{5}$

5. **(c)** The electric potential V at a point due to an infinitely long charged wire with linear charge density λ at a perpendicular distance r from wire is given by
$V = -\int \vec{E}.d\vec{r} = \int \frac{2k\lambda}{r} dr = 2k\lambda \ln r + c$
Potential is a scalar quantity, total potential
$V = V_x + V_y + V_z$
Net potential due to all wire
$$V = 2k\lambda \ln\sqrt{x^2+y^2} + 2k\lambda \ln\sqrt{y^2+z^2} + 2k\lambda \ln\sqrt{z^2+x^2} + c$$
For equipotential surface, V = constant
$\Rightarrow \sqrt{(x^2+y^2)(y^2+z^2)(z^2+x^2)} = \text{constant}$
$\therefore (x^2+y^2)(y^2+z^2)(z^2+x^2) = \text{constant}$

6. **(c)** For a dipole,
$E_A = \frac{2kP}{r^3} = E_0$ & $V_A = \frac{kP}{r^2} = V_0$
$\therefore E_B = \frac{kP}{(2r)^3} = \frac{kP}{8r^3} = \frac{E_0}{16}$ & $V_B = \frac{k\vec{P}.\hat{r}}{r^2} = 0$

7. **(36)** Given, Radius of half ring, R = 10 cm
Linear charge density, $\lambda = 4\text{n Cm}^{-1}$
$V = \frac{KQ}{R} = \frac{K\lambda\pi R}{R}$
$\Rightarrow V = K\lambda\pi = 9 \times 10^9 \times 4 \times 10^{-9}\pi$
$\Rightarrow V = 36\pi\ V = x\pi V \quad \therefore x = 36$

8. **(c)** Let V_A and V_B be potential due to spheres having radii a and b respectively.
Potential will be same at surface of spheres, $V_A = V_B$
$\therefore \frac{q_A}{4\pi\varepsilon_0 a} = \frac{q_B}{4\pi\varepsilon_0 b}$
$\Rightarrow \frac{q_A}{q_B} = \frac{a}{b}$

9. **(a)** Static charge can be developed due to air friction. So, it causes combustion. So, metallic chains is used to discharge excess charge to the ground.

10. **(27)**

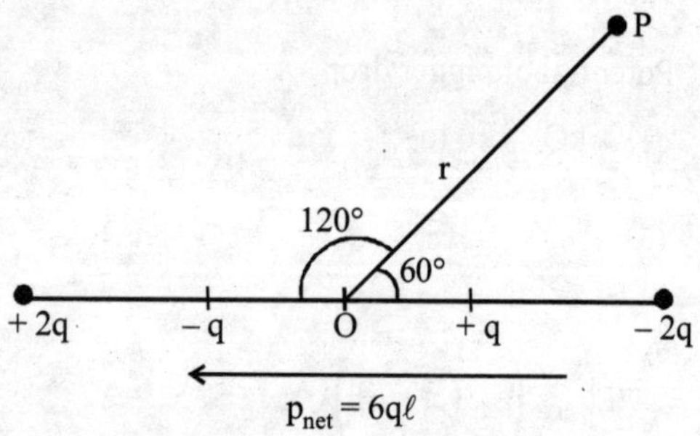

Dipole moment due to charge –q and +q separated by distance $2\ell = q(2\ell)$
Dipole moment due to charge –2q and +2q separated by $4\ell = 2q(4\ell)$
Net dipole moment = $6q\ell$
Potential at P
$V_p = \frac{K\vec{p}.\vec{r}}{r^3} = \frac{9 \times 10^9 (6q\ell)}{r^2}\cos(120°)$
$= -(27)\left(\frac{q\ell}{r^2}\right) \times 10^9 V = -\alpha\left(\frac{q\ell}{r^2}\right) \times 10^9 V$ (given)
$\Rightarrow \alpha = 27$

11. **(b)** The electrostatic potential due to an electronic dipole at a distance *r* given as $V_P \propto \frac{1}{r^2}$

12. **(c)** Electric potential, $V = \frac{KQ}{r}$
$\therefore V_P - V_Q = \Delta V = \frac{KQ}{r_1} - \frac{KQ}{r_2}$

$$r_1 = \sqrt{(\sqrt{3})^2 + (\sqrt{3})^2} = \sqrt{6}\text{m}$$

$$r_2 = \sqrt{(\sqrt{6})^2 + 0} = \sqrt{6}\text{m} \therefore \; r_1 - r_2 = 0 \;\therefore\; \Delta V = 0$$

13. **(8)** Radius of the nucleus, $R = 9 \times 10^{-13} \times 10^{-2}$ m

Electric potential at the surface of an atomic nucleus =

$$\frac{k.Ze}{R} = \frac{9\times10^{9}\times50\times1.6\times10^{-19}}{9\times10^{-13}\times10^{-2}} = 8\times10^{6}\text{ V}$$

14. **(d)** Electric lines of force are always perpendicular to equipotential surface $\therefore$ angle between force (F) and displacement(s) = 90°

$\therefore$ Work done, $W = FS\cos\theta = 0$

15. **(160)** Let charge on each drop = q

Charge on the bigger drop = 64q

Radius of small drop = r

Radius of big drop = R

Potential on each small drop $V = \frac{kq}{r} = 10\text{mV}$ (i)

$\because$ Volume remains same

$$\therefore 64\times\frac{4}{3}\pi r^3 = \frac{4}{3}\pi R^3 \Rightarrow R = 4r$$

Potential of bigger drop

$$V' = \frac{kQ}{R} = \frac{k64q}{4r} = 16\frac{kq}{r} = 16\times10\text{mV} = 160\text{mV}.$$

16. **(5)**

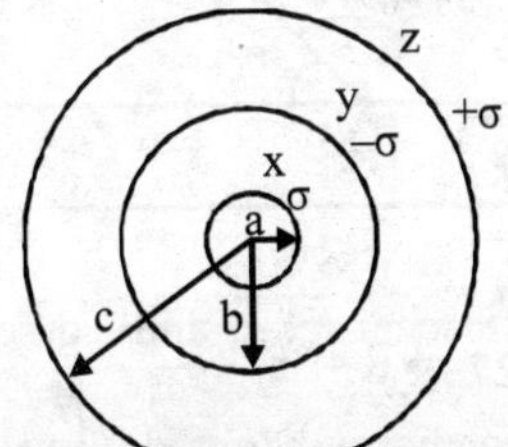

$q_x = \sigma 4\pi a^2$; $q_y = -\sigma 4\pi b^2$; $q_z = \sigma 4\pi c^2$

Given, potential of X = potential of Z

$$V_x = V_z \quad \left(\because V = \frac{q}{4\pi\varepsilon_0 r}\right)$$

$$\frac{q_x}{4\pi\varepsilon_0 a} + \frac{q_y}{4\pi\varepsilon_0 b} + \frac{q_z}{4\pi\varepsilon_0 c} = \frac{q_x}{4\pi\varepsilon_0 c} + \frac{q_y}{4\pi\varepsilon_0 c} + \frac{q_z}{4\pi\varepsilon_0 c}$$

$$\frac{\sigma4\pi a^2}{a} - \frac{\sigma4\pi b^2}{b} + \frac{\sigma4\pi c^2}{c} = \frac{4\pi\sigma[a^2 - b^2 + c^2]}{c}$$

$c(a - b + c) = a^2 - b^2 + c^2$

$c(a - b) = a^2 - b^2$

$c = a + b = 5$ cm

17. **(c)** Electric potential at a point P due to a point charge, $(\because K = 9\times10^9)$

$$V_P = \frac{KQ}{r}$$

$$\Rightarrow 50 = \frac{9\times10^9\times5\times10^{-9}}{r}$$

$$\Rightarrow r = \frac{45}{50} = \frac{9}{10} = 0.9\text{m} = 90\text{cm}$$

18. **(a)** For a uniformly charged thin spherical shell,

$$V_{inside} = V_{surface} = \frac{kQ}{R} = \text{constant}$$

$$V_{outside} = \frac{kQ}{r}$$

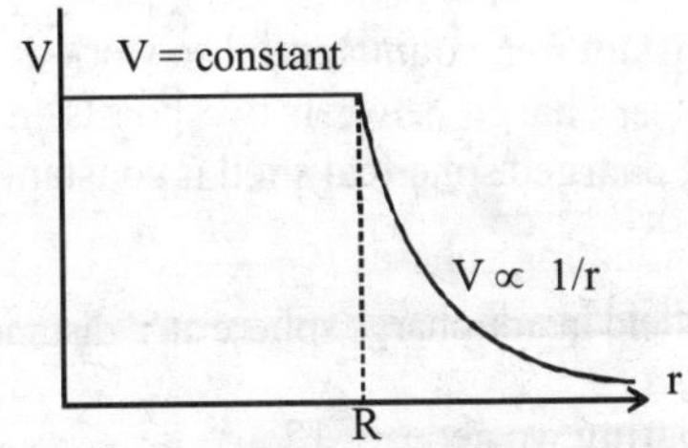

19. **(c)** For $r < R \Rightarrow V = \frac{1}{4\pi\epsilon_0}\frac{Q}{R} \Rightarrow V = \text{constant}$

for $r > R \Rightarrow V = \frac{1}{4\pi\epsilon_0}\frac{Q}{r} \Rightarrow V \propto \frac{1}{r}$

So, graph will look like

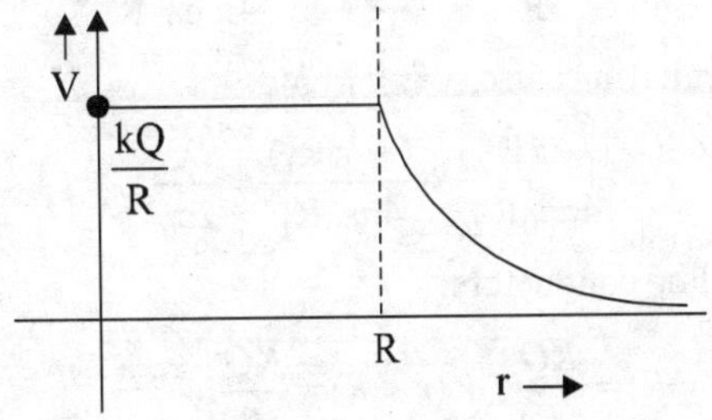

20. **(a)** For a group of positive charge, net potential at any point is never zero but electric field can be zero.

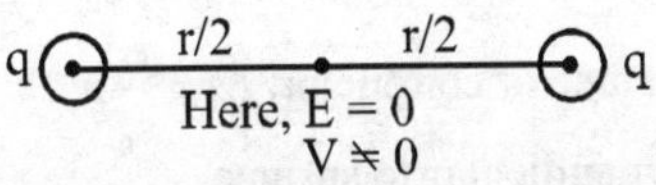

21. **(d)**

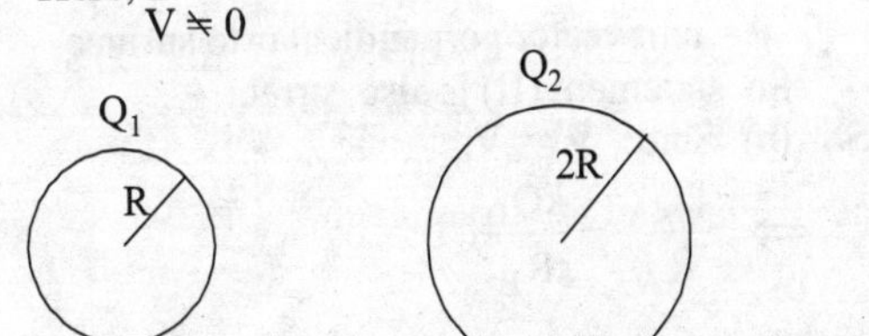

$Q_1 = \sigma(4\pi R^2) = 4\pi R^2\sigma$ $\qquad$ $Q_2 = \sigma 4\pi(2R)^2 = 16\pi R^2\sigma$

After connected by a thin conducting wire the potential of both sphere will be equal.

$\therefore V_1 = V_2$

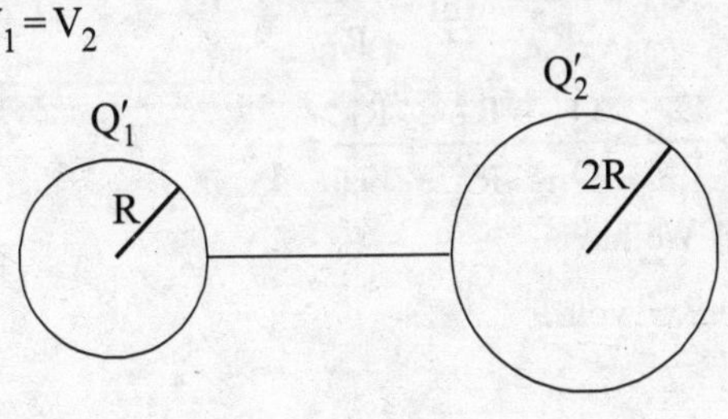

$$\frac{Q_1'}{4\pi\varepsilon_0 R} = \frac{Q_2'}{4\pi\varepsilon_0 (2R)}$$

$\therefore Q_2' = 2Q_1'$

Initial total charge = Final total charge

$$Q_1' + Q_2' = Q_1 + Q_2 \therefore \frac{Q_2'}{2} + Q_2' = 20\pi R^2\sigma$$

$$\frac{3}{2}Q_2' = 20\pi R^2\sigma$$

$$\therefore \sigma'\left[4\pi(2R)^2\right] = \frac{2}{3}\cdot\left(20\pi R^2\sigma\right) \therefore \frac{\sigma'}{\sigma} = \frac{5}{6}$$

22. **(12)** $E = -\frac{dV}{dr} = -\frac{d}{dr}(2ar^2 + b) = -4ar$

But electric field inside charge sphere at 'r' distance is $\frac{\rho r}{3\in}$.

So, by comparing we get $\rho = -12\,a\in$

23. **(b)**

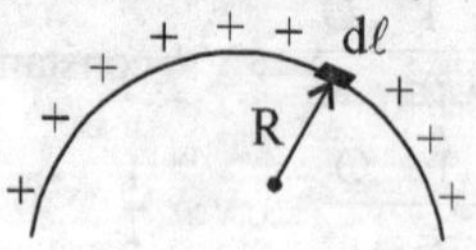

$$dV = \frac{1}{4\pi\in_0}\frac{\lambda d\ell}{R};\ V = \int dV = \frac{1}{4\pi\in_0}\frac{\lambda\ell}{R}$$

Potential at centre, $V = V_2 + V_1$

$$\Rightarrow \quad V = \frac{(\lambda\cdot\pi R_2)}{4\pi\varepsilon_0 R_2} + \frac{(\lambda\cdot\pi R_1)}{4\pi\varepsilon_0 R_1} = \frac{\lambda}{2\varepsilon_0}$$

24. **(a)** For conductors

$$V(r=R) = \frac{KQ}{R};\ V(r<R) = \frac{KQ}{R}$$

So, potential inside the surface is same as the potential on surface. So statement (I) is correct.

Just outside the surface of conductor, $\vec{E} = \frac{\sigma}{\varepsilon_0}\hat{n}$

$\hat{n}$ = unit vector perpendicular to surface

So, statement (II) is also correct.

25. **(b)** Since, $V_A = V_B$

$$\Rightarrow \frac{kQ_A}{R_A} = \frac{kQ_B}{R_B}.$$

Here, $K = \frac{1}{4\pi\in_o}$

$$\Rightarrow \frac{Q_A}{Q_B} = \frac{R_A}{R_B} = \frac{1}{2}$$

$$E_A = \frac{kQ_A}{R_A^2};\ E_B = \frac{kQ_B}{R_B^2}$$

$$\Rightarrow \frac{E_A}{E_B} = \frac{Q_A}{Q_B}\times\frac{R_B^2}{R_A^2} = \frac{R_B}{R_A} = \frac{2}{1}$$

26. **(d)** We have

$V = 3x^2$ volt

$$E_x = \frac{-\partial V}{\partial x} = -6x \Rightarrow E_{x/x=1} = -6$$

$$E_y = -\frac{\partial V}{\partial y} = 0$$

$$E_z = -\frac{\partial V}{\partial z} = 0$$

So, $|\vec{E}| = \sqrt{(-6)^2 + 0^2 + 0^2} = 6Vm^{-1}$

In vector form,

$\vec{E} = -6\hat{i}V/m$ (Along –ve x-axis)

27. **(d)**

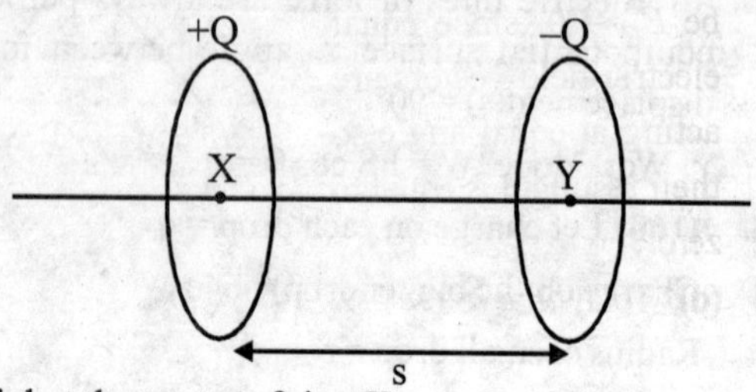

Potential at the centre of ring X

$$V_X = \frac{Q}{4\pi\in_0 a} - \frac{Q}{4\pi\in_0\sqrt{a^2+s^2}}$$

Potential at the centre of ring Y

$$V_Y = \frac{-Q}{4\pi\in_0 a} + \frac{Q}{4\pi\in_0\sqrt{a^2+s^2}}$$

$$V_X - V_Y = \frac{2Q}{4\pi\in_0 a} - \frac{2Q}{4\pi\in_0\sqrt{a^2+s^2}}$$

$$= \frac{Q}{2\pi\in_0}\left(\frac{1}{a} - \frac{1}{\sqrt{s^2+a^2}}\right)$$

28. **(243)** $V_i = V_f$

$$\Rightarrow \frac{4}{3}\pi R^3 = \frac{4}{3}\pi(r)^3\times 27 \Rightarrow R = 3r$$

Now, $U_{shell} = \frac{Q^2}{8\pi\in_0 R} \Rightarrow U \propto \frac{Q^2}{R}$

So, $\frac{U_B}{U_s} = \left(\frac{27Q}{Q}\right)^2\times\frac{r}{3r} = 243$

29. **(128)** Potential, $V = \frac{Kq}{r}$

If charge on each drop = q and radius = r

$$\therefore V = \frac{Kq}{r} = 2\text{ V}$$

$$\frac{4}{3}\pi R^3 = 512\times\frac{4}{3}\pi r^3$$

$$\Rightarrow R = (512\times r^3)^{1/3} = 8r$$

Potential of bigger drop,

$$V=\frac{K(512)q}{8r}=\frac{512}{8}\frac{Kq}{r}=\frac{512}{8}\times 2\text{ V}=128\text{ V}.$$

30. **(c)** Potential at the centre, $V_C=\frac{KQ_{net}}{R}$

$\because Q_{net}=0 \quad \therefore V_C=0$

Let E be electric field produced by each charge at the centre, then resultant electric field will be $E_C=0$, since equal electric field vectors are acting at equal angle so their resultant is equal to zero.

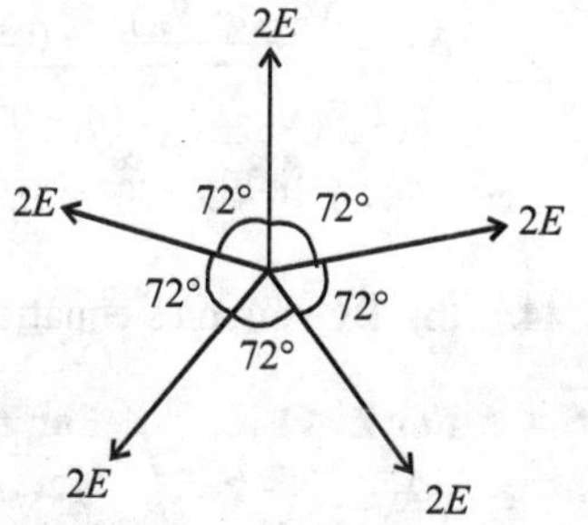

31. **(d)** Total charge $Q_1+Q_2=Q'_1+Q'_2$

$=12\mu C-3\mu C=9\mu C$

Two isolated conducting spheres S_1 and S_2 are now connected by a conducting wire.

$$\therefore V_1=V_2\Rightarrow\frac{KQ'_1}{\frac{2}{3}R}=\frac{KQ'_2}{\frac{R}{3}}$$

$Q'_1=2Q'_2\Rightarrow 2Q'_2+Q'_2=9\mu C$

$\therefore Q'_1=6\mu C$ and $Q'_2=3\mu C$

32. **(a)** We have given two metallic hollow spheres of radii R and $4R$ having charges Q_1 and Q_2 respectively.

Potential on the surface of inner sphere (at A)

$$V_A=\frac{kQ_1}{R}+\frac{kQ_2}{4R}$$

Potential on the surface of outer sphere (at B)

$$V_B=\frac{kQ_1}{4R}+\frac{kQ_2}{4R}$$

$\left(\text{Here, k}=\frac{1}{4\pi\varepsilon_0}\right)$

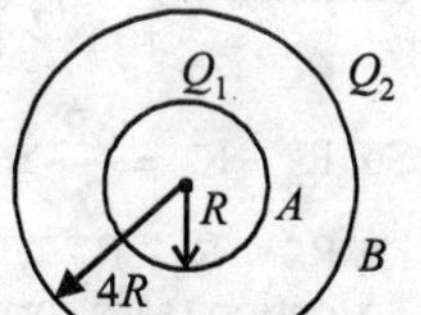

Potential difference,

$$\Delta V=V_A-V_B=\frac{3}{4}\cdot\frac{kQ_1}{R}=\frac{3}{16\pi\epsilon_0}\cdot\frac{Q_1}{R}$$

33. **(d)** Let σ be the surface charge density of the shells.

Charge on the inner shell, $Q_1=\sigma 4\pi r^2$

Charge on the outer shell, $Q_2=\sigma 4\pi R^2$

$\therefore$ Total charge, $Q=\sigma 4\pi(r^2+R^2)$

$$\Rightarrow\sigma=\frac{Q}{4\pi(r^2+R^2)}$$

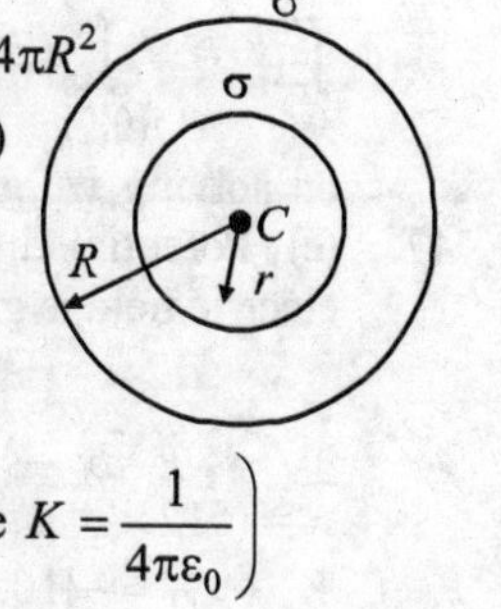

Potential at the common centre,

$$V_C=\frac{KQ_1}{r}+\frac{KQ_2}{R}\qquad\left(\text{where }K=\frac{1}{4\pi\varepsilon_0}\right)$$

$$=\frac{K\sigma 4\pi r^2}{r}+\frac{K\sigma 4\pi R^2}{R}=K\sigma 4\pi(r+R)$$

$$=\frac{KQ4\pi(r+R)}{4\pi(r^2+R^2)}=\frac{1}{4\pi\varepsilon_0}\frac{(r+R)Q}{(r^2+R^2)}$$

34. **(b)** The electric potential at the bisector is zero and electric field is antiparallel to the dipole moment.

$$\therefore V=0\text{ and }\vec{E}=\frac{1}{4\pi\varepsilon_0}\left(\frac{-\vec{P}}{d^3}\right)$$

35. **(c)** Potential at any point of the charged ring

$$V_P=\frac{Kq}{\sqrt{R^2+Z^2}}$$

$R=3a,\ Z=4a$

$\ell=\sqrt{R^2+Z^2}=5a$

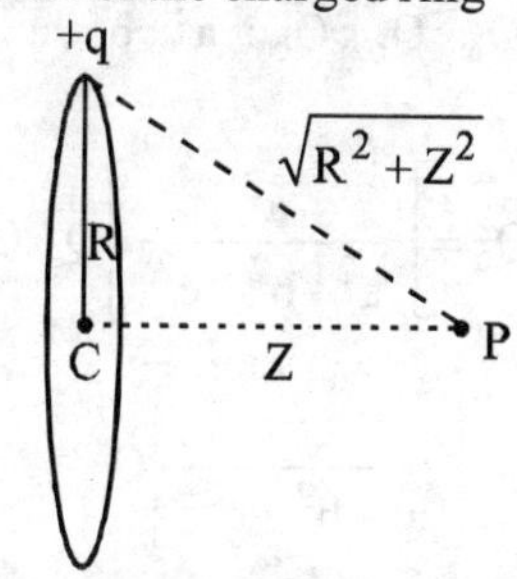

The minimum velocity (v_0) should just sufficient to reach the point charge at the center, therefore

$$\frac{1}{2}mv_0^2=q\,[V_C-V_P]=q\left[\frac{Kq}{3a}-\frac{Kq}{5a}\right]$$

$$\Rightarrow v_0^2=\frac{4Kq^2}{15ma}=\frac{4}{15}\frac{1}{4\pi\varepsilon_0}\frac{q^2}{ma}$$

$$\Rightarrow v_0=\sqrt{\frac{2}{m}}\left(\frac{2q^2}{15\times 4\pi\varepsilon_0 a}\right)^{\frac{1}{2}}$$

36. **(d)** When charge Q is on inner solid conducting sphere

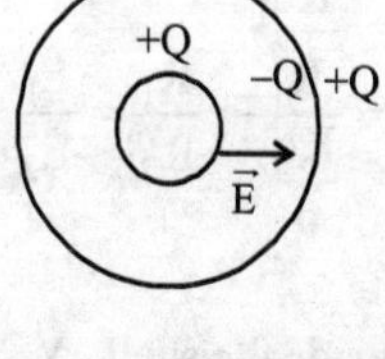

Electric field between spherical surface, $\vec{E}=\frac{KQ}{r^2}$

Now when a charge $-4Q$ is given to hollow shell Electric field between surface remain unchanged because Q_{inside} remain same. So, $\vec{E}=\frac{KQ}{r^2}$

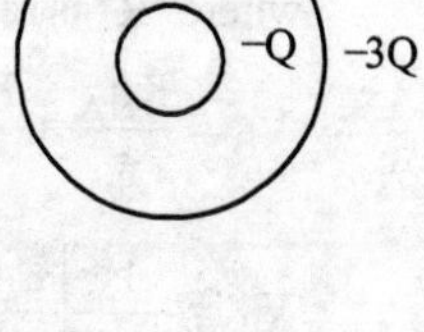

Now as $dV=-\vec{E}.d\vec{r}$ and $E=$ cons.

So, $dV=$ cons.

So, potential will be V only.

37. **(c)** Given, $\vec{E}=(Ax+B)\hat{i}$ or $E=20x+10$

Using $V=-\int Edx$, we have

$$V_2 - V_1 = -\int_1^{-5}(20x+10)\,dx = -180\text{ V}$$

or $V_1 - V_2 = 180$ V

38. **(b)** Electric potential is constant inside a charged spherical shell and outside it vary inversely with distance.

39. **(d)** Potential at point P, $V = \frac{kQ_a}{a} + \frac{kQ_b}{b} + \frac{kQ_c}{c}$

Since surface charge densities are equal to one another i.e., $\sigma_a = \sigma_b = \sigma_c$

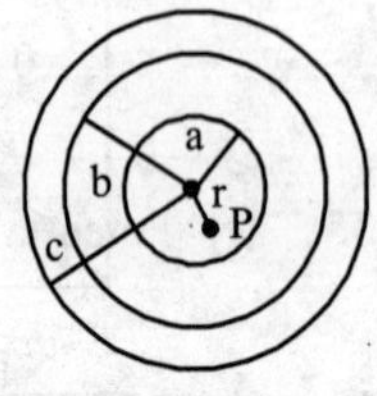

$\therefore Q_a : Q_b : Q_c :: a^2 : b^2 : c^2$

$$\therefore Q_a = \left[\frac{a^2}{a^2+b^2+c^2}\right]Q,\quad Q_b = \left[\frac{b^2}{a^2+b^2+c^2}\right]Q$$

$$Q_c = \left[\frac{c^2}{a^2+b^2+c^2}\right]Q \qquad \therefore V = \frac{Q}{4\pi\epsilon_0}\left[\frac{(a+b+c)}{a^2+b^2+c^2}\right]$$

40. **(d)** Let at a distance 'x' from point B, both the dipoles produce same potential

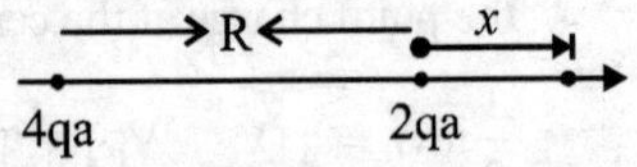

$$\therefore \frac{4qa}{(R+x)^2} = \frac{2qa}{(x^2)}$$

$$\Rightarrow \sqrt{2}x = R + x \Rightarrow x = \frac{R}{\sqrt{2}-1}$$

Therefore distance from A at which both of them produce the same potential

$$= \frac{R}{\sqrt{2}-1} + R = \frac{\sqrt{2}R}{\sqrt{2}-1}$$

41. **(b)** $\frac{V_1}{V_2} = \frac{\frac{Kq_1}{R_1}}{\frac{Kq_2}{R_2}} = \frac{\frac{Kq_1}{R_1^2}}{\frac{Kq_2}{R_2^2}} \times \frac{R_1}{R_2} = \frac{E_1}{E_2} \times \frac{R_1}{R_2} = \left(\frac{R_1}{R_2}\right)^2$

42. **(b)** Potential outside the shell, $V_{outside} = \frac{KQ}{r}$

where r is distance of point from the centre of shell

Potential inside the shell, $V_{inside} = \frac{KQ}{R}$

where 'R" is radius of the shell

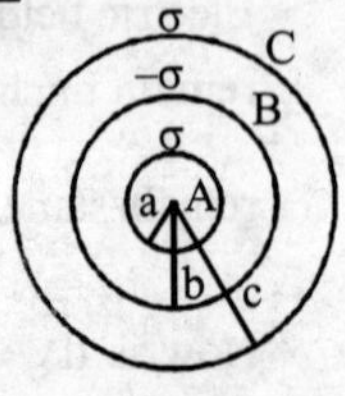

$$V_B = \frac{Kq_A}{r_b} + \frac{Kq_B}{r_b} + \frac{Kq_C}{r_c}$$

$$V_B = \frac{1}{4\pi\epsilon_0}\left[\frac{\sigma 4\pi a^2}{b} - \frac{\sigma 4\pi b^2}{b} + \frac{\sigma 4\pi c^2}{c}\right]$$

$$V_B = \frac{\sigma}{\epsilon_0}\left[\frac{a^2 - b^2}{b} + c\right]$$

43. **(b)** $V_{avg} = \frac{V_{max} + V_{min}}{2}$

$$= \frac{589.8 + 589}{2} = 589.4\text{V}$$

$$\Delta V = \frac{V_{max} - V_{min}}{2} = 0.4\text{V}$$

$\therefore V' = V_{avg} - \Delta V_{cos}\theta$

$= 589.4 - 0.4\cos 60° = 589.2$V

44. **(b)** By Poisson's equation $\frac{\partial^2 V}{\partial z^2} = \frac{-\rho}{\epsilon_0}$

For Z ≤ 1	For Z ≥ 1
$\frac{\partial^2 V}{\partial z^2} = -10$	$\frac{\partial^2 V}{\partial z^2} = 0$
$\Rightarrow \frac{-\rho}{\epsilon_0} = -10$	$\Rightarrow \frac{-\rho}{\epsilon_0} = 0$
$\Rightarrow \rho = 10\epsilon_0$	$\Rightarrow \rho = 0$

45. **(a)** $\frac{KQ}{R} = V_0$

Now, $V = \frac{KQ}{r}$, $r \ge R$

$V = \frac{KQ}{2R^3}(3R^2 - r^2)$, $r < R$

when $r = R_1$, $V = \frac{3V_0}{2}$

$$\therefore \frac{3V_0}{2} = \frac{3}{2}\frac{KQ}{R} - \frac{KQR_1^2}{R^3} \Rightarrow \frac{3V_0}{2} = \frac{3}{2}V_0 - \frac{KQ}{R^3}\cdot R_1^2$$

$\Rightarrow R_1 = 0$

Similarly, $R_2 = \frac{R}{\sqrt{2}}, R_3 = \frac{4R}{3}, R_4 = 4R$

So, $R_4 - R_3 = \frac{8R}{3} > \frac{R}{\sqrt{2}}$

i.e. $R_2 < R_4 - R_3$

So, option (a) is correct.

46. **(a)** As we know, $dV = -\int_{\vec{r}_1}^{\vec{r}_2} \vec{E}.\vec{dr}$

Potential at the point x = 2m, y = 2m is given by :

$$\int_0^V dV = -\int_{(0,0)}^{(2,2)}(25dx + 30dy) = -\int_0^2 25\,dx - \int_0^2 30\,dy$$

on solving we get, V = – 110 volt.

47. **(c)** Potential difference between any two points in an electric field is given by,

$$dV = -\vec{E}\cdot\vec{dx} \Rightarrow \int_{V_O}^{V_A} dV = -\int_0^2 30x^2\,dx$$

$$V_A - V_O = -[10x^3]_0^2 = -80\text{ J/C}$$

48. **(a)** The potential of uncharged body is less than that of the charged conductor and more than at infinity because in direction of field potential decreases.

49. **(c)** $q = 1\mu C = 1\times10^{-6}C, r = 4\text{ cm} = 4\times10^{-2}\text{ m}$
There will be no potential due to induced charges.

Potential $V = \frac{kq}{r} = \frac{9\times10^9\times10^{-6}}{4\times10^{-2}} = 2.25\times10^5\text{ V.}$

Induced electric field $E = -\frac{kq}{r^2}$

$[\because \vec{E}_i + \vec{E}_{1\mu c} = 0 \Rightarrow \vec{E}_i = -\vec{E}_{1\mu c}]$

$= -\frac{9\times10^9\times1\times10^{-6}}{16\times10^{-4}} = -5.625\times10^6\text{ V/m}$

50. **(d)** $V(x) = 4x^2$

$E = -\frac{dV}{dx} = -8x$

We will have flux through surface 1 and 2 only.

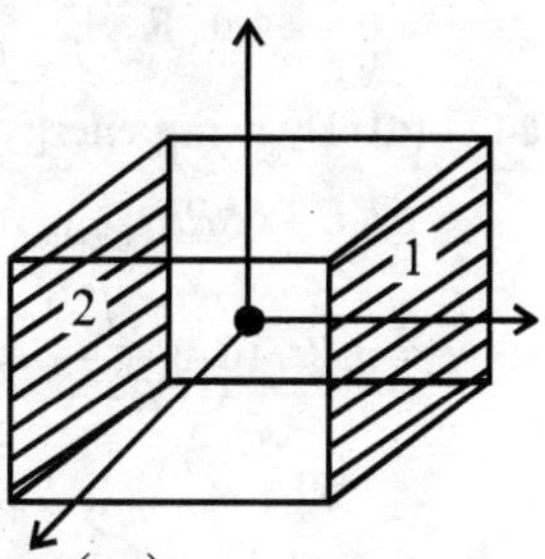

$\phi = \phi_1 + \phi_2$

$= \left(-8\times\frac{1}{2}\right)\times1\times1 + \left(-8\times\frac{-1}{2}\right)\times1\times(-1)$

$= -4-4 = -8 \quad [\because \phi = \vec{E}\cdot\vec{A} = EA\cos\theta]$

By Gauss law,

$\phi = \frac{q_{in}}{\epsilon_0} \Rightarrow q_{in} = \phi\,\epsilon_0 = -8\,\epsilon_0$

51. **(c)** Electric field

$E = -\frac{d\phi}{dr} = -2ar$ ....(i)

By Gauss's theorem

$E = \frac{1}{4\pi\varepsilon_0}\frac{q}{r^2}$ ....(ii)

From (i) and (ii),

$q = -8\pi\varepsilon_0 ar^3 \Rightarrow dq = -24\pi\varepsilon_0 ar^2\,dr$

Charge density, $\rho = \frac{dq}{4\pi r^2 dr} = -6\varepsilon_0 a$

52. **(c)** The distance of point $A(\sqrt{2},\sqrt{2})$ from the origin,

$r_A = \sqrt{(\sqrt{2})^2+(\sqrt{2})^2}$

$= \sqrt{4} = 2$ units.

The distance of point $B(2, 0)$ from the origin,

$r_B = \sqrt{(2)^2+(0)^2} = 2$ units.

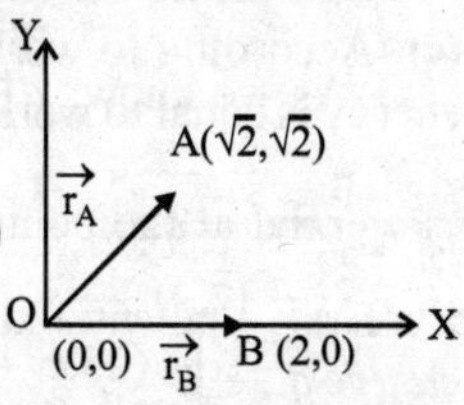

$V_A - V_B = \frac{1}{4\pi\epsilon_0}\cdot\frac{10^{-3}}{r_A} - \frac{1}{4\pi\epsilon_0}\cdot\frac{10^{-3}}{r_B}$

$= \frac{10^{-3}}{4\pi\epsilon_0}\left(\frac{1}{r_A}-\frac{1}{r_B}\right) = \frac{10^{-3}}{4\pi\epsilon_0}\left(\frac{1}{2}-\frac{1}{2}\right) = \frac{Q}{4\pi\epsilon_0}\times0 = 0.$

53. **(a)** As shown in the figure, the resultant electric fields before and after interchanging the charges will have the same magnitude, but opposite directions.

As potential is a scalar quantity, So the potential will be same in both cases.

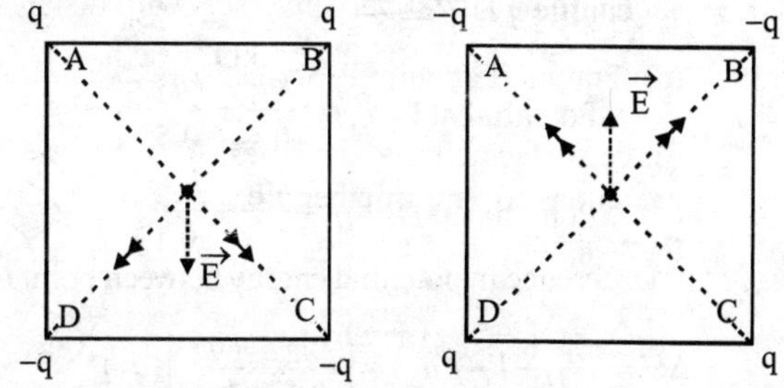

54. **(a)** Given, potential $V(x) = \frac{20}{x^2-4}$ volt

Electric field $E = -\frac{dV}{dx} = -\frac{d}{dx}\left(\frac{20}{x^2-4}\right)$

$\Rightarrow E = +\frac{40x}{(x^2-4)^2}$

At $x = 4\ \mu$m,

$E = +\frac{40\times4}{(4^2-4)^2} = +\frac{160}{144} = +\frac{10}{9}$ volt / μm.

Positive sign indicates that $\vec{E}$ is in +ve x-direction.

55. **(a)**

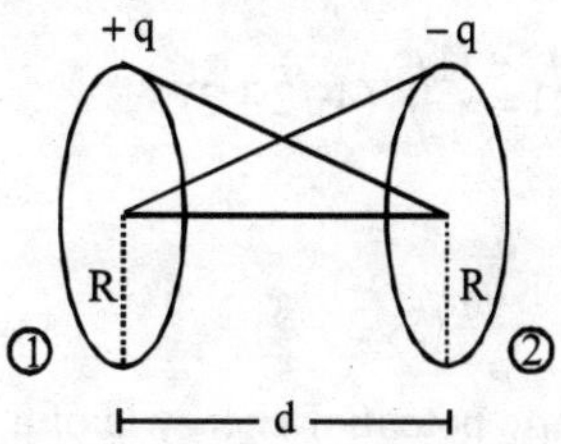

Potential at the center of ring of charge $+q$ = potential due to iteself + potential due to other ring of charge $-q$.

$\Rightarrow V_1 = \frac{1}{4\pi\varepsilon_0}\left[\frac{q}{R} - \frac{q}{\sqrt{R^2+d^2}}\right]$

Potential at the centre of ring of charge $-q$ = potential due to itself + potential due to other ring of charge $+q$.

$\Rightarrow V_2 = \frac{1}{4\pi\varepsilon_0}\left[\frac{-q}{R} + \frac{q}{\sqrt{R^2+d^2}}\right]$

$\Delta V = V_1 - V_2$

$= \frac{1}{4\pi\varepsilon_0}\left[\frac{q}{R}+\frac{q}{R}-\frac{q}{\sqrt{R^2+d^2}}-\frac{q}{\sqrt{R^2+d^2}}\right]$

$= \frac{q}{2\pi\varepsilon_0}\left[\frac{1}{R}-\frac{1}{\sqrt{R^2+d^2}}\right]$

56. **(c)** Electric potential due to charge Q at point P is

$V_1 = \frac{1}{4\pi\varepsilon_o}\frac{Q}{R/2} = \frac{1}{4\pi\varepsilon_o}\frac{2Q}{R}$

Electric potential due to charge q at point p is $V_2 = \frac{1}{4\pi\varepsilon_o}\frac{q}{R}$

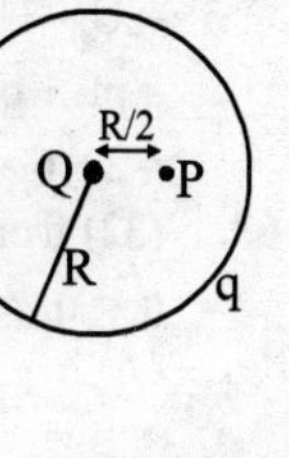

$\therefore$ The net electric potential at point P is

$$V = V_1 + V_2 = \frac{1}{4\pi\varepsilon_o}\frac{2Q}{R} + \frac{1}{4\pi\varepsilon_o}\frac{q}{R}$$

57. **(a)** The potential at a point at a distance r from a point charge q is $V = \frac{kq}{r}$

$\therefore$ Potential at C, $V_C = \frac{kq_1}{0.4} + \frac{kq_2}{0.5}$

$\therefore$ Potential at D, $V_D = \frac{kq_1}{0.4} + \frac{kq_2}{0.1}$

$\therefore$ Difference in potential energy between point C and D is

$$\Delta U = (V_D - V_C)(q_3) = \left(\frac{kq_2}{0.1} - \frac{kq_2}{0.5}\right)(q_3)$$

$$= 8kq_2q_3 = \frac{8q_2q_3}{4\pi\varepsilon_0} = \frac{kq_2q_3}{4\pi\varepsilon_0} \text{ (given)} \quad \therefore \; k = 8.$$

58. **(12)** Given, $p = 6 \times 10^{-6}$ Cm, $E = 10^6$ v/m

The expression of work done in rotating a dipole

$W = \Delta U = -pE(\cos\theta_f - \cos\theta_i) = -pE(\cos 180° - \cos 0°)$

$W = 2pE = 2 \times 6 \times 10^{-6} \times 10^6 = 12$ J

59. **(d)** For the given configuration (1) and (2),

$$U_1 = \frac{4Kq_0^2}{a} + \frac{2Kq_0^2}{\sqrt{2}a} = \frac{Kq_0^2}{a}(4+\sqrt{2})$$

$$U_2 = \frac{Kq_0^2}{\left(\frac{a}{\sqrt{2}}\right)}(4+\sqrt{2}) = \frac{Kq_0^2}{a}(4\sqrt{2}+2)$$

$$\therefore \Delta U = U_2 - U_1 = \frac{Kq_0^2}{a}(3\sqrt{2}-2)$$

60. **(d)** The electrostatic potential energy U of a system of two point charges is given by

$$U = \frac{1}{4\pi\varepsilon_0}\frac{q_1q_2}{r}$$

$$r = \sqrt{(x_2-x_1)^2 + (y_2-y_1)^2 + (z_2-z_1)^2} = 14 \text{ cm}$$

$$\therefore U = \frac{9\times10^9 \times 7\times10^{-6} \times (-4)\times10^{-6}}{14\times10^{-2}} = -1.8 \text{ J}$$

61. **(a)** The potential energy of an electric dipole in a uniform electric field is given by

$U = -PE\cos\theta$

The work done in rotating the dipole from an initial angle θ_i to a final angle θ_f

$W_{ext} = \Delta U = U_f - U_i = -PE\cos 180° + PE\cos 0°$

$W_{ext} = 2PE = 2 \times (4 \times 10^{-6})(18) \times 10^4 = 144 \times 10^{-2} = 14.4$ mJ

62. **(18)** Dipole moment, $p = 6 \times 10^{-6}$ Cm

Electric field, $E = 1.5 \times 10^3$ NC^{-1}

The work done in rotating the electric dipole $= \Delta U$

$= U_f - U_i = (-pE\cos(180°)) - (-pE\cos(0°))$

$= pE + pE = 2pE$ $\qquad (\because U = -pE\cos\theta)$

$= 2 \times 6 \times 10^{-6} \times 1.5 \times 10^3 = 18$mJ

63. **(32)** from law of conservation of charge

$q_i = q_f \Rightarrow 8 \times (4\pi E_0 R) \times 12 = (4\pi E R') \times V_f$

$\Rightarrow 96R = V_f R'$ .... (i)

And, $8 \times \frac{4}{3}\pi R^3 = \frac{4}{3}\pi R'^3$

$8 = \left(\frac{R'}{R}\right)^3$

$R' = 2R$ ....(ii)

From (i) & (ii), we get

So, $96R = V_f \times 2R \Rightarrow V_f = 48$ Volt

$$E_f = \frac{1}{2}C_f V_f^2 = \frac{1}{2} \times (4\pi\varepsilon_0 R')V_f^2$$

$$= \frac{1}{2} \times (4\pi\varepsilon_0 \times 2R) \times 48^2$$

$$= \left(\frac{1}{2} \times 4\pi\varepsilon_0 R \times 12^2\right) \times \frac{48^2 \times 2}{12^2} = 32E$$

64. **(d)** By using energy conservation,

$$\Delta KE + (\Delta PE)_{\text{Electro}} + (\Delta PE)_{\text{gravitational}} = 0$$

$$\frac{1}{2}mV^2 + \left(k\frac{Qq}{R+y} - k\frac{Qq}{R}\right) + (-mgy) = 0$$

$$\Rightarrow \frac{1}{2}mV^2 = mgy + kQq\left(\frac{1}{R} - \frac{1}{R+y}\right)$$

$$\Rightarrow V^2 = 2gy + \frac{2kQq}{m}\frac{y}{R(R+y)}$$

$$\text{or, } V^2 = 2y\left[\frac{qQ}{4\pi\varepsilon_0 R(R+y)m} + g\right]$$

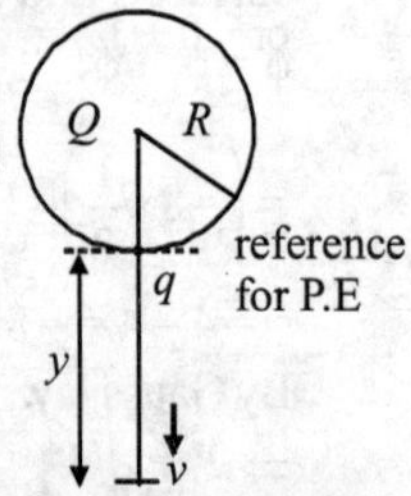

65. **(d)** Change in potential energy, $\Delta U = q(V_f - V_i)$

Potential of $-q$ is same at initial and final point of the path.

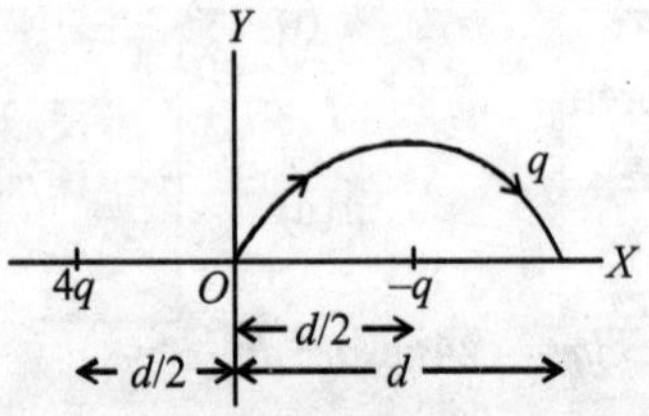

$$\Delta U = q\left(\frac{k4q}{3d/2} - \frac{k4q}{d/2}\right) = -\frac{4q^2}{3\pi\varepsilon_0 d}$$

–ve sign shows the energy of the charge is decreasing.

66. **(c)** According to work energy theorem, gain in kinetic energy is equal to work done in displacement of charge.

$$\therefore \frac{1}{2}mv^2 = q\Delta V$$

Here, ΔV = potential difference between two positions of charge q.

For same q and ΔV.

$$v \propto \frac{1}{\sqrt{m}}$$

Mass of hydrogen ion $m_H = 1$; Mass of helium ion $m_{He} = 4$

$$\therefore \frac{v_H}{v_{He}} = \sqrt{\frac{4}{1}} = 2:1.$$

67. **(c)** Using conservation of energy

$$U_i = U_F + \frac{1}{2}mv^2;\ \frac{kq_1q_2}{r_1} = \frac{kq_1q_2}{r_2} + \frac{1}{2}mv^2$$

$$\Rightarrow \frac{1}{2}mv^2 = kq_1q_2\left[\frac{1}{r_1} - \frac{1}{r_2}\right] \Rightarrow v^2 = \frac{2kq_1q_2}{m}\left[\frac{1}{r_1} - \frac{1}{r_2}\right]$$

$$= \frac{2\times 9\times 10^9 \times 10^{-12}}{4\times 10^{-6}\times 10^{-3}}\left[1 - \frac{1}{9}\right] = 4\times 10^{+6}$$

$v = 2\times 10^3$ m/s

68. **(d)** $U = \frac{1}{4\pi\epsilon_0}\left[\frac{q(-q)}{d} + \frac{qQ}{\left(D+\frac{d}{2}\right)} + \frac{(-q)Q}{\left(D-\frac{d}{2}\right)}\right]$

$$= \frac{1}{4\pi\epsilon_0}\left[-\frac{q^2}{d} - \frac{qQd}{D^2}\right], \text{Ignoring } \frac{d^2}{4}$$

69. **(b)** Using, $[K+U]_i = [K+U]_f$

or $0 + Vq = \frac{1}{2}mv^2 + V'q$ or $\frac{1}{2}mv^2 = (V - V')q$

$$= -q\int_{r_0}^{r} Edr = q\int_{r_0}^{r}\frac{\lambda}{2\pi\epsilon_0 r}dr = \frac{\lambda q}{2\pi\epsilon_0}\ln\left(\frac{r}{r_0}\right)$$

$$\Rightarrow v \propto \sqrt{ln\left(\frac{r}{r_0}\right)}$$

70. **(b)** Net electrostatic energy for the system

$$U = K\left[\frac{q^2}{a} + \frac{Qq}{a} + \frac{Qq}{a\sqrt{2}}\right] = 0$$

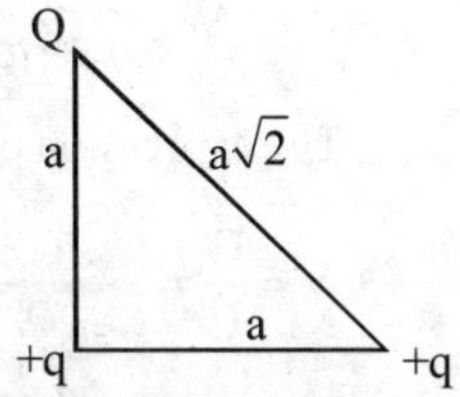

$$\Rightarrow q = -Q\left[1 + \frac{1}{\sqrt{2}}\right]$$

$$\Rightarrow Q = \frac{-q\sqrt{2}}{\sqrt{2}+1}$$

71. **(b)** Potential at origin

$$v = \frac{KQ}{2} + \frac{KQ}{2} + \frac{KQ}{\sqrt{20}} + \frac{KQ}{\sqrt{20}}$$

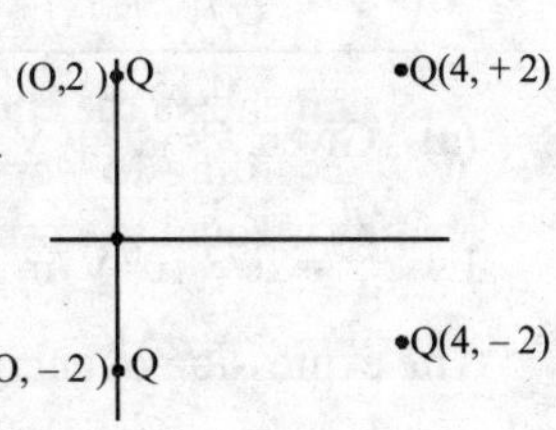

$$= KQ\left(1 + \frac{1}{\sqrt{5}}\right)$$

and potential at $\infty = 0$

$\therefore$ Work required to put a fifth charge Q at origin W = VQ

$$= \frac{Q^2}{4\pi\varepsilon_0}\left(1 + \frac{1}{\sqrt{5}}\right)$$

72. **(c)** The work done in moving a charge along an equipotential surface is always zero.
The direction of electric field is perpendicular to the equipotential surface.

73. **(c)** $V_B - V_A = -\int_A^B \vec{E}.d\vec{r} = -\int_0^R \frac{\rho r}{3\epsilon_0}dr = -\frac{\rho R^2}{6\epsilon_0}$

$$\therefore \Delta U = q(V_B - V_A) = -\frac{\rho R^2 \rho}{6\epsilon_0}$$

For $r < R$, $\vec{E} = \frac{\rho r}{3\epsilon_0}\hat{r}$

74. **(d)** Initial potential of the charge,

$$V_A = \frac{2kq}{a} - \frac{2kq}{a\sqrt{5}}$$

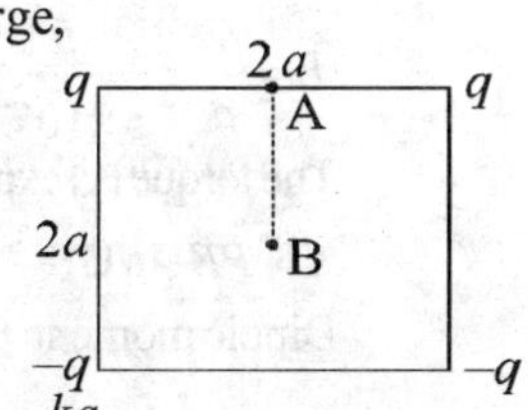

$$\Rightarrow V_A = \frac{1}{4\pi\varepsilon_0}\frac{2q}{a}\left(1 - \frac{1}{\sqrt{5}}\right)$$

(Here potential due to each $q = \frac{kq}{a}$ and potential due to each $-q = \frac{-kq}{a\sqrt{5}}$)

Final potential of the charge, $V_B = 0$

($\because$ Point B is equidistant from all the four charges)

$\therefore$ Using work energy theorem,

$(KE)_B = (W_{AB})_{\text{electric}} = Q(V_A - V_B)$

$$= \frac{2qQ}{4\pi\varepsilon_0 a}\left[1 - \frac{1}{\sqrt{5}}\right] = \left(\frac{1}{4\pi\varepsilon_0}\right)\frac{2Qq}{a}\left[1 - \frac{1}{\sqrt{5}}\right]$$

75. **(c)** Work done, $W_{PQ} = q(V_Q - V_P)$

$= (-100\times 1.6\times 10^{-19})(-4-10) = +2.24\times 10^{-16}$J

76. **(a)** Gain in kinetic energy = work done by potential difference

$$eV = \frac{1}{2}mv^2 \Rightarrow v = \sqrt{\frac{2eV}{m}}$$

$$= \sqrt{\frac{2\times 1.6\times 10^{-19}\times 20}{9.1\times 10^{-31}}} = 2.65\times 10^6 \text{ m/s}$$

77. **(d)** $\frac{1}{2}mv^2 = \frac{kQq}{r} \Rightarrow \frac{1}{2}m(2v)^2 = \frac{kqQ}{r'} \Rightarrow r' = \frac{r}{4}$

78. **(a)** By using

$$W = q(V_B - V_A) \therefore V_B - V_A = \frac{2\text{J}}{20\text{C}} = 0.1\text{J/C} = 0.1\text{V}$$

79. **(15)** Capacitance $C = \frac{K\varepsilon_0 A}{d}$

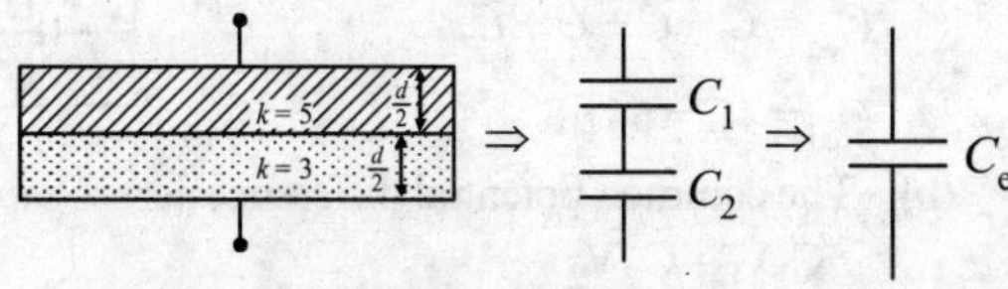

$$C_1 = \frac{5\times 4\times 10^{-4}\times 8.85\times 10^{-12}}{\frac{1.77}{2}\times 10^{-3}} = 20 \text{ pF}$$

$$C_2 = \frac{3\times 4\times 10^{-4}\times 8.85\times 10^{-12}}{\frac{1.77}{2}\times 10^{-3}} = 12 \text{ pF}$$

Capacitance C_1 and C_2 are in series

$$C_{eq} = \frac{C_1C_2}{C_1+C_2} = \frac{12\times20}{12+20} = 7.5 \text{ pF}$$

Now another capacitor of capacitance 7.5 pF is connected in parallel with this capacitor.

$\therefore (C_{eq})_{final} = 7.5 + 7.5 = 15$ pF

80. **(a)** Given : $q = 2\times10^{-6}$ c, $d = 2\times10^{-6}$ m, $(\Delta V) = 5$ V

D (Between plates) $= 0.5\times10^{-3}$ m

$$E = \frac{v}{d} = \frac{5}{5\times10^{-4}} = 10^4 \text{ V/m}$$

The torque (τ) experienced by a dipole in an electric field

$\tau = PE\sin\theta$

Dipole moment $(P) = qa = 2\times10^{-6}\times5\times10^{-7}$

$= 1\times10^{-12}$ C-m

$$\therefore \tau = 1\times10^{-12}\times10^4\times\frac{1}{2} = 5\times10^{-9} \text{ N-m.}$$

81. **(5)** Given :

Q (Charge on parallel plate capacitor) $= 5\times10^{-6}$ C

Q_i (induced) $= 4\times10^{-6}$ C

$$Q_{in} = Q\left(1-\frac{1}{K}\right)$$

$$\Rightarrow 4\times10^{-6} = 5\times10^{-6}\left(1-\frac{1}{K}\right) \Rightarrow 1-\frac{1}{K} = \frac{4}{5} \Rightarrow K = 5.$$

82. **(c)** Given : $K = 4$

When the dielectric is filled, $C_{new} = CK$

$C_1 = 4C$

$C_1 = 4\times5 = 20$ μF

$C_2 = C_3 = 5$ μF

C_1 and C_2 are in series which is parallel to C_3. So

$$C_{eq} = \frac{C_1C_2}{C_1+C_2} + C_3 = \frac{20\times5}{20+5} + 5 = 4 + 5 = 9 \text{ μF}$$

Correct Option (c).

83. **(64)** Applying equipotential surface concept

Given : $C = 16$ μF

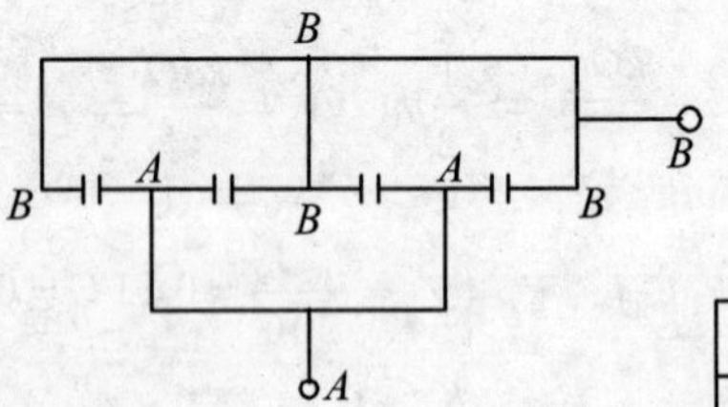

Redrawing

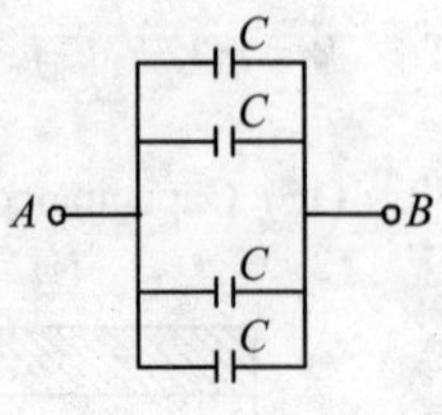

$C_{eq} = C + C + C + C$

$C_{eq} = 4C = 64$ μF.

84. **(b)** The common potential difference across capacitors is

$$V_C = \frac{C_1V_1 + C_2V_2}{C_1+C_2}$$

Here, $V_C = 20$ V, $C_1 = 100$pF, $V_1 = 60$V

$V_2 = 0$

$$\therefore \quad 20 = \frac{100\times60}{100+C_2} \qquad \therefore \quad C_2 = 200\text{pF}$$

85. **(b)** The capacitance of parallel plate capacitor in air is

$$C_0 = \frac{\varepsilon_0 A}{d}$$

For first configuration,

$C = 2\varepsilon_2 C_0$, $C' = 2\varepsilon_1 C_0$

$$\therefore \quad C_1 = \frac{CC'}{C+C'} = \frac{4\varepsilon_2\varepsilon_1 C_0^2}{2C_0(\varepsilon_2+\varepsilon_1)} = \frac{2\varepsilon_2\varepsilon_1 C_0}{(\varepsilon_2+\varepsilon_1)}$$

For second configuration,

$$C = \frac{\varepsilon_1\varepsilon_0 A}{2d} = \frac{\varepsilon_1 C_0}{2},\ C' = \frac{\varepsilon_2 C_0}{2}$$

$$C_2 = C' + C = (\varepsilon_1+\varepsilon_2)\frac{C_0}{2}$$

$$\therefore \quad \frac{C_1}{C_2} = \frac{2\varepsilon_2\varepsilon_1 C_0}{(\varepsilon_2+\varepsilon_1)} \times \frac{2}{(\varepsilon_1+\varepsilon_2)C_0} = \frac{4\varepsilon_1\varepsilon_2}{(\varepsilon_1+\varepsilon_2)^2}$$

86. **(c)** Given, $C_1 = 6$ μF, $C_2 = 12$ μF

Voltage, $V_0 = 5$V

At equilibrium, charge, Q = constant

$\Rightarrow \quad 6V_C + 12V_C = 30 + 0 \quad (\because Q = CV)$

$$\therefore \quad V_C = \frac{30}{18} = \frac{5}{3} V$$

$\therefore$ Final charges on the capacitors are

$$\Rightarrow \quad q_1 = C_1V_C = \frac{6\times5}{3} = 10\mu C$$

$$\Rightarrow \quad q_2 = C_2V_C = \frac{12\times5}{3} = 20\mu C$$

87. **(1320)** Given,

Radius of circular plate, r = 10 cm

Current, I = 0.15 A

Rate of change of potential difference,

$$\frac{V}{t} = 7\times10^8 \text{ V/S}$$

The expression of potential difference across capacitor is

$$V = \frac{Q}{C} = \frac{It}{\left(\frac{\epsilon_0 A}{d}\right)} = \frac{Itd}{\epsilon_0(\pi r^2)} \Rightarrow \quad d = \frac{\epsilon_0(\pi r^2)}{I}\left(\frac{V}{t}\right)$$

$$= \frac{(9\times10^{-12})\left(\frac{22}{7}\right)(0.1)^2}{0.15}(7\times10^8)\text{m} = 1320\mu\text{m}$$

88. **(a)** Given, C = 1 μF, V = 20 V, d = 1 μm

$$E = \frac{V}{d} = 20\times10^6 \text{ V/m}$$

The expression of energy density is

$$u = \frac{1}{2}\varepsilon_0 E^2 = \frac{1}{2}\times8.85\times10^{-12}\times(20\times10^6)^2$$

$u = 1.77\times10^3$ J/m^3 $\approx 1.8\times10^3$ J/m^3

89. **(d)** When capacitors are connected in parallel, the voltage across each capacitor is the same.

$V_1 = V_2 = V$

The charge stored in a capacitor, q = CV

$q_1 = C_1V$, $q_2 = C_2V$

The energy stored in a capacitor is given by

$U=\frac{1}{2}CV^2$

As $q_1<q_2$ $\therefore$ $C_1<C_2$ & $U_1<U_2$

90. **(a)** Initially
$l=3$ cm, $b=1$ cm; $d=3\mu m$; $A_1=l\times b=3\times 1$
Initial capacitance of parallel plate capacitor,

$$C_1=\frac{\epsilon_0 A_1}{d}=\frac{\epsilon_0\times 3\times 1}{3}=\epsilon_0 \text{ units}$$

For option C, $l=6$ cm, $b=5$ cm, $d=3$ μm

$$C_2=\frac{\epsilon_0 A_2}{d_2}=\frac{\epsilon_0\times 6\times 5}{3}=10\,\epsilon_0 \text{ units}$$

For option E, $l=5$ cm, $b=2$ cm, $d=1$ μm

$$C_2=\frac{\epsilon_0 A_2}{d}=\frac{\epsilon_0\times 5\times 2}{1}=10\,\epsilon_0 \text{ units}$$

91. **(a)** For parallel plate capacitor,

Energy density, $u=\frac{U}{V}=\frac{1}{2}\epsilon_0 E^2$

$$\therefore U=\frac{1}{2}\epsilon_0 E^2 V=\frac{1}{2}\epsilon_0 E^2 Ad$$

92. **(c)** Given, capacitance of capacitor, $C=40\mu F$
Voltage supply, $V=100$ V. Initially, $q=CV$
On introducing dielectric, $q'=KCV$
$\Delta q=(KC-C)V$
$=40\times 10^{-6}\times 100=4\times 10^{-3}=4$ mC
Change in energy,

$$\Delta U=\frac{1}{2}C'V^2-\frac{1}{2}CV^2=\frac{1}{2}(K-1)CV^2$$

$$=\frac{1}{2}CV^2(2-1)=\frac{1}{2}\times 40\times 10^{-6}\times 10000=0.2\text{ J}$$

93. **(a)** Capacitance of capacitor with air between plate,

$$C=\frac{\varepsilon_0 A}{d}$$

All capacitor are in parallel combination.

$\therefore$ $C_{eq}=C_1+C_2+C_3$

$$\Rightarrow C_{eq}=\frac{A\varepsilon_0}{3d}+\frac{A\varepsilon_0}{3(2d)}+\frac{A\varepsilon_0}{3(3d)}$$

$$=\frac{A\varepsilon_0}{3d}\left(1+\frac{1}{2}+\frac{1}{3}\right)$$

$$\Rightarrow C_{eq}=\frac{11A\varepsilon_0}{18d}$$

94. **(a)** Capacitance if air is used as medium, $C_0=\frac{A\epsilon_0}{d}$

Capacitance if slab is used, $C=\frac{A\epsilon_0}{t+\frac{d}{k}}$

$$\frac{A\epsilon_0}{d}=\frac{A\varepsilon_0}{\left(0.2+\frac{d}{k}\right)}$$

$$\Rightarrow 0.6=0.2+\frac{0.6}{k}\quad \therefore k=\frac{3}{2}=1.50$$

95. **(0.0095)** $C_0=10\,\mu F, A=4cm^2, K_1=2, K_2=3, F=8N$

$$\therefore C=K_{eq}C_0=\left(\frac{K_1+K_2}{2}\right)C_0=\frac{5}{2}\times 10=25\mu F$$

Now, $F=\frac{Q^2}{2A\varepsilon_0}\Rightarrow \alpha=(2A\varepsilon_0 F)^{\frac{1}{2}}$

$\therefore$ Voltage supply, $V=\frac{Q}{C}=\frac{(2A\varepsilon_0 F)^{\frac{1}{2}}}{C}$

$$=\frac{(2\times 4\times 10^{-4}\times 8.834\times 10^{-12}\times 8)^{\frac{1}{2}}}{25\times 10^{-6}}$$

$=0.0095$V

96. **(86)** For parallel combination, $E=\frac{1}{2}(C_1+C_2+C_3)V^2$

$$\therefore E=\frac{1}{2}(25+30+45)(100)^2 \quad \text{...(i)}$$

For series combination, $\frac{9}{x}E=\frac{1}{2}C_{eq}V^2$

$$\therefore \frac{9}{x}E=\frac{1}{2}\frac{1}{\left(\frac{1}{25}+\frac{1}{30}+\frac{1}{45}\right)}(100)^2 \quad \text{...(ii)}$$

From eqn(i) and (ii), we get $x=86$

97. **(a)** $V_c=\frac{Q_{net}}{C_{net}}=\frac{C_1V_1+C_2V_2}{C_1+C_2}=\frac{CV+2CV}{2C}=\frac{3V}{2}$

Decrease in energy $\Delta E=E_2-E_1$

$$\Delta E=\frac{1}{2}CV^2+\frac{1}{2}C(2V)^2-\frac{1}{2}2C\left(\frac{3V}{2}\right)^2=\frac{CV^2}{4}$$

98. **(2)** Charge drawn by capacitor without dielectric, $Q=CV$

$$\Rightarrow Q_1=\frac{\epsilon_0 AV}{d}\quad\left(\because C=\frac{\epsilon_0 A}{d}\right)$$

Charge drawn by capacitor on introducing dielectric

$$\Rightarrow Q_2=\frac{\epsilon_0 AV}{d-t+\frac{t}{K}}$$

We have given $\frac{Q_2-Q_1}{Q_1}\times 100=25\Rightarrow\frac{Q_2}{Q_1}-1=\frac{1}{4}$

$$\Rightarrow Q_2=1.25Q_1\ \Rightarrow\frac{\epsilon_0 AV}{d-t+\frac{t}{K}}=(1.25)\frac{\epsilon_0 AV}{d}$$

$$\Rightarrow 1.25\left(3+\frac{2}{K}\right)=5\Rightarrow K=2$$

99. **(2)**

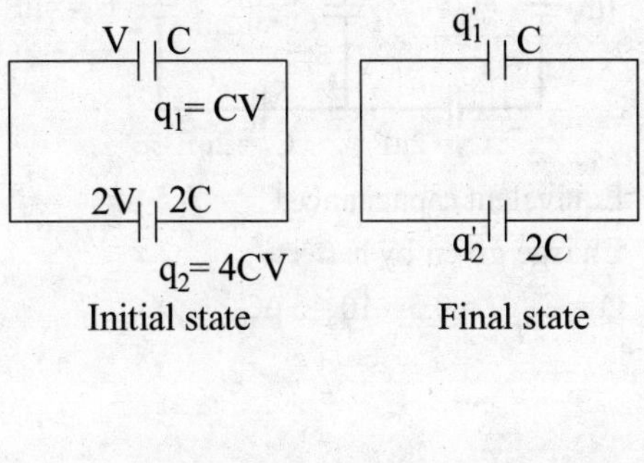

From the change conservation

$$q_1' + q_2' = q_1 + q_2$$

$$q_1' + q_2' = 5\,CV \qquad .....(1)$$

$$\frac{q_1'}{C} = \frac{q_2'}{2C} \qquad .....(2)$$

From equation (1) and (2)

$$q_1' = \frac{5}{3}CV,\ q_2' = \frac{10}{3}CV$$

Loss of energy, $\Delta E = \frac{\Delta q_1^2}{2C_1} + \frac{\Delta q_2^2}{2C_2}$

$$= \frac{\left(CV - \frac{5}{3}CV\right)^2}{2C} + \frac{\left(4CV - \frac{10}{3}CV\right)^2}{2(2C)}$$

$$= \frac{4}{18}CV^2 + \frac{2}{18}CV^4 = \frac{1}{3}CV^2 = \frac{2}{3}E = \frac{x}{3}E \text{ (given)}$$

$\therefore\ x = 2$

100. (5) Charge of on the capacitor, Q = CV

Charge on C_1 is $Q_1 = 2 \times 10 = 20\mu C$

Charge on C_2 is $Q_2 = x \times 10 = 10x\mu C$

Charge on C_3 is $Q_3 = 3 \times 10 = 30\mu C$

Total charge, $Q = Q_1 + Q_2 + Q_3$

$20 + 10x + 30 = 100 \Rightarrow x = 5$

101. (b) Initially

$Q_1 = CV = 2V$ $\qquad (\because C = 2F \text{ given})$

$$E_1 = \frac{1}{2}\frac{Q_1^2}{C} = \frac{1}{2}\frac{(2V)^2}{2} = V^2$$

Finally

Charge will be divided equally

$$\therefore Q_2 = \frac{Q_1}{2} = \frac{2V}{2} = V$$

$$E_2 = 2\left(\frac{1}{2}\frac{Q_2^2}{C}\right) = \frac{V^2}{2} \qquad \therefore \frac{E_2}{E_1} = \frac{1}{2}$$

102. (c) From the Gauss's law

$$\phi = \frac{q_{in}}{\epsilon_0} = \frac{Q}{\epsilon_0} \qquad \because Q = CV$$

$$\therefore \phi = \frac{CV}{\epsilon_0}$$

103. (4)

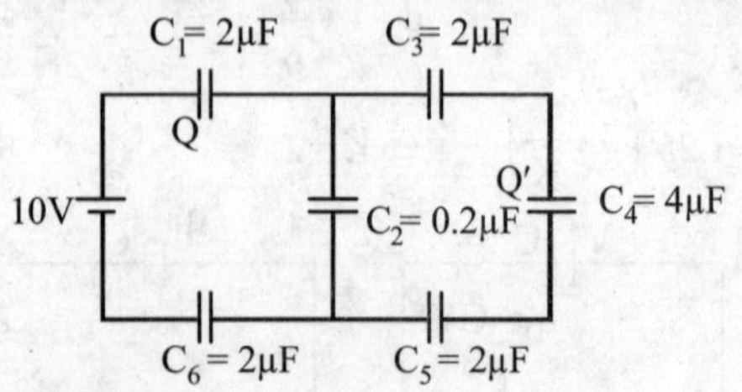

Equivalent capacitance $C_{eq} = 0.5\ \mu F$

Charge given by battery

$Q = C_{eq}V = 0.5 \times 10 = 5\ \mu C$

Charge on C_4,

$$Q' = \frac{5\mu C \times 0.8}{0.8 + 0.2} = 4\mu C$$

104. (c) The circuit can be reduced to

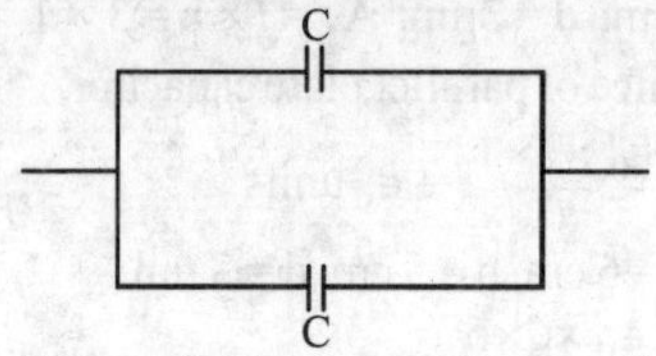

The equivalent capacitance of the combination is

$C_{eq} = C + C = 2C$

105. (6) $C = 600 \times 10^{-12}$ F; V = 200 V

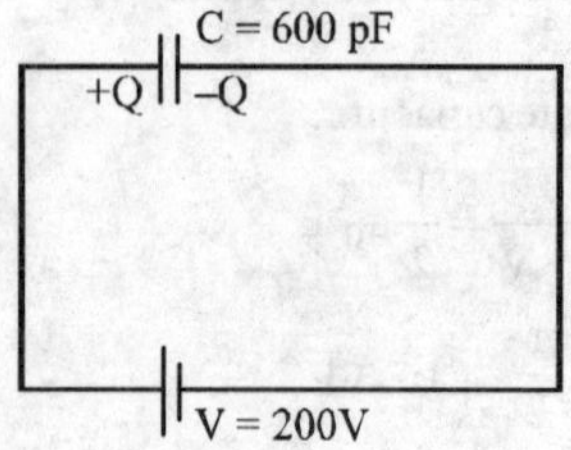

$Q = CV = 600 \times 10^{-12} \times 200 = 12 \times 10^{-8}$ C

Initial energy $= \frac{1}{2}CV^2$

$$= \frac{1}{2} \times 600 \times 10^{-12} \times (200)^2 = 12\mu J$$

When connected to another uncharged capacitor

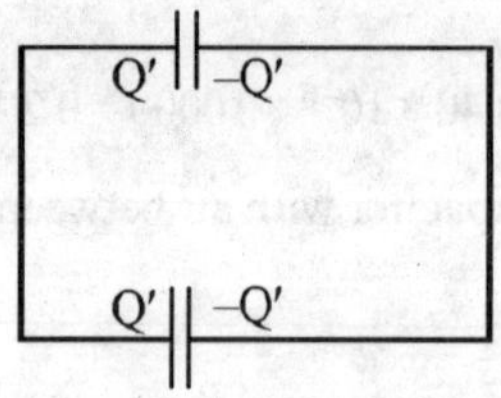

Charge will be equally distributed on identical capacitor

$$Q' = \frac{Q}{2} = \frac{12 \times 10^{-8}}{2} = 6 \times 10^{-8} C$$

Final energy $= 2 \times \frac{Q'^2}{2C} = \frac{Q'^2}{C}$

$$= \frac{\left(6 \times 10^{-8}\right)^2}{600 \times 10^{-12}} = 6\mu J$$

Energy lost in the process = Initial energy – Final energy

$= (12 - 6)\ \mu J = 6\mu J$

106. (3) Capacitance of parallel plate capacitor with dielectric slab between plates

$$C = \frac{\epsilon_0 A}{d - \left(t - \frac{t}{K}\right)}$$

Here, t = thickness of the dielectric material

For thickness, $x = \frac{d}{3}$

$$C_1 = \frac{\epsilon_0 A}{\left(\frac{d}{3K}+\frac{2d}{3}\right)} = \frac{\epsilon_0 A}{\frac{d}{12}+\frac{2d}{3}} = \frac{\epsilon_0 A}{\frac{d}{12}(1+8)}$$

$$= \frac{\epsilon_0 A}{d} \times \left(\frac{12}{9}\right) = \frac{4\epsilon_0 A}{3d} = 2\mu F$$

For thickness, $x = \frac{2d}{3}$

$$C_2 = \frac{\epsilon_0 A}{\left(\frac{2d/3}{K}+\frac{d}{3}\right)} = \frac{\epsilon_0 A}{\frac{2d}{12}+\frac{d}{3}} = \frac{2\epsilon_0 A}{d}$$

$$\therefore \frac{C_1}{C_2} = \frac{\frac{4\epsilon_0 A}{3d}}{\frac{2\epsilon_0 A}{d}} = \frac{4}{3} \times \frac{1}{2} \Rightarrow C_2 = \frac{3}{2} C_1$$

$$\Rightarrow C_2 = \frac{3}{2} \times 2 = 3\mu F$$

107. (5) Plate area, $A = 200\ cm^2 = 200 \times 10^{-4}\ m^2$

The equivalent capacitance of the combination is

$$C_{eq} = \frac{\varepsilon_0 A}{(d-c)} = \frac{\varepsilon_0 \times 200 \times 10^{-4}}{4 \times 10^{-3}} = 5\varepsilon_0$$

108. (a) Both the spheres will hold the same amount of charge. It is because, the two spheres has equal capacitance. The capacitance of a sphere depends only on its radius. It does not matter whether the sphere is hollow or solid.

109. (55) If battery is disconnected, then, voltage becomes $\frac{1}{K}$ times

If battery is not disconnected, then voltage remains constant.

So, common potential $= \frac{C_1V_1 + C_2V_2}{C_1 + C_2}$

$$= \frac{C \times \frac{100}{K} + C \times 100}{C + C}, \text{ where } C = C_1 = C_2$$

$$= \frac{\frac{100}{10} + 100}{2} = 55\,V$$

110. (225) The loss of energy in this process,

$C_1 = C_2 = C = 900 \times 10^{-6}\ F$

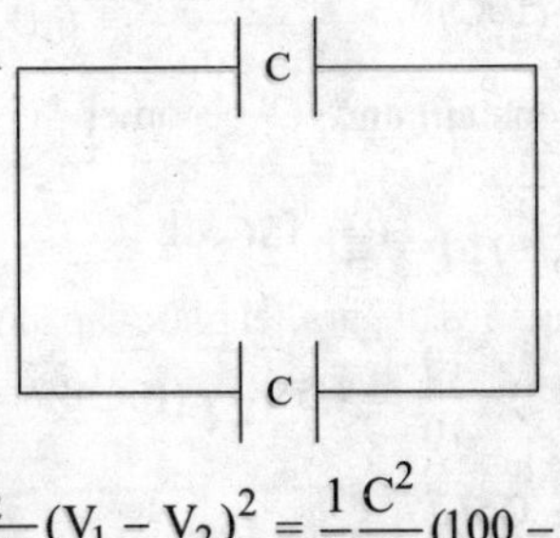

$$= \frac{1}{2}\frac{C_1C_2}{C_1+C_2}(V_1 - V_2)^2 = \frac{1}{2}\frac{C^2}{2C}(100-0)^2$$

$$= \frac{1}{2}\frac{900 \times 10^{-6}}{2} \times 10^4 = \frac{9}{4}\text{ joule} = 2.25\text{ joule}$$

111. (c) Given that

Area, $A = 40\ cm^2 = 40 \times 10^{-4}\ m^2$

Plates separation

$d = 2\ mm$

$= 2 \times 10^{-3}\ m$

Thickness, $t = 1\ mm$

$= 1 \times 10^{-3}\ m$

Dielectric constant, $K = 5$

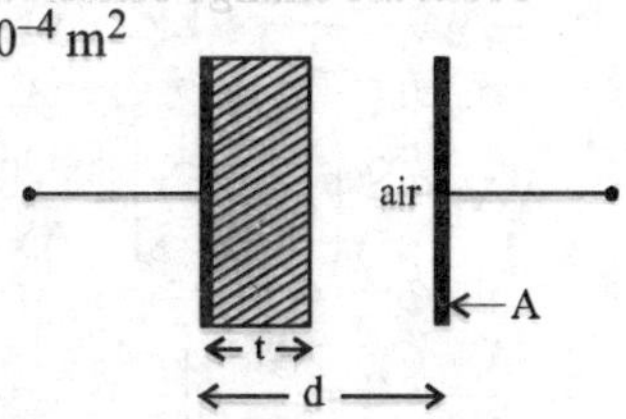

Here two capacitors in series,

$$\frac{1}{C_{eq}} = \frac{1}{C_1} + \frac{1}{C_2} = \frac{1}{\frac{K\epsilon_0 A}{t}} + \frac{1}{\frac{\epsilon_0 A}{d-t}}$$

$$= \frac{t}{K\epsilon_0 A} + \frac{d-t}{\epsilon_0 A} = \frac{1 \times 10^{-3}}{5\epsilon_0 \times 40 \times 10^{-4}} + \frac{1 \times 10^{-3}}{\epsilon_0 \times 40 \times 10^{-4}}$$

$$\frac{1}{C_{eq}} = \frac{1}{20\epsilon_0} + \frac{1}{4\epsilon_0}$$

$$C_{eq} = \frac{20 \times 4\epsilon_0}{24} = \frac{10\epsilon_0}{3}\ F$$

112. (6) Initialy capacitor by air medium of thikness d,

$$C = \frac{\epsilon_0 A}{d} = 5\mu F$$

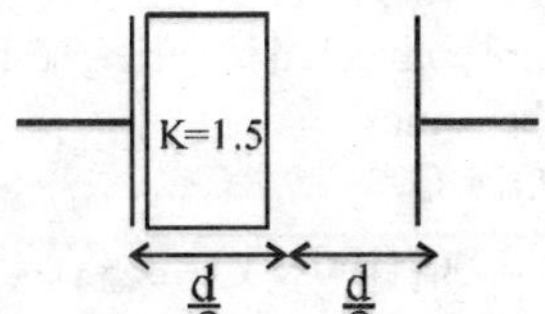

$$\frac{1}{C_{eq}} = \frac{\frac{d}{2}}{\epsilon_0 A} + \frac{\frac{d}{2}}{\epsilon_0 AK}$$

$$C_{eq} = \left(\frac{2K}{K+1}\right)\frac{\epsilon_0 A}{d} = \left(\frac{2 \times 1.5}{1.5+1}\right)5 = 6\mu F$$

113. (105) Initially, $C_0 = \frac{\epsilon_0 A}{d} = 15pF$

Finally, $C = \frac{K\epsilon_0 A}{2d} = \frac{3.5}{2} \times 15pF = \frac{105}{4}pF$

114. (c) Electric field between plates given by, $E = \frac{q_1 - q_2}{2A\epsilon_0}$

(Here, $q_1 > q_2$)

The, the potential difference will be

$$V = Ed = \frac{q_1 - q_2}{2A\epsilon_0}d = \frac{q_1 - q_2}{2C} \qquad \left(\because C = \frac{\epsilon_0 A}{d}\right)$$

115. (a) Equivalent capacitance in series,

$$C_{eq} = \frac{C(KC)}{C+KC} = \frac{KC}{K+1}$$

$$24 = \frac{K40}{K+1} \qquad [\because C_{eq} = 24\mu F]$$

$$24(K+1) = 40K \therefore K = 1.5$$

116. (a) We have, $x + y + \frac{3d}{4} = d$

$$\Rightarrow x + y = \frac{d}{4}$$

$$\Delta V = Ex + \frac{E}{k} \times \frac{3d}{4} + Ey$$

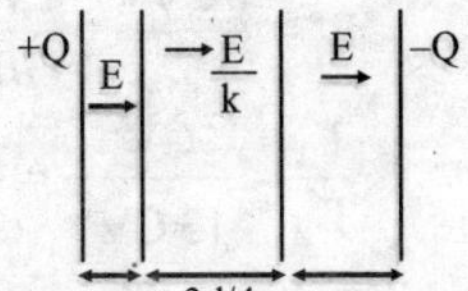

$\Rightarrow \Delta V = \frac{3Ed}{4k} + E(x+y) \Rightarrow \Delta V = E\left[\frac{3d}{4k} + \frac{d}{4}\right]$

$\Delta V = \frac{\sigma}{\epsilon_0}\left[\frac{3d + dk}{4k}\right] = \frac{Qd}{A\epsilon_0}\left[\frac{3+k}{4k}\right]$

$\frac{Q}{\Delta V} = C = \frac{A\epsilon_0}{d}\left[\frac{4k}{3+k}\right] = \frac{4kC_0}{k+3} \qquad \left[\because C_0 = \frac{A\epsilon_0}{d}\right]$

117. (240) $C_{eff} = \frac{\epsilon_0 A_1}{d} + \frac{K\epsilon_0 A_2}{d}$

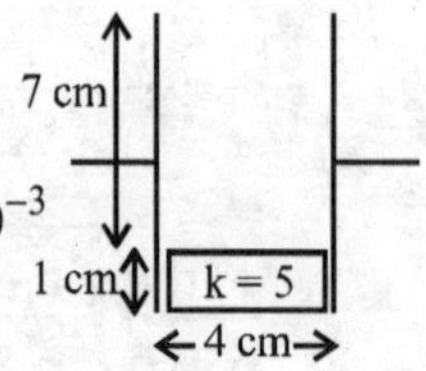

$C_{eff} = \left[\frac{\varepsilon_0(7\times 4)}{4/10} + \frac{5\varepsilon_0(1\times 4)}{4/10}\right]\times 10^{-3}$

$\Rightarrow \quad C_{eff} = 1.2\varepsilon_0$

Energy $= \frac{1}{2}C_{eff}V^2 = \frac{1}{2}(1.2)\varepsilon_0(20)(20) = 240\varepsilon_0 J$

118. (60) Capacitances of the capacitors,

$C_1 = \frac{\epsilon_0 \epsilon_{r_1} A}{t_1}, \; C_2 = \frac{\epsilon_0 \epsilon_{r_2} A}{t_2}$

Let V be the voltage of conducting foil. As the capacitors are connected in series, so charge on the capacitors should be same

$Q_1 = Q_2$

$\Rightarrow \; C_1(100 - V) = C_2 V \quad (\because Q = CV)$

$\Rightarrow \; \frac{\epsilon_0 \epsilon_{r_1} A}{t_1}(100 - V) = \frac{\epsilon_0 \epsilon_{r_2} AV}{t_2}$

$\Rightarrow \; \frac{3\times(100 - V)}{0.5\times 10^{-3}} = \frac{4\times V}{1\times 10^{-3}}$

$\Rightarrow \; 600 - 6V = 4V \Rightarrow V = 60\,V$

119. (a) Given that all capacitors are connected in parallel so, the equivalent capacitance will be $C_{eq} = C_1 + C_2 + C_3 + C_4$ $= 1 + 2 + 4 + 3 = 10\,\mu F$

Voltage of battery V = 20 V

We have, $Q = CV = 10\,\mu F \times 20 = 200\,\mu C$

120. (c) When switch is closed then both capacitor will be parallel.

$C_{eq} = 2C$

Energy $E_1 = \frac{1}{2}C_{eq}V^2 = \frac{1}{2}\times 2C \times V^2$

$E_1 = CV^2$

When switch is opened charge on right capacitor remain CV while potential on left capacitor remain same.

Dielectric K = 5

$C' = KC$

$C' = 5C$

$E_2 = \frac{1}{2}C'V^2 + \frac{q^2}{2C'} \Rightarrow E_2 = \frac{1}{2}(5C)V^2 + \frac{(CV)^2}{2(5C)}$

$E_2 = \frac{5CV^2}{2} + \frac{CV^2}{10} \Rightarrow E_2 = \frac{13CV^2}{5}$

$\frac{E_1}{E_2} = \frac{CV^2}{\frac{13CV^2}{5}} = \frac{5}{13} \Rightarrow \frac{E_1}{E_2} = \frac{5}{13}$

121. (6)

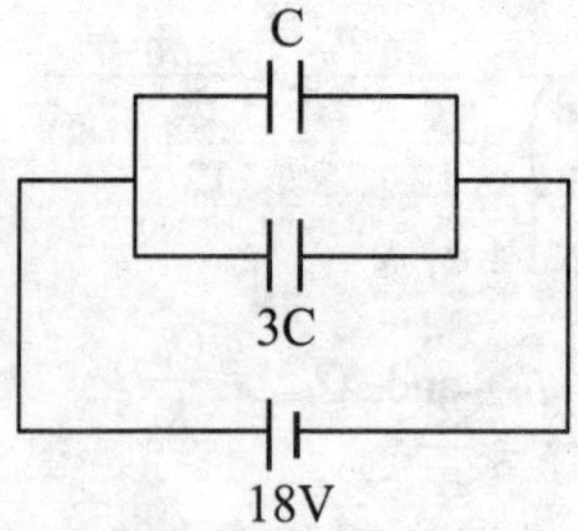

Initial charge on C = 18 CV

initial charge on 3C = 54 CV

Let final common potential difference = V′

$Q_{final} = Q_{initial}$

$9CV' + 3CV' = 18CV + 54CV$

$\Rightarrow 12CV' = 72CV \Rightarrow V' = 6V$

122. (a) Capacitance of isolated conducting sphere is given as $C_1 = 4\pi\varepsilon_0 R_1$

By enclosing it inside another sphere of radius R_2, new capacitance, $C_2 = \frac{4\pi\varepsilon_0 R_1 R_2}{(R_2 - R_1)}$

Given, $C_2 = nC_1 = \frac{4\pi\varepsilon_0 R_1 R_2}{(R_2 - R_1)} = n\times 4\pi\varepsilon_0 R_1$

$\Rightarrow \frac{R_2}{(R_2 - R_1)} = n \Rightarrow \frac{\frac{R_2}{R_1}}{\left(\frac{R_2}{R_1} - 1\right)} = n \Rightarrow \frac{R_2}{R_1} = n\frac{R_2}{R_1} - n$

$\Rightarrow \frac{R_2}{R_1} = \frac{n}{(n-1)}$

123. (a) Clearly, C_1 and C_2 are in parallel as shown

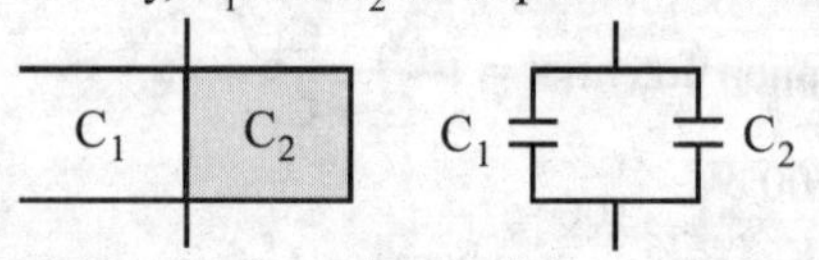

So, $C_{eq} = C_1 + C_2 = \frac{\frac{A}{2}\varepsilon_0}{d} + \frac{\frac{A}{2}\varepsilon_0\varepsilon_r}{d} = \frac{A\varepsilon_0}{2d}[1 + \varepsilon_r]$

$\therefore C_{eq} = \frac{C_0}{2}(1 + \varepsilon_r)$

124. (a) $U_i = \frac{1}{2}(10C)V^2$

Here, V = constant and 'C' becomes 1.5 times

So, $U_f = \frac{1}{2}C_f V_f^2 = \frac{1}{2}\times 15C \times V^2$

$= 15\times\frac{1}{2}CV^2 \times \frac{10}{10} = 1.5\times\frac{1}{2}\times(10C)V^2 = 1.5\,U_i$

$\frac{\Delta U}{U}\times 100 = \frac{0.5U_i}{U_i}\times 100 = 50\%$

125. (d) We have

$U = \frac{Q^2}{2C} = \frac{Q_0^2 e^{-2t/\tau}}{2C} = U_0 e^{-2t/\tau}$

as, $U_{t=t_1} = \frac{U_0}{2}$

$U_0 e^{-2t_1/\tau} = \frac{U_0}{2}$

$e^{-2t_i/\tau} = \frac{1}{2} \Rightarrow t_1 = \frac{\tau}{2} ln2$ and, $Q_{t=t_2} = \frac{Q_0}{8}$

$Q_0 e^{-t_2/\tau} = \frac{Q_0}{8} \Rightarrow e^{-t_2/\tau} = \frac{1}{8} \Rightarrow t_2 = \tau\, ln8 \Rightarrow t_2 = 3\tau ln2$

So, $\frac{t_1}{t_2} = \frac{\frac{\tau}{2} ln2}{3\tau ln2} = \frac{1}{6}$

126. (100) $q_i = q_f$

$\Rightarrow 5 \times 30 = (5+10)V' \Rightarrow V' = 10$ Volt

So, $Q_{C_2} = C_2 V' = 10 \times 10\,\mu C = 100\,\mu C$

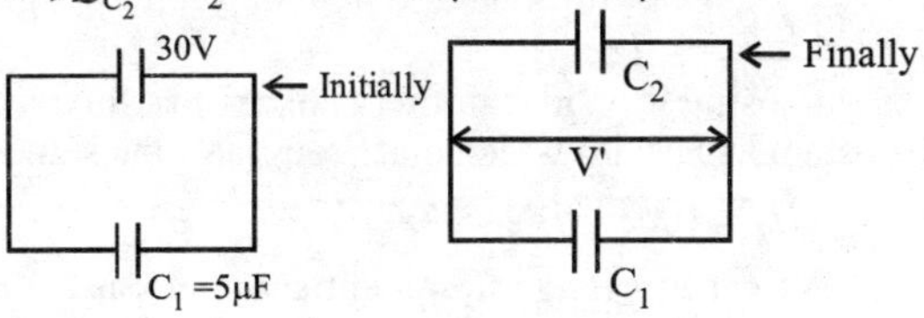

127. (a) Force on charge particle will be given by

$F = qE$

$F \propto E$

$$F_2 = F_1\left(\frac{E_2}{E_1}\right) = F_1\left(\frac{\frac{\sigma}{2\epsilon_0}}{\frac{\sigma}{\epsilon_0}}\right) = \frac{F_1}{2} = \frac{10}{2} = 5N$$

128. (a) We have, $q_i = C_1 V$

Let V' be the final voltage.

Then, $q_f = (C_1 + C_2)V'$

as $q_i = q_f \quad \therefore\ C_1 V = (C_1 + C_2)V'$

$V' = \frac{C_1 V}{C_1 + C_2}$

$q_{C_2} = V' C_2 = \frac{C_1 C_2 V}{C_1 + C_2}$

129. (a) $\frac{1}{C_{eq}} = \frac{1}{10} + \frac{1}{15} + \frac{1}{20} = \frac{6+4+3}{60} \therefore C_{eq} = \frac{60}{13}\mu F$

So, $Q = C_{eq} V = \frac{60}{13} \times 13\mu C = 60\mu C$

In series, charge across each capacitor is same and is equal to net charge.

So, $Q_{15\mu F} = 60\mu C$

130. (c) We have, $C_i = \frac{A\epsilon_0}{d} = 4\mu F$

$$C_f = \frac{A\epsilon_0}{d - t + \frac{t}{k}} = \frac{A\epsilon_0}{d - \frac{d}{2} + \frac{d}{2\times 3}} = \frac{A\epsilon_0}{d\left(1 - \frac{1}{2} + \frac{1}{6}\right)}$$

$$= \frac{4\mu F}{\frac{2}{3}} = 6\mu F$$

131. (a) Dielectric constant of conductor $= \infty$. So $K = \infty$

Now, $C_f = \frac{A\epsilon_0}{d - t + t/k} = \frac{A\epsilon_0}{d - \frac{d}{2} + \frac{d}{\infty}}$

$= \frac{A\epsilon_0}{\frac{d}{2}} = 2\frac{A\epsilon_0}{d} = 2C_i$. So, $\frac{C_f}{C_i} = \frac{2}{1}$

132. (d) $q = CV$

$\Rightarrow q = \frac{kA\epsilon_0}{d}.Ed \Rightarrow k = \frac{q}{A\epsilon_0 E}$

$$= \frac{7\times 10^{-6}}{30\pi \times 10^{-4} \times \frac{1}{4\pi \times 9 \times 10^9} \times 3.6 \times 10^7} = \frac{7}{3} = 2.33$$

133. (a) $U = \frac{q^2}{2C}$. As C = constant, $U \propto q^2$

So, $\frac{U_2}{U_1} = \left(\frac{q_2}{q_1}\right)^2 \Rightarrow q_1^2 = \frac{U_1}{U_2} q_2^2$

$\Rightarrow q_1^2 = \frac{U}{1.44U} \times (q_1 + 2)^2 \Rightarrow \left(\frac{q_1}{q+2}\right)^2 = \frac{1}{1.44}$

$\Rightarrow \frac{q_1}{q_1 + 2} = \frac{1}{1.2}$

$\Rightarrow 1.2q_1 = q_1 + 2 \Rightarrow 0.2\,q_1 = 2$

$\Rightarrow q_1 = 10\,C$

134. (b) Given,

Voltage of battery, $V_0 = 20V$

Voltage across capacitor, $V = 2V$

$V = V_0(1 - e^{-t/RC}) \Rightarrow 2 = 20\left(1 - e^{-t/RC}\right)$

$\Rightarrow \frac{1}{10} = 1 - e^{-t/RC} \Rightarrow e^{-t/RC} = \frac{9}{10} \Rightarrow e^{t/RC} = \frac{10}{9}$

$\Rightarrow \frac{t}{RC} = \ell n\left(\frac{10}{9}\right) \Rightarrow C = \frac{t}{R\ell n\left(\frac{10}{9}\right)}$

$\Rightarrow C = \frac{10^{-6}}{10 \times 0.105} = 0.95\mu F$

135. (100)

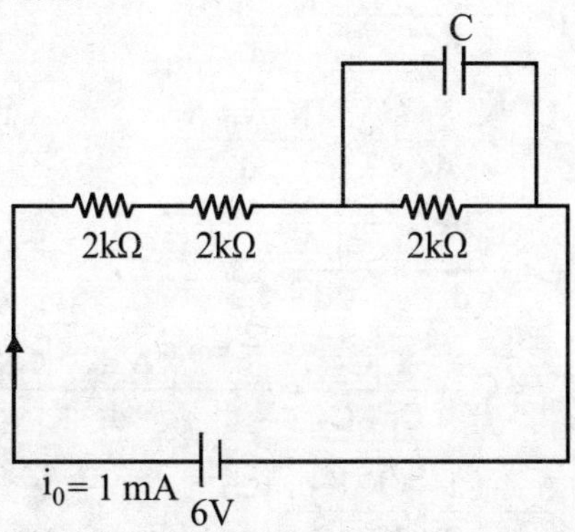

Potential difference across each resistor $= 2V$ as each of 2Ω resistance and current is same

Charge $q = CV$

$= 50 \times 10^{-6} \times 2 = 100 \times 10^{-6} = 100\,\mu C$

136. (4) Change in energy stored

$=\frac{1}{2}C_2V^2-\frac{1}{2}C_1V^2=\frac{1}{2}KCV^2-\frac{1}{2}CV^2$

$=\frac{1}{2}(2-1)CV^2$ [Putting $K=2$]

$=\frac{1}{2}\times 200\times 10^{-6}\times 200\times 200=4J$

137. (a) Equivalent circuit

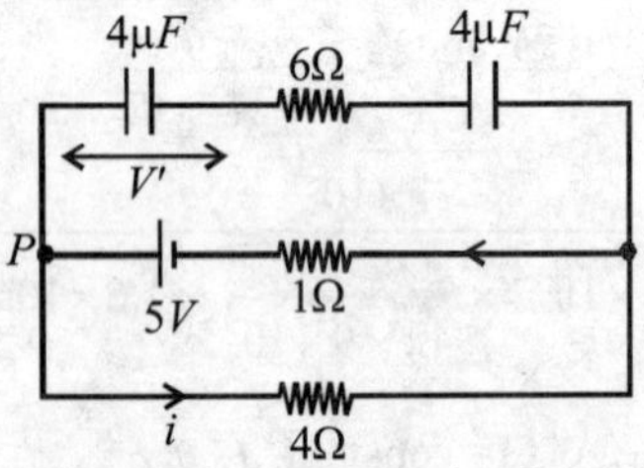

Capacitor blocks the DC current. So current will flow in lower loop only.

So, $V_{PQ}=\frac{5}{4+1}\times 4=4V$

Charge on $4\mu F=V'C=\left(\frac{4\times 4}{4+4}\right)\times 4=8\mu C$

138. (c) $V=\frac{qd}{AK\in_0}$

$i=\frac{V}{R}=\frac{qd}{AK\in_0\rho\frac{d}{A}}$ $\left(\because R=\rho\frac{d}{A}\right)$

$i_{max}=\frac{q_{max}}{\rho K\in_0}\Rightarrow i_{max}=\frac{2\times 10^{-12}\times 40}{200\times 50\times 8.85\times 10^{-12}}$

$=0.9\ mA$

139. (a) Equivalent circuit is

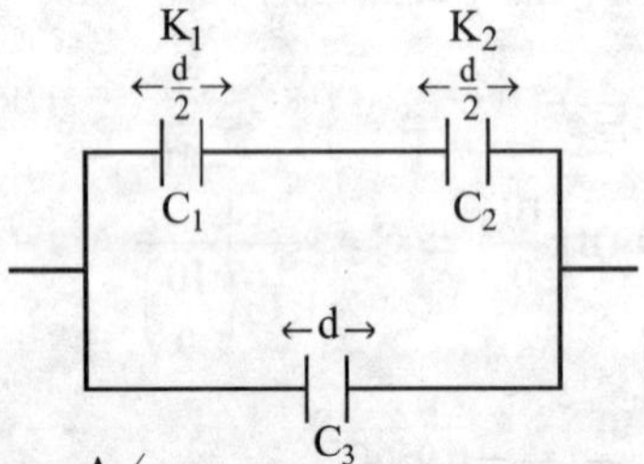

$C_1=\frac{K_1\in_0 A/2}{d/2}=\frac{K_1\in_0 A}{d}$

$C_2=\frac{K_2\in_0 A/2}{d/2}=\frac{K_2\in_0 A}{d}$

$C_3=\frac{\in_0 A/2}{d}=\frac{\in_0 A}{2d}$

$C_{eq}=C_3+\frac{C_1C_2}{C_1+C_2}=\frac{\frac{A}{2}\in_0}{d}+\frac{A\in_0}{d}\frac{K_1K_2}{K_1+K_2}$

$=\frac{A\in_0}{d}\left(\frac{1}{2}+\frac{K_1K_2}{K_1+K_2}\right)$

140. (2) At, $t=0$, potential difference across resistance

$V=\frac{Q}{C}=\frac{30}{3}=10\text{ V}$

From $V=RI$

$\Rightarrow\ I=\frac{10}{R}=\frac{10}{5\times 10^6}=2\times 10^{-6}\text{ A}=2\ \mu\text{A}=x\mu\text{A}$

$\therefore\ x=2$

141. (161) Given : $A=100\text{ m}^2$, $d=10$ m, $t=5$ m and $k=10$

Capacitance, $C=\frac{\varepsilon_0 A}{d-t+t/k}=\frac{8.85\times 10^{-12}\times 100}{10-5+5/10}$

$=1.609\times 10^{-10}=160.9\times 10^{12}=161\text{ pF}=x\text{pF}\ \therefore x=161$

142. (864) Initial energy stored in the capacitor,

$U_i=\frac{1}{2}CV^2=1008\text{ pJ}$

After inserting porcelin plate, $k=7$

$U_f=\frac{1}{2}\frac{CV^2}{k}=144\text{ pJ}$

Constant M.E. with which plate oscillates back and forth $=U_i-U_f=1008-144=864\text{ pJ}$

143. (2) We can rearrange four identical plates as follows

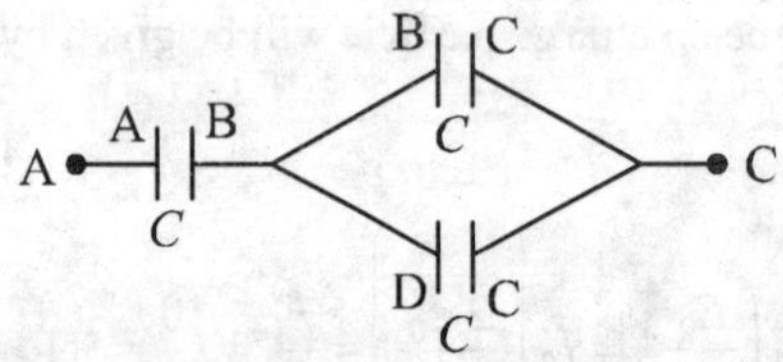

$C_{eq}=\frac{C\times 2C}{C+2C}=\frac{2C}{3}\Rightarrow C_{eq}=\frac{2C}{3}=\frac{2}{3}\frac{\varepsilon_0 A}{d}=\frac{2}{3}\frac{\varepsilon_0\ell\times b}{d}$

$C_{eq}=\frac{2}{3}\frac{\varepsilon_0\times 2}{d}\times\frac{3}{2}=2\frac{\varepsilon_0}{d}=\frac{x\varepsilon_0}{d}\ \therefore x=2$

144. (16) When battery is connected,

$Q_1=C_1V=2\times 10=20\ \mu C$

When battery is removed and the capacitor is connected

$Q_1=C_1V+C_2V\Rightarrow 20=(2+8)V\quad\therefore V=2V$

Therefore charge in $C_2=C_2V=16\mu C$

145. (3) Capacitance of the set-up

$C=\frac{\varepsilon_0 A}{d-t+\frac{t}{k}}=\frac{\varepsilon_0\times 2}{1-0.5+\frac{0.5}{3.2}}=\frac{\varepsilon_0\times 2}{\frac{2.1}{3.2}}=3.04\,\varepsilon_0\approx 3\,\varepsilon_0$

146. (c) From figure, $v_1\cos\alpha=v_2\cos\beta$

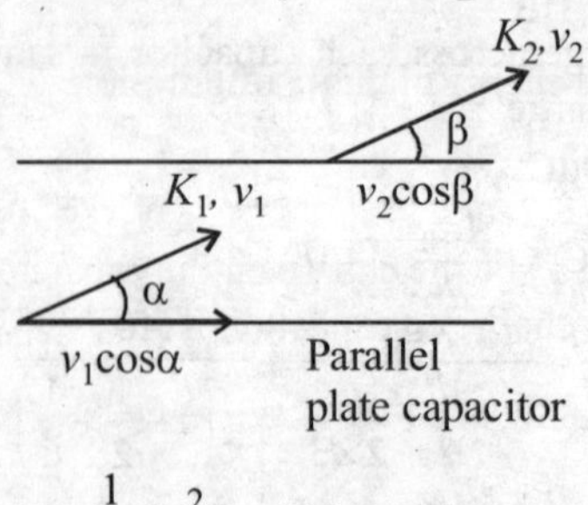

$\therefore\frac{K_1}{K_2}=\frac{\frac{1}{2}mv_1^2}{\frac{1}{2}mv_2^2}=\frac{v_1^2}{v_2^2}=\frac{\cos^2\beta}{\cos^2\alpha}$

147. (c) For series combination,

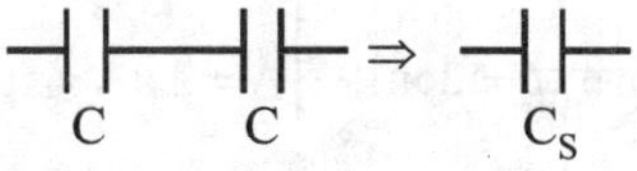

$$\frac{1}{C_S}=\frac{1}{C}+\frac{1}{C}\Rightarrow \boxed{C_S=\frac{C}{2}}$$

For parallel combination,

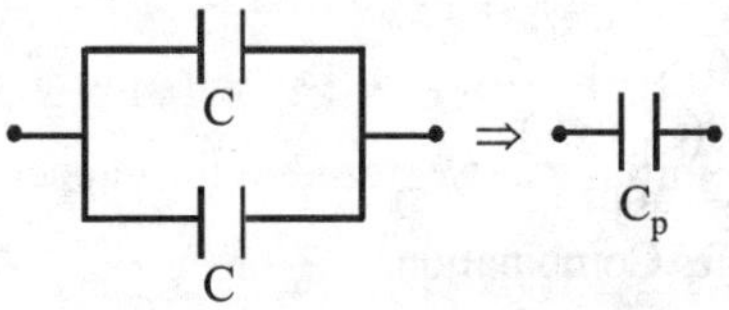

$$C_p=C+C\Rightarrow C_p=2C\Rightarrow \frac{C_S}{C_p}=\frac{\left(\frac{C}{2}\right)}{2C}=\frac{1}{4}=1:4$$

148. (None) When C_1 and C_2 capacitors are connected in series, their equivalent capacitance is

$$C'=\frac{C_1C_2}{C_1+C_2}$$

When C_1 and C_2 are connected in parallel.
Their equivalent capacitance is $C''=C_1+C_2$
According to question

$$\frac{C_1C_2}{C_1+C_2}=\frac{4}{15}(C_1+C_2)\Rightarrow 4(C_1+C_2)^2=15C_1C_2$$

Put $\frac{C_2}{C_1}=x \Rightarrow 4(x+1)^2=15x \Rightarrow x \simeq 0.91$

None of the values of x satisfies the condition.

149. (b) When capacitors C and $2C$ capacitance are charged to V and 2V respectively.

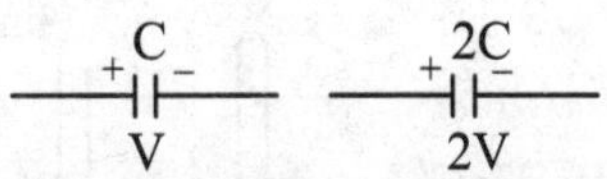

$Q_1=CV$ $Q_2=2C\times 2V=4CV$

When connected in parallel
By conservation of charge

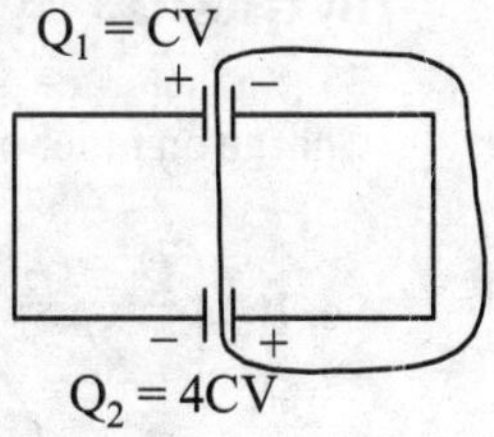

$$4CV-CV=(C+2C)V_{\text{common}}$$

$$\Rightarrow V_{\text{common}}=\frac{3CV}{3C}=V$$

Therefore final energy of this configuration,

$$U_f=\left(\frac{1}{2}CV^2+\frac{1}{2}\times 2CV^2\right)=\frac{3}{2}CV^2$$

150. (c) Let q_1 and q_2 be the charge on the capacitors of 2μF and 4μF. Then charge on capacitor of 5μF

$$Q=q_1+q_2$$

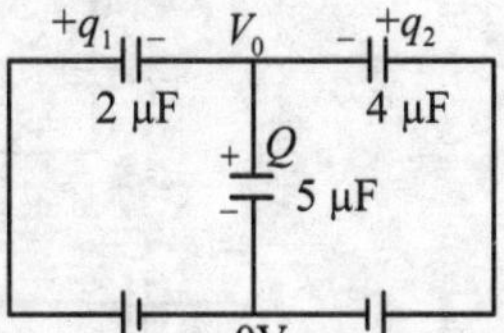

$$\Rightarrow 5V_0=2(6-V_0)+4(6-V_0)$$

$$\Rightarrow 5V_0=12-2V_0+24-4V_0$$

$$\Rightarrow 11V_0=36\Rightarrow V_0=\frac{36}{11}V\Rightarrow Q=5V_0=\frac{180}{11}\mu C$$

$$=16.36\mu C$$

151. (d) When two capacitors with capacitance C_1 and C_2 at potential V_1 and V_2 connected to each other by wire, charge begins to flow from higher to lower potential till they acquire common potential. Here, some loss of energy takes place which is given by.

Heat loss, $H=\frac{C_1C_2}{2(C_1+C_2)}(V_1-V_2)^2$

In the equation, put $V_2=0$, $V_1=V_0$

$$C_1=C,\ C_2=\frac{C}{2}$$

$$\text{Loss of heat}=\frac{C\times\frac{C}{2}}{2\left(C+\frac{C}{2}\right)}(V_0-0)^2=\frac{C}{6}V_0^2$$

$$\Delta H=\frac{1}{6}CV_0^2$$

152. (b) According to question,

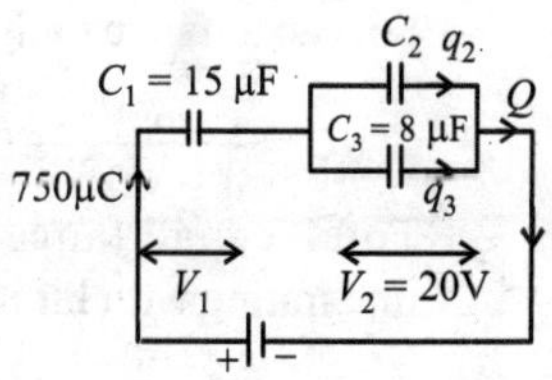

$Q=750\mu C=q_2+q_3$

Capacitors C_2 and C_3 are in parallel hence,
Voltage across C_2 = voltage across $C_3=20$ V
Change on capacitor C_3,

$$q_3=C_3\times V_3=8\times 20=160\mu C$$

$$\therefore q_2=750\mu C-160\mu C=590\mu C$$

153. (4) Given, $C_1=5$ μF and $V_1=220$ Volt
When capacitor C_1 fully charged it is disconnected from the supply and connected to uncharged capacitor C_2.
$C_2=2.5$ μF, $V_2=0$
Energy change during the charge redistribution,

$$\Delta U=U_i-U_f=\frac{1}{2}\frac{C_1C_2}{C_1+C_2}(V_1-V_2)^2$$

$$=\frac{1}{2}\times\frac{5\times 2.5}{(5+2.5)}(220-0)^2\ \mu J$$

$$=\frac{5}{2\times 3}\times 22\times 22\times 100\times 10^{-6}\ J$$

$$=\frac{5\times 11\times 22}{3}\times 10^{-4}\ J=\frac{55\times 22}{3}\times 10^{-4}\ J$$

$$=\frac{1210}{3}\times 10^{-4}\ J=\frac{1210}{3}\times 10^{-3}\ J\simeq 4\times 10^{-2}\ J$$

According to questions, $\frac{x}{100}=4\times 10^{-2}$ $\therefore x=4$

154. (a) Given,
Capacitance of capacitor, $C_1=10$ μF
Potential difference before removing the source voltage, $V_1=50$ V

If C_2 be the capacitance of uncharged capacitor, then common potential is

$$V=\frac{C_1V_1+C_2V_2}{C_1+C_2}\Rightarrow 20=\frac{10\times 50+0}{20+C_2}\Rightarrow C_2=15\mu F$$

155. (c) In parallel combination, $C_{eq}=C_1+C_2=10\,\mu F$

When connected across 1 V battery, then

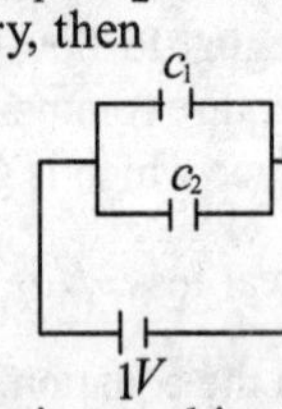

$$\frac{U_1}{U_2}=\frac{\left(\frac{1}{2}C_1V^2\right)}{\left(\frac{1}{2}C_2V^2\right)}=\frac{1}{4}\Rightarrow\frac{C_1}{C_2}=\frac{1}{4}$$

$\therefore C_2=8\,\mu F$ and $C_1=2\,\mu F$

Now C_1 and C_2 are connected in series combination,

$$\therefore C_{\text{equivalent}}=\frac{C_1C_2}{C_1+C_2}=\frac{2\times 8}{2+8}=\frac{16}{10}=1.6\mu F$$

156. (a) Consider an infinitesimal strip of capacitor of thickness dx at a distance x as shown.

Capacitance of parallel plate capacitor of area A is given by $C=\frac{\varepsilon_0 A}{t}$

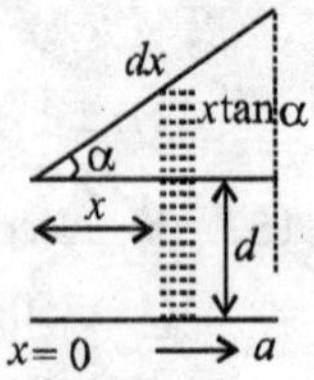

[Here t = seperation between plates]

So, capacitance of thickness dx will be

$$\therefore dC=\frac{\varepsilon_0 adx}{d+x\tan\alpha}$$

Total capacitance of system can be obtained by integrating with limits x = 0 to x = a

$$\therefore C_{eq}=\int dC=a\varepsilon_0\int_{x=0}^{x=a}\frac{dx}{x\tan\alpha+d}$$

$$=a\varepsilon_0\int_{x=0}^{x=a}\frac{dx}{d\left(1+\frac{x}{d}\tan\alpha\right)}$$

[By Binomial expansion]

$$\Rightarrow C_{eq}=\frac{a\varepsilon_0}{d}\int_0^a\left(1-\frac{x\tan\alpha}{d}\right)dx=\frac{a\varepsilon_0}{d}\left(x-\frac{x^2\tan\alpha}{2d}\right)_0^a$$

$$\Rightarrow C_{eq}=\frac{a^2\varepsilon_0}{d}\times\left(1-\frac{a\tan\alpha}{2d}\right)=\frac{\varepsilon_0 a^2}{d}\left(1-\frac{\alpha a}{2d}\right)$$

157. (6) In the first condition, electrostatic energy is

$$U_i=\frac{1}{2}CV_0^2=\frac{1}{2}\times 60\times 10^{-12}\times 400=12\times 10^{-9}\,J$$

In the second condition $U_F=\frac{1}{2}C'V'^2$

$$U_f=\frac{1}{2}2C.\left(\frac{V_0}{2}\right)^2\qquad\left(\because C'=2C, V'=\frac{V_0}{2}\right)$$

$$=\frac{1}{4}\times 60\times 10^{-12}\times(20)^2\quad=6\times 10^{-9}\,J$$

Energy lost $=U_i-U_f=12\times 10^{-9}J-6\times 10^{-9}J=6\,nJ$

158. (d) $V'=\frac{CV+(nC)V}{kC+nC}$

$[\because q_i=q_f\Rightarrow(C+nC)V=V'(KC+nC)]$

$$=\frac{(n+1)V}{k+n}$$

159. (d) $V_1+V_2=10$

and $4V_1=6V_2$

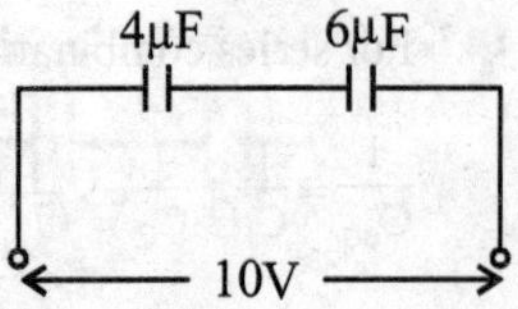

On solving above equations, we get

$V_1=6\,V$

Charge on 4 μF,

$q=CV_1=4\times 6=24\,\mu C$

160. (a) For parallel combination

$q_1=10(C_1+C_2)$

$q_1=500\,\mu C$

$500=10(C_1+C_2)$

$C_1+C_2=50\mu F$ ....(i)

For Series Combination–

$$q_2=10\frac{C_1C_2}{(C_1+C_2)}$$

$$80=10\frac{C_1C_2}{50}\quad\text{[From equation (i)]}$$

$C_1C_2=400$ ....(ii)

From equation (i) and (ii)

$C_1=10\mu F\quad C_2=40\mu F$

161. (b) $W=U_f-U_i=\frac{q^2}{2}\left(\frac{1}{C_f}-\frac{1}{C_i}\right)$

$$=\frac{(5\times 10^{-6})^2}{2}\left(\frac{1}{2}-\frac{1}{5}\right)\times 10^6=3.75\times 10^{-6}J$$

162. (b) Capacitance of a capacitor with a dielectric of dielectric constant k is given by

$$C=\frac{k\in_0 A}{d}\quad\because E=\frac{V}{d}\quad\therefore C=\frac{k\in_0 AE}{V}$$

$$15\times 10^{-12}=\frac{k\times 8.86\times 10^{-12}\times 10^{-4}\times 10^{6}}{500}$$

$k=8.5$

163. (b) $V=\frac{Q}{C}$

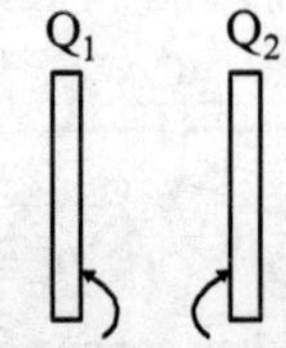

By Gauss law

Charge on inner plates $=\left|\frac{Q_1-Q_2}{2}\right|\quad\frac{Q_1-Q_2}{2}\quad-\frac{(Q_1-Q_2)}{2}$

$$\text{So, } V=\left(\frac{Q_1-Q_2}{2C}\right)=\left(\frac{4-2}{2\times 1}\right)=1\,V$$

164. (c) $E=\frac{\sigma}{\varepsilon_0}=\frac{Q}{A\varepsilon_0}$

$\therefore Q=\varepsilon_0.E.A=8.85\times 10^{-12}\times 100\times 1=8.85\times 10^{-10}C$

165. (a)

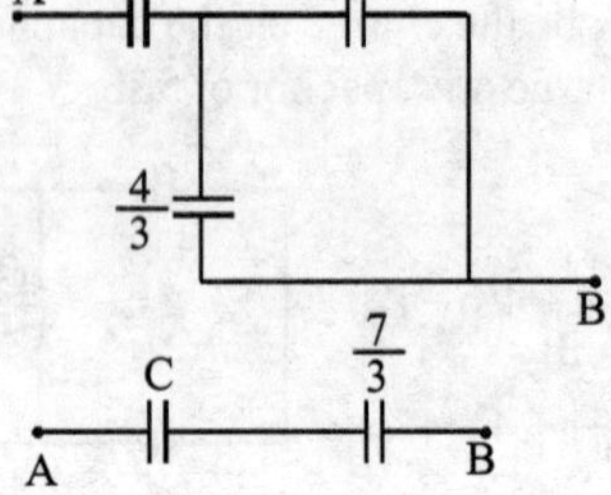

For series combination

$$\frac{1}{C_{eq}}=\frac{1}{C_1}+\frac{1}{C_2}\Rightarrow\frac{\frac{7C}{3}}{\frac{7}{3}+C}=\frac{1}{2}\Rightarrow 14C=7+3C\Rightarrow C=\frac{7}{11}\mu F$$

166. (d)

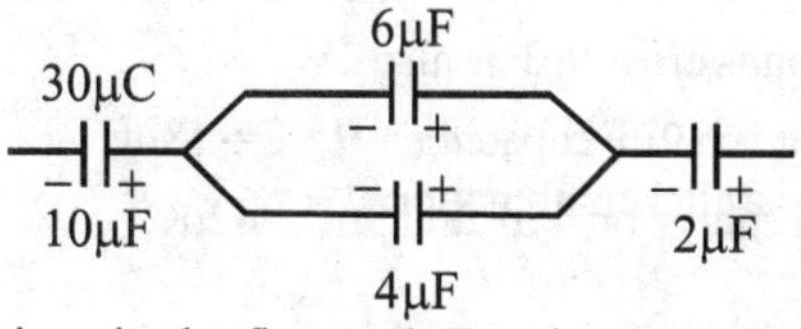

As given in the figure, 6μF and 4μF are in parallel. Now using charge conservation.

Charge on 6μF capacitor $=\frac{6}{6+4}\times 30=18\mu C$

Since charge is asked on right plate therefore is $+18\mu C$

167. (b) $W=-\Delta U$
$=U_i-U_f$

$$=U_i-\frac{U_i}{K}=U_i\left(1-\frac{1}{K}\right)=\frac{1}{2}CV^2\left(1-\frac{1}{K}\right)=508\text{ pJ}$$

168. (c) Let dielectric constant of material used be K.

$$\frac{k_1\epsilon_0 A_1}{d}+\frac{k_2\epsilon_0 A_2}{d}+\frac{k_3\epsilon_0 A_3}{d}=\frac{k\epsilon_0 A}{d}$$

$$\Rightarrow\frac{10\epsilon_0 A/3}{d}+\frac{12\epsilon_0 A/3}{d}+\frac{14\epsilon_0 A/3}{d}=\frac{K\epsilon_0 A}{d}$$

$$\Rightarrow\frac{\epsilon_0 A}{d}\left(\frac{10}{3}+\frac{12}{3}+\frac{14}{3}\right)=\frac{K\epsilon_0 A}{d}\quad\therefore K=12$$

169. (b) From figure, $\frac{y}{x}=\frac{d}{a}\Rightarrow y=\frac{d}{a}x$

$$dy=\frac{d}{a}(dx)$$

$$\Rightarrow\frac{1}{dC}=\frac{y}{K\varepsilon_0 adx}+\frac{(d-y)}{\varepsilon_0 adx}$$

$$\frac{1}{dC}=\frac{1}{(dx)a\varepsilon_0}\left(\frac{y}{K}+d-y\right)$$

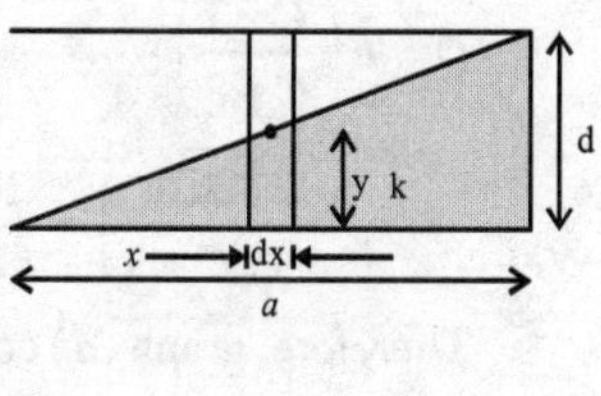

$$\int dC=\int\frac{\varepsilon_0 adx}{\frac{y}{K}+d-y}$$

$$\Rightarrow C=\varepsilon_0 a.\frac{a}{d}\int_0^d\frac{dy}{d+y\left(\frac{1}{K}-1\right)}$$

$$=\frac{\varepsilon_0 a^2}{\left(\frac{1}{K}-1\right)d}\left[\ell n\left(d+y\left(\frac{1}{K}-1\right)\right)\right]_0^d$$

$$=\frac{K\epsilon_0 a^2}{(1-K)d}\ell n\left(\frac{d+d\left(\frac{1}{K}-1\right)}{d}\right)$$

$$=\frac{K\epsilon_0 a^2}{(1-K)d}\ell n\left(\frac{1}{K}\right)=\frac{K\epsilon_0 a^2\ell nK}{(K-1)d}$$

170. (a) $Q_i=Q_{Plate}\left(1-\frac{1}{K}\right)$

$$=KCV\left(1-\frac{1}{K}\right)=CV(K-1)$$

$$=90\text{ pF}\times 20\times\left(\frac{5}{3}-1\right)=1.2\text{ nC}$$

171. (d) The simplified circuit of the circuit given in question as follows:

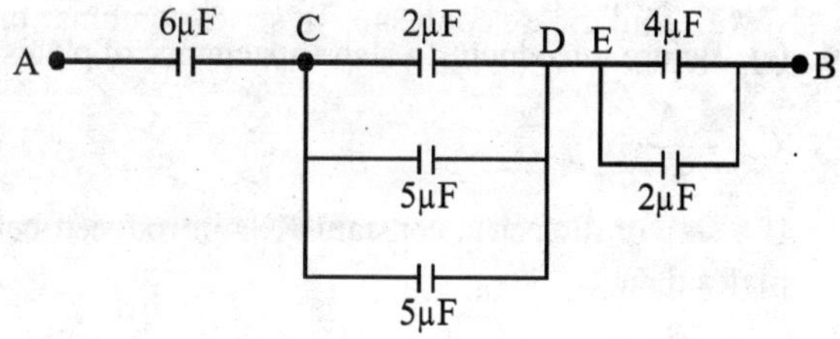

The equivalent capacitance between C & D, capacitors of 2μF, 5μF and 5 μF are in parallel.

$\therefore$ $C_{CD}=2+5+5=12$ μF ($\because$ In parallel grouping $C_{eq}=C_1+C_2+....+C_n$)

Similarly equivalent capacitance between E & B, $C_{EB}=4+2=6\mu F$

Now equivalent capacitance between A & B

$$\frac{1}{C_{eq}}=\frac{1}{6}+\frac{1}{12}+\frac{1}{6}=\frac{5}{12}$$

$\Rightarrow C_{eq}=\frac{12}{5}=2.4$ μF ($\because$ In series grouping, $\frac{1}{C_{eq}}=\frac{1}{C_1}+\frac{1}{C_2}+.........+\frac{1}{C_n}$)

172. (d) $C'=\frac{C_2C_3}{C_2+C_3}=\frac{3\times 6}{3+6}=2\mu F$

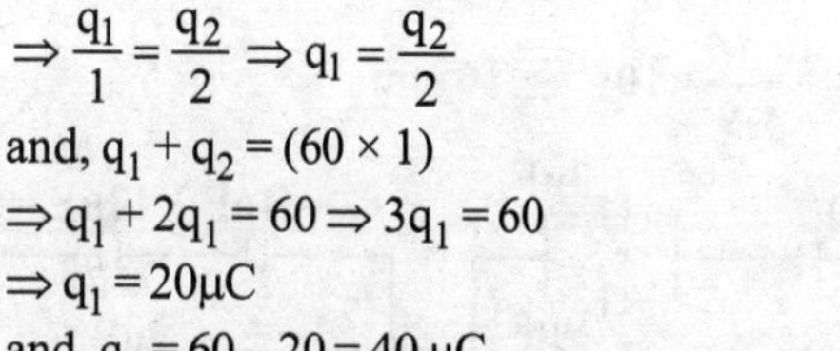

As $V=V'\Rightarrow\frac{q_1}{C_1}=\frac{q_2}{C'}$

$$\Rightarrow\frac{q_1}{1}=\frac{q_2}{2}\Rightarrow q_1=\frac{q_2}{2}$$

and, $q_1+q_2=(60\times 1)$

$\Rightarrow q_1+2q_1=60\Rightarrow 3q_1=60$

$\Rightarrow q_1=20\mu C$

and, $q_2=60-20=40$ μC.

173. (b) To get a capacitance of 2μF arrangement of capacitors of capacitance 1μF as shown in figure 8 capacitors of 1μF in parallel with four such branches in series i.e., 32 such capacitors are required.

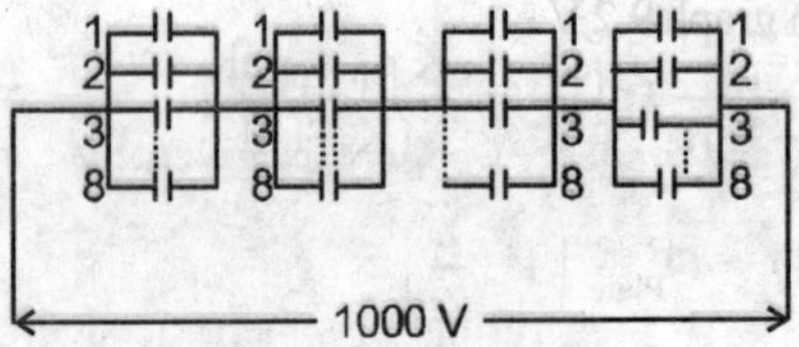

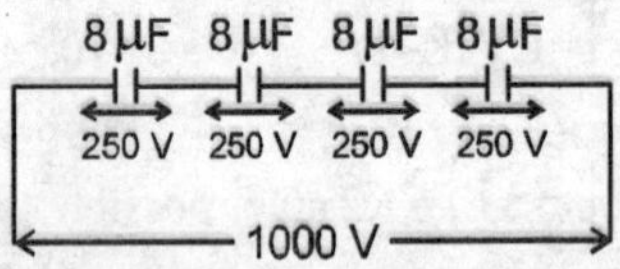

$$\frac{1}{C_{eq}}=\frac{1}{8}+\frac{1}{8}+\frac{1}{8}+\frac{1}{8} \quad \therefore C_{eq}=2\,\mu F$$

174. (c) Before introducing a slab capacitance of plates

$$C=\frac{\varepsilon_0 A}{3}=C_{eq}$$

If a slab of dielectric constant K is introduced between plates then

$$C_1=\frac{K\varepsilon_0 A}{3} \text{ then}$$

$C_1'=\frac{\varepsilon_0 A}{2.4}$. As V and q do not change. So C_{eq} will also not change.

C_1 and C_1' are in series hence,

$$\frac{\varepsilon_0 A}{3}=\frac{K\frac{\varepsilon_0 A}{3}\cdot\frac{\varepsilon_0 A}{2.4}}{K\frac{\varepsilon_0 A}{3}+\frac{\varepsilon_0 A}{2.4}}$$

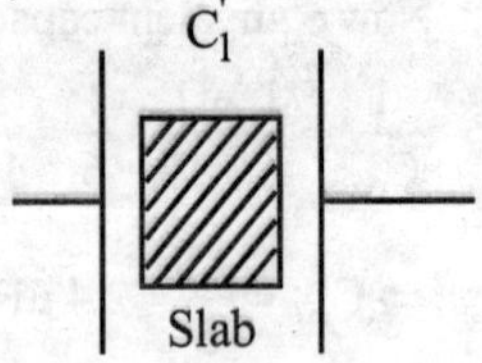

$3K=2.4K+3$

$0.6K=3$

Hence, the dielectric constant of slab is given by,

$$K=\frac{30}{6}=5$$

175. (d) Energy of sphere $=\frac{Q^2}{2C}$

$$\Rightarrow 4.5=\frac{16\times10^{-12}}{2C} \Rightarrow C=\frac{16\times10^{-12}}{9}=4\pi\varepsilon_0 R$$

(capacity of spherical conductor)

$$R=\frac{16\times10^{-12}}{9}\times\frac{1}{4\pi\varepsilon_0} \quad \left[\because \frac{1}{4\pi\varepsilon_0}=9\times10^9\right]$$

$$=9\times10^9\times\frac{16}{9}\times10^{-12}=16\,\text{mm}$$

176. (a)

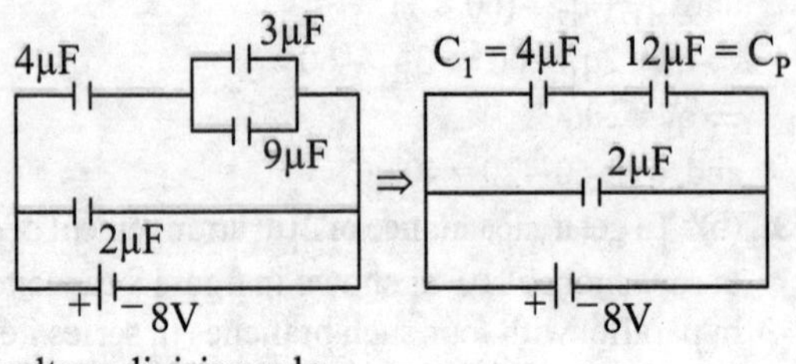

By voltage division rule,

Charge on C_1 is $q_1=\left[\left(\frac{12}{4+12}\right)\times8\right]\times4=24\mu C$

The voltage across C_P is $V_P=\frac{4}{4+12}\times8=2V$

$\therefore$ Voltage across 9µF is also 2V

$\therefore$ Charge on 9µF capacitor $=9\times2=18\mu C$

$\therefore$ Total charge on 4 µF and 9µF = 42µC

$$\therefore E=\frac{KQ}{r^2}=9\times10^9\times\frac{42\times10^{-6}}{30\times30}=420\,NC^{-1}$$

177. (d) To get effective capacitance of 6 µF two capacitors of 4 µF each connected in sereies and one of 4 µF capacitor in parallel with them.

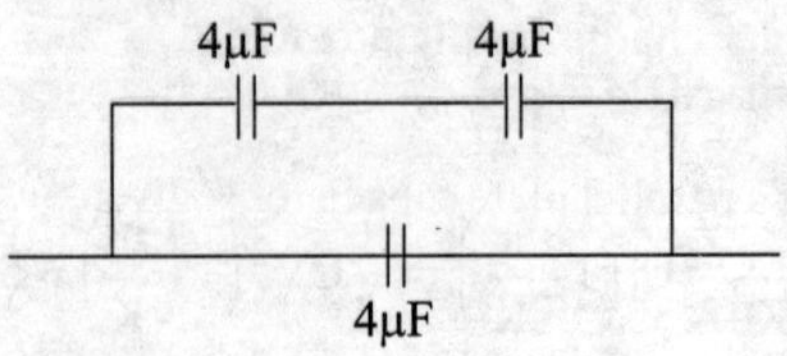

Two capacitances in series

$$\therefore \frac{1}{C}=\frac{1}{C_1}+\frac{1}{C_2}=\frac{1}{4}+\frac{1}{4}=\frac{1}{2}$$

1 capacitor in parallel

$$\therefore C_{eq}=C_3+C=4+2=6\,\mu F$$

178. (d)

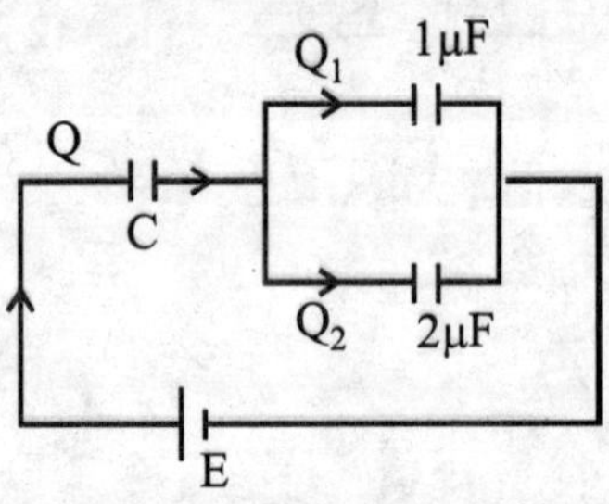

From figure, $Q_2=\frac{2}{2+1}Q=\frac{2}{3}Q$

$$Q=E\left(\frac{C\times3}{C+3}\right)$$

$$\therefore Q_2=\frac{2}{3}\left(\frac{3CE}{C+3}\right)=\frac{2CE}{C+3}$$

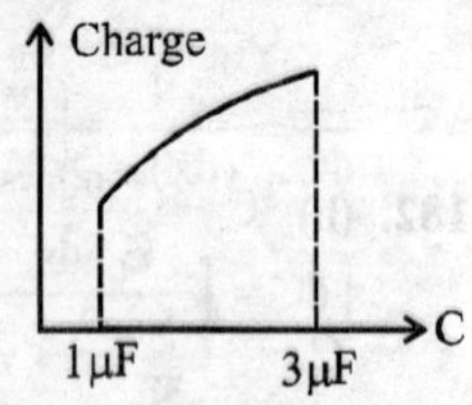

Therefore, graph (*d*) correctly dipicts.

179. (a) Initially,

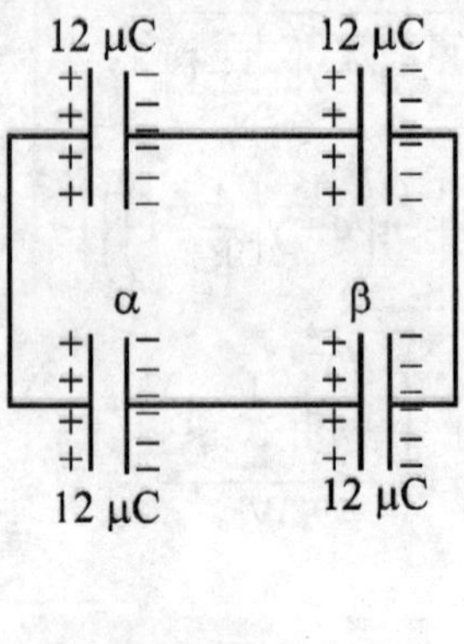

$(Q_\alpha + Q_\beta)_i = (-12 + 12)\ \mu C = 0$

Finally,

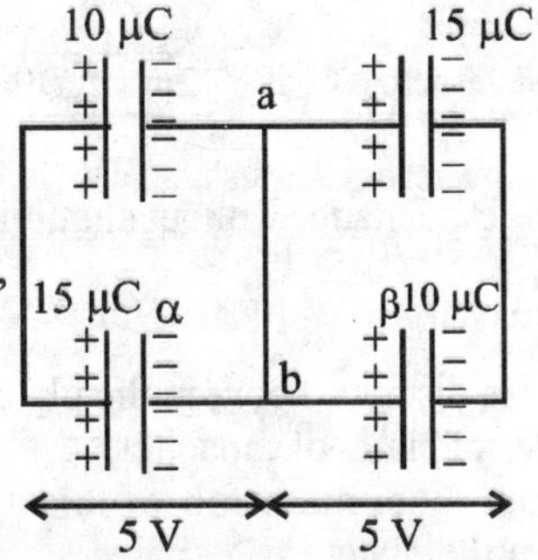

$(Q_\alpha + Q_\beta)_f = -15 + 10 = -5\ \mu C$

So, 5 μC charge from b → a

so that net charge on plate α and β is zero.

180. (a) Electric field in presence of dielectric between the two plates of a parallel plate capacitor is given by, $E = \dfrac{\sigma}{K\varepsilon_0}$

Then, charge density, $\sigma = K\varepsilon_0 E$

$= 2.2 \times 8.85 \times 10^{-12} \times 3 \times 10^4 \approx 6 \times 10^{-7}\ C/m^2$

181. (c)

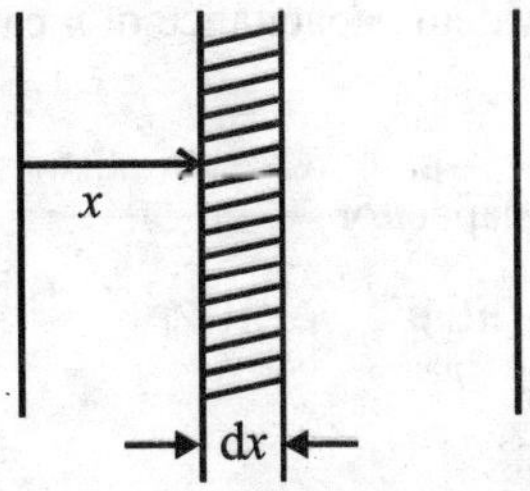

Let us take an elemental capacitor of width 'dx'. Then,

$$dC_{eq} = \frac{(K_0+\lambda x)A \in_0 d}{dx\ d} = \frac{(K_0+\lambda x)C_0 d}{dx}, C_0 = \frac{A \in_0}{d}$$

$$\int \frac{1}{dc_{eq}} = \int_0^d \frac{dx}{(K_0+\lambda x)C_0 d} \Rightarrow \frac{1}{C_{eq}} = \frac{1}{\lambda C_0 d}[\ln(K_0+\lambda x)]_0^d$$

$$\Rightarrow \frac{1}{C_{eq}} = \frac{1}{\lambda C_0 d}\left[\ln\left(\frac{K_0+\lambda d}{K_0}\right)\right] \Rightarrow C_{eq} = \frac{\lambda C_0 d}{\ln\left(1+\dfrac{\lambda d}{K_0}\right)}$$

182. (b) $C_0 = \dfrac{K \in_0 A}{d}$

$$C = \frac{2K \in_0 A}{3d} + \frac{2K \in_0 A}{3d} = \frac{4}{3}\frac{K \in_0 A}{d}$$

$$\therefore \frac{C}{C_0} = \frac{\frac{4}{3}\frac{K \in_0 A}{d}}{\frac{K \in_0 A}{d}} = \frac{4}{3}$$

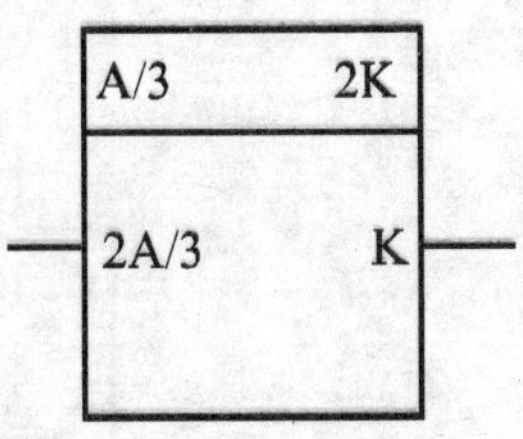

183. (c) As, $C = \dfrac{Q}{V} = \dfrac{It}{V}$

$$\Rightarrow \frac{V}{t} = \frac{I}{C} = \frac{2}{1 \times 10^{-6}} = 2 \times 10^6\ V/s$$

184. (d) At $t = \tau$

$$V = \frac{V_0}{e} = \frac{25}{2.71} = 9.2V$$

From graph 9.2 V lies between 100 sec and 150 sec

185. (d) Capacitance of sphere is given by :

$C = 4\pi \in_0 r$

If, C = 1F then radius of sphere needed:

$$r = \frac{C}{4\pi \in_0} = \frac{1}{4\pi \times 8.85 \times 10^{-12}} \Rightarrow r = \frac{10^{12}}{4\pi \times 8.85} = 9 \times 10^9 m$$

9×10^9 m is very large, it is not possible to obtain such a large sphere. Infact earth has radius 6.4×10^6 m only and capacitance of earth is 711μF.

186. (d) Equivalent capacitance of n_2 number of capacitors each of capacitance C_2 in parallel, $C_1 = n_2 C_2$

Equivalent capacitance of n_1 number of capacitors each of capacitances C_1 in series.

$$C_s = \frac{C_1}{n_1}$$

According to question, total energy stored in both the combinations are same

i.e., $\dfrac{1}{2}\left(\dfrac{C_1}{n_1}\right)(4V)^2 = \dfrac{1}{2}(n_2 C_2)V^2 \quad \therefore C_2 = \dfrac{16 C_1}{n_1 n_2}$

187. (c) Initial energy of capacitor, $E_1 = \dfrac{q_1^2}{2C}$

Final energy of capacitor,

$$E_2 = \frac{1}{2}E_1 = \frac{q_1^2}{4C} = \frac{(q_1/\sqrt{2})^2}{2C} = \frac{q_2^2}{2C}$$

$\therefore t_1$ = time for the charge to reduce to $\dfrac{1}{\sqrt{2}}$ of its initial value

and t_2 = time for the charge to reduce to $\dfrac{1}{4}$ of its initial value

We have, $q_2 = q_1 e^{-t/CR} \Rightarrow \ln\left(\dfrac{q_2}{q_1}\right) = -\dfrac{t}{CR}$

$$\therefore \ln\left(\frac{1}{\sqrt{2}}\right) = \frac{-t_1}{CR} \quad ...(1)$$

$$\text{and } \ln\left(\frac{1}{4}\right) = \frac{-t_2}{CR} \quad ...(2)$$

By (1) and (2), $\dfrac{t_1}{t_2} = \dfrac{\ln\left(\frac{1}{\sqrt{2}}\right)}{\ln\left(\frac{1}{4}\right)} = \dfrac{1}{2}\dfrac{\ln\left(\frac{1}{2}\right)}{2\ln\left(\frac{1}{2}\right)} = \dfrac{1}{4}$

188. (c)

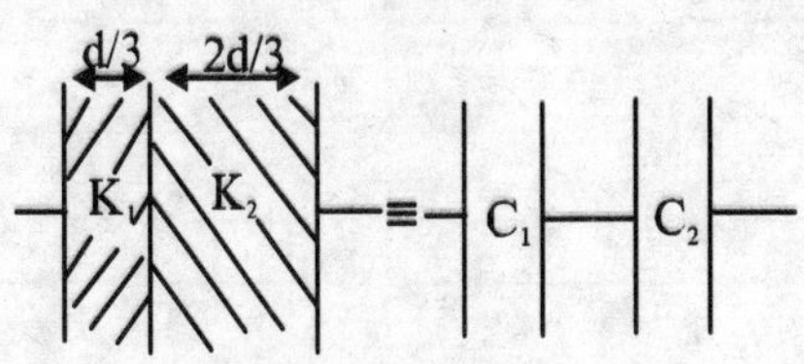

The capacitance with air between the plates

$$C = \frac{\varepsilon_0 A}{d} = 9\text{pF}$$

On introducing two dielectric between the plates, the given capacitance is equal to two capacitances connected in series where

$$C_1 = \frac{k_1 \epsilon_0 A}{d/3} = \frac{3 \epsilon_0 A}{d/3} = \frac{3 \times 3 \epsilon_0 A}{d} = \frac{9 \epsilon_0 A}{d}$$

and $C_2 = \frac{k_2 \epsilon_0 A}{2d/3} = \frac{3k_2 \epsilon_0 A}{2d} = \frac{3 \times 6 \epsilon_0 A}{2d} = \frac{9 \epsilon_0 A}{d}$

The equivalent capacitance C_{eq} is

$$\frac{1}{C_{eq}} = \frac{1}{C_1} + \frac{1}{C_2} = \frac{d}{9 \epsilon_0 A} + \frac{d}{9 \epsilon_0 A} = \frac{2d}{9 \epsilon_0 A}$$

$$\therefore C_{eq} = \frac{9}{2} \frac{\epsilon_0 A}{d} = \frac{9}{2} \times 9\ pF = 40.5 pF$$

189. (a) The potential energy of a charged capacitor is given by $U = \frac{Q^2}{2C}$.

When a dielectric slab is introduced between the plates the energy is given by $\frac{Q^2}{2KC}$, where K is the dielectric constant.

Again, when the dielectric slab is removed slowly its energy increases to initial potential energy. Thus, work done is zero.

190. (b) So, we will get (n–1) capacitor in parallel

$$\therefore \; C_{eq} = C_1 + C_2 + ... + C_{n-1} = (n-1)\,C$$

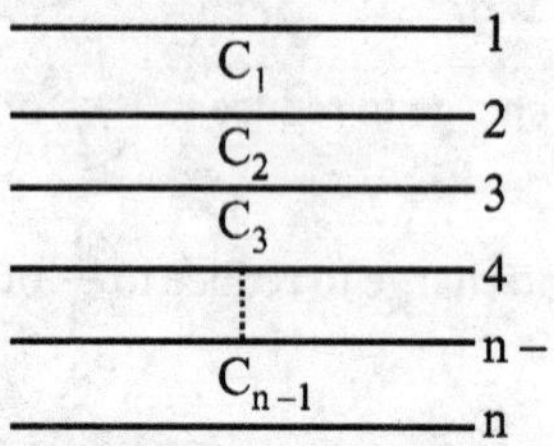

191. (c) Applying conservation of energy,

Electric potential energy of capacitor = heat absorbed

$$\frac{1}{2}CV^2 = m.s\,\Delta T; \; V = \sqrt{\frac{2m.s.\Delta T}{C}}$$

192. (b) The capacitance without aluminium foil is

$$C = \frac{\varepsilon_0 A}{d}$$

Here, d is distance between the plates of a capacitor

A = Area of plates of capacitor

When an aluminium foil of thickness t is introduced between the plates.

Capacitance, $C' = \frac{\varepsilon_0 A}{d - t}$

If thickness of foil is negligible $d - t \sim d$. Hence, $C = C'$.

193. (d) The work done is stored in the form of potential energy which is given by

$$U = \frac{1}{2}\frac{Q^2}{C} \quad \therefore U = \frac{1}{2} \times \frac{\left(8 \times 10^{-18}\right)^2}{100 \times 10^{-6}} = 32 \times 10^{-32}\text{ J}$$

194. (b) In parallel, equivalent capacitance of n capacitor of capacitance C

$C' = nC$

Energy stored in this capacitor

$$E = \frac{1}{2}C'V^2 \Rightarrow E = \frac{1}{2}(nC)V^2 = \frac{1}{2}nCV^2$$

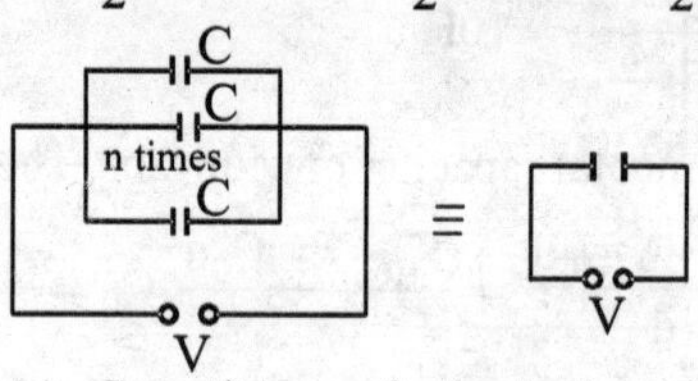

195. (a) Capacitance of spherical conductor = $4\pi\varepsilon_0 R$

Here, R is radius of conductor

$$\therefore C = 4\pi \epsilon_0 R = \frac{1}{9 \times 10^9} \times 1 = 1.1 \times 10^{-10} F$$

Current Electricity

3

1. **(d)** $I = \frac{dq}{dt}$

$q = \int i\, dt$

$\Rightarrow \int_1^2 (0.02t + 0.01)\, dt$

$q = \left[0.02\frac{t^2}{2} + 0.01\, t\right]_1^2$

$= \left[\frac{0.02}{2} \times 4 + 0.01 \times 2\right] - \left[0.02 \times \frac{1}{2} + 0.01 \times 1\right]$

$= 0.01\,(3) + 0.01\,(1) = 0.04\,\text{C}.$

2. **(d)**

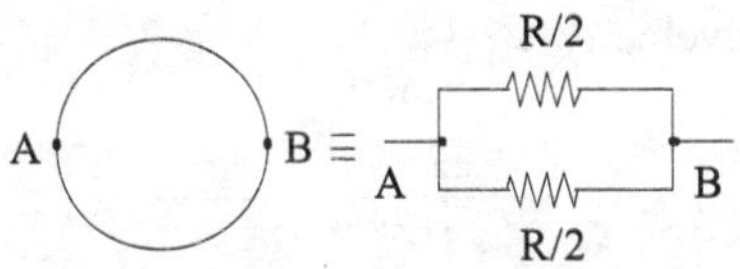

$L = 25\text{m}, A = 5\ \text{mm}^2 = 5 \times 10^{-6}\ \text{m}^2$

$\rho = 2 \times 10^{-6}\ \Omega\text{m}$

$R = \frac{\rho L}{A} = \frac{2 \times 10^{-6} \times 25}{5 \times 10^{-6}} = 10\Omega$

For parallel combination,

$R_{eq} = \frac{R_1 R_2}{R_1 + R_2} = \frac{\left(\frac{R}{2}\right)\left(\frac{R}{2}\right)}{\frac{R}{2} + \frac{R}{2}}$

$= \frac{R}{4} = \frac{10}{4} = 2.5\Omega$

3. **(b)** When the key is closed at t = 0, the capacitor is initially unchanged.
Initially capacitor behave as a short circuit (zero potential across it)
So current will be maximum. $I_0 = \frac{E}{R}$
Charge on capacitor will zero.
Potential difference across capacitor will be zero.
As time progresses, the capacitor starts charging and the current decreases exponentially.

4. **(a)** Wire bound resistors have weak dependence of resistivity with temperature.

5. **(1027)** Given, Resistance of room temperature, $R_0 = 50\Omega$
Temperature coefficient of material, $\alpha = 2.4 \times 10^{-4}\ °\text{C}^{-1}$

$R = R_0(1 + \alpha\Delta T)$

$\Rightarrow 62 = 50\,[1 + 2.4 \times 10^{-4}\,\Delta T]$

$\Rightarrow 0.24 = 2.4 \times 10^{-4}\,\Delta T \Rightarrow \Delta T = 1000°\text{C}$

$\Rightarrow T - 27° = 1000°\text{C}$

$\Rightarrow T = 1027°\text{C}$

6. **(748)** We know that
Resistance $R = R_0(1 + \alpha\Delta T)$

$\Rightarrow \frac{\Delta R}{R_0} = \alpha\Delta T$

Taking temperatures 0°C and 100°C at a time

$\frac{R - R_0}{R_0} = \alpha\Delta T$

$\frac{10.2 - 10}{10} = \alpha(100 - 0)$ ...(i)

$\alpha = \frac{0.2}{1000}$

Taking temperatures 0°C and t°C we get

$\Rightarrow \frac{10.95 - 10}{10} = \alpha(t - 0)$

$\frac{0.95}{10} = \frac{0.2}{1000} \times t$ [using (i)]

$\therefore \frac{t}{100} = \frac{0.95}{0.2} = 475°\text{C}$

$\Rightarrow t = 475 + 273 = 748\ \text{K}$

7. **(16)** Resistance $R = \frac{\rho l}{A} \Rightarrow R = \frac{\rho l}{\pi r^2} \Rightarrow R \propto \frac{l}{r^2}$

Volume of wire remains constant,
$V_i = V_f$

$\pi r^2 l = \pi \frac{r^2}{4} l'$

$l' = 4l$

$\frac{R'}{R} = \left(\frac{4l}{\frac{r^2}{4}}\right)\frac{r^2}{l} = 16 \Rightarrow R' = 16R = xR$ (given)

Hence, x = 16

8. **(b)** Using $R_t = R_0\,(1 + \alpha\Delta t)$
Given $R_0 = 8\Omega$, $R_{100} = 10\Omega$
$\therefore R_{100} = R_0\,(1 + \alpha\Delta T)$
Also, $R_{400} = R_0\,(1 + \alpha\Delta T')$

$\therefore 10 = 8\,(1 + \alpha \times 100) \Rightarrow 100\alpha = \frac{1}{4}$

$\therefore R_{400} = 8\,(1 + 400\alpha) = 8\,(1 + 1) = 16\Omega$

9. **(32)** Resistance

$R = \frac{\rho \ell}{A} = \frac{\rho V}{A^2}$ or, $R \propto \frac{1}{(r^2)^2}$

$\therefore \frac{R_A}{R_B} = \frac{r_B^4}{r_A^4} \Rightarrow \frac{R_A}{2} = \left[\frac{4 \times 10^{-3}}{2 \times 10^{-3}}\right]^4$

$\therefore R_A = 32\ \Omega$

10. (22) Current, $i = \frac{dq}{dt}$

$$\therefore q = \int_1^2 i\,dt = \int_1^2 \left(3t^2 + 4t^3\right)\cdot dt = \left|t^3 + t^4\right|_1^2$$

$= 2^3 - 1 + 2^4 - 1 = 22\,C$

11. (c) Resistance, $R_0 = 60\Omega$, $R_T = \frac{220}{2.75} = 80\Omega$

$R_T = R_0(1 + \alpha\Delta T)$

$80 = 60[1 + 2\times10^{-4}(T - 27)]$

$T \approx 1694°C$

12. (d) Charge, $q = \int_{t_1}^{t_2} I.dt = \int_0^{20}(20 + 3t)dt$

$$= 20t + \left.\frac{3t^2}{2}\right|_0^{20} = 1000C$$

13. (b) Equivalent resistance

$R_{eq} = R_1 + R_2 + R_3$

Resistivity depends on temperature.

The resistivity of the material is given by

$\rho_T = \rho_0(1 + \alpha\,\Delta T)$

14. (a) Let

Initial length $= l_1$

Final length $= l_2$

Initial area $= A_1$

Final area $= A_2$

$\therefore$ Volume remains same

$\therefore A_1 l_1 = A_2 l_2 \Rightarrow A_1 l_1 = A_2 \frac{l_1}{4}$

$\Rightarrow 4A_1 = A_2$

Initial resistance, $R_1 = \frac{\rho l_1}{A_1} = 160\Omega$ (given)

Final resistance, $R_2 = \frac{\rho l_2}{A_2}$

$\therefore \frac{R_2}{R_1} = \frac{l_2 A_1}{A_2 l_1} = \frac{l_1}{4}\frac{A_1}{4A_1 l_1}$

$\Rightarrow R_2 = \frac{1}{16}R_1 = \frac{1}{16}\times160 = 10\Omega$

15. (15) For same supply voltage

$i_0 R_0 = i_{100} R_{100}$

$\Rightarrow 2R_0 = 1.2R_0[1 + 100\alpha]$ ... (i)

$\Rightarrow 1 + 100\alpha = \frac{5}{3} \Rightarrow 100\alpha = \frac{2}{3} \Rightarrow \alpha = \frac{1}{150}$

At 50°C temperature

$i_{50} R_{50} = i_0 R_0$

$\Rightarrow i_{50} = \frac{i_0 R_0}{R_{50}} = \frac{2\times R_0}{R_0(1 + 50\alpha)}$

$$= \frac{2}{1 + 50\times\frac{1}{150}} = \frac{6}{4} = 1.5A = 15\times10^2\,mA$$

16. (b) Under the effect of applied electric field, the free electrons get accelerated due to electric field and they move from lower to higher potential. However, electron collides with lattice ion during passage and follow curved path.

17. (3)

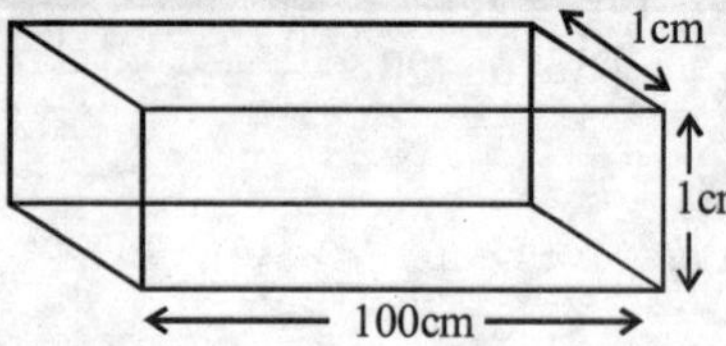

Given,

Specific resistance, $\rho = 3\times10^{-7}\,\Omega\,m$

Resistance, $R = \rho\frac{\ell}{A} = \frac{3\times10^{-7}\times\left(1\times10^{-2}\right)}{100\times1\times10^{-4}}$

$= 3\times10^{-7}\,\Omega$

18. (25) Current, $I = 2A$; Cross sectional Area, $A = 25\,mm^2 = 25\times10^{-6}\,m^2$

Number of free electron, $n = 2\times10^{28}$

Charge on electron, $e = 1.6\times10^{-19}\,C$

Drift velocity $V_d = \frac{I}{neA}$

$$= \frac{2}{2\times10^{28}\times1.6\times10^{-19}\times25\times10^{-6}}$$

$= 25\times10^{-6}\,ms^{-1}$

19. (125) Number density of free electrons

$n = 8\times10^{28}\,m^{-3}$

Area $= 2\times10^{-6}\,m^2$ Current, $I = 3.2\,A$

Drift speed $V_d = \frac{I}{neA} = \frac{3.2}{8\times10^{28}\times1.6\times10^{-19}\times2\times10^{-6}}$

$= 125\times10^{-6}\,m/s$

20. (25) Let ℓ = initial length of the wire

A = cross section area of the wire

ρ = resistivity of material of the wire

Initial resistance, $R = \rho\frac{\ell}{A}$

New resistance,

$R' = \rho\frac{1.2\ell}{0.96A} = 1.25\rho\frac{\ell}{A} = 1.25R$

% change $= \frac{R' - R}{R} = \frac{1.25R - R}{R}\times100 = 25\%$

21. (b) Resistivity (ρ) is given by

$\rho = \frac{m}{ne^2\tau}$

With increase in temperature, number density n increases and relaxation time τ decreases. But n is dominant over τ. Hence, resistivity (ρ) decreases.

22. (a) We have

$V_d = \frac{eV\tau}{mL} \Rightarrow V_d$ is independent of cross sectional area.

23. (d) Given that

$Q = (\alpha t - \beta t^2 + \gamma t^3)$

$i = \frac{dQ}{dt} = (\alpha - 2\beta t + 3\gamma t^2)$ ...(i)

For minimum value of current

$\frac{di}{dt} = (6\gamma t - 2\beta) = 0 \quad \Rightarrow t = \frac{\beta}{3\gamma}$

Put this value in equation (i)

$$i = (\alpha - 2\beta t + 3\gamma t^2) = \alpha - 2\beta \times \frac{\beta}{3\gamma} + 3\gamma \times \left(\frac{\beta}{3\gamma}\right)^2$$

$$= \left(\alpha - \frac{\beta^2}{3\gamma}\right)$$

24. **(b)** Given that
Current, I = 2A
Voltage, V = 3.4 V
Using Ohm's Law, V = IR
Here resistivity, $\rho = 1.7 \times 10^{-8}\ \Omega$-m

$R = \frac{3.4}{2} = 1.7\Omega$

$1.7 = \frac{\rho L}{A} \quad \left(\because R = \frac{\rho L}{A}\right)$

$L = \frac{1.7(A)}{\rho};\ L^2 = \frac{1.7 A.L}{\rho} = \frac{1.7 V}{\rho}$

M = density × volume

$\text{Volume, } V = \frac{8.92 \times 10^{-3}}{8.92 \times 10^3} = 10^{-6}\ m^3$

$L^2 = \frac{1.7}{\rho}(10^{-6}) = \frac{1.7 \times 10^{-6}}{1.7 \times 10^{-8}} = 100$

$\therefore L = 10m$

25. **(c)**

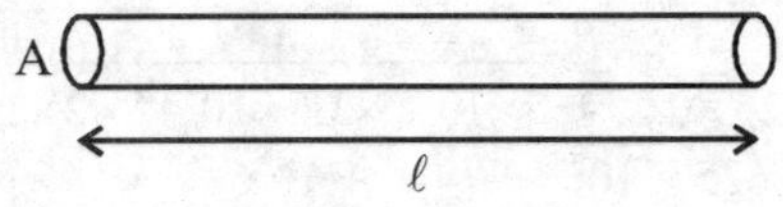

Let resistance of a wire R and length ℓ.

$R = \frac{\rho \ell}{A} = 5\Omega$

$\therefore$ Volume of wire is constant in stretching

$V_i = V_f \quad \Rightarrow \quad A_i \ell_i = A_f \ell_f$

$A\ell = A'(5\ell) \Rightarrow A' = \frac{A}{5}$

$$R_f = \frac{\rho \ell_f}{A_f} = \frac{\rho(5\ell)}{\left(\frac{A}{5}\right)} = 25\left(\frac{\rho \ell}{A}\right) = 25 \times 5 = 125\ \Omega$$

26. **(2)** $R = \frac{\rho \ell}{A} = \frac{\rho \ell}{\pi(b^2 - a^2)}$ [$\because$ the C.S.A is $\pi(b^2 - a^2)$]

$$= \frac{2.4 \times 10^{-8} \times \pi}{\pi(4^2 - 2^2) \times 10^{-6}} = \frac{2.4}{12} \times 10^{-2} = 0.2 \times 10^{-2} = 2 \times 10^{-3}\Omega$$

$= n \times 10^{-3}\Omega$ (given)

So, n = 2

27. **(44)** Let $L_i = l_i$. Then, $L_f = 1.2\,l_i$
As volume is constant,
So resistance ∝ (length)^2

$$\frac{R_f}{R_i} = \left(\frac{1.2 l_i}{l_i}\right)^2 = 1.44 \Rightarrow R_f = 1.44 R_i$$

$$\text{So, } \frac{\Delta R}{R_i} \times 100 = \frac{1.44 R_i - R_i}{R_i} \times 100 = 44\%$$

28. **(b)** Consider of the length of X part is ℓ then the length of Y part will be $1 - \ell$.

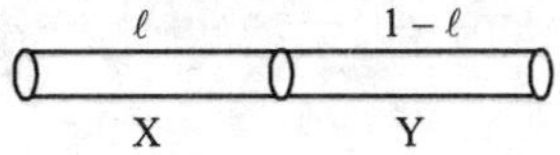

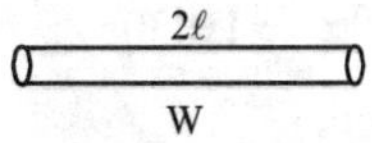

$\frac{R_X}{R_Y} = \frac{\ell_X}{\ell_Y} \qquad \left(\because R = \frac{\rho \ell}{A}\right)$

When wire is stretched to double of its length, then resistance becomes 4 times.

$R_W = 4R_X = 2R_Y$

$\frac{R_X}{R_Y} = \frac{1}{2} \quad \text{So } \frac{\ell_X}{\ell_Y} = \frac{1}{2}.$

29. **(a)** Resistance, $R_1 = \rho \frac{L_1}{A_1}$

$$R_2 = \rho\left(\frac{3L_1}{A_1/3}\right) = 9\rho \frac{L_1}{A_1}$$

$$\therefore \frac{R_2}{R_1} = \frac{9\rho \frac{L_1}{A_1}}{\rho \frac{L_1}{A_1}} = 9$$

30. **(b)** Drift velocity, $v_d = \left(\frac{e\tau}{m}\right)E$

Here, τ = relaxation time
E = electric field

$\Rightarrow v_d = \left(\frac{e\tau}{m}\right)\left(\frac{V}{\ell}\right) \quad \left(\because E = \frac{V}{\ell}\right)$

$\therefore v_d \propto \frac{\tau V}{\ell}$

Here, ℓ = length of conductor
V = Potential difference applied across the wire.
As temperature increases, relaxation time decreases, hence v_d decreases.
Also, $v_d \propto V$, so drift velocity depends on the applied potential difference.
Drift velocity is also given by

$v_d = \frac{I}{neA}$,

it is not mentioned that current is at steady state neither it is mentioned that n is constant for given conductor. So it can't be said that v_d is inversely proportional to A.
First and fourth statements are correct.

31. **(144)** We have
R = slope of V–I curve = tan 45° = 1Ω

As, $R = \frac{\rho l}{A} \Rightarrow \rho = \frac{RA}{l} = \frac{1 \times \pi \times (1.2 \times 10^{-2})^2}{31.4 \times 10^{-2}}$

$= 144 \times 10^{-3}\ \Omega\text{cm}$

32. **(a)** We know that

$R = R_0(1 + \alpha\Delta T)$

So, $2 = R_0(1 + \alpha \times 10)$ ...(i)

$3 = R_0(1 + \alpha \times 30)$ ...(ii)

Dividing (i) by (ii), we get

$\frac{3}{2} = \frac{1+30\alpha}{1+10\alpha} \Rightarrow 3 + 30\alpha = 2 + 60\alpha \Rightarrow 1 = 30\alpha$

$\Rightarrow \alpha = \frac{1}{30} = 0.033\ °C^{-1}$

33. **(48)** I = JA

$= 4 \times 10^6 \left[\pi R^2 - \pi\left(\frac{R}{2}\right)^2\right] = 4\pi \times 10^6 \left[\frac{3R^2}{4}\right]$

$= \pi \times 10^6 \times 3 \times 16 \times 10^{-6} = 48\,\pi\,A$

34. **(12)** We have

I = JA

$= 10^6 \times \pi\,[4^2 - 2^2] \times 10^{-6} = \pi \times 12\,A = x\,\pi\,A$

So, $x = 12$

35. **(c)** $R = \frac{\rho l}{A}$

$\frac{\Delta R}{R} = \frac{\Delta l}{l} - \frac{\Delta A}{A}$

[**Note:** This is not relative error case, so –ve sign comes with $\frac{\Delta A}{A}$]

Now, V = constant

Al = constant

$\Rightarrow \frac{\Delta A}{A} + \frac{\Delta l}{l} = 0 \Rightarrow -\frac{\Delta A}{A} = \frac{\Delta l}{l}$

So, $\frac{\Delta R}{R} = \frac{\Delta l}{l} + \frac{\Delta l}{l} \quad \left[\because \frac{\Delta l}{l} = -\frac{\Delta A}{A}\right]$

$= \frac{2\Delta l}{l} = 2 \times 0.4 = 0.8\%$

36. **(300)** If length is increased by n times then, resistance increased by n^2 times

So, $R_f = 4R_i$

$\Delta R = R_f - R_i = 3R_i$

$\%\,\Delta R = \frac{3R_i}{R_i} \times 100 = 300\%$

37. **(a)** Current is constant in the conductor. I = constant

Resistance of element of conductor, $dR = \frac{\rho dx}{\pi r^2}$

$dV = idR = \frac{i\rho dx}{\pi r^2}$

$E = \frac{dV}{dx} = \frac{i\rho}{\pi r^2}$

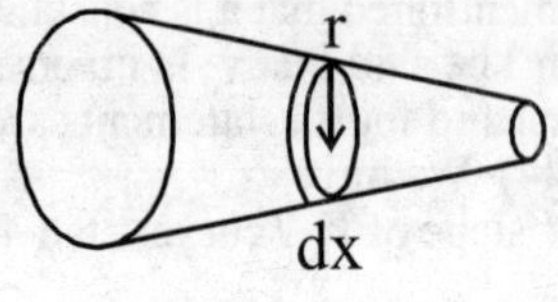

Drift velocity,

$V_d = \frac{eE\tau}{m}$ As $V_d \propto E$ and $E \propto \frac{1}{r^2}$

So, if r decreases, E will increase and hence V_d

38. **(1)** From graph (figure 1) voltage at t = 3.2s = 6V

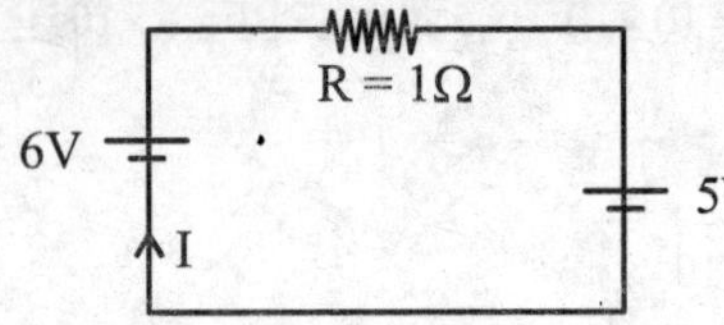

Current, $I = \frac{V}{R} = \frac{6-5}{1} = 1A$

39. **(d)** In parallel combination, equivalent resistance,

$R = \frac{R_1 R_2}{R_1 + R_2} = \frac{\ell}{A} \cdot \frac{\rho_1 \rho_2}{\rho_1 + \rho_2} \left[\because R = \rho\frac{\ell}{A}\right]$

or, $R = \frac{25 \times 10^{-2}}{3 \times 10^{-6}} \times \frac{1.7 \times 2.6 \times 10^{-16}}{4.3 \times 10^{-8}} \quad \therefore R = 0.858\text{m}\Omega$

40. **(4)** $l_1 = l_2 = l_{eq}$

$A_1 = A_2 = A$ and $A_{eq} = 2A$

$R_{eq} = \frac{R_1 R_2}{R_1 + R_2} \Rightarrow \frac{\rho(l)}{2A} = \frac{\left(\frac{\rho_1 l}{A}\right)\left(\frac{\rho_2 l}{A}\right)}{\frac{\rho_1 l}{A} + \frac{\rho_2 l}{A}}$

$\Rightarrow \frac{\rho}{2} = \frac{\rho_1 \rho_2}{\rho_1 + \rho_2} \Rightarrow \frac{\rho}{2} = \frac{\rho_1 \rho_2}{(\rho_1 + \rho_2)} = \frac{6 \times 3}{(6+3)} = 2$

$\therefore \rho = 4\ \Omega\ \text{cm}$

41. **(c)** From formula, drift velocity, $I = neV_dA$

$\Rightarrow n = \frac{I}{AeV_d} = \frac{10}{5 \times 10^{-6} \times 1.6 \times 10^{-19} \times 2 \times 10^{-3}}$

$= 625 \times 10^{25}$

42. **(d)** Resistance of the conducting wire, $R_1 = \frac{\rho l}{A}$

When length of the wire is doubled, it length becomes $l' = 2l$.

New area of cross-section, $A' = \frac{A}{2}$

$R_2 = \frac{\rho \times (2l)}{\left(\frac{A}{2}\right)} = 4\frac{\rho l}{A}$

From the Ohm's law, $V = IR$

$\therefore$ Current, $I = \frac{V}{R_2} = \frac{V}{\left(\frac{4\rho l}{A}\right)} \Rightarrow I = \frac{1}{4}\frac{VA}{\rho l}$

43. **(b)** Given, $i = \alpha_0 t + \beta t^2$

Put $\alpha_0 = 20$ and $\beta = 8$, we get $i = 20t + 8t^2$

Current, $i = \frac{dq}{dt} \Rightarrow \int dq = \int i dt \Rightarrow q = \int_0^{15} (20t + 8t^2)dt$

$\Rightarrow q = \left(\frac{20t^2}{2} + \frac{8t^3}{3}\right)_0^{15} = 20 \times \left(\frac{15^2 - 0^2}{2}\right) + \frac{8}{3}(15^3 - 0^3)$

$\Rightarrow q = 10 \times (15)^2 + \frac{8(15)^3}{3} \Rightarrow q = 2250 + 9000 = 11250\,\text{C}$

44. **(5)** Given : Conductivity of wire, $\sigma = 5\times10^7$ S/m

Radius of wire, $r = 0.5$ mm $= 5\times10^{-4}$ m

Electric field, $E = 10\times10^{-3}$ V/m

$J = \sigma E = 10\times10^{-3}\times5\times10^7 \Rightarrow J = 5\times10^5$

Since, $J = \frac{i}{A} \Rightarrow \frac{i}{A} = 5\times10^5$

$\Rightarrow i = 5\times10^5\times\pi r^2$

$= 5\times10^5\times\pi\times(5\times10^{-4})^2 = 125\pi\times10^{-3}$ A

$\therefore i = 125\pi$ mA $\quad \therefore i = 5^3\pi$ mA $= x^3\pi$ mA $\quad \therefore x = 5$

45. **(300)** Given,
Charge, $q = 20$ C, potential difference, $\Delta V = 15$ V
Work done, $W = q\Delta V = 20\times15 = 300$ J

46. **(b)** Resistance, $R = \frac{\rho\ell}{A} = \frac{\rho\ell^2}{(\text{Vol.})}$ or $R \propto \ell^2$

or, $\frac{R_1}{R_2} = \frac{\ell_1^2}{\ell_2^2}$

As length increases 25%, if $\ell_1 = \ell$ then

$\ell_2 = \ell + \ell\times\frac{25}{100} = 1.25\ell$

$\therefore \frac{R}{R_2} = \frac{\ell^2}{(1.25\ell)^2}$ or, $R_2 = 1.5625R$

$\therefore$ % increase in resistance $R \simeq 56\%$

47. **(b)** $\rho_M = 98\times10^{-8}$; $\rho_A = 2.65\times10^{-8}$

$\rho_C = 1.724\times10^{-8}$; $\rho_T = 5.65\times10^{-8}$

$\therefore \rho_M > \rho_T > \rho_A > \rho_C$

48. **(d)** Charge mobility

$(\mu) = \frac{V_d}{E}$ [Where V_d = drift velocity]

and resistivity $(\rho) = \frac{E}{j} = \frac{EA}{I} \Rightarrow E = \frac{I(\rho)}{A}$

$\Rightarrow \mu = \frac{V_d}{E} = \frac{V_d A}{I\rho}$

$$= \frac{1.1\times10^{-3}\times\pi\times\left(5\times10^{-3}\right)^2}{5\times1.7\times10^{-8}} \Rightarrow \mu = 1.0\frac{m^2}{Vs}$$

49. **(b)** Equation of straight line from graph

$y = -mx + c \Rightarrow \ln R = -m\left(\frac{1}{T^2}\right) + c$

here, m & c are constants

$$R = e^{\left[-m\left(\frac{1}{T^2}\right)+c\right]} = e^{-m\left(\frac{1}{T^2}\right)}\times e^c \Rightarrow R(T) = R_0 e^{\frac{-T_0^2}{T^2}}$$

50. **(a)** $dR = \frac{(\rho)(dx)}{4\pi x^2}$

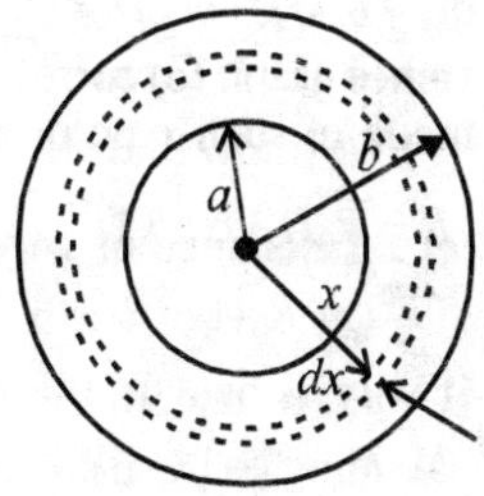

$R = \int dR = \rho\int_a^b \frac{dx}{4\pi x^2}$

$= \frac{\rho}{4\pi}\left[\frac{-1}{x}\right]_a^b$

$\therefore R = \left(\frac{\rho}{4\pi}\right).\left(\frac{1}{a} - \frac{1}{b}\right)$

51. **(c)** $\rho = \frac{m}{ne^2\tau} = \frac{9.1\times10^{-31}}{8.5\times10^{28}\times(1.6\times10^{-19})^2\times25\times10^{-15}}$

$= 10^{-8}\,\Omega$-m

52. **(a)** Clearly, from graph

Current, $I = \frac{dq}{dt} = 0$ [Since at t = 4s, q is constant]

53. **(a)** Using, $I = neAv_d$

$\therefore$ Drift speed $v_d = \frac{I}{neA}$

$= \frac{1.5}{9\times10^{28}\times1.6\times10^{-19}\times5\times10^{-6}} = 0.02$ mms^{-1}

54. **(c)** Resistance, $R = \frac{\rho\ell}{A}$

$R = \rho\frac{\ell}{A}\times\frac{\ell}{\ell} = \frac{\rho\ell^2}{V}$

Since resistivity and volume remains constant therefore % change in resistance

$\frac{\Delta R}{R} = \frac{2\Delta\ell}{\ell} = 2\times(0.5) = 1\%$

55. **(a)** Given, $R_1 = 100\,\Omega$, $r' = r/2$, $R_2 = ?$

Resistance of wire, $R = \frac{\rho l}{A}$

Hence, $R = \frac{\rho V}{A^2}$ [$\because$ Area × length = volume]

Since, $\rho \to$ constant, $V \to$ constant

$R \propto \frac{1}{A^2}$ or $R \propto \frac{1}{r^4}$ [$\because A = \pi r^2$]

$\frac{R_2}{R_1} = 16 \Rightarrow R_2 = 16\times100 = 1600\,\Omega$

56. **(b)** $V = IR = (neAv_d)\rho\frac{\ell}{A} \quad \therefore \rho = \frac{VA}{neAV_d l} = \frac{V}{neV_d l}$

Here V = potential difference
l = length of wire
n = no. of electrons per unit volume of conductor.
e = no. of electrons
Placing the value of above parameters we get resistivity

$$\rho = \frac{5}{8\times10^{28}\times1.6\times10^{-19}\times2.5\times10^{-4}\times0.1} = 1.6\times10^{-5}\,\Omega m$$

57. **(c)** $i = neAV_d$ and $V_d \propto \sqrt{E}$ (Given)

or, $i \propto \sqrt{E}$ or, $i^2 \propto E$ or, $i^2 \propto V$

Hence graph (c) correctly dipicts the *V-I* graph for a wire made of such type of material.

58. **(c)** Resistance of wire $(R) = \rho\frac{l}{A}$

If wire is bent in the middle then $l' = \frac{l}{2}, A' = 2A$

$\therefore$ New resistance, $R' = \rho\frac{l'}{A'} = \frac{\rho\frac{l}{2}}{2A} = \frac{\rho l}{4A} = \frac{R}{4}$.

59. **(a)** Resistance of wire

$R = \frac{\rho l}{A} = \frac{\rho l^2}{V}$ ($\because V = A\ell$)

Hence, $R = \rho\frac{\ell^2}{V}$ = constant $\times \ell^2$

$\therefore$ Fractional change in resistance

$\frac{\Delta R}{R} = 2\frac{\Delta l}{l} \Rightarrow 100\times\frac{\Delta R}{R} = 200\times\left(\frac{d\ell}{\ell}\right)$

$\because \frac{d\ell}{\ell}\times 100 = 0.1\%$

$\therefore$ % change in $R = 2\times 0.1\% = 0.2\%$

$\therefore$ Resistance will increase by 0.2%.

60. **(a)** Let *j* be the current density.

Then $j\times 2\pi r^2 = I \Rightarrow j = \frac{I}{2\pi r^2}$

$\therefore E = \rho j = \frac{\rho I}{2\pi r^2}$

Now, $V'_{BC} = -\int_{a+b}^{a}\vec{E}\cdot\vec{dr} = -\int_{a+b}^{a}\frac{\rho I}{2\pi r^2}dr$

$= -\frac{\rho I}{2\pi}\left[-\frac{1}{r}\right]_{a+b}^{a} = \frac{\rho I}{2\pi a} - \frac{\rho I}{2\pi\,(a+b)}$

On applying superposition principle as mentioned we get

$\Delta V_{BC} = 2\times\Delta V'_{BC} = \frac{\rho I}{\pi a} - \frac{\rho I}{\pi(a+b)}$

61. **(c)** As shown in solution (56), $E = \frac{\rho I}{2\pi r^2}$

62. **(d)** Resistance of a metal conductor at temperature t°C is given by $R_t = R_0(1+\alpha t)$,

R_0 is the resistance of the wire at 0°C

and α is the temperature coefficient of resistance.

Resistance at 50°C, $R_{50} = R_0(1+50\alpha)$ .. (i)

Resistance at 100°C, $R_{100} = R_0(1+100\alpha)$ ... (ii)

From (i), $R_{50} - R_0 = 50\alpha R_0$ ... (iii)

From (ii), $R_{100} - R_0 = 100\alpha R_0$ ... (iv)

Dividing (iii) by (iv), we get

$\frac{R_{50}-R_0}{R_{100}-R_0} = \frac{1}{2}$

Here, $R_{50} = 5\Omega$ and $R_{100} = 6\Omega$

$\therefore \frac{5-R_0}{6-R_0} = \frac{1}{2}$ or, $6 - R_0 = 10 - 2R_0$ or, $R_0 = 4\Omega$.

63. **(d)** Let d_A and d_B are the diameter of wire *A* and *B* respectively.

Let ρ_B and ρ_A be the resistivity of wire *A* and *B*. We have given

$\rho_B = 2\rho_A, d_B = 2d_A$

If both resistances are equal

$R_B = R_A$

$\Rightarrow \frac{\rho_B\ell_B}{A_B} = \frac{\rho_A\ell_A}{A_A} \therefore \frac{\ell_B}{\ell_A} = \frac{\rho_A}{\rho_B}\times\frac{d_B^2}{d_A^2} = \frac{\rho_A}{2\rho_A}\times\frac{4d_A^2}{d_A^2} = 2$

64. **(b)** Given,

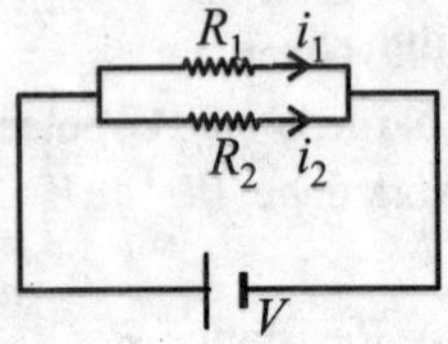

$\frac{\ell_1}{\ell_2} = \frac{4}{3}$ and $\frac{r_1}{r_2} = \frac{2}{3}$

$R_1 = \frac{\rho\ell_1}{\pi r_1^2}$; $R_2 = \frac{\rho\ell_2}{\pi r_2^2}$

When wires are in parallel to the circuit potential difference across each wire is same

$i_1R_1 = i_2R_2$

$\therefore \frac{i_1}{i_2} = \frac{R_2}{R_1} = \frac{\rho\ell_2}{\pi r_2^2}\times\frac{\pi r_1^2}{\rho\ell_1} = \frac{\ell_2}{\ell_1}\times\frac{r_1^2}{r_2^2} = \frac{3}{4}\times\frac{4}{9} = \frac{1}{3}$

65. **(d)** The resistance of each segment of wire is, $r = \frac{R}{6}$

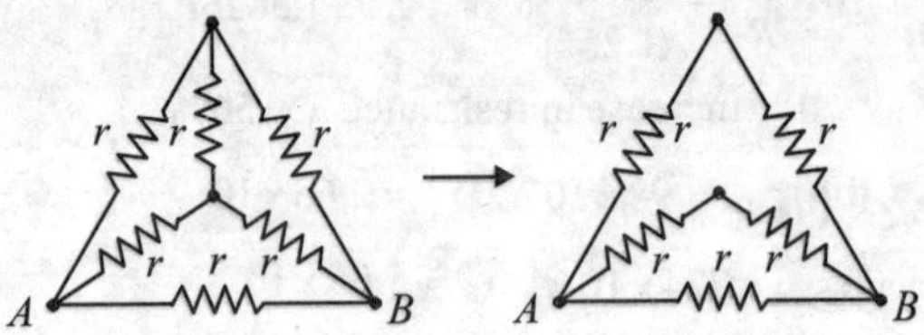

Now, equivalent resistance between *A* and *B* can be written as

$\frac{1}{R_{AB}} = \frac{1}{2r} + \frac{1}{2r} + \frac{1}{r} = \frac{2}{r}$

$\therefore R_{AB} = \frac{R}{12} = \frac{R}{n}$ (given)

$\therefore n = 12$.

66. **(a)** Solving from option, option (1) satisfies the condition

$\frac{1}{R_P} = \frac{1}{10} + \frac{1}{15} = \frac{3+2}{30} = \frac{1}{6}$

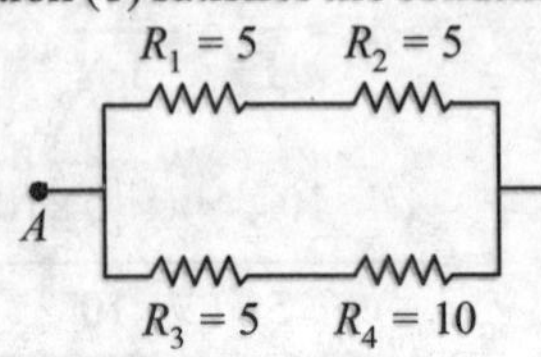

67. **(5)** The equivalent circuit diagram is

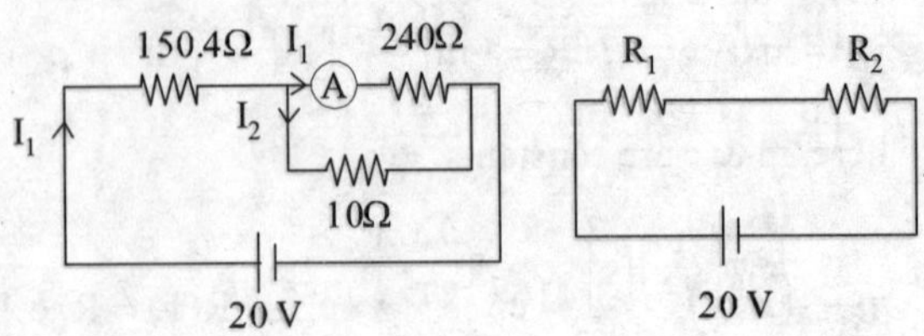

Here, $R_1 = 150.4\,\Omega$

$$R_2 = \frac{R_a R_s}{R_a + R_s} = \frac{240 \times 10}{240 + 10} = 9.6\,\Omega$$

$$R_{eq} = R_1 + R_2 = 150.4 + 9.6 = 160\,\Omega$$

$$\therefore \quad I = \frac{V}{R_{eq}} = \frac{20}{160} = \frac{1}{8}A$$

$\therefore$ Reading of ammeter,

$$I_1 = \left(\frac{R_s}{R_a + R_s}\right) I = \left(\frac{10}{240 + 10}\right) \times \frac{1}{8} = 5mA$$

68. **(c)**

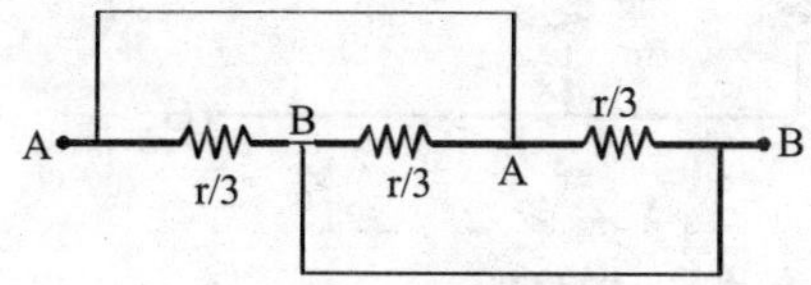

The given circuit can be redrawn as

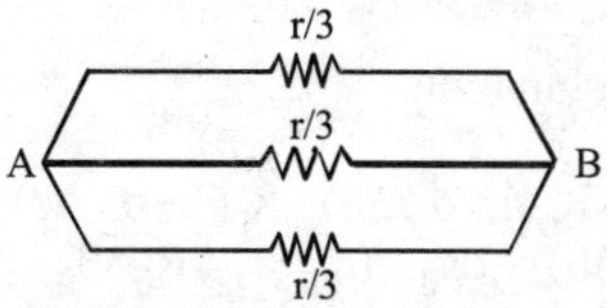

Use formula for equivalent resistance in parallel combination

$$\frac{1}{R_{eq}} = \frac{1}{R_1} + \frac{1}{R_2} + \frac{1}{R_3}$$

$$R_{eq} = \frac{r/3}{3} = r/9$$

69. **(d)** The formula for resistance is given by $R = \frac{\rho \ell}{A}$

So, $R \propto \ell$

Side length of triangle is $\left(\frac{1}{3}\right)^{rd}$ of its total length and $\left(\frac{1}{4}\right)^{th}$ of that of for square.

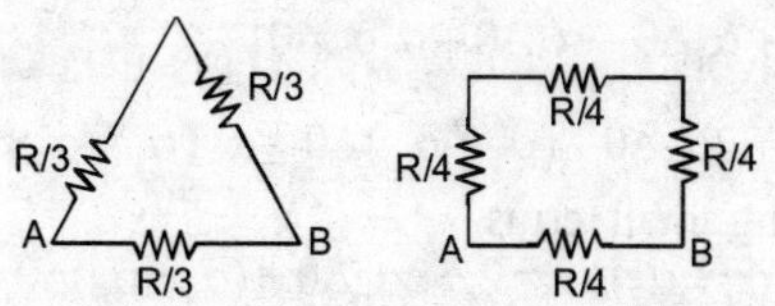

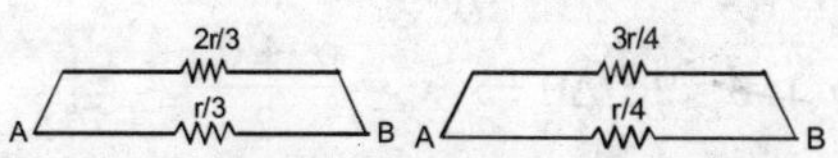

$$\left(R_{eq}\right)_1 = \frac{2r/3 \times r/3}{2r/3 + r/3}, \qquad \left(R_{eq}\right)_2 = \frac{3r/4 \times r/4}{3r/4 \times r/4}$$

$$\left(R_{eq}\right)_1 = 2r/9, \qquad \left(R_{eq}\right)_2 = 3r/16$$

$$\therefore \frac{\left(R_{eq}\right)_1}{\left(R_{eq}\right)_2} = \frac{2r/9}{3r/16} = \frac{32}{27}$$

70. **(2)** Resistance, $R = \rho \frac{\ell}{A} \Rightarrow R \propto \ell$

$\therefore R_1 = R_2 = R_3 = 3\Omega$

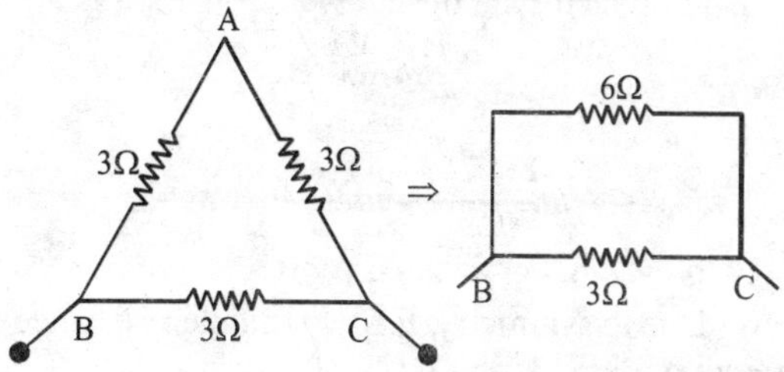

$$\therefore R_{BC} = \frac{6 \times 3}{6 + 3} = 2\Omega$$

71. **(16)**

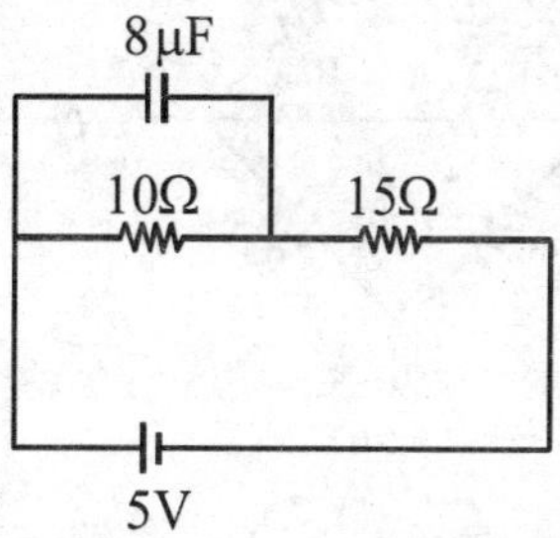

At steady state, the capacitor is fully charged, means no current flows through the capacitor

$$V_{10\Omega} = V_{capacitor} = 10 \times \frac{5}{25} = 2$$

The charge stored in a capacitor is given by

$Q = C\,V_{capacitor}$

$Q = (8 \times 10^{-6}) \times 2 = 16\mu C$

72. **(1)** At steady state, capacitor behaves as open wire , so equivalent circuit diagram is

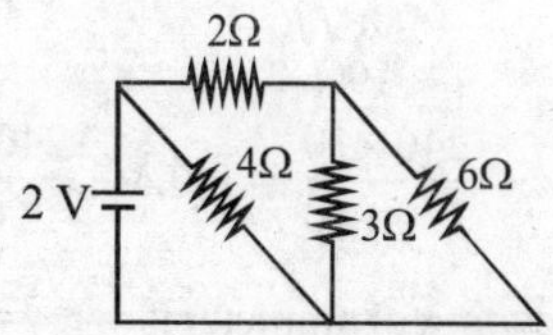

$$R_{eq} = [(2 + 6 \,\|\, 3) \,\|\, 4] = \left[\left(2 + \frac{6 \times 3}{6 + 3}\right) \| \, 4\right] = 2\Omega$$

$$\therefore \text{Current, } I = \frac{V}{R_{eq}} = \frac{2}{2} = 1A$$

73. **(d)**

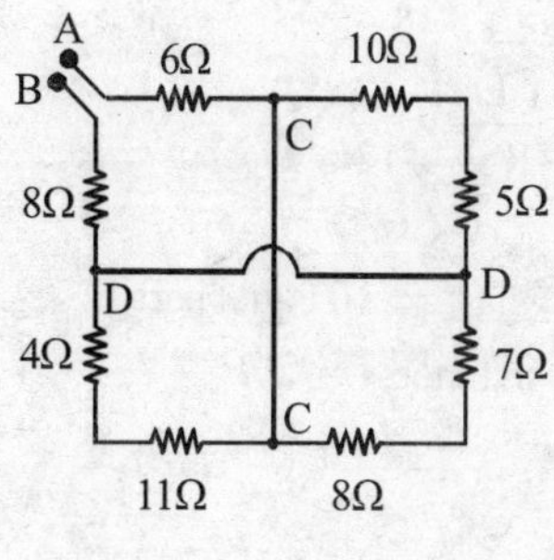

The given circuit can be redrawn as

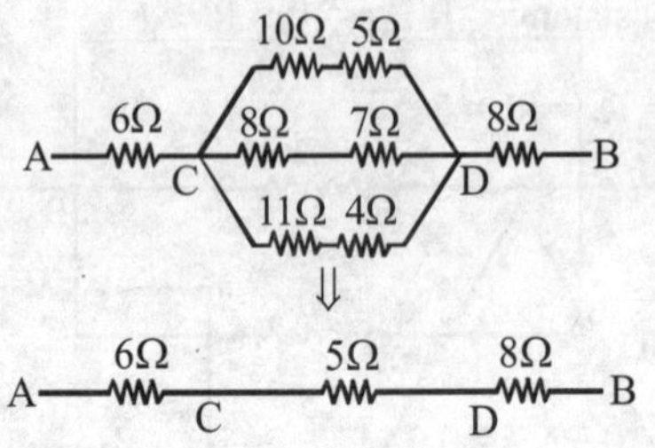

$\Rightarrow \quad R_{eq} = 6\Omega + 5\Omega + 8\Omega = 19\Omega$

74. (b) Due to symmetry, the two middle resistance can be removed.
Hence, the new circuit will be :

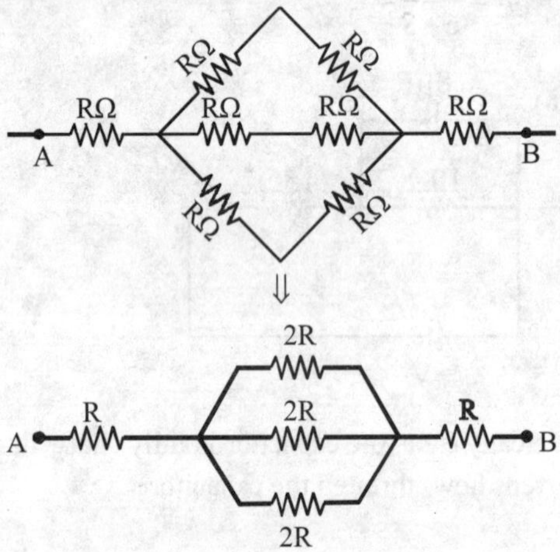

New resistance $= 2R + \frac{2R}{3} = \frac{8R}{3}$

75. (160) Equivalent resistance is,

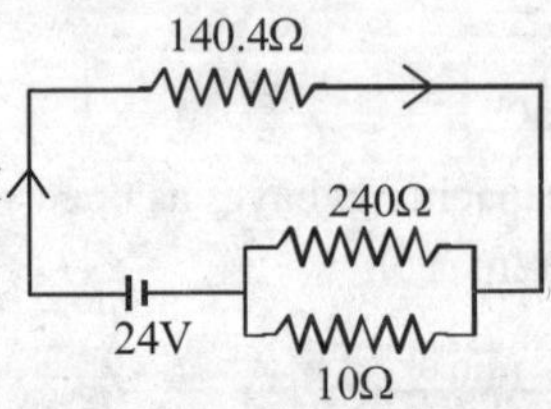

$R = 140.4 + \frac{240 \times 10}{240 + 10} = 140.4 + \frac{2400}{250} = 150\Omega$

Hence, reading of ammeter is $= \frac{V}{R} = \frac{24}{150} = 160\,\text{mA}$

76. (d) $R_{eq} = R_1 + (R_2 \| R_3 \| R_4)$

$= R_1 + \frac{R_2 R_3 R_4}{R_2 R_3 + R_3 R_4 + R_4 R_2}$

$= 10 + \frac{8 \times 4 \times 8}{8 \times 4 + 4 \times 8 + 8 \times 8} = 12\Omega$

and current, $I = \frac{E}{R_{eq}} = \frac{12}{12} = 1A$

77. (5) 

$\Rightarrow$ 10 equal part

Each part has resistance, r = 2Ω

$\therefore R_{eq} = (r \| r) \times 5$

$= \frac{r \cdot r}{r + r} \times 5 = \frac{r}{2} \times 5 = \frac{2}{2} \times 5 = 5\Omega$

78. (1) Resistance 6Ω is short circuited

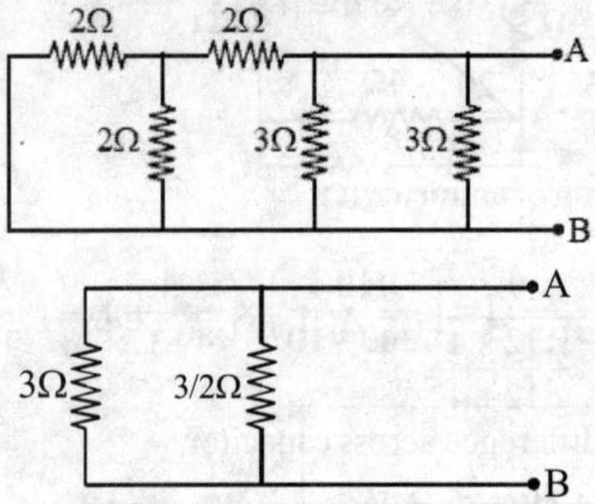

$\therefore \quad R_{eq} = \frac{\frac{3}{2} \times 3}{\frac{3}{2} + 3} = 1\Omega$

79. (b) In series combination

$R_{eq} = R_1 + R_2$

$2R_0(1 + \alpha_s \Delta\theta) = R_0(1 + \alpha_1 \Delta\theta) + R_0(1 + \alpha_2 \Delta\theta)$

$\Rightarrow 2R_0(1 + \alpha_s \Delta\theta) = 2R_0 + (\alpha_1 + \alpha_2) R_0 \Delta\theta$

$\Rightarrow \alpha_s = \frac{\alpha_1 + \alpha_2}{2}$

For parallel combination $\frac{1}{R_p} = \frac{1}{R_1} + \frac{1}{R_2}$

$\Rightarrow \frac{1}{\frac{R}{2}(1 + \alpha_p \Delta\theta)} = \frac{1}{R(1 + \alpha_1 \Delta\theta)} + \frac{1}{R(1 + \alpha_2 \Delta\theta)}$

$\Rightarrow \frac{2}{1 + \alpha_p \Delta\theta} = \frac{1}{1 + \alpha_1 \Delta\theta} + \frac{1}{1 + \alpha_2 \Delta\theta}$

$\Rightarrow \frac{2}{1 + \alpha_p \Delta\theta} = \frac{1 + \alpha_2 \Delta\theta + 1 + \alpha_1 \Delta\theta}{(1 + \alpha_1 \Delta\theta)(1 + \alpha_2 \Delta\theta)}$

$\Rightarrow 2\left[(1 + \alpha_1 \Delta\theta)(1 + \alpha_2 \Delta\theta)\right]$

$= \left[2 + (\alpha_1 + \alpha_2)\Delta\theta\right]\left[1 + \alpha_p \Delta\theta\right]$

$\Rightarrow 2\left[1 + \alpha_1 \Delta\theta + \alpha_2 \Delta\theta + \alpha_1 \alpha_2 \Delta\theta\right]$

$= 2 + 2\alpha_p \Delta\theta + (\alpha_1 + \alpha_2)\Delta\theta + \alpha_p(\alpha_1 + \alpha_2)\Delta\theta^2$

Neglecting small terms

$\Rightarrow 2 + 2(\alpha_1 + \alpha_2)\Delta\theta = 2 + 2\alpha_p \Delta\theta + (\alpha_1 + \alpha_2)\Delta\theta$

$\Rightarrow (\alpha_1 + \alpha_2)\Delta\theta = 2\alpha_p \Delta\theta \Rightarrow \alpha_p = \frac{\alpha_1 + \alpha_2}{2}$

80. (c) Equivalent resistance in series grouping,

$R_{eq} = R_1 + R_2 + \ldots + R_n$

$R_{eq} = 3300 + 100 + 100 + 100 + 100 + 100 + 100 + 100$

$i = \frac{4}{4000} = \frac{1}{1000} A$

$V_0 = i.R = \frac{1}{1000} \times 500 = 0.5V$

81. (81) At steady state, capacitor behaves as open circuit.

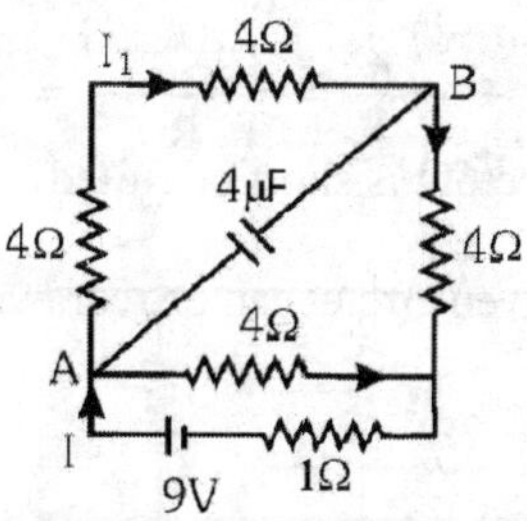

$$\therefore I=\frac{V}{R_{eq}}=\frac{9}{1+\frac{12\times4}{12+4}}=\frac{9}{4}A;\ I_1=\left(\frac{4}{4+12}\right)\cdot\frac{9}{4}=\frac{9}{16}A$$

$\therefore$ Potential difference across capacitor,

$$V=I_1\times8=\frac{9}{16}\times8=\frac{9}{2}V$$

$$\therefore\ U=\frac{1}{2}CV^2=\frac{1}{2}\times4\times\frac{81}{4}\mu J=\frac{81}{2}\mu J=\frac{x}{2}\mu J\ \text{(given)}$$

$\therefore\ x=81$

82. (b)

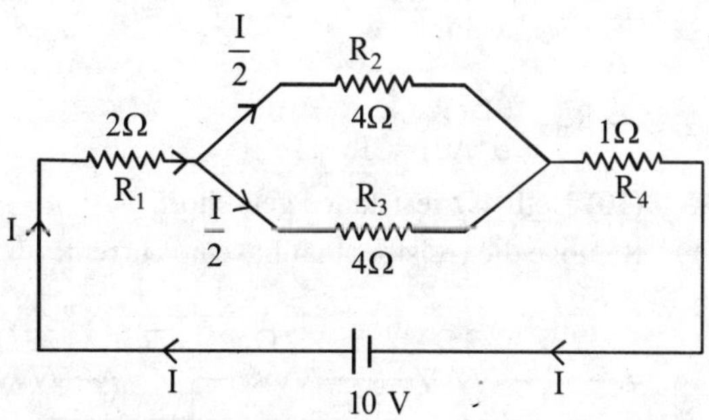

$R_{eq}=2\,\Omega+2\,\Omega+1\,\Omega=5\,\Omega$

Current, $I=\frac{V}{R_{eq}}=\frac{10}{5}=2A$

Current in resistance $R_3=\frac{I}{2}=1A$

83. (36) At steady state, capacitor bahaves as an open circuit and current flows in circuit as shown in the figure below.

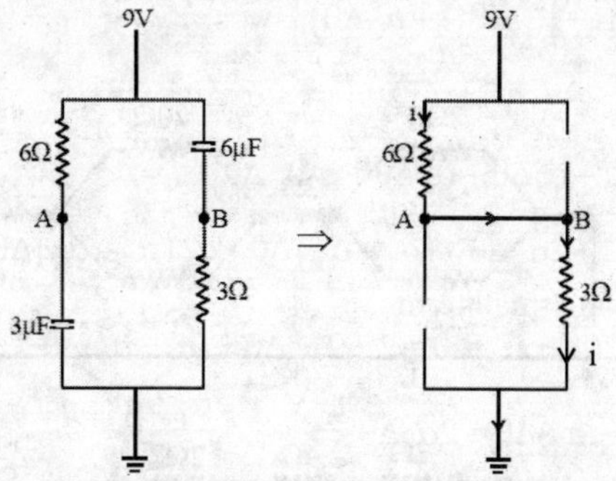

$R_{eq}=9\Omega$

Current, $i=\frac{9V}{9\Omega}=1A$

Potential difference across 6μF capacitor is

$\Delta V=6i=1\times6=6V$

$\therefore$ Charge stored in 6μF capacitor,

$Q=C\Delta V=6\times6\times10^{-6}C=36\,\mu C$

84. (30)

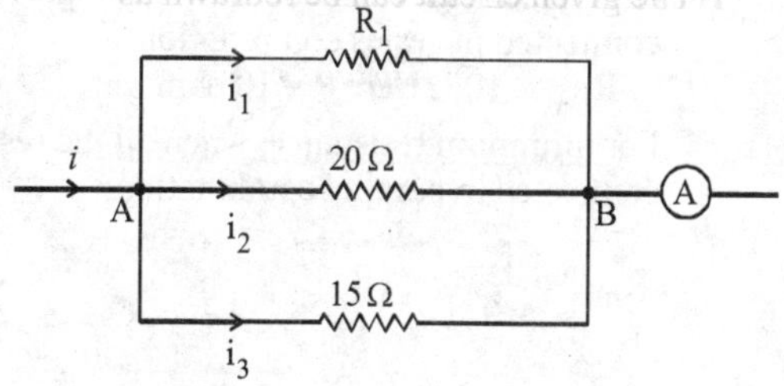

Given, $i_2=0.3\,A, i=0.9A$

So, $V_{AB}=i_2\times R_2=20\times0.3=6V$

$$i_3=\frac{V_{AB}}{R_3}=\frac{6}{15}=\frac{2}{5}A$$

$$\therefore\ i_1=i-(i_2+i_3)=\frac{9}{10}-\left(\frac{3}{10}+\frac{2}{5}\right)=\frac{1}{5}A$$

$$\therefore\ R_1=\frac{V_{AB}}{i_1}=\frac{6}{\frac{1}{5}}=30\Omega$$

85. (a) As wire is cut into 5 equal parts so resistance of each part $=\frac{R}{5}$

In parallel grouping, $\frac{1}{R_p}=\frac{1}{R_1}+\frac{1}{R_2}+............+\frac{1}{R_n}$

$\therefore$ Total resistance $=\frac{1}{5}\times\frac{R}{5}=\frac{R}{25}$

86. (a)

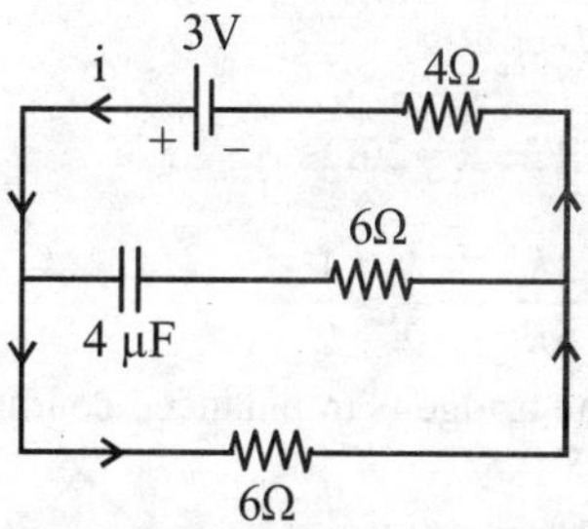

Current flowing in the circuit in steady state

$$I=\frac{V}{r_{eq}}=\frac{3}{6+4}=\frac{3}{10}$$

[∵ In steady state current through capacitor is zero]

Potential difference on 6Ω resistance

$$V=R\times I=6\times\frac{3}{10}=1.8\text{ volt}$$

Capacitor will have same potential so charge,

$q=CV=(4\,\mu F).(1.8\text{ volt})=7.2\,\mu C.$

87. (b) The circuit can be reduced to

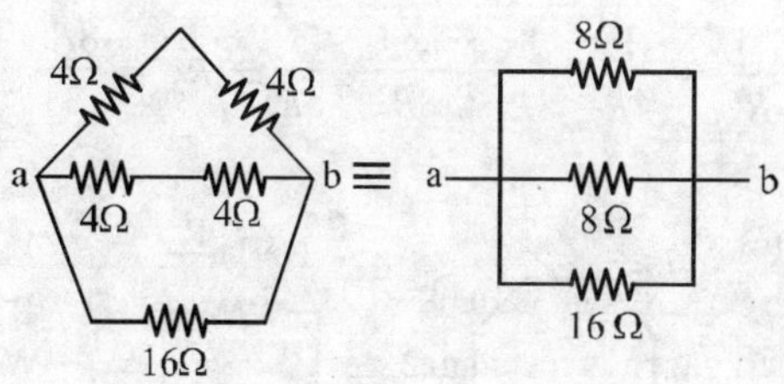

$$\Rightarrow R_{eq}=\frac{16\times4}{16+4}=\frac{16}{5}\Omega=3.2\Omega$$

88. **(100)** For maximum resistance, when all the resisters are connected in series combination.

$\therefore R_{max} = 10R$, Here R = 10 ohm

For minimum resistance, when all the resistance are connected in parallel combination.

$$R_{min} = \frac{R}{10}$$

$$\therefore \frac{R_{max}}{R_{min}} = \frac{10R \times 10}{R} = 100$$

89. **(a)** Since, $V_A = V_D; V_C = V_B$

All resistors are in parallel.

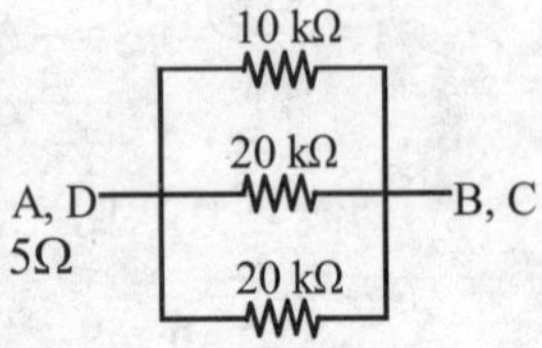

$$\frac{1}{R_{eq}} = \frac{1}{10} + \frac{1}{20} + \frac{1}{20} \Rightarrow R_{eq} = 5\Omega$$

90. **(20)** Current through 5Ω resistance,

$$\Rightarrow i_1 = \frac{2V}{5\Omega} = \frac{2}{5}A$$

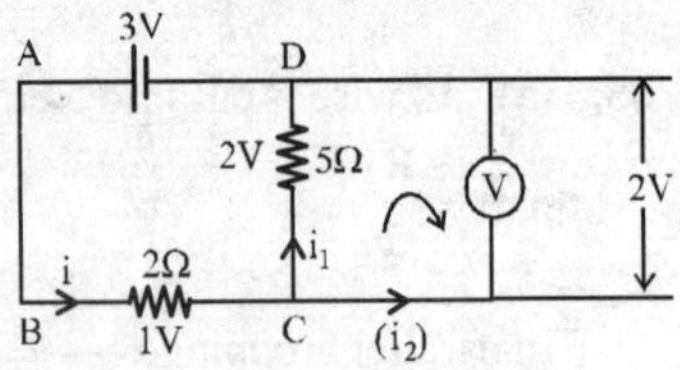

Current through 2Ω resistance,

$$i = \frac{1V}{2\Omega} = \frac{1}{2}A$$

$\therefore$ Current flowing through voltmeter

$$i_2 = i - i_1 = \frac{1}{2} - \frac{2}{5} = \frac{1}{10}A$$

For voltmeter, $V = i_2 R$

$$\Rightarrow 2 = \left(\frac{1}{10}\right) R \Rightarrow R = 20\Omega$$

91. **(c)** As, $\frac{R}{2R} = \frac{3R}{6R} \Rightarrow \frac{1}{2} = \frac{1}{2}$

$\therefore$ Wheatstone bridge is in balanced condition.

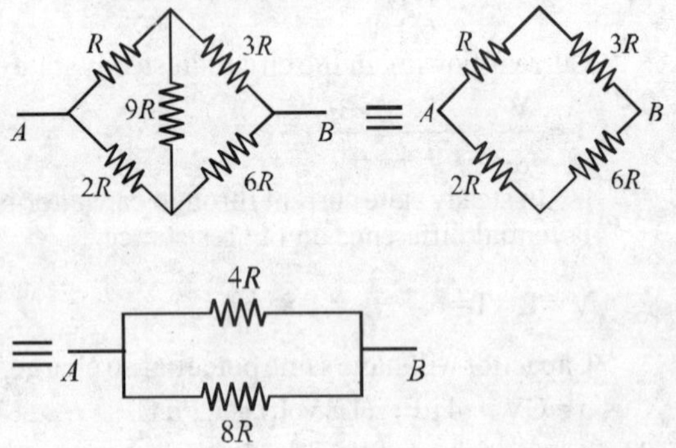

$$\frac{1}{R_{eq}} = \frac{1}{4R} + \frac{1}{8R} = \frac{2+1}{8R} = \frac{3}{8R} \Rightarrow R_{eq} = \frac{8R}{3}$$

92. **(c)**

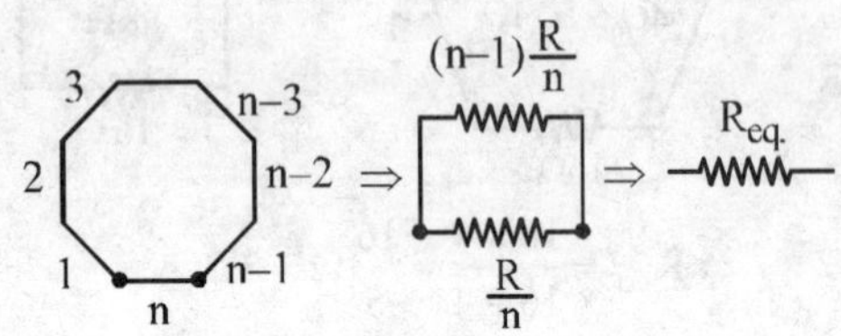

$$R_{eq} = \frac{\frac{R}{n}.(n-1)\frac{R}{n}}{\frac{R}{n}+(n-1)\frac{R}{n}} = \frac{\frac{R^2}{n^2}(n-1)}{R} = \frac{R(n-1)}{n^2}$$

93. **(a)** The given circuit can be redrawn as

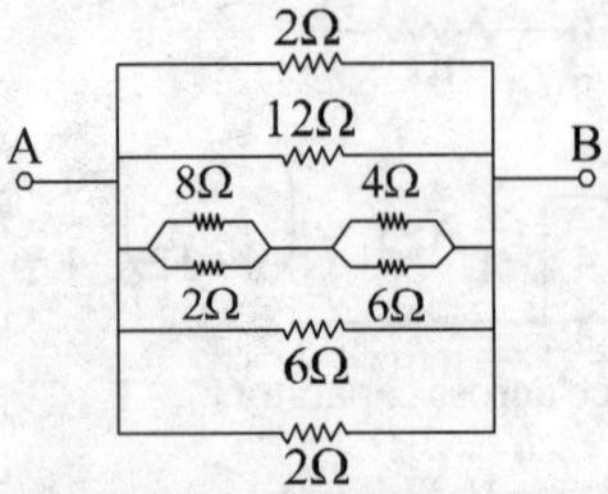

The net resistance between A and B

$$\frac{1}{R_{net}} = \frac{1}{2} + \frac{1}{12} + \frac{1}{4} + \frac{1}{6} + \frac{1}{2} = \frac{6+1+3+2+6}{12} = \frac{18}{12} = \frac{3}{2}$$

$$\Rightarrow R_{net} = \frac{2}{3}\Omega$$

94. **(10)** Both 4Ω resistance gets short.

Remove the resistors that have no current.

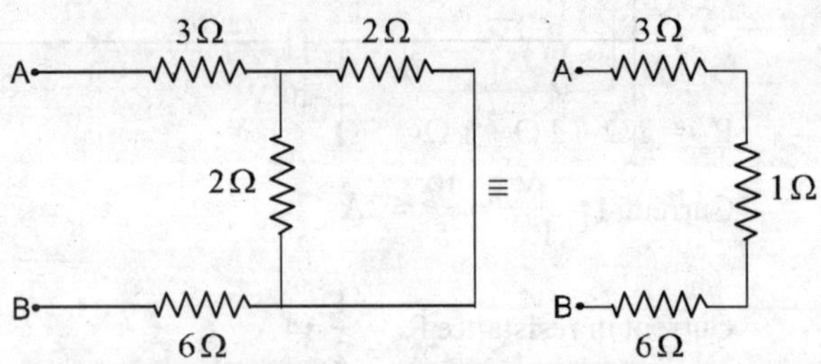

$R_{eq} = 3 + 1 + 6$

$R_{eq} = 10\,\Omega$

95. **(d)**

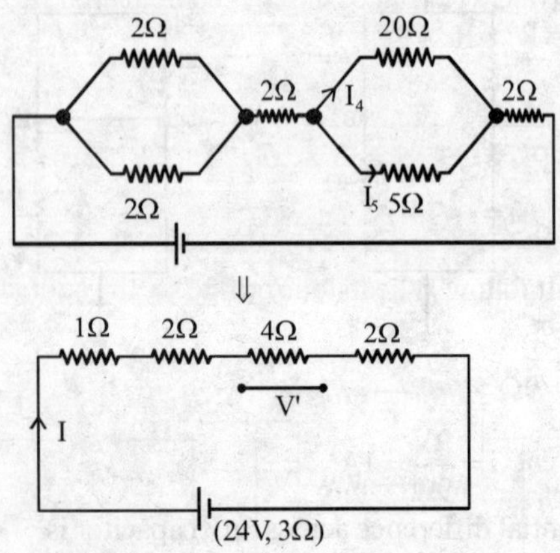

So, $R_{eq.} = 12\ \Omega$

As $I = \frac{V}{R_{eq}} \Rightarrow I = \frac{24}{12} A\ 2A$

So, $V' = 2 \times 4 = 8$ volt

$I_4 = \frac{V'}{20} = \frac{8}{20} = \frac{2}{5} A$

$I_5 = \frac{V'}{5} = \frac{8}{5} A$

96. (a)

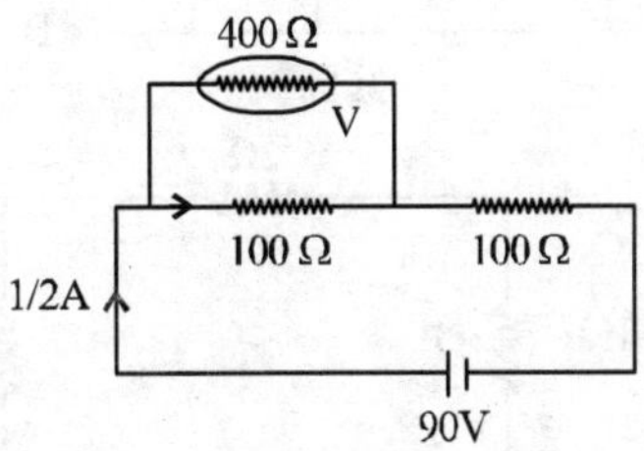

$$R_{eq} = \frac{400 \times 100}{500} + 100 = 180\Omega$$

$$i = \frac{V}{R_{eq}} \Rightarrow i = \frac{90}{180} = \frac{1}{2} A$$

$$\text{Reading} = \frac{1}{2} \times \left(\frac{400}{500} \times 100\right) = 40 \text{ volt}$$

97. (2) Equivalent circuit of the above circuit is as shown below

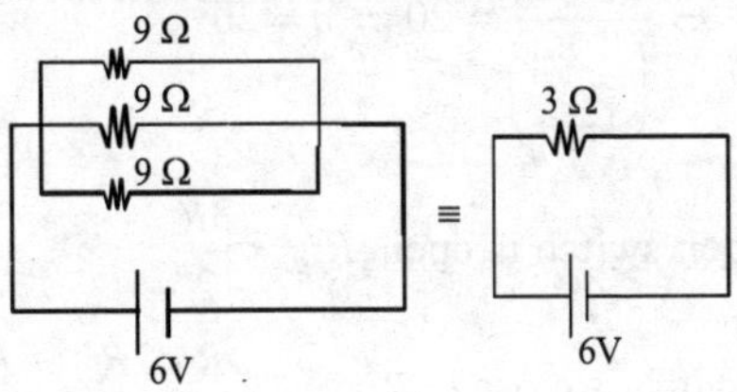

$I = \frac{6}{3} = 2A$

98. (b)

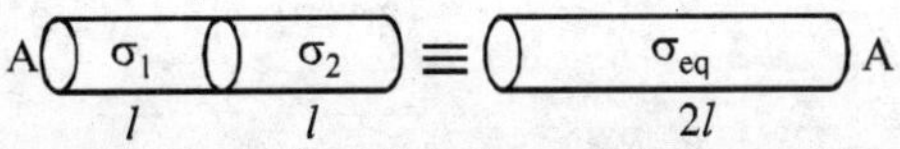

Let length of wire be 'ℓ' and area of wire as 'A'
For equivalent wire length = 2ℓ; area will be A
Equivalent resistance in series combination will be given as
$R_{eq} = R_1 + R_2$

$$\Rightarrow \frac{2\ell}{\sigma_{eq} A} = \frac{\ell}{\sigma_1 A} + \frac{\ell}{\sigma_1 A} \quad \left[\because R = \frac{\ell}{\sigma A}\right]$$

$$\Rightarrow \sigma_{eq} = \frac{2\sigma_1\sigma_2}{\sigma_1 + \sigma_2}$$

99. (a) It follow Wheatstone bridge so, the equivalent circuit will be

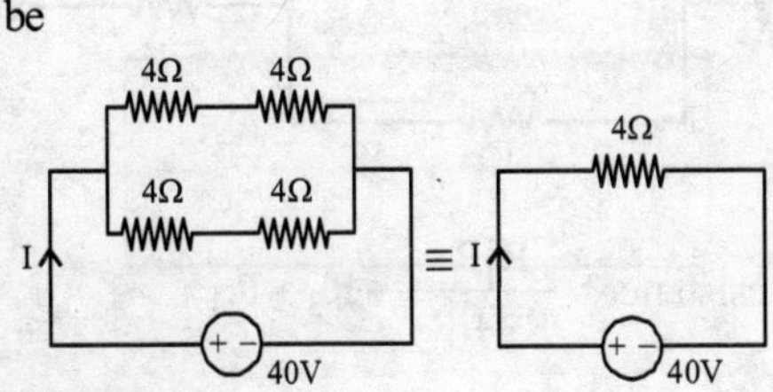

Applying KVL
$-40 + 4I = 0 \Rightarrow I = 10$ A

100. (4) R_{eq} of 8 wires $= \frac{\rho\ell}{\pi\left(\frac{d}{2}\right)^2 \times 8} = \frac{\rho\ell}{2\pi d^2}$

R_{eq} of single wire $= \frac{\rho \times 2\ell}{\pi \frac{D^2}{4}} = \frac{8\rho\ell}{\pi D^2}$

So, $\frac{\rho\ell}{2\pi d^2} = \frac{8\rho\ell}{\pi D^2}$

$\Rightarrow D^2 = 16\, d^2$

$\Rightarrow D = 4d$

101. (d)

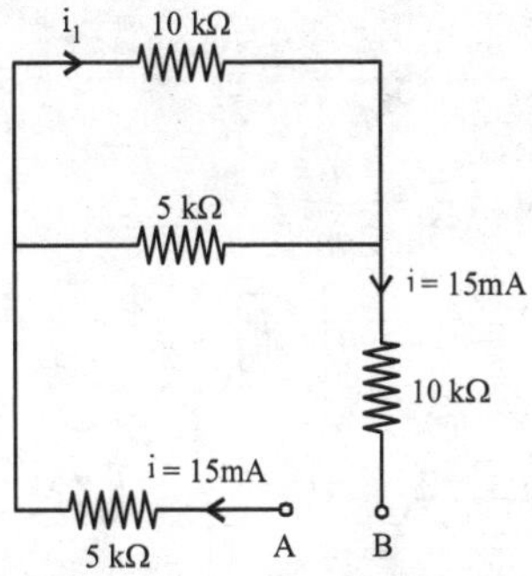

$i_1 = \frac{5}{10+5} \times 15 \text{ mA} = 5 \text{ mA}$

Applying Kirchhoff's voltage law,
$V_A - 5i - 10i_1 - 10\,i = V_B$
$V_A - V_B = 75 + 50 + 150 = 275$ V

102. (b) As $R = \frac{\rho l}{A} \Rightarrow R_i = 14\Omega = \frac{\rho l_0}{\pi r_0^2}$

$$R_f = \frac{\rho l_0}{\pi\left(\frac{r_0}{3}\right)^2} = 9\frac{\rho l_0}{\pi r_0^2} = 9R_i = 9 \times 14 = 126\Omega$$

$$R_{eq} = \frac{R_f}{7} = \frac{126}{7} = 18\Omega$$

103. (10) We have $i = \frac{15}{1+2} = 5A$

So, $V_B - V_A = 2i = 2 \times 5 = 10V.$

104. (8)

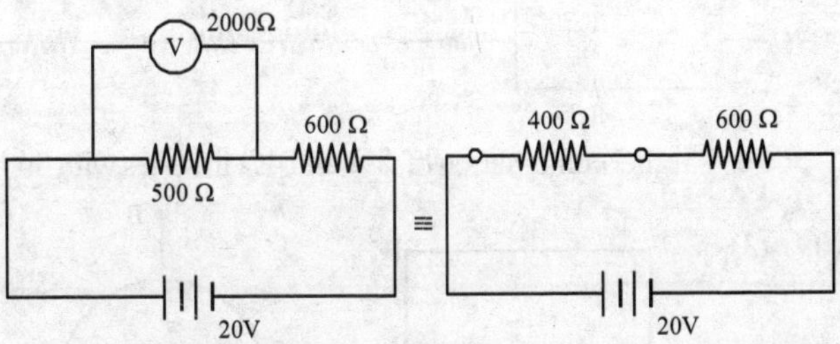

Voltage measured by voltmeter

$= \text{voltage across } 400\Omega = \frac{400}{400+600} \times 20 = 8V$

105. (8)

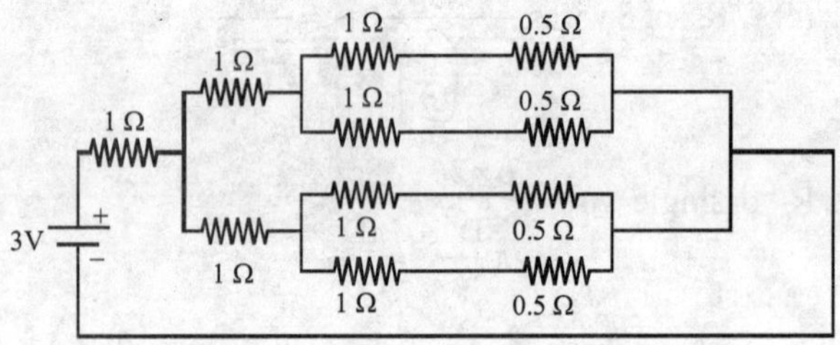

From above figure, $R_{eq}=\frac{15}{8}\Omega$

So, $I=\frac{V}{R}=\frac{3}{15/8}=\frac{8}{5}A=\frac{a}{5}A$ So, $a=8$.

106. (3) We have $R_{eq}=\frac{ma}{3}+\frac{a}{2m}$

For $R_{\min}$, $\frac{dR_{eq}}{dm}=0\Rightarrow\frac{a}{3}-\frac{a}{2m^2}=0$

$\Rightarrow\frac{a}{3}=\frac{a}{2m^2}\Rightarrow m^2=\frac{3}{2}\Rightarrow m=\sqrt{\frac{3}{2}}$

107. (c)

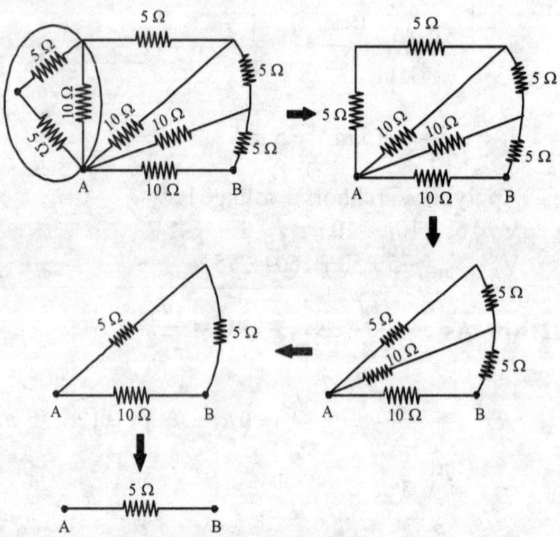

108. (b) As $R_{eq}=\frac{22}{3}\Omega=7.33\Omega$

So, the best arrangement will be

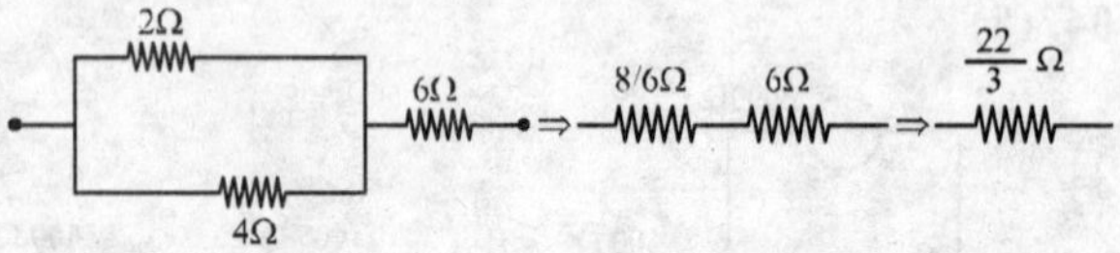

i.e., Parallel combination of 2Ω and 4Ω in series with 6Ω

109. (3)

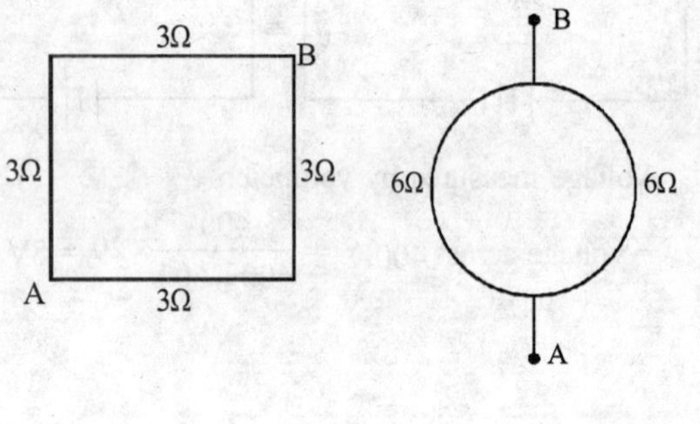

$$\frac{1}{R_{eq}}=\frac{1}{6}+\frac{1}{6}\Rightarrow R_{eq}=3\Omega$$

110. (d) Current does not pass through 5Ω as potential difference across it is zero. Then circuit becomes

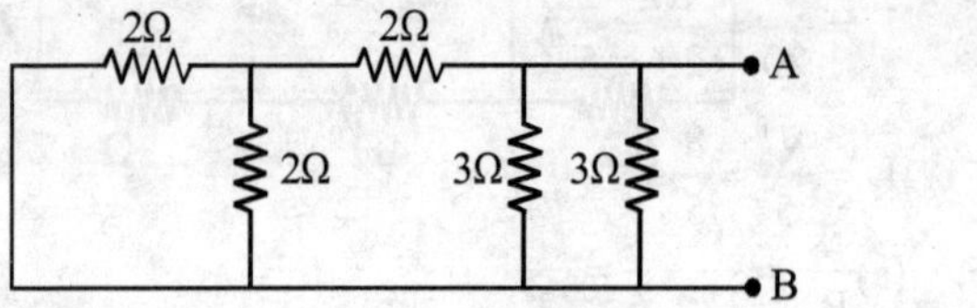

$2\Omega \parallel 2\Omega$ $\quad$ $3\Omega \parallel 3\Omega$

$R_{eq}=\frac{3\times1.5}{3+1.5}=1\Omega$

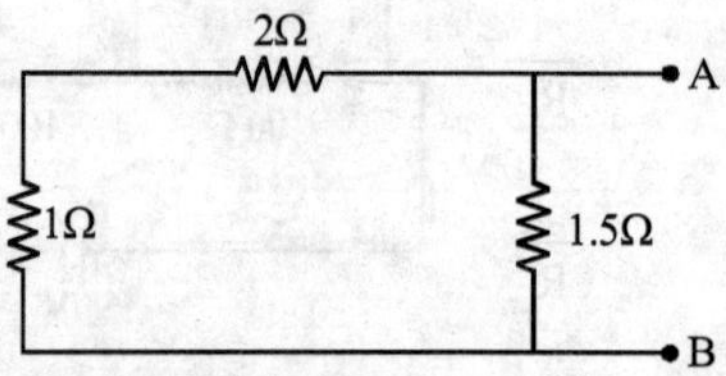

111. (20) In series, current, $i_1=\frac{20}{10+10n}=\frac{2}{1+n}$

In parallel, current, $i_2=\frac{20}{\frac{10}{n}+10}=\frac{2n}{1+n}$

$$\frac{i_2}{i_1}=20\Rightarrow\frac{\left(\frac{2n}{1+n}\right)}{\left(\frac{2}{1+n}\right)}=20\Rightarrow n=20$$

112. (9) When switch is open, $R_{eq}=\frac{3R}{2}$

When switch is closed, $R_{eq}=2\times\frac{R\times2R}{3R}=\frac{4R}{3}$

$\frac{R_{\text{open}}}{R_{\text{closed}}}=\frac{3R}{2}\times\frac{3}{4R}=\frac{9}{8}=\frac{x}{8}$ (given) $\quad\therefore x=9$

113. (b) $R_{net}=\frac{R_1R_2}{R_1+R_2}$

$$R_{\text{net}}=\left(\frac{\rho_1\rho_2}{\rho_1+\rho_2}\right)\frac{l}{A}\Rightarrow l=\frac{(\rho_1+\rho_2)AR_{\text{net}}}{\rho_1\rho_2}$$

Putting the required value we get $l=97$ m

114. (d)

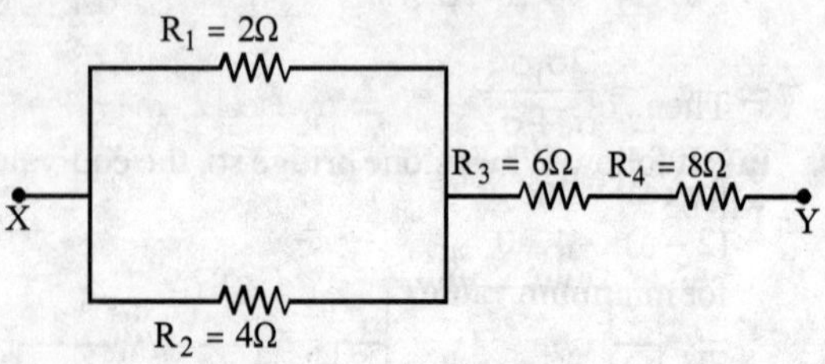

Net Resistance $=\frac{R_1R_2}{R_1+R_2}+R_3+R_4$

$=\frac{2\times4}{2+4}+14=\frac{46}{3}\Omega$

115. (10) When switch S_1 and S_2 are closed

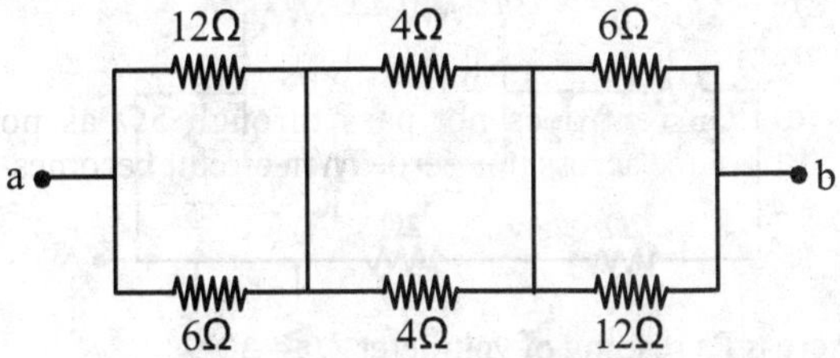

$$\frac{1}{R_1}=\frac{1}{12}+\frac{1}{6}=\frac{1+2}{12}=\frac{1}{4} \quad \therefore R_1=4\Omega$$

$$\frac{1}{R_2}=\frac{1}{4}+\frac{1}{4}=\frac{1+1}{4} \therefore R_2=\frac{4}{2}=2\Omega$$

$$\frac{1}{R_3}=\frac{1}{6}+\frac{1}{12}=\frac{2+1}{12} \therefore R_3=\frac{12}{3}=4\Omega$$

$\therefore R_{ab}=R_1+R_2+R_3=4+2+4=10\Omega.$

116. (70) Resistance 50 Ω and 20 Ω are in parallel.

$$\therefore \frac{1}{R}=\frac{1}{50}+\frac{1}{20}=\frac{20+50}{1000} \Rightarrow R=\frac{1000}{70}$$

So, equivalent resistance of circuit

$$R_{eq}=10+\frac{1000}{70}$$

By voltage division rule

$$V_{10}=\frac{V\times R_{10}}{R_{eq}}=\frac{170\times 10}{10+\frac{1000}{70}}=70\text{V}=x\text{V} \quad \therefore x=70\text{ V}$$

117. (4) Equivalent resistance in series, $s=R_1+R_2$

Equivalent resistance in parallel, $p=\frac{R_1R_2}{R_1+R_2}$

Given $(R_1+R_2)=n\left(\frac{R_1R_2}{R_1+R_2}\right)$

$\Rightarrow R_1^2+R_2^2+2R_1R_2=nR_1R_2$

$$\Rightarrow \frac{R_1^2}{R_1R_2}+\frac{R_2^2}{R_1R_2}+2=n \Rightarrow \frac{R_1}{R_2}+\frac{R_2}{R_1}+2=n$$

Let $\frac{R_1}{R_2}=\alpha$

Then, $\alpha+\frac{1}{\alpha}+2=n \Rightarrow \alpha^2+\alpha(2-n)+1=0$

for real value of 'α'

$(2-n)^2-4\geq 0$

for minimum value $(2-n)^2-4=0$

$\Rightarrow n=0$ or $n=4$

$\Rightarrow n=4 \quad [\because n\neq 0]$

118. (3) In the given circuit, three $3k\Omega$ resistors are connected in parallel, this combination is in series with two resistors of $5k\Omega$ and $1k\Omega$.

Thus, equivalent resistance of the circuit

$$R_{eq}=5k\Omega+1k\Omega+1k\Omega=7k\Omega$$

$$i=\frac{E}{R_{eq}}=\frac{21}{7}\text{mA}=3\text{ mA}$$

119. (c) The diagram can be drawn as

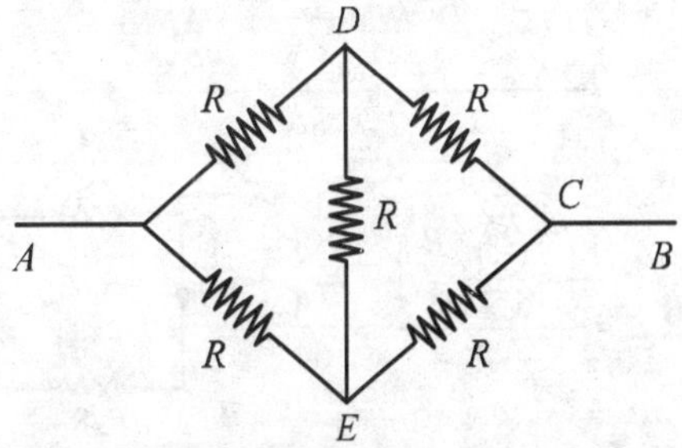

This is a wheatstone bridge. So, resistance along arm *DE* is ignored.

$$\text{So, } \frac{1}{R_{eq}}=\frac{1}{2R}+\frac{1}{2R} \Rightarrow R_{eq}=R$$

120. (b) From circuit diagram,

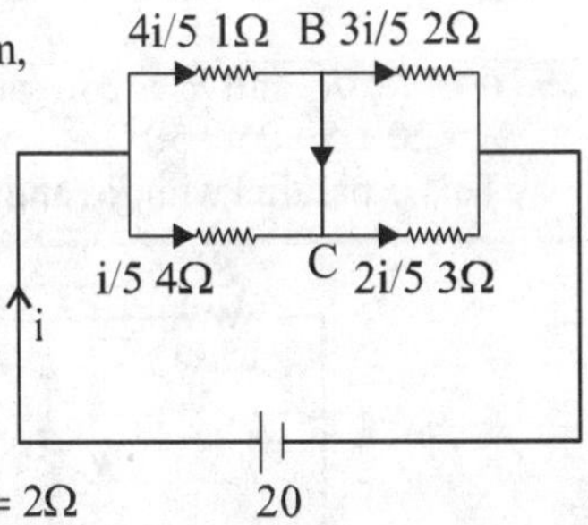

$$\frac{1}{R_1}=\frac{1}{1}+\frac{1}{4} \Rightarrow R_1=\frac{4}{5}$$

$$\frac{1}{R_2}=\frac{1}{2}+\frac{1}{3} \Rightarrow R_2=\frac{6}{5}$$

$$R_{eff}=R_1+R_2=\frac{4}{5}+\frac{6}{5}=2\Omega$$

$$i=\frac{V}{R_{eff}}=\frac{20}{2}=10A$$

$$\therefore \quad I_{BC}=\frac{4i}{5}-\frac{3i}{5}=\frac{i}{5}=2A$$

121. (30)

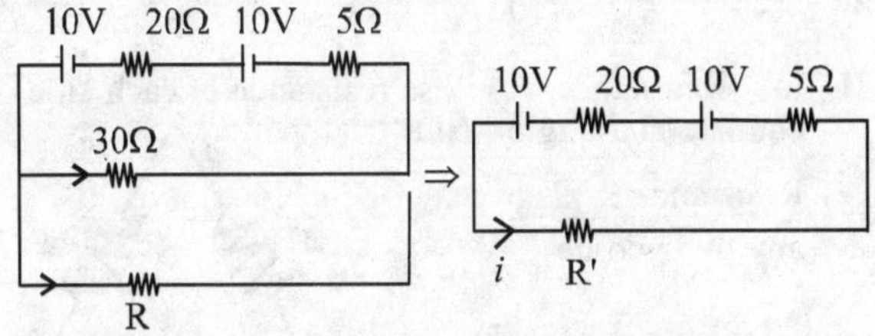

The resistance of 30Ω is in parallel with *R*. Their effective resistance

$$\frac{1}{R'}=\frac{1}{30}+\frac{1}{R} \Rightarrow R'=\frac{30R}{30+R}$$

Now, $10-i\times 20=0$

$i=0.5\text{ A}$

$$\text{as, } i=\frac{E}{R_{eq}} \Rightarrow 0.5=\frac{10+10}{R'+25}$$

$\Rightarrow 0.5 = \dfrac{20}{\dfrac{30R}{R+30}+25} \Rightarrow R = 30\Omega$

122. (c) $V_{1\Omega} = \dfrac{0.5}{2+0.5} \times 1 = 0.2V$

$\therefore$ Current in 1Ω resistor,

$I = \dfrac{0.2}{1} = 0.2A$

123. (b) $R_{eq} = \dfrac{\left(\dfrac{7R}{8}\right)\left(\dfrac{R}{8}\right)}{R} = \dfrac{7R}{64}$

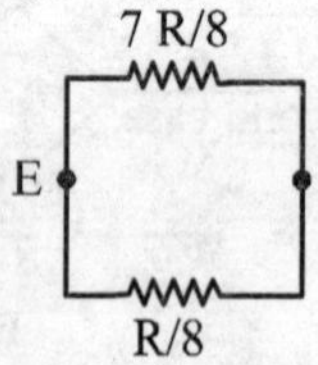

124. (c) When length becomes double its resistance becomes

$R = 4 \times 3 = 12\,\Omega$ $(\because R \propto l^2)$

$R_{eq} = \dfrac{2 \times 10}{12} = \dfrac{5}{3}\Omega$

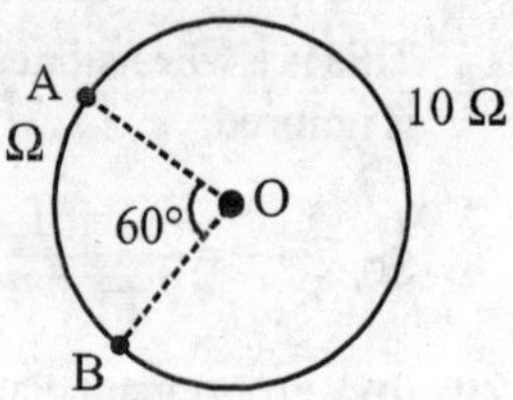

125. (c) R_3, R_4 and R_5 are in series so their equivalent $R = 20 + 5 + 25 = 50\,\Omega$

This is parallel with R_2, and so net resistance of the circuit

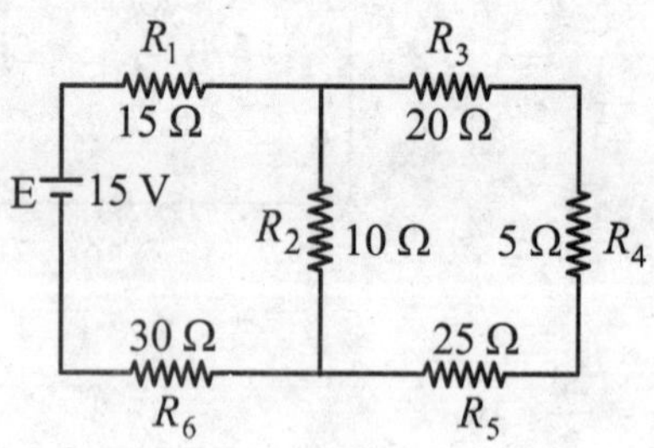

$R_{eq} = \left(\dfrac{10 \times 50}{10+50}\right) + 15 + 30 = \dfrac{160}{3}\Omega$

So, $i = \dfrac{E}{R_{eq}} = \dfrac{15}{(100/3)} = \dfrac{9}{32}A$

126. (a) Resistance, $R \propto l$ so resistance of each side of the equilateral triangle $= 6\,\Omega$

Resistance R_{eq} between any two vertices

$\dfrac{1}{R_{eq}} = \dfrac{1}{12} + \dfrac{1}{6} \Rightarrow R_{eq.} = 4\Omega$

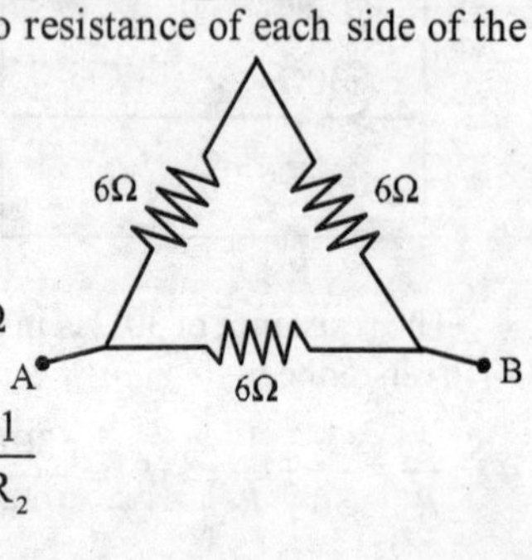

127. (b) Using, $\dfrac{1}{R_{eq}} = \dfrac{1}{R_1} + \dfrac{1}{R_2}$

$0.95\,R = \dfrac{RR_v}{R+R_v}$ (measured value 5% less then internal resistance of voltmeter)

or, $0.95 \times 30 = 0.05\,R_v$

$\therefore R_v = 19 \times 30 = 570\,\Omega$

128. (a)

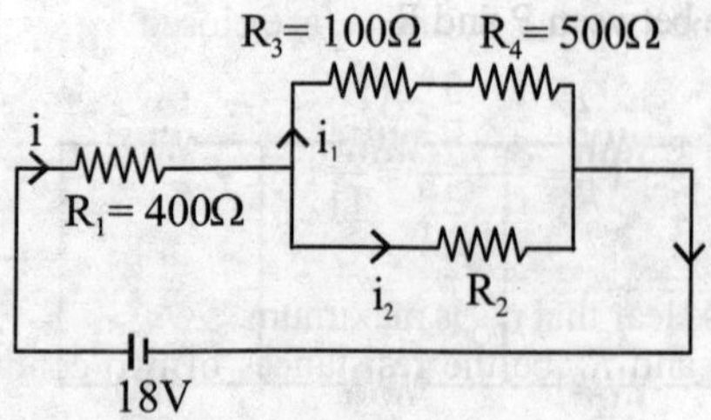

Across R_4 reading of voltmeter, $V_4 = 5V$

Current, $i_1 = \dfrac{V_4}{R_4} = 0.01A$

$V_3 = i_1 R_3 = 1V$

$\therefore \quad V_3 + V_4 = 6V = V_2$

$\Rightarrow \quad V_1 + V_3 + V_4 = 18V$

$\Rightarrow V_1 = 12\,V$

$i = \dfrac{V_1}{R_1} = 0.03A$

$i = i_1 + i_2 \Rightarrow i_2 = i - i_1, = 0.03 - 0.01A = 0.02\,A$

$\therefore R_2 = \dfrac{V_2}{i_2} = \dfrac{6}{0.02} = 300\Omega$

129. (a) Current will follow the path ACCCCB so we will get our final circuit as shown below

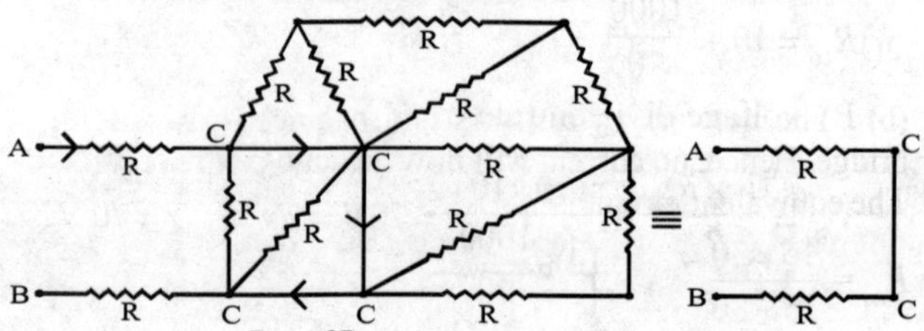

$R_{eq} = 2R$

130. (a) In steady state, flow of current through capacitor will be zero.

Current through the circuit,

$i = \dfrac{E}{r + r_2}$

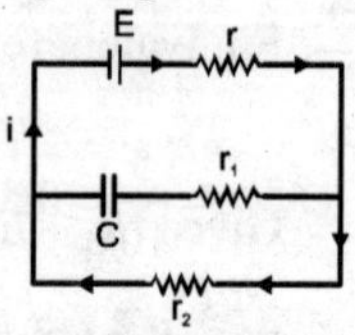

Potential difference across capacitor

$V_c = \dfrac{Q}{C} = E - ir = E - \left(\dfrac{E}{r+r_2}\right)r$

$\therefore \quad Q = CE\dfrac{r_2}{r + r_2}$

131. (b) The potential difference in each loop is zero.

$\therefore$ No current will flow or current in each resistance is zero.

132. (b) Resistance between P and Q

$r_{PQ} = r \,||\left(\dfrac{r}{3} + \dfrac{r}{2}\right) = \dfrac{r \times \dfrac{5}{6}r}{r + \dfrac{5}{6}r} = \dfrac{5}{11}r$

Resistance between Q and R

$r_{QR} = \dfrac{r}{2} \,||\, (r + \dfrac{r}{3}) = \dfrac{\dfrac{r}{2} \times \dfrac{4}{3}r}{\dfrac{r}{2} + \dfrac{4}{3}r} = \dfrac{4}{11}r$

Resistance between P and R

$$r_{PR} = \frac{r}{3} \| \left(\frac{r}{2}+r\right) = \frac{\frac{r}{3}\times\frac{3}{2}r}{\frac{r}{3}+\frac{3}{2}r} = \frac{3}{11}r$$

Hence, it is clear that r_{PQ} is maximum.

133. (d) Let R_1 and R_2 be the resistances of two conductors, then

$R_1 = R_0[1+\alpha_1\Delta t]$

$R_2 = R_0[1+\alpha_2\Delta t]$

Here, R_0 is the resistance of conductor at 0°C

In Series, $R = R_1 + R_2$

$$= R_0[2+(\alpha_1+\alpha_2)\Delta t] = 2R_0\left[1+\left(\frac{\alpha_1+\alpha_2}{2}\right)\Delta t\right]$$

$$\therefore \alpha_{eq} = \frac{\alpha_1+\alpha_2}{2}$$

In Parallel, $\frac{1}{R} = \frac{1}{R_1}+\frac{1}{R_2} = \frac{1}{R_0[1+\alpha_1\Delta t]}+\frac{1}{R_0[1+\alpha_2\Delta t]}$

$$= \frac{1}{\frac{R_0}{2}(1+\alpha_{eq}\Delta t)} = \frac{1}{R_0(1+\alpha_1\Delta t)}+\frac{1}{R_0(1+\alpha_2\Delta t)}$$

$$2(1-\alpha_{eq}\Delta t) = (1-\alpha_1\Delta t)(1-\alpha_2\Delta t) \qquad \therefore \alpha_{eq} = \frac{\alpha_1+\alpha_2}{2}$$

134. (b) The network of resistors is a balanced wheatstone bridge. Hence, no current will flow through centre resistor. The equivalent circuit is

$$R_{eq} = \frac{15\times 30}{15+30} = 10\,\Omega$$

$$\Rightarrow I = \frac{V}{R} = \frac{5}{10} = 0.5\text{ A}$$

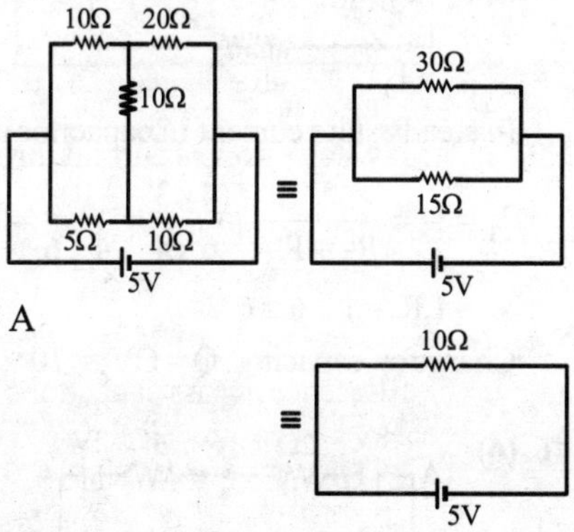

135. (a)

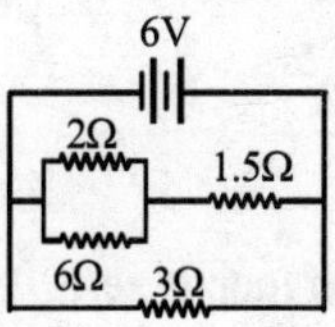

hence $R_{eq} = 3/2$;

$$\therefore I = \frac{6}{3/2} = 4\text{A}$$

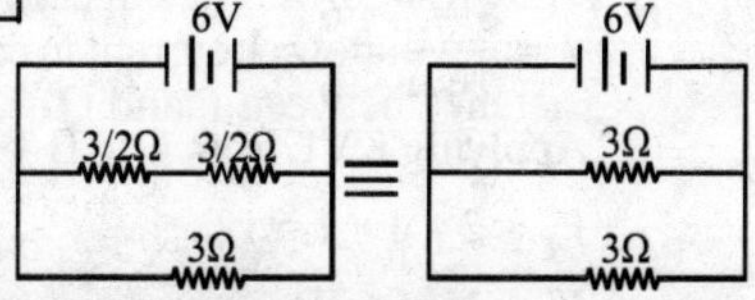

136. (b) In the given circuit, resistance of 3Ω is in parallel with series combination of two 3Ω resistance.

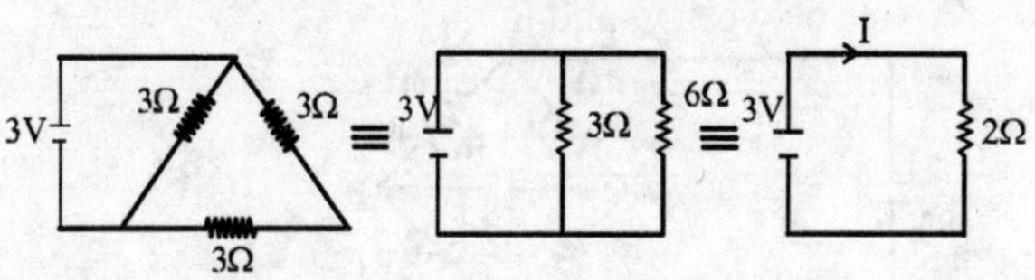

$$R_p = \frac{3\times 6}{3+6} = \frac{18}{9} = 2\Omega$$

Using ohm's law $V = IR \Rightarrow I = \frac{V}{R} = \frac{3}{2} = 1.5A$

137. (4) For first case,

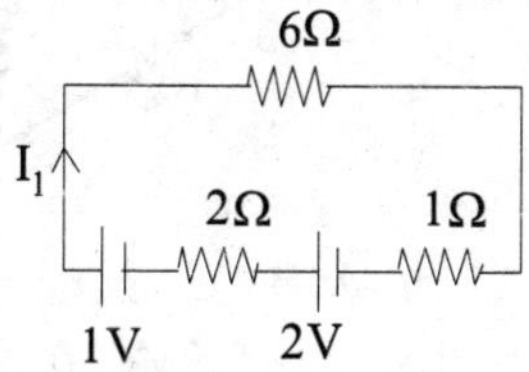

$$\text{Current, } I_1 = \frac{E_{net}}{R_{net}} = \frac{(2+1)}{(6+1+2)} = \frac{1}{3}\text{A}$$

For second case,

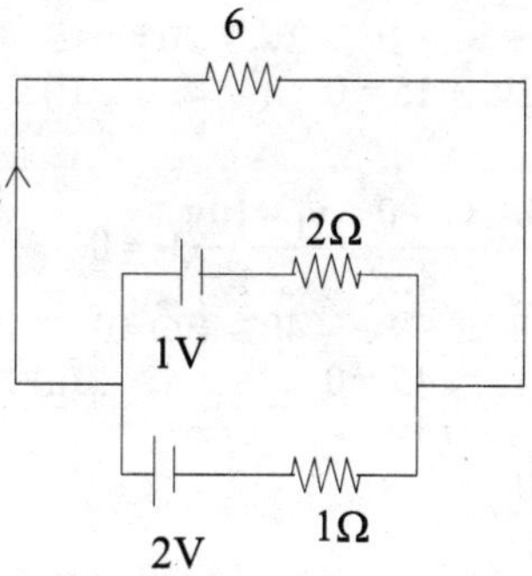

$$E_{eq} = \frac{\frac{\varepsilon_1}{r_1}+\frac{\varepsilon_2}{r_2}}{\frac{1}{r_1}+\frac{1}{r_2}} = \frac{\frac{1}{2}+\frac{2}{1}}{\frac{1}{2}+\frac{1}{1}} = \frac{5}{3}$$

$$R_{eq} = \frac{2\times 1}{3}+6 = \frac{20}{3}\Omega$$

$$\therefore I_2 = \frac{E_{eq}}{R_{eq}} = \frac{1}{4}$$

$$\therefore \frac{I_1}{I_2} = \frac{4}{3}$$

138. (2)

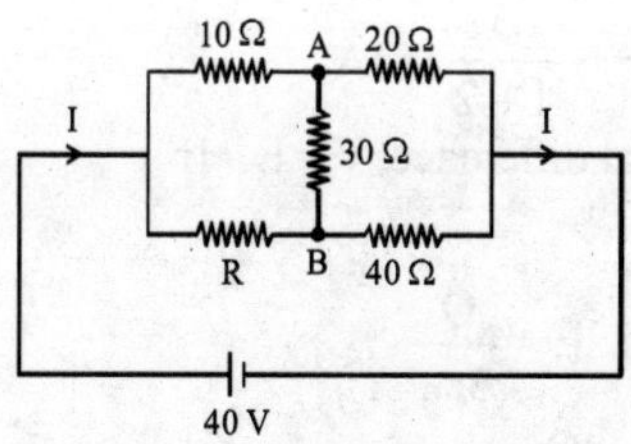

$V_A = V_B \Rightarrow$ the bridge wheats stone is balanced

$$\therefore \frac{P}{Q} = \frac{R}{S}$$

$$\Rightarrow \frac{10}{R} = \frac{20}{40} \Rightarrow R = 20\Omega \quad \therefore \quad I = \frac{E}{R_{eq}} = \frac{40}{20} = 2\text{A}$$

139. (d) Equivalent emf of two nonideal batteries connected in parallel

$$E_{eq} = \frac{\frac{E}{r_1}+\frac{E}{r_2}}{\frac{1}{r_1}+\frac{1}{r_2}} = E; \quad r_{eq} = \frac{r_1 r_2}{r_1+r_2} < r_1 \text{ or } r_2$$

140. (25)

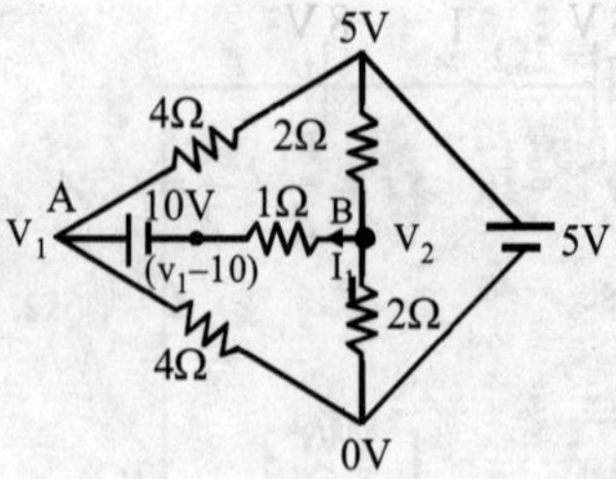

At B

$$\frac{v_2-5}{2}+\frac{v_2-0}{2}+\frac{v_2-v_1+10}{1}=0$$

$\Rightarrow \quad v_2-5+v_2+2v_2-2v_1+20=0$

$\Rightarrow \quad 4v_2-2v_1+15=0$ ...(i)

At A

$$\Rightarrow \quad \frac{v_1-5}{4}+\frac{v_1-0}{4}+\frac{v_1-10v_2}{1}=0$$

$\Rightarrow \quad v_1-5+v_1+4v_1-40-4v_2=0$

$\Rightarrow \quad 6v_1-4v_2-45=0$ ...(ii)

From eq (i) and (ii),

$4v_1-30=0$

$$\Rightarrow \quad v_1=\frac{15}{2} \;\&\; 4v_2-15+15=0 \Rightarrow v_2=0$$

$$\therefore \quad I_1=\frac{v_2-v_1+10}{1}$$

$$=\frac{0-7.5+10}{1}=2.5\text{A}=\frac{n}{10}\text{A}$$

$\therefore \quad n=25$

141. (a) The circuit can be redrawn as

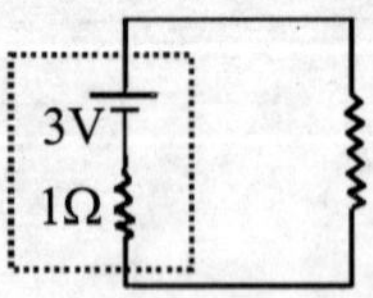

$$\text{Current, } i=\frac{E}{R_{eq}}=\frac{3}{1+2}=1\text{A}$$

Terminal potential difference, $V=E-Ir$

$=3-1\times1=2\text{V}$

142. (4) Electric field, $E=\dfrac{Q}{A\varepsilon_0}$

$$\Rightarrow E \propto Q \Rightarrow \frac{E_2}{E_1}=\frac{Q_2}{Q_1}$$

$$\Rightarrow \frac{E_0/3}{E_0}=\frac{Q_2}{Q_0} \Rightarrow Q_2=\frac{Q_0}{3}$$

Now, $Q=Q_0e^{-\frac{t}{RC}}$

$$\text{Also } \frac{Q_0}{3}=Q_0e^{-\frac{t}{RC}} \Rightarrow ln\left(\frac{1}{3}\right)=\frac{-t}{RC}$$

$$\Rightarrow R=\frac{t}{C\ln 3}=4\Omega$$

143. (c) All the cells are connected in series,

$$I=\frac{E_{eq}}{r_{eq}}=\frac{8\times5}{8\times0.2}=25\text{A}$$

Potential, $V=E-ir=5-0.2\times25=0\text{ V}$

144. (d)

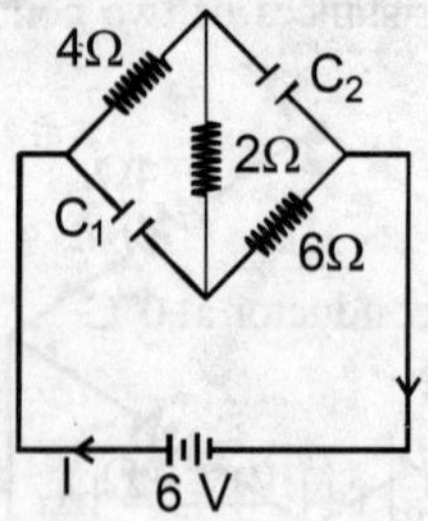

Equivalent Resistance in steady state $R_{eq}=4+2+6=12\Omega$

$$\text{Current, } I=\frac{V}{R_{eq}}=\frac{4\times3}{6\times4}=\frac{6}{12}=0.5\text{A}$$

Potential across C_1 is $V_1=(4+2)I=6\times0.5=3\text{V}$

Potential across C_2 is $V_2=(6+2)I=8\times0.5=4\text{V}$

$$\frac{q_1}{q_2}=\frac{C_1V_1}{C_2V_2}=\frac{1}{2}$$

145. (60)

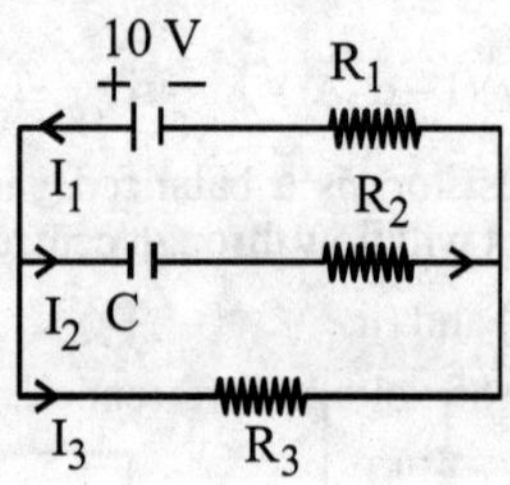

In steady state current in capacitor is zero i.e., $I_2=0$

$$I_1=I_3=\frac{V}{R_1+R_3}=\frac{4\times3}{6\times4}=\frac{10}{4+6}=1\text{A}$$

$V_c=I_3R_3=1\times6=6\text{ V}$

Charge on capacitor, $Q=CV_c=10\times6=60\ \mu\text{C}$

146. (6)

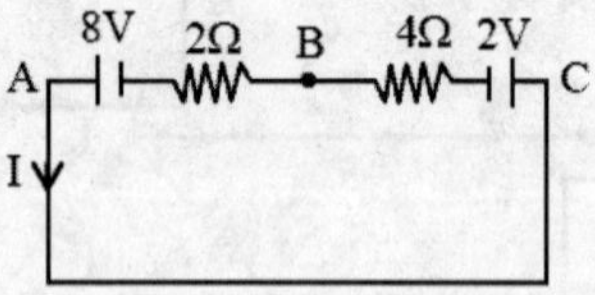

$$I=\frac{8-2}{2+4}=\frac{6}{6}=1A$$

Applying KVL from C to B

$V_C-2-4\times1=V_B$

$V_C-V_B=6\text{ V}$

147. (400) Considering circuit diagram

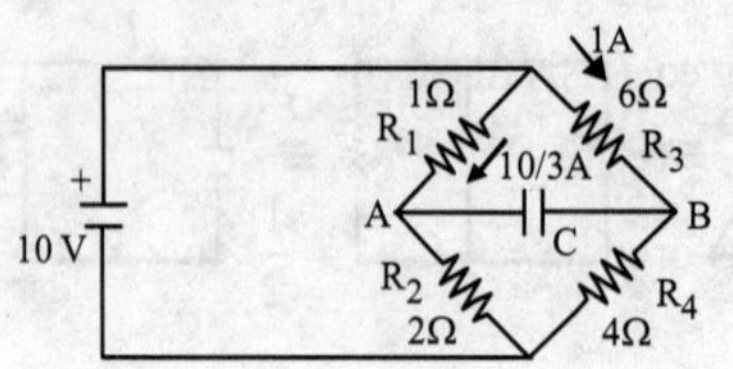

$$V_A + \frac{10}{3}(1) - 6(1) = V_B$$

$$\Rightarrow V_A - V_B = 6 - \frac{10}{3} = \frac{8}{3} V$$

$$Q = C(V_A - V_B) = 150 \times \frac{8}{3} = 400 \mu C$$

148. (3)

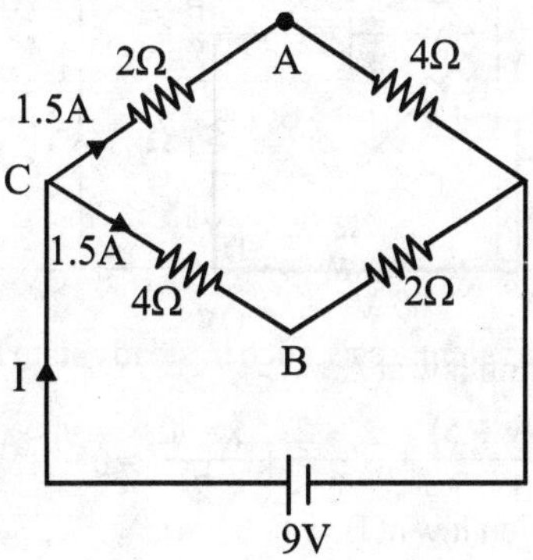

$$R_{eq} = \frac{6 \times 6}{6+6} = 3\Omega$$

In the circuit, $I = \frac{9}{3} = 3A$ ($\because V = IR_{eq}$)

$V_C - V_A = 2 \times 1.5 = 3V$ .....(i)

$V_C - V_B = 4 \times 1.5 = 6V$ .....(ii)

from equation (i) and (ii)

The magnitude of voltage difference between the point A and B,

$V_A - V_B = 6 - 3 = 3\,V$

149. (75) For the given circuit, current

$$I_1 = \frac{V}{R_{eq}} = \frac{12}{3+9} = 1A$$

$$I_2 = \frac{V}{R'_{eq}} = \frac{12}{4+2} = 2A$$

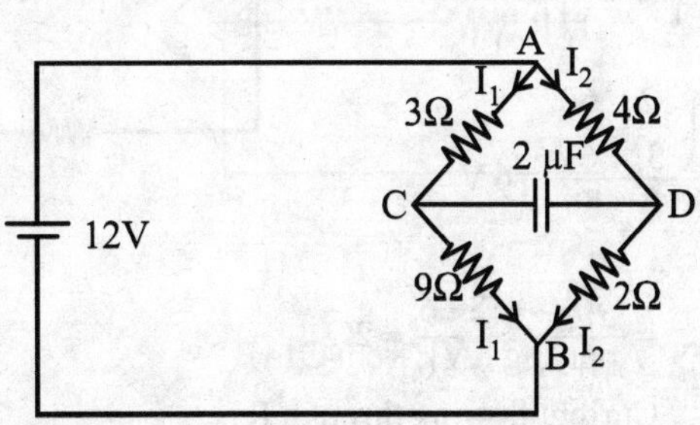

Now using Kirchhoff rule,

$V_A - V_C = 3I_1 = 3V$ ...(i)

$V_A - V_D = 2 \times 4 = 8V$ ... (ii)

Subtracting eq. (i) from eq. (ii)

$V_C - V_D = 5V \Rightarrow V = 5V$

$\therefore$ Energy stored $U = \frac{1}{2}CV^2 = \frac{1}{2} \times 6 \times 5^2 = 75 \mu J$

150. (1)

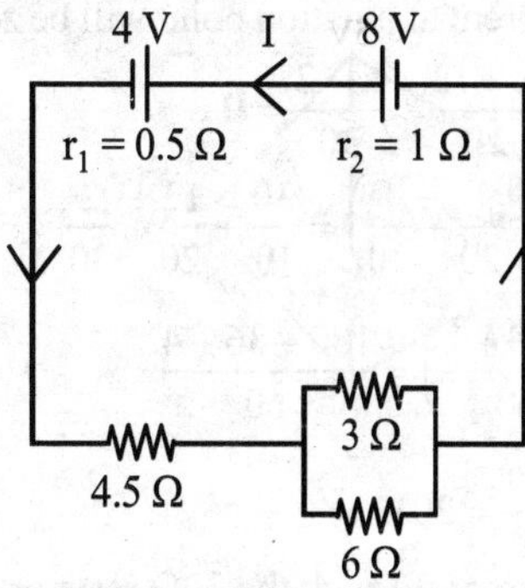

Effective emf,

$E_2 - E_1 = 8 - 4 = 4V$

$$R_{eq} = 0.5\,\Omega + 1\,\Omega + 4.5\,\Omega + \frac{6 \times 3}{3+6}\,\Omega = 8\,\Omega$$

Current, $I = \frac{4}{8} = 0.5A$

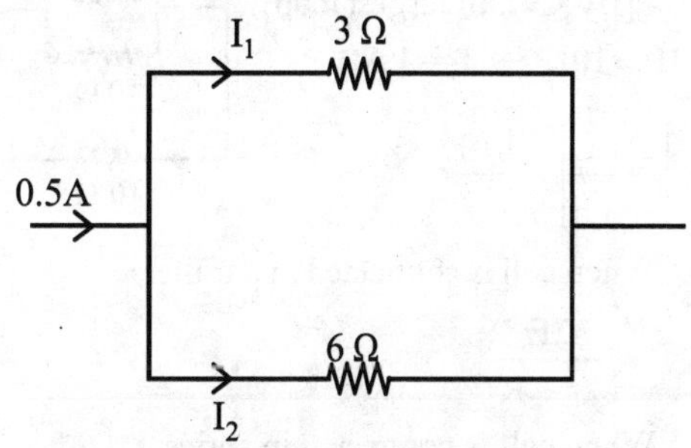

Current through 3Ω resistance,

$$\therefore I_1 = \left(\frac{6}{3+6}\right) \times 0.5 = \frac{2}{3} \times 0.5 = \frac{1}{3} A$$

$$\Rightarrow I_1 = \frac{x}{3} = \frac{1}{3} \therefore x = 1$$

151. (5) Reading of voltmeter, $V = 1.5\,V$

Net emf, $E = 1.5\,V + 1.5\,V = 3\,V$

External resistance, $R = 10\Omega$

Let r be the internal resistance of each cell.

Using, $V = IR$

$V = I \times 10$

$$\Rightarrow 1.5 = \left(\frac{3}{10+2r}\right) \times 10 \Rightarrow r = 5\Omega$$

152. (b) Resistance of the coil of galvanometer, $R_G = 2\Omega$.

Capacitor behave as open circuit.

At steady state, current in the circuit is

$$i = \frac{4}{6+2+8} = \frac{1}{4} A$$

Voltage across C_1 is

$$V_1 = V_{AC} = i(R_{AB} + R_G) = i(6+2) = \frac{1}{4} \times 8 = 2V$$

Voltage across C_2 is

$$V_2 = V_{BD} = i(R_G + R_{CD}) = i(2+8) = \frac{1}{4} \times 10 = 2.5V$$

$$\Rightarrow \frac{V_1}{V_2} = \frac{2}{2.5} = \frac{4}{5}$$

153. (a) Sum of current at junction point will be zero:

$$\frac{x-30}{10}+\frac{x-12}{20}+\frac{x-2}{30}=0$$

$$\Rightarrow x\left(\frac{1}{10}+\frac{1}{20}+\frac{1}{30}\right)=\frac{30}{10}+\frac{12}{20}+\frac{2}{30}$$

$$\Rightarrow x\left(\frac{6+3+2}{60}\right)=\frac{180+36+4}{60}$$

$$\Rightarrow x=\frac{220}{11}=20V$$

Hence, current through the 20Ω resistor will be

$$I=\frac{x-12}{20}=\frac{20-12}{20}=\frac{2}{5}=0.4A$$

154. (2) We have $10I_1 = 10I_2 = 10V$

$\Rightarrow I_1 = I_2 = 1A$

and, $-10+20-10I_3=0$

[By KVL in largest loop]

$\Rightarrow 10 = 10I_3 \Rightarrow I_3 = 1A$

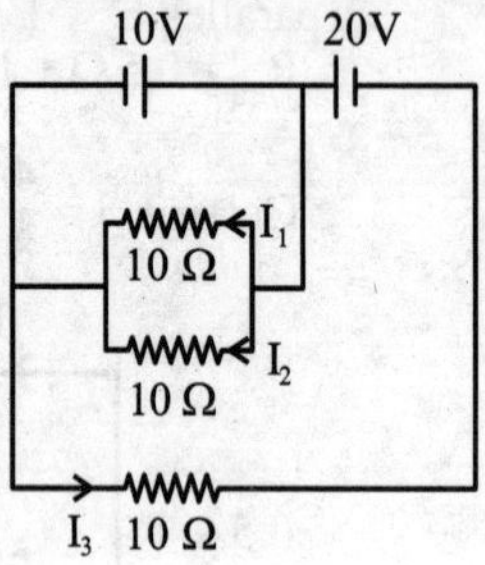

So, $\left|\frac{I_1+I_3}{I_2}\right|=\frac{1+1}{1}=2$

155. (5) When cell is connected in parallel

$$i=\frac{2E}{2r+R}$$

When cell is connected in series

$$i=\frac{E}{\frac{r}{2}+R}=\frac{2E}{r+2R}\ \text{So,}\ \frac{2E}{2r+R}=\frac{2E}{r+2R}$$

$\Rightarrow 2r+R=r+2R \Rightarrow r=R \Rightarrow r=5Ω$

156. (1) As capacitor behaves like open switch to DC. So, circuit will look like as shown below

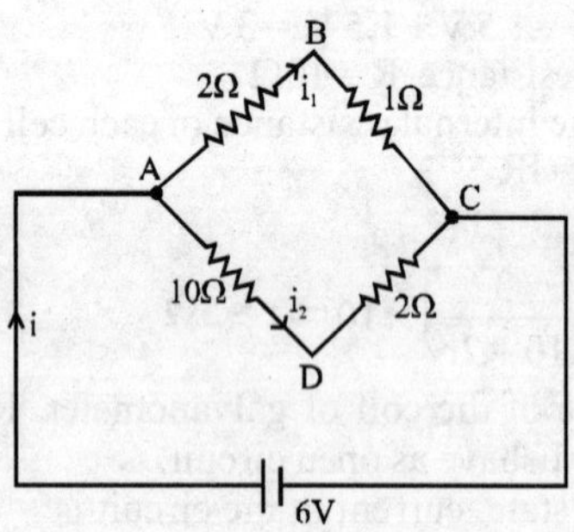

Here, $R_{eq}=\frac{3\times12}{3+12}=\frac{36}{15}=\frac{12}{5}Ω$

So, $i=\frac{6}{\frac{12}{5}}=\frac{5}{2}=2.5A$

and, $i_1=2.5\times\frac{12}{15}=2A$; $i_2=2.5\times\frac{3}{15}=0.5A$

Applying KVL between B and D, we get

$V_B-V_D=-2i_1+10i_2=-2\times2+10\times0.5=1A$

157. (1.5)

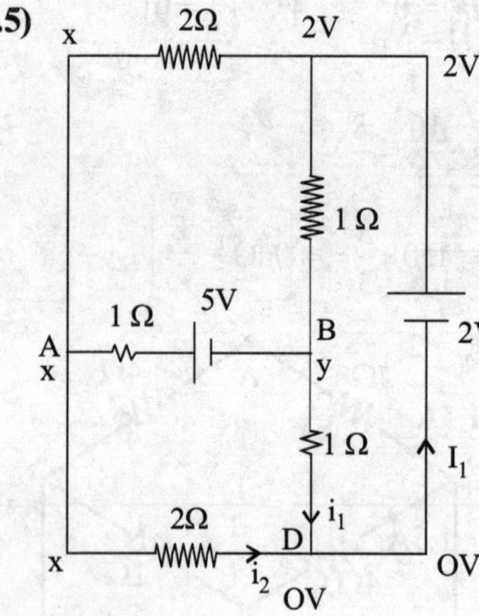

Junction law at A,

$$\frac{x-(y+5)}{1}+\frac{x-2}{2}+\frac{x-0}{2}=0 \quad ...(1)$$

Junction law at B,

$$\frac{y+5-x}{1}+\frac{y-0}{1}+\frac{y-2}{1}=0 \quad ...(2)$$

On solving equation (1) and Equation (2)

$x = 3$ & $y = 0$

At D junction

$I_1 = i_1 + i_2$

$$I_1=\frac{y-0}{1}+\frac{x-0}{2}=\frac{0-0}{1}+\frac{3-0}{2}$$

$I_1 = 1.5A$

158. (1)

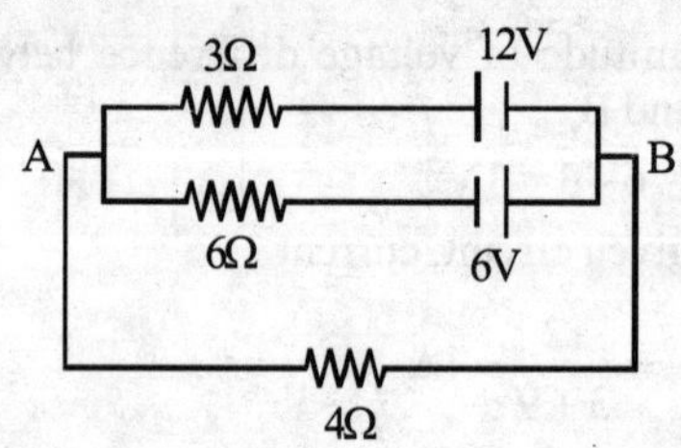

$$E_{eq}=\frac{\left(\frac{E_1}{r_1}+\frac{E_2}{r_2}\right)}{1/R_{eq}}$$

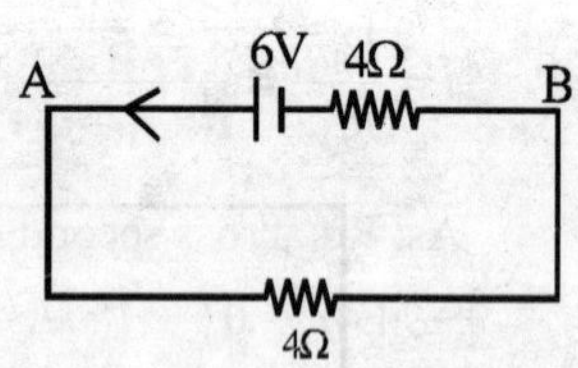

$$=\frac{\left(\frac{12}{3}-\frac{6}{6}\right)}{\frac{1}{3}+\frac{1}{6}}=6V$$

$$R_{eq}=\frac{r_1r_2}{r_1+r_2}=\frac{3\times6}{3+6}=2Ω$$

∴ Current flowing through R is

$$I=\frac{E_{eq}}{R_{eq}+R}=\frac{6}{2+4}=1A$$

159. (a)

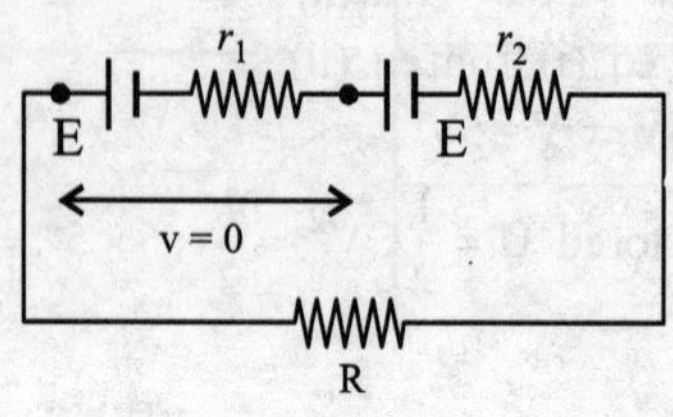

We have, $E - ir_1 = 0$ $\quad [\because V_{r_1} = 0]$

$\Rightarrow E = ir_1$

Now, $i = \frac{2E}{r_1 + r_2 + R}$

$\Rightarrow \frac{E}{r_1} = \frac{2E}{r_1 + r_2 + R}$ $\quad \left[\because i = \frac{E}{r_1}\right]$

$\Rightarrow r_1 + r_2 + R = 2r_1 \Rightarrow \quad R = r_1 - r_2$

160. (4) $E_{eq} = \frac{\frac{E_1}{r_1} + \frac{E_2}{r_2} + \frac{E_3}{r_3}}{\frac{1}{r_1} + \frac{1}{r_2} + \frac{1}{r_3}} = \frac{\frac{2}{1} + \frac{4}{1} + \frac{4}{1}}{\frac{1}{1} + \frac{1}{1} + \frac{1}{1}} = \frac{12}{3} = 4V$

$r_{eq} = \frac{1}{3} k\Omega$

So, $V_o = E_{eq} - ir_{eq} = \frac{12}{3} - 0 \times \frac{1}{3} = 4$ V

161. (a) In series

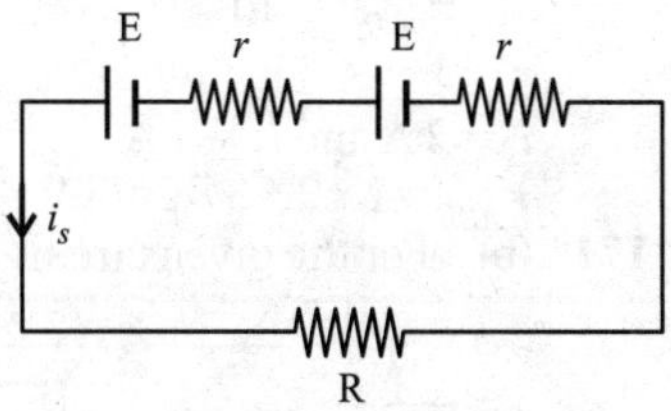

$i_s = \frac{2E}{2r + 2}$

In parallel

$i_p = \frac{E}{\frac{r}{2} + 2} = \frac{2E}{r + 4}$

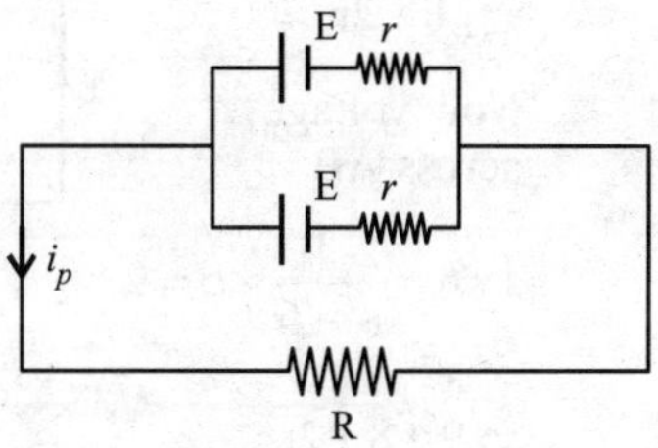

as, $i_s = i_p$

$\Rightarrow \frac{2E}{2r + 2} = \frac{2E}{r + 4}$

$\Rightarrow r = 2\Omega$

162. (a) We have,

$i = \frac{E_{net}}{R_{net}} = \frac{2E}{R + r_1 + r_2}$

As. P.d across second cell = 0

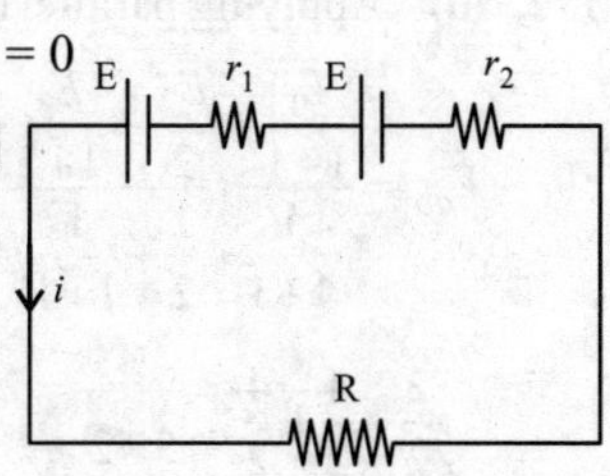

$\Rightarrow E - ir_2 = 0 \; [\because V = E - iR]$

$\Rightarrow E - \frac{2Er_2}{R + r_1 + r_2} = 0$

$\Rightarrow R + r_1 + r_2 - 2r_3 = 0$

$\Rightarrow R = r_2 - r_1$

163. (c)

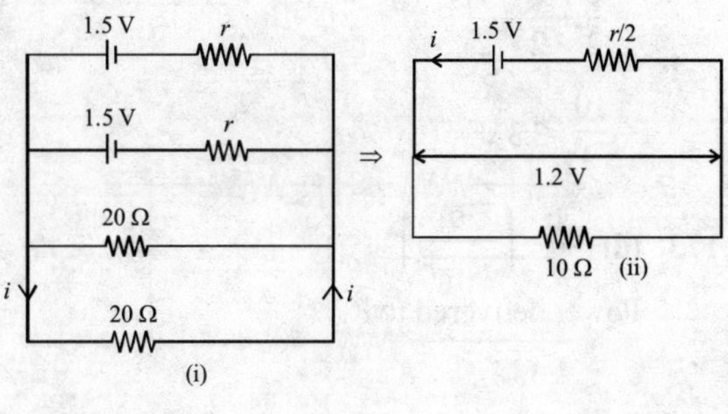

Here, $i = \frac{1.2}{10}$ A 0.12 A

From figure (ii), $1.2 = 1.5 - i\frac{r}{2}$

$1.2 = 1.5 - (0.12)\frac{r}{2} \Rightarrow 0.06r = 0.3$

$\Rightarrow r = \frac{0.30}{0.06} = 5\Omega$

164. (a) $i_r = \frac{25}{5 + R}$

in parallel, $E_{eq} = E = 5V$

$r_{eq} = \frac{r}{5} = \frac{1}{5}\Omega$

$i_2 = \frac{5}{R + \frac{1}{5}}$

putting $i_1 = i_2$,

we get, $4R = 4 \Rightarrow R = 1\Omega$.

165. (15) Here, $I = \frac{E}{R + r}$

$\therefore$ Terminal voltage $V = IR = \frac{ER}{R + r}$

When potential difference, V = 1.25V and $R_L = 5\Omega$, then

$1.25 = \frac{E(5)}{5 + r}$ ...(i)

When potential difference V = 1 V and $R_L = 2\Omega$, then

$1 = \frac{E(2)}{2 + r}$ ...(ii)

From eq. (i) and (ii)

we get $E = \frac{3}{2} = \frac{x}{10}$ $\quad \therefore$ value of x = 15

166. (c)

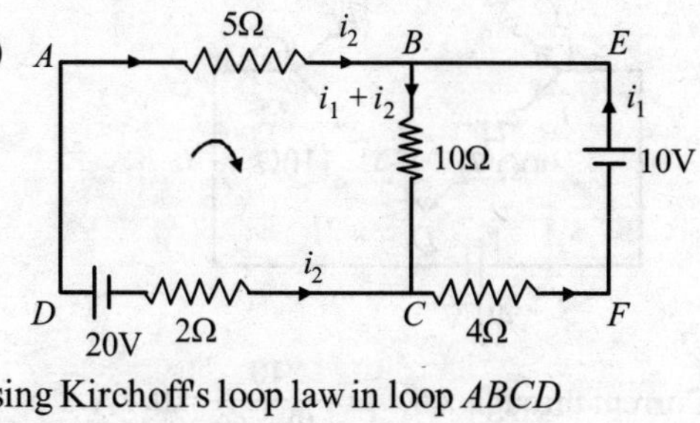

Using Kirchoff's loop law in loop *ABCD*

$-5i_2 - 10(i_1 + i_2) - 2i_2 + 20 = 0$

$\Rightarrow -10i_1 - 17i_2 + 20 = 0$ ...(i)

Using Kirchoff's loop law in loop *BEFC*

$\Rightarrow -10 + 4i_1 + 10(i_1 + i_2) = 0$

$\Rightarrow 14i_1 + 10i_2 - 10 = 0$ ...(ii)

Multiplying equation (i) by 10, we have

$(10i_1 + 17i_2 = 20) \times 10$

$\Rightarrow 100i_1 - 170i_2 = 200$ ...(iii)

Multiplying equation (ii) by 17, we have

$(14i_1 + 10i_2 = 10) \times 17$

$\Rightarrow 238i_1 - 170i_2 = 170$ ...(iv)

On solving equations (iii) and (iv), we get

$-138i_1 = 30 \Rightarrow i_1 = -\frac{30}{138} = -0.217$ A

i_1 is negative it means current flows from positive to negative terminal in the battery.

167. (d)

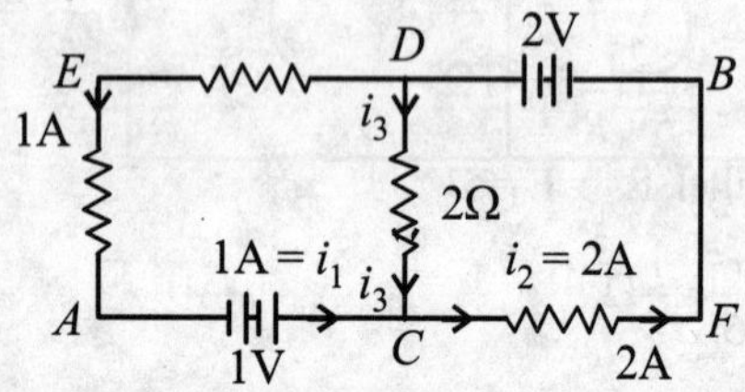

Let us assume the potential at $A = V_A = 0$

Using Kirchoff's junction rule at C, we get

$i_1 + i_3 = i_2$

$1\text{A} + i_3 = 2\text{ A} \Rightarrow i_3 = 1A$

Now using Kirchoff's loop law along $ACDB$

$V_A + 1 + i_3(2) - 2 = V_B$

$\Rightarrow V_B - V_A = 1 + 1 \times 2 - 2 = 1$ volt

168. (c) The equivalent circuit can be drawn as

As $V_B = V_D$.

So arm BD will be removed

Voltage across AC = 8 V

Resistance $R_{AC} = 4 + 4$
$= 8\,\Omega$

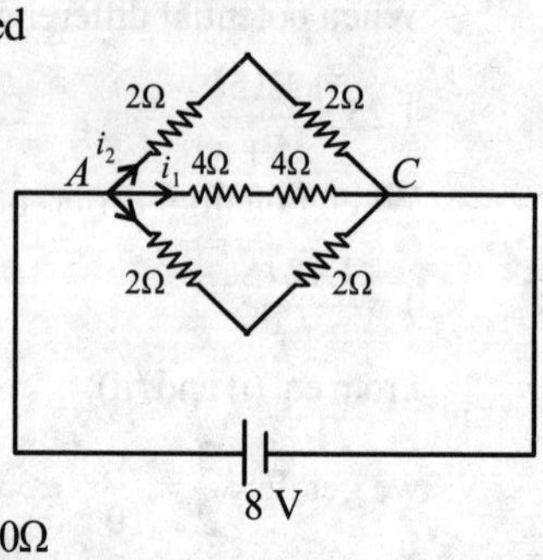

$i_1 = \frac{V}{R_{AC}} = \frac{8}{4+4}$

$= 1$A

169. (2)

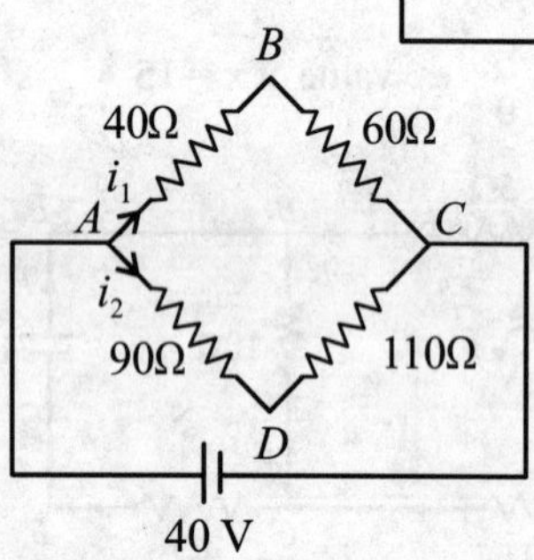

Current through AB, $i_1 = \frac{40}{40+60} = 0.4$ A

Current through AD, $i_2 = \frac{40}{90+110} = \frac{1}{5}$ A

Using KVL in BAD loop

$V_B + i_1(40) - i_2(90) = V_D$

$\Rightarrow V_B - V_D = \frac{1}{5}(90) - \frac{4}{10}(40)$

$\Rightarrow V_B - V_D = 18 - 16 = 2$ V

170. (8)

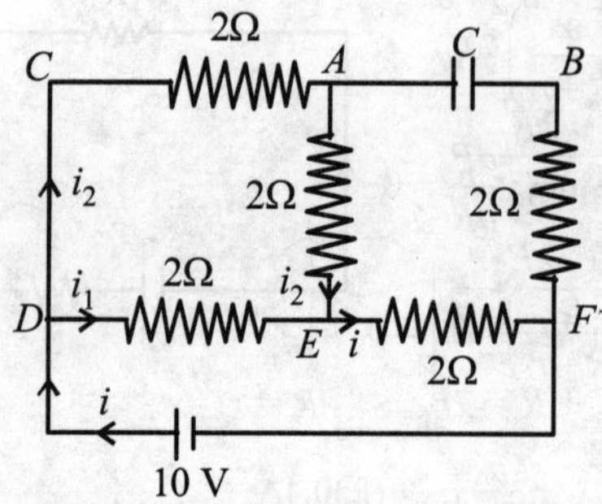

As capacitor is fully charged no current will flow through it. We have the current distribution as shown in the figure.

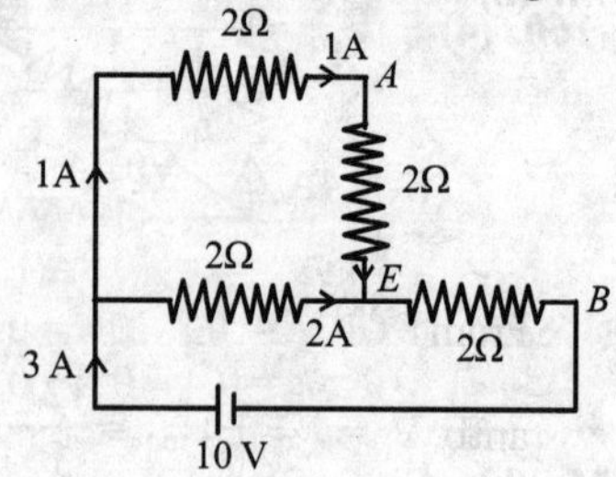

Equivalent resistance,

$$R_{eq} = \left(\frac{4\times 2}{4+2}\right) + 2$$

Net current,

$$i = \frac{10}{\frac{4}{3}+2} = \frac{10\times 3}{10} = 3A$$

$i_1 = 2$ A and $i_2 = 1$ A

$V_{AEB} = 1 \times 2 + 3 \times 2 = 8$ V

171. (b) For the given circuit

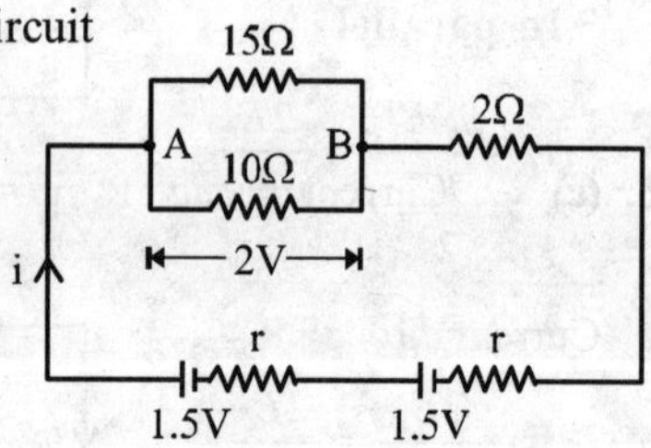

$$i = \frac{3}{8+2r}$$

Now voltage across AB

$$i \times 6 = \frac{3}{8+2r} \times 6 = 2$$

$\Rightarrow 9 = 8 + 2r$

$\Rightarrow r = \frac{1}{2}\Omega$

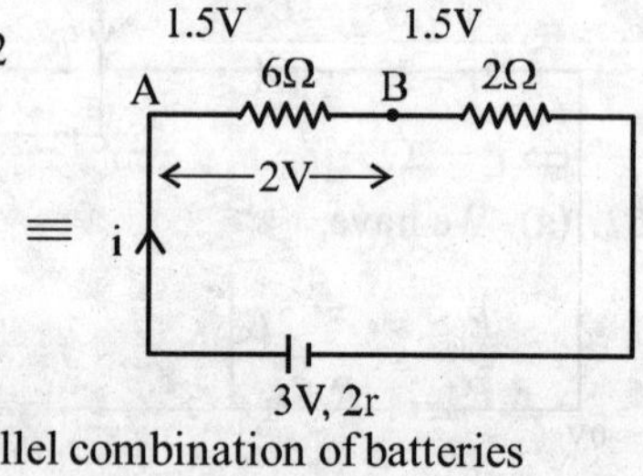

172. (d) Applying parallel combination of batteries

$$V_{ab} = \frac{\frac{E_1}{1+1} + \frac{E_2}{2} + \frac{E_3}{1+1}}{\frac{1}{1+1} + \frac{1}{2} + \frac{1}{1+1}}$$

$$= \frac{\frac{2}{2} + \frac{4}{2} + \frac{4}{2}}{\frac{1}{2} + \frac{1}{2} + \frac{1}{2}} = \frac{5\times 2}{3}$$

$$= \frac{10}{3} = 3.3\text{ V}$$

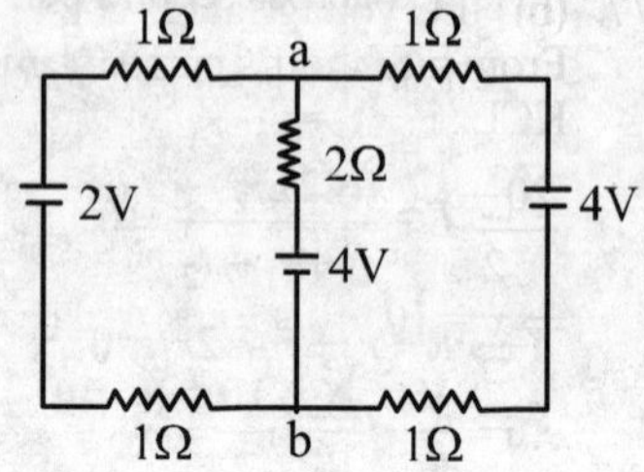

173. (d) $i = \left(\frac{\varepsilon}{R+r}\right)$

Power delivered to R.

$P = i^2R = \left(\frac{\varepsilon}{R+r}\right)^2 R$

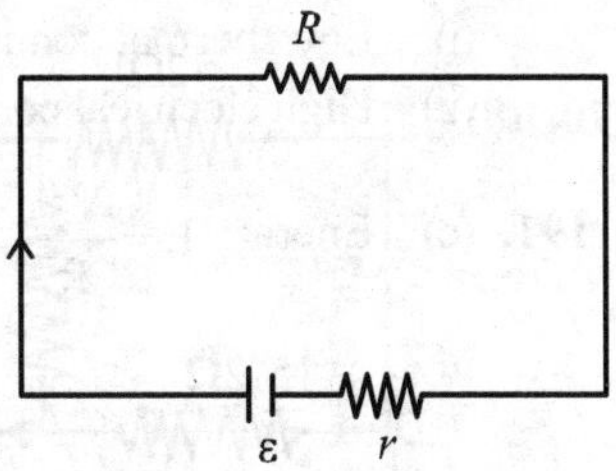

P to be maximum, $\frac{dP}{dR} = 0$

or $\frac{d}{dR}\left[\left(\frac{\varepsilon}{R+r}\right)^2 R\right] = 0$

or $R = r$

174. (b)

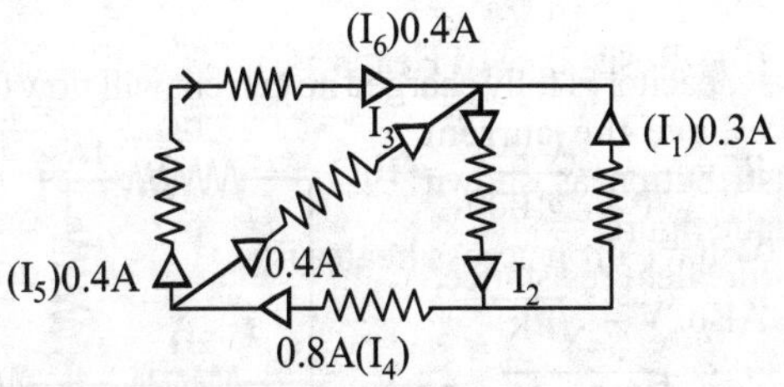

From KCL, $I_3 = 0.8 - 0.4 = 0.4$ A

$I_2 = 0.4 + 0.4 + 0.3 = 1.1$ A

and $I_6 = 0.4$ A

175. (b) Given, $E_1 = 1V, E_2 = 2V, E_3 = 3V, r_1 = 1\Omega$, $r_2 = 1\Omega$ and $r_3 = 1\Omega$. Here, $V_{AB} = V_{CD}$ because $i = 0$

$$V_{AB} = V_{CD} = \frac{\frac{E_1}{r_1} + \frac{E_2}{r_2} + \frac{E_3}{r_3}}{\frac{1}{r_1} + \frac{1}{r_2} + \frac{1}{r_3}} = \frac{\frac{1}{1} + \frac{2}{1} + \frac{3}{1}}{\frac{1}{1} + \frac{1}{1} + \frac{1}{1}} = \frac{6}{3} = 2V$$

176. (c) $\therefore$ Current through R_1, $i_1 = \frac{10-0}{20} = 0.5$ A

Current through R_2, $i_2 = \frac{0-0}{20} = 0$

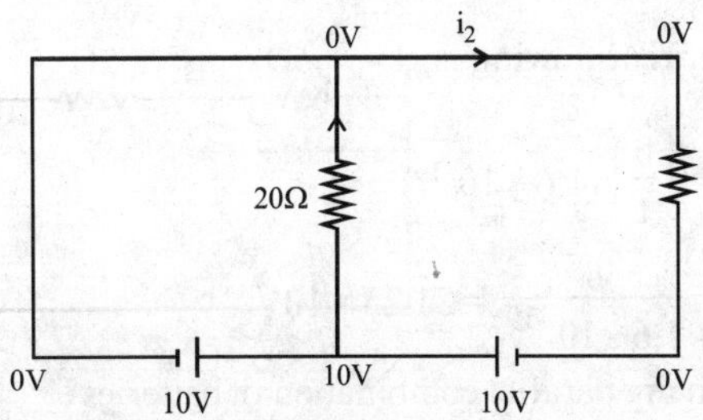

177. (b) Let voltage at C = xV

From kirchhoff's current law,

KCL : $i_1 + i_2 = i$

$\frac{20-x}{2} + \frac{10-x}{4} = \frac{x-0}{2}$

$\Rightarrow x = 10$

$\therefore i = \frac{V}{R} = \frac{X}{R} = \frac{10}{2} = 5$A

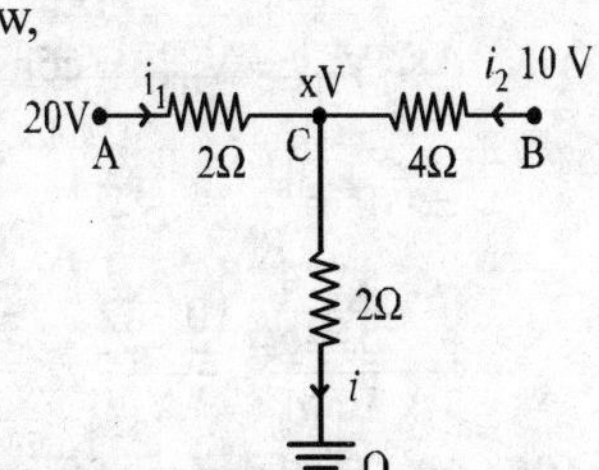

178. (b) Using Kirchhoff's law at *P* we get

$\frac{V-12}{1} + \frac{V-13}{2} + \frac{V-0}{10} = 0$

[Let potential at P, Q, U = 0 and at R = V]

$\Rightarrow \frac{V}{1} + \frac{V}{2} + \frac{V}{10} = \frac{12}{1} + \frac{13}{2} + \frac{0}{10}$

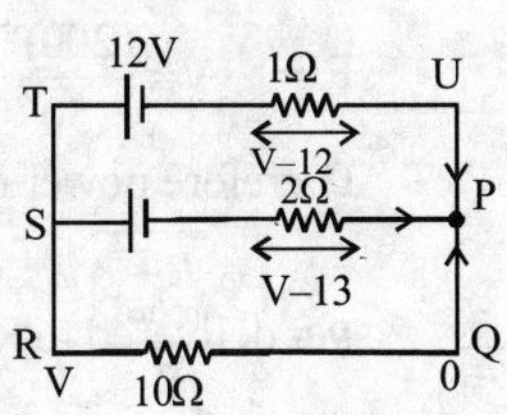

$\Rightarrow \frac{10+5+1}{10}V = \frac{24+13}{2} \Rightarrow V\left(\frac{16}{10}\right) = \frac{37}{2}$

$\Rightarrow V = \frac{37 \times 10}{16 \times 2} = \frac{370}{32} = 11.56$ volt

179. (a) From KVL, $-6 + 3I_1 + 1(I_1 - I_2) = 0$

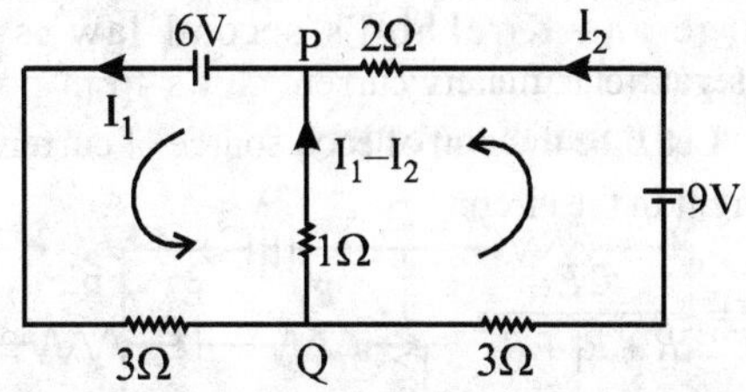

$6 = 3I_1 + I_1 - I_2$; $4I_1 - I_2 = 6$ ...(i)

$-9 + 2I_2 - (I_1 - I_2) + 3I_2 = 0$

$-I_1 + 6I_2 = 9$ ...(ii)

On solving (i) and (ii)

$I_1 = 0.13$A

Direction Q to P, since $I_1 > I_2$.

180. (c) As 200 V battery is charging

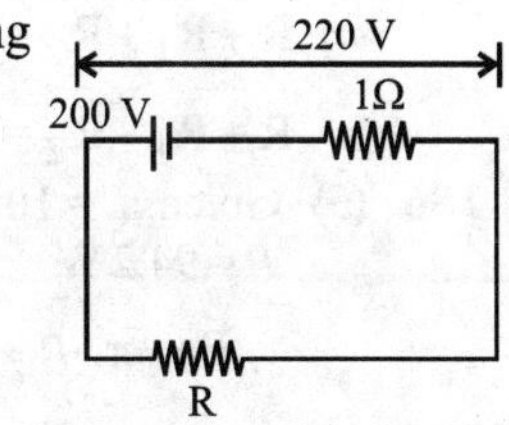

So, $220 = 200 + i.1$

$i = 20$ A

So, $20 \times R = 220$ V

$\Rightarrow R = 11\,\Omega$

181. (c) $\frac{100}{R+r} = \frac{90}{R} \Rightarrow \frac{R+r}{R} = \frac{10}{9} \Rightarrow 1 + \frac{0.5}{R} = \frac{10}{9}$

$\Rightarrow \frac{0.5}{R} = \frac{1}{9} \therefore R = 4.5\,\Omega$

182. (c) Applying Kirchoff's second law in *AB P_2P_1A*, we get

$-2i + 5 - 10\,i_1 = 0$

$2i + 10i_1 = 5$ .....(i)

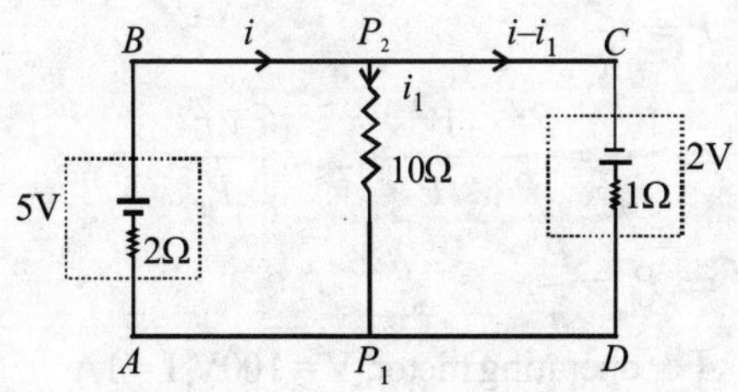

Again applying Kirchoff's second law in $P_2\,CDP_1P_2$ we get,

$10\,i_1 + 2 - i + i_1 = 0$

$2i - 22i_1 = 4$ ....(ii)

From (i) and (ii)

$32i_1 = 1 \Rightarrow i_1 = \frac{1}{32}$ A from P_2 to P_1

183. (a) Energy in capacitor $= \frac{1}{2}CV^2$

Work done by battery $= QV = CV^2$

where C = Capacitance of capacitor
V = Potential difference,

$$\text{Required ratio} = \frac{\frac{1}{2}CV^2}{CV^2} = \frac{1}{2}$$

184. **(d)** Kirchhoff's first law is based on conservation of charge and Kirchhoff's second law is based on conservation of energy.

185. **(a)** Let E be the emf of each source of current

Current in the circuit

$$I = \frac{2E}{R + R_1 + R_2}$$

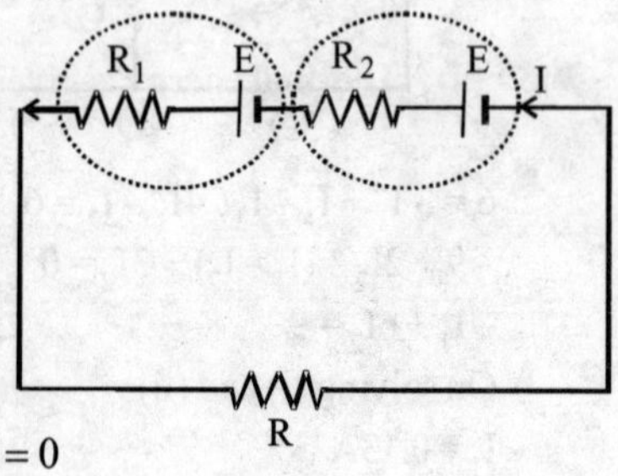

Potential difference across cell having internal resistance R_2

$V = E - iR_2 = 0$

$$E - \frac{2E}{R + R_1 + R_2} \cdot R_2 = 0$$

$\Rightarrow R + R_1 + R_2 - 2R_2 = 0$

$\Rightarrow R + R_1 - R_2 = 0 \Rightarrow R = R_2 - R_1$

186. **(5)** Given, $L = 10$ cm, $d = 0.5$ mm, $T = 1727°C = 2000$ K, $P = 94.2$ W

$\therefore$ Power, $P = \varepsilon\sigma A T^4 = \varepsilon\sigma(\pi dL)T^4$

$$\Rightarrow 94.2 = \varepsilon \times (6\times10^{-8})(3.14)(0.5)(10^{-3}) \times (10\times10^{-2})(2000)^4$$

$$\therefore \varepsilon = \frac{94.2}{(94.2)(16)} = \frac{5}{8} = \frac{x}{8} \text{ (given)}$$

$\therefore x = 5.$

187. **(c)** When the resistances are connected in series

$R_s = R_1 + R_2 + R_3 + ... + R_n$

$$P = \frac{V^2}{R}$$

$$\Rightarrow \frac{V^2}{P_s} = \frac{V^2}{P} + \frac{V^2}{P} + ... + \frac{V^2}{P_n}$$

$$\Rightarrow P_s = \frac{P}{n}.$$

188. **(c)** For operating motor, V = 100V, I = 1A

$P_{input} = VI = 100 \times 1 = 100$ W

$$\therefore \quad \text{Efficiency, } \eta = \frac{P_{out}}{P_{input}} = 0.916$$

$P_{out} = 91.6$ W

Loss in power = 100 – 91.6 = 8.4 J/s = 2 cal/s

189. **(c)** Energy stored in mobile phone battery when it is fully charged,

$= vq = 4.2 \times 5800 \times 3600 \times 10^{-3}$ J = 87.696 kJ

190. **(c)** The conversion of heat energy into electrical energy is given by See-back effect, and according to it, the material should have

(i) Low thermal conductivity
(ii) High electrical conductivity

191. **(c)** Energy, $E = \frac{V^2}{R}t = \frac{V^2 tA}{\rho l} \Rightarrow l \propto t$

$$= \frac{t_2}{t_1} = \frac{15}{20} = \frac{\ell_2}{\ell_1} \therefore \ell_2 = \frac{3}{4}\ell_1$$

Thus length should be decreased by $\frac{3}{4}$ times.

192. **(5)** Resistance of Heater is,

$$R_h = \frac{V^2}{P} = \frac{(100)^2}{1000} = 10\Omega$$

Voltage drop across heater is,

Also, $V = \sqrt{PR}$

$V = \sqrt{62.5 \times 10} = 25$ V

Current through 10Ω resistance

$$i_1 = \frac{75}{10} = 7.5\text{A},$$

Current through heater is

$$i_h = \frac{25}{10} = 2.5\text{ A}.$$

Using Kirchoff's law

$i_R = i_1 - i_h = 5$

$$R = \frac{V}{i_R} = \frac{25}{5} = 5\Omega$$

193. **(a)** Power (P) = V.I.

$\Rightarrow 110 = (220)(I) \Rightarrow I = 0.5$ A

Now, current flowing is, $I = \frac{n.e}{t}$

$$\Rightarrow 0.5 = \left(\frac{n}{t}\right)(1.6\times10^{-19})$$

$$\Rightarrow \frac{n}{t} = \frac{0.5}{1.6\times10^{-19}} = 31.25\times10^{17}$$

194. **(b)** Power, $P = \frac{V^2}{R}$

As, $V_{5\Omega} = V_{10\Omega}$ = constant

$$\Rightarrow P \propto \frac{1}{R}$$

$$\therefore \quad \frac{P_{5\Omega}}{P_{10\Omega}} = \frac{10}{5} = \frac{2}{1}$$

195. **(a)** Resistance of filament of electric bulb

$$R = \frac{V^2}{P} = \frac{(200)^2}{50} = \frac{40000}{50} = 800\Omega$$

Therefore power dissipation of the bulb,

$$P = \frac{(V_{applied})^2}{R} = \frac{(100)^2}{800} = 12.5 \text{ watt}$$

196. (c) Power,

$P = I^2 R$

Here, I = current, R = Resistance

$P_i = I_i^2 R$

$P_f = (0.8I_i)^2 R = 0.64 I_i^2 R$

$$\%\text{ decrease in power} = \frac{P_i - P_f}{P_i} \times 100 = (1 - 0.64) \times 100$$

$$= 0.36 \times 100 = 36\%$$

197. (3) Given,

Emf of the battery, V = 2V

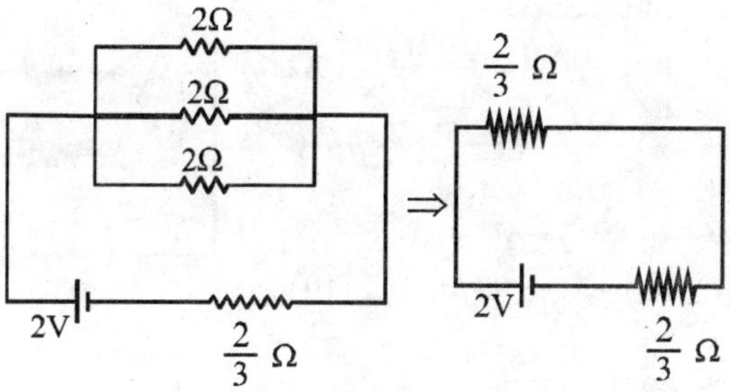

$$\therefore \quad R_{eq} = 2 \times \frac{2}{3} = \frac{4}{3}\Omega$$

$$\text{Power, } P = \frac{V^2}{R_{eq}} = \frac{4}{4/3} = 3\text{ W}$$

198. (d) $\frac{v^2}{R} = W$ ...(i)

$$\frac{v^2}{\frac{1}{2}\left(\frac{R}{2}\right)} = W' \quad \text{...(ii)}$$

From (i) & (ii), we get

$W' = 4W$

199. (d) Power $P = I^2 R$ or, $P \propto R$ [$\because$ I = Constant]

For different combinations of resistances equivalent resistance

$$R_A = \frac{3R}{2}, R_B = \frac{2R}{3}, R_C = \frac{R}{3}, R_D = 3R$$

Therefore increasing order of power dissipation is $P_C < P_B < P_A < P_D$

200. (16) Power $P = VI = I^2 R = \frac{V^2}{R}$ and $R = \frac{\rho l}{A}$

If wire is cut in two equal half then $R = \frac{R}{2}$

$$\text{Initial Power } P_1 = \frac{V_0^2}{R}$$

$$\text{Final Power } P_2 = \frac{V_0^2}{R'} \times 2 \Rightarrow \frac{V_0^2}{R} \times 4$$

$$\therefore \frac{P_2}{P_1} = 4 = \frac{\sqrt{x}}{1} \quad \therefore x = 16$$

201. (a) In parallel combination

$$\text{Heat produced, } H_1 = \frac{V^2}{\left(\frac{R}{2}\right)} t = \frac{2V^2 t}{R}$$

In series combination

$$\text{Heat produced, } H_2 = \left(\frac{V^2}{2R}\right) t \quad \therefore \frac{H_1}{H_2} = 4$$

202. (b) We have $H = i^2 Rt \Rightarrow H \propto i^2$

$$\Rightarrow \frac{H_1}{H_2} = \left(\frac{4}{16}\right)^2 = \frac{1}{16} \Rightarrow H_2 = 16H_1 = 16H$$

203. (a) Thermal energy is given by

$$H = P \times t = \frac{V^2}{R} \times t$$

Here, voltage V is same.

$$\therefore \quad \frac{H_1}{H_2} = \frac{\frac{V^2 t}{R}}{\frac{V^2 t}{3R}} = 3:1$$

204. (a) They have high resistivity, low-temperature coefficient of resistance ability to attain good resistance values in small size.

205. (14) For Bulb 1

$$R_1 = \frac{V^2}{P} = \frac{220^2}{100} = 484\Omega$$

For Bulb 2

$$R_2 = \frac{V^2}{P} = \frac{220^2}{60} = 484\left(\frac{10}{6}\right)\Omega$$

$$I = \frac{220}{484 + 484 \times \frac{10}{6}} \qquad \left[\because I = \frac{E}{R_1 + R_2}\right]$$

$P_1 = I^2 R_1 = 14.06\text{ W}$

206. (c) **Statement 1** - R = 80 Ω

$R_1 = R_2 = R_3 = R_4 = 20\,\Omega$

$$\text{In parallel } R_{eq} = \frac{20}{4} = 5\,\Omega$$

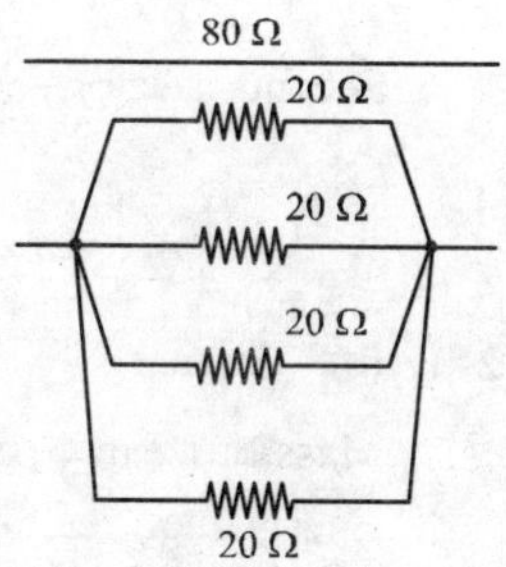

Statement 2 –

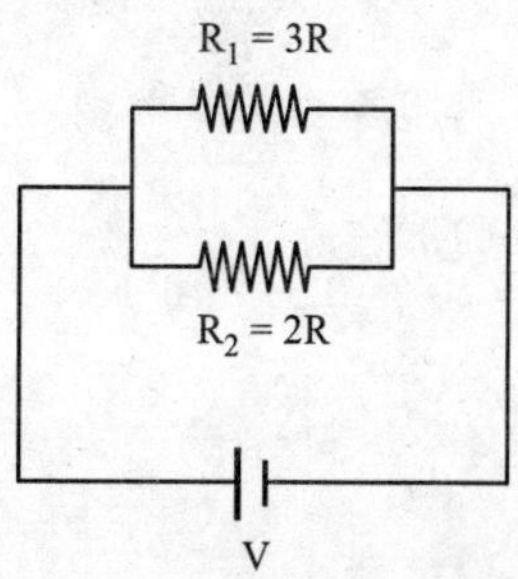

$$P_{th} = \frac{v^2}{R} \Rightarrow P \propto \frac{1}{R}$$

$$\text{So, } \frac{P_1}{P_2} = \left(\frac{R_2}{R_1}\right) = \frac{2}{3} \text{ (where P is power)}$$

207. (15) Thermal energy, $H = \frac{V^2}{R}t$

So, $H = \frac{V^2}{R_1} \times 20$ ...(i)

$H = \frac{V^2}{R_2} \times 60$ ...(ii)

Dividing (ii) by (i), we get

$1 = \frac{R_2}{R_1} \times \frac{1}{3} \Rightarrow R_2 = 3R_1$

Now, as $H = \frac{V^2}{R_{eq}} \times t \Rightarrow \frac{V^2}{R_1} \times 20 = \frac{V^2 \times t}{R_{eq}}$ [from (i)]

$\Rightarrow \frac{20}{R_1} = \frac{t}{\frac{3R_1}{4}}$ $\left[\because R_{eq} = \frac{R_1 . 3R_1}{R_1 + 3R_1}\right]$

$\Rightarrow t = 15$ minutes

208. (b) Given

$\frac{1}{2} . \frac{\Delta U}{\Delta t} = P_{Bulb} \times N$, where $N =$ no. of bulb

$\Rightarrow \frac{9 \times 10^4 \times 10 \times 40}{2 \times 3600} = 100 \times N \left[\because \frac{\Delta U}{\Delta t} = \frac{mgh}{3600}\frac{J}{S}\right]$

$\Rightarrow N = \frac{36 \times 10^6}{72 \times 10^4} \Rightarrow N = \frac{1}{2} \times 100 \Rightarrow N = 50$

209. (975) As, $P = Vi$

$\Rightarrow 5 = 25i \Rightarrow i = 0.2A$

Now, $i = \frac{V_R}{R} \Rightarrow 0.2 = \frac{220 - 25}{R}$

$R = \frac{195}{0.2} = 975\,\Omega$

210. (4) First case $P_1 = \frac{V^2}{R} = \frac{(240)^2}{36}$

Second case $P_2 = \frac{V^2}{R/2} \times 2 = \frac{4V^2}{R} = \frac{4 \times (240)^2}{36}$

$\therefore \frac{P_1}{P_2} = \frac{1}{4} = \frac{1}{x} \Rightarrow x = 4$

211. (b) Given,

Mass of ice $m = \rho A \ell = 10^3 \times 10^{-4} \times 1 = 10^{-1}$ kg

Energy required to melt the ice

$\Delta Q_1 = ms\Delta t + mL = m(S\Delta t + L)$

$= 10^{-1}(2 \times 10^3 \times 10 + 3.33 \times 10^5) = 3.53 \times 10^4$ J

Heat produced in resistance

$\Delta Q_2 = i^2 Rt = \left(\frac{1}{2}\right)^2 (4 \times 10^3)(t)$

According to Question, $\Delta Q_1 = \Delta Q_2$

$\Rightarrow 3.53 \times 10^4 = \left(\frac{1}{2}\right)^2 \times 4 \times 10^3 \times t$

$\Rightarrow t = 35.3$ sec

212. (3840) Rate of energy dissipated

$= \frac{192J}{S} = i^2 R \Rightarrow 192 = 16 \times R \Rightarrow R = 12\Omega$

Energy $= i^2 Rt = (8)^2 \times 12 \times 5 = 3840\,J$

213. (c) Resistances $4\Omega, 8\Omega, 4 + 8\Omega, 2 + 4\Omega$ are in parallel

$\frac{1}{R_{eq}} = \frac{1}{4} + \frac{1}{8} + \frac{1}{12} + \frac{1}{6} = \frac{6+3+2+4}{24} = \frac{15}{24}$

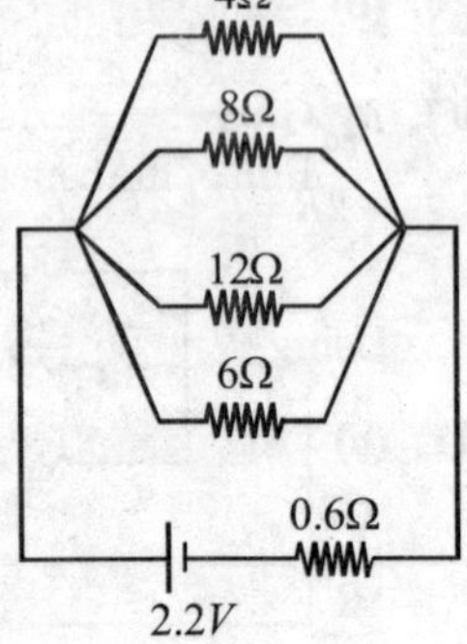

$\Rightarrow R_{eq} = \frac{24}{15} 1.6\Omega$

Total resistance

$R = 1.6 + 0.6 = 2.2\Omega$

$P = \frac{V^2}{R} = \frac{(2.2)^2}{2.2} = 2.2W$

214. (a) $\Delta Q = \Delta U + \Delta W$

$100 = \Delta U + 90 \Rightarrow \Delta U = 10$ watt

So, $t = \frac{25 \times 100}{10} = 250$ sec

215. (a) Given, Power of electric bulb, P = 500W

$\Rightarrow I = \frac{P}{V} = \frac{500}{100} = 5$ A

As current remains same in series, using ohm's law

$V = I \times R_{eq}$

$\Rightarrow 200 = 5 \times R_{eq} \Rightarrow R_{eq} = 40$

$\therefore R + 20 = 40 \Rightarrow R = 20\,\Omega$

216. (2500) From, $H = i^2 Rt$

$10 \times 10^{-3} = (2 \times 10^{-3})^2 \times R \times 1$

$\therefore R = 2500\,\Omega$

217. (d) Given : Power, $P = 1$ kW $= 1000$ W $= P_{output}$

$R = 2\Omega$, $V = 220$ V

Current, $I = \frac{P}{V} = \frac{1000}{220}$

$P_{loss} = I^2 R = \left(\frac{1000}{220}\right)^2 \times 2$

$P_{in} = P_{output} + P_{loss}$

$\therefore$ Efficiency $= \frac{1000}{1000 + P_{loss}} \times 100 = 96\%$.

218. (b) Maximum power in external resistance is generated when it is equal to internal resistance of battery i.e., P_R maximum when $r = R$

The maximum Joule heating in R will take place for, the resistance of small element

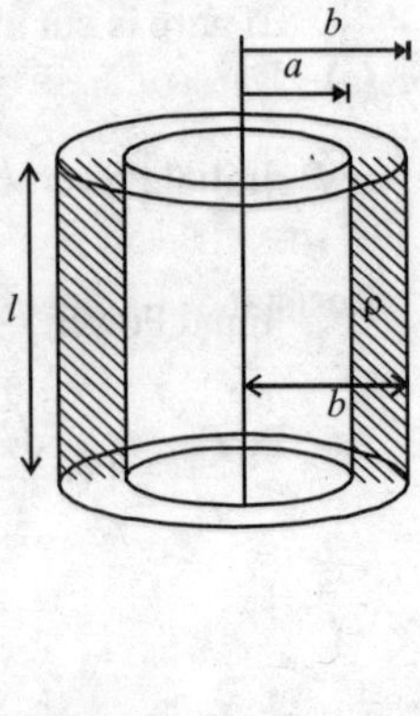

$dR = \frac{\rho dr}{2\pi rl} \Rightarrow R = \frac{\rho}{2\pi l}\int_a^b \frac{dr}{r}$

$\Rightarrow R = \frac{\rho}{2\pi l} \ln \frac{b}{a}$

219. (d) Net Power, P
$= 15 \times 45 + 15 \times 100 + 15 \times 10 + 2 \times 1000$
$= 15 \times 155 + 2000$ W

Power, $P = VI \Rightarrow I = \frac{P}{V}$

$\therefore \; I_{main} = \frac{15 \times 155 + 2000}{220} = 19.66\ A \approx 20 A$

220. (b) Equivalent resistance,

$R_{eq} = \frac{4R \times 4R}{4R + 4R} + R + \frac{6R \times 12R}{6R + 12R} + R$
$= 2R + R + 4R + R = 8R.$

Using, $P = \frac{V^2}{R_{eq}} \Rightarrow 4 = \frac{16^2}{8R} \quad \therefore\ R = \frac{16^2}{4 \times 8} = 8\,\Omega$

221. (b) $\Delta Q = mS\Delta T + mL$

$\frac{V^2}{R} \times t = mS\Delta T + mL$

$\Rightarrow \frac{200^2}{20} \times t = 1 \times 4200 \times 80 + 1 \times 2260 \times 10^3$

$\Rightarrow t = 1298$ sec $\Rightarrow t \simeq 22$ minutes

222. (a) As $R = \frac{V^2}{P}$, so $R_1 = \frac{220^2}{25}$ and $R_2 = \frac{220^2}{100}$

Current flown $i = \frac{220}{R_1 + R_2}$

$P_1 = i^2 R_1 = \frac{220^2}{\left(\frac{220^2}{25} + \frac{220^2}{100}\right)^2} \times \frac{220^2}{25} = 16$ W

Similarly, $P_2 = i^2 R_2 = 4$ W

223. (b) When two resistances are connected in series, $R_{eq} = 2R$

Power consumed, $P = \frac{\varepsilon^2}{R_{eq}} = \frac{\varepsilon^2}{2R}$

In parallel condition, $R_{eq} = R/2$.

New power, $P' = \frac{\varepsilon^2}{(R/2)}$

or $P' = 4P = 240$ W ($\because$ P = 60 W)

224. (a) Power, $P = I^2R$
$4.4 = 4 \times 10^{-6} \times R \Rightarrow R = 1.1 \times 10^6 \Omega$
When supply of 11 V is connected

Power, $P' = \frac{V^2}{R} = \frac{11 \times 11}{1.1 \times 10^6} = 11 \times 10^{-5}$ W

225. (a) Rate of heat i.e., Power developed in the wire $= P = \frac{V^2}{R}$

Resistance of the wire of length, L $R_1 = \frac{\rho L}{A} = \frac{\rho L}{\pi r^2}$

$\therefore$ Power, $P_1 = \frac{V^2}{R_1}$

Resistance of the wire when length is halved i.e., $L/2$ and radius is doubled

$R_2 = \frac{\rho \frac{L}{2}}{\pi (2r)^2} = \frac{\rho L}{\pi 8 r^2} = \frac{R_1}{8}$

$\therefore$ Power, $P_2 = \frac{V^2}{\frac{R_1}{8}} = \frac{8V^2}{R_1}$ or, $P_2 = 8P_1$

i.e., power increased 8 times of previous or original wire.

226. (a) Current through resistance r is

$i = \frac{V}{R + \frac{Rr}{R + r}} \times \left(\frac{R}{R + r}\right) = \frac{V}{R + 2r}$

So, $P = i^2 R = \frac{V^2 r}{(R + 2r)^2}$

for H_{max}, P is maximum

$\frac{dp}{dr} = 0 \Rightarrow \frac{(R + 2r)^2 - 2(R + 2r).2r}{(R + 2r)^4} = 0$

$\Rightarrow (R + 2r)(R + 2r - 4r) = 0$

$\Rightarrow r = -\frac{R}{2}$ or $r = \frac{R}{2} \Rightarrow r = \frac{R}{2}$ so, $f = \frac{1}{2}$

227. (c) Total power consumed by electrical appliances in the building, $P_{total} = 2500$W
Watt = Volt $\times$ ampere
$\Rightarrow\ 2500 = V \times I \Rightarrow 2500 = 220\, I$

$\Rightarrow\ I = \frac{2500}{220} = 11.36 \approx 12$A

(Minimum capacity of main fuse)

228. (c) Current in each bulb $= \frac{\text{Power}}{\text{Voltage}} = \frac{100}{220} = 0.45$A

Current through ammeter = $0.45 \times 3 = 1.35$ A

229. (d)

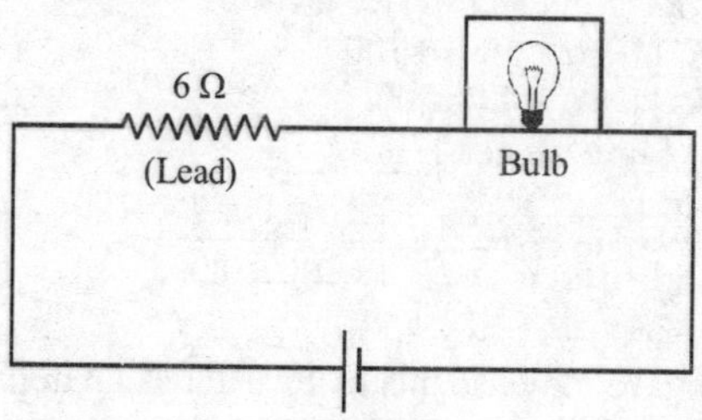

120 V

Power of bulb = 60 W (given)

Resistance of bulb $= \frac{120 \times 120}{60} = 240\Omega \quad \left[\because P = \frac{V^2}{R}\right]$

Power of heater = 240W (given)

Resistance of heater $= \frac{120 \times 120}{240} = 60\Omega$

Voltage across bulb before heater is switched on,

$$V_1 = \frac{240}{246} \times 120 = 117.73 \text{ volt}$$

Voltage across bulb after heater is switched on,

$$V_2 = \frac{48}{54} \times 120 = 106.66 \text{ volt}$$

Hence decrease in voltage
$V_1 - V_2 = 117.073 - 106.66 = 10.04$ volt (approximately)

230. (c) $R_{25W} = \frac{220^2}{25} = 1936\Omega$

$$R_{100W} = \frac{220^2}{100} = 484\Omega$$

$$V_{25W} = \frac{1936}{2420} \times 440 = 252\text{ V} > 220\text{ V}$$

So, 25 watt bulb will fuse.

231. (c) Resistors 4 Ω, 6 Ω and 12 Ω are connected in parallel, its equivalent resistance (R) is given by

$$\frac{1}{R} = \frac{1}{4} + \frac{1}{6} + \frac{1}{12} \Rightarrow R = \frac{12}{6} = 2\Omega$$

Again R is connected to 1.5 V battery whose internal resistance $r = 1\ \Omega$.
Equivalent resistance now,
$R' = 2\Omega + 1\Omega = 3\Omega$

$$\text{Current, } I_{\text{total}} = \frac{V}{R'} = \frac{1.5}{3} = \frac{1}{2}\text{A}$$

$$I_{\text{total}} = \frac{1}{2} = 3x + 2x + x = 6x \Rightarrow x = \frac{1}{12}$$

$\therefore$ Current through 4Ω resistor = $3x$

$$= 3 \times \frac{1}{12} = \frac{1}{4} A$$

Therefore, rate of Joule heating in the 4Ω resistor

$$= I^2R = \left(\frac{1}{4}\right)^2 \times 4 = \frac{1}{4} = 0.25\text{W}$$

232. (b) Let resistance of bulb filament be R_0 at 0°C using $R = R_0(1 + \alpha\,\Delta t)$ we have
$R_1 = R_0[1 + \alpha \times 100] = 100$ ....(1)
$R_2 = R_0[1 + \alpha \times T] = 200$ ....(2)
On dividing we get

$$\frac{200}{100} = \frac{1 + \alpha T}{1 + 100\alpha} \Rightarrow 2 = \frac{1 + 0.005T}{1 + 100 \times 0.005}$$

$\Rightarrow T = 400°C$

Note : We may use this expression as an approximation because the difference in the answers is appreciable. For accurate results one should use $R = R_0 e^{\alpha \Delta T}$

233. (c) The resistance of the electric bulb is

$$R = \frac{V^2}{P} = \frac{(220)^2}{100}$$

The power consumed when operated at 110 V is

$$P' = \frac{V^2}{R} \Rightarrow P = \frac{(110)^2}{(220)^2/100} = \frac{100}{4} = 25 \text{ W}$$

234. (b) Heat generated,

$$H = \frac{V^2 t}{R}$$

After cutting equal length of heater coil will become half.
As $R \propto \ell$

Resistance of half of the coil = $\frac{R}{2}$

$$H' = \frac{V^2 t}{\frac{R}{2}} = 2H$$

$\therefore$ As R reduces to half, 'H' will be doubled.

235. (b) Power, $P = Vi = \frac{V^2}{R}$

$\therefore$ Resistance of tungsten filament when in use

$$R_{\text{hot}} = \frac{V^2}{P} = \frac{200 \times 200}{100} = 400\Omega$$

Resistance when not in use *i.e.*, cold resistance

$$R_{\text{cold}} = \frac{400}{10} = 40\Omega$$

236. (a) Thermistors are usually made of metal oxides with high temperature coefficient of resistivity.

237. (a) Heat supplied in time t for heating 1L water from 10°C to 40°C
$\Delta Q = mC_p \times \Delta T = 1 \times 4180 \times (40 - 10) = 4180 \times 30$

$$\text{But } \Delta Q = P \times t = 836 \times t \Rightarrow t = \frac{4180 \times 30}{836} = 150\text{s}$$

238. (b) **Case 1 :** Initial power dissipation,

$$P_1 = \frac{V^2}{R}$$

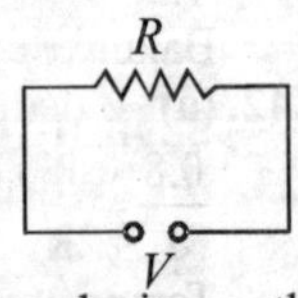

Case 2 : When wire is cut into two equal pieces, the resistance of each piece is $\frac{R}{2}$. When they are connected in parallel

$$\text{Equivalent resistance, } R_{eq} = \frac{R/2}{2} = \frac{R}{4}$$

Power dissipated,

$$P_2 = \frac{V^2}{R/4} = 4\left(\frac{V^2}{R}\right) = 4P_1$$

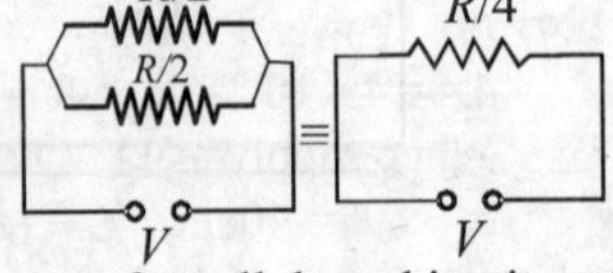

239. (b) The equivalent resistance of parallel combination of 2Ω and R is $R_{eq} = \frac{2 \times R}{2 + R}$

$$\therefore \text{Power dissipation } P = \frac{V^2}{R_{eq}} \quad \therefore\ 150 = \frac{(15)^2}{R_{eq}}$$

$$\Rightarrow 150 = \frac{225 \times (R + 2)}{2R} \Rightarrow \frac{2R}{2 + R} = \frac{3}{2}$$

$\Rightarrow 4R = 6 + 3R \Rightarrow R = 6\Omega$

240. (c) Solving resistance in series and parallel,

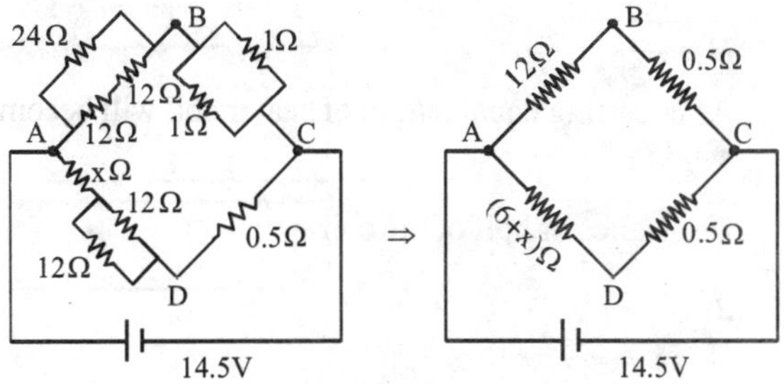

Using balanced condition of Wheatstone Bridge

$$\frac{12}{6+x}=\frac{0.5}{0.5}\Rightarrow x=6\Omega$$

241. (b)

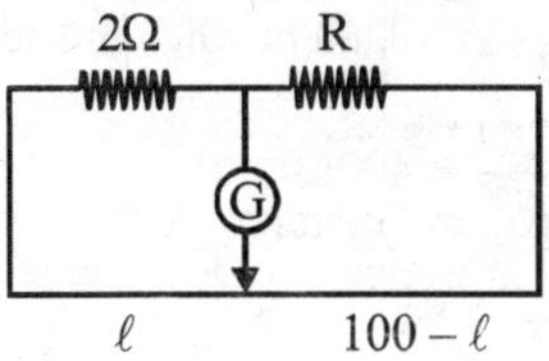

Condition for Wheatstone Bridge is, $\frac{R_1}{R_2}=\frac{\ell}{100-\ell}$

In Ist case $\frac{2}{R}=\frac{40}{60}\Rightarrow R=3\Omega$

on shunting 3Ω with 2Ω resistance $R'=\frac{2\times3}{2+3}=1.2\Omega$

Applying condition of Wheatstone Bridge

$$\frac{2}{1.2}=\frac{\ell}{100-\ell}\Rightarrow \ell=\frac{200}{3.2}=62.5\text{cm}$$

Balance length changes by 62.5 – 40 = 22.5 cm

242. (a) If deflection through galvanometer is zero.

$$\frac{0.8}{1}=\frac{R}{3}\Rightarrow R=2.4\text{ m}\Omega$$

Temperature fall is 20°C

Resistance, $\Delta R = R\,\alpha\,\Delta t$

$$\Rightarrow \alpha=\frac{\Delta R}{R\Delta t}=\frac{-0.6}{3\times20}=-1\times10^{-2}\,{}^{\circ}\text{C}^{-1}$$

243. (d)

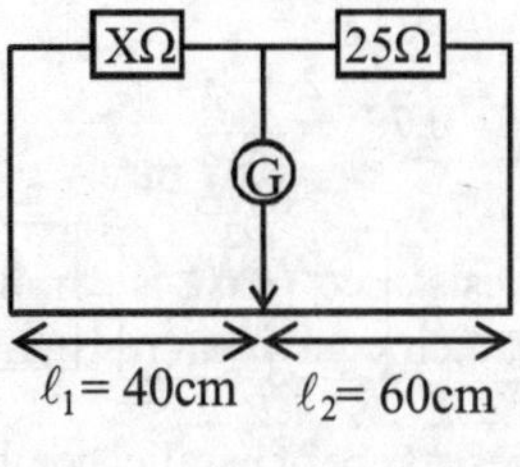

From the balancing condition of meter bridge

In first case

$$\frac{X}{r\ell_1}=\frac{25}{r\ell_2} \quad ...(i)$$

In second case

$$\frac{X}{2r\ell'_1}=\frac{25}{2r\ell'_2} \quad ...(ii)$$

From (i) and (ii)

$\ell'_1=\ell_1=40$ cm

244. (200)

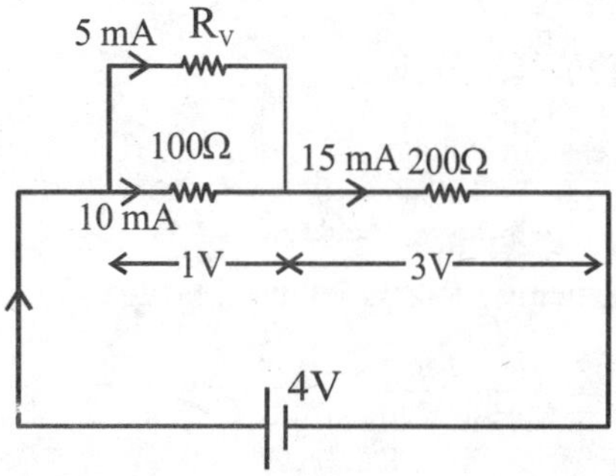

$3=200\times i\Rightarrow i=15\text{ mA}$

$1=100\times i_1\Rightarrow i_1=10\text{ mA}$

∴ Current through voltmeter, $i-i_1=15-10=5\text{ mA}$

$$\therefore R_v=\frac{1}{5\text{mA}}=\frac{1}{5\times10^{-3}}=200\,\Omega$$

245. (c) For null point in balanced condition, $\frac{P}{Q}=\frac{\ell_1}{100-\ell_1}$

$$\therefore \frac{4.5}{60}=\frac{R}{40}$$

Also, $R=\frac{\rho\ell}{A}=\frac{\rho\ell}{\pi r^2}\Rightarrow 4.5\times40=\rho\times\frac{0.1}{\pi\times7\times10^{-8}}\times60$

$\therefore \rho=66\times10^{-7}\,\Omega\text{m}$

246. (d) Specific resistance is independent of dimension of wire so, it will remain unchanged i.e., S_1

247. (d) From Kirchhoff's voltage law,

$V_1+V_2-V_3=0\Rightarrow V_1+V_2=V_3$

248. (2) The given circuit can be redrawn as

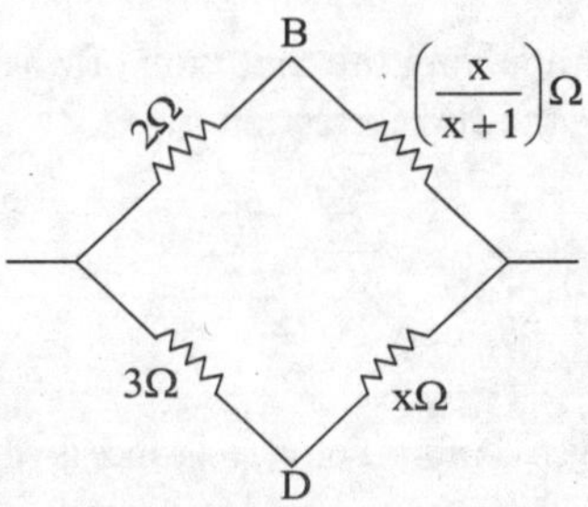

As potential difference between B and D is zero, the given circuit is wheatstone bridge,

$$\frac{2}{3}=\frac{\frac{x}{x+1}}{x}\Rightarrow\frac{2}{3}=\frac{1}{x+1}\Rightarrow x=0.5=\frac{1}{2}=\frac{1}{n}$$

$\therefore n=2$

249. (30) **Case I:** $\frac{R_1+R_2}{10}=\frac{60}{40}=\frac{3}{2}\Rightarrow R_1+R_2=15$

Case II: $\frac{R_1R_2}{(R_1+R_2)\times 3}=\frac{40}{60}\Rightarrow\frac{R_1R_2}{45}=\frac{2}{3}$

$\Rightarrow R_1R_2=30\,\Omega^2$

250. **(20)** Given,

Resistance in the left gap, $P=4\Omega$

From the balancing condition of meter bridge

$$\frac{P}{Q}=\frac{l_1}{100-l_1}=\frac{40\text{ cm}}{100\text{ cm}-40\text{ cm}}$$

$$\frac{4}{Q}=\frac{40}{60}\Rightarrow\frac{4}{Q}=\frac{2}{3}\quad\Rightarrow Q=6\,\Omega$$

When unknown resistance x is connected in series with P, then

$$\frac{4+x}{6}=\frac{80}{100-80}=\frac{80}{20}\quad\Rightarrow\frac{4+x}{6}=4$$

$\Rightarrow\ 4+x=24\ \Rightarrow x=20\,\Omega$

251. **(a)** Balanced wheatstone bridge in circuit, so there is no current in 5Ω resistor so it can be removed from the circuit.

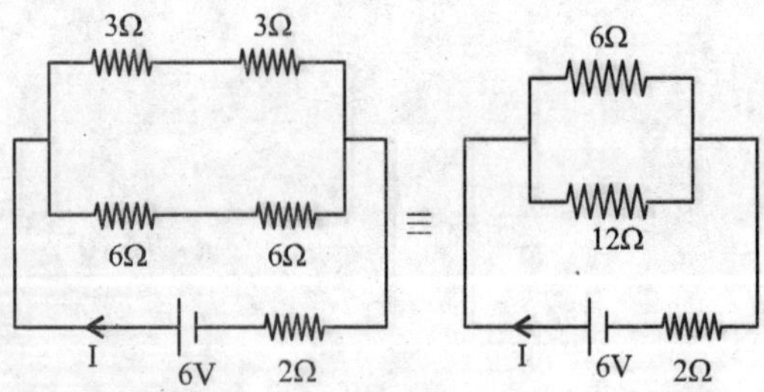

The equivalent resistance will be

$$R_{eq}=\frac{6\times 12}{6+12}+2=6\,\Omega$$

$$\therefore\ \ I=\frac{V}{R_{eq}}=\frac{6}{6}=1\text{A}$$

252. **(19)** After applying the condition of wheatstone bridge and adding the end correction

$$\frac{15}{43+2}=\frac{R}{102-45}\Rightarrow\frac{15}{45}=\frac{R}{57}$$

$\therefore R=19\Omega$

253. **(10)** The given combination of resistance is wheatstone bridge

So, $6\times R=4\times 3\Rightarrow R=2\Omega$

Therfore, the effective resistance of circuit is

$$R_{eq}=\frac{6\times 9}{6+9}=\frac{54}{15}=\frac{18}{5}\Omega$$

$$\text{So, } i=\frac{36}{R_{eq}}=\frac{36}{\frac{18}{5}}=10A$$

254. **(48)** In balanced condition,

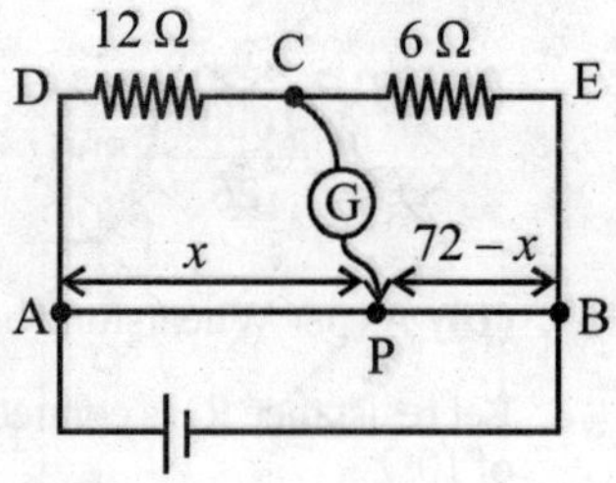

$$\frac{x}{12}=\frac{72-x}{6}$$

$\Rightarrow x=2(72-x)$

or, $3x=144$

$\therefore\quad x=\frac{144}{3}=48\text{ cm}$

255. **(d)**

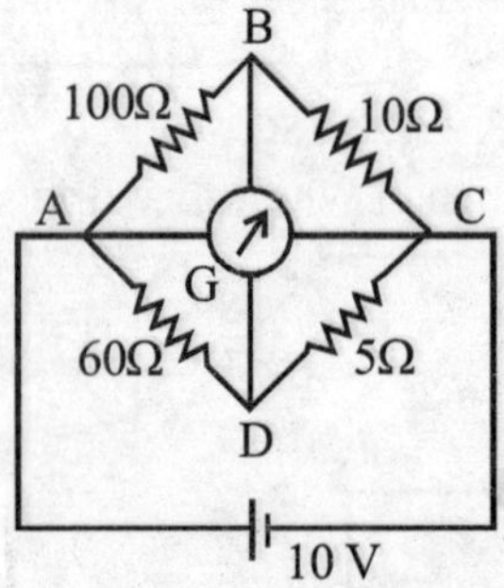

Current through the galvanometer,

$$I_G=\frac{V_B-V_D}{R_{BD}}$$

$$\frac{V_B-V_A}{100}+\frac{V_B-V_D}{15}+\frac{V_B-V_C}{10}=0$$

$$\Rightarrow\frac{V_B-10}{100}+\frac{V_B-V_D}{15}+\frac{V_B-0}{10}=0$$

$$\Rightarrow\frac{V_B-10}{20}+\frac{V_B-V_D}{3}+\frac{V_B}{2}=0$$

$$\Rightarrow 3V_B-30+20V_B-20V_D+30V_B=0$$

$$\Rightarrow 53V_B-20V_D=30 \qquad \text{...(i)}$$

Similarly,

$$\frac{V_D-10}{60}+\frac{V_D-V_B}{15}+\frac{V_D-0}{5}=0$$

$$\Rightarrow V_D-10+4V_D-4V_B+12V_D=0$$

$$\Rightarrow -4V_B+17V_D=10 \qquad \text{...(ii)}$$

Solving equations (i) and (ii), we get

$V_B=0.86$ V and $V_D=0.79$ V

$$\therefore I_G=\frac{V_B-V_D}{R_{BD}}=\frac{0.86-0.79}{15}=4.87\text{ mA A}$$

256. **(d)** The voltmeter of resistance 10kΩ is parallel to the resistance of 400Ω. So, their equivalent resistance is

$$\frac{1}{R'}=\frac{1}{10\,k\Omega}+\frac{1}{400\Omega}=\frac{1}{10000}+\frac{1}{400}$$

$$\Rightarrow\frac{1}{R'}=\frac{1+25}{10000}=\frac{26}{10000}\quad\Rightarrow R'=\frac{10000}{26}\Omega$$

Using Ohm's law, current in the circuit

$$I=\frac{\text{Voltage}}{\text{Net Resistance}}=\frac{6}{\frac{10000}{26}+800}$$

Potential difference measured by voltmeter

$$V = IR' = \frac{6}{\frac{10000}{26}+800} \times \frac{10000}{26} \Rightarrow V = \frac{150}{77} = 1.95 \text{ volt}$$

257. **(10)** As per Wheatstone bridge balance condition $\frac{P}{Q} = \frac{S}{R}$

Let resistance R' is connected in parallel with resistance S of 10Ω

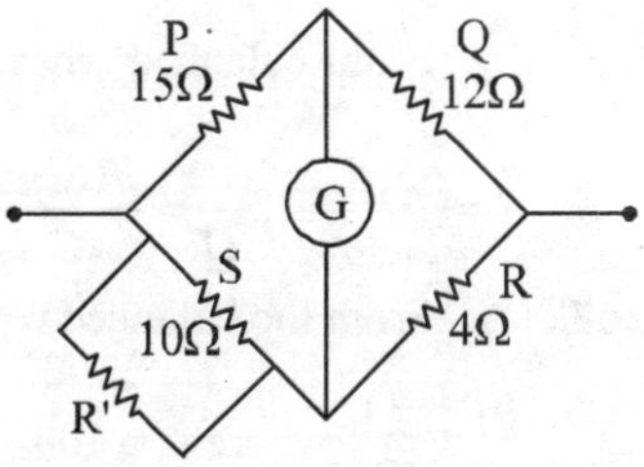

$$\therefore \frac{15}{12} = \frac{10R'}{(10+R')\times 4}$$

$$\Rightarrow 5 = \frac{10R'}{10+R'}$$

$$\Rightarrow 50 + 5R' = 10R'$$

$$\therefore R' = \frac{50}{5} = 10\Omega$$

258. **(c)** We have given

$$\frac{dR}{d\ell} \propto \frac{1}{\sqrt{\ell}} \Rightarrow \frac{dR}{d\ell} = k \times \frac{1}{\sqrt{\ell}} \quad \text{(where k is constant)}$$

$$dR = k\frac{d\ell}{\sqrt{\ell}}$$

Let R_1 and R_2 be the resistance of AP and PB respectively. Using wheatstone bridge principle

$$\therefore \frac{R'}{R'} = \frac{R_1}{R_2} \text{ or } R_1 = R_2$$

$$\text{Now, } \int dR = k\int \frac{d\ell}{\sqrt{\ell}} \quad \therefore R_1 = k\int_0^{\ell} \ell^{-1/2} d\ell = k.2.\sqrt{\ell}$$

$$R_2 = k\int_{\ell}^{1} \ell^{-1/2} d\ell = k.(2 - 2\sqrt{\ell})$$

Putting $R_1 = R_2$

$$k2\sqrt{\ell} = k(2 - 2\sqrt{\ell}) \quad \therefore 2\sqrt{\ell} = 1 \Rightarrow \sqrt{\ell} = \frac{1}{2}$$

i.e., $\ell = \frac{1}{4}\text{m} \Rightarrow 0.25 \text{ m}$

259. **(b)** Given, Emf of cell, $\varepsilon = 0.5$ V

Rheostat resistance, $R_h = 2\Omega$

Potential gradient is

$$\frac{dV}{dL} = \left(\frac{6}{2+4}\right) \times \frac{4}{L}$$

Let null point be at ℓ cm when cell of emf, $\varepsilon = 0.5$ V is used.

$$\text{thus } \varepsilon_1 = 0.5\text{V} = \left(\frac{6}{2+4}\right) \times \frac{4}{L} \times \ell \quad \text{... (i)}$$

For resistance $R_h = 6\Omega$ new potential gradient is $\left(\frac{6}{4+6}\right) \times \frac{4}{L}$ and at null point

$$\left(\frac{6}{4+6}\right)\left(\frac{4}{L}\right) \times \ell = \varepsilon_2 \quad \text{... (ii)}$$

Dividing equation (i) by (ii) we get

$$\frac{0.5}{\varepsilon_2} = \frac{10}{6} \text{ thus } \varepsilon_2 = 0.3\text{V}$$

260. **(d)** Initially, $\frac{P}{Q} = \frac{R_1}{X}$ ... (i)

After interchanging P and Q

$$\frac{Q}{P} = \frac{R_2}{X} \quad \text{... (ii)}$$

From (i) and (ii), we get

$$1 = \frac{R_1R_2}{X^2} \Rightarrow X = \sqrt{R_1R_2} = \sqrt{400 \times 405} = 402.5\,\Omega$$

261. **(c)** Initially at null deflection $\frac{R_1}{R_2} = \frac{2}{3}$ ...(i)

Finally at null deflection, when null point is shifted

$$\frac{R_1+10}{R_2} = 1 \Rightarrow R_1 + 10 = R_2 \quad \text{...(ii)}$$

Solving equations (i) and (ii) we get

$$\frac{2R_2}{3} + 10 = R_2$$

$$10 = \frac{R_2}{3} \Rightarrow R_2 = 30\Omega \;\&\; R_1 = 20\Omega$$

Now if required resistance is R then $\frac{\frac{30\times R}{30+R}}{30} = \frac{2}{3} \Rightarrow R = 60\Omega$

262. **(c)** $R_1 + R_2 = 1000$

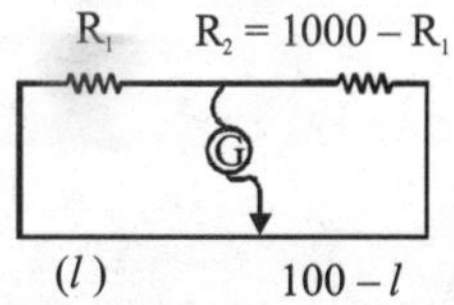

$\Rightarrow R_2 = 1000 - R_1$

On balancing condition,

$R_1(100 - l) = (1000 - R_1)l$ ...(i)

On Interchanging resistance balance point shifts left by 10 cm.

On balancing condition,

$(1000 - R_1)(110 - l) = R_1(l - 10)$

or, $R_1(l - 10) = (1000 - R_1)(110 - l)$ ...(ii)

Dividing eqn (i) by (ii)

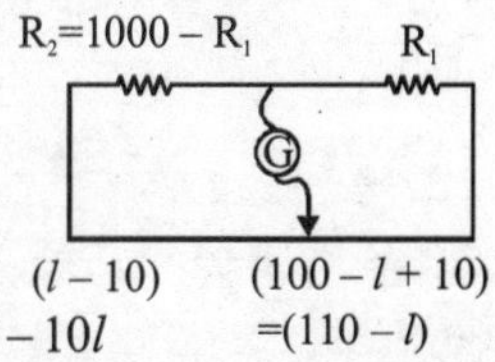

$$\frac{100-l}{l-10} = \frac{l}{110-l}$$

$\Rightarrow (100 - l)(110 - l) = l(l - 10)$

$\Rightarrow 11000 - 100l - 110l + l^2 = l^2 - 10l$

$\Rightarrow 11000 = 200l$ or, $l = 55$

Putting the value of 'l' in eqn (i)

$R_1(100 - 55) = (1000 - R_1)\,55$

$\Rightarrow R_1(45) = (1000 - R_1)\,55$

$\Rightarrow R_1(9) = (1000 - R_1)\,11$

$\Rightarrow 20R_1 = 11000$

$\therefore R_1 = 550\text{K}\Omega$

263. (d) There is no change in null point, if the cell and the galvanometer are exchanged in a balanced wheatstone bridge.

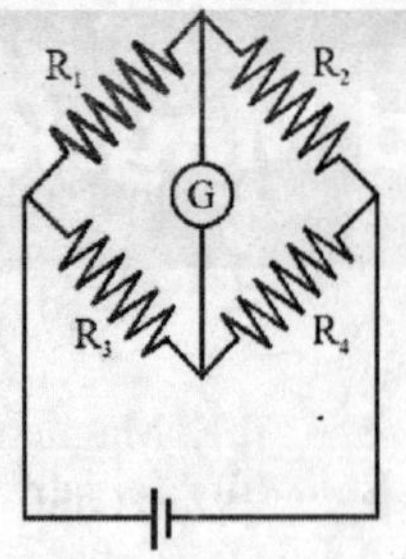

On balancing condition $\frac{R_1}{R_3} = \frac{R_2}{R_4}$

After exchange

On balancing condition

$\frac{R_1}{R_2} = \frac{R_3}{R_4}$

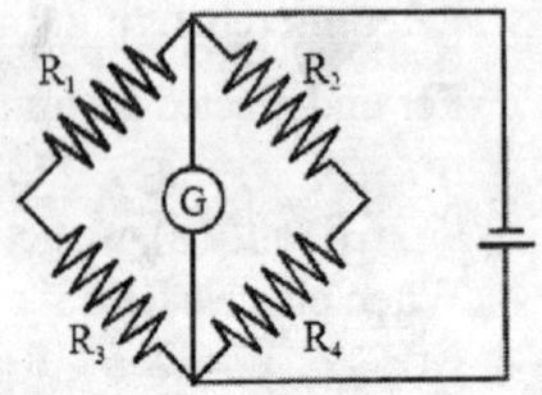

264. (d) In balance position of bridge, $\frac{P}{Q} = \frac{l}{(100-l)}$

Initially neutral position is 60 cm. from A, so

$\frac{4}{60} = \frac{Q}{40} \Rightarrow Q = \frac{16}{6} = \frac{8}{3}\Omega$

Now, when unknown resistance R is connected in series to P, neutral point is 80 cm from A then,

$\frac{4+R}{80} = \frac{Q}{20} \quad \Rightarrow \frac{4+R}{80} = \frac{8}{60}$

$R = \frac{64}{6} - 4 = \frac{64-24}{6} = \frac{40}{6}\Omega$

Hence, the value of unknown resistance R is $= \frac{20}{3}\Omega$

265. (b) Given, balance point from one end, $\ell_1 = 20$ cm

From the condition for balance of metre bridge, we have

$\frac{55}{R} = \frac{\ell_1}{100-\ell_1} \Rightarrow \frac{55}{R} = \frac{20}{80} \Rightarrow R = 220\Omega$

266. (b) From balanced wheat stone bridge $\frac{P}{Q} = \frac{R}{S}$ where

$S = \frac{S_1S_2}{S_1+S_2} \therefore \frac{P}{Q} = \frac{R(S_1+S_2)}{S_1S_2}$

267. (c) From the balanced wheat stone bridge

$\frac{R_1}{R_2} = \frac{\ell_1}{\ell_2}$ where $\ell_2 = 100 - \ell_1$

In the first case $\frac{X}{Y} = \frac{20}{80}$; $Y = 4X$

In the second case

$\frac{4X}{Y} = \frac{\ell}{100-\ell} \Rightarrow \frac{4X}{4X} = \frac{\ell}{100-\ell} \Rightarrow \ell = 50$ cm

268. (d) $i_g \times G = (i - i_g)S \therefore S = \frac{i_g \times G}{i - i_g} = \frac{1 \times 0.81}{10-1} = 0.09\Omega$

4 Moving Charges and Magnetism

1. **(d)** The radius of circular path for a charged particle moving perpendicularly in a magnetic field is

$$R=\frac{mv}{qB}$$

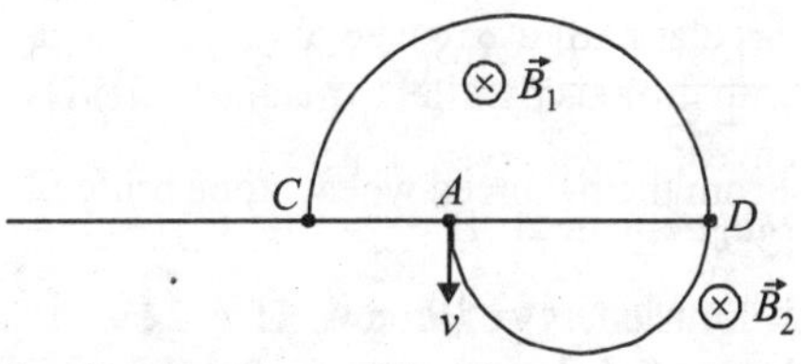

The displacement of the charged particle is
$AC = CD - AD = 2R_1 - 2R_2$

$$=\frac{2mv}{qB_1}-\frac{2mv}{qB_2}=\frac{2mv}{qB_1}\left[1-\frac{B_1}{B_2}\right].$$

2. **(d)** The radius of circular arc for a charge particle enters in magnetic field perpendicularly is

$$r=\frac{mv}{Bq}=\frac{\sqrt{2mE}}{Bq}$$

$$E=\left(\frac{q^2B^2}{2m}\right)r^2 \qquad \therefore E\propto r^2$$

Hence, r vs E graph is

3. **(0 or 10)**

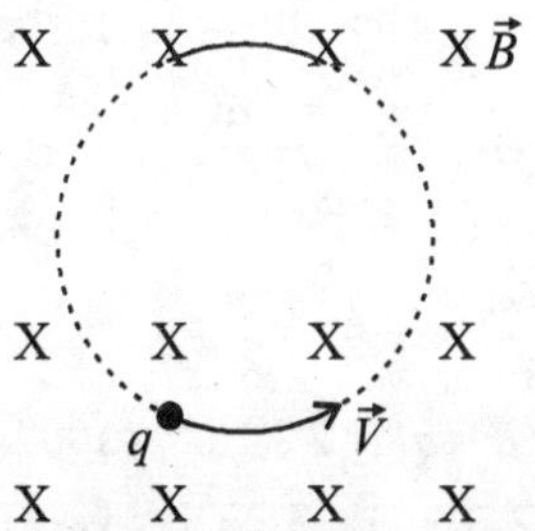

When a charged particle moves perpendicular to a magnetic field, it undergoes circular motion, and the time it takes to complete one full revolution is Time period.

$$T=\frac{2\pi m}{qB}=\frac{2\pi\times 16\times 10^{-9}\ \text{kg}}{1.6\times 10^{-6}\times 6.28}$$

$T = 0.01$ seconds
NTA Answer is 10.
Correct Answer is 0 (nearest integer).

4. **(a)** The radius of curvature of a charged particle in magnetic field is

$$r=\frac{mv}{qB}=\frac{p}{qB}\Rightarrow r\propto\frac{1}{q}\ \text{(For same momentum p)}$$

In first case, $q_o > q_H \Rightarrow r_o < r_H$
In second case, $q_e = q_p \Rightarrow r_e = r_p$
$\therefore$ Assertion is true reason is false.

5. **(b)** Magnetic force, $\vec{F}=q(\vec{v}\times\vec{B})=0$
$\Rightarrow qv\,B\sin\theta = 0 \therefore \theta = 0°$ or $180°$

6. **(2)** Given, Speed of proton, $v = 2\times 10^5\ \text{ms}^{-1}$
Radius of circular path, $R = 2$ cm
For undeflected motion of proton,

$$v=\frac{E}{B}\Rightarrow B=\frac{E}{v}$$

When electric field is switched off

$$R=\frac{mv}{eB}=\frac{mv^2}{eE}$$

$$\Rightarrow E=\frac{mv^2}{eR}=\frac{1.6\times10^{-27}\times4\times10^{10}}{1.6\times10^{-19}\times2\times10^{-2}}=2\times10^4\ \text{N/C}$$

7. **(c)** In the uniform magnetic field, radius of the circular path is given by;

$$R=\frac{mv}{qB}=\frac{\sqrt{2m(K.E)}}{qB}$$

$$\Rightarrow R\propto\frac{\sqrt{m}}{q};\ (\because \text{KE in same})$$

$$\therefore \frac{R_{deutron}}{R_{proton}}=\sqrt{\frac{m_d}{m_p}}\times\frac{q_p}{q_d} \quad \because m_d=2m_p \text{ and } q_d=q_p$$

$$\therefore \frac{R_d}{R_p}=\sqrt{\frac{2m_p}{m_p}}\times\frac{q_p}{q_p}=\sqrt{2}\ \therefore R_d:R_p=\sqrt{2}:1$$

8. **(4)** Given,
Mass of electron, $m = 9\times10^{-31}$ kg
Kinetic energy of electron, = 5eV
Magnetic field, $B = 3\times10^{-6}$ T
As the electron is moving undeflected, net force should be zero.
$\therefore qE = qvB \Rightarrow E = vB$

$$\because KE=\frac{1}{2}mv^2 \Rightarrow v=\sqrt{\frac{2KE}{m}}$$

$$\therefore E=\sqrt{\frac{2\times KE}{m}}\times B=\sqrt{\frac{2\times5\times1.6\times10^{-19}}{9\times10^{-31}}}\times3\times10^{-6}$$

$$=\sqrt{\frac{16}{9}\times10^{12}}\times3\times10^{-6}=4\ \text{N/C}$$

9. **(c)** Electrostatic force, $\vec{F}_1=q\vec{E}$

Magnetic force, $\vec{F}_2=q\left(\vec{V}\times\vec{B}\right)$

10. **(b)** Electron is moving along the direction of magnetic field, i.e., $\vec{v} \parallel \vec{B}$ so force on electron due to magnetic field is zero i.e., $F_{net} = qvB \sin\theta = qvB \sin 0° = 0$

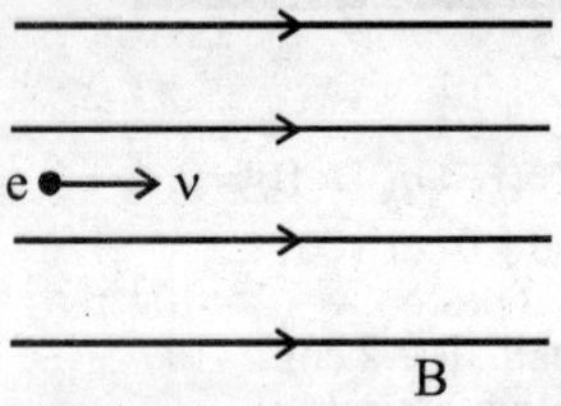

So it will move along the axis with uniform velocity.

11. **(5)** Magnetic force acting on electron in a uniform magnetic field is given by

$\vec{F} = q(\vec{V} \times \vec{B})$

$\Rightarrow 5e\hat{k} = e(3\hat{i} + 5\hat{j}) \times (B_0\hat{i} + 2B_0\hat{j})$

$\Rightarrow 5e\hat{k} = e(6B_0\hat{k} - 5B_0\hat{k}) \qquad \Rightarrow B_0 = 5T$

12. **(d)** The radius of circular path is

$$R = \frac{mv}{qB} = \frac{p}{qB} = \frac{\sqrt{2m(KE)}}{qB} = \frac{\sqrt{2mqV}}{qB}$$

$$\Rightarrow m \propto R^2 \qquad \therefore \frac{m_1}{m_2} = \left(\frac{R_1}{R_2}\right)^2$$

13. **(32)** $q = 4\mu C$, $\vec{v} = 4 \times 10^6 \hat{j}$ m/s, $\vec{B} = 2\hat{k}T$

The force acting on charge is

$\vec{F} = q(\vec{v} \times \vec{B})$

$= 4 \times 10^{-6}(4 \times 10^6 \hat{j} \times 2\hat{k})$

$= 4 \times 10^{-6} \times 8 \times 10^6 \hat{i}$

$\vec{F} = 32\hat{i}$ N $= x\hat{i}$ N(given)

$\therefore \; x = 32$

14. **(c)** Net force F_{net} on particle must be zero

i.e., $q\vec{E} + q\vec{V} \times \vec{B} = 0$

Possible cases are (i) $\vec{E} \,\&\, \vec{B} = 0$

(ii) $\vec{V} \times \vec{B} = 0 \Rightarrow \vec{B} \neq 0, \vec{E} = 0$;

(iii) $q\vec{E} = -q\vec{V} \times \vec{B}$ (i.e., $E \neq 0$ and $B \neq 0$)

15. **(a)** Magnetic force

$\vec{F} = -e(\vec{v} \times \vec{B})$ counters centripetal force

As magnetic force is $\perp$ to velocity, path of electron must be a circle.

Force will be along –ve y-axis.

16. **(b)** $\vec{F} = q(\vec{v} \times \vec{B}) = qvB\sin\theta$

As $\vec{v}$ is parallel to $\vec{B}$, angle between $\vec{v}$ and $\vec{B}$ is 0°.

$\therefore F = 0$

Hence, electron will continue to move with constant velocity.

17. **(c)** Let v_1 is component of velocity parallel to the magnetic field and v_2 is the component of velocity perpendicular to the magnetic field.

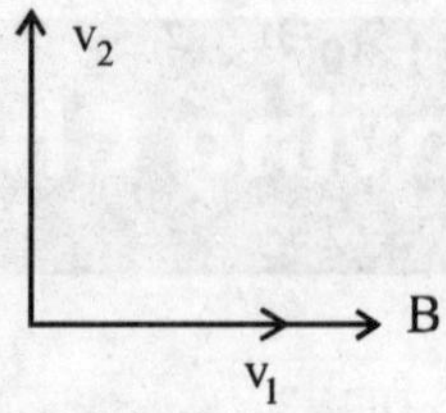

Due to component v_1,

magnetic force $F = qv_1B\sin\theta = 0$ ($\because \theta = 0$)

So v_1 remains unchanged, but due to component v_2 magnetic force act towards centre i.e. moving it along circular path with forward motion. So path is helical with the axis parallel to magnetic field B.

18. **(40)** Given,

Magnetic field, $B = \frac{\pi}{2} \times 10^{-3}$ T

Kinetic energy of proton, KE = 2 eV

Mass of proton, $m = 1.6 \times 10^{-27}$ kg

Charge on proton, $e = 1.6 \times 10^{-19}$C

Pitch = $v_{\parallel}$ × time period of one rotation

$$= v\cos 60° \times \frac{2\pi m}{eB} \quad (\because \theta = 60°)$$

$$\because \; K.E. = \frac{1}{2}mv^2 \Rightarrow v = \sqrt{\frac{2KE}{m}}$$

$$\therefore \quad \text{Pitch} = \sqrt{\frac{2KE}{m}} \cos 60° \times \frac{2\pi m}{eB}$$

$$= \sqrt{\frac{2 \times 2 \times 1.6 \times 10^{-19}}{1.6 \times 10^{-27}}} \times \cos 60° \times \frac{2\pi \times 1.6 \times 10^{-27}}{1.6 \times 10^{-19} \times \frac{\pi}{2} \times 10^{-3}}$$

$$= 2 \times 10^4 \times \frac{1}{2} \times 4 \times 10^{-5} = 4 \times 10^{-1} \text{ m} = 40 \text{ cm}$$

19. **(144)** Given, Charge, $q = 2\,\mu C = 2 \times 10^{-6}$ C

Magnetic field, $B = 4 \times 10^{-3}$ T

radius $r = 3$ cm $= 3 \times 10^{-2}$ m

$$r = \frac{mv}{qB} = \frac{\sqrt{2Km}}{qB}, m = \frac{r^2q^2B^2}{2K}$$

Kinetic energy $K = qV = 2 \times 10^{-6} \times 100$

$$m = \frac{(3 \times 10^{-2})^2 \times (2 \times 10^{-6})^2 \times (4 \times 10^{-3})^2}{2 \times 10^{-6} \times 100} = 144 \times 10^{-18} \text{kg}$$

20. **(b)** Given that uniform magnetic field, $\vec{B} = (2\hat{i} + 3\hat{j})$T

Acceleration $\vec{a} = (\alpha\hat{i} - 4\hat{j})$m/s^2

We know that

$F = q(\vec{v} \times \vec{B}) \Rightarrow m\vec{a} = q(\vec{v} \times \vec{B})$

Here, $\vec{a} \perp \vec{B}$, so, $\vec{a}.\vec{B} = 0$

$(\alpha\hat{i} - 4\hat{j})(2\hat{i} + 3\hat{j}) = 0 \Rightarrow 2\alpha - 12 = 0 \Rightarrow \alpha = 6$

21. **(a)** Given, $\vec{E} = E\hat{k}$

$\vec{B} = B\hat{j}$, B = 12 mT = 12×10^{-3} T. Energy = 728 eV

$$\text{Energy} = \frac{1}{2}mv^2$$

$728\text{ eV} = \frac{1}{2} \times 9.1 \times 10^{-31} \times v^2$

$728 \times 1.6 \times 10^{-19} = \frac{1}{2} \times 9.1 \times 10^{-31} \times v^2$

$= v = \sqrt{\frac{2 \times 728 \times 1.6 \times 10^{-9}}{9.1 \times 10^{-31}}}$

we have, $v = 16 \times 10^6$ m/s

$E = vB \Rightarrow E = 16 \times 10^6 \times 12 \times 10^{-3}$

$\Rightarrow E = 192 \times 10^3$ V/m $= 192$ k Vm^{-1}

22. (b) As $R = \frac{mv}{qB} = \frac{\sqrt{2m(K.E)}}{qB} \Rightarrow q \propto \frac{\sqrt{m}}{R}$

$\Rightarrow \frac{q_1}{q_2} = \sqrt{\frac{m_1}{m_2}} \times \frac{R_2}{R_1} = \sqrt{\frac{9}{4}} \times \frac{5}{6} = \frac{3}{2} \times \frac{5}{6} = \frac{5}{4}$

23. (c) $f = \frac{1}{T} = \frac{eB}{2\pi m}$ $\left[\because T = \frac{2\pi m}{qB}\right]$

$= \frac{1.6 \times 10^{-19} \times 10^{-4}}{2\pi \times 9 \times 10^{-31}} = 2.8 \times 10^6 \text{Hz}$

24. (c) Electric field can change speed and kinetic energy as $\vec{F}$ is parallel or antiparallel to $\vec{v}$.
But, magnetic field cannot change speed and kinetic energy as magnetic force is perpendicular to velocity.

25. (10) We know that

$r = \frac{mV}{qB} = \frac{\sqrt{2m(KE)}}{qB}$

$= \frac{\sqrt{2 \times 24 \times 1.67 \times 10^{-27} \times 5000 \times 1.6 \times 10^{-19}}}{1.6 \times 10^{-19} \times 0.5} = 0.1 \text{ m} = 10 \text{ cm}$

26. (2) We have,

$r = \frac{mv}{qB} = \frac{\sqrt{2m(K.E)}}{qB}$

$r \propto \sqrt{\frac{m}{q}} \Rightarrow \frac{r_d}{r_p} = \sqrt{\frac{m_d}{m_p}} \times \frac{q_p}{q_d} = \sqrt{\frac{2}{1}} \times \frac{1}{1} = \frac{\sqrt{2}}{1}$

27. (c) As $R = \frac{mv}{qB}$

$\Rightarrow R \propto \frac{m}{q} \Rightarrow \frac{R_\alpha}{R_P} = \left(\frac{m_\alpha}{m_P}\right) \times \frac{q_P}{q_\alpha} \Rightarrow \frac{R_\alpha}{R_P} = \left(\frac{4}{1}\right) \times \frac{1}{2} = 2$

28. (a) In uniform magnetic field

$\vec{F} = q(\vec{v} \times \vec{B})$ So, $\vec{F} \perp \vec{v} \Rightarrow \vec{a} \perp \vec{v}$.

So, speed and kinetic energy remains same.

29. (d) Radius of circular path in uniform magnetic field is given by, $r = \frac{mV}{qB} = \frac{\sqrt{2m(K.E.)}}{qB}$

As $K.E$ = constant

So, $r \propto \frac{\sqrt{m}}{q}$

So, $r_p : r_d : r_\alpha = \frac{\sqrt{m}}{e} : \frac{\sqrt{2m}}{e} : \frac{\sqrt{4m}}{2e} = 1 : \sqrt{2} : 1$

30. (b) Given,

$\frac{q_1}{q_2} = \frac{1}{2}$ & $\frac{v_1}{v_2} = \frac{2}{3}$

Radius of circular path,

$R = \frac{mv}{qB}$ Where,

m = mass of charged particle

B = magnetic field

$\therefore \frac{R_1}{R_2} = \frac{\frac{mv_1}{q_1 B}}{\frac{mv_2}{q_2 B}} = \frac{v_1}{v_2} \times \frac{q_2}{q_1} = \frac{2}{3} \times \frac{2}{1} = \frac{4}{3}$

31. (b) Radius of circular path in uniform magnetic field is

$r = \frac{\sqrt{2mE}}{qB} \Rightarrow r \propto \frac{\sqrt{m}}{q}$

$m_\alpha = 2m_d$ and $q_\alpha = 2q_d$

$\therefore \frac{r_d}{r_\alpha} = \frac{\sqrt{m_d}}{q_d} \times \frac{2q_d}{\sqrt{2m_d}} = \sqrt{2}$

32. (d) Radius of circular path in uniform magnetic field is

$r = \frac{\sqrt{2mK}}{qB}$ $(\because p = mV = \sqrt{2mK})$

$\Rightarrow \frac{r^2 q^2 B^2}{2m} = K$

$K_p = \frac{r_p^2 q_p^2 B^2}{2m_p}$ and $K_\alpha = \frac{r_\alpha^2 q_\alpha^2 B^2}{2m_\alpha}$

$\therefore \frac{K_p}{K_\alpha} = \frac{r_p^2 q_p^2 m_\alpha}{r_\alpha^2 q_\alpha^2 m_p} = \left(\frac{2}{1}\right)^2 \left(\frac{1}{2}\right)^2 \left(\frac{4}{1}\right)$ or, $\frac{K_p}{K_\alpha} = 4:1$

33. (b) Force on a charge particle of charge q moving with velocity $\vec{v}$ in the magnetic field of strength $\vec{B}$ is given by

$\vec{F}_m = q\vec{v} \times \vec{B}$

Since, force on a point charge by magnetic field is always perpendicular to $\vec{v}$. So, work done by magnetic force on the point charge is zero.

34. (c) $\vec{E} = 300\hat{j}$ V/cm $= 3 \times 10^4 \hat{j}$V/m

$\vec{V} = 6 \times 10^6 \hat{i}$

$\vec{B}$ must be in +z axis.

$q\vec{E} + q\vec{V} \times \vec{B} = 0$

$E = VB$

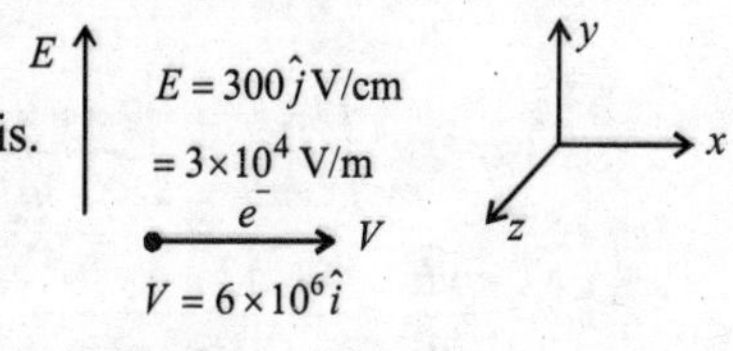

$\therefore B = \frac{E}{V} = \frac{3 \times 10^4}{6 \times 10^6} = 5 \times 10^{-3} T$

Hence, magnetic field $B = 5 \times 10^{-3}$ T along +z direction.

35. (c) If $r \le d$, particle will not hit screen

$r = \frac{mv}{qB_0}$ $\left[\because \frac{mv^2}{r} = qvB_0\right]$

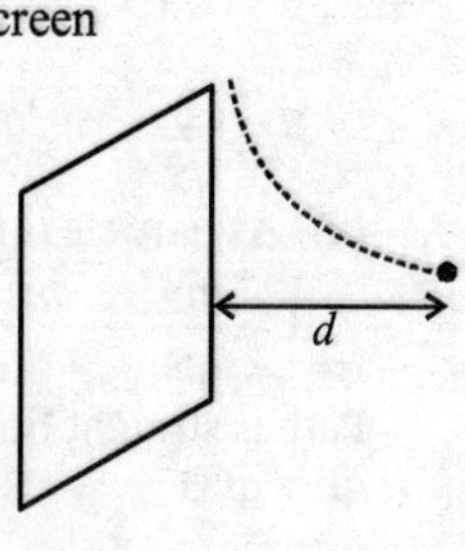

Hence, minimum value of v for which the particle will not hit the screen.

$v = \frac{qB_0 d}{m}$

36. **(b)** [Given: $q = 1\mu C = 1\times10^{-6}C$;
$\vec{V} = (2\hat{i} + 3\hat{j} + 4\hat{k})$ m/s and
$\vec{B} = (5\hat{i} + 3\hat{j} - 6\hat{k}) \times 10^{-3}$ T

$$\vec{F} = q(\vec{V}\times\vec{B}) = 10^{-6}\times10^{-3}\begin{vmatrix}\hat{i} & \hat{j} & \hat{k}\\ 2 & 3 & 4\\ 5 & 3 & -6\end{vmatrix}$$

$$= (-30\hat{i} + 32\hat{j} - 9\hat{k})\times10^{-9}\text{ N}$$

$$\therefore \vec{F} = (-30\hat{i} + 32\hat{j} - 9\hat{k})$$

37. **(d)** Pitch $= (v\cos\theta)T$ and $T = \dfrac{2\pi m}{qB}$

$$\therefore \text{ Pitch} = (V\cos\theta)\frac{2\pi m}{qB}$$

$$= (4\times10^5\cos 60°)\frac{2\pi}{0.3}\left(\frac{1.67\times10^{-27}}{1.69\times10^{-19}}\right) = 4\text{ cm}$$

38. **(c)** Time period of one revolution of proton, $T = \dfrac{2\pi m}{qB}$

Here, m = mass of proton
q = charge of proton
B = magnetic field.
Linear distance travelled in one revolution,
$p = T(v\cos\theta)$ (Here, v = velocity of proton)
$\therefore$ Length of region, $l = 10\times(v\cos\theta)T$

$$\Rightarrow l = 10\times v\cos60°\times\frac{2\pi m}{qB}$$

$$\Rightarrow l = \frac{10\pi mv}{qB} = \frac{10\times3.14\times1.67\times10^{-27}\times4\times10^5}{1.6\times10^{-19}\times0.3}$$

$$\Rightarrow l = 0.44\text{ m}$$

39. **(a)**

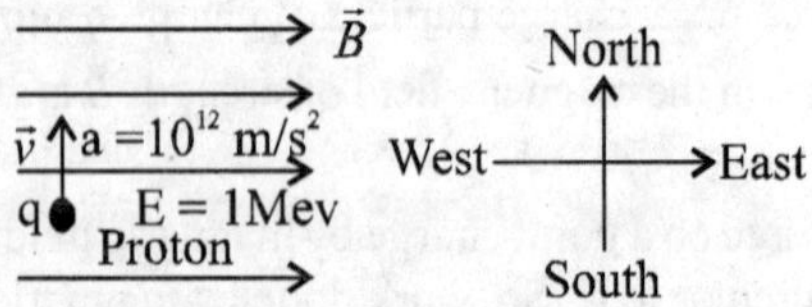

As we know, magnetic force $F = qvB = ma$

$$\therefore\ a = \left(\frac{qvB}{m}\right) \text{ perpendicular to velocity.}$$

$$\therefore \text{ Also } v = \sqrt{\frac{2KE}{m}} = \sqrt{\frac{2\times e\times10^6}{m}}$$

$$\therefore\ a = \frac{qvB}{m} = \frac{eB}{m}\sqrt{\frac{2\times e\times10^6}{m}}$$

$$\therefore\ 10^{12} = \left(\frac{1.6\times10^{-19}}{1.67\times10^{-27}}\right)^{\frac{3}{2}}.\sqrt{2}\times10^3 B$$

$$\therefore B \simeq \frac{1}{\sqrt{2}}\times10^{-3}T = 0.71\text{ mT (approx)}$$

40. **(d)** As particle is moving along a circular path

$$\therefore \text{R} = \frac{\text{mv}}{\text{qB}} \qquad ...(i)$$

Path is straight line, then
qE = qvB

$$\text{E} = \text{vB} \Rightarrow \text{v} = \frac{\text{E}}{\text{B}} \qquad ....(ii)$$

From equation (i) and (ii)

$$\text{m} = \frac{\text{qB}^2\text{R}}{\text{E}} = \frac{1.6\times10^{-19}\times(0.5)^2\times0.5\times10^{-2}}{100}$$

$$\therefore \text{m} = 2.0\times10^{-24}\text{ kg}$$

41. **(b)** The radius of the circular path is,

$$\text{R} = \frac{\text{mv}}{\text{Bq}} = \frac{\sqrt{2\text{mE}_\text{k}}}{\text{Bq}}$$

$$\Rightarrow \text{R} = \frac{\sqrt{2\times9.1\times10^{-31}\times100\times1.6\times10^{-19}}}{1.5\times10^{-3}\times1.6\times10^{-19}}$$

$$\Rightarrow \text{R} = 2.25\text{ cm}$$

$$\text{Now, } \sin\theta = \frac{2}{2.25} = \frac{8}{9}$$

$$\Rightarrow \cos\theta = \frac{\sqrt{17}}{9} \text{ and } \tan\theta = \frac{8}{\sqrt{17}}$$

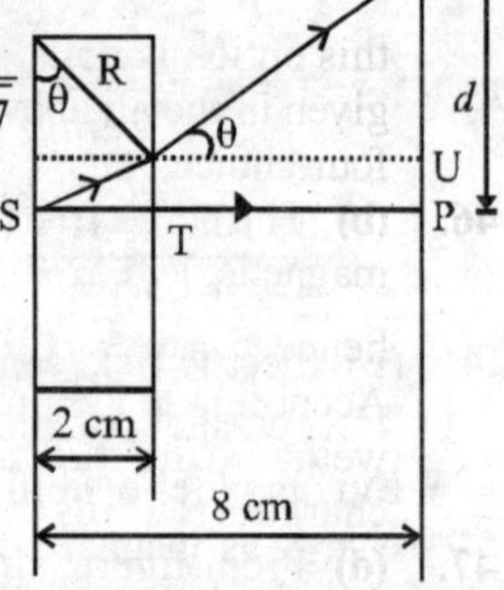

$$\text{PU} = \text{R}(1-\cos\theta)$$

$$= 2.25\times\left(1 - \frac{\sqrt{17}}{9}\right)$$

$$\Rightarrow \text{PU} = 1.22\text{ cm}$$

$$\tan\theta = \frac{\text{QU}}{6}$$

$$\Rightarrow \text{QU} = 6\times\frac{8}{\sqrt{17}} = 11.64\text{ cm}$$

$$\therefore \text{PQ} = \text{PU} + \text{QU} = 1.22 + 11.64 = 12.86\text{ cm}$$

42. **(b)** As mvr = qvB $\Rightarrow$ $\text{r} = \dfrac{\text{mv}}{\text{qB}} = \dfrac{\sqrt{2\text{mK.E.}}}{\text{qB}}$

$$\left[\text{As}: \frac{1}{2}\text{mv}^2 = \text{K.E.}\right.$$

$$\left.\Rightarrow\ \text{m}^2\text{v}^2 = 2\text{mK.E.} \Rightarrow\ \text{mv} = \sqrt{2\text{mK.E.}}\right]$$

For proton, electron and α-particle,
$\text{m}_\text{He} = 4\text{m}_\text{p}$ and $\text{m}_\text{p} >> \text{m}_\text{e}$
Also $\text{q}_\text{He} = 2\text{q}_\text{p}$ and $\text{q}_\text{p} = \text{q}_\text{e}$

$\therefore$ As KE of all the particles is same then, $\text{r}\ \alpha\ \dfrac{\sqrt{\text{m}}}{\text{q}}$

$$\therefore \text{r}_\text{He} = \text{r}_\text{p} > \text{r}_\text{e} \Rightarrow \text{r}_\text{e} < \text{r}_\text{p} = \text{r}_\text{He}$$

43. **(a)** Radius of the circular path will be $\text{r} = \dfrac{\text{mv}}{\text{qB}}$

$$\Rightarrow \text{r} = \frac{\sqrt{2\text{mKE}}}{\text{qB}} \quad (\because\ \text{p} = \text{mv} = \sqrt{2\text{mKE}})$$

$$\because\ \text{KE} = \text{q}\Delta\text{V}$$

$$\therefore\ \text{r} = \frac{\sqrt{2\text{mq}\Delta\text{V}}}{\text{qB}} \Rightarrow \text{r} \propto \sqrt{\frac{\text{m}}{\text{q}}} \qquad \therefore\ \frac{\text{r}_\text{p}}{\text{r}_\alpha} = \frac{1}{\sqrt{2}}$$

44. **(d)** Radius of the path is given by $\text{r} = \dfrac{\text{mv}}{\text{qB}}$

$$\text{r} = \frac{\sqrt{2\text{mk}}}{\text{eB}} \quad \left(\because\ \text{p} = \text{mv} = \sqrt{2\text{mk}}\right)$$

$= \frac{\sqrt{2meV}}{eB}$ $\quad (\because k = eV)$

$r = \frac{\sqrt{\frac{2m}{e}V}}{B} = \frac{\sqrt{\frac{2\times 9.1\times 10^{-31}}{1.6\times 10^{-19}}(500)}}{100\times 10^{-3}}$

$\Rightarrow r = \frac{\sqrt{\frac{9.1}{0.16}\times 10^{-10}}}{10^{-1}} = \frac{3}{.4}\times 10^{-4} = 7.5\times 10^{-4}\,\text{m}$

45. **(None)** Assuming particle enters from (0, d)

$r = \frac{mv}{qB},\ d = \frac{r}{2}$

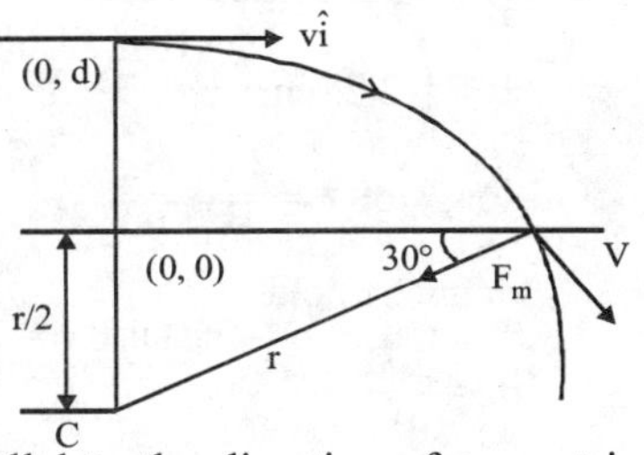

$a = \frac{qVB}{m}\left[\frac{-\sqrt{3}i - j}{2}\right]$

this option is not given in the all above four choices.

46. **(b)** The force is parallel to the direction of current in magnetic field,

hence $F = q(v\times \vec{B})$

According to Fleming's left hand rule, we have, the direction of motion of charge is towards the wire.

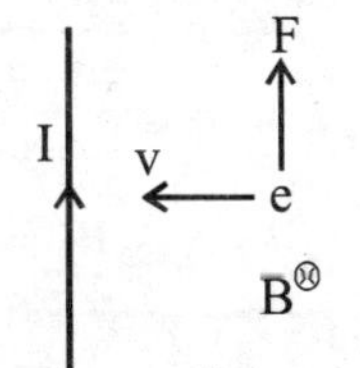

47. **(d)** From figure, $\sin\alpha = d/R$

And we know, $\frac{mv^2}{R} = qvB$

$\Rightarrow R = \frac{mv}{qB}$

$\therefore \sin\alpha = \frac{dqB}{mv}$

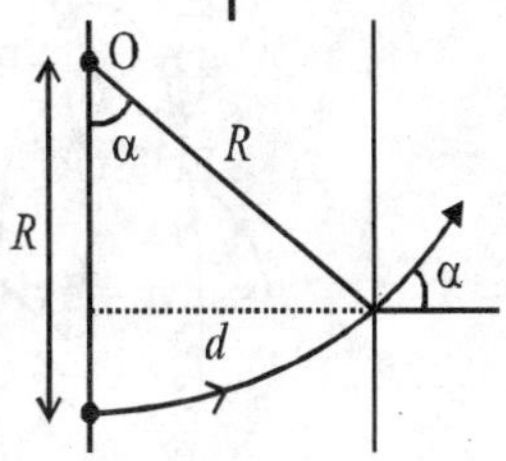

$\sin\alpha = Bd\sqrt{\frac{q}{2mV}}\left[\because qV = \frac{1}{2}mv^2\right]$

48. **(d)** When a charged particle enters the magnetic field in perpendicular direction then it experience a force in perpendicular direction.

i.e. $F = Bqv\sin\theta$

Due to which it moves in a circular path with constant speed as well as kinetic energy.

49. **(c)** As velocity $v = \frac{E}{B} = \frac{7.7\times 10^3}{0.14} = 55$ km/s

50. **(a)** The charge experiences both electric and magnetic force.

Electric force, $F_e = qE$

Magnetic force, $F_m = q(\vec{v}\times\vec{B})$

$\therefore$ Net force, $\vec{F} = q\left[\vec{E} + \vec{v}\times\vec{B}\right]$

$$= q\left[3\hat{i} + \hat{j} + 2\hat{k} + \begin{vmatrix}\hat{i} & \hat{j} & \hat{k}\\ 3 & 4 & 1\\ 1 & 1 & -3\end{vmatrix}\right]$$

$$= q\left[3\hat{i} + \hat{j} + 2\hat{k} + \hat{i}(-12-1) - \hat{j}(-9-1) + k(3-4)\right]$$

$$= q\left[3\hat{i} + \hat{j} + 2\hat{k} - 13\hat{i} + 10\hat{j} - \hat{k}\right] = q\left[-10\hat{i} + 11\hat{j} + \hat{k}\right]$$

$F_y = 11q$

51. **(b)** As velocity is not changing, charge particle must go undeflected, then

$qE = qvB \Rightarrow v = \frac{E}{B}$

Also, $\vec{E}\times\vec{B} \parallel \vec{V}$

and $\left|\frac{\vec{E}\times\vec{B}}{B^2}\right| = \frac{E\,B\sin\theta}{B^2} = \frac{E\,B\sin 90°}{B^2} = \frac{E}{B} = v$

52. **(b)** When a charged particle enters a magnetic field at a direction perpendicular to the direction of motion, the path of the motion is circular. In circular motion the direction of velocity changes at every point (the magnitude remains constant).

Therefore, the momentum will change at every point, but kinetic energy will remain constant as it is given by $\frac{1}{2}mv^2$ and v^2 is the square of the magnitude of velocity which does not change.

53. **(b)** The charged particle will move along the lines of electric field (and magnetic field). Magnetic field will exert no force. The force by electric field will be along the lines of uniform electric field. Hence the particle will move in a straight line.

54. **(c)** Equating magnetic force to centripetal force,

$\frac{mv^2}{r} = qvB\sin 90° \Rightarrow \frac{mv}{r} = Bq \Rightarrow v = \frac{Bqr}{m}$

Time to complete one revolution, $T = \frac{2\pi r}{v} = \frac{2\pi m}{qB}$

55. **(b)** Due to electric field, it experiences force and accelerates i.e. its velocity decreases.

56. **(b)** The workdone, $dW = Fds\cos\theta$

The angle between force and displacement is 90°. Therefore work done is zero.

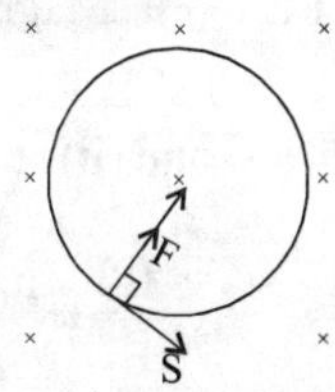

57. **(a)** When a moving charged particle is subjected to a perpendicular magnetic field, then it describes a circular path of radius.

$r = \frac{p}{qB}$

where q = Charge of the particle

p = Momentum of the particle

B = Magnetic field

Here p, q and B are constant for electron and proton, therefore the radius will be same.

58. **(a)** The time period of a charged particle of charge q and mass m moving in a magnetic field (B) is $T = \frac{2\pi m}{qB}$

Clearly time period is independent of speed of the particle.

59. **(c)** Resultant magnetic field at the centre of square loop

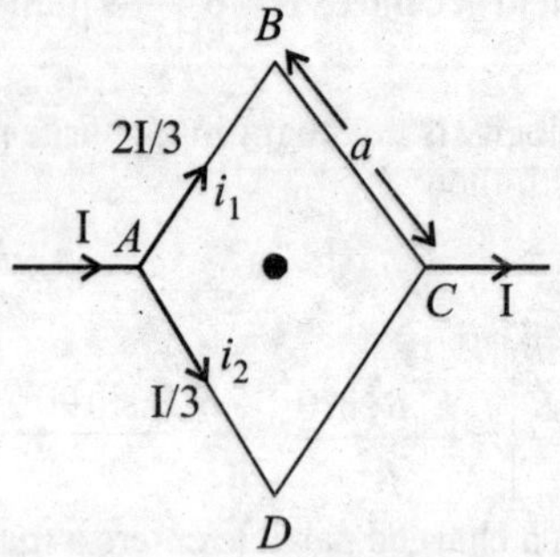

$$\vec{B} = \vec{B}_{AB} + \vec{B}_{BC} + \vec{B}_{CD} + \vec{B}_{DA}$$

$$\vec{B} = \left[\frac{-\mu_0(2I/3)}{4\pi(a/2)}\sqrt{2} - \frac{\mu_0(2I/3)}{4\pi(a/2)}\sqrt{2} + \frac{\mu_0(I/3)}{4\pi(a/2)}\sqrt{2} + \frac{\mu_0(I/3)}{4\pi(a/2)}\sqrt{2}\right]\hat{k}$$

$$\Rightarrow \vec{B} = \left[\frac{-2\sqrt{2}\mu_0 I}{3\pi a} + \frac{\sqrt{2}\mu_0 I}{3\pi a}\right]\hat{k} \quad \therefore \vec{B} = \frac{-\sqrt{2}\mu_0 I}{3\pi a}\hat{k}$$

60. **(1)** Magnetic field due to wires AB & CD = 0

$$B_0 = B_{BC} - B_{AD} = \frac{\mu_0 I}{4R_2} - \frac{\mu_0 I}{4R_1} = \frac{\mu_0 I}{4}\left(\frac{1}{R_2} - \frac{1}{R_1}\right)$$

Here, $R_1 = 6\pi\,m$, $R_2 = 4\pi\,m$

$$\therefore B_0 = \frac{4\pi\times10^{-7}\times12}{4}\left(\frac{1}{4\pi} - \frac{1}{6\pi}\right)$$

$$= 12\pi\times10^{-7}\left(\frac{1}{12\pi}\right) = 1\times10^{-7}\,T = k\times10^{-7}\,T \text{ (given)}$$

$\therefore \quad K = 1$

61. **(c)** Magnetic field (i) at the centre of the loop $B_1 = \frac{\mu_0 i}{2R}$

and (ii) at the axis at point P $B_2 = \frac{\mu_0 i R^2}{2x^3}$

$$\therefore \frac{B_2}{B_1} = \frac{\mu_0 i R^2}{2x^3}\bigg|\frac{\mu_0 i}{2R} = \frac{R^3}{x^3} = \left(\frac{4}{5}\right)^3 = \frac{64}{125}$$

62. **(8)** Given,

Magnetic field, $B = 2.9\times10^{-4}$T

Radius, r = 10 cm

Number of turns, N = 200

Current, I = 0.29 A

The expression of magnetic field of solenoid is

$$B = \mu_0\left(\frac{N}{\ell}\right)I$$

$$\Rightarrow \ell = \frac{\mu_0 NI}{B} = \frac{\left(4\pi\times10^{-7}\right)(200)(0.29)}{2.9\times10^{-4}}\,m = 8\pi\,cm$$

63. **(b)** For long straight current carrying wire,
Maximum possible magnetic field is

$$B_{max} = B_{surface} = \frac{\mu_0 I}{2\pi a}$$

For inside, $B_{inside} = \frac{B_{max}}{2}$

$$\frac{\mu_0 Ir}{2\pi a^2} = \frac{\mu_0 I}{4\pi a} \Rightarrow r = \frac{a}{2}$$

For outside,

$$B_{outside} = \frac{B_{max}}{2} \Rightarrow \frac{\mu_0 I}{2\pi r} = \frac{\mu_0 I}{4\pi a} \Rightarrow r = 2a$$

The distance from the centre of wire is $\left[\frac{a}{2}, 2a\right]$

64. **(b)**

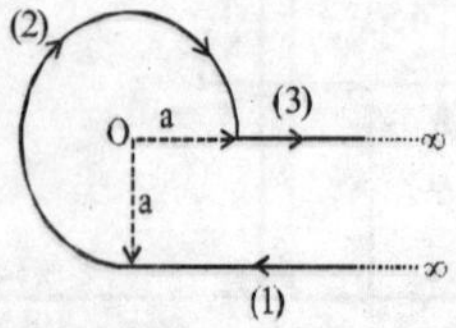

Magnetic field

$$B_1 = \frac{\mu_0 I}{4\pi a}\otimes$$

$$B_2 = \frac{\mu_0}{4\pi}\frac{I}{a}\left(\frac{3\pi}{2}\right)\otimes$$

$$B_3 = 0$$

$$\therefore \quad B_0 = B_1 + B_2 + B_3 = \frac{\mu_0}{4\pi}\frac{I}{a}\left(1 + \frac{3\pi}{2}\right)\otimes$$

65. **(c)**

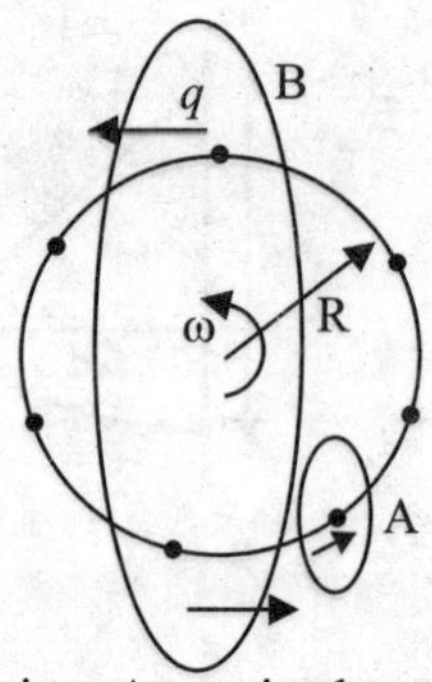

For given Amperian loops A and B,

$$I_A = \frac{Nq}{T} = \frac{Nq}{\frac{2\pi}{\omega}} = \frac{Nq\omega}{2\pi}$$

From loop B, Incoming current is equal to outgoing current. So, $I_B = 0$

$$\therefore I_A - I_B = \frac{Nq\omega}{2\pi}$$

66. **(a)**

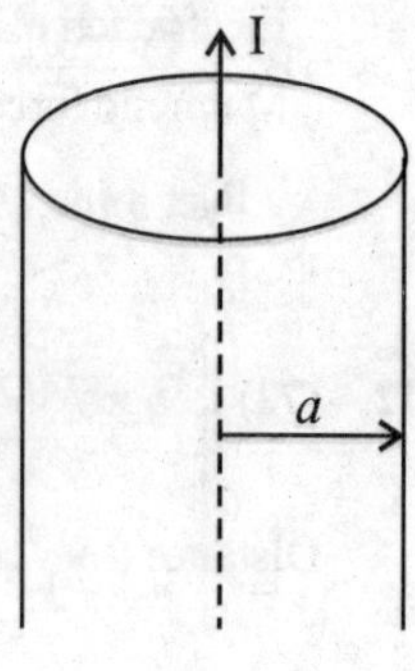

For given straight circular wire,

$$B_{in} = \left(\frac{\mu_0 I}{2\pi R^2}\right)r$$

$\therefore B_{in} \propto r$ for $r < R$

$$B_{out} = \frac{\mu_0 I}{2\pi r}$$

$\therefore B_{net} \propto \frac{1}{r}$ for $r > R$

67. (250) Given, Current, I = 1.5A
Time period of revolution of electron, T = 75 ns
For charged particle moving on circle in magnetic field,

$$T=\frac{2\pi m}{qB}=\frac{2\pi m}{q(\mu_0 nI)}$$

$$\Rightarrow 75\times10^{-9}=\frac{2\pi(9\times10^{-31})}{1.6\times10^{-19}\times4\pi\times10^{-7}\times n\times1.5}$$

$\therefore n=250$

68. (8)

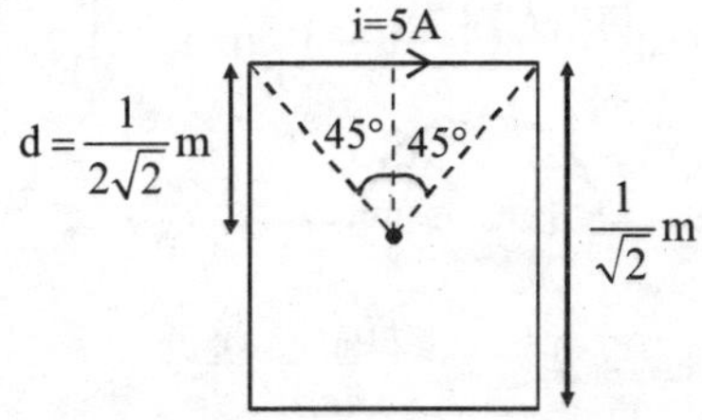

The net magnetic field at the centre of square loop is

$$B_{net}=4\,B_{\text{single side}}=\frac{\mu_0 i}{4\pi d}(\sin\theta_1+\sin\theta_2)$$

$$=\frac{4\times10^{-7}\times5\times2}{\frac{1}{2\sqrt{2}}}\times\frac{1}{\sqrt{2}}=8\times10^{-6}$$

$=p\times10^{-6}$ (given) $\therefore p=8$

69. (c) Magnetic field due to straight wire, $B=\frac{\mu_0 I}{2\pi r}$
Magnetic field at 2a is,

$$B_{2a}=\frac{\mu_0 I}{2\pi(2a)}=\frac{\mu_0 I}{4\pi a}$$

Inside the wire, $B_{a/2}=\left(\frac{\mu_0 I}{2\pi a^2}\right)\cdot\frac{a}{2}=\frac{\mu_0 I}{4\pi a}$

$$\therefore \frac{B_{a/2}}{B_{2a}}=\frac{1}{1}=1:1$$

70. (500) Magnetic field inside a solenoid, $B=\mu_0 ni$

$$\Rightarrow \mu_0\left(\frac{m}{\ell}\right)i=B \qquad \left[\because n=\frac{m}{\ell}\right]$$

$$\Rightarrow m=\frac{B\ell}{\mu_0 i}=\frac{6.28\times10^{-3}\times0.5}{12.56\times10^{-7}\times5}\quad \therefore m=500$$

71. (5) Magnetic field at point A

$$B_A=\frac{\mu_0 i}{2\pi r}+\frac{\mu_0(2i)}{2\pi(3r)}=\frac{5\mu_0 i}{6\pi r}$$

Magnetic field at point C

$$B_C=\frac{\mu_0(2i)}{2\pi r}+\frac{\mu_0 i}{2\pi(3r)}=\frac{7\mu_0 i}{6\pi r}$$

$$\therefore \frac{B_A}{B_C}=\frac{5}{7}=\frac{x}{7}\qquad \therefore x=5$$

72. (72) Length of each side, $a=\frac{4\pi}{6}$

Distance from centre, $r=\frac{\sqrt{3}}{2}a=\frac{\sqrt{3}}{2}\times\frac{4\pi}{6}=\frac{\pi}{\sqrt{3}}$

Magnetic field at centre, $B=6\left(\frac{\mu_0 I}{4\pi r}\right)(\sin\theta_1+\sin\theta_2)$

$$\Rightarrow B=\frac{6\times4\pi\times10^{-7}\times4\pi\sqrt{3}}{4\pi\times\left(\frac{\pi}{\sqrt{3}}\right)}(\sin30°+\sin30°)$$

$\Rightarrow B=6\times4\times3\times10^{-7}=72\times10^{-7}T=x\times10^{-7}$ T (given)
$\therefore x=72$

73. (20)

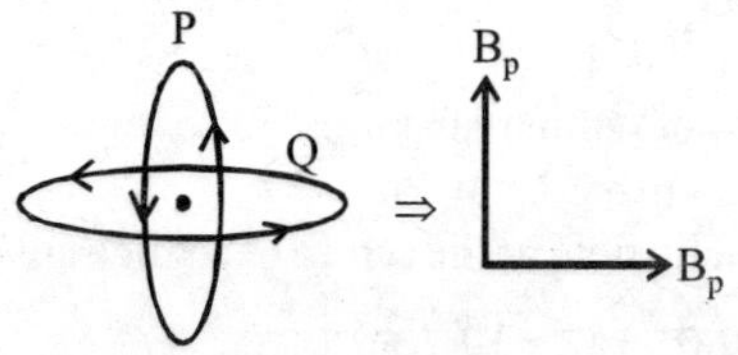

Magnetic field due to coil P

$$B_P=\frac{\mu_0 Ni_1}{2r}=\frac{\mu_0\times1\times100}{2\pi}=2\times10^{-3}T$$

Magnetic field due to coil Q

$$B_Q=\frac{\mu_0 Ni_2}{2r}=\frac{\mu_0\times2\times100}{2\pi}=4\times10^{-3}T$$

$$B_{net}=\sqrt{B_P^2+B_Q^2+2B_PB_Q\cos90°}=\sqrt{B_P^2+B_Q^2}$$

$$=\sqrt{(2\times10^{-3})^2+(4\times10^{-3})^2}=\sqrt{20}\times10^{-3}T$$

$=\sqrt{20}$ mT $=\sqrt{x}$ mT (given)
$\therefore$ x=20

74. (c)

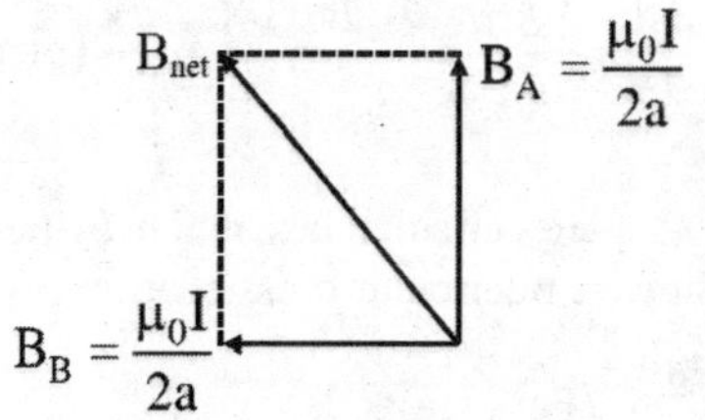

$$B_{net}=\sqrt{B_A^2+B_B^2}=\sqrt{2}B_A$$

$$\because B_{net}=\frac{\sqrt{2}\mu_0 I}{2a}=\frac{\mu_0 I}{\sqrt{2}a}$$

75. (40) The magnetic field at the centre is

$$B=4\times\frac{\mu_0 i}{4\pi r}(\sin45°+\sin45°)\quad\left(\because r=\frac{1}{2}m\right)$$

$$B=4\times\frac{\mu_0 i}{4\pi(1/2)}\left(\frac{1}{\sqrt{2}}+\frac{1}{\sqrt{2}}\right)$$

$$=4\times10^{-7}\times5\times2\times\sqrt{2}=40\sqrt{2}\times10^{-7}T$$

$=x\sqrt{2}\times10^{-7}$T (given)
$\therefore$ x=40

76. (3) Here magnetic field at the centre (O) of a wire loop formed by two semicircular wires of radii R_1 and R_2

$$\vec{B}_{centre}=\frac{\mu_0 i}{2R_2}\left(\frac{\pi}{2\pi}\right)\otimes+\frac{\mu_0 i}{2R_1}\left(\frac{\pi}{2\pi}\right)\otimes$$

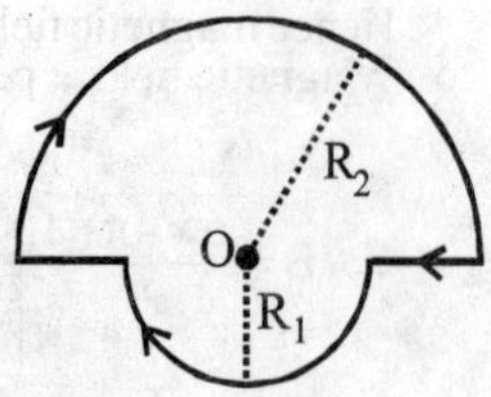

$$= \left(\frac{\mu_0 i}{4R_2} + \frac{\mu_0 i}{4R_1}\right) \otimes$$

$$= \frac{4\pi \times 10^{-7} \times 4}{4 \times 4\pi} + \frac{4\pi \times 10^{-7} \times 4}{4 \times 2\pi}$$

$= 3 \times 10^{-7}\,T = \alpha \times 10^{-7}\,T$ (given)

$\therefore \alpha = 3$

77. (2) Given,

Current, $I = 14A$

Radius of semicircular arc,

$R = 2.2\text{ cm} = 2.2 \times 10^{-2}\text{m}$

Magnetic field at the centre of semicircular arc

$$B = \frac{\mu_0 I}{4R} = \frac{4\pi \times 10^{-7} \times 14}{4 \times 2.2 \times 10^{-2}}$$

$$= \frac{4 \times 22 \times 10^{-7} \times 14}{4 \times 7 \times 2.2 \times 10^{-2}} = 2 \times 10^{-4}\,T$$

78. (8) Magnetic field at centre $(B_1) = \frac{\mu_0 I}{2r}$

Magnetic field on axis $(B_2) = \frac{\mu_0 I r^2}{2\left(r^2 + x^2\right)^{3/2}}$

Value of $x = r$ (given)

$$B_2 = \frac{\mu_0 I}{4\sqrt{2}r}$$

$$\frac{B_1}{B_2} = \frac{\mu_0 I}{2r} \times \frac{4\sqrt{2}r}{\mu_0 I} = \frac{2\sqrt{2}}{1} = \frac{\sqrt{8}}{1} = \frac{\sqrt{x}}{1}\text{(given)}$$

$\therefore x = 8$

79. (c) Using Ampere's circuital law, magnetic field for long straight wire of circular cross-section is given by

$B = \frac{\mu_0 I r}{2\pi a^2}$ (For $r < a$) $\therefore B \propto r$

For $r > a$

$B = \frac{\mu_0 I}{2\pi r}$ $\therefore B \propto \frac{1}{r}$

80. (628) Given,

Radius of circular wires, $r = 20\text{ cm} = 0.2\text{ m}$

Current, $i = \sqrt{2}A$

Magnetic field due to circular wire X = magnetic field due to circular wire Y

$$\therefore B_x = B_y = \frac{\mu_0 i}{2r}$$

$$= \frac{4\pi \times 10^{-7}}{2 \times 0.2} \times \sqrt{2}T \qquad [\because \mu_0 = 4\pi \times 10^{-7}]$$

Net magnetic field is

$$B_N = \sqrt{B_x^2 + B_y^2} = \frac{4\pi \times 10^{-7} \times \sqrt{2}}{2 \times 0.2} \times \sqrt{2}T = 2\pi \times 10^{-6}T$$

$= 200\pi \times 10^{-8}\,T = 2 \times 314 \times 10^{-8}\,T$

$= 628 \times 10^{-8}T$

81. (a) The magnetic field at the point P is $B_P = B_{AB} + B_{BC} + B_{CD}$

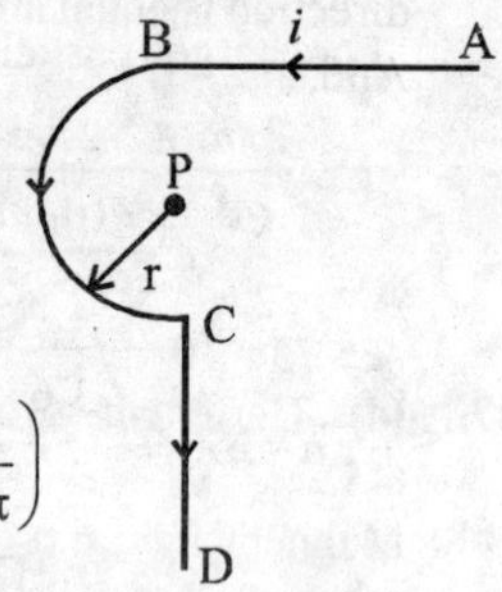

$$B_{AB} = \frac{\mu_0 i}{4\pi r}$$

$$B_{BC} = \frac{\mu_0 i}{4\pi r} \times \pi = \frac{\mu_0 i}{4r}$$

$$B_{CD} = 0$$

$$B_P = \left(\frac{\mu_0 i}{4r} + \frac{\mu_0 i}{4\pi r}\right) = \frac{\mu_0 i}{2r}\left(\frac{1}{2} + \frac{1}{2\pi}\right)$$

82. (c)

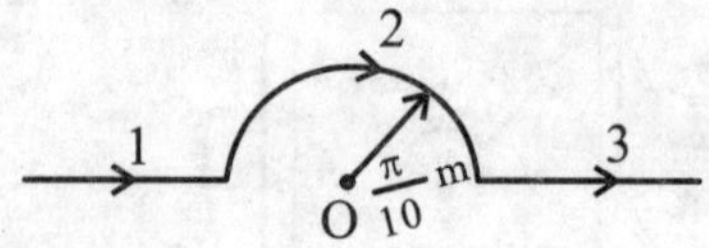

$$\vec{B}_{net} = \vec{B}_1 + \vec{B}_2 + \vec{B}_3 = \left(0 + \frac{\mu_0 I}{4r} + 0\right) \otimes$$

$$= \left(\frac{\mu_0 \times 3 \times 10}{4 \times \pi}\right) \otimes = 10^{-7} \times 3 \times 10 \otimes = 3\,\mu T$$

83. (c) We have,

$$B_1 = \frac{N\mu_0 I}{2r_1} \text{ and } B_2 = \frac{n\mu_0 I}{2r_2}$$

Now, $N \times 2\pi r_1 = n \times 2\pi r_2 \Rightarrow N r_1 = n r_2 \Rightarrow \frac{r_2}{r_1} = \frac{N}{n}$

So, $\frac{B_1}{B_2} = \frac{N r_2}{r_1 n} = \frac{N}{n}\left(\frac{r_2}{r_1}\right) = \frac{N^2}{n^2}$

84. (c) The magnetic moment associated with circular coil,

$M = NIA$

$M_A = M_B$

$\therefore N_A I_A A_A = N_B I_B A_B$

$\therefore N_A I_A \pi(0.1)^2 = N_B I_B \pi(0.2)^2$ $(\because A = \pi r^2)$

$\therefore N_A I_A = 4 N_B I_B$

85. (d)

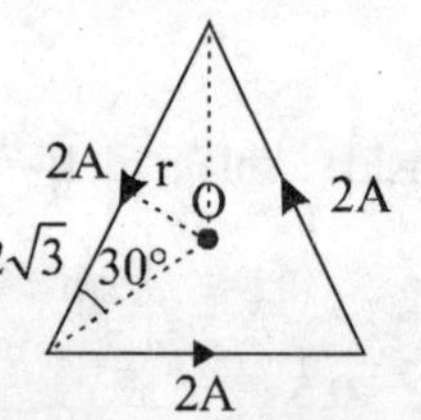

From Figure

$$\Rightarrow \frac{r}{2\sqrt{3}} = \tan 30° \Rightarrow r = 2\text{ cm}$$

Magnetic field at the centroid of triangle,

$$B = 3 \times \frac{\mu_0 i}{4\pi r}(\sin 60° + \sin 60°) \Rightarrow B = 3 \times \frac{\mu_0 i}{2\pi r}\sin 60°$$

$$= 3 \times \frac{2 \times 10^{-7} \times 2}{2 \times 10^{-2}} \times \frac{\sqrt{3}}{2} = 3\sqrt{3} \times 10^{-5}\,T$$

86. (c)

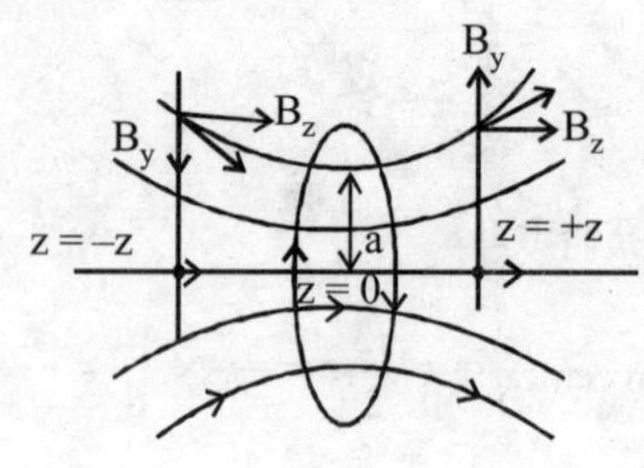

We can see from magnetic field lines drawing that B_y direction is equal and opposite on z = +z and z = –z
And, at $z = 0 \Rightarrow B_y = 0$ So, curves will look like.

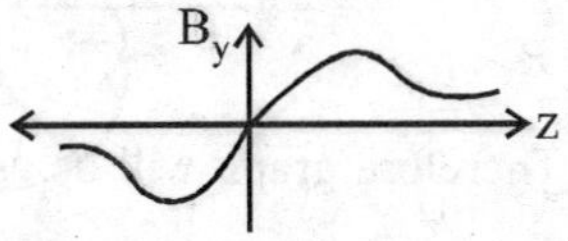

87. **(d)** Current in AB and ED will not generate magnetic field at O.
Magnetic field at point 'O' due to current in BC and ET are outward.

$$\therefore\ B_0 = \frac{\mu_0 I}{4\pi \frac{a}{2}} + \frac{\mu_0 I}{4\pi \frac{a}{2}} = 2 \times \frac{\mu_0 I}{4\pi \frac{a}{2}} = \frac{\mu_0 I}{\pi a}.$$

88. **(c)** Initial length = Final length
$\Rightarrow 4 \times 2\pi R_i = 2\pi R_f \Rightarrow R_f = 4R_i$

As $B = \frac{\mu_0 i}{2R} \times n \Rightarrow B \propto \frac{n}{R}$

$$\Rightarrow \frac{B_2}{B_1} = \frac{n_2}{n_1} \times \frac{R_1}{R_2} = \frac{1}{4} \times \frac{1}{4} = \frac{1}{16}$$

So, $B_2 = \frac{B_1}{16} = \frac{32}{16} = 2T$

89. **(c)** (A)

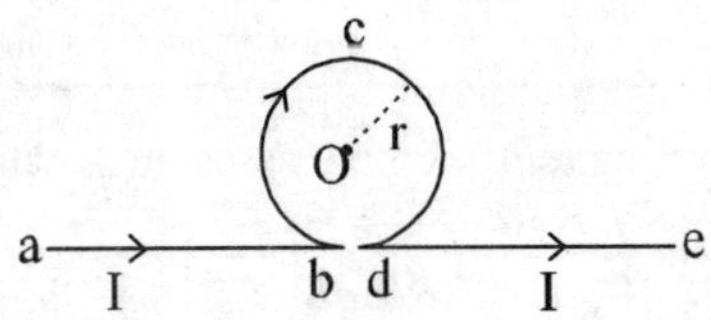

$B_{ab} = \frac{\mu_0 I}{4\pi r}$ (out of the plane)

$B_{bcd} = \frac{\mu_0 I}{4\pi r}(2\pi)$ (into the plane)

$B_{de} = \frac{\mu_0 I}{4\pi r}$ (out of the plane)

Hence magnetic field at O is
$B_0 = B_{ab} + B_{bcd} + B_{de}$

$$B_0 = -\frac{\mu_0 I}{4\pi r} + \frac{\mu_0 I}{4\pi r}(2\pi) - \frac{\mu_0 I}{4\pi r}$$

$$B_0 = \frac{\mu_0 I}{4\pi r}(2\pi - 2)$$

$$B_0 = \frac{\mu_0 I}{2\pi r}(\pi - 1) \quad \text{......(III)}$$

(B)

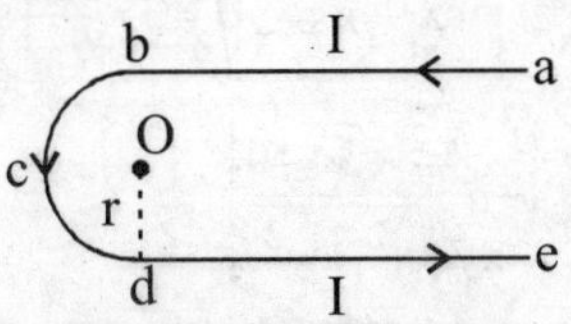

$B_{ab} = \frac{\mu_0 I}{4\pi r}$ (out of the plane)

$B_{bcd} = \frac{\mu_0 I}{4\pi r}(\pi)$ (out of the plane)

$B_{de} = \frac{\mu_0 I}{4\pi r}$ (out of the plane)

Hence magnetic field at O is
$B_0 = B_{ab} + B_{bcd} + B_{de}$

$$B_0 = \frac{\mu_0 I}{4\pi r} + \frac{\mu_0 I}{4\pi r}(\pi) + \frac{\mu_0 I}{4\pi r}$$

$$B_0 = \frac{\mu_0 I}{4\pi r}(\pi + 2) \quad \text{.....(I)}$$

(C)

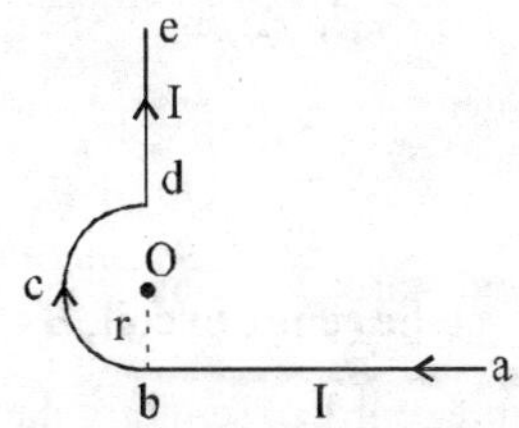

$B_{ab} = \frac{\mu_0 I}{4\pi r}$ (into the plane)

$B_{bcd} = \frac{\mu_0 I}{4\pi r}(\pi)$ (into the plane)

$B_{de} = 0$ (at the axis)
Hence, magnetic field at O is
$B_0 = B_{ab} + B_{bcd} + B_{de}$

$$B_0 = \frac{\mu_0 I}{4\pi r}(1 + \pi) \quad \text{.....(IV)}$$

(D)

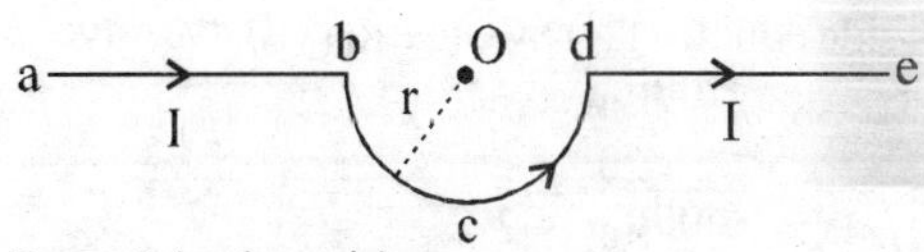

$B_{ab} = 0$ (at the axis)

$B_{bcd} = \frac{\mu_0 I}{4\pi r}(\pi)$ (out of the plane)

$B_{de} = 0$ (at the axis)
Hence magnetic field at O is
$B_0 = B_{ab} + B_{bcd} + B_{de}$

$$= 0 + \frac{\mu_0 I}{4\ r} + 0$$

$$B_0 = \frac{\mu_0 I}{4r} \quad \text{.....(II)}$$

90. **(b)** Magnetic field inside solenoid is given as
$B = \mu_0 n I$
$= 4\pi \times 10^{-7} \times 70 \times 10^2 \times 2 = 56\pi \times 10^{-4} T = 176 \times 10^{-4} T$

91. **(11)**

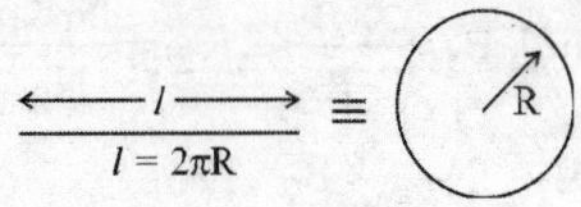

As, length of wire = $2\pi R$

$$\Rightarrow \frac{314}{100} = 2\pi R \quad \Rightarrow R = 0.5\,m$$

Magnetic Moment = I × A = 14 × πR^2

$$= 14 \times (3.14) \times \frac{1}{4} = 10.99 \approx 11.00\ A - m^2$$

92. **(3)** Magnetic field at centre of ring/coil is given as,

$$B_{centre} = \frac{N\mu_0 I}{2R}$$

$$\Rightarrow 37.68 \times 10^{-4} = \frac{100 \times 4\pi \times 10^{-7} \times I}{2 \times 5 \times 10^{-2}} \quad \Rightarrow I = 3A$$

93. **(c)** $B_1 = \frac{\mu_0 I}{2R}$ ...(i)

$B_2 = \frac{\mu_0 I R^2}{2(R^2 + 3R^2)^{3/2}}$ ...(ii)

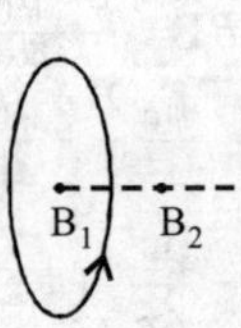

$= \frac{1}{8}\left(\frac{\mu_0 I}{2R}\right) = \frac{B_1}{8}$

Dividing (i) by (ii), we get

$\Rightarrow \frac{B_1}{B_2} = \frac{8}{1}$

94. **(b)** The magnetic field at the centre of coil, $B = \frac{\mu_0 N i}{2R}$

Given, for coil x

Number of turns, $N_x = 200$

Radius, $R_x = 20$ cm

$\Rightarrow B_x = \frac{\mu_0 N_x i}{2R_x} = \frac{200\mu_0 i}{2 \times 20}$ ..... (i)

For coil y

Number of turns, $N_y = 400$

Radius, $R_y = 20$ cm

$\Rightarrow B_y = \frac{\mu_0 N_y i}{2R_y} = \frac{400\mu_0 i}{2 \times 20}$ ..... (ii)

Divide equation (i) by equation (ii), we have

$$\Rightarrow \frac{B_x}{B_y} = \frac{\frac{200\mu_0 i}{2\times 20}}{\frac{400\mu_0 i}{2\times 20}} = \frac{1}{2}$$

95. **(b)** From the given figure. Magnetic moment will be,

$\vec{M}$ = current $(\vec{I}) \times$ Area $(\vec{A})$

$\vec{M} = -I\pi(0.5)^2\hat{k} + I\pi(0.3)^2\hat{k}$

$\vec{M} = -7 \times \frac{22}{7}\left(\frac{25}{100} - \frac{9}{100}\right)\hat{k} = -22\left(\frac{16}{100}\right)\hat{k}$

$\vec{M} = -3.52\,\hat{k}\ Am^2 = -\frac{7}{2}\hat{k}\ Am^2$

96. **(b)** Magnetic field at centre is given as

$B = \frac{N\mu_0 i}{2R}$

So, $B_1 = \frac{N_1\mu_0 i}{2R_1}$ and, $B_2 = \frac{N_2\mu_0 i}{R_2}, R_2 = \frac{N_1 R_1}{N_2}$

$\frac{B_2}{B_1} = \frac{N_2}{N_1}\frac{R_1}{R_2} = \frac{N_2}{N_1} \times \frac{N_2}{N_1} \Rightarrow \frac{B_2}{B_1} = \frac{25}{4}$

97. **(d)** Let us consider a circular strip of length 'dr' at distance 'r' from centre

Magnetic field at centre due to this strip is given as

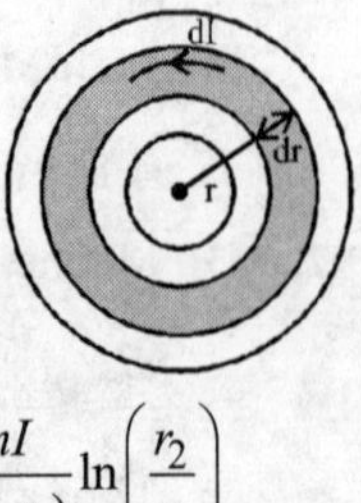

$dB = \frac{\mu_0 I}{2r}dn = \frac{\mu_0 I}{2r}\cdot\frac{n}{r_2 - r_1}\cdot dr$

So, $B_{net} = \int dB = \frac{\mu_0 nI}{2(r_2 - r_1)}\int_{r_1}^{r_2}\frac{dr}{r} = \frac{\mu_o nI}{2(r_2 - r_1)}\ln\left(\frac{r_2}{r_1}\right)$

98. **(d)** $\vec{B} = O, r < R$

$= \frac{\mu_0 I}{2r}, r \geq R$

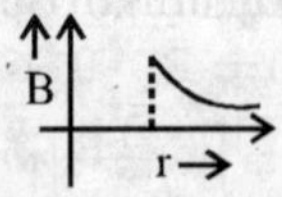

So, $B = 0, r < R$

$\propto \frac{1}{r}, r \geq R$. Therefore graph will be as such shown in figure above.

99. **(b)** Let us draw an amperian loop of radius 'r' as shown then.

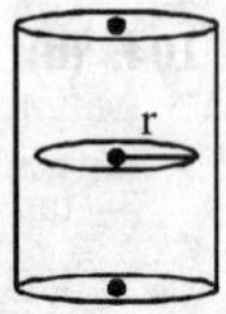

By Ampere's law

$\int \vec{B}.\vec{dr} = \mu_0 I_{in}$

$\Rightarrow B \times 2\pi r = \mu_0 \frac{I}{\pi R^2} \times \pi r^2 \Rightarrow B = \frac{\mu_0 I r}{2\pi R^2} \Rightarrow B \propto r$

100. **(a)** Magnetic field due to solenoid is given by

$B = \mu_0 nI \ \Rightarrow B \propto nI \Rightarrow \frac{B_2}{B_1} = \frac{n_2 I_2}{n_1 I_1}$

$\Rightarrow B_2 = \frac{n_2 I_2}{n_1 I_1} \times B_1 \ \Rightarrow B_2 = \frac{1}{2} \times 2 \times B_1 \ \therefore B_2 = B_1 = B$

101. **(c)** $B_{centre} = \frac{\mu_0 I}{2r}$

$$B_{r/2} = \frac{\mu_0 I}{2}\frac{r^2}{\left[r^2 + \left(\frac{r}{2}\right)^2\right]^{3/2}} = \frac{\mu_0 I}{2}\frac{r^2}{\left[\frac{5r^2}{4}\right]^{3/2}}$$

$$= \frac{\mu_0 I}{2r}\left(\frac{4}{5}\right)^{3/2} = B_{centre}\left(\frac{2}{\sqrt{5}}\right)^3$$

102. **(a)** $B_1 = \frac{\mu_0 I}{4\pi y}[\sin 90 + \sin\theta_1]$

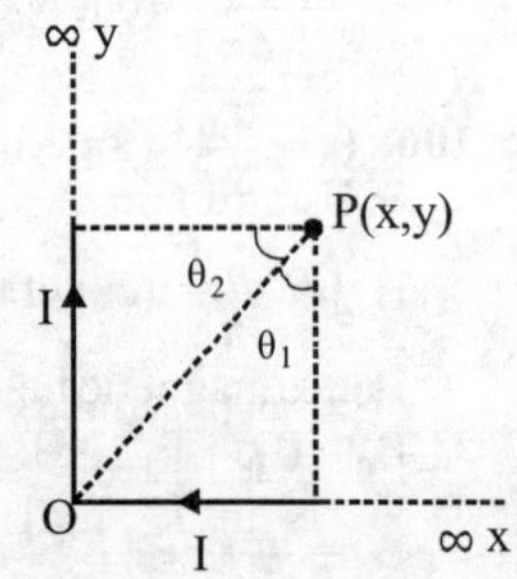

$= \frac{\mu_0 I}{4\pi y}\left(1 + \frac{x}{\sqrt{x^2 + y^2}}\right)$ ...(i)

$B_2 = \frac{\mu_0}{4\pi}\frac{I}{x}(\sin 90 + \sin\theta_2)$

$= \frac{\mu_0}{4\pi}\frac{I}{y}\left(1 + \frac{y}{\sqrt{x^2 + y^2}}\right)$ ...(ii)

$B_{Net} = B_1 + B_2$

$$B = \frac{\mu_0 I}{4\pi}\left[\frac{1}{y} + \frac{x}{y\sqrt{x^2 + y^2}} + \frac{1}{x} + \frac{y}{x\sqrt{x^2 + y^2}}\right]$$

$$\Rightarrow B = \frac{\mu_0 I}{4\pi}\left[\frac{x + y}{xy} + \frac{x^2 + y^2}{xy\sqrt{x^2 + y^2}}\right]$$

$$= \frac{\mu_0 I}{4\pi}\left[\frac{x + y}{xy} + \frac{\sqrt{x^2 + y^2}}{xy}\right]$$

$$\Rightarrow B = \frac{\mu_0 I}{4\pi xy}\left[\sqrt{x^2 + y^2} + (x + y)\right]$$

103. (a) No. of turns in dr width $= \dfrac{N}{b-a}dr$

$$B = \int dB = \int_a^b \left(\frac{N}{b-a}\right) dr \frac{\mu_0 I}{2x}$$

$$\therefore B = \frac{N\mu_0 I}{2(b-a)} \ell n\left(\frac{b}{a}\right)$$

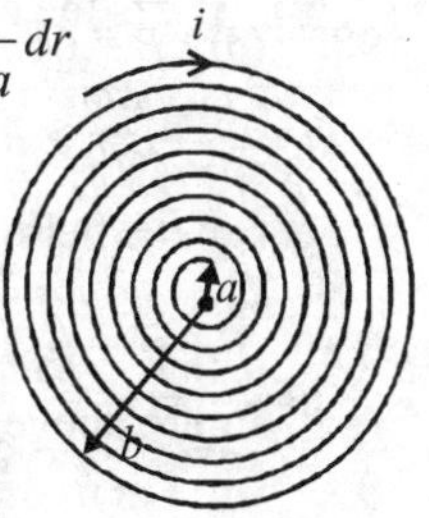

104. (d) $B = 3\left[\dfrac{\mu_0 i}{4\pi d}(\sin 60° + \sin 60°)\right]$

$\tan 60° = \dfrac{a/2}{d}$

put $a = 9 \times 10^{-2}\ m \Rightarrow d = \dfrac{9\times 10^{-2}}{2\sqrt{3}}m$

$B = 3 \times 10^{-5}\ T.$

$\vec{B}$ is into the plane of triangle.

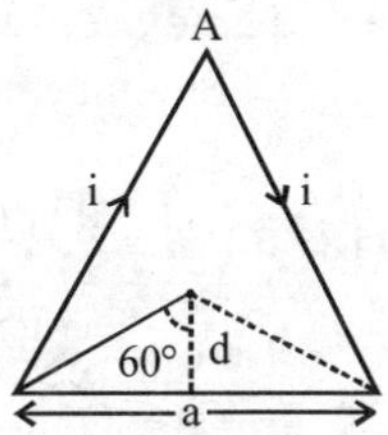

105. (3) No. of turn in triangle

$N_1 = \dfrac{24a}{3a} = 8$

No. of turns in square

$N_2 = \dfrac{24a}{4a} = 6$

$\dfrac{M_1}{M_2} - \dfrac{N_1 I A_1}{N_2 I A_2}$ [As $i_1 = i_2 = I$]

$= \dfrac{8\times\dfrac{\sqrt{3}}{4}\times a^2}{6\times a^2}$

$\dfrac{M_1}{M_2} = \dfrac{1}{\sqrt{3}} = \dfrac{1}{\sqrt{y}}$ so, $y = 3$

106. (a) For x < a

Using ampere's circuital law

$\int B.dl = \mu_o$ (current enclosed)

$\Rightarrow B_1(2\pi x) = \mu_o\left(\dfrac{i_o}{\pi a^2}\right)\pi x^2$

$\Rightarrow B_1(2\pi x) = \dfrac{\mu_o i_o x^2}{a^2}$

$\Rightarrow B_1 = \dfrac{\mu_o i_o x}{2\pi a^2}$ ...(i)

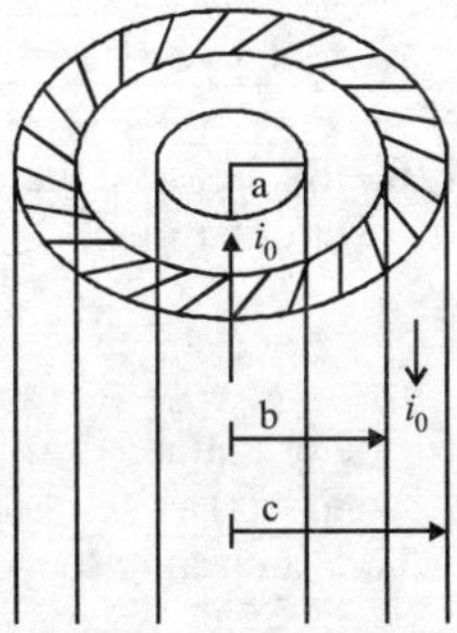

For a < x < b

$B_2(2\pi x) = \mu_o i_o \Rightarrow B_2 = \dfrac{\mu_o i_o}{2\pi x}$ ...(ii)

$$\therefore \frac{B_1}{B_2} = \frac{\mu_o i_o \dfrac{x}{2\pi a^2}}{\dfrac{\mu_o i_o}{2\pi x}} = \frac{x^2}{a^2}$$

107. (d) magnetic filed at centre (B_c) is

$B_c = \dfrac{\mu_0 i}{2a}$

magnetic field at distance r from centre B_r on axis is

$$B_r = \frac{\mu_0 i a^2}{2(a^2+r^2)^{3/2}}$$

fractional change in magnetic field

$$= \frac{\dfrac{\mu_0 i}{2a} - \dfrac{\mu_0 i a^2}{2(a^2+r^2)^{2/3}}}{\dfrac{\mu_0 i}{2a}} = 1 - \frac{1}{\left[1+\left(\dfrac{r^2}{a^2}\right)\right]^{3/2}}$$

$$\approx 1 - \left[1 - \frac{3}{2}\frac{r^2}{a^2}\right] = \frac{3}{2}\frac{r^2}{a^2}$$

Note : $\left(a + \dfrac{r^2}{a^2}\right)^{-3/2} \approx \left(1 - \dfrac{3}{2}\dfrac{r^2}{a}\right)$

108. (c) $(B_{surface})_A = \dfrac{\mu_0 I}{2\pi a}$

$(B_{surface})_B = \dfrac{\mu_0 I}{2\pi b}$

As $a < b$ So, $(B_{surface})_A > (B_{surface})_B$

109. (a) Magnetic field at P,

$B_P = 2$ (Magnetic field due to straight wire) + (Magnetic field due to semicircle)

$$B_P = 2\left(\frac{\mu_0 I}{4\pi r}\right) + \frac{\mu_0 I}{4r} \quad \therefore B_P = \frac{\mu_0 I}{2\pi r} + \frac{\mu_0 I}{4r} = \frac{\mu_0 I}{4\pi r}(2+\pi)$$

110. (d) Length of the circular path, $l = 2\pi r$

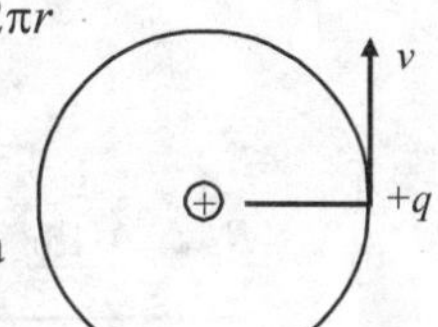

Current, $i = \dfrac{q}{T} = \dfrac{qv}{2\pi r}$

Magnetic moment M = Current × Area

$= i \times \pi r^2 = \dfrac{qv}{2\pi r} \times \pi r^2$

$M = \dfrac{1}{2} q \cdot v \cdot r$

Radius of circular path in magnetic field, $r = \dfrac{mv}{qB}$

$\therefore M = \dfrac{1}{2} qv \times \dfrac{mv}{qB} \Rightarrow M = \dfrac{mv^2}{2B}$

Direction of $\vec{M}$ is opposite of $\vec{B}$ therefore

$\vec{M} = \dfrac{-mv^2\vec{B}}{2B^2}$

111. (d) Given : $I_A = 2$ A, $R_A = 2$ cm, $\theta_A = 2\pi - \dfrac{\pi}{2} = \dfrac{3\pi}{2}$

$I_B = 3$ A, $R_B = 4$ cm, $\theta_B = 2\pi - \dfrac{\pi}{3} = \dfrac{5\pi}{3}$

Using, magnetic field, $B = \dfrac{\mu_0 I\theta}{4\pi R}$

$$\frac{B_A}{B_B} = \frac{I_A}{I_B} \times \frac{\theta_A R_B}{\theta_B R_A} = \frac{2\times\dfrac{3\pi}{2}\times 4}{3\times\dfrac{5\pi}{3}\times 2} = \frac{6}{5}$$

112. (c) Magnetic field due to one side of hexagon

$$B=\frac{\mu_0 I}{4\pi\frac{\sqrt{3}a}{2}}(\sin 30°+\sin 30°)$$

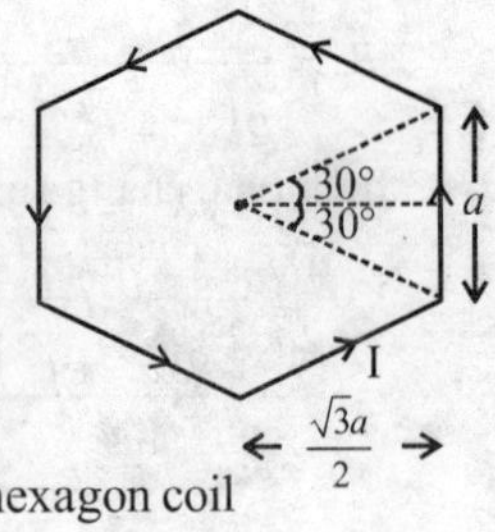

$$\Rightarrow B=\frac{\mu_0 I}{2\sqrt{3}a}\left(\frac{1}{2}+\frac{1}{2}\right)=\frac{\mu_0 I}{2\sqrt{3}a\pi}$$

Now, magnetic field due to one hexagon coil

$$B=6\times\frac{\mu_0 I}{2\sqrt{3}a\pi}$$

Again magnetic field at the centre of hexagonal shape coil of 50 turns,

$$B=50\times 6\times\frac{\mu_0 I}{2\sqrt{3}a\pi}\qquad\left[\because a=\frac{10}{100}=0.1\text{ m}\right]$$

$$\text{or, } B=\frac{150\mu_0 I}{\sqrt{3}\times 0.1\times\pi}=500\sqrt{3}\frac{\mu_0 I}{\pi}$$

113. (a) Let a be the radius of the wire

Magnetic field at point A (inside)

$$B_A=\frac{\mu_0 ir}{2\pi a^2}=\frac{\mu_0 i\frac{a}{3}}{2\pi a^2}=\frac{\mu_0 i}{\pi a^2}\frac{a}{6}=\frac{\mu_0 i}{6\pi a}$$

Magnetic field at point B (outside)

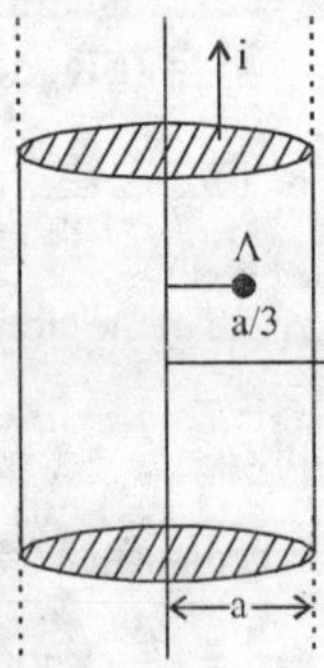

$$B_B=\frac{\mu_0 i}{2\pi(2a)}$$

$$\therefore \frac{B_A}{B_B}=\frac{\frac{\mu_0 i}{6\pi a}}{\frac{\mu_0 i}{2\pi(2a)}}=\frac{4}{6}=\frac{2}{3}$$

114. (b) Magnetic field inside the solenoid is given by

$B=\mu_0 nI$ .... (i)

Here, n = number of turns per unit length

The path of charge particle is circular. The maximum possible radius of electron $=\frac{R}{2}$

$$\therefore \frac{mV_{max}}{qB}=\frac{R}{2}$$

$$\Rightarrow V_{max}=\frac{qBR}{2m}=\frac{eR\mu_0 nI}{2m}\text{ (using (i))}$$

115. (a) $B_0=B_1+B_2+B_3+B_4$

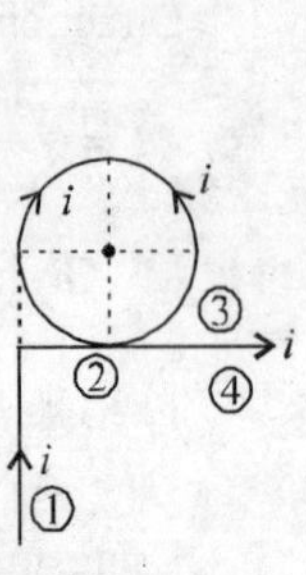

$$=-\frac{\mu_0 I}{4\pi R}[\sin 90°-\sin 45°]+\frac{\mu_0 I}{2R}+\frac{\mu_0 I}{4\pi R}[\sin 45°+\sin 90°]$$

$$=-\frac{\mu_0 I}{4\pi R}\left(1-\frac{1}{\sqrt{2}}\right)+\frac{\mu_0 I}{2R}+\frac{\mu_0 I}{4\pi R}\left(1+\frac{1}{\sqrt{2}}\right)$$

$$B_0=\frac{\mu_0 I}{2\pi R}\left(\pi+\frac{1}{\sqrt{2}}\right)$$

116. (a) $\vec{B}=\vec{B_1}+\vec{B_2}=\frac{\mu_0}{2\pi}\cdot\left(\frac{i}{d}\hat{k}+\frac{i}{d}\left(-\hat{k}\right)\right)=0$

117. (b) If q is the charge on the ring, then

$$i=\frac{q}{T}=\frac{q\omega}{2\pi}$$

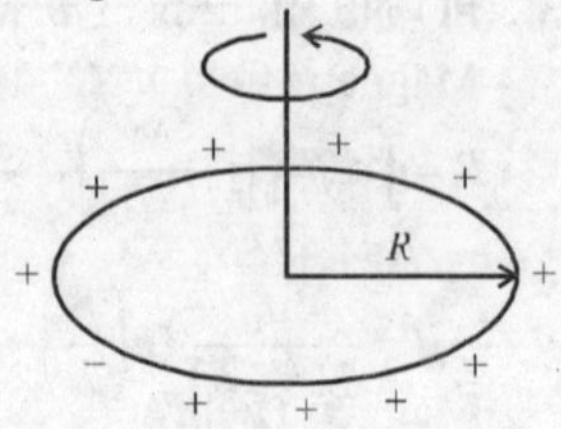

Magnetic field,

$$B=\frac{\mu_0 i}{2R}=\frac{\mu_0\left(\frac{q\omega}{2\pi}\right)}{2R}$$

$$\text{or }\ 3.8\times 10^{-9}=\left(\frac{\mu_0}{4\pi}\right)\frac{q\omega}{R}=\left(10^{-7}\right)\frac{q\times 40\pi}{0.10}$$

$\therefore\ q=3\times 10^{-5}$C.

118. (b) $B=\frac{\mu_0}{4\pi}\frac{i}{r}(\sin\alpha+\sin\beta)$

Here $r=\sqrt{5^2-3^2}=4$ cm

$\alpha=\beta=37°$

$\therefore B=10^{-7}\times\frac{5}{4}\times 2\sin 37°=1.5\times 10^{-5}$ T

119. (a) $\tan 30°=\frac{r}{a/2}$

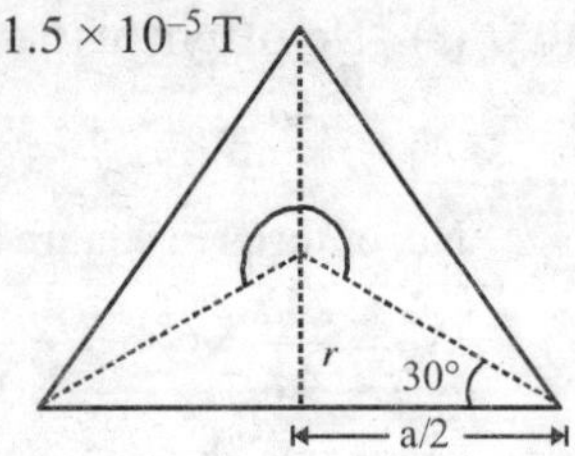

$$\therefore r=\frac{a}{2\sqrt{3}}$$

$$B_0=3\left[\frac{\mu_0 I}{4\pi r}(\sin 60°+\sin 60°)\right]$$

$$=\frac{3\mu_0 I}{4\pi\left(\frac{a}{2\sqrt{3}}\right)}\times(2)\left(\frac{\sqrt{3}}{2}\right)=\frac{9}{2}\left(\frac{\mu_0 I}{\pi a}\right)$$

$$=\frac{9\times 2\times 10^{-7}\times 10}{1}=18\,\mu\text{T}$$

120. (d) Let a be the area of the square and r be the radius of circular loop.

$$2\pi r=4a\Rightarrow r=\left(\frac{2a}{\pi}\right)$$

For square

m = (I) a^2

For circular loop

$$M=(I)\pi r^2\Rightarrow M=(I)(\pi)\left(\frac{4a^2}{\pi^2}\right)$$

$$M=\frac{4Ia^2}{\pi}\Rightarrow M=\frac{4m}{\pi}\quad(\because m=Ia^2)$$

121. (c) Let i be the current in each wire.

Magnetic field at 'O' due to LP and QM will be zero.

i.e., $B_0=B_{PS}+B_{QN}$

$\therefore$ Net magnetic field $B_0=\frac{\mu_0 i}{4\pi d}+\frac{\mu_0 i}{4\pi d}$

$$\Rightarrow 10^{-4}=\frac{\mu_0 i}{2\pi d}=\frac{2\times 10^{-7}\times i}{4\times 10^{-2}}$$

$\therefore$ i = 20 A and the direction of magnetic field is perpendicular into the plane

122. **(d)** There will be no magnetic field at O due to wire PQ and RS

Magnetic field at 'O' due to arc QR

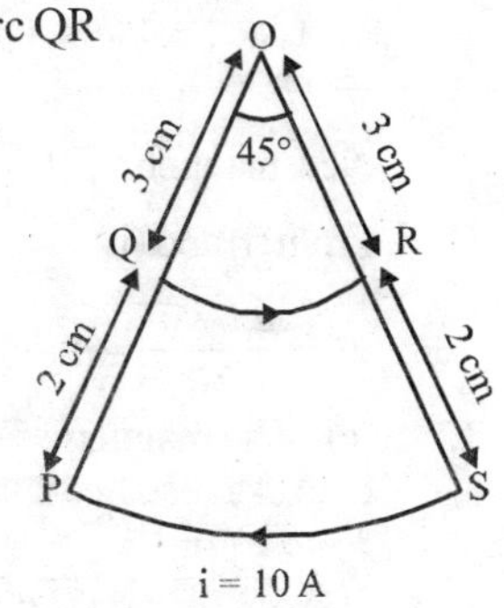

$$= \frac{\mu_0}{4\pi}\frac{\left(\frac{\pi}{4}\right).I}{r_1}$$

Magnetic field at 'O' due to are PS

$$= \frac{\mu_0}{4\pi}\frac{\left(\frac{\pi}{4}\right).I}{r_2}$$

∴ Net magnetic field at 'O'

$$B = \frac{\mu_0}{4\pi}(\pi/4)\times 10\left[\frac{1}{(3\times10^{-2})} - \frac{1}{(5\times10^{-2})}\right]$$

$$= \frac{\pi}{3}\times10^{-5}\,T \approx 1\times10^{-5}\,T$$

123. **(d)**

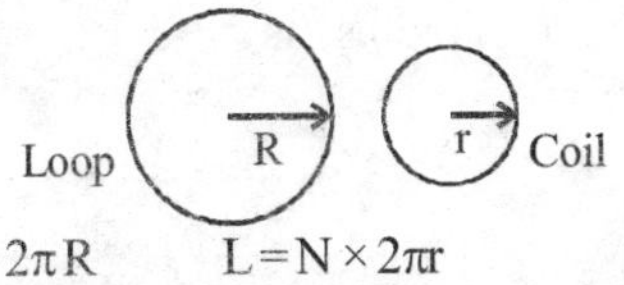

$L = 2\pi R \qquad L = N\times 2\pi r$

$R = Nr \Rightarrow r = \frac{R}{N}$

$$B_{Loop} = \frac{\mu_0 i}{2R},\ B_{coil} = \frac{\mu_0 Ni}{2r} = \frac{\mu_0 Ni}{2\left(\frac{R}{N}\right)} = \frac{\mu_0 N^2 i}{2R}\ \therefore \frac{B_L}{B_C} = \frac{1}{N^2}$$

124. **(c)** Magnetic field at the centre of loop, $B_1 = \frac{\mu_0 I}{2R}$

Dipole moment of circular loop is m = IA

$m_1 = I.A = I.\pi R^2$ {R = Radius of the loop}

If magnetic moment is doubled (keeping current constant) R becomes $\sqrt{2}R$

$$B_2 = \frac{\mu_0 I}{2(\sqrt{2}R)} \qquad \therefore \frac{B_1}{B_2} = \frac{\frac{\mu_0 I}{2R}}{\frac{\mu_0 I}{2(\sqrt{2}R)}} = \sqrt{2}$$

125. **(b)** Point *P* is situated at the mid-point of the line joining the centres of the circular wires which have same radii (R).

The magnetic fields ($\vec{B}$) at P due to the currents in the loops are in same direction.

Magnitude of magnetic field at point, *P*

$$B = 2\left\{\frac{\mu_0 NIR^2}{2\left(R^2 + \frac{R^2}{4}\right)^{3/2}}\right\} = \frac{8\mu_0 NI}{5^{3/2}R}$$

126. **(b)** **Case (a) :**

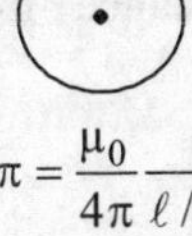

$$B_A = \frac{\mu_0}{4\pi}\frac{I}{R}\times 2\pi = \frac{\mu_0}{4\pi}\frac{I}{\ell/2\pi}\times 2\pi \quad (\because 2\pi R = \ell)$$

$$= \frac{\mu_0}{4\pi}\frac{I}{\ell}\times(2\pi)^2$$

Case (b) :

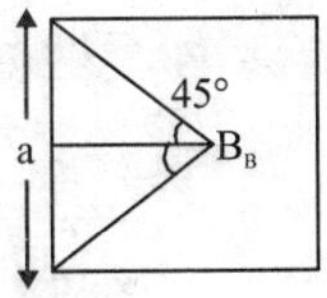

$$B_B = 4\times\frac{\mu_0}{4\pi}\frac{I}{a/2}[\sin 45° + \sin 45°]$$

$$= 4\times\frac{\mu_0}{4\pi}\times\frac{I}{\ell/8}\times\frac{2}{\sqrt{2}} = \frac{\mu_0}{4\pi}\frac{I}{\ell}\times\frac{64}{\sqrt{2}} = \frac{\mu_0 I}{4\pi\ell}32\sqrt{2} \qquad [4a = \ell]$$

$$\Rightarrow \frac{B_A}{B_B} = \frac{\pi^2}{8\sqrt{2}}$$

127. **(b)** For loop $B = \frac{\mu_0 nI}{2a}$

where, a is the radius of loop.

Then, $B_1 = \frac{\mu_0 I}{2a}$

Now, for coil $B = \frac{\mu_0 I}{4\pi}.\frac{2nA}{x^3}$

at the centre x = radius of loop

$$B_2 = \frac{\mu_0}{4\pi}.\frac{2\times3\times(I/3)\times\pi(a/3)^2}{(a/3)^3} = \frac{\mu_0.3I}{2a}$$

$$\therefore \frac{B_1}{B_2} = \frac{\mu_0 I/2a}{\mu_0.3I/2a}$$

$$B_1 : B_2 = 1:3$$

128. **(a)** Magnetic field at any point lies on axial position of current carrying conductor B = 0

129. **(b)** Given : Radius = R

Distance $x = 2\sqrt{2}R$

$$\frac{B_{centre}}{B_{axis}} = \left(1 + \frac{x^2}{R^2}\right)^{3/2} = \left(1 + \frac{(2\sqrt{2}R)^2}{R^2}\right)^{3/2} = (9)^{3/2} = 27$$

130. **(a)** The magnetic field due to a disc is given as

$$B = \frac{\mu_0\omega Q}{2\pi R} \quad \text{i.e., } B \propto \frac{1}{R}$$

131. **(d)** Let *R* be the radius of semicircular ring. Let an elementary length *dl* is cut for finding magnetic field. So,

Current in a small element, $dI = \frac{d\theta}{\pi}I$

Magnetic field due to the element

$$dB = \frac{\mu_0}{4\pi}\frac{2dI}{R} = \frac{\mu_0 I}{2\pi^2 R}$$

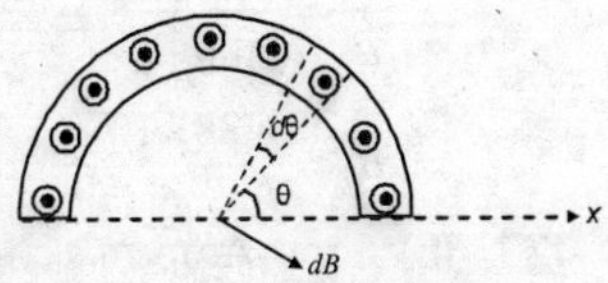

The component *dB* sin θ, of the field is cancelled by another opposite component.

Therefore, $B_{net} = 2\int dB\cos\theta = \frac{\mu_0 I}{\pi^2 R}\int_0^{\pi/2}\cos\theta d\theta = \frac{\mu_0 I}{\pi^2 R}$

132. **(a)** The magnetic field varies inversely with the distance for a long conductor. That is, $B \propto \frac{1}{d}$

so, graph in option (a) is the correct one.

133. (c) The magnetic field is

$B = \frac{\mu_0}{4\pi}\frac{2I}{r} = 10^{-7} \times \frac{2 \times 100}{4}$

$= 5 \times 10^{-6}\ T$ [∵ Since power line will behave as infinite conductor]

Current flows from east to west. Point is below the power line, using right hand thumb rule, the magnetic field is directed towards south.

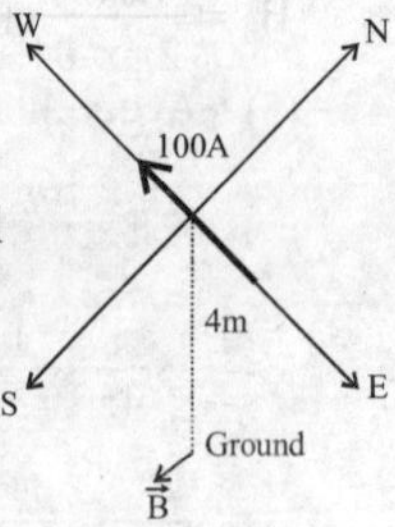

134. (d) There is no current inside the pipe. From Ampere's circuital law $\oint \vec{B}\cdot\vec{dl} = \mu_0 I \quad [\because I = 0]$

$\therefore B = 0$

135. (c) The direction of magnetic field induction due to current through AB and CD at P are indicated as B_1 and B_2. The magnetic fields at a point P, equidistant from AOB and COD will have directions perpendicular to each other, as they are placed normal to each other.

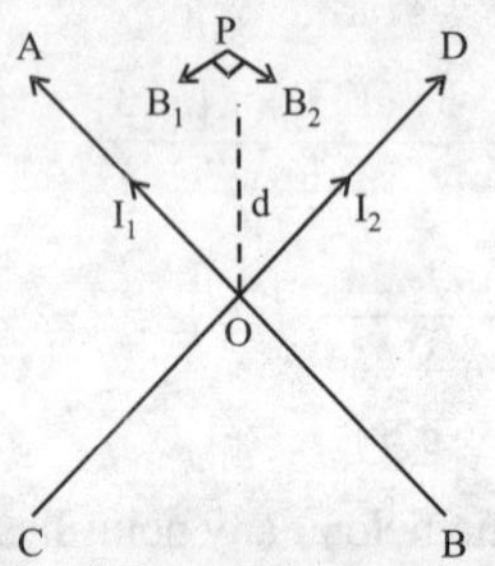

Magnetic field at P due to current through AB,

$B_1 = \frac{\mu_0 I_1}{2\pi d}$

Magnetic field at P due to current through CD,

$B_2 = \frac{\mu_0 I_2}{2\pi d}$

$\therefore$ Resultant field, $B = \sqrt{B_1^2 + B_2^2}$

$\therefore B = \sqrt{\left(\frac{\mu_0}{2\pi d}\right)^2 \left(I_1^2 + I_2^2\right)}$

or, $B = \frac{\mu_0}{2\pi d}\left(I_1^2 + I_2^2\right)^{1/2}$

136. (a) Magnetic field due to long solenoid is given by $B = \mu_0 nI$

In first case $B_1 = \mu_0 n_1 I_1$

In second case, $B_2 = \mu_0 n_2 I_2$

$\therefore \frac{B_2}{B_1} = \frac{\mu_0 n_2 i_2}{\mu_0 n_1 i_1} \Rightarrow \frac{B_2}{6.28 \times 10^{-2}} = \frac{100 \times \frac{i}{3}}{200 \times i}$

$\Rightarrow B_2 = \frac{6.28 \times 10^{-2}}{6} = 1.05 \times 10^{-2}\ \text{Wb/m}^2$

137. (b) From Ampere's circuital law

$\int \vec{B}\cdot\vec{dl} = \mu_0 i \Rightarrow B \times 2\pi r = \mu_0 i$

Here i is zero, for $r < R$, whereas R is the radius

$\therefore B = 0$

138. (b) Magentic field at the centre of a circular coil of radius R carrying current i is $B = \frac{\mu_0 i}{2R}$

The circumference of the first loop $= 2\pi R$. If it is bent into n circular coil of radius r'.

$n \times (2\pi r') = 2\pi R$

$\Rightarrow nr' = R$ ...(i)

New magnetic field, $B' = \frac{n \cdot \mu_0 i}{2r'}$ ...(ii)

From (i) and (ii),

$B' = \frac{n\mu_0 i \cdot n}{2R} = n^2 B$

139. (c) The magnetic field at a point on the axis of a circular loop at a distance x from centre is,

$B = \frac{\mu_0 i\, a^2}{2(x^2 + a^2)^{3/2}}$

Magnetic field at the centre of loop is

$B' = \frac{\mu_0 i}{2a} \Rightarrow B' = \frac{B \cdot (x^2 + a^2)^{3/2}}{a^3}$

Put $x = 4$ & $a = 3 \Rightarrow B' = \frac{54(5^3)}{3 \times 3 \times 3} = 250\ \mu T$

140. (a) Magnetic field induction at the centre of current carrying circular coil of radius r is

$B = \frac{\mu_0}{4\pi}\frac{I}{R} \times 2\pi$

Here $B_A = \frac{\mu_0}{4\pi}\frac{I}{R} \times 2\pi$

and $B_B = \frac{\mu_0}{4\pi}\frac{2I}{2R} \times 2\pi \Rightarrow \frac{B_A}{B_B} = \frac{I/R}{2I/2R} = 1$

141. (d) Given, $\tau = 80\sqrt{3}$ Nm, $\theta = 60°$

The torque experienced by a magnetic dipole is

$\tau = \text{MB}\sin\theta = \text{MB}\sin 60° = \frac{\sqrt{3}}{2}\text{MB} = 80\sqrt{3}$

$\therefore$ MB = 160

The potential energy of the dipole is

$U = -\text{MB}\cos\theta = -\text{MB}\cos 60° = -160 \times 1/2 = -80$ J

142. (48) The magnetic force on a current carrying wire placed in magnetic field is $F = I\ell B$

Here, $I = 8A$, $B = 0.15T$, $\ell = 4\ \text{cm} = 4 \times 10^{-2}$ m

$\therefore F = 8 \times \frac{4}{100} \times 0.15 = 48 \times 10^{-3}\ \text{N} = 48$ mN

143. (d)

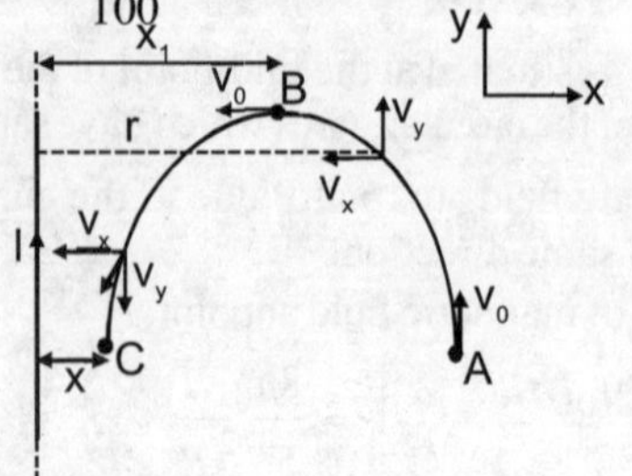

A → B

$\vec{V} = -v_x\hat{i} + v_y\hat{j}$

The magnetic field at a distance r from an infinitely long, straight wire is $\vec{B} = \frac{\mu_0 I}{2\pi r}$

$\vec{B} = \frac{\mu_0 I}{2\pi r}(-\hat{k})$

A charge q moving with velocity in magnetic field experiences a Lorentz force.

$$\vec{F} = q(\vec{v} \times \vec{B}) = \frac{\mu_0 q}{2\pi r}\left[-v_x \hat{j} - v_y \hat{i}\right]$$

$$a_x = -\frac{\mu_0 I q}{2\pi m}.\frac{v_y}{r} \qquad a_y = -\frac{\mu_0 I q}{2\pi m}.\frac{v_x}{r}$$

$$\frac{v_x dv_x}{dr} = -\frac{\mu_0 I q}{2\pi m}\frac{v_y}{r} \qquad \frac{v_x dv_x}{v_y} = -\frac{\mu_0 I q}{2\pi m}\frac{dr}{r}$$

$$\int_0^{v_0} \frac{v_x dv_x}{\sqrt{v_0^2 - v_x^2}} = -\frac{\mu_0 I q}{2\pi m}\int_a^{x_1}\frac{dr}{r}$$

Let, $z^2 = v_0^2 - v_x^2$

$2zdz = -2v_x dv_x$

$zdz = -v_x dv_x$

$$\frac{v_x dv_x}{\sqrt{v_0^2 - v_x^2}} = \frac{-zdz}{z} = -dz$$

then integral becomes

$$-\int_{v_0}^{0} dz = -\frac{\mu_0 I q}{2\pi m}\ln\frac{x_1}{a}$$

$$v_0 = -\frac{\mu_0 I q}{2\pi m}\ln\frac{x_1}{a}$$

$$x_1 = ae^{-\frac{2\pi m v_0}{\mu_0 I q}} \quad \text{.......... (1)}$$

For B→C

$$\vec{v} = -v_x\hat{i} - v_y\hat{j} \quad ; \quad \vec{B} = \frac{\mu_0 I}{2\pi r}(-\hat{k})$$

$$\vec{F} = q\left(\vec{v} \times \vec{B}\right) = \frac{\mu_0 I q}{2\pi r}\left(-v_x\hat{j} - v_y\hat{i}\right)$$

$$a_x = +\frac{\mu_0 I q}{2\pi m}\frac{v_y}{r} \quad ; \quad a_y = -\frac{\mu_0 I q}{2\pi m}.\frac{v_x}{r}$$

$$\frac{v_x dv_x}{dr} = \frac{\mu_0 I q}{2\pi m}\frac{v_y}{r}$$

$$\int_{v_0}^{0}\frac{v_x dv_x}{\sqrt{v_0^2 - v_x^2}} = \frac{\mu_0 I q}{2\pi m}\int_{x_1}^{x}\frac{dr}{r}$$

$$\frac{\mu_0 I q}{2\pi m}\ln\frac{x}{x_1} = -\int_{v_0}^{0} dz = -v_0$$

$$x = x_1 e^{-\frac{2\pi m v_0}{\mu_0 I q}} \quad \text{.......... (ii)}$$

From equation (i) and (ii)

$$X = ae^{-\frac{4\pi m v_0}{\mu_0 I q}}$$

144. (1)

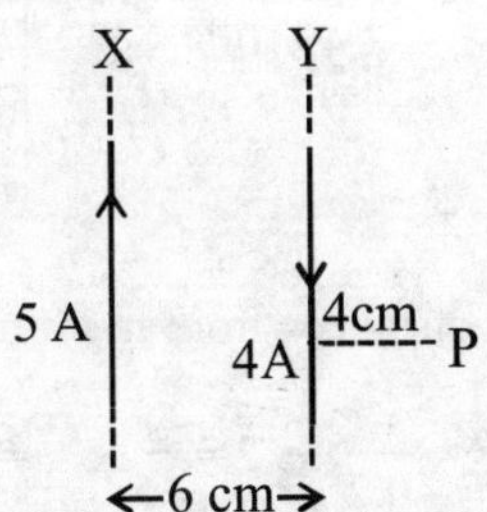

The net magnetic field at point P is $B_P = B_y - B_x$

$$B_p = \frac{\mu_0(4)}{2\pi \times .04} - \frac{\mu_0(5)}{2\pi \times 0.1} = \frac{100\mu_0}{4\pi} = 100 \times 10^{-7} = 1 \times 10^{-5}\text{T}$$

145. (5) At equilibrium, $F_{magnetic} = mg$

$$\Rightarrow \quad iLB = mg \text{ and } L = \frac{AR}{\rho}$$

$$\therefore B = \frac{mg\rho}{iAR} = \frac{0.5 \times 10 \times 2 \times 10^{-6}}{2 \times 10 \times 10^{-6} \times 1} = 0.5\text{T} = 5 \times 10^{-1}\text{T}$$

146. (50) Force on segment parallel to x-axis will cancel each other. Hence F_{net} will be due to portion parallel to y-axis.

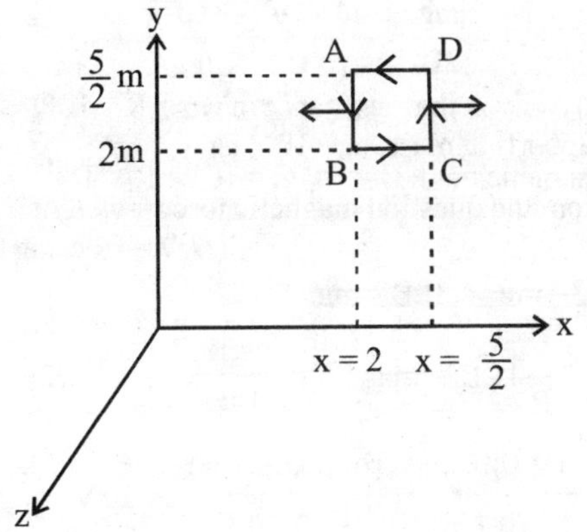

$\vec{F}_{BC} + \vec{F}_{DA} = 0$

$\vec{F}_{AB} = i\ell B = 0.5 \times 0.5(0.2 \times 5) = 0.25\text{ N}$

$\vec{F}_{CD} = 0.5 \times 0.5(0.2 \times 6) = 0.3\text{ N}$

$\vec{F}_{net} = 0.05\text{ N} = 50\text{ mN}$

147. (a)

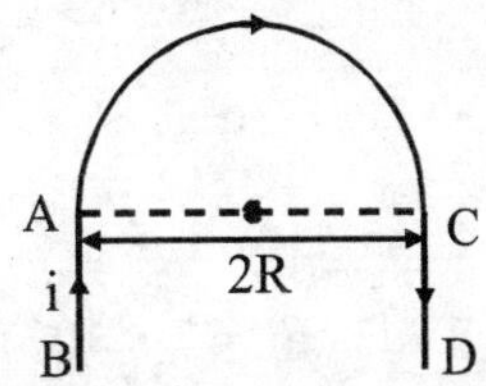

Force due to AB and CD will cancel each other.

Magnetic field is in $+\hat{k}$ direction

Effective length of semicircular loop, $\vec{\ell}_{eff} = 2R\hat{i}$

$\vec{F} = i\vec{\ell} \times \vec{B}$

$\Rightarrow \vec{F} = i\vec{\ell}_{eff}B$

$\Rightarrow \vec{F} = i\left(2R\hat{i} \times B_0\hat{k}\right) = -2iBR\hat{j}$

148. (b) Given,

Magnetic field, $B = 2 \times 10^{-3}$T

Current in rectangular loop, I = 5A

Area of rectangular loop, $A = \ell \times b = 0.2 \times 0.1\text{ m}^2$

Magnetic moment of rectangular coil, $\vec{M} = i\vec{A}$

$= 5 \times (0.2) \times (0.1)\left(-\hat{i}\right) = 0.1\left(-\hat{i}\right)$

Torque, $\vec{\tau} = \vec{M} \times \vec{B} = 0.1(-\hat{i}) \times (2 \times 10^{-3})(\hat{j})$

$= 2 \times 10^{-4}\ (-\hat{k})\text{N} - \text{m}$

149. (35) Magnetic field, $B_H = 3.5 \times 10^{-5}$ T, Current, $i = \sqrt{2}$A

Force on the conductor, $F = i\ell B \sin\theta$

$$\frac{F}{\ell} = iB \sin 45° = \sqrt{2} \times 3.5 \times 10^{-5} \times \frac{1}{\sqrt{2}} = 35 \times 10^{-6} \text{ N/m}$$

150. (160) Magnetic field midway between the wires, B_p

$$= \left(\frac{\mu_0 i}{2\pi a}\right) \times 2 = \frac{4\pi \times 10^{-7} \times 10}{\pi \times \left(\frac{5}{2} \times 10^{-2}\right)} = 16 \times 10^{-5} = 160 \mu T$$

151. (2) For equilibrium, force $Mg = I\ell B$

$$\therefore I = \frac{mg}{\ell B} = \frac{40 \times 10^{-3} \times 10}{50 \times 10^{-2} \times 0.4} = 2A$$

152. (d) Given that resistance of wire, $R = 10\Omega$
Supporting mass, $m = 10^{-3}$ kg
Magnetic field, $B = 10^3$ G $= 10^3 \times 10^{-4}$T
From the question magnetic force = weight
($\because V = IR$ using Ohm's Law)

$F_m = mg \;\; \therefore ILB = mg$

$$\therefore \left(\frac{V}{R}\right) LB = mg \;\therefore V = \frac{mgR}{LB}$$

$$= \frac{(1 \times 10^{-3}\text{ kg})(10 \text{ m/s}^2)(10\Omega)}{(0.1\text{ m})(10^3 \times 10^{-4}\text{T})} = 10\text{ V}$$

153. (b) Magnetic force per unit length on wire PQ due to wire RS is given by

$$\frac{F}{\ell} = \frac{\mu_0}{4\pi}\frac{2I_1I_2}{r} \quad \therefore F_1 \propto I_1 I_2 \propto (I)^2$$

When ammeter current reading is changed from I to 2I.
$F_2 \propto (2I)^2$
$\therefore \; F_1 : F_2 = 1 : 4$

154. (68)

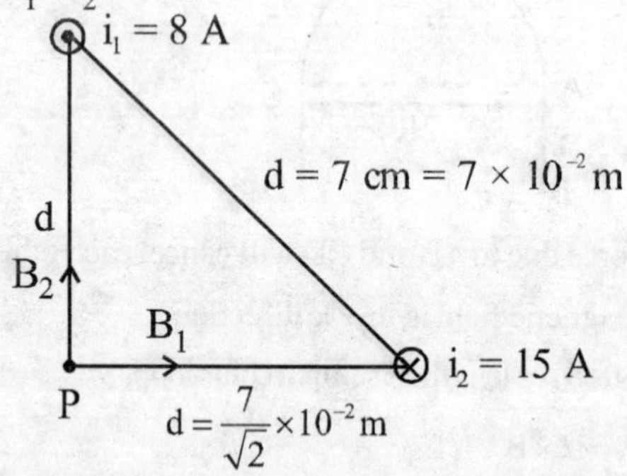

Magnetic fields due to both wires will be perpendicular to each other.

$$B_1 = \frac{\mu_0 i_1}{2\pi d} \quad B_2 = \frac{\mu_0 i_2}{2\pi d}$$

$$B_{net} = \sqrt{B_1^2 + B_2^2} = \frac{\mu_0}{2\pi d}\sqrt{i_1^2 + i_2^2}$$

$$= \frac{4\pi \times 10^{-7}}{2\pi \times (7/\sqrt{2}) \times 10^{-2}} \times \sqrt{8^2 + 15^2}$$

$= 68 \times 10^{-6}$ T

155. (a) We have

$$\frac{F}{\ell} = \frac{\mu_0}{2\pi}\frac{I_1I_2}{d} \Rightarrow F = \frac{\mu_0}{2\pi}\frac{I_1I_2\ell}{d}$$

So, $F \propto \frac{I_1I_2}{d}$

$$\frac{F_2}{F_1} = \frac{\left(\frac{2I_1.2I_2}{d/2}\right)}{\left(\frac{I_1I_2}{d}\right)} = 8 \Rightarrow F_2 = 8F_1$$

156. (9)

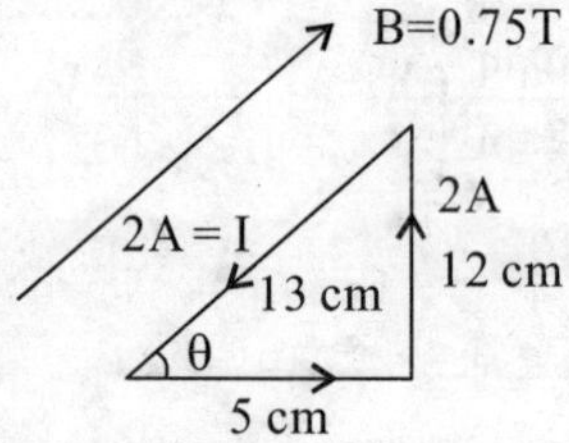

$$\vec{F} = I(\vec{L} \times \vec{B})$$

Force on 5 cm side is

$|\vec{F}| = ILB \sin\theta$

$$= (2)(5 \times 10^{-2}) \times \frac{3}{4} \times \frac{12}{13} = \frac{9}{130}\text{N} = \frac{x}{130}\text{N (given)}$$

So, $x = 9$

157. (a) Given, length of wire X, $\ell_1 = 50$ cm $= 0.5$ m
Length of wire Y, $\ell_2 = 5$m
Distance between two wire, $r = 5$ cm $= 5 \times 10^{-2}$m
Force of interaction $= I_1 \ell_1 B_{12}$

$$= \frac{\mu_0 I_1 I_2}{2\pi r} \quad \left(\because B_{12} = \frac{\mu_0 I_2}{2\pi r}\right)$$

$$= \frac{4\pi \times 10^{-7} \times 6 \times 0.5}{2\pi \times 5 \times 10^{-2}} = 1.2 \times 10^{-5}\text{ N towards X}$$

158. (a) Force, $F = ILB \sin\theta$
From figure, $mg \sin 45° = ILB \cos 45°$

$$\therefore I = \left(\frac{m}{L}\right)\frac{g}{B}$$

$$= \frac{(0.45)(10)}{0.15} = 30A$$

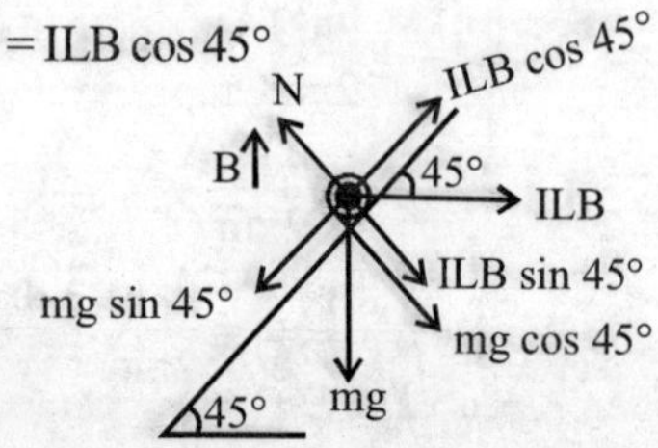

159. (c) F_M on CD $= BI\ell_{eff}$
$= 0.5 \times (10) \times (5 \sin 60 \times 10^{-2}) = 0.216$ N

160. (b) Here, current cannot be in same direction in both wire because $B_p = 0$. So current should be in opposite direction.

$B_p = B_{1p} + B_{2p}$

$$B_p = 2\frac{\mu_0 I}{2\pi r}$$

$$300 \times 10^{-6} = 2 \times 2 \times 10^{-7} \times \frac{I}{4 \times 10^{-2}}$$

$3 \times 10^{-4} = 10^{-5} \times I$

$I = 30\,A$

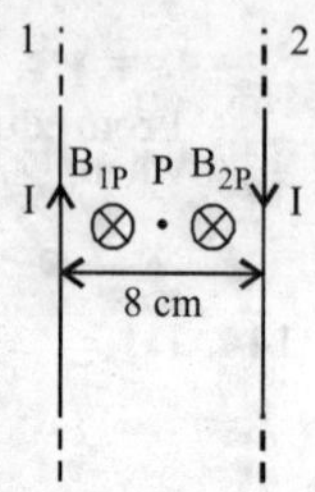

161. (c) We know that

$$\frac{F}{l} = \frac{\mu_0}{2\pi}\frac{I_1I_2}{d}$$

$$\Rightarrow 2 \times 10^{-6} = 2 \times 10^{-7} \times \frac{x^2}{0.2} \Rightarrow 2 = x^2 \Rightarrow x = \sqrt{2} = 1.4$$

162. (c) We have,

$$B_p = \frac{\mu_0}{2\pi}\left[\frac{I_1}{r_1} - \frac{I_2}{r_2}\right](-\hat{k}) = \frac{2\times10^{-7}}{10^{-2}}\left[\frac{4}{4} - \frac{2}{6}\right](-\hat{k})$$

$$= 2\times10^{-5}\left[1 - \frac{1}{3}\right](-\hat{k}) = \frac{4}{3}\times10^{-5}(-\hat{k})$$

So, $\vec{F} = q(\vec{V}\times\vec{B})$

$$= 3\pi\left[(2\hat{i}+3\hat{j})\times\left(-\frac{4}{3}\times10^{-5}\hat{k}\right)\right] = 3\pi\left[\frac{8}{3}\hat{j} - 4\hat{i}\right]\times10^{-5}$$

$$= [8\pi\hat{j} - 12\pi\hat{i}]\times10^{-5} = 4\pi[2\hat{j} - 3\hat{i}]\times10^{-5}$$

$$= 4\pi\times10^{-5}[-3\hat{i}+2\hat{j}]. = 4\pi\times10^{-5}[-x\hat{i}+2\hat{j}]$$ So, $x = 3$

163. (5) We have

$$\frac{F}{l} = \frac{\mu_0}{2\pi}\frac{i_1 i_2}{r} \Rightarrow r = \frac{\mu_0}{2\pi} i_1 i_2 \times \frac{l}{F}$$

$$\Rightarrow r = 2\times10^{-7}\times25\times\frac{0.1}{10^{-5}}$$

$$= 50\times10^{-7}\times10^{4} = 50\times10^{-3}\,\text{m} = 5\,\text{cm}$$

164. (a) $\vec{\tau} = \vec{M}\times\vec{B} = MB\hat{n}$

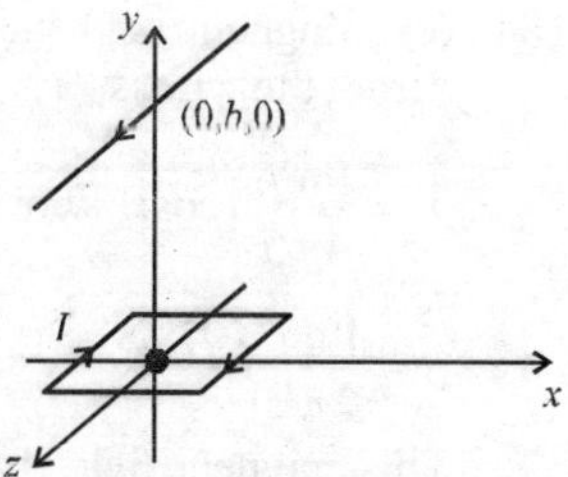

$$|\vec{\tau}| = I(2a)^2 \times \frac{\mu_0 I}{2\pi b}$$

$$= \frac{2\mu_0 I^2 a^2}{\pi b}$$

165. (b) Magnetic moment of loop *ABCD*,

M_1 = area of loop × current

$\vec{M}_1 = (abI)\hat{k}$ (Here, ab = area of rectangle)

Magnetic moment of loop *DEFA*,

$\vec{M}_2 = (abI)\hat{j}$

Net magnetic moment,

$\vec{M} = \vec{M}_1 + \vec{M}_2 \Rightarrow \vec{M} = abI(\hat{k}+\hat{j})$

$$\Rightarrow |\vec{M}| = \sqrt{2}abI\left(\frac{\hat{j}}{\sqrt{2}} + \frac{\hat{k}}{\sqrt{2}}\right)$$

166. (c) Torque on circular loop, $\tau = -MB\sin\theta$

where, M = magnetic moment

B = magnetic field

Now, using $\tau = I\alpha$

$$\therefore \tau = -MB\sin\theta = I\alpha \Rightarrow -\pi a^2 IB\theta = \frac{ma^2\alpha}{2}$$

($\because M = IA$ and moment of inertia of circular loop, $I = \frac{ma^2}{2}$)

$$\Rightarrow \alpha = -\left(\frac{2\pi IB}{m}\right)\theta$$

$$\Rightarrow \omega = \sqrt{\frac{2\pi IB}{m}} \Rightarrow \frac{2\pi}{T} = \sqrt{\frac{2\pi IB}{m}} \Rightarrow T = \sqrt{\frac{2\pi m}{IB}}$$

167. (d) As net force on the third wire C is zero.

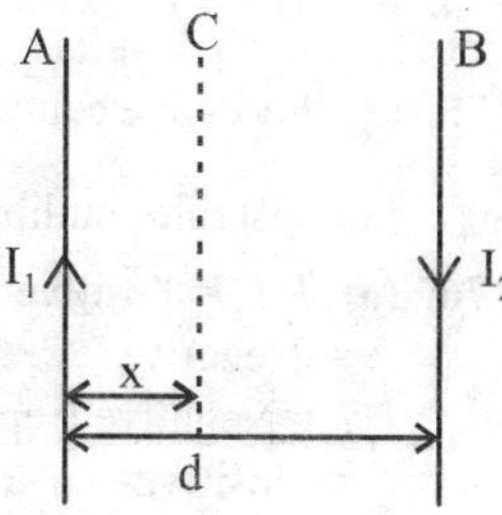

$$\Rightarrow F = \frac{\mu_0 I_1}{2\pi x} + \frac{\mu_0 I_2}{2\pi(d-x)} = 0$$

$$\Rightarrow \frac{\mu_0 I_1}{2\pi x} = \frac{\mu_0 I_2}{2\pi(x-d)}$$

$$\Rightarrow I_1 x - I_1 d = I_2 x$$

$$x = \frac{I_1 d}{I_1 - I_2} \text{ or } \frac{I_1 d}{I_2 - I_1}$$

Two cases may be possible if $I_1 > I_2$ or $I_2 > I_1$

168. (d) $\tau = MB\sin45° = N(iA)B\sin45°$

$$= 100\times3(5\times2.5)\times10^{-4}\times1\times\frac{1}{\sqrt{2}} = 0.27\,\text{N-m}$$

169. (c) $F = \frac{\mu_0}{2\pi}\left(\frac{i_1 i_2}{a} - \frac{i_1 i_2}{2a}\right)\times a = \frac{\mu_0 i_1 i_2}{4\pi}$

170. (b) $|\vec{\tau}| = |\vec{\mu}\times\vec{B}|$ $[\mu = NIA]$

$= NIA\times B\sin90°$ $[A = \pi r^2]$

$\Rightarrow \tau = NI\pi r^2 B$

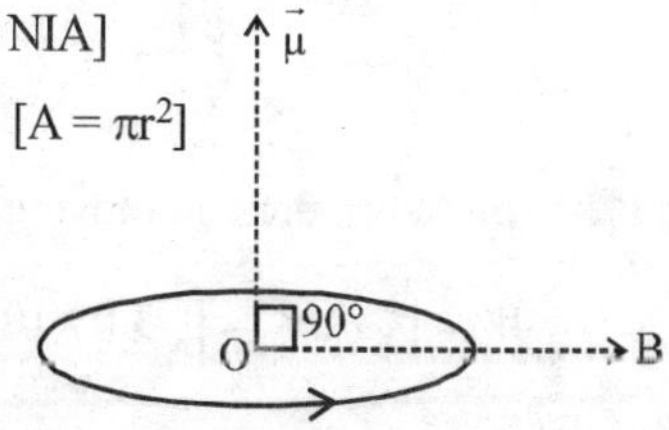

171. (d) Force on one pole,

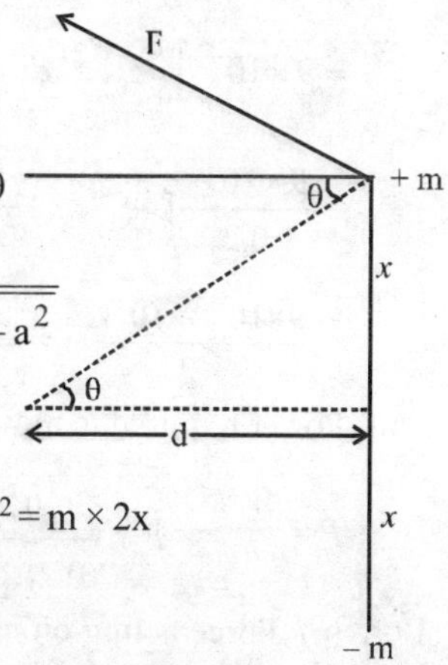

$$F = m\times\frac{\mu_0 I}{2\pi\sqrt{d^2+x^2}}$$

Total force, $F_{total} = 2F\sin\theta$

$$= 2\times\frac{\mu_0 Im}{2\pi\sqrt{d^2+a^2}}\times\frac{x}{\sqrt{d^2+a^2}}$$

$$= \frac{\mu_0 Imx}{\pi(d^2+a^2)}$$

Magnetic moment, $M = I\pi a^2 = m\times2x$

$$\therefore m = \frac{I\pi a^2}{2x}$$

or, Total force, $F_{total} = \frac{\mu_0 I^2 a^2}{2(d^2+a^2)} = \frac{\mu_0 I^2 a^2}{2d^2}$ $[\because d >> a]$

Clearly $F_{total} \propto \frac{a^2}{d^2}$

172. (a) Magnetic moment, $\mu = IA = \frac{qv}{2\pi r}(\pi r^2)$

or, $\mu = \frac{qr\omega}{2\pi r}(\pi r^2) = \frac{1}{2}qr^2\omega$

173. (c) Magnetic moment of current carrying rectangular loop of area A is given by M = NIA

magnetic moment of current carrying coil is a vector and its direction is given by right hand thumb rule, for stable equilibrium of rectangular loop, $\vec{B}$ and $\vec{M}$ at centre due to current in loop are always parallel.

Hence, (c) corresponds to stable equilibrium.

174. (c) $\vec{F}_1 = \vec{F}_2 = 0$ because of action and reaction pair.

175. (a) For stable equilibrium $\vec{M} \| \vec{B}$

For unstable equilibrium $\vec{M} \| (-\vec{B})$

176. (a) $I_1 I_2$ = Positive (attract)
F = Negative
$I_1 I_2$ = Negative (repell)
F = Positive
Hence, option (a) is the correct answer.

177. (c) For small arc length
$2T \sin\theta = BIR\, 2\theta$ (As F = BIL and L = R2θ)
For small θ, $\sin\theta \approx \theta$
$\therefore T = BIR$

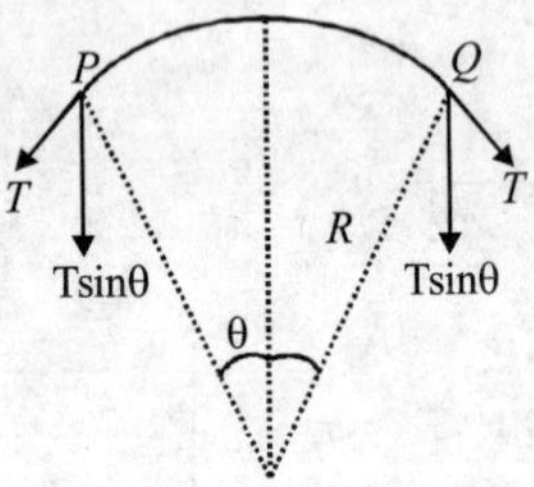

178. (b) Work done in moving the conductor is,

$$W = \int_0^2 F dx = \int_0^2 3.0 \times 10^{-4} e^{-0.2x} \times 10 \times 3 dx$$

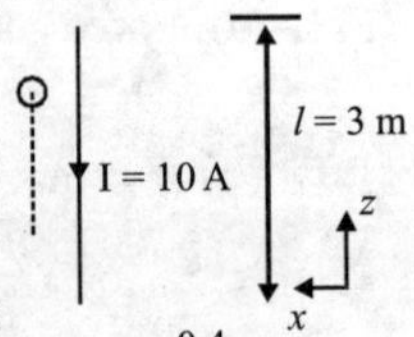

$$= 9 \times 10^{-3} \int_0^2 e^{-0.2x} dx$$

$$= \frac{9 \times 10^{-3}}{0.2} [-e^{-0.2 \times 2} + 1] = \frac{9 \times 10^{-3}}{0.2} \times [1 - e^{-0.4}]$$

$$= \frac{9 \times 10^{-3} \times (0.33)}{2} = \frac{2.97 \times 10^{-3}}{2}$$

Power required to move the conductor is,

$$P = \frac{W}{t} \Rightarrow P = \frac{2.97 \times 10^{-3}}{(0.2) \times 5 \times 10^{-3}} = 2.97 \text{ W}$$

179. (a) Force acting on conductor B due to conductor A is given by relation

$$F = \frac{\mu_0 I_1 I_2 l}{2\pi r}$$

l-length of conductor B
r-distance between two conductors

$$\therefore F = \frac{4\pi \times 10^{-7} \times 10 \times 2 \times 2}{2 \times \pi \times 0.1} = 8 \times 10^{-5} \text{ N}$$

180. (a) The magnetic field at O due to current in DA is

$$B_1 = \frac{\mu_o}{4\pi}\frac{I}{a} \times \frac{\pi}{6} \text{ (directed vertically upwards)}$$

The magnetic field at O due to current in BC is

$$B_2 = \frac{\mu_o}{4\pi}\frac{I}{b} \times \frac{\pi}{6} \text{ (directed vertically downwards)}$$

The magnetic field due to current AB and CD at O is zero.
Therefore the net magnetic field is

$B = B_1 - B_2$ (directed vertically upwards)

$$= \frac{\mu_o}{4\pi}\frac{I}{a}\frac{\pi}{6} - \frac{\mu_o}{4\pi}\frac{I}{b} \times \frac{\pi}{6} = \frac{\mu_o I}{24}\left(\frac{1}{a} - \frac{1}{b}\right) = \frac{\mu_o I}{24ab}(b-a)$$

181. (d) $\vec{F} = I(\vec{\ell} \times \vec{B})$
The force on AD and BC due to current I_1 is zero. This is because the directions of current element $I\vec{d\ell}$ and magnetic field $\vec{B}$ are parallel.

182. (a) Force acting between two long conductor carrying current,

$$F = \frac{\mu_0}{4\pi}\frac{2I_1 I_2}{d} \times \ell \quad \text{...(i)}$$

Where d = distance between the conductors
ℓ = length of conductor

$$\text{In second case, } F' = -\frac{\mu_0}{4\pi}\frac{2(2I_1)I_2}{3d}\ell \quad \text{..(ii)}$$

From equation (i) and (ii), we have

$$\therefore \frac{F'}{F} = \frac{-2}{3}$$

183. (b) When current is passed through a spring then current flows parallel in the adjacent turns in the same direction. As a result the various turn attract each other and spring get compress.

184. (c) Magnetic field due to current in wire 1 at point P distant r from the wire is

$$B = \frac{\mu_0}{4\pi}\frac{i_1}{r}[\cos\theta + \cos\theta]$$

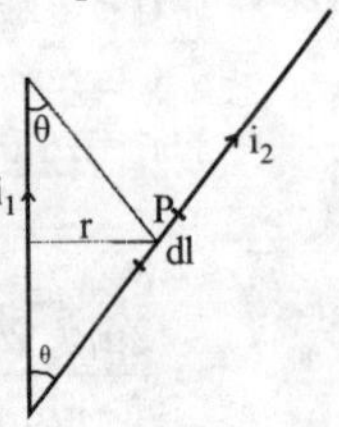

$$B = \frac{\mu_0}{2\pi}\frac{i_1 \cos\theta}{r}$$

This magnetic field is directed perpendicular to the plane of paper, inwards.
The force exerted due to this magnetic field on current element $i_2\, dl$ is
$dF = i_2\, dl\, B \sin 90°$

$$\therefore dF = i_2 dlB \Rightarrow dF = i_2 dl\left(\frac{\mu_0}{2\pi}\frac{i_1 \cos\theta}{r}\right)$$

$$= \frac{\mu_0}{2\pi r} i_1 i_2\, dl \cos\theta$$

185. (a) Surface energy released voltage sensitivity

$$= \frac{\theta}{V} = \frac{NAB}{KR}$$

$$\frac{V_s(M_1)}{V_s(M_2)} = \left(\frac{N_1A_1B_1}{N_2A_2B_2}\right)\frac{R_2}{R_1} = \frac{15 \times 3.6 \times 0.25}{21 \times 1.8 \times 0.5} \times \frac{7}{5} = \frac{1}{1}$$

186. (c) Here,
I_g = full scale deflection current of galvanometer
G = resistance of galvanometer
I = total current to be measured
R = required shunt resistance

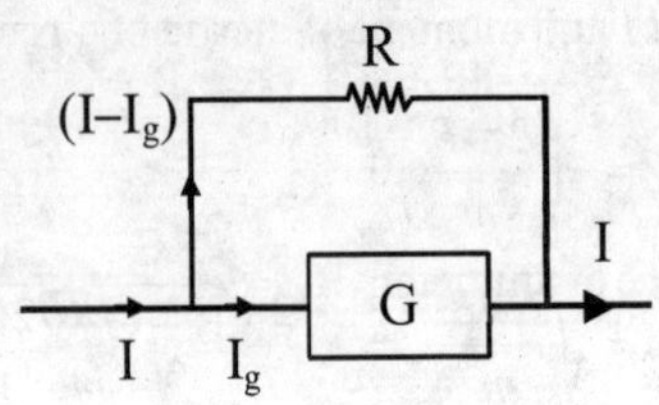

$I_g G = (I - I_g) R$

$20 \times 10^{-3} \times 30 = (3 - 0.02) \times R$

$$R = \left(\frac{0.6}{2.98}\right) = \frac{30}{x}$$

$$x = \left(\frac{2.98 \times 30}{0.6}\right) = 149$$

187. (a) A moving coil galvanometer works on the principle that when a current-carrying coil is placed in a magnetic field, it experiences a torque which is given by z = BINA This torque causes the coil to rotate against restoring torque $\therefore \tau = C\theta$

(A) $\tau = C\theta \Rightarrow [ML^2T^2]$

(B) The expression of curent sensitivity is

$$C.S. = \frac{\theta}{I} = \frac{BNA}{C}$$

$$V.S. = \frac{BNA}{RC}$$ [R also depends on 'N']

$\therefore$ If N increases, then R also increases. So, V.S. = constant

188. (b) Given,

Resistance of galvanometer,

$G = 200\,\Omega$

Current through galvanometer,

$i_g = 20\,\mu A$

$$i = i_g\left(\frac{G}{S} + 1\right)$$

$$\Rightarrow 20 \times 10^{-3} = 20 \times 10^{-6}\left(\frac{200}{S} + 1\right)$$

$$10^3 = \frac{200}{S} + 1 \Rightarrow \frac{200}{S} = 999$$

$\Rightarrow S \approx 0.2\,\Omega$

189. (c) $i_g = \dfrac{V}{(R+G)} = \dfrac{10}{400+100} = 20 \times 10^{-3}$ A

For ammeter,

Let shunt resistance = S

$\therefore V_G = V_S \Rightarrow i_g R = (i - i_g) S$

$\Rightarrow 20 \times 10^{-3} \times 100 = 10S \therefore S = 20 \times 10^{-2}\,\Omega$

190. (c) $V = I_g(R + R_g)$

$$\Rightarrow R = \frac{V}{I_g} - R_g = \frac{100}{5 \times 10^{-3}} - 50 = 19950\Omega$$

191. (d)

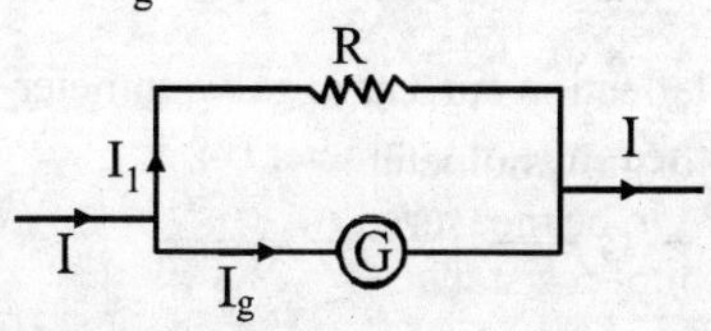

$$I_1 R = I_g G \Rightarrow \frac{95}{100} IR = \frac{5I}{100} G \Rightarrow R = \frac{G}{19}$$

$$\text{Resistance of ammeter, } R_a = \frac{RG}{R+G} = \frac{\frac{G^2}{19}}{\frac{20G}{19}} = \frac{G}{20}$$

192. (4) Torque, $\tau = NIAB\sin\theta$

Torque is also given as, $\tau = K\theta$

on comparing, $NIAB\sin\theta = K\theta$

$$K = \frac{NIAB\sin\theta}{\theta}$$

$$K = \frac{100 \times 0.01 \times 0.01 \times 2 \times 10^{-4}}{0.05} = 4 \times 10^{-5} \text{ N-m/rad.}$$

$= x \times 10^{-5}$ N-m/rad (given) $\therefore x = 4$

193. (c) $I_g = 3mA, I = 8A, G = 10\Omega$

Since S and G are in parallel,

$\therefore V_G = V_S \Rightarrow GI_g = S(I - I_g) \Rightarrow 10 \times 3 = 5(8000 - 3)$

$\therefore S = 3.75 \times 10^{-3}\,\Omega$

194. (d) For moving coil galvanometer,

Let x = current/division

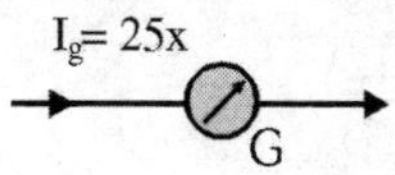

After applying shunt

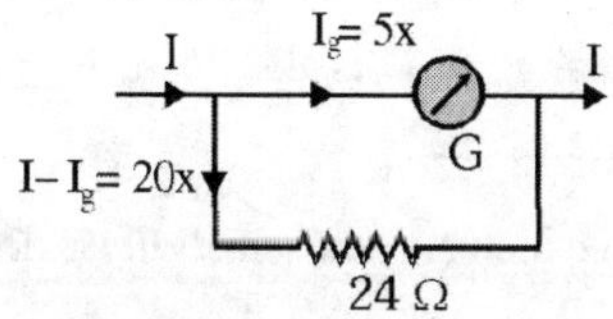

$\therefore V_G = V_S$

Now, $5x \times G = 20x \times 24$

$G = 4 \times 24 = 96\Omega$

195. (c) Electric current,

$i \propto \theta$ (angle of deflection)

$$\therefore \frac{i_2}{i_1} = \frac{\theta_2}{\theta_1} \Rightarrow \frac{i_2}{200\,\mu A} = \frac{\pi/10}{\pi/3} = \frac{3}{10} \quad \left[\because 60° = \frac{\pi}{3}\right]$$

$\therefore i_2 = 60\,\mu A$

196. (c) Here, $I = 10 \times 10^{-3}A$; $I_g = 1 \times 10^{-3}A$

For voltmeter,

$$R = \frac{V}{I_g} - G = \frac{50}{10^{-3}} - 54 \approx 50k\Omega$$

(A) is correct.

For ammeter,

$$r = \frac{I_g G}{I - I_g} = \frac{10^{-3} \times 54}{(10-1) \times 10^{-3}} = 6\Omega(C)$$

197. (50) When a resistance of 5Ω is shunted with a MCG

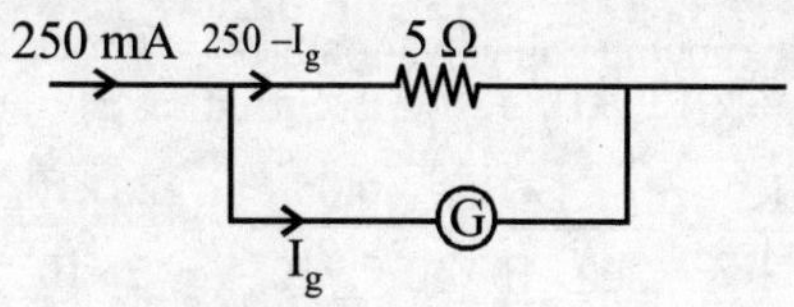

$$I_g = \frac{250\text{mA} \times 5}{5 + R_G} \quad \text{.......(i)}$$

When a resistance of 1050Ω is shunted with a MCG

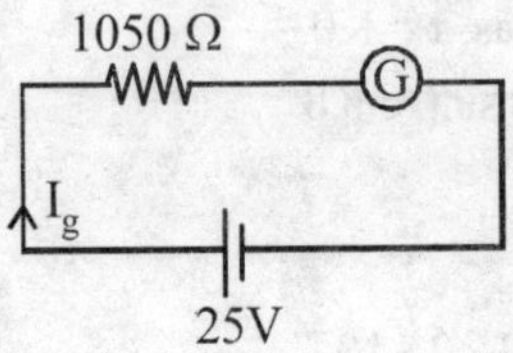

$$I_g = \frac{25}{1050 + R_G} \qquad ......(ii)$$

From (i) and (ii)

$$\frac{25}{1050 + R_G} = \frac{5}{4(5 + R_G)}$$

$100(5 + R_G) = 1050 \times 5 + R_G \times 5 \Rightarrow 95 R_G = 4750$

$\therefore R_G = 50\Omega$

198. (a) Current sensitivity,

$I_s = \frac{NBA}{C}$ and

Voltage sensitivity $V_s = \frac{NBA}{CG} \Rightarrow V_s = \frac{I_s}{G}$,

As G is constant,

$\therefore V_s \propto I_s$

$\therefore$ Percentage change in V_s is also 25%.

199. (d) Current sensitive $= \frac{BNA}{k}$

So, if N is doubled then current sensitivity is doubled. Voltage sensitivity,

$V = \frac{BNA}{Rk}$, as N is doubled R is also doubled. So, no change in voltage sensitivity.

200. (b) Due to motion or oscillates of the coil, the eddy current develops in the metallic material which oppose the motion of the coil thus bringing the coil to rest.

201. (a) For measuring potential difference, we always use voltmeter having high resistance. Because higher is the resistance, lesser is the current drawn by voltmeter which results in correct voltage drop across the load.

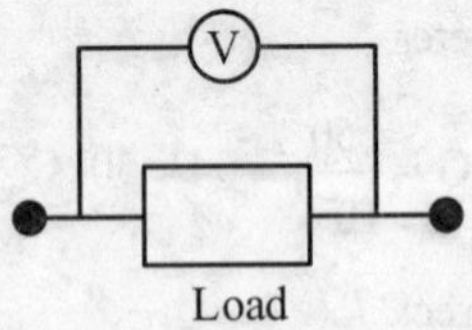

202. (d) We have

$C.S = \frac{NBA}{K}$ and $V.S = \frac{NBA}{KR}$

As current sensitivity increases by 50%. So, no. of turns increases by 50%. As no. of turn increases by 50%, so resistance also increases by 50%. Therefore, V.S remains constant.

203. (b) In moving coil galvanometer

$BINA = K\theta$

$$A = \frac{K\theta}{NIB} = \frac{4\times10^{-5}\times0.05}{200\times10\times10^{-3}\times10^{-2}} = \frac{20\times10^{-7}}{2\times10^{-2}}$$

$$= \frac{10\times10^{-7}}{10^{-2}} = 10^{-4}\,m^2 = 1\,cm^2$$

204. (d) Current $I = \left(\frac{K}{NAB}\right)\theta$

where, K = torsional constant and B = magnetic field

$$\therefore \frac{\theta}{I} = \frac{NAB}{K} \Rightarrow \frac{\theta}{I} \propto B \text{ and } \frac{\theta}{I} \propto \frac{1}{K}$$

205. (b) We have, $S(i - i_g) = i_g G$

$$\Rightarrow S\left(1 - \frac{i_g}{i}\right) = \frac{i_g}{i}G$$

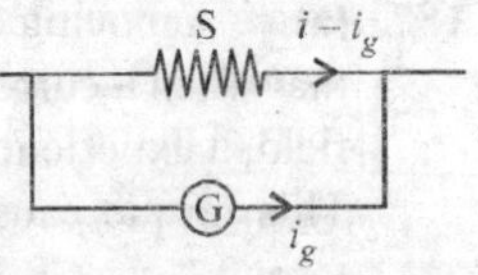

$$\Rightarrow \frac{G}{S} = \frac{i}{i_g} - 1$$

$$\Rightarrow \frac{i}{i_g} = \frac{G}{S} + 1 = \frac{72}{8} + 1 = 10 \Rightarrow \frac{i_g}{i} = \frac{1}{10}. \text{ So, } \%\frac{i_g}{i} = 10\%$$

206. (d) By current division rule

$$I_g = I \times \frac{S}{G+S} \Rightarrow I = \frac{G+S}{S} I_g \qquad ...(i)$$

Now, figure of merit $(k) = \frac{I_g}{\theta}$

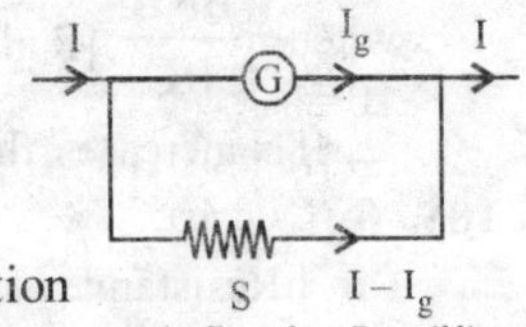

$\Rightarrow I_g = K\theta$

as, n = actual number of deflection

So, $I_g = kn$, [$\because$ k represent current per deflection] ...(ii)

So, $I = \frac{Kn}{S}(G + S)$. [From (i) & (ii)]

207. (c)

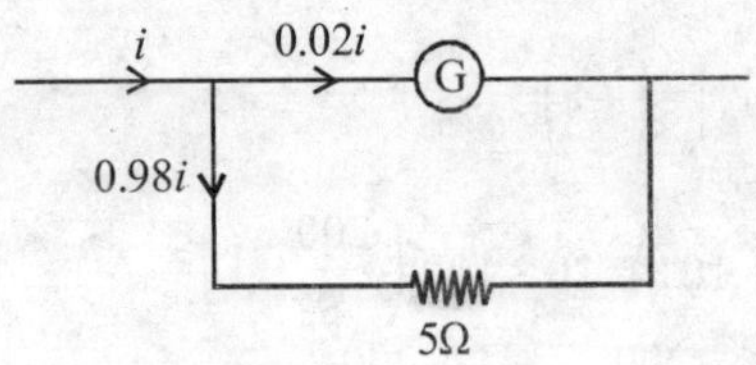

$0.02i R_g = 0.98i \times 5$

$R_g = 245\Omega$.

208. (d) Given, current sensitivity = 2 div/mA

Full scale current,

$$I_{max} = \frac{50}{2} = 25mA$$

$$\therefore \text{ Resistance, } R = \frac{V}{I} = \frac{50\,mV}{25mA} = 2\Omega$$

209. (d) Galvanometer of resistance (G) converted into a voltmeter of range 0-1 V.

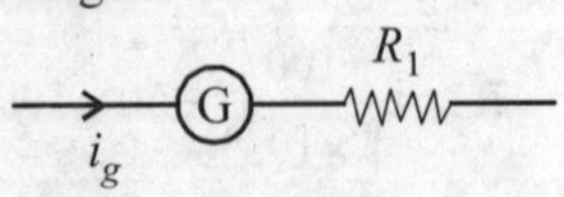

$$V = 1 = i_g(G + R_1) \qquad ...(i)$$

To increase the range of voltmeter 0-2 V

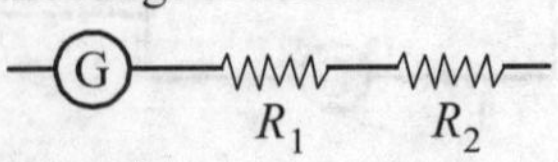

$$2 = i_g(R_1 + R_2 + G) \qquad ...(ii)$$

Dividing eq. (i) by (ii),

$$\Rightarrow \frac{1}{2} = \frac{G + R_1}{G + R_1 + R_2} \Rightarrow G + R_1 + R_2 = 2G + 2R_1$$

$\therefore R_2 = G + R_1$

210. (d) Given,
Current passing through galvanometer, $I = 6$ mA
Deflection, $\theta = 2°$
Figure of merit of galvanometer

$$= \frac{I}{\theta} = \frac{6 \times 10^{-3}}{2} = 3 \times 10^{-3} \text{ A/div.}$$

211. (20) Given,
Area of galvanometer coil, $A = 3 \times 10^{-4}\,\text{m}^2$
Number of turns in the coil, $N = 500$
Current in the coil, $I = 0.5$ A

Torque $\tau = |\vec{M} \times \vec{B}| = NiAB\sin(90°) = NiAB$

$$\Rightarrow B = \frac{\tau}{NiA} = \frac{1.5}{500 \times 0.5 \times 3 \times 10^{-4}} = 20T$$

212. (c) $i_g = 20 \times 50 = 1000\,\mu A = 1$ mA
Using, $V = i_g(G + R)$, we have
$2 = 10^{-3}(100 + R_1)$
$R_1 = 1900\,\Omega$
when, $V = 10$ volt
$10 = 10^{-3}(100 + R_2 + R_1)$
$\Rightarrow 10000 = (100 + R_2 + 1900) \Rightarrow R_2 = 8000\,\Omega$

213. (b) In an ammeter, $i_g = i_0 \dfrac{R_A}{R_A + G}$

and for voltmeter,
$V = i_g(G + R_V) = Gi_0$
On solving above equations, we get
$R_A R_V = G^2$

$$\text{and } \frac{R_A}{R_V} = \left(\frac{i_g}{i_0 - i_g}\right)^2$$

214. (None) $v = i_g(R + G) \Rightarrow 5 = 10^{-4}(2 \times 10^6 + x)$
$x = -195 \times 10^4\Omega$. As the resistance of given galvanometer cannot be negative. Hence the given condition is impossible.

215. (c) $V = i_g(G + R) = 4 \times 10^{-3}(50 + 5000) = 20$V

216. (d) $C\theta = NBiA \sin 90°$

$$\text{or } 10^{-6}\left(\frac{\pi}{180}\right) = 175B(10^{-3}) \times 10^{-4}$$

$\therefore B = 10^{-3}$T

217. (d) Using, $i_g = i\dfrac{S}{S + G}$

$$0.002 = 0.5\frac{S}{S + 50}$$

On solving, we get

$$S = \frac{100}{498} \simeq 0.2\,\Omega$$

218. (b) When key K_1 is closed and key K_2 is open

$$i_g = \frac{E}{220 + R_g} = C\theta_0 \quad \text{...(i)}$$

When both the keys are closed

$$I_g = \left(\frac{E}{220 + \dfrac{5R_g}{5 + R_g}}\right) \times \frac{5}{(R_g + 5)} = \frac{C\theta_0}{5}$$

$$\Rightarrow \frac{5E}{225R_g + 1100} = \frac{C\theta_0}{5} \quad \text{...(ii)}$$

Dividing (i) by (ii), we get

$$\Rightarrow \frac{225R_g + 1100}{1100 + 5R_g} = 5$$

$\Rightarrow 5500 + 25R_g = 225R_g + 1100$
$\Rightarrow 200R_g = 4400 \Rightarrow R_g = 22\Omega$

219. (b) Galvanometer has 25 divisions $I_g = 4 \times 10^{-4} \times 25 = 10^{-2}$A

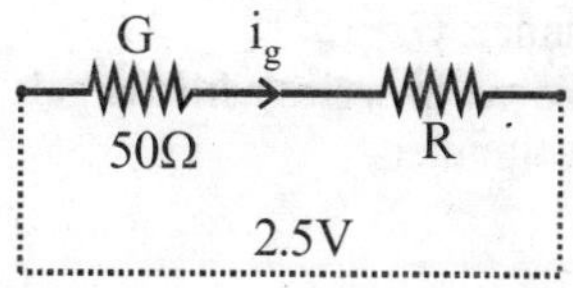

$V = I_g R_{net}$

$V = I_g(G + R)$
$2.5 = (50 + R)\,10^{-2}\ \therefore R = 200\Omega$

220. (c) Deflection current $= (Ig)_{max} = nk = 0.005 \times 30$
Where, n = Number of divisions = 30 and k = 0.005 amp/division
$= 15 \times 10^{-2} = 0.15$
$V = I_g[20 + R]$
$\Rightarrow 15 = 0.15\,[20 + R] \Rightarrow 100 = 20 + R \Rightarrow R = 80\,\Omega$

221. (d) Given,

Resistance of galvanometer, $G = 100\Omega$
Current, $i_g = 1\,mA$
A galvanometer can be converted into voltmeter by connecting a large resistance R in series with it.
Total resistance of the combination $= G + R$
According to Ohm's law, $V = i_g(G + R)$
$\therefore 10 = 1 \times 10^{-3}(100 + R_0)$
$\Rightarrow R_0 = 10000 - 100 = 9900\,\Omega = 9.9\,\text{k}\Omega$

222. (b) Figure of merit of a galvanometer is the current required to produce a deflection of one division in the galvanometer i.e., figure of merit $= \dfrac{I}{\theta}$

$$I = \frac{\varepsilon}{R + G} \qquad G = \frac{1}{9}K\Omega$$

$$\frac{1}{2} = \frac{\varepsilon}{R + \dfrac{GS}{G + S}} \times \frac{S}{S + G} \Rightarrow \frac{1}{2} = \frac{\varepsilon S}{R(S + G) + GS}$$

$$S = \frac{RG \times \dfrac{1}{2}}{\varepsilon - \dfrac{(R + G)I}{2}}$$

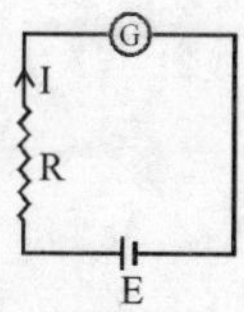

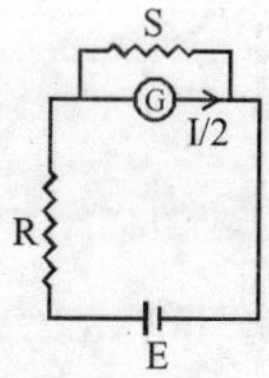

$$S = \frac{11 \times 10^3 \times \dfrac{1}{2} \times 10^2 \times 270 \times 10^{-6}}{6 - \left(\dfrac{6}{2}\right)} = 110\Omega$$

223. (d) According to question, current through galvanometer, $I_g = 1$ mA
Current through shunt, $(I - I_g) = 2$ A
Galvanometer resistance $R_g = 25\Omega$
Resistance of shunt, S = ?
$I_g R_g = (I - I_g)S$

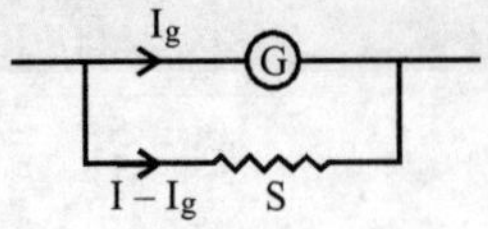

$\Rightarrow S = \frac{10^{-3} \times 25}{2}$

$S \simeq 1.25 \times 10^{-2}\Omega$

224. (c) Given : Current through the galvanometer, $i_g = 5 \times 10^{-3} A$
Galvanometer resistance, $G = 15\Omega$
Let resistance R to be put in series with the galvanometer to convert it into a voltmeter.
$V = i_g(R + G)$
$10 = 5 \times 10^{-3}(R + 15)$
$\therefore \; R = 2000 - 15 = 1985 = 1.985 \times 10^3 \,\Omega$

225. (c) $I_g G = (I - I_g)S$
$\therefore 10^{-3} \times 100 = (10 - 10^{-3}) \times S$
$\therefore S \approx 0.01\Omega$

226. (d) As we know, $I = \frac{V}{R} = \frac{5}{50} = 0.1$ A

$I' = 0.099$ A
When Galvanometer is connected

$R_{eq} = 50 + \frac{100S}{100 + S} = \frac{V}{I'}$

$\Rightarrow \frac{100S}{100 + S} = \frac{5}{0.099} - 50$

$\Rightarrow \frac{100S}{100 + S} = 50.50 - 50 \Rightarrow \frac{100S}{100 + S} = 0.5$

$\Rightarrow 100S = 50 + 0.55 \Rightarrow 99.5S = 50$

$\Rightarrow S = r_s = \frac{50}{99.05} = 0.5\Omega$

So, shunt of resistance = 0.5Ω is connected in parallel with the galvanometer.

227. (a) According to Ohm's Law, $I = \frac{V}{R}$

$I_g = \frac{V}{R + G}$

where, I_g-Galvanometer current, G-Galvonometer resistance

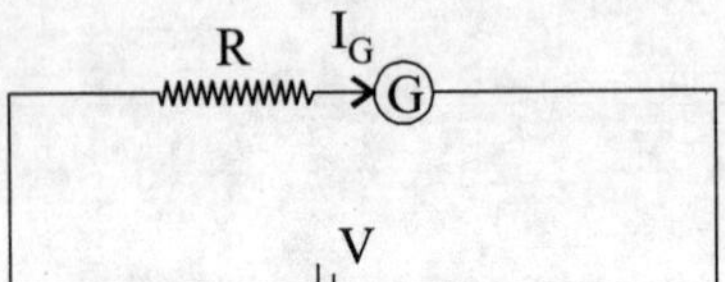

When shunt of resistance S is connected parallel to the Galvanometer then $G = \frac{GS}{G + S}$

$\therefore I = \frac{V}{R + \frac{GS}{G + S}}$

Equal potential difference is given by
$I'_g G = (I - I'_g)S$
$I'_g(G + S) = IS$

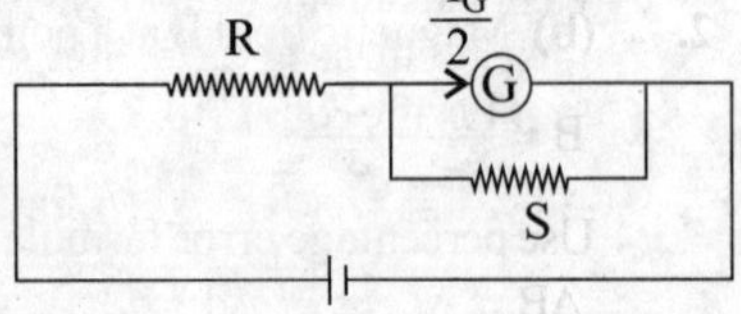

$\Rightarrow \frac{I_g}{2} = \frac{IS}{G + S}$

$\Rightarrow \frac{V}{2(R + G)} = \frac{V}{R + \frac{GS}{G + S}} \times \frac{S}{G + S}$

$\Rightarrow \frac{1}{2(R + G)} = \frac{S}{R(G + S) + GS}$

$\Rightarrow R(G + S) + GS = 2S(R + G)$

$\Rightarrow RG + RS + GS = 2S(R + G)$

$\Rightarrow RG = 2S(R + G) - S(R + G) \Rightarrow RG = S(R + G)$

228. (b) To measure AC voltage across a resistance a moving coil galvanometer is used.

229. (d) Statements I is false and Statement II is true

For ammeter, shunt resistance, $S = \frac{I_g G}{I - I_g}$

Therefore for I to increase, S should decrease, So additional S can be connected across it.

230. (d) The current that will given full scale deflection in the absence of the shunt is nearly equal to the current through the galvanometer when shunt is connected i.e. I_g

As $I_g = \frac{IS}{G + S} = \frac{5.5 \times 1}{120 + 1} = 0.045$ ampere.

231. (c) Resistance of Galvanometer,

$G = \frac{\text{Current sensitivity}}{\text{Voltage sensitivity}} \Rightarrow G = \frac{10}{2} = 5\Omega$

Here i_g = Full scale deflection current = $\frac{150}{10} = 15$ mA

V = voltage to be measured = 150 volts
($\because$ each division reads 1 volt)

$\Rightarrow R = \frac{150}{15 \times 10^{-3}} - 5 = 9995\Omega$

5 Magnetism and Matter

1. (c) Magnetic monopoles cannot exist as magnetic field lines requires north and south pole to form closed loops.

2. (b) Magnetic field B at a point P due to a magnetic dipole

$B = \frac{\mu_0}{4\pi}\frac{m(2\ell)}{d^3}$

Use percentage error formula

$\frac{\Delta B}{B} = \frac{\Delta \ell}{\ell} + 3 \times \left(\frac{\Delta d}{d}\right) = 1 + 3 \times 1 = 4\%$

% uncertainity in B = 4%

3. (28) Given,

Magnetic moment of straight wire $= m \times l = 44\ Am^2$

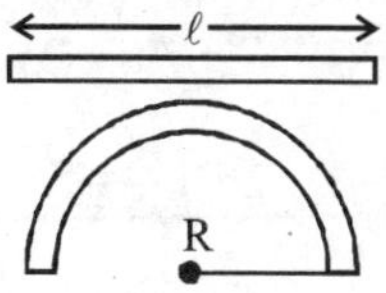

Initial length of wire = length of arc $= \pi R = l$

Effective length of semicircular arc, $l' = 2R \rightarrow l' = \frac{2l}{\pi}$

Magnetic moment of arc

$= m \times l' = m \times \frac{2\ell}{\pi} = \frac{44 \times 2}{\pi} = \frac{88}{\pi} = 28\ Am^2$

4. (5) External work done is given by:

$W_{ext} = \Delta U = U_2 - U_1$

$= -MB \cos 90 + MB \cos 0 = MB = NIAB$

$= 200 \times 100 \times 10^{-6} \times 2.5 \times 10^{-4} \times 1 = 5\mu J$

5. (100) Work done is equal to change in potential energy,

$W = U_f - U_i$

$W = -MB \cos 180° + MB \cos 90°$

$\Rightarrow W = (100 \times 5 \times 10^{-3} \times 1 \times 10^{-3}) \times 0.2\ J = 100\ \mu J$

6. (b) $U = -MB \cos\theta$

At stable equilibrium

$U = -MB \cos 0^0 = -MB$

At unstable equilibrium

$U = -MB \cos 180° = +MB$

$W = \Delta U = 2MB$

$= 2\ (0.5)\ 8 \times 10^{-2} = 8 \times 10^{-2} J$

7. (6) The magnetic potential due to magnetic dipole at axial point is

$V = \frac{\mu_0}{4\pi}\frac{M}{r^2} \Rightarrow 1.5 \times 10^{-5} = 10^{-7} \times \frac{M}{(20 \times 10^{-2})^2}$

$\Rightarrow M = \frac{1.5 \times 10^{-5} \times 20 \times 20 \times 10^{-4}}{10^{-7}}$

$M = 1.5 \times 4 = 6\ Am^2$

8. (a) Work done in rotating a dipole in magnetic field B from angle 'θ_1' to 'θ_2' is given as

$W = MB(\cos\theta_1 - \cos\theta_2)$

$= 5 \times 0.4\ (\cos 0° - \cos 180°) = 5 \times 0.4 \times 2 = 4J$

9. (b) As $\vec{M} = I\vec{A}$

$\Rightarrow |\vec{M}| = \frac{e}{\frac{2\pi R}{v}}\pi R^2 \qquad \left[\because I = \frac{Q}{T} = \frac{e}{\frac{2\pi R}{v}}\right]$

$\Rightarrow |\vec{M}| = \frac{1}{2} evR \Rightarrow |\vec{M}| = \frac{mvR}{1} \cdot \frac{e}{2m}$

$\Rightarrow |\vec{M}| = \frac{eL}{2m} \Rightarrow \vec{M} = -\frac{e\vec{L}}{2M}$

[$\because$ Here $\vec{M}$ and $\vec{L}$ will always be opposite]

10. (a) Work done in rotating a dipole from angle θ_1 to θ_2 is given by

$W = MB(\cos\theta_1 - \cos\theta_2) = MB(\cos 0° - \cos 60°)$

$= \frac{MB}{2} = \frac{2 \times 10^5 \times 14 \times 10^{-5}}{2} = 14J$

11. (1) Torque on a dipole

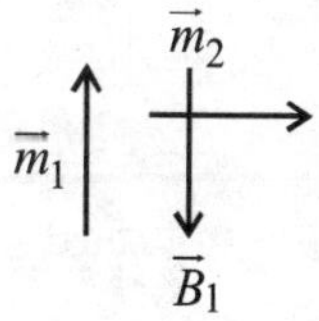

$T = \vec{M} \times \vec{B}$

Torque on m_2.

$= \vec{M}_2 \times \vec{B}_1$

$= 1 \times \frac{\mu_0 1}{4\pi(1)^3} = 10^{-7}\ N.m$

12. (3) Given,

Side of triangle, $\ell = 10\ cm = 10 \times 10^{-2}\ m$

Magnetic field, $B = 20\ mT = 20 \times 10^{-3}\ T$

Current, $I = 0.2\ A$

Area of equilateral triangle, $A = \frac{\sqrt{3}}{4}\ell^2$

Torque, $\tau = MB\sin 90°$

$= IAB = \frac{I\sqrt{3}\ell^2}{4}B = \frac{0.2 \times \sqrt{3} \times (10 \times 10^{-2})^2}{4} \times 20 \times 10^{-3}$

$\therefore\ \tau = \sqrt{3} \times 10^{-5}\ Nm$

13. (8) Given : Magnetic moment, $M = 9.85 \times 10^{-2} A/m^2$, moment of inertia, $I = 5 \times 10^{-6}\ kg\ m^2$

10 oscillations in 5 seconds

$\therefore$ Time period $= \frac{5}{10} = 0.5s$

$T = 2\pi\sqrt{\frac{I}{MB}} \Rightarrow \left(\frac{T}{2\pi}\right)^2 = \frac{I}{MB}$

$\Rightarrow\ B = \frac{I \times 4\pi^2}{M \times T^2} = \frac{5 \times 10^{-6} \times 4 \times 9.85}{9.85 \times 10^{-2} \times (0.5)^2}$

$\therefore\ B = 80 \times 10^{-4} = 8m\ T$

14. (c) Here, $\theta = 30°$, $\tau = 0.018$ N-m, $B = 0.06$ T

Torque on a bar magnet :

$\tau = MB\sin\theta \Rightarrow 0.018 = M \times 0.06 \times \sin 30°$

$\Rightarrow 0.018 = M \times 0.06 \times \frac{1}{2} \Rightarrow M = 0.6\ A\text{-}m^2$

Position of stable equilibrium ($\theta = 0°$)

Position of unstable equilibrium ($\theta = 180°$)
Minimum work required to rotate bar magnet from stable to unstable equilibrium

$\Delta U = U_f - U_i = -MB\cos 180° - (-MB\cos 0°)$

$W = 2MB = 2 \times 0.6 \times 0.06 = 7.2 \times 10^{-2}$ J

15. **(a)** Given,
Moment of inertia of circular coil, $I = 0.8$ kg m^2
Magnetic moment of circular coil, $M = 20$ Am^2

Rotational kinetic energy of circular coil, KE $= \frac{1}{2}I\omega^2$

Here, ω = angular speed of coil
Potential energy of bar magnet $= -MB\cos\phi$
From energy conservation

$$\frac{1}{2}I\omega^2 = U_{in} - U_f = 0 - (-MB\cos 60°)$$

$$[\because U_i = -MB\cos 90° = 0]$$

$$\Rightarrow \frac{MB}{2} = \frac{1}{2}I\omega^2 \Rightarrow \frac{20 \times 4}{2} = \frac{1}{2}(0.8)\omega^2$$

$$\Rightarrow 100 = \omega^2 \Rightarrow \omega = 10 \text{ rad/sec}$$

16. **(b)** $B_1 = \frac{\mu_0}{4\pi}\frac{2M}{(d/2)^3}$

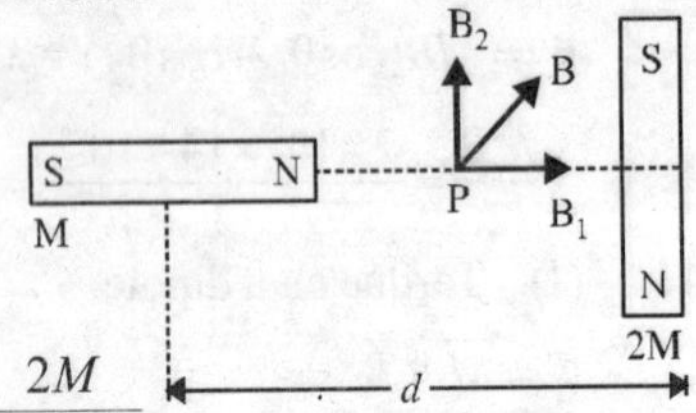

and $B_2 = \frac{\mu_0}{4\pi}\frac{2M}{(d/2)^3}$

$$\therefore \quad \tan\theta = \frac{B_2}{B_1} = \frac{\frac{\mu_0}{4\pi}\frac{2M}{(d/2)^3}}{\frac{\mu_0}{4\pi}\frac{2M}{(d/2)^3}} = 1$$

or $\theta = 45°$

The resultant field is 45° from B_1. The angle between $\vec{B}$ and $\vec{v}$ zero, so force on the particle is zero.

17. **(c)** Work done, W = 2 M·B
$= 2 \times 10^{-2} \times 1\cos(0.125) = 0.02$ J

18. **(c)** Potential energy of dipole, U = – MB cos θ
Torque experienced by dipole τ = MB sin θ
Torque will be maximum (τ_{max}) when θ = 90° then potential energy U = 0

19. **(b)**

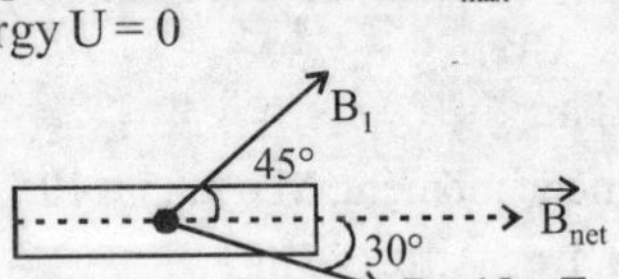

Clearly, $\tan 30° = \frac{B_1 \sin 75}{B_2 + B_1\cos 75°}$

Putting $B_2 = 15$ mT in above equation, we get $B_1 \simeq 11$ mT

20. **(c)** $\vec{M}$ (mag. moment/volume) $= \frac{NiA}{A\ell}$

$$= \frac{Ni}{\ell} = \frac{(500)15}{25 \times 10^{-2}} = 30000 \text{ Am}^{-1}$$

21. **(c)** $B_{eq} = \frac{\mu_0}{4\pi}\frac{M}{(l^2 + d^2)^{3/2}}$

$$= 10^{-7} \times \frac{4}{\left[(3 \times 10^{-2})^2 + (2)^2\right]^{3/2}} = 5 \times 10^{-8}T$$

22. **(b)** $M = IA = \frac{QA}{T} = \frac{\sigma A^2}{\frac{2\pi R}{V}} = \frac{\sigma \times \pi^2 R^4}{\frac{2\pi R}{\omega R}} = \frac{\sigma\pi\omega R^4}{2}$

23. **(d)** A magnetic needle kept in non uniform magnetic field experience a force and torque due to unequal forces acting on poles.

24. **(b)** Initially, time period of magnet

$$T = 2\pi\sqrt{\frac{I}{MB}} = 2s \text{ where } I = \frac{1}{12}m\ell^2$$

When the magnet is cut into three pieces the pole strength will remain the same and Moment of inertia of each part,

$$(I') = \frac{1}{12}\left(\frac{m}{3}\right)\left(\frac{\ell}{3}\right)^2 \times 3 = \frac{I}{9}$$

We have, Magnetic moment (M)
= Pole strength (m) × ℓ
∴ New magnetic moment,

$$M' = m \times \left(\frac{\ell}{3}\right) \times 3 = m\ell = M$$

New time period, $T' = 2\pi\sqrt{\frac{I'}{M'B}}$

$$= 2\pi\sqrt{\frac{I}{9MB}} \Rightarrow T' = \frac{T}{\sqrt{9}} = \frac{2}{3}s.$$

25. **(a)** Workdone to turn a magnetic needle from angle θ_1 to θ_2 is given by

$W = MB(\cos\theta_1 - \cos\theta_2)$

$$\therefore W = MB(\cos 0° - \cos 60°) = MB\left(1 - \frac{1}{2}\right) = \frac{MB}{2}$$

$$\therefore \text{ Torque, } \tau = MB\sin\theta = MB\sin 60° = \sqrt{3}\frac{MB}{2} = \sqrt{3}W$$

26. **(d)** The magnetic field lines of bar magnet form closed lines. As shown in the figure, the magnetic lines of force are directed from south to north inside a bar magnet. Outside the bar magnet magnetic field lines directed from north to south pole.

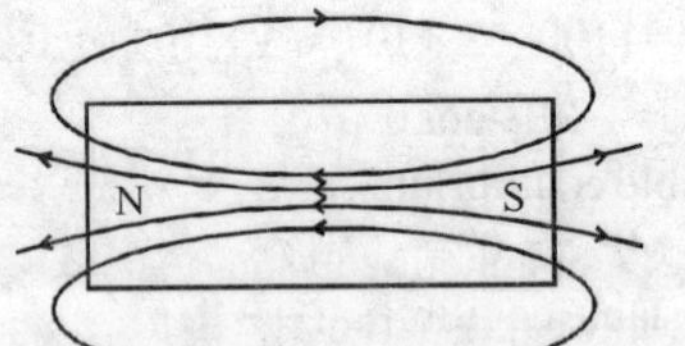

27. **(a)** The percentage increase in magnetic field B is

$$\%\Delta B = \frac{B_{new} - B_{old}}{B_{old}} \times 100\%$$

$$= \frac{\mu ni - \mu_0 ni}{\mu_0 ni} \times 100\% = \frac{(\mu - \mu_0)}{\mu_0} \times 100\%$$

$$= \frac{(\mu_0\mu_r - \mu_0)}{\mu_0} \times 100\% = (\mu_r - 1) \times 100\%$$

$$= \chi \times 100\% \qquad [\because \chi = \mu_r - 1]$$

$$= 1.2 \times 10^{-3}\% = \frac{6}{5} \times 10^{-3}\%.$$

28. **(a)** Relative permeability

$$\mu_r = (1 + \chi) \Rightarrow \chi = (\mu_r - 1)$$

Also, $\mu = \mu_0 \mu_r \Rightarrow \mu_r = \frac{\mu}{\mu_0}$

$\therefore$ Magnetic susceptibility $\chi = \left(\frac{\mu}{\mu_0} - 1\right)$

29. (d) When paramagnetic substances are placed in magnetic field, they align themselves along the directions of external magnetic field. For paramagnetic substance $\boxed{\chi > 0}$

30. (10) Coercivity is given by:

$$\Rightarrow H_0 = \frac{B}{\mu_0}$$

$$= \frac{\mu_0 ni}{\mu_0} = ni \qquad \left[\because n = \frac{N}{\ell}\right]$$

$$5 \times 10^3 = \frac{150}{30} \times 100 \times i \Rightarrow i = 10A$$

31. (d) (A) Magnetic susceptibility is independent on magnetising field

(B) Magnetic field due to a current carrying wire for x < a:

$$B = \frac{\mu_0 ir}{2\pi a^2} \Rightarrow B \propto r$$

(C) Magnetic field due to a current carrying wire for x > a:

$$B = \frac{\mu_0 i}{2\pi a} \Rightarrow B \propto \frac{1}{a}$$

(D) Magnetic field inside solenoid varies with distance as,

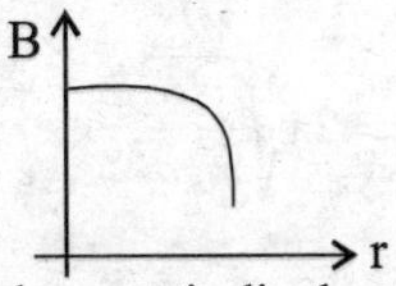

32. (a) The induced magnetic dipole moment in diamagnetic sample is always opposite to the magnetizing field. Diamagnetic property is independent on temperature.

33. (6) Given, magnetic intensity, $H = 2.4 \times 10^3$ A/m

$$H = ni = \frac{Ni}{\ell} \Rightarrow 2.4 \times 10^3 = \frac{60}{15 \times 10^{-2}} i$$

$$i = \frac{2.4 \times 15 \times 10}{60} = \frac{36}{6} = 6A$$

34. (b) Diamagnetic substances tend to move from stronger to weaker part of the field.

Magnetic susceptibility of diamagnetic substance is negative.

35. (2) Number of turns $n = \frac{8}{10^{-2}}$ per metre

Magnetic intensity,

$$H = \frac{B}{\mu_0} = \frac{\mu_0 ni}{\mu_0} = ni \qquad (\because B = \mu_0 ni)$$

$$i = \frac{H}{n} = \frac{1.6 \times 10^3}{\left(\frac{8}{10^{-2}}\right)} = 2A$$

36. (b) Given that number of turns, n = 1200

Current, I = 2A

Magnetic field at centre inside the solenoid is given by, $B = \mu_0 nI$

So magnetic intensity at centre of the solenoid,

$$H = \frac{B}{\mu_0} = nI = \left(\frac{1200}{2}\right)(2) \qquad (\because B = \mu_0 nI)$$

$$H = 1.2 \times 10^3 \text{ Am}^{-1}$$

37. (a) $\chi = 1.2 \times 10^{-5}$

$$\mu_r = 1 + \chi = 1 + 1.2 \times 10^{-5}$$

Now, $B = \mu_0 \mu_r nI \Rightarrow \frac{\Delta B}{B} = \frac{\Delta \mu_r}{\mu_r}$

$$\Rightarrow \frac{\Delta B}{B} = \frac{\mu_r - 1}{1} = \mu_r - 1 = X = 1.2 \times 10^{-5}$$

38. (c) $\mu_r = 1 + \chi = 1 + 99 = 100$

$$\mu = \mu_0 \mu_r = \mu_0 \times 100 = 4\pi \times 10^{-7} \times 100 = 4\pi \times 10^{-5}$$

39. (a) According to Curie's law, magnetic susceptibility is inversely proportional to temperature for a fixed value of external magnetic field i.e. $X = \frac{C}{T}$

The same is applicable for ferromagnet & the relation is given as

$$X = \frac{C}{T - T_C} \quad (T_c \text{ is curie's Temperature})$$

Dimagnetism is due to non-cooperative behaviour of orbiting electrons when exposed to external magnetic field.

40. (a) As temperature increases, domains disintegrate so ferromagnetism decreases and above Curie's temperature it becomes paramagnet. Also, at high temperature the domain area of ferromagnetic substance decreases.

41. (b) Given,

Magnetic susceptibility $\chi_m = 499$

Relation between absolute permeability and magnetic susceptibility (χ) is, $\mu = \mu_0(1 + \chi) = 4\pi \times 10^{-7} \times 500$

$= 2\pi \times 10^{-4}$ H/m

42. (a) Magnetic susceptibility(χ) is independent of temperature for diamagnetic material.

Magnetization is directly proportional to magnetising field and take place in direction opposite to it.

43. (250) $I = \frac{M}{V}$, I = Magnetisation intensity

$$\chi H = \frac{M}{V}, \; M = \text{Magnetic moment}$$

For solenoid ($H = ni$) and as $\mu_r = 1 + \chi$

So, $(\mu_r - 1)niV = M$

$M \propto \mu_r - 1$

$$\therefore \frac{M_2}{M_1} = \frac{749}{499} \Rightarrow M_2 = \frac{749}{499} M_1$$

$$\text{So, } \frac{M_2 - M_1}{M_1} = \frac{\frac{250}{499} M_1}{M_1} = \frac{250}{499} = \frac{x}{499} \text{(given)}$$

44. (d) Soft ferromagnetic materials are materials which can be easily magnetised and demagnetised by external magnetic field. When external field is applied, the domains in which magnetic dipole are aligned in the direction of external magnetic field will increase in size. While the domain in which magnetic dipole aligned opposite to magnetic field decrease in size.

45. (c) In a ferromagnetic material, below the curie temperature, domain is defined as the region in which magnetic moment is aligned in such a way to produce saturated magnetization.

46. (b) Given, Volume of iron rod, $V = 10^{-3}\ m^3$
Relative permeability, $\mu_r = 1000$
Number of turns per unit length, $n = 10$
Magnetic moment of an iron core solenoid,
$M = (\mu_r - 1) \times NiA$

$$\Rightarrow M = (\mu_r - 1) \times Ni\frac{V}{l} \Rightarrow M = (\mu_r - 1) \times \frac{N}{l} iV$$

$$\Rightarrow M = 999 \times \frac{10}{10^{-2}} \times 0.5 \times 10^{-3} = 499.5 \approx 500.$$

47. (d) For paramagnetic material, according to Curies law,

$$\chi \propto \frac{1}{T}$$

For two temperatures T_1 and T_2, $\chi_1 T_1 = \chi_2 T_2$

But $\chi = \frac{I}{B}$ $\quad \therefore \frac{I_1}{B_1} T_1 = \frac{I_2}{B_2} T_2$

$$\Rightarrow \frac{6}{0.4} \times 4 = \frac{I_2}{0.3} \times 24 \Rightarrow I_2 = \frac{0.3}{0.4} = 0.75 \text{ A/m}$$

48. (b) When magnetic field is applied to a diamagnetic substance, it produces magnetic field in opposite direction so net magnetic field inside the cavity of sphere will be zero. So, field inside the paramagnetic substance kept inside the cavity is zero.

49. (a) According to Curie law for paramagnetic substance,

$$\chi \propto \frac{1}{T_C} \Rightarrow \frac{\chi_1}{\chi_2} = \frac{T_{C_2}}{T_{C_1}} \Rightarrow \frac{2.8 \times 10^{-4}}{\chi_2} = \frac{300}{350}$$

$$\chi_2 = \frac{2.8 \times 350 \times 10^{-4}}{300} = 3.266 \times 10^{-4}$$

50. (d) Magnetic susceptibility, $\chi = \frac{I}{H}$

where, $I = \frac{\text{Magnetic moment}}{\text{Volume}} = \frac{20 \times 10^{-6}}{10^{-6}} = 20\ \text{N/m}^2$

Now, $\chi = \frac{20}{60 \times 10^3} = \frac{1}{3} \times 10^{-3} = 3.3 \times 10^{-4}$

51. (b) Corecivity, $H = \frac{B}{\mu_0}$ and $B = \mu_0 ni$

or, $H = \frac{N}{\ell} i = \frac{100}{0.2} \times 5.2 = 2600$ A/m $\quad \left(\because n = \frac{N}{\ell}\right)$

52. (b) Magnetic field inside the superconductor is zero. as $\chi = -1$ for super conductors
$\therefore \mu_r = 1 + \chi = 1 - 1 = 0$
So, $B = \mu H = 0$

53. (b) Diamagnetic materials does not allow magnetic field lines to passes through them.
Bar B represents diamagnetic materials.

54. (d) In magnetic dipole
Interaction energy, $U = -\vec{M} \cdot \vec{B}$ and $B \propto \frac{1}{r^3}$

$$\therefore F = -\frac{dU}{dr} \propto \frac{1}{r^4}$$

55. (b) For a diamagnetic material, the value of μ_r is slightly less than one. For any material, the value of $\in_r$ is always greater than 1.

56. (b) Ferromagnetic substance has magnetic domains, whereas paramagnetic substances have magnetic dipoles which get attracted to a magnetic field. Ferromagnetic material magnetised strongly in the direction of magnetism field. Hence, N_1 will be attracted. Paramagnetic substance attract weekly in the direction of field. Hence, N_2 will weakly attracted. Diamagnetic substances do not have magnetic dipole but in the presence of external magnetic field due to their orbital motion of electrons these substances are repelled. Hence, N_3 will be repelled.

57. (a) The temperature above which a ferromagnetic substance becomes paramagnetic is called Curie's temperature.

58. (b) As, $T = 2\pi\sqrt{\frac{I}{MB}}$

$$\Rightarrow T^2 \propto \frac{I}{M} \Rightarrow M \propto \frac{I}{T^2}$$

$$\Rightarrow \frac{M_1}{M_2} = \left(\frac{I_1}{I_2}\right)\left(\frac{T_2}{T_1}\right)^2 = \frac{3}{2} \times \left(\frac{4}{3}\right)^2 = \frac{3}{2} \times \frac{16}{9} = \frac{8}{3}$$

59. (a) Let I_1 and I_2 be the moment of inertia in first and second case respectively.

$I_1 = 2MR^2$

$$I_2 = MR^2 + \frac{MR^2}{2} = \frac{3}{2}MR^2$$

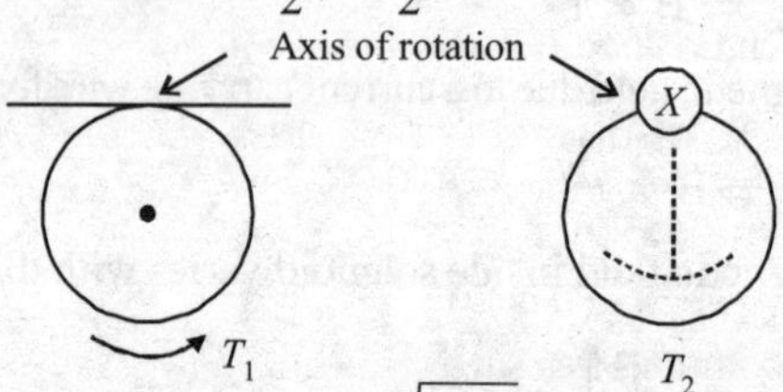

Time period, $T = 2\pi\sqrt{\frac{I}{mgd}}$

$T \propto \sqrt{I}$

$$\therefore \frac{T_1}{T_2} = \sqrt{\frac{I_1}{I_2}} = \sqrt{\frac{2MR^2}{\frac{3}{2}MR^2}} = \frac{2}{\sqrt{3}}$$

60. (a) Using, time period,

$$T = 2\pi\sqrt{\frac{I}{MB}}$$

Where, M = magnetic moment, I moment of inertia and B = magnetic field

$$T_h = 2\pi\sqrt{\frac{mR^2}{(2MB)}}$$

$$T_c = 2\pi\sqrt{\frac{1/2mR^2}{MB}}$$

Clearly, $T_h = T_c$

61. (c) Given : Magnetic moment, $M = 6.7 \times 10^{-2}\ Am^2$
Magnetic field, $B = 0.01\ T$
Moment of inertia, $I = 7.5 \times 10^{-6}\ \text{Kgm}^2$
Using,

$$T = 2\pi\sqrt{\frac{I}{MB}} = 2\pi\sqrt{\frac{7.5 \times 10^{-6}}{6.7 \times 10^{-2} \times 0.01}} = \frac{2\pi}{10} \times 1.06\ s$$

Time taken for 10 complete oscillations
$t = 10T = 2\pi \times 1.06 = 6.6568 \approx 6.65$ s

6 Electromagnetic Induction

1. (b)

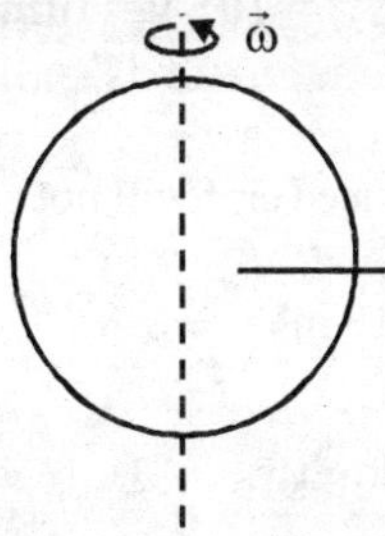

The magnetic flux at any instant of time t is

$\phi = BAN.\cos(\omega t)$

$\therefore$ Induced emf, $\varepsilon = \frac{-d\phi}{dt} = BA\omega N.\sin(\omega t)$

When B is parallel to plane of coil, $\omega t = \frac{\pi}{2}$

$\therefore \phi = 0, \varepsilon = BA\omega N$

2. (5) emf, $\varepsilon = -\frac{d\phi}{dt} = NAB\omega\sin(\omega t)$

emf is maximum when $\sin\omega t = 1$

$\varepsilon_{max} = NAB\omega$

$= 200 \times 0.2 \times 0.01 \times \pi = \frac{2\pi}{5}V = \frac{2\pi}{\beta}V$ (given) $\therefore \beta = 5$

3. (a) Given,

Initial Magnetic field, $B_i = 5000T$

Final Magnetic field, $B_f = 3000T$

Radius, $r = \frac{0.02}{2} = 0.01$ m

From Faraday's law of electromagnetic induction

$$e = -N\left(\frac{\Delta\phi}{\Delta t}\right) = -N\frac{(\Delta B)A}{\Delta t} = \frac{N(B_i - B_f)\pi r^2}{\Delta t}$$

$$\Rightarrow 22 = N\frac{2000\pi(0.01)^2}{2} \Rightarrow N = 70$$

4. (2) From Faraday's law of Induction

Induced

emf, $\varepsilon = -\left(\frac{d\phi}{dt}\right) = -\frac{d}{dt}(5t^2 - 36t + 1) = -(10t - 36)$

At $t = 2, \varepsilon = -(10 \times 2 - 36) = 16$ V

Induced current, $i = \frac{\varepsilon}{R} = \frac{16}{8} = 2$ A

5. (2)

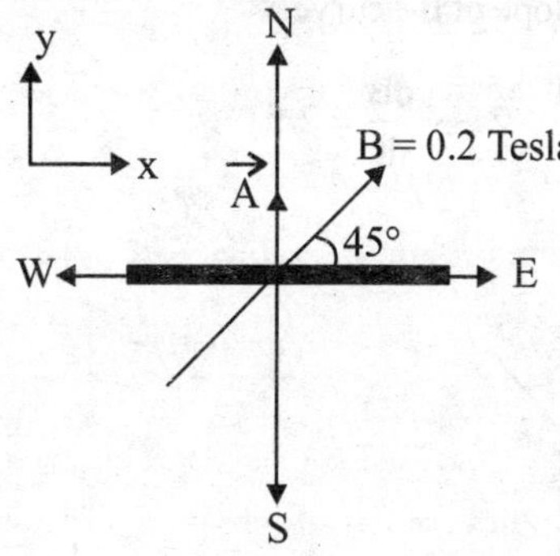

From above diagram,

$\vec{A} = (0.1)^2\hat{j}$; $\vec{B} = \frac{0.2}{\sqrt{2}}\hat{i} + \frac{0.2}{\sqrt{2}}\hat{j}$

By Faraday's law,

Induced emf, $|e| = \frac{\Delta\phi}{\Delta t} = \frac{\vec{B}\cdot\vec{A} - 0}{1} = \sqrt{2} \times 10^{-3}V$

$= \sqrt{x} \times 10^{-3}V$ (given)

$\therefore x = 2$

6. (c) Average emf, $e = \frac{\text{Change in flux}}{\text{Time}} = -\frac{\Delta\phi}{\Delta t}$

$$= -\frac{0 - (4 \times (2.5 \times 2)\cos 60°)}{10} = +1V$$

7. (60) $\Delta Q = -\frac{\Delta\phi}{R} = -\left(\frac{\phi_2 - \phi_1}{R}\right)$ $[\because \phi = NBA]$

$$\therefore \Delta Q = \frac{2NBA}{R} = \frac{2 \times 100 \times 1.5 \times 24 \times 10^{-4}}{12}$$

$= 6 \times 10^{-2}$ C $= 60$ mC

8. (50) Given,

Magnetic field, $B = 0.4T$

Radius of the loop, $r = 2$ cm

Rate of expansion of radius, $\frac{dr}{dt} = 1$ mm/s

Magnetic flux $\phi = BA = (0.4)\pi r^2$

$$\varepsilon = \left|\frac{-d\phi}{dt}\right| = \left|\frac{BdA}{dt}\right| = 0.4\pi\left(2r\frac{dr}{dt}\right)$$

$= 0.4 \times 2 \times \pi \times 2 \times 10^{-2} \times 10^{-3}V$

$= 16\pi\mu V = 50.24\mu V$

9. (8) Given,

Area of metallic plate, $A = 4\ m^2$

Induced emf,

$$\varepsilon = \left|\frac{d\phi}{dt}\right| = \frac{d(BA)}{dt} = \frac{AdB}{dt}$$

But $\frac{dB}{dt}$ = Slope of B-t curve

$$= \tan\theta = \frac{10}{5} = 2 \Rightarrow \frac{dB}{dt} = 2mT$$

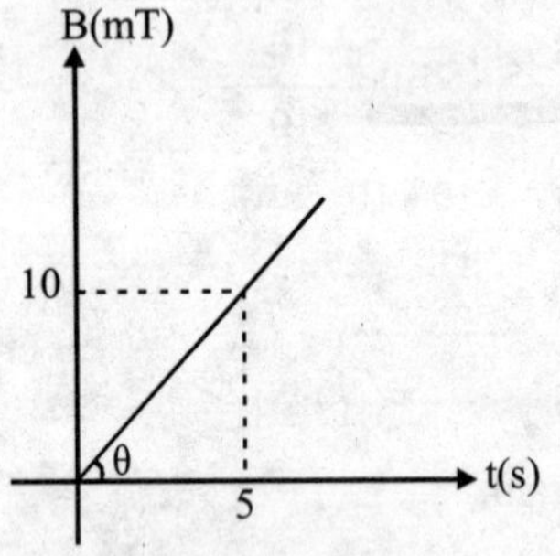

$\therefore \varepsilon = 4 \times 2 = 8\ mV$

10. (d) When coil is moving inside a uniform magnetic field either with uniform or non-uniform speed, it doesn't changes flux, so, no emf is induced.
Magnetic flux, $\phi = BA\cos\theta$
In statement C, θ is changing, hence emf will induce.
In statement D, area is changing, hence emf will induce.

11. (d) We know that
$\phi_B = BA\cos\theta$
So, if 'B', 'A' and θ changes then magnetic flux also get changed.
By reversing the direction of magnetic field only the direction of magnetic flux or magnetic field lines will change, there will be no change in its magnitude.

12. (44) If plane of coil makes angle 60° with field, then angle between $\vec{B}$ and $\vec{A}$ at that instant is 30°

$$e = -\frac{Nd\phi}{dt} = -N\frac{d}{dt}(BA\cos\omega t) = NBA\,\omega\sin\omega t$$

$$= 600 \times 0.4 \times 70 \times 10^{-4} \times 2\pi \times \frac{500}{60}\sin 30° \simeq 44\,V$$

13. (c) As $\phi = BA\cos\theta$

So, $\phi = \mu_0\mu_r \frac{N}{l} IA\cos 0°$

$$\Rightarrow 4\pi \times 10^{-6} = 4\pi \times 10^{-7} \times \mu_r \times \frac{400}{0.4} \times 0.4 \times 2 \times 10^{-4}$$

$\Rightarrow \mu_r = 125$

14. (10) Given, magnetic field, $B = 0.8\ T$

Rate of shrinking of radius, $\frac{dr}{dt} = 2cm^{-1}$

Magnetic flux, $\phi = BA = B(\pi r^2)$

$$EMF = \left|\frac{-d\phi}{dt}\right| = \frac{d}{dt}(B\pi r^2) = 2B\pi r\frac{dr}{dt}$$

$= 2 \times \pi \times 0.1 \times 0.8 \times 2 \times 10^{-2}$
$= 2\pi \times 1.6 = 10.06$ [round off 10.06 = 10]

15. (b) As $\varepsilon|_{t=0.5sec} = -\frac{d\phi}{dt}$

$$= -A\frac{dB}{dt} \quad [\because \theta = 0° \Rightarrow \cos\theta = 1]$$

$$= -\pi \times \left(\frac{10}{\sqrt{\pi}}\right)^2 \times 10^{-4} \times \frac{0-0.5}{0.5} = 10^{-2}\,V = 10\,mV$$

As $\frac{dB}{dt}$ = constant $\Rightarrow$ Induced emf will not change with time. So, $e|_{0.5sec} = e|_{0.25sec} = 10\ mV$

16. (12) Given,

Magnetic field, $B = (3t^3\hat{j} + 3t^2\hat{k})$

Magnetic flux, $\phi = \vec{B}.\vec{A}$

$$= (3t^3\hat{j} + 3t^2\hat{k}).(\pi(1)^2\hat{k}) = 3t^2\pi$$

Induced emf, $\varepsilon = \left|\frac{d\phi}{dt}\right| = \frac{d(3t^2\pi)}{dt} = 6t\pi$

$\therefore \quad \varepsilon_{t=2} = 6 \times 2 \times \pi = 12\pi$

17. (2) The magnetic flux given by $\phi = \frac{2}{3}(9 - t^2)$

when the flux becomes zero then $\phi = \frac{2}{3}(9 - t^2) = 0$

$t = 3$ sec
The emf will be

$$e = \frac{-d\phi}{dt} = -\frac{2}{3}(0 - 2t) = \frac{4t}{3}$$

Heat produced in 3 sec $= \int_0^t \frac{e^2}{r}dt = \int_0^3 \frac{16t^2}{9 \times 8}dt = 2\ J$

18. (250) We have magnetic flux given as,
$\phi = 8t^2 - 9t + 5$

Inuced emf $= -\frac{d\phi}{dt} = -(16t - 9)$

At $t = 0.25$ s
Emf $= -[(16 \times 0.25) - 9] = 5V$

So, Current $= \frac{\text{Emf}}{\text{Resistance}} = \frac{5V}{20\Omega}$

$$\Rightarrow \frac{1}{4}A = \frac{1000}{4}mA = -250\ mA$$

19. (d) $P = \frac{e^2}{R} = \frac{\left(-NA\frac{dB}{dt}\right)^2}{\rho l} \times A_C,$

A = area of coil
A_c = area of cross section of wire used in coil
$P \propto NA_c$

$$\frac{P_2}{P_1} = \frac{\frac{N}{2} \times 4A_c}{NA_c} = 2; \ P_2 = 2P_1$$

20. **(a)** $\phi = 5t^3 + 4t^2 + 2t - 5$

$|e| = 15t^2 + 8t + 2 \qquad \left[\because |e| = \frac{d\phi}{dt}\right]$

$R = 5\Omega$

So, $i = \frac{|e|}{R} = \frac{15t^2 + 8t + 2}{5} = 3t^2 + 1.6t + 0.4$

$= 3 \times 2^2 + 1.6 \times 2 + 0.4 = 15.6\,\text{A} \qquad (\because t = 2\text{ sec})$

21. **(440)** $e = -\frac{d\phi}{dt} = -\frac{d}{dt}(\text{NBA} \cos \omega t) = \text{NBA}\,\omega \sin \omega t$

So, $|e|_{max} = \text{NBA}\,\omega$

$= 1000 \times 0.07 \times 1 \times 2\pi \times 1 = 140\,\pi \simeq 440\,\text{V}$

22. **(a)** From Lenz law direction of induced current can be found, induced current will oppose the decrease in outward magnetic field, by producing its own outward magnetic field.

23. **(60)** Given,

Magnetic field, $B = 3 \times 10^{-2}$T

Angular speed of coil, $\omega = 50$ rad s^{-1}

Number of turns in coil, n = 20

Maximum emf, $\varepsilon = \text{N}\omega\text{AB}$

$\Rightarrow \varepsilon = 20 \times 50 \times \pi \times (0.08)^2 \times 3 \times 10^2 = 60.28 \times 10^2$

Rounded off to nearest integer = 60 V.

24. **(60)** Given,

Magnetic flux, $\phi = 10t^2 + 20t$

Resistance of circular loop, $R = 2\Omega$

Emf induced in the coil,

$\varepsilon = \left|-\frac{d\phi}{dt}\right| = 20t + 20$

Current in the circuit,

$i = \frac{\varepsilon}{R} = \frac{120}{2} = 60\,\text{mA}$

25. **(c)** From figure,

$B_A = 3\text{T}$

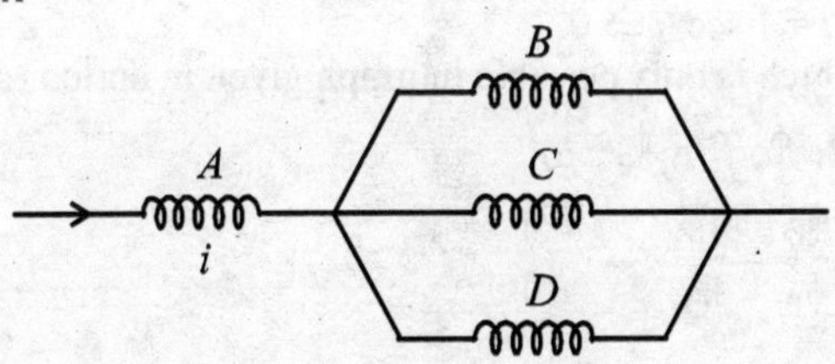

$I_B + I_C + I_D = I_A$

By symmetry, $I_B = I_C = I_D = i_0$ (let) $\Rightarrow i_0 = \frac{i}{3}$

We have $B_A = \mu_0 ni$

and, $B_C = \mu_0 n i_0 = \frac{\mu_0 ni}{3} = \frac{B_A}{3} = 1\text{T}$

26. **(5)**

For coil C_1, No. of turns $N_1 = 500$ and radius, $r = 1$ cm.

For coil C_2, No. of turns $N_2 = 200$ and radius, $R = 20$ cm

$I = (5t^2 - 2t + 3) \Rightarrow \frac{dI}{dt} = (10t - 2)$

$\phi_{\text{small}} = BA = \left(\frac{\mu_0 I N_2}{2R}\right)(\pi r^2)$

Induced emf in small coil,

$e = \frac{d\phi}{dt} = \left(\frac{\mu_0 N_2}{2R}\right)\pi r^2 N_1 \frac{di}{dt} = \left(\frac{\mu_0 N_1 N_2 \pi r^2}{2R}\right)(10t - 2)$

At $t = 1$ s

$e = \left(\frac{\mu_0 N_1 N_2 \pi r^2}{2R}\right)8 = 4\frac{\mu_0 N_1 N_2 \pi r^2}{R}$

$= \frac{4(4\pi)10^{-7} \times 200}{20} \times 500 \times \frac{10^{-4}}{10^{-2}}\pi$

$= 80 \times \pi^2 \times 10^{-7} \times 10 \times 10^2 \times 10^{-2}$

$= 8 \times 10^{-4}$ volt $= 0.8$ mV$= \frac{4}{x} \Rightarrow x = 5.$

27. **(a)** As we know, emf $\varepsilon = NAB\omega \cos \omega t$, Here $N = 1$

Average power,

$< P > = < \frac{\varepsilon^2}{R} > = < \frac{A^2 B^2 \omega^2 \cos^2 \omega t}{R} >$

$= \frac{A^2 B^2 \omega^2}{R}\left(\frac{1}{2}\right)$

Therefore average power loss in the loop due to Joule heating

$< P > = \frac{\pi^2 a^2 b^2 B^2}{2R}(\omega^2)$

28. **(b)** $e = -\frac{d\phi}{dt} = -\frac{d}{dt}(BA \cos 0°) = -A\frac{dB}{dt}$

$i = \frac{e}{R} = A\frac{dB}{dt} \times \frac{A'}{\rho l'}$

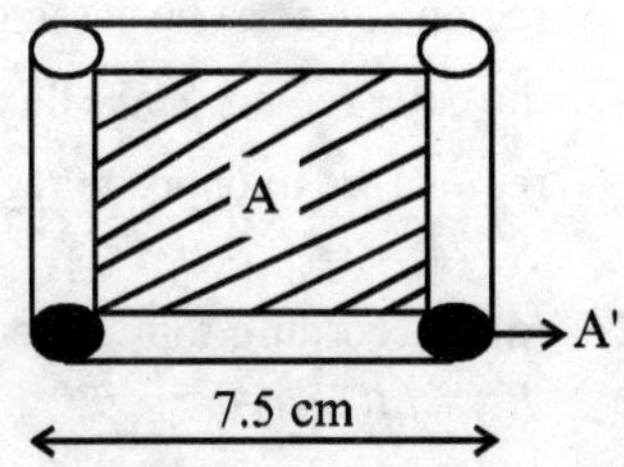

$= l^2 \frac{dB}{dt} \times \frac{\pi r^2}{\rho l'}$

Here, $l = 7.5 \times 10^{-2}$m

$l' = 0.30$m

$r = 2 \times 10^{-3}$m

$\frac{dB}{dt} = 0.032$ T/s

Putting above value; we get

$i = 0.61$A

29. **(15)**

Here, $B = 3.0 \times 10^{-5}$ T, $R = 10$ cm $= 0.1$ m

$\omega = \frac{2\pi}{2T} = \frac{\pi}{0.2}$

Flux as a function of time $\phi = \vec{B} \cdot \vec{A} = AB \cos(\omega t)$

Emf induced, $e = \frac{-d\phi}{dt} = AB\omega \sin(\omega t)$

Max. value of Emf $= AB\omega = \pi R^2 B\omega$

$= 3.14 \times 0.1 \times 0.1 \times 3 \times 10^{-5} \times \frac{\pi}{0.2}$

$= 15 \times 10^{-6}$ V $= 15$ μV

30. **(a)** According to question, $dB = 1000 - 500 = 500$ Gauss $= 500 \times 10^{-4} T$

Time $dt = 5$ s

Using faraday law

$$\text{Induced } EMF,\ e = \left|-\frac{d\phi}{dt}\right| = \left|A\frac{dB}{dt}\right|$$

$$\frac{dB}{dt} = \frac{1000 - 500}{5} \times 10^{-4} = 10^{-2} \text{ T/sec}$$

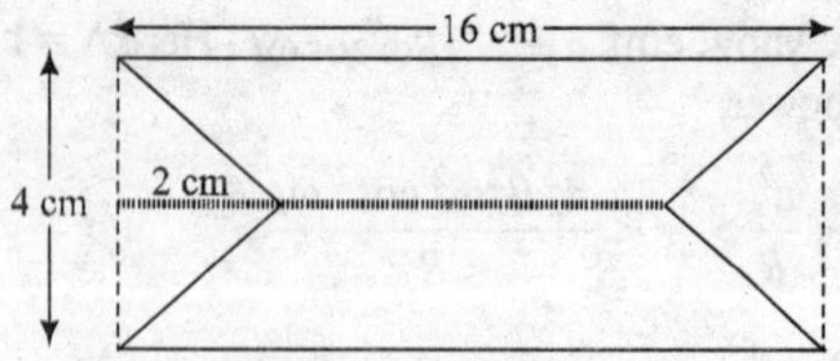

Area, A = ar of Δ –2 ar of Δ = (16 × 4 – 2 × Area of triangle) cm²

$= \left(64 - 2 \times \frac{1}{2} \times 2 \times 4\right) \text{cm}^2 = 56 \times 10^{-4} \text{ m}^2$

$$\therefore \varepsilon_{\text{induced}} = \left|A\frac{dB}{dt}\right| = 56 \times 10^{-4} \times 10^{-2} = 56 \times 10^{-6} V = 56 \mu V$$

31. **(d)** As magnetic field lines form close loop, hence every magnetic field line creating magnetic flux through the inner region (ϕ_i) must be passing through the outer region. Since flux in two regions are in opposite region.

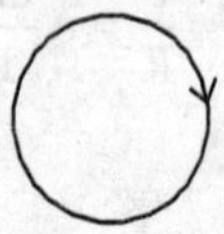

$\therefore \quad \phi_i = -\phi_0$

32. **(d)** According to question,

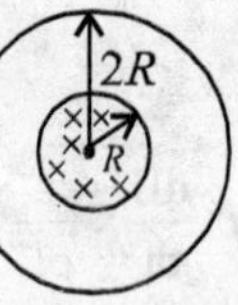

$I(t) = I_0 t(1 - t)$

$\therefore I = I_0 t - I_0 t^2$

$\phi = B.A$

$\phi = (\mu_0 nI) \times (\pi R^2)$

$(\because B = \mu_0 \text{nI} \textit{ and } A = \pi R^2)$

$$V_R = \frac{-d\phi}{dt}$$

$V_R = \mu_0 n \pi R^2 (I_0 - 2I_0 t) \Rightarrow V_R = 0$ at $t = \frac{1}{2} s$

after $t > \frac{1}{2}$ sec, V_R is – Ve

i.e. current will change direction after $t > \frac{1}{2}$ sec.

33. **(175.00)**

Flux through the loop ABCDEFA,

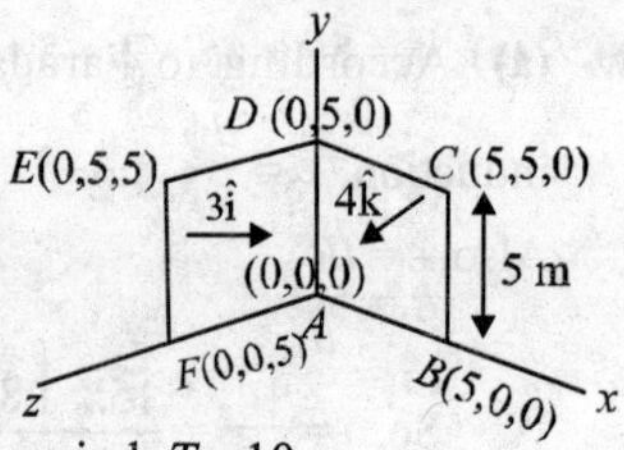

$\phi = \vec{B}.\vec{A}$

$= (3\hat{i} + 4\hat{k}).(25\hat{i} + 25\hat{k})$

$\Rightarrow \phi = (3 \times 25) + (4 \times 25)$

$= 175$ weber

34. **(b)** We have given, time period, $T = 10$s

$\therefore$ Angular velocity, $\omega = \frac{2\pi}{10} = \frac{\pi}{5}$

Magnetic flux, $\phi(t) = BA \cos \omega t$

$$\text{Emf induced, } E = \frac{-d\phi}{dt} = BA\omega \sin \omega t = BA\omega \sin(\omega t)$$

Induced emf, $|\varepsilon|$ is maximum when $\omega t = \frac{\pi}{2}$

$$\Rightarrow t = \frac{\frac{\pi}{2}}{\frac{\pi}{5}} = 2.5 \text{ s}$$

For induced emf to be minimum i.e zero

$$\omega t = \pi \Rightarrow t = \frac{\pi}{\frac{\pi}{5}} = 5 \ s$$

$\therefore$ Induced emf is zero at t = 5 s

35. **(a)** $\phi = BA = (\mu_0 n_i)A = \mu_0 n (kt\, e^{-\alpha t})A$

$$e = -\frac{d\phi}{dt} = -\mu_0 nAk \frac{d}{dt}(te^{-\alpha t})$$

$$= -\mu_0 nAk[t(-\alpha)e^{-\alpha t} + e^{-\alpha t} \times 1] = -\mu_0 nAk[e^{-\alpha t}(1 - \alpha t)]$$

$$i = \frac{e}{R} = \frac{-\mu_0 nAk}{R}[e^{-\alpha t}(1 - \alpha t)]$$

At $t = 0$, $i \Rightarrow$ –ve

at $t = 1$ sec, $i \Rightarrow 0$

Which is only possible in graph given in option (a).

36. **(a)** $\phi_Q \propto i_P, \phi_P \propto i_Q$

$$\therefore \frac{\phi_Q}{\phi_P} = \frac{i_P}{i_Q}$$

$$\phi_P = \frac{i_Q}{i_P}\phi_Q = \frac{2}{3} \times 10^{-3} = 6.67 \times 10^{-4} \text{ Wb}$$

37. **(None)**

$$\text{Net charge } Q = \frac{\Delta\phi}{R} = \frac{1}{10}A(B_f - B_i)$$

$$= \frac{1}{10} \times 3.5 \times 10^{-3}\left(0.4 \sin\frac{\pi}{2} - 0\right)$$

$$= \frac{1}{10}(3.5 \times 10^{-3})(0.4 - 0) = 1.4 \times 10^{-4}$$

No option matches, So it should be a None.

38. **(a)** According to Faraday's law of electromagnetic induction, $\varepsilon = \frac{d\phi}{dt}$

Also, $\varepsilon = iR$

$\therefore \quad iR = \frac{d\phi}{dt} \quad \Rightarrow \quad \int d\phi = R\int idt$

Magnitude of change in flux $(d\phi) = R \times$ area under current vs time graph

or, $d\phi = 100 \times \frac{1}{2} \times \frac{1}{2} \times 10 = 250$ Wb

39. **(b)** Electric flux is given by

$\phi = B.A$

$\phi = B_0 \pi r^2 e^{-t/\tau} \qquad \left(\because B = B_0 e^{-t/\tau}\right)$

Induced E.m.f. $\varepsilon = \frac{d\phi}{dt} = \frac{-B_0 \pi r^2}{\tau} e^{-t/\tau}$

$\text{Heat} = \int_0^\infty \frac{\varepsilon^2}{R} = \frac{\pi^2 r^4 B_0^2}{2\tau R}$

40. **(a)** Inside the sphere field varies linearly i.e., $E \propto r$ with distance and outside varies according to $E \propto \frac{1}{r^2}$

Hence the variation is shown by curve (a).

41. **(a)** **Given:** No. of turns $N = 1000$

Face area, $A = 4\ cm^2 = 4 \times 10^{-4}\ m^2$

Change in magnetic field,

$\Delta B = 10^{-2}\ wbm^{-2}$

Time taken, $t = 0.01s = 10^{-2}$ sec

Emf induced in the coil $e = ?$

Applying formula,

Induced emf, $e = \frac{-d\phi}{dt} = N\left(\frac{\Delta B}{\Delta t}\right) A \cos\theta$

$= \frac{1000 \times 10^{-2} \times 4 \times 10^{-4}}{10^{-2}} = 400\ \text{mV}$

42. **(a)** As we know, Magnetic flux, $\phi = B.A$

[Trick: In such case we assume that current is flowing through bigger coil and flux due to magnetic field of bigger coil on smaller coil is determined.]

$\frac{\mu_0 (2)(20 \times 10^{-2})^2}{2[(0.2)^2 + (0.15)^2]^{1.5}} \times \pi (0.3 \times 10^{-2})^2$

On solving $= 9.216 \times 10^{-11} = 9.2 \times 10^{-11}$ weber

43. **(d)** Because of the Lenz's law.

An emf will be induced in A*l* which will oppose its cause, which is oscillating coil.

44. **(c)** As $e = \frac{\Delta\phi}{\Delta t}$ or $Ri = \frac{\Delta\phi}{\Delta t} \quad (\because e = Ri)$

$\Rightarrow \Delta\phi = R(i.\Delta t)$

$= R \times$ area under $i - t$ graph

$= 10 \times \frac{1}{2} \times 4 \times 0.1 = 2$ Weber

45. **(b)** Electric flux, $\phi = 10t^2 - 50t + 250$

Induced emf, $e = -\frac{d\phi}{dt} = -(20t - 50)$

$e_{t=3} = -10\ V$

46. **(b)** $\frac{\Delta\phi}{\Delta t} = \frac{(W_2 - W_1)}{t}$

$R_{tot} = (R + 4R)\Omega = 5R\Omega$

$i = \frac{nd\phi}{R_{tot} dt} = \frac{-n(W_2 - W_1)}{5Rt}$

($\because W_2$ & W_1 are magnetic flux)

47. **(1)** Given : $R = 100\pi\Omega, X_L = 2\pi \times 50 = 100\pi\Omega$

$Z = \sqrt{R^2 + (X_L)^2} = \sqrt{R^2 + (\omega L)^2}$

$= \sqrt{(100\pi)^2 + (2\pi \times 50 \times 1)^2}$

$= \sqrt{(100\pi)^2 + (100\pi)^2} = \sqrt{2} \times 100\pi$

$I_{rms} = \frac{V}{z} = \frac{100\pi}{\sqrt{2} \times 100\pi} = \frac{1}{\sqrt{2}}$

$I_{max} = \sqrt{2} I_{rms} = \sqrt{2} \times \frac{1}{\sqrt{2}} = 1$ Ampere

Correct Answer : 1.

48. **(10)**

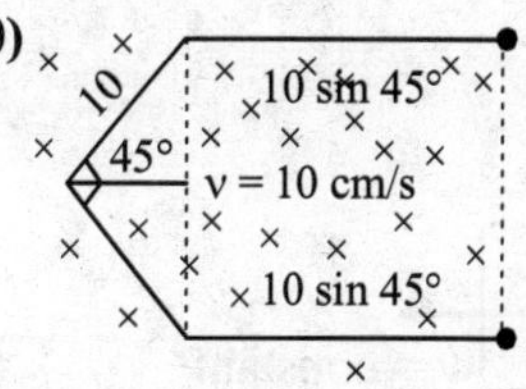

Only bent portion of centre will generate emf.

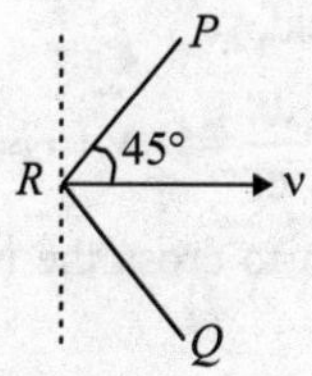

So EMF :

$\varepsilon = Bvl$ for PR

$\Rightarrow \varepsilon = 2 \times \left(\frac{1}{\sqrt{2}} \times \frac{10}{100} \times \frac{10}{100} \times \sin 45°\right)$

$\Rightarrow \varepsilon = 10^{-2}\ V \Rightarrow \varepsilon = 10$ mV.

49. **(d)** Magnetic energy per unit volume

$= \frac{B^2}{2\mu_r u_0} \Rightarrow U = \frac{B^2}{4\mu_0} V = \frac{B^2}{4\mu_0} A\ell \left[\because \mu_r = 2\right]$

50. (c) The flux associated with coil-1 is

$$\phi_1 = L_1 I_1 + M_{12} I_2$$

The emf induced in coil-1 is

$$\varepsilon_1 = -\frac{d\phi_1}{dt} = -L_1\frac{dI_1}{dt} - M_{12}\frac{dI_2}{dt}$$

51. (c) Given, Magnetic field, B = 0.4 T
radius, R = 20 cm
angular velocity, $\omega = 10\pi$ rad/s
The expression for potential difference developed between the axis of the disc and the rim is

$$E = \frac{1}{2}B\omega R^2 = \frac{1}{2}\times 0.4\times 10\pi\times(0.2)^2 = 0.2512\text{ V}$$

52. (1) The expression of the induced EMF of a conducting bar moving with velocity v

$E = \ell vB$

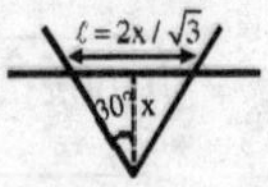

$$\Rightarrow \quad E = \frac{2x}{\sqrt{3}}\times vB \text{ and } x = vt$$

$$\Rightarrow \quad E = \frac{2}{\sqrt{3}}v^2 Bt \qquad \Rightarrow E \propto t^1$$

53. (a) Self-inductance is the property of a coil that opposes changes in the current flowing through it by inducing emf within coil.
The formula for self-inductance of coil is given by

$$L = \frac{\mu_0 N^2 A}{2\pi R}$$

54. (d)

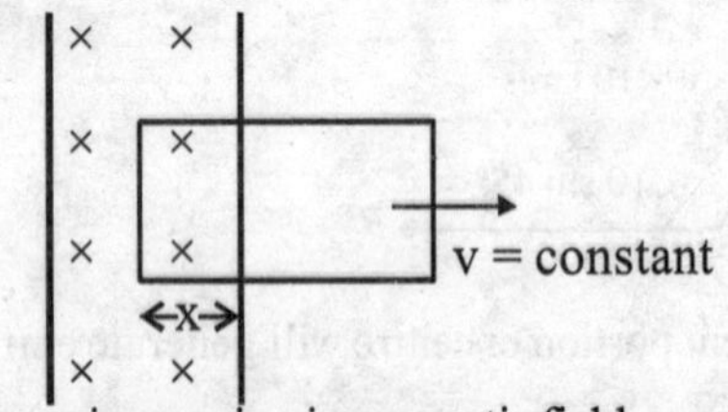

For a wire moving in magnetic field,

Motional emf; $\varepsilon = \frac{d}{dt}(Bvx) = Bv\frac{dx}{dt} = B\ell v = \text{constant}$

55. (c) Time taken by square loop to cross the magnetic field $= \frac{50}{2} = 25s$

At t = 10 seconds, whole loop is inside magnetic field therefore flux will not change

$\therefore$ Induced emf, $e = \frac{d\phi}{dt} = 0$

56. (3) Given,
Area of square loop, $A = 3.6\times 10^{-3}\text{ m}^2$
Magnetic field, B = 0.5 T
Resistance of square loop, R = 100 Ω
Induced emf, $E = NB\ell v$

Current in the loop $I = \frac{E}{R} = \frac{NB\ell v}{R}$

Magnetic force on left arm of loop,

$$F = N(I\ell B) = \frac{N^2B^2\ell^2 v}{R}$$

Work done in pulling the loop out of the field,

$$W = F\times\ell = \frac{N^2B^2\ell^3}{R}\left(\frac{\ell}{t}\right)$$

$$\Rightarrow W = \frac{(10\times 10)(0.5)^2\times(3.6\times 10^{-3})^2}{100\times 1} \quad (\because A = \ell^2)$$

$$\Rightarrow W = 3.24\times 10^{-6}\text{ J}$$

57. (a) Induced Emf is given as,

$$E_L = -L\frac{di}{dt}$$

$$\Rightarrow 0.1 = \frac{(L)[+2-(-2)]}{0.2} \Rightarrow L = 5\text{mH}$$

58. (4) Given, $I = 3t + 8$, $\varepsilon = 12$ mV
$\therefore$ The marnitude of induced emf is

$$|\varepsilon| = \left|\frac{dI}{dt}\right|$$

$$\Rightarrow \quad 12 = L\left|\frac{d}{dt}(3t+8)\right| = L\times 3$$

$\therefore$ L = 4mH

59. (0) Potential difference at end A of the rod,

$$V_0 - V_A = \frac{B\omega\ell^2}{2}$$

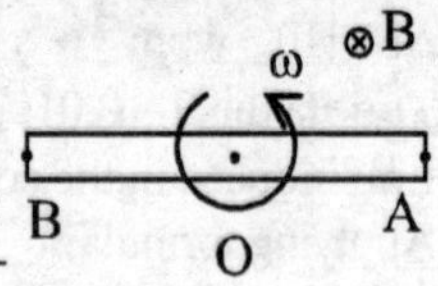

and at end B, $V_0 - V_B = \frac{B\omega\ell^2}{2}$

So $V_A = A_B \therefore V_A - V_B = 0$

60. (216)

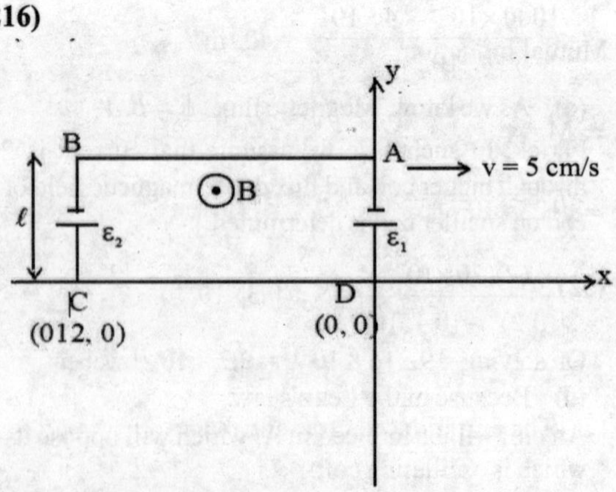

Let B_0 be the magnetic field at origin

$$\frac{dB}{dx} = -\frac{10^{-3}}{10^{-2}}$$

$$\int_{B_0}^{B} dB = -\int_0^x 10^{-1}dx \Rightarrow B - B_0 = -10^{-1}x$$

$B = \left(B_0 - \frac{x}{10}\right)$

Motional emf in both ABCD = 0

Motional emf in AD, $\varepsilon_1 = B_0 \ell v$

On rod BC, magnetic field

$$B = \left(B_0 - \frac{(-12\times10^{-2})}{10}\right)$$

$$\text{Motional emf in rod BC} = \varepsilon_2 = \left(B_0 + \frac{12\times10^{-2}}{10}\right)\ell \times v$$

$\varepsilon_{eq} = \varepsilon_2 - \varepsilon_1 = 300 \times 10^{-7}V$

Emf induced due to variation in magnetic field

$$(\varepsilon_{eq})' = A\frac{dB}{dt} = 60\times10^{-7}\,V$$

$$\text{Net Emf } (\varepsilon_{eq})_{net} = \varepsilon_{eq} + (\varepsilon_{eq})' = 360\times10^{-7}\,V$$

$$\therefore \text{Power} = \frac{(\varepsilon_{eq})^2_{net}}{R} = 216\times10^{-9}\,W$$

61. (128)

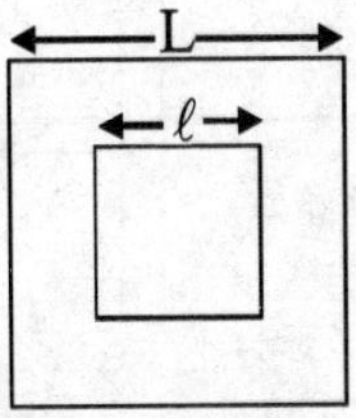

Magnetic flux linked with inner loop

$\phi = B_{center} \cdot l^2$

$$= 4\times\frac{\mu_0 i}{4\pi\frac{L}{2}}(\sin45°+\sin45°)l^2 \quad \Rightarrow \phi = 2\sqrt{2}\frac{\mu_0 i}{\pi L}l^2$$

$$\text{Mutual Induction, } M = \frac{\phi}{i} = \frac{2\sqrt{2}\mu_0 l^2}{\pi L} = 2\sqrt{2}\frac{\mu_0}{\pi}$$

$$\Rightarrow M = 2\sqrt{2}\times\frac{4\pi}{\pi}\times10^{-7} = 8\sqrt{2}\times10^{-7}\,H$$

$$= \sqrt{128}\times10^{-7}\,H = \sqrt{x}\times10^{-7}\,H \text{ (given)} \Rightarrow x = 128$$

62. (32) $B_v = B\sin30° = .5\times10^{-4}\sin30° = \frac{1}{4}\times10^{-4}T$

$$\omega = 2\pi\times f = \frac{2\pi}{60}\times1200 \text{ rad/s}$$

The emf induced cross the blades

$$\varepsilon = \frac{1}{2}B_v\omega\ell^2$$

$$= \frac{1}{2}\times\frac{1}{4}\times10^{-4}\times\frac{2\pi}{60}\times1200\times(0.8)^2 = 32\pi\times10^{-5}V$$

$= n\pi\times10^{-5}V$

$\therefore$ n = 32

63. (3) $B_H = 0.60\times10^{-4}\,Wb/m^2$

Motional Induced emf, $e = B_H V \ell$

$= 0.60\times10^{-4}\times10\times5$

$= 3\times10^{-3}V$

64. (2) Electric flux, $\phi = Mi = Mi_0\sin\omega t \quad [\because i = i_0\sin\omega t]$

$$\therefore e = -M\frac{di}{dt} = -0.002(i_0\omega\cos\omega t)$$

$e_{max} = i_0\omega(0.002) = (5)(50\pi)(0.002) \qquad \because \cos\omega t = 1$

$$e_{max} = \frac{\pi}{2}V = \frac{\pi}{\alpha}V \text{ (given)} \quad \therefore \alpha = 2$$

65. (d) Induced emf, $\varepsilon = -L\frac{dI}{dt} \Rightarrow 20 = -L\frac{(0-2)}{10^{-3}}$

Hence, inductance of the coil.

L = 10 mH

66. (88) Here ω = 210 rpm

$$= 210\times\frac{2\pi}{60}\text{rad/s} \Rightarrow \omega = 7\pi \text{ rad/s}$$

Length of metallic rod, ℓ = 20 cm = 0.2m

Magnetic field, B = 0.2 T

$$\text{emf developed across rod is} = \frac{1}{2}B\omega\ell^2$$

$$= \frac{1}{2}\times0.2\times7\pi\times(0.2)^2 = 88\text{ mV}$$

67. (c) When bar magnet is dropped through a metallic cylindrical pipe, magnetic flux linked with cylindrical pipe increases and eddy currents are produced. This oppose the motion of magnet through a metallic cylindrical pipe. As a result it takes more time to come down as compared to a non-magnetic bar with same geometry and mass.

68. (150) Given,

Speed of metallic cube, $v = 2\,ms^{-1}$

Magnetic field, B = 0.5 T

Length of metallic cube, ℓ = 15 cm = 0.15 m

Potential difference, $\Delta V = (v\times B)\ell$

$\Rightarrow \Delta V = (2\times1/2)\,0.15$

$\Rightarrow \Delta V = 150\,mV$

69. (18) Magnetic field $F = i\ell B \quad (\because \varepsilon = iR)$

$$= \left(\frac{\varepsilon}{R}\right)\ell B = \left(\frac{vB\ell}{R}\right)\ell B = \frac{vB^2\ell^2}{R} = \frac{4}{5}\times\left(\frac{15}{100}\right)^2\times1^2$$

$$= \frac{4}{5}\times\frac{225}{10^4} = 18\times10^{-3}N \qquad (\because \varepsilon = vB\ell)$$

70. (44) Given,

Number of turns in solenoid, n = 50

Amplitude of current, $I_0 = 2.5\,A$

Length of square loop ℓ = 2 cm

Area of square loop, $A = 2\times2 = 4\,cm^2$

Magnetic field due to solenoid, $B = \mu_0 nI$

Flux through solenoid, $\phi = B.A = \mu_0 nI\times A$

$$\text{Induced emf, } e = \mu_0 nA\times\frac{dI}{dt} \quad (\because I = I_0\sin\omega t)$$

$= \mu_0 nA\times I_0\,\omega\cos\omega t$

Amplitude of induced emf, $e_0 = \mu_0 n A \times I_0 \omega$

$= 4\pi \times 10^{-7} \times \dfrac{50}{10^{-2}} \times 4 \times 10^{-4} \times 2.5 \times 700$

$[\because \omega = 700 \text{ rad s}^{-1} \text{ given}]$

$= 44 \times 10^{-4}\text{V}$

71. **(a)** Length of metal rod, $\ell = 10$ cm $= 0.10$ m

Induced emf $= B\ell v$

$0.08 = 0.4 \times 0.10 \times v$

$\Rightarrow v = \dfrac{0.08}{0.4 \times 0.10} \quad \therefore v = 2$ m/s

72. **(4)** For coil (1),

Radius, $r_1 = 1$cm, Number of turns, $N_1 = 10$

For coil (2),

Radius, $r_2 = 1000$ cm, Number of turns, $N_2 = 200$

Using $\phi_{1,2} = MI$

$N_2 \vec{B}.N_1 \vec{A_1} = MI \quad \left(\because B = \dfrac{\mu_0 I}{2r_2}\right)$

$\Rightarrow N_1 N_2 \dfrac{\mu_0 I}{2r_2}.\pi r_1^2 = MI$

The mutual inductance of this arrangement will be

$\Rightarrow M = \dfrac{10 \times 200 \times 4\pi \times 10^{-7} \times \pi \times (0.01)^2}{2 \times 10}$

$\Rightarrow M = 4 \times 10^{-8}$H

73. **(b)** When metal ball is passing through magnetic field, eddy current will produce and it will oppose the motion, so it will take more time.

74. **(30)** Given, $\dfrac{dI}{dt} = -1\dfrac{A}{sec}$

$V_A - IR - L\dfrac{dI}{dt} - 12 = V_B$

$V_A - 2 \times 12 - 6(-1) - 12 = V_B$

$V_{AB} = V_A - V_B = 36 - 6 = 30$ volt

75. **(1584)**

Given, area of rectangular coil, $A = 14 \times 10^{-2}$ m^2

Magnetic field, B = 3T

Number of turns, n = 100

$\varepsilon_{max} = NAB\omega$

$= 100 \times 14 \times 10^{-2} \times 3 \times \dfrac{360 \times 2\pi}{60} = 1584$ V

76. **(c)**

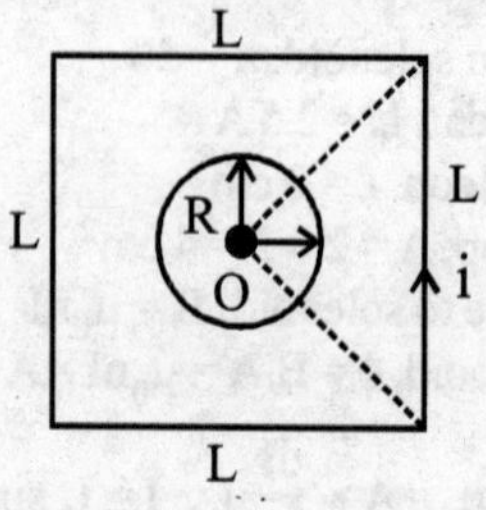

$\therefore$ At the centre of square loop

Magnetic field, $B = \left(4 \times \dfrac{\mu_0}{4\pi\left(\frac{L}{2}\right)}\left(\dfrac{1}{\sqrt{2}} + \dfrac{1}{\sqrt{2}}\right)\right)$

Magnetic flux, $\phi = (BA) = \left(\dfrac{4\mu_0 i}{4\pi\left(\frac{L}{2}\right)}\sqrt{2}\right)\pi R^2$

$\phi = Mi \Rightarrow M = \dfrac{2\sqrt{2}\mu_0 R^2}{L}$

77. **(b)** $\ell = 5$ cm

t = 1 sec

$\therefore V = \dfrac{0.05}{1} = 0.05$m / s.

And, $i = \dfrac{B\ell V}{R} \Rightarrow i = \dfrac{40 \times 0.05 \times 0.05}{10} = 0.01$A

So, $F = Bi\ell = 40 \times 0.01 \times 0.05 \Rightarrow F = 0.02$ N

$\therefore \quad W = 0.02 \times \ell = 0.02 \times .05 \Rightarrow W = 1 \times 10^{-3}$ J

78. **(d)** Given, length, $\ell = 1$m

Velocity, v = 8 m/s

Magnetic Field, B = 2T

Induced emf across the ends = $Bv\ell$

$= 2 \times 8 \times 1 = 16$ V

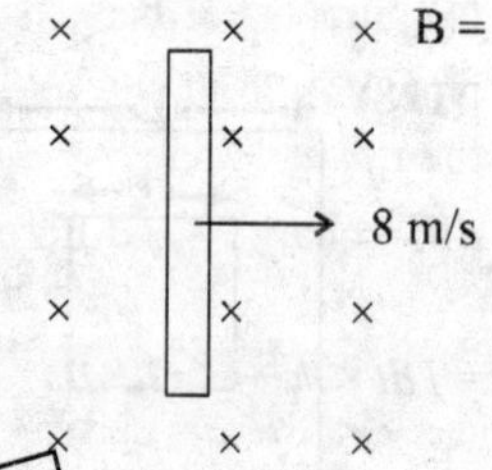

79. **(c)**

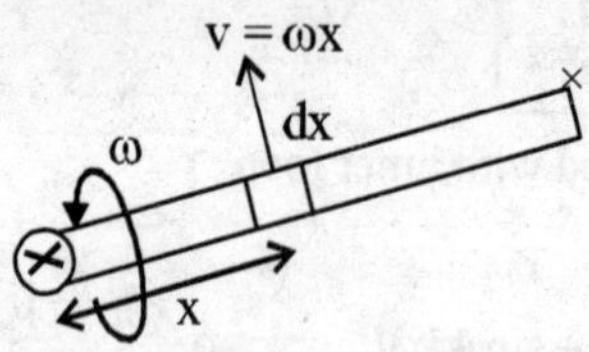

$\Rightarrow \int d\varepsilon = \int B(\omega x)dx \Rightarrow \varepsilon = B\omega\int_0^L x dx = \dfrac{B\omega L^2}{2}$

80. **(c)** $\phi_2 \propto I_1$

$\phi_2 = M_{21}I_1$

$M_{21} = \dfrac{\phi_2}{I_1} = \dfrac{B_1 A_2}{I_1}$

Now, $B_1 = \dfrac{\mu_0 I_1}{4\pi L/2}[\sin 45° + \sin 45°] \times 4$

$= \dfrac{\mu_0 I_1}{2\pi L} \times \dfrac{8}{\sqrt{2}} = 2\sqrt{2}\dfrac{\mu_0 I_1}{\pi L}$

So, $M_{21} = \dfrac{\frac{2\sqrt{2}\mu_0 I_1}{\pi L}\ell^2}{I_1} = \dfrac{\mu_0 \ell^2}{\pi L} \times 2\sqrt{2}$

81. **(16)** The moving rod will cut the vertical component of magnetic field perpendicularly.

So, $e = B_v l V$

$= B_H \times lV \quad [\because Dip = 45° \Rightarrow B_v = B_H]$

$= 4 \times 10^{-3} \times 0.2 \times 20 = 16 \times 10^{-3}\text{V} = 16$ mV.

82. **(400)** 10 Ω, 20 mH

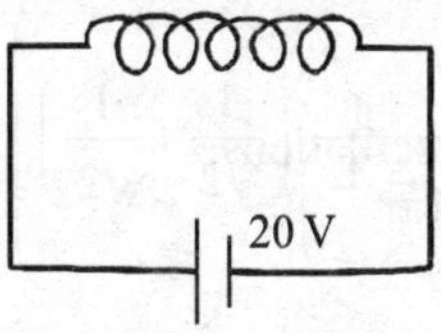

$$|e| = \frac{Ldi}{dt} = \frac{L(i_{initial} - 0)}{t} = \frac{LV}{Rt}$$

$$\therefore |e| = \frac{20 \times 10^{-3} \times 20}{10 \times 100 \times 10^{-6}} = 400\,V$$

83. **(d)** If current in both inductor is in same direction then $L_{eq} = L_1 + L_2 + 2M$ and, if current in both inductor is in opposite direction, then $L_{eq} = L_1 + L_2 - 2M$

84. **(b)** emf induced between the two ends is given by

$$e = \frac{1}{2}B_H \omega l^2 = \frac{1}{2} \times 0.2 \times 10^{-4} \times 5 \times 1^2$$

$$= 0.1 \times 10^{-4} \times 5 = 5 \times 10^{-5} = 50\,\mu V$$

85. **(4)** Suppose at any instant, current through inductor is 'I'. Now if in inductor when current change by dI in dt time then, work done by battery is given by

dW = charge × P.difference

$$= I\,dt \times L\frac{dI}{dt} = LI\,dI$$

Now, $I = 2\sin(t^2)A$

If $I = 0 \Rightarrow t = 0$

If $I = 2A \Rightarrow 2 = 2\sin(t^2) \Rightarrow \sin t^2 = \sin \pi/2 \Rightarrow t = \sqrt{\frac{\pi}{2}}$.

So, amount of total energy spent

$U = \int dw = \int LIdI$

$= \int 2 \times 2\sin(t^2) \times 4t\cos(t^2)\,dt$

$$= 8\int_0^{\sqrt{\pi}/2} t\sin(2t^2)\,dt = 8 \times \frac{1}{2} = 4J$$

86. **(a)** Magnetic field at centre due to each side

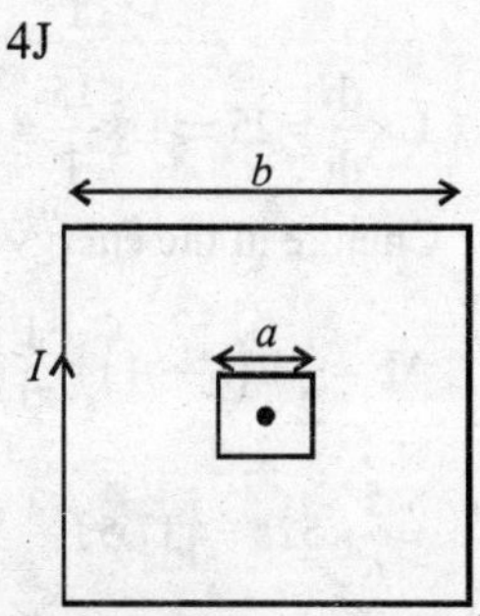

$$= \frac{\mu_0 I}{4\pi \frac{b}{2}}(\sin 45° + \sin 45°)$$

Field due to all 4 sides

$$= B = \frac{\mu_0}{4\pi}\frac{I}{b/2}(\sin 45° + \sin 45°)4$$

$$\Rightarrow B = \left[\frac{\mu_0}{4\pi}\frac{I}{b/2} \times 2\sin 45\right] \times 4 = 2\sqrt{2}\frac{\mu_0}{\pi}\frac{I}{b}$$

Flux through small square will be

$$\Rightarrow \phi = 2\sqrt{2}\frac{\mu_0}{\pi}\frac{I}{b} \times a^2$$

Coefficient of mutual induction, $Mi_b = \phi_a$ is given as

$$M = \frac{\phi}{I} = \frac{2\sqrt{2}\mu_0 a^2}{\pi b} = \frac{\mu_0}{4\pi}8\sqrt{2}\frac{a^2}{b}$$

87. **(c)** Given,

Magnetic field, $B = 1T$

At $t = 0$s, the centre of circular ring will touch the boundary.

$\therefore$ Induced emf $= Blv$, l = length $\perp_r$ to $\vec{B}$ and $\vec{V}$.

$= 1.(2R).1 = 2V$.

88. **(3)** Steady state current in the inductor when T_1 and T_2 are connected, $I = \frac{6}{6} = 1A$ then current through inductor remains same when T_1 and T_3 are connected. So potential difference across $r = 3\Omega$

$V = Ir = 1 \times 3 = 3V$

89. **(d)** From question, $U = 25\%$ of U_0

$$U = \frac{1}{2}LI^2 = \frac{1}{4} \times \frac{1}{2}LI_0^2 \Rightarrow I = \frac{I_0}{2}$$

$$\text{Also, } I = I_0\left(1 - e^{-\frac{Rt}{L}}\right) \Rightarrow \frac{I_0}{2} = I_0\left(1 - e^{-\frac{Rt}{L}}\right)$$

$$\Rightarrow e^{\frac{-Rt}{L}} = \frac{1}{2} \Rightarrow (t)\frac{R}{L} = \ln 2 \quad \therefore t = \frac{L}{R}\ln 2$$

90. **(b)**

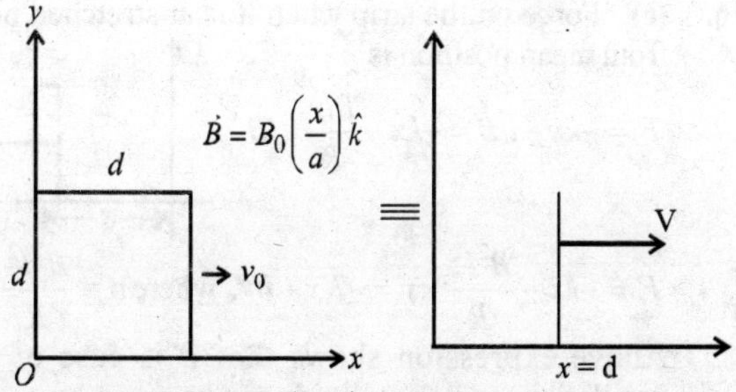

So, Emf induced in the loop.

$$\varepsilon = B_0 \times \left(\frac{d}{a}\right) \times d \times v_0 - 0 = \frac{B_0 v_0 d^2}{a}$$

91. **(a)** *Emf* induced $\varepsilon_{ind} = (B_v)LV$ and $B_v = B_{Total}\sin 60°$

$$= (2.5 \times 10^{-4})(\sin 60°) \times 10 \times 180 \times \frac{5}{18}$$

$$= 2.5 \times 10^{-4} \times \frac{\sqrt{3}}{2} \times 500 = 108.25\,mV$$

92. **(d)** Magnetic field at a distance r from the wire

$$B = \frac{\mu_0 I}{2\pi r}$$

Magnetic flux for small displacement dr,

$\phi = B \cdot A = Bldr$

$[\because A = l\,dr$ and $B.A = BA\cos 0°]$

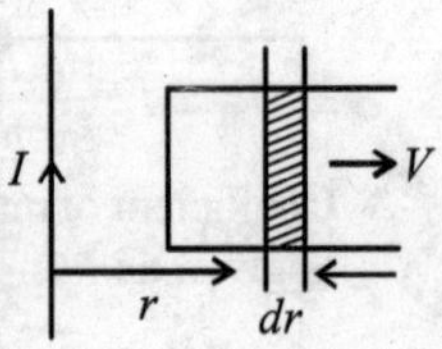

$\Rightarrow \phi = \frac{\mu_0 I}{2\pi r} l\,dr$

Emf, $e = \frac{d\phi}{dt} = \frac{\mu_0 Il}{2\pi r} \cdot \frac{dr}{dt} \Rightarrow e = \frac{\mu_0}{2\pi} \cdot \frac{Ivl}{r}$

Induce current in the loop, $i = \frac{e}{R} = \frac{\mu_0}{2\pi} \cdot \frac{Ivl}{Rr}$

93. **(10)** Given $dI = 0.25 - 0 = 0.25$ A
$dt = 0.025$ ms
Induced voltage
$E_{ind} = 100$ v
Self-inductance, L = ?

Using, $E_{ind} = \frac{\Delta\phi}{\Delta t} \Rightarrow 100 = \frac{L(0.25-0)}{.025\times10^{-3}}$

$\Rightarrow L = 0.01\text{ H} = 10\text{ mH}$

94. **(b)** Induced emf,
$e = Bv\ell = 1\times10^{-2}\times0.05 = 5\times10^{-4}$ V
Equivalent resistance,

$R = \frac{4\times2}{4+2} + 1.7 = \frac{4}{3} + 1.7 \simeq 3\,\Omega$

Current, $i = \frac{e}{R} = \frac{5\times10^{-4}}{3} \simeq 170\,\mu A$

95. **(d)** Inductance $= \frac{\mu_0 N^2 A}{L}$

96. **(c)** Force on the strip when it is at stretched position x from mean position is

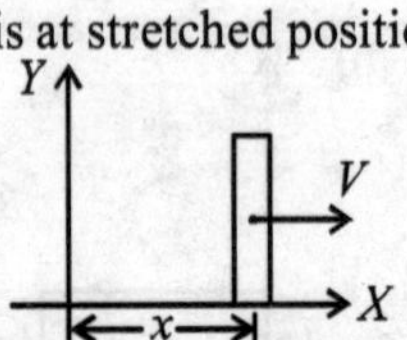

$F = -kx - ilB = -kx - \frac{Blv}{R} \times lB$

$F = -kx - \frac{B^2l^2}{R} \times v = -Kx - bV$, where $b = \frac{B^2l^2}{R}$

Above expression shows that it is case of damped oscillation, so its amplitude can be given by

$\Rightarrow A = A_0 e^{-\frac{bt}{2m}}$

$\Rightarrow \frac{A_0}{e} = A_0 e^{-\frac{bt}{2m}}$ [as per question $A = \frac{A_0}{e}$]

$\Rightarrow t = \frac{2m}{\left(\frac{B^2l^2}{R}\right)} = \frac{2\times50\times10^{-3}\times10}{0.01\times0.01}$

Given, $m = 50\times10^{-3}$ kg
B = 0.1 T
$l = 0.1$ m
R = 10 Ω
k = 0.5 N

Time period, $T = 2\pi\sqrt{\frac{m}{k}} \simeq 2$ s

so, required number of oscillations,

$N = \frac{10000}{2} = 5000$

97. **(a)** Induced emf, $\varepsilon = Bv\ell$
$= 0.3\times10^{-4}\times5\times10$
$= 1.5\times10^{-3}$ V

98. **(d)** The rate of mutual inductance is given by

$M = \mu_0 n_1 n_2 \pi r_1^2$ ...(i)

The rate of self inductance is given by

$L = \mu_0 n_1^2 \pi r_1^2$ ...(ii)

Dividing (i) by (ii)

$\Rightarrow \frac{M}{L} = \frac{n_2}{n_1}$

99. **(c)** $L = \mu_0 n^2 Al$
$L \propto l$

$\therefore \frac{L_2}{L_1} = \frac{l_2}{l_1}$

$\Rightarrow L_2 = 3L_1$ $[\because l_2 = 3l_1]$

100. **(a)** Potential difference between two faces perpendicular to x-axis
$= l\,V.B = 2\times10^{-2}\,(6\times0.1) = 12$ mV

101. **(c)** $dQ = \rho\,dx$

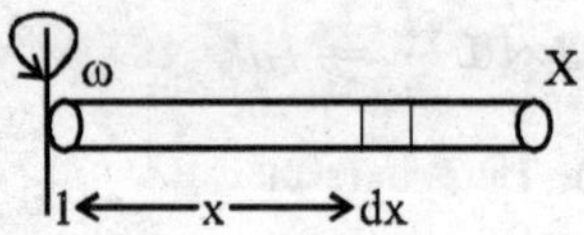

$dI = \frac{dQ}{2\pi}.\omega$

$dM = dI\times A = \frac{\omega}{2\pi}.\frac{\rho_0}{\ell}.x\pi x^2\,dx$

$\Rightarrow M = \frac{\rho_0}{\ell} n\pi \int_0^\ell x^3\,dx = \frac{\pi}{4}.n\rho_0\ell^3$

102. **(b)** According to faraday's law of electromagnetic induction, $e = \frac{-d\phi}{dt}$

$L\times\frac{di}{dt} = 25 \Rightarrow L\times\frac{15}{1} = 25$ or $L = \frac{5}{3}$ H

Change in the energy of the inductance,

$\Delta E = \frac{1}{2}L\left(i_1^2 - i_2^2\right) = \frac{1}{2}\times\frac{5}{3}\times(25^2 - 10^2)$

$= \frac{5}{6}\times525 = 437.5$ J

103. **(a)** Given: Capacitance, $C = 0.2\,\mu F = 0.2\times10^{-6}$ F
Inductance $L = 0.5$ mH $= 0.5\times10^{-3}$ H
Current I = ?
Using energy conservation

$\frac{1}{2}CV^2 = \frac{1}{2}CV_1^2 + \frac{1}{2}LI^2$

$$\Rightarrow \frac{1}{2}\times 0.2\times 10^{-6}\times 10^{2}+0$$

$$=\frac{1}{2}\times 0.2\times 10^{-6}\times 5^{2}+\frac{1}{2}\times 0.5\times 10^{-3}I^{2}$$

$$\therefore \quad I=\sqrt{3}\times 10^{-1}\text{A}=0.17\text{A}$$

104. (d) According to Faraday's law of electromagnetic induction,

Induced emf, $e=\dfrac{Ldi}{dt}$

$$50=L\left(\frac{5-2}{0.1\sec}\right)\Rightarrow L=\frac{50\times 0.1}{3}=\frac{5}{3}=1.67\text{ H}$$

105. (b) Time in which left arm reach at x = 10 cm

$$t=\frac{0.1}{10}=\frac{1}{100}=0.01\sec$$

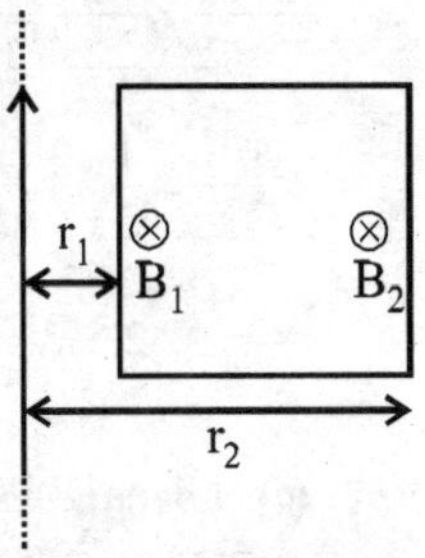

$r_1=10\times 0.01=0.1$ m

$e=lV(B_1-B_2)$

$$=0.1\times 10\left(\frac{\mu_0 i}{2\pi r_1}-\frac{\mu_0 i}{2\pi r_2}\right)$$

$$=0.1\times 10\times 2\times 10^{-7}\times 1\left(\frac{1}{0.1}-\frac{1}{2\times 0.1}\right)$$

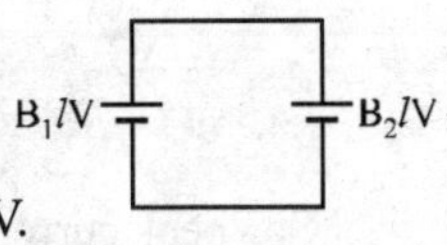

$$=2\times 10^{-7}\left(\frac{2-1}{2\times 0.1}\right)=10^{-6}\text{V}=1\,\mu\text{V}.$$

106. (d) Here, induced e.m.f.

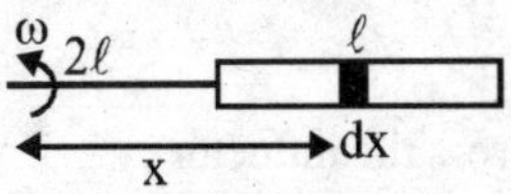

$de=(\omega x)Bdx$

$$e=\int_{2\ell}^{3\ell}(\omega x)Bdx=B\omega\frac{[(3\ell)^2-(2\ell)^2]}{2}=\frac{5B\ell^2\omega}{2}$$

107. (d) Self inductance of a long solenoid is given by

$$L=\frac{\mu_0 N^2 A}{l}$$

Magnetic field at the centre of solenoid

$$B=\frac{\mu_0 NI}{l}$$

So both the statements are correct and statement 2 is not correct explanation of statement 1.

108. (c) $e=Bl V$

$=5\times 10^{-5}\times 2\times 1.5$

$=15\times 10^{-5}\text{V}=0.15\text{mV}$

109. (b) Due to the movement of resistor R, an emf equal to Blv will be induced in it as shown in figure clearly,

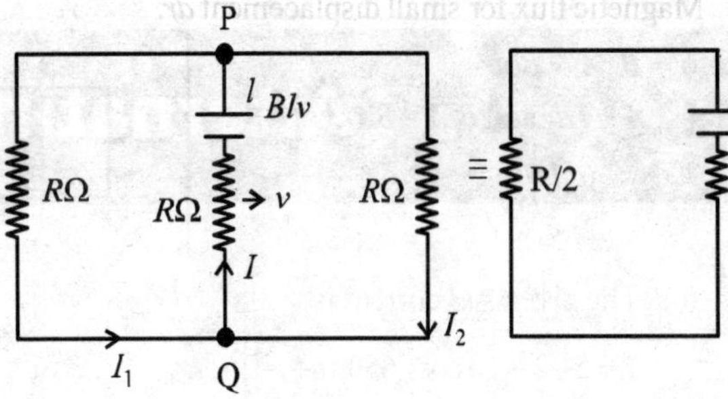

$$R_{eq}=\frac{R}{2}+R=\frac{3R}{2}$$

$$I=\frac{e}{R_{eq}}=\frac{B\ell v}{\frac{3R}{2}}=\frac{2B\ell v}{3R}$$

we get $I_1=I_2=\dfrac{Blv}{3R}$ $\quad[\because I_1=I_2]$

110. (d) Given, Area of cross-section of pipe,

$A=10\text{ cm}^2$

Length of pipe, $\ell=20$ cm

$$M=\frac{\mu_0 N_1 N_2 A}{\ell}=\frac{4\pi\times 10^{-7}\times 300\times 400\times 10\times 10^{-4}}{0.2}$$

$$\Rightarrow M=\frac{\mu_0 N_1 N_2 A}{\ell}=2.4\pi\times 10^{-4}\text{H}$$

111. (c) We will sit on one tube and observe the motion of other. Relative velocity of the tube of width l, $=v-(-v)=2v$ ∴ Induced emf. $=B.l\,(2v)$

112. (b) Given, length of conductor $\ell=1$m,

Angular speed, $\omega=5$ rad/s,

Magnetic field, $B=0.2\times 10^{-4}\,T$

EmF generated between two ends of conductor

$$\varepsilon=\frac{B\omega l^2}{2}=\frac{0.2\times 10^{-4}\times 5\times 1}{2}=50\mu V$$

113. (b) Mutual inductance depends on the relative position and orientation of the two coils.

114. (d) As the side BC is outside the field, no emf is induced across BC. Further, sides AB and CD are not cutting any flux. So, they will not centribute in flux.

Only side AD is cutting the flux 50 emf will be induced due to AD only.

The induced emf is

$$e=\frac{-d\phi}{dt}=-\frac{d(\vec{B}\cdot\vec{A})}{dt}=\frac{-d(BA\cos 0^\circ)}{dt}$$

$$\therefore e=-B\frac{dA}{dt}=-B\frac{d(\ell\times x)}{dt}$$

$$\therefore e=-B\ell\frac{dx}{dt}=-B\ell v$$

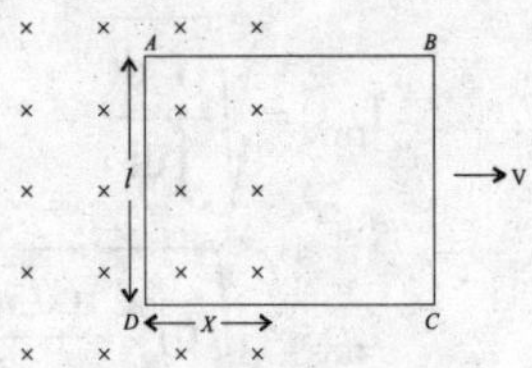

7 Alternating Current

1. **(c)** The given AC current is

$$i = 5\sqrt{2} + 10\cos\left(650\pi t + \frac{\pi}{6}\right)$$

$$\Rightarrow i^2 = 50 + 100\cos^2\left(650\pi t + \frac{\pi}{6}\right) + (2)(5\sqrt{2})(10)\cos\left(650\pi t + \frac{\pi}{6}\right)$$

$$\Rightarrow \langle i^2 \rangle = 50 + \frac{100}{2} + 0 = 100$$

$\therefore$ The r.m.s. value of current is

$$i_{rms} = \sqrt{\langle i^2 \rangle} = \sqrt{100} = 10 \text{ amp.}$$

2. **(c)** Given ; $i_0 = 100\sqrt{2}$, $\omega = 100\pi$

$$i_{rms} = \frac{i_0}{\sqrt{2}} = 100 \text{ A} \qquad f = \frac{\omega}{2\pi} = \frac{100\pi}{2\pi} = 50 \text{ Hz.}$$

3. **(a)** The power in AC circuit is $P = V_{rms} . I_{rms} \cos\phi$

For purely resistive (bulb) circuit, $\cos\phi = 1$

$\therefore \quad P = v_{rms}\, i_{rms}$

$\therefore \quad \Rightarrow i_{rms} = \frac{100}{220}$

$\therefore \quad$ Peak current, $i_0 = \sqrt{2}\, i_{rms} = 0.64$ A

4. **(c)** $i_{rms} = \sqrt{\langle I^2 \rangle} = \sqrt{\langle (I_A \sin\omega t + I_B \cos\omega t)^2 \rangle}$

$$\sqrt{\frac{I_A^2 + I_B^2}{2}}$$

5. **(22)** Capacitive reactance is,

$$X_C = \frac{1}{\omega c} = \frac{1}{100 \times 2 \times 10^6} = 5000\Omega$$

Maximum current is, $i_o = \frac{V_o}{X_c} = \frac{110\sqrt{2}}{5000}$

RMS value of current is given by, $i_{rms} = \frac{i_o}{\sqrt{2}}$

$$i_{rms} = \frac{110\sqrt{2}}{5000\sqrt{2}} = \frac{110}{5} \text{ mA} = 22 \text{ mA}$$

6. **(8)** Given, $i = \left[6 + \sqrt{56}\sin(100\pi t + \frac{\pi}{3})\right]$ A

$$I_{rms} = \sqrt{\frac{\int i^2 dt}{\int dt}}$$

$$I_{rms} = \sqrt{(6)^2 + \frac{(\sqrt{56})^2}{2}} = \sqrt{36 + 28} = \sqrt{64} \;\therefore I_{rms} = 8\text{A}$$

7. **(b)** Rising half to peak

t = T/6

$$t = \frac{2\pi}{6\omega} = \frac{\pi}{3\omega} = \frac{\pi}{300\pi} = \frac{1}{300} = 3.3 \text{ ms}$$

Here $\omega = 100\,\pi$ rad / s

8. **(b)** The average power in one cycle is

$$P_{avg} = V_{rms} I_{rms} \cos(\Delta\phi)$$

$$= \frac{100}{\sqrt{2}} \times \frac{100 \times 10^{-3}}{\sqrt{2}} \times \cos\left(\frac{\pi}{3}\right) \left[\because V_{rms} = \frac{V_0}{\sqrt{2}}, I_{rms} = \frac{I_0}{\sqrt{2}}\right]$$

$$= \frac{10^4}{2} \times \frac{1}{2} \times 10^{-3}$$

$$= \frac{10}{4} = 2.5 \text{ W}$$

9. **(c)** Given, $E = 440 \sin 100\pi t$, $\quad L = \frac{\sqrt{2}}{\pi} H$

As, $X_L = \omega L = 100\pi \frac{\sqrt{2}}{\pi} = 100\sqrt{2}\,\Omega$

So, peak current, $I_0 = \frac{E_0}{X_L} = \frac{440}{100\sqrt{2}} = 2.2\sqrt{2}$A

We know AC ammeter reads RMS value therefore reading will be I_{rms}, $I_{rms} = \frac{I_0}{\sqrt{2}} = 2.2$A

10. **(d)** Voltage amplitude across the inductor

$$v_0 = i_0 x_L = i_0(\omega L)$$

$$= (5)(49\pi)(30 \times 10^{-3}) = 23.1\text{V}$$

Voltage will lead current by 90°.

Therefore the equation for the voltage across the inductor $V = 23.1 \sin(49\,\pi t + 60°)$

11. **(b)** As $H = i_{rms}^2 Rt$

$$H \propto i_{rms}^2 R$$

$$\frac{H_1}{H_2} = \left(\frac{i_{rms_1}}{i_{rms2}}\right)^2 \times \frac{R_1}{R_2} = \left(\frac{4}{\frac{4}{\sqrt{2}}}\right)^2 \times \frac{3}{2} = 2 \times \frac{3}{2} = \frac{3}{1}$$

12. **(d)** We have

$\omega = 120\pi$

and, $T = \frac{2\pi}{\omega} = \frac{2\pi}{120\pi} = \frac{\pi}{60\pi} = \frac{1}{60}$

So, req. time $= \frac{T}{4} = \frac{1}{240}$ s

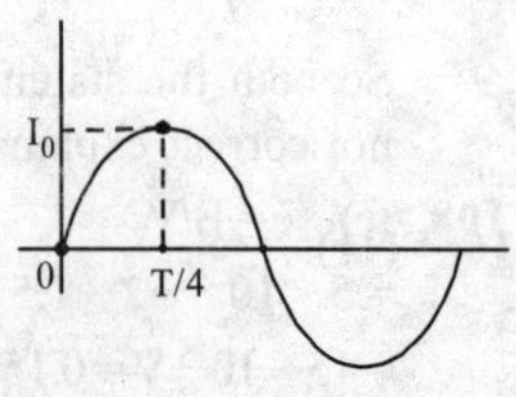

13. **(242)** When voltage is minimum, current is maximum. So, circuit is purely inductive i.e. $R = 0$

So, $Z = X_L$ Then, $i_0 = \frac{V_0}{X_L}$

$$\Rightarrow i_0 = \frac{V_0}{2\pi f L} \Rightarrow i_0 = \frac{\sqrt{2}V_{rms}}{2\pi f L}$$

$$\Rightarrow i_0 = \frac{\sqrt{2}\times 220}{2\pi\times 50\times 0.2} = \frac{220\sqrt{2}}{20\pi} = \frac{11\sqrt{2}}{\pi} = \frac{\sqrt{242}}{\pi}$$

$\therefore\ a = 242$

14. **(a)** When $I = I_0$

$I_0 \sin(\omega t_1 + \phi) = I_0 \Rightarrow \sin(\omega t_1 + \phi) = 1$

$$\omega t_1 + \phi = \frac{\pi}{2} \quad ...(i)$$

when $I = \frac{I_0}{\sqrt{2}} \Rightarrow I_0 \sin(\omega t_2 + \phi) = \frac{I_0}{\sqrt{2}}$

$$\Rightarrow \sin(\omega t_2 + \phi) = \frac{1}{\sqrt{2}} \Rightarrow \omega t_2 + \phi = \frac{\pi}{4} \quad ...(ii)$$

Substracting (ii) from (i), we get

$$\omega(t_1 - t_2) = \frac{\pi}{4} \Rightarrow 2\pi f(t_1 - t_2) = \frac{\pi}{4}$$

$$\Rightarrow t_1 - t_2 = \frac{1}{8f} = \frac{1}{400}\ s = 2.5\ ms.$$

15. **(a)** From $I_{rms} = \sqrt{\frac{\int I^2 dt}{T}}$

$$\Rightarrow I^2_{rms} = \frac{\int_0^T (I_1 \sin\omega t + I_2 \cos\omega t)^2\, dt}{T}$$

$$(I_{rms})^2 = \frac{\int_0^T [I_1^2 \sin^2\omega t + I_2^2\cos^2\omega t + 2I_1I_2\sin(\omega t)\times\cos(\omega t)]dt}{T}$$

$$\Rightarrow I_{rms} = \sqrt{\frac{I_1^2}{2} + \frac{I_2^2}{2} + 0}$$

$$= \sqrt{\frac{I_1^2 + I_2^2}{2}} \quad \left[\because \frac{\int_0^T I^2 \sin^2\omega t\, dt}{T} = \frac{1}{2} \text{ and } \int_0^T I\sin\omega t = 0\right]$$

16. **(11)** $i_{rms} = \sqrt{i^2_{1rms} + i^2_{2rms}} = \sqrt{\left(\frac{\sqrt{42}}{\sqrt{2}}\right)^2 + 10^2}$

$= \sqrt{121} \Rightarrow i_{rms} = 11A$

17. **(d)** As $V(t) = 220 \sin 100\pi t$ so, $I(t) = \frac{220}{50}\sin 100\pi t$

i.e., $I = I_m \sin(100\pi t)$ For $I = I_m$

$t_1 = \frac{\pi}{2}\times\frac{1}{100\pi} = \frac{1}{200}$ sec. and for $I = \frac{I_m}{2}$

$$\Rightarrow \frac{I_m}{2} = I_m \sin(100\pi t_2)$$

$$\Rightarrow \frac{\pi}{6} = 100\pi t_2 \Rightarrow t_2 = \frac{1}{600}s$$

$$\therefore t_{req} = \frac{1}{200} - \frac{1}{600} = \frac{2}{600} = \frac{1}{300}s = 3.3\text{ ms}$$

18. **(b)** In a pure inductive circuit current always lags behind the emf by $\frac{\pi}{2}$.

If $V(t) = v_0 \sin\omega t$

then $I = I_0 \sin\left(\omega t - \frac{\pi}{2}\right)$

Now, given $V(t) = 100 \sin(500t)$

and $I_0 = \frac{E_0}{\omega L} = \frac{100}{500\times 0.02}$ $[\because L = 0.02H]$

$$I = 10\sin\left(500t - \frac{\pi}{2}\right) \Rightarrow I = -10\cos(500t)$$

19. **(c)** We know that power consumed in a.c. circuit is given by,

$P = E_{rms}.I_{rms}\cos\phi$

Here, $E = E_0 \sin\omega t$

$I = I_0 \sin\left(\omega t - \frac{\pi}{2}\right)$

This means the phase difference, is $\phi = \frac{\pi}{2}$

$$\because \cos\phi = \cos\frac{\pi}{2}\ \therefore P = E_{rms}.I_{rms}.\cos\frac{\pi}{2} = 0$$

20. **(b)** $\phi = \vec{B}.\vec{A}$; $\phi = BA\cos\omega t$

$$\varepsilon = -\frac{d\phi}{dt} = \omega BA\sin\omega t\ ;\ i = \frac{\omega BA}{R}\sin\omega t$$

$$P_{inst} = i^2R = \left(\frac{\omega BA}{R}\right)^2 \times R\sin^2\omega t$$

$$P_{avg} = \frac{\int_0^T P_{inst}\times dt}{\int_0^T dt} = \frac{(\omega BA)^2}{R}\frac{\int_0^T \sin^2\omega t dt}{\int_0^T dt} = \frac{1}{2}\frac{(\omega BA)^2}{R}$$

$$\therefore P_{avg} = \frac{(\omega B\pi r^2)^2}{8R} \quad \left[A = \frac{\pi r^2}{2}\right]$$

21. **(a)** D.C. ammeter measure average value of current. In AC current, average value of current in complete cycle is zero. Hence reading will be zero.

22. **(1)** Given: $X_L = 100\,\Omega, X_C = 50\,\Omega, R = 50\,\Omega, V_{rms} = 10\,V$

$P = V_{rms} I_{rms} \cos\phi$

$\cos\phi = \frac{R}{z}$, $I_{rms} = \frac{V_{rms}}{z}$

$\Rightarrow P = V_{rms} \times \frac{V_{rms}}{z} \times \frac{R}{z}$

$\Rightarrow P = V_{rms}^2 \times \frac{R}{z^2}$

$\Rightarrow z = \sqrt{R^2 + (x_L - x_C)^2} \Rightarrow z = 50\sqrt{2}\ \Omega$

$\therefore P = 100 \times \frac{50}{2500 \times 2} = 1$ W.

$\therefore x = 199.$

23. **(5)** Phasor diagram for V_{in} is

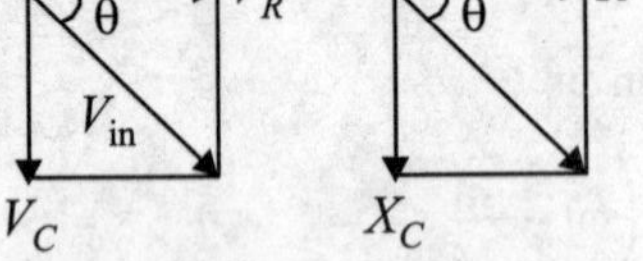

Phasor diagram for $(V_B - V_A)$ is

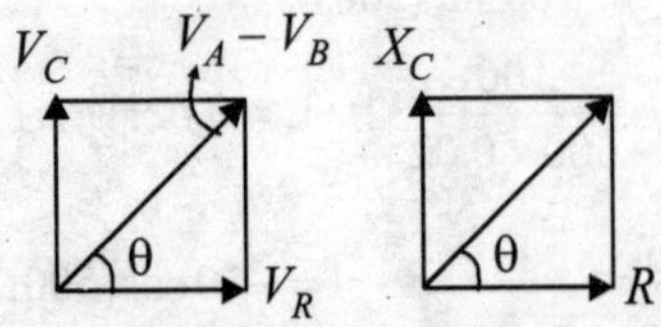

From above Phasor diagrams,

$\theta + \theta = 90° \Rightarrow \theta = 45°$

$\therefore \tan\theta = \frac{X_C}{R} \Rightarrow X_C = R \Rightarrow \frac{1}{\omega C} = R$

$\therefore \omega = \frac{1}{RC} = \frac{1}{100 \times 10^3 \times 100 \times 10^{-12}} = 10^5$ rad/s

$= 10^x$ rad/s (given)

$\therefore x = 5$

24. **(c)** A choke coil is simply an LR circuit having large inductance, but small resistance. The voltage across resistor (mercury-tube) is

$V_R = IR = \frac{R}{\sqrt{R^2 + \omega^2 L^2}} V$

$\therefore$ The voltage across tube is reduced by a factor $\left(R / \sqrt{R^2 + \omega^2 L^2}\right)$

25. **(2)** Resonance condition in LCR circuit is observed at maximum current.

$\therefore$ Angular frequency, $\omega = \frac{1}{\sqrt{LC}}$

$= \frac{1}{\sqrt{25 \times 10^{-9} \times 100 \times 10^{-3}}} = 2 \times 10^4\ \text{rad s}^{-1}$

26. **(3)** Given: rate of change of current, $\frac{dI}{dt} = -8$A/s

$R = 12\Omega$

From Kirchhoff's voltage law,

$\varepsilon - \frac{LdI}{dt} - IR = 0$

$12 - 3 \times (-8) - 1 \times 12 = 0$

$\therefore I = 3A$

27. **(b)** In LCR circuit,

Initially, $I_0 = \frac{E}{R}$

Finally, $I_0' = \frac{E}{2R} = \frac{I_0}{2}$

28. **(a)** Capacitance of capacitor, $C = \frac{\varepsilon_0 A}{d}$

On introducing dielectric between the plates,

$C' = \frac{K\varepsilon_0 A}{d}$

$\therefore$ Capacitance increases

Reactance $Z = \sqrt{R^2 + X_C^2}$

Capacitive reactance, $X_C = \frac{1}{\omega C}$ decreases

$\therefore$ Current, $I = \frac{V}{2}$ increases

So, current increases & thus bulb will glow more brighter.

29. **(10)** In First case

Current, $I = \frac{V}{R} = \frac{20}{R}$

$\Rightarrow R = \frac{20}{5} = 4\Omega$

In second case

Current, $i = \frac{20}{Z}$

$\Rightarrow 4 = \frac{20}{\sqrt{R^2 + X_L^2}} \Rightarrow \sqrt{R^2 + X_L^2} = 5$

$\Rightarrow R^2 + X_L^2 = 25 \Rightarrow X_L = 3\Omega$

$\Rightarrow L = \frac{3}{2\pi f} = \frac{1}{2 \times 50} = \frac{1000}{100}$ mH

$\Rightarrow L = 10$ mH

30. **(4)** Given,

Capacitive reactance, $X_c = 4\sqrt{3}\Omega$

Resistance, $R = 4\,\Omega$

Reactance is given by

$Z = \sqrt{R^2 + X_c^2}$

$\Rightarrow Z = \sqrt{4^2 + \left(4\sqrt{3}\right)^2} = 8\Omega$

RMS voltage

$$V_{rms} = \frac{V_0}{\sqrt{2}} = \frac{8\sqrt{2}}{\sqrt{2}} = 8V$$

RMS current

$$I_{rms} = \frac{V_{rms}}{Z} = \frac{8}{8} = 1A$$

Power dissipated, $P = I^2_{rms} \times R = 1 \times 4 = 4W$

31. **(b)** At resonance reactance, $Z = R$

$$\therefore \text{Current } I = \frac{V}{R} = \text{maximum}$$

When resistance, $R' = \frac{R}{2}$

Current, $I' = \frac{V}{R'} = \frac{2V}{R} = 2I$

32. **(a)**

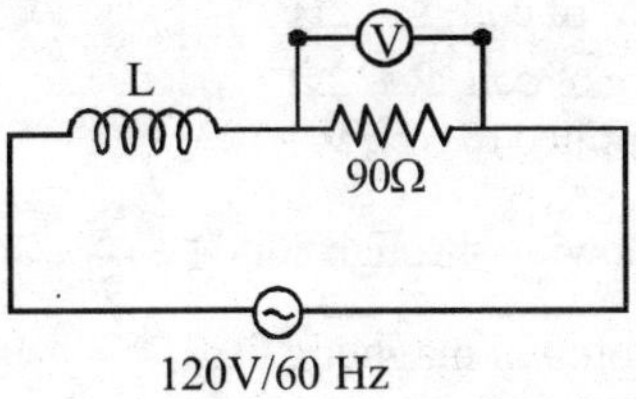

Using ohm's law, $V = I_{rms}R$

$36 = I_{rms}R$

$$36 = \frac{120}{\sqrt{X_L^2 + R^2}} \times R \qquad \left[\because I_{rms} = \frac{V}{\sqrt{X_L^2 + R^2}}\right]$$

$$36 = \frac{120 \times 90}{\sqrt{X_L^2 + 90^2}}$$

$$\sqrt{X_L^2 + 90^2} = 300$$

$$X_L = 286.18 \qquad [\because X_L = \omega L]$$

$$\omega L = 286.18 \Rightarrow L = \frac{286.18}{376.8} \qquad L = 0.76\,H$$

33. **(c)** Current is $I_m = \frac{V_m}{\sqrt{R^2 + (X_L - X_C)^2}}$

At resonance $X_L = X_C$

$$\therefore I_m = \frac{V_m}{R}$$

Statement I is correct.

Current in purely resistive circuit is $I = \frac{V}{R}$.

Statement II is also correct.

34. **(250)** Using Ohm's law,

$$R = \frac{V}{I} = \frac{100}{5} = 20\,\Omega$$

Impedence is given as:

$$Z = \sqrt{X_L^2 + R^2} = \sqrt{3 \times 400 + 400} = 40\,\Omega$$

Power dissipated in circuit is,

$$P = \left(\frac{V_{rms}}{Z}\right)^2 \times R = \left(\frac{\frac{200}{\sqrt{2}}}{40}\right)^2 \times 20 = 250\,W$$

35. **(10)** In resonance frequency is given by:

$$f_0 = \frac{1}{2\pi\sqrt{LC}}$$

$$f_0 = \frac{1}{2\pi\sqrt{100 \times 10^{-3} \times 2.5 \times 10^{-9}}} = \frac{1}{2\pi\sqrt{25 \times 10^{-11}}}$$

$$= 10 \times 10^3\,Hz$$

36. **(50)** $(V_{rms})_c = i_{rms}X_c = \frac{V_{rms}}{\sqrt{(X_L - X_c)^2 + R^2}} X_c$

Now, $X_c = \frac{1}{\omega C} = 500\Omega \;\&\; X_L = \omega L = 100\Omega$

$$\therefore (V_{rms})_c = \frac{50}{\sqrt{(500 - 100)^2 + 300^2}} \times 500 = 50\,V$$

37. **(c)** Voltage across inductor, $V_L = IX_L = I\omega L$

$$\therefore I = \frac{V_L}{\omega L} = \frac{V_L}{2\pi f L} = \frac{31.4}{2\pi \times 50 \times 10 \times 10^{-3}} = 10A$$

38. **(d)** In an ac circuit, the instantaneous current = 0 when the instantaneous voltage is maximum.

This is possible when phase difference is $\frac{\pi}{2}$ between current and voltage and this can be achieved when the source is connected to combination of an inductor (L) and capacitor (C).

39. **(d)** **A** – For purely capacitive circuit V lags by 90° from I hence option (I) is correct.

B – For purely inductive circuit V lead by 90° from I hence option (IV) is correct

C – In LCR series circuit at resonance $X_L = X_C$. Hence circuit is purely resistive so option (II) is correct

D – In LCR series circuit V is at some angle from I hence (III) is correct

40. **(c)** Since resonance frequency is unchanged $\omega' = \omega$

$$\frac{1}{\sqrt{L'C'}} = \frac{1}{\sqrt{LC}} \Rightarrow L'C' = LC$$

$$\Rightarrow L'(4C) = LC \Rightarrow L' = \frac{L}{4}$$

New Inductance should be decreased by

$$L - L' = L - \frac{L}{4} = \frac{3L}{4}$$

41. **(d)** Average power dissipated

$$\langle P \rangle = I_{rms} V_{rms} \cos\phi$$

$$= \frac{I_o}{\sqrt{2}} \frac{V_o}{\sqrt{2}} \cos 60^\circ$$

$$= \frac{20}{\sqrt{2}} \times \frac{10}{\sqrt{2}} \times \cos 60^\circ = 50\,W$$

42. (c) Given, E = 25 sin (1000 t)

$$P.F. = \cos\theta = \frac{1}{\sqrt{2}} = \frac{R}{\sqrt{R^2+\omega_1^2 L^2}} \quad (\because X_L = \omega L)$$

$R = \omega_1 L$ (Given)

We can say that angular frequency of second emf is double of first alternating emf

$\omega_2 = 2\omega_1$, given

$$P.F = \frac{R}{\sqrt{R^2+\omega_2^2 L^2}} = \frac{R}{\sqrt{R^2+4\omega_1^2 L^2}} = \frac{R}{\sqrt{5R^2}} = \frac{1}{\sqrt{5}}$$

43. (1) Given, $L = \frac{100}{\pi}\times 10^{-3}$ H, $C = \frac{10^{-3}}{\pi}$ F

and R = 10 Ω

$$X_c = \frac{1}{\omega C} = \frac{\pi}{2\pi\times 50\times 10^{-3}} = 10\Omega$$

$$X_L = \omega L = 2\pi\times 50\times\frac{100}{\pi}\times 10^{-3} = 10\,\Omega$$

$\because X_C = X_L$ So, circuit represents resonance.

$\therefore$ Power factor $= \frac{R}{Z} = \frac{R}{R} = 1$

44. (25) Energy stored in an inductor $E = \frac{1}{2}LI^2$

Rate of energy storing $= \frac{dE}{dt} = LI\frac{dI}{dt}$ and for $R-L$ circuit

$$I = \frac{E}{R}\left(1-e^{-t\frac{R}{L}}\right) \quad \therefore \frac{dI}{dt} = \frac{E}{L}e^{\frac{tR}{L}}$$

$$\frac{dE}{dt} = \frac{E^2}{R}\left(1-e^{-\frac{tR}{L}}\right)\left(e^{-t\frac{R}{L}}\right)$$

Time at which rate of power storing will be max,

$$t = \frac{L}{R\ln 2}$$

$$\therefore \frac{dE}{dt} = \frac{E^2}{R}\left(1-\frac{1}{2}\right)\times\frac{1}{2}$$

$$\Rightarrow \frac{E^2}{4R} = \frac{E^2}{100} = \frac{E^2}{2\times 50}$$

$a = 2, b = 50$

$$\therefore \frac{b}{a} = \frac{50}{2} = 25$$

45. (a) For resonance, $\phi = 0$, hence both inductor and capacitor must be present. Also power factor is zero for pure inductor or pure capacitor hence both the component consume zero average power.
Therefore both the statements I and II are false.

46. (d) Current, $I = \frac{V}{Z}$

When frequency of an a.c. source increases impedance (Z) first decreases and then increases.

$\therefore$ Current will first increase, attains its maximum value and then decreases.

Power factor, $\cos\phi = \frac{R}{Z}$

At Resonance $Z = R$

$\therefore \cos\phi = 1$

47. (c) Capacitive reactance,

$$X_c = \frac{1}{\omega C} = \frac{1}{(2\pi f)C} \quad \therefore X_c \propto \frac{1}{f}$$

X_C vs f graph is rectangular hyperbola

Inductive reactance

$X_L = \omega L = (2\pi f)L$

$\therefore X_L \propto f$

$\therefore X_L$ vs f graph is straight line.

48. (625) Given,

Inductance of coil, L = 2H

Resistance of coil, R = 2Ω

Voltage applied to coil, V = 10 V

Current flowing through coil $I = \frac{V}{R} = \frac{5}{2} = 2.5$ A

Energy stored in magnetic field

$$u = \frac{1}{2}LI^2 = \frac{1}{2}\times 2\times(2.5)^2$$

$\Rightarrow u = 625\times 10^{-2}$ J

49. (c) Power will be maximum when impedance is minimum,

$$Z = \left[R^2+(X_L-X_C)^2\right]^{\frac{1}{2}}$$

At resonance, $X_L = X_C \Rightarrow Z_{min} = R$

Voltage and current in same phase. Power through the LCR circuit is given by

$$P_{max} = \frac{V^2}{Z^2}R = \frac{V^2}{R} = i^2R$$

50. (a) Given, capacitance, C = 150 μF

Maximum value of emf, $E_0 = 36$

Maximum value of current,

$$I_0 = \frac{E_0}{X_c} = \frac{E_0}{\frac{1}{\omega c}} = E_0\omega c$$

$\Rightarrow I_0 = 36\times 120\pi\times 150\times 10^{-6} = 2.03 \simeq 2$A

51. (40) In LCR circuit power will maximum at the condition of resonance .

$X_L = X_C$

$$79.6 = \frac{1}{\omega C} \quad \therefore C = \frac{1}{2\pi\times 50\times 79.6} = 40\mu F$$

52. (242) As $\phi = 0$, we have resonance condition

So, Z = R

Thus, $I_{rms} = \frac{V_{rms}}{R} = \frac{220}{20} = 11$A

$I_0 = \sqrt{2}\,I_{rms} = 11\sqrt{2} = \sqrt{242}$

53. (a) We have

$$X_L = \omega L \Rightarrow X_L = 2\pi f L = 2\pi \times \frac{628}{2\pi} \times 5 \times 10^{-3} = 3.14\,\Omega$$

54. (25) We have

$$I_0 = \frac{V_0}{Z} = \frac{2500}{\sqrt{R^2 + (X_L - X_C)^2}} = \frac{2500}{\sqrt{80^2 + 60^2}} = \frac{2500}{100}$$

$= 25\text{A}$

55. (b) For LR circuit, the power factor is

$$P_1 = \frac{R}{Z} \Rightarrow P_1 = \frac{R}{\sqrt{R^2 + X_L^2}} = \frac{R}{R\sqrt{2}} \text{ (as } X_L = R)$$

$$P_1 = \frac{1}{\sqrt{2}}$$

$$P_2 = \frac{R}{\sqrt{R^2 + (X_L - X_L)^2}},\ P_2 = 1 \quad (\text{as } X_C = X_L)$$

$$\frac{P_1}{P_2} = \frac{1}{\sqrt{2}}$$

56. (a) Given,

Capacitive reactance, $X_C = 100\,\Omega$

Inductive reactance, $X_L = 200\,\Omega$

Resistance, $R = 100\,\Omega$

Impedance, $Z = \sqrt{R^2 + (X_L - X_C)^2}$

$$Z = \sqrt{100^2 + (200 - 100)^2} = 100\sqrt{2}\,\Omega$$

RMS value of current,

$$i_{rms} = \frac{V_{rms}}{Z} = \frac{200\sqrt{2}}{100\sqrt{2}} = 2\text{A}$$

57. (a)

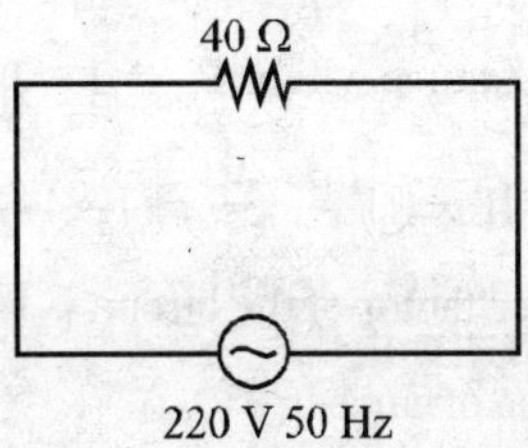

Here, $Z = R$

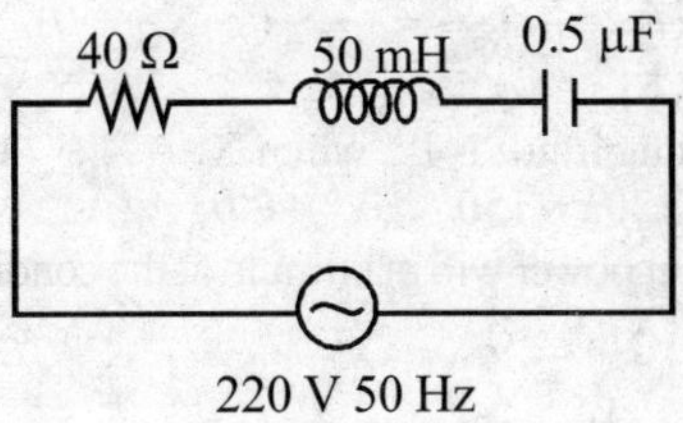

Here, $Z = \sqrt{R^2 + (X_L - X_C)^2}$

and, $X_L \neq X_C$

So, $Z > R$

So, rms current in (b) can never be larger than (a).

58. (3872) For maximum current, $X_L = X_C$

i.e., $\frac{1}{2\pi f C} = 2\pi f L$

$$\Rightarrow C = \frac{1}{4\pi^2 f^2 L} = \frac{1}{4 \times \pi^2 \times 49 \times 10^6 \times 2 \times 10^{-6}}$$

$$\Rightarrow C = \frac{1}{3872}\text{F}$$

$x = 3872$

59. (100) Given,

$C = 62.5\text{ nF} = 62.5 \times 10^{-9}\text{ F}$

$R = 50\,\Omega$

$f = 2\text{ kHz} = 2000\text{ Hz}$

$$f = \frac{1}{2\pi\sqrt{LC}}$$

$$2000 = \frac{1}{2\pi\sqrt{L \times 62.5 \times 10^{-9}}}$$

$$L = \frac{1}{4\pi^2 \times 2000^2 \times 62.5 \times 10^{-9}} = 0.1\text{ H} = 100\text{ mH}$$

60. (8) The power factor of circuit define as

$$\cos\phi = \frac{R}{Z} = \frac{R}{\sqrt{R^2 + (X_C - X_L)^2}}$$

$$\cos\phi = \frac{80}{\sqrt{(80)^2 + (60)^2}} \Rightarrow \cos\phi = \frac{80}{100} \Rightarrow \frac{8}{10}$$

So, $x = 8$

61. (c) Given circuit is R – L growth circuit

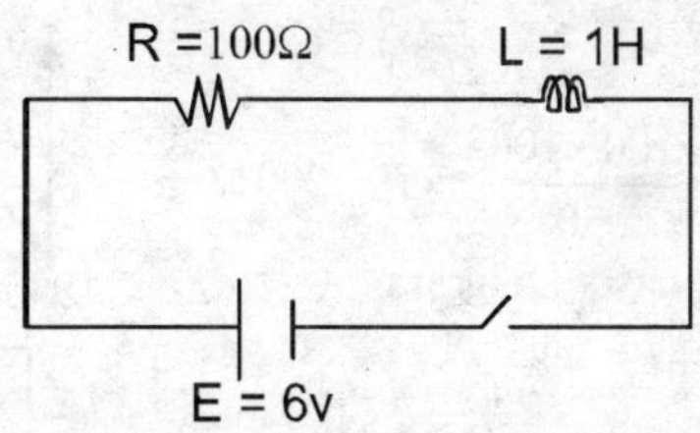

$$i = \frac{E}{R}(1 - e^{-t/\tau}) = i_0(1 - e^{-t/\tau}), \quad \text{when } i = \frac{i_0}{2}$$

$$\Rightarrow \frac{E}{2R} = \frac{E}{R}(1 - e^{-t/\tau})$$

Solving we get, $t = \tau \ln 2$

$$\Rightarrow t = \frac{L}{R}\ln 2 = \frac{1}{100}0.693 = 0.00693 = 7\text{ ms}$$

$$\text{Now, } i(15\text{ms}) = \frac{E}{R}\left(1 - e^{-\frac{15}{10}}\right) = \frac{E}{R}(1 - e^{-3/2})$$

$$\Rightarrow i = \frac{6}{100}\left(1 - \frac{1}{4}\right) = \frac{3}{4} \times \frac{6}{100} = \frac{18}{400} = \frac{9}{200}\text{A}$$

$$\text{As, } U = \frac{1}{2}LI^2 = \frac{1}{2} \times 1 \times \left(\frac{9}{200}\right)^2 \approx 10^{-3}\text{J} = 1\text{mJ}$$

62. (10) Here, inductance, $L = 50\text{ mH} = 50 \times 10^{-3}\text{ H}$

Capacitance, $C = 500 \times 10^{-6}\text{ F}$

Energy stored in capacitor given by

$$=\frac{1}{2}CV^2=\frac{1}{2}\times 500\times 10^{-6}\times 10^4=\frac{5}{2}J$$

Current will be maximum when whole energy capacitor becomes energy of inductor.

$$\frac{1}{2}LI^2=\frac{5}{2}\Rightarrow I=\sqrt{\frac{5}{L}}=\sqrt{\frac{5}{50\times 10^{-3}}}=10\text{ A}$$

63. **(d)** Element X should be resistance with R

$$=\frac{E}{I}=\frac{100}{5}=20\,\Omega$$

Element Y should be inductive with $X_L = 20\,\Omega$

When X and Y are connected in series

$$I_0=\frac{E_0}{Z}=\frac{E_0}{\sqrt{X_L^2+R^2}}=\frac{100}{20\sqrt{2}}=\frac{5}{\sqrt{2}}A$$

The rms value of the current will be,

$$I_{rms}=\frac{I_0}{\sqrt{2}}=\frac{5}{2}A$$

64. **(250)** Band width, $\Delta\omega=\omega_2-\omega_1=\frac{R}{L}=232-212=\frac{R}{L}$

$$\Rightarrow\ 20=\frac{R}{L}\Rightarrow L=\frac{R}{20}=\frac{5}{20}=250\text{ mH}$$

65. **(1)** Just after closing the switch S, inductor behaves like an open circuit.

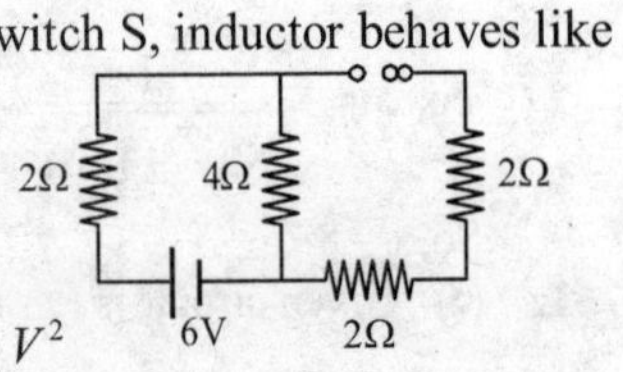

$$I=\frac{6}{2+4}=1\text{ A}$$

66. **(3)** Power $P=\frac{V^2}{R}\Rightarrow R=\frac{V^2}{P}$

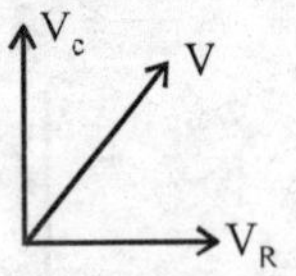

$$\Rightarrow\ R=\frac{100\times 100}{50}\Rightarrow R=200\Omega$$

By phasor Diagram

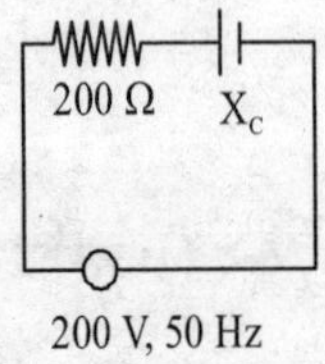

$$V_R^2+V_C^2=V^2$$

$$\Rightarrow\ (100)^2+V_C^2=(200)^2$$

$$\Rightarrow\ V_C=100\sqrt{3}$$

Now, $I=\frac{100}{200}=\frac{1}{2}A$

As, $V_C=I\times X_C$; $V_C=100\sqrt{3}$ and $I=\frac{1}{2}A$

So, $X_C=200\sqrt{3}$

$$\Rightarrow\ 200\sqrt{3}=\frac{1}{\omega C}\Rightarrow C=\frac{1}{200\sqrt{3}\times 2\pi\times 50}=\frac{50}{\pi\sqrt{3}}\mu F$$

So, $x=3$

67. **(10)** In steady state, capacitor is fully charged. Therefore current through it zero.

Charge $q=CV_{100\Omega}$

$$=(1.1\times 10^{-6})\left(\frac{10}{R+r}R\right)\quad(\because V_{100}=IR_{100})$$

$$=1.1\times 10^{-6}\left(\frac{10}{110}\times 100\right)=10\,\mu C$$

68. **(c)** In LCR circuit,

At resonance $X_L = X_C$

$$\Rightarrow\ \omega_0 L=\frac{1}{\omega_0 C}\Rightarrow\omega_0=\frac{1}{\sqrt{LC}}$$

$$\omega_0=\frac{1}{\sqrt{LC}}=\frac{1}{\sqrt{(0.01)(1\times 10^{-6})}}=10^4\text{ rad/sec}$$

At frequency 60% lower than resonant frequency,

$$\omega'=\frac{40}{100}\times 10^4=4000\text{ rad/sec}$$

$$\text{Current, } I=\frac{V_0}{\sqrt{R^2+(X_C'-X_L')^2}}$$

$$=\frac{50}{\sqrt{(10)^2+\left(\frac{1}{4000\times 1\times 10^{-6}}-4000\times 0.01\right)^2}}$$

$$=\frac{50}{\sqrt{100+\left(\frac{1000}{4}-40\right)^2}}=238\text{ mA}$$

69. **(44)** Here circuit can be reduced as

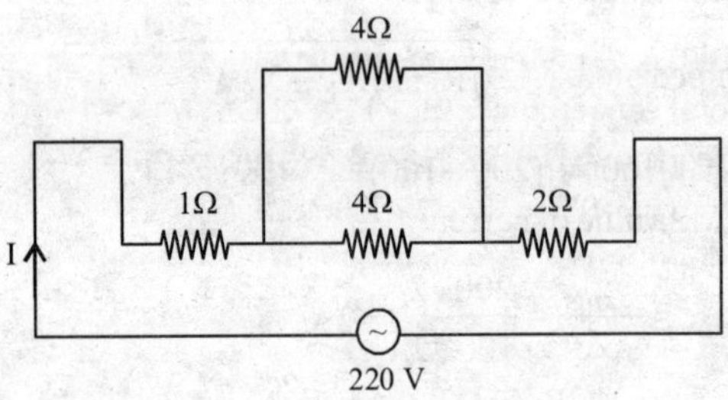

Then equivalent resistance will be $R_{eq}=1+2+2=5\,\Omega$

From Ohm's law $V=IR\Rightarrow I=\frac{220}{5}\Rightarrow I=44\text{ A}$

70. **(b)** Given that power factor of the circuit is P_1, when $X_L = R$

$$P_1=\frac{R}{\sqrt{R^2+(X_L-X_C)^2}}=\frac{1}{\sqrt{2}}\quad(\because X_C=0)$$

Power factor of the circuit is P_2, when $X_L = X_C$

$$P_2=\frac{R}{\sqrt{R^2+(X_L-X_C)^2}}=1$$

So, we have $\frac{P_1}{P_2}=\frac{1}{\sqrt{2}}$

71. **(c)** Resonant frequency, $f_r=\frac{1}{2\pi\sqrt{LC}}$

$$f_r\propto\frac{1}{\sqrt{C}}$$

When another capacitor is added in series, C_{eq} decreases. So, f_r increases.

72. **(b)** Metal detector works on the principle of transmitting an electromagnetic signal and analyses a return signal from the target. So it works on the principle of resonance in AC circuit.

73. **(500)** For maximum power

power factor $= \cos\theta = 1 \therefore \frac{R}{Z} = 1$

$R^2 = Z^2$

$R^2 = (X_L - X_C)^2 + R^2$

$X_L = X_C$

$$\frac{70}{11} = \frac{1}{100\pi \times C} \Rightarrow C = \frac{11}{7000\pi} = 500 \times 10^{-6} F = 500\mu F$$

74. **(c)** If $\omega \downarrow\downarrow \Rightarrow X_L \downarrow\downarrow$. So to the left of ω_r the circuit and, $X_C \uparrow\uparrow$ is mainly capacitive.
So option (a) is correct and option (b) is incorrect.
At resonance i.t. at $\omega = \omega_r$, we have $Z = R$
So option (c) is correct and option (d) is incorrect.

75. **(5)** Here $\Delta\phi = 0$. So, $Z = R$

$$\text{So, } f_{\text{ac source}} = f_{\text{resonant}} = \frac{1}{2\pi\sqrt{LC}}$$

$$= \frac{1}{2\pi\sqrt{10^{-7}}} = 500 \text{ Hz} = 5 \times 10^2$$

76. **(100)** $C = 0.01\ \mu F/km = 0.01 \times 10^{-6} F/km$
Total capacitance $(C_{total}) = C \times L = 0.01 \times 10^{-6} \times 100$
Frequency (f) 0.5 KHz $= 0.5 \times 10^3$Hz
$w = 2\pi \times (0.5 \times 10^3) = 1000\pi$ rad/s
For minimum impedence in an LCR circuit, use need:
$X_L = X_C$

$$L = \left(\frac{1}{(1000\pi)^2 \times 10^{-6}}\right)$$

$$L = \frac{1}{\pi^2} = \frac{1}{10} H$$

L = 100 mH

77. **(15)** $Z = \sqrt{R^2 + (X_L - X_C)^2} = \sqrt{120^2 + (10-100)^2}$

$= \sqrt{120^2 + 90^2} = 150\Omega$

$[\because X_L = \omega L = 100 \times 100 \times 10^{-3} = 10\Omega]$

$$X_C = \frac{1}{\omega_c} = \frac{1}{100 \times 100 \times 10^{-6}} = 100\Omega$$

$$\text{So, } i_{rms} = \frac{V_{rms}}{Z} = \frac{20}{150} = \frac{2}{15} A$$

Now, Heat lost $= ms\Delta T = i_{rms}^2 R\Delta t$

$$\Rightarrow \Delta t = \frac{(ms)\Delta T}{i_{rms}^2 R} = \frac{2 \times 16}{\frac{4}{225} \times 120} = 15 \text{ sec}$$

78. **(0)** We have
$X_L = \omega L \Rightarrow X_L = 2\pi f L$

Now, $V_L = 2V_R \Rightarrow iX_L = 2iR \Rightarrow X_L = 2R$
So, $2R = 2\pi f L \Rightarrow R = \pi f L \Rightarrow 5 = \pi \times 50 \times L$

$$\Rightarrow L = \frac{1}{10\pi} H = \frac{1000}{10\pi} mH = \frac{1}{0.01\pi} mH \text{ So, } K = 0.01 \approx 0$$

79. **(0)** This is a case of parallel RLC circuit

$$\text{So, } \frac{1}{Z} = \sqrt{\left(\frac{1}{X_L} - \frac{1}{X_C}\right)^2}$$

At resonance, $X_L = X_C$. So $Z \to \infty$

$$I = \frac{V}{Z} \to 0$$

80. **(b)** Wattless current flow in a circuit only when circuit is resistanceless i.e. circuit is purely capacitive or inductive.

81. **(22)** $Z = \sqrt{R^2 + (x_L - x_C)^2}$

$$= \sqrt{10^2 + \left(2\pi \times 50 \times 0.1 - \frac{1}{2\pi \times 50 \times 10^{-4}}\right)} \approx 10\Omega$$

$$\text{So, } i = \frac{E}{Z} = \frac{220}{10} = 22A$$

82. **(a)** $X_L = \omega L = 3000 \times 10^{-2}\Omega = 30\,\Omega$

$$X_C = \frac{1}{\omega c} = \frac{1}{3000 \times 25 \times 10^{-6}} = \frac{1000}{75} = \frac{40}{3}\Omega$$

$$\text{So, } X_L - X_C = \left(30 - \frac{40}{3}\right)\Omega = \frac{50}{3}\Omega$$

Given, $R = 100\,\Omega$

$$\text{As, } \tan\phi = \frac{|X_L - X_C|}{R} = \frac{50}{3 \times 100} = \frac{1}{6}$$

$$\phi = \tan^{-1}\left(\frac{1}{6}\right) = \tan^{-1}(0.17)$$

The reactance can only be zero if circuit contain both inductor and capacitor, and $X_L = X_C$.

83. **(c)** The reactance can only be zero if circuit contain both inductor and capacitor, and $X_L = X_C$. When a circuit is purely inductive or capacitive then

$$\phi = \pm\frac{\pi}{2}, \text{ then } P_{avg} = V_{rms} I_{rms} \cos\left(\pm\frac{\pi}{2}\right) = 0$$

Also when circuit has only capacitor and inductor, then $P_{avg} = 0$ because their is no resistance to waste the energy.

84. **(c)** We know in steady state, inductor behaves as a conductor. So, equivalent circuit is

$$\frac{1}{R'} = \frac{1}{R_2} + \frac{1}{R_3} + \frac{1}{R_4} = \frac{1}{3} + \frac{1}{3} + \frac{1}{3} \Rightarrow R' = 1\Omega$$

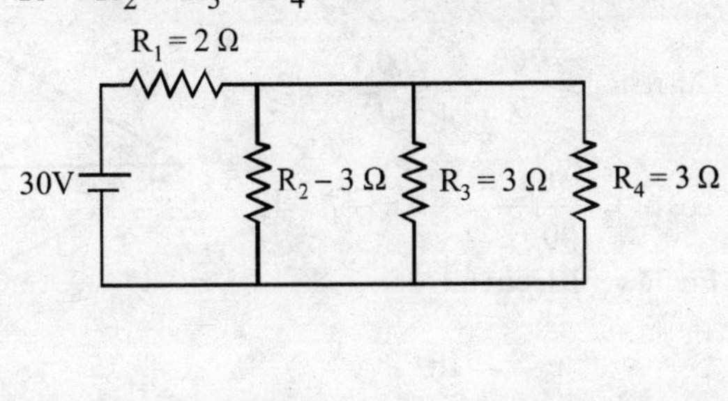

Now total resistance of circuit $R_{eq} = R' + R_1 = 2\Omega + 1\Omega = 3\Omega$

$\Rightarrow$ Current, $i = \frac{30}{3} = 10A$

85. **(c)** Impedance in LCR circuit

$$Z = \sqrt{(X_L - X_C)^2 + R^2} \quad \because X_L = X_C = R \quad \therefore Z = R$$

86. **(2)** At very high frequency inductive reactance X_L

$X_L = 2\pi fL$ is very high while $X_c = \frac{1}{2\pi fC}$ i.e. capacitive reactance is very small. So inductor acts as open circuit while capacitor as a conducting wire. Final circuit will look like

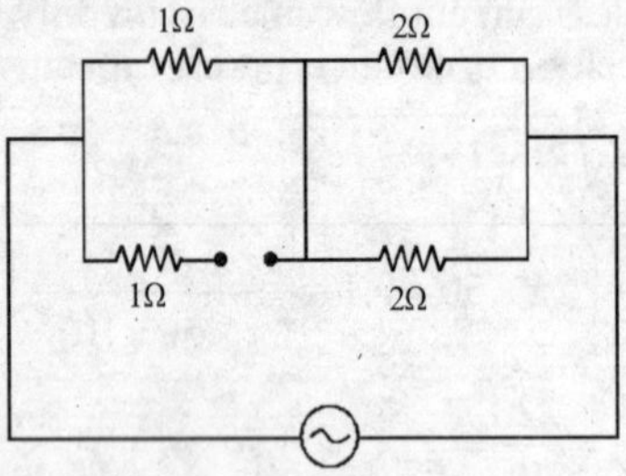

Equivalent resistance or impedance

$$Z_{eq} = 1 + \frac{2 \times 2}{2+2} = 2\,\Omega$$

87. **(1)** Power factor for RL circuit

$$\cos\phi = \frac{R}{\sqrt{R^2 + X_L^2}} = \frac{R}{\sqrt{R^2 + 9R^2}} = \frac{1}{\sqrt{10}}$$

Power factor for LCR circuit

$$\cos\phi' = \frac{R}{\sqrt{R^2 + \left(X_L^2 - X_C^2\right)}} = \frac{R}{\sqrt{R^2 + R^2}} = \frac{1}{\sqrt{2}}$$

$$\frac{\cos\phi'}{\cos\phi} = \frac{\sqrt{10}}{\sqrt{2}} = \frac{\sqrt{5}}{1} \quad \therefore x = 1$$

88. **(a)** Power will be maximum at resonance

At resonance $X_L = X_C \Rightarrow L\omega = \frac{1}{C\omega}$

$$\Rightarrow 250\pi = \frac{1}{2\pi(50)C} \Rightarrow C = 4 \times 10^{-6}\,F = 4\,\mu F$$

89. **(None)** For upper part of RC circuit

$$Z = \sqrt{(X_c)^2 + R^2} = \sqrt{\left(\frac{1}{\omega C}\right)^2 + R^2}$$

$$= \sqrt{\left(\frac{1}{100 \times 100 \times 10^{-6}}\right)^2 + 100^2} = \sqrt{(100)^2 + (100)^2} = 100\sqrt{2}$$

Current $I = \frac{200}{Z} = \frac{200}{100\sqrt{2}} = \sqrt{2}$

$\cos\phi_1 = \frac{100}{100\sqrt{2}} = \frac{1}{\sqrt{2}} \Rightarrow \phi_1 = 45°$

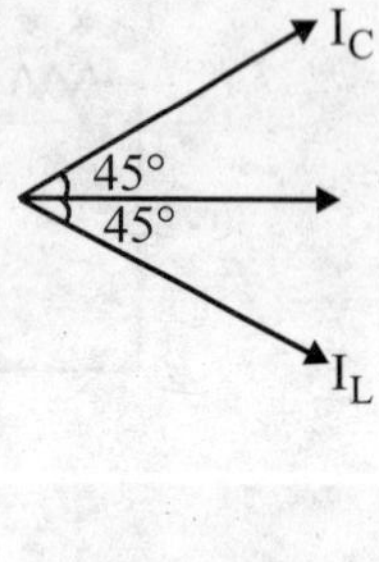

For lower RL circuit

$$Z = \sqrt{(\omega L)^2 + R^2} = \sqrt{(100 \times 0.5)^2 + 50^2} = 50\sqrt{2}$$

Current $I = \frac{200}{Z} = \frac{200}{50\sqrt{2}} = 2\sqrt{2}$

$\cos\phi_2 = \frac{50}{50\sqrt{2}} = \frac{1}{\sqrt{2}} \Rightarrow \phi_2 = 45°$

Net current $I = \sqrt{I_C^2 + I_L^2} = \sqrt{2+8} = \sqrt{10} = 3.16$ A

90. **(74)** Maximum value of current,

$$I_{max} = \frac{V}{R} = \frac{20V}{10K\Omega} = 2mA$$

For LR – circuit

$I = I_{max} e^{-Rt/L}$

$$\Rightarrow I = 2 \times e^{\frac{-10 \times 10^3 \times 1 \times 10^{-6}}{10 \times 10^{-3}}} \Rightarrow I = 2e^{-1}$$

$$\Rightarrow I = 2 \times 0.37\text{ mA} \Rightarrow I = \frac{74}{100}\text{ mA}$$

Hence value of x = 74

91. **(500)** The first circuit is purely resistive circuit and second circuit is LCR circuit. We have given,

$P_1 = P_2$

$$\therefore \left(\frac{V^2}{R}\right)_1 = \left(\frac{V^2}{Z}\right)_2 \Rightarrow R = Z$$

$$\Rightarrow R = \sqrt{\left(\omega L - \frac{1}{\omega C}\right)^2 + (R)^2} \Rightarrow 5 = \sqrt{\left(\omega L - \frac{1}{\omega C}\right)^2 + 25}$$

$$\Rightarrow 25 = \left[\omega(0.1) - \frac{1}{\omega\left(40 \times 10^{-6}\right)}\right]^2 + 25$$

$$\Rightarrow \omega^2(0.1) = \frac{1}{40 \times 10^{-6}} \Rightarrow \omega^2 = \frac{1}{4} \times 10^6$$

$\therefore \quad \omega = 500$ rad/s

92. **(c)** The correct phasor diagram for such circuit is (c), in capacitor, current lead voltage by $\frac{\pi}{2}$.

93. **(3)** For current leads the voltage by 45°

$$\tan 45° = \frac{X_C - X_L}{R}$$

$$\Rightarrow X_C - X_L = R \Rightarrow \frac{1}{\omega C} - \omega L = R \Rightarrow \frac{1}{\omega C} - 300 \times 0.03 = 1$$

$$\frac{1}{\omega C} = 10 \Rightarrow C = \frac{1}{10\omega} = \frac{1}{10 \times 300} \Rightarrow C = \frac{1}{3} \times 10^{-3}$$

Hence, value of x = 3.

94. **(c)** Inductive reactance,

$X_L = \omega L = 2\pi fL$

where f = frequency

$$I = \frac{\varepsilon}{X_L} = \frac{\varepsilon}{2\pi fL}$$

$\therefore$ When frequency is halved, $f' = \frac{f}{2}$ inductive reactance gets halved and current gets doubled.

95. (3.906) Given, voltage applied, $V_0 = 250$ V

RMS value of voltage,

$$V_{rms} = \frac{V_0}{\sqrt{2}} = \frac{250}{\sqrt{2}}$$

At resonance, LCR series circuit behaves as pure resistive circuit.

Power dissipated at resonance

$$= \frac{V_{rms}^2}{R} = \frac{(250)^2}{2 \times 8} = 3.906 \times 10^3 \text{ W}$$

96. (900) At resonance, $X_L - X_C = 0$

Power drawn by circuit, $P = \frac{V^2}{R}$

$$\Rightarrow R = \frac{V^2}{P} = \frac{(120)^2}{16} = 900\Omega.$$

97. (0.8, 250) $I_{rms} = \frac{V}{R} = \frac{15}{60} = \frac{1}{4}$ A

Now, $20 = I_{rms} \times X_L \Rightarrow 20 = \frac{1}{4} \times X_L \Rightarrow X_L = 80\ \Omega$

$$X_L = L\omega \Rightarrow L = \frac{X_L}{\omega} = \frac{80}{100} \Rightarrow L = 0.8 \text{ H}$$

Again, $10 = X_C \times I_{rms}$

$$\Rightarrow X_C = \frac{10}{I_{rms}} = \frac{10}{\frac{1}{4}} = 40\Omega$$

And, $X_C = \frac{1}{C\omega} \Rightarrow C = \frac{1}{X_C\omega} = \frac{1}{40 \times 100}$

$$\Rightarrow C = \frac{1}{4000} F = 250\ \mu F$$

98. (c) $\cos 45° = \frac{R}{X_L}$ ....(i)

$\cos 45° = \frac{R}{X_C}$ ....(ii)

So, from (i) & (ii) we get $X_L = X_C$ i.e. circuit in resonance. LCR circuit is in resonance, behaves as resistive circuit.

$$\therefore I_{rms} = \frac{V_{rms}}{R} = \frac{220}{110} = 2 \text{ A}$$

99. (33) As current is decreasing. So inductor will support current.

Here, $L = 50$ mH $= 50 \times 10^{-3}$ H; $I = 1$ A, $R = 2\Omega$

$$V_P - L\frac{dl}{dt} - 30 + RI = V_Q$$

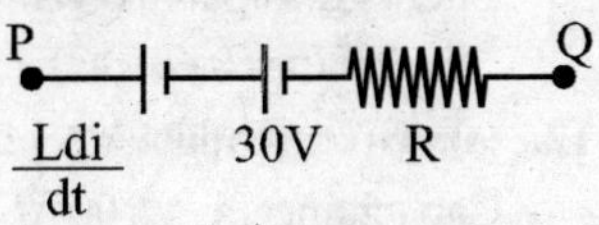

$$\Rightarrow V_P - V_Q = 50 \times 10^{-3} \times 10^2 + 30 - 1 \times 2 = 5 + 30 - 2 = 33 \text{ V.}$$

100. (400)

$P = V_{rms} I_{rms} \cos\phi$

$$\Rightarrow 400 = \frac{V_{rms}^2}{Z} \times 0.8$$

$$\Rightarrow Z = \frac{250^2}{400} \times 0.8 = 125\ \Omega$$

$$\cos\phi = \frac{R}{Z}$$

$R = Z\cos\phi = 125 \times 0.8 = 100\ \Omega$

Now, $Z^2 = X_L^2 + R^2$

$\Rightarrow 125^2 = X_L^2 + 100^2$

$\Rightarrow X_L = 75\ \Omega$

At resonance

$X_L = X_C$

$$\Rightarrow 75 = \frac{1}{\omega C} \Rightarrow C = \frac{1}{2\pi \times 75 \times 50} \Rightarrow C = \frac{400}{3\pi}\mu F$$

101. (a) Potential energy stored in the inductor

$$U = \frac{1}{2}LI^2$$

During growth of current,

$$I = I_{max}\left(1 - e^{-Rt/L}\right)$$

For U to be $\frac{U_{max}}{n}$; i has to be $\frac{I_{max}}{\sqrt{n}}$

$$\therefore \quad \frac{I_{max}}{\sqrt{n}} = I_{max}(1 - e^{-Rt/L})$$

$$\Rightarrow e^{-Rt/L} = 1 - \frac{1}{\sqrt{n}} = \frac{\sqrt{n}-1}{\sqrt{n}} \Rightarrow -\frac{Rt}{L} = \ln\left(\frac{\sqrt{n}-1}{\sqrt{n}}\right)$$

$$\Rightarrow t = \frac{L}{R}\ln\left(\frac{\sqrt{n}}{\sqrt{n}-1}\right)$$

102. (d) Here, $R = 100, X_L = L\omega = 0.1803 \times 750 \times 2\pi = 850\Omega$,

$$X_C = \frac{1}{C\omega} = \frac{1}{10^{-5} \times 2\pi \times 750} = 21.23\Omega$$

Impedance $Z = \sqrt{R^2 + (X_L - X_C)^2}$

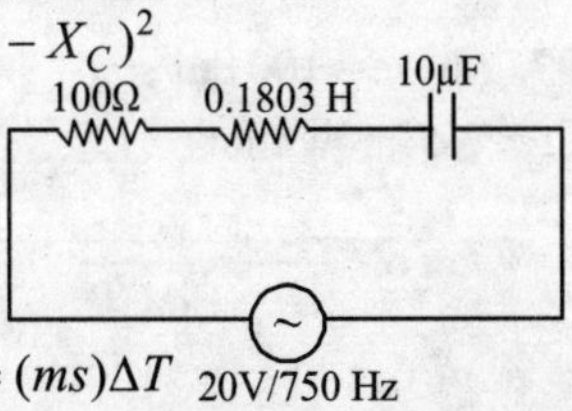

$= \sqrt{100^2 + (850 - 21.23)^2}$

$= 834.77 \simeq 835\ \Omega$

$$H = i_{rms}^2 Rt = \left(\frac{V_{rms}}{|Z|}\right)^2 Rt = (ms)\Delta T$$

$$\Rightarrow \frac{20}{835} \times \frac{20}{835} \times 100t = (2) \times 10$$

$\because V_{rms} = 20$ V and $\Delta T = 10°$C

$\therefore$ Time, $t = 348.61$ s.

103. (a) Given,

Reactance of inductance coil, $Z = 100\Omega$

Frequency of AC signal, f = 1000 Hz

Phase angle, $\phi = 45°$

$$\tan\phi = \frac{X_L}{R} = \tan 45° = 1 \Rightarrow X_L = R$$

Reactance, $Z = 100 = \sqrt{X_L^2 + R^2}$

$$\Rightarrow 100 = \sqrt{R^2 + R^2} \Rightarrow \sqrt{2}R = 100 \Rightarrow R = 50\sqrt{2}$$

$$\therefore X_L = 50\sqrt{2} \Rightarrow L\omega = 50\sqrt{2} \quad (\because X_L = \omega L)$$

$$\Rightarrow L = \frac{50\sqrt{2}}{2\pi \times 1000} = \frac{25\sqrt{2}}{\pi}\text{ mH} \quad (\because \omega = 2\pi f)$$

$$= 1.1 \times 10^{-2}\text{ H}$$

104. (b) We have, $i = i_0(1 - e^{-t/e}) = \frac{\varepsilon}{R}(1 - e^{-t/e})$

Charge, $q = \int_0^\tau i dt = \frac{\varepsilon}{R}\int_0^\tau (1 - e^{-t/\tau})dt$

$$= \frac{E}{R}\frac{\tau}{e} = \frac{E}{R} \times \frac{(L/R)}{e} = \frac{EL}{2.7R^2}$$

105. (d) $I = I_0\left(1 - e^{-\frac{Rt}{L}}\right)$ Here $R = R_L + r = 1\Omega$

$$\Rightarrow 0.8I_0 = I_0\left(1 - e^{-\frac{t}{.01}}\right) \Rightarrow 0.8 = 1 - e^{-100t}$$

$$\Rightarrow e^{-100t} = 0.2 = \left(\frac{1}{5}\right)$$

$$\Rightarrow 100t = \ln 5 \Rightarrow t = \frac{1}{100}\ln 5 = 0.016\text{ s}$$

106. (c) $i^2R = Li\frac{di}{dt} \Rightarrow \frac{di}{dt} = \frac{i}{\tau}$

$$e^{-Rt/L} = \frac{1}{2}$$

$$\Rightarrow t = \tau\, ln2 = 2ln2 \left[\text{as}\, \tau = \frac{L}{R} = \frac{20}{10} = 2\right]$$

107. (d) $\omega = 100$ rad/s

We know that

$$\tan\phi = \frac{X_C}{R} = \frac{1}{\omega CR}$$

$$\Rightarrow \tan 45° = \frac{1}{\omega CR} \Rightarrow \omega CR = 1$$

LHS: $\omega CR = 100 \times 10 \times 10^{-6} \times 10^3 = 1$

108. (d) Long time after switch is closed, the inductor will be idle so, the equivalent diagram will be as below

$$I = \frac{\varepsilon}{\left(\frac{R \times R}{R + R}\right)} = \frac{2\varepsilon}{R} = \frac{2 \times 15}{5} = 6\text{ A}$$

109. (None) Capacitive reactance,

$$X_c = \frac{1}{\omega C}$$

$$= \frac{2}{10^{-6} \times \sqrt{3} \times 100} = \frac{2 \times 10^4}{\sqrt{3}}$$

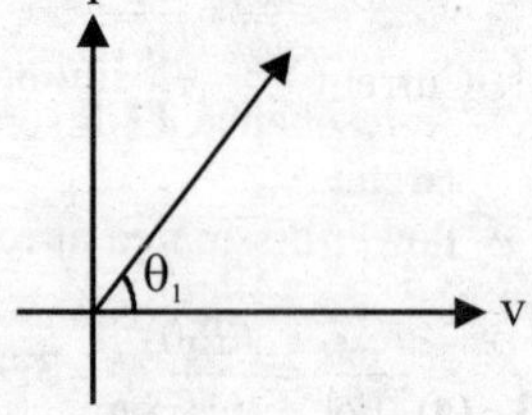

$$\tan\theta_1 = \frac{X_C}{R_2} = \frac{10^3}{\sqrt{3}}$$

θ_1 is close to 90°

For L-R circuit

$$X_L = \omega L = 100 \times \frac{\sqrt{3}}{10} = 10\sqrt{3}$$

$R_1 = 10$

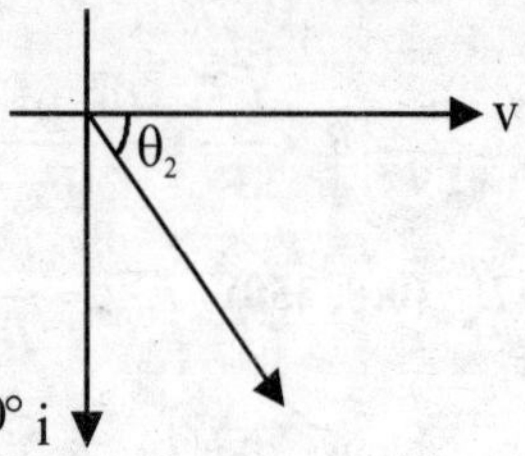

$$\tan\theta_2 = \frac{X_L}{R_1} \tan\theta_2 = \sqrt{3}$$

$$\Rightarrow \theta_2 = \tan^{-1}(\sqrt{3}) \Rightarrow \theta_2 = 60°$$

So, phase difference comes out 90° + 60° = 150°

110. (b)

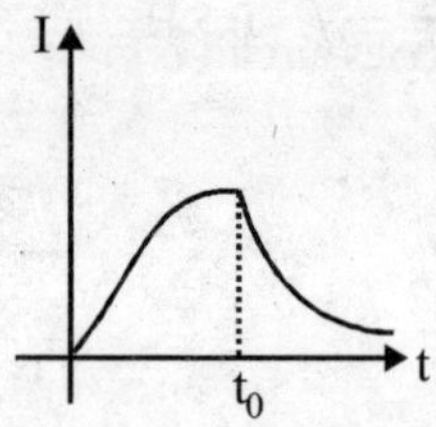

The current will grow for the time t = 0 to $t = t_0$ and after that decay of current takes place.

111. (c) Given: $R = 60\Omega$, f = 50 Hz, $\omega = 2\pi f = 100\pi$ and V = 24v

$C = 120\,\mu f = 120 \times 10^{-6} f$

$$X_C = \frac{1}{\omega C} = \frac{1}{100\pi \times 120 \times 10^{-6}} = 26.52\Omega$$

$$X_L = \omega L = 100\pi \times 20 \times 10^{-3} = 2\pi\Omega$$

$$X_C - X_L = 20.24 \approx 20$$

$$Z = \sqrt{R^2 + (X_C - X_L)^2} \Rightarrow Z = 20\sqrt{10}\Omega$$

$$\cos\phi = \frac{R}{Z} = \frac{60}{20\sqrt{10}} = \frac{3}{\sqrt{10}}$$

$$P_{avg} = VI\cos\phi, I = \frac{V}{Z} = \frac{V^2}{Z}\cos\phi = 8.64\text{ watt}$$

Energy dissipated (Q) in time t = 60s is

$$Q = P.t = 8.64 \times 60 = 5.17 \times 10^2\text{J}$$

112. (a) Given, Inductance, $L = 40$ mH

Capacitance, $C = 100\,\mu F$

Impedance, $Z = X_C - X_L$

$$\Rightarrow\ Z=\frac{1}{\omega C}-\omega L \quad \left(\because X_c=\frac{1}{\omega C}\text{ and } X_L=\omega L\right)$$

$$=\frac{1}{314\times100\times10^{-6}}-314\times40\times10^{-3}=19.28\Omega$$

Here $X_C > X_L$. So current will lead.

$$\text{Current, } i=\frac{V_0}{Z}\sin(\omega t+\pi/2)\left[\because \tan\phi=\frac{X_C-X_C}{R} \text{ and } R=0\right]$$

$$\Rightarrow\ i=\frac{10}{19.28}\cos\omega t=0.52\cos(314\,t)$$

113. (a) For series connection of a resistor and inductor, time variation of current is $I=I_0(1-e^{-t/t_c})$

Here, $t_c=\dfrac{L}{R}$

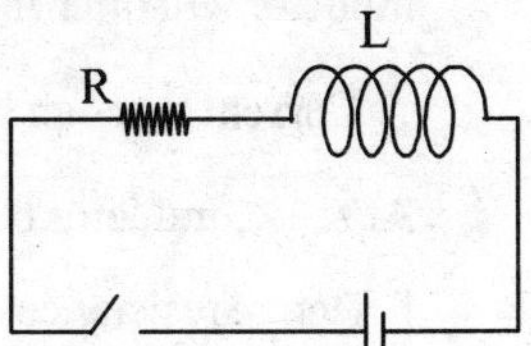

$$q=\int_0^{t_c} idt$$

$$\Rightarrow q=\int\frac{E}{R}\left(1-e^{-t/t_c}\right)dt \Rightarrow q=\frac{E}{R}\left[t+t_c e^{-t/t_c}\right]_0^{t_c}$$

$$\Rightarrow q=\frac{E}{R}\left[t_c+\frac{t_c}{e}-t_c\right]\Rightarrow q=\frac{E}{R}\frac{L}{\text{Re}}\ \therefore q=\frac{EL}{R^2e}$$

114. (a) The current (I) in LR series circuit is given by

$$I=\frac{V}{R}\left(1-e^{-\frac{tR}{L}}\right);\text{ At } t=\infty,$$

$$I_\infty=\frac{20}{5}\left(1-e^{\frac{-\infty}{L/R}}\right)=4 \qquad \text{...(i)}$$

At $t=40$s,

$$I_{40}=4\left(1-e^{\frac{-40\times5}{10\times10^{-3}}}\right)=4(1-e^{-20,000}) \qquad \text{...(ii)}$$

Dividing (i) by (ii) we get

$$\Rightarrow\ \frac{I_\infty}{I_{40}}=\frac{1}{1-e^{-20,000}}\simeq1.06$$

115. (b) As we know, average power $P_{avg}=V_{rms}I_{rms}\cos\theta$

$$=\left(\frac{V_0}{\sqrt2}\right)\left(\frac{I_0}{\sqrt2}\right)\cos\theta=\left(\frac{100}{\sqrt2}\right)\left(\frac{20}{\sqrt2}\right)\cos45^\circ\ (\because\theta=45^\circ)$$

$$P_{avg}=\frac{1000}{\sqrt2}\text{ watt}$$

Wattless current $I=I_{rms}\sin\theta$

$$=\frac{I_0}{\sqrt2}\sin\theta=\frac{20}{\sqrt2}\sin45^\circ=10A$$

116. (b) Given,

$V_0=283$ volt, $\omega=320$, $R=5\,\Omega$, $L=25$ mH, $C=1000\,\mu F$

$x_L=\omega L=320\times25\times10^{-3}=8\,\Omega$

$$x_C=\frac{1}{\omega C}=\frac{1}{320\times1000\times10^{-6}}=3.1\,\Omega$$

Total impedance of the circuit:

$$Z=\sqrt{R^2+(X_L-X_C)^2}=\sqrt{25+(4.9)^2}=7\,\Omega$$

Phase difference between the voltage and current

$$\tan\phi=\frac{X_L-X_C}{R}\Rightarrow\tan\phi=\frac{4.9}{5}\approx1\Rightarrow\phi=45^\circ$$

117. (b) Here

$$i=\frac{e}{\sqrt{R^2+X_L^2}}=\frac{e}{\sqrt{R^2+\omega^2L^2}}=\frac{e}{\sqrt{R^2+4\pi^2v^2L^2}}$$

$$\Rightarrow10=\frac{220}{\sqrt{64+4\pi^2(50)^2L^2}}\quad\left[\because R=\frac{V}{I}=\frac{80}{10}=8\right]$$

On solving we get

L = 0.065 H

118. (b) $I(0)=\dfrac{15}{0.15\times10^3}=0.1A$

So, $I(t)=0.1\,e^{\frac{-t}{L/R}}$

$$I(t)=0.1\,e^{\frac{-0.15\times10^{+3}\times10^{-3}}{0.03}}=0.67mA$$

119. (c) Applying Kirchhoff's law of voltage in closed loop

$$-V_R-V_C=0\ \Rightarrow\ \frac{V_R}{V_C}=-1$$

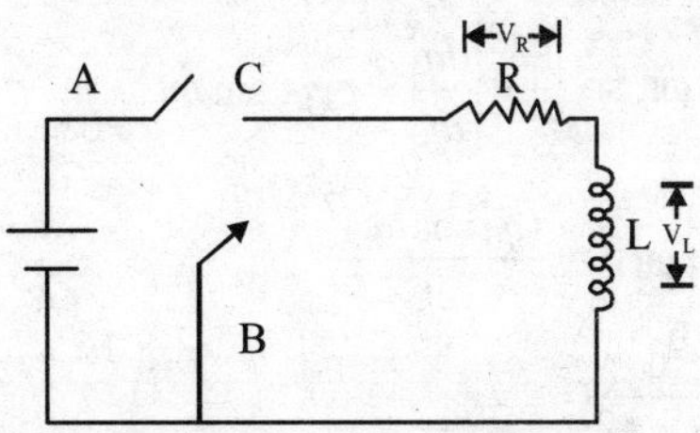

120. (c) Given, $V_L:V_C:V_R=1:2:3$

$V=100$ V, let $V_L=x$. Then, $V_C=2x$ and $V_R=3x$

$V_R=?$

As we know,

$$V=\sqrt{V_R^2+(V_L-V_C)^2}=\sqrt{9x^2+x^2}=\sqrt{10}x$$

$$\text{So, } x=\frac{V}{\sqrt{10}}=\frac{100}{\sqrt{10}}$$

So $V_R=300/\sqrt{10}=90$V

121. (c) Charge on he capacitor at any time t is given by $q=CV(1-e^{-t/\tau})$

at $t=2\tau$

$q=CV(1-e^{-2})$

122. (d) Power factor $_{(old)}$

$$=\frac{R}{\sqrt{R^2+X_L^2}}=\frac{R}{\sqrt{R^2+(2R)^2}}=\frac{R}{\sqrt5R}$$

Power factor$_{(new)}$

$$= \frac{R}{\sqrt{R^2+(X_L-X_C)^2}} = \frac{R}{\sqrt{R^2+(2R-R)^2}} = \frac{R}{\sqrt{2}R}$$

$$\therefore \frac{\text{New power factor}}{\text{Old power factor}} = \frac{\frac{R}{\sqrt{2}R}}{\frac{R}{\sqrt{5}R}} = \sqrt{\frac{5}{2}}$$

123. (d) Impedance (Z) of the series LCR circuit is

$$Z = \sqrt{R^2+(X_L-X_C)^2}$$

At resonance, $X_L = X_C$

Therefore, $Z_{minimum} = R$

124. (c) Voltage E of the ac source, $E = \sqrt{(V_C - V_L)^2 + V_R^2}$

$E = V_C - V_L = 400\text{ V} - 300\text{ V} = 100\text{ V}$

125. (d) In case of series RLC circuit,

Equation of voltage is given by

$$V^2 = V_R^2 + (V_L - V_C)^2$$

Here, $V = 220\ V$; $V_L = V_C = 300$ V

$\therefore V_R = \sqrt{V^2} = 220\text{V}$

Current $i = \frac{V}{R} = \frac{220}{100} = 2.2\text{A}$

126. (a) Energy will be equal when

$$\therefore \frac{1}{2}Li^2 = \frac{1}{2}\frac{q^2}{C}$$

As, $q = q_0 \cos \omega t$, so $|i| = \left|\frac{dq}{dt}\right| = \omega q_0 \sin \omega t$

$$\Rightarrow \frac{1}{2}L(\omega q_0 \sin \omega t)^2 = \frac{(q_0 \cos \omega t)^2}{2C}$$

$$\Rightarrow \tan^2 \omega t = \frac{1}{LC\omega^2}$$

$$\Rightarrow \tan^2 \omega t = \frac{\omega^2}{\omega^2} \qquad \left[\because \omega^2 = \frac{1}{LC}\right]$$

$$\Rightarrow \tan \omega t = 1 \Rightarrow \omega t = \frac{\pi}{4}$$

$$\Rightarrow t = \frac{\pi}{4\omega} = \frac{\pi}{4}\sqrt{LC}$$

127. (b) We have, $V = V_0(1 - e^{-t/RC})$

$$\Rightarrow 120 = 200(1 - e^{-t/RC});\ e^{-t/RC} = \frac{200-120}{200} = \frac{80}{200}$$

$$\frac{-5}{R \times 2 \times 10^{-6}} = \ell n(0.4) \qquad [\because \tau = RC]$$

$\Rightarrow R = 2.71 \times 10^6\ \Omega$

128. (c) Time constant for parallel combination $= 2RC$

Time constant for series combination $= \frac{RC}{2}$

In first case :

$$V = V_0 e^{-\frac{t_1}{2RC}} = \frac{V_0}{2} \qquad ....(1)$$

In second case :

In series grouping, equivalent capacitance $= \frac{C}{2}$

$$V = V_0 e^{-\frac{t_2}{(RC/2)}} = \frac{V_0}{2} \qquad ....(2)$$

From (1) and (2)

$$\frac{t_1}{2RC} = \frac{t_2}{(RC/2)}$$

$$\Rightarrow t_2 = \frac{t_1}{4} = \frac{10}{4} = 2.5 \text{ sec.}$$

129. (b) At $t = 0$, no current will flow through L and R_1 as inductor will offer infinite resistance.

$\therefore$ Current through battery, $i = \frac{V}{R_2}$

At $t = \infty$, inductor behave as conducting wire

Effective resistance, $R_{eff} = \frac{R_1 R_2}{R_1 + R_2}$

$\therefore$ Current through battery $= \frac{V}{R_{eff}} = \frac{V(R_1+R_2)}{R_1 R_2}$

130. (c) Growth in current in branch containing L and R_2 when switch is closed is given by

$$i = \frac{E}{R_2}[1 - e^{-R_2 t/L}]$$

$$\Rightarrow \frac{di}{dt} = \frac{E}{R_2}\cdot\frac{R_2}{L}\cdot e^{-R_2 t/L} = \frac{E}{L}e^{-\frac{R_2 t}{L}}$$

Hence, potential drop across L.

$$V_L = \frac{Ldi}{dt} = \left(\frac{E}{L}e^{-R_2 t/L}\right)L$$

$$= Ee^{-R_2 t/L} = 12e^{-\frac{2t}{400 \times 10^{-3}}} = 12e^{-5t}\text{V}$$

131. (c) Across resistor, $I = \frac{V}{R} = \frac{100}{1000} = 0.1\ A$

At resonance,

$$X_L = X_C = \frac{1}{\omega C} = \frac{1}{200 \times 2 \times 10^{-6}} = 2500\ \Omega$$

Voltage across L is

$I X_L = 0.1 \times 2500 = 250$ V

132. (a) Initially, when steady state is achieved,

$$i = \frac{E}{R}$$

Let E is short circuited at $t = 0$. Then

At $t = 0$

Maximum current, $i_0 = \frac{E}{R} = \frac{100}{100} = 1A$

When battery is disconnected

$$i = i_0\ e^{-\frac{R}{L}t}$$

$\Rightarrow i=\frac{E}{R}e^{-\frac{R}{L}t}=1\times e^{\frac{-100\times10^{-3}}{100\times10^{-3}}}=\frac{1}{e}$

133. (a) Phase difference for R–L circuit lies between $\left(0,\ \frac{\pi}{2}\right)$

134. (b) Given, Resistance of circuit, $R=12\,\Omega$
Imedance of circuit, $Z=15\,\Omega$

Power factor $=\cos\phi=\frac{R}{Z}=\frac{12}{15}=\frac{4}{5}=0.8$

135. (a) Current in inductor circuit is given by,

$i=i_0\left(1-e^{-\frac{Rt}{L}}\right)$

$\Rightarrow \frac{i_0}{2}=i_0(1-e^{-\frac{Rt}{L}})\Rightarrow e^{-\frac{Rt}{L}}=\frac{1}{2}$

Taking log on both the sides,

$-\frac{Rt}{L}=\log 1-\log 2\Rightarrow t=\frac{L}{R}\log 2$

$=\frac{300\times10^{-3}}{2}\times 0.69\Rightarrow t=0.1$ sec.

136. (d) For maximum power, $X_L=X_C$, which yields

$C=\frac{1}{(2\pi n)^2 L}=\frac{1}{4\pi^2\times 50\times 50\times 10}$

$\Rightarrow C=0.1\times10^{-5}F=1\mu F$

137. (d) In a series LCR circuit voltage across the inductor and capacitor are in opposite phase
$\therefore$ Net voltage difference across
$LC=50-50=0$

138. (a) Resonant frequency, $f_r=\frac{1}{2\pi\sqrt{LC}}$
For resonant frequency to remain same
$LC=$ constant

$\therefore LC=L'C'\Rightarrow LC=L'\times 2C\Rightarrow L'=\frac{L}{2}$

139. (b) Net impedance of circuit $Z=\sqrt{X_L^2+R^2}$

Power factor, $\cos\phi=\frac{R}{Z}$

$\Rightarrow \cos\phi=\frac{R}{\sqrt{R^2+\omega^2L^2}}$

140. (d) All three inductors are connected in parallel. The equivalent inductance L_p is given by

$\frac{1}{L_p}=\frac{1}{L_1}+\frac{1}{L_2}+\frac{1}{L_3}=\frac{1}{3}+\frac{1}{3}+\frac{1}{3}=\frac{3}{3}=1$

$\therefore L_p=1\,\text{H}$

141. (b) Efficiency of transformer $\eta=\frac{V_sI_s}{V_pI_p}$

$\Rightarrow 0.8=\frac{240 I_s}{4000}\Rightarrow I_s=13.33\text{A}$

142. (45) Voltage $V=230$ V
$P_i=2300\times5$ watt

Efficienfy, $\eta=\frac{P_0}{P_i}$

$P_0=0.9\,P_i$
$P_0=2300\times5\times0.9=230\times I_2$ $[\because P=IV]$
$I_2=45\text{A}$

143. (b) For LC oscillation,
Maximum current, $I=Q_0.\omega$

$=\frac{CV}{\sqrt{LC}}=V\sqrt{\frac{C}{L}}$ $\left[\because \omega=\frac{1}{\sqrt{LC}}\ \&\ Q_0=CV\right]$

$=12\sqrt{\frac{100\times10^{-6}}{6.4\times10^{-3}}}=1.5\text{A}$

144. (c) In transformer,

$\frac{N_P}{N_S}=\frac{V_P}{V_S}=\frac{I_S}{I_P}\ \therefore \frac{V_P}{V_S}=\frac{N_P}{N_S}\Rightarrow \frac{230}{V_S}=\frac{10}{1}$

$\therefore\ V_S=23\text{V}$

Power consumed $=\frac{V_S^2}{R}=\frac{23\times23}{46}=11.5$ W

145. (9) Inductance, $L=75\times10^{-3}$ H
Capacitance, $C=1.2\times10^{-6}$F
Maximum charge to the capacitor $Q_{max}=2.7\times10^{-6}$C
Maximum energy stored in inductor = maximum energy stored in capacitance

$\frac{1}{2}Li_{max}^2=\frac{1}{2}\frac{Q_{max}^2}{C}$; $i_{max}=\sqrt{\frac{1}{LC}}Q_{max}$

$=\frac{2.7\times10^{-6}}{\sqrt{75\times10^{-3}\times1.2\times10^{-6}}}=9\text{mA}$

146. (240) Given,
Voltage across the primary,
$V_p=12\times10^3$ volts
Voltage across the secondary, $V_s=120$ volts
Power across the secondary,
$P_s=60\text{ kW}=V_s\times I_s$

$\Rightarrow I_s=\frac{60\times10^3}{120}=5\times10^2\text{A}$ Resistive load

$R_S=\frac{V_s}{I_s}=\frac{120}{5\times10^2}=24\times10^{-2}=240\times10^{-3}\,\Omega$
$=240\text{ m}\Omega$

147. (a) The resonance frequency of LC oscillations circuit is

$\omega=\frac{1}{\sqrt{L'C'}}\Rightarrow L'\to 2L$

$C'\to 8C$

$\omega=\frac{1}{\sqrt{2L\times8C}}=\frac{1}{4\sqrt{LC}},\ \omega_0=\frac{1}{\sqrt{LC}}$

$\omega=\frac{\omega_0}{4}$ So, $x=\frac{1}{4}$

148. (c) $\dfrac{(8\times10^3)^2}{R_P}=80\times10^3\quad\left[\because P=\dfrac{V^2}{R}\right]$

$\Rightarrow R_P=800\,\Omega$

Similarly, $\dfrac{(160)^2}{R_S}=80\times10^3\Rightarrow R_S=0.32\,\Omega$

149. (a) AC Generator:- It converts mechanical energy into electrical energy.
Galvanometer:- It shows deflection when current passes through it, so it is used to show presence of current in the wire.
Transformer:- It is used to step up or step down the voltage.
Metal detector:- It contain inductor coil and use principle of induction and resonance in AC circuit.

150. (a) Primary voltage $V_P=220$ V

Secondary voltage $V_S=\dfrac{P}{i}=\dfrac{60}{0.11}=545.45$ V

Hence step-up transformer.

151. (440) As we know,

$$\frac{N_P}{N_S}=\frac{V_P}{V_S}$$

Since, $N_S=24$, $V_P=220$ V and $V_S=12$ V

$$\frac{N_P}{24}=\frac{220}{12}\Rightarrow N_P=\frac{220\times24}{12}=440$$

152. (a) When first pulse is applied, the potential across capacitor

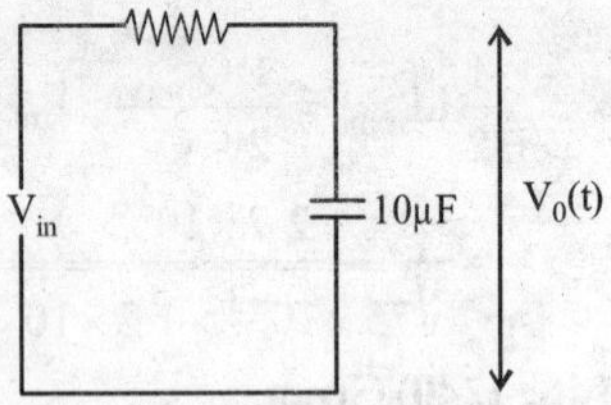

$$V_0(t)=V_{in}\left(1-e^{\frac{-t}{RC}}\right)$$

At $t=5\mu s=5\times10^{-6}\,s$

$$V_0(t)=5\left(1-e^{\frac{-5\times10^{-6}}{10^3\times10\times10^{-9}}}\right)=5(1-e^{-0.5})=2V$$

When no pulse is applied, capacitor will discharge.
Now, $V_{in}=0$ means discharging.

$$V_0(t)=2e^{\frac{-t}{RC}}=2e^{-0.5}=1.21\text{ V}$$

Now for next 5 μs, $V_0(t)=5(1-e^{-t/\tau})+1.21e^{-t/\tau}$

$$V_0(t)=5-3.79e^{\frac{-\tau}{RC}}$$

When $t=5\mu s$

$V_0(t)=5-3.79e^{-0.5}$
$=5-3.79\times2.71^{-0.5}\simeq3V$

Hence, graph (a) correctly depicts.

153. (c) Power output $(V_2I_2)=2.2$ kW

$\therefore V_2=\dfrac{2.2\text{kW}}{(10\text{A})}=220$ volts

$\therefore$ Input voltage for step-down transformer

$$\frac{V_1}{V_2}=\frac{N_1}{N_2}=2$$

$V_{input}=2\times V_{output}=2\times220=440$ V

Also $\dfrac{I_1}{I_2}=\dfrac{N_2}{N_1}$

$\therefore\quad I_1=\dfrac{1}{2}\times10=5$A

154. (b) Efficiency, $\eta=\dfrac{P_{out}}{P_{in}}=\dfrac{V_sI_s}{V_pI_p}$

$\Rightarrow0.9=\dfrac{230\times I_s}{2300\times5}\Rightarrow I_s=0.9\times50=45$A

Output current = 45A

155. (c) When the capacitor is completely charged, the total energy in the LC circuit is with the capacitor and that energy is given by

$$U_{max}=\frac{1}{2}\frac{Q^2}{C}$$

When half energy is with the capacitor in the form of electric field between the plates of the capacitor we get

$$\frac{U_{max}}{2}=\frac{1}{2}\frac{q'^2}{C}$$

Here q' is the charge on the plate of capacitor when energy is shared equally.

$$\therefore\quad\frac{1}{2}\times\frac{1}{2}\frac{Q^2}{C}=\frac{1}{2}\frac{q'^2}{C}\Rightarrow q'=\frac{Q}{\sqrt{2}}$$

156. (a) Laminated core provide less area of cross-section for the current to flow. Because of this, resistance of the core increases and current decreases there by decreasing the energy loss due to eddy current.

157. (b) Number of turns in primary $N_p=140$
Number of turns in secondary $N_s=280$, $I_p=4A$, $I_s=?$

Using transformation ratio for a transformer $\dfrac{I_s}{I_p}=\dfrac{N_p}{N_s}$

$$\Rightarrow\frac{I_s}{4}=\frac{140}{280}\Rightarrow I_s=2\,A$$

8 Electromagnetic Waves

1. **(c)** The radiation pressure is

$$P_{rad} = \frac{2I}{c} = \frac{2P}{4\pi r^2 c}$$

Here, P = 450W, r = 2m, c = 3 × 10^8 m/s

$$\therefore \quad P_{rad} = \frac{2 \times 450}{4\pi \times (2)^2 \times 3 \times 10^8} = \frac{150}{8\pi \times 10^8}$$

$= 5.97 \times 10^{-8} \approx 6 \times 10^{-8}$ pascals

2. **(b)**

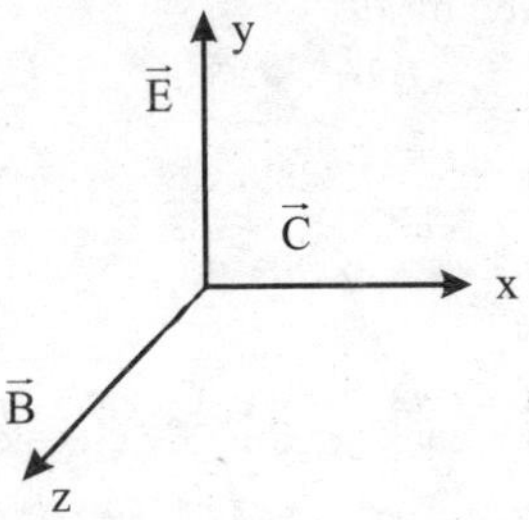

Direction of propogation of an electromagnetic wave is $= \vec{E} \times \vec{B}$

3. **(a)** Given,

$$\vec{B} = \left(\frac{\sqrt{3}}{2}\hat{i} + \frac{1}{2}\hat{j}\right) 30 \sin\left[\omega\left(t - \frac{z}{c}\right)\right]$$

Using the relations

$\vec{E} = \vec{B} \times \vec{c}$ and $E = B_0 c$

Here, direction of $\vec{E}$ is $\left(\frac{\sqrt{3}}{2}(-\hat{j}) + \frac{1}{2}\hat{i}\right)$

$E_0 = 30c$

$$\vec{E} = \left(\frac{1}{2}\hat{i} - \frac{\sqrt{3}}{2}\hat{j}\right) 30c \sin\left[\omega\left(t - \frac{z}{c}\right)\right]$$

4. **(b)** The formula for energy density of an EM wave,

Energy density $= \frac{1}{2} \epsilon_0 E^2$

Energy $= \frac{1}{2} \epsilon_0 E^2 \times$ volume

Here, $(\text{Energy})_1 = (\text{Energy})_2$

$$\frac{1}{2}\epsilon_0 E_1^2 \pi R_1^2 \times L_1 = \frac{1}{2}\epsilon_0 E_2^2 \pi R_2^2 \times L_2$$

$\Rightarrow E_1 R_1 = E_2 R_2$

$\Rightarrow 100 \times R_1 = E_2 \times \frac{R_1}{2} \Rightarrow E_2 = 200\,N/C$

$\therefore E' = 200 \sin(\omega t - kx)\, NC^{-1}$

5. **(d)** For EM wave,

E = BC

9.3 = B × 3 × 10^8

$$B = \frac{9.3}{3 \times 10^8} = 3.1 \times 10^{-8}\,T$$

6. **(100)** The displacement current in a capacitor is given by

$$I_d = \frac{CdV}{dt}$$

$$\Rightarrow \frac{dV}{dt} = \frac{I_d}{C} = \frac{0.25 \times 10^{-3}}{2.5 \times 10^{-6}} = 100\,V/s$$

7. **(d)** The relation between the electric field E and magnetic field B is given by $C = \frac{E_0}{B_0}$

$$\vec{C} = 3\hat{i} + 4\hat{j} \Rightarrow \hat{C} = \frac{3\hat{i} + 4\hat{j}}{5}$$

$$\hat{E} = \frac{4\hat{i} - 3\hat{j}}{5};\ \hat{B} = \hat{K} \times \hat{E} = -\hat{K}$$

$$B_0 = \frac{E_0}{C} = \frac{57}{3 \times 10^8}$$

8. **(1200)** For a capacitor, $I_d = I_c = 6A$

$\therefore$ Displacement current through small area is

$$I'_d = \left(\frac{A_0}{A}\right) I_d = \frac{3.2}{16} \times 6 = 1200\text{mA}$$

9. **(b)** $\lambda = 4\,\text{mm} = 4 \times 10^{-3}\,\text{m}$

$$K = \frac{2\pi}{4 \times 10^{-3}} = \frac{\pi}{2} \times 10^3\,\text{m}^{-1}$$

$$\omega = VXK = 3 \times 10^8 \times \frac{\pi}{2} \times 10^3 = \frac{3\pi}{2} \times 10^{11}$$

$\hat{E} \times \hat{B} = \hat{C}$ and $B_0 = \frac{E_0}{C} = 2 \times 10^{-7}\,T$

$$B_Z = 2 \times 10^{-7} \sin\left[\frac{\pi}{2} \times 10^3 (x - 3 \times 10^8 t)\right] \hat{K} T$$

10. **(d)** $\nu = \frac{C}{\sqrt{\mu_r \epsilon_r}}$

$$\Rightarrow 1.5 \times 10^8 = \frac{3 \times 10^8}{\sqrt{2 \times \epsilon_r}} \Rightarrow 2\epsilon_r = 4 \Rightarrow \epsilon_r = 2$$

11. **(b)** $C = \frac{\omega}{k} = \frac{0.5 \times 10^{11}}{1.5 \times 10^3} = \frac{10^8}{3}$

Using $\Rightarrow E_0 = 3.5 \times 10^{-7} \times 3 \times 10^8 = 105$

$E_z = 105 \sin(1.5 \times 10^3 x + 0.5 \times 10^{11} t)\ Vm^{-1}$

Hence option (b) is correct.

12. **(a)** Intensity is given as:

$$I = \frac{1}{2}\varepsilon_0 E_0^2 c = \frac{1}{2} \times 9 \times 10^{-12} \times (600)^2 \times 3 \times 10^8$$

$$\Rightarrow \frac{9}{2} \times 36 \times 3 = 486\ \text{w/m}^2$$

13. (d) Displacement current, $I_d = C\frac{dv}{dt} = C\frac{d}{dt}(V_0 \sin\omega t)$

$= (CV_0\omega \cos\omega t)$

$\therefore \quad (I_d)_{max} = CV_0\omega$

$= (12 \times 10^{-6})(40)(2\pi \times 4 \times 10^3) \simeq 12A$

14. (d) Given, $\vec{E} = \hat{i}40\cos\omega\left(t - \frac{z}{c}\right)$

$\vec{E}$ is along +x direction; $\vec{v}$ is along + z direction

So direction of $\vec{B}$ will be along +y and magnitude of B will be $\frac{E}{c}$ since, $\vec{c} = \vec{E} \times \vec{B}$

$\therefore \vec{B} = \frac{40}{c}\cos\omega\left(t - \frac{z}{c}\right)\hat{j}$

15. (d) Conduction current $I = \frac{V}{X_c}$

$I = V\omega C = 230 \times 300 \times 200 \times 10^{-12}\ \mu A = 13.8\mu A$

Conduction current and displacement current is same.

16. (b) Wavelength of wave is given as:

$\lambda = \frac{c}{f} = \frac{3 \times 10^8}{60 \times 10^6} = 5$

$\therefore \quad \lambda = 5m$

17. (a) Given,

Amplitude of oscillating electric field,

$E_0 = 50$ V/m $\qquad U_E = \frac{1}{2} \epsilon_0 E_0^2$

Average energy density,

$U_E = \frac{1}{2} \times 8.85 \times 10^{-12} \times (50)^2$

$= 1.106 \times 10^{-8}\ J/m^3$

18. (b) Given $\vec{E} = E_0 \cos(\omega t - kz)\hat{i}$

$\vec{B} = \frac{\hat{k} \times \vec{E}}{\omega} = \frac{(\hat{k} \times \hat{i})kE_0 \cos(\omega t - kz)}{\omega}$

$\vec{B} = \frac{E_0}{C}\cos(\omega t - kz)\hat{j}$

19. (b) Total energy transferred, $E = 6.48 \times 10^5$ J

Momentum, $p = \frac{E}{c} = \frac{6.48 \times 10^5}{3 \times 10^8} = 2.16 \times 10^{-3}$ kgm/s

$(\because c = 3 \times 10^8\ m/s)$

20. (d) Ampere's law, $\oint \vec{B}.d\vec{\ell} = \mu_0 I$

Gauss's law of electrostatics

$\oint \vec{E}.d\vec{a} = \frac{1}{\varepsilon_0}\int \rho dV$

21. (c) Ampere – Maxwell law is $\oint \vec{B}.\vec{dl} = \mu_0 i_c + \mu_0\varepsilon_0 \frac{d\phi_E}{dt}$

Faraday law is $\oint \vec{E}.\vec{dl} = \frac{d\phi_B}{dt}$

Gauss' law for electricity is $\oint \vec{E}.\vec{dA} = \frac{Q}{\varepsilon_0}$

Gauss' law for magnetism is $\oint \vec{B}.\vec{dA} = 0$

22. (c) The speed of electromagnetic wave is

$c = \frac{E}{B} \Rightarrow B = \frac{E}{c}$

$\therefore \ B = \frac{9.6}{3 \times 10^8}$

$B = 3.2 \times 10^{-8}$ T

Direction of B is given by

$\hat{B} = \hat{v} \times \hat{E}$

$= \hat{i} \times \hat{j} = \hat{k}$

$\therefore \ \vec{B} = 3.2 \times 10^{-8}\hat{k}$ T

23. (b) The intensity of wave, $I = \frac{1}{2}\varepsilon_0 E_0^2 \times c$

$I = \frac{1}{2} \times 8.85 \times 10^{-12} \times 4 \times 10^4 \times 3 \times 10^8 = 53.1\ W/m^2$

24. (c) Radiation pressure $= \frac{1}{v} = \frac{1.\mu}{c}$ $\qquad \left[\because v = \frac{c}{\mu}\right]$

$= \frac{6 \times 10^8 \times 3}{3 \times 10^8} = 6\ N/m^2$

25. (c) For time varying condition Maxwell's equation,

$\oint \vec{E}.\vec{dl} = -\frac{d\phi_B}{dt}$

26. (a) Source of time varying magnetic field may be accelerated or retarded charge which produces varying electric and magnetic fields.

27. (b) Direction of propagation of EM wave will be in the direction of $\vec{E} \times \vec{B}$ i.e., negative y-axis direction.

28. (c) Given : $\vec{E} = 20\sin\omega\left(t - \frac{x}{C}\right)\hat{j}$N/C

$\therefore E_0 = 20$

Volume, $V = 5 \times 10^{-4}\ m^3$

Energy density of an EM wave, $u = \frac{1}{2} \epsilon_0 E_0^2$

$\therefore$ Energy stored $= \left(\frac{1}{2} \epsilon_0 E_0^2\right) \times V$

$= \frac{1}{2} \times 8.85 \times 10^{-12} \times (20)^2 \times (5 \times 10^{-4})$

$= 200 \times 8.85 \times 10^{-12} \times 5 \times 10^{-4}$

$= 8.85 \times 10^{-13}$ J

29. (b) Given, electric field, $\vec{E} = 6.6\hat{j}$

Frequency of electromagnetic wave,

$\nu = 20$ MHz

Speed of wave,

$\vec{c} = 3 \times 10^8\hat{i}$

From the relation

Magnetic field $|\vec{B}| = \frac{|\vec{E}|}{c} = \frac{6.6}{3\times10^8} = 2.2\times10^{-8}$ T

Using $\hat{E}\times\hat{B}=\hat{c}$

The magnetic field is along z-axis.

$\therefore\ \vec{B} = 2.2\times10^{-8}\hat{k}T$

30. (c) Let electric field of the wave is given as

$E = E_0 \sin(\omega t - kx)$

Here $\omega = 2\pi f$; $k = \frac{2\pi}{\lambda}$

Now, energy density of electromagnetic wave is

$u = \frac{1}{2}\varepsilon_0 E^2 = \frac{1}{2}\varepsilon_0 E_0^2 \sin^2$

$= \frac{\varepsilon_0 E_0^2}{4}[1-\cos(2\omega t - 2kx)]$

Corresponding frequency will be $f' = \frac{2\omega}{2\pi} = 2f$

31. (b) Given,

Amplitude of magnetic field in an electromagnetic wave, $B = 6\times10^{-7}$T

Relation between magnetic field (B) and electric field (E) of an electromagnetic wave is given by

$\frac{E}{B} = c$

$\Rightarrow E = Bc$

$= 6\times10^{-7}\times3\times10^8$ $\quad(\because c = 3\times10^8$ m/s$)$

$\Rightarrow E = 180$ Vm^{-1}

32. (a) $\frac{\text{Energy density due to electric field }(u_E)}{\text{Energy density due to magnetic field }(u_B)}$

$= \frac{\frac{1}{2}\in_0 E_{rms}^2}{\left(\frac{B_{rms}^2}{2\mu_0}\right)}$

$= \left(\frac{E_{rms}}{B_{rms}}\right)^2 .\mu_0 \in_0 \quad \left[\because C = \frac{E}{B} \,\&\, C^2 = \frac{1}{\mu_0 \in_0}\right]$

$= \frac{C^2}{C^2} = 1$

33. (d) The relation is given by

$U_E = \frac{1}{2}\in_0 E^2$ and $U_B = \frac{B^2}{2\mu_0}$

34. (b) We have $\frac{U_E}{U_T} = \frac{U_E}{U_E+U_B} = \frac{U_E}{2U_E} = \frac{1}{2}$

$\left[\because U_E = U_B = \frac{1}{2}\in_0 E_0^2 = \frac{B_0^2}{2\mu_0}\right]$

35. (5) $\mu = \frac{C}{V} = \frac{C}{0.2C} = 5$

As $V = \frac{1}{\sqrt{\mu_0\mu_r \in_0 \in_r}} \Rightarrow \frac{C}{\mu} = \frac{C}{\sqrt{\mu_r \in_r}}$

$\Rightarrow \mu = \sqrt{\mu_r \in_r} \Rightarrow \mu^2 = \mu_r \in_r$

$\Rightarrow \frac{\mu}{\mu_r} = \frac{\in_r}{\mu} \Rightarrow \frac{5}{1} = \frac{\in_r}{\mu}$

36. (b) Given that

Power, p = 20 mW

$= 20\times10^{-3}$ W

time = 300 ns = 300×10^{-9} s

Momentum $= \frac{\text{Energy}}{c} = \frac{\text{Power}\times\text{time}}{c}$

$= \frac{(20\times10^{-3}\text{ W})(300\times10^{-9}\text{ s})}{3\times10^8\text{ m/s}} = 2\times10^{-17}$ kg m/s

37. (4) Given that power, $P_0 = 24$W

Force $= \int PdA\cos\theta$

$= \frac{2I}{C}\int dA\cos\theta = \frac{2I}{C}\pi R^2 = 2\frac{p_0}{4\pi R^2}\cdot\frac{\pi R^2}{C}$

$= \frac{p_0}{2C} = \frac{24}{2\times3\times10^8} = 4\times10^{-8}$ N

38. (b) Energy emitted per unit time $= \frac{5}{100}\times100 = 5$ W

At a distance 5 m from the source,

Intensity $= \frac{\text{Power}}{4\pi(5)^2} = \frac{5}{4\pi(5)^2}$

The intensity is due to both electric and magnetic field,

Intensity due to electric field $= \frac{1}{2}\times\frac{5}{4\pi\times5^2} = \frac{1}{40\pi}$ W/m^2

39. (a) Statement-I is correct as electromagnetic wave are neutral.

Statement-II is wrong.

as, $E_0 = \frac{1}{\sqrt{\mu_0\varepsilon_0}}B_0$

40. (a) Poynting vector is given by $\vec{S} = \vec{E}\times\vec{H}$

Energy transport = negative z direction $(-\hat{k})$

Electric field = positive y direction $(+\hat{j})$

$(-\hat{k}) = (+\hat{j})\times[\hat{i}]$

Hence according to vector cross product magnetic field should be in positive x direction.

41. (a) By Gauss law of Electrostatics

$\oint \vec{E}.\vec{dA} = \frac{q_{in}}{\in_0}$

By Faraday's law $\oint \vec{E}.\vec{dl} = \frac{-d\phi_B}{dt}$

By Gauss law in magnetism $\oint \vec{B}.\vec{dA} = 0$

By Ampere - Maxwell law

$\oint \vec{B}.\vec{dl} = \mu_0 i_c + \mu_0 \in_0 \frac{d\phi_E}{dt}$

42. (a) For an EM wave

$\vec{B} = \frac{1}{\omega}(\vec{K}\times\vec{E})$

43. **(a)** For an EM wave, we have $c=\frac{\omega}{k}=\frac{E_0}{B_0}$

44. **(84)** Power of bulb,

$P' = 10\% \text{ of } 110 \text{ W} = \frac{10}{100}\times 110 = 11 \text{ W}$

$\text{Intensity(I)} = \frac{\text{Pressure(P)}}{\text{Area(A)}}$

The change in average intensities,

$I_1 - I_2 = \frac{P'}{4\pi r_1^2} - \frac{P'}{4\pi r_2^2}$

$= \frac{11}{4\pi}\left[\frac{1}{1}-\frac{1}{25}\right] = \frac{11}{4\pi}\frac{24}{25} = \frac{264}{\pi}\times 10^{-2} = 84\times 10^{-2} \text{ W/m}^2$

So, $a = 84$

45. **(d)** $\frac{I}{C}\times \text{area} = \text{force}$

$\Rightarrow \frac{I}{C}\times 36\times 10^{-4} = 7.2\times 10^{-9}$

$\Rightarrow I = \frac{7.2\times 10^{-9}\times 3\times 10^8}{36\times 10^{-9}\times 10} = \frac{6\times 10^{-1}}{10^{-3}}$

$\Rightarrow I = 6\times 10^2 \frac{W}{m^2} \Rightarrow I = 0.06\frac{W}{cm^2}$

46. **(b)** In a plane electromagnetic wave, electric field, magnetic field and direction of propagation of wave must be perpendicular to each other.

For EM wave

$\frac{E}{B} = c$

Here, E = amplitude of electric field,

B = amplitude of magnetic field

The energy in electromagnetic wave is equally divided between electric and magnetic fields.

47. **(a)** The magnetic field of a plane electromagnetic wave is given by $\vec{B} = 2\times 10^{-8}\sin(0.5\times 10^3 x + 1.5\times 10^{11} t)\hat{j}T$

The amplitude of the electric field would be

$|E| = c|B| = 3\times 10^8 \times 2\times 10^{-8}$

$\Rightarrow |E| = 6 \text{ Vm}^{-1}$ along x-axis.

48. **(d)** Compare with $B_y = B_0 \sin(kx - \omega t)$

$\omega = 4\times 10^8, K = 5$

Here, $B_0 = 5\times 10^{-6}$

$\omega = 4\times 10^8, K = 5$

$v = \text{Speed of wave} = \frac{4\times 10^8}{5} = 8\times 10^7 \left[\because v = \frac{\omega}{k}\right]$

$E_0 = vB_0 = 40\times 10^1$

$= 4\times 10^2 \text{ V/m}$

49. **(b)** $I_o = \frac{V_o}{X_C} \Rightarrow I_o = V_o\omega C \Rightarrow I_{rms} = V_{rms}\omega C$

$\Rightarrow C = \frac{I_{rms}}{V_{rms}\omega} = \frac{6.9\times 10^{-6}}{230\times 600} = 5\times 10^{-11} = 50 \text{ pF}$

50. **(a)** As $E_y = 540\sin\pi\times 10^4(x - ct) \text{ Vm}^{-1}$

So, $E_0 = 540 \text{ Vm}^{-1}$

$B_0 = \frac{E_0}{C} = \frac{540}{3\times 10^8} = 18\times 10^{-7} \text{ T}$

51. **(b)** We know that,

$B_0 = \frac{E_0}{v} = \frac{300}{\frac{w}{k}} = \frac{300\times 5\pi\times 10^3}{3\pi\times 10^{11}} = 5\times 10^{-6}T$

52. **(43)** $I = \frac{B^2 C}{2\mu_0} \Rightarrow B = \sqrt{\frac{2\mu_0 I}{C}}$

$= \sqrt{\frac{2\times 4\pi\times 10^{-7}\times 0.22}{3\times 10^8}} = 4.29\times 10^{-8}\,T \simeq 43\times 10^{-9}\,\text{T}$

53. **(8)** Displacement current, $I_d = \epsilon_0 \frac{d\phi_E}{dt}$

$I_d = \epsilon_0 \frac{d}{dt}(EA)$

$= \epsilon_0 \frac{d}{dt}\left(\frac{q}{A\epsilon_0}.A\right)\left[\because E = \frac{q}{A\epsilon_0}\right] = \frac{dq}{dt} = C\frac{dV}{dt}$

$= \frac{A\epsilon_0}{d}.\frac{dV}{dt} = \frac{40\times 10^{-4}\times 8.85\times 10^{-12}\times 10^6}{d}$

Now, $I_d = \frac{3.54\times 10^{-8}}{d} \Rightarrow d = \frac{3.54\times 10^{-8}}{4.425\times 10^{-6}}$

$\Rightarrow d = 8\times 10^{-3} \text{ m} \Rightarrow d = 8 \text{ mm}$

54. **(b)** $V = \frac{E_0}{B_0} = \frac{2.25}{1.5\times 10^{-8}} = \frac{2.25\times 10^8}{1.5} = 1.5\times 10^8 \text{ m/s}$

$t = \frac{\text{Total distance}}{\text{velocity of work}} = \frac{6\times 10^3}{1.5\times 10^8} = 4\times 10^{-5} \text{ sec}$

55. **(b)** We have,

$B_0 = \frac{E_0}{c} = \frac{60}{3\times 10^8} = 2\times 10^{-7}\,T$

$\hat{E}\times\hat{B}$ must be in direction of propogation

So, $\hat{B}$ direction is $\hat{K}$

$K = \frac{2\pi}{\lambda} = \frac{2\pi}{8}\times 10^3 = \frac{\pi}{4}\times 10^3 \text{ m}^{-1}$

So, $E_y = 60\sin\left[\frac{\pi}{4}\times 10^3(x - ct)\right]\hat{j} \text{ V/m}$

$B_z = 2\times 10^{-7}\sin\left[\frac{\pi}{4}\times 10^3(x - ct)\right]\hat{k} \text{ T}$

56. **(c)** By Maxwell's law, time varying electric field produce time - varying magnetic field and vice-versa. So statement I is correct and, $V = \frac{c}{\sqrt{\mu_r \epsilon_r}}$. So statement II is incorrect.

57. **(c)** Let $\vec{E} = \vec{E_1} + \vec{E_2}$, where

$\vec{E_1} = 301.6\sin(kz - \omega t)(-\hat{a}_x)$ and

$\vec{E_2} = 452.4\sin(kz - \omega t)(\hat{a}_y)$

Then, $\vec{H}_1 = \frac{301.6}{c\mu_0}\sin(kz-\omega t)(-\hat{a}_y)$

$$\left[\because H_{10} = \frac{E_{10}}{\mu_0 c} \text{ and } \hat{E} = -\hat{i} \text{ and } \hat{E}\times\hat{B} = \hat{K} \Rightarrow \hat{B} = -\hat{j}\right]$$

$$\vec{H}_2 = \frac{452.4}{\mu_0 c}\sin(kz-\omega t)(-\hat{a}_x)$$

$$\left[\because H_{20} = \frac{E_{20}}{\mu_0 c} \text{ and, } \hat{E} = \hat{j} \text{ and } \hat{E}\times\hat{B} = \hat{K} \Rightarrow \hat{B} = -\hat{i}\right]$$

So, $\vec{H} = \vec{H}_1 + \vec{H}_2$

$$= -0.8\sin(kz-\omega t)\hat{a}_y - 1.2\sin(kz-\omega t)\hat{a}_x$$

58. **(b)** Intensity of EM wave is given by

$$I = \frac{1}{2}\epsilon_0 E^2 C$$

$$= \frac{1}{2}\times 8.85\times 10^{-12}\times(56.5)^2\times 3\times 10^8 = 4.24\ Wm^{-2}$$

59. **(a)** We know that $v = \frac{c}{\sqrt{\mu_r \epsilon_r}}$

$$\Rightarrow 2\times 10^8 = \frac{3\times 10^8}{\sqrt{1\times\epsilon_r}} \Rightarrow \sqrt{\epsilon_r} = 1.5 \Rightarrow \epsilon_r = 1.5^2 = 2.25$$

60. **(None)** $c = \frac{1}{\sqrt{\mu_0\epsilon_0}}$, and $V = \frac{1}{\sqrt{\mu\epsilon}} = \frac{1}{\sqrt{\mu_0\mu_r\epsilon_0\epsilon_r}}$

$$= \frac{c}{\sqrt{\mu_r\epsilon_r}}$$

Now, $V = E/B \Rightarrow E = VB = \frac{c}{\sqrt{\mu_r\epsilon_r}}\times 4.5\times 10^{-2}$

$$= \frac{3\times 10^8}{\sqrt{1.61\times 6.44}}\times 4.5\times 10^{-2} = 4192546.584 \simeq 4.2\times 10^6$$

61. **(b)** Intensity of *EM* wave is given as

$$I = \frac{B_0^2 c}{2\mu_0} \Rightarrow \frac{P}{A}\times\eta = \frac{B_0^2 c}{2\mu_0} \Rightarrow \frac{P}{4\pi r^2}\times\eta = \frac{B_0^2 c}{2\mu_0}$$

$$\Rightarrow B_0 = \sqrt{\frac{2\mu_0 P\eta}{c\times 4\pi}}\times\frac{1}{r}$$

$$= \sqrt{\frac{2\times 10^{-7}\times 200\times 3.5\times 10^{-2}}{3\times 10^8}}\times\frac{1}{4} = 1.71\times 10^{-8}\ T$$

62. **(a)** Speed $= \frac{\omega}{k} = \frac{2\times 10^{10}}{200} = 10^8\ \text{m/s}$

Refractive index, $\mu = \frac{c}{v} = \frac{3\times 10^8}{10^8} = 3$

Refractive index, $\mu = \frac{c}{v} = \sqrt{\frac{\mu\epsilon}{\mu_0\epsilon_0}} \Rightarrow \mu = \sqrt{\mu_r\epsilon_r}$

$$\left[\because C = \frac{1}{\sqrt{\mu_0\epsilon_0}} \text{ and } V = \frac{1}{\sqrt{\mu\epsilon}}\right]$$

$$\Rightarrow 3 = \sqrt{\epsilon_r(1)} \Rightarrow \epsilon_r = 9 \quad (\because \mu_r = 1)$$

63. **(500)** $E = 50\sin\left(\omega t - \frac{\omega}{c}.x\right) \therefore E_0 = 50\ \text{NC}^{-1}$

Energy density $= \frac{1}{2}\epsilon_0 E_0^2$

Energy = energy density × volume

$$\Rightarrow \frac{1}{2}\epsilon_0 E_0^2.V = 5.5\times 10^{-12}$$

$$\Rightarrow \frac{1}{2}8.8\times 10^{-12}\times(50)^2\times V = 5.5\times 10^{-12}$$

$$\Rightarrow V = \frac{5.5\times 2}{2500\times 8.8} = 500\ \text{cm}^3$$

64. **(c)** Magnetic force on charge q is $q(\vec{v}\times\vec{B})$

Force on q_1

$$\vec{F}_{q_1} = 4\pi\left[0.5c\hat{i}\times B_o\left(\frac{\hat{i}+\hat{j}}{2}\right)\cos\left(K\frac{\pi}{K}-0\right)\right]$$

Force on q_2

$$\vec{F}_{q_2} = 2\pi\left[0.5c\hat{i}\times B_o\left(\frac{\hat{i}+\hat{j}}{2}\right)\cos\left(K\frac{3\pi}{K}-0\right)\right]$$

Using $\cos\pi = -1$ & $\cos 3\pi = -1$ we get, $\frac{F_1}{F_2} = 2$

65. **(c)** Velocity of wave (v)

$$v = \frac{\omega}{K} = \frac{10\times 10^{10}}{500} = 2\times 10^8 \Rightarrow v = \frac{2c}{3}$$

66. **(2)** Given,

Electric field, E = 6V/m

Relation between electric field and magnetic field for an electromagnetic wave is given by

$$|B| = \frac{|E|}{c} = \frac{6}{3\times 10^8} = 2\times 10^{-8}T$$

$\therefore x = 2$

67. **(354)** Intensity of EM wave

$$I = \frac{1}{2}\epsilon_0 E_0^2.c$$

Radiation pressure for perfectly reflecting surface

$$P = \frac{2I}{c} = \frac{2}{c}\times\frac{1}{2}\epsilon_0 E_0^2 c = \epsilon_0 E_0^2$$

$$= 8.85\times 10^{-12}\times 200^2 = 8.85\times 10^{-8}\times 4 = \frac{354}{10^9}$$

68. **(c)** Given,

Relative permittivity, $\epsilon_r = 81$

Velocity of light in distilled water is

$$v = \frac{c}{\sqrt{\mu_r\epsilon_r}} \Rightarrow v = \frac{3\times 10^8}{\sqrt{1\times 81}} = 3.3\times 10^7\text{m/s}$$

69. **(a)** Average intensity, $I_{avg} = \frac{B_0^2 C}{2\mu_0}$ & $\frac{1}{\mu_0} = \epsilon_0 c^2$

$$\Rightarrow I = \frac{B_0^2}{2}\epsilon_0 c^3$$

$\therefore$ Peak value of magnetic field, $B_0 = \sqrt{\frac{2I}{\epsilon_0 c^3}}$

or, $B_0 = \sqrt{\frac{2\times 0.092}{8.85\times 10^{-12}\times(3\times 10^8)^3}} = 2.77\times 10^{-8}\text{T}$

70. **(d)** The direction of electric field is along +x axis. The magnetic field is acting along +z axis.

The direction of EM wave is given by direction of $\vec{E} \times \vec{B}$.

Unit vector in direction $\vec{E} \times \vec{B} = \frac{\vec{E} \times \vec{B}}{|\vec{E} \times \vec{B}|}$

$= \frac{E_0 \hat{i} \times B_0 \hat{k}}{E_0 B_0 \sin 90°}$ $(\because \vec{A} \times \vec{B} = AB \sin \theta)$

$= \hat{i} \times \hat{k} = -\hat{j}$

71. **(a)** From, $|E_0| = |B_0| C$ and direction will be in $(\vec{B} \times \hat{k})$ direction i.e., in direction $(\hat{k} \times \hat{i})$ i.e., $\hat{j}$ direction

$\Rightarrow |E_0| = 2 \times 10^{-8} \times 3 \times 10^8$ V/m

$\therefore |E_0| = 6$ V/m

So, $\vec{E} = 6$ V/m $\hat{j}$

72. **(25)** Given : $F = 2.5 \times 10^{-6}$ N, $A = 30$ cm², $I = ?$

Using, $P = \frac{I}{c}$ or $\frac{F}{A} = \frac{I}{c}$

$\Rightarrow F = \frac{I}{c} A \Rightarrow I = \frac{F \times c}{A}$

$\Rightarrow I = \frac{2.5 \times 10^{-6} \times 3 \times 10^8}{30} = 25$ W cm^{-2}

73. **(3)** Electric field intensity,

$I = \frac{P}{A} = \frac{P}{4\pi r^2} = \frac{1}{2} c \varepsilon_0 E^2$

$\therefore E \propto \frac{\sqrt{P}}{r}$ or $E_1 = K\frac{\sqrt{P_1}}{r_1}$ and $E_2 = K\frac{\sqrt{P_2}}{r_2}$

$[r_1 = r_2$ given$]$

$\therefore E_2 = E_1 \sqrt{\frac{P_2}{P_1}} = E\sqrt{\frac{60}{100}} = E\sqrt{\frac{3}{5}}$

74. **(6)** Conduction current,

$i_c = \frac{V}{R} = \frac{V_0}{R} \sin \omega t = \frac{AV_0 \sin(2\pi f t)}{\rho d}$ $\left(\because R = \rho \frac{d}{A}\right)$

Displacement current,

$i_d = \frac{dq}{dt} = \frac{cdV}{dt} = \frac{k \varepsilon_0 A}{d} V_0 \omega \cos \omega t$

$\therefore \frac{j_c}{j_d} = \frac{i_{c/A}}{i_{d/A}} = \frac{\frac{AV_0 \sin(2\pi f t)}{\rho d}}{\frac{\varepsilon_r \varepsilon_0 AV_0 \omega \cos(2\pi f t)}{d}} = \frac{\tan(2\pi f t)}{\rho(2\pi f)\varepsilon_r \varepsilon_0}$

$= \frac{\tan\left(2\pi \times 9 \times 10^2 \times \frac{1}{800}\right)}{0.25 \times 2\pi \times 9 \times 10^2 \times \frac{1}{800} \times \frac{1}{4\pi \times 9 \times 10^9} \times 80}$

Solving we get,

$\frac{j_c}{j_d} = 10^6$ $\quad \therefore j_c = 10^6 j_d \quad \therefore x = 6.$

75. **(b)** The energy in electromagnetic waves is divided equally between the electric and magnetic fields.

$\therefore U_e = U_m$

Energy density of electric field, $U_e = \frac{1}{2} \varepsilon_0 E_0^2$

Energy density of magnetic field, $U_B = \frac{1}{2} \frac{B_0^2}{\mu_0}$

Now, $\frac{B_0^2}{2\mu_0} = \frac{E_0^2}{2C^2 \mu_0} = \frac{E_0^2}{2\mu_0} \times \mu_0 \varepsilon_0 = \frac{1}{2} \varepsilon_0 E_0^2$

76. **(a)** $\vec{E} \times \vec{B}$ is along positive y-direction.

$\frac{|\vec{E}|}{|\vec{B}|} = c = 3 \times 10^8$

Given, magnetic field, $\vec{B} = 8 \times 10^{-8} \hat{z} T$

In electromagnetic wave, amplitude of electric field (E) and magnetic field (B) are related by the relation, $\frac{E}{B} = c$

$\Rightarrow E = cB = 3 \times 10^8 \times 8 \times 10^{-8} = 24$ V/m.

Since, electromagnetic wave is of transverse nature. Hence if magnetic field is along z-axis, the propagation is along y-axis. Using vector algebra, electric field must be along negative x-axis.

$\therefore \vec{E} = -24\hat{x}$ V/m.

77. **(15)** Given : Frequency of wave $\nu = 5$ GHz $= 5 \times 10^9$ Hz

Relative permittivity of the medium, $\varepsilon_r = 2$

and relative permeability of the medium, $\mu_r = 2$

Since speed of light in a medium is given by,

$v = \frac{1}{\sqrt{\mu\varepsilon}} = \frac{1}{\sqrt{\mu_r \mu_0 \varepsilon_r \varepsilon_0}} = \frac{c}{\sqrt{\mu_r \varepsilon_r}}$

where c is speed of light is vacuum.

$\therefore v = \frac{3 \times 10^8}{\sqrt{4}} = \frac{30 \times 10^7}{2}$ m/s $= 15 \times 10^7$ m/s

78. **(667)** Wavelength of EM wave λ_v in vacuum

$= \frac{c}{f} = \frac{3 \times 10^8}{3 \times 10^9} = 0.1$ m

$\therefore$ Wavelength of EM wave λ_m in medium $= \frac{0.1}{\mu}$

Where μ is refractive index given by,

$\mu = \sqrt{\mu_r \varepsilon_r}$

Assuming material to be non-magnetic,

$\mu_r = 1$

$\therefore \mu = \sqrt{1 \times 2.25} = 1.5$

$\lambda_m = \frac{0.1}{1.5} = \frac{1}{15}$ m $= 6.67$ cm $= 667 \times 10^{-2}$ cm.

79. **(2)** $I = \frac{P}{4\pi r^2} = \frac{1}{2} \varepsilon_0 E_0^2 \cdot c \Rightarrow E_0 = \left(\frac{P}{2\pi \varepsilon_0 c r^2}\right)^{\frac{1}{2}}$

$= \left(\frac{8 \times \mu_0 c^2}{2\pi c r^2}\right)^{\frac{1}{2}} = \left(\frac{4 \times \mu_0 c}{\pi \times 10^2}\right)^{\frac{1}{2}} = \frac{2}{10}\sqrt{\frac{\mu c}{\pi}} \frac{N}{C}$

$\therefore x = 2.$ $\left[\because \varepsilon_0 = \frac{1}{\mu_0 c^2}\right]$

80. **(137)** $I=\frac{\eta P}{4\pi r^2}=\frac{1}{2}\varepsilon_0 E_0^2\cdot c$

$\Rightarrow \frac{0.0125\times 1000}{4\times 3.14\times 2^2}=\frac{1}{2}\times 8.85\times 10^{-12}\times 3\times 10^8 E_0^2$

$\Rightarrow E_0 = 13.69$ V/m $= 136.89\times 10^{-1}$ V/m

$\therefore$ $x = 137$ (Rounding off to nearest integer)

81. **(a)** Relation between electric field E_0 and magnetic field B_0 of an electromagnetic wave is given by

$c=\frac{E_0}{B_0}$ (Here, c = Speed of light)

$\Rightarrow E_0 = B_0\times c = 1.2\times 10^{-7}\times 3\times 10^8 = 36$

As the wave is propagating along –ve x-direction, magnetic field is along z-direction

and $(\hat{E}\times\hat{B})\parallel\hat{c}$

$\therefore \vec{E}$ should be along –ve y-direction. So, electric field

$=[-36\sin(0.5\times 10^3 x+1.5\times 10^{11}t)\hat{j}]\frac{V}{m}$

82. **(c)** In electromagnetic wave, $\frac{E_0}{B_0}=c$

$\therefore$ Maximum value of magnetic field, $B_0=\frac{E_0}{c}$

$F_{max}=qVB_{max}\sin 90^\circ=\frac{qV_0E_0}{c}$

(Given $V_0 = 0.1\,c$ and $E_0 = 30$)

$=\frac{1.6\times 10^{-19}\times 0.1\times 3\times 10^8\times 30}{3\times 10^8}=4.8\times 10^{-19}$N

83. **(a)** $\vec{E}=E_0(\hat{x}+\hat{y})\sin(kz-\omega t)$

Direction of propagation of EM wave $=+\hat{k}$

Unit vector in the direction of electric field, $\hat{E}=\frac{\hat{i}+\hat{j}}{\sqrt{2}}$

The direction of electromagnetic wave is perpendicular to both electric and magnetic field.

$\therefore \hat{k}=\hat{E}\times\hat{B}\Rightarrow\hat{k}=\left(\frac{\hat{i}+\hat{j}}{\sqrt{2}}\right)\times\hat{B}\Rightarrow\hat{B}=\frac{-\hat{i}+\hat{j}}{\sqrt{2}}$

$\therefore \vec{B}=\frac{E_0}{c}(-\hat{x}+\hat{y})\sin(kz-\omega t)$

84. **(d)** Given : $\vec{B}=3\times 10^{-8}\sin[200\pi(y+ct)]\hat{i}T$

$\therefore B_0=3\times 10^{-8}$

$E_0=cB_0\Rightarrow E_0=3\times 10^8\times 3\times 10^{-8}=9$ V/m

Direction of wave propagation

$(\bar{E}\times\bar{B})\parallel\bar{c}$

$\hat{B}=\hat{i}$ and $\hat{c}=-\hat{j}$ $\qquad\therefore\hat{E}=-\hat{k}$

$\therefore\vec{E}=E_0\sin[200\pi(y+ct)](-\hat{k})$ V/m

or, $\bar{E}=-9\sin[200\pi(y+ct)]\hat{k}$ V/m

85. **(c)** As $B_0=\frac{E_0}{c}=\frac{E_0}{1/\sqrt{\mu_0\varepsilon_0}}=E_0\sqrt{\mu_0\varepsilon_0}$

As the electromagnetic wave is propagating along x direction and electric field is along y direction.

$\therefore\hat{E}\times\hat{B}\parallel\hat{c}$ (Here, $\hat{c}$ = direction of propagation of wave)

$\therefore\vec{B}$ should be in $\hat{k}$ direction.

$\therefore$ B $= E_0\sqrt{\mu_0\varepsilon_0}\cos(\omega t-kx)\,\hat{k}$

At t = 0, B $= E_0\sqrt{\mu_0\varepsilon_0}\cos(kx)\,\hat{k}$

86. **(b)** Energy density $=\frac{1}{2}\frac{B^2}{\mu_0}$

$\Rightarrow B=\sqrt{2\times\mu_0\times\text{Energy density}}$

$\therefore B=\sqrt{2\times 4\pi\times 10^{-7}\times 1.02\times 10^{-8}}=160\times 10^{-9}=160\,\text{nT}$

87. **(a)** Electromagnetic wave will propagate perpendicular to the direction of Electric and Magnetic fields

$\hat{c}=\hat{E}\times\hat{B}$

Here unit vector $\hat{c}$ is perpendicular to both $\hat{E}$ and $\hat{B}$

Given, $\vec{E}=\hat{k}$, $\vec{B}=2\hat{i}-2\hat{j}$

$\therefore\ \hat{c}=\hat{E}\times\hat{B}=\frac{1}{\sqrt{2}}\begin{vmatrix}\hat{i}&\hat{j}&\hat{k}\\0&0&1\\1&-1&0\end{vmatrix}=\frac{\hat{i}+\hat{j}}{\sqrt{2}}\Rightarrow\hat{c}=\frac{\hat{i}+\hat{j}}{\sqrt{2}}$

88. **(d)** Given: $\vec{E}_1=E_0\hat{j}\cos(\omega t-kx)$

i.e., Travelling in +*ve* x-direction $\vec{E}\times\vec{B}$ should be in x-direction

$\therefore$ $\vec{B}$ is in $\hat{K}$

$\therefore\ \vec{B}_1=\frac{E_0}{c}\cos(\omega t-kx)\hat{k}\ \left(\because B_0=\frac{E_0}{c}\right)$

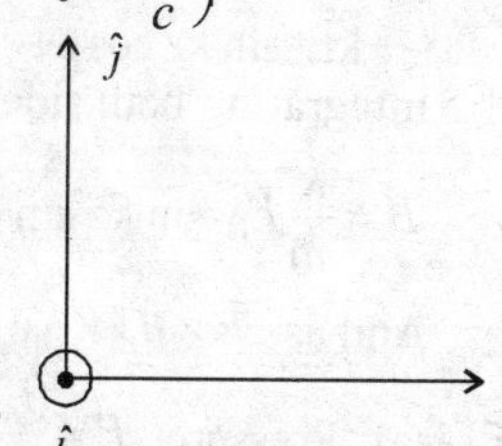

$\vec{E}_2=E_0\hat{k}\cos(\omega t-ky)$

$\vec{B}_2=\frac{E_0}{c}\hat{i}\cos(\omega t-ky)$

$\therefore$ Travelling in +*ve* y-axis

$\vec{E}\times\vec{B}$ should be in y-axis

$\therefore$ Net force $\vec{F}=q\vec{E}+q(\vec{v}\times\vec{B})$

$q(\vec{E}_1+\vec{E}_2)+q(0.8c\hat{j}\times(\vec{B}_1+\vec{B}_2))$

If t = 0 and x = y = 0

$\vec{E}_1=E_0\hat{j}\qquad\vec{E}_2=E_0\hat{k}$

$\vec{B}_1=\frac{E_0}{c}\hat{k}\qquad\vec{B}_2=\frac{E_0}{c}\hat{i}$

$\therefore\vec{F}_{net}=qE_0(\hat{j}+\hat{k})+q\times 0.8c\times\frac{E_0}{c}\hat{j}\times(\hat{k}+\hat{i})$

$=qE_0(\hat{j}+\hat{k})+0.8qE_0(\hat{i}-\hat{k})=qE_0(0.8\hat{i}+\hat{j}+0.2\hat{k})$

89. **(d)** Amplitude of electric field (E) and Magnetic field (B) of an electromagnetic wave are related by the relation

$\frac{E}{B}=c\Rightarrow E=Bc$

$\Rightarrow E = 5\times 10^{-8}\times 3\times 10^8 = 15\,N/C \Rightarrow\vec{E}=15\hat{i}\,V/m$

90. **(b)** Given, $\vec{B} = 3\times10^{-8}\sin(1.6\times10^3 x + 48\times10^{10} t)$

Using, $E_0 = B_0 \times c = 3\times10^{-8}\times3\times10^8 = 9\ V/m$

$\therefore$ Electric field, $\vec{E} = 9\sin(1.6\times10^3 x + 48\times10^{10} t)\hat{k}\ V/m$

91. **(c)** At $t = 0, z = \dfrac{\pi}{k}$

$\therefore \quad \vec{E} = \dfrac{E_0}{\sqrt{2}}(\hat{i}+\hat{j})\cos[\pi] = -\dfrac{E_0}{\sqrt{2}}(\hat{i}+\hat{j})$

$\vec{F}_E = q\vec{E}$

Force due to electric field will be in the direction $\dfrac{-(\hat{i}+\hat{j})}{\sqrt{2}}$

Force due to magnetic field is in direction $q(\vec{v}\times\vec{B})$ and $\vec{v}$ anti $\|^r$ $\hat{k}$. Therefore, it is parallel to $\vec{E}$.

$\Rightarrow \quad \vec{F}_{net} = \vec{F}_E + \vec{F}_B$ is antiparallel to $\dfrac{\hat{i}+\hat{j}}{\sqrt{2}}$

92. **(c)** $\vec{k} = 6\hat{j} - 8\hat{k}$

Direction of propagation, $\vec{S} = -\hat{k} = \left(\dfrac{-3\hat{j}+4\hat{k}}{5}\right)$

93. **(b)** $B_0 = \dfrac{E_0}{c} = \dfrac{60}{3\times10^8} = 20\times10^{-8}\,T = 2\times10^{-7}\,T$

$K = \dfrac{\omega}{V} = \dfrac{2\pi f}{V} = \dfrac{2\pi\times23.9\times10^9}{3\times10^8} = 500$

Therefore, $\vec{B} = B_0\sin(kz - \omega t)$

$= 2\times10^{-7}\sin(0.5\times10^3 z - 1.5\times10^{11} t)\hat{i}$

94. **(a)** $\nabla\times E = \dfrac{-\partial B}{\partial t} \quad \dfrac{\partial E}{\partial Z} = \dfrac{-\partial B}{\partial t} \quad \dfrac{\partial B}{\partial t} = \dfrac{-\partial E}{\partial z}$

$= +kE_0 \sin kz \cos\omega t$

Integrating both sides of given equation w.r.t. t, we get

$B = \dfrac{k}{\omega}E_0 \sin kz \sin\omega t = \dfrac{E_0}{C}\sin kz \sin\omega t$

And as, $\vec{E}\times\vec{B} \,\|\, \vec{V}$ so, $\vec{B}$ will be in $\hat{j}$ direction

95. **(c)** Pressure, $P = \dfrac{I}{c} \Rightarrow \dfrac{F}{A} = \dfrac{I}{c}$

$\Rightarrow F = \dfrac{IA}{c} = \dfrac{\Delta p}{\Delta t} \Rightarrow \Delta p = \dfrac{I}{c}A\Delta t$

$= \dfrac{(25\times25)\times10^4\times10^{-4}\times40\times60}{3\times10^8}$ N-s $= 5\times10^{-3}$ N-s

96. **(a)** $B_0 = \sqrt{B_0^2 + B_1^2} = \sqrt{30^2 + 2^2}\times10^{-6} = 30\times10^{-6}\,T$

$\therefore E_0 = cB = 3\times10^8\times30\times10^{-6} = 9\times10^3$ V/m

$E_{r.ms} = \dfrac{E_0}{\sqrt{2}} = \dfrac{9}{\sqrt{2}}\times10^3$ V/m

Force on the charge,

$F = EQ = \dfrac{9}{\sqrt{2}}\times10^3\times10^{-4} \simeq 0.64N$

97. **(b)** As we know,

$|\vec{B}| = \dfrac{|\vec{E}|}{C} = \dfrac{6.3}{3\times10^8} = 2.1\times10^{-8}\,T$

and $\hat{E}\times\hat{B} = \hat{c}$

$\hat{j}\times\hat{B} = \hat{i}$ [$\because$ EM wave travels along +(ve) x-direction.]

$\therefore \hat{B} = \hat{k}$ or $\vec{B} = 2.1\times10^{-8}\hat{k}T$

98. **(b)** $F = (1+r)\dfrac{IA}{C} = \dfrac{(1+0.25)\times50\times1}{3\times10^8} \simeq 20\times10^{-8}\,N$

99. **(a)** The relation between amplitudes of electric and magnetic field in free space is given by

$B_0 = \dfrac{E_0}{c} = \dfrac{6}{3\times10^8} = 2\times10^{-8}\,T$

Propagation direction $= \hat{E}\times\hat{B}$

$\hat{i} = \hat{j}\times\hat{B} \Rightarrow \hat{B} = \hat{k}$

$\therefore$ The magnetic field component will be along z direction.

100. **(d)** $I = \dfrac{B_0^2}{2\mu_0}\cdot c \Rightarrow \dfrac{B_0^2}{2} = \dfrac{I\mu_0}{c}$

$\Rightarrow B_{rms} = \sqrt{\dfrac{I\mu_0}{c}} = \sqrt{\dfrac{10^8\times4\pi\times10^{-7}}{3\times10^8}} \simeq 6\times10^{-4}\,T$

Which is closest to 10^{-4}.

101. **(c)** The speed of electromagnetic wave in free space is given by

$c = \dfrac{1}{\sqrt{\mu_0 \in_0}}$ ...(i)

In medium, $v = \dfrac{1}{\sqrt{k \in_0 \mu_0}}$ ...(ii)

Dividing equation (i) by (ii), we get

$\therefore \quad \dfrac{c}{v} = \sqrt{k} = n$

$\dfrac{1}{2}\in_0 E_0^2 c = \text{intensity} = \dfrac{1}{2}\in_0 kE^2 v$

$\therefore E_0^2 c = kE^2 v \Rightarrow \dfrac{E_0^2}{E^2} = \dfrac{kV}{c} = \dfrac{n^2}{n} \Rightarrow \dfrac{E_0}{E} = \sqrt{n}$

similarly

$\dfrac{B_0^2 c}{2\mu_0} = \dfrac{B^2 v}{2\mu_0} \Rightarrow \dfrac{B_0}{B} = \dfrac{1}{\sqrt{n}}$

102. **(d)** EM wave intensity

$\Rightarrow I = \dfrac{\text{Power}}{\text{Area}} = \dfrac{1}{2}\varepsilon_0 E_0^2 c$

[where E_0 = maximum electric field]

$\Rightarrow \dfrac{27\times10^{-3}}{10\times10^{-6}} = \dfrac{1}{2}\times9\times10^{-12}\times E_0^2\times3\times10^8$

$\Rightarrow E_0 = \sqrt{2}\times10^3\,kV/m = 1.4\,kV/m$

103. **(b)** Using, formula $E_0 = B_0 \times C$

$= 100\times10^{-6}\times3\times10^8 = 3\times10^4$ N/C

Here we assumed that

$B_0 = 100\times10^{-6}$ is in tesla (T) units

104. **(c)** Velocity of EM wave is given by $v = \dfrac{1}{\sqrt{\mu\in}}$

Velocity in air $= \dfrac{\omega}{k} = c$

Velocity in medium = $\frac{c}{2}$

Here, $\mu_1 = \mu_2 = 1$ as medium is non-magnetic

$$\therefore \frac{\frac{1}{\sqrt{\epsilon_{r_1}}}}{\frac{1}{\sqrt{\epsilon_{r_2}}}} = \frac{c}{\left(\frac{c}{2}\right)} = 2 \Rightarrow \frac{\epsilon_{r_1}}{\epsilon_{r_2}} = \frac{1}{4}$$

105. (a) Speed of EM wave in force space (c) $= \frac{E_0}{B_0}$

or $\vec{E} = cB_0 \sin(kx + \omega t)\hat{k}$ V/m

106. (b) As we know, $\vec{E}.\vec{B} = 0 \because [\vec{E} \perp \vec{B}]$

and $\vec{E} \times \vec{B}$ should be along Z-direction

As $(-2\hat{i} - 3\hat{j}) \times (3\hat{i} - 2\hat{j}) = 13\hat{k}$

Hence, option (b) is the correct answer.

107. (c) $E_0 = cB_0$ and $c = \frac{1}{\sqrt{\mu_0 \varepsilon_0}}$

Electric energy density $= \frac{1}{2}\varepsilon_0 E_0{}^2 = \mu_E$

Magnetic energy density $= \frac{1}{2}\frac{Bo^2}{\mu_0} = \mu_B$

Thus, $\mu_E = \mu_B$

Energy is equally divided between electric and magnetic field.

108. (c) Average energy density $u_E = \frac{1}{4}\varepsilon_0 E^2$

$$\Rightarrow u_E = \frac{1}{4} \times 8.8 \times 10^{-12} \times (4)^2 = 35.2 \times 10^{-12}\ \text{J/m}^3$$

109. (b) From question,

$B_0 = 20\,nT = 20 \times 10^{-9} T$

($\because$ velocity of light in vacuum $c = 3 \times 10^8$ ms^{-1})

$\vec{E}_0 = \vec{B}_0 \times \vec{c}$

$|\vec{E}_0| = |\vec{B}| \cdot |\vec{c}| = 20 \times 10^{-9} \times 3 \times 10^8 = 6$ V/m.

110. (d) Electromagnetic waves do not required any medium to propagate. They can travel in vacuum. They are transverse in nature like light. They carry both energy and momentum. A changing electric field produces a changing magnetic field and vice-versa. Which gives rise to a transverse wave known as electromagnetic wave.

111. (b) Direction of polarization is always $\|^r$ to $\vec{E}$

and $\vec{k} \parallel \vec{E} \times \vec{B}$

112. (c) In an electromagnetic wave, electric field and magnetic field are perpendicular to the direction of propagation of wave. The vector equation for the electric field is $\vec{E} = E_0 \cos\left(\omega t - \frac{2\pi}{\lambda} y\right)\hat{z}$

113. (a) Frequency remains unchanged during refraction

Velocity of EM wave in vacuum

$$V_{\text{vacuum}} = \frac{1}{\sqrt{\mu_0 \epsilon_0}} = c$$

$$v_{\text{med}} = \frac{1}{\sqrt{\mu_0 \epsilon_0 \times 4}} = \frac{c}{2}$$

$$\frac{\lambda_{\text{med}}}{\lambda_{\text{vacuum}}} = \frac{v_{\text{med}}}{v_{\text{vacuum}}} = \frac{c/2}{c} = \frac{1}{2}$$

$\therefore$ Wavelength is halved and frequency remains unchanged

114. (a) The phenomenon of polarisation is shown only by transverse waves. The vibration of electromagnetic wave are restricted through polarization in a direction perpendicular to wave propagation.

115. (c) The ascending order of wavelength of EMW is

γ-rays < X-rays < UV rays < Visible rays < IR rays < Micro waves < Radio waves

$\therefore \lambda_4 < \lambda_2 < \lambda_3 < \lambda_1$

116. (b) Infra-red > Ultraviolet > X-ray > γ-ray
$\longleftarrow$ wavelength (λ)

117. (c) Order of wavelength is as follows

$\lambda_1 < \lambda_2 < \lambda_3 < \lambda_4$

118. (b) Increasing order of wavelength

X-ray	1 nm to 10^{-3} nm
Ultra Violet	400 nm to 1 nm
Intra red	1 mm to 700 nm
Micro wave	0.1 m to 1 mm

119. (a) Optical communication is performed with EM waves having wavelength less than that of microwave.

As frequency of infrared is more than frequency of microwave.

$\therefore$ Infrared EM waves are more energetic than microwave (used in Radar)

120. (b) X rays are emitted when target metal is bombarded with high energy electron.

121. (c) X-rays is used to detect inner shell of electrons and γ-rays is used for radioactive decay and radio waves used for rapid acceleration & deceleration or electron.

122. (c)

Microwave	$\rightarrow$	Aircraft navigation
UV rays	$\rightarrow$	Lasik eye surgery
Infra-red	$\rightarrow$	Physiotherapy
X-rays	$\rightarrow$	Treatment of cancer

123. (b) (a) UV rays – used in water purification.

(b) x-rays are used for diagnosing fracture.

(c) Microwaves are used for mobile and radar communication.

(d) Infrared waves show less scattering effect therefore used in foggy days.

124. (a) Microwave is used in Radar system

X-rays is used in studying crystal structure due to high penetrating power.

125. (b) The wave having highest energy, has lowest wavelength as,

$E_\gamma > E_{X\text{-rays}} > E_{visible} > E_{microwave}$

$\Rightarrow \lambda_\gamma < \lambda_{X\text{-rays}} < \lambda_{visible} < \lambda_{microwave}$

126. **(d)** Energy sequence of radiations is

$$E_{\gamma\text{-Rays}} > E_{\text{X-Rays}} > E_{\text{microwave}} > E_{\text{AM Radiowaves}}$$

$$\therefore \lambda_{\gamma\text{-Rays}} < \lambda_{\text{X-Rays}} < \lambda_{\text{microwave}} < \lambda_{\text{AM Radiowaves}}$$

From the above sequence, we have

(a) Microwave $\rightarrow 10^{-3}$ m (iv)

(b) Gamma Rays $\rightarrow 10^{-15}$ m (ii)

(c) AM Radio wave $\rightarrow$ 100 m (i)

(d) X-Rays $\rightarrow 10^{-10}$ m (iii)

127. **(b)** The orderly arrangement of different parts of EM wave in decreasing order of wavelength is as follows:

$$\lambda_{\text{radiowaves}} > \lambda_{\text{microwaves}} > \lambda_{\text{visible}} > \lambda_{\text{X-rays}}$$

128. **(c)** Optical Fibre Communication – Infrared Light

Radar – Radio Waves

Sonar – Ultrasound

Mobile Phones – Microwaves

129. **(c)** E, Decreases $\longrightarrow$

γ-rays X-rays uv-rays Visible rays (VIBGYOR) IR rays Microwaves Radio waves

Radio wave < yellow light < blue light < X-rays (Increasing order of energy)

130. **(c)** Microwave oven acts on the principle of giving vibrational energy to water molecules.

131. **(d)**

(1) Infrared rays are used to treat muscular strain because these are heat rays.

(2) Radio waves are used for broadcasting because these waves have very long wavelength ranging from few centimeters to few hundred kilometers.

(3) X-rays are used to detect fracture of bones because they have high penetrating power but they can't penetrate through denser medium like dones.

(4) Ultraviolet rays are absorbed by ozone of the atmosphere.

132. **(c)** Gamma rays < X-rays < Ultra violet < Visible rays < Infrared rays < Microwaves < Radio waves.

133. **(d)** Vibration of atoms and molecules 700 nm to 1 mm Radioactive decay of the nucleus < 10^{-3} nm. Magnetron value 1 mm to 0.1 m.

134. **(a)** $E = \dfrac{hc}{\lambda} \Rightarrow \lambda = \dfrac{hc}{E}$

$$\Rightarrow \lambda = \frac{6.6\times10^{-34}\times3\times10^{8}}{11\times1000\times1.6\times10^{-19}} = 12.4\text{ Å}$$

Which belongs to X-ray region.

135. **(a)** Frequency range of γ-ray,
$b = 10^{18} - 10^{23}$ Hz

Frequency range of X-ray,
$a = 10^{16} - 10^{20}$ Hz

Frequency range of ultraviolet ray,
$c = 10^{15} - 10^{17}$ Hz

$\therefore a < b;\ b > c$

Ray Optics and Optical Instruments

9

1. **(c)** Applying magnification formula,

$$m = -\frac{v}{u} = -\left(\frac{v}{-u}\right) = \frac{v}{u}$$

$$\Rightarrow \frac{1}{4} = \frac{v}{u} \Rightarrow u = 4v$$

$\Rightarrow v + u = 40$ (Given)

$\Rightarrow 5v = 40$

$\therefore v = 8$ cm, $u = 32$ cm

Applying mirror formula,

$$\frac{1}{v} + \frac{1}{u} = \frac{1}{f}$$

$$\Rightarrow \frac{1}{8} - \frac{1}{32} = \frac{1}{f} \Rightarrow \frac{4-1}{32} = \frac{1}{f} \Rightarrow f = \frac{32}{3} = 10.7 \text{ cm}$$

2. **(b)**

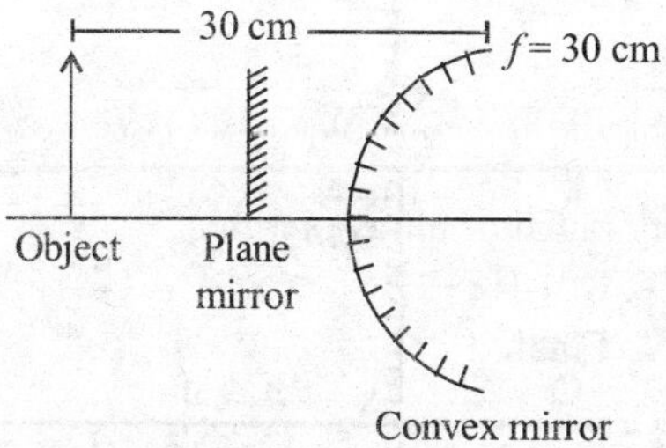

$u = -30$ cm, $f = 30$ cm

For Convex mirror

$$\frac{1}{v} + \frac{1}{u} = \frac{1}{f} \Rightarrow \frac{1}{v} - \frac{1}{30} = \frac{1}{30}$$

$$\Rightarrow \frac{1}{v} = \frac{2}{30} = \frac{1}{15} \Rightarrow v = 15 \text{ cm}$$

Image formed 15 cm in positive direction from the convex mirror.

So, distance between object and image $= 30 + 15 = 45$ cm.

Image formed by convex mirror is at 45 cm from object so plane mirror should be placed midway at 22.5 cm from object so that both of their images may coinside.

Therefore distance between both mirrors $= 30 - 22.5 = 7.5$ cm

Correct Answer : Option (b).

3. **(a)** Given, $u = -40$ cm, $m_1 = \frac{1}{2}$, $m_2 = \frac{1}{3}$

$$m_1 = \frac{1}{2} = \frac{f}{f-u} \Rightarrow \frac{1}{2} = \frac{f}{f-(-40)}$$

$$\Rightarrow f + 40 = 2f \Rightarrow f = 40 \text{ cm}$$

$$\text{Now, } m_2 = \frac{1}{3} = \frac{40}{40-u}$$

$$\Rightarrow 40 - u = 120 \Rightarrow u = -80$$

Hence, the object is to be shifted by 40 cm away from the mirror.

4. **(45)** Using magnification formula,

$$m = -\frac{1}{3} \quad m = -\frac{V}{U} = -\frac{1}{3}$$

$$\Rightarrow \frac{-V}{-U} = \frac{-1}{3} \Rightarrow V = \frac{U}{3}$$

Distance between object and image :

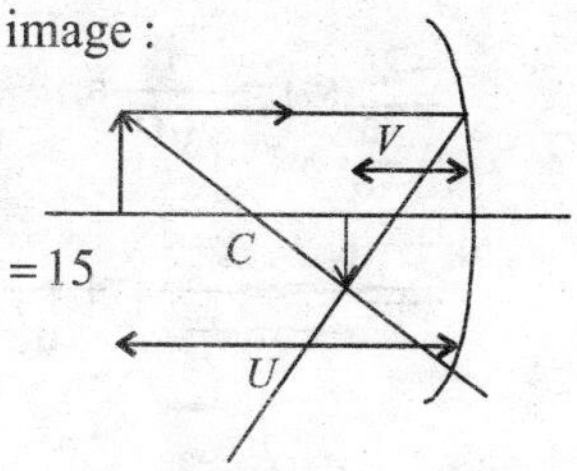

$U - V = 30$

$$\Rightarrow U - \frac{U}{3} = 30 \Rightarrow U = 45, V = 15$$

$$\Rightarrow \frac{1}{f} = \frac{1}{V} + \frac{1}{U} = -\frac{1}{15} - \frac{1}{45}$$

$$\Rightarrow f = \frac{45}{4}$$

$\therefore x = 45.$

5. **(a)** In case of concave mirror

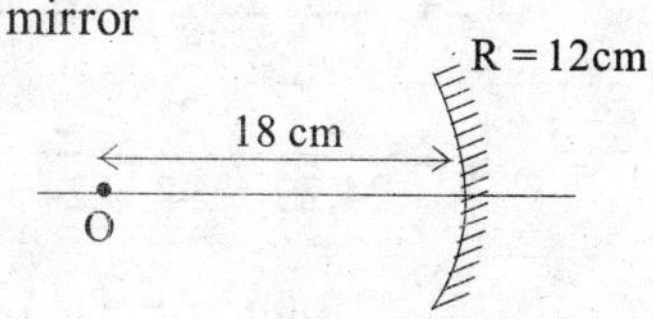

$$\text{Using } m = \frac{f}{u-f}$$

$$\therefore m_1 = \frac{6}{18-6} = \frac{1}{2}$$

Again for convex mirror

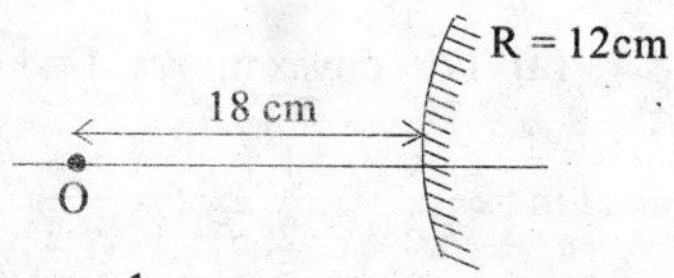

$$m_2 = \frac{6}{18+6} = \frac{1}{4} \quad \therefore \frac{m_2}{m_1} = \frac{1}{2}$$

6. **(d)** The expression of magnification for mirror is $m = -\frac{v}{u}$

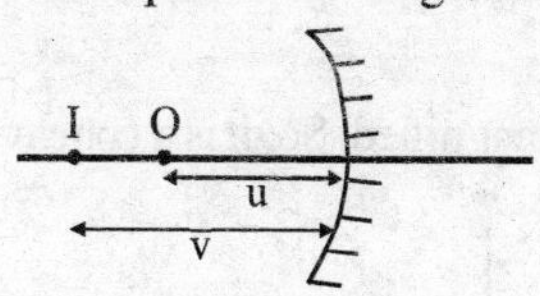

$m = -3 = -\frac{v}{u}$ and $v - u = -20$ cm

$\Rightarrow u = -10$ cm, $u = -30$ cm

From mirror formula $\frac{1}{f} = \frac{1}{u} + \frac{1}{v}$

$$f = \frac{vu}{v+u} = \frac{(-30)(-10)}{-30-10} = -\frac{15}{2} \text{cm}$$

$\therefore$ $R = 2f = -15$ cm

7. **(8)** From the mirror formula,

$$v = \frac{uf}{u-f} = \frac{-24 \times 1}{-24-1} = \frac{24}{25} \text{m}$$

Magnification,

$$m=\frac{-v}{u}=-\frac{24}{25(-24)}=\frac{1}{25}$$

$$v_1=-m^2 . v_0=\frac{-1}{(25)^2}\cdot 25=\frac{-1}{25}$$

From mirror formula, $\frac{1}{v}+\frac{1}{u}=\frac{1}{f}$

Diff.

$$\frac{-1}{v^2}\left(\frac{dv}{dt}\right)-\frac{1}{u^2}\left(\frac{du}{dt}\right)=0 \qquad \left[\frac{dv}{dt}=v_I;+\frac{du}{dt}=v_0\right]$$

$$\frac{+2}{v^3}.(v_I)^2-\frac{1}{v^2}\cdot a_I+\frac{2}{u^3}\cdot(v_0)^2-\frac{1}{u^2}\cdot a_0=0$$

$$\Rightarrow \frac{a_I}{v^2}=\frac{2}{v^3}\cdot v_I^2+\frac{2}{u^3}\cdot v_0^2$$

$$\Rightarrow a_I=\frac{2}{v}\cdot v_I^2+\frac{2v^2}{u^3}\cdot v_0^2$$

$$=\frac{2\cdot 25}{24}\cdot\frac{1}{25}\cdot\frac{1}{25}-\frac{2}{(24)^3}\cdot\frac{24}{25}\cdot\frac{24}{25}\cdot 25\cdot 25$$

$$\therefore a_I=\frac{2}{24.25}-\frac{2}{24}=\frac{2}{24}\cdot\frac{-24}{25}=\frac{-2}{25}$$

$$\Rightarrow 100a=100|a_I|=\frac{2}{25}\times 100=8$$

8. **(a)** For convex mirror, $f=+15$cm

$$m=-\frac{v}{u}=+\frac{1}{2}\Rightarrow v=-\frac{u}{2}$$

By mirror formula,

$$\frac{1}{v}+\frac{1}{u}=\frac{1}{f}\Rightarrow\frac{2}{-u}+\frac{1}{u}=\frac{1}{f}$$

$u=-f=-15$ cm

9. **(c)** Since, image is magnified. So, it is a concave mirror.

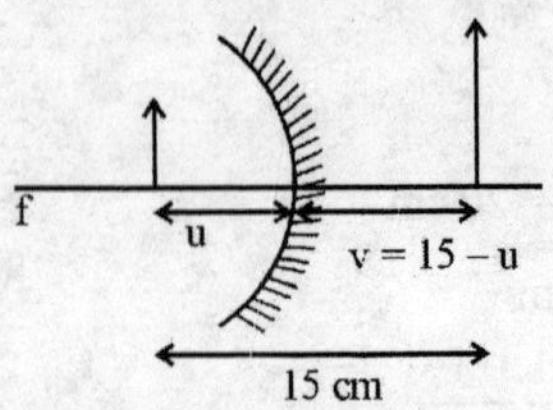

$$\therefore\ m=2=\frac{-v}{u}$$

$$2=\frac{-(15-u)}{-u}$$

$2u=15-u$

$3u=15\ \Rightarrow\ u=5$ cm

$v=15-u=15-5=10$ cm

By mirror formula,

$$\frac{1}{f}=\frac{1}{v}+\frac{1}{u}$$

$$=\frac{1}{10}+\frac{1}{(-5)}=\frac{1-2}{10}=\frac{-1}{10}$$

$f=-10$ cm

10. **(b)**

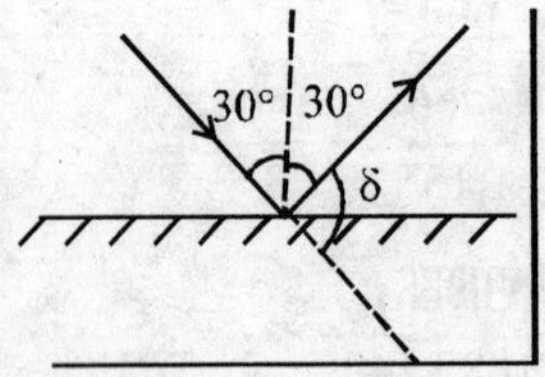

From the figure, it is clear

$\delta=180°-2\times 30°=120°$

11. **(b)**

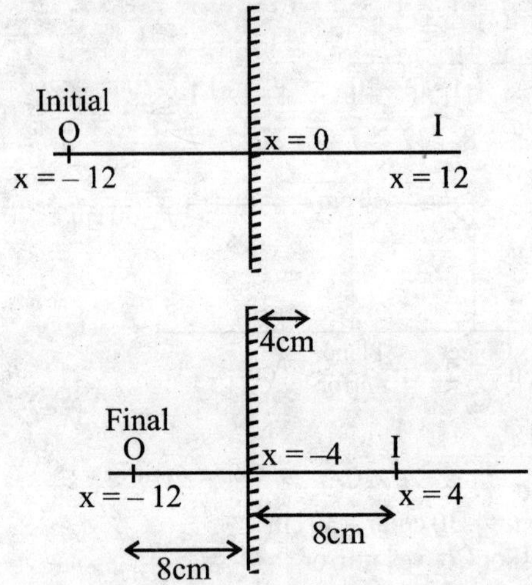

$\therefore$ Shifting of image will be 8 cm towards mirror.

12. **(18)** The first image in mirror A is at a distance 12 cm from mirror B. Now that is the second image in mirror B which is the third object for mirror A at a distance 18 cm from the mirror A.

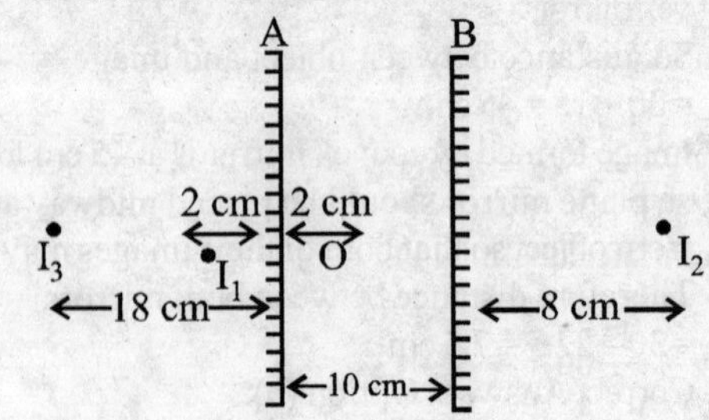

The desired image is 18 cm from mirror A.

13. **(32)**

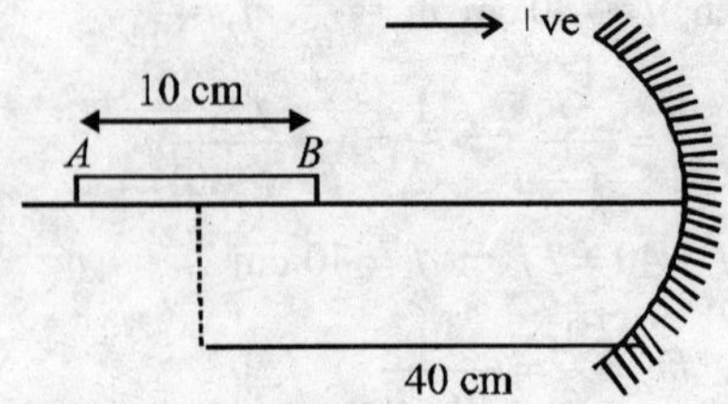

$u_A=-45$ cm, $f=-20$ cm

$$\frac{1}{v_A}+\frac{1}{u_A}=\frac{1}{f} \Rightarrow \frac{1}{v_A}-\frac{1}{45}=\frac{-1}{20}$$

$$\frac{1}{v_A}=\frac{1}{45}-\frac{1}{20} \Rightarrow v_A=-36 \text{ cm} \quad \text{And } u_B=-35 \text{ cm}$$

$$\frac{1}{v_B}+\frac{1}{u_B}=\frac{1}{f} \Rightarrow \frac{1}{v_B}-\frac{1}{35}=\frac{-1}{20}$$

$$\frac{1}{v_B}=\frac{1}{35}-\frac{1}{20} \Rightarrow v_B=-\frac{700}{15}=-\frac{140}{3} \text{ cm}$$

$\therefore \quad v_A-v_B=$ length of image

$$=\left(-36+\frac{140}{3}\right) \text{cm}=\frac{-108+140}{3} \text{ cm}=\frac{32}{3} \text{ cm} \quad \therefore x=32$$

14. **(c)** For A, u = –15 cm, f = – 20 cm

$$\text{So, } \frac{1}{V_A}+\frac{1}{-15}=\frac{1}{-20} \Rightarrow \frac{1}{V_A}=\frac{1}{15}-\frac{1}{20}$$

$$\Rightarrow \frac{1}{V_A}=\frac{20-15}{300} \Rightarrow \frac{1}{V_A}=\frac{5}{300}$$

$$\Rightarrow V_A=\frac{300}{5}=60 \text{ cm}$$

For B, u = – 25 cm, f = – 20 cm

$$\text{So, } \frac{1}{V_B}+\frac{1}{-25}=\frac{1}{-20} \Rightarrow \frac{1}{V_B}=-\frac{1}{20}+\frac{1}{25}$$

$$\Rightarrow \frac{1}{V_B}=\frac{-5}{500} \Rightarrow V_B=-100 \text{ cm}$$

So, distance between image of A and B are (100 – (–60)) cm i.e. 160 cm

15. **(32)** Using the formula,

$$\frac{1}{v}+\frac{1}{u}=\frac{1}{f} \quad \text{...(1)}$$

$$\frac{-1}{120}-\frac{1}{40}=\frac{1}{f}, f=-30 \text{ cm}$$

Differentiate equation (1),

$$\frac{-1}{v^2}dv-\frac{1}{u^2}du=-\frac{1}{f^2}df$$

$$\text{Also } dv=du=\frac{1}{20} \text{cm} \therefore \frac{\frac{1}{20}}{(120)^2}+\frac{\frac{1}{20}}{(40)^2}=\frac{df}{(30)^2}$$

$$df=\frac{1}{32} \text{ cm} \therefore K=32$$

16. **(400)** After 10 sec.

u = –80 cm, f = –100 cm

By mirror formula

$$\frac{1}{v}+\frac{1}{u}=\frac{1}{f} \Rightarrow v=400 \text{ cm}$$

17. **(210)**

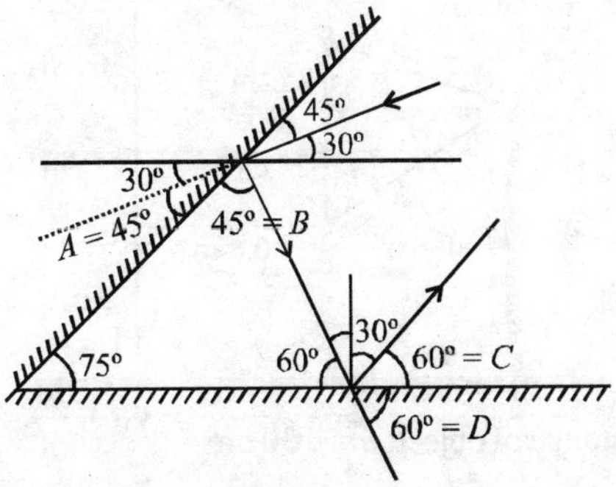

We have, $\delta=\delta_1+\delta_2$

$=(A+B)+(C+D)$

$=45°+45°+60°+60°=210°$

18. **(b)** From figure shortest distance between the images

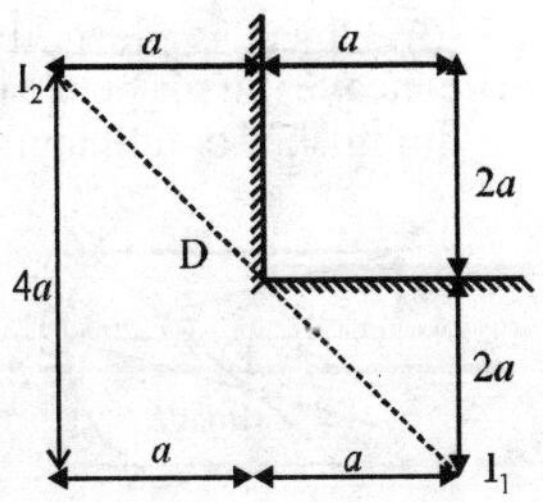

$$D=\sqrt{(4a)^2+(2a)^2}$$

$=2\sqrt{5}\,a$

$=4.6\,a$

19. **(a)**

20. **(d)**

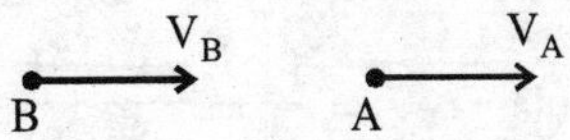

Mirror will be convex mirror

$V_i=-m^2 v_0$

V_i is velocity of image w.r.t mirror

V_o is velocity of object w.r.t mirror

$V_o=40$ m/s

$$\text{Magnification } m=\frac{f}{f-u}=\frac{10}{10+190}=\frac{1}{20}$$

$$v_i=-\frac{1}{20^2}\times 40=-0.1 \text{ m/s}$$

21. **(a)** For convex mirror, focus is behind the mirror. So, its focal length (f) is positive.

$$\therefore \quad f=+\frac{r}{2}$$

22. **(a)** Magnification (m) is given by

$$m=\frac{\text{height of the image}}{\text{height of the object}}=\frac{-25 \text{ cm}}{-100 \text{ cm}}=\frac{1}{4}$$

Image is diminished and magnification is positive.

$\therefore$ It is possible if object is placed in front of convex mirror.

23. **(150.00)** The distance between the extreme points where man can see the image of light source in the mirror,

$PQ = 2 \times (50+25)$ cm = 150 cm

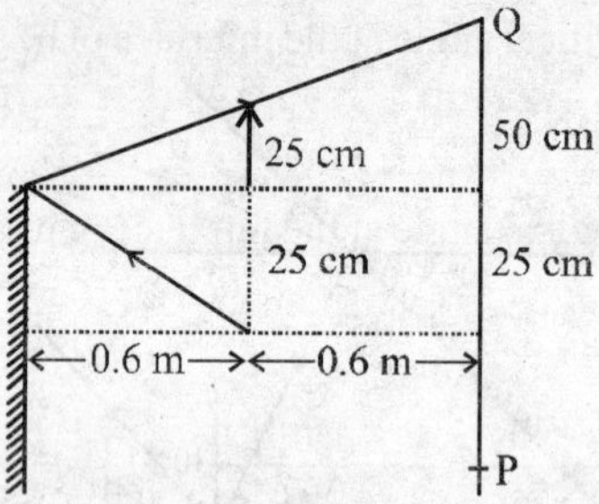

24. **(1)**

Distance of object, $u = -30$ cm

Distance of image, $v = 10$ cm

Magnification, $m = \frac{-v}{u} = \frac{(-10)}{-30} = \frac{1}{3}$

Speed of image $= m^2 \times$ speed of object $= \frac{1}{9} \times 9 = 1$ cm s^{-1}

25. **(d)** Object is placed beyond radius of curvature (R) of concave mirror hence image formed is real, inverted and diminished or unmagnified.

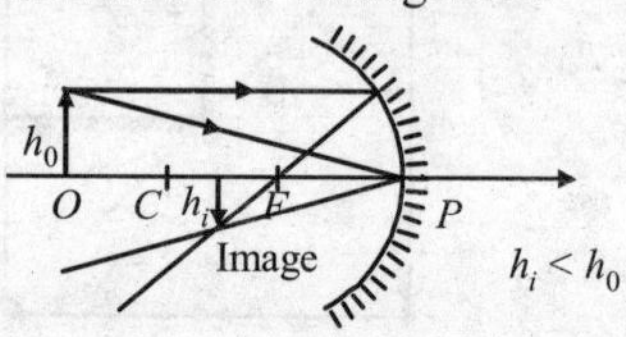

26. **(c)** $+5 = -\frac{v}{u} \Rightarrow v = -5u$

From mirror formula, $\frac{1}{v} + \frac{1}{u} = \frac{1}{f} \Rightarrow \frac{1}{-5u} + \frac{1}{u} = \frac{1}{0.4}$

$\therefore$ u = 0.32 m

27. **(b)**

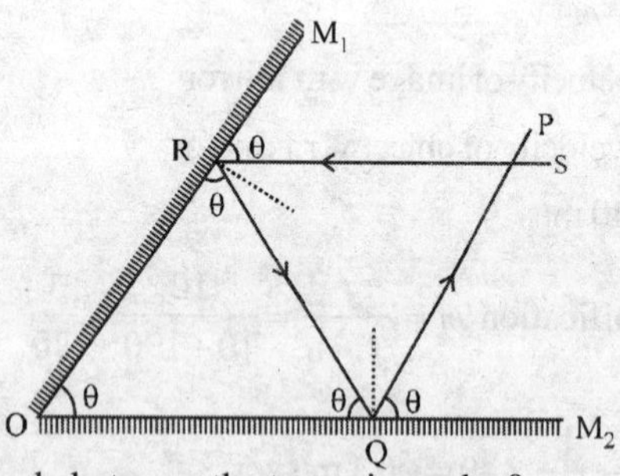

Let angle between the two mirrors be θ.

Ray PQ ∥ mirror M_1 and RS ∥ mirror M_2

$\therefore$ $M_1RS = \angle ORQ = \angle M_1OM_2 = \theta$

Similarly, $\angle M_2QP = \angle OQR = \angle M_2OM_1 = \theta$

$\therefore$ In ΔORQ, $3\theta = 180° \Rightarrow \theta = \frac{180°}{3} = 60°$

28. **(c)** Using mirror formula, magnification is given by

$m = \frac{f}{u-f} = \frac{-1}{1-\frac{u}{f}}$

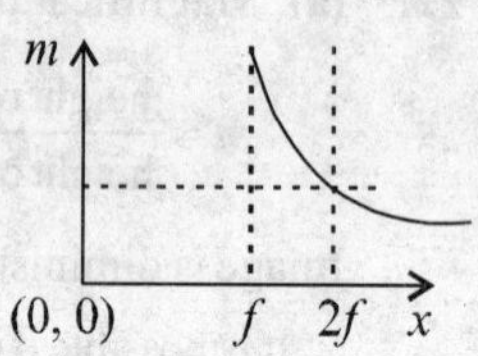

At focus magnification is ∞

And at $u = 2f$, magnification is 1.

Hence graph (c) correctly depicts 'm' versus distance of object 'x' graph.

29. **(c)** Convex morror is used as a shaving mirror.

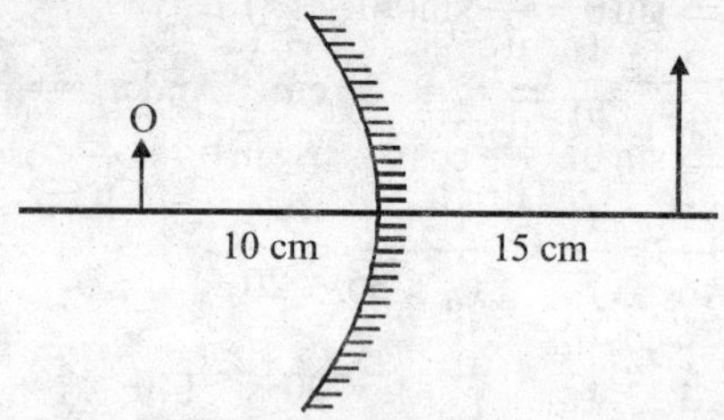

From question : $v = 15$ cm, $u = -10$ cm

Radius of curvature, $R = 2f = ?$

Using mirror formula, $\frac{1}{v} + \frac{1}{u} = \frac{1}{f}$

$\frac{1}{15} + \frac{1}{(-10)} = \frac{1}{f} \Rightarrow f = -30$ cm

Therefore radius of curvature,

$R = 2f = -60$ cm

30. **(a)** From mirror formula

$\frac{1}{v} + \frac{1}{u} = \frac{1}{f}$

Differentiating the above equation, we get

$\frac{dv}{dt} = -\frac{v^2}{u^2}\left(\frac{du}{dt}\right)$

Also,

$\frac{v}{u} = \frac{f}{u-f} \Rightarrow \frac{dv}{dt} = -\left(\frac{f}{u-f}\right)^2 \frac{du}{dt}$

$\Rightarrow \frac{dv}{dt} = \left(\frac{0.2}{2.8-0.2}\right)^2 \times 15 \Rightarrow \frac{dv}{dt} = \frac{1}{15}$ m/s

31. **(b)** The number of images formed is given by

$n = \frac{360}{\theta} - 1 \Rightarrow \frac{360}{\theta} - 1 = 3 \Rightarrow \theta = \frac{360°}{4} = 90°$

32. **(a)** When two plane mirrors are inclined at each other at an angle θ then the number of the images (n) of a point object kept between the plane mirrors is

$n = \frac{360°}{\theta} - 1$, (if $\frac{360°}{\theta}$ is even integer)

$\therefore$ Number of images formed $= \frac{360°}{60°} - 1 = 5$

33. **(c)** $\mu_g = \frac{3}{2} = 1.5$, $\mu_{air} = 1$

$\mu = \frac{C}{v}$, here we can say. It does not depend on mass density.

So (A) is correct but (R) is not correct.

34. **(c)**

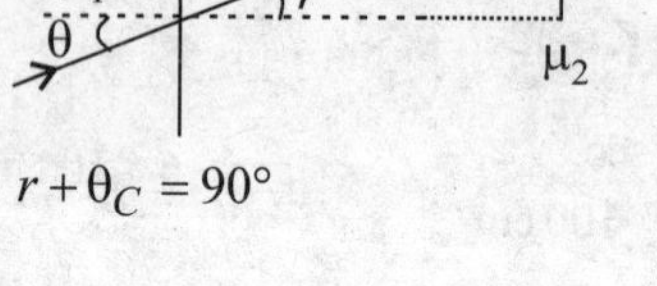

$r + \theta_C = 90°$

Using Snell's law,

$$\mu_1 \sin\theta = \mu_2 \sin r$$

$$\Rightarrow \sin\theta = \frac{\mu_2}{\mu_1}\sin(90-\theta_C)$$

$$\Rightarrow \sin\theta = \frac{\mu_2}{\mu_1}\cos\theta_C \Rightarrow \sin\theta_C = \frac{\mu_1}{\mu_2}$$

$$\Rightarrow \cos\theta_C = \sqrt{1-\frac{\mu_1^2}{\mu_2^2}}$$

$$\Rightarrow \sin\theta = \frac{\mu_2}{\mu_1}\sqrt{1-\frac{\mu_1^2}{\mu_2^2}}$$

$$\Rightarrow \sin\theta = \sqrt{\frac{\mu_2^2-\mu_1^2}{\mu_1^2}} = \sqrt{\frac{\frac{25}{16}-1}{1}}$$

$$\Rightarrow \sin\theta = \frac{3}{4} \Rightarrow \theta = \sin^{-1}\left(\frac{3}{4}\right).$$

35. (80)

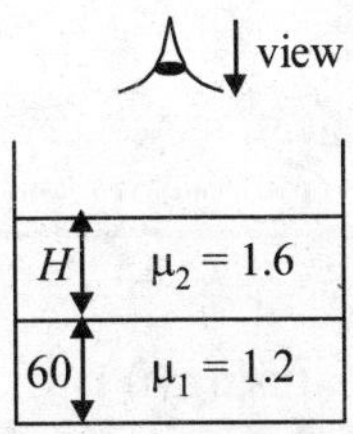

Here, Real depth, $d = H + 60$, Shift $= 40$ cm

The apparent depth of container bottom is

$$d' = \frac{H}{1.6} + \frac{60}{1.2}$$

$\therefore$ Shift $= d - d'$

$$\Rightarrow H + 60 - \frac{H}{1.6} - \frac{60}{1.2} = 40 \Rightarrow \frac{6}{16}H = 30$$

$\therefore H = 80$ cm.

36. (6) The velocity of light can be expressed in terms of the permeability (μ) and permitivity ($\in$).

$$V = \frac{1}{\sqrt{\mu\in}} = \frac{1}{\sqrt{\mu_0\mu_r}} \times \frac{1}{\sqrt{\in_0\in_r}}$$

$$= \frac{1}{\sqrt{\mu_r \in_r}} \times \frac{1}{\sqrt{\mu_0\in_0}} = \frac{C}{\sqrt{\mu_r\in_r}} = \frac{C}{\sqrt{\frac{10}{\pi}\times\frac{1}{0.0885}}}$$

$$= \frac{C}{\sqrt{36}} = \frac{C}{6} \Rightarrow V = \frac{C}{6}$$

$\therefore C = 6V$

Velocity of light in vacuum is greater by 6 times the velocity of light in medium.

Answer is 6.

37. (a) The wavelength of light in air, $\lambda_a = \frac{c}{f_a} = \frac{3\times10^8}{5\times10^{14}}$

$$= 6\times10^{-7}\,\text{m}$$

$\therefore$ The wavelength of light in medium is

$$\lambda_m = \frac{\lambda_a}{\mu} = \frac{6\times10^{-7}}{2} = 300\,\text{nm}$$

38. (d) Critical angle at the interface of n_1 and n_2 is

$$\sin\theta_{1C} = \frac{n_1}{n_2}$$

Critical angle at the interface of n_1 and n_3 is

$$\sin\theta_{2C} = \frac{n_1}{n_3} \qquad \therefore \quad \sin\theta_{2C} - \sin\theta_{1C} = \frac{1}{2}$$

$$\Rightarrow \frac{n_1}{n_3} - \frac{n_1}{n_2} = \frac{1}{2} \Rightarrow \frac{n_1}{n_2}\left(\frac{n_2}{n_3}-1\right) = \frac{1}{2}$$

$$\Rightarrow \frac{n_1}{n_2}\left(\frac{2}{5}-1\right) = \frac{1}{2} \;\therefore\; \frac{n_1}{n_2} = \frac{-5}{6}$$

$$\therefore \quad \theta_{1C} = \sin^{-1}\left(\frac{n_1}{n_2}\right) = \theta_{1C} = \sin^{-1}\left(-\frac{5}{6}\right)$$

39. (6)

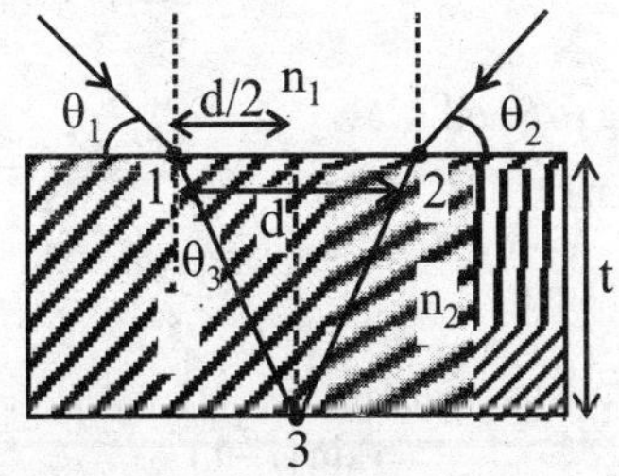

By Snell's law,

$$n_1 \sin(90-\theta_1) = n_2 \sin\theta_3$$

$$\Rightarrow n_1\cos\theta_1 = n_2\sin\theta_3 \Rightarrow \cos\theta_1 = \frac{n_2}{n_1}\sin\theta_3$$

$$\Rightarrow \frac{n_2}{2n_1} = \frac{n_2}{n_1}\sin\theta_3 \Rightarrow \sin\theta_3 = \frac{1}{2}$$

$\therefore \theta_3 = 30°$

$$\text{Now, } \tan 30° = \frac{d}{2(t)}$$

$$\therefore \; t = \frac{d\sqrt{3}}{2} = \frac{4\sqrt{3}\times\sqrt{3}}{2}\,\text{cm} = 6\,\text{cm}$$

40. (c) To determine the minimum refractive index (μ) required for the coin to be visible, we need to analyze the refraction at the curved liquid air interface.

The critical angle and refractive index relation is given by

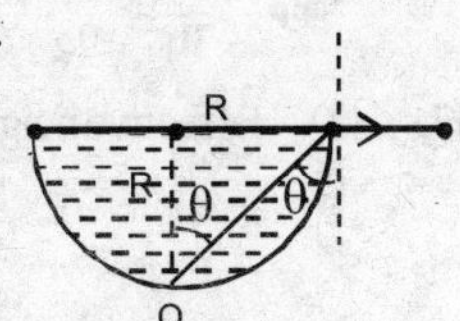

$$\sin C = \frac{1}{\mu}$$

for $\mu \to$ least, $\quad C \to$ maximum

$\theta = C = 45°$

$$\therefore \quad \mu = \frac{1}{\sin 45°} = \sqrt{2}°$$

41. (d) Focal length of mirror is independent of medium.

42. (d) When a light ray passes through a parallel-sided glass palced in air, it undergoes refraction at both interfaces but emerges parallel to the original direction, with a lateral shift

$$x = \frac{h\sin(i-r)}{\cos r}$$

43. **(a)** Critical angle is given by $\sin\theta_c = \dfrac{\mu_r}{\mu_d}$

$$\therefore \sin\theta_c = \frac{\mu_r}{\mu_d} = \frac{1}{\sqrt{2}}$$

$$\Rightarrow \frac{1}{\sqrt{2}} = \frac{\mu_r}{\mu_d}$$

$$\Rightarrow \frac{\mu_d}{\mu_r} = \frac{\sqrt{2}}{1}$$

44. **(2)** Given, thickness of glass slab

$t = 4\sqrt{3}$ cm, $\mu = \sqrt{2}$ and

$i = \theta_c$

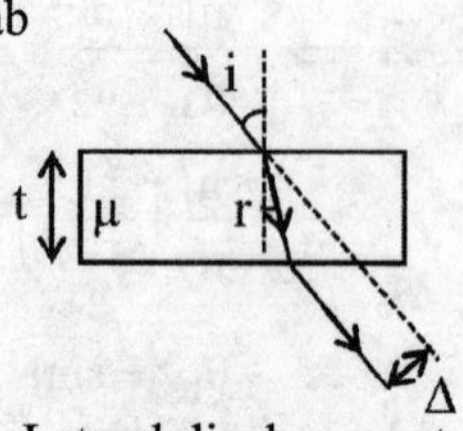

Lateral displacement

$$\Rightarrow i = \sin^{-1}\left(\frac{1}{\mu}\right) = \sin^{-1}\left(\frac{1}{\sqrt{2}}\right)$$

$\therefore i = 45°$

And according to Snell's law

$\mu_1 \sin i = \mu_2 \sin r$

$1 \sin 45° = \sqrt{2} \sin r$

$\therefore r = 30°$

Lateral displacement $\Delta = \dfrac{t\sin(i-r)}{\cos r}$

$$\Rightarrow \Delta = \frac{4\sqrt{3}\times\sin 15°}{\cos 30°} = \frac{4\sqrt{3}\times 0.25}{\frac{\sqrt{3}}{2}}$$

$\therefore \Delta = 2\text{cm}$

45. **(31)** Apparent depth

$$h_{app} = \frac{h_1}{\mu_1} + \frac{h_2}{\mu_2} = \frac{6}{3/2} + \frac{6}{8/5} = 4 + \frac{15}{4} = \frac{31}{4}\text{ cm}$$

Therefore, $\alpha = 31$

46. **(b)** Apparent depth.

$$d_{app} = \frac{d_1}{n_1} + \frac{d_2}{n_2} = \frac{d}{2}\left[\frac{1}{n_1} + \frac{1}{n_2}\right];\ d_{app} = \frac{d}{2}\left[\frac{n_1+n_2}{n_1 n_2}\right]$$

47. **(3)** Effective height of the bird as seen by the fish, $Y = y + \mu y'$

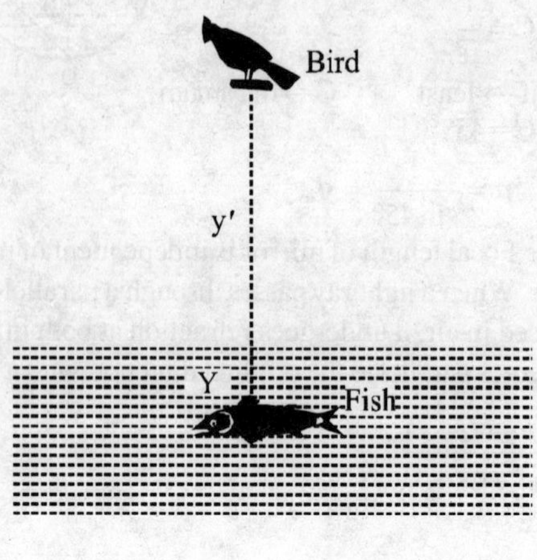

Differentiating both sides w.r.t.t

$$\frac{dY}{dt} = \frac{dy}{dt} + \mu\frac{dy'}{dt}$$

Given $\dfrac{dY}{dt} = 12\text{ ms}^{-1}$; $\dfrac{dy}{dt} = 8\text{ ms}^{-1}$

$$\therefore 12 = 8 + \frac{4}{3}\frac{dy}{dt}$$

$$\therefore \frac{dy'}{dt} = \frac{4\times 3}{4} = 3\text{ ms}^{-1}$$

48. **(b)** Refractive index, $\mu = \dfrac{d_{real}}{d_{apparent}}$

When viewed from one side

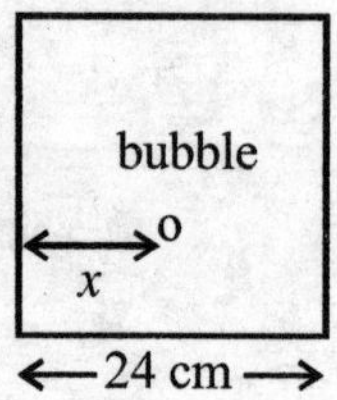

$$\mu = \frac{x}{12} \quad ...(i)$$

When viewed from other side

$$\mu = \frac{24-x}{4} \quad ...(ii)$$

$$\Rightarrow \frac{x}{12} = \frac{24-x}{4}$$

$$\Rightarrow \frac{x}{12} = 6 - \frac{x}{4} \Rightarrow \frac{x}{12} + \frac{x}{4} = 6$$

$$\Rightarrow \frac{4x}{12} = 6 \Rightarrow x = 18$$

$$\therefore \mu = \frac{18}{12} = 1.5 \quad \text{(using (i))}$$

49. **(98)** Given,

Actual depth d = 8 cm

Refractive index, $\mu = \dfrac{4}{3}$

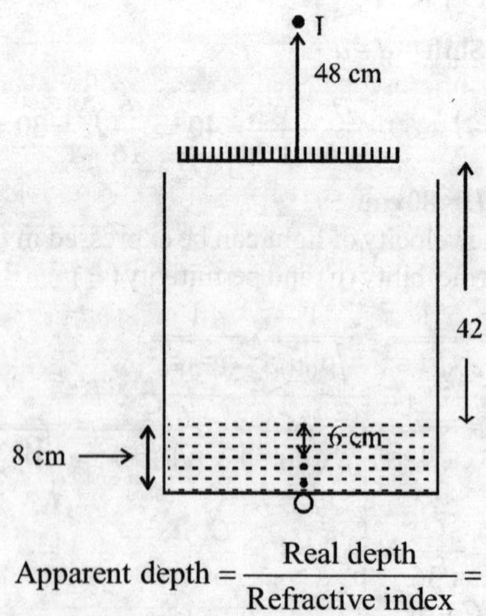

$$\text{Apparent depth} = \frac{\text{Real depth}}{\text{Refractive index}} = \frac{d}{\mu} = \frac{8}{\frac{4}{3}} = 6$$

Distance between object (O) and Image (I) = 48 + 50 = 98 cm

50. **(b)** From the Snell's law of refraction

$\mu_1 \sin 45° = \mu_2 \sin 30°$

Here, μ_1 and μ_2 are the refractive index of two medium.

$\therefore \mu_1 \times \frac{1}{\sqrt{2}} = \mu_2 \times \frac{1}{2} \Rightarrow \frac{\mu_1}{\mu_2} = \frac{1}{\sqrt{2}} \quad \because \mu \propto \frac{1}{\lambda}$

$\therefore \frac{\mu_1}{\mu_2} = \frac{\lambda_2}{\lambda_1} = \frac{1}{\sqrt{2}} \Rightarrow \lambda_2 = \frac{\lambda_1}{\sqrt{2}}$

Frequency of light wave doesn't change with medium.

51. (50)

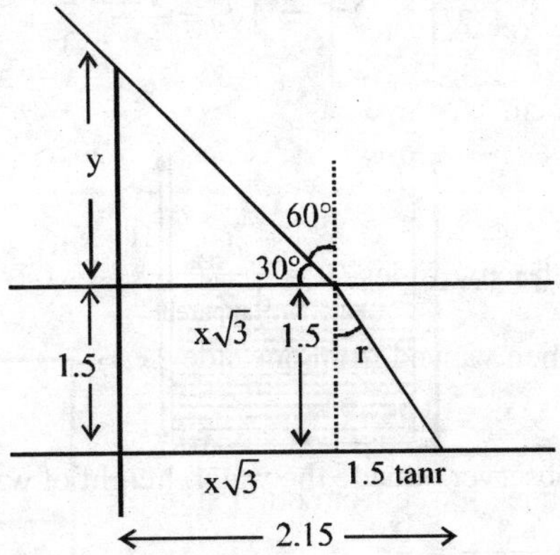

Using Snell's law

$1 \sin 60° = \frac{4}{3} \sin r \Rightarrow \sin r = \frac{3\sqrt{3}}{8}$

$\cos r = \sqrt{1 - \left(\frac{3\sqrt{3}}{8}\right)^2} = \frac{\sqrt{37}}{8}$

$\therefore \tan r = \frac{\sin r}{\cos r} = \frac{3\sqrt{3}}{\sqrt{37}}$

From the diagram

$y\sqrt{3} + 1.5 \tan r = 2.15 \Rightarrow y\sqrt{3} = 2.15 - 1.5 \times \frac{3\sqrt{3}}{\sqrt{37}}$

$y = \frac{2.15}{\sqrt{3}} - \frac{1.5 \times 3}{\sqrt{37}} = 1.241 - 0.739 = 0.502 \approx 0.50 \text{ meter}$

$x = 50 \text{ cm}$

52. (a) As microscope need to be raised by 30 cm to focus the object again. So, apparent shift is 30 cm.

As, apparent shift $= t\left(1 - \frac{1}{\mu}\right)$

$\Rightarrow \quad 30 = t\left(1 - \frac{3}{5}\right) \Rightarrow 30 = \frac{2t}{5} \Rightarrow t = 75 \text{ cm}$

53. (52)

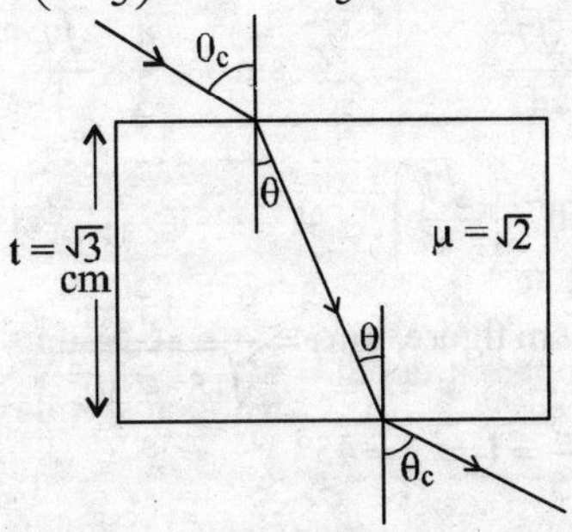

$\sin \theta_c = \frac{\mu_{air}}{\mu_{glass}}$; $\mu_{air} = 1$; $\mu_{glass} = \sqrt{2}$

$\sin \theta_c = \frac{1}{\sqrt{2}} \Rightarrow \theta_c = 45°$

$\sin \theta_c = \mu \sin \theta$

$\frac{1}{\sqrt{2}} = \sqrt{2} \sin \theta \Rightarrow \theta = 30°$

The lateral displacement of the ray

$x = t \sin(i - r) \sec r = t \sin(\theta_c - \theta) \sec \theta$

$x = \sqrt{3} \sin(45° - 30°) \sec 30°$

$x = \sqrt{3}(0.26)\left(\frac{2}{\sqrt{3}}\right)$; $x = 0.52 \text{ cm}$

$x = 52 \times 10^{-2} \text{ cm}$

54. (15) $\vec{A} = 4\sqrt{3}\hat{i} - 3\sqrt{3}\hat{j} - 5\hat{k}$

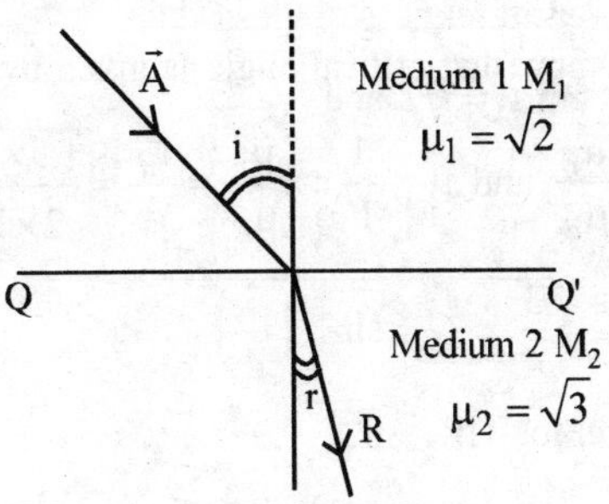

As incident vector A makes i angle with normal z-axis and refracted vector R makes r angle with normal z-axis with help of direction cosine

$i = \cos^{-1}\left(\frac{A_z}{A}\right) = \cos^{-1}\left(\frac{5}{\sqrt{(4\sqrt{3})^2 + (3\sqrt{3})^2 + 5^2}}\right)$

$\cos^{-1}\left(\frac{5}{10}\right) \Rightarrow i = 60°$

By Snell's law, we have

$\sqrt{2} \sin 60° = \sqrt{3} \times \sin r \Rightarrow r = 45°$

Difference between i and $r = 60° - 45° = 15°$

55. (c) $i = 45°, D = 15°$

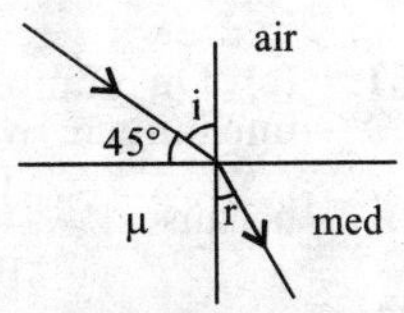

$D = i - r$

$15° = 45° - r \Rightarrow r = 30°$

$n_1 \sin 1 = n_2 \sin r$ (from snells's law)

$1 \sin 45° = \mu \sin 30°$

$\frac{1}{\sqrt{2}} = \mu \frac{1}{2} \Rightarrow \mu = \sqrt{2} = 1.414$

56. (a) Refractive index, $\frac{\mu_2}{\mu_{air}} = \frac{c}{v_2}$ or, $\therefore \frac{\sqrt{\mu_{r_2} \varepsilon_{r_2}}}{(1)} = \frac{c}{v_2}$

$\therefore \sqrt{(1)(9)} = \frac{c}{v_2} \qquad \therefore v_2 = \frac{c}{3}$

57. (a) $n_d \sin i_c = n_r \sin 90°$ ($\because$ From Snell's law)

$\sin i_c = \frac{n_r}{n_d} = \frac{v_d}{v_r} \qquad \left(\therefore v = \frac{c}{n}\right)$

$\sin i_c = \frac{1.5 \times 10^8}{2 \times 10^8} = \frac{1.5}{2}$; $\sin i_c = \frac{3}{4}$

$\tan i_c = \frac{3}{\sqrt{4^2 - 3^2}} \Rightarrow \frac{3}{\sqrt{7}}$

The critical angle between them, $i_c = \tan^{-1}\left(\frac{3}{\sqrt{7}}\right)$

58. (a) $t_2 - t_1 = 5 \times 10^{-10}$

$$\Rightarrow \frac{d}{v_B} - \frac{d}{v_A} = 5\times10^{-10} \Rightarrow \frac{d\mu_B}{C} - \frac{d\mu_A}{C} = 5\times10^{-10}$$

$$\Rightarrow d\left(\frac{\mu_B}{C} - \frac{\mu_A}{C}\right) = 5\times10^{-10}$$

$$\Rightarrow d\left(\frac{2\mu_A}{C} - \frac{\mu_A}{C}\right) = 5\times10^{-10}$$

$$\Rightarrow d = \frac{5\times10^{-10}}{\frac{\mu_A}{C}} = 5\times10^{-10} v_A \text{ m}$$

59. (d) We know that critical angle is given by

$$\sin C = \frac{\mu_r}{\mu_d} \text{ and } \mu \propto \frac{1}{V} \Rightarrow \frac{\mu_r}{\mu_d} = \frac{v_d}{v_r} = \frac{1.5\times10^{10}}{2\times10^{10}} = \frac{3}{4}$$

So, $\sin C = \frac{3}{4} \Rightarrow C = \sin^{-1}\left(\frac{3}{4}\right)$

Therefore, for TIR

$\theta > C$

$\Rightarrow \theta > \sin^{-1}\left(\frac{3}{4}\right)$

60. (d) Let $2i$ be angle of incidence

By Snell's law

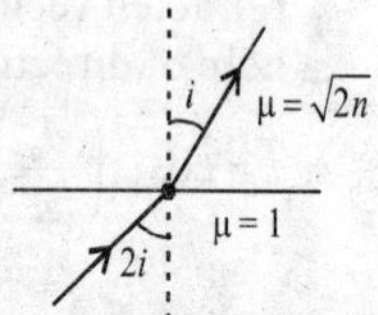

$\Rightarrow 1 \sin 2i = \sqrt{2n} \sin i$

$\Rightarrow 2 \sin i \cos i = \sqrt{2n} \sin i$

$\Rightarrow \cos i = \sqrt{\frac{n}{2}} \Rightarrow i = \cos^{-1}\left(\sqrt{\frac{n}{2}}\right)$

$\Rightarrow 2i = 2\cos^{-1}\left(\sqrt{\frac{n}{2}}\right)$

61. (c) On going from rare to denser, frequency remain unchanged, whereas speed and wavelength decreases because $V \propto \frac{1}{\mu}$ and $\lambda \propto \frac{1}{\mu}$.

62. (9) We have, $\sin c = \frac{1}{\mu} = \frac{3}{4}$

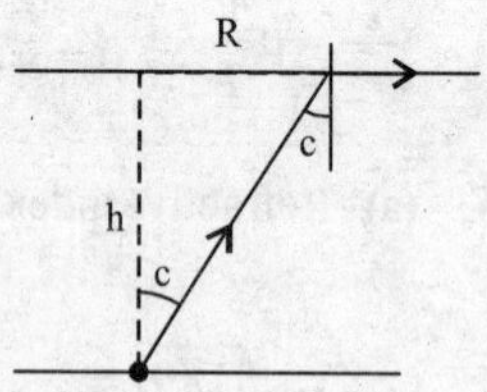

So, $\tan c = \frac{3}{\sqrt{7}}$

Also, $\tan c = \frac{R}{h} \Rightarrow \frac{3}{\sqrt{7}} = \frac{R}{\sqrt{7}} \Rightarrow R = 3\text{m}$

So, area of illumination $= \pi \times 3^2 = 9\pi m^2$

63. (c) Given, $V_A - V_B = 2.6 \times 10^7$ m/s

$$V_B = \frac{C}{\mu_B} = \frac{3\times10^8}{1.47} = 2.04 \times 10^8 \text{ m/s} = 20.4 \times 10^7 \text{ m/s}$$

So,

$V_A = (V_B + 2.6 \times 10^7)$ m/s $= (20.4 + 2.6) \times 10^7$ m/s $= 23 \times 10^7$ m/s

As, $\frac{\mu_B}{\mu_A} = \frac{V_A}{V_B} \Rightarrow \frac{\mu_B}{\mu_A} = \frac{23\times10^7}{20.4\times10^7} = 1.13$

64. (12) The formula of lateral shift in glass slab is given by

$$l = t\sin i\left[1 - \frac{\cos i}{\sqrt{\mu^2 - \sin^2 i}}\right] \Rightarrow 4\sqrt{3} = t\sin 60°\left[1 - \frac{\cos 60°}{\sqrt{3 - \frac{3}{4}}}\right]$$

$$\Rightarrow 4\sqrt{3} = t \times \frac{\sqrt{3}}{2}\left[1 - \frac{\frac{1}{2}}{\frac{3}{2}}\right] \Rightarrow 4\sqrt{3} = \frac{\sqrt{3}t}{2} \times \frac{2}{3}$$

$\Rightarrow t = 12$ cm

65. (a)

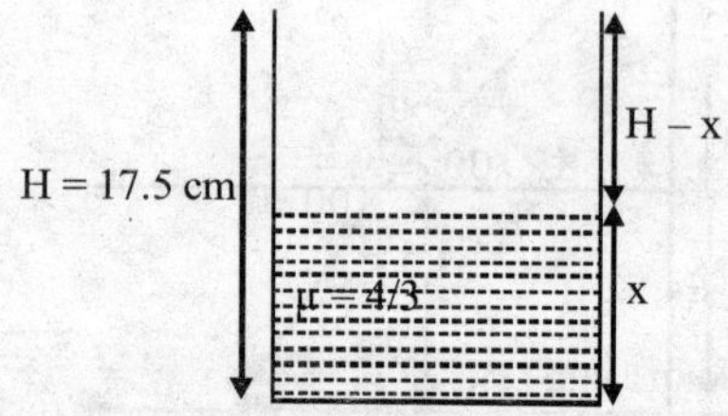

For an observer outside the water, height of water

$$= \frac{x}{\mu_w} = \frac{x}{(4/3)} = \frac{3x}{4}$$

Now, $\frac{3x}{4} = \frac{17.5}{2} \Rightarrow x = \frac{35}{3} = 11.7$ cm

66. (a) At maximum angle θ, the ray goes along the surface. For all angles less than θ, total internal reflection occurs.

At point B

$\frac{4}{3} \times \sin\theta'' = 1 \times \sin 90° \Rightarrow \theta'' = \sin^{-1}\left(\frac{3}{4}\right)$

$\theta' = \left(\frac{\pi}{2} - \theta''\right)$

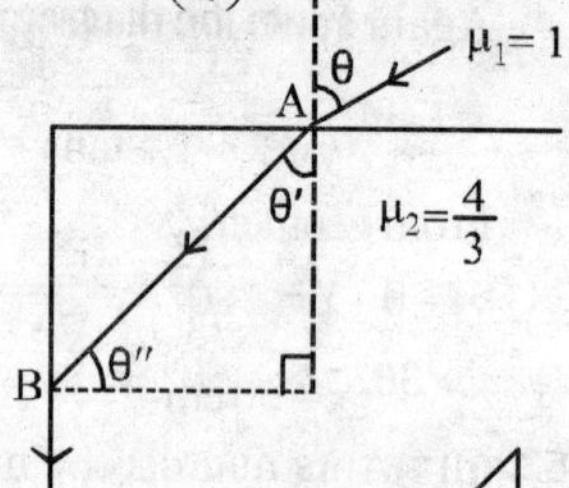

At point A,

$1 \times \sin\theta = \frac{4}{3} \times \sin\theta'$

$\Rightarrow \sin\theta = \frac{4}{3} \times \sin\left(\frac{\pi}{2} - \theta''\right)$

$\Rightarrow \sin\theta = \frac{4}{3}\cos\theta''$

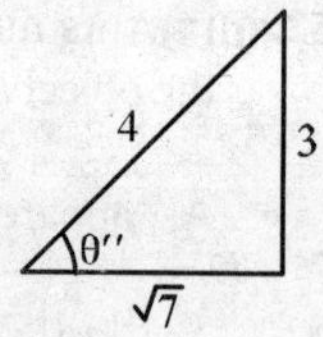

From the figure,

$\cos\theta'' = \frac{\sqrt{7}}{4}$ $\quad\therefore \sin\theta = \frac{4}{3} \times \frac{\sqrt{7}}{4}$

or, $\theta = \sin^{-1}\left(\frac{\sqrt{7}}{3}\right)$

67. (158) From figure, $\sin i = \frac{15}{\sqrt{15^2 + 30^2}}$

$\tan r = \frac{15}{15} = 1 \Rightarrow r = 45°$

From Snell's law, $\mu \times \sin i = 1 \times \sin r$

$$\Rightarrow \mu \times \frac{15}{\sqrt{15^2 + 30^2}} = 1 \times \sin 45° = \frac{1}{\sqrt{2}}$$

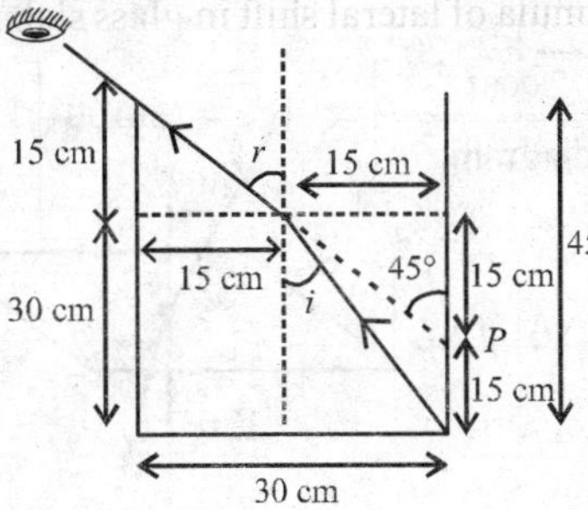

$$\therefore \mu = \frac{\frac{1}{\sqrt{2}}}{\frac{15}{\sqrt{1125}}} = 158 \times 10^{-2} = \frac{N}{100}$$

Hence, value of $N \simeq 158$.

68. (90.00) In the figure, QR is the reflected ray and QS is refracted ray. CQ is normal.

Apply Snell's law at P

$1 \sin 60° = \sqrt{3} \sin r$

$\Rightarrow \sin r = \frac{1}{2}$

$\Rightarrow r = 30°$

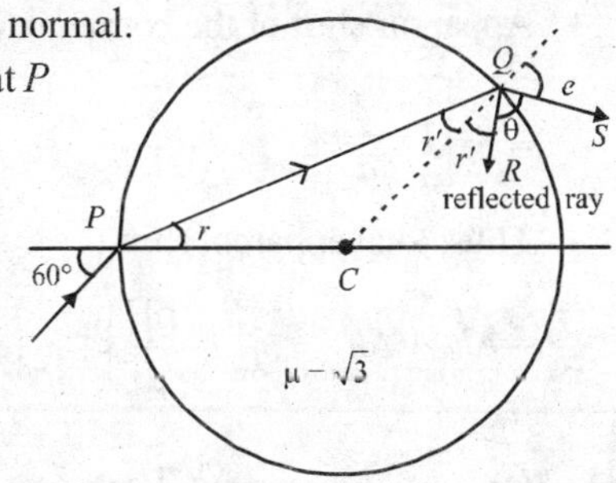

From geometry,

$CP = CQ =$ radius

$\therefore r' = 30°$

Again apply snell's law at Q,

$\sqrt{3} \sin r' = 1 \sin e \Rightarrow \frac{\sqrt{3}}{2} = \sin e \Rightarrow e = 60°$

From geometry

$r' + \theta + e = 180°$ (As angles lies on a straight line)

$\Rightarrow 30° + \theta + 60° = 180° \Rightarrow \theta = 90°$.

69. (d) Apparent depth,

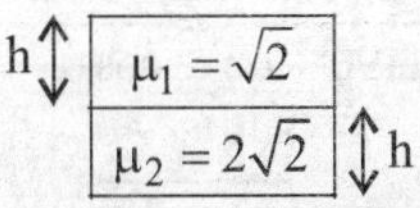

$$D = \frac{t_1}{\mu_1} + \frac{t_2}{\mu_2} = \frac{h}{\sqrt{2}} + \frac{h}{2\sqrt{2}} = \frac{3h}{2\sqrt{2}} = \frac{3h\sqrt{2}}{4}$$

70. (c) Given,

Refractive index, $\mu = \frac{4}{3}$

$\frac{4}{3} \sin \theta = 1 \sin 90° \Rightarrow \sin\theta = \frac{3}{4}$

$\cos \theta = \frac{\sqrt{7}}{4}$

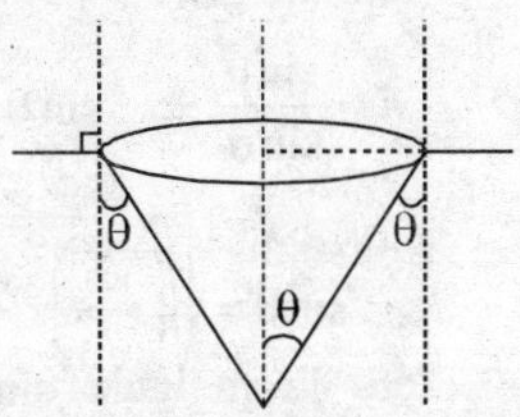

Solid angle, $\Omega = 2\pi(1 - \cos\theta) = 2\pi(1 - \sqrt{7}/4)$

Fraction of energy transmitted

$$= \frac{2\pi(1 - \cos\theta)}{4\pi} = \frac{1 - \sqrt{7}/4}{2} = 0.17$$

Percentage of light emerges out of surface $= 0.17 \times 100 = 17\%$

71. (b) As, $n = \sqrt{\mu_r \varepsilon_r} = \sqrt{\frac{4}{3} \times 3} = 2$

And, $\sin \theta_c = \frac{1}{n} = \frac{1}{2}$ $\therefore$ Critical angle, $\theta_C = 30°$

72. (c) If v is the distance of image formed by mirror, then

$$\frac{1}{v} + \frac{1}{u} = \frac{1}{f} \Rightarrow \frac{1}{v} + \frac{1}{-5} = \frac{1}{-20} \quad \therefore v = \frac{20}{3} \text{ cm}$$

Distance of this image from water surface

$= \frac{20}{3} + 5 = \frac{35}{3}$ cm ; Using, $\frac{RD}{AD} = \mu$

$$\therefore AD = d = \frac{RD}{\mu} = \frac{(35/3)}{1.33} = 8.8 \text{ cm}$$

73. (d) From the given figure

As $\sin 60° = \mu \sin 30°$

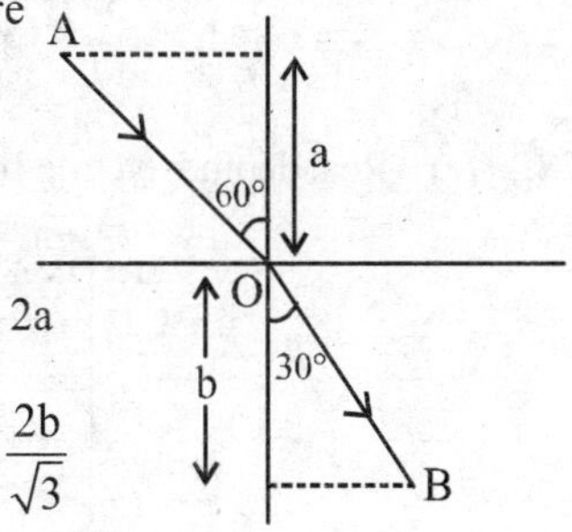

$\Rightarrow \mu = \frac{\sin 60°}{\sin 30°} = \sqrt{3}$

$\frac{a}{AO} = \cos 60° \Rightarrow AO = 2a$

$\frac{b}{BO} = \cos 30° \Rightarrow BO = \frac{2b}{\sqrt{3}}$

Optical path length $= AO + \mu BO$

$= 2a + (\sqrt{3})\frac{2b}{\sqrt{3}} = 2a + 2b$

74. (d) Using Snell's law of refraction,

$1 \times \sin 40° = 1.31 \sin \theta$

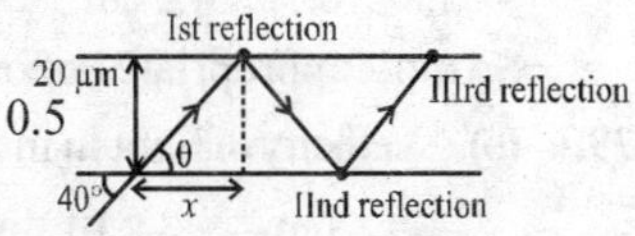

$\Rightarrow \sin \theta = \frac{0.64}{1.31} = 0.49 \approx 0.5$

$\Rightarrow \theta = 30°$

$x = 20$ μm $\times \cot \theta$

$\therefore$ Number of reflections

$$= \frac{2}{20 \times 10^{-6} \times \cot\theta} = \frac{2 \times 10^6}{20 \times \sqrt{3}} = 57735 \approx 57000$$

75. (c) $I_{\text{glass slab}} = \frac{96}{100} I_{\text{air}} \Rightarrow \frac{1}{2}\epsilon_0 \epsilon_r E^2 V = \frac{96}{100} \times \frac{1}{2} \epsilon_0 E_0^2 C$

$$\Rightarrow \epsilon_r E^2 V = \frac{96}{100} E_0^2 C \quad \Rightarrow E = \sqrt{\frac{96}{100} \times E_0^2 \times \frac{C}{\epsilon_r V}}$$

Putting $E_0 = 30, \mu = \frac{C}{V}$ and $\sqrt{\epsilon_r} = \mu$ $[\because \mu_r = 1]$

$E = 24\, V/m$

76. (d)

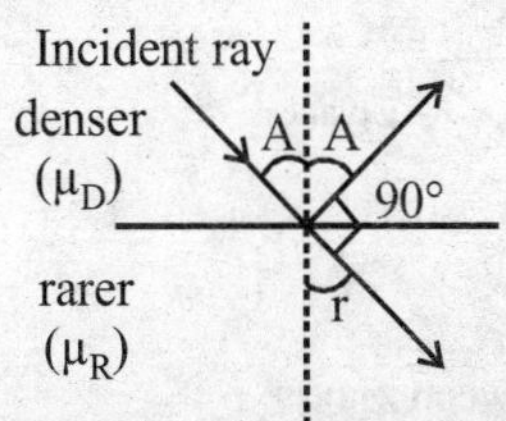

From Snell's law, $\frac{\mu_R}{\mu_D} = \frac{\sin i}{\sin r}$ ..... (i)

If $\angle i = A$ and $\angle r = (90° - A)$

We also know that, $\sin \theta_C = \frac{\mu_R}{\mu_D}$

From equation (i), $\sin\theta_C = \dfrac{\sin A}{\sin(90° - A)}$

$\Rightarrow \sin\theta_C = \dfrac{\sin A}{\cos A}$

$\Rightarrow \quad \sin\theta_C = \tan A \Rightarrow A = \tan^{-1}(\sin\theta_C)$

77. **(d)** $\sin C = \dfrac{1}{\mu} = \dfrac{R}{\sqrt{R^2 + h^2}} = \dfrac{3}{4}$

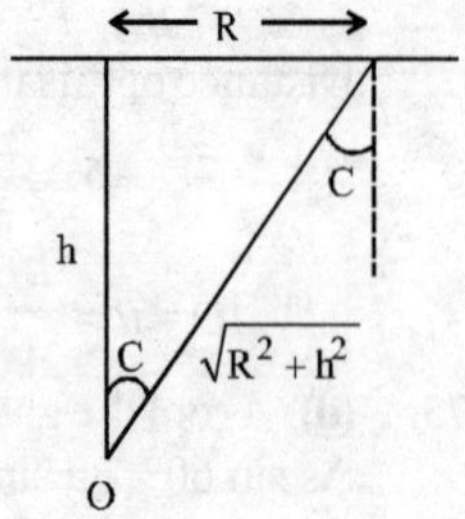

$\Rightarrow \quad 16R^2 = 9R^2 + 9h^2$

$\Rightarrow 7R^2 = 9h^2$

$\Rightarrow \quad R = \dfrac{3}{\sqrt{7}}h = \dfrac{3}{\sqrt{7}} \times 15\,\text{cm}$

78. **(c)** Real depth = 5 cm + 1 cm = 6 cm

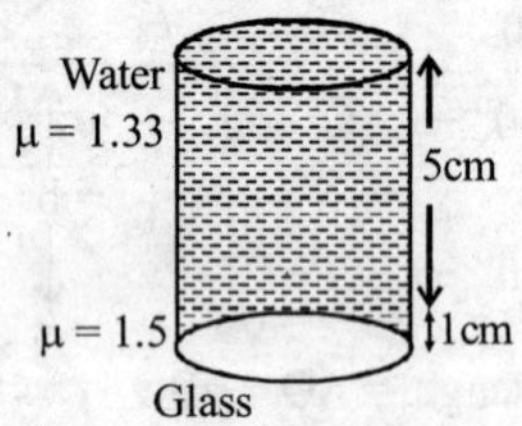

Apparent depth = $\dfrac{d_1}{\mu_1} + \dfrac{d_2}{\mu_2} = \dfrac{5}{1.33} + \dfrac{1}{1.5}$

$\simeq 3.8 + 0.7 \simeq 4.5$ cm $\therefore$ Shift = 6 cm – 4.5 cm $\cong$ 1.5 cm

So most appropriate option will be (c).

79. **(b)** Velocity of light in medium

$$V_{med} = \frac{3\,\text{cm}}{0.2\text{ns}} = \frac{3\times10^{-2}\,\text{m}}{0.2\times10^{-9}\,\text{s}} = 1.5\times10^8\,\text{m/s}$$

Refractive index of the medium

$$\mu = \frac{V_{air}}{V_{med}} = \frac{3\times10^8}{1.5\times10^8\,\text{m/s}} = 2$$

As $\mu = \dfrac{1}{\sin C}$ $\therefore \sin C = \dfrac{1}{\mu} = \dfrac{1}{2} = 30°$

Condition of TIR is angle of incidence i must be greater than critical angle. Hence ray will suffer TIR in case of (B) (i = 40° > 30°) only.

80. **(a)**

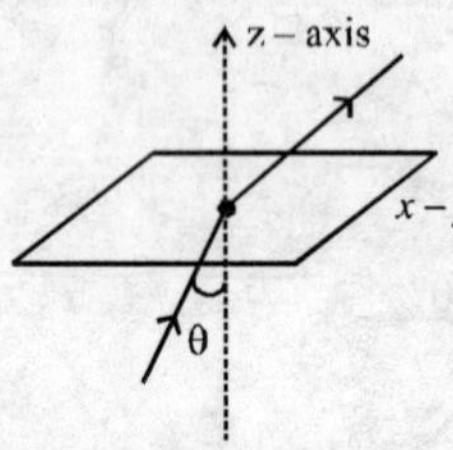

Let θ be the angle between z-axis and incident ray.

$$\therefore \cos\theta = \frac{\hat{k}.(6\sqrt{3}\,\hat{i} + 8\sqrt{3}\,\hat{j} - 10\hat{k})}{\sqrt{108+192+100}}; \ \cos\theta = \frac{-10}{20}$$

$\theta = \cos^{-1}\left(\dfrac{-1}{2}\right) = 120°$

So, the ray diagram should be

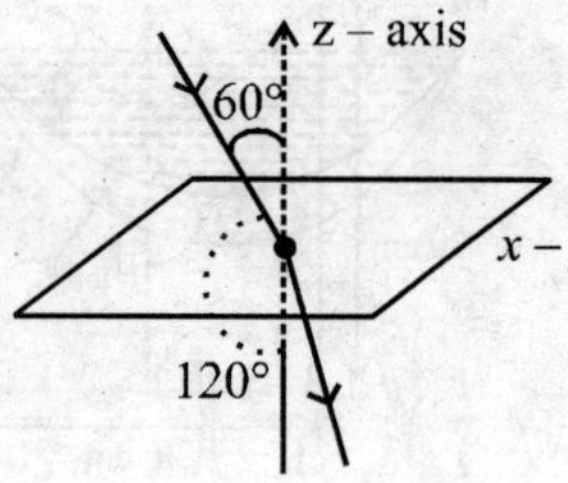

Clearly, $i = 180 - 120$

$= 60°$

By Snell's law

$\Rightarrow \sqrt{2}\sin 60 = \sqrt{3}\sin r$

$\Rightarrow \sin r = \dfrac{1}{\sqrt{2}} \Rightarrow r = 45°$

81. **(b)** Apparent shift of the bottom due to water,

$$\Delta h_1 = h_1\left[1 - \frac{1}{\mu_1}\right]$$

Apparent shift of the bottom due to kerosene, Δh_2

$$= h_2\left[1 - \frac{1}{\mu_2}\right]$$

Thus, total apparent shift :

$$= \Delta h_1 + \Delta h_2 = h_1\left(1 - \frac{1}{\mu_1}\right) + h_2\left(1 - \frac{1}{\mu_2}\right)$$

82. **(c)**

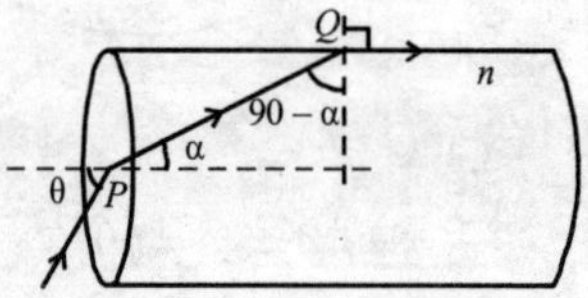

Applying Snell's law for medium inside the cylinder and air at Q we get

$$n = \frac{\sin 90°}{\sin(90° - \alpha)} = \frac{1}{\cos\alpha} \Rightarrow \cos\alpha = \frac{1}{n}$$

$$\therefore \ \sin\alpha = \sqrt{1 - \cos^2\alpha} = \sqrt{1 - \frac{1}{n^2}} = \frac{\sqrt{n^2 - 1}}{n} \quad ...(i)$$

Applying Snell's Law for air and medium inside the cylinder at P we get

$$n = \frac{\sin\theta}{\sin\alpha} \Rightarrow \ \sin\theta = n \times \sin\alpha = \sqrt{n^2 - 1}; \ [\text{from (i)}]$$

$$\therefore \ \sin\theta = \sqrt{\left(\frac{2}{\sqrt{3}}\right)^2 - 1} = \sqrt{\frac{4}{3} - 1} = \frac{1}{\sqrt{3}}$$

or $\theta = \sin^{-1}\left(\dfrac{1}{\sqrt{3}}\right)$

83. **(a)** In an optical fibre, light is sent through the fibre without any loss by the phenomenon of total internal reflection. Total internal reflection of light waves confine the light rays inside the optical fiber.

84. (d) Using Lens Maker's formula

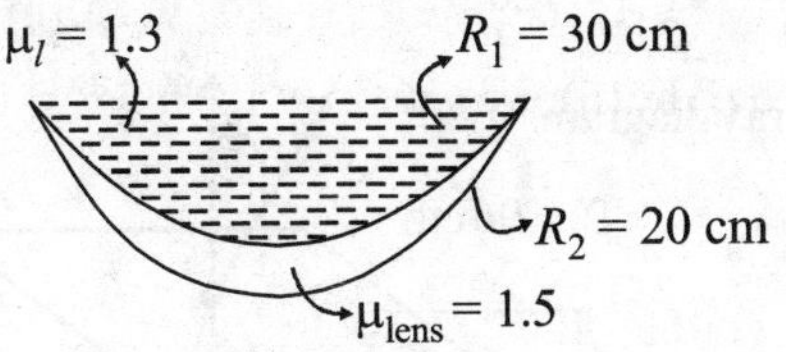

$$\frac{1}{f}=\left(\frac{1.3-1}{1}\right)\left(\frac{1}{\infty}-\frac{1}{-30}\right)$$

$$=\left(\frac{1.5-1}{1}\right)\left(\frac{1}{-30}-\frac{1}{-30}\right)$$

$$=\frac{0.3}{30}+\frac{0.5}{60}=\frac{1}{100}+\frac{1}{120}=\frac{6+5}{600}=\frac{11}{600}$$

$$\therefore f=\frac{600}{11}\text{ cm}$$

85. (d) Equivalent focal length of the convex and concave lens system,

$$\frac{1}{f}=\frac{1}{f_1}+\frac{1}{f_2}=\frac{1}{30}+\frac{1}{-20}=\frac{2-3}{60}=-\frac{1}{60}$$

$\therefore f=-60$ cm

Now using lens formula,

$$\frac{1}{v}-\frac{1}{u}=\frac{1}{f}$$

$$\Rightarrow \frac{1}{v}-\frac{1}{-20}=\frac{1}{-60}\Rightarrow\frac{1}{v}=-\frac{1}{20}-\frac{1}{60}=\frac{-3-1}{60}$$

86. (b) By lens maker's formula,

$$\frac{1}{f}=\left[\frac{\mu_L}{\mu_m}-1\right]\left[\frac{1}{R_1}-\frac{1}{R_2}\right]$$

For air, $\mu_m=1,\ \mu_L=1.6,\ f=12$ cm

$$\therefore\frac{1}{12}=[1.6-1]\left[\frac{1}{R_1}-\frac{1}{R_2}\right]\Rightarrow\frac{1}{12}=\frac{6}{10}\left[\frac{1}{R_1}-\frac{1}{R_2}\right]$$

$$\Rightarrow\left[\frac{1}{R_1}-\frac{1}{R_2}\right]=\frac{10}{72}\qquad\text{...(i)}$$

For water, $\mu_m=1.28$

$$\frac{1}{f}=\left[\frac{1.6}{1.28}-1\right]\left[\frac{10}{72}\right]=\frac{32}{128}\times\frac{10}{72}\qquad\text{[Using (i)]}$$

$$\Rightarrow\frac{1}{f}=\frac{1}{4}\times\frac{10}{72}$$

$\therefore f=28.8$ cm $=288$ mm.

87. (d) The equivalent focal length of combination of two lens placed at a distance d is

$$\frac{1}{f_{eq}}=\frac{1}{f_1}+\frac{1}{f_2}-\frac{d}{f_1f_2}$$

Here, $f_1=30$ cm $=0.3$ m, $f_2=10$ cm $=0.1$ m, $d=10$ cm $=0.1$ m

$$\therefore\frac{1}{f_{eq}}=\frac{1}{0.3}+\frac{1}{0.1}-\frac{0.1}{(0.3)(0.1)}$$

$$\Rightarrow\frac{1}{f_{eq}}=\frac{1}{0.1}$$

$$\therefore\text{ Power, }P=\frac{1}{f_{eq}}=10D.$$

88. (4) For refraction through a curved surface,

$$\frac{\mu_2}{v}-\frac{\mu_1}{u}=\frac{\mu_2-\mu_1}{R}$$

Here, $\mu_1=1,\ \mu_2=1.5$

$u=-x,\ v=-200$ cm, $R=50$ cm

$$\Rightarrow\frac{1.5}{200}-\frac{1}{-x}=\frac{1.5-1}{50}$$

$$\Rightarrow\frac{1}{x}=\frac{1}{100}-\frac{3}{400}$$

$\Rightarrow x=400$ cm

$\therefore x=4$ m

89. (c) By lens maker's formula,

$$\frac{1}{f}=(\mu-1)\left(\frac{1}{R_1}-\frac{1}{R_2}\right)$$

Here, f = 12 cm, $R_1=10$ cm, $R_2=-15$ cm

$$\frac{1}{12}=(\mu-1)\left(\frac{1}{10}-\frac{1}{-15}\right)$$

$$\frac{1}{12}=(\mu-1)\left(\frac{3+2}{30}\right)$$

$$\therefore\quad\mu=\frac{3}{2}$$

90. (b) Here $\frac{1}{f_1}=\frac{1}{f_2}$

or, $(\mu-1)\left(\frac{1}{R_1}-\frac{1}{-R_2}\right)=(\mu-1)\left(\frac{2}{R}\right)$

$$\frac{1}{R_1}+\frac{1}{R_2}=\frac{2}{R}=\frac{2}{\frac{1}{6}}=12$$

Therefore the possible combination of R_1 and R_2 is $\frac{1}{5}$ cm and $\frac{1}{7}$ cm

91. (b) From Lens formula, $\frac{1}{v}-\frac{1}{u}=\frac{1}{f}$

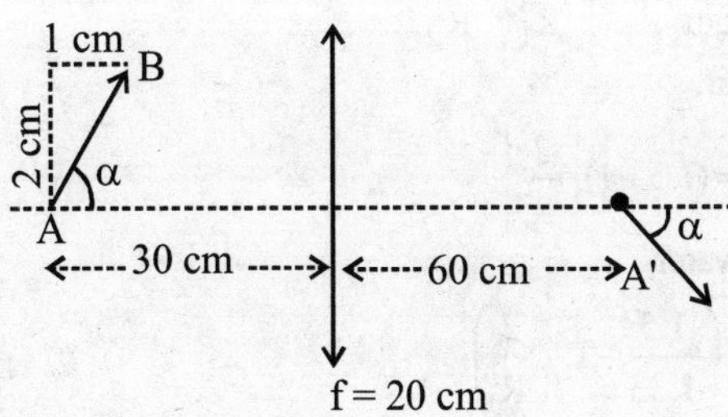

Location of image of A

$\frac{1}{v}-\frac{1}{u}=\frac{1}{f}\Rightarrow\frac{1}{v}-\frac{1}{-30}=\frac{1}{20}\Rightarrow\frac{1}{v}=\frac{1}{60}\Rightarrow v=60\text{cm}$

$\therefore$ magnification $m=\frac{v}{u}=\frac{60}{30}=2$

Since size of object is small wrt the location.

$\therefore dv=m^2du\Rightarrow dv=4\times 1=4$ cm

$h_i=mh_0\Rightarrow h_i(dy)=2\times 2=4$ cm

$\therefore$ Angle made with principle axis = –45°

92. **(b)** Using $\frac{\mu_2}{v}-\frac{\mu_1}{u}=\frac{\mu_2-\mu_1}{R}$

or, $\frac{1.5}{v}-\frac{1}{(-0.2)}=\frac{1.5-1}{0.4}$

$\Rightarrow\frac{1.5}{v}=\frac{0.5}{0.4}-\frac{1}{0.2}\Rightarrow\frac{1.5}{v}=-\frac{1.5}{0.4}$

$\therefore\quad v=-0.4$ m

i.e., 0.4m left to the spherical surface.

93. **(d)** The focal length of lenses are

$f_{L_1}=f_{L_2}=f$ and $f_{L_3}=f_{L_4}=2f$

$\therefore\quad f_{L_1}:f_{L_3}=1:2$

94. **(b)** For refraction at a spherical surface,

$\frac{\mu_2}{v}-\frac{\mu_1}{u}=\frac{\mu_2-\mu_1}{R}$

For B,

$\frac{1.5}{V}+\frac{1}{\frac{R}{2}}=\frac{0.5}{-R}\Rightarrow\frac{1.5}{V}=-\frac{1}{2R}-\frac{2}{R}$

$\Rightarrow\frac{1.5}{V}=\frac{-5}{2R}\Rightarrow V_B=-0.6\,R$

Similarly, for A,

$\frac{1.5}{V}+\frac{2}{3R}=\frac{0.5}{-R}\Rightarrow\frac{1.5}{V}=-\frac{1}{2R}-\frac{2}{3R}\Rightarrow\frac{1.5}{V}=-\frac{7}{6R}$

$\therefore V_A=-\frac{9}{7}R$

$\therefore$ Distance between images

$=2R-\left(0.6R+\frac{9}{7}R\right)=0.114R$

95. **(b)** By lens maker's formula,

$\frac{1}{f}=\left(\frac{\mu_\ell}{\mu_s}-1\right)\left[\frac{1}{R_1}-\frac{1}{R_2}\right]$

For air,

$\frac{1}{24}=(1.5-1)\left[\frac{2}{R}\right]$ .... (i)

For water,

$\frac{1}{f'}=\left(\frac{1.5}{1.33}-1\right)\left(\frac{2}{R}\right)$

$\frac{1}{f'}=\left(\frac{1.5\times 3}{4}-1\right)\frac{2}{R}$ .... (ii)

Dividing eq (i) by (ii), we get

$\frac{f'}{24}=4\quad\therefore\quad f'=96\,\text{cm}$

96. **(c)**

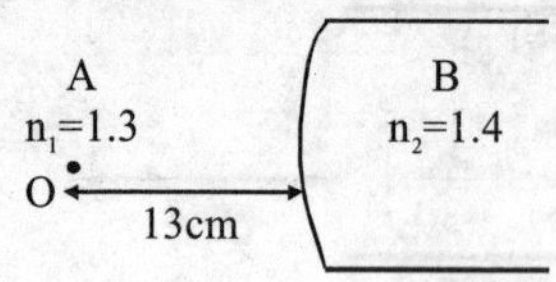

For refraction at a spherical surface

$\frac{n_2}{v}-\frac{n_1}{u}=\frac{n_2-n_1}{R}$

$\Rightarrow\frac{1.4}{v}-\frac{1.3}{-13}=\frac{0.1}{R}$

$\Rightarrow\frac{1.4}{v}=\frac{1-R}{10R}$

From magnification formula, $m=\frac{v/n_2}{u/n_1}$

$\Rightarrow\quad -2\times\frac{(-13)}{1.3}=\frac{10R}{1-R}$

$\Rightarrow\quad R=\frac{2}{3}$ cm

97. **(a)** For drone camera, size = $(2\times 2)\times 10^{-4}\,m^2$

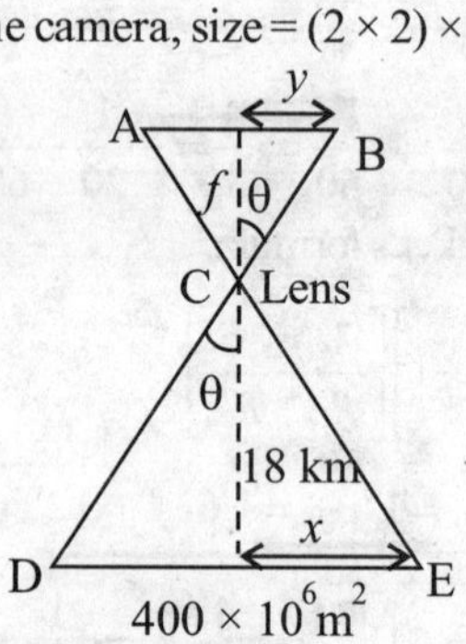

From similar ΔABC & ΔCDE,

$\frac{y}{x}=\frac{f}{18\times 10^3}\Rightarrow\frac{2\times 10^{-2}}{20\times 10^3}=\frac{f}{18\times 10^3}$

$\therefore f=18\times 10^{-3}\,m=18\,mm=1.8\,cm$

98. **(a)** The focal length of a lens is $F=\frac{1}{P}m$

$\therefore F_1=\frac{1}{P_1}=\frac{1}{2.5}m\qquad(\because P_1=2.5\,D\text{ given})$

$\therefore F_2=\frac{1}{P_2}=\frac{1}{2.6}m\qquad(\because P_2=2.6\,D\text{ given})$

$\therefore$ Relative decrease in focal length is

$\frac{F_1-F_2}{F_1}=\frac{\frac{2}{5}-\frac{5}{13}}{\frac{2}{5}}=1-\frac{25}{26}=\frac{1}{26}=0.04$

99. (b) For silvered lens,

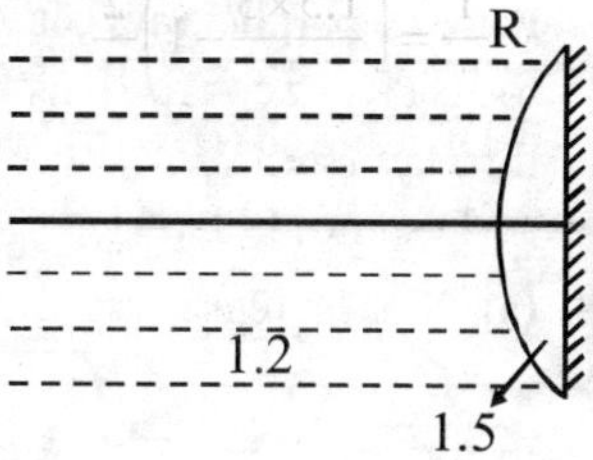

$$-\frac{1}{F_e}=\frac{2}{f_l}-\frac{1}{f_m}$$

$$\Rightarrow -\frac{1}{-0.2}=2\left[\frac{1.5}{1.2}-1\right]\frac{1}{R}$$

$$\therefore R=\frac{2\times0.3\times0.2}{1.2}=0.10\text{m}$$

100. (b) By lens maker's formula,

$$\frac{1}{f}=\left(\frac{\mu_2}{\mu_1}-1\right)\left(\frac{1}{R_1}-\frac{1}{R_2}\right)$$

In air, $R_1=2$ cm

$$\therefore\ \frac{1}{f_1}=(1.5-1)\left[\frac{1}{2}-0\right]\Rightarrow f_1=4\text{cm}$$

In water, $R_1=3$ cm

$$\therefore\ \frac{1}{f_2}=\left(\frac{1.5}{1.2}-1\right)\left(\frac{1}{3}-0\right)\Rightarrow\ \frac{1}{f_2}=\frac{0.3}{1.2}\times\frac{1}{3}$$

$$\Rightarrow f_2=12\text{ cm}\ \therefore\ f_1:f_2=4:12=1:3$$

101. (a)

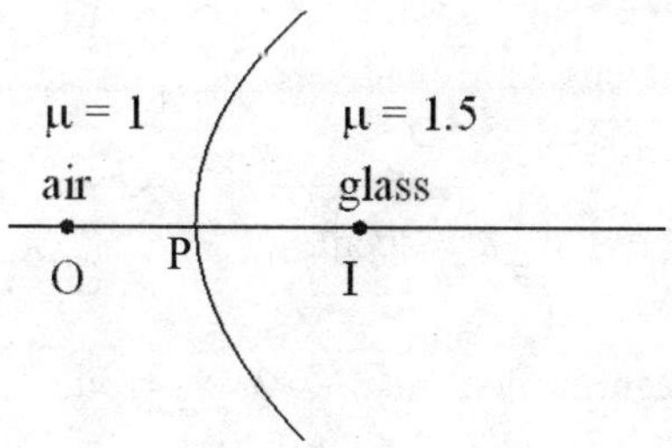

Distances are measured from P and are taken as positive in the direction of incident ray of light

$PO=u=-x$

$PI=v=x$ $\quad [\because PO=PI]$

Using formula of refraction on a spherical surfaces

$$\frac{\mu_2}{v}-\frac{\mu_1}{u}=\frac{\mu_2-\mu_1}{R}$$

$$\frac{1.5}{x}+\frac{1}{x}=\frac{1}{2R}\Rightarrow\frac{5}{2x}=\frac{1}{2R}\ \therefore\ x=5R$$

102. (b) The expression of the focal length is given by

$$\frac{1}{f}=\left(\frac{\mu_2}{\mu_1}-1\right)\left(\frac{1}{R_1}-\frac{1}{R_2}\right)$$

Since, the second surface is silvered, light passing through the lens will reflect back, effectively making it work like a combination of two lens and a concave mirror

$$\frac{1}{F_{eff}}=\frac{2}{F_{lens}}+\frac{1}{F_{minor}}$$

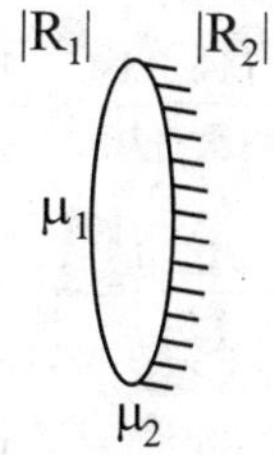

$$\frac{1}{f_{eq}}=\frac{2}{f_L}-\frac{1}{f_m}$$

$$f_m=-\frac{|R_2|}{2}$$

$$\frac{1}{f_L}=\left(\frac{\mu_2}{\mu_1}-1\right)\left(\frac{1}{|R_1|}+\frac{1}{|R_2|}\right)$$

$$\frac{1}{f_{eq}}=2\left(\frac{\mu_2-\mu_1}{\mu_1}\right)\left(\frac{|R_1|+|R_2|}{|R_1||R_2|}\right)+\frac{2}{|R_2|}$$

$$=\frac{2}{|R_2|}\left[\frac{(\mu_2-\mu_1)(|R_1|+|R_2|)+\mu_1|R_1|}{\mu_1|R_1|}\right]$$

$$=\frac{2}{|R_2|}\left[\frac{\mu_2|R_1|+\mu_2|R_2|-\mu_1|R_1|-\mu_1|R_2|+\mu_1|R_1|}{\mu_1|R_1|}\right]$$

$$\frac{1}{f_{eq}}=\frac{2[\mu_2|R_1|+\mu_2|R_2|-\mu_1|R_2|]}{\mu_1|R_1||R_2|}$$

For same size of image

$u=2f$

$$u=\frac{\mu_1|R_1||R_2|}{\mu_2(|R_1|+|R_2|)-\mu_1|R_2|}$$

103. (d) For original lens,

$$\frac{1}{f_1}=(\mu-1)\frac{2}{R}=P_1=4D$$

For remaining lens,

$$P_2=\frac{1}{f_2}=(\mu-1)\frac{1}{R}=\frac{P_1}{2}=2D$$

104. (b)

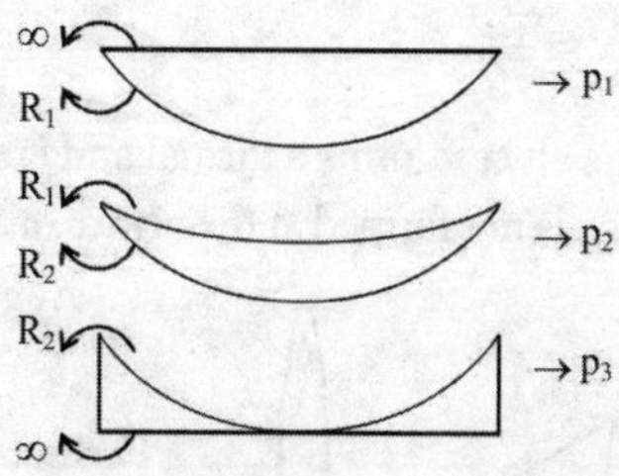

$$\frac{1}{f_{eq}}=\frac{1}{f_1}+\frac{1}{f_2}+\frac{1}{f_3}$$

$$p_{eq}=p_1+p_2+p_3$$

$$p_1 = \left(\frac{4}{3}-1\right)\left(\frac{1}{\infty}-\frac{1}{-|R_1|}\right) \Rightarrow p_1 = \left(\frac{1}{3|R_1|}\right)$$

$$p_2 = \left(\frac{1}{2}\right)\left(\frac{1}{-|R_1|}+\frac{1}{|R_2|}\right) = \frac{1}{2}\left(\frac{1}{|R_2|}-\frac{1}{|R_1|}\right)$$

$$p_3 = \left(\frac{1}{3}\right)\left(\frac{1}{-|R_2|}-\frac{1}{\infty}\right) = -\frac{1}{3|R_2|}$$

$$p_{eq} = \frac{1}{3}\left(\frac{1}{|R_1|}-\frac{1}{|R_2|}\right)-\frac{1}{2}\left(\frac{1}{|R_1|}-\frac{1}{|R_2|}\right)$$

$$= -\frac{1}{6}\left(\frac{1}{|R_1|}-\frac{1}{|R_2|}\right)$$

105. (d) $P_{eq} = 2P_\ell + P_m$

$$-\frac{1}{f_{eq}} = \frac{2}{f_\ell} - \frac{1}{f_m}$$

But, $\frac{1}{f_\ell} = (\mu-1)\left[\frac{1}{R}-\left(\frac{1}{-R}\right)\right]$

$$\Rightarrow \frac{1}{f_\ell} = \frac{2(\mu-1)}{R}$$

$$\therefore \frac{-1}{f_{eq}} = \frac{4(\mu-1)}{R} - \frac{2}{-R} = \frac{1}{R}(4\mu-4+2) = \frac{1}{R}(4\mu-2)$$

$$\Rightarrow \frac{1}{f_{eq}} = \frac{-1}{R}(4\mu-2) \Rightarrow f_{eq} = -\frac{R}{(4\mu-2)}$$

$\therefore$ Object distance,

$$v = -2f_{eq} = 2\left(\frac{R}{4\mu-2}\right) = \frac{R}{(2\mu-1)}$$

106. (a) Using lens formula

$$\frac{1}{v}-\frac{1}{u}=\frac{1}{f}$$

When object is at centre of curvature of lens, $u=-2f$

$$\therefore \frac{1}{v}-\frac{1}{-2f}=\frac{1}{-f}$$

For concave lens focal length is negative.

$$\Rightarrow \frac{1}{v}=\frac{-3}{2f} \Rightarrow v = \frac{-2f}{3}$$

Concave lens always forms a virtual and erect image. Hence, image is not formed at the centre of curvature.

107. (c)

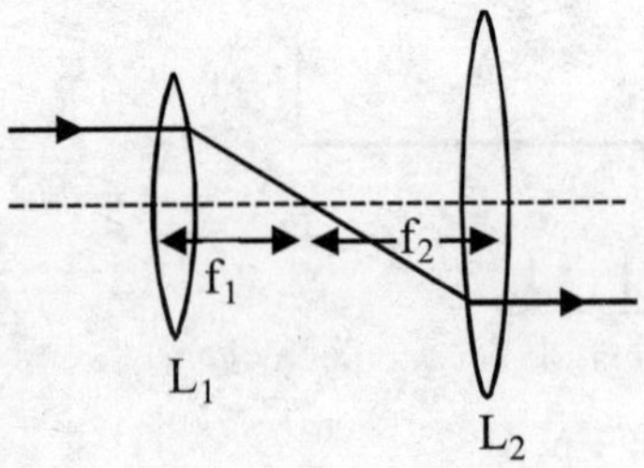

Paraxial parallel rays converges at focus (f_1) and ray from focus (f_2) of convex lens becomes parallel after passing through lens.

$\therefore$ Distance $L_1 L_2 = f_1 + f_2 = 25$ cm

108. (a) For Ist lens given $f_1 = 10$, $u = -30$, $v = ?$

using, $\frac{1}{v}-\frac{1}{u}=\frac{1}{f} \Rightarrow v = \frac{uf}{u+f} = \frac{-30\times 10}{-30+10} = 15$

For 2nd lens given $f_1 = -10$, $u = 10$, $v = ?$

using, $\frac{1}{v}-\frac{1}{u}=\frac{1}{f} \Rightarrow v = \frac{uf}{u+f} = \frac{10\times -10}{10-10} = \infty$

for 3rd lens given $f = 30$, $u = -\infty$, $v = ?$

using, $\frac{1}{v}-\frac{1}{u}=\frac{1}{f}$ So v will be 30.

109. (c) Using lens makers formula,

$$\frac{1}{f} = \left(\frac{\mu_1}{\mu_a}-1\right)\left(\frac{1}{R_1}-\frac{1}{R_2}\right)$$

$$\Rightarrow \frac{1}{20} = (\mu-1)\left(\frac{1}{15}-\frac{1}{(-30)}\right) \Rightarrow \mu-1 = \frac{1}{2} \Rightarrow \mu = 1.5$$

110. (a) Equivalent Power, $P_{eq} = P_1 + P_2 + P_3 + ...$

Since all lenses are identical

$\therefore 5P = 25D \Rightarrow P = 5D$

$$\therefore \frac{1}{f} = 5 \Rightarrow f = \frac{1}{5}m = 20cm$$

111. (15) Magnification, $m = \frac{v}{u} = 3 \Rightarrow v = 3u$

Given, $v + u = 20$

$3u - u = 20 \Rightarrow u = 10$ cm.

using lens formula, $\frac{1}{u}-\frac{1}{v}=\frac{1}{f}$

$$\frac{1}{-30}-\frac{1}{-10}=\frac{1}{f} \Rightarrow \frac{2}{30}=\frac{1}{f} \Rightarrow f = 15cm$$

112. (200) Given,

Radius of convex curved surface, R = 20 cm

Distance of object from curved surface, u = 100 cm

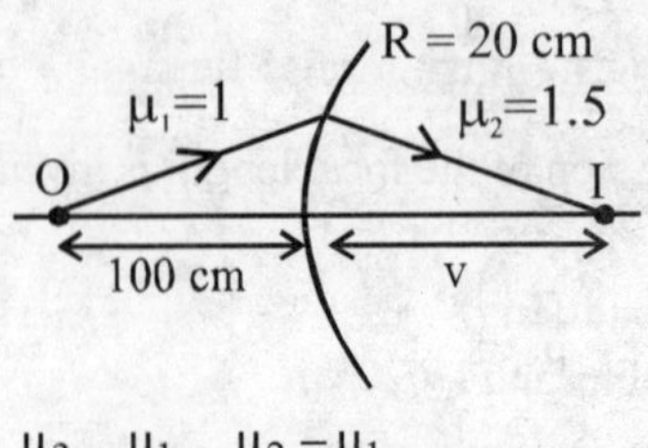

$$\frac{\mu_2}{v}-\frac{\mu_1}{u}=\frac{\mu_2-\mu_1}{R}$$

$$\Rightarrow \frac{1.5}{v}-\frac{1}{-100}=\frac{1.5-1}{20} \Rightarrow \frac{1.5}{v}=\frac{1}{40}-\frac{1}{100}$$

$\Rightarrow v = 100$ cm

Distance from object = 100 + 100 = 200 cm

113. (10) Magnification, $m=\frac{v}{u}=-2$

$v=-2u$

$v-u=45 \Rightarrow -2u-u=45$

$u=-15\text{ cm} \Rightarrow u=-15\text{ cm}$

$v=30\text{ cm}$

$\frac{1}{f}=\frac{1}{v}-\frac{1}{u}=\frac{1}{30}+\frac{1}{15} \Rightarrow f=+10\text{ cm}$

114. (d) By lens maker formula,

$\frac{1}{f}=\left(\frac{\mu_2}{\mu_1}-1\right)\left(\frac{1}{R_1}-\frac{1}{R_2}\right)$

$\frac{1}{20}=(1.5-1)\left(\frac{2}{R}\right) \Rightarrow R=20\text{ cm}$

$\frac{1}{f'}=\left(\frac{1.5}{1.6}-1\right)\left(\frac{2}{R}\right)$

$[\because R_1=R, R_2=-R]$

$=\frac{-1}{16}\times\frac{2}{20}$

$f'=-160\text{ cm}$

115. (5) Power of lens, $P=\frac{1}{f}$

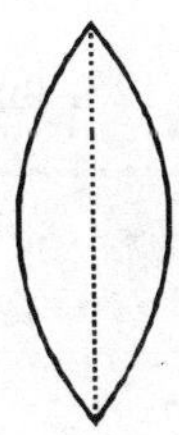

Let power of each part is P_1

$\therefore \quad P_1+P_1=P=\frac{1}{f} \Rightarrow 2P_1=\frac{1}{0.1}=10$

$\Rightarrow \quad P_1=5D$

Hence the power of each lens after cut is 5D

116. (100)

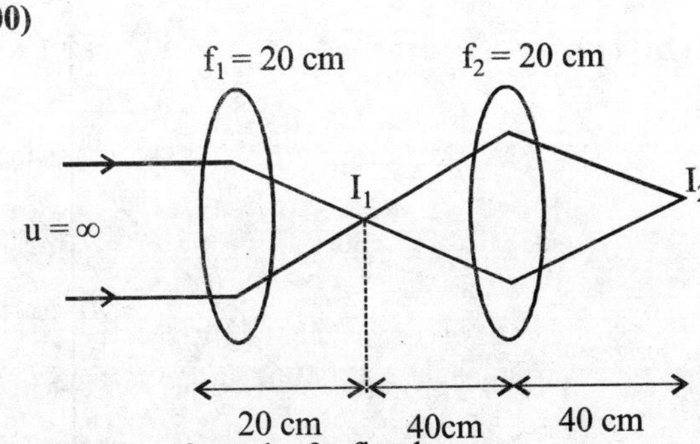

Using lens formula, for first lens

$\frac{1}{v}-\frac{1}{u}=\frac{1}{f} \Rightarrow \frac{1}{v}-\frac{1}{\infty}=\frac{1}{f} \quad \therefore v=f$

For second lens

$u=-40\text{ cm}$ and $f=20\text{ cm}$

$\Rightarrow \frac{1}{v}-\frac{1}{u}=\frac{1}{f} \Rightarrow \frac{1}{v}-\frac{1}{(-40)}=\frac{1}{20}$

$\Rightarrow \frac{1}{v}=\frac{1}{20}-\frac{1}{40}$

$\therefore v=40\text{ cm}$

Distance of final image

$=20\text{ cm}+40\text{ cm}+40\text{ cm}=100\text{ cm}$

117. (4) From the lens maker's formula

$\frac{1}{f}=\left(\frac{\mu_\ell}{\mu_m}-1\right)\left(\frac{1}{R_1}-\frac{1}{R_2}\right)$

Here $R_1=R_2=R=20\text{ cm}$

$\therefore P=\frac{1}{f}=\left(\frac{\mu_\ell}{\mu_m}-1\right)\left(\frac{1}{R}-\frac{1}{-R}\right)$

When lens is in air

$\mu_\ell=1.8, \mu_{air}=1$

$\therefore P=(1.8-1)\left(\frac{1}{20}+\frac{1}{20}\right)$

When lens is immersed in liquid

$\mu_\ell=1.8, \mu_m=1.5$

$\therefore P'=\left(\frac{1.8}{1.5}-1\right)\left(\frac{1}{20}+\frac{1}{20}\right);$

Dividing $\frac{P}{P'}=\frac{0.8}{1.2-1}=4$

118. (34) For first lens

$\frac{1}{v}-\frac{1}{u}=\frac{1}{f}$

$\Rightarrow \frac{1}{v}-\frac{1}{-6}=\frac{1}{24}$

$\Rightarrow \frac{1}{v}+\frac{1}{6}=\frac{1}{24}=\frac{1}{24}-\frac{1}{6}=-\frac{1}{8}$.

$\Rightarrow v=-8\text{ cm}$

For second lens

$\frac{1}{v}-\frac{1}{-18}=\frac{1}{9} \Rightarrow \frac{1}{v}=\frac{1}{9}-\frac{1}{18}$

$\Rightarrow \frac{1}{v}=\frac{2-1}{18} \Rightarrow v=18\text{ cm}$

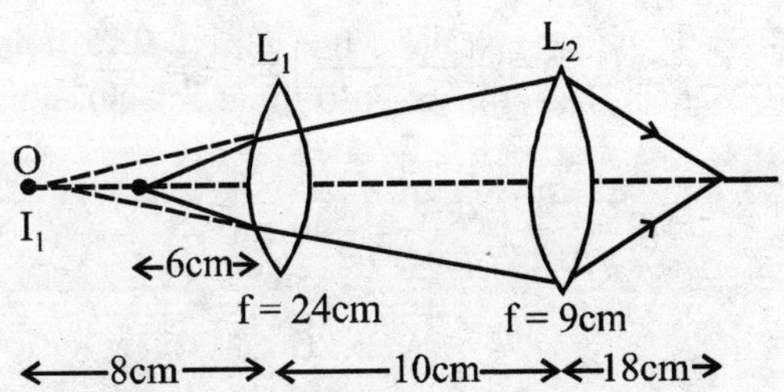

Distance between object and its final image

$=6+10+18=34\text{ cm}$

119. (30)

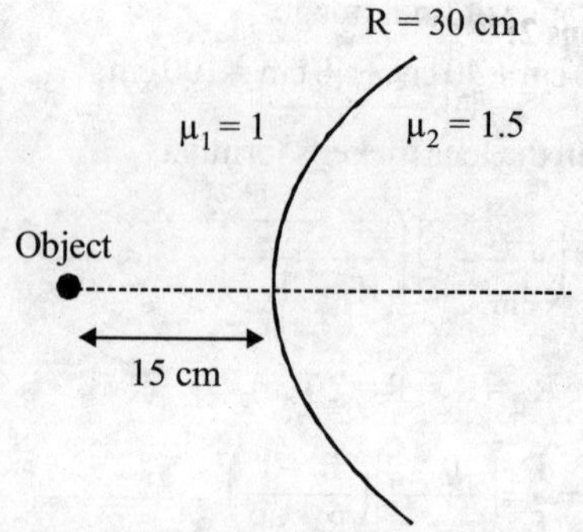

$$\frac{\mu_2}{v}-\frac{\mu_1}{u}=\frac{\mu_2-\mu_1}{R}$$

$$\Rightarrow \frac{1.5}{v}-\frac{1}{-15}=\frac{1.5-1}{30}=\frac{1}{60}\Rightarrow\frac{1.5}{v}+\frac{1}{15}=\frac{1}{60}$$

$$\Rightarrow \frac{1.5}{v}=\frac{1}{60}-\frac{1}{15}=\frac{-1}{20}\Rightarrow\frac{1.5}{v}=-\frac{1}{20}\Rightarrow v=-30\text{cm}$$

The distance of image from pole of the surface is 30 cm.

120. (d) The distance for the distinct vision, $u=-25$ cm

$P=2D=2\text{m}^{-1}$

$$\Rightarrow\frac{1}{f}=\frac{2}{100}\text{cm}^{-1}$$

$$\frac{1}{v}-\frac{1}{u}=\frac{1}{f}\Rightarrow\frac{1}{v}-\left(-\frac{1}{25}\right)=\frac{2}{100}\Rightarrow\frac{1}{v}=\frac{1}{50}-\frac{1}{25}$$

The least distance of distinct vision for this person.

$\Rightarrow v=-50$ cm

121. (30) How I_1 is image formed by lens and I_2 is image formed by mirror.

f = 10 cm

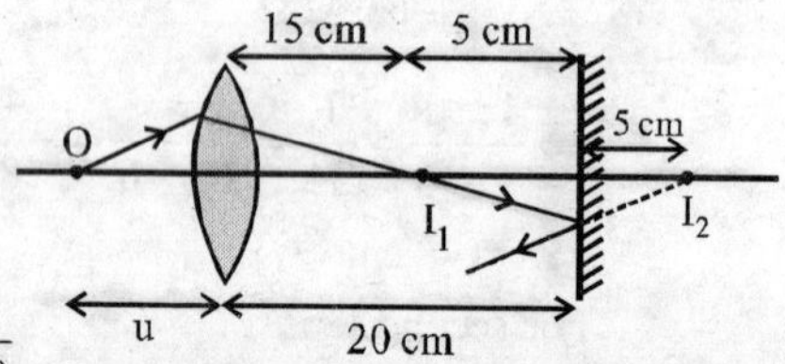

$$\frac{1}{v}-\frac{1}{u}=\frac{1}{f}$$

$$\frac{1}{15}-\frac{1}{-u}=\frac{1}{10}$$

$$\Rightarrow\frac{1}{u}=\frac{1}{10}-\frac{1}{15}\Rightarrow u=30\text{ cm}$$

122. (120) For plano-concave lens $(R_1=\infty, R_2=30\text{ cm})$

$$\frac{1}{f_1}=(1.75-1)\left(\frac{1}{\infty}-\frac{1}{+30}\right)\Rightarrow\frac{1}{f_1}=\frac{0.75}{-30}$$

$$\Rightarrow f_1=\frac{-30}{0.75}=\frac{-30}{\frac{3}{4}}=-40\text{ cm}$$

For plano-convex lens $(R_1=30\text{cm}, R_2=\infty)$

$$\frac{1}{f_2}=(1.75-1)\left(\frac{1}{30}-\frac{1}{\infty}\right)\Rightarrow\frac{1}{f_2}=\frac{0.75}{30}\Rightarrow f_2=40\text{ cm}$$

So, image from L_1 will be virtual and on the left of L_1 at focal length 40 cm. So the object for L_2 will be 80 cm from L_2 which is 2f. Final image is formed at 80 cm from L_2 on the right.

So, x = 120

123. (b) Chromatic aberration is a phenomenon in which light rays passing through a lens focus at different points, depending on their wavelength.

124. (54) By len's maker formulae, we have

$$\frac{1}{f_{H_2O}}=\left(\frac{\mu_g}{\mu_{H_2O}}-1\right)\left(\frac{2}{R}\right)=\frac{1}{8}\left(\frac{2}{R}\right)=\frac{1}{4R}$$

$$\frac{1}{f_{air}}=(\mu_g-1)\left(\frac{2}{R}\right)=\frac{0.5\times2}{R}=\frac{1}{R}$$

So, $f_{H_2O}=4f_{air}=72$ cm

So change in focal length = 72 – 18 = 54 cm

125. (b) $\frac{1}{f}=({}_m\mu_g-1)\left(\frac{1}{R_1}-\frac{1}{R_2}\right)$

$$\Rightarrow 1.25=\left(\frac{1.5}{\mu_m}-1\right)\left(\frac{1}{0.2}+\frac{1}{0.4}\right)$$

$$\Rightarrow\frac{1.25}{7.5}=\frac{1.5}{\mu_m}-1\Rightarrow\frac{7}{6}=\frac{1.5}{\mu_m}\Rightarrow\mu_m=\frac{9}{7}$$

126. (10)

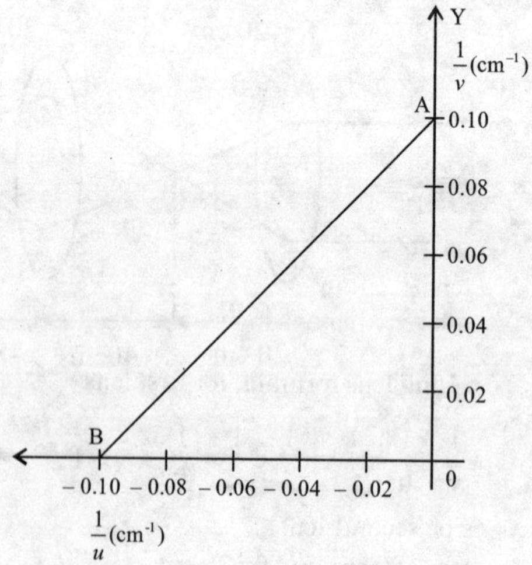

For point B, $\frac{1}{u}=-0.10\text{ cm}^{-1}$, $\frac{1}{v}=0$

$\therefore$ Thus, $u=-10$ cm, $v=\infty$

i.e, f = 10 cm

$$\Rightarrow\frac{1}{10\text{ cm}}=(1.5-1)\left(\frac{2}{R}\right)=\frac{1}{R}\Rightarrow R=10\text{ cm}$$

127. (b) u = – 240 cm

v = 12 cm

frani lens formula

$$f = \frac{uv}{u - v} = \frac{-240 \times 12}{-240 - 12} = \frac{80}{7}$$

Normal shift by glass

$$= t\left(1 - \frac{1}{u}\right) = \frac{1}{3} \text{cm}$$

Now, the image distance should be as

$$v' = \left(v - \frac{1}{3}\right) = 12 - \frac{1}{3} = \frac{35}{3} \text{cm}$$

from lens formula

$$-\frac{1}{u'} = \frac{1}{f} - \frac{1}{v'} = \frac{7}{80} - \frac{3}{35}$$

$$= \frac{49 - 48}{560}$$

$u' = -560$ cm

$u' = -5.6$ m

Shift in position of object

$= 5.6 - 2.4 = 3.2$m

128. (225) At surface 1

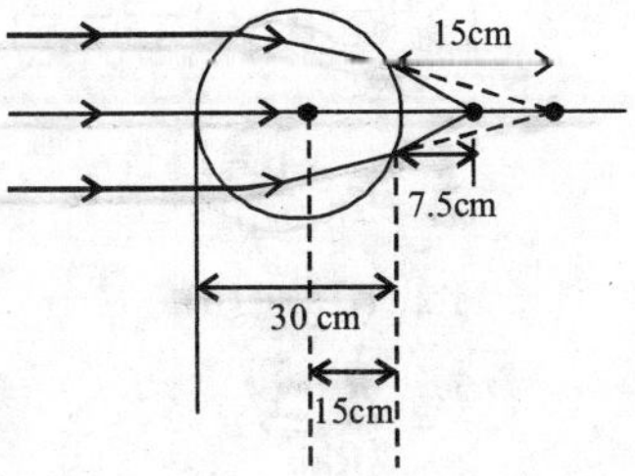

$$\frac{1.5}{v_1} - \frac{1}{\infty} = \frac{1.5 - 1}{15}$$

$$\frac{1.5}{v_1} = \frac{1}{30} \Rightarrow v = 45 \text{ cm}$$

For surface 2

$$\frac{1}{v} - \frac{1.5}{15} = \frac{1 - 1.5}{-15} \Rightarrow \frac{1}{v} - \frac{1}{10} = \frac{1}{30}$$

$$\frac{1}{v} = \frac{1}{30} + \frac{1}{10} = \frac{1 + 3}{30}$$

$\therefore v = \frac{30}{4} \Rightarrow v = 7.5$ cm. So required distance $= (15 + 7.5)$cm $= 22.5$ cm $= 225$ mm.

129. (a) If a lens is cut into two half along principal axis, its focal length remain unchanged only the intensity of image get reduced.

So, power of L_1 will remain unaffected as $P = \frac{1}{f}$.

130. (10) By lens maker formula,

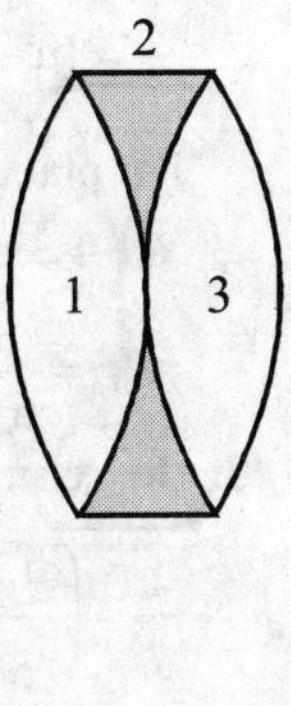

$$\frac{1}{f} = ({}_g\mu_a - 1)\left(\frac{1}{R_1} - \frac{1}{R_2}\right)$$

For Lens 1:

$$\frac{1}{f_1} = (1.5 - 1)\left(\frac{1}{R} - \frac{1}{-R}\right)$$

$$\Rightarrow \frac{1}{f_1} = 0.5 \times \frac{2}{R}$$

$$\Rightarrow f_1 = R \Rightarrow R = 15 \text{ cm}$$

For Lens 2:

$$\frac{1}{f_2} = (1.25 - 1)\left(\frac{1}{-R} - \frac{1}{+R}\right)$$

$$= -0.25 \times \frac{2}{R} = -\frac{0.5}{R} = \frac{-1}{2R}$$

$$\therefore f_2 = -2R$$

For Lens 3:

Similarly like lens 1, $f_3 = R$

$$\text{So, } \frac{1}{f_{net}} = \frac{1}{f_1} + \frac{1}{f_2} + \frac{1}{f_3} = \frac{1}{R} + \frac{1}{-2R} + \frac{1}{R}$$

$$= \frac{3}{2R} = \frac{3}{2 \times 15} = \frac{1}{10} \quad \therefore f_{net} = 10 \text{ cm}$$

131. (c) $u = -f$

Using lens formula

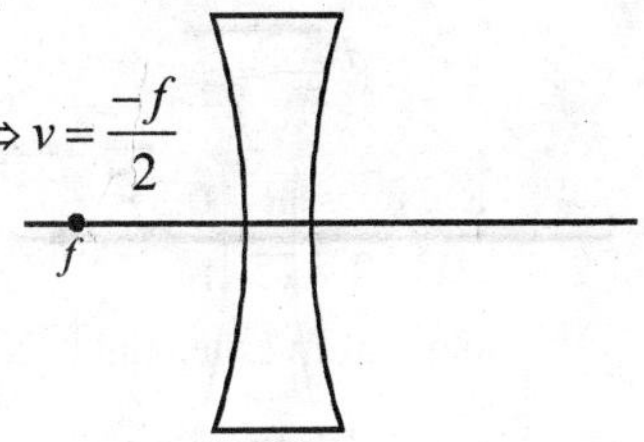

$$\frac{1}{v} - \frac{1}{u} = \frac{1}{-f} \Rightarrow \frac{1}{v} = -\frac{2}{f} \Rightarrow v = \frac{-f}{2}$$

$$m = \frac{v}{u} = \frac{1}{2}$$

$$\text{Distance} = \frac{f}{2}$$

132. (a) Using lens formula

$$\frac{1}{v} - \frac{1}{u} = \frac{1}{f} \Rightarrow \frac{1}{V_1} - \frac{1}{-30} = \frac{1}{+10} \Rightarrow V_1 = 15 \text{ cm.}$$

So, $u_2 = (15 - 5)$ cm $= 10$ cm

$$\frac{1}{V_2} - \frac{1}{+10} = \frac{1}{-10} \Rightarrow \frac{1}{V_2} = 0 \Rightarrow V_2 = \infty$$

So, $u_3 = \infty$

Then, $V_3 = 30$ cm

$OV_3 = 75$ cm

133. (b) Focal length of plano-convex lens

$$\frac{1}{f_1} = (\mu_1 - 1)\left(\frac{1}{R}\right) \quad [\because R_1 = \infty \text{ and } R_2 = -R]$$

Focal length of plano-concave lens

$$\frac{1}{f_2} = (\mu_2 - 1)\left(\frac{-1}{R}\right) \quad [\because R_1 = -R \text{ and } R_2 = \infty]$$

Focal length of combination

$$\frac{1}{f_{eq}} = \frac{1}{f_1} + \frac{1}{f_2} = \frac{(\mu_1 - 1) - (\mu_2 - 1)}{R}$$

$$\Rightarrow \frac{1}{f_{eq}} = \frac{(\mu_1 - \mu_2)}{R} \Rightarrow \frac{R}{f_{eq}} = (\mu_1 - \mu_2)$$

134. (50)

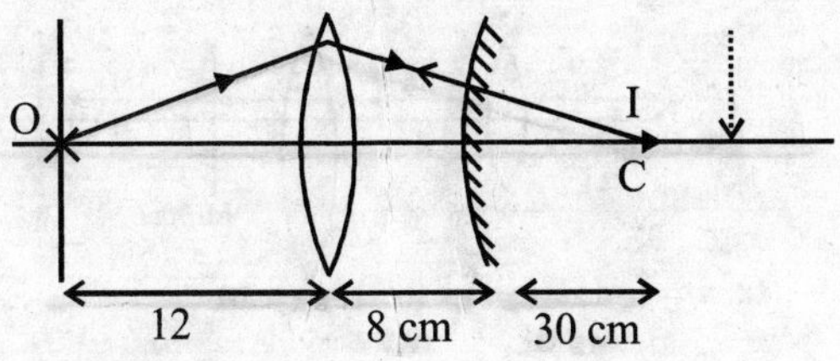

Image will coincide with object, if light fall perpendicularly to mirror. Which is possible if light converge at C of mirror. In the absence of mirror, light will be converging at centre of curvature. Distance $= 12 + 8 + 30 = 50$ cm

135. (d) In given case, medium 1 has refractive index 1.25 and medium 2 has refractive index 1.4.

From the refraction formula

$$\frac{n_2}{v} - \frac{n_1}{u} = \frac{n_2 - n_1}{R} \Rightarrow \frac{1.4}{v} - \frac{1.25}{-40} = \frac{1.4 - 1.25}{(-25)}$$

$$\Rightarrow \frac{1.4}{v} = \frac{-0.15}{25} - \frac{1.25}{40} \Rightarrow v = -37.58 \text{cm}$$

136. (d) From figure,

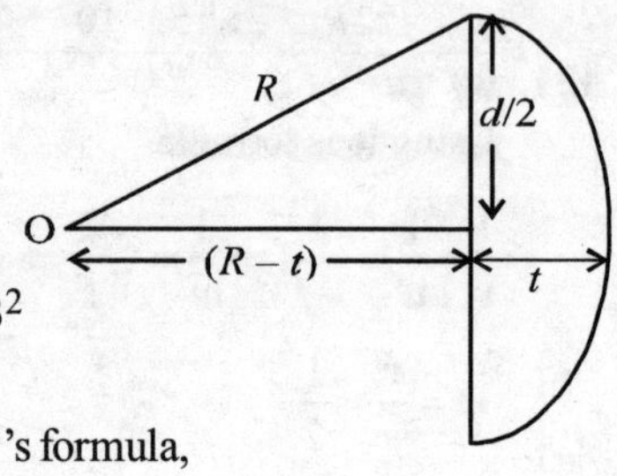

$t = 0.3$ cm

$\frac{d}{2} = 3$ cm

So, $R^2 = (R - 0.3)^2 + 3^2$

$\Rightarrow R = 15$ cm

Now from Lens maker's formula,

$$\frac{1}{f} = (\mu - 1)\left(\frac{1}{R_{flat}} - \frac{1}{R_2}\right)$$

$$\Rightarrow \frac{1}{f} = (\mu - 1) \times \frac{1}{R} \qquad [\because R_{flat} = \infty \text{ and } R_2 = -R]$$

$$\Rightarrow \frac{1}{f} = \frac{(1.5 - 1)}{15} \Rightarrow f = 30 \text{ cm}$$

137. (30) Refraction formula (refraction from curved surface)

$$\frac{\mu_2}{v} - \frac{\mu_1}{u} = \frac{\mu_2 - \mu_1}{R}, \ |u| = \frac{3}{2} \times |v| = \frac{3}{2} \times 10 = 15 \text{ cm}$$

$$\left[\because \lambda \propto \frac{1}{\mu}\right]$$

$\mu_1 = 1$ (in air) $\qquad \mu_2 = 1.5$ (inside the surface)

$$\frac{1.5}{+10} - \frac{1}{-15} = \frac{1.5 - 1}{+R} \Rightarrow \frac{1.5}{10} + \frac{1}{15} = \frac{0.5}{R} \Rightarrow R = \frac{30}{13} \text{ m}$$

$\therefore x = 30.$

138. (15) $m = \frac{f}{f + u}$; As, $m_1 = -m_2$

$$\frac{f}{f - 10} = \frac{-f}{f - 20} \Rightarrow f = 15 \text{ cm}$$

139. (a) Focal length of the convex lens, $f = 0.5$ m

Object is at $2f$ so, image (I_1) will also be at $2f$.

Image of I_1 i.e., I_2 will be 1 m behind mirror.

Now I_2 will be object for lens.

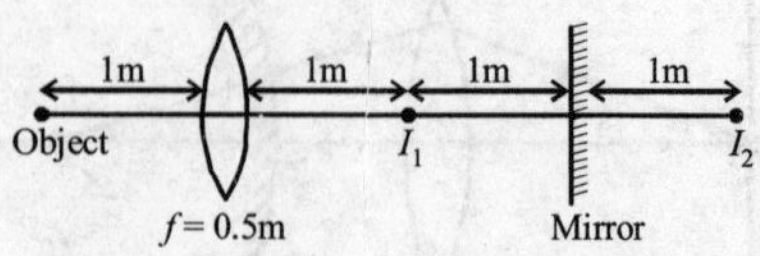

$\therefore u = (-1) + (-1) + (-1) = -3$ m

Using lens formula, $\frac{1}{v} - \frac{1}{u} = \frac{1}{f}$

$$\Rightarrow \frac{1}{v} = \frac{1}{f} + \frac{1}{u} = \frac{1}{+0.5} + \frac{1}{-3} \text{ or } v = \frac{3}{5} = 0.6 \text{ m}$$

Hence, distance of image from mirror $= 2 + 0.6 = 2.6$ m and real.

140. (d) Given, using lens maker's formula

$$\frac{1}{f} = (\mu - 1)\left(\frac{1}{R_1} - \frac{1}{R_2}\right)$$

Here, $R_1 = R_2 = R$ (For double convex lens)

$$\therefore \frac{1}{f} = (\mu - 1)\left(\frac{1}{R} - \frac{1}{-R}\right)$$

$$\Rightarrow P = \frac{1}{f} = (\mu - 1)\frac{2}{R} \qquad ...(i)$$

For plano convex lens,

$R_1 = R'$, $R_2 = \infty$

Using lens maker's formula again, we have

$$1.5P = (\mu - 1)\left(\frac{1}{R'} - \frac{1}{\infty}\right)$$

$$\Rightarrow \frac{3}{2}P = \frac{\mu - 1}{R'} \qquad ...(ii)$$

From (i) and (ii),

$$\frac{3}{2} = \frac{R}{2R'} \Rightarrow R' = \frac{R}{3}$$

141. (476.19) Given,

Distance between an object and screen, $D = 100$ cm

Distance between the two position of lens, $d = 40$ cm

Focal length of lens,

$$f = \frac{D^2 - d^2}{4D} = \frac{100^2 - 40^2}{4(100)} = \frac{(100 + 40)(100 - 40)}{4(100)} = 21 \text{ cm}$$

Power, $P = \frac{1}{f} = \frac{100}{21} = \frac{N}{100}$

$\therefore N = 476.19.$

142. (60) Given : $\mu = 1.5$; $R_{curved} = 30$ cm

Using, Lens-maker formula

$$\frac{1}{f} = (\mu - 1)\left(\frac{1}{R_1} - \frac{1}{R_2}\right)$$

For plano-convex lens

$R_1 \to \infty$ then $R_2 = -R$

$$\therefore f = \frac{R}{\mu - 1} = \frac{30}{1.5 - 1} = 60 \text{ cm}$$

143. (b) Using lens maker's formula

$$\frac{1}{f} = \left(\frac{\mu_g}{\mu_a} - 1\right)\left[\frac{1}{R_1} - \frac{1}{R_2}\right]$$

Here, μ_g and μ_a are the refractive index of glass and air respectively

$$\Rightarrow \frac{1}{f}=(1.5-1)\left(\frac{1}{R_1}-\frac{1}{R_2}\right) \quad ...(i)$$

When immersed in liquid

$$\frac{1}{f_l}=\left(\frac{\mu_g}{\mu_l}-1\right)\left(\frac{1}{R_1}-\frac{1}{R_2}\right)$$

[Here, μ_l = refractive index of liquid]

$$\Rightarrow \frac{1}{f_l}=\left(\frac{1.5}{1.42}-1\right)\left(\frac{1}{R_1}-\frac{1}{R_2}\right) \quad ...(ii)$$

Dividing (i) by (ii)

$$\Rightarrow \frac{f_l}{f}=\frac{(1.5-1)1.42}{0.08}=\frac{1.42}{0.16}=\frac{142}{16}\approx 9$$

144. (d) From the equation of line

$m=k_1v+k_2 \quad (\because y=mx+c)$

$$\Rightarrow \frac{v}{u}=k_1v+k_2 \quad \left(\because m=\frac{v}{u}\right)$$

$$\Rightarrow \frac{1}{u}=k_1+\frac{k_2}{v} \quad \text{(Dividing both sides by v)}$$

$$\Rightarrow \frac{k_2}{v}=\frac{1}{u}-k_1 \Rightarrow \frac{k_2}{v}-\frac{1}{u}=-k_1$$

Comparing with lens formula $\frac{1}{v}-\frac{1}{u}=\frac{1}{f}$, we get

$$k_1=\frac{1}{-f}\Rightarrow f=-\frac{1}{k_1}=-\frac{1}{\text{slope of m - v graph}}$$

$$\therefore f=\frac{-1}{\text{slope of } m-v \text{ graph}}=-\frac{b}{c}$$

145. (a) $\frac{1}{f_{lens}}=(\mu_\ell-1)\left(\frac{1}{R_1}-\frac{1}{R_2}\right)$

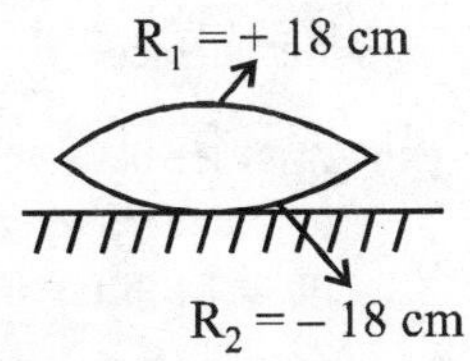

$$=(1.5-1)\left(\frac{1}{-18}-\frac{1}{-18}\right)=\frac{1}{18}$$

$\therefore f_{lens}=18$ cm

For liquid in between lens

$$\frac{1}{f_{liq}}=(\mu-1)\left(\frac{1}{R_1}-\frac{1}{R_2}\right)$$

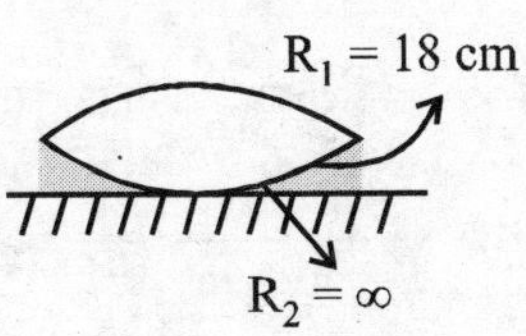

$$=(\mu-1)\left(\frac{1}{-18}-\frac{1}{\infty}\right)=\left(\frac{\mu-1}{-18}\right)$$

$$\therefore \frac{1}{f}=\frac{1}{f_{liq}}+\frac{1}{f_{lens}}$$

$$\Rightarrow \frac{1}{27\text{cm}}=\frac{1}{18\text{cm}}-\frac{(\mu-1)}{18\text{cm}} \quad \Rightarrow \mu=\frac{4}{3}$$

146. (b) We will have 3 phenomena one by one

(i) Refraction from lens.

(ii) Reflection from mirror

(iii) Refraction from lens

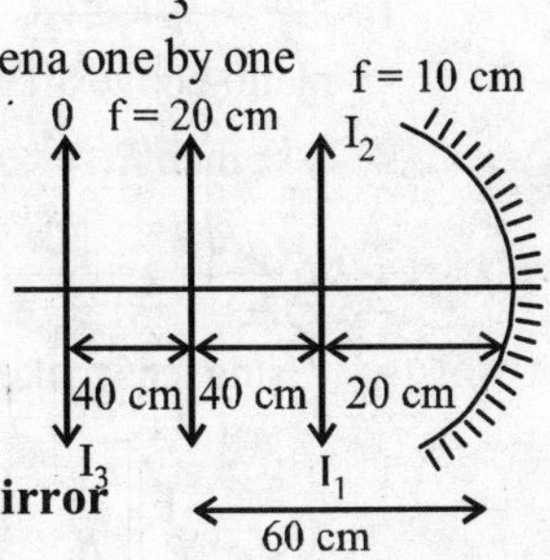

Ist refraction from lens

$u=-40$ cm, $f=+20$ cm

$\Rightarrow V=+40$ cm (I_1), $m_1=-1$

Reflection from concave mirror

$u=-20$ cm, $f=-10$ cm

$\Rightarrow V=-20$ cm (I_2) and $m_2=-1$

2nd refraction from lens

$u=-40$ cm, $f=+20$ cm

$\Rightarrow V=40$ cm (I_3) and $m_3=-1$

So, $M_{net}=-1\times-1\times-1=-1$

$\therefore$ Final image is formed at distance 40 cm from the convergent mirror and is of same size as the object

147. (c) For lens

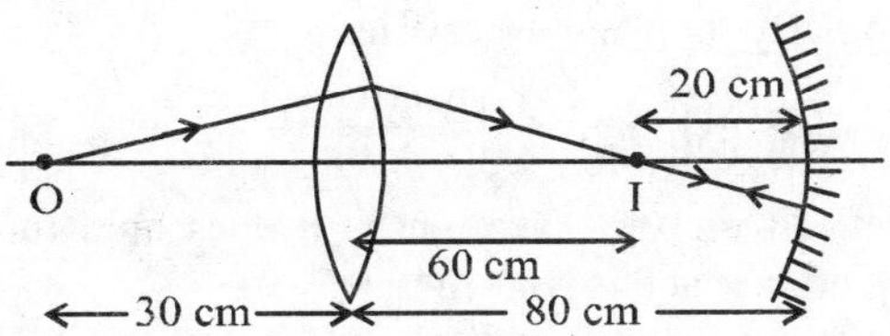

$$\frac{1}{v}-\frac{1}{u}=\frac{1}{f}\Rightarrow \frac{1}{v}-\frac{1}{-30}=\frac{1}{20}$$

$\therefore \quad v=+60$ cm

According to the condition, image formed by lens should be the centre of curvature of the mirror, and so $2f'=20$ or $f'=10$ cm

148. (d) By lens's formula, $\frac{1}{V}-\frac{1}{u}=\frac{1}{f}$

For first lens, $[u_1=-20]$

$$\frac{1}{V_1}-\frac{1}{-20}=\frac{1}{5}\Rightarrow V_I=\frac{20}{3}$$

Image formed by first lense will behave as an object for second lens

so, $u_2=\frac{20}{3}-2=\frac{14}{3}$; $\frac{1}{V_2}-\frac{1}{\frac{14}{3}}=\frac{1}{-5}\Rightarrow V_2=70\text{cm}$

149. (a) According to lens maker's formula,

$$\frac{1}{f}=(\mu_{rel}-1)\left(\frac{1}{R_1}-\frac{1}{R_2}\right)$$

Focal length of lens will change due to change in refractive index μ_{rel}. So, image will be formed at new position. But, position of screen remain unchanged. Hence image disappears.

150. (d) By lens formula

$$\frac{1}{v}-\frac{1}{u}=\frac{1}{f}\Rightarrow \frac{1}{v}-\frac{1}{(-20)}=\frac{10}{3}$$

$$\frac{1}{v}=\frac{10}{3}-\frac{1}{20}\Rightarrow \frac{1}{v}=\frac{197}{60}\Rightarrow v=\frac{60}{197}$$

Magnification of lens (m) is given by

$$m=\left(\frac{v}{u}\right)=\frac{\left(\frac{60}{197}\right)}{20}$$

velocity of image wrt. to lens is given by

$v_{I/L}=m^2v_{O/L}$

direction of velocity of image is same as that of object

$v_{O/L} = 5$ m/s

$$v_{I/L} = \left(\frac{60\times 1}{197\times 20}\right)^2 (5) = 1.16\times 10^{-3} \text{ m/s towards the lens}$$

151. (b) From lens maker's formula,

$$\frac{1}{f} = (\mu-1)\left(\frac{1}{R_1}-\frac{1}{R_2}\right)$$

$$\frac{1}{f_1} = (\mu_1-1)\left(\frac{1}{\infty}-\frac{1}{-R}\right) = \frac{1}{2f_2}$$

Similarly, for plano-concave lens

$$\frac{1}{f_2} = (\mu_2-1)\left(\frac{1}{-R}-\frac{1}{\infty}\right) = -\frac{(\mu_2-1)}{R}$$

We will ignore – ve sign as it does hold too much importance at this point of time.

Dividing $\frac{1}{f_1}$ by $\frac{1}{f_2}$ we get,

$$\frac{(\mu_1-1)}{R} = \frac{(\mu_2-1)}{2R} \Rightarrow 2\mu_1-\mu_2 = 1$$

152. (d) using, $\frac{\mu_2}{v}-\frac{\mu_1}{u} = \frac{\mu_2-\mu_1}{R}$

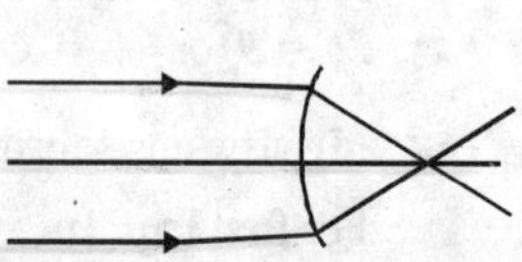

R = 7.8 mm

$\mu_1 = 1\ \mu_2 = 1.34$

$$\Rightarrow \frac{1.34}{V}-\frac{1}{\infty} = \frac{1.34-1}{7.8}\ [\because u=\infty]$$

$\therefore$ V = 30.7 mm = 3.07 cm ≃ 3.1 cm

153. (d)

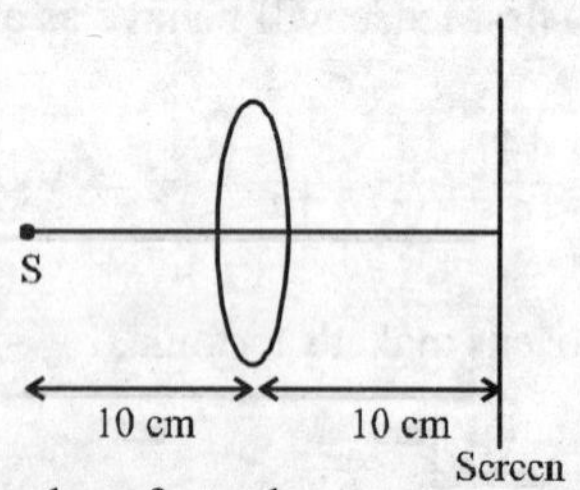

Using lens formula

$$\frac{1}{v}-\frac{1}{u} = \frac{1}{f} \Rightarrow \frac{1}{10}-\frac{1}{-10} = \frac{1}{f} \Rightarrow f = 5\text{cm}$$

Shift due to slab, $= t\left(1-\frac{1}{\mu}\right)$ in the direction of incident ray

or, $d = 1.5\left(1-\frac{2}{3}\right) = 0.5$

Now, u = – 9.5

Again using lens formulas $\frac{1}{v}-\frac{1}{-9.5} = \frac{1}{5}$

$$\Rightarrow \frac{1}{v} = \frac{1}{5}-\frac{2}{19} = \frac{9}{95} \quad \text{or,} \quad v = \frac{95}{9} = 10.55\text{cm}$$

Thus, screen is shifted by a distance

d = 10.55 – 10 = 0.55 cm away from the lens.

154. (a) For minimum spherical aberration separation,

$d = f_1 - f_2 = 2$ cm

Resultant focal length = F = 10 cm

Using $\frac{1}{F} = \frac{1}{f_1}+\frac{1}{f_2}-\frac{d}{f_1 f_2}$ and solving, we get f_1, f_2 as 18 cm and 20 cm respectively.

155. (a) Given, focal length of lens (f) = 15 cm

object is placed at a distance (u) = – 20 cm

By lens formula,

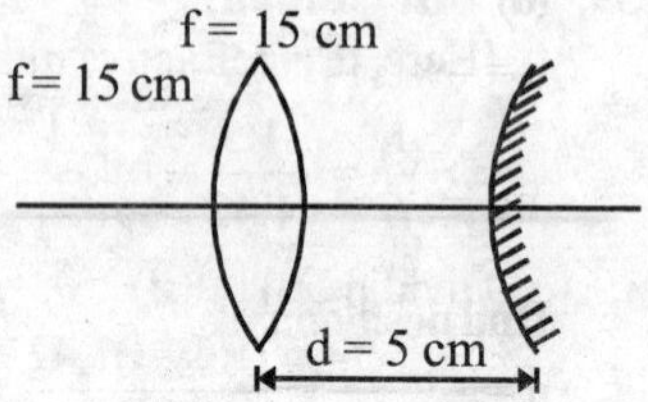

$$\frac{1}{f} = \frac{1}{v}-\frac{1}{u},\ u = -20 \text{ cm},\ f = 15 \text{ cm}$$

$$\frac{1}{v} = \frac{1}{f}+\frac{1}{u} = \frac{1}{15}-\frac{1}{20}$$

$$\Rightarrow \frac{1}{v} = \frac{4-3}{60}$$

v = 60 cm

for mirror, u = 55 cm

for the mirror to form image at 'O'

$$u = R = 2f \Rightarrow f = \frac{R}{2} = 27.5\text{cm}$$

156. (b) Given, radius of hemispherical glass R = 10 cm

$$\therefore \text{Focal length } f = \frac{10}{2} = -5 \text{ cm}$$

u = (10 – 6) = –4 cm.

By using mirror formula,

$$\frac{1}{v}+\frac{1}{u} = \frac{1}{f} \Rightarrow \frac{1}{v}+\frac{1}{-4} = \frac{1}{-5} \Rightarrow v = 20 \text{ cm.}$$

Apparent height, $h_a = h_r\frac{\mu_r}{\mu_i} = 30\times\frac{1}{1.5} = 20$ cm below flat surface.

157. (d) $P = \frac{1}{f}$

$$|P| = P_1 + P_2 = \frac{1}{f}+\frac{1}{f} = \frac{2}{f}$$

As there is reflection happening, the power will have negative sign so

$$P = -\frac{2}{f} \Rightarrow f_{eff} = -\frac{f}{2};\ \text{Now, } \frac{1}{v}-\frac{1}{u} = \frac{1}{f}$$

$$\frac{1}{\frac{-a}{3}}+\frac{1}{-a} = \frac{1}{\frac{-f}{2}} \Rightarrow \frac{-3}{a}+\frac{-1}{a} = \frac{-2}{f} \Rightarrow a = 2f$$

158. (b) By Lens maker's formula for convex lens

$$\Rightarrow \frac{f_m}{f} = \frac{(\mu-1)}{\left(\frac{\mu}{\mu_m}-1\right)}$$

$$\Rightarrow \frac{f_1}{f} = \frac{\left(3/2-1\right)}{\left(\frac{3/2}{4/2}-1\right)} = 4$$

$\Rightarrow f_1 = 4f$

$$\frac{f_2}{f} = \frac{\left(\frac{3}{2}-1\right)}{\left(\frac{3/2}{5/2}-1\right)} = -5$$

$f_2 < 0$

159. (d) **Ist position:**

$u = -x, v = +y$

$$m_1 = \frac{v}{u} = \frac{+y}{-x}$$

IInd position:

$u = -y, v = +x$

$$m_2 = \frac{v}{u} = \frac{+x}{-y}$$

Here, $\frac{m_1}{m_2} = \frac{3}{2} \Rightarrow \frac{y}{x} = \sqrt{\frac{3}{2}}$ ...(i)

Also, $y - x = 10$ ...(ii)

solving (i) & (ii), we get

$y = 44.5$ cm and $x = 54.5$ cm

So, $d = x + y = 99$ cm

160. (d) If side of object square = ℓ

and side of image square = ℓ'

From question, $\frac{\ell'^2}{\ell^2} = 9$ or $\frac{\ell'}{\ell} = 3$

i.e., magnification $m = 3$

$u = -40$ cm

$v = 3 \times 40 = 120$ cm

$f = ?$

From formula, $\frac{1}{v} - \frac{1}{u} = \frac{1}{f} \Rightarrow \frac{1}{120} - \frac{1}{-40} = \frac{1}{f}$

or, $\frac{1}{f} = \frac{1}{120} + \frac{1}{40} = \frac{1+3}{120}$ $\therefore$ $f = 30$ cm

161. (d) The focal length of the lens

$$\frac{1}{f} = \frac{1}{v} - \frac{1}{u} = \frac{1}{12} + \frac{1}{240} = \frac{20+1}{240} = \frac{21}{240} \Rightarrow f = \frac{240}{21}\text{ cm}$$

When glass plate is interposed between lens and film, so shift produced will be

$$\text{Shift} = t\left(1-\frac{1}{\mu}\right) \Rightarrow 1\left(1-\frac{1}{3/2}\right) = 1 \times \frac{1}{3}$$

Now image should be form at

$$v' = 12 - \frac{1}{3} = \frac{35}{3}\text{ cm}$$

Now the object distance u.

Using lens formula again

$$\frac{1}{f} = \frac{1}{v'} - \frac{1}{u} \Rightarrow \frac{1}{u} = \frac{1}{v'} - \frac{1}{f}$$

$$\Rightarrow \frac{1}{u} = \frac{3}{35} - \frac{21}{240} = \frac{1}{5}\left[\frac{3}{7} - \frac{21}{48}\right] \Rightarrow \frac{1}{u} = \frac{1}{5}\left[\frac{48-49}{7 \times 16}\right]$$

$\Rightarrow u = -7 \times 16 \times 5 = -560$ cm $= -5.6$ m

162. (a) From the Cauchy

Formula, $\mu = A + \frac{B}{\lambda^2} + \frac{C}{\lambda^4}$ $\therefore \mu \propto \frac{1}{\lambda}$

As, $\lambda_{blue} < \lambda_{red} \therefore \mu_{blue} > \mu_{red}$

From lens maker's formula

and $\frac{1}{f} = (\mu - 1)\left(\frac{1}{R_1} - \frac{1}{R_2}\right) \Rightarrow \frac{1}{f_B} > \frac{1}{f_R} \Rightarrow f_R > f_B.$

163. (c) When two thin lenses are in contact coaxially, power of combination is given by

$P = P_1 + P_2 = (-15+5)D = -10D.$

Also, $P = \frac{1}{f} \Rightarrow f = \frac{1}{P} = \frac{1}{-10}$ metre

$$\therefore f = -\left(\frac{1}{10} \times 100\right)\text{ cm} = -10\text{ cm}.$$

164. (b) According to lens maker's formula in air

$$\frac{1}{f_a} = \left({}_a\mu_g - 1\right)\left(\frac{1}{R_1} - \frac{1}{R_2}\right)$$

$$\Rightarrow \frac{1}{f_a} = \left(\frac{1.5}{1} - 1\right)\left(\frac{1}{R_1} - \frac{1}{R_2}\right) \quad \text{.... (i)}$$

Using lens maker's formula in liquid medium,

$$\frac{1}{f_m} = \left(\frac{\mu_g}{\mu_m} - 1\right)\left(\frac{1}{R_1} - \frac{1}{R_2}\right)$$

$$\Rightarrow \frac{1}{f_m} = \left(\frac{1.5}{1.6} - 1\right)\left(\frac{1}{R_1} - \frac{1}{R_2}\right) \quad \text{.... (ii)}$$

Dividing (i) by (ii),

$$\frac{f_m}{f_a} = \frac{(1.5-1)}{\left(\frac{1.5}{1.6}-1\right)} = -8;\ P_a = -5 = \frac{1}{f_a} \Rightarrow f_a = -\frac{1}{5}$$

$$\Rightarrow f_m = -8 \times f_a = -8 \times -\frac{1}{5} = \frac{8}{5}$$

$$P_m = \frac{\mu}{f_m} = \frac{1.6}{8} \times 5 = 1\text{D}$$

165. (c) The focal length(F) of the final mirror is

$$\frac{1}{F} = \frac{2}{f_\ell} + \frac{1}{f_m}$$

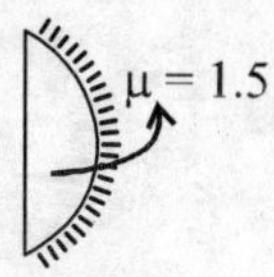

Using lens maker's formula

Here $\frac{1}{f_\ell} = (\mu - 1)\left(\frac{1}{R_1} - \frac{1}{R_2}\right)$ (Here, $R_1 = \infty \Rightarrow R_2 = 30\,cm$)

$= (1.5 - 1)\left[\frac{1}{\infty} - \frac{1}{-30}\right] = \frac{1}{60}$

$\frac{1}{F} = 2 \times \frac{1}{60} + \frac{1}{30/2} = \frac{1}{10} \quad \therefore F = 10\,cm$

Real image will be equal to the size of the object if the object distance

$u = 2F = 20\,cm$

166. (a) For minimum deviation through a prism,

$i = e, r_1 = r_2 = \frac{A}{2}$ and $\delta_{min} = 2i - A$

Also, refracted ray becomes parallel to the base of prism.

167. (45) Refractive index of prism material

$$\mu = \frac{\sin\left(\frac{A + \delta_m}{2}\right)}{\sin\left(\frac{A}{2}\right)}$$

$$\sqrt{2} = \frac{\sin\left(\frac{60° + \delta_m}{2}\right)}{\sin\left(\frac{60°}{2}\right)}$$

$\delta_m = 2i - A [\because i = e] \quad \therefore i = 45°$

168. (b) The condition for dispersion without deviation is

$\delta_{net} = 0;\ (\mu_1 - 1)A_1 - (\mu_2 - 1)A_2 = 0$

$(1.54 - 1)4 - (1.72 - 1)A_2 = 0;\ A_2 = 3°$

169. (b) The refractive index (μ) of a prism is given by

$$\mu = \frac{\sin\left(\frac{A + \delta_{min}}{2}\right)}{\sin\frac{A}{2}}$$

Given, $\delta_{min} = A, \mu = \sqrt{3}$

$$\Rightarrow \sqrt{3} = \frac{\sin A}{\sin\frac{A}{2}} = \frac{2\sin\frac{A}{2}\cos\frac{A}{2}}{\sin\frac{A}{2}} \Rightarrow \cos\frac{A}{2} = \frac{\sqrt{3}}{2}$$

$\therefore A = 60°$

170. (60) Given, $\frac{\delta_{min}}{A} = 1$

$\Rightarrow \frac{2i - A}{A} = 1 \Rightarrow i = A$

Using Snell's law,

$1 \times \sin i = \mu \sin r$

$\Rightarrow \sin i = \mu \sin\left(\frac{A}{2}\right) \Rightarrow \sin A = \mu \sin\left(\frac{A}{2}\right)$

$\Rightarrow 2\sin\frac{A}{2}\cos\frac{A}{2} = \sqrt{3}\sin\left(\frac{A}{2}\right) \Rightarrow \cos\left(\frac{A}{2}\right) = \frac{\sqrt{3}}{2}$

$\therefore A = 60°$

171. (a) As $\lambda_{red} > \lambda_{yellow} > \lambda_{violet}$

Also deviation $\propto \mu \propto \frac{1}{\lambda}$

$\therefore$ Light ray with longer wavelength bends less.

172. (c) Refraction index of the material of the prism

$$\mu = \frac{\sin\left(\frac{A + \delta_m}{2}\right)}{\sin\left(\frac{A}{2}\right)}$$

$$\Rightarrow \cot\frac{A}{2} = \frac{\cos\frac{A}{2}}{\sin\frac{A}{2}} = \frac{\sin\left(\frac{A + \delta_m}{2}\right)}{\sin\frac{A}{2}} \quad \left[\because \mu = \cot\frac{A}{2}\right]$$

$$\Rightarrow \sin\left(\frac{\pi}{2} - \frac{A}{2}\right) = \sin\left(\frac{A + \delta_m}{2}\right) \Rightarrow \frac{\pi}{2} - \frac{A}{2} = \frac{A}{2} + \frac{\delta_m}{2}$$

$\Rightarrow \delta_m = \pi - 2A$

173. (a) Using $\mu = \frac{\sin\left(\frac{A + \delta_{min}}{2}\right)}{\sin\frac{A}{2}}$

$$\Rightarrow \cot\frac{A}{2} = \frac{\sin\left(\frac{A + \delta_{min}}{2}\right)}{\sin\frac{A}{2}} \Rightarrow \cos\frac{A}{2} = \sin\left(\frac{A + \delta_{min}}{2}\right)$$

$\Rightarrow \frac{A + \delta_{min}}{2} = \frac{\pi}{2} - \frac{A}{2}$ or, $\delta_{min} = \pi - 2A$

174. (30) Given, refractive index, $\mu = \sqrt{2}$

As $\mu = \frac{\sin\left(\frac{\delta_{min} + A}{2}\right)}{\sin\left(\frac{A}{2}\right)}$ (Here δ_{min} = minimum deviation)

$$\sqrt{2} = \frac{\sin\frac{(\delta_{min} + 60)}{2}}{\sin\left(\frac{60}{2}\right)} \Rightarrow \frac{1}{\sqrt{2}} = \sin\left(\frac{\delta_{min} + 60}{2}\right)$$

$\therefore \frac{\delta_{min} + 60}{2} = 45 \Rightarrow \delta_{min} = 30$

175. (45) Refractive index

$$\mu = \frac{\sin\left(\frac{A + \delta\sin}{2}\right)}{\sin\left(\frac{A}{2}\right)} \Rightarrow \sqrt{2} = \frac{\sin\left(\frac{60 + \delta_{min}}{2}\right)}{\sin 30°}$$

$\Rightarrow \frac{1}{2} = \sin\left(\frac{60 + \delta_{min}}{2}\right) \Rightarrow 45° = \frac{60 + \delta_{min}}{2}$

$\Rightarrow \delta_{min} = 30°$

$\delta = i + e - A$

Here, e = i

So, $\delta_{min} = 2e - A \Rightarrow 2e = \delta_{min} + A$

$$e = \frac{\delta_{min} + A}{2} = \frac{30° + 60°}{2} = 45°$$

176. (a) We known that

$$\mu = \frac{\frac{\sin(A+\delta_m)}{2}}{\sin\left(\frac{A}{2}\right)} \Rightarrow \cot\frac{A}{2} = \frac{\sin\left(\frac{A+\delta_m}{1}\right)}{\sin\left(\frac{A}{2}\right)}$$

$$\Rightarrow \cos\left(\frac{A}{2}\right) = \sin\left(\frac{A+\delta_m}{2}\right)$$

$$\Rightarrow \sin\left(\frac{\pi}{2} - \frac{A}{2}\right) = \sin\left(\frac{A}{2} + \frac{\delta_m}{2}\right)$$

$$\Rightarrow \frac{\pi}{2} - \frac{A}{2} = \frac{A}{2} + \frac{\delta_m}{2} \Rightarrow \frac{\pi}{2} - A = \frac{\delta m}{2}$$

$$\therefore \delta_m = \pi - 2A$$

177. (5) $PA = .1 \sin 60° = .1\frac{\sqrt{3}}{2}$

[As $i = A = 60°$]

Minimum deviation is given by

$\delta_{min} = i + e - A \Rightarrow \delta_{min} = 2i - A$

[At minimum deviation $i = e$]

$= 2 \times 60° - 60° = 60°$

$$\mu = \frac{\sin\left(\frac{\delta_{min} + A}{2}\right)}{\sin\left(\frac{A}{2}\right)}$$

Putting values of δ_{min} & A

We get

$$\mu = \frac{\sin\left(\frac{60+60}{2}\right)}{\sin\frac{60}{2}} = \sqrt{3}$$

velocity of light in prism $= \frac{C}{\sqrt{3}} = \frac{3\times10^8}{\sqrt{3}}\ m/s$

time taken to travel $PA = \frac{PA}{V} = \frac{.1\sqrt{3}\times\sqrt{3}}{3\times10^8\times2}$

$= 5 \times 10^{-10}$ sec

178. (a) According to prism formula,

$$\mu = \frac{\sin\left(\frac{A+\delta_m}{2}\right)}{\sin\left(\frac{A}{2}\right)}$$

Here, δ_m = angle of minimum deviation

A = angle of prism

We have given, $\delta_m = A$

$$\therefore \quad \mu = \frac{\sin A}{\sin A/2} \Rightarrow \mu = 2\cos\frac{A}{2} \Rightarrow A = 2\cos^{-1}\left(\frac{\mu}{2}\right)$$

179. (60) Given $i = 2r_1 = A$

And at minimum deviation $r_1 = r_2 = \frac{A}{2}$

From Snell's law,

$\mu_1 \sin i = \mu_2 \sin r_1 \therefore 1.\sin i = \sqrt{3}\sin r_1$

$$\Rightarrow 1\sin A = \sqrt{3}\sin\frac{A}{2} \Rightarrow 2\sin\frac{A}{2}\cos\frac{A}{2} = \sqrt{3}\sin\frac{A}{2}$$

$$\Rightarrow \cos\frac{A}{2} = \frac{\sqrt{3}}{2} \Rightarrow \frac{A}{2} = 30° \text{ or, } A = 60°$$

Hence angle of prism = 60°

180. (c) For minimum deviation:

$$r_1 = r_2 = \frac{A}{2} = 30°$$

by Snell's law $\mu_1 \sin i = \mu_2 \sin r$

$$1 \times \sin i = \sqrt{3} \times \frac{1}{2} = \frac{\sqrt{3}}{2} \Rightarrow i = 60$$

181. (c) Angle of prism, A = 30°, i = 60°, angle of deviation, δ = 30°

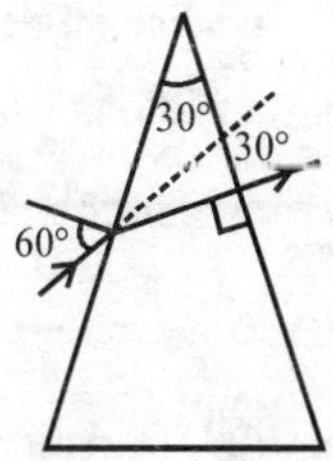

Using formula, $\delta = i + e - A$

$\Rightarrow e = \delta + A - i$

$= 30° + 30° - 60° = 0°$

$\therefore$ Emergent ray will be perpendicular to the face

So it will make angle 90° with the face through which it emerges.

182. (c) $r_2 < QC$

$$r_2 < \sin^{-1}\left(1/\mu\right)$$

$$\sin r^2 < \frac{1}{\mu}$$

$$\text{Sin } Q = \mu \sin r_1$$

$$r_1 = \sin^{-1}\left(\frac{\sin Q}{\mu}\right)$$

$$\sin(A - r_1) < \frac{1}{\mu}$$

$$\sin\left(A - \left(\sin^{-1}\left(\frac{\sin Q}{\mu}\right)\right)\right) < \frac{1}{\mu}$$

$$A - \sin^{-1}\left(\frac{\sin Q}{\mu}\right) < \sin^{-1}\left(\frac{1}{\mu}\right)$$

$$A - \sin^{-1}\left(\frac{1}{\mu}\right) < \sin^{-1}\left(\frac{\sin Q}{\mu}\right)$$

$$\left[\sin\left(A - \sin^{-1}\left(\frac{1}{\mu}\right)\right)\right] < \frac{\sin Q}{\mu}$$

$$\left[\sin^{-1}\left[\mu\left[\sin\left(A - \sin^{-1}\left(1/\mu\right)\right)\right]\right]\right] < Q$$

183. (d) For light to come out through face 'AC', total internal reflection must not take place.
i.e., $\theta < c \Rightarrow \sin\theta < \sin c$

$\Rightarrow \sin\theta < \frac{1}{\mu} \Rightarrow \mu < \frac{1}{\sin\theta} \Rightarrow \mu < \frac{1}{\sin 45°}$

$\Rightarrow \mu < \sqrt{2} \quad \Rightarrow \quad \mu < 1.414$

184. (a) When angle of prism is small,
Angle of deviation, $D = (\mu - 1)A$
Since $\lambda_b < \lambda_r$
$\Rightarrow \mu_r < \mu_b \qquad \Rightarrow D_1 < D_2$

185. (b) For total internal reflection
Incident angle $(i) >$ critical angle (i_c),
$\therefore \sin i > \sin i_c$

$\Rightarrow \sin 45° > \sin i_c \quad \Rightarrow \sin i_c = \frac{1}{n} \quad \therefore \sin 45° > \frac{1}{n}$

$\Rightarrow \frac{1}{\sqrt{2}} > \frac{1}{n} \quad \Rightarrow n > \sqrt{2}$

186. (d) To redirect the light that enters the telescope to the eyepiece or camera. The primary mirror of a reflecting telescopes gathers the light and reflects towards the secondary mirror which then reflect the light towards the eyepiece allowing the observer to see image.
It has advantage of a large focal length in a short telescope.

187. (a) $\because f_0 + f_e = 30$
And magnification, $m = \frac{f_0}{f_e}$

$2 = \frac{f_0}{f_e} \Rightarrow f_0 = 2f_e \Rightarrow f_0 + \frac{f_0}{2} = 30 \therefore f_0 = 20$ cm

188. (25) In case of simple microscope,
Magnification, $m = 1 + \frac{D}{f_0}$ or, $6 = 1 + \frac{D}{f_0}$

$\Rightarrow \quad 5 = \frac{25}{f_0} \quad \therefore f_0 = 5$cm

As total magnification double using an eyepiece along with the given lens i.e, case of compound microscope,

Magnification, $m = \frac{\ell . D}{f_0 . f_e}$

or, $12 = \frac{60 \times 25}{5 . f_e} \quad \therefore f_e = 25$cm

189. (d) Image size is bigger than object size, the angular size of the image is equal to the angular size of object.

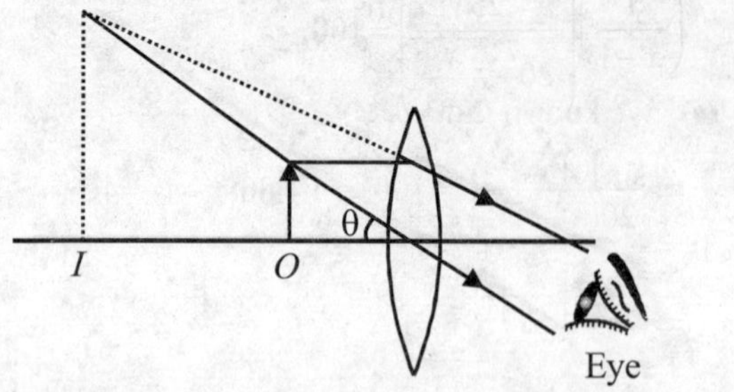

190. (50) Given : Length of compound microscope, $L = 10$ cm
Focal length of objective $f_0 = 1$ cm and of eye-piece, $f_e = 5$ cm
$u_0 = f_e = 5$ cm
Final image formed at infinity (∞), $v_e = \infty$
$v_0 = 10 - 5 = 5$

Using lens formula, $\frac{1}{v} - \frac{1}{u} = \frac{1}{f}$

$\frac{1}{v_0} - \frac{1}{u_0} = \frac{1}{f_0} \Rightarrow \frac{1}{5} - \frac{1}{u_0} = \frac{1}{1} \Rightarrow u_0 = -\frac{5}{4}$ cm

$\Rightarrow \frac{5}{4} = \frac{N}{40} \Rightarrow N = \frac{200}{4} = 50$ cm.

191. (4.48) According to question, final image i.e., $v_2 = 25$ cm, $f_0 = 1$ cm, magnification, $m = m_1 m_2 = 100$

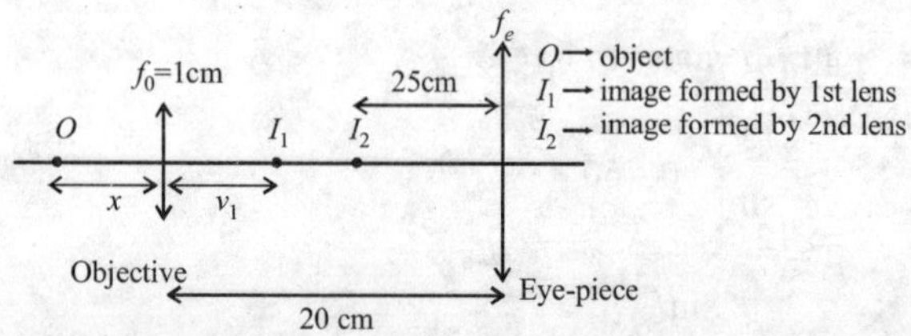

Using lens formula,

For first lens or objective $= \frac{1}{v_1} - \frac{1}{-x} = \frac{1}{1} \Rightarrow v_1 = \frac{x}{x-1}$

Also magnification $|m_1| = \left|\frac{v_1}{u_1}\right| = \frac{1}{x-1}$

For 2nd lens or eye-piece, this is acting as object

$\therefore u_2 = -(20 - v_1) = -\left(20 - \frac{x}{x-1}\right)$ and $v_2 = -25$ cm

Angular magnification $|m_A| = \left|\frac{D}{u_2}\right| = \frac{25}{|u_2|}$

Total magnification $m = m_1 m_A = 100$

$$\left(\frac{1}{x-1}\right)\left(\frac{25}{20-\frac{x}{x-1}}\right) = 100$$

$$\Rightarrow \frac{25}{20(x-1)-x} = 100 \Rightarrow 1 = 80(x-1) - 4x$$

$$\Rightarrow 76x = 81 \Rightarrow x = \frac{81}{76} \Rightarrow u_2 = -\left(20 - \frac{\frac{81}{76}}{\frac{81}{76}-1}\right) = \frac{-19}{5}$$

Again using lens formula for eye-piece

$$\frac{1}{-25} - \frac{1}{-\frac{19}{5}} = \frac{1}{f_e} \Rightarrow f_e = \frac{25\times 19}{106} \approx 4.48\,\text{cm}$$

192. (d) For telescope

Tube length (L) $= f_o + f_e = 60$

and magnification $(m) = \frac{f_o}{f_e} = 5 \Rightarrow f_0 = 5f_e$

$\therefore\ f_o = 50$ cm and $f_e = 10$ cm

Hence focal length of eye-piece, $f_e = 10$ cm

193. (a) According question, $M = 375$

$L = 150$ mm, $f_0 = 5$ mm and $f_e = ?$

Using, magnification, $M \simeq \frac{L}{f_0}\left(1 + \frac{D}{f_e}\right)$

$$\Rightarrow 375 = \frac{150}{5}\left(1 + \frac{250}{f_e}\right) \quad (\because D = 25\,\text{cm} = 250\,\text{mm})$$

$$\Rightarrow 12.5 = 1 + \frac{250}{f_e} \Rightarrow f_e = \frac{250}{11.5} = 21.7 \approx 22\,mm$$

194. (b) $M = \frac{\theta_i}{\theta_o},\ 20 = \frac{h/d_i}{h/d_o} \Rightarrow 20 = \frac{d_o}{d_i} \Rightarrow d_i = \frac{d_o}{20}$

195. (c) $M.P = \frac{f_0}{f_e} = \frac{150}{5} = 30$

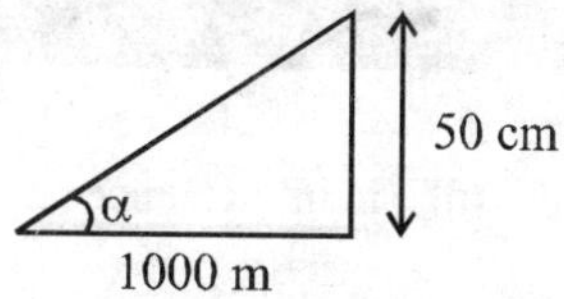

$\tan\alpha = \frac{50}{1000} = \frac{1}{20}$

According to the question,

angle formed by the image of the tower is θ. We can write

$\tan\beta = \tan\theta$ ...(i)

$\tan\beta = (M.P)\tan\alpha$ [$\because$ α and β are small]

$= 30 \times \frac{1}{20} = \frac{3}{2} = 1.5$

from (i), $\tan\theta = 1.5$

$\theta = \tan^{-1}(1.5) = 56.31°$

196. (d) Given, Focal length of objective, $f_0 = 30$ cm

focal length of eye lens, $f_e = 3.0$ cm

Magnifying power, M = ?

Magnifying power of the Galilean telescope,

$$M_D = \frac{f_0}{f_e}\left(1 - \frac{f_e}{D}\right) = \frac{30}{3}\left(1 - \frac{3}{25}\right) \quad [\because D = 25\,\text{cm}]$$

$$= 10 \times \frac{22}{25} = 8.8\,\text{cm}$$

197. (c) One side of mirror is opaque and another side is reflecting this is not in case of lens hence, it is easier to provide mechanical support to large size mirrors than large size lenses. Reflecting telescopes are based on the same principle except that the formation of images takes place by reflection instead of refraction.

198. (a) To find the refractive index of glass using a travelling microscope, a vernier scale is provided on the microscope

199. (c) A real, inverted and enlarged image of the object is formed by the objective lens of a compound microscope.

10 Wave Optics

1. **(d)** In interference pattern, $I_1 : I_2 = 1 : 9$

$\therefore \quad I_1 = I, I_2 = 9I$

$$\therefore \quad \frac{I_{max}}{I_{min}} = \frac{\left(\sqrt{I_1}+\sqrt{I_2}\right)^2}{\left(\sqrt{I_1}-\sqrt{I_2}\right)^2} = \frac{(4)^2}{(2)^2} = \frac{16}{4} = 4$$

2. **(24)** In interference, $I_1 = 4I, I_2 = 9I$

$$\therefore \quad I_{max} = \left(\sqrt{I_1}+\sqrt{I_2}\right)^2 = \left(\sqrt{4I}+\sqrt{9I}\right)^2 = 25I$$

$$\therefore \quad I_{min} = \left(\sqrt{I_1}-\sqrt{I_2}\right)^2 = \left(\sqrt{4I}-\sqrt{9I}\right)^2 = I$$

$\therefore \quad I_{max} - I_{min} = 24\,I = xI$ (given)

$\therefore \quad x = 24$

3. **(a)** Let the angle made by light with x, y and z axis are α, β and γ respectively.
The direction of propagation of light is perpendicular to the wave front and is symmetric about x, y and z axis.

$\therefore \quad \cos\alpha = \cos\beta = \cos\gamma.$

Also $\cos^2\alpha + \cos^2\beta + \cos^2\gamma = 1$ [Sum of direction cosines]

$$\therefore \alpha = \cos^{-1}\frac{1}{\sqrt{3}}$$

4. **(54)** From the concept of a thin film interference.
Maxima condition,

$$2\mu t = n\lambda \Rightarrow t = \frac{n\lambda}{2\mu} \Rightarrow t = \frac{\lambda}{2\mu}, \frac{2\lambda}{2\mu}, ...$$

Minima condition, $2\mu t = (2n-1)\lambda/2$

$$\Rightarrow t = \frac{(2n-1)\lambda}{4\mu} \quad \Rightarrow t = \frac{\lambda}{4\mu}, \frac{3\lambda}{4\mu}, ...$$

$$\therefore \Delta t = \frac{2\lambda}{4\mu}$$

$$\text{Rate of evaporation} = \frac{A(\Delta t)}{\text{time}} = \frac{(\pi r^2)(2\lambda)}{4\mu \times \text{time}}$$

$= 54\pi \times 10^{-13}\ m^3/s$

5. **(c)**

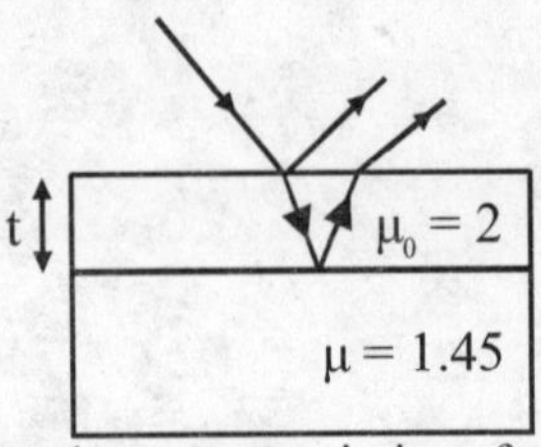

For maximum transmission of green light, reflected green should be minimum

$$\therefore \text{Path difference, } \Delta x = 2\mu_0 t = n\lambda \Rightarrow t = \frac{n\lambda}{2\mu_0}$$

$$\therefore t_{min} = \frac{\lambda}{2\mu_0} = \frac{550}{2\times 2} = 137.5\text{nm} \quad (\because \text{given } \lambda = 550 \text{ nm})$$

6. **(8)** Maximum intensity is,

$I_{max} = (\sqrt{I}+\sqrt{4I})^2 = 9I$ ...(i)

Minimum intensity is,

$I_{min} = (\sqrt{4I}-\sqrt{I})^2 = I$ ...(ii)

Substract (i) and (ii),

Hence, $I_{max} - I_{min} = 9I - I = 8I$

7. **(d)**

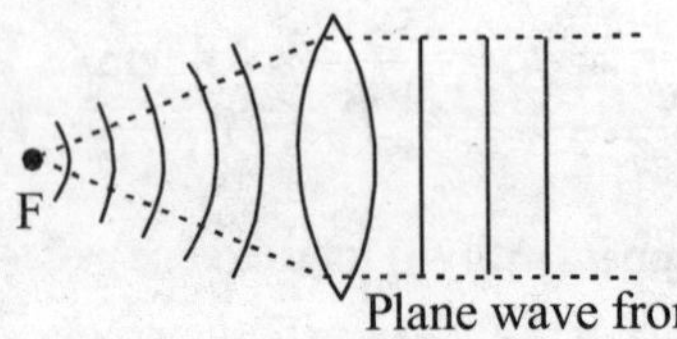

Plane wave front

As, the light emerges parallel. So, plane wavefront is formed.

8. **(c)** As medium changes, optical path changes.

From $\Delta\phi = \frac{2\pi}{\lambda}\Delta x$

Hence phase difference changes.

9. **(24)** We know that,

$I_{net} = I_1 + I_2 + 2\sqrt{I_1}\sqrt{I_2}\cos\phi$

As, I_{max} for $\phi = 0$ and I_{min} for $\phi = \pi$

$$I_{max} = \left(\sqrt{I_1}+\sqrt{I_2}\right)^2 = \left(\sqrt{9I}+\sqrt{4I}\right)^2 = 25I$$

$$I_{min} = \left(\sqrt{I_1}-\sqrt{I_2}\right)^2 = \left(\sqrt{9I}-\sqrt{4I}\right)^2 = I$$

So, $I_{max} - I_{min} = 25I - I = 24I$

10. **(2)** $I_A = I + 4I + 2\sqrt{I}\sqrt{4I}\cos\frac{\pi}{2}$

$I_A = 5I$

and $I_B = I + 4I + 2\sqrt{I}\sqrt{4I}\cos\frac{\pi}{3}$

$= 5I + 2I = 7I$

$I_B - I_A = 2I$

11. **(b)** We have given, $\frac{I_1}{I_2} = \frac{1}{4}$

$\Rightarrow I_2 = 4I_1$

$$I_{max} = \left(\sqrt{I_1}+\sqrt{I_2}\right)^2 = 9I_1$$

$$I_{min} = \left(\sqrt{I_1}-\sqrt{I_2}\right)^2 = I_1$$

$$\therefore \quad \frac{I_{max}+I_{min}}{I_{max}-I_{min}} = \frac{9I_1+I_1}{9I_1-I_1} = \frac{10}{8} = \frac{5}{4} = \frac{2\alpha+1}{\beta+3}$$

$\alpha = 2 \quad \beta = 1$

$$\therefore \quad \frac{\alpha}{\beta} = \frac{2}{1} = 2$$

12. (b) $I_P = I + 9I + 2\sqrt{I.9I}\cos{\pi}/_{2} = 10I$

$I_Q = I + 9I + 2\sqrt{I.9I}\cos\pi$
$= I + 9I + 2.3I \times -1 = 4I$
So, $I_P - I_Q = 10I - 4I = 6I$

13. (b) We have

$$\frac{I_{max}}{I_{min}} = \frac{(\sqrt{I_1}+\sqrt{I_2})^2}{(\sqrt{I_1}-\sqrt{I_2})^2} = \frac{(\sqrt{4}+\sqrt{1})^2}{(\sqrt{4}-\sqrt{1})^2} = \frac{9}{1}$$

By componendo and dividendo

$$\frac{I_{max}+I_{min}}{I_{max}-I_{min}} = \frac{9+1}{9-1} = \frac{10}{8} = \frac{5}{4}$$

So, $x = 4$

14. (d) $\frac{I_{max}}{I_{min}} = \frac{(\sqrt{I_1}+\sqrt{I_2})^2}{(\sqrt{I_1}-\sqrt{I_2})^2} = \frac{(\sqrt{9}+\sqrt{4})^2}{(\sqrt{9}-\sqrt{4})^2} = \frac{25}{1}$

15. (d) Maximum intensity in interference pattern

$$I_{max} = \left(\sqrt{I_1}+\sqrt{I_2}\right)^2 = \left(2\sqrt{I_0}\right)^2 = 4I_0.$$

16. (2) The difference in the number of waves when yellow light propagates through air and vacuum columns of the same thickness is one

So, thickness $t = n\lambda_{vac} = (n+1)\lambda_{air}$

or, $n\lambda = (n+1)\frac{\lambda}{\mu_{air}} \Rightarrow n = \frac{1}{\mu_{air}-1} = \frac{10^4}{3}$

$\therefore$ Thickness of air column

$$t = n\lambda = \frac{10^4}{3} \times 6000Å = 2\text{ mm}$$

17. (d) Given : $\frac{I_2}{I_1} = 2x$

So if $I_2 = 2x$ then $I_1 = 1$

$$\therefore \frac{I_{max}-I_{min}}{I_{max}+I_{min}} = \frac{(\sqrt{I_2}+\sqrt{I_1})^2-(\sqrt{I_2}-\sqrt{I_1})^2}{(\sqrt{I_2}+\sqrt{I_1})^2+(\sqrt{I_2}-\sqrt{I_1})^2}$$

$$= \frac{(\sqrt{2x}+1)^2-(\sqrt{2x}-1)^2}{(\sqrt{2x}+1)^2+(\sqrt{2x}-1)^2} = \frac{4\sqrt{2x}}{4x+2} = \frac{2\sqrt{2x}}{2x+1}$$

18. (b) For (A)

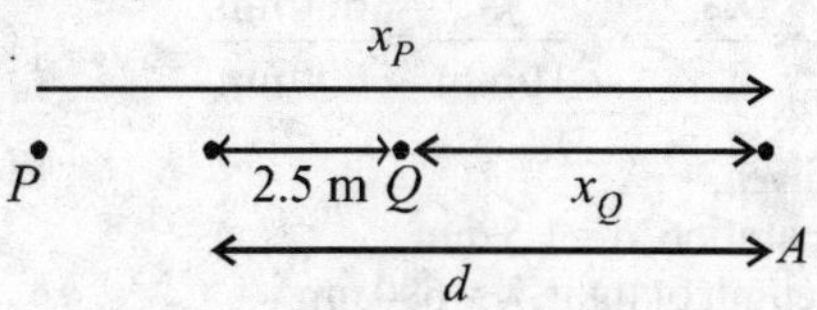

$x_P - x_Q = (d+2.5) - (d-2.5) = 5$ m

Phase difference $\Delta\phi$ due to path difference

$$= \frac{2\pi}{\lambda}(\Delta x) = \frac{2\pi}{20}(5) = \frac{\pi}{2}.$$

At A, Q is ahead of P by path, as wave emitted by Q reaches before wave emitted by P.

$\therefore$ Total phase difference at A $\frac{\pi}{2} - \frac{\pi}{2} = 0$

(due to P being ahead of Q by 90°)

$$I_A = I_1 + I_2 + 2\sqrt{I_1}\sqrt{I_2}\cos\Delta\phi$$
$$= I + I + 2\sqrt{I}\sqrt{I}\cos(0) = 4I$$

For C, Path difference, $x_Q - x_P = 5$ m

Phase difference $\Delta\phi$ due to path difference

$$= \frac{2\pi}{\lambda}(\Delta x) = \frac{2\pi}{20}(5) = \frac{\pi}{2}$$

Total phase difference at $C = \frac{\pi}{2} + \frac{\pi}{2} = \pi$

$$I_{net} = I_1 + I_2 + 2\sqrt{I_1}\sqrt{I_2}\cos(\Delta\phi)$$
$$= I + I + 2\sqrt{I}\sqrt{I}\cos(\pi) = 0$$

For B,

Path difference, $x_P - x_Q = 0$

Phase difference, $\Delta\phi = \frac{\pi}{2}$

(due to P being ahead of Q by 90°)

$$I_B = I + I + 2\sqrt{I}\sqrt{I}\cos\frac{\pi}{2} = 2I$$

Therefore intensities of radiation at A, B and C will be in the ratio

$I_A : I_B : I_C = 4I : 2I : 0 = 2:1:0.$

19. (d) Initially, $S_2L = 2$ m

$$S_1L = \sqrt{2^2 + \left(\frac{3}{2}\right)^2} = \frac{5}{2} = 2.5\text{ m}$$

Path difference, $\Delta x = S_1L - S_2L = 0.5\text{ m} = \frac{\lambda}{2}$.

So we have minima here

At L', we have maxima

So, $\Delta x = \lambda$

$d - 2 = \lambda \quad \Rightarrow d - 2 = 1$

$\Rightarrow d = 3$ mm

20. (c) Optical path for first ray which travels a path L_1 through a medium of refractive index $n_1 = n_1L_1$

Optical path for second ray which travels a path L_2 through a medium of refractive index $n_2 = n_2L_2$

Path difference $= n_1L_1 - n_2L_2$

Now, phase difference

$$= \frac{2\pi}{\lambda} \times \text{path difference} = \frac{2\pi}{\lambda} \times (n_1L_1 - n_2L_2)$$

21. (c) Given amplitude ratio of waves is $\frac{a_2}{a_1} = \frac{3}{1}$

$$\therefore \frac{I_{max}}{I_{min}} = \left(\frac{a_2+a_1}{a_2-a_1}\right)^2 = \left(\frac{\frac{a_2}{a_1}+1}{\frac{a_2}{a_1}-1}\right)^2 = \left(\frac{3+1}{3-1}\right)^2 = \left(\frac{4}{2}\right)^2 = \frac{4}{1} = 4$$

22. **(b)** As we know, $\frac{I_{max}}{I_{min}} = \left(\frac{A_1 + A_2}{A_1 - A_2}\right)^2$

and $\sqrt{\frac{I_1}{I_2}} = \frac{A_1}{A_2}$

$\frac{I_{max}}{I_{min}} = 16 \Rightarrow \frac{A_{max}}{A_{min}} = 4 \Rightarrow \frac{A_1 + A_2}{A_1 - A_2} = \frac{4}{1}$

Using componendo and dividendo.

$\frac{A_1}{A_2} = \frac{5}{3} \Rightarrow \frac{I_1}{I_2} = \left(\frac{5}{3}\right)^2 = \frac{25}{9}$

23. **(b)**

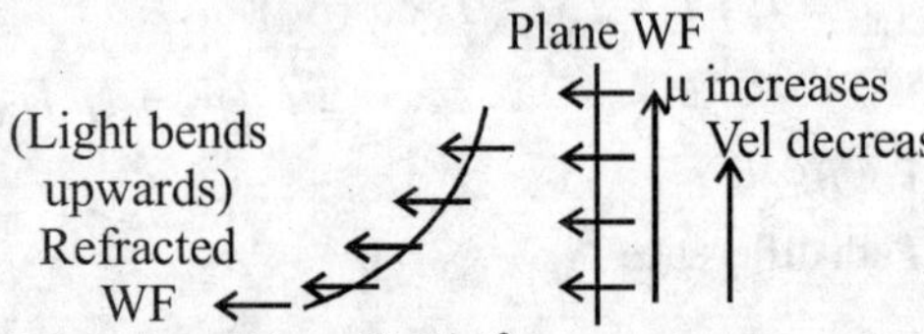

24. **(d)** Intensity ∝ (amplitude)²

$\frac{I_1}{I_2} = \frac{16}{9} = \frac{a_1^2}{a_2^2}$

$\Rightarrow a_1 = 4;\ a_2 = 3$

Therefore the ratio of intensities of bright and dark parts

$\frac{I_{Bright}}{I_{Dark}} = \frac{(a_1 + a_2)^2}{(a_1 - a_2)^2} = \frac{(4+3)^2}{(4-3)^2} = \frac{49}{1}$

25. **(d)** $\frac{\text{(Maximum intensity) coherent interference}}{\text{(Maximum intensity) in coherent interference}}$

$= \frac{n^2 I_0}{n I_0} = n$

26. **(b)** $\beta = \frac{\lambda D}{d}$

Now, for small angle α

$\sin\alpha = \frac{d}{D} \Rightarrow \frac{1}{\sin\alpha} = \frac{D}{d} \qquad \therefore \beta = \frac{\lambda}{\sin\alpha} \approx \frac{\lambda}{\alpha}$

27. **(b)** When light beam is moving and as it enters the medium, the refractive index will decrease from the axis towards the periphery of the beam.

Therefore, the beam will cover less distance as one move from the axis to the periphery and hence the beam will converge.

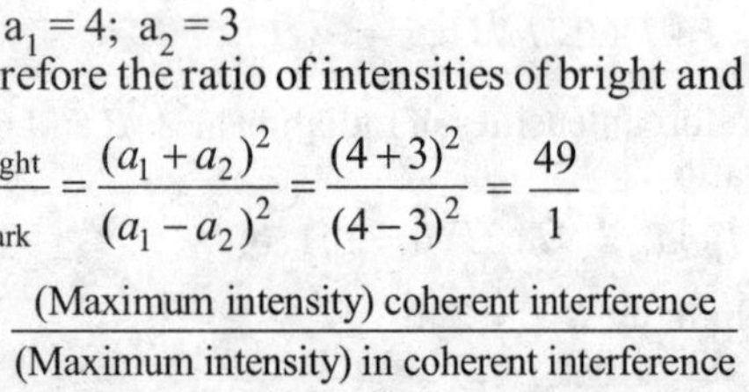

28. **(d)** Initially the parallel beam is cylindrical. Therefore, the wavefront will be planar.

29. **(a)** The speed of light (v) in a medium of refractive index (μ) is given by

$\mu = \frac{c}{v}$, where c is the speed of light in vacuum

$\therefore v = \frac{c}{\mu} = \frac{c}{\mu_0 + \mu_2(I)}$

As I is decreasing with increasing radius, it is maximum on the axis of the beam. Therefore, v is minimum on the axis of the beam.

30. **(d)** To demonstrate the phenomenon of interference we require two sources of light which emit radiation of same frequency and having a definite phase relationship (a phase relationship that does not change with time).

31. **(b)** In Young's double slit experiment, if one slit is covered by red filter and another by a green filter, no interference pattern is observed on the screen.

32. **(15)** Given : $d = 1.5\text{ mm} = 1.5 \times 10^{-3}\text{ m}$

$D = 2\text{ m}$

$\lambda = 400\text{ nm} = 400 \times 10^{-9}\text{ m}$

$\Delta\theta = \frac{\lambda}{d} \Rightarrow 20 \times \Delta\theta = \frac{2\lambda}{a}$

$\Rightarrow \frac{20\lambda D}{d} = \frac{2\lambda D}{a} \Rightarrow \frac{10}{d} = \frac{1}{a}$

$\therefore a = \frac{d}{10} = \frac{1.5 \times 10^{-1}}{10}\text{ cm} = 15 \times 10^{-3}\text{ cm}$

Value of x is 15.

Answer is 15.

33. **(c)** Given, $d_1 = 0.2\text{ mm}, d_2 = 0.4\text{ mm}$

$\beta = \frac{D\lambda}{d} \propto \frac{1}{d}$

$\beta_{new} = \frac{\beta}{2}$

Hence, β is half so 50% decrement.

34. **(b)** In YDSE,

I ∝ width of slit

$\therefore I_1 = I_0, I_2 = 2I_0$

$\frac{I_{max}}{I_{min}} = \frac{\left(\sqrt{I_1} + \sqrt{I_2}\right)^2}{\left(\sqrt{I_1} - \sqrt{I_2}\right)^2} = \frac{\left(\sqrt{2}+1\right)^2}{\left(\sqrt{2}-1\right)^2} \Rightarrow \frac{3+2\sqrt{2}}{3-2\sqrt{2}}$

35. **(11)** In Young's Double-slit experiment (YDSE), the position of nth-bright fringe (constructive interference) from the central maximum is given by

$y_n = \frac{n\lambda D}{d}$

Here n, D and d are same

$y_n \propto \lambda$

$\therefore \frac{y_2}{y_1} = \frac{\lambda_2}{\lambda_1} \Rightarrow \frac{y_2}{10\text{mm}} = \frac{660\text{nm}}{600\text{nm}} \Rightarrow y_2 = 11\text{ mm}$

36. **(a)** Given,

Slit separation, d = 1.5 mm

Wavelength of light, λ = 690 nm

Distance between screen and slits, D = 0.72 m

In YDSE,

Fringe width, $\beta = \left(\frac{\lambda}{\mu}\right) \times \frac{D}{d} = \frac{690 \times 10^{-9} \times 0.72}{1.44 \times 1.5 \times 10^{-3}}$

$= 0.23\text{ mm}$

37. **(d)** In YDSE, for bright fringes

$y = \frac{n\lambda D}{d}$; Here, $\lambda_1 = 480\text{ nm}, \lambda_2 = 600\text{ nm}$

For coincide of two bright fringes at two different wavelengths λ_1 and λ_2 is $y_1 = y_2$

$\therefore \frac{n_1\lambda_1 D}{d} = \frac{n_2\lambda_2 D}{d} \Rightarrow n_1 480 = n_2 600$

For first coincide of bright fringes for λ_2, $n_2 = 1$

$\therefore n_{1,min} = 5$

38. **(c)** In YDSE,

$I \propto (width)^2$

$$\therefore \frac{I_{max}}{I_{min}} = \left(\frac{\sqrt{I_1}+\sqrt{I_2}}{\sqrt{I_1}-\sqrt{I_2}}\right)^2 = \frac{9}{4} \Rightarrow \frac{\sqrt{I_1}+\sqrt{I_2}}{\sqrt{I_1}-\sqrt{I_2}} = \frac{3}{2}$$

$$\Rightarrow \frac{(x+1)d}{(x-1)d} = \frac{3}{2} \Rightarrow 3x - 3 = 2x + 2$$

$\therefore x = 5$

39. **(b)** In YDSE,

Fringe width, $\beta = \frac{\lambda D}{d}$

As, $\lambda_R > \lambda_B \therefore \beta_R > \beta_B$

40. **(a)** Fringe width is given by $\beta = \frac{\lambda D}{d} \Rightarrow \beta \propto \lambda$

In denser medium, wavelength λ decreases and hence fringe width decreases and fringe come closer

Also, $\mu = \frac{c}{v} \Rightarrow v = \frac{c}{\mu}$

Hence, speed of light reduces.

Frequency remains unchanged during refraction.

41. **(200)** Intensity at a point is given by

$$I = I_0 \cos^2\left(\frac{\Delta\phi}{2}\right)$$

$\Rightarrow$ According to question

$$\frac{I_0}{4} = I_0 \cos^2\left(\frac{\Delta\phi}{2}\right)$$

$$\Delta\phi = \frac{2\pi}{3}$$

$$\therefore \quad \Delta\phi = \frac{2\pi}{\lambda}\left(\frac{yd}{D}\right) = \frac{2\pi}{3}$$

$$\Rightarrow \quad y = \frac{\lambda D}{3d} = \frac{600\times10^{-9}\times1}{3\times10^{-3}} = 2\times10^{-4}\,m$$

42. **(4)** Wavelength of light, $\lambda = 500$ nm

Refractive index, $\mu = 1.5$

Path difference,

$\therefore \Delta x = n\lambda \Rightarrow 4\pi(\mu-1)t = n\lambda$

$\Rightarrow (1.5-1)t = 4\times500\times10^{-9}$ m

$\Rightarrow \frac{t}{2} = 2000\times10^{-9}$ m

$\Rightarrow t = 4000\times10^{-9}$ m

$\Rightarrow t = 4\ \mu m$

43. **(10)** The position of 3rd maxima,

$$X_3 = \frac{3\lambda D}{d} = \frac{3\times5000\times10^{-10}\times200\times10^{-2}}{0.3\times10^{-3}} = 10mm$$

44. **(9)** For overlap, $n_2\lambda_2 = n_1\lambda_1$

$$\Rightarrow \frac{n_2}{n_1} = \frac{\lambda_1}{\lambda_2} = \frac{450}{650} = \frac{9}{13}$$

Therefore $n = 9$

45. **(125)** Intensity at screen,

$$I = I_0 + I_0 + 2\sqrt{I_0 I_0}\cos\phi = 2I_0(1+\cos\phi)$$

Intensity is maximum at centre $I_{max} = 4I_0$

Now, $\frac{I_{max}}{2} = 2I_0(1+\cos\phi)$

$$\frac{4I_0}{2} = 2I_0(1+\cos\phi)$$

$\cos\phi = 0 \quad \therefore \phi = \frac{\pi}{2}$

We know that

$$\frac{2\pi}{\lambda} d\sin\theta = \frac{\pi}{2} \Rightarrow \frac{2}{\lambda} d \times \frac{y}{D} = \frac{1}{2}$$

$$y = \frac{\lambda D}{4d} = \frac{5\times10^{-7}\times1}{4\times10^{-3}} = 125\times10^{-6}\ m$$

46. **(0.20)** For first dark fringe at P,

Path difference, $\Delta x = \frac{\lambda}{2}$

$$\Rightarrow 2\sqrt{D^2+d^2} - 2D = \frac{\lambda}{2} \Rightarrow d^2 = \frac{D\lambda}{2} + \frac{\lambda^2}{16}$$

$$d^2 = \frac{0.2\times400\times10^{-9}}{2} + \frac{4\times10^{-14}}{4} \approx 400\times10^{-10}$$

$\therefore d = 20\times10^{-5} = 0.20$ mm

47. **(d)** Path difference, $\Delta x = \frac{7\lambda}{4}$

$\therefore$ Phase difference, $\phi = \frac{2\pi}{\lambda}\Delta x = \frac{2\pi}{\lambda}\times\frac{7\lambda}{4} = \frac{7\pi}{2}$

Intensity at a point,

$$I = I_{max}\cos^2\left(\frac{\phi}{2}\right)$$

$$\therefore \frac{I}{I_{max}} = \cos^2\left(\frac{\phi}{2}\right) = \cos^2\left(\frac{7\pi}{2\times2}\right) = \cos^2\left(\frac{7\pi}{4}\right)$$

$$= \cos^2\left(2\pi - \frac{\pi}{4}\right) = \cos^2\frac{\pi}{4} = \frac{1}{2}$$

48. **(c)** Intensity at a point in Young's double slit experiment is given by

$$I = I_1 + I_2 + 2\sqrt{I_1 I_2}\cos\phi$$

Here $I_1 = I_2 = I_0$ (say)

At P

$$\therefore I_P = I_0 + I_0 + 2I_0\cos\frac{\pi}{3} = 2I_0 + 2I_0\times\frac{1}{2} = 3I_0$$

At Q

$$I_Q = I_0 + I_0 + 2I_0\cos 90° = 2I_0$$

$$\frac{I_P}{I_Q} = \frac{3}{2}$$

49. **(c)** Fringe width, $\beta_1 = 2mm = 2\times10^{-3}$ m
$\lambda_1 = 400$ nm ; $\lambda_2 = 600$ nm
$\lambda_1 = 400\times10^{-9}$m, $\lambda_2 = 600\times10^{-9}$

Fringe width $(\beta) = \frac{D\lambda}{d} \Rightarrow \frac{\beta_2}{\beta_1} = \frac{\lambda_2}{\lambda_1}$

$\Rightarrow \frac{\beta_2}{2\times10^{-3}} = \frac{600\times10^{-9}}{400\times10^{-9}} = \frac{3}{2} \Rightarrow \beta_2 = 3mm$

50. **(462)** Distance between two slits, d = 2.5 mm,
Distance between slits and screen, D = 150 cm
Wavelength $\lambda_1 = 7000$Å
Wavelength, $\lambda_2 = 5500$Å

Fringe width, $\beta = \frac{\lambda D}{d}$

Let n^{th} bright fringe of λ_1 match with m^{th} bright fringe of λ_2.

$\Rightarrow n\beta_1 = m\beta_2$

$\Rightarrow \frac{n\lambda_1 D}{d} = \frac{m\lambda_2 D}{d}$

$\Rightarrow n\lambda_1 = m\lambda_2$

$\Rightarrow \frac{n}{m} = \frac{\lambda_2}{\lambda_1} = \frac{5500}{7000} \Rightarrow \frac{n}{m} = \frac{11}{14}$

Distance where bright fringe will match

$= n\beta_1 = \frac{n\lambda_1 D}{d} = \frac{11\times7000\times150m\times10^{-2}\times10^{-1}}{0.25m\times10^{-2}}$

$= 462\times10^{-5}$

51. **(4)** Due to insertion of thin plate, there will be an extra path difference of $t(\mu-1)$

As $\Delta\gamma = \gamma_f - \gamma_i$, $\Delta\gamma$ = path difference

$\Rightarrow \Delta\gamma = \frac{\Delta x_f D}{d} - \frac{\Delta x_i D}{d} \Rightarrow x\beta_0 = \frac{D}{d}(\Delta x_f - \Delta x_i)$

$\Rightarrow x\beta_0 = \frac{D}{d}[t(\mu-1)-0] \Rightarrow x\frac{\lambda D}{d} = \frac{D}{d}t(\mu-1)$

$\Rightarrow x = \frac{t(\mu-1)}{\lambda} \Rightarrow x = \frac{10^{-5}\times0.2}{500\times10^{-9}} \Rightarrow x = 4$

52. **(48)** Let us suppose that Y is shortest distance from centre where bright fringe of the two wavelength coincide. Let m^{th} order bright fringe of 800 nm coincide with n^{th} order bright fringe of 600 nm.

Then, $Y = \frac{n\lambda_1 D}{d} = \frac{m\lambda_2 D}{d}$

$[\because Y_n = \frac{n\lambda D}{d}, Y_n = \text{position of } n^{th} \text{ maxima}]$

$\Rightarrow n\lambda_1 = m\lambda_2$

$\Rightarrow n\times800 = m\times600 \Rightarrow \frac{n}{m} = \frac{3}{4}$

i.e. 3rd bright fringe of 800 nm will coincide with 4th bright fringe of 600 nm.

So, $Y = \frac{3\times800\times10^{-9}\times7}{35\times10^{-5}} = 0.048m = 48mm$

53. **(10)** Path difference at P be Δx
$\Delta x = (\mu_2-\mu_1)t = (1.55-1.51)0.1$ mm
$= 0.04\times10^{-4}$
$\Delta x = 4\times10^{-6} = 4\ \mu m$
Here, y is the distance of central maxima from geometric center

$y = \frac{\Delta x D}{d} = 4\times10^{-6}\frac{D}{d}$

fringe width $(\beta) = \frac{\lambda D}{d} = 4\times10^{-7}\frac{D}{d}$

$\therefore$ Central bright fringe spot will shift by 'x'

Number of shift $= \frac{y}{\beta} = \frac{4\times10^{-6}\ D/d}{4\times10^{-7}\ D/d} = 10$

54. **(3)** Resultant intensity in Young's double slit experiment

$I = 4I_0\cos^2\left(\frac{\Delta\phi}{2}\right)$

For path difference $\frac{\lambda}{4}$ phase difference,

$\Delta\phi = \frac{2\pi}{\lambda}\times\frac{\lambda}{4} = \frac{\pi}{4}$ $\quad\therefore I_1 = 4I_0\cos^2\left(\frac{\pi}{4}\right) = 2I_0$

For path difference $\frac{\lambda}{3}$

$I_2 = 4I_0\cos^2\left(\frac{2\pi}{\lambda}\times\frac{\lambda}{3}\right) = I_0 \Rightarrow \frac{I_1+I_2}{I_0} = 3$

55. **(c)** Given, wavelength of light $\lambda = 800$ nm
D = 5 cm
From the condition of first minima

Path difference $= A_2P - A_1P = \frac{\lambda}{2}$

$\Rightarrow \sqrt{D^2+a^2} - D = \frac{\lambda}{2} \Rightarrow D\left(1+\frac{a^2}{D^2}\right)^{1/2} - D = \frac{\lambda}{2}$

From the binomial expansion

$\Rightarrow D\left(1+\frac{1}{2}\times\frac{a^2}{D^2}\right) - D = \frac{\lambda}{2}$

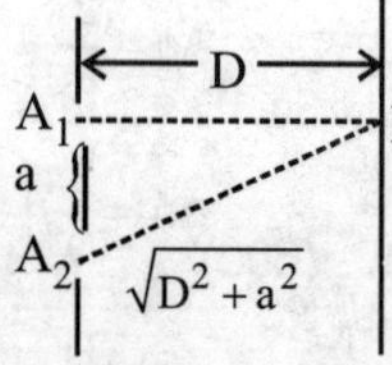

$\Rightarrow \frac{a^2}{2D} = \frac{\lambda}{2} \Rightarrow a = \sqrt{\lambda.D}$

$= \sqrt{800\times10^{-6}\times50}$

$\Rightarrow a = 0.2$ mm

56. **(a)** Given, the distance between slits and screen, D = 1 m
Wavelength of monochromatic light, $\lambda = 600$ nm $= 600\times10^{-9}$, m & n = 5
The separation between the slits is

$d = \frac{n\lambda D}{y_{th}}$ (For bright fringes)

$\Rightarrow \frac{5\times600\times10^{-9}\times1}{d} = 5\times10^{-2}$

$\Rightarrow d = \frac{5\times600\times10^{-9}\times1}{5\times10^{-2}} = 60\times10^{-6}$ m

$\Rightarrow d = 60\ \mu m$

57. **(630)** Fringe width, $\beta = \frac{D}{d}\lambda \quad \therefore \beta \propto \lambda$

$\therefore\ \lambda_2 = \frac{8.1}{7.2}\lambda_1 = \frac{9}{8}\lambda_1$

$\therefore$ wavelength of second light, $\lambda_2 = \frac{9}{8}\lambda_1 = \frac{9}{8} \times 560 = 630\,\text{nm}$

58. **(b)** Fringes width, $\beta = 12\,\text{mm}$

Refractive index of water, $\mu = \frac{4}{3}$

The fringes width is given by,

$$\beta = \frac{D\lambda}{d} \quad \text{..... (i)}$$

Here, λ is wavelength of light.

D is distance between screen and source.

d is distance between coherent source.

If the entire arrangement is placed in water then fringes width becomes $\beta' = \frac{D\lambda'}{d}$ ..... (ii)

Dividing equation (ii) by (i), we have

$$\Rightarrow \frac{\beta'}{\beta} = \frac{\lambda'}{\lambda}$$

$$\Rightarrow \beta' = \frac{12 \times 3}{4} \qquad \left(\because\ \mu = \frac{\lambda}{\lambda'}\right)$$

$$\Rightarrow \beta' = 9\ \text{mm}$$

59. **(d)** Width $\propto$ Amplitude^2 $\propto$ I

$$So, \frac{I_{max}}{I_{min}} = \frac{\left(\sqrt{I_1} + \sqrt{I_2}\right)^2}{\left(\sqrt{I_1} - \sqrt{I_2}\right)^2} = \frac{\left(\sqrt{9} + \sqrt{16}\right)^2}{\left(\sqrt{9} - \sqrt{16}\right)^2} = \frac{49}{1}$$

60. **(d)** As $\beta = \frac{\lambda D}{d}$

$$\beta \propto \frac{\lambda}{d} \Rightarrow \frac{\beta_2}{\beta_1} = \frac{\lambda_2}{\lambda_1} \times \frac{d_1}{d_2} = \frac{6}{5} \times \frac{d_1}{2d_1} = \frac{3}{5}$$

So, $\beta_2 = \frac{3}{5}\beta_1 = \frac{3}{5} \times 0.5\,\text{mm} = 0.3\,\text{mm}$

61. **(600)** $\beta = \frac{\lambda D}{d}$

$$\Delta\beta = \lambda \frac{\Delta D}{d}$$

$$\lambda = \frac{\beta d}{\Delta D} = \frac{3 \times 10^{-5} \times 10^{-3}}{5 \times 10^{-2}} = 0.6 \times 10^{-6} = 600 \times 10^{-9}\,\text{m}$$

$= 600\ nm$

62. **(b)**

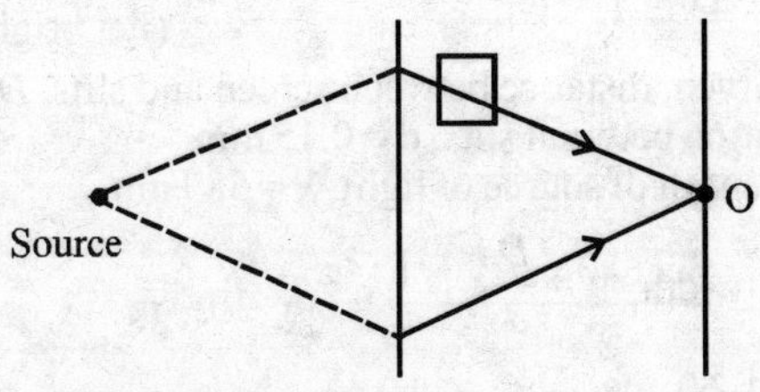

Path difference at O $= t(\mu - 1)$

If the intensity at O remains (maximum) unchanged, path difference must be $n\lambda$

So, $t(\mu - 1) = n\lambda$

$\Rightarrow x\lambda(1.5 - 1) = n\lambda$

$\Rightarrow x = 2n$

For $n = 1, x = 2$.

63. **(450)** Given, $\frac{d}{2}$ = position of Ist dark fringe

$$\Rightarrow \frac{d}{2} = \frac{\beta}{2} \Rightarrow d = \frac{\lambda D}{d}$$

$$\Rightarrow \lambda = \frac{d^2}{D} = \frac{(0.6 \times 10^{-3})^2}{0.8} = 450 \times 10^{-9}\,\text{m} = 450\,\text{nm}$$

64. **(1)** We have given

Amplitude $\propto$ slit width (w)

Also intensity $\propto$ (Amplitude)^2 $\propto$ (Slit width)^2

$$\frac{I_2}{I_1} = \left(\frac{w_2}{w_1}\right)^2 = \left(\frac{3w_1}{w_1}\right)^2 = \left(\frac{3}{1}\right)^2 = 9 (\because w_2 = 3w_1)$$

$$\Rightarrow I_2 = 9I_1$$

$$\Rightarrow \frac{I_{min}}{I_{max}} = \left(\frac{\sqrt{I_2} - \sqrt{I_1}}{\sqrt{I_1} + \sqrt{I_2}}\right)^2 = \left(\frac{3-1}{3+1}\right)^2 = \frac{1}{4} = \frac{x}{4}$$

$$\Rightarrow x = 1.00$$

65. **(5)** $y_4 = \frac{2.4cm}{2} = 1.2cm$

Position of 4th bright fringe $= \frac{4\lambda D}{d}$

$$\frac{4\lambda D}{d} = 1.2\ cm \quad \text{...(i)}$$

Putting $D = 1.5, d = .3 \times 10^{-3}\ m$ in equation (i)

$$\Rightarrow \lambda = \frac{1.2 \times 10^{-2} \times .3 \times 10^{-3}}{1.5 \times 4},$$

Putting λ in $v = \frac{c}{\lambda}$ we get

$$v = \frac{c}{\lambda} = \frac{3 \times 10^8 \times 1.5 \times 4}{1.2 \times 10^{-2} \times .3 \times 10^{-3}} \Rightarrow v = 5 \times 10^{14}\,\text{Hz}$$

66. **(300)** Position of 1st bright fringe is given by

$$y_n = n\frac{\lambda D}{d}$$

For red, $y_1 = \frac{D\lambda_{red}}{d} = 3.5\ \text{mm}$

For violet, $y_2 = \frac{D}{d}\lambda_{violet} = 2\ \text{mm}$

$y_1 - y_2 = 3.5 - 2 = 1.5\ \text{mm}$

$1.5\ \text{mm} = \frac{D}{d}\left(\lambda_{red} - \lambda_{violet}\right)$

Put $D = 1.5\ m$ and d = .3 mm

$\lambda_{red} - \lambda_{violet} = 3 \times 10^{-7} = 300\ \text{nm}$

67. **(b)** Fringe width, $\beta = \frac{\lambda D}{d}$

Here, λ = wavelength of light

D = Distance between sources and screen
d = distance between two slits
As $\lambda_{blue} < \lambda_{orange}$
So, $\beta_{blue} < \beta_{orange}$

68. **(b)** Fringe width $\beta = \frac{D\lambda}{d}$

$$\beta = \frac{D\lambda}{(d_0 + a_0 \sin \omega t)} \quad [\because d = d_0 + a_0 \sin \omega t]$$

$$\beta_{max} - \beta_{min} = \frac{D\lambda}{d_0 - a_0} - \frac{D\lambda}{d_0 + a_0}$$

$$= D\lambda\left[\frac{d_0 + a_0 - d_0 + a_0}{(d_0 + a_0)(d_0 - a_0)}\right] = \frac{2a_0 D\lambda}{d_0^2 - a_0^2}$$

69. **(b)** Given :
$d = 0.5$ mm $= 0.5 \times 10^{-3}$ m and $D = 0.5$ m

Fringe width $\beta = \frac{\lambda D}{d} = \frac{5890 \times 10^{-10} \times 0.5}{0.5 \times 10^{-3}} = 589 \times 10^{-6}$ m

Hence, distance between the first and third bright fringe $= 2\beta = 2 \times 589 \times 10^{-6}$ m $= 1178 \times 10^{-6}$ m

70. **(600)** Given,
Separation between two slits, $d = 1$ mm
Distance of the screen, $D = 10$ m
Fringe width,

$$\beta = \frac{D\lambda}{d} \Rightarrow 6 \times 10^{-3} = \frac{10\lambda}{10^{-3}}$$

$\therefore \lambda = 600$ nm.

71. **(a)** Fringe width,

$$\beta = \frac{\lambda \cdot D}{d}$$

Here, λ = wavelength of light
D = Distance of screen from source
d = Distance between the slits
$\because \lambda_R > \lambda_V$
As wavelength of light decreases, fringe width will decrease and fringe line will come closer.

72. **(a)** Fringe width, $\beta = \frac{\lambda D}{d}$
Where D is the distance between slit and screen, d is the distance between two slits, λ is the wavelength of light.

$$\therefore \quad \beta = \frac{500 \times 10^{-9} \times 1}{2 \times 10^{-3}} = 250 \times 10^{-6} = 0.25 \text{ mm}$$

73. **(9)** In young's double slit experiment, intensity at a point is given by

$$I = I_0 \cos^2 \frac{\phi}{2} \quad ...(i)$$

where, ϕ = phase difference,

Using phase difference, $\phi = \frac{2\pi}{\lambda} \times$ path difference

For path difference λ, phase difference $\phi_1 = 2\pi$

For path difference, $\frac{\lambda}{6}$, phase difference $\phi_2 = \frac{\pi}{3}$

Using equation (i),

$$\frac{I_1}{I_2} = \frac{\cos^2\left(\frac{\phi_1}{2}\right)}{\cos^2\left(\frac{\phi_2}{2}\right)} = \frac{\cos^2\left(\frac{2\pi}{2}\right)}{\cos^2\left(\frac{\pi}{6}\right)}$$

$$\Rightarrow \frac{K}{I_2} = \frac{1}{\frac{3}{4}} = \frac{4}{3} \Rightarrow I_2 = \frac{3K}{4} = \frac{9K}{12} \quad \therefore n = 9.$$

74. **(b)** Given : Wavelength of light, $\lambda = 500$ nm
Distance between the slits, $d = 0.05$ mm
Angular width of the fringe formed,

$$\theta = \frac{\lambda}{d} = \frac{500 \times 10^{-9}}{0.05 \times 10^{-3}} = 0.01 \text{ rad} = 0.57°.$$

75. **(a)** Path difference, $\Delta P = d \sin\theta = d\theta$
d = distance between slits = 1 mm = 10^{-3} mm
D = distance between the slits and screen = 100 cm = 1 m
y = distance between central bright fringe and observed fringe = 1.27 mm

$$\therefore \Delta P = \frac{dy}{D} = \frac{10^{-3} \times 1.270 \times 10^{-3}}{1 \text{ m}} = 1.27 \ \mu\text{m}$$

76. **(d)** Let n_1 fringes are visible with light of wavelength λ_1 and n_2 with light of wavelength λ_2. Then

$$\frac{n_1 D\lambda_1}{d} = \frac{n_2 D\lambda_2}{d}$$

$$\Rightarrow \frac{n_2}{n_1} = \frac{\lambda_1}{\lambda_2} \Rightarrow n_2 = \frac{700}{400} \times 16 = 28$$

77. **(750)** Fringe width, $\beta = \frac{\lambda D}{d}$ where, λ = wavelength, D = distance of screen from slits, d = distance between slits
ATQ

$$15 \times \frac{\lambda_1 D}{d} = 10 \times \frac{\lambda_2 D}{d} \Rightarrow 15\lambda_1 = 10\lambda_2$$

$\Rightarrow \lambda_2 = 1.5\lambda_1 = 1.5 \times 500$ nm $\Rightarrow \lambda_2 = 750$ nm

78. **(a)** Given, Path difference, $\Delta x = \frac{\lambda}{8}$

Phase differences, $\Delta\phi = \frac{2\pi}{\lambda}\Delta x = \frac{2\pi}{\lambda} \times \frac{\lambda}{8} = \frac{\pi}{4}$

$$I = I_0 \cos^2\left(\frac{\Delta\phi}{2}\right)$$

$$\Rightarrow \frac{I}{I_0} = \cos^2\left(\frac{\frac{\pi}{4}}{2}\right) = \cos^2\left(\frac{\pi}{8}\right) \Rightarrow \frac{I}{I_0} = 0.853$$

79. **(c)** Given, distance between screen and slits, $D = 1.5$ m
Seperation between slits, d = 0.15 mm
Wavelength of source of light, $\lambda = 589$ nm

Fringe-width, $w = \frac{D}{d}\lambda$

$$= \frac{1.5}{0.15 \times 10^{-3}} \times 589 \times 10^{-9} \, m$$

$= 589 \times 10^{-2}$ mm $= 5.89$ mm ≈ 5.9 mm

80. (c) Given, $\Delta x = \beta$

or $\frac{D(\mu-1)t}{d} = \frac{D\lambda}{d}$ $\quad \therefore \ t = \frac{\lambda}{(\mu-1)}$

81. (None) Shift $= n\beta$ (given)

$\therefore \ D\frac{(\mu-1)t}{a} = \frac{n\lambda D}{a} \left[\because \text{Shift} = \frac{D(\mu-1)t}{a}\right]$

or $t = \frac{n\lambda}{(\mu-1)}$

82. (a) Path difference $= d\sin\theta \approx d\theta$

$= 0.1 \times \frac{1}{40}$ mm $= 2500$ nm

For bright fringe, path difference must be integral multiple of λ.

$\therefore 2500 = n\lambda_1 = m\lambda_2$

$\therefore \lambda_1 = 625$ nm (for $n = 4$), $\lambda_2 = 500$ nm (for $m = 5$)

83. (a) Here, $x_1 = 2d$ and $x_2 = \sqrt{5}d$

For, first minima, $\Delta x = \frac{\lambda}{2}$

$\therefore \Delta x = x_2 - x_1 = \sqrt{5}d - 2d = \frac{\lambda}{2} \Rightarrow d = \frac{\lambda}{2(\sqrt{5}-2)}$

84. (d) For 'n' number of maximas

$d\sin\theta = n\lambda$

$0.32 \times 10^{-3} \sin 30° = n \times 500 \times 10^{-9}$

$\therefore \quad n = \frac{0.32 \times 10^{-3}}{500 \times 10^{-9}} \times \frac{1}{2} = 320$

Hence total no. of maximas observed in angular range $-30° \le \theta \le 30°$

$= 320 + 1 + 320 = 641$

85. (d) For common maxima, $n_1\lambda_1 = n_2\lambda_2$

$\Rightarrow \quad \frac{n_1}{n_2} = \frac{\lambda_2}{\lambda_1} = \frac{520 \times 10^{-9}}{650 \times 10^{-9}} = \frac{4}{5}$

For λ_1

$y = \frac{n_1\lambda_1 D}{d}$, $\lambda_1 = 650$ nm

$y = \frac{4 \times 650 \times 10^{-9} \times 1.5}{0.5 \times 10^{-3}}$ or, $y = 7.8$ mm

86. (c) $2I_0 = 4I_0 \cos^2\left(\frac{\Delta\phi}{2}\right)$ here, $\Delta\phi = \frac{\pi}{2}$

But, $\Delta\phi = \frac{2\pi}{\lambda}\Delta x$ $\quad$ so, $\Delta x = \frac{\lambda}{4}$

So, $Y = \frac{\Delta x D}{d} = \frac{\lambda}{4} \cdot \frac{D}{d} = \frac{\beta}{4}$

87. (c) β (width of central maxima) $= 2\left(\frac{D\lambda}{a}\right)$,

a = slit width

$\beta_{interfer} = \frac{D\lambda}{d} \times n$ $\quad \therefore \frac{2D\lambda}{a} = \frac{D\lambda \times n}{d}$

$\Rightarrow \frac{2}{a} = \frac{n}{d} \Rightarrow n = \frac{2d}{a} = \frac{2 \times 6.1a}{a} \simeq 12$

88. (d) It will be concentric circles because locus of all the point having same path difference lies on concentric circle.

89. (a) Fringe width $\beta = \frac{D}{d}\lambda$

And number of fringes observed in the field of view is obtained by $\frac{d}{\lambda}$

90. (d) Let a_1 be the amplitude of light from first slit and a_2 be the amplitude of light from second slit.

$a_1 = a$, Then $a_2 = 2a$

Intensity $I \propto$ (amplitude)^2

$\therefore \quad I_1 = a_1^2 = a^2$

$I_2 = a_2^2 = 4a^2 = 4I_1$

$I_r = a_1^2 + a_2^2 + 2a_1a_2\cos\phi = I_1 + I_2 + 2\sqrt{I_1I_2}\cos\phi$

$I_r = I_1 + 4I_1 + 2\sqrt{4I_1^2}\cos\phi$

$\Rightarrow \quad I_r = 5I_1 + 4I_1\cos\phi \quad$ ...(i)

Now, $I_{max} = (a_1 + a_2)^2 = (a + 2a)^2 = 9a^2$

$I_{max} = 9I_1 \Rightarrow I_1 = \frac{I_{max}}{9}$

Substituting in equation (i)

$I_r = \frac{5I_{max}}{9} + \frac{4I_{max}}{9}\cos\phi \Rightarrow I_r = \frac{I_{max}}{9}[5 + 4\cos\phi]$

$\Rightarrow I_r = \frac{I_{max}}{9}\left[5 + 8\cos^2\frac{\phi}{2} - 4\right]$

$\Rightarrow I_r = \frac{I_{max}}{9}\left[1 + 8\cos^2\frac{\phi}{2}\right] = \frac{I_m}{9}\left(1 + 8\cos^2\frac{\phi}{2}\right)$

91. (c) We know that,

$$\frac{I_{max}}{I_{min}} = \frac{\left(\sqrt{I_1} + \sqrt{I_2}\right)^2}{\left(\sqrt{I_2} - \sqrt{I_1}\right)^2} \Rightarrow \left(\frac{\sqrt{\frac{I_1}{I_2}} + 1}{1 - \sqrt{\frac{I_1}{I_2}}}\right)^2$$

$$\therefore \frac{I_{max}}{I_{min}} = \frac{\left(\sqrt{\frac{1}{25}} + 1\right)^2}{\left(\sqrt{\frac{1}{25}} - 1\right)^2} \left(\frac{\omega_1}{\omega_2} = \frac{1}{25} \text{ given and } \frac{\omega_1}{\omega_2} = \frac{I_1}{I_2}\right)$$

On solving we get,

$$\frac{I_{max}}{I_{min}} = \frac{\frac{36}{25}}{\frac{16}{25}} = \frac{9}{4} = 9 : 4$$

92. (b) As $\sin\theta = \frac{n\lambda}{d}$ and $\sin\theta$ cannot be > 1

$\therefore 1 = \frac{n\lambda}{1.8\lambda} \Rightarrow n = 1.8$

Hence maximum number of possible interference maximas, 0, ±1 i.e. 3

93. (c) Shift $= \frac{\beta(\mu-1)}{\lambda}t_1 - \frac{\beta(\mu-1)}{\lambda}t_2 = \frac{\beta(\mu-1)}{\lambda}(t_1 - t_2)$

94. **(a)** Path difference at P, $\Delta x_1 = 0$

$\therefore$ Phase difference at P will be

$$\Delta\phi_1 = \frac{2\pi}{\lambda}\Delta x_1 = \frac{2\pi}{\lambda}\times 0 = 0°$$

Resultant Intensity at P

$$I_1 = I_0 + I_0 + 2I_0\cos 0° = 4I_0$$

Path difference at Q, $\Delta x_2 = \frac{\lambda}{4}$

$\therefore$ Phase difference at Q, $\Delta = \frac{2\pi}{\lambda}\cdot\frac{\lambda}{4} = \left(\frac{\pi}{2}\right)$

Resultant intensity at Q.

$$I_2 = I_0 + I_0 + 2I_0\cos\frac{\pi}{2} = 2I_0$$

Thus, $\frac{I_1}{I_2} = \frac{4I_0}{2I_0} = \frac{2}{1}$

95. **(a)** For coherent sources, intensity at mid point $I_1 \propto (a+a)^2 \Rightarrow I_1 \propto (2a)^2$

For incoherent sources, intensity of mid point is $I_2 \propto 2a^2$

$\therefore \frac{I_1}{I_2} = \frac{2}{1}$

96. **(b)** Let λ be the wavelength of unknown light. Third bright fringe of known light coincides with the 4th bright fringe of the unknown light.

$$\therefore \frac{3\lambda_1 D}{d} = \frac{4\lambda D}{d} \Rightarrow \frac{3(590)D}{d} = \frac{4\lambda D}{d}$$

$$\Rightarrow \lambda = \frac{3}{4}\times 590 = 442.5\,\text{nm}$$

97. **(a)** For path difference of λ, the phase difference is 2π

For path difference of $\frac{\lambda}{6}$, the phase difference is

$$\frac{2\pi\times\lambda/6}{\lambda} = \frac{\pi}{3}$$

Resultant intensity

$$I = I_0\cos^2\left(\frac{\pi/3}{2}\right) = I_0\cos^2\left(\frac{\pi}{6}\right) = \frac{3}{4}I_0$$

So, $\frac{I}{I_0} = \frac{3}{4}$

98. **(d)** In case of double hole experiment shape is hyperbolic, but in case of double slit experiment shape is straight line, if source is monochromatic.

99. **(b)** For constructive interference path difference (As $\sin\theta \le 1$)

$d\sin\theta = n\lambda$

Given $d = 2\lambda$

$\therefore 2\lambda\sin\theta = n\lambda \Rightarrow \sin\theta = \frac{n}{2}$

$n = 0, 1, -1, 2, -2$ hence five maxima are possible.

100. **(c)** The amplitude of resultant wave is

$$E = \sqrt{E_1^2 + E_2^2 + 2E_1E_2\cos\phi}$$

Here, $E_1 = E_2 = E_0$, $\phi = \frac{\pi}{3}$

$$\therefore E = \sqrt{(E_0)^2 + (E_0)^2 + 2(E_0)(E_0)\cos\frac{\pi}{3}}$$

$$= \sqrt{2E_0^2 + E_0^2} = \sqrt{3}E_0 = 1.73E_0.$$

101. **(a)** Two crossed polarisers : P_1 and P_2 are at 90° to each other - No light passes through.

Now, finding the angle between P_2 and P_3, so that transmitted intensity is Maximum.

Through P_2, $I_1 = I_0\sin^2\left(\frac{\pi}{2} - \theta\right)$

Angle between P_1 and $P_3 = \theta$

Then angle between P_3 and $P_2 = 90° - \theta$

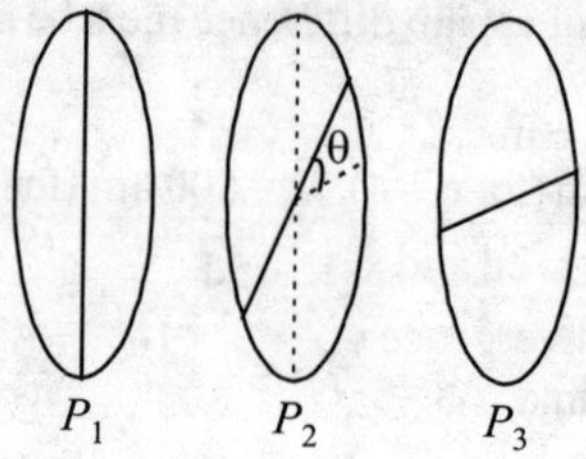

$$I_1 = I_0\cos^2\theta$$

Through P_3, $I_{net} = (I_0\cos^2\theta)\sin^2\theta$

$I_{net} = \frac{I_0}{4}[\sin(2\theta)]^2$, for max $I_{net}: \theta = 45°$

So angle between P_2 and $P_3 = \frac{\pi}{4}$

Correct Ans. (a).

102. **(6)** $\theta_1 = \sin^{-1}\left(\frac{2\lambda}{a}\right)$

$$\theta_2 = \sin^{-1}\left(\frac{3\lambda}{a}\right)$$

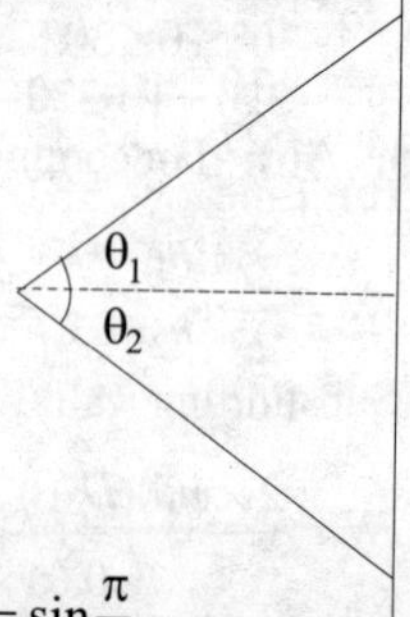

$\because \theta_1 + \theta_2 = 30°$

$$\Rightarrow \sin^{-1}\left(\frac{2\lambda}{a}\right) + \sin^{-1}\left(\frac{3\lambda}{a}\right) = \frac{\pi}{6}$$

$$\Rightarrow \frac{2\lambda}{a}\sqrt{1-\left(\frac{3\lambda}{a}\right)^2} + \frac{3\lambda}{a}\sqrt{1+\left(\frac{2\lambda}{a}\right)^2} = \sin\frac{\pi}{6}$$

Putting $\lambda = 628$ nm and solving we get, $a = 6.07\mu$m

103. **(b)**

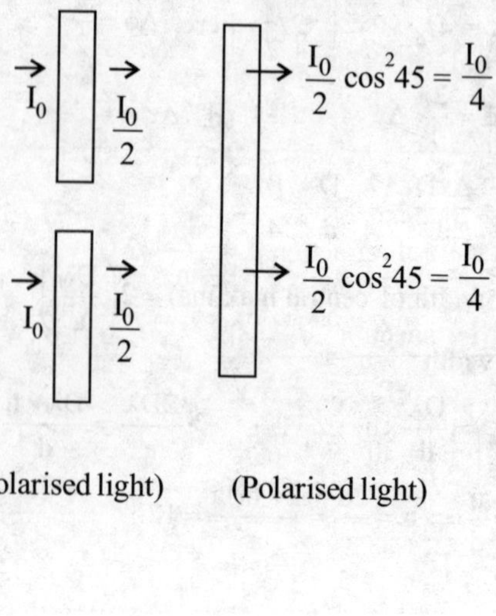

(Unpolarised light) (Polarised light)

Path difference, $x = \frac{\lambda}{3}$

$\therefore$ Intensity, $I = 4\left(\frac{I_0}{4}\right)\cos^2\left(\frac{\pi}{\lambda}x\right)$

$= I_0 \cos^2\left(\frac{\pi}{\lambda}\cdot\frac{\lambda}{3}\right) = I_0 \cos^2\frac{\pi}{3} = \frac{I_0}{4}$

104. (6) Given,
Wavelength of monochromatic light, $\lambda = 600$ nm
Width of slit, a = 0.4 mm
For second order minima

$\sin\theta \simeq \theta \simeq \frac{2\lambda}{a}$

$= \frac{2\times600\times10^{-9}}{4\times10^{-4}} = 3\times10^{-3}$ rad

Total angular divergence $= 2\times3\times10^{-3} = 6\times10^{-3}$ rad

105. (2) $10W_d = W_s$

$10\left(\frac{\lambda D}{d}\right) = \frac{2\lambda D}{a}$

$a = \frac{d}{5} = \frac{10\times10^{-4}\text{ m}}{5} = 2\times10^{-4}$

106. (275) Distance of nth order maxima,

$x_n = \frac{n\lambda D}{d}$

$\therefore x_1 = \frac{\lambda D}{d} = \frac{550\times10^{-9}\times100\times10^{-2}}{0.2\times10^{-3}} = 275\times10^{-5}$ m

$= x\times10^{-5}$ m (given)

$\therefore x = 275$

107. (b) Linear width is given as:

$W = \frac{2\lambda}{a}f$

Substituting values

$W = \frac{2\times6000\times10^{-10}\times20\times10^{-2}}{0.01\times10^{-3}} = 24\times10^{-3} = 24\text{mm}$

108. (a) Condition for minima is $d\sin\theta = n\lambda$
For Ist minima $d\sin\theta = \lambda$

$\sin\theta = \frac{\lambda}{d} = \frac{2}{4} = \frac{1}{2} \quad \therefore \theta = 30°$

Angular spread $= 2\theta = 60°$

109. (a) From Brewster's law
$\mu = \tan i_p$

$\Rightarrow \mu = \tan 60° = \sqrt{3}$

Using Snell's law of refraction $\mu = \frac{\sin i}{\sin r}$

$\Rightarrow \sin r = \frac{\sin i}{\mu} = \frac{\sin 60°}{\sqrt{3}} = \frac{\sqrt{3}}{2}\times\frac{1}{\sqrt{3}}$

$\Rightarrow r = \sin^{-1}\left(\frac{1}{2}\right) = 30°$

110. (a) Width of 1st secondary maxima $= \frac{\lambda}{d}D$

$= \frac{400\times10^{-9}}{0.2\times10^{-3}}\times100\times10^{-2} = 2$ mm

111. (a) When intensity I_0 is passed through a polaroid A,

$I' = \frac{I_0}{2}$

Intensity of emergent light after passed through polaroid B,

$= I'\cos^2 45° = \frac{I_0}{2}\cos^2 45° = \frac{I_0}{4}$

112. (2) In diffraction, for n^{th} minima
$\Delta y = d\sin\theta = n\lambda$

$\because$ θ is small $\Rightarrow \sin\theta \approx \theta = \frac{y}{D}$

$\therefore d\frac{y}{D} = n\lambda$

$\Rightarrow y_n = \frac{n\lambda D}{d}$ (Position of nth minima)

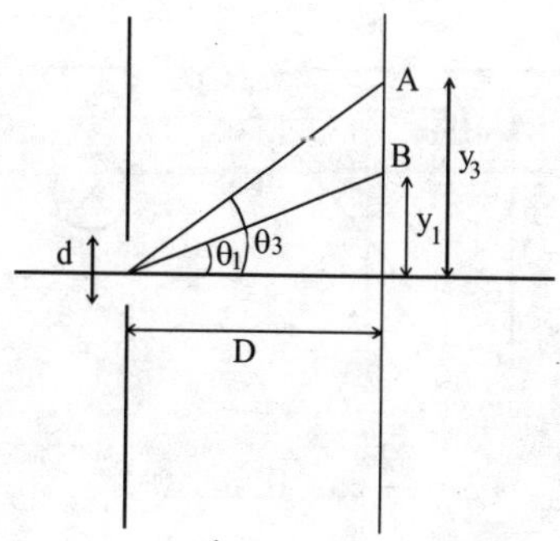

B → 1st minima, A → 3rd minima,

$\therefore y_3 = \frac{3\lambda D}{d},\ y_1 = \frac{\lambda D}{d}$

$\Delta y = y_3 - y_1 = \frac{2\lambda D}{d}$

$3\times10^{-3} = \frac{2\times6000\times10^{-10}\times0.5}{d}$

$d = \frac{2\times6000\times10^{-10}\times0.5}{3\times10^{-3}}$

$d = 2\times10^{-4}$ m

113. (d) Let I_0 = intensity of unpolarised light incident on first polaroid.

I_1 = Intensity of light transmitted from 1st polaroid $= \frac{I_0}{2}$

θ = angle between 1st and 2nd polaroid and φ = angle between 2nd and 3rd polaroid

$\theta + \phi = 90°$ ($\because$ 1st and 3rd polaroid are crossed)

$\Rightarrow \phi = 90° - \theta$

I_2 = Intensity from 2nd polaroid $= I_1\cos^2\theta = \frac{I_0}{2}\cos^2\theta$

I_3 = Intensity from 3rd polaroid $= I_2 \cos^2 \phi$

$I_3 = I_1 \cos^2\theta \cos^2\phi = \frac{I_0}{2}\cos^2\theta\cos^2\phi$

$\phi = 90 - \theta$

$I_3 = \frac{I_0}{2}\cos^2\theta\sin^2\theta = \frac{I_0}{2}\left[\frac{2\sin\theta\cos\theta}{2}\right]^2$

$I_3 = \frac{I_0}{8}\sin^2 2\theta$

I_3 will be maximum when $\sin 2\theta = 1$

$2\theta = 90°$ $\therefore \theta = 45°$

114. (30) For first minima, $a\sin\theta = \lambda$
where θ = angle of diffraction

$\Rightarrow \sin\theta = \frac{\lambda}{a} = \frac{5000\times10^{-10}}{1\times10^{-6}} = \frac{1}{2}$

$\therefore \theta = 30°$

115. (b) Wavelength of monochromatic light,
$\lambda = 600\times10^{-9}$ m
As for first minima $a\sin\theta = \lambda$
$\Rightarrow a\sin30° = 600\times10^{-9}$
$\Rightarrow a = 1200\times10^{-9}$ m $= 1.2\ \mu$m

116. (30) $\longrightarrow \quad \longrightarrow \frac{I_0}{2}\text{Cos}^2\theta \longrightarrow \frac{I_0}{2}\cos^2\theta\sin^2\theta$

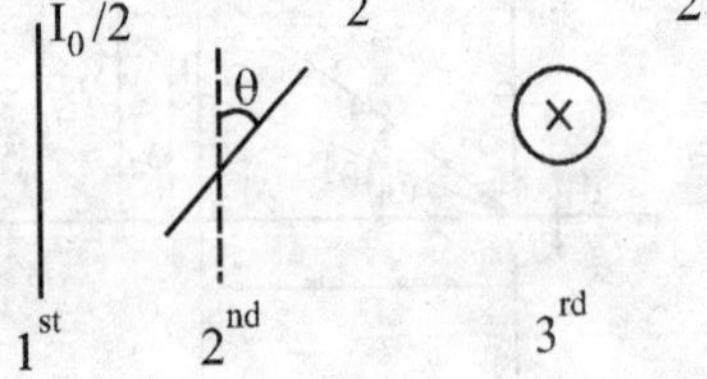

$I_{net} = 3 = \frac{32}{2}\cos^2\theta.\sin^2\theta$

$\frac{3}{4} = 4\sin^2\theta.\cos^2\theta = (\sin 2\theta)^2$

$\frac{\sqrt{3}}{2} = \sin(2\theta)$ Hence, $\theta = 30°$

117. (a)

118. (c)

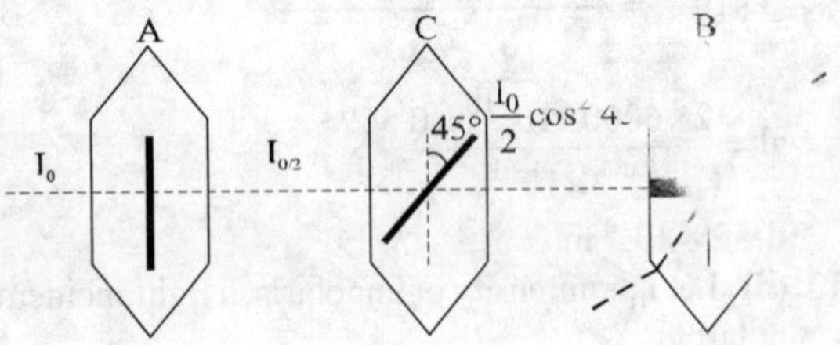

$\frac{I_0}{2}\cos^2 45°\cos^2 45° = \frac{I_0}{2}\times\frac{1}{4} = \frac{I_0}{8}$

119. (24) Given, intensity of light, $I = 256\frac{W}{m^2}$

By first polaroid P_1 intensity will be halved.

After P_2, $I_2 = \frac{I}{2}\cos^2 60° = \frac{I}{8}$

After P_3, $I_3 = \frac{I}{8}\cos^2 30° = \frac{I}{8}\left(\frac{\sqrt{3}}{2}\right)^2 = \frac{3I}{32}$

Intensity $= \frac{256\times3}{32} = 24\ \frac{W}{m^2}$

120. (7) We have, $\mu_1 = \sqrt{2.8\times1} = \sqrt{2.8}$ $\left[\because \mu = \sqrt{\varepsilon_r\mu_r}\right]$

and, $\mu_2 = \sqrt{6.8\times1} = \sqrt{6.8}$
$i + r = 90°$
$\Rightarrow r = 90° - i \Rightarrow \sin r = \sin(90 - i) \Rightarrow \sin r = \cos i$
By Snells law, $\mu_1\sin i = \mu_2\sin r$
$\Rightarrow \mu_1\sin i = \mu_2\cos i$

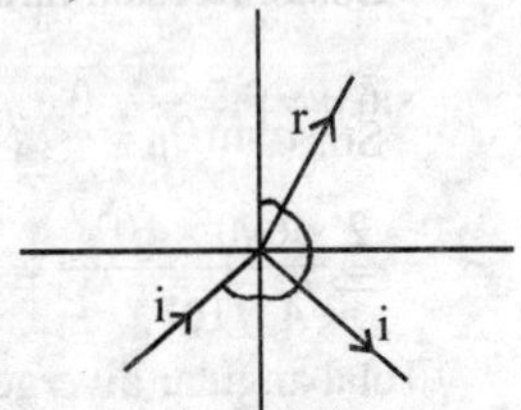

$\Rightarrow \tan i = \frac{\mu_2}{\mu_1} = \sqrt{\frac{6.8}{2.8}}$

$\Rightarrow \tan i = \left(\frac{2.8+4}{2.8}\right)^{1/2} \Rightarrow i = \tan^{-1}\left(1+\frac{10}{7}\right)^{1/2}$

121. (b)

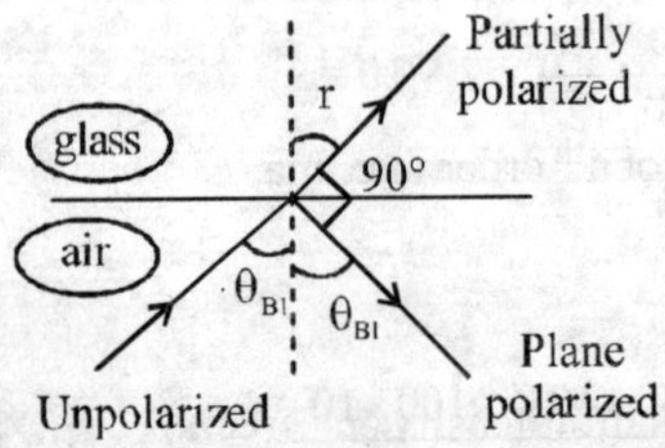

By Brewster's law
$\tan\theta_{B1} = \mu g$ ....(i)
By Brewster's law
$\tan\theta_{B2} = \frac{1}{\mu g}$
$\Rightarrow \cot\theta_{B2} = \mu g$ ...(ii)
$\Rightarrow \theta_{B2} = \cot^{-1}(\mu g)$
From (i) & (ii), we get
$\tan\theta_{B1} = \cot\theta_{B2}$
$\Rightarrow \theta_{B1} = \frac{\pi}{2} - \theta_{B2}$
$\Rightarrow \theta_{B2} = \frac{\pi}{2} - \theta_{B1}$

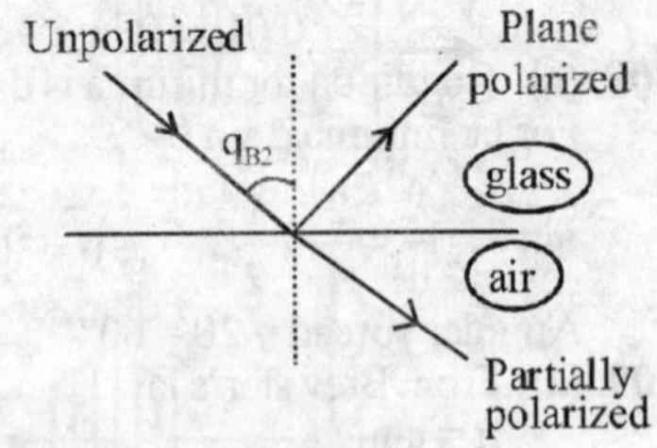

So, (I) is correct.
and (II) is incorrect.

122. (c) The intensity of the light given by, $I = I_0\cos^2\theta$

$I_1 = \frac{1}{2}(2I_0)\cos^2 0 = I_0$

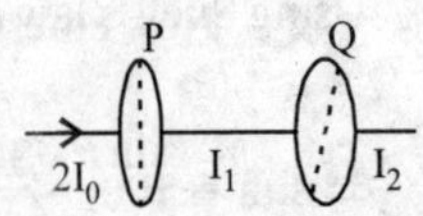

$I_2 = I_1\cos^2 30° = I_0.\frac{3}{4} = \frac{3I_0}{4}$

Hence, the intensity of the emergent light is $\frac{3I_0}{4}$.

123. (d) We know that when light strike the interface of two media at Brewster's angle, then reflected light will be plane-polarized with its $\vec{E}$ vector vibrating in a single plane. Now, if there will be no electric field vector then there will no '$\vec{E}$' vibrating in reflected light. So there will be no reflected light.

124. (3) Condition for maxima is

$$a\sin\theta = \left(n+\frac{1}{2}\right)\lambda$$

So, for Ist maxima $a\sin\theta = \frac{3}{2}\lambda$

So, $a\sin\theta_1 = \frac{3}{2}\lambda_1$

$$\Rightarrow a.\frac{y_1}{L} = \frac{3}{2}\lambda_1 \Rightarrow y_1 = \frac{3}{2}\frac{\lambda_1 L}{a}$$

Similarly, $y_2 = \frac{3}{2}\frac{\lambda_2}{a}L$

So, $\Delta y = y_2 - y_1 = \frac{3}{2}\frac{L}{a}(\lambda_2 - \lambda_1)$

$$= \frac{3}{2}\times\frac{2}{0.5\times10^{-3}}\times(5\times10^{-9}) = 3\times10^{-5}$$

125. (30) Initialy polaroids have angle of 0° between them. From the law of Malus,

$$I = \frac{I_0}{2}\cos^2\theta$$

Here I = resultant intensity on screen

$$\therefore \frac{I}{2}\cos^2\theta = \frac{3I}{8}$$

$$\Rightarrow \cos^2\theta = \frac{3}{4} \Rightarrow \cos\theta = \frac{\sqrt{3}}{2} \Rightarrow \theta = 30°$$

126. (75) Assuming initially axis of polarizer and analyzer are parallel

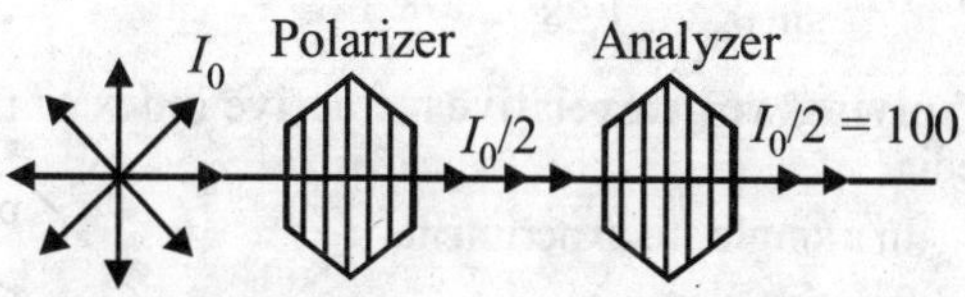

When the angle between axis of polarizer and analyzer is 30°.

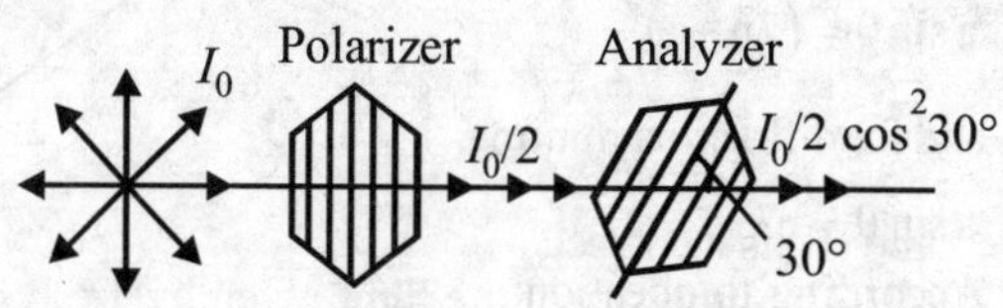

Now emerging intensity $= \frac{I_0}{2}\cos^2 30°$

$$= 100\left(\frac{\sqrt{3}}{2}\right)^2 = 100\times\frac{3}{4} = 75 \text{ Lumens}$$

127. (b) Given :
Intensity, $I_0 = 3.3$ Wm^{-2}
Area, $A = 3\times10^{-4}$ m^2
Angular speed, $\omega = 31.4$ rad/s

Average energy $= I_0 A < \cos^2\theta > T$

$$\because <\cos^2\theta> = \frac{1}{2}$$

$$\therefore \text{Average energy} = \frac{(3.3)(3\times10^{-4})}{2}\times\frac{2\pi}{\omega} \simeq 10^{-4}\text{ J}$$

128. (198) For obtaining secondary minima at a point path difference should be integral multiple of wavelength

$$\therefore d\sin\theta = n\lambda \Rightarrow \sin\theta = \frac{n\lambda}{d}$$

For n to be maximum $\sin\theta = 1$

$$n = \frac{d}{\lambda} = \frac{6\times10^{-5}}{6\times10^{-7}} = 100$$

Total number of minima on one side = 99
Total number of minima = 198.

129. (b) According to question, the intensity of light coming out of the analyser is just 10% of the original intensity (I_0)
Using, $I = I_0\cos^2\theta$

$$\Rightarrow \frac{I_0}{10} = I_0\cos^2\theta \Rightarrow \frac{1}{10} = \cos^2\theta$$

$$\Rightarrow \cos\theta = \frac{1}{\sqrt{10}} = 0.316 \Rightarrow \theta \approx 71.6°$$

Therefore, the angle by which the analyser need to be rotated further to reduce the output intensity to be zero $\phi = 90° - \theta = 90° - 71.6° = 18.4°$

130. (c) $I = \left(\frac{I_0}{2}\right)\cos^2 30°\cos^2 60°$

$$= \frac{I_0}{2}\times\frac{3}{4}\times\frac{1}{4}$$

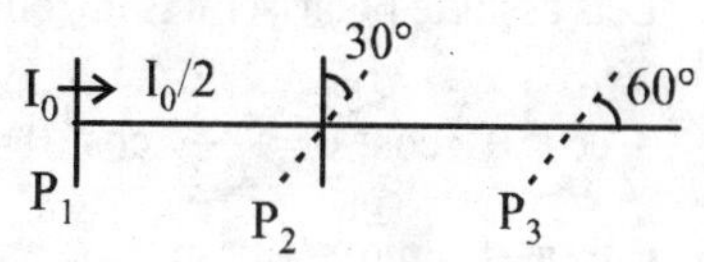

$$\therefore \frac{I_0}{I} = \frac{32}{3} = 10.67$$

131. (b) Angular width between first and second diffraction minima $\theta \simeq \frac{\lambda}{a}$ and angular width of fringe due to double slit is $\theta' = \frac{\lambda}{d}$.

$$\text{So, number of fringes} = \frac{\theta}{\theta'} = \frac{\left(\frac{\lambda}{a}\right)}{\left(\frac{\lambda}{d}\right)} = \left(\frac{d}{a}\right) = \frac{19.44}{4.05} = 4.81 \simeq 5$$

132. (a) d sin θ = λ
Here, θ = 30°

$$\frac{d}{2} = \lambda$$

$$\lambda = 5000\text{ A°} \quad \left[\therefore d = 1\times10^{-6}\text{m}\right]$$

Fringe width, $\beta = \frac{\lambda D}{d'}$

$$10^{-2} = \frac{5000 \times 10^{-10} \times 0.5}{d'}$$

$d' = 25 \times 10^{-6}$ m

$\boxed{d' = 25\ \mu m}$

133. (c) Axis of transmission of A & B are parallel.

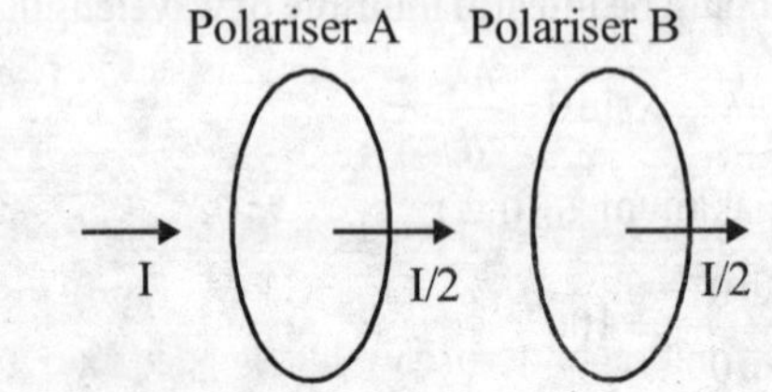

After introducing polariser C between A and B,

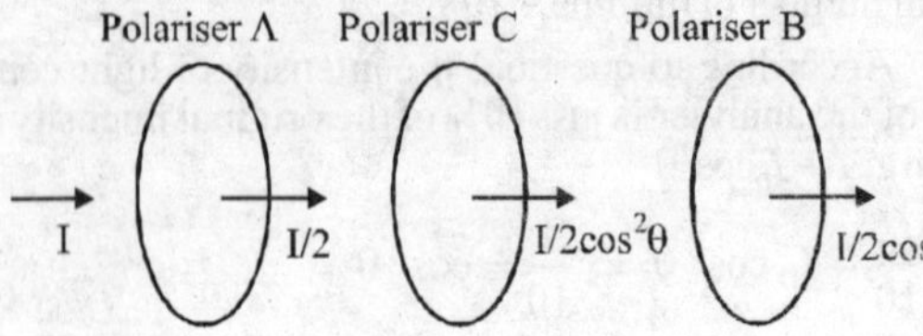

$$\frac{I}{2}\cos^4\theta = \frac{I}{8} \Rightarrow \cos^4\theta = \frac{1}{4}$$

$$\Rightarrow \cos\theta = \frac{1}{\sqrt{2}} \text{ or, } \theta = 45°$$

134. (a) Polariser A and B have same alignment of transmission axis.

Lets assume polariser c is introduced at θ angle

$$\frac{1}{2}\cos^2\theta \times \cos^2\theta = \frac{1}{3} \Rightarrow \cos^4\theta = \frac{2}{3} \Rightarrow \cos\theta = \left(\frac{2}{3}\right)^{1/4}$$

135. (b) $a = 0.1\text{ mm} = 10^{-4}\text{ m}$,

$\lambda = 6000 \times 10^{-10}\text{ m} = 6 \times 10^{-7}\text{ m}, D = 0.5\text{ m}$

for 3rd dark band, $a \sin\theta = 3\lambda$

or $\sin\theta = \frac{3\lambda}{a} = \frac{x}{D}$

The distance of the third dark band from the central bright band

$$x = \frac{3\lambda D}{a} = \frac{3 \times 6 \times 10^{-7} \times 0.5}{10^{-4}} = 9\text{ mm}$$

136. (b) $b \sin\theta = m\lambda$

for second diffraction minimum,

$$b\sin\theta_2 = 2\lambda \Rightarrow \sin\theta_2 = \frac{2\lambda}{b}$$

$$\sin\theta_2 = \tan\theta_2 = \frac{x^2}{D} = \frac{3}{100}$$

$$\frac{2\lambda}{b} = \frac{3}{100} \Rightarrow \frac{\lambda}{b} = \frac{3}{200}$$

for the first minimum

$$b\sin\theta_1 = 1 \times \lambda \Rightarrow x_c = D\tan\theta_1 = \frac{D\lambda}{b}$$

Now, length of central maximum

$l = 2x_c$

$$l = 2\frac{D\lambda}{b} = 2 \times 100 \times \frac{3}{200}$$

$\boxed{l = 3\text{ cm}}$

137. (a) Given geometrical spread = a

$$\text{Diffraction spread} = \frac{\lambda}{a} \times L = \frac{\lambda L}{a}$$

The sum $b = a + \frac{\lambda L}{a}$

For b to be minimum

$$\frac{db}{da} = 0, \quad \therefore \quad \frac{d}{da}\left(a + \frac{\lambda L}{a}\right) = 0 \Rightarrow a = \sqrt{\lambda L}$$

$$b_{min} = \sqrt{\lambda L} + \sqrt{\lambda L} = 2\sqrt{\lambda L} = \sqrt{4\lambda L}$$

138. (d) According to malus law, intensity of emerging beam is given by,

$I = I_0 \cos^2\theta$

Now, $I_{A'} = I_A \cos^2 30°$

$I_{B'} = I_B \cos^2 60°$

As $I_{A'} = I_{B'}$

$$\Rightarrow I_A \times \frac{3}{4} = I_B \times \frac{1}{4}; \frac{I_A}{I_B} = \frac{1}{3}$$

139. (c) Here, $\sin\theta_{ic} / \sin\theta_{iB} = 1.28$

As we know,

$$\mu = \frac{\sin\theta_{iB}}{\sin\left(\frac{\pi}{2} - \theta_{iB}\right)}$$

where, θ_{iB} is Brewster's angle of incidence,

$$\text{And, } \mu = \frac{1}{\sin\theta_{ic}} = \frac{1}{1.28} = 0.8$$

On solving we get, relative refractive index of the two media.

140. (b) In a single slit experiment,

For diffraction maxima,

$$a\sin\theta = (2n+1)\frac{\lambda}{2}$$

and for diffraction minima,

$a \sin\theta = n\lambda$

According to question,

$$(2 \times 1 + 1)\frac{\lambda}{2} = 1 \times 6600 \qquad (\because \lambda_R = 6600\text{Å})$$

$$\Rightarrow \lambda = \frac{6600 \times 2}{3} \Rightarrow \lambda = 4400\text{Å}$$

141. **(c)** Relation between intensities

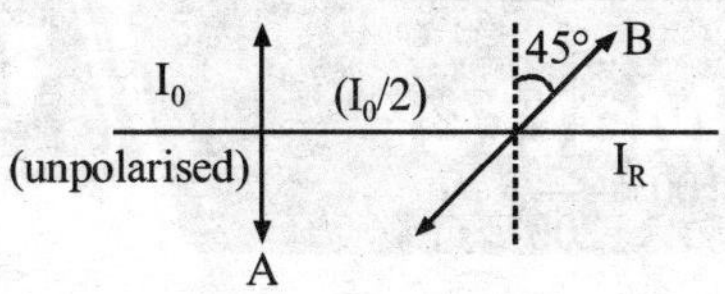

$$I_r = \left(\frac{I_0}{2}\right)\cos^2(45°) = \frac{I_0}{2} \times \frac{1}{2} = \frac{I_0}{4}$$

142. **(c)** Wavelength of radio waves is greater than microwaves hence frequency of radio waves is less than microwaves.

The degree of diffraction is greater whose wavelength is greater.

143. **(c)** For first minimum, $d\sin\theta = \lambda$

$$\Rightarrow \quad d = \frac{\lambda}{\sin\theta} = \frac{5000 \times 10^{-8}\text{ cm}}{\sin 30°}$$

$$= \frac{5000 \times 10^{-8}\text{ cm}}{1/2} = 10 \times 10^{-5}\text{ cm}$$

144. **(a)** For $I = \frac{I_0}{2}$

and $I = I_0 \cos^2\theta = \frac{I_0}{2}$

$\therefore \theta = 45°$

Therefore the angle through which either polaroids turned is 135° (= 180° – 45°)

145. **(b)** When viewed through a polaroid which is rotated then the light from a clear blue portion of the sky shows a rise and fall of intensity. The light coming from the sky is polarised due to scattering of sunlight by particles in the atmosphere.

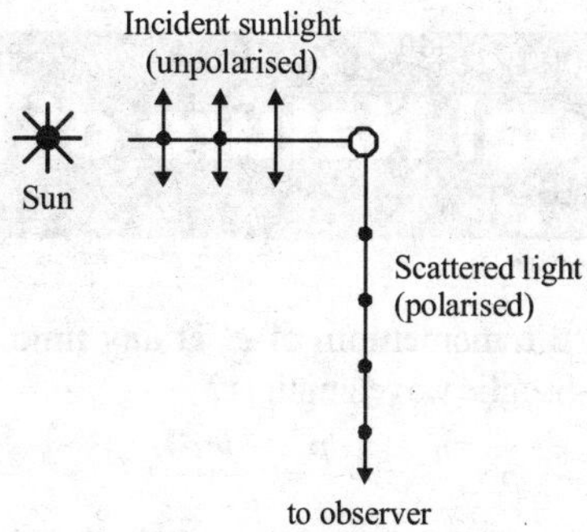

146. **(d)** The electron beam will be diffracted and the maxima is obtained at $y = 0$.

Also, the diffraction pattern, should be wider than the slit width.

147. **(a)** $I = I_0\left(\frac{\sin\phi}{\phi}\right)^2$ and $\phi = \frac{\pi}{\lambda}(b\sin\theta)$

When the slit width is doubled, the amplitude of the wave at the centre of the screen is doubled, so the intensity at the centre is increased by a factor 4.

148. **(b)** From the law of Malus, $I = I_0\cos^2\theta$

When an unpolarised light is converted into plane polarised light by passing through polariod, its intensity become half.

$\therefore$ Intensity of polarized light $= \frac{I_0}{2}$

$\Rightarrow$ Intensity of untransmitted light $= I_0 - \frac{I_0}{2} = \frac{I_0}{2}$

149. **(d)** From the Brewster's law, angle of incidence for total polarization is given by $\tan\theta = n$

$\Rightarrow \theta = \tan^{-1} n$

Where n is the refractive index of the glass.

Dual Nature of Radiation and Matter 11

1. **(2)** Let the momentum of e^- at any time t is p and its De-broglie wavelength is λ.

Then, $p = \frac{h}{\lambda} \Rightarrow \frac{dp}{dt} = \frac{-h}{\lambda^2}\frac{d\lambda}{dt}$

$\Rightarrow ma = F = -\frac{h}{\lambda^2}\frac{d\lambda}{dt}$ [Here m = mass of e^-]

Where, –ve sign represents decrease in λ with time

$\Rightarrow ma = \frac{-h}{(h/p)^2}\frac{d\lambda}{dt} \Rightarrow a = -\frac{p^2}{mh}\frac{d\lambda}{dt} = -\frac{mv^2}{h}\frac{d\lambda}{dt}$

$\Rightarrow \frac{d\lambda}{dt} = -\frac{ah}{mv^2}$ ...(i)

Here, $a = \frac{qE}{m} = \frac{e}{m}\frac{\sigma}{2\varepsilon_0}$ and $v = u + at = 0 + at = at$

Substituting values of a and v in equation (i)

$\frac{d\lambda}{dt} = -\frac{2h\varepsilon_0}{\sigma e t^2} \Rightarrow \frac{d\lambda}{dt} \propto \frac{1}{t^2}$ $\therefore n = 2$

2. **(d)** Magnetic field does not work

$\therefore$ Speed will not change, so De-Broglie wavelength remains same.

3. **(b)** de-Broglie wavelength of the particle is

$\lambda = \frac{h}{mv} = \frac{h}{\sqrt{2mK}} \Rightarrow \lambda^2 = \frac{h^2}{2m}\left(\frac{1}{K}\right)$

$\therefore$ The graph of $\frac{1}{K}$ versus λ is

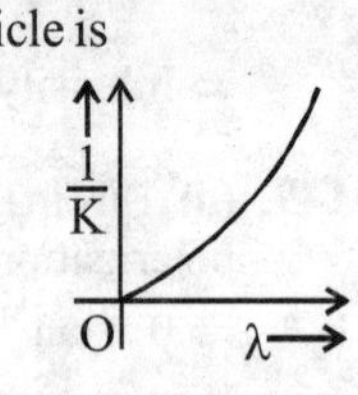

4. **(a)** Photoelectric equation $E = W + K_{max}$

K.E. $= E - W$

$\lambda = \frac{h}{P}$ (For electrons)

$\lambda_e = \frac{h}{\sqrt{2m\,\text{K.E.}}}, E = \frac{hc}{\lambda_i}, W = \frac{hc}{\lambda_0}$

λ_0 is threshold frequency

$\frac{h^2}{2m\lambda_e^2} = \frac{hc}{\lambda_i} - \frac{hc}{\lambda_0} \Rightarrow \lambda_e = \sqrt{\frac{h}{2mc\left(\frac{1}{\lambda_i} - \frac{1}{\lambda_0}\right)}}$

5. **(c)** The energy of photon is

$E_{photon} = \frac{hc}{\lambda} = E$; $E_{proton} = \frac{1}{2}m_p v^2 = E$

$\frac{\lambda_{proton}}{\lambda_{photon}} = \frac{h/p}{hc/E} = \frac{h/\sqrt{2m_pE}}{hc/E} = \frac{E}{c\sqrt{2m_pE}}$

$\therefore \frac{\lambda_{proton}}{\lambda_{photon}} = \frac{1}{c}\sqrt{\frac{E}{2m_p}}$

6. **(d)** By principle of homogeneity,

$[E^2] = [M^2L^4T^{-4}]$

$[P^2C^2] = [MLT^{-1}]^2[LT^{-1}]^2 = [M^2L^4T^{-4}]$

$[m^2C^4] = [M^2L^4T^{-4}]$

$E^2 = P^2C^2 + m^2C^4$

7. **(a)** Velocity of electron after time t is

$\vec{v}' = v_0\hat{i} - \frac{E_0 e}{m}t\hat{k}$; $v' = \sqrt{v_0^2 + \frac{E_0^2e^2t^2}{m^2}}$

Initial de-Broglie wavelenth, $\lambda_0 = \frac{h}{mv_0}$

de-Broglie wavelength after time t is

$\lambda' = \frac{h}{mv'} = \frac{h}{mv_0\sqrt{1 + \frac{E_0^2e^2t^2}{v_0^2m^2}}} = \frac{\lambda_0}{\sqrt{1 + \frac{E_0^2e^2t^2}{v_0^2m^2}}}$

8. **(b)** The formula for de-Broglie wave is given by

$\lambda = \frac{h}{p} = \frac{6.63\times10^{-34}}{10^{-30}\times2.21\times10^6} = 3\times10^{-10}$m

Hence, particle will behave as *X*-ray.

9. **(b)** de-Broglie wavelength is given by

$\lambda = \frac{h}{mv} = \frac{h}{\sqrt{2mk}}$ $\therefore \lambda\alpha = \frac{1}{\sqrt{m}}$

$\because$ $m_e < m_P < m_\alpha$ $\therefore \lambda_\alpha < \lambda_p < \lambda_e$

10. **(a)** de-Broglie wavelength

$\lambda = \frac{h}{p} = \frac{h}{\sqrt{2m\,K}}$

Here, h = constant

λ is same for both electron and proton

$\because m_eK_e = m_pK_p \Rightarrow \frac{K_p}{K_e} = \frac{m_e}{m_p} = \frac{1}{1836}$

11. **(d)** Let De-Broglie wavelength of proton & electron be λ

We know that $\lambda = \frac{h}{p}$

$\Rightarrow p_p = p_e$

kinetic energy is,

$K = \frac{p^2}{2m} \Rightarrow K \propto \frac{1}{m}$

since $m_e < m_P$

then, $K_p < K_e$

12. **(d)** De-Broglie wavelength $\lambda = \frac{h}{p} = \frac{h}{mv}$

For α-particle, $2\lambda = \frac{h}{4mv_\alpha}$ ...(i)

For proton, $\lambda = \frac{h}{mv_p}$ ...(ii)

Divide (ii) by (i)

$\frac{2\lambda}{\lambda} = \frac{h}{4mv_\alpha} \times \frac{mv_p}{h} \Rightarrow \frac{v_p}{v_\alpha} = \frac{8}{1} = 8:1$

13. **(d)** Energy of photon,

$E_p = \frac{hc}{\lambda_p} \Rightarrow \lambda_p = \frac{hc}{E_p}$

de-Broglie wavelength of electron,

$\lambda_e = \frac{h}{m_e v_e} = \frac{hv_e}{2K_e}$ $\left[\because K_e = \frac{1}{2}m_e v_e^2\right]$

Given $v_e = 0.25$ c; $\lambda_e = \frac{h \times 0.25c}{2K_e} = \frac{hc}{8K_e}$

Also, $\lambda_p = \lambda_e$

$\frac{hc}{E_p} = \frac{hc}{8K_e}$; $\frac{K_e}{E_p} = \frac{1}{8}$

14. **(c)** De-Broglie wavelength,

$\lambda = \frac{h}{p}$ $\left(\because K.E = \frac{p^2}{2m}\right)$

$\lambda = \frac{h}{\sqrt{2mK.E}} = \frac{h}{\sqrt{2mE}}$

$\lambda' = \frac{h}{\sqrt{2m\left(\frac{E}{4}\right)}} = \frac{2h}{\sqrt{2mE}} = 2\lambda$

15. **(a)** De-Broglie wavelength,

$\lambda = \frac{h}{m\upsilon} = \frac{h}{\sqrt{2mK}} = \frac{h}{\sqrt{2mq\Delta V}}$

$\therefore \frac{\lambda_\alpha}{\lambda_p} = \sqrt{\frac{2m_p V_p q_p}{2m_\alpha V_\alpha q_\alpha}}$

$\because V_p = 2V$ and $V_\alpha = 4V$ (given)

$m_\alpha = 4m_p$ and $q_\alpha = 2q_p$

$\therefore \frac{\lambda_\alpha}{\lambda_p} = \sqrt{\frac{1 \times 2 \times 1}{4 \times 4 \times 2}} = \frac{1}{4} \Rightarrow \lambda_p : \lambda_\alpha = 4:1$

16. **(a)** de-Broglie wavelength, $\lambda = \frac{h}{\sqrt{2mKE}}$

$\therefore \lambda \propto \frac{1}{\sqrt{m}}$ $\therefore \frac{\lambda_p}{\lambda_e} = \sqrt{\frac{m_e}{m_p}} = \sqrt{\frac{m_e}{1849m_e}} = 1:43$

17. **(d)** De Broglie wavelength is $\lambda = \frac{h}{mv}$

$\lambda_p = \lambda_e \Rightarrow \frac{h}{m_p v_p} = \frac{h}{m_e v_e}$

$m_e v_e = m_p v_p \Rightarrow p_e = p_v$ $\therefore \frac{p_p}{p_e} = \frac{1}{1}$

18. **(c)** $\lambda = \frac{h}{\sqrt{2mK}}$

$\lambda_e = \frac{h}{\sqrt{2 \times \frac{m}{1840} \times 4K}}, \lambda_\alpha = \frac{h}{\sqrt{2 \times 4m \times 2K}}, \lambda\rho = \frac{h}{\sqrt{2mK}}$

$\lambda_\alpha < \lambda_p < \lambda_e$

19. **(a)** $KE = \frac{P^2}{2m} = \frac{h^2}{2m\lambda^2}$ $\left(\because P = \frac{h}{\lambda}\right)$

$\Rightarrow KE \propto \frac{1}{m}; \frac{KE_p}{KE_\alpha} = \frac{m_\alpha}{m_p} = 4:1$ $(\because m_\alpha = 4m_p)$

20. **(b)** When electron is accelerated through a potential difference V_1

Kinetic energy, $KE = \frac{P^2}{2m}$, $\left(\because \text{Momentum, } P = \frac{h}{\lambda}\right)$

$KE = eV_1 = \frac{\left(\frac{h}{\lambda}\right)^2}{2m}$

When potential is changed to V_2, then

$KE = eV_2 = \frac{\left(\frac{h}{1.5\lambda}\right)^2}{2m}$ $\therefore \frac{V_1}{V_2} = (1.5)^2 = \frac{9}{4}$

21. **(c)** We have $\lambda = \frac{h}{mV} = \frac{h}{\sqrt{2mK.E.}} = \frac{h}{\sqrt{2mqv}}$

So, $\frac{\lambda_\alpha}{\lambda_p} = \frac{\frac{h}{\sqrt{2m_\alpha q_\alpha V}}}{\frac{h}{\sqrt{2m_p q_p V}}}$ $(\because m_\alpha = 4m_p, q_\alpha = 2q_p)$

$\Rightarrow \frac{\lambda_\alpha}{\lambda_p} = \sqrt{\frac{1}{8}} \Rightarrow m = 8$

22. **(d)** When electron is accelerated through potential difference V, then the de-Broglie wavelength could be

$\Rightarrow \lambda = \frac{h}{\sqrt{2m(KE)}} = \frac{h}{\sqrt{2meV}}$ $(\because kE = eV)$

$\therefore \lambda \propto \frac{1}{\sqrt{V}}$ $\therefore \frac{\lambda}{\lambda_0} = \sqrt{\frac{20}{40}}$ $\therefore \lambda = \frac{\lambda_0}{\sqrt{2}}$

23. **(b)** De-Broglie wavelength is given as

$\lambda_D = \frac{h}{p} = \frac{h}{\sqrt{2mK}}$ $\therefore \lambda \propto \frac{1}{\sqrt{m}}$ [$\because$ K = constant]

As $m_\alpha > m_p > m_e$. So, $\lambda_e > \lambda_p > \lambda_\alpha$

24. **(b)** The linear momenta given by,

$p = \sqrt{2mE} = \sqrt{2mqV}$ $(\because E = qV)$

$\frac{p_\alpha}{p_p} = \sqrt{\frac{m_\alpha q_\alpha}{m_p q_p}} = \sqrt{\frac{4m_p.2q_p}{m_p q_p}} = \sqrt{\frac{4}{1} \times \frac{2}{1}} = \frac{2\sqrt{2}}{1}$

25. **(d)** Wavelength, $\lambda = \frac{h}{p} = \frac{h}{mv}$

$\Rightarrow \lambda = \frac{h}{\sqrt{2m(K.E)}}$ $\therefore \lambda = \frac{h}{\sqrt{2mqV}}$

Putting the values of m and q and solving we get

$$\lambda = \frac{1.22}{\sqrt{V}} \text{nm}$$

So x stands for $\sqrt{V}$.

26. **(d)** As $v = u + at$

$$v = v_0 + \frac{eE_0 t}{m}$$

$$\text{Now, } \lambda = \frac{h}{mv} = \frac{h}{m\left(v_0 + \frac{eE_0 t}{m}\right)}$$

$$= \frac{h}{mv_0\left(1 + \frac{eE_0 t}{mv_0}\right)} = \frac{\lambda_0}{1 + \frac{eE_0 t}{mv_0}}$$

27. **(c)**

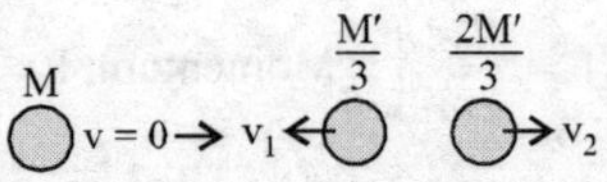

$|\vec{P}_1| = |\vec{P}_2|$ (from the conservation of momentum)

Here $\vec{p}$ is momentum.

So, $\lambda = \frac{h}{P}$ Here wavelength will be same.

28. **(d)** Kinetic energy gained by a charged particle accelerated by a potential V is qV

KE = qV

$$\Rightarrow \frac{p^2}{2m} = qV \quad \Rightarrow p = \sqrt{2mqV}$$

$$p = \frac{h}{\lambda} \Rightarrow \lambda = \frac{h}{\sqrt{2mqV}}$$

$$\text{now } \frac{\lambda_p}{\lambda_d} = \sqrt{\frac{m_d V_d}{m_p V_p}} \Rightarrow \frac{1}{\sqrt{2}} = \sqrt{\frac{2V_d}{V_p}} \Rightarrow \frac{V_p}{V_d} = 4$$

29. **(a)** We have $\lambda_e = \frac{h}{P} = \frac{h}{\sqrt{2mK_e}}$

$$\text{Now, } K_p = \frac{hc}{\lambda_p}$$

$$\text{As, } K_p = K_e \ \therefore \lambda_e = \frac{h\sqrt{\lambda_p}}{\sqrt{2mc}} \Rightarrow \lambda_e^2 \propto \lambda_p$$

30. **(b)** $\lambda = \frac{h}{p} \Rightarrow \lambda = \frac{h}{\sqrt{2mK.E}} \Rightarrow \lambda \propto \frac{1}{\sqrt{m}}$

$$\text{So, } \frac{\lambda_\alpha}{\lambda_c} = \sqrt{\frac{m_c}{m_\alpha}} = \sqrt{\frac{12}{4}} = \frac{\sqrt{3}}{1}$$

31. **(b)** $\lambda_e = \lambda_{\text{photon}} \Rightarrow \frac{h}{mv} = \frac{h}{P_{ph}} \Rightarrow P_{ph} = mv$

$$\frac{E_e}{E_{ph}} = \frac{\frac{1}{2}mv^2}{P_{ph}c} = \frac{mv.v}{2P_{ph}c} = \frac{v}{2c}$$

32. **(b)** As $\lambda = \frac{h}{p} = \frac{h}{\sqrt{2m(K.E)}}$

As $K.E$ = constant

$$\text{So, } \lambda \propto \frac{1}{\sqrt{m}}$$

As, $m_e < m_p < m_n < m_\alpha$. So, $\lambda_e > \lambda_p > \lambda_n > \lambda_\alpha$

33. **(a)** De-Broglie wavelength is given by

$$\lambda = \frac{h}{p} = \frac{h}{\sqrt{2mk.E}}$$

$$\text{For gas k.E} = \frac{3kT}{2}$$

$$\therefore \lambda = \frac{h}{\sqrt{2 \times m \times \frac{3}{2}kT}} = \frac{h}{\sqrt{3mkT}}$$

$$= \frac{6.6 \times 10^{-34}}{\sqrt{3 \times 9 \times 10^{-31} \times 1.38 \times 10^{-23} \times 300}}$$

$$\Rightarrow \lambda = 6.26 \times 10^{-9} \text{ m} = 6.26\text{nm}$$

34. **(a)** de-Broglie wavelength of proton, $\lambda_P = \frac{h}{P_P}$

de-Broglie wavelength of electron, $\lambda_e = \frac{h}{P_e} \quad \therefore P_P = P_e$

Kinetic energy $(K.E) = \frac{P^2}{2m} \quad \therefore$ For same P, $K.E = \frac{1}{m}$

$\therefore K_P < K_e$ As $m_P > m_e$

35. **(c)** $\Delta x.\Delta p \geq \frac{h}{4\pi} \Rightarrow \Delta x = \frac{h}{4\pi m \Delta v}$

$$v = \sqrt{\frac{3KT}{m}} \Rightarrow v \propto \frac{1}{\sqrt{m}}$$

$$\text{Also } \Delta v \propto \frac{1}{\sqrt{m}} \Rightarrow \Delta x \propto \frac{1}{m \cdot \frac{1}{\sqrt{m}}} \Rightarrow \Delta x \propto \frac{1}{\sqrt{m}}$$

$$\frac{\Delta x_e}{\Delta x_p} = \sqrt{\frac{m_p}{m_e}}$$

36. **(b)** De-Broglie wavelength

$$\lambda = \frac{h}{p} = \frac{h}{\sqrt{2mE}} \quad (\because p = \sqrt{2mE})$$

$$\Rightarrow \lambda^2 = \frac{h^2}{2mE} \Rightarrow \lambda \propto \frac{1}{\sqrt{E}}$$

$$\therefore \frac{\lambda_2}{\lambda_1} = \sqrt{\frac{E_1}{E_2}}$$

$$\text{But } \lambda_2 = \frac{75}{100}\lambda_1 \Rightarrow \lambda_2 = \frac{3}{4}\lambda_1$$

$$\therefore \frac{E_1}{E_2} = \left(\frac{3\lambda_1}{4\lambda_1}\right)^2 = \left(\frac{3}{4}\right)^2 \Rightarrow E_2 = \frac{16}{9}E_1$$

Extra energy given $E_2 - E_1 = \frac{16}{9}E_1 - E_1 = \frac{7}{9}E_1$

37. **(d)** de-Broglie wavelength, $\lambda = \frac{h}{P}$

Here momentum is same for both the particles of mass 3 m and m, so ratio of de-Broglie wavelength,

$$\frac{\lambda_{3m}}{\lambda_m} = \frac{1}{1}$$

38. **(d)** Given : $v_{\text{particle}} = 4v_{\text{electron}}$ and, $\lambda_{\text{particle}} = 2\lambda_{\text{electron}}$

Using $\lambda = \frac{h}{p} \Rightarrow \lambda p$ = constant

$$\therefore \quad \lambda_{\text{particle}} \times p_{\text{particle}} = \lambda_{\text{electron}} \times p_{\text{electron}}$$

$$\Rightarrow \lambda_{\text{particle}} \times m_{\text{particle}} \times v_{\text{particle}} = \lambda_{\text{electron}} \times m_{\text{electron}} \times v_{\text{electron}}$$

$$\therefore \quad m_{\text{particle}} v_{\text{particle}} = \frac{m_{\text{electron}} v_{\text{electron}}}{2}$$

$$\Rightarrow \quad m_{\text{particle}} = \frac{m_{\text{electron}}}{8}$$

39. **(b)** de-Broglie wavelength, $\lambda = \frac{h}{P}$

$$\lambda_P = \frac{h}{\sqrt{2m_P \times (q_P \times V)}} \text{ and } \lambda_\alpha = \frac{h}{\sqrt{2m_\alpha \times (2q_P \times V)}}$$

$$\therefore \frac{\lambda_P}{\lambda_\alpha} = \sqrt{\frac{m_\alpha}{m_P} \times 2} = \sqrt{4 \times 2} = 2\sqrt{2} = 2.8.$$

40. **(d)** de-Broglie wavelength, $\lambda = \frac{h}{p}$

$$\lambda_e = \frac{h}{m_e \times v} \text{ and } \lambda_p = \frac{h}{m_p \times v}$$

$$\therefore \frac{\lambda_e}{\lambda_p} = \frac{m_p}{m_e} = \frac{1836 m_e}{m_e} = 1836.$$

41. **(b)** de-Broglie wavelength,

$$\lambda = \frac{h}{mv}$$

Given,

$$\lambda_P = \lambda_\alpha$$

$$\therefore \frac{h}{m_p v_p} = \frac{h}{m_\alpha v_\alpha} \Rightarrow m_p v_p = m_\alpha v_\alpha$$

$$\Rightarrow m_p v_p = 4 m_p v_\alpha \qquad (\because m_\alpha = 4m_p)$$

$$\Rightarrow \frac{v_P}{v_\alpha} = 4$$

42. **(c)** de-Broglie wavelength, $\lambda = \frac{h}{P} = \frac{h}{\sqrt{2m(\text{KE})}}$

$$\therefore \lambda \propto \frac{1}{\sqrt{m}}$$

As $m_{\text{He}^{++}} > m_P > m_e \Rightarrow \lambda_e > \lambda_P > \lambda_{\text{He}^{++}}$

43. **(a)** Rms speed of gas molecule, $V_{rms} = \sqrt{\frac{3kT}{m}}$

de Broglie wavelength, $\lambda = \frac{h}{p} = \frac{h}{\sqrt{2mk}}$

$$\therefore \lambda = \frac{h}{\sqrt{2m \times \frac{1}{2} m V_{rms}^2}} = \frac{h}{\sqrt{m^2 \times \frac{3kT}{m}}} = \frac{h}{\sqrt{3mkT}}$$

Substituting the respective values we get

$$\lambda = \frac{6.63 \times 10^{-34}}{\sqrt{3 \times 4.64 \times 10^{-26} \times 1.38 \times 10^{-23} \times 400}} = 0.24\text{Å}$$

44. **(d)**

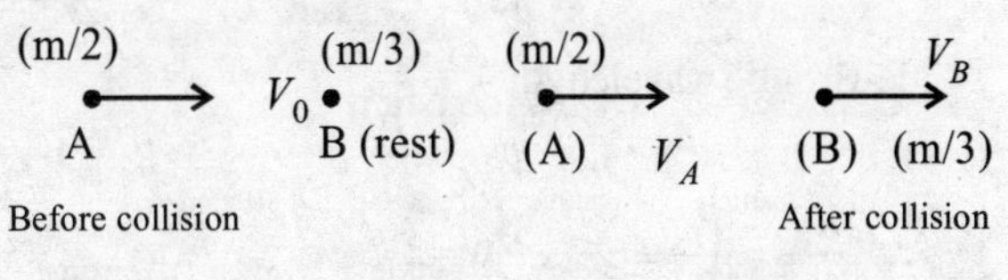

Applying momentum conservation

$$\frac{m}{2} \times V_0 + \frac{m}{3} \times (0) = \frac{m}{2} V_A + \frac{m}{3} V_B$$

$$= \frac{V_0}{2} = \frac{V_A}{2} + \frac{V_B}{3} \qquad \text{...(i)}$$

Since, collision is elastic

$$e = 1 = \frac{V_B - V_A}{V_0} \Rightarrow V_0 = V_B - V_A \qquad \text{...(ii)}$$

On solving equations (i) and (ii) : $V_A = \frac{V_0}{5}$

Now, de-Broglie wavelength of A before collision :

$$\lambda_0 = \frac{h}{m_A V_0} = \frac{h}{\left(\frac{m}{2}\right) V_0} \Rightarrow \lambda_0 = \frac{2h}{m V_0}$$

Final de-Broglie wavelength :

$$\lambda_f = \frac{h}{m_A V_0} = \frac{h}{\frac{m}{2} \times \frac{V_0}{5}} \Rightarrow \lambda_f = \frac{10h}{m V_0}$$

$$\therefore \Delta\lambda = \lambda_f - \lambda_0 = \frac{10h}{mV_0} - \frac{2h}{mV_0}$$

$$\Rightarrow \Delta\lambda = \frac{8h}{mv_0} \Rightarrow \Delta\lambda = 4 \times \frac{2h}{mv_0}$$

$$\therefore \Delta\lambda = 4\lambda_0$$

45. **(d)** de-Broglie wavelength $\lambda = \frac{h}{mv} \Rightarrow m = \frac{h}{\lambda v}$.

Clearly, $m \propto \frac{1}{\lambda v}$

If λ and v be the wavelength and velocity of electron and λ' and v' be the wavelength and velocity of the particle then

$$\Rightarrow \frac{m'}{m} = \frac{v\lambda}{v'\lambda'} = \frac{1}{5} \times \frac{1}{1.878 \times 10^{-4}}$$

$$\Rightarrow m = 9.7 \times 10^{-28} \text{ kg} \qquad [\because m = 9.1 \times 10^{-31} \text{ kg}]$$

46. **(c)** As per question, when KE of particle E, wavelength λ and when KE becomes $E + \Delta E$ wavelength becomes $\lambda/2$

Using, $\lambda = \frac{h}{\sqrt{2mKE}}$

$$\frac{\lambda}{2} = \frac{h}{\sqrt{2m(KE + \Delta E)}}$$

$$\Rightarrow \frac{\lambda}{\lambda/2} = \sqrt{\frac{KE + \Delta E}{KE}} \qquad \Rightarrow 4 = \frac{KE + \Delta E}{KE}$$

$$\Rightarrow \quad 4KE - KE = \Delta E \qquad \Rightarrow \quad \Delta E = 3\,KE = 3\,E$$

47. **(d)** Acceleration of electron in electric field, $a = \frac{eE}{m}$

Using equation

$$v = u + at \;\Rightarrow\; v = 0 + \frac{eE}{m} t$$

$\Rightarrow \; v = \frac{eEt}{m}$ ..... (i)

De-broglie wavelength λ is given by

$\lambda = \frac{h}{mv} = \frac{h}{m\left(\frac{eEt}{m}\right)} \Rightarrow \lambda = \frac{h}{eEt}$ [using (i)]

Differentiating w.r.t. t

$$\frac{d\lambda}{dt} = \frac{d\left(\frac{h}{eEt}\right)}{dt} \Rightarrow \frac{d\lambda}{dt} = \frac{-h}{eEt^2}$$

48. **(c)** Given, Initial velocity, $u = v_0\hat{i} + v_0\hat{j}$

Acceleration, $a = \frac{qE_0}{m} = \frac{eE_0}{m}$

Using $v = u + at$

$v = v_0\hat{i} + v_0\hat{j} + \frac{eE_0}{m}t\hat{k}$

$\therefore |\vec{v}| = \sqrt{2v_0^2 + \left(\frac{eE_0 t}{m}\right)^2}$

de-Broglie wavelength, $\lambda = \frac{h}{p}$

$\Rightarrow \lambda = \frac{h}{mv}$ $(\because p = mv)$

Initial wavelength, $\lambda_0 = \frac{h}{mv_0\sqrt{2}}$

Final wavelength,

$$\lambda = \frac{h}{\sqrt{2v_0^2 + \left(\frac{eE_0 t}{m}\right)^2} \times m}$$

So, $\frac{\lambda}{\lambda_0} = \frac{1}{\sqrt{1 + \left(\frac{eE_0 t}{\sqrt{2}mv_0}\right)^2}} \Rightarrow \lambda = \frac{\lambda_0}{\sqrt{1 + \frac{e^2E_0^2t^2}{2m^2v_0^2}}}$

49. **(b)** $P_1 - P_2 = P$

As $P \propto \frac{1}{\lambda}$ or $\frac{1}{\lambda_x} - \frac{1}{\lambda_y} = \frac{1}{\lambda}$

$\Rightarrow \frac{\lambda_y - \lambda_x}{\lambda_x\lambda_y} = \frac{1}{\lambda} \Rightarrow \lambda = \frac{\lambda_x\lambda_y}{\lambda_y - \lambda_x}$

50. **(a)** From the de-Broglie relation,

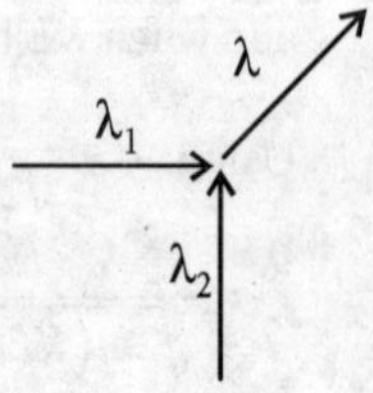

$p_1 = \frac{h}{\lambda_1}$

$p_2 = \frac{h}{\lambda_2}$

Momentum of the final particle (p_f) is given by

$\therefore \; p_f = \sqrt{p_1^2 + p_2^2}$ $\left[\because p_f = p_i \text{ and } p_i = \sqrt{p_1^2 + p_2^2}\right]$

$\Rightarrow \frac{h}{\lambda} = \sqrt{\frac{h^2}{\lambda_1^2} + \frac{h^2}{\lambda_2^2}} \Rightarrow \frac{1}{\lambda^2} = \frac{1}{\lambda_1^2} + \frac{1}{\lambda_2^2}$

51. **(c)** de Broglie wavelength (λ) is given by

$K = qV$

$\lambda = \frac{h}{p} = \frac{h}{\sqrt{2mK}} = \frac{h}{\sqrt{2mqV}}\left(\because p = \sqrt{2mK}\right)$

Substituting the values we get

$\therefore \; \frac{\lambda_A}{\lambda_B} = \frac{\sqrt{2m_Bq_BV_B}}{\sqrt{2m_Aq_AV_A}} = \sqrt{\frac{4m.q.2500}{m.q.50}}$

$= 2\sqrt{50} = 2 \times 7.07 = 14.14$

52. **(d)** de-Broglie wavelength,

$\lambda_e = \frac{h}{mv} = 10^{-3}\left(\frac{3\times10^8}{6\times10^{14}}\right)$ $\left[\because \lambda = \frac{c}{f}\right]$

$\Rightarrow v = \frac{6.63\times10^{-34}\times6\times10^{14}}{9.1\times10^{-31}\times3\times10^5} \Rightarrow v = 1.45\times10^6$ m/s

53. **(d)** Using, $\lambda = \frac{h}{p}$ {given: $\lambda = 7.5\times10^{-12}$}

$\Rightarrow P = \frac{h}{\lambda}$

Minimum energy required,

$$KE = \frac{P^2}{2m} = \frac{(h/\lambda)^2}{2m} = \frac{\left\{\frac{6.6\times10^{-34}}{7.5\times10^{-12}}\right\}^2}{2\times9.1\times10^{-31}} J = 25\,keV$$

54. **(c)** According to question, $\lambda_p = \lambda_\alpha$

Using, $\lambda = \frac{h}{p} = \frac{h}{mv}$

So, $\frac{h}{m_p \times v_p} = \frac{h}{m_\alpha \times v_\alpha} \Rightarrow \frac{v_p}{v_\alpha} = \frac{m_\alpha}{m_p} = \frac{4m_p}{m_p}$

($\because$ mass of α-particle is 4 times of mass of proton)

So, $\frac{v_p}{v_\alpha} = \frac{4}{1}$; i.e., 4 : 1

55. **(d)** From question, $m_A = m$; $m_B = \frac{m}{2}$

$u_A = u \quad u_B = 0$

Let after collision velocity of $A = V_1$ and velocity of $B = V_2$

Applying law of conservation of momentum,

$mu = mv_1 + \left(\frac{m}{2}\right)v_2$ or $2u = 2v_1 + v_2$ ....(i)

By law of collision

$e = \frac{v_2 - v_1}{u - 0}$ or $u = v_2 - v_1$ ....(ii)

[$\because$ collision is elastic, $e = 1$]

using eqns (i) and (ii)

$v_1 = \frac{u}{3}$ and $v_2 = \frac{4}{3}u$

de-Broglie wavelength $\lambda = \frac{h}{p}$

$$\therefore \; \frac{\lambda_A}{\lambda_B} = \frac{P_B}{P_A} = \frac{\frac{m}{2}\times\frac{4}{3}u}{m\times\frac{u}{3}} = 2$$

56. **(c)** de-Broglie wavelength,

$$\lambda = \frac{h}{P} = \frac{h}{mv} = \frac{h}{\sqrt{2mqV}}$$

or, $$\lambda = \frac{6.6\times10^{-34}}{\sqrt{2\times9.1\times10^{-31}\times1.6\times10^{-19}\times50}} = 1.7\text{ Å}$$

57. **(d)** As $m_{\text{dust particle}} >> m_{\text{subatomic particle}}$

So, its de-Broglie wavelength is very small. So to analyse its wave nature is very difficult. Secondly, dust particle motion is not linear.

58. **(d)** Energy in joule (E) = charge × potential diff. in volt

$E_{electron} = q_e V$ and $E_{proton} = q_p 4V$

de-Broglie wavelength $\lambda = \frac{h}{P} = \frac{h}{\sqrt{2mE}}$

$$\lambda_e = \frac{h}{\sqrt{2m_e eV}} \text{ and } \lambda_P = \frac{h}{\sqrt{2m_P e4V}} \quad (\because q_e = q_P)$$

$$\therefore \frac{\lambda_e}{\lambda_P} = \frac{\frac{h}{\sqrt{2m_e eV}}}{\frac{h}{\sqrt{2m_P e4V}}} = \sqrt{\frac{2m_P e4V}{2m_e eV}} = 2\sqrt{\frac{m_P}{m_e}}$$

59. **(d)** de-Broglie wavelength,

$$\lambda = \frac{h}{p} = \frac{h}{mv} \quad \text{.... (i)}$$

but $K.E = \frac{1}{2}mv^2 \Rightarrow K.E = \frac{(mv)^2}{2m}$

$\Rightarrow mv = \sqrt{2m\,K.E}$

$$\lambda = \frac{h}{\sqrt{2m\,K.E}} \Rightarrow \lambda \propto \frac{1}{\sqrt{K.E}}$$

So, if $K.E.$ is doubled, wavelength becomes $\frac{\lambda}{\sqrt{2}}$

60. **(d)** Momentum imparted by the laser pulse

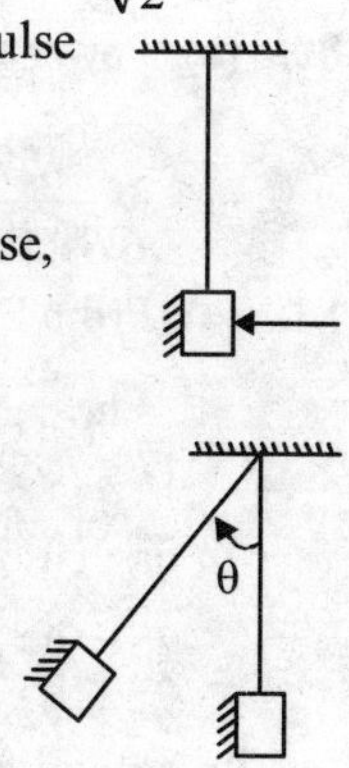

$$\Delta P = \frac{2E}{C}$$

The mirror's velocity V after the pulse,

$$mV = \frac{2E}{C} \Rightarrow V = \frac{2E}{mC}$$

From work energy theorem

W_g = Change in Kinetic energy

$$\Rightarrow -mgl(1-\cos\theta) = 0 - \frac{1}{2}mv^2 \quad \text{...(i)}$$

$$\Rightarrow gl\left(2\sin^2\frac{\theta}{2}\right) = \frac{v^2}{2}$$

$\because \theta$ is small

$$\therefore gl2\left(\frac{\theta}{2}\right)^2 = \frac{1}{2}\frac{4E^2}{m^2c^2} \quad \text{(On putting value of } v\text{)}$$

$$\Rightarrow gl\theta^2 = \frac{4E^2}{m^2c^2} \Rightarrow \theta = \frac{2E}{mc\sqrt{gl}}.$$

61. **(b)** $eV_s = h\nu - \phi \Rightarrow V_s = \frac{h\nu - \phi}{e}$

It depends only on the frequency of the incident light, not its intensity.

$$I = \frac{nh\nu}{A}$$

$\Rightarrow I \uparrow\uparrow \Rightarrow n \uparrow\uparrow \Rightarrow$ increased photocurrent.

62. **(b)** By Einstein photoelectric equation,

$$(KE)_{max} = \frac{hc}{\lambda} - \phi$$

For photoelectrons emission, $(KE)_{max} > 0$

$$\Rightarrow \frac{hc}{\lambda} > \phi$$

$$\Rightarrow \lambda < \frac{hc}{\phi} \Rightarrow \lambda < \frac{1242}{3}\text{ nm} = 414\text{ nm}$$

So, it is blue light.

63. **(c)** From photoelectric equation $KE_{max} = \frac{hc}{\lambda} - \phi$

Momentum $p = \sqrt{2mK_{max}}$

$$\therefore p = \sqrt{2m\left(\frac{hc}{\lambda} - \phi\right)}$$

$$d_{AB} = 2R = 2\left[\frac{p}{eB}\right]$$

$$d_{AB} = \frac{2\sqrt{2m\left(\frac{hc}{\lambda} - \phi\right)}}{eB} = \frac{\sqrt{8m\left(\frac{hc}{\lambda} - \phi\right)}}{eB}$$

64. **(c)** Einstein's photoelectric equation is,

$$\frac{hc}{\lambda} = W + eV_s \quad \text{.......... (i)}$$

$$\frac{hc}{\lambda} = W + (K_{max}) \quad \text{.......... (ii)}$$

Comparing equations (i) and (ii), we get

$$\therefore V_s = \frac{K_{max}}{e}$$

65. **(d)** In photoelectric effect, the maximum negative electrode potential at which emission of electrons stops, is known as stopping potential and its variation with frequency is given as

$V_0 = \left(\frac{h}{e}\right)\nu - \phi$, which is a linear variation.

66. **(b)** EM wave carry both energy and momentum. The mass of a photon is zero.

67. **(c)** From Einstein photoelectric equation,

$h\nu = \phi + KE_{max}$

$\because KE_{max} = eV_0$; $V_0 = \left(\frac{h}{e}\right)\nu - \frac{\phi}{e}$

$\therefore V_0$ v/s ν is linear $\therefore$ Slope $= \frac{h}{e}$

68. (5) Given,
For source S_1 wavelength of light, $\lambda_1 = 600$ nm
For source S_2 wavelength of light, $\lambda_2 = 300$ nm
Number of photons, $n_1 = 2 \times 10^{15}$
The power of a light source,

$$P = \frac{n}{t}\left(\frac{hc}{\lambda}\right) \Rightarrow P \propto \frac{n}{\lambda};\ \frac{P_1}{P_2} = \left(\frac{\lambda_2}{\lambda_1}\right)\frac{n_1}{n_2}$$

$$2 = \left(\frac{300}{600}\right) \times \frac{2\times10^{15}}{n_2}$$

$$n_2 = \frac{1}{2}\times10^{15} = 5\times10^{14} \text{ Photon/sec}$$

69. (d) The Einstein photoelectric equation is
$eV_s = E - \phi$.
$2\,eV = E - 2.14\,eV$
$E = 4.14\,eV$
The wavelength of em-wave is

$$\lambda = \frac{hc}{E(\text{inev})} = \frac{1242}{4.14} = 300\text{nm}$$

70. (c) By Einstein's photoelectric equation,

$$\frac{hc}{\lambda} = \phi + K_1 \Rightarrow \frac{hc}{\lambda} = 1 + 2 = 3\text{eV} \quad ...(i)$$

Again, $\frac{hc}{\lambda/2} = 1 + K_2$
Using eq (i)
$2 \times 3 = 1 + K_2 \quad \therefore K_2 = 5\text{eV}$

71. (b) Given, wavelength of incident light, $\lambda = 550$ nm

Energy of incident photon, $E = \frac{1240}{\lambda} = \frac{1240}{550} \simeq 2.25$ eV

As E > work function for cesium
Hence, photo electric effect is possible in Cs only

72. (b) Given, work function of metal surface, $\phi_0 = 3.13$ eV
Energy of incident photon, E = 4.13 eV
According to Einstein's photoelectric equation

$E = \phi_0 + K.E_{max}$
$\Rightarrow 4.13 = 3.13 + K.E_{max}$
$\Rightarrow K.E_{max} = 4.13 - 3.13 = 1\text{eV}$

73. (b) Given,
Force, $F = 2.4 \times 10^{-4}$ N
Light energy flux, $I = 360$ W/cm^2

Pressure, $P = \frac{I}{C} = \frac{F}{A}$

$$A = \frac{FC}{I} = \frac{2.4\times10^{-4}\times3\times10^{8}\times10^{-4}}{360}$$

$\Rightarrow A = 2 \times 10^{-2}\,m^2 = 0.02\,m^2$

74. (b) Photoelectric effect cannot be explained by wave nature of light.

75. (c) According to photoelectric equation,

$$eV_0 = \frac{hc}{\lambda} - \phi$$

$$\Rightarrow eV_0 = \frac{1240}{300}\text{eV} - 2.13\text{ eV}$$

$\Rightarrow eV_0 = 4.13\text{ eV} - 2.13\text{ eV}$
$\Rightarrow$ So, $V_0 = 2V$

76. (a) From the graph, $V_0 = \frac{h}{e}f - \frac{h}{e}f_0$

$\Rightarrow \text{Slope} = \frac{h}{e} \quad \therefore$ Statement-I is correct

Now, $(f_0)_1 < (f_0)_2$
$\Rightarrow$ Work function of M_1 is less than work function of M_2
$\therefore$ Statement-II is incorrect
$\therefore$ Option (a) is correct.

77. (c) $KEmax = h\nu - \phi_0 = eV_0$
Also, stopping potential depends only on frequency and independent of intensity.

78. (c) Given lights are of same wavelength and stopping potential is independent on intensity.
Hence stopping potential will remain same.
Intensity $I_2 > I_1$, hence saturation current corresponding to I_2 will be greater than that corresponding to I_1.

79. (d) Intensity of light $I = \frac{nh\nu}{A}$
where n = no. of photons per unit time.

$n = \frac{IA}{h\nu}$ so on increasing frequency ν, n decreases taking intensity constant. So assertion is incorrect.

Maximum kinetic energy, $k_{max} = h\nu - \phi$
So on increasing ν, kinetic energy increases so reason is correct.

80. (a) Power emitted, $P = nhf \Rightarrow n = \frac{P}{hf}$

$$n = \frac{2\times10^{-3}}{6.6\times10^{-34}\times6\times10^{14}} = 5\times10^{15}$$

81. (a) From Einstein's photoelectric equation $K_{max} = E - \phi_0$

$\phi_0 = \frac{hc}{\lambda_0}$ and $K_{max} = eV_s$

In first case

$$\therefore 8e = \frac{hc}{\lambda} - \frac{hc}{\lambda_0} \quad ....(i) \quad \left(\because E = \frac{hc}{\lambda}\right)$$

In second case

$$2e = \frac{hc}{3\lambda} - \frac{hc}{\lambda_0} \quad ....(ii)$$

Solving (i) & (ii) we get
$\Rightarrow \lambda_0 = 9\lambda$

82. (c) Initially light has frequency 1.5 times the threshold frequency $f = \frac{3}{2}f_0$

When frequency is halved $f' = \frac{3}{4} f_0$

$\therefore f' < f_0$

For photoelectric emission incident frequency should be more than threshold frequency. There will be no emission of photoelectrons.

Hence, photoelectric current = 0

83. **(b)** Work function, $W_e = 3eV$

$$\frac{hc}{\lambda_{max}} = W_e = 3eV$$

$$\Rightarrow \lambda_{max} = \frac{1240 \text{ nm eV}}{3} = 414 \text{nm}$$

84. **(d)** From the Einstein photoelectric equation

$E_k = h\nu - \phi$

$\tan\theta = h$

85. **(a)** $\lambda_1 = 300\text{nm}, \lambda_2 = 500\text{nm}$

Power, $P = \frac{E}{t} \Rightarrow P \propto E$

$$\therefore \frac{P_1}{P_2} = \frac{E_1}{E_2} = \frac{n_1 hc/\lambda_1}{n_2 hc/\lambda_2}$$

$$\frac{200}{200} = \frac{n_1}{n_2} \cdot \frac{500}{300}$$

$\therefore n_1 : n_2 = 3:5$

86. **(d)** The photoelectric current will remain same as amount of energy incident on cell is same.

87. **(d)** Work function, $\phi_0 = h\nu_0$

$6.63 \times 1.6 \times 10^{-19} = 6.63 \times 10^{-34} \nu_0$

$$\Rightarrow \nu_0 = \frac{1.6 \times 10^{-19}}{10^{-34}} = 1.6 \times 10^{15} \text{ Hz}$$

88. **(b)** Energy $E = \frac{hc}{\lambda} \Rightarrow \lambda = \frac{hc}{E} = \frac{hc}{\phi} = \frac{1242}{\phi}$

$$\lambda_A = \left(\frac{1242}{9}\right) = 138 \text{nm}$$

$$\lambda_B = \left(\frac{1242}{4.5}\right) = 276 \text{nm}$$

89. **(a)** From equation, $KE_{max} = hf - hf_0$

out of the rays given, UV rays have maximum frequency hence are most effective for emission of electrons from a metallic surface.

90. **(4125)** Net energy absorbed,

$$E = E_1 - E_2 = \frac{hc}{\lambda_1} - \frac{hc}{\lambda_2} = hc\left(\frac{1}{\lambda_1} - \frac{1}{\lambda_2}\right)$$

$$= 6.6 \times 10^{-34} \times 3 \times 10^8 \left(\frac{1}{500 \times 10^{-9}} - \frac{1}{600 \times 10^{-9}}\right)$$

$= 6.6 \times 10^{-20}$ J

$$= \frac{6.6 \times 10^{-20}}{1.6 \times 10^{-19}} eV = 4.125 \times 10^{-1} eV$$

$= 4125 \times 10^{-4}$ eV

91. **(d)** From the Einstein's photoelectric equation

$$eV_0 = \frac{hc}{\lambda} - \phi_0 = \frac{hc}{\lambda} - \frac{hc}{\lambda_0} \quad ...(i)$$

$$\& \frac{eV_0}{4} = \frac{hc}{2\lambda} - \frac{hc}{\lambda_0} \quad ...(ii)$$

$$\Rightarrow \frac{1}{4}\left(\frac{hc}{\lambda} - \frac{hc}{\lambda_0}\right) = \frac{hc}{2\lambda} - \frac{hc}{\lambda_0}$$

$$\Rightarrow \frac{1}{\lambda_0} - \frac{1}{4\lambda_0} = \frac{1}{2\lambda} - \frac{1}{4\lambda} \Rightarrow \frac{3}{4\lambda_0} = \frac{1}{4\lambda}$$

$\Rightarrow \lambda_0 = 3\lambda$

92. **(c)** From the figure, it is clear that threshold frequency for the metal is $\nu = 5 \times 10^{14}$

$\therefore$ Work function $\phi = h\nu$

$= 6.6 \times 10^{-34} \times 5 \times 10^{14}$ ($\because h = 6.6 \times 10^{-34}$ Js)

$= 33 \times 10^{-20}$ J

$$\Rightarrow \phi = \frac{33 \times 10^{-20}}{1.6 \times 10^{-19}} = 2.07eV$$

93. **(a)** Intensity of light $\propto$ number of photons $\propto$ no of photo electrons $\propto$ photo current

Here, A is correct

$KE_{max} = h\nu - \phi$

Here ν is the frequency of incident light.

KE_{max} depends on frequency

So, C is correct.

94. **(c)** From the Einstein's photoelectric equation,

$eV_s = hf - \phi$

$$\Rightarrow V_s = \left\{\frac{h}{e}\right\} f + \left\{\frac{-\phi}{e}\right\}$$

Slope $\left(\frac{h}{e}\right)$ is independent of nature of metal

$\therefore$ slope $(V_s)^{Gold}$ = slope $(V_s)^{Aluminium}$

95. **(a)** We have

$$\frac{1}{2} mv^2 = hf - hf_0$$

$$\text{So, } \frac{1}{2} mv_1^2 = h(2f_0) - hf_0 \Rightarrow \frac{1}{2} mv_1^2 = hf_0 \quad ...(i)$$

$$\frac{1}{2} mv_2^2 = h(5f_0) - hf_0 \Rightarrow \frac{1}{2} mv_2^2 = 4hf_0 \quad ...(ii)$$

Dividing (i) by (ii), we get

$$\left(\frac{v_1}{v_2}\right)^2 = \frac{1}{4} \Rightarrow \frac{v_1}{v_2} = \frac{1}{2}$$

96. **(c)** As $P = IA$

$$\Rightarrow P = \frac{nh\nu}{At} A \Rightarrow P = \frac{n}{t} h\nu$$

$$\Rightarrow \nu = \frac{P}{h(n/t)} = \frac{15000}{10^{16} \times 6 \times 10^{-34}} = 2.5 \times 10^{21}$$

As $\nu = 2.5 \times 10^{21}$ sec^{-1} $\Rightarrow$ Radiation lies in gamma region.

97. **(d)** As $E=\frac{1240}{350}=3.54$ eV, and $\phi_A=4.8$ eV & $\phi_B=2.2$ eV

So, $\phi_A > E \Rightarrow$ So, 'A' will not emit photoelectrons

and $\phi_B < E \Rightarrow$ So, 'B' will emit photoelectrons

98. **(d)** UV rays have wavelength less than 5500 A.

So, irrespective of power, photoelectrons will be emitted.

99. **(d)** Stopping potential $V_s=\frac{KE_{max}}{e}=\frac{\frac{hC}{\lambda}-\phi}{e}$

Stopping potential does not depend on intensity or power of light used, it only depends on frequency or wavelength of incident light.

So, both statements I and II are correct.

100. **(c)** We have

$$V_{stopp}=\frac{h\nu}{e}-\frac{\phi}{e}=\frac{1}{e}(h\nu-\phi)$$

So, stopping potential depends both on work function and frequency of radiation. So, (A) is false.

Increasing intensity, increases total number of photoelectrons emitted per second, there by increases saturation current.

So, (B) is true.

The maximum energy of photoelectrons depends only on frequency and work function so, (C) is false.

Photoelectric effect can be explained only by particle theory of light. So, (D) is false.

101. **(b)** By Einstein photoelectric equation,

$$E=\frac{hc}{\lambda}-\phi \quad ...(i)$$

$$\text{and, } 2E=\frac{hc}{\lambda'}-\phi \quad ...(ii)$$

Substracting (i) from (ii), we get

$$E=hc\left(\frac{1}{\lambda'}-\frac{1}{\lambda}\right) \Rightarrow \lambda'=\frac{hc\lambda}{E\lambda+hc}$$

102. **(c)**. According to question,

$$\frac{1}{2}mv_1^2=4\phi \quad ...(i)$$

$$\frac{1}{2}mv_2^2=9\phi \quad ...(ii)$$

Dividing (i) by (ii), we get

$$\frac{v_1}{v_2}=\frac{2}{3}$$

103. **(b)** From the Einstein's photoelectric equation,

$$\frac{1}{2}mv_{max}^2=hf-\phi$$

Here, f = frequency of incident light

ϕ = work function

Photoelectric effect can be explained by particle nature of light.

104. **(b)** Given that

Wavelength of light, $\lambda=900$ nm $=900\times10^{-9}$ m

Intensity, $I=100$ Wm^{-2}

Area, $A=1$ cm$^2=10^{-4}$ m^2

Time, t = 1 second

The number of photons in one second

$$=\frac{\text{Energy of all photons}(E')}{\text{Energy of one photon }(E)} \quad ..... (i)$$

Energy of all photons, (E') = Power (P) × time (t)

= Intensity (I) × Area (A) × time (t)

$= 100\times10^{-4}\times1=0.01$ J

Energy of each photon, (E) $=\frac{hc}{\lambda}$

$$=\frac{6.6\times10^{-34}\times3\times10^8}{900\times10^{-9}\times1.6\times10^{-19}}=1.375\text{ eV}$$

Putting this value in equation (i), we have

$$\text{The number of photons}=\frac{0.01}{1.375\times1.6\times10^{-19}}=4.5\times10^{16}$$

105. **(c)** As $K.Emax=\frac{hc}{\lambda}-\phi$

$$\text{So, } K_1=\frac{1230}{800}-\phi \quad ...(i) \quad \left[\because \frac{hc}{\lambda}=\frac{1240}{\lambda(nm)}\right]$$

$$2K_1=\frac{1230}{500}-\phi \quad ...(ii)$$

Multiplying (i) by 2 and substracting it from (ii)

$$\left(\frac{1230}{400}-\frac{1230}{500}\right)eV=\phi$$

$\Rightarrow \phi=0.615$ eV

106. **(a)** As we know

$$I=\frac{E}{At}=\frac{nh\nu}{At}\Rightarrow\frac{n}{t}=\frac{IA\lambda}{hc}\Rightarrow\frac{n}{t}=\frac{\lambda}{hc}\Rightarrow\frac{n}{t}\alpha P\lambda$$

$$\Rightarrow\left(\frac{n}{t}\right)_2=\left(\frac{n}{t}\right)_1\times\frac{p_2\lambda_2}{p_1\lambda 1}=9\times10^{20}\times\frac{P}{P}\times\frac{800}{600}=12\times10^{20}$$

107. **(b)** For emission of electrons from metal surface energy of incident radiation ≥ work function

as, $hf=hf_0+K.E$

at $f=f_0 \Rightarrow K.E=0$

108. **(d)** For maximum *K.E.*, we will take

higher frequency i.e, $\left(f=\frac{9\times10^{15}}{2\pi}Hz\right)$

$$(K.E.)_{max}=hf-\phi=\left(\frac{4.14\times10^{-15}\times9\times10^{15}}{2\pi}-2.50\right)eV$$

$= 5.93 - 2.50 = 3.43eV$, nearest is 3.42 *eV*.

109. **(b)** $K.E_{max}=\frac{hc}{\lambda}-\frac{hc}{\lambda_0}$ where λ_0 = cut off frequency

$$\text{So, } k_1=\frac{hc}{\lambda_1}-\frac{hc}{\lambda_0}\Rightarrow3k_1=\frac{3hc}{\lambda_1}-\frac{3hc}{\lambda_0}$$

$$\Rightarrow3k_1=\frac{hc}{\lambda_2}-\frac{3hc}{\lambda_0}$$

$k_2 = \frac{hc}{\lambda_2} - \frac{hc}{\lambda_0}$

Substracting (ii) from (i), we get

$3k_1 - k_2 = \frac{-2hc}{\lambda_0} \Rightarrow k_2 - 3k_1 = \frac{2hc}{\lambda_0}$

$\Rightarrow k_2 - 3k_1 = 2\phi \Rightarrow k_2 > 3k_1 \Rightarrow k_1 < \frac{k_2}{3}$

110. (a) K.E of $e^- = \frac{1}{2}mv^2 = \frac{1}{2}m\left(\frac{qBR}{m}\right)^2 = \frac{q^2B^2R^2}{2m}$

$[\because R = \frac{mv}{qB}]$

So, K.E of $e^- = \frac{(1.6\times10^{-19}\times2\times10^{-3}\times2\times10^{-3})^2}{2\times9.1\times10^{-31}}$

$\simeq 1.40eV$

So, $\phi = E - K = 2.76eV - 1.40eV = 1.36eV$

111. (35) As $K.E._{max} = e\,V_s = 0.42\,eV$

Now,

$h\nu = h\nu_0 + (K.E)_{max}$

$\Rightarrow \frac{12400}{6630} = h\nu_0 + 0.42 \Rightarrow 1.87 - 0.42 = h\nu_0$

$\Rightarrow \nu_0 = \frac{1.45\times1.6\times10^{-19}}{6.63\times10^{-34}} \Rightarrow \nu_0 = 3.5\times10^{14}$

$\therefore \nu_0 = 35\times10^{13}$ Hz

112. (2) Case I: $h\nu = h\nu_0 + \frac{1}{2}mv_1^2$

$h(2\nu_0) = h\nu_0 + \frac{1}{2}mv_1^2$

$h\nu_0 = \frac{1}{2}mv_1^2$ ...(i)

Case II: $h(5\nu_0) = h\nu_0 + \frac{1}{2}mv_2^2$

$4h\nu_0 = \frac{1}{2}mv_2^2$ ...(ii)

Dividing (ii) by (i), we get

$4 = \left(\frac{v_2}{v_1}\right)^2 \Rightarrow \frac{v_2}{v_1} = 2 \Rightarrow v_2 = 2v_1$

113. (b) $K.E_{max} = h\nu - \phi$

$\frac{\frac{1}{2}mV_{max1}^2}{\frac{1}{2}mV_{max2}^2} = \frac{3.8-0.6}{1.4-0.6} \Rightarrow \left(\frac{V_{max1}}{V_{max2}}\right)^2 = \frac{3.2}{0.8}$

$\therefore \frac{V_{max1}}{V_{max2}} = 2$

114. (d) Kinetic energy of electron depends on frequency of incident light & increasing intensity only increases no. of photons not frequency.

115. (b) Given wavelength of neon lemp $\lambda_1 = 670.5$nm

Stopping potential $V_1 = 0.48V$.

From the Einstein's photoelectric equation

$KE_{max} = \frac{hc}{\lambda_1} + \phi \Rightarrow eV_1 = \frac{hc}{\lambda_1} + \phi$

$\Rightarrow e(0.48) = \frac{1240}{670.5} + \phi$ ...(i)

For second cae wavelength $\lambda_2 = 474.6$ nm stlopping potential $= V_2$.

$e(V_2) = \frac{1240}{474.6} + \phi$ ...(ii)

Subtrating equation (i) from (ii) we get

$e(V_2 - 0.48) = 1240\left(\frac{1}{474.6} - \frac{1}{670.5}\right)eV$

$\Rightarrow V_2 = 0.48 + 1240\left(\frac{670.5-474.6}{474.6\times670.5}\right)$ volts

$\Rightarrow V_2 = 0.48 + 0.76 = 1.25V$

116. (a) From photoelectric effect maximum kinetic energy $= eV_s$

$eV_s = \frac{hc}{\lambda} - \phi \Rightarrow eV_{s_1} = \frac{1240}{280} - 2.5 = 1.93eV$

$\Rightarrow V_{s_1} = 1.93V$

and, $eV_{s_2} = \frac{1240}{400} - 2.5 = 0.6eV$

$\Rightarrow V_{s_1} - V_{s_2} = 1.93 - 0.6 = 1.33V$

117. (910) From de-Broglie wavelength $\lambda = \frac{h}{P}$

For photon, $\lambda_1 = \frac{h}{P} = \frac{6.6\times10^{-34}}{10^{-27}}$

For particle, $\lambda_2 = \frac{h}{mv} = \frac{6.6\times10^{-34}}{9.1\times10^{-31}\times10^6}$

$\therefore \frac{\lambda_1}{\lambda_2} = \frac{9.1\times10^{-25}}{10^{-27}} = 910$

118. (10) Here,

$E_{k_\alpha} = E_k - E_L$

or, $\frac{hc}{\lambda_{k_\alpha}} = E_k - E_L \Rightarrow E_L = E_k - \frac{hc}{\lambda_{k_\alpha}}$

$\Rightarrow E_L = 27.5\text{ KeV} - \frac{12.42\times10^{-7}}{0.071\times10^{-9}}$

$\Rightarrow E_L = (27.5 - 17.5)\text{keV} = 10\text{ keV}$

119. (122) From the Einstein's photoelectric equation,

$\frac{hc}{\lambda} = \phi + KE_{max} \Rightarrow \frac{1240}{500} = 1.25 + KE_{max}$

$\Rightarrow KE_{max} = 1.23$ eV

Radius of the circular path,

$R = \frac{mv}{qB} = \frac{\sqrt{2m\,KE}}{qB}$ $\left(\because P = mv = \sqrt{2m\,KE}\right)$

$\Rightarrow B = \frac{\sqrt{2m\,KE}}{qR}$

$\Rightarrow B = \frac{\sqrt{2\times9.1\times10^{-31}\times1.18\times1.6\times10^{-19}}}{1.6\times10^{-19}\times30\times10^{-2}}$

$\Rightarrow B = 0.122\times10^{-4}$

$\Rightarrow B = 12.2\times10^{-6} \Rightarrow B = 12.2\,\mu T = 122\times10^{-7}T$

120. (b) From the Einstein's photoelectric equation

$KE = h\nu - W \Rightarrow eV = \frac{hc}{\lambda} - W$

For first case

$e(3V_0) = \frac{hc}{\lambda_0} - W$ ... (i)

For second case

$eV_0 = \frac{hc}{2\lambda_0} - W$ ... (ii)

From equation (i) and (ii)

$W = \frac{hc}{4\lambda_0}$

Also, Work function, $W = \frac{hc}{\lambda_{th}}$

$\Rightarrow \frac{hc}{4\lambda_0} = \frac{hc}{\lambda_{th}} \Rightarrow \lambda_{th} = 4\lambda_0$

121. (c) As de-Broglie wavelengths are equal

$\lambda_e = \lambda_{Ph}$

$\Rightarrow \frac{h}{p_e} = \frac{h}{p_{ph}} \quad \left(\because \lambda = \frac{h}{p}\right)$

Momentum of photon, $p = \sqrt{2mk_e} = \frac{E_{ph}}{c}$

$\Rightarrow 2mk_e = \frac{(E_{ph})^2}{c^2} \Rightarrow \frac{k_e}{E_{ph}} = \frac{E_{ph}}{c^2}\left(\frac{1}{2m}\right)$

$= \frac{p_{ph}}{c}\left(\frac{1}{2m}\right) \quad (\because E_{photon} = pc)$

$= \frac{p_e}{c}\left(\frac{1}{2m}\right) \quad (\because p_{photon} = p_{electron})$

$= \frac{mv}{c}\frac{1}{2m} = \frac{v}{2c}$

122. (d) Energy of the photon emitted from the transition of electron from n = 3 to n = 2 is given by

$E_p = 13.6\left[\frac{1}{R_1^2} - \frac{1}{R_2^2}\right] eV = 13.6\left[\frac{1}{4} - \frac{1}{9}\right]$

$\Rightarrow E_p = 1.89$ eV

For Gold plate

Using Einstein's photoelectric equation

$E_p = \phi + K.E_{max}$ ...(i)

$\Rightarrow \phi = E_p - KE_{max} \Rightarrow \phi = E_p - \frac{1}{2}mv^2$

Radius of electron, $R = \frac{mv}{qB}$

$\Rightarrow v = \frac{RqB}{m} = \frac{7\times10^{-3}\times1.6\times10^{-19}\times5\times10^{-4}}{9.1\times10^{-31}} = 6.15\times10^5$

$K.E. = \frac{1}{2}mV^2$

$\Rightarrow K.E. = \frac{1}{2}\times\frac{9.1\times10^{-31}\times(6.15\times10^5)^2}{1.6\times10^{-19}} eV = 1.077 eV$

$\phi = 1.89 - 1.077$ [using equation (i)]

$\Rightarrow \phi = 0.813$ eV

123. (c) According to Einstein's photoelectric equation,

$K_{max} = h\nu - \phi_0 \Rightarrow eV_0 = h\nu - \phi$

Where, V_0 = Stopping potential,

ϕ = Work function,

ν = Frequency of incident light

∴ Stopping potential depends on frequency.

124. (1) From $KE_{max} = \frac{1}{2}mv^2 = E - \phi$

$\Rightarrow V^2 = \frac{2}{m}(E - \phi)$

From question, $E_1 = 2\phi$ and $E_2 = 10\phi$

$\Rightarrow V_1^2 = \frac{2}{m}(\phi)$ ...(i)

$V_2^2 = \frac{2}{m}(9\phi)$ ...(ii)

From (i) & (ii),

$\frac{V_1^2}{V_2^2} = \frac{\frac{2\phi}{m}}{\frac{18\phi}{m}} \Rightarrow \frac{V_1}{V_2} = \frac{1}{3}$ so $\frac{x}{y} = \frac{1}{3}$ ∴ $x = 1$

125. (c) From Einstein's photoelectric equation,

$\frac{hc}{\lambda} = \phi + eV$

$\frac{hc}{\lambda_1} - eV_1 = \frac{hc}{\lambda_2} - eV_2$ [∵ for same surface ϕ is same]

$\Rightarrow \frac{1240}{491} - 0.71 = \frac{1240}{\lambda_2} - 1.43$

$\Rightarrow \frac{1240}{\lambda_2} = \frac{1240}{491} + 0.72 \Rightarrow \lambda_2 = 382$ nm

126. (a) We know that

$E = e\Delta V$

Here, V = voltage,

$\frac{hc}{\lambda} = e\Delta V$ e = charge on electron

$\Rightarrow \lambda = \frac{hc}{e\Delta V}$

[put, $h = 6.63\times10^{-34}$ Js,

$C = 3\times10^8$ m/s, $e = 1.6\times10^{-19}$ C]

We get

$\lambda_{min} = \frac{1240}{\Delta V}(nm) = \frac{1240}{1.24\times10^6} = 10^{-3}$ nm

127. (50) From Bragg's equation $2d\sin\theta = \lambda$ and de-Broglie wavelength, $\lambda = \frac{h}{P} = \frac{h}{\sqrt{2mE}}$

$2d\sin\theta = \lambda = \frac{h}{\sqrt{2mE}}$

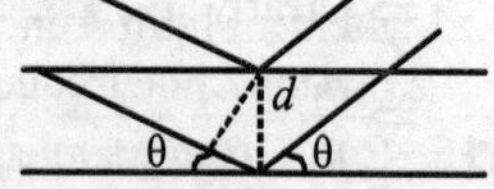

$\Rightarrow 2\times10^{-10}\times\frac{\sqrt{3}}{2} = \frac{6.6\times10^{-34}}{\sqrt{2mE}} \Rightarrow E = 50$ eV

[∵ $\theta = 60°$, $d = 1Å = 1\times10^{-10}$ m and $m = 9.1\times10^{-31}$ kg]

128. **(2)** From the Einstein's photoelectric equation
Energy of photon
= Kinetic energy of photoelectrons + Work function
⇒ Kinetic energy = Energy of Photon – Work Function
Let ϕ_0 be the work function of metal and v_1 and v_2 be the velocity of photoelectrons. Using Einstein's photoelectric equation we have

$$\frac{1}{2}mv_1^2 = 4-\phi_0 \quad ...(i)$$

$$\frac{1}{2}mv_2^2 = 2.5-\phi_0 \quad ...(ii)$$

$$\Rightarrow \frac{\frac{1}{2}mv_1^2}{\frac{1}{2}mv_2^2} = \frac{4-\phi_0}{2.5-\phi_0}$$

$$\Rightarrow (2)^2 = \frac{4-\phi_0}{2.5-\phi_0} \Rightarrow 10-4\phi_0 = 4-\phi_0$$

$$\phi_0 = 2eV$$

129. **(a)** Graph of V_s and f given at B (5.5, 0)
$h\nu = \phi + eV_s$ at B, $V_s = 0$
So, $h\nu = \phi$

$$\phi = h\nu \text{ joule or } \phi = \frac{h\nu}{e} \text{ eV (for } V=0)$$

$$\therefore \phi = \frac{6.62\times10^{-34}\times5.5\times10^{14}}{1.6\times10^{-19}} \text{eV} = 2.27 \text{eV}$$

130. **(d)** As stopping potential is independent of intensity. So there will be no change in curve.

131. **(d)** Using equation, $= \frac{hc}{\lambda} - \phi$

$$KE_{max} = \frac{hc}{\lambda} - \phi = \frac{hc}{500} - \phi \quad ...(i)$$

$$\text{Again, } 3KE_{max} = \frac{hc}{200} - \phi \quad ...(ii)$$

Dividing equation (ii) by (i),

$$\frac{3KE_{max}}{KE_{max}} = \frac{3}{1} = \frac{\frac{hc}{200}-\phi}{\frac{hc}{500}-\phi}$$

Putting the value of $hc = 1237.5$ and solving we get, work function, $\phi = 0.61$ eV.

132. **(a)** IA = P

$$\text{and, } I = \frac{Nh\nu}{At}$$

$$\frac{N}{t} = \frac{IA}{h\nu} \Rightarrow \frac{N}{t} = \frac{P}{E}$$

$$n = \frac{P}{E} = \frac{P}{h\nu} = \frac{P\lambda}{hc} \quad (\because E = h\nu \text{ and } \nu = \frac{c}{\lambda})$$

⇒ Clearly $n \propto \lambda$

$$\Rightarrow \frac{n_1}{n_2} = \frac{\lambda_1}{\lambda_2} = \frac{1}{500}$$

133. **(9)** When radiation of wavelength A, λ_A is used to illuminate, stopping potential $V_A = V$

$$\frac{hc}{\lambda} = \phi + eV \quad ...(i)$$

When radiation of wavelength B, λ_B is used to illuminate, stopping potential, $V_B = \frac{V}{4}$

$$\frac{hc}{3\lambda} = \phi + \frac{eV}{4} \quad ...(ii)$$

From eq. (i) – (ii),

$$\frac{hc}{\lambda}\left(1-\frac{1}{3}\right) = \frac{3}{4}eV \Rightarrow \frac{hc}{\lambda}\frac{2}{3} = \frac{3}{4}eV \Rightarrow eV = \frac{8}{9}\frac{hc}{\lambda}$$

$$\frac{hc}{\lambda} = \phi + \frac{8}{9}\frac{hc}{\lambda}$$

$$\therefore \phi = \frac{hc}{9\lambda} = \frac{hc}{n\lambda}, \text{ so, } n = 9.$$

134. **(a)** Using Einstein's photoelectric equation,
$E = \phi_0 + KE_{max} \Rightarrow \phi_0 = E - (K.E)_{max}$

$$p = \sqrt{2mKE} \Rightarrow KE = \frac{p^2}{2m}$$

$$r = \frac{p}{eB} \Rightarrow p = reB$$

$$K_{max} = \frac{r^2e^2B^2}{2m} \qquad KE_{max} = \frac{12420}{\lambda} - \phi_0$$

$$\Rightarrow \phi_0 = \frac{12420}{6561} - \frac{r^2e^2B^2}{2m}(In\,\text{eV})$$

$$= 1.89(eV) - \frac{(10^{-4})(1.6\times10^{-19})^2\,9\times10^{-8}}{2\times9.07\times10^{-31}\times e}$$

$$= 1.89(eV) - \frac{(10^{-4})(1.6\times10^{-19})^2\,9\times10^{-8}}{2\times9.07\times10^{-31}\times1.6\times10^{-19}}$$

$= (1.89 - 0.79)\,eV = 1.1\,eV$

135. **(a)** de-Broglie wavelength (λ),

$$\text{Momentum, } mv = \frac{h}{\lambda} = p = \sqrt{2m(KE)}$$

$$\therefore \lambda = \frac{h}{\sqrt{2mKE}} \Rightarrow \lambda \propto \frac{1}{\sqrt{KE}}$$

$$\therefore \frac{\lambda_A}{\lambda_B} = \sqrt{\frac{K_B}{K_A}} = \sqrt{\frac{T_A-1.5}{T_A}} \quad \text{(as given)}$$

$$\text{Also, } \frac{\lambda_A}{\lambda_B} = \frac{1}{2}$$

On solving we get, $T_A = 2\,eV$
$\therefore KE_B = T_A - 1.5 = 2 - 1.5 = 0.5\,eV$
∴ Work function of metal B is
$\phi_B = E_B - KE_B = 4.5 - 0.5 = 4\,eV$

136. **(11.00)** Energy of proton

$$E = \frac{hc}{\lambda} = \frac{1240}{310} = 4eV > 2eV\ [= \phi]$$

(so emission of photoelectron will take place)
$= 4\times1.6\times10^{-19} = 6.4\times10^{-19}$ joule

$$\frac{N}{t} = \frac{6.4\times10^{-5}\times1}{6.4\times10^{-19}} = 10^{14} \left[\because I = \frac{Nh\nu}{At}\right]\left[\frac{N}{t} = \frac{IA}{h\nu}\right]$$

No. of photoelectrons emitted per second

$= \frac{10^{14}}{10^3} = 10^{11}$ ($\because$ 1 in 10^3 photons ejects an electron)

$\therefore$ Value of $X = 11.00$

137. (b) $f_0 = 4 \times 10^{14}$ Hz

$W_0 = hf_0 = 6.63 \times 10^{-34} \times (4 \times 10^{14})$ J

$= \frac{(6.63\times10^{-34})\times(4\times10^{14})}{1.6\times10^{-19}} = 1.66$ eV

138. (a) $KE_{max} = E - \phi_0$

(where E = energy of incident light ϕ_0 = work function)

$= \frac{hc}{\lambda} - \frac{hc}{\lambda_0} = 1237\left[\frac{1}{260} - \frac{1}{380}\right] = \frac{1237\times120}{380\times260} = 1.5$eV

139. (a) Energy of photon (E) is given by

$E = \frac{hc}{\lambda}$

Number of photons of wavelength λ emitted in t second from laser of power P is given by

$n = \frac{Pt\lambda}{hc} \Rightarrow n = \frac{P\times\lambda}{hc} = \frac{2\times10^{-3}\times5\times10^{-7}}{19.8\times10^{-26}}$ ($\because t = 1$S)

$\Rightarrow n = 5 \times 10^{15}$

140. (c) Here $\omega = 2\pi \times 6 \times 10^{14}$ or $f = 6 \times 10^{14}$Hz

Wavelength $\lambda = \frac{C}{f} = \frac{3\times10^8}{6\times10^{14}} = 0.5\times10^{-6}\,m = 5000$A°

Now $E = \frac{12374}{5000} = 2.48\,eV$

Using $E = w + eV_s$

$2.48 = 2 + eV_s$ or $V_s = 0.48$ V

141. (d) $hv = \phi + \frac{V_0}{2}e$ ....(i)

$\frac{hv}{2} = \phi + V_0 e$ ...(ii)

On solving, Both the equation $\phi = \frac{3}{2}hv$

$hv_0 = -\frac{3}{2}hv \Rightarrow v_0 = -\frac{3}{2}v$

142. (d) Using, wavelength, $\lambda = \frac{12375}{\Delta E}$

or, $\lambda = \frac{1237.5}{4.9} \simeq 250$nm

143. (c) Using, intensity $I = \frac{nE}{At}$

n = no. of photoelectrons

$\Rightarrow 16\times10^{-3} = \left(\frac{n}{t}\right)\times\frac{10\times1.6\times10^{-19}}{10^{-4}}$ or, $\frac{n}{t} = 10^{12}$

So, effective number of photoelectrons ejected per unit time $= 10^{12} \times 10/100 = 10^{11}$

$K.E_{max} = (10-5)$eV $= 5$eV

144. (a) From Einstein's photoelectric equation,

$\frac{hc}{\lambda_1} = \phi + \frac{1}{2}m(2v)^2$ ....(i)

and $\frac{hc}{\lambda_2} = \phi + \frac{1}{2}mv^2$ ....(ii)

As per question, maximum speed of photoelectrons in two cases differ by a factor 2

From eqn. (i) & (ii)

$\Rightarrow \frac{\frac{hc}{\lambda_1} - \phi}{\frac{hc}{\lambda_2} - \phi} = 4 \Rightarrow \frac{hc}{\lambda_1} - \phi = \frac{4hc}{\lambda_2} - 4\phi \Rightarrow \frac{4hc}{\lambda_2} - \frac{hc}{\lambda_1} = 3\phi$

$\Rightarrow \phi = \frac{1}{3}hc\left(\frac{4}{\lambda_2} - \frac{1}{\lambda_1}\right) = \frac{1}{3}\times1240\left(\frac{4\times350-540}{350\times540}\right) = 1.8$ eV

145. (d) $B = B_0[\sin(3.14\times10^7 ct) + \sin(6.28\times10^7 ct)]$

Maximum KE will associate with maximum frequency.

$f_{max} = \frac{6.28\times10^7 c}{2\pi} = 10^7\times3\times10^8$ Hz $= 3\times10^{15}$ Hz

$KE_{max} = hf_{max} - \phi$

$= \left[\frac{6.63\times10^{-34}\times3\times10^{15}}{1.6\times10^{-19}} - 4.7\text{eV}\right] = 7.72$ eV

146. (c) In X-ray tube, $\lambda_{min} = \frac{hc}{eV}$

$ln\ \lambda_{min} = \left(\frac{hc}{e}\right) - lnV$

Clearly, log λ_{min} versus log V graph slope is negative hence option (c) correctly depicts.

147. (a) Given, $\lambda = 660$ nm, Power = 0.5 kW, t = 60 ms

Power $P = \frac{nhc}{\lambda t} \Rightarrow n = \frac{P\lambda t}{hc}$

$= 0.5\times10^3\times\frac{660\times10^{-9}\times60\times10^{-3}}{6.6\times10^{-34}\times3\times10^8} = 100\times10^{18} = 10^{20}$

148. (c) $h\frac{c}{\lambda} - hv_0 = \frac{1}{2}mv^2$ $\therefore \frac{4}{3}\frac{hc}{\lambda} - hv_0 = \frac{1}{2}mV^2$

$\therefore \frac{V^2}{v^2} = \frac{\frac{4}{3}v - v_0}{v - v_0}$ $\therefore V = v\sqrt{\frac{\frac{4}{3}v - v_0}{v - v_0}}$

$\therefore V > v\sqrt{\frac{4}{3}}$

149. (a) From Einstein's photoelectric equation

$K.E._\lambda = \frac{hc}{\lambda} - \phi$ ...(i)

(for monochromatic light of wavelength λ)

where ϕ is work function

$K.E._{\lambda/2} = \frac{hc}{\lambda/2} - \phi$ ...(ii)

(for monochromatic light of wavelength λ/2)
From question,

$$K.E._{\lambda/2} = 3(K.E._{\lambda}) \Rightarrow \frac{hc}{\lambda/2} - \phi = 3\left(\frac{hc}{\lambda} - \phi\right)$$

$$\frac{2hc}{\lambda} - \phi = 3\frac{hc}{\lambda} - 3\phi \Rightarrow 2\phi = \frac{hc}{\lambda} \Rightarrow \phi = \frac{hc}{2\lambda}$$

150. (b) Given, $\lambda_1 = 4972Å$ and $\lambda_2 = 6216Å$

and $I = 3.6 \times 10^{-3} Wm^{-2}$
Intensity associated with each wavelength

$$= \frac{3.6\times10^{-3}}{2} = 1.8 \times 10^{-3} Wm^{-2}$$

work function $\phi = h\nu$

$$= \frac{hc}{\lambda} = \frac{(6.62\times10^{-34})(3\times10^{8})}{\lambda} = \frac{12.4\times10^{3}}{\lambda} ev$$

for different wavelengths

$$\phi_1 = \frac{12.4\times10^3}{\lambda_1} = \frac{12.4\times10^3}{4972} = 2.493 \text{ eV} = 3.984 \times 10^{-19} \text{ J}$$

$$\phi_2 = \frac{12.4\times10^3}{\lambda_2} = \frac{12.4\times10^3}{6216} = 1.994 \text{ eV} = 3.184 \times 10^{-19} \text{ J}$$

Work function for metallic surface $\phi = 2.3$ eV (given)
$\phi_2 < \phi$
Therefore, ϕ_2 will not contribute in this process.
Now, no. of electrons per m^2-s = no. of photons per m^2-s

$$\text{no. of electrons per } m^2\text{-s} = \frac{1.8\times10^{-3}}{3.984\times10^{-19}} \times 10^{-4}$$

$(\because 1\,cm^2 = 10^{-4} m^2) = 0.45 \times 10^{12}$

So, the number of photo electrons liberated in 2 sec.
$= 0.45 \times 10^{12} \times 2 = 9 \times 10^{11}$

151. (c) According to photo-electric equation :
$K.E_{max} = h\nu - h\nu_0$ (Work function)
Some sort of energy is used in ejecting the photoelectrons.

152. (c) By Einstein photoelectric equation,
$K_{max} = eV_0 = h\nu - h\nu_0$
When ν is doubled, K_{max} and V_0 become more than double.

153. (d) We know that $eV_0 = K_{max} = h\nu - \phi$
where, ϕ is the work function.
X-rays have higher frequency (ν) than ultraviolet rays. Therefore as ν increases K.E and V_0 both increases.
The kinetic energy ranges from zero to maximum because of loss of energy due to subsequent collisions before getting ejected.

154. (a) Wavelength of incident light, $\lambda = 400$ nm $hc = 1240$ eV.nm
$K.E = 1.68$ eV
Using Einstein's photoelectric equation

$$\frac{hc}{\lambda} - W = K.E \Rightarrow W = \frac{hc}{\lambda} - K.E$$

$$\Rightarrow W = \frac{1240}{400} - 1.68 = 3.1 - 1.68 = 1.41 \text{ eV}$$

155. (a) Energy of a photon of frequency ν is given by $E = h\nu$.

Also, $E = mc^2 \quad \therefore mc^2 = h\nu \Rightarrow mc = \frac{h\nu}{c} \Rightarrow p = \frac{h\nu}{c}$

156. (a) Work function, $\phi = 6.2\text{ eV} = 6.2 \times 1.6 \times 10^{-19}$ J
Stopping potential, $V = 5$ volt
From the Einstein's photoelectric equation

$$\frac{hc}{\lambda} - \phi = eV_0 \Rightarrow \lambda = \frac{hc}{\phi + eV_0}$$

$$= \frac{6.6\times10^{-34}\times3\times10^{8}}{1.6\times10^{-19}(6.2+5)} \approx 10^{-7} \text{ m}$$

This range lies in ultra violet range.

157. (b) The photoelectric emission is an instantaneous process without any apparent time lag. It is known that emission starts in the time of the order of 10^{-9} second. So, the approximate time taken by a photoelectron to come out after the photon strikes is 10^{-10} second.

158. (b) $I \propto i$

$$I \propto E \qquad \therefore i \propto E \Rightarrow i \propto \frac{1}{\lambda}$$

159. (a) $I \propto \frac{1}{r^2}; \frac{I_1}{I_2} = \left(\frac{r_2}{r_1}\right)^2 = \frac{1}{4}$

$I_2 \rightarrow 4$ times I_1

When intensity becomes 4 times, no. of photoelectrons emitted would increase by 4 times, since number of electrons emitted per second is directly proportional to intensity.

160. (b) Momentum of photon of energy E is $= \frac{E}{c}$

When a photon hits a perfectly reflecting surface, it reflects black in opposite direction with same energy and momentum.

$$\therefore \text{Change in momentum} = \frac{E}{C} - \left(\frac{-E}{C}\right) = \frac{2E}{C}$$

This is equal to momentum transferred to the surface.

161. (d) From the Einstein photoelectric equation $K.E. = h\nu - \phi$
Here, ϕ = work function of metal
h = Plank's constant
slope of graph of $K.E.$ & ν is h (Plank's constant) which is same for all metals.

162. (a) Work function of metal (ϕ) is given by

$$\phi = \frac{hc}{\lambda} \Rightarrow \lambda = \frac{hc}{\phi} \Rightarrow \lambda = \frac{6.63\times10^{-34}\times3\times10^{8}}{4\times1.6\times10^{-19}} = 310 \text{nm}$$

163. (a) Let work function be W and v_1 and v_2 be the velocity of electrons for frequencies f_1 and f_2.
Using Einstein's photo electric equation for one photodiode, we get

$$hf_1 - W = \frac{1}{2}mv_1^2 \quad ....(i)$$

Using Einstein's photo electric equation for another photodiode we get,

$$hf_2 - W = \frac{1}{2}mv_2^2 \quad ....(ii)$$

Subtracting (ii) from (i) we get

$$(hf_1 - W) - (hf_2 - W) = \frac{1}{2}mv_1^2 - \frac{1}{2}mv_2^2$$

$$\Rightarrow h(f_1 - f_2) = \frac{m}{2}(v_1^2 - v_2^2) \quad \therefore v_1^2 - v_2^2 = \frac{2h}{m}(f_1 - f_2)$$

164. (c) We know that work function,

$$E = h\upsilon = \frac{hC}{\lambda} \quad \therefore \frac{E_{Na}}{E_{Cu}} = \frac{\lambda_{Cu}}{\lambda_{Na}} \Rightarrow \frac{\lambda_{Na}}{\lambda_{Cu}} = \frac{E_{Cu}}{E_{Na}} = \frac{4.5}{2.3} \approx \frac{2}{i}$$

Atoms

12

1. **(156)** Given, Distance of closer approach, $r_{min} = 4.5 \times 10^{-14}$ m
Atomic number, z = 80
In alpha particle scattering experiment

$$\frac{1}{2}mv^2 = \frac{1}{4\pi \epsilon_0}\frac{Z(e)(2e)}{r_{min}}$$

$$v = \sqrt{\frac{4Ze^2}{4\pi \epsilon_0 r_{min} m}}$$

$$\sqrt{\frac{4 \times 9 \times 10^9 \times 80}{6.72 \times 10^{-27} \times 4.5 \times 10^{-14}}} \times 1.6 \times 10^{-19}$$

$= 9.759 \times 10^{25} \times 1.6 \times 10^{-19} = 156 \times 10^5$ m/s

2. **(a)** Rutherford atomic model states that most of mass of atom and all its positive charge is concentrated in tiny nucleus & electron revolve around it.
On the other hand, according to Thomson atomic model, atom is spherical cloud of positive charge with electron embedded in it. Hence, Statement I is true but statement II false.

3. **(3)** Impact parameter (d) = $\alpha \cot^2\left(\frac{\theta}{2}\right)$, θ = Scattering angle

$$\Rightarrow \frac{d_1}{d_2} = \frac{\cot^2\left(\frac{\theta_1}{2}\right)}{\cot^2\left(\frac{\theta_2}{2}\right)} = \frac{\cot^2(30°)}{\cot^2(45°)} = \frac{3}{1}$$

As, $\frac{d_1}{d_2} = x$. So, $x = 3$.

4. **(c)** In Rutherford model, e^- revolve around the nucleus in circular path. Now, according to classical theory of electromagnetism, an accelerating charge radiates energy in form of EM waves and here motion of e^- is circular so it is also accelerated, therefore it will radiate energy and finally collapse to nucleus.

5. **(d)** $y \propto \frac{1}{\sin^4 \frac{\theta}{2}}$

at $\theta = 0 \Rightarrow y \to \infty$
at $\theta = \pi \Rightarrow y$ is a small finite value.

6. **(c)** As α-particles are doubly ionised helium He^{++} i.e. Positively charged and nucleus is also positively charged and we know that like charges repel each other.

7. **(b, c)** Work done to stop the α particle is equal to K.E.

$$\therefore qV = \frac{1}{2}mv^2 \Rightarrow q \times \frac{K(Ze)}{r} = \frac{1}{2}mv^2$$

$$\Rightarrow r = \frac{2(2e)K(Ze)}{mv^2} = \frac{4KZe^2}{mv^2} \Rightarrow r \propto \frac{1}{v^2} \text{ and } r \propto \frac{1}{m}.$$

8. **(a)** Distance of closest approach

$$\frac{1}{4\pi \epsilon_0}\frac{Ze2e}{r_0} = \frac{1}{2}mv^2 \Rightarrow r_0 = \frac{Ze(2e)}{4\pi\varepsilon_0\left(\frac{1}{2}mv^2\right)}$$

Energy, $\frac{1}{2}mv^2 = 5 \times 10^6 \times 1.6 \times 10^{-19}$ J

$$\therefore \quad r_0 = \frac{9 \times 10^9 \times (92 \times 1.6 \times 10^{-19})(2 \times 1.6 \times 10^{-19})}{5 \times 10^6 \times 1.6 \times 10^{-19}}$$

$\Rightarrow r_h = 5.2 \times 10^{-14} m = 5.3 \times 10^{-12}$ cm

9. **(d)** The energy of transition in hydrogen atom is

$$\Delta E = 13.6Z^2\left[\frac{1}{n_1^2} - \frac{1}{n_2^2}\right]$$

Here, $n_1 = 1, n_2 = 3, \Delta E = 108.8$ eV

$$\Rightarrow (13.6)Z^2\left[\frac{1}{1} - \frac{1}{9}\right] = 108.8$$

$$\Rightarrow \frac{(13.6)(8)}{9}(Z^2) = 108.8$$

$\therefore Z = 3.$

10. **(b)** For largest wavelength of Lyman series, $n_1 = 1, n_2 = 2$

$$\Rightarrow \frac{1}{\lambda_1} = R\left[\frac{1}{1} - \frac{1}{4}\right] = \frac{3R}{4}$$

$$\therefore \lambda_1 = \frac{4}{3R}$$

For largest wavelength of Balmer series, $n_1 = 2, n_2 = 3$

$$\Rightarrow \frac{1}{\lambda_2} = R\left[\frac{1}{4} - \frac{1}{9}\right] = \frac{5R}{36}$$

$\therefore \lambda_2 = \frac{36}{5R}$

$\therefore \frac{\lambda_1}{\lambda_2} = \frac{5}{27}.$

11. (c) $E = h\nu$

$ML^2T^{-2} = [h] \times [T^{-1}] \Rightarrow [h] = [ML^2T^{-1}]$

$L = MVR = [ML^2T^{-1}]$

Hence, the dimension of both is same

$L = \frac{nh}{2\pi}$ L is integral multiple of $\frac{h}{2\pi}$.

12. (d) $r_n = \frac{n^2 \times r}{z}$

r = Bohr radius, for $Li^{2+}: z = 3$, for $He^+: z = 2, n = 5$

$r = r \cdot \frac{n^2}{z}$

For Li^{2+} $r_5 = r \cdot \frac{25}{3}$

For He^+ $r_5 = r \cdot \frac{25}{2}$

$\therefore \frac{r_{Li^{2+}}}{r_{He^{2+}}} = \frac{2}{3}.$

13. (c) Bohr's model is applicable for hydrogen and hydrogen-like atoms. In this model, repulsive force between electrons is neglected.

14. (2) In hydrogen atom, the energy of transition is

$E = 13.6\left(\frac{1}{n_1^2} - \frac{1}{n_1^2}\right)$

$\Rightarrow 2.86 = 13.6\left(\frac{1}{n^2} - \frac{1}{5^2}\right)$

$\Rightarrow \frac{1}{n^2} = 0.21 + \frac{1}{2.5}$

$\therefore n = 2$

15. (c) Radius $r = r_0 \frac{n^2}{z}$ and $z = 3$ for Li^{+2} and $n = 1$

$\therefore r = r_0 \frac{1^2}{3} = \frac{r_0}{3} = \frac{1}{3}$ $\therefore x = 3$

16. (b) From relation $E \propto \frac{Z}{n^2}$

And $Z_H = 1$ $Z_{He^+} = 2$ $Z_{Li^{+2}} = 3$

For 1st excited state $\Rightarrow n = 2$ and for 2nd excited state $\Rightarrow n = 3$

statements only A & B are correct.

17. (a) The expression of number of spectral lines emitted

$= \frac{(n_2 - n_1)(n_2 - n_1 + 1)}{2} = \frac{(4-1)(4-1+1)}{2} = 6$

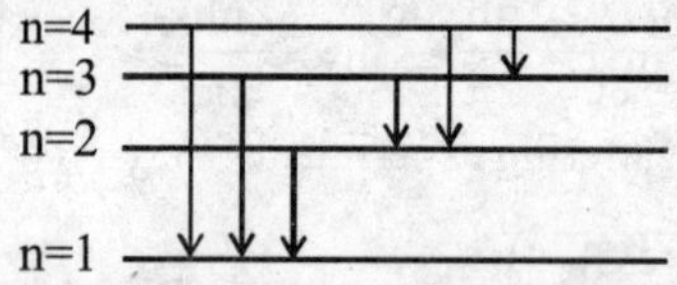

18. (b) Frequency of revolution is inversely proportional to the cube of principle quantum number.

$f \propto \frac{1}{n^3}$

19. (a)

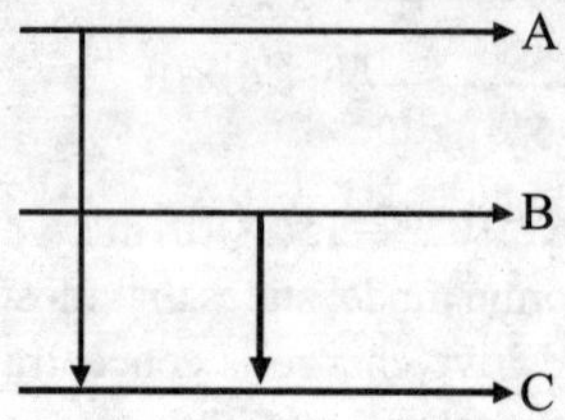

By Bohr's theory, $\Delta E = E_2 - E_1 = \frac{hc}{\lambda}$

$\therefore E_A - E_C = \frac{hc}{2000Å}$ ...(i)

and $E_B - E_C = \frac{hc}{6000Å}$ ...(ii)

Now, $E_A - E_B = (E_A - E_C) - (E_B - E_C)$

$\frac{hc}{\lambda_{AB}} = \frac{hc}{2000} - \frac{hc}{6000}$

$\frac{1}{\lambda_{AB}} = \frac{1}{3000Å}$

$\therefore \lambda_{AB} = 3000Å$

20. (d) For motion of electron perpendicular to magnetic field B,

$r = \frac{mv}{eB}$ & $mvr = \frac{nh}{2\pi}$

$\therefore (eBr)r = \frac{nh}{2\pi}$

$\Rightarrow r = \sqrt{\frac{nh}{2\pi eB}}$

For first excited state, n = 2

$\therefore r = \sqrt{\frac{h}{\pi eB}}$

21. **(a)** Given, orbital radius of election in ground state, $r_1 = 5.3 \times 10^{-11}$ m

Orbital radius of election in third excited state, $r_4 = 8.48 \times 10^{-10}$ m

de-Broglie wavelength, $\lambda = \frac{h}{mv}$

From Bohr's model, $mvr = \frac{nh}{2\pi} \Rightarrow mv = \frac{nh}{2\pi r}$

$$\therefore \quad \lambda = \frac{2\pi rh}{nh} \Rightarrow \lambda \propto \frac{r}{n}$$

$\because$ For third excited state, $n_4 = 4$ ground and for ground state, $n_1 = 1$

$$\therefore \quad \frac{\lambda_1}{\lambda_4} = \frac{r_1 n_4}{n_1 r_4} = \frac{5.3 \times 10^{-11} \times 4}{1 \times 84.8 \times 10^{-11}} \Rightarrow \frac{\lambda_1}{\lambda_4} = \frac{1}{4}$$

22. **(d)** Energy, $\Delta E = 10.2\text{eV}$

So, electron is excited to second state.

Hence, number of spectral lines = 1

23. **(a)** Wavelength of H-atom is

$$\frac{1}{\lambda} = Rz^2\left(\frac{1}{n_1^2} - \frac{1}{n_2^2}\right)$$

Shortest wavelength for Balmer series:

$$\frac{1}{\lambda_B} = Rz^2\left(\frac{1}{2^2} - \frac{1}{\infty}\right) \quad \text{...(i)}$$

Shortest wavelength for Lyman series:

$$\frac{1}{\lambda_L} = Rz^2\left(\frac{1}{1^2} - \frac{1}{\infty}\right) \quad \text{...(ii)}$$

Dividing eq. (i) and (ii),

$\lambda_B : \lambda_L = 4 : 1$

24. **(16)** Radius of H-atom is,

$$r = 0.529\frac{n^2}{Z} \Rightarrow 8.48 = 0.529\frac{n^2}{1}$$

$\Rightarrow n^2 = 16 \Rightarrow n = 4$

Energy of orbit is,

$$E = \frac{-13.6Z^2}{n^2}\text{eV} \Rightarrow E \propto \frac{1}{n^2}$$

$$\Rightarrow E' = \frac{E}{16} = \frac{E}{x}\text{(given)}$$

$\therefore x = 16$

25. **(d)** Longest wavelength in Paschen's series occurs for $n_2 = 4$ to $n_1 = 3$

$$\frac{1}{\lambda} = R\left[\frac{1}{n_1^2} - \frac{1}{n_2^2}\right]$$

$$\frac{1}{\lambda} = R\left[\frac{1}{(3)^2} - \frac{1}{(4)^2}\right] = R\left[\frac{16-9}{16 \times 9}\right]$$

$$\Rightarrow \lambda = \frac{16 \times 9}{7R} = \frac{16 \times 9}{7 \times 1.097 \times 10^7} = 1.876 \times 10^{-6}\text{ m}$$

26. **(122)** Energy of H-atom, $E = \frac{hc}{\lambda}$

$$\Rightarrow 10.2\text{ eV} = \frac{hc}{\lambda} = \frac{1245\text{ nm} - \text{eV}}{\lambda}$$

$$\lambda = \frac{1245\text{ eV} - \text{nm}}{10.2\text{eV}} = 122.06\text{nm}$$

27. **(d)**

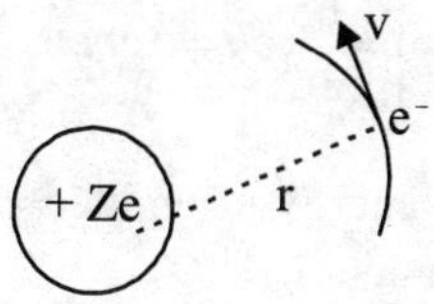

For electron revolution, $F_{centripetal} = \frac{mv^2}{r}$

$$\Rightarrow \frac{KZe^2}{r^2} = \frac{mv^2}{r} \Rightarrow mv^2 = \frac{KZe^2}{r}$$

$$\therefore K = \frac{1}{2}mv^2 = \frac{KZe^2}{2r} \text{ and } U = -\frac{KZe^2}{r}$$

$$\therefore E = K + U = -\frac{KZe^2}{2r}$$

$$\therefore \frac{E}{U} = \frac{1}{2}$$

28. **(a)** Centripetal force, $F_C = \frac{mv^2}{r}$

$$\Rightarrow \frac{Kq_1q_2}{r^2} = \frac{mv^2}{r} \Rightarrow mv^2r^2 = Kq_1q_2r$$

$$\Rightarrow \frac{m^2v^2r^2}{m} = Kq_1q_2r$$

$$\therefore L \propto \sqrt{r} \quad [\because L = mvr]$$

29. **(6588)** Lyman Series: $n_1 = 1, n_2 = 2, 3 \ldots\ldots \infty$

For shortest wavelength, $n_1 = 1, n_2 = \infty$

$$\frac{1}{\lambda_{shortest}} = R_H\left(\frac{1}{n_1^2} - \frac{1}{n_2^2}\right)$$

$\therefore R_H = \dfrac{1}{915}$

Balmer Series : $n_1 = 2, n_2 = 3, 4 \ldots\ldots \infty$

For longest wavelength, $n_1 = 2, n_2 = 3$

$$\therefore \frac{1}{\lambda_{longest}} = R_H\left(\frac{1}{2^2} - \frac{1}{3^2}\right)$$

$$\Rightarrow \lambda_{longest} = \frac{36 \times 915}{5} = 6588A°$$

30. **(7)** As per question, $\Delta E = 13.6\left(\dfrac{1}{n_1^2} - \dfrac{1}{n_2^2}\right)$

or $\quad \Delta E = 13.6\left(\dfrac{1}{2^2} - \dfrac{1}{3^2}\right) = \dfrac{68}{36}\text{ eV} = 1.9\text{ eV}$

Here due to recoil, the energy of emitted photon will decrease by very small amount.

$$\therefore \%\text{ change in } \lambda = \frac{\lambda' - \lambda}{\lambda} \times 100\%$$

$$= \frac{\Delta E}{C^2 \times 2m} \times 100$$

$$= \frac{1.9 \times 1.6 \times 10^{-19} \times 100}{(3 \times 10^8)^2 \times 2 \times 1.6 \times 10^{-27}} = 1.05 \times 10^{-7}\%$$

$= 1 \times 10^{-n}\%$ (given) $\quad \therefore n = 7$

31. **(c)** Angular momentum as per Bohr's second postulate,

$$L = mvr = \frac{nh}{2\pi} = \frac{4h}{2\pi} = \frac{2h}{\pi}$$

32. **(c)** Energy in balmer series is lower for 3 to 2 transition

$$E = 13.6\left(\frac{1}{n_1^2} - \frac{1}{n_2^2}\right) = 13.6\left(\frac{1}{4} - \frac{1}{9}\right) = 1.9\text{eV}$$

33. **(32)** Frequency is given as:

$$f = K\left(\frac{1}{n_f^2} - \frac{1}{n_i^2}\right)$$

For $n_i = 2$ to $n_f = 1$

$$f_1 = K\left(\frac{1}{1} - \frac{1}{4}\right) \Rightarrow 3 \times 10^{15} = \frac{3}{4}K \qquad \therefore K = 4 \times 10^{15}$$

For $n_i = 3$ to $n_f = 1$

$$f_2 = K\left(\frac{1}{1} - \frac{1}{9}\right) \Rightarrow f_2 = 4 \times 10^{15} \times \frac{8}{9} = \frac{32}{9} \times 10^{15}\text{ Hz}$$

$= \dfrac{x}{9} \times 10^{15}$ Hz (given) $\therefore x = 32$

34. **(b)** For first member of Lyman series, $n_1 = 1, n_2 = 2$

$$\therefore \frac{1}{\lambda} = \frac{13.6z^2}{hc}\left[\frac{1}{1^2} - \frac{1}{2^2}\right] \qquad \ldots\ldots \text{(i)}$$

For second member of Lyman series $n_1 = 1, n_2 = 3$

$$\therefore \frac{1}{\lambda'} = \frac{13.6z^2}{hc}\left[\frac{1}{1^2} - \frac{1}{3^2}\right] \qquad \ldots\ldots \text{(ii)}$$

On dividing (i) and (ii)

$$\lambda' = \frac{27}{32}\lambda$$

35. **(c)** For orbiting electron

$$K.E = \frac{|PE|}{2}$$

$$\therefore \frac{KE}{|PE|} = \frac{1}{2}$$

36. **(6)** Energy of excited state of hydrogen atom given as

$$E_n = -\frac{13.6}{n^2} = -0.85$$

$$n^2 = \frac{13.6}{8.5} = 16 \Rightarrow n = 4$$

No. of transition

$$= \frac{n(n-1)}{2} = \frac{4(4-1)}{2} = 6$$

37. **(b)** Magnetic moment = current × Area

$$\mu_n = \frac{ev}{2\pi r} \times \pi r^2$$

$$\propto \left(\frac{1}{n}\right)n^2 \qquad \left(\because v \propto \frac{1}{n}\right)\left(\because r \propto n^2\right)$$

$\mu_n \propto n \propto n^x$ (given) $\therefore x = 1$

38. **(17)** Using $\Delta E = -13.6\left(\dfrac{1}{n_2^2} - \dfrac{1}{n_1^2}\right)$eV

$n_1 = 1, n_2 = 2$

$$\Delta E = -13.6\left(\frac{1}{4} - 1\right)$$

$\Delta E = 10.2$ eV

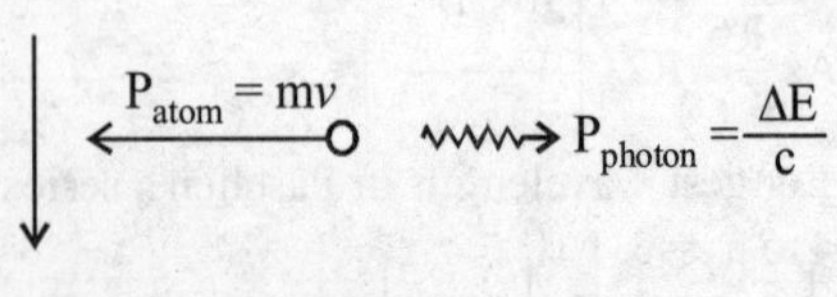

By conservation of momentum,

$$P_{atom} = P_{photon}$$

$$\Rightarrow mv = \frac{\Delta E}{c}$$

$$\Rightarrow v = \frac{\Delta E}{mc} = \frac{10.2 \times 1.6 \times 10^{-19}}{1.6 \times 10^{-27} \times 3 \times 10^{8}} = \frac{17}{5} m/s$$

$$= \frac{x}{5} m/s \text{ (given)}$$

$\therefore$ x = 17

39. (121) For minimum potential difference electron has to make transition from n = 3 to n = 2 state to observe a Balmer series, but first electron has to reach ground state (n = 1) to n = 3.

$\therefore$ Energy of bombarding electron is

$$\Delta E = 13.6\left(\frac{1}{n_1^2} - \frac{1}{n_2^2}\right) eV = 13.6\left(1 - \frac{1}{9}\right) eV \simeq 12.1 eV$$

$\therefore$ The minimum potential difference,

$$V = \frac{\Delta E}{e} = 12.1V = \frac{121}{10}V = \frac{\alpha}{10}V \text{ (given)}$$

$\therefore \alpha = 121$

40. (b) Radius, $r \propto \frac{n^2}{Z}$ where n = no. of orbit and Z = atomic number

$$\therefore \frac{r_4}{r_3} = \frac{4^2}{3^2} \Rightarrow r_4 = \frac{16}{9}R$$

41. (144) Longest wavelength in Paschen series corresponds to transition between n = 3 and n = 4

$$\frac{1}{\lambda} = RZ^2\left(\frac{1}{3^2} - \frac{1}{4^2}\right) = RZ^2\left(\frac{1}{9} - \frac{1}{16}\right) = \frac{7RZ^2}{9 \times 16}$$

$$\Rightarrow \lambda = \frac{144}{7R} = \frac{\alpha}{7R} \text{ (given) for } Z = 1 \quad \therefore \alpha = 144$$

42. (5)

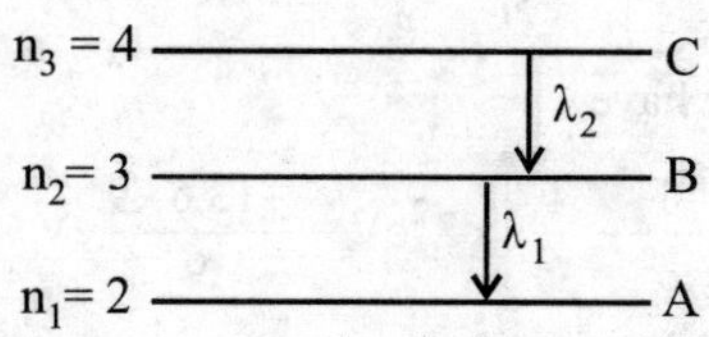

$$\text{As } \frac{1}{\lambda} = RZ^2\left[\frac{1}{n_1^2} - \frac{1}{n_2^2}\right]$$

$$\frac{1}{\lambda_1} = R(1)^2\left[\frac{1}{(2)^2} - \frac{1}{(3)^2}\right] = R\left(\frac{5}{36}\right) \quad \text{... (i)}$$

$$\frac{1}{\lambda_2} = R(1)^2\left[\frac{1}{(3)^2} - \frac{1}{(4)^2}\right] = R\left(\frac{7}{144}\right) \quad \text{... (ii)}$$

Divide equation (ii) by (i)

$$\frac{\lambda_1}{\lambda_2} = \frac{7/144}{5/36} = \frac{7}{20} = \frac{7}{4n} \text{ (given)} \quad \therefore n = 5$$

43. (2) Radius $r \propto \frac{n^2}{z}$

Where n = no. of orbit and z = atomic number

$$\frac{r_{He^+}}{r_{Be^{3+}}} = \frac{2^2 \times 4}{2 \times 4 \times 4} = \frac{1}{2} \quad \therefore \frac{r_2}{r_1} = \frac{2}{1} = \frac{x}{1} \text{ (given)}$$

$\therefore x = 2$

44. (c) If we assume electron in hydrogen atom takes energy 12.09 eV from the incoming radiation, the maximum excited state of electron will be n = 3. So, number of spectral lines is $\frac{3(3-1)}{2} = 3$.

Here we assume some part of energy 12.5eV – 12.09 eV – 0.41 eV get lost due to collision.

45. (3) As number of spectral lines emitted is 6

$$\frac{n(n-1)}{2} = 6 \Rightarrow n^2 - n - 12 = 0$$

$$\Rightarrow (n-4)(n+3) \Rightarrow n = 4$$

For ground state, n = 1

$$\therefore \Delta E = E_4 - E_1 \Rightarrow h\nu = 13.6\left(\frac{1}{1^2} - \frac{1}{4^2}\right)$$

$$\therefore \nu = 13.6\left(\frac{1}{1^2} - \frac{1}{4^2}\right) \times \frac{1}{4.25 \times 10^{-15}} = 3 \times 10^{15} Hz$$

46. (c) For first excited state, n = 2

$$\text{Energy } E_n = \frac{-13.6Z^2}{n^2}$$

$$\therefore E_2 = \frac{-13.6 \times 4}{4} = -13.6eV$$

47. (c) From the Bohr's quantization rule

$$\text{Angular momentum, } L = \frac{nh}{2\pi} \Rightarrow L_1 = \frac{1.h}{2\pi} = L$$

(n = 1 for first orbit)

$$L_2 = \frac{2h}{2\pi} = 2L \text{ (n = 2 for second orbit)}$$

Hence, change in angular momentum = $L_2 - L_1$ = 2L – L = L

48. **(27)** For H-atom

$$\frac{1}{\lambda}=R\left[\frac{1}{{n_1}^2}-\frac{1}{{n_2}^2}\right]$$

For Balmer series, $n_1=2$

$$\frac{1}{\lambda}=R\left[\frac{1}{4}-\frac{1}{{n_2}^2}\right]$$

For H_α, $n_2=3$

$$\frac{1}{\lambda_{H_\alpha}}=R\left[\frac{1}{4}-\frac{1}{9}\right]=\frac{5R}{36} \quad ...(i)$$

For H_β, $n_2=4$

$$\frac{1}{\lambda_{H_\beta}}=R\left[\frac{1}{4}-\frac{1}{16}\right]=\frac{3R}{16} \quad ...(ii)$$

Divide equation (i) and (ii)

$$\frac{1/\lambda_{H_\alpha}}{1/\lambda_{H_\beta}}=\frac{5R/36}{3R/16}\Rightarrow\frac{\lambda_{H_\alpha}}{\lambda_{H_\beta}}=\frac{27}{20}=\frac{x}{20} \text{ (given)}$$

The value of x is 27.

49. **(b)** $\Delta E=\frac{hc}{\lambda}\Rightarrow\lambda\propto\frac{1}{\Delta E}$

For minimum wavelength, energy gap must be maximum.

ΔE is maximum for transition from n = 3 to n = 1.

50. **(425)** Given,

Radius of hydrogen atom, $r_0 = 0.51$Å

For fifth orbit, n = 5

$$r_n = r_0\frac{n^2}{z}\Rightarrow r_n = 0.51\times\frac{25}{3}\text{Å} = 4.25\times10^{-10}\text{m}$$

$= 425\times10^{-12}$ m

51. **(16)** Binding energy of system is given by

$$\Rightarrow\frac{ke^2}{2r}=12.8\text{eV}$$

$$\Rightarrow\frac{9\times10^9\times\left(1.6\times10^{-19}\right)^2}{2r}=12.8\times1.6\times10^{-19}$$

$$\left(\because K=\frac{1}{4\pi\varepsilon_0}=9\times10^9\ \text{Nm}^2/\text{C}^2 \text{ and electronic charge} = 1.6\times10^{-19}\text{C}\right)$$

$$\Rightarrow r=\frac{9\times10^9\times1.6\times10^{-19}}{12.8\times2}\Rightarrow r=\frac{9\times10^{-10}}{16}\text{m}$$

$$=\frac{9}{x}\times10^{-10}\text{m (given)}$$

The value of x is 16.

52. **(828)** In the ground state energy = – 13.6 eV

Total energy = ground state energy + incident energy

$$\frac{-13.6\text{eV}}{n^2}=-13.6+12.75\Rightarrow\frac{-13.6\text{eV}}{n^2}=-0.85$$

$$\Rightarrow n^2=16\Rightarrow n=4$$

$$\text{Angular momentum}=\frac{nh}{2\pi}=\frac{4h}{2\pi}=\frac{2h}{\pi}$$

$$\text{Angular momentum}=\frac{2}{\pi}\times4.14\times10^{-15}=\frac{828\times10^{-17}}{\pi}\text{eVs}$$

53. **(d)** Energy released = $E_4 - E_2$

$$=-13.6\left(\frac{1}{4^2}-\frac{1}{2^2}\right)\times Z^2\text{eV}$$

$$=-13.6\left(\frac{1}{16}-\frac{1}{4}\right)\times16\text{eV}$$

$$=-13.6\left(-\frac{12}{64}\right)\times16\text{eV}$$

$$=13.6\times\frac{3}{16}\times16\text{eV}=40.8\text{eV}$$

54. **(27)** We have,

$$\frac{1}{\lambda}=R_H\left(\frac{1}{n_1^2}-\frac{1}{n_2^2}\right)$$

$$\text{So, }\frac{1}{\lambda_1}=R_H\left(\frac{1}{1^2}-\frac{1}{3^2}\right)=R_H\left(\frac{8}{9}\right)\Rightarrow\lambda_1=\frac{9}{8R_H}$$

$$\frac{1}{\lambda_2}=R_H\left(\frac{1}{1^2}-\frac{1}{2^2}\right)=R_H\left(\frac{3}{4}\right)\Rightarrow\lambda_2=\frac{4}{3R_H}$$

$$\therefore\frac{\lambda_1}{\lambda_2}=\frac{\frac{9}{8R_H}}{\frac{4}{3R_H}}=\frac{27}{32}$$

55. **(b)** We have

$$R_2=R\Rightarrow2^2R_1=R\Rightarrow R_1=\frac{R}{4}$$

$$R_3=R_1\times3^2=9R_1=\frac{9R}{4}=2.25R$$

56. **(136)** We have,

$$\Delta E=-13.6\left[\frac{1}{\infty^2}-\frac{1}{3^2}\right]\times Z^2\text{eV}=\frac{+13.6\times9}{9}\text{eV}$$

$$=13.6\text{ eV}=136\times10^{-1}\text{ eV}$$

57. **(d)** $v_n\propto\frac{Z}{n}$ (for hydrogen atom, Z = 1)

$$\therefore v_n\propto\frac{1}{n}\quad\therefore\frac{v_3}{v_7}=\frac{7}{3}$$

Speed of electron in 3rd orbit will be

$\therefore v_3 = \frac{7}{3} v_7 \quad (\because v_7 = 3.6 \times 10^6 \text{ m/s})$

$= \frac{7}{3} \times 3.6 \times 10^6 \text{ m/s} = 8.4 \times 10^6 \text{ m/s}$

58. **(27)** Second excited state → first excited state
$n = 3 \rightarrow n = 2$

$$\frac{hc}{\lambda_0} = 13.6\left(\frac{1}{2^2} - \frac{1}{3^2}\right) \quad ...(i)$$

Third excited state → second orbit
$n = 4 \rightarrow n = 2$

$$\frac{hc}{\lambda'} = 13.6\left(\frac{1}{2^2} - \frac{1}{4^2}\right) \quad \left(\because \lambda' = \frac{20\lambda_0}{x}\right)$$

$$\frac{hc}{(20\lambda_0/x)} = 13.6\left(\frac{1}{2^2} - \frac{1}{4^2}\right) \quad ...(ii)$$

Divide equation (ii) by (i),

$$\frac{x}{20} = \frac{\frac{1}{2^2} - \frac{1}{4^2}}{\frac{1}{2^2} - \frac{1}{3^2}} = \frac{\frac{1}{4} - \frac{1}{16}}{\frac{1}{4} - \frac{1}{9}} = \frac{\frac{4-1}{16}}{\frac{9-4}{36}} = \frac{\frac{3}{16}}{\frac{5}{36}} \quad \therefore x = 27$$

59. **(d)** As $E(\text{eV}) = \frac{1240}{\lambda\,(\text{nm})} = \frac{1240}{124.1} \simeq 10 \text{ eV}$

For only transition (D), the energy gap is 10 eV
So, option (d) is correct

60. **(a)** $E_{photon} = E_4 - E_1 = \frac{hc}{\lambda} = -13.6\left(\frac{1}{4^2} - \frac{1}{1^2}\right)$

$\frac{hc}{\lambda} = \left[1 - \frac{1}{16}\right](13.6)$ So, $\lambda = 94.1$ nm

61. **(a)** From Bohr's atomic model,

$$E_n = \frac{-13.6}{n^2}\text{eV} \Rightarrow \frac{E_2 - E_1}{E_\infty - E_1} = \frac{13.6\left(1 - \frac{1}{4}\right)}{13.6} = \frac{3}{4}$$

62. **(5)** For first line of Lyman series

$$\frac{1}{\lambda} = R\left(1 - \frac{1}{4}\right) = R\left(\frac{3}{4}\right) \Rightarrow \lambda = \frac{4}{3R} \quad ..... (i)$$

For 3rd line of Paschen series

$$\frac{1}{\lambda_3} = R\left(\frac{1}{3^2} - \frac{1}{6^2}\right) = \frac{R}{9} \times \frac{3}{4}$$

For 2nd line of Balmer series

$$\frac{1}{\lambda_2} = R\left(\frac{1}{2^2} - \frac{1}{4^2}\right) = \frac{R}{4} \times \frac{3}{4}$$

Thus, $a\lambda = \lambda_3 - \lambda_2 = \frac{12}{R} - \frac{16}{3R} = \frac{20}{3R}$

Putting in equation (i) we get

$$a\left(\frac{4}{3R}\right) = \frac{20}{3R} \Rightarrow a = 5$$

63. **(a)** By Bohr' quantization rule

$$L_n = \frac{nh}{2\pi} \Rightarrow mvr = \frac{nh}{2\pi} \Rightarrow mv = \frac{nh}{2\pi r} \Rightarrow p = \frac{nh}{2\pi r}$$

64. **(5)** When electron moves from $n_2 \rightarrow n_1$ in hydrogen atom

$$\Delta E = \left|13.6\left(\frac{1}{n_1^2} - \frac{1}{n_2^2}\right)\right|$$

$$\frac{13.6\left(\frac{1}{2^2} - \frac{1}{3^2}\right)}{13.6\left(\frac{1}{2^2} - 0\right)} = \frac{x}{x+4} \Rightarrow \frac{\frac{1}{4} - \frac{1}{9}}{\frac{1}{4}} = \frac{x}{x+4}$$

$$\Rightarrow \frac{5}{9} = \frac{x}{x+4}$$

$\Rightarrow 5x + 20 = 9x$

$\Rightarrow 4x = 20 \Rightarrow x = 5$

65. **(b)** $E_1 = \frac{-Rch}{(1)^2}$

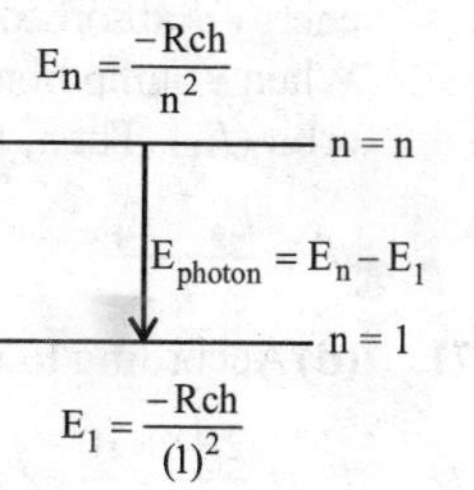

$$\Rightarrow \frac{-Rch}{(n)^2} + \frac{Rch}{1} = \frac{hc}{\lambda}$$

$$\Rightarrow \frac{-R}{n^2} + R = \frac{1}{\lambda}$$

$$\Rightarrow R - \frac{1}{\lambda} = \frac{R}{n^2} \Rightarrow \frac{\lambda R - 1}{\lambda} = \frac{R}{n^2}$$

$$n^2 = \frac{\lambda R}{\lambda R - 1} \Rightarrow n = \sqrt{\frac{\lambda R}{\lambda R - 1}}$$

66. **(286)** As, $E_n = \frac{P.E_n}{2} = -\frac{1.08}{2} = -0.544$

So, $\Delta E = E_f - E_i = -0.544 - \left(-\frac{13.6}{2^2}\right) = 3.4 - 0.544$

$\approx 2.86eV = 286 \times 10^{-2} eV$

67. **(b)** Change in energy
$\Delta E = E_f - E_2$
$\Rightarrow 12.09 = E_f - (-13.6)$
$\Rightarrow E_f = -1.51\text{eV}$

$$\Rightarrow \frac{-13.6}{n^2} = -1.51$$

$\Rightarrow n^2 = 9 \Rightarrow n = 3$

So, $\Delta L = L_f - L_i$

$$= \frac{h}{2\pi}(3-1) = \frac{2h}{2\pi} = \frac{h}{\pi} = \frac{6.63\times10^{-34}}{3.14} = 2.11\times10^{-34}\text{ Js}$$

68. **(114)** We have

$$\Delta E = 13.6\left(1-\frac{1}{3^2}\right)\times Z^2$$

$$\Rightarrow \Delta E = 13.6\times\frac{8}{9}\times9 \Rightarrow \frac{12400}{\lambda(\text{in } \mathring{A})} = 13.6\times 8eV$$

$$\Rightarrow \lambda(\text{in } \mathring{A}) = \frac{12400}{13.6\times8} \simeq 114\mathring{A}$$

69. **(b)** $\Delta E = -13.6\left(\frac{1}{n^2}\right) - \left(-\frac{13.6}{1^2}\right)$

$$\Rightarrow 10.2 = -13.6\left(\frac{1}{n^2}-1\right) \Rightarrow 10.2 = 13.6\left(1-\frac{1}{n^2}\right)$$

$\Rightarrow n = 2$ So, $L_{n=2} = \frac{2h}{2\pi}$ and $L_{(n=1)} = \frac{h}{\pi}$

Then, $\Delta L = L_{(n=2)} - L_{(n=1)} = \frac{2h}{2\pi} - \frac{h}{2\pi}$

$$= \frac{h}{2\pi} = 1.05\times10^{-34}\text{ J-s}$$

70. **(d)** When e^- jumps from lower to higher energy orbit, energy is absorbed not emitted. So, statement I is incorrect. When e^- jump from higher energy orbit (E_2) to lower energy orbit (E_1). Then, frequency of emitted photon is

$$f = \frac{E_2 - E_1}{h}$$

71. **(d)** According to Bohr, speed of e^- in n^{th} orbit is given by

$$V_n = \frac{2.18\times10^6}{n}\times Z \;\therefore\; V_n \propto Z$$

$$\frac{(V_3)_{He^+}}{(V_3)_H} = \frac{Z_{He^+}}{Z_H} = \frac{2}{1}$$

72. **(b)** $E_n = -\frac{13.6}{n^2}$, $P.E_n = 2\times -\frac{13.6}{n^2}$ and $K.E_n = \frac{13.6}{n^2}$

Clearly, when n increases

'E' increases, *'PE'* increases and *'K.E'* decreases.

73. **(c)** Initially, potential energy of H^+ ion & electron is zero due to large separation.

Initially

Energy = P.E + K,E = 0 + 2.6ev = 2.6ev

Final energy of hydrogen atom in 1st exited state = –3.4 eV

Energy lost = 2.6 – (–3.4) = 6eV

which is emitted as photon of wavelength

$$\lambda = \frac{12400}{6} = 2060\text{Å}$$

$$\therefore\; v = \frac{3\times10^8}{2060\times10^{-10}} = 1.45\times10^{15}\text{ Hz} = 1.45\times10^9\text{ MHz}$$

74. **(15)** Number of wavelengths possible $= \frac{n(n-1)}{2}$

$$= \frac{6\times(6-1)}{2} = \frac{6\times5}{2} = 15$$

75. **(d)** $v_0 = cRz^2\left(\frac{1}{n_1^2}-\frac{1}{n_2^2}\right)$

$$v_1 = cRz^2\left(\frac{1}{1^2}-\frac{1}{3^2}\right) \quad ...(i)$$

Put $v_1 = 2.92\times10^{15}$ Hz

$$v_2 = cRz^2\left(\frac{1}{1^2}-\frac{1}{2^2}\right) \quad ...(ii)$$

Dividing (ii) by (i), we get $v_2 = 2.46\times10^{15}$ Hz

76. **(112)** Given,

Radius of circular orbit, r = 0.5Å

Speed of electron, $v = 2.2\times10^6$ m/s

Current, $i = \frac{q}{t} = \frac{q}{2\pi r/v}$ $\quad\left(\because v = \frac{2\pi r}{t}\right)$

$$= \frac{1.6\times10^{-19}\times2.2\times10^6}{2\pi\times0.5\times10^{-10}} = 1.12\text{ mA} = 112\times10^{-2}\text{ mA}$$

77. **(b)** $E_n = -\frac{mZ^2e^4}{8\epsilon_0 n^2h^2}$

$$E_n \propto m \Rightarrow \frac{I.E_2}{I.E_1} = \frac{m_2}{m_1} \Rightarrow I.E_2 = \frac{207m_e}{m_e}\times I.E_1$$

$\Rightarrow I.E_2 = 207\times(13.6) = 2815.2$ eV

78. **(3)** Given : $U(r) = U_0r^4$

And from $F(r) = -\frac{dU(r)}{dr} = \frac{d(U_0r^4)}{dr} = -4U_0r^3$

$$\frac{mv_n^2}{r_n} = 4U_0r_n^3 \Rightarrow v \propto r_n^2 \quad ...(i)$$

From Bohr's postulate,

$$mvr_n = n\frac{h}{2\pi} \quad ...(ii)$$

From equations (i) and (ii),

$$r_n \propto n^{\frac{1}{3}} \propto n^{\frac{1}{\alpha}} \quad \therefore \alpha = 3.$$

79. **(15)** For Balmer series, $\frac{1}{\lambda} = R\left(\frac{1}{2^2}-\frac{1}{n^2}\right)$ where $n = 3, 4, 5$.

For first line, $\frac{1}{\lambda_1} = R\left(\frac{1}{4}-\frac{1}{9}\right) = \frac{5R}{36}$

For second line, $\frac{1}{\lambda_2} = R\left(\frac{1}{4}-\frac{1}{16}\right) = \frac{12R}{64}$

For third line, $\frac{1}{\lambda_3} = R\left(\frac{1}{4} - \frac{1}{25}\right) = \frac{21R}{100}$

The required ratio

$\Rightarrow \frac{\lambda_1}{\lambda_3} = \frac{21R}{100} \times \frac{36}{5R} = 1.512 = 15 \times 10^{-1}$

80. **(c)** The wavelength of spectral line of the third member of Lyman series is given by

$\frac{1}{\lambda_1} = R\left[\frac{1}{1} - \frac{1}{16}\right] \Rightarrow \lambda_1 = \frac{16}{15R}$ ...(i)

The wavelength of spectral line of the first member of Paschen series is given by

$\frac{1}{\lambda_2} = R\left[\frac{1}{9} - \frac{1}{16}\right] \Rightarrow \lambda_2 = \frac{144}{7R}$ ...(ii)

Dividing (ii) by (i) we get

$\frac{\lambda_1}{\lambda_2} = \frac{16}{15R} \times \frac{7R}{144} = \frac{7}{135}$

81. **(c)** Momentum, $P = \frac{\Delta E}{C}$

$\Rightarrow mv = \frac{\Delta E}{c} \Rightarrow v = \frac{\Delta E}{mc} = \frac{\left(\frac{13.6}{1^2} - \frac{13.6}{5^2}\right) \times 1.6 \times 10^{-19}}{1.67 \times 10^{-27} \times 3 \times 10^8}$

$= \frac{13.06 \times 1.6 \times 10^{-19}}{3 \times 10^8 \times 1.67 \times 10^{-27}}$ $\therefore v \simeq 4.17$ m/s

82. **(b)** From $\frac{1}{\lambda} = R\left[\frac{1}{n_2^2} - \frac{1}{n_1^2}\right]$

$\Rightarrow \frac{1}{\lambda} = R\left[\frac{1}{1^2} - \frac{1}{2^2}\right] = R \times \left(\frac{3}{4}\right)$

$\Rightarrow \lambda = \frac{4}{3R} = \frac{4}{3 \times 1.09 \times 10^7} = 121.8$ nm

83. **(c)** A corresponds to transition from $n = \infty$ to $n = 1$.
It is a series limit of Lymen series.
B corresponds to transition from $n = 5$ to $n = 2$.
It is a third member of Balmer series.
C corresponds to transition from $n = 5$ to $n = 3$, thus, it is second member of Paschen series.

84. **(b)** Highest frequency transition will be of maximum energy difference.

Energy difference, $\Delta E = 13.6\left[\frac{1}{n_1^2} - \frac{1}{n_2^2}\right] = h\nu$

Energy difference is maximum if $n_1 = 1$ and $n_2 = 2$.

85. **(51)** $M_H = 931\ \text{MeV/c}^2 \simeq 5\ m_{particle}$

Before collision **After collision**

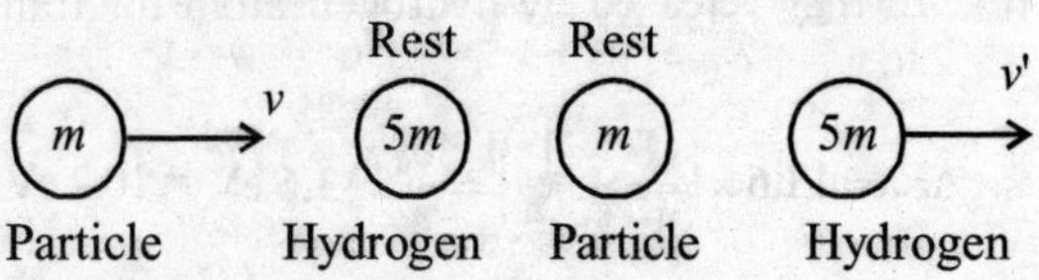

From linear momentum conservation, $P_i = P_f$

$mv + 0 = 0 + 5mv' \Rightarrow v' = \frac{v}{5}$

Loss of KE $= KE_i - KE_f = \frac{1}{2}mv^2 - \frac{1}{2}(5m)\left(\frac{v}{5}\right)^2$

$= \frac{1}{2}mv^2\left(1 - \frac{1}{5}\right) = \frac{4}{5}\left(\frac{mv^2}{2}\right) = \frac{4}{5}KE_i$

Now, Loss in K.E = Gain in P.E

$\Rightarrow \frac{4}{5}K.E_i = 10.2eV$

$K.E_i = 12.75\text{eV} = \frac{N}{4}\text{eV} \Rightarrow N = 51$

The value of $N = 51$.

86. **(10553.14)** From Bohr's formula for hydrogen atom,

$\frac{1}{\lambda} = R\left(\frac{1}{n_1^2} - \frac{1}{n_2^2}\right)$, $R = 1.097 \times 10^7\ \text{m}^{-1}$

For Lyman series :

$\frac{1}{\lambda_{min.}} = R(1) = R$

$\because n_2 = \infty$ and $n_1 = 1$

$\frac{1}{\lambda_{max.}} = R\left\{1 - \frac{1}{4}\right\} = \frac{3R}{4}$ $\because n_1 = 2,\ n_1 = 1$

$\therefore \lambda_{max.} - \lambda_{min.} = \frac{4}{3R} - \frac{1}{R} = \frac{1}{3R} = 304$ (Given)

For Paschen series :

$\lambda'_{min.} = R\left(\frac{1}{9}\right)$ and $\lambda'_{max.} = R\left(\frac{1}{9} - \frac{1}{16}\right) = \frac{7R}{16 \times 9}$

$\lambda'_{max.} - \lambda'_{min.} = \frac{16 \times 9}{7R} - \frac{9}{R} = \frac{81}{7R}$

or, $\lambda'_{max.} - \lambda'_{min.} = \frac{81}{7R} = \frac{81 \times 3}{7 \times 3R} = \frac{81 \times 3}{7} \times 304$

$\left(\because \frac{1}{3R} = 304\text{Å}\right)$

$\therefore$ For Pachen series, $\lambda'_{max.} - \lambda'_{min.} = 10553.14$ Å

87. **(b)** Total energy of electron in n^{th} orbit of hydrogen atom

$E_n = -\frac{Rhc}{n^2}$

Total energy of electron in $(n + 1)^{th}$ level of hydrogen atom

$E_{n+1} = -\frac{Rhc}{(n+1)^2}$

When electron makes a transition from $(n + 1)^{th}$ level to n^{th} level

Change in energy,

$\Delta E = E_{n+1} - E_n$

$h\nu = Rhc \cdot \left[\frac{1}{n^2} - \frac{1}{(n+1)^2}\right]$ $(\because E = h\nu)$

$$\nu = R\cdot c\left[\frac{(n+1)^2-n^2}{n^2(n+1)^2}\right] \Rightarrow \nu = R\cdot c\left[\frac{1+2n}{n^2(n+1)^2}\right]$$

For $n >> 1$

$$\Rightarrow \nu = R\cdot c\left[\frac{2n}{n^2\times n^2}\right] = \frac{2Rc}{n^3}$$

$$\Rightarrow \nu \propto \frac{1}{n^3}$$

88. **(b)** $9R_H = R_H\left(\frac{1}{1^2}-\frac{1}{(\infty)^2}\right)Z^2$

$\Rightarrow \quad Z = 3$

and, $\frac{1}{\lambda} = RZ^2\left[\frac{1}{n_1^2}-\frac{1}{n_2^2}\right] \quad \because (Z=3)$

$$\therefore \quad \frac{1}{\lambda} = 9R\left(1-\frac{1}{9}\right)$$

$$\Rightarrow \quad \lambda = \frac{1}{8R} = \frac{1}{8\times 10973731.6}$$

$\Rightarrow \lambda = 11.39$ nm

89. **(486)** The wavelength of the spectral line of hydrogen spectrum is given by formula

$\frac{1}{\lambda} = R\left(\frac{1}{n_f^2}-\frac{1}{n_i^2}\right)$ Where, R = Rydberg constant

For the first member of Balmer series $n_f = 2, n_i = 3$

$$\therefore \frac{1}{\lambda} = R\left(\frac{1}{2^2}-\frac{1}{3^2}\right) \quad \text{...(i)}$$

For second member of Balmer series, $n_f = 2, n_i = 4$

$$\text{So, } \frac{1}{\lambda'} = R\left[\frac{1}{4}-\frac{1}{16}\right] \quad \text{...(ii)}$$

Dividing (i) by (ii), we get

$$\Rightarrow \frac{\lambda'}{\lambda} = \frac{5\times 16}{9\times 4\times 3}$$

$$\Rightarrow \lambda' = \frac{5\times 4\times 656.1}{9\times 3}(nm) = 486\,nm$$

90. **(b)** For first excited state n = 2

Time period $T \propto n^3$ [$\because r \propto n^2$ and $V \propto \frac{1}{n}$ and $T = \frac{2\pi r}{V}$]

$$\Rightarrow \frac{T_2}{T_1} = \frac{n'^3}{n^3} = \frac{2^3}{1^3}$$

$\therefore T_2 = 8T_1 = 8 \times 1.6 \times 10^{-16}$

$$\therefore \text{Frequency, } \nu = \frac{1}{T_2} = \frac{1}{8\times 1.6\times 10^{-16}} \approx 7.8\times 10^{14}\text{ Hz}$$

91. **(b)** $E = E_1 + E_2$

$$13.6\times(2)^2\times\left(\frac{1}{1^2}-\frac{1}{n^2}\right) = \frac{1240}{108.5}+\frac{1240}{30.4}$$

On solving, n = 5

92. **(a)** $\frac{1}{\lambda_1} = R\left(\frac{1}{3^2}-\frac{1}{4^2}\right) = \frac{7R}{16\times 9}$

And $\frac{1}{\lambda_2} = R\left(\frac{1}{2^2}-\frac{1}{3^2}\right) = \frac{5R}{36}$

Now $\frac{\lambda_1}{\lambda_2} = \frac{(5R/36)}{7R/(16\times 9)} = \frac{20}{7}$

93. **(d)** $v = \frac{c}{137n} = \frac{c}{137\times 3}$

$$\lambda = \frac{h}{p} = \frac{h}{mv} = \frac{h}{\left(\frac{m\times c}{3\times 137}\right)} = \frac{h}{mc}\times(3\times 137) = 9.7\text{ Å}$$

94. **(d)** Spectral lines obtained on account of transition from nth orbit to various lower orbits is $\frac{n(n-1)}{2}$

$$\Rightarrow 6 = \frac{n(n-1)}{2} \Rightarrow n = 4$$

$$\Rightarrow \frac{1}{\lambda} = Z^2\left(\frac{13.6eV}{hc}\right)\left(\frac{1}{n_2^2}-\frac{1}{n_1^2}\right)$$

$$= \frac{(13.4)\times(3)^2}{1242}\left[1-\frac{1}{16}\right]eV$$

$$\Rightarrow \lambda = \frac{1242\times 16}{(13.4)\times(9)(15)}\text{nm} \simeq 10.8\text{nm}$$

95. **(b)** $\frac{1}{\lambda_1} = R\left(\frac{1}{2^2}-\frac{1}{3^2}\right) = \frac{5R}{36}$

$$\frac{1}{\lambda_2} = R\left(\frac{1}{2^2}-\frac{1}{4^2}\right) = \frac{3R}{16}$$

$$\therefore \frac{\lambda_2}{\lambda_1} = \frac{80}{108} \Rightarrow \lambda_2 = \frac{80}{108}\lambda_1 = \frac{80}{108}\times 660 = 488.9\text{ nm.}$$

96. **(c)** $E_n = -13.6\frac{Z^2}{n^2}$

For He^+, $E_2 = \frac{-13.6(2)^2}{2^2} = -13.60$ eV

Ionization energy $= 0 - E_2 = 13.60$ eV

97. **(d)** Energy released by hydrogen atom for transition n = 2 to n = 1

$$\therefore \Delta E_1 = 13.6\times\left(\frac{1}{1^2}-\frac{1}{2^2}\right) = \frac{3}{4}\times 13.6\text{ eV} = 10.2\text{ eV}$$

This energy is absorbed by He^+ ion in transition from n = 2 to n = n_1 (say)

$$\therefore \Delta E_2 = 13.6\times 4\times\left(\frac{1}{4}-\frac{1}{n_1^2}\right) = 10.2\text{ eV}$$

$\Rightarrow n_1 = 4$

So, possible transition is n = 2 → n = 4

98. **(c)** Energy of photon $= \frac{hc}{\lambda} = \frac{12500}{980} = 12.75\text{eV}$

Energy of electron in n^{th} orbit is given by

$$En = \frac{-13.6}{n^2}$$

$$\Rightarrow E_n - E_1 = -13.6\left[\frac{1}{n^2}\frac{-1}{1^2}\right]$$

$$\Rightarrow 12.75 = 13.6\left[\frac{1}{1^2}\frac{-1}{n^2}\right] \Rightarrow n = 4$$

∴ Electron will excite to n = 4

We know that radius, $r \propto n^2$

∴ Radius of atom will be $16a_0$

99. **(d)** When electron jumps from M → L shell

$$\frac{1}{\lambda} = K\left(\frac{1}{2^2}-\frac{1}{3^2}\right) = \frac{K\times 5}{36} \quad \text{..... (i)}$$

When eletron jumps from N → L shell

$$\frac{1}{\lambda'} = K\left(\frac{1}{2^2}-\frac{1}{4^2}\right) = \frac{K\times 3}{16} \quad \text{....(ii)}$$

solving equation (i) and (ii) we get $\lambda' = \frac{20}{27}\lambda$

100. **(a)** Wavelength of emitted photon from n^{th} state to the ground state,

$$\frac{1}{\Lambda_n} = RZ^2\left(\frac{1}{1^2}-\frac{1}{n^2}\right) \Rightarrow \Lambda_n = \frac{1}{RZ^2}\left(1-\frac{1}{n^2}\right)^{-1}$$

Since n is very large, using binomial theorem

$$\Lambda_n = \frac{1}{RZ^2}\left(1+\frac{1}{n^2}\right) \Rightarrow \Lambda_n = \frac{1}{RZ^2}+\frac{1}{RZ^2}\left(\frac{1}{n^2}\right)$$

As we know, $\lambda_n = \frac{2\pi r}{n} = 2\pi\left(\frac{n^2h^2}{4\pi^2 mZe^2}\right)\frac{1}{n} \propto n$

$$\Lambda_n \approx A + \frac{B}{\lambda_n^2}$$

101. **(d)** $h\nu_L = E_\infty - E_1$ ...(i)

$h\nu_f = E_\infty - E_5$ ...(ii)

$$E \propto \frac{z^2}{n^2} \Rightarrow \frac{E_5}{E_1} = \left(\frac{1}{5}\right)^2 = \frac{1}{25}$$

$$\text{Eqn (i)/(ii)} \Rightarrow \frac{h\nu_L}{h\nu_f} = \frac{E_1}{E_5}$$

$$\Rightarrow \frac{\nu_L}{\nu_f} = \frac{25}{1} \Rightarrow \nu_f = \frac{\nu_L}{25}$$

102. **(b)** Energy required to remove e^- from singly ionized helium atom $= \frac{(13.6)Z^2}{1^2} = 54.4\text{ eV}$ $(\because Z = 2)$

Energy required to remove e^- from helium atom = x eV

According to question, 54.4 eV = 2.2x ⇒ x = 24.73 eV

Therefore, energy required to ionize helium atom = (54.4 + 24.73) eV = 79.12 eV

103. **(b)** From energy level diagram, using $\Delta E = \frac{hc}{\lambda}$

For wavelength $\lambda_1, \Delta E = -E-(-2E) = \frac{hc}{\lambda_1}$

$$\therefore \quad \lambda_1 = \frac{hc}{E}$$

For wavelength $\lambda_2, \Delta E = -E-\left(-\frac{4E}{3}\right) = \frac{hc}{\lambda_2}$

$$\therefore \quad \lambda_2 = \frac{hc}{\left(\frac{E}{3}\right)} \qquad \therefore r = \frac{\lambda_1}{\lambda_2} = \frac{1}{3}$$

104. **(c)** Speed of electron in first orbit (n = 1) of hydrogen atom (z = 1), $v = \frac{e^2}{2\varepsilon_0 h}$

radius of Bohr's first orbit,

$$r = \frac{h^2\varepsilon_0}{\pi me^2} \Rightarrow \varepsilon_0 = \frac{r\pi me^2}{h^2} \quad \text{....... (i)}$$

Acceleration of electron,

$$\frac{v^2}{r} = \frac{e^4}{4\varepsilon_0^2 h^2}\times\frac{\pi me^2}{h^2\varepsilon_0}$$

$$= \frac{e^4\times\pi me^2}{4h^4\varepsilon_0^3} \quad \text{....... (ii)}$$

eliminating ε_0 from eq (ii),

$$= \frac{e^4\,\pi me^2 h^6}{4h^4 r^3\pi^3 m^3 e^6} \quad \text{[Using eq}^n\text{ (i)]}$$

$$= \frac{h^2}{4\pi^2 m^2 r^3}$$

105. **(d)** Magnetic field at the centre of nucleus of H-atom,

$$B = \frac{\mu_0 I}{2r} \quad \text{..... (i)}$$

According to Bohr's model, radius of orbit $r \propto n^2$

from eq. (i) we can also write as $B \propto n^{-2}$

106. **(b)** A hydrogen atom makes a transition from n = 2 to n = 1

$$\Delta E = Rhcz^2\left[\frac{1}{n_1^2}-\frac{1}{n_2^2}\right] = Rhc(1)^2\left[1-\frac{1}{4}\right]$$

$$\lambda = Rhc\left[\frac{3}{4}\right] \quad ...(i)$$

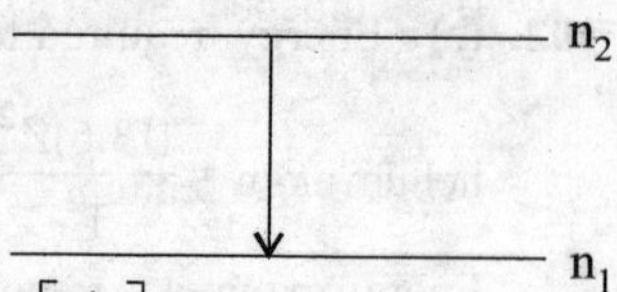

For ionized lithium

$$\Delta E = Rhc(3)^2\left[\frac{1}{n^2}\right] = Rhc9\left[\frac{1}{n^2}\right] \quad ...(ii)$$

From equations (i) and (ii), we get

$$\Rightarrow Rhc\left[\frac{3}{4}\right] = Rc9\left[\frac{1}{n^2}\right]$$

$$\Rightarrow \frac{3}{4} = \frac{9}{n^2} \Rightarrow n = \sqrt{12} = 2\sqrt{3}$$

$\therefore$ The least quantum number must be 4.

107. (c) Kinetic energy of electron is K.E. $\propto \left(\frac{Z}{N}\right)^2$

When the electron makes transition from excited state to ground state, then n decreases and kinetic energy increases.

Total energy = – KE

$\therefore$ Total energy also decreases.

Potential energy is lowest for ground state.

108. (b) de-Broglie wavelength of electron $\lambda = \frac{h}{mV}$

As we know, $V \propto \frac{1}{n}$

So, $\lambda \propto n$

$\lambda_4 = 4\lambda_1$

λ_1 is the de-Broglie wavelength of the electron in the ground state.

109. (b) Radius of circular path followed by electron is given by,

$$r = \frac{mv}{qB} = \frac{\sqrt{2meV}}{eB} \Rightarrow V = \frac{B^2r^2e}{2m} = 0.8V$$

$\therefore$ $(KE)_{max} = eV = 0.8eV$

For transition between 3 to 2.

$$E = 13.6\left(\frac{1}{4}-\frac{1}{9}\right) = \frac{13.6\times5}{36} = 1.88eV$$

Work function = E–$(KE)_{max}$

= 1.88 eV – 0.8 eV = 1.08 eV ≈ 1.1eV

110. (c) Wave number $\frac{1}{\lambda} = RZ^2\left[\frac{1}{n_1^2}-\frac{1}{n^2}\right] \Rightarrow \lambda \propto \frac{1}{Z^2}$

$\therefore \lambda Z^2$ = constant

By question $n_1 = 2$ and $n_2 = 1$

Then, $\lambda_1 = \lambda_2 = 4\lambda_3 = 9\lambda_4$

111. (b) For first excited state, n = 2 and for Li^{++} Z = 3

$$E_n = \frac{13.6}{n^2}\times Z^2 = \frac{13.6}{4}\times 9 = 30.6\text{ eV}$$

112. (d) $\Delta E = h\nu$

$$\nu = \frac{\Delta E}{h} = k\left[\frac{1}{(n-1)^2}-\frac{1}{n^2}\right] = \frac{k(2n-1)}{n^2(n-1)^2} \approx \frac{2k}{n^3} \text{ or } \nu \propto \frac{1}{n^3}$$

113. (a) $E = \frac{hc}{\lambda} \Rightarrow \lambda = \frac{hc}{E} = \frac{6.62\times10^{-34}\times3\times10^8}{12.5\times1.6\times10^{-19}} = 993\,A°$

$$\frac{1}{\lambda} = R\left(\frac{1}{n_1^2}-\frac{1}{n_2^2}\right)$$

(where Rydberg constant, R = 1.097×10^7)

or, $\frac{1}{993\times10^{-10}} = 1.097\times10^7\left(\frac{1}{1^2}-\frac{1}{n_2^2}\right)$

Solving we get $n_2 = 3$

Spectral lines

Total number of spectral lines = 3

Two lines in Lyman series for (n = 3 to n = 1) & (n = 2 to n = 1) and one line in Balmer series for (n = 3 to n = 2).

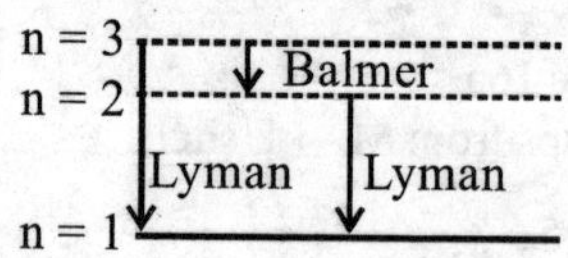

114. (c) According to the question,

$$2\pi r = n\lambda = \frac{nh}{p} = \frac{nh}{mv}$$

$$\text{or } mvr = \frac{nh}{2\pi} \text{ or } mv = \frac{nh}{2\pi r}$$

$$F = qvB = \frac{mv^2}{r} \quad \text{or, } qB = \frac{mv}{r} = \frac{nh}{2\pi r.r}$$

$$\text{or, } r^2 = \frac{nh}{2\pi qB} \quad \text{or, } r = \sqrt{\frac{nh}{2\pi qB}}$$

i.e., $r \propto n^{1/2}$

115. (d) For ground state, the principal quantum no. $(n) = 1$. Principal quantum number 4 belongs to 3rd excited state. The possible number of the spectral lines from a state n to ground state is

$$= \frac{n(n-1)}{2} = \frac{4(4-1)}{2} = 6$$

116. (d) The energy of the system of two atoms of diatomic molecule $E = \frac{1}{2}I\omega^2$, where I = moment of inertia

ω = Angular velocity = $\frac{L}{I}$, L = Angular momentum

$$I = (m_1r_1^2 + m^2r_2^2) = \left(\frac{m_1m_2}{m_1+m_2}\right)r$$

$$\text{Thus, } E = \frac{1}{2}(m_1r_1^2 + m_2r_2^2)\omega^2 \quad ...(i)$$

$$E=\frac{1}{2}(m_1r_1^2+m_2r_2^2)\frac{L^2}{I^2}$$

$L=n\hbar$

(According to Bohr's Hypothesis)

$$E=\frac{1}{2}(m_1r_1^2+m_2r_2^2)\frac{L^2}{(m_1r_1^2+m_2r_2^2)^2}$$

$$\Rightarrow \quad E=\frac{1}{2}\frac{L^2}{(m_1r_1^2+m_2r_2^2)} \Rightarrow E=\frac{(m_1+m_2)n^2\hbar^2}{2r^2m_1m_2}$$

117. (a) Velocity of electron in n^{th} orbit of hydrogen atom is given by : $V_n=\frac{2\pi KZe^2}{nh}$

Substituting the values we get,

$$V_n=\frac{2.2\times10^6}{n}\text{m/s} \quad \text{or} \quad V_n\propto\frac{1}{n}$$

As principal quantum number increases, velocity decreases.

118. (b) If $n_1=n$ and $n_2=(n+1)$ $\Rightarrow \frac{1}{\lambda}=R\left(\frac{1}{n^2}-\frac{1}{(n+1)^2}\right)$

Maximum wavelength $\lambda_{max}=\frac{n^2(n+1)^2}{(2n+1)R}$

Therefore, for large n, $\lambda_{max}\propto n^3$

119. (b) Energy of excitation (ΔE) is

$$\Delta E=13.6\,z^2\left(\frac{1}{n_1}-\frac{1}{n_2}\right)eV$$

$$\Rightarrow \Delta E=13.6\,(3)^2\left(\frac{1}{1^2}-\frac{1}{3^2}\right)=108.8\,eV$$

120. (c) It is given that transition from the state $n=4$ to $n=3$ in a hydrogen like atom result in ultraviolet radiation. For infrared radiation $\left(\frac{1}{n_1^2}-\frac{1}{n_2^2}\right)$ should be less. The only option is $5\rightarrow4$.

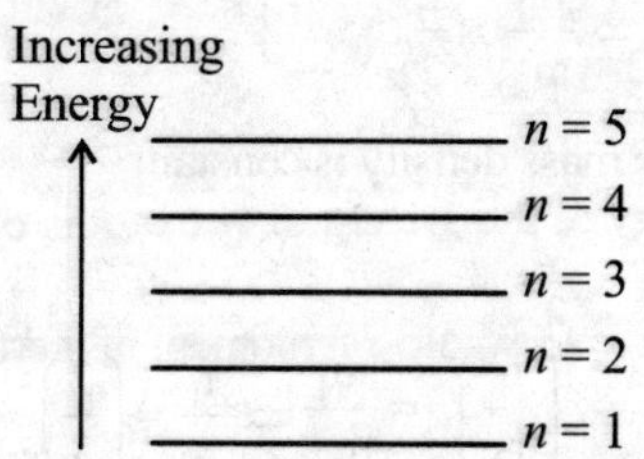

121. (b) Given,

Centripetal force $=\frac{k}{r}$

Then

$$\frac{k}{r}=\frac{mv^2}{r}\Rightarrow k=mv^2$$

$$\Rightarrow \quad T_n=\frac{1}{2}mv^2=\frac{1}{2}k$$

T_n is independent of n

Also,

Angular momentum, $L=\frac{nh}{2\pi}$

$$\Rightarrow \quad mvr_n=\frac{nh}{2\pi} \quad (\because L=mvr)$$

$$\Rightarrow \quad r_n=\frac{nh}{2\pi\sqrt{km}} \quad \left(\begin{matrix}\because m^2v^2=km \\ \text{or } mv=\sqrt{km}\end{matrix}\right)$$

Clearly, $r_n\propto n$

122. (d) We have to find the frequency of emitted photons. For emission of photons electron should makes a transition from higher energy level to lower energy level. so, option (a) and (b) are incorrect.

Clearly, $E_{n2}-E_{n1}>E_{n6}-E_{n2}$

So, option (d) is correct.

123. (b) Eenrgy of radiation that corresponds to energy difference between two energy levels n_1 and n_2 is given as

$$E=Rhc\left[\frac{1}{n_1^2}-\frac{1}{n_2^2}\right] \qquad \therefore E\propto\left(\frac{1}{n_1^2}-\frac{1}{n_2^2}\right)$$

E will be maximum for the transition for which $\left[\frac{1}{n_1^2}-\frac{1}{n_2^2}\right]$ is maximum. Here n_2 is the higher energy level.

Clearly, $\left[\frac{1}{n_1^2}-\frac{1}{n_2^2}\right]$ is maximum for the third transition, *i.e.*, $2\rightarrow1$. I transition is showing the absorption of energy.

124. (c) The wavelength of spectrum is given by

$$\frac{1}{\lambda}=Rz^2\left(\frac{1}{n_1^2}-\frac{1}{n_2^2}\right) \quad \text{where } R=\frac{1.097\times10^7}{1+\frac{m}{M}}$$

where m = mass of electron

M = mass of nucleus.

Thus, wavelength involved in the spectrum of hudrogen like atom depends upon masses of nucleus. The mass number of hydrogen and deuterium is 1 and 2 respectively, so spectrum of deuterium will be different from hydrogen.

125. (c) The energy required to remove the electron from the nth orbit of hydrogen is given by

$$E_n=\frac{13.6}{n^2}\text{ eV/atom}$$

For $n=2$, $E_n=\frac{13.6}{4}=3.4\,eV$

Therefore the energy required to remove electron from $n=2$ is 3.4 eV.

Nuclei

13

1. **(d)** Mass density of nucleus is independent of mass number A.

2. **(b)** $\rho = \frac{\text{Mass of nucleus}}{\text{Volume of nucleus}}$

$$\rho = \frac{M}{V} = \frac{m_n \times A}{\frac{4}{3}\pi R^3} = \frac{m_n \times A}{\frac{4}{3}\pi A R_0^3}$$

This shows that the density is independent of the mass number A.

$\therefore R = R_0 A^{1/3} \Rightarrow R \propto A^{1/3}$

3. **(27)** Radius of nucleus, $R = R_0 A^{1/3}$

$\therefore R^3 \propto A$

$$\left(\frac{R_1}{R_2}\right)^3 = \frac{A_1}{A_2}$$

$$\left(\frac{4.8}{4}\right)^3 = \frac{64}{A} \Rightarrow A = \frac{64}{1.44 \times 1.2} = \frac{1000}{27}$$

on comparing A with $\frac{1000}{x}$ we get $x = 27$.

4. **(a)** Radius of Nucleus, $R = R_0 A^{1/3}$

Here, A = mass number of nuclei

According to question,

$$R_1 = \frac{R_2}{2}$$

$$\Rightarrow R_0 (A_1)^{1/3} = \frac{R_0}{2}(A_2)^{1/3} \Rightarrow A_1 = \frac{192}{8} = 24$$

5. **(4)** Radius of nucleus, $R = R_0 A^{1/3}$

Volume of nucleus having mass number A_1

$$V_1 = \frac{4}{3}\pi R_1^3 = \frac{4}{3}\pi\left(R_0 A_1^{1/3}\right)^3$$

$$= \frac{4}{3}\pi R_0^3 A_1$$

Volume of nucleus having mass number A_2

$$V_2 = \frac{4}{3}\pi R_2^3 = \frac{4}{3}\pi\left(R_0 A_2^{1/3}\right)^3$$

$$\Rightarrow V_2 = \frac{4}{3}\pi R_0^3 A_2 = \frac{4}{3}\pi R_0^3 4A_1 = 4V_1 \quad (\because A_2 = 4A_1 \text{ given})$$

$$\therefore \frac{V_2}{V_1} = 4$$

6. **(2)** As density of two parts is same

$$\frac{m_1}{\frac{4}{3}\pi r_1^3} = \frac{m_2}{\frac{4}{3}\pi r_2^3}$$

$$\Rightarrow \frac{m_1}{m_2} = \frac{r_1^3}{r_2^3} = \frac{1}{2} \quad \left(\because \frac{r_1}{r_2} = \frac{1}{2^{1/3}} \text{ given}\right)$$

Using conservation of linear momentum

$$m_1 v_1 = m_2 v_2$$

$$\frac{v_1}{v_2} = \frac{m_2}{m_1} = \frac{2}{1}$$

7. **(b)** Nuclear density is independent of A. Hence, Assertion is False.

The radius R of nucleus is related to its mass number by $R = R_0 A^{1/3}$ (Here, R_0 = constant)

8. **(c)** As nuclear density $= \frac{Au}{\frac{4}{3}\pi R^3}$. Also, $R = R_0 A^{\frac{1}{3}}$

$\Rightarrow$ Nuclear density $\Rightarrow R^3 = R_0^3 A$

$$= \frac{Au}{\frac{4}{3}\pi R_0^3 A} = \frac{3u}{4\pi R_0^3}$$

$\Rightarrow$ Nuclear density is independent of A.

9. **(2)** Given, $\frac{v_1}{v_2} = \frac{3}{2}$

Apply conservation of momentum

$$m_1 v_1 = m_2 v_2 \Rightarrow \frac{m_1}{m_2} = \frac{2}{3}$$

Since, Nuclear mass density is constant

$\rho_1 = \rho_2$

$$\frac{m_1}{\frac{4}{3}\pi r_1^3} = \frac{m_2}{\frac{4}{3}\pi r_2^3} \Rightarrow \left(\frac{r_1}{r_2}\right)^3 = \frac{m_1}{m_2} \Rightarrow \frac{r_1}{r_2} = \left(\frac{2}{3}\right)^{\frac{1}{3}}$$

$$= \left(\frac{x}{3}\right)^{\frac{1}{3}} \quad \text{(Given)}$$

So, $x = 2$

10. **(11)** Density of nuclei,

$$\rho = \frac{A \times 1.6 \times 10^{-27}}{\frac{4}{3}\pi \times \left(1.5 \times 10^{-15}\right)^3 \times A} = 0.113 \times 10^{18}\ \text{kg/m}^3$$

Density of water $= \rho_w = 10^3\ \text{kg/m}^3$

So, $\frac{\rho}{\rho_w} = 11.3 \times 10^{13}\ \text{kg/m}^3 \approx 11 \times 10^{13}\ \text{kg/m}^3$

11. **(b)** The radius of nucelus is given by $R = R_0 A^{1/3}$

Volume, $V = \frac{4}{3}\pi R^3 = \frac{4}{3}\pi R_0^3 A$

Mass = m × A

$$\text{Density, } (\rho) = \frac{\text{Mass}}{\text{Volume}} = \frac{m \times A}{\frac{4}{3}\pi R_0^3 A} = \frac{m}{\frac{4}{3}\pi R_0^3} \Rightarrow \rho \propto A^0$$

Here m = mass of protons or neutrons
So density is indepoendent of A.

12. **(c)** Radius of nucleus, $R = R_0 A^{\frac{1}{3}}$ ...(i)

$$\text{Density of nucleus} = \frac{\text{Mass of nucleus}}{\text{volume of nucleus}}$$

$\rho = \frac{m \times A}{\frac{4}{3}\pi R^3}$ Here m = mass of proton or neutron

from equation (i), we have

$$\rho = \frac{m \times A}{\frac{4}{3}\pi R_0^3 A} \Rightarrow \rho \propto A^0$$

Hence density of nucleus is independent of mass number.

13. **(b)** As radius of nucleus varies as

$$R = R_0 A^{1/3} \Rightarrow \frac{R}{R_0} = A^{1/3}$$

$$\Rightarrow \ln\frac{R}{R_0} = \frac{1}{3}\ln A$$

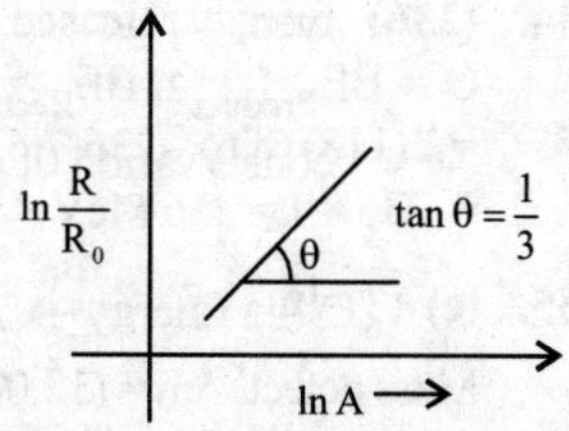

Comparing it with straight line equation $y = mx + c$, we get

$m = \frac{1}{3}$ and $c = 0$.

So, graph will be very much like as shown below.

14. **(d)** Density of nucleus, $\rho = \frac{\text{Mass}}{\text{Volume}} = \frac{mA}{\frac{4}{3}\pi R^3}$

$$\Rightarrow \rho = \frac{mA}{\frac{4}{3}\pi (R_0 A^{1/3})^3} \quad (\because R = R_0 A^{1/3})$$

Here m = mass of a nucleon

$$\therefore \rho = \frac{3 \times 1.67 \times 10^{-27}}{4 \times 3.14 \times (1.3 \times 10^{-15})^3} \quad (\text{Given, } R_0 = 1.3 \times 10^{-15})$$

$$\Rightarrow \rho = 2.38 \times 10^{17}\ \text{kg/m}^3$$

15. **(a)** Nuclear density is independent of atomic number.

16. **(c)** Let heavy nucleus breaks into two nuclei of mass m_1 and m_2 and move away with velocities V_1 and V_2 respectively.

According to question, $\frac{V_1}{V_2} = \frac{8}{27}$

$m_1 V_1 = m_2 V_2$ (Law of momentum conservation)

$$\Rightarrow \frac{m_1}{m_2} = \frac{V_2}{V_1} = \frac{27}{8} = \frac{\rho \times \frac{4}{3}\pi R_1^3}{\rho \times \frac{4}{3}\pi R_2^3} \quad \left(\because \text{density } \rho = \frac{\text{mass}}{\text{volume}}\right)$$

$$\Rightarrow \left(\frac{R_1}{R_2}\right) = \left(\frac{27}{8}\right)^{\frac{1}{3}} = \left(\frac{3}{2}\right)^{3 \times \frac{1}{3}} \quad \therefore \frac{R_1}{R_2} = \frac{3}{2}$$

17. **(b)** Nucleus consist of neutron and proton.

18. **(b)** Radius of a nucleus,

$R = R_0 (A)^{1/3}$

Here, R_0 is a constant
A = atomic mass number

$$\therefore \frac{R_1}{R_2} = \left(\frac{A_1}{A_2}\right)^{1/3} = \left(\frac{27}{125}\right)^{1/3} = \frac{3}{5}$$

$$\Rightarrow R_2 = \frac{5}{3} \times 3.6 = 6 \text{ fermi}$$

19. **(b)** The reaction of neutron with uranium is fission reaction.

$2H_2 + O_2 \rightarrow 2H_2O$ is a chemical reaction.

${}_1^2H + {}_1^2H \rightarrow {}_2^3He + {}_0^1n$ (Fusion with positive Q-value)

${}_1^1H + {}_1^3H \rightarrow {}_1^2H + {}_1^2H$ (Fusion with negative Q-value)

20. **(c)** $\underset{1.1\text{MeV}}{{}_1H^2} + {}_1H^2 \rightarrow \underset{7.0\text{MeV}}{{}_1He^4}$

$E_B = BE_{\text{reactant}} - BE_{\text{product}}$
$= 1.1 \times 2 + 1.1 \times 2 - 7 \times 4 = -23.6$ MeV
$= Q = 23.6$ MeV

21. **(a)** Theoretical equation for β^{-1} decay

$$n \rightarrow p + e^- + \bar{v}$$

22. **(b)** The binding energy per nucleon (BE/A) is a measure of the stability of a nucleus. Nuclear force is a short range force.

23. **(5)** ${}^4He + {}^4He + {}^4He \rightarrow {}^{12}C + Q.$

$\therefore \quad Q = (3m_{He} - m_C)C^2$

$= (3\times 4.0026 - 12)(3\times 10^8)^2$
$= 7.266$ MeV

$\text{Power} = \frac{N}{t}Q$

$\Rightarrow \frac{N}{t} = \frac{\text{Power}}{Q} = \frac{5.808\times 10^{30}}{7.266\times 10^6 \times 1.6\times 10^{-19}}$

$\Rightarrow \frac{N}{t} = 5\times 10^{42} = n\times 10^{42}$ (given)

Hence, n = 5

24. **(c)** For each fusion reaction, ^1_1H nucleus are used.

The energy released per one Nuclei of $^1_1H = \frac{26.7}{4}$ MeV

Number of moles in 2kg of hydrogen $= \frac{2000}{1}$

∴ Energy released in fusion of 2 kg hydrogen (E_H)

$= \frac{2000}{1}\times\frac{26.7}{4}\text{MeV}\times N_A$

∴ Energy released in fission of 2 kg Uranium (E_U)

$= \frac{2000}{235}\times 200\text{MeV}\times N_A$

$\therefore \frac{E_H}{E_U} = 235\times\frac{26.7}{4\times 200} \approx 7.62$

25. **(a)** Using law of conservation of linear momentum

$\vec{p}_i = \vec{p}_f$

$0 = m_1u_1 + m_2u_2 \Rightarrow \frac{u_1}{u_2} = \frac{-m_2}{m_1} = \frac{-1}{2}$

Negative sign shows they move in opposite direction

26. **(b)** Binding energy, $E = \Delta mc^2$

$\Rightarrow 18\times 10^8 \text{ J} = \Delta m\times(3\times 10^8)^2$

$\Rightarrow \Delta m = \frac{18\times 10^8}{9\times 10^{16}}$

$\Rightarrow \Delta m = 2\times 10^{-8}\text{ kg} = 20\ \mu g$

27. **(c)** Atomic number in LHS = 92
Atomic number in RHS = 56 + 36 = 92
Mass number in LHS = 236
Mass number in RHS = 141 + 92 = 233
∴ 236 – 233 = 3 neutrons are released

28. **(a)** Binding energy
$B.E. = \Delta mC^2 = (M_r - M_p)C^2$
$= (5M_p + 7M_n - M_o)C^2$

29. **(737)** $3^4_2He \to ^{12}_6C + Q$

∴ Mass defect, $\Delta M = (3m_{He} - m_c) = (3\times 4.002603 - 12)$
$= 0.007809$u

∴ Energy released, $Q = \Delta MC^2 = 0.007809uc^2$
$= 0.007809\times 931\text{ MeV} = 727\times 10^{-2}$ MeV

30. **(d)** Nuclear fission between neutron $\left(^1_0n\right)$ and uranium isotope $^{235}_{92}U$

$^{235}_{92}U + ^1_0n \to ^{144}_{56}Ba + ^{89}_{36}Kr + 3^1_0n$

31. **(c)** According to Bohr's IInd postulate, angular momentum is an integral multiple of $\hbar$. It is given as :

$L = n\hbar = \frac{nh}{2\pi}$

Central forces are those forces whose magnitude depends upon the distance between objects and the direction of force is along the line joining the objects. Hence nuclear force is a non-central force.
Non-central force depends upon spin and do not obey inverse square law.

32. **(c)** (X) → (Y) + (Z) + (P)
M M/3 M/3 M/3

Total Kinetic energy = ΔMc^2

$\Delta Mc^2 = 3\times\frac{1}{2}\frac{M}{3}v^2 \Rightarrow v = c\sqrt{\frac{2\Delta M}{M}}$

33. **(d)** The chain reaction in hydrogen bomb is

$_3Li^6 + _0n^1 \to _2He^4 + _1H^3$

$_1H^2 + _1H^3 \to _2He^4 + _0n^1$

$_3Li^6 + _1H^2 \to 2(_2He^4)$

Energy released in process
$Q = \Delta mc^2 = (M_{reactant} - M_{product})c^2$
$= [M(Li) + M(_1H^2) - 2\times M(_2He^4)]\times 931.5$ MeV
$= [6.01690 + 2.01471 - 2\times 4.00388]\times 931.5$ MeV
$= 22.22$ MeV

34. **(236)** Energy released in fission reaction,
$Q = BE_{Product} - BE_{Rectant}$
$= 2(118)(8.6) - 236(7.6)$
$= 236\times 1 = 236$ MeV

35. **(c)** $_6C^{13} + \text{Energy} \to _6C^{12} + _0n^1$
Mass defect, $\Delta m = (12.000000 + 1.008665) - 13.003354$
$= 0.005311$ u
∴ Energy required $= \Delta m\times 931.5$
$= 0.005311\times 931.5\text{ MeV} = 4.95$ MeV

36. **(d)** For the given nuclear reaction, amount of energy released $Q = (m_A - m_B - m_D)\times 931.5\ MeV$
$= (238.05079 - 234.04363 - 4.00260)\times 931.5$
or, $Q = 4.25\ MeV$

37. **(d)** The constancy of the binding energy in the range $30 < A < 170$ is a consequence of the fact that the nuclear force is short- ranged.
Binding energy per nucleon is almost same for nuclei of mass number ranging 30 to 170.

38. (a) Surface energy per nucleon,

$$b_s \propto \frac{r^2}{A} \propto \frac{A^{2/3}}{A} \propto \frac{1}{A^{1/3}} = \frac{a_1}{A^{1/3}} \quad (\because r = r_0 A^{1/3})$$

A is incorrect

Contribution to binding energy by columbic forces is given by

$$b_c = \frac{-a_2 Z(Z-1)}{A^{1/3}}$$

B is incorrect

Volume energy, $b_v \propto A$

C is correct

For (D), if we consider only surface energy contribution then option is correct.

For (E) only 3 interactions contribute to surface energy.

39. (121) Initial binding energy = 242 × 7.6 MeV

Final binding enrgy = 2 × 121 × 8.1 MeV
= 242 × 8.1 MeV

Total gain in binding energy = Final binding energy – initial binding energy
= 242 (8.1 – 7.6) = 121 MeV

40. (c) Given, $m_P = 1.0073u$; $m_N = 1.0087u$; $m_{He} = 4.0015u$

B.E of Helium = $(2m_P + 2m_N - m_{He})\, c^2$
$= (2 \times 1.0073 + 2 \times 1.0087 - 4.0015) \times 931.5 = 28.4\,\text{MeV}$

41. (6) Difference in B.E

$$= |n_A \times (B.E \cdot P.N)_A - n_B \times (B.E \cdot P.N)_B|$$

$$= |34 \times 1.2 - 26 \times 1.8|\,\text{MeV} = 6\,\text{MeV}$$

42. (6) No. of mole $(n) = \frac{120}{240} = \frac{1}{2}$

No. of molecules $(N) = \frac{1}{2} \times N_A = \frac{1}{2} \times 6 \times 10^{23}$

Energy released $= \frac{1}{2} \times 6 \times 10^{23} \times 200$
$= 6 \times 10^{25}$ MeV

43. (26) Energy released in the given process = Binding energy of product – Binding energy of reactants
= 7.6 × 4 – (1.1 × 2) × 2
= 30.4 – 4.4 = 26 MeV

44. (d) Energy released per fission
= (105 × 6.4) + (115 × 6.4) – 200 × 5.6 = 176 MeV.

45. (27.16) Given,

Mass of proton, $m_p = 1.00726$ u

Mass of neutron, $m_n = 1.00866$ u

Mass of Aluminium nucleus, $m_{Al} = 27.18846$ u

Mass defect, $\Delta m = (Zm_p + (A - Z)m_n) - M_{Al}$

$\Delta m = [13 \times 1.00726 + 14 \times 1.00866 - 27.18846]$
$= 27.21562 - 27.18846 = 0.02716$

Binding Energy $= \Delta mc^2 = 27.16 \times 10^{-3}$ J

46. (d) Energy of γ ray photon, $E = h\nu$

Momentum, $P = mv = \frac{h}{\lambda} = \left(\frac{h\nu}{c}\right)$

Total momentum is conserved,

$$\vec{P}_\gamma + \vec{P}_{Nu} = 0 \Rightarrow P_\gamma = P_{Nu}$$

$$\Rightarrow \quad P_{Nu} = \frac{h\nu}{c}$$

Kinetic energy of nuclei,

$$\frac{1}{2}mv^2 = \frac{(P_{Nu})^2}{2m}$$

Loss of energy $= \frac{1}{2}mv^2 + h\nu = \frac{1}{2}\frac{(P_{Nu})^2}{m} + h\nu$

$$= \frac{1}{2m}\left(\frac{h\nu}{c}\right)^2 + h\nu \qquad \left(\because (P_{Nu})^2 = \frac{h\nu}{c}\right)$$

$$= h\nu\left(1 + \frac{h\nu}{2mc^2}\right)$$

47. (d) ${}^7_3\text{Li} + {}^1_1\text{H} \longrightarrow 2\left({}^4_2\text{He}\right)$

$\Delta m = [m_{Li} + m_H] - 2[M_{He}]$

Energy released $= \Delta mc^2$

In use of 1 g Li energy released $= \frac{\Delta mc^2}{m_{Li}}$

In use of 20 g energy released $= \frac{\Delta mc^2}{m_{Li}} \times 20$ g

$$= \frac{[(7.016 + 1.0079) - 2 \times 4.0026]u \times c^2}{7.016 \times 1.6 \times 10^{-24}} \times 20\text{ g}$$

$$= \left(\frac{0.0187 \times 1.6 \times 10^{-19} \times 10^9}{7.016 \times 1.6 \times 10^{-24}} \times 20\right) = 480 \times 10^{10}\,\text{J}$$

$\because 1\text{ J} = 2.778 \times 10^{-7}$ kWh

$\therefore$ Energy released $= 480 \times 10^{10} \times 2.778 \times 10^{-7}$
$= 1.33 \times 10^6$ kWh

48. (c) For the momentum and energy conservation, mass defect (Δm) should be positive. Since some energy is lost in every process. $(m_p + m_n) > m_d$

49. (d) Mass defect,

$\Delta m = (50m_p + 70m_n) - (m_{Sn})$
$= (50 \times 1.00783 + 70 \times 1.008) - (119.902199) = 1.096$ U

Binding energy $= (\Delta m)C^2 = (\Delta m) \times 931 = 1020.56$ MeV

$$\frac{\text{Binding energy}}{\text{Nucleon}} = \frac{1020.5631}{120} = 8.5\text{ MeV}$$

50. **(b)** Power output of the reactor,

$$P=\frac{\text{energy}}{\text{time}}$$

$$=\frac{2000}{235}\times\frac{6.023\times10^{23}\times200\times1.6\times10^{-13}}{30\times24\times60\times60}\simeq60\,\text{MW}$$

51. **(None)** The given nuclear fission is:

$Ne^{20}\rightarrow2He^4+C^{12}$

If $(B.E)_f > B.E_i$ (heat is released)

If $(B.E)_f < (B.E)_i$ (heat is supplied)

B.E of Nucleon = 8.03 MeV

So, $(B.E)_i=(8.03)\times20$ MeV

$=16.06\text{ MeV}\times10\Rightarrow160.6$ MeV

B.E of He^4 and C^{12} are 7.07 MeV and 87.86 MeV.

$(B.E)_f=[(7.07)\times4]2+[7.86\times12]$

$=56.56+94.32=150.88$ MeV

Difference between *initial* and final binding energies $(160.60-150.88)=9.72$ MeV

$\therefore$ Energy of 9.72 MeV has to be supplied

52. **(c)** Power level of reactor, $P=\frac{E}{\Delta t}=\frac{\Delta mc^2}{\Delta t}$

mass of the fuel consumed per hour in the reactor,

$$\frac{\Delta m}{\Delta t}=\frac{P}{c^2}=\frac{10^9}{(3\times10^8)^2}=4\times10^{-2}\text{ gm}$$

53. **(c)** ${}_1H^2+{}_1H^2\rightarrow{}_2He^4$

Total binding energy of two deuterium nuclei = 1.1×4 = 4.4 MeV

Binding energy of a $({}_2He^4)$ nuclei = $4\times7=28$ MeV

Energy released in this process = $28-4.4=23.6$ MeV

54. **(a)** ${}_0^1n\longrightarrow{}_1^1H+{}_{-1}e^0+\bar{v}+Q$

The mass defect during the process

$\Delta m=m_n-m_H-m_e=1.6725\times10^{-27}-(1.6725\times10^{-27}+9\times10^{-31})$

$=-9\times10^{-31}$ kg

The energy released during the process

$E=\Delta mc^2$

$E=9\times10^{-31}\times9\times10^{16}=81\times10^{-15}$ joules

$$=\frac{81\times10^{-15}}{1.6\times10^{-19}}=0.511\text{MeV}$$

55. **(c)** Initial momentum of system, $p_i=0$

Let p_1 and p_2 be the momentum of broken nuclei of masses m_1 and $5m_1$ respectively.

From the conservation of momentum

$p_i=p_f\Rightarrow0=p_1-p_2$

$p_1=p_2$

From de-Broglie relation, wavelength

$$\lambda_1=\frac{h}{p_1}\text{ and }\lambda_2=\frac{h}{p_2}$$

$\therefore\ \lambda_1=\lambda_2=\lambda.$

56. **(c)** In nuclear fission, the binding energy per nucleon of daughter nuclei is always greater than the parent nucleus because fission happen only to get more BE/nucleon.

57. **(b)** Mass defect, $\Delta M=\left[(M+\Delta m)-\left(\frac{M}{2}+\frac{M}{2}\right)\right]$

$=[M+\Delta m-M]=\Delta m$

Energy released, $Q=\Delta Mc^2=\Delta mc^2$ ...(i)

From the law of conservation of momentum

$$(M+\Delta m)\times0=\frac{M}{2}v_1-\frac{M}{2}\times v_2\Rightarrow v_1=v_2$$

$$\text{Now, }Q=\frac{1}{2}\left(\frac{M}{2}\right)v_1^2+\frac{1}{2}\left(\frac{M}{2}\right)v_2^2$$

$$=\frac{M}{2}v_1^2\ (\because v_1=v_2)\quad\text{...(ii)}$$

From equation (i) and (ii), we get

$$\left(\frac{M}{2}\right)v_1^2=\Delta mc^2$$

$$\Rightarrow v_1^2=\frac{2\Delta mc^2}{M}\Rightarrow v_1=c\sqrt{\frac{2\Delta m}{M}}$$

58. **(d)** We know that energy is released when heavy nuclei undergo fission or light nuclei undergo fusion. Therefore statement (1) is correct.

The second statement is false because for heavy nuclei the binding energy per nucleon decreases with increasing Z and for light nuclei, B.E/nucleon increases with increasing Z.

59. **(c)** Number of protons in oxygen isotope, $Z=8$

Number of neutrons = $17-8=9$

Binding energy

$=[ZM_P+(A-Z)M_N-M]c^2$

$=[8M_P+9M_N-M]c^2=[8M_P+9M_N-M_o]c^2$

60. **(c)** ${}_3^7Li+{}_1^1p\longrightarrow{}_4^8Be+{}_0^0\gamma$

We see that both atomic number and mass number are equal in both sides, so emitted particle should be massless gamma photons.

61. **(b)** Given,

Binding energy per nucleon of ${}_3^7Li=5.60$ MeV

Binding energy per nucleon of ${}_2^4He=7.06$ MeV

Let E be the energy of proton, then

$E+7\times5.6=2\times[4\times7.06]$

$\Rightarrow E=56.48-39.2=17.28$MeV

62. **(b)** Given :
$$\frac{v_1}{v_2}=\frac{2}{1}$$
From conservation of momentum $m_1v_1 = m_2v_2$
$$\Rightarrow\left(\frac{m_1}{m_2}\right)=\left(\frac{v_2}{v_1}\right)=\frac{1}{2}$$
We know that mass of nucleus, $m \propto A$
Nuclear size $R \propto A^{1/3} \propto m^{1/3}$
$$\frac{R_1}{R_2}=\left(\frac{m_1}{m_2}\right)^{1/3} \Rightarrow \left(\frac{R_1}{R_2}\right)=\left(\frac{1}{2}\right)^{1/3}$$

63. **(c)** Mass of α particle, $m_\alpha = 4\,u$
Mass of nucleus after fission, $m_n = 234u$
From conservation of linear momentum,
$$238\times 0 = 4\,u + 234\,v \Rightarrow v = -\frac{4}{234}u$$
$\therefore$ Recoil speed $=|v|=\frac{4}{234}u$

64. **(d)** The average kinetic energy per molecule at temperature T is $=\frac{3}{2}kT$
Where k = Boltzmann's constant
This kinetic energy should be able to provide the repulsive potential energy
$$\therefore\ \frac{3}{2}kT = 7.7\times10^{-14}$$
$$\Rightarrow T=\frac{2\times7.7\times10^{-14}}{3\times1.38\times10^{-23}}=3.7\times10^{9}K$$

Semiconductor Electronics : Materials, Devices and Simple Circuits

14

1. **(b)** Here D_1 is reverse biased and D_2 is forward biased. Therefore current flows through D_2 and 5 V drop on resistor.
So, $V_{out} = 0$

2. **(c)**

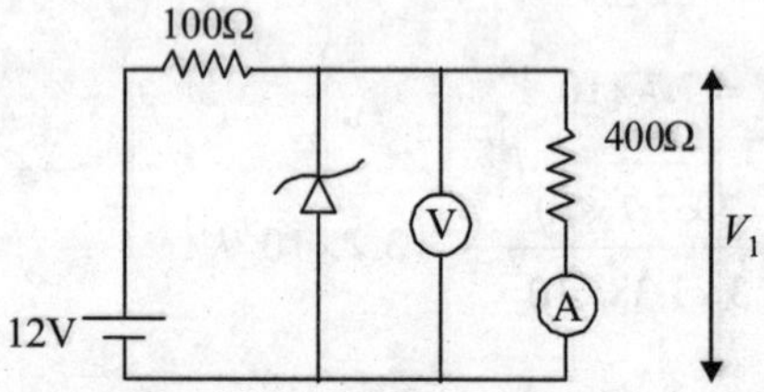

The voltage across the load is

$$V_1 = \frac{400}{100+400} \times 12V = \frac{4}{5} \times 12 = \frac{48}{5} V$$

Here, $V_z = 4$ V

$\therefore V_1 > V_z$

So, Zener breakdown will be take place and voltage across 400Ω will be 4V

$\therefore$ Reading of ammeter is

$$I = \frac{4}{400} A = \frac{1}{100A} = 10mA.$$

3. **(c)**
(A) *n* type semiconductor holes are minority carries and e^- are majority carriers.
(B) Dopant are pentavalent atom.
(C) $n_e \cdot n_h = n_1^2$ for intrinsic semiconductor
(E) In *n* type semiconductor primary source of holes generation are thermal excitation.

4. **(a)**

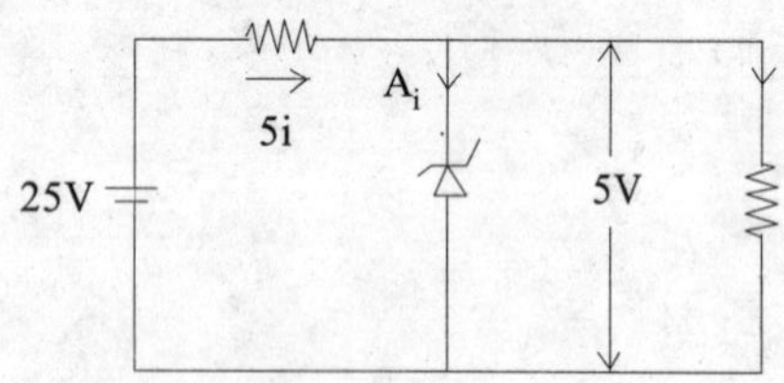

From the circuit diagram,

$$5i = \frac{V - V_Z}{R} = \frac{25-5}{400} = \frac{20}{400} = \frac{1}{20} A$$

$$\therefore \; i = \frac{1}{100} A = 10 \text{ mA} = \text{Load current}$$

Also, load voltage $V_L = 5$ V

$$\therefore \text{Load resistance } R_L = \frac{V_L}{i} = \frac{5}{10 \times 10^{-3}} \Omega = 500\Omega$$

5. **(d)** Due to the presence of the diode, current flows only from B to A through the resistor,
As a result, V_{AB} will only have negative values.

$V = V_0 \sin\omega t$

Input →

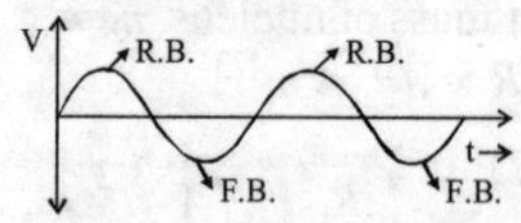

Output →

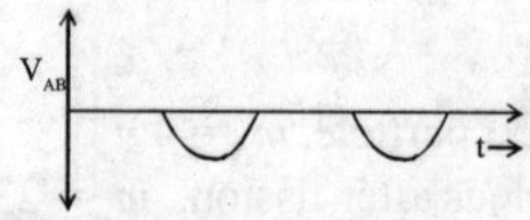

6. **(d)** A solar cell consists of a large junction compared to a photodiode without any external biasing.
LED is made of highly dopped p-n junction and increase in its forward current does not continuously increase light intensity.

7. **(c)** The diodes are in forward bias, so they allow current to flow.
$R_{eq} = 10\,\Omega$

$$\therefore I = \frac{V}{R} = \frac{5}{10} = 0.5A$$

8. **(a)** During forward biasing, diode does not offer any resistance.

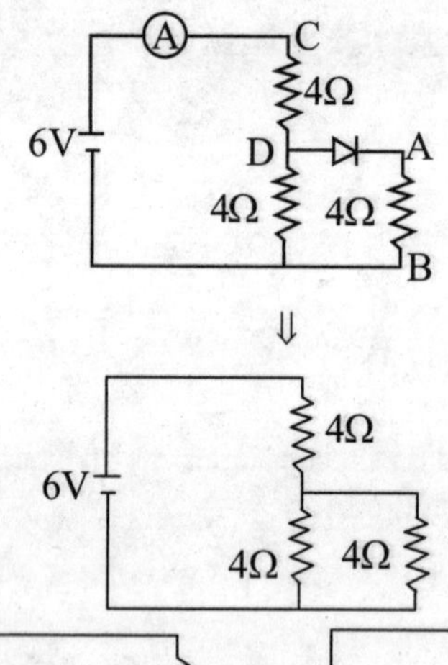

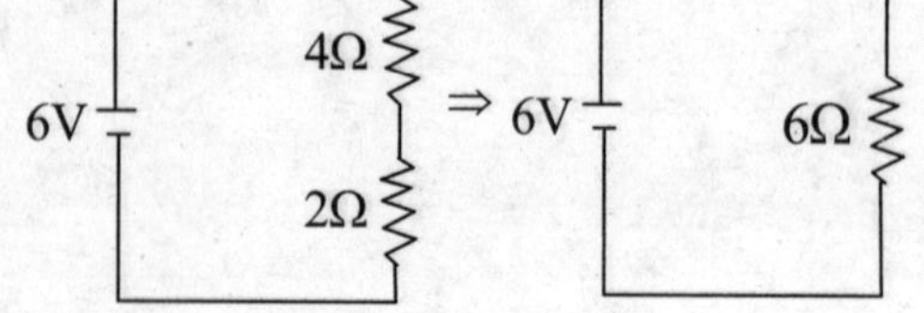

Current in ammeter, $I = \frac{E}{R} = \frac{6}{6} = 1A$

Current in AB $= \frac{I}{2} = \frac{1}{2} A = 0.5A$

$V_{AB} = 0.5 \times 4 = 2$ volt

$V_{CD} = 1 \times 4 = 4$ volt

9. **(c)** In (B), (C), (E) the P side is connected at higher potential than the N side of the diode.

10. **(c)** Given,

B and gap of semiconducting material Eg = 1.42 eV

Using, $Eg = \frac{1240}{\lambda}$

$\Rightarrow \lambda = \frac{1240}{1.42} = 875$ nm (Approx)

11. **(c)** The graph shows reverse baised characteristics, so it should be zener diode which is used as voltage regulator.

12. **(5)** Length of wire is,

d = 20 cm

Voltage of diode,

$V_d = 1.8 + 3.2 = 5V$

Length of PQ is,

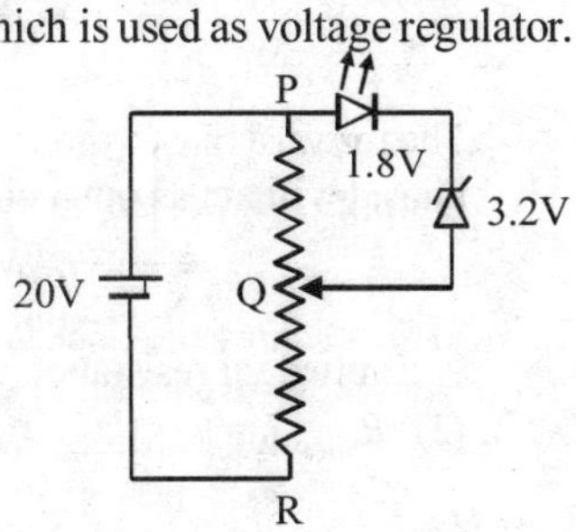

$l' = \frac{V_d l}{V} = \frac{20}{20} \times 5 = 5$ cm

13. **(c)** Energy is, $E = \frac{hc}{\lambda}$

$E = \frac{1242\text{nm-eV}}{\lambda} \Rightarrow \lambda = \frac{1242}{6} = 207$ nm

14. **(c)** In the given circuit upper diode is forward biased and lower diode is reverse biased circuit is as shown in figure.

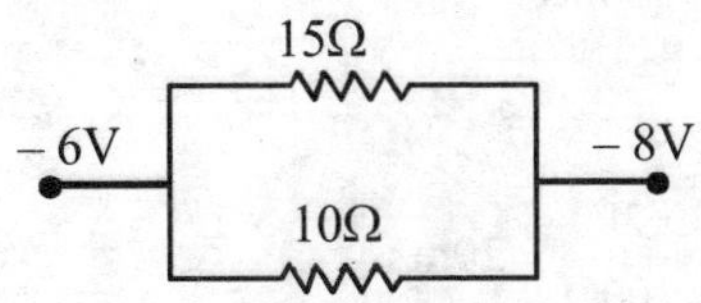

Clearly 15Ω and 10Ω are in parallel

$\therefore R_{eq} = \frac{15 \times 10}{15 + 10} = \frac{15 \times 10}{25} = 6\Omega$

15. **(b)** Only circuit given in option (b) is forward biased so it can be used for the measurement of dynamic resistance of p-n junction diode.

16. **(c)**

17. **(c)** Energy band gap, $E_g = \frac{hc}{\lambda} = \frac{1240 \text{ nm-eV}}{660 \text{ nm}} = \frac{15}{8}$eV

$\therefore x = 15$

18. **(c)** From ohm's law, $I = \frac{V}{R}$

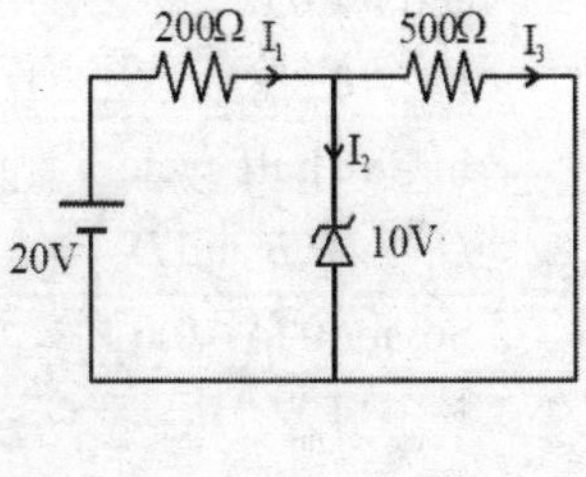

$I_3 = \frac{10}{500} = \frac{1}{50}$;

$I_1 = \frac{10}{200} = \frac{1}{20}$

$I_2 = I_1 - I_3 = \left(\frac{1}{20} - \frac{1}{50}\right) = \left(\frac{3}{100}\right) = 30$ mA

19. **(a)**

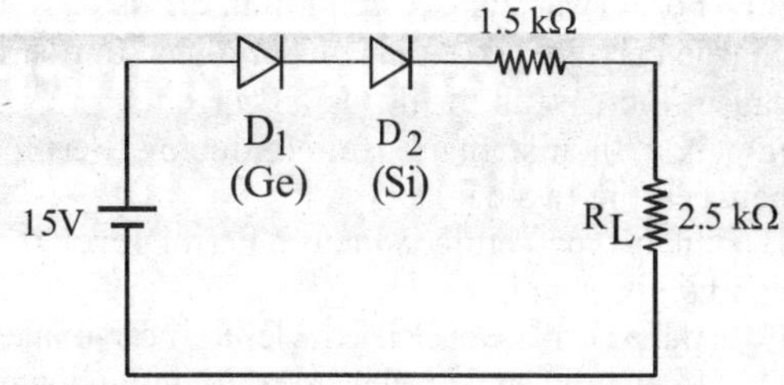

$i = \frac{15 - 0.7 - 0.3}{2.5 + 1.5} = 3.5$mA

$V_L = IR_L = 3.5 \times 2.5$ volt $= 8.75$ volt

20. **(b)** $V_z = 3V, R_1 = 1000\,\Omega, R_2 = 2000\,\Omega$

$i_{battery} = \frac{10 - 3}{R_1} = \frac{10 - 3}{1000} = 7$ mA

$i_{2k\Omega} = \frac{V_z}{R_2} = \frac{3}{2000} = 1.5$ mA

$i_z = (7 - 1.5)$ mA $= 5.5$ mA

21. **(d)** P end should be at lower potential and N end at higher potential for reverse biased.

22. **(a)** In the circuit D_1 and D_3 are forward biased and D_2 is reverse biased.

So, D_1 and D_3 will be short circuit and D_2 will be open circuit.

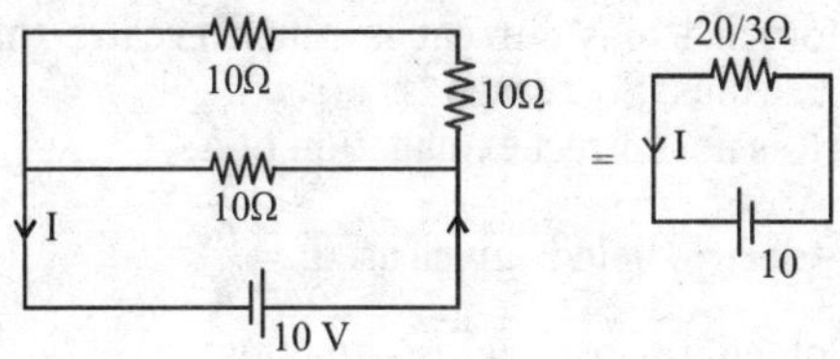

$\therefore I = \frac{10}{20/3} = \frac{3}{2} A = 1.5A$

23. **(c)**

24. **(d)** Diode D_2 is reversed biased.

$\therefore I_3 = 0$

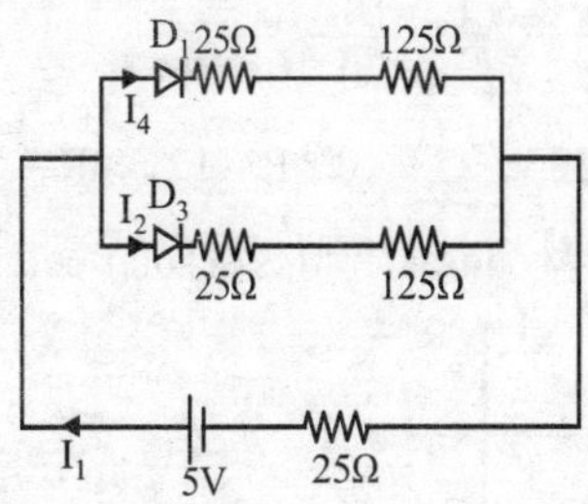

Equivalent resistance of circuit,

$R_{eq} = \frac{150 \times 150}{300} + 25 = 100\Omega$

$\therefore$ Current, $I_1 = \frac{V}{R} = \frac{5}{100} = 0.05A$ and $I_2 = I_4 = \frac{0.05}{2} = 0.025A$

$\therefore \frac{I_1}{I_2} = 2$

25. **(c)** For forward biased condition, diffusion of majority charge carriers takes place from p-side to n-side which causes increase in the diffusion current.

26. **(c)** A. In instrinsic semiconductor, Fermi level is between the two bands.
B. In n-type semiconductor, Fermi level is close to conduction band.
C. In p-type semiconductor, fermi level is near to valence band.
D. In metal, fermi level inside the conduction band.

27. **(a)** Zener diode behave as voltage regulator in reverse bias and as a normal p-n junction diode in forward bias.

28. **(c)** When temperature increases, more electrons excite to conduction band and hence conductivity increases, therefore resistance decreases.

29. **(c)** LED is a heavily doped p-n junction. LED emits light only when it is forward biased. Energy of light emitted from LED may be slightly less or equal to band gap.

30. **(c)** Photodiodes are used in reverse bias condition. For P-N junction diode current in forward bias (for $V \geq V_0$) is always greater than current in reverse bias (for $V \leq V_z$). Hence Assertion if false but Reason is true.

31. **(d)** When P-N junction is formed an electric field is generated from N-side to P-side due to which barrier potential arise & majority charge carrier cannot flow through the junction due to barrier potential so current is zero unless we apply forward bias voltage.

32. **(b)** Photodiodes are always operated in reverse bias. So, assertion is correct.
In forward bias current is much greater than that in reverse bias. So, reason is correct.
But R is not correct explanation of A.

33. **(1)** Electric field is given as, $E = \frac{V}{d}$

$= \frac{\text{Potential barrier across junction}}{\text{Width of depletion layer}}$

$= \frac{0.6\text{V}}{6\times10^{-6}\text{m}} = 1\times10^5 \text{ V/m} = 1\times10^5 \text{N/C}$

34. **(480)** Given, zener current, $I = 25\text{ mA} = 25\times10^{-3}\text{ A}$
$V_z = 8\text{ V}$
Apply K.V.L, we have
$\varepsilon - IR - V_z = 0;\ 20 - IR - 8 = 0$
$IR = 12 \Rightarrow 25\times10^{-3} R = 12 \Rightarrow R = \frac{12}{25\times10^{-3}} = 480\ \Omega$

35. **(b)** I-V characterstics of solar cell is as shown below.

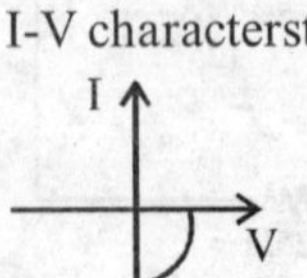

36. **(9)** Maximum voltage across 4 kΩ resistance is $= 120\text{ V} - 60\text{ V} = 60\text{ V}$

Current through 4 kΩ resistance, $I_1 = \frac{60}{4000} = 0.015\text{ A}$

The current through 10 kΩ will be $I_2 = \frac{60}{10000} = 0.006$

Hence, maximum current through zener diode is
$I_Z = I_1 - I_2 = 0.009\text{ A} = 9\text{ mA}$

37. **(3)** $E(\text{eV}) = \frac{12400}{\lambda(\text{Å})} = \frac{12400}{4000} \simeq 3\text{ eV}$

38. **(25)** Left diode is formed biased and right diode is reverse biased.

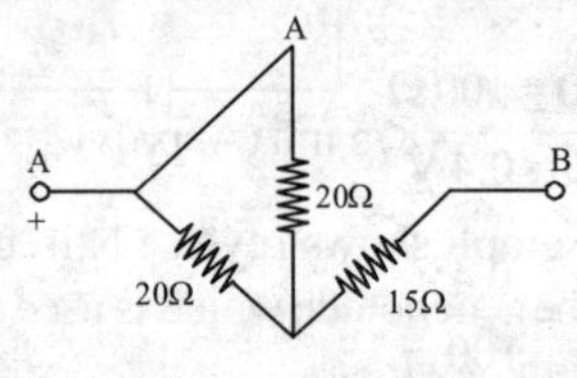

The forward biased diode will act as wire while the reverse biased will act as open circuit

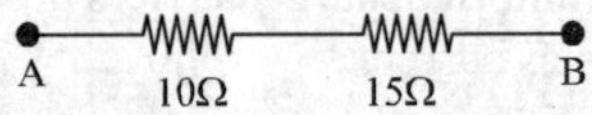

∴ Equivalent resistance = 10 + 15 = 25 Ω

39. **(2)** $R_{max}, I_z = I_{zm}$

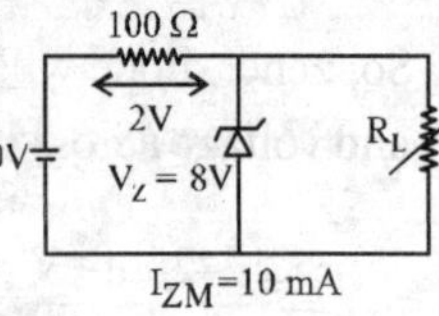

$I = I_z + I_L$

$\Rightarrow \frac{2}{100}\times10^3 = 10 + I_L$

$\Rightarrow I_L = 10\text{ mA}$

So, $V_L = I_L (R_L)_{\min}$

$(R_L)_{\min} = \frac{V_L}{I_L} = \frac{8}{10\text{ mA}} = 800\ \Omega$

For $R_{min}, I_z = 0$
So, $I_L = I$

$I_L = 20\text{ mA}\left[\because I = \frac{2V}{100\Omega} = 20mA\right]$

So, $V_L = I_L (R_L)_{\min}$
$8V = 20\text{ mA}\times(R_L)_{\min} \Rightarrow (R_L)_{\min} = 400\Omega$

So, $\frac{(R_L)_{\max}}{(R_L)_{\min}} = \frac{800\Omega}{400\Omega} = 2$

40. **(d)** In +ve half cycle

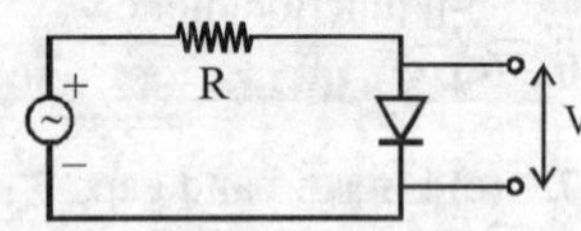

$D_1 \rightarrow F.B$ and $D_2 \rightarrow R.B$
So, circuit will be
So, for $0 - 0.6\ V$
$V_{out} = V_{in}$
and $V > 0.6\text{ V}$
$V_{out} = 0.6\text{ V}$
In – ve half cycle
$D_1 \rightarrow R.B$ and $D_2 \rightarrow F.B$
So, for 0 to $-0.6\ V$
$V_{out} = V_{in}$
and for $V < -0.6\ V$
$V_{out} = -0.6V$

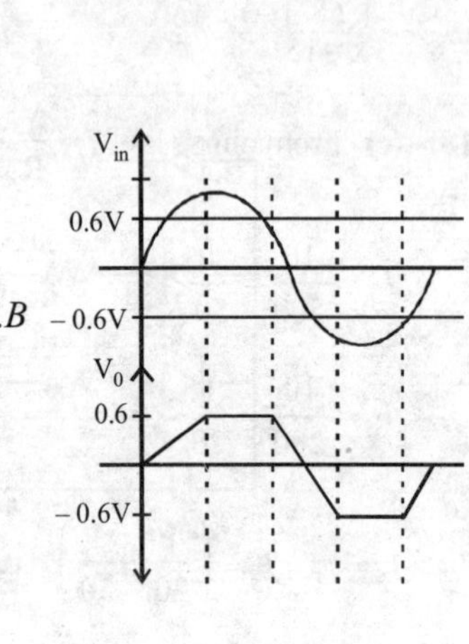

41. **(4)** As D_1 is *FB* and D_2 is *RB*

So, D_2 will be open circuited and D_1 will be replaced by $0.6V$ battery.

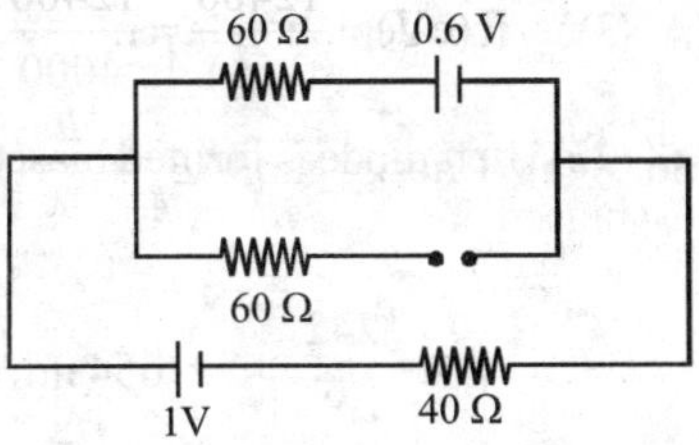

$R_{eq} = 60 + 40 = 100\,\Omega$

$V_{net} = 1 - 0.6 = 0.4$ V

So, $i = \dfrac{V_{net}}{R_{eq}} = \dfrac{0.4}{100} A = 4$ mA

42. **(d)** Photodiode is used in reverse bias because very small change in minority charge carriers produces high value of reverse bias current.

43. **(5)** $V_{\text{zener diode}} = 10 \times \dfrac{1}{1+0.8} = \dfrac{10}{1.8} = 5.56V > V_z$

So, zener diode will act as a constant voltage regulator, and voltage across it will be equal to $V_z = 5V$.

$\therefore I_L = \dfrac{V_Z}{R_L} = \dfrac{5V}{1k\Omega} = 5mA$

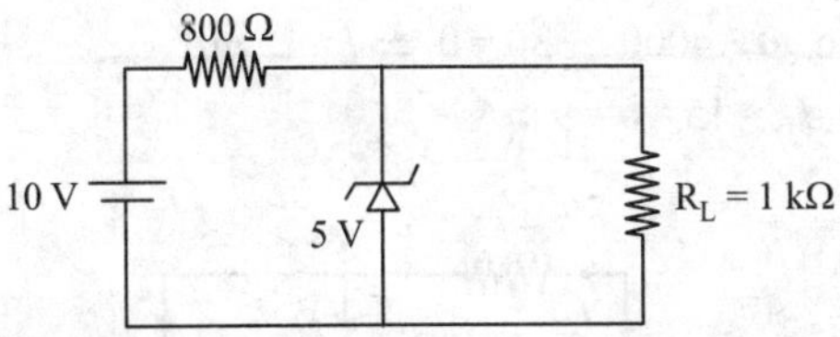

44. **(c)** In the given circuit

Diode D_1 & D_2 are in forward bias whereas diode D_3 is in reverse bias. The circuit can be redrawn as shown in figure

$$\frac{1}{R'} = \frac{1}{R_1 + R_2} + \frac{1}{R_3 + R_4}$$

$$\Rightarrow \frac{1}{R'} = \frac{1}{30\Omega + 130\Omega} + \frac{1}{30\Omega + 130\Omega}$$

$$\Rightarrow \frac{1}{R'} = \frac{1}{160} + \frac{1}{160} \Rightarrow R' = \frac{160}{2}$$

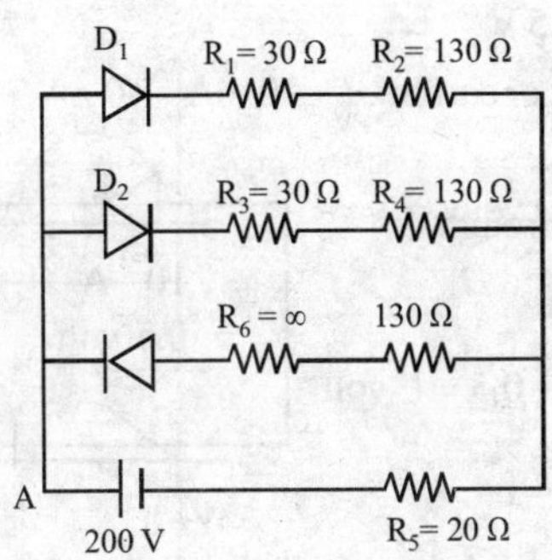

$$R_{eq} = R' + R_5 = \frac{160}{2} + 20 = 100\Omega$$

$$\text{Current, } I = \frac{V}{R_{eq}} = \frac{200}{100} = 2\text{A}$$

45. **(a)** When pentavalent atom loses an electron, equal +ve charge appears on the atom. So overall neutral.

46. **(20)** Given, Power rating of zener diode, $P_z = 2W$.

For highest value of unregulated voltage $14V$, voltage across zener diode must be $14V - 10V = 4V$.

Using $P_z = V_z I. \Rightarrow I = \dfrac{2}{10} = 0.2A$

Using Ohm's law

$\Delta V_s = IR_s. \Rightarrow R_s = \dfrac{4}{0.2} = 20\Omega$

47. **(120)**

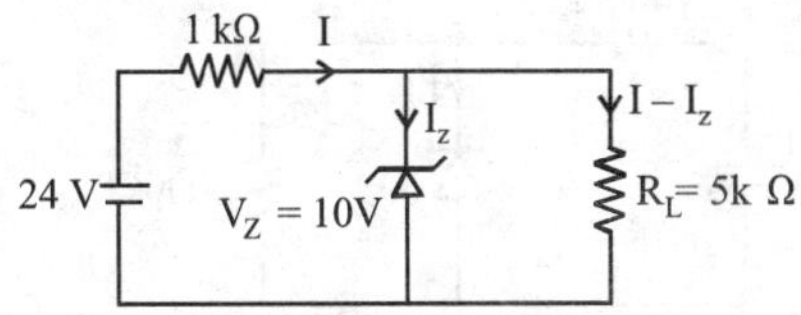

As zener diode provides constant output voltage, so

$$I - I_Z = \frac{10V}{5k\Omega} = 2\text{mA}$$

$$I = \frac{V - V_Z}{R} = \frac{14V}{1k\Omega} = 14\text{mA}$$

$I - I_Z = 2$ mA $\Rightarrow I_Z = 14 - 2 = 12$ mA

$\therefore$ Power, $P = I_z V_z = 12$ mA $\times$ 10V $= 120$ mW

48. **(5)** For a doped semi-conductor in thermal equilibrium

$$n_e n_h = n_i^2 \Rightarrow n_e = \frac{n_i^2}{n_h} = \frac{\left(1.5 \times 10^{16}\right)^2}{4.5 \times 10^{22}}$$

$$= \frac{1.5 \times 1.5 \times 10^{32}}{4.5 \times 10^{22}} = 5 \times 10^9 \text{m}^{-3}$$

49. **(500)** Voltage across load, $R_L = 5V$

$$\therefore i_2 = \frac{V_L}{R_L} = \frac{5}{R_L}$$

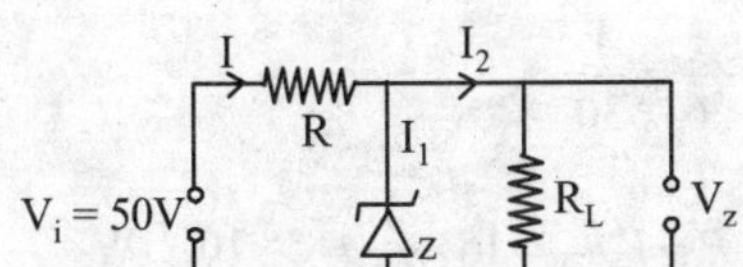

Also voltage across resistance $R = 50 - 5 = 45$ V

$$\therefore R = \frac{V}{I} = \frac{45}{I_1 + I_2} \Rightarrow R = \frac{45}{90mA + \frac{5}{R_L}}$$

Current in zener diode is maximum when $R_L \to \infty$ ($I_2 \to 0$ So, $I_1 = I$)

or, resistance, $R = \frac{45}{90mA} = 500\ \Omega$

50. **(a)** Band gap $= \frac{hc}{\lambda_0} = \frac{1242\ ev - nm}{621\ nm} = 2eV$

$\because \lambda_0 =$ threshold wavelength

51. **(a)** In the given circuit, the zener diode is used as a voltage regulating device.
Hence, the voltage across 2000Ω is 50V.

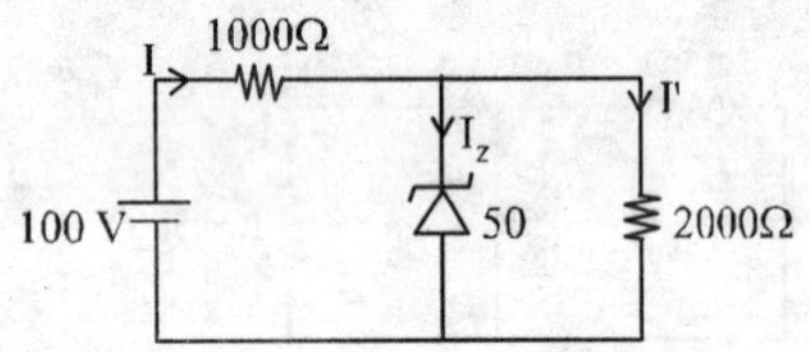

Current flowing through 1000Ω, $I = \frac{100 - 50}{1000} = 50mA$

Current flowing through 2000Ω, $I' = \frac{50}{2000} = 25mA$

Using Kirchoff's current law,
$I = I_z + I' \Rightarrow I_z = I - I' = 50 - 25 = 25mA$

52. **(192)** Power $P = Vi$ or, $0.5 = 8i$

$$\therefore i = \frac{0.5}{8} = \frac{1}{16} A$$

$$E = V + iR_p \Rightarrow 20 = 8 + iR_p \Rightarrow R_P = \frac{12}{i} = 12 \times 16 = 192\ \Omega$$

53. **(5)** Current flowing through 35 Ω resistance,

$$i = \frac{V}{R_s} = \frac{22 - 15}{35} = \frac{1}{5} A$$

Current flowing through R_L, $i' = \frac{15}{90} = \frac{1}{6}$ A

Current flowing through zener diode,

$$i_z = \frac{1}{5} - \frac{1}{6} = \frac{1}{30}\ A$$

Power, $P = V_z I_z = 15 \times \frac{1}{30} = 5 \times 10^{-1}$ W

54. **(25)** Current through 2 kΩ resistance is

$$I = \frac{5}{2 \times 10^3} = 2.5 \times 10^{-3}\ A \quad \Rightarrow \quad I = 25 \times 10^{-4} A$$

55. **(b)** As Zener diode is heavily doped *p-n* diode so it has narrow depletion layer.

56. **(a)** Energy gap, $\Delta E_g = \frac{hc}{\lambda} \Rightarrow \lambda = \frac{hc}{\Delta E_g}$

$\Rightarrow \lambda = \frac{1242}{1.9}$ nm $= 654$ nm of red colour

57. **(20)** As diode D_2 is reverse baised, no current will flow through it.

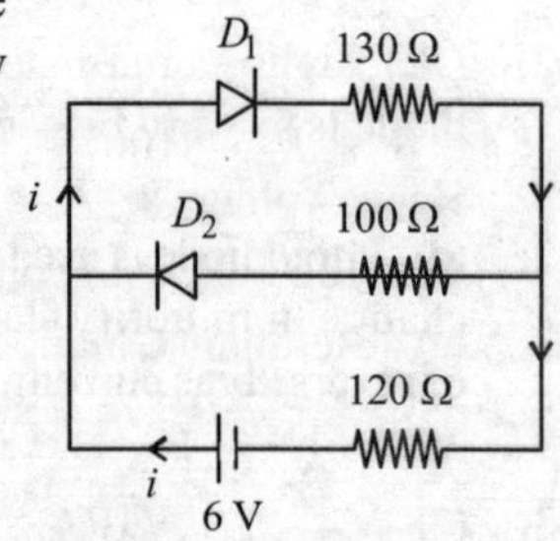

Current through resistance of 120 Ω,

$$i = \frac{V}{R_{net}} = \frac{6}{50 + 130 + 120}$$

$$= \frac{6000}{300} \times 10^{-3}\ A = 20\ mA$$

58. **(09.00)** Current passing through the diode $I_D = I - I_1$

$$I_1 = \frac{30}{5,000} = 6\ mA$$

And $90 + 4000I + 30 = 0 \Rightarrow I = 15$ mA
$\therefore\ I_D = 15 - 6 = 9$ mA

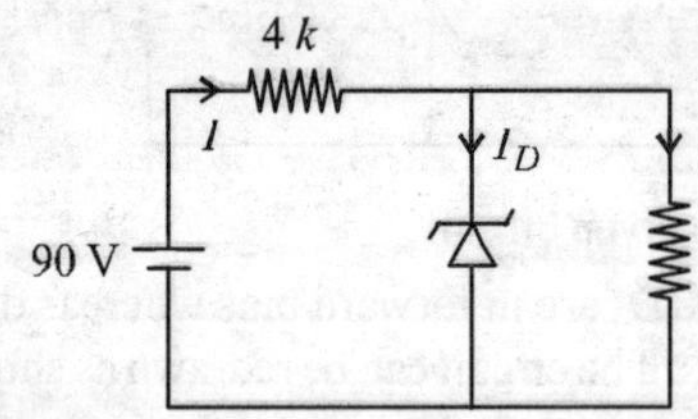

59. **(d)** I-V characteristic of a photodiode is as follows :

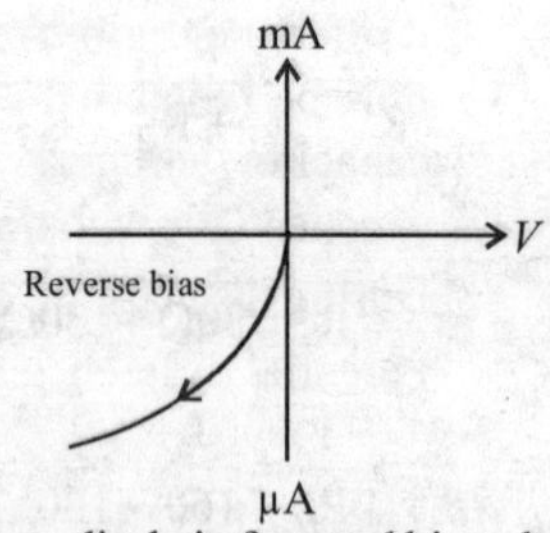

On increasing the biasing voltage of a photodiode, the magnitude of photocurrent first increases and then attains a saturation.

60. **(c)** According to question, when diode is forward biased,
$V_{diode} = 0.5$ V
Safe limit of current, $I = 10$ mA $= 10^{-2}$ A
$R_{min} = ?$
Voltage through resistance

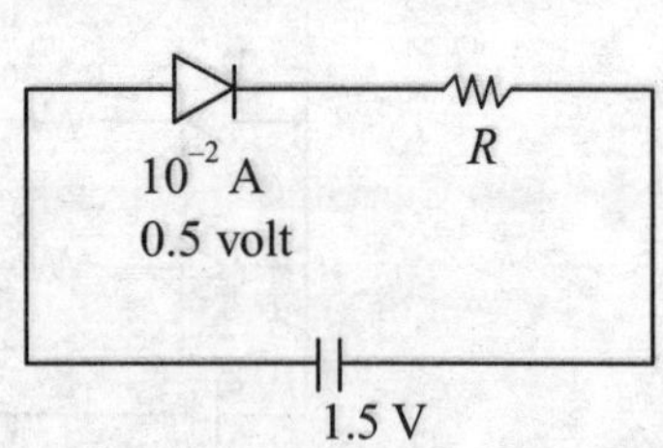

$V_R = 1.5 - 0.5 = 1$ volt
$iR = 1 = V_R$

$$\therefore R_{min} = \frac{1}{i} = \frac{1}{10^{-2}} = 100\ \Omega$$

61. (d) Given,

Wavelength of photon, $\lambda = 400$ nm

A photodiode can detect a wavelength corresponding to the energy of band gap. If the signal is having wavelength greater than this value, photodiode cannot detect it.

$\therefore$ Band gap $E_g = \frac{hc}{\lambda} = \frac{1237.5}{400} = 3.09$ eV

62. (12) Right hand diode is reversed biased and left hand diode is forward biased.

Hence Voltage at 'A'

$V_A = 12.7 - 0.7 = 12$ volt

63. (c) Resultant Circuit will be

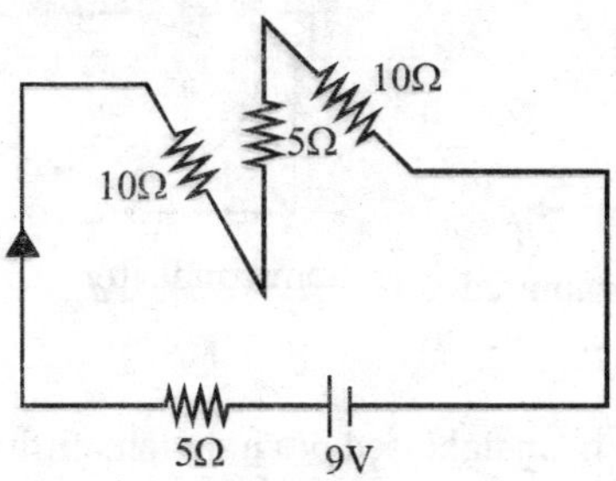

$R_{resultant} = 30\Omega$

$I = \frac{9}{30} = \frac{3}{10}$

$\boxed{I = 0.3A}$

64. (c) In case I diode is reverse biased, so no current flows

$\therefore\ Q_A = CV$

In case II, current will flow as diode is forward biased. So, it offers negligible resistance to the flow of current and thus be replaced by short circuit. Now, the charge of capacitor will leak through the resistance and decay exponentially with time.

During discharging of capacitor

Potential difference across the capacitor at any instant

$V' = Ve^{-\frac{t}{CR}}$

But $t = CR$

$V' = Ve^{-1} = \frac{V}{e}$ $\quad\therefore$ Charge $Q_B = CV' = \frac{CV}{e}$

65. (40) Current in the circuit, $I = \frac{12-8}{400} = 10^{-2} A$

Power dissipited in each diode, $P = VI$

$\Rightarrow P = 4 \times 10^{-2} = 40\ mW$

66. (a) The given circuit has two $10k\Omega$ resistances in parallel, so we can reduce this parallel combination to a single equivalent resistance of $5k\Omega$.

Diode is in forward bias. So it will behave like a conducting wire.

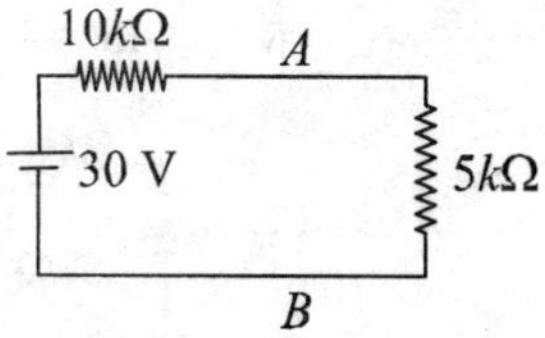

$V_A - V_B = \frac{30}{5+10} \times 5 = 10\,V$

67. (d) Current in load resistance,

$i_L = \frac{6}{4 \times 10^3} = 1.5 \times 10^{-3}$ A $= 1.5$ mA

For $V = 16$ volt,

$i_L = \frac{(16-6)}{2 \times 10^3} = 5$ mA

$\therefore\ i_Z = i_s - i_L = 5 - 1.5 = 3.5$ mA

68. (c)

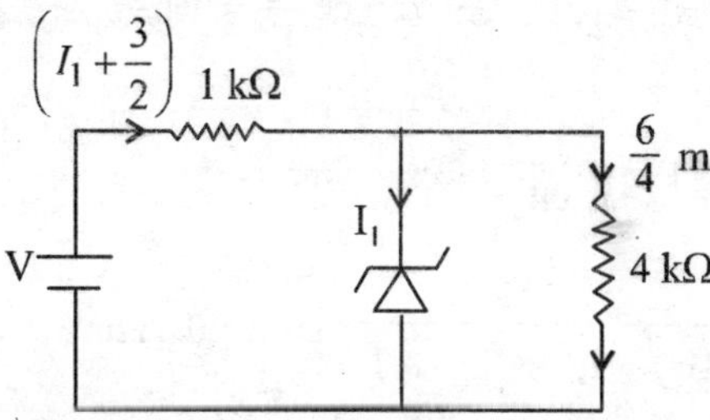

For voltage, V = 8V

Current, $I_1 = \left(8 - 6 - \frac{3}{2}\right) = \frac{1}{2} = 0.5$ mA

For voltage, V = 16V

Current, $I_2 = \left(16 - 6 - \frac{3}{2}\right) = 8.5$ mA

69. (a) P.D. across 800Ω resistors = 5.6 V

so, $I_{800\Omega} = \frac{5.6}{800} A = 7$ mA

Now, P.D. across 200Ω resistors $= 9 - 5.6$ V $= 3.4$ V

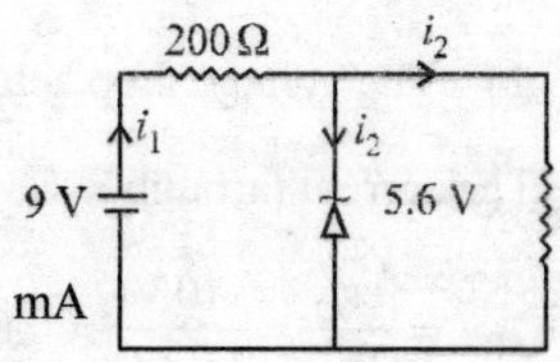

so, $I_{200\Omega} = \frac{9-5.6}{200} = 17$ mA

so, current through zener diode $= I_2 = 17 - 7 = 10$ mA

70. (a) Since voltage across zener diode does not reach to breakdown voltage therefore its resistance will be infinite & current through it is 0.

71. (b) As D_2 is reversed biased, so no current through 75Ω resistor.

now $R_{eq} = 150 + 50 + 100 = 300\ \Omega$

So, required current $I = \frac{\text{Battery Voltage}}{300}$

$I = \frac{6}{300} = 0.02$

72. (a) The voltage across zener diode is constant

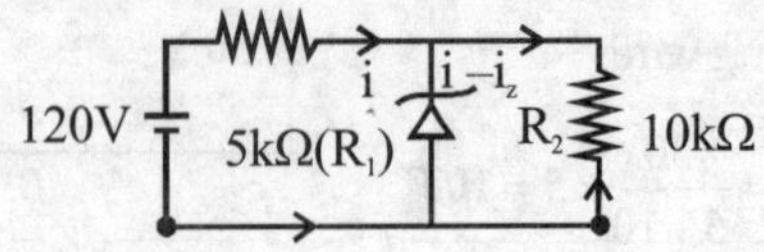

$$i_{(R_2)}=\frac{V}{R}=\frac{50}{10\times10^3}=5\times10^{-3}\text{ A}$$

$$i_{(R_1)}=\frac{V}{R}=\frac{120-50}{5\times10^3}=\frac{70}{5\times10^3}14\times10^{-3}\text{ A}$$

$\therefore i_{zenerdiode}=14\times10^{-3}-5\times10^{-3}=9\times10^{-3}\text{ A}=9\text{ mA}$

73. (c) As we know, current density,

$$j=\sigma E=nev_d \Rightarrow \sigma=ne\frac{v_d}{E}=ne\mu$$

$$\frac{1}{\sigma}=\rho=\frac{1}{n_e e\mu_e}=\text{Resistivity}$$

$$=\frac{1}{10^{19}\times1.6\times10^{-19}\times1.6} \text{ or } \rho=0.4\,\Omega\text{m}$$

74. (d) Initially Ge and Si are both forward biased so current will effectively pass through Ge diode $\therefore V_0=12-0.3=11.7\text{ V}$

And if "Ge" is revesed then current will flow through "Si" diode

$\therefore V_0=12-0.7=11.3\text{ V}$

Clearly, V_0 changes by $11.7-11.3=0.4\text{ V}$

75. (c) Clearly from fig. given in question, Silicon diode is in forward bias.

$\therefore$ Potential barrier across diode $\Delta V=0.7$ volts

$$\text{Current, } I=\frac{V-\Delta V}{R}=\frac{3-0.7}{200}=\frac{2.3}{200}=11.5\text{mA}$$

76. (b) The voltage drop across R_2 is $V_{R_2}=V_Z=10\text{V}$

The current through R_2 is

$$I_{R_2}=\frac{V_{R_2}}{R_2}=\frac{10\text{ V}}{1500\Omega}=0.667\times10^{-2}\text{ A}$$

$=6.67\times10^{-3}\text{ A}=6.67\text{ mA}$

The voltage drop across R_1 is

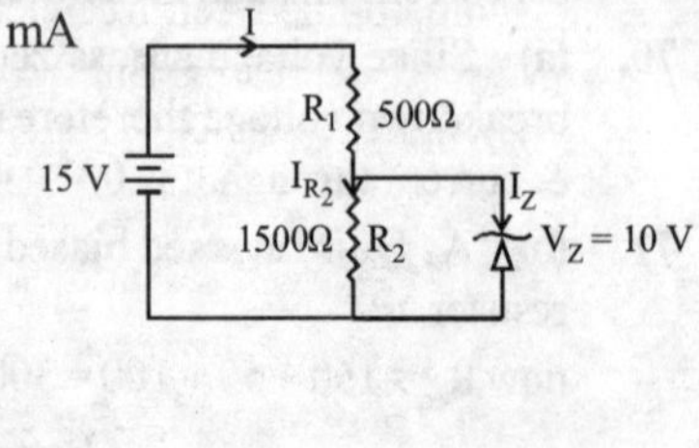

$V_{R_1}=15\text{V}-V_{R_2}$

$=15\text{ V}-10\text{ V}=5\text{ V}$

The current through R_1 is

$$I_{R_1}=\frac{V_{R_1}}{R_1}=\frac{5\text{ V}}{500\Omega}=10^{-2}\text{ A}=10\times10^{-3}\text{ A}=10\text{mA}$$

The current through the zener diode is

$$I_Z=I_{R_1}-I_{R_2}=(10-6.67)\text{mA}=3.3\text{mA}$$

77. (a) The conductivity of semiconductor

$\sigma=e(\eta_e\mu_e+\eta_h\mu_h)=1.6\times10^{-19}(5\times10^{18}\times2+5\times10^{19}\times0.01)$

$=1.6\times1.05=1.68\,(\Omega\text{m})^{-1}$.

78. (c) Graph (i) is for a simple diode.

Graph (ii) is showing the V Break down used for zener diode.

Graph (iii) is for solar cell which shows cut-off voltage and open circuit current.

Graph (iv) shows the variation of resistance h and hence current with intensity of light.

79. (a)

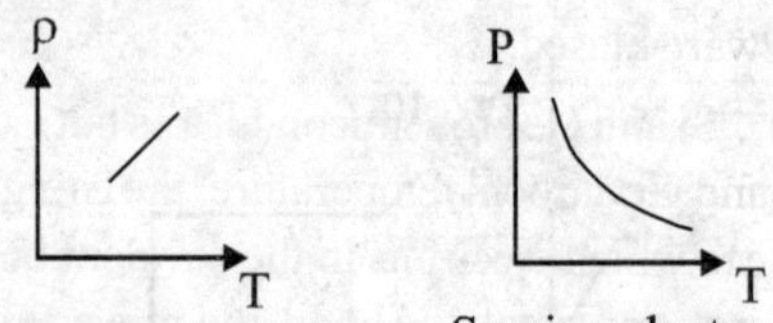

Metal (for limited range of temperature) Semiconductor $\rho=\rho_o e^{\frac{-Eg}{K_B T}}$

80. (d) Electrons in an unbiased *p-n* junction, diffuse from *n* -region i.e. higher electron concentration to *p*-region i.e. low electron concentration region.

81. (a)

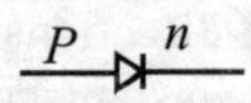

For forward bias, p-side must be at higher potential than *n*-side. $\Delta V=(+)Ve$.

82. (d) Energy band gap range is given by,

$$E_g=\frac{hc}{\lambda}$$

For visible region $\lambda=\left(4\times10^{-7}\sim7\times10^{-7}\right)\text{m}$

$$E_{gL}=\frac{6.6\times10^{-34}\times3\times10^8}{7\times10^{-7}} \text{ and}$$

$$E_{gH}=\frac{6.6\times10^{-34}\times3\times10^8}{4\times10^{-7}}=3.09\text{eV}\simeq3\text{eV}$$

$$E_{gL}=\frac{19.8\times10^{-26}}{7\times10^{-7}}=\frac{2.8\times10^{-19}}{1.6\times10^{-19}}\Rightarrow E_{gL}=1.75\text{ eV}$$

83. (d) Here, $R=4\text{ k}\Omega=4\times10^3\,\Omega$

$V_i=60\text{ V}$

Zener voltage $V_z=10\text{ V}$

$R_L=2\text{ k}\Omega=2\times10^3\,\Omega$

$$\text{Load current, } I_L=\frac{V_Z}{R_L}=\frac{10}{2\times10^3}=5\text{ mA}$$

Current through R, $I = \frac{V_i - V_Z}{R}$

$= \frac{60-10}{4\times10^3} = \frac{50}{4\times10^3} = 12.5\text{ mA}$

Fom circuit diagram,

$I = I_Z + I_L \Rightarrow 12.5 = I_Z + 5 \Rightarrow I_Z = 12.5 - 5 = 7.5\text{ mA}$

84. **(a)** For same value of current higher value of voltage is required for higher frequency. Hence (a) is correct answer.

85. **(d)** Temperature coefficient of resistance is negative for pure semiconductor. And no. of charge carriers in conduction band increases with increase in temperature.

86. **(a)** The current will flow through R_L when the diode is forward biased.

87. **(a)** Si and Ge are semiconductors but C is an insulator. In Si and Ge at room temperature, the energy band gap is low due to which electrons in the covalent bonds gains kinetic energy and break the bond and move to conduction band. As a result, hole is created in valence band. So, the number of free electrons is significant in Si and Ge.

88. **(c)** A crystal structure is made up of a unit cell arranged in a particular way; which is periodically repeated in three dimensions on a lattice. The spacing between unit cells in various directions is called its lattice constants. As lattice constants increases the band-gap (E_g), also increases which means more energy would be required by electrons to reach the conduction band from the valence band. Automatically E_c and E_v decreases.

89. **(b)** Van der Waal's bonding is attributed to the attractive forces between molecules of a liquid. The conductivity of semiconductors (covalent bonding) and insulators (ionic bonding) increases with increase in temperature.

Solid which is formed by covalent bond is not transparent to visible light and its conductivity increase with temperature.

90. **(c)** Relation between drift velocity and current is

$I = nAeV_d$

$\frac{I_e}{I_h} = \frac{n_e eAv_e}{n_h eAv_h} \Rightarrow \frac{7}{4} = \frac{7}{5} \times \frac{v_e}{v_h} \Rightarrow \frac{v_e}{v_h} = \frac{5}{4}$

91. **(b)** D_2 is forward biased.

D_1 is reversed biased. So, it will act like an open circuit.

So effective resistance of the circuit

$R = 4 + 2 = 6\Omega \qquad \therefore i = \frac{E}{R} = \frac{12}{6} = 2\text{ A}$

92. **(d)** p-side connected to low potential and n-side is connected to high potential.

93. **(d)** Band gap = energy of photon of wavelength 2480 nm.

So, Band gap, $E_g = \frac{hc}{\lambda}$

$= \left(\frac{6.63\times10^{-34}\times3\times10^8}{2480\times10^{-9}}\right) \times \frac{1}{1.6\times10^{-19}} eV = 0.5\text{ eV}$

94. **(a)** In forward biasing, the p type is connected to positive terminal and n type is connected with negative terminal. So holes from p region and electron from n region are pushed towards the junction which reduces the width of depletion layer. Also, distance between diffused holes and electrons decrease, which decrease electric field hence barrier potential.

95. **(c)** Copper is a conductor and in conductor resistance decreases with decrease in temperature. Germanium is a semiconductor. In semi-conductor resistance increases with decrease in temperature.

96. **(b)** When the temperature increases, certain bounded electrons become free which tend to promote conductivity. Simultaneously number of collisions between electrons and positive kernels increases which decrease the relaxation time.

97. **(a)** In reverse biasing the width of depletion region increases, and current flowing through diode is zero. Thus, electric field is zero at middle of depletion region.

98. **(c)** Pure silicon, at OK, will contain all the electrons in bounded state. The conduction band will be empty. So there will be no free electrons (in conduction band) and holes (in valence band).Therefore no electrons from valence band are able to shift to conduction band due to thermal agitation. Pure silicon will act as insulator.

99. **(c)** Specific resistance (resistivity) is given by

$\rho = \frac{m}{ne^2\tau}$

where n = no. of free electrons per unit volume

and τ = average relaxation time

For a conductor with rise in temperature n increases. Increase in temperature results increase in number of collision between free electrons due to which relaxation time T decreases. But the decrease in τ is more dominant than increase in n resulting an increase in the value of ρ.

For a semiconductor with rise in temperature, n increases and τ decreases. But the increase in n is more dominant than decrease in τ resulting in a decrease in the value of ρ.

100. **(c)** In insulators, valence band is completely filled while conduction band is empty. The energy band gap is maximum in insulators.

101. (a) Redrawing diagram

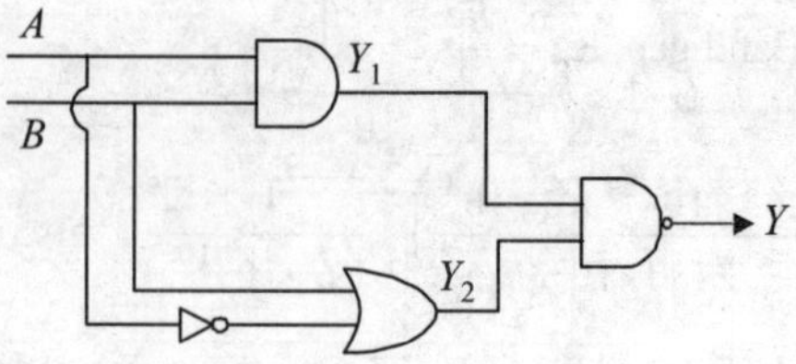

$$Y_1 = A \cdot B,\ Y_2 = \overline{A} + B$$

$$Y = \overline{Y_1 \cdot Y_2} = \overline{Y_1} + \overline{Y_2}$$

$$Y = \overline{A \cdot B} + \overline{\overline{A} + B}$$

$$Y = \overline{A} + \overline{B} + A \cdot \overline{B}$$

Preparing table based on diagram.

A	B	Y
0	0	1
1	0	1
0	1	1
1	1	0

102. (b) Simplifying the expression step by step and drawing the diagram.

(A)

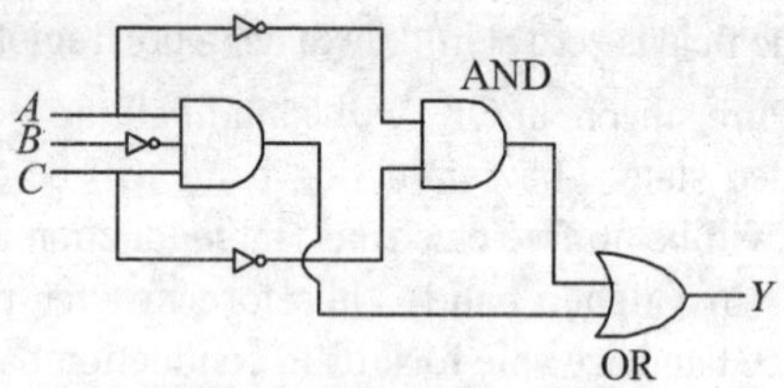

(B)

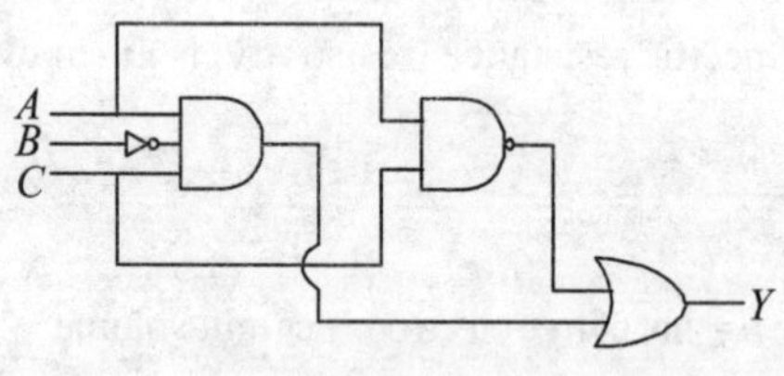

$\because \overline{A} \cdot \overline{C} + \overline{A + C} \equiv$ NOR gate.

103. (b)

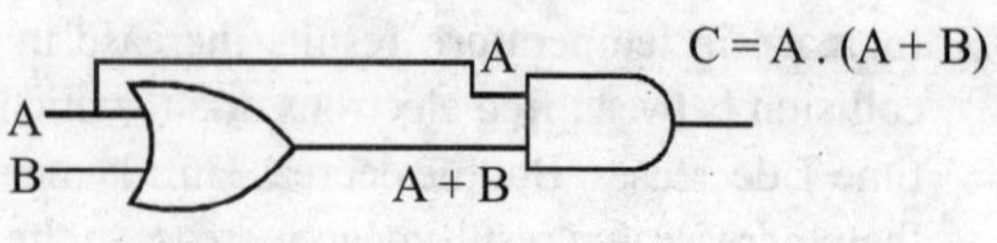

The output of the logic circuit is

C = A.(A + B) = A.A + A.B = A + AB = A. (1 + B) = A.1 = A

$\therefore$ Output C only depends on input A and independent of input B

104. (b) For option (b), Y = A . (A + B) = A. A + A.B = A + A.B = A.(1 + B) = A.1 = A

$\therefore$ Output Y is same as that of input A and independent of input B.

105. (b)

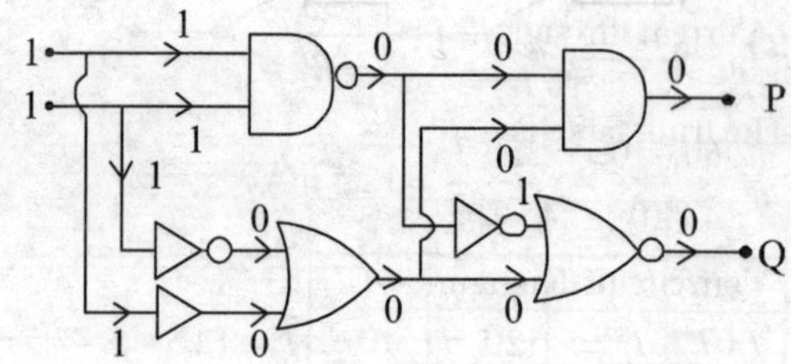

106. (a)

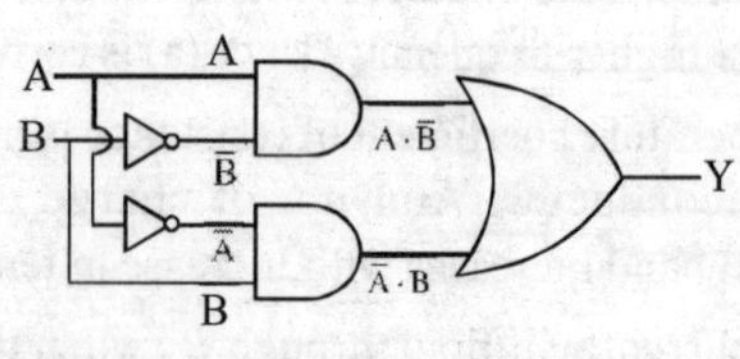

The output of the logic circuit is

$Y = A.\overline{B} + \overline{A}.B$

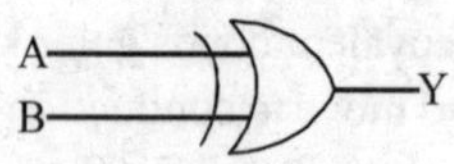

XOR (Exclusive OR)

$\therefore$ Truth table of the circuit is

A	B	Y
0	0	0
0	1	1
1	0	1
1	1	0

107. (a) The truth table of given logic circuit is

A	B	Y
0	0	0
0	1	1
1	0	1
1	1	1

$\therefore$ It is an OR Gate

108. (a)

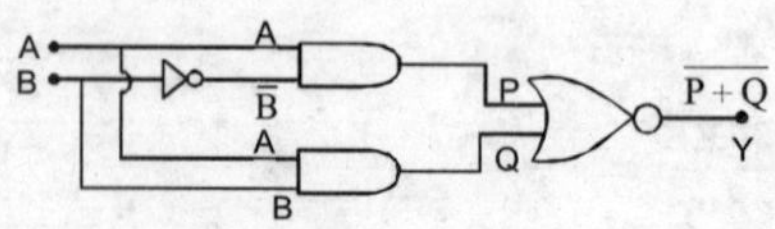

$P = A.\overline{B}$ $\qquad$ $Q = A.B$

$Y = \overline{P + Q} = \overline{A.\overline{B} + A.B}$

Using the complement law

$A + \overline{A} = 1$

$Y = \overline{A.(B + \overline{B})} = \overline{A.1}$

$Y = \overline{A}$

109. (c)

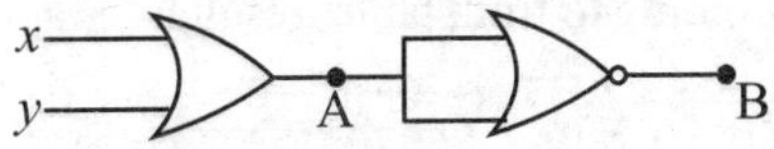

The truth table is

x	y	A	B
0	0	0	1
0	1	1	0
1	0	1	0
1	1	1	0

110. (a)

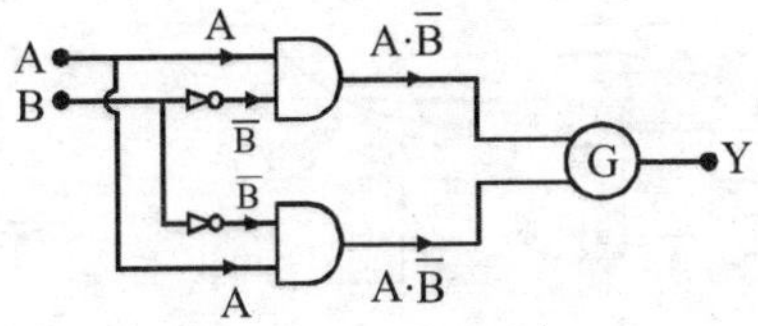

For NOR gate; $y = \overline{A \cdot \overline{B} + A \cdot \overline{B}} = \overline{A\overline{B}} = \overline{A} + B$

∴ Truth table is

A	B	Y
0	0	1
0	1	1
1	0	0
1	1	1

111. (a) Output, $Y = \overline{AB + \overline{A}\overline{B}}$

For x

$A = 0, B = 1$

$\therefore Y = 1$

for Y $A = 1, B = 0$

$\therefore Y = 1$

112. (c)

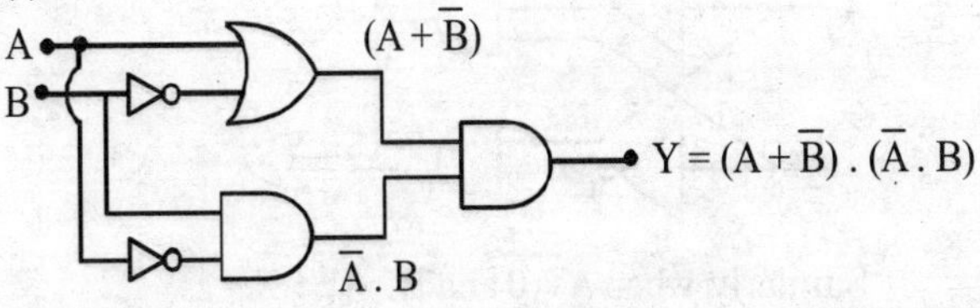

Substituting different values of A and B, we get following truth table

A	B	Y
0	0	0
0	1	0
1	0	0
1	1	0

Hence, output is always zero.

113. (b)

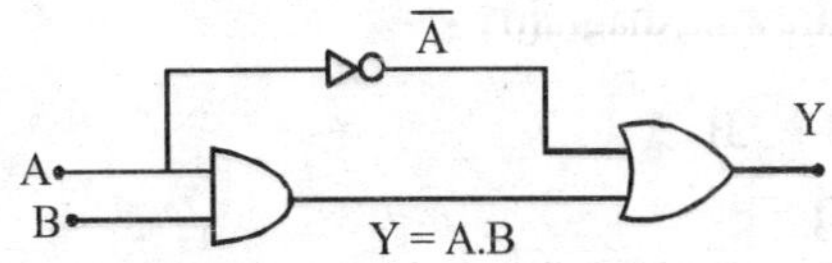

A	B	$\overline{A}$	$Y = A \cdot B$	Y
0	0	1	0	1
0	1	1	0	1
1	0	0	0	0
1	1	0	1	1

114. (b) For bulb to glow, there sould be potential drop across the bulb.

If $A = 1, B = 0, C = 0$ & $D = 0$

∴ We get one end of the bulb at high (1) and other end low (0).

So, bulb will glow.

115. (b) Output, $Y = (A + B) + (B \cdot 1)$

$= A + B + B = A + B$

$\therefore Y = 0 \Rightarrow A = 0$ and $B = 0$

116. (b) Output $Y = \overline{\overline{A}.\overline{B}} = \overline{\overline{A + B}}$ (By De-Morgan Law)

$\therefore Y = A + B$

This Boolean expression represents OR gate.

117. (d)

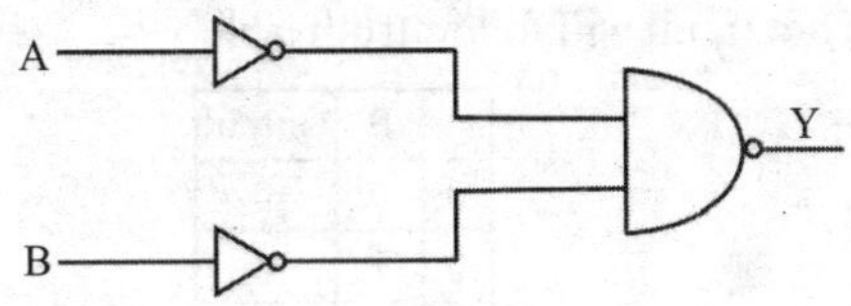

Output, $Y = \overline{\overline{A}.\overline{B}}. = \overline{\overline{A}} + \overline{\overline{B}} = A + B$

Hence, circuit is equivalent to OR gate.

118. (c) Output, $Y = \overline{(A + \overline{B}) + (\overline{A} + B)}$

For $A = 0$; $\overline{A} = 1$

$A = 1$; $\overline{A} = 0$

$B = 0$; $\overline{B} = 1$

$B = 1$; $\overline{B} = 0$

It will give $Y = 0$ in all cases

119. (c) The given circuit is

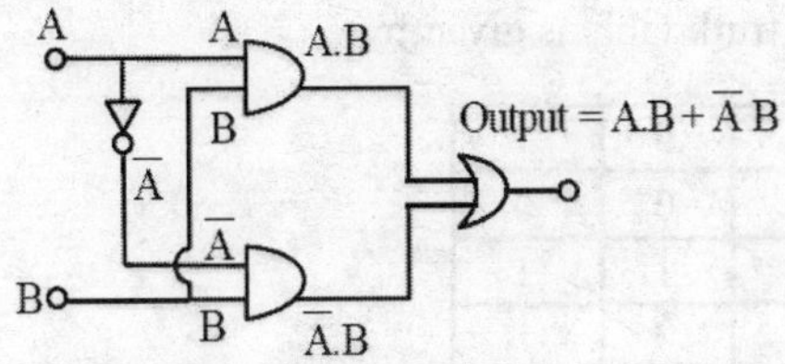

Output, $Y = A.B + \bar{A}.B$

$= (A + \bar{A}).B$

$Y = 1.B$

$Y = B$

120. (b) Boolean expression of given circuit

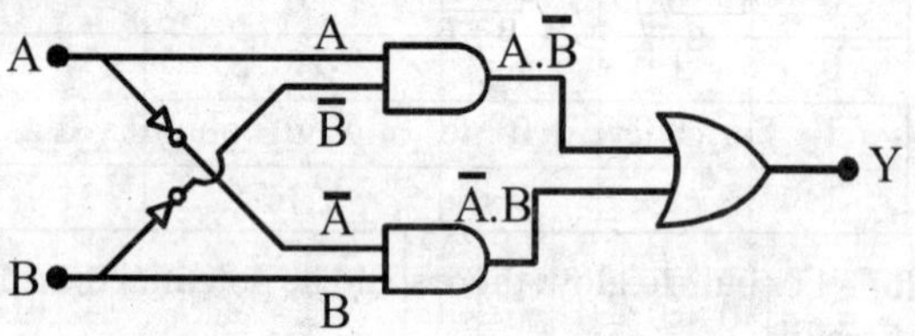

$Y = A\cdot\bar{B} + \bar{A}\cdot B = A \oplus B$. This is XOR gate.

121. (d) Output for the given circuit, $Y = \overline{\bar{A}\cdot B} = A + \bar{B}$

Therefore the correct option for output is (d)

122. (a)

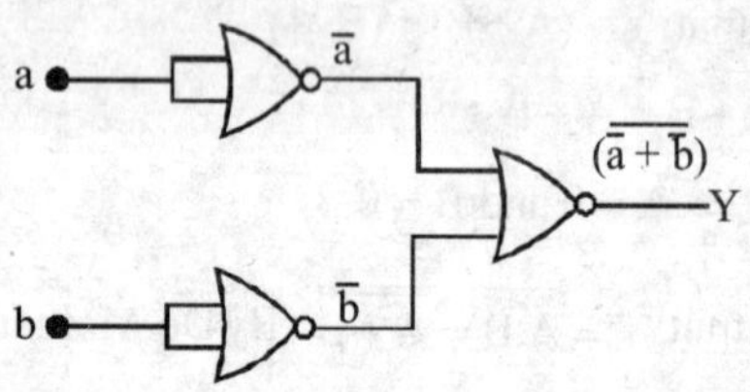

$Y = \overline{\bar{a} + \bar{b}} = a.b$

The circuit will follow truth table

A	B	Output
0	0	0
0	1	0
1	0	0
1	1	1

Hence it is AND gate.

123. (a)

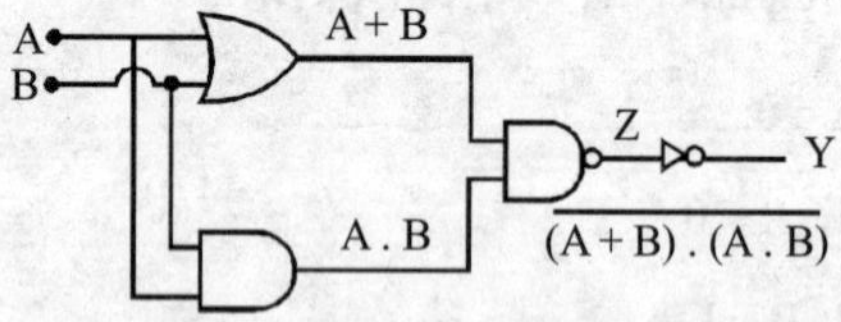

$Y = \bar{\bar{Z}} = (A+B).(A.B) = A.A.B + AB.B$

$Y = AB + AB = A.B$

∴ It is an AND Gate.

124. (d) $Y = \overline{\bar{A}.\bar{B}} = A + B$

The truth table is given by

A	B	Y
0	0	0
0	1	1
1	0	1
1	1	1

According to truth table, resultant graph is

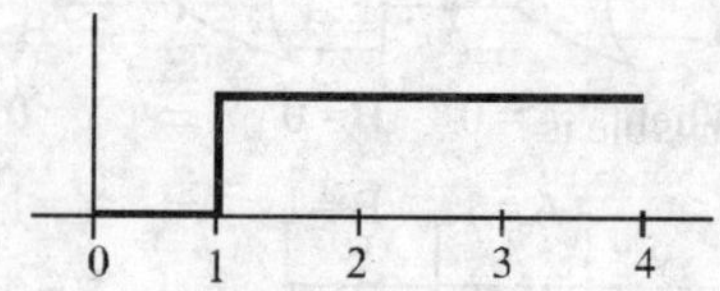

125. (b) It is clear from the given circuit, when both A and B is open current flows through LED. When either of A and B or both of A & B is closed. LED does not glow. Hence it is "NOR" gate

126. (d)

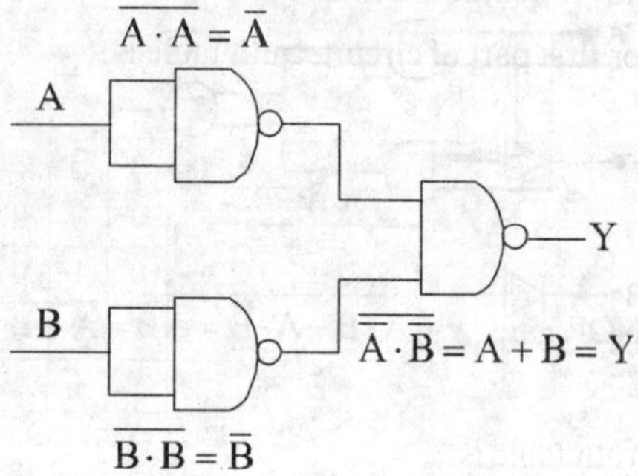

	A	B	Y=A+B
$t_1 \to$	1	0	1
$t_2 \to$	1	1	1
$t_3 \to$	1	1	1
$t_4 \to$	0	1	1
$t_5 \to$	0	0	0
$t_6 \to$	0	1	1

It represent OR gate.

127. (d) $Y = \overline{A\cdot\overline{(A\cdot B)}\cdot B\cdot\overline{(A\cdot B)}}$

$= A\cdot(\overline{AB}) + B\cdot(\overline{AB}) = A\cdot(\bar{A}+\bar{B}) + B\cdot(\bar{A}+\bar{B})$

$= A\bar{B} + B\bar{A}$ = XOR gate

128. (b) When $A = 0$ and $B = 0 \Rightarrow X = 0$

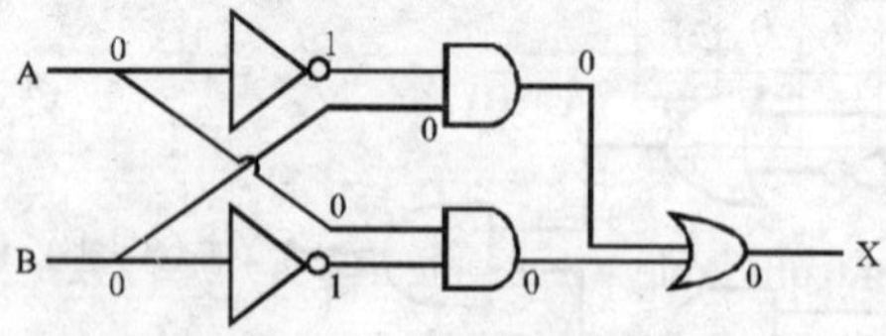

Similarly, when $A = 0$ and $B = 1 \Rightarrow X = 1$

when $A = 1$ and $B = 0 \Rightarrow X = 1$

when $A = 1$ and $B = 1 \Rightarrow X = 0$

129. (b) The truth table for given circuit is

A_1	B_1	Y
0	0	1
0	1	1
1	0	1
1	1	0

So, we have $Y=\overline{A\cdot B}$ i.e. circuit is equivalent to NAND GATE.

130. (a) When $A=0,\quad B=0 \Rightarrow Y=0$

$A=1,\quad B=1 \Rightarrow Y=1$

$A=0,\quad B=1 \Rightarrow Y=0$

$A=1,\quad B=0 \Rightarrow Y=0$

So, this must be AND gate.

131. (c) Let '*O*' denotes O volt

and '*I*' denotes *E* volt

For first part of circuit, truth table is

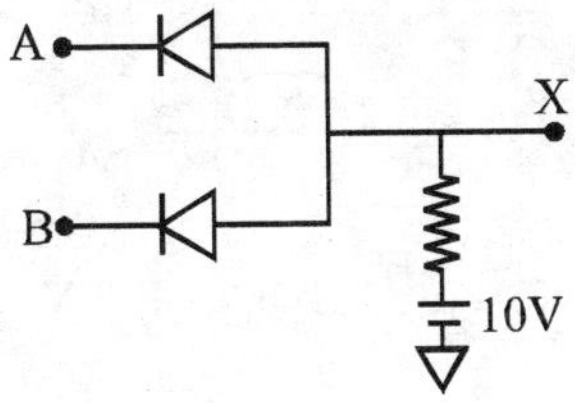

A	B	X
0	0	0
0	1	0
1	0	0
1	1	1

Clearly first part is AND gate

for second part of circuit, truth table is

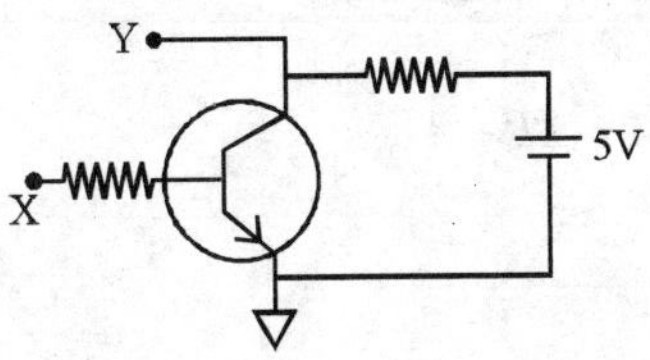

X	Y
0	1
1	0

So, second part of circuit is NOT gate.

So, overall truth table is

A	B	X	Y
0	0	0	1
0	1	0	1
1	0	0	1
1	1	1	0

From truth table we can clearly say $\gamma=\overline{AB}$ and overall circuit behaves as NAND gate.

132. (b) We have truth table as shown

A	B	Y
0	0	1
0	1	1
1	0	1
1	1	0

Which resembles to NAND gate so, correct option is (b).

133. (a) $\overline{\overline{A+A}+\overline{B+B}}=\overline{\overline{A}+\overline{B}}=\overline{\overline{A.B}}=A.B\Rightarrow$ AND Gate

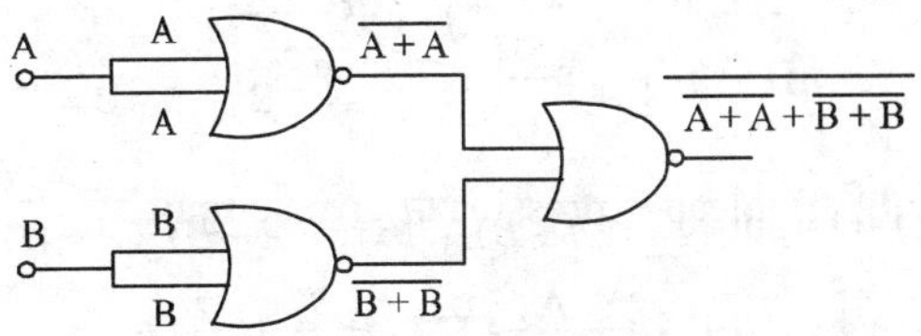

134. (c) For different values of A & B output y is as follows

A	B	P	Q	Y
0	0	0	0	1
0	1	0	1	1
1	0	0	1	1
1	1	1	1	0

Output $y=\overline{(A\cdot B)\cdot A+B}$ matches with option (c).

135. (c) For different values of A & B input, obtain output Z.

A	*B*	*X*	*Y*	*Z*
1	1	0	0	0
1	0	0	1	0
0	1	1	0	0
0	0	1	1	1

Therefore the given circuit represents NOR gate.

136. (d)

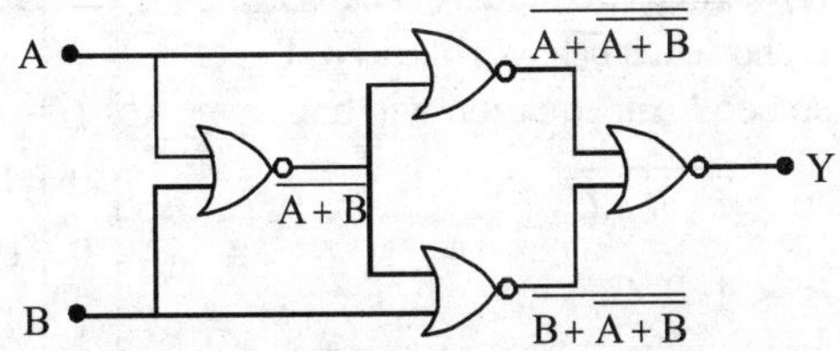

$$y=\overline{\overline{\left(A+\overline{A+B}\right)}+\overline{\left(B+\overline{A+B}\right)}}=\left(A+\overline{A+B}\right).\left(B+\overline{A+B}\right)$$

$\therefore$ Truth table

A	B	y
0	0	1
0	1	0
1	0	0
1	1	1

137. (b) The equivalent circuit is AND gate.

Output $y=\overline{\overline{A}+\overline{B}}=A.B$

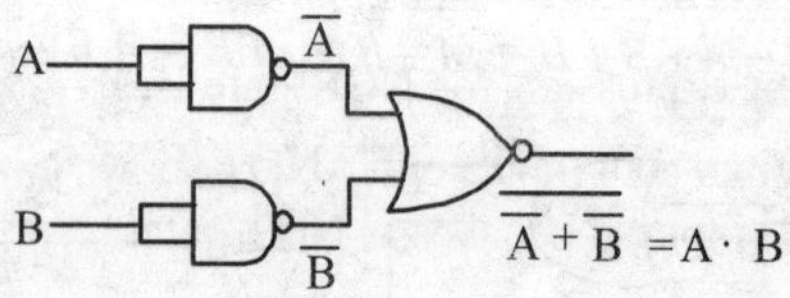

138. (b) Output, $Y=(A+B)\cdot(\overline{AB})=(A+B)\cdot(\overline{A}+\overline{B})$

$=\overline{A}A+\overline{A}B+A\overline{B}+\overline{B}B=O+\overline{A}B+A\overline{B}+O$

or, $Y=\overline{A}B+A\overline{B}$ which is output of *XOR* gate.

139. (0) Output $Y=0$

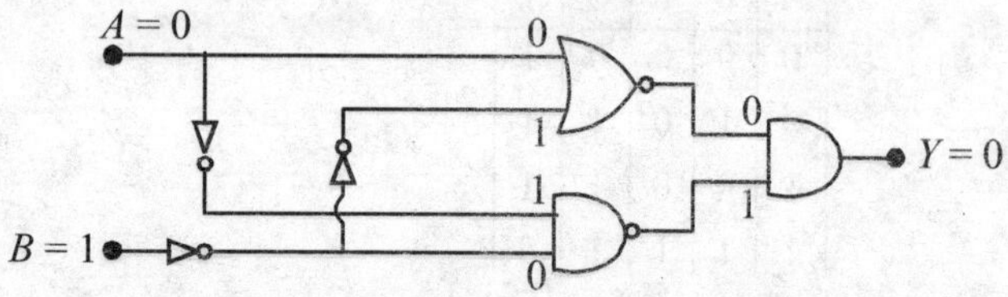

140. (c) Output $Y=\overline{\overline{\overline{A}}\,\overline{\overline{B}}}=\overline{\overline{A}}+\overline{\overline{B}}=\overline{A+B}$

Hence the given logic gate is equivalent to NOR gate.

141. (c) Truth table for the given logic circuit :

A	B	$C=A\cdot\overline{B}$	$D=\overline{A}\cdot B$	$C+D$	$\overline{C+D}=Y$
0	0	0	0	0	1
0	1	0	1	1	0
1	0	1	0	1	0
1	1	0	0	0	1

142. (c) When two inputs of NAND gate is shorted, it behaves like a NOT gate so boo Lean equation will be

$y=\overline{\overline{A}+\overline{B}+\overline{C}}$

$y=A\cdot B\cdot C$

Thus, whole arrangement behaves like a AND gate.

A	B	C	
0	0	0	0
1	0	0	0
0	1	0	0
0	0	1	0
1	1	0	0
1	0	1	0
0	1	1	0
1	1	1	1

143. (a)

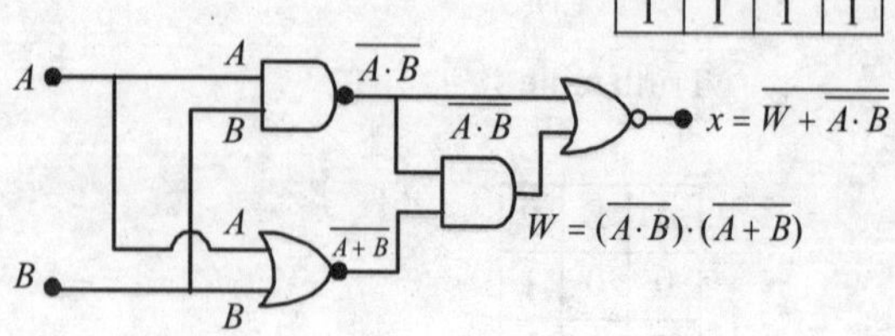

A	B	$\overline{A\cdot B}$	$\overline{A+B}$	$W=(\overline{A\cdot B})\cdot(\overline{A+B})$	$Q=W+\overline{A\cdot B}$	$\overline{Q}=x$
1	0	1	0	0	1	0
0	1	1	0	0	1	0
1	1	0	0	0	0	1
0	0	1	1	1	0	0

144. (d)

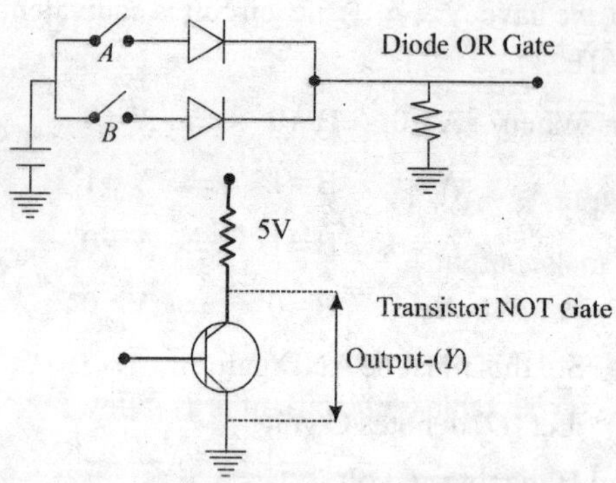

OR + NOT $\rightarrow$ NOR Gate

Hence Boolean relation at the output stage – *Y* for the circuit,

$Y=\overline{A+B}=\overline{A}.\overline{B}$

145. (a)

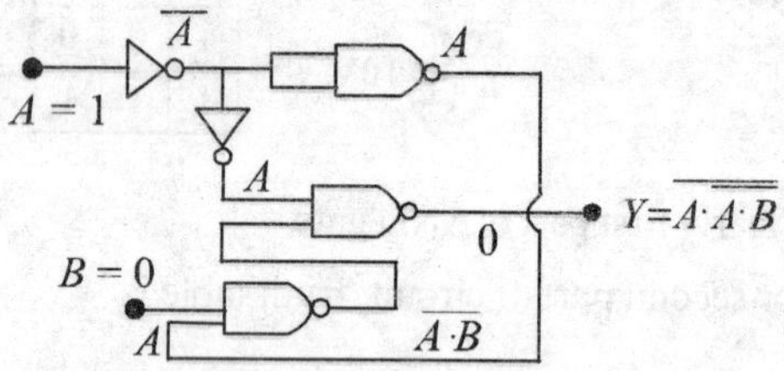

$Y=\overline{\overline{AB}.A}=\overline{\overline{AB}}+\overline{A}=AB+\overline{A}$

For $A=1, B=0$

$Y=(1)\times 0+0$

$\Rightarrow Y=0+0=0$

146. (d) A logic gate is reversible if we can recover input data from the output. Hence NOT gate.

147. (b) Truth table $\rightarrow$

The output is of OR-gate

A	B	$\overline{A}$	$\overline{B}$	$\overline{\overline{A}.\overline{B}}$
0	0	1	1	0
0	1	1	0	1
1	0	0	1	1
1	1	0	0	1

148. (c)

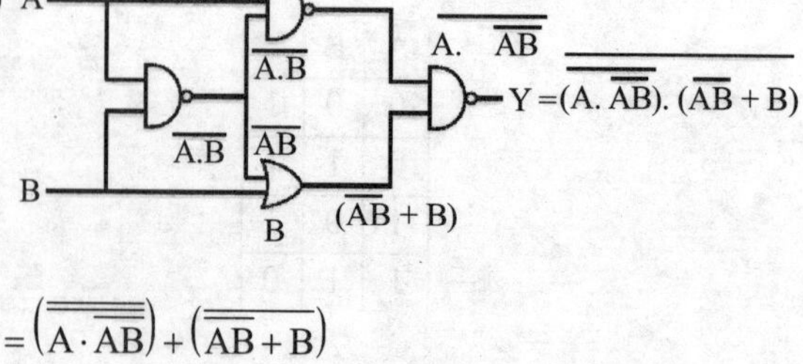

$Y=\left(\overline{A\cdot\overline{AB}}\right)+\left(\overline{\overline{AB}+B}\right)$

$=A\cdot\overline{AB}+AB\cdot\overline{B}=A\cdot(\overline{A}+\overline{B})+AB\cdot\overline{B}=A\overline{B}$

149. (c) From the given logic circuit,

$P = \bar{x} + y$

$Q = \overline{\bar{y}.x} = y + \bar{x}$

Output, $R = \overline{P + Q}$

To make output 1

$P + Q$ must be '0'

So, $x = 1, y = 0$

150. (c) Truth table of the circuit is as follows

x	y	$\bar{x}$	$a = x.y$	$b = \bar{x}.y$	$z = \overline{a.b}$
0	0	1	0	0	1
0	1	1	0	1	1
1	0	0	0	0	1
1	1	0	1	0	1

151. (a) In case of an 'OR' gate the input is zero when all inputs are zero. If any one input is ' 1', then the output is '1'.

152. (a) It represents OR-Gate.

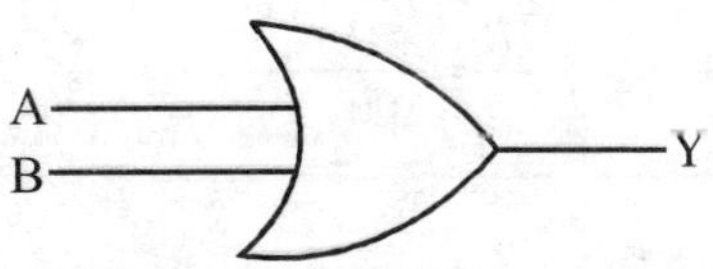

A	B	A + B = Y
0	0	0
0	1	1
1	0	1
1	1	1

153. (c) Truth table is as shown :

A	B	$\bar{A}$	$\bar{B}$	$\bar{A}+\bar{B}$	$\overline{\bar{A}+\bar{B}}$
0	0	1	1	1	0
0	1	1	0	1	0
1	0	0	1	1	0
1	1	0	0	0	1

Thus the combination of two NOT gates and one NOR gate is equivalent to a AND gate.

154. (a) NOR gate is the combination of NOT and OR gate.

Boolean expression for NOR gate is $Y = \overline{A + B}$

155. (b) When both inputs of NAND gate are jointed to form a single input, it behaves as NOT gate

OR + NOT = NOR.

$\left(\overline{A + B}\right)$ = NOR gate

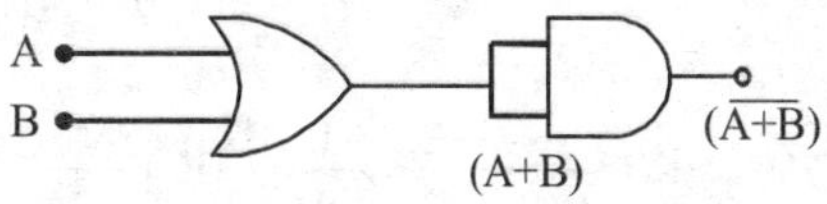

156. (a) The final boolean expression of these gates is,

$X = \overline{\left(\bar{A}.\bar{B}\right)} = \bar{\bar{A}} + \bar{\bar{B}} = A + B \Rightarrow$ OR gate

It means OR gate is formed.

157. (d) The final boolean expression

$Y = \overline{(\bar{A} + \bar{B})} = \bar{\bar{A}}.\bar{\bar{B}} = A \cdot B$.

Thus, it is an AND gate for which truth table is

A	B	Y
0	0	0
0	1	0
1	0	0
1	1	1

158. (d)

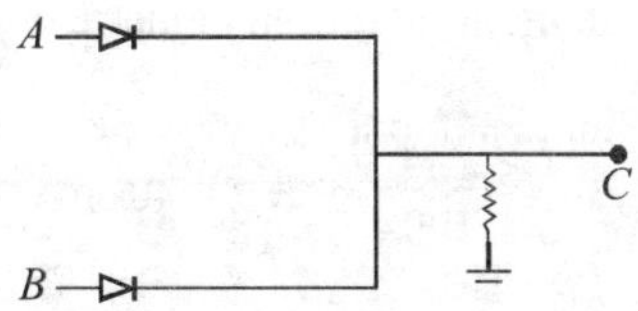

The truth table for the above circuit is :

A	B	C
1	1	1
1	0	1
0	1	1
0	0	0

when either A or B conducts, the gate conducts. It means $C = A + B$ which is for OR gate.

15 Experimental Skills-XII

1. **(5)** Figure of merit of galvanometer is $K = \frac{I}{\theta} = \frac{E}{\theta(R+G)}$

Given, $E = 2V \Rightarrow K = \frac{2}{\theta(R+G)} \Rightarrow \frac{1}{\theta} = \frac{K}{2}R + \frac{KG}{2}$

$\therefore$ Slope $= \frac{K}{2} = \frac{2-\frac{1}{2}}{6-0}$ $\qquad \therefore K = 5 \times 10^{-1}$ A/div

2. **(c)** A galvanometer is converted into ammeter by connecting a very low resistances in parallel to it. To convert galvanometer into voltmeter, a very high resistance should be added in series.

3. **(2)** From the balancing condition of metre bridge

$\frac{R_1}{R_2} = \frac{\ell}{100-\ell} \Rightarrow \frac{2}{3} = \frac{\ell}{100-\ell} \Rightarrow \ell = 40$ cm

When XΩ resistor is added to 3Ω balancing condition

$\frac{2}{\left(\frac{3X}{3+X}\right)} = \frac{40+22.5}{60-22.5} = \frac{62.5}{37.5} = \frac{5}{3} \Rightarrow \frac{6}{5} = \frac{3x}{3+x}$

$\Rightarrow 6 + 2X = 5X \Rightarrow X = 2$

4. **(2400)** At D, we have null point

So, $\frac{S}{30} = \frac{R}{70} \Rightarrow \frac{S}{30} = \frac{5600}{70} \Rightarrow S = \frac{3}{7} \times 5600 = 2400\ \Omega$

5. **(b)** 1/3rd deflection means $= \frac{I}{I_g} = 3$ or $I_g = \frac{I}{3}$

$I_g G = (I - I_g)S$

$\Rightarrow I_g(G+S) = IS$

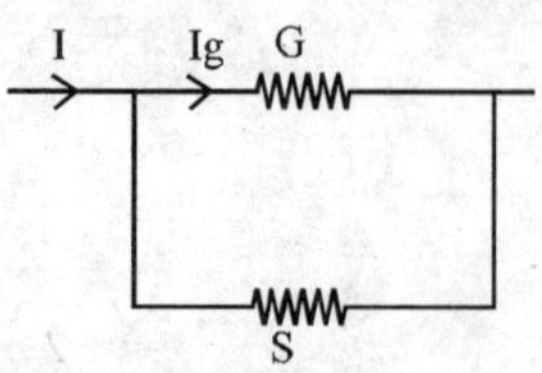

$\Rightarrow \frac{I_g}{I} = \frac{S}{G+S} \Rightarrow \frac{1}{3} = \frac{S}{G+S}$

$\Rightarrow G + S = 3S \Rightarrow G = 2S$

6. **(c)** Ammeter is always connected in series and voltmeter is always connected in parallel in a circuit.

7. **(40)** For the given meter bridge

$\frac{R}{S} = \frac{\ell_1}{100-\ell_1}$ Where, ℓ_1 = balancing length

$\Rightarrow \frac{R}{S} = \frac{25}{75} = \frac{1}{3}$ ...(i)

New resistance,

$R' = \frac{\rho\frac{\ell}{2}}{\frac{A}{4}} = \rho\frac{\ell \times 2}{A}$ $\qquad \left(\because R = \rho\frac{\ell}{A}\right)$

$\Rightarrow R' = 2R$

$\frac{R'}{S} = \frac{\ell_2}{100-\ell_2} \Rightarrow \frac{2R}{S} = \frac{\ell_2}{100-\ell_2}$

$\Rightarrow 2 \times \frac{1}{3} = \frac{\ell_2}{100-\ell_2}$ Using (i)

$\Rightarrow \ell_2 = 40$ cm

8. **(a)** When $V = V_0 \simeq 0$, then $I = 1000$ mA, and when $I = 0$ $= 1A$

then, $E = V + Ir \qquad \Rightarrow V = E$

$E = 0 + Ir \qquad \Rightarrow E = 1.5$ V

$E = r\ [\because I = 1]$ So, $E = 1.5$ V and $r = 1.5\ \Omega$

9. **(c)** For a balanced bridge $\frac{R_1}{R_2} = \frac{l_2}{l_1} \Rightarrow \frac{R}{X} = \frac{l}{100-l}$

Using the above expression $X = \frac{R(100-l)}{l}$

for observation (1) $X = \frac{1000 \times 40}{60} = \frac{2000}{3}\Omega$

for observation (2) $X = \frac{100 \times 87}{13} = \frac{8700}{13}\Omega$

for observation (3) $X = \frac{10 \times 98.5}{1.5} = \frac{1970}{3}\Omega$

for observation (4) $X=\frac{1\times 99}{1}=99\Omega$

Clearly we can see that the value of x calculated in observation (4) is inconsistent than other.

10. **(a)** Bridge wire in a sensitive meter bridge wire should be of high resistivity and low temperature coefficient.

11. **(b)** LC = 1 MSD – 1 VSD

300 msd = 15 cm

$1 \text{ msd} = \frac{15}{300} \text{ cm} = 0.05 \text{ cm}$

25 vsd = 24 msd

$1 \text{ vsd} = \frac{24}{25} \text{ msd}$

LC = 1 msd – 1 vsd

$LC = 1 \text{ msd} - \frac{24}{25} \text{ msd} = \frac{1}{25} \text{ msd}$

$LC = \frac{1}{25}\times 0.05 = 0.002 \text{ cm.}$

Correct option (b).

12. **(b)** By Newton's formula for focal length of a lens, $(u-f)(v+f)=f^2$

$\therefore$ v versus u graph is

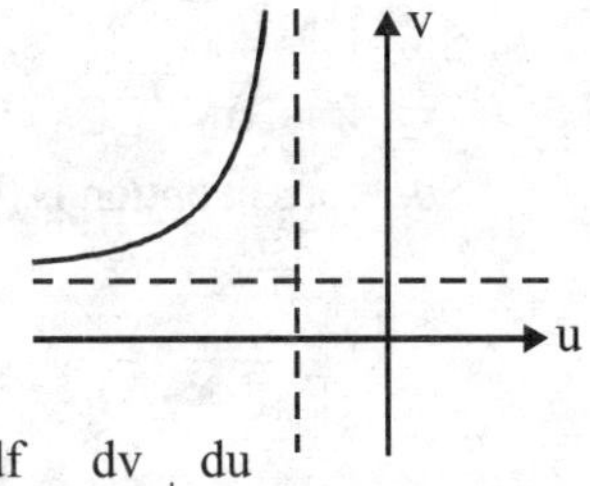

13. **(b)** Using, $\frac{1}{f}=\frac{1}{v}-\frac{1}{u} \Rightarrow \frac{df}{f^2}=\frac{dv}{v^2}+\frac{du}{u^2}$

For small change df = Δf $\quad \therefore \Delta f = f^2\left[\frac{\Delta v}{v^2}+\frac{\Delta u}{u^2}\right]$

14. **(20)** Using $\frac{1}{f}=\frac{1}{v}-\frac{1}{u} \Rightarrow \frac{1}{f+20}-\frac{1}{-(f+20)}=\frac{1}{f}$

$\frac{2}{f+20}=\frac{1}{f} \Rightarrow f = 20\text{cm}$

By Newton formula

$f^2 = xy = 20\times 20 \quad \therefore f = 20 \text{ cm}$

15. **(c)** Least count, L.C. = 0.2 cm

Distance of object u = (100 ± 0.2) – (80 ± 0.2) = (20 ± 0.4) cm

Distance of image v = (180 ± 0.2) – (100 ± 0.2) = (80 ± 0.4)cm

From lens formula,

$\frac{1}{v}-\frac{1}{u}=\frac{1}{f} \Rightarrow \frac{1}{f}=\frac{1}{80}-\frac{1}{-20} \therefore f = 16 \text{ cm}$

Also $\frac{\Delta f}{f^2}=\frac{\Delta v}{v^2}+\frac{\Delta u}{u^2}$

The % error in the estimation of focal length is

$\Rightarrow \left|\frac{\Delta f}{f}\times 100\right| = \left|\frac{\Delta v}{v^2}\times 100\right| + \left|\frac{\Delta u}{u^2}\times 100\right|$

$\Rightarrow \left|\frac{\Delta f}{f}\times 100\right| = \left|\left(\frac{\Delta v}{v^2}+\frac{\Delta u}{u^2}\right)\times f \times 100\right|$

$=\left(\frac{0.4}{400}+\frac{0.4}{6400}\right)\times 16\times 100 \Rightarrow \left|\frac{\Delta f}{f}\times 100\right| = 1.70\%$

16. **(15)** By Newton's formula

$vu = f^2 \Rightarrow f^2 = 225 \Rightarrow f = 15 \text{ cm}$

17. **(c)** We know that $i + e - A = \delta$

$35° + 79° - A = 40° \quad \therefore \quad A = 74°$

But $\mu = \frac{\sin\left(\frac{A+\delta_m}{2}\right)}{\sin A/2} = \frac{\sin\left(\frac{74+\delta_m}{2}\right)}{\sin\frac{74}{2}}$

$=\frac{5}{3}\sin\left(37°+\frac{\delta_m}{2}\right)$

μ_{max} can be $\frac{5}{3}$. That is μ_{max} is less than $\frac{5}{3}=1.67$

But δ_m will be less than 40°

18. **(c)** For the prism as the angle of incidence (i) increases, the angle of deviation (δ) first decreases goes to minimum value and then increases.

19. **(d)** For the graph to intersect y = x line. The value of $|v|$ and $|u|$ must be equal.

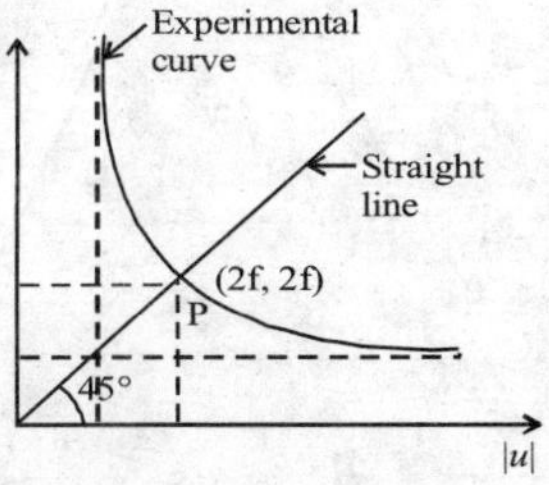

From lens formula

$\frac{1}{v}-\frac{1}{u}=\frac{1}{f}$

When $u=-2f, v=2f$

Also $v=\frac{f}{1+\frac{f}{u}}$

As $|u|$ increases, v decreases for $|u|>f$. The graph between $|v|$ and $|u|$ is shown in the figure. A straight line passing through the origin and making an angle of 45° with the x-axis meets the experimental curve at $P(2f, 2f)$.

20. **(c)** From the lens formula $\frac{1}{f}=\frac{1}{v}-\frac{1}{u}$

This graph suggest that when

$u=-f, v=+\infty$

When u is at $-\infty$,

$v=f$.

When the object is moved further away from the lens, v decreases but remains positive.

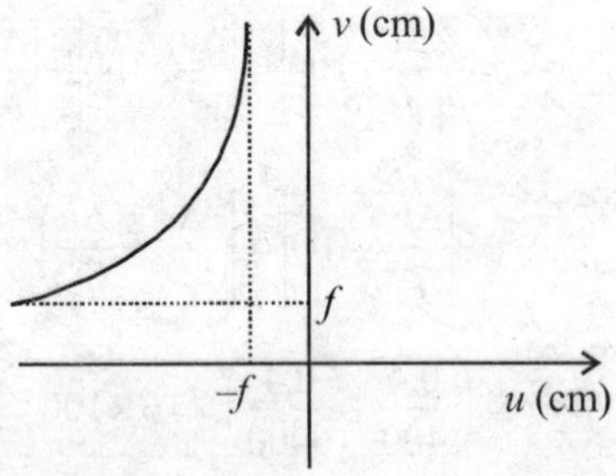

21. **(125)** Power $P_{ZM}=I_{ZM}V_Z$

$$I_{ZM}=\frac{P_{ZM}}{V_Z}=\frac{0.4}{10}=0.04A$$

$$\text{So, } R_P=\frac{E_{max}-V_z}{I_{ZM}}=\frac{15-10}{0.04}=\frac{5}{0.04}=125\Omega$$

22. **(14)** Potential barrier will reduce electron's kinetic energy by $q\ V_{barrier}$

So, $k_f-k_i=qV_{Barrier}$

$$\Rightarrow \frac{1}{2}\times 9.1\times 10^{-31}\times\left(36\times 10^{10}-V_i^2\right)=1.6\times 10^{-19}\times 0.4$$

$$\Rightarrow (36\times 10^{10}-V_i^2)=1.40\times 10^{11}\Rightarrow 2.2\times 10^{11}=V_i^2$$

$$\Rightarrow V_i=4.67\times 10^5\text{ m/s}\Rightarrow V_i=\frac{14}{3}\times 10^5\text{ m/s}$$

23. **(b)** $R=\frac{\Delta V}{\Delta i}\Rightarrow \frac{R_1}{R_2}=\frac{\Delta V_1}{\Delta i_1}\times\frac{\Delta i_2}{\Delta V_2}$

$$=\frac{2.1-2}{10-5}\times\frac{250-200}{4.2-4}=\frac{0.1}{5}\times\frac{50}{0.2}=\frac{5}{1}$$

24. **(b)** Multimeter shows deflection in both cases i.e. before and after reversing the probes if the chosen component is capacitor.

25. **(b)** Forward bias resistance $=\frac{\Delta V}{\Delta I}=\frac{0.1}{10\times 10^{-3}}=10\,\Omega$

Reverse bias resistance $=\frac{10}{10^{-6}}=10^7\,\Omega$

Ratio of resistances $=\frac{\text{Forward bias resistance}}{\text{Reverse bias resistance}}=10^{-6}$

26. **(b)** $V=iR+V_D\Rightarrow V_D=V-iR=V-100i$

Maximum power dissipated across the zener diode is given by

$P=V_Di=(V-100i)\,i\le 1$

$100i^2-Vi+1\ge 0$

Current should be real,

So, $V^2>400$

$V>20\text{ V}$

So, correct option is (b).

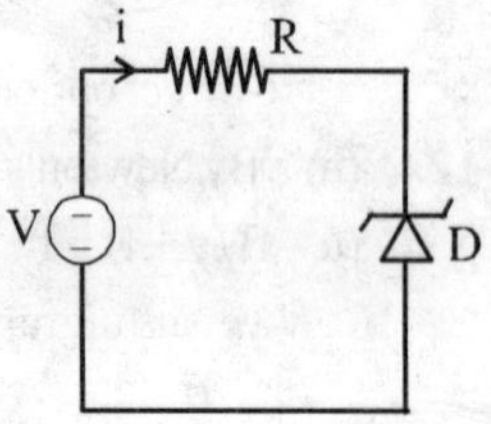